An Introduction to Medicinal Chemistry

an introduction to
MEDICINAL CHEMISTRY

sixth edition

GRAHAM L. PATRICK

OXFORD
UNIVERSITY PRESS

OXFORD

UNIVERSITY PRESS

Great Clarendon Street, Oxford, OX2 6DP,
United Kingdom

Oxford University Press is a department of the University of Oxford.
It furthers the University's objective of excellence in research, scholarship,
and education by publishing worldwide. Oxford is a registered trade mark of
Oxford University Press in the UK and in certain other countries

Third edition 2005
Fourth edition 2009
Fifth edition 2013

Impression: 1

io 07817606

Published in the United States of America by Oxford University Press
198 Madison Avenue, New York, NY 10016, United States of America

British Library Cataloguing in Publication Data

Data available

Library of Congress Control Number: 2017932606

ISBN 978–0–19–874969–1

Printed in Italy by L.E.G.O. S.p.A.

Preface

This text is aimed at undergraduates and postgraduates who have a basic grounding in chemistry and are studying a module or degree in medicinal chemistry. It attempts to convey, in a readable and interesting style, an understanding about drug design and the molecular mechanisms by which drugs act in the body. In so doing, it highlights the importance of medicinal chemistry in all our lives and the fascination of working in a field which overlaps the disciplines of chemistry, biochemistry, physiology, microbiology, cell biology, and pharmacology. Consequently, the book is of particular interest to students who might be considering a future career in the pharmaceutical industry.

Following the success of the first five editions, as well as useful feedback from readers, there has been some reorganization and updating of chapters, especially those in section E. A chapter on cardiovascular agents has also been added.

Following the introductory chapter, the book is divided into five parts:

- Part A contains five chapters that cover the structure and function of important drug targets such as receptors, enzymes, and nucleic acids. Students with a strong background in biochemistry will already know this material, but may find these chapters a useful revision of the essential points.

- Part B covers pharmacodynamics in Chapters 7–10 and pharmacokinetics in Chapter 11. Pharmacodynamics is the study of how drugs interact with their molecular targets, and the consequences of those interactions. Pharmacokinetics relates to the issues involved in a drug reaching its target in the first place.

- Part C covers the general principles and strategies involved in discovering and designing new drugs and developing them for the marketplace.

- Part D looks at particular 'tools of the trade' which are invaluable in drug design, i.e. QSAR, combinatorial synthesis, and computer-aided design.

- Part E covers a selection of specific topics within medicinal chemistry—antibacterial, antiviral, and anticancer agents, cholinergics and anticholinesterases, adrenergics, opioid analgesics, anti-ulcer agents, and cardiovascular agents. To some extent, those chapters reflect the changing emphasis in medicinal chemistry research. Antibacterial agents, cholinergics, adrenergics, and opioids have long histories and much of the early development of these drugs relied heavily on random variations of lead compounds on a trial and error basis. This approach was wasteful but it led to the recognition of various design strategies which could be used in a more rational approach to drug design. The development of the anti-ulcer drug cimetidine (Chapter 25) represents one of the early examples of the rational approach to medicinal chemistry. However, the real revolution in drug design resulted from giant advances made in molecular biology and genetics which have provided a detailed understanding of drug targets and how they function at the molecular level. This, allied to the use of molecular modelling and X-ray crystallography, has revolutionized drug design. The development of protease inhibitors as antiviral agents (Chapter 20), kinase inhibitors as anticancer agents (Chapter 21), and the statins as cholesterol-lowering agents (Case study 1) are prime examples of the modern approach.

G. L. P.
December 2016

About the book

The sixth edition of *An Introduction to Medicinal Chemistry* and its accompanying companion website contains many learning features. This section illustrates each of these learning features and explains how they will help you to gain a deeper understanding of this fascinating subject.

Emboldened keywords

Terminology is emboldened and defined in an extensive glossary at the end of the book, helping you to become familiar with the language of medicinal chemistry.

Glossary

3D QSAR QSAR studies which relate the biological activities of a series of compounds to their steric and electrostatic fields determined by molecular modelling software.

Abzyme An antibody with catalytic properties.

ADME Refers to drug abs drug metabolism, and d

Adrenal medulla A gland

Adrenaline A catecholam neurotransmitter, and w

Boxes

Boxes are used to present in-depth material and to explore how the concepts of medicinal chemistry are applied in practice.

582   **Chapter 21** Anticancer agents

BOX 21.7 General synthesis of gefitinib and related analogues

A general synthesis for gefitinib and its analogues starts from a quinazolinone starting material which acts as the central scaffold for the molecule. The synthesis is then a case of introducing the two important substituents. Selective demethylation reveals a phenol which is then

subsequent reagents. Chlori the carbonyl group, and the substituted by an aniline to substituent. Deprotection of tion with an alkyl halide int

Key points

Summaries at the end of major sections within chapters highlight key concepts and provide a useful basis for revision.

KEY POINTS

- Pharmaceutical companies tend to concentrate on developing drugs for diseases which are prevalent in developed countries, and aim to produce compounds with better properties than existing drugs.

- A molecular target is chosen which is believed to influence a particular disease when affected by a drug. The greater the selectivity that can be achieved, the less chance of side effects.

Unfortunately, this comple difficult and the compoun from their natural source– cient process. As a result, t designing simpler analogu

Many natural products structures which no chem sizing. For example, the a (Fig. 12.6) is a natural unstable looking trioxane structures to have appeare

Questions

End-of-chapter questions allow you to test your own understanding and apply concepts presented in the chapter.

QUESTIONS

1. How would you convert penicillin G to 6-aminopenicillanic acid (6-APA) using chemical reagents? Suggest how you would make ampicillin from 6-APA.

2. Penicillin is produced biosynthetically from cysteine and valine. If the biosynthetic pathway could accept different amino acids, what sort of penicillin analogues might be formed if valine was replaced by alanine, phenylalanine,

8. The following structure i sort of properties do you cefoxitin itself?

Further reading

Selected references allow you to easily research those topics that are of particular interest to you.

FURTHER READING

Abraham, D. J. (ed.) (2003) Narcotic analgesics. in *Burger's medicinal chemistry and drug discovery*, 6th edn. Chapter 7, John Wiley and Sons, New York.

Corbett, A. D., et al. (2006) 75 Years of opioid research: the exciting but vain quest for the Holy Grail. *British Journal of Pharmacology*, **147**, S153–62.

Pouletty, P. (2002) Drug addi and medically treatable dis *Discovery*, **1**, 731–6.

Roberts, S. M., and Price, B. buprenorphine, a potent an *chemistry—the role of orga*

Appendices

There are several appendices provided at the end of the book, providing further information which you may find useful. Appendix 1 shows the structures of common amino acids, with the standard genetic code given in Appendix 2. Statistical data for QSAR is provided in Appendix 3, while further information relating to the action of nerves, and microorganisms, are given in Appendices 4 and 5, respectively. Appendix 6 lists trade names and the drug(s) to which they correspond, while trade names corresponding to specific drugs in the main index are shown in brackets. Appendix 7 shows the likely hydrogen bonding interactions for different functional groups. Related appendices on the website give information on properties such as molecular weight, log P, the number of hydrogen bonding groups and rotatable bonds, molecular weight, and polar surface area for several clinically important drugs.

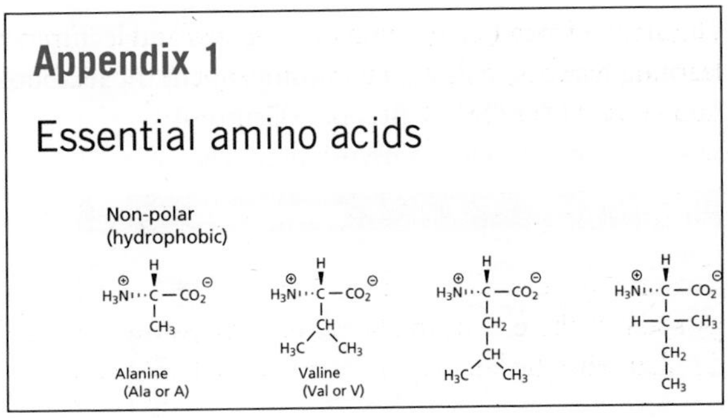
Appendix 1
Essential amino acids
Non-polar
(hydrophobic)
Alanine
(Ala or A)
Valine
(Val or V)

Links

Links have been added to the text which alert the reader to relevant articles and molecular modelling exercises on the accompanying website for the textbook. These exercises involve the use of Spartan and/or ChemBio3D molecular modelling software, as well as Excel.

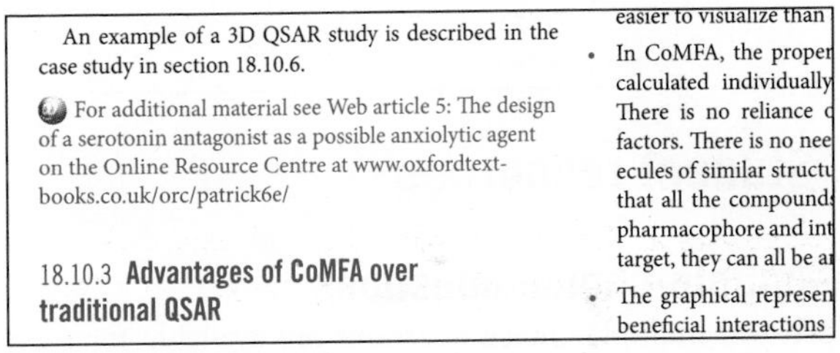

An example of a 3D QSAR study is described in the case study in section 18.10.6.

For additional material see Web article 5: The design of a serotonin antagonist as a possible anxiolytic agent on the Online Resource Centre at www.oxfordtextbooks.co.uk/orc/patrick6e/

18.10.3 **Advantages of CoMFA over traditional QSAR**

easier to visualize than
- In CoMFA, the proper calculated individually There is no reliance c factors. There is no nee ecules of similar structu that all the compound pharmacophore and int target, they can all be a
- The graphical represen beneficial interactions

Case Studies

Case Studies help you to link the underlying theory to its pharmaceutical applications and appreciate the real-world applications of the science.

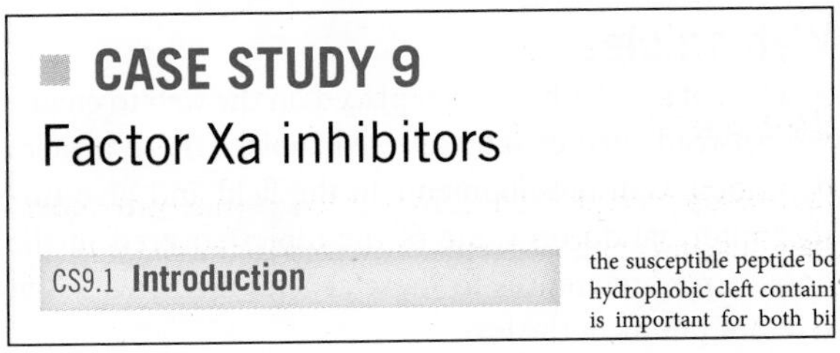

CASE STUDY 9

Factor Xa inhibitors

CS9.1 **Introduction**

the susceptible peptide bc hydrophobic cleft contain is important for both bi

About the Online Resource Centre

Online Resource Centres provide students and lecturers with ready-to-use teaching and learning resources. They are free-of-charge, and designed to complement the textbook.

You will find the Online Resource Centre at:

www.oxfordtextbooks.co.uk/orc/patrick6e/

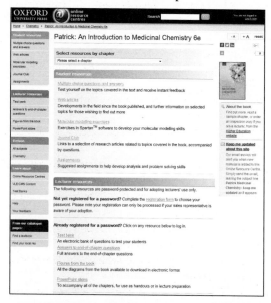

Student resources

Multiple-choice questions

Self-test multiple-choice questions are available for each chapter allowing you to test your knowledge and understanding of key concepts as you progress through the book.

Web articles

A series of articles have been placed on the web to enable you to read further into selected topics. These articles describe recent developments in the field and give further information on some of the topics covered in the book. Cross-references to these articles are provided at relevant points in the text.

Molecular modelling exercises

A series of molecular modelling exercises have been added to the website aimed at students using Spartan or ChemBio3D molecular modelling software. Alerts are provided in the book to molecular modelling exercises related to specific topic areas.

Journal Club

Suggested papers are provided along with questions and answer guidance, to help you to critically analyse the research literature.

Assignments

Suggested assignments are provided to help you develop your analysis and problem-solving skills.

Lecturer resources

For registered adopters of the book

Test Bank

A bank of multiple-choice questions, with links to relevant sections in the book, which can be downloaded and customized for your teaching.

Answers

Answers to end-of-chapter questions in the book.

Figures from the book

All of the figures from the textbook are available to download electronically for use in lectures and handouts.

PowerPoint® slides

PowerPoint® slides are provided to accompany selected topics from the book.

Acknowledgements

The author and Oxford University Press would like to thank the following people who have given advice on the various editions of this textbook:

Dr Lee Banting, School of Pharmacy and Biomedical Sciences, University of Portsmouth, UK

Dr Don Green, Department of Health and Human Sciences, London Metropolitan University, UK

Dr Mike Southern, Department of Chemistry, Trinity College, University of Dublin, Ireland

Dr Mikael Elofsson (Assistant Professor), Department of Chemistry, Umeå University, Sweden

Dr Ed Moret, Faculty of Pharmaceutical Sciences, Utrecht University, The Netherlands

Professor John Nielsen, Department of Natural Sciences, Royal Veterinary and Agricultural University, Denmark

Professor H. Timmerman, Department of Medicinal Chemistry, Vrije Universiteit, Amsterdam, The Netherlands

Professor Nouri Neamati, School of Pharmacy, University of Southern California, USA

Professor Kristina Luthman, Department of Chemistry, Gothenburg University, Sweden

Professor Taleb Altel, College of Pharmacy, University of Sarjah, United Arab Emirates

Professor Dirk Rijkers, Faculty of Pharmaceutical Sciences, Utrecht University, The Netherlands

Dr Sushama Dandekar, Department of Chemistry, University of North Texas, USA

Dr John Spencer, Department of Chemistry, School of Life Sciences, University of Sussex, UK

Dr Angeline Kanagasooriam, School of Physical Sciences, University of Kent at Canterbury, UK

Dr A. Ganesan, School of Chemistry, University of Southampton, UK

Dr Rachel Dickens, Department of Chemistry, University of Durham, UK

Dr Gerd Wagner, School of Chemical Sciences and Pharmacy, University of East Anglia, UK

Dr Colin Fishwick, School of Chemistry, University of Leeds, UK

Professor Paul O'Neil, Department of Chemistry, University of Liverpool, UK

Professor Trond Ulven, Department of Chemistry, University of Southern Denmark, Denmark

Professor Jennifer Powers, Department of Chemistry and Biochemistry, Kennesaw State University, USA

Professor Joanne Kehlbeck, Department of Chemistry, Union College, USA

Dr Robert Sindelar, Faculty of Pharmaceutical Sciences, University of British Columbia, Canada

Professor John Carran, Department of Chemistry, Queen's University, Canada

Professor Anne Johnson, Department of Chemistry and Biology, Ryerson University, Canada

Dr Jane Hanrahan, Faculty of Pharmacy, University of Sydney, Australia

Dr Ethel Forbes, School of Science, University of the West of Scotland, UK

Dr Zoe Waller, School of Pharmacy, University of East Anglia, UK

Dr Susan Matthews, School of Pharmacy, University of East Anglia, UK

Professor Ulf Nilsson, Organic Chemistry, Lund University, Sweden

Dr Russell Pearson, School of Physical and Geographical Sciences, Keele University, UK

Dr Rachel Codd, Sydney Medical School, The University of Sydney, Australia

Dr Marcus Durrant, Department of Chemical and Forensic Sciences, Northumbria University, UK

Dr Alison Hill, College of Life and Environmental Sciences, University of Exeter, UK

Dr Connie Locher, School of Biomedical, Biomolecular and Chemical Sciences, University of Western Australia, Australia

Associate Professor Jon Vabeno, Department of Pharmacy, University of Tromso, Norway

Dr Celine Cano, Northern Institute for Cancer Research, Newcastle University, UK

Professor Steven Bull, Department of Chemistry, University of Bath, UK

Professor John Marriott, School of Pharmacy, University of Birmingham, UK

Associate Professor Jonathan Watts, University of Southampton, UK

Associate Professor Alexander Zelikin, Aarhus University, Denmark

Prof. Dr Iwan de Esch, Division of Medicinal Chemistry, VU University Amsterdam, The Netherlands

Dr Patricia Ragazzon, School of Environment & Life Sciences, University of Salford, UK

Dr David Adams, School of Pharmacy and Biomedical Sciences, University of Central Lancashire, UK

The author would like to express his gratitude to Dr John Spencer of the University of Sussex for co-authoring Chapter 16, the preparation of several web articles, and for preparation of Journal Club to accompany the sixth edition. Much appreciation is due to Nahoum Anthony and Dr Rachel Clark of the Strathclyde Institute for Pharmaceutical and Biomedical Sciences at the University of Strathclyde, for their assistance with creating Figures 2.9, Box 8.2 Figures 1 and 3, Figures 17.9, 17.44, 20.15, 20.22, 20.54, and 20.55 from pdb files, some of which were obtained from the RSCB Protein Data Bank. Dr James Keeler of the Department of Chemistry, University of Cambridge, kindly generated the molecular models that appear on the book's Online Resource Centre. Thanks also to Dr Stephen Bromidge of GlaxoSmithKline for permitting the description of his work on selective 5-HT2C antagonists, and for providing many of the diagrams for that web article. Many thanks to Cambridge Scientific, Oxford Molecular, and Tripos for their advice and assistance in the writing of Chapter 17. Finally, thanks are due to Dr Des Nichol, Dr Jorge Chacon, Dr Ciaran Ewins, Dr Callum McHugh, and Dr Fiona Henriquez for their invaluable support at the University of the West of Scotland.

Brief contents

List of boxes   xxvi
Acronyms and abbreviations   xxviii

1 Drugs and drug targets: an overview   1

PART A Drug targets

2 Protein structure and function   17
3 Enzymes: structure and function   30
4 Receptors: structure and function   44
5 Receptors and signal transduction   61
6 Nucleic acids: structure and function   77

PART B Pharmacodynamics and pharmacokinetics

7 Enzymes as drug targets   93
8 Receptors as drug targets   109
9 Nucleic acids as drug targets   128
10 Miscellaneous drug targets   144
11 Pharmacokinetics and related topics   162
   Case study 1: Statins   187

PART C Drug discovery, design, and development

12 Drug discovery: finding a lead   197
13 Drug design: optimizing target interactions   223
14 Drug design: optimizing access to the target   256
15 Getting the drug to market   284
   Case study 2: The design of ACE inhibitors   302
   Case study 3: Artemisinin and related antimalarial drugs   309
   Case study 4: The design of oxamniquine   315

PART D Tools of the trade

16 Combinatorial and parallel synthesis   325
17 Computers in medicinal chemistry   349
18 Quantitative structure–activity relationships (QSAR)   395
   Case study 5: Design of a thymidylate synthase inhibitor   419

PART E Selected topics in medicinal chemistry

19 Antibacterial agents   425
20 Antiviral agents   490
21 Anticancer agents   543
22 Cholinergics, anticholinergics, and anticholinesterases   620
23 Drugs acting on the adrenergic nervous system   654
24 The opioid analgesics   678
25 Anti-ulcer agents   705
26 Cardiovascular drugs   735
   Case study 6: Steroidal anti-inflammatory agents   766
   Case study 7: Current research into antidepressant agents   776
   Case study 8: The design and development of aliskiren   781
   Case study 9: Factor Xa inhibitors   788
   Case study 10: Reversible inhibitors of HCV NS3-4A protease   795

Appendix 1 Essential amino acids   801
Appendix 2 The standard genetic code   802
Appendix 3 Statistical data for QSAR   803
Appendix 4 The action of nerves   807
Appendix 5 Microorganisms   811
Appendix 6 Trade names and drugs   813
Appendix 7 Hydrogen bonding interactions   822

Glossary   824
General further reading   845
Index   847

Detailed contents

List of boxes   xxvi
Acronyms and abbreviations   xxviii

1 Drugs and drug targets: an overview   1

1.1 What is a drug?   1
1.2 Drug targets   3
   1.2.1   Cell structure   3
   1.2.2   Drug targets at the molecular level   4
1.3 Intermolecular bonding forces   5
   1.3.1   Electrostatic or ionic bonds   5
   1.3.2   Hydrogen bonds   6
   1.3.3   Van der Waals interactions   8
   1.3.4   Dipole–dipole and ion–dipole interactions   8
   1.3.5   Repulsive interactions   9
   1.3.6   The role of water and hydrophobic interactions 10
1.4 Pharmacokinetic issues and medicines   11
1.5 Classification of drugs   11
1.6 Naming of drugs and medicines   12

PART A Drug targets

2 Protein structure and function   17

2.1 The primary structure of proteins   17
2.2 The secondary structure of proteins   18
   2.2.1   The α-helix   18
   2.2.2   The β-pleated sheet   18
   2.2.3   The β-turn   18
2.3 The tertiary structure of proteins   19
   2.3.1   Covalent bonds: disulphide links   21
   2.3.2   Ionic or electrostatic bonds   21
   2.3.3   Hydrogen bonds   21
   2.3.4   Van der Waals and hydrophobic interactions   22
   2.3.5   Relative importance of bonding interactions   23
   2.3.6   Role of the planar peptide bond   23
2.4 The quaternary structure of proteins   23
2.5 Translation and post-translational modifications   25
2.6 Proteomics   26
2.7 Protein function   26
   2.7.1   Structural proteins   26
   2.7.2   Transport proteins   27
   2.7.3   Enzymes and receptors   27
   2.7.4   Miscellaneous proteins and protein–protein interactions   28

3 Enzymes: structure and function   30

3.1 Enzymes as catalysts   30
3.2 How do enzymes catalyse reactions?   31
3.3 The active site of an enzyme   31
3.4 Substrate binding at an active site   32

3.5 The catalytic role of enzymes   32
   3.5.1   Binding interactions   32
   3.5.2   Acid–base catalysis   33
   3.5.3   Nucleophilic groups   34
   3.5.4   Stabilization of the transition state   35
   3.5.5   Cofactors   35
   3.5.6   Naming and classification of enzymes   37
   3.5.7   Genetic polymorphism and enzymes   37
3.6 Regulation of enzymes   38
3.7 Isozymes   40
3.8 Enzyme kinetics   41
   3.8.1   The Michaelis–Menten equation   41
   3.8.2   Lineweaver–Burk plots   42
Box 3.1 The external control of enzymes by nitric oxide   39

4 Receptors: structure and function   44

4.1 Role of the receptor   44
4.2 Neurotransmitters and hormones   44
4.3 Receptor types and subtypes   47
4.4 Receptor activation   47
4.5 How does the binding site change shape?   47
4.6 Ion channel receptors   49
   4.6.1   General principles   49
   4.6.2   Structure   50
   4.6.3   Gating   51
   4.6.4   Ligand-gated and voltage-gated ion channels   51
4.7 G-protein-coupled receptors   52
   4.7.1   General principles   52
   4.7.2   Structure   53
   4.7.3   The rhodopsin-like family of G-protein-coupled receptors   53
   4.7.4   Dimerization of G-coupled receptors   55
4.8 Kinase receptors   55
   4.8.1   General principles   55
   4.8.2   Structure of tyrosine kinase receptors   56
   4.8.3   Activation mechanism for tyrosine kinase receptors   56
   4.8.4   Tyrosine kinase receptors as targets in drug discovery   57
      4.8.4.1   The ErbB family of tyrosine kinase receptors   57
      4.8.4.2   Vascular endothelial growth factor receptors   58
      4.8.4.3   Platelet-derived growth factor receptor   58
      4.8.4.4   Stem cell growth factor receptor   58
      4.8.4.5   Anaplastic lymphoma kinase (ALK)   58
      4.8.4.6   The RET receptor   58
      4.8.4.7   Hepatocyte growth factor receptor or c-MET receptor   58

4.9 Intracellular receptors 59

4.10 Regulation of receptor activity 59

4.11 Genetic polymorphism and receptors 60

5 Receptors and signal transduction 61

5.1 Signal transduction pathways for G-protein-coupled receptors 61
  5.1.1 Interaction of the receptor–ligand complex with G-proteins 61
  5.1.2 Signal transduction pathways involving the α-subunit 62

5.2 Signal transduction involving G-proteins and adenylate cyclase 63
  5.2.1 Activation of adenylate cyclase by the α_s-subunit 63
  5.2.2 Activation of protein kinase A 64
  5.2.3 The G_i-protein 65
  5.2.4 General points about the signalling cascade involving cyclic AMP 66
  5.2.5 The role of the βγ-dimer 66
  5.2.6 Phosphorylation 66

5.3 Signal transduction involving G-proteins and phospholipase C_β 68
  5.3.1 G-protein effect on phospholipase C_β 68
  5.3.2 Action of the secondary messenger: diacylglycerol 68
  5.3.3 Action of the secondary messenger: inositol triphosphate 68
  5.3.4 Resynthesis of phosphatidylinositol diphosphate 70

5.4 Signal transduction involving kinase receptors 70
  5.4.1 Activation of signalling proteins and enzymes 70
  5.4.2 The MAPK signal transduction pathway 71
  5.4.3 Activation of guanylate cyclase by kinase receptors 71
  5.4.4 The JAK-STAT signal transduction pathway 72
  5.4.5 The PI3K/Akt/mTOR signal transduction pathway 73

5.5 The hedgehog signalling pathway 74

6 Nucleic acids: structure and function 77

6.1 Structure of DNA 77
  6.1.1 The primary structure of DNA 77
  6.1.2 The secondary structure of DNA 77
  6.1.3 The tertiary structure of DNA 80
  6.1.4 Chromatins 82
  6.1.5 Genetic polymorphism and personalized medicine 82

6.2 Ribonucleic acid and protein synthesis 82
  6.2.1 Structure of RNA 82
  6.2.2 Transcription and translation 83
  6.2.3 Small nuclear RNA 85
  6.2.4 The regulatory role of RNA 85

6.3 Genetic illnesses 85

6.4 Molecular biology and genetic engineering 87

PART B Pharmacodynamics and pharmacokinetics

7 Enzymes as drug targets 93

7.1 Inhibitors acting at the active site of an enzyme 93
  7.1.1 Reversible inhibitors 93
  7.1.2 Irreversible inhibitors 94

7.2 Inhibitors acting at allosteric binding sites 96

7.3 Uncompetitive and non-competitive inhibitors 96

7.4 Transition-state analogues: renin inhibitors 97

7.5 Suicide substrates 98

7.6 Isozyme selectivity of inhibitors 99

7.7 Medicinal uses of enzyme inhibitors 99
  7.7.1 Enzyme inhibitors used against microorganisms 99
  7.7.2 Enzyme inhibitors used against viruses 101
  7.7.3 Enzyme inhibitors used against the body's own enzymes 101
  7.7.4 Enzyme modulators 103

7.8 Enzyme kinetics 104
  7.8.1 Lineweaver–Burk plots 104
  7.8.2 Comparison of inhibitors 106

Box 7.1 A cure for antifreeze poisoning 94

Box 7.2 Irreversible inhibition for the treatment of obesity 96

Box 7.3 Suicide substrates 100

Box 7.4 Designing drugs to be isozyme selective 101

Box 7.5 Action of toxins on enzymes 102

Box 7.6 Kinase inhibitors 104

8 Receptors as drug targets 109

8.1 Introduction 109

8.2 The design of agonists 109
  8.2.1 Binding groups 109
  8.2.2 Position of the binding groups 111
  8.2.3 Size and shape 112
  8.2.4 Other design strategies 112
  8.2.5 Pharmacodynamics and pharmacokinetics 112
  8.2.6 Examples of agonists 113
  8.2.7 Allosteric modulators 113

8.3 The design of antagonists 114
  8.3.1 Antagonists acting at the binding site 114
  8.3.2 Antagonists acting outwith the binding site 117

8.4 Partial agonists 118

8.5 Inverse agonists 119

8.6 Desensitization and sensitization 119

8.7 Tolerance and dependence 121

8.8   Receiver types and subtypes | 122

8.8   Receptor types and subtypes   122
8.9   Affinity, efficacy, and potency   124
Box 8.1   An unexpected agonist   113
Box 8.2   Estradiol and the estrogen receptor   116

9   Nucleic acids as drug targets   128

9.1   Intercalating drugs acting on DNA   128
9.2   Topoisomerase poisons: non-intercalating   129
9.3   Alkylating and metallating agents   131
    9.3.1   Nitrogen mustards   132
    9.3.2   Nitrosoureas   132
    9.3.3   Busulfan   132
    9.3.4   Cisplatin   133
    9.3.5   Dacarbazine and procarbazine   134
    9.3.6   Mitomycin C   135
9.4   Chain cutters   136
9.5   Chain terminators   137
9.6   Control of gene transcription   138
9.7   Agents that act on RNA   139
    9.7.1   Agents that bind to ribosomes   139
    9.7.2   Antisense therapy   139

10   Miscellaneous drug targets   144

10.1   Transport proteins as drug targets   144
10.2   Structural proteins as drug targets   144
    10.2.1   Viral structural proteins as drug targets   144
    10.2.2   Tubulin as a drug target   145
        10.2.2.1   Agents which inhibit tubulin polymerization   145
        10.2.2.2   Agents which inhibit tubulin depolymerization   146
10.3   Biosynthetic building blocks as drug targets   147
10.4   Biosynthetic processes as drug targets: chain terminators   148
10.5   Protein–protein interactions   148
10.6   Lipids as a drug target   152
    10.6.1   'Tunnelling molecules'   152
    10.6.2   Ion carriers   155
    10.6.3   Tethers and anchors   156
10.7   Carbohydrates as drug targets   157
    10.7.1   Glycomics   157
    10.7.2   Antigens and antibodies   158
    10.7.3   Cyclodextrins   160
Box 10.1   Antidepressant drugs acting on transport proteins   145
Box 10.2   Targeting transcription factor–coactivator interactions   149
Box 10.3   Cyclodextrins as drug scavengers   159

11   Pharmacokinetics and related topics   162

11.1   The three phases of drug action   162
11.2   A typical journey for an orally active drug   162
11.3   Drug absorption   163
11.4   Drug distribution   165
    11.4.1   Distribution round the blood supply   165
    11.4.2   Distribution to tissues   165
    11.4.3   Distribution to cells   165
    11.4.4   Other distribution factors   165
    11.4.5   Blood–brain barrier   166
    11.4.6   Placental barrier   166
    11.4.7   Drug–drug interactions   166
11.5   Drug metabolism   167
    11.5.1   Phase I and phase II metabolism   167
    11.5.2   Phase I transformations catalysed by cytochrome P450 enzymes   167
    11.5.3   Phase I transformations catalysed by flavin-containing monooxygenases   170
    11.5.4   Phase I transformations catalysed by other enzymes   170
    11.5.5   Phase II transformations   171
    11.5.6   Metabolic stability   172
    11.5.7   The first pass effect   176
11.6   Drug excretion   176
11.7   Drug administration   177
    11.7.1   Oral administration   178
    11.7.2   Absorption through mucous membranes   178
    11.7.3   Rectal administration   178
    11.7.4   Topical administration   178
    11.7.5   Inhalation   179
    11.7.6   Injection   179
    11.7.7   Implants   180
11.8   Drug dosing   180
    11.8.1   Drug half-life   181
    11.8.2   Steady state concentration   181
    11.8.3   Drug tolerance   182
    11.8.4   Bioavailability   182
11.9   Formulation   182
11.10   Drug delivery   183
Box 11.1   Metabolism of an antiviral agent   175

Case study 1: Statins   187

■ CS1.1   Cholesterol and coronary heart disease   187
■ CS1.2   The target enzyme   188
■ CS1.3   The discovery of statins   190
■ CS1.4   Mechanism of action for statins: pharmacodynamics   192
■ CS1.5   Binding interactions of statins   192
■ CS1.6   Other mechanisms of action for statins   193
■ CS1.7   Other targets for cholesterol-lowering drugs   194

PART C Drug discovery, design, and development

12 Drug discovery: finding a lead   197
12.1 Choosing a disease   197
12.2 Choosing a drug target   197
   12.2.1 Drug targets   197
   12.2.2 Discovering drug targets   197
   12.2.3 Target specificity and selectivity between species   199
   12.2.4 Target specificity and selectivity within the body   199
   12.2.5 Targeting drugs to specific organs and tissues 200
   12.2.6 Pitfalls   200
   12.2.7 Multi-target drugs   201
12.3 Identifying a bioassay   203
   12.3.1 Choice of bioassay   203
   12.3.2 *In vitro* tests   203
   12.3.3 *In vivo* tests   203
   12.3.4 Test validity   204
   12.3.5 High-throughput screening   204
   12.3.6 Screening by NMR   205
   12.3.7 Affinity screening   205
   12.3.8 Surface plasmon resonance   205
   12.3.9 Scintillation proximity assay   206
   12.3.10 Isothermal titration calorimetry   206
   12.3.11 Virtual screening   207
12.4 Finding a lead compound   207
   12.4.1 Screening of natural products   207
      12.4.1.1 The plant kingdom 207
      12.4.1.2 Microorganisms   208
      12.4.1.3 Marine sources   209
      12.4.1.4 Animal sources   209
      12.4.1.5 Venoms and toxins 210
   12.4.2 Medical folklore   210
   12.4.3 Screening synthetic compound 'libraries' 210
   12.4.4 Existing drugs   211
      12.4.4.1 'Me too' and 'me better' drugs 211
      12.4.4.2 Enhancing a side effect 211
   12.4.5 Starting from the natural ligand or modulator 214
      12.4.5.1 Natural ligands for receptors 214
      12.4.5.2 Natural substrates for enzymes 214
      12.4.5.3 Enzyme products as lead compounds 214
      12.4.5.4 Natural modulators as lead compounds 215
   12.4.6 Combinatorial and parallel synthesis   215
   12.4.7 Computer-aided design of lead compounds 215
   12.4.8 Serendipity and the prepared mind   215
   12.4.9 Computerized searching of structural databases 217
   12.4.10 Fragment-based lead discovery   217
   12.4.11 Properties of lead compounds   219
12.5 Isolation and purification   220
12.6 Structure determination   220
12.7 Herbal medicine   220

Box 12.1 Recently discovered targets: the caspases 198
Box 12.2 Pitfalls in choosing particular targets   200
Box 12.3 Early tests for potential toxicity   201
Box 12.4 Selective optimization of side activities (SOSA) 213
Box 12.5 Natural ligands as lead compounds   214
Box 12.6 Examples of serendipity   216
Box 12.7 The use of NMR spectroscopy in finding lead compounds 217
Box 12.8 Click chemistry *in situ*   219

13 Drug design: optimizing target interactions 223
13.1 Structure–activity relationships   223
   13.1.1 Binding role of alcohols and phenols   224
   13.1.2 Binding role of aromatic rings   225
   13.1.3 Binding role of alkenes   226
   13.1.4 The binding role of ketones and aldehydes 226
   13.1.5 Binding role of amines   226
   13.1.6 Binding role of amides   228
   13.1.7 Binding role of quaternary ammonium salts 229
   13.1.8 Binding role of carboxylic acids   229
   13.1.9 Binding role of esters   230
   13.1.10 Binding role of alkyl and aryl halides 230
   13.1.11 Binding role of thiols and ethers   231
   13.1.12 Binding role of other functional groups 231
   13.1.13 Binding role of alkyl groups and the carbon skeleton 231
   13.1.14 Binding role of heterocycles   232
   13.1.15 Isosteres   233
   13.1.16 Testing procedures   234
   13.1.17 SAR in drug optimization   234
13.2 Identification of a pharmacophore   235
13.3 Drug optimization: strategies in drug design   236
   13.3.1 Variation of substituents   236
      13.3.1.1 Alkyl substituents 236
      13.3.1.2 Substituents on aromatic or heteroaromatic rings 237
      13.3.1.3 Synergistic effects 238
   13.3.2 Extension of the structure   239
   13.3.3 Chain extension/contraction   239
   13.3.4 Ring expansion/contraction   239
   13.3.5 Ring variations   241
   13.3.6 Ring fusions   242
   13.3.7 Isosteres and bio-isosteres   243
   13.3.8 Simplification of the structure   244
   13.3.9 Rigidification of the structure   247
   13.3.10 Conformational blockers   248
   13.3.11 Structure-based drug design and molecular modelling 248
   13.3.12 Drug design by NMR spectroscopy   250
   13.3.13 The elements of luck and inspiration 250
   13.3.14 Designing drugs to interact with more than one target 252
      13.3.14.1 Agents designed from known drugs 252
      13.3.14.2 Agents designed from non-selective lead compounds 253

Box 13.1 Converting an enzyme substrate to an inhibitor by extension tactics 240

Box 13.2 Simplification 245

Box 13.3 Rigidification tactics in drug design 249

Box 13.4 The structure-based drug design of crizotinib 251

14 Drug design: optimizing access to the target 256

14.1 Optimizing hydrophilic/hydrophobic properties 256

  14.1.1 Masking polar functional groups to decrease polarity 257
  14.1.2 Adding or removing polar functional groups to vary polarity 257
  14.1.3 Varying hydrophobic substituents to vary polarity 257
  14.1.4 Variation of N-alkyl substituents to vary pK_a 258
  14.1.5 Variation of aromatic substituents to vary pK_a 258
  14.1.6 Bio-isosteres for polar groups 258

14.2 Making drugs more resistant to chemical and enzymatic degradation 259

  14.2.1 Steric shields 259
  14.2.2 Electronic effects of bio-isosteres 259
  14.2.3 Steric and electronic modifications 260
  14.2.4 Metabolic blockers 260
  14.2.5 Removal or replacement of susceptible metabolic groups 261
  14.2.6 Group shifts 261
  14.2.7 Ring variation and ring substituents 262

14.3 Making drugs less resistant to drug metabolism 263

  14.3.1 Introducing metabolically susceptible groups 263
  14.3.2 Self-destruct drugs 263

14.4 Targeting drugs 264

  14.4.1 Targeting tumour cells: 'search and destroy' drugs 264
  14.4.2 Targeting gastrointestinal infections 265
  14.4.3 Targeting peripheral regions rather than the central nervous system 265
  14.4.4 Targeting with membrane tethers 265

14.5 Reducing toxicity 266

14.6 Prodrugs 266

  14.6.1 Prodrugs to improve membrane permeability 267
    14.6.1.1 Esters as prodrugs 267
    14.6.1.2 N-Methylated prodrugs 268
    14.6.1.3 Trojan horse approach for transport proteins 268
  14.6.2 Prodrugs to prolong drug activity 269
  14.6.3 Prodrugs masking drug toxicity and side effects 270
  14.6.4 Prodrugs to lower water solubility 270
  14.6.5 Prodrugs to improve water solubility 270
  14.6.6 Prodrugs used in the targeting of drugs 271
  14.6.7 Prodrugs to increase chemical stability 272
  14.6.8 Prodrugs activated by external influence (sleeping agents) 273

14.7 Drug alliances 273

  14.7.1 'Sentry' drugs 273

14.7.2 Localizing a drug's area of activity 274
  14.7.3 Increasing absorption 274

14.8 Endogenous compounds as drugs 274

  14.8.1 Neurotransmitters 274
  14.8.2 Natural hormones, peptides, and proteins as drugs 275
  14.8.3 Antibodies as drugs 276

14.9 Peptides and peptidomimetics in drug design 277

  14.9.1 Peptidomimetics 278
  14.9.2 Peptide drugs 280

14.10 Oligonucleotides as drugs 280

Box 14.1 The use of bio-isosteres to increase absorption 259

Box 14.2 Shortening the lifetime of a drug 264

Box 14.3 Identifying and replacing potentially toxic groups 267

Box 14.4 Varying esters in prodrugs 269

Box 14.5 Prodrugs masking toxicity and side effects 271

Box 14.6 Prodrugs to improve water solubility 272

15 Getting the drug to market 284

15.1 Preclinical and clinical trials 284

  15.1.1 Toxicity testing 284
  15.1.2 Drug metabolism studies 285
  15.1.3 Pharmacology, formulation, and stability tests 287
  15.1.4 Clinical trials 287
    15.1.4.1 Phase I studies 288
    15.1.4.2 Phase II studies 288
    15.1.4.3 Phase III studies 289
    15.1.4.4 Phase IV studies 289
    15.1.4.5 Ethical issues 290

15.2 Patenting and regulatory affairs 291

  15.2.1 Patents 291
  15.2.2 Regulatory affairs 293
    15.2.2.1 The regulatory process 293
    15.2.2.2 Fast tracking and orphan drugs 294
    15.2.2.3 Good laboratory, manufacturing, and clinical practice 294
    15.2.2.4 Analysis of cost versus benefits 295

15.3 Chemical and process development 295

  15.3.1 Chemical development 295
  15.3.2 Process development 297
  15.3.3 Choice of drug candidate 299
  15.3.4 Natural products 299

Box 15.1 Drug metabolism studies and drug design 286

Box 15.2 Synthesis of ebalzotan 296

Box 15.3 Synthesis of ICI D7114 297

Case study 2: The design of ACE inhibitors 302

Box CS2.1 Synthesis of captopril and enalaprilat 307

Case study 3: Artemisinin and related antimalarial drugs 309

  ■ **CS3.1 Introduction** 309

  ■ **CS3.2 Artemisinin** 309

  ■ **CS3.3 Structure and synthesis of artemisinin** 310

■ CS3.4 Structure–activity relationships 310

■ CS3.5 Mechanism of action 311

■ CS3.6 Drug design and development 313

Box CS3.1 Clinical properties of artemisinin and analogues 313

Case study 4: The design of oxamniquine 315

■ CS4.1 Introduction 315

■ CS4.2 From lucanthone to oxamniquine 315

■ CS4.3 Mechanism of action 319

■ CS4.4 Other agents 319

Box CS4.1 Synthesis of oxamniquine 320

PART D Tools of the trade

16 Combinatorial and parallel synthesis 325

16.1 Combinatorial and parallel synthesis in medicinal chemistry projects 325

16.2 Solid-phase techniques 326

16.2.1 The solid support 326

16.2.2 The anchor/linker 327

16.2.3 Examples of solid-phase syntheses 329

16.3 Planning and designing a compound library 330

16.3.1 'Spider-like' scaffolds 330

16.3.2 Designing 'drug-like' molecules 330

16.3.3 Synthesis of scaffolds 331

16.3.4 Substituent variation 331

16.3.5 Designing compound libraries for lead optimization 331

16.3.6 Computer-designed libraries 332

16.4 Testing for activity 333

16.4.1 High-throughput screening 333

16.4.2 Screening 'on bead' or 'off bead' 333

16.5 Parallel synthesis 334

16.5.1 Solid-phase extraction 334

16.5.2 The use of resins in solution-phase organic synthesis (SPOS) 336

16.5.3 Reagents attached to solid support: catch and release 336

16.5.4 Microwave technology 337

16.5.5 Microfluidics in parallel synthesis 337

16.6 Combinatorial synthesis 340

16.6.1 The mix and split method in combinatorial synthesis 340

16.6.2 Structure determination of the active compound(s) 341

16.6.2.1 Tagging 341

16.6.2.2 Photolithography 343

16.6.3 Dynamic combinatorial synthesis 343

Box 16.1 Examples of scaffolds 332

Box 16.2 Dynamic combinatorial synthesis of vancomycin dimers 346

17 Computers in medicinal chemistry 349

17.1 Molecular and quantum mechanics 349

17.1.1 Molecular mechanics 349

17.1.2 Quantum mechanics 349

17.1.3 Choice of method 350

17.2 Drawing chemical structures 350

17.3 3D structures 350

17.4 Energy minimization 351

17.5 Viewing 3D molecules 351

17.6 Molecular dimensions 353

17.7 Molecular properties 353

17.7.1 Partial charges 353

17.7.2 Molecular electrostatic potentials 354

17.7.3 Molecular orbitals 355

17.7.4 Spectroscopic transitions 355

17.7.5 The use of grids in measuring molecular properties 356

17.8 Conformational analysis 358

17.8.1 Local and global energy minima 358

17.8.2 Molecular dynamics 358

17.8.3 Stepwise bond rotation 359

17.8.4 Monte Carlo and the Metropolis method 360

17.8.5 Genetic and evolutionary algorithms 362

17.9 Structure comparisons and overlays 363

17.10 Identifying the active conformation 364

17.10.1 X-ray crystallography 364

17.10.2 Comparison of rigid and non-rigid ligands 365

17.11 3D pharmacophore identification 366

17.11.1 X-ray crystallography 367

17.11.2 Structural comparison of active compounds 367

17.11.3 Automatic identification of pharmacophores 367

17.12 Docking procedures 368

17.12.1 Manual docking 368

17.12.2 Automatic docking 369

17.12.3 Defining the molecular surface of a binding site 369

17.12.4 Rigid docking by shape complementarity 370

17.12.5 The use of grids in docking programs 372

17.12.6 Rigid docking by matching hydrogen bonding groups 373

17.12.7 Rigid docking of flexible ligands: the FLOG program 373

17.12.8 Docking of flexible ligands: anchor and grow programs 373

17.12.8.1 Directed Dock and Dock 4.0 374

17.12.8.2 FlexX 374

17.12.8.3 The Hammerhead program 376

17.12.9 Docking of flexible ligands: simulated annealing and genetic algorithms 377

17.13 Automated screening of databases for lead compounds and drug design 378

17.14 Protein mapping 378

17.14.1 Constructing a model protein: homology modelling 378

17.14.2 Constructing a binding site: hypothetical pseudoreceptors 380

17.15 *De novo* drug design 381
    17.15.1 General principles of *de novo* drug design 381
    17.15.2 Automated *de novo* drug design 383
        17.15.2.1 LUDI 383
        17.15.2.2 SPROUT 387
        17.15.2.3 LEGEND 389
        17.15.2.4 GROW, ALLEGROW, and SYNOPSIS 390

17.16 Planning compound libraries 390
17.17 Database handling 392
Box 17.1 Energy minimizing apomorphine 352
Box 17.2 Study of HOMO and LUMO orbitals 356
Box 17.3 Finding conformations of cyclic structures by molecular dynamics 359
Box 17.4 Identification of an active conformation 365
Box 17.5 Constructing a receptor map 382
Box 17.6 Designing a non-steroidal glucocorticoid agonist 391

18 Quantitative structure–activity relationships (QSAR) **395**

18.1 Graphs and equations 395
18.2 Physicochemical properties 396
    18.2.1 Hydrophobicity 397
        18.2.1.1 The partition coefficient (P) 397
        18.2.1.2 The substituent hydrophobicity constant (π) 398
        18.2.1.3 P versus π 399
    18.2.2 Electronic effects 400
    18.2.3 Steric factors 402
        18.2.3.1 Taft's steric factor (E_s) 403
        18.2.3.2 Molar refractivity 403
        18.2.3.3 Verloop steric parameter 403
    18.2.4 Other physicochemical parameters 404

18.3 Hansch equation 404
18.4 The Craig plot 404
18.5 The Topliss scheme 406
18.6 Bio-isosteres 409
18.7 The Free–Wilson approach 409
18.8 Planning a QSAR study 409
18.9 Case study 410
18.10 3D QSAR 413
    18.10.1 Defining steric and electrostatic fields 413
    18.10.2 Relating shape and electronic distribution to biological activity 414
    18.10.3 Advantages of CoMFA over traditional QSAR 415
    18.10.4 Potential problems of CoMFA 415
    18.10.5 Other 3D QSAR methods 416
    18.10.6 Case study: inhibitors of tubulin polymerization 416

Box 18.1 Altering log P to remove central nervous system side effects 399
Box 18.2 Insecticidal activity of diethyl phenyl phosphates 402

Box 18.3 Hansch equation for a series of antimalarial compounds 405
Case study 5: Design of a thymidylate synthase inhibitor **419**

PART E Selected topics in medicinal chemistry

19 Antibacterial agents **425**

19.1 History of antibacterial agents 425
19.2 The bacterial cell 427
19.3 Mechanisms of antibacterial action 427
19.4 Antibacterial agents which act against cell metabolism (antimetabolites) 428
    19.4.1 Sulphonamides 428
        19.4.1.1 The history of sulphonamides 428
        19.4.1.2 Structure–activity relationships 428
        19.4.1.3 Sulphanilamide analogues 428
        19.4.1.4 Applications of sulphonamides 429
        19.4.1.5 Mechanism of action 430
    19.4.2 Examples of other antimetabolites 432
        19.4.2.1 Trimethoprim 432
        19.4.2.2 Sulphones 432

19.5 Antibacterial agents which inhibit cell wall synthesis 433
    19.5.1 Penicillins 433
        19.5.1.1 History of penicillins 433
        19.5.1.2 Structure of benzylpenicillin and phenoxymethylpenicillin 434
        19.5.1.3 Properties of benzylpenicillin 434
        19.5.1.4 Mechanism of action for penicillin 435
        19.5.1.5 Resistance to penicillin 438
        19.5.1.6 Methods of synthesizing penicillin analogues 440
        19.5.1.7 Structure–activity relationships of penicillins 441
        19.5.1.8 Penicillin analogues 441
        19.5.1.9 Synergism of penicillins with other drugs 447
    19.5.2 Cephalosporins 448
        19.5.2.1 Cephalosporin C 448
        19.5.2.2 Synthesis of cephalosporin analogues at position 7 449
        19.5.2.3 First-generation cephalosporins 450
        19.5.2.4 Second-generation cephalosporins 451
        19.5.2.5 Third-generation cephalosporins 452
        19.5.2.6 Fourth-generation cephalosporins 452
        19.5.2.7 Fifth-generation cephalosporins 453
        19.5.2.8 Resistance to cephalosporins 453
    19.5.3 Other β-lactam antibiotics 454
        19.5.3.1 Carbapenems 454
        19.5.3.2 Monobactams 455
    19.5.4 β-Lactamase inhibitors 455
        19.5.4.1 Clavulanic acid 455
        19.5.4.2 Penicillanic acid sulphone derivatives 457

	19.5.4.3 Olivanic acids	457
	19.5.4.4 Avibactam	457
19.5.5	Other drugs which act on bacterial cell wall biosynthesis	458
	19.5.5.1 D-Cycloserine and bacitracin	458
	19.5.5.2 The glycopeptides: vancomycin and vancomycin analogues	459

19.6 Antibacterial agents which act on the plasma membrane structure — 464

19.6.1	Valinomycin and gramicidin A	464
19.6.2	Polymyxin B	464
19.6.3	Killer nanotubes	464
19.6.4	Cyclic lipopeptides	464

19.7 Antibacterial agents which impair protein synthesis: translation — 466

19.7.1	Aminoglycosides	466
19.7.2	Tetracyclines	468
19.7.3	Chloramphenicol	472
19.7.4	Macrolides	473
19.7.5	Lincosamides	474
19.7.6	Streptogramins	475
19.7.7	Oxazolidinones	475
19.7.8	Pleuromutilins	476

19.8 Agents that act on nucleic acid transcription and replication — 476

19.8.1	Quinolones and fluoroquinolones	476
19.8.2	Aminoacridines	478
19.8.3	Rifamycins	479
19.8.4	Nitroimidazoles and nitrofurantoin	479
19.8.5	Inhibitors of bacterial RNA polymerase	479

19.9 Miscellaneous agents — 480

19.10 Drug resistance — 482

19.10.1	Drug resistance by mutation	483
19.10.2	Drug resistance by genetic transfer	483
19.10.3	Other factors affecting drug resistance	483
19.10.4	The way ahead	484

Box 19.1	Sulphonamide analogues with reduced toxicity	429
Box 19.2	Treatment of intestinal infections	430
Box 19.3	Clinical properties of benzylpenicillin and phenoxymethylpenicillin	435
Box 19.4	*Pseudomonas aeruginosa*	438
Box 19.5	The isoxazolyl penicillins	444
Box 19.6	Clinical aspects of β-lactamase-resistant penicillins	444
Box 19.7	Ampicillin prodrugs	446
Box 19.8	Clinical aspects of broad-spectrum penicillins	447
Box 19.9	Synthesis of 3-methylated cephalosporins	451
Box 19.10	Clinical aspects of cephalosporins	454
Box 19.11	Clinical aspects of miscellaneous β-lactam antibiotics	456
Box 19.12	Clinical aspects of cycloserine, bacitracin, and vancomycin	464
Box 19.13	Clinical aspects of drugs acting on the plasma membrane	465
Box 19.14	Clinical aspects of aminoglycosides	468
Box 19.15	Clinical aspects of tetracyclines and chloramphenicol	472
Box 19.16	Clinical aspects of macrolides, lincosamides, streptogramins, oxazolidinones, and pleuromutilins	477
Box 19.17	Synthesis of ciprofloxacin	479
Box 19.18	Clinical aspects of quinolones and fluoroquinolones	480
Box 19.19	Clinical aspects of rifamycins and miscellaneous agents	482
Box 19.20	Organoarsenicals as antiparasitic drugs	487

20 Antiviral agents — **490**

20.1 Viruses and viral diseases — 490

20.2 Structure of viruses — 490

20.3 Life cycle of viruses — 491

20.4 Vaccination — 492

20.5 Antiviral drugs: general principles — 493

20.6 Antiviral drugs used against DNA viruses — 494

20.6.1	Inhibitors of viral DNA polymerase	494
20.6.2	Inhibitors of tubulin polymerization	498
20.6.3	Antisense therapy	498

20.7 Antiviral drugs acting against RNA viruses: the human immunodeficiency virus (HIV) — 498

20.7.1	Structure and life cycle of HIV	498
20.7.2	Antiviral therapy against HIV	500
20.7.3	Inhibitors of viral reverse transcriptase	500
	20.7.3.1 Nucleoside reverse transcriptase inhibitors	500
	20.7.3.2 Non-nucleoside reverse transcriptase inhibitors	501
20.7.4	Protease inhibitors	504
	20.7.4.1 The HIV protease enzyme	504
	20.7.4.2 Design of HIV protease inhibitors	505
	20.7.4.3 Saquinavir	507
	20.7.4.4 Ritonavir and lopinavir	508
	20.7.4.5 Indinavir	512
	20.7.4.6 Nelfinavir	513
	20.7.4.7 Palinavir	514
	20.7.4.8 Amprenavir and darunavir	514
	20.7.4.9 Atazanavir	514
	20.7.4.10 Tipranavir	515
	20.7.4.11 Alternative design strategies for antiviral drugs targeting the HIV protease enzyme	516
20.7.5	Inhibitors of other targets	517

20.8 Antiviral drugs acting against RNA viruses: flu virus — 519

20.8.1	Structure and life cycle of the influenza virus	519
20.8.2	Ion channel disrupters: adamantanes	521

20.8.3   Neuraminidase inhibitors   522
　　20.8.3.1 Structure and mechanism of neuraminidase   522
　　20.8.3.2 Transition-state inhibitors: development of zanamivir (Relenza)   524
　　20.8.3.3 Transition-state inhibitors: 6-carboxamides   525
　　20.8.3.4 Carbocyclic analogues: development of oseltamivir (Tamiflu)   526
　　20.8.3.5 Other ring systems   528
　　20.8.3.6 Resistance studies   529

20.9   Antiviral drugs acting against RNA viruses: cold virus   530

20.10   Antiviral drugs acting against RNA viruses: hepatitis C   531
　20.10.1 Inhibitors of HCV NS3-4A protease   532
　　20.10.1.1 Introduction   532
　　20.10.1.2 Design of boceprevir and telaprevir   532
　　20.10.1.3 Second-generation protease inhibitors   534
　20.10.2 Inhibitors of HCV NS5B RNA-dependent RNA polymerase   535
　20.10.3 Inhibitors of HCV NS5A protein   535
　20.10.4 Other targets   538

20.11   Broad-spectrum antiviral agents   539
　20.11.1 Agents acting against cytidine triphosphate synthetase   539
　20.11.2 Agents acting against S-adenosylhomocysteine hydrolase   539
　20.11.3 Ribavirin   540
　20.11.4 Interferons   540
　20.11.5 Antibodies and ribozymes   540

20.12   Bioterrorism and smallpox   541

Box 20.1 Clinical aspects of viral DNA polymerase inhibitors   497

Box 20.2 Clinical aspects of antiviral drugs used against HIV   501

Box 20.3 Clinical aspects of reverse transcriptase inhibitors   503

Box 20.4 Clinical aspects of protease inhibitors   516

Box 20.5 Clinical aspects of antiviral agents used in the treatment of hepatitis C   538

21   Anticancer agents   543

21.1 Cancer: an introduction   543
　21.1.1   Definitions   543
　21.1.2   Causes of cancer   543
　21.1.3   Genetic faults leading to cancer: proto-oncogenes and oncogenes   543
　　21.1.3.1 Activation of proto-oncogenes   543
　　21.1.3.2 Inactivation of tumour suppressor genes (anti-oncogenes)   544
　　21.1.3.3 The consequences of genetic defects   544

21.1.4   Abnormal signalling pathways   544
21.1.5   Insensitivity to growth-inhibitory signals   545
21.1.6   Abnormalities in cell cycle regulation   545
21.1.7   Apoptosis and the p53 protein   547
21.1.8   Telomeres   548
21.1.9   Angiogenesis   549
21.1.10 Tissue invasion and metastasis   550
21.1.11 Treatment of cancer   550
21.1.12 Resistance   552

21.2 Drugs acting directly on nucleic acids   553
　21.2.1   Intercalating agents   553
　21.2.2   Non-intercalating agents which inhibit the action of topoisomerase enzymes on DNA   555
　　21.2.2.1 Podophyllotoxins   555
　　21.2.2.2 Camptothecins   555
　21.2.3   Alkylating and metallating agents   555
　　21.2.3.1 Nitrogen mustards   556
　　21.2.3.2 Cisplatin and cisplatin analogues: metallating agents   558
　　21.2.3.3 CC 1065 analogues   558
　　21.2.3.4 Other alkylating agents   558
　21.2.4   Chain cutters   559
　21.2.5   Antisense therapy   559

21.3 Drugs acting on enzymes: antimetabolites   560
　21.3.1   Dihydrofolate reductase inhibitors   560
　21.3.2   Inhibitors of thymidylate synthase   561
　21.3.3   Inhibitors of ribonucleotide reductase   563
　21.3.4   Inhibitors of adenosine deaminase   564
　21.3.5   Inhibitors of DNA polymerases   564
　21.3.6   Purine antagonists   565

21.4 Hormone-based therapies   567
　21.4.1   Glucocorticoids, estrogens, progestins, and androgens   567
　21.4.2   Luteinizing hormone-releasing hormone receptor agonists and antagonists   568
　21.4.3   Anti-estrogens   568
　21.4.4   Anti-androgens   568
　21.4.5   Aromatase inhibitors   570

21.5 Drugs acting on structural proteins   572
　21.5.1   Agents which inhibit tubulin polymerization   572
　21.5.2   Agents which inhibit tubulin depolymerization   573

21.6 Inhibitors of signalling pathways   575
　21.6.1   Inhibition of farnesyl transferase and the Ras protein   575
　21.6.2   Protein kinase inhibitors   577
　　21.6.2.1 Kinase inhibitors of the epidermal growth factor receptor (EGFR)   579
　　21.6.2.2 Kinase inhibitors of Abelson tyrosine kinase, c-KIT, PDGFR, and SRC   582
　　21.6.2.3 Inhibitors of cyclin-dependent kinases (CDKs)   586
　　21.6.2.4 Kinase inhibitors of the MAPK signal transduction pathway   587
　　21.6.2.5 Kinase inhibitors of PI3K-PIP$_3$ pathways   588

21.6.2.6 Kinase inhibitors of anaplastic
lymphoma kinase (ALK) 589

21.6.2.7 Kinase inhibitors of RET and
KIF5B-RET 590

21.6.2.8 Kinase inhibitors of Janus kinase 590

21.6.2.9 Kinase inhibitors of vascular
endothelial growth factor receptor
(VEGFR) 591

21.6.2.10 Multi-receptor tyrosine kinase
inhibitors 591

21.6.2.11 Kinase inhibition involving
protein–protein binding
interactions 595

21.6.3 Receptor antagonists of the hedgehog
signalling pathway 595

21.7 Miscellaneous enzyme inhibitors 596

21.7.1 Matrix metalloproteinase inhibitors 596

21.7.2 Proteasome inhibitors 597

21.7.3 Histone deacetylase inhibitors 600

21.7.4 Inhibitors of poly ADP ribose polymerase 602

21.7.5 Other enzyme targets 603

21.8 Agents affecting apoptosis 603

21.9 Miscellaneous anticancer agents 604

21.9.1 Synthetic agents 605

21.9.2 Natural products 606

21.9.3 Protein therapy 608

21.9.4 Modulation of transcription
factor–coactivator interactions 608

21.10 Antibodies, antibody conjugates, and gene therapy 609

21.10.1 Monoclonal antibodies 609

21.10.2 Antibody–drug conjugates 611

21.10.3 Antibody-directed enzyme prodrug therapy
(ADEPT) 612

21.10.4 Antibody-directed abzyme prodrug therapy
(ADAPT) 614

21.10.5 Gene-directed enzyme prodrug therapy
(GDEPT) 614

21.10.6 Other forms of gene therapy 615

21.11 Photodynamic therapy 615

21.12 Viral therapy 616

Box 21.1 Clinical aspects of intercalating agents 554

Box 21.2 Clinical aspects of non-intercalating agents
inhibiting the action of topoisomerase enzymes
on DNA 556

Box 21.3 Clinical aspects of alkylating and metallating
agents 559

Box 21.4 Clinical aspects of antimetabolites 565

Box 21.5 Clinical aspects of hormone-based therapies 571

Box 21.6 Clinical aspects of drugs acting on
structural proteins 575

Box 21.7 General synthesis of gefitinib and related
analogues 582

Box 21.8 General synthesis of imatinib and analogues 586

Box 21.9 Design of sorafenib 592

Box 21.10 Clinical aspects of kinase inhibitors 593

Box 21.11 Clinical aspects of antibodies and
antibody–drug conjugates 609

Box 21.12 Gemtuzumab ozogamicin: an antibody–drug
conjugate 613

22 Cholinergics, anticholinergics,
and anticholinesterases 620

22.1 The peripheral nervous system 620

22.2 Motor nerves of the peripheral nervous system 620

22.2.1 The somatic motor nervous system 621

22.2.2 The autonomic motor nervous system 621

22.2.3 The enteric system 622

22.2.4 Defects in motor nerve transmission 622

22.3 The cholinergic system 622

22.3.1 The cholinergic signalling system 622

22.3.2 Presynaptic control systems 623

22.3.3 Cotransmitters 623

22.4 Agonists at the cholinergic receptor 623

22.5 Acetylcholine: structure, SAR, and receptor
binding 624

22.6 The instability of acetylcholine 626

22.7 Design of acetylcholine analogues 627

22.7.1 Steric shields 627

22.7.2 Electronic effects 627

22.7.3 Combining steric and electronic effects 628

22.8 Clinical uses for cholinergic agonists 628

22.8.1 Muscarinic agonists 628

22.8.2 Nicotinic agonists 628

22.9 Antagonists of the muscarinic cholinergic receptor 629

22.9.1 Actions and uses of muscarinic antagonists 629

22.9.2 Muscarinic antagonists 629

22.9.2.1 Atropine and hyoscine 629

22.9.2.2 Structural analogues of atropine
and hyoscine 631

22.9.2.3 Simplified analogues of atropine 631

22.9.2.4 Quinuclidine muscarinic agents 633

22.9.2.5 Other muscarinic antagonists 633

22.10 Antagonists of the nicotinic cholinergic receptor 635

22.10.1 Applications of nicotinic antagonists 635

22.10.2 Nicotinic antagonists 635

22.10.2.1 Curare and tubocurarine 635

22.10.2.2 Decamethonium and
suxamethonium 636

22.10.2.3 Steroidal neuromuscular
blocking agents 637

22.10.2.4 Atracurium and mivacurium 637

22.10.2.5 Other nicotinic antagonists 638

22.11 Receptor structures 639

22.12 Anticholinesterases and acetylcholinesterase 640

22.12.1 Effect of anticholinesterases 640

22.12.2 Structure of the acetylcholinesterase enzyme 640

22.12.3 The active site of acetylcholinesterase 640

22.12.3.1 Crucial amino acids within
the active site 641

22.12.3.2 Mechanism of hydrolysis 641

22.13 Anticholinesterase drugs — 642
  22.13.1 Carbamates — 642
    22.13.1.1 Physostigmine — 642
    22.13.1.2 Analogues of physostigmine — 644
  22.13.2 Organophosphorus compounds — 645
    22.13.2.1 Nerve agents — 645
    22.13.2.2 Medicines — 646
    22.13.2.3 Insecticides — 646

22.14 Pralidoxime: an organophosphate antidote — 647

22.15 Anticholinesterases as 'smart drugs' — 648
  22.15.1 Acetylcholinesterase inhibitors — 648
  22.15.2 Dual-action agents acting on the acetylcholinesterase enzyme — 649
  22.15.3 Multi-targeted agents acting on the acetylcholinesterase enzyme and the muscarinic M_2 receptor — 650

Box 22.1 Clinical applications for muscarinic antagonists — 634

Box 22.2 Muscarinic antagonists for the treatment of COPD — 634

Box 22.3 Mosses play it smart — 652

23 Drugs acting on the adrenergic nervous system — 654

23.1 The adrenergic nervous system — 654
  23.1.1 Peripheral nervous system — 654
  23.1.2 Central nervous system — 654

23.2 Adrenergic receptors — 654
  23.2.1 Types of adrenergic receptor — 654
  23.2.2 Distribution of receptors — 655

23.3 Endogenous agonists for the adrenergic receptors — 656

23.4 Biosynthesis of catecholamines — 656

23.5 Metabolism of catecholamines — 657

23.6 Neurotransmission — 657
  23.6.1 The neurotransmission process — 657
  23.6.2 Cotransmitters — 657
  23.6.3 Presynaptic receptors and control — 658

23.7 Drug targets — 659

23.8 The adrenergic binding site — 659

23.9 Structure–activity relationships — 660
  23.9.1 Important binding groups on catecholamines — 660
  23.9.2 Selectivity for α- versus β-adrenoceptors — 661

23.10 Adrenergic agonists — 662
  23.10.1 General adrenergic agonists — 662
  23.10.2 α_1-, α_2-, β_1-, and β_3-Agonists — 662
  23.10.3 β_2-Agonists and the treatment of asthma — 663

23.11 Adrenergic receptor antagonists — 666
  23.11.1 General α/β-blockers — 666
  23.11.2 α-Blockers — 666
  23.11.3 β-Blockers as cardiovascular drugs — 667
    23.11.3.1 First-generation β-blockers — 667
    23.11.3.2 Structure–activity relationships of aryloxypropanolamines — 668

23.11.3.3 Selective β_1-blockers (second-generation β-blockers) — 669
23.11.3.4 Short-acting β-blockers — 669

23.12 Other drugs affecting adrenergic transmission — 672
  23.12.1 Drugs that affect the biosynthesis of adrenergics — 672
  23.12.2 Drugs inhibiting the uptake of noradrenaline into storage vesicles — 672
  23.12.3 Release of noradrenaline from storage vesicles — 673
  23.12.4 Reuptake inhibitors of noradrenaline into presynaptic neurons — 673
  23.12.5 Inhibition of metabolic enzymes — 675

Box 23.1 Clinical aspects of adrenergic agents — 656

Box 23.2 Synthesis of salbutamol — 664

Box 23.3 Synthesis of aryloxypropanolamines — 668

Box 23.4 Clinical aspects of β-blockers — 670

24 The opioid analgesics — 678

24.1 History of opium — 678

24.2 The active principle: morphine — 678
  24.2.1 Isolation of morphine — 678
  24.2.2 Structure and properties — 679

24.3 Structure–activity relationships — 679

24.4 The molecular target for morphine: opioid receptors — 682

24.5 Morphine: pharmacodynamics and pharmacokinetics — 682

24.6 Morphine analogues — 684
  24.6.1 Variation of substituents — 684
  24.6.2 Drug extension — 684
  24.6.3 Simplification or drug dissection — 686
    24.6.3.1 Removing ring E — 686
    24.6.3.2 Removing ring D — 686
    24.6.3.3 Removing rings C and D — 687
    24.6.3.4 Removing rings B, C, and D — 688
    24.6.3.5 Removing rings B, C, D, and E — 689
  24.6.4 Rigidification — 690

24.7 Agonists and antagonists — 693

24.8 Endogenous opioid peptides and opioids — 695
  24.8.1 Endogenous opioid peptides — 695
  24.8.2 Analogues of enkephalins and δ-selective opioids — 696
  24.8.3 Binding theories for enkephalins — 697
  24.8.4 Inhibitors of peptidases — 699
  24.8.5 Endogenous morphine — 699

24.9 The future — 700
  24.9.1 The message-address concept — 700
  24.9.2 Receptor dimers — 700
  24.9.3 Selective opioid agonists versus multi-targeted opioids — 701
  24.9.4 Peripheral-acting opioids — 701

24.10 Case study: design of nalfurafine — 701

Box 24.1 Clinical aspects of morphine — 679

Box 24.2 Synthesis of *N*-alkylated morphine analogues — 685

Box 24.3 Opioids as antidiarrhoeal agents | 690
Box 24.4 Synthesis of the orvinols | 692
Box 24.5 A comparison of opioids and their effects on opioid receptors | 695
Box 24.6 Design of naltrindole | 698

25 Anti-ulcer agents | 705

25.1 Peptic ulcers | 705
  25.1.1 Definition | 705
  25.1.2 Causes | 705
  25.1.3 Treatment | 705
  25.1.4 Gastric acid release | 705
25.2 H_2 antagonists | 706
  25.2.1 Histamine and histamine receptors | 707
  25.2.2 Searching for a lead | 708
    25.2.2.1 Histamine | 708
    25.2.2.2 N^{α}-Guanylhistamine | 708
  25.2.3 Developing the lead: a chelation bonding theory | 711
  25.2.4 From partial agonist to antagonist: the development of burimamide | 711
  25.2.5 Development of metiamide | 713
  25.2.6 Development of cimetidine | 716
  25.2.7 Cimetidine | 717
    25.2.7.1 Biological activity | 717
    25.2.7.2 Structure and activity | 718
    25.2.7.3 Metabolism | 718
  25.2.8 Further studies of cimetidine analogues | 719
    25.2.8.1 Conformational isomers | 719
    25.2.8.2 Desolvation | 720
    25.2.8.3 Development of the nitroketeneaminal binding group | 720
  25.2.9 Further H_2 antagonists | 722
    25.2.9.1 Ranitidine | 722
    25.2.9.2 Famotidine and nizatidine | 723
    25.2.9.3 H_2 antagonists with prolonged activity | 724
  25.2.10 Comparison of H_1 and H_2 antagonists | 724
  25.2.11 H_2 receptors and H_2 antagonists | 725
25.3 Proton pump inhibitors | 725
  25.3.1 Parietal cells and the proton pump | 725
  25.3.2 Proton pump inhibitors | 726
  25.3.3 Mechanism of inhibition | 727
  25.3.4 Metabolism of proton pump inhibitors | 728
  25.3.5 Design of omeprazole and esomeprazole | 728
  25.3.6 Other proton pump inhibitors | 731
25.4 *Helicobacter pylori* and the use of antibacterial agents | 732
  25.4.1 Discovery of *Helicobacter pylori* | 732
  25.4.2 Treatment | 732
25.5 Traditional and herbal medicines | 733
Box 25.1 Synthesis of cimetidine | 718
Box 25.2 Synthesis of omeprazole and esomeprazole | 731

26 Cardiovascular drugs | 735

26.1 Introduction | 735
26.2 The cardiovascular system | 735
26.3 Antihypertensives affecting the activity of the RAAS system | 737
  26.3.1 Introduction | 737
  26.3.2 Renin inhibitors | 737
  26.3.3 ACE inhibitors | 738
  26.3.4 Angiotensin receptor antagonists | 739
  26.3.5 Mineralocorticoid receptor antagonists | 741
  26.3.6 Dual-action agents | 742
26.4 Endothelin receptor antagonists as antihypertensive agents | 742
  26.4.1 Endothelins and endothelin receptors | 742
  26.4.2 Endothelin antagonists | 742
  26.4.3 Dual-action agents | 743
26.5 Vasodilators | 744
  26.5.1 Modulators of soluble guanylate cyclase | 744
  26.5.2 Phosphodiesterase type 5 inhibitors | 746
  26.5.3 Neprilysin inhibitors | 747
  26.5.4 Prostacyclin agonists | 747
  26.5.5 Miscellaneous vasodilators | 747
26.6 Calcium entry blockers | 748
  26.6.1 Introduction | 748
  26.6.2 Dihydropyridines | 750
  26.6.3 Phenylalkylamines | 751
  26.6.4 Benzothiazepines | 752
26.7 Funny ion channel inhibitors | 753
26.8 Lipid-regulating agents | 754
  26.8.1 Statins | 754
  26.8.2 Fibrates | 754
  26.8.3 Dual- and pan-PPAR agonists | 755
  26.8.4 Antisense drugs | 756
  26.8.5 Inhibitors of transfer proteins | 756
  26.8.6 Antibodies as lipid-lowering agents | 756
26.9 Antithrombotic agents | 757
  26.9.1 Anticoagulants | 758
    26.9.1.1 Introduction | 758
    26.9.1.2 Direct thrombin inhibitors | 758
    26.9.1.3 Factor Xa inhibitors | 759
  26.9.2 Antiplatelet agents | 760
    26.9.2.1 Introduction | 760
    26.9.2.2 PAR-1 antagonists | 760
    26.9.2.3 $P2Y_{12}$ antagonists | 761
    26.9.2.4 GpIIb/IIIa antagonists | 763
  26.9.3 Fibrinolytic drugs | 763
Box 26.1 Synthesis of dihydropyridines | 749

Case study 6: Steroidal anti-inflammatory agents | 766
  ■ CS6.1 Introduction to steroids | 766
  ■ CS6.2 Orally active analogues of cortisol | 767
  ■ CS6.3 Topical glucocorticoids as anti-inflammatory agents | 768

Case study 7: Current research into antidepressant agents   776

■ CS7.1 Introduction   776

■ CS7.2 The monoamine hypothesis   776

■ CS7.3 Current antidepressant agents   776

■ CS7.4 Current areas of research   777

■ CS7.5 Antagonists for the 5-HT$_7$ receptor   777

Case study 8: The design and development of aliskiren   781

■ CS8.1 Introduction   781

■ CS8.2 Reaction catalysed by renin   781

■ CS8.3 From lead compound to peptide inhibitors   781

■ CS8.4 Peptidomimetic strategies   783

■ CS8.5 Design of non-peptide inhibitors   783

■ CS8.6 Optimization of the structure   785

Case study 9: Factor Xa inhibitors   788

■ CS9.1 Introduction   788

■ CS9.2 The target   788

■ CS9.3 General strategies in the design of factor Xa inhibitors   789

■ CS9.4 Apixaban: from hit structure to lead compound   789

■ CS9.5 Apixaban: from lead compound to final structure   790

■ CS9.6 The development of rivoraxaban   793

■ CS9.7 The development of edoxaban   794

Case study 10: Reversible inhibitors of HCV NS3-4A protease   795

■ CS10.1 Introduction   795

■ CS10.2 Identification of a lead compound   795

■ CS10.3 Modifications of the lead compound   796

■ CS10.4 From hexapeptide to tripeptide   797

■ CS10.5 From tripeptide to macrocycle (BILN-2061)   798

■ CS10.6 From BILN-2061 to simeprevir   799

Appendix 1 Essential amino acids   801
Appendix 2 The standard genetic code   802
Appendix 3 Statistical data for QSAR   803
Appendix 4 The action of nerves   807
Appendix 5 Microorganisms   811
Appendix 6 Trade names and drugs   813
Appendix 7 Hydrogen bonding interactions   822

Glossary   824
General further reading   845
Index   847

List of boxes

General interest

3.1 The external control of enzymes by nitric oxide — 39

7.1 A cure for antifreeze poisoning — 94

7.2 Irreversible inhibition for the treatment of obesity — 96

7.3 Suicide substrates — 100

7.4 Designing drugs to be isozyme selective — 101

7.5 Action of toxins on enzymes — 102

7.6 Kinase inhibitors — 104

8.1 An unexpected agonist — 113

8.2 Estradiol and the estrogen receptor — 116

10.1 Antidepressant drugs acting on transport proteins — 145

10.2 Targeting transcription factor–coactivator interactions — 149

10.3 Cyclodextrins as drug scavengers — 159

11.1 Metabolism of an antiviral agent — 175

12.1 Recently discovered targets: the caspases — 198

12.2 Pitfalls in choosing particular targets — 200

12.3 Early tests for potential toxicity — 201

12.4 Selective optimization of side activities (SOSA) — 213

12.5 Natural ligands as lead compounds — 214

12.6 Examples of serendipity — 216

12.7 The use of NMR spectroscopy in finding lead compounds — 217

12.8 Click chemistry *in situ* — 219

13.1 Converting an enzyme substrate to an inhibitor by extension tactics — 240

13.2 Simplification — 245

13.3 Rigidification tactics in drug design — 249

13.4 The structure-based drug design of crizotinib — 251

14.1 The use of bio-isosteres to increase absorption — 259

14.2 Shortening the lifetime of a drug — 264

14.3 Identifying and replacing potentially toxic groups — 267

14.4 Varying esters in prodrugs — 269

14.5 Prodrugs masking toxicity and side effects — 271

14.6 Prodrugs to improve water solubility — 272

15.1 Drug metabolism studies and drug design — 286

16.1 Examples of scaffolds — 332

17.1 Energy minimizing apomorphine — 352

17.2 Study of HOMO and LUMO orbitals — 356

17.3 Finding conformations of cyclic structures by molecular dynamics — 359

17.4 Identification of an active conformation — 365

17.5 Constructing a receptor map — 382

17.6 Designing a non-steroidal glucocorticoid agonist — 391

18.1 Altering log *P* to remove central nervous system side effects — 399

18.2 Insecticidal activity of diethyl phenyl phosphates — 402

18.3 Hansch equation for a series of antimalarial compounds — 405

19.1 Sulphonamide analogues with reduced toxicity — 429

19.2 Treatment of intestinal infections — 430

19.5 The isoxazolyl penicillins — 444

19.7 Ampicillin prodrugs — 446

19.20 Organoarsenicals as antiparasitic drugs — 487

21.9 Design of sorafenib — 592

21.12 Gemtuzumab ozogamicin: an antibody–drug conjugate — 613

22.3 Mosses play it smart — 652

24.3 Opioids as antidiarrhoeal agents — 690

24.6 Design of naltrindole — 698

Synthesis

15.2 Synthesis of ebalzotan — 296

15.3 Synthesis of ICI D7114 — 297

16.2 Dynamic combinatorial synthesis of vancomycin dimers — 346

19.9 Synthesis of 3-methylated cephalosporins — 451

19.17 Synthesis of ciprofloxacin — 479

21.7 General synthesis of gefitinib and related analogues — 582

21.8 General synthesis of imatinib and analogues — 586

23.2 Synthesis of salbutamol — 664

23.3 Synthesis of aryloxypropanolamines — 668

24.2 Synthesis of *N*-alkylated morphine analogues — 685

24.4 Synthesis of the orvinols — 692

25.1 Synthesis of cimetidine — 718

25.2 Synthesis of omeprazole and esomeprazole — 731

26.1 Synthesis of dihydropyridines 749
CS2.1 Synthesis of captopril and enalaprilat 307
CS4.1 Synthesis of oxamniquine 320

Clinical correlation

19.3 Clinical properties of benzylpenicillin and phenoxymethylpenicillin 435

19.4 *Pseudomonas aeruginosa* 438

19.6 Clinical aspects of β-lactamase-resistant penicillins 444

19.8 Clinical aspects of broad-spectrum penicillins 447

19.10 Clinical aspects of cephalosporins 454

19.11 Clinical aspects of miscellaneous β-lactam antibiotics 456

19.12 Clinical aspects of cycloserine, bacitracin, and vancomycin 464

19.13 Clinical aspects of drugs acting on the plasma membrane 465

19.14 Clinical aspects of aminoglycosides 468

19.15 Clinical aspects of tetracyclines and chloramphenicol 472

19.16 Clinical aspects of macrolides, lincosamides, streptogramins, oxazolidinones, and pleuromutilins 477

19.18 Clinical aspects of quinolones and fluoroquinolones 480

19.19 Clinical aspects of rifamycins and miscellaneous agents 482

20.1 Clinical aspects of viral DNA polymerase inhibitors 497

20.2 Clinical aspects of antiviral drugs used against HIV 501

20.3 Clinical aspects of reverse transcriptase inhibitors 503

20.4 Clinical aspects of protease inhibitors 516

20.5 Clinical aspects of antiviral agents used in the treatment of hepatitis C 538

21.1 Clinical aspects of intercalating agents 554

21.2 Clinical aspects of non-intercalating agents inhibiting the action of topoisomerase enzymes on DNA 556

21.3 Clinical aspects of alkylating and metallating agents 559

21.4 Clinical aspects of antimetabolites 565

21.5 Clinical aspects of hormone-based therapies 571

21.6 Clinical aspects of drugs acting on structural proteins 575

21.10 Clinical aspects of kinase inhibitors 593

21.11 Clinical aspects of antibodies and antibody–drug conjugates 609

22.1 Clinical applications for muscarinic antagonists 634

22.2 Muscarinic antagonists for the treatment of COPD 634

23.1 Clinical aspects of adrenergic agents 656

23.4 Clinical aspects of β-blockers 670

24.1 Clinical aspects of morphine 679

24.5 A comparison of opioids and their effects on opioid receptors 695

CS3.1 Clinical properties of artemisinin and analogues 313

CS6.1 Clinical aspects of glucocorticoids 774

Acronyms and abbreviations

Note: Abbreviations for amino acids are given in Appendix 1

5-HT5	hydroxytryptamine (serotonin)
7-ACA7	aminocephalosporinic acid
6-APA6	aminopenicillanic acid
ACE	angiotensin-converting enzyme
ACh	acetylcholine
AChE	acetylcholinesterase
ACP	acyl carrier protein
ACT	artemisinin combination therapy
ADAPT	antibody-directed abzyme prodrug therapy
ADEPT	antibody-directed enzyme prodrug therapy
ADH	alcohol dehydrogenase
ADME	absorption, distribution, metabolism, excretion
ADP	adenosine 5′-diphosphate
AGO	argonaute protein
AIC	5-aminoimidazole-4-carboxamide
AIDS	acquired immune deficiency syndrome
Akt	protein kinase B
ALK	anaplastic lymphoma kinase
AME	aminoglycoside modifying enzyme
AML	acute myeloid leukaemia
AMP	adenosine 5′-monophosphate
AT	angiotensin
ATP	adenosine 5′-triphosphate
AUC	area under the curve
BiTE	bi-specific T-cell engager
BuChE	butyrlcholinesterase
BTK	Bruton's tyrosine kinase
cAMP	cyclic AMP
β-CCE	carboline-3-carboxylate
CCK	cholecystokinin
CDKs	cyclin-dependent kinases
CETP	cholesteryl ester transfer protein
cGMP	cyclic GMP
CHO cells	Chinese hamster ovarian cells
CKIs	cyclin-dependent kinase inhibitors
c-KIT	mast/stem cell growth factor receptor
Clog P	calculated logarithm of the partition coefficient
c-MET receptor	hepatocyte growth factor receptor
CML	chronic myeloid leukaemia
CMV	cytomegalovirus
CNS	central nervous system
CoA	coenzyme A
CoMFA	comparative molecular field analysis
COMT	catechol O-methyltransferase
COPD	chronic obstructive pulmonary disease
COX	cyclooxygenase
CSD	Cambridge Structural Database
CYP	enzymes that constitute the cytochrome P450 family
D-receptor	dopamine receptor
dATP	deoxyadenosine triphosphate
DCC	dicyclohexylcarbodiimide
dCTP	deoxycytosine triphosphate
DG	diacylglycerol
dGTP	deoxyguanosine triphosphate
DHFR	dihydrofolate reductase
Dhh	desert hedgehog
DMAP	dimethlaminopyridine
DNA	deoxyribonucleic acid
DOR	delta opioid receptor
dsDNA	double-stranded DNA
dsRNA	double-stranded RNA
dTMP	deoxythymidylate monophosphate
dTTP	deoxythymidylate triphosphate
dUMP	deoxyuridylate monophosphate
EC_{50}	concentration of drug required to produce 50% of the maximum possible effect
E_s	Taft's steric factor
EGF	epidermal growth factor
EGFR	epidermal growth factor receptor
EMEA	European Agency for the Evaluation of Medicinal Products
EPC	European Patent Convention
EPO	European Patent Office
EPO	erythropoietin
ErbB	epidermal growth factor receptor
ERK	see MAPK
ET	endothelin
FDA	US Food and Drug Administration
FdUMP	fluorodeoxyuracil monophosphate
FGF	fibroblast growth factor
FGFR	fibroblast growth factor receptor

FH$_4$	tetrahydrofolate
F	oral bioavailability
F	inductive effect of an aromatic substituent in QSAR
F-SPE	fluorous solid-phase extraction
FLOG	Flexible Ligands Orientated on Grid
FPGS	folylpolyglutamate synthetase
FPP	farnesyl diphosphate
FT	farnesyl transferase
FTI	farnesyl transferase inhibitor
G-protein	guanine nucleotide binding protein
GABA	γ-aminobutyric acid
GAP	GTPase activating protein
GCP	Good Clinical Practice
GDEPT	gene-directed enzyme prodrug therapy
GDP	guanosine diphosphate
GEF	guanine nucleotide exchange factors
GGTase	geranylgeranyltransferase
GH	growth hormone
GIT	gastrointestinal tract
GLP	Good Laboratory Practice
GMC	General Medical Council
GMP	Good Manufacturing Practice
GMP	guanosine monophosphate
GnRH	gonadotrophin-releasing hormone
gp	glycoprotein
GRB2	growth factor receptor bound protein 2
gt	genotype
GTP	guanosine triphosphate
h-PEPT	human intestinal proton-dependent oligopeptide transporter
H-receptor	histamine receptor
HA	haemagglutinin
HAART	highly active antiretroviral therapy
HAMA	human anti-mouse antibodies
HBA	hydrogen bond acceptor
HBD	hydrogen bond donor
HCV	hepatitis C virus
HDL	high density lipoprotein
HERG	human ether-a-go-go related gene
HER	human epidermal growth factor receptor
HGFR	hepatocyte growth factor receptor
HIF	hypoxia-inducible factor
HIV	human immunodeficiency virus

HMG-SCoA	3-hydroxy-3-methylglutaryl-coenzyme A
HMGR	3-hydroxy-3-methylglutaryl-coenzyme A reductase
HOMO	highest occupied molecular orbital
HPLC	high-performance liquid chromatography
HPMA	*N*-(2-hydroxypropyl)methacrylamide
HPT	human intestinal di-/tripeptide transporter
HRV	human rhinoviruses
HSV	herpes simplex virus
HTS	high-throughput screening
IC$_{50}$	concentration of drug required to inhibit a target by 50%
ICMT1	isoprenylcysteine carboxylmethyltransferase
If	funny ion channels
IGF-1R	insulin growth factor 1 receptor
Ihh	Indian hedgehog
IND	Investigational exemption to a New Drug application
IP$_3$	inositol triphosphate
IPER	International Preliminary Examination Report
IRB	Institutional Review Board
ISR	International Search Report
ITC	isothermal titration calorimetry
IUPAC	International Union of Pure and Applied Chemistry
IV	intravenous
JAK	Janus kinase
K_D	dissociation binding constant
K_i	inhibition constant
K_M	Michaelis constant
KOR	kappa opioid receptor
LAAM	L-α-acetylmethadol
LD$_{50}$	lethal dose required to kill 50% of a test sample of animals
LDH	lactate dehydrogenase
LDL	low density lipoprotein
LH	luteinizing hormone
LHRH	luteinizing hormone-releasing hormones
LipE	lipophilic efficiency
log *P*	logarithm of the partition coefficient
LDL	low density lipoprotein
LUMO	lowest unoccupied molecular orbital
M-receptor	muscarinic receptor
MAA	Marketing Authorization Application

MAB	monoclonal antibody
MAO	monoamine oxidase
MAOI	monoamine oxidase inhibitor
MAOS	microwave assisted organic synthesis
MAP	mitogen-activated protein
MAPK	mitogen-activated protein kinases
MCHR	melanin-concentrating hormone receptor
MDR	multidrug resistance
MDRTB	multidrug-resistant tuberculosis
MEP	molecular electrostatic potential
miRNA	micro RNA
miRNP	micro RNA protein
MMAE	monomethyl auristatin E (vedotin)
MMP	matrix metalloproteinase
MMPI	matrix metalloproteinase inhibitor
MOR	mu opioid receptor
MR	molar refractivity
mRNA	messenger RNA
MRSA	methicillin-resistant *Staphylococcus aureus*
mRTKI	multi-receptor tyrosine kinase inhibitors
MTP	microsomal triglyceride transfer protein
MTDD	multi-target drug discovery
mTOR	mechanistic or mammalian target of rapamycin
mTORC	mechanistic or mammalian target of rapamycin complex
mTRKI	multi-tyrosine receptor kinase inhibitor
MWt	molecular weight
N-receptor	nicotinic receptor
NA	neuraminidase or noradrenaline
NAD^+/NADH	nicotinamide adenine dinucleotide
$NADP^+$/NADPH	nicotinamide adenine dinucleotide phosphate
NAG	*N*-acetylglucosamine
NAM	*N*-acetylmuramic acid
NCE	new chemical entity
NDA	new drug application
NEP	neutral endopeptidase
NHS	National Health Service
NICE	National Institute for Health and Clinical Excellence
NMDA	*N*-methyl-D-aspartate
NME	new molecular entity
NMR	nuclear magnetic resonance
NNRTI	non-nucleoside reverse transcriptase inhibitor
NO	nitric oxide
NOR	nociceptin opioid receptor
NOS	nitric oxide synthase
NRTI	nucleoside reverse transcriptase inhibitor
NS	non-structural
NSAID	non-steroidal anti-inflammatory drug
NSCLC	non-small-cell lung carcinoma
NVOC	nitroveratryloxycarbonyl
ORL1	opioid receptor-like receptor
P	partition coefficient
P_2Y receptor	purinergic G-protein-coupled receptor
PABA	*p*-aminobenzoic acid
PAR	protease activated receptor
PARP	poly ADP ribose polymerase
PBP	penicillin binding protein
PCP	phencyclidine, otherwise known as 'angel dust'
PCT	patent cooperation treaty
PD-1 receptor	programmed cell death 1 receptor
PDB	protein data bank
PDE	phosphodiesterase
PDGF	platelet-derived growth factor
PDGFR	platelet-derived growth factor receptor
PDK1	phosphoinositide dependent kinase 1
PDT	photodynamic therapy
PEG	polyethylene glycol
PGE	prostaglandin E
PGF	prostaglandin F
PGI_2	prostacyclin
PH	Pleckstrin homology
PI3K	phosphoinositide 3-kinases
PIP_2	phosphatidylinositol diphosphate
PIP_3	phosphatidylinositol (3,4,5)-triphosphate
PI	protease inhibitor
piRNA	piwi-interacting RNA
PKA	protein kinase A
PKB	protein kinase B
PKC	protein kinase C
PLC	phospholipase C
PLS	partial least squares
PPAR	peroxisome proliferator-activated receptor
PPBI	protein–protein binding inhibitor
PPI	proton pump inhibitor

PPts	pyridinium 4-toluenesulphonate		SPA	scintillation proximity assay
PTase	palmitoyl transferase		SPE	solid-phase extraction
PTCH	patched receptor		SPOS	solution phase organic synthesis
QSAR	quantitative structure–activity relationships		SPR	surface plasmon resonance
r	regression or correlation coefficient		ssDNA	single-stranded DNA
R	resonance effect of an aromatic substituent in QSAR		SSRI	selective serotonin reuptake inhibitor
			ssRNA	single-stranded RNA
RAAS	renin–angiotensin–aldosterone system		STAT	signal transducer and activator of transcription
RANK	receptor activator of nuclear factor-kappa B		TB	tuberculosis
RCE1	ras converting enzyme 1		TCA	tricyclic antidepressants
RES	reticuloendothelial system		TFA	trifluoroacetic acid
RET	rearranged during transcription		TGF-α	transforming growth factor α
RFC	reduced folate carrier		TGF-β	transforming growth factor β
RISC	RNA induced silencing complex		THF	tetrahydrofuran
RMSD	root mean square distance		TM	transmembrane
rRNA	ribosomal RNA		TNF	tumour necrosis factor
RNA	ribonucleic acid		TNFR	tumour necrosis factor receptor
RNAi	RNA interference		TNT	trinitrotoluene
s	standard error of estimate or standard deviation		TRAIL	TNF-related apoptosis-inducing ligand
SAR	structure–activity relationships		TRIPS	trade related aspects of intellectual property rights
SCAL	safety-catch acid-labile linker		tRNA	transfer RNA
SCF	stem cell factor		T-VEC	talimogene laherparepvec
SCFR	mast/stem cell growth factor receptor		UTI	urinary tract infection
SCID	severe combined immunodeficiency disease		vdW	van der Waals
sGC	soluble guanylate cyclase		VEGF	vascular endothelial growth factor
SH	src homology		VEGFR	vascular endothelial growth factor receptor
Shh	sonic hedgehog		VIP	vasoactive intestinal peptide
siRNA	small interfering RNA		VOC-Cl	vinyloxycarbonyl chloride
SKF	Smith-Kline and French		VRE	vancomycin-resistant enterococci
Smo	Smoothened receptor		VRSA	vancomycin-resistant Staphylococci aureus
SNRI	selective noradrenaline reuptake inhibitors		VZV	varicella-zoster viruses
siRNA	Small inhibitory RNA		WHO	World Health Organization
snRNA	Small nuclear RNA		WTO	World Trade Organization
SOP	standard operating procedure			
SOS	son of sevenless protein			

1 Drugs and drug targets: an overview

1.1 What is a drug?

Medicinal chemistry involves the design and synthesis of a pharmaceutical agent that has a desired biological effect on the human body or some other living system. Such a compound could also be called a 'drug', but this is a word that many scientists dislike because of the way it is viewed by society. With media headlines such as 'Drugs Menace' or 'Drug Addiction Sweeps City Streets', this is hardly surprising. However, it suggests that a distinction can be drawn between drugs that are used in medicine and drugs that are abused. But is this really true? Can we draw a neat line between 'good drugs' like penicillin and 'bad drugs' like heroin? If so, how do we define what is meant by a good or a bad drug in the first place? Where would we place a so-called social drug like cannabis in this divide? What about nicotine, or alcohol?

The answers we get depend on who we ask. As far as the law is concerned, the dividing line is defined in black and white. As far as the party-going teenager is concerned, the law is an ass. As far as we are concerned, the questions are irrelevant. Trying to divide drugs into two categories—safe or unsafe, good or bad—is futile and could even be dangerous.

First, let us consider the so-called 'good' drugs used in medicines. How 'good' are they? If a drug is to be truly 'good' it would have to do what it is meant to do, have no toxic or unwanted side effects, and be easy to take.

How many drugs fit these criteria?

The short answer is 'none'. There is no pharmaceutical compound on the market today that can completely satisfy all these conditions. Admittedly, some come quite close to the ideal. **Penicillin**, for example, has been one of the safest and most effective antibacterial agents ever discovered. Yet it too has drawbacks. It cannot treat all known bacterial infections, and, as the years have gone by, more and more bacterial strains have become resistant. Moreover, some individuals can experience severe allergic reactions to the compound.

Penicillin is a relatively safe drug, but there are some drugs that are distinctly dangerous. **Morphine** is one such example. It is an excellent analgesic, yet it suffers from serious side effects such as tolerance, respiratory depression, and addiction. It can even kill if taken in excess. **Barbiturates** are also known to be dangerous. At Pearl Harbor, American casualties were given barbiturates as general anaesthetics before surgery. However, a poor understanding of how barbiturates are stored in the body led to many patients receiving a fatal overdose. In fact, it is thought that more casualties died at the hands of the anaesthetists at Pearl Harbor than died of their wounds.

To conclude, the 'good' drugs are not as perfect as one might think.

What about the 'bad' drugs then? Is there anything good that can be said about them? Surely there is nothing we can say in defence of the highly addictive drug **heroin**?

Well, let us look at the facts about heroin. It is one of the best painkillers known to medicine. In fact, it was named heroin at the end of the nineteenth century because it was thought to be the 'heroic' drug that would banish pain for good. Heroin went on the market in 1898, but had to be withdrawn from general distribution 5 years later when its addictive properties became evident. However, heroin is still used in medicine today—under strict control, of course. The drug is called **diamorphine** and it is the drug of choice for treating patients dying of cancer. Not only does diamorphine reduce pain to acceptable levels, it also produces a euphoric effect that helps to counter the depression faced by patients close to death. Can we really condemn such a drug as being all 'bad'?

By now, it should be evident that the division between 'good' and 'bad' drugs is a woolly one and is not really relevant to our discussion of medicinal chemistry. All drugs have their good and bad points. Some have more good points than bad and vice versa, but, like people, they all have their own individual characteristics. So how are we to define a drug in general?

One definition could be to classify drugs as 'compounds which interact with a biological system to produce a biological response'. This definition covers all the drugs we have discussed so far, but it goes further. There are chemicals which we take every day and which have a biological effect on us. What are these everyday drugs?

One is contained in the cups of tea, coffee, and cocoa that we consume. All of these beverages contain the stimulant **caffeine**. Whenever you take a cup of coffee, you are a drug user. We could go further. Whenever you crave a cup of coffee, you are a drug addict. Even children are not immune. They get their caffeine 'shot' from Coke or Pepsi. Whether you like it or not, caffeine is a drug. When you take it, you experience a change of mood or feeling.

So too, if you are a worshipper of the 'nicotine stick'. The biological effect is different. In this case you crave sedation or a calming influence, and it is the **nicotine** in the cigarette smoke which induces that effect. **Alcohol** is another example of a 'social' drug and, as such, causes society more problems than all other drugs put together. One only has to study road accident statistics to appreciate that fact. If alcohol was discovered today, it would probably be restricted in exactly the same way as **cocaine**. Considered in a purely scientific way, alcohol is a most unsatisfactory drug. As many will testify, it is notoriously difficult to judge the correct dose required to gain the beneficial effect of 'happiness' without drifting into the higher dose levels that produce unwanted side effects such as staggering down the street. Alcohol is also unpredictable in its biological effects. Either happiness or depression may result, depending on the user's state of mind. On a more serious note, **addiction** and **tolerance** in certain individuals have ruined the lives of addicts and relatives alike.

Our definition of a drug can also be used to include less obvious compounds; for example poisons and toxins. They too interact with a biological system and produce a biological response—a bit extreme perhaps, but a response all the same. The idea of poisons acting as drugs may not appear so strange if we consider penicillin. We have no problem in thinking of penicillin as a drug, but if we were to look closely at how penicillin works, then it acts as a poison. It interacts with bacteria (the biological system) and kills them (the biological response). Fortunately for us, penicillin has no such effect on human cells.

Even those drugs which do not act as poisons have the potential to become poisons—usually if they are taken in excess. We have already seen this with morphine. At low doses it is a painkiller. At high doses, it is a poison which kills by suppressing breathing. Therefore, it is important that we treat all medicines as potential poisons and treat them with respect.

There is a term used in medicinal chemistry known as the **therapeutic index**, which indicates how safe a particular drug is. The therapeutic index is a measure of the drug's beneficial effects at a low dose, versus its harmful effects at a high dose. To be more precise, the therapeutic index compares the dose level required to produce toxic effects in 50% of patients to the dose level required to produce the maximum therapeutic effects in 50% of patients. A high therapeutic index means that there is a large safety margin between beneficial and toxic doses. The values for cannabis and alcohol are 1000 and 10 respectively, which might imply that cannabis is safer and more predictable than alcohol. Indeed, a cannabis preparation (**nabiximols**) has now been approved to relieve the symptoms of multiple sclerosis. However, this does not suddenly make cannabis safe. For example, the favourable therapeutic index of cannabis does not indicate its potential toxicity if it is taken over a long period of time (chronic use). For example, the various side effects of cannabis include panic attacks, paranoid delusions, and hallucinations. Clearly, the safety of drugs is a complex matter and it is not helped by media sensationalism.

If useful drugs can be poisons at high doses or over long periods of use, does the opposite hold true? Can a poison be a medicine at low doses? In certain cases, this is found to be so.

Arsenic is well known as a poison, but arsenic-derived compounds are used as antiprotozoal and anticancer agents. **Curare** is a deadly poison which was used by the native people of South America to tip their arrows such that a minor arrow wound would be fatal, yet compounds based on the **tubocurarine** structure (the active principle of curare) are used in surgical operations to relax muscles. Under proper control and in the correct dosage, a lethal poison may well have an important medical role. Alternatively, lethal poisons can be the starting point for the development of useful drugs. For example, ACE inhibitors are important cardiovascular drugs that were developed, in part, from the structure of a snake venom.

Since our definition covers any chemical that interacts with any biological system, we can include all the pesticides used in agriculture as drugs. They interact with the biological systems of harmful bacteria, fungi, and insects to produce a toxic effect that protects plants.

Even food can act like a drug. Junk foods and fizzy drinks have been blamed for causing hyperactivity in children. It is believed that junk foods have high concentrations of certain amino acids which can be converted in the body to neurotransmitters—chemicals that pass messages between nerves. In excess, these chemical messengers overstimulate the nervous system, leading to the disruptive behaviour observed in susceptible individuals. Allergies due to food additives and preservatives are also well recorded.

Some foods even contain toxic chemicals. Broccoli, cabbage, and cauliflower all contain high levels of a chemical that can cause reproductive abnormalities in rats. Peanuts and maize sometimes contain fungal toxins, and it is thought that fungal toxins in food were responsible for one of the biblical plagues. Basil contains over 50 compounds that are potentially carcinogenic, and other herbs contain some of the most potent carcinogens known. Carcinogenic compounds have also been identified in radishes, brown mustard, apricots, cherries, and plums. Such unpalatable facts might put you off your dinner, but take comfort—these chemicals are present in such small quantities that the risk is insignificant. Therein lies a great truth, which was recognized as long ago as the fifteenth century when it was stated that 'Everything is a poison, nothing is a poison. It is the dose that makes the poison'.

Almost anything taken in excess will be toxic. You can make yourself seriously ill by taking 100 aspirin tablets or a bottle of whisky or 9 kg of spinach. The choice is yours!

To conclude, drugs can be viewed as actual or potential poisons. An important principle is that of **selective toxicity**. Many drugs are effective because they are toxic to 'problem cells', but not normal cells. For example, antibacterial, antifungal, and antiprotozoal drugs are useful in medicine when they show a selective toxicity to microbial cells, rather than mammalian cells. Clinically effective anticancer agents show a selective toxicity for cancer cells over normal cells. Similarly, effective antiviral agents are toxic to viruses rather than normal cells.

Having discussed what drugs are, we shall now consider why, where, and how they act.

KEY POINTS

- Drugs are compounds that interact with a biological system to produce a biological response.

- No drug is totally safe. Drugs vary in the side effects they might have.

- The dose level of a compound determines whether it will act as a medicine or as a poison.

- The therapeutic index is a measure of a drug's beneficial effect at a low dose versus its harmful effects at a higher dose. A high therapeutic index indicates a large safety margin between beneficial and toxic doses.

- The principle of selective toxicity means that useful drugs show toxicity against foreign or abnormal cells, but not against normal host cells.

1.2 Drug targets

Why should chemicals, some of which have remarkably simple structures, have such an important effect on such

a complicated and large structure as a human being? The answer lies in the way that the human body operates. If we could see inside our bodies to the molecular level, we would see a magnificent array of chemical reactions taking place, keeping the body healthy and functioning.

Drugs may be mere chemicals, but they are entering a world of chemical reactions with which they interact. Therefore, there should be nothing odd in the fact that they can have an effect. The surprising thing might be that they can have such *specific* effects. This is more a result of *where* they act in the body—the drug targets.

1.2.1 Cell structure

Since life is made up of cells, then quite clearly drugs must act on cells. The structure of a typical mammalian cell is shown in Fig. 1.1. All cells in the human body contain a boundary wall called the **cell membrane** which encloses the contents of the cell—the **cytoplasm**. The cell membrane seen under the electron microscope consists of two identifiable layers, each of which is made up of an ordered row of phosphoglyceride molecules such as **phosphatidylcholine (lecithin)** (Fig. 1.2). The outer layer of the membrane is made up of phosphatidylcholine whereas the inner layer is made up of phosphatidylethanolamine, phosphatidylserine, and phosphatidylinositol. Each phosphoglyceride molecule consists of a small polar head-group, and two long hydrophobic (water-hating) chains.

In the cell membrane, the two layers of phospholipids are arranged such that the hydrophobic tails point towards each other and form a fatty, hydrophobic centre, while the ionic head-groups are placed at the inner and outer surfaces of the cell membrane (Fig. 1.3). This is a stable structure because the ionic, hydrophilic head-groups

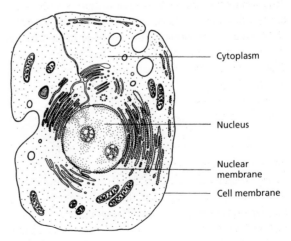

FIGURE 1.1 A typical mammalian cell. Taken from J. Mann, *Murder, magic, and medicine*, Oxford University Press (1992), with permission.

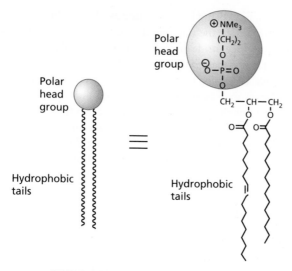

FIGURE 1.2 Phosphoglyceride structure.

interact with the aqueous media inside and outside the cell, whereas the hydrophobic tails maximize hydrophobic interactions with each other and are kept away from the aqueous environments. The overall result of this structure is to construct a fatty barrier between the cell's interior and its surroundings.

The membrane is not just made up of phospholipids, however. There are a large variety of proteins situated in the cell membrane (Fig. 1.3). Some proteins lie attached to the inner or the outer surface of the membrane. Others are embedded in the membrane with part of their structure exposed to one surface or both. The extent to which these proteins are embedded within the cell membrane structure depends on the types of amino acid present. Portions of protein that are embedded in the cell membrane have a large number of hydrophobic amino acids, whereas those portions that stick out from the surface have a large number of hydrophilic amino acids. Many surface proteins also have short chains of carbohydrates attached to them and are thus classed as **glycoproteins**.

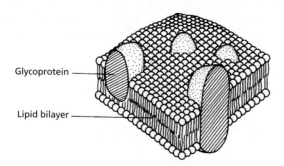

FIGURE 1.3 Cell membrane. Taken from J. Mann, *Murder, magic, and medicine*, Oxford University Press (1992), with permission.

These carbohydrate segments are important to cell–cell recognition (section 10.7).

Within the cytoplasm there are several structures, one of which is the **nucleus**. This acts as the 'control centre' for the cell. The nucleus contains the genetic code—the DNA—which acts as the blueprint for the construction of all the cell's proteins. There are many other structures within a cell, such as the mitochondria, the Golgi apparatus, and the endoplasmic reticulum, but it is not the purpose of this book to look at the structure and function of these organelles. Suffice it to say that different drugs act on molecular targets at different locations in the cell.

1.2.2 **Drug targets at the molecular level**

We shall now move to the molecular level, because it is here that we can truly appreciate how drugs work. The main molecular targets for drugs are proteins (enzymes, receptors, and transport proteins), and nucleic acids (DNA and RNA). These are large molecules (**macromolecules**) having molecular weights measured in the order of several thousand atomic mass units. They are much bigger than a typical drug, which has a molecular weight in the order of a few hundred atomic mass units.

The interaction of a drug with a macromolecular target involves a process known as binding. There is usually a specific area of the macromolecule where this takes place, and this is known as the **binding site** (Fig. 1.4). Typically, this takes the form of a hollow or canyon on the surface of the macromolecule allowing the drug to sink into the body of the larger molecule. Some drugs react with the binding site and become permanently attached via a covalent bond that has a bond strength of 200–400 kJ mol⁻¹. However, most drugs interact through weaker forms of interaction known as **intermolecular bonds**. These include electrostatic or ionic bonds, hydrogen bonds, van der Waals interactions, dipole–dipole interactions, and hydrophobic interactions. (It is also possible for these interactions to take place *within* a molecule, in which case they are called **intramolecular bonds**; see for example protein structure, sections 2.2 and 2.3). None of these bonds is as strong as the covalent bonds that make up the skeleton of a molecule, and so they can be formed, then broken again. This means that an equilibrium takes place between the drug being bound and unbound to its target. The binding forces are strong enough to hold the drug for a certain period of time to let it have an effect on the target, but weak enough to allow it to depart once it has done its job. The length of time the drug remains at its target will depend on the number of intermolecular bonds involved in holding it there. Drugs having a large number of interactions are likely to remain bound longer than those that have only a few. The relative strength of the different intermolecular binding forces

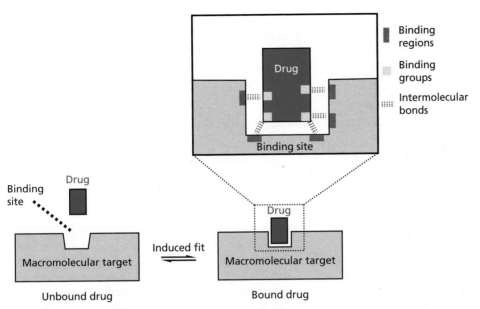

FIGURE 1.4 The equilibrium of a drug being bound and unbound to its target.

is also an important factor. Functional groups present in the drug can be important in forming intermolecular bonds with the target binding site. If they do so, they are called **binding groups**. However, the carbon skeleton of the drug also plays an important role in binding the drug to its target through van der Waals interactions. As far as the target binding site is concerned, it too contains functional groups and carbon skeletons which can form intermolecular bonds with 'visiting' drugs. The specific regions where this takes place are known as **binding regions**. The study of how drugs interact with their targets through binding interactions and produce a pharmacological effect is known as **pharmacodynamics**. Let us now consider the types of intermolecular bond that are possible.

1.3 Intermolecular bonding forces

There are several types of intermolecular bonding interactions, which differ in their bond strengths. The number and types of these interactions depend on the structure

of the drug and the functional groups that are present (section 13.1 and Appendix 7). Thus, each drug may use one or more of the following interactions, but not necessarily all of them.

1.3.1 Electrostatic or ionic bonds

An ionic or electrostatic bond is the strongest of the intermolecular bonds (20–40 kJ mol^{-1}) and takes place between groups having opposite charges such as a carboxylate ion and an aminium ion (Fig. 1.5). The strength of the interaction is inversely proportional to the distance between the two charged atoms, and it is also dependent on the nature of the environment, being stronger in hydrophobic environments than in polar environments. Usually, the binding sites of macromolecules are more hydrophobic in nature than the surface, and so this enhances the effect of an ionic interaction. The drop-off in ionic bonding strength with separation is less than in other intermolecular interactions, so if an ionic interaction is possible, it is likely to be the most important initial interaction as the drug enters the binding site.

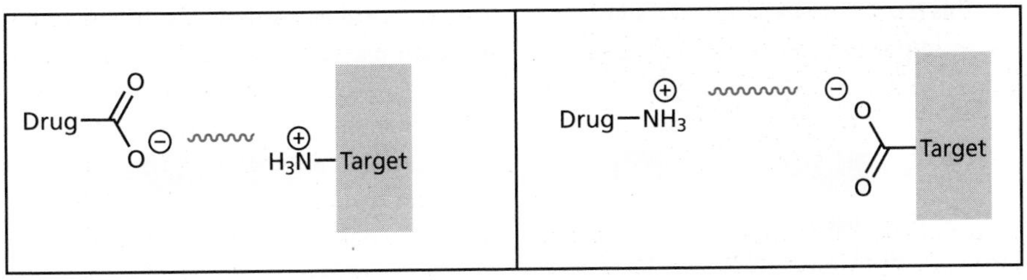

FIGURE 1.5 Electrostatic (ionic) interactions between a drug and the binding site.

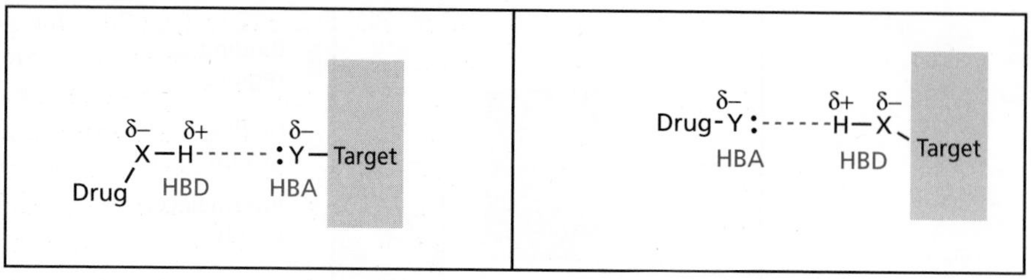

$$\delta- \quad \delta+ \qquad \delta-$$
$$\overset{\text{X}-\text{H}------:\text{Y}-\text{Target}}{\underset{\text{Drug}\quad\text{HBD}\qquad\text{HBA}}{}}$$

$$\delta- \qquad \delta+ \ \delta-$$
$$\overset{\text{Drug-Y}:------\text{H}-\text{X}}{\underset{\text{HBA}\qquad\text{HBD}\quad\text{Target}}{}}$$

FIGURE 1.6 Hydrogen bonding shown by a dashed line between a drug and a binding site (X, Y = oxygen or nitrogen; HBD = hydrogen bond donor, HBA = hydrogen bond acceptor).

1.3.2 Hydrogen bonds

A hydrogen bond can vary substantially in strength, and normally takes place between an electron-rich heteroatom and an electron-deficient hydrogen (Fig. 1.6). The electron-rich heteroatom has to have a lone pair of electrons and is usually oxygen or nitrogen.

The electron-deficient hydrogen is usually linked by a covalent bond to an electronegative atom, such as oxygen or nitrogen. As the electronegative atom (X) has a greater attraction for electrons, the electron distribution in the covalent bond (X–H) is weighted towards the more electronegative atom, and so the hydrogen gains a slight positive charge. Such a hydrogen atom can act as a **hydrogen bond donor (HBD)**. The electron-rich heteroatom that receives the hydrogen bond is known as the **hydrogen bond acceptor (HBA)**. Some functional groups can provide both hydrogen bond donors and hydrogen bond acceptors (e.g. OH, NH_2). When such a group is present in a binding site, it is possible that it might bind to one ligand as a hydrogen bond donor and to another as a hydrogen bond acceptor. This characteristic is given the term **hydrogen bond flip-flop**.

Hydrogen bonds have been viewed as a weak form of electrostatic interaction, because the heteroatom is slightly negative and the hydrogen is slightly positive. However, there is more to hydrogen bonding than an attraction between partial charges. Unlike other intermolecular interactions, an interaction of orbitals takes place between the two molecules (Fig. 1.7). The orbital containing the lone pair of electrons on heteroatom Y interacts with the atomic orbitals normally involved in the covalent bond between X and H. This results in a weak form of sigma (σ) bonding and has an important directional consequence

that is not evident in electrostatic bonds. The optimum orientation is where the X–H bond points directly to the lone pair on Y, such that the angle formed between X, H, and Y is 180°. This is observed in very strong hydrogen bonds. However, the angle can vary between 130° and 180° for moderately strong hydrogen bonds, and can be as low as 90° for weak hydrogen bonds. The lone pair orbital of Y also has a directional property, depending on its hybridization. For example, the nitrogen of a pyridine ring is sp^2 hybridized and so the lone pair points directly away from the ring, and in the same plane (Fig. 1.8). The best location for a hydrogen bond donor would be the region of space indicated in the figure.

The strength of a hydrogen bond can vary widely, but most hydrogen bonds in drug–target interactions are moderate in strength, varying from 16 to 60 kJ mol^{-1}—approximately 10 times less than a covalent bond. The bond distance reflects this, and hydrogen bonds are typically 1.5–2.2 Å compared with 1.0–1.5 Å for a covalent bond. The strength of a hydrogen bond depends on the strengths of the hydrogen bond acceptor and the hydrogen bond donor. A good hydrogen bond acceptor has to be electronegative and have a lone pair of electrons. Nitrogen and oxygen are the most common atoms involved as hydrogen bond acceptors in biological systems. Nitrogen has one lone pair of electrons and can act as an acceptor for one hydrogen bond; oxygen has two lone pairs of electrons and can act as an acceptor for two hydrogen bonds (Fig. 1.9).

Several drugs and macromolecular targets contain a sulphur atom, which is also electronegative. However, sulphur is a weak hydrogen bond acceptor because its lone pairs are in third-shell orbitals, which are larger and more diffuse than second-shell orbitals. This means that

FIGURE 1.7 Orbital overlap in a hydrogen bond.

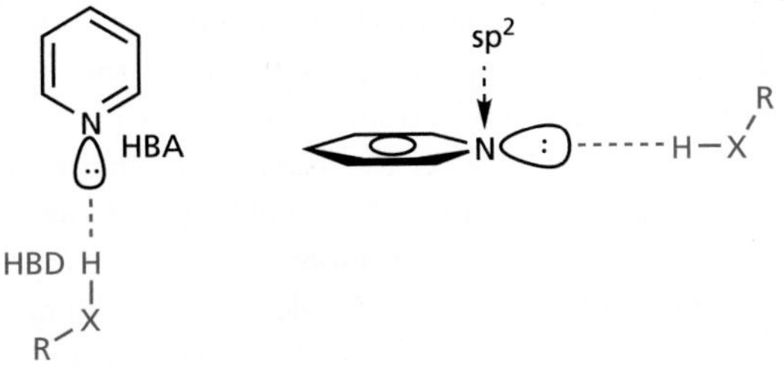

FIGURE 1.8 Directional influence of hybridization on hydrogen bonding.

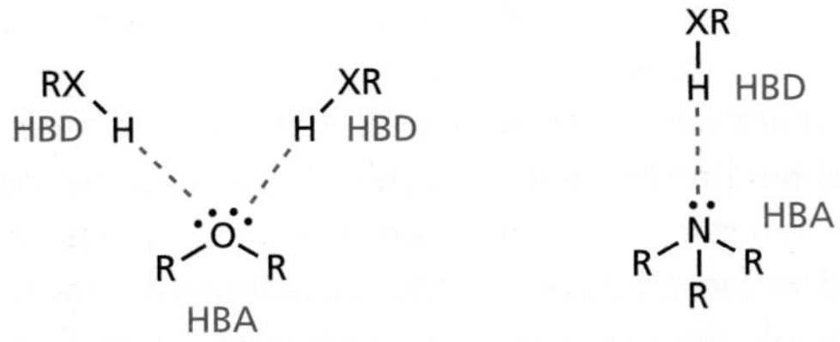

FIGURE 1.9 Oxygen and nitrogen acting as hydrogen bond acceptors (HBD = hydrogen bond donor, HBA = hydrogen bond acceptor).

the orbitals concerned interact less efficiently with the small 1s orbital of a hydrogen atom.

Fluorine, which is present in several drugs, is more electronegative than either oxygen or nitrogen. It also has three lone pairs of electrons, and this might suggest that it would make a good hydrogen bond acceptor. In fact, it is rather a weak hydrogen bond acceptor. It has been suggested that fluorine is so electronegative that it clings on tightly to its

lone pairs of electrons, making them incapable of hydrogen bond interactions. This is in contrast to a fluoride ion which is a very strong hydrogen bond acceptor.

Any feature that affects the electron density of the hydrogen bond acceptor is likely to affect its ability to act as a hydrogen bond acceptor; the greater the electron density of the heteroatom the greater its strength as a hydrogen bond acceptor. For example, the oxygen of a negatively charged carboxylate ion is a stronger hydrogen bond acceptor than the oxygen of the uncharged carboxylic acid (Fig. 1.10). Phosphate ions can also act as good hydrogen bond acceptors. Most hydrogen bond acceptors present in drugs and binding sites are neutral functional groups such as ethers, alcohols, phenols, amides, amines, and ketones. These groups will form moderately strong hydrogen bonds.

It has been proposed that the pi (π) systems present in alkynes and aromatic rings are regions of high electron density and can act as hydrogen bond acceptors. However, the electron density in these systems is diffuse, and so the hydrogen bonding interaction is much weaker than those involving oxygen or nitrogen. As a result, aromatic rings and alkynes are only likely to be significant hydrogen bond acceptors if they interact with a strong hydrogen bond donor such as an alkylammonium ion (NHR_3^+).

More subtle effects can influence whether an atom is a good hydrogen bond acceptor or not. For example, the nitrogen atom of an aliphatic tertiary amine is a better hydrogen bond acceptor than the nitrogen of an amide or an aniline (Fig. 1.11). In the latter functional groups,

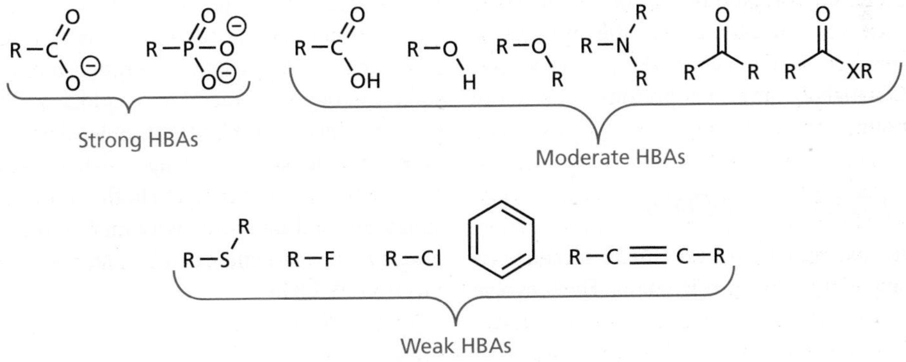

FIGURE 1.10 Relative strengths of hydrogen bond acceptors (HBAs).

Tertiary amine—good HBA   Amide—N acts as poor HBA   Aniline—N acts as poor HBA

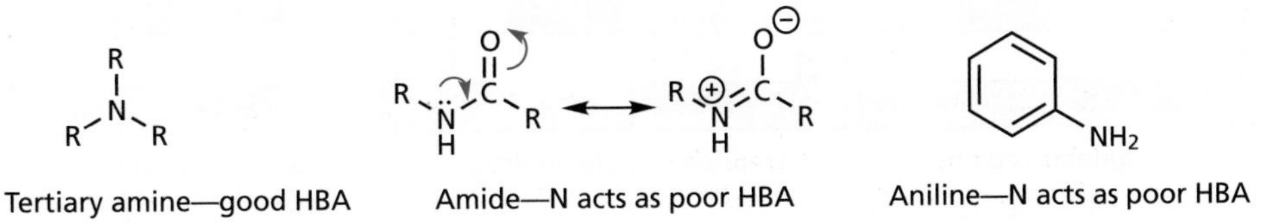

FIGURE 1.11 Comparison of different nitrogen-containing functional groups as hydrogen bond acceptors (HBAs).

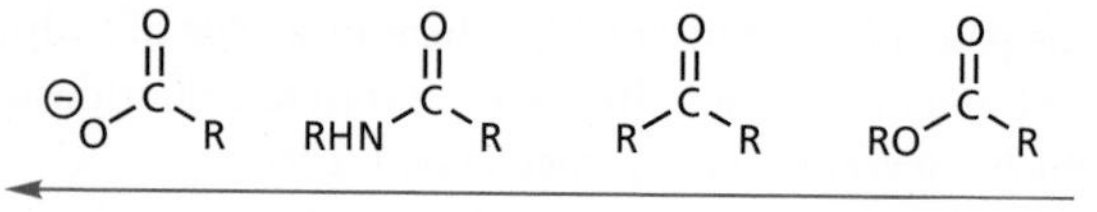

FIGURE 1.12 Comparison of carbonyl oxygens as hydrogen bond acceptors.

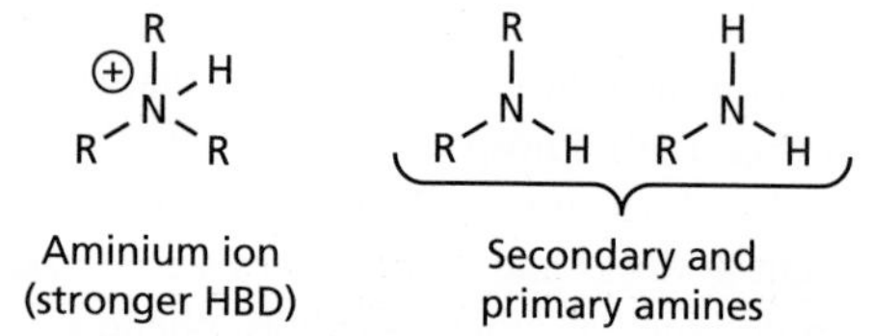

Aminium ion
(stronger HBD)

Secondary and
primary amines

FIGURE 1.13 Comparison of hydrogen bond donors (HBDs).

the lone pair of the nitrogen can interact with neighbouring pi systems to form various resonance structures. As a result, it is less likely to take part in a hydrogen bond.

Similarly, the ability of a carbonyl group to act as a hydrogen bond acceptor varies depending on the functional group involved (Fig. 1.12).

It has also been observed that an sp³ hybridized oxygen atom linked to an sp² carbon atom rarely acts as an HBA. This includes the alkoxy oxygen of esters, and the oxygen atom present in aromatic ethers or furans.

Good hydrogen bond donors contain an electron-deficient proton linked to oxygen or nitrogen. The more electron-deficient the proton, the better it will act as a hydrogen bond donor. For example, a proton attached to a positively charged nitrogen atom acts as a stronger hydrogen bond donor than the proton of a primary or secondary amine (Fig. 1.13). Because the nitrogen is positively charged, it has a greater pull on the electrons surrounding it, making attached protons even more electron-deficient.

1.3.3 Van der Waals interactions

Van der Waals interactions are very weak interactions that are typically 2–4 kJ mol⁻¹ in strength. They involve interactions between hydrophobic regions of different molecules, such as aliphatic substituents or the overall carbon skeleton. The electronic distribution in neutral, non-polar regions is never totally even or symmetrical, and there are always transient areas of high and low electron densities leading to temporary dipoles. The dipoles in one molecule can induce dipoles in a neighbouring molecule, leading to weak interactions between the two molecules (Fig. 1.14). Thus, an area of high electron density on one molecule can have an attraction for an area of low electron density on another molecule. The strength of these interactions falls off rapidly the further the two molecules are apart, decreasing to the seventh power of the separation. Therefore, the drug has to be close to the target binding site before the interactions become important. Van der Waals interactions are also referred to as **London forces**. Although the interactions are individually weak, there may be many such interactions between a drug and its target, and so the overall contribution of van der Waals interactions is often crucial to binding. Hydrophobic forces are also important when the non-polar regions of molecules interact (section 1.3.6).

1.3.4 Dipole–dipole and ion–dipole interactions

Many molecules have a permanent dipole moment resulting from the different electronegativities of the atoms and functional groups present. For example, a ketone has a dipole moment due to the different electronegativities of the carbon and oxygen making up the carbonyl bond. The binding site also contains functional groups, so it is inevitable that it too will have various local dipole moments. It is possible for the dipole moments of the drug and the binding site to interact as a drug approaches, aligning the drug such that the dipole moments are parallel and in opposite directions (Fig. 1.15). If this positions the drug such that other intermolecular interactions can take place between the drug and the binding site, then the alignment is beneficial to both binding and activity. If not, then binding and activity may be weakened. An example of such an effect can be found in anti-ulcer drugs (section 25.2.8.3).

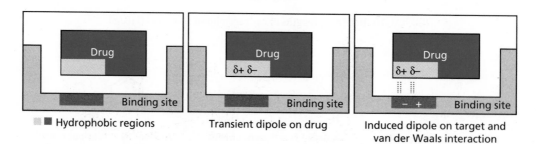

FIGURE 1.14 Van der Waals interactions between hydrophobic regions of a drug and a binding site.

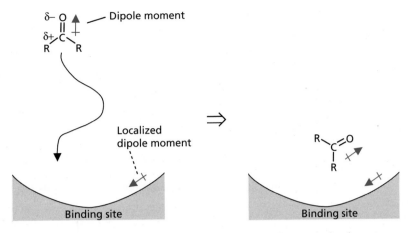

FIGURE 1.15 Dipole–dipole interactions between a drug and a binding site.

The strength of dipole–dipole interactions reduces with the cube of the distance between the two dipoles. This means that dipole–dipole interactions fall away more quickly with distance than electrostatic interactions, but less quickly than van der Waals interactions.

An ion–dipole interaction is where a charged or ionic group in one molecule interacts with a dipole in a second molecule (Fig. 1.16). This is stronger than a dipole–dipole interaction, and falls off less rapidly with separation (decreasing relative to the square of the separation).

Interactions involving an induced dipole moment have been proposed. There is evidence that an aromatic ring can interact with an ionic group such as a quaternary ammonium ion. Such an interaction is feasible if the positive charge of the quaternary ammonium group distorts the π electron cloud of the aromatic ring to produce a dipole moment, where the face of the aromatic ring is electron-rich and the edges are electron-deficient (Fig. 1.17). This is also called a **cation-pi interaction**. An important neurotransmitter called **acetylcholine** forms this type of interaction with its binding site (section 22.5).

1.3.5 Repulsive interactions

So far we have concentrated on attractive forces, which increase in strength the closer the molecules approach each other. Repulsive interactions are also important. Otherwise, there would be nothing to stop molecules

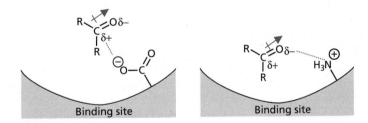

FIGURE 1.16 Ion–dipole interactions between a drug and a binding site.

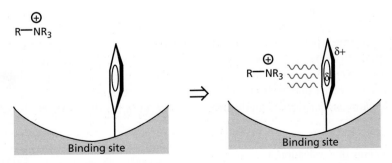

FIGURE 1.17 Induced dipole interaction between an alkylammonium ion and an aromatic ring.

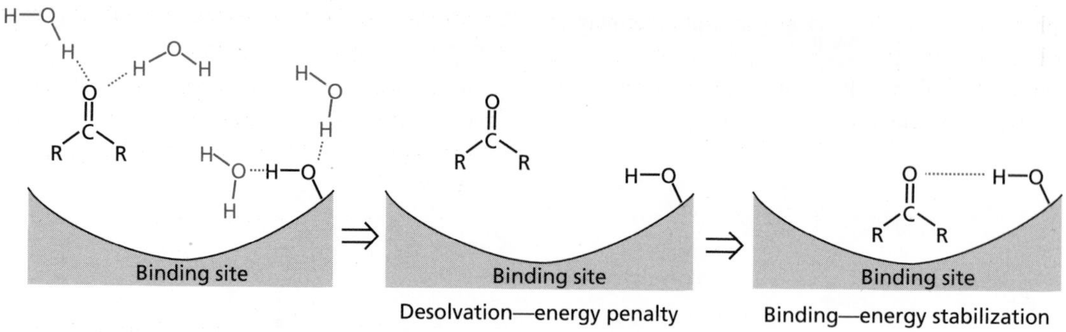

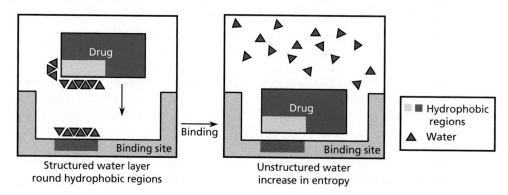

FIGURE 1.18 Desolvation of a drug and its target binding site prior to binding.

Desolvation—energy penalty

Binding—energy stabilization

Structured water layer round hydrophobic regions

Unstructured water increase in entropy

Hydrophobic regions

Water

FIGURE 1.19 Hydrophobic interactions.

trying to merge with each other! If molecules come too close, their molecular orbitals start to overlap and this results in repulsion. Other forms of repulsion are related to the types of groups present in both molecules. For example, two charged groups of identical charge are repelled.

1.3.6 The role of water and hydrophobic interactions

A crucial feature that is often overlooked when considering the interaction of a drug with its target is the role of water. The macromolecular targets in the body exist in an aqueous environment, and the drug has to travel through that environment in order to reach its target. Therefore, both the drug and the macromolecule are solvated with water molecules before they meet each other. The water molecules surrounding the drug and the target binding site have to be stripped away before the interactions described above can take place (Fig. 1.18). This requires energy, and if the energy required to desolvate both the drug and the binding site is greater than the stabilization energy gained by the binding interactions, then the drug may be ineffective. In certain cases, it has even proved beneficial to remove a polar binding group from a drug in order to lower its energy of desolvation. For example, a polar binding group was removed during the development of the antiviral drug **ritonavir** (section 20.7.4.4).

Sometimes polar groups are added to a drug to increase its water solubility. If this is the case, it is important that such groups are positioned in such a way that they protrude from the binding site when the drug binds; in other words they are solvent-accessible or solvent-exposed. In this way, the water that solvates this highly polar group does not have to be stripped away, and there is no energy penalty when the drug binds to its target. Examples of this can be seen in sections 21.6.2.1, 26.9.1.2, and Case study 5.

It is not possible for water to solvate the non-polar or hydrophobic regions of a drug or its target binding site. Instead, the surrounding water molecules form stronger than usual interactions with each other, resulting in an ordered layer of water next to the non-polar surface. This represents a negative entropy due to the increase in order. When the hydrophobic region of a drug interacts with a hydrophobic region of a binding site, these water molecules are freed and become less ordered (Fig. 1.19). This leads to an increase in entropy and a gain in binding energy.[1] The interactions involved are small, at 0.1–0.2 kJ mol^{-1} for each square angstrom of hydrophobic surface, but overall they can be substantial. Sometimes, a hydrophobic region in the drug may not be sufficiently close to a

---

[1]The free energy gained by binding (ΔG) is related to the change in entropy (ΔS) by the equation $\Delta G = \Delta H - T\Delta S$. If entropy increases, ΔS is positive which makes ΔG more negative. The more negative the value of ΔG, the more likely binding will take place.

hydrophobic region in the binding site, and water may be trapped between the two surfaces. The entropy increase is not so substantial in that case, and there is a benefit in designing a better drug that fits more snugly.

1.4 Pharmacokinetic issues and medicines

Pharmacodynamics is the study of how a drug binds to its target binding site and produces a pharmacological effect. However, a drug capable of binding to a particular target is not necessarily going to be useful as a clinical agent or medicine. For that to be the case, the drug not only has to bind to its target, it has to reach it in the first place. For an orally administered drug, that involves a long journey with many hazards to be overcome. The drug has to survive stomach acids, then digestive enzymes in the intestine. It has to be absorbed from the gut into the blood supply, then it has to survive the liver where enzymes try to destroy it (drug metabolism). It has to be distributed round the body and not get mopped up by fat tissue. It should not be excreted too rapidly or else frequent doses will be required to maintain activity. On the other hand, it should not be excreted too slowly or its effects could linger on longer than required. The study of how a drug is absorbed, distributed, metabolized, and excreted (known as ADME in the pharmaceutical industry) is called **pharmacokinetics**. Pharmacokinetics has sometimes been described as 'what the body does to the drug' as opposed to pharmacodynamics—'what the drug does to the body'.

There are many ways in which medicinal chemists can design a drug to improve its pharmacokinetic properties, but the methods by which a drug is formulated and administered are just as important. Medicines are not just composed of the active pharmaceutical agent. For example, a pill contains a whole range of chemicals which are present to give structure and stability to the pill, and also to aid the delivery and breakdown of the pill at the desired part of the gastrointestinal tract.

KEY POINTS

- Drugs act on molecular targets located in the cell membrane of cells or within the cells themselves.
- Drug targets are macromolecules that have a binding site into which the drug fits and binds.
- Most drugs bind to their targets by means of intermolecular bonds.
- Pharmacodynamics is the study of how drugs interact with their targets and produce a pharmacological effect.
- Electrostatic or ionic interactions occur between groups of opposite charge.

- Hydrogen bonds occur between an electron-rich heteroatom and an electron-deficient hydrogen.
- The hydrogen involved in a hydrogen bond is called the hydrogen bond donor. The electronegative atom that interacts with the hydrogen in a hydrogen bond is called the hydrogen bond acceptor.
- Van der Waals interactions take place between non-polar regions of molecules and are caused by transient dipole–dipole interactions.
- Ion–dipole and dipole–dipole interactions are a weak form of electrostatic interaction.
- Hydrophobic interactions involve the displacement of ordered layers of water molecules which surround hydrophobic regions of molecules. The resulting increase in entropy contributes to the overall binding energy.
- Polar groups have to be desolvated before intermolecular interactions take place. This results in an energy penalty.
- The pharmacokinetics of a drug relate to its absorption, distribution, metabolism, and excretion in the body.

1.5 Classification of drugs

There are four main ways in which drugs might be classified or grouped.

By pharmacological effect. Drugs can be classified depending on the biological or pharmacological effect that they have; for example analgesics, antipsychotics, antihypertensives, anti-asthmatics, and antibiotics. This is useful if one wishes to know the full scope of drugs available for a certain ailment, but it means that the drugs included are numerous and highly varied in structure. This is because there are a large variety of targets at which drugs could act in order to produce the desired effect. It is, therefore, not possible to compare different painkillers and expect them to look alike or to have some common mechanism of action.

The chapters on antibacterial, antiviral, anticancer, anti-ulcer, and cardiovascular drugs (Chapters 19, 20, 21, 25, and 26) illustrate the variety of drug structures and mechanisms of action that are possible when drugs are classified according to their pharmacological effect.

By chemical structure. Many drugs which have a common skeleton are grouped together; for example penicillins, barbiturates, opiates, steroids, and catecholamines. In some cases, this is a useful classification since the biological activity and mechanism of action is the same for the structures involved; for example, the antibiotic activity of penicillins. However, not all compounds with similar chemical structure have the same biological action. For example, steroids share a similar tetracyclic structure, but they have very different effects in the body. In this text, various groups of structurally related drugs are discussed; for example, penicillins, cephalosporins, sulphonamides, opioids, and glucocorticoids (sections 19.4–19.5, Chapter 24, and Case study 6). These are examples of compounds with a similar structure and similar mechanism of action. However, there are exceptions. Most sulphonamides are used as antibacterial agents, but there are a few which have totally different medical applications.

By target system. Drugs can be classified according to whether they affect a certain target system in the body. An example of a target system is where a neurotransmitter is synthesized, released from its neuron, interacts with a protein target, and is either metabolized or reabsorbed into the neuron. This classification is a bit more specific than classifying drugs by their overall pharmacological effect. However, there are still several different targets with which drugs could interact in order to interfere with the system, and so the drugs included in this category are likely to be quite varied in structure due to the different mechanisms of action that are involved. In Chapters 22 and 23, we look at drugs that act on target systems—the cholinergic and the adrenergic system respectively.

By target molecule. Some drugs are classified according to the molecular target with which they interact. For example, anticholinesterases (sections 22.12–22.15) are drugs which act by inhibiting the enzyme acetylcholinesterase. This is a more specific classification since we have now identified the precise target at which the drugs act. In this situation, we might expect some structural similarity between the agents involved and a common mechanism of action, although this is not an inviolable assumption. However, it is easy to lose the wood for the trees and to lose sight of why it is useful to have drugs which switch off a particular enzyme or receptor. For example, it is not intuitively obvious why an anticholinesterase agent could be useful in treating Alzheimer's disease or glaucoma.

1.6 Naming of drugs and medicines

The vast majority of chemicals that are synthesized in medicinal chemistry research never make it to the market place and it would be impractical to name them all. Instead, research groups label them with a code which usually consists of letters and numbers. The letters are specific to the research group undertaking the work, and the number is specific for the compound. Thus, Ro31-8959, ABT-538, and MK-639 were compounds prepared by Roche, Abbott, and Merck pharmaceuticals respectively. If the compounds concerned show promise as therapeutic drugs, they are taken forward to pre-clinical trials then clinical studies, by which time they are often named. For example, the above compounds showed promise as anti-HIV drugs and were named **saquinavir**, **ritonavir**, and **indinavir** respectively. Finally, if the drugs prove successful and are marketed as medicines, they are given a proprietary,

brand, or trade name which only the company can use. For example, the above compounds were marketed as **Fortovase®**, **Norvir®**, and **Crixivan®** respectively (note that brand names always start with a capital letter and have the symbol R or TM to indicate that they are registered brand names). The proprietary names are also specific for the preparation or formulation of the drug. For example, Fortovase® (or Fortovase™) is a preparation containing 200 mg of saquinavir in a gel-filled, beige-coloured capsule. If the formulation is changed, then a different name is used. For example, Roche sell a different preparation of saquinavir called **Invirase®** which consists of a brown/green capsule containing 200 mg of saquinavir as the mesylate salt. When a drug's patent has expired, it is possible for any pharmaceutical company to produce and sell that drug as a generic medicine. However, they are not allowed to use the trade name used by the company that originally invented it. European law requires that generic medicines are given a **recommended International Non-proprietary Name** (rINN) which is usually identical to the name of the drug. In Britain, such drugs were given a **British Approved Name** (BAN), but these have now been modified to fall in line with rINNs. rINNs generally have a suffix which indicates the therapeutic area for the named drug. For example, saquinavir, ritonavir, and indinavir all end with the suffix -vir indicating that they are antiviral agents.

Since the naming of drugs is progressive, early research papers in the literature may only use the original letter/number code since the name of the drug had not been allocated at the time of publication.

Throughout this text, the names of the active constituents are used rather than the trade names, although the trade name may be indicated if it is particularly well known. For example, it is indicated that **sildenafil** is **Viagra®** and that **paclitaxel** is **Taxol®**. If you wish to find out the trade name for a particular drug, these are listed in the index. If you wish to 'go the other way', Appendix 6 contains trade names and directs you to the relevant compound name. Only those drugs covered in the text are included and if you cannot find the drug you are looking for, you should refer to other textbooks or formularies such as the British National Formulary (see General further reading).

KEY POINTS

- Drugs can be classified by their pharmacological effect, their chemical structure, their effect on a target system, or their effect on a target structure.

- Clinically useful drugs have a trade (or brand) name, as well as a recommended international non-proprietary name.

- Most structures produced during the development of a new drug are not considered for the clinic. They are identified by simple codes that are specific to each research group.

QUESTIONS

1. The hormone adrenaline interacts with proteins located on the surface of cells and does not cross the cell membrane. However, larger steroid molecules such as estrone cross cell membranes and interact with proteins located in the cell nucleus. Why is a large steroid molecule able to cross the cell membrane when a smaller molecule such as adrenaline cannot?

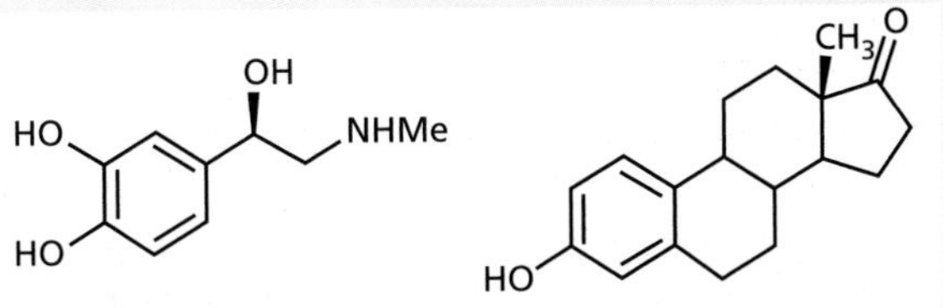

Adrenaline   Estrone

Structure I

2. Valinomycin is an antibiotic which is able to transport ions across cell membranes and disrupt the ionic balance of the cell. Find out the structure of valinomycin and explain why it is able to carry out this task.

3. Archaea are microorganisms which can survive in extreme environments such as high temperature, low pH, or high salt concentration. It is observed that the cell membrane phospholipids in these organisms (see structure I) are markedly different from those in eukaryotic cell membranes. What differences are present and what function might they serve?

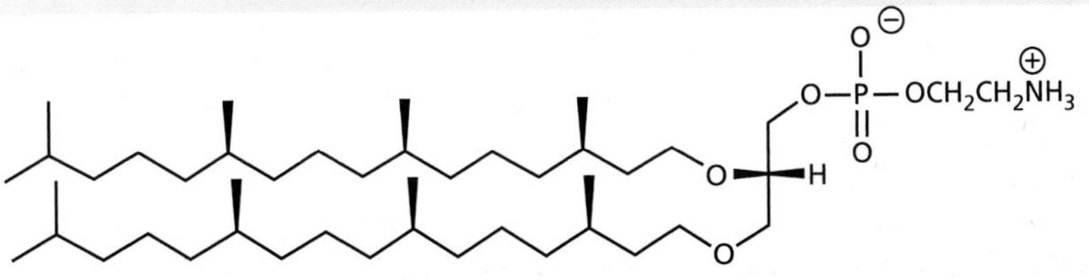

4. Teicoplanin is an antibiotic which 'caps' the building blocks used in the construction of the bacterial cell wall, such that they cannot be linked up. The cell wall is a barrier surrounding the bacterial cell membrane, and the building blocks are anchored to the outside of this cell membrane prior to their incorporation into the cell wall. Teicoplanin contains a very long alkyl substituent which plays no role in the capping mechanism. However, if this substituent is absent, activity drops. What role do you think this alkyl substituent might serve?

5. The Ras protein is an important protein in signalling processes within the cell. It exists freely in the cell cytoplasm, but must become anchored to the inner surface of the cell membrane in order to carry out its function. What kind of modification to the protein might take place to allow this to happen?

6. Cholesterol is an important constituent of eukaryotic cell membranes and affects the fluidity of the membrane. Consider the structure of cholesterol (shown below) and suggest how it might be orientated in the membrane.

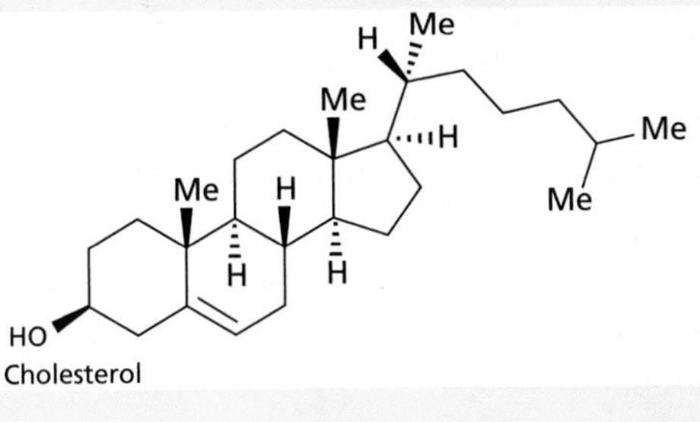

7. Most unsaturated alkyl chains in phospholipids are *cis* rather than *trans*. Consider the *cis*-unsaturated alkyl chain in the phospholipid shown in Fig. 1.2. Redraw this chain to give a better representation of its shape and compare it with the shape of its *trans*-isomer. What conclusions can you make regarding the packing of such chains in the cell membrane, and the effect on membrane fluidity?

8. The relative strength of carbonyl oxygens as hydrogen bond acceptors is shown in Fig. 1.12. Suggest why the order is as shown.

9. Consider the structures of adrenaline, estrone, and cholesterol and suggest what kind of intermolecular interactions are possible for these molecules and where they occur.

10. Using the index and Appendix 8 (on the website), identify the structures and trade names for the following drugs—amoxicillin, ranitidine, gefitinib, atracurium.

Multiple-choice questions are available on the Online Resource Centre at www.oxfordtextbooks.co.uk/orc/patrick6e/

Cholesterol

FURTHER READING

Kubinyi, H. (2001) Hydrogen bonding: The last mystery in drug design? in Testa, B., van de Waterbeemd, H., Folkers, G., and Guy, R. (eds), *Pharmacokinetic optimization in drug research*. Wiley-VCH, Weinheim.

Mann, J. (1992) *Murder, magic, and medicine*, Chapter 1. Oxford University Press, Oxford.

Page, C., Curtis, M., Sutter, M., Walker, M., and Hoffman, B. (2002) Drug names and drug classification systems. in *Integrated pharmacology 2nd edn*, Chapter 2. Mosby, Elsevier, Maryland Heights, MO.

Sneader, W. (2005) *Drug discovery: a history.* John Wiley and Sons, Chichester.

WEBSITES

International non-proprietary names, World Health Organization. www.who.int/medicines/services/inn/en/

Brand names of some commonly used drugs. www.mwrckmanuals.com/professional/appendices/brand-names-of-some-commonly-used-drugs?starting with=a

Titles for general further reading are listed on p.845.

PART A

Drug targets

Medicinal chemistry is the study of how novel drugs can be designed and developed. This process is helped immeasurably by a detailed understanding of the structure and function of the molecular targets that are present in the body.

The major drug targets are normally large molecules (macromolecules), such as proteins and nucleic acids. Knowing the structures, properties, and functions of these macromolecules is crucial if we are to design new drugs. There are a variety of reasons for this.

Firstly, it is important to know what functions different macromolecules have in the body and whether targeting them is likely to have a beneficial effect in treating a particular disease. There is no point designing a drug to inhibit a digestive enzyme if one is looking for a new analgesic.

Secondly, a knowledge of macromolecular structure is crucial if one is to design a drug that will bind effectively to the target. Knowing the target structure and its functional groups will allow the medicinal chemist to design a drug that contains complementary functional groups that will bind the drug to the target.

Thirdly, a drug must not only bind to the target, it must bind to the correct region of the target. Proteins and nucleic acids are extremely large molecules in comparison to a drug and if the drug binds to the wrong part of the macromolecule, it may not have any effect. An appreciation of the target's structure and function will guide the medicinal chemist in this respect.

Finally, an understanding of how a macromolecule operates is crucial if one is going to design an effective drug that will interfere with that process. For example, understanding the mechanism of how enzymes catalyse reactions has been extremely important in the design of many important drugs, for example the protease inhibitors used in HIV therapy (section 20.7).

Proteins are the most important drug targets used in medicinal chemistry and so it should be no surprise that the major focus in Part A (Chapters 2–5) is devoted to them. However, there are some important drugs which interact with nucleic acids. The structure and function of these macromolecules are covered in Chapter 6.

If you have a background in biochemistry, much of the material in this section may already be familiar to you, and you may wish to move directly to Part B. Alternatively, you may find the material in Part A useful revision.

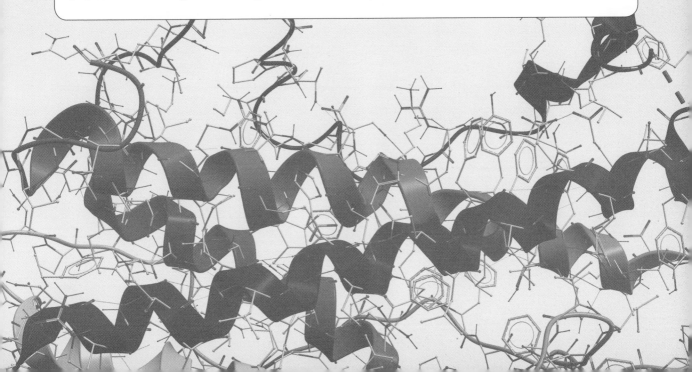

2 Protein structure and function

The vast majority of drugs used in medicine are targeted on proteins such as receptors, enzymes, and transport proteins. Therefore, it is important to understand protein structure in order to understand drug action on proteins. Proteins have four levels of structure—primary, secondary, tertiary, and quaternary.

2.1 The primary structure of proteins

The primary structure is the order in which the individual amino acids making up the protein are linked together through peptide bonds (Fig. 2.1). The 20 common amino acids found in humans are listed in Table 2.1, with the three-letter and one-letter codes often used to represent

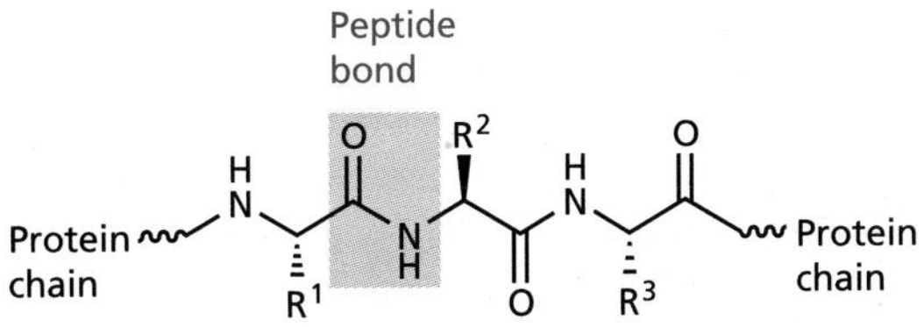

FIGURE 2.1 Primary structure of proteins (R^1, R^2, and R^3 = amino acid side chains).

them. The structures of the amino acids are shown in Appendix 1. The primary structure of **Met-enkephalin** (one of the body's own painkillers) is shown in Fig. 2.2.

The peptide bond in proteins is planar in nature as a result of the resonance structure shown in Fig. 2.3. This gives the peptide bond a significant double bond character which prevents rotation. As a result, bond rotation in the protein backbone is only possible for the bonds on

TABLE 2.1 The 20 common amino acids found in humans.

Synthesized in the human body			Essential to the diet		
Amino acid	Codes		Amino acid	Codes	
	3-letter	1-letter		3-letter	1-letter
Alanine	Ala	A	Histidine	His	H
Arginine	Arg	R	Isoleucine	Ile	I
Asparagine	Asn	N	Leucine	Leu	L
Aspartic acid	Asp	D	Lysine	Lys	K
Cysteine	Cys	C	Methionine	Met	M
Glutamic acid	Glu	E	Phenylalanine	Phe	F
Glutamine	Gln	Q	Threonine	Thr	T
Glycine	Gly	G	Tryptophan	Trp	W
Proline	Pro	P	Valine	Val	V
Serine	Ser	S			
Tyrosine	Tyr	Y			

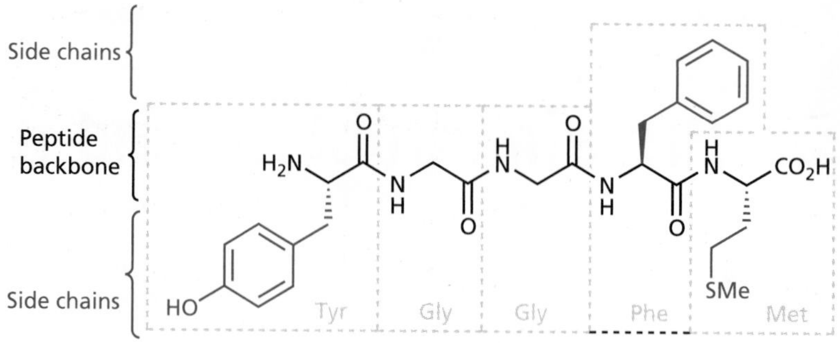

FIGURE 2.2 Met-enkephalin. The short hand notation for this peptide is H-Tyr-Gly-Gly-Phe-Met-OH or YGGFM.

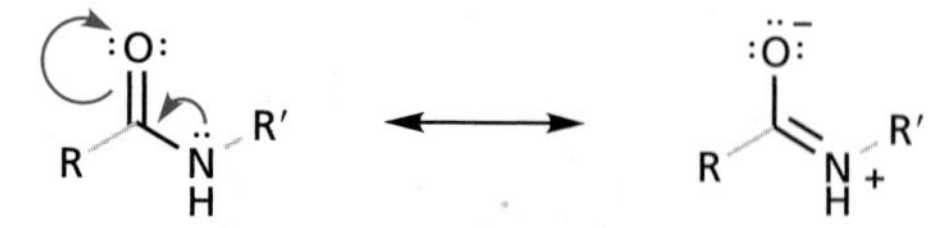

FIGURE 2.3 The planar peptide bond (free bond rotation allowed for coloured bonds only).

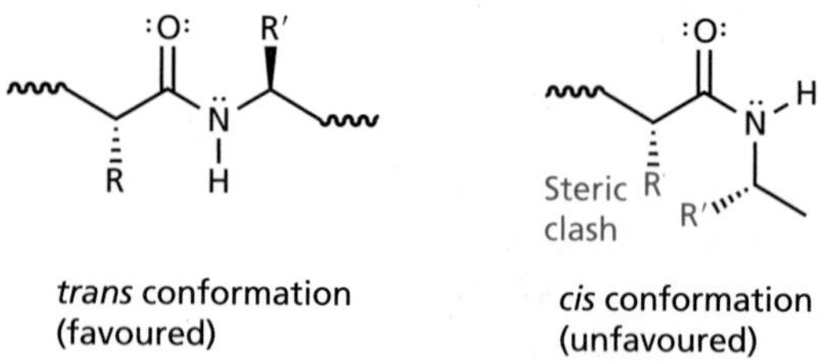

trans conformation (favoured)

cis conformation (unfavoured)

FIGURE 2.4 *Trans* and *cis* conformations of the peptide bond.

either side of each peptide bond. This has an important consequence for protein tertiary structure (section 2.3.6).

There are two possible conformations for the peptide bond (Fig. 2.4). The *trans* conformation is the one that is normally present in proteins, because the *cis* conformation leads to a steric clash between the residues. However, the *cis* conformation is possible for peptide bonds next to a proline residue.

2.2 The secondary structure of proteins

The secondary structure of proteins consists of regions of ordered structure adopted by the protein chain. In structural proteins such as wool and silk, secondary structures are extensive and determine the overall shape and properties of such proteins. However, there are also regions of secondary structure in most other proteins. There are three main secondary structures—the α-helix, β-pleated sheet, and β-turn.

2.2.1 The α-helix

The α-helix results from coiling of the protein chain such that the peptide bonds making up the backbone are able to form hydrogen bonds between each other. These hydrogen bonds are directed along the axis of the helix, as shown in Fig. 2.5. The side chains of the component amino acids stick out at right angles from the helix, thus minimizing steric interactions and further stabilizing the structure. Other less common types of helices can occur in proteins, such as the 3(10)-helix which is more stretched than the ideal α-helix, and the π-helix which is more compact and extremely rare.

Ⓦ Test your understanding and practise your molecular modelling with Exercise 2.1 on the Online Resource Centre: at www.oxfordtextbooks.co.uk/orc/patrick6e/

2.2.2 The β-pleated sheet

The β-pleated sheet is a layering of protein chains one on top of another, as shown in Fig. 2.6. Here too, the structure is held together by hydrogen bonds between the peptide chains. The side chains are situated at right angles to the sheets, once again to reduce steric interactions. The chains in β-sheets can run in opposite directions (antiparallel) or in the same direction (parallel) (Fig. 2.7).

2.2.3 The β-turn

A β-turn allows the polypeptide chain to turn abruptly and go in the opposite direction. This is important in allowing the protein to adopt a more globular compact shape. A hydrogen bonding interaction between the first and third peptide bond of the turn is important in stabilizing the turn (Fig. 2.8). Less abrupt changes in the direction of the polypeptide chain can also take place through longer loops, which are less regular in their structure, but are often rigid and well defined.

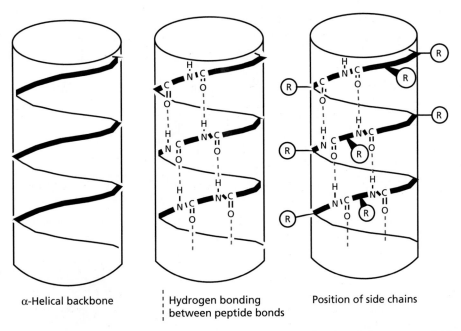

FIGURE 2.5 The α-helix for proteins showing intramolecular hydrogen bonds and the position of side chains.

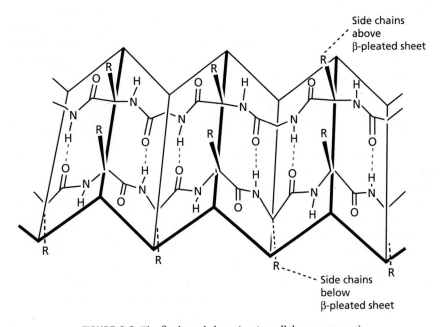

FIGURE 2.6 The β-pleated sheet (antiparallel arrangement).

2.3 The tertiary structure of proteins

The tertiary structure is the overall three-dimensional shape of a protein. Structural proteins are quite ordered in shape, whereas globular proteins, such as enzymes and receptors (Chapters 3 and 4), fold up to form more complex structures. The tertiary structure of enzymes and receptors is crucial to their function and also to their interaction with drugs; therefore it is important to appreciate the forces that control tertiary structure.

Globular proteins often contain regions of ordered secondary structure, the extent of which varies from protein to protein. For example, **cyclin-dependent kinase 2** (a protein that catalyses phosphorylation reactions) has several regions of α-helices and β-pleated sheets (Fig. 2.9), whereas the digestive enzyme **chymotrypsin**

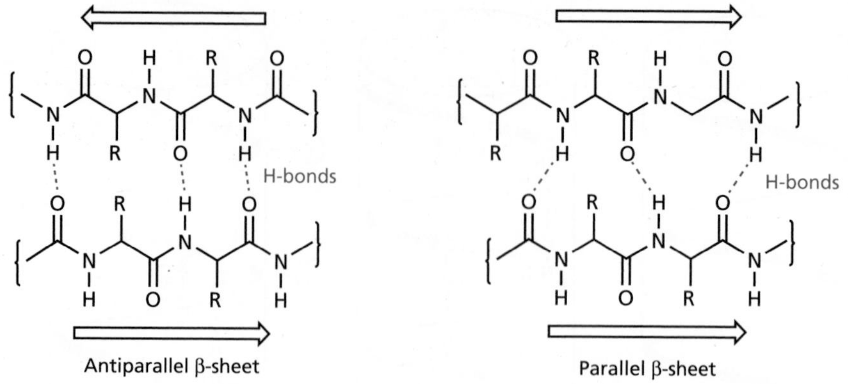

FIGURE 2.7 Hydrogen bonding in antiparallel and parallel β-sheets (the arrows are pointing to the *C*-terminal end of the chain).

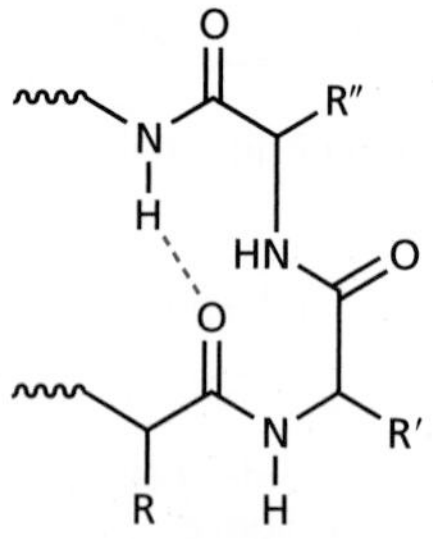

FIGURE 2.8 The β-turn showing hydrogen bonding between the first and third peptide bond.

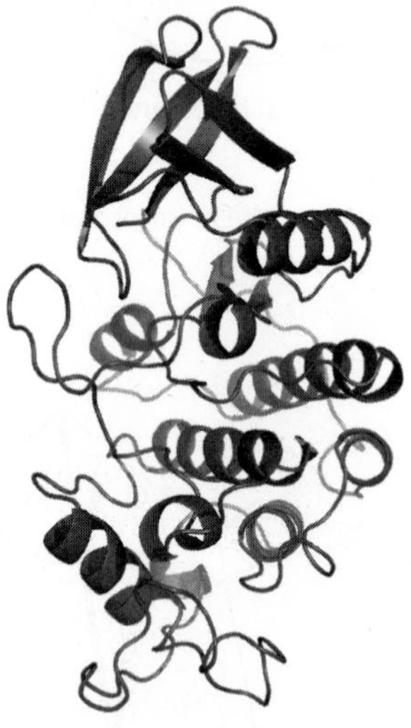

has very little secondary structure. Nevertheless, the protein chains in both cyclin-dependent kinase 2 and chymotrypsin fold up to form a complex, but distinctive, globular shape. How does this come about?

At first sight, the three-dimensional structure of cyclin-dependent kinase 2 looks like a ball of string after the cat has been at it. In fact, the structure shown is a very precise shape which is taken up by every molecule of this protein, and which is determined by the protein's primary structure[1]. Indeed, in the laboratory, it is possible to synthesize proteins which automatically adopt the same three-dimensional structure and function as the naturally occurring protein. The HIV-1 protease enzyme is an example (section 20.7.4.1).

This poses a problem. Why should a chain of amino acids take up such a precise three-dimensional shape? At first sight, it does not make sense. If we place a length of string on the table, it does not fold itself up into a precise complex shape. So why should a chain of amino acids do such a thing?

The answer lies in the fact that a protein is not just a bland piece of string. That is because it contains a large

FIGURE 2.9 The pdb file (1hcl) for human cyclin-dependent kinase 2 (CDK2) where cylinders represent α-helices and arrows represent β-sheets. A pdb file contains the 3D structural information for a protein and can be downloaded from the Brookhaven protein data bank. Each protein structure file is given a code, for example, 1hcl.

number of different chemical functional groups, which include the peptide bonds of the polypeptide backbone, as well as a variety of functional groups in the amino acid side chains. These can interact with each other, such that there is either an attractive or a repulsive interaction. Thus, the protein will twist and turn to minimize the unfavourable interactions and maximize the favourable ones until the most stable shape or conformation is found—the tertiary structure (Fig. 2.10).

With the exception of disulphide bonds, the attractive interactions involved in tertiary structure are the same as the **intermolecular bonds** described in section 1.3. The latter occur between different molecules, whereas the bonds controlling protein tertiary structure occur within

---

[1] Some proteins contain species known as **cofactors** (e.g. metal ions or small organic molecules) which also have an effect on tertiary structure.

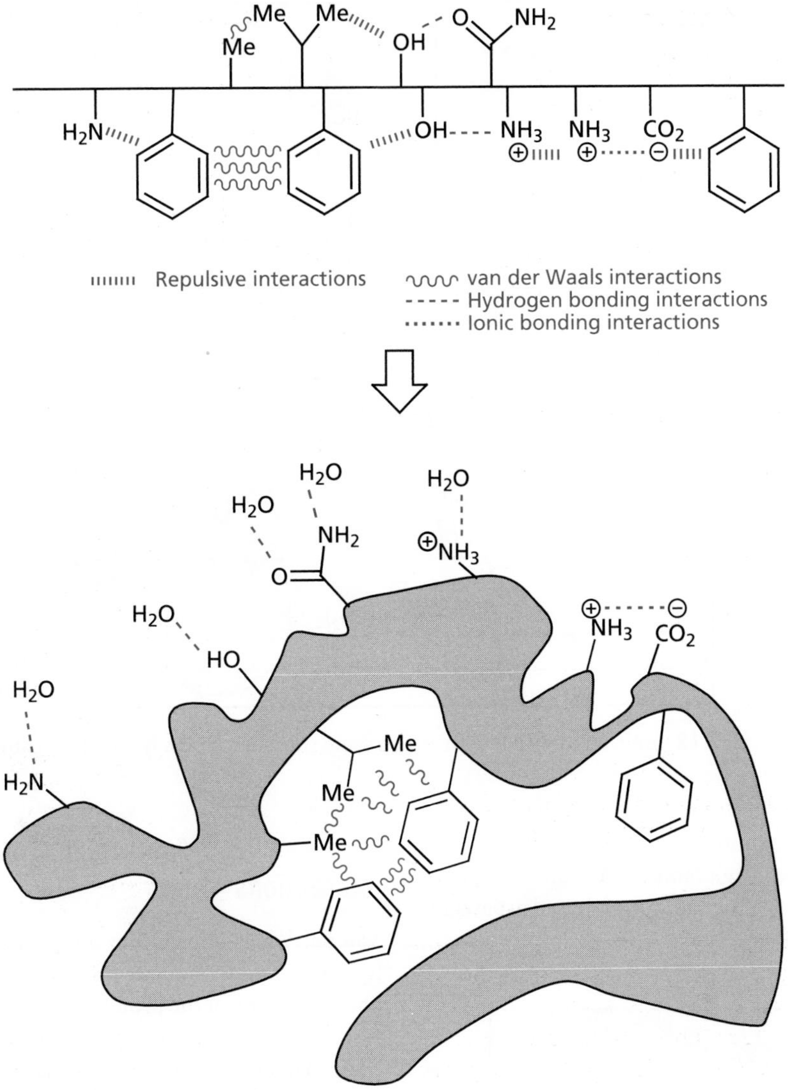

FIGURE 2.10 Tertiary structure formation as a result of intramolecular interactions.

the same molecule, and so they are called **intramolecular bonds**. Nevertheless, the principles described in section 1.3 are the same.

Test your understanding and practise your molecular modelling with Exercise 2.2 on the Online Resource Centre: at www.oxfordtextbooks.co.uk/orc/patrick6e/

2.3.1 Covalent bonds: disulphide links

Cysteine has a residue containing a thiol group capable of forming a covalent bond in protein tertiary structure. When two such residues are close together, a covalent disulphide bond can be formed as a result of oxidation. A covalent bridge is thus formed between two different parts of the protein chain (Fig. 2.11). It should be noted that the two cysteine residues involved in this bond

formation may be far apart from each other in the primary structure of the protein, but are brought close together as a result of protein folding.

2.3.2 Ionic or electrostatic bonds

An ionic bond or salt bridge can be formed between the carboxylate ion of an acidic residue such as aspartic acid or glutamic acid, and the aminium ion of a basic residue such as lysine, arginine, or histidine (Fig. 2.12). This is the strongest of the intramolecular bonds.

2.3.3 Hydrogen bonds

Hydrogen bonds can be viewed as a weak form of ionic interaction as they involve interactions between atoms having partial charges. They can be formed between

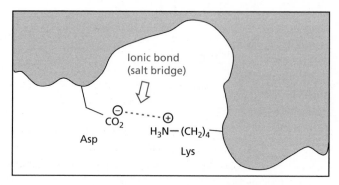

FIGURE 2.11 The formation of a disulphide covalent bond between two cysteine side chains.

FIGURE 2.12 Ionic bonding between an aspartate side chain and a lysine side chain.

a large number of amino acid residues such as serine, threonine, aspartic acid, glutamic acid, glutamine, lysine, arginine, histidine, tryptophan, tyrosine, and asparagine. Two examples are shown in Fig. 2.13.

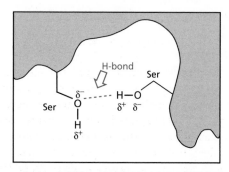

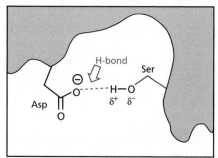

FIGURE 2.13 Hydrogen bonding between amino acid side chains.

2.3.4 Van der Waals and hydrophobic interactions

Van der Waals interactions are weaker interactions than hydrogen bonds, and can take place between two hydrophobic regions of the protein. For example, they can take place between two alkyl groups (Fig. 2.14). The amino acids alanine, valine, leucine, isoleucine, phenylalanine, and proline all have hydrophobic side chains capable of interacting with each other by van der Waals interactions. The side chains of other amino acids such as methionine, tryptophan, threonine, and tyrosine contain polar functional groups, but the side chains also have a substantial

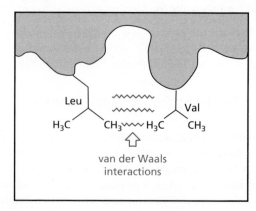

FIGURE 2.14 Van der Waals interactions between amino acid side chains.

hydrophobic character and so van der Waals interactions are possible for these amino acids as well. Hydrophobic interactions (section 1.3.6) are also important in the coming together of hydrophobic residues.

2.3.5 Relative importance of bonding interactions

We might expect the relative importance of the bonding interactions in protein tertiary structure to follow the same order as their strengths: covalent, ionic, hydrogen bonding, and finally van der Waals. In fact, the opposite is generally true. Usually the most important bonding interactions are those due to van der Waals interactions and hydrogen bonding, while the least important interactions are those due to covalent and ionic bonding.

There are two reasons for this. Firstly, in most proteins there are more possible opportunities for van der Waals and hydrogen bonding interactions than for covalent or ionic bonding. We only need to consider the relative number of amino acids present in a typical globular protein to see why. The only amino acid that can form a covalent disulphide bond is cysteine, whereas there are many more amino acids that can interact with each other through hydrogen bonding and van der Waals interactions.

Having said that, there *are* examples of proteins with a large number of disulphide bridges, where the relative importance of the covalent link to tertiary structure is more significant. Disulphide links are also more significant in small polypeptides such as the peptide hormones **vasopressin** and **oxytocin** (Fig. 2.15). Nevertheless, in most proteins, disulphide links play a minor role in controlling tertiary structure.

As far as ionic bonding is concerned, there is only a limited number of amino acids with residues capable of forming ionic bonds, and so these, too, are outnumbered by the number of residues capable of forming hydrogen bonds or van der Waals interactions.

There is a second reason why van der Waals interactions are normally the most important form of bonding in tertiary structure. Proteins do not exist in a vacuum; they are surrounded by water. Water is a highly polar compound that interacts readily with polar, hydrophilic

amino acid residues capable of forming hydrogen bonds (Fig. 2.16). The remaining non-polar, hydrophobic amino acid residues cannot interact favourably with water, so the most stable tertiary structure will ensure that most of the hydrophilic groups are on the surface where they can interact with water, and most of the hydrophobic groups are in the centre where they avoid water and interact with each other. Since the hydrophilic amino acids form hydrogen bonds with water, the number of ionic and hydrogen bonds contributing to the tertiary structure is reduced and this leaves hydrophobic and van der Waals interactions to largely determine the three-dimensional shape of the protein.

For the reasons stated above, the centre of the protein must be hydrophobic and non-polar. This has important consequences. For example, it helps to explain why enzymes catalyse reactions that should be impossible in the aqueous environment of the human body. Enzymes contain a hollow or canyon on their surface called an **active site**. As the active site protrudes into the centre of the protein, it tends to be hydrophobic in nature and can provide a non-aqueous environment for the reaction taking place (Chapter 3).

Many other types of protein contain similar hollows or clefts that act as **binding sites** for natural ligands. They, too, are more hydrophobic than the surface, and so van der Waals and hydrophobic interactions play an important role in the binding of the ligand. An understanding of these interactions is crucial to the design of effective drugs that will target these binding sites.

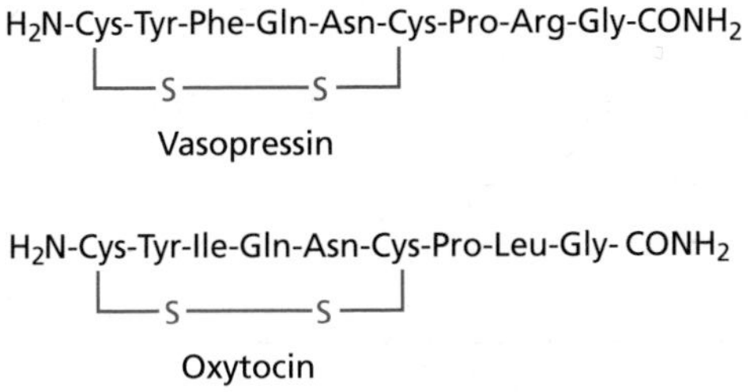

2.3.6 Role of the planar peptide bond

Planar peptide bonds indirectly play an important role in tertiary structure. Bond rotation in peptide bonds is hindered, with the *trans* conformation generally favoured, and so the number of possible conformations that a protein can adopt is significantly restricted, making it more likely that a specific conformation is adopted. Polymers without peptide bonds do not fold into a specific conformation, because the entropy change required to form a highly ordered structure is extremely unfavourable. Peptide bonds can also form hydrogen bonds with amino acid side chains and play a further role in determining tertiary structure.

2.4 The quaternary structure of proteins

Only proteins that are made up of a number of protein subunits have quaternary structure. For example, **haemoglobin** is made up of four protein molecules—two identical alpha subunits and two identical beta subunits

H$_2$N-Cys-Tyr-Phe-Gln-Asn-Cys-Pro-Arg-Gly-CONH$_2$

Vasopressin

H$_2$N-Cys-Tyr-Ile-Gln-Asn-Cys-Pro-Leu-Gly-CONH$_2$

Oxytocin

FIGURE 2.15 Vasopressin and oxytocin.

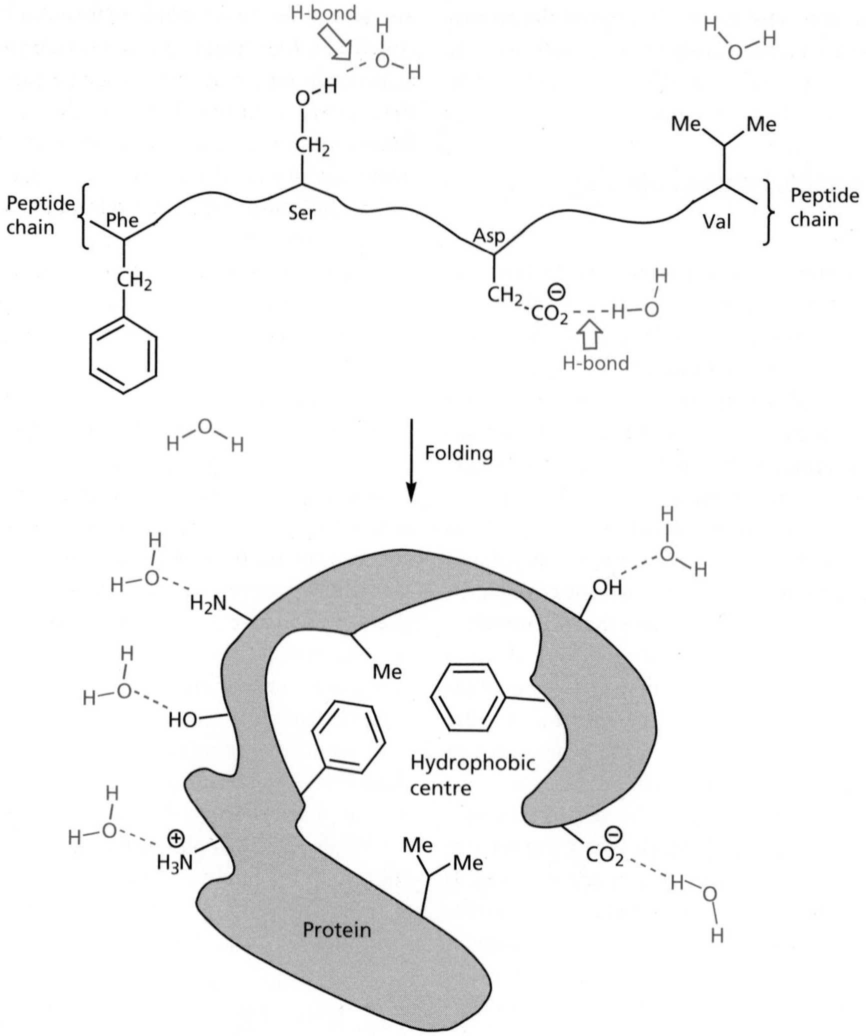

FIGURE 2.16 Bonding interactions with water.

(not to be confused with the alpha and beta terminology used in secondary structure). The quaternary structure of haemoglobin is the way in which these four protein units associate with each other.

Since this must inevitably involve interactions between the exterior surfaces of proteins, ionic bonding can be more important to quaternary structure than it is to tertiary structure. Nevertheless, hydrophobic and van

der Waals interactions have a role to play. It is not possible for a protein to fold up such that all its hydrophobic groups are placed towards the centre. Some of these groups may be stranded on the surface. If they form a small hydrophobic area on the protein surface, there is a distinct advantage for two protein molecules to form a dimer such that the two hydrophobic areas face each other rather than be exposed to an aqueous environment

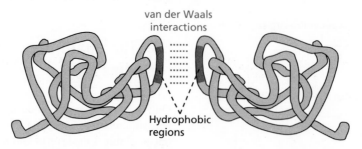

FIGURE 2.17 Quaternary structure involving two protein subunits.

(Fig. 2.17). It is also possible for protein molecules to interlock in a quaternary structure (section 24.9.2).

2.5 Translation and post-translational modifications

The process by which a protein is synthesized in the cell is called **translation** (section 6.2.2). Many proteins are modified following translation (Fig. 2.18), and these modifications can have wide-ranging effects. For example, the *N*-terminals of many proteins are acetylated, making these proteins more resistant to degradation. Acetylation of proteins also has a role to play in the control of transcription, cell proliferation, and differentiation (section 21.7.3).

The fibres of **collagen** are stabilized by the hydroxylation of proline residues, and insufficient hydroxylation results in scurvy (caused by a deficiency of vitamin C).

The glutamate residues of **prothrombin**, a clotting protein, are carboxylated to form γ-carboxyglutamate structures. In cases of vitamin K deficiency, carboxylation does not occur and excessive bleeding results. The serine, threonine, and tyrosine residues of many proteins are phosphorylated and this plays an important role in signalling pathways within the cell (sections 5.2–5.4).

Many of the proteins present on the surface of cells are linked to carbohydrates through asparagine residues. Such carbohydrates are added as post-translational modifications and are important to cell–cell recognition, disease processes, and drug treatments (section 10.7). The proteins concerned are called **glycoproteins** or **proteoglycans**, and are members of a larger group of molecules called **glycoconjugates**.

Several proteins are cleaved into smaller proteins or peptides following translation. For example, the **enkephalins** are small peptides which are derived from proteins in this manner (section 24.8). Active enzymes are sometimes formed by cleaving a larger protein precursor.

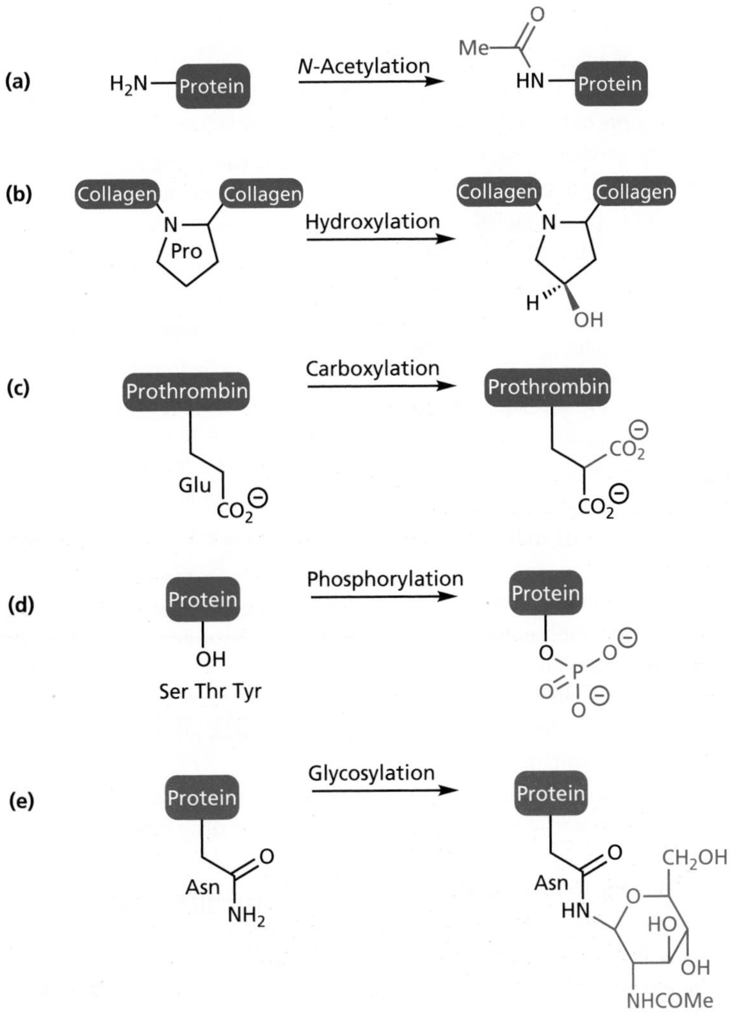

FIGURE 2.18 Examples of post-translational modifications carried out on proteins.

Often this serves to protect the cell from the indiscriminate action of an enzyme. For example, digestive enzymes are stored in the pancreas as inactive protein precursors and are only produced once the protein precursor is released into the intestine. In blood clotting, the soluble protein **fibrinogen** is cleaved to insoluble **fibrin** when the latter is required (section 26.9). Some polypeptide hormones are also produced from the cleavage of protein precursors. Finally, the cleavage of a viral polyprotein into constituent proteins is an important step in the life cycle of retroviruses, and has proved a useful target for several drugs currently used to combat AIDS and hepatitis C (sections 20.7.4 and 20.10.1).

2.6 Proteomics

A lot of publicity has been rightly accorded to the Human Genome Project, which has now been completed. The science behind this work is called **genomics** and involves the identification of the genetic code in humans and other species. The success of this work was a major breakthrough that heralded the start of a new era in medicinal chemistry research. However, it is important to appreciate that identifying the human genome only marked the start of a more prolonged process. As we shall see in Chapter 6, DNA is the blueprint for the synthesis of proteins, and so a lot of research has now gone into identifying the proteins present in each cell of the body and, more importantly, how they interact with each other—an area of science known as **proteomics**. Proteomics is far more challenging than genomics because of the complexity of interactions that can take place between proteins (see Chapter 5). Moreover, the pattern and function of proteins present in any cell depend on the type of cell involved, and whether it is in the normal or diseased state. Nevertheless, progress has been made in identifying the structure and function of proteins, several of which are proving to be novel drug targets. This has been no easy task, and it is made all the more difficult by the fact that one cannot simply derive the structure of proteins based on the known gene sequences. This is because different proteins can be derived from a single gene, and proteins are often modified following their synthesis (section 2.5). There are roughly 40 000 genes, whereas a typical cell contains hundreds of thousands of different proteins. Moreover, knowing the structure of a protein does not necessarily indicate what its function or interactions might be.

Identifying the proteins present in a cell usually involves analysing the contents of the cell and separating out the proteins using a technique known as two-dimensional gel electrophoresis. Mass spectrometry can then be used to study the molecular weight of each protein. Assuming a pure sample of protein is obtained, its primary structure can be identified by traditional sequencing techniques. The analysis of secondary and tertiary structure is trickier. If the protein can be crystallized, then it is possible to determine its structure by X-ray crystallography. Not all proteins can be crystallized, though, and even if they are, it is possible that the conformation in the crystal form is different from that in solution. In recent years nuclear magnetic resonance (NMR) spectroscopy has been successful in identifying the tertiary structure of some proteins.

There then comes the problem of identifying what role the protein has in the cell and whether it would serve as a useful drug target. If it does show promise as a target, the final problem is to discover or design a drug that will interact with it.

KEY POINTS

- The order in which amino acids are linked together in a protein is called the primary structure.

- The secondary structure of a protein refers to regions of ordered structure within the protein, such as α-helices, β-pleated sheets, or β-turns.

- The overall three-dimensional shape of a protein is called its tertiary structure.

- Proteins containing two or more subunits have a quaternary structure which defines how the subunits are arranged with respect to each other.

- Secondary, tertiary, and quaternary structures are formed to maximize favourable intramolecular and intermolecular bonds and to minimize unfavourable interactions.

- Amino acids with polar residues are favoured on the outer surface of a protein because this allows hydrogen bonding interactions with water. Amino acids with non-polar residues are favoured within the protein because this maximizes van der Waals and hydrophobic interactions.

- Many proteins undergo post-translational modifications.

- Proteomics is the study of the structure and function of novel proteins discovered through genomics.

2.7 Protein function

We are now ready to discuss the various types of protein which act as drug targets.

2.7.1 Structural proteins

Structural proteins do not normally act as drug targets. However, the structural protein **tubulin** is an exception. Tubulin molecules polymerize to form small tubes called **microtubules** in the cell's cytoplasm (Fig. 2.19). These

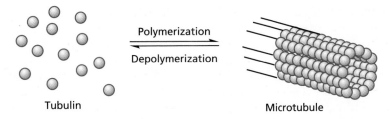

FIGURE 2.19 Polymerization of tubulin.

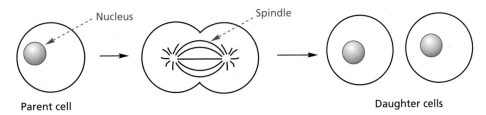

FIGURE 2.20 Cell division.

microtubules have various roles within the cell, including the maintenance of shape, exocytosis, and release of neurotransmitters. They are also involved in the mobility of cells. For example, inflammatory cells called **neutrophils** are mobile cells which normally protect the body against infection. However, they can also enter joints, leading to inflammation and arthritis.

Tubulin is also crucial to cell division. When a cell is about to divide, its microtubules depolymerize to give tubulin. The tubulin is then repolymerized to form a structure called a **spindle**, which then serves to push apart the two new cells and to act as a framework on which the chromosomes of the original cell are transferred to the nuclei of the daughter cells (Fig. 2.20). Drugs that target tubulin and inhibit this process are useful anticancer agents (section 10.2.2).

The structural proteins of viruses are important to the survival of the virus outside their host cell. Some of these proteins are proving to be interesting drug targets for the design of new antiviral agents, and are discussed in more detail in sections 20.7.5 and 20.9.

2.7.2 Transport proteins

Transport proteins are present in the cell membrane and act as the cell's 'smugglers'—smuggling the important chemical building blocks of amino acids, sugars, and nucleic acid bases across the cell membrane, such that the cell can synthesize its proteins, carbohydrates, and nucleic acids. They are also important in transporting important neurotransmitters (section 4.2) back into the

neuron that released them, so that the neurotransmitters only have a limited period of activity. But why is this smuggling operation necessary? Why can't these molecules pass through the membrane by themselves? Quite simply, the molecules concerned are polar structures and cannot pass through the hydrophobic cell membrane.

The transport proteins can float freely within the cell membrane because they have hydrophobic residues on their outer surface which interact favourably with the hydrophobic centre of the cell membrane. The portion of the transport protein that is exposed on the outer surface of the cell membrane contains a binding site that can bind a polar molecule such as an amino acid, stow it away in a hydrophilic pocket, and ferry it across the membrane to release it on the other side (Fig. 2.21).

Transport proteins are not all identical; there are specific transport proteins for the different molecules that need to be smuggled across the membrane. The binding sites for these transport proteins vary in structure such that they can recognize and bind their specific guest. There are several important drugs which target transport proteins (section 10.1).

2.7.3 Enzymes and receptors

The most important drug targets in medicinal chemistry are enzymes and receptors. Individual chapters are devoted to the structure and function of these proteins—Chapters 3 and 4 respectively.

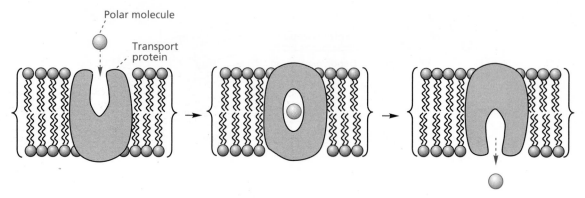

FIGURE 2.21 Transport proteins.

2.7.4 Miscellaneous proteins and protein–protein interactions

There are many situations in cell biology where proteins are required to interact with each other in order to produce a particular cellular effect. We have already seen an example of this in the polymerization of tubulin proteins to form microtubules (section 2.7.1). The structures of many important drug targets such as ion channels, enzymes, and receptors consist of two or more protein subunits associated with each other. The signal transduction processes described in Chapter 5 show many instances where a variety of proteins such as receptors, signal proteins, and enzymes associate with each other in order to transmit a chemical signal into the cell. The actions of insulin are mediated through a protein–protein interaction (section 4.8.3). The control of gene expression involves the prior assembly of a variety of different proteins (section 4.9 and Box 8.2). An important part of the immune response involves proteins called antibodies interacting with foreign proteins (section 10.7.2). Cell–cell recognition involves protein–protein interactions—a process which is not only important in terms of the body's own proteins, but in the mechanism by which viruses invade human cells (sections 20.7.1, 20.8.1, and 20.9). Important processes that have an influence on tumour growth such as angiogenesis and apoptosis (section 21.1) involve the association of proteins. Proteins called chaperones help to stabilize partially folded proteins during translation through protein–protein interactions. They are also important in the process by which old proteins are removed to the cell's recycling centre. Chaperones are particularly important when the cell experiences adverse environmental conditions which might damage proteins. It has been found that the synthesis of chaperones increases in tumour cells, and this may reflect some of the stresses experienced in such

cells; for example, lack of oxygen, pH variation, and nutrient deprivation. Inhibiting chaperones could well lead to more damaged proteins and cell death. There are current studies looking into methods of inhibiting a chaperone protein called HSP90 (HSP stands for heat shock protein). Inhibition might prevent the synthesis of important receptors and enzymes involved in the process of cell growth and division, and provide a new method of treating tumour cells. The inhibition of an enzyme acting as a chaperone protein is also being considered as a potential therapy for the treatment of Alzheimer's disease (section 22.15.2).

Protein–protein interactions are not limited to human biochemistry. Interfering with these interactions in other species could lead to novel antibacterial, antifungal, and antiviral agents. For example, HIV protease is an important enzyme in the life cycle of the HIV virus, and is an important target for antiviral agents (section 20.7.4). The enzyme consists of two identical proteins which bind together to produce the active site. Finding a drug that will prevent this association would be a novel method of inhibiting this enzyme.

To conclude, there is a lot of research currently underway looking at methods of inhibiting or promoting protein–protein interactions (section 10.5).

KEY POINTS

- Transport proteins, enzymes, and receptors are common drug targets.
- Transport proteins transport essential polar molecules across the hydrophobic cell membrane.
- Tubulin is a structural protein which is crucial to cell division and cell mobility.
- Many cell processes depend on the interactions of proteins with each other.

QUESTIONS

1. Draw the full structure of L-alanyl-L-phenylalanyl-glycine.

2. What is unique about glycine compared to other naturally occurring amino acids?

3. Identify the intermolecular/intramolecular interactions that are possible for the side chains of the following amino acids: serine, phenylalanine, glycine, lysine, aspartic acid, and aspartate.

4. The chains of several cell membrane-bound proteins wind back and forth through the cell membrane, such that some parts of the protein structure are extracellular, some parts are intracellular, and some parts lie within the cell membrane. How might the primary structure of a protein help in distinguishing the portions of the protein embedded within the cell membrane from those that are not?

5. What problems might you foresee if you tried to synthesize L-alanyl-L-valine directly from its two component amino acids?

6. The tertiary structure of many enzymes is significantly altered by the phosphorylation of serine, threonine, or tyrosine residues. Identify the functional groups that are involved in these phosphorylations and suggest why phosphorylation affects tertiary structure.

7. What is the one-letter code for the polypeptide Glu-Leu-Pro-Asp-Val-Val-Ala-Phe-Lys-Ser-Gly-Gly-Thr?

Multiple-choice questions are available on the Online Resource Centre at www.oxfordtextbooks.co.uk/orc/patrick6e/

FURTHER READING

Ball, P. (2009) Proteins unravelled. *Chemistry World*, December, 58–62.

Berg, C., Neumeyer, K., and Kirkpatrick, P. (2003) Teriparatide. *Nature Reviews Drug Discovery*, **2**, 257–8.

Dobson, C. M. (2003) Protein folding and disease: a view from the first Horizon symposium. *Nature Reviews Drug Discovery*, **2**, 154–60.

Ezzell, C. (2002) Proteins rule. *Scientific American*, **286**, April, 7–33 (proteomics).

Harris, J. M., and Chess, R. B. (2003) Effect of pegylation on pharmaceuticals. *Nature Reviews Drug Discovery*, **2**, 214–21.

Stevenson, R. (2002) Proteomic analysis honoured. *Chemistry in Britain*, **38**, November, 21–3.

Teague, S. J. (2003) Implications of protein flexibility for drug discovery. *Nature Reviews Drug Discovery*, **2**, 527–41.

Titles for general further reading are listed on p.845.

3 Enzymes: structure and function

In this chapter we discuss the structure and function of enzymes. Drug action at enzymes is discussed in Chapter 7 and in other chapters throughout the text.

3.1 Enzymes as catalysts

Enzymes are proteins which act as the body's catalysts—agents that speed up a chemical reaction without being consumed themselves. Without them, the cell's chemical reactions would either be too slow or not take place at all. An example of an enzyme-catalysed reaction is the reduction of **pyruvic acid** to **lactic acid**, which takes place when muscles are over-exercised, and is catalysed by an enzyme called **lactate dehydrogenase** (Fig. 3.1).

Note that the reaction is shown as an equilibrium. It is, therefore, more correct to describe an enzyme as an agent that speeds up the approach to equilibrium, because the enzyme speeds up the reverse reaction just as

efficiently as the forward reaction. The final equilibrium concentrations of the starting materials and products are unaffected by the presence of an enzyme.

How do enzymes affect the rate of a reaction without affecting the equilibrium? The answer lies in the existence of a high-energy **transition state** that must be formed before the starting material (the **substrate**) can be converted to the product. The difference in energy between the transition state and the substrate is the **activation energy**, and it is the size of this activation energy that determines the rate of a reaction, rather than the difference in energy between the substrate and the product (Fig. 3.2). An enzyme acts to lower the activation energy by helping to stabilize the transition state. The energy of both the substrate and products is unaffected, and therefore the equilibrium ratio of substrate to product is unaffected. We can relate energy to the rate and equilibrium constants with the following equations:

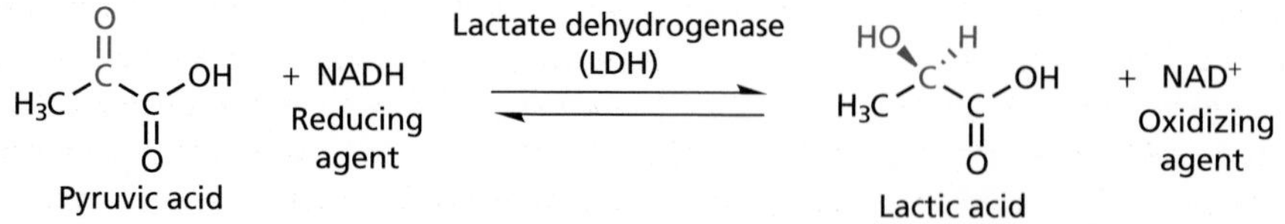

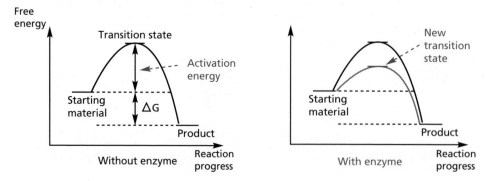

FIGURE 3.1 Reaction catalysed by lactate dehydrogenase.

FIGURE 3.2 Graphs demonstrating the stabilization of a reaction's transition state by an enzyme.

$$\text{Energy difference} = \Delta G = -RT \ln K$$

where K is the equilibrium constant (= [products]/[reactants]), R is the gas constant (= 8.314 J mol^{-1} K^{-1}), and T is the temperature.

$$\text{Rate constant} = k = Ae^{-E/RT}$$

where E is the activation energy and A is the frequency factor. Note that the rate constant k does not depend on the equilibrium constant K.

We have stated that enzymes catalyse reactions, but we have still to explain how.

3.2 How do enzymes catalyse reactions?

The factors involved in enzyme catalysis are summarized below and will be discussed in more detail in sections 3.2–3.5.

- Enzymes provide a reaction surface and a suitable environment.
- Enzymes bring reactants together and position them correctly so that they easily attain their transition-state configurations.
- Enzymes weaken bonds in the reactants.
- Enzymes may participate in the mechanism.
- Enzymes form stronger interactions with the transition state than with the substrate or the product.

An enzyme catalyses a reaction by providing a surface to which a substrate can bind, resulting in the weakening of high-energy bonds. The binding also holds the substrate in the correct orientation to increase the chances of reaction. The reaction takes place, aided by the enzyme, to give a product which is then released (Fig. 3.3). Note again that it is a reversible process. Enzymes can catalyse both forward and backward reactions. The final equilibrium mixture will, however, be the same, regardless of whether we supply the enzyme with substrate or product.

Substrates bind to, and react at, a specific area of the enzyme called the **active site**—usually quite a small part of the overall protein structure.

3.3 The active site of an enzyme

The active site of an enzyme (Fig. 3.4) has to be on or near the surface of the enzyme if a substrate is to reach it. However, the site could be a groove, hollow, or gully allowing the substrate to sink into the enzyme. Normally the active site is more hydrophobic in character than the surface of the enzyme, providing a suitable environment for many reactions that would be difficult or impossible to carry out in an aqueous environment.

Because of the overall folding of the enzyme, the amino acid residues that are close together in the active site may be far apart in the primary structure. Several amino acids in the active site play an important role in enzyme function, and this can be demonstrated by comparing the primary structures of the same enzyme from different organisms. Here, the primary structure differs from species to species as a result of mutations happening over millions of years, and the variability is proportional to how far apart the organisms are on the evolutionary ladder. However, there are certain amino acids that remain constant, no matter the source of the enzyme. These are amino acids that are crucial to the enzyme's function and are often present in the active site. If one of these amino acids should be altered through mutation, the enzyme could become useless and the cell bearing this mutation would have a poor chance of survival. Thus, the mutation would not be preserved. The only exception to this would be if the mutation introduced an amino acid which could either perform the same task as the original amino acid, or improve substrate binding. This consistency of amino acids in the active site can often help scientists determine which amino acids are present in an active site, if this is not known already.

Amino acids present in the active site can have one of two roles:

- binding—the amino acid residue is involved in binding the substrate or a cofactor to the active site;

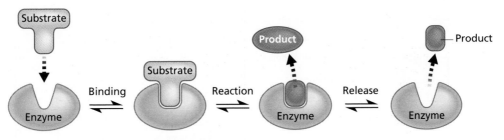

FIGURE 3.3 The process of enzyme catalysis.

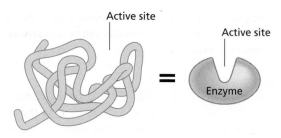

FIGURE 3.4 The active site of an enzyme.

- catalytic—the amino acid is involved in the mechanism of the reaction.

We shall study these in turn.

3.4 **Substrate binding at an active site**

The interactions which bind substrates to the active sites of enzymes include ionic bonds, hydrogen bonds, dipole–dipole and ion–dipole interactions, as well as van der Waals and hydrophobic interactions (section 1.3). These binding interactions are the same bonding interactions responsible for the tertiary structure of proteins, but their relative importance differs. Ionic bonding plays a relatively minor role in protein tertiary structure compared with hydrogen bonding or van der Waals interactions, but it can play a crucial role in the binding of a substrate to an active site.

As intermolecular bonding forces are involved in substrate binding, it is possible to look at the structure of a substrate and postulate the probable interactions that it will have with its active site. As an example, consider **pyruvic acid**—the substrate for **lactate dehydrogenase** (Fig. 3.5).

If we look at the structure of pyruvic acid, we can propose three possible interactions by which it might bind to its active site—an ionic interaction involving the ionized carboxylate group, a hydrogen bond involving the ketonic oxygen, and a van der Waals interaction involving

the methyl group. If these postulates are correct, it means that within the active site there must be **binding regions** containing suitable amino acids that can take part in these intermolecular interactions. Lysine, serine, and phenylalanine residues respectively would fit the bill. A knowledge of how a substrate binds to its active site is invaluable in designing drugs that will target specific enzymes (Chapter 7).

3.5 **The catalytic role of enzymes**

We now move on to consider the mechanism of enzymes, and how they catalyse reactions. In general, enzymes catalyse reactions by providing binding interactions, acid–base catalysis, nucleophilic groups, and cofactors.

3.5.1 **Binding interactions**

In the past, it was thought that a substrate fitted its active site in a similar way to a key fitting a lock (**Fischer's lock and key hypothesis**). Both the enzyme and the substrate were seen as rigid structures, with the substrate (the key) fitting perfectly into the active site (the lock) (Fig. 3.6). However, this scenario does not explain how some enzymes can catalyse a reaction on a range of different substrates. It implies, instead, that an enzyme has an optimum substrate that fits it perfectly, whereas all other substrates fit less perfectly. This, in turn, would imply that the catalysed reaction is only efficient for the optimum substrate. As this is not the case for many enzymes, the lock and key analogy must be invalid.

It is now proposed that the substrate is not quite the ideal shape for the active site, and that it forces the active site to change shape when it enters—a kind of moulding process. This theory is known as **Koshland's theory of induced fit** since the substrate induces the active site to take up the ideal shape to accommodate it (Fig. 3.6).

For example, a substrate such as **pyruvic acid** might interact with specific binding regions in the active site of lactate dehydrogenase via one hydrogen bond, one ionic bond, and one van der Waals interaction (Fig. 3.7). However, if the fit is not perfect, the three bonding

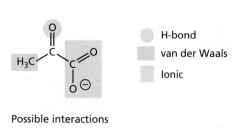

Possible interactions

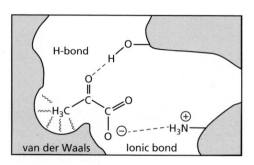

FIGURE 3.5 Binding interactions between pyruvic acid and lactate dehydrogenase.

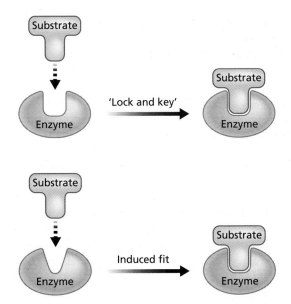

FIGURE 3.6 The 'lock and key' and 'induced fit' hypotheses for substrate-enzyme binding.

interactions are not ideal either. For example, the binding groups may be slightly too far away from the corresponding binding regions in the active site. In order to maximize the strength of these bonds, the enzyme changes shape such that the amino acid residues involved in the binding move closer to the substrate.

This theory of induced fit helps to explain why some enzymes can catalyse reactions involving a wide range of substrates. Each substrate induces the active site into a shape that is ideal for it, and as long as the moulding process does not distort the active site so much that the

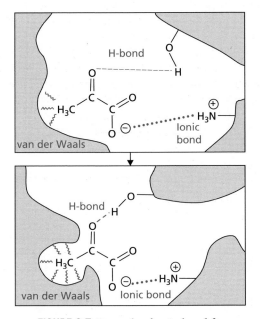

FIGURE 3.7 Example of an induced fit.

reaction mechanism proves impossible, the reaction can proceed. The range of substrates that can be accepted depends on the substrates being the correct size to fit the active site, and having the correct binding groups in the correct relative positions.

But note this. The substrate is not a passive spectator to the moulding process going on around it. As the enzyme changes shape to maximize bonding interactions, the same thing can happen to the substrate. It, too, may alter shape. Bond rotation may occur to fix the substrate in a particular conformation—and not necessarily the most stable one. Bonds may even be stretched and weakened. Consequently, this moulding process designed to maximize binding interactions may force the substrate into the ideal conformation for the reaction to follow and may also weaken the very bonds that have to be broken.

Once bound to an active site, the substrate is now held ready for the subsequent reaction. Binding has fixed the 'victim' (the substrate) so that it cannot evade attack, and this same binding has weakened its defences (the bonds) so that reaction is easier (a lower activation energy).

There is another point relating to substrate binding. The binding interactions with the active site must be sufficiently strong to hold the substrate for the subsequent reaction, but they cannot be too strong. If they were, the product might also be bound strongly and fail to depart the active site. This would block the active site of the enzyme and prevent it from catalysing another reaction. Therefore, a balance must be struck.

Finally, it is important to realize that the enzyme also binds the transition state involved in the enzyme-catalysed reaction. Indeed, the binding interactions involved are stronger than those binding the substrate, which means that the transition state is stabilized relatively more than the substrate. This results in a lower activation energy compared to the non-catalysed reaction (see section 3.5.4).

3.5.2 Acid–base catalysis

Acid–base catalysis is often provided by the amino acid **histidine**, which contains an imidazole ring as part of its side chain. The imidazole ring acts as a weak base, which means that it exists in equilibrium between its protonated and free base forms (Fig. 3.8), allowing it to accept or donate protons during a reaction mechanism. This is important, as there are often very few water molecules present in an active site to carry out this role. Histidine is not the only amino acid residue that can provide acid–base catalysis. For example, a **glutamic acid** residue acts as a proton source in the reaction mechanism of the enzyme HMG-CoA reductase (Case study 1), while **aspartic acid** and **aspartate** residues act as proton donors and proton acceptors respectively in other enzyme-catalysed reactions (sections 7.4 and 20.7.4.1). **Tyrosine** acts as a

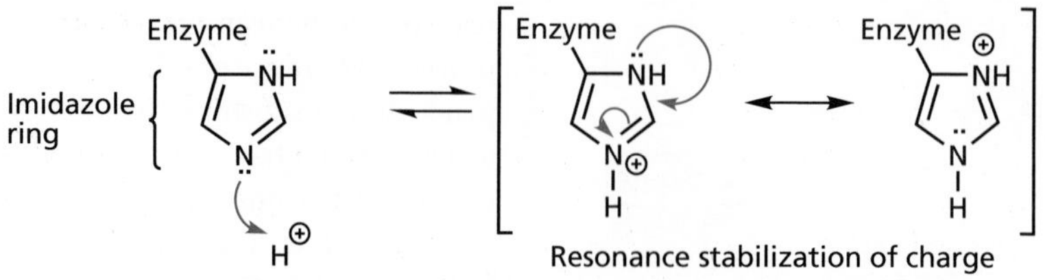

FIGURE 3.8 Histidine acting as a weak base.

proton source in the mechanism by which the enzyme 17β-hydroxysteroid type 1 catalyses the conversion of **estrone** to **estradiol** (Box 13.1).

For additional material see Web article 1: Steroids as novel anticancer agents on the Online Resource Centre at www.oxfordtextbooks.co.uk/orc/patrick6e/

3.5.3 Nucleophilic groups

The amino acids **serine** and **cysteine** are present in the active sites of some enzymes. These amino acids have nucleophilic residues (OH and SH respectively) which are able to participate in the reaction mechanism. They do this by reacting with the substrate to form intermediates that would not be formed in the uncatalysed reaction.

These intermediates offer an alternative reaction pathway that may avoid a high-energy transition state and hence increase the rate of the reaction.

Normally, an alcoholic OH group such as the one on serine is not a good nucleophile. However, there is usually a histidine residue close by to catalyse the reaction. For example, the mechanism by which chymotrypsin hydrolyses peptide bonds (Fig. 3.9) involves a **catalytic triad** of amino acids—serine, histidine, and aspartic acid. Serine and histidine participate in the mechanism as a nucleophile and acid–base catalyst respectively. The aspartate group interacts with the histidine ring and serves to activate and orient it correctly for the mechanism.

The presence of a nucleophilic serine residue means that water is not required in the initial stages of the

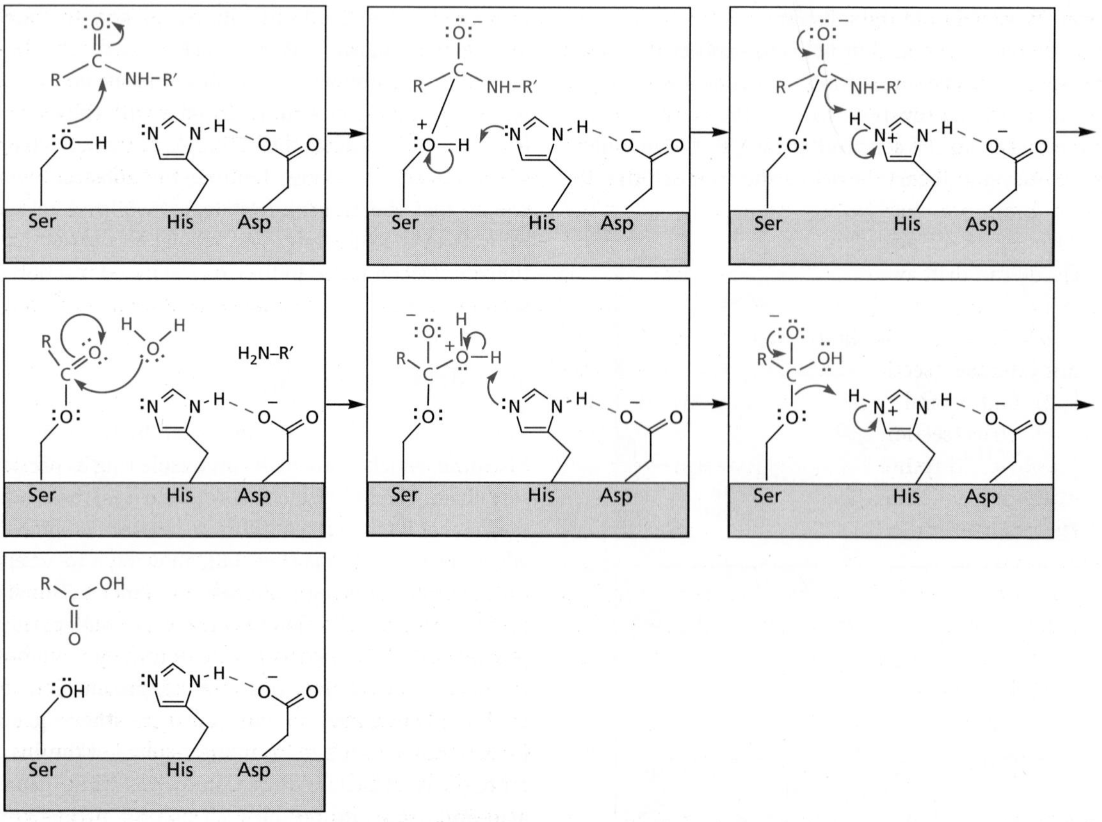

FIGURE 3.9 Hydrolysis of peptide bonds catalysed by the enzyme chymotrypsin.

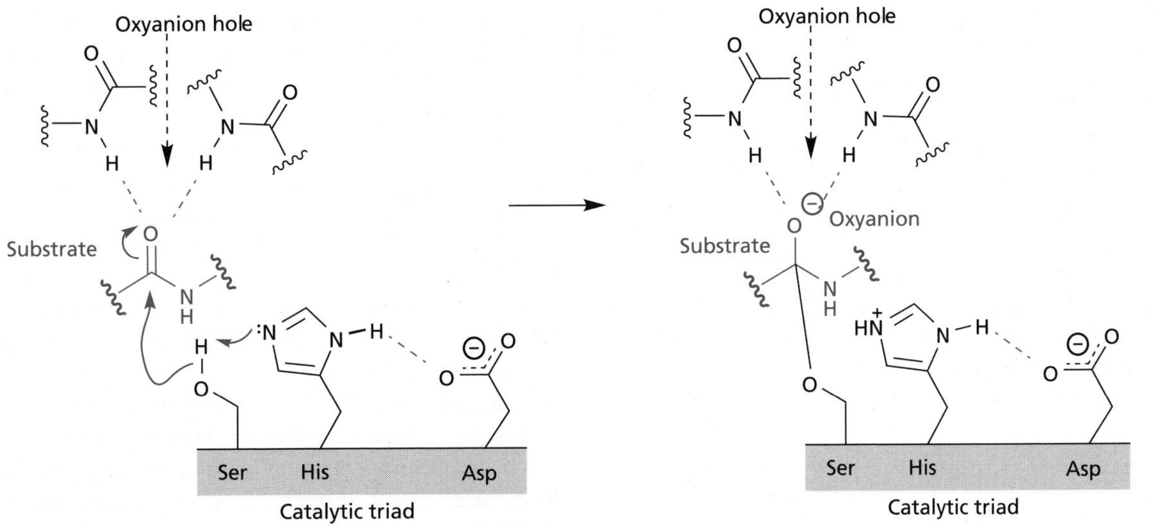

FIGURE 3.10 The stabilizing role of the oxyanion hole in enzyme-catalysed peptide bond hydrolysis.

mechanism. This is important, because water is a poor nucleophile and may also find it difficult to penetrate the occupied active site. Secondly, a water molecule would have to drift into the active site, and search out the carboxyl group before it could attack it. This would be something similar to a game of blind man's bluff. The enzyme, on the other hand, can provide a serine OH group, positioned in exactly the right spot to react with the substrate. Therefore, the nucleophile has no need to search for its substrate. The substrate has been delivered to it.

Water is eventually required to hydrolyse the acyl group attached to the serine residue. However, this is a much easier step than the hydrolysis of a peptide link, as esters are more reactive than amides. Furthermore, the hydrolysis of the peptide link means that one half of the peptide can drift away from the active site and leave room for a water molecule to enter. A similar enzymatic mechanism is involved in the action of the enzyme **acetylcholinesterase** (section 22.12.3), **pancreatic lipase** (Box 7.2), and a viral protease enzyme carried by the hepatitis C virus (section 20.10).

The amino acid **lysine** has a primary amine group on its side chain which should make it a better nucleophilic group than serine or cysteine. However, the group is generally protonated at physiological pH, which precludes it acting as a nucleophile. Having said that, some enzymes have a lysine residue located in a hydrophobic pocket, which means that it is not protonated and can, indeed, act as a nucleophilic group.

3.5.4 Stabilization of the transition state

As stated earlier, binding interactions serve to stabilize the transition state of an enzyme-catalysed reaction,

and thus lower the activation energy for the reaction. One example of this involves what is known as the **oxyanion hole**, which is present in enzymes such as chymotrypsin. This region is occupied by the oxyanion intermediate that is formed during the reaction mechanism described in Fig. 3.9. In the vicinity of the oxyanion hole, there are peptide bonds that form hydrogen bonds with the oxyanion and help to stabilize its negative charge (Fig. 3.10). These hydrogen bonds are also possible with the carbonyl group of the substrate, but the interaction is stronger with the oxyanion because of the negative charge. A negatively charged oxygen atom is a stronger hydrogen bond acceptor than a neutral oxygen atom (section 1.3.2). Hydrogen bonding also stabilizes the developing negative charge in the transition state leading to the oxyanion. As a result, the activation energy of that stage of the reaction mechanism is lowered.

3.5.5 Cofactors

Many enzymes require additional non-protein substances called cofactors for the reaction to take place. Deficiency of cofactors can arise from a poor diet and leads to loss of enzyme activity and subsequent disease (e.g. scurvy). Cofactors are either metal ions (e.g. zinc) or small organic molecules called **coenzymes** (e.g. NAD^+, pyridoxal phosphate). Most coenzymes are bound by ionic bonds and other non-covalent bonding interactions, but some are bound covalently and are called **prosthetic groups**. Coenzymes are derived from water-soluble vitamins and act as the body's chemical reagents. For example, **lactate dehydrogenase** requires the coenzyme **nicotinamide adenine dinucleotide (NAD^+)** (Fig. 3.11) in order to

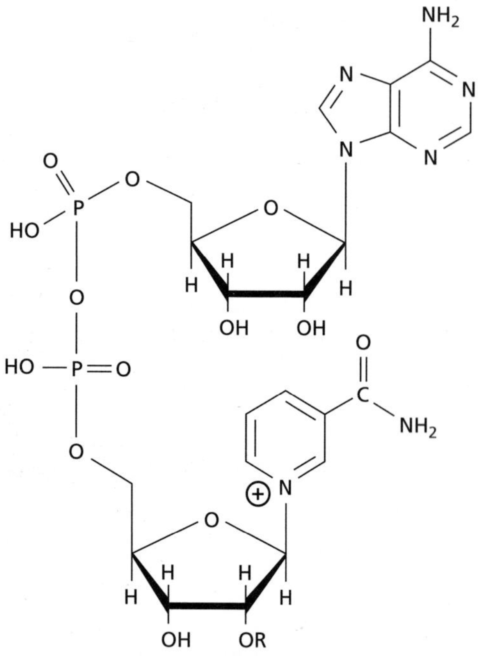

FIGURE 3.11 Nicotinamide adenine dinucleotide (R = H) and nicotinamide adenine dinucleotide phosphate (R = phosphate).

A large number of enzymes contain a metal ion cofactor and are termed **metalloenzymes**. For example, the cytochrome P450 enzymes involved in drug metabolism (section 11.5.2) contain an iron ion which can interchange between two different oxidation states (Fe^{2+} and Fe^{3+}). This allows cytochrome enzymes to catalyse oxidation and reduction reactions.

Other examples are the kinase enzymes responsible for catalysing phosphorylation reactions, and which serve as important drug targets (section 21.6.2). These contain a magnesium ion which helps to bind the coenzyme ATP to the active site, such that it can act as a phorphorylating agent.

Several protease enzymes (responsible for catalysing the hydrolysis of peptide bonds) contain a zinc ion as a cofactor and are categorized as **zinc metalloproteases**. Examples include matrix metalloproteinases (section 21.7.1) and angiotensin-converting enzyme (section 26.3.3 and Case study 2).

The role of the zinc ion in aiding catalysis can be illustrated in the reaction catalysed by **histone deacetylase**—a reaction that removes an acetyl group from the side chain of an *N*-acetylated lysine residue (section 21.7.3). Zinc is present in the active site as Zn^{2+} and is held in place by three amino acids which act as ligands. The reaction mechanism is similar to the mechanism shown in Fig. 3.10 and involves the formation of an oxyanion with the aid of histidine and aspartate residues (Fig. 3.13). However, in this case, there is no serine residue to act as a nucleophile. Instead, a water molecule acts as the nucleophile. One of the catalytic roles of zinc is to activate both the substrate and water by accepting both molecules as ligands. This also helps to position the water correctly for the reaction. Nucleophilic addition to the carbonyl group takes place with histidine removing a proton from water as it reacts. An oxyanion intermediate is formed, which interacts strongly with zinc. Zinc also stabilizes the developing negative charge on the transition state leading to the oxyanion intermediate, thus lowering the activation energy for that stage of the reaction mechanism.

catalyse the dehydrogenation of **lactic acid** to **pyruvic acid**. NAD^+ is bound to the active site along with lactic acid, and acts as the oxidizing agent. During the reaction, it is converted to its reduced form (NADH) (Fig. 3.12). Conversely, NADH can bind to the enzyme and act as a reducing agent when the enzyme catalyses the reverse reaction.

$NADP^+$ and NADPH are phosphorylated analogues of NAD^+ and NADH respectively and carry out redox reactions by the same mechanism. NADPH is used almost exclusively for reductive biosynthesis, whereas NADH is used primarily for the generation of ATP.

A knowledge of how the coenzyme binds to the active site allows the possibility of designing enzyme inhibitors that will fit the same region (see Case study 5 and section 21.6.2; see also web article 1).

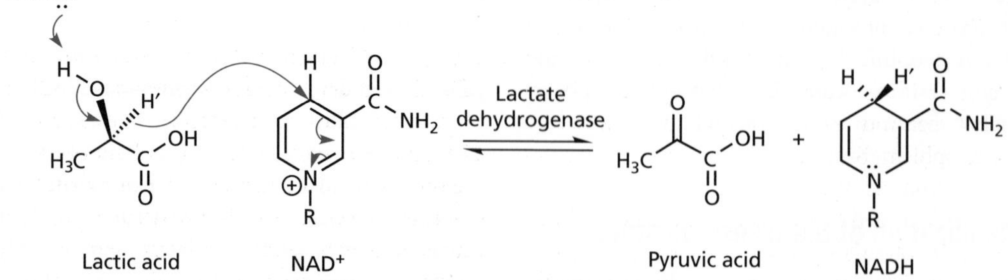

FIGURE 3.12 NAD^+ acting as a coenzyme.

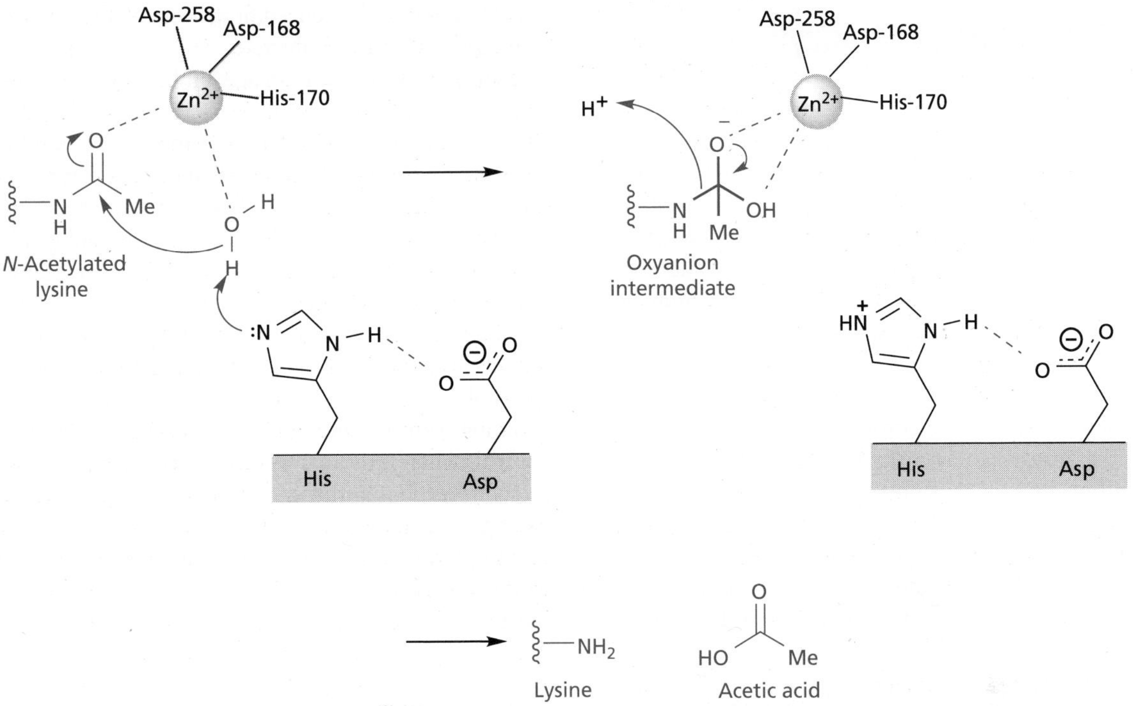

FIGURE 3.13 The role of zinc in enzyme catalysis.

3.5.6 Naming and classification of enzymes

The name of an enzyme reflects the type of reaction it catalyses, and has the suffix -ase to indicate that it is an enzyme. For example, an oxidase enzyme catalyses an oxidation reaction. It is important to appreciate that enzymes can catalyse the forward and back reactions of an equilibrium reaction. This means that an oxidase enzyme can catalyse reductions as well as oxidations. The reaction catalysed depends on the nature of the substrate; that is whether it is in the reduced or oxidized form.

Enzymes are classified according to the general class of reaction they catalyse and are coded with an EC number (Table 3.1).

3.5.7 Genetic polymorphism and enzymes

There are often subtle differences in the structure and properties of an enzyme between different individuals. This is due to the fact that the DNA that codes for proteins (Chapter 6) is not identical from person to person. On average, there is a difference of one base pair in every thousand between individuals, and this is known as genetic polymorphism. Since the nucleic acid bases act as the code for amino acids in proteins, a difference at this level may result in a different amino acid being introduced into the protein. Often, this has no observable effect on protein function, but not always. Some polymorphisms can adversely affect the proper functioning of an

TABLE 3.1 Classification of enzymes

EC number	Enzyme class	Type of reaction
E.C.1.x.x.x	Oxidoreductases	Oxidations and reductions
E.C.2.x.x.x	Transferases	Group transfer reactions
E.C.3.x.x.x	Hydrolases	Hydrolysis reactions
E.C.4.x.x.x	Lyases	Addition or removal of groups to form double bonds
E.C.5.x.x.x	Isomerases	Isomerizations and intra-molecular group transfers
E.C.6.x.x.x	Ligases	Joining two substrates at the expense of ATP hydrolysis

Note: EC stands for Enzyme Commission, a body set up by the International Union of Biochemistry (as it then was) in 1955.

enzyme and lead to genetic disease. Others can have an influence on drug therapy. For example, individuals differ in their ability to metabolize drugs as a result of this phenomenon (section 11.5.6). Polymorphism can alter the sensitivity of an enzyme towards a drug, making the latter less effective. This is a particular problem in anticancer, antibacterial, and antiviral therapies, where drug resistance can develop through the survival of cells containing less sensitive enzymes (Chapters 19–21).

3.6 **Regulation of enzymes**

Virtually all enzymes are controlled by agents which can either enhance or inhibit catalytic activity. Such control reflects the local conditions within the cell. For example, the enzyme **phosphorylase *a*** catalyses the breakdown of **glycogen** (a polymer of glucose monomers) to **glucose-1-phosphate** subunits (Fig. 3.14). It is stimulated by **adenosine 5′-monophosphate (AMP)** and inhibited by glucose-1-phosphate. Thus, rising levels of the product (glucose-1-phosphate) act as a self-regulating 'brake' on the enzyme.

But how does this control take place?

The answer is that many enzymes have an **allosteric binding site**, which is quite separate from the active site (Fig. 3.15). This is where the agents controlling the activity of the enzyme bind. When this occurs, an induced fit takes place which alters not only the allosteric binding site, but the active site as well. Agents that inhibit the

enzyme produce an induced fit that makes the active site unrecognizable to the substrate.

We might wonder why an agent inhibiting the enzyme has to bind to a separate, allosteric binding site and not to the active site itself. After all, if the agent could bind to the active site, it would directly block the natural substrate from entering. There are two explanations for this.

Firstly, many of the enzymes that are under allosteric control are at the start of a biosynthetic pathway (Fig. 3.16). A biosynthetic pathway involves a series of enzymes, all working efficiently to produce a final product. Eventually, the cell will have enough of the required material and will need to stop production. The most common control mechanism is known as **feedback control**, where the final product controls its own synthesis by inhibiting the first enzyme in the biochemical pathway. When there are low levels of final product in the cell, the first enzyme in the pathway is not inhibited and works normally. As the levels of final product increase, more and more of the enzyme is blocked and the rate

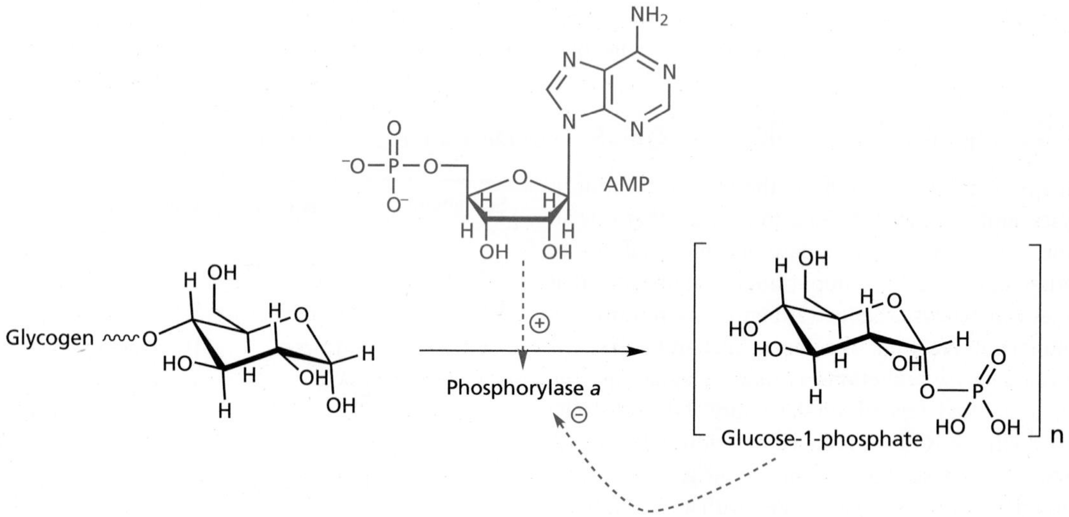

FIGURE 3.14 Internal control of the catalytic activity of phosphorylase a by glucose-1-phosphate and AMP.

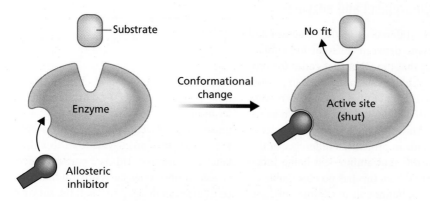

FIGURE 3.15 Allosteric inhibition of an enzyme.

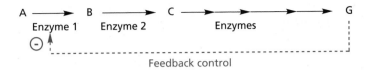

FIGURE 3.16 Feedback control of enzyme 1 by final product G.

of synthesis drops off in a graded fashion. Crucially, the final product has undergone many transformations from the original starting material and so it is no longer recognized by the active site of the first enzyme. A separate allosteric binding site is therefore needed which recognizes the final product. The biosynthesis of noradrenaline in section 23.4 is an example of a biosynthetic pathway under feedback control.

Secondly, binding of the final product to the active site would not be a very efficient method of feedback control, as the product would have to compete with the enzyme's substrate. If levels of the latter increased, then the inhibitor would be displaced and feedback control would fail.

Many enzymes can also be regulated externally (Box 3.1). We shall look at this in more detail in Chapter 5, but, in essence, cells receive chemical messages from their

---

BOX 3.1 The external control of enzymes by nitric oxide

The external control of enzymes is usually initiated by external chemical messengers which do not enter the cell. However, there is an exception to this. It has been discovered that cells can generate the gas **nitric oxide** by the reaction sequence shown in Fig. 1, catalysed by the enzyme **nitric oxide synthase**.

Because nitric oxide is a gas, it can easily diffuse through cell membranes into target cells. There, it activates enzymes called **cyclases** to generate **cyclic GMP** from **GTP** (Fig. 2). Cyclic GMP then acts as a secondary messenger to influence other reactions within the cell. By this process, nitric oxide has an influence on a diverse range of physiological processes including blood pressure, neurotransmission, and immunological defence mechanisms.

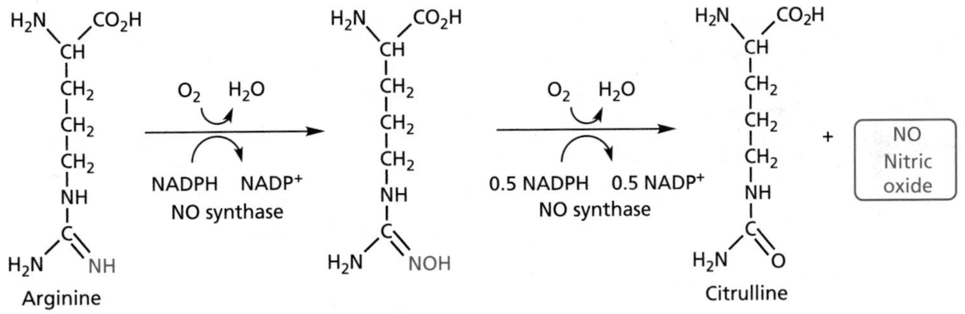

FIGURE 1 Synthesis of nitric oxide.

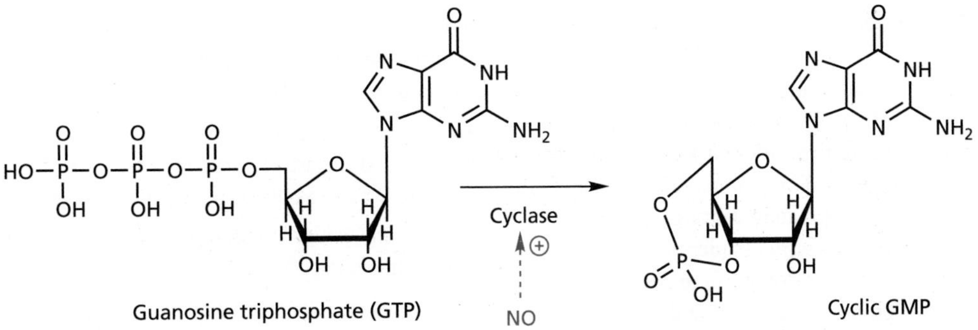

FIGURE 2 Activation of cyclase enzymes by nitric oxide (NO).

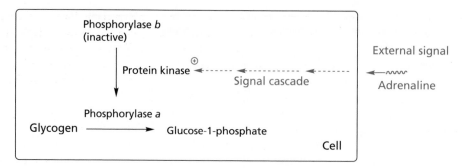

FIGURE 3.17 External control of phosphorylase a.

environment which trigger a cascade of signals within the cell. These in turn ultimately activate a set of enzymes known as **protein kinases**. The protein kinases play an important part in controlling enzyme activity within the cell by phosphorylating amino acids such as **serine**, **threonine**, or **tyrosine** in target enzymes—a covalent modification. For example, the hormone **adrenaline** is an external messenger which triggers a signalling sequence resulting in the activation of a protein kinase enzyme. Once activated, the protein kinase phosphorylates an inactive enzyme called **phosphorylase b** (Fig. 3.17). This enzyme now becomes active and is called **phosphorylase a**. It catalyses the breakdown of glycogen and remains active until it is dephosphorylated back to phosphorylase b.

In this case, phosphorylation of the target enzyme leads to activation. Other enzymes may be deactivated by phosphorylation. For example, **glycogen synthase**—the enzyme that catalyses the *synthesis* of glycogen from glucose-1-phosphate—is inactivated by phosphorylation and activated by dephosphorylation. The latter is effected by the hormone **insulin**, which triggers a different signalling cascade from that of adrenaline.

Protein–protein interactions can also play a role in the regulation of enzyme activity. For example, signal proteins in the cell membrane are responsible for regulating the activity of membrane-bound enzymes (section 5.2).

3.7 **Isozymes**

Enzymes having a quaternary structure are made up of a number of polypeptide subunits, and the combination of these subunits can differ in different tissues. Such variations are called isozymes. For example, there are five different isozymes of mammalian **lactate dehydrogenase** (LDH)—a tetrameric enzyme made up of four polypeptide subunits. There are two different types of subunits involved, which are labelled 'H' and 'M'. The former predominates in the LDH present in heart muscle, while the latter predominates in the LDH present in skeletal muscle. Since there are two different types of subunit,

five different isozymes are possible—HHHH, HHHM, HHMM, HMMM, MMMM. Isozymes differ in their properties. For example, the M_4 isozyme in skeletal muscle catalyses the conversion of pyruvic acid to lactic acid, and is twice as active as the H_4 isozyme in heart muscle. The H_4 isozyme catalyses the reverse reaction and is inhibited by excess pyruvic acid, whereas the M_4 isozyme is not.

KEY POINTS

- Enzymes are proteins that act as the body's catalysts by binding substrates and participating in the reaction mechanism.

- The active site of an enzyme is usually a hollow or cleft in the protein. There are important amino acids present in the active site that either bind substrates or participate in the reaction mechanism.

- Binding of a substrate to an active site involves intermolecular bonds.

- Substrate binding involves an induced fit where the shape of the active site alters to maximize binding interactions. The binding process also orientates the substrate correctly and may weaken crucial bonds in the substrate to facilitate the reaction mechanism.

- The amino acid histidine is often present in active sites and acts as an acid–base catalyst. Glutamic acid, aspartic acid, and tyrosine also act as acid–base catalysts in some enzymes.

- The amino acids serine and cysteine act as nucleophiles in the reaction mechanisms of some enzymes. In some enzymes, lysine can act as a nucleophile.

- Cofactors are metal ions or small organic molecules (coenzymes) which are required by many enzymes. Coenzymes can be viewed as the body's chemical reagents.

- Prosthetic groups are coenzymes which are bound covalently to an enzyme.

- Enzymes are regulated by internal and/or external control.

- External control involves regulation initiated by a chemical messenger from outside the cell, and which ultimately involves the phosphorylation of enzymes.

- Allosteric inhibitors bind to a different binding site from the active site and alter the shape of the enzyme, such that the active site is no longer recognizable. Allosteric inhibitors are often involved in the feedback control of biosynthetic pathways.

- Isozymes are variations of the same enzyme. They catalyse the same reaction but differ in their primary structure, substrate specificity, and tissue distribution.

- The amino acid sequence in enzymes may differ between individuals due to genetic polymorphism. This may or may not result in a difference in enzyme activity.

3.8 Enzyme kinetics

3.8.1 The Michaelis–Menten equation

The Michaelis–Menten equation holds for an enzyme (E) which combines with its substrate (S) to form an enzyme–substrate complex (ES). The enzyme–substrate complex can then either dissociate back to E and S, or go on to form a product (P). It is assumed that formation of the product is irreversible.

$$E + S \underset{k_2}{\overset{k_1}{\rightleftharpoons}} ES \xrightarrow{k_3} E + P$$

where k_1, k_2, and k_3 are rate constants.

For enzymes such as these, plotting the rate of enzyme reaction versus substrate concentration [S] gives a curve as shown in Fig. 3.18. At low substrate concentrations the rate of reaction increases almost proportionally to the substrate concentration, whereas at high substrate concentration the rate becomes almost constant and approaches a maximum rate (rate$_{max}$), which is independent of substrate

concentration. This reflects a situation where there is more substrate present than active sites available, and so increasing the amount of substrate will have little effect.

The Michaelis–Menten equation relates the rate of reaction to the substrate concentration for the curve in Fig. 3.16.

$$\text{rate} = \text{rate}_{max} \frac{[S]}{[S] + K_M}$$

The derivation of this equation is not covered here but can be found in most biochemistry textbooks. The constant K_M is known as the **Michaelis constant** and is equal to the substrate concentration at which the reaction rate is half of its maximum value. This can be demonstrated as follows. If $K_M = [S]$, then the Michaelis–Menten equation becomes:

$$\text{rate} = \text{rate}_{max} \frac{[S]}{[S] + [S]} = \text{rate}_{max} \frac{[S]}{2[S]} = \text{rate}_{max} \times \frac{1}{2}$$

The K_M of an enzyme is significant because it measures the concentration of substrate at which half the active sites in the enzyme are filled. This in turn provides a measure of the substrate concentration required for significant catalysis to occur.

K_M is also related to the rate constants of the enzyme-catalysed reaction:

$$K_M = \frac{k_2 + k_3}{k_1}$$

Consider now the situation where there is rapid equilibration between S and ES, and a slower conversion to product P. This means that the substrate binds to the active site and departs several times before it is finally converted to product.

$$E + S \underset{k_2 \text{ fast}}{\overset{k_1 \text{ fast}}{\rightleftharpoons}} ES \xrightarrow[\text{slow}]{k_3} E + P$$

Under these conditions, the dissociation rate (k_2) of ES is much greater than the rate of formation of product (k_3). k_3 now becomes insignificant relative to k_2 and the equation simplifies to the following:

$$K_M = \frac{k_2 + k_3}{k_1} = \frac{k_2}{k_1}$$

In this situation, K_M effectively equals the dissociation constant of ES, and can be taken as a measure of how strongly the substrate binds to the enzyme.

$$[ES] \rightleftharpoons [E] + [S] \qquad \text{dissociation constant} = \frac{[E][S]}{[ES]}$$

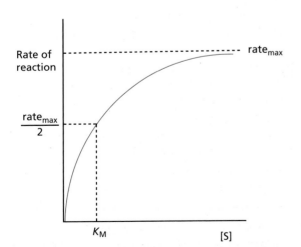

FIGURE 3.18 Reaction rate versus substrate concentration.

A high value of K_M indicates weak binding because the equilibrium is pushed to the right; a low K_M indicates strong binding because the equilibrium is to the left. K_M is also dependent on the particular substrate involved and on environmental conditions such as pH, temperature, and ionic strength.

The maximum rate is related to the total concentration of enzyme ($[E]_{total} = [E] + [ES]$) as follows:

$$\text{rate}_{max} = k_3 [E]_{total}$$

A knowledge of the maximum rate and the enzyme concentration allows the determination of k_3. For example, the enzyme **carbonic anhydrase** catalyses the formation of hydrogen carbonate and does so at a maximum rate of 0.6 moles of hydrogen carbonate molecules formed per second for a solution containing 10^{-6} moles of the enzyme. Altering the above equation, k_3 can be determined as follows:

$$k_3 = \frac{\text{rate}_{max}}{[E]_{total}} = \frac{0.6 \ \text{moles s}^{-1}}{10^{-6} \ \text{moles}} = 600000 \ \text{s}^{-1}$$

Therefore, each enzyme is catalysing the formation of 600 000 hydrogen carbonate molecules per second. The turnover number is the time taken for each catalysed reaction to take place, i.e. $1/600\,000 = 1.7 \ \mu s$.

3.8.2 Lineweaver–Burk plots

A problem related to Michaelis–Menten kinetics is the fact that there may not be sufficient data points to determine whether the curve of the Michaelis–Menten plot has reached a maximum value or not. This means that values for the maximum rate and K_M are likely to be inaccurate. More accurate values for these properties can be obtained by plotting the reciprocals of the rate and the

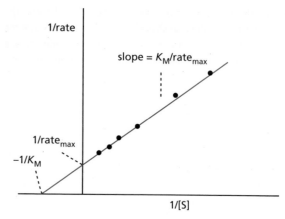

FIGURE 3.19 Lineweaver–Burk plot.

substrate concentration to give a Lineweaver–Burk plot (Fig. 3.19):

$$\frac{1}{\text{rate}} = \frac{K_M}{\text{rate}_{max}} \cdot \frac{1}{S} + \frac{1}{\text{rate}_{max}} \qquad y = m.x + c$$

The maximum rate can then be obtained from the intersect of the line with the y-axis, while K_M can be obtained from the slope of the line, or the intersect with the x-axis.

KEY NOTES

- The Michaelis–Menten equation relates the rate of an enzyme-catalysed reaction to substrate concentration.

- The Michaelis constant is equal to the substrate concentration at which the rate of the enzyme catalysed reaction is half of its maximum value.

- A Lineweaver–Burk plot provides more accurate values for the maximum rate and K_M.

QUESTIONS

1. Enzymes can be used in organic synthesis. For example, the reduction of an aldehyde is carried out using aldehyde dehydrogenase. Unfortunately, this reaction requires the use of the cofactor NADH, which is expensive and is used up in the reaction. If ethanol is added to the reaction, only catalytic amounts of cofactor are required. Why?

2. Acetylcholine is the substrate for the enzyme acetylcholinesterase. Suggest what sort of binding interactions could be involved in holding acetylcholine to the active site.

Acetylcholine

3. The ester bond of acetylcholine is hydrolysed by acetylcholinesterase. Suggest a mechanism by which the enzyme catalyses this reaction.

4. Suggest how binding interactions might make acetylcholine more susceptible to hydrolysis.

5. 17β-Hydroxysteroid dehydrogenase type 1 (17β-HSD1) is an enzyme that catalyses the conversion of estrone to estradiol in the presence of the cofactor NADH. The initial rate data for the enzyme-catalysed reaction in the absence of an inhibitor is as follows:

Substrate concentration (10^{-2} mol dm^{-3})	5	10	25	50	100
Initial rate (10^{-1} mol dm^{-3} s^{-1})	28.6	51.5	111	141	145

Create a Michaelis–Menten plot and a Lineweaver–Burk plot. Use both plots to calculate the values of K_M and the maximum rate of reaction. Identify which plot is likely to give the more accurate results and explain why this is the case.

6. Lactate dehydrogenase has a 1000-fold selectivity for lactate as a substrate over malate. However, if a mutation occurs that alters an active site glutamine residue to an arginine residue, the enzyme shows a 10000-fold selectivity for malate over lactate. Explain this astonishing transformation.

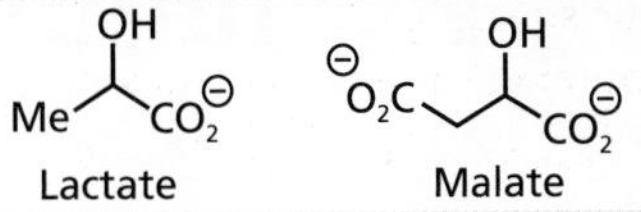

Lactate   Malate

7. Why should the hydrogen bonding identified in Fig. 3.10 be stronger with the transition state than with the substrate?

Multiple-choice questions are available on the Online Resource Centre at www.oxfordtextbooks.co.uk/orc/patrick6e/

FURTHER READING

Broadwith, P. (2010) Enzymes do the twist. *Chemistry World*, February, p.30.

Teague, S. J. (2003) Implications of protein flexibility for drug discovery. *Nature Reviews Drug Discovery*, **2**, 527–41.

Titles for general further reading are listed on p.845.

4 Receptors: structure and function

In this chapter we discuss the structure and function of receptors. Drug action at receptors is discussed in Chapter 8 and in other chapters throughout the text.

4.1 Role of the receptor

Receptors are proteins which are by far the most important drug targets in medicine. They are implicated in ailments such as pain, depression, Parkinson's disease, psychosis, heart failure, asthma, and many other problems. What are these receptors and what do they do?

In a complex organism, there has to be a communication system between cells. After all, it would be pointless if individual heart cells were to contract at different times. The heart would then be a wobbly jelly and totally useless in its function as a pump. Communication is essential to ensure that all heart muscle cells contract at the same time. The same is true for all the organs and tissues of the body if they are to operate in a coordinated and controlled fashion.

Control and communication come primarily from the brain and spinal column (the **central nervous system**), which receives and sends messages via a vast network of nerves (Fig. 4.1). The detailed mechanism by which nerves transmit messages along their length need not concern us here (see Appendix 4). It is sufficient for our purposes to think of the message as being an electrical pulse which travels down the nerve cell (**neuron**) towards the target, whether that be a muscle cell or another neuron. If that was all there was to it, it would be difficult to imagine how drugs could affect this communication system. However, there is one important feature that is crucial to our understanding of drug action. Neurons do not connect directly to their target cells. They stop just short of the cell surface. The distance is minute, about 100 Å, but it is a space that the electrical 'pulse' is unable to jump.

Therefore, there has to be a method of carrying the message across the gap between the nerve ending and the target cell. The problem is solved by the release of a chemical messenger called a **neurotransmitter** from the nerve cell (Fig. 4.2). Once released, this chemical messenger diffuses across the gap to the target cell, where it binds and interacts with a specific protein (receptor) embedded in the cell membrane. This process of binding leads to a series or cascade of secondary effects, which results either in a flow of ions across the cell membrane or in the switching on (or off) of enzymes inside the target cell. A biological response then results, such as the contraction of a muscle cell or the activation of fatty acid metabolism in a fat cell.

The first person to propose the existence of receptors was Langley in 1905. Up until that point, it was thought that drugs acted to prevent the release of neurotransmitter from the neuron, but Langley was able to show that certain target cells responded to the drug nicotine, even when the neurons supplying those cells were dead.

So far, we have talked about cellular communication involving neurons and neurotransmitters, but cells also receive chemical messages from circulating **hormones**. Once again, receptors are responsible for binding these messengers and triggering a series of secondary effects.

We shall consider these secondary effects and how they result in a biological action in Chapter 5, but for the moment, the important thing to note is that the communication system depends crucially on a chemical messenger. As a chemical process is involved, it should be possible for other chemicals (drugs) to interfere or interact with the process.

4.2 Neurotransmitters and hormones

There are a large variety of messengers which interact with receptors, and they vary significantly in structure and complexity. Some neurotransmitters are simple molecules, such as monoamines (e.g. **acetylcholine**,

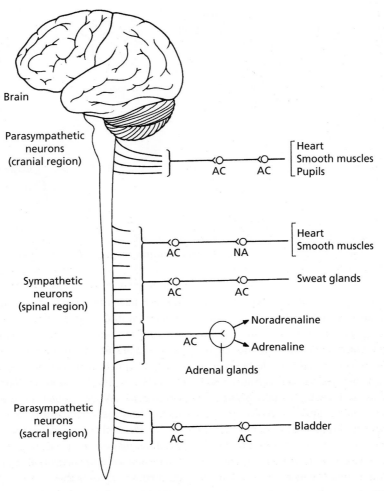

FIGURE 4.1 The central nervous system (AC = acetylcholine; NA = noradrenaline). Taken from J. Mann, *Murder, magic, and medicine*, Oxford University Press (1992), with permission.

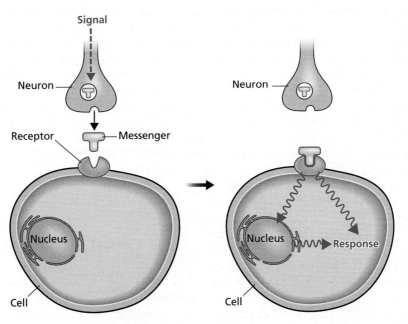

FIGURE 4.2 Neurotransmitters act as chemical messengers that bind to receptors and trigger reactions within a cell.

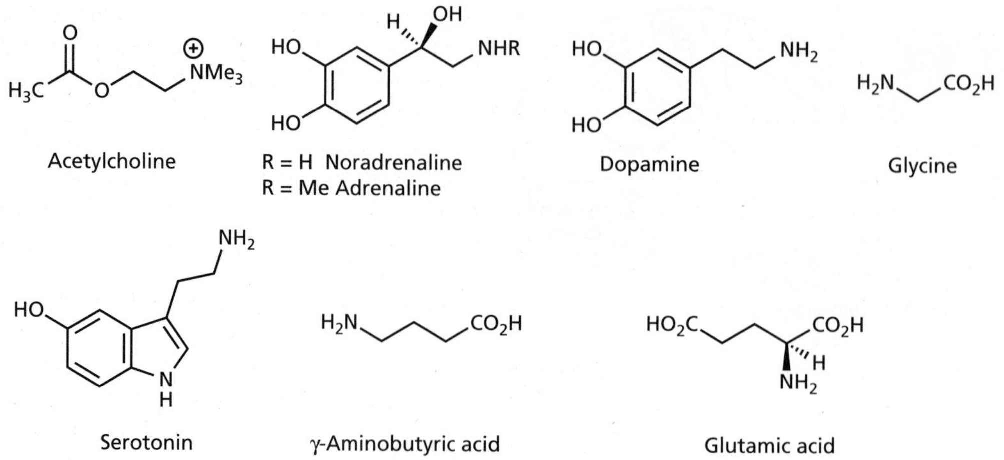

Acetylcholine

R = H Noradrenaline
R = Me Adrenaline

Dopamine

Glycine

Serotonin

γ-Aminobutyric acid

Glutamic acid

FIGURE 4.3 Examples of neurotransmitters and the hormone, adrenaline.

noradrenaline, **dopamine**, and **serotonin**) or amino acids (e.g. γ-**aminobutyric acid** (GABA), **glutamic acid**, and **glycine**) (Fig. 4.3). Even the calcium ion can act as a chemical messenger. Other chemical messengers are more complex in structure and include lipids such as **prostaglandins**; purines such as **adenosine** or **ATP** (Chapter 6); neuropeptides such as **endorphins** and **enkephalins** (section 24.8); peptide hormones such as **angiotensin** or **bradykinin**; and even enzymes such as **thrombin**.

In general, a neuron releases mainly one type of neurotransmitter, and the receptor which awaits it on the target cell will be specific for that messenger. However, that does not mean that the target cell has only one type of receptor protein. Each target cell has a large number of nerves communicating with it and they do not all use the same neurotransmitter (Fig. 4.4). Therefore, the target

cell will have other types of receptors specific for those other neurotransmitters. It may also have receptors waiting to receive messages from chemical messengers that have longer distances to travel. These are the hormones released into the circulatory system by various glands in the body. The best known example of a hormone is **adrenaline**. When danger or exercise is anticipated, the adrenal medulla gland releases adrenaline into the bloodstream where it is carried round the body, preparing it for vigorous exercise.

Hormones and neurotransmitters can be distinguished by the route they travel and by the way they are released, but their action when they reach the target cell is the same. They both interact with a receptor and a message is received. The cell responds to that message and adjusts its internal chemistry accordingly, and a biological response results.

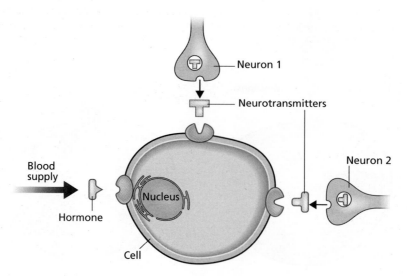

FIGURE 4.4 Target cell containing various receptors specific to different types of messenger.

4.3 Receptor types and subtypes

Receptors are identified by the specific neurotransmitter or hormone which activates them. Thus, the receptor activated by **dopamine** is called the **dopaminergic receptor**, the receptor activated by **acetylcholine** is called the **cholinergic receptor**, and the receptor activated by **adrenaline** or **noradrenaline** is called the **adrenergic receptor** or **adrenoceptor**.

However, not all receptors activated by the same chemical messenger are exactly the same throughout the body. For example, the adrenergic receptors in the lungs are slightly different from the adrenergic receptors in the heart. These differences arise from slight variations in amino acid composition, and if the variations are in the binding site, it allows medicinal chemists to design drugs which can distinguish between them. For example, adrenergic drugs can be designed to be 'lung' or 'heart' selective. In general, there are various **types** of a particular receptor and various **subtypes** of these, which are normally identified by numbers or letters. Having said that, some of the early receptors that were discovered were named after the natural products which bound to them; for example the muscarinic and nicotinic types of cholinergic receptor (section 22.4).

Some examples of receptor types and subtypes are given in Fig. 4.16. The identification of many of these subtypes is relatively recent and the current emphasis in medicinal chemistry is to design drugs that are as selective as possible for receptor types and subtypes so that the drugs are tissue selective and have fewer side effects.

4.4 Receptor activation

A receptor is a protein molecule usually embedded within the cell membrane with part of its structure exposed on the outside of the cell. The protein surface is a complicated shape containing hollows, ravines, and ridges, and somewhere within this complicated geography there is an area that has the correct shape to accept the incoming messenger. This area is known as the **binding site** and is analogous to the active site of an enzyme (section 3.3). When the chemical messenger fits into this site, it 'switches on' the receptor molecule and a message is received (Fig. 4.5). However, there is an important difference between enzymes and receptors in that the chemical messenger does not undergo a chemical reaction. It fits into the binding site of the receptor protein, passes on its message, and then leaves unchanged. If no reaction takes place, what *has* happened? How does the chemical messenger tell the receptor its message and how is this message conveyed to the cell? The first thing to note is that when the messenger fits the binding site of the protein receptor, it causes the binding site to change shape. This is known as an **induced fit**. This, in turn, has wider ramifications since there is a knock-on effect which causes the overall protein to change shape. But how does an induced fit happen, and what is the significance of the receptor changing shape?

4.5 How does the binding site change shape?

As we have seen, the binding site of a receptor changes shape when a chemical messenger fits into it. This is not a moulding process in which the binding site wraps itself around the messenger. Instead, the induced fit is brought about by the intermolecular binding interactions that take place between the messenger and the binding site. This is exactly the same process that occurs when a substrate binds to the active site of an enzyme (section 3.5.1), but in this situation no catalysed reaction follows binding.

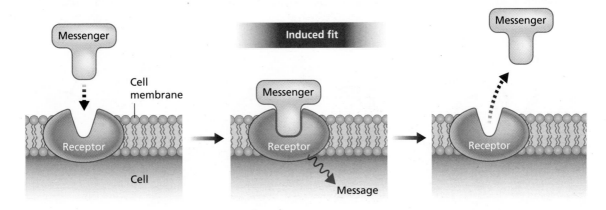

FIGURE 4.5 Binding of a chemical messenger to a protein receptor.

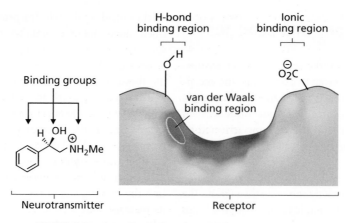

FIGURE 4.6 A hypothetical receptor and neurotransmitter.

To illustrate how binding interactions result in an induced fit, let us consider a hypothetical neurotransmitter and a hypothetical binding site, as shown in Fig. 4.6. The neurotransmitter has an aromatic ring that can take part in van der Waals interactions, an alcohol OH group that can take part in hydrogen bonding interactions, and a charged nitrogen centre that can take part in ionic or electrostatic interactions. These functional groups are the messenger's **binding groups**.

The hypothetical binding site contains three **binding regions** which contain functional groups that are complimentary to the binding groups of the messenger. The messenger fits into the binding site such that intermolecular interactions take place between the messenger's binding groups and the receptor's binding regions (Fig. 4.7). However, the fit is not perfect. In the diagram, there are good van der Waals and hydrogen bond interactions, but the ionic interaction is not as strong as it could be. The ionic binding region is close enough to have a weak interaction with the messenger, but not close enough for the optimum interaction. The receptor protein therefore alters shape to bring the carboxylate group closer to the positively charged nitrogen, and to obtain a stronger interaction. As a result, the

shape of the binding site is altered and an induced fit has taken place.

The illustration shown here is a simplification of the induced fit process and, in reality, both the messenger and the binding site take up different conformations or shapes to maximize the bonding forces between them. As with enzyme–substrate binding, there is a fine balance involved in receptor–messenger binding. The bonding forces must be large enough to change the shape of the binding site, but not so strong that the messenger is unable to leave. Most neurotransmitters bind quickly to their receptors, then 'shake themselves loose' once their message has been received.

We have now seen how a chemical messenger can cause an induced fit in the binding site of a receptor protein. However, this induced fit has a knock-on effect which alters the overall shape of the protein. It is this overall shape change that is crucial to the activation of a receptor, and in its ability to trigger an amazing 'domino effect' which affects the cell's internal chemistry. This domino effect involves several different proteins and enzymes, and ultimately produces an observed biological effect. The process by which this takes place is called **signal transduction** and is covered in more detail in

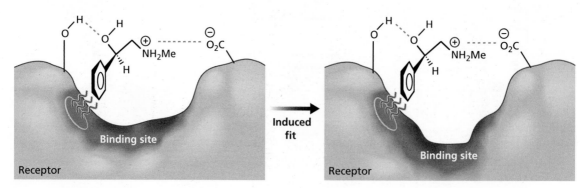

FIGURE 4.7 Binding of a hypothetical neurotransmitter to a binding site resulting in an induced fit.

Chapter 5. Signal amplification is an important feature of this process since it means that a relatively small number of neurotransmitter molecules can have a dramatic effect on the cell's internal chemistry. In this chapter, we shall focus on the structure of different receptors and the process by which they are activated and trigger the signal transduction process.

There are three main types (or families) of membrane-bound receptors:

- ion channel receptors
- G-protein-coupled receptors
- kinase-linked receptors.

We shall consider each of these in turn in sections 4.6–4.8.

KEY POINTS

- Most receptors are membrane-bound proteins which contain an external binding site for hormones or neurotransmitters. Binding results in an induced fit that changes the receptor conformation. This triggers a series of events that ultimately results in a change in cellular chemistry.

- Neurotransmitters and hormones do not undergo a reaction when they bind to receptors. They depart the binding site unchanged once they have passed on their message.

- The interactions that bind a chemical messenger to the binding site must be strong enough to allow the chemical message to be received, but weak enough to allow the messenger to depart.

- Binding groups are the functional groups present on a messenger molecule which are used for binding it to the receptor binding site.

- Binding regions are regions of the receptor binding site which contain functional groups capable of forming intermolecular bonds to the binding groups of a messenger molecule.

4.6 **Ion channel receptors**

4.6.1 General principles

Some neurotransmitters operate by controlling ion channels. What are these ion channels and why are they necessary? Let us look again at the structure of the cell membrane.

As described in section 1.2.1, the membrane is made up of a bilayer of phospholipid molecules, and so the middle of the cell membrane is 'fatty' and hydrophobic. Such a barrier makes it difficult for polar molecules or ions to move in or out of the cell. Yet it is important that these species should cross. For example, the movement of sodium and potassium ions across the membrane is crucial to the function of nerves (Appendix 4). It seems an intractable problem, but once again the ubiquitous proteins provide the answer by forming ion channels.

Ion channels are complexes made up of protein subunits which traverse the cell membrane (Fig. 4.8). The centre of the complex is hollow and is lined with polar amino acids to give a hydrophilic tunnel or pore.

Ions can cross the fatty barrier of the cell membrane by moving through these hydrophilic channels or tunnels. But there has to be some control. In other words, there has to be a 'lock gate' that can be opened or closed as required. It makes sense that this lock gate should be controlled by a receptor protein sensitive to an external chemical messenger, and this is exactly what happens. In fact, the receptor protein is an integral part of the ion channel complex and is one or more of the constituent protein subunits. In the resting state, the ion channel is closed (i.e. the lock gate is shut). However, when a chemical messenger binds to the external binding site of the receptor protein, it causes an induced fit which causes the protein to change shape. This in turn causes the overall protein complex to change shape, opening up the lock

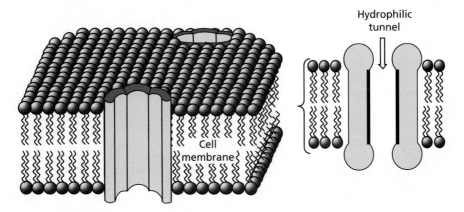

FIGURE 4.8 The structure of an ion channel. The bold lines show the hydrophilic sides of the channel.

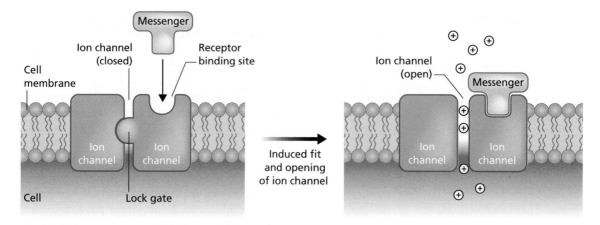

FIGURE 4.9 Lock-gate mechanism for opening ion channels.

gate and allowing ions to pass through the ion channel (Fig. 4.9). We shall look at this in more detail in section 4.6.3.

The operation of an ion channel explains why the relatively small number of neurotransmitter molecules released by a neuron is able to have such a significant biological effect on the target cell. By opening a few ion channels, several thousand ions are mobilized for each neurotransmitter molecule involved. Moreover, the binding of a neurotransmitter to an ion channel results in a rapid response, measured in a matter of milliseconds. This is why the synaptic transmission of signals between neurons usually involves ion channels.

Ion channels are specific for certain ions. For example there are different cationic ion channels for sodium (Na$^+$), potassium (K$^+$), and calcium (Ca^{2+}) ions. There are also anionic ion channels for the chloride ion (Cl$^-$). The ion selectivity of different ion channels is dependent on the amino acids lining the ion channel, and it is interesting to note that the mutation of just one amino acid in this area is sufficient to change a cationic selective ion channel to one that is selective for anions.

4.6.2 Structure

The protein subunits that make up an ion channel are actually **glycoproteins** (sections 2.5 and 10.7.1) but we will refer to them here as proteins. The protein subunits in an ion channel are not identical. For example, the ion channel controlled by the nicotinic cholinergic receptor is made up of five subunits of four different types (two α, β, γ, δ); the ion channel controlled by the glycine receptor is made up of five subunits of two different types (3 × α, 2 × β) (Fig. 4.10).

The receptor protein in the ion channel controlled by glycine is the α-subunit. Three such subunits are present, all capable of interacting with glycine. However, the situation is slightly more complex in the nicotinic

ion channel controlled by the neurotransmitter acetylcholine. Most of the binding site is on the α-subunit, but there is some involvement from neighbouring subunits. In this case, the ion channel complex as a whole might be viewed as the receptor.

Let us now concentrate on the individual protein subunits. Although there are various types of these, they all fold up in a similar manner such that the protein chain

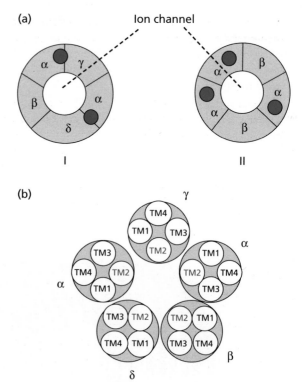

FIGURE 4.10 (a) Pentameric structure of ion channels (transverse view): I, ion channel controlled by a nicotinic cholinergic receptor; II, ion channel controlled by a glycine receptor. The coloured circles indicate ligand binding sites. (b) Transverse view of I including transmembrane regions.

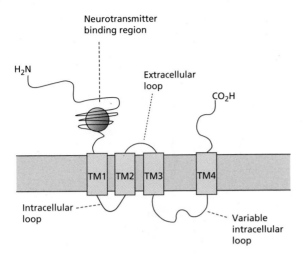

FIGURE 4.11 Structure of the four transmembrane (4-TM) receptor subunit.

traverses the cell membrane four times. This means that each subunit has four transmembrane (TM) regions which are hydrophobic in nature. These are labelled TM1–TM4. There is also a lengthy *N*-terminal extracellular chain which (in the case of the α-subunit) contains the ligand-binding site (Fig. 4.11).

The subunits are arranged such that the second transmembrane region of each subunit faces the central pore of the ion channel (Fig. 4.10). We shall see the significance of this when we look at the next section.

4.6.3 Gating

When the receptor binds a ligand it changes shape (the induced fit). This has a knock on effect on the protein complex, causing the ion channel to open—a process called **gating** (Fig. 4.12).

The binding of a neurotransmitter to its binding site causes a conformational change in the receptor, which eventually opens up the central pore and allows ions to flow. This conformational change is quite complex, involving several knock-on effects from the initial binding process. This must be so, as the binding site is quite far from the lock gate. Studies have shown that the lock gate is made up of five kinked α-helices (the 2-TM region) where one helix is contributed by each of the five protein subunits. In the closed state the kinks are pointing towards each other. The conformational change induced by ligand binding causes each of these helices to rotate such that the kink points the other way, thus opening up the pore (Fig. 4.13).

4.6.4 Ligand-gated and voltage-gated ion channels

The ion channels that we have discussed so far are called **ligand-gated ion channels** as they are controlled by chemical messengers (ligands). There are other types of ion channel which are not controlled by ligands, but are sensitive instead to the potential difference that exists across a cell membrane—the **membrane potential**. These ion channels are present in the axons of excitable cells (i.e. neurons) and are called **voltage-gated ion channels**. They are crucial to the transmission of a signal along individual neurons and are important drug targets for local anaesthetics. A description of these ion channels is given in Appendix 4. Voltage-gated ion channels also play an important role in controlling blood pressure and heart rate (section 26.6.1).

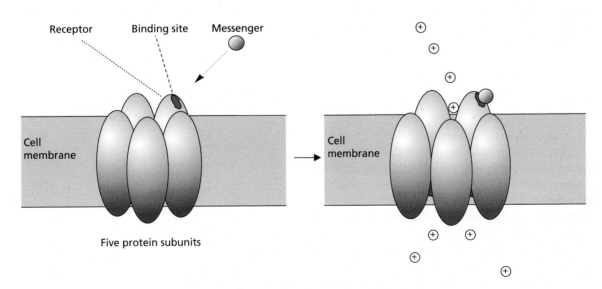

FIGURE 4.12 Opening of an ion channel (gating).

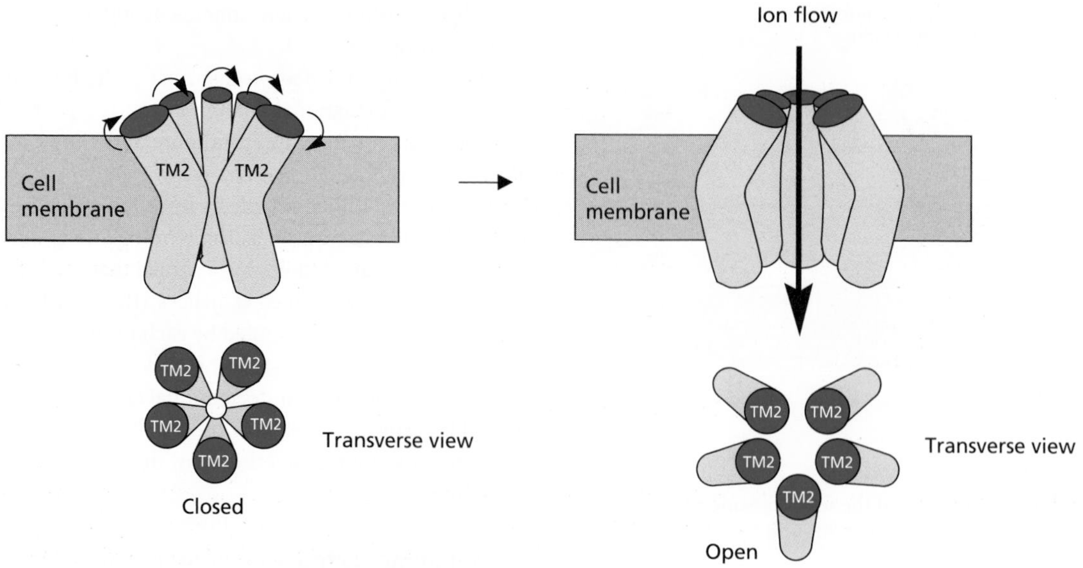

FIGURE 4.13 Opening of the 'lock gate' in an ion channel.

- Receptors controlling ion channels are an integral part of the ion channel. Binding of a messenger induces a change in shape, which results in the rapid opening of the ion channel.

- Receptors controlling ion channels are called ligand-gated ion channel receptors. Many consist of five protein subunits with the receptor binding site being present on one or more of the subunits.

- Binding of a neurotransmitter to an ion channel receptor causes a conformational change in the protein subunits such that the second transmembrane domain of each subunit rotates to open the channel.

4.7 G-protein-coupled receptors

4.7.1 General principles

The G-protein-coupled receptors (GPCRs) are some of the most important drug targets in medicinal chemistry. Indeed, some 30% of all drugs on the market act by binding to these receptors. They include the **muscarinic receptor** (section 22.11) **adrenergic receptors** (section 23.2) and **opioid receptors** (section 24.4).

The response from activated G-protein-coupled receptors is measured in seconds. This is slower than the response of ion channels, but faster than the response of kinase-linked receptors (section 4.8), which takes a matter of minutes. There are a large number of different G-protein-coupled receptors interacting with important neurotransmitters such as acetylcholine, dopamine, histamine, serotonin, glutamate, and noradrenaline. Other G-protein-coupled receptors are activated by peptide and protein hormones such as the enkephalins and endorphins.

G-Protein-coupled receptors are membrane-bound proteins that can activate proteins called **G-proteins** (Fig. 4.14). The G-proteins then fragment and act as **signal proteins** to activate or deactivate membrane-bound enzymes (sections 5.1–5.3). Consequently, activation of a GPCR by a chemical messenger triggers a process that influences the reactions that take place within the cell.

The GPCR protein is embedded within the membrane, with the binding site for the chemical messenger exposed on the outer surface. On the inner surface, there is another binding site which is normally closed (Fig. 4.14, frame 1). When the chemical messenger binds to its binding site, the GPCR protein changes shape, opening up the binding site on the inner surface. This new binding site is recognized by the G-protein, which then binds (Fig. 4.14, frame 2). The G-protein is attached to the inner surface of the cell membrane and is made up of three protein subunits, but once it binds to the GPCR, the complex is destabilized and fragments to a monomer and a dimer (Fig. 4.14, frame 3). These then interact with membrane-bound enzymes to continue the signal transduction process (sections 5.1–5.3).

There are several different G-proteins, which are recognized by different types of receptor. Some of the activated subunits from these G-proteins have an inhibitory effect on a membrane-bound enzyme, while others have a stimulatory effect. Nevertheless, the mechanism by which the G-protein is activated by fragmentation is

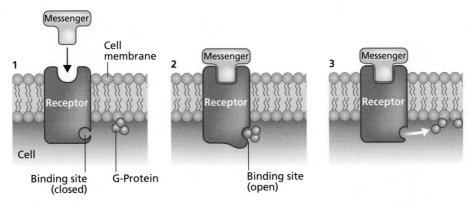

FIGURE 4.14 Activation of a G-protein-coupled receptor and G-protein.

the same. There is also a substantial amplification of the signal in this process, as one activated receptor activates several G-proteins.

4.7.2 Structure

The G-protein-coupled receptors fold up within the cell membrane such that the protein chain winds back and forth through the cell membrane seven times (Fig. 4.15). Each of the seven transmembrane sections is hydrophobic and helical in shape, and it is usual to assign these helices with roman numerals (I, II, etc.) starting from the *N*-terminus of the protein. Owing to the number of transmembrane regions, the G-proteins are also called **7-TM receptors**. The binding site for the G-protein is situated on the intracellular side of the protein, and involves part of the *C*-terminal chain as well as part of the variable intracellular loop (so called because the length of this loop varies between different types of receptor). As one might expect, the binding site for the neurotransmitter or hormone messenger is on the extracellular portion of the protein. The exact position of the binding site varies from receptor to receptor. For example, the binding site for the adrenergic receptor is in a deep binding pocket between the transmembrane helices, whereas the binding site for the glutamate receptor involves the *N*-terminal chain, and is situated above the surface of the cell membrane.

4.7.3 The rhodopsin-like family of G-protein-coupled receptors

The G-protein-coupled receptors include the receptors for some of the best-known chemical messengers in medicinal chemistry (e.g. glutamic acid, GABA, noradrenaline, dopamine, acetylcholine, serotonin, prostaglandins, adenosine, endogenous opioids, angiotensin, bradykinin, and thrombin). Considering the structural variety of the chemical messengers involved, it is remarkable that the overall structures of the GPCRs are so similar. Nevertheless, despite their similar overall structure, the amino acid

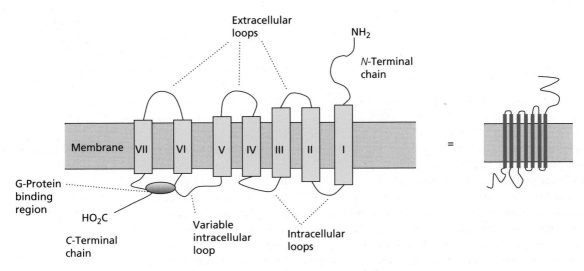

FIGURE 4.15 Structure of G-protein-coupled receptors.

sequences of the receptors vary quite significantly. This implies that the GPCRs have evolved over millions of years from an ancient common ancestral protein. Comparing the amino acid sequences of the GPCRs allows us to construct an evolutionary tree and to group the receptors of this superfamily into various sub-families, which are defined as class A (rhodopsin-like receptors), class B (secretin-like receptors), and class C (metabotropic glutamate-like and pheromone receptors). The most important of these, as far as medicinal chemistry is concerned, is the rhodopsin-like family—so called because the first receptor of this family to be studied in detail was the rhodopsin receptor itself, a receptor involved in the visual process. A study of the evolutionary tree of rhodopsin-like receptors throws up some interesting observations (Fig. 4.16).

First of all, the evolutionary tree illustrates the similarity between different kinds of GPCRs based on their relative positions on the tree. Thus, the muscarinic, α-adrenergic, β-adrenergic, histamine, and dopamine receptors have evolved from a common branch of the evolutionary tree and have greater similarity to each other than to any receptors arising from an earlier evolutionary branch (e.g. the **angiotensin receptor**). Such receptor similarity may prove a problem in medicinal chemistry. Although the receptors are distinguished by different neurotransmitters or hormones in the body, a drug may not manage to make that distinction. Therefore, it is important to ensure that any new drug aimed at one kind of receptor (e.g. the dopamine receptor) does not interact with a similar kind of receptor (e.g. the muscarinic receptor).

Receptors have further evolved to give receptor **types** and **subtypes** which recognize the same chemical messenger, but are structurally different. For example, there are two types of adrenergic receptor (α and β), each of which has various subtypes (α_1, α_{2A}, α_{2B}, α_{2C}, β_1, β_2, β_3). There are two types of cholinergic receptor—nicotinic (an ion channel receptor) and muscarinic (a 7-TM receptor). Five subtypes of the muscarinic cholinergic receptor have been identified.

The existence of receptor subtypes allows the possibility of designing drugs that are selective for one receptor subtype over another. This is important, because one receptor subtype may be prevalent in one part of the body (e.g. the gut) while a different receptor subtype is prevalent in another part (e.g. the heart). Therefore, a drug that is designed to interact selectively with the receptor subtype in the gut is less likely to have side effects on the heart. Even if the different receptor subtypes are present in the same part of the body, it is still important to make drugs as selective as possible, because different receptor subtypes frequently activate different signalling systems, leading to different biological results.

A closer study of the evolutionary tree reveals some curious facts about the origins of receptor subtypes. As one might expect, various receptor subtypes have diverged from a common evolutionary branch (e.g. the dopamine subtypes D2, D3, D4). This is known as **divergent evolution** and there should be a close structural similarity between these subtypes. However, receptor subtypes are also found in separate branches of the tree. For example, the dopamine receptor subtypes ($D1_A$, $D1_B$, and D5) have developed from a different evolutionary branch. In other words, the ability of a receptor to bind dopamine has developed in different evolutionary branches—an example of **convergent evolution**.

Consequently, there may sometimes be greater similarities between receptors which bind different ligands and have evolved from the same branch of the tree,

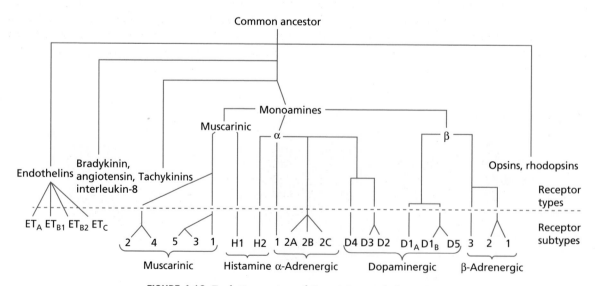

FIGURE 4.16 Evolutionary tree of G-protein-coupled receptors.

than there are between the various subtypes of receptors which bind the same ligand. For example, the histamine H$_1$ receptor resembles a muscarinic receptor more closely than it does the histamine H$_2$ receptor. Again, this has important consequences in drug design because there is an increased possibility that a drug aimed at a muscarinic receptor may also interact with a histamine H$_1$ receptor and lead to unwanted side effects.

As these receptors are membrane-bound, it is not easy to crystallize them for X-ray crystallographic studies. However, the X-ray crystal structures of the β$_2$- and β$_1$-adrenoceptors have now been determined.

4.7.4 Dimerization of G-coupled receptors

There is strong evidence that some G-coupled receptors can exist as dimeric structures containing identical or different types of receptor—homodimers or heterodimers respectively. The presence of these receptor dimers appears to vary between different tissues and this has important consequences for drug design. An agent that is selective for one type of receptor would not normally affect other types. However, if receptor heterodimers are present, a 'communication' is possible between the component receptors such that an agent interacting with one half of the dimer may affect the activity of the other half. This is discussed further in section 24.9 with respect to opioid receptors.

KEY POINTS

- G-protein-coupled receptors activate signal proteins called G-proteins. Binding of a messenger results in the opening of a binding site for the signal protein. The latter binds and fragments, with one of the subunits departing to activate a membrane-bound enzyme.

- The G-protein-coupled receptors are membrane-bound proteins with seven transmembrane sections. The *C*-terminal chain lies within the cell and the *N*-terminal chain is extracellular.

- The location of the binding site differs between different G-protein-coupled receptors.

- The rhodopsin-like family of G-protein-coupled receptors includes many receptors that are targets for currently important drugs.

- Receptor types and subtypes recognize the same chemical messenger, but have structural differences; this makes it possible to design drugs that are selective for one type (or subtype) of receptor over another.

- Receptor subtypes can arise from divergent or convergent evolution.

- It is possible for some G-protein coupled receptors to exist as dimeric structures.

4.8 Kinase receptors

4.8.1 General principles

Kinase receptors are a superfamily of receptors which activate enzymes directly and do not require a G-protein (Fig. 4.17). **Tyrosine kinase receptors** are important examples of such receptors, and are proving to be highly important targets for novel anticancer drugs (section 21.6.2). The protein concerned plays the dual role of receptor and enzyme. The receptor protein is embedded within the cell membrane, with part of its structure exposed on the outer surface of the cell and part exposed on the inner surface. The outer surface contains the binding site for the chemical messenger, and the inner surface has an active site that is closed in the resting state. When a chemical messenger binds to the receptor, it causes the protein to change shape. This results in the active site being opened up, allowing the protein to act as an enzyme within the cell. The reaction which is catalysed is a phosphorylation reaction where tyrosine residues on a protein substrate are phosphorylated. An enzyme that

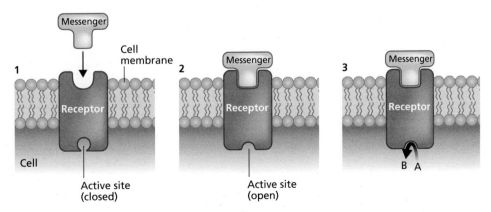

FIGURE 4.17 Enzyme activation.

catalyses phosphorylation reactions is known as a kinase enzyme and so the protein is referred to as a tyrosine kinase receptor. ATP is required as a cofactor to provide the necessary phosphate group. The active site remains open for as long as the messenger molecule is bound to the receptor, and so several phosphorylation reactions can occur, resulting in an amplification of the signal. A curiosity of this enzyme-catalysed reaction is that the substrate for the reaction is the receptor itself. This is explained more fully in section 4.8.3.

Tyrosine kinase receptors can be a single protein, as described above, or a protein complex where one protein acts as the receptor and another acts as the kinase. They are activated by a large number of polypeptide hormones, growth factors, and cytokines. Loss of function of these receptors can lead to developmental defects or hormone resistance. Over-expression can result in malignant growth disorders.

4.8.2 Structure of tyrosine kinase receptors

The basic structure of a single protein tyrosine kinase receptor consists of an extracellular region (the *N*-terminal chain) that includes the binding site for the chemical messenger, a single hydrophobic region that traverses the membrane as an α-helix of seven turns (just sufficient to traverse the membrane), and a *C*-terminal chain on the inside of the cell membrane (Fig. 4.18). The *C*-terminal region contains the catalytic active site for kinase activity. Examples of tyrosine kinase receptors like this include the receptors for the hormones **epidermal growth factor** and **vascular endothelial growth factor**.

4.8.3 Activation mechanism for tyrosine kinase receptors

A specific example of a tyrosine kinase receptor is the **epidermal growth factor receptor** (EGFR). As the name suggests, the endogenous ligand for EGFR is **epidermal growth factor** (EGF), which is a **bivalent ligand** that can bind to two EGF receptors at the same time. This results in **receptor dimerization** as well as activation of enzymatic activity. The dimerization process is important because the active site on each half of the receptor dimer catalyses the phosphorylation of accessible tyrosine residues on the other half (Fig. 4.19). If dimerization did not occur, no phosphorylation would take place. Note that these phosphorylations occur on the intracellular portion of the receptor protein chain. The relevance of these phosphorylation reactions will be explained in section 5.4.1. The important point to grasp at this stage is that an external chemical messenger has managed to convey its message to the interior of the cell without itself being altered or having to enter the cell.

Dimerization and auto-phosphorylation is a common theme for tyrosine kinase receptors. However, some of the receptors in this family already exist as dimers or tetramers and only require binding of the ligand. For example, the **insulin** receptor is a heterotetrameric complex (Fig. 4.20).

Some tyrosine kinase receptors bind ligands and dimerize in a similar fashion to the ones described above, but do not have inherent catalytic activity in their *C*-terminal chain. However, once they have dimerized, they can bind and activate a tyrosine kinase enzyme from the cytoplasm. The **growth hormone receptor** is an example of this type of receptor (Fig. 4.21). **Growth hormone** (GH) is an example of a family of chemical messengers called the **cytokines**. These include important polypeptide and protein chemical messengers, such as interleukins, interferons, erythropoietin (EPO), thrombopoietin, prolactin, and colony-stimulating factor. There are a corresponding variety of cytokine receptors that respond to these cytokines. Several of these cytokine receptors bind intracellular kinases known as **Janus kinases**. There are four members of the JAK family labelled JAK1, JAK2, JAK3, and TYK4, and the types of Janus kinase present in the activated receptor complex depend on the individual receptor involved (see also section 5.4.4).

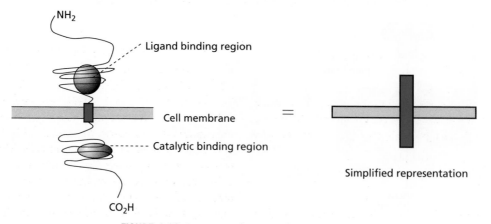

FIGURE 4.18 Structure of tyrosine kinase receptors.

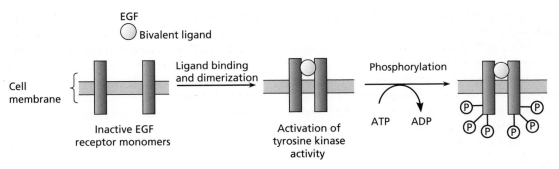

FIGURE 4.19 Activation mechanism for the epidermal growth factor receptor (EGFR).

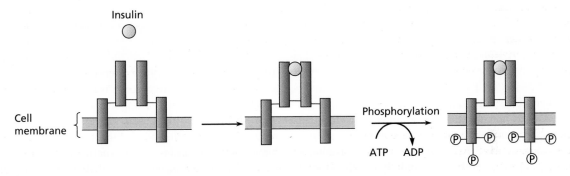

FIGURE 4.20 Ligand binding and activation of the insulin receptor.

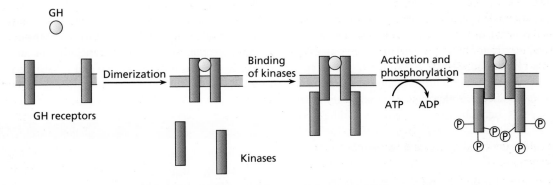

FIGURE 4.21 Activation of the growth hormone (GH) receptor.

4.8.4 Tyrosine kinase receptors as targets in drug discovery

Because of their key role in controlling cell growth and division, tyrosine kinase receptors are important targets for novel anticancer drugs. The following are some examples.

4.8.4.1 The ErbB family of tyrosine kinase receptors

The ErbB family of receptors are structurally related receptors that include the epidermal growth factor receptor (EGFR), which is also labelled as ErbB-1 or HER1 (HER1 refers to the human form of the receptor). Other receptors in the family are HER2 (or ErbB-2), Her 3 (or ErbB-3), and Her 4 (or ErbB-4). EGFR responds not only to EGF, but also to the chemical messenger **transforming growth factor α (TGF-α)**. Abnormal ErbB-2 receptors have been observed in breast cancer cells. These receptors are inherently active despite the lack of a ligand. Other cancer cells overproduce TGF-α, resulting in overactivation of EGFR. A number of kinase inhibitors that target the ErbB family of receptors are discussed in section 21.6.2.1.

4.8.4.2 Vascular endothelial growth factor receptors

The vascular endothelial growth factor receptor (**VEGFR**) is present in the cell membranes of blood vessel endothelial cells, and plays an important role in angiogenesis (section 21.1.9). There are three subtypes of vascular endothelial growth factor receptor, numbered 1, 2, and 3. There are also five varieties of vascular endothelial growth factor (VEGF) labelled A–E. A variety of different kinase inhibitors have been designed that inhibit VEGFR, in particular VEGFR-2 (section 21.6.2.9).

4.8.4.3 Platelet-derived growth factor receptor

There are two types of platelet-derived growth factor receptor (PDGFR) labelled alpha and beta. There are also four types of platelet-derived growth factors (PDGF) labelled A–D. PDGF can dimerize to form homodimers or heterodimers depending on the types of PDGF involved. Once formed, the PDGF dimer binds to its receptor and promotes receptor dimerization. Some cancer cells overproduce PDGF, resulting in overactivation of PDGFR. PDGFR inhibitors include **imatinib**, which is discussed in section 21.6.2.2.

4.8.4.4 Stem cell growth factor receptor

The mast/stem cell growth factor receptor (SCFR) is also known as c-KIT or CD117. It is present in several stem cells and is activated by stem cell factor (SCF)—an example of a cytokine messenger. Activation of the receptor plays an important role in the formation of sperm, melanin, and blood cells. Following dimerization, the receptor triggers a signal transduction process leading to cell growth, cell division, and cell differentiation. The receptor kinase is inhibited by imatinib (section 21.6.2.2).

4.8.4.5 Anaplastic lymphoma kinase (ALK)

Anaplastic lymphoma kinase (ALK) is part of the insulin receptor superfamily. It is also labelled as **CD246**. ALK is thought to play an important part in the embryonic development of the brain and neural systems, but its natural ligand is uncertain. Animal experiments indicate that ALK is present in only small quantities following birth. However, it has been detected in tumour cell lines involving anaplastic large cell lymphoma. Such tumours can result from fusion genes that produce a fusion protein involving ALK linked to another protein. In these fusion proteins, ALK is constantly active, resulting in excessive cell growth and division. ALK inhibitors have proved useful in the treatment of non-small-cell lung cancer (section 21.6.2.6).

4.8.4.6 The RET receptor

The letters *RET* refer to the gene that codes for the RET receptor. The letters stand for REarranged during Transcription. The resulting tyrosine kinase receptor responds to neurotrophic factors, which are important to the growth, survival, and development of neurons. Point mutations and fusion genes involving the *RET* gene have resulted in RET proteins that are overactive and are associated with tumours of the thyroid, parathyroid, and adrenal glands. Kinase inhibitors for RET have been approved for the treatment of thyroid cancers (section 21.6.2.7).

4.8.4.7 Hepatocyte growth factor receptor or c-MET receptor

The hepatocyte growth factor receptor (HGFR) is also known as the c-MET receptor. It is normally only active in stem cells and plays an important role in embryonic and organ development. It is also involved in wound repair. The natural ligands for the c-MET receptor are **hepatocyte growth factor** and **scatter factor**. Deregulated MET receptors are associated with most solid tumours, and result in tumour growth, angiogenesis, and metastasis. This is observed in a large number of different types of cancer affecting different organs, and is the result of point mutations or fusion proteins where c-MET becomes inherently active. **Crizotinib** is an anticancer agent that inhibits both ALK and the c-MET receptor (section 21.6.2.6). **Cabozantinib** is an anticancer agent that inhibits both RET and c-MET (section 21.6.2.7).

KEY POINTS

- Tyrosine kinase receptors have an extracellular binding site for a chemical messenger, and an intracellular enzymatic active site which catalyses the phosphorylation of tyrosine residues in protein substrates.

- Ligand binding to the epidermal growth factor (EGF) receptor results in dimerization and opening of kinase active sites. The active site on one half of the dimer catalyses the phosphorylation of tyrosine residues present on the *C*-terminal chain of the other half.

- The insulin receptor is a preformed heterotetrameric structure which acts as a tyrosine kinase receptor.

- The growth hormone receptor dimerizes on binding its ligand, then binds and activates tyrosine kinase enzymes from the cytoplasm.

- Several tyrosine kinase receptors are important targets for anticancer drugs.

4.9 **Intracellular receptors**

Not all receptors are located in the cell membrane. Some receptors are within the cell, and are defined as intracellular receptors. There are about 50 members of this group and they are particularly important in directly regulating gene transcription. As a result, they are often called **nuclear hormone receptors** or **nuclear transcription factors**. The chemical messengers for these receptors include steroid hormones, thyroid hormones, and retinoids. In all these cases, the messenger has to pass through the cell membrane in order to reach its receptor, and so it has to be hydrophobic in nature. The response time resulting from the activation of the intracellular receptors is measured in hours or days, and is much slower than the response times of the membrane-bound receptors.

The intracellular receptors all have similar general structures. They consist of a single protein containing a ligand binding site at the C-terminus and a binding region for DNA near the centre (Fig. 4.22). The DNA binding region contains nine cysteine residues, eight of which are involved in binding two zinc ions. The zinc ions play a crucial role in stabilizing and determining the conformation of the DNA binding region. As a result, the stretches of protein concerned are called the **zinc finger domains**.

The DNA binding region for each receptor can identify particular nucleotide sequences in DNA. For example, the zinc finger domains of the **estrogen receptor** recognize the sequence 5′-AGGTCA-3′, where A, G, C, and T are adenine, guanine, cytosine, and thymine.

The mechanism by which intracellular receptors work is also very similar (Fig. 4.23). Once the chemical messenger (ligand) has crossed the cell membrane, it seeks out its receptor and binds to it at the ligand binding site. An induced fit takes place which causes the receptor to change shape. This in turn leads to a dimerization of the ligand–receptor complex. The dimer then binds to a protein called a **coactivator** and finally the whole complex binds to a particular region of the cell's DNA. As there are two receptors in the complex and two DNA binding regions, the complex recognizes two identical sequences of nucleotides in the DNA, separated by a short distance. For example, the estrogen ligand–receptor dimer binds to a nucleotide sequence of 5′-AGGTCANNNTGAC-CT-3′ where N can be any nucleic acid base. Depending on the complex involved, binding of the complex to DNA either triggers or inhibits the start of transcription, and affects the eventual synthesis of a protein.

4.10 **Regulation of receptor activity**

The role of allosteric binding sites in regulating the activity of enzymes was covered in section 3.6. Allosteric binding sites also play a role in regulating or modulating the activity of various receptors. These include ligand-gated ion channels such as the nicotinic and the γ-aminobutyric acid receptors, and several G-protein-coupled receptors such as the muscarinic, adenosine, and dopamine receptors. Structures that interact with these sites are called **allosteric modulators** and can either enhance or decrease the effect of the chemical messenger on the receptor (sections 8.2.7 and 8.3.2).

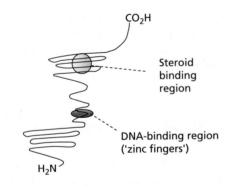

FIGURE 4.22 Structure of intracellular receptors.

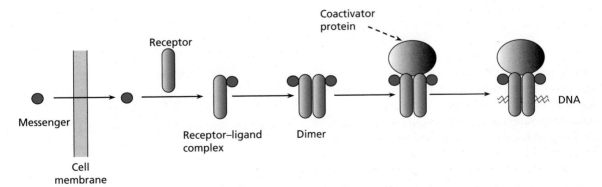

FIGURE 4.23 From messenger to control of gene transcription.

4.11 **Genetic polymorphism and receptors**

Genetic polymorphism was discussed in section 3.5.7 with respect to enzymes. Polymorphism is also responsible for receptors having subtle differences in structure and activity between individuals. In some cases, this can lead to diseases such as cancer (section 21.1.3).

KEY NOTES

• Intracellular receptors are located within the cell and are important in controlling transcription.

• The chemical messengers for intracellular receptors must be sufficiently hydrophobic to pass through the cell membrane.

• The binding of a ligand with an intracellular receptor results in dimerization and the formation of a transcription factor complex which binds to a specific nucleotide sequence on DNA.

QUESTIONS

1. Explain the distinction between a binding site and a binding region.

2. Consider the structures of the neurotransmitters shown in Fig. 4.3 and suggest what type of binding interactions could be involved in binding them to a receptor binding site. Identify possible amino acids in the binding site that have side chains which could take part in each of these binding interactions.

3. There are two main types of adrenergic receptor—the α- and β-adrenoceptors. Noradrenaline shows a slight selectivity for the α-receptor, whereas isoprenaline shows selectivity for the β-adrenoceptor. Adrenaline shows no selectivity and binds equally well to both the α- and

β-adrenoceptors. Suggest an explanation for these differences in selectivity.

Noradrenaline   Isoprenaline

4. Suggest why the transmembrane regions of many membrane-bound proteins are α-helices.

Multiple-choice questions are available on the Online Resource Centre at www.oxfordtextbooks.co.uk/orc/patrick6e/

FURTHER READING

Alexander, S. P. H., Mathie, A., and Peters, J. A. (2006) Guide to receptors and channels. *British Journal of Pharmacology*, **147**, Supplement 3.

Chalmers, D. T., and Behan, D. P. (2002) The use of constitutively active GPCRs in drug discovery and functional genomics. *Nature Reviews Drug Discovery*, **1**, 599–608.

Christopoulis, A. (2002) Allosteric binding sites on cell surface receptors: novel targets for drug discovery. *Nature Reviews Drug Discovery*, **1**, 198–210.

Kenakin, T. (2002) Efficacy at G-protein-coupled receptors. *Nature Reviews Drug Discovery*, **1**, 103–10.

Kobilka, B., and Schertler, G. F. X. (2008) New G-protein-coupled receptor crystal structures: insights and limitations. *Trends in Pharmacological Sciences*, **29**, 79–83.

Maehle, A.-H., Prull, C.-R., and Halliwell, R. F. (2002) The emergence of the drug receptor theory. *Nature Reviews Drug Discovery*, **1**, 637–41.

Palczewski, K. (2010) Oligomeric forms of G protein-coupled receptors (GCPRs). *Trends in Biochemical Sciences*, **35** (11), 595–600.

Sansom, C. (2010) Receptive receptors. *Chemistry World*, August, 52–5.

van Rijn, R. M., Whistler, J. L., and Waldhoer, M. (2010) Opioid-receptor-heteromer-specific trafficking and pharmacology. *Current Opinion in Pharmacology*, **20**, 73–9.

Zhan-Guo, G., and Jacobson, K. A. (2006) Allosterism in membrane receptors. *Drug Discovery Today*, **11**, 191–202.

Receptors and signal transduction

In Chapter 4, we discussed the structure and function of receptors. In this chapter, we consider what happens once a receptor has been activated. The interaction of a receptor with its chemical messenger is only the first step in a complex chain of events involving several secondary messengers, proteins, and enzymes that ultimately leads to a change in cell chemistry. These events are referred to as **signal transduction**. Unfortunately, a full and detailed account of these processes would fill a textbook in itself, and so the following account is focused mainly on the signal transduction processes that result from activation of G-protein-coupled receptors and kinase receptors. The signal transduction pathways following activation of G-protein-coupled receptors are of particular interest as 30% of all drugs on the market interact with these kinds of receptor. The transduction pathways for kinase receptors are also of great interest as they offer exciting new targets for novel drugs, particularly in the area of anticancer therapy (section 21.6.2). An understanding of the pathways and the various components involved, help to identify suitable drug targets.

5.1 Signal transduction pathways for G-protein-coupled receptors

G-protein-coupled receptors activate a signalling protein called a G-protein, which then initiates a signalling cascade involving a variety of enzymes. The sequence of events leading from the combination of receptor and ligand (the chemical messenger) to the final activation of a target enzyme is quite lengthy and so we shall look at each stage of the process in turn.

5.1.1 Interaction of the receptor–ligand complex with G-proteins

The first stage in the process is the binding of the chemical messenger or ligand to the receptor, followed by the binding of a G-protein to the receptor–ligand complex (Fig. 5.1). G-proteins are membrane-bound proteins situated at the inner surface of the cell membrane, and are made up of three protein subunits (α, β, and γ). The α-subunit has a binding pocket which can bind guanyl nucleotides (hence the name G-protein) and which binds **guanosine diphosphate (GDP)** when the G-protein is in the resting state. There are several types of G-protein (e.g. G_s, G_i/G_o, G_q/G_{11}) and several subtypes of these. Specific G-proteins are recognized by specific receptors. For example, G_s is recognized by the β-adrenoceptor, but not the α-adrenoceptor. However, in all cases, the G-protein acts as a molecular 'relay runner' carrying the message received by the receptor to the next target in the signalling pathway.

We shall now look at what happens in detail.

Firstly, the receptor binds its neurotransmitter or hormone (Fig. 5.1, frame 1). As a result, the receptor changes shape and exposes a new binding site on its inner surface (Fig. 5.1, frame 2). The newly exposed binding site now recognizes and binds a specific G-protein. Note that the cell membrane structure is a fluid structure and so it is possible for different proteins to 'float' through it. The binding process between the receptor and the G-protein causes the latter to change shape, which, in turn, changes the shape of the guanyl nucleotide binding site. This weakens the intermolecular bonding forces holding GDP, and so GDP is released (Fig. 5.1, frame 3).

However, the binding pocket does not stay empty for long because it is now the right shape to bind **GTP** (**guanosine triphosphate**). Therefore, GTP replaces GDP (Fig. 5.1, frame 4).

Binding of GTP results in another conformational change in the G-protein (Fig. 5.1, frame 5), which weakens the links between the protein subunits such that the α-subunit (with its GTP attached) splits off from the β- and γ-subunits (Fig. 5.1, frame 6). Both the α-subunit and the $\beta\gamma$-dimer then depart the receptor.

The receptor–ligand complex is able to activate several G-proteins in this way before the ligand departs and switches off the receptor. This leads to an amplification of the signal.

Both the α-subunit and the $\beta\gamma$-dimer are now ready to enter the second stage of the signalling mechanism. We shall first consider what happens to the α-subunit.

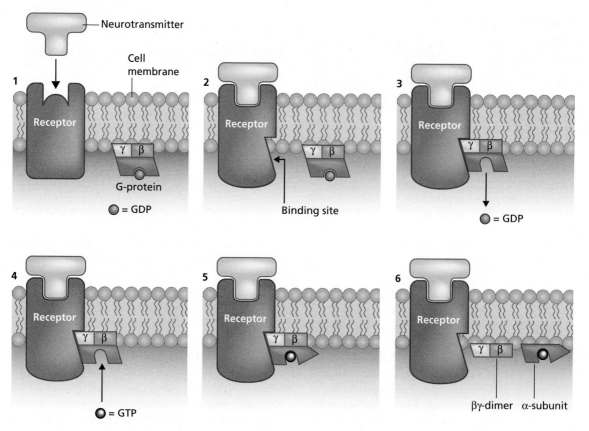

FIGURE 5.1 Activation of G-protein-coupled receptors and their interaction with G-proteins.

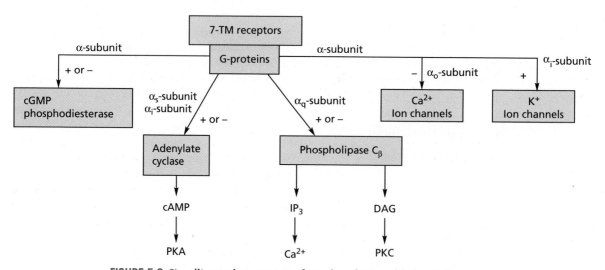

FIGURE 5.2 Signalling pathways arising from the splitting of different G-proteins.

5.1.2 Signal transduction pathways involving the α-subunit

The first stage of signal transduction (i.e. the splitting of a G-protein) is common to all of the 7-TM receptors. However, subsequent stages depend on what type of G-protein is involved and which specific α-subunit is formed (Fig. 5.2). Different α-subunits—there are at least 20 of them—have different targets and different effects:

- α_s stimulates adenylate cyclase.
- α_i inhibits adenylate cyclase and may also activate potassium ion channels.
- α_o activates receptors that inhibit neuronal calcium ion channels.
- α_q activates phospholipase C_β.

We do not have the space to study all these pathways in detail. Instead, we shall concentrate on two—the activation of **adenylate cyclase** and the activation of **phospholipase C_β**.

5.2 Signal transduction involving G-proteins and adenylate cyclase

5.2.1 Activation of adenylate cyclase by the α_s-subunit

The α_s-subunit binds to a membrane-bound enzyme called adenylate cyclase (or adenylyl cyclase) and 'switches' it on (Fig. 5.3). This enzyme now catalyses the synthesis of a molecule called cyclic AMP (cAMP) (Fig. 5.4). cAMP is an example of a **secondary messenger** which

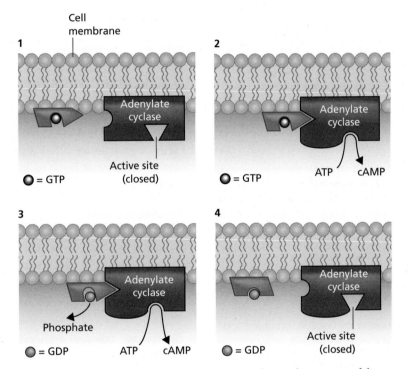

FIGURE 5.3 Interaction of α_s-subunit with adenylate cyclase and activation of the enzyme.

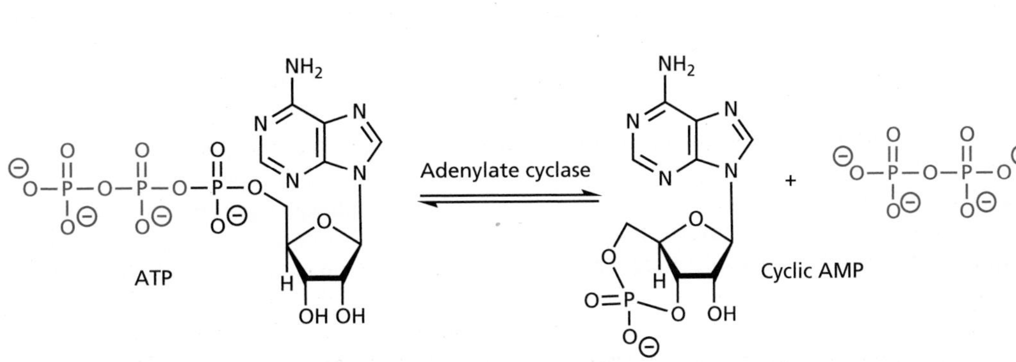

FIGURE 5.4 Synthesis of cyclic AMP.

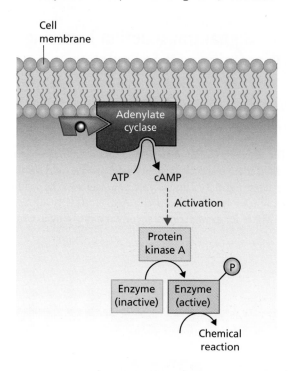

FIGURE 5.5 Activation of protein kinase A by cyclic AMP (℗ = phosphate).

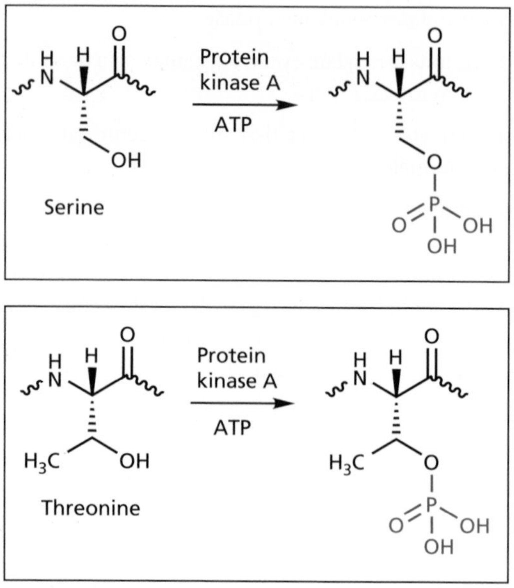

FIGURE 5.6 Phosphorylation of serine and threonine residues in protein substrates.

moves into the cell's cytoplasm and carries the signal from the cell membrane into the cell itself. The enzyme will continue to be active as long as the α_s-subunit is bound, and this results in the synthesis of several hundred cyclic AMP molecules, representing another substantial amplification of the signal. However, the α_s-subunit has intrinsic GTPase activity (i.e. it can catalyse the hydrolysis of its bound GTP to GDP) and so it deactivates itself after a certain time period and returns to the resting state. The α_s-subunit then departs the enzyme and recombines with the $\beta\gamma$-dimer to reform the G_s-protein, while the enzyme returns to its inactive conformation.

5.2.2 Activation of protein kinase A

cAMP now proceeds to activate an enzyme called protein kinase A (PKA) (Fig. 5.5). PKA belongs to a group of enzymes called the **serine-threonine kinases** which catalyse the phosphorylation of serine and threonine residues in protein substrates (Fig. 5.6).

Protein kinase A catalyses the phosphorylation and activation of further enzymes with functions specific to the particular cell or organ in question; for example, lipase enzymes in fat cells are activated to catalyse the breakdown of fat. The active site of a protein kinase has to be capable of binding the region of the protein substrate which is to be phosphorylated, as well as the cofactor ATP which provides the necessary phosphate group.

There may be several more enzymes involved in the signalling pathway between the activation of PKA and the activation (or deactivation) of the target enzyme. For example, the enzymes involved in the breakdown and synthesis of glycogen in a liver cell are regulated as shown in Fig. 5.7.

Adrenaline is the initial hormone involved in the regulation process and is released when the body requires immediate energy in the form of **glucose**. The hormone initiates a signal at the β-**adrenoceptor** leading to the synthesis of cAMP and the activation of PKA by the mechanism already discussed. The catalytic subunit of PKA now phosphorylates three enzymes within the cell, as follows:

- An enzyme called **phosphorylase kinase** is phosphorylated and activated. This enzyme then catalyses the phosphorylation of an inactive enzyme called **phosphorylase b** which is converted to its active form, **phosphorylase a**. Phosphorylase a now catalyses the breakdown of glycogen by splitting off glucose-1-phosphate units.

- **Glycogen synthase** is phosphorylated to an inactive form, thus preventing the synthesis of glycogen.

- A molecule called **phosphorylase inhibitor** is phosphorylated. Once phosphorylated, it acts as an inhibitor for the **phosphatase** enzyme responsible for the conversion of phosphorylase a back to phosphorylase b. The lifetime of phosphorylase a is thereby prolonged.

The overall result of these different phosphorylations is a coordinated inhibition of glycogen synthesis and

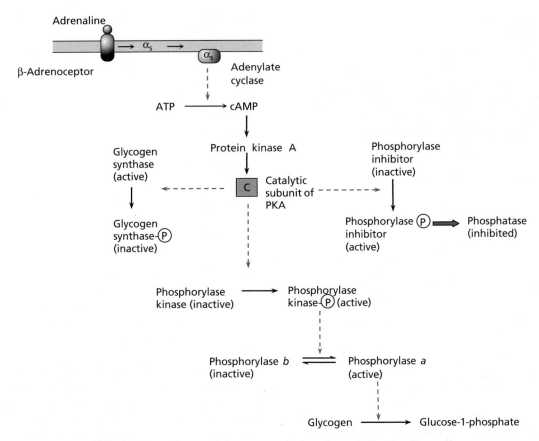

FIGURE 5.7 Regulation of glycogen synthesis and metabolism in a liver cell.

enhancement of glycogen metabolism to generate glucose in liver cells. Note that the effect of adrenaline on other types of cell may be quite different. For example, adrenaline activates β-adrenoceptors in fat cells leading to the activation of protein kinases as before. This time, however, phosphorylation activates **lipase** enzymes which then catalyse the breakdown of fat to act as another source of glucose.

5.2.3 The G_i-protein

We have seen how the enzyme adenylate cyclase is activated by the α_s-subunit of the G_s-protein. Adenylate cyclase can also be inhibited by a different G-protein—the G_i-protein. The G_i-protein interacts with different receptors from those that interact with the G_s-protein, but the mechanism leading to inhibition is the same as that leading to activation. The only difference is that the α_i-subunit released binds to adenylate cyclase and inhibits the enzyme rather than activates it.

Receptors that bind G_i-proteins include the **muscarinic M_2 receptor** of cardiac muscle, α_2-**adrenoceptors** in smooth muscle, and **opioid receptors** in the central nervous system.

The existence of G_i- and G_s-proteins means that the generation of the secondary messenger cAMP is under the dual control of a brake and an accelerator, and this explains the process by which two different neurotransmitters can have opposing effects at a target cell. A neurotransmitter which stimulates the production of cAMP forms a receptor–ligand complex which activates a G_s-protein, whereas a neurotransmitter which inhibits the production of cAMP forms a receptor–ligand complex which activates a G_i-protein. For example, **noradrenaline** interacts with the β-**adrenoceptor** to activate a G_s-protein, whereas **acetylcholine** interacts with the muscarinic receptor to activate a G_i-protein.

As there are various different types of receptor for a particular neurotransmitter, it is actually possible for that neurotransmitter to activate cAMP in one type of cell but inhibit it in another. For example, noradrenaline interacts with the β-adrenoceptor to activate adenylate cyclase, because the β-adrenoceptor binds the G_s-protein. However, noradrenaline interacts with the α_2-adrenoceptor to inhibit adenylate cyclase, because this receptor binds the G_i-protein. This example illustrates that it is the receptor that determines which G-protein is activated, and not the neurotransmitter or hormone.

It is also worth pointing out that enzymes such as adenylate cyclase and the kinases are never fully active or inactive. At any one time, a certain proportion of these enzymes are active and the role of the G_s- and G_i-proteins is to either increase or decrease that proportion. In other words, the control is graded rather than all or nothing.

5.2.4 General points about the signalling cascade involving cyclic AMP

The signalling cascade involving the G_s-protein, cAMP, and PKA appears very complex and you might wonder whether a simpler signalling process would be more efficient. There are several points worth noting about the process as it stands.

- Firstly, the action of the G-protein and the generation of a secondary messenger explains how a message delivered to the outside of the cell surface can be transmitted to enzymes within the cell—enzymes that have no direct association with the cell membrane or the receptor. Such a signalling process avoids the difficulties involved in a messenger molecule (which is commonly hydrophilic) having to cross a hydrophobic cell membrane.

- Secondly, the process involves a molecular 'relay runner' (the G-protein) and several different enzymes in the signalling cascade. At each of these stages, the action of one protein or enzyme results in the activation of a much larger number of enzymes. Therefore, the effect of one neurotransmitter interacting with one receptor molecule results in a final effect several factors larger than one might expect. For example, each molecule of **adrenaline** is thought to generate 100 molecules of cAMP and each cAMP molecule starts off an amplification effect of its own within the cell.

- Thirdly, there is an advantage in having the receptor, the G-protein, and adenylate cyclase as separate entities. The G-protein can bind to several different types of receptor–ligand complexes. This means that different neurotransmitters and hormones interacting with different receptors can switch on the same G-protein leading to activation of adenylate cyclase. Therefore, there is an economy of organization involved in the cellular signalling chemistry, as the adenylate cyclase signalling pathway can be used in many different cells, and yet respond to different signals. Moreover, different cellular effects will result depending on the type of cell involved (i.e. cells in different tissues will have different receptor types and subtypes, and the signalling system will switch on different target enzymes). For example, **glucagon** activates G_s-linked receptors in the liver leading to gluconeogenesis, **adrenaline** activates G_s-linked β_2-adrenoceptors in

fat cells leading to lipolysis, and **vasopressin** interacts with G_s-linked vasopressin (V_2) receptors in the kidney to affect sodium/water resorption. Adrenaline acts on $G_{i/o}$-linked α_2-adrenoceptors leading to contraction of smooth muscle, and **acetylcholine** acts on $G_{i/o}$-linked M_2 receptors leading to relaxation of heart muscle. All these effects are mediated by the cAMP signalling pathway.

- Finally, the dual control of 'brake/accelerator' provided by the G_s- and G_i-proteins allows fine control of adenylate cyclase activity.

5.2.5 The role of the βγ-dimer

If you've managed to follow the complexity of the G-protein signalling pathway so far, well done. Unfortunately, there's more! You may remember that when the G-protein binds to a receptor–ligand complex, it breaks up to form an α-subunit and a $\beta\gamma$-dimer. Until recently, the $\beta\gamma$-dimer was viewed merely as an anchor for the α-subunit to ensure that it remained bound to the inner surface of the cell membrane. However, it has now been found that the $\beta\gamma$-dimers from both the G_i- and the G_s-proteins can themselves activate or inhibit adenylate cyclase. There are actually six different types (or isozymes) of adenylate cyclase, and activation or inhibition depends on the isozyme involved. Moreover, adenylate cyclase is not the only enzyme that can be controlled by the $\beta\gamma$-dimer. The $\beta\gamma$-dimer is more promiscuous than the α-subunits and can affect several different targets, leading to a variety of different effects. This sounds like a recipe for anarchy. However, there is some advantage in the dimer having a signalling role, since it adds an extra subtlety to the signalling process. For example, it is found that higher concentrations of the dimer are required to result in any effect compared to the α-subunit. Therefore, regulation by the dimers becomes more important when a greater number of receptors are activated.

By now it should be clear that the activation of a cellular process is more complicated than the interaction of one type of neurotransmitter interacting with one type of receptor. In reality, the cell is receiving a whole myriad of signals from different chemical messengers via various receptors and receptor–ligand interactions. The final signal depends on the number and type of G-proteins activated at any one time, as well as the various signal transduction pathways that these proteins initiate.

5.2.6 Phosphorylation

As we have seen above, phosphorylation is a key reaction in the activation or deactivation of enzymes.

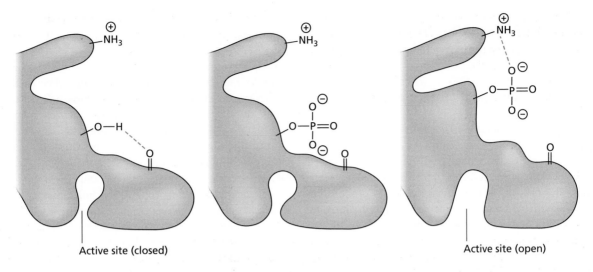

FIGURE 5.8 Conformational changes in a protein, induced by phosphorylation.

Phosphorylation requires ATP as a source for the phosphate group and occurs on the phenolic group of tyrosine residues when catalysed by **tyrosine kinases**, and on the alcohol groups of serine and threonine residues when catalysed by **serine-threonine kinases**. These functional groups are all capable of participating in hydrogen bonding, but if a bulky phosphate group is added to the OH group, hydrogen bonding is disrupted. Furthermore, the phosphate group is usually ionized at physiological pH and so phosphorylation introduces two negatively charged oxygens. These charged groups can now form strong ionic bonds with a suitably positioned positively charged group in the protein, causing the enzyme to change its tertiary structure. This change in shape results in the exposure or closure of the active site (Fig. 5.8).

Phosphorylation by kinase enzymes also accounts for the desensitization of G-protein-linked receptors. Phosphorylation of serine and threonine residues occurs on the intracellular C-terminal chain after prolonged ligand binding. Since the C-terminal chain is involved in G-protein binding, phosphorylation changes the conformation of the protein in that region and prevents the G-protein from binding. Thus the receptor–ligand complex is no longer able to activate the G-protein.

- G-proteins consist of three protein subunits, with the α-subunit bound to GDP. There are several types of G-protein.

- Receptor–ligand binding opens a binding site for the G-protein. On binding, GDP is exchanged for GTP, and the G-protein fragments into an α-subunit (bearing GTP) and a $\beta\gamma$-dimer.

- G-proteins are bound and split for as long as the chemical messenger is bound to the receptor, resulting in a signal amplification.

- An α_s-subunit binds to adenylate cyclase and activates it such that it catalyses the formation of cAMP from ATP. The reaction proceeds for as long as the α_s-subunit is bound, representing another signal amplification. An α_i-subunit inhibits adenylate cyclase.

- The α-subunits eventually hydrolyse bound GTP to GDP and depart adenylate cyclase. They combine with their respective $\beta\gamma$-dimers to reform the original G-proteins.

- cAMP acts as a secondary messenger within the cell and activates PKA. PKA catalyses the phosphorylation of serine and threonine residues in other enzymes, leading to a biological effect determined by the type of cell involved.

- The signalling cascade initiated by receptor–ligand binding results in substantial signal amplification and does not require the original chemical messenger to enter the cell.

- The overall activity of adenylate cyclase is determined by the relevant proportions of G_s- and G_i-proteins that are split, which in turn depends on the types of receptors that are being activated.

- The $\beta\gamma$-dimer of G-proteins has a moderating role on the activity of adenylate cyclase and other enzymes when it is present in relatively high concentration.

- Tyrosine kinases are enzymes which phosphorylate the phenol group of tyrosine residues in enzyme substrates. Serine-threonine kinases phosphorylate the alcohol groups of serine and threonine in enzyme substrates. In both cases, phosphorylation results in conformational changes that affect the activity of the substrate enzyme.

- Kinases are involved in the desensitization of receptors.

5.3 Signal transduction involving G-proteins and phospholipase C_β

5.3.1 G-protein effect on phospholipase C_β

Certain receptors bind G_s- or G_i-proteins and initiate a signalling pathway involving adenylate cyclase (section 5.2). Other 7-TM receptors bind a different G-protein called a **G_q-protein** which initiates a different signalling pathway. This pathway involves the activation or deactivation of a membrane-bound enzyme called phospholipase C_β (PLC_β). The first part of the signalling mechanism is the interaction of the G-protein with a receptor–ligand complex as described previously in Fig. 5.1. This time, however, the G-protein is a G_q-protein rather than a G_s- or G_i-protein, and so an α_q-subunit is released. Depending on the nature of the released α_q-subunit, phospholipase C is activated or deactivated. If activated, phospholipase C_β catalyses the hydrolysis of **phosphatidylinositol diphosphate** (PIP_2) (an integral part of the cell membrane structure) to generate the two secondary messengers **diacylglycerol** (DG) and **inositol triphosphate** (IP_3) (Figs. 5.9 and 5.10).

5.3.2 Action of the secondary messenger: diacylglycerol

Diacylglycerol is a hydrophobic molecule and remains in the cell membrane once it is formed (Fig. 5.11). There, it activates an enzyme called **protein kinase C** (PKC) which moves from the cytoplasm to the cell membrane and then catalyses the phosphorylation of serine and threonine residues of enzymes within the cell. Once phosphorylated, these enzymes are activated and catalyse specific reactions within the cell. These induce effects such as tumour propagation, inflammatory responses, contraction or relaxation of smooth muscle, the increase or decrease of neurotransmitter release, the increase or decrease of neuronal excitability, and receptor desensitizations.

5.3.3 Action of the secondary messenger: inositol triphosphate

Inositol triphosphate is a hydrophilic molecule and moves into the cytoplasm (Fig. 5.12). This messenger works by mobilizing calcium ions from calcium stores in the endoplasmic reticulum. It does so by binding to

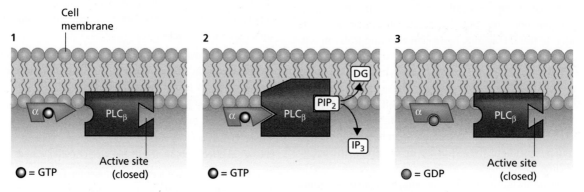

FIGURE 5.9 Activation of phospholipase C_β by an α_q-subunit.

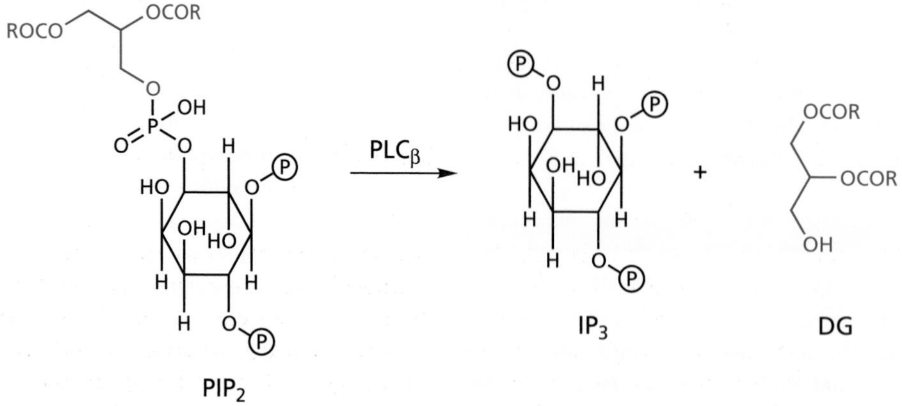

FIGURE 5.10 Hydrolysis of PIP_2 to inositol triphosphate (IP_3) and diacylglycerol (DG) (Ⓟ = phosphate).

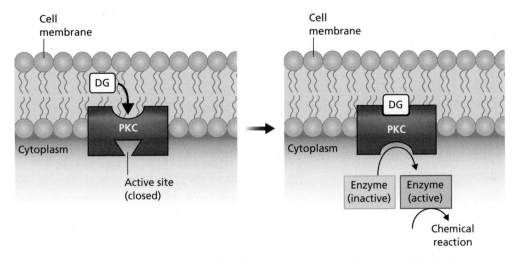

FIGURE 5.11 Activation of protein kinase C (PKC) by diacylglycerol (DG).

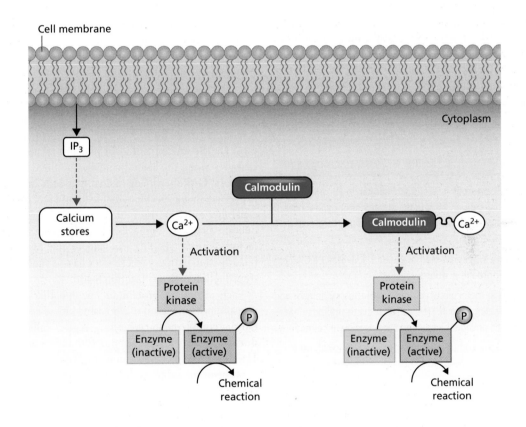

FIGURE 5.12 Signal transduction initiated by inositol triphosphate (IP$_3$).

a receptor and opening up a calcium ion channel. Once the ion channel is open, calcium ions flood the cell and activate calcium-dependent protein kinases which in turn phosphorylate and activate cell-specific enzymes. The released calcium ions also bind to a calcium binding protein called **calmodulin**, which then activates calmodulin-dependent protein kinases that phosphorylate and activate other cellular enzymes. Calcium has effects on contractile proteins and ion channels, but it is not possible to cover these effects in detail in this text. Suffice it to say that the release of calcium is crucial to a large variety of cellular functions including smooth muscle and cardiac muscle contraction, secretion from exocrine glands, transmitter release from nerves, and hormone release.

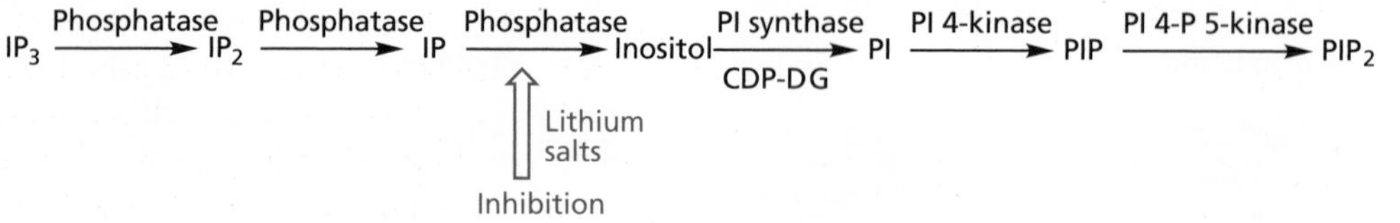

IP$_3$ $\xrightarrow{\text{Phosphatase}}$ IP$_2$ $\xrightarrow{\text{Phosphatase}}$ IP $\xrightarrow{\text{Phosphatase}}$ Inositol $\xrightarrow[\text{CDP-DG}]{\text{PI synthase}}$ PI $\xrightarrow{\text{PI 4-kinase}}$ PIP $\xrightarrow{\text{PI 4-P 5-kinase}}$ PIP$_2$

Lithium salts

Inhibition

FIGURE 5.13 Resynthesis of PIP$_2$ from IP$_3$ (CDP-DG = cytidine diphosphate-diacylglycerol).

5.3.4 Resynthesis of phosphatidylinositol diphosphate

Once IP$_3$ and DG have completed their tasks, they are recombined to form phosphatidylinositol diphosphate (PIP$_2$). Oddly enough, they cannot be linked directly and both molecules have to undergo several metabolic steps before resynthesis can occur. For example, IP$_3$ is dephosphorylated in three steps to inositol which is then used as one of the building blocks for the resynthesis of PIP$_2$ (Fig. 5.13). It is thought that **lithium salts** control the symptoms of manic depressive illness by interfering with this complex synthesis. They do so by inhibiting the monophosphatase enzyme responsible for the final dephosphorylation leading to inositol.

Note that PIP$_2$ is also involved in the PI3K signal transduction process (section 5.4.5).

> **KEY POINTS**
>
> - G$_q$-proteins are split in a similar manner to G$_s$- and G$_i$-proteins. The α_q-subunit affects the activity of phospholipase C which catalyses the hydrolysis of PIP$_2$ to form the secondary messengers IP$_3$ and DG.
>
> - DG remains in the cell membrane and activates PKC which is a serine-threonine kinase.
>
> - IP$_3$ is a polar molecule which moves into the cytoplasm and mobilizes calcium ions. The latter activate protein kinases both directly and via the calcium binding protein calmodulin.
>
> - IP$_3$ and DG are combined in a series of steps to reform PIP$_2$. Lithium salts are believed to interfere with this process.

5.4 Signal transduction involving kinase receptors

5.4.1 Activation of signalling proteins and enzymes

We saw in section 4.8 that the binding of a chemical messenger to a kinase-linked receptor activates kinase activity such that a phosphorylation reaction takes place on the receptor itself. In the case of a tyrosine kinase, this involves the phosphorylation of tyrosine residues. We now continue that story.

Once phosphorylation has taken place, the phosphotyrosine groups and the regions around them act as binding sites for various signalling proteins or enzymes. Each phosphorylated tyrosine region can bind a specific signalling protein or enzyme. Some of these signalling proteins or enzymes become phosphorylated themselves once they are bound, and act as further binding sites for yet more signalling proteins (Fig. 5.14).

Not all of the phosphotyrosine binding regions can be occupied by signalling proteins at the one time, and so the type of signalling that results depends on which signalling proteins *do* manage to bind to the kinase receptors available. There is no room in an introductory text to consider what each and every signalling protein does, but most are the starting point for phosphorylation (kinase) cascades along the same principles as the cascades initiated by G-proteins (Fig. 5.15). It is also important to appreciate that a specific tyrosine kinase receptor has the capability to trigger more than one type of signal transduction pathway. For example, the epidermal growth factor receptor (EGFR) can trigger the MAPK signal transduction pathway (section 5.4.2) and the PI3K signal transduction pathway (section 5.4.5).

Each signalling protein or enzyme that binds to a phosphorylated region of a tyrosine kinase receptor contains a complementary binding region called the **SH2 domain**, which involves about 100 amino acids. About 115 such proteins have been identified.

Some of the proteins that are bound are enzymes. For example, some kinase receptors bind a specific subtype of

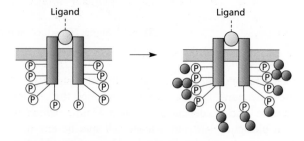

FIGURE 5.14 Binding of signalling proteins or enzymes (indicated by dark circles) to activated kinase-linked receptors (Ⓟ = phosphate).

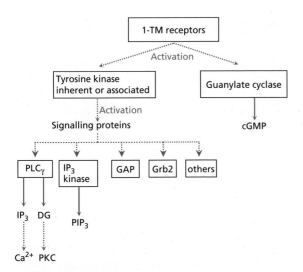

FIGURE 5.15 Signalling pathways from 1-TM receptors.

phospholipase C (PLC$_\gamma$) (Fig. 5.15)—an enzyme which catalyses phospholipid breakdown leading to the generation of IP$_3$ and subsequent calcium release. This is the same process described in section 5.3.3.

Other proteins bind to the kinase receptor and act as chemical 'adaptors'. These serve to transfer a signal from the receptor to a wide variety of other proteins, including many involved in cell division and differentiation.

We shall now look at some examples of signal transduction pathways resulting from tyrosine kinase receptors.

5.4.2 The MAPK signal transduction pathway

The MAPK signal transduction pathway is initiated by a number of receptors including the epidermal growth factor receptor (EGFR) (Fig. 5.16). The pathway involves a kinase signalling cascade that stimulates transcription of particular genes involved in cell growth and division.

An adaptor protein called **Grb2** contains the SH2 domain that allows it to bind to a phosphorylated region of EGFR. Grb2 also contains two SH3 domains that bind a protein called **SoS**. Once bound, SoS is activated and interacts with a membrane protein called **Ras**. Ras contains a bound molecule of **GDP**, but interaction with SoS causes GDP to be replaced with GTP. Ras is now activated and triggers the activation of an intracellular serine-threonine kinase called **Raf**. In turn, Raf activates another serine-threonine kinase called **MEK** (also called MAPKK). Finally, MEK activates **mitogen-activated protein kinase (MAPK or ERK)**. Once activated, MAPK phosphorylates and activates proteins called **transcription factors** which enter the nucleus and initiate gene expression resulting in various responses including cell division. Many cancers can arise from malfunctions of this signalling cascade if the kinases involved become

permanently activated, despite the absence of the initial receptor signal. Alternatively, some cancer cells overexpress kinases, and, as a result, the cell becomes supersensitive to signals that stimulate growth and division. Consequently, inhibiting kinase receptors or targeting their signalling pathways is proving to be an important strategy in designing new anticancer drugs (sections 21.6.1 and 21.6.2.4).

The Ras signal protein described above is an example of a class of signal proteins called the **small G-proteins**, so called because they are about two-thirds the size of the G-proteins described in sections 5.1–5.3. There are several subfamilies of small G-proteins (Ras, Rho, Arf, Rab, and Ran), and they can be viewed as being similar to the α-subunit of the larger G-proteins. Like the α-subunits, they are able to bind either GDP in the resting state, or GTP in the activated state. Unlike their larger cousins, the small G-proteins are not activated by direct interaction with a receptor, but are activated downstream of receptor activation through intermediary proteins which are classed as **guanine nucleotide exchange factors** (GEFs). For example, activation of Ras in Fig. 5.16 requires the prior involvement of the adaptor protein Grb2 and the guanine nucleotide exchange factor SoS following receptor activation. Like the α-subunits, small G-proteins can autocatalyse the hydrolysis of bound GTP to give bound GDP, resulting in a return to the resting state. However, this process can be accelerated by helper proteins known as **GTPase activating proteins (GAPs)**. This means that the activity level of small G-proteins is under simultaneous brake and accelerator control involving GAP and GEF respectively.

The small G-proteins are responsible for stimulating cell growth and differentiation through different signal transduction pathways. Many cancers are associated with defects in small G-proteins such as the Ras protein. *Ras* is the gene coding for the Ras protein and is one of the genes most commonly mutated in human tumours. There are three Ras proteins in mammalian cells; H-, K-, and N-Ras. Mutations can occur which result in the inability of these proteins to autocatalyse the hydrolysis of bound GTP. As a result, they remain permanently activated, leading, in turn, to permanent cell growth and division—see also section 21.6.1.

5.4.3 Activation of guanylate cyclase by kinase receptors

Some kinase receptors have the ability to catalyse the formation of **cyclic GMP** from **GTP**. Therefore, they are both receptor and enzyme (guanylate cyclase). The membrane-bound receptor/enzyme spans the cell membrane and has a single transmembrane segment. It has an extracellular receptor binding site and an intracellular

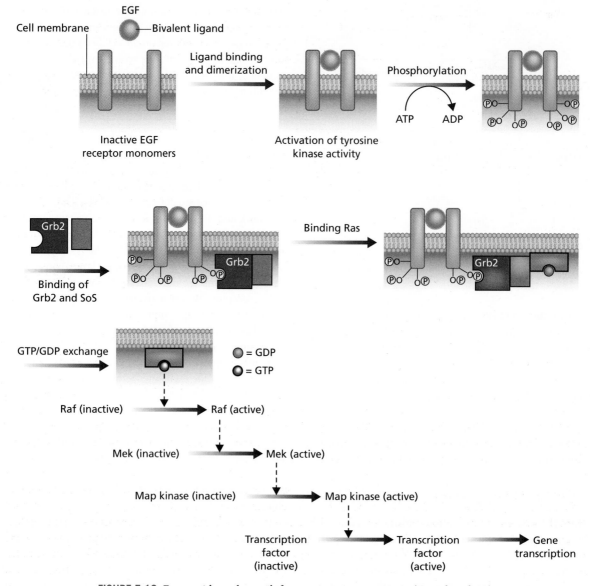

FIGURE 5.16 From epidermal growth factor to gene transcription ($\mathbb{P}$ = phosphate).

guanylate cyclase active site. Its ligands are α-**atrial natriuretic peptide** and **brain natriuretic peptide**. Cyclic GMP appears to open sodium ion channels in the kidney, promoting the excretion of sodium.

5.4.4 The JAK-STAT signal transduction pathway

Janus kinase enzymes (JAK) serve as the tyrosine kinase components of several cytokine receptors (section 4.8.3) (Fig. 5.17). These enzymes contain two regions which have the amino acid sequence that one would expect for the catalytic site. However, only one of these regions actually has a catalytic function. The other appears only to regulate the activity of the functioning catalytic region.

As a result, this second region is known as a pseudokinase region. The presence of two similar regions with different functions has resulted in the enzyme being named after Janus—the two-faced Roman god of doors and new beginnings.

Once the receptor complex has been formed, the two Janus kinase enzymes are brought close together and catalyse phosphorylation of tyrosine residues on each other. Further phosphorylations then take place on both JAK and the receptor protein to provide binding sites for proteins containing the SH2 domain. These proteins then serve as substrates for JAK and become phosphorylated in turn. The protein substrates in question include **Shc**, **Grb2**, and **PI3K**, but of particular importance are the **STAT** proteins (**Signal Transducer and Activator of**

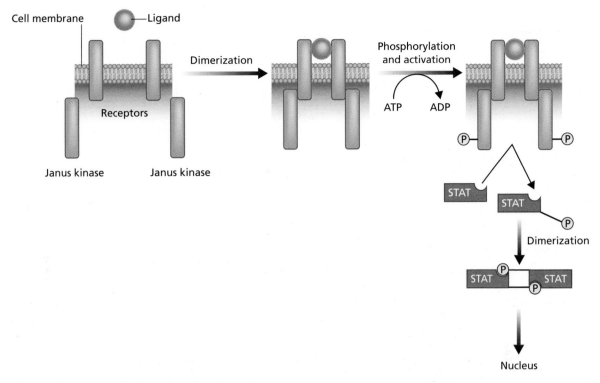

FIGURE 5.17 Activation of Janus kinases and subsequent signal transduction.

Transcription). There are at least seven types of STAT protein, with each type showing a preference for specific cytokine receptors. Once phosphorylated, the STAT proteins are released into the cytoplasm and dimerize to form a transcription factor. Dimerization takes place because the phosphorylated tyrosine residue in one STAT protein binds to a binding site in the other STAT protein, and vice versa. The STAT dimer then moves to the nucleus and binds to DNA where it regulates gene expression in a remarkably fast process measured in minutes. STATs are eventually deactivated by degradation or by phosphatase enzymes. Because of their role in controlling gene transcription, JAK kinase inhibitors are currently being studied for the treatment of autoimmune disease and cancer (section 21.6.2.8).

5.4.5 The PI3K/Akt/mTOR signal transduction pathway

Phosphoinositide 3-kinases (PI3K) are membrane-bound kinases that can be activated downstream of G-protein-coupled receptors or tyrosine kinase receptors (Fig. 5.18). As far as GPCRs are concerned, the $\beta\gamma$-dimers formed from G-protein fragmentation can activate PI3K. As far as tyrosine kinase receptors are concerned, IP_3 can be phosphorylated and activated by binding directly to the receptor or by binding to activated Ras.

There are several classes of PI3K, but the one that has been studied the most is class 1. The kinases in class 1 can be further subdivided into class 1A (which includes three subclasses—α, β, and γ) and class 1B. Class 1A kinases

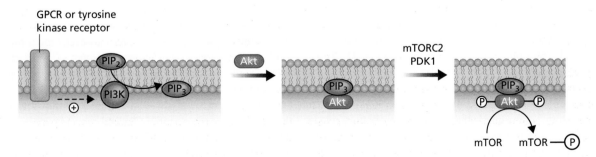

FIGURE 5.18 Activation of Akt by PIP_3.

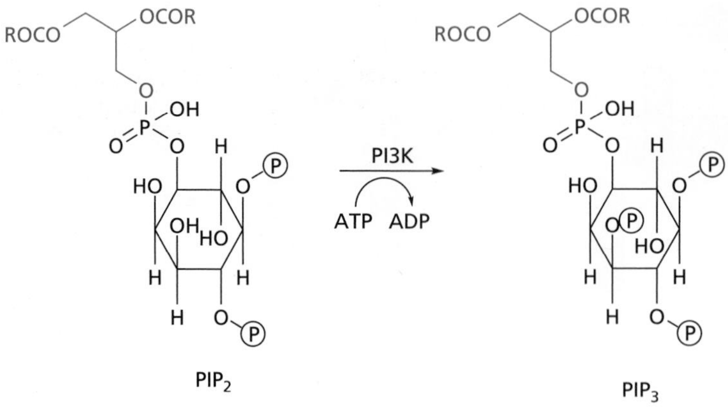

FIGURE 5.19 Phosphorylation of PIP_2 by PI3K.

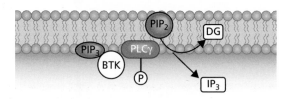

FIGURE 5.20 Activation of Bruton's tyrosine kinase.

are activated by tyrosine kinase receptors, and class 1B kinases are activated by G-protein-coupled receptors.

The various PI3Ks are protein dimers that contain a regulatory subunit and a catalytic subunit. The regulatory subunit binds the dimer to receptors and other regulatory proteins, while the catalytic subunit (identified as p110α, β, γ, or δ) contains the kinase active site.

Once activated, the PI3K enzyme catalyses the phosphorylation of **phosphatidylinositol 4,5-bisphosphate (PIP$_2$)** to form **phosphatidylinositol (3,4,5)-triphosphate (PIP$_3$)** (Fig. 5.19).

PIP_3 remains bound to the cell membrane and acts as a docking site for cytoplasmic proteins containing a region called the **Pleckstrin homology domain** (PH domain). One such protein is **Akt** (also known as **protein kinase B** or **PKB**) (Fig. 5.18). Akt is a serine-threonine kinase which plays a key role in cellular survival and the inhibition of apoptosis, making it an important factor in several cancers.

Once Akt is bound to PIP_3, it becomes a substrate for two serine-threonine kinases called **mTORC2** and **PDK1** (phosphoinositide-dependent kinase 1). These enzymes activate Akt by phosphorylating a serine and a threonine residue. Akt now phosphorylates a number of intracellular substrates such as **mTOR**. mTOR is another serine-threonine kinase that plays a key regulatory role in a number of processes including metabolism, transcription, cell growth, and cell division. Inhibitors of mTOR are described in section 21.6.2.11. Inhibitors of other parts of

the PI3K signal transduction process are also being studied as potential anticancer agents (section 21.6.2.5).

Another protein containing the PH domain is **Bruton's tyrosine kinase** (BTK)—a component of B-cell receptor signal transduction pathways (Fig. 5.20). Once it becomes bound to PIP_3, BTK is activated and proceeds to phosphorylate **phospholipase Cγ (PLC$_\gamma$)**. PLC$_\gamma$ then hydrolyses PIP_2 to form inositol triphosphate and diacylglycerol (compare section 5.3.1). This stimulates signal transduction pathways that regulate B-cell proliferation and activation. BTK is the target for the anticancer drug **ibrutinib** (section 21.6.2.5).

5.5 The hedgehog signalling pathway

The hedgehog signalling pathway is present in stem cells and plays an important role in cellular differentiation and organ development (Fig. 5.21). It is activated by proteins known as **hedgehogs**. There are three types of hedgehog protein identified as **Sonic hedgehog (Shh)**, **Indian hedgehog (Ihh)**, and **Desert hedgehog (Dhh)**. When secreted, the hedgehog proteins bind to a membrane-bound protein receptor called **Patched** (Ptch) containing 12 transmembrane regions.

The Patched receptor normally suppresses a receptor called **Smoothened (Smo)**, which contains seven transmembrane regions, just like GPCRs. However, Smo is not a GPCR because it does not activate G-proteins. Therefore, it is defined as a GPCR-like receptor. When a hedgehog protein binds to Ptch, it prevents Ptch from supressing Smo. Smo can then initiate a signal transduction pathway that affects gene transcription. Several cancers have been identified where the hedgehog pathway is abnormally active, and so antagonists that block Smo have been developed as anticancer agents (section 21.6.3).

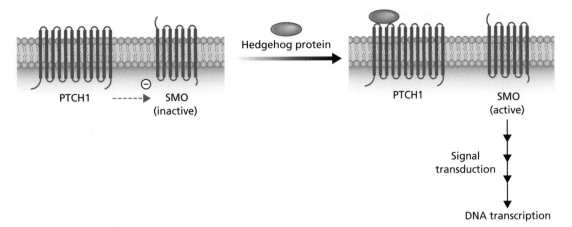

FIGURE 5.21 The hedgehog signalling pathway.

KEY POINTS

- The phosphorylated tyrosine residues on activated kinase receptors act as binding sites for various signalling proteins and enzymes which are activated in turn.

- Small G-proteins are similar in nature to G-proteins, binding GDP in the resting state, and GTP in the activated state.

They are single proteins activated by guanine nucleotide exchange factors.

- Some kinase receptors have an intracellular active site capable of catalysing the formation of cyclic GMP from GTP.

QUESTIONS

1. A model binding site for ATP was created for EGF receptor kinase, which demonstrates how ATP is bound. Structure I is known to inhibit the binding of ATP. Suggest how structure I might bind.

'ribose' pocket

2. Small G-proteins like Ras have an autocatalytic property. What does this mean and what consequences would there be (if any) should that property be lost?

3. Farnesyl transferase is an enzyme which catalyses the attachment of a long hydrophobic chain to the Ras protein. What do you think is the purpose of this chain and what would be the effect if the enzyme was inhibited?

4. Consider the signal transduction pathways shown in Fig. 5.16 and identify where signal amplification takes place.

5. The enzyme cAMP phosphodiesterase hydrolyses cAMP to AMP. What effect would an inhibitor of this enzyme have on glucose-1-phosphate production (Fig. 5.7)?

6. An enzyme was produced by genetic engineering where several of the serine residues were replaced by glutamate residues. The mutated enzyme was permanently active, whereas the natural enzyme was only active in the presence of a serine-threonine protein kinase. Give an explanation.

7. Suggest why tyrosine kinases phosphorylate tyrosine residues in protein substrates, but not serine or threonine residues.

8. Antibodies have been generated to recognize the extracellular regions of growth-factor receptors. Binding of the antibody to the receptor should block the growth factor from reaching its binding site and block its signal. However, it has been observed that antibodies can sometimes trigger the same signal as the growth factor. Why should this occur? Consult section 10.7.2 to see the structure of an antibody.

Multiple-choice questions are available on the Online Resource Centre at www.oxfordtextbooks.co.uk/orc/patrick6e/

FURTHER READING

Cohen, P. (2002) Protein kinases—the major drug targets of the twenty-first century? *Nature Reviews Drug Discovery*, **1**, 309–15.

Flower, D. (2000) Throwing light on GPCRs. *Chemistry in Britain*, November, 25.

George, S. R., O'Dowd, B. F., and Lee, S. P. (2002) G-protein-coupled receptor oligomerization and its potential for drug discovery. *Nature Reviews Drug Discovery*, **1**, 808–20.

Kenakin, T. (2002) Efficacy at G-protein-coupled receptors. *Nature Reviews Drug Discovery*, **1**, 103–10.

Neubig, R. R., and Siderovski, D. P. (2002) Regulators of G-protein signalling as new central nervous system drug targets. *Nature Reviews Drug Discovery*, **1**, 187–97.

Schwarz, M. K., and Wells, T. N. C. (2002) New therapeutics that modulate chemokine networks. *Nature Reviews Drug Discovery*, **1**, 347–58.

Takai, Y., Sasaki, T., and Matozaki, T. (2001) Small GTP-binding proteins. *Physiological Reviews*, **81**, 153–208.

Vlahos, C. J., McDowell, S. A., and Clerk, A. (2003) Kinases as therapeutic targets for heart failure. *Nature Reviews Drug Discovery*, **2**, 99–113.

Titles for general further reading are listed on p.845.

6

Nucleic acids: structure and function

In this chapter we discuss the structure and function of nucleic acids. Drug action at nucleic acids is discussed in Chapter 9 and in other chapters throughout the text. Although most drugs act on protein structures, there are several examples of important drugs which act directly on nucleic acids. There are two types of nucleic acid—DNA (deoxyribonucleic acid) and RNA (ribonucleic acid). We first consider the structure of DNA.

6.1 Structure of DNA

Like proteins, DNA has a primary, secondary, and tertiary structure.

6.1.1 The primary structure of DNA

The primary structure of DNA is the way in which the DNA building blocks are linked together. Whereas proteins have over 20 building blocks to choose from, DNA has only four—the nucleosides **deoxyadenosine**, **deoxyguanosine**, **deoxycytidine**, and **deoxythymidine** (Fig. 6.1). Each nucleoside is constructed from two components—a **deoxyribose** sugar and a nucleic acid base. The sugar is the same in all four nucleosides and only the nucleic acid base is different. The four possible bases

are two bicyclic purines (**adenine** and **guanine**) and two smaller pyrimidine structures (**cytosine** and **thymine**) (Fig. 6.2).

The nucleoside building blocks are joined together through phosphate groups which link the 5′-hydroxyl group of one nucleoside to the 3′-hydroxyl group of the next (Fig. 6.3). With only four types of building block available, the primary structure of DNA is far less varied than the primary structure of proteins. As a result, it was long thought that DNA had only a minor role to play in cell biochemistry, since it was hard to see how such an apparently simple molecule could have anything to do with the mysteries of the genetic code. The solution to this mystery lies in the secondary structure of DNA.

6.1.2 The secondary structure of DNA

Watson and Crick solved the secondary structure of DNA by building a model that fitted all the known experimental results. The structure consists of two DNA chains arranged together in a double helix of constant diameter (Fig. 6.4). The double helix has a major groove and a minor groove, which are of some importance to the action of several anticancer agents acting as intercalators (section 9.1).

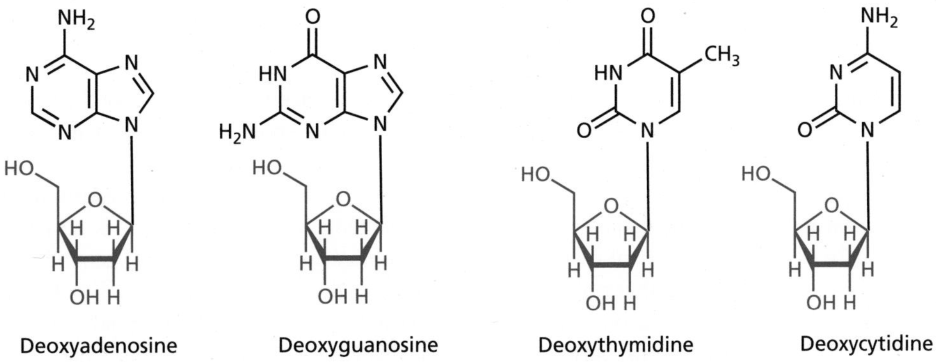

FIGURE 6.1 Nucleosides—the building blocks of DNA.

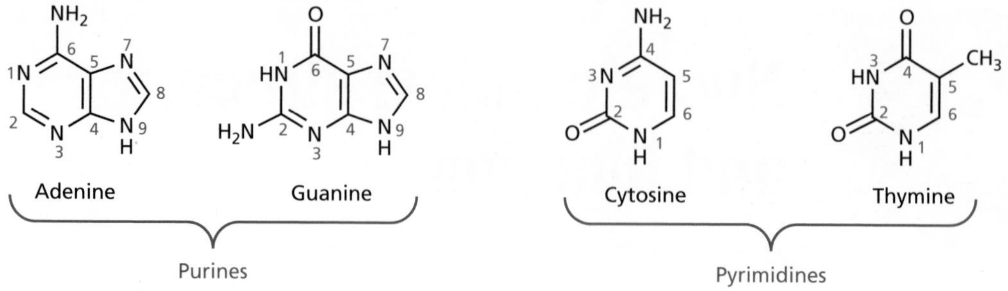

FIGURE 6.2 The nucleic acid bases for DNA.

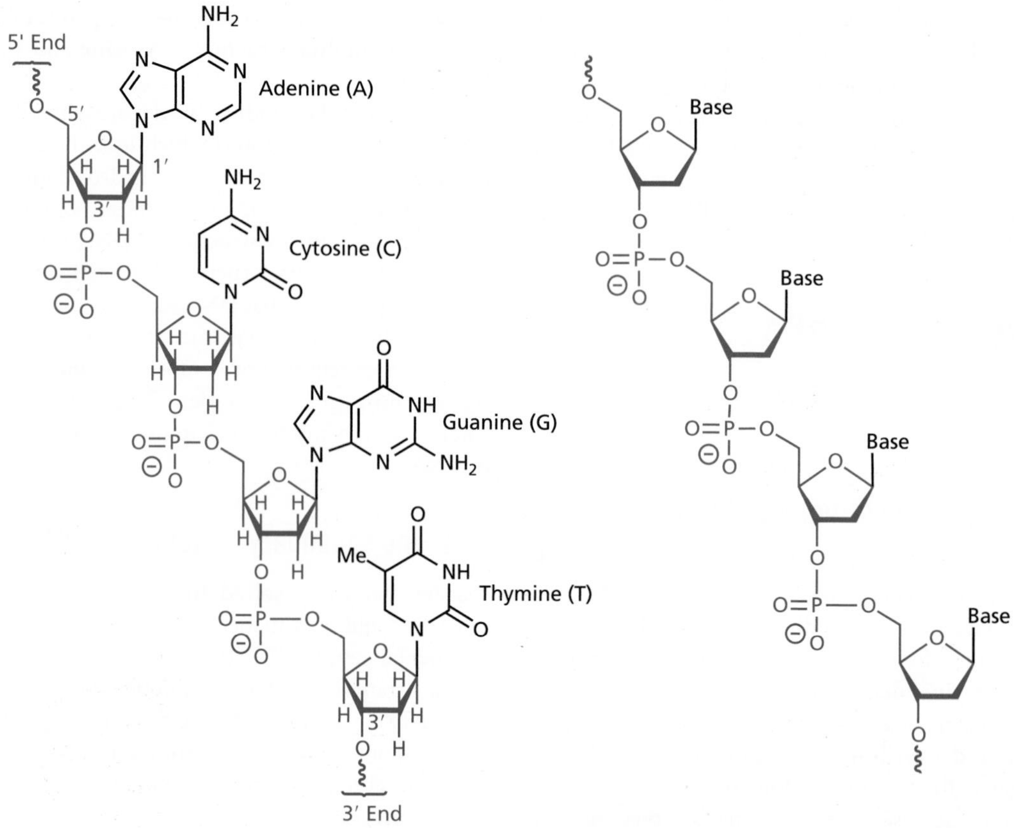

FIGURE 6.3 Linkage of nucleosides through phosphate groups.

The structure relies crucially on the pairing up of nucleic acid bases between the two chains. Adenine pairs with thymine via two hydrogen bonds, whereas guanine pairs with cytosine via three hydrogen bonds. Thus, a bicyclic purine base is always linked with a smaller monocyclic pyrimidine base to allow the constant diameter of the double helix. The double helix is further stabilized by the fact that the base pairs are stacked one on top of each other, allowing hydrophobic interactions between the faces of the heterocyclic rings. The polar sugar–phosphate

backbone is placed to the outside of the structure and can form favourable polar interactions with water.

The fact that adenine always binds to thymine, and cytosine always binds to guanine, means that the chains are complementary to each other. It is now possible to see how **replication** (the copying of genetic information) is feasible. If the double helix unravels, a new chain can be constructed on each of the original chains (Fig. 6.5). In other words, each of the original chains acts as a template for the construction of a new and identical double

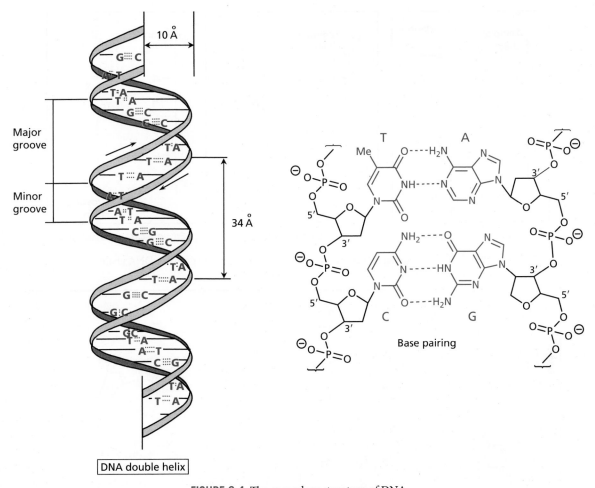

FIGURE 6.4 The secondary structure of DNA.

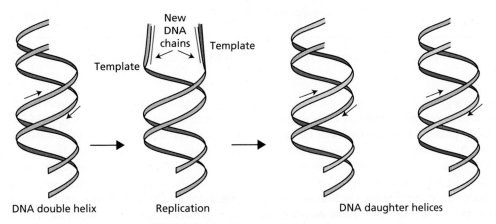

FIGURE 6.5 Replication of DNA chains.

helix. The mechanism by which this takes place is shown in Figs. 6.6 and 6.7. The template chain has exposed bases which can base pair by hydrogen bonding with individual nucleotides in the form of triphosphates. Once a nucleotide has base paired, an enzyme-catalysed reaction takes place where the new nucleotide is spliced on to the growing complementary chain with the loss of a diphosphate group—the latter acting as a good leaving group. Note that the process involves each new nucleotide reacting with the 3′ end of the growing chain.

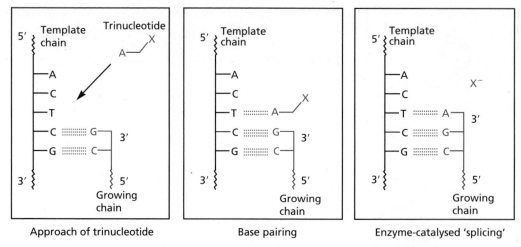

FIGURE 6.6 Base pairing of a trinucleotide and extension of the growing DNA chain.

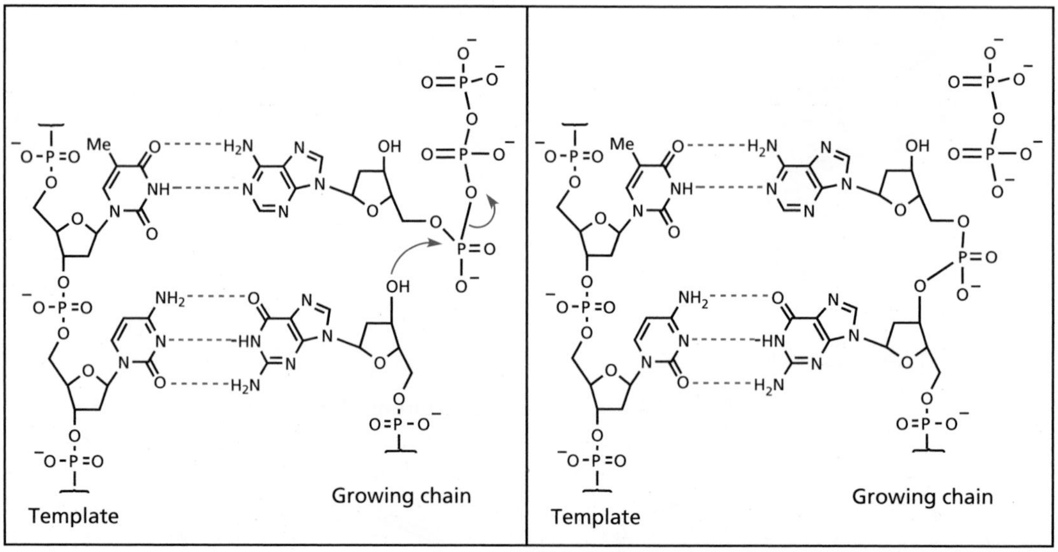

FIGURE 6.7 Mechanism by which a nucleotide is linked to the growing DNA chain.

We can now see how genetic information is passed on from generation to generation, but it is less obvious how DNA codes for proteins. How can only four nucleotides code for over 20 amino acids? The answer lies in the **triplet code**. In other words, an amino acid is coded not by one nucleotide, but by a set of three. There are 64 (4^3) ways in which four nucleotides can be arranged in sets of three—more than enough for the task required. Appendix 2 shows the standard genetic code for the various triplets. We shall look at how this code is interpreted to produce a protein in section 6.2.

6.1.3 The tertiary structure of DNA

The tertiary structure of DNA is often neglected or ignored, but it is important to the action of the quinolone group of antibacterial agents (section 9.2) and to several anticancer agents (sections 9.1–9.2). DNA is an extremely long molecule, so long in fact that it would not fit into the nucleus of the cell if it existed as a linear molecule. It has to be coiled into a more compact three-dimensional shape which *can* fit into the nucleus—a process known as **supercoiling.** This process requires the action of a family of enzymes called **topoisomerases**, which can catalyse the seemingly impossible act of passing one stretch of DNA helix across another stretch. They do this by temporarily cleaving one or both strands of the DNA helix to create a temporary gap, then resealing the strand(s) once the crossover has taken place. Supercoiling allows the efficient storage of DNA, but the DNA has to be uncoiled again if replication and transcription (section 6.2.2) are to take place. If uncoiling did not take place, the unwinding process (catalysed by **helicase** enzymes) that takes place during replication and transcription would lead to increased tension due to increased supercoiling of the remaining DNA double helix. You can demonstrate the principle of this by pulling apart the strands of rope or sisal. The same topoisomerase enzymes are responsible for catalysing the uncoiling process, so inhibition of these enzymes would effectively block transcription and replication.

Topoisomerase II is a mammalian enzyme that is crucial to the effective replication of DNA. The enzyme binds to parts of DNA where two regions of the double helix are in near proximity (Fig. 6.8). The enzyme binds to one of these DNA double helices, and tyrosine residues are used to nick both strands of the DNA (Fig. 6.9). This results in a temporary covalent bond between the enzyme and the resulting 5′ end of each strand, thus stabilizing the DNA. The strands are now pulled in opposite directions to form a gap through which the intact DNA region can be passed. The enzyme then reseals the strands and departs.

Topoisomerase I is similar to topoisomerase II in that it relieves the torsional stress of supercoiled DNA during replication, transcription, and the repair of DNA. The difference is that it cleaves a single strand of double-stranded DNA, whereas topoisomerase II cleaves both strands. The enzyme catalyses a reversible transesterification reaction similar to that shown in Fig. 6.9, but the tyrosine residue of the enzyme is linked to the 3′ phosphate end of the DNA strand rather than the 5′ end. This creates a 'cleavable complex' with a single-strand break. Relaxation of torsional strain takes place either by allowing the

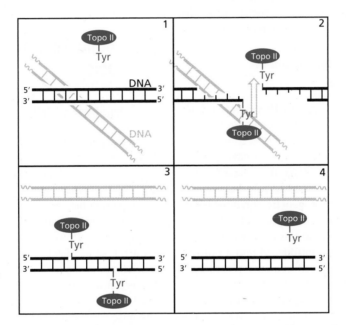

FIGURE 6.8 Method by which topoisomerase II catalyses the crossover of DNA strands. Note that the same enzyme bonds covalently to each DNA strand.

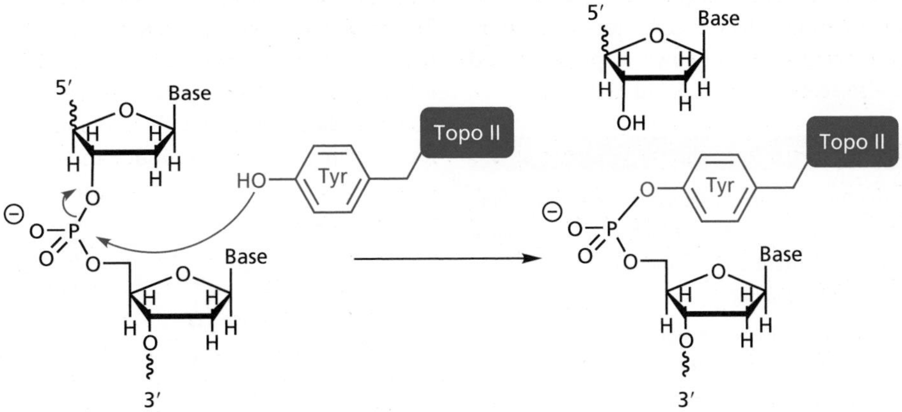

FIGURE 6.9 Mechanism by which topoisomerase II splits a DNA chain.

intact strand to pass through the nick or by free rotation of the DNA about the uncleaved strand. Once the torsional strain has been relieved, the enzyme rejoins the cleaved strand of DNA and departs.

Topoisomerase IV is a bacterial enzyme that carries out the same process as the mammalian enzyme topoisomerase II, and is an important target for the fluoroquinolone antibacterial agents (section 9.2).

6.1.4 Chromatins

So far we have focused on the structure of DNA. However, DNA is not an isolated macromolecule within the nucleus of the cell. It is associated with a variety of proteins such as histones in a structure called a chromatin (Fig. 21.5). The histones and associated DNA form a structure called a **nucleosome** which occurs regularly along the length of the chromatin and plays a crucial role in the regulation of DNA transcription (section 21.7.3).

6.1.5 Genetic polymorphism and personalized medicine

The process of replication is not 100% perfect and occasionally a mutation can occur. If the mutation does not prove fatal, it will be carried on from generation to generation. This leads to different individuals having subtly different gene sequences. On average, there is a difference of one base pair in every thousand base pairs between individuals, and this is known as **genetic polymorphism**. Since the nucleic acid bases act as the code for amino acids in proteins, a difference at this level may result in a different amino acid being introduced into a protein, which may or may not have an effect on that protein's activity or function (sections 3.5.7 and 4.11). Genetic polymorphism has important consequences with respect to the susceptibility of individuals to disease, and also to the kinds of drug therapies that are best suited for individuals. A detailed knowledge of a patient's genome opens up the possibility of predicting and preventing disease, as well as choosing the ideal drug therapy for that patient should a disease occur. This is known as **personalized medicine** (see also sections 15.1.4.4 and 21.1.11).

KEY POINTS

- The primary structure of DNA consists of a sugar–phosphate backbone with nucleic acid bases attached to each sugar moiety. The sugar is deoxyribose and the bases are adenine, thymine, cytosine, and guanine.

- The secondary structure of DNA is a double helix where the nucleic acid bases are stacked in the centre, and paired up such that adenine pairs with thymine, and cytosine pairs with guanine. Hydrogen bonding is responsible for the base

pairing and there are van der Waals interactions between the stacks of bases. Polar interactions occur between the sugar phosphate backbone and surrounding water.

- The DNA double helix is coiled up into a tertiary structure. The coiling and uncoiling of the double helix requires topoisomerase enzymes.

- The copying of DNA from one generation to the next is known as replication. Each strand of a parent DNA molecule acts as the template for a new daughter DNA molecule.

- The genetic code consists of nucleic acid bases, which are read in sets of three during the synthesis of a protein. Each triplet of bases codes for a specific amino acid.

- Knowing a patient's genome opens up the possibility of predicting disease and identifying the best therapies for that individual. This is known as personalized medicine.

6.2 Ribonucleic acid and protein synthesis

6.2.1 Structure of RNA

The primary structure of RNA is the same as that of DNA, with two exceptions: **ribose** (Fig. 6.10) is the sugar component rather than **deoxyribose**, and **uracil** (Fig. 6.10) replaces thymine as one of the bases.

Base pairing between nucleic acid bases can occur in RNA, with adenine pairing to uracil, and cytosine pairing to guanine. However, the pairing is between bases within the same chain, and it does not occur for the whole length of the molecule (e.g. Fig. 6.11). Therefore, RNA is not a double helix, but it does have regions of helical secondary structure.

Because the secondary structure is not uniform along the length of the RNA chain, more variety is allowed in RNA tertiary structure. There are three main types of RNA molecules with different cellular functions. The three are **messenger RNA** (mRNA), **transfer RNA** (tRNA), and **ribosomal RNA** (rRNA). These three molecules are crucial to the process by which protein synthesis takes place. Although DNA contains the genetic code for proteins, it cannot produce these proteins directly.

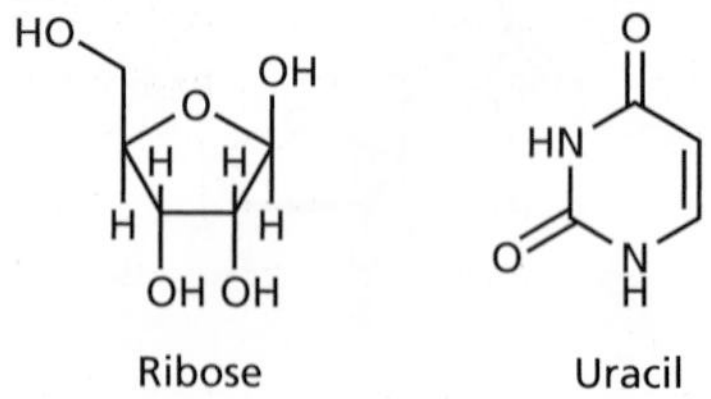

FIGURE 6.10 Ribose and uracil.

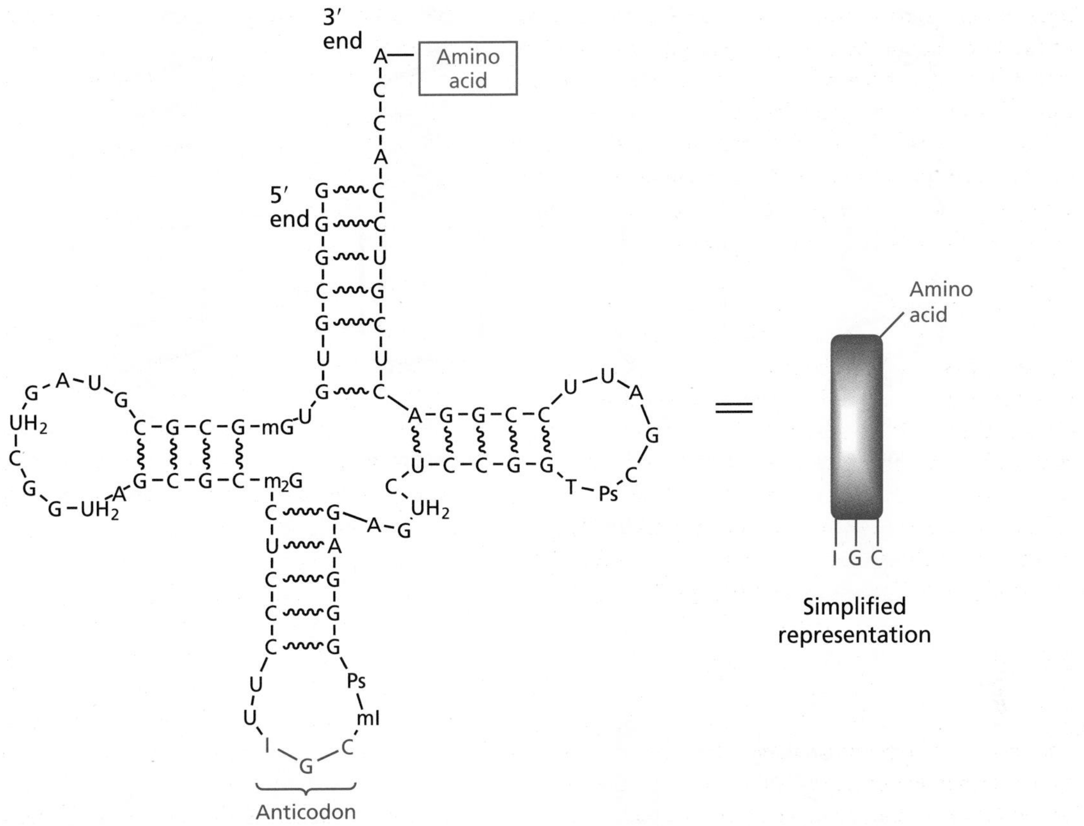

FIGURE 6.11 Yeast alanine transfer RNA. The wiggly lines indicate base pairing (mI = methylinosine; I = inosine; UH_2 = dihydrouridine; T = ribothymidine; Ps = pseudouridine; mG = methylguanosine; m_2G = dimethylguanosine).

Instead, RNA takes on that role, acting as the crucial 'middle man' between DNA and proteins. This has been termed the '**central dogma**' of molecular biology.

The bases adenine, cytosine, guanine, and uracil are found in mRNA and are predominant in rRNA and tRNA. However, tRNA also contains a number of less common nucleic acids—see for example Fig. 6.11.

6.2.2 Transcription and translation

A molecule of mRNA represents a copy of the genetic information required to synthesize a single protein. Its role is to carry the required code out of the nucleus to a cellular organelle called the **endoplasmic reticulum**. This is where protein production takes place on bodies called **ribosomes**. The segment of DNA which is copied is called a gene, and the process involved is called **transcription**. The DNA double helix unravels and the stretch that is exposed acts as a template on which the mRNA can be built (Fig. 6.12). Once complete, the mRNA departs the nucleus to seek out a ribosome, while the DNA re-forms its double helix.

Ribosomal RNA is the most abundant of the three types of RNA and is the major component of ribosomes. These can be looked upon as the production sites for protein synthesis—a process known as **translation**. The ribosome binds to one end of the mRNA molecule, then travels along it to the other end, allowing the triplet code to be read and catalysing the construction of the protein molecule one amino acid at a time (Fig. 6.13). There are two segments to the mammalian ribosome, known as the 60S and 40S subunits. These combine to form an 80S ribosome. In bacterial cells, the ribosomes are smaller and consist of 50S and 30S subunits combining to form a 70S ribosome. The terms 50S, etc., refer to the sedimentation properties of the various structures. These are related qualitatively to size and mass, but not quantitatively—that is why a 60S and a 40S subunit can combine to form an 80S ribosome.

rRNA is the major component of each subunit, making up two thirds of the ribosome's mass. The 40S subunit contains one large rRNA molecule along with several proteins, whereas the 60S subunit contains three different-sized rRNAs, again with accompanying proteins. The secondary structure of rRNA includes extensive stretches of base pairing (**duplex regions**), resulting in a well-defined tertiary structure. At one time, it was thought that rRNA only played a structural role, and that the proteins were acting as enzymes to catalyse translation. The rRNA molecules

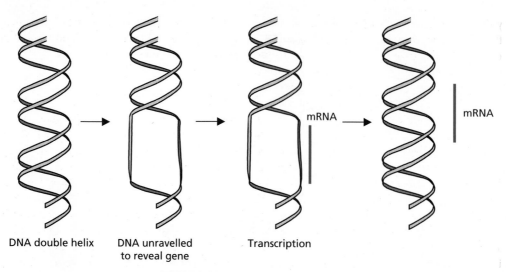

DNA double helix   DNA unravelled   Transcription
                    to reveal gene

FIGURE 6.12 Formation of mRNA.

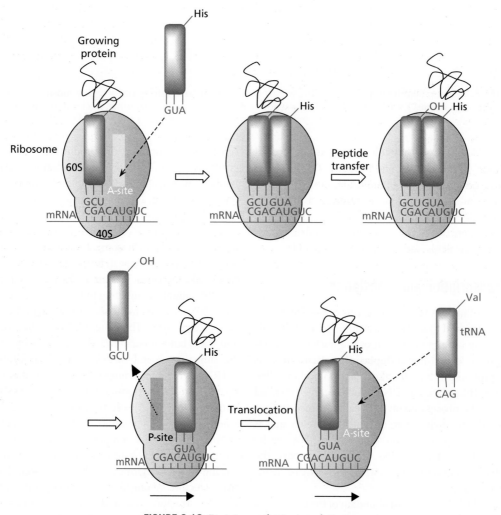

FIGURE 6.13 Protein synthesis: translation.

certainly do have a crucial structural role, but it is now known that they, rather than the ribosomal proteins, have the major catalytic role. Indeed, the key sites in the ribosome where translation takes place are made up almost entirely of rRNA. The proteins are elongated structures which meander through the ribosome structure and are thought to have a fine-tuning effect on the translation process.

Transfer RNA is the crucial adaptor unit which links the triplet code on mRNA to a specific amino acid. This means there has to be a different tRNA for each amino acid. All the tRNAs are clover-leaf in shape, with two different binding regions at opposite ends of the molecule (see Fig. 6.11). One binding region is for the amino acid, where a specific amino acid is covalently linked to a terminal adenosyl residue. The other is a set of three nucleic acid bases (**anticodon**) which will base pair with a complementary triplet on the mRNA molecule. A tRNA having a particular anticodon will always have the same amino acid attached to it.

Let us now look at how translation takes place in more detail. As rRNA travels along mRNA, it reveals the triplet codes on mRNA one by one. For example, in Fig. 6.13 the triplet code CAU is revealed along with an associated binding site called the A site. The A stands for aminoacyl and refers to the attached amino acid on the incoming tRNA. Any tRNA molecule can enter this site but it is accepted only if it has the necessary anticodon capable of base pairing with the exposed triplet on mRNA. In this case, tRNA having the anticodon GUA is accepted and brings with it the amino acid histidine. The peptide chain that has been created so far is attached to a tRNA molecule which is bound to the P binding site (standing for peptidyl). A grafting process then takes place, catalysed by rRNA, where the peptide chain is transferred to histidine (Fig. 6.14). The tRNA occupying the P binding site now departs and the ribosome shifts along mRNA to reveal the next triplet (a process called **translocation**), and so the process continues until the whole strand is read. The new protein is then released from the ribosome,

which is now available to start the process again. The overall process of transcription and translation is summarized in Fig. 6.15.

6.2.3 Small nuclear RNA

After transcription, mRNA molecules are frequently modified before translation takes place. This involves a splicing operation where the middle section of mRNA (the **intron**) is excised and the ends of the mRNA molecule (the **exons**) are spliced together (Fig. 6.16).

Splicing requires the aid of an RNA-protein complex called a **spliceosome**. The RNA molecules involved in this complex are called **small nuclear RNAs** (snRNAs). As the name indicates, these are small RNA molecules with fewer than 300 nucleotides that occur in the nucleus of the cell. The role of the snRNAs in the spliceosome is to base pair with particular segments of mRNA such that the mRNA can be manipulated and aligned properly for the splicing process. Splice sites are recognized by their nucleotide sequences, but, on occasion, a mutation in DNA may introduce a new splice site somewhere else on mRNA. This results in faulty splicing, an altered mRNA, and a defective protein. About 15% of genetic diseases are thought to be due to mutations that result in defective splicing.

6.2.4 The regulatory role of RNA

In recent years, it has been discovered that RNA can have a regulatory role in gene expression. This will be discussed in section 9.7.2.

6.3 Genetic illnesses

A number of genetic illnesses are due to genetic abnormalities that result in the non-expression of particular proteins, or the expression of defective proteins. For example, **albinism** is a condition where the skin, hair, and

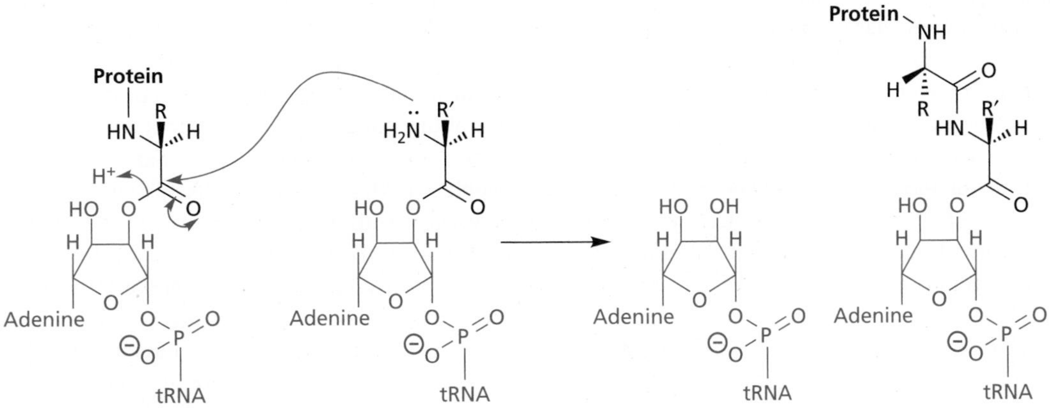

FIGURE 6.14 Mechanism by which a growing protein is transferred to the next amino acid.

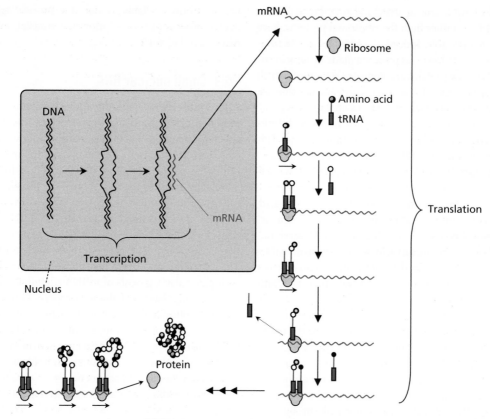

FIGURE 6.15 Transcription and translation.

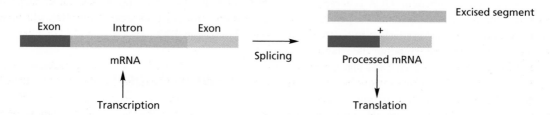

FIGURE 6.16 Splicing messenger RNA (mRNA).

eyes lack pigment; it is associated with a deficiency of an enzyme called **tyrosinase**. This is a copper-containing enzyme that catalyses the first two stages in the synthesis of the pigment **melanin**. Over 90 mutations of the tyrosinase gene have been identified which lead to the expression of inactive enzyme. Mutations in the triplet code result in one or more amino acids being altered in the resulting protein, and if these amino acids are important to the activity of the enzyme, activity is lost. Mutations which alter amino acids in the active site are the ones most likely to result in loss of activity.

Phenylketonuria is a genetic disease caused by the absence or deficiency of an enzyme called **phenylalanine hydroxylase**. This enzyme normally converts phenylalanine to tyrosine. In its absence, the blood levels of

phenylalanine rise substantially, along with alternative metabolic products such as phenylpyruvate. If left untreated, this disease results in severe mental retardation.

Haemophilias are inherited genetic diseases in which one of the blood coagulation factors is deficient. This results in uncontrolled bleeding after an injury. In the past, people with this disease were likely to die in their youth. Nowadays, with the proper treatment, affected individuals should have a normal life expectancy. Treatment in severe cases involves regular intravenous infusion with the relevant coagulation factor. In less severe cases, transfusions can be used when an injury has taken place. The coagulation factors used to be typically derived from blood plasma, but this meant that people with haemophilia were susceptible to infection from infected blood samples. For example, during the period 1979–1985 more than 1200 people in the UK were infected with HIV as a result of taking infected blood products. For the same reason, they were also prone to viral infections caused by hepatitis B and C. During the 1990s, recombinant DNA technology (section 6.4) successfully produced blood coagulation factors, and these are now the agents of choice as they eliminate the risk of infection. Unfortunately, some patients produce an immune response to the infused factor, which can preclude their use. At present, clinical trials are under way to test whether gene therapy can be used as a treatment. This involves the introduction of a gene which will code for the normal coagulation factor so that it can be produced naturally in the body (section 6.4).

Muscular dystrophy is another genetic disease that affects 1 in every 3500 males and is characterized by the absence of a protein called **dystrophin**. This has an important structural role in cells, and its absence results in muscle deterioration. Gene therapy is also being considered for this disease.

Many cancers are associated with genetic defects which result in molecular signalling defects in the cell. This is covered more fully in Chapter 21.

6.4 Molecular biology and genetic engineering

Over the last few years, rapid advances in molecular biology and genetic engineering have had important repercussions for medicinal chemistry. It is now possible to clone specific genes and to include these genes into the DNA of fast-growing cells such that the proteins encoded by these genes are expressed in the modified cell. As the cells are fast growing, this leads to a significant quantity of the desired protein, which permits its isolation, purification, and structural determination. Before

these techniques became available, it was extremely difficult to isolate and purify many proteins from their parent cells due to the small quantities present. Even if one was successful, the low yields inherent in the process made an analysis of the protein's structure and mechanism of action very difficult. Advances in molecular biology and recombinant DNA techniques have changed all that.

Recombinant DNA technology allows scientists to manipulate DNA sequences to produce modified DNA or completely novel DNA. The technology makes use of natural enzymes called **restriction enzymes** and **ligases** (Fig. 6.17). The restriction enzymes recognize a particular sequence of bases in each DNA molecule and split a specific sugar–phosphate bond in each strand of the double helix. With some restriction enzymes, the break is not a clean one; there is an overlap between the two chains, resulting in a tail of unpaired bases on each side of the break. The bases on each tail are complementary and can still recognize each other, so they are described as 'sticky' ends. The same process is carried out on a different molecule of DNA and the molecules from both processes are mixed together. As these different molecules have the same sticky ends, they recognize each other such that base pairing takes place in a process called **annealing**. Treatment with the ligase enzyme then repairs the sugar–phosphate backbone and a new DNA molecule is formed.

If the DNA molecule of interest does not have the required sequence recognized by the restriction enzyme, a synthetic DNA linker that *does* contain the sequence can be added to either end of the molecule using a ligase enzyme. This is then treated with the restriction enzyme as before (Fig. 6.18).

There are many applications for this technology, one of which is the ability to amplify and express the gene for a particular human protein in bacterial cells. In order to do this, it is necessary to introduce the gene to the bacterial cell. This is done by using a suitable **vector** which will carry the gene into the cell. There are two suitable

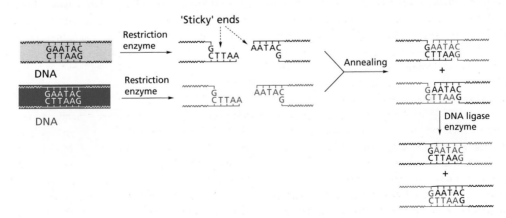

FIGURE 6.17 Recombinant DNA technology.

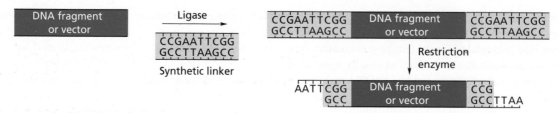

FIGURE 6.18 Attaching sequences recognized by restriction enzymes.

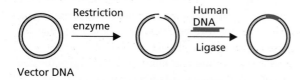

FIGURE 6.19 Inserting a human gene into a plasmid by recombinant DNA technology.

vectors—**plasmids** and **bacteriophages**. Plasmids are segments of circular DNA which are transferred naturally between bacterial cells and allow the sharing of genetic information. Because the DNA is circular, the DNA representing a human gene can be inserted into the vector's DNA by the same methods described above (Fig. 6.19). Bacteriophages (phages for short) are viruses which infect bacterial cells. There are a variety of these, but the same recombinant DNA techniques can be used to insert human DNA into viral DNA.

Whichever vector is used, the modified DNA is introduced into the bacterial cell where it is cloned and amplified (Fig. 6.20). For example, once a phage containing modified nucleic acid infects a bacterial cell, the phage takes over the cell's biochemical machinery to produce multiple copies of itself and its nucleic acid.

Human genes can be introduced to bacterial cells such that the gene is incorporated into bacterial DNA and expressed as if it was the bacterial cell's own DNA. This allows the production of human proteins in much greater quantity than would be possible by any other

means. Such proteins could then be used for medicinal purposes as described below. Modified genes can also be introduced and expressed to produce modified proteins to see what effect a mutation would have on the structure and function of a protein.

The following are some of the applications of genetic engineering to the medical field.

Harvesting important proteins. The genes for important hormones or growth factors such as **insulin** and **human growth factor** have been included in fast-growing unicellular organisms. This allows the harvesting of these proteins in sufficient quantity that they can be marketed and administered to patients who are deficient in these important hormones. Genetic engineering has also been crucial in the production of monoclonal antibodies (section 14.8.3).

Genomics and the identification of new protein drug targets. Nowadays, it is relatively easy to isolate and identify a range of signalling proteins, enzymes, and receptors by cloning techniques. This has led to the identification of a growing number of isozymes and receptor subtypes which offer potential drug targets for the future. The **Human Genome Project** involved the mapping of human DNA (completed in 2000) and has led to the discovery of new proteins previously unsuspected. These, too, may offer potential drug targets. The study of the structure and function of new proteins discovered from genomics is called **proteomics** (section 2.6).

Study of the molecular mechanism of target proteins. Genetic engineering allows the controlled

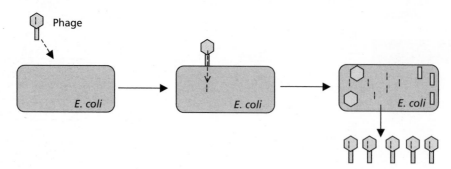

FIGURE 6.20 Infecting *Escherichia coli* with a phage.

mutation of proteins such that specific amino acids are altered. This allows researchers to identify which amino acids are important to enzyme activity or to receptor binding. In turn, this leads to a greater understanding of how enzymes and receptors operate at the molecular level.

Somatic gene therapy. Somatic gene therapy involves the use of a carrier virus to smuggle a healthy gene into cells in the body where the corresponding gene is defective. Once the virus has infected the cell, the healthy gene is inserted into the host DNA where it undergoes transcription and translation. This approach has great therapeutic potential for cancers, AIDS, and genetic abnormalities such as cystic fibrosis. However, the approach is still confined to research labs and there is still a long way to go before it is used clinically. There are several problems still to be tackled, such as how to target the viruses specifically to the defective cells, how to insert the gene into DNA in a controlled manner, how to regulate gene expression once it is in DNA, and how to avoid immune responses to the carrier virus. Progress in this field was set back significantly in 1999 as a result of a fatality to a teenage volunteer during a clinical trial in the USA. This was attributed to an over-reactive immune response to the carrier virus used in the trial. Consequently, there are now studies looking into the use of artificial viruses which would be less likely to cause an immune response. Non-viral delivery systems are also being studied, involving caged molecules called cyclodextrins. In addition, lipids, polyaminoesters, glycine polymers, and carbon buckyballs are being investigated as carriers.

KEY POINTS

- The primary structure of RNA is similar to that of DNA, but it contains ribose instead of deoxyribose. Uracil is used as a base in place of thymine and other bases may be present in smaller quantities.

- Base pairing and sections of helical secondary structure are possible within the structure of RNA.

- There are three main types of RNA—messenger RNA, transfer RNA, and ribosomal RNA.

- Transcription is the process by which a segment of DNA is copied as mRNA. mRNA carries the genetic information required for the synthesis of a protein from the nucleus to the endoplasmic reticulum.

- rRNA is the main constituent of ribosomes where protein synthesis takes place. A ribosome moves along mRNA revealing each triplet of the genetic code in turn.

- tRNA interprets the coded message in mRNA. It contains an anticodon of three nucleic acid bases which binds to a complementary triplet on mRNA. Each tRNA carries a specific amino acid, the nature of which is determined by the anticodon.

- The process of protein synthesis is called translation. The growing protein chain is transferred from one tRNA to the amino acid on the next tRNA and is only released once the complete protein molecule has been synthesized.

- Genetic engineering has been used in the production of important hormones for medicinal purposes, the identification of novel drug targets, the study of protein structure and function, and gene therapy.

QUESTIONS

1. Proflavine is a topical antibacterial agent which intercalates bacterial DNA and was used to treat wounded soldiers in the Far East during the Second World War. What role (if any) is played by the tricyclic ring and the primary amino groups? The drug cannot be used systemically. Suggest why this is the case.

Proflavine

2. The following compounds are antiviral drugs which mimic natural nucleosides. What nucleosides do they mimic?

3. Adenine is an important component of several important biochemicals. It has been proposed that adenine was synthesized early on in the evolution of life when the Earth's atmosphere consisted of gases such as hydrogen cyanide and methane. It has also been possible to synthesize adenine from hydrogen cyanide. Consider the structure of adenine and identify how cyanide molecules might act as the building blocks for this molecule.

4. The genetic code involves three nucleic acid bases coding for a single amino acid (the triplet code). Therefore, a mutation to a particular triplet should result in a different amino acid. However, this is not always the case. For any triplet represented by XYZ, which mutation is least likely to result in a change in amino acid—X, Y, or Z?

5. The amino acids serine, glutamate, and phenylalanine were found to be important binding groups in a receptor binding site (see Appendix 1 for structures). The triplet codes for these amino acids in the mRNA for this receptor were AGU, GAA, and UUU respectively. Explain what effect the following mutations might have, if any:

AGU to ACU; AGU to GGU; AGU to AGC
GAA to GAU; GAA to AAA; GAA to GUA
UUU to UUC; UUU to UAU; UUU to AUU

⦿ Multiple-choice questions are available on the Online Resource Centre at www.oxfordtextbooks.co.uk/ orc/patrick6e/

FURTHER READING

Aldridge, S. (2003) The DNA story. *Chemistry in Britain*, April, 28–30.

Breaker, R. R. (2004) Natural and engineered nucleic acids as tools to explore biology. *Nature*, **432**, 838–45.

Broad, P. (2009) Biology's Nobel molecule factory. *Chemistry World*, Nov., 42–4 (ribosomes).

Burke, M. (2003) On delivery. *Chemistry in Britain*, February, 36–8.

Dorsett, Y., and Tuschl, T. (2004). siRNAs: applications in functional genomics and potential as therapeutics. *Nature Reviews Drug Discovery*, **3**, 318–29.

Fletcher, H., and Hickey, I. (2012) *Instant notes genetics*. Garland Science.

Johnson, I. S. (2003) The trials and tribulations of producing the first genetically engineered drug. *Nature Reviews Drug Discovery*, **2**, 747–51.

Judson, H. F. (1979) *The eighth day of creation*. Simon and Schuster, New York.

Langer, R. (2003) Where a pill won't reach. *Scientific American*, April, 32–9.

Lewcock, A. (2010) Medicine made to measure. *Chemistry World*, July, 56–61.

Lindpaintner, K. (2002) The impact of pharmacogenetics and pharmacogenomics on drug discovery. *Nature Reviews Drug Discovery*, **1**, 463–69.

Nicholl, D. S. T. (2008) *An introduction to genetic engineering*, 3rd edn. Cambridge University Press, Cambridge.

Opalinska, J. B., and Gewirtz, A. M. (2002) Nucleic-acid therapeutics: basic principles and recent applications. *Nature Reviews Drug Discovery*, **1**, 503–14.

Petricoin, E. F., et al. (2002) Clinical proteomics. *Nature Reviews Drug Discovery*, **1**, 683–95.

Stark, H., et al. (2001) Arrangement of RNA and proteins in the spliceosomal U1 small nuclear ribonucleoprotein particle. *Nature*, **409**, 539–42.

Titles for general further reading are listed on p.845.

Pharmacodynamics and pharmacokinetics

The role of the medicinal chemist is to design and synthesize new drugs. In order to carry out this role, it is important to identify the particular target for a specific drug and to establish how the drug interacts with that target to produce a biological effect. In Chapters 2–6 of Part A we looked at the structure and function of various drug targets that are present in living systems. In Part B we shall look at the general mechanisms by which drugs can produce a pharmacological or biological effect. This is an area of study known as pharmacodynamics.

Drugs are normally small molecules with a molecular weight of less than 500 atomic mass units, and so they are much smaller than their macromolecular targets. As a result, they interact directly with only a small portion of the macromolecule. This is called a binding site. The binding site usually has a defined shape into which a drug must fit if it is to have an effect; therefore, it is important that the drug is the correct size and shape. However, there is more to drug action than just a good 'fit'. Once an active drug enters a binding site, a variety of intermolecular bonding interactions are set up which hold it there and lead to further effects, culminating, eventually, in a biological effect. For this to occur, the drug must have the correct functional groups and molecular skeleton capable of participating in these interactions.

Optimizing the interactions that a structure has with its target is clearly important if we are to design an effective drug. Having said that, there are examples of compounds which interact extremely well with their target but are useless in a clinical sense. That is because the compounds involved fail to reach their target in the body once they have been administered. There are various ways in which drugs can be administered, but, generally, the aim is to get the drug into the bloodstream such that it can be carried to its particular target. Following the administration of a drug, there are a wide variety of hurdles and problems that have to be overcome. These include the efficiency with which a drug is absorbed into the bloodstream, how rapidly it is metabolized and excreted, and to what extent it is distributed round the body. This is an area of study known as pharmacokinetics and we shall consider this in Chapter 11.

As this is a medicinal chemistry textbook, the focus is very much on the design of drugs to optimize their pharmacokinetic and pharmacodynamic properties. However, it is important to appreciate that formulation and drug delivery is an extremely important area of research in developing new and improved medicines (a brief overview is given in sections 11.7.1, 11.9, and 11.10). Indeed, drug action has been categorized into three phases, which occur in the following order: pharmaceutical, pharmacokinetic, and pharmacodynamic. The pharmaceutical phase includes the disintegration of a pill or capsule in the gastrointestinal tract, the release of the drug contained within, and its dissolution. This is followed by the pharmacokinetic and pharmacodynamics phases as described above.

Part B includes Case study 1, which is a study on the clinically important statins used to lower cholesterol levels. It illustrates some of the principles of enzyme inhibitors mentioned in Chapter 7.

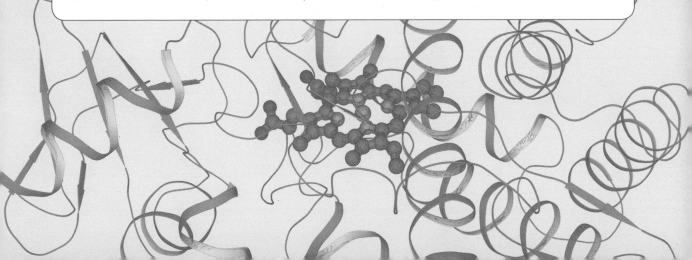

Enzymes as drug targets

Many important drugs act as enzyme inhibitors. In other words, they hinder or prevent enzymes acting as catalysts for a particular reaction. We covered the structure and function of enzymes in Chapter 3. In this chapter, we concentrate on how drugs target enzymes and inhibit their action.

7.1 Inhibitors acting at the active site of an enzyme

7.1.1 Reversible inhibitors

In Chapter 3, we emphasized the importance of binding interactions between an enzyme and its substrate. If there are no interactions holding a substrate to the active site, then the substrate will drift in and drift back out again before there is a chance for it to react. Therefore, the more binding interactions there are, the stronger the substrate will bind, and the better the chance of reaction. But there is a catch! What happens if a strongly bound substrate gives a product that also binds strongly to the active site (Fig. 7.1)?

The answer is that the enzyme becomes clogged up and is unable to accept any more substrate. Therefore, the binding interactions holding the substrate or the product to the enzyme must be properly balanced. They must be sufficiently strong to hold the substrate in the active site

long enough for the reaction to occur, but weak enough to allow the product to leave. This bonding balancing act can be turned to great advantage if the medicinal chemist wishes to inhibit a particular enzyme, or switch it off altogether. A molecule can be designed which is similar to the natural substrate or product, and can fit the active site, but which binds more strongly. It may not undergo any reaction when it is in the active site, but as long as it stays there, it blocks access to the natural substrate and prevents the enzymatic reaction (Fig. 7.2). This is known as **competitive inhibition**, as the drug is competing with the natural substrate for the active site. The longer the inhibitor is present in the active site, the greater the inhibition. Therefore, if a medicinal chemist knows the position and nature of different binding regions within an active site, it is possible to design molecules that will fit that active site, bind strongly, and act as inhibitors.

Competitive inhibitors bind to the active site through intermolecular bonds and so the binding is reversible, allowing an equilibrium to occur between bound drug and unbound drug—a kind of 'yoyo' effect where the drug binds to the active site, is released, then binds again. This means that the inhibition caused by the drug is reversible. If the concentration of substrate increases, it competes more effectively with the drug for the active site, and so inhibition by the drug will be less effective (Box 7.1).

There are many examples of useful drugs that act as competitive inhibitors. For example, the **sulphonamides**

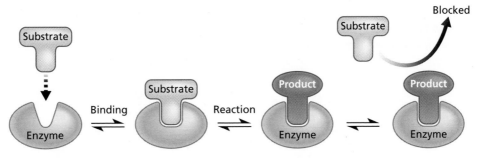

FIGURE 7.1 Example of an enzyme being 'clogged up' if the product remains bound.

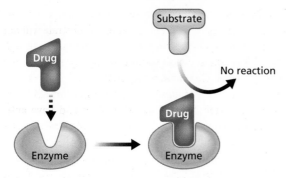

FIGURE 7.2 Competitive inhibition.

act as antibacterial agents by inhibiting a bacterial enzyme in this fashion (section 19.4.1.5). Many diuretics used to control blood pressure are competitive inhibitors, as are some antidepressants (section 23.12.5). Other examples include the statins (Case study 1), ACE inhibitors (Case study 2), and protease inhibitors (section 20.7.4). Indeed, the majority of clinically useful enzyme inhibitors are of this nature.

As stated above, competitive inhibitors frequently bear some resemblance to the natural substrate, allowing them to be recognized by the active site. Some of these inhibitors may have additional features which allow them to form extra binding interactions to regions of the active site that are not occupied by the substrate. This allows them to bind more strongly and to be more effective inhibitors. The statins described in Case study 1 are a good example of this.

Although competitive inhibitors often bear some resemblance to the substrate, this is not always the case. As long as the drug has the right shape to fit the active site,

and has functional groups that can interact with the binding regions available, it can still bind to the active site and inhibit the enzyme. Therefore, it is possible for drugs with a totally different skeleton to the substrate to act as competitive inhibitors. Such drugs may bind to a combination of binding regions within the active site, some of which are used by the substrate, and some of which are not.

It should also be remembered that the product of an enzyme-catalysed reaction is bound to the active site before it is finally released, and so it is possible to have enzyme inhibitors which resemble the structure of the product more closely than the substrate. Other drugs are designed to mimic the transition state of the enzyme-catalysed reaction (section 7.4).

Finally, some competitive inhibitors bind to the active site, but do not compete with the substrate. How can this occur? The answer lies in the fact that the active sites of several enzymes bind a substrate *and* an enzyme **cofactor.** Therefore, it is possible to have competitive inhibitors that occupy the binding region normally occupied by the cofactor, and so the competition is with the cofactor rather than the substrate. The kinase inhibitors described in section 21.6.2 are a good example of this. Many of these agents compete with the cofactor ATP for the active site of kinase enzymes, and not the protein substrate. The competitive nature of the inhibition is illustrated in resistant tumour cells, where a mutated enzyme shows greater affinity for ATP over the inhibitors (Box 21.11).

7.1.2 Irreversible inhibitors

Irreversible enzyme inhibitors can form a covalent bond to a key amino acid in the active site and permanently block

---

BOX 7.1 A cure for antifreeze poisoning

Competitive inhibitors can generally be displaced by increasing the level of natural substrate. This feature has been useful in the treatment of accidental poisoning by antifreeze. The main constituent of antifreeze is **ethylene glycol,** which is oxidized in a series of enzymatic reactions to the toxic compound **oxalic acid**. Blocking the synthesis of oxalic acid leads to recovery.

The first step in this enzymatic process is the oxidation of ethylene glycol by **alcohol dehydrogenase (ADH)**. Ethylene

glycol is acting here as a substrate, but we can view it as a competitive inhibitor because it is competing with the natural substrate for the enzyme. If the levels of natural substrate are increased, it will compete far better with ethylene glycol and prevent it from reacting. Toxic oxalic acid will no longer be formed and the unreacted ethylene glycol is eventually excreted from the body. The cure, then, is to administer high doses of the natural substrate—alcohol!

HO␣␣␣␣␣␣␣␣␣␣␣␣␣␣␣␣␣␣O␣␣␣H

␣␣␣␣␣␣␣␣␣␣␣ADH␣␣␣␣␣␣␣␣␣␣␣␣␣Enzymes␣␣␣␣␣␣COOH
␣␣␣␣␣␣␣␣␣␣——————→␣␣␣␣␣␣␣␣——————→␣␣␣␣␣|
␣␣␣␣␣␣OH␣␣␣␣␣␣␣␣␣␣␣␣␣␣␣␣␣OH␣␣␣␣␣␣␣␣␣COOH

Ethylene␣␣␣␣␣␣␣␣␣␣␣␣␣␣␣␣␣␣␣␣␣␣␣␣␣␣␣␣␣␣Oxalic acid
glycol

the affected enzyme (Fig. 7.3). The most effective **irreversible inhibitors** are those that contain an electrophilic functional group (X) capable of reacting with a nucleophilic group present on an amino acid side chain. Invariably, the amino acid affected is either **serine** or **cysteine.** This is because these amino acids contain nucleophilic functional groups in their side chains (OH and SH respectively) that are often involved in enzyme-catalysed reactions (section 3.5.3). Electrophilic functional groups used in irreversible inhibitors include alkyl halides, epoxides, α, β-unsaturated ketones, acrylamides, or strained lactones and lactams (Fig. 7.4). The highly toxic **nerve agents** (section 22.13.2.1) contain electrophilic fluorophosphonate groups, and are irreversible inhibitors of mammalian enzymes.

Not all irreversible inhibitors are highly toxic, though, and several are used clinically. For example, **penicillins** (section 19.5.1) contain a β-lactam group that irreversibly inhibits an enzyme that is crucial to bacterial cell wall synthesis. **Disulfiram (Antabuse)** (Box 12.6) is an irreversible inhibitor of the enzyme alcohol dehydrogenase and is used to treat alcoholism. The **proton pump inhibitors** described in section 25.3 are irreversible inhibitors and are used as anti-ulcer agents. The anti-obesity drug **orlistat** is also an irreversible inhibitor (Box 7.2). Having said that, it is generally better to inhibit an enzyme with a reversible inhibitor rather than an irreversible inhibitor. As irreversible inhibitors have reactive functional groups, there is a risk that they might react with other proteins or nucleic acids and cause toxic side effects. For example, covalently linking a drug to a protein could trigger an immune response.

Irreversible enzyme inhibitors are not competitive inhibitors. Increasing the concentration of substrate will not reverse their inhibition since the inhibitors cannot be displaced from the active site. This can cause problems if the build-up of a particular substrate leads to toxic side effects. For example, the **monoamine oxidase inhibitors** (MAOIs) block the metabolism of **noradrenaline** and have antidepressant activity (section 23.12.5). Unfortunately, the metabolism of substrates other than noradrenaline is also inhibited, leading to a build-up of those substrates and serious side effects. More modern MAOIs have been designed as reversible inhibitors in order to avoid this problem.

The potential problems associated with irreversible inhibitors have generally discouraged research teams from designing such agents. However, in recent years, there has been a resurgence of interest in designing irreversible inhibitors that are more selective and have fewer side effects. A successful strategy is to design an agent that initially binds reversibly to a target binding site using intermolecular bonding interactions. Only if the drug is held long enough in the active site by those interactions, will an irreversible reaction take place with the binding site. In order to achieve this, a mildly reactive electrophilic group is incorporated into the drug at a position where it will be close to a nucleophilic amino acid residue when the drug binds by reversible binding. A mildly reactive electrophilic group is less likely to react with the first nucleophilic group it meets, cutting down on potential side effects. Therefore, the priorities are to choose an electrophilic group with the right level of reactivity, and to ensure that it is correctly positioned in the drug. For

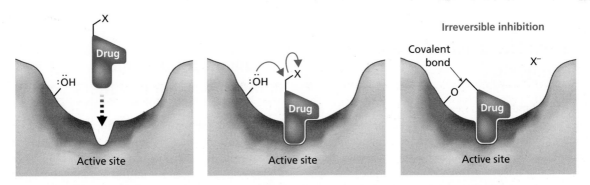

FIGURE 7.3 Irreversible inhibition of an enzyme with an alkylating agent. (X = halogen leaving group.)

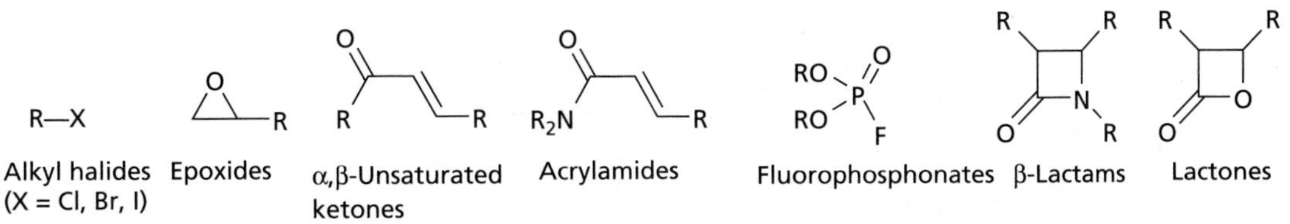

FIGURE 7.4 Examples of electrophilic functional groups.

BOX 7.2 Irreversible inhibition for the treatment of obesity

Fat in the diet is composed mainly of triglycerides which are digested in the small intestine to fatty acids and 2-mono-glycerides. The digestion products are then absorbed and act as the building blocks for fat biosynthesis in the body. The enzyme **pancreatic lipase** is responsible for catalysing the digestion of fats, and so inhibition of this enzyme will result in reduced absorption of glycerides and fatty acids

from the gut. Consequently, less fat will be synthesized in the body. **Orlistat** is an anti-obesity drug that acts as an irreversible inhibitor of pancreatic lipase due to the presence of an electrophilic four-membered lactone group. This acylates a serine residue in the active site, which is part of a catalytic triad of serine, histidine, and aspartic acid (compare section 3.5.3).

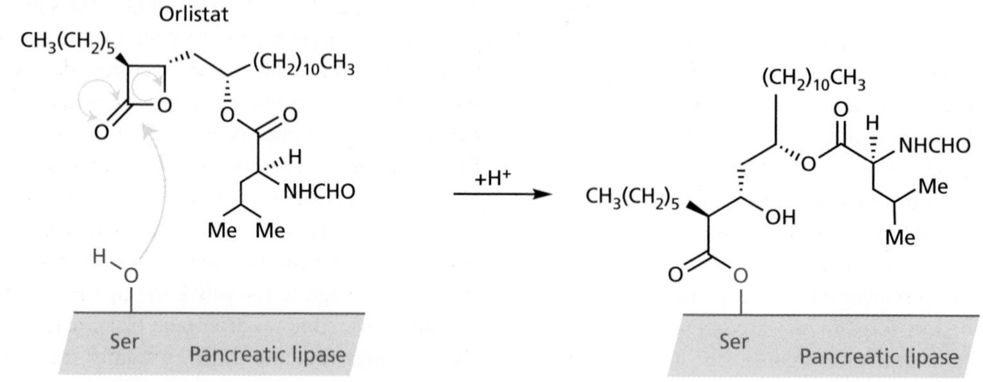

example, an acrylamide group is less reactive than an α, β-unsaturated ketone, and has been incorporated into a number of kinase inhibitors such that it will react with a specific cysteine residue in a target binding site. Several of these agents have been approved as anticancer agents (sections 21.6.2.1 and 21.6.2.5).

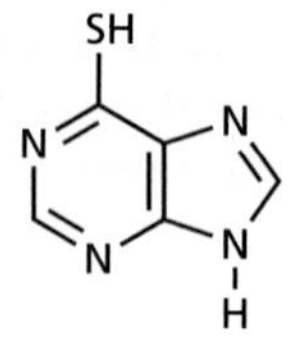

7.2 Inhibitors acting at allosteric binding sites

Allosteric binding sites were discussed in section 3.6, and are a means by which enzyme activity can be controlled by natural inhibitors. When an allosteric inhibitor binds to its binding site, the resulting induced fit also deforms the active site such that it becomes unrecognizable to the substrate. Drugs can be designed to mimic this natural control of the enzyme. If the drug binds through intermolecular interactions, the inhibition is reversible. If the drug contains a reactive group allowing it to form a covalent bond to the allosteric binding site, irreversible inhibition results.

The drug **6-mercaptopurine** (Fig. 7.5), used in the treatment of leukaemia, is an example of an allosteric inhibitor. It inhibits the first enzyme involved in the synthesis of purines (section 6.1.1) and blocks purine synthesis. This in turn blocks DNA synthesis.

FIGURE 7.5 6-Mercaptopurine.

7.3 Uncompetitive and non-competitive inhibitors

Uncompetitive inhibitors are inhibitors that bind reversibly to an enzyme when the substrate is already bound to the active site. In other words, the inhibitor binds to the enzyme–substrate complex. In this situation, increasing the substrate concentration will not overcome inhibition. Indeed, the level of inhibition is dependent on sufficient substrate being present to form the enzyme–substrate complex. Therefore, uncompetitive inhibitors are less effective at low substrate concentrations. Uncompetitive inhibitors are not very common.

In theory, a non-competitive inhibitor binds to an allosteric binding site and inhibits the enzyme-catalysed reaction, without affecting the strength of substrate binding. This would occur if the induced fit arising from the binding of the allosteric inhibitor distorts the active site sufficiently to prevent the catalytic mechanism, but has no effect on the substrate binding process. In practice, this ideal situation is extremely rare, if it even occurs at all. It is almost inevitable that any active site distortion affecting the catalytic process will also affect substrate binding. Therefore, those inhibitors which inhibit the catalytic process, whilst still allowing substrates to bind, normally cause some inhibition of substrate binding. This is known as **mixed inhibition** since it is neither pure competitive inhibition, nor pure non-competitive inhibition.

7.4 Transition-state analogues: renin inhibitors

An understanding of an enzyme mechanism can help medicinal chemists design more powerful inhibitors. For example, it is possible to design inhibitors which bind so strongly to the active site (using non-covalent forces) that they are effectively irreversible inhibitors—a bit like inviting someone for dinner and finding that they have moved in on a permanent basis. One way of doing this is to design a drug that resembles the transition state for the catalysed reaction. Such a drug should bind more strongly than either the substrate or the product. Such compounds are known as **transition-state analogues or inhibitors**.

The use of transition-state analogues has been particularly effective in the development of **renin inhibitors**

(Fig. 7.6). Renin is a protease enzyme which is responsible for hydrolysing a specific peptide bond in the protein **angiotensinogen** to form **angiotensin I**. Angiotensin I is further converted to **angiotensin II** (see Case study 2 and section 26.3), which acts to constrict blood vessels and retain fluid in the kidneys, both of which lead to a rise in blood pressure. Therefore, an inhibitor of renin should act as an antihypertensive agent (i.e. lower blood pressure) by preventing the first stage in this process.

Renin contains two aspartyl residues and a bridging water molecule in the active site which are crucial to the mechanism by which an amide bond in the substrate is hydrolysed (Fig. 7.7). In the first stage of this mechanism, a tetrahedral intermediate is formed. In order to form this intermediate, the reaction mechanism has to proceed through a high-energy transition state, and it is this transition state that we wish to mimic with a transition-state analogue. However, it is not possible to isolate such a high-energy species in order to study its structure, so how can one design a drug to mimic it? The answer is to base the design of the drug on the reaction intermediate. The rationale for this is as follows. Since the intermediate is less stable than the substrate, it is presumed that it is closer in character to the transition state. This in turn implies that the transition state is more tetrahedral in character than planar. Therefore, drugs based on the structure of the tetrahedral intermediate are more likely to mimic the transition state.

The intermediate itself is reactive and easily cleaved. Therefore, an analogue has to be designed which binds just as strongly, but is stable to hydrolysis. This can be done by

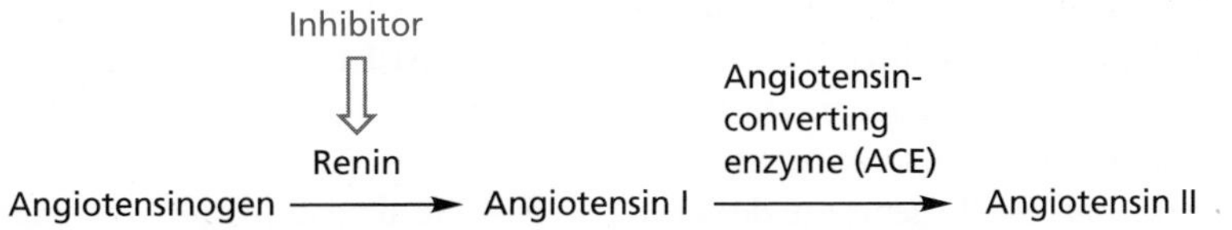

FIGURE 7.6 Inhibition of renin to block the synthesis of angiotensin I and angiotensin II.

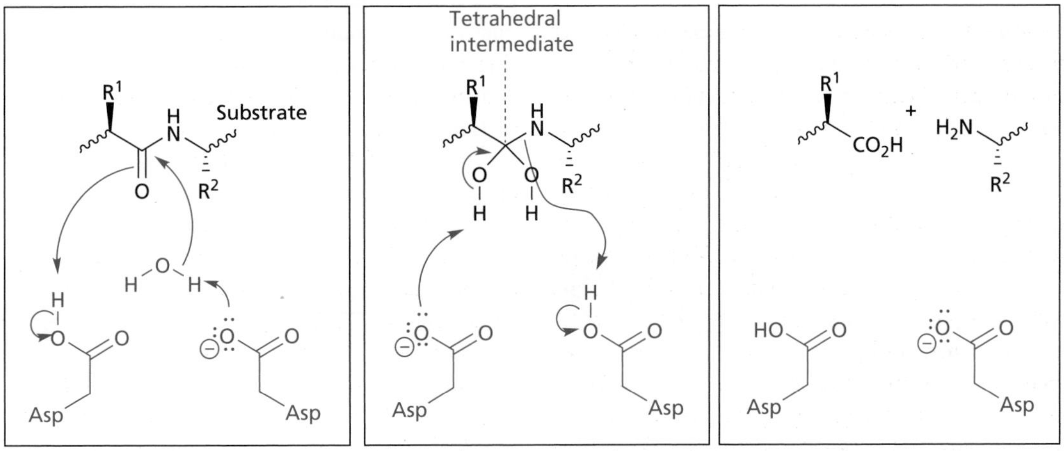

FIGURE 7.7 Mechanism of renin-catalysed hydrolysis.

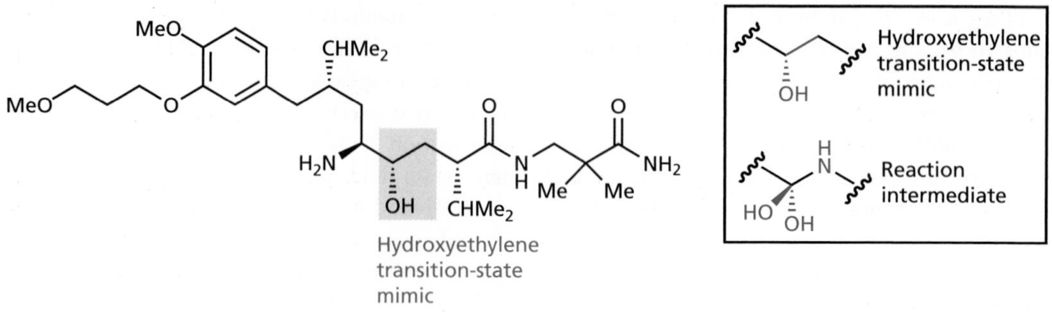

FIGURE 7.8 Aliskiren.

introducing a feature that mimics the tetrahedral structure of the intermediate, but has no leaving group for the second part of the reaction mechanism. A variety of mimics have been tried and a hydroxyethylene moiety has proved effective (e.g. **aliskiren**; Fig. 7.8). The hydroxyethylene group has the required tetrahedral geometry and one of the two hydroxyl groups required for good binding. It is also stable to hydrolysis because there is no leaving group present. Aliskiren was approved by the US FDA in 2007 for the treatment of hypertension (see also Case study 8).

Similar strategies have been used successfully to design antiviral agents which act as transition-state analogue inhibitors for the **HIV protease enzyme** (section 20.7.4). The **statins** can also be viewed as transition-state analogues (Case study 1).

7.5 Suicide substrates

Transition-state analogues can be viewed as *bona fide* visitors to an enzyme's active site that become stubborn squatters once they have arrived. Other apparently harmless visitors can turn into lethal assassins once they have bound to their target enzyme. Such agents are designed to undergo an enzyme-catalysed transformation which converts them into a highly reactive species that forms a covalent bond to the active site.

One example of a suicide substrate is **clavulanic acid**, which is used clinically in antibacterial medications (e.g. **Augmentin**) to inhibit the bacterial β-**lactamase** enzyme (section 19.5.4.1). This enzyme is responsible for the

penicillin resistance observed in several bacterial strains, because it catalyses the hydrolysis of the penicillin β-lactam ring. The mechanism involves a serine residue in the active site acting as a nucleophile to form an intermediate where serine is covalently linked via an ester group to the ring-opened penicillin. The ester group is then hydrolysed to release the inactivated penicillin and free up the active site, such that the catalytic process can be repeated (Fig. 7.9).

Clavulanic acid also fits the active site of β-lactamase, and the β-lactam ring is opened by the serine residue in the same manner. However, the acyl–enzyme intermediate then reacts further with another enzymatic nucleophilic group (possibly NH_2) to bind the drug irreversibly to the enzyme (Fig. 7.10). The mechanism requires the loss or gain of protons at various stages, and an amino acid such as histidine in the active site would be capable of acting as a proton donor/acceptor (compare sections 3.5.2 and 22.12.3.2).

Drugs that operate in this way are often called **mechanism-based inhibitors** or **suicide substrates**, because the enzyme is committing suicide by reacting with them (see also Box 7.3). The great advantage of this approach is that the alkylating agent is generated at the site where it is meant to act and is, therefore, highly selective for the target enzyme. If the alkylating group had not been disguised in this way, the drug would have alkylated the first nucleophilic group it met in the body and would have shown little or no selectivity. The uses of alkylating agents and the problems associated with them are discussed in sections 9.3 and 21.2.3.

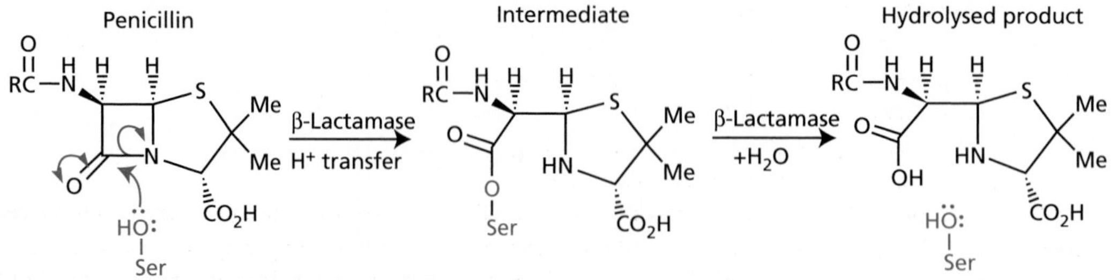

FIGURE 7.9 Reaction catalysed by bacterial β-lactamase enzymes.

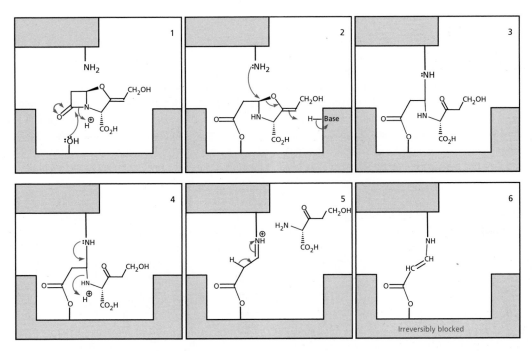

FIGURE 7.10 Clavulanic acid acting as a suicide substrate.

The main use for suicide substrates has been in labelling specific enzymes for diagnostic purposes. The substrates can be labelled with radioactive isotopes and reacted with their target enzyme in order to locate the enzyme in tissue preparations. However, some clinically useful agents do act as suicide substrates, such as clavulanic acid described above. Some monoamine oxidase inhibitors are also thought to be suicide substrates (Box 7.4). Another interesting example of a suicide substrate is **5-fluorodeoxyuracil monophosphate** (5-FdUMP). The anticancer agent **5-fluorouracil** is used to treat cancers of the breast, liver, and skin, and is converted to 5-FdUMP in the body. This then acts as a suicide substrate for the enzyme thymidylate synthase (section 21.3.2). In this case, the covalent bond is formed between the suicide substrate and the enzyme cofactor, but the overall effect is the same.

7.6 Isozyme selectivity of inhibitors

Identification of isozymes that predominate in some tissues, but not others, allows the possibility of designing tissue-selective enzyme inhibitors (Box 7.4).

For example, the non-steroidal anti-inflammatory drug (NSAID) **indometacin** (Fig. 7.11) is used to treat inflammatory diseases such as rheumatoid arthritis, and works by inhibiting the enzyme **cyclooxygenase**. This enzyme is involved in the biosynthesis of **prostaglandins**— agents which are responsible for the pain and inflammation of rheumatoid arthritis. Inhibiting the enzyme lowers prostaglandin levels and alleviates the symptoms

of the disease. However, the drug also inhibits the synthesis of beneficial prostaglandins in the gastrointestinal tract and the kidney. It has been discovered that cyclooxygenase has two isozymes, COX-1 and COX-2. Both isozymes carry out the same reactions, but COX-1 is the isozyme that is active under normal healthy conditions. In rheumatoid arthritis, the normally dormant COX-2 becomes activated and produces excess inflammatory prostaglandins. Therefore, drugs such as **valdecoxib**, **rofecoxib**, and **celecoxib** were developed to be selective for the COX-2 isozyme, so that only the production of inflammatory prostaglandins is reduced. Selectivity is possible by taking advantage of the fact that an isoleucine group is present in the binding site of COX-1, whereas the corresponding group in COX-2 is valine. Rofecoxib was authorized in 1999, but had to be withdrawn in 2004 as it was linked to an increased risk of heart attack and stroke when taken over a period of 18 months or so.

7.7 Medicinal uses of enzyme inhibitors

7.7.1 Enzyme inhibitors used against microorganisms

Inhibitors of enzymes have been extremely successful in the war against infection. If an enzyme is crucial to a microorganism, then switching it off will clearly kill the cell or prevent it from growing. Ideally, the enzyme chosen

BOX 7.3 Suicide substrates

Suicide substrates are agents which are converted to highly reactive species when they undergo an enzyme-catalysed reaction. They form covalent bonds to the enzyme and cause irreversible inhibition. In some cases, this can cause toxicity. For example, the diuretic agent **tienilic acid** had to be withdrawn from the market because it was found to act as a suicide substrate for the cytochrome P450 enzymes involved in drug metabolism (section 11.5.2). Unfortunately, the metabolic reaction carried out by these enzymes converted tienilic acid to a thiophene sulphoxide which proved highly electrophilic. This encouraged a Michael reaction leading to alkylation of a thiol group in the enzyme's active site and irreversible inhibition. Loss of water from the thiophene sulphoxide restored the thiophene ring in the inhibitor.

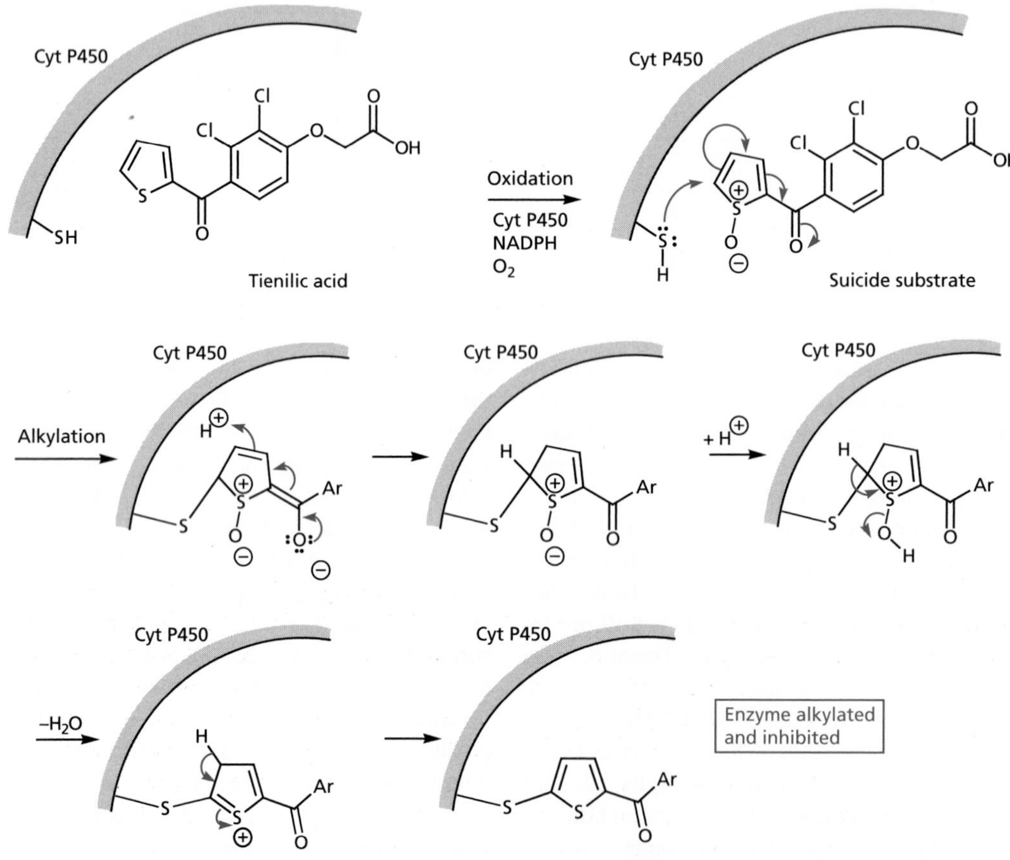

Irreversible inhibition of cytochrome P450 by tienilic acid.

should be one that is not present in our own bodies. Fortunately, such enzymes exist because of the significant biochemical differences between bacterial cells and our own. Nature, of course, is well ahead in this game. For example, many fungal strains produce metabolites that act as inhibitors of bacterial enzymes, but have no effect on fungal enzymes. This gives fungi an advantage over their microbiological competitors when competing for nutrients. It has also provided medicine with important antibiotics such as **penicillin** and **cephalosporin** C.

Although it is preferable to target enzymes that are unique to the foreign invader, it is also possible to selectively target bacterial enzymes that have equivalent mammalian counterparts, as long as there are significant differences between them. Such differences are perfectly feasible. Although the enzymes in both species may have derived from a common ancestral protein, they have evolved and mutated separately over several million years. Identifying these differences allows the medicinal chemist to design drugs that will bind and act selectively against the bacterial enzyme. Chapter 19 covers antibacterial agents such as the **sulphonamides**, **penicillins**, and **cephalosporins**, all of which act by inhibiting enzymes. Synthetic enzyme inhibitors such as the **fluoroquinolones** are also covered in this chapter.

7.7.2 Enzyme inhibitors used against viruses

Enzyme inhibitors are also extremely important in the battle against viral infections (e.g. herpesvirus and HIV). Successful antiviral drugs include **aciclovir** for herpes,

and drugs such as **zidovudine** and **saquinavir** for HIV (see Chapter 20).

7.7.3 Enzyme inhibitors used against the body's own enzymes

Drugs that act on the body's own enzymes are important in medicine. There are many examples discussed elsewhere in this text and Table 7.1 indicates several of these with a cross-reference to the relevant section in the book.

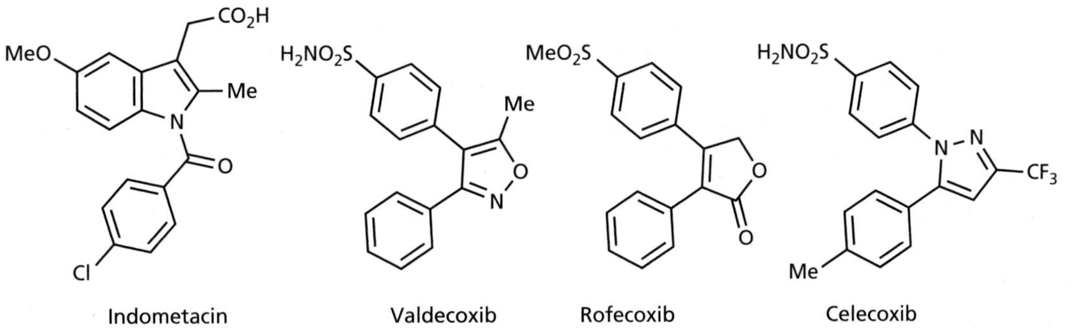

FIGURE 7.11 Cyclooxygenase inhibitors.

BOX 7.4 Designing drugs to be isozyme selective

Designing drugs to be isozyme selective means that they can be designed to act on different diseases despite acting on the same enzyme. This is because isozymes differ in substrate specificity and are distributed differently in the body. **Monoamine oxidase (MAO)** is one of the enzymes responsible for the metabolism of important neurotransmitters such as **dopamine, noradrenaline**, and **serotonin** (section 4.2), and exists in two isozymic forms (MAO-A and MAO-B). These isozymes differ in substrate specificity, tissue distribution, and primary structure, but carry out the same reaction by the same mechanism (Fig. 1). MAO-A is selective for noradrenaline and serotonin, whereas MAO-B is selective for dopamine. MAO-A inhibitors such as **clorgiline** are used clinically as antidepressants, while MAO-B inhibitors such as **selegiline** are

administered with **levodopa** for the treatment of Parkinson's disease (Fig. 2). MAO-B inhibition protects levodopa from metabolism. Clorgiline and selegiline are thought to act as suicide substrates where they are converted by the enzyme to reactive species that react with the enzyme and form covalent bonds. The amine and alkyne functional groups present in both drugs are crucial to this process.

FIGURE 1 Reaction catalysed by MAO.

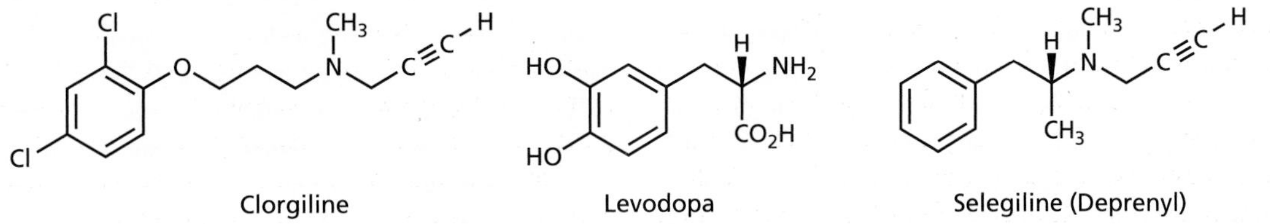

FIGURE 2 Clorgiline, levodopa, and selegiline.

TABLE 7.1 Enzyme inhibitors that act against enzymes in the body.

Drug	Target enzyme	Field of therapy	Relevant section
Aspirin	Cyclooxygenase	Anti-inflammatory	13.1.9
Captopril and enalapril	Angiotensin-converting enzyme (ACE)	Antihypertension	26.3.3 and Case study 2
Simvastatin	HMG-CoA reductase	Lowering cholesterol levels	Case study 1
Phenelzine	Monoamine oxidase	Antidepressant	23.12.5
Clorgiline, Moclobemide	Monoamine oxidase-A	Antidepressant	Box 7.4, 23.12.5
Selegiline	Monoamine oxidase-B	Parkinson's disease	Box 7.4
Methotrexate, permetrexed, pralatrexate	Dihydrofolate reductase	Anticancer	21.3.1
5-Fluorouracil, raltitrexid	Thymidylate synthase	Anticancer	21.3.2
Gefitinib, imatinib, etc.	Tyrosine kinases	Anticancer	21.6.2
Sildenafil	Phosphodiesterase enzyme (PDE5)	Treatment of male erectile dysfunction. Vasodilator in cardiovascular medicine	12.4.4.2 26.5.2
Allopurinol	Xanthine oxidase	Treatment of gout	
Hydroxycarbamide	Ribonucleotide reductase	Anticancer	21.3.3
Pentostatin	Adenosine deaminase	Antileukaemia	21.3.4
Cytarabine, gemcitabine, fludarabine	DNA polymerases	Anticancer	21.3.5
Omeprazole, lansoprazole, pantoprazole, rabeprazole	Proton pump	Anti-ulcer	25.3
Physostigmine, donepezil, tacrine, organophosphates	Acetylcholinesterase	Myasthenia gravis, glaucoma, Alzheimer's disease	22.12–22.15
Various structures	Matrix metalloproteinase	Potential anticancer agents	21.7.1
Racecadotril	Enkephalinase	Treatment of diarrhoea	24.8.4
Zileutin	5-Lipoxygenase	Anti-asthmatic	
Bortezomib	Proteasome	Anticancer	21.7.2
Vorinostat	Histone deacetylase	Anticancer	21.7.3
Lonafarnib	Farnesyl transferase	Anticancer	21.6.1
Sacubitril	Neprilysin	Vasodilator	26.5.3
Aliskiren	Renin	Antihypertensive	26.3.2 and Case study 8
Apixaban	Factor Xa	Anticoagulant	26.9.1.3 and Case study 9

BOX 7.5 Action of toxins on enzymes

The toxicity of several poisons, toxins, and heavy metals result from their action on enzymes. Heavy metals such as lead, cadmium, and mercury have teratogenic effects leading to babies being born with malformed limbs. The worst case of mercury poisoning was in Japan, where a local population ate fish contaminated with methylmercury (MeHg$^+$) that had been used as an agricultural fungicide. The compound inac-tivates enzymes by reacting with the thiol groups (R–SH) of cysteine residues to form covalent bonds (R–S–HgMe).

Mercury poisoning can also affect enzymes in the central nervous system leading to strange behaviour. For example, mercury nitrate was used by hat makers to soften and shape animal furs, and inevitably some of the chemical was absorbed through the skin. So many in the trade were poisoned

(Continued)

BOX 7.5 Action of toxins on enzymes (*Continued*)

in this way that their peculiar manner of behaviour led to the phrase 'mad as a hatter'.

The poison arsenite (AsO_3^{3-}) reacts with the thiol groups of an enzyme cofactor called dihydrolipoate, which is a prosthetic group (section 3.5.5) in some enzymes (Fig. 1). It is possible to reverse the poisoning by administering reagents with adjacent thiol groups that displace the arsenic from the cofactor. **2,3-Dimercaptopropanol** was developed after World War 1 as an antidote to an arsenic-based chemical weapon called **lewisite.**

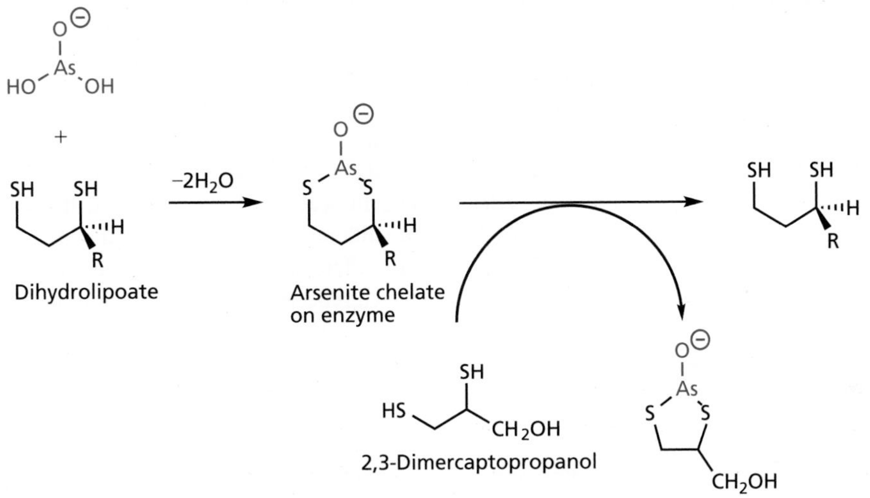

FIGURE 1 Mechanism of arsenite poisoning and its treatment.

The search continues for new enzyme inhibitors, especially those that are selective for a specific isozyme, or act against recently discovered enzymes. Some current research projects include investigations into inhibitors of **factor Xa** (Case study 9), histone deacetylase (section 21.7.3), proteasome inhibitors (section 21.7.2), and **caspases**. The caspases are implicated in the processes leading to cell death, and inhibitors of caspases may have potential in the treatment of stroke victims (Box 12.1). A vast amount of research is also taking place on **kinase inhibitors** (Box 7.6). The kinase enzymes catalyse the phosphorylation of proteins and play an important role in signalling pathways within cells (see also Chapter 5 and section 21.6.2).

For additional material see Web article 1: Steroids as novel anticancer agents on the Online Resource Centre at www.oxfordtextbooks.co.uk/orc/patrick6e/

7.7.4 Enzyme modulators

Modulators are agents that bind to the allosteric binding site of an enzyme and modulate its activity by making it more sensitive to low levels of substrate. Alternatively, they convert the enzyme from an inactive conformation to an active conformation. Compared to enzyme inhibitors, there are far fewer drugs acting as enzyme modulators. However, one example is the vasodilator **riociguat** which stimulates an enzyme called **soluble guanylate cyclase** (section 26.5.1).

KEY POINTS

- Enzyme inhibition is reversible if the drug binds through intermolecular interactions. Irreversible inhibition results if the drug reacts with the enzyme and forms a covalent bond.

- Competitive inhibitors bind to the active site and compete with either the substrate or the cofactor.

- Allosteric inhibitors bind to an allosteric binding site which is different from the active site. They alter the shape of the enzyme such that the active site is no longer recognizable.

- Transition-state analogues are enzyme inhibitors designed to mimic the transition state of an enzyme-catalysed reaction mechanism. They bind more strongly than either the substrate or the product.

- Suicide substrates are molecules that act as substrates for a target enzyme, but which are converted into highly reactive species as a result of the enzyme-catalysed reaction mechanism. These species react with amino acid residues present in the active site to form covalent bonds and act as irreversible inhibitors.

- Drugs that selectively inhibit isozymes are less likely to have side effects, and will be more selective in their effect.

- Enzyme inhibitors are used in a wide variety of medicinal applications.

7.8 Enzyme kinetics

Studies of enzyme kinetics are extremely useful in determining the properties of an enzyme inhibitor. In this section, we will look at how Lineweaver–Burk plots are used to determine what type of inhibition is occurring, as well as important quantitative measurements related to that inhibition.

7.8.1 Lineweaver–Burk plots

The Lineweaver–Burk plot (described in section 3.8.2) can be used to determine whether the inhibitor of an enzyme-catalysed reaction is competitive, uncompetitive or non-competitive (Figs. 7.12b and 7.14). The reciprocals of the reaction rate and the substrate concentration are plotted, with and without an inhibitor being present. This generates straight lines having the following equation, where the slope (m) corresponds to K_M/rate_{max}, and the intersection with the y-axis (c) corresponds to $1/\text{rate}_{max}$.

$$\frac{1}{\text{rate}} = \frac{K_M}{\text{rate}_{max}} \cdot \frac{1}{[S]} + \frac{1}{\text{rate}_{max}} \qquad y = m.x + c$$

In the case of competitive inhibition, the lines cross the y-axis at the same point (i.e. the maximum rate of the enzyme-catalysed reaction is unaffected), but the slopes are different (i.e. the values of the **Michaelis constant** K_M are different). The fact that the maximum rate is unaffected reflects the fact that the inhibitor and substrate are competing for the same active site, and that increasing the substrate concentration sufficiently will overcome the inhibition. The increase in the slope that results from

---

BOX 7.6 Kinase inhibitors

Kinases play a crucial role in the processes by which signal transduction is communicated from a membrane-bound receptor into a cell (section 5.4). However, when this system goes wrong, it can lead to excess cell growth and division resulting in cancer. Therefore, a large amount of research has gone into developing kinase inhibitors for the treatment of various cancers (section 21.6.2). A flurry of kinase inhibitors have been approved as novel anticancer drugs over the last 10 years. However, there are other therapeutic areas where kinase inhibitors could be useful. In 2013, **tofacitinib** was

the first kinase inhibitor to be approved for the treatment of rheumatoid arthritis. The agent inhibits the **Janus kinase enzyme** (sections 4.8.4 and 5.4.4) by binding to the region of the active site normally occupied by the cofactor ATP. There are two particularly important hydrogen bonding interactions involving the pyrimidine ring system, where one nitrogen acts as a hydrogen bond acceptor, and an NH proton acts as a hydrogen bond donor. One of the methyl substituents fits into a hydrophobic pocket and forms van der Waals interactions with the amino acid residues present there.

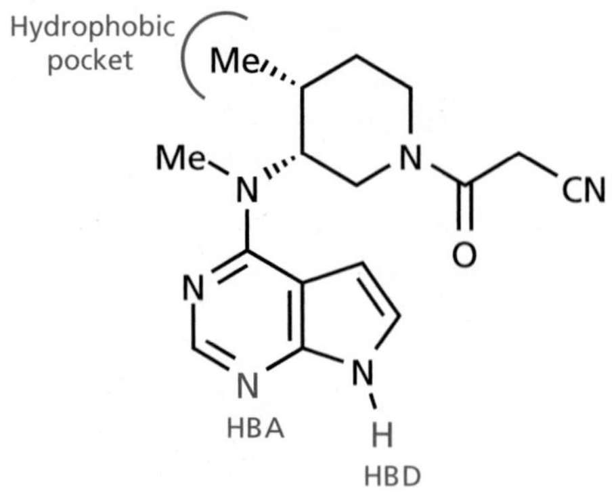

Tofacitinib.

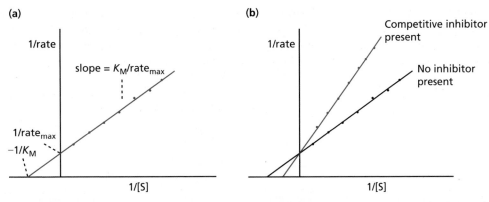

FIGURE 7.12 (a) Lineweaver–Burk plot. (b) Lineweaver–Burk plots with and without a competitive inhibitor present.

adding an inhibitor is a measure of how strongly the inhibitor binds to the enzyme and decreases the rate of the enzyme-catalysed reaction. In the presence of a competitive inhibitor, the apparent value of K_M is increased by a constant α (the **degree of inhibition**) that depends on the concentration of inhibitor present;

$$K_M(\text{app}) = \alpha K_M$$

The degree of inhibition (α) can be determined by rearranging this equation as follows:

$$\alpha = \frac{K_M(\text{app})}{K_M}$$

A useful measure of inhibition is the apparent **inhibition constant** K_i, which is a measure of the equilibrium between the enzyme–inhibitor complex and the uncomplexed enzyme and inhibitor.

$$EI \rightleftharpoons E + I \qquad K_i = \frac{[E][I]}{[EI]}$$

We can write an expression linking the apparent inhibition constant K_i to the inhibitor concentration [I] and the degree of inhibition α;

$$K_i = \frac{[I]}{\alpha - 1}$$

Replacing α with $K_M(\text{app})/K_M$, then rearranging the equation gives the straight line equation shown in Fig. 7.13. A plot of this line will give the Michaelis constant K_M as the intersect with the y-axis, while the slope corresponds to K_M/K_i. From this, one can get the value of K_i.

To create this plot, a series of Lineweaver–Burk plots is first created in order to get values of $K_M(\text{app})$ at different inhibitor concentrations. The plot of $K_M(\text{app})$ versus [I] is then drawn, allowing K_i to be calculated from the slope of the line. The lower the value of K_i, the more potent the inhibitor.

In the case of an uncompetitive inhibitor (Fig. 7.14), the inhibitor binds to the enzyme–substrate complex rather than the free enzyme. Enzyme inhibition studies result in Lineweaver–Burk plots where the lines are parallel and cross the y-axis at different points, indicating that the maximum rate for the enzyme has been reduced. For a reversible, non-competitive inhibitor, the lines have the same intercept point on the x-axis (i.e. K_M is unaffected), but have different slopes and different intercepts on the y-axis. Therefore, the maximum rate for the enzyme has been reduced.

Lineweaver–Burk plots are extremely useful in determining the nature of inhibition, but they have their limitations and are not applicable to enzymes that are under allosteric control.

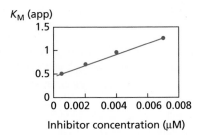

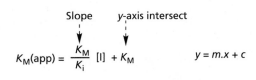

FIGURE 7.13 Plot of $K_M(\text{app})$ versus inhibitor concentration [I].

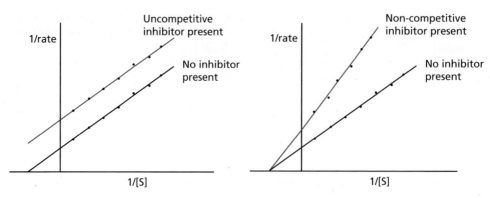

FIGURE 7.14 Lineweaver–Burk plots with and without an uncompetitive inhibitor or non-competitive inhibitor.

Test your understanding and practise your enzyme kinetics with exercise 7.1 on the Online Resource Centre at www.oxfordtextbooks.co.uk/orc/patrick6e/

7.8.2 Comparison of inhibitors

When comparing the activity of enzyme inhibitors, the IC_{50} value is often quoted. This is the concentration of inhibitor required to reduce the activity of the enzyme by 50%. Compounds with a high IC_{50} are less powerful inhibitors than those with a low IC_{50}, as a higher concentration of the former is required to attain the same level of inhibition.

K_i values are also reported in enzyme inhibition studies and it can be shown that $IC_{50} = K_i + [E]_{total}/2$. If the concentration of inhibitor required to inhibit the enzyme by 50% is much greater than the concentration of enzyme, then K_i is much larger than $[E]_{total}$ and this equation approximates to $IC_{50} = K_i$.

IC_{50} and K_i values are measured in assays involving isolated enzymes. However, it is often useful to carry out enzyme inhibition studies where the enzyme is present in whole cells or tissues. In these studies, a cellular effect resulting from enzyme activity is monitored. EC_{50} values represent the concentration of inhibitor required to reduce that particular cellular effect by 50%. It should be noted that the effect being measured may be several stages downstream from the enzyme reaction concerned.

KEY POINTS

- The Michaelis–Menten equation relates the rate of an enzyme-catalysed reaction to substrate concentration.
- Lineweaver–Burk plots are derived from the Michaelis–Menten equation and are used to determine whether inhibition is competitive, uncompetitive, or non-competitive.
- The activity of different enzymes can be compared by measuring values of EC_{50}, K_i, or IC_{50}.

QUESTIONS

1. It is known that the amino acid at position 523 of the cyclooxygenase enzyme is part of the active site. In the isoenzyme COX-1, this amino acid is isoleucine, whereas in COX-2, it is valine. Suggest how such information could be used in the design of drugs that selectively inhibit COX-2.

2. Neostigmine is an inhibitor of acetylcholinesterase. The enzyme attempts to catalyse the same reaction on neostigmine as it does with acetylcholine. However, a stable intermediate is formed which prevents completion of the

process and which results in a molecule being covalently linked to the active site. Identify the stable intermediate and explain why it is stable.

3. The human immunodeficiency virus contains a protease enzyme that is capable of hydrolysing the peptide bond of L-Phe-L-Pro. Structure I was designed as a transition-state inhibitor of the protease enzyme. What is a transition-state inhibitor and how does structure I fit the description of a transition-state inhibitor? What is meant by IC_{50} 6500 nM?

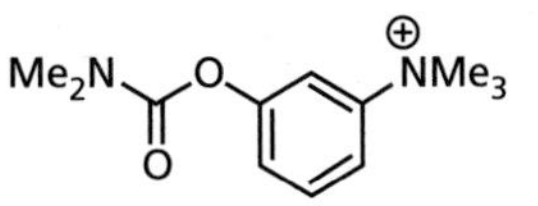

Neostigmine

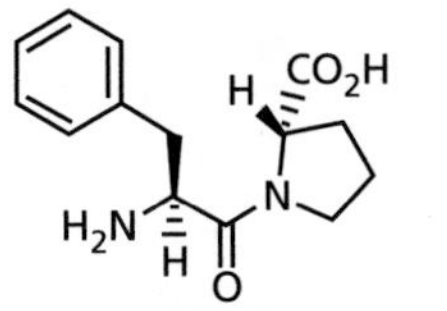

L-Phe-L-Pro

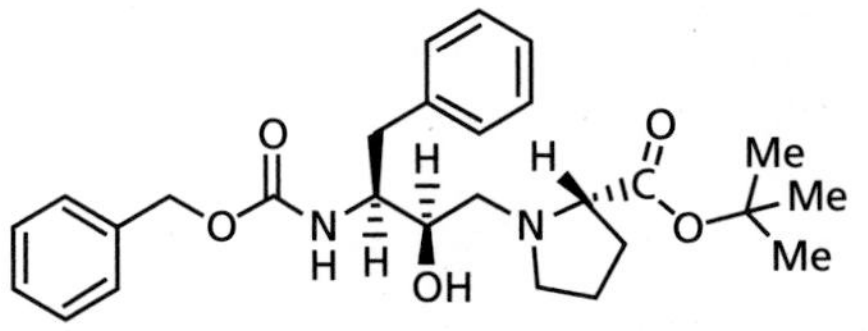

(I) IC$_{50}$ 6500 nM

4. Why should a transition state be bound more strongly to an enzyme than a substrate or a product?

5. The methylation of cytosine residues in DNA plays a role in the regulation of transcription and is catalysed by the enzyme DNA methyltransferase. The mechanism is as follows.

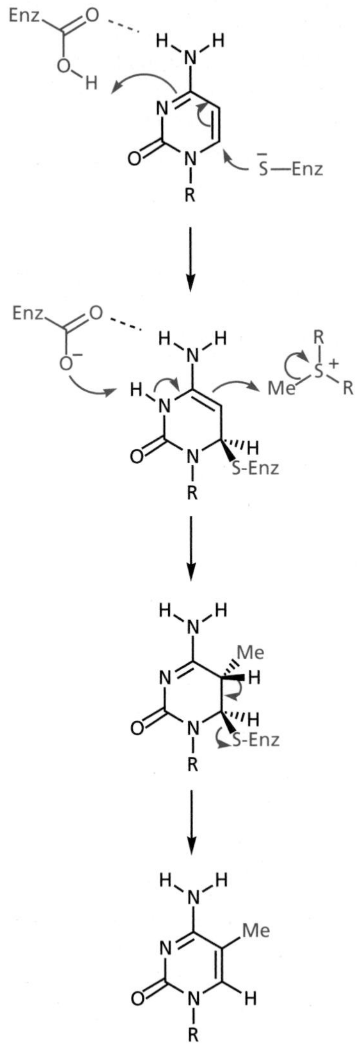

5-Azacytidine and 5-fluoro-2′-deoxycytidine are mechanism-based inhibitors of DNA methyltransferase. Explain why.

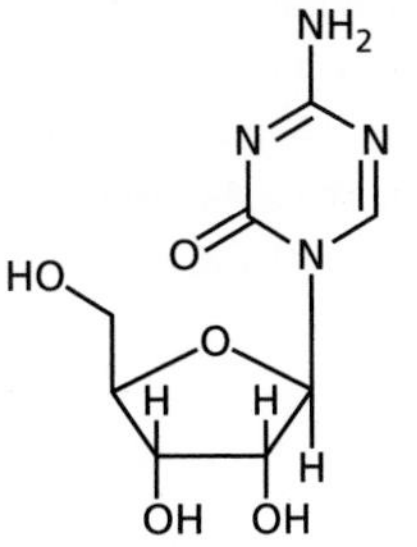

5-Azacytidine 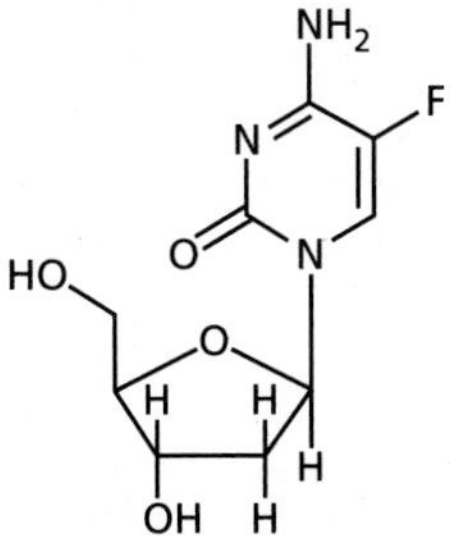   5-Fluoro-2′-deoxycytidine

6. 17β-Hydroxysteroid dehydrogenase type 1 (17β-HSD1) is an enzyme that catalyses the conversion of estrone to estradiol in the presence of the cofactor NADH. The initial rate data for the enzyme-catalysed reaction in the absence of an inhibitor is given in Q5 of Chapter 3. EM-1745 is an inhibitor of the enzyme. The following data was determined with EM-1745 present at a concentration of 4nM. Using this and the data recorded in Chapter 3, determine whether the compound acts as a competitive, uncompetitive, or mixed inhibitor. Calculate the value of K$_i$.

Substrate concentration (10^{-2} mol dm^{-3})	5	10	20	40	83.3
Initial rate (10^{-1} mol dm^{-3} s^{-1})	1.0	2.0	3.45	6.25	10.0

7. The quinazoline structure shown is an inhibitor of the enzyme scytalone dehydratase. One of the binding interactions between the inhibitor and the active site is a hydrogen bond to a water molecule which acts as a hydrogen-bonding bridge to two tyrosine residues. Explain why analogue I is three times less active, whereas analogue II is 20 times more active.

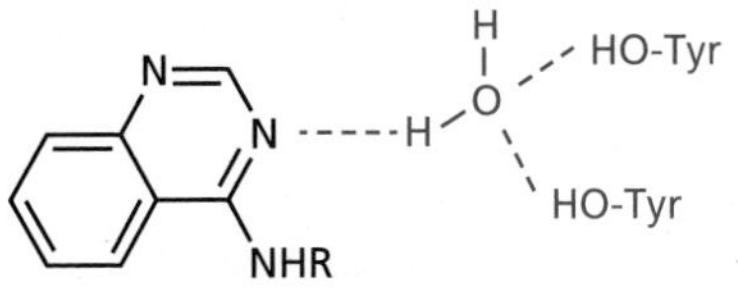

Quinazoline

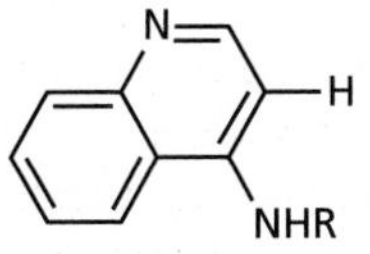

Analogue I 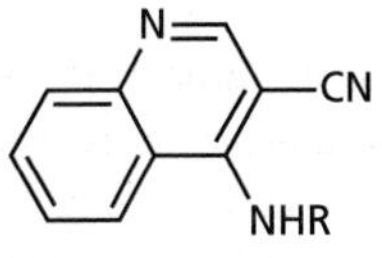   Analogue II

8. Cytidine deaminase is an enzyme which converts cytidine to uridine. Suggest a mechanism by which this reaction might occur, considering that a highly conserved water molecule is present in the active site. Zebularine is a natural product which is converted into a highly potent transition-state inhibitor of the enzyme ($K_i = 1.2$ pM) when it is in the active site. Suggest what the structure of the TS inhibitor might be and why it is so effective. Explain why 3,4-dihydrozebularine has a binding affinity that is only 30 μM.

Zebularine

3,4-Dihydrozebularine

Multiple-choice questions are available on the Online Resource Centre at www.oxfordtextbooks.co.uk/orc/patrick6e/

FURTHER READING

Clark, J. D., Flanagan, M. E., and Telliez, J.-B. (2014) Discovery and development of Janus Kinase (JAK) inhibitors for inflammatory diseases, *Journal of Medicinal Chemistry*, **57**, 5023–38.

Flower, R. J. (2003) The development of COX-2 inhibitors. *Nature Reviews Drug Discovery*, **2**, 179–91.

Lowe, D. (2010) In the pipeline. *Chemistry World*, September, p18. (Kinases as drug targets.)

Mitchell, J. A. and Warner, T. D. (2006) COX isoforms in the cardiovascular system: understanding the activities of non-steroidal anti-inflammatory drugs. *Nature Reviews Drug Discovery*, **5**, 75–86.

Siragy, H. M., Kar, S., and Kirkpatrick, P. (2007) Aliskiren. *Nature Reviews Drug Discovery*, **6**, 779–80.

Teague, S. J. (2003) Implications of protein flexibility for drug discovery. *Nature Reviews Drug Discovery*, **2**, 527–41.

Titles for general further reading are listed on p.845.

Receptors as drug targets

8.1 Introduction

The structures and functions of various receptors were described in Chapter 4. Receptors and their chemical messengers are crucial to the communication systems of the body. Such communication is clearly essential to the normal workings of the body, and when it goes wrong, a huge variety of ailments can arise, such as depression, heart problems, schizophrenia, and muscle fatigue to name just a few. What sort of things *could* go wrong though?

One problem would be if too many chemical messengers were released. The target cell could (metaphorically) start to overheat. Alternatively, if too few messengers were sent out, the cell could become sluggish. It is at this point that drugs can play a role by either acting as replacement messengers, or blocking receptors from receiving their natural messengers. Drugs that mimic the natural messengers and activate receptors are known as **agonists**. Drugs that block receptors are known as **antagonists**. The latter compounds still bind to the receptor, but they do not activate it. However, since they are bound, they prevent the natural messenger from binding.

What determines whether a drug acts as an agonist or an antagonist, and is it possible to predict whether a new drug will act as one or the other? To answer these questions, we have to move down to the molecular level and consider what happens when a small molecule such as a drug or a neurotransmitter interacts with a receptor protein.

In sections 4.4–4.5, we looked at a hypothetical receptor and neurotransmitter. We saw that a chemical messenger caused the receptor to change shape—a process

known as an induced fit. It is this induced fit which activates the receptor and leads to the 'domino' effect of signal transduction—the method by which the message carried by the chemical messenger is transferred into the cell (Chapters 4 and 5).

8.2 The design of agonists

We are now at the stage of understanding how drugs might be designed in such a way that they mimic the natural chemical messengers. Assuming that we know what binding regions are present in the receptor site and where they are located, we can design drugs to interact with the receptor in the same way. Let us look at this more closely and consider the following requirements in turn.

- The drug must have the correct binding groups.
- The drug must have these binding groups correctly positioned.
- The drug must be the right size for the binding site.

8.2.1 Binding groups

If we know the structure of the natural chemical messenger, and can identify the functional groups that form important interactions with the binding site, then we might reasonably predict which of a series of molecules would interact in the same way. For example, consider the hypothetical neurotransmitter shown in Fig. 8.1. The important binding groups are indicated in blue—an

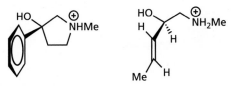

Hypothetical
neurotransmitter

FIGURE 8.1 A hypothetical neurotransmitter and possible agonists (binding groups shown in colour).

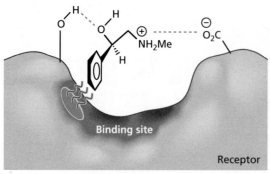

(a) Three interactions

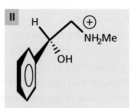

(b) Two interactions

FIGURE 8.2 A comparison of interactions involving (a) the hypothetical neurotransmitter and (b) its mirror image with a hypothetical binding site.

aromatic ring, alcohol, and aminium ion. These interact with the binding site through van der Waals interactions, hydrogen bonding, and ionic bonding respectively (Fig. 8.2a). Consider now the other structures in Fig. 8.1. They all look different, but they all contain functional groups which could interact in the same way. Therefore, they may well be potential agonists that will activate the receptor.

What about the structures in Fig. 8.3? They lack one or more of the required binding groups and should, therefore, have poor activity. We would expect them to drift into the binding site then drift back out again, binding only weakly, if at all.

Of course, we are making an assumption here; namely that all three binding groups are essential. It might be argued that a compound such as structure II in Fig. 8.3

might be effective even though it lacks a suitable hydrogen bonding group. Why, for example, could it not bind initially by van der Waals interactions alone and then alter the shape of the receptor protein via ionic bonding?

In fact, this seems unlikely when we consider that neurotransmitters appear to bind, pass on their message, and then leave the binding site relatively quickly. In order to do that, there must be a fine balance in the binding interactions between the receptor and the neurotransmitter. They must be strong enough to bind the neurotransmitter effectively such that the receptor changes shape. However, the binding interactions cannot be too strong or else the neurotransmitter would not be able to leave and the receptor would not be able to return to its original shape. Therefore, it is reasonable to assume that a neurotransmitter needs all of its binding interactions to be effective.

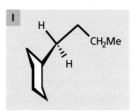

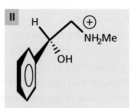

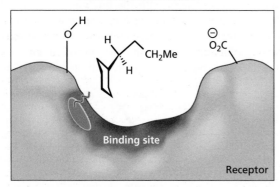

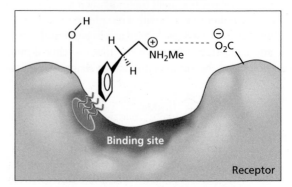

FIGURE 8.3 Weaker binding to the hypothetical receptor by structures that possess fewer than the required binding groups.

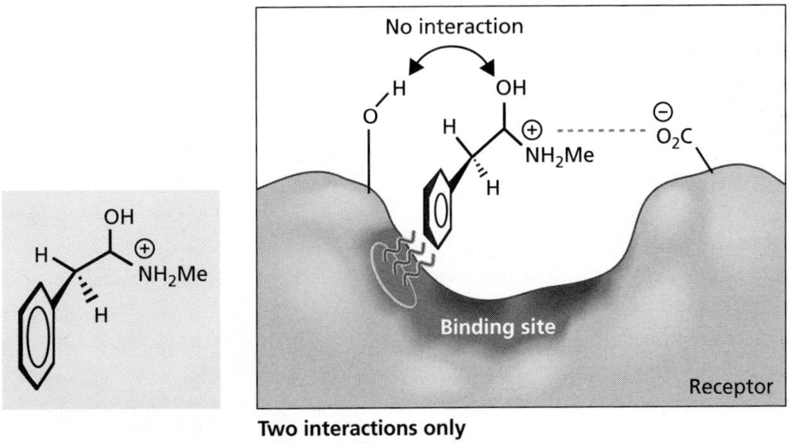

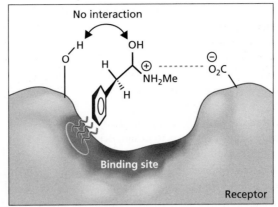

Two interactions only

FIGURE 8.4 Weaker binding to the hypothetical receptor by a molecule containing binding groups in incorrect positions.

The lack of even one of these interactions would lead to a significant loss in activity.

8.2.2 Position of the binding groups

The molecule may have the correct binding groups, but if they are in the wrong relative positions, they will not be able to form bonds at the same time. As a result, bonding would be too weak to be effective.

A molecule such as the one shown in Fig. 8.4 obviously has one of its binding groups (the hydroxyl group) in the wrong position, but there are more subtle examples of molecules that do not have the correct arrangement of binding groups. For example, the mirror image of our hypothetical neurotransmitter would not bind strongly to the binding site (Fig. 8.5). The structure has the same formula and the same constitutional structure as our original structure. It will have the same physical properties and undergo the same chemical reactions, but it is not the same shape. It is a non-superimposable mirror image and it cannot interact with all the binding regions of the receptor binding site at the same time (Fig. 8.2b).

Compounds which exist as non-superimposable mirror images are termed **chiral** or **asymmetric.** There are only two detectable differences between the two mirror images (or **enantiomers)** of a chiral compound. They rotate plane polarized light in opposite directions and they interact differently with other chiral systems, such as enzymes and receptors. This has very important consequences for the pharmaceutical industry.

Pharmaceutical agents are usually synthesized from simple starting materials using simple, achiral (symmetrical) chemical reagents. These reagents are incapable of distinguishing between the two mirror images of a chiral compound. As a result, most chiral drugs used to be synthesized as a mixture of both mirror images—a **racemate**. However, we have seen from our own simple

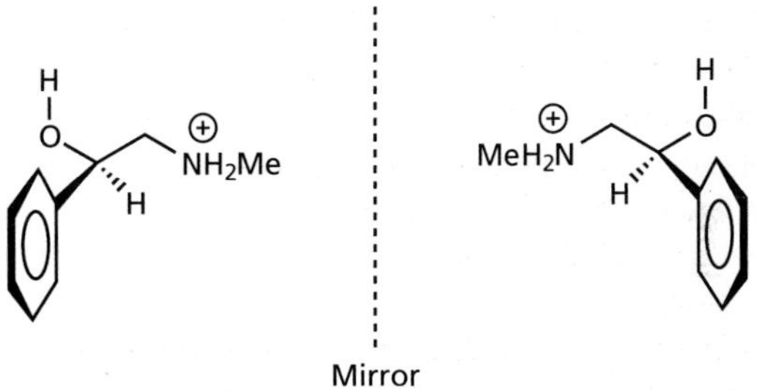

FIGURE 8.5 Mirror image of a hypothetical neurotransmitter.

example that only one of these enantiomers is going to interact properly with a target receptor. What happens to the other enantiomer?

At best, it floats about in the body doing nothing. At worst, it interacts with a totally different target and results in an undesired side effect. Even if the 'wrong' enantiomer does not do any harm, it seems to be a great waste of time, money, and effort to synthesize drugs that are only going to be 50% efficient. That is why one of the biggest areas of chemical research in recent years has been in **asymmetric synthesis**—the selective synthesis of a single enantiomer of a chiral compound.

Of course, nature has been at it for millions of years, having chosen to work predominantly with the 'left-handed' enantiomer of amino acids[1]. Therefore, enzymes

---

[1] Naturally occurring asymmetric amino acids exist in mammals as the one enantiomer, termed the L-enantiomer. This terminology is historical, and defines the absolute configuration of the asymmetric carbon present at the head-group of the amino acid. The current terminology for asymmetric centres is to define them as R or S according to a set of rules known as the Cahn–Ingold–Prelog rules. The L-amino acids exist as the (S)-configuration (except for cysteine which is R), but the older terminology still dominates here. Experimentally, the L-amino acids are found to rotate plane-polarized light anticlockwise or to the left. It should be noted that D-amino acids can occur naturally in bacteria (see for example section 19.5.5).

are made up of left-handed amino acids and are present as a single mirror image. As a consequence, they catalyse **enantiospecific** reactions—reactions which give only one enantiomer. Moreover, the enantiomers of asymmetric enzyme inhibitors can be distinguished by the target enzyme which means that one enantiomer is more potent than the other.

The importance of having binding groups in the correct position has led medicinal chemists to design drugs based on what is considered to be the important **pharmacophore** of the messenger molecule. In this approach, it is assumed that the correct positioning of the binding groups is what decides whether the drug will act as a messenger or not, and that the rest of the molecule serves as a scaffold to hold the groups in those positions. Therefore, the activity of apparently disparate structures at a receptor can be explained if they all contain the correct binding groups at the correct positions. Totally novel structures or molecular frameworks could then be designed to obey this rule, leading to a new series of drugs. There is, however, a limiting factor to this, which will now be discussed.

8.2.3 Size and shape

It is possible for a compound to have the correct binding groups in the correct positions, and yet fail to interact effectively if it has the wrong size or shape. As an example, consider the structure shown in Fig. 8.6 as a possible ligand for our hypothetical receptor.

The structure has a *meta*-methyl group on the aromatic ring and a long alkyl chain attached to the nitrogen atom. Both of these features would prevent this molecule from binding effectively to the binding site shown.

The *meta*-methyl group acts as a steric shield and prevents the structure from sinking deep enough into the binding site for effective binding. Similarly, the long alkyl chain on the nitrogen atom makes that part of the molecule too long for the space available to it. A thorough understanding of the space available in the binding site is therefore necessary when designing drugs to fit it. Having said that, it is important to appreciate that there is a level of flexibility in the binding site. A potential agonist may appear too large, but a slightly different induced fit might occur which allows the molecule to fit and bind, yet still activate the receptor. In exceptional cases, there can be quite significant alterations in the induced fit (Box 8.1).

8.2.4 Other design strategies

The discussion above describes how agonists can be designed from a knowledge of the structure, shape, and binding interactions of the natural messenger. However,

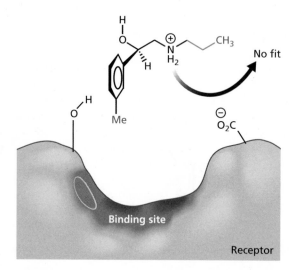

FIGURE 8.6 Failed interaction of a structure with a binding site because of steric factors.

there are several agonists that are quite different in structure from the natural messenger. How are these designed? It has to be remembered that the binding site is bristling with amino acid residues and peptide links, all of which might be capable of interacting with a visiting molecule by different types of intermolecular bonds. In other words, there may be other binding regions present than just those used by the natural messenger (Fig. 8.7). A drug that has the same three binding interactions described above, plus an extra binding interaction, would be expected to bind more strongly and be a more potent agonist if the correct induced fit still takes place for receptor activation. Moreover, the loss of one of the key binding groups required by the natural messenger could be compensated by the presence of other binding groups capable of interacting with different binding regions.

8.2.5 Pharmacodynamics and pharmacokinetics

The study of how molecules interact with targets such as receptors or enzymes to produce a pharmacological effect is called pharmacodynamics. Such studies can typically be carried out on the pure target protein, or on isolated cells or tissues which bear the target protein (*in vitro* studies), but it is important to appreciate that designing a drug to interact effectively with a protein *in vitro* does not guarantee a clinically useful drug. Studies should also be carried out concurrently to ensure that promising-looking drugs are active in whole organisms (*in vivo* studies). This is a field known as pharmacokinetics and is covered in Chapters 11 and 14. It is also important to identify at an early stage whether the structures being studied might be prone to toxic or unacceptable

BOX 8.1 An unexpected agonist

Glucocorticoid steroids such as **cortisol** are used clinically as anti-inflammatory agents and act as agonists at the glucocorticoid receptor (Case study 6). These have the correct size, shape, and binding groups to fit the binding site, and produce the required induced fit for receptor activation. Recently, it has been discovered that **cortivazol** acts as an agonist, yet it lacks one of the important binding groups (the ketone) and has two extra rings that should make it too big for the binding site. A crystal structure of the receptor–ligand complex was studied and it was found that a different induced fit had occurred from normal. This had resulted in a new channel being opened up in the binding site that could accommodate the extra rings. Moreover, extra interactions with the rings compensated for the loss of the usually crucial ketone group. Normally, a different induced fit would be expected to result in antagonist activity, but, in this case, the receptor was still activated (see also Box 17.6).

Cortisol

Cortivazol; R = Ac
Deacetylcortivazol; R = H

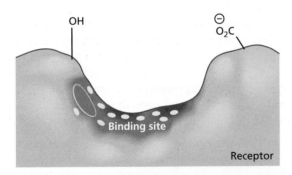

FIGURE 8.7 The hypothetical binding site showing extra binding regions (in white) that are not used by the natural chemical messenger.

side effects, so that time is not wasted taking a candidate drug all the way to clinical trials only for it to be rejected (Box 12.3).

8.2.6 Examples of agonists

There are numerous examples of drugs that act as agonists at various target receptors. In this textbook, you will find a description of cholinergic agonists which are used in the treatment of glaucoma and myasthenia gravis (section 22.8). There is a description of the adrenergic agonists used as anti-asthmatic agents in section 23.10.3, while Chapter 24 covers the opioid analgesics which act as agonists. The glucocorticoids that are used as anti-inflammatory agents also act as agonists (Case study 6; see also Box 8.1), and prostacyclin agonists are used in cardiovascular medicine (section 26.5.4).

Other examples of agonists used in the clinic are dopamine agonists used in the treatment of Parkinson's disease, and serotonin agonists used in the treatment of migraine. Agonists designed to act on the estrogen receptor are used as contraceptives. There are many more examples.

8.2.7 Allosteric modulators

Some drugs have an indirect agonist effect by acting as allosteric modulators. By binding to an allosteric site on a target receptor, they mimic the action of endogenous modulators and enhance the action of the natural or endogenous chemical messenger (section 4.10). For example, the **benzodiazepines** used as sleep medicines target the allosteric binding site of the **$GABA_A$ receptor**. **Cinacalcet** (Fig. 8.8) is used to treat thyroid problems, and is an allosteric modulator for a G-protein-coupled receptor known as the **calcium-sensing receptor**. **Galantamine** acts as an enzyme inhibitor in the treatment of Alzheimer's disease (section 22.15), but is also an allosteric modulator of the **nicotinic receptor**.

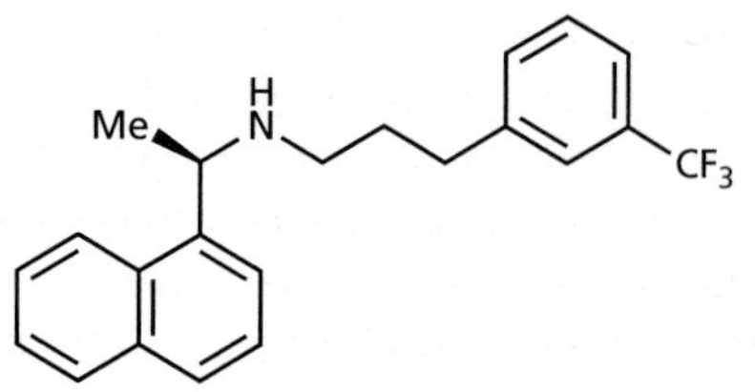

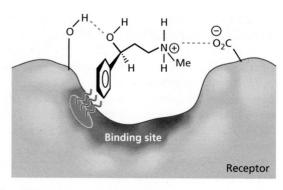

FIGURE 8.8 Cinacalcet.

8.3 The design of antagonists

8.3.1 Antagonists acting at the binding site

We have seen how it might be possible to design drugs (agonists) to mimic natural chemical messengers, and how these would be useful in treating a shortage of the natural ligand. However, suppose that we have too many messengers operating in the body. How could a drug counteract that? The answer would be to design a drug (an antagonist) that will bind to the binding site, but will not activate the receptor. Since it is bound, it will prevent the normal ligand from binding and activating the receptor.

There are several strategies in designing antagonists, but one way is to design a drug which is the right shape to bind to the receptor binding site, but which either fails to change the shape of the binding site or distorts it in the wrong way. Consider the following scenario.

The compound shown in Fig. 8.9 fits the binding site perfectly and, as a result, does not cause any change of shape. Therefore, there is no biological effect and the binding site is blocked to the natural neurotransmitter.

Another strategy is to find different binding regions within the binding site that are not used by the natural

FIGURE 8.9 Compound acting as an antagonist at the binding site.

chemical messenger (Fig. 8.7). Drugs could be designed to interact with some of these extra binding regions such that the resultant binding produces a quite different induced fit from that obtained when the natural messenger binds—an induced fit that fails to activate the receptor.

Extra binding regions do not necessarily have to be within the part of the binding site occupied by the natural messenger. It is quite common to find antagonists that are larger than the natural messenger and which access extra binding regions beyond the reach of the usual messenger. Many antagonists are capable of binding to both the normal binding site and these neighbouring regions.

To illustrate this, we will once more consider our hypothetical neurotransmitter and its receptor, but this time we will represent the binding site in a different way, as if we were looking at it from above and drawing a map of where the binding regions are located (Fig. 8.10). This kind of representation is frequently used to simplify binding site diagrams, but it is important to appreciate that

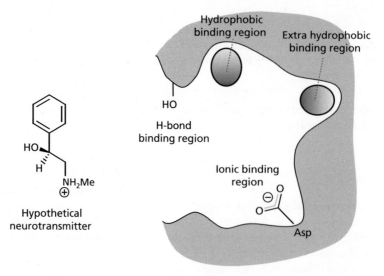

FIGURE 8.10 'Map' of the hypothetical binding site.

the binding site is a three-dimensional shape and that the interactions involved are also in three dimensions.

The three important binding regions are still present, but our 'map' shows an extra hydrophobic region which could act as a potential binding region.

Binding of the hypothetical neurotransmitter results in the correct induced fit required for receptor activation (Fig. 8.11). Note that the extra binding region is not within range of the messenger molecule.

We could now design a molecule which would bind to all four of these binding regions (Fig. 8.12). This molecule will bind more strongly than the natural messenger due to the extra binding interaction. If the binding produces the same induced fit, then we have designed a more potent agonist, but, in this case, the induced fit is significantly

different and so the receptor is not activated. Therefore, the molecule acts as an antagonist; it binds to the receptor, but fails to activate it. Moreover, by occupying the binding site, it prevents the normal messenger from binding.

Antagonists which bind very strongly to a target binding site are often used to label receptors. Such antagonists are synthesized with a radioactive isotope incorporated into their structure, allowing them to be detected more easily.

For additional material see Web article 2: Antagonists as molecular labels on the Online Resource Centre at www.oxfordtextbooks.co.uk/orc/patrick6e/

To sum up, if we know the shape and characteristics of a receptor binding site, then we should be able to design

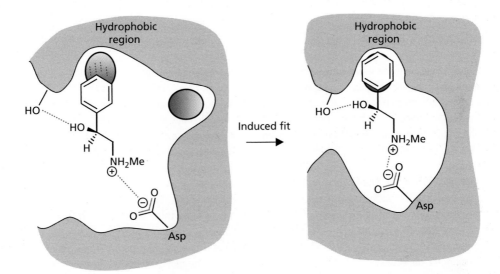

FIGURE 8.11 Binding of the natural chemical messenger resulting in an induced fit that activates the receptor.

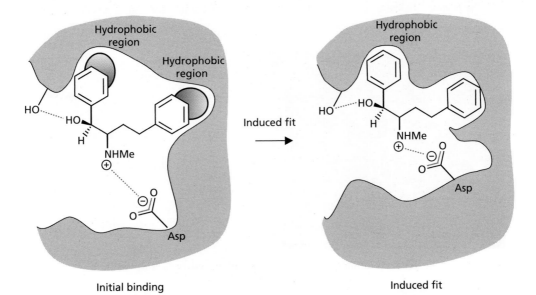

FIGURE 8.12 Binding of an antagonist leading to a different induced fit.

drugs to act as agonists *or* antagonists. Unfortunately, determining the layout of a receptor binding site is not as straightforward as it sounds. For many years, the only feasible approach was to synthesize a large number of compounds, identify those that fitted the binding site and those that did not, then propose what the binding site might look like from those results—a bit like a three-dimensional jigsaw. Nowadays, the use of genetic engineering, X-ray crystallography of protein targets, and computer-based molecular modelling allows a more accurate representation of proteins and their binding sites (Chapter 17). This has heralded new approaches to developing new drugs such as *de novo* **drug design** and **structure-based drug design** (section 17.15 and Case study 5; see also Box 8.2). Some of these studies can reveal surprising

results, where the binding of a particular drug causes a different kind of induced fit from normal, resulting in the exposure of new potential binding regions (Box 8.1).

There are many examples in this book of antagonists that act at the binding site of a receptor. These include the histamine H_2 antagonists used for the treatment of ulcers (Chapter 25), the adrenergic antagonists used in cardiovascular medicine (section 23.11.3), serotonin antagonists as potential CNS-active drugs (Case study 7), and the cholinergic antagonists used as neuromuscular blockers (section 22.10.2). Another example is **raloxifene** which acts as an antagonist of the **estrogen receptor** (Box 8.2). This compound is an example of an antagonist which binds to the same binding regions as the natural ligand, as well as an extra binding region.

BOX 8.2 Estradiol and the estrogen receptor

17β-Estradiol is a steroid hormone which affects the growth and development of a number of tissues. It does so by crossing cell membranes and interacting with the binding site of an estrogen intracellular receptor. Estradiol uses its alcohol and phenol groups to form hydrogen bonds with three amino acids in the binding site, while the hydrophobic skeleton of the molecule forms van der Waals and hydrophobic interactions with other regions (Fig. 1). The binding pocket is hydrophobic in nature and quite spacious, except for the region where the phenol ring binds. This is a narrow slot and will only accept a planar aromatic ring. Due to these constraints, the binding of estradiol's phenolic ring determines the orientation for the rest of the molecule.

The binding of estradiol induces a conformational change in the receptor which sees a helical section known as H12 folding across the binding site like a lid (Fig. 2). This not

only seals estradiol into its binding site, it also exposes a hydrophobic region called the activating function (AF-2) region which acts as a binding site for a coactivator protein. Since dimerization has also taken place, there are two of these regions available and the coactivator binds to both to complete the nuclear transcription factor. This now binds to a specific region of DNA and switches on the transcription of a gene, resulting in the synthesis of a protein.

Raloxifene (Fig. 3) is an antagonist of the estrogen receptor and is used for the treatment of hormone-dependent breast cancer. It is a synthetic agent which binds to the binding site without activating the receptor, and presents estradiol from binding. The molecule has two phenol groups which mimic the phenol and alcohol group of estradiol. The skeleton is also hydrophobic and matches the hydrophobic character of estradiol's tetracyclic skeleton. So why does raloxifene not act

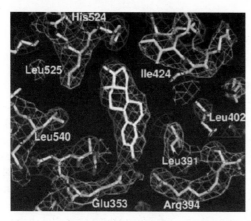

FIGURE 1 Binding mode of estradiol with the estrogen receptor.

(Continued)

BOX 8.2 Estradiol and the estrogen receptor (*Continued*)

as an agonist? The answer lies in a side chain. This side chain contains an amino group which is protonated and forms a hydrogen bond to Asp-351—an interaction that does not take place with estradiol. In doing so, the side chain protrudes from the binding pocket and prevents the receptor helix H12 folding over as a lid. As a result, the AF-2 binding region is not exposed, the coactivator cannot bind and the transcription factor cannot be formed. Hence, the side chain is crucial to antagonism. It must contain an amine group of the correct basicity such that it ionizes and forms the interaction with Asp-351, and it must be of the correct length and flexibility to place the amine in the correct position for binding.

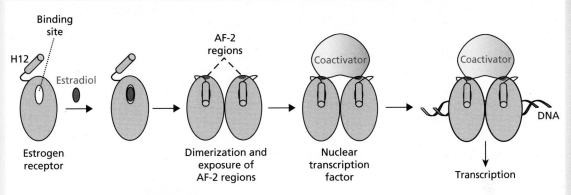

FIGURE 2 Control of transcription by the estrogen receptor.

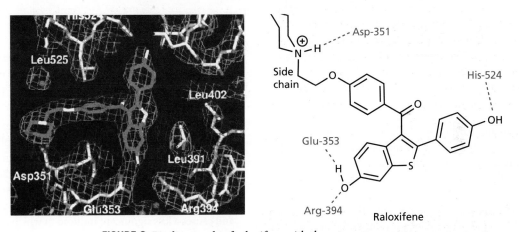

FIGURE 3 Binding mode of raloxifene with the estrogen receptor.

8.3.2 Antagonists acting outwith the binding site

There are examples of antagonists which do not bind to the binding site used by the natural chemical messenger. How do these antagonists work? There are two possible explanations.

Allosteric modulators. Some receptors have allosteric binding sites. These are binding sites which are located on a different part of the receptor surface from the binding site, and which bind natural molecules called **modulators** that 'modulate' the activity of receptors by either enhancing it (section 8.2.7) or diminishing it. If activity is diminished, the modulator is acting indirectly as an antagonist. The mechanism by which this takes place could be viewed in a similar way to the allosteric inhibition of enzymes (section 3.6). The modulator binds to the allosteric binding site and causes it to change shape—an induced fit. This has a 'knock-on' effect which alters the shape of the normal binding site. If the site becomes too distorted, then it is no longer able to bind the normal chemical messenger, or binds it less effectively. Therefore, it is possible to

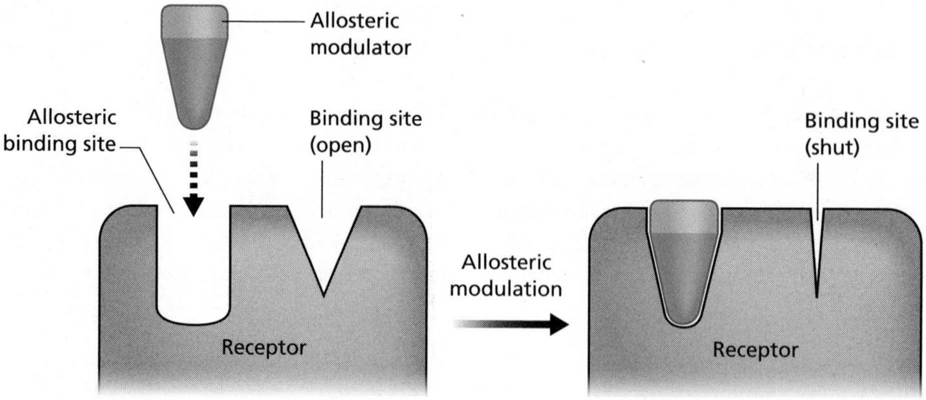

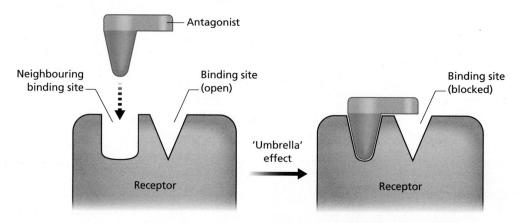

FIGURE 8.13 Principle by which an allosteric antagonist distorts a binding site.

FIGURE 8.14 Antagonism by the 'umbrella effect'.

design an antagonist that will bind to the allosteric binding site rather than to the normal binding site (Fig. 8.13).

Antagonism by the 'umbrella' effect. Some antagonists are thought to bind to regions of the receptor which are close to the normal binding site. Although they do not bind directly to the binding site, the molecule acts as a 'shield' or as an 'umbrella', preventing the normal messenger from accessing the binding site (Fig. 8.14).

8.4 Partial agonists

Frequently a drug is discovered which cannot be defined either as a pure antagonist or a pure agonist. The compound acts as an agonist and produces a biological effect, but that effect is not as great as one would get with a full agonist. Therefore, the compound is called a **partial agonist**. There are several possible explanations for this.

- A partial agonist obviously must bind to a receptor in order to have an agonist effect. However, it may be binding in such a way that the conformational change induced is not ideal, and the subsequent effects of receptor activation are decreased. For example, a receptor may be responsible for the opening of an ion channel. The normal chemical messenger causes an induced fit that results in the ion channel fully opening up. A partial agonist, however, binds to the receptor and causes a less significant induced fit which results in only a slight distortion of the receptor. As a result, the ion channel is only partially opened (Fig. 8.15).

- The partial agonist may be capable of binding to a receptor in two different ways by using different binding regions in the binding site. One method of binding activates the receptor (an agonist effect), but the other does not (an antagonist effect). The balance of agonism versus antagonism would then depend on the relative proportions of molecules binding by either method. This theory was used to explain the activity of partial agonists observed during the development of the anti-ulcer drug cimetidine (section 25.2). An alternative explanation is that a partial agonist has the ability to stabilize two different conformations of the receptor—one which is active and one which is not (Fig. 8.16).

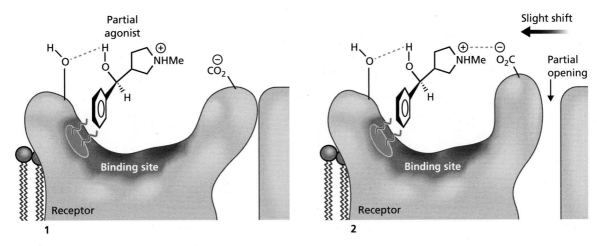

FIGURE 8.15 Partial agonism.

- Receptors which bind the same chemical messenger are not all the same. The partial agonist may be capable of distinguishing between different receptor types or subtypes, acting as an agonist at one subtype, but as an antagonist at another subtype.

Examples of partial agonists in the opioid and antihistamine fields are discussed in Chapters 24 and 25, respectively.

8.5 Inverse agonists

Many antagonists which bind to a receptor binding site are in fact more properly defined as inverse agonists. For example, some of the sartans used as antihypertensive agents act as inverse agonists (section 26.3.4). An inverse agonist has the same effect as an antagonist in that it binds to a receptor, fails to activate it, and prevents the normal chemical messenger from binding. However, there is more to an inverse agonist than that. Some receptors (e.g. the GABA[1], serotonin, and dihydropyridine receptors) are found to have an inherent activity, even in the absence of the chemical messenger. They are said to be **constitutionally active**. An inverse agonist is capable of preventing this activity as well.

The discovery that some receptors have an inherent activity has important implications for receptor theory. It suggests that these receptors do not have a 'fixed' inactive conformation, but are continually changing shape such that there is an equilibrium between the active conformation and different inactive conformations. In that equilibrium, most of the receptor population is in an inactive conformation but a small proportion of the receptors is in the active conformation. The action of agonists and antagonists is then explained by how that equilibrium is affected by binding preferences (Fig. 8.16).

[1] GABA = γ-aminobutyric acid

If an agonist is introduced (frame B), it binds preferentially to the active conformation and stabilizes it, shifting the equilibrium to the active conformation and leading to an increase in the biological activity associated with the receptor.

In contrast, it is proposed that an antagonist binds equally well to all receptor conformations (both active and inactive) (frame C). In the absence of the natural ligand, the receptor's equilibrium is unaffected and there is no change in biological activity. The subsequent introduction of an agonist has no effect either, because all the receptor binding sites are already occupied by the antagonist. Antagonists such as these will have some structural similarity to the natural agonist.

An inverse agonist is proposed to have a binding preference for an inactive conformation. This stabilizes the inactive conformation and shifts the equilibrium away from the active conformation, leading to a drop in inherent biological activity (frame D). An inverse agonist need have no structural similarity to an agonist, as it could be binding to a different part of the receptor.

A partial agonist has a slight preference for the active conformation over any of the inactive conformations. The equilibrium is shifted to the active conformation but not to the same extent as with a full agonist, and so the increase in biological activity is less (frame E). Moreover, the binding of the natural ligand may be suppressed.

8.6 Desensitization and sensitization

Desensitization can occur by a number of mechanisms. Some drugs bind relatively strongly to a receptor, switch it on, but then subsequently block the receptor after a certain period of time. Thus, they are acting as agonists,

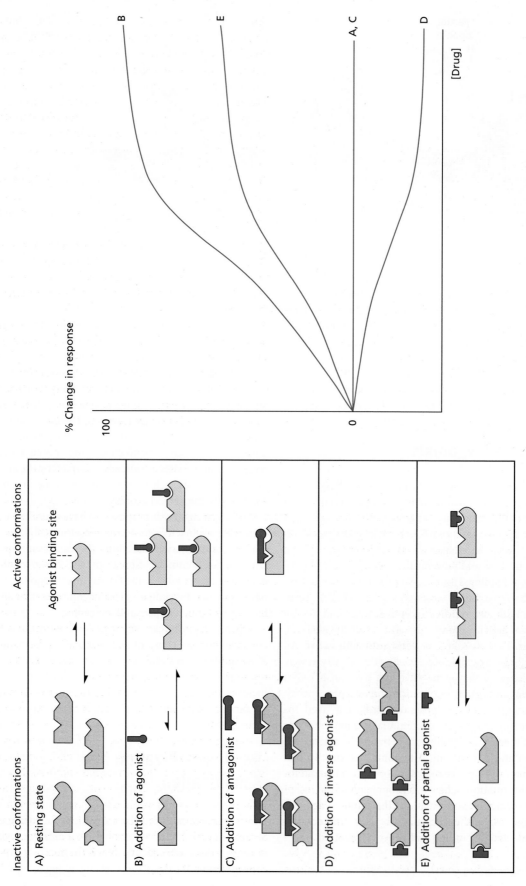

FIGURE 8.16 Equilibria between active and inactive receptor conformations, and the effect of agonists, antagonists, inverse agonists, and partial agonists in the absence of the natural ligand.

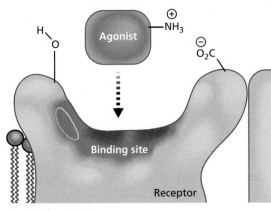

1. Resting state

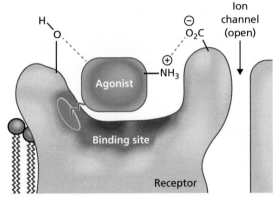

2. Induced fit

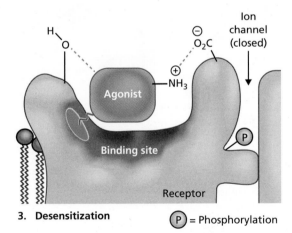

3. Desensitization   (P) = Phosphorylation

FIGURE 8.17 Desensitization of a receptor following prolonged binding of an agonist.

then antagonists. The mechanism of how this takes place is not clear, but it is believed that prolonged binding of the agonist to the receptor results in phosphorylation of hydroxyl or phenolic groups in the receptor. This causes the receptor to alter shape to an inactive conformation, despite the binding site being occupied by the agonist. In the case of an ion channel, this would mean that the

channel is closed (Fig. 8.17). In the case of a G-protein-coupled receptor, the binding site for the G-protein is closed. This altered tertiary structure is then maintained as long as the binding site is occupied by the agonist. When the drug eventually leaves, the receptor is dephosphorylated and returns to its original resting shape.

On even longer exposure to a drug, the receptor–drug complex may be removed completely from the cell membrane by a process called **endocytosis.** Here, the relevant portion of the membrane is 'nipped out', absorbed into the cell, and metabolized. Receptor endocytosis may also occur after short exposures to a ligand, but, in this situation, the receptor is often recycled back to the cell membrane in a resensitization process.

Finally, prolonged activation of a receptor may result in the cell reducing its synthesis of the receptor protein. Consequently, it is generally true that the best agonists bind swiftly to the receptor to pass on their message, and then leave quickly.

Antagonists, in contrast, tend to be slow to add and slow to leave. Prolonged exposure of a target receptor to an antagonist may lead to the opposite of desensitization (i.e. sensitization). This is where the cell synthesizes more receptors to compensate for the receptors that are blocked. This is known to happen when some β-blockers are given over long periods (section 23.11.3).

8.7 Tolerance and dependence

As mentioned above, depriving a target receptor of its natural ligand by administering an antagonist may induce that cell to synthesize more receptors. By doing so, the cell gains a greater sensitivity for its natural ligand. This process can explain the phenomena of tolerance and dependence (Fig. 8.18).

Tolerance is a situation where higher levels of a drug are required to get the same biological response. If a drug is acting to suppress the binding of a chemical messenger, then the cell may respond by increasing the number of receptors. This would require increasing the dose to regain the same level of antagonism.

If the drug is suddenly stopped, then all the receptors suddenly become available. There is now an excess of receptors, which makes the cell supersensitive to normal levels of messenger. This would be equivalent to receiving an overdose of an agonist. The resulting biological effects would explain the distressing withdrawal symptoms resulting from the cessation of certain drugs. These withdrawal symptoms would continue until the number of receptors returned to their original level. During this period, the patient may be tempted to take the drug again in order to feel 'normal' and will have then acquired a dependence on the drug.

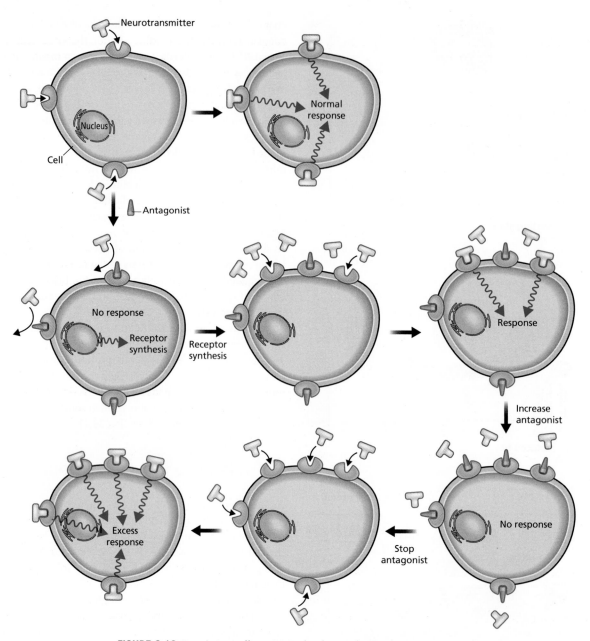

FIGURE 8.18 Increasing cell sensitivity by the synthesis of more receptors.

Tolerance and dependence also occur with agonists such as the opioids. In this situation, increased doses are required due to receptor desensitization. Increased levels of agonist are required to activate the receptors that are still available.

8.8 Receptor types and subtypes

The receptors for a particular chemical messenger are not all identical. There are various types and subtypes of receptor, and it is found that these are not evenly distributed throughout the different organs and tissues in the body. This means that designing a drug which is selective for a particular type or subtype of receptor leads to selectivity of action against a particular organ in the body (see also sections 4.3 and 4.7.3)

Some examples of receptor types and subtypes are given in Table 8.1. The identification of many of these subtypes is relatively recent, and the current emphasis in medicinal chemistry is to design drugs which are as selective as possible so that the drugs are tissue selective and have fewer side effects. For example, there are five types of dopaminergic receptor. All clinically effective antipsychotic

TABLE 8.1 Some examples of receptor types and subtypes

Receptor	Types	Subtypes	Examples of agonist therapies	Examples of antagonist therapies
Cholinergic (Chapter 22)	Nicotinic (N) Muscarinic (M)	Nicotinic (various) M_1–M_5	Stimulation of GIT motility (M_1) Glaucoma (M)	Neuromuscular blockers and muscle relaxants (N) Peptic ulcers (M_1) Motion sickness (M)
Adrenergic (adrenoceptors) (Chapter 23)	Alpha (α_1, α_2) Beta (β)	$\alpha_{1A}\alpha_{1B}\alpha_{1D}$ α_{2A}–α_{2C} (β_1, β_2, β_3)	Anti-asthmatics (β_2)	β-Blockers (β_1)
Dopamine		D_1–D_5	Parkinson's disease	Antidepressant (D_2/D_3)
Histamine (Chapter 25)		H_1–H_3	Vasodilation (limited use)	Treatment of allergies, anti-emetics, sedation (H_1) Anti-ulcer agents (H_2)
Opioid and opioid-like (Chapter 24)		μ, κ, δ, ORL1	Analgesics (κ)	Antidote to morphine overdose
5-Hydroxytryptamine (serotonin)	5-HT_1–5-HT_7	5-HT_{1A}, 5HT_{1B}, 5$HT_{1D–1F}$ 5$HT_{2A–2C}$ 5-HT_{5A} 5-HT_{5B}	Antimigraine (5-HT_{1D}) Stimulation of GIT motility (5-HT_4)	Anti-emetics (5-HT_3)
Estrogen (section 21.4)			Contraception	Breast cancer (Tamoxifen)

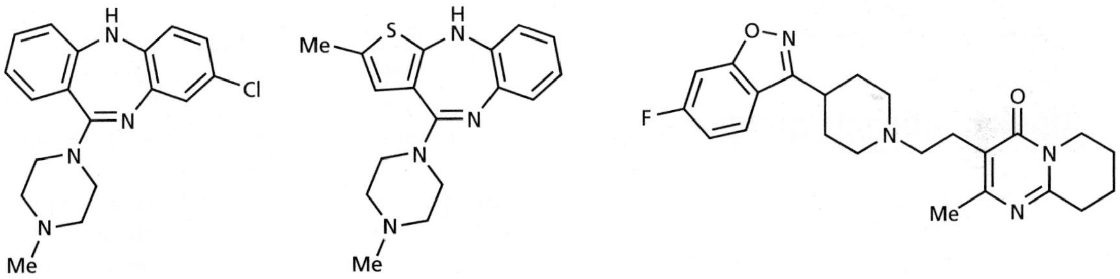

Clozapine   Olanzapine   Risperidone

FIGURE 8.19 Antipsychotic agents.

agents (e.g. **clozapine**, **olanzapine**, and **risperidone**; Fig. 8.19) antagonize the dopaminergic receptors D_2 and D_3. However, blockade of D_2 receptors may lead to some of the side effects observed, and a selective D_3 antagonist may have better properties as an antipsychotic.

Other examples of drugs targeting specific receptor subtypes include:

- muscarinic (M_2) agonists for the treatment of heart irregularities;
- agonists targeting peroxisome proliferator-activated receptors (PPARαs) for lowering triglyceride levels in the blood supply;
- N-methyl-D-aspartate (NMDA) antagonists for the treatment of stroke;
- cannabinoid (CB_1) antagonists for the treatment of memory loss.

KEY POINTS

- Agonists are compounds that mimic the natural ligand for the receptor.
- Antagonists are agents that bind to the receptor, but which do not activate it. They block binding of the natural ligand.
- Agonists may have a similar structure to the natural ligand.

- Antagonists bind differently from the natural ligand such that the receptor is not activated.

- Antagonists can bind to regions of the receptor that are not involved in binding the natural ligand. In general, antagonists tend to have more binding interactions than agonists and bind more strongly.

- Partial agonists induce a weaker effect than a full agonist.

- Inverse agonists act as antagonists, but also eliminate any resting activity associated with a receptor.

- Desensitization may occur when an agonist is bound to its receptor for a long period of time. Phosphorylation of the receptor results in a change of conformation.

- Sensitization can occur when an antagonist is bound to a receptor for a long period of time. The cell synthesizes more receptors to counter the antagonist effect.

- Tolerance is a situation where increased doses of a drug are required over time to achieve the same effect.

- Dependence is related to the body's ability to adapt to the presence of a drug. On stopping the drug, withdrawal symptoms occur as a result of abnormal levels of target receptor.

- There are several receptor types and subtypes which vary in their distribution round the body. They also vary in their selectivity for agonists and antagonists.

- Pharmacodynamics is the study of how drugs interact with their targets to produce a pharmacological effect. Pharmacokinetics is the study of factors that affect the ability of drugs to reach their targets *in vivo*.

8.9 Affinity, efficacy, and potency

The **affinity** of a drug for a receptor is a measure of how strongly that drug binds to the receptor. **Efficacy** is a measure of the maximum biological effect that a drug can produce as a result of receptor binding. It is important to appreciate the distinction between affinity and efficacy. A compound with high affinity does not necessarily have high efficacy. For example, an antagonist can bind with high affinity but have no efficacy. The **potency** of a drug refers to the amount of drug required to achieve a defined biological effect—the smaller the dose required, the more potent the drug. It is possible for a drug to be potent (i.e. active in small doses) but have a low efficacy.

Affinity can be measured using a process known as **radioligand labelling**. A known antagonist (or ligand) for the target receptor is labelled with radioactivity and is added to cells or tissue such that it can bind to the receptors present. Once an equilibrium has been reached, the unbound ligands are removed by washing, filtration, or centrifugation. The extent of binding can then be measured by detecting the amount of radioactivity present in the cells or tissue, and the amount of radioactivity that was removed. The equilibrium constant for bound versus unbound radioligand is defined as the **dissociation binding constant (K_d)**.

$$L + R \rightleftharpoons \underset{\substack{\text{Receptor–ligand} \\ \text{complex}}}{LR} \qquad K_d = \frac{[L] \times [R]}{[LR]}$$

[L] and [LR] can be found by measuring the radioactivity of unbound ligand and bound ligand respectively, after correction for any background radiation. However, it is not possible to measure [R], and so we have to carry out some mathematical manipulations to remove [R] from the equation.

The total number of receptors present must equal the number of receptors occupied by the ligand ([LR]) and those that are unoccupied ([R]), i.e.

$$[R_{tot}] = [R] + [LR]$$

This means that the number of receptors unoccupied by a ligand is

$$[R] = [R_{tot}] - [LR].$$

Substituting this into the first equation and rearranging leads to the **Scatchard equation**, where both [LR] and [L] are measurable:

$$\frac{[\text{Bound ligand}]}{[\text{Free ligand}]} = \frac{[LR]}{[L]} = \frac{R_{tot} - [LR]}{K_d}$$

We are still faced with the problem that K_d and R_{tot} cannot be measured directly. However, these terms can be determined by drawing a graph based on a number of experiments where different concentrations of a known radioligand are used. [LR] and [L] are measured in each case and a **Scatchard plot** (Fig. 8.20) is drawn which compares the ratio [LR]/[L] versus [LR]. This gives a straight line, and the point where it meets the x-axis represents the total number of receptors available (R_{tot}) (line A; Fig. 8.21). The slope is a measure of the radioligand's affinity for the receptor and allows K_d to be determined.

We are now in the position to determine the affinity of a novel drug.

This is done by repeating the radioligand experiments in the presence of the unlabelled drug. The drug competes with the radioligand for the receptor's binding sites and is called a **displacer**. The stronger the affinity of the drug, the more effectively it will compete for binding sites and the less radioactivity will be measured for [LR]. This will result in a different line in the Scatchard plot.

If the drug competes directly with the radiolabelled ligand for the same binding site on the receptor, then the

(a)

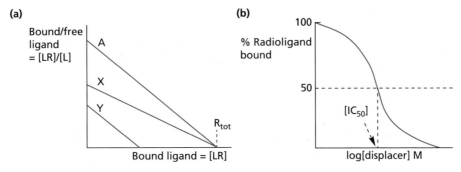

(b)

FIGURE 8.20 (a) Scatchard plot. (A = radioligand only, X = radioligand + competitive ligand, Y = radioligand + non-competitive ligand). (b) The displacement or inhibition curve.

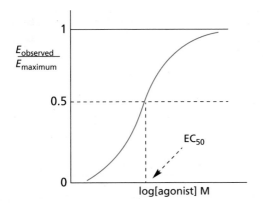

FIGURE 8.21 Measurement of EC_{50}.

slope is decreased but the intercept on the *x*-axis remains the same (line X in the graph). In other words, if the radioligand concentration is much greater than the drug it will bind to all the receptors available.

Agents that bind to the receptor at an allosteric binding site do not compete with the radioligand for the same binding site and so cannot be displaced by high levels of radioligand. However, by binding to an allosteric site they make the normal binding site unrecognizable to the radioligand, and so there are fewer receptors available. This results in a line with an identical slope to line A, but crossing the *x*-axis at a different point, thus indicating a lower total number of available receptors (line Y).

The data from these displacement experiments can be used to plot a different graph which compares the percentage of the radioligand that is bound to a receptor versus the concentration of the drug (or displacer). This results in a sigmoidal curve termed the **displacement** or **inhibition curve**, which can be used to identify the IC_{50} **value** for the drug (i.e. the concentration of compound that prevents 50% of the radioactive ligand being bound).

The **inhibitory** or **affinity constant** (K_i) for the drug is the same as the IC_{50} value if non-competitive interactions are involved. For compounds that *are* in competition with the radioligand for the binding site, the inhibitory

constant depends on the level of radioligand present and is defined as

$$K_i = \frac{IC_{50}}{1 + [L]_{tot} / K_d}$$

where K_d is the dissociation constant for the radioactive ligand and $[L]_{tot}$ is the concentration of radioactive ligand used in the experiment.

Efficacy is determined by measuring the maximum possible effect resulting from receptor–ligand binding. Potency can be determined by measuring the concentration of drug required to produce 50% of the maximum possible effect (EC_{50}) (Fig. 8.21). The smaller the value of EC_{50}, the more potent the drug. In practice, pD_2 is taken as the measure of potency where $pD_2 = -\log[EC_{50}]$.

A **Schild analysis** is used to determine the dissociation constant (K_d) of competitive antagonists (Fig. 8.22). An agonist is first used at different concentrations to activate the receptor, and an observable effect is measured at each concentration. The experiment is then repeated several times in the presence of different concentrations of antagonist. Comparing the effect ratio ($E_{observed}/E_{maximum}$) versus the log of the agonist concentration (log[agonist]) produces a series of sigmoidal curves where the EC_{50} of the agonist increases with increasing antagonist concentration. In other words, greater concentrations of agonist are required to compete with the antagonist. A **Schild plot** is then constructed, which compares the log of the reciprocal of the dose ratio versus the log of the antagonist concentration. The **dose ratio** is the agonist concentration required to produce a specified level of effect when no antagonist is present, compared to the agonist concentration required to produce the same level in the presence of antagonist. The line produced from these studies can be extended to the *x*-axis to find pA_2 ($= -\log K_d$), which represents the affinity of the competitive antagonist.

Schild plots can be used to determine whether different agonists are showing similar selectivities towards

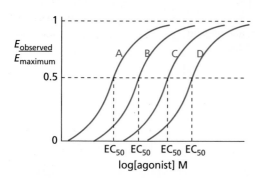
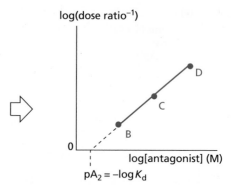

FIGURE 8.22 Schild analysis (A = no antagonist present, B–D = increasing concentrations of antagonist present).

different types of receptor. The pA_2 values of a non-selective antagonist acting on a population of the different receptor types are determined in the presence of each of the agonists. If the pA_2 values are similar, it is indicative that the agonists are showing similar receptor selectivity.

cell. Potency relates to how effective a drug is in producing a cellular effect.

- Affinity can be measured from Scatchard plots derived from radioligand displacement experiments.

- Efficacy is determined by the EC_{50} value—the concentration of agent required to produce 50% of the maximum possible effect resulting from receptor activation.

- A Schild analysis is used to determine the dissociation constant of competitive antagonists.

KEY POINTS

- Affinity is a measure of how strongly a drug binds to a receptor. Efficacy is a measure of the effect of that binding on the

QUESTIONS

1. Structure I is an agonist which binds to the cholinergic receptor and mimics the action of the natural ligand acetylcholine. Structure II on the other hand shows no activity and does not bind to the receptor. Suggest why this might be the case.

2. Isoprenaline undergoes metabolism to give the inactive metabolite shown above. Suggest why this metabolite is inactive.

3. Salbutamol is an anti-asthmatic agent which acts as an adrenergic agonist. Do you think it is likely to show any selectivity between the α- or β-adrenoceptors? Explain your answer.

4. Propranolol is an adrenergic antagonist. Compare the structure of propranolol with noradrenaline and identify which features are similar in both molecules. Suggest why this molecule might act as an antagonist rather than an agonist, and whether it might show any selectivity between the different types of adrenergic receptor.

5. If you were asked to design drugs which acted as selective antagonists of the dopamine receptor, what structures might you consider synthesizing?

6. Tamoxifen acts as an antagonist for the estrogen receptor. Suggest how it might bind to the receptor in order to do this.

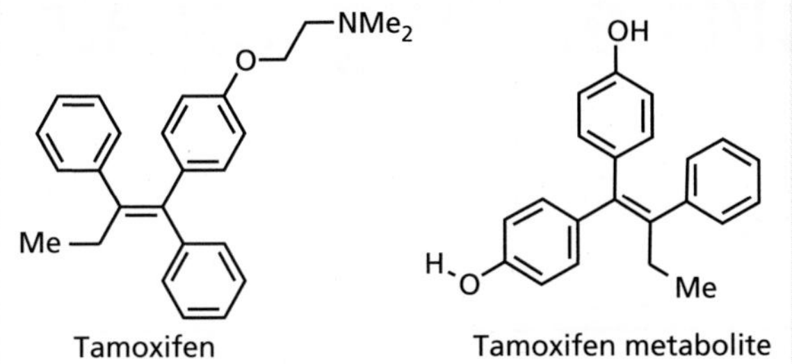

Tamoxifen   Tamoxifen metabolite

7. The tamoxifen metabolite shown acts as an estrogen agonist rather than an antagonist. Why?

8. The ability of the opioid antagonist naloxone to antagonize the opioid agonists normorphine, Met-enkephalin, and metkephamid was determined on tissue containing different types of opioid receptor. From the data provided, prepare Schild plots for the opioid antagonist naloxone in the presence of each of the opioid agonists and determine the pA_2 values in each case. Identify whether any of the opioid agonists show a similar selectivity for opioid receptors. ([I] is the concentration of naloxone present).

Normorphine and naloxone

[I] mol l^{-1}	1×10^{-6}	1×10^{-7}	3.162×10^{-8}
Dose ratio	0.0018	0.0178	0.1122

Metkephamid and naloxone

[I] mol l^{-1}	3.16×10^{-7}	1×10^{-7}	3.162×10^{-8}
Dose ratio	0.0562	0.1585	0.7943

Met-enkephalin and naloxone

[I] mol l^{-1}	3.16×10^{-7}	1×10^{-7}	3.162×10^{-8}
Dose ratio	0.0398	0.2512	0.8913

Multiple-choice questions are available on the Online Resource Centre at www.oxfordtextbooks.co.uk/orc/patrick6e/

FURTHER READING

Chalmers, D. T., and Behan, D. P. (2002) The use of constitutively active GPCRs in drug discovery and functional genomics. *Nature Reviews Drug Discovery*, **1**, 599–608.

Christopoulis, A. (2002) Allosteric binding sites on cell surface receptors: novel targets for drug discovery. *Nature Reviews Drug Discovery*, **1**, 198–210.

Kreek, M. J., LaForge, K. S., and Butelman, E. (2002) Pharmacotherapy of addictions. *Nature Reviews Drug Discovery*, **1**, 710–25.

Maehle, A.-H., Prull, C.-R., and Halliwell, R. F. (2002) The emergence of the drug receptor theory. *Nature Reviews Drug Discovery*, **1**, 637–41.

Pouletty, P. (2002) Drug addictions: towards socially accepted and medically treatable diseases. *Nature Reviews Drug Discovery*, **1**, 731–6.

Schlyer, S., and Horuk, R. (2006) I want a new drug: G-protein-coupled receptors in drug development. *Drug Discovery Today*, **11**, 481–93.

Zhan-Guo, G., and Jacobson, K. A. (2006) Allosterism in membrane receptors. *Drug Discovery Today*, **11**, 191–202.

Titles for general further reading are listed on p.845.

9 Nucleic acids as drug targets

Although proteins are the target for the majority of clinically useful drugs, there are many important drugs which target nucleic acids, especially in the areas of antibacterial and anticancer therapy (see sections 19.7, 19.8, and 21.2). In this chapter, we concentrate on the mechanism of action of some of these drugs. Further information and clinical aspects are covered in Chapters 19 and 21. The structure and function of nucleic acids was discussed in Chapter 6.

We shall first consider the drugs that interact with DNA. In general, we can group these under the following categories:

- intercalating agents
- topoisomerase poisons (non-intercalating)
- alkylating agents
- chain cutters
- chain terminators.

9.1 Intercalating drugs acting on DNA

Intercalating drugs are compounds that contain planar or heteroaromatic features which slip between the base pair layers of the DNA double helix. Some of these drugs prefer to approach the helix via the major groove; others prefer access via the minor groove. Once they are inserted between the nucleic acid base pairs, the aromatic/heteroaromatic rings are held there by van der Waals interactions with the base pairs above and below. Several intercalating drugs also contain ionized groups which can interact with the charged phosphate groups of the DNA backbone, thus strengthening the interaction. Once the structures have become intercalated, a variety of other processes may take place which prevent replication and transcription, leading finally to cell death. The following are examples of drugs that are capable of intercalating DNA.

Proflavine (Fig. 9.1) is an example of a group of antibacterial compounds called the **aminoacridines**, which were used during the First and Second World Wars to treat deep surface wounds. They proved highly effective in preventing infection and reduced the number of fatalities resulting from wound infections. Proflavine is completely ionized at pH 7 and interacts directly with bacterial DNA. The flat tricyclic ring intercalates between the DNA base pairs, and interacts with them by van der Waals forces, while the aminium cations form ionic bonds with the negatively charged phosphate groups on

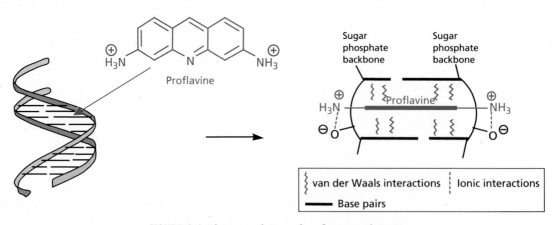

FIGURE 9.1 The intercalation of proflavine with DNA.

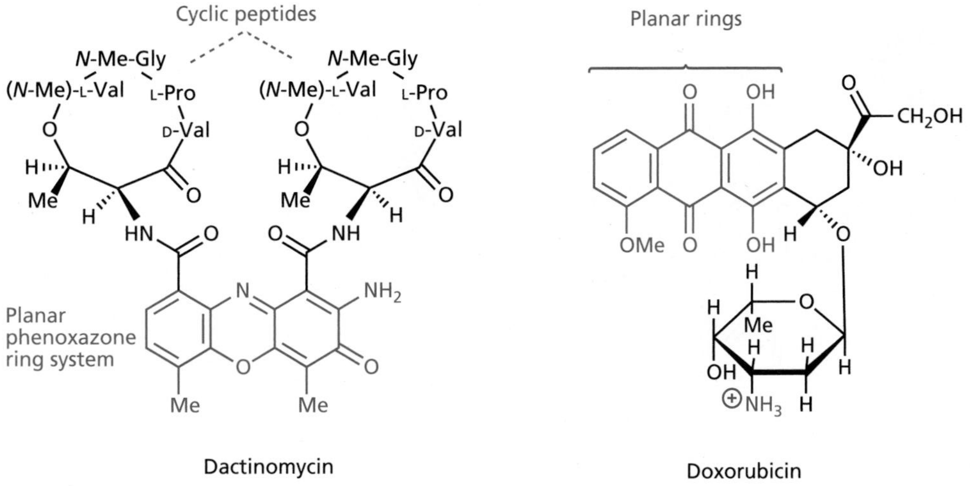

FIGURE 9.2 Dactinomycin and doxorubicin.

the sugar phosphate backbone. Once inserted, proflavine deforms the DNA double helix and prevents the normal functions of replication and transcription.

Dactinomycin (Fig. 9.2) (previously called actinomycin D) is a naturally occurring antibiotic that was first isolated from *Streptomyces parvullus* in 1953, and was shown to be an effective anticancer agent in children. It contains two cyclic pentapeptides, but the important feature is a flat, tricyclic, heteroaromatic structure which slides into the double helix via the minor groove. It appears to favour interactions with guanine–cytosine base pairs and, in particular, between two adjacent guanine bases on alternate strands of the helix. The molecule is further held in position by hydrogen bond interactions between the nucleic acid bases of DNA and the cyclic pentapeptides positioned on the outside of the helix. The 2-amino group of guanine plays a particularly important role in this interaction. The resulting bound complex is very stable and prevents the unwinding of the double helix. This, in turn, prevents DNA-dependent RNA polymerase from catalysing the synthesis of messenger RNA (mRNA) and thus prevents transcription.

Doxorubicin (Fig. 9.2) is one of the most effective anticancer drugs ever discovered, and belongs to a group of naturally occurring antibiotics called the **anthracyclines**. It was first isolated from *Streptomyces peucetius* in 1967, and contains a tetracyclic system where three of the rings are planar. The drug approaches DNA via the major groove of the double helix and intercalates using the planar tricyclic system. The charged amino group attached to the sugar is also important, as it forms an ionic bond with the negatively charged phosphate groups of the DNA backbone. This is supported by the fact that structures lacking the aminosugar have poor activity. Intercalation prevents the normal action of an enzyme

called **topoisomerase II**—an enzyme that is crucial to replication and mitosis. The mechanism by which this enzyme works is described in section 6.1.3 and includes the formation of a DNA–enzyme complex where the enzyme is covalently linked to the DNA. When doxorubicin is intercalated into DNA it stabilizes this DNA–enzyme complex and stalls the process. Agents such as doxorubicin are referred to as topoisomerase II poisons rather than inhibitors since they do not prevent the enzyme functioning directly. Other mechanisms of action for doxorubicin and its analogues have also been proposed—see section 21.2.1.

Bleomycins (Fig. 9.3) are complex natural products that were isolated from *Streptomyces verticillus* in 1962, and are some of the few anticancer drugs not to cause bone marrow depression. Their structure includes a bithiazole ring system which intercalates with DNA. Once the structure has become intercalated, the nitrogen atoms of the primary amines, pyrimidine ring, and imidazole ring chelate a ferrous ion which then interacts with oxygen and is oxidized to a ferric ion, leading to the generation of superoxide or hydroxyl radicals. These highly reactive species abstract hydrogen atoms from DNA, which results in the DNA strands being cut, particularly between purine and pyrimidine nucleotides. Bleomycin also appears to prevent the enzyme **DNA ligase** from repairing the damage caused.

9.2 Topoisomerase poisons: non-intercalating

The following structures are classed as poisons rather than inhibitors because they stabilize the normally transient cleavable complex that is formed between

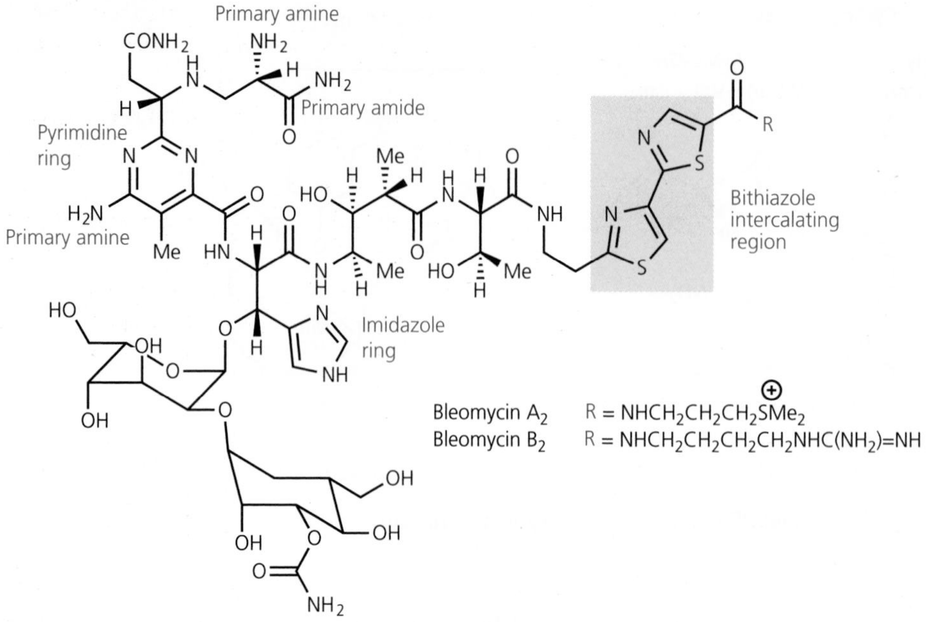

FIGURE 9.3 Bleomycins.

DNA and topoisomerase enzymes, thus inhibiting the rejoining of the DNA strand or strands (section 6.1.3). We have already mentioned topoisomerase poisons in section 9.1 where we discussed the anthracyclines. In this section, we look at topoisomerase poisons which do not intercalate into the DNA structure. However, since DNA is part of the target complex, we can view these poisons as targeting DNA as well as the topoisomerase enzyme.

The anticancer agents **etoposide** and **teniposide** (Fig. 9.4) belong to a group of compounds called the **podophyllotoxins,** and are semi-synthetic derivatives of **epipodophyllotoxin**—an isomer of a naturally occurring agent called **podophyllotoxin**. Both agents act as topoisomerase poisons. DNA strand breakage is also thought

to occur by a free radical process involving oxidation of the 4′-phenolic group and the production of a semiquinone free radical. Evidence supporting this comes from the fact that the 4′-methoxy structures are inactive. The presence of the glucoside sugar moiety also increases the ability to induce breaks.

Camptothecin (Fig. 9.5) is a natural product which was extracted from a Chinese bush (*Camptotheca acuminata*) in 1966. It stabilizes the cleavable complex formed between DNA and the enzyme **topoisomerase I** (section 6.1.3). As a result, single-strand breaks accumulate in the DNA. These can be repaired if the drug departs, but if replication is taking place when the drug–enzyme–DNA complex is present, an irreversible double-strand break takes place which leads to cell death. Semi-synthetic

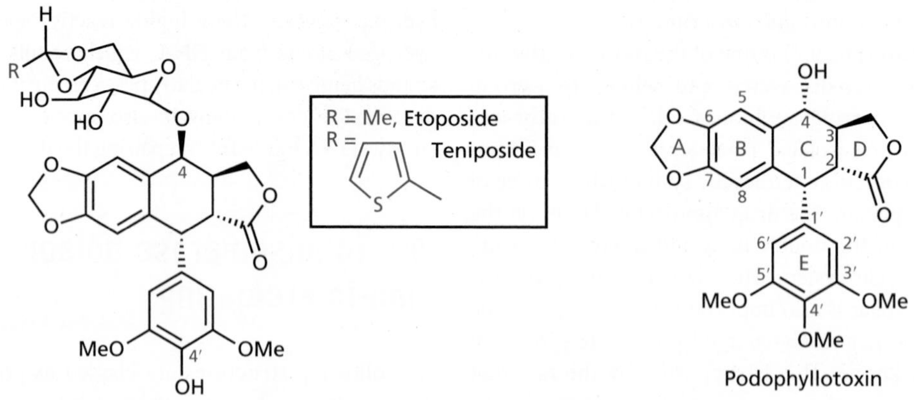

FIGURE 9.4 Podophyllotoxins.

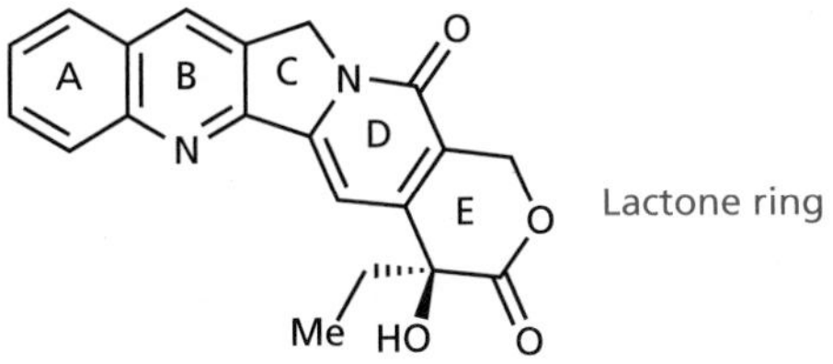

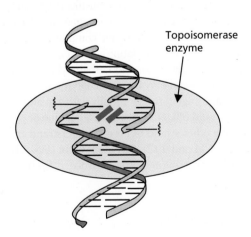

FIGURE 9.5 Camptothecin.

analogues of camptothecin have been developed as clinically useful anticancer agents (section 21.2.2.2).

The antibacterial **quinolones** and **fluoroquinolones** (section 19.8.1) are synthetic agents that inhibit the replication and transcription of bacterial DNA by stabilizing the complex formed between DNA and bacterial topoisomerases. Inhibition arises by the formation of a ternary complex involving the drug, the enzyme, and bound DNA (Fig. 9.6). The binding site for the fluoroquinolones only appears once the enzyme has 'nicked' the DNA strands, and the strands are ready to be crossed over. At that point, four fluoroquinolone molecules are bound in a stacking arrangement such that their aromatic rings are coplanar. The carbonyl and carboxylate groups of the fluoroquinolones interact with DNA by hydrogen bonding, while the fluoro substituent at position 6, the substituent at C-7, and the carboxylate ion are involved in binding interactions with the enzyme.

KEY POINTS

- Intercalating drugs contain planar aromatic or heteroaromatic ring systems which can slide between the base pairs of the DNA double helix.

- The anthracyclines are intercalating drugs that act as topoisomerase II poisons, stabilizing the cleavage complex formed between the enzyme and DNA.

- Bleomycins are intercalating drugs which form complexes with ferrous ions. These complexes generate reactive oxygen species that cleave the strands of DNA.

- Etoposide and teniposide are non-intercalating drugs that act as topoisomerase II poisons.

- Camptothecin is a non-intercalating drug that acts as a topoisomerase I poison. It stabilizes an enzyme–DNA complex where a single strand of DNA has been cleaved.

9.3 Alkylating and metallating agents

Alkylating agents are highly electrophilic compounds that react with nucleophiles to form strong covalent bonds. There are several nucleophilic groups present on the nucleic acid bases of DNA which can react with electrophiles—in particular the N-7 of guanine (Fig. 9.7).

Drugs with two alkylating groups can react with a nucleic acid base on each chain of DNA to cross-link the strands such that replication or transcription is disrupted. Alternatively, the drug could link two nucleophilic groups on the same chain such that the drug is attached like a limpet to the side of the DNA helix. That portion of DNA then becomes masked from the enzymes required to catalyse DNA replication and transcription.

Miscoding due to alkylated guanine units is also possible. The guanine base usually exists as the keto tautomer, allowing it to base pair with cytosine. Once alkylated,

FIGURE 9.6 Complex formed between DNA, the topoisomerase enzyme, and fluoroquinolones; R^6 = F for fluoroquinolones.

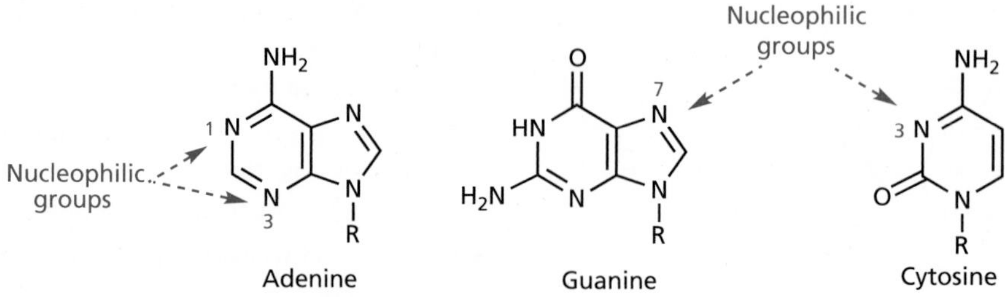

FIGURE 9.7 Nucleophilic groups on adenine, guanine, and cytosine.

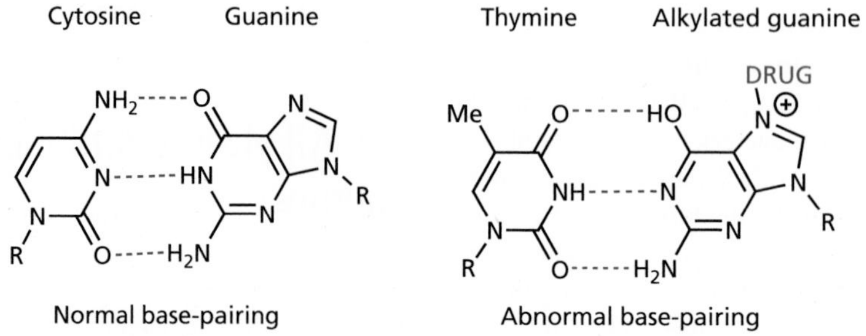

FIGURE 9.8 Normal and abnormal base pairing of guanine.

however, guanine prefers the enol tautomer and is more likely to base pair with thymine (Fig. 9.8). Such miscoding ultimately leads to an alteration in the amino acid sequence of proteins, which, in turn, can lead to disruption of protein structure and function.

Unfortunately, alkylating agents can alkylate nucleophilic groups on proteins, as well as DNA, which means they have poor selectivity and have toxic side effects. They can even lead to cancer in their own right. Nevertheless, alkylating drugs are still useful in the treatment of cancer (section 21.2.3). Examples of how some of these drugs alkylate DNA are now given (see also Case study 4 for an example of an antiparasitic drug that alkylates DNA).

9.3.1 Nitrogen mustards

The nitrogen mustards get their name because they are related to the sulphur-containing mustard gases used during the First World War. In 1942, the nitrogen mustard compound **chlormethine** (Fig. 9.9) was the first alkylating agent to be used medicinally, although full details were not revealed until after the war due to secrecy surrounding all nitrogen mustards. The nitrogen atom is able to displace a chloride ion intramolecularly to form the highly electrophilic aziridinium ion. This is an example of a **neighbouring group effect**, also called **anchimeric assistance**. Alkylation of DNA can then take place. As the process can be repeated, cross-linking between chains or within the one chain will occur. Monoalkylation of DNA guanine units is also possible if the second alkyl halide reacts with water, but cross-linking

is the major way in which these drugs inhibit replication and act as anticancer agents.

Analogues of chlormethine have been designed to improve selectivity and to reduce side effects (section 21.2.3.1). Other agents such as **cyclophosphamide** have been designed as prodrugs, and are converted into the alkylating drug once they have been absorbed into the blood supply (section 21.2.3.1).

9.3.2 Nitrosoureas

The anticancer agents **lomustine** and **carmustine** (Fig. 9.10) were discovered in the 1960s, and are chloroethylnitrosoureas which decompose spontaneously in the body to form two active compounds—an alkylating agent and a carbamoylating agent (Fig. 9.11). The organic isocyanate that is formed carbamoylates lysine residues in proteins and may inactivate DNA repair enzymes. The alkylating agent reacts initially with a guanine moiety on one strand of DNA, then with a guanine or cytosine unit on the other strand to produce interstrand cross-linking (Figs. 9.11 and 9.12). **Streptozotocin** (Fig. 9.10) is a naturally occurring nitrosourea isolated from *Streptomyces achromogenes*.

9.3.3 Busulfan

Busulfan (Fig. 9.13) was synthesized in 1950 as part of a systematic search for novel alkylating agents. It is an anticancer agent which causes interstrand cross-linking between guanine units. The sulphonate groups are good

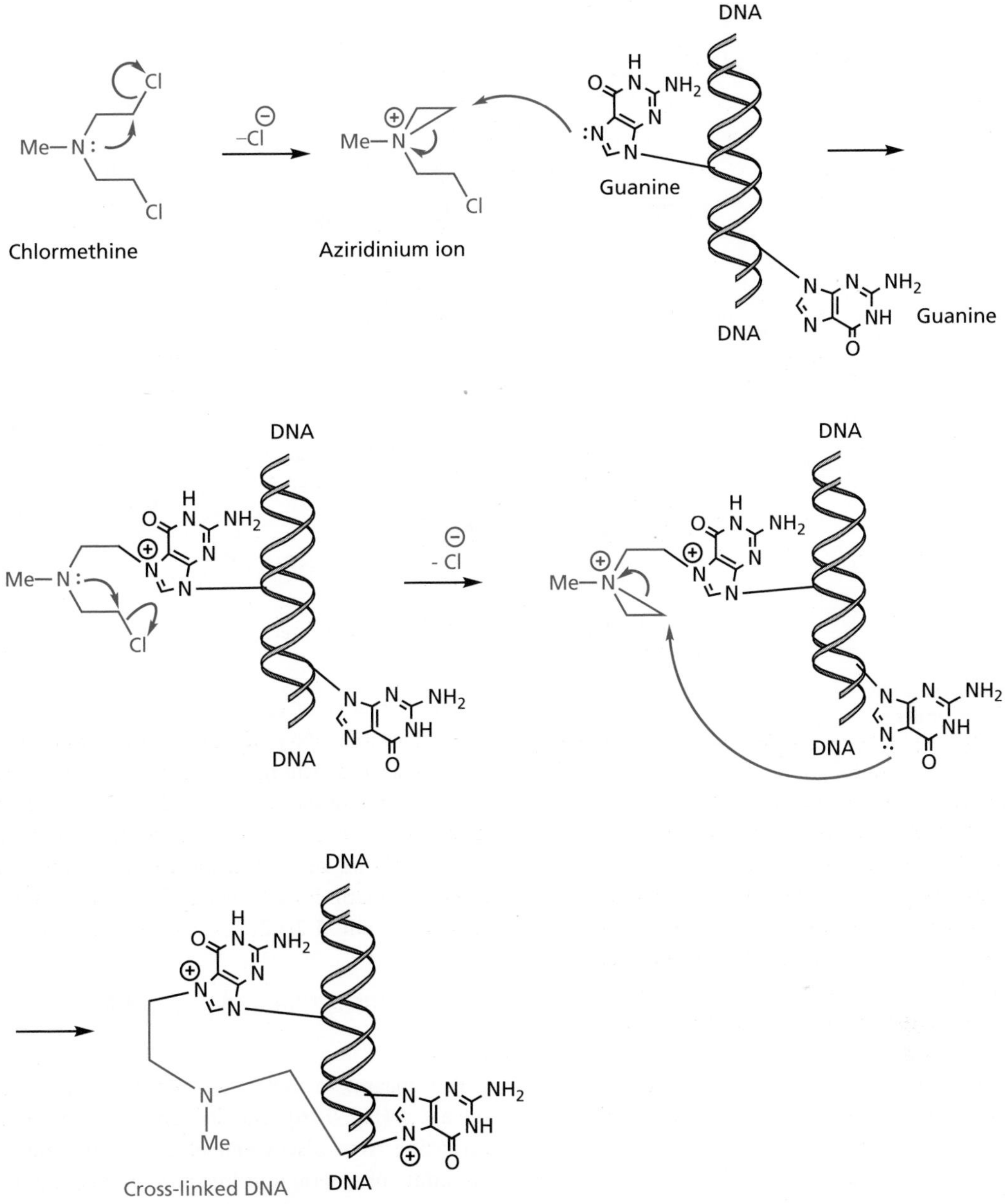

FIGURE 9.9 Alkylation of DNA by chlormethine.

leaving groups and play a similar role to the chlorines in the nitrogen mustards. However, the mechanism involves a direct S_N2 nucleophilic substitution of the sulphonate groups and does not involve any intermediates such as the aziridinium ion.

9.3.4 Cisplatin

Cisplatin (Fig. 9.14) is one of the most frequently used anticancer drugs in medicine. Its discovery was fortuitous in the extreme, arising from research carried out in the 1960s to investigate the effects of an electric current on bacterial growth. During these experiments, it

was discovered that bacterial cell division was inhibited. Further research led to the discovery that an electrolysis product from the platinum electrodes was responsible for the inhibition and the agent was eventually identified as *cis*-diammonia dichloroplatinum(II), now known as cisplatin.

The structure consists of a central platinum atom, covalently linked to two chloro substituents, while the two ammonia molecules act as ligands. The overall structure is neutral and unreactive. Once cisplatin enters cells, however, it enters an environment which has a low concentration of chloride ions. This leads to aquation where the chloro substituents of cisplatin are displaced by

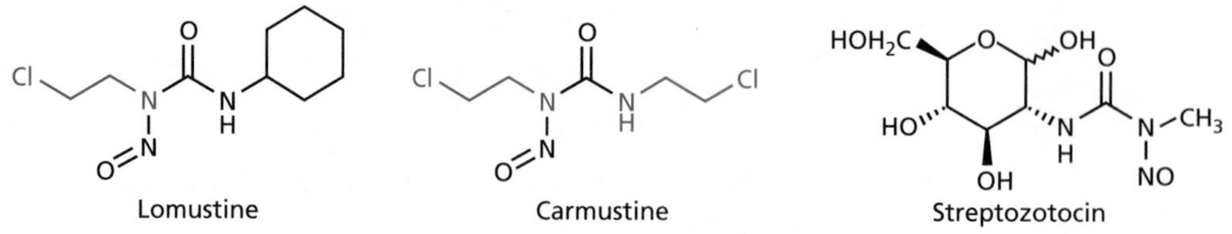

FIGURE 9.10 Nitrosourea alkylating agents.

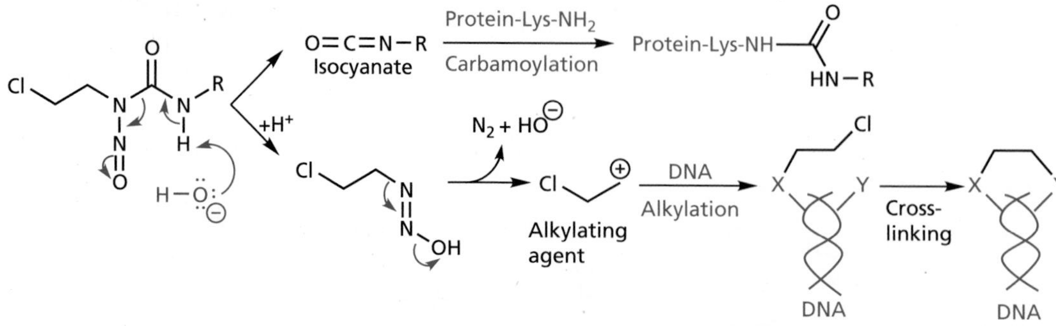

FIGURE 9.11 Mechanisms of action for nitrosoureas.

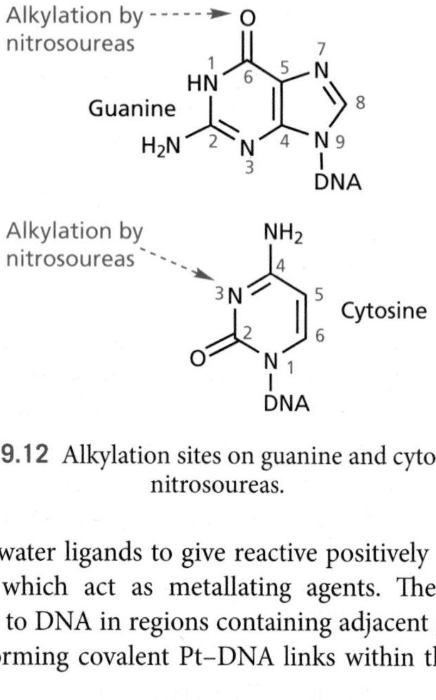

FIGURE 9.12 Alkylation sites on guanine and cytosine for nitrosoureas.

neutral water ligands to give reactive positively charged species which act as metallating agents. These bind strongly to DNA in regions containing adjacent guanine units, forming covalent Pt–DNA links within the same strand (intrastrand cross-linking). It is likely that this takes place to the N-7 and O-6 positions of adjacent guanine molecules. The hydrogen bonds that are normally involved in base-pairing guanine to cytosine are disrupted by the cross-links, leading to localized unwinding of the DNA helix and inhibition of transcription. Derivatives of cisplatin have been developed with reduced side effects (section 21.2.3.2).

9.3.5 Dacarbazine and procarbazine

Dacarbazine and procarbazine (Fig. 9.15) are prodrugs which generate a methyldiazonium ion as the alkylating agent (Fig. 9.16). The antitumour properties of procarbazine were discovered in the 1960s following the screening of several hundred compounds that had been prepared as potential antidepressants.

Dacarbazine is activated by *N*-demethylation in the liver—a reaction catalysed by cytochrome P450 enzymes

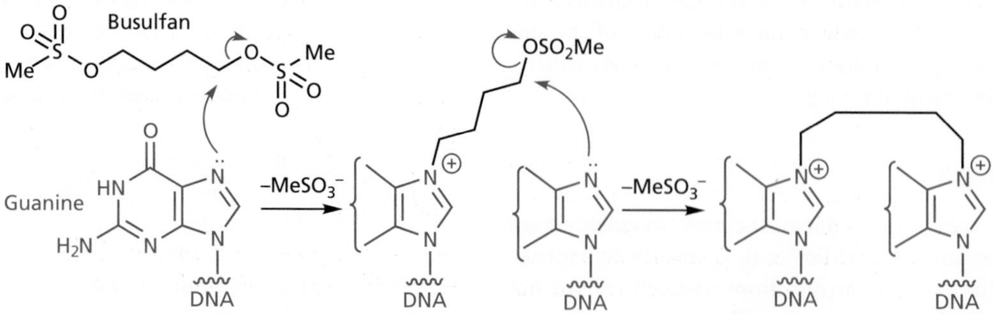

FIGURE 9.13 Cross-linking mechanism involving busulfan.

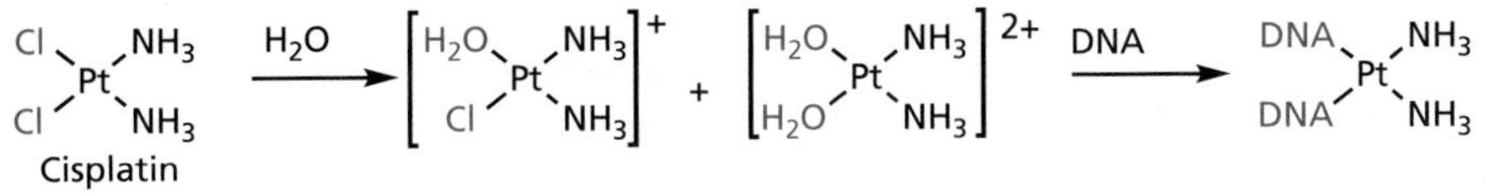

FIGURE 9.14 Activation of cisplatin and intrastrand cross-linking of DNA.

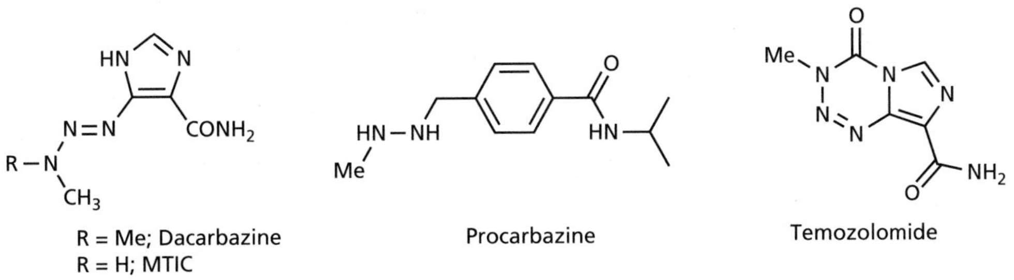

R = Me; Dacarbazine
R = H; MTIC

Procarbazine

Temozolomide

FIGURE 9.15 Dacarbazine, procarbazine, and temozolomide.

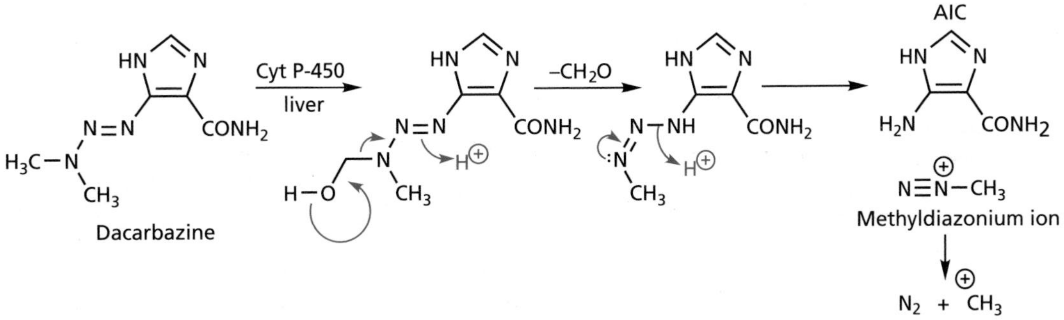

FIGURE 9.16 Mechanism of action of dacarbazine.

(section 11.5.2) (Fig. 9.16). Formaldehyde is then lost to form a product which spontaneously degrades to form 5-aminoimidazole-4-carboxamide (AIC) and the methyldiazonium ion. Reaction of this ion with RNA or DNA results in methylation, mainly at the 7-position of guanine. DNA fragmentation can also occur. AIC has no cytotoxic effect and is present naturally as an intermediate in purine synthesis. **Temozolomide** (Fig. 9.15) also acts as a prodrug and is hydrolysed in the body to form 5-(3-methyltriazen-1-yl)imidazole-4-carboxamide (MTIC), which decomposes in a similar fashion to form the methyldiazonium ion. The O-6 oxygen atom of a guanine group is particularly methylated by this agent.

9.3.6 Mitomycin C

Mitomycin C (Fig. 9.17) was discovered in the 1950s, and is a naturally occurring compound obtained from the microorganism *Streptomyces caespitosus*. It is one of the most toxic anticancer drugs in clinical use, and acts as a prodrug, being converted to an alkylating agent within the body. The process by which this takes place is initiated by an enzyme-catalysed reduction of the quinone

ring system to a hydroquinone. Loss of methanol and opening of the three-membered aziridine ring then takes place to generate the alkylating agent. Guanine residues on different DNA strands are then alkylated, leading to interstrand cross-linking, and the inhibition of DNA replication and cell division. Since a reduction step is involved in the mechanism, it has been proposed that this drug should be more effective against tumours in an oxygen-starved (hypoxic) environment, such as the centre of solid tumour masses.

KEY POINTS

- Alkylating agents contain electrophilic groups that react with nucleophilic centres on DNA. If two electrophilic groups are present, interstrand and/or intrastrand cross-linking of the DNA is possible.

- Nitrogen mustards react with guanine groups on DNA to produce cross-linking.

- Nitrosoureas have a dual mechanism of action whereby they alkylate DNA and carbamoylate proteins.

- Cisplatin is an alkylating agent which causes intrastrand cross-linking.

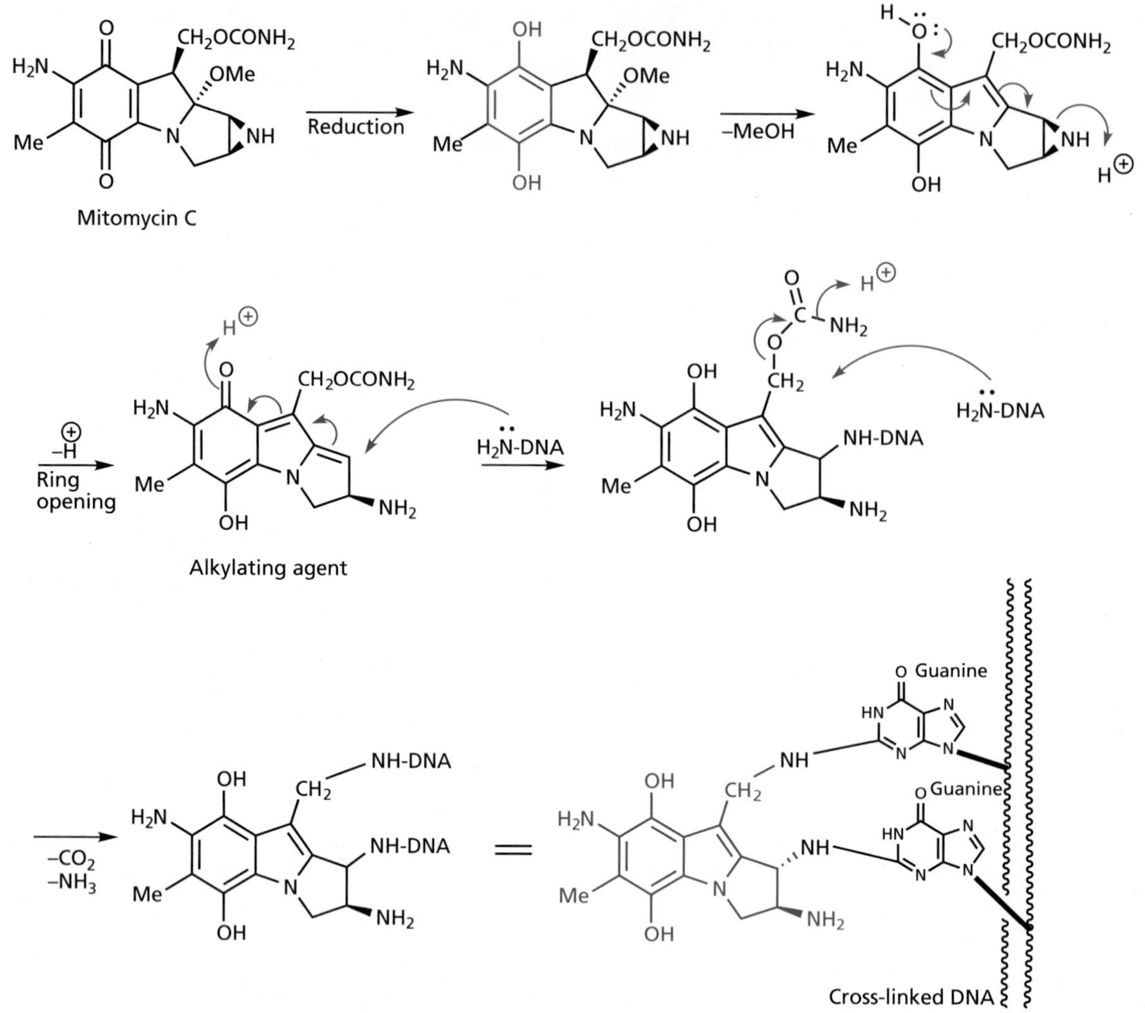

FIGURE 9.17 DNA cross-linking by mitomycin C.

- Dacarbazine and procarbazine are prodrugs that are activated by enzymes to produce a methyldiazonium ion which acts as an alkylating agent.
- Mitomycin C is a natural product that is converted to an alkylating agent by enzymatic reduction. Interstrand cross-linking takes place between guanine groups.

9.4 Chain cutters

'Chain cutters' cut the strands of DNA and prevent the enzyme **DNA ligase** from repairing the damage. They appear to act by creating radicals on the DNA structure. These radicals react with oxygen to form peroxy species, and the DNA chain fragments. The bleomycins (section 9.1) and the podophyllotoxins (section 9.2) are examples of drugs that can act in this way, as are the **nitroimidazoles** and **nitrofurantoin** which target bacterial DNA and are used as antibacterial agents (section 19.8.4). Another example is the antitumour agent **calicheamicin γ**$^\text{I}$ (Fig. 9.18), which was isolated from a bacterium. This compound binds to the minor groove of DNA and cuts the DNA chain by the mechanism shown in Fig. 9.19. The driving force behind the reaction mechanism is the formation of an aromatic ring from the unusual enediyne system. The reaction starts with a nucleophile attacking the trisulphide group. The thiol which is freed then undergoes an intramolecular Michael addition with a reactive α,β-unsaturated ketone. The resulting intermediate then cycloaromatizes (a reaction known as the **Bergman cyclization**) to produce an aromatic diradical species which snatches two hydrogens from DNA. As a result, the DNA becomes a diradical. Reaction with oxygen then leads to chain cutting.

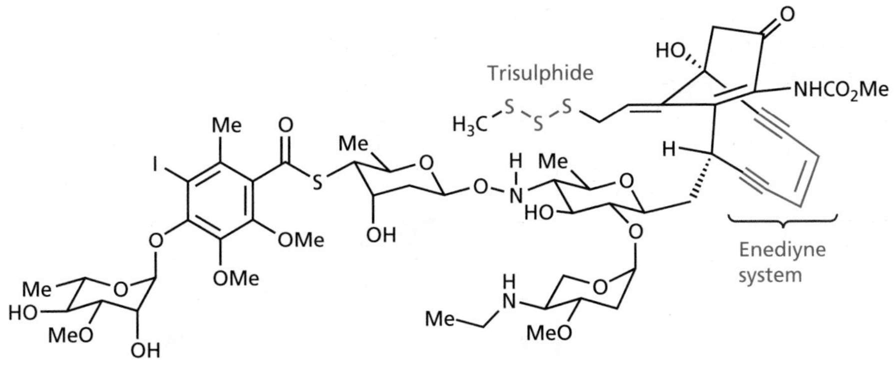

FIGURE 9.18 Calicheamicin γ^I.

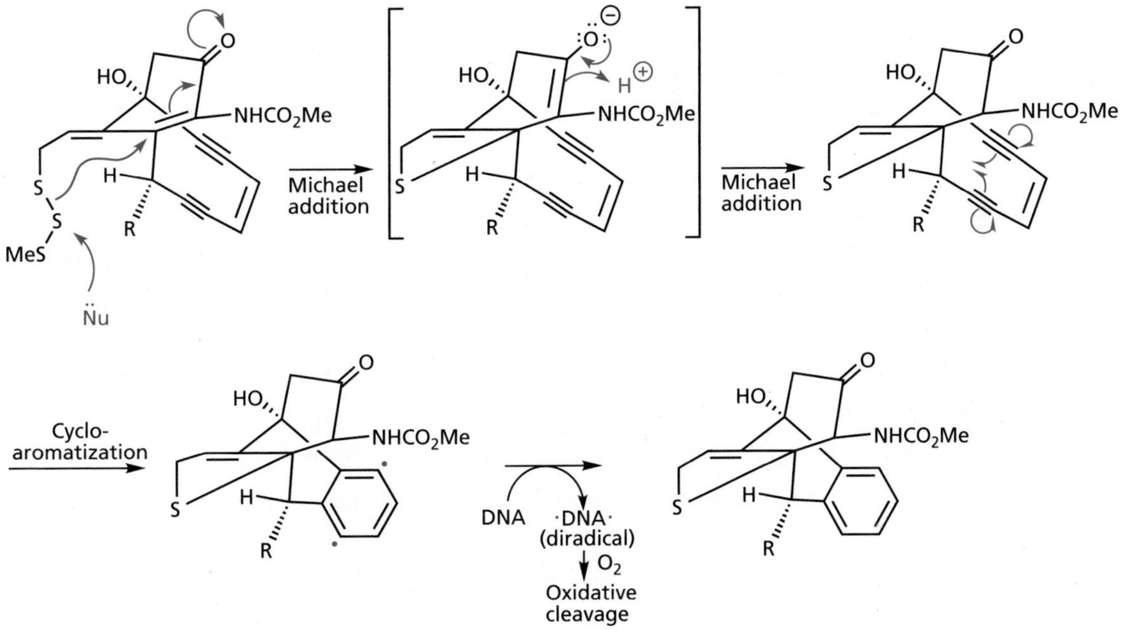

FIGURE 9.19 Mechanism of action of calicheamicin γ^I.

9.5 Chain terminators

Chain terminators are drugs which act as 'false substrates' and are incorporated into the growing DNA chain during replication. Once they have been added, the chain can no longer be extended and chain growth is terminated. The drugs which act in this way are 'mistaken' for the nucleotide triphosphates that are the authentic building blocks for DNA synthesis. The mechanism by which these nucleotides are added to the end of the growing DNA chain is shown in Fig. 9.20 and involves the loss of a diphosphate group—a process catalysed by the enzyme **DNA polymerase**. Before each building block is linked to the chain, it has to be 'recognized' by the complementary nucleic acid base on the template chain. This involves base

pairing between a nucleic acid base on the template and the nucleic acid base on the nucleotide.

Chain terminators, therefore, have to satisfy three conditions. Firstly, they have to be recognized by the DNA template by interacting with a nucleic acid base on the template strand. Secondly, they should have a triphosphate group such that they can undergo the same enzyme-catalysed reaction mechanism as the normal building blocks. Thirdly, their structure must make it impossible for any further building blocks to be added.

Aciclovir (Fig. 9.21) is an important antiviral drug that was discovered in the 1970s, and acts as a chain terminator, satisfying all three requirements. Firstly, it contains a guanine base which means that it can base pair to cytosine moieties on the template chain. Secondly, although it does not contain a triphosphate group, this is added to the

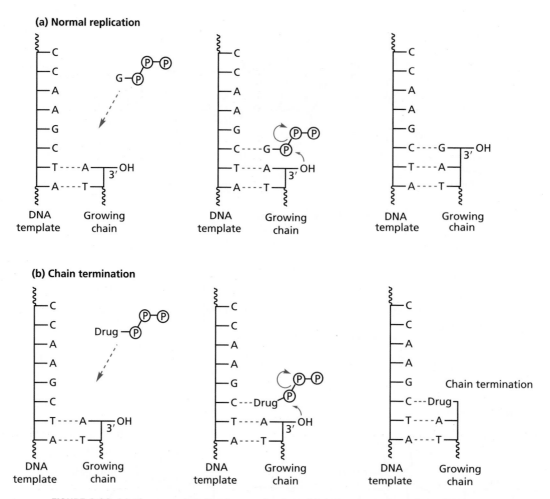

FIGURE 9.20 (a) The normal replication mechanism. (b) A drug acting as a chain terminator.

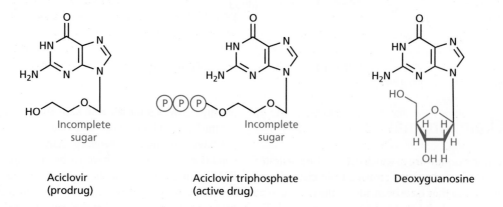

FIGURE 9.21 Structure of aciclovir, aciclovir triphosphate, and deoxyguanosine.

molecule in virally infected cells. Thirdly, the sugar unit is incomplete and lacks the required OH group normally present at position 3′—compare the structure of deoxyguanosine in Fig. 9.21. Therefore, the nucleic acid chain cannot be extended any further. Several other structures acting in a similar fashion are used in antiviral therapies and are described in sections 20.6.1 and 20.7.3.1.

9.6 Control of gene transcription

Various research groups are looking into the design of synthetic molecules that can bind to DNA by recognizing nucleic acid base pairs, and, by doing so, control gene transcription. It has been found that 'hairpin' polyamide

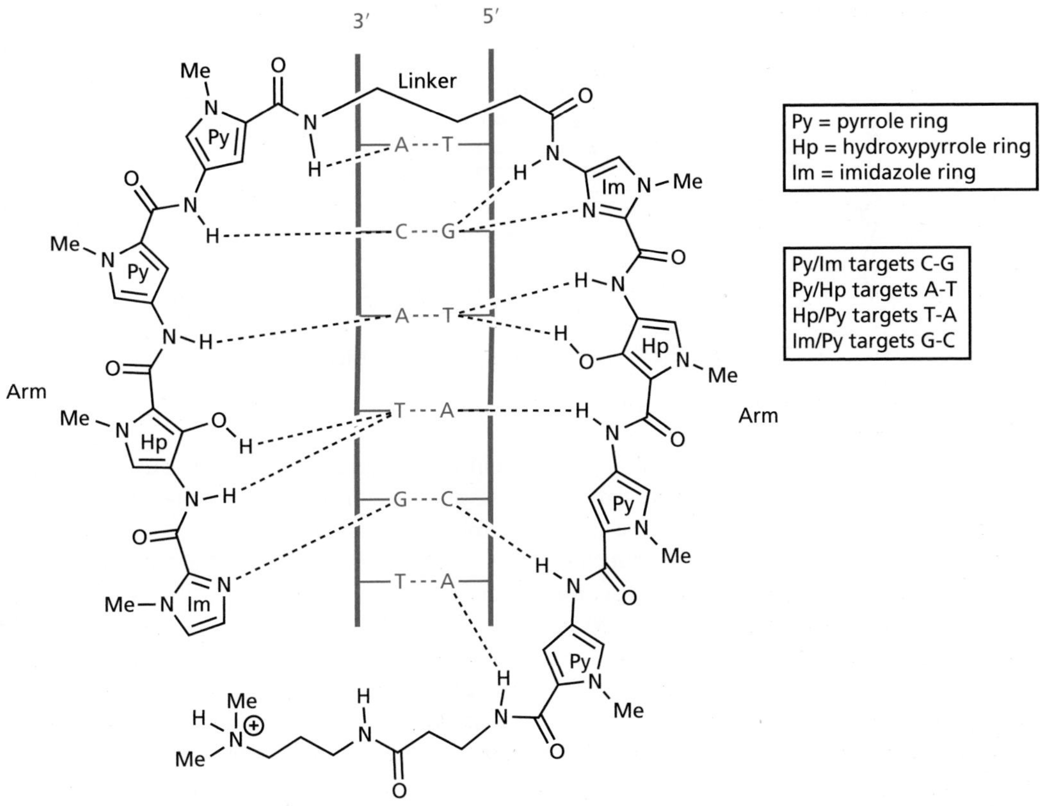

FIGURE 9.22 Synthetic polyamides capable of recognizing and binding to a particular sequence of nucleic acid base pairs.

structures containing heterocyclic rings have this capacity and bind in the minor groove of DNA (Fig. 9.22). The molecule is made up of two arms connected by means of a linker unit. The molecule attaches itself to DNA like a clamp with each arm binding to one of the DNA strands. The binding interactions are through hydrogen bonding to the base pairs of DNA, and involve both the heterocyclic rings and the amide bonds. Polyamides containing eight heterocyclic rings bind with an affinity and specificity that is comparable to naturally occurring DNA-binding proteins. Experiments have shown that it is feasible for these drugs to cross the cell membrane and to inhibit transcription by binding to the regulatory element of a gene—in other words where a transcription factor would normally bind. Binding at this specific region is achieved by designing the drug to recognize the base pair sequences in that region. This is possible by using particular patterns of pyrrole, hydroxypyrrole, and imidazole rings on each arm of the molecule. Binding to the regulatory element of the gene is crucial, since polyamides that bind to the coding region of the gene do not appear to prevent transcription. Presumably, they are displaced during the transcription process. However, it may be possible to attach an alkylating agent to the molecule such that a covalent bond is formed and the gene is 'knocked out'.

It may also be possible to design polyamides that activate the transcription process, rather than switch it off. Initial work has involved linking the polyamide to a peptide. The polyamide acts as the binding unit for DNA, while the attached peptide acts as the activating unit for transcription. It will be interesting to see whether any of these approaches leads to a clinically useful drug.

9.7 Agents that act on RNA

9.7.1 Agents that bind to ribosomes

A large number of clinically important antibacterial agents prevent protein synthesis in bacterial cells by binding to ribosomes and inhibiting the translation process. These are described in section 19.7.

9.7.2 Antisense therapy

A great deal of research has been carried out into the possibility of using **oligonucleotides** to block the coded messages carried by mRNA. This is an approach known as **antisense therapy** and has great potential. The rationale is as follows (Fig. 9.23). Assuming that the primary

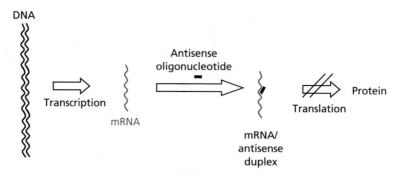

FIGURE 9.23 The principles of antisense therapy.

sequence of an mRNA molecule is known, an oligonucleotide can be synthesized containing nucleic acid bases that are complementary to a specific stretch of the mRNA molecule. As the oligonucleotide has a complementary base sequence, it is called an **antisense oligonucleotide**. When mixed with mRNA, the antisense oligonucleotide recognizes its complementary section in mRNA, interacts with it, and forms a duplex structure such that the bases pair up by hydrogen bonding. This section now acts as a barrier to the translation process and blocks protein synthesis.

There are several advantages to this approach. First of all, it can be highly specific. Statistically, an oligonucleotide of 17 nucleotides should be specific for a single mRNA molecule and block the synthesis of a single protein. The number of possible oligonucleotides containing 17 nucleotides is 4^{17}, assuming four different nucleic acid bases. Therefore, the chances of the same segment being present in two different mRNA molecules is remote. Secondly, because one mRNA leads to several copies of the same protein, inhibiting mRNA should be more efficient than inhibiting the resulting protein. Both these factors should allow the antisense drug to be used in low doses and result in fewer side effects than conventional protein inhibition.

However, there are several difficulties involved in designing suitable antisense drugs. mRNA is a large molecule with a secondary and tertiary structure. Care has to be taken to choose a section that is exposed. There are also problems relating to the poor absorption of nucleotides and their susceptibility to metabolism.

Nevertheless, antisense oligonucleotides are potential antiviral and anticancer agents, as they should be capable of preventing the biosynthesis of 'rogue' proteins and have fewer side effects than currently used drugs. Design strategies aimed at solving many of the pharmacokinetic problems of oligonucleotides are described in section 14.10. The first antisense oligonucleotide to be approved for the market was the antiviral agent **fomivirsen** (Vitravene) in 1998 (section 20.6.3). **Mipomersen** is another

antisense oligonucleotide that has reached the market. It was approved in 2013 as a treatment for lowering cholesterol levels (section 26.8.4).

Antisense oligonucleotides are also being considered for the treatment of genetic diseases such as **muscular dystrophy** and **β-thalassaemia**. Abnormal mRNA is sometimes produced as a result of a faulty splicing mechanism (section 6.2.3). Designing an antisense molecule which binds to the faulty splice might disguise that site and prevent the wrong splicing mechanism taking place.

It has been discovered in recent years that a significant proportion of the RNA produced in cells is involved in regulating gene expression. Some molecules of RNA interact directly with specific genes to prevent transcription, while other molecules intercept mRNA and prevent it taking its message to ribosomes for protein translation. This is known as **RNA interference (RNAi)**.

The interception of mRNA is promoted by short segments of double-stranded RNA typically containing 19–25 base pairs. There are three main classes of these involved in eukaryotic cells, namely **micro-RNAs (miRNAs)**, **small interfering RNAs (siRNAs)**, and **piwi-interacting RNAs (piRNAs)**. It is not feasible to describe all of these in a textbook of this nature, and so we will restrict ourselves to a more detailed description of how miRNAs are formed and trigger RNA interference.

The process by which miRNAs are produced starts off in the nucleus with the transcription of a long, single-stranded RNA molecule (pri-miRNA). This folds up such that some segments of the molecule base pair with each other (compare transfer RNA, Fig. 6.11). An endonuclease enzyme called **Drosha** then excises a segment of base-paired RNA, along with its connecting hairpin loop, to form a structure called pre-miRNA (Fig. 9.24). Pre-miRNA exits the nucleus into the cytoplasm by means of a transport protein, and is intercepted by a multiprotein complex which includes another endonuclease enzyme

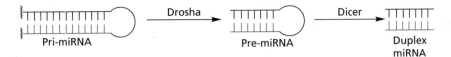

FIGURE 9.24 Cleavage of RNA to produce micro-RNAs (miRNA).

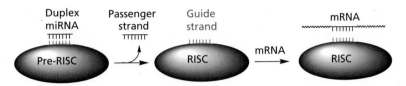

FIGURE 9.25 Activation of RISC.

called **Dicer**. Once pre-miRNA is bound to the protein complex, Dicer removes the hairpin loop to produce a short, double-stranded length of RNA corresponding to the duplex form of miRNA.

The miRNA duplex remains bound to the protein complex and interacts with another protein called the **argonaute** (Ago) protein. One strand of the duplex (the passenger strand) is discarded, while the other (the guide strand) is incorporated into Ago. The resulting protein complex is now known as **RISC (RNA induced silencing complex)**. The guide strand of miRNA can now base pair to a specific mRNA molecule if the latter contains a complementary sequence of nucleic acid bases. This brings mRNA and RISC together (Fig. 9.25), and the enzyme complex then cleaves the mRNA.

A similar type of mechanism is involved in the processing and action of siRNAs.

The control of gene expression by RNA interference is important both to the normal development of the cell and to the development of tumours. Therefore, research has been carried out to design drugs that will take advantage of these mechanisms. For example, siRNAs have been shown to regulate HIV-1 expression in cultured cells and have the potential to be used in gene therapy for the treatment of AIDS. One of the advantages of these mechanisms over conventional antisense therapy is a greater efficiency in suppressing translation. One siRNA molecule can be responsible for the cleavage of several mRNA molecules through the RISC pathway.

However, there are many difficulties still to be overcome. If siRNAs are to be effective as drugs, they will have to be metabolically stable (section 14.10) and there are also difficulties in ensuring that they:

- reach their target cells
- are taken up into the target cell.

One method that is being tried to solve these problems is to encapsulate the siRNA into small, stable nucleic acid–lipid particles that remain stable in the bloodstream, and are then taken up by target cells. For example, experiments have shown that it is possible to deliver siRNA molecules to liver cells by this method. If siRNA molecules could be designed to 'knock out' the mRNA that codes for low density lipoproteins (LDLs), this could be an effective way of lowering cholesterol levels. LDLs play an important role in transporting cholesterol round the body (see Case study 1).

KEY POINTS

- Calicheamicin is a natural product which reacts with nucleophiles to produce a diradical species. Reaction with DNA ultimately leads to cutting of the DNA chains.

- Aciclovir and related antiviral agents act as prodrugs which are converted to incomplete or unnatural nucleotides that act as DNA chain terminators.

- Synthetic agents are being designed which can bind to the regulatory elements of DNA in order to control gene transcription.

- Antisense therapy involves the use of oligonucleotides which are complementary to small sections of mRNA. They form a duplex with mRNA and prevent translation.

- Small inhibitory RNA molecules can inhibit protein synthesis by binding to mRNA and then either blocking translation or cleaving mRNA.

QUESTIONS

1. Puromycin is an antibiotic which inhibits the translation of proteins. When inhibition is taking place, partially constructed proteins are found to be present in the cytoplasm. These proteins are covalently linked to the drug.

   Suggest a mechanism by which this drug causes inhibition.

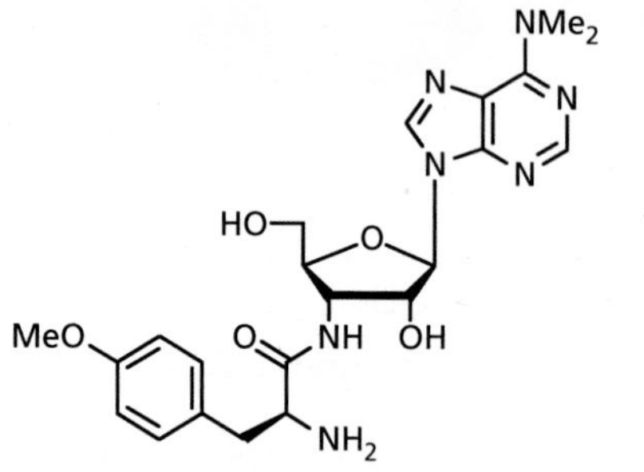

Puromycin

2. Alkylating agents have been observed to cause breaks in the DNA chain as shown below. Suggest a mechanism.

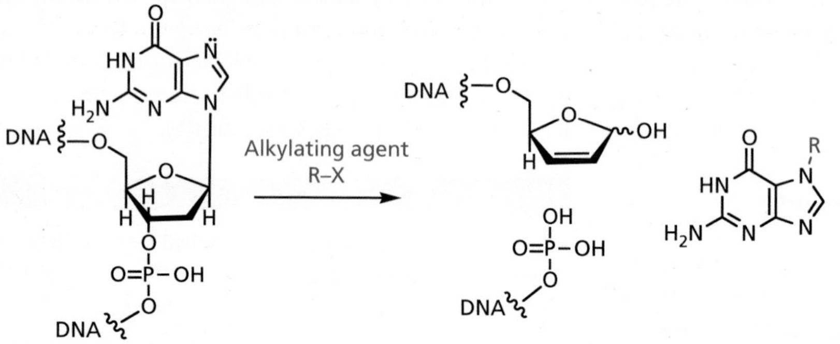

3. The following structure is an important antiviral agent. Suggest what mode of action it may have and the mechanism by which it works.

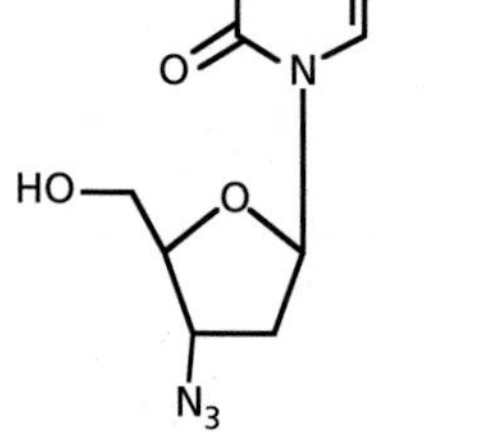

4. Propose a mechanism showing how the anticancer drug temozolomide acts as a prodrug for MTIC.

Multiple-choice questions are available on the Online Resource Centre at www.oxfordtextbooks.co.uk/orc/patrick6e/

FURTHER READING

Aldridge, S. (2003) The DNA story. *Chemistry in Britain*, April, 28–30.

Avitabile, C., Cimmino, C., and Romanelli, A. (2014) Oligonucleotide analogues as modulators of the expression and function of noncoding RNAs (ncRNAs): emerging therapeutics applications. *Journal of Medicinal Chemistry*, **57**, 10220–40.

Burke, M. (2003) On delivery. *Chemistry in Britain*, February, 36–8.

Dorsett, Y., and Tuschl, T. (2004) siRNAs: applications in functional genomics and potential as therapeutics. *Nature Reviews Drug Discovery*, **3**, 318–29.

Fletcher, H., and Hickey, I. (2012) *Instant notes genetics*. Garland Science.

Fortune, J. M., and Osheroff, N. (2000) Topoisomerase II as a target for anticancer drugs. *Progress in Nucleic Acid Research*, **64**, 221–53.

Johnson, I. S. (2003) The trials and tribulations of producing the first genetically engineered drug. *Nature Reviews Drug Discovery*, **2**, 747–51.

Judson, H. F. (1979) *The eighth day of creation*. Simon and Schuster, New York.

Kelland, L. (2007) The resurgence of platinum-based cancer chemotherapy. *Nature Reviews Cancer*, **7**, 573–84.

Langer, R. (2003) Where a pill won't reach. *Scientific American*, April, 32–9.

Lindpaintner, K. (2002) The impact of pharmacogenetics and pharmacogenomics on drug discovery. *Nature Reviews Drug Discovery*, **1**, 463–9.

Opalinska, J. B., and Gewirtz, A. M. (2002) Nucleic acid therapeutics: basic principles and recent applications. *Nature Reviews Drug Discovery*, **1**, 503–14.

Petricoin, E. F., et al. (2002) Clinical proteomics. *Nature Reviews Drug Discovery*, **1**, 683–95.

Sansom, C. (2009) Temozolomide—birth of a blockbuster. *Chemistry World*, July, 48–51.

Wang, D., and Lippard, S. J. (2005) Cellular processing of platinum anticancer drugs. *Nature Reviews Drug Discovery*, **4**, 307–20.

Winter, P. C., Hickey, G. I., and Fletcher, H. L. (1998) *Instant notes genetics*. Bios Scientific Publishers, Oxford.

Titles for general further reading are listed on p.845.

10 Miscellaneous drug targets

In Chapters 7–9 we looked at the most common drug targets in medicinal chemistry (i.e. enzymes, receptors, and nucleic acids). In this chapter, we shall look at other important drug targets to illustrate the variety of ways in which drugs can act.

10.1 Transport proteins as drug targets

Transport proteins were described in section 2.7.2. They have a binding site which 'recognizes' and binds a specific guest molecule, but it is sometimes possible to fool a transport protein into accepting a drug which resembles the usual guest. If that drug remains strongly bound to the transport protein, it will prevent the protein from carrying out its normal role. Some important drugs operate in this way. For example, **cocaine** and the **tricyclic antidepressants** bind to transport proteins and prevent neurotransmitters such as **noradrenaline** or **dopamine** from re-entering nerve cells (section 23.12.4). This results in an increased level of the neurotransmitter at nerve synapses, and has the same effect as adding drugs that mimic the neurotransmitter. Other antidepressant drugs act on the transport proteins for serotonin (Box 10.1). Drugs which inhibit the reuptake of neurotransmitters may affect more than one type of neurotransmitter. For example, several antidepressant drugs inhibit more than one type of transport protein (section 23.12.4). Another example is the anti-obesity drug **sibutramine** (Fig. 10.1), which acts centrally to inhibit the reuptake of serotonin, noradrenaline, and, to a lesser extent, dopamine. It is thought that the increase in serotonin levels dulls the appetite. Sibutramine was introduced in 1997 and is chemically related to the amphetamines. However, it was withdrawn in 2010 due to side effects. Another transport protein that is currently being investigated as a drug target is the **microsomal triglyceride transfer protein** (section 26.8.5). Inhibiting this protein can help to lower plasma cholesterol levels.

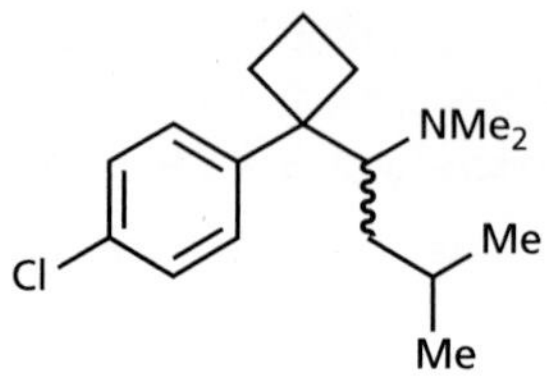

FIGURE 10.1 Sibutramine.

Transport proteins can also be targeted as a means of transporting polar drugs across the cell membrane and into the cell (see Case study 1, and sections 14.6.1.3 and 23.12.4).

10.2 Structural proteins as drug targets

In general, there are not many drugs which target structural proteins. However, some antiviral drugs have been designed to act against viral structural proteins, and there are established anticancer agents which target the structural protein, tubulin.

10.2.1 Viral structural proteins as drug targets

Viruses consist of a nucleic acid encapsulated within a protein coat called a **capsid**. If a virus is to multiply within a host cell, this protein coat has to be dismantled in order to release the nucleic acid into the cell. Drugs have been designed which bind to the structural proteins that make up the capsid, and which prevent the uncoating process. The drugs concerned show potential as antiviral agents against the cold virus (section 20.9).

Capsid proteins are also important in the mechanism by which viruses infect host cells. The viral proteins interact with host cell proteins which are present in the cell membranes, and this triggers processes which allow the virus to enter the cell. Drugs which bind to viral proteins and inhibit this protein–protein interaction can therefore

BOX 10.1 Antidepressant drugs acting on transport proteins

The antidepressant drugs **fluoxetine (Prozac)**, **citalopram**, and **escitalopram** (Fig. 1) selectively block the transport protein responsible for the uptake of a neurotransmitter called **serotonin** from nerve synapses, and are called **selective serotonin reuptake inhibitors** (SSRIs) (see also Case study 7). A lack of serotonin in the brain has been linked with depression and, by blocking its uptake, the serotonin that is released has a longer duration of action. Fluoxetine and citalopram are chiral molecules which are marketed as racemates. The *S*-enantiomer of citalopram is more active than the *R*-enantiomer and is now marketed as escitalopram. Replacing a racemic drug with a more effective enantiomer is known as **chiral switching** (section 15.2.1).

Other examples of clinically important SSRIs include **sertraline**, **paroxetine**, and **fluvoxamine** (Fig. 2).

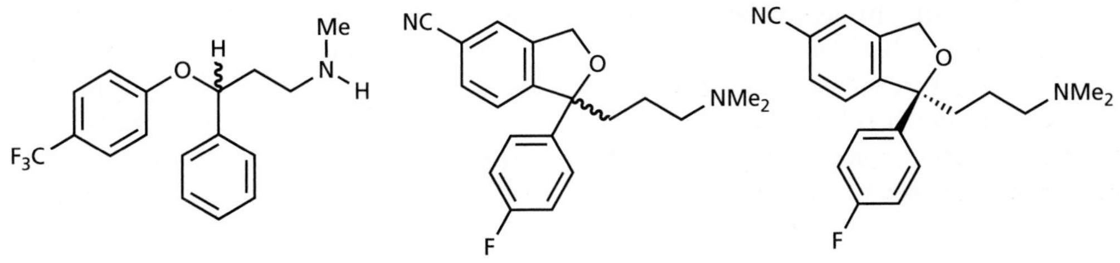

Fluoxetine (Prozac) Citalopram *S*-Citalopram (Escitalopram)

Figure 1 Antidepressant drugs acting to block the uptake of serotonin.

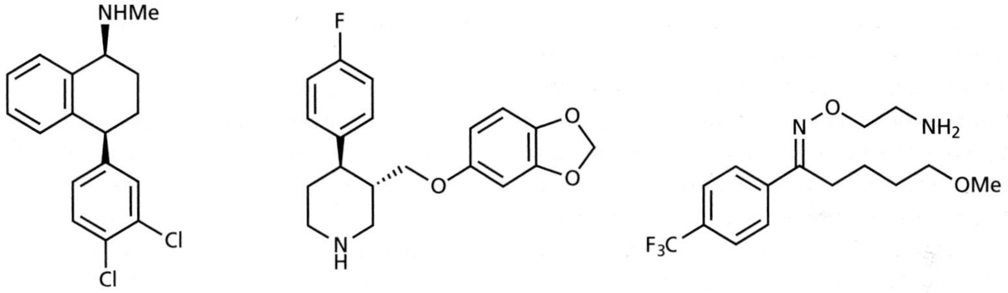

Sertraline Paroxetine Fluvoxamine

Figure 2 Further examples of SSRIs.

act as antiviral agents. **Enfuvirtide** is an example of an antiviral agent working in this way (section 20.7.5) that was approved in March 2003.

10.2.2 Tubulin as a drug target

In section 2.7.1, we described the role of the structural protein tubulin in cell division—a process which involves the polymerization and depolymerization of microtubules using tubulin proteins as building blocks. A variety of drugs interfere with this process by either binding to tubulin and inhibiting the polymerization process, or binding to the microtubules to stabilize them and thus inhibit depolymerization. Either way, the balance between polymerization and depolymerization is disrupted, which leads to a toxic effect and the inability of the cell to divide. Drugs that target tubulin have been found to be useful anticancer and anti-inflammatory agents, and some of the most important are described below.

10.2.2.1 Agents which inhibit tubulin polymerization

Colchicine (Fig. 10.2) is an example of a drug that binds to tubulin and prevents its polymerization. It can be used in the treatment of gout by reducing the mobility

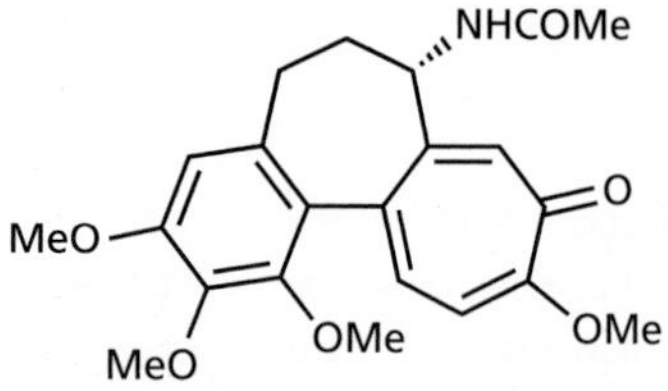

FIGURE 10.2 Colchicine.

of neutrophils into joints. Unfortunately, colchicine has many side effects, and so it is restricted to the treatment of acute attacks of this disease.

The **Vinca alkaloids vincristine, vinblastine, vindesine**, and **vinorelbine** (Fig. 10.3) bind to tubulin to prevent polymerization and are useful anticancer agents. A range of other natural products have also been found to prevent the polymerization of microtubules and are currently being studied as potential anticancer agents (section 21.5.1).

10.2.2.2 Agents which inhibit tubulin depolymerization

Paclitaxel (Taxol) and the semi-synthetic analogues **docetaxel** and **cabazitaxel** (Fig. 10.4) are important anticancer agents that inhibit tubulin depolymerization (section 21.5.2). Paclitaxel itself was isolated from the bark of yew trees (*Taxus* spp.) and identified in 1971 following a screening programme for new anticancer agents carried out by the US National Cancer Institute. Obtaining sufficient paclitaxel was initially a problem, as the bark from two yew trees was required to supply sufficient paclitaxel for one patient! A full synthesis of paclitaxel was achieved in 1994, but was impractical for large-scale production because it involved 30 steps and gave a low overall yield. Fortunately, it has been possible

to carry out a semi-synthetic synthesis (section 15.3.4) using a related natural product which can be harvested from the yew needles without damaging the tree. The semi-synthetic route involves docetaxel as an intermediate. The term **taxoids** is used generally for paclitaxel and its derivatives.

Tubulin is actually made up of two separate proteins and the taxoids are found to bind to the β-subunit of tubulin. In contrast to the drugs described in section 10.2.2.1, the binding of paclitaxel accelerates polymerization and stabilizes the resultant microtubules, which means that depolymerization is inhibited. As a result, the cell division cycle is halted.

The benzoyl and acetyl substituents, at positions 2 and 4 respectively, play an important role in this binding interaction, as do the side chain and the oxetane ring. These groups dominate the 'lower' or 'southern' half of the molecule (as the structure is normally presented), and so the variations that are possible in this region are restricted when making analogues. In contrast, it is possible to carry out more variations in the 'northern' half of the molecule. This can affect the *in vivo* efficacy of the molecule, allowing modification of aqueous solubility and pharmacokinetic properties. **BMS 188797** and **BMS 184476** (Fig. 10.5) are two taxoids which have recently been developed and have reached clinical trials.

More substantial variations have resulted in a second generation of taxoids where potency has been increased by 2–3 orders of magnitude. For example, it was possible to replace the aromatic rings of paclitaxel with other hydrophobic groups. Having a suitable acyl group at position 10 has also been found to increase activity against drug-resistant strains of cancers. Such compounds have the ability not only to bind to tubulin, but to inhibit the P-glycoprotein efflux pump. This is a protein which is present in the cell membrane of cancer cells and can

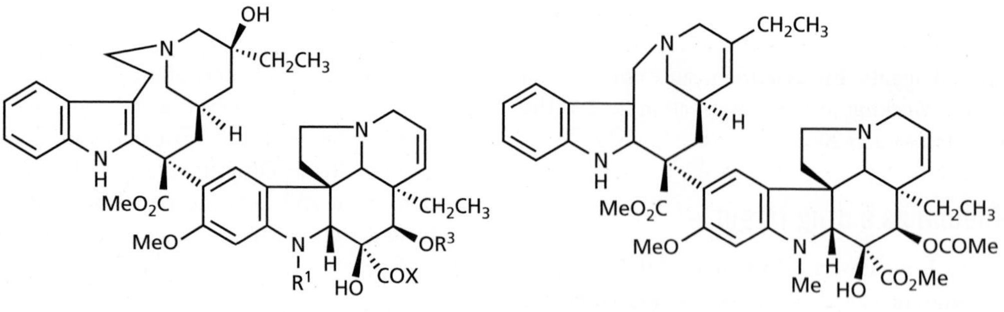

Vinblastine   (R^1=Me; X=OMe; R^3= COMe)
Vincristine   (R^1=CHO; X=OMe; R^3= COMe)
Vindesine   (R^1=Me; X=NH$_2$; R^3= H)

Vinorelbine

FIGURE 10.3 The Vinca alkaloids.

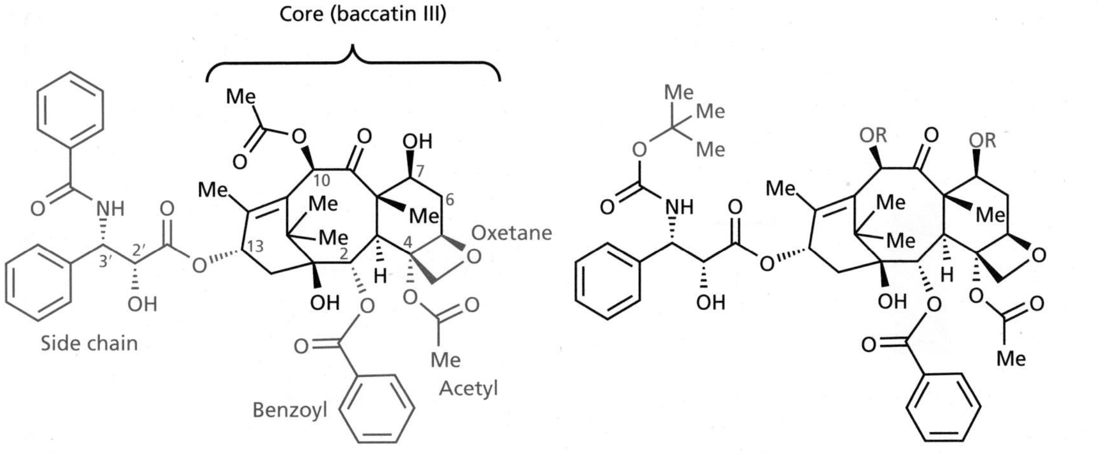

FIGURE 10.4 Paclitaxel (Taxol) with important binding groups in colour, as well as docetaxel (Taxotere) and cabazitaxel.

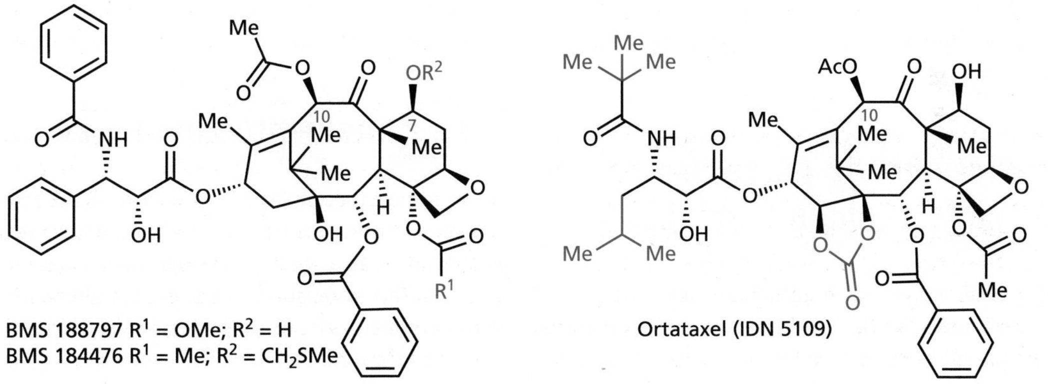

FIGURE 10.5 Analogues of paclitaxel.

pump drugs out of the cell before they get the chance to work effectively. Further work has demonstrated that acylating the 7-hydroxy group with hydrophobic groups is also effective in blocking efflux.

Finally, the addition of a methyl substituent at C-2′ has been found to increase activity by inhibiting rotation of the C-2′–C-3′ bond. The first orally active taxoid structure **ortataxel** (Fig. 10.5) has now been developed and has entered clinical trials.

Since the discovery of paclitaxel a variety of other natural products have been found to have a similar mechanism of action and are currently being studied as potential anticancer agents (section 21.5.2).

KEY POINTS

- Transport proteins transport polar molecules across the hydrophobic cell membrane. Drugs can be designed to take advantage of this transport system in order to gain access to cells, or to block the transport protein.

- Drugs that target viral structural proteins can prevent viruses entering host cells. They can also inhibit the uncoating process.

- Tubulin is a structural protein which is crucial to cell division and cell mobility, and which is the target for several anticancer drugs.

- The Vinca alkaloids bind to tubulin and inhibit the polymerization process.

- Paclitaxel and its derivatives bind to tubulin and accelerate polymerization by stabilizing the resulting microtubules.

10.3 Biosynthetic building blocks as drug targets

The target for the antibacterial agent **vancomycin** is rather unique in that it is a biosynthetic building block. Essentially, vancomycin 'caps' the building block and

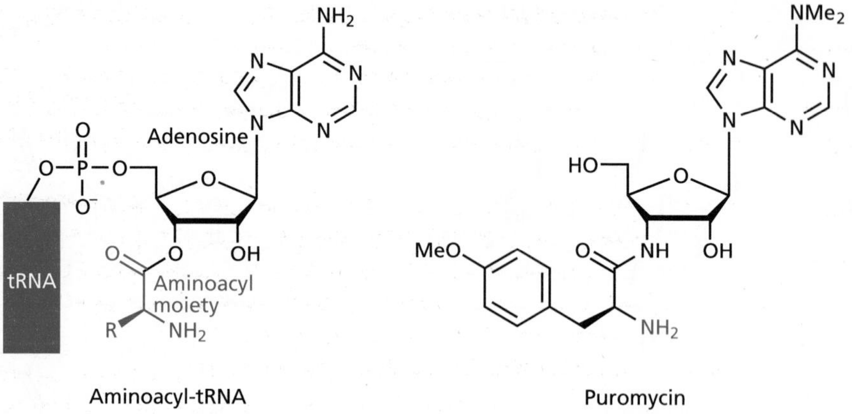

FIGURE 10.6 Comparison of puromycin and aminoacyl-tRNA.

prevents its incorporation into the growing bacterial cell wall. There is a small peptide chain on the building block which can bind to vancomycin by hydrogen bond interactions. Indeed, vancomycin is acting like a receptor by providing a binding site for the building block (see section 19.5.5.2).

10.4 Biosynthetic processes as drug targets: chain terminators

In section 9.5, we looked at antiviral drugs which act as chain terminators for the synthesis of new DNA. **Puromycin** is an antibiotic which can be viewed in the same light, except that it terminates the growth of protein chains during translation. It is able to carry out this role because it mimics the terminus of an aminoacyl-tRNA molecule (Fig. 10.6). Aminoacyl-tRNA is the molecule which brings an amino acid to the ribosome such that it can be added to the growing protein chain (section 6.2.2).

Because puromycin resembles the aminoacyl and adenosine moieties of aminoacyl-tRNA, it is able to enter the A site of the ribosome and prevent aminoacyl-tRNA

molecules from binding. It has the amino group required for the transfer reaction, and so the peptide chain is transferred from tRNA in the P binding site to puromycin in the A binding site. Puromycin departs the ribosome carrying a stunted protein along with it (Fig. 10.7).

10.5 Protein–protein interactions

Many important cellular processes involve the association of two or more proteins (section 2.7.4), and so several research teams are trying to develop drugs that might interfere with this association. Such drugs could be useful in a variety of medicinal fields. For example, a drug that prevents protein–protein interactions as part of a signal transduction process (Chapter 5) could inhibit cell growth and cell division, and hence be a useful anticancer agent. An agent which prevents the formation of transcription factor complexes could prevent the transcription of specific genes (Box 10.2). There is also evidence that the abnormal protein structures observed in Alzheimer's disease result from protein–protein interactions (section 22.15).

One way of inhibiting protein–protein interactions is to use antibodies (section 10.7.2), and these agents have

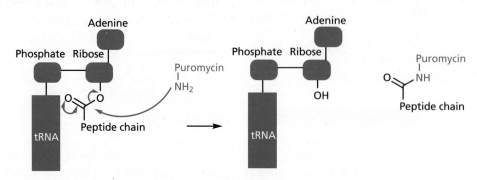

FIGURE 10.7 Transfer of peptide chain to puromycin.

been particularly successful in preventing protein–protein interactions for a family of extracellular proteins called **integrins**. The integrins are adhesive proteins which are important to processes such as blood clotting, inflammation, cell protection, and the immune response.

Indeed, **daclizumab** is an antibody which is used as an immunosuppressant in kidney transplants while **abciximab** is an antibody fragment which inhibits blood clotting following angioplasty procedures aimed at unblocking coronary arteries. Successful though antibodies may

---

BOX 10.2 Targeting transcription factor–coactivator interactions

The transcription of a gene is initiated by a protein complex that is formed between a transcription factor and a coactivator protein (Box 8.2). A drug that inhibits the interactions between these proteins would prevent formation of the complex, prevent transcription, and be potentially useful in treating some cancers. The crucial interactions between two proteins can often involve a relatively short α-helical segment. For example, the interaction between the ESX transcription factor and its coactivator protein Sur-2 involves an eight amino acid α-helix present on the transcription factor (Fig. 1). One of these eight amino acids is a tryptophan residue (Trp) which plays a particularly important binding role, and so one research group screened a number of chemical libraries for compounds containing indole rings that could mimic this residue. This led to the discovery of a lead compound called **adamanolol** (Fig. 2), which was found to inhibit the interaction between the proteins.

Structure–activity studies showed that:

- The indole ring system was essential and mimics the tryptophan residue.

- The adamantane ring is important and is thought to mimic a cluster of isoleucine and leucine residues that are on the α-helix. It may also bind to a hydrophobic pocket in the coactivator protein.
- The isopropyl group can be replaced with bulky substituents. These substituents enforce a configuration around the urea linker where the molecule forms a helix-like shape with the adamantane and indole rings in close proximity.

From these results, a more active water-soluble agent called **wrenchnolol** was designed—so named since it resembles the shape of a wrench. The molecule has two hydrophobic 'jaws' and a polar 'handle'. The non-polar components are clustered on one face of the molecule with the polar handle angled away, resulting in an amphiphilic molecule that mimics the amphiphilic α–helix of the transcription factor. The hydrophobic jaws make contact with the Sur-2 protein and mimic the amino acid residues of tryptophan, leucine, and isoleucine.

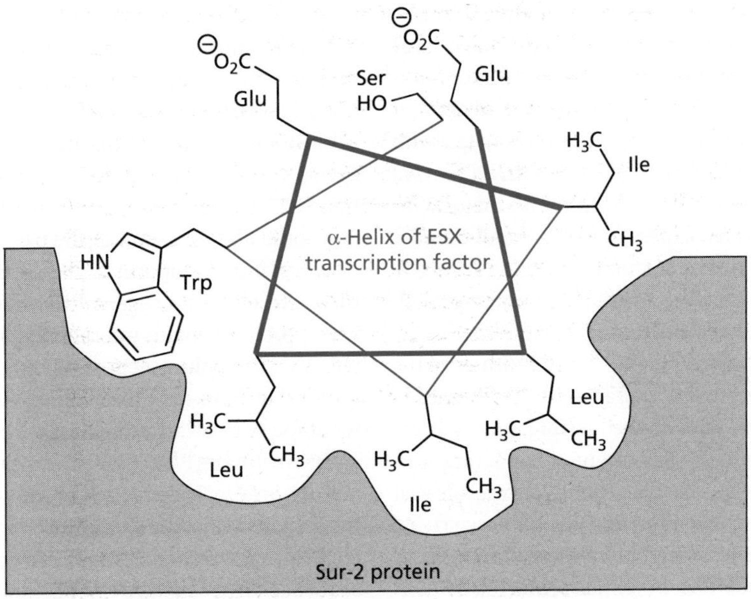

FIGURE 1 Interaction of an α-helix of the ESX transcription factor with the protein Sur-2.

(Continued)

BOX 10.2 Targeting transcription factor–coactivator interactions (*Continued*)

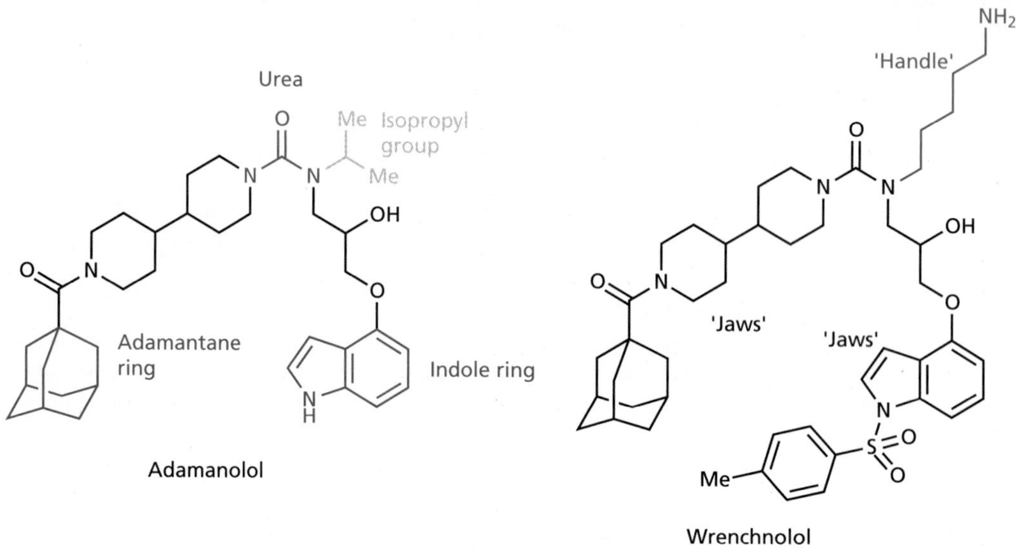

FIGURE 2 Structures of adamanolol and wrenchnolol.

be, they are limited in application to extracellular proteins, and so it would be advantageous to design drug-sized molecules which could have the same action on protein targets both extracellularly and intracellularly.

Finding a drug to do this might seem a tall order. Drugs, after all, are small molecules in comparison to a protein, and protein–protein interactions involve large surface areas of the associated proteins. The idea of binding a drug to a protein surface in order to ward off another protein seems a bit optimistic. It might be equated with landing a spacecraft on the moon and expecting it to ward off meteorites. Fortunately, it has been found that the interactions between proteins often involve a small number of particularly important interactions involving relatively small areas. For example, the binding of human growth factor with its receptor certainly involves large surface areas of both proteins, where 31 amino acid residues of the human growth factor protein interact with 33 residues of the receptor. However, 85% of the binding energy is associated with 8 residues of the hormone interacting with 9 residues of the receptor. Therefore, it is conceivable that a drug could be designed to bind to some of these crucial residues and hinder the association of these proteins.

However, there are other potential problems to consider. The protein surfaces involved in protein–protein interactions are often relatively flat and do not contain the kind of binding sites that we are used to with enzymes and receptors. Therefore, identifying a particular feature

on a protein surface that could be 'recognized' by a drug might be difficult. A final problem is that drugs which inhibit protein–protein interactions are likely to be larger than the average-sized drug. This might hinder drugs from passing through cell membranes in order to reach their intracellular targets.

Despite these problems, there is active research in finding drugs that can inhibit protein–protein interactions. Such drugs are known as **protein–protein binding inhibitors** (PPBIs). PPBIs have potential as anticancer agents (sections 21.8 and 21.9.4), antiviral agents (section 20.7.5), analgesics, and anti-inflammatory agents, and could also be useful in the treatment of autoimmune diseases and osteoporosis. It is worth pointing out that there are already drugs on the market which interfere with protein–protein interactions, mainly those that interact with tubulin (section 10.2.2). Drugs have also been found that bind to various integrins to prevent their interaction with other proteins. One example is the clinical agent **tirofiban** (Fig. 10.8) which is used as an anticoagulant by preventing protein–protein binding between an integrin and the blood clotting agent fibrinogen (section 26.9.2.4). It is thought that the drug mimics a tripeptide sequence (Arg-Gly-Asp) in fibrinogen that plays an important role in the binding process between the two proteins. When the drug binds to integrin, it prevents this interaction taking place, and so one could view the drug as an ultra-simplified analogue of the fibrinogen protein!

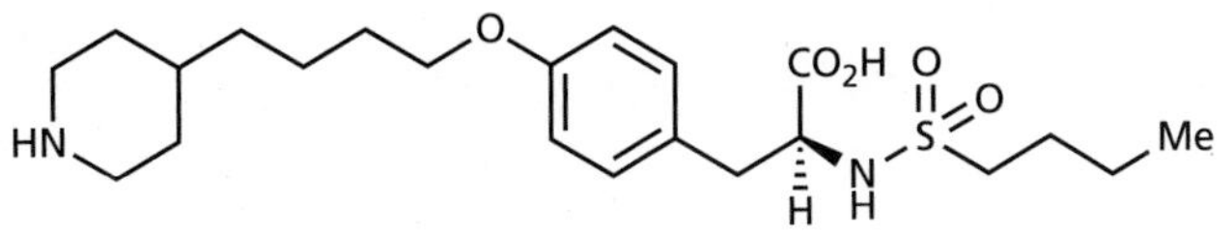

FIGURE 10.8 Tirofiban.

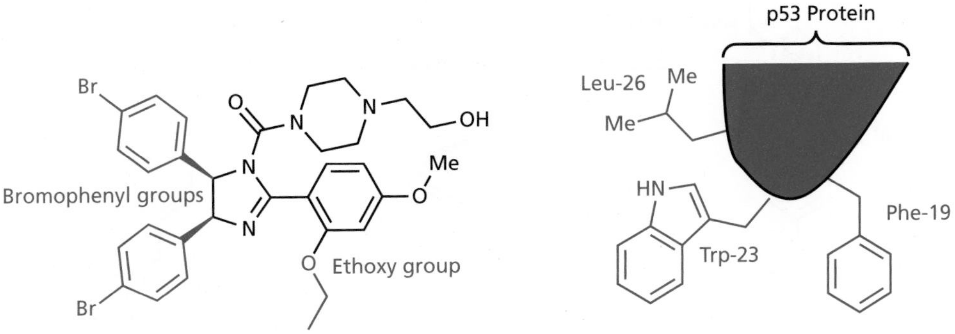

FIGURE 10.9 Nutlin-2 mimicking the three amino acid 'finger' residues of p53.

An important example of a protein–protein interaction involves the proteins p53 and MDM2 (or HDM2[1]). The former protein (p53) is produced in cells that are damaged or are under stress, and serves to restrict cell growth or even induce cell death (section 21.1.7). This activity is important to the health and survival of an organism, because it suppresses the growth of defective cells such as tumour cells. MDM2 is a protein which down-regulates the activity of p53 by binding or interacting with it. In some tumour cells, a genetic defect results in excess levels of MDM2, which means that p53 can no longer function, allowing tumour cells to multiply. Therefore, drugs which prevent this interaction could be useful anticancer drugs. **Nutlin-2** (Fig. 10.9) is an example of a series of structurally related compounds which are capable of preventing this protein–protein interaction. It binds to a region of MDM2 that is normally involved in the protein–protein interaction with p53, and mimics three amino acid residues present on p53 (Leu-26, Trp-23, Phe-19). These three amino acid residues normally fit like three fingers into complementary pockets on the MDM2 surface. The ethoxy group and the two bromophenyl groups of nutlin-2 mimic these three fingers.

One easy way of designing a PPBI is to identify a peptide that will mimic a crucial peptide binding region for one of the proteins. This peptide would then be recognized by the complementary protein and bind with it, thus preventing protein–protein binding. However, peptides have many disadvantages as drugs (section 14.9), and non-peptide drugs are preferable. To that end, medicinal chemists have attempted to design peptide mimics. In order to achieve that goal, molecules need to be designed with substituents that will mimic the side chains of amino acids. The substituents also need to be attached to a stable molecular scaffold in such a way that they are positioned in the same relative positions as amino acid residues in common protein features, i.e. α-helices, β-sheets, β-turns, and loops. A lot of work has been carried out designing drugs to mimic β-turns, but more recently, researchers have been turning their attention to structures that mimic α-helices—an extremely important area since α-helices play crucial roles in many protein–protein interactions (see also section 21.8).

An example of this research involves **terphenyl structures** (Fig. 10.10). The three aromatic rings that are directly linked together in these compounds are not coplanar. Instead, they are at different angles with respect to each other and mimic the twist of the α-helix. These rings act as the scaffold onto which different substituents can be placed to mimic amino acid side chains. The *meta*-substituent and the two *ortho*-substituents shown in Fig. 10.10a mimic the side chains of amino acids which would be at the first, fourth, and seventh positions of an α-helix. This structure has been shown to act as an antagonist for the protein **calmodulin**, but by varying the nature of the substituents, one can obtain structures that are recognized by different proteins. For example, the terphenyl structure shown in Fig. 10.10b binds to a protein called BCl-x$_L$. This protein plays an important role in apoptosis—the process by which cells are destroyed (sections 21.1.7 and 21.8). Another terphenyl structure bearing three aliphatic residues has been shown to bind to a viral protein that is crucial to the process by which HIV enters a host cell, and so the terphenyl structure can inhibit that process (section 20.7.5).

[1]Note: MDM2 is produced in mice and is used for research. HDM2 is the human version of MDM2.

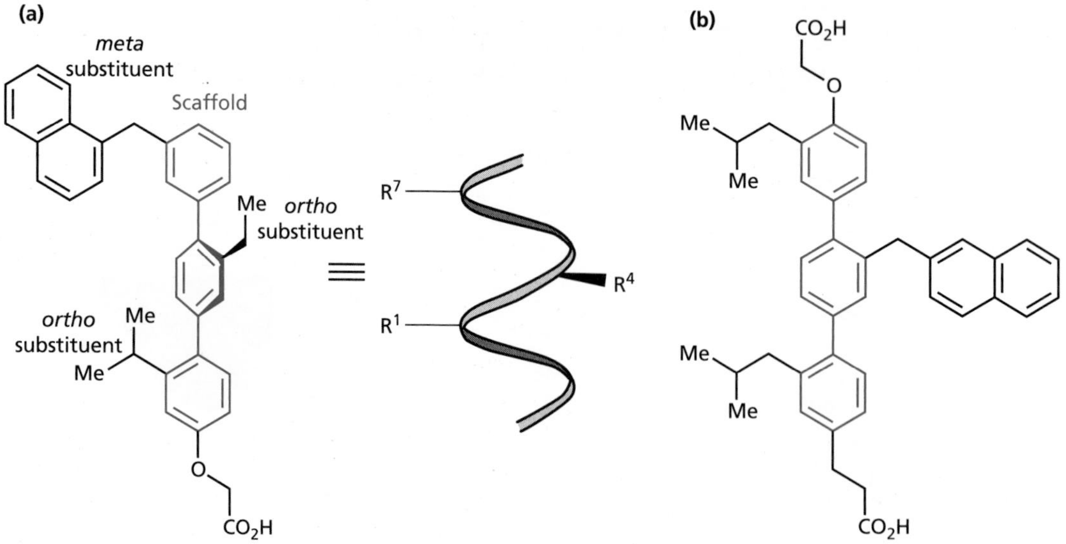

FIGURE 10.10 (a) Terphenyl-based structure mimicking an α-helix. (b) Terphenyl structure that binds to the protein BCl-x$_L$.

Drugs which mimic β-sheets are also being investigated. Such drugs have potential as antiviral agents in the treatment of AIDS. One of the important viral proteins in the life cycle of HIV is a protease enzyme which is made up of two identical proteins, interacting with each other by means of an antiparallel β-sheet (section 20.7.4.1). A drug which could mimic this feature might prevent dimerization of the protein and prevent it from functioning. Other antiviral drugs are being designed to target a variety of other protein–protein interactions involving HIV, especially those involved in the process of cell entry (section 20.7.5).

A different approach to inhibiting protein–protein interactions is to use an oligonucleotide. Oligonucleotide–protein interactions are common in the biological world, and it has been shown that it is possible to obtain oligonucleotides that bind to specific protein targets with a high degree of selectivity. Such oligonucleotides are called **aptamers** (derived from the Latin *aptus*, to fit, and the Greek *meros*, part or region). A procedure called **SELEX** has been developed that allows researchers to find an aptamer that will bind to virtually any protein target. A library of oligonucleotides is synthesized using mixed combinatorial synthesis (Chapter 16). Each oligonucleotide is 20–40 nucleotides in length, and the library contains in the order of 10^{15} potential aptamers. The library is tested against a particular protein target and aptamers that bind to the target are selected and amplified through cloning. Further cycles of selection and amplification can then be carried out to find the aptamer with the greatest selectivity and binding strength. This approach has been successful in generating a clinically useful aptamer called **pegaptanib**, which binds to a hormone called **vascular endothelial growth factor** (VEGF) and prevents it from binding to its receptor (VEGFR). Activation of this receptor is important to the formation of new blood vessels (sections 4.8.4.2 and 21.1.9 and Box 21.11), and pegaptanib was approved in 2004 for the treatment of an eye disease where there is an overproduction of blood vessels. The aptamer is linked to polyethylene glycol (PEG) to improve the half-life of the agent (section 11.10).

The antibody **bevacizumab** works in a similar manner by binding to VEGF, and is used as an anticancer agent (sections 21.1.9 and Box 21.12).

10.6 Lipids as a drug target

The number of drugs that interact with lipids is relatively small and, in general, they all act in the same way—by disrupting the lipid structure of cell membranes. For example, it has been proposed that general anaesthetics work by interacting with the lipids of cell membranes to alter the structure and conducting properties of the membranes. Another agent which is thought to disrupt cell membrane structure is the anticancer agent **cephalostatin I**, which is thought to span the phospholipid bilayer. Finally, **daptomycin** is an antibiotic which disrupts multiple functions of the bacterial cell membrane (section 19.6.4).

10.6.1 'Tunnelling molecules'

The antifungal agent **amphotericin B** (Fig. 10.11) (used topically against athletes foot and systemically against life-threatening fungal diseases) interacts with the lipids

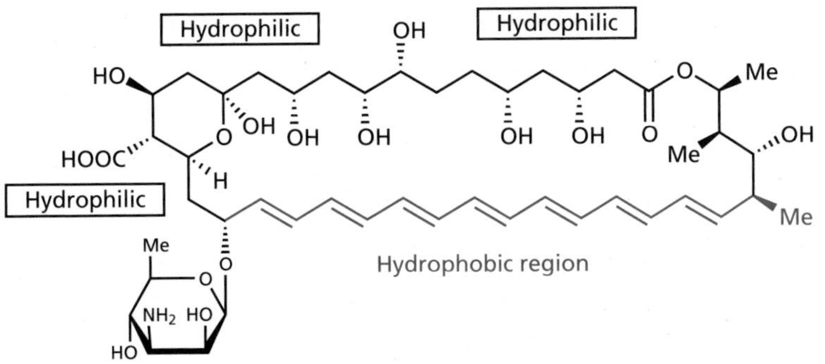

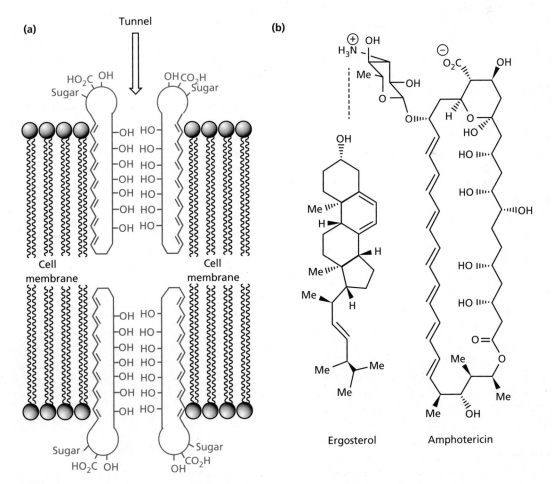

FIGURE 10.11 Amphotericin B.

and sterols of fungal cell membranes to build 'tunnels' through the membrane. Once in place, the contents of the cell are drained away and the cell is killed.

Amphotericin B is a fascinating molecule in that one half of the structure is made up of double bonds and is hydrophobic, whereas the other half contains a series of hydroxyl groups and is hydrophilic. It is a molecule of extremes and is ideally suited to act on the cell membrane in the way that it does. Several amphotericin molecules cluster together such that the alkene chains face outwards to interact favourably with the hydrophobic centre of the cell membrane. The tunnel resulting from this cluster is lined with the hydroxyl groups and so it is hydrophilic, allowing the polar contents of the cell to drain away (Fig. 10.12a). Amphotericin B is a natural product derived from a microorganism (*Streptomyces nodosus*). Recently,

FIGURE 10.12 (a) Ion channel pore through the cell membrane formed by amphotericin (ergosterol not shown). (b) Interaction between amphotericin and ergosterol in the ion pore channel.

Val-Gly-Ala-Leu-Ala-Val-Val-Val-Trp-Leu-Trp-Leu-Trp-Leu-Trp-NH-CH$_2$-CH$_2$-OH

FIGURE 10.13 Gramicidin A.

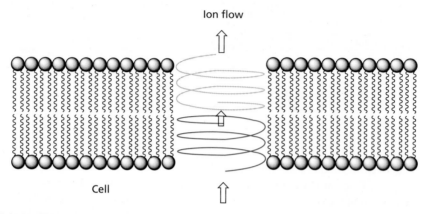

FIGURE 10.14 Gramicidin helices aligned end-to-end to traverse the cell membrane.

it has been established that each molecule of amphotericin forms a hydrogen bonding interaction with a molecule of **ergosterol** in order to create the ion pore channel. Therefore, the ion pore is actually made up of both amphotericin and ergosterol. Ergosterol is the fungal equivalent of cholesterol and is an important constituent of the fungal cell membrane. The crucial interaction involves the charged aminium group on the carbohydrate ring of amphotericin (Fig. 10.12b).

The antibiotic **gramicidin A** (Fig. 10.13) is a peptide containing 15 amino acids which is thought to coil into a helix such that the outside of the helix is hydrophobic and interacts with the membrane lipids, while the inside of the helix contains hydrophilic groups, thus allowing the passage of ions. Therefore, gramicidin A could also be viewed as an escape tunnel through the cell membrane. In fact, one molecule of gramicidin would not be long enough to traverse the membrane, and it has

been proposed that two gramicidin helices align themselves end-to-end in order to achieve the length required (Fig. 10.14).

Magainins (section 12.4.1.4) are 23-residue polypeptide antibiotics which form helical structures that also disrupt the permeability of cell membranes. However, the helices are thought to associate only with the head groups of the cell membrane, then cause segments of the lipid membrane to bend back on themselves to form a toroidal structure or wormhole (Fig. 10.15). The magainin helices remain associated with the head groups of the cell membrane to stabilize the pores that are formed.

Work is currently in progress to design cyclic peptides which will self-assemble in bacterial cell membranes to form tubules. These tubules have been labelled as 'killer nanotubes' (Fig. 10.16). Once formed, the nanotubes would allow molecules to leach out from the cell and cause cell death. The cyclic peptides concerned are

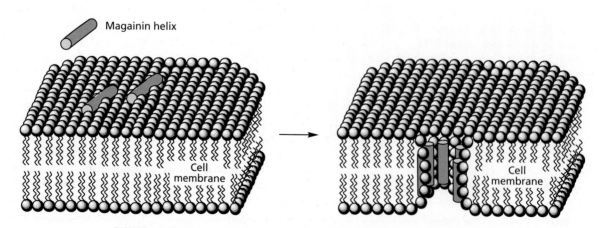

FIGURE 10.15 The wormhole or toroidal model for magainin antibiotic action.

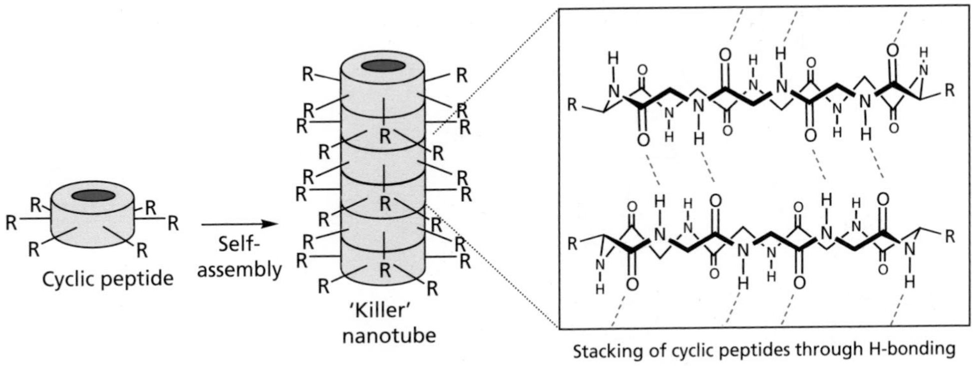

FIGURE 10.16 Self-assembly of 'killer nanotubes'.

designed to have 6–8 alternating D- and L-amino acids such that the amide groups are perpendicular to the plane of the cyclic structure, with the side chains pointing outwards in the same plane. This means that the side chains do not interfere with the stacking process while the amide groups in each cyclic peptide form hydrogen bonds to the cyclic peptides above and below it, thus promoting the stacking process. Modifying the types of residues present has been successful in introducing selectivity *in vitro* for bacterial cells versus red blood cells. For example, the inclusion of a basic amino acid such as lysine is useful for selectivity. Lysine has a primary amino group which can become protonated and gain a positive charge. This encourages the structures to target bacterial membranes, because the latter tend to have a negative charge on their surface. *In vivo* studies have also been carried out successfully on mice.

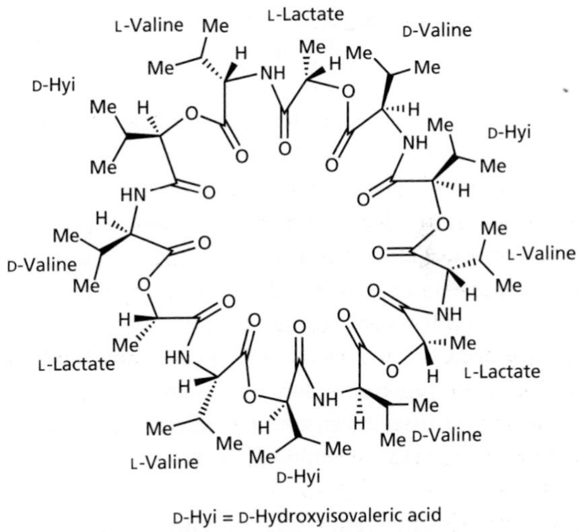

10.6.2 Ion carriers

Valinomycin (Fig. 10.17) is a cyclic structure obtained from *Streptomyces* fermentation. It contains three molecules of L-valine, three molecules of D-valine, three molecules of L-lactic acid, and three molecules of D-hydroxyisovalerate. These four components are linked in an ordered fashion such that there is an alternating sequence of ester and amide linking bonds around the cyclic structure. This is achieved by the presence of a lactic or hydroxyisovaleric acid unit between each of the six valine units. Further ordering can be observed by noting that the L and D portions of valine alternate around the cycle, as do the lactate and hydroxyisovalerate units.

Valinomycin acts as an ion carrier and could be looked upon as an inverted detergent. As it is cyclic, it forms a doughnut-type structure where the polar carbonyl oxygens of the ester and amide groups face inwards, while the hydrophobic side chains of the valine and hydroxyisovalerate units point outwards. This is clearly favoured because the hydrophobic side chains can interact via van

D-Hyi = D-Hydroxyisovaleric acid

FIGURE 10.17 Valinomycin.

der Waals interactions with the fatty lipid interior of the cell membrane, while the polar hydrophilic groups are clustered together in the centre of the doughnut to produce a hydrophilic environment. This hydrophilic centre is large enough to accommodate an ion and it is found that a 'naked' potassium ion (i.e. one with no surrounding water molecules) fits the space and is complexed by the amide carboxyl groups (Fig. 10.18).

Valinomycin can therefore collect a potassium ion from the inner surface of the membrane, carry it across the membrane, and deposit it outside the cell, thus disrupting the ionic equilibrium of the cell (Fig. 10.19). Normally, cells contain a high concentration of potassium ions and a low concentration of sodium ions. The fatty cell membrane prevents passage of ions between the cell and its environment, and ions can only pass through the cell membrane aided by specialized and controlled ion transport systems. Valinomycin introduces an uncontrolled ion transport system which proves fatal.

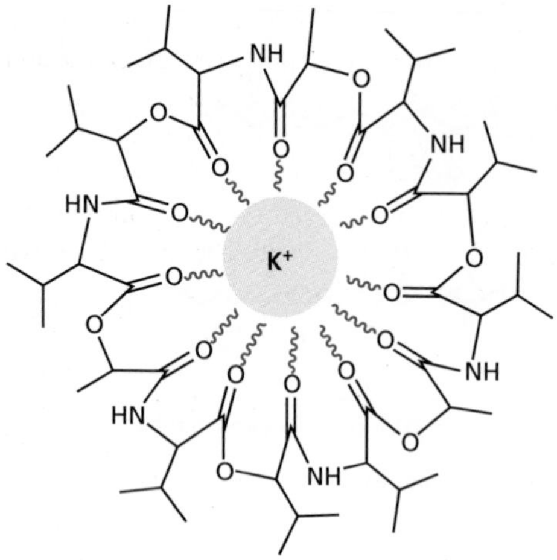

FIGURE 10.18 Potassium ion in the hydrophilic centre of valinomycin.

Nigericin R^1–R^4 = Me; R^5 = R^6 = H; X =

Monensin A R^1 = Et; R^2–R^4 = H; R^5 = R^6 = Me; X =

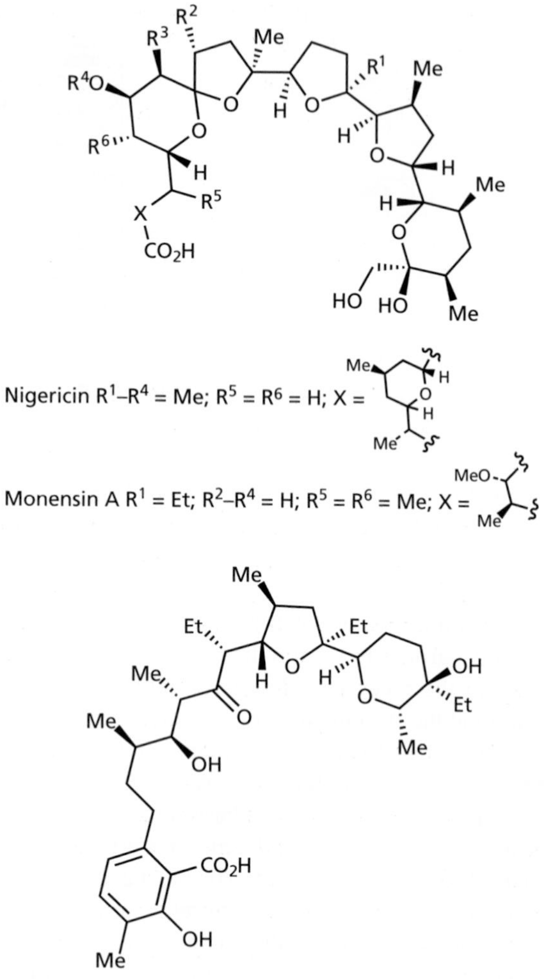

Lasalocid A

FIGURE 10.20 Ionophores used in veterinary medicine.

Valinomycin is specific for potassium ions over sodium ions, and one might be tempted to think that sodium ions would be too small to be properly complexed. The real reason is that sodium ions do not lose their surrounding water molecules very easily and would have to be transported as the hydrated ion. As such, they are too big for the central cavity of valinomycin.

The ionophores **nigericin**, **monensin A**, and **lasalocid A** (Fig. 10.20) function in much the same way as valinomycin and are used in veterinary medicine to control the levels of bacteria in the rumen of cattle and the intestines of poultry.

The polypeptide antibiotic **polymyxin B** (section 19.6.2) acts like valinomycin, but it causes the leakage of small molecules (e.g. nucleosides) from the cell, rather than ions.

10.6.3 **Tethers and anchors**

Several drugs contain hydrophobic groups that are designed to anchor the drug to the membranes of cells and organelles. These drugs are not targeting the membrane itself, but are tethered such that they interact more easily with molecular targets which are also tethered to the membrane—see sections 14.4.4 and 19.5.5.2.

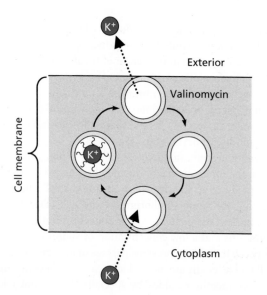

FIGURE 10.19 Valinomycin disrupts the ionic equilibrium of a cell.

KEY POINTS

- 'Tunnelling' molecules and ion carriers act on the plasma membrane and result in the uncontrolled movement of ions across the cell membrane leading to cell death.

- Cyclic peptides are being designed which will self-assemble to form nanotubes in the cell membranes of bacteria.

- Tethering drugs to a membrane is a useful method of targeting them against structures that are attached to membranes.

10.7 Carbohydrates as drug targets

10.7.1 Glycomics

The term **glycomics** is used to describe the study of carbohydrates, either as drugs or as drug targets. Carbohydrates are polyhydroxy structures, many of which have the general formula $C_nH_{2n}O_n$. Examples of some simple carbohydrate structures include **glucose, fructose**, and **ribose** (Fig. 10.21). These are called **monosaccharides** because they can be viewed as the monomers required to make more complex polymeric carbohydrates. For example, glucose monomers are linked together to form the natural polymers **glycogen, cellulose** (Fig. 10.22), or **starch**.

Until relatively recently, carbohydrates were not considered useful drug targets. The main roles for carbohydrates in the cell were seen as energy storage (e.g. glycogen) or structural (e.g. starch and cellulose). It is now known that carbohydrates have important roles to play in various cellular processes such as cell recognition, cell regulation, and cell growth. Various disease states are associated with these cellular processes. For example, bacteria and viruses have to recognize host cells before they can infect them, and so the carbohydrate molecules involved in cell recognition are crucial to that process

(sections 20.3, 20.7.1, and 20.8.1). Designing drugs to bind to these carbohydrates may well block the ability of bacteria and viruses to invade host cells. Alternatively, vaccines or drugs may be developed based on the structure of these important carbohydrates (section 20.8.3).

It has also been observed that autoimmune diseases and cancers are associated with changes in the structure of cell surface carbohydrates (section 21.1.10). Understanding how carbohydrates are involved in cell recognition and cell regulation may well allow the design of novel drugs to treat these diseases (section 21.10).

Many of the important cell recognition roles played by carbohydrates are not acted out by pure carbohydrates, but by carbohydrates linked to proteins (**glycoproteins** or **proteoglycans**) or lipids (**glycolipids**). Such molecules are called **glycoconjugates**. Usually, the lipid or protein portion of the molecule is embedded within the cell membrane with the carbohydrate portion hanging free on the outside, like the streamer of a kite. This allows the carbohydrate portion to serve the role of a molecular tag that labels and identifies the cell. The tag may also play the role of a receptor, binding other molecules or cells.

There is actually good sense in having a carbohydrate as a molecular tag rather than a peptide or a nucleic acid, because more structural variations are possible for carbohydrates than for other types of structure. For example, two molecules of alanine can only form one possible dipeptide, as there is only one way in which they can be linked (Fig. 10.23). However, because of the different hydroxyl groups on a carbohydrate, there are 11 possible disaccharides that can be formed from two glucose molecules (Fig. 10.24). This allows nature to create an almost infinite number of molecular tags based on different

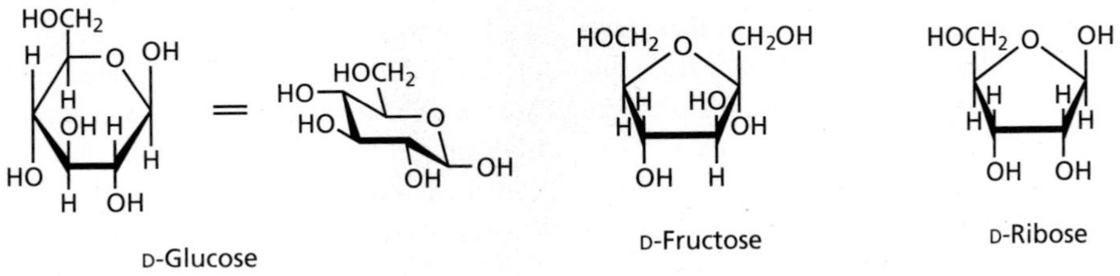

FIGURE 10.21 Examples of monosaccharides.

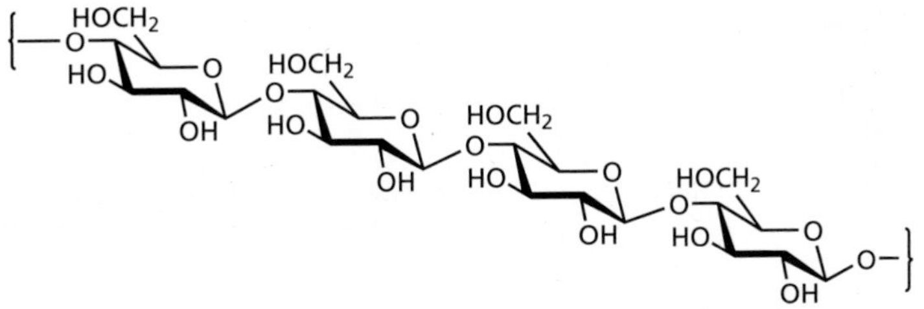

FIGURE 10.22 Cellulose, where glucosyl units are linked β-1,4.

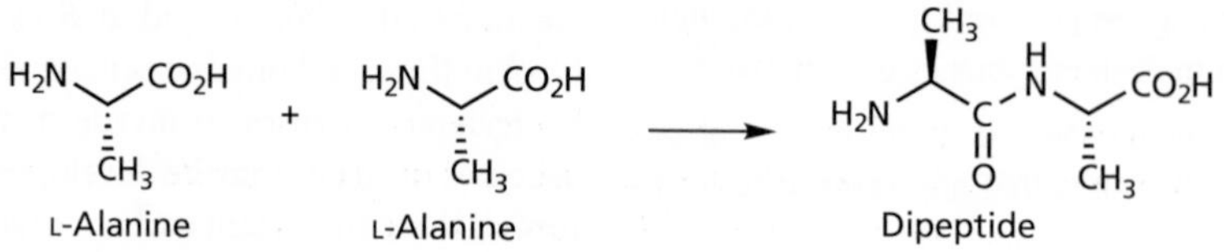

FIGURE 10.23 Dipeptide formed from linking two L-alanines.

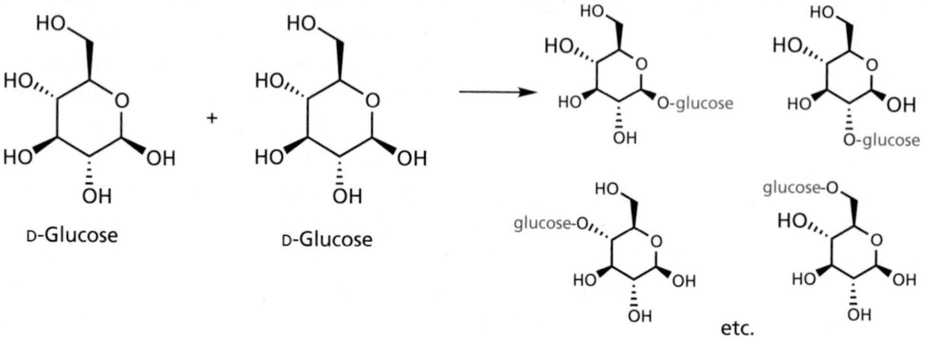

FIGURE 10.24 Variety of carbohydrate structures formed from two glucose molecules.

numbers and types of sugar units. Indeed, it has been calculated that 15 million possible structures can be derived from combining just four carbohydrate monomers.

For additional material see Web article 3: **Glycosphingolipids** on the Online Resource Centre at www.oxfordtextbooks.co.uk/orc/patrick6e/

10.7.2 Antigens and antibodies

The molecular tags that act as cell recognition molecules commonly act as **antigens** if that cell is introduced into a different individual. In other words, they identify that cell as being foreign. For example, bacteria have their own cell recognition molecules which are different from our own. When we suffer a bacterial infection, the immune system recognizes foreign molecular tags and produces **antibodies** which bind to them and trigger an immune response aimed at destroying the invader.

Antibodies are Y-shaped molecules that are made up of two heavy and two light peptide chains (Fig. 10.25). At the *N*-terminals of these chains, there is a highly variable region of amino acids which differs from antibody to antibody. It is this region which recognizes particular antigens. Once an antigen is recognized, the antibody binds to it and recruits the body's immune response to destroy the foreign cell (Fig. 10.26). All cells (including our own) have antigens on their outer surface. They act as a molecular signature for different cells, allowing the body to distinguish between its own cells and 'foreigners'. Fortunately, the body does not normally produce antibodies against its own cells and so we are safe from attack. However, antibodies will be produced against cells from other individuals, and this poses a problem when it comes to organ transplants and blood transfusions. Therefore, it is important to get as close a match as possible between donor and recipient. Immunosuppressant drugs may also be required to allow transplants to be accepted. Another problem can

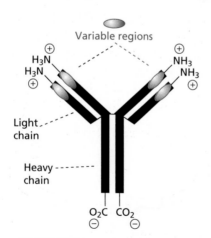

FIGURE 10.25 Structure of an antibody.

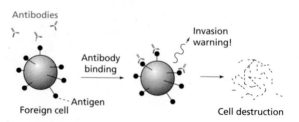

FIGURE 10.26 Role of antibodies in cell destruction.

BOX 10.3 Cyclodextrins as drug scavengers

Sugammadex is a cyclodextrin which has been designed to scavenge the steroidal neuromuscular blocking agents **rocuronium** and **vecuronium** in order to reduce their lifetime in the blood supply. This, in turn, results in faster recovery times for patients who have undergone surgery.

Sugammadex consists of eight identical carbohydrate molecules. The faces of the carbohydrate rings form the interior of the macrocycle creating a relatively hydrophobic environment, while the hydroxyl and carboxylate groups interact with water. This makes the cyclodextrin water soluble.

The dimensions of the cyclodextrin cavity are such that the steroid is neatly encapsulated inside the cyclodextrin ring. The cavity diameter of sugammadex is 7.5–8.3 Å which matches the molecular width of rocuronium (about 7.5 Å). The carboxylate groups help to lock the steroid into the cyclodextrin by forming ionic interactions with the quaternary ammonium ion of the drug.

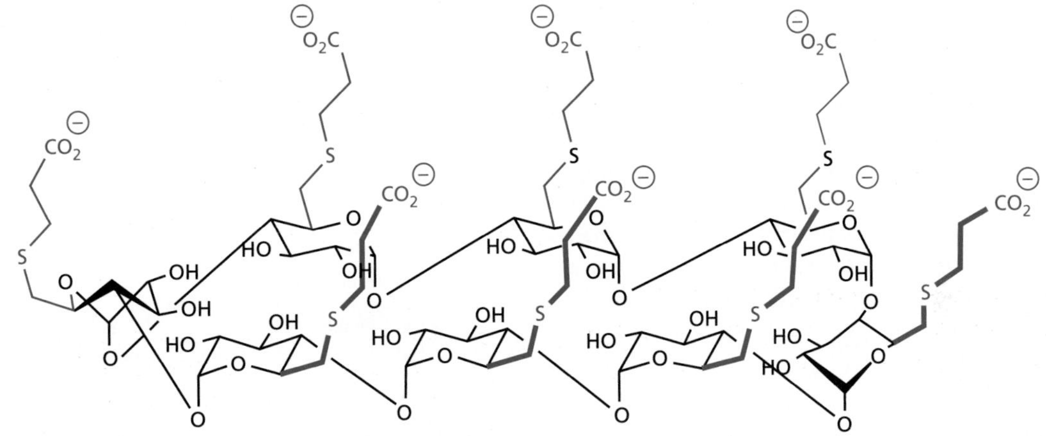

FIGURE 1 Structure of sugammadex.

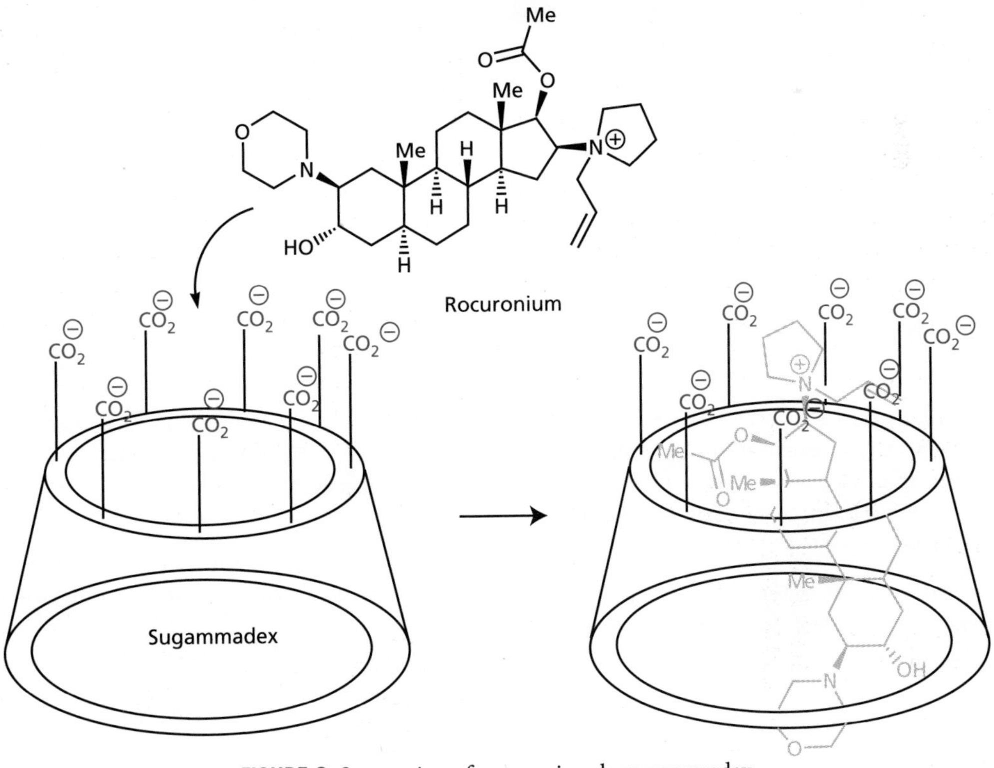

FIGURE 2 Scavenging of rocuronium by sugammadex.

arise when proteins are being used as drugs, as these are large enough to stimulate the immune response.

There has been a lot of progress in using antibodies in the treatment of cancer by producing antibodies which will target antigens that are overexpressed on the surface of cancer cells. They can either be used by themselves to mark cancer cells out for destruction, or as a means of delivering anticancer drugs to cancer cells. This is covered in more detail in sections 14.8.3 and 21.10. Antibodies have also been used in the treatment of autoimmune and inflammatory diseases (section 14.8.3).

10.7.3 Cyclodextrins

Cyclodextrins are macrocyclic structures made up of carbohydrate building blocks. As the interior of cyclodextrins is relatively hydrophobic and can accommodate drug-sized molecules, cyclodextrins have been extensively studied as a means of drug delivery for hydrophobic drugs. Moreover, a novel application has recently been approved for a cyclodextrin called **sugammadex**, which acts as a 'scavenger' for the neuromuscular agents **rocuronium** and **vecuronium** (Box 10.3).

For additional material see Web article 4: Inside the 'doughnut': the versatile chemistry of cyclodextrins on the Online Resource Centre at www.oxfordtextbooks. co.uk/orc/patrick6e/

KEY POINTS

* Carrier proteins transport essential polar molecules across the hydrophobic cell membrane. Drugs can be designed to take advantage of this transport system in order to gain access to cells, or to block the carrier protein.

* Tubulin is a structural protein which is crucial to cell division and cell mobility, and which is the target for several anticancer and anti-inflammatory drugs.

* Viral capsid proteins are promising targets for new antiviral agents.

* Drugs are being designed to inhibit protein–protein interactions. The drugs concerned mimic features of protein secondary structure such as α-helices.

* General anaesthetics target the phospholipid bilayer of cell membranes.

* Several antifungal and antibacterial agents act on the cell membrane of cells. Some agents form tunnels through the cell membrane, while others act as ion carriers. In both situations, an uncontrolled passage of ions or small molecules takes place across the cell membrane leading to cell death.

* Carbohydrates are of increasing importance as drugs or drug targets in developing new therapies for infection, cancer, and autoimmune disease.

* Carbohydrates are more challenging to synthesize than peptides, but offer a greater variety of potential novel structures.

* Antibodies are proteins which are important to the body's immune response and which can identify foreign cells or macromolecules, marking them out for destruction. They have been used therapeutically and can also be used to carry drugs to specific targets.

QUESTIONS

1. The carboxylate groups in sugammadex play an important binding role in locking rocuronium into the central cavity of the cyclodextrin, but they also have an important role in allowing the drug access to the cavity. Suggest possible reasons for this.

2. The carboxylate groups in sugammadex are linked to the carbohydrate rings by a four atom linker chain. Suggest whether a shorter or longer chain would make any difference, and whether there are any advantages in having the linker chain used.

Multiple-choice questions are available on the Online Resource Centre at www.oxfordtextbooks.co.uk/ orc/patrick6e/

FURTHER READING

Berg, C., Neumeyer, K., and Kirkpatrick, P. (2003) Teriparatide. *Nature Reviews Drug Discovery*, **2**, 257–8.

Buolamwini, J. K., et al. (2005) Small molecule antagonists of the MDM2 oncoprotein as anticancer agents. *Current Cancer Drug Targets*, **5**, 57–68.

Dwek, R. A., et al. (2002) Targeting glycosylation as a therapeutic approach. *Nature Reviews Drug Discovery*, **1**, 65–75.

Farina, V. (ed.) (1995) *The chemistry and pharmacology of taxol and its derivatives.* Elsevier, Amsterdam.

Le, G. T., et al. (2003) Molecular diversity through sugar scaffolds. *Drug Discovery Today*, **8**, 701–9.

Maeder, T. (2002) Sweet medicines. *Scientific American*, **287** (1) (July), 24–31.

Ng, E. W. M., et al. (2006) Pegaptanib, a targeted anti-VEGF aptamer for ocular vascular disease. *Nature Reviews Drug Discovery*, **5**, 123–32.

Ojima, I., et al. (2005) Design, synthesis, and structure–activity relationships of novel taxane-based multidrug resistance reversal agents. *Journal of Medicinal Chemistry*, **48**, 2218–28.

Palacios, D. S., et al. (2011) Organic synthesis toward small-molecule probes and drugs special feature: synthesis-enabled functional group deletions reveal key underpinnings of amphotericin B ion channel and antifungal activities. *Proceedings of the National Academy of Sciences USA*, **108**, 6733–8.

Shimogawa, H., et al. (2004) A wrench-shaped synthetic molecule that modulates a transcription factor–coactivator interaction. *Journal of the American Chemical Society*, **126**, 3461–71.

Toogood, P. L. (2002) Inhibition of protein–protein association by small molecules: approaches and progress. *Journal of Medicinal Chemistry*, **45**, 1543–58.

Vassilev, L. T. (2005) p53 activation by small molecules: application in oncology. *Journal of Medicinal Chemistry*, **48**, 4491–9.

Wong, C. (2003) *Carbohydrate-based drug discovery.* John Wiley and Sons, Chichester.

Yin, H., et al. (2005) Terphenyl-based Bak BH3 α-helical proteomimetics as low-molecular-weight antagonists of Bcl-x$_L$. *Journal of the American Chemical Society*, **127**, 10191–6.

Pharmacokinetics and related topics

11.1 The three phases of drug action

There are three phases involved in drug action. The first of these is the **pharmaceutical phase**. For an orally administered drug, this includes the disintegration of a pill or capsule in the gastrointestinal tract (GIT), the release of the drug, and its dissolution. The pharmaceutical phase is followed by the **pharmacokinetic phase**, which includes absorption from the GIT into the blood supply, and the various factors that affect a drug's survival and progress as it travels to its molecular target. The final **pharmacodynamic phase** involves the mechanism by which a drug interacts with its molecular target and the resulting pharmacological effect.

In previous chapters, we have focused on drug targets and drug design, where the emphasis is on the pharmacodynamic aspects of drug action; for example, optimizing the binding interactions of a drug with its target. However, the compound with the best binding interactions for a target is not necessarily the best drug to use in medicine. This is because a drug has to reach its target in the first place if it is to be effective. Therefore, when carrying out a drug design programme, it is important to study pharmacokinetics alongside pharmacodynamics. The four main topics to consider in pharmacokinetics are absorption, distribution, metabolism, and excretion (often abbreviated to ADME).

11.2 A typical journey for an orally active drug

The preferred method of drug administration is the oral route, and so we shall consider some of the hurdles and hazards faced by such a drug in order to reach its eventual target. When a drug is swallowed, it enters the **gastrointestinal tract** (GIT), which comprises the mouth, throat, stomach, and the upper and lower intestines. A certain amount of the drug may be absorbed through the mucosal membranes of the mouth, but most passes down into the stomach where it encounters gastric juices and hydrochloric acid. These chemicals aid in the digestion of food and will treat a drug in a similar fashion if it is susceptible to breakdown and is not protected within an acid-resistant pill or capsule. For example, the first clinically useful penicillin was broken down in the stomach and had to be administered by injection. Other acid-labile drugs include the **local anaesthetics** and **insulin**. If the drug *does* survive the stomach, it enters the upper intestine where it encounters digestive enzymes that serve to break down food. Assuming the drug survives this attack, it then has to pass through the cells lining the gut wall. This means that it has to pass through a cell membrane on two occasions, first to enter the cell and then to exit it on the other side. Once the drug has passed through the cells of the gut wall, it can enter the blood supply relatively easily as the cells lining the blood vessels are loose fitting and there are pores through which most drugs can pass. In other words, drugs enter blood vessels by passing between cells rather than through them.

The drug is now transported in the blood to the body's 'customs office'—the liver. The liver contains enzymes which are ready and waiting to intercept foreign chemicals, and modify them such that they are more easily excreted—a process called drug metabolism (section 11.5). Following this, the drug has to be carried by the blood supply round the body to reach its eventual target, which may require crossing further cell membranes—always assuming that it is neither excreted before it gets there, nor diverted to parts of the body where it is not needed.

It can be seen that stringent demands are made on any orally administered drug. It must be stable to both chemical and enzymatic attack. It must also have the correct physicochemical properties to allow it to reach its target in therapeutic concentrations. This includes efficient absorption, effective distribution to target tissues, and an acceptable rate of excretion. We will now look more closely at the various stages.

11.3 **Drug absorption**

In order to be absorbed efficiently from the GIT, a drug must have the correct balance of water versus fat solubility. If the drug is too polar (hydrophilic), it will fail to pass through the fatty cell membranes of the gut wall (section 1.2.1). On the other hand, if the drug is too fatty (hydrophobic), it will be poorly soluble in the gut and will dissolve in fat globules. This means that there will be poor surface contact with the gut wall, resulting in poor absorption.

It is noticeable how many drugs contain an amine functional group. There are good reasons for this. Amines are often involved in a drug's binding interactions with its target. However, they are also an answer to the problem of balancing the dual requirements of water and fat solubility. Amines are weak bases, and it is found that many of the most effective drugs contain amine groups having a pK_a value in the range 6–8. In other words, they are partially ionized at the slightly acidic and alkaline pHs present in the intestine and blood respectively, and can easily equilibrate between their ionized and non-ionized forms. This allows them to cross cell membranes in the non-ionized form, while the presence of the ionized form gives the drug good water solubility and permits good binding interactions with its target binding site (Fig. 11.1).

The extent of ionization at a particular pH can be determined by the **Henderson–Hasselbalch equation**:

$$pH = pK_a + \log\frac{[RNH_2]}{[RNH_3^+]}$$

where $[RNH_2]$ is the concentration of the free base and $[RNH_3^+]$ is the concentration of the ionized amine. K_a is the equilibrium constant for the equilibrium shown in Fig. 11.1, and the Henderson–Hasselbalch equation can be derived from the equilibrium constant:

$$K_a = \frac{[H^+][RNH_2]}{[RNH_3^+]}$$

$$\text{Therefore } pK_a = -\log\frac{[H^+][RNH_2]}{[RNH_3^+]}$$

$$= -\log[H^+] - \log\frac{[RNH_2]}{[RNH_3^+]}$$

$$= pH - \log\frac{[RNH_2]}{[RNH_3^+]}$$

$$\text{Therefore } pH = pK_a + \log\frac{[RNH_2]}{[RNH_3^+]}$$

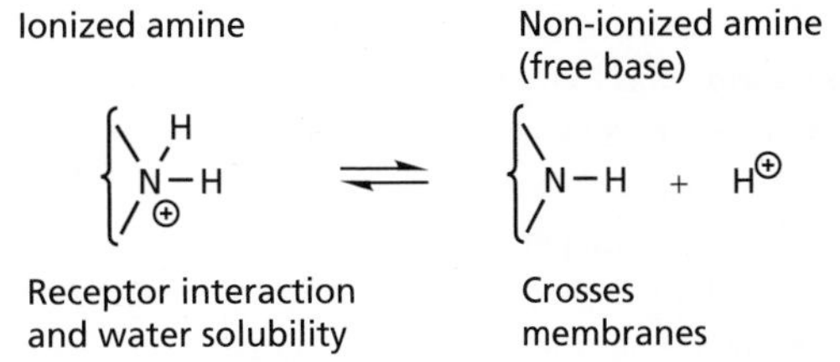

FIGURE 11.1 Equilibrium between the ionized and non-ionized form of an amine.

Note that when the concentration of the ionized and unionized amines are identical (i.e. when $[RNH_2] = [RNH_3^+]$), the ratio $[RNH_2]/[RNH_3^+]$ is 1. Since $\log 1 = 0$, the Henderson–Hasselbalch equation will simplify to pH $= pK_a$. In other words, when the amine is 50% ionized, pH $= pK_a$. Therefore, drugs with a pK_a of 6–8 are approximately 50% ionized at blood pH (7.4) or the slightly acidic pH of the intestines.

The hydrophilic/hydrophobic character of the drug is the crucial factor affecting absorption through the gut wall; in theory the molecular weight of the drug should be irrelevant. For example, **ciclosporin** is successfully absorbed through cell membranes, although it has a molecular weight of about 1200. In practice, however, larger molecules tend to be poorly absorbed because they are likely to contain a large number of polar functional groups. As a rule of thumb, orally absorbed drugs tend to obey what is known as Lipinski's **rule of five**. The rule of five was derived from an analysis of compounds from the World Drugs Index database, aimed at identifying features that were important in making a drug orally active. It was found that the factors concerned involved numbers that are multiples of 5:

- a molecular weight less than 500
- no more than 5 hydrogen bond donor (HBD) groups
- no more than 10 hydrogen bond acceptor (HBA) groups
- a calculated **log P** value less than +5 (log P is a measure of a drug's hydrophobicity—section 14.1).

The rule of five has been an extremely useful rule of thumb for many years, but it is neither quantitative nor foolproof. For example, orally active drugs such as **atorvastatin**, **rosuvastatin**, **ciclosporin**, and **vinorelbine** do not obey the rule of five. It has also been demonstrated that a high molecular weight does not in itself cause poor oral bioavailability. Another source of debate concerns the calculation of the number of HBAs. In Lipinski's original paper, the number of HBAs corresponded to the total number of oxygen and nitrogen atoms present in a structure. This was done for simplicity's sake, but most medicinal chemists would discount weak HBAs such as amide nitrogens (see also section 1.3.2 and Appendix 7). Therefore, it is better to view Lipinski's rules as a set of guidelines rather than rules. Lipinski himself stated that a compound was likely to be orally active as long as it did not break more than one of his 'rules'.

Further research has been carried out to find guidelines that are independent of molecular weight. Work carried out by Veber et al. in 2002 demonstrated the rather surprising finding that molecular flexibility plays an important role in oral bioavailability; the more flexible the molecule, the less likely it is to be orally active. In order to measure flexibility, one can count the number of freely rotatable bonds that result in significantly different conformations. Bonds to simple substituents such as methyl or alcohol groups are not included in this analysis as their rotation does not result in significantly different conformations.

Veber's studies also demonstrated that the polar surface area of the molecule could be used as a factor instead of the number of hydrogen bonding groups. These findings led to the following parameters for predicting acceptable oral activity. Either:

- a polar surface area ≤ 140 Å and ≤ 10 rotatable bonds;

or

- ≤ 12 hydrogen bond donors and acceptors in total and ≤ 10 rotatable bonds.

Some researchers set the limit of rotatable bonds to ≤ 7 since the analysis shows a marked improvement in oral bioavailability for such molecules.

These rules are independent of molecular weight and open the way to studying larger structures that have been 'shelved' up to now. Unfortunately, structures having a molecular weight larger than 500 are quite likely to have more than 10 rotatable bonds. However, the new rules suggest that rigidifying the structures to reduce the number of rotatable bonds would be beneficial. Rigidification tactics are described in section 13.3.9 as a strategy to improve a drug's pharmacodynamic properties, but these same tactics could also be used to improve pharmacokinetic properties. Appendix 9 (available on the website) provides information on MWt, log P, HBDs, HBAs, rotatable bonds, and polar surface area for several of the drugs covered in this text.

Polar drugs that break the above rules are usually poorly absorbed and have to be administered by injection.

Nevertheless, some highly polar drugs *are* absorbed from the digestive system as they are able to 'hijack' **transport proteins** present in the membranes of cells lining the gut wall (sections 2.7.2 and 10.1). These transport proteins normally transport the highly polar building blocks required for various biosynthetic pathways (e.g. amino acids and nucleic acid bases) across cell membranes. If the drug bears a structural resemblance to one of these building blocks, then it, too, may be smuggled across. For example, **levodopa** is transported by the transport protein for the amino acid phenylalanine, while **fluorouracil** is transported by transport proteins for the nucleic acid bases thymine and uracil. The antihypertensive agent **lisinopril** is transported by transport proteins for dipeptides. The anticancer agent **methotrexate** and the antibiotic **erythromycin** are also absorbed by means of transport proteins.

Other highly polar drugs can be absorbed into the blood supply if they have a low molecular weight (less than 200), as they can then pass through small pores between the cells lining the gut wall.

Occasionally, polar drugs with high molecular weight can cross the cells of the gut wall without actually passing through the membrane. This involves a process known as **pinocytosis** where the drug is engulfed by the cell membrane and a membrane-bound vesicle is pinched off to carry the drug across the cell (Fig. 11.2). The vesicle then fuses with the membrane to release the drug on the other side of the cell.

Sometimes, drugs are deliberately designed to be highly polar so that they are *not* absorbed from the GIT. These are usually antibacterial agents targeted against gut infections. Making them highly polar ensures that the drug reaches the site of infection in higher concentration (Box 19.2).

Finally, it should be noted that the absorption of some drugs can be affected adversely by interactions with food or other drugs in the gut (section 11.7.1).

Other drug administration routes may involve an absorption process and this is discussed in section 11.7.

@ For additional information see Web article 25: Looking at medicinal chemistry post Lipinski on the Online Resource Centre at www.oxfordtextbooks.co.uk/orc/patrick6e/

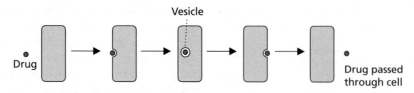

FIGURE 11.2 Pinocytosis.

11.4 **Drug distribution**

Once a drug has been absorbed, it is rapidly distributed around the blood supply, then more slowly distributed to the various tissues and organs. The rate and extent of distribution depends on various factors, including the physical properties of the drug itself.

11.4.1 **Distribution round the blood supply**

The vessels carrying blood round the body are called **arteries**, **veins**, and **capillaries** (section 26.2). The heart is the pump that drives the blood through these vessels. The major artery carrying blood from the heart is called the **aorta**, and as it moves further from the heart, it divides into smaller and smaller arteries—similar to the limbs and branches radiating from the trunk of a tree. Eventually, the blood vessels divide to such an extent that they become extremely narrow—equivalent to the twigs of a tree. These blood vessels are called capillaries, and it is from these vessels that oxygen, nutrients, and drugs can escape in order to reach the tissues and organs of the body. At the same time, waste products such as cell breakdown products and carbon dioxide are transferred from the tissues into the capillaries to be carried away and disposed of. The capillaries now start uniting into bigger and bigger vessels, resulting in the formation of veins which return the blood to the heart.

Once a drug has been absorbed into the blood supply, it is rapidly and evenly distributed throughout the blood supply within a minute—the time taken for the blood volume to complete one circulation. However, this does not mean that the drug is evenly distributed around the body—the blood supply is richer to some areas of the body than to others.

11.4.2 **Distribution to tissues**

Drugs do not stay confined to the blood supply. If they did, they would be of little use since their targets are the cells of various organs and tissues. The drug has to leave the blood supply in order to reach those targets. The body has an estimated 10 billion capillaries with a total surface area of 200 m^2. They probe every part of the body, such that no cell is more than 20–30 μm away from a capillary. Each capillary is very narrow, not much wider than the red blood cells that pass through it. Its walls are made up of a thin single layer of cells packed tightly together. However, there are pores between the cells which are 90–150 Å in diameter—large enough to allow most drug-sized molecules to pass though, but not large enough to allow the **plasma proteins** present in blood to escape. Therefore, drugs do not have to cross cell membranes in order to leave the blood system and can be freely and rapidly distributed into the aqueous fluid surrounding the various tissues and organs of the body. Having said that, some drugs bind to plasma proteins in the blood. Since the plasma proteins cannot leave the capillaries, the proportion of drug bound to these proteins is also confined to the capillaries and cannot reach its target.

11.4.3 **Distribution to cells**

Once a drug has reached the tissues, it can immediately be effective if its target site is a receptor situated in a cell membrane. However, there are many drugs that have to enter the individual cells of tissues in order to reach their target. These include local anaesthetics, enzyme inhibitors, and drugs which act on nucleic acids or intracellular receptors. Such drugs must be hydrophobic enough to pass through the cell membrane, unless they are smuggled through by carrier proteins or taken in by pinocytosis. Since many drugs contain an amine functional group, the same principles described in section 11.3 apply. The drug must pass through the cell membrane as the free base, but, once it is inside the cell, the amine may become protonated to allow a strong interaction with the target binding site. Experiments have demonstrated this with a number of drugs such as **verapamil** (section 26.6.3).

11.4.4 **Other distribution factors**

The concentration levels of free drug circulating in the blood supply rapidly fall away after administration as a result of the distribution patterns described above, but there are other factors at work. Drugs that are excessively hydrophobic are often absorbed into fatty tissues and removed from the blood supply. This fat solubility can lead to problems. For example, obese patients undergoing surgery require a larger than normal volume of general anaesthetic because the gases used are particularly fat soluble. Unfortunately, once surgery is over and the patient has regained consciousness, the anaesthetics stored in the fat tissues will be released and may render the patient unconscious again. **Barbiturates** were once seen as potential intravenous anaesthetics which could replace the anaesthetic gases. Unfortunately, they, too, are fat soluble and it is extremely difficult to estimate a sustained safe dosage. The initial dose can be estimated to allow for the amount of barbiturate taken up by fat cells, but further doses eventually lead to saturation of the fat depot and result in a sudden, and perhaps fatal, increase of barbiturate levels in the blood supply.

Ionized drugs may become bound to various macromolecules and be removed from the blood supply. Drugs may also be bound reversibly to blood plasma proteins such as **albumin**, thus lowering the level of free drug.

Therefore, only a small proportion of the drug that has been administered may actually reach the desired target.

11.4.5 Blood–brain barrier

The blood–brain barrier is an important barrier that drugs have to negotiate if they are to enter the brain. The blood capillaries feeding the brain are lined with tight-fitting cells which do not contain pores (unlike capillaries elsewhere in the body). Moreover, the capillaries are coated with a fatty layer formed from nearby cells, providing an extra fatty barrier through which drugs have to cross. Therefore, drugs entering the brain have to dissolve through the cell membranes of the capillaries and also through the fatty cells coating the capillaries. As a result, polar drugs such as **penicillin** do not easily enter the brain.

The existence of the blood–brain barrier makes it possible to design drugs which will act at various parts of the body (e.g. the heart) and have no activity in the brain, thus reducing any central nervous system (CNS) side effects. This is done by increasing the polarity of the drug such that it does not cross the blood–brain barrier. On the other hand, drugs that are intended to act in the brain must be designed such that they *are* able to cross the blood–brain barrier. This means that they are limited in the number of polar groups that they can have. Alternatively, polar groups may have to be temporarily masked in order to allow passage through the blood–brain barrier (see prodrugs; section 14.6). Having said that, some polar drugs can cross the blood–brain barrier with the aid of carrier proteins, while others (e.g. **insulin**) can cross by the process of pinocytosis previously described. The ability to cross the blood–brain barrier has an important bearing on the analgesic activity of opioids (section 24.5). Research is also being carried out to find ways of increasing the permeability of the blood–brain barrier using techniques such as ultrasound or drugs such as **sildenafil**.

11.4.6 Placental barrier

The placental membranes separate a mother's blood from the blood of her fetus. The mother's blood provides the fetus with essential nutrients and carries away waste products, but these chemicals must pass through the placental barrier. As food and waste products can pass through the placental barrier, it is perfectly feasible for drugs to pass through as well. Drugs such as **alcohol**, **nicotine**, and **cocaine** can all pass into the fetal blood supply. Fat-soluble drugs will cross the barrier most easily, and drugs such as **barbiturates** will reach the same levels in fetal blood as in maternal blood. Such levels may have unpredictable effects on fetal

development. They may also prove hazardous once the baby is born. Before birth, drugs and other toxins can be removed from fetal blood by the maternal blood and detoxified. Once the baby is born, it may have the same levels of drugs in its blood as the mother, but it does not have the same ability to detoxify or eliminate them. As a result, drugs will have a longer lifetime and may have fatal effects.

11.4.7 Drug–drug interactions

Drugs such as **warfarin** and **methotrexate** are bound to albumin and plasma proteins in the blood, and are unavailable to interact with their targets. When another drug is taken which can compete for plasma protein binding (e.g. **sulphonamides**), then a certain percentage of previously bound drug is released, increasing the concentration of the drug and its effect.

KEY POINTS

- Pharmacodynamics is the study of how drugs interact with a molecular target to produce a pharmacological effect, whereas pharmacokinetics is the study of how a drug reaches its target in the body and how it is affected on that journey.
- The four main issues in pharmacokinetics are absorption, distribution, metabolism, and excretion.
- Orally taken drugs have to be chemically stable to survive the acidic conditions of the stomach, and metabolically stable to survive digestive and metabolic enzymes.
- Orally taken drugs must be sufficiently polar to dissolve in the GIT and blood supply, but sufficiently fatty to pass through cell membranes.
- Most orally taken drugs obey Lipinski's rule of five and have no more than seven rotatable bonds.
- Highly polar drugs can be orally active if they are small enough to pass between the cells of the gut wall, are recognized by carrier proteins, or are taken across the gut wall by pinocytosis.
- Distribution round the blood supply is rapid. Distribution to the interstitial fluid surrounding tissues and organs is also rapid if the drug is not bound to plasma proteins.
- Some drugs have to enter cells in order to reach their target.
- A certain percentage of a drug may be absorbed into fatty tissue and/or bound to macromolecules.
- Drugs entering the CNS have to cross the blood–brain barrier. Polar drugs are unable to cross this barrier unless they make use of carrier proteins or are taken across by pinocytosis.
- Some drugs cross the placental barrier into the fetus and may harm development or prove toxic in newborn babies.

11.5 Drug metabolism

When drugs enter the body, they are subject to attack from a range of metabolic enzymes. The role of these enzymes is to degrade or modify the foreign structure, such that it can be more easily excreted. As a result, most drugs undergo some form of metabolic reaction, resulting in structures known as **metabolites**. Very often these metabolites lose the activity of the original drug, but, in some cases, they may retain a certain level of activity. In exceptional cases, the metabolite may even be more active than the parent drug. Some metabolites can possess a different activity from the parent drugs, resulting in side effects or toxicity. A knowledge of drug metabolism and its possible consequences can aid the medicinal chemist in designing new drugs which do not form unacceptable metabolites. Equally, it is possible to take advantage of drug metabolism to activate drugs in the body. This is known as a prodrug strategy (see section 14.6). It is now a requirement to identify all the metabolites of a new drug before it can be approved. The structure and stereochemistry of each metabolite has to be determined and each metabolite must be tested for biological activity (section 15.1.2).

11.5.1 Phase I and phase II metabolism

The body treats drugs as foreign substances and has methods of getting rid of such chemical invaders. If the drug is polar, it will be quickly excreted by the kidneys (section 11.6). However, non-polar drugs are not easily excreted and the purpose of drug metabolism is to convert such compounds into more polar molecules that *can* be easily excreted.

Non-specific enzymes (particularly **cytochrome P450 enzymes** in the liver) are able to add polar functional groups to a wide variety of drugs. Once the polar functional group has been added, the overall drug is more polar and water soluble, and is more likely to be excreted when it passes through the kidneys. An alternative set of enzymatic reactions can reveal masked polar functional groups which might already be present in a drug. For example, there are enzymes which can demethylate a methyl ether to reveal a more polar hydroxyl group. Once again, the more polar product (metabolite) is excreted more efficiently.

These reactions are classed as phase I reactions and generally involve oxidation, reduction, and hydrolysis (see Figs. 11.3–11.9). Most of these reactions occur in the liver, but some (such as the hydrolysis of esters and amides) can also occur in the gut wall, blood plasma, and other tissues. Some of the structures most prone to oxidation are *N*-methyl groups, aromatic rings, the terminal positions of alkyl chains, and the least hindered positions of alicyclic rings. Nitro, azo, and carbonyl groups

are prone to reduction by **reductases**, while amides and esters are prone to hydrolysis by **peptidases** and **esterases** respectively. For many drugs, two or more metabolic reactions might occur, resulting in different metabolites; other drugs may not be metabolized at all. A knowledge of the metabolic reactions that are possible for different functional groups allows the medicinal chemist to predict the likely metabolic products for any given drug, but only drug metabolism studies will establish whether these metabolites are really formed.

Drug metabolism has important implications when it comes to using chiral drugs, especially if the drug is to be used as a racemate. The enzymes involved in catalysing metabolic reactions will often distinguish between the two enantiomers of a chiral drug, such that one enantiomer undergoes different metabolic reactions from the other. As a result, both enantiomers of a chiral drug have to be tested separately to see what metabolites are formed. In practice, it is usually preferable to use a single enantiomer in medicine, or design the drug such that it is not asymmetric (section 13.3.8).

A series of metabolic reactions classed as phase II reactions also occur, mainly in the liver (see Figs. 11.10–11.16). Most of these reactions are **conjugation reactions**, whereby a polar molecule is attached to a suitable polar 'handle' that is already present on the drug or has been introduced by a phase I reaction. The resulting conjugate has greatly increased polarity, thus increasing its excretion rate in urine or bile even further.

Both phase I and phase II reactions can be species specific, which has implications for *in vivo* metabolic studies. In other words, the metabolites formed in an experimental animal may not necessarily be those formed in humans. A good knowledge of how metabolic reactions differ from species to species is important in determining which test animals are relevant for drug metabolism tests. Both sets of reactions can also be regioselective and stereoselective. This means that metabolic enzymes can distinguish between identical functional groups or alkyl groups located at different parts of the molecule (regioselectivity) as well as between different stereoisomers of chiral molecules (stereoselectivity).

11.5.2 Phase I transformations catalysed by cytochrome P450 enzymes

The enzymes that constitute the cytochrome P450 family are the most important metabolic enzymes and are located in liver cells. They are **haemoproteins** (containing haem and iron) and they catalyse a reaction that splits molecular oxygen, such that one of the oxygen atoms is introduced into the drug and the other ends up in water (Fig. 11.3). As a result they belong to a general class of enzymes called the **monooxygenases**.

$$Drug-H + O_2 + NADPH + H^+ \xrightarrow{\substack{\text{Cytochrome P450} \\ \text{enzymes}}} Drug-OH + NADP^+ + H_2O$$

FIGURE 11.3 Oxidation by cytochrome P450 enzymes.

There are at least 33 different cytochrome P450 enzymes, grouped into four main families, CYP1–CYP4. Within each family there are various subfamilies designated by a letter, and each enzyme within that subfamily is designated by a number. For example, CYP3A4 is enzyme 4 in the sub family A of the main family 3. Most drugs in current use are metabolized by five primary CYP enzymes (CYP3A, CYP2D6, CYP2C9, CYP1A2, and CYP2E1). The isozyme CYP3A4 is particularly important in drug metabolism and is responsible for the metabolism of most drugs. The reactions catalysed by cytochrome P450

enzymes are shown in Figs. 11.4 and 11.5 and can involve the oxidation of carbon, nitrogen, phosphorus, sulphur, and other atoms.

Oxidation of carbon atoms can occur if the carbon atom is either exposed (i.e. easily accessible to the enzyme) or activated (Fig. 11.4). For example, methyl substituents on the carbon skeleton of a drug are often easily accessible and are oxidized to form alcohols, which may be oxidized further to carboxylic acids. In the case of longer-chain substituents, the terminal carbon and the penultimate carbon are the most exposed carbons in

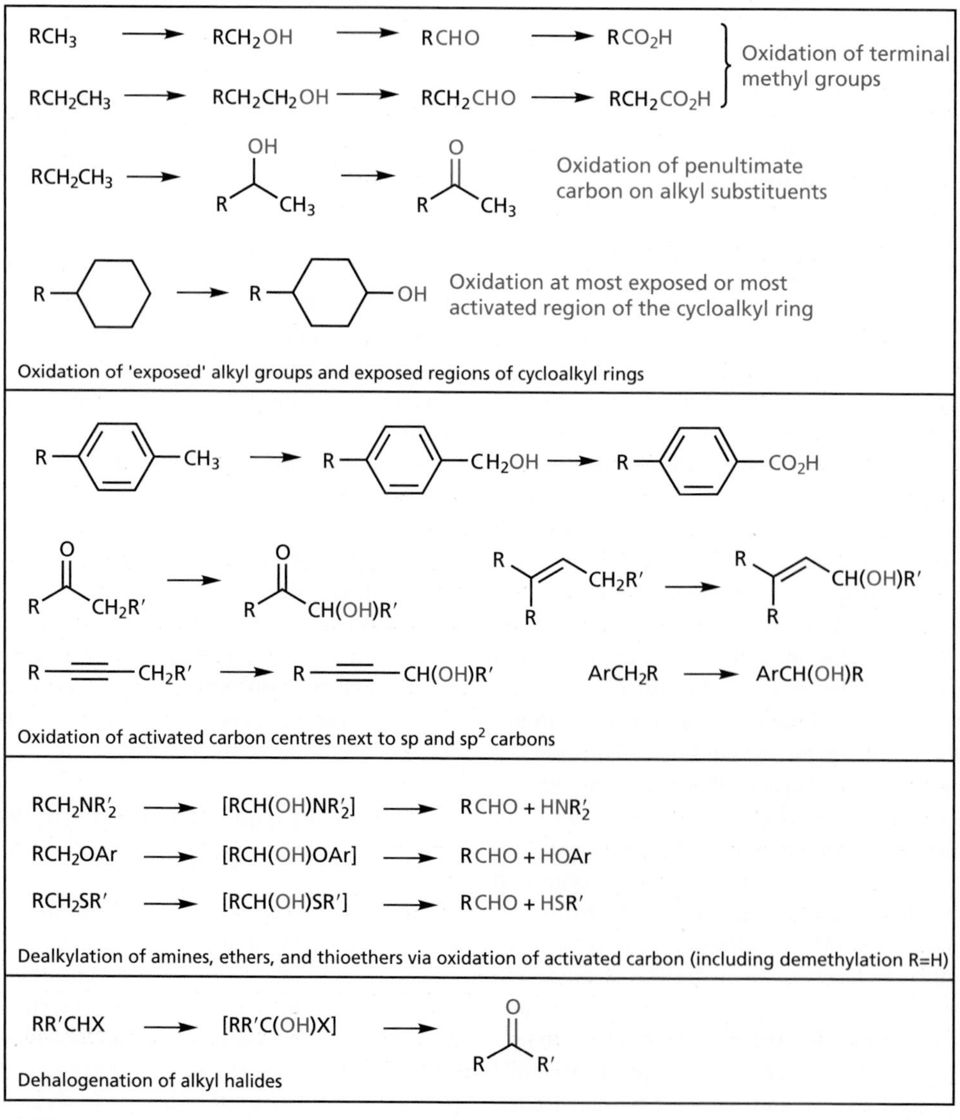

FIGURE 11.4 Oxidative reactions catalysed by cytochrome P450 enzymes on saturated carbon centres.

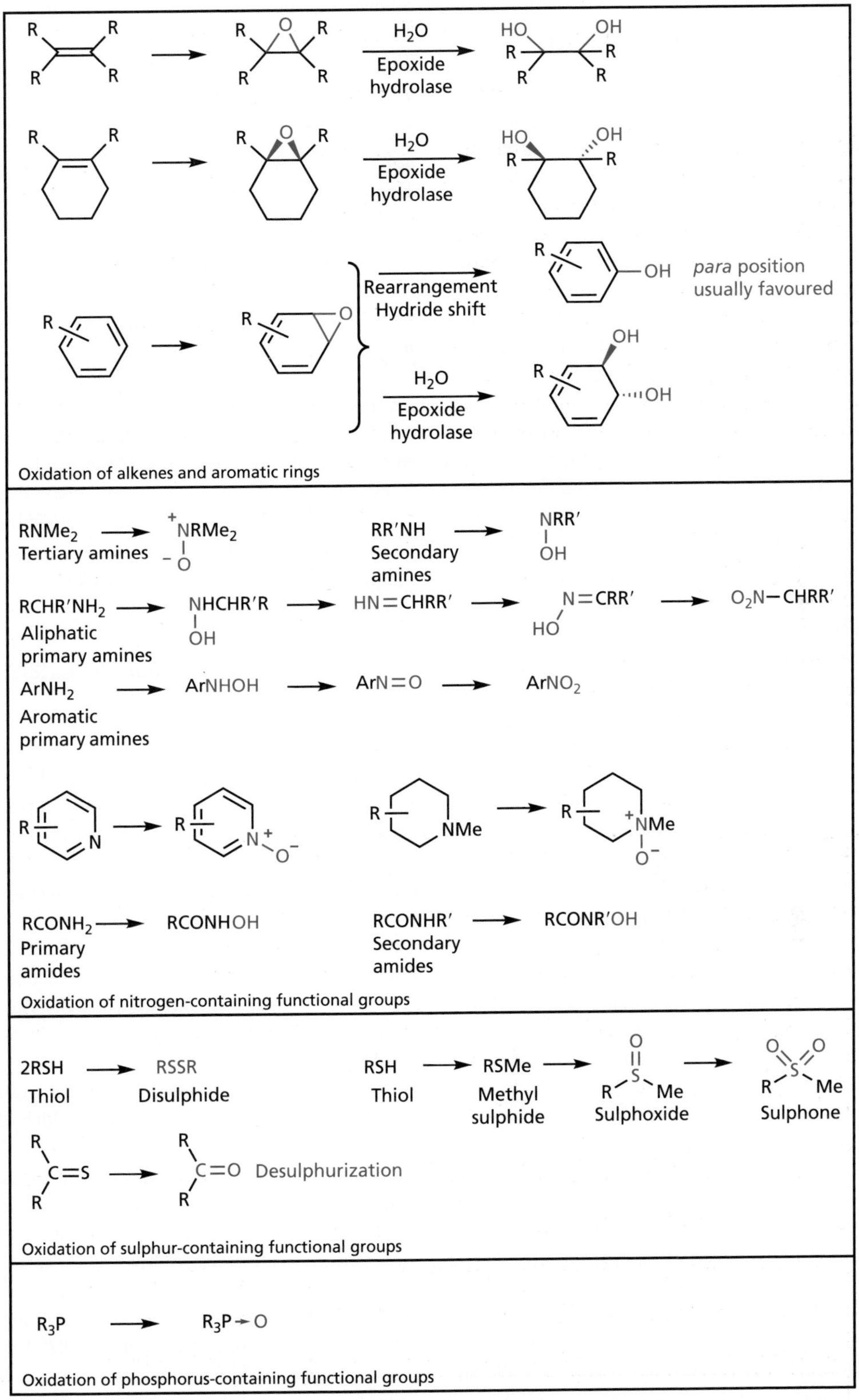

FIGURE 11.5 Oxidative reactions catalysed by cytochrome P450 enzymes on heteroatoms and unsaturated carbon centres.

the chain and are both susceptible to oxidation. If an aliphatic ring is present, the most exposed region is the part most likely to be oxidized.

Activated carbon atoms next to an sp^2 carbon centre (i.e. allylic or benzylic positions) or an sp carbon centre (i.e. a propynylic position) are more likely to be oxidized than exposed carbon atoms (Fig. 11.4). Carbon atoms which are alpha to a heteroatom are also activated and prone to oxidation. In this case, hydroxylation results in an unstable metabolite that is immediately hydrolysed resulting in the dealkylation of amines, ethers, and thioethers, or the dehalogenation of alkyl halides. The aldehydes which are formed from these reactions generally undergo further oxidation to carboxylic acids by aldehyde dehydrogenases (section 11.5.4). Tertiary amines are found to be more reactive to oxidative dealkylation than secondary amines because of their greater basicity, while O-demethylation of aromatic ethers is faster than O-dealkylation of larger alkyl groups. O-Demethylation is important to the analgesic activity of **codeine** (section 24.5).

Cytochrome P450 enzymes can catalyse the oxidation of unsaturated sp^2 and sp carbon centres present in alkenes, alkynes, and aromatic rings (Fig. 11.5). In the case of alkenes, a reactive epoxide is formed which is deactivated by the enzyme **epoxide hydrolase** to form a diol. In some cases, the epoxide may evade the enzyme. If this happens, it can act as an alkylating agent and react with nucleophilic groups present in proteins or nucleic acids, leading to toxicity. The oxidation of an aromatic ring results in a similarly reactive epoxide intermediate which can have several possible fates. It may undergo a rearrangement reaction involving a hydride transfer to form a phenol, normally at the *para* position. Alternatively, it may be deactivated by epoxide hydrolase to form a diol, or react with **glutathione S-transferase** to form a conjugate (section 11.5.5). If the epoxide intermediate evades these enzymes, it may act as an alkylating agent and prove toxic. Electron-rich aromatic rings are likely to be epoxidized more quickly than those with electron-withdrawing substituents, and this has consequences for drug design.

Tertiary amines are oxidized to N-oxides as long as the alkyl groups are not sterically demanding. Primary and secondary amines are also oxidized to N-oxides, but these are rapidly converted to hydroxylamines and beyond. Aromatic primary amines are also oxidized in stages to aromatic nitro groups—a process which is related to the toxicity of aromatic amines, as highly electrophilic intermediates are formed which can alkylate proteins or nucleic acids. Aromatic primary amines can also be methylated in a phase II reaction (section 11.5.5) to a secondary amine which can then undergo phase I oxidation to produce formaldehyde and primary hydroxylamines. Primary and secondary amides can be oxidized to hydroxylamides. These functional groups have also been linked with toxicity and carcinogenicity. Thiols can be oxidized to disulphides. There is evidence that thiols can be methylated to methyl sulphides, which are then oxidized to sulphides and sulphones.

For additional information see Web article 5: The design of a serotonin antagonist as a possible anxiolytic agent on the Online Resource Centre at www.oxfordtextbooks.co.uk/orc/patrick6e/

11.5.3 Phase I transformations catalysed by flavin-containing monooxygenases

Another group of metabolic enzymes present in the endoplasmic reticulum of liver cells consists of the **flavin-containing monooxygenases.** These enzymes are chiefly responsible for metabolic reactions involving oxidation at nucleophilic nitrogen, sulphur, and phosphorus atoms, rather than at carbon atoms. Several examples are given in Fig. 11.6. Many of these reactions are also catalysed by cytochrome P450 enzymes.

11.5.4 Phase I transformations catalysed by other enzymes

There are several oxidative enzymes in various tissues around the body that are involved in the metabolism of endogenous compounds, but can also play a role in drug metabolism (Fig. 11.7). For example, **monoamine oxidases** are involved in the deamination of catecholamines (section 23.5), but have been observed to oxidize some drugs. Other important oxidative enzymes include alcohol dehydrogenases and aldehyde dehydrogenases. The aldehydes formed by the action of alcohol dehydrogenases on primary alcohols are usually not observed, as they are converted to carboxylic acids by aldehyde dehydrogenases.

Reductive phase I reactions are less common than oxidative reactions, but reductions of aldehyde, ketone, azo, and nitro functional groups have been observed in specific drugs (Fig. 11.8). Many of the oxidation reactions described for heteroatoms in Figs. 11.5–11.7 are reversible, and are catalysed by reductase enzymes. Cytochrome P450 enzymes are involved in catalysing some of these reactions. Remember: enzymes can catalyse a reaction in both directions, depending on the nature of the substrate. So although cytochrome P450 enzymes are predominantly oxidative enzymes, it is possible for them to catalyse some reductions.

The hydrolysis of esters and amides is a common metabolic reaction, catalysed by **esterases** and **peptidases** respectively (Fig. 11.9). These enzymes are present in various organs of the body including the liver. Amides

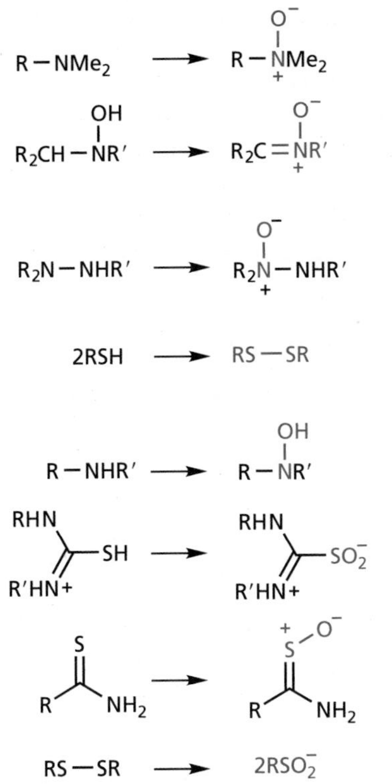

FIGURE 11.6 Phase I reactions catalysed by flavin monooxygenases.

tend to be hydrolysed more slowly than esters. The presence of electron-withdrawing groups can increase the susceptibility of both amides and esters to hydrolysis.

11.5.5 Phase II transformations

Most phase II reactions are **conjugation reactions** catalysed by transferase enzymes. The resulting conjugates are usually inactive, but there are exceptions to this rule. Glucuronic acid conjugation is the most common of these reactions. Phenols, alcohols, hydroxylamines, and carboxylic acids form **O-glucuronides** by reaction with **UDFP-glucuronate** such that a highly polar glucuronic acid molecule is attached to the drug (Fig. 11.10). The resulting conjugate is excreted in the urine, but may also be excreted in the bile if the molecular weight is over 300.

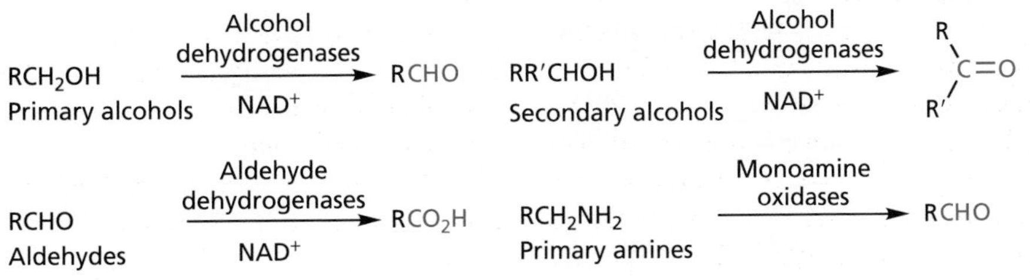

A variety of other functional groups such as sulphonamides, amides, amines, and thiols (Fig. 11.11) can react to form *N*- or *S*-glucuronides. *C*-Glucuronides are also possible in situations where there is an activated carbon centre next to carbonyl groups.

Another form of conjugation is sulphate conjugation (Fig. 11.12). This is less common than glucuronidation and is restricted mainly to phenols, alcohols, arylamines, and *N*-hydroxy compounds. The reaction is catalysed by **sulphotransferases** using the cofactor **3′-phosphoadenosine 5′-phosphosulphate** as the sulphate source. Primary and secondary amines, secondary alcohols, and phenols form stable conjugates, whereas primary alcohols form reactive sulphates which can act as toxic alkylating agents. Aromatic hydroxylamines and hydroxylamides also form unstable sulphate conjugates that can be toxic.

Drugs bearing a carboxylic acid group can become conjugated to amino acids by the formation of a peptide link. In most animals, glycine conjugates are generally formed, but L-glutamine is the most common amino acid used for conjugation in primates. The carboxylic acid present in the drug is first activated by formation of a coenzyme A thioester which is then linked to the amino acid (Fig. 11.13).

Electrophilic functional groups such as epoxides, alkyl halides, sulphonates, disulphides, and radical species can react with the nucleophilic thiol group of the tripeptide **glutathione** to give glutathione conjugates which can be subsequently transformed to **mercapturic acids** (Fig. 11.14). The glutathione conjugation reaction can take place in most cells, especially those in the liver and kidney, and is catalysed by **glutathione transferase.** This conjugation reaction is important in detoxifying potentially dangerous environmental toxins or electrophilic alkylating agents formed by phase I reactions (Fig. 11.15). Glutathione conjugates are often excreted in the bile, but are more usually converted to mercapturic acid conjugates before excretion.

Not all phase II reactions result in increased polarity. Methylation and acetylation are important phase II reactions which usually *decrease* the polarity of the drug (Fig. 11.16). An important exception is the methylation of pyridine rings, which leads to polar quaternary salts. The functional groups that are susceptible to

FIGURE 11.7 Phase I oxidative reactions catalysed by miscellaneous enzymes.

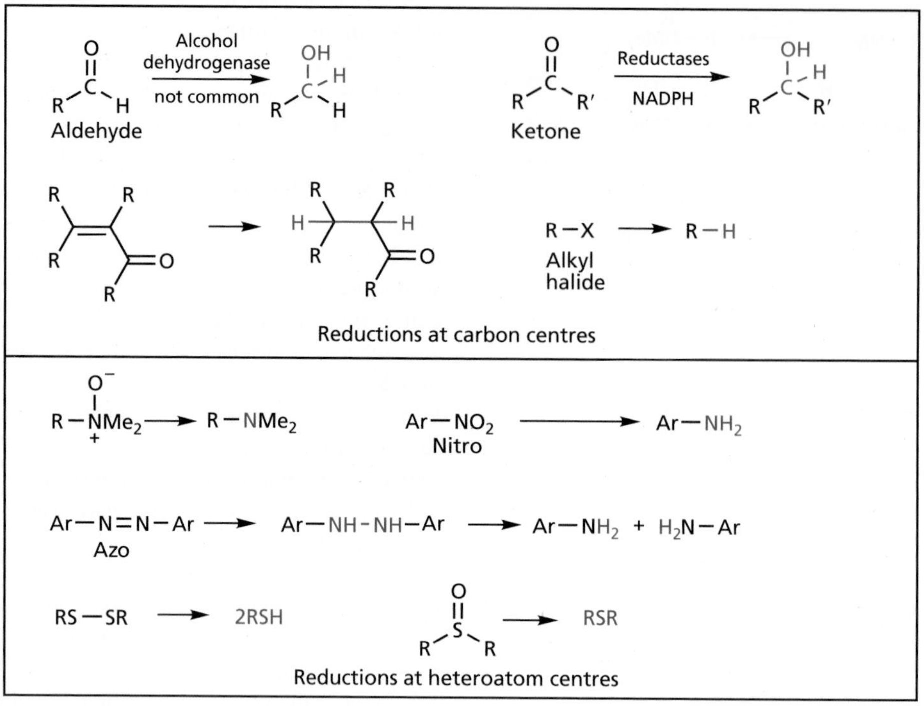

FIGURE 11.8 Phase I reductive reactions.

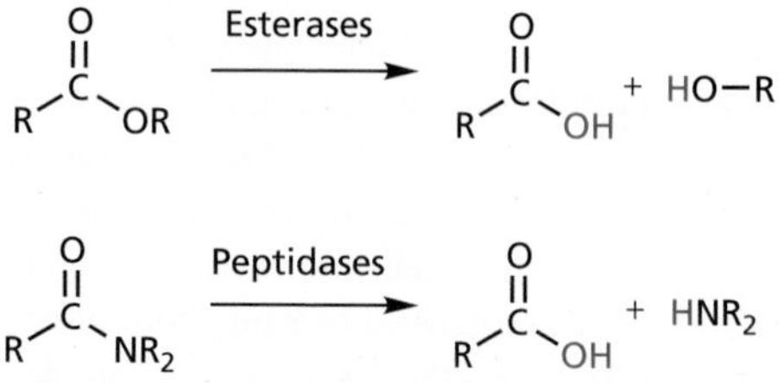

FIGURE 11.9 Hydrolysis of esters and amides.

methylation are phenols, amines, and thiols. Primary amines are also susceptible to acetylation. The enzyme cofactors involved in contributing the methyl group or acetyl group are **S-adenosyl methionine** and **acetyl SCoA** respectively. Several methyltransferase enzymes are involved in the methylation reactions. The most important enzyme for *O*-methylations is **catechol *O*-methyltransferase**, which preferentially methylates the *meta* position of catechols (section 23.5). It should be pointed out, however, that methylation occurs less frequently than other conjugation reactions and is more important in biosynthetic pathways or the metabolism of endogenous compounds.

It is possible for drugs bearing carboxylic acids to become conjugated with **cholesterol.** Cholesterol conjugates can also be formed with drugs bearing an ester group by means of a transesterification reaction. Some drugs with an alcohol functional group form conjugates with fatty acids by means of an ester link.

11.5.6 Metabolic stability

Ideally, a drug should be resistant to drug metabolism because the production of metabolites complicates drug therapy (see Box 11.1). For example, the metabolites formed will usually have different properties from the

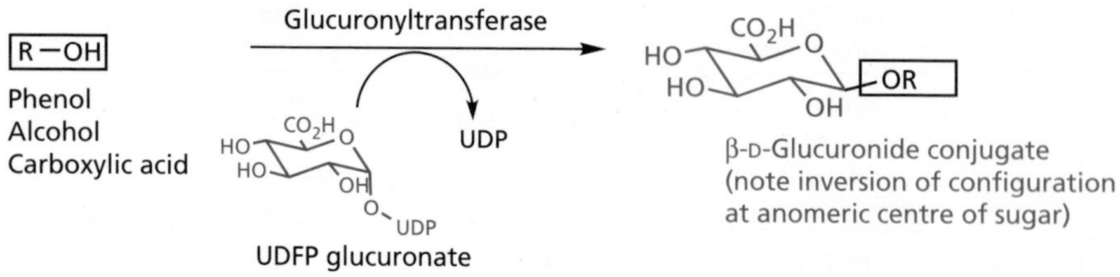

FIGURE 11.10 Glucuronidation of alcohols, phenols, and carboxylic acids.

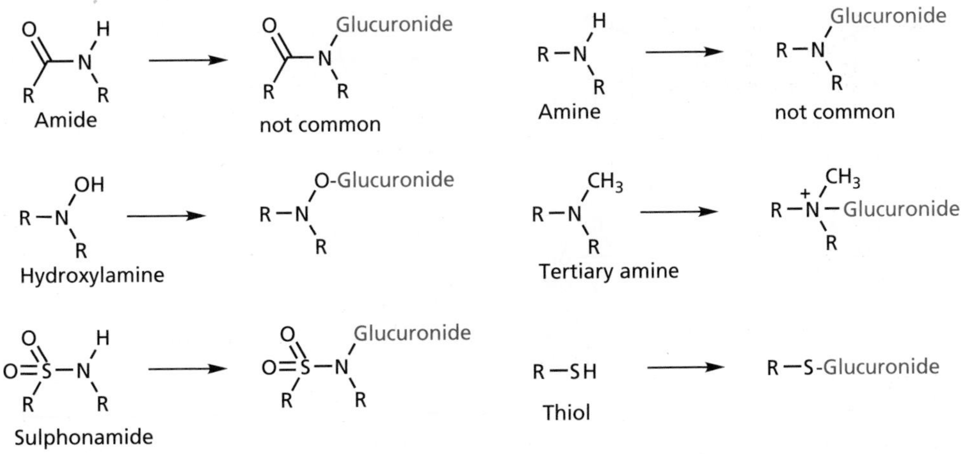

FIGURE 11.11 Glucuronidation of miscellaneous functional groups.

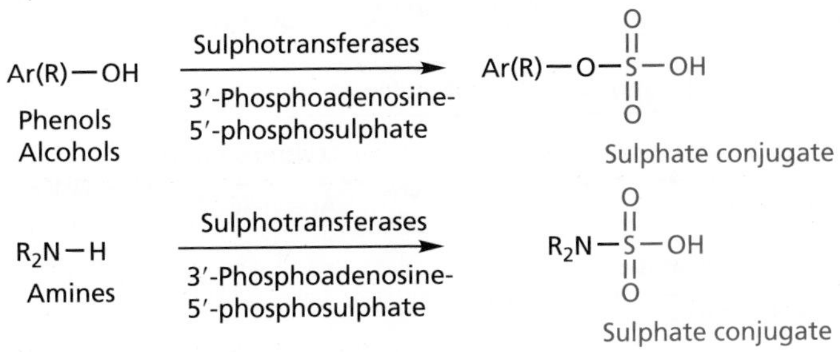

FIGURE 11.12 Examples of sulphoconjugation phase II reactions.

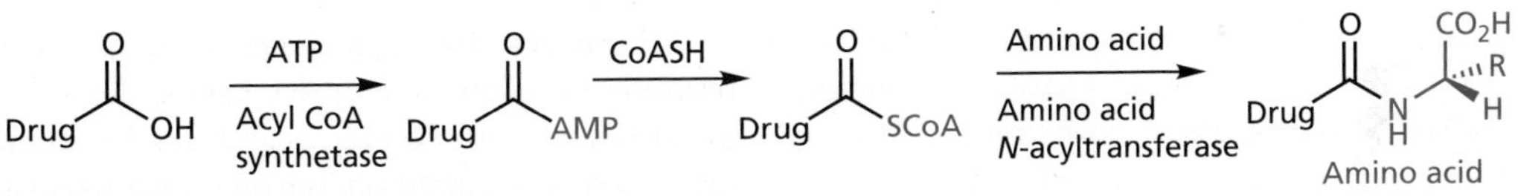

FIGURE 11.13 Formation of amino acid conjugates.

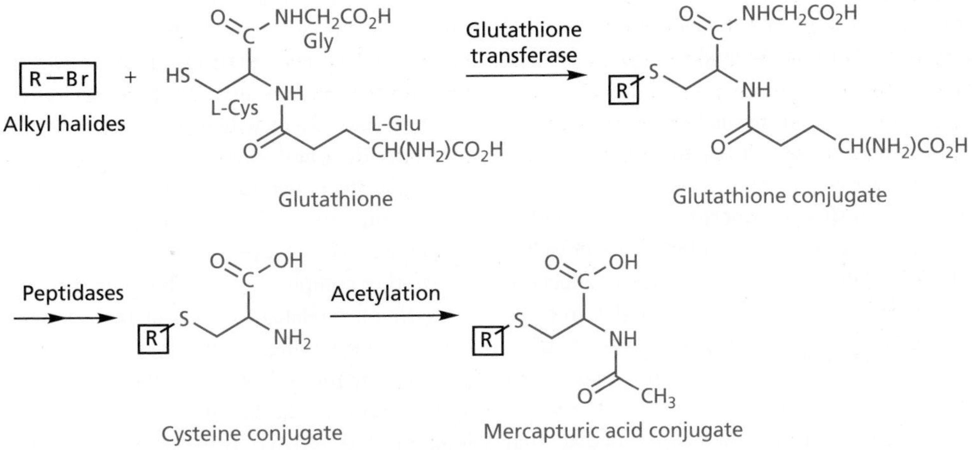

FIGURE 11.14 Formation of glutathione and mercapturic acid conjugates from an alkyl halide.

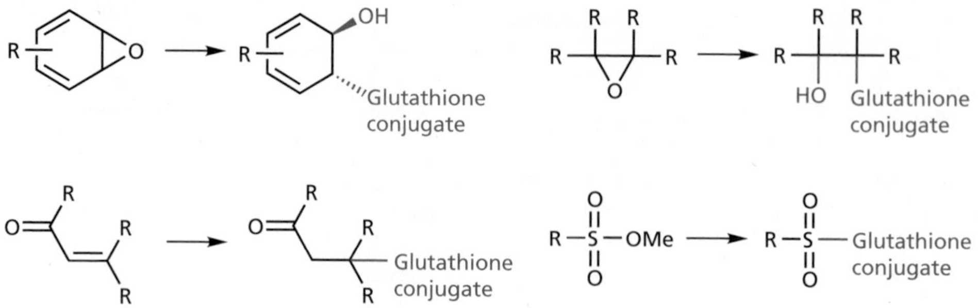

FIGURE 11.15 Formation of glutathione conjugates (Glu-Cys-Gly) with electrophilic groups.

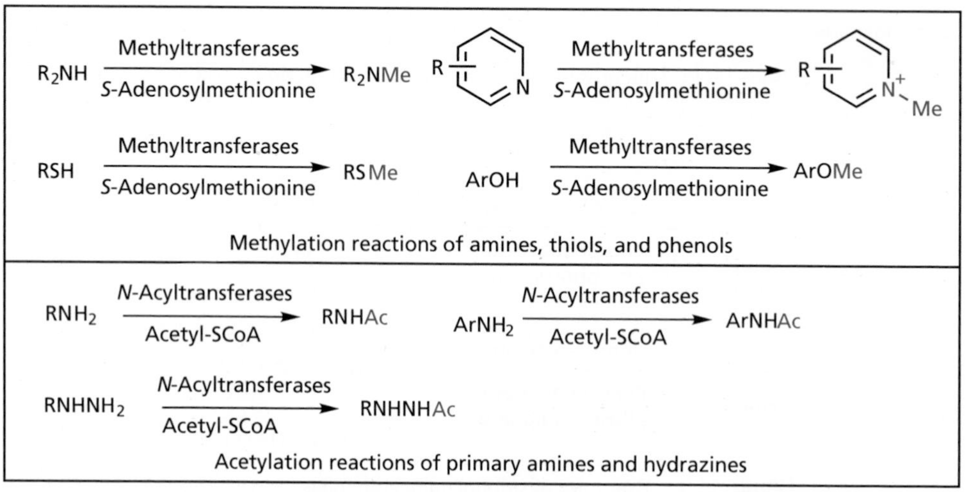

FIGURE 11.16 Methylation and acetylation.

original drug. In some cases, activity may be lost. In others, the metabolite may prove to be toxic. For example, the metabolites of **paracetamol** cause liver toxicity, and the carcinogenic properties of some polycyclic hydrocarbons are due to the formation of epoxides.

Another problem arises from the fact that the activity of metabolic enzymes varies from individual to individual. This is especially true of the cytochrome P450 enzymes with at least a 10-fold variability for the most important isoform CYP3A4. Individuals may even lack particular isoforms. For example, 8% of Americans lack the CYP2D6 isoform, which means that drugs normally metabolized by this enzyme can rise to toxic levels. Examples of drugs that are normally metabolized by this isozyme are **desipramine**, **haloperidol**, and **tramadol**. Some prodrugs require metabolism by CYP2D6 in order to be effective. For example, the analgesic effects of **codeine** are due to its metabolism by CYP2D6 to morphine. Therefore, codeine is ineffective in patients lacking this isozyme. The profile of these enzymes in different patients can vary, resulting in a difference in the way a drug is metabolized. As a result, the amount of drug that can be safely administered also varies.

Differences across populations can be quite significant, resulting in different countries having different recommended dose levels for particular drugs. For example, the rate at which the antibacterial agent **isoniazid** is acetylated and deactivated varies amongst populations. Asian populations acylate the drug at a fast rate, whereas 45–65% of Europeans and North Americans have a slow rate of acylation. **Pharmacogenomics** is the study of genetic variations between individuals and the effect that has on individual responses to drugs. In the future, it is possible that 'fingerprints' of an individual's genome may allow better prediction of which drugs would be suitable for that individual, and which drugs might produce unacceptable side effects—an example of **personalized medicine**. This in turn may avoid drugs having to be withdrawn from the market as a result of rare toxic side effects.

Another complication involving drug metabolism and drug therapy relates to the fact that cytochrome P450 activity can be affected by other chemicals. For example, certain foods have an influence. Brussels sprouts and cigarette smoke enhance activity, whereas grapefruit juice inhibits activity. This can have a significant effect on the activity of drugs metabolized by cytochrome P450

BOX 11.1 Metabolism of an antiviral agent

Indinavir is an antiviral agent used in the treatment of HIV and is prone to metabolism, resulting in seven different metabolites (Fig. 1). Studies have shown that the CYP3A subfamily of cytochrome P450 enzymes is responsible for six of these metabolites. The metabolites concerned arise from *N*-dealkylation of the piperazine ring, *N*-oxidation of the pyridine ring, *para*-hydroxylation of the phenyl ring, and hydroxylation of the indane ring. The seventh metabolite is a glucuronide conjugate of the pyridine ring. All these reactions occur individually to produce five separate metabolites. The remaining two metabolites arise from two or more metabolic reactions taking place on the same molecule.

The major metabolites are those resulting from *N*-dealkylation. As a result, research has been carried out to try and design indinavir analogues that are resistant to this reaction.

For example, structures having two methyl substituents on the activated carbon next to pyridine have been effective in blocking *N*-dealkylation (Fig. 2).

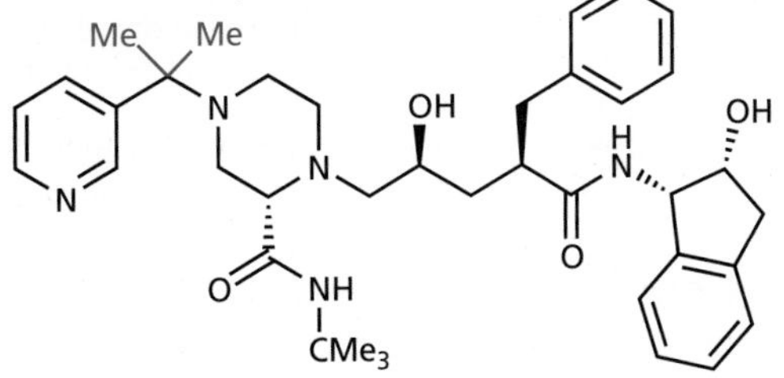

FIGURE 2 Analogue of indinavir resistant to *N*-dealkylation.

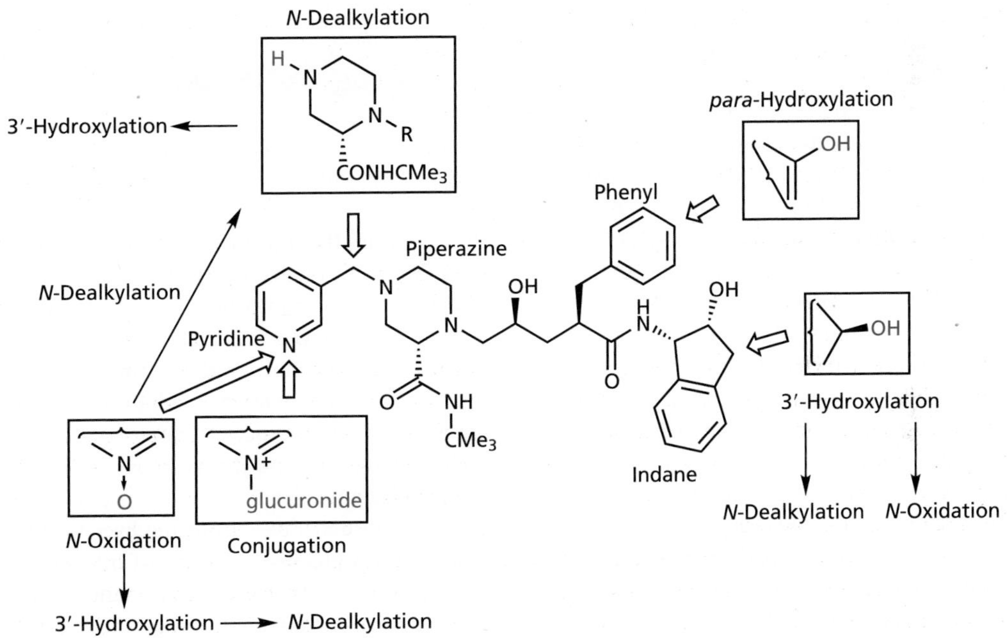

FIGURE 1 Metabolism of indinavir.

enzymes. For example, the immunosuppressant drug **ciclosporin** and the dihydropyridine hypotensive agents are more efficient when taken with grapefruit juice, because their metabolism is reduced. However, serious toxic effects can arise if the antihistamine agent **terfenadine** is taken with grapefruit juice. Terfenadine is actually a prodrug and is metabolized to the active agent **fexofenadine** (Fig. 11.17). If metabolism is inhibited by grapefruit

juice, terfenadine persists in the body and can cause serious cardiac toxicity. As a result, fexofenadine itself is now favoured over terfenadine and is marketed as **Allegra**.

Certain drugs are also capable of inhibiting or promoting cytochrome P450 enzymes, leading to a phenomenon known as **drug–drug interactions** where the presence of one drug affects the activity of another. For example, several antibiotics can act as cytochrome P450 inhibitors

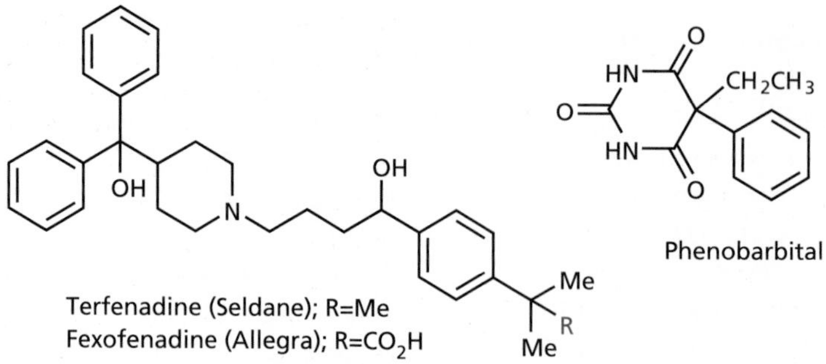

FIGURE 11.17 Drugs which are metabolized by cytochrome P450 enzymes or affect the activity of cytochrome P450 enzymes.

and will slow the metabolism of drugs metabolized by these enzymes. Other examples are the drug–drug interactions that occur between the anticoagulant **warfarin** (section 26.9.1.1) and the barbiturate **phenobarbital** (Fig. 11.17), or between warfarin and the anti-ulcer drug **cimetidine** (section 25.2.7.3).

Phenobarbital stimulates cytochrome P450 enzymes and accelerates the metabolism of warfarin, making it less effective. On the other hand, cimetidine inhibits cytochrome P450 enzymes, thus slowing the metabolism of warfarin. Such drug–drug interactions affect the plasma levels of warfarin and could cause serious problems if the levels move outwith the normal therapeutic range.

Herbal medicine is not immune from this problem either. **St. John's wort** is a popular remedy used for mild to moderate depression. However, it promotes the activity of cytochrome P450 enzymes and decreases the effectiveness of contraceptives and warfarin.

Because of the problems caused by cytochrome P450 activation or inhibition, new drugs are usually tested to check whether they have any effect on cytochrome P450 activity, or are, themselves, metabolized by these enzymes. Indeed, an important goal in many projects is to ensure that such properties are lacking.

Drugs can be defined as hard or soft with respect to their metabolic susceptibility. In this context, **hard drugs** are those that are resistant to metabolism and remain unchanged in the body. **Soft drugs** are designed to have a predictable, controlled metabolism where they are inactivated to non-toxic metabolites and excreted. A group is normally incorporated which is susceptible to metabolism, but will survive sufficiently long to allow the drug to achieve what it is meant to do before it is metabolized and excreted. Drugs such as these are also called **antedrugs**.

11.5.7 The first pass effect

Drugs that are taken orally pass directly to the liver once they enter the blood supply. Here, they are exposed to drug metabolism before they are distributed around the rest of the body, and so a certain percentage of the drug is transformed before it has the chance to reach its target. This is known as the first pass effect. Drugs that are administered in a different fashion (e.g. injection or inhalation) avoid the first pass effect and are distributed around the body before reaching the liver. Indeed, a certain proportion of the drug may not pass through the liver at all, but may be taken up in other tissues and organs en route.

11.6 Drug excretion

Drugs and their metabolites can be excreted from the body by a number of routes. Volatile or gaseous drugs are excreted through the lungs. Such drugs pass out of the capillaries that line the air sacs (alveoli) of the lungs, then diffuse through the cell membranes of the alveoli into the air sacs, from where they are exhaled. Gaseous **general anaesthetics** are excreted in this way and move down a concentration gradient from the blood supply into the lungs. They are also administered through the lungs, in which case the concentration gradient is in the opposite direction and the gas moves from the lungs to the blood supply.

The **bile duct** travels from the liver to the intestines and carries a greenish fluid called **bile** which contains bile acids and salts that are important to the digestion process. A small number of drugs are diverted from the blood supply back into the intestines by this route. Since this happens from the liver, any drug eliminated in this way has not been distributed round the body. Therefore, the amount of drug distributed is less than that absorbed. However, once the drug has entered the intestine, it can be reabsorbed, so it has another chance.

It is possible for as much as 10–15% of a drug to be lost through the skin in sweat. Drugs can also be excreted through saliva and breast milk, but these are minor excretion routes compared with the kidneys. There are concerns, however, that mothers may be passing on drugs such as **nicotine** to their baby through breast milk.

The **kidneys** are the principal route by which drugs and their metabolites are excreted (Fig. 11.18). The kidneys

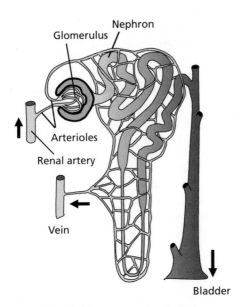

FIGURE 11.18 Excretion by the kidneys.

filter the blood of waste chemicals and these chemicals are subsequently removed in the urine. Drugs and their metabolites are excreted by the same mechanism.

Blood enters the kidneys by means of the **renal artery**. This divides into a large number of capillaries, each one of which forms a knotted structure called a **glomerulus** that fits into the opening of a duct called a **nephron**. The blood entering these glomeruli is under pressure, and so plasma is forced through the pores in the capillary walls into the nephron, carrying with it any drugs and metabolites that might be present. Any compounds that are too big to pass through the pores, such as plasma proteins and red blood cells, remain in the capillaries with the remaining plasma. Note that this is a filtration process, so it does not matter whether the drug is polar or hydrophobic: all drugs and drug metabolites will be passed equally efficiently into the nephron. However, this does not mean that every compound will be *excreted* equally efficiently, because there is more to the process than simple filtration.

The filtered plasma and chemicals now pass through the nephron on their route to the bladder. However, only a small proportion of what starts that journey actually finishes it. This is because the nephron is surrounded by a rich network of blood vessels carrying the filtered blood away from the glomerulus, permitting much of the contents of the nephron to be reabsorbed into the blood supply. Most of the water that was filtered into the nephron is quickly reabsorbed through pores in the nephron cell membrane which are specific for water molecules, and bar the passage of ions or other molecules. These pores are made up of protein molecules called **aquaporins**. As water is reabsorbed, drugs and other agents are concentrated in the nephron and a concentration gradient

is set up. There is now a driving force for compounds to move back into the blood supply down the concentration gradient. However, this can only happen if the drug is sufficiently hydrophobic to pass through the cell membranes of the nephron. This means that hydrophobic compounds are efficiently reabsorbed back into the blood, whereas polar compounds remain in the nephron and are excreted. This process of excretion explains the importance of drug metabolism to drug excretion. Drug metabolism creates polar metabolites which are less likely to be reabsorbed from the nephrons.

Some drugs are actively transported from blood vessels into the nephrons. This process is called **facilitated transport**, and is important in the excretion of penicillins (section 19.5.1.9).

KEY POINTS

- Drugs are exposed to enzyme-catalysed reactions which modify their structure. This is called drug metabolism and can take place in various tissues. However, most reactions occur in the liver.

- Orally taken drugs are subject to the first pass effect.

- Drugs administered by methods other than the oral route avoid the first pass effect.

- Phase I metabolic reactions typically involve the addition or exposure of a polar functional group. Cytochrome P450 enzymes present in the liver carry out important phase I oxidation reactions. The types of cytochrome P450 enzymes present vary between individuals, leading to varying rates of drug metabolism.

- The activity of cytochrome P450 enzymes can be affected by food, chemicals, and drugs, resulting in drug–drug interactions and possible side effects.

- Phase II metabolic reactions involve the addition of a highly polar molecule to a functional group. The resulting conjugates are more easily excreted.

- Drug excretion can take place through sweat, exhaled air, or bile, but most excretion takes place through the kidneys.

- The kidneys filter blood such that drugs and their metabolites enter nephrons. Non-polar substances are reabsorbed into the blood supply, but polar substances are retained in the nephrons and excreted in the urine.

11.7 Drug administration

There are a large variety of ways in which drugs can be administered and many of these avoid some of the problems associated with oral administration. The main routes are: oral, sublingual, rectal, epithelial, inhalation, and injection. The method chosen will depend on the target organ and the pharmacokinetics of the drug.

11.7.1 Oral administration

Orally administered drugs are taken by mouth. This is the preferred option for most patients, so there is more chance that the patient will comply with the drug regime and complete the course. However, the oral route places the greatest demands on the chemical and physical properties of the drug, as described in the previous sections of this chapter.

Drugs given orally can be taken as pills, capsules, or solutions. Drugs taken in solution are absorbed more quickly and a certain percentage may even be absorbed through the stomach wall. For example, approximately 25–33% of **alcohol** is absorbed into the blood supply from the stomach; the rest is absorbed from the upper intestine. Drugs taken as pills or capsules are mostly absorbed in the upper intestine. The rate of absorption is partly determined by the rate at which the pills and capsules dissolve. This, in turn, depends on such factors as particle size and crystal form. In general, about 75% of an orally administered drug is absorbed into the body within 1–3 hours. Specially designed pills and capsules can remain intact in the stomach to help protect acid-labile drugs from stomach acids. The containers then degrade once they reach the intestine.

Care has to be taken if drugs interact with food. For example, **tetracycline** binds strongly to calcium ions, which inhibits absorption, so foods such as milk should be avoided. Some drugs bind other drugs and prevent absorption. For example, **colestyramine** (used to lower cholesterol levels) binds to **warfarin** and also to the thyroid drug **levothyroxine sodium**, so these drugs should be taken separately.

11.7.2 Absorption through mucous membranes

Some drugs can be absorbed through the mucous membranes of the mouth or nose, thus avoiding the digestive and metabolic enzymes encountered during oral administration. For example, heart patients take **glyceryl trinitrate** (Fig. 11.19) by placing it under the tongue (sublingual administration). The opiate analgesic **fentanyl** (Fig. 11.19) has been given to children in the form of a lollipop, and is absorbed through the mucous membranes of the mouth. The Incas absorbed **cocaine** sublingually by chewing coca leaves.

Nasal decongestants are absorbed through the mucous membranes of the nose. Cocaine powder is absorbed in this way when it is sniffed, as is **nicotine** in the form of snuff. Nasal sprays have been used to administer analogues of peptide hormones such as **antidiuretic hormone**. These drugs would be quickly degraded if taken orally.

Eye drops are used to administer drugs directly to the eye and thus reduce the possibility of side effects elsewhere in the body. For example, the eye condition known as glaucoma is treated in this way. Nevertheless, some absorption into the blood supply can still occur and some asthmatic patients suffer bronchospasms when taking **timolol** eye drops.

11.7.3 Rectal administration

Some drugs are administered rectally as **suppositories**, especially if the patient is unconscious, vomiting, or unable to swallow. However, there are several problems associated with rectal administration: the patient may suffer membrane irritation, and, although drug absorption is efficient, it can be unpredictable. It is not the most popular of methods with patients either!

11.7.4 Topical administration

Topical drugs are those which are applied to the skin. For example, steroids are applied topically to treat local skin irritations. It is also possible for some of the drug to be absorbed through the skin (**transdermal absorption**) and to enter the blood supply, especially if the drug is lipophilic. **Nicotine patches** work in this fashion, as do hormone replacement therapies for **estrogen**. Drugs are absorbed by this method at a steady rate, and avoid the acidity of the stomach, or the enzymes in the gut or gut wall. Other drugs that have been applied in this way include the analgesic **fentanyl** and the antihypertensive agent **clonidine**. Once applied, the drug is slowly released from the patch and absorbed through the skin into the blood supply over several days. As a result, the level of drug remains relatively constant over that period.

A technique known as **iontophoresis** is being investigated as a means of topical administration. Two miniature electrode patches are applied to the skin and linked

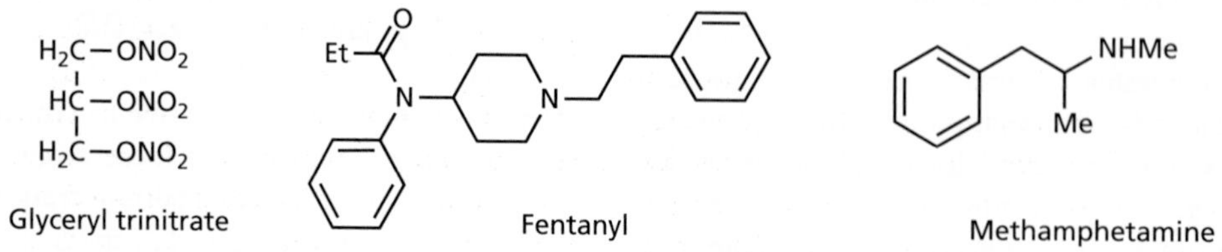

FIGURE 11.19 Glyceryl trinitrate, fentanyl, and methamphetamine.

to a reservoir of the drug. A painless pulse of electricity is applied, which has the effect of making the skin more permeable to drug absorption. By timing the electrical pulses correctly, the drug can be administered such that fluctuations in blood levels are kept to a minimum. Similar devices are being investigated which use ultrasound to increase skin permeability.

11.7.5 Inhalation

Drugs administered by inhalation avoid the digestive and metabolic enzymes of the GIT or liver. Once inhaled, the drugs are absorbed through the cell linings of the respiratory tract into the blood supply. Assuming the drug is able to pass through the hydrophobic cell membranes, absorption is rapid and efficient because the blood supply is in close contact with the cell membranes of the lungs. For example, **general anaesthetic gases** are small, highly lipid-soluble molecules which are absorbed almost as fast as they are inhaled.

Non-gaseous drugs can be administered as **aerosols**. This is how anti-asthmatic drugs are administered and it allows them to be delivered to the lungs in far greater quantities than if they were given orally or by injection. In the case of anti-asthmatics, the drug is made sufficiently polar that it is poorly absorbed into the bloodstream. This localizes it in the airways and lowers the possibility of side effects elsewhere in the body (e.g. action on the heart). However, a certain percentage of an inhaled drug is inevitably swallowed and can reach the blood supply by the oral route. This may lead to side effects. For example, tremor is a side effect of the anti-asthmatic **salbutamol** as a result of the drug reaching the blood supply.

Several drugs of abuse are absorbed through inhalation or smoking (e.g. **nicotine, cocaine, marijuana, heroin**, and **methamphetamine** (Fig. 11.19)). Smoking is a particularly hazardous method of taking drugs. A normal cigarette is like a mini-furnace producing a complex mixture of potentially carcinogenic compounds, especially from the tars present in tobacco. These are not absorbed into the blood supply but coat the lung tissue, leading to long-term problems such as lung cancer. The tars in cannabis are considerably more dangerous than those in tobacco. If cannabis is to be used in medicine, safer methods of administration are desirable (i.e. inhalers).

11.7.6 Injection

Drugs can be introduced into the body by intravenous, intramuscular, subcutaneous, or intrathecal injection. Injection of a drug produces a much faster response than oral administration because the drug reaches the blood supply more quickly. The levels of drug administered are also more accurate because absorption by the oral route

has a level of unpredictability due to the first pass effect. Injecting a drug, however, is potentially more hazardous. For example, some patients may have an unexpected reaction to a drug and there is little that can be done to reduce the levels once the drug has been injected. Such side effects would be more gradual and treatable if the drug was given orally. Furthermore, sterile techniques are essential when giving injections to avoid the risks of bacterial infection or of transmitting hepatitis or AIDS from a previous patient. Finally, there is a greater risk of receiving an overdose when injecting a drug.

The **intravenous** route involves injecting a solution of the drug directly into a vein. This method of administration is not particularly popular with patients, but it is a highly effective method of administering drugs in accurate doses and it is the fastest of the injection methods. However, it is also the most hazardous method of injection. Since its effects are rapid, the onset of any serious side effects or allergies is also rapid. It is important, therefore, to administer the drug as slowly as possible and to monitor the patient closely. An intravenous drip allows the drug to be administered in a controlled manner such that there is a steady level of drug in the system. The local anaesthetic **lidocaine** is given by intravenous injection. Drugs that are dissolved in oily liquids cannot be given by intravenous injection as this may result in the formation of blood clots.

The **intramuscular** route involves injecting drugs directly into muscle, usually in the arm, thigh, or buttocks. Drugs administered in this way do not pass round the body as rapidly as they would if given by intravenous injection, but they are still absorbed faster than by oral administration. The rate of absorption depends on various factors such as the diffusion of the drug, blood supply to the muscle, the solubility of the drug, and the volume of the injection. Local blood flow can be reduced by adding adrenaline to constrict blood vessels. Diffusion can be slowed by using a poorly absorbed salt, ester, or complex of the drug (see also section 14.6.2). The advantage of slowing down absorption is in prolonging activity. For example, oily suspensions of steroid hormone esters are used to slow absorption. Drugs are often administered by intramuscular injection when they are unsuitable for intravenous injection, and so it is important to avoid injecting into a vein.

Subcutaneous injection involves injecting the drug under the surface of the skin. Absorption depends on factors such as how fast the drug diffuses, the level of blood supply to the skin, and the ability of the drug to enter the blood vessels. Absorption can be slowed by the same methods described for intramuscular injection. Drugs which can act as irritants should not be administered in this way as they can cause severe pain and may damage local tissues.

Intrathecal injection means that the drug is injected into the spinal cord. Antibacterial agents that do not normally cross the blood–brain barrier are often administered in this way. Intrathecal injections are also used to administer **methotrexate** in the treatment of childhood leukaemia in order to prevent relapse in the CNS.

Intraperitoneal injection involves injecting drugs directly into the abdominal cavity. This is very rarely used in medicine, but it is a method of injecting drugs into animals during preclinical tests.

11.7.7 Implants

Continuous osmotically driven minipumps for **insulin** have been developed which are implanted under the skin. The pumps monitor the level of insulin in the blood and release the hormone as required to keep levels constant. This avoids the problem of large fluctuations in insulin levels associated with regular injections.

Gliadel is a wafer that has been implanted into the brain to administer anticancer drugs direct to brain tumours, thus avoiding the blood–brain barrier.

Polymer-coated, drug-releasing stents have been used to keep blood vessels open after a clot clearing procedure called angioplasty.

Investigations are underway into the use of implantable microchips which could detect chemical signals in the body and release drugs in response to these signals.

KEY POINTS

- Oral administration is the preferred method of administering drugs, but it is also the most demanding on the drug.

- Drugs administered by methods other than the oral route avoid the first pass effect.

- Drugs can be administered such that they are absorbed through the mucous membranes of the mouth, nose, or eyes.

- Some drugs are administered rectally as suppositories.

- Topically administered drugs are applied to the skin. Some drugs are absorbed through the skin into the blood supply.

- Inhaled drugs are administered as gases or aerosols to act directly on the respiratory system. Some inhaled drugs are absorbed into the blood supply to act systemically.

- Polar drugs which are unable to cross cell membranes are given by injection.

- Injection is the most efficient method of administering a drug, but it is also the most hazardous. Injection can be intravenous, intramuscular, subcutaneous, or intrathecal.

- Implants have been useful in providing controlled drug release such that blood concentrations of the drug remain as level as possible.

11.8 Drug dosing

Because of the number of pharmacokinetic variables involved, it can be difficult to estimate the correct dose regimen for a drug (i.e. the amount of drug used for each dose and the frequency of administration). There are other issues to consider as well. Ideally, the blood levels of any drug should be constant and controlled, but this would require a continuous, intravenous drip which is clearly impractical for most drugs. Therefore, drugs are usually taken at regular time intervals. This means that the doses taken have to be such that blood levels of the drug are within a maximum and minimum level, where they are not too high to be toxic, yet not too low to be ineffective. In general, the concentration of free drug in the blood (i.e. not bound to plasma protein) is a good indication of the availability of that drug at its target site. This does not mean that blood concentration levels are the same as the concentration levels at the target site. However, any variations in blood concentration will result in similar fluctuations at the target site. Thus, blood concentration levels can be used to determine therapeutic and safe dosing levels for a drug.

Figure 11.20 shows two dose regimens. Dose regimen A quickly reaches the therapeutic level but continues to rise to a steady state which is toxic. Dose regimen B involves half the amount of drug provided with the same frequency. The time taken to reach the therapeutic level is certainly longer, but the steady state levels of the drug remain between the therapeutic and toxic levels—the **therapeutic window**.

Dose regimens involving regular administration of a drug work well in most cases, especially if the size of each dose is less than 200 mg and doses are taken once or twice a day. However, there are certain situations where timed doses are not suitable. The treatment of diabetes with **insulin** is a case in point. Insulin is normally secreted continuously by the pancreas, so the injection of insulin at timed intervals is unnatural and can lead to a whole range of physiological complications.

Other dosing complications include differences of age, sex, and race. Diet, environment, and altitude also have an influence. Obese people present a particular problem, as it can be very difficult to estimate how much of a drug will be stored in fat tissue and how much will remain free in the blood supply. The precise time when drugs are taken may be important, because metabolic reaction rates can vary throughout the day.

Drugs can interact with other drugs. For example, some drugs used for diabetes are bound by plasma protein in the blood supply and are therefore not free to react with their targets. However, drugs such as **aspirin** may displace them from plasma protein, leading

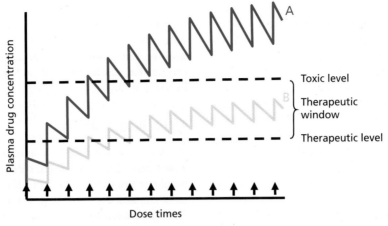

FIGURE 11.20 Dosing regimens.

to a drug overdose. Aspirin has this same effect on anticoagulants.

Problems can also occur if a drug inhibits a metabolic reaction and is taken with a drug normally metabolized by that reaction. The latter is more slowly metabolized than normal, increasing the risk of an overdose. For example, the antidepressant drug **phenelzine** inhibits the metabolism of amines and should not be taken with drugs such as **amphetamines** or **pethidine**. Even amine-rich foods can lead to adverse effects, implying that cheese and wine parties are hardly the way to cheer the victim of depression. Other examples were described in section 11.5.6.

When one considers all these complications, it is hardly surprising that individual variability to drugs can vary by as much as a factor of 10.

11.8.1 Drug half-life

The half-life ($t_{1/2}$) of a drug is the time taken for the concentration of the drug in blood to fall by half. The removal or elimination of a drug takes place through both excretion and drug metabolism, and is not linear with time. Therefore, drugs can linger in the body for a significant period of time. For example, if a drug has a half-life of 1 hour, then there is 50% of it left after 1 hour. After 2 hours, there is 25% of the original dose left, and after 3 hours, 12.5% remains. It takes 7 hours for the level to fall below 1% of the original dose. Some drugs, such as the

opioid analgesic **fentanyl**, have short half-lives (45 minutes), whereas others such as **diazepam** (**Valium**) have a half-life measured in days. In the latter case, recovery from the drug may take a week or more.

11.8.2 Steady state concentration

Drugs are metabolized and eliminated as soon as they are administered, so it is necessary to provide regular doses in order to maintain therapeutic levels in the body. Therefore, it is important to know the half-life of the drug in order to calculate the frequency of dosing required to reach and maintain these levels. In general, the time taken to reach a **steady state concentration** is six times the drug's half-life. For example, the concentration levels of a drug with a half-life of 4 hours, supplied at 4-hourly intervals, are shown in Table 11.1 and Fig. 11.21.

Note that there is a fluctuation in level in the period between each dose. The level is at a maximum after each dose, and falls to a minimum before the next dose is provided. It is important to ensure that the level does not drop below the therapeutic level, but does not rise to such a level that side effects are induced. The time taken to reach steady state concentration is not dependent on the size of the dose, but the blood level achieved at steady state is. Therefore, the levels of drug present at steady state concentration depend on the size of each dose given, as well as the frequency of dosing. During

TABLE 11.1 Fluctuation of drug concentration levels on regular dosing

Time of dosing (h)	0	4	8	12	16	20	24
Max level (µg/ml)	1.0	1.5	1.75	1.87	1.94	1.97	1.98
Min level (µg/ml)	0.5	0.75	0.87	0.94	0.97	0.98	0.99

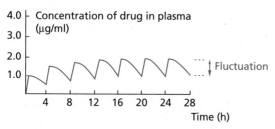

FIGURE 11.21 Graphical representation of fluctuation of drug concentration levels on regular dosing.

clinical trials, blood samples are taken from patients at regular time intervals to determine the concentration of the drug in the blood. This helps determine the proper dosing regimen in order to get the ideal blood levels.

The **area under the plasma drug concentration curve** (AUC) represents the total amount of drug that is available in the blood supply during the dosing regimen.

11.8.3 Drug tolerance

With certain drugs, it is found that the effect of the drug diminishes after repeated doses, and it is necessary to increase the size of the dose in order to achieve the same results. This is known as drug tolerance. There are several mechanisms by which drug tolerance can occur. For example, the drug can induce the synthesis of metabolic enzymes which result in increased metabolism of the drug. **Pentobarbital** (Fig. 11.22) is a barbiturate sedative which induces enzymes in this fashion.

Alternatively, the target may adapt to the presence of a drug. Occupancy of a target receptor by an antagonist may induce cellular effects which result in the synthesis of more receptor (section 8.7). As a result, more drug will be needed in the next dose to antagonize all the receptors.

Physical dependence is usually associated with drug tolerance. Physical dependence is a state in which a patient becomes dependent on the drug in order to feel normal. If the drug is withdrawn, uncomfortable **withdrawal symptoms** may arise which can only be alleviated by re-taking the drug. These effects can be explained in part by the effects which lead to drug tolerance. For example, if cells have synthesized more receptors to counteract the presence of an antagonist, the removal of the antagonist means that the body will have too many receptors. This results in a 'kickback' effect where the cell

becomes oversensitive to the normal neurotransmitter or hormone, and this is what produces withdrawal symptoms. These will continue until the excess receptors have been broken down by normal cellular mechanisms—a process that may take several days or weeks (see also sections 8.6–8.7).

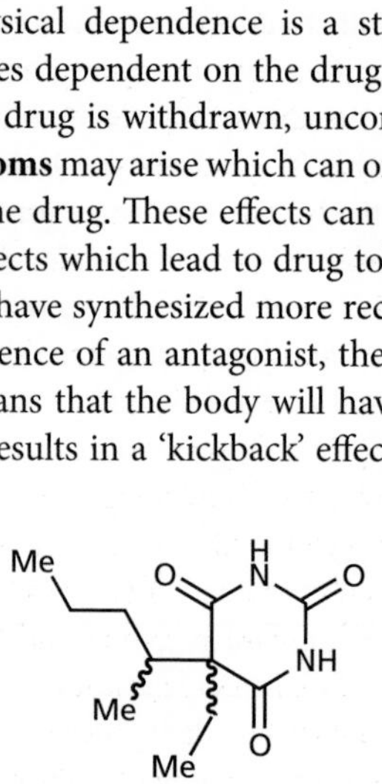

11.8.4 Bioavailability

Bioavailability refers to how quickly and how much of a particular drug reaches the blood supply once all the problems associated with absorption, distribution, metabolism, and excretion have been taken into account. **Oral bioavailability (F)** is the fraction of the ingested dose that survives to reach the blood supply. This is an important property when it comes to designing new drugs and should be considered alongside the pharmacodynamics of the drug (i.e. how effectively the drug interacts with its target).

11.9 Formulation

The way a drug is formulated can avoid some of the problems associated with oral administration. Drugs are normally taken orally as tablets or capsules. A tablet is usually a compressed preparation that contains 5–10% of the drug, 80% of fillers, disintegrants, lubricants, glidants, and binders, and 10% of compounds which ensure easy disintegration, disaggregation, and dissolution of the tablet in the stomach or the intestine—a process which is defined as the **pharmaceutical phase** of drug action. The disintegration time can be modified for a rapid effect or for sustained release. Special coatings can make the tablet resistant to the stomach acids such that it only disintegrates in the duodenum as a result of enzyme action or pH. Pills can also be coated with sugar, varnish, or wax to disguise taste. Some tablets are designed with an osmotically active bilayer core surrounded by a semi-permeable membrane with one or more laser drilled pores in it. The osmotic pressure of water entering the tablet pushes the drug through the pores at a constant rate as the tablet moves through the digestive tract. Therefore, the rate of release is independent of varying pH or gastric motility. Several drugs such as **hydromorphone**, **albuterol**, and **nifedipine** have been administered in this way.

A capsule is a gelatinous envelope enclosing the active substance. Capsules can be designed to remain intact for some hours after ingestion in order to delay absorption. They may also contain a mixture of slow-release and fast-release particles to produce rapid and sustained absorption in the same dose.

The drug itself needs to dissolve in aqueous solution at a controlled rate. Such factors as particle size and crystal

FIGURE 11.22 Pentobarbital.

form can significantly affect dissolution. Fast dissolution is not always ideal. For example, slow dissolution rates can prolong the duration of action or avoid initially high plasma levels.

Formulation can also play an important role in preventing drugs being abused. For example, a tablet preparation (**Oxecta**) of the opioid analgesic **oxycodone** was approved in 2011 as an orally active opioid analgesic and includes deterrents to abuse. For example, chemicals are present that prevent the drug being dissolved in solvent and injected. Other chemicals cause a burning sensation in the nose, which discourages drug abusers crushing the tablets and snorting the powder. Finally, other chemicals are present which produce non-toxic, but very unpleasant, effects if too many pills are taken orally.

11.10 Drug delivery

The various aspects of drug delivery could fill a textbook in themselves, so any attempt to cover the topic in a single section is merely tickling the surface, let alone scratching it! However, it is worth appreciating that there are various methods by which drugs can be physically protected from degradation and/or targeted to treat particular diseases such as cancer and inflammation. One approach is to use a prodrug strategy (section 14.6), which involves chemical modifications to the drug. Another approach covered in this section is the use of water-soluble macromolecules to help the drug reach its target. The macromolecules concerned are many and varied and include synthetic polymers, proteins, liposomes, and antibodies. The drug itself may be covalently linked to the macromolecule or encapsulated within it. The following are some illustrations of drug delivery systems.

Antibodies were described in section 10.7.2 and have long been seen as a method of targeting drugs to cancer cells. Methods have been devised for linking anticancer drugs to antibodies to form **antibody–drug conjugates** that remain stable on their journey through the body, but release the drug at the target cell. A lot of research has been carried out on these conjugates, and this is discussed in detail in section 21.10.2. However, there are problems associated with antibodies. The amount of drug that can be linked to the protein is quite limited and there is the risk of an immune reaction where the body identifies the antibody as foreign and tries to reject it.

A similar approach is to link the drugs to synthetic polymers such as polyethylene glycol (PEG), polyglutamate, or N-(2-hydroxypropyl)methacrylamide (HPMA) to form polymer–drug conjugates (Fig. 11.23). Again the amount of drug that can be linked is limited, but a variety of anticancer–polymer conjugates are currently undergoing clinical trials. Such conjugates help to protect the

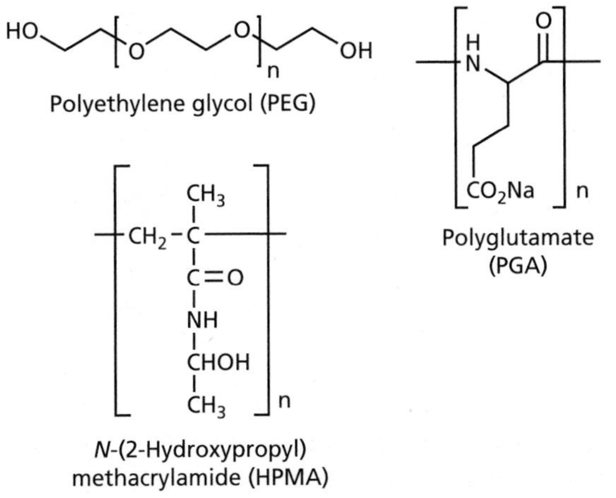

FIGURE 11.23 Synthetic polymers used for polymer–drug conjugates.

lifetime of the drug by decreasing the rates of metabolism and excretion. **Pegaptanib** is a preparation that was approved for treating a vascular disease in the eye and consists of an oligonucleotide drug linked to PEG (section 10.5). Pegylation has also been used to design a peripherally acting opioid that is unable to cross the blood–brain barrier (section 24.9.4)

Protein-based polymers are being developed as drug delivery systems for the controlled release of ionized drugs. For example, the cationic drugs **Leu-enkephalin** or **naltrexone** could be delivered using polymers with anionic carboxylate groups. Ionic interactions between the drug and the protein result in folding and assembly of the protein polymer to form a protein–drug complex and the drug is then released at a slow and constant rate. The amount of drug carried could be predetermined by the density of carboxylate binding sites present and the accessible surface area of the vehicle. The rate of release could be controlled by varying the number of hydrophobic amino acids present. The greater the number of hydrophobic amino acids present, the weaker the affinity between the carboxylate binding groups and the drug. Once the drug is released, the protein carrier would be metabolized like any normal protein.

A physical method of protecting drugs from metabolic enzymes in the bloodstream and allowing a steady slow release of the drug is to encapsulate the drug within small vesicles called **liposomes**, and then inject them into the blood supply (Fig. 11.24). These vesicles or globules consist of a bilayer of fatty phospholipid molecules (similar to a cell membrane) and will travel round the circulation, slowly leaking their contents. Liposomes are known to be concentrated in malignant tumours and this provides a possible method of delivering antitumour drugs to these cells. It is also found that liposomes can fuse with the plasma membranes of a variety of cells, allowing the delivery of drugs or DNA into these cells. As a result, they

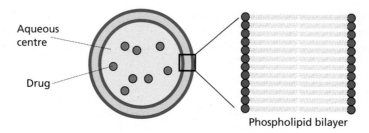

FIGURE 11.24 Liposome containing a drug.

may be useful for gene therapy. The liposomes can be formed by sonicating a suspension of a phospholipid (e.g. phosphatidylcholine) in an aqueous solution of the drug.

Another future possibility for targeting liposomes is to incorporate antibodies into the liposome surface such that specific tissue antigens are recognized. Liposomes have a high drug-carrying capacity, but it can prove difficult to control the release of drug at the required rate. Slow leakage is a problem if the liposome is carrying a toxic anticancer drug such as **doxorubicin**. The liposomes can also be trapped by the **reticuloendothelial system** (RES) and removed from the blood supply. The RES is a network of cells which can be viewed as a kind of filter. One answer to this problem has been to attach PEG polymers to the liposome (see also section 14.8.2). The tails of the PEG polymers project out from the liposome surface and act as a polar outer shell which protects and shields the liposome from both destructive enzymes and the reticuloendothelial system. This significantly increases its lifetime and reduces leakage of its passenger drug. **DOXIL** is a PEGylated liposome containing doxorubicin which is used successfully in anticancer therapy as a once-monthly infusion.

The use of injectable **microspheres** has been approved for the delivery of human growth hormone. The microspheres containing the drug are made up of a biologically degradable polymer and slowly release the hormone over a 4-week period.

A large number of important drugs have to be administered by injection because they are either susceptible to digestive enzymes or cannot cross the gut wall. This includes the ever-growing number of therapeutically useful peptides and proteins being generated by biotechnology companies using recombinant DNA technology. Drug delivery systems which could deliver these drugs orally would prove a huge step forward in medicine. For example, liposomes are currently being studied as possible oral delivery systems. Another approach currently being investigated is to link a therapeutic protein to a hydrophobic polymer such that it is more likely to be absorbed. However, it is important that the conjugate breaks up before the drug enters the blood supply or else

it would have to be treated as a new drug and undergo expensive preclinical and clinical trials. **Hexyl-insulin monoconjugate 2** consists of a polymer linked to a lysine residue of insulin. It is currently being investigated as an oral delivery system for **insulin**.

Biologically erodible microspheres have also been designed to stick to the gut wall such that absorption of the drug within the sphere through the gut wall is increased. This has still to be used clinically, but has proved effective in enhancing the absorption of insulin and **plasmid DNA** in test animals. In a similar vein, drugs have been coated with bioadhesive polymers designed to adhere to the gut wall so that the drug has more chance of being absorbed. The use of anhydride polymers has the added advantage that these polymers are capable of crossing the gut wall and entering the bloodstream, taking their passenger drug with them. **Emisphere Technologies Inc.** have developed derivatives of amino acids and shown that they can enhance the absorption of specific proteins. It is thought that the amino acid derivatives interact with the protein and make it more lipophilic so that it can cross cell membranes directly.

Drug delivery systems are being investigated which will carry oligonucleotides such as DNA, antisense molecules, and siRNAs (section 9.7.2). For example, nucleic acid–lipid particles are being investigated as a means of delivering oligonucleotides into liver cells. Such particles are designed to have a positive charge on their exterior since this encourages adsorption to the negatively charged cell membranes of target cells. Another method of carrying and delivering oligonucleotides is to incorporate them into viruses that are capable of infecting cells. However, there are risks associated with this approach and there have been instances of fatalities during clinical trials. Therefore, nanotechnology is being used to construct artificial viruses which will do the job more safely. Clinical trials have demonstrated that it is possible to use engineered viruses to target drugs to tumour cells.

Other areas of research include studies of crown ethers, nanoparticles, nanospheres, nanowires, nanomagnets, biofuel cells, hydrogel polymers, and superhydrophobic materials as methods of delivering drugs.

- Drugs should be administered at the correct dose levels and frequency to ensure that blood concentrations remain within the therapeutic window.

- The half-life of a drug is the time taken for the blood concentration of the drug to fall by half. A knowledge of the half-life is required to calculate how frequently doses should be given to ensure a steady state concentration.

- Drug tolerance is where the effect of a drug diminishes after repeated doses. In physical dependence a patient becomes dependent on a drug and suffers withdrawal symptoms on stopping the treatment.

- Formulation refers to the method by which drugs are prepared for administration, whether by solution, pill, capsule, liposome, or microsphere. Suitable formulations can protect drugs from particular pharmacokinetic problems.

QUESTIONS

1. Benzene used to be a common solvent in organic chemistry, but is no longer used because it is a suspect carcinogen. Benzene undergoes metabolic oxidation by cytochrome P450 enzymes to form an electrophilic epoxide which can alkylate proteins and DNA. Toluene is now used as a solvent in place of benzene. Toluene is also oxidized by cytochrome P450 enzymes, but the metabolite is less toxic and is rapidly excreted. Suggest what the metabolite might be, and why the metabolism of toluene is different from that of benzene.

2. The prodrug of the antipsychotic drug **fluphenazine** shown below has a prolonged period of action when it is given by intramuscular injection, but not when it is given by intravenous injection. Suggest why this is the case.

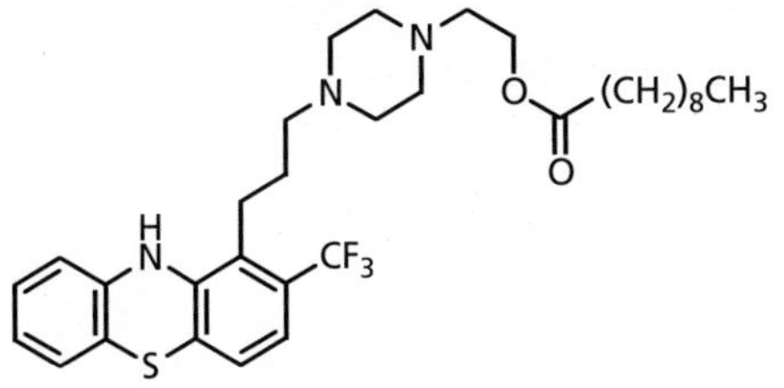

Fluphenazine prodrug

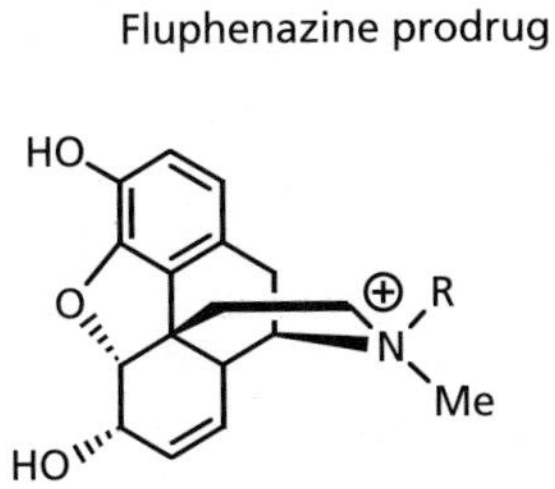

Morphine; R = H
Quaternary salt; R = Me

3. Morphine binds strongly to opioid receptors in the brain to produce analgesia. *In vitro* studies on opioid receptors show that the quaternary salt of morphine also binds strongly. However, the compound is inactive *in vivo* when injected intravenously. Explain this apparent contradiction.

4. The phenol group of morphine is important in binding morphine to opioid receptors and causing analgesia. Codeine

has the same structure as morphine but the phenol group is masked as a methyl ether. As a result, codeine binds poorly to opioid receptors and should show no analgesic activity. However, when it is taken *in vivo*, it shows useful analgesic properties. Explain how this might occur.

5. The pK_a of histamine is 5.74. What is the ratio of ionized to unionized histamine (a) at pH 5.74 (b) at pH 7.4?

6. A drug contains an ionized carboxylate group and shows good activity against its target in *in vitro* tests. When *in vivo* tests were carried out, the drug showed poor activity when it was administered orally, but good activity when it was administered by intravenous injection. The same drug was converted to an ester, but proved inactive *in vitro*. Despite that, it proved to be active *in vivo* when it was administered orally. Explain these observations.

7. **Atomoxetine** and **methylphenidate** are used in the treatment of attention deficit hyperactivity disorder. Suggest possible metabolites for these structures.

8. Suggest metabolites for the proton pump inhibitor **omeprazole**.

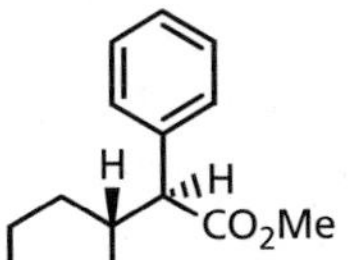

Methylphenidate 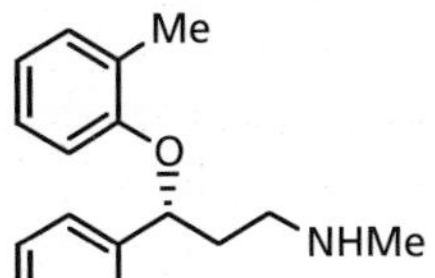   Atomoxetine

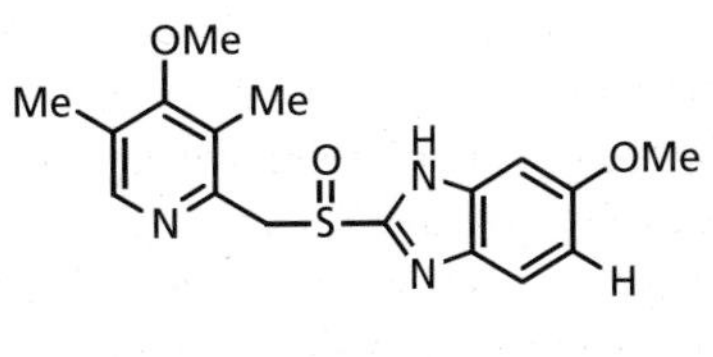

Omeprazole

9. A drug has a half-life of 4 hours. How much of the drug remains after 24 hours?

10. Salicylic acid is absorbed more effectively from the stomach than from the intestines, whereas quinine is absorbed more effectively from the intestines than from the stomach. Explain these observations.

🌐 Multiple-choice questions are available on the Online Resource Centre at www.oxfordtextbooks.co.uk/orc/patrick6e/

FURTHER READING

Cairns, D. (2012) *Essentials of pharmaceutical chemistry*, 4th edn. Pharmaceutical Press, London.

Duncan, R. (2003) The dawning era of polymer therapeutics. *Nature Reviews Drug Discovery*, **2**, 347–60.

Goldberg, M., and Gomez-Orellana, I. (2003) Challenges for the oral delivery of macromolecules. *Nature Reviews Drug Discovery*, **2**, 257–8.

Guengerich, F. P. (2002) Cytochrome P450 enzymes in the generation of commercial products. *Nature Reviews Drug Discovery*, **1**, 359–66.

King, A. (2011) Breaking through the barrier. *Chemistry World*, June, 36–9.

Langer, R. (2003) Where a pill won't reach. *Scientific American*, April, 32–9.

LaVan, D. A., Lynn, D. M., and Langer, R. (2002) Moving smaller in drug discovery and delivery. *Nature Reviews Drug Discovery*, **1**, 77–84.

Lindpaintner, K. (2002) The impact of pharmacogenetics and pharmacogenomics on drug discovery. *Nature Reviews Drug Discovery*, **1**, 463–9.

Lipinski, C. A., Lombardo, F., Dominy, B. W., and Feeney, P. J. (1997) Experimental and computational approaches to estimate solubility and permeability in drug discovery and development settings. *Advanced Drug Delivery Reviews*, **23**, 3–25 (rule of five).

Mastrobattista, E., et al. (2006) Artificial viruses: a nanotechnological approach to gene delivery. *Nature Reviews Drug Discovery*, **5**, 115–21.

Nicholson, J. K., and Wilson, I. D. (2003) Understanding global systems biology: metabonomics and the continuum of metabolism. *Nature Reviews Drug Discovery*, **2**, 668–76.

Pardridge, W. M. (2002) Drug and gene targeting to the brain with molecular Trojan horses. *Nature Reviews Drug Discovery*, **1**, 131–9.

Roden, D. M., and George, A. L. (2002) The genetic basis of variability in drug responses. *Nature Reviews Drug Discovery*, **1**, 37–44.

Roses, A. D. (2002) Genome-based pharmacogenetics and the pharmaceutical industry. *Nature Reviews Drug Discovery*, **1**, 541–9.

Saltzman, W. M., and Olbricht, W. L. (2002) Building drug delivery into tissue engineering. *Nature Reviews Drug Discovery*, **1**, 177–86.

Stevenson, R. (2003) Going with the flow. *Chemistry in Britain*, November, 18–20 (aquaporins).

Veber, D. F., et al. (2002) Molecular properties that influence the oral bioavailability of drug candidates. *Journal of Medicinal Chemistry*, **45**, 2615–23.

Willson, T. M., and Kliewer, S. A. (2002) PXR, CAR and drug metabolism. *Nature Reviews Drug Discovery*, **1**, 259–66.

Titles for general further reading are listed on p.845.

CASE STUDY 1
Statins

Statins are an important group of cholesterol-lowering drugs that act as enzyme inhibitors. The market for cholesterol-lowering drugs is the largest in the pharmaceutical sector and is dominated by the statins, with substantial rewards for the companies that produce them. In 2002, **atorvastatin** and **simvastatin** recorded revenues of about 7 billion and 5.3 billion dollars respectively. In this Case study, we shall look at how these drugs were discovered, and how they interact with their target at the molecular level. First, we shall consider the role of cholesterol in coronary heart disease and how the inhibition of an enzyme can lower cholesterol levels.

CS1.1 Cholesterol and coronary heart disease

Cholesterol (Fig. CS1.1a) is an important constituent of cell membranes and is also the biosynthetic precursor for steroid hormones. Therefore it is vital to the normal, healthy functioning of cells, and can be obtained both from the diet and biosynthesis in cells. Problems arise if too much cholesterol is present in the diet since this can lead to cardiovascular disease.

As cholesterol is a fatty molecule, it cannot dissolve in blood and so it has to be transported round the body by particles known as **low-density lipoprotein (LDL)** (Fig. CS1.1b) or **high-density lipoprotein (HDL)**. LDLs are particles about 22 nm in diameter having a mass of 3 million Daltons. Each particle contains a lipoprotein of 4536 amino acid residues that encircles a variety of fatty acids, keeping them soluble in the aqueous environment of the blood supply. The particle also contains a polyunsaturated fatty acid called **linoleate**, several phospholipids, and a large number of cholesterol molecules. LDL serves to transport cholesterol and triglycerides from the liver to the peripheral tissues. When a cell requires cholesterol, it produces LDL receptors which are placed in the cell membrane. LDL binds to these receptors and is then endocytosed into the cell where it releases cholesterol into the cytoplasm.

HDLs are lipoprotein particles about 8–11 nm in diameter that carry fatty acids and cholesterol from tissues back to the liver, where they are removed from the blood supply. They are called high-density lipoproteins because they contain a higher proportion of protein than the LDLs. When they travel round the body, they steadily increase in size as they pick up cholesterol from the tissues.

Mortality from coronary heart disease has been shown to be associated with high levels of LDL or low levels of HDL. Inevitably, LDLs transport cholesterol to the arteries and, if cholesterol is retained there, it can lead to the formation of fatty plaques which narrow the arteries, resulting in an increased risk of atherosclerosis. If a clot forms and blocks an artery supplying blood to heart muscle, it leads to a heart attack. If the clot blocks an artery serving the brain, a stroke results. Thus, lowering the levels of LDL and/or increasing HDL should reduce the risk of heart attacks and strokes. When the statins were first

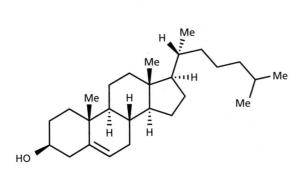

(a) Cholesterol

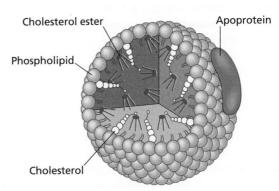

(b) Particle of low-density lipoprotein

FIGURE CS1.1 Cholesterol and a particle of low-density lipoprotein.

FIGURE CS1.2 The biosynthesis of cholesterol. The rate-limiting step is coloured blue.

FIGURE CS1.3 Reaction catalysed by 3-hydroxy-3-methylglutaryl-coenzyme A reductase (HMGR or HMG-CoA reductase).

designed, the aim was to lower the levels of cholesterol that were synthesized in the body. The statins certainly do this, but as we shall see later, it is the subsequent effects on LDL plasma levels that are more important in the protective actions of statins against cardiovascular disease.

CS1.2 The target enzyme

Cholesterol is synthesized within cells, and one way of lowering cholesterol levels in the blood is to block this synthesis. This can be achieved by finding a drug which will inhibit one of the enzymes involved in cholesterol biosynthesis (Fig. CS1.2). However, there are more than 30 enzymes involved in the biosynthetic pathway, so how does one decide which of these enzymes is the best target? The choice can be narrowed down by targeting the enzyme that catalyses the rate-limiting step for the overall process, as this provides the most effective inhibition of the biosynthetic pathway. The enzyme catalysing the rate-limiting step is a reductase enzyme called **3-hydroxy-3-methylglutaryl-coenzyme A reductase** (HMGR or HMG-CoA reductase). The reaction involved is the conversion of 3-hydroxy-3-methylglutaryl-coenzyme A (HMG-CoA) to mevalonate with the aid of NADPH as a cofactor (section 3.5.5) (Fig. CS1.3).

HMGR consists of four protein subunits and is one of the most highly regulated enzymes known. The enzyme's activity can be decreased in a number of ways if too much cholesterol is produced. Firstly, high levels of cholesterol within the cell trigger a signal transduction process which activates a protein kinase responsible for phosphorylating HMGR and inactivating it. Secondly, the rate at which the enzyme is synthesized through transcription and translation is controlled by intracellular

levels of cholesterol. Finally, the rate at which the enzyme is degraded appears to be influenced by cholesterol levels.

There are two active sites present in the tetrameric structure and each one is located between two of the monomers. The portion of the active site which binds the substrate HMG-CoA is predominantly on one monomer, while the portion that binds the cofactor NADPH is situated on a neighbouring monomer.

As far as the reaction pathway is concerned, it involves a reductive cleavage involving two hydride transfers (Fig. CS1.4). The hydride is provided by the cofactor NADPH, and so two NADPH molecules are required for the cleavage reaction.

The HMGR enzyme is found to be highly flexible in its three-dimensional structure and this has an important role to play in the binding and activity of statins.

Different amino acid residues in the active site have important roles to play in the enzyme-catalysed reaction, either in binding the substrate or in the mechanism of the reaction.

As far as binding goes, a positively charged lysine residue (Lys-735) forms an ionic bond with the negatively charged carboxylate group of HMG-CoA. Other residues such as Ser-684 and Asp-690 interact with the alcohol group by hydrogen bonding, while Lys-691 is involved in a hydrogen bonding interaction with the carbonyl group (Fig. CS1.5). The coenzyme A moiety is also bound by different interactions into a narrow hydrophobic slot within the active site.

Other amino acids play an important part in the mechanism of the enzyme-catalysed reaction (Fig. CS1.6). A histidine residue (His-866) acts as an acid catalyst and provides the proton required by coenzyme A to depart as a leaving group. A special mention needs to be made regarding the binding role of Lys-691. We have already

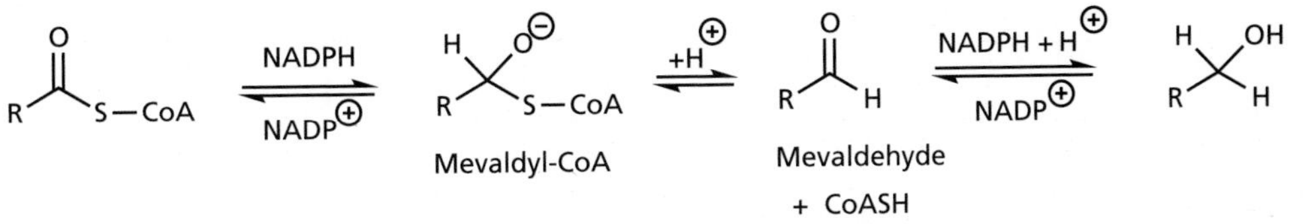

FIGURE CS1.4 Reaction pathway catalysed by 3-hydroxy-3-methylglutaryl-coenzyme A reductase (HMGR or HMG-CoA reductase).

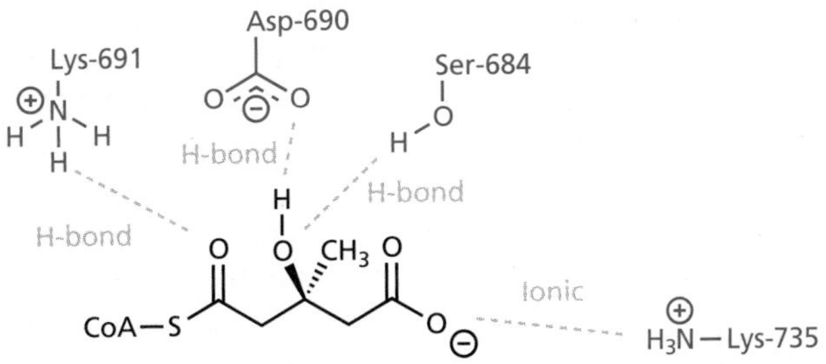

FIGURE CS1.5 Binding interactions for 3-hydroxy-3-methylglutaryl-coenzyme A (HMG-CoA).

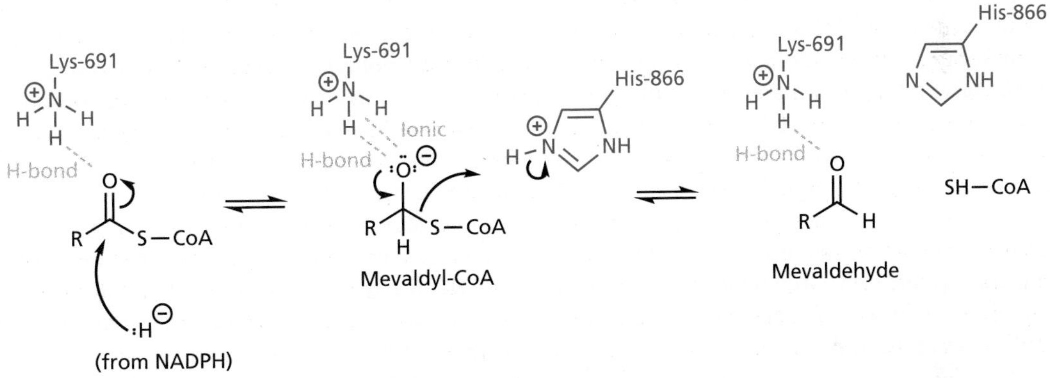

FIGURE CS1.6 Mechanism of the first reduction.

seen that this forms a hydrogen bond to the substrate, but it also plays a particularly important role in stabilizing the negatively charged oxygen of mevaldyl-CoA through hydrogen bonding and ionic interactions. This not only helps to stabilize the intermediate but also stabilizes the transition state leading to it. Consequently, the activation energy for the first step in the mechanism is lowered, allowing the reaction to occur more easily.

An uncharged glutamic acid residue (Glu-559) is also involved as an acid catalyst and provides a proton for the final stage where mevaldehyde is reduced to mevalonate (Fig. CS1.7).

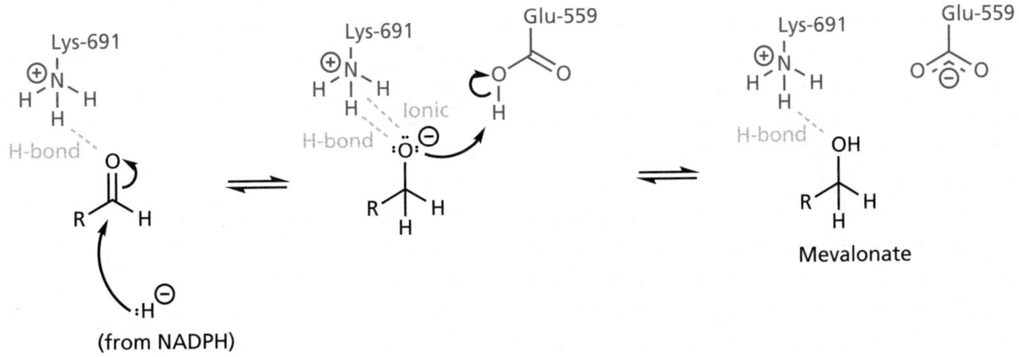

FIGURE CS1.7 Mechanism of the second reduction.

FIGURE CS1.8 Hydrogen bonding network involving Lys-691, Asp-767, and Glu-559.

It is rather unusual to have an uncharged glutamic acid residue present in an active site. The reason that it is not ionized is that there is a neighbouring aspartate residue (Asp-767) which affects the pK_a of the glutamic acid residue. The aspartate residue also helps to stabilize the ionic form of Lys-691 through the hydrogen bonding network shown in Fig. CS1.8.

CS1.3 The discovery of statins

Once the HMGR enzyme was identified as a potential target, researchers set out to find a lead compound that would inhibit it. They started by concentrating their attention on compounds produced by microorganisms. This might appear odd, but the rationale was that microorganisms are constantly involved in chemical warfare with each other, and so a microorganism that produces a chemical that is toxic to another microorganism gains an advantage in the never-ending fight for survival (section 12.4.1.2). It seemed likely that microorganisms lacking HMGR might produce HMGR inhibitors which would be toxic to microbes that require HMGR in order to produce important sterols.

CS1.3.1 Type I statins

Compactin (Mevastatin) (Fig. CS1.9) was the first potent statin to be found that inhibited HMGR, and it can be viewed as the lead compound for this group of drugs. It is a natural product that was isolated from *Penicillium citrinum* in the 1970s, following an investigation of

FIGURE CS1.10 General structure for the Type I statins (* represents an asymmetric centre).

6000 microbes by the Japanese pharmaceutical scientist Akira Endo. Studies showed that it was a highly potent inhibitor and had a 10 000-fold higher affinity for the enzyme than the natural substrate. Although it entered clinical trials, the drug never reached the market. The reason for this has never been fully revealed, but it is likely that adverse toxic effects were observed during preclinical trials.

In 1978, Merck isolated a closely related structure called **mevinolin** from the fermentation broth of *Aspergillus terreus*. This was also a potent inhibitor and clinical trials began in 1980. The drug was marketed in 1987 as **lovastatin** (Fig. CS1.9) and it revolutionized the treatment of hypercholesterolaemia (high cholesterol levels).

Other statins soon followed (Fig. CS1.9). **Simvastatin** is a semi-synthetic structure prepared from lovastatin and was first approved in 1988. **Pravastatin** is derived from compactin by biotransformation and reached the market in 1991.

These statins represent the first generation of statins and have been classified as **Type I statins**. They are all derived directly or indirectly from fungal metabolites, and share a similar structure which contains a polar 'head group' and a hydrophobic moiety which includes a bicyclic decalin ring (Fig. CS1.10).

Observant readers will notice that the structures for lovastatin and simvastatin contain a lactone ring, and not

Compactin IC_{50} = 23 nM
(Mevastatin)

Lovastatin (R = H) IC_{50} = 24 nM
Simvastatin (R = Me) IC_{50} = 11 nM

Pravastatin IC_{50} = 1900 nM

FIGURE CS1.9 Type I statins.

the acyclic polar head group shown in Fig. CS1.10. However, the lactone rings observed in these structures are hydrolysed by enzymes in the body to produce the polar 'head group', and this structure represents the active drug. Lovastatin and simvastatin are therefore termed as prodrugs (section 14.6).

Although the Type I statins have been extremely effective in lowering cholesterol levels, they do suffer from side effects. They are also difficult to synthesize due to the number of asymmetric centres associated with the decalin ring, and so further work was carried out to find statins with improved activity and reduced side effects, and which would be easier to synthesize. This resulted in a second generation of statins known as the Type II statins.

CS1.3.2 Type II statins

In contrast to type I statins, type II statins are synthetic structures that contain a different (and larger) hydrophobic moiety from the decalin ring system present in Type I statins (Fig. CS1.11). The hydrophobic moieties present in Type II statins may be larger than the decalin system,

but they are easier to synthesize since they contain no asymmetric centres (*simplification*; section 13.3.8). **Fluvastatin** was marketed in 1994, **atorvastatin** in 1997, **cerivastatin** in 1998, and **rosuvastatin** in 2003. The structures share a number of common structural features, and can be viewed as 'me too' or 'me better' drugs (section 12.4.4.1). In 2001, atorvastatin became the biggest selling drug in history. It is the most commonly prescribed statin and has remained the biggest selling drug in the world for several years, bringing in nearly £10 billion in sales for Pfizer during 2010 alone.

Of these structures, cerivastatin is the most hydrophobic, while pravastatin and rosuvastatin are the least hydrophobic. Studies have shown that statins with a lower hydrophobic character are more selective for liver cells, where most cholesterol synthesis takes place, and that such statins have fewer side effects. Side effects are thought to be caused by the inhibition of HMGR in other tissues, particularly muscle cells, where a condition known as **myalgia** can occur. This is a type of muscle pain or weakness that is particularly prevalent among individuals who take statins and who exercise vigorously.

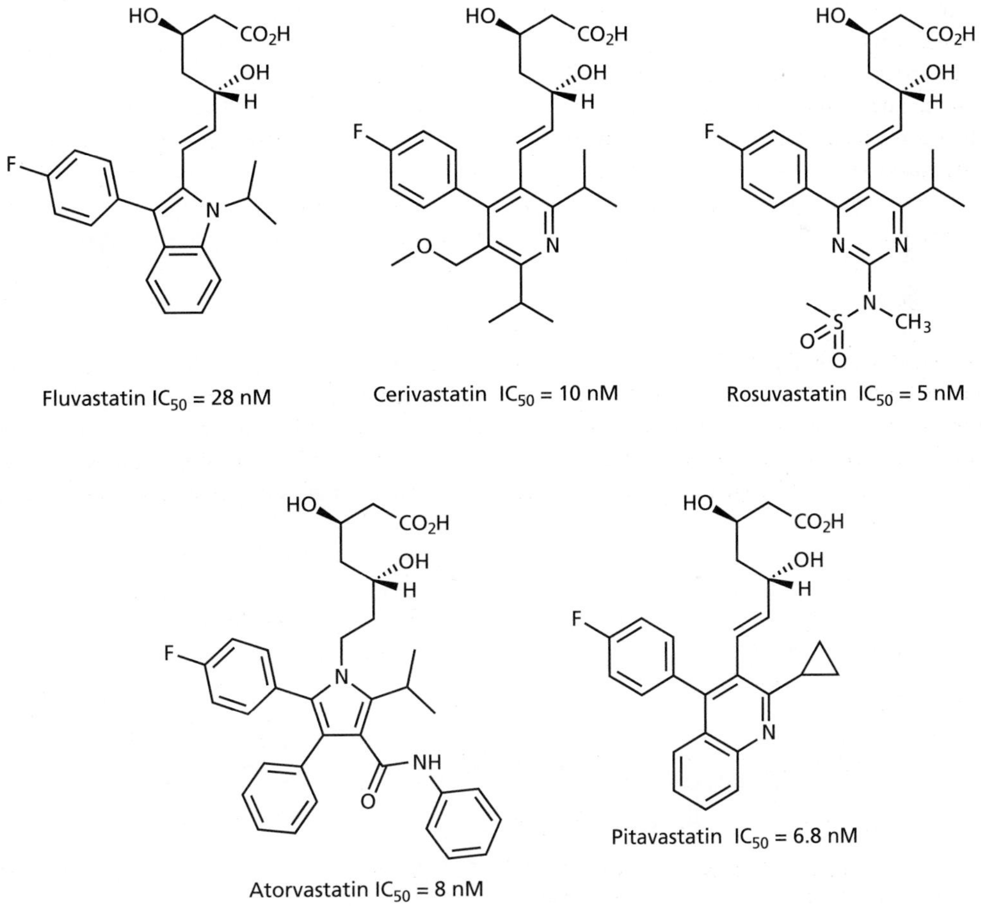

Fluvastatin IC_{50} = 28 nM

Cerivastatin IC_{50} = 10 nM

Rosuvastatin IC_{50} = 5 nM

Atorvastatin IC_{50} = 8 nM

Pitavastatin IC_{50} = 6.8 nM

FIGURE CS1.11 Type II statins..

A severe form of muscle toxicity is a condition known as **rhabdomyolysis** which can be fatal. Indeed, cerivastatin was withdrawn in 2001 following a large number of reported cases of rhabdomyolysis, which included 50 fatalities caused by kidney failure.

The most potent statin currently available is rosuvastatin. This structure contains a sulphonamide group which was introduced to lower the drug's hydrophobic character. Coincidentally, the introduction of this group resulted in enhanced binding interactions as described in section CS1.5 below.

The selectivity associated with less hydrophobic statins is a result of the way they access cells. The less hydrophobic statins do not diffuse easily through cell membranes and require transport proteins to reach effective levels within cells (see also sections 2.7.2 and 14.1). Liver cells possess a transport protein which can carry statins across the cell membrane, whereas muscle cells do not.

CS1.4 Mechanism of action for statins: pharmacodynamics

The statins work by acting as competitive inhibitors (section 7.1.1). They mimic the natural substrate and compete with it in order to bind to the active site. Unlike the natural substrate, they do not undergo an enzyme-catalysed reaction and they bind more strongly. How can we explain all of this?

Both the polar head group and the hydrophobic moieties are important to the action of statins. All the statins share the same polar head group and it is this group which mimics the natural substrate (HMG-SCoA). This can be seen more clearly if we redraw the structure of HMG-SCoA as shown in Fig. CS1.12 and compare it with a general structure for the statins. The head group of the statins can, therefore, mimic the natural substrate and bind to the active site using the same binding interactions. We now need to explain why statins bind more strongly than the natural substrate and why they are resistant to the enzyme-catalysed reaction.

- Firstly, the statins contain an extra hydrophobic region which can form additional hydrophobic interactions

FIGURE CS1.13 Structural comparison of mevaldyl CoA with statins.

with a hydrophobic binding region present in the enzyme. This allows the statins to bind more strongly.

- Secondly, the statins are resistant to the enzyme-catalysed reaction since the coenzyme A moiety in the substrate (which acts as a leaving group) has been replaced with a hydrophobic group that cannot act as a leaving group.

There is one other interesting feature about the statins. They are actually more similar to the first intermediate in the enzyme-catalysed mechanism—mevaldyl CoA—than to the substrate (Fig. CS1.13). Assuming that mevaldyl CoA is less stable than the substrate, this implies that the statins bear some resemblance to the transition state leading to mevaldyl CoA. Consequently, they would be expected to have a stronger binding interaction than the natural substrate and are likely to be acting as transition-state analogues (section 7.4). We shall now look in more detail at the binding interactions of the statins.

CS1.5 Binding interactions of statins

The binding interactions of the substrate with the enzyme have been studied by X-ray crystallography, as have the binding interactions of the statins[1] (see also section 13.3.11).

The polar head group of the statins binds in a similar fashion to the substrate, as described previously (Fig. CS1.5). As far as the hydrophobic region is concerned, we might be tempted to think that it would bind to the same

---

[1] These studies were actually carried out using the catalytic portion of the HMGR enzyme, rather than the whole enzyme.

FIGURE CS1.12 Structural comparison of HMG-SCoA with statins.

region of the active site as coenzyme A. However, studies carried out on the enzyme–substrate complex show that the binding pocket for coenzyme A is narrow and could not possibly accommodate the bulky hydrophobic groups that are present in statins. There is also no other hydrophobic region into which the statins could bind, and so they should really be inactive compounds. The fact that they do bind to the enzyme reflects a marked flexibility that is inherent to the enzyme.

Let us return to look more closely at how the substrate binds to HMGR. When the substrate binds, an alpha-helical section of the protein folds over the active site, shielding it from water and creating a narrow hydrophobic binding region for the coenzyme A portion of the substrate (Fig. CS1.14a). When a statin binds, the enzyme alters shape in a different manner. Movement of flexible C-terminal alpha helices exposes a shallow but different hydrophobic binding region next to the active site that can accommodate the hydrophobic moiety present in the statin (Fig. CS1.14b). Thus, the statins are effective inhibitors because they can take advantage of the enzyme's flexibility and essentially create their own binding site.

Comparing the binding interactions of type I and type II statins with the enzyme, it is found that the methylethyl group in a type II statin binds to the same part of the shallow hydrophobic region as the decalin ring of a type I statin. Type II statins have additional interactions which include van der Waals interactions with the hydrophobic side chains of amino acids such as leucine, valine, and alanine. A particularly important interaction involves the fluorophenyl group of type II statins and an arginine residue in the binding region (Fig. CS1.15). First of all, there is a polar interaction between this residue and the fluoro substituent. Secondly, the planar guanidinium group of the residue is stacked over the phenyl ring allowing additional interactions.

Atorvastatin and rosuvastatin can form an extra hydrogen bonding interaction with the enzyme that does not occur with other statins. This involves a serine residue which acts as hydrogen bond donor to the carbonyl oxygen atom of atorvastatin (Fig. CS1.15) or to the sulphone oxygen of rosuvastatin.

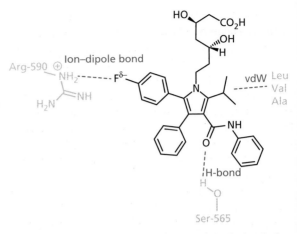

FIGURE CS1.15 Binding interactions for the hydrophobic moiety of atorvastatin with HMGR (3-hydroxy-3-methylglutaryl-coenzyme A reductase).

Rosuvastatin is unique among the statins in having an extra binding interaction between the sulphone group of the drug and Arg-568, making it the most strongly bound statin.

CS1.6 Other mechanisms of action for statins

The action of statins is not purely down to inhibition of HMGR. Inhibition certainly causes a decrease in the levels of mevalonate and cholesterol, but this in turn leads to up-regulation of the transcription and translation processes, leading to new HMGR which should counteract the inhibition. The fact that statins are still effective is due to other factors. In particular, the lowering of cholesterol levels in liver cells causes an increase in the synthesis of hepatic LDL receptors which are then incorporated into the cell membrane. These receptors are responsible for clearing LDL-cholesterol from the plasma and it is this that is crucial to the effectiveness of the statins.

FIGURE CS1.14 Crystal structure of HMG-CoA reductase with (a) substrate (pdb 1DQA) and (b) rosuvastatin (pdb 1HWL) bound to the active site.

CS1.7 **Other targets for cholesterol-lowering drugs**

We mentioned earlier that over 30 enzymes are involved in the biosynthesis of cholesterol. Early attempts to find cholesterol-lowering drugs studied the inhibition of enzymes catalysing the later steps in the biosynthetic pathway. There is sense in this, because inhibiting an enzyme late on in a biosynthesis is likely to have a more selective action. In other words, levels of the final product are lowered without affecting the biosynthesis of other compounds which share part of the same biosynthetic pathway. Although the inhibitors were effective, it led to an accumulation of unused substrate which proved insoluble and toxic. When HMG-CoA reductase is inhibited, the substrate is water soluble and easily metabolized. Therefore, it does not build up to toxic levels.

FURTHER READING

Bottorff, M., and Hansten, P. (2000) Long-term safety of hepatic hydroxymethyl glutaryl coenzyme A reductase inhibitors. *Archives of Internal Medicine*, **160**, 2273–80.

Jain, K. S., et al. (2007) The biology and chemistry of hyperlipidemia. *Bioorganic and Medicinal Chemistry*, **15**, 4674–99.

Istvan, E. S., et al. (2000) Crystal structure of the catalytic portion of human HMG-CoA reductase: insights into regulation of activity and catalysis. *The EMBO Journal*, **19**, 819–30.

Istvan, E. S., and Deseinhofer, J. (2001) Structural mechanisms for statin inhibition of HMG-CoA reductase. *Science*, **292**, 1160–4.

Istvan, E. (2003) Statin inhibition of HMG-CoA reductase: a 3-dimensional view. *Atherosclerosis Supplements*, **4**, 3–8.

Tobert, J. A. (2003) Lovastatin and beyond: the history of the HMG-CoA reductase inhibitors. *Nature Reviews Drug Discovery*, **2**, 517–26.

PART C

Drug discovery, design, and development

Drug discovery, design, and development: the past

Before the twentieth century, medicines consisted mainly of herbs and potions, and it was not until the mid-nineteenth century that the first serious efforts were made to isolate and purify the **active principles** of those remedies (i.e. the pure chemicals responsible for the medicinal properties). The success of these efforts led to the birth of many of the pharmaceutical companies we know today. Since then, many naturally occurring drugs have been obtained and their structures determined (e.g. morphine from opium, cocaine from coca leaves, quinine from the bark of the cinchona tree).

These natural products sparked off a major synthetic effort where chemists made literally thousands of analogues in an attempt to improve on what nature had provided. Much of this work was carried out on a trial and error basis, but the results obtained revealed several general principles behind drug design. Many of these principles are described in Chapters 13 and 14.

An overall pattern for drug discovery and drug development also evolved, but there was still a high element of trial and error involved in the process. The mechanism by which a drug worked at the molecular level was rarely understood and drug research very much focused on what is known as the **lead**[1] **compound**—an active principle isolated from a natural source or a synthetic compound prepared in the laboratory.

Drug discovery, design, and development: the present

In recent years, medicinal chemistry has undergone a revolutionary change. Rapid advances in the biological sciences have resulted in a much better understanding of how the body functions at the cellular and the molecular level. As a result, most research projects in the pharmaceutical industry or university sector now begin by identifying a suitable target in the body and designing a drug to interact with that target. An understanding of the structure and function of the target, as well as the mechanism by which it interacts with potential drugs, is crucial to this approach. Generally, we can identify

the following stages in drug discovery, design, and development:

Drug discovery: finding a lead (Chapter 12)

- Choose a disease!
- Choose a drug target.
- Identify a bioassay.
- Find a 'lead compound'.
- Isolate and purify the lead compound.
- Determine the structure of the lead compound.

Drug design (Chapters 13 and 14)

- Identify structure–activity relationships (SARs).
- Identify the pharmacophore.
- Improve target interactions (pharmacodynamics).
- Improve pharmacokinetic properties.

Drug development (Chapter 15)

- Patent the drug.
- Carry out preclinical trials (drug metabolism, toxicology, formulation and stability tests, pharmacology studies, etc.).
- Design a manufacturing process (chemical and process development).
- Carry out clinical trials.
- Register and market the drug.
- Make money!

Many of these stages run concurrently and are dependent on each other. For example, preclinical trials are usually carried out in parallel with the development of a manufacturing process. Even so, the discovery, design, and development of a new drug can take 15 years or more, involve the synthesis of over 10 000 compounds, and cost in the region of $800 million (£450 million).

There are three case studies in this section covering the discovery and design of clinically important agents. Case study 2 covers the design of ACE inhibitors, which are important cardiovascular drugs that act as antihypertensive agents. Case study 3 describes the discovery of the antimalarial agent artemisinin, and the design of analogues based on an understanding of its mechanism of action. Case study 4 is an example of how traditional drug design strategies were used in the design of important drugs that are used against the tropical disease of bilharzia.

---

[1] Pronounced 'leed'.

Drug discovery: finding a lead

In this chapter, we shall look at what happens when a pharmaceutical company or university research group initiates a new medicinal chemistry project through to the identification of a lead compound.

12.1 Choosing a disease

How does a pharmaceutical company decide which disease to target when designing a new drug? Clearly, it would make sense to concentrate on diseases where there is a need for new drugs. However, pharmaceutical companies have to consider economic factors as well as medical ones. A huge investment has to be made towards the research and development of a new drug. Therefore, companies must ensure that they get a good financial return for their investment. As a result, research projects tend to focus on diseases that are important in the developed world, because this is the market best able to afford new drugs. A great deal of research is carried out on ailments such as migraine, depression, ulcers, obesity, flu, cancer, and cardiovascular disease. Less is carried out on the tropical diseases of the developing world. Only when such diseases start to make an impact on western society do the pharmaceutical companies sit up and take notice. For example, there has been a noticeable increase in antimalarial research as a result of the increase in tourism to more exotic countries, and the spread of malaria into the southern states of the USA (see also Case study 3). Moreover, pharmaceutical companies are becoming more involved in partnerships with governments and philanthropic organizations such as the **Wellcome Trust**, the **Bill and Melinda Gates Foundation,** and **Medicines for Malaria Venture** in order to study diseases such as tuberculosis, malaria, and dengue.

Choosing which disease to tackle is usually a matter for a company's market strategists. The science becomes important at the next stage.

12.2 Choosing a drug target

12.2.1 Drug targets

Once a therapeutic area has been identified, the next stage is to identify a suitable drug target (e.g. receptor, enzyme, or nucleic acid). An understanding of which biomacromolecules are involved in a particular disease state is clearly important (see Box 12.1). This allows the medicinal research team to identify whether agonists or antagonists should be designed for a particular receptor, or whether inhibitors should be designed for a particular enzyme. For example, agonists of serotonin receptors are useful for the treatment of migraine, while antagonists of dopamine receptors are useful as antidepressants. Sometimes it is not known for certain whether a particular target will be suitable or not. For example, **tricyclic antidepressants,** such as **desipramine** (Fig. 12.1), are known to inhibit the uptake of the neurotransmitter **noradrenaline** from nerve synapses by inhibiting the carrier protein for noradrenaline (section 23.12.4). However, these drugs also inhibit uptake of a separate neurotransmitter called **serotonin,** and the possibility arose that inhibiting serotonin uptake might also be beneficial. A search for **selective serotonin uptake inhibitors** was initiated, which led to the discovery of the best-selling antidepressant drug **fluoxetine (Prozac)** (Fig. 12.1), but when this project was initiated it was not known for certain whether serotonin uptake inhibitors would be effective or not.

12.2.2 Discovering drug targets

If a drug or a poison produces a biological effect, there must be a molecular target for that agent in the body. In the past, the discovery of drug targets depended on finding the drug first. Many early drugs such as the analgesic **morphine** are natural products derived from plants, and just happen to interact with a molecular target in the human body. As this involves coincidence more than

BOX 12.1 Recently discovered targets: the caspases

The **caspases** are examples of recently discovered enzymes which may prove useful as drug targets. They are a family of protease enzymes that catalyse the hydrolysis of important cellular proteins, and which have been found to play a role in inflammation and cell death. Cell death is a natural occurrence in the body, and cells are regularly recycled. Therefore, caspases should not necessarily be seen as 'bad' or 'undesirable' enzymes. Without them, cells could be more prone to unregulated growth, resulting in diseases such as cancer.

The caspases catalyse the hydrolysis of particular target proteins such as those involved in DNA repair and the regulation of cell cycles. By understanding how these enzymes operate, there is the possibility of producing new therapies for a variety of diseases. For example, agents which promote the activity of caspases and lead to more rapid cell death might be useful in the treatment of diseases such

as cancer, autoimmune disease, and viral infections. For example, **carboplatin** is an anticancer agent that promotes caspase activity. Alternatively, agents which inhibit caspases and reduce the prevalence of cell death could provide novel treatments for trauma, neurodegenerative disease, and strokes. It is already known that the active site of caspases contains two amino acids that are crucial to the mechanism of hydrolysis—cysteine, which acts as a nucleophile, and histidine, which acts as an acid–base catalyst. The mechanism is similar to that used by acetylcholinesterase (section 22.12.3.2).

Caspases recognize an aspartate residue within protein substrates and cleave the peptide link next to it. Selective inhibitors have been developed which include aspartate or a mimic of it, but it remains to be seen whether such inhibitors have a clinical role.

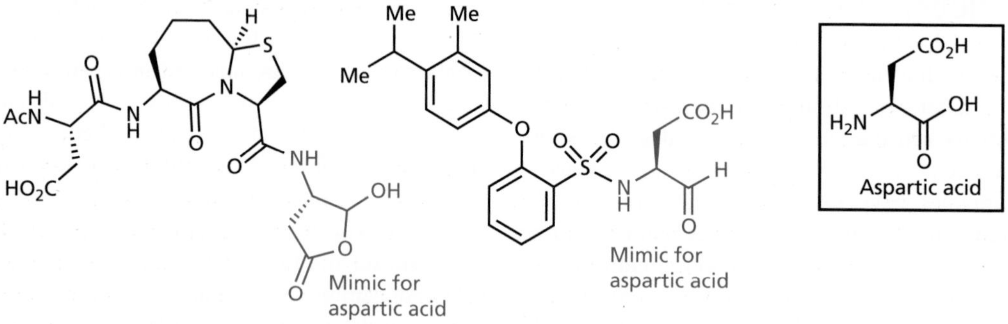

FIGURE 1 Selective caspase inhibitors.

design, the detection of drug targets was very much a hit and miss affair. Later, the body's own chemical messengers started to be discovered and pointed the finger at further targets. For example, since the 1970s a variety of peptides and proteins have been discovered which act as the body's own analgesics (enkephalins and endorphins). Another example is the rather surprising discovery that

nitric oxide acts as a chemical messenger (Box 3.1 and sections 22.3.2 and 26.5.1). Despite this, relatively few of the body's messengers were identified, either because they were present in such small quantity or because they were too short lived to be isolated. Indeed, many chemical messengers still remain undiscovered today. This, in turn, means that many of the body's potential drug

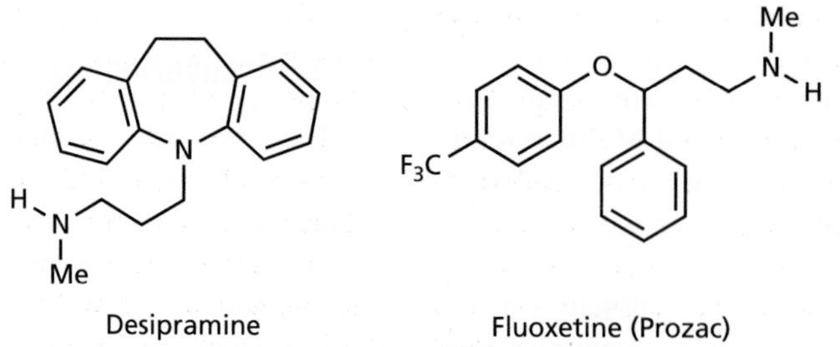

Desipramine   Fluoxetine (Prozac)

FIGURE 12.1 Antidepressant drugs.

targets remain hidden. Or at least it did! The advances in genomics and proteomics have changed all that. The various genome projects which have mapped the DNA of humans and other life forms, along with the newer field of proteomics (section 2.6), are revealing an ever increasing number of new proteins which are potential drug targets for the future. In many cases, the natural chemical messengers for these proteins are unknown, and medicinal chemistry is faced with new targets, but with no lead compounds to interact with them. Such targets have been defined as **orphan receptors**. The challenge is now to find a chemical which will interact with each of these targets in order to find out what their function is and whether they will be suitable as drug targets. This was one of the main driving forces behind the development of **combinatorial** and **parallel synthesis** (Chapter 16).

12.2.3 Target specificity and selectivity between species

Target specificity and selectivity is a crucial factor in modern medicinal chemistry research. The more selective a drug is for its target, the less chance there is that it will interact with different targets and have undesirable side effects.

In the field of antimicrobial agents, the best targets to choose are those that are unique to the microbe and are not present in humans. For example, **penicillin** targets an enzyme involved in bacterial cell wall biosynthesis. Mammalian cells do not have a cell wall, so this enzyme is absent in human cells and penicillin has minimal side effects (section 19.5). In a similar vein, sulphonamides inhibit a bacterial enzyme not present in human cells (section 19.4.1.5), while several agents used to treat AIDS inhibit an enzyme called **retroviral reverse transcriptase** which is unique to the infectious agent HIV (section 20.7.3).

Other cellular features that are unique to microorganisms could also be targeted. For example, the microorganisms which cause sleeping sickness in Africa are propelled by means of a tail-like structure called a **flagellum**. This feature is not present in mammalian cells, so designing drugs that bind to the proteins making up the flagellum to prevent it working could be potentially useful in treating that disease.

Having said all that, it is still possible to design drugs against targets which are present both in humans and microbes, as long as the drugs show selectivity against the microbial target. Fortunately, this is perfectly feasible. An enzyme which catalyses a reaction in a bacterial cell differs significantly from the equivalent enzyme in a human cell. The enzymes may have been derived from an ancient common ancestor, but several million years of evolution have

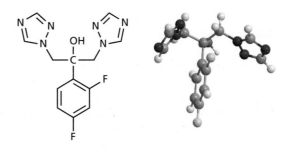

FIGURE 12.2 Fluconazole.

resulted in significant structural differences. For example, the antifungal agent **fluconazole** (Fig. 12.2) inhibits a fungal demethylase enzyme involved in steroid biosynthesis. This enzyme is also present in humans, but the structural differences between the two enzymes are significant enough that the antifungal agent is highly selective for the fungal enzyme. Other examples of bacterial or viral enzymes which are sufficiently different from their human equivalents are **dihydrofolate reductase** (section 19.4.2) and **viral DNA polymerase** (section 20.6.1).

12.2.4 Target specificity and selectivity within the body

Selectivity is also important for drugs acting on targets within the body. Enzyme inhibitors should only inhibit the target enzyme and not some other enzyme. Receptor agonists/antagonists should ideally interact with a specific kind of receptor (e.g. the adrenergic receptor) rather than a variety of different receptors. However, nowadays medicinal chemists aim for even higher standards of target selectivity. Ideally, enzyme inhibitors should show selectivity between the various isozymes of an enzyme (isozymes are the structural variants of an enzyme that result from different amino acid sequences or quaternary structure—section 3.7). For example, there are three different isoforms of **nitric oxide synthase** (NOS)—the enzyme responsible for generating the chemical messenger **nitric oxide** (Box 3.1). Selective inhibitors for one of these isoforms (nNOS) could potentially be useful in treating cerebral palsy and other neurodegenerative diseases.

Receptor agonists and antagonists should not only show selectivity for a particular receptor (e.g. an adrenergic receptor) or even a particular receptor type (e.g. the β-adrenergic receptor), but also for a particular receptor subtype (e.g. the $β_2$-adrenergic receptor). One of the current areas of research is to find antipsychotic agents with fewer side effects. Traditional antipsychotic agents act as antagonists of dopamine receptors. However, it has been found that there are five dopamine receptor subtypes and that traditional antipsychotic agents antagonize two of these (D_3 and D_2). There is good evidence that the D_2 receptor is responsible for the undesirable Parkinsonian type side effects of current drugs, and so research is now underway to find a selective D_3 antagonist.

12.2.5 Targeting drugs to specific organs and tissues

Targeting drugs against specific receptor subtypes often allows drugs to be targeted to specific organs or to specific areas of the brain. This is because the various receptor subtypes are not uniformly distributed around the body, but are often concentrated in particular tissues. For example, the β-adrenergic receptors in the heart are predominantly β_1, whereas those in the lungs are β_2. This makes it feasible to design drugs that will work on the lungs with a minimal side effect on the heart, and vice versa.

Attaining subtype selectivity is particularly important for drugs that are intended to mimic neurotransmitters. Neurotransmitters are released close to their target receptors and once they have passed on their message, they are quickly deactivated and do not have the opportunity to 'switch on' more distant receptors. Therefore, only those receptors which are fed by 'live' nerves are switched on.

In many diseases, there is a 'transmission fault' to a particular tissue or in a particular region of the brain. For example, in Parkinson's disease, **dopamine** transmission is deficient in certain regions of the brain, although it is functioning normally elsewhere. A drug could be given to mimic dopamine in the brain. However, such a drug acts like a hormone rather than as a neurotransmitter because it has to travel round the body in order to reach its target. This means that the drug could potentially 'switch on' all the dopamine receptors around the body and not just the ones that are suffering the dopamine deficit. Such drugs would have a large number of side effects, so it is important to make the drug as selective as possible for the particular type or subtype of dopamine receptor affected in the brain. This would target the drug more effectively to the affected area and reduce side effects elsewhere in the body.

12.2.6 Pitfalls

A word of caution! It is possible to identify whether a particular enzyme or receptor plays a role in a particular ailment. However, the body is a highly complex system. For any given function, there are usually several messengers, receptors, and enzymes involved in the process. For example, there is no one simple cause for hypertension (high blood pressure). This is illustrated by the variety of receptors and enzymes which can be targeted in its treatment. These include β_1-adrenoceptors, calcium ion channels, angiotensin-converting enzyme (ACE), potassium ion channels, and angiotensin II receptors (Chapters 23 and 26).

As a result, more than one target may need to be addressed for a particular ailment (Box 12.2). For example, most of the current therapies for asthma involve a combination of a bronchodilator (β_2-agonist) and an anti-inflammatory agent such as a corticosteroid.

BOX 12.2 Pitfalls in choosing particular targets

Drugs are designed to interact with a particular target because that target is believed to be important to a particular disease process. Occasionally, though, a particular target may not be as important to a disease as was first thought. For example, the dopamine D_2 receptor was thought to be involved in causing nausea. Therefore, the D_2 receptor antagonist **metoclopramide** was developed as an anti-emetic agent. However, it was found that more potent D_2 antagonists were less effective, implying that a different receptor might be more important in producing nausea. Metoclopramide also antagonizes the 5-hydroxytryptamine (5-HT_3) receptor, so antagonists for this receptor were studied, which led to the development of the anti-emetic drugs **granisetron** and **ondansetron**.

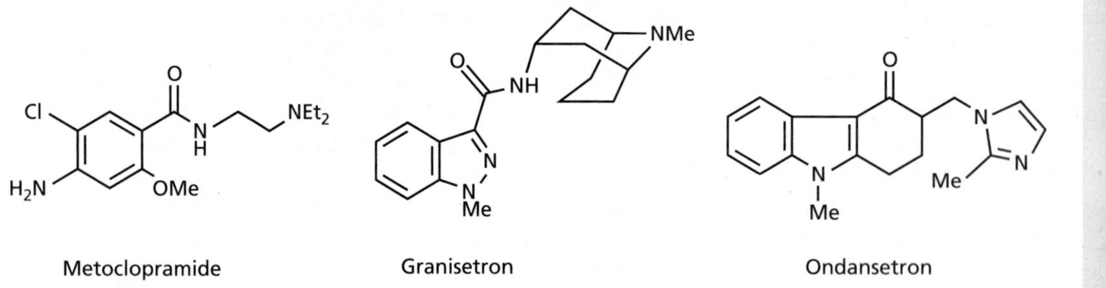

FIGURE 1 Anti-emetic agents.

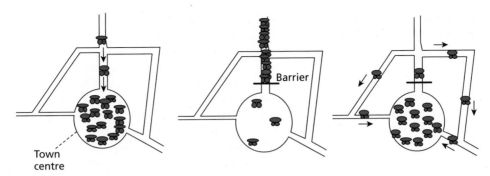

FIGURE 12.3 Avoiding the jam.

Sometimes, drugs designed against a specific target become less effective over time. Because cells have a highly complex system of signalling mechanisms, it is possible that the blockade of one part of that system could be bypassed. This could be compared to blocking the main road into town to try and prevent congestion in the town centre. To begin with, the policy works, but in a day or two commuters discover alternative routes, and congestion in the centre becomes as bad as ever (Fig. 12.3).

12.2.7 Multi-target drugs

In certain diseases and afflictions, there can be an advantage in 'hitting' a number of different targets selectively, as this can be more beneficial than hitting just one. Combination therapy is normally used to achieve this by administering two or more drugs showing selectivity against the different targets. This is particularly the case in the treatment of cancer (Chapter 21) and HIV infection (Box 20.2). However, combination therapies are also used in a variety of other situations (sections 19.4.2.1, 19.5.4, and 20.10). The disadvantage of combination therapies is the number of different medications and the associated dose regimens. Therefore, there are benefits in designing a single drug that can act selectively at different targets in a controlled manner—a **multi-target-directed ligand**. Many research projects now set out to discover new drugs with a defined profile of activity against a range of specific targets. For example, a research team may set out to find a drug that has agonist activity for one receptor subtype and antagonist activity at another. A further requirement may be that the drug neither acts as a substrate for cytochrome p450 enzymes, nor affects the activity of those metabolic enzymes (section 11.5). Tests are also carried out to ensure that the drug does not interact with targets that could lead to toxicity (Box 12.3). A current area of research is in designing dual-action drugs to treat depression (section 23.12.4 and Case study 7). Dual and triple action drugs are also being studied for the treatment of Alzheimer's disease (section 22.15).

BOX 12.3 Early tests for potential toxicity

In vivo and *in vitro* tests are often carried out at an early stage to find out whether lead compounds or candidate drugs are likely to have certain types of toxicity. One such test is to see whether compounds inhibit HERG potassium ion channels in the heart. HERG stands for the gene that codes for this protein, the so called **Human Ether-a-go-go Related Gene**! Who makes up these names? Several promising drugs have had to be withdrawn at a very late stage in their development because they were found to inhibit the HERG potassium ion channels. Inhibition can result in disruption in the normal rhythm of the heart, leading to fibrillation, heart failure, and death. The gastric agent **cisapride** (Fig. 1) and the antihistamine **terfenadine** (section 11.5.6) both had to be withdrawn from the market because of this problem. A large variety

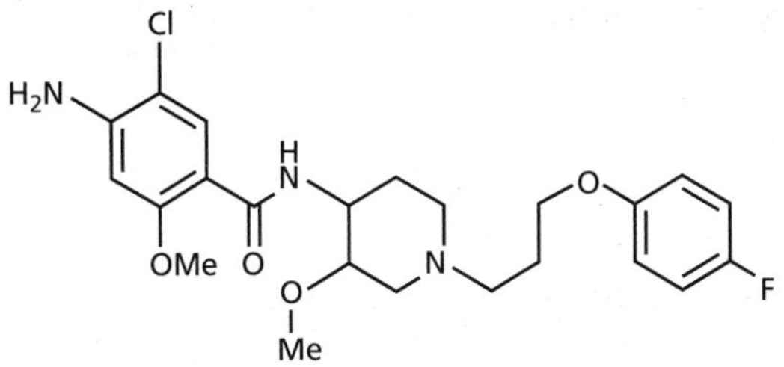

FIGURE 1 Cisapride.

of other structures have been found to have this unwanted effect, and so tests to detect this property are best done as

(Continued)

BOX 12.3 Early tests for potential toxicity (*Continued*)

early as possible in order to remove this property as part of the drug optimization process.

The **Ames test** is another early test that is worth carrying out in order to detect potential mutagenicity or carcinogenicity in new compounds. It involves the use of a mutated bacterial strain of *Salmonella typhimurium* that lacks the ability to synthesize histidine and can only grow in a medium containing that amino acid. The test involves growing the mutant strain in a medium that contains a small amount of histidine, as well as the test compound. Since there is only a small amount of histidine present, the mutant bacteria will soon stop growing and dividing. However, some of the mutant bacteria will 'back mutate' to the original wild-type strain. These cells are now able to synthesize their own histidine and will keep growing. The bacterial colonies that are present on the plate are subcultured onto plates lacking histidine to detect the presence of the wild-type strains, allowing a measure of the mutation rate. Any mutagenic or carcinogenic drug that is present in the original medium will increase the mutation rate, relative to a reference culture containing no drug.

Many research groups now concentrate on 'taming' Ames and HERG liabilities at an early stage of drug development. For example, structure I (Fig. 2) is an antagonist for the **melanin-concentrating hormone receptor** (MCHR)—a receptor that has been identified as an important target for novel anti-obesity drugs. Unfortunately, structure I blocks HERG ion channels and has Ames liability (i.e. it has mutagenic properties). A library of analogues was prepared by parallel synthesis (Chapter 16), which identified structure II as a potent antagonist having no Ames liability. Further work led to structure III, which lacked the Ames liability and has a greatly reduced capacity to block HERG ion channels.

Another example where studies were carried out to avoid interactions with the HERG ion channels was in the development of the antiviral agent **maraviroc** (section 20.7.5).

Microbioassay tests are also been developed to test for drug toxicity. These involve the use of microfluidic systems on microchips. Cells from different organs are grown in microchannels on the microchip and then tiny volumes of drug solution are passed through the microchip to see what effect they have.

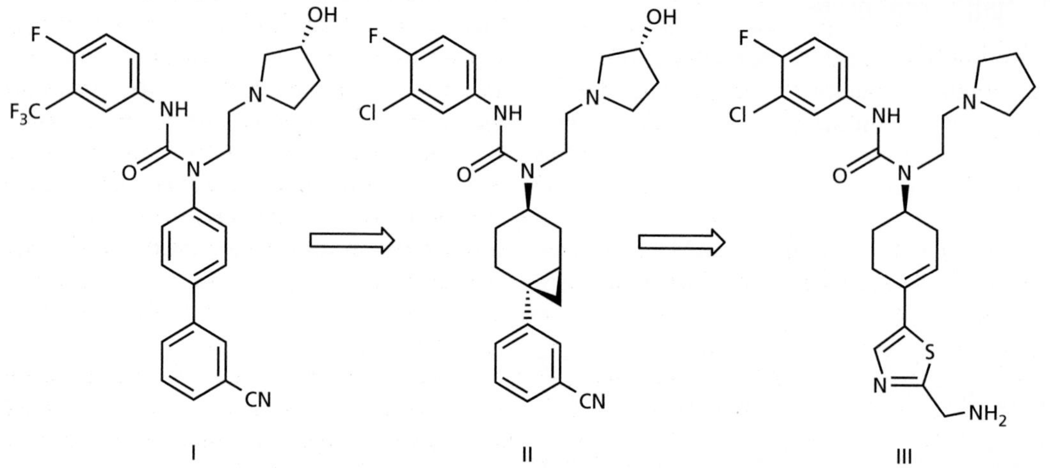

FIGURE 2 The development of agents to remove undesirable properties.

A less selective example is **olanzapine** (Fig. 12.4). This drug binds to more than a dozen receptors for serotonin, dopamine, muscarine, noradrenaline, and histamine. This kind of profile would normally be unacceptable, but olanzaprine has been highly effective in the treatment of schizophrenia, probably because it blocks both serotonin and dopamine receptors. Drugs which interact with a large range of targets are called **promiscuous ligands** or **dirty drugs**. Such drugs can act as lead compounds for the development of more selective multi-targeted ligands (see also section 22.15.3).

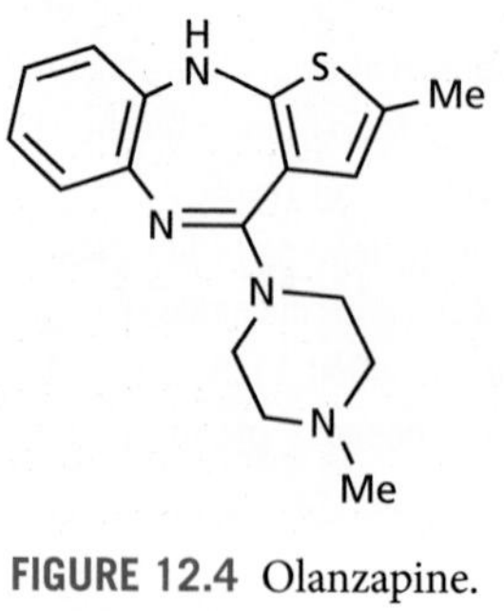

FIGURE 12.4 Olanzapine.

12.3 **Identifying a bioassay**

12.3.1 **Choice of bioassay**

Choosing the right bioassay or test system is crucial to the success of a drug research programme. The test should be simple, quick, and relevant, as there are usually a large number of compounds to be analysed. Human testing is not possible at such an early stage, so the test has to be done *in vitro* (i.e. on isolated cells, tissues, enzymes, or receptors) or *in vivo* (on animals). In general, *in vitro* tests are preferred over *in vivo* tests because they are cheaper, easier to carry out, less controversial, and they can be automated. However, *in vivo* tests are often needed to check whether drugs have the desired pharmacological activity, and whether they have acceptable pharmacokinetic properties. In modern medicinal chemistry, a variety of tests are usually carried out both *in vitro* and *in vivo* to determine not only whether the candidate drugs are acting at the desired target, but also whether they have activity at other undesired targets (Box 12.3). The direction taken by projects is then determined by finding drugs that have the best balance of good activity at the desired target and minimal activity at other targets. In this way, there is less likelihood of millions of dollars being wasted developing a drug that will either fail clinical trials or be withdrawn from the market with all the associated litigation that involves—a '**fail fast, fail cheap**' strategy.

12.3.2 *In vitro* tests

In vitro tests do not involve live animals. Instead, specific tissues, cells, or enzymes are used. Enzyme inhibitors can be tested on the pure enzyme in solution. In the past, it could be a major problem to isolate and purify sufficient enzyme to test, but nowadays genetic engineering can be used to incorporate the gene for a particular enzyme into fast-growing cells such as yeast or bacteria. These then produce the enzyme in larger quantities, making isolation easier. For example, **HIV protease** (section 20.7.4.1) has been cloned and expressed in the bacterium *Escherichia coli*. A variety of experiments can be carried out on this enzyme to determine whether an enzyme inhibitor is competitive or non-competitive, and to determine IC_{50} values (section 7.8).

Receptor agonists and antagonists can be tested on isolated tissues or cells which express the target receptor on their surface. Sometimes these tissues can be used to test drugs for physiological effects. For example, bronchodilator activity can be tested by observing how well compounds inhibit contraction of isolated tracheal smooth muscle. Alternatively, the affinity of drugs for receptors (how strongly they bind) can be measured by radioligand studies (section 8.9). Many *in vitro* tests have been designed by genetic engineering where the gene coding for a specific receptor is identified, cloned, and expressed in fast-dividing cells such as bacterial, yeast, or tumour cells. For example, **Chinese Hamster Ovarian cells** (CHO cells) are commonly used for this purpose, as they express a large amount of the cloned receptor on their cell surface. *In vitro* studies on whole cells are useful because there are none of the complications of *in vivo* studies where the drug has to survive metabolic enzymes or cross barriers such as the gut wall. The environment surrounding the cells can be easily controlled and both intracellular and intercellular events can be monitored, allowing measurement of efficacy and potency (section 8.9). Primary cell cultures (i.e. cells that have not been modified) can be produced from embryonic tissues; transformed cell lines are derived from tumour tissue. Cells grown in this fashion are all identical.

Antibacterial drugs are tested *in vitro* by measuring how effectively they inhibit or kill bacterial cells in culture. It may seem strange to describe this as an *in vitro* test, as bacterial cells are living microorganisms. However, *in vivo* antibacterial tests are defined as those that are carried out on animals or humans to test whether antibacterial agents combat infection.

The lack of a suitable *in vitro* test can actually prevent progress being made in a particular field of medicinal chemistry. For example, research into finding novel antiviral agents for the treatment of hepatitis C was hindered for many years because of the lack of a suitable *in vitro* test. An *in vitro* test has now been developed, but the time needed to achieve that goal was one of the main reasons why novel antiviral agents for the treatment of hepatitis C have only appeared on the market since 2011, whereas antiviral drugs for the treatment of HIV appeared in the 1980s.

In vitro tests are also used to test for the pharmacokinetic properties of compounds. For example, the **Caco-2 cell monolayer absorption** model is used to assess how well a drug is likely to be absorbed from the gastrointestinal tract. Microsomes and hepatocytes extracted from liver cells contain cytochrome P450 enzymes, and can be used to assess the likely metabolism of drug candidates, as well as identifying possible drug–drug interactions. Another *in vitro* assay using artificial membranes has been developed as a simple and rapid measure of how effectively drugs will cross the blood–brain barrier.

12.3.3 *In vivo* tests

In vivo tests on animals often involve inducing a clinical condition in the animal to produce observable symptoms. The animal is then treated to see whether the drug alleviates the problem by eliminating the observable symptoms.

For example, the development of non-steroidal inflammatory drugs was carried out by inducing inflammation on test animals, then testing drugs to see whether they relieved the inflammation.

Transgenic animals are often used in *in vivo* testing. These are animals whose genetic code has been altered. For example, it is possible to replace some mouse genes with human genes. The mouse produces the human receptor or enzyme and this allows *in vivo* testing against that target. Alternatively, the mouse's genes could be altered such that the animal becomes susceptible to a particular disease (e.g. breast cancer). Drugs can then be tested to see how well they prevent that disease.

There are several problems associated with *in vivo* testing. It is slow and expensive, and it also causes animal suffering. There are the many problems of pharmacokinetics (Chapter 11), and so the results obtained may be misleading and difficult to rationalize if *in vivo* tests are carried out in isolation. For example, how can one tell whether a negative result is due to the drug failing to bind to its target or not reaching the target in the first place? Thus, *in vitro* tests are usually carried out first to determine whether a drug interacts with its target, and *in vivo* tests are then carried out to test pharmacokinetic properties.

Certain *in vivo* tests might turn out to be invalid. It is possible that the observed symptoms might be caused by a different physiological mechanism than the one intended. For example, many promising anti-ulcer drugs which proved effective in animal testing were ineffective in clinical trials. Finally, different results may be obtained in different animal species. For example, **penicillin methyl ester prodrugs** (Box 19.7) are hydrolysed in mice or rats to produce active penicillins, but are not hydrolysed in rabbits, dogs, or humans. Another example involves **thalidomide** which is teratogenic in rabbits and humans, but has no such effect in mice.

Despite these issues, *in vivo* testing is still crucial in identifying the particular problems that might be associated with using a drug *in vivo* and which cannot be picked up by *in vitro* tests.

12.3.4 **Test validity**

Sometimes the validity of testing procedures is easy and clear-cut. For example, an antibacterial agent can be tested *in vitro* by measuring how effectively it kills bacterial cells. A local anaesthetic can be tested *in vitro* on how well it blocks action potentials in isolated nerve tissue. In other cases, the testing procedure is more difficult. For example, how do you test a new antipsychotic drug? There is no animal model for this condition and so a simple *in vivo* test is not possible. One way round this problem is to propose which receptor or receptors might be involved in a

medical condition, and to carry out *in vitro* tests against these in the expectation that the drug will have the desired activity when it comes to clinical trials. One problem with this approach is that it is not always clear-cut whether a specific receptor or enzyme is as important as one might think to the targeted disease (see Box 12.2).

12.3.5 **High-throughput screening**

Robotics and the miniaturization of *in vitro* tests on genetically modified cells has led to a process called high-throughput screening (HTS) which is particularly effective in identifying potential new lead compounds. This involves the automated testing of large numbers of compounds versus a large number of targets; typically, several thousand compounds can be tested at once in 30–50 biochemical tests. It is important that the test should produce an easily measurable effect which can be detected and measured automatically. This effect could be cell growth, an enzyme-catalysed reaction which produces a colour change, or displacement of radioactively labelled ligands from receptors.

Receptor antagonists can be studied using modified cells which contain the target receptor in their cell membrane. Detection is possible by observing how effectively the test compounds inhibit the binding of a radiolabelled ligand. Another approach is to use yeast cells which have been modified such that activation of a target receptor results in the activation of an enzyme which, when supplied with a suitable substrate, catalyses the release of a dye. This produces an easily identifiable colour change.

In general, positive hits are compounds which have an activity in the range 30 μM–1 nM. Unfortunately, HTS can generate many false-positive hits, and there is a high failure rate between the number of hits and those compounds which are eventually identified as authentic lead compounds. One of the main causes of false hits is what are known as **promiscuous inhibitors**. These are agents which appear to inhibit a range of different target proteins and show very poor selectivity. It is believed that agents working in this manner come together in solution to form molecular aggregates which adsorb target proteins onto their surface, resulting in the inhibition observed. The effect is more pronounced if mixtures of compounds are being tested in solution, such as those prepared by combinatorial syntheses. This kind of inhibition is of no use to drug design and it is important to eliminate these agents early on as potential lead compounds, such that time is not wasted resynthesizing and investigating them. One way of finding out whether promiscuous inhibition is taking place is to add a detergent to the test solution. This reverses and prevents the phenomenon.

Other false hits include agents which are chemically reactive and carry out a chemical reaction with the target protein, such as the alkylation or acylation of a susceptible nucleophilic group. This results in an irreversible inhibition of the protein since the agent becomes covalently linked to the target. Although there are important drugs which act as irreversible inhibitors, the emphasis in HTS is to find reversible inhibitors which interact with their targets through intermolecular binding interactions. For that reason, known alkylating or acylating agents should not be included in HTS, or, if they are, they should not be considered as potential lead compounds. Examples of reactive groups include alkyl halides, acid chlorides, epoxides, aldehydes, α-chloroketones, and trifluoromethyl ketones.

12.3.6 Screening by NMR

Nuclear magnetic resonance (NMR) spectroscopy is an analytical tool which has been used for many years to determine the molecular structure of compounds. More recently, it has been used to detect whether a compound binds to a protein target. In NMR spectroscopy, a compound is radiated with a short pulse of energy which excites the nuclei of specific atoms such as hydrogen, carbon, or nitrogen. Once the pulse of radiation has stopped, the excited nuclei slowly relax back to the ground state giving off energy as they do so. The time taken by different nuclei to give off this energy is called the **relaxation time** and this varies depending on the environment or position of each atom in the molecule. Therefore, a different signal will be obtained for each atom in the molecule, and a spectrum is obtained which can be used to determine the structure.

The size of the molecule also plays an important role in the length of the relaxation time. Drugs are generally small molecules and have long relaxation times, whereas large molecules such as proteins have short relaxation times. Therefore it is possible to delay the measurement of energy emission such that only small molecules are detected. This is the key to the detection of binding interactions between a protein and a test compound.

First of all, the NMR spectrum of the drug is taken, then the protein is added and the spectrum is re-run, introducing a delay in the measurement such that the protein signals are not detected. If the drug fails to bind to the protein, then its NMR spectrum will still be detected. If the drug binds to the protein, it essentially becomes part of the protein. As a result, its nuclei will have a shorter relaxation time and no NMR spectrum will be detected.

This screening method can also be applied to a mixture of compounds arising from a natural extract or from a combinatorial synthesis. If any of the compounds present bind to the protein, its relaxation time is shortened and so signals due to that compound will disappear from the spectrum. This will show that a component of the mixture is active and determine whether it is worthwhile separating the mixture or not.

There are several advantages in using NMR as a detection system:

- It is possible to screen 1000 small-molecular-weight compounds a day with one machine.
- The method can detect weak binding which would be missed by conventional screening methods.
- It can identify the binding of small molecules to different regions of the binding site (section 12.4.10).
- It is complimentary to HTS. The latter may give false-positive results, but these can be checked by NMR to ensure that the compounds concerned are binding in the correct binding site (section 12.4.10).
- The identification of small molecules which bind weakly to part of the binding site allows the possibility of using them as building blocks for the construction of larger molecules that bind more strongly (section 12.4.10).
- Screening can be done on a new protein without needing to know its function.

Disadvantages include the need to purify the protein and to obtain it in a significant quantity (at least 200 mg).

12.3.7 Affinity screening

A nice method of screening mixtures of compounds for active constituents is to take advantage of the binding affinity of compounds for the target. This not only detects the presence of such agents, but picks them out from the mixture. For example, the vancomycin family of antibacterial agents has a strong binding affinity for the dipeptide D-Ala-D-Ala (section 19.5.5.2). D-Ala-D-Ala was linked to sepharose resin, and the resin was mixed with extracts from various microbes which were known to have antibacterial activity. If an extract lost antibacterial activity as a result of this operation, it indicated that active compounds had bound to the resin. The resin could then be filtered off and, by changing the pH, the compounds could be released from the resin for identification.

12.3.8 Surface plasmon resonance

Surface plasmon resonance (SPR) is an optical method of detecting when a ligand binds to its target. The procedure is patented by Pharmacia Biosensor as **BIAcore** and makes use of a dextran-coated, gold-surfaced glass chip (Fig. 12.5). A ligand that is known to bind to the target is immobilized by linking it covalently to the dextran matrix, which is in a flow of buffer solution. Monochromatic, plane-polarized light is shone at an angle of incidence (α)

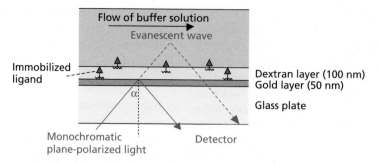

FIGURE 12.5 Surface plasmon resonance. The word evanescent means 'passing out of sight'.

from below the glass plate and is reflected back at the interface between the dense gold-coated glass and the less dense buffer solution. However, a component of the light called the **evanescent wave** penetrates a distance of about one wavelength into the buffer/dextran matrix. Normally, all of the light including the evanescent wave is reflected back, but if the gold film is very thin (a fraction of the evanescent wavelength), and the angle of incidence is exactly right, the evanescent wave interacts with free oscillating electrons called **plasmons** in the metal film. This is the surface plasmon resonance. Energy from the incident light is then lost to the gold film. As a result, there is a decrease in the reflected light intensity, which can be measured.

The angle of incidence when SPR occurs depends crucially on the refractive index of the buffer solution close to the metal film surface. This means that if the refractive index of the buffer changes, the angle of incidence at which SPR takes place changes as well.

If the macromolecular target for the immobilized ligand is now introduced into the buffer flow, some of it will be bound by the immobilized ligand. This leads to a change of refractive index in the buffer solution close to the metal-coated surface, which can be detected by measuring the change in the angle of incidence required to get SPR. The technique allows the detection of ligand–target binding and can also be used to measure rate and equilibrium binding constants.

Suppose, now, we want to test whether a novel compound is binding to the target. This can be tested by introducing the novel compound into the buffer flow along with the target. If the test compound *does* bind to the target, less target will be available to bind to the immobilized ligands, so there will be a different change in both the refractive index and the angle of incidence.

12.3.9 Scintillation proximity assay

Scintillation proximity assay (SPA) is a visual method of detecting whether a ligand binds to a target. It involves the immobilization of the target by linking it covalently to beads which are coated with a scintillant. A solution

of a known ligand labelled with iodine-125 is then added to the beads. When the labelled ligand binds to the immobilized target, the ^{125}I acts as an energy donor and the scintillant-coated bead acts as an energy acceptor, resulting in an emission of light that can be detected. In order to find out whether a novel compound interacts with the target, the compound is added to the solution of the labelled ligand and the mixture is added to the beads. Successful binding by the novel compound will mean that less of the labelled ligand will bind, resulting in a reduction in the emission of light.

12.3.10 Isothermal titration calorimetry

Isothermal titration calorimetry (ITC) is a technique that is used to determine the thermodynamic properties of binding between a drug and its protein target, in particular the binding affinity and enthalpy change. Two identical glass cells are used which are filled with buffer solution. One of the cells acts as the reference cell, while the other acts as the sample cell and contains the protein target in solution. The reference cell is heated slightly to a constant temperature. The sample cell is heated to the same temperature through an automatic feedback system, whereby any temperature difference between the two cells is detected and power is applied to the sample cell to equalize the temperature. Once the apparatus has stabilized, a constant level of power is being used to maintain the two cells at the same constant temperature.

The drug is now added to the sample cell and binds to the protein target. If the binding interaction is exothermic, heat energy is generated within the sample cell and so less external power is needed to maintain the cell temperature. If the interaction is endothermic, the opposite holds true and more external power has to be applied to maintain the temperature. The external power required to maintain the temperature of the sample cell is measured with respect to time, with power 'spikes' occurring every time the drug is injected into the cell. Measurement of these spikes allows the determination of the thermodynamic properties of binding.

12.3.11 Virtual screening

Virtual screening involves the use of computer programs to assess whether known compounds are likely to be lead compounds for a particular target. There is no guarantee that 'positive hits' from a virtual screening will, in fact, be active, and the compounds still have to be screened experimentally, but the results from a virtual screening can be used to make experimental screening methods more efficient. In other words, if there are several thousand compounds available for testing, virtual screening can be used to identify those compounds which are most likely to be active, and so those are the structures which would be given priority for actual screening. Virtual screening can involve a search for pharmacophores known to be required for activity. Alternatively, compounds can be docked into target binding sites (sections 17.11–17.13).

KEY POINTS

- Pharmaceutical companies tend to concentrate on developing drugs for diseases which are prevalent in developed countries, and aim to produce compounds with better properties than existing drugs.

- A molecular target is chosen which is believed to influence a particular disease when affected by a drug. The greater the selectivity that can be achieved, the less chance of side effects.

- A suitable bioassay must be devised which will demonstrate whether a drug has activity against a particular target. Bioassays can be carried out *in vitro* or *in vivo*, and usually a combination of tests is used.

- HTS involves the miniaturization and automation of *in vitro* tests such that a large number of tests can be carried out in a short period of time.

- Compounds can be tested by NMR spectroscopy for their affinity to a macromolecular target. The relaxation times of ligands bound to a macromolecule are shorter than when they are unbound.

- SPR, SPA, and ITC are three visual methods of detecting whether ligands bind to macromolecular targets.

- Virtual screening can be used to identify the compounds most likely to be active in experimental screening.

12.4 Finding a lead compound

Once a target and a testing system have been chosen, the next stage is to find a lead compound—a compound which shows the desired pharmacological activity. The level of activity may not be very great and there may be undesirable side effects, but the lead compound provides a start for the drug design and development process. There are various ways in which a lead compound might be discovered as described in the following sections.

12.4.1 Screening of natural products

Natural products are a rich source of biologically active compounds. Many of today's medicines are either obtained directly from a natural source or were developed from a lead compound originally obtained from a natural source. Usually, the natural source has some form of biological activity, and the compound responsible for that activity is known as the **active principle**. Such a structure can act as a lead compound. Most biologically active natural products are **secondary metabolites** with quite complex structures and several chiral centres. This has an advantage in that they are extremely novel compounds. Unfortunately, this complexity also makes their synthesis difficult and the compounds usually have to be extracted from their natural source—a slow, expensive, and inefficient process. As a result, there is usually an advantage in designing simpler analogues (section 13.3.8).

Many natural products have radically new chemical structures which no chemist would dream of synthesizing. For example, the antimalarial drug **artemisinin** (Fig. 12.6) is a natural product with an extremely unstable looking trioxane ring—one of the most unlikely structures to have appeared in recent years (see also Case study 3).

The study of medicines derived from natural sources is known as **pharmacognosy**, and includes both crude extracts and purified active principles.

12.4.1.1 The plant kingdom

Plants have always been a rich source of lead compounds (e.g. **morphine**, **cocaine**, **digitalis**, **quinine**, **tubocurarine**, **nicotine**, and **muscarine**). Many of these lead compounds are useful drugs in themselves (e.g. morphine and quinine), and others have been the basis for synthetic drugs (e.g. local anaesthetics developed from cocaine). Plants still remain a promising source of new drugs and will continue to be so. Clinically useful drugs which have recently been isolated from plants include the anticancer agent **paclitaxel (Taxol)** from the yew tree, the antimalarial agent artemisinin from a Chinese plant (Fig. 12.6), and the Alzheimer's drug **galantamine** from daffodils (section 22.15.1).

Plants provide a bank of rich, complex, and highly varied structures which are unlikely to be discovered from other sources. Furthermore, evolution has already carried out a screening process that favours compounds which provide plants with an 'edge' when it comes to survival. For example, biologically potent compounds

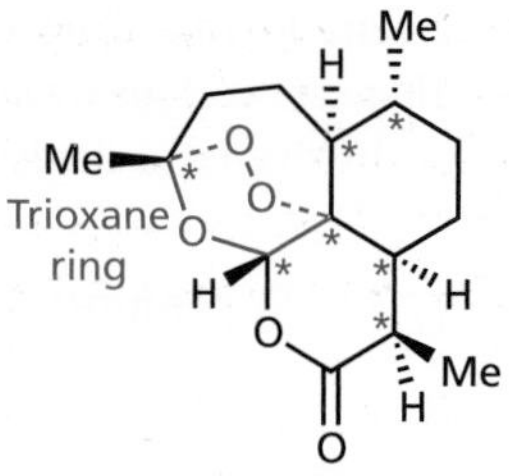

Artemisinin

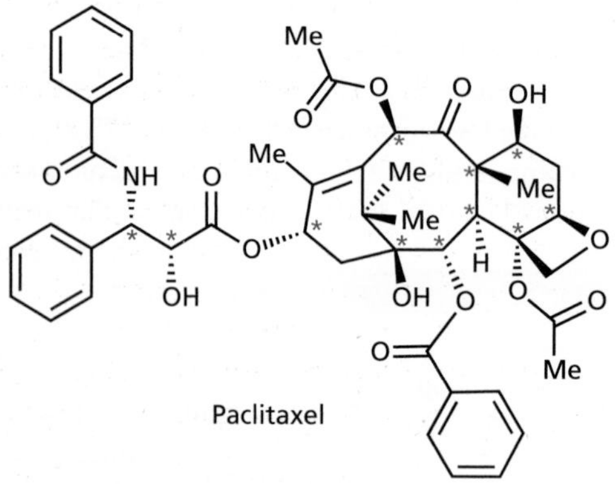

Paclitaxel

FIGURE 12.6 Plant natural products as drugs (the asterisks indicate chiral centres).

can deter animals or insects from eating the plants that contain them. Considering the debt medicinal chemistry owes to the natural world, it is sobering to think that very few plants have been fully studied and the vast majority have not been studied at all. The rainforests of the world are particularly rich in plant species which have still to be discovered, let alone studied. Who knows how many exciting new lead compounds await discovery for the fight against cancer, AIDS, or any of the other myriad of human afflictions? This is one reason why the destruction of rainforests and other ecosystems is so tragic; once these ecosystems are destroyed, unique plant species are lost to medicine for ever. For example, **silphion**—a plant that was cultivated near Cyrene in North Africa and was famed as a contraceptive agent in ancient Greece—is now extinct. It is certain that many more useful plants have become extinct without medicine ever being aware of them.

12.4.1.2 Microorganisms

Microorganisms such as bacteria and fungi have also provided rich pickings for drugs and lead compounds. These organisms produce a large variety of antimicrobial agents which have evolved to give their hosts an advantage over their competitors in the microbiological world. The screening of microorganisms became highly popular

after the discovery of **penicillin**. Soil and water samples were collected from all round the world in order to study new fungal or bacterial strains, leading to an impressive arsenal of antibacterial agents such as the **cephalosporins**, **tetracyclines**, **aminoglycosides**, **rifamycins**, **chloramphenicol**, and **vancomycin** (Chapter 19). Although most of the drugs derived from microorganisms are used in antibacterial therapy, some microbial metabolites have provided lead compounds in other fields of medicine. For example, **asperlicin**—isolated from *Aspergillus alliaceus*—is a novel antagonist of a peptide hormone called **cholecystokinin** (CCK) which is involved in the control of appetite. CCK also acts as a neurotransmitter in the brain and is thought to be involved in panic attacks. Analogues of asperlicin may therefore have potential in treating anxiety (see also Box 13.2).

Other examples include the fungal metabolite **lovastatin**, which was the first of the clinically useful statins found to lower cholesterol levels (Case study 1), and another fungal metabolite called **ciclosporin** (Fig. 12.7) which is used to suppress the immune response after organ transplants. **Lipstatin** (Fig. 12.7) is a natural product which was isolated from *Streptomyces toxytricini*. It inhibits pancreatic lipase, and was the lead compound for the anti-obesity compound **orlistat** (Box 7.2). Finally, a fungal metabolite called **rasfonin** (isolated from a fungus in New Zealand) promotes cell death (apoptosis) in

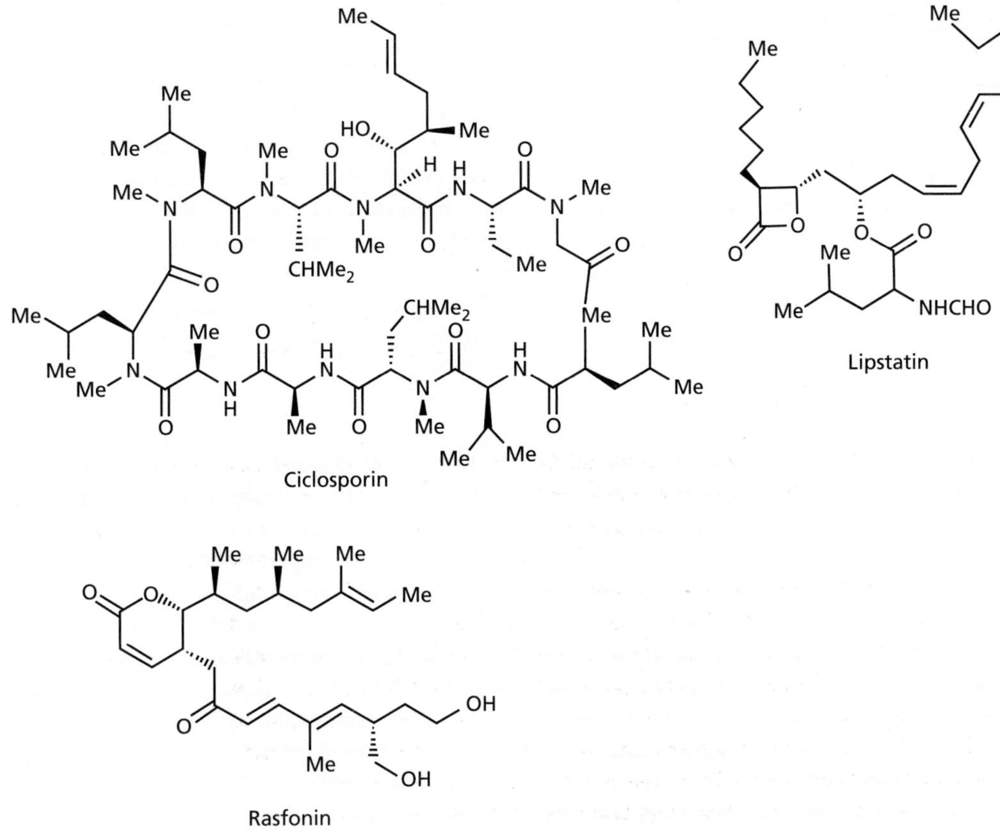

Ciclosporin

Lipstatin

Rasfonin

FIGURE 12.7 Lead compounds from microbiological sources.

cancer cells, but not normal cells. It represents a promising lead compound for novel anticancer agents.

12.4.1.3 Marine sources

In recent years, there has been great interest in finding lead compounds from marine sources. Coral, sponges, fish, and marine microorganisms have a wealth of biologically potent chemicals with interesting inflammatory, antiviral, and anticancer activity. For example, **curacin A** (Fig. 12.8) is obtained from a marine cyanobacterium, and shows potent antitumour activity. Other antitumour agents derived from marine sources include **eleutherobin**, **bryostatins**, **dolastatins**, **cephalostatins**, and **halichondrin B** (sections 21.5.2 and 21.9.2). In 2010, a simplified analogue of halichondrin B was approved for the treatment of breast cancer.

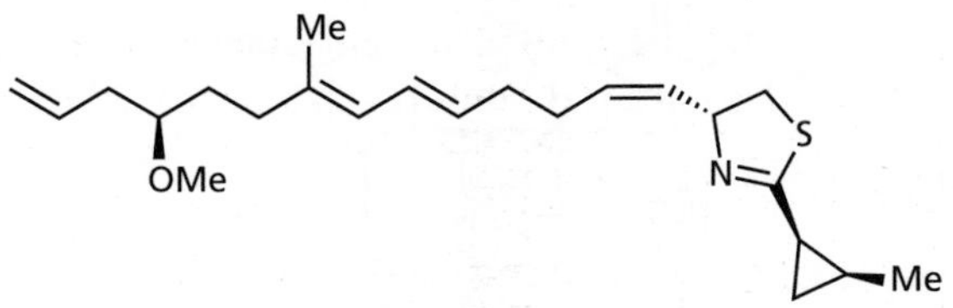

12.4.1.4 Animal sources

Animals can sometimes be a source of new lead compounds. For example, a series of antibiotic polypeptides known as the **magainins** were extracted from the skin of the African clawed frog *Xenopus laevis*. These agents protect the frog from infection and may provide clues to the development of novel antibacterial and antifungal agents in human medicine. Another example is a potent analgesic compound called **epibatidine** (Fig. 12.9), obtained from the skin extracts of the Ecuadorian poison frog.

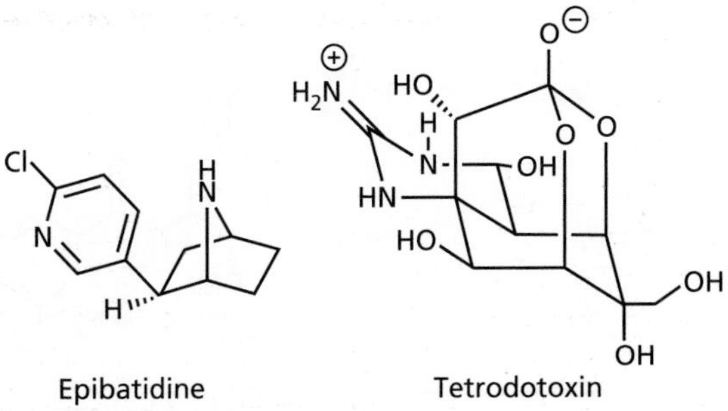

FIGURE 12.8 Curacin A.

Epibatidine

Tetrodotoxin

FIGURE 12.9 Natural products as drugs.

12.4.1.5 Venoms and toxins

Venoms and toxins from animals, plants, snakes, spiders, scorpions, insects, and microorganisms are extremely potent because they often have very specific interactions with a macromolecular target in the body. As a result, they have proved important tools in studying receptors, ion channels, and enzymes. Many of these toxins are polypeptides (e.g. α-**bungarotoxin** from cobras). However, non-peptide toxins such as **tetrodotoxin** from the puffer fish (Fig. 12.9) are also extremely potent.

Venoms and toxins have been used as lead compounds in the development of novel drugs. For example, **teprotide**, a peptide isolated from the venom of the Brazilian viper, was a lead compound for the development of the antihypertensive agents **cilazapril** and **captopril** (Case study 2).

The neurotoxins from *Clostridium botulinum* are responsible for serious food poisoning (**botulism**), but they have a clinical use as well. They can be injected into specific muscles (such as those controlling the eyelid) to prevent muscle spasm. These toxins prevent cholinergic transmission (Chapter 22) and could well prove a lead for the development of novel anticholinergic drugs.

Finally, **conotoxin** is a peptide toxin derived from the marine cone snail, and has very powerful analgesic properties in humans. A synthetic form of conotoxin called **ziconotide** was approved in 2004 for the treatment of chronic pain.

12.4.2 Medical folklore

In the past, ancient civilizations depended greatly on local flora and fauna for their survival. They would experiment with various berries, leaves, and roots to find out what effects they had. As a result, many brews were claimed by the local healer or shaman to have some medicinal use. More often than not, these concoctions were useless or downright dangerous, and if they worked at all, it was because the patient willed them to work—a **placebo effect**. However, some of these extracts may indeed have a real and beneficial effect, and a study of medical folklore can give clues as to which plants might be worth studying in more detail. **Rhubarb** root has been used as a purgative for many centuries. In China, it was called 'The General' because of its 'galloping charge'! The most significant chemicals in rhubarb root are anthraquinones, which were used as the lead compounds in the design of the laxative—**dantron** (Fig. 12.10).

The ancient records of Chinese medicine also provided the clue to the novel antimalarial drug **artemisinin** mentioned in section 12.4.1 (see also Case study 3). The therapeutic properties of the opium poppy (active principle **morphine**) were known in Ancient Egypt, as were those of the *Solanaceae* plants in ancient Greece (active principles **atropine** and **hyoscine**; section 22.9.2). The snakeroot plant was well regarded in India (active principle **reserpine**; Fig. 12.10), and herbalists in medieval England used extracts from the willow tree (active principle **salicin**; Fig. 12.10) and foxglove (active principle **digitalis**—a mixture of compounds such as digitoxin, digitonin, and digitalin). The Aztec and Mayan cultures of South America used extracts from a variety of bushes and trees including the ipecacuanha root (active principle **emetine**; Fig. 12.10), coca bush (active principle **cocaine**), and cinchona bark (active principle **quinine**).

12.4.3 Screening synthetic compound 'libraries'

The thousands of compounds which have been synthesized by the pharmaceutical companies over the years

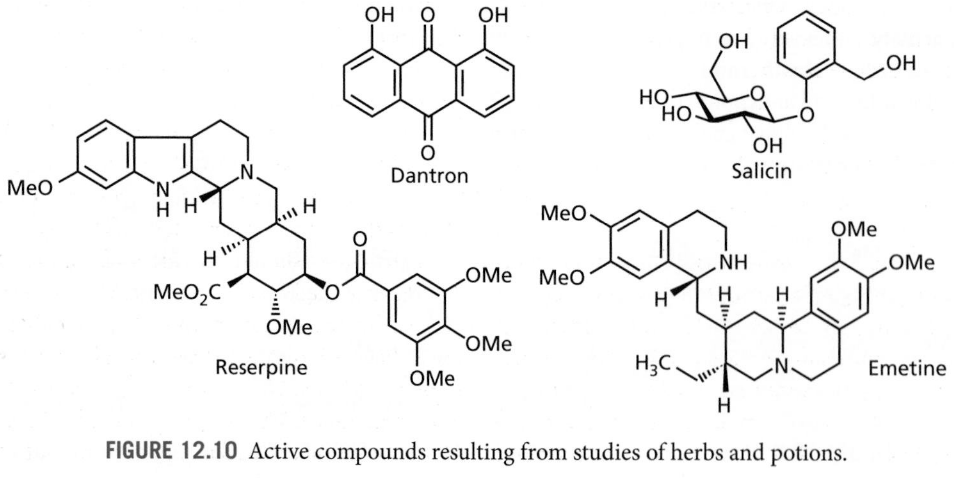

FIGURE 12.10 Active compounds resulting from studies of herbs and potions.

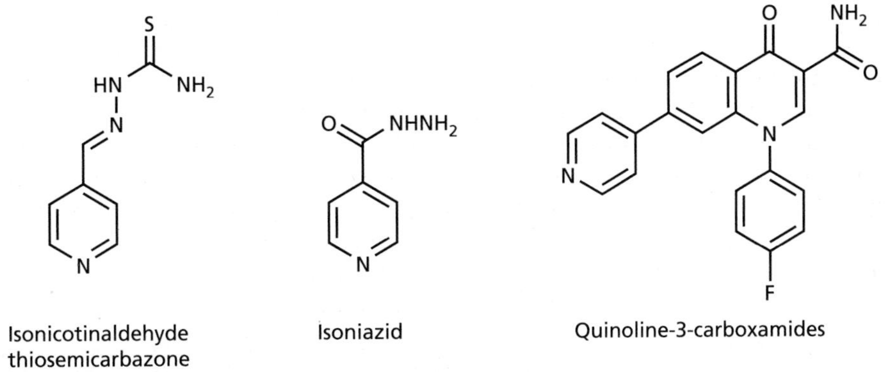

FIGURE 12.11 Pharmaceutically active compounds discovered from synthetic intermediates.

are another source of lead compounds. The vast majority of these compounds have never made the market place, but they have been stored in compound 'libraries' and are still available for testing. Pharmaceutical companies often screen their library of compounds whenever they study a new target. However, it has to be said that the vast majority of these compounds are merely variations on a theme; for example 1000 or so different penicillin structures. This reduces the chances of finding a novel lead compound.

Pharmaceutical companies often try to diversify their range of structures by purchasing novel compounds prepared by research groups elsewhere—a useful source of revenue for hard-pressed university departments! These compounds may never have been synthesized with medicinal chemistry in mind, but there is always the chance that they may have useful biological activity.

It can also be worth testing synthetic intermediates. For example, a series of thiosemicarbazones was synthesized and tested as antitubercular agents in the 1950s. This included **isonicotinaldehyde thiosemicarbazone**, the synthesis of which involved the hydrazide structure **isoniazid** (Fig. 12.11) as a synthetic intermediate. It was subsequently found that isoniazid had greater activity than the target structure. Similarly, a series of **quinoline-3-carboxamide** intermediates (Fig. 12.11) were found to have antiviral activity.

12.4.4 Existing drugs

12.4.4.1 'Me too' and 'me better' drugs

Many companies use established drugs from their competitors as lead compounds in order to design a drug that gives them a foothold in the same market area. The aim is to modify the structure sufficiently such that it avoids patent restrictions, retains activity, and ideally has improved therapeutic properties. For example, the antihypertensive drug **captopril** was used as a lead compound by various

companies to produce their own antihypertensive agents (Fig. 12.12, see also Case study 2).

Although often disparaged as 'me too' drugs, they can often offer improvements over the original drug ('me better' drugs). For example, modern penicillins are more selective, more potent, and more stable than the original penicillins. Newer statins that lower cholesterol levels also have improved properties over older ones (Case study 1). It should also be noted that it is not unusual for companies to be working on similar looking structures for a particular disease at the same time. The first of these drugs to reach the market gets all the kudos, but it is rather unfair to call the drugs that follow it as 'me too' drugs, since they were designed and developed independently.

12.4.4.2 Enhancing a side effect

An existing drug usually has a minor property or an undesirable side effect which could be of use in another area of medicine. As such, the drug could act as a lead compound on the basis of its side effects. The aim would then be to enhance the desired side effect and to eliminate the major biological activity. This has been described as the SOSA approach—**selective optimization of side activities**. Choosing a known drug as the lead compound for a side effect has the advantage that the compound is already 'drug-like' and it should be more feasible to develop a clinically useful drug with the required pharmacodynamic and pharmacokinetic properties. Many of the 'hits' obtained from HTS do not have a 'drug-like' structure and it may require far more effort to optimize them. Indeed, it has been argued that modifications of known drug structures should provide lead compounds in several areas of medicinal chemistry. Many research groups are now screening compounds that are either in clinical use or have reached late-stage clinical trials to see whether they have side effects that would make them suitable lead compounds. The John Hopkins Clinical Compound Library is one such source of these compounds.

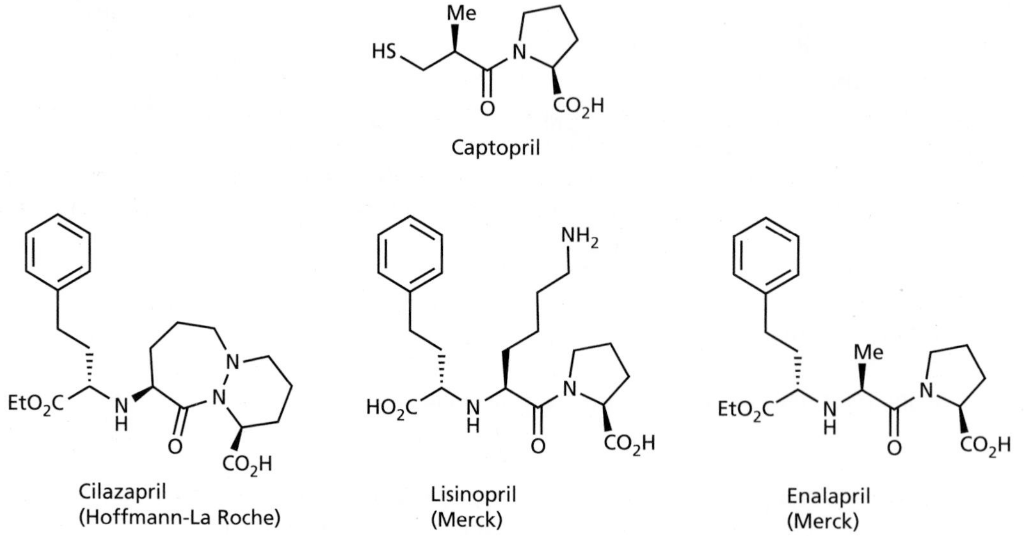

FIGURE 12.12 Captopril and 'me too' drugs.

For example, most sulphonamides have been used as antibacterial agents. However, some sulphonamides with antibacterial activity could not be used clinically because they had convulsive side effects brought on by **hypoglycaemia** (lowered glucose levels in the blood). Clearly, this is an undesirable side effect for an antibacterial agent, but the ability to lower blood glucose levels would be useful in the treatment of diabetes. Therefore, structural alterations were made to the sulphonamides concerned in order to eliminate the antibacterial activity and to enhance the hypoglycaemic activity. This led to the antidiabetic agent **tolbutamide** (Fig. 12.13). Another example was the discovery that the anticoagulant **warfarin** is also a weak inhibitor of a viral enzyme that is important in the life cycle of HIV. Warfarin was used as the lead compound in the development of an anti-HIV drug called **tipranavir** (section 20.7.4.10).

In some cases, the side effect may be strong enough that the drug can be used without modification. For example, the anti-impotence drug **sildenafil (Viagra)** (Fig. 12.13) was originally designed as a vasodilator to treat angina and hypertension (section 26.5.2). During clinical trials, it was found that it acted as a vasodilator more effectively in the penis than in the heart, resulting in increased erectile function. The drug is now used to treat erectile dysfunction and sexual impotence. Another example is the antidepressant drug **bupropion**. Patients taking this drug reported that it helped them give up smoking, and so the drug is now marketed as an antismoking aid (**Zyban**) (section 23.12.4). **Astemizole** (Fig. 12.13) is a medication used in the treatment of allergy, but has been found to be a potent antimalarial agent.

The moral of the story is that a drug used in one field of medicinal chemistry could be the lead compound

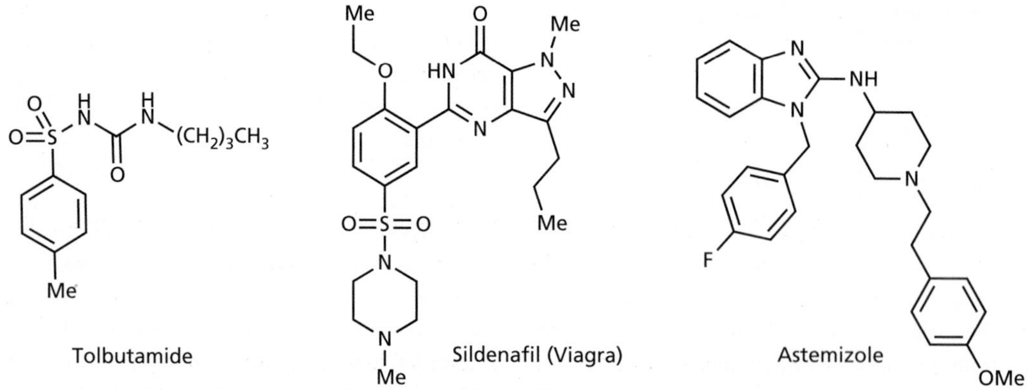

FIGURE 12.13 Tolbutamide, sildenafil (Viagra), and astemizole.

BOX 12.4 Selective optimization of side activities (SOSA)

Several drugs have been developed by enhancing the side effect of another drug (Fig. 1). **Chlorpromazine** is used as a neuroleptic agent in psychiatry, but was developed from the antihistamine agent **promethazine**. This might appear an odd thing to do, but it is known that promethazine has sedative side effects, and so medicinal chemists modified the structure to enhance the sedative effects at the expense of antihistamine activity. Similarly, the development of sulphonamide diuretics such as **chlorothiazide** arose from the observation that **sulphanilamide** has a diuretic effect in large doses (due to its action on an enzyme called **carbonic anhydrase**).

Sometimes, slight changes to a structure can result in significant changes in pharmacological activity. For example, **minaprine** (Fig. 2) is an antidepressant agent that acts as a serotonin agonist. Adding a phenolic substituent resulted in **4-hydroxyminaprine**, which is a potent dopamine agonist, whereas adding a cyano substituent gave **bazinaprine**, which is a potent inhibitor of the enzyme **monoamine oxidase-A**. Minaprine also binds weakly to muscarinic receptors, and modifications were successfully carried out to give structure I, having potent activity for the muscarinic receptor and negligible activity for dopamine and serotonin receptors. Minaprine also has weak affinity for the cholinesterase enzyme, and modifications led to structure II with over 1000-fold increased affinity.

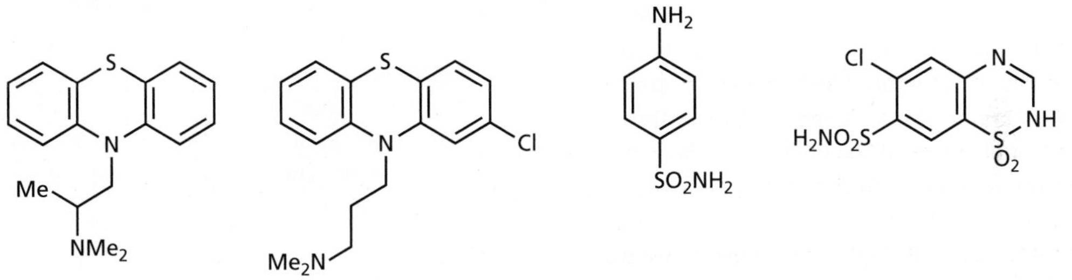

Promethazine Chlorpromazine Sulphanilamide Chlorothiazide

FIGURE 1 Drugs developed by enhancing a side effect.

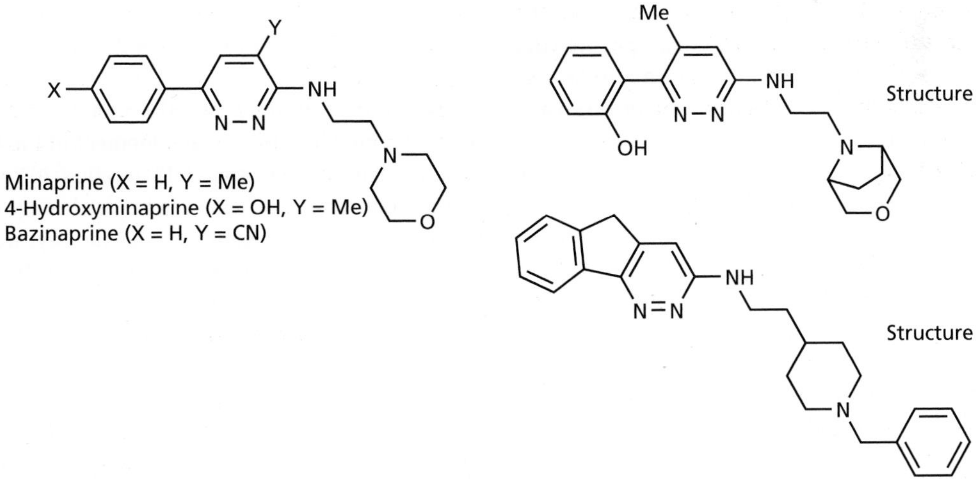

FIGURE 2 Structures with different pharmacological properties derived from the lead compound minaprine.

in another field (Box 12.4). Furthermore, one can fall into the trap of thinking that a structural group of compounds all have the same type of biological activity. The sulphonamides are generally thought of as antibacterial agents, but we have seen that they can also have other properties.

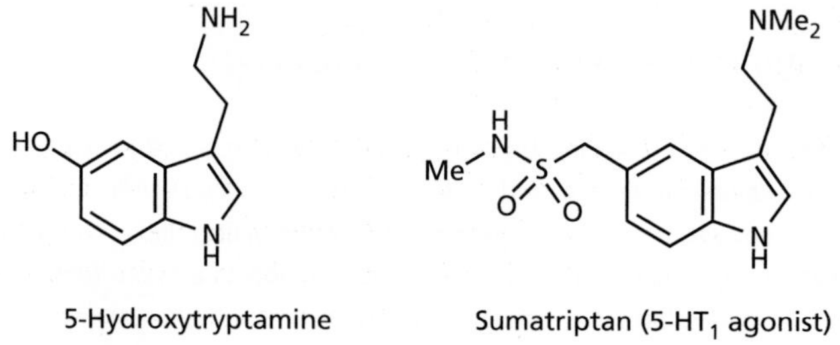

5-Hydroxytryptamine   Sumatriptan (5-HT$_1$ agonist)

FIGURE 12.14 5-Hydroxytryptamine and sumatriptan.

12.4.5 Starting from the natural ligand or modulator

12.4.5.1 Natural ligands for receptors

The natural ligand of a target receptor has sometimes been used as a lead compound. The natural ligands **adrenaline** and **noradrenaline** were the starting points for the development of adrenergic β-agonists such as **salbutamol**, **dobutamine**, and **xamoterol** (section 23.10), and 5-hydroxytryptamine (5-HT) was the starting point for the development of the 5-HT$_1$ agonist **sumatriptan** (Fig. 12.14).

The natural ligand of a receptor can also be used as the lead compound in the design of an antagonist. For example, **histamine** was used as the original lead compound in the development of the H$_2$ histamine antagonist **cimetidine** (section 25.2). Turning an agonist into an antagonist is frequently achieved by adding extra binding groups to the lead structure. Other examples include the development of the adrenergic antagonist **pronethalol** (section 23.11.3.1), the H$_2$ antagonist **burimamide** (section 25.2.4), and the 5-HT$_3$ antagonists **ondansetron** and **granisetron** (Box 12.2)

Sometimes the natural ligand for a receptor is not known (an **orphan receptor**) and the search for it can be a major project in itself. If the search is successful, however, it opens up a brand-new area of drug design (see Box 12.5). For example, the identification of the opioid receptors for **morphine** led to a search for endogenous opioids (natural body painkillers), which eventually led to the discovery of **endorphins** and **enkephalins**, and their use as lead compounds (section 24.8).

12.4.5.2 Natural substrates for enzymes

The natural substrate for an enzyme can be used as the lead compound in the design of an enzyme inhibitor. For example, **enkephalins** have been used as lead compounds for the design of enkephalinase inhibitors. **Enkephalinases** are enzymes which metabolize enkephalins, and their inhibition should prolong the activity of enkephalins (section 24.8.4).

BOX 12.5 Natural ligands as lead compounds

The discovery of **cannabinoid** receptors in the early 1990s led to the discovery of two endogenous cannabinoid messengers—**arachidonylethanolamine** (**anandamide**) and **2-arachidonyl glycerol**. These have now been used as lead compounds to develop agents that interact with cannabinoid receptors. Such agents may prove useful in suppressing nausea during chemotherapy or in stimulating appetite in patients with AIDS.

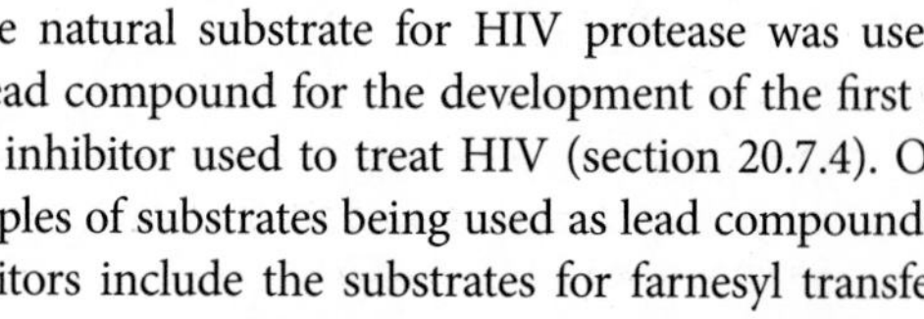

FIGURE 1 Anandamide.

The natural substrate for HIV protease was used as the lead compound for the development of the first protease inhibitor used to treat HIV (section 20.7.4). Other examples of substrates being used as lead compounds for inhibitors include the substrates for farnesyl transferase (section 21.6.1), matrix metalloproteinase (section 21.7.1), HCV NS3-4A protease (section 20.10.1), renin (Case study 8), and 17β-hydroxysteroid dehydrogenase type 1.

For additional material see Web article 1: Steroids as novel anticancer agents (*web article on 17β-HSD type 1*) on the Online Resource Centre at www.oxfordtextbooks.co.uk/orc/patrick6e/

12.4.5.3 Enzyme products as lead compounds

It should be remembered that enzymes catalyse a reaction in both directions, and so the product of an enzyme-catalysed reaction can also be used as a lead compound for an enzyme inhibitor. For example, the design of the carboxypeptidase inhibitor **L-benzylsuccinic acid** was based on the products arising from the carboxypeptidase-catalysed hydrolysis of peptides (see Case study 2). Similarly, the development of some antiviral agents for the treatment of hepatitis C began with a lead compound that was the product of the enzyme-catalysed reaction (Case study 10).

12.4.5.4 Natural modulators as lead compounds

Many receptors and enzymes are under allosteric control (sections 3.6 and 8.3.2). The natural or endogenous chemicals that exert this control (modulators) could also serve as lead compounds. For example, ATP is a natural antagonist for a platelet receptor called the $P2Y_{12}$ receptor. It was used as the lead compound in the development of the antiplatelet agent **cangrelor** (section 26.9.2.3).

In some cases, a modulator for an enzyme or receptor is suspected but has not yet been found. For example, the **benzodiazepines** are synthetic compounds that modulate the receptor for γ-**aminobutyric acid** (GABA) by binding to an allosteric binding site. The natural modulators for this allosteric site were not known at the time benzodiazepines were synthesized, but endogenous peptides called **endozepines** have since been discovered which bind to the same allosteric binding site, and which may serve as lead compounds for novel drugs having the same activity as the benzodiazepines.

12.4.6 Combinatorial and parallel synthesis

The growing number of potentially new drug targets arising from genomic and proteomic projects has meant that there is an urgent need to find new lead compounds to interact with them. Unfortunately, the traditional sources of lead compounds have not managed to keep pace and in the last decade or so, research groups have invested greatly in combinatorial and parallel synthesis in order to tackle this problem. Combinatorial synthesis is an automated solid-phase procedure aimed at producing as many different structures as possible in as short a time as possible. The reactions are carried out on very small scale, often in a way that will produce mixtures of compounds in each reaction vial. In a sense, combinatorial synthesis aims to mimic what plants do, i.e. produce a pool of chemicals, one of which may prove to be a useful lead compound. Combinatorial synthesis has developed so swiftly that it is almost a branch of chemistry in itself and a separate chapter is devoted to it (Chapter 16). Parallel synthesis involves the small scale synthesis of large numbers of compounds at the same time using specialist miniaturized equipment. The synthesis can be carried out in solution or solid phase, and each reaction vial contains a distinct product (Chapter 16). Nowadays, parallel synthesis is generally preferred over combinatorial synthesis in order to produce smaller, more focused compound libraries.

12.4.7 Computer-aided design of lead compounds

A detailed knowledge of a target binding site significantly aids in the design of novel lead compounds intended to bind with that target. In cases where enzymes or receptors can be crystallized, it is possible to determine the structure of the protein and its binding site by **X-ray crystallography**. Molecular modelling software programs can then be used to study the binding site, and to design molecules which will fit and bind to the site—*de novo* drug design (section 17.15).

In some cases, the enzyme or receptor cannot be crystallized and so X-ray crystallography cannot be carried out. However, if the structure of an analogous protein has been determined, this can be used as the basis for generating a computer model of the protein. This is covered in more detail in section 17.14. NMR spectroscopy has also been effective in determining the structure of proteins and can be applied to proteins that cannot be studied by X-ray crystallography.

12.4.8 Serendipity and the prepared mind

Frequently, lead compounds are found as a result of serendipity (i.e. chance). However, it still needs someone with an inquisitive nature or a prepared mind to recognize the significance of chance discoveries and to take advantage of these events. The discovery of **cisplatin** (section 9.3.4) and **penicillin** (section 19.5.1.1) are two such examples, but there are many more (see Box 12.6).

Sometimes, the research carried out to improve a drug can have unexpected and beneficial spin offs. For example, **propranolol** (Fig. 12.15) and its analogues are effective β-blockers (antagonists of β-adrenergic receptors) (section 23.11.3). However, they are also lipophilic, which means that they can cross the blood–brain barrier and cause CNS side effects. To counteract this, more hydrophilic analogues were designed by decreasing the size of the aromatic ring system and adding a hydrophilic amide group. One of the compounds made was **practolol**. As expected, this compound had fewer CNS side effects, but, more importantly, it was found to be a selective antagonist for the β-receptors of the heart over β-receptors in other organs—a result that was highly desirable, but not the one that was being looked for at the time.

Frequently, new lead compounds have arisen from research projects carried out in a totally different field of medicinal chemistry. This emphasizes the importance of keeping an open mind, especially when testing for biological activity. For example, we have already described the development of the antidiabetic drug **tolbutamide** (section 12.4.4.2), based on the observation that some antibacterial sulphonamides could lower blood glucose levels.

BOX 12.6 Examples of serendipity

During the Second World War, an American ship carrying **mustard gas** exploded in an Italian harbour. It was observed that many of the survivors who had inhaled the gas lost their natural defences against microbes. Further study showed that their white blood cells had been destroyed. It is perhaps hard to see how a drug that weakens the immune system could be useful. However, there is one disease where this *is* the case—leukaemia. Leukaemia is a form of cancer which results in the excess proliferation of white blood cells, so a drug that kills these cells is potentially useful. As a result, a series of mustard-like drugs were developed based on the structure of the original mustard gas (sections 9.3.1 and 21.2.3.1).

Another example involved the explosives industry, where it was quite common for workers to suffer severe headaches. These headaches resulted from dilatation of blood vessels in the brain, caused by handling **trinitroglycerine.** Once again, it is hard to see how such a drug could be useful. Certainly, the dilatation of blood vessels in the brain may not be particularly beneficial, but dilating the blood vessels in the heart is useful in cardiovascular medicine. As a result, trinitroglycerine (or **glyceryl trinitrate** as it is called in medical circles) is used as a spray or sublingual tablet for the prophylaxis and treatment of angina. The agent acts as a prodrug for the generation of **nitric oxide**, which causes vasodilation (see also section 26.5.1).

Workers in the rubber industry found that they often acquired a distaste for **alcohol**! This was caused by an antioxidant used in the rubber manufacturing process which found its way into workers' bodies and prevented the normal oxidation of alcohol in the liver. As a result, there was a build-up of **acetaldehyde**, which was so unpleasant that workers preferred not to drink. The antioxidant became the lead compound for the development of **disulfiram (Antabuse)**—used for the treatment of chronic alcoholism.

The following are further examples of lead compounds arising as a result of serendipity:

* **Clonidine** was originally designed to be a nasal vasoconstrictor for use in nasal drops and shaving soaps. Clinical trials revealed that it caused a marked fall in blood pressure, and so it became an important antihypertensive instead.

* **Imipramine** was synthesized as an analogue of chlorpromazine (Box 12.4), and was initially to be used as an antipsychotic. However, it was found to alleviate depression and this led to the development of a series of compounds classified as the **tricyclic antidepressants** (section 23.12.4).

* **Aminoglutethimide** was prepared as a potential anti-epileptic drug, but is now used as an anticancer agent (section 21.4.5).

* The anti-impotence drug **sildenafil (Viagra)** (Fig. 12.13) was discovered by chance from a project aimed at developing a new heart drug (see also section 26.5.2).

* **Isoniazid** (Fig. 12.11) was originally developed as an antituberculosis agent. Patients taking it proved remarkably cheerful and this led to the drug becoming the lead compound for a series of antidepressant drugs known as the **monoamine oxidase inhibitors** (MAOIs) (section 23.12.5).

* **Chlorpromazine** (Box 12.4) was synthesized as an antihistamine for possible use in preventing surgical shock, and was found to make patients relaxed and unconcerned. This led to the drug being tested in people with manic depression, where it was found to have tranquillizing effects. As a result, it was marketed as the first of the neuroleptic drugs (major tranquillizers) used for schizophrenia.

* **Ciclosporin A** (Fig. 12.7) suppresses the immune system and is used during organ and bone marrow transplants to prevent the immune response rejecting the donor organs. The compound was isolated from a soil sample as part of a study aimed at finding new antibiotics. Fortunately, the compounds were more generally screened and the immunosuppressant properties of ciclosporin A were identified.

* In a similar vein, the anticancer alkaloids **vincristine** and **vinblastine** (section 10.2.2) were discovered by chance when searching for compounds that could lower blood sugar levels. Vincristine is used in the treatment of Hodgkin's disease.

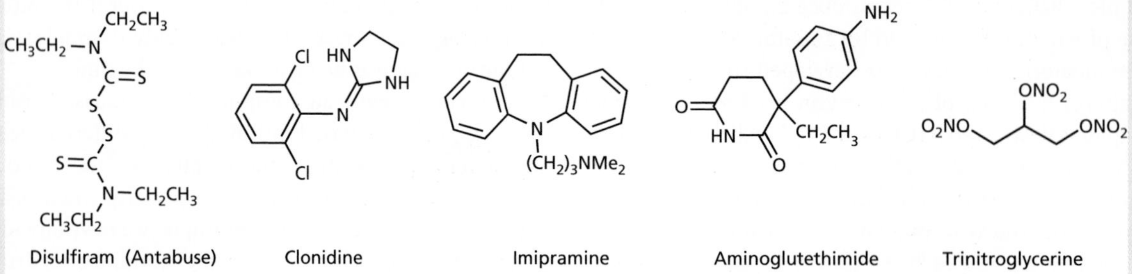

Disulfiram (Antabuse)   Clonidine   Imipramine   Aminoglutethimide   Trinitroglycerine

FIGURE 1 Drugs discovered by serendipity.

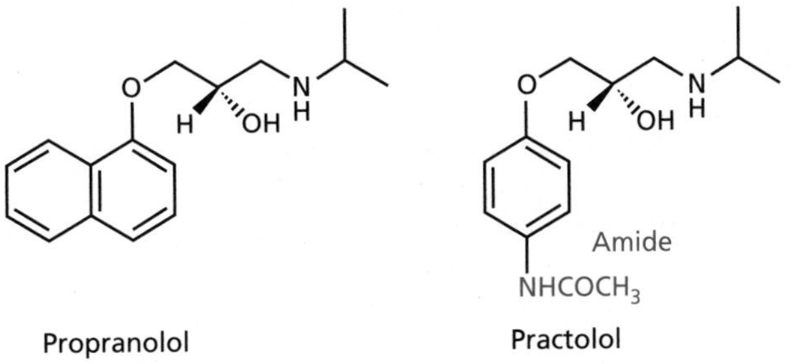

FIGURE 12.15 Propranolol and practolol.

12.4.9 Computerized searching of structural databases

New lead compounds can be found by carrying out computerized searches of structural databases. In order to carry out such a search, it is necessary to know the desired **pharmacophore** (sections 13.2 and 17.11). Alternatively, docking experiments can be carried out if the structure of the target binding site is known (section 17.12). This type of database searching is also known as **database mining** and is described in section 17.13.

12.4.10 Fragment-based lead discovery

So far we have described methods by which a lead compound can be discovered from a natural or synthetic source, but all these methods rely on an active compound being present. Unfortunately, there is no guarantee that this will be the case. Recently, NMR spectroscopy has been used to *design* a lead compound rather than to discover one (see Box 12.7). In essence, the method sets out to find small molecules (**epitopes**) which will bind to specific, but different, regions of a protein's binding site.

BOX 12.7 The use of NMR spectroscopy in finding lead compounds

NMR spectroscopy was used in the design of high-affinity ligands for the FK506 binding protein—a protein involved in the suppression of the immune response. Two optimized epitopes (A and B) were discovered, which bound to different regions of the binding site. Structure C was then synthesized, where the two epitopes were linked by a propyl link. This compound had higher affinity than either of the individual epitopes and represents a lead compound for further development.

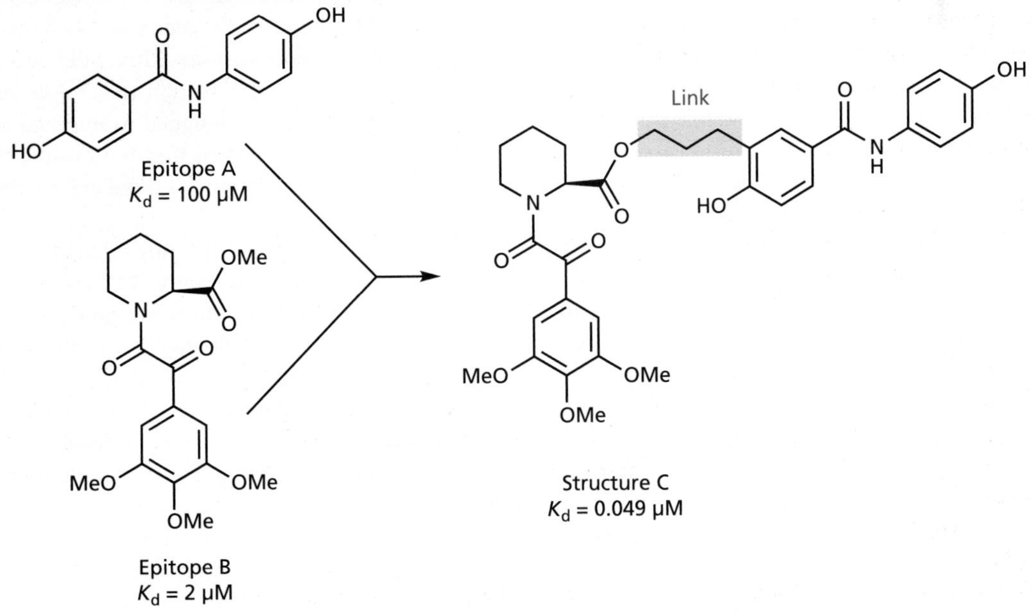

FIGURE 1 Design of a ligand for the FK506 binding protein. (K_d is defined in section 8.9.)

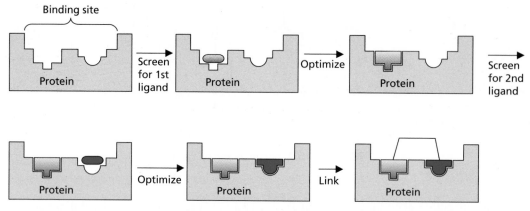

FIGURE 12.16 Epitope mapping.

These molecules will have no activity in themselves since they only bind to one part of the binding site, but if a larger molecule is designed which links these epitopes together, then a lead compound may be created which *is* active and which binds to the whole of the binding site (Fig. 12.16).

Lead discovery by NMR is also known as SAR by NMR (SAR = structure–activity relationships) and can be applied to proteins of known structure which are labelled with ^{15}N or ^{13}C, such that each amide bond in the protein has an identifiable peak.

A range of low-molecular-weight compounds is screened to see whether any of them bind to a specific region of the binding site. Binding can be detected by observing a shift in any of the amide signals, which will not only show that binding is taking place, but will also reveal which part of the binding site is occupied. Once a compound (or ligand) has been found that binds to one region of the binding site, the process can be repeated to find a ligand that will bind to a different region. This is usually done in the presence of the first ligand to ensure that the second ligand does, in fact, bind to a distinct region.

Once two ligands (or epitopes) have been identified, the structure of each can be optimized to find the best ligand for each of the binding regions, then a molecule can be designed where the two ligands are linked together.

There are several advantages to this approach. Since the individual ligands are optimized for each region of the binding site, a lot of synthetic effort is spared. It is much easier to synthesize a series of small-molecular-weight compounds to optimize the interaction with specific parts of the binding site, than it is to synthesize a range of larger molecules to fit the overall binding site. A high level of diversity is also possible, as various combinations of fragments could be used. A further advantage is that it is more likely to find epitopes that will bind to a particular region of a binding site, than to find a lead compound that will bind to the overall binding site.

Moreover, fragments are more likely to be efficient binders, having a high binding energy per unit molecular mass. Finally, some studies have demonstrated a 'super-additivity' effect where the binding affinity of the two linked fragments is much greater than one might have expected from the binding affinities of the two independent fragments.

The method described above involves the linking of fragments. Another strategy is to 'grow' a lead compound from a single fragment—a process called **fragment evolution**. This involves the identification of a single fragment that binds to part of the binding site, then finding larger and larger molecules which contain that fragment, but which bind to other parts of the binding site as well.

A third strategy is known as **fragment self-assembly** and is a form of dynamic combinatorial chemistry (section 16.6.3). Fragments are chosen that can bind to different regions of the binding site, then react with each other to form a linked molecule *in situ*. This could be a reversible reaction as described in section 16.6.3. Alternatively, the two fragments can be designed to undergo an irreversible linking reaction when they bind to the binding site. This has been called '**click chemistry *in situ***' (see Box 12.8).

NMR spectroscopy is not the only method of carrying out fragment-based lead discovery. It is also possible to identify fragments that bind to target proteins using the techniques of X-ray crystallography, *in vitro* bioassays, and mass spectrometry. X-ray crystallography, like NMR, provides information about how the fragment binds to the binding site, and does so in far greater detail. However, it can be quite difficult obtaining crystals of protein–fragment complexes because of the low affinity of the fragments. Recently, a screening method called **CrystalLEAD** has been developed which can quickly screen large numbers of compounds, and detect ligands by monitoring changes in the electron density map of protein–fragment complexes, relative to the unbound protein.

BOX 12.8 Click chemistry *in situ*

A femtomolar inhibitor for the acetylcholinesterase enzyme was obtained by fragment self-assembly within the active site of the enzyme. One of the molecular fragments contained an azide group while the other contained an alkyne group. In the presence of the enzyme, both fragments were bound to the active site, and were positioned close enough to each other for an irreversible 1,3 dipolar cycloaddition to take place, forming the inhibitor *in situ* (Fig. 1). This type of reaction has been called 'click chemistry *in situ*'.

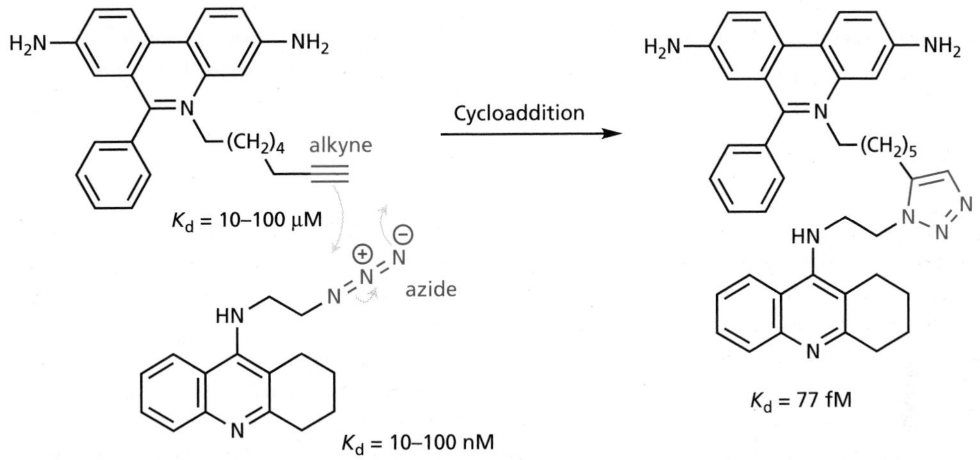

FIGURE 1 'Click' chemistry by means of a cycloaddition reaction. (K_d is defined in section 8.9.)

Finally, it is possible to use fragment-based strategies as a method of optimizing lead compounds that may have been obtained by other means. The strategy is to identify distinct fragments within the lead compound and then to optimize these fragments by the procedures already described. Once the ideal fragments have been identified, the full structure is synthesized incorporating the optimized fragments. This can be a much quicker method of optimization than synthesizing analogues of the larger lead compound.

For additional material see Web article 17: Fragment-based drug discovery on the Online Resource Centre at www.oxfordtextbooks.co.uk/orc/patrick6e/

12.4.11 Properties of lead compounds

Some of the lead compounds that have been isolated from natural sources have sufficient activity to be used directly in medicine without serious side effects; for example morphine, quinine, and paclitaxel. However, most lead compounds have low activity and/or unacceptable side effects, which means that a significant amount of structural modification is required (see Chapters 13 and 14). If the aim of the research is to develop an orally active compound, certain properties of the lead compound should be taken into account. Most orally active drugs obey the rules laid down in Lipinski's rule of five or Veber's parameters (section 11.3). A study of known orally active drugs and the lead compounds from which they were derived demonstrated that the equivalent rules for a lead compound should be more stringent. This is because the structure of the lead compound almost certainly has to be modified and increased, both in terms of size and hydrophobicity. The suggested properties for a lead compound are that it should have a molecular weight of 100–350 amu and a Clog P value of 1–3. (Clog P is a measure of how hydrophobic a compound is; section 14.1). In general, there is an average increase in molecular weight of 80 amu, and an increase of 1 in Clog P when going from a lead compound to the final drug. Studies also show that a lead compound generally has fewer aromatic rings and hydrogen bond acceptors compared to the final drug. Such considerations can be taken into account when deciding which lead compound to use for a research project if several such structures are available. Another approach in making this decision is to calculate the **binding** or **ligand 'efficiency'** of each potential lead compound. This can be done by dividing the free energy of binding for each molecule by the number of non-hydrogen atoms present in the structure. The better the ligand efficiency, the lower the molecular weight of the final optimized structure is likely to be. Moreover, if you have a choice of lead compounds, the most suitable one is not necessarily the most potent.

For fragment-based lead discovery (section 12.4.10), a rule of three has been suggested for the fragments used:

- a molecular weight less than 300;
- no more than 3 hydrogen bond donors;
- no more than 3 hydrogen bond acceptors;
- a Clog P of no more than 3;
- no more than 3 rotatable bonds;
- a polar surface area no more than 60 Å^2.

12.5 Isolation and purification

If the lead compound (or **active principle**) is present in a mixture of compounds from a natural source or a combinatorial synthesis (Chapter 16), it has to be isolated and purified. The ease with which the active principle can be isolated and purified depends very much on the structure, stability, and quantity of the compound. For example, Fleming recognized the antibiotic qualities of **penicillin** and its remarkable non-toxic nature to humans, but he disregarded it as a clinically useful drug because he was unable to purify it. He could isolate it in aqueous solution, but whenever he tried to remove the water, the drug was destroyed. It was not until the development of new experimental procedures such as freeze-drying and chromatography that the successful isolation and purification of penicillin and other natural products became feasible. A detailed description of the experimental techniques involved in the isolation and purification of compounds is outwith the scope of this textbook, and can be obtained from textbooks covering the practical aspects of chemistry.

12.6 Structure determination

It is sometimes hard for present-day chemists to appreciate how difficult structure determinations were before the days of NMR and IR spectroscopy. A novel structure, which may now take a week's work to determine, would have provided two or three decades of work in the past. For example, the microanalysis of **cholesterol** was carried out in 1888 to get its molecular formula, but its chemical structure was not fully established until an X-ray crystallographic study was carried out in 1932.

In the past, structures had to be degraded to simpler compounds, which were further degraded to recognizable fragments. From these scraps of evidence, a possible structure was proposed, but the only sure way of proving the proposal was to synthesize the structure and to compare its chemical and physical properties with those of the natural compound.

Today, structure determination is a relatively straightforward process and it is only when the natural product is obtained in minute quantities that a full synthesis is required to establish its structure. The most useful analytical techniques are **X-ray crystallography** and **NMR spectroscopy**. The former technique comes closest to giving a 'snapshot' of the molecule, but requires a suitable crystal of the sample. The latter technique is used more commonly as it can be carried out on any sample, whether it be a solid, oil, or liquid. There are a large variety of different NMR experiments that can be used to establish the structures of quite complex molecules. These include various two-dimensional NMR experiments which involve a comparison of signals from different types of nuclei in the molecule (e.g. carbon and hydrogen). Such experiments allow the chemist to build up a picture of the molecule atom by atom, and bond by bond.

In cases where there is not enough sample for an NMR analysis, mass spectrometry can be helpful. The fragmentation pattern can give useful clues about a structure, but does not prove it. A full synthesis is still required as final proof.

12.7 Herbal medicine

We have described how useful drugs and lead compounds can be isolated from natural sources, so where does this place herbal medicine? Are there any advantages or disadvantages in using herbal medicines instead of the drugs developed from their active principles? There are no simple answers to this. Herbal medicines contain a large variety of different compounds, several of which may have biological activity, so there is a significant risk of side effects and even toxicity. The active principle is also present in small quantity, so the herbal medicine may be expected to be less active than the pure compound. Herbal medicines such as **St. John's wort** can also interact with prescribed medicines (section 11.5.6), and, in general, there is a lack of regulation or control over their use. Another example is **Ginkgo** which is often used to treat memory problems. However, it also has anticoagulant properties and should not be used alongside other drugs having similar properties; for example warfarin, aspirin, or ibuprofen. Having said all that, several of the issues identified above may actually be advantageous. If the herbal extract contains the active principle in small quantities, there is an inbuilt safety limit to the dose levels received. Different compounds within the extract may also have roles to play in the medicinal properties of the plant and enhance the effect of the active principle—a phenomenon known as **synergy**. Alternatively, some plant extracts have a wide variety of different active principles which act together to produce a beneficial effect. The **aloe plant** (the 'wand of heaven') is an example of this. It is a cactus-like plant found in the deserts of Africa and Arizona and has long been revered for its curative properties. Supporters of herbal medicine have proposed the use of aloe preparations to treat burns, irritable bowel syndrome, rheumatoid arthritis, asthma, chronic leg ulcers, itching, eczema, psoriasis, and acne, thus avoiding the undesirable side effects of long-term steroid use. The preparations are claimed to contain analgesic, anti-inflammatory, antimicrobial, and many other agents, which all contribute to the overall effect. Trying to isolate each active principle would detract from this. On the other hand, critics have stated that many of the beneficial effects claimed for aloe preparations have not been proven and that, although the effects may be useful in some ailments, they are not very effective.

KEY POINTS

- A lead compound is a structure which shows a useful pharmacological activity and can act as the starting point for drug design.

- Natural products are a rich source of lead compounds. The agent responsible for the biological activity of a natural extract is known as the active principle.

- Lead compounds have been isolated from plants, trees, microorganisms, animals, venoms, and toxins. A study of medical folklore indicates plants and herbs which may contain novel lead compounds.

- Lead compounds can be found by screening synthetic compounds obtained from combinatorial syntheses and other sources.

- Existing drugs can be used as lead compounds for the design of novel structures in the same therapeutic area. Alternatively, the side effects of an existing drug can be enhanced to design novel drugs in a different therapeutic area.

- The natural ligand, substrate, product, or modulator for a particular target can act as a lead compound.

- The ability to crystallize a molecular target allows the use of X-ray crystallography and molecular modelling to design lead compounds which will fit the relevant binding site.

- Serendipity has played a role in the discovery of new lead compounds.

- A knowledge of an existing drug's pharmacophore allows the computerized searching of structural databases to identify possible new lead compounds which share that pharmacophore. Docking experiments are also used to identify potential lead compounds.

- NMR spectroscopy can be used to identify whether small molecules (epitopes) bind to specific regions of a binding site. Epitopes can be optimized then linked together to give a lead compound.

- If a lead compound is present in a natural extract or a combinatorial synthetic mixture, it has to be isolated and purified such that its structure can be determined. X-ray crystallography and NMR spectroscopy are particularly important in structure determination.

- Herbal medicines contain different active principles that may combine to produce a beneficial effect. However, toxic side effects and adverse interactions may occur when taken in combination with prescribed medicines.

QUESTIONS

1. What is meant by target specificity and selectivity? Why is it important?

2. What are the advantages and disadvantages of natural products as lead compounds?

3. Fungi have been a richer source of antibacterial agents than bacteria. Suggest why this might be so.

4. Scuba divers and snorkelers are advised not to touch coral. Why do you think this might be? Why might it be of interest to medicinal chemists?

5. You are employed as a medicinal chemist and have been asked to initiate a research programme aimed at finding a drug which will prevent a novel tyrosine kinase receptor from functioning. There are no known lead compounds that have this property. What approaches can you make to establish a lead compound? (Consult section 4.8 to find out more about protein kinase receptors.)

6. A study was set up to look for agents that would inhibit the kinase active site of the epidermal growth factor receptor (section 4.8). Three assay methods were used: an assay

carried out on a genetically engineered form of the protein that was water soluble and contained the kinase active site; a cell assay that measured total tyrosine phosphorylation in the presence of epidermal growth factor; and an *in vivo* study on mice that had tumours grafted onto their backs. How do you think these assays were carried out to measure the effect of an inhibitor? Why do you think three assays were necessary? What sort of information did they provide?

Multiple-choice questions are available on the Online Resource Centre at www.oxfordtextbooks.co.uk/orc/patrick6e/

FURTHER READING

Abad-Zapatero, C., and Metz, J. T. (2005) Ligand efficiency indices as guideposts for drug discovery. *Drug Discovery Today*, **10**, 464–9.

Bleicher, K. H., et al. (2003) Hit and lead generation: beyond high-throughput screening. *Nature Reviews Drug Discovery*, **2**, 369–78.

Blundell, T. L., Jhoti, H., and Abell, C. (2002) High-throughput crystallography for lead discovery in drug design. *Nature Reviews Drug Discovery*, **1**, 45–54.

Bolognesi, M. L., et al. (2009) Alzheimer's disease: new approaches to drug discovery. *Current Opinion in Chemical Biology*, **13**, 303–8.

Cavalli, A., et al. (2008) Multi-target-directed ligands to combat neurodegenerative diseases. *Journal of Medicinal Chemistry*, **51**, 347–72.

Clardy, J., and Walsh, C. (2004) Lessons from natural molecules. *Nature*, **432**, 829–37.

Di, L., et al. (2003) High throughput artificial membrane permeability assay for blood–brain barrier. *European Journal of Medicinal Chemistry*, **38**, 223–32.

Engel, L. W., and Straus, S. E. (2002) Development of therapeutics: opportunities within complementary and alternative medicine. *Nature Reviews Drug Discovery*, **1**, 229–37.

Gershell, L. J., and Atkins, J. H. (2003) A brief history of novel drug discovery technologies. *Nature Reviews Drug Discovery*, **2**, 321–7.

Honma, T. (2003) Recent advances in *de novo* design strategy for practical lead identification. *Medicinal Research Reviews*, **23**, 606–32.

Hopkins, A. L., and Groom, C. R. (2002) The druggable genome. *Nature Reviews Drug Discovery*, **1**, 727–30.

Keseru, G. M., Erlanson, D. A., Ferenczy, G. G., et al. (2016) Design principle for fragment libraries: maximising the value of learnings from Pharma fragment-based drug discovery (FBDD) programs for use in academia. *Journal of Medicinal Chemistry*, DOI: 10.1021/acs.jmedchem.6b00197.

Lewis, R. J., and Garcia, M. L. (2003) Therapeutic potential of venom peptides. *Nature Reviews Drug Discovery*, **2**, 790–802.

Lindsay, M. A. (2003) Target discovery. *Nature Reviews Drug Discovery*, **2**, 831–8.

Lipinski, C., and Hopkins, A. (2004) Navigating chemical space for biology and medicine. *Nature*, **432**, 855–61.

Lowe, D. (2009) In the pipeline. *Chemistry World*, Nov., 20 (screening assays).

Megget, K. (2011) Of mice and men. *Chemistry World*, April, 42–5.

Pellecchia, M., Sem, D. S., and Wuthrich, K. (2002) NMR in drug discovery. *Nature Reviews Drug Discovery*, **1**, 211–9.

Perks, B. (2011) Extreme potential. *Chemistry World*, June, 48–51.

Phillipson, J. D. (2007) Phytochemistry and pharmacognosy. *Phytochemistry*, **68**, 2960–72.

Rees, D. C., et al. (2004) Fragment-based lead discovery. *Nature Reviews Drug Discovery*, **3**, 660–72.

Rishton, G. B. (2003) Nonleadlikeness and leadlikeness in biochemical screening. *Discovering Drugs Today*, **8**, 86–96.

Sauter, G., Simon, R., and Hillan, K. (2003) Tissue microarrays in drug discovery. *Nature Reviews Drug Discovery*, **2**, 962–72.

Shuker, S. B., Hajduk, P. J., Meadows, R. P., and Fesik, S. W. (1996) Discovering high-affinity ligands for proteins: SAR by NMR. *Science*, **274**, 1531–4.

Srivastava, A. S., et al. (2005) Plant-based anticancer molecules. *Bioorganic Medicinal Chemistry*, **13**, 5892–908.

Stockwell, B. R. (2004) Exploring biology with small organic molecules. *Nature*, **432**, 846–54.

Su, J., et al. (2007) SAR study of bicyclo[4.1.0]heptanes as melanin-concentrating hormone receptor R1 antagonists: taming hERG. *Bioorganic and Medicinal Chemistry*, **15**, 5369–85.

Walters, W. P., and Namchuk, M. (2003) Designing screen: how to make your hits a hit. *Nature Reviews Drug Discovery*, **2**, 259–66.

Wermuth, C. G. (2006) Selective optimization of side activities: the SOSA approach. *Drug Discovery Today*, **11**, 160–4.

Titles for general further reading are listed on p.845.

13 Drug design: optimizing target interactions

In Chapter 12, we looked at the various methods of discovering a lead compound. Once it *has* been discovered, the lead compound can be used as the starting point for drug design. There are various aims in drug design. The eventual drug should have a good selectivity and level of activity for its target, and have minimal side effects. It should be easily synthesized and chemically stable. Finally, it should be non-toxic and have acceptable pharmacokinetic properties. In this chapter, we concentrate on design strategies that can be used to optimize the interaction of the drug with its target in order to produce the desired pharmacological effect; in other words its **pharmacodynamic** properties. In Chapter 14, we look at the design strategies that can improve the drug's ability to reach its target and have an acceptable lifetime—in other words its **pharmacokinetic** properties. Although these topics are in separate chapters, it would be wrong to think that they are tackled separately during drug optimization. For example, it would be foolish to spend months or years perfecting a drug that interacts perfectly with its target, but has no chance of reaching that target because of adverse pharmacokinetic properties. Pharmacodynamics and pharmacokinetics should have equal priority in influencing drug design strategies and determining which analogues are synthesized.

13.1 Structure–activity relationships

Once the structure of a lead compound is known, the medicinal chemist moves on to study its structure–activity relationships (SAR). The aim is to identify those parts of the molecule that are important to biological activity and those that are not. If it is possible to crystallize the target with the lead compound bound to the binding site, the crystal structure of the complex could be solved by X-ray crystallography, then studied with molecular modelling software to identify important binding interactions.

However, this is not possible if the target structure has not been identified or cannot be crystallized. It is then necessary to revert to the traditional method of synthesizing a selected number of compounds that vary slightly from the original structure, then studying what effect that has on the biological activity.

One can imagine the drug as a chemical knight going into battle with an affliction. The drug is armed with a variety of weapons and armour, but it may not be obvious which weapons are important to the drug's activity, or which armour is essential to its survival. We can only find this out by removing some of the weapons and armour to see if the drug is still effective. The weapons and armour involved are the various structural features in the drug that can either act as binding groups with the target binding site (section 1.3), or assist and protect the drug on its journey through the body (Chapter 14). Recognizing functional groups and the sort of intermolecular bonds that they can form is important in understanding how a drug might bind to its target.

Let us imagine that we have isolated a natural product with the structure shown in Fig. 13.1. We shall name it glipine. There are a variety of functional groups present in the structure, and the diagram shows the potential binding interactions that are possible with a target binding site.

It is unlikely that all of these interactions take place, so we have to identify those that do. By synthesizing analogues (such as the examples shown in Fig. 13.2) where one particular functional group of the molecule is removed or altered, it is possible to find out which groups are essential and which are not. This involves testing all the analogues for biological activity and comparing them with the original compound. If an analogue shows a significantly lowered activity, then the group that has been modified must have been important. If the activity remains similar, then the group is not essential.

The ease with which this task is carried out depends on how easily we can synthesize the necessary analogues. It may be possible to modify some lead compounds directly

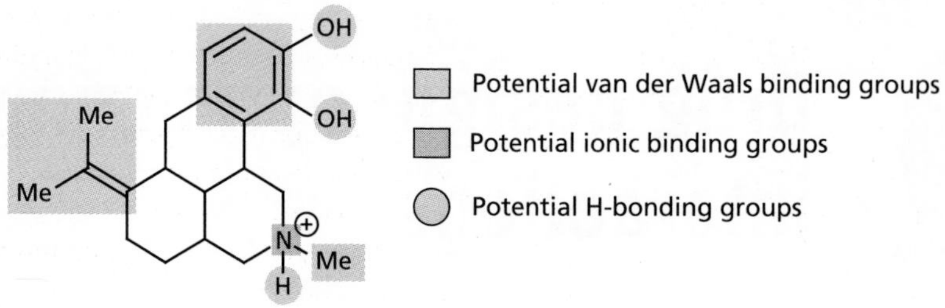

FIGURE 13.1 Glipine.

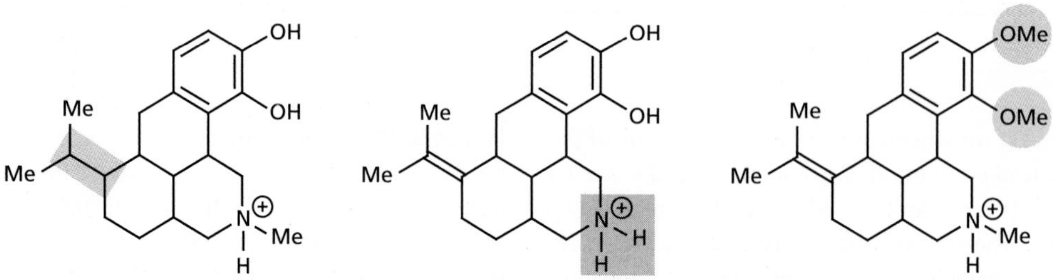

FIGURE 13.2 Modifications of glipine.

to the required analogues, whereas the analogues of other lead compounds may best be prepared by total synthesis. Let us consider the binding interactions that are possible for different functional groups, and the analogues that could be synthesized to establish whether they are involved in binding or not (see also section 1.3 and Appendix 7).

13.1.1 Binding role of alcohols and phenols

Alcohols and phenols are functional groups which are commonly present in drugs and are often involved in hydrogen bonding. The oxygen can act as a hydrogen bond acceptor, and the hydrogen can act as a hydrogen bond donor (Fig. 13.3). The directional preference for hydrogen bonding is indicated by the arrows in the figure, but it is important to realize that slight deviations are possible (section 1.3.2). One or all of these interactions may be important in binding the drug to the binding site. Synthesizing a methyl ether or an ester analogue would be relevant in testing this, as it is highly likely that

the hydrogen bonding would be disrupted in either analogue. Let us consider the methyl ether first.

There are two reasons why the ether might hinder or prevent the hydrogen bonding of the original alcohol or phenol. The obvious explanation is that the proton of the original hydroxyl group is involved as a hydrogen bond donor and, by removing it, the hydrogen bond is lost (Frames 1 and 2 in Fig. 13.4). However, suppose the oxygen atom is acting as a hydrogen bond acceptor (Frame 3, Fig. 13.4)? The oxygen is still present in the ether analogue, so could it still take part in hydrogen bonding? Well, it may, but possibly not to the same extent. The extra bulk of the methyl group should hinder the close approach that was previously attainable and is likely to disrupt hydrogen bonding (Frame 4, Fig. 13.4). The hydrogen bonding may not be completely prevented, but we could reasonably expect it to be weakened.

An ester analogue cannot act as a hydrogen bond donor either. There is still the possibility of it acting as a hydrogen bond acceptor, but the extra bulk of the acyl

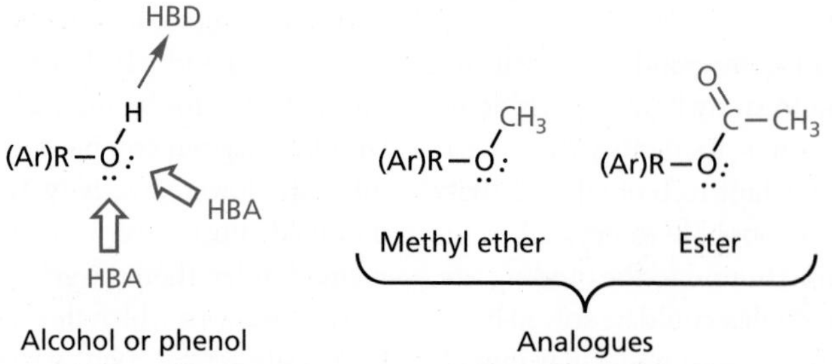

FIGURE 13.3 Possible hydrogen bonding interactions for an alcohol or phenol.

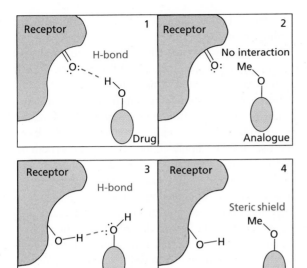

FIGURE 13.4 Possible hydrogen bond interactions for an alcohol/phenol in comparison with an ether analogue.

carbonyl oxygen is potentially a more effective hydrogen bond acceptor; however, it is in a different position relative to the rest of the molecule and may be poorly positioned to form an effective hydrogen bond interaction with the target binding region.

It is relatively easy to acetylate alcohols and phenols to their corresponding esters, and this was one of the early reactions that was carried out on natural products such as morphine (sections 24.3 and 24.5). Alcohols and phenols can also be converted easily to ethers.

In this section, we considered the OH group of alcohols and phenols. It should be remembered that the OH group of a phenol is linked to an aromatic ring, which can also be involved in intermolecular interactions (section 13.1.2).

13.1.2 Binding role of aromatic rings

Aromatic rings are planar, hydrophobic structures, commonly involved in van der Waals interactions with flat hydrophobic regions of the binding site. An analogue containing a cyclohexane ring in place of the aromatic ring is less likely to bind so well, as the ring is no longer flat. The axial protons can interact weakly, but they also serve as buffers to keep the rest of the cyclohexane ring at a distance (Fig. 13.6). The binding region for the aromatic ring may also be a narrow slot rather than a planar surface. In that scenario, the cyclohexane ring would be incapable of fitting into it, because it is a bulkier structure.

Although there are methods of converting aromatic rings to cyclohexane rings, they are unlikely to be

group is even greater than the methyl group of the ether, and this too should hinder the original hydrogen bonding interaction. There is also a difference between the electronic properties of an ester and an alcohol. The carboxyl group has a weak pull on the electrons from the neighbouring oxygen, giving the resonance structure shown in Fig. 13.5. Because the lone pair is involved in such an interaction, it will be less effective as a hydrogen bond acceptor. Of course, one could then argue that the

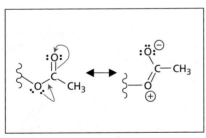

Electronic factor

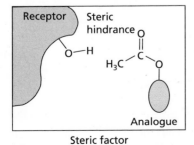

Steric factor

FIGURE 13.5 Factors by which an ester group can disrupt the hydrogen bonding of the original hydroxyl group.

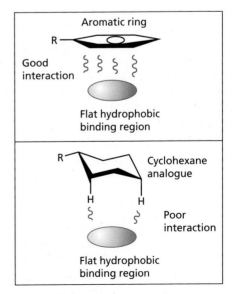

FIGURE 13.6 Binding comparison of an aromatic ring with a cyclohexyl ring.

successful with most lead compounds, and so such analogues would normally be prepared using a full synthesis.

Aromatic rings could also interact with an aminium or quaternary ammonium ion through induced dipole interactions or hydrogen bonding (sections 1.3.4 and 1.3.2). Such interactions would not be possible for the cyclohexyl analogue.

13.1.3 Binding role of alkenes

Like aromatic rings, alkenes are planar and hydrophobic, so they too can interact with hydrophobic regions of the binding site through van der Waals interactions. The activity of the equivalent saturated analogue would be worth testing, since the saturated alkyl region is bulkier and cannot approach the relevant region of the binding site so closely (Fig. 13.7). Alkenes are generally easier to reduce than aromatic rings, so it may be possible to prepare the saturated analogue directly from the lead compound.

13.1.4 The binding role of ketones and aldehydes

A ketone group is not uncommon in many of the structures studied in medicinal chemistry. It is a planar group that can interact with a binding site through hydrogen bonding where the carbonyl oxygen acts as a hydrogen bond acceptor (Fig. 13.8). Two such interactions are possible, as two lone pairs of electrons are available on the carbonyl oxygen. The lone pairs are in sp^2 hybridized orbitals which are in the same plane as the functional group. The carbonyl group also has a significant dipole

moment and so a dipole–dipole interaction with the binding site is also possible.

It is relatively easy to reduce a ketone to an alcohol and it may be possible to carry out this reaction directly on the lead compound. This significantly changes the geometry of the functional group from planar to tetrahedral. Such an alteration in geometry may well weaken any existing hydrogen bonding interactions and will certainly weaken any dipole–dipole interactions, as both the magnitude and orientation of the dipole moment will be altered (Fig. 13.9). If it was suspected that the oxygen present in the alcohol analogue might still be acting as a hydrogen bond acceptor, then the ether or ester analogues could be studied, as described in section 13.1.1. Reactions are available that can reduce a ketone completely to an alkane and remove the oxygen, but they are unlikely to be practical for many of the lead compounds studied in medicinal chemistry.

Aldehydes are less common in drugs because they are more reactive and are susceptible to metabolic oxidation to carboxylic acids. However, they could interact in the same way as ketones, and similar analogues could be studied.

13.1.5 Binding role of amines

Amines are extremely important functional groups in medicinal chemistry and are present in many drugs. They may be involved in hydrogen bonding, either as a hydrogen bond acceptor or a hydrogen bond donor (Fig. 13.10). The nitrogen atom has one lone pair of electrons and can act as a hydrogen bond acceptor for one

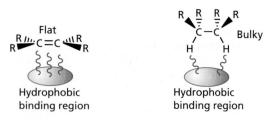

FIGURE 13.7 Binding comparison of an alkene with an alkane.

FIGURE 13.8 Binding interactions that are possible for a carbonyl group.

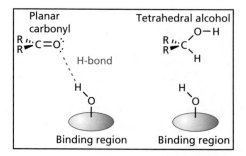

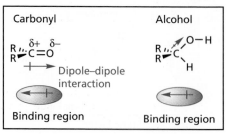

FIGURE 13.9 Effect on binding interactions following the reduction of a ketone or aldehyde.

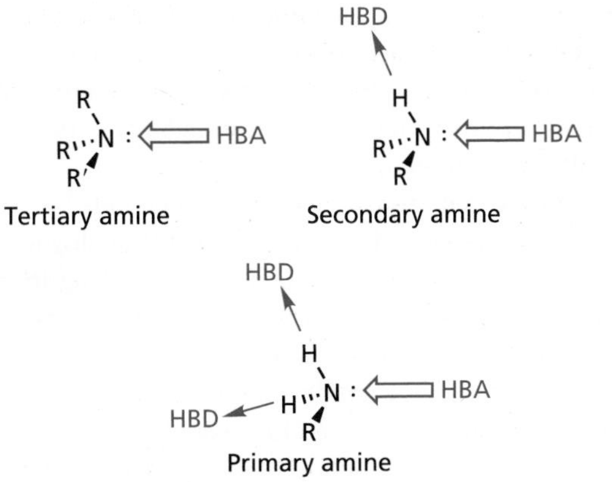

FIGURE 13.10 Possible binding interactions for amines.

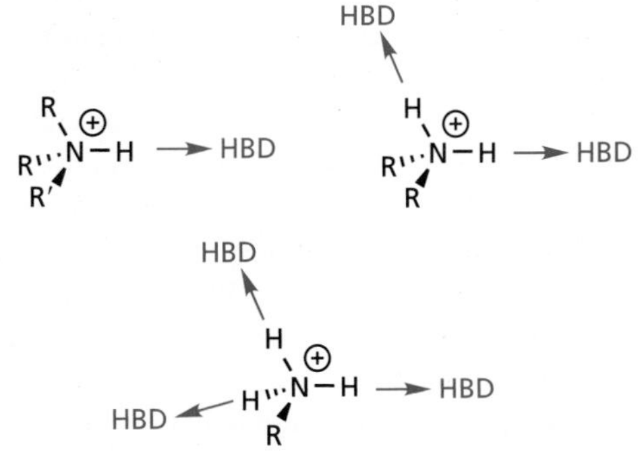

FIGURE 13.11 Possible hydrogen bonding interactions for ionized amines.

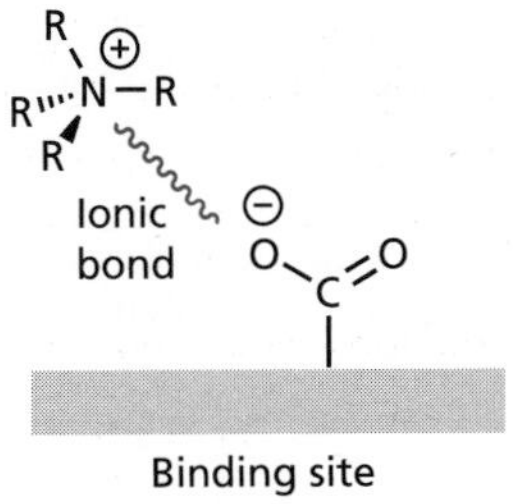

FIGURE 13.12 Ionic interaction between an ionized amine and a carboxylate ion (R = H, alkyl, or aryl).

hydrogen bond. Primary and secondary amines have N–H groups and can act as hydrogen bond donors. Aromatic and heteroaromatic amines act only as hydrogen bond donors, because the lone pair interacts with the aromatic or heteroaromatic ring.

In many cases, the amine may be protonated when it interacts with its target binding site, which means that it is ionized and cannot act as a hydrogen bond acceptor. However, it can still act as a hydrogen bond donor and will form stronger hydrogen bonds than if it was not ionized (Fig. 13.11). Alternatively, a strong ionic interaction may take place with a carboxylate ion in the binding site (Fig. 13.12).

To test whether ionic or hydrogen bonding interactions are taking place, an amide analogue could be studied. This will prevent the nitrogen acting as a hydrogen bond acceptor, as the nitrogen's lone pair will interact with the neighbouring carbonyl group (Fig. 13.13). This interaction also prevents protonation of the nitrogen and rules out the possibility of ionic interactions. You might argue that the right-hand structure in Fig. 13.13a has a positive charge on the nitrogen and could still take part in an ionic interaction. However, this resonance structure represents one extreme and is never present as a distinct entity. The amide group as a whole is neutral, and so lacks the net positive charge required for ionic bonding.

It is relatively easy to form secondary and tertiary amides from primary and secondary amines respectively, and it may be possible to carry out this reaction directly on the lead compound. A tertiary amide lacks the N–H group of the original secondary amine and would test whether this is involved as a hydrogen bond donor. The secondary amide formed from a primary amine still has an N–H group present, but the steric bulk of the acyl group should hinder it acting as a hydrogen bond donor.

Tertiary amines cannot be converted directly to amides, but if one of the alkyl groups is a methyl group, it is often possible to remove it with vinyloxycarbonyl chloride (VOC-Cl) to form a secondary amine, which could then be converted to the amide (Fig. 13.14). This demethylation reaction is extremely useful and has been used to good effect in the synthesis of morphine analogues (see Box 24.2 for the reaction mechanism).

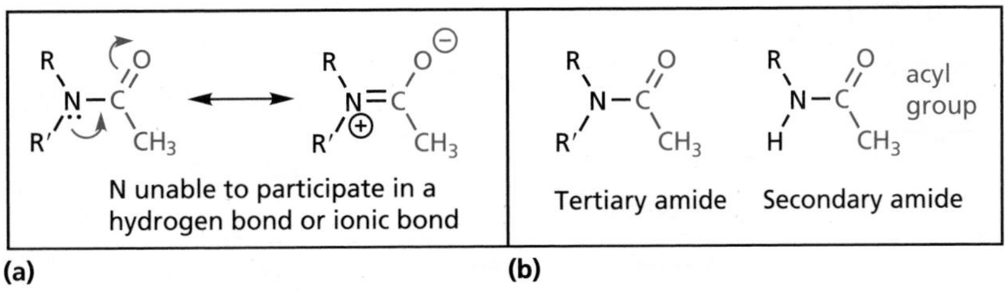

FIGURE 13.13 (a) Interaction of the nitrogen lone pair with the neighbouring carbonyl group in amides. (b) Secondary and tertiary amides.

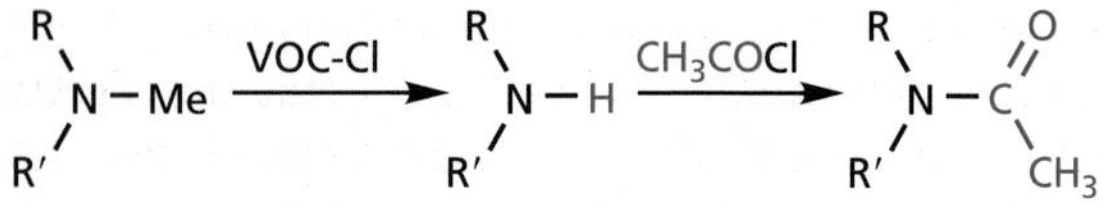

FIGURE 13.14 Demethylation of a tertiary amine and formation of a secondary amide.

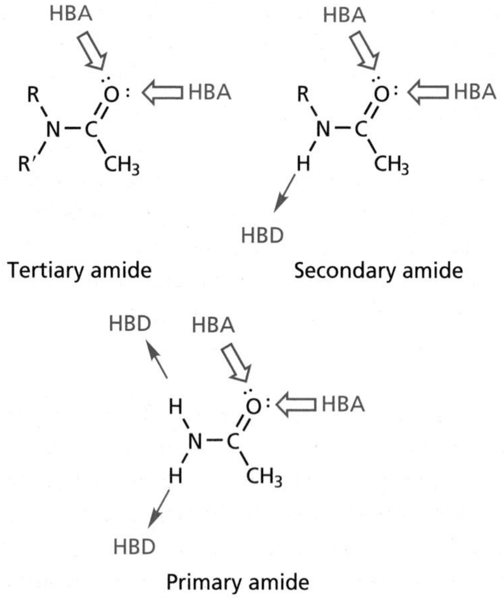

FIGURE 13.15 Possible hydrogen bonding interactions for amides.

13.1.6 Binding role of amides

Many of the lead compounds currently studied in medicinal chemistry are peptides or polypeptides consisting of amino acids linked together by peptide or amide bonds (section 2.1). Amides are likely to interact with binding sites through hydrogen bonding (Fig. 13.15). The carbonyl oxygen atom can act as a hydrogen bond acceptor and has the potential to form two hydrogen bonds. Both the lone pairs involved are in sp^2 hybridized orbitals which are located in the same plane as the amide group. The nitrogen cannot act as a hydrogen bond acceptor because the lone pair interacts with the neighbouring carbonyl group (Fig. 13.13a). Primary and secondary amides have an N–H group, which allows the possibility of this group acting as a hydrogen bond donor.

The most common type of amide in peptide lead compounds is the secondary amide. Suitable analogues that could be prepared to test out possible binding interactions are shown in Fig. 13.16. All the analogues, apart from the primary and secondary amines, could be used to check whether the amide is acting as a hydrogen bond donor. The alkenes and amines could be tested to see whether the amide is acting as a hydrogen bond acceptor. However, there are traps for the unwary. The amide group is planar and does not rotate because of its partial double bond character. The ketone, the secondary amine and the tertiary amine analogues have a single bond at the equivalent position which *can* rotate. This would alter the relative positions of any binding groups on either side of the amide group and lead to a loss of binding, even if the amide itself was not involved in binding. Therefore, a loss of activity would not necessarily mean that the amide is important as a binding group. With these groups, it would only be safe to say that the amide group is not essential if activity is retained. Similarly, the primary amine and carboxylic acid may be found to have no activity, but this might be due to the loss of important binding groups in one half of the molecule. These particular analogues would only be worth considering if the amide group is peripheral to the molecule (e.g. R–NHCOMe or R–CONHMe) and not part of the main skeleton.

The alkene would be a particularly useful analogue to test because it is planar, cannot rotate, and cannot act as a hydrogen bond donor or hydrogen bond acceptor. However, the synthesis of this analogue may not be simple. In fact, it is likely that all the analogues described would have to be prepared using a full synthesis. Amides are relatively stable functional groups and, although several of the analogues described might be attainable directly from the lead compound, it is more likely that the lead

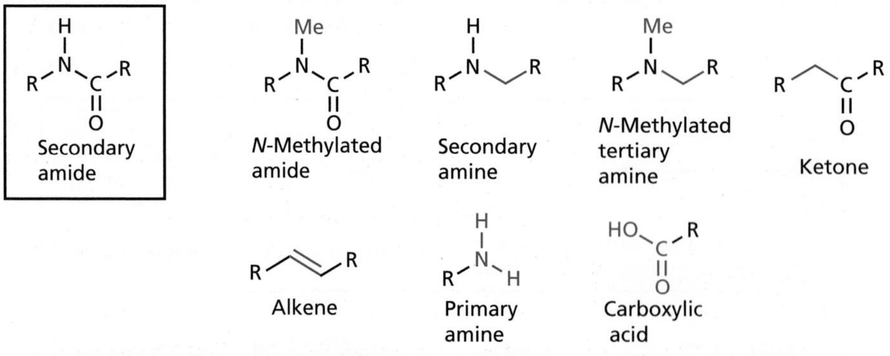

FIGURE 13.16 Possible analogues to test the binding interactions of a secondary amide.

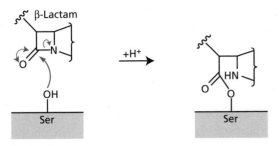

FIGURE 13.17 β-Lactam ring acting as an acylating agent.

compound would not survive the forcing conditions required.

Amides which are within a ring system are called lactams. They, too, can form intermolecular hydrogen bonds as described above. However, if the ring is small and suffers ring strain, the lactam can undergo a chemical reaction with the target leading to the formation of a covalent bond. The best examples of this are the penicillins which contain a four-membered β-lactam ring. This acts as an acylating agent and irreversibly inhibits a bacterial enzyme by acylating a serine residue in the active site (Fig. 13.17) (section 19.5.1.4).

13.1.7 Binding role of quaternary ammonium salts

Quaternary ammonium salts are ionized and can interact with carboxylate groups by ionic interactions (Fig. 13.18). Another possibility is an **induced dipole interaction** between the quaternary ammonium ion and any aromatic rings in the binding site. The positively charged nitrogen can distort the π electrons of the aromatic ring such that a dipole is induced, whereby the face of the ring is slightly negative and the edges are slightly positive. This allows an interaction between the slightly negative faces of the aromatic rings and the positive charge of the quaternary ammonium ion. This is also known as a **π-cation interaction**.

The importance of these interactions could be tested by synthesizing an analogue that has a tertiary amine group rather than the quaternary ammonium group. Of course, it is possible that such a group could ionize by becoming protonated, then interact in the same way. Converting the amine to an amide would prevent this possibility. The neurotransmitter **acetylcholine** has a quaternary ammonium group which is thought to bind to the binding site of its target receptor by ionic bonding and/or induced dipole interactions (section 22.5).

13.1.8 Binding role of carboxylic acids

The carboxylic acid group is reasonably common in drugs. It can act as a hydrogen bond acceptor or as a hydrogen bond donor (Fig. 13.19). Alternatively, it may exist as the carboxylate ion. This allows the possibility of an ionic interaction and/or a strong hydrogen bond where the carboxylate ion acts as a hydrogen bond acceptor. The carboxylate ion is also a good ligand for metal ion cofactors present in several enzymes; for example zinc metalloproteinases (section 21.7.1 and Case study 2).

In order to test the possibility of such interactions, analogues such as esters, primary amides, primary alcohols, and ketones could be synthesized and tested (Fig. 13.20). None of these functional groups can ionize, so a loss of activity could imply that an ionic bond is important. The primary alcohol could shed light on whether the carbonyl oxygen is involved in hydrogen bonding, whereas the ester and ketone could indicate whether the hydroxyl group of the carboxylic acid is involved in hydrogen

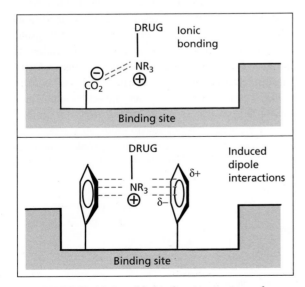

FIGURE 13.18 Possible binding interactions of a quaternary ammonium ion.

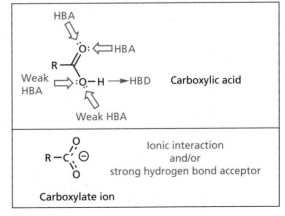

FIGURE 13.19 Possible binding interactions for a carboxylic acid and carboxylate ion.

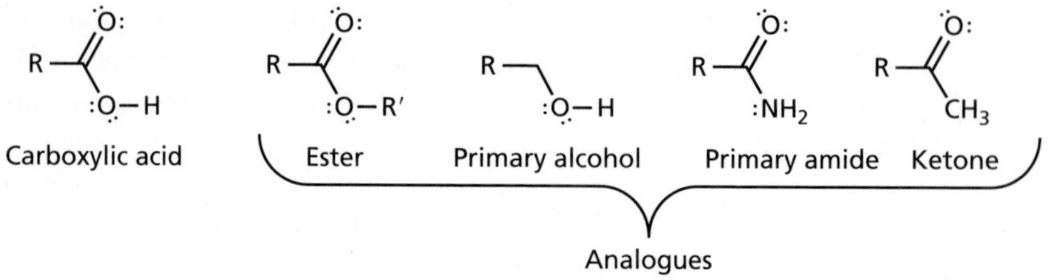

FIGURE 13.20 Analogues to test the binding interactions for a carboxylic acid.

bonding. It may be possible to synthesize the ester and amide analogues directly from the lead compound, but the reduction of a carboxylic acid to a primary alcohol requires harsher conditions and this sort of analogue would normally be prepared by a full synthesis. The ketone would also have to be prepared by a full synthesis.

13.1.9 Binding role of esters

An ester functional group has the potential to interact with a binding site as a hydrogen bond acceptor only (Fig. 13.21). The carbonyl oxygen is more likely to act as the hydrogen bond acceptor than the alkoxy oxygen (section 1.3.2), as it is sterically less hindered and has a greater electron density. The importance or otherwise of the carbonyl group could be judged by testing an equivalent ether, which would require a full synthesis.

Esters are susceptible to hydrolysis *in vivo* by metabolic enzymes called **esterases**. This may pose a problem if the lead compound contains an ester that is important to binding, as it means the drug might have a short lifetime *in vivo*. Having said that, there are several drugs that *do* contain esters and are relatively stable to metabolism,

thanks to electronic factors that stabilize the ester or steric factors that protect it.

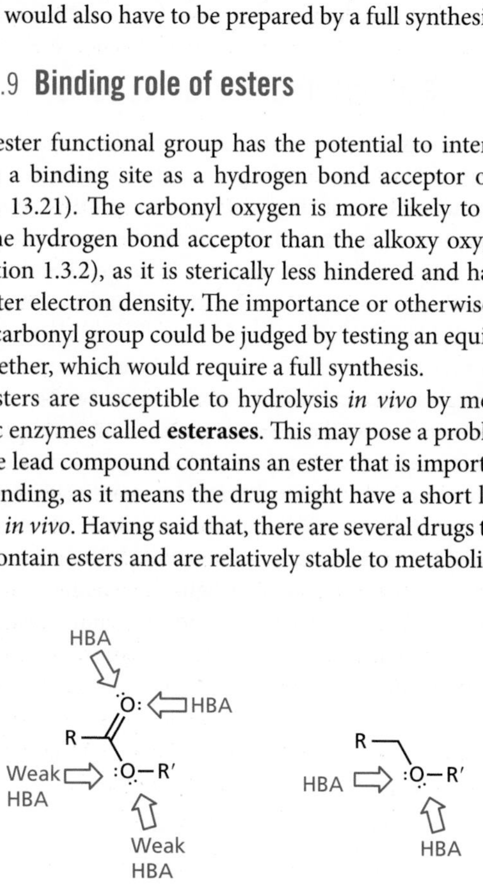

Esters that are susceptible to metabolic hydrolysis are sometimes used deliberately to mask a polar functional group such as a carboxylic acid, alcohol, or phenol in order to achieve better absorption from the gastrointestinal tract. Once in the blood supply, the ester is hydrolysed to release the active drug. This is known as a **prodrug** strategy (section 14.6).

Special mention should be made of the ester group in aspirin. Aspirin has an anti-inflammatory action resulting from its ability to inhibit an enzyme called **cyclooxygenase** (COX) which is required for **prostaglandin** synthesis. It is often stated that aspirin acts as an acylating agent, and that its acetyl group is covalently attached to a serine residue in the active site of COX (Fig. 13.22). However, this theory has been disputed and it is stated that aspirin acts, instead, as a prodrug to generate salicylic acid, which then inhibits the enzyme through non-covalent interactions.

13.1.10 Binding role of alkyl and aryl halides

Alkyl halides involving chlorine, bromine, or iodine tend to be chemically reactive, since the halide ion is a good leaving group. As a result, a drug containing an alkyl halide is likely to react with any nucleophilic group that it encounters and become permanently linked to that group by a covalent bond—an alkylation reaction (Fig. 13.23). This poses a problem, as the drug is likely to alkylate a large variety of macromolecules which have nucleophilic groups, especially amine groups in proteins and nucleic acids. It is possible to moderate the reactivity to some extent, but selectivity is still a problem and leads to severe side effects. These drugs are, therefore, reserved

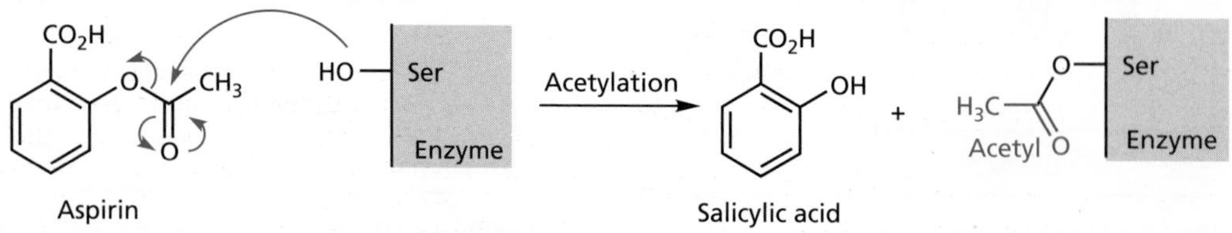

FIGURE 13.21 Possible binding interactions for an ester and an ether.

FIGURE 13.22 The disputed theory of aspirin acting as an acylating agent.

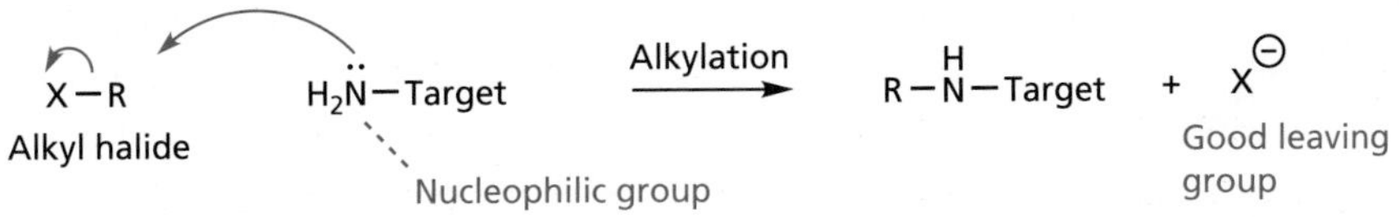

FIGURE 13.23 Alkylation of macromolecular targets by alkyl halides.

for life-threatening diseases such as cancer (sections 9.3 and 21.2.3). Alkyl fluorides, on the other hand, are not alkylating agents, because the strong C–F bond is not easily broken. Fluorine is commonly used to replace a proton as it is approximately the same size, but has different electronic properties. It may also protect the molecule from metabolism (sections 13.3.7 and 14.2.4).

Aryl halides do not act as alkylating agents and pose less of a problem in that respect. As the halogen substituents are electron-withdrawing groups, they affect the electron density of the aromatic ring and this may have an influence on the binding of the aromatic ring. The halogen substituents chlorine and bromine are hydrophobic in nature and may interact favourably with hydrophobic pockets in a binding site. Hydrogen bonding is likely to be weak if it occurs at all. Although halide ions are strong hydrogen bond acceptors, halogen substituents are generally poor hydrogen bond acceptors. Having said that, an ion–dipole interaction involving an aryl fluoride is important in the binding of some statins to their target binding site (Case study 1).

Aliphatic and aromatic analogues lacking the halogen substituent could be prepared by a full synthesis to test whether the halogen has any importance towards the activity of the lead compound.

13.1.11 Binding role of thiols and ethers

The thiol group (S–H) is known to be a good ligand for d-block metal ions and has been incorporated into several drugs designed to inhibit enzymes containing a zinc cofactor; for example, the zinc metalloproteinases (sections 21.7.1 and 26.3.3 and Case study 2). If the lead compound has a thiol group, the corresponding alcohol could be tested as a comparison. This would have a far weaker interaction with zinc.

An ether group (R′OR) might act as a hydrogen bond acceptor through the oxygen atom (Fig. 13.21). This could be tested by increasing the size of the neighbouring alkyl group to see whether it diminishes the ability of the group to take part in hydrogen bonding. Analogues where the oxygen is replaced with a methylene (CH_2) isostere should show significantly decreased binding affinity.

The oxygen atom of an aromatic ether is generally a poor hydrogen bond acceptor (section 1.3.2).

13.1.12 Binding role of other functional groups

In some drugs, sulphonamides can have a binding role where the oxygen atoms act as hydrogen bond acceptors and the NH proton acts as a hydrogen bond donor. In certain circumstances, the sulphonamide group may be ionized and participate in an ionic interaction with the binding site (section 26.4.2)

Lead compounds may contain a wide variety of other functional groups that have no direct binding role, but could be important in other respects. Some may influence the electronic properties of the molecule (e.g. nitro groups or nitriles). Others may restrict the shape or conformation of a molecule (e.g. alkynes) (Box 13.3). Functional groups may also act as metabolic blockers (e.g. aryl halides) (section 14.2.4).

13.1.13 Binding role of alkyl groups and the carbon skeleton

The alkyl substituents and carbon skeleton of a lead compound are hydrophobic and may bind with hydrophobic regions of the binding site through van der Waals interactions. The relevance of an alkyl substituent to binding can be determined by synthesizing an analogue which lacks the substituent. Such analogues generally have to be synthesized using a full synthesis if the substituents are attached to the carbon skeleton of the molecule. However, if the alkyl group is attached to nitrogen or oxygen, it may be possible to remove the group from the lead compound, as shown in Fig. 13.24. The analogues obtained may then be expected to have less activity if the alkyl group was involved in important hydrophobic interactions.

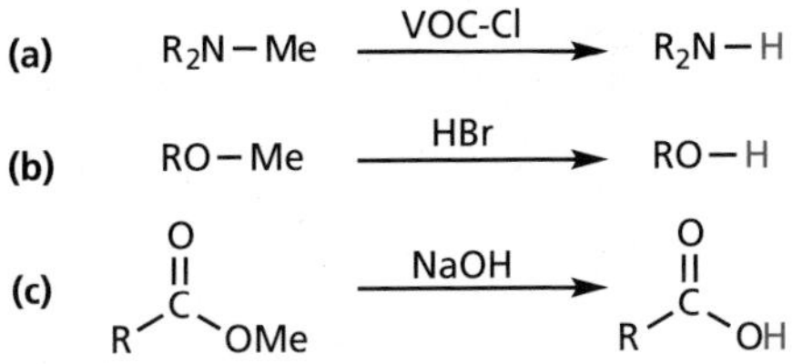

FIGURE 13.24 (a) N-Demethylation of a tertiary amine with vinyloxycarbonyl chloride (see Box 24.2 for mechanism). (b) Demethylation of a methyl ether using hydrogen bromide where nucleophilic substitution leads to an alcohol (or phenol) plus bromomethane. (c) Hydrolysis of an ester using sodium hydroxide where OH replaces OMe.

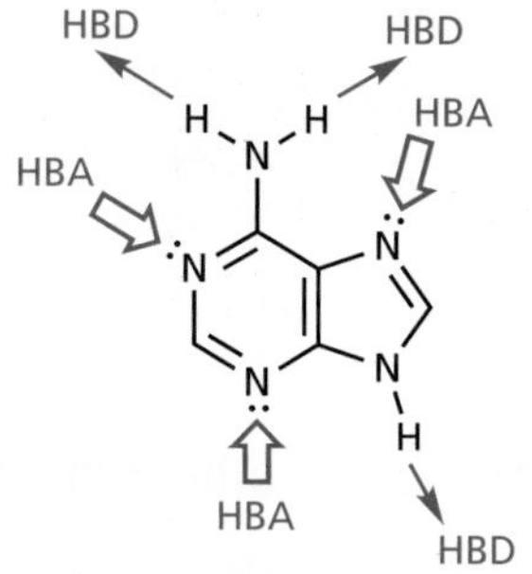

FIGURE 13.25 Possible hydrogen bonding interactions for adenine.

13.1.14 **Binding role of heterocycles**

A large diversity of heterocycles are found in lead compounds. Heterocycles are cyclic structures that contain one or more heteroatoms such as oxygen, nitrogen, or sulphur. Nitrogen-containing heterocycles are particularly prevalent. The heterocycles can be aliphatic or aromatic in character, and have the potential to interact with binding sites through a variety of bonding forces. For example, the overall heterocycle can interact through van der Waals and pi–pi interactions, while the individual heteroatoms present in the structure could interact by hydrogen bonding or ionic bonding.

As far as hydrogen bonding is concerned, there is an important directional aspect. The position of the heteroatom in the ring and the orientation of the ring in the binding site can be crucial in determining whether or not a good interaction takes place. For example, adenine can take part in six hydrogen bonding interactions, three as a hydrogen bond donor and three as a hydrogen bond acceptor. The ideal directions for these interactions are

shown in Fig. 13.25. Van der Waals or pi–pi interactions are also possible to regions of the binding site above and below the plane of the ring system.

Heterocycles can be involved in quite intricate hydrogen bonding networks within a binding site. For example, the anticancer drug **methotrexate** contains a diaminopteridine ring system that interacts with its binding site as shown in Fig. 13.26.

If the lead compound contains a heterocyclic ring, it is worth synthesizing analogues containing a benzene ring or different heterocyclic rings to explore whether all the heteroatoms present are really necessary.

A complication with heterocycles is the possibility of **tautomers**. This played an important role in determining the structure of DNA (section 6.1.2). The structure of DNA consists of a double helix with base pairing between two sets of heterocyclic nucleic acid bases. Base pairing involves three hydrogen bonds between the base pair guanine and cytosine, and two hydrogen bonds between the base pair adenine and thymine (Fig. 13.27). The rings involved in the base pairing are coplanar, allowing the optimum orientation for the hydrogen bond donors and hydrogen bond acceptors. This in turn means that the base pairs are stacked above each other, allowing van der Waals interactions between the faces of each base pair. However, when Watson and Crick originally tried to devise a model for DNA, they incorrectly assumed that the preferred tautomers for the nucleic acid bases were as shown in the right-hand part of Fig. 13.27. With these tautomers, the required hydrogen bonding is not possible and would not explain the base pairing observed in the structure of DNA.

In a similar vein, knowing the preferred tautomers of heterocycles can be important in understanding how

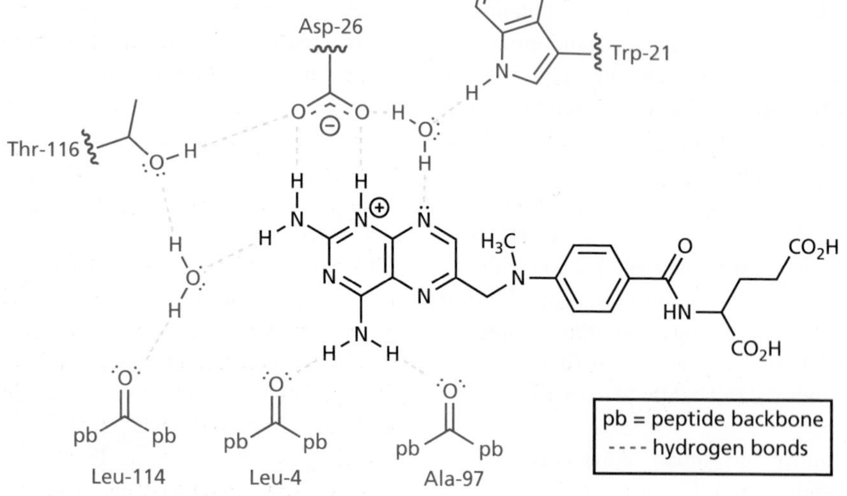

FIGURE 13.26 Binding interactions for the diaminopteridine ring of methotrexate in its binding site.

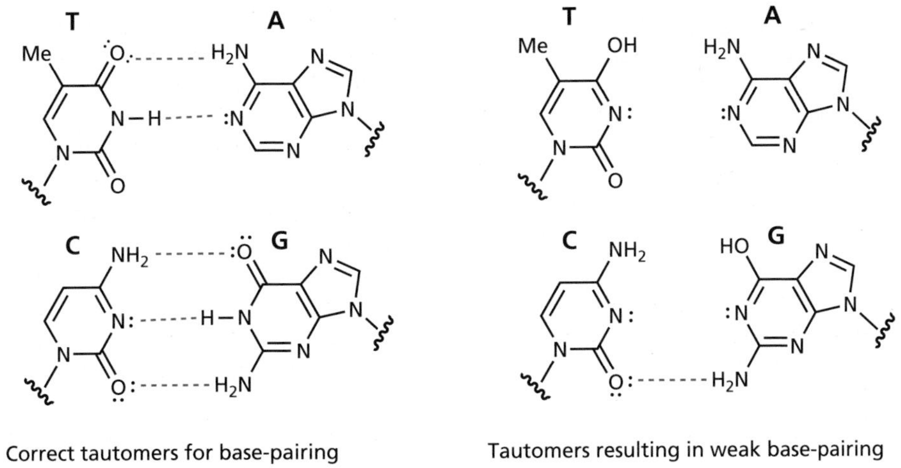

Correct tautomers for base-pairing   Tautomers resulting in weak base-pairing

FIGURE 13.27 Base pairing in DNA and the importance of tautomers.

drugs interact with their binding sites. This is amply illustrated in the design of the anti-ulcer agent **cimetidine** (section 25.2).

With heterocyclic compounds, it is possible for a hydrogen bond donor and a hydrogen bond acceptor to be part of a conjugated system. Polarization of the electrons in the conjugated system permits π-**bond cooperativity**, where the strength of the hydrogen bond donor is enhanced by the hydrogen bond acceptor and vice versa. This has also been called **resonance-assisted hydrogen bonding**. This type of hydrogen bonding is possible for the hydrogen bond donors and acceptors for the nucleic acid base pairs (Fig. 13.28).

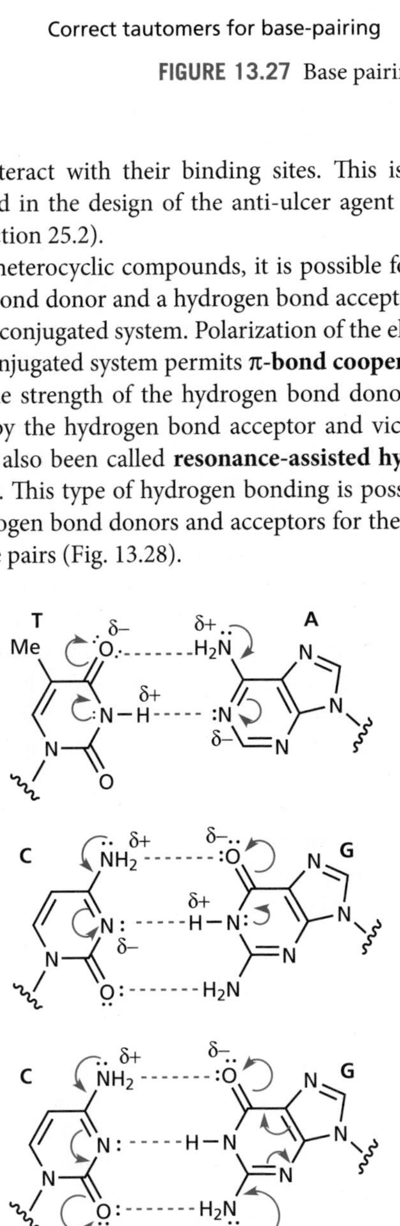

FIGURE 13.28 π-Bond cooperativity in hydrogen bonding.

Note that not all heteroatoms in heterocyclic systems are able to act as good hydrogen bond acceptors. If a heteroatom's lone pair of electrons is part of an aromatic sextet of electrons, it is not available to form a hydrogen bond.

13.1.15 Isosteres

Isosteres are atoms or groups of atoms which share the same valency and which have chemical or physical similarities (Fig. 13.29).

For example, SH, NH_2, and CH_3 are isosteres of OH, whereas S, NH, and CH_2 are isosteres of O. Isosteres can be used to determine whether a particular group is an important binding group or not, by altering the character of the molecule in as controlled a way as possible. Replacing O with CH_2, for example, makes little difference to the size of the analogue, but will have a marked effect on its polarity, electronic distribution, and bonding. Replacing OH with the larger SH may not have such an influence on the electronic character, but steric factors become more significant.

Isosteric groups could be used to determine whether a particular group is involved in hydrogen bonding. For example, replacing OH with CH_3 would completely eliminate hydrogen bonding, whereas replacing OH with NH_2 would not.

The β-blocker **propranolol** has an ether linkage (Fig. 13.30). Replacement of the OCH_2 segment with the isosteres CH=CH, SCH_2, or CH_2CH_2 eliminates activity, whereas replacement with $NHCH_2$ retains activity (though reduced). These results show that the ether oxygen is important to the activity of the drug and suggests that it is involved in hydrogen bonding with the receptor.

The use of isosteres in drug design is described in section 13.3.7.

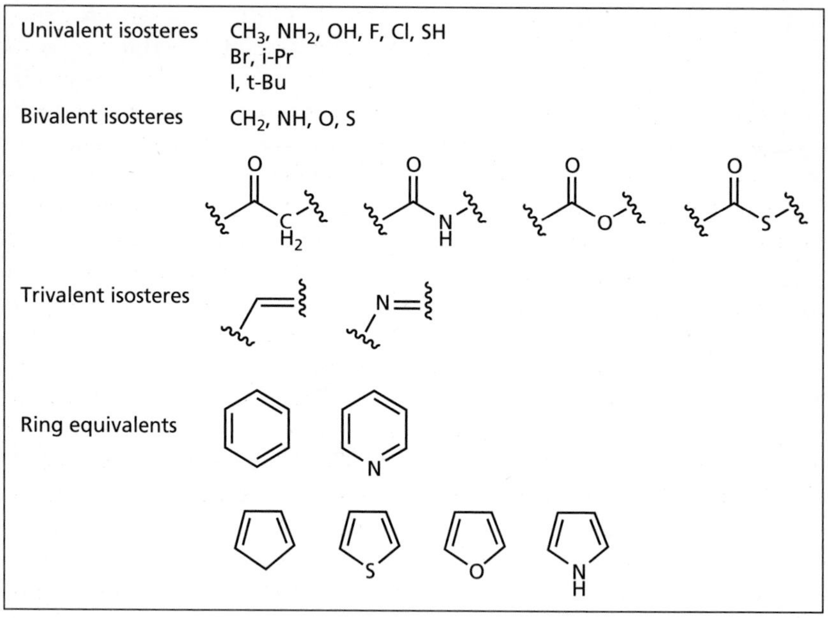

Univalent isosteres	CH$_3$, NH$_2$, OH, F, Cl, SH
	Br, i-Pr
	I, t-Bu
Bivalent isosteres	CH$_2$, NH, O, S

FIGURE 13.29 Examples of classic isosteres.

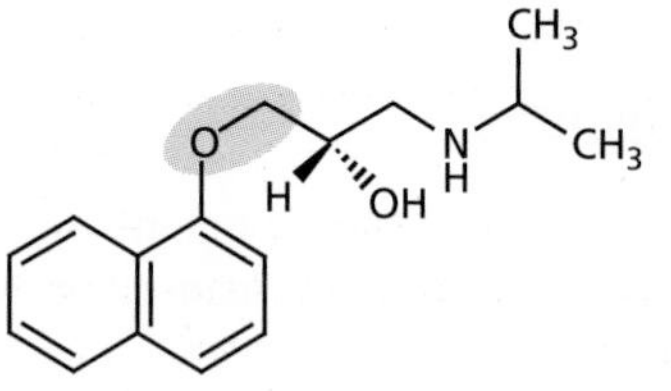

FIGURE 13.30 Propranolol.

13.1.16 Testing procedures

When investigating structure–activity relationships for drug–target binding interactions, biological testing should involve *in vitro* tests; for example inhibition studies on isolated enzymes or binding studies on membrane-bound receptors in whole cells. The results then show conclusively which binding groups are important in drug–target interactions. If *in vivo* testing is carried out, the results are less clear-cut because loss of activity may be due to the inability of the drug to reach its target rather than reduced drug–target interactions. However, *in vivo* testing may reveal functional groups that are important in protecting or assisting the drug in its passage through the body. This would not be revealed by *in vitro* testing.

NMR spectroscopy can also be used to test structure–activity relationships, as described in section 12.4.10.

As mentioned in the introduction, there is little point in designing a drug that has optimum interactions with its target if it has undesirable pharmacokinetic properties. Calculating a structure's hydrophobicity can provide an indication as to whether it is likely to suffer from pharmacokinetic problems. This is because hydrophobic drugs have been found to be more prone to adverse pharmacokinetic properties. For example, they are more likely to interact with other protein targets, resulting in unwanted side effects. They are generally less soluble, show poor permeability, and are more likely to produce toxic metabolites. The hydrophobic nature of a drug can be calculated by its Clog D value (section 14.1). In recent years, several research groups have optimized drugs by optimizing **lipophilic efficiency** (LipE), where LipE = pK_i (or pIC$_{50}$) − Clog D. Drugs with a good level of lipophilic efficiency will have high activity (pK_i or pIC$_{50}$) and low hydrophobic character. Optimizing LipE involves a parallel optimization of potency and hydrophobicity. This quantitative method of optimizing both the pharmacodynamic and pharmacokinetic properties has been called **property-based drug design** and was used in the structure-based drug design of **crizotinib** (Box 13.4).

13.1.17 SAR in drug optimization

In this section, we have focused on SAR studies aimed at identifying important binding groups in a lead compound. SAR studies are also used in drug optimization, where the aim is to find analogues with better activity and selectivity. This involves further modifications of the lead compound to identify whether these are beneficial or detrimental to activity. This is covered in section 13.3 where the different strategies of optimizing drugs are discussed.

13.2 Identification of a pharmacophore

Once it is established which groups are important for a drug's activity, it is possible to move on to the next stage—the identification of the **pharmacophore**. The pharmacophore summarizes the important binding groups which are required for activity, and their relative positions in space with respect to each other. For example, if we discover that the important binding groups for our hypothetical drug glipine are the two phenol groups, the aromatic ring, and the nitrogen atom, then the pharmacophore is as shown in Fig. 13.31. Structure I shows the two-dimensional (2D) pharmacophore and structure II shows the three-dimensional (3D) pharmacophore. The latter specifies the relative positions of the important groups in space. In this case, the nitrogen atom is 5.063 Å from the centre of the phenolic ring and lies at an angle of 18° from the plane of the ring. Note that it is not necessary to show the specific skeleton connecting the important groups. Indeed, there are benefits in not doing so, as it is easier to compare the 3D pharmacophores from different structural classes of compound to see if they are similar. Three-dimensional pharmacophores can be defined using molecular modelling (section 17.11), which allows the definition of 'dummy bonds', such as the one in Fig. 13.31 between nitrogen and the centre of the aromatic ring. The centre of the ring can be defined by a dummy atom called a **centroid** (not shown).

An even more general type of 3D pharmacophore is the one shown as structure III (Fig. 13.31)—a bonding-type pharmacophore. Here the bonding characteristics of each functional group are defined, rather than the group itself. Note also that the groups are defined as points in space. This includes the aromatic ring, which is defined by the centroid. All the points are connected by pharmacophoric triangles to define their positions. This allows the comparison of molecules which may have the same pharmacophore and binding interactions, but which use different functional groups to achieve these interactions. In this case, the phenol groups can act as hydrogen bond donors or acceptors, the aromatic ring can participate in van der Waals interactions, and the amine can act as a hydrogen bond acceptor (or as an ionic centre if it is protonated). We shall return to the concept and use of 3D pharmacophores in sections 17.11 and 18.10.

Identifying 3D pharmacophores is relatively easy for rigid cyclic structures such as the hypothetical glipine. With more flexible structures, it is not so straightforward because the molecule can adopt a large number of shapes or conformations which place the important binding groups in different positions relative to each other. Normally, only one of these conformations is recognized and bound by the binding site. This conformation is known as the **active conformation**. In order to identify the 3D pharmacophore, it is necessary to know the active conformation. There are various ways in which this might be done. Rigid analogues of the flexible compound could be synthesized and tested to see whether activity is retained (section 13.3.9). Alternatively, it may be possible to crystallize the target with the compound bound to the binding site. X-ray crystallography could then be used to identify the structure of the complex as well as the active conformation of the bound ligand (section 17.10). Finally, progress has been made in using NMR spectroscopy to solve the active conformation of isotopically labelled molecules bound to their binding sites (see also Case study 10).

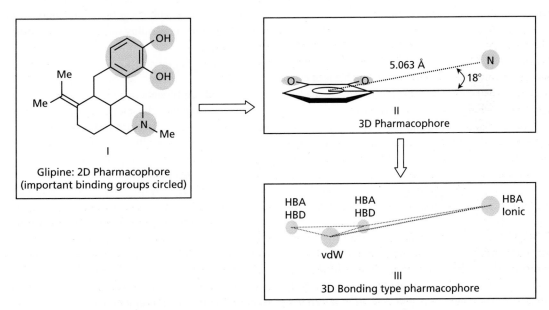

FIGURE 13.31 Pharmacophore for the fictitious structure glipine.

We finish this section with a warning! A drawback with pharmacophores is their unavoidable emphasis on functional groups as the crucial binding groups. In many situations this is certainly true, but in other situations, it is not. It is not uncommon to find compounds that have the correct pharmacophore, but show disappointing activity and poor binding. It is important to realize that the overall skeleton of the molecule is involved in interactions with the binding site through van der Waals and hydrophobic interactions. The strength of these interactions can sometimes be crucial in whether a drug binds effectively or not, and the 3D pharmacophore does not take this into account. The pharmacophore also does not take into account the size of a molecule and whether it will fit the binding site. Finally, a functional group that is part of the pharmacophore may not be so crucial if an agent can form an alternative binding interaction with the binding site. For example, the phenol group is an important part of the analgesic pharmacophore for **morphine** and closely related analogues, but is less important for analgesics such as the **oripavines**. Other analgesics such as **pethidine** and **methadone** lack the phenol group entirely (Chapter 24).

KEY POINTS

- SARs define the functional groups or regions of a lead compound which are important to its biological activity.

- Functional groups such as alcohols, amines, esters, amides, carboxylic acids, phenols, and ketones can interact with binding sites by means of hydrogen bonding.

- Functional groups such as aminium ions, quaternary ammonium salts, and carboxylate groups can interact with binding sites by ionic bonding. In some cases, a sulphonamide group can become ionized and form an ionic bond.

- Functional groups such as alkenes and aromatic rings can interact with binding sites by means of van der Waals interactions. π-π interactions are also possible with aromatic rings.

- Alkyl substituents and the carbon skeleton of the lead compound can interact with hydrophobic regions of binding sites by means of van der Waals interactions.

- Interactions involving dipole moments or induced dipole moments may play a role in binding a lead compound to a binding site.

- Reactive functional groups such as alkyl halides may lead to irreversible covalent bonds being formed between a lead compound and its target.

- The relevance of a functional group to binding can be determined by preparing analogues where the functional group is modified or removed in order to see whether activity is affected by such a change.

- Some functional groups can be important to the activity of a lead compound for reasons other than target binding. They may play a role in the electronic or stereochemical properties of the compound, or they may have an important pharmacokinetic role.

- Replacing a group in the lead compound with an isostere (a group having the same valency) makes it easier to determine whether a particular property such as hydrogen bonding is important.

- *In vitro* testing procedures should be used to determine the SAR for target binding.

- The pharmacophore summarizes the groups which are important in binding a lead compound to its target, as well as their relative positions in three dimensions.

13.3 Drug optimization: strategies in drug design

Once the important binding groups and pharmacophore of the lead compound have been identified, it is possible to synthesize analogues that contain the same pharmacophore. But why is this necessary? If the lead compound has useful biological activity, why bother making analogues? The answer is that very few lead compounds are ideal. Most are likely to have low activity, poor selectivity, and significant side effects. They may also be difficult to synthesize, so there is an advantage in finding analogues with improved properties. We look now at strategies that can be used to optimize the interactions of a drug with its target in order to gain better activity and selectivity.

13.3.1 Variation of substituents

Varying easily accessible substituents is a common method of fine tuning the binding interactions of a drug.

13.3.1.1 Alkyl substituents

Certain alkyl substituents can be varied more easily than others. For example, the alkyl substituents of ethers, amines, esters, and amides are easily varied as shown in Fig. 13.32. In these cases, the alkyl substituent already present can be removed and replaced by another substituent. Alkyl substituents which are part of the carbon skeleton of the molecule are not easily removed, and it is usually necessary to carry out a full synthesis in order to vary them.

If alkyl groups are interacting with a hydrophobic pocket in the binding site, then varying the length and bulk of the alkyl group (e.g. methyl, ethyl, propyl, butyl, isopropyl, isobutyl, or *t*-butyl) allows one to probe the depth and width of the pocket. Choosing a substituent that will fill the pocket will then increase the binding interactions (Fig. 13.33).

FIGURE 13.32 Methods of modifying an alkyl group.

Larger alkyl groups may also confer selectivity on the drug. For example, in the case of a compound that interacts with two different receptors, a bulkier alkyl substituent may prevent the drug from binding to one of those receptors and so cut down side effects (Fig. 13.34). For example, **isoprenaline** is an analogue of **adrenaline** where a methyl group was replaced by an isopropyl group, resulting in selectivity for adrenergic α-receptors over adrenergic β-receptors (section 23.11.3).

13.3.1.2 Substituents on aromatic or heteroaromatic rings

If a drug contains an aromatic or heteroaromatic ring, the position of substituents can be varied to find better binding interactions, resulting in increased activity (Fig. 13.35).

For example, the best anti-arrhythmic activity for a series of benzopyrans was found when the sulphonamide substituent was at position 7 of the aromatic ring (Fig. 13.36).

Changing the position of one substituent may have an important effect on another. For example, an

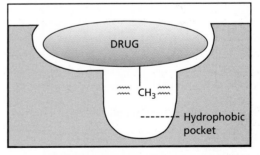

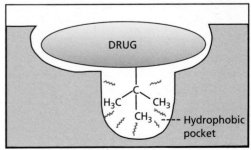

FIGURE 13.33 Variation of an alkyl substituent to fill a hydrophobic pocket.

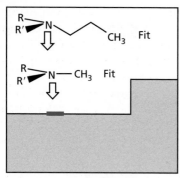

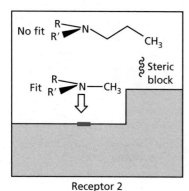

■ Binding region for N

FIGURE 13.34 Use of a larger alkyl group to confer selectivity on a drug.

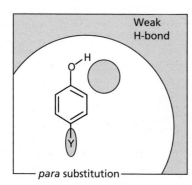

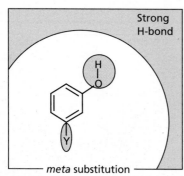

◯ Binding region (H-bond) ◯ Binding region (for Y)

FIGURE 13.35 Varying the position of a substituent on an aromatic ring.

FIGURE 13.36 Benzopyrans.

electron-withdrawing nitro group will affect the basicity of an aromatic amine more significantly if it is at the *para* position rather than the *meta* position (Fig. 13.37). At the *para* position, the nitro group will make the amine a weaker base and less liable to protonate. This would decrease the amine's ability to interact with ionic binding groups in the binding site, and decrease activity.

If the substitution pattern is ideal, then we can try varying the substituents themselves. Substituents have different steric, hydrophobic, and electronic properties, and so varying these properties may have an effect on binding and activity. For example, activity might be improved by having a more electron-withdrawing substituent, in which case a chloro substituent might be tried in place of a methyl substituent.

The chemistry involved in these procedures is usually straightforward, so these analogues are made as a matter of course whenever a novel drug structure is developed. Furthermore, the variation of substituents is open to **quantitative structure–activity relationship** (QSAR) studies, as described in Chapter 18.

13.3.1.3 Synergistic effects

Finally, a warning! When varying substituents, it is normal to study analogues where only one substituent is

meta (inductive electron-withdrawing effect)

para (electron-withdrawing effect due to resonance *and* inductive effects)

FIGURE 13.37 Electronic effects of different substitution patterns on an aromatic ring.

added or altered at a time. In that way, one can identify those substituents that are good for activity and those that are not. However, it does not take into account the synergistic effect that two or more substituents may have on activity. For example, two substituents that are individually bad for activity may actually be beneficial for activity when they are both present. The design of the anticancer drug **sorafenib** provides an illustration of this effect (Box 21.10).

13.3.2 Extension of the structure

The strategy of extension involves the addition of another functional group or substituent to the lead compound in order to probe for extra binding interactions with the target. Lead compounds are capable of fitting the binding site and have the necessary functional groups to interact with some of the important binding regions present. However, it is possible that they do not interact with all the binding regions available. For example, a lead compound may bind to three binding regions in the binding site but fail to use a fourth (Fig. 13.38). Therefore, why not add extra functional groups to probe for that fourth region?

Extension tactics are often used to find extra hydrophobic regions in a binding site by adding various alkyl or arylalkyl groups. These groups can be added to functional groups such as alcohols, phenols, amines, and carboxylic acids should they be present in the drug, as long as this does not disrupt important binding interactions that are already present. Alternatively, they could be built into the building blocks used in the synthesis of various analogues. By the same token, substituents containing polar functional groups could be added to probe for extra hydrogen bonding or ionic interactions. A good example of the use of extension tactics to increase binding interactions involves the design of the ACE inhibitor **enalaprilat** from the lead compound **succinyl proline**; see Case study 2, Figs CS2.8–2.9.

Extension strategies are used to strengthen the binding interactions and activity of a receptor agonist or an enzyme inhibitor, but they can also be used to convert an agonist into an antagonist. This will happen if the extra binding interaction results in a different induced fit from that required to activate the receptor. As a result, the antagonist binds to an inactive conformation of the receptor and blocks access to the endogenous agonist. The strategy has also been used to alter an enzyme substrate into an inhibitor (Box 13.1).

The extension tactic has been used successfully to produce more active analogues of morphine (sections 24.6.2 and 24.6.4) and more active adrenergic agents (sections 23.9–23.11). It was also used to improve the activity and selectivity of the protein kinase inhibitor **imatinib** (section 21.6.2.2). Other examples of the extension strategy can be found in Case studies 2, 5, 6, and 7, and Box 17.6, as well as sections 19.7.7, 20.7.4, and 26.4.2.

An unusual example of an extension strategy is where a substituent was added to an enzyme substrate such that extra binding interactions took place with a neighbouring cofactor in the binding site. This resulted in the analogue acting as an inhibitor, rather than a substrate (Box 13.1).

13.3.3 Chain extension/contraction

Some drugs have two important binding groups linked together by a chain, in which case it is possible that the chain length is not ideal for the best interaction. Therefore, shortening or lengthening the chain length is a useful tactic to try (Fig. 13.39, see also Box 13.1, section 24.6.2, and Case study 2).

13.3.4 Ring expansion/contraction

If a drug has one or more rings that are important binding groups, it is generally worth synthesizing analogues where one of these rings is expanded or contracted. The principle behind this approach is much the same

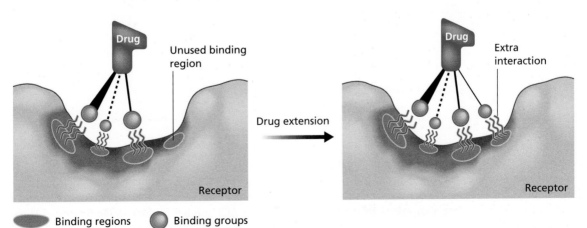

FIGURE 13.38 Extension of a drug to provide a fourth binding group.

BOX 13.1 Converting an enzyme substrate to an inhibitor by extension tactics

The enzyme **17β-hydroxysteroid dehydrogenase type 1** catalyses the conversion of **estrone** to the female steroid hormone **estradiol**, with the cofactor **NADH** acting as the reducing agent for the reaction (Fig. 1; see also Chapter 3, Figs. 3.11–3.12). Inhibition of this enzyme may prove useful in the treatment of estradiol-dependent tumours since the levels of estradiol present in the body would be lowered.

The cofactor NADH is bound next to estrone in the active site, and so it was reasoned that a direct bonding interaction between an estrone analogue and NADH would lock the analogue into the active site and block access to estrone itself. Therefore, the analogue would act as an enzyme inhibitor. Various substituents were added at position 16 to achieve this goal since crystallographic and molecular modelling studies had shown that such substituents would

be ideally placed for an interaction with the cofactor. This led to a structure (Fig. 2) which showed promising activity as an inhibitor. The amide group interacts with the primary amide of NADH by hydrogen bonding, while the pyridine ring interacts with the phosphate groups of the cofactor. A more conventional extension strategy was to add an ethyl group at C-2, which allowed additional van der Waals interactions with a small hydrophobic pocket in the active site. It was also observed that two protons acted as steric blockers and prevented NADH reducing the ketone group of the analogue.

For additional material see Web article 1: Steroids as novel anticancer agents on the Online Resource Centre at www.oxfordtextbooks.co.uk/orc/patrick6e/

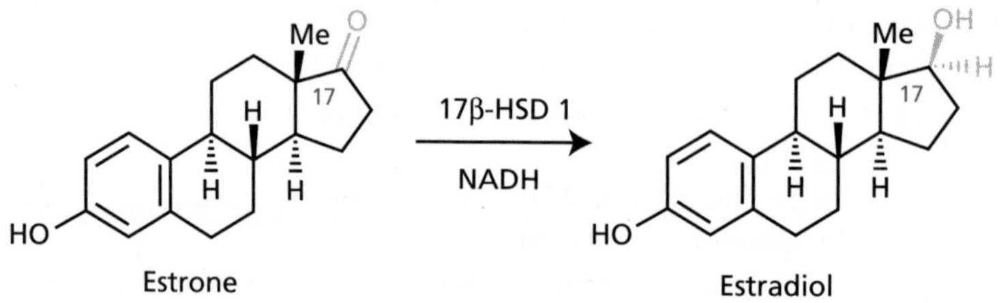

FIGURE 1 The enzyme-catalysed conversion of estrone to estradiol.

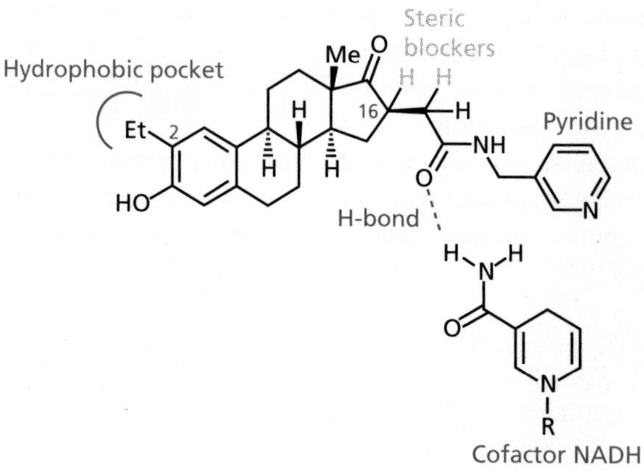

FIGURE 2 Extra binding interactions resulting from the extension strategy.

as varying the substitution pattern of an aromatic ring. Expanding or contracting a ring may put other rings in different positions relative to each other, and may lead to better interactions with specific regions in the binding site (Fig. 13.40).

Varying the size of a ring can also bring substituents into a good position for binding. For example, during the development of the antihypertensive agent **cilazaprilat** (another ACE inhibitor), the bicyclic structure I showed promising activity (Fig. 13.41). The important binding

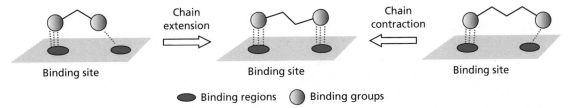

FIGURE 13.39 Chain contraction and chain extension.

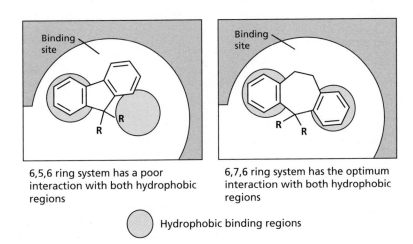

6,5,6 ring system has a poor interaction with both hydrophobic regions

6,7,6 ring system has the optimum interaction with both hydrophobic regions

Hydrophobic binding regions

FIGURE 13.40 Ring expansion.

groups were the two carboxylate groups and the amide group. By carrying out various ring contractions and expansions, cilazaprilat was identified as the structure having the best interaction with the binding site.

Another example of a ring expansion can be seen in the design and development of the cardiovascular agent **ivabradine** (section 26.7). An example of a beneficial ring contraction can be seen in the design of the antiviral agent **simeprevir**, where a 15-membered macrocycle was contracted to a 14-membered macrocycle (Case study 10).

13.3.5 **Ring variations**

A popular strategy used for compounds containing an aromatic or heteroaromatic ring is to replace the original ring with a range of other heteroaromatic rings of different ring size and heteroatom positions. For example, several non-steroidal anti-inflammatory agents (NSAIDs) have been reported, all consisting of a central ring with 1,2-biaryl substitution. Different pharmaceutical companies have varied the central ring to produce a range of active compounds (Fig. 13.42).

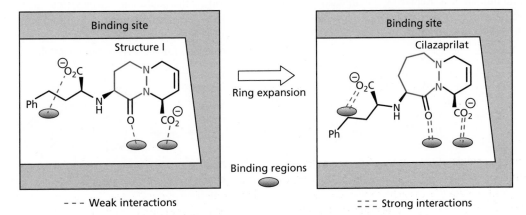

FIGURE 13.41 Development of cilazaprilat.

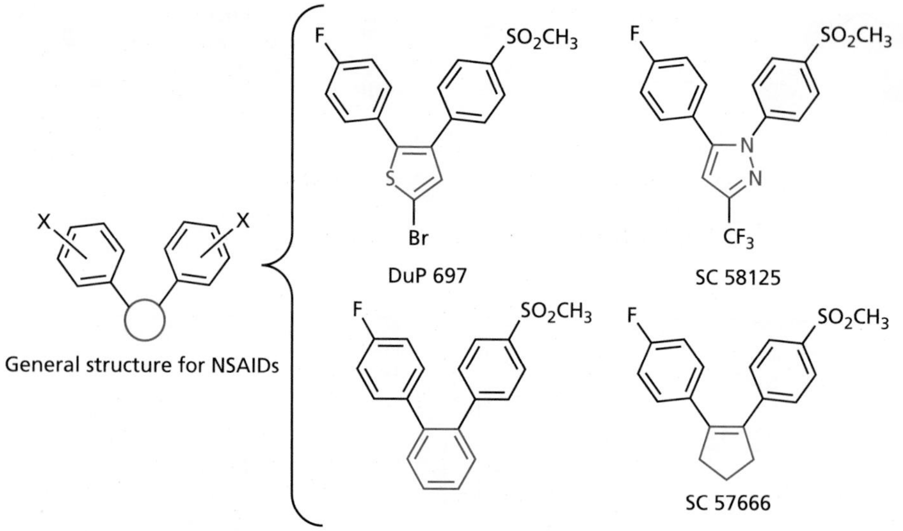

FIGURE 13.42 Non-steroidal anti-inflammatory drugs (NSAIDS).

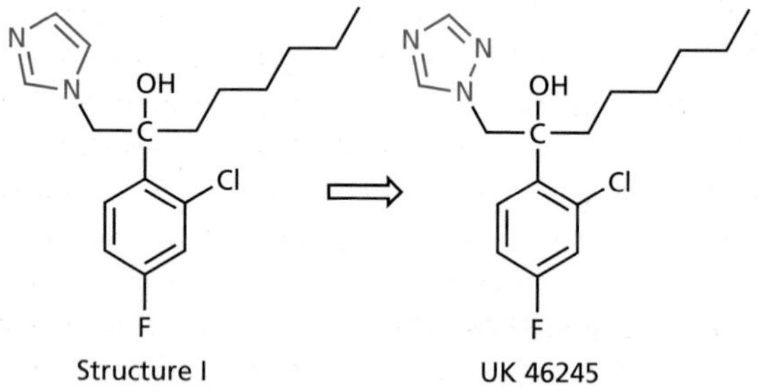

FIGURE 13.43 Development of UK 46245.

One advantage of altering an aromatic ring to a heteroaromatic ring is that it introduces the possibility of an extra hydrogen bonding interaction with the binding site, should a suitable binding region be available (*extension strategy*). For example, structure I (Fig. 13.44) was the lead compound for a project looking into novel antiviral agents. Replacing the aromatic ring with a pyridine ring resulted in an additional binding interaction with the target enzyme. Further development led eventually to the antiviral agent **nevirapine** (Fig. 13.44 and section 20.7.3.2).

Admittedly, a lot of these changes are merely ways of avoiding patent restrictions ('**me too' drugs**), but there can often be significant improvements in activity, as well as increased selectivity and reduced side effects ('**me better' drugs**). For example, the antifungal agent (I) (Fig. 13.43) acts against an enzyme present in both fungal and human cells. Replacing the imidazole ring of structure (I) with a 1,2,4-triazole ring to give UK 46245 resulted in better selectivity against the fungal form of the enzyme.

13.3.6 Ring fusions

Extending a ring by ring fusion can sometimes result in increased interactions or increased selectivity. One of the major advances in the development of the selective β-blockers was the replacement of the aromatic ring in **adrenaline** with a naphthalene ring system (**pronethalol**) (Fig. 13.45). This resulted in a compound that was able to distinguish between two very similar receptors—the α- and β-receptors for adrenaline. One possible explanation

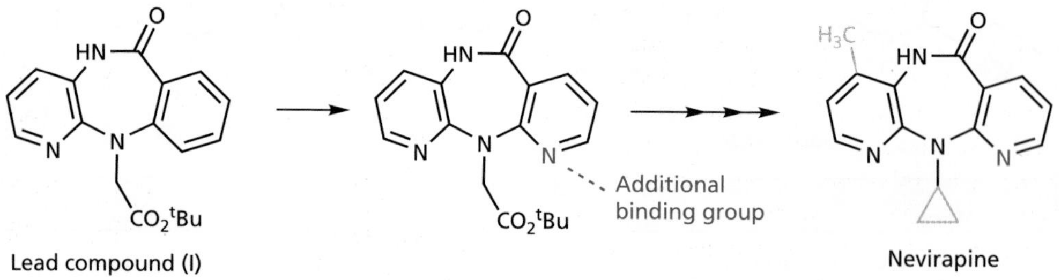

FIGURE 13.44 Development of nevirapine.

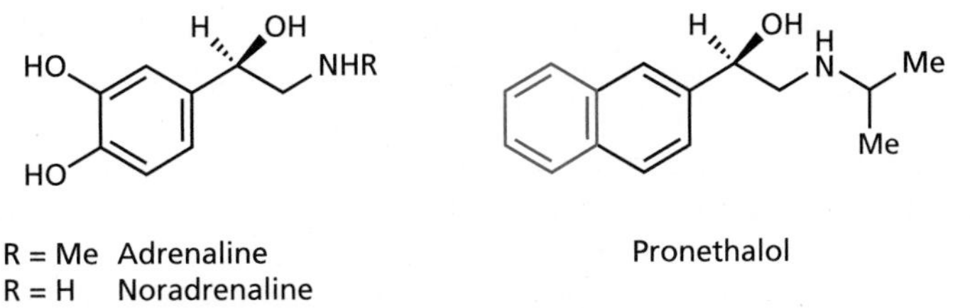

FIGURE 13.45 Structures of adrenaline, noradrenaline, and pronethalol.

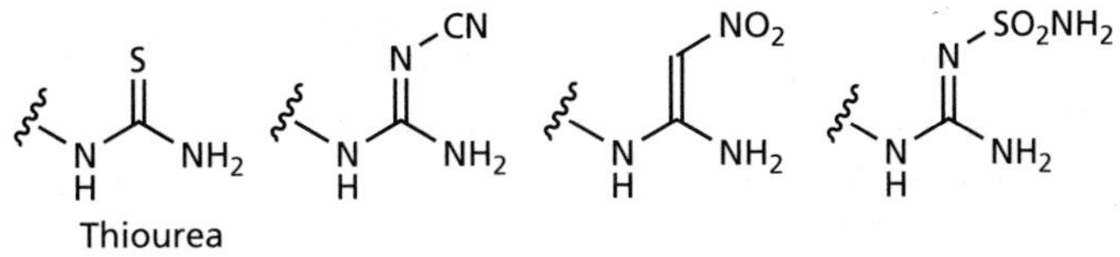

FIGURE 13.46 Non-classical isosteres for a thiourea group.

for this could be that the β-receptor has a larger van der Waals binding area for the aromatic system than the α-receptor, and can interact more strongly with pronethalol than with adrenaline. Another possible explanation is that the naphthalene ring system is sterically too big for the α-receptor, but is just right for the β-receptor.

13.3.7 Isosteres and bio-isosteres

Isosteres (section 13.1.15) have often been used in drug design to vary the character of the molecule in a rational way with respect to features such as size, polarity, electronic distribution, and bonding. Some isosteres can be used to determine the importance of size towards activity, whereas others can be used to determine the importance of electronic factors. For example, fluorine is often used as an isostere of hydrogen since it is virtually the same size. However, it is more electronegative and can be used to vary the electronic properties of the drug without having any steric effect.

The presence of fluorine in place of an enzymatically labile hydrogen can also disrupt an enzymatic reaction, as C–F bonds are not easily broken. For example, the antitumour drug **5-fluorouracil** described in section 21.3.2 is accepted by its target enzyme because it appears little different from the normal substrate—**uracil**. However, the mechanism of the enzyme-catalysed reaction is totally disrupted, as the fluorine has replaced a hydrogen which is normally lost during the enzyme mechanism.

Several non-classical isosteres have been used in drug design as replacements for particular functional groups. Non-classical isosteres are groups which do not obey the steric and electronic rules used to define classical isosteres, but which have similar physical and chemical properties. For example, the structures shown in Fig. 13.46 are non-classical isosteres for a thiourea group. They are all planar groups of similar size and basicity.

The term **bio-isostere** is used in drug design and includes both classical and non-classical isosteres. A bio-isostere is a group that can be used to replace another group while retaining the desired biological activity. For example, a cyclopropyl group has been used as a bio-isostere for an alkene group in prodrugs (section 14.6.1.1) and opioid antagonists (section 24.6.2). Bio-isosteres are often used to replace a functional group that is important for target binding, but is problematic in one way or another. For example, the thiourea group was present as an important binding group in early histamine antagonists, but was responsible for toxic side effects. Replacing it with bio-isosteres allowed the important binding interactions to be retained for histamine antagonism, but avoided the toxicity problems (section 25.2.6). Further examples of the use of bio-isosteres are given in sections 14.1.6, 14.2.2, 20.7.4, and 20.10.3. It is important to realize that bio-isosteres are specific for a particular group of compounds and their target. Replacing a functional group with a bio-isostere is not guaranteed to retain activity for every drug at every target.

As stated above, bio-isosteres are commonly used in drug design to replace a problematic group while

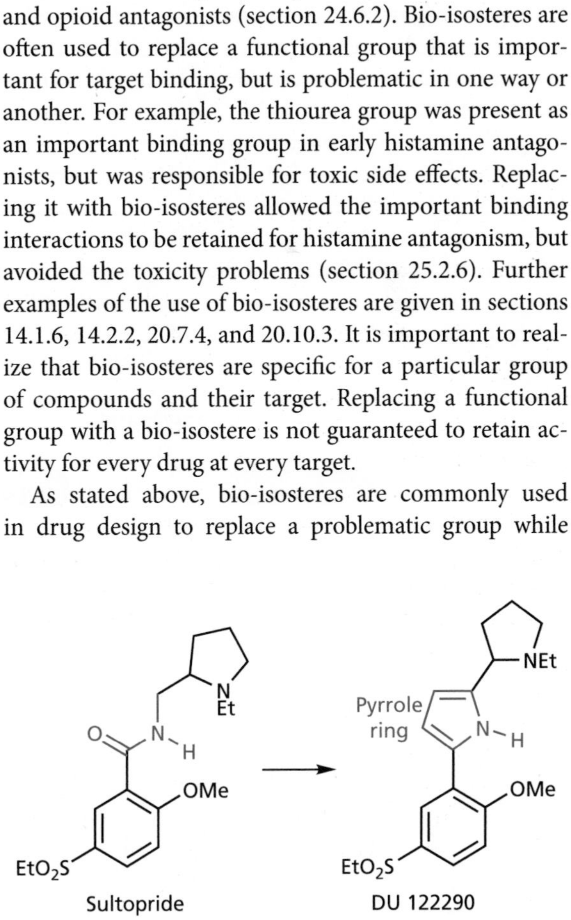

FIGURE 13.47 Introducing a pyrrole ring as a bio-isostere for an amide group.

retaining activity. In some situations, the use of a bio-isostere can actually increase target interactions and/or selectivity. For example, a pyrrole ring has frequently been used as a bio-isostere for an amide. Carrying out this replacement on the dopamine antagonist **sultopride** led to increased activity and selectivity towards the dopamine D_3-receptor over the dopamine D_2-receptor (Fig. 13.47). Such agents show promise as antipsychotic agents which lack the side effects associated with the D_2-receptor.

Introducing a bio-isostere to replace a problematic group often involves introducing further functional groups that might form extra binding interactions with the target binding site (section 13.3.2). For example, a ten-fold increase in activity was observed for an antiviral agent when an *N*-acylsulphonamide was used as a bio-isostere for a carboxylic acid (Fig. 13.48). The *N*-acylsulphonamide group introduces the possibility of further hydrogen bonding and van der Waals interactions with the binding site (see also Case study 10).

Transition-state isosteres are a special type of isostere used in the design of transition-state analogues. These are drugs that are used to inhibit enzymes (section 7.4). During an enzymatic reaction, a substrate goes through a transition state before it becomes product. It is proposed that the transition state is bound more strongly than either the substrate or the product, so it makes sense to design drugs based on the structure of the transition state rather than the structure of the substrate or the product. However, the transition state is inherently unstable and so transition-state isosteres are moieties that are used to mimic the crucial features of the transition state, but which are stable to the enzyme-catalysed reaction. For example, the transition state of an amide hydrolysis is

thought to resemble the tetrahedral reaction intermediate shown in Fig. 13.49. This is a geminal diol, which is inherently unstable. The hydroxyethylene moiety shown is a transition-state isostere because it shares the same tetrahedral geometry, retains one of the hydroxyl groups, and is stable to hydrolysis. Further examples of the use of transition-state isosteres are given in sections 20.7.4, 20.8.3, and 21.3.4, and Case studies 1 and 2.

13.3.8 Simplification of the structure

Simplification is a strategy which is commonly used on the often complex lead compounds arising from natural sources (see Box 13.2). Once the essential groups of such a drug have been identified by SAR, it is often possible to discard the non-essential parts of the structure without losing activity. Consideration is given to removing functional groups which are not part of the pharmacophore, simplifying the carbon skeleton (for example removing rings), and removing asymmetric centres.

This strategy is best carried out in small stages. For example, consider our hypothetical natural product glipine (Fig. 13.50). The essential groups have been highlighted, and we might aim to synthesize simplified compounds in the order shown. These still retain the essential groups making up the pharmacophore.

Chiral drugs pose a particular problem. The easiest and cheapest method of synthesizing a chiral drug is to make the racemate. However, both enantiomers then have to be tested for their activity and side effects, doubling the number of tests that have to be carried out. This is because different enantiomers can have different activities. For example, compound **UH-301** (Fig. 13.51) is inactive as a racemate, whereas its enantiomers have opposing agonist

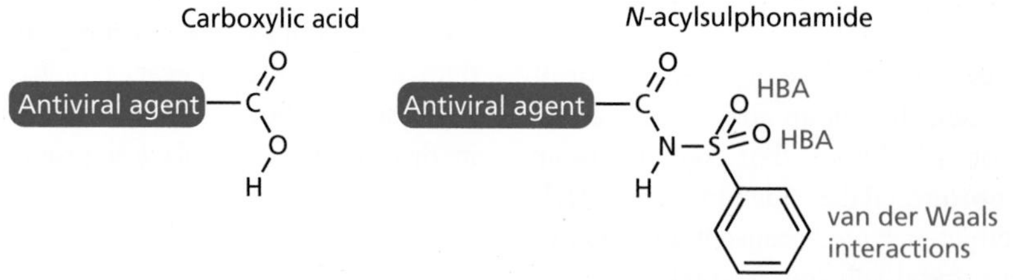

FIGURE 13.48 Extra binding interactions that might be possible when using an *N*-acylsulphonamide as a bio-isostere for a carboxylic acid.

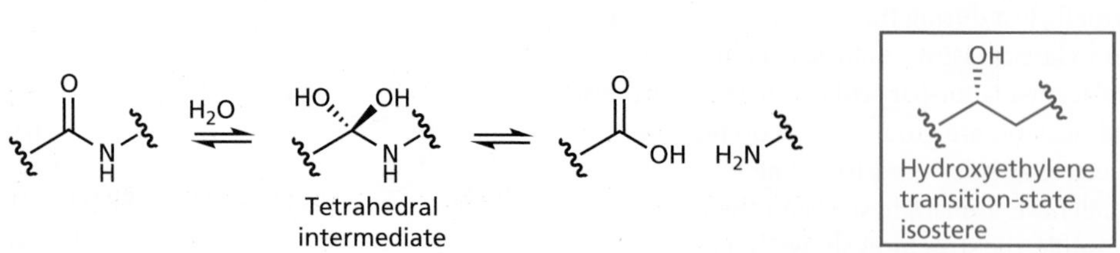

FIGURE 13.49 Example of a transition-state isostere designed to resemble a tetrahedral intermediate formed during an enzyme-catalysed reaction. The transition state is believed to resemble the tetrahedral intermediate.

BOX 13.2 Simplification

Simplification tactics have been used successfully with the alkaloid **cocaine**. Cocaine has local anaesthetic properties, and its simplification led to the development of local anaesthetics which could be easily synthesized in the laboratory. One of the earliest was **procaine** (**Novocaine**), discovered in 1909 (Fig. 1). Simplification tactics have also proved effective in the design of simpler morphine analogues (section 24.6.3).

Simplification tactics were also used in the development of **devazepide** from the microbial metabolite **asperlicin**. The benzodiazepine and indole skeletons inherent in asperlicin are important to activity and have been retained. Both asperlicin and devazepide act as antagonists of a neuropeptide chemical messenger called **cholecystokinin** (CCK) which has been implicated in causing panic attacks. Therefore, antagonists may be of use in treating such attacks.

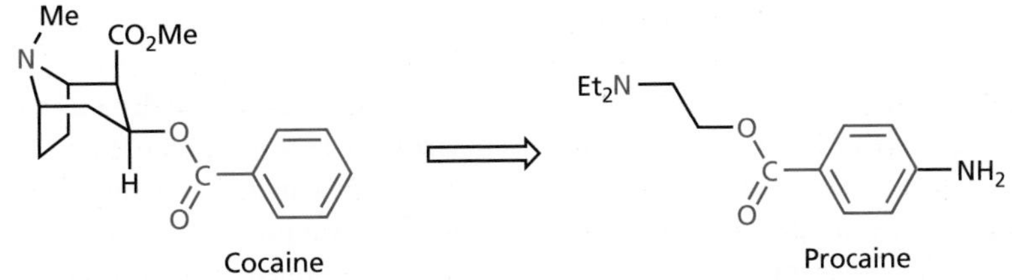

FIGURE 1 Simplification of cocaine (pharmacophore shown in colour).

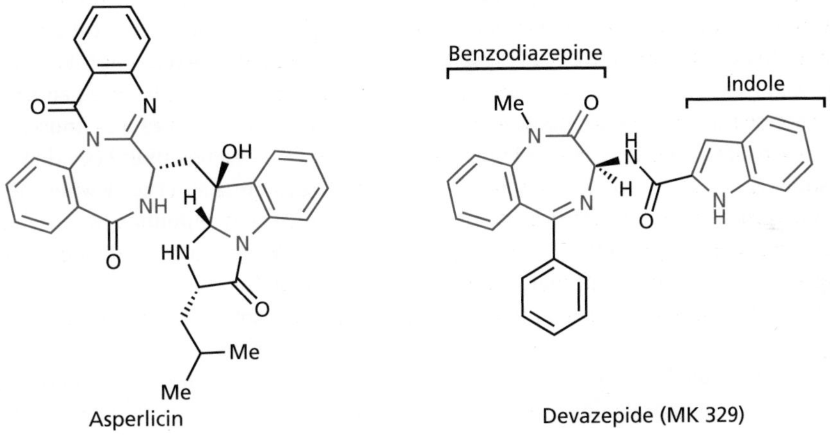

FIGURE 2 Simplification of asperlicin.

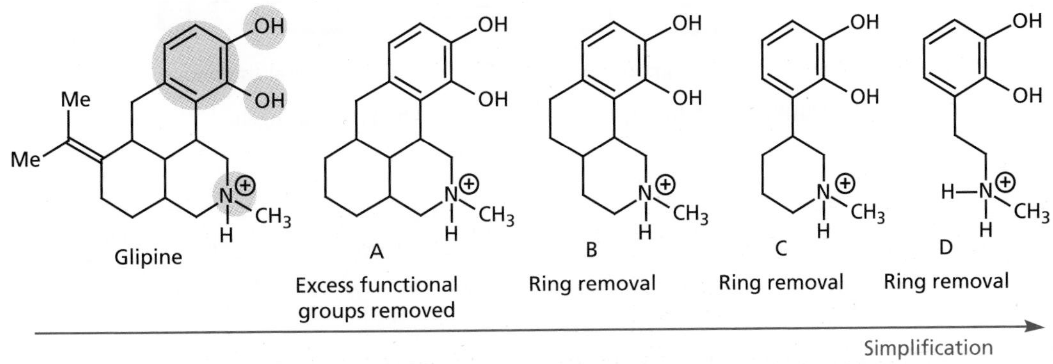

FIGURE 13.50 Glipine analogues.

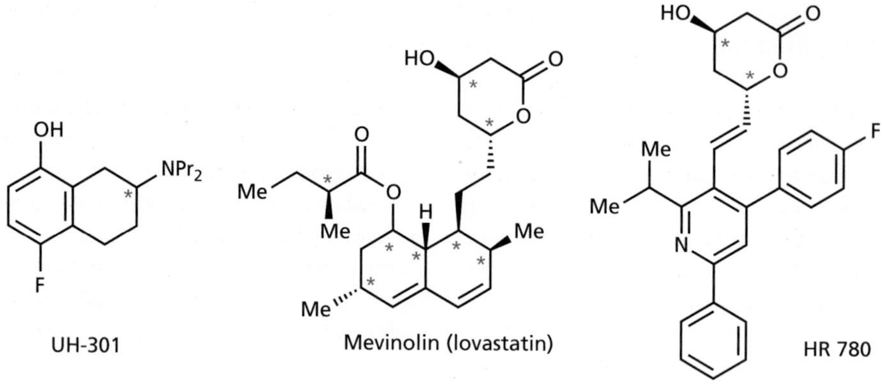

FIGURE 13.51 UH-301, mevinolin, and HR 780.

and antagonist activity at the serotonin receptor (5-HT$_{1A}$). Another notorious example is **thalidomide** where one of the enantiomers is teratogenic (section 21.9.1).

The use of racemates is discouraged and it is preferable to use a pure enantiomer. This could be obtained by separating the enantiomers of the racemic drug, or carrying out an asymmetric synthesis. Both options inevitably add to the cost of the synthesis, and so designing a structure that lacks some or all of the asymmetric centres can be advantageous and represents a simplification of the structure. For example, the cholesterol-lowering agent **mevinolin** has eight asymmetric centres, but a second generation of cholesterol-lowering agents has been developed which contain far fewer (e.g. **HR 780**; Fig. 13.51, see also Case study 1).

Various tactics can be used to remove asymmetric carbon centres. For example, replacing the carbon centre with nitrogen has been effective in many cases (Fig. 13.52). An illustration of this can be seen in the design of thymidylate synthase inhibitors described in Case study 5. However, it should be noted that the introduction of an amine in this way may well have significant effects on the

pharmacokinetics of the drug in terms of log P, basicity, polarity, etc.; see Chapters 11 and 14.

Another tactic is to introduce symmetry where originally there was none. For example, the muscarinic agonist (II) was developed from (I) in order to remove asymmetry (Fig. 13.53). Both structures have the same activity.

Simplification strategies have been applied extensively in many areas of medicinal chemistry, some of which are described in this text; for example, antiprotozoal agents (Case studies 3 and 4), local anaesthetics (section 17.9, Box 13.2), antibacterial agents (section 19.5.5.2), antiviral agents (section 20.7.4.8), anticancer agents (sections 21.2.1, 21.2.3.3, and 21.5.2), muscarinic antagonists (section 22.9.2.3), and opioids (section 24.6.3). Simplification strategies are also crucial when developing drugs from peptide lead compounds, where the aim is to reduce the size of the structure to the equivalent of a dipeptide or tripeptide (see Case study 10).

The advantage of simpler structures is that they are easier, quicker, and cheaper to synthesize in the laboratory. Usually the complex lead compounds obtained from natural sources are impractical to synthesize and have to be

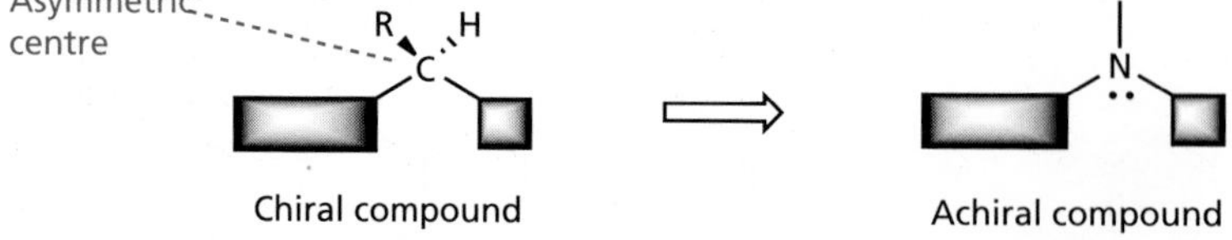

FIGURE 13.52 Replacing an asymmetric carbon with nitrogen.

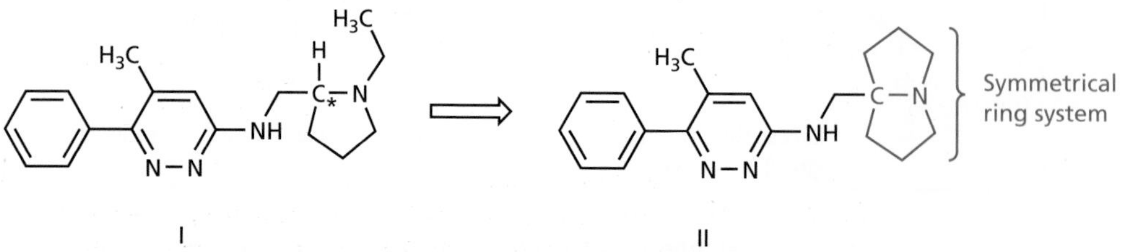

FIGURE 13.53 Introducing symmetry.

extracted from the source material—a slow, tedious, and expensive business. Removing unnecessary functional groups can also be advantageous in removing side effects if these groups interact with other targets, or are chemically reactive. There are, however, potential disadvantages in oversimplifying molecules. Simpler molecules are often more flexible and can sometimes bind differently to their targets compared to the original lead compound, resulting in different effects. It is best to simplify in small stages, checking that the desired activity is retained at each stage. Oversimplification may also result in reduced activity, reduced selectivity, and increased side effects. We shall see why in the next section (section 13.3.9).

13.3.9 Rigidification of the structure

Rigidification has often been used to increase the activity of a drug or to reduce its side effects. In order to understand why this tactic can work, let us consider again our hypothetical neurotransmitter from Chapter 5 (Fig. 13.54). This is quite a simple, flexible molecule with several rotatable bonds that can lead to a large number of conformations or shapes. One of these conformations is recognized by the receptor and is known as the **active conformation**. The other conformations are unable to interact efficiently with the receptor and are inactive conformations. However, it is possible that a different receptor exists which *is* capable of binding one of these alternative conformations. If this is the case, then our model neurotransmitter could switch on two different receptors

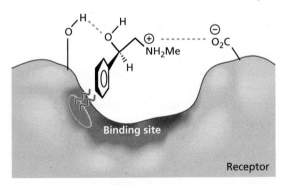

FIGURE 13.54 Active conformation of a hypothetical neurotransmitter.

and give two different biological responses, one which is desired and one which is not.

The body's own neurotransmitters are highly flexible molecules (section 4.2), but, fortunately, the body is efficient at releasing them close to their target receptors, then quickly inactivating them so that they do not make the journey to other receptors. This is not the case for drugs. They have to be sturdy enough to travel throughout the body and will interact with all the receptors that are prepared to accept them. The more flexible a drug molecule is, the more likely it will interact with more than one receptor and produce other biological responses (side effects). Too much flexibility is also bad for oral bioavailability (section 11.3).

The strategy of rigidification is to make the molecule more rigid, such that the active conformation is retained and the number of other possible conformations is decreased. This should reduce the possibility of other receptor interactions and side effects. This same strategy should also increase activity. By making the drug more rigid, it is more likely to be in the active conformation when it approaches the target binding site and should bind more readily. This is also important when it comes to the thermodynamics of binding. A flexible molecule has to adopt a single active conformation in order to bind to its target, which means that it has to become more ordered. This results in a decrease in entropy and, since the free energy of binding is related to entropy by the equation $\Delta G = \Delta H - T\Delta S$, any decrease in entropy will adversely affect ΔG. This in turn lowers the binding affinity (K_i), which is related to ΔG by the equation $\Delta G = -RT\ln K_i$. A totally rigid molecule, on the other hand, is already in its active conformation, and there is no loss of entropy involved in binding to the target. If the binding interactions (ΔH) are exactly the same as for the more flexible molecule, the rigid molecule will have the better overall binding affinity.

Incorporating the skeleton of a flexible drug into a ring is the usual way of locking a conformation. For our model compound, the analogue shown in Fig. 13.55 would be suitably rigid.

A ring was used to rigidify the acyclic pentapeptide shown in Fig. 13.56. This is a highly flexible molecule that acts as an inhibitor of a proteolytic enzyme. It was decided to rigidify the structure by linking the asparagine

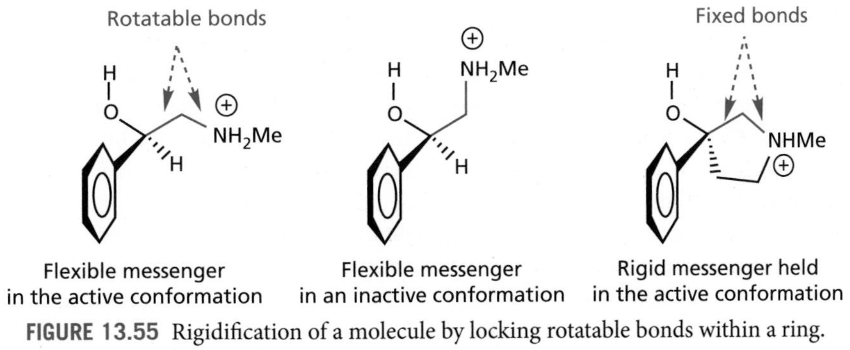

FIGURE 13.55 Rigidification of a molecule by locking rotatable bonds within a ring.

residue with the aromatic ring of the phenylalanine residue to form a macrocyclic ring. The resulting structure showed a 400-fold increase in activity. Macrocyclic rings have an advantage over smaller rings because they rigidify the molecule, but still allow a level of flexibility that increases the chances of the active conformation being adopted. Another example of macrocycles being used in rigidification can be found in Case study 10.

Similar rigidification tactics have been useful in the development of the antihypertensive agent **cilazapril** (Fig. 12.12) from **captopril**, and the development of the sedative **etorphine** (section 24.6.4). Other examples of rigidification can be seen in sections 25.2.8.1, 26.3.3, and 26.7.

Locking a rotatable bond into a ring is not the only way a structure can be rigidified. A flexible side chain can be partially rigidified by incorporating a rigid functional group such as a double bond, alkyne, amide, or aromatic ring (see Box 13.3).

Rigidification also has potential disadvantages. Rigidified structures may be more complicated to synthesize. There is also no guarantee that rigidification will retain the active conformation; it is perfectly possible that rigidification will lock the compound into an inactive conformation. Another disadvantage involves drugs acting on targets which are prone to mutation. If a mutation alters the shape of the binding site, then the drug may no longer be able to bind, whereas a more flexible drug may adopt a different conformation that *could* bind.

📀 For additional material see Web article 5: The design of a serotonin antagonist as a possible anxiolytic agent on the Online Resource Centre at www.oxfordtextbooks.co.uk/orc/patrick6e/

13.3.10 Conformational blockers

We have seen how rigidification tactics can restrict the number of possible conformations for a compound. Another tactic that has the same effect is the use of conformational blockers. In certain situations, a quite simple substituent can hinder the free rotation of a single bond. For example, introducing a methyl substituent to the dopamine (D_3) antagonist (I in Fig. 13.57) gives structure II and results in a dramatic reduction in affinity. The explanation lies in a bad steric clash between the new methyl group and an *ortho* proton on the neighbouring ring which prevents both rings being in the same plane. Free rotation around the bond between the two rings is no longer possible and so the structure adopts a conformation where the two rings are at an angle to each other. In structure I, free rotation around the connecting bond allows the molecule to adopt a conformation where the aromatic rings are coplanar—the active conformation for the receptor. In this case, a conformational blocker

'rejects' the active conformation. Examples of a conformational blocker favouring the active conformation can be seen with 4-methylhistamine (section 25.2.2.2), the design of a serotonin antagonist (see Web article 5), and the development of the anticancer agent imatinib (section 21.6.2.2). In the last case, conformational restraint not only increased activity, but also introduced selectivity between two similar target binding sites.

Rigidification is also possible through intramolecular hydrogen bonding, which may help to stabilize particular conformations (Fig. 13.58).

13.3.11 Structure-based drug design and molecular modelling

So far we have discussed the traditional strategies of drug design. These were frequently carried out with no knowledge of the target structure, and the results obtained were useful in providing information about the target binding site. Clearly, if a drug has an important binding group, there must be a complementary binding region present in the binding site of the receptor or enzyme.

If the macromolecular target can be isolated and crystallized, then it may be possible to determine the structure using X-ray crystallography. Unfortunately, this does not reveal where the binding site is, and so it is better to crystallize the protein with a known inhibitor or antagonist (ligand) bound to the binding site. X-ray crystallography can then be used to determine the structure of the complex and this can be downloaded to a computer. Molecular modelling software is then used to identify where the ligand is and thus identify the binding site. Moreover, by measuring the distances between the atoms of the ligand and neighbouring atoms in the binding site, it is possible to identify important binding interactions between the ligand and the binding site. Once this has been done, the ligand can be removed from the binding site *in silico*, and novel lead compounds can be inserted *in silico* to see how well they fit. (The term *in silico* indicates that the virtual process concerned is being carried out on a computer using molecular modelling software.) Regions in the binding site which are not occupied by the lead compound can be identified and used to guide the medicinal chemist as to what modifications and additions can be made to design a new drug that occupies more of the available space and binds more strongly. The drug can then be synthesized and tested for activity. If it proves active, the target protein can be crystallized with the new drug bound to the binding site, and then X-ray crystallography and molecular modelling can be used again to identify the structure of the complex to see if binding took place as expected. This approach is known as structure-based drug design. Examples of the use of structure-based drug design can be found in Case studies

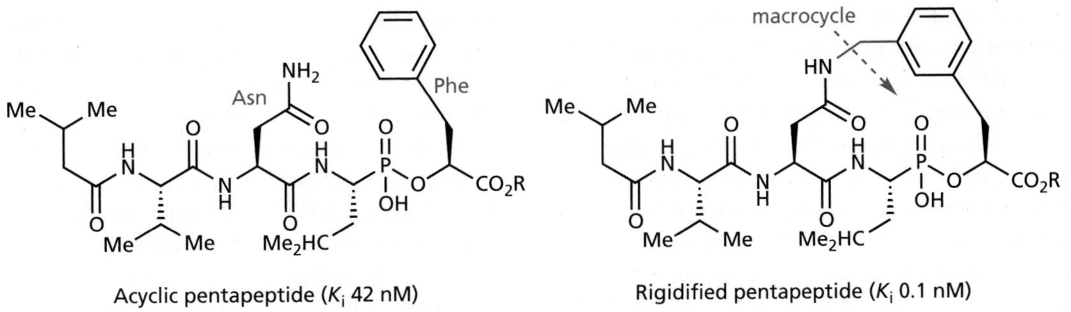

Acyclic pentapeptide (K_i 42 nM)   Rigidified pentapeptide (K_i 0.1 nM)

FIGURE 13.56 Rigidification of an acyclic pentapeptide.

2 and 5, Box 13.4, and Web article 5, as well as sections 14.9.1, 20.7.3.2, 20.7.4, 20.9, and 21.6.2.

A related process is known as *de novo* drug design (section 17.15). This involves the design of a novel drug structure, based on a knowledge of the binding site alone.

This is quite a demanding exercise, but there are examples where *de novo* design has successfully led to a novel lead compound which can then be the starting point for structure-based drug design (see Case study 5 and section 20.7.4.4).

BOX 13.3 Rigidification tactics in drug design

The diazepine (I) is an inhibitor of platelet aggregation and binds to its target receptor by means of a guanidine functional group and a diazepine ring system. These binding groups are linked together by a highly flexible chain. Structures (II)

and (III) are examples of active compounds where the connecting chain between the guanidine group and the bicyclic system has been partially rigidified by the introduction of rigid functional groups.

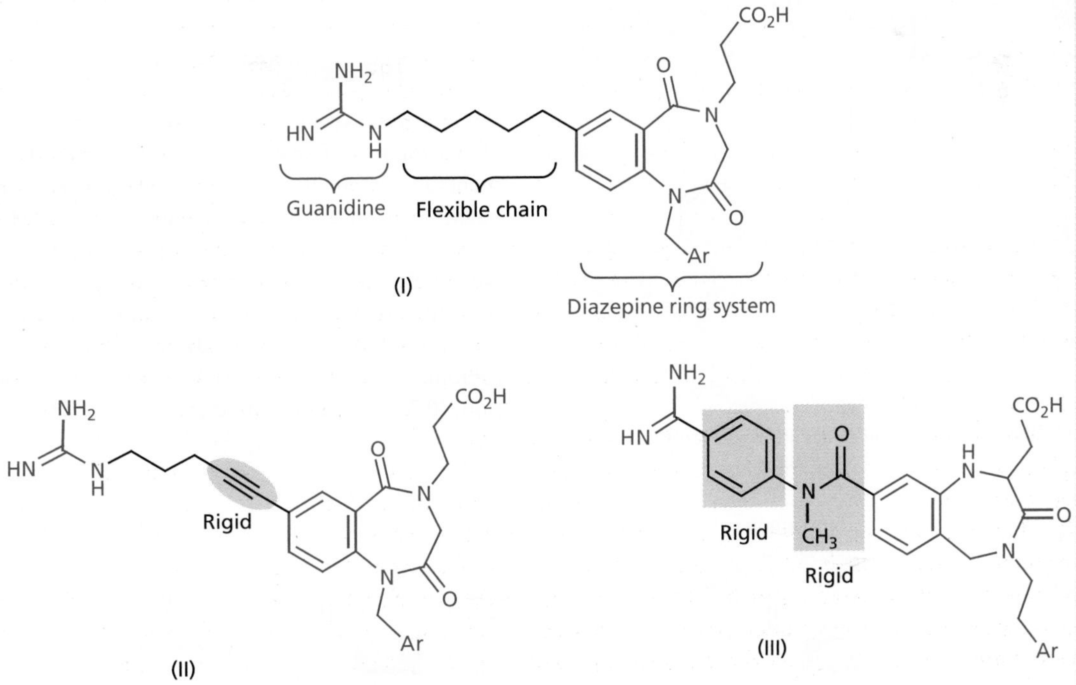

FIGURE 1 Rigidification of flexible chains.

Structure-based drug design cannot be used in all cases. Sometimes the target for a lead compound may not have been identified and, even if it has, it may not be possible to crystallize it. This is particularly true for membrane-bound proteins. One way round this is to identify a protein which is thought to be similar to the target protein, and which *has* been crystallized and studied by X-ray crystallography. The structural and mechanistic information obtained from that analogous protein can then be used to design drugs for the target protein (see Case studies 2 and 5).

Molecular modelling can also be used to study different compounds which are thought to interact with the same target. The structures can be compared and the important pharmacophore identified (section 17.11), allowing the design of novel structures containing the same pharmacophore. Compound databanks can be searched for those pharmacophores to identify novel lead compounds (section 17.13).

There are many other applications of molecular modelling in medicinal chemistry, some of which are described in Chapter 17. However, a point of caution is worth making at this stage. Molecular modelling studies tackle only one part of a much bigger problem—the design of an effective drug. True, one might design a compound that binds perfectly to a particular enzyme or receptor *in silico*, but that becomes pointless if the drug cannot be synthesized or never reaches the target protein in the body.

There have also been various examples where a binding site has altered shape in an unpredictable way to accommodate ligands that would not normally be expected to bind. Examples include binding sites for the **statins** (Case study 1) and an anti-inflammatory steroid (Box 8.1). Another example involves the dimeric structure of **galantamine** which has been studied as an inhibitor of the enzyme **acetylcholinesterase** (section 22.15.2).

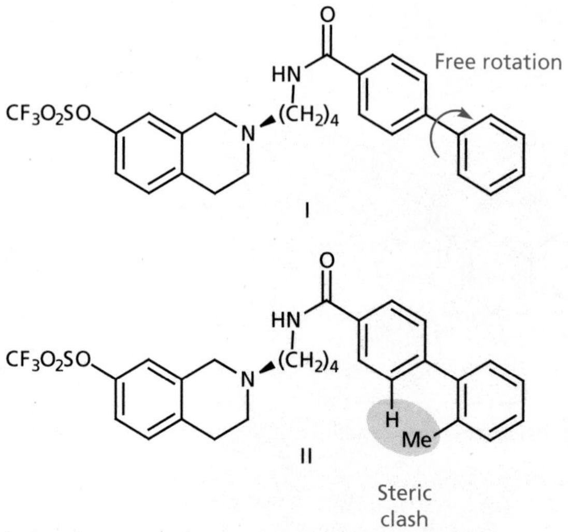

13.3.12 Drug design by NMR spectroscopy

The use of NMR spectroscopy in designing lead compounds has already been discussed in section 12.4.10. This can also be seen as a method of drug design since the focus is not only on designing a lead compound, but in designing a *potent* lead compound. Usually, drug design aims to optimize a lead compound once it has been discovered. In the NMR method, the component parts (**epitopes**) are optimized first to maximize binding interactions, then linked together to produce the final compound.

NMR is also being increasingly used to identify the structure of target proteins that cannot be crystallized and studied by X-ray crystallography. Once the structure has been identified, molecular modelling techniques can be used for drug design as described in section 13.3.11.

13.3.13 The elements of luck and inspiration

It is true to say that drug design has become more rational, but the role of chance or the need for hard-working, mentally alert medicinal chemists has not yet been eliminated. Most of the drugs currently on the market were developed by a mixture of rational design, trial and error, hard graft, and pure luck. There are a growing number of drugs that were developed by rational design such as the ACE inhibitors (Case study 2), thymidylate synthase inhibitors (Case study 5), HIV protease inhibitors (section 20.7.4), neuraminidase inhibitors (section 20.8.3),

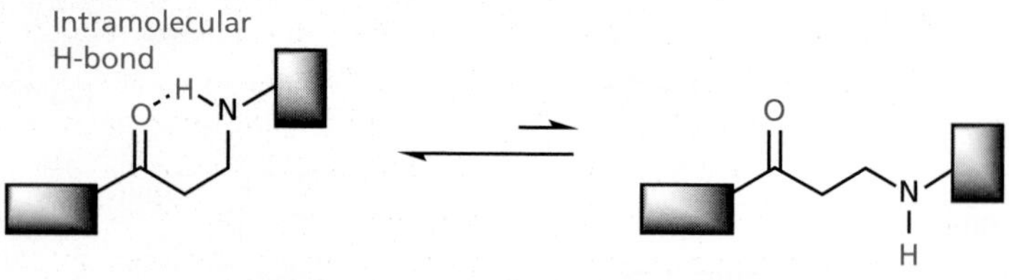

FIGURE 13.57 Introducing rigidity by conformational blocking.

FIGURE 13.58 Rigidification involving an intramolecular hydrogen bond.

BOX 13.4 The structure-based drug design of crizotinib

Structure-based drug design is normally used to observe the binding interactions of a ligand, then to identify modifications that will result in better interactions and greater activity. This approach was used in the design of a recently approved anticancer agent called crizotinib, and included a substantial modification which totally altered the scaffold of the molecule. **PHA-665752** was the starting point for this research, and had been obtained from structure-based drug design of a previous lead compound. However, it had a large molecular weight and was too hydrophobic to be orally active. The structure was co-crystallized with the target enzyme and the crucial binding interactions were identified. These

included the dihydroindolone ring system which formed two important hydrogen bonds (HBD and HBA), as well as the dichloroaromatic ring. As a result of this study, it was noted that much of the scaffold connecting these binding groups was redundant, and so a much simpler, less hydrophobic skeleton was designed which would position the important binding groups in a similar but more efficient manner. The thought process behind this design involved a ring fusion, ring cleavage, and chain contraction. When the novel structures were synthesized, they were found to bind as predicted, and further structure-based drug design was used in the optimization process leading to crizotinib.

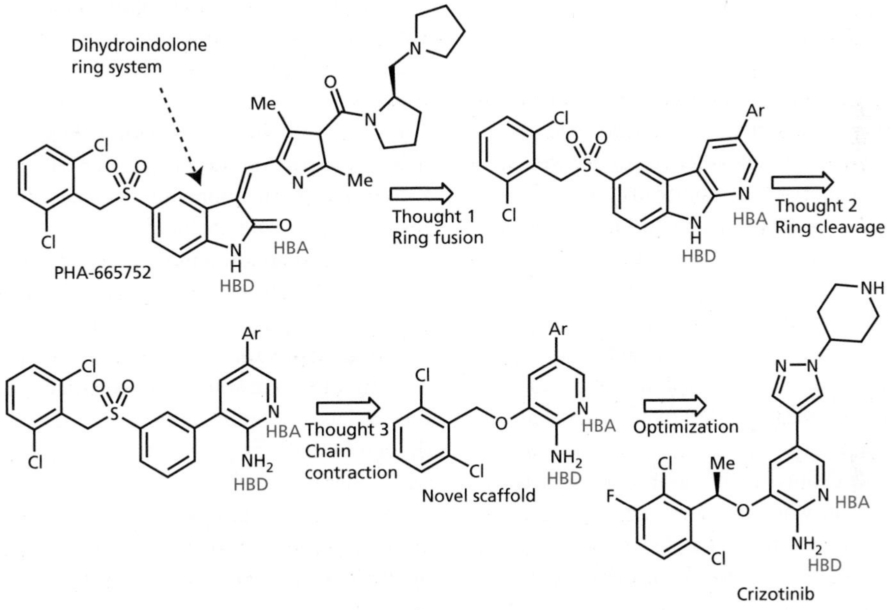

FIGURE 1 Design process in the development of crizotinib.

pralidoxime (section 22.14), and **cimetidine** (section 25.2), but they are still in the minority.

Frequently, the development of drugs is helped by watching the literature to see what works on related compounds and what doesn't, then trying out similar alterations to one's own work. It is often a case of groping in the dark, with the chemist asking whether the addition of a group at a certain position will have a steric, electronic, or interactive effect. Even when drug design is carried out on rational lines, good fortune often has a role to play—for example the discovery of the β-blocker **propranolol** (section 23.11.3).

Finally, there are some cases where the use of logical step-by-step modifications to a structure fails to result in significantly improved activity. In such cases, there may be some advantage in synthesizing a large range of structures with different substituents or modifications in the hope of striking lucky. This is illustrated in the development of the anticancer agent **sorafenib** (Box 21.9). The breakthrough here was the discovery of an active structure which contained two substituents that were known to be bad for activity when only one or other was present. When both were present, however, there was a beneficial synergistic effect.

13.3.14 Designing drugs to interact with more than one target

Many diseases require a cocktail of drugs interacting with different targets to provide suitable treatments. A better approach would be to design agents that interact with two or more targets in a controlled fashion in order to reduce the number of drugs that have to be taken. This is known as **multi-target drug discovery** (MTDD) (section 12.2.7). There have been two approaches to designing such multi-target-directed ligands. One is to design agents from known drugs and pharmacophores such that the new agent has the combined properties of the drugs involved. The other approach is to start from a lead compound which has activity against a wide range of targets, and then modify the structure to try and narrow the activity down to the desired targets.

13.3.14.1 Agents designed from known drugs

One strategy is to link two known drugs to form a dimeric structure. The advantage of this approach is that there is a good chance that the resulting dimer will have a similar selectivity and potency to the original individual drugs for both intended targets. The disadvantage is the increased number of functional groups and rotatable bonds that result, as this may have detrimental effects on whether the resulting dimer is orally active or not. There is also the problem that linking one drug to another may block each individual component binding to its target binding site. Nevertheless, the design of dimers has been successful in a number of fields.

Dimers can be defined as homodimeric or heterodimeric depending on whether the component drugs are the same or not. Homodimeric and heterodimeric opioid ligands have been synthesized to take advantage of the fact that opioid receptors form homodimeric and heterodimeric arrays in certain tissues of the body (section 24.9.2).

There is also a potential use for dimers in the treatment of Alzheimer's disease. The **acetylcholinesterase** enzyme has an active site and a peripheral binding site, both of which play a role in the symptoms of the disease. Dimers have been designed that can interact with both of these sites and act as **dual-action agents** (section 22.15.2). Research is also being carried out to design triple-action agents that will interact with the two binding sites in the acetylcholinesterase enzyme plus a totally different target that is also involved in the symptoms or development of the disease.

Enzyme inhibitors have also been designed that contain structural components of the substrate and cofactor of **17β-hydroxysteroid dehydrogenase type 1** (see Web article 1).

A second strategy for designing dual-action drugs is to consider the pharmacophores of two different drugs, and to then design a hybrid structure where the two pharmacophores are merged. Such drugs are called **hybrid drugs**. One example of this is **ladostigil** (Fig. 13.59), which is a hybrid structure of the acetylcholinesterase inhibitor **rivastigmine** and the monoamine oxidase inhibitor **rasagiline**. The feature in blue indicates the structural features of ladostigil that are present in both component drugs.

A third strategy is to design a **chimeric drug** that contains key pharmacophore features from two different drugs. For example, a structure containing features of **2-methoxyestradiol** and **colchicine** has been synthesized as a potential anticancer agent (Fig. 13.60). Although both of the parent structures have anticancer activity, they have serious drawbacks. 2-Methoxyestradiol is rapidly metabolized, while colchicine has toxic side effects. The chimeric structure also has anticancer activity, but improved pharmacokinetic properties.

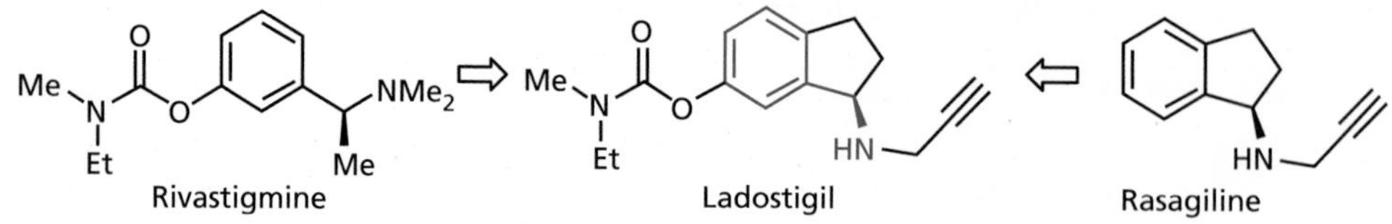

FIGURE 13.59 Design of the hybrid drug ladostigil.

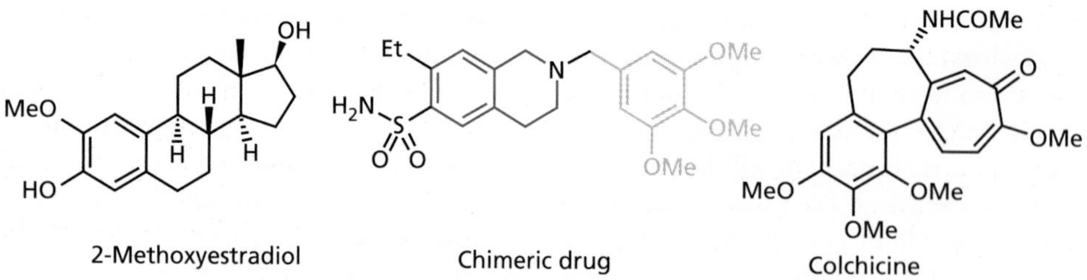

FIGURE 13.60 Design of a chimeric drug.

13.3.14.2 Agents designed from non-selective lead compounds

The second approach to designing multi-target drugs is to identify a lead compound that already shows the ability to interact with a wide variety of targets. Such an agent is termed a **promiscuous ligand** or a **dirty drug**. Linear polyamines have been suggested as ideal lead compounds in this approach as they have several amine groups that can act as good binding groups to protein targets. Moreover, the flexibility of the structure means that an active conformation is likely to exist for a large number of protein targets. The challenge is then to modify the structure such that it shows selectivity towards the desired targets. This approach has been used in the design of an agent which shows activity both as an acetylcholinesterase inhibitor and a muscarinic antagonist (section 22.15.3). Such agents may be useful in the treatment of Alzheimer's disease. Multi-tyrosine receptor kinase inhibitors have also been developed as anticancer agents (section 21.6.2.10).

KEY POINTS

- Drug optimization aims to maximize the interactions of a drug with its target binding site in order to improve activity and selectivity, and to minimize side effects. Designing a drug that can be synthesized efficiently and cheaply is another priority.

- The length and size of alkyl substituents can be modified to fill up hydrophobic pockets in the binding site or to introduce selectivity for one target over another. Alkyl groups attached to heteroatoms are most easily modified.

- Aromatic substituents can be varied in character and/or ring position.

- Extension is a strategy where extra functional groups are added to the lead compound such that they interact with extra binding regions in the binding site.

- Chains connecting two important binding groups can be modified in length in order to maximize the interactions of each group with the corresponding binding regions.

- Ring systems can be modified to maximize binding interactions through strategies such as expansion, contraction, variation, or fusion with other rings.

- Classical and non-classical isosteres are frequently used in drug optimization.

- Simplification involves removing functional groups from the lead compound that are not part of the pharmacophore. Unnecessary parts of the carbon skeleton or asymmetric centres can also be removed in order to design drugs that are easier and cheaper to synthesize. Oversimplification can result in molecules that are too flexible, resulting in decreased activity and selectivity.

- Rigidification is applicable to flexible lead compounds. The aim is to reduce the number of conformations available while retaining the active conformation. Locking rotatable rings into ring structures or introducing rigid functional groups are common methods of rigidification.

- Conformational blockers are groups which are introduced into a lead compound to reduce the number of conformations that the molecule can adopt.

- Structure-based drug design makes use of X-ray crystallography and computer-based molecular modelling to study how a lead compound and its analogues bind to a target binding site.

- NMR studies can be used to determine protein structure and to design novel drugs.

- Serendipity plays a role in drug design and optimization.

- Multi-target-directed ligands can be designed by linking or merging established drugs, or by modifying a lead compound that interacts with a large number of targets.

QUESTIONS

1. DU 122290 was developed from sultopride (Fig. 13.47) and shows improved activity and selectivity. Suggest possible reasons for this.

2. Methotrexate inhibits the enzyme dihydrofolate reductase. The pteridine ring system of methotrexate binds to the binding site as shown in Fig. 13.26. Suggest how dihydrofolate (the natural substrate for the enzyme) might bind.

Dihydrofolate

3. A lead compound containing a methyl ester was hydrolysed to give a carboxylic acid. An *in vivo* bioassay suggested that the ester was active and the acid was inactive. However, an *in vitro* bioassay suggested that the ester was inactive and the acid was active. Explain these contradictory results.

4. A lead compound contains an aromatic ring. The following structures were made as analogues. Structures I and II were similar in activity to the lead compound, whereas structure III showed a marked increase in activity. Explain these results and describe the strategies involved.

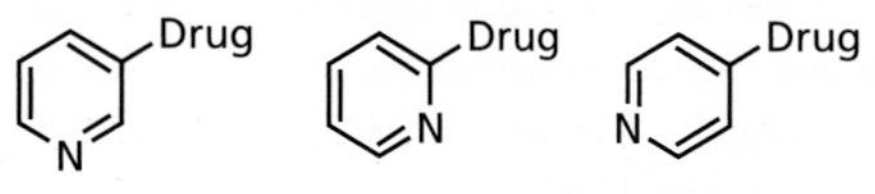

I   II   III

5. The pharmacophore of cocaine is shown in Box 13.2. Identify possible cyclic analogues which are simpler than cocaine and which would be expected to retain activity.

6. Procaine (Box 13.2) has been a highly successful local anaesthetic and yet there are three bonds between the important ester and amine binding groups, compared to four in cocaine. This might suggest that these groups are too close together in procaine. In fact, this is not the case. Suggest why not.

7. The aromatic amine on procaine is not present in cocaine. Comment on its possible role.

8. Explain how you would apply the principles of rigidification to structure IV below in order to improve its pharmacological properties. Give two specific examples of rigidified structures.

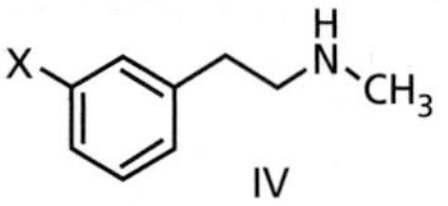

IV

9. Combretastatin is an anticancer agent discovered from an African plant. Analogue V is more active than combretastatin whereas analogue VI is less active. What strategy was used in designing analogues V and VI? Why is analogue V more active and analogue VI less active than combretastatin?

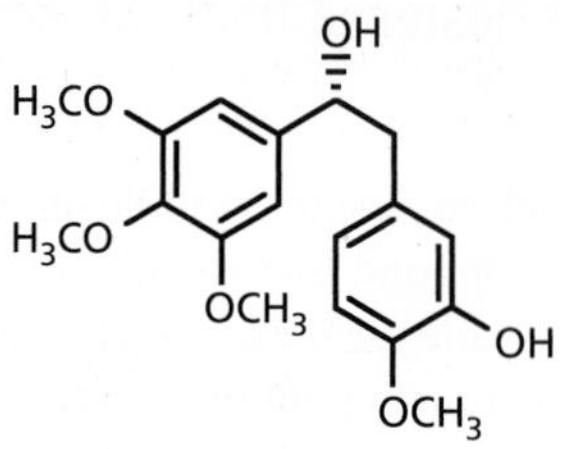

Combretastatin

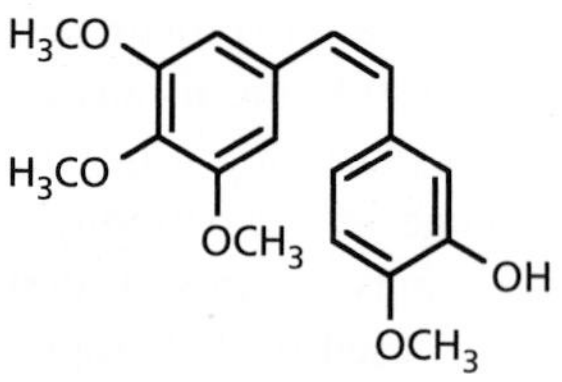

V

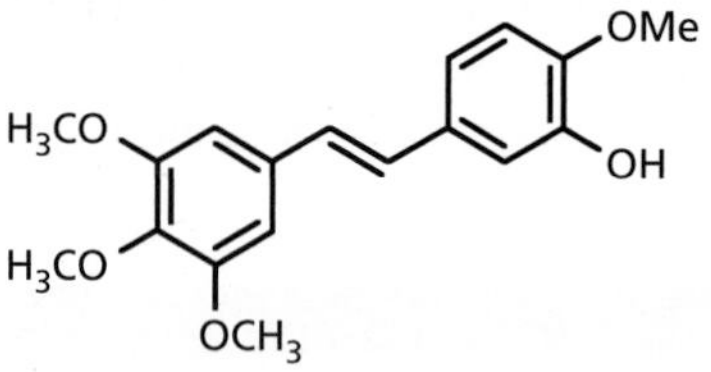

VI

10. Structure VII is a serotonin antagonist. A methyl group has been introduced into analogue VIII, resulting in increased activity. What role does the methyl group play and what is the term used for such a group? Explain why increased activity arises.

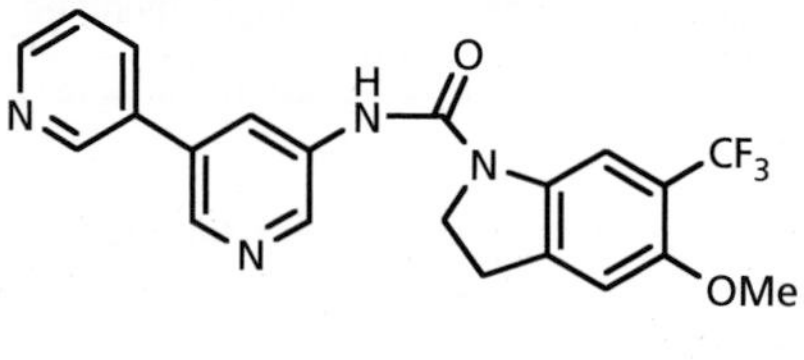

VII

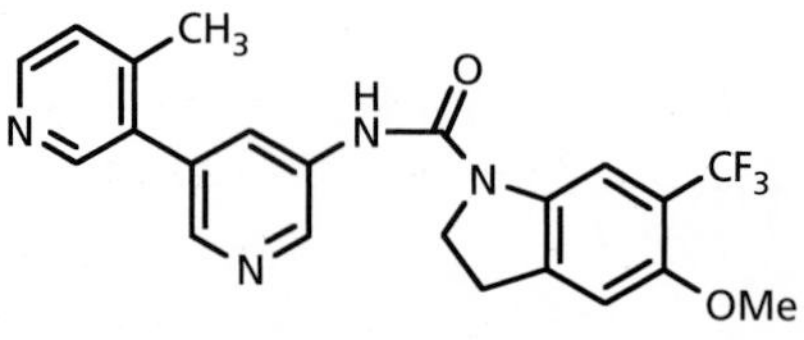

VIII

11. Explain what kind of drug design strategies were carried out in the design of enalaprilat (Case study 2).

12. Salicylamides are inhibitors for an enzyme called scytalone dehydratase. SAR shows that there are three important hydrogen bonding interactions. Explain whether you think quinazolines could act as a bio-isostere for salicylamides.

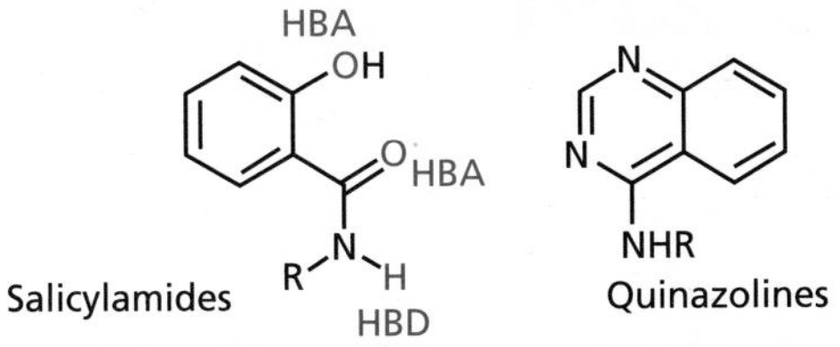

Salicylamides   Quinazolines

13. Structure IX (X = NH) is an inhibitor of a metalloenzyme called thermolysin and forms interactions as shown. Explain why the analogue (X = O) has reduced binding affinity by a factor of 1000, and why the analogue (X = CH_2) has roughly the same binding affinity.

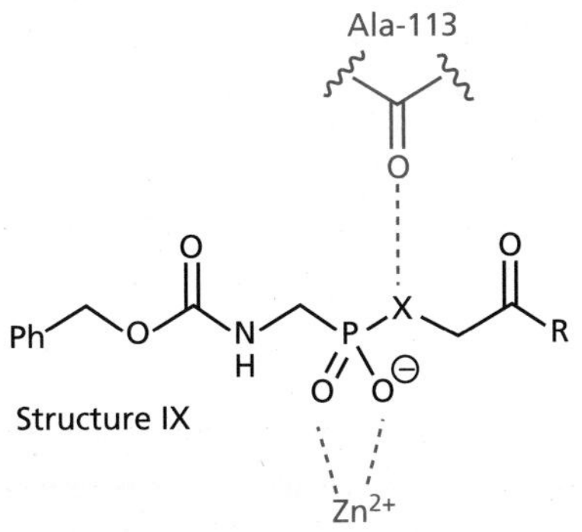

Structure IX

14. Suggest why the oxygen atoms in the following structures are poor hydrogen bond acceptors.

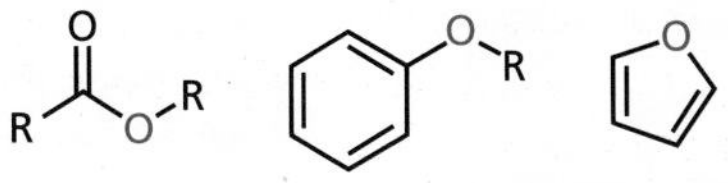

15. Compare the ability of the nitrogen atoms in the following structures to act as hydrogen bond acceptors.

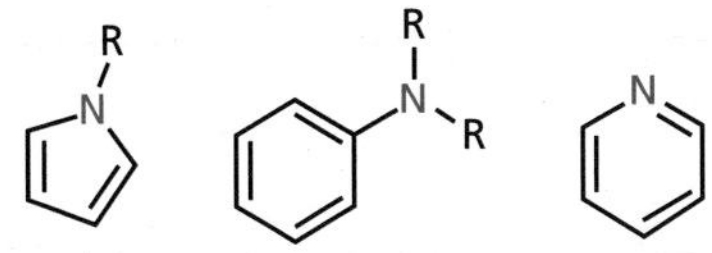

16. Explain why a benzimidazole group could be considered a bio-isostere for an N-phenyl amide, and why using a benzimidazole group might increase binding affinity.

🌐 Multiple-choice questions are available on the Online Resource Centre at www.oxfordtextbooks.co.uk/orc/patrick6e/

FURTHER READING

Acharya, K. R., et al. (2003) ACE revisited: a new target for structure-based drug design. *Nature Reviews Drug Discovery*, **2**, 891–902.

Cavalli, A., et al. (2008) Multi-target-directed ligands to combat neurodegenerative diseases. *Journal of Medicinal Chemistry*, **51**, 347–72.

Cui, J. J., et al. (2011) Structure-based drug design of crizotinib. *Journal of Medicinal Chemistry*, **54**, 6342–63.

Hruby, V. J. (2002) Designing peptide receptor agonists and antagonists. *Nature Reviews Drug Discovery*, **1**, 847–58.

Kubinyi, H. (2001) Hydrogen bonding: the last mystery in drug design? in Testa, B., van de Waterbeemd, H., Folkers, G., and Guy, R. (eds), *Pharmacokinetic optimization in drug research*. Wiley-VCH, Weinheim.

Luca, S., et al. (2003) The conformation of neurotensin bound to its G-protein-coupled receptor. *Proceedings of the National Academy of Sciences of the USA*, **100**, 10706–11 (active conformation by NMR).

Morphy, R., et al. (2004) From magic bullets to designed multiple ligands. *Drug Discovery Today*, **9**, 641–51.

Morphy, R., and Rankovic, Z. (2005) Designed multiple ligands. An emerging drug discovery paradigm. *Journal of Medicinal Chemistry*, **48**, 6523–43.

Pellecchia, M., Sem, D. S., and Wuthrich, K. (2002) NMR in drug discovery. *Nature Reviews Drug Discovery*, **1**, 211–9.

Rees, D. C., et al. (2004) Fragment-based lead discovery. *Nature Reviews Drug Discovery*, **3**, 660–72.

Titles for general further reading are listed on p.845.

14 Drug design: optimizing access to the target

In Chapter 13, we looked at drug design strategies aimed at optimizing the binding interactions of a drug with its target. However, the compound with the best binding interactions is not necessarily the best drug to use in medicine. The drug needs to overcome many barriers if it is to reach its target in the body (Chapter 11). In this chapter, we shall study design strategies which can be used to counter such barriers, and which involve modification of the drug itself. There are other methods of aiding a drug in reaching its target, which include linking the drug to polymers or antibodies, or encapsulating it within a polymeric carrier. These topics are discussed in sections 11.10 and 21.10. In general, the aim is to design drugs that will be absorbed into the blood supply, will reach their target efficiently, be stable enough to survive the journey, and will be eliminated in a reasonable period of time. This all comes under the banner of a drug's pharmacokinetics.

14.1 Optimizing hydrophilic/ hydrophobic properties

The relative hydrophilic/hydrophobic properties of a drug are crucial in influencing its solubility, absorption, distribution, metabolism, and excretion (ADME). Drugs which are too polar or hydrophilic do not easily cross the cell membranes of the gut wall. One way round this is to inject them, but they cannot be used against intracellular targets since they will not cross cell membranes. They are also likely to have polar functional groups which will make them prone to plasma protein binding, metabolic phase II conjugation reactions, and rapid excretion (Chapter 11). Very hydrophobic drugs fare no better. If they are administered orally, they are likely to be dissolved in fat globules in the gut and will be poorly absorbed. If they are injected, they are poorly soluble in blood and are likely to be taken up by fat tissue, resulting in low circulating levels. It has also been observed that toxic metabolites are more likely to be formed from hydrophobic drugs.

The hydrophobic character of a drug can be measured experimentally by testing the drug's relative distribution in an *n*-octanol/water mixture. Hydrophobic molecules will prefer to dissolve in the *n*-octanol layer of this two-phase system, whereas hydrophilic molecules will prefer the aqueous layer. The relative distribution is known as the partition coefficient (*P*) and is obtained from the following equation:

$$P = \frac{\text{Concentration of drug in octanol}}{\text{Concentration of drug in aqueous solution}}$$

Hydrophobic compounds have a high *P* value, whereas hydrophilic compounds have a low *P* value. In fact, log *P* values are normally used as a measure of hydrophobicity. Other experimental procedures to determine log *P* include high-performance liquid chromatography (HPLC) and automated potentiometric titration procedures. It is also possible to calculate log *P* values for a given structure using suitable software programs. Such estimates are referred to as **Clog *P*** values to distinguish them from experimentally derived log *P* values.

Many drugs can exist as an equilibrium between an ionized and an un-ionized form. However, log *P* measures only the relative distribution of the un-ionized species between water and octanol. The relative distribution of all species (both ionized and un-ionized) is given by **log *D***.

In general, the hydrophilic/hydrophobic balance of a drug can be altered by changing easily accessible substituents. Such changes are particularly open to a quantitative approach known as QSAR (quantitative structure–activity relationships), discussed in Chapter 18.

As a postscript, the hydrophilic/hydrophobic properties of a drug are not the only factors that influence drug absorption and oral bioavailability. Molecular flexibility

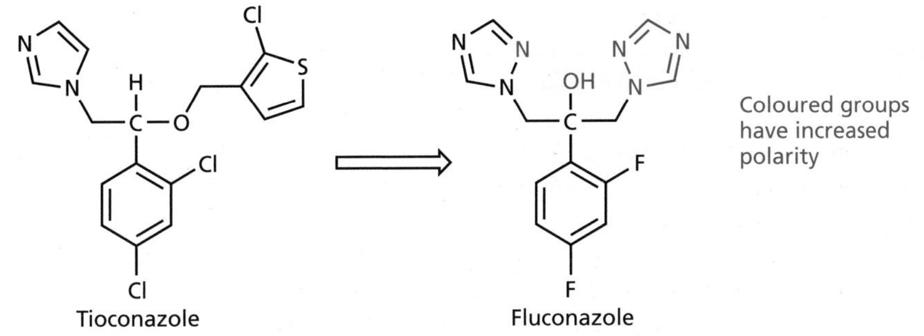

FIGURE 14.1 Increasing polarity in antifungal agents.

also has an important role in oral bioavailability (section 11.3), and so the tactics of rigidification described in section 13.3.9 can be useful in improving drug absorption.

14.1.1 Masking polar functional groups to decrease polarity

Molecules can be made less polar by masking a polar functional group with an alkyl or acyl group. For example, an alcohol or a phenol can be converted to an ether or ester, a carboxylic acid can be converted to an ester or amide, and primary and secondary amines can be converted to amides or to secondary and tertiary amines. Polarity is decreased not only by masking the polar group, but by the addition of an extra hydrophobic alkyl group—larger alkyl groups having a greater hydrophobic effect. One has to be careful in masking polar groups, though, as they may be important in binding the drug to its target. Masking such groups would decrease binding interactions and lower activity. If this is the case, it is often useful to mask the polar group temporarily such that the mask is removed once the drug is absorbed (section 14.6).

14.1.2 Adding or removing polar functional groups to vary polarity

A polar functional group could be added to a drug to increase its polarity. For example, the antifungal agent **tioconazole** is only used for skin infections because it is non-polar and poorly soluble in blood. Introducing a polar hydroxyl group and more polar heterocyclic rings led to the orally active antifungal agent **fluconazole** with improved solubility and enhanced activity against systemic infection (i.e. in the blood supply) (Fig. 14.1). Another example can be found in Case study 1 where a polar sulphonamide group was added to **rosuvastatin** to make it more hydrophilic and more tissue selective. Finally, nitrogen-containing heterocycles (e.g. morpholine or pyridine) are often added to drugs in order to increase their polarity and water solubility. This is because the

nitrogen is basic in character, and it is possible to form water-soluble salts. Examples of this tactic can be seen in the design of **gefitinib** (section 21.6.2.1) and a thymidylate synthase inhibitor (Case study 5). If a polar group is added in order to increase water solubility, it is preferable to add it to the molecule in such a way that it is still exposed to surrounding water when the drug is bound to the target binding site. This means that energy does not have to be expended in desolvation (section 1.3.6).

The polarity of an excessively polar drug can be lowered by removing polar functional groups. This strategy has been particularly successful with lead compounds derived from natural sources (e.g. alkaloids or endogenous peptides). It is important, though, not to remove functional groups which are important to the drug's binding interactions with its target. In some cases, a drug may have too many essential polar groups. For example, the antibacterial agent shown in Fig. 14.2 has good *in vitro* activity but poor *in vivo* activity because of the large number of polar groups. Some of these groups can be removed or masked, but most of them are required for activity. As a result, the drug cannot be used clinically.

14.1.3 Varying hydrophobic substituents to vary polarity

Polarity can be varied by the addition, removal, or variation of suitable hydrophobic substituents. For

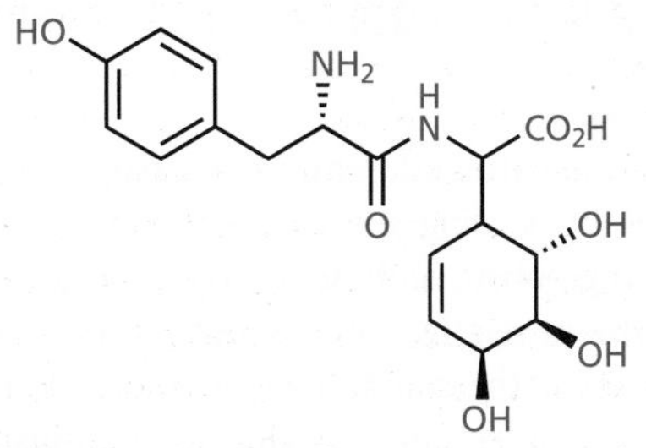

FIGURE 14.2 Excess polarity (coloured) in a drug.

example, extra alkyl groups could be included within the carbon skeleton of the molecule to increase hydrophobicity if the synthetic route permits. Alternatively, alkyl groups already present might be replaced with larger groups. If the molecule is not sufficiently polar, then the opposite strategy can be used (i.e. replacing large alkyl groups with smaller alkyl groups, or removing them entirely). Sometimes there is a benefit in increasing the size of one alkyl group and decreasing the size of another. This is called a **methylene shuffle** and has been found to modify the hydrophobicity of a compound. The addition of halogen substituents also increases hydrophobicity. Chloro or fluoro substituents are commonly used, and, less commonly, a bromo substituent.

14.1.4 Variation of *N*-alkyl substituents to vary pK_a

Drugs with a pK_a outside the range 6–9 tend to be too strongly ionized and are poorly absorbed through cell membranes (section 11.3). The pK_a can often be altered to bring it into the preferred range. For example, this can be done by varying any *N*-alkyl substituents that are present. However, it is sometimes difficult to predict how such variations will affect the pK_a. Extra *N*-alkyl groups or larger *N*-alkyl groups have an increased electron-donating effect which should increase basicity, but increasing the size or number of alkyl groups increases the steric bulk around the nitrogen atom. This hinders water molecules from solvating the ionized form of the base and prevents stabilization of the ion. This in turn decreases the basicity of the amine. Therefore, there are two different effects acting against each other. Nevertheless, varying alkyl substituents is a useful tactic to try.

A variation of this tactic is to 'wrap up' a basic nitrogen within a ring. For example, the benzamidine structure (I in Fig. 14.3) has antithrombotic activity, but the amidine group present is too basic for effective absorption. Incorporating the group into an isoquinoline ring system (**PRO 3112**) reduced basicity and increased absorption.

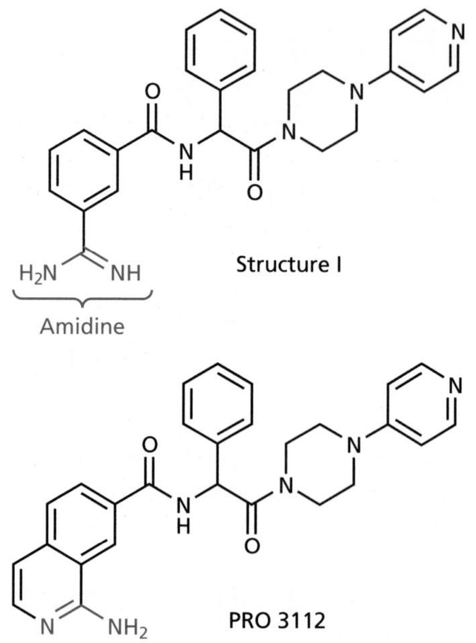

14.1.5 Variation of aromatic substituents to vary pK_a

The pK_a of an aromatic amine or carboxylic acid can be varied by adding electron-donating or electron-withdrawing substituents to the ring. The position of the substituent relative to the amine or carboxylic acid is important if the substituent interacts with the ring through resonance (section 18.2.2). An illustration of this can be seen in the development of **oxamniquine** (Case study 4).

FIGURE 14.3 Varying basicity in antithrombotic agents.

14.1.6 Bio-isosteres for polar groups

The use of bio-isosteres has already been described in section 13.3.7 in the design of compounds with improved target interactions. Bio-isosteres have also been used as substitutes for important functional groups that are required for target interactions, but which pose pharmacokinetic problems. For example, a carboxylic acid is a highly polar group which can ionize and hinder absorption of any drug containing it. One way of getting round this problem is to mask it as an ester prodrug (section 14.6.1.1). Another strategy is to replace it with a bio-isostere which has similar physicochemical properties, but which offers some advantage over the original carboxylic acid. Several bio-isosteres have been used for carboxylic acids, but among the most popular is a 5-substituted tetrazole ring (Fig. 14.4). Like carboxylic acids, tetrazoles contain an acidic proton and are ionized at pH 7.4. They are also planar in structure. However, they have an advantage in that the tetrazole anion is 10 times more lipophilic than a carboxylate anion. Drug absorption is enhanced as a result (see Box 14.1). They are also resistant to many of the metabolic reactions that occur on carboxylic acids. *N*-Acylsulphonamides have

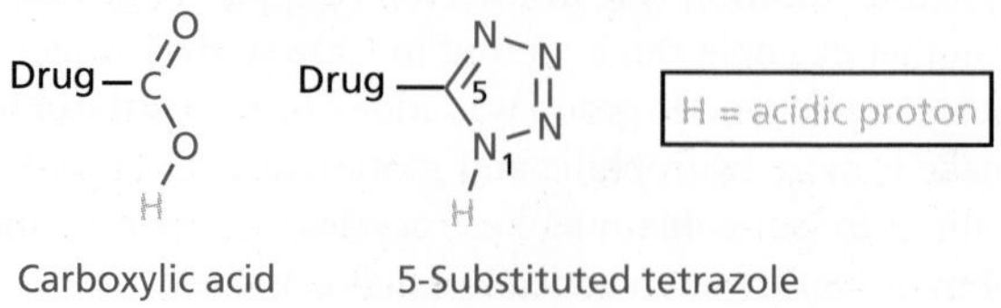

FIGURE 14.4 5-Substituted tetrazole ring as a bio-isostere for a carboxylic acid.

BOX 14.1 The use of bio-isosteres to increase absorption

The biphenyl structure (Structure I) was shown by Du Pont to inhibit the receptor for angiotensin II and had potential as an antihypertensive agent. However, the drug had to be injected as it showed poor absorption through the gut wall. Replacing the carboxylic acid with a tetrazole ring led to **losartan**, which was launched in 1994 (section 26.3.4).

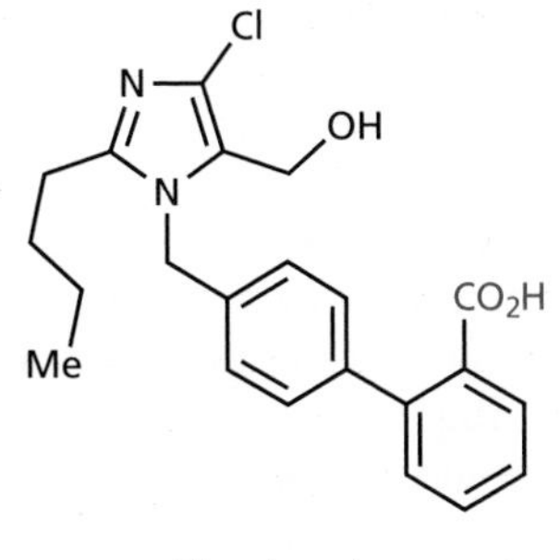

Structure I   Losartan

FIGURE 1 Development of losartan.

also been used as bio-isosteres for carboxylic acids (section 13.3.7).

Phenol groups are commonly present in drugs but are susceptible to metabolic conjugation reactions. Various bio-isosteres involving amides, sulphonamides, or heterocyclic rings have been used where an N–H group mimics the phenol O–H group.

14.2 Making drugs more resistant to chemical and enzymatic degradation

There are various strategies that can be used to make drugs more resistant to hydrolysis and drug metabolism, and thus prolong their activity.

14.2.1 Steric shields

Some functional groups are more susceptible to chemical and enzymatic degradation than others. For example, esters and amides are particularly prone to hydrolysis. A common strategy that is used to protect such groups is to add steric shields, designed to hinder the approach of a nucleophile or an enzyme to the susceptible group. These usually involve the addition of a bulky alkyl group close to the functional group. For example, the *t*-butyl group in the antirheumatic agent **D 1927** serves as a

steric shield and blocks hydrolysis of the terminal peptide bond (Fig. 14.5). Steric shields have also been used to protect penicillins from lactamases (section 19.5.1.8), and to prevent drugs interacting with cytochrome P450 enzymes (section 22.7.1).

For additional material see Web article 5: The design of a serotonin antagonist as a possible anxiolytic agent on the Online Resource Centre at www.oxfordtextbooks.co.uk/orc/patrick6e/

14.2.2 Electronic effects of bio-isosteres

Another popular tactic used to protect a labile functional group is to stabilize the group electronically using a bio-isostere. Isosteres and non-classical isosteres are frequently used as bio-isosteres (see also sections 13.1.15, 13.3.7, and 14.1.6). For example, replacing

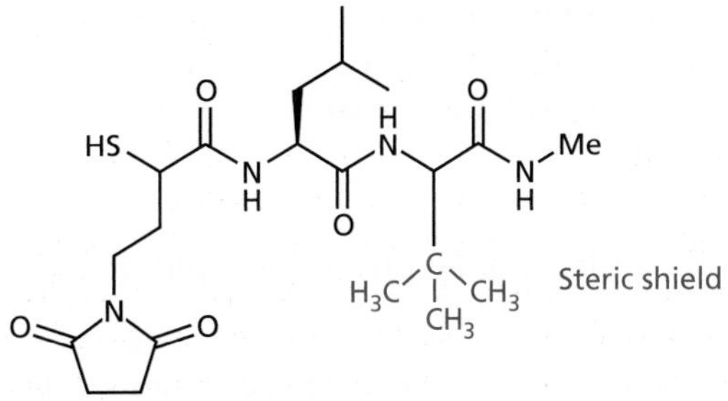

FIGURE 14.5 The use of a steric shield to protect the antirheumatic agent D 1927.

FIGURE 14.6 Isosteric replacement of a methyl group with an amino group.

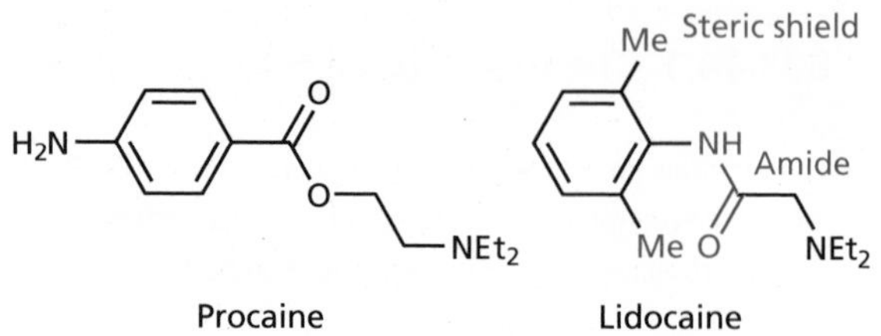

FIGURE 14.7 Steric and electronic modifications which make lidocaine a longer lasting local anaesthetic compared to procaine.

the methyl group of an ethanoate ester with NH_2 results in a urethane functional group which is more stable than the original ester (Fig. 14.6). The NH_2 group is the same valency and size as the methyl group and so it has no steric effect. However, it has totally different electronic properties as it can feed electrons into the carboxyl group and stabilize it from hydrolysis. The cholinergic agonist **carbachol** is stabilized in this way (section 22.7.2), as is the cephalosporin **cefoxitin** (section 19.5.2.4).

Alternatively, a labile ester group could be replaced by an amide group (NH replacing O). Amides are more resistant to chemical hydrolysis, due again to the lone pair of the nitrogen feeding its electrons into the carbonyl group and making it less electrophilic.

It is important to realize that bio-isosteres are often specific to a particular area of medicinal chemistry. Replacing an ester with a urethane or an amide may work in one category of drugs, but not another. One must also appreciate that bio-isosteres are different from isosteres. It is the retention of important biological activity that determines whether a group is a bio-isostere, not the valency. Therefore, non-isosteric groups can be used as bio-isosteres. For example, a pyrrole ring was used as a bio-isostere for an amide bond in the development of the dopamine antagonist **Du 122290** from **sultopride** (section 13.3.7). Similarly, thiazolyl rings were used as bio-isosteres for pyridine rings in the development of **ritonavir** (section 20.7.4.4).

One is not confined to the use of bio-isosteres to increase stability. Groups or substituents having an inductive electronic effect have frequently been incorporated into molecules to increase the stability of a labile functional group. For example, electron-withdrawing groups were incorporated into the side chain of penicillins to increase their resistance to acid hydrolysis (section 19.5.1.8). The inductive effects of groups can also determine the ease with which ester prodrugs are hydrolysed (Box 14.4).

14.2.3 **Steric and electronic modifications**

Steric hindrance and electronic stabilization have often been used together to stabilize labile groups. For example, **procaine** (Fig. 14.7) is a good, but short-lasting, local anaesthetic because its ester group is quickly hydrolysed.

Changing the ester group to the less reactive amide group reduces susceptibility to chemical hydrolysis. Furthermore, the presence of two *ortho*-methyl groups on the aromatic ring helps to shield the carbonyl group from attack by nucleophiles or enzymes. This results in the longer acting local anaesthetic **lidocaine**. Further successful examples of steric and electronic modifications are demonstrated by **oxacillin** (Box 19.5) and **bethanechol** (section 22.7.3).

14.2.4 **Metabolic blockers**

Some drugs are metabolized by the introduction of polar groups at particular positions in their skeleton. For example, steroids can be oxidized at position 6 of the tetracyclic skeleton to introduce a polar hydroxyl group (Fig. 14.8). The introduction of this group allows the formation of polar conjugates which can be quickly eliminated from the system. By introducing a methyl group at position 6, metabolism is blocked and the activity of the steroid is prolonged. The oral contraceptive **megestrol acetate** is an agent which contains a 6-methyl blocking group.

On the same lines, a popular method of protecting aromatic rings from metabolism at the *para*-position is to introduce a fluoro substituent. For example, **CGP 52411** (Fig. 14.9) is an enzyme inhibitor which acts on the kinase active site of the epidermal growth factor receptor

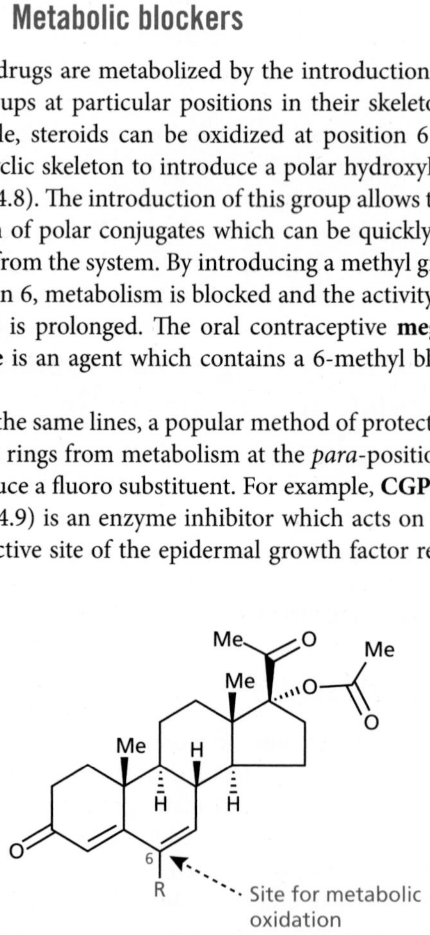

FIGURE 14.8 Metabolically susceptible steroid (R = H), metabolite (R = OH), and megestrol acetate (R = Me).

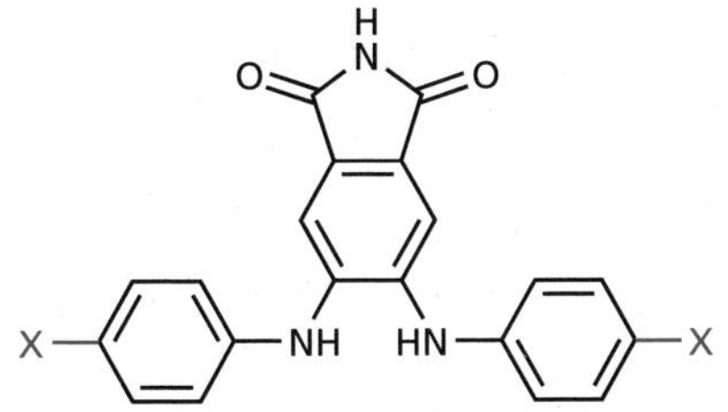

FIGURE 14.9 The use of fluorine substituents as metabolic blockers. X = H, CGP 52411; X = OH, metabolite; X = F, CGP 53353.

(section 4.8). It went forward for clinical trials as an anti-cancer agent and was found to undergo oxidative metabolism at the *para*-position of the aromatic rings. Fluoro-substituents were successfully added in the analogue **CGP 53353** to block this metabolism. This tactic was also applied successfully in the design of **gefitinib** (section 21.6.2.1). Fluorine has now been used extensively to block metabolism in a variety of structural situations.

Another approach which is actively being explored is to replace a hydrogen atom with a deuterium isotope. The covalent bond between carbon and deuterium is twice as strong as that between carbon and hydrogen, and this might help to block metabolic mechanisms.

For additional information see Web article 28: Use of fluorine in medicinal chemistry on the Online Resource Centre at www.oxfordtextbooks.co.uk/orc/patrick6e/

14.2.5 Removal or replacement of susceptible metabolic groups

Certain substituents are particularly susceptible to metabolic enzymes. For example, methyl groups on aromatic rings are often oxidized to carboxylic acids (section 11.5.2). These acids can then be quickly eliminated from the body. Other common metabolic reactions include aliphatic and aromatic *C*-hydroxylations, *N*- and *S*-oxidations, *O*- and *S*-dealkylations, and deaminations (section 11.5).

Susceptible groups can sometimes be removed or replaced by groups that are stable to oxidation, in order to prolong the lifetime of the drug. For example, the aromatic methyl substituent of the antidiabetic **tolbutamide** was

replaced by a chloro substituent to give **chlorpropamide**, which is much longer lasting (Fig. 14.10). This tactic was also used in the design of **gefitinib** (section 21.6.2.1). An alternative strategy which is often tried is to replace the susceptible methyl group with CF_3, CHF_2, or CH_2F. The fluorine atoms alter the oxidation potential of the methyl group and make it more resistant to oxidation.

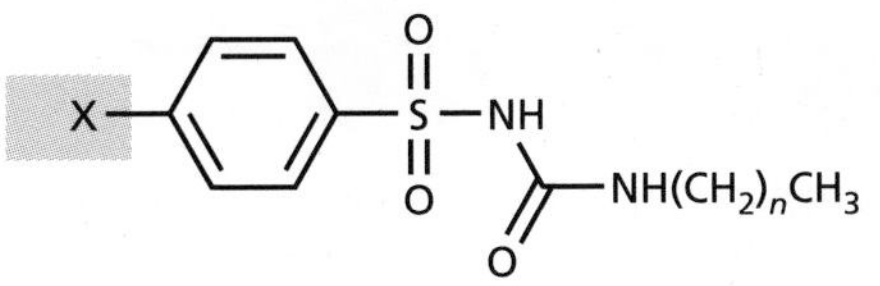

Another example where a susceptible metabolic group is replaced is seen in section 19.5.2.3 where a susceptible ester in cephalosporins is replaced with metabolically stable groups to give **cephaloridine** and **cefalexin**.

14.2.6 Group shifts

Removing or replacing a metabolically vulnerable group is feasible if the group concerned is not involved in important binding interactions with the binding site. If the group *is* important, then we have to use a different strategy.

There are two possible solutions. We can either mask the vulnerable group on a temporary basis by using a prodrug (section 14.6) or we can try shifting the vulnerable group within the molecular skeleton. The latter tactic was used in the development of **salbutamol** (Fig. 14.11). Salbutamol was introduced in 1969 for the treatment of asthma, and is an analogue of the neurotransmitter **noradrenaline**—a catechol structure containing two *ortho*-phenolic groups.

One of the problems faced by catechol compounds is metabolic methylation of one of the phenolic groups. Since both phenol groups are involved in hydrogen bonds to the receptor, methylation of one of the phenol groups disrupts the hydrogen bonding and makes the compound inactive. For example, the noradrenaline analogue (I in Fig. 14.12) has useful anti-asthmatic activity, but the effect is of short duration because the compound is rapidly metabolized to the inactive methyl ether (II in Fig. 14.12).

Removing the OH or replacing it with a methyl group prevents metabolism, but also prevents the important hydrogen bonding interactions with the binding site. So

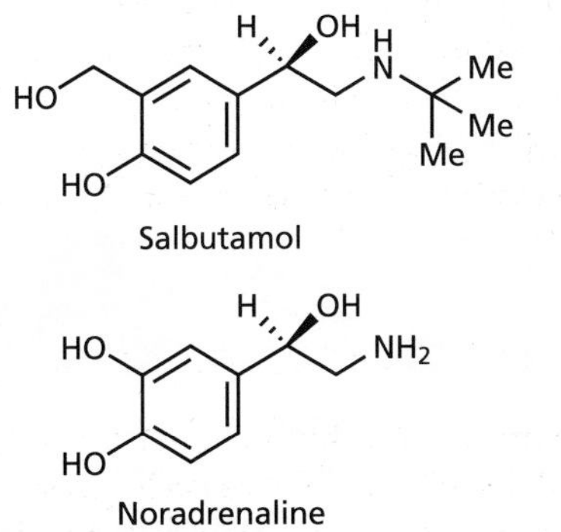

Salbutamol

Noradrenaline

FIGURE 14.11 Salbutamol and noradrenaline.

FIGURE 14.10 Tolbutamide (X = Me; *n* = 3) and chlorpropamide (X = Cl; *n* = 2).

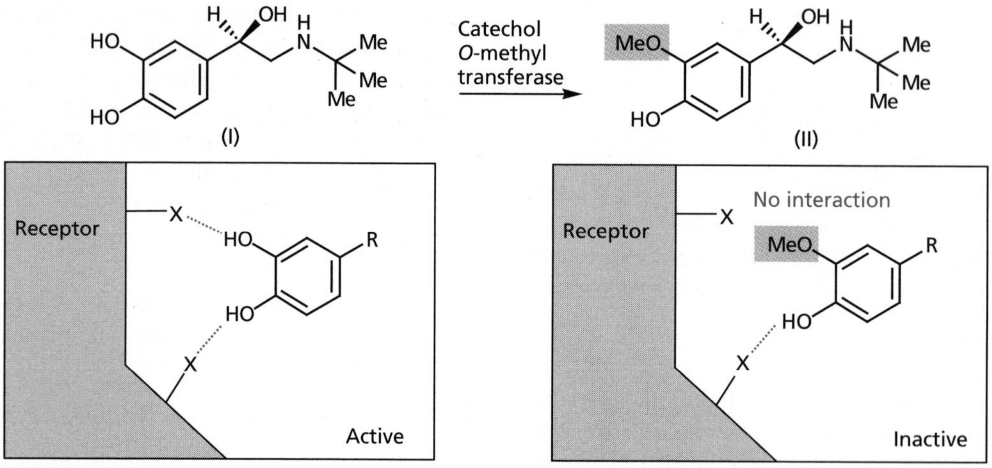

FIGURE 14.12 Metabolic methylation of a noradrenaline analogue. X denotes an electronegative atom.

how can this problem be solved? The answer was to move the vulnerable hydroxyl group out from the ring by one carbon unit. This was enough to make the compound unrecognizable to the metabolic enzyme, but not to the receptor binding site.

Fortunately, the receptor appears to be quite lenient over the position of this hydrogen bonding group and it is interesting to note that a hydroxyethyl group is also acceptable (Fig. 14.13). Beyond that, activity is lost because the OH group is out of range, or the substituent is too large to fit. These results demonstrate that it is better to consider a binding region within the receptor binding site as an available volume, rather than imagining it as being fixed at one spot. A drug can then be designed such that the relevant binding group is positioned in any part of that available volume. Another example of a successful group shift strategy can be seen in Case study 7.

Shifting an important binding group that is metabolically susceptible cannot be guaranteed to work in every situation. It may well make the molecule unrecognizable both to its target and to the metabolic enzyme.

14.2.7 Ring variation and ring substituents

Certain ring systems may be susceptible to metabolism, and so varying the ring might improve metabolic stability. This can be done by adding a nitrogen into the ring to lower the electron density of the ring system. For example, the imidazole ring of the antifungal agent **tioconazole** mentioned in section 14.1.2 is susceptible to metabolism, but replacement with a 1,2,4-triazole ring as in **fluconazole** results in improved stability (Fig. 14.1).

Electron-rich aromatic rings such as phenyl groups are particularly prone to oxidative metabolism, but can be stabilized by replacing them with nitrogen-containing heterocyclic rings such as pyridine or pyrimidine. Alternatively, electron-withdrawing substituents could be added to the aromatic ring to lower the electron density (see Web article 5).

Ring variation can also help to stabilize metabolically susceptible aromatic or heteroaromatic methyl substituents. Such substituents could be replaced with more stable substituents as described in section 14.2.5, but sometimes the methyl substituent has to be retained for good

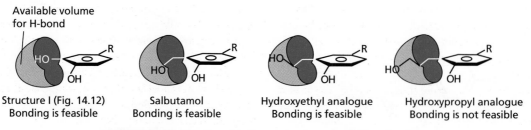

FIGURE 14.13 Viewing a binding region as an available volume.

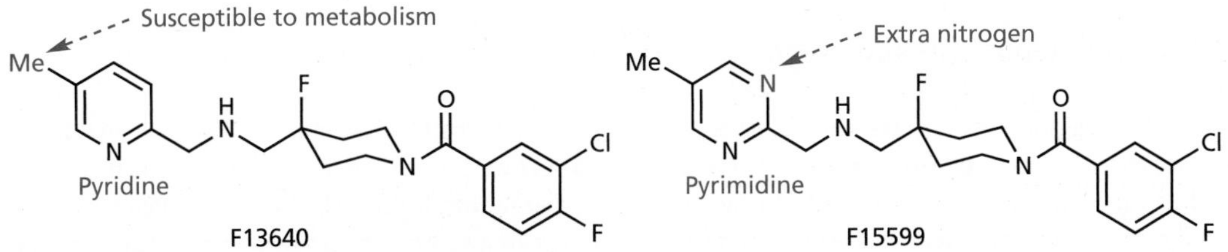

FIGURE 14.14 Stabilizing an aromatic or heteroaromatic methyl substituent by adding a nitrogen to the ring.

activity. In such cases, introducing a nitrogen atom into the aromatic/heteroaromatic ring can be beneficial, since lowering the electron density in the ring also helps to make the methyl substituent more resistant to metabolism. For example, **F13640** underwent phase II clinical trials as an analgesic (Fig. 14.14). The methyl substituent on the pyridine ring is susceptible to oxidation and is converted to a carboxylic acid which is inactive. The methyl group plays an important binding role and has to be present. Therefore, the pyridine ring was changed to a pyrimidine ring resulting in a compound (**F15599**) that has increased metabolic stability without affecting binding affinity.

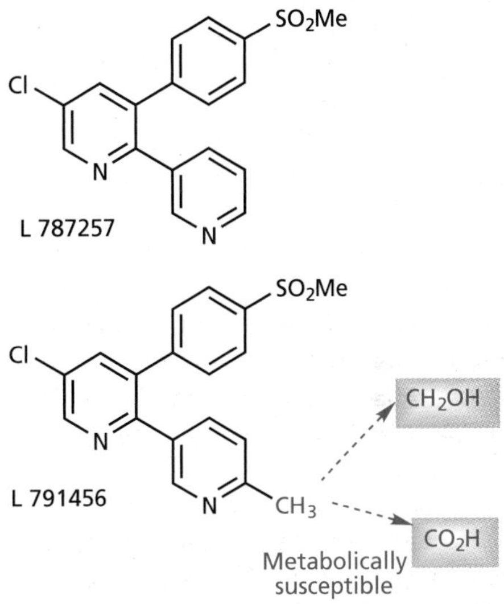

14.3 Making drugs less resistant to drug metabolism

So far, we have looked at how the activity of drugs can be prolonged by inhibiting their metabolism. However, a drug that is extremely stable to metabolism and is very slowly excreted can pose just as many problems as one that is susceptible to metabolism. It is usually desirable to have a drug that does what it is meant to do, then stops doing it within a reasonable time. If not, the effects of the drug could last too long and cause toxicity and lingering side effects. Therefore, designing drugs with decreased chemical and metabolic stability can sometimes be useful.

14.3.1 Introducing metabolically susceptible groups

Introducing groups that are susceptible to metabolism is a good way of shortening the lifetime of a drug (see Box 14.2). For example, a methyl group was introduced to the anti-arthritic agent **L 787257** to shorten its lifetime. The methyl group of the resulting compound (**L 791456**) was metabolically oxidized to a polar alcohol as well as to a carboxylic acid (Fig. 14.15).

FIGURE 14.15 Adding a metabolically labile methyl group to shorten a drug's lifetime.

Another example is the analgesic **remifentanil** (section 24.6.3.4), where ester groups were incorporated to make it a short-lasting agent. The beta-blocker **esmolol** was also designed to be a short-acting agent by introducing an ester group (section 23.11.3.4).

14.3.2 Self-destruct drugs

A self-destruct drug is one which is chemically stable under one set of conditions, but becomes unstable and spontaneously degrades under another set of conditions. The advantage of a self-destruct drug is that inactivation does not depend on the activity of metabolic enzymes, which could vary from patient to patient. The best example of a self-destruct drug is the neuromuscular blocking agent **atracurium**, which is stable at acid pH but self-destructs when it meets the slightly alkaline conditions of the blood (section 22.10.2.4). This means that the drug has a short duration of action, allowing anaesthetists to control its blood levels during surgery by providing it as a continuous, intravenous drip.

BOX 14.2 Shortening the lifetime of a drug

Anti-asthmatic drugs are usually taken by inhalation to re- duce the chances of side effects elsewhere in the body. However, a significant amount is swallowed and can be ab- sorbed into the blood supply from the gastrointestinal tract. Therefore, it is desirable to have an anti-asthmatic drug which is potent and stable in the lungs, but which is rapidly metabolized in the blood supply. **Cromakalim** has useful anti- asthmatic properties, but has cardiovascular side effects if

it gets into the blood supply. Structures **UK 143220** and **UK 157147** were developed from cromakalim so that they would be quickly metabolized. UK 143220 contains an ester which is quickly hydrolysed by esterases in the blood to produce an inactive carboxylic acid, while UK 157147 contains a phenol group which is quickly conjugated by metabolic conjugation enzymes and eliminated. Both these compounds were con- sidered as clinical candidates.

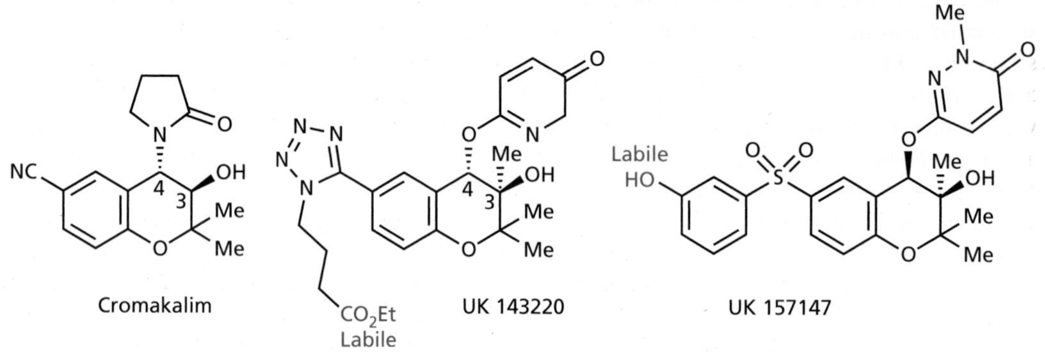

FIGURE 1 Metabolically labile analogues of cromakalim.

KEY POINTS

- The polarity or pK_a of a lead compound can be altered by varying alkyl substituents or functional groups, allowing the drug to be absorbed more easily.

- Drugs can be made more resistant to metabolism by in- troducing steric shields to protect susceptible functional groups. It may also be possible to modify the functional group itself to make it more stable as a result of electronic factors.

- Metabolically stable groups can be added to block metabo- lism at certain positions.

- Groups which are susceptible to metabolism may be modi- fied or removed to prolong activity, as long as the group is not required for drug–target interactions.

- Metabolically susceptible groups which are necessary for drug–target interactions can be shifted in order to make them unrecognizable by metabolic enzymes, as long as they are still recognizable to the target.

- Varying a heterocyclic ring in the lead compound can some- times improve metabolic stability.

- Drugs which are slowly metabolized may linger too long in the body and cause side effects.

- Groups which are susceptible to metabolic or chemical change can be incorporated to reduce a drug's lifetime.

14.4 Targeting drugs

One of the major goals in drug design is to find ways of targeting drugs to the exact locations in the body where they are most needed. The principle of targeting drugs can be traced back to Paul Ehrlich who developed anti- microbial drugs that were selectively toxic for microbial cells over human cells. Drugs can also be made more se- lective to distinguish between different targets within the body, as discussed in Chapter 13. Here, we discuss other tactics related to the targeting of drugs.

14.4.1 Targeting tumour cells: 'search and destroy' drugs

A major goal in cancer chemotherapy is to target drugs efficiently against tumour cells rather than normal cells. One method of achieving this is to design drugs which make use of specific molecular transport systems. The idea is to attach the active drug to an important 'building block' molecule that is needed in large amounts by the rapidly dividing tumour cells. This could be an amino acid or a nucleic acid base (e.g. uracil mustard; section 21.2.3.1). Of course, normal cells require these building blocks as well, but tumour cells often grow more quickly than normal cells and require the building blocks more urgently. Therefore, the uptake is greater in tumour cells.

A more recent idea has been to attach the active drug (or a poison such as **ricin**) to **monoclonal antibodies** which can recognize antigens unique to the tumour cell. Once the antibody binds to the antigen, the drug or poison is released to kill the cell. The difficulties in this approach include the identification of suitable antigens and the production of antibodies in significant quantity. Nevertheless, the approach has great promise for the future and is covered in more detail in section 21.10.2. Another tactic which has been used to target anticancer drugs is to administer an enzyme–antibody conjugate where the enzyme serves to activate an anticancer prodrug, and the antibody directs the enzyme to the tumour. This is a strategy known as **ADEPT** and is covered in more detail in section 21.10.3. Other targeting strategies include **ADAPT** and **GDEPT** covered in sections 21.10.4 and 21.10.5 respectively. Antibodies are also being studied as a means of targeting viruses (section 20.11.5).

14.4.2 Targeting gastrointestinal infections

If a drug is to be targeted against an infection of the gastrointestinal tract, it must be prevented from being absorbed into the blood supply. This can be done by using a fully ionized drug which is incapable of crossing cell membranes. For example, highly ionized sulphonamides are used against gastrointestinal infections (Box 19.2).

14.4.3 Targeting peripheral regions rather than the central nervous system

It is often possible to target drugs such that they act peripherally and not in the central nervous system. By increasing the polarity of drugs, they are less likely to cross the blood–brain barrier (section 11.4.5), and this means they are less likely to have central nervous system side effects. Achieving selectivity for the central nervous system over the peripheral regions of the body is not so straightforward. In order to achieve that, the drug would have to be designed to cross the blood–brain barrier efficiently, whilst being metabolized rapidly to inactive metabolites in the peripheral system.

14.4.4 Targeting with membrane tethers

Several drug targets are associated with cell membranes, and one way of targeting drugs to these targets is to attach membrane tethers to the drug such that the molecule is anchored in the membrane close to the target. The antibacterial agent **teicoplanin** is one such example and is discussed in section 19.5.5.2. Another membrane-tethered drug has been designed to inhibit the enzyme β-**secretase**, with the ultimate aim of treating Alzheimer's disease (AD). This enzyme generates the proteins that are responsible for the toxic protein aggregates found in the brains of AD sufferers, and does so mainly in cellular organelles called **endosomes**. A peptide transition-state inhibitor has been linked to a sterol such that it is taken into endosomes by endocytosis. The sterol then acts as the membrane tether to lock the drug in position, such that it targets β-secretase in endosomes rather than β-secretase in other locations. Potential agents for AD treatment are also being targeted to mitochondria where AD leads to the generation of radicals and oxidation reactions that are damaging to the cell. **MitoQ** (Fig. 14.16) is an agent undergoing clinical trials which contains an antioxidant prodrug linked to a hydrophobic triphenylphosphine moiety. The latter group aids the drug's entry into mitochondria, then tethers it to the phospholipid bilayers of the mitochondria membrane. The quinone ring system is rapidly reduced to the active quinol form which can then act as an antioxidant to neutralize free radicals. A different approach for targeting antioxidant drugs to mitochondria has been to modify known antibacterial agents (e.g. **gramicidin S**) such that they act as antioxidants rather than antibacterial agents. The

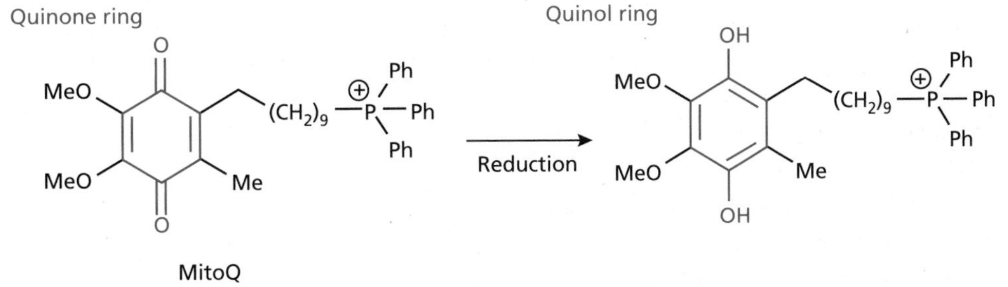

FIGURE 14.16 MitoQ acting as a prodrug.

rationale here is that the mitochondrial membrane is similar in nature to bacterial cell membranes, and so antibacterial agents may show selectivity for mitochondrial membranes over cell membranes.

14.5 Reducing toxicity

It is often found that a drug fails clinical trials because of toxic side effects. This may be due to toxic metabolites, in which case the drug should be made more resistant to metabolism as described earlier (section 14.2). It is also worth checking to see whether there are any functional groups present which are particularly prone to producing toxic metabolites. For example, it is known that functional groups such as aromatic nitro groups, aromatic amines, bromoarenes, hydrazines, hydroxylamines, or polyhalogenated groups are often metabolized to toxic products (see section 11.5 for typical metabolic reactions; see also Box 14.3).

Side effects might also be reduced or eliminated by varying apparently harmless substituents. For example, the halogen substituents of the antifungal agent **UK 47265** were varied in order to find a compound that was less toxic to the liver. This led to the successful antifungal agent **fluconazole** (Fig. 14.17).

Varying the position of substituents can sometimes reduce or eliminate side effects. For example, the dopamine antagonist **SB 269652** inhibits cytochrome P450 enzymes as a side effect. Placing the cyano group at a different position prevented this inhibition (Fig. 14.18).

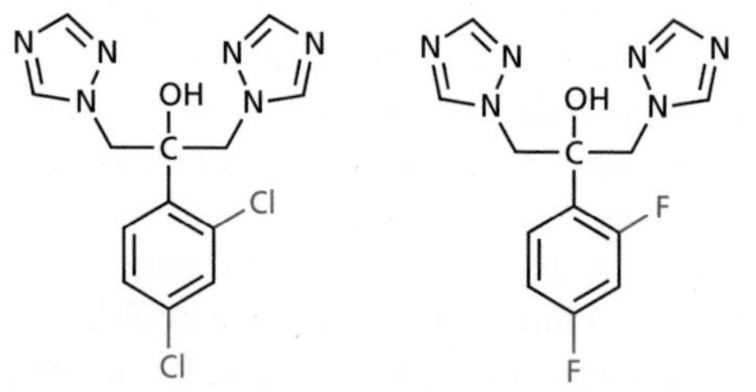

KEY POINTS

- Strategies designed to target drugs to particular cells or tissues are likely to lead to safer drugs with fewer side effects.
- Drugs can be linked to amino acids or nucleic acid bases to target them against fast-growing and rapidly dividing cells.
- Drugs can be targeted to the gastrointestinal tract by making them ionized or highly polar such that they cannot cross the gut wall.
- The central nervous system side effects of peripherally acting drugs can be eliminated by making the drugs more polar so that they do not cross the blood–brain barrier.
- Drugs with toxic side effects can sometimes be made less toxic by varying the nature or position of substituents, or by preventing their metabolism to a toxic metabolite.

14.6 Prodrugs

Prodrugs are compounds which are inactive in themselves, but which are converted in the body to the active drug. They have been useful in tackling problems such as acid sensitivity, poor membrane permeability, drug toxicity, bad taste, and short duration of action. Usually, a metabolic enzyme is involved in converting the prodrug to the active drug, and so a good knowledge of drug metabolism and the enzymes involved allows the medicinal chemist to design a suitable prodrug which turns drug metabolism into an advantage rather than a problem. Prodrugs have been designed to be activated by a variety of metabolic enzymes. Ester prodrugs which are hydrolysed by esterase enzymes are particularly common, but prodrugs have also been designed which are activated by N-demethylation, decarboxylation, and the hydrolysis of amides and phosphates. Not all prodrugs are activated by metabolic enzymes, however. For example, photodynamic therapy involves the use of an external light source to activate prodrugs. When designing prodrugs, it is important to ensure that the prodrug is effectively converted to the active drug once it has been absorbed into the blood supply, but it is also important to ensure that any groups that are cleaved from the molecule are non-toxic.

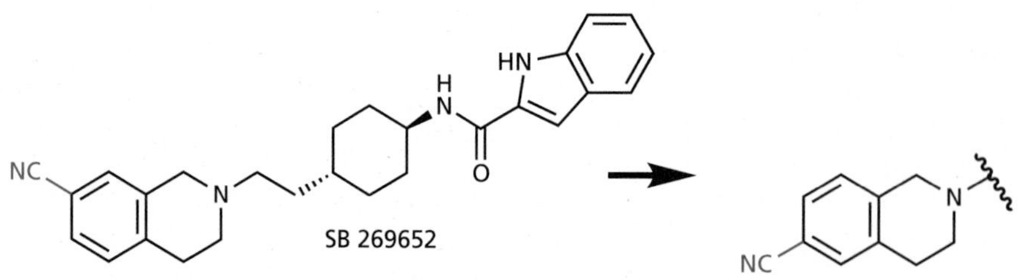

FIGURE 14.17 Varying aromatic substituents to reduce toxicity.

FIGURE 14.18 Varying substituent positions to reduce side effects.

BOX 14.3 Identifying and replacing potentially toxic groups

Replacing potentially toxic functional groups is often carried out at an early stage in drug development, even if there is no direct evidence of actual toxicity. For example, the presence of a 1,4-diamino-substituted aromatic ring is seen as a significant risk factor for toxicity. The presence of the two amino groups activates the aromatic ring and makes it electron rich such that it can undergo metabolism to form a 1,4-diiminoquinone structure. Diiminoquinones can act as electrophilic agents and react with nucleophilic groups such as lysine in proteins, or guanine in nucleic acids.

One strategy that can be used to combat this problem is to make the ring less electron rich by introducing a nitrogen atom. In other words, the benzene ring is replaced with a pyridine ring. This strategy was used in the development of the anticancer agent **sonidegib** (section 21.6.3). Another strategy that was used in the development of the anticancer agent **ceritinib** was to carry out a 'ring reversal'. This involved reversing a piperidine ring in the lead compound such that the nitrogen was no longer linked to the aromatic ring. In addition, a methyl substituent was added to act as a metabolic blocker.

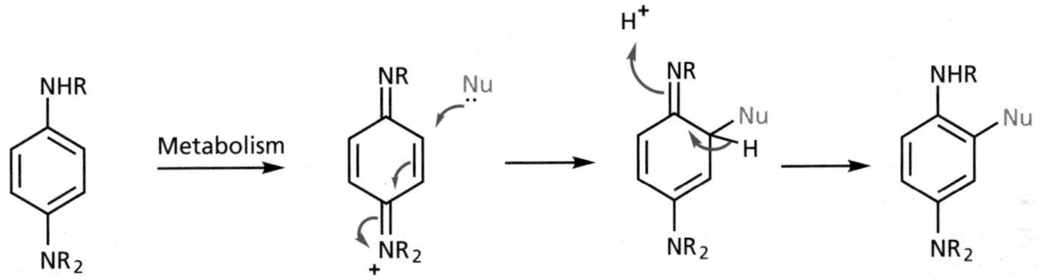

1,4-Diamino-substituted aromatic ring   1,4-Diiminoquinone ring

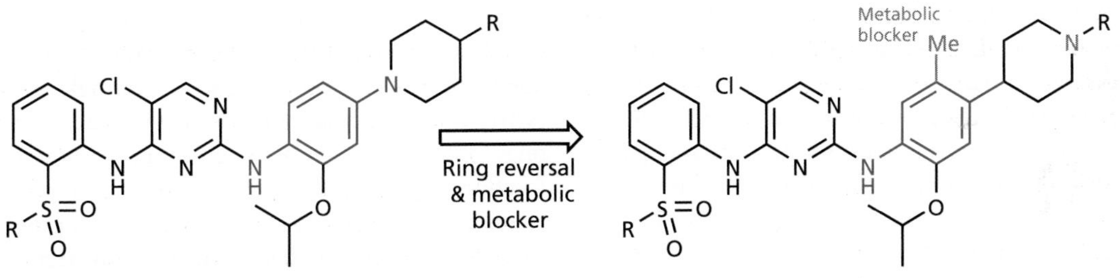

Ring reversal & metabolic blocker

14.6.1 Prodrugs to improve membrane permeability

14.6.1.1 Esters as prodrugs

Prodrugs have proved very useful in temporarily masking an 'awkward' functional group which is important to target binding, but which hinders the drug from crossing the cell membranes of the gut wall. For example, a carboxylic acid functional group may have an important role to play in binding a drug to its binding site via ionic or hydrogen bonding. However, the very fact that it is an ionizable group may prevent it from crossing a fatty cell membrane. The answer is to protect the acid function as an ester. The less polar ester can cross fatty cell membranes and, once it is in the bloodstream, it is hydrolysed

back to the free acid by esterases in the blood. Examples of ester prodrugs used to aid membrane permeability include many of the ACE inhibitors (section 26.3.3 and Case study 2), **sacubitril** (section 26.5.3), and **pivampicillin**, which is a penicillin prodrug (Box 19.7).

Not all esters are hydrolysed equally efficiently, and a range of esters may need to be tried to find the best one (Box 14.4). It is possible to make esters more susceptible to hydrolysis by introducing electron-withdrawing groups to the alcohol moiety (e.g. OCH_2CF_3, OCH_2CO_2R, $OCONR_2$, OAr). The inductive effect of these groups aids the hydrolytic mechanism by stabilizing the alkoxide leaving group (Fig. 14.19). Care has to be taken, however, not to make the ester too reactive in case it becomes chemically unstable and is hydrolysed by the acid conditions of

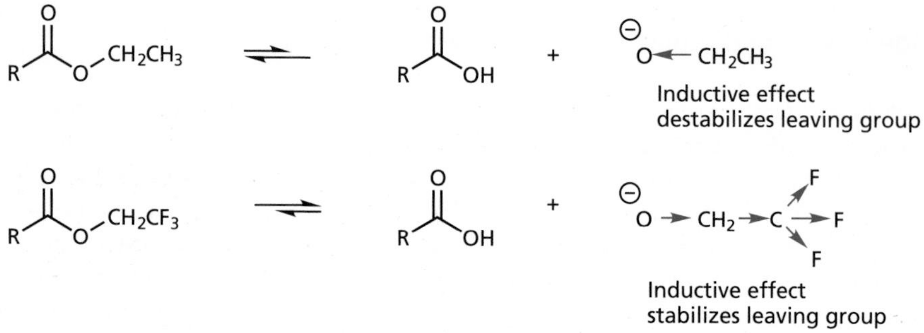

FIGURE 14.19 Inductive effects on the stability of leaving groups.

the stomach or the more alkaline conditions of the intestine before it reaches the blood supply. To that end, it may be necessary to make the ester more stable. For example, cyclopropanecarboxylic acid esters have been studied as potential prodrugs because the cyclopropane ring has the ability to stabilize the carbonyl group of a neighbouring ester (Fig. 14.20). In this respect, it is acting as a bio-isostere for a double bond (see also section 13.3.7). A conjugated double bond stabilizes a neighbouring carbonyl group due to interaction of the π-systems involved. It is proposed that the σ bonds of a cyclopropane ring are correctly orientated to allow a hyperconjugative interaction that has a similar stabilizing effect on a neighbouring carbonyl group. The interaction proposed involves hyperconjugative donation to the antibonding π orbital of the carbonyl group.

There are some instances where esterases fail to hydrolyse an ester prodrug because the ester is shielded by a bulky group. For example, this is often the case if the ester is linked to a multicylic ring system. In such cases, an **extended ester** can be the answer. This involves esterifying the drug with a group that contains a second ester or carbonate group, which will be positioned further away from the ring system. Consequently, it will be more accessible to esterase enzymes. Enzyme-catalysed hydrolysis of the more accessible ester then leads to a product which is designed to be chemically unstable and will spontaneously degrade to give the active drug without the need for enzyme intervention. Examples of extended esters as prodrugs can be found in some penicillins (Box 19.7) and antihypertensive agents (section 26.3.4).

14.6.1.2 *N*-Methylated prodrugs

N-Demethylation is a common metabolic reaction in the liver, so polar amines can be *N*-methylated to reduce polarity and improve membrane permeability. Several hypnotics and anti-epileptics take advantage of this reaction, for example **hexobarbitone** (Fig. 14.21).

14.6.1.3 Trojan horse approach for transport proteins

Another way round the problem of membrane permeability is to design a prodrug which can take advantage of transport proteins in the cell membrane, such as the ones responsible for carrying amino acids into a cell (section 2.7.2). A well-known example of such a prodrug is **levodopa** (Fig. 14.22). Levodopa is a prodrug for the neurotransmitter **dopamine** and has been used in the treatment of Parkinson's disease—a condition due primarily to a deficiency of that neurotransmitter in the brain. Dopamine itself cannot be used, since it is too polar to cross the blood–brain barrier. Levodopa is even more polar and seems an unlikely prodrug, but it is also an amino acid, and so it is recognized by the transport proteins for amino acids which carry it across the cell membrane. Once in the brain, a decarboxylase enzyme removes the acid group and generates dopamine.

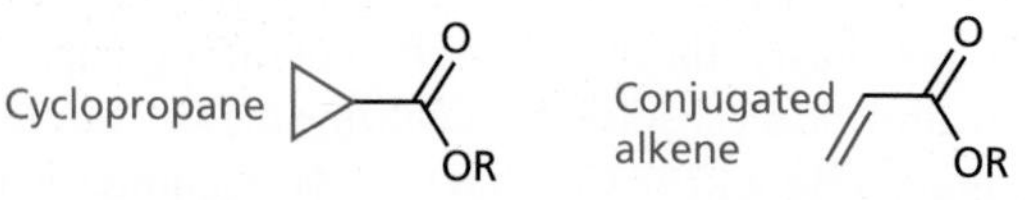

FIGURE 14.20 Cyclopropane carboxylic acid esters as prodrugs and bio-isosteres for α,β-unsaturated esters.

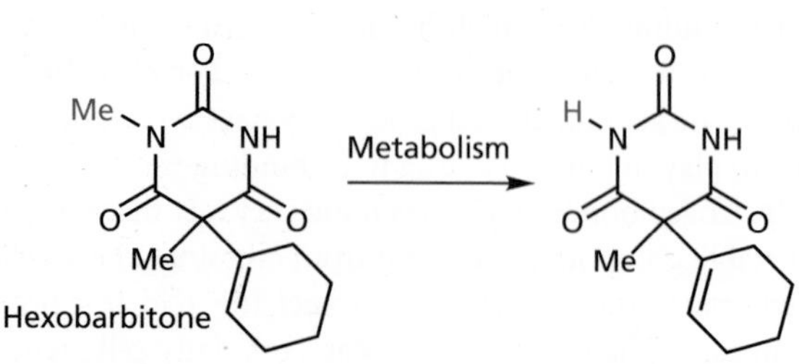

FIGURE 14.21 *N*-Demethylation of hexobarbitone.

Varying esters in prodrugs

The protease inhibitor **candoxatrilat** has to be given in-travenously because it is too polar to be absorbed from the gastrointestinal tract. Different esters were tried as prodrugs to get round this problem. It was found that an ethyl ester was absorbed but was inefficiently hydrolysed. A more activated ester was required, and a 5-indanyl ester proved to be the best. The 5-indanol released on hydroly-sis is non-toxic (Fig. 1).

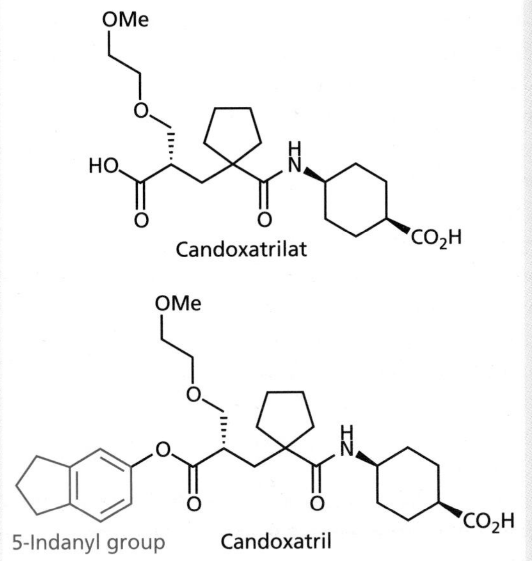

FIGURE 1 Protease inhibitors.

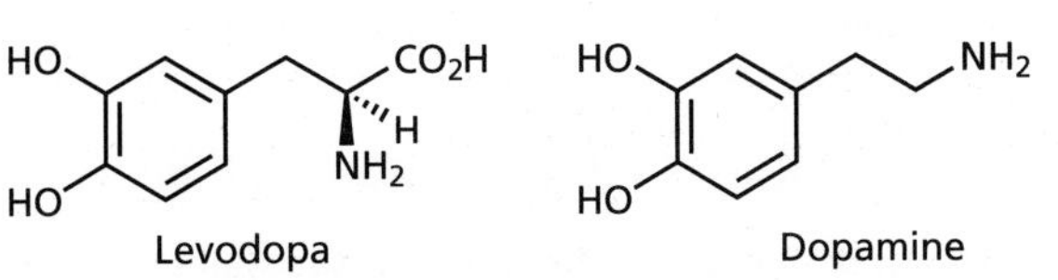

FIGURE 14.22 Levodopa and dopamine.

14.6.2 Prodrugs to prolong drug activity

Sometimes prodrugs are designed to be converted slowly to the active drug, thus prolonging a drug's activity. For example, **6-mercaptopurine** (Fig. 14.23) suppresses the body's immune response and is, therefore, useful in pro-tecting donor grafts. Unfortunately, the drug tends to be eliminated from the body too quickly. The prodrug **aza-thioprine** has the advantage that it is slowly converted to 6-mercaptopurine by being attacked by **glutathione** (section 11.5.5), allowing a more sustained activity. The rate of conversion can be altered, depending on the elec-tron-withdrawing ability of the heterocyclic group. The greater the electron-withdrawing power, the faster the breakdown. The NO_2 group is therefore present to ensure an efficient conversion to 6-mercaptopurine, since it is strongly electron-withdrawing on the heterocyclic ring.

There is a belief that the well-known sedatives **Valium** (Fig. 14.24) and **Librium** might be prodrugs and are ac-tive because they are metabolized by *N*-demethylation to **nordazepams**. Nordazepam itself has been used as a sedative, but loses activity quite quickly as a result of metabolism and excretion. Valium, if it is a prodrug for nordazepam, demonstrates again how a prodrug can be used to produce a more sustained action.

Another approach to maintaining a sustained level of drug over long periods is to deliberately associate a very lipophilic group to the drug. This means that most of the drug is stored in fat tissue, from where it is stead-ily and slowly released into the bloodstream. The anti-malarial agent **cycloguanil pamoate** (Fig. 14.25) is one such agent. The active drug is bound ionically to an anion containing a large lipophilic group, and is only released into the blood supply following slow dissociation of the ion complex.

Similarly, lipophilic esters of the antipsychotic drug **fluphenazine** are used to prolong its action. The prodrug is given by intramuscular injection and slowly diffuses from fat tissue into the blood supply where it is rapidly hydrolysed (Fig. 14.26).

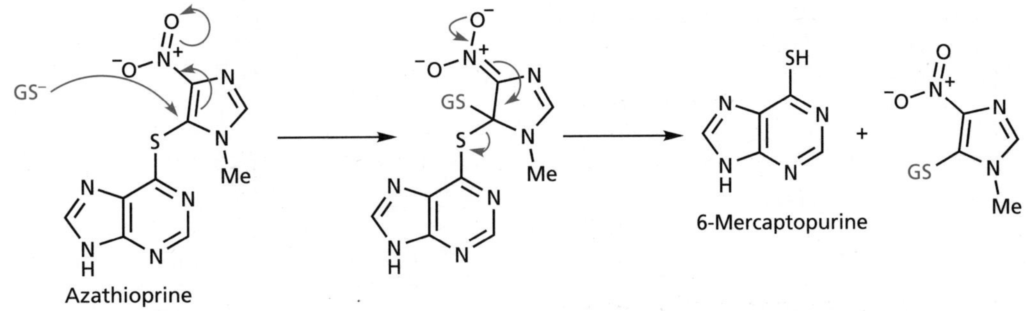

FIGURE 14.23 Azathioprine acts as a prodrug for 6-mercaptopurine (GS = glutathione).

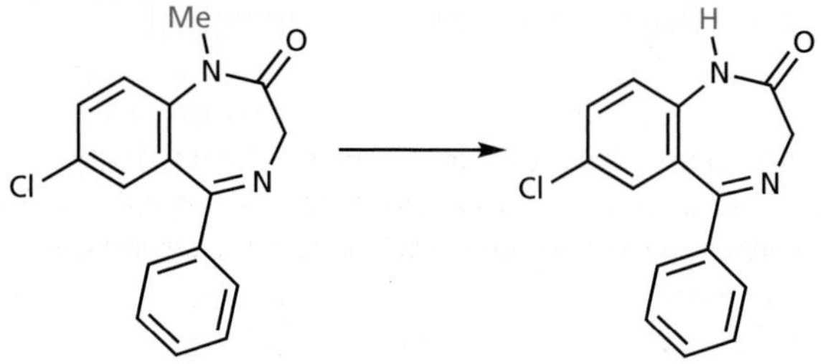

FIGURE 14.24 Valium (diazepam) as a possible prodrug for nordazepam.

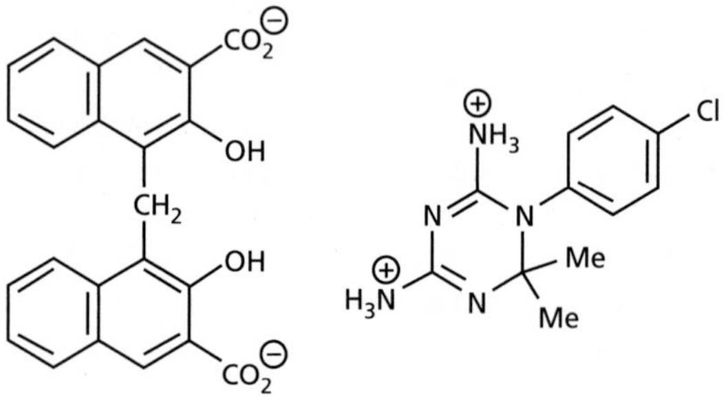

FIGURE 14.25 Cycloguanil pamoate.

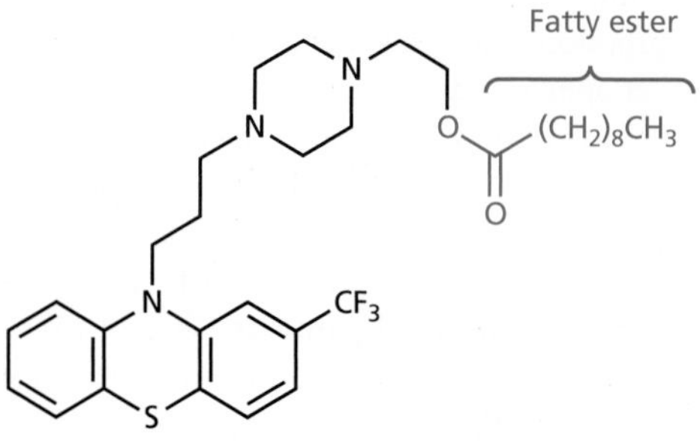

FIGURE 14.26 Fluphenazine decanoate.

14.6.3 Prodrugs masking drug toxicity and side effects

Prodrugs can be used to mask the side effects and toxicity of drugs (Box 14.5). For example, **salicylic acid** is a good painkiller, but causes gastric bleeding due to the free phenolic group. This is overcome by masking the phenol as an ester (**aspirin**) (Fig. 14.27). The ester is later hydrolysed to free the active drug.

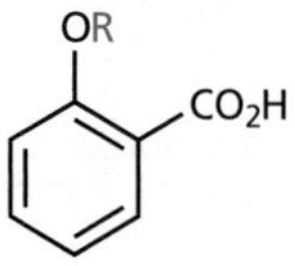

FIGURE 14.27 Aspirin (R = COCH$_3$) and salicylic acid (R = H).

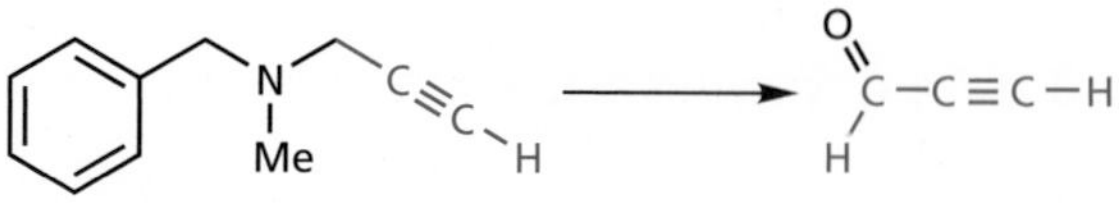

FIGURE 14.28 Pargyline as a prodrug for propiolaldehyde.

Prodrugs can be used to give a slow release of drugs that would be too toxic to give directly. **Propiolaldehyde** is useful in the aversion therapy of alcohol, but is not used itself because it is an irritant. The prodrug **pargyline** can be converted to propiolaldehyde by enzymes in the liver (Fig. 14.28).

Cyclophosphamide is a successful, non-toxic prodrug which can be safely taken orally. Once absorbed, it is metabolized in the liver to a toxic alkylating agent which is useful in the treatment of cancer (section 21.2.3.1).

Many important antiviral drugs such as **aciclovir** and **penciclovir** are non-toxic prodrugs which show selective toxicity towards virally infected cells. This is because they are activated by a viral enzyme which is only present in infected cells (sections 9.5 and 20.6.1). In a similar vein, the antischistosomal agent **oxamniquine** is converted to an alkylating agent by an enzyme which is only present in the parasite (Case study 4).

14.6.4 Prodrugs to lower water solubility

Some drugs have a revolting taste! One way to avoid this problem is to reduce their water solubility to prevent them dissolving on the tongue. For example, the bitter taste of the antibiotic **chloramphenicol** can be avoided by using the palmitate ester (Fig. 14.29). This is more hydrophobic because of the masked alcohol and the long chain fatty group that is present. It does not dissolve easily on the tongue and is quickly hydrolysed once swallowed.

14.6.5 Prodrugs to improve water solubility

Prodrugs have been used to increase the water solubility of drugs (Box 14.6). This is particularly useful for drugs which are given intravenously, as it means that higher

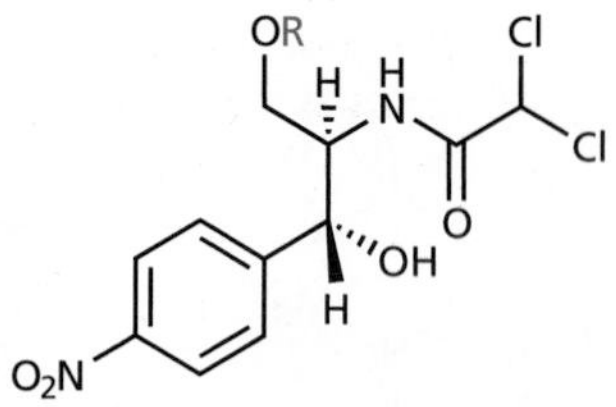

FIGURE 14.29 Chloramphenicol (R = H) and chloramphenicol prodrugs: chloramphenicol palmitate (R = CO(CH$_2$)$_{14}$CH$_3$); chloramphenicol succinate (R = CO(CH$_2$)$_2$CO$_2$H).

BOX 14.5 Prodrugs masking toxicity and side effects

LDZ is an example of a diazepam prodrug which avoids the drowsiness side effects associated with **diazepam**. These side effects are associated with the high initial plasma levels of diazepam following administration. The use of a prodrug avoids this problem. An aminopeptidase enzyme hydrolyses the prodrug to release a non-toxic lysine moiety, and the resulting amine spontaneously cyclizes to the diazepam (as shown below).

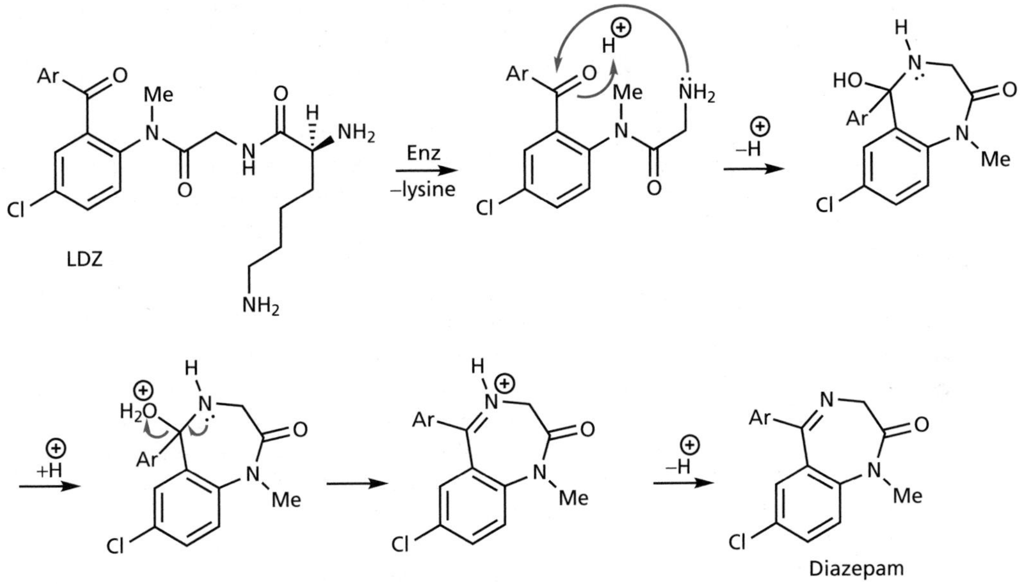

FIGURE 1 LDZ as a diazepam prodrug.

concentrations and smaller volumes can be used. For example, the succinate ester of **chloramphenicol** (Fig. 14.29) increases the latter's water solubility due to the extra carboxylic acid that is present. Once the ester is hydrolysed, chloramphenicol is released along with succinic acid, which is naturally present in the body.

Prodrugs designed to increase water solubility have proved useful in preventing the pain associated with some injections, which is caused by the poor solubility of the drug at the site of injection. For example, the antibacterial agent **clindamycin** is painful when injected, but this is avoided by using a phosphate ester prodrug which has much better solubility because of the ionic phosphate group (Fig. 14.30).

14.6.6 Prodrugs used in the targeting of drugs

Methenamine (Fig. 14.31) is a stable, inactive compound when the pH is more than 5. At a more acidic pH, however, the compound spontaneously degrades to generate **formaldehyde,** which has antibacterial properties. This is useful in the treatment of some urinary tract infections. The normal pH of blood is slightly alkaline (7.4) and so methenamine passes round the body unchanged. However, once it is excreted into the infected urinary tract, it encounters urine which can be acidic as a result of certain bacterial infections. Consequently, methenamine degrades to generate formaldehyde just where it is needed.

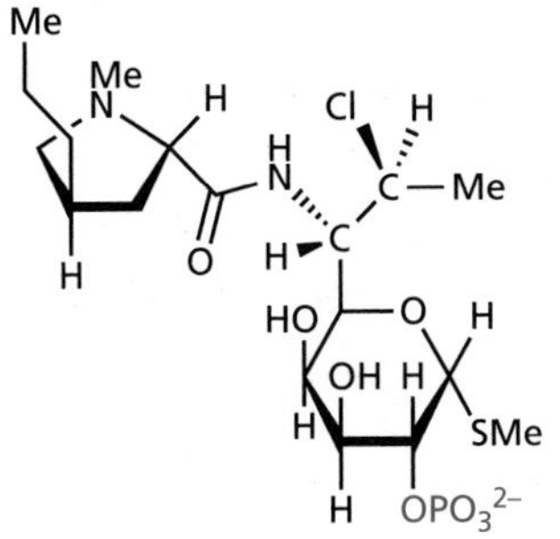

FIGURE 14.30 Clindamycin phosphate.

Polar prodrugs have been used to improve the absorption of non-polar drugs from the gut. Drugs have to have some water solubility if they are to be absorbed, otherwise they dissolve in fatty globules and fail to interact effectively with the gut wall. The steroid **estrone** is one such drug. By using a lysine ester prodrug, water solubility and absorption are increased. Hydrolysis of the prodrug releases the active drug, and the amino acid lysine as a non-toxic by-product.

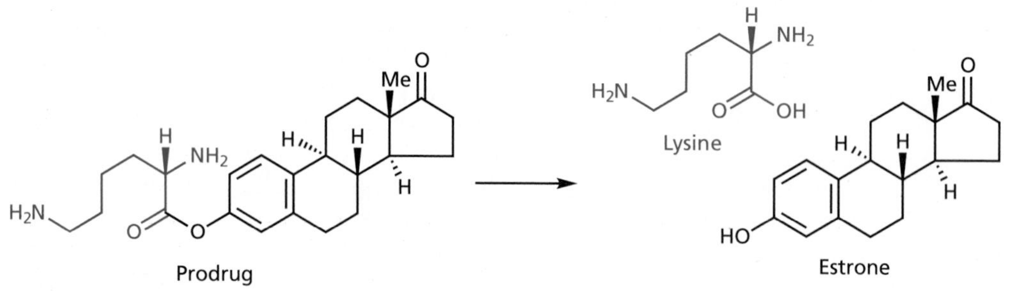

FIGURE 1 The lysine ester of estrone to improve water solubility and absorption.

Prodrugs of sulphonamides have also been used to target intestinal infections (Box 19.2). Other examples of prodrugs used to target infections are the antischistosomal drug **oxamniquine** (Case study 4) and the antiviral drugs described in sections 9.5 and 20.6.1.

The targeting of prodrugs to tumour cells by antibody-related strategies was mentioned in section 14.4.1 and is described in more detail in section 21.10. Antibody–drug conjugates can also be viewed as prodrugs and are described in that section.

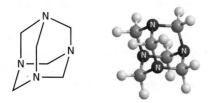

FIGURE 14.31 Methenamine.

Finally, the **proton pump inhibitors** are prodrugs which are activated by the acid conditions of the stomach (section 25.3).

14.6.7 **Prodrugs to increase chemical stability**

The antibacterial agent **ampicillin** decomposes in concentrated aqueous solution as a result of intramolecular attack of the side chain amino group on the lactam ring (section 19.5.1.8). **Hetacillin** (Fig. 14.32) is a prodrug which locks up the offending nitrogen in a ring and prevents this reaction. Once the prodrug has been administered, hetacillin slowly decomposes to release ampicillin and acetone.

In the field of antiviral agents, cyclopropane carboxylic acid esters (section 14.6.1.1) are being studied as potential prodrugs for aciclovir, in order to prolong chemical stability in solution.

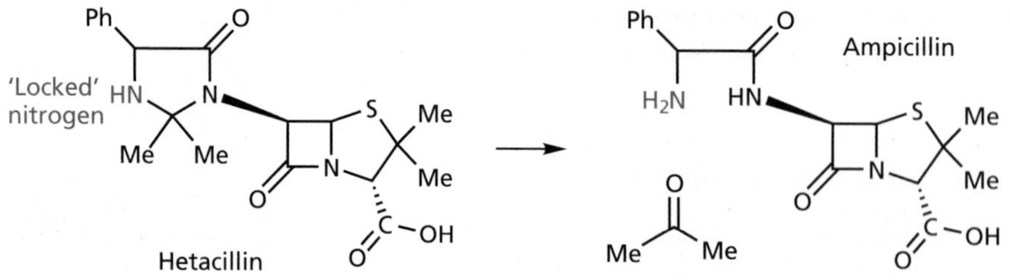

FIGURE 14.32 Hetacillin and ampicillin.

14.6.8 Prodrugs activated by external influence (sleeping agents)

Conventional prodrugs are inactive compounds which are normally metabolized in the body to the active form. A variation of the prodrug approach is the concept of a 'sleeping agent'. This is an inactive compound which is only converted to the active drug by some form of external influence. The best example of this approach is the use of photosensitizing agents such as **porphyrins** or **chlorins** in cancer treatment—a strategy known as **photodynamic therapy**. Given intravenously, these agents accumulate within cells and have some selectivity for tumour cells. By themselves, the agents have little effect, but if the cancer cells are irradiated with light, the porphyrins are converted to an excited state and react with molecular oxygen to produce highly toxic singlet oxygen. This is covered in section 21.11.

KEY POINTS

- Prodrugs are inactive compounds which are converted to active drugs in the body—usually by drug metabolism.

- Esters are commonly used as prodrugs to make a drug less polar, allowing it to cross cell membranes more easily. The nature of the ester can be altered to vary the rate of hydrolysis.

- Extended esters can be used if a simple ester is shielded by a bulky group.

- Introducing a metabolically susceptible *N*-methyl group can sometimes be advantageous in reducing polarity.

- Prodrugs with a similarity to important biosynthetic building blocks may be capable of crossing cell membranes with the aid of transport proteins.

- The activity of a drug can be prolonged by using a prodrug which is converted slowly to the active drug.

- The toxic nature of a drug can be reduced by using a prodrug which is slowly converted to the active compound, preferably at the site of action.

- Prodrugs which contain metabolically susceptible polar groups are useful in improving water solubility. They are particularly useful for drugs which have to be injected, or for drugs which are too hydrophobic for effective absorption from the gut.

- Prodrugs which are susceptible to pH or chemical degradation can be effective in targeting drugs or increasing stability in solution prior to injection.

- Prodrugs which are activated by light are the basis for photodynamic therapy.

14.7 Drug alliances

Some drugs are found to affect the activity or pharmacokinetic properties of other drugs, and this can be put to good use. The following are some examples.

14.7.1 'Sentry' drugs

In this approach, a second drug is administered with the principal drug in order to guard or assist it. Usually, the second drug inhibits an enzyme that metabolizes the principal drug. For example, **clavulanic acid** inhibits the enzyme β-**lactamase** and is therefore able to protect penicillins from that particular enzyme (sections 7.5 and 19.5.4.1).

The antiviral preparation **Kaletra**, used in the treatment of AIDS, is a combination of two drugs called **ritonavir** and **lopinavir**. Although the former has antiviral activity, it is principally present to protect lopinavir, which is metabolized by the metabolic cytochrome P450 enzyme (CYP3A4). Ritonavir is a strong inhibitor of this enzyme and so the metabolism of lopinavir is decreased allowing lower doses to be used for therapeutic plasma levels (section 20.7.4.4).

Another example is to be found in the drug therapy of Parkinson's disease. The use of **levodopa** as a prodrug for **dopamine** has already been described (section 14.6.1.3). To be effective, however, large doses of levodopa (3–8 g per day) are required, and over a period of time these dose levels lead to side effects such as nausea and vomiting. Levodopa is susceptible to the enzyme **dopa decarboxylase** and, as a result, much of the levodopa administered is decarboxylated to dopamine before it reaches the central nervous system (Fig. 14.33). This build-up of dopamine in the peripheral blood supply leads to the observed nausea and vomiting.

The drug **carbidopa** has been used successfully as an inhibitor of dopa decarboxylase and allows smaller doses of levodopa to be used. Furthermore, since it is a highly polar compound containing two phenolic groups, a hydrazine moiety, and an acidic group, it is unable to cross the blood–brain barrier, and so cannot prevent the

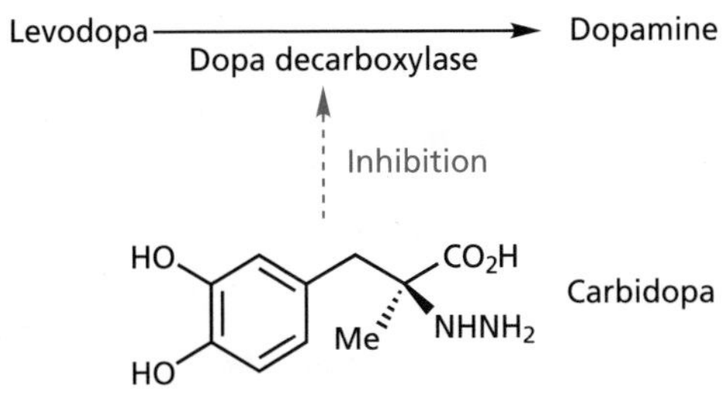

FIGURE 14.33 Inhibition of levodopa decarboxylation.

conversion of levodopa to dopamine in the brain. Carbidopa is marketed as a mixture with levodopa and is called **co-careldopa**.

Several important peptides and proteins could be used as drugs if it were not for the fact that they are quickly broken down by **protease** enzymes. One way round this problem is to inhibit the protease enzymes. **Candoxatril** (Box 14.4) is a protease inhibitor which has some potential in this respect and is under clinical evaluation.

Further examples of enzyme inhibitors that are used to block the metabolism of other drugs include **cilastatin** (Box 19.11), **tipiracil** (section 21.3.2), and **cobicistat** (section 20.7.5).

Finally, the action of penicillins can be prolonged if they are administered alongside **probenecid** (section 19.5.1.9). This agent slows the rate at which penicillins are excreted in the kidneys.

14.7.2 Localizing a drug's area of activity

Adrenaline is an example of a drug which has been used to localize the area of activity for another drug. When injected with the local anaesthetic **procaine**, adrenaline constricts the blood vessels in the vicinity of the injection, and so prevents procaine being rapidly removed from the area by the blood supply.

14.7.3 Increasing absorption

Metoclopramide (Fig. 14.34) is administered alongside analgesics in the treatment of migraine. Its function is to increase gastric motility, leading to faster absorption of the analgesic and quicker pain relief.

KEY POINTS

- A sentry drug is a drug which is administered alongside another drug to enhance the latter's activity.
- Many sentry drugs protect their partner drug by inhibiting an enzyme which acts on the latter.
- Other drugs have been used to localize the site of action of local anaesthetics, and to increase the absorption of drugs from the gastrointestinal tract.

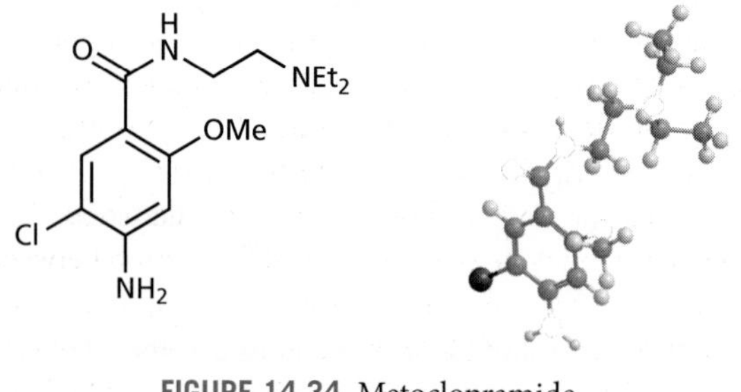

FIGURE 14.34 Metoclopramide.

14.8 Endogenous compounds as drugs

Endogenous compounds are molecules which occur naturally in the body. Many of these could be extremely useful in medicine. For example, the body's hormones are natural chemical messengers, so why not use them as medicines instead of synthetic drugs that are foreign to the body? In this section, we look at important molecules such as neurotransmitters, hormones, peptides, proteins, and antibodies, to see how feasible it is to use them as drugs.

14.8.1 Neurotransmitters

Many non-peptide neurotransmitters are simple molecules which can easily be prepared in the laboratory, so why are these not used commonly as drugs? For example, if there is a shortage of dopamine in the brain, why not administer more dopamine to make up the balance?

Unfortunately, this is not possible for a number of reasons. Many neurotransmitters are not stable enough to survive the acid of the stomach, and would have to be injected. Even if they were injected, there is little chance that they would survive to reach their target receptors. The body has efficient mechanisms which inactivate neurotransmitters as soon as they have passed on their message from nerve to target cell. Therefore, any neurotransmitter injected into the blood supply would be swiftly inactivated by enzymes, or taken up by cells via transport proteins. Even if they were not inactivated or removed, they would be poor drugs indeed, leading to many undesirable side effects. For example, the shortage of neurotransmitter may only be at one small area in the brain; the situation may be normal elsewhere. If we gave the natural neurotransmitter, how would we stop it producing an overdose of transmitter at these other sites? Of course this is a problem with all drugs, but it has been discovered that the receptors for a specific neurotransmitter are not all identical. There are different types and subtypes of a particular receptor, and their distribution around the body is not uniform. One subtype of receptor may be common in one tissue, whereas a different subtype is common in another tissue. The medicinal chemist can design synthetic drugs which take advantage of that difference, ignoring receptor subtypes which the natural neurotransmitter would not. In this respect, the medicinal chemist has actually improved on nature.

We cannot even assume that the body's own neurotransmitters are perfectly safe, and free from the horrors of tolerance and addiction associated with drugs such as **heroin**. It is quite possible to be addicted to one's own neurotransmitters and hormones. Some people are addicted to exercise, and are compelled to exercise long hours each day in order to feel good. The very process of exercise leads to the release of hormones and neurotransmitters which can produce a 'high', and this drives susceptible people to exercise more and more. If they stop exercising, they suffer withdrawal symptoms such as deep depression. The same phenomenon probably drives mountaineers into attempting feats which they know might well lead to their death. The thrill of danger produces hormones and neurotransmitters which in turn produce a 'high'. This may also explain why some individuals choose to become mercenaries and risk their lives travelling the globe in search of wars to fight.

To conclude, many of the body's own neurotransmitters are known and can be easily synthesized, but they cannot be effectively used as medicines.

14.8.2 Natural hormones, peptides, and proteins as drugs

Unlike neurotransmitters, natural hormones have potential in drug therapy as they normally circulate round the body and behave like drugs. Indeed, **adrenaline** is commonly used in medicine to treat (among other things) severe allergic reactions (section 23.10.1). Most hormones are peptides and proteins, and some naturally occurring peptide and protein hormones are already used in medicine. These include **insulin, calcitonin, erythropoietin, human growth factor, interferons**, and **colony stimulating factors**.

The availability of many protein hormones owes a great deal to genetic engineering (section 6.4). It is extremely tedious and expensive to obtain substantial quantities of these proteins by other means. For example, isolating and purifying a hormone from blood samples is impractical because of the tiny quantities of hormone present. It is far more practical to use **recombinant DNA techniques**, whereby the human genes for the protein are cloned and then incorporated into the DNA of fast-growing bacterial, yeast, or mammalian cells. These cells then produce sufficient quantities of the protein.

Using these techniques, it is possible to produce 'cut down' versions of important body proteins and polypeptides which can also be used therapeutically. For example, **teriparatide** is a polypeptide which has been approved for the treatment of osteoporosis, and was produced by recombinant DNA technology using a genetically modified strain of the bacterium *Escherichia coli*. It consists of 34 amino acids that represent the *N*-terminal end of **human parathyroid hormone** (consisting of 84 amino acids). Another recombinant protein that has been approved is **etanercept**, which is used for the treatment of rheumatoid arthritis. More than 80 polypeptide drugs

have reached the market as a result of the biotechnology revolution, with more to come. Another example is **abatacept** which was approved in 2005 for the treatment of rheumatoid arthritis. This disease is caused by T-cells binding and interacting with susceptible cells to cause cell damage and inflammation. The binding process involves a protein–protein interaction between a T-cell protein and a protein in the membrane of the susceptible cell. Abatacept is an agent which mimics the T-cell protein and binds to the susceptible cell before the T-cell does, thus preventing the damage and inflammation that would result from such an interaction. Abatacept was prepared by taking the extracellular portion of the T-cell protein and linking it to part of an antibody. Therefore, it is classed as a **fusion protein**. **Belatacept** is a very similar fusion protein that was approved in 2011 as an immunosuppressant used for extending graft survival.

Recombinant enzymes have also been produced. For example, **glucarpidase** is a carboxypeptidase enzyme which was approved in 2012. It is administered to cancer patients with failed kidneys when they are taking the anticancer drug **methotrexate**. The enzyme serves to metabolize methotrexate and prevent it from reaching toxic levels. Another recombinant enzyme that has been recently approved for the treatment of cancer patients is **rasburicase**. This is a recombinant version of the enzyme urate oxidase which catalyses the conversion of uric acid to allantoin. Uric acid can build up to toxic levels as a result of cell death from chemotherapy—a condition called **tumour lysis syndrome**. If untreated, this can result in kidney failure. The agent has also been investigated as a treatment for gout.

Other examples of recombinant enzymes used in the clinic include **imiglucerase, taliglucerase alfa**, and **velaglucerase alfa** for the treatment of **Gaucher's disease**, which is a hereditary disease caused by a deficiency in the enzyme glucocerebrosidase, and **agalsidase beta** for the treatment of **Fabry disease**. **Alglucosidase alfa** is a recombinant enzyme that is used to treat **Pompe disease**, where the patient suffers a deficiency in alpha-glucosidase, while **elsosulfase alfa** was approved in 2014 for the treatment of **Morquio syndrome**, where the patient suffers a deficiency in *N*-acetylgalactosamine-6-sulphatase.

Collagenase clostridium histolyticum is a bacterial enzyme produced by *Clostridium histolyticum* that catalyses the degradation of collagen. It has been approved for the treatment of a condition where patients cannot straighten their fingers due to a build-up of collagen in the palms of the hand. **Ocriplasmin** was given approval in 2012 for the treatment of vitreomacular adhesion. This is a condition where there is abnormal adhesion between the vitreous gel and retina of the eye. Ocriplasmin is a truncated version of plasmin—a serine protease enzyme that dissolves the proteins responsible for this condition.

In 2015, **sebelipase alfa** was approved as an orphan drug for patients deficient in the enzyme lysosomal acid lipase (**Wolman disease**). This deficiency means that fats are not properly metabolized, leading to a build-up of fat in the digestive system, the internal organs, and blood vessels. Infants suffering from this enzyme deficiency rarely survive beyond their first year due to poor absorption through the fat-coated digestive system. It is estimated that less than 0.002% of the population in Europe suffers from this deficiency, and sebelipase alfa is the first licensed treatment. The recombinant enzyme is extracted from the egg white produced by genetically modified chickens and is the first recombinant protein to be produced in that way for human medicine.

Asfatase alfa is another recombinant enzyme that was approved in 2015 to treat a rare condition where patients have a deficit of the enzyme tissue-non-specific alkaline phosphatase—an enzyme that plays an important role in bone mineralization. Such patients suffer from bone diseases such as rickets.

Despite all these successes, many endogenous peptides and proteins have proved ineffective. This is because peptides and proteins suffer serious drawbacks such as susceptibility to digestive and metabolic enzymes, poor absorption from the gut, and rapid clearance from the body. Furthermore, proteins are large molecules which could possibly induce an adverse immunological response where the body produces antibodies against the therapeutic agent.

Solutions to some of these problems are appearing, though. It has been found that linking the polymer **polyethylene glycol** (PEG) to a protein can increase the latter's solubility and stability, as well as decreasing the likelihood of an immune response (Fig. 14.35). PEGylation, as it is called, also prevents the removal of small proteins from the blood supply by the kidneys or the reticuloendothelial system. The increased size of the PEGylated protein means that it is not filtered into the kidney nephrons and remains in the blood supply.

The PEG molecules surrounding the protein can be viewed as a kind of hydrophilic, polymeric shield which both protects and disguises the protein. The PEG polymer has the added advantage that it shows little toxicity. The enzymes L-**asparaginase** and **adenosine deaminase** have

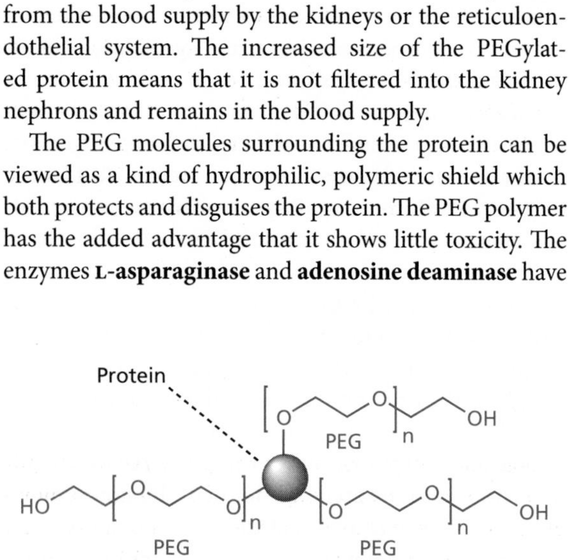

been treated in this way to give protein–PEG conjugates called **pegaspargase** and **pegademase**, which have been used for the treatment of leukaemia and **severe combined immunodeficiency** (SCID) syndrome respectively. SCID is an immunological defect associated with a lack of **adenosine deaminase**. The conjugates have longer plasma half-lives than the enzymes alone and are less likely to produce an immune response. **Interferon** has similarly been PEGylated to give a preparation called **peginterferon α2b** which is used for the treatment of hepatitis C.

Pegvisomant is the PEGylated form of **human growth hormone antagonist** and is used for the treatment of a condition known as acromegaly which results in abnormal enlargement of the skull, jaw, hands, and feet due to the excessive production of growth hormone. **Pegfilgrastim** is the PEGylated form of **filgrastim** (**recombinant human granulocyte-colony stimulating factor**), and is used as an anticancer agent. **Pegloticase** is a recombinant porcine-like uricase that has been PEGylated and was approved in 2010 for the treatment of gout. It metabolizes uric acid to allantoin. Compared to rasburicase (see above), PEGylation increases the half-life from 8 hours to 10 or 12 days and decreases the immunogenicity of the protein.

PEGylation has also been used to protect liposomes for drug delivery (section 11.10).

14.8.3 Antibodies as drugs

Biotechnology companies are producing an ever increasing number of antibodies and antibody-based drugs with the aid of genetic engineering and monoclonal antibody technology.

Because antibodies can recognize the chemical signature of a particular cell or macromolecule, they have great potential in targeting cancer cells or viruses. Alternatively, they could be used to carry drugs or poisons to specific targets (see sections 14.4.1, 20.11.5 and 21.10). Antibodies that recognize a particular antigen are generated by exposing a mouse to the antigen so that the mouse produces the desired antibodies (known as **murine antibodies**). However, the antibodies themselves are not isolated. Antibodies are produced by cells called **B lymphocytes**, and it is a mixture of B lymphocytes that is isolated from the mouse. The next task is to find the B lymphocyte responsible for producing the desired antibody. This is done by fusing the mixture with immortal (cancerous) human B lymphocytes to produce cells called **hybridomas**. These are then separated and cultured. The culture that produces the desired antibody can be identified by its ability to bind to the antigen, and is then used to produce antibody on a large scale. Since all the cells in this culture are identical, the antibodies produced are also identical and are called **monoclonal antibodies**.

FIGURE 14.35 PEGylated protein.

There was great excitement when this technology appeared in the 1980s, which spawned an expectation that antibodies would be the magic bullet to tackle many diseases. Unfortunately, the early antibodies failed to reach the clinic, because they triggered an immune response in patients which resulted in antibodies being generated against the antibodies! In hindsight, this is not surprising; the antibodies were mouse-like in character and were identified as 'foreign' by the human immune system, resulting in the production of human anti-mouse antibodies (the **HAMA response**).

In order to tackle this problem, **chimeric antibodies** have been produced which are part human (66%) and part mouse in origin, to make them less 'foreign'. Genetic engineering has also been used to generate **humanized antibodies** which are 90% human in nature. In another approach, genetic engineering has been used to insert the human genes responsible for antibodies into mice, such that the mice (transgenic mice) produce human antibodies rather than murine antibodies when they are exposed to the antigen. As a result of these efforts, a variety of antibodies have reached the clinic and are being used as antiviral and anticancer agents (sections 20.11.5 and 21.10.1), as well as lipid-lowering agents (section 26.8.6) and drug antidotes (section 26.9.1.2). Others are being used as immunosuppressants, For example, **omalizumab** is an example of a recombinant humanized monoclonal antibody which targets **immunoglobulin E** (IgE) and was approved in 2003 for the treatment of allergic asthmatic disease. It is known that exposure to allergens results in increased levels of IgE, which triggers the release of many of the chemicals responsible for the symptoms of asthma. Omalizumab works by binding to IgE and prevents it from acting in this way. More recently, **reslizumab** was approved in 2016 for the treatment of severe asthma. Another example of an immunosuppressant is **belimumab**, which was approved in 2011 for the treatment of lupus—an autoimmune disease. The antibody downgrades the immune response by inhibiting B-cell activating factor.

A number of antibodies have been marketed that target the cytokines and cytokine receptors that play an important role in inflammation. **Adalimumab** was the first fully humanized antibody to reach the market, and was approved for the treatment of rheumatoid arthritis. It works by binding to **tumour necrosis factor** (TNF-α) which is overproduced in arthritis and causes chronic inflammation. By binding to the cytokine, the antibody prevents it interacting with its receptor. The antibody can also tag cells that are producing the chemical messenger, leading to the cell's destruction by the body's immune system. **Infliximab** is another monoclonal antibody that targets TNF-α, but this is a chimeric monoclonal antibody and there is greater chance of the body developing

an immune response against it during long-term use. **Tocilizumab** is a humanized monoclonal antibody that was approved in 2010, and targets a cytokine receptor rather than a cytokine itself. To be specific, it targets the interleukin-6 receptor and prevents interleukin 6 from binding. **Secukinumab** was approved in 2015 for the treatment of psoriasis and acts by targeting the cytokine interleukin 17, while **mepolizumab** was approved in 2015 for the treatment of eosinophilic asthma and targets interleukin 5.

Vedolizumab was approved in 2014 for the treatment of ulcerative colitis and Crohn's disease. It binds to an integrin and blocks it from promoting inflammation. **Natalizumab** was approved in 2004 for the treatment of multiple sclerosis.

Ranibizumab is a fragment of the monoclonal antibody **bevacizumab** used in cancer therapy (section 21.10.1), and is used for the treatment of a condition that results in age-related vision loss. **Denosumab** is a fully humanized monoclonal antibody that was approved in 2009 for the treatment of osteoporosis, while **raxibacumab** is a human monoclonal antibody that was approved in 2012 for the treatment of inhaled anthrax.

Work on the large-scale production of antibodies has also been continuing. They have traditionally been produced using hybridoma cells in bioreactors, but more recently companies have been looking at the possibility of using transgenic animals in order to collect antibodies in milk. Another possibility is to harvest transgenic plants which produce the antibody in their leaves or seeds.

A different approach to try and prevent antibodies producing an immune response has been to treat them with PEG (section 14.8.2). Unfortunately, this tends to be counterproductive, as it prevents the antibody acting out its role as a targeting molecule. However, controlling the PEGylation such that it only occurs on the thiol group of cysteine residues could be beneficial, as it would limit the number of PEG molecules attached and make it more likely that the antibody remains functional.

An interesting idea under investigation involves coating the inside of nanotubes with antibodies that can recognize infectious agents such as viruses. It is hoped that such nanotubes could be administered to trap and remove viruses from the blood supply.

14.9 Peptides and peptidomimetics in drug design

Endogenous peptides and proteins serve as highly important lead compounds for the design of novel drugs. Current examples include renin inhibitors (Case study 8), protease inhibitors (section 20.7.4), LHRH agonists (section 21.4.2), matrix metalloproteinase

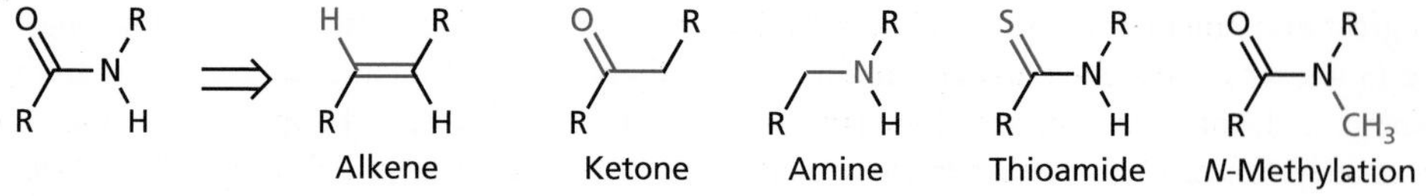

FIGURE 14.36 Examples of functional groups that might be used to replace a peptide bond.

inhibitors (section 21.7.1), and enkephalin analogues (section 24.8.2). Peptides will continue to be important lead compounds because many of the new targets in medicinal chemistry involve peptides as receptor ligands or as enzyme substrates; for example the protein kinases. Consequently, drugs which are designed from these lead compounds are commonly peptide-like in nature. The pharmacokinetic properties of these 'first-generation' drugs are often unsatisfactory, and so various strategies have been developed to try and improve bioavailability and attain more acceptable levels in the blood supply. This usually involves strategies aimed at disguising or reducing the peptide nature of the lead compound to generate a structure which is more easily absorbed from the gastrointestinal tract, and is more resistant to digestive and metabolic enzymes. Such analogues are known as **peptidomimetics**.

14.9.1 Peptidomimetics

One approach that is used to increase bioavailability is to replace a chemically or enzymatically susceptible peptide bond with a functional group that is either more stable to hydrolytic attack by peptidase enzymes, or binds less readily to the relevant active sites. For example, a peptide bond might be replaced by an alkene (Fig. 14.36). If the compound retains activity, then the alkene represents a **bio-isostere** for the peptide link. An alkene has the advantage that it mimics the double bond nature of a peptide bond and is not a substrate for peptidases. However, the peptide bonds in lead compounds are often involved in hydrogen bond interactions with the target binding site, where the NH acts as a hydrogen bond donor and the carbonyl C=O acts as a hydrogen bond acceptor. Replacing both of these groups may result in a significant drop in binding strength. Therefore, an alternative approach might be to replace the amide with a ketone or an amine, such that only one possible interaction is lost. The problem now is that the double bond nature of the original amide group is lost, resulting in greater chain flexibility and a possible drop in binding affinity (see section 13.3.9). A thioamide group is another option. This group retains the planar shape of the amide, and the NH moiety can still act as a hydrogen bond donor. The sulphur is a poor hydrogen bond acceptor, but this could be advantageous if the original

carbonyl oxygen forms a hydrogen bond to the active site of peptidase enzymes.

A different approach is to retain the amide, but to protect or disguise it. One strategy that has been used successfully is to methylate the nitrogen of the amide group. The methyl group may help to protect the amide from hydrolysis by acting as a steric shield, or prevent an important hydrogen bonding interaction taking place between the NH of the original amide and the active site of the peptidase enzyme that would normally hydrolyse it.

A second strategy is to replace an L-amino acid with the corresponding D-enantiomer (Fig. 14.37). Such a move alters the relative orientation of the side chain with respect to the rest of the molecule and can make the molecule unrecognizable to digestive or metabolic enzymes, especially if the side chain is involved in binding interactions. The drawback to this strategy is that the resulting peptidomimetic may become unrecognizable to the desired target as well.

A third strategy is to replace natural amino acid residues with unnatural ones. This is a tactic that has worked successfully in structure-based drug design where the binding interactions of the peptidomimetic and a protein target are studied by X-ray crystallography and molecular modelling. The idea is to identify binding subsites in the target binding site into which various amino acid side chains fit and bind. The residues are then replaced by groups which are designed to fit the subsites better, but which are not found on natural amino acids. This increases the binding affinity of the peptidomimetic to the target binding site, and, at the same time, makes it less recognizable to digestive and metabolic enzymes.

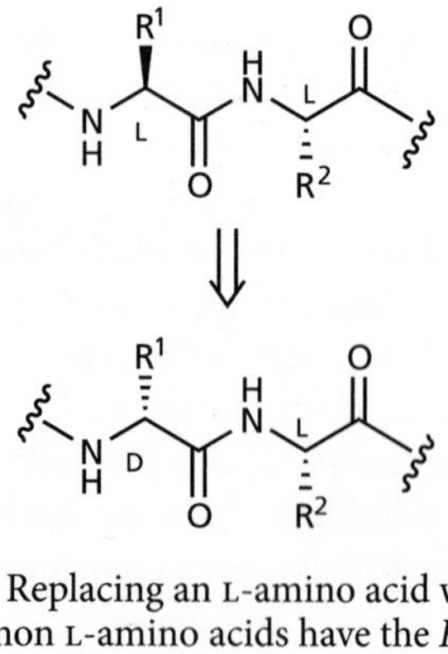

FIGURE 14.37 Replacing an L-amino acid with a D-amino acid. The common L-amino acids have the *R*-configuration except for L-cysteine which has the *S*-configuration.

For example, the lead compound for the antiviral drug **saquinavir** contained an L-proline residue that occupied a hydrophobic subsite of a viral protease enzyme. The proline residue was replaced by a decahydroisoquinoline ring which filled the hydrophobic subsite more fully, resulting in better binding interactions (Fig. 14.38; see also section 20.7.4.3).

It is even possible to design extended groups which fill two different subsites (Fig. 14.39). This means that the peptidomimetic can be pruned to a smaller molecule. The resulting decrease in molecular weight often leads to better absorption (see also Case study 8 and sections 20.7.4.6 and 20.7.4.7).

Peptidomimetics are often hydrophobic in nature, and this can pose a problem because poor water solubility may result in poor oral absorption. Water solubility can be increased by increasing the polarity of residues. For example, an aromatic ring could be replaced by a pyridine ring. However, it is important that this group is not involved in any binding interactions with the target and remains exposed to the surrounding water medium when the peptidomimetic is bound (Fig. 14.40). Otherwise, it would have to be desolvated and this would carry an energy penalty that would result in a decreased binding affinity.

Another potential problem with peptide lead compounds is that they are invariably flexible molecules with a large number of freely rotatable bonds. Flexibility has been shown to be detrimental to oral bioavailability (section 11.3), and so rigidification tactics (section 13.3.9) may well be beneficial.

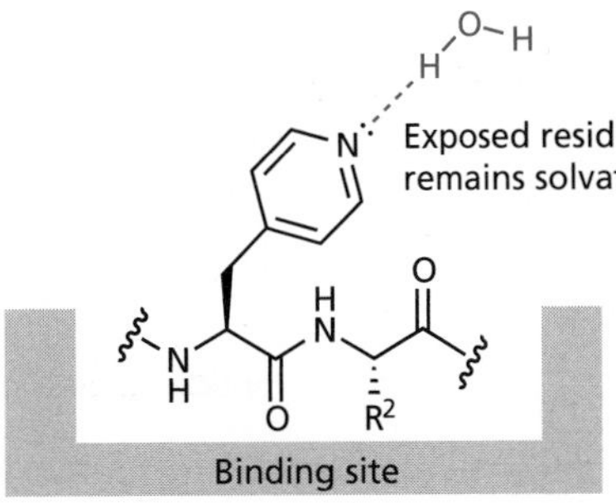

FIGURE 14.40 Altering exposed residues to increase water solubility.

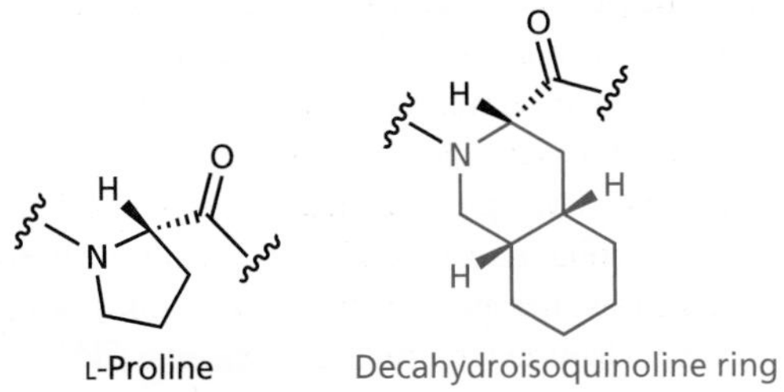

The structure-based design of various peptidomimetic enzyme inhibitors is described in sections 20.7.4 and 20.10.1, as well as Case study 10. These examples illustrate many of the principles described above. Crystal structures of peptide lead compounds bound to their target binding site are invaluable in identifying what kind of modifications are likely to be successful in achieving a particular strategy. Crystal structures also allow docking studies to be carried out to see whether planned analogues are likely to bind or not. It is even possible to use docking alongside virtual screening to modify peptide lead compounds (section 17.13).

Finally, there is current research into designing structures which mimic particular features of protein secondary structure such as α-helices, β-sheets and β-turns (section 10.5). The goal here is to design a stable molecular scaffold that contains substituents capable of mimicking the side chains of amino acids as they would be positioned in common protein features such as α-helices. This might be useful in designing peptidomimetics that mimic peptide neurotransmitters or peptide hormones. For example, it is found that such messengers adopt a helical conformation when they bind to their receptor. 1,1,6-Trisubstituted indanes have been designed to mimic three consecutive amino acid side chains in an α-helix (Fig. 14.41).

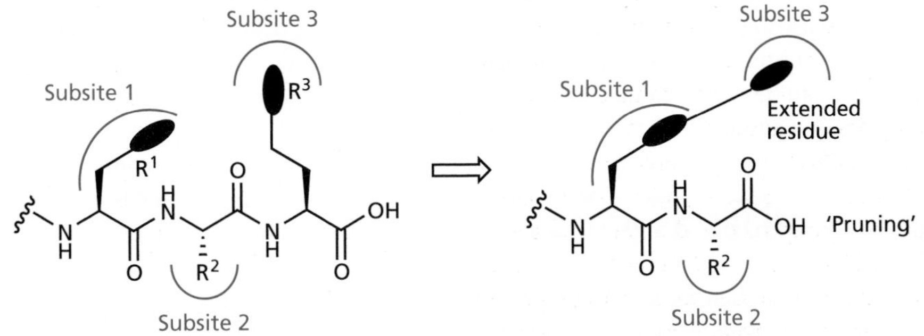

FIGURE 14.38 Replacing a natural residue with an unnatural one.

FIGURE 14.39 Extended residues.

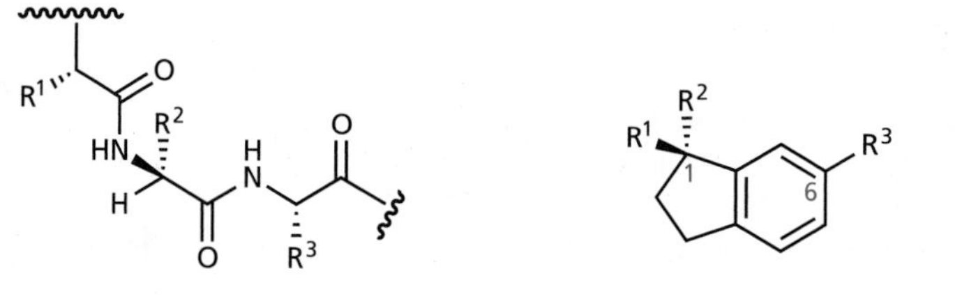

FIGURE 14.41 Trisubstituted indanes as a peptidomimetic for a tripeptide sequence in an α-helix.

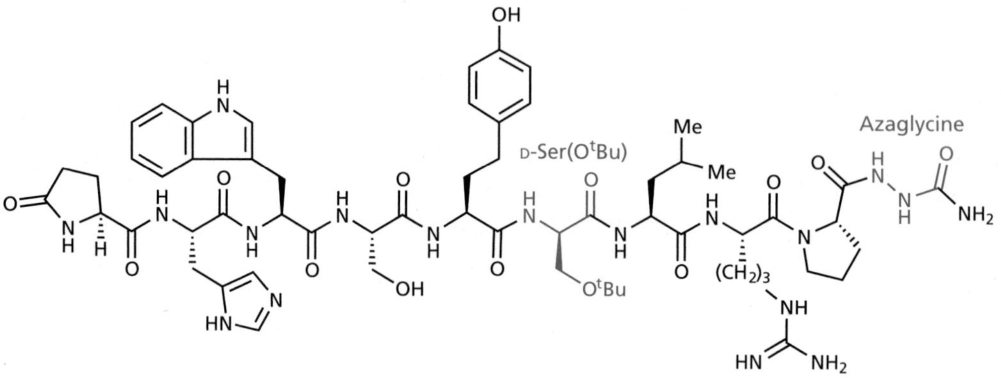

FIGURE 14.42 Goserelin (Zoladex). Moieties in blue increase metabolic resistance and receptor affinity.

14.9.2 Peptide drugs

As stated above, there is often a reluctance to use peptides as drugs because of the many pharmacokinetic difficulties that can be encountered, but this does not mean that peptide drugs have no role to play in medicinal chemistry. For example, the immunosuppressant **ciclosporin** can be administered orally (section 11.3). Another important peptide drug is **goserelin** (Fig. 14.42), which is administered as a subcutaneous implant and is used against breast and prostate cancers, earning $700 million dollars a year for its maker (section 21.4.2). In 2003, **enfuvirtide (Fuzeon)** was approved as the first of a new class of anti-HIV drugs (section 20.7.5). It is a polypeptide of 36 amino acids which is injected subcutaneously and offers another weapon in the combination therapies used against HIV. **Teriparatide**, which was mentioned in section 14.8.2, is also administered by subcutaneous injection. Peptide drugs can be useful if one chooses the right disease and method of administration.

14.10 Oligonucleotides as drugs

Oligonucleotides are being studied as **antisense drugs** and **aptamers**. The rationale and therapeutic potential

of these agents are described in sections 9.7.2 and 10.5. However, there are disadvantages to the use of oligonucleotides as drugs, as they are rapidly degraded by enzymes called **nucleases**. They are also large and highly charged, and are not easily absorbed through cell membranes. Attempts to stabilize these molecules, and to reduce their polarity, have involved modifying the phosphate linkages in the sugar–phosphate backbone. For example, phosphorothioates and methylphosphonates have been extensively studied, and oligonucleotides containing these linkages show promise as therapeutic agents (Fig. 14.43). An antisense oligonucleotide with such a modified backbone has been approved as an antiviral drug (section 20.6.3), as has one for the treatment of high cholesterol levels (section 26.8.4). Alterations to the sugar moiety have also been tried. For example, placing a methoxy group at position 2′, or using the α-anomer of a deoxyribose sugar, increases resistance to nucleases. Bases have also been modified to improve and increase the number of hydrogen bonding interactions with target nucleic acids.

The biopharmaceutical company Genta has developed an antisense drug called **oblimersen** which consists of 18 deoxynucleotides linked by a phosphorothioate backbone. It binds to the initiation codon of the messenger RNA molecule carrying the genetic instructions for **Bcl-2**. Bcl-2 is a protein which suppresses cell death

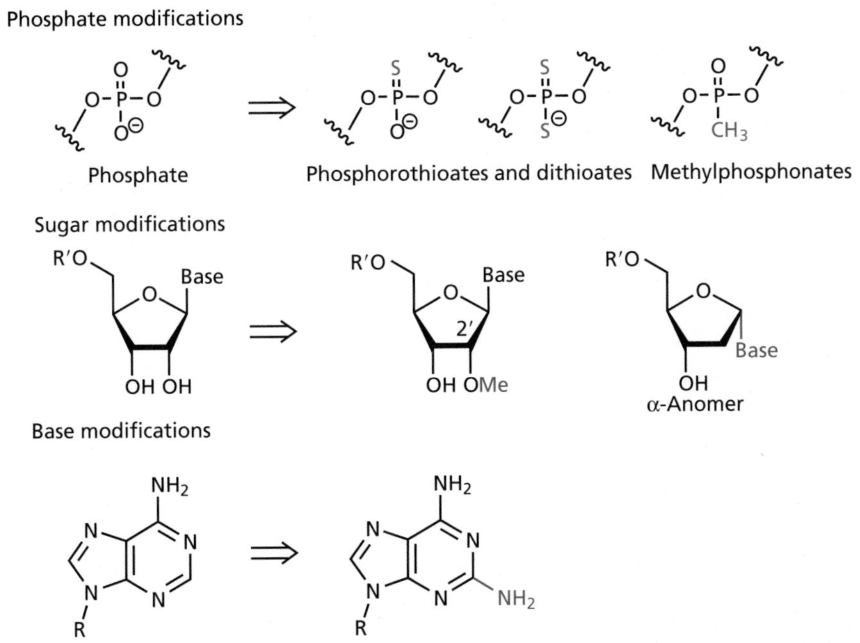

Phosphate modifications

Phosphate ⟹ Phosphorothioates and dithioates Methylphosphonates

Sugar modifications

OH OH ⟹ OH OMe OH
α-Anomer

Base modifications

FIGURE 14.43 Modifications on oligonucleotides.

(apoptosis) (section 21.1.7), and so suppressing its synthesis will increase the chances of apoptosis taking place when chemotherapy or radiotherapy is being used for the treatment of cancer. The drug is currently undergoing Phase III clinical trials in combination with the anticancer drugs **docetaxel** and **irenotecan**.

Phosphorothioate oligonucleotides are also being investigated which will target the genetic instructions for **Raf** and **PKCγ**, two proteins which are involved in signal transduction pathways (section 5.4). These oligonucleotides also have potential as anticancer drugs.

KEY POINTS

• Neurotransmitters are not effective as drugs as they have a short lifetime in the body, and have poor selectivity for the various types and subtypes of a particular target.

• Hormones are more suitable as drugs and several are used clinically. Others are susceptible to digestive or metabolic enzymes, and show poor absorption when taken orally. Adverse immune reactions are possible.

• Peptides and proteins generally suffer from poor absorption or metabolic susceptibility. Peptidomimetics are compounds that are derived from peptide lead compounds, but have been altered to disguise their peptide character.

• Many of the body's hormones are peptides and proteins and can be produced by recombinant DNA techniques. However, there are several disadvantages in using such compounds as drugs.

• Antibodies are proteins which are important to the body's immune response, and can identify foreign cells or macromolecules, marking them out for destruction. They have been used therapeutically and can also be used to carry drugs to specific targets.

• Oligonucleotides are susceptible to metabolic degradation, but can be stabilized by modifying the sugar–phosphate backbone so that they are no longer recognized by relevant enzymes.

• Antisense molecules have been designed to inhibit the mRNA molecules that code for the proteins which suppress cell death.

QUESTIONS

1. Suggest a mechanism by which methenamine (Fig. 14.31) is converted to formaldehyde under acid conditions.

2. Suggest a mechanism by which ampicillin (Fig. 14.32) decomposes in concentrated solution.

3. Carbidopa (Fig. 14.34) protects levodopa from decarboxylation in the peripheral blood supply, but is too polar to cross the blood–brain barrier into the central nervous system. Carbidopa is reasonably similar in structure to levodopa, so why can it not mimic levodopa and cross the blood–brain barrier by means of a transport protein?

4. Acetylcholine (Fig. 4.3) is a neurotransmitter that is susceptible to chemical and enzymatic hydrolysis. Suggest strategies that could be used to stabilize the ester group of acetylcholine, and show the sort of analogues which might have better stability.

5. Decamethonium is a neuromuscular blocking agent which requires both positively charged nitrogen groups to be present. Unfortunately, it is slowly metabolized and lasts too long in the body. Suggest analogues which might be expected to be metabolized more quickly and lead to inactive metabolites.

6. Miotine has been used in the treatment of a muscle-wasting disease, but there are side effects because a certain amount of the drug enters the brain. Suggest how one might modify the structure of miotine to eliminate this side effect.

7. The oral bioavailability of the antiviral drug aciclovir is only 15–30%. Suggest why this may be the case, and how one might increase the bioavailability of this drug.

Decamethonium

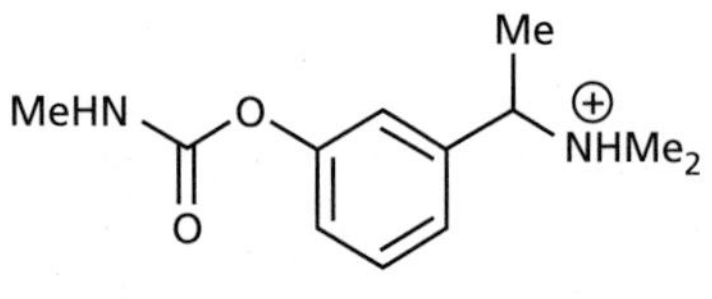

Miotine

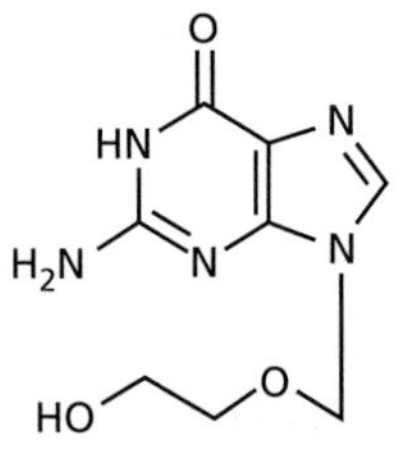

Aciclovir

8. CGP 52411 is a useful inhibitor of a protein kinase enzyme. Studies on structure–activity relationships demonstrate that substituents on the aromatic rings such as Cl, Me, or OH are bad for activity. Drug metabolism studies also show that *para*-hydroxylation occurs to produce inactive metabolites. How would you modify the structure to protect it from metabolism?

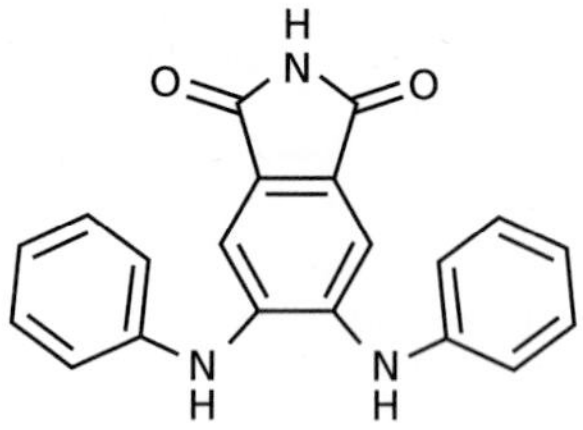

CGP52411

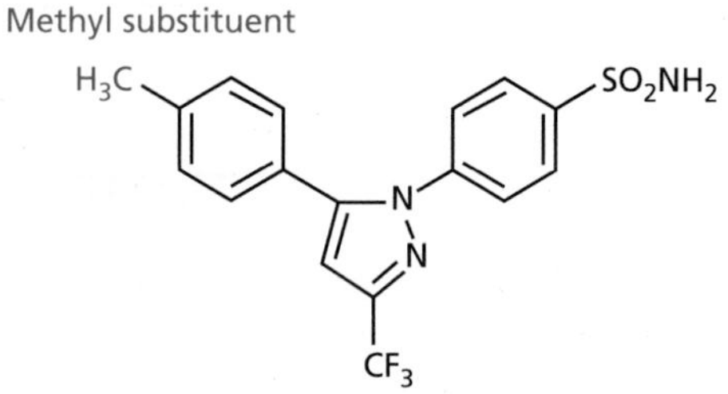

Celecoxib

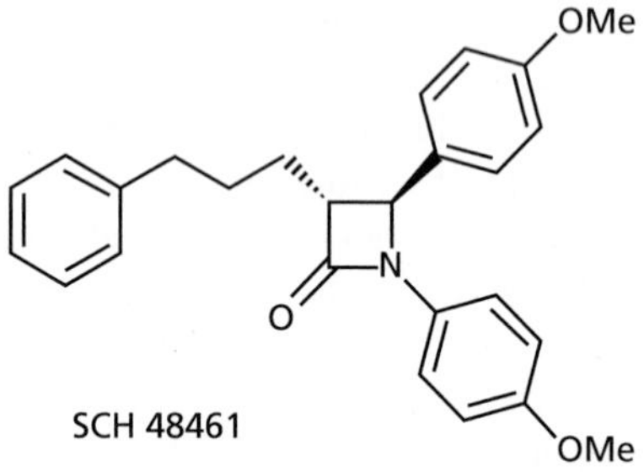

SCH 48461

9. Celecoxib is a COX-2 inhibitor and contains a methyl substituent on the phenyl ring. It is known that inhibitory activity increases if this methyl substituent is not present, or if it is replaced with a chloro substituent. However, neither of these analogues were used clinically. Why not?

10. SCH 48461 has been found to lower cholesterol levels by inhibiting cholesterol absorption. Unfortunately, it is susceptible to metabolism. Identify the likely metabolic reactions which this molecule might undergo, and what modifications could be made to reduce metabolic susceptibility.

Multiple-choice questions are available on the Online Resource Centre at www.oxfordtextbooks.co.uk/orc/patrick6e/

FURTHER READING

Berg, C., Neumeyer, K., and Kirkpatrick, P. (2003) Teriparatide. *Nature Reviews Drug Discovery*, **2**, 257–8.

Bolgnesi, M. L., et al. (2009) Alzheimer's disease: new approaches to drug discovery. *Current Opinions in Chemical Biology*, **13**, 303–8.

Burke, M. (2002) Pharmas market. *Chemistry in Britain*, June, 30–2 (antibodies).

Duncan, R. (2003) The dawning era of polymer therapeutics. *Nature Reviews Drug Discovery*, **2**, 347–60.

Ezzell, C. (2001) Magic bullets fly again. *Scientific American*, October, 28–35 (antibodies).

Harris, J. M., and Chess, R. B. (2003) Effect of pegylation on pharmaceuticals. *Nature Reviews Drug Discovery*, **2**, 214–21.

Herr, R. J. (2002) 5-Substituted-1H-tetrazoles as carboxylic acid isosteres: medicinal chemistry and synthetic methods. *Bioorganic and Medicinal Chemistry*, **10**, 3379–93.

Matthews, T., et al. (2004) Enfuvirtide: the first therapy to inhibit the entry of HIV-1 into host CD4 lymphocytes. *Nature Reviews Drug Discovery*, **3**, 215–25.

Moreland, L., Bate, G., and Kirkpatrick, P. (2006) Abatacept. *Nature Reviews Drug Discovery*, **5**, 185–6.

Opalinska, J. B., and Gewirtz, A. M. (2002) Nucleic-acid therapeutics: basic principles and recent applications. *Nature Reviews Drug Discovery*, **1**, 503–14.

Pardridge, W. M. (2002) Drug and gene targeting to the brain with molecular Trojan horses. *Nature Reviews Drug Discovery*, **1**, 131–9.

Reichert, J. M., and Dewitz, M. C. (2006) Anti-infective monoclonal antibodies: perils and promise of development. *Nature Reviews Drug Discovery*, **5**, 191–5.

Rotella, D. P. (2002) Phosphodiesterase 5 inhibitors: current status and potential applications. *Nature Reviews Drug Discovery*, **1**, 674–82.

Titles for general further reading are listed on p.845.

Getting the drug to market

The methods by which lead compounds are discovered were discussed in Chapter 12. In Chapters 13 and 14, we looked at how lead compounds can be optimized to improve their target interactions and pharmacokinetic properties. In this chapter, we look at the various issues that need to be tackled before a promising-looking drug candidate reaches the clinic and goes into full-scale production. This final phase is significantly more expensive in terms of time and money than either lead discovery or drug design, and many drugs will fall by the wayside. On average, for every 10 000 structures synthesized during drug design, 500 will reach animal testing, 10 will reach phase I clinical trials, and only 1 will reach the market place. The average overall development cost of a new drug was recently estimated as $800 million or £444 million.

Three main issues are involved in getting the drug to the market. Firstly, the drug has to be tested to ensure that it is safe and effective, and can be administered in a suitable fashion. This involves preclinical and clinical trials covering toxicity, drug metabolism, stability, formulation, and pharmacological tests. Secondly, there are the various patenting and legal issues. Thirdly, the drug has to be synthesized in ever-increasing quantities for testing and eventual manufacture. This is a field known as chemical and process development. Many of these issues have to be tackled in parallel.

15.1 Preclinical and clinical trials

15.1.1 Toxicity testing

One of the first priorities for a new drug is to test if it has any toxicity. This often starts with *in vitro* tests on genetically engineered cell cultures and/or *in vivo* testing on transgenic mice to examine any effects on cell reproduction and to identify potential carcinogens. Any signs of carcinogenicity would prevent the drug being taken any further.

The drug is also tested for acute toxicity by administering sufficiently large doses *in vivo* to produce a toxic effect or death over a short period of time. Different animal species are used in the study and the animals are dissected to test whether particular organs are affected. Further studies on acute toxicity then take place over a period of months, where the drug is administered to laboratory animals at a dose level expected to cause toxicity but not death. Blood and urine samples are analysed over that period, and then the animals are killed so that tissues can be analysed by pathologists for any sign of cell damage or cancer.

Finally, long-term toxicology tests are carried out over a period of years at lower dose levels to test the drug for chronic toxic effects, carcinogenicity, special toxicology, mutagenicity, and reproduction abnormalities.

The toxicity of a drug used to be measured by its LD_{50} value (the lethal dose required to kill 50% of a group of animals). The ratio of LD_{50} to ED_{50} (the dose required to produce the desired effect in 50% of test animals) is known as the **therapeutic ratio** or **therapeutic index**. A therapeutic ratio of 10 indicates an LD_{50}:ED_{50} ratio of 10:1. This means that a tenfold increase in the ED_{50} dose would result in a 50% death rate. The dose–response curves for a drug's therapeutic and lethal effects can be compared to determine whether the therapeutic ratio is safe or not (Fig. 15.1). Ideally, the curves should not overlap on the *x*-axis, which means that the more gradual the two slopes, the riskier the drug will be. The graph provided (Fig. 15.1) shows the therapeutic and lethal dose–response curves for a sedative. Here, a 50 mg dose of the drug will act as a sedative for 95% of the test animals, but will be lethal for 5%. Such a drug would be unacceptable, even though it is effective in 95% of cases treated.

A better measure of a drug's safety is to measure the ratio of the lethal dose for 1% of the population to the effective dose for 99% of the population. A sedative drug with the ratio LD_1:ED_{99} of 1 would be safer than the one shown in Fig. 15.1.

However, LD values and therapeutic ratios are not the best indicators of a drug's toxicity, as they fail to register any non-lethal or long-term toxic effects. Therefore,

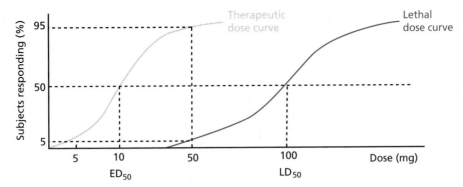

FIGURE 15.1 Comparison of therapeutic and lethal dose curves.

toxicity testing should include a large variety of different *in vitro* and *in vivo* tests designed to reveal different types of toxicity. This is not foolproof, however, and a new and unexpected toxic effect may appear during later clinical trials which will require the development of a new test. For example, when **thalidomide** was developed, nobody appreciated that drugs could cause fetal deformities, and so there was no test for this. Moreover, even if there had been such tests available, only *in vivo* tests on rabbits would have detected the potential risk.

Many promising drugs fail toxicity testing—a frustrating experience indeed for the drug design teams. For example, the antifungal agent **UK 47265** (Fig. 14.17) was an extremely promising antifungal agent, but *in vivo* tests on mice, dogs, and rats showed that it had liver toxicity and was potentially **teratogenic**. The design team had to synthesize more analogues and finally discovered the clinically useful drug **fluconazole** (section 14.5). A variety of other drugs have had to be withdrawn at a late stage in their development because they were found to have potentially serious heart effects, which could result in heart failure caused by inhibition of calcium ion channels known as **HERG K$^+$ ion channels**. As a result of this, *in vivo* and *in vitro* tests are now carried out at an early stage in drug development to detect this kind of activity (Box 12.3).

It should also be borne in mind that it is rare for a drug to be 100% pure. There are bound to be minor impurities present arising from the synthetic route used, and these may well have an influence on the toxicity of the drug. The toxicity results of a drug prepared by one synthetic route may not be the same for the same drug synthesized by a different route, and so it is important to establish the manufacturing synthesis as quickly as possible (section 15.3).

Another aim of toxicity testing is to discover what dose levels are likely to be safe for future clinical trials. Animal toxicity tests do not, however, always highlight potential problems, and the toxic properties in test animals may differ from those ultimately observed in humans. For example, clinical trials were started for the antiviral agent **fialuridine** (Fig. 15.2) after it had passed toxicity tests on animals. However, the clinical trials had to be stopped when it was found that the drug had severe liver and kidney toxicity. Half the patients (15 in total) suffered liver failure, resulting in five deaths, and the two survivors requiring liver transplants. It was later found that the drug was incorporated into mitochondrial DNA—something that was not observed in the animal toxicity tests.

Having said that, it is unlikely that the thorny problem of animal testing will disappear for a long time. There are so many variables involved in a drug's interaction with the body that it is impossible to anticipate them all. One has also to take into account that the drug will be metabolized to other compounds, all with their own range of biological properties. It appears impossible, therefore, to predict whether a potential drug will be safe by *in vitro* tests alone. Therein lies the importance of animal experiments. Only animal tests can test for the unexpected. Unless we are prepared to volunteer ourselves as guinea-pigs, animal experiments will remain an essential feature of preclinical trials for many years to come.

15.1.2 Drug metabolism studies

The body has an arsenal of metabolic enzymes that can modify foreign chemicals in such a way that they are rapidly excreted (sections 11.5–11.6). The structures formed

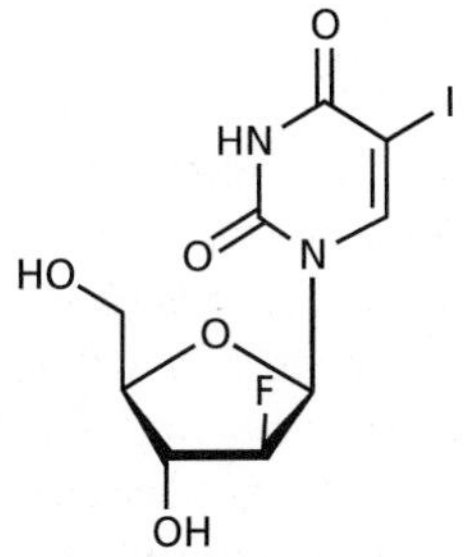

FIGURE 15.2 Fialuridine.

from these reactions are called drug metabolites, and it is important to find out what metabolites are formed from any new drug. The structure and stereochemistry of each metabolite has to be determined and the metabolite tested to see what sort of biological activity it might have. This is a safety issue, since some metabolites might prove toxic and others may have side effects that will affect the dose levels that can be used in clinical trials. Ideally, any metabolites that are formed should be inactive and quickly excreted. However, it is quite likely that they will have some form of biological activity (see Box 15.1).

In order to carry out such studies, it is necessary to synthesize the drug with an isotopic label such as deuterium (^{2}H or D), carbon-13 (^{13}C), tritium (^{3}H or T), or carbon-14 (^{14}C). This makes it easier to detect any metabolites that might be formed. Metabolites containing radioisotopes, such as ^{3}H and ^{14}C, can be detected at small levels by measuring their β radiation. Metabolites containing stable heavy isotopes, such as deuterium, can be detected by mass spectrometry or, in the case of ^{13}C, NMR spectroscopy.

Normally, a synthesis is carried out to include the isotopic label at a specific position in the molecule. It may be possible to use the established synthetic route for the drug, but, in many cases, a different route may have to be developed in order to incorporate the label in an efficient manner. Usually, it is preferable to include the label at the latest possible stage of the synthesis. It is not necessary to label every single molecule of the drug, as detection methods are sensitive enough to detect the label even if only a small proportion of the molecules are labelled.

Deuterium or tritium can be very easily incorporated into any molecule containing an exchangeable proton, such as those of an alcohol, carboxylic acid, or phenol. This is done simply by shaking a solution of the drug with D_2O or T_2O. Unfortunately, the label is just as easily lost as a result of proton exchange with water during *in vivo* testing. Therefore, it is best to carry out a synthesis that places the label on the carbon skeleton of the drug. Nevertheless, there is always the possibility that deuterium or tritium could be lost through a metabolic reaction, such that the metabolite is not detected.

Introducing a carbon isotope often means devising a different synthetic route from the normal one. The effort is often worthwhile, though, as there is less chance of the isotope being lost as a result of a metabolic reaction. Having said that, it is not impossible for an isotope to be lost in this way. For example, labelling an *N*-methyl group is asking for problems, as *N*-demethylation is a well-known metabolic reaction.

Once a labelled drug has been synthesized, a variety of *in vitro* and *in vivo* tests can be carried out. *In vivo* tests are carried out by administering the labelled drug to a test animal in the normal way, then taking blood and urine samples for analysis to see if any metabolites have been formed. For radiolabelled drugs, this can be done by using high-performance liquid chromatography (HPLC) with a radioactivity detector. It is important to

---

BOX 15.1 Drug metabolism studies and drug design

Drug metabolism studies can sometimes be useful in drug design. On several occasions, it has been found that an active drug *in vivo* is inactive *in vitro*. This is often a sign that the structure is not really active at all, but is being converted to the active drug by metabolism. The story of **oxamniquine** (Case study 4) illustrates this. Another example was the discovery that the antihypertensive structure I (Fig. 1) was less active in vitro than it was in vivo, implying that it was being converted into an active metabolite. Further studies led to the discovery that the active metabolite was **cromakalim**, which proved superior to structure I as an antihypertensive agent.

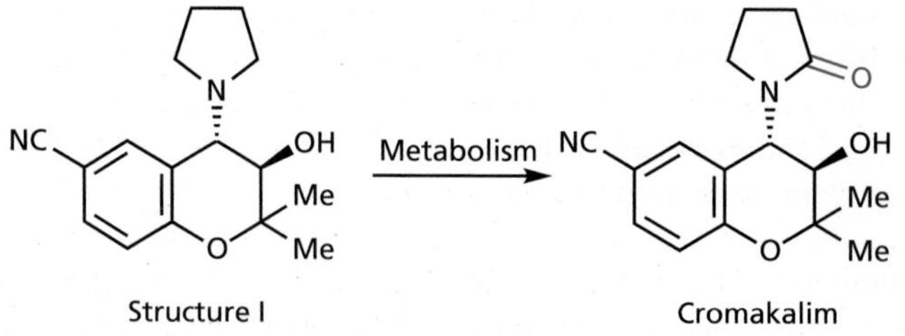

FIGURE 1 Discovery of cromakalim.

choose the correct animal for these studies, as there are significant metabolic differences across different species. *In vivo* drug metabolism tests are also carried out as part of phase I clinical trials to see whether the drug is metabolized differently in humans from any of the test animals.

In vitro drug metabolism studies can also be carried out using perfused liver systems, liver microsomal fractions, or pure enzymes. Many of the individual cytochrome P450 enzymes that are so important in drug metabolism are now commercially available.

15.1.3 **Pharmacology, formulation, and stability tests**

Although the pharmacology of the drug may have been studied during the drug discovery and drug design stages, it is usually necessary to carry out more tests to see whether the drug has activity at targets other than the intended one, and to gain a better insight into the drug's mechanism of action. These studies also determine a dose–response relationship and define the drug's duration of action.

Formulation studies involve developing a preparation of the drug which is both stable and acceptable to the patient. For orally taken drugs, this usually involves incorporating the drug into a tablet or a capsule. It is important to appreciate that a tablet contains a variety of other substances apart from the drug itself, and studies have to be carried out to ensure that the drug is compatible with these other substances. Preformulation involves the characterization of a drug's physical, chemical, and mechanical properties in order to choose what other ingredients should be used in the preparation. Formulation studies then consider such factors as particle size, salt forms, crystal polymorphism, solvates, pH, and solubility, as all of these can influence bioavailability and, hence, the activity of a drug. The drug must be combined with inactive additives by a method which ensures that the quantity of drug present is consistent in each dosage unit. The dosage should have a uniform appearance, with an acceptable taste, tablet hardness, or capsule disintegration.

It is unlikely that these studies will be complete by the time clinical trials commence. This means that simple preparations are developed initially for use in phase I clinical trials (section 15.1.4.1). These typically consist of hand-filled capsules containing a small amount of the drug and a diluent. Proof of the long-term stability of these formulations is not required, as they will be used in a matter of days. Consideration has to be given to what is called the **drug load**—the ratio of the active drug to the total contents of the dose. A low drug load may cause homogeneity problems. A high drug load may pose flow problems or require large capsules if the compound has a low bulk density.

By the time phase III clinical trials are reached (section 15.1.4.3), the formulation of the drug should have been developed to be close to the preparation that will ultimately be used in the market. A knowledge of stability is essential by this stage and conditions must have been developed to ensure that the drug is stable in the preparation. If the drug proves unstable, it will invalidate the results from clinical trials since it would be impossible to know what the administered dose actually was. Stability studies are carried out to test whether temperature, humidity, ultraviolet light, or visible light have any effect, and the preparation is analysed to see if any degradation products have been formed. It is also important to check whether there are any unwanted interactions between the preparation and the container. If a plastic container is used, tests are carried out to see whether any of the ingredients become adsorbed on to the plastic, and whether any plasticizers, lubricants, pigments, or stabilizers leach out of the plastic into the preparation. Even the adhesives for the container label need to be tested to ensure that they do not leach through the plastic container into the preparation. Despite extensive testing, there is always the possibility that an unexpected problem might occur that can result in contamination of drugs. For example, several batches of paracetamol had to be withdrawn from the market in 2009 because they had a musty smell and caused nausea and stomach pains. It was discovered that the batches were contaminated with the breakdown product of a fungicide that had been applied to the wooden pallets used to transport packaging materials.

15.1.4 **Clinical trials**

Once the preclinical studies described above have been completed, the company decides whether to proceed to clinical trials. Usually, this will happen if the drug has the desired effect in animal tests, demonstrates a distinct advantage over established therapies, and has acceptable pharmacokinetics, few metabolites, a reasonable half-life, and no serious side effects. Clinical trials are the province of the clinician rather than the scientist, but this does not mean that the research team can wash its hands of the candidate drug and concentrate on other things. Many promising drug candidates fail this final hurdle, and further analogues may need to be prepared before a clinically acceptable drug is achieved. For example, a study carried out for the period 1990–2002 showed that there was an average failure rate of 90% for the drugs that reached clinical trials. Clinical trials involve testing the drug on volunteers and patients, so the procedures involved must be ethical and beyond reproach. These trials can take 5–7 years to carry out, involve hundreds to thousands of patients, and be extremely expensive. There are four phases of clinical trials.

15.1.4.1 Phase I studies

Phase I studies take about a year and involve 100–200 volunteers. They are carried out on healthy human volunteers to provide a preliminary evaluation of the drug's safety, its pharmacokinetics, and the dose levels that can be administered, but they are not intended to demonstrate whether the drug is effective or not.

The drug is tested at different dose levels to see what levels can be tolerated. For each dose level, 6–12 subjects are given the active drug and 2–4 subjects are given a placebo. Normally, the initial dose is a tenth of the highest safest dose used in animal testing. Pharmacokinetic studies are then carried out in order to follow the drug and its metabolites. After a full safety assessment has been made, a higher dose is given, and this is continued until mild adverse effects are observed. This indicates the maximum tolerated dose and further studies will then concentrate on smaller doses.

During the study, volunteers do not take medication, caffeine, alcohol, or cigarettes. This is to avoid any complications that might arise due to drug–drug interactions (section 11.4.7). As a result, these effects may appear in later phase studies. Studies are carried out early on, however, to determine whether there are any interactions between the drug and food. This is essential in order to establish when the dose should be taken relative to meals.

Another study involves 4–8 healthy volunteers being given a radiolabelled drug in order to follow the absorption, distribution, and excretion of the drug. These studies also determine how the drug is metabolized in humans.

Studies may be carried out on special age groups. For example, drugs intended for Alzheimer's disease are tested on healthy, elderly volunteers to test the drug's pharmacokinetics in that particular population. Studies may also be carried out to test whether there are interactions with any other drugs likely to be taken by such a cohort. For example, a drug for Alzheimer's disease will be used mostly on elderly patients who are likely to be taking drugs such as diuretics or anticoagulants. Special studies may be carried out on volunteers with medical conditions that will affect the pharmacokinetics of the drug. These include patients with abnormal rates of metabolism, liver or kidney problems, inflammatory bowel disease, or other gastrointestinal diseases.

Bioavailability refers to the fraction of administered drug that reaches the blood supply in a set period of time. This can vary depending on a variety of factors, such as the crystal form of the drug, whether the drug is administered as a tablet or a capsule, or a variation in the constituents of a tablet or capsule. As a result, it is important to check that bioavailability remains the same should there be any alteration to the manufacturing, formulation, or storage processes. Such checks are called **bioequivalence** studies. For example, bioequivalence studies are required when different dosage forms are used in the early and late phases of clinical trials. Powder-filled capsules are frequently used in phase I, whereas tablets are used in phases II and III. Therefore, it is necessary to establish that these formulations show bioequivalence in healthy volunteers. In addition, it has to be demonstrated that dissolution of both formulations is similar.

In situations where the drug is potentially toxic and is to be used on a life-threatening disease such as AIDS or cancer, volunteer patients are used for phase I studies rather than healthy volunteers.

The decision on whether to proceed to phase II can be difficult, as only a limited amount of safety data is available. Any adverse effects that are observed may or may not be due to the drug. For example, abnormal liver function in a healthy patient may be due to the drug or to alcohol. Nevertheless, evidence of a serious adverse effect will usually result in clinical trials being terminated.

15.1.4.2 Phase II studies

Phase II studies generally last about 2 years and may start before phase I studies are complete. They are carried out on patients to establish whether the drug has the therapeutic property claimed, to study the pharmacokinetics and short-term safety of the drug, and to define the best dose regimen. Phase II trials can be divided into early and late studies (IIa and IIb respectively).

Initial trials (phase IIa) involve a limited number of patients to see if the drug has any therapeutic value at all, and to see if there are any obvious side effects. If the results are disappointing, clinical trials may be terminated at this stage.

Later studies (phase IIb) involve a larger number of patients. They are usually carried out as double-blind placebo-controlled studies. This means that the patients are split into two groups where one group receives the drug and the other group receives a placebo. In a double-blind study, neither the doctor nor the patient knows whether a placebo or drug is administered. In the past, it has been found that investigators can unwittingly 'give the game away' if they know which patient is getting the actual drug. The studies demonstrate whether the patients receiving the drug show an improvement relative to the patients receiving the placebo. The placebo effect can be particularly marked for patients involved in trials for novel antidepressants or anxiolytics. Different dosing levels and regimes are also determined to find the most effective. Most phase II trials require 20–80 patients per dose group to demonstrate efficacy.

Some form of rescue medication may be necessary for those patients taking a placebo. For example, it would be unethical to continue asthmatic patients on a placebo if they suffer a severe asthmatic attack. A conventional drug would be given and its use documented. The study would then compare how frequently the placebo group needed to use the rescue medicine compared with those taking the new drug. With life-threatening diseases such as AIDS or cancer, the use of a placebo is not ethical and an established drug is used as a standard comparison.

The **endpoint** is the measure that is used to determine whether a drug is successful or not. It can be any parameter that is relevant, measurable, sensitive, and ethically acceptable. Examples of endpoints include blood assays, blood pressure, tumour regression, and the disappearance of an invading pathogen from tissues or blood. Less defined endpoints include perception of pain, use of rescue medications, and level of joint stiffness.

15.1.4.3 Phase III studies

Phase III studies normally take about 3 years and can be divided into phases IIIa and IIIb. These studies may begin before phase II studies are completed. The drug is tested in the same way as in phase II, using double-blind procedures, but on a much larger sample of patients. Patients taking the drug are compared with patients taking a placebo or another available treatment. Comparative studies of this sort must be carried out without bias and this is achieved by randomly selecting the patients—those who will receive the new drug and those who will receive the alternative treatment or placebo. Nevertheless, there is always the possibility of a mismatch between the two groups with respect to factors such as age, race, sex, or disease severity, and so the greater the number of patients in the trial the better.

Phase IIIa studies establish whether the drug is really effective or whether any beneficial effects are psychological. They also allow further 'tweaking' of dose levels to achieve the optimum dose. Any side effects not previously detected may be picked up with this larger sample of patients. If the drug succeeds in passing phase IIIa, it can be registered. Phase IIIb studies are carried out after registration, but before approval. They involve a comparison of the drug with those drugs that are already established in the field.

In certain circumstances where the drug shows a clear beneficial effect early on, the phase III trials may be terminated earlier than planned. Some patients in the phase III studies will be permitted to continue taking the drug if it has proved effective, and will be monitored to assess the long-term safety of the drug. However, serious side effects observed during phase III may result in early termination of the clinical trials and the abandonment of further development. For example, the development of Pfizer's **torcetrapib** (a cholesterol-lowering agent) was terminated in 2006 when it was discovered that there was a statistically increased risk of death associated with its use. The drug had been developed over a period of 16 years at a cost of $800 million and represented one of the costliest failures in pharmaceutical history.

15.1.4.4 Phase IV studies

The drug is now placed on the market and can be prescribed, but it is still monitored for effectiveness and for any rare or unexpected side effects. In a sense, this phase is a never-ending process as unexpected side effects may crop up many years after the introduction of the drug. In the UK, the medicines committee runs a voluntary yellow card scheme where doctors and pharmacists report suspected adverse reactions to drugs. This system has revealed serious side effects for a number of drugs after they had been put on the market. For example, the β-blocker **practolol** had to be withdrawn after several years of use because some patients suffered blindness and even death. The toxic effects were unpredictable and are still not understood, and so it has not been possible to develop a test for this effect.

The diuretic agent **tienilic acid** (Fig. 15.3) had to be withdrawn from the market because it damaged liver cells in 1 out of every 10 000 patients. The anti-inflammatory agent **phenylbutazone** (Fig. 15.3) can cause a rare, but fatal, side effect in 22 patients out of every million treated with the drug! Such a rare toxic effect would clearly not be detected during phase III trials. A more recent example is **cerivastatin** (Fig. 15.3), which was marketed as a potent anticholesterol drug (Case study 1). Unfortunately, it had to be withdrawn in 2001 as a result of adverse drug–drug interactions which resulted in muscle damage and the deaths of 40 people worldwide. **Rofecoxib** (VIOXX) (Fig. 7.11) was used to treat rheumatoid arthritis for five years before a clinical trial carried out after its release showed that it was associated with increased risks of heart attack and stroke. The drug was withdrawn voluntarily by Merck in 2004, but in the 5 years it had been on the market, rofecoxib had been prescribed to 1.3 million patients in the USA and to 700 000 patients in 80 other countries. Annual profits from the drug had reached $1.2 billion, which represented 18% of Merck's net income. The loss of this income was so serious that the company's share price dropped 27% in value in a single day. Not only that, the company was faced with a lengthy litigation battle as thousands of patients sought compensation for alleged personal injuries as a result of taking the drug.

The withdrawal of drugs could potentially be avoided if the genomes of individual patients were 'fingerprinted' to establish who might be at risk from rare toxic effects—a process known as **personalized medicine** (see also section 21.1.11). For example, genetic fingerprinting has been used to determine doses for the anticoagulant drug **warfarin** (Fig. 11.7) for different individuals. Similarly, genetic fingerprinting has been used to identify those patients most likely to respond to the anticancer drug **panitumumab** (Box 21.11). It may also be possible to re-establish drugs that have previously been withdrawn if a biomarker can be developed that identifies those patients at risk. For example, the anti-inflammatory agent **lumiracoxib** (Fig. 15.3) is a selective inhibitor for the **cyclooygenase-2** enzyme and was introduced to the European market in 2006, only to be withdrawn a year later as a result of severe liver toxicity in a small number of patients. A genetic biomarker has now been identified that can determine which patients are at risk to these side effects, and there is a possibility that the drug may be reintroduced.

15.1.4.5 Ethical issues

In phases I–III of clinical trials, the permission of the patient is mandatory. However, ethical problems can still arise. For example, unconscious patients and mentally ill patients cannot give consent, but might benefit from the improved therapy. Should one include them or not? The ethical problem of including children in clinical trials is also a thorny issue, and so most clinical trials exclude them. As most licensed drugs have been licensed for adults, it means that around 40% of medicines given to children have never actually been tested on that age group. When it comes to prescribing for children, clinicians are faced with the problem of deciding what dose levels to use, and simple arithmetic mistakes made by tired health staff can have tragic consequences. Furthermore, children are not small adults. It is not a simple matter of modifying dose levels based purely on relative body weight. The pharmacodynamic and pharmacokinetic properties of a drug are significantly different in a child compared with an adult. For example, drug metabolism varies considerably with the age and development of a child. Adverse side effects also differ. The **grey baby syndrome** associated with **chloramphenicol** is one such example (Box 19.15).

Regulatory and professional bodies are actively addressing these issues in different countries worldwide. For example, the **Medicines for Children Research Network** was set up in England with the aim of carrying out high quality clinical studies on children for both established drugs and new chemical entities. In 2005, the **British National Formulary for Children** was published and there is pressure for European-wide regulation for the testing and prescribing of children's medicines. The European Medicines Agency now offers a license extension for new drugs to companies who have included testing on children. A newly created licence called the **Paediatric Use Marketing Authorisation** (PUMA) has also been introduced for established drugs.

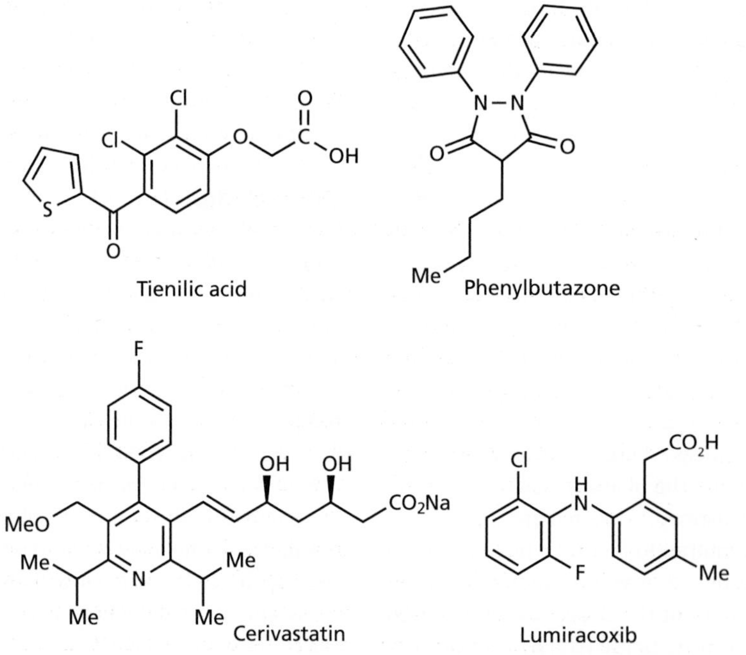

FIGURE 15.3 Drugs which have been removed from the market due to rare toxic side effects.

- Toxicity tests are carried out *in vitro* and *in vivo* on drug candidates to assess acute and chronic toxicity. During animal studies, blood and urine samples are taken for analysis. Individual organs are analysed for tissue damage or abnormalities. Toxicity testing is important in defining what the initial dose level should be for phase I clinical trials.

- Drug metabolism studies are carried out on animals and humans to identify drug metabolites. The drug candidate is labelled with an isotope to aid the detection of metabolites.

- Pharmacology tests are carried out to determine a drug's mechanism of action and to determine whether it acts at targets other than the intended one.

- Formulation studies aim to develop a preparation of the drug which can be administered during clinical trials and beyond. The drug must remain stable in the preparation under a variety of environmental conditions.

- Clinical trials involve four phases. In phase I, healthy volunteers are normally used to evaluate the drug's safety, its pharmacokinetics, and the dose levels that can safely be administered. Phase II studies are carried out on patients to assess whether the drug is effective, to give further information on the most effective dose regimen, and to identify side effects. Phase III studies are carried out on larger numbers of patients to ensure that results are statistically sound, and to detect less common side effects. Phase IV studies are ongoing and monitor the long term use of the drug in specific patients, as well as the occurrence of rare side effects.

15.2 Patenting and regulatory affairs

15.2.1 Patents

Having spent enormous amounts of time and money on research and development, a pharmaceutical company quite rightly wants to reap the benefit of all its hard work. To do so, it needs to have the exclusive rights to sell and manufacture its products for a reasonable period of time, and at a price which will not only recoup its costs, but will generate sufficient profits for further research and development. Without such rights, a competitor could synthesize the same product without suffering the expense involved in designing and developing it.

Patents allow companies the exclusive right to the use and profits of a novel pharmaceutical for a limited term. In order to gain a patent, the company has to first submit or file the patent. This should reveal what the new pharmaceutical is, what use it is intended for, and how

it can be synthesized. This is no straightforward task. Each country has its own patents, so the company has to first decide in which countries it is going to market its new drug and then file the relevant patents. Patent law is also very precise and varies from country to country. Therefore, submitting a patent is best left to the patent attorneys and lawyers who are specialists in the field. The cost and effort involved in obtaining patents from different countries can be reduced in two ways. Firstly, a patent application can be made to the **European Patent Office** (EPO). If it is approved, a European patent is granted which can then be converted to country-specific patents relevant to the individual countries belonging to the **European Patent Convention** (EPC). There are 27 such countries and the applicant can decide how many of the 27 individual patents should be taken out. A second approach is to file an international application which designates one or more of the 122 countries who have signed up to the **Patent Cooperation Treaty** (PCT). An **International Search Report** (ISR) and **International Preliminary Examination Report** (IPER) can be obtained, which can then be used when applying for patents from individual countries. No PCT or international patents are awarded, but the reports received help the applicant decide which patent applications to individual countries are likely to succeed.

Once a patent has been filed, the patent authorities decide whether the claims are novel and whether they satisfy the necessary requirements for that patent body. One universal golden rule is that the information supplied has not previously been revealed, either in print or by word of mouth. As a result, pharmaceutical companies only reveal their work after the structures involved have been safely patented.

It is important that a patent is filed as soon as possible. Such is the competition between the pharmaceutical companies that it is highly likely that a novel agent discovered by one company may be discovered by a rival company only weeks or months later. This means that patents are filed as soon as a novel agent or series of agents is found to have significant activity. Usually, the patent is filed before the research team has had the chance to start all the extensive preclinical tests that need to be carried out on novel drugs. It may not even have synthesized all the possible structures it is intending to make. Therefore, the team is in no position to identify which specific compound in a series of structures is likely to be the best drug candidate. As a result, most patents are designed to cover a series of compounds belonging to a particular structural class, rather than one specific structure. Even if a specific structure has been identified as the best drug candidate, it is best to write the patent to cover a series of analogues. This prevents a rival company making a close analogue of the specified structure and selling it in competition. All the structures that are to be protected by the patent should be specified, but only a representative few need to be described in detail.

Patents in most countries run for 20 years after the date of filing. This sounds a reasonable time span, but it has to be remembered that the protection period starts from the time of filing, not from when the drug comes onto the market. A significant period of patent protection is lost due to the time required for preclinical tests, clinical trials, and regulatory approval. This often involves a period of 6–10 years. In some cases, this period may threaten to be even longer, in which case the company may decide to abandon the project as the duration of patent protection would be deemed too short to make sufficient profits. This illustrates the point that not all patents lead to a commercially successful product.

The income obtained from a successful drug is so important to a company's financial viability that **pay-for-delay deals** have become a growing trend in the pharmaceutical sector for drugs that are nearing the end of their patent lifetime. These involve a pharmaceutical company making a deal with a manufacturer that specializes in producing off-patent drugs or **generics**. The generic manufacturer receives a lump sum if it agrees to delay manufacturing the generic version for an agreed time period (typically a year), allowing the inventor to gain several months of additional income. One reason for this trend was the fact that the patents of many profitable small drug molecules ran out during 2010–14. As there were fewer drugs reaching the market to take their place, this became known as the **patent cliff** and caused many large pharmaceutical companies to rethink their business strategy. Many firms now outsource projects to other companies, and concentrate more on research into novel drug delivery processes or combination therapies using existing drugs.

Patents can be taken out to cover specific products, the medicinal use of the products, the synthesis of the products, or preferably all three of these aspects. Taking out a patent which only covers the synthesis of a novel product offers poor patent protection. A rival company could quite feasibly develop a different synthesis to the same structure and then sell it legally.

One of the issues that affected the patent area in recent years was **chiral switching**. In the period 1983–87, 30% of approved drugs were pure enantiomers, 29% were racemates, and 41% were achiral. Nowadays, most of the drugs reaching the market are either achiral or pure enantiomers. The problem with racemates is that each enantiomer usually has a different level of activity. Moreover, the enantiomers often differ in the way they are metabolized and in their side effects. Consequently, it is better to market the pure enantiomer rather than the racemate. The issue of chiral switching related mostly to racemic

drugs that had been on the market for several years and were approaching the end of their patent life. By switching to the pure enantiomer, companies could argue that it was a new invention and take out a new patent. Timing was important and the company ideally wanted to have the pure enantiomer reaching the market just as the patent on the original racemate expired. However, they had to prove that the pure enantiomer was an improvement on the original racemate and that they could not reasonably have been expected to know that when the racemate was originally patented. A full appreciation of how different enantiomers can have different biological properties was realized in the 1980s, and so chiral switches have normally been carried out on racemic drugs that reached the market before that date. It is not possible to patent a new drug in its racemic form today then expect to market it later as the pure enantiomer, as the issue of stereoisomerism is now an established fact.

Bupivacaine is an example of an established chiral drug that has undergone chiral switching. It is a long-lasting local anaesthetic used as a spinal and epidural anaesthetic for childbirth and hip replacements, and acts by blocking sodium ion channels in nerve axons. Unfortunately, it also affects the heart, which prevents it being used for intravenous injections. The *S*-enantiomer of bupivacaine is called **levobupivacaine** (Fig. 15.4). It has less severe side effects and is a safer local anaesthetic. Other examples of drugs that have undergone chiral switching include **salbutamol** (section 23.10.3) and **omeprazole** (section 25.3). **Armodafinil** (Fig. 15.4) is another example of chiral switching and was approved in 2007 for the treatment of excess sleepiness. It is an enantiomer of the racemic drug **modafinil.**

New patents can also be taken out on an existing drug if it is marketed as a new salt, or if the formulation is altered in a significant way. Finally, there may be scope for pharmaceutical companies to develop patented drugs in veterinary medicine, rather than human medicine.

A looming threat to the patents and profits of the pharmaceutical industry relates to rules called the **Trade Related Aspects of Intellectual Property Rights** (TRIPS), which are set out by the **World Trade Organization** (WTO). Under these rules a country can grant a generics company a compulsory license to produce a patented drug if that country faces a national health crisis. In 2012, India used these rules to allow one of its generics companies to produce the patented anticancer drug **sorafenib** at a fraction of its current cost, because it was seen as a life-saving drug. This might set a precedent which would result in any novel life-saving drug being prone to similar compulsory licenses. If so, there is a risk that the pharmaceutical industry would cut back its commitment to the design of novel anticancer drugs.

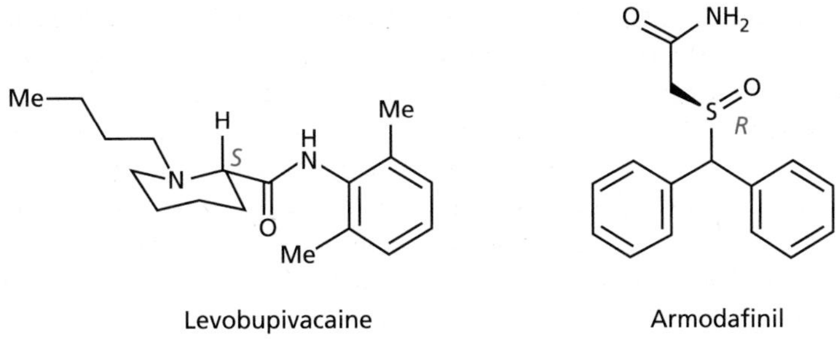

FIGURE 15.4 Levobupivacaine and armodafinil.

15.2.2 Regulatory affairs

15.2.2.1 The regulatory process

Regulatory bodies such as the **Food and Drug Administration** (**FDA**) in the USA and the **European Agency for the Evaluation of Medicinal Products** (**EMEA**) in Europe come into play as soon as a pharmaceutical company believes it has a useful drug. Before clinical trials can begin, the company has to submit the results of its scientific and preclinical studies to the relevant regulatory authority. In the USA, this takes the form of an **Investigational Exemption to a New Drug Application** (**IND**), which is a confidential document submitted to the FDA. The IND should contain information regarding the chemistry, manufacture, and quality control of the drug, as well as information on its pharmacology, pharmacokinetics, and toxicology. The FDA assesses this information and then decides whether clinical trials can begin. Dialogue then continues between the FDA and the company throughout the clinical trials. Any adverse results must be reported to the FDA who will discuss with the company whether the trials should be stopped.

If the clinical trials proceed smoothly, the company applies to the regulatory authority for marketing approval. In the USA, this involves the submission of a **New Drug Application** (**NDA**) to the FDA; in Europe, the equivalent submission is called a **Marketing Authorisation Application** (**MAA**). An NDA or MAA is typically 400–700 volumes in size with each volume containing 400 pages! The application has to state what the drug is intended to do, along with scientific and clinical evidence for its efficacy and safety. It should also give details of the chemistry and manufacture of the drug, as well as the controls and analysis which will be in place to ensure that the drug has a consistent quality. Any advertising and marketing material must be submitted to ensure that it makes accurate claims and that the drug is being promoted for its intended use. The labelling of a drug preparation must also be approved to ensure that it instructs physicians about the mechanism of action of the drug, the medical situations for which it should be used, and the correct dosing levels and frequency. Possible side effects, toxicity,

or addictive effects should be detailed, as well as special precautions which might need to be taken (e.g. avoiding drugs that interact with the preparation).

The FDA has inspectors who will visit clinical investigators to ensure that their records are consistent with those provided in the NDA, and that patients have been adequately protected. An approval letter is finally given to the company and the product can be launched, but the FDA will continue to monitor the promotion of the product, as well as further information regarding any unusual side effects.

Once an NDA is approved, any modifications to a drug's manufacturing synthesis or analysis must be approved. In practice, this means that the manufacturer will stick with the route described in the NDA and perfect that, rather than consider alternative routes.

An abbreviated NDA can be filed by manufacturers who wish to market a generic variation of an approved drug whose patent life has expired. The manufacturer is only required to submit chemistry and manufacturing information, and demonstrate that the product is comparable with the product already approved.

The term **new chemical entity** (**NCE**) or **new molecular entity** (**NME**) refers to a novel drug structure. In the 1960s, about 70 NCEs reached the market each year, but this had dropped to less than 30 per year by 1971. In part, this was due to more stringent testing regulations that were brought in after the thalidomide disaster. Another factor was the decreasing number of lead compounds available at the time. Since the 1970s, there has been an emphasis on understanding the mechanism of a disease and designing new drugs in a scientific fashion. Although the approach has certainly been more scientific than the previous trial and error approach, the number of NCEs reaching the market was still low. In 2002, only 18 NCEs were approved by the FDA and 13 by the EMEA. In recent years, the situation has improved, helped by a significant increase in the number of biologics that have reached the market. In 2014, 41 new therapeutics were approved, including 11 biologics. Biologics are drugs such as antibodies and recombinant proteins.

As far as small molecular weight drugs are concerned, the last five years has seen a flurry of antiviral agents for the treatment of hepatitis C, and kinase inhibitors for the treatment of various cancers.

15.2.2.2 Fast tracking and orphan drugs

The regulations of many regulatory bodies include the possibility of **fast tracking** certain types of drug, so that they reach the market as quickly as possible. Fast tracking is made possible by requiring a smaller number of phase II and phase III clinical trials before the drug is put forward for approval. Fast tracking is carried out for drugs that show promise for diseases where no current therapy exists, and for drugs that show distinct advantages over existing ones in the treatment of life-threatening diseases such as cancer. An example of a fast tracked drug is **oseltamivir (Tamiflu)** for the treatment of flu (section 20.8.3.4).

Orphan drugs are drugs that are effective against relatively rare medical problems. In the USA, an orphan drug is defined as one that is used for less than 200 000 people. Because there is a smaller market for such drugs, pharmaceutical companies may be less likely to reap huge financial benefits and may decide not to develop and market an orphan drug. Therefore, financial and commercial incentives are given to firms in order to encourage the development and marketing of such drugs. The attitude of pharmaceutical companies towards orphan drugs has changed following the therapeutic and financial success of **imatinib**—an orphan drug which was designed against a specific form of cancer (section 21.6.2.2). Although there are not many patients for a specific orphan disease, it is estimated that tens of millions of patients in Europe suffer from some form of orphan disease, and so there is a significant market to be tapped.

15.2.2.3 Good laboratory, manufacturing, and clinical practice

Good Laboratory Practice (GLP) and **Good Manufacturing Practice (GMP)** are scientific codes of practice for a pharmaceutical company's laboratories and production plants. They detail the scientific standards that are necessary, and the company must prove to regulatory bodies that it is adhering to these standards.

GLP regulations apply to the various research laboratories involved in pharmacology, drug metabolism, and toxicology studies. GMP regulations apply to the production plant and chemical development laboratories. They encompass the various manufacturing procedures used in the production of the drug, as well as the procedures used to ensure that the product is of a consistently high quality.

As part of GMP regulations, the pharmaceutical company is required to set up an independent quality control unit which monitors a wide range of factors including employee training, the working environment, operational procedures, instrument calibration, batch storage, labelling, and the quality control of all solvents, intermediates, and reagents used in the process. The analytical procedures which are used to test the final product must be defined, as well as the specifications that have to be met. Each batch of drug that is produced must be sampled to ensure that it passes those specifications. Written operational instructions must be in place for all special equipment (e.g. freeze dryers), and standard operating procedures (SOPs) must be written for the use, calibration, and maintenance of equipment.

Detailed and accurate paperwork on the above procedures must be available for inspection by the regulatory bodies. This includes calibration and maintenance records, production reviews, batch records, master production records, inventories, analytical reports, equipment cleaning logs, batch recalls, and customer complaints. Although record keeping is crucial, it is possible that the extra paperwork involved can stifle innovations in the production process.

Investigators involved in clinical research must demonstrate that they can carry out the work according to **Good Clinical Practice (GCP)** regulations. The regulations require proper staffing, facilities, and equipment for the required work, and each test site involved must be approved. There must also be evidence that a patient's rights and well-being are properly protected. In the USA, approval is given by the **Institutional Review Board (IRB)**. While the work is in progress, regulatory authorities may carry out data audits to ensure that no research misconduct is taking place (e.g. plagiarism, falsification of data, poor research procedures, etc.). In the UK, the **General Medical Council** or the **Association of the British Pharmaceutical Industry** can discipline unethical researchers. Problems can arise due to the pressures which are often placed on researchers to obtain their results as speedily as possible. This can lead to hasty decisions, which result in mistakes and poorly thought-out procedures. There have also been individuals who have deliberately falsified results or have cut corners. Sometimes personal relationships can prove a problem. The investigator can be faced with a difficult dilemma between doing the best for the patient, and maintaining good research procedures. Patients may also mislead clinicians if they are desperate for a new cure and falsify their actual condition in order to take part in the trial. Other patients have been known to get their drugs analysed to see if they are getting a placebo or the drug.

15.2.2.4 Analysis of cost versus benefits

A medicine that is successfully licensed and reaches the market has one other hurdle to negotiate—a cost versus benefit analysis carried out by individual government authorities. For example, the UK's **National Institute for Health and Clinical Excellence** (**NICE**) determines whether novel drugs should be used by the National Health Service. The decisions of NICE have a significant economic impact on world-wide pharmaceutical sales, as more than 60 other countries adopt the NICE guidelines rather than carry out their own cost/benefit analysis.

KEY POINTS

- Patents are taken out as soon as a useful drug has been identified. They cover a structural class of compounds rather than a single structure.

- A significant period of the patent is lost as a result of the time taken to get a drug through clinical trials to the market place.

- Patents can cover structures, their medicinal use, and their method of synthesis.

- Regulatory bodies are responsible for approving the start of clinical trials and the licensing of new drugs for the market place.

- Drugs may be fast tracked if they show promise in a field which is devoid of a current therapy.

- Special incentives are given to companies to develop orphan drugs. These are drugs that are effective in rare diseases.

- Pharmaceutical companies are required to abide by professional codes of practice known as Good Laboratory Practice, Good Manufacturing Practice, and Good Clinical Practice.

15.3 **Chemical and process development**

15.3.1 **Chemical development**

Once a compound goes forward for preclinical tests, it is necessary to start the development of a large-scale synthesis as soon as possible. This is known as chemical development and is carried out in specialist laboratories. To begin with, a quantity of the drug may be obtained by scaling up the synthetic route used by the research laboratories. In the longer term, however, such routes often prove unsuitable for large-scale manufacture. There are several reasons for this. During the drug discovery/design phase, the emphasis is on producing as many different compounds in as short a period of time as possible.

The yield is unimportant as long as sufficient material is obtained for testing. The reactions are also done on a small scale, which means that the cost is trivial, even if expensive reagents or starting materials are used. Hazardous reagents, solvents, or starting materials can also be used because of the small quantities involved.

The priorities in chemical development are quite different. A synthetic route has to be devised which is straightforward, safe, cheap, efficient, high yielding, has the minimum number of synthetic steps, and will provide a consistently high-quality product which meets predetermined specifications of purity.

During chemical development, the conditions for each reaction in the synthetic route are closely studied and modified in order to get the best yields and purity. Different solvents, reagents, and catalysts may be tried. The effects of temperature, pressure, reaction time, excess reagent or reactant, concentration, and method of addition are studied. Consideration is also given to the priorities required for scale up. For example, the original synthesis of aspirin from salicylic acid involved acetylation with acetyl chloride (Fig. 15.5). Unfortunately, the by-product from this is hydrochloric acid, which is corrosive and environmentally hazardous. A better synthesis involves acetic anhydride as the acylating agent. The by-product formed here is acetic acid, which does not have the unwanted properties of hydrochloric acid and can also be recycled.

Therefore, the final reaction conditions for each stage of the synthesis may be radically different from the original conditions, and it may even be necessary to abandon the original synthesis and devise a completely different route (Box 15.3).

Once the reaction conditions for each stage have been optimized, the process needs to be scaled up. The priorities here are cost, safety, purity, and yield. Expensive or hazardous solvents or chemicals should be avoided and replaced by cheaper, safer alternatives. Experimental procedures may have to be modified. Several operations carried out on a research scale are impractical on large scale. These include the use of drying agents, rotary evaporators, and separating funnels. Alternative large-scale procedures for these operations are, respectively, removing water as an azeotrope, distillation, and stirring the different phases.

There are several stages in chemical development. In the first stage, about a kilogram of drug is required for short-term toxicology and stability tests, analytical research, and pharmaceutical development. Often, the original synthetic route will be developed quickly and scaled up in order to produce this quantity of material, as time is of the essence. The next stage is to produce about 10 kg for long-term toxicology tests, as well as for formulation studies. Some of the material may also be used for phase I clinical trials. The third stage involves a further

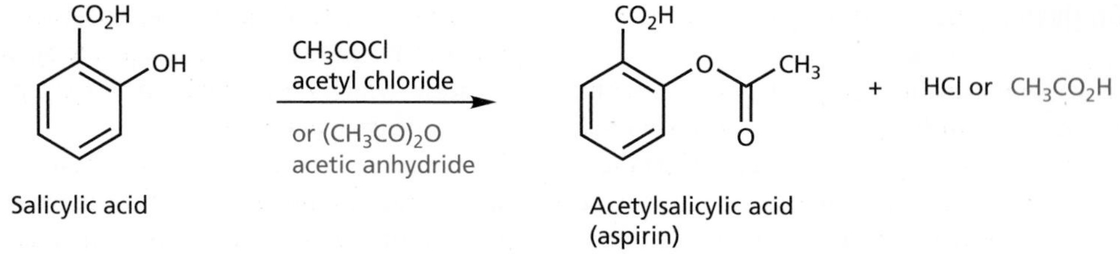

FIGURE 15.5 Synthesis of aspirin.

scale up to the pilot plant, where about 100 kg is prepared for phase II and phase III clinical trials.

Because of the time scales involved, the chemical process used to synthesize the drug during stage 1 may differ markedly from that used in stage 3. However, it is important that the quality and purity of the drug remains as constant as possible for all the studies carried out. Therefore, an early priority in chemical development is to optimize the final step of the synthesis and to develop a purification procedure which will consistently give a high-quality product. The **specifications** of the final product are defined and determine the various analytical tests and purity standards required. These define predetermined limits for a range of properties such as melting point, colour of solution, particle size, polymorphism, and pH. The product's chemical and stereochemical purities must also be defined, and the presence of any impurities or solvent should be identified and quantified if they are present at a level greater than 1%. Acceptable limits for different compounds are proportional to their toxicity. For example, the specifications for ethanol, methanol, mercury, sodium, and lead are 2%, 0.05%, 1 ppm, 300 ppm, and 2 ppm respectively. Carcinogenic compounds such as benzene or chloroform should be completely absent, which means, in practice, that they must not be used as solvents or reagents in the final stages of the synthesis.

All future batches of the drug must meet these specifications. Once the final stages have been optimized, future

BOX 15.2 Synthesis of ebalzotan

Ebalzotan is an antidepressant drug produced by Astra which works as a selective serotonin (5-HT$_{1A}$) antagonist. The original synthesis from structure I involved 6 steps and included several expensive and hazardous reagents, resulting in a paltry overall yield of 3.7%. Development of the route involved the replacement of 'problem' reagents and optimization of the reaction conditions, leading to an increase in the overall yield to 15%. Thus, the expensive and potentially toxic reducing agent sodium cyanoborohydride was replaced with hydrogen gas over a palladium catalyst. In the demethylation step, BBr$_3$, which is corrosive, toxic, and expensive, was replaced with the cheaper and less toxic HBr.

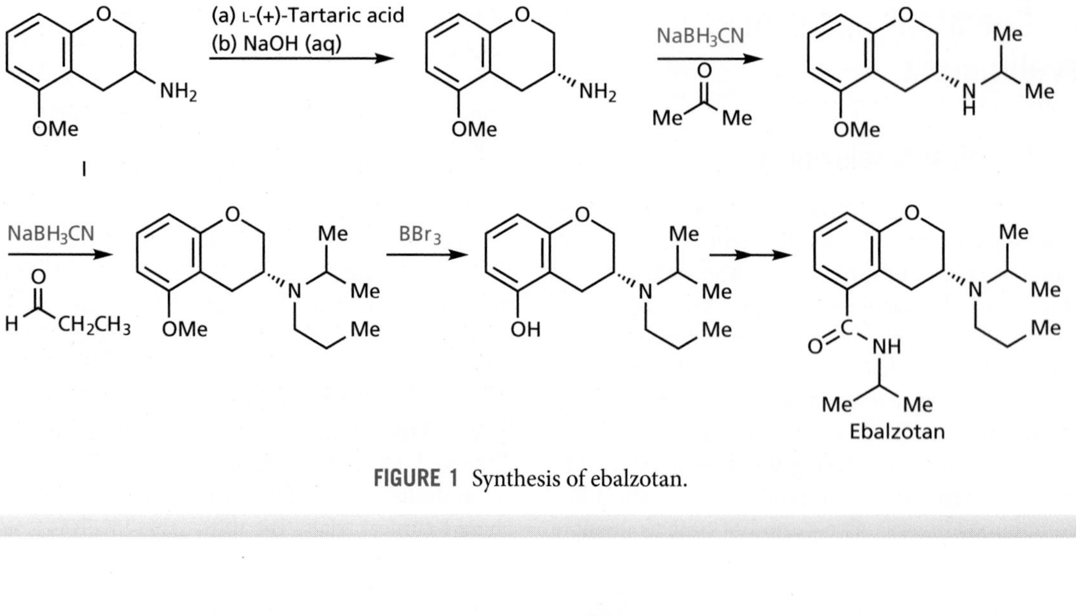

FIGURE 1 Synthesis of ebalzotan.

development work can then look to optimize or alter the earlier stages of the synthesis (Box 15.2).

In some development programmes, the structure originally identified as the most promising clinical prospect may be supplanted by another structure that demonstrates better properties. The new structure may be a close analogue of the original compound, but such a change can have radical effects on chemical development and require totally different conditions to maximize the yields for each synthetic step.

15.3.2 Process development

Process development aims to ensure that the number of reactions in the synthetic route is as small as possible and that all the individual stages in the process are integrated with each other, such that the full synthesis runs smoothly and efficiently on a production scale. The aim is to reduce the number of operations to a minimum. For example, rather than isolating each intermediate in the synthetic sequence, it is better to move it in solution directly from one reaction vessel to the next for the subsequent step. Ideally, the only purification step carried out is on the final product.

Environmental and safety issues are extremely important. Care is taken to minimize the risk of chemicals escaping into the surrounding environment, and chemical recycling is carried out as much as possible. The use of 'green technology' such as electrochemistry, photochemistry, ultrasound, or microwaves may solve potential environmental problems.

Keeping costs low is a high priority and it is more economic to run the process such that a small number of large batches are produced rather than a large number of small batches. Extreme care must be taken over safety. Any accident in a production plant has the potential to be a major disaster, so there must be strict adherence to safety procedures, and close monitoring of the process when it is running. However, the overriding priority is that the final product should still be produced with a consistently high purity in order to meet the required specifications.

Process development is very much aimed at optimizing the process for a specific compound (Box 15.3). If the original structure is abandoned in favour of a different analogue, the process may have to be rethought completely.

The regulatory authorities require that every batch of a drug product is analysed to ensure that it meets the required specifications, and that all impurities present at more than 1% are characterized, identified, and quantified. The identification of impurities is known as **impurity profiling** and typically involves their isolation by preparative HPLC, followed by NMR spectroscopy or mass spectrometry to identify their structure.

BOX 15.3 Synthesis of ICI D7114

ICI D7114 is an agonist at adrenergic β_3-receptors and was developed for the treatment of obesity and non-insulin-dependent diabetes. The original synthetic route used in the research laboratory is shown in Fig. 1.

The overall yield was only 1.1% and there were various problems in applying the route to the production scale. The first reaction involves hydroquinone and ethylene dibromide, both of which could react twice to produce side products. Moreover, ethylene dibromide is a carcinogen, and toxic vinyl bromide is generated as a volatile side product during the reaction. A chromatographic separation was required after the second stage to remove a side product—a process which is best avoided on large scale. The use of high-pressure hydrogenation at 20 times atmospheric pressure to give structure III was not possible on the plant scale, as the equipment available could only achieve 500 kPa (about five times atmospheric pressure). Finally, the product has an asymmetric centre and so it was necessary to carry out a resolution. This involved forming a salt with a chiral acid and carrying out eight crystallizations—a process that would be totally unsuitable at production scale.

The revised synthetic route shown in Fig. 2 avoided these problems and improved the overall yield to 33%. To avoid the possibility of any dialkylated side product being formed, *para*-benzyloxyphenol was used as starting material. Ethylene dimesylate was used in place of the carcinogenic ethylene dibromide. As a result, vinyl bromide was no longer generated as a side product. The alkylated product (IV) was not isolated, but was treated *in situ* with benzylamine, thus cutting down the number of operations involved. Hydrogenolysis of the benzyl ether group was carried out in the presence of methanesulphonic acid, the latter helping to prevent hydrogenolysis of the *N*-benzyl group. This gave structure II which was one of the intermediates in the original synthesis. Alkylation gave structure V and an asymmetric reaction was carried out with the epoxide to avoid the problem of resolution. The product from this reaction (VI) could be hydrogenated to the final product *in situ* without the need to isolate VI, again cutting down the number of operations.

(Continued)

BOX 15.3 Synthesis of ICI D7114 (*Continued*)

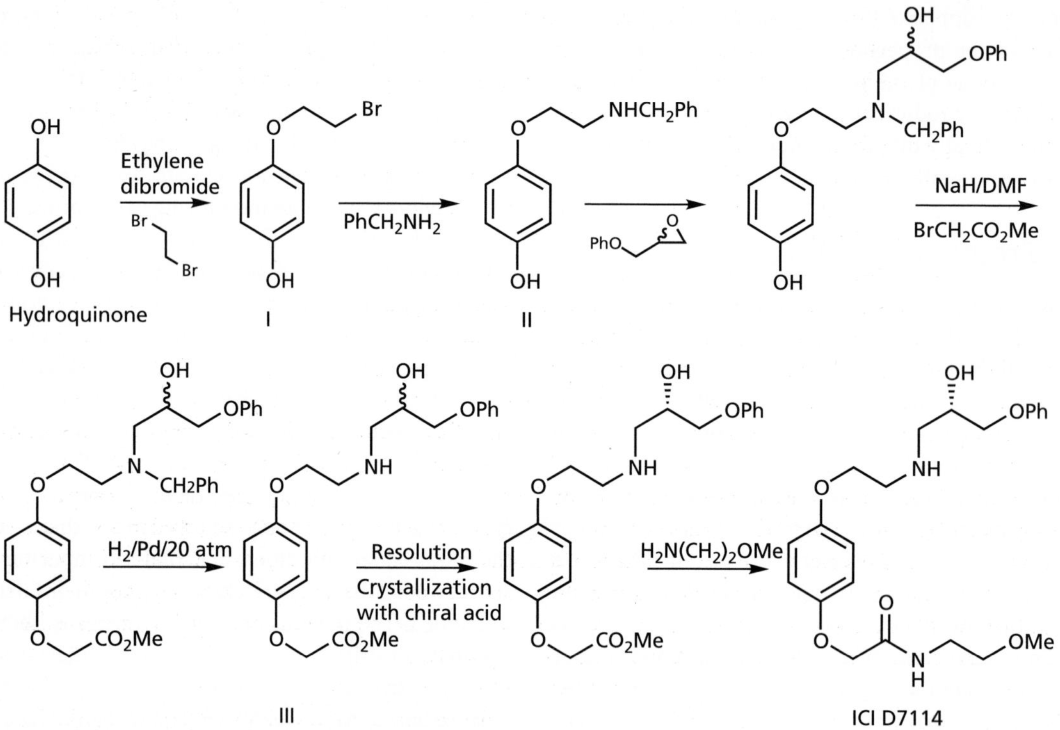

FIGURE 1 Research synthesis of ICI D7114.

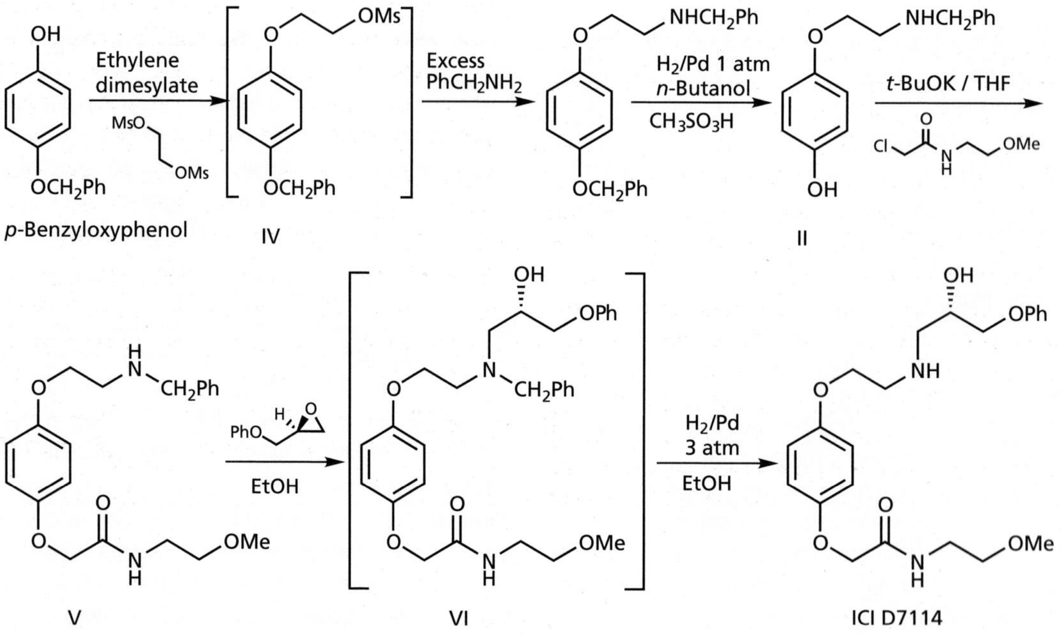

FIGURE 2 Revised synthetic route to ICI D7114.

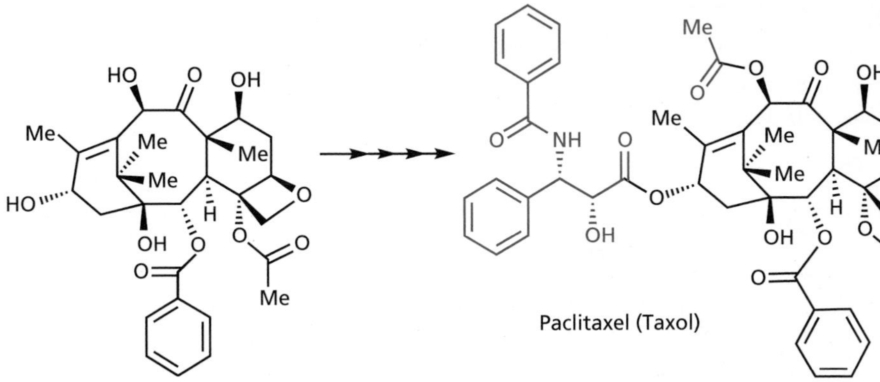

FIGURE 15.6 Semi-synthetic synthesis of paclitaxel (Taxol).

15.3.3 **Choice of drug candidate**

The issues surrounding chemical and process development can affect the choice of which drug candidate is taken forward into drug development. If it is obvious that a particular structure is going to pose problems for large-scale production, an alternative structure may be chosen which poses fewer problems, even if it is less active.

15.3.4 **Natural products**

Not all drugs can be fully synthesized. Many natural products have quite complex structures that are too difficult and expensive to synthesize on an industrial scale. These include drugs such as **penicillin, morphine**, and **paclitaxel (Taxol)**. Such compounds can only be harvested from their natural source—a process which can be tedious, time consuming, and expensive, as well as being wasteful on the natural resource. For example, four mature yew trees have to be cut down to obtain enough paclitaxel to treat one patient! Furthermore, the number of structural analogues that can be obtained from harvesting is severely limited.

Semi-synthetic procedures can sometimes get round these problems. This often involves harvesting a biosynthetic intermediate from the natural source, rather than the final compound itself. The intermediate can then be converted to the final product by conventional synthesis. This approach can have two advantages. Firstly, the intermediate may be more easily extracted in higher yield than the final product itself. Secondly, it may allow the possibility of synthesizing analogues of the final product. The semi-synthetic penicillins are an illustration of this approach (section 19.5.1.6). Another more recent, example is that of paclitaxel. It is manufactured by extracting **10-deacetylbaccatin III** from the needles of the yew tree, then carrying out a four-stage synthesis (Fig. 15.6).

KEY POINTS

- Chemical development involves the development of a synthetic route which is suitable for the large-scale synthesis of a drug.

- The priorities in chemical development are to develop a synthetic route which is straightforward, safe, cheap, efficient, has the minimum number of synthetic steps, and will provide a consistently good yield of high-quality product that meets predetermined purity specifications.

- An early priority in chemical development is to define the purity specifications of the drug and to devise a purification procedure which will satisfy these requirements.

- Process development aims to develop a production process which is safe, efficient, economic, environmentally friendly, and provides a product having consistent yield and quality to satisfy purity specifications.

- Drugs derived from natural sources are usually produced by harvesting the natural source or through semi-synthetic methods.

QUESTIONS

1. Discuss whether the doubly labelled atropine molecule shown below is suitable for drug metabolism studies.

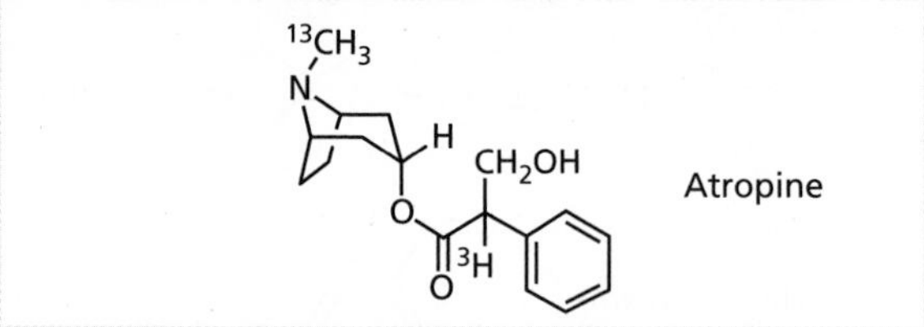

   Atropine

2. What is meant by a placebo and what sort of issues need to be considered in designing a suitable placebo?

3. Usually, a 'balancing act' of priorities is required during chemical development. Explain what this means.

4. Discuss whether chemical development is simply a scale-up exercise.

5. The following synthetic route was used for the initial synthesis of fexofenadine (R = CO_2H)—an analogue of terfenadine (R = CH_3). The synthesis was suitable for the large-scale synthesis of terfenadine, but not for fexofenadine. Suggest why not. (Hint: consider the electronic effects of R.)

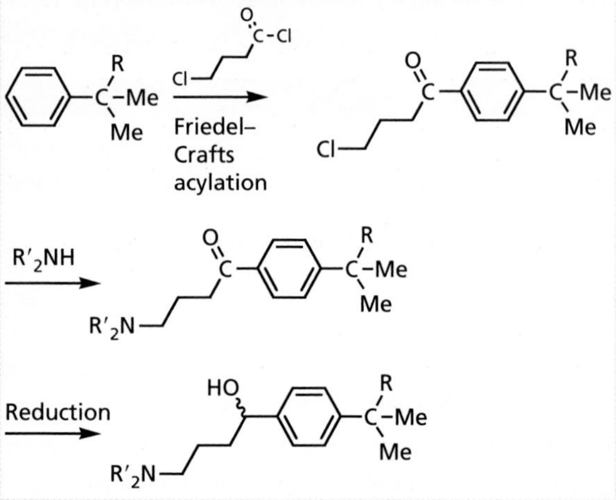

6. The following reaction was carried out using ethanol or water as solvents but gave poor yields in both cases. Suggest why this might be the case, and how these problems could be overcome.

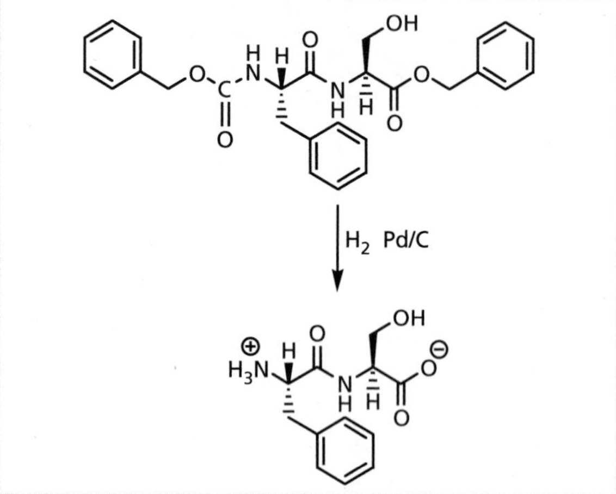

7. The following reaction was carried out with heating under reflux at 110 °C. However, the yield was higher when the condenser was set for distillation. Explain.

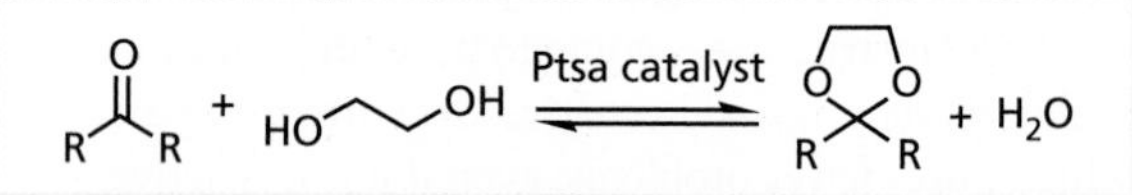

8. What considerations do you think have to be taken into account when choosing a solvent for scale up? Would you consider diethyl ether or benzene as a suitable solvent?

9. Phosphorus tribromide was added to an alcohol to give an alkyl bromide, but the product was contaminated with an ether impurity. Explain how this impurity might arise and how the reaction conditions could be altered to avoid the problem.

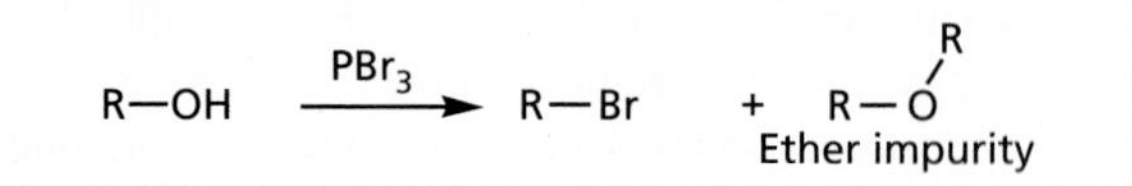

   Ether impurity

🔘 Multiple-choice questions are available on the Online Resource Centre at www.oxfordtextbooks.co.uk/orc/patrick6e/

FURTHER READING

Preclinical trials

Cavagnaro, J. A. (2002) Preclinical safety evaluation of biotechnology-derived pharmaceuticals. *Nature Reviews Drug Discovery*, 1, 469–75.

Lindpaintner, K. (2002) The impact of pharmacogenetics and pharmacogenomics on drug discovery. *Nature Reviews Drug Discovery*, 1, 463–9.

Matfield, M. (2002) Animal experimentation: the continuing debate. *Nature Reviews Drug Discovery*, 1, 149–52.

Nicholson, J. K., Connelly, J., Lindon, J. C., and Holmes, E. (2002) Metabonomics: a platform for studying drug toxicity and gene function. *Nature Reviews Drug Discovery*, 1, 153–61.

Pritchard, J. F. (2003) Making better drugs: decision gates in non-clinical drug development. *Nature Reviews Drug Discovery*, **2**, 542–53.

Ulrich, R., and Friend, S. H. (2002) Toxicogenomics and drug discovery: will new technologies help us produce better drugs? *Nature Reviews Drug Discovery*, **1**, 84–8.

Chemical and process development

Delaney, J. (2009) Spin-outs: protecting your assets. *Chemistry World*, July, 54–5.

Lee, S., and Robinson, G. (1995) *Process development: fine chemicals from grams to kilograms.* Oxford University Press, Oxford.

Repic, O. (1998) *Principles of process research and chemical development in the pharmaceutical industry.* John Wiley and Sons, Chichester.

Saunders, J. (2000) *Top drugs: top synthetic routes.* Oxford University Press, Oxford.

Patenting

Agranat, I., Caner, H., and Caldwell, J. (2002) Putting chirality to work: the strategy of chiral switches. *Nature Reviews Drug Discovery*, **1**, 753–68.

Southall, N. T., et al. (2006) Kinase patent space visualization using chemical replacements. *Journal of Medicinal Chemistry*, **49**, 2103–9.

Webber, P. M. (2003) Protecting your inventions: the patent system. *Nature Reviews Drug Discovery*, **2**, 823–30.

Regulatory affairs

Engel, L. W., and Straus, S. E. (2002) Development of therapeutics: opportunities within complementary and alternative medicine. *Nature Reviews Drug Discovery*, **1**, 229–37.

Haffner, M. E., Whitley, J., and Moses, M. (2002) Two decades of orphan product development. *Nature Reviews Drug Discovery*, **1**, 821–5.

Houlton, S. (2010) Recalling pharma. *Chemistry World*, July, 18–9.

Maeder, T. (2003) The orphan drug backlash. *Scientific American*, May, 71–7.

Perks, B. (2011a) Faking it. *Chemistry World*, January, 56–9.

Perks, B. (2011b) Orphans come in from the cold. *Chemistry World*, September, 60–3.

Reichert, J. M. (2003) Trends in development and approval times for new therapeutics in the United States. *Nature Reviews Drug Discovery*, **2**, 695–702.

Clinical trials

Issa, A. M. (2002) Ethical perspectives on pharmacogenomic profiling in the drug development process. *Nature Reviews Drug Discovery*, **1**, 300–8.

Lewcock, A. (2010) Medicine made to measure. *Chemistry World*, July, 56–61.

Schreiner, M. (2003) Paediatric clinical trials: redressing the balance. *Nature Reviews Drug Discovery*, **2**, 949–61.

Sutcliffe, A. G., and Wong, I. C. K. (2006) Rational prescribing for children. *British Medical Journal*, **332**, 1464–5.

Titles for general further reading are listed on p.845.

CASE STUDY 2
The design of ACE inhibitors

ACE inhibitors are important pharmaceutical agents that are used in the treatment of hypertension. ACE stands for **angiotensin-converting enzyme**, which is a key component of the renin–angiotensin–aldosterone system—also known as the RAAS cascade (section 26.3). The RAAS cascade produces a potent vasoconstricting hormone called **angiotensin II**, and ACE is the enzyme that catalyses its formation from the biosynthetic precursor **angiotensin I** (Fig. CS2.1). Therefore, drugs that inhibit ACE have proved to be potent antihypertensives.

The design of ACE inhibitors demonstrates how it is possible to design drugs for a protein target in a rational manner even if the structure of the target has not been determined. The **angiotensin-converting enzyme** (ACE) is a membrane-bound enzyme which has been difficult to isolate and study. It is a member of a group of enzymes called the **zinc metalloproteinases** and catalyses the hydrolysis of a dipeptide fragment from the end of the decapeptide **angiotensin I** to give the octapeptide **angiotensin II** (Fig. CS2.2).

Although the enzyme ACE could not be isolated, the design of ACE inhibitors was helped by studying the structure and mechanism of another zinc metalloproteinase that could—an enzyme called **carboxypeptidase.** This enzyme splits the terminal amino acid from a peptide chain as shown in Fig. CS2.3 and is inhibited by **L-benzylsuccinic acid.**

The active site of carboxypeptidase (Fig. CS2.4) contains a charged arginine unit (Arg-145) and a zinc ion, which are both crucial in binding the substrate peptide. The peptide binds such that the terminal carboxylic acid is bound ionically to the arginine unit, while the carbonyl group of the terminal peptide bond is bound to the zinc ion. There is also a hydrophobic pocket called the S1′ pocket which can accept the side chain of the terminal amino acid. Aromatic rings are found to bind strongly to this pocket and this explains the specificity of the enzyme towards peptide substrates containing an aromatic amino acid at the *C*-terminus (Phe in the example shown). Hydrolysis of the terminal peptide bond then takes place

Angiotensinogen $\xrightarrow{\text{Renin}}$ Angiotensin I $\xrightarrow{\text{ACE}}$ Angiotensin II (Potent vasoconstrictor)

FIGURE CS2.1 Biosynthesis of angiotensin II.

Asp-Arg-Val-Tyr-Ile-His-Pro-Phe-His-Leu $\xrightarrow{\text{ACE}}$ Asp-Arg-Val-Tyr-Ile-His-Pro-Phe + His-Leu

Angiotensin I   Angiotensin II

FIGURE CS2.2 Reaction catalysed by angiotensin-converting enzyme (ACE).

Peptide $\sim$ aa³-aa²-aa¹$-CO_2H$ $\xrightarrow{\text{Carboxypeptidase}}$ Peptide $\sim$ aa³-aa² $- CO_2H$ + aa¹

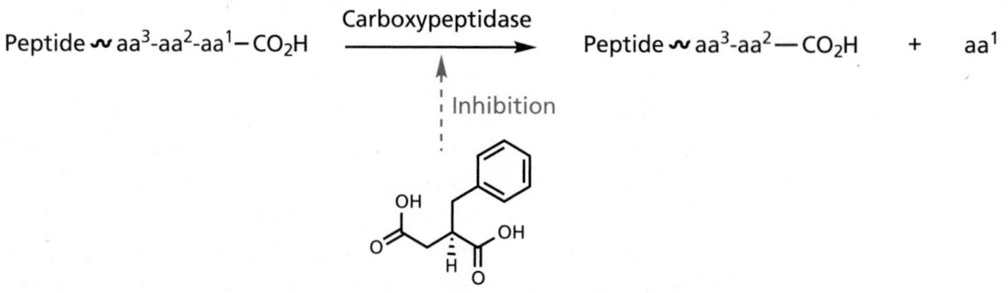

L-Benzylsuccinic acid

FIGURE CS2.3 Hydrolysis of a terminal amino acid from a peptide chain by the carboxypeptidase enzyme. The asymmetric centre of L-benzylsuccinic acid has the *R* configuration.

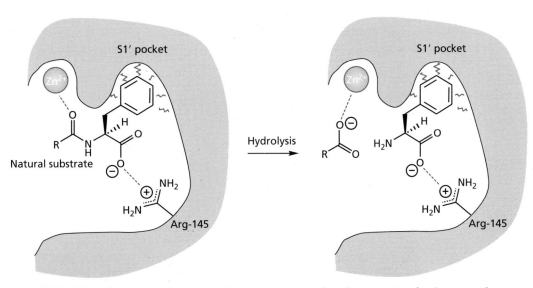

FIGURE CS2.4 Binding site interactions for a substrate bound to the active site of carboxypeptidase.

aided by the zinc ion, which plays a crucial role in the mechanism by polarizing the carbonyl group and making the amide group more susceptible to hydrolysis (see also section 3.5.5 on the mechanism of action of zinc metalloproteinases).

The design of the carboxypeptidase inhibitor L-benzylsuccinic acid was based on the hydrolysis products arising from this enzymatic reaction. The benzyl group was included to occupy the S1′ pocket, while the adjacent carboxylate anion was present to form an ionic interaction with Arg-145. The second carboxylate was present to act as a ligand to the zinc ion, mimicking the carboxylate ion of the other hydrolysis product.

L-Benzylsuccinic acid binds as shown in Fig. CS2.5. However, it cannot be hydrolysed as there is no peptide bond present, and so the enzyme is inhibited for as long as the compound stays attached.

An understanding of the above mechanism and inhibition helped in the design of ACE inhibitors. First of all, it was assumed that the active site contained the same zinc ion and arginine group. However, as ACE splits a dipeptide unit from the peptide chain rather than a single amino acid, these groups are likely to be further apart, and so an analogous inhibitor to benzylsuccinic acid would be a succinyl-substituted amino acid. The next step was to choose which amino acid to use. Unlike carboxypeptidase, ACE shows no specificity for peptide substrates containing any particular C-terminal amino acid, and so the binding pocket for the C-terminal side chain must be different for the two enzymes. The relevant pockets would be S2′ for ACE (not shown in the diagrams) and S1′ for the carboxypeptidase enzyme. As there was no selectivity

shown for peptide substrates, it was decided to study peptides that acted as ACE inhibitors, and to identify whether any C-terminal amino acids were commonly present in these structures. The amino acid proline is the C-terminal amino acid in a known ACE inhibitor called **teprotide** (Fig. CS2.6)—a nonapeptide which was isolated from the venom of the Brazilian pit viper. Although teprotide is a reasonably potent inhibitor, it is susceptible to digestive enzymes and is orally inactive. Other ACE inhibitors found in snake venoms also contain the terminal proline group, implying that the ring might be involved in some binding interaction with the binding site. **Succinyl proline** was the end result of this design philosophy.

Succinyl proline was found to be a weak, but specific, inhibitor of ACE, and it was proposed that both carboxylate groups were ionized—one interacting with the arginine group and one with the zinc ion (Fig. CS2.7). It was now argued that there must be pockets available to accommodate amino acid side chains on either side of the reaction centre (pockets S1 and S1′). The strategy of extension (section 13.3.2) was now used to find a group that would fit the S1′ pocket and increase the binding affinity. A methyl group fitted the bill and resulted in an increase in activity (Fig. CS2.8). The next step was to see whether there was a better group than the carboxylate ion to interact with zinc, and it was discovered that a thiol group led to dramatically increased activity. This resulted in **captopril**, which was the first non-peptide ACE inhibitor to become commercially available. The stereochemistry of the methyl substituent in both SQ13 297 and captopril is important for activity since the opposite enantiomers show 100-fold less activity.

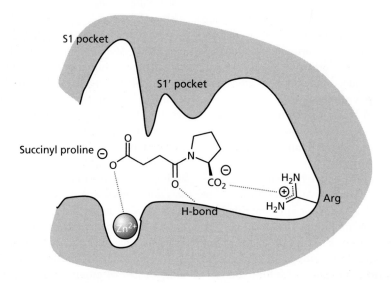

FIGURE CS2.5 Inhibition by L-benzylsuccinic acid (the *R*-enantiomer).

FIGURE CS2.6 Angiotensin-converting enzyme (ACE) inhibitors.

Succinyl proline; IC_{50} 628 μM

Glu-Trp-Pro-Arg-Pro-Gln-Ile-Pro

Teprotide; IC_{50} 0.9 μM

FIGURE CS2.7 Binding site interaction for succinyl proline in the active site of angiotensin-converting enzyme (ACE).

The most common side effects associated with captopril are rashes and loss of taste, which are thought to be associated with the thiol group. Therefore, other workers sought to find an ACE inhibitor that was as potent as captopril, but which lacked the thiol group. This meant re-introducing the carboxylate group as the zinc ligand, despite the fact that it is a far weaker binding group for zinc. To compensate for this, groups were introduced that could form extra binding interactions with the active site (*extension strategy*; section 13.3.2). Firstly, it was decided to extend the chain length of succinyl proline in order to introduce an NH group. The rationale was that

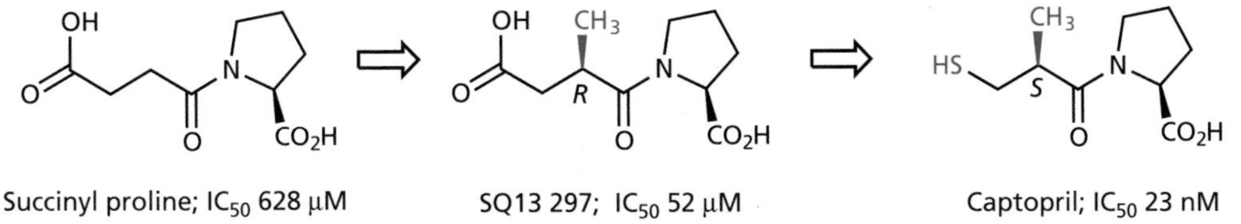

FIGURE CS2.8 Development of captopril.

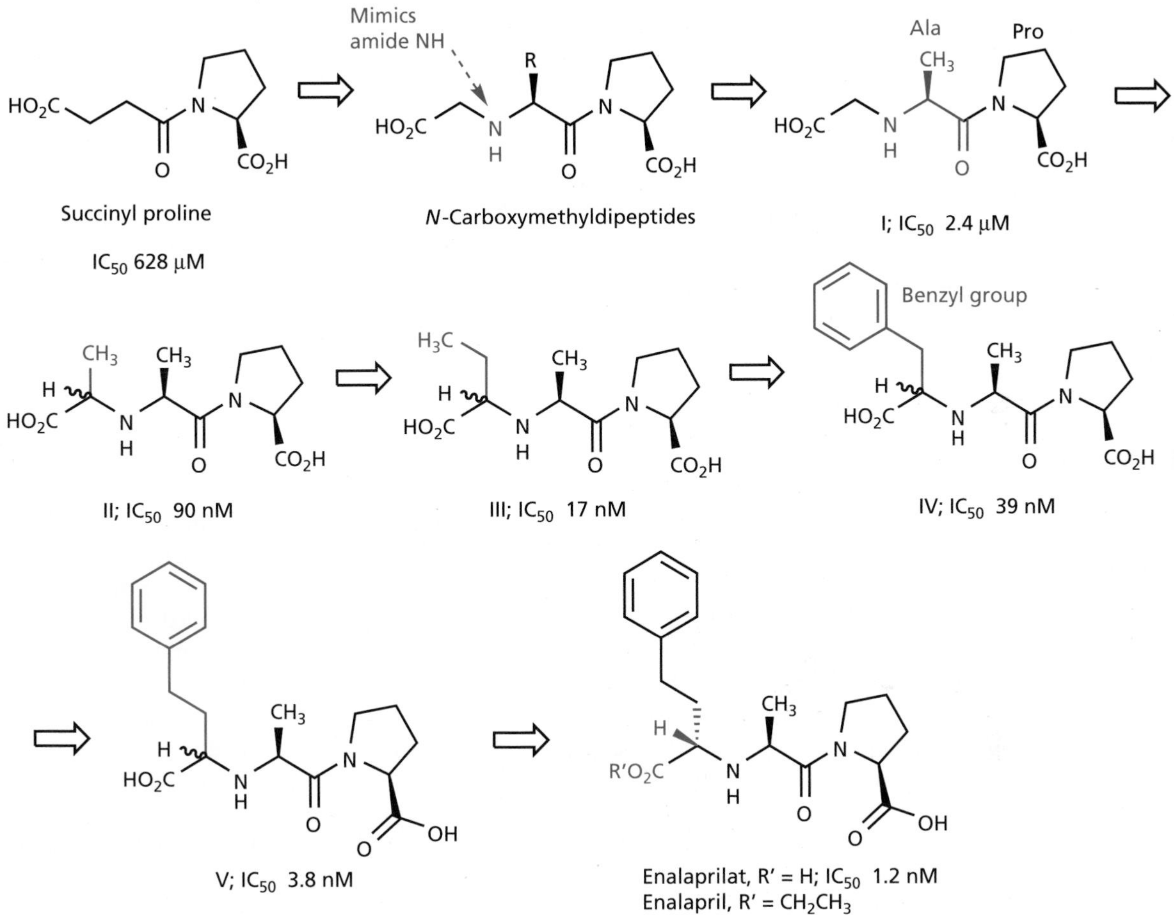

FIGURE CS2.9 Development of enaprilat.

the NH group would mimic the amide NH of the peptide link that would normally be hydrolysed by the enzyme (Fig. CS2.9). It seemed reasonable to assume that this group could be involved in some kind of hydrogen bonding with the active site. Introducing the NH group meant that a second amino acid had now been introduced into the structure, and so a series of *N*-carboxymethyl dipeptides was studied. Incorporating L-alanine introduced the methyl substituent that is present in captopril (structure I, Fig. CS2.9). The activity of this compound was better than succinyl proline, but greatly inferior to captopril. Therefore, it was decided to 'grow' a substituent

from the penultimate carbon atom in the structure to search for the binding pocket S1 shown in Fig. CS2.7. This pocket would normally accept the phenylalanine side chain of angiotensin I and should be hydrophobic in nature. Therefore, methyl and ethyl substituents were introduced (structures II–III; Fig. CS2.9). The analogues showed increased activity, with the ethyl analogue proving as effective as captopril. Activity dropped slightly with the introduction of a benzyl group, but a chain extension (section 13.3.3) led to a dramatic increase in activity, such that structure V proved to be more active than captopril.

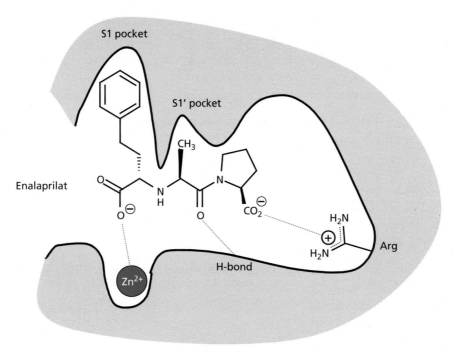

FIGURE CS2.10 Enalaprilat.

The addition of a new substituent to structures II–V meant that a new asymmetric centre had been introduced, and all these structures had been tested as mixtures of diastereomers—the *R,S,S* and *S,S,S* diastereomers. Structure V was now separated by chromatography and the *S,S,S* diastereomer was found to be 700 times more active than the *R,S,S* diastereomer. This structure was named **enalaprilat** and is proposed to bind to the active site as shown in Fig. CS2.10. **Enalapril** is the ethyl ester prodrug of enalaprilat and is used clinically (*prodrugs*; section 14.6.1.1). The prodrug is absorbed more easily from the gut than enalaprilat itself, and is converted to enalaprilat by esterase enzymes.

Lisinopril (Fig. CS2.11) is another successful ACE inhibitor which is similar to enalaprilat, but where the methyl substituent has been extended to an aminobutyl substituent—the side chain for the amino acid lysine. In 2003, a crystal structure of ACE complexed with lisinopril was finally determined by X-ray crystallography (Fig. CS2.12). This provided a detailed picture of the three-dimensional structure of ACE, and how lisinopril binds to the active site. In fact, there is a marked difference in structure between ACE and carboxypeptidase A, which means that ACE inhibitors do not bind as originally thought. For example, the ionic interaction originally thought to involve an arginine residue involves a lysine residue instead. Now that an accurate picture has been obtained of the active site and the manner in which lisinopril binds, it is possible that a new generation of ACE inhibitors will be designed with improved binding characteristics using structure-based drug design (section 13.3.11).

FIGURE CS2.11 Lisinopril.

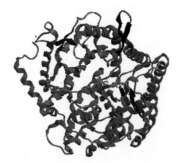

FIGURE CS2.12 Crystal structure of lisinopril bound to ACE (pdb 1O86).

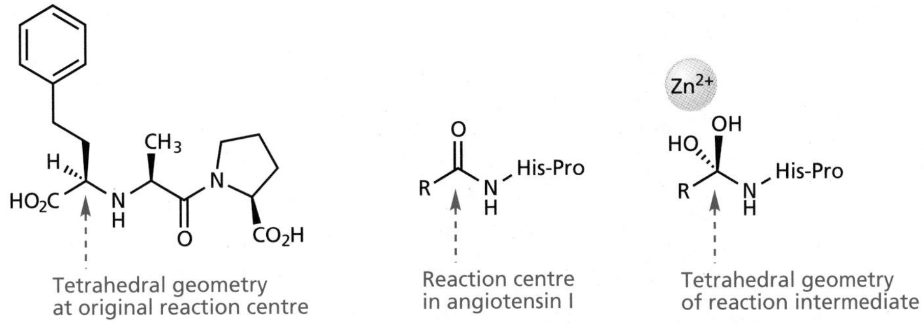

FIGURE CS2.13 Comparison of enalaprilat with angiotensin I and the reaction intermediate formed during enzyme-catalysed hydrolysis.

Both enalaprilat and lisinopril have a tetrahedral geometry at what would normally be the reaction centre for the enzyme-catalysed reaction (Fig. CS2.13). As a result, they have been described as transition-state analogues (section 7.4). This is because the transition state for the enzyme-catalysed reaction would be expected to be similar in nature to the tetrahedral intermediate formed during the hydrolysis mechanism.

BOX CS2.1 Synthesis of captopril and enalaprilat

There are several methods of synthesizing captopril. One method is to take the *t*-butyl ester of proline and react it with the carboxylic acid (I) in the presence of a coupling agent to form the amide (II) (Fig. 1). The *t*-butyl and acetyl protecting groups can then be removed in the presence of acid and base respectively to give captopril.

Enalaprilat can be synthesized by reacting the ketoacid (III) with L-alanyl-L-proline. Two diastereomers are formed which are separated by chromatography.

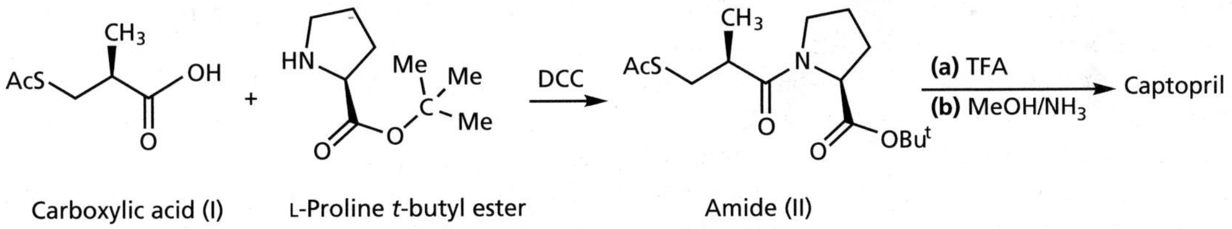

FIGURE 1 Synthesis of captopril (DCC = dicyclohexylcarbodiimide).

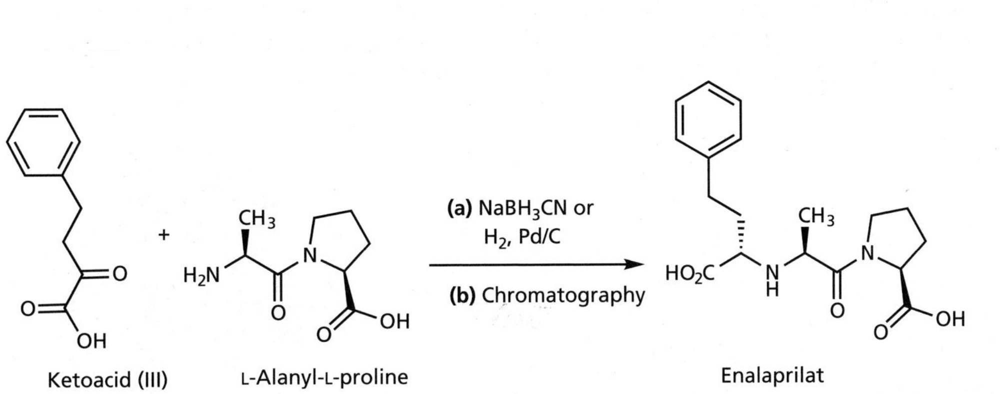

FIGURE 2 Synthesis of enalaprilat.

FURTHER READING

Acharya, K. R., et al. (2003) ACE revisited: a new target for structure-based drug design. *Nature Reviews Drug Discovery*, **2**, 891–902.

Ganellin, C. R., and Roberts, S. M. (eds.) (1994) Angiotensin-converting enzyme (ACE) inhibitors and the design of cilazapril. *Medicinal chemistry—the role of organic research in drug research*, 2nd edn. Chapter 9, Academic Press, London.

Natesh, R., et al. (2003) Crystal structure of the human angiotensin-converting enzyme–lisinopril complex. *Nature*, **420**, 551–4.

Ondetti, M. A., Rubin, B., and Cushman, D. W. (1977) Design of specific inhibitors of angiotensin-converting enzyme: new class of orally active antihypertensive agents. *Science*, **196**, 441–4 (captopril).

Patchett, A. A., et al. (1980) A new class of angiotensin-converting enzyme inhibitors. *Nature*, **288**, 280–3.

Saunders, J. (2000) Inhibitors of angiotensin converting enzyme as effective antihypertensive agents. *Top drugs: top synthetic routes*. Chapter 1, Oxford University Press, Oxford.

Zaman, M. A., Oparil, S., and Calhoun, D. A. (2002) Drugs targeting the renin–angiotensin–aldosterone system. *Nature Reviews Drug Discovery*, **1**, 621–36.

CASE STUDY 3
Artemisinin and related antimalarial drugs

CS3.1 Introduction

Malaria is an ancient disease that has resulted in millions of deaths and much human misery. It is caused by a protozoal parasite which is carried by mosquitoes and is transmitted between mosquitoes and humans by mosquito bites. The malarial parasite is a microorganism belonging to the *Plasmodium* genus, of which there are four species—*P. vivax, P. falciparum, P. ovale,* and *P. malariae. Plasmodium falciparum* is the most dangerous of these and can result in death. The disease is currently associated with tropical countries, but, in the past, it was present in Europe and North America. Campaigns were carried out in the 1950s and 60s to try and eradicate mosquitoes by spraying their breeding grounds with DDT. These efforts, along with the use of quinine-based drugs (Fig. CS3.1), successfully reduced the prevalence of malaria.

Quinine was the first of the antimalarial agents to be used and is still effective today. However, it can cause adverse reactions such as ringing in the ears and partial deafness. Therefore, its use is currently limited to the treatment of malaria rather than as a **prophylactic.** A prophylactic is a protective agent that is administered to prevent a disease occurring. The agent that largely replaced quinine as the antimalarial drug of choice was **chloroquine**, which has far fewer side effects. This was introduced in the 1950s and at one point it was thought that the disease would be conquered. Unfortunately, from 1961 onwards, the parasite has developed resistance to chloroquine such that the drug is no longer effective in many malarial infected areas of the world, especially in sub-Saharan Africa. It is, therefore, crucial that new antimalarial therapies are discovered that can combat these drug-resistant strains. An added urgency comes from the belief that global warming might result in the return of the disease to North America and Europe. This is particular worrying with respect to the potentially fatal *Plasmodium falciparum*. Resistance appears to be a result of the parasite having a cell membrane protein which can pump the drug out of the cell. Fortunately, a new drug has been discovered in recent years that has been found to be active against these drug-resistant strains—artemisinin.

CS3.2 Artemisinin

For over 2000 years, Chinese herbalists have used concoctions or teas (called **qinghao**) obtained from an abundant Chinese plant called *Artemisia annua*. The herb was first described as a remedy for haemorrhoids in 168 BC, and the first mention of it as an antimalarial preparation was in 340 AD. Further references to the plant were made in 1596, when it was used for the treatment of chills and fever resulting from malaria.

In 1972, the active principle of the plant was isolated and identified as **artemisinin** (or **qinghaosu**). The compound caused great excitement because it was found to be effective against the particularly dangerous chloroquine-resistant *Plasmodium falciparum*, and also acted more quickly against chloroquine-sensitive strains. The **Walter Reed Army Institute of Research** in the USA was particularly interested in this new compound.

FIGURE CS3.1 Quinine and quinine-based antimalarial drugs.

Historically, more military casualties have resulted from malaria than from battle action. For example, during the Burmese campaign of the Second World War, a huge number of British and Indian soldiers were incapacitated by malaria, and had to be withdrawn from action. Many of the politically unstable countries in the world today are malarial infected areas, and so there is an obvious interest among the military to find novel antimalarial drugs for their troops. Unfortunately, the only known source of *Artemisia annua* was in China, and the Chinese communist authorities were understandably reluctant to grant US army scientists free access into China. Negotiations were certainly not helped by American negotiators appearing in full dress uniform. As a result, the Americans were denied access to Chinese supplies and were forced to carry out a worldwide search to see if they could find an artemisinin-producing plant in a different country. Ironically enough, a suitable plant was eventually found growing in the American capital!

CS3.3 Structure and synthesis of artemisinin

The multicyclic structure of artemisinin (Fig. CS3.2) contains seven asymmetric centres and an unusual, unstable-looking trioxane ring that includes an endoperoxide

group. Despite the unstable appearance of the molecule, it is stable to heat and light. Once the compound was identified, the next stage was to synthesize a range of analogues to investigate structure–activity relationships (section 13.1).

CS3.4 Structure–activity relationships

Artemisinin is a complex structure and although it has been fully synthesized, this is not a practical method of obtaining it, or for producing a variety of different analogues. Consequently, analogues were prepared from artemisinin itself—a semi-synthetic approach. This was done by first reducing the lactone group of artemisinin to give **dihydroartemisinin** (Fig. CS3.3). This contains an alcoholic group which can then be alkylated to give various ethers such as **artemether** and **arteether.**

Esterifications can also be carried out on dihydroartemisinin, and a particularly important ester is **sodium artesunate** from the reaction of artemisinin with succinic acid (Fig. CS3.4).

Dihydroartemisinin, artemether, arteether, and sodium artesunate are all more active than artemisinin itself, and so the lactone carbonyl group of artemisinin is not crucial to its antimalarial activity. A variety of other

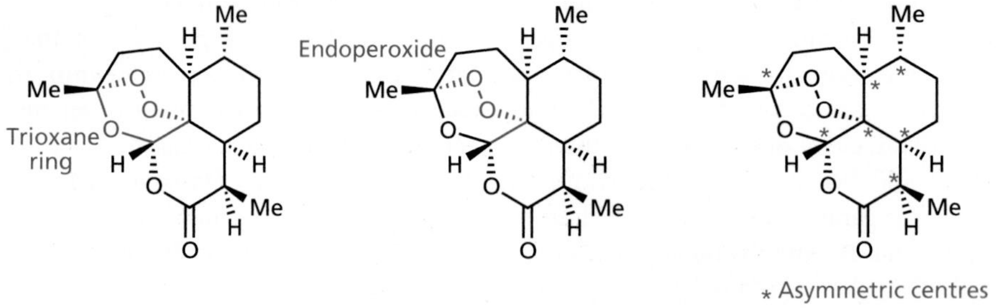

* Asymmetric centres

FIGURE CS3.2 Structure of artemisinin.

Artemisinin   Dihydroartemisinin   Artemether; R = CH_3
Arteether; R = CH_2CH_3

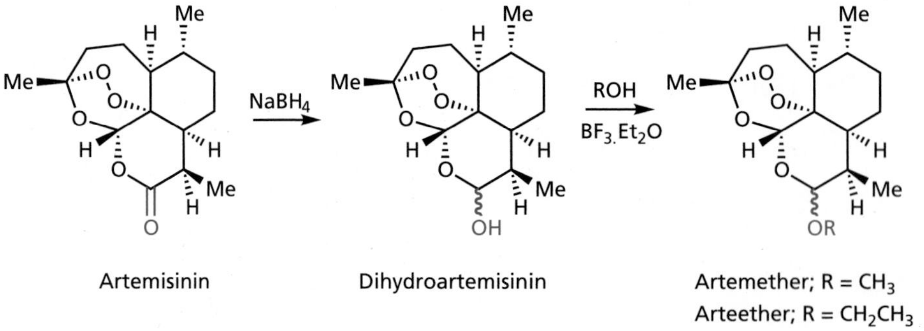

FIGURE CS3.3 Preparation of artemether and arteether.

artemisinin analogues have also been studied. For example, **deoxyartemisinin** (Fig. CS3.5) is a metabolite of
artemisinin and is 300–1000-fold less active. **Deoxodeoxyartemisinin** is also poorly active, whereas **deoxoartemisinin** has a similar activity to arteether.

The results from these and other structures led to the
conclusion that the endoperoxide group in the trioxane
ring was the essential group required for antimalarial activity, and that this represented the **pharmacophore** (section 13.2) for antimalarial activity (see also Box CS3.1).

CS3.5 **Mechanism of action**

Artemisinin has a totally different mechanism of action from the quinine-based drugs and has, therefore,

proved effective against chloroquine-resistant strains of
malaria. The secret behind its action lies in the endoperoxide group. This acts as a molecular trigger for a kind
of 'scattergun' action which causes severe damage within
the parasite cell and ultimately leads to its death. As the
group is acting as a 'trigger', something has to pull the
trigger. This turns out to be iron ions, and, in particular,
ferrous ions. In the presence of these ions, a reduction
of the endoperoxide group takes place which generates
two possible radical species (Fig. CS3.6). Further reactions take place to generate a series of other cytotoxic free
radicals and reactive electrophiles which alkylate, cross-
link, and oxidize vital biomolecules within the parasite.
Cell death is the result.

This explains the action of artemisinin on protozoal
cells, but why does it not kill human cells as well? In

FIGURE CS3.4 Synthesis of sodium artesunate.

FIGURE CS3.5 Analogues of artemisinin.

FIGURE CS3.6 Activation of artemisinin by ferrous ions.

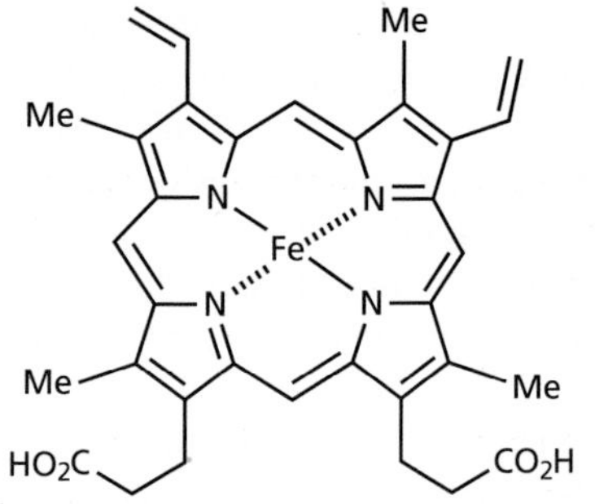

FIGURE CS3.7 Structure of heme.

When the malarial parasite infects red blood cells, it breaks down haemoglobin as a food source to provide itself with amino acids. This, of course, releases heme into the parasite cell. The ferrous ion present in heme can now react with artemisinin leading to the parasite's demise. Therefore, artemisinin and its analogues can be viewed as **prodrugs** (section 14.6) which are activated as a result of the parasite's own destructive tendencies to haemoglobin—poetic justice really!

A lot of research has been carried out to investigate the detailed radical mechanisms that follow on from the two radical products shown in Fig. CS3.6. The story is quite complex, but there is evidence that a particularly important mechanistic route for high antimalarial activity is the formation of a C-4 radical via 1,5-hydrogen atom abstraction (Fig. CS3.8). This produces the major metabolite that is observed for artemisinin, and also generates a highly reactive ferric hydroxide species which can go on to cause havoc within the cell.

Support for this theory comes from the activities of the simplified artemisinin analogues shown in Fig. CS3.9. Structure II is twice as active as artemisinin *in vitro*, whereas structures I and III are 100-fold less active. The 1,5-hydrogen shift shown in Fig. CS3.8 is not possible for structures I and III, where the crucial hydrogen atom has been replaced with an α-methyl group. These compounds still react with the ferrous ion, but the 1,5-hydrogen shift is blocked. There is some evidence that the β-alkyl group at position 4 of structure II enhances activity, possibly by stabilizing the radical at position 4.

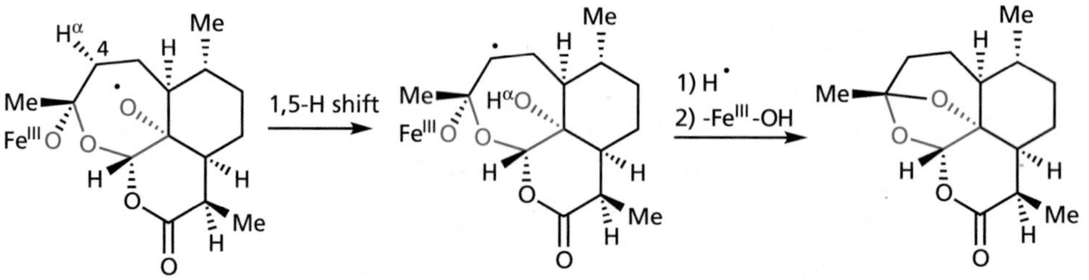

particular, why does the drug not destroy red blood cells which are rich in iron-containing **haemoglobin**—the protein responsible for carrying oxygen from the lungs to the rest of the body.

To answer these questions, we need to consider the life cycle of the parasite. This is quite a complex process involving both humans and mosquitoes, but part of the parasite's life cycle in humans involves the invasion of red blood cells. As mentioned above, red blood cells contain haemoglobin. This is a protein that contains an iron (II)-centred porphyrin called **heme** (Fig. CS 3.7). The porphyrin and the ferrous ion are buried deep within the protein and are effectively shielded. This explains why artemisinin is not toxic to normal, uninfected red blood cells. The ferrous iron, which would trigger its destructive capability, is 'hidden from view'.

FIGURE CS3.8 Generation of a C-4 radical by 1,5-hydrogen atom abstraction.

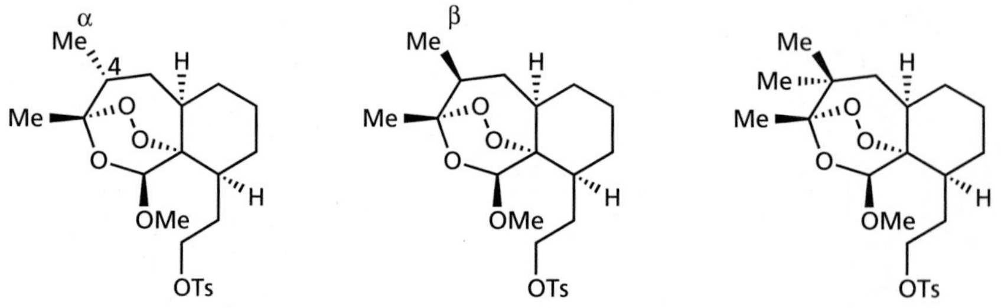

FIGURE CS3.9 Simplified analogues of artemisinin.

BOX CS3.1 Clinical properties of artemisinin and analogues

Artemisinin has proved highly effective in treating malaria, but there are problems related with its use. First of all, it is not water soluble and it has to be administered by intramuscular injection. It is also found that malaria recurs in up to 25% of patients treated after 1 month. Artemether and arteether are more hydrophobic than artemisinin and can be administered more easily in the field by injection in oil. They are also more potent. Sodium artesunate is also used clinically. Due to the ionized carboxylate group, sodium artesunate is water soluble and can be administered by intravenous injection.

Currently, artemisinin, artemether, and sodium artesunate are used clinically. These compounds are now considered to be an essential component of **artemisinin combination therapy** (ACT) against drug-resistant malaria. They show brisk and potent activity while cross-resistance with the more traditional antimalarial drugs is unlikely, due to the different mechanism of action.

Drawbacks for these drugs include a short plasma half-life, which is typically less than 1 hour, and rapid elimination. This means that the drug is cleared from the system within a day of administration, leaving the longer-lived drugs of the combination therapy to continue the battle alone. This increases the risk of drug-resistant parasites emerging.

CS3.6 Drug design and development

Since artemisinin is poorly soluble in both water and oil, early research was aimed at producing analogues which would be more soluble in one or other of these media. Dihydroartemisinin was found to be twice as active as artemisinin itself and was the gateway to the synthesis of a range of ethers and esters (Figs. CS3.4 and CS3.5). Many of these were found to have enhanced activity as well as better solubility. The most interesting of these are artemether and arteether which, being more hydrophobic in nature, are more soluble in oil. Among the esters, the most interesting compound is sodium artesunate, which is ionized and water soluble.

Research has also been carried out with the aim of designing an antimalarial agent that can be synthesized easily and which has the same mechanism of action as the lead compound, artemisinin. As with many lead compounds of complex structure, the strategy of simplification (section 13.3.8) has been used. Artemisinin has a complex tetracyclic structure with seven asymmetric centres, which makes it far too complex to synthesize economically in the laboratory. A variety of simpler structures retaining the trioxane ring have been synthesized—one of the most interesting of these is **fenozan**, which has a tricyclic ring system as its core and two asymmetric centres (Fig. CS3.10). This structure shows comparable activity to arteether and sodium artesunate against some malarial strains.

Other simplified structures having comparable activity to artemisinin or its semisynthetic analogues include

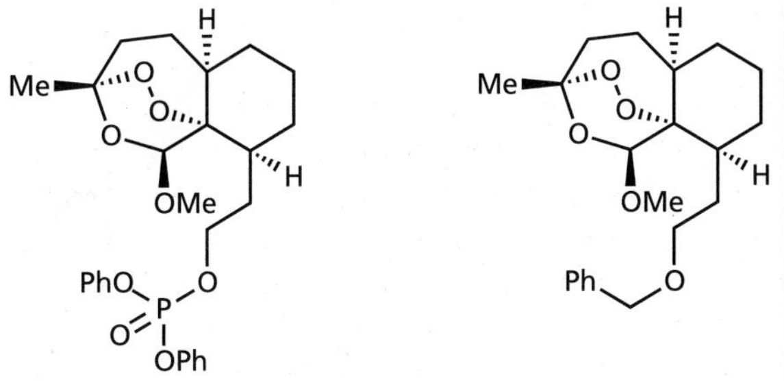

FIGURE CS3.11 Trioxanes having comparable activity to artemisinin.

Fenozan

Spiroalkyl trioxanes

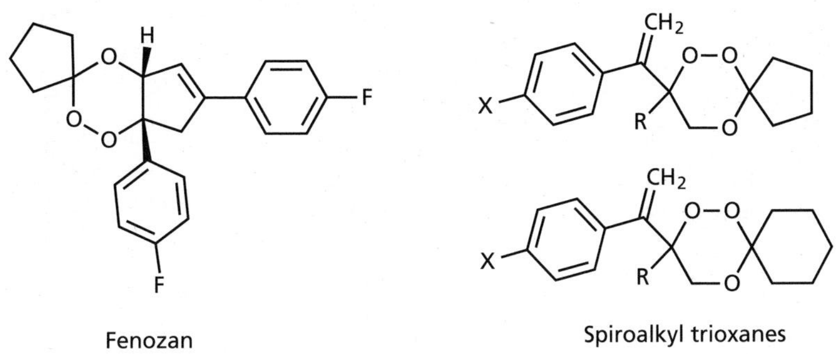

FIGURE CS3.10 Fenozan and spiroalkyl trioxanes.

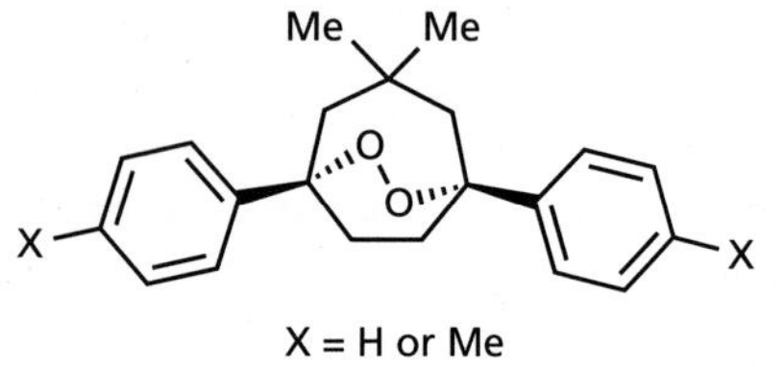

FIGURE CS3.12 Symmetrical analogues of artemisinin.

bicyclic spiroalkyl trioxanes (Fig. CS3.10), which are as active as artemisinin in mice experiments, and the trioxanes shown in Fig. CS3.11, which have comparable activity to artemisinin *in vitro*.

Simple, symmetrical endoperoxides have also been synthesized (Fig. CS3.12). These have been designed to take advantage of the proposed 1,5-H abstraction mechanism described in Fig. CS3.8. The advantage of a symmetrical artemisinin is that degradation can occur in the same manner regardless of which oxygen reacts with iron. The potency of this compound is about a seventh of artemisinin *in vitro*, but this is still considered to be high.

Yingzhaosu A (Fig. CS3.13) is a naturally occurring endoperoxide which was isolated in 1979 from a traditional Chinese herbal remedy for fever (made from the plant *Artabotrys uncinatus*) and shows antimalarial activity. However, the plant is a rare ornamental vine, and extraction of the natural compound is difficult and erratic. A synthesis was devised to produce a synthetic analogue of the structure, resulting in the discovery of **arteflene**, which is half as active as artemisinin.

To date, none of the simplified structures described have found widespread use as clinical agents, but there would be clear benefits in having a simple synthetic structure with the same mechanism of action as artemisinin, and which could be produced efficiently and cheaply for a market that cannot afford expensive drugs.

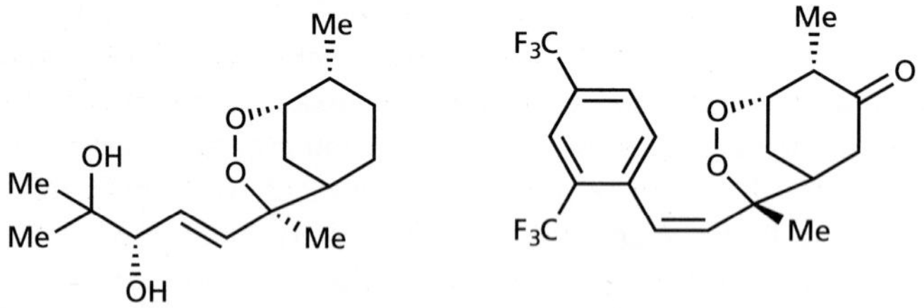

FIGURE CS3.13 Yingzhaosu A and arteflene.

FURTHER READING

Cumming, J. N., Ploypradith, P., and Posner, G. H. (1997) Antimalarial activity of artemisinin (Qinghaosu) and related trioxanes: mechanism(s) of action. *Advances in Pharmacology*, **37**, 253–97.

Davies, J. (2010) Cultivating the seeds of hope. *Chemistry World*, June, 50–3.

Drew, M. G. B., et al. (2006) Reactions of artemisinin and arteether with acid: implications for stability and mode of antimalarial action. *Journal of Medicinal Chemistry*, **49**, 6065–73.

Olliaro, P. L., et al. (2001) Possible modes of action of the artemisinin-type compounds. *Trends in Parasitology*, **17**, 122–6.

Posner, G. H., and O'Neill, P. M. (2004) Knowledge of the proposed chemical mechanism of action and cytochrome P450 metabolism of antimalarial trioxanes like artemisinin allows rational design of new antimalarial peroxides. *Accounts of Chemical Research*, **37**, 397–404.

Posner, G.H. (2007) Malaria-infected mice are cured by a single dose of novel artemisinin derivatives. *Journal of Medicinal Chemistry*, **50**, 2516–19.

Wu, Y. (2002) How might qinghaosu (artemisinin) and related compounds kill the intraerythrocytic malaria parasite? A chemist's view. *Accounts of Chemical Research*, **35**, 255–9.

■ CASE STUDY 4
The design of oxamniquine

CS4.1 Introduction

The development of **oxamniquine** (Fig. CS4.1) by Pfizer pharmaceuticals is a nice example of how traditional strategies can be used in the development of a drug where the molecular target is unknown. It also demonstrates that strategies can be used in any order and may be used more than once.

Oxamniquine is an important drug in developing countries, and is used in the treatment of **schistosomiasis (bilharzia)**. After malaria, this disease is the most endemic parasitic disease in the world, affects an estimated 200 million people, and is responsible for almost 500 000 deaths each year. Urinary infection can cause bladder cancer while intestinal infection can result in liver damage. The disease is caused by small flatworms called **schistosomes** which are contracted by swimming or wading in infected water. The parasites can rapidly penetrate human skin in the larval form, and, once they are in the blood supply, the larvae develop into adult flatworms. The females then produce eggs that become trapped in organs and tissues, leading to inflammation and a long debilitating disease that can last up to 20 years. In severe cases, the disease can be fatal. There are three pathogenic species of the parasite—*Schistosoma mansoni*, *S. haematobium*, and *S. japonicum*.

In the early 1960s, the only drugs available were the tricyclic structure **lucanthone** (Fig. CS4.1), and antimonials such as **stibocaptate** (Fig. CS4.2). However, stibocaptate and lucanthone have serious drawbacks as therapeutics. Stibocaptate is orally inactive, while both drugs require frequent dosing regimens and produce toxic side effects.

For example, lucanthone has to be administered 3–5 times per day, and can cause nausea and vomiting. More seriously, it can result in severe toxic effects on the heart and the central nervous system. Finally, it is not effective against all three of the pathogenic strains.

CS4.2 From lucanthone to oxamniquine

In 1964, Pfizer initiated a project aimed at developing an orally active, non-toxic agent that would be effective as a single dose against all three pathogenic strains, and which would be affordable for patients in developing countries. This research ultimately led to the discovery of oxamniquine (Fig. CS4.1), which met all but one of those goals.

Lucanthone was chosen as the lead compound because it was orally active, and it was decided to try *simplifying* (section 13.3.8) the structure to see whether the tricyclic system was really necessary. Several compounds were made—the most interesting structure was one where the two rings seen on the left in Fig. CS4.1 had been removed.

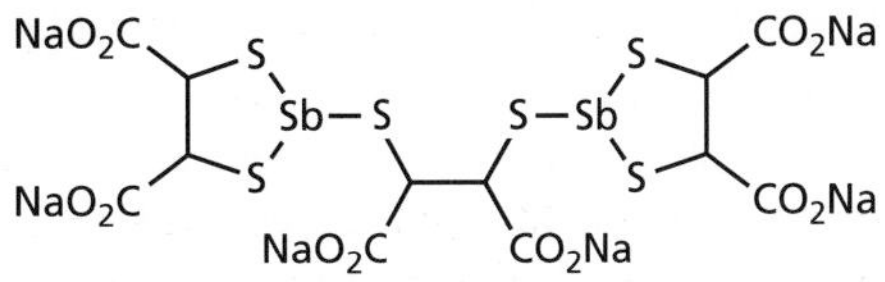

FIGURE CS4.2 Stibocaptate.

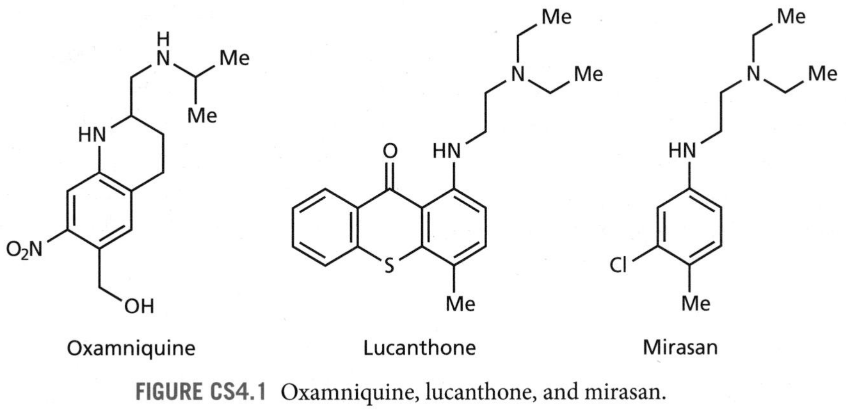

Oxamniquine   Lucanthone   Mirasan

FIGURE CS4.1 Oxamniquine, lucanthone, and mirasan.

This gave a compound called **mirasan** (Fig. CS4.1), which retained the right-hand aromatic ring containing the methyl and β-aminoethylamino side chains *para* to each other. *Varying substituents* (section 13.3.1.2) showed that an electronegative chloro substituent, positioned where the sulphur atom had been, was beneficial to activity. Mirasan was active against the bilharzia parasite in mice, but not in humans.

It was now reasoned that the β-aminoethylamino side chain was important in binding the drug to a target binding site and would adopt a particular conformation in order to bind efficiently (*active conformation*, section 13.2). This conformation would be only one of many conformations available to a flexible molecule such as mirasan, and so there would only be a limited chance of it being adopted at any one time. Therefore, it was decided to restrict the number of possible conformations by incorporating the side chain into a ring (*rigidification*, section 13.3.9). This would increase the chances of the molecule having the correct conformation when it approached its target binding site. There was the risk that the active conformation itself would be disallowed by this tactic and so, rather than incorporate all of the side chain into a ring, compounds were first designed such that only portions of the chain were included.

The bicyclic structure (I in Fig. CS4.3) contains one of the side chain bonds fixed in a ring to prevent rotation round that bond. It was found that this gave a dramatic improvement in activity. The compound was still not active in humans, but, unlike mirasan, it was active in monkeys. This gave hope that the chemists were on the right track. Further *rigidification* led to structure II in Fig. CS4.3, where two of the side chain bonds were constrained. This compound showed even more activity in mouse studies and it was decided to concentrate on this compound.

By now, the structure of the compound had been altered significantly from mirasan. When this is the case, it is advisable to check whether past results still hold true. For example, does the chloro group still have to be *ortho* to the methyl group? Can the chloro group be changed for something else? Novel structures may fit the binding site slightly differently from the lead compound, such that the binding groups are no longer in the optimum positions for binding.

Therefore, structure II was modified by *varying substituents* and *substitution patterns* on the aromatic ring (section 13.3.1.2), and by *varying alkyl substituents* on the amino groups (section 13.3.1.1). Chains were also *extended* (section 13.3.3) to search for other possible binding regions.

The results and possible conclusions were as follows:

- It was possible to vary substituents on the aromatic ring, but the substitution pattern itself could not be altered and was essential for activity. Altering the substitution pattern presumably places the essential binding groups out of position with respect to their binding regions.

- Replacing the chloro substituent with more electronegative substituents improved activity, with the nitro group being the best substituent. Therefore, an electron-deficient aromatic ring is beneficial to activity. One possible explanation for this could be the effect of the aromatic ring on the basicity of the cyclic nitrogen atom. A strongly electron-deficient aromatic ring would pull the cyclic nitrogen's lone pair of electrons into the ring, thus reducing its basicity (Fig. CS4.4). This, in turn, might alter the pK_a of the drug such that it is less ionized and is able to pass through the cell membranes of the gut and target cells more easily (see sections 14.1.4 and 14.1.5). The electronic effect of a

FIGURE CS4.3 Bicyclic structures I and II (restricted bonds in colour).

FIGURE CS4.4 Effect of aromatic substituents on pK_a.

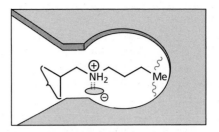

FIGURE CS4.5 Relative activity of amino side chains.

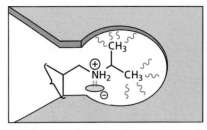

FIGURE CS4.7 Branching of the alkyl chain.

substituent on a distant functional group is a useful strategy in drug design (see sections 13.3.1.2, 14.1.5, and 19.5.1.8).

- The best activities were found if the amino group on the side chain was secondary rather than primary or tertiary (Fig. CS4.5).

- The alkyl group on this nitrogen could be increased up to four carbon units with a corresponding increase in activity. Longer chains led to a reduction in activity. The latter result might imply that large substituents are too bulky and prevent the drug from fitting the binding site. Acyl groups eliminated activity altogether, emphasizing the importance of this nitrogen atom. Most likely, it is ionized and interacts with the binding site through an ionic bond (Fig. CS4.6).

- Branching of the alkyl chain increased activity. A possible explanation is that branching increases van der Waals interactions to a hydrophobic region of the binding site (Fig. CS4.7). Alternatively, the lipophilicity of the drug might be increased, allowing easier passage through cell membranes.

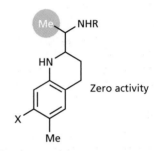

FIGURE CS4.8 Addition of a methyl group.

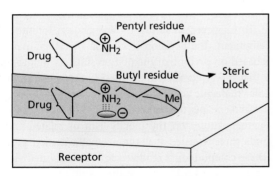

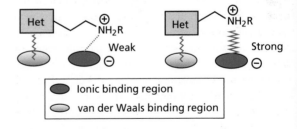

FIGURE CS4.9 Effect of extension of the side chain.

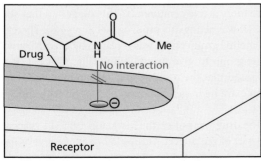

FIGURE CS4.6 Proposed ionic binding interaction.

- Putting a methyl group on the side chain eliminated activity (Fig. CS4.8). A methyl group is a bulky group compared with a proton and it is possible that it prevents the side chain taking up the correct binding conformation—*conformational blocking* (section 13.3.10).

- Extending the length of the side chain by an extra methylene group eliminated activity (Fig. CS4.9). This tactic was tried in case the binding groups were not far enough apart for optimum binding. This result suggests the opposite.

Asymmetric centre

III

IV

Piperazine ring

V

FIGURE CS4.10 The optimum structure (III) and the tricyclic structures (IV) and (V) (restricted bonds in colour).

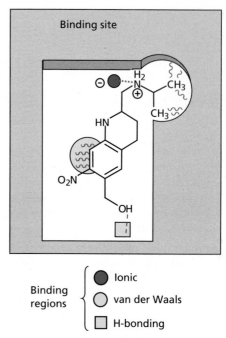

Binding regions { Ionic / van der Waals / H-bonding }

FIGURE CS4.11 Proposed binding interactions for oxamniquine to a binding site.

The optimum structure based on these results was structure III (Fig. CS4.10). It has one asymmetric centre and, as one might expect, the activity was much greater for one enantiomer than it was for the other.

The tricyclic structure IV (Fig. CS4.10) was also constructed. In this compound, the side chain is fully incorporated into a ring structure, drastically restricting the number of possible conformations (*rigidification*). As mentioned earlier, there was a risk that the active conformation would no longer be allowed, but, in this case, good activity was still obtained. The same *variations* as above were carried out to show that a secondary amine was essential (R = H) and that an electronegative group on the aromatic ring was required. However, some conflicting results were obtained compared with the previous results for structure III. A chloro substituent on the aromatic ring was better than a nitro group, and it could be in either of the two possible *ortho* positions relative to the methyl group. These results demonstrate that optimizing substituents in one structure does not necessarily mean that they will be optimum in another. One possible explanation for the chloro substituent being better than the nitro is that a less electronegative substituent is required to produce the optimum pK_a or basicity for membrane permeability (section 14.1.5).

Adding a further methyl group to the aromatic ring to give structure (V) (Fig. CS4.10) increased activity. It

was proposed that the bulky methyl group could interact with the piperazine ring, causing it to twist out of the plane of the other two rings (*conformational blocker*). The resulting increase in activity suggests that a better fitting conformation is obtained for the binding site.

Compound V was three times more active than structure III. However, structure III was chosen for further development. The decision to choose III rather than V was based on preliminary toxicity results, as well as the fact that it was cheaper to synthesize. Structure III is a simpler molecule and, in general, simpler molecules are easier and cheaper to synthesize.

Further studies on the metabolism of related compounds then revealed that the aromatic methyl group on these compounds is oxidized to a hydroxymethylene group (section 11.5.2) and that the resulting metabolites were more active compounds. This suggested that structure III was acting as a prodrug (section 14.6). Therefore, the methyl group on III was replaced by a hydroxymethylene group to give oxamniquine (Fig. CS4.11). It was proposed that the new hydroxyl group may be involved in an extra hydrogen bonding interaction with the binding site.

The drug was put on the market in 1975, 11 years after the start of the project. Oxamniquine is effective as a single oral dose for treating infections of *Schistosoma mansoni*. Side effects are relatively mild compared to those of lucanthone, the most frequent being dizziness, drowsiness, and headache, which can last for a few hours after administration. Although the drug is not effective against all three strains of the parasite, it met all the other goals of the project and proved to be a highly successful drug. It is still used today in countries such as Brazil. The contribution that the drug made to tropical medicine earned Pfizer the Queen's Award for Technological Achievement in 1979.

CS4.3 **Mechanism of action**

When oxamniquine was being developed, its mechanism of action and target binding site were unknown. Oxamniquine is now known to inhibit nucleic acid synthesis in schistosomal cells. The mechanism of action is thought to involve prior activation of the drug by a **sulphotransferase** enzyme that is present in parasitic cells, but not in mammalian cells. Once oxamniquine is bound to the active site of the schistosomal enzyme, the hydroxyl

group is converted to a sulphate ester (Fig. CS4.12). The ester is a much better leaving group than the original hydroxyl group, and so the molecule is now set up to dissociate. This is aided by the *para*-substituted nitrogen which can feed its lone pair of electrons into the aromatic ring. The structure that is formed is an alkylating agent which alkylates the DNA of the parasite and prevents DNA replication. This theory fits in nicely with the SAR results previously described, which emphasize the importance of the hydroxyl group, the aromatic amine, and the electron-deficient aromatic ring. It also explains why the agent is selectively toxic for the parasite rather than mammalian cells. Therefore, oxamniquine is acting as a prodrug (section 14.6.6), which is activated by the parasitic sulphotransferase enzyme.

CS4.4 **Other agents**

The knowledge that the CH_2OH group is crucial to the activity and mechanism of action of oxamniquine led to

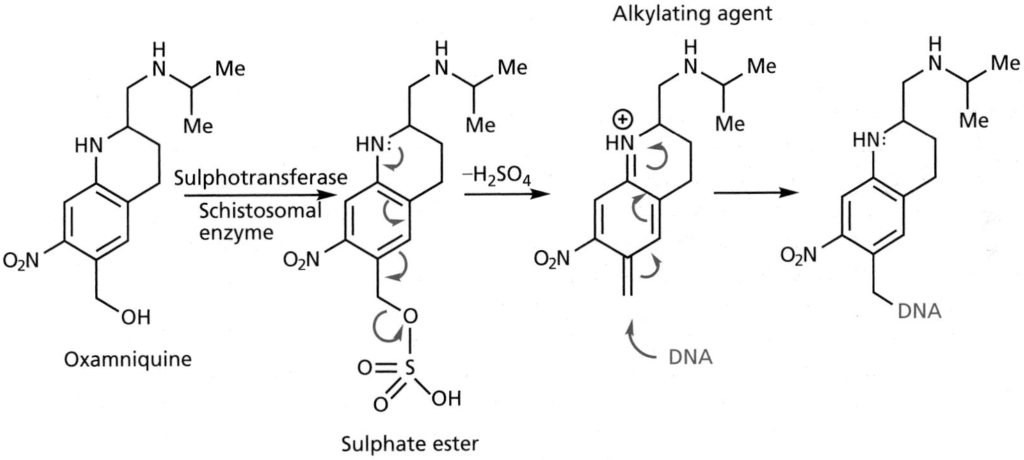

FIGURE CS4.12 Mechanism by which oxamniquine might dissociate to form an alkylating agent.

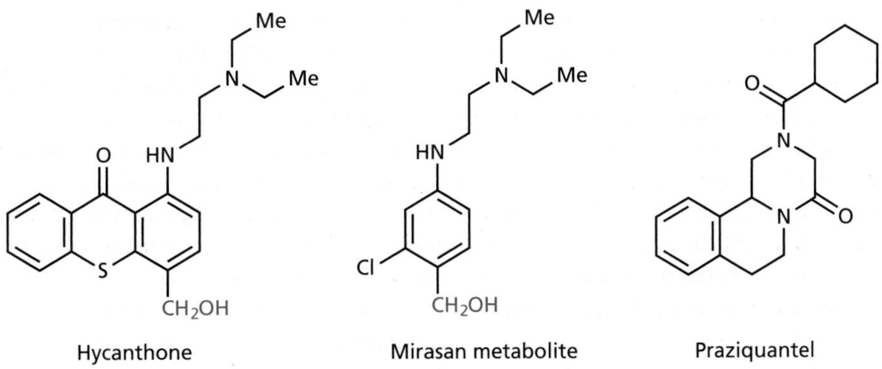

FIGURE CS4.13 Hycanthone, mirasan metabolite, and praziquantel.

BOX CS4.1   Synthesis of oxamniquine

One method of synthesizing oxamniquine is to start from the quinoline structure (I; Fig. 1). The methyl substituent on the heterocyclic ring is selectively chlorinated and the alkyl chloride (II) undergoes a nucleophilic substitution with 2-aminopropane to form structure III. Reduction with hydrogen gas over a nickel catalyst gives the tetrahydroquinoline IV, which is nitrated to give a mixture of isomers. These are separated and the desired isomer is then hydroxylated in the presence of the fungus *Aspergillus sclerotiorum*. Microbial enzymes catalyse the oxidation reaction.

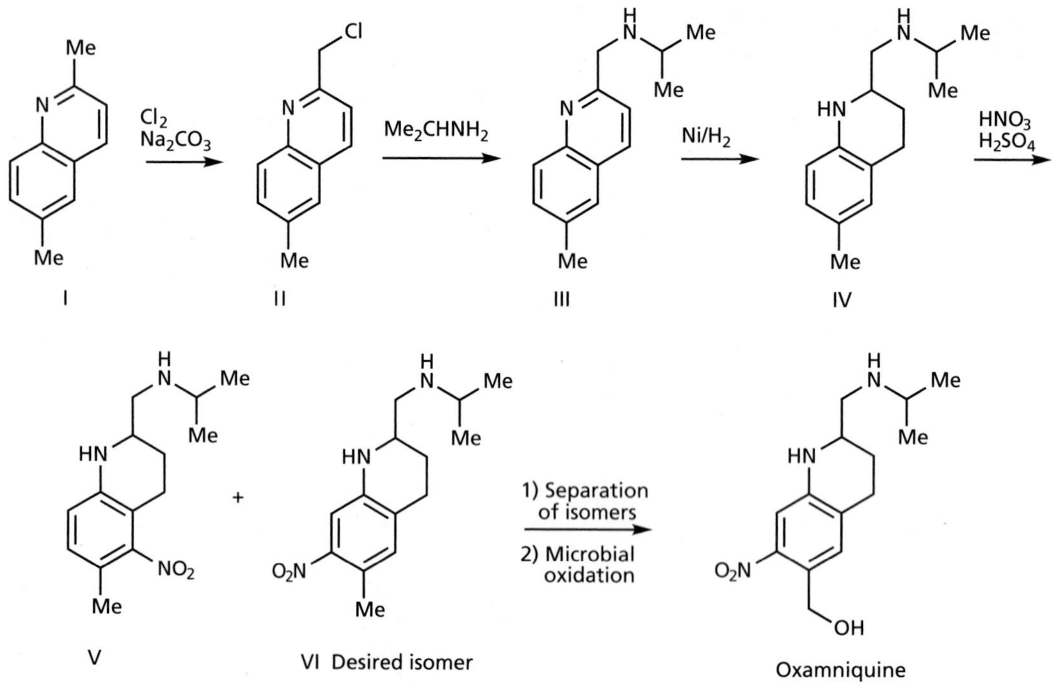

FIGURE 1 Synthesis of oxamniquine.

further investigations on lucanthone and mirasan. Studies showed that lucanthone was being oxidized *in vivo* to a metabolite called **hycanthone** (Fig. CS4.13), which then acts as a prodrug in a similar manner to oxamniquine. Hycanthone was shown to be more active than lucanthone and replaced it for the treatment of schistosomal infection. It was widely used in the 1970s, at which time it was considered the best treatment available. However, it suffered the same toxic side effects as lucanthone and was suspected of being mildly carcinogenic. It was gradually withdrawn from the market to be replaced by safer and more potent agents such as oxamniquine.

Mirasan was also found to be oxidized *in vivo* to a metabolite (Fig. CS4.13) that proved to be active in a range of species. However, the metabolite was never used clinically. It is likely that the observed inactivity of mirasan in monkeys is due to it being resistant to metabolic oxidation in that species.

Praziquantel (Fig. CS4.13) is now the recommended treatment for schistosomiasis in the UK, as it is active against all three parasitic strains. Unfortunately, it is more expensive than oxamniquine which limits its use in less affluent nations. New agents would be desirable, but schistosomicides are not economically attractive to the pharmaceutical industry.

FURTHER READING

Cioli, D., et al. (1995) Antischistosomal drugs: past, present... and future? *Pharmacology & Therapeutics*, **68**, 35–85.

Filho, S. B., et al. (2002) Synthesis and evaluation of new oxamniquine derivatives. *International Journal of Pharmaceutics*, **233**, 35–41.

Filho, R. P., et al. (2007) Design, synthesis, and in vivo evaluation of oxamniquine methacrylate and acrylamide prodrugs. *Bioorganic and Medicinal Chemistry*, **15**, 1229–36.

Roberts, S. M., and Price, B. J. (eds) (1985) Oxamniquine: a drug for the tropics. *Medicinal chemistry—the role of organic research in drug research*. Chapter 14, Academic Press, London.

PART D

Tools of the trade

In Part D, we shall study three topics which are invaluable tools in the discovery and design of drugs. The topics covered are combinatorial and parallel synthesis, molecular modelling, and quantitative structure–activity relationships. It should be emphasized that these are not the only topics that could be considered as 'tools of the trade'. For example, a detailed knowledge of organic synthesis is clearly vital in order to synthesize new compounds and for drug design. There is no point designing drugs that are impossible to synthesize. However, organic synthesis is a major area in itself and a single chapter would do no justice to it, especially when there are many excellent undergraduate texts covering the subject in depth. Instead, examples of synthetic routes used to produce important drugs have been included as Boxes within the case studies and the various chapters of Part E. To read more about drug synthesis, see the companion volume to this text *An introduction to drug synthesis*, Oxford University Press (2015), by the same author.

Combinatorial and parallel synthesis are the topics covered in Chapter 16. These are methods of rapidly preparing large numbers of compounds in an automated or semi-automated fashion. The techniques were developed to meet the urgent need for new lead compounds for the ever-increasing number of novel targets discovered by genomic and proteomic projects. Parallel synthesis, in particular, is now an effective method of producing large numbers of analogues for drug discovery, studies into structure–activity relationships, and drug optimization.

In Chapter 17, we look at some of the operations that can be carried out using computers, and which aid the drug design process. Computers and molecular modelling software packages have now become an integral part of the drug design process and have been instrumental in a more scientific approach to medicinal chemistry. Molecular modelling is key to structure-based drug design and *de novo* drug design. The website for this volume contains a variety of molecular modelling exercises that can be tried out if you have access to the software packages Spartan or ChemBio3D.

In Chapter 18, we look at quantitative structure–activity relationships (QSAR). This topic has been around for many years and it is a well-established tool in medicinal chemistry. QSAR attempts to relate the physicochemical properties of compounds to their biological activity in a quantitative fashion by the use of equations. In traditional QSAR, this typically involves synthesizing a series of analogues with different substituents, and studying how the physicochemical properties of the substituents affect the biological activities of the analogues. Typically, the hydrophobic, steric, and electronic properties of each substituent are considered when setting up a QSAR equation. With the advent of computers and suitable software programs, traditional QSAR studies have been largely superseded by three-dimensional quantitative structure–activity relationships (3D QSAR), where the physicochemical properties of the complete molecule are calculated and then related to biological activity.

All of these tools are extremely important in medicinal chemistry research, but they are meant to be used in combination with each other and should also be used when they are appropriate. Critics have sometimes argued that none of these tools has ever resulted in a clinically useful drug per se. That is not the point. One cannot build a house with just a hammer, so it is unrealistic to suggest that a drug can be discovered by using a single scientific tool. The case study in this section (Case study 5) illustrates an early example of *de novo* drug design. The website contains an article (Web article 5) that illustrates how the tools mentioned in Chapters 16, 17, 18 are applied alongside more traditional drug design strategies. There are also exercises on QSAR that involve the use of Excel.

16 Combinatorial and parallel synthesis

Combinatorial and parallel synthesis have become established tools in drug discovery and drug development, allowing the use of a defined reaction route to produce a large number of compounds in a short period of time. The full set of compounds produced in this way is called a **compound library**. Reactions are usually carried out on small scale and the process can be automated or semi-automated, allowing reactions to be carried out in several reaction vessels at the same time and under identical conditions, but using different reagents for each vessel. The compact nature of the apparatus means that the process can be carried out within a normal fume cupboard.

16.1 Combinatorial and parallel synthesis in medicinal chemistry projects

In the past, medicinal chemistry involved the identification of a lead compound having a useful activity which was then modified to develop a clinically useful drug. Identification of the molecular target for the drug, and the mechanism by which it worked often took many years to establish. Today, most medicinal chemistry projects start with an identifiable target, and the emphasis is on discovering a lead compound that will interact with

this target. This reversal of priorities came about as a result of the human genome project and the proteomic revolution that followed. Once the genome was mapped, a vast number of previously unknown proteins were identified, all of which could be considered as potential drug targets. Pharmaceutical companies were faced with the problem of identifying the function of each target and finding a lead compound to interact with it. Before the advent of combinatorial chemistry and parallel synthesis, the need to find a lead compound was becoming the limiting factor in the whole process. Now, with the aid of these techniques, research groups can rapidly synthesize and screen thousands of structures in order to find new lead compounds, identify structure–activity relationships, and find analogues with good activity and minimal side effects (Fig. 16.1).

The procedures used in combinatorial synthesis are designed to produce mixtures of different compounds within each reaction vessel, whereas those used in parallel synthesis produce a single product in each vessel. In general, parallel synthesis is favoured because it is easier to identify the structures that are synthesized. However, there is still scope for combinatorial chemistry in finding lead compounds, especially since this procedure can generate significantly more structures in a set period of time, thus increasing the chances of finding a lead compound. Both methods generally involve the use of solid-phase techniques which are discussed in the next section.

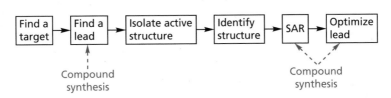

FIGURE 16.1 The stages in drug development requiring synthesis of large numbers of compounds (SAR = structure–activity relationships).

16.2 Solid-phase techniques

Solid-phase techniques can be used to carry out reactions where the starting material is linked to a solid support such as a resin bead. Several reactions can then be carried out in sequence on the attached molecule. The final structure is then detached from the solid support. There are several advantages to this:

- since the starting material, intermediates, and final product are bound to a solid support, excess reagents or unbound by-products from each reaction can easily be removed by washing the resin with a suitable solvent;

- large excesses of reagents can be used to drive the reactions to completion (greater than 99%) because of the ease with which excess reagent can be removed;

- intermediates in a reaction sequence are bound to the bead and do not need to be purified;

- the polymeric support can be regenerated and reused if appropriate cleavage conditions and suitable anchor/linker groups are chosen (see section 16.2.2);

- automation is possible;

- if a combinatorial synthesis is being carried out, a range of different starting materials can be bound to separate beads. The beads can be mixed together such that all the starting materials are treated with another reagent in a single experiment. The starting materials and products are still physically distinct, as they

are bound to separate beads. The individual beads can be separated at the end of the experiment to give individual products. In solution chemistry, mixing all the starting materials together is a recipe for disaster, with polymerizations and side reactions producing a tarry mess.

The essential requirements for solid-phase synthesis are:

- a cross-linked insoluble polymeric support which is inert to the synthetic conditions (e.g. a resin bead);

- an anchor or linker covalently linked to the resin— the anchor has a reactive functional group that can be used to attach a substrate;

- a bond linking the substrate to the linker, which will be stable to the reaction conditions used in the synthesis;

- a means of cleaving the product or the intermediates from the linker;

- protecting groups for functional groups not involved in the synthetic route.

16.2.1 The solid support

The first successful example of solid-phase synthesis was the Merrifield peptide synthesis. The resin involved consists of polystyrene beads where the styrene is partially cross-linked with 1% divinylbenzene. The beads are derivatized with a chloromethyl group (the anchor/linker) to which amino acids can be coupled via an ester group

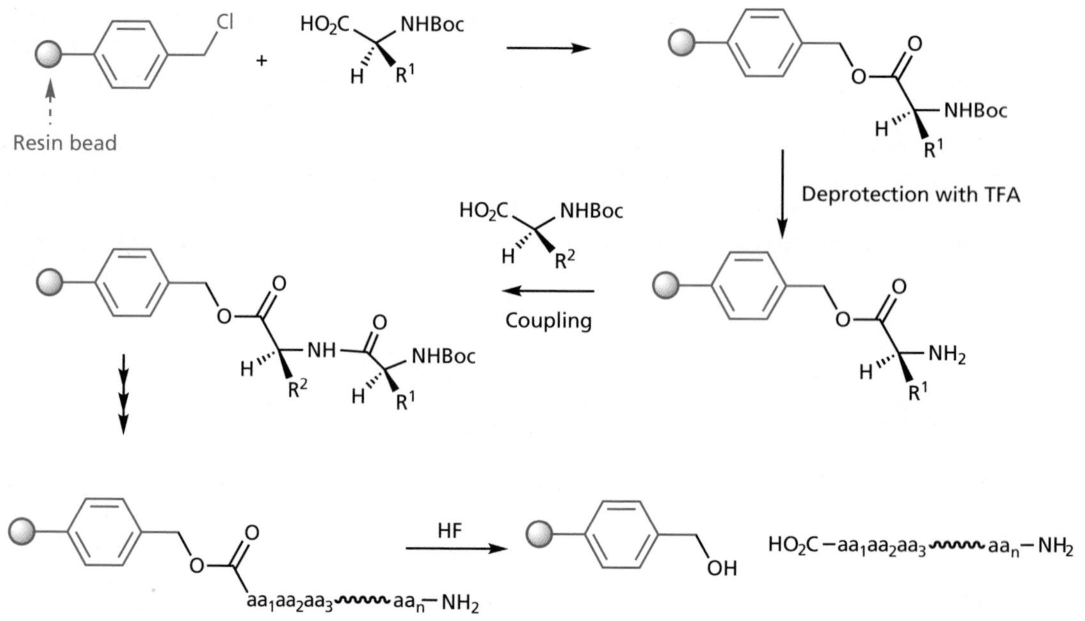

FIGURE 16.2 Peptide synthesis on a solid support (Boc = *tert*-butyloxycarbonyl = *t*-BuO-CO; TFA = trifluoroacetic acid).

(Fig. 16.2). This ester group is stable to the reaction conditions used in peptide synthesis, but can be cleaved at the end of the synthesis using vigorous acidic conditions (hydrofluoric acid).

One disadvantage of polystyrene beads is the fact that they are hydrophobic and the growing peptide chain is hydrophilic. As a result, the growing peptide chain is not solvated and often folds in on itself to form internal hydrogen bonds. This, in turn, hinders access of further amino acids to the exposed end of the growing chain. To address this, more polar solid phases were developed such as **Sheppard's polyamide resin**. Other resins have been developed to be more suitable for the synthesis of non-peptides. For example, **TentaGel resin** is 80% polyethylene glycol grafted to cross-linked polystyrene, and provides an environment similar to ether or tetrahydrofuran. Regardless of the polymer that is used, the bead should be capable of swelling in solvent while remaining stable. Swelling is important because most of the reactions involved in solid-phase synthesis take place in the interior of the bead rather than on the surface. It is wrong to think of resin beads as being like miniature marbles with an impenetrable surface. Each bead is a polymer and swelling involves unfolding of the polymer chains such that solvent and reagents can move between the chains into the heart of the polymer (Fig. 16.3).

Although beads are the common shape for the solid support, a range of other shapes such as pins have been designed to maximize the surface area available for reaction, and, hence, maximize the amount of compound linked to the solid support. Functionalized glass surfaces have also been used and are suitable for oligonucleotide synthesis.

16.2.2 The anchor/linker

The anchor/linker is a molecular unit covalently attached to the polymer chain making up the solid support. It contains a reactive functional group with which the starting material in the proposed synthesis can react and, hence,

become attached to the resin. The resulting link must be stable to the reaction conditions used throughout the synthesis, but be easily cleaved to release the final compound once the synthesis is complete (Fig. 16.4). As the linkers are distributed along the length of the polymer chain, most of them will be in the interior of the polymer bead, emphasizing the importance of the bead swelling if the starting material is to reach them.

Different linkers are used depending on:

- the functional group which will be present on the starting material;
- the functional group which is desired on the final product once it is released.

Resins having different linkers are given different names (Fig. 16.5). For example, the **Wang resin** has a linker which is suitable for the attachment and release of carboxylic acids. It can be used in peptide synthesis by linking an *N*-protected amino acid to the resin by means of an ester link. This ester link remains stable to coupling and deprotection steps in the peptide synthesis, and can then be cleaved using trifluoroacetic acid (TFA) to release the final peptide from the bead (Fig. 16.6). One problem with the Wang resin is that the first amino acid linked to the resin is prone to racemization. The **Barlos resin** contains a trityl linker and was designed to avoid this problem. The final product can be cleaved under very mild conditions (e.g. HOAc/TFE/CH_2Cl_2 or TFA/CH_2Cl_2) due to the high stability of the trityl cations that are formed. Molecules can also be linked to the resin by means of an alcohol group.

Starting materials with a carboxylic acid (RCO_2H) can be linked to the **Rink resin** via an amide link. Once the reaction sequence is complete, treatment with TFA releases the product with a primary amide group, rather than the original carboxylic acid ($R'CONH_2$; Fig. 16.7).

Primary and secondary alcohols (ROH) can be linked to a dihydropyran-functionalized resin. Linking the alcohol is done in the presence of pyridinium 4-toluenesulphonate (PPts) in dichloromethane. Once the reaction

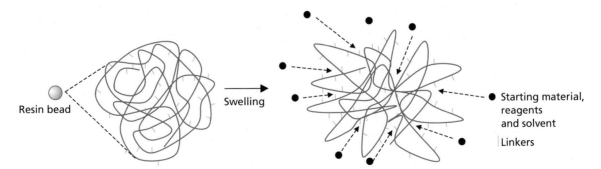

FIGURE 16.3 Swelling of a resin bead allowing access of reagents and solvent.

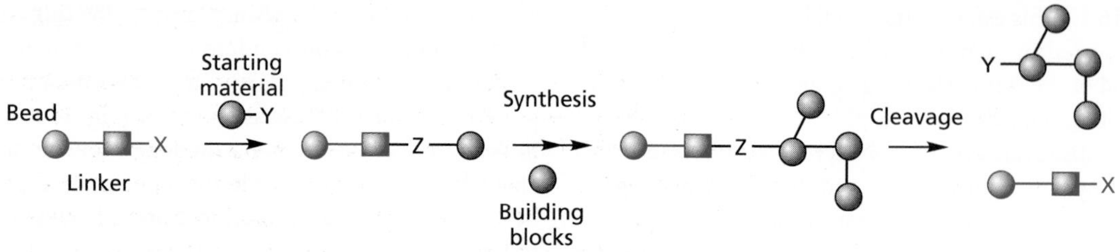

FIGURE 16.4 The principles of an anchor/linker. X, Y, and Z are functional groups.

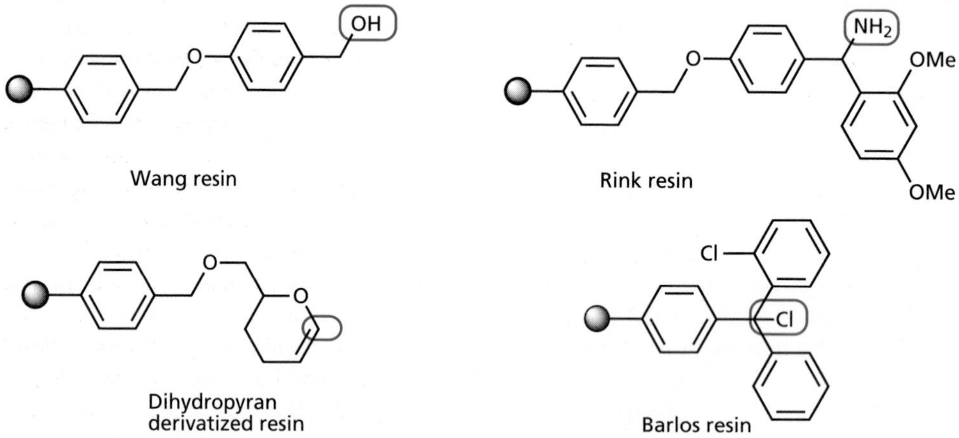

FIGURE 16.5 Types of resin with the linkage point circled.

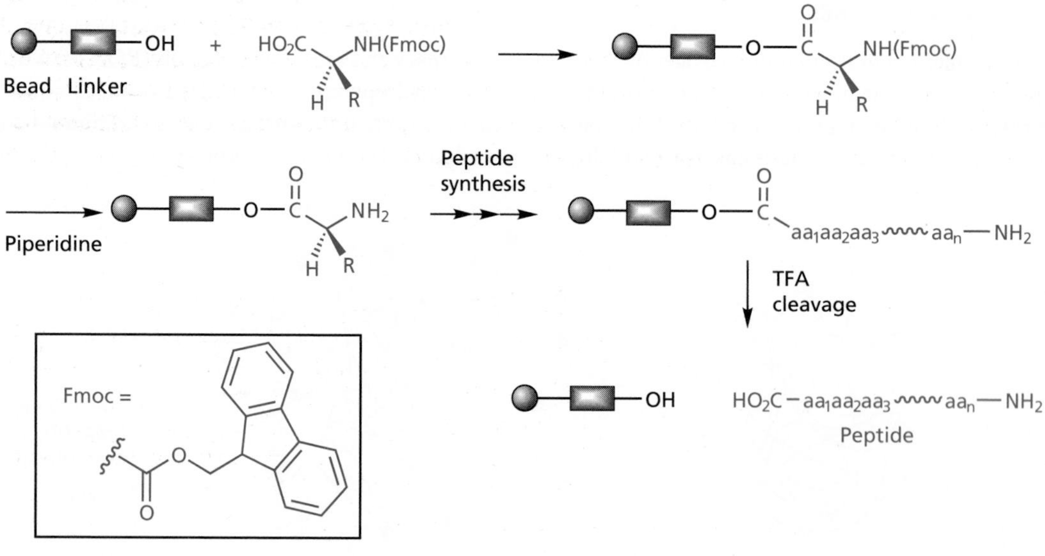

FIGURE 16.6 Peptide synthesis with a Wang resin—the structure of the linker is shown in Fig. 16.5.

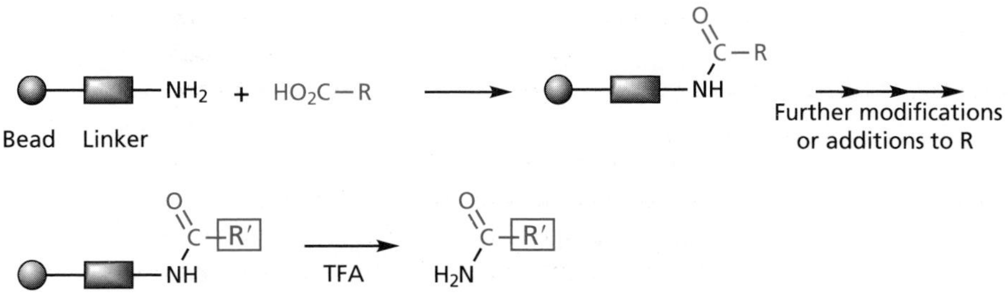

FIGURE 16.7 Solid-phase synthesis with a Rink resin (R contains functional groups which allows further modifications of the molecule to give R′). The structure of the linker is shown in Fig. 16.5.

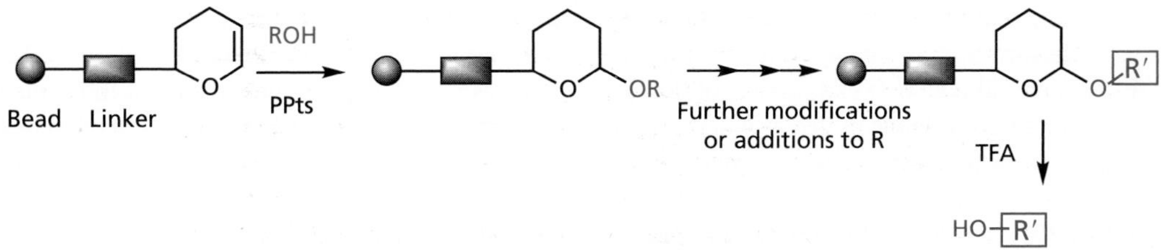

FIGURE 16.8 Solid-phase synthesis with a dihydropyran-functionalized resin (R contains functional groups which allows further modifications of the molecule to give R′).

sequence has been completed, cleavage can be carried out using TFA (Fig. 16.8).

16.2.3 Examples of solid-phase syntheses

Solid-phase synthesis was pioneered by Merrifield, and most of the early work involved peptide synthesis. However, peptides pose particular problems as drugs in terms of their pharmacokinetic properties (section 14.8.2), and so a large amount of research was carried out to extend solid-phase synthetic methods to the synthesis of small non-peptide molecules. The first move away from natural peptides was to use the same peptide coupling procedures, but with non-natural amino acids. Peptides could also be modified once they were built by reactions such as *N*-methylation. *N*-Substituted glycine units were used to produce structures known as **peptoids** where the

side chain is attached to the nitrogen rather than the α-carbon. Some of these have been shown to be ligands for various important receptors and show increased metabolic stability.

A disadvantage with all the above structures is the fact that they are linear, flexible molecules linked together by a regular molecular backbone. The real interest in solid-phase synthesis began when it became possible to produce heterocyclic structures. Heterocycles are less susceptible to metabolism, and have better pharmacokinetic properties. They are more rigid, and diversity is possible by varying the substituents around the heterocyclic 'core'.

1,4-Benzodiazepines have been synthesized by linking a selection of amino acids to resin beads through the carboxylic acid group (Fig. 16.9). Reaction with a variety of imines gave the adducts shown. Treatment with TFA released the adducts which then cyclized to give the final

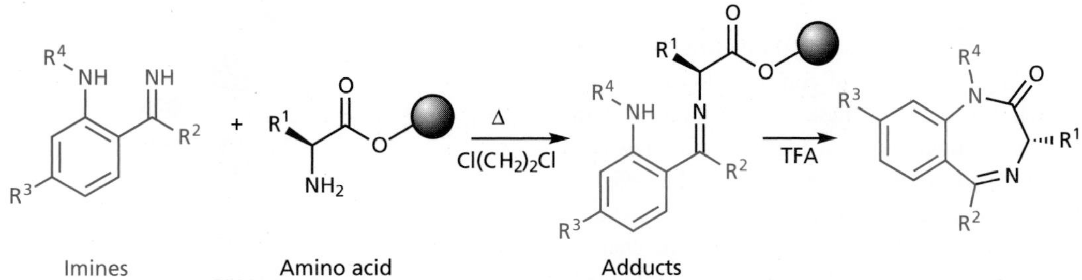

FIGURE 16.9 Benzodiazepine synthesis involving a cyclo-release strategy.

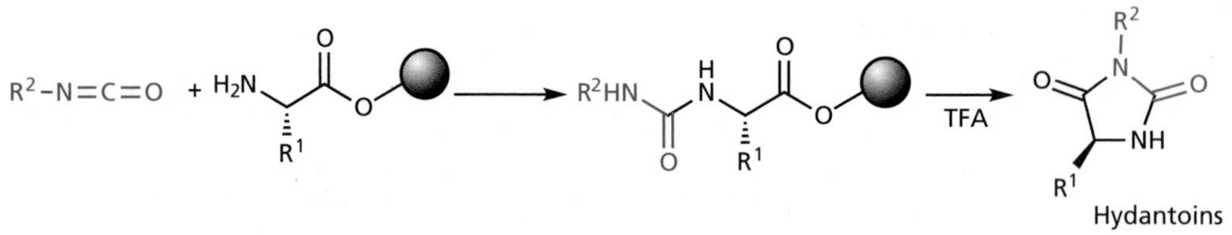

FIGURE 16.10 Synthesis of hydantoins.

products. The advantage of this synthesis lies in the fact that the functional group released from the resin takes part in the final cyclization and does not remain as an extra, and possibly redundant, group. The final product has four variable substituents spread evenly around the bicyclic ring system. This allows exploration of conformational space around the whole molecule when searching for binding interactions with a drug target (see section 16.3.1).

A similar strategy was employed for the synthesis of hydantoins (Fig. 16.10), and a large variety of heterocyclic compounds have now been synthesized using solid-phase methods.

The range of reactions which can be carried out on solid phase has also been extended: most common reactions are now feasible, including moisture-sensitive and organometallic reactions. For example, aldol condensations, DIBAL reductions, Wittig reactions, LDA reductions, Heck couplings, Stille couplings, and Mitsunobu reactions are all possible. Automated or semi-automated synthesizers can cope with 6, 12, 42, 96, or 144 reaction vials depending on the instrument and the size of the reaction tubes used. The addition of solvent, starting materials, and reagents can be carried out automatically using syringes. Automated work-up procedures such as the removal of solvent, washing, and liquid–liquid separations are also possible. Reactions can be stirred and carried out under inert atmospheres, and the reactions can be heated or cooled as required.

16.3 Planning and designing a compound library

The techniques of solid-phase synthesis have been used to produce large quantities of compounds from a particular reaction sequence. These can be stored as compound libraries, then accessed to search for new lead compounds capable of interacting with novel or existing drug targets. It is important that the molecules in these libraries are structurally diverse to increase the chances of success, and so some thought has to be put into planning and designing a compound library.

16.3.1 'Spider-like' scaffolds

A compound library is generated from a specific sequence of reactions, so there is a danger that it will contain a large number of very similar molecules. Therefore, care has to be taken about the type of molecule synthesized, the synthetic route employed, and the types of substituents involved, in order to achieve structural diversity. In general, it is best to synthesize 'spider-like' molecules, so called because they consist of a central body (called the **centroid** or **scaffold**) from which various 'arms' (substituents) radiate (Fig. 16.11). These arms contain different functional groups which are used to probe a binding site for binding regions once the spider-like molecule has entered (Fig. 16.12). The chances of success are greater if the 'arms' are evenly spread around the scaffold, as this allows a more thorough exploration of the three-dimensional space (**conformational space**) around the molecule. The molecules made in the synthesis are planned in advance to ensure that they contain different functional groups on their arms, placed at different distances from the central scaffold.

16.3.2 Designing 'drug-like' molecules

The 'spider-like' approach increases the chances of finding a lead compound which will interact with a target

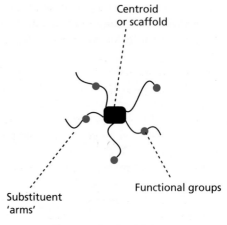

FIGURE 16.11 'Spider-like' molecule.

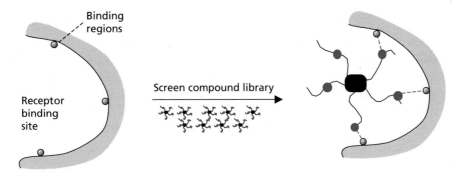

FIGURE 16.12 Probing for an interaction.

binding site, but it is also worth remembering that compounds with good binding interactions do not necessarily make good medicines. There are also the pharmacokinetic issues to take into account (Chapter 11), and so it is worthwhile introducing certain restrictions to the types of molecule that will be produced in order to increase the chances that the lead compound will be orally active. In general, the chances of oral activity are increased if the structure obeys Lipinski's rule of five or Veber's parameters (section 11.3). However, allowance has to be made for the fact that any lead compounds identified are almost certainly going to require substantial optimization, in which case more stringent guidelines should be applied (section 12.4.11). Other restrictions should be considered. For example, groups such as esters should be avoided because they are easily metabolized. Scaffolds or substituents likely to result in toxic compounds should also be avoided; for example alkylating groups or aromatic nitro groups.

16.3.3 Synthesis of scaffolds

Most scaffolds are constructed using the synthetic route employed for the solid-phase synthesis, and this also determines the number and variety of substituents that can be attached to the scaffold. The ideal scaffold should be small in order to allow a wide variation of substituents (see Box 16.1). It should also have its substituents widely dispersed round its structure (spider-like) rather than restricted to one part of the structure (tadpole-like) if the conformational space around it is to be fully explored (Fig. 16.13). Finally, the synthesis should allow each of the substituents to be varied independently of each other.

Scaffolds can be flexible (e.g. a peptide backbone) or rigid (a cyclic system). They may contain heteroatoms that are capable of forming useful bonding interactions with the binding site, or they may not. Some scaffolds are already common in medicinal chemistry (e.g.

benzodiazepine, hydantoin, tetrahydroisoquinoline, benzenesulphonamide, and biaryls) and are associated with a diverse range of activities. Such scaffolds are termed **privileged scaffolds**.

16.3.4 Substituent variation

The variety of substituents chosen in a combinatorial synthesis depends on their availability and the diversity required. This would include such considerations as structure, size, shape, lipophilicity, dipole moment, electrostatic charge, and functional groups present. It is usually best to identify which of these factors should be diversified before commencing the synthesis.

16.3.5 Designing compound libraries for lead optimization

If a compound library is being planned for drug optimization, the variations planned should take into account several factors such as the biological and physical properties of the compound, its binding interactions with the target, and the potential problems of particular substituents. For example, if the binding interactions of a target receptor with its usual ligand are known, this knowledge can be used to determine what size of compounds it would be best synthesizing, the types of functional

'Spider' scaffold with dispersed substituents   'Tadpole' scaffold with restricted substituents

FIGURE 16.13 Dispersed and restricted substituents.

BOX 16.1 Examples of scaffolds

Benzodiazepines, hydantoins, β-lactams, and pyridines are examples of extremely good scaffolds. They all have small molecular weights, and there are various synthetic routes available which produce the substitution patterns required to fully explore the conformational space about them. For example, it is possible to synthesize benzodiazepines such that there are variable substituents round the whole structure.

Peptide scaffolds are flexible scaffolds which have the capacity to form hydrogen bonds with target binding sites. They are easy to synthesize and a large variety of different substituents are possible by using the amino acid building blocks. Further substitution is possible on the terminal amino and carboxylic acid functions. The substituents are widely distributed along the peptide chain allowing a good exploration of conformational space. If we consider Lipinski's rule of five, the peptide scaffold should ideally be restricted to di- and tripeptides in order to keep the molecular weight below 500. It is interesting to note that the orally active antihypertensive agents **captopril** and **enalapril** are dipeptide-like, whereas larger peptides such as the **enkephalins** are not orally active. Oral activity has also been a problem with HIV protease inhibitors having molecular weights over 500 (section 20.7.4).

Some of the scaffolds shown in Fig. 1 have various disadvantages. Although **glucose** has a small molecular weight and the possibility of five variable substituents, it contains multiple hydroxyl groups. Attaching different substituents to similar groups usually requires complex protection and deprotection strategies. Nevertheless, the potential of sugar-based drugs is so great that a lot of progress has been made in developing solid-phase syntheses based on sugar scaffolds.

Steroids might appear attractive as scaffolds. However, the molecular weight of the steroid skeleton itself (314) limits the size and number of the substituents which can be added if we wish to keep the overall molecular weight below 500.

The indole scaffold shown suffers a disadvantage in having its variable substituents located in the same region of the molecule, preventing a full exploration of conformational space (i.e. it is a 'tadpole-like' scaffold).

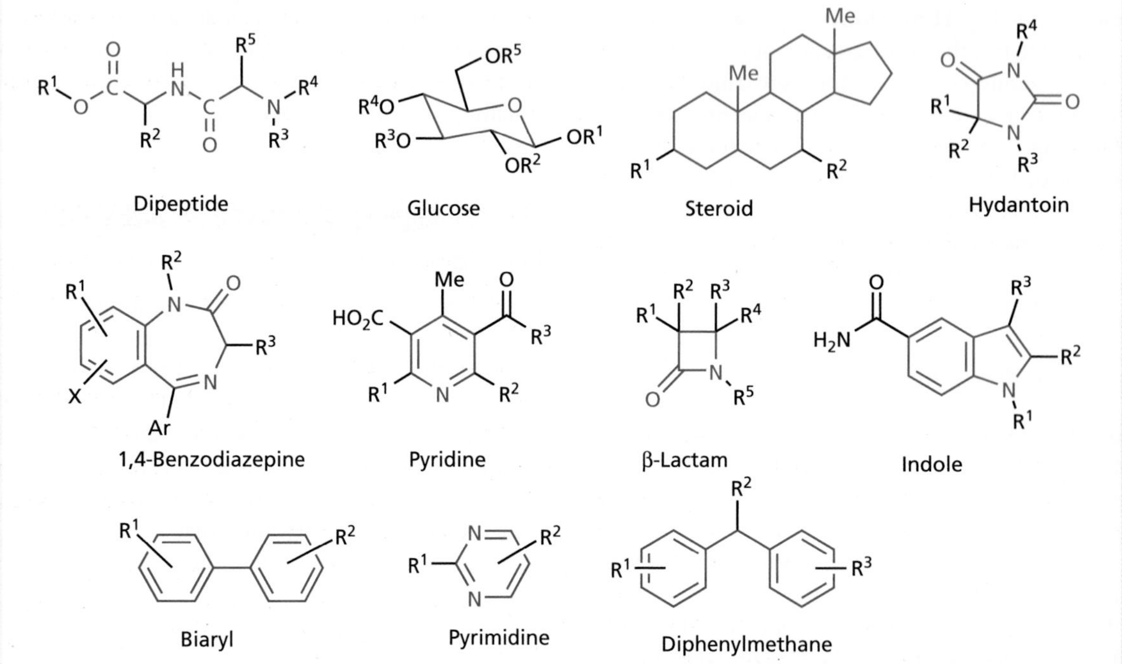

FIGURE 1 Examples of scaffolds.

groups that ought to be present, and their relative positions. For example, if the target is a zinc-containing protease (e.g. angiotensin-converting enzyme), a library of compounds containing a carboxylic acid or thiol group would be relevant.

16.3.6 Computer-designed libraries

It has been claimed that half of all known drugs involve only 32 scaffolds. Furthermore, it has been stated that a relatively small number of moieties account for the large majority of side chains in known drugs. This may imply that it is possible to define 'drug-like molecules' and use computer software programs to design more focused compound libraries. Descriptors used in this approach include log *P*, molecular weight, number of hydrogen bond donors, number of hydrogen bond acceptors, number of rotatable bonds, aromatic density, the degree of branching in the structure, and the presence or absence of specific functional groups. One can also choose to filter out compounds that do not obey the rules mentioned in section 16.3.2. Computer programs can also be used to identify the structures which should be synthesized in order to maximize the number of different pharmacophores produced (section 17.16).

16.4 Testing for activity

We shall now look in more detail at how the structures in a compound library are tested for biological activity.

16.4.1 High-throughput screening

Because solid-phase syntheses can produce a large quantity of structures in a very short period of time, biological testing has to be carried out quickly and automatically. The process is known as high-throughput screening (HTS) and was developed before combinatorial and parallel synthetic methods were devised. Indeed, the existence of HTS was one of the driving forces that led to the development of these synthetic procedures. Since biological testing was so rapid and efficient, the pharmaceutical companies soon ran out of novel structures to test, and the synthesis of new structures became the limiting factor in the whole process of drug discovery. Combinatorial and parallel syntheses have solved that problem and the number of new compounds synthesized each year has increased dramatically. In fact, there are now so many compounds being produced that the focus is on making HTS even more efficient. Traditionally, compounds are automatically tested and analysed on a plate containing 96 small wells, each with a capacity of 0.1 ml. There is now a move to use test plates of similar size but containing 1536 wells, where the test volumes are only 1–10 µl. Moreover, methods such as fluorescence and chemiluminescence are being developed which will allow the simultaneous identification of active wells. Further miniaturization of open systems is unlikely because of the problems of evaporation involving small volumes less than 1 µl. However, miniaturization using closed systems is on the horizon. The next major advance will be in the science of **microfluidics**, which involves the manipulation of tiny volumes of liquids in confined space. Microfluidic circuits on a chip can be used to control

fluids electronically, allowing separation of an analytical sample using capillary electrophoresis. Companies are now developing machines that combine ultra-small-scale synthesis (section 16.5.5) and miniaturized analysis. A single 10×10 cm^2 silicon wafer can be microfabricated to support 10^5 separate syntheses/bioassays on a nanolitre scale!

16.4.2 Screening 'on bead' or 'off bead'

Sometimes structures can be tested for biological activity when they are still attached to the solid phase. 'On bead' screening assays involve interactions with targets which are tagged with an enzyme, fluorescent probe, radionuclide, or chromophore. A positive interaction results in a recognizable effect such as fluorescence or a colour change. These screening assays are rapid, and 10^8 beads can be readily screened. Active beads can then be picked out by micromanipulation and the structure of the active compound determined.

A false negative might be obtained if the solid phase sterically interferes with the assay. If such interference is suspected, it is better to release the drug from the solid phase before testing. This avoids the uncertainty of false negatives. On the other hand, there are cases where the compounds released prove to be insoluble in the test assay and give a negative result, whereas they give a positive result when attached to the bead.

KEY POINTS

- Solid-phase synthetic methods have proved valuable in producing compounds for lead discovery, structure–activity relationships and drug optimization.

- Parallel synthesis involves the synthesis of a different compound in each reaction vial, and is useful in all aspects of medicinal chemistry where synthesis is required.

- Combinatorial synthesis involves the synthesis of mixtures of compounds in each reaction vial, and has been useful in discovering lead compounds.

- Solid-phase synthesis has several advantages. Intermediates do not need to be isolated or purified. Reactants and reagents can be used in excess to drive the reaction to completion. Impurities and excess reagents or reactants are easily removed.

- In combinatorial syntheses, different compounds are linked to different solid-phase surfaces such that they are physically separated, allowing them to undergo reactions and work-up procedures in the same reaction vessel.

- The solid support consists of a polymeric surface and a linker molecule which allows a starting material to be covalently linked to the support.

- Different linkers are used depending on the functional group present on the starting material and the functional group which is desired on the product.

- A scaffold is the core structure of a molecule round which variations are possible through the use of different substituents.

- Spider-like scaffolds allow substituent variation around the whole molecule, making it possible to explore all the conformational space around the scaffold. This increases the possibility of finding a lead compound which will bind to a target binding site.

- The chances of identifying an orally active lead compound can be improved by applying Lipinski's rule of five when planning compound libraries. More rigid guidelines may be used if the lead compound is likely to undergo substantial optimization.

- A privileged scaffold is one which is commonly present in known drugs.

- Computer software is available to assist in the planning of compound libraries.

- High-throughput screening automatically analyses the biological activity of large numbers of samples against defined targets. The analysis requires only small quantities of each sample.

- Screening can be carried out on compounds attached to resin beads, or on compounds which have been released into solution.

16.5 Parallel synthesis

In parallel synthesis, a reaction is carried out in a series of wells such that each well contains a single product. This method is a 'quality rather than quantity' approach and is often used for focused lead optimization studies. For parallel synthesis to be fast and efficient, it is necessary to remove or simplify the bottlenecks associated with classical organic synthesis. These include laborious work-ups, extractions, solvent evaporations, and purifications. A typical medicinal chemist may synthesize one or two new entities a week using traditional methods of organic synthesis. With parallel synthesis, that same researcher can synthesize a dozen or more pure molecules,

thus increasing the synthetic output, and speeding up the lead optimization process. Parallel synthesis can be carried out on solid phase with all the advantages that entails (section 16.2). However, parallel syntheses can also be carried out in solution (**solution-phase organic synthesis (SPOS)**). We will now focus on methods that increase the efficiency of SPOS.

The synthesis of an amide typically involves the reaction of a carboxylic acid with an amine in the presence of a coupling reagent such as dicyclohexylcarbodiimide (DCC) (Fig. 16.14). Conventionally, a work-up procedure has to be carried out once the reaction is complete. This involves washing the organic solution with aqueous acid to remove unreacted amine. Once the aqueous and organic layers are separated, the organic layer is then washed with an aqueous base in order to remove unreacted acid. The organic and basic layers are separated, and then the organic layer is treated with a drying agent such as magnesium sulphate. The drying agent is filtered off and then the solvent is removed to afford the crude amide. Purification then has to be carried out by crystallization or chromatography. In order to synthesize a small 12-component amide library by reacting different carboxylic acids with the same amine, one would have to repeat all of these steps and this would prove both time consuming and equipment intensive.

Equipment miniaturization for parallel synthesis means that it is possible to house a mini-parallel synthesis laboratory in a fume cupboard for each chemist (Fig. 16.15). Small footprint work stations allow reactions to be carried out in parallel, followed by simultaneous evaporations on a normal heater–stirrer unit. Multiple parallel or sequential automated chromatography units can facilitate purification, and microwave reactors can dramatically speed up reaction times. In this way, all 12 amides in our library can be made at the same time in parallel. A variety of useful techniques can also be used to minimize the work-up procedures, as described below.

16.5.1 Solid-phase extraction

Solid-phase extraction (SPE) can be used to avoid the 'hassle' of carrying out liquid–liquid extractions to remove acidic or basic compounds or impurities. For example, solutions of the twelve crude amides that have

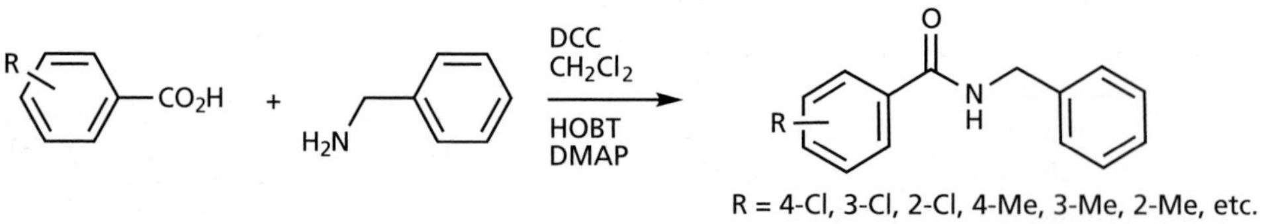

R = 4-Cl, 3-Cl, 2-Cl, 4-Me, 3-Me, 2-Me, etc.

FIGURE 16.14 Coupling reaction of a carboxylic acid and an amine to give an amide library (DCC = dicyclohexylcarbodiimide, DMAP = dimethylaminopyridine, HOBT = hydroxybenzotriazole).

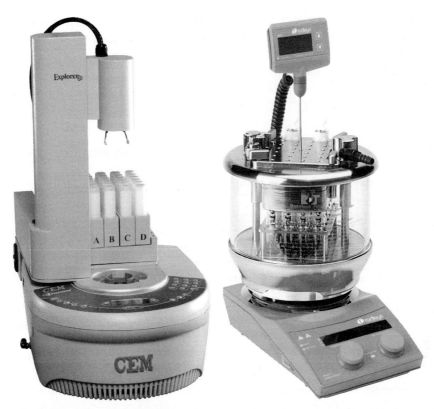

FIGURE 16.15 Laboratory stations for microwave-assisted organic reactions (CEM Explorer-24) and parallel synthesis (Radley's Greenhouse).

been prepared can be taken up from their wells at the same time using a multi-channel pipettor and applied to a battery of silica columns. An acidic column (SCX column) removes basic impurities, while a basic column (SAX column) removes acid impurities. Once the solutions have passed through the columns, the solvents are concentrated in parallel to yield essentially pure amide.

Another method of removing excess amine from a reaction is to use **fluorous solid-phase extraction** (**F-SPE**). This consists of silica columns where the silica has been linked to alkyl chains containing a large number of fluoro substituents. The highly fluorinated silica has a high affinity for fluorinated molecules and can be used to separate fluorinated compounds from non-fluorinated compounds. For example, consider the reaction shown in Fig. 16.16 where an isocyanate is treated with an amine to give a urea product. The amine is used in excess in order to

drive the reaction to completion, but the amine left over has to be removed. In order to do this, a fluorinated isocyanate is added which reacts with the excess amine to produce a fluorinated urea (Fig. 16.17). The crude solution is passed through an F-SPE column which acts as a scavenger resin to retain the highly fluorinated urea, and allow the desired unfluorinated urea product to pass through.

Sometimes an aqueous work-up cannot be avoided. For example, an aqueous work-up is required following a Grignard reaction which means that the aqueous and organic phases have to be separated. Fortunately, there are efficient methods of carrying out such separations in parallel.

One such method is to use a **lollipop phase separator**. A pin is inserted into a mixture of the two phases, and the mixture is rapidly cooled in a dry ice/acetone bath at −78 °C. The aqueous phase freezes onto the pin to form

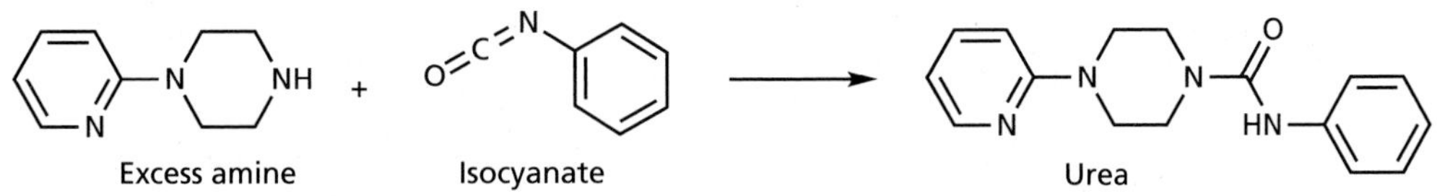

FIGURE 16.16 Reaction of an isocyanate with excess amine to produce a urea.

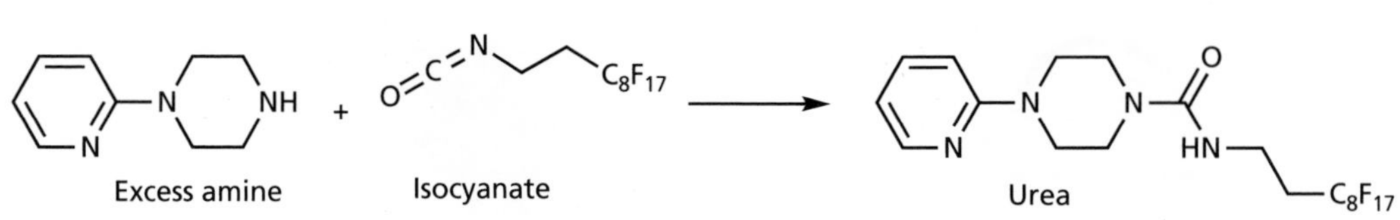

FIGURE 16.17 Removal of excess amine by reaction with a fluorinated isocyanate followed by F-SPE.

a 'lollipop'. The pin and its lollipop can then be removed from the reaction vial, leaving the organic phase behind. Up to 96 such separations may be performed in parallel with specially designed units.

Another method is to use **phase separation columns**, which can be used to separate a dense chlorinated organic layer from an aqueous phase. The lower organic layer passes through a hydrophobic frit by gravity, whereas the upper aqueous layer is retained on the frit. It is important not to apply pressure, otherwise the aqueous phase may be forced through the frit as well.

16.5.2 The use of resins in solution-phase organic synthesis (SPOS)

By carrying out a parallel synthesis in solution, it is easy to monitor the reaction by ^{1}H NMR spectroscopy or by thin layer chromatography. Work-up procedures can be greatly simplified by the use of a variety of resins. Since resins are solid-supported, there is little interaction between different types, allowing a variety of resins to be used in the same reaction. Thus, it is common to have a reaction cocktail which includes nucleophilic and electrophilic resins, or acidic and basic resins, without any problems arising.

Reactions are carried out such that one of the reagents—usually the cheaper and more readily available—is used in excess to drive the reaction to completion (A in Fig. 16.18). The crude reaction mixture will comprise the product AB and excess starting material A. The crude mixture is treated with a solid-supported scavenger resin that is capable of reacting with the excess reagent (A). As a result, excess reagent becomes attached to the resin and can be removed by filtering the resin. Removal of the solvent then leaves the pure product (AB).

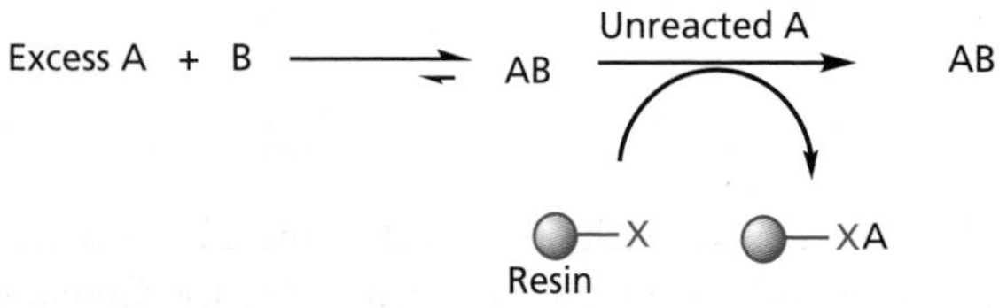

FIGURE 16.18 The use of scavenger resins in solution-phase organic synthesis.

16.5.3 Reagents attached to solid support: catch and release

It is possible to attach a reagent to a solid support. This has the advantage that the reagent or its by-product can be easily removed at the end of the reaction. For example, the coupling agent used for amide synthesis can be attached to a resin instead of being present in solution (Fig. 16.19). The reaction involves a carboxylic acid starting material reacting with the coupling reagent to form an intermediate which is still linked to the resin. Thus, the carboxylic acid is taken out of solution—the 'catch' phase. The resin-bound intermediate now reacts with the amine, and the amide product is released back into solution. The urea by-product which is formed remains bound to the resin and is easily removed when the resin is finally filtered. Acidic and basic resins can also be added to remove reagents and excess starting materials as described above.

The formation of a sulphonamide library shown in Fig. 16.20 makes use of a variety of different resins. The reaction involves an amine being treated with an excess of a sulphonyl chloride. A basic catalyst is required for the reaction and triethylamine is normally used in a conventional synthesis. However, this is quite a smelly, volatile compound and would have to be removed once the reaction was complete. Instead of triethylamine, a resin-bound base such as morpholine (PS-morpholine) can be used.

Following the reaction, nucleophilic and electrophilic scavenger resins are added. The nucleophilic resin PS-trisamine reacts with excess sulphonyl chloride to remove it from solution, while the electrophilic resin PS-isocyanate removes unreacted amine (Fig. 16.21). Filtration, to remove the resins, leaves the pure sulphonamide in solution.

Solid-supported reagents can be used in a variety of very common synthetic reactions. For example, a solid-supported borohydride can be used to reduce carbonyl groups (Fig. 16.22). In some reactions, it is also possible to reduce the toxicity and odour of reagents and their by-products. For example, the normal Swern oxidation involves the formation of dimethyl sulphide as a by-product—a compound which has a pungent cabbage odour! This is avoided by using a solid-supported reagent instead (Fig. 16.23).

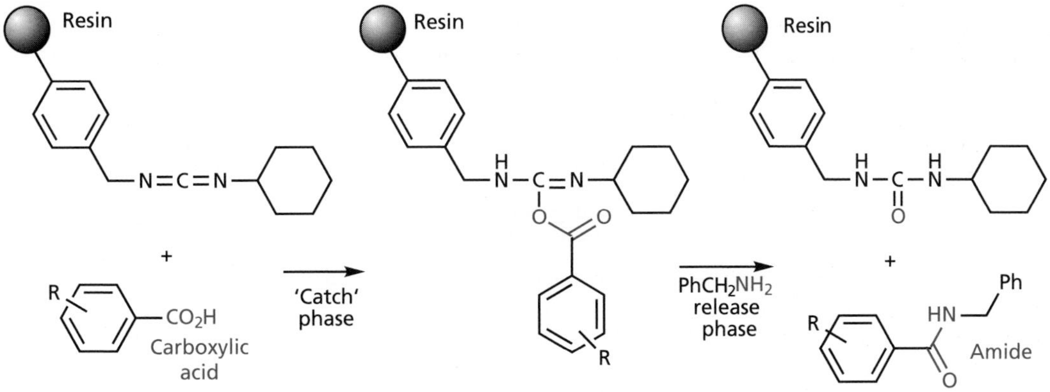

FIGURE 16.19 'Catch and release' during a coupling reaction.

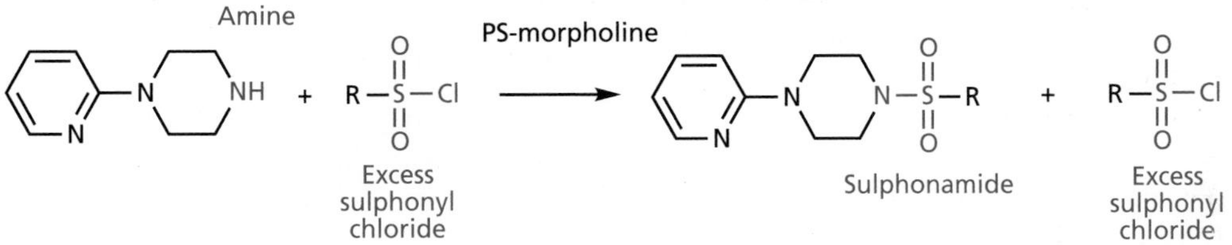

FIGURE 16.20 Formation of a sulphonamide library.

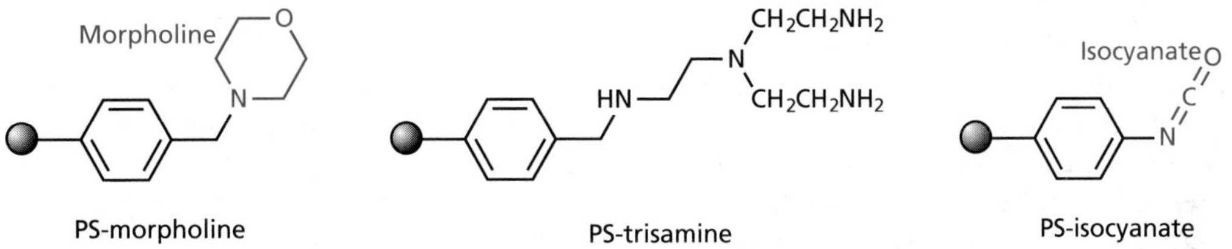

FIGURE 16.21 PS-morpholine, PS-trisamine, and PS-isocyanate.

16.5.4 Microwave technology

Drug discovery is a very expensive process and **microwave assisted organic synthesis** (MAOS) is proving to be a very useful tool for accelerating syntheses and making the process more efficient. There are many examples of thermal reactions that take several hours to complete using heaters or oil baths, but which are carried out in minutes using microwave conditions. There is a much greater efficiency of energy transfer using microwave technology which accounts for the faster reaction times. Moreover, yields can sometimes be dramatically improved with less decomposition and side reactions. Specially designed microwave units are now commonly employed in library syntheses (Fig. 16.15). Examples of reactions that have been carried out using microwave technology include the formation of amides from acids and amines without the need for coupling agents (Fig. 16.24), metal-catalysed

Suzuki couplings which can be performed even on usually unreactive aryl chlorides (Fig. 16.25), and metal-mediated reductions and aminations (Figs. 16.26). The reduction shown in Fig. 16.26 took 24 h using conventional heating, and only 15 min using microwave heating.

For additional material see Web article 21: Microwave technology applied to medicinal chemistry on the Online Resource Centre at www.oxfordtextbooks.co.uk/orc/patrick6e/

16.5.5 Microfluidics in parallel synthesis

The science of **microfluidics** involves the manipulation of tiny volumes of liquids in a confined space. Companies are devising microreactors that can be used to carry out parallel syntheses on microchips (Fig. 16.27) using a continuous flow of reactants in microfluidic channels.

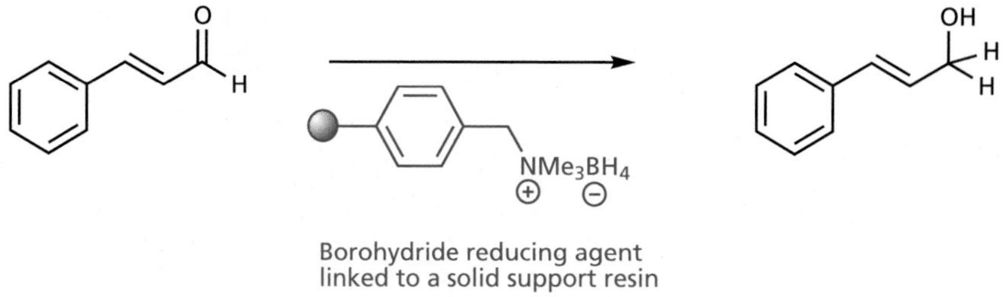

FIGURE 16.22 Reduction of an aldehyde with a solid-supported borohydride reagent.

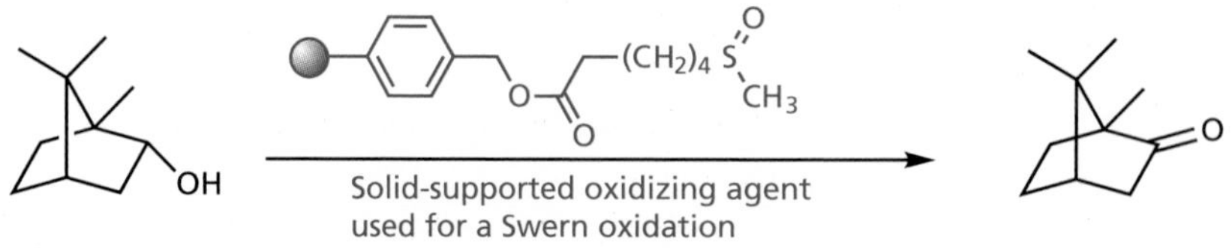

FIGURE 16.23 Swern oxidation using a solid-supported reagent.

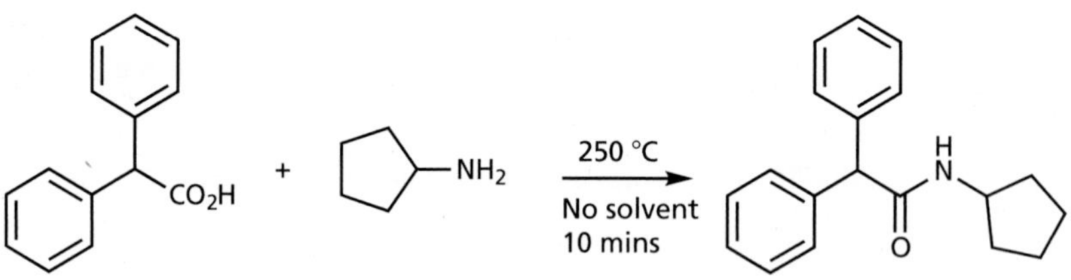

FIGURE 16.24 Amide formation using microwave technology.

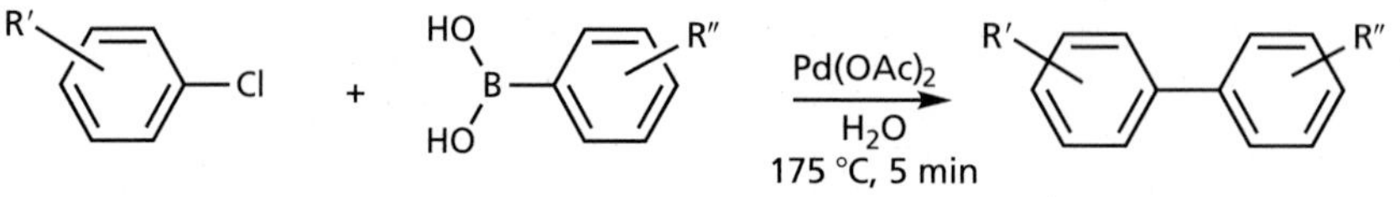

FIGURE 16.25 A Suzuki coupling carried out under microwave conditions.

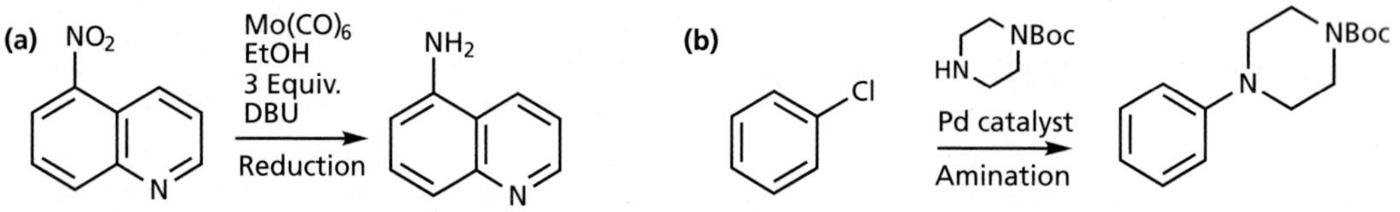

FIGURE 16.26 Microwave-assisted transition metal-mediated reactions. (a) Reductions and (b) aminations.

The channels are designed such that various reactants are mixed and reacted as they flow through the microchip. Several reactions have already been successfully carried out on microscale and it is found that reaction times can sometimes be shortened from hours to minutes. Some reactions take place in higher yield and with less side products. It is also possible to control the temperature of each reaction extremely accurately. Another advantage of microreactors is the potential to handle a vast number of parallel reactions on microchips. The channels through each chip can be fabricated to allow all possible mixing combinations of the various reactants, either on separate microchips or on a three-dimensional microchip. The example in Fig 16.27 is a simple illustration of how a microreactor system could be set up to create a mini-library from the reaction of A or B with C or D.

For additional material see Web article 20: Modern chemistry techniques in medicinal chemistry on the Online Resource Centre at www.oxfordtextbooks.co.uk/orc/patrick6e/

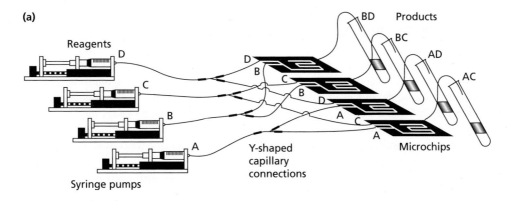

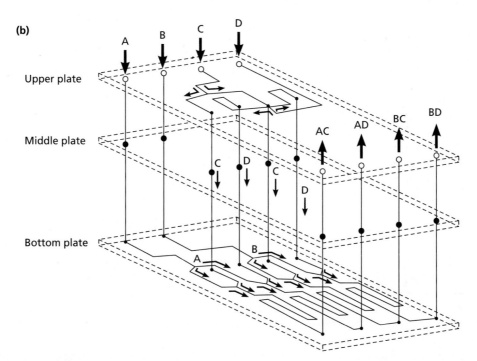

FIGURE 16.27 Parallel synthesis on a microchip. Parallel synthesis of four products using (a) four separate 2D microchips, and (b) a 3D microchip. (Reprinted by permission from Macmillan Publishers Ltd: *Nature Reviews Drug Discovery* (**5**, 210–18) 2006.)

KEY POINTS

- In parallel synthesis, a reaction or series of reactions is carried out in a series of wells to produce a range of analogues. Each reaction well contains a single product.

- Parallel synthesis can be carried out on solid phase or in solution.

- Parallel synthesis allows the synthesis of a large number of easily identifiable analogues which can be quickly and easily tested, speeding up the optimization process.

- Solid-phase extraction is often used in parallel synthesis for work-up. It involves the use of columns to remove impurities and excess reagents.

- An aqueous phase can be separated from an organic phase using phase separation columns or by freezing the aqueous phase onto a solid surface.

- Catch and release strategies involve reagents which are linked to a solid support. Reactants are taken out of solution when they react with the reagent, and are then released when a subsequent reaction takes place.

- Solid-supported reagents are easily removed at the end of a reaction. The potential toxicity of the reagent or its by-product is reduced when attached to a solid support.

- Microwave technology can prove advantageous over conventional heating.

16.6 Combinatorial synthesis

In combinatorial synthesis, mixtures of compounds are deliberately produced in each reaction vessel, allowing chemists to produce thousands and even millions of novel structures in the time that it would take to synthesize a few dozen by conventional means. This method of synthesis goes against the grain of conventional organic synthesis where chemists set out to produce a single identifiable structure which can be purified and characterized. The structures in each reaction vessel of a combinatorial synthesis are not separated and purified, but are tested for biological activity as a whole. If there is no activity, then there is no need to continue studies on that mixture and it is stored. If activity *is* observed, then one or more components in the mixture are active, although false positives can sometimes be an issue (section 12.3.5). Overall, there is an economy of effort, as a negative result for a mixture of 100 compounds saves the effort of synthesizing, purifying, and identifying each component of that mixture. On the other hand, identifying the active component of an active mixture is not straightforward.

In a sense, combinatorial synthesis can be looked upon as the synthetic equivalent of nature's chemical pool. Through evolution, nature has produced a huge number and variety of chemical structures, some of which are biologically active. Traditional medicinal chemistry dips into that pool to pick out the **active principles** and develop them. Combinatorial synthesis produces pools of purely synthetic structures that we can dip into for active compounds. The diversity of structures from the natural pool is far greater than that likely to be achieved by combinatorial synthesis, but isolating, purifying, and identifying new agents from natural sources is a relatively slow process, and there is no guarantee that a lead compound will be discovered against a specific drug target. The advantage of combinatorial chemistry is the fact that it produces new compounds faster than those derived from natural sources, and can produce a diversity not found in the traditional banks of synthetic compounds held by pharmaceutical companies.

A few words of caution should be made here with regards to negative assays. There is always the possibility that a combinatorial mixture does not contain all the structures expected. This can happen if some of the starting materials or intermediates in the synthesis do not react as expected. A negative assay would then lead to the conclusion that these compounds are inactive when they are not actually present. This could mean that an active compound is missed. Assays might also be affected adversely if the individual components of a mixture interact with each other or have conflicting activities.

16.6.1 The mix and split method in combinatorial synthesis

A combinatorial synthesis is designed to produce a mixture of products in each reaction vessel, starting with a wide range of starting materials and reagents. This does not mean that all possible starting materials are thrown together in the one reaction flask. If this was done, a black tarry mess would result. Instead, molecular structures are synthesized on solid supports such as beads. Each individual bead may contain a large number of such molecules, but all the molecules on that bead are identical—'the one-bead-one-compound concept'. Different beads have different structures attached and can be mixed together in a single vial such that the molecules attached to the beads undergo the same reaction. In this way, each vial contains a mixture of structures, but each structure is physically distinct from the others because it is attached to a different bead.

Planning has to go into designing a combinatorial synthesis to minimize the effort involved and to maximize the number of different structures obtained. The strategy of **mix and split** is a crucial part of this. As an example, suppose we wish to synthesize all the possible dipeptides of five different amino acids. Using orthodox chemistry, we would synthesize these one at a time. There are 25 possible dipeptides, and so we would have to carry out 25 separate experiments (Fig. 16.28).

Using a mix and split strategy, the same products can be obtained with far less effort (Fig. 16.29). First of all, the beads are split between five reaction vials. The first amino acid is attached to the beads, using a different amino acid for each vial. The beads from all five flasks are collected, mixed together, then split back into the five

Gly	25 separate	Gly-Gly	Ala-Gly	Phe-Gly	Val-Gly	Ser-Gly
Ala	procedures	Gly-Ala	Ala-Ala	Phe-Ala	Val-Ala	Ser-Ala
Phe	→	Gly-Phe	Ala-Phe	Phe-Phe	Val-Phe	Ser-Phe
Val		Gly-Val	Ala-Val	Phe-Val	Val-Val	Ser-Val
Ser		Gly-Ser	Ala-Ser	Phe-Ser	Val-Ser	Ser-Ser

FIGURE 16.28 The possible dipeptides that can be synthesized from five different amino acids. Each procedure involves protection, coupling, and deprotection stages.

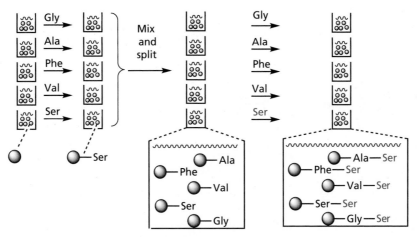

FIGURE 16.29 Synthesis of five different dipeptides using the mix and split strategy. Note that the addition of each amino acid involves protection, coupling, and deprotection steps.

vials. This means that each vial now has the same mixture. The second amino acid is now coupled to the amino acid on each bead, with a different amino acid being used for each vial. Each vial now contains five different dipeptides with no two vials containing the same dipeptide. Each of the five mixtures can now be tested for activity. If the results are positive, the emphasis is on identifying which of the dipeptides is active. If there is no activity present, the mixture can be ignored.

In studies such as these, one can generate large numbers of mixtures, many of which are inactive. However, these mixtures are not discarded. Although they may not contain a lead compound on this particular occasion, they may provide the necessary lead compound for a different target in medicinal chemistry. Therefore, all the mixtures (both active and inactive) resulting from a combinatorial synthesis are stored as **compound libraries**. The example above produced 25 compounds in five mixtures. However, combinatorial synthesis can be used to produce several thousand compounds.

16.6.2 **Structure determination of the active compound(s)**

The direct structural determination of components in a compound mixture is no easy task, but advances have been made in obtaining interpretable mass, NMR, Raman, infrared, and ultraviolet spectra on products attached to a single resin bead. Peptides can be sequenced while still attached to the bead. Each 100 μm bead contains about 100 pmol of peptide, which is enough for microsequencing. With non-peptides, the structural determination of an active compound can be achieved by deconvolution methods. Alternatively, **tagging** procedures can be used during the synthesis.

For additional material see Web article 6: Deconvolution on the Online Resource Centre at www.oxford-textbooks.co.uk/orc/patrick6e/

16.6.2.1 Tagging

In this process, two molecules are built up on the same bead. One of these is the intended structure, the other is a molecular tag (usually a peptide or oligonucleotide) which will act as a code for each step of the synthesis. For this to work, the bead must have a multiple linker capable of linking both the target structure and the molecular tag. A starting material is added to one part of the linker, and an encoding amino acid or nucleotide to another part. After each subsequent stage of the combinatorial synthesis, an amino acid or nucleotide is added to the growing tag to indicate what reagent was used. One example of a multiple linker is called the **Safety-Catch Acid-labile Linker (SCAL)** (Fig. 16.30), which includes lysine and tryptophan. Both these amino acids have a free amino group.

The target structure is constructed on the amino group of the tryptophan moiety and, after each stage of the synthesis, an amino acid is added to the growing peptide tag on the lysine moiety. Figure 16.31 illustrates the procedure for a synthesis involving three reagents. By the end of the process, the tripeptide tag contains amino acids which define the identity of the variable groups R, R′, and R″ in the target structure.

The non-peptide target structure can be cleaved by reducing the two sulphoxide groups in the safety-catch linker, then treating with acid. Under these conditions, the tripeptide sequence remains attached to the bead and can be sequenced on the bead to identify the structure of the compound that was released.

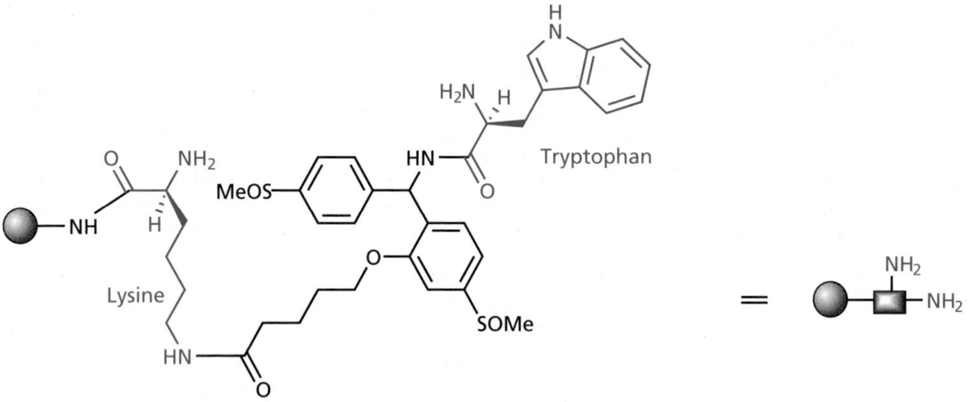

FIGURE 16.30 SCAL (Safety-Catch Acid-labile Linker).

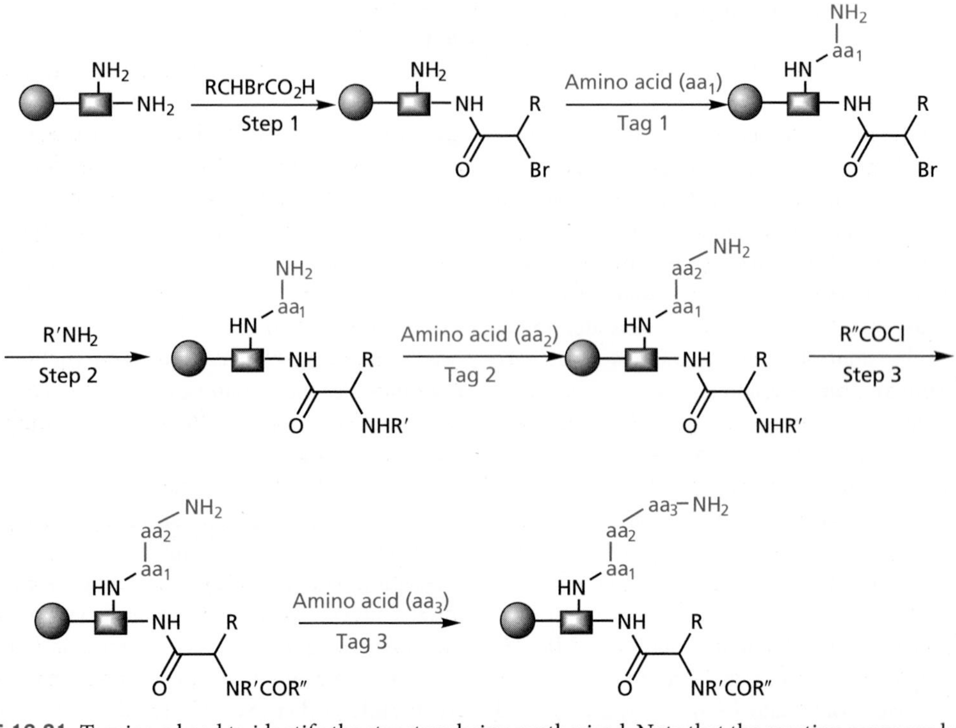

FIGURE 16.31 Tagging a bead to identify the structure being synthesized. Note that the reaction sequence has been simplified here to illustrate the principle of tagging. Amino acids are *N*-protected when coupled and the protecting group is removed before the next coupling. Coupling agents are also present. An orthogonal protection strategy is also required to distinguish between the amino groups of the safety-catch acid-labile linker.

The same strategy can be used with an oligonucleotide as the tagging molecule. The oligonucleotide can be amplified by replication and the code read by DNA sequencing.

There are drawbacks to tagging processes. They are time consuming and require elaborate instrumentation. Building the coding structure itself also adds extra restraints on the protection strategies that can be employed and may impose limitations on the reactions that can be used. In the case of oligonucleotides, their inherent instability can prove a problem. Another possible problem

with tagging is the possibility of an unexpected reaction taking place, resulting in a different structure from that expected. Nevertheless, the tagging procedure is still valid since it identifies the starting materials and the reaction conditions used at each stage. When these are repeated on a larger scale, any unusual reactions would be discovered.

The tagging methods described above require the use of a specific molecular tag to represent each reagent used in the synthesis. Moreover, the resultant molecular tag has to be sequenced at the end of the synthesis. A more efficient method of tagging and identifying the final product is to use some form of encryption or '**bar code**'. For example, it is possible to identify which one of seven possible reagents has been used in the first stage of a synthesis with the use of only three molecular labels (A–C). This is achieved by adding different combinations of the three tags to set up a triplet code on the bead. Thus, adding just one of the tags (A, B, or C) will allow the identification of three of the reagents. Adding two of the tags at the same time allows the identification of another three reagents, and adding all three tags at the same time allows the identification of a seventh reagent. The presence (1) or absence (0) of the tag forms a triplet code; the presence of a single molecular tag (A, B, or C) gives three triplet codes (100, 010, and 001); the presence of two different tags is indicated by another three triplet codes (110, 101, 011); and the presence of all three tags is represented by 111. The tags are linked to the bead by means of a photocleavable bond, so irradiating the bead releases all the tags. These can then be passed through a gas chromatograph and identified by their retention time.

Three different molecular tags could now be used to represent seven reagents in the second stage, and so on. All the tags used to represent the second reagent would have longer retention times than the tags used to represent the first reagent. Similarly, all subsequent tags would have longer retention times that any previous tag. Once the synthesis is complete, all the tags are released simultaneously and passed through the gas chromatograph. The 'bar code' is then read from the chromatograph in one go, not only identifying the reagents used, but the order in which they were used (Fig. 16.32).

16.6.2.2 Photolithography

Photolithography is a technique that permits miniaturization and spatial resolution such that specific products are synthesized on a plate of immobilized solid support. In the synthesis of peptides, the solid support surface contains an amino group protected by the photolabile **nitroveratryloxycarbonyl** (**NVOC**) protecting group (Fig. 16.33). Using a mask, part of the surface is exposed to light resulting in deprotection of the exposed region. The plate is then treated with a protected amino acid and the coupling reaction takes place only on the region of the plate which has been deprotected. The plate is then washed to remove excess amino acid. The process can be repeated on a different region using a different mask, and so different peptide chains can be built on different parts of the plate; the sequences are known from the record of masks and reagents used.

Incubation of the plate with a protein receptor can then be carried out to detect active compounds which bind to the binding site of the receptor. A convenient method to assess such interactions is to incubate the plate with a fluorescently tagged receptor. Only those regions of the plate which contain active compounds will bind to the receptor and fluoresce. The fluorescence intensity can be measured using fluorescence microscopy and is a measure of the affinity of the compound for the receptor. Alternatively, testing can be carried out such that active compounds are detected by radioactivity or chemiluminescence.

The photodeprotection described above can be achieved in high resolution. At a 20-μm resolution, plates can be prepared with 250 000 separate compounds per square centimetre.

16.6.3 **Dynamic combinatorial synthesis**

Dynamic combinatorial synthesis is an exciting development which has been used as an alternative to the classic mix and split combinatorial syntheses in the search for new lead compounds. The aim of dynamic combinatorial synthesis is to synthesize all the different compounds in one flask at the same time, screen them *in situ* as they are being formed, and thus identify the most active compound in a much faster time period (Box 16.2). How can this be achieved? There are several important principles which are followed.

- The best way of screening the compounds is to have the desired target present in the reaction flask along with the building blocks. This means that any active compounds can bind to the target as soon as they are formed. The trick is then to identify which products are binding.

- The reactions involved should be reversible. If this is the case, a huge variety of products are constantly being formed in the flask then breaking back down into their constituent building blocks. The advan-

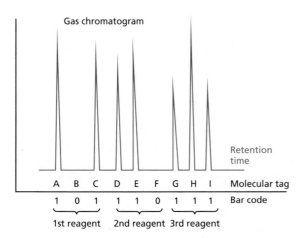

FIGURE 16.32 Identification of reagents and order of use by bar coding.

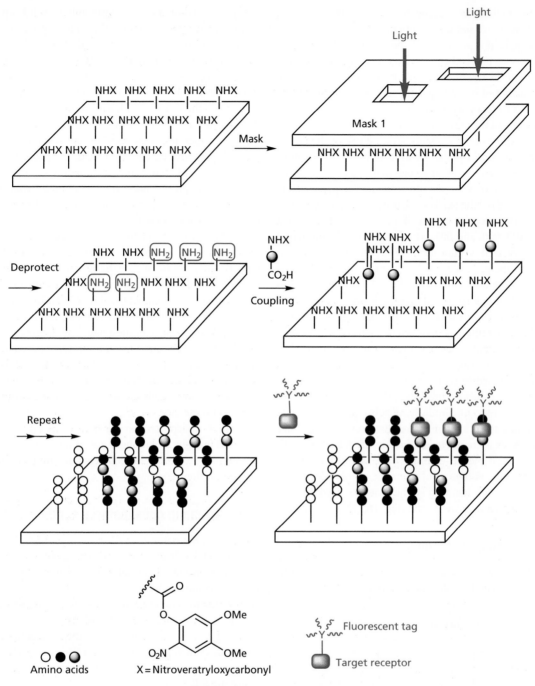

FIGURE 16.33 Photolithography.

tage of this may not seem obvious, but it allows the possibility of 'amplification' where the active compound is present to a greater extent than the other possible products. By having the target present, active compounds become bound and are effectively removed from the equilibrium mixture. The equilibrium is now disturbed such that more of the active product is formed. Thus, the target serves not only to screen for active compounds, but also to amplify them.

- In order to identify the active compounds, it is necessary to 'freeze' the equilibrium reaction such that it no longer takes place. This can be done by carrying out a further reaction which converts all the equilibrium products into stable compounds that cannot revert back to starting materials.

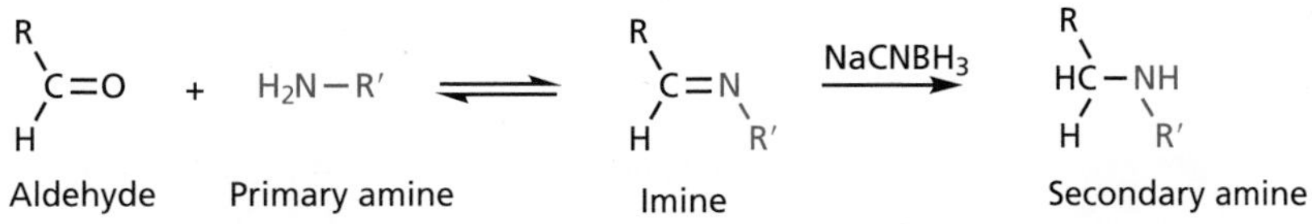

FIGURE 16.34 Example of dynamic combinatorial synthesis.

A simple example of dynamic combinatorial synthesis involved the reversible formation of imines from aldehydes and primary amines (Fig. 16.34). A total of three aldehydes and four amines were used in the study (Fig. 16.35), allowing the possibility of 12 different imines in the equilibrium mixture.

The building blocks were mixed together with the target enzyme **carbonic anhydrase** and allowed to interact. After a suitable period of time, sodium cyanoborohydride was added to reduce all the imines present to secondary amines so that they could be identified (Fig. 16.34). The mixture was separated by reverse-phase HPLC, allowing each product to be quantified and identified. These results were compared with those obtained

when the experiment was carried out in the absence of carbonic anhydrase, making it possible to identify which products had been amplified. In this experiment, the sulphonamide shown in Fig. 16.36 was significantly amplified, which demonstrated that the corresponding imine was an active compound.

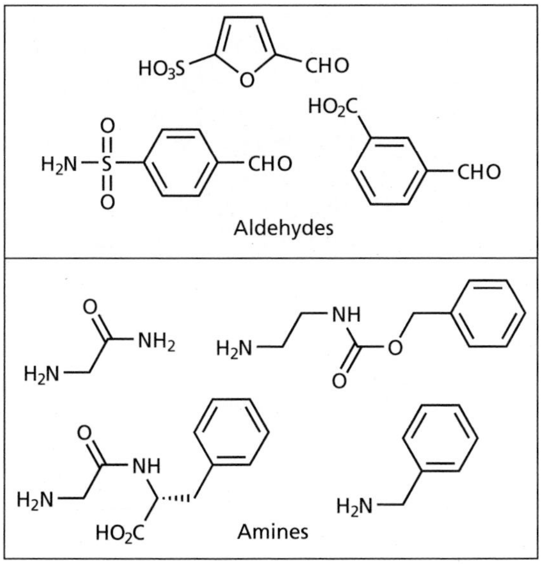

The above example illustrates a simple case involving one reaction and two sets of building blocks, but it is feasible to have more complex situations. For example, a molecule with two or more functional groups could be present to act as a scaffold on to which various substituents could be added from the building blocks available (Fig. 16.37). The use of a central scaffold has another benefit; it helps the amplification process. If the number of scaffold molecules present is equal to the number of target molecules, then the number of products formed cannot be greater than the number of targets available. If any of these products binds to the target, the effect on the equilibrium will be greater than if there were more products than targets available.

There are certain limitations to dynamic combinatorial chemistry:

- Conditions must be chosen such that the target does not react chemically with any of the building blocks, or is unstable under the reaction conditions used.

- The target is normally in an aqueous environment, so the reactions have to be carried out in aqueous solution.

- The reactions themselves have to undergo fast equilibration rates to allow amplification.

- It is important to avoid using building blocks that are too reactive, as this would bias the equilibrium towards particular products and confuse the identification of the amplified product.

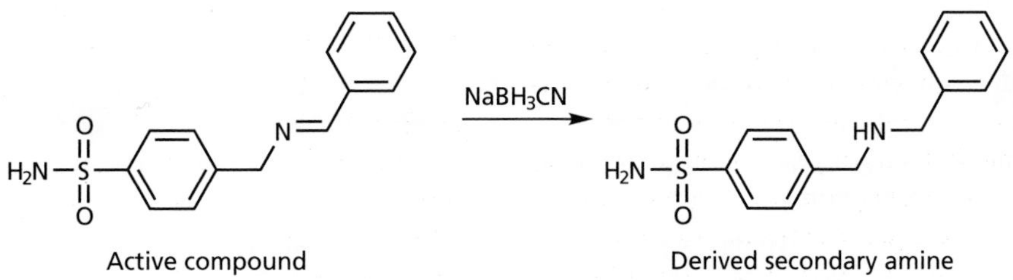

FIGURE 16.35 Aldehyde and amine building blocks used in the dynamic combinatorial synthesis of imines.

FIGURE 16.36 Amplified imine and the amine obtained from reduction.

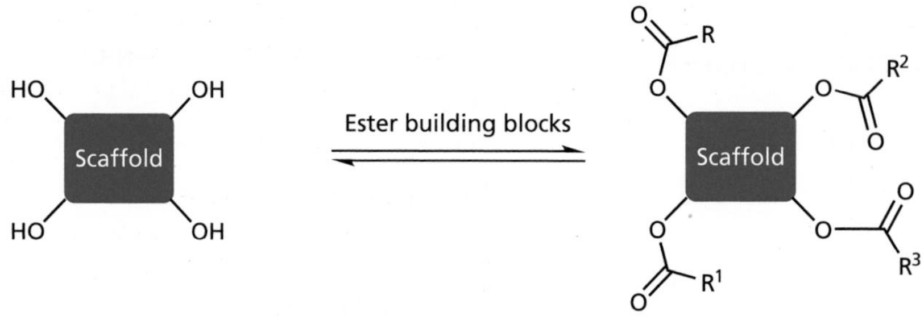

FIGURE 16.37 Use of a scaffold molecule.

BOX 16.2 Dynamic combinatorial synthesis of vancomycin dimers

Vancomycin is an antibiotic that works because it masks the building blocks required for bacterial cell wall synthesis (section 19.5.5.2). Binding takes place specifically between the antibiotic and a peptide sequence (L-Lys-D-Ala-D-Ala) which is present in the building block. It is also known that this binding promotes dimerization of the vancomycin–target complex, which suggests that covalently linked vancomycin dimers might be more effective antibacterial agents than vancomycin itself. A dynamic combinatorial synthesis was carried out to synthesize a variety of different vancomycin dimers covalently linked by bridges of different lengths. The vancomycin monomers used had been modified such that they contained long-chain alkyl substituents with double bonds at the end. Reaction between the double bonds in the presence of a catalyst then led to bridge formation through a reaction known as olefin metathesis (Fig. 1).

The tripeptide target was present to accelerate the rate of bridge formation and to promote formation of vancomycin dimers having the ideal bridge length. As shown in Fig. 2, the vancomycin monomers bind the tripeptide which encourages the self-assembly of non-covalently linked dimers. Once formed, those dimers having the correct length of substituent are more likely to react together to form the covalent bridge (Fig. 2).

Having established the optimum length of bridge, another experiment was carried out on eight vancomycin monomers which had the correct length of 'tether' but varied slightly in their structure. The mixture of 36 possible products was analysed by mass spectrometry to indicate the relative proportion of each dimer formed. Eleven of the 36 compounds were then synthesized separately and it was found that their antibacterial activity matched their level of amplification, i.e. the compounds present in greater quantities had the greater activity.

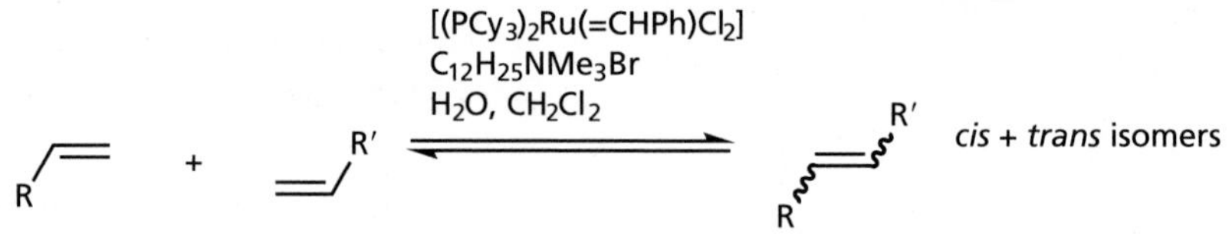

FIGURE 1 Olefin metathesis.

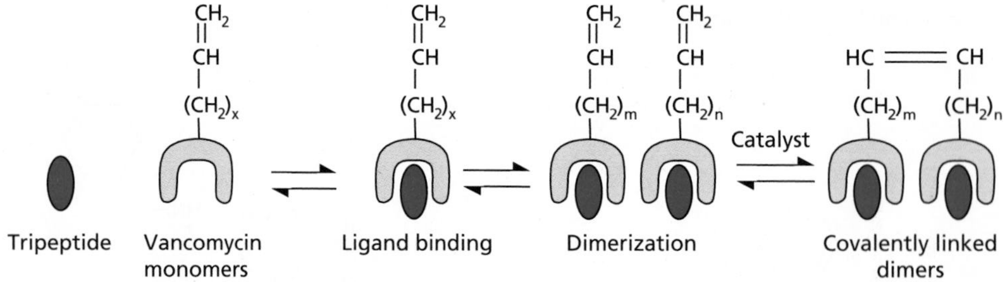

FIGURE 2 Formation of covalently linked dimers.

- Most combinatorial syntheses are carried out using automated or semi-automated synthesizers.

- The mix and split method allows the efficient synthesis of large numbers of compounds with a minimum number of operations.

- The compounds synthesized in a combinatorial synthesis are stored as compound libraries.

- Tagging involves the construction of a tagging molecule on the same solid support as the target molecule. Tagging molecules are normally peptides or oligonucleotides. After each stage of the target synthesis, the peptide or oligonucleotide is extended and the amino acid or nucleotide used defines the reactant or reagent used in that stage.

- Photolithography is a technique involving a solid support surface containing functional groups protected by photolabile groups. Masks are used to reveal defined areas of the plate to light, thus removing the protecting groups and allowing a reactant to be linked to the solid support. A record of the masks and reagents used determines what reactions have been carried out at different regions of the plate.

- Combinatorial synthesis has been used for the synthesis of peptides, peptoids, and heterocyclic structures. Most organic reactions are feasible.

- Dynamic combinatorial chemistry involves the formation of a mixture of compounds in the presence of a target under equilibrium conditions. Binding of a product with the target amplifies that product in the equilibrium mixture.

QUESTIONS

1. Identify three stages of the drug discovery, design, and development process where combinatorial chemistry or parallel synthesis is of importance.

2. A pharmaceutical laboratory wishes to synthesize all the possible dipeptides containing the amino acids tyrosine, lysine, phenylalanine, and leucine. Identify the number of possible dipeptides and explain how the lab would carry this out using combinatorial techniques.

3. What particular precautions have to be taken with the amino acids tyrosine and lysine in the above synthesis?

4. Identify the advantages and disadvantages of the following structures as scaffolds.

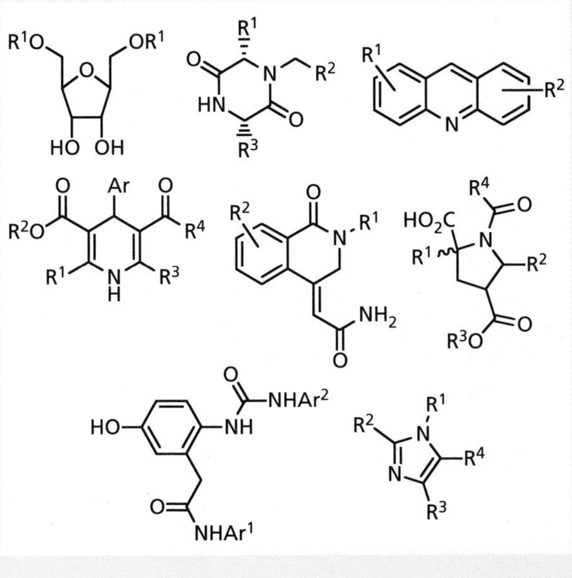

5. You wish to carry out the combinatorial synthesis shown in Fig. 16.31 using bar coding techniques rather than the conventional tagging scheme shown in the figure. You have nine molecules suitable for tagging purposes (A–I), seven bromo acids (B1–B7), seven amines (A1–A7), and seven acid chlorides (C1–C7). Construct a suitable coding system for the synthesis.

6. Based on your coding scheme from Question 5, what product is present on the bead if the released tags resulted in the gas chromatograph shown in Fig. 16.32.

Multiple-choice questions are available on the Online Resource Centre at www.oxfordtextbooks.co.uk/orc/patrick6e/

FURTHER READING

Beck-Sickinger, A., and Weber, P. (2002) *Combinatorial strategies in biology and chemistry*. John Wiley and Sons, New York.

Bhalay, G., Dunstan, A., and Glen, A. (2000) Supported reagents: opportunities and limitations. *Synlett*, 1846–59.

Braeckmans, K., et al. (2002) Encoding microcarriers: present and future technologies. *Nature Reviews Drug Discovery*, **1**, 447–56.

Dittrich, P. S., and Manz, A. (2006) Lab-on-a-chip: microfluidics in drug discovery. *Nature Reviews Drug Discovery*, **5**, 210–18.

Dobson, C. M. (2004) Chemical space and biology. *Nature*, **432**, 824–8.

Dolle, R. E. (2003) Comprehensive survey of combinatorial library synthesis: 2002. *Journal of Combinatorial Chemistry*, **5**, 693–753.

Geysen, H. M., et al. (2003) Combinatorial compound libraries for drug discovery: an ongoing challenge. *Nature Reviews Drug Discovery*, **2**, 222–30.

Guillier, F., Orain, D., and Bradley, M. (2000) Linkers and cleavage strategies in solid-phase organic synthesis and combinatorial chemistry. *Chemical Reviews*, **100**, 2091–158.

Houlton, S. (2002) Sweet synthesis. *Chemistry in Britain*, April, 46–9.

Kappe, C. O. (2004) Controlled microwave heating in modern organic synthesis. *Angewandte Chemie International Edition*, **43**, 6250–84.

Keseru, G. M., et al. (2016) Design principle for fragment libraries: maximizing the value of learnings from Pharma fragment-based drug discovery (FBDD) programs for use in academia. *Journal of Medicinal Chemistry*, DOI: 10.1021/acs/jmedchem.6b00197.

Le, G. T., et al. (2003) Molecular diversity through sugar scaffolds. *Drug Discovery Today*, **8**, 701–9.

Ley, S. V., and Baxendale, I. R. (2002) New tools and concepts for modern organic synthesis. *Nature Reviews Drug Discovery*, **1**, 573–86.

Mavandadi, F., and Pilotti, Å. (2006) The impact of microwave-assisted organic synthesis in drug discovery. *Drug Discovery Today*, **11**, 165–74.

Nicolaou, K. C., et al. (2000) Target-accelerated combinatorial synthesis and discovery of highly potent antibiotics effective against vancomycin-resistant bacteria. *Angewandte Chemie International Edition*, **39**, 3823–8.

Ramstrom, O., and Leh, J.-M. (2002) Drug discovery by dynamic combinatorial libraries. *Nature Reviews Drug Discovery*, **1**, 26–36.

Reader, J. C. (2004) Automation in medicinal chemistry. *Current Topics in Medicinal Chemistry*, **4**, 671–86.

Titles for general further reading are listed on p.845; see also articles in Journal of Combinatorial Chemistry.

17 Computers in medicinal chemistry

Computers are an essential tool in modern medicinal chemistry and are important in both drug discovery and drug development. Rapid advances in computer hardware and software have meant that many of the operations which were once the exclusive province of the expert can now be carried out on ordinary laboratory computers with little specialist expertise in the molecular or quantum mechanics involved. In this chapter, we shall look at examples of how computers are used in medicinal chemistry. However, it has to be appreciated that it is not possible to do full justice to the subject in a single chapter. For example, a full coverage would include details of the mathematics and equations used in different algorithms, and that is not possible here. Readers with an interest in gaining a more detailed appreciation of how software programs work at the mathematical level are encouraged to read more specialized textbooks and journal articles (see Further reading at the end of the chapter).

17.1 Molecular and quantum mechanics

The various operations which are carried out in molecular modelling involve the use of programs or **algorithms** which calculate the structure and property data for the molecule in question. For example, it is possible to calculate the energy of a particular arrangement of atoms (conformation), modify the structure to create an energy minimum, and calculate properties such as charge, dipole moment, and heat of formation. The computational methods that are used to calculate structure and property data can be split into two categories—molecular mechanics and quantum mechanics.

17.1.1 Molecular mechanics

In molecular mechanics, equations are used which follow the laws of classical physics and apply them to nuclei

without consideration of the electrons. In essence, the molecule is treated as a series of spheres (the atoms) connected by springs (the bonds). Equations derived from classical mechanics are used to calculate the different interactions and energies (**force fields**) resulting from bond stretching, angle bending, non-bonded interactions, and torsional energies. Torsional energies are associated with atoms that are separated from each other by three bonds. The relative orientation of these atoms is defined by the dihedral or torsion angle—see for example Fig. 17.16.

These calculations require data or parameters which are stored in tables within the program and which describe interactions between different sets of atoms. The energies calculated by molecular mechanics have no meaning as absolute quantities, but are useful when comparing different conformations of the same molecule. Molecular mechanics is fast and is less intensive on computer time than quantum mechanics. However, it cannot calculate electronic properties because electrons are not included in the calculations.

17.1.2 Quantum mechanics

Quantum mechanics uses quantum physics to calculate the properties of a molecule by considering the interactions between the electrons and nuclei of the molecule. Unlike molecular mechanics, atoms are not treated as solid spheres. In order to make the calculations feasible, various approximations have to be made.

- Nuclei are regarded as motionless. This is reasonable since the motion of the electrons is much faster in comparison. As electrons are considered to be moving around fixed nuclei, it is possible to describe electronic energy separately from nuclear energy.

- It is assumed that the electrons move independently of each other, so the influence of other electrons and nuclei is taken as an average.

Quantum mechanical methods can be subdivided into two broad categories—*ab initio* and semi-empirical. The former is more rigorous and does not require any stored parameters or data. However, it is expensive on computer time and is restricted to small molecules. Semi-empirical methods compute for valence electrons only. They are quicker, though less accurate, and can be carried out on larger molecules. There are various forms of semi-empirical software (i.e. programs such as MINDO/3, MNDO, MNDO-d, AM1, and PM3). These methods are quicker because they use further approximations and make use of stored parameters.

17.1.3 Choice of method

The method of calculation chosen depends on what calculation needs to be done, as well as the size of the molecule. As far as size of molecule is concerned, *ab initio* calculations are limited to molecules containing tens of atoms, semi-empirical calculations on molecules containing hundreds of atoms, and molecular mechanics on molecules containing thousands of atoms.

Molecular mechanics is useful for the following operations or calculations:

- energy minimization;
- identifying stable conformations;
- energy calculations for specific conformations;
- generating different conformations;
- studying molecular motion.

Quantum mechanical methods are suitable for calculating the following:

- molecular orbital energies and coefficients;
- heat of formation for specific conformations;
- partial atomic charges calculated from molecular orbital coefficients;

- electrostatic potentials;
- dipole moments;
- transition state geometries and energies;
- bond dissociation energies.

17.2 Drawing chemical structures

Chemical drawing packages do not require the calculations described in section 17.1, but they are often integrated into molecular modelling programs. Various software packages such as ChemDraw, ChemWindow, and Symyx Draw are available which can be used to construct diagrams quickly and to a professional standard. For example, the diagrams in this book have mostly been prepared using the ChemDraw package.

Some drawing packages are linked to other items of software which allow quick calculations of various molecular properties. For example, the following properties for **adrenaline** were obtained using ChemDraw Ultra; the structure's correct IUPAC chemical name, molecular formula, molecular weight, exact mass, and theoretical elemental analysis. It was also possible to get calculated predictions of the compound's ^{1}H and ^{13}C NMR chemical shifts, melting point, freezing point, log *P* value, molar refractivity, and heat of formation (Fig. 17.1).

17.3 3D structures

Molecular modelling software allows the chemist to construct a three-dimensional (3D) molecular structure on the computer. There are several software packages available such as ChemBio3D, Hyperchem, Discovery

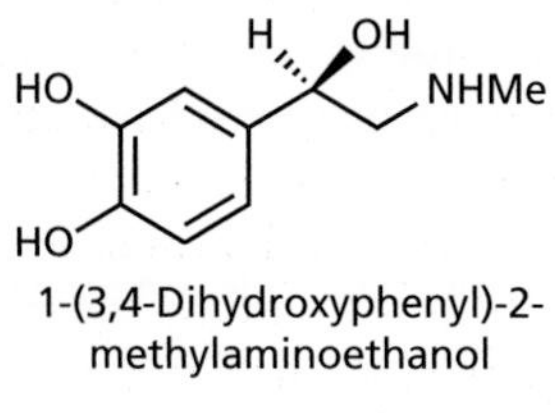

1-(3,4-Dihydroxyphenyl)-2-methylaminoethanol

Calculated properties
$C_9H_{13}NO_3$
Exact mass: 183.09
MWt.: 183.20
C, 59.00; H, 7.15; N, 7.65; O, 26.20

Predicted properties
Log *P* = −0.61–0.63
Molar refractivity 48.66–49.08 cm³/mol
b.pt. 618.55 K; freezing point 539.03 K
Heat of formation −451.22 kJ/mol

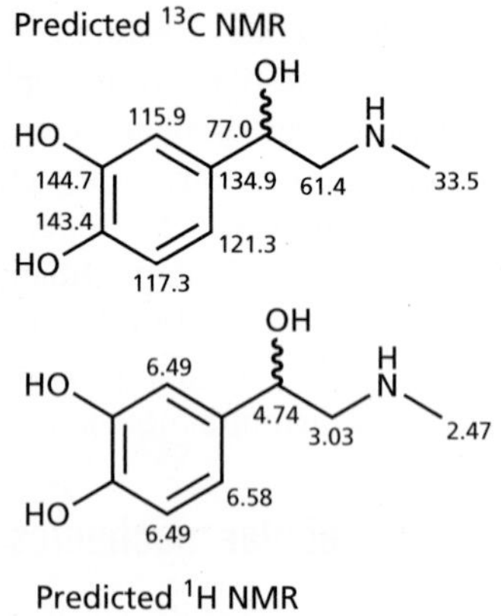

Predicted ^{13}C NMR

Predicted ^{1}H NMR

FIGURE 17.1 Structure and calculated/predicted properties of adrenaline.

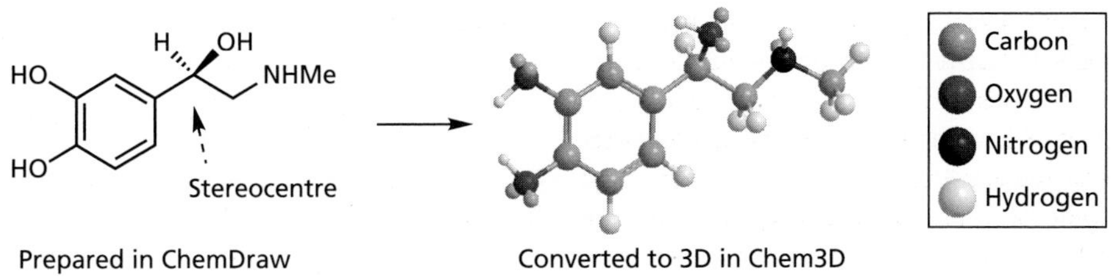

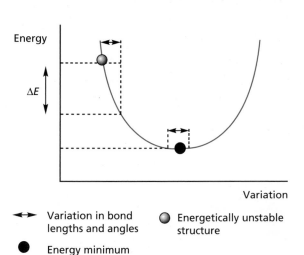

FIGURE 17.2 Conversion of a 2D drawing to a 3D model.

Studio Pro, and Spartan. The 3D model can be made by constructing the molecule atom by atom, and bond by bond. It is also possible to automatically convert a 2D drawing into a 3D structure, and most molecular modelling packages have this facility. For example, the 2D structure of adrenaline in Fig. 17.2 was drawn in ChemDraw, then copied and pasted into ChemBio3D, resulting in the automatic construction of the 3D model shown. The 3D structures of a large number of small molecules can also be accessed and downloaded from the **Cambridge Structural Database** (CSD). This database contains over 200 000 molecules which have been crystallized and their structure determined by X-ray crystallography.

17.4 Energy minimization

Whichever software program is used to create a 3D structure, a process called energy minimization should be carried out once the structure is built. This is because the construction process may have resulted in unfavourable bond lengths, bond angles, or torsion angles. Unfavourable non-bonded interactions may also be present (i.e. atoms from different parts of the molecule occupying the same region of space). The energy minimization process is usually carried out by a molecular mechanics program which calculates the energy of the starting molecule, then varies the bond lengths, bond angles, and torsion angles to create a new structure. The energy of the new structure is calculated to see whether it is energetically more stable or not. If the starting structure is inherently unstable, a slight alteration in bond angle or bond length will have a large effect on the overall energy of the molecule resulting in a large energy difference (ΔE; Fig. 17.3). The program will recognize this and carry out more changes, recognizing those which lead to stabilization and those which do not. Eventually, a structure will be found where structural variations result in only slight changes in energy—an energy minimum. The program will interpret this as the most stable structure and will stop at that stage (Box 17.1).

17.5 Viewing 3D molecules

Once a structure has been energy minimized, it can be rotated in various axes to study its shape from different angles. It is also possible to display the structure in different formats (i.e. cylindrical bonds, wire frame, ball and stick, space filling; Fig. 17.4).

There is another format known as the ribbon format which is suitable for portraying regions of protein secondary structure such as α-helices. This often simplifies the highly complex-looking structure of a protein, allowing easier visualization of its secondary and tertiary structure. The ball and stick model of an α-helical decapeptide consisting of 10 alanine units is shown in Fig. 17.5, along with the same molecule displayed as a ribbon.

Test your understanding and practise your molecular modelling with Exercises 17.1 and 17.2 on the Online Resource Centre at www.oxfordtextbooks.co.uk/orc/patrick6e/

FIGURE 17.3 Energy minimization.

BOX 17.1 Energy minimizing apomorphine

A 2D structure of **apomorphine** was converted to a 3D structure using ChemBio3D. However, the catechol ring was found to be non-planar with different lengths of C–C bond.

Energy minimization corrected the deformed aromatic ring, resulting in the desired planarity and the correct length of bonds.

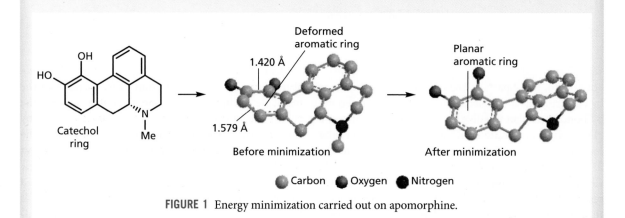

FIGURE 1 Energy minimization carried out on apomorphine.

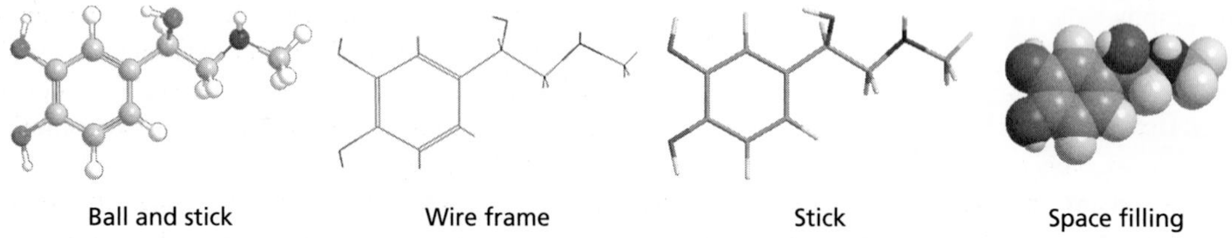

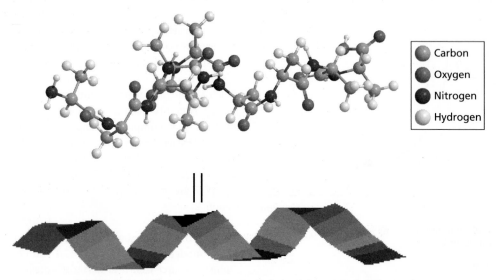

FIGURE 17.4 Different methods of visualizing molecules.

FIGURE 17.5 Ribbon representation of a helical decapeptide (Chem3D).

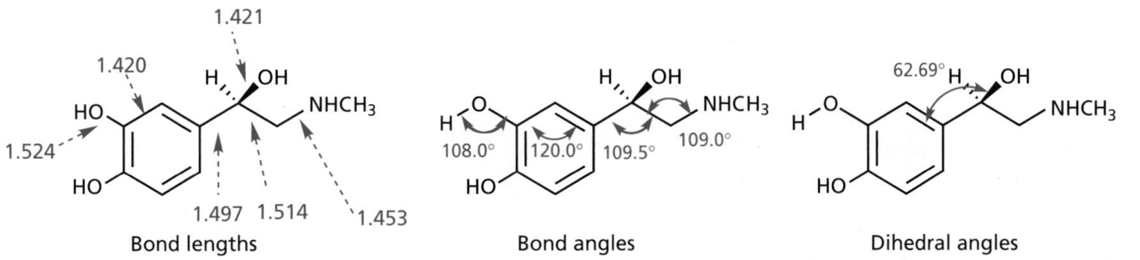

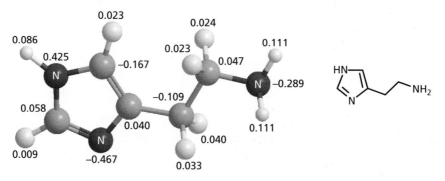

FIGURE 17.6 Molecular dimensions for adrenaline (Chem3D).

17.6 Molecular dimensions

Once a 3D model of a structure has been constructed, it is a straightforward procedure to measure all of its bond lengths, bond angles, and torsion (or dihedral) angles. These values can be read from tables or by highlighting the relevant atoms and bonds on the structure itself. The various bond lengths, bond angles, and torsion angles measured for **adrenaline** are illustrated in Fig. 17.6. It is also a straightforward process to measure the separation between any two atoms in a molecule (see molecular modelling Exercise 17.1).

17.7 Molecular properties

Various properties of the 3D structure can be calculated once it has been built and minimized. For example, the steric energy is automatically measured as part of the minimization process and takes into account the various strain energies within the molecule, such as bond stretching or bond compression, deformed bond angles, deformed torsion angles, non-bonded interactions arising from atoms too close to each other in space, and un-favourable dipole–dipole interactions. The steric energy is useful when comparing different conformations of the same structure, but the steric energies of different molecules should not be compared.

Other properties for the structure can be calculated, such as the predicted heat of formation, dipole moment, charge density, electrostatic potential, electron spin density, hyperfine coupling constants, partial charges, polarizability, and infrared vibrational frequencies. Some of these are described in the following sections.

17.7.1 Partial charges

It is important to realize that the valence electrons in molecules are not fixed to any one particular atom and can move around the molecule as a whole. As the electrons are likely to spend more of their time nearer electronegative atoms than electropositive atoms, this distribution is not uniform and results in some parts of the molecule being slightly positive and other parts being slightly negative. For example, the partial charges for **histamine** are shown in Fig. 17.7.

The calculation of partial charges has important consequences in the way we view ions. Conventionally, we consider charges to be fixed on a particular atom (unless delocalization is possible). For example, the histamine ion is normally drawn showing the positive charge on the terminal nitrogen atom (Fig. 17.8). In fact, calculation of partial charges shows that some of the positive charge is localized on the hydrogens attached to the terminal nitrogen. This has important consequences for the way we think of ionic interactions between a drug and

FIGURE 17.7 Partial charges for histamine.

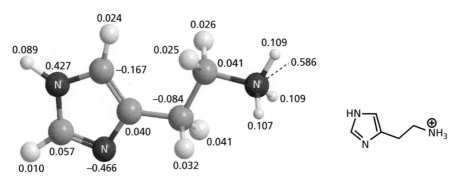

FIGURE 17.8 Charge distribution on the histamine ion.

its binding site. It implies that charged areas in the binding site and the drug are more diffuse than one might think. This, in turn, suggests that we have wider scope in designing novel drugs. For example, in the classical viewpoint of charge distribution, a certain molecule might be considered to have its charged centre too far away from the corresponding 'centre' in the binding site. If these charged areas are actually more diffuse, then this is not necessarily true.

It is worth pointing out, however, that such calculations are carried out on structures in isolation from their environment. Histamine is in an aqueous environment in the body and would be surrounded by water molecules, which would solvate the charge and consequently have an effect on charge distribution. Furthermore, water has a high dielectric constant, which means that electrostatic interactions are more effectively masked than in a hydrophobic environment.

Partial charges can also be represented by dot clouds. The size of each cloud represents the amount of charge, and the clouds can be coloured to show what sort of charge it is.

Test your understanding and practise your molecular modelling with Exercise 17.3 on the Online Resource Centre at www.oxfordtextbooks.co.uk/orc/patrick6e/

17.7.2 Molecular electrostatic potentials

Another way to consider charge distribution is to view the molecule as a whole rather than as individual atoms and bonds. This allows one to identify *areas* of the molecule which are electron rich or electron poor. This is particularly important in the **3D QSAR** technique of **CoMFA** described in section 18.10. It can also be useful in identifying how compounds with different structures might line up to interact with corresponding electron-rich and electron-poor areas in a binding site.

Molecular electrostatic potentials (MEPs) can be calculated using quantum mechanics by considering the

molecular orbitals. The MEP for histamine shown in Fig. 17.9 was calculated using the semi-empirical method AM1. Another method of calculating MEPs is described in section 17.7.5.

An example of how electrostatic potentials have been used in drug design can be seen in the design of the **cromakalim** analogue (II; Fig. 17.10), where the cyanoaromatic ring was replaced by a pyridine ring. This was part of a study looking into analogues of cromakalim which would have similar antihypertensive properties, but which might have different pharmacokinetics. In order to retain activity, it was important that any replacement heteroaromatic ring was similar in character

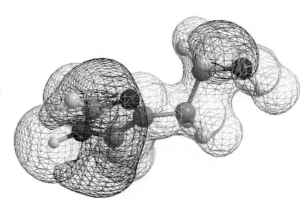

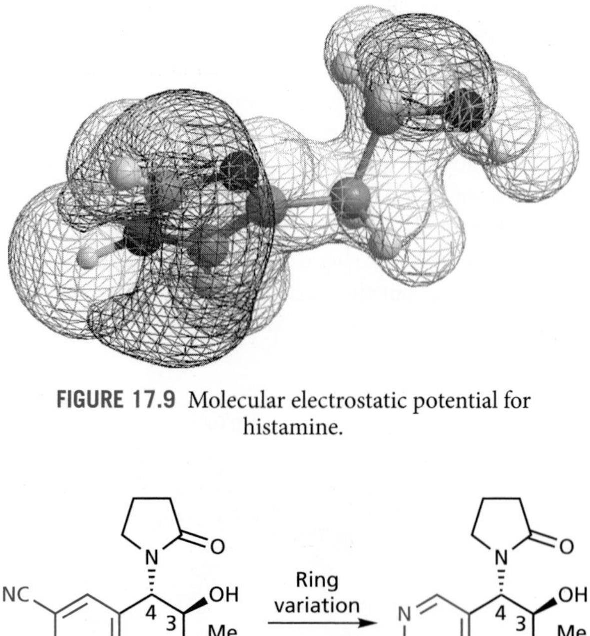

FIGURE 17.9 Molecular electrostatic potential for histamine.

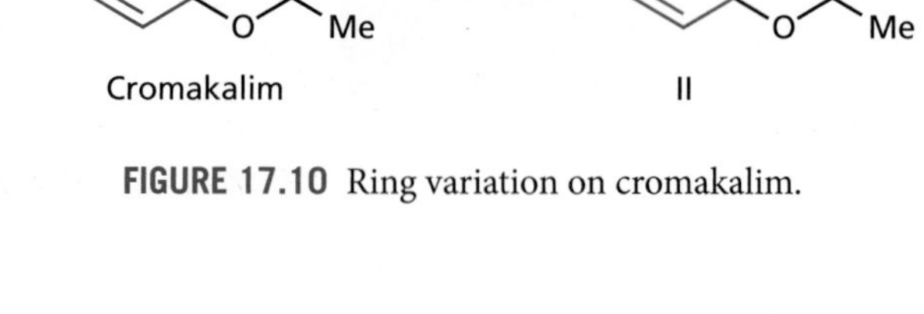

Cromakalim

II

FIGURE 17.10 Ring variation on cromakalim.

to the original aromatic ring. Consequently, the MEPs of various bicyclic systems were calculated and compared with the parent bicyclic system (III; Fig. 17.11). In order to simplify the analysis, the study was carried out in 2D within the plane of the bicyclic systems, and maps were created showing areas of negative potential (Fig. 17.12). The contours represent the various levels of the MEP, and can be taken to indicate possible hydrogen bonding regions around each molecule. The analysis demonstrated that the bicyclic system (IV) had similar electrostatic properties to (III), resulting in the choice of structure (II) as an analogue.

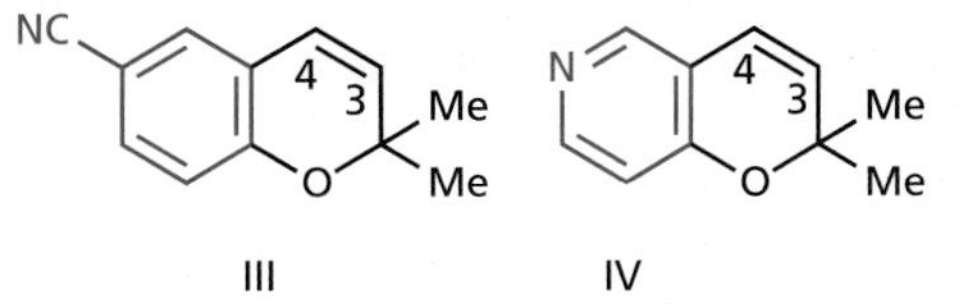

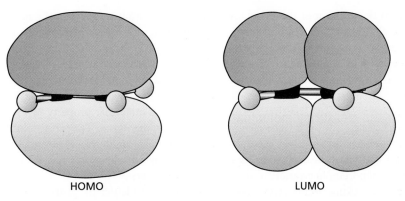

FIGURE 17.11 Bicyclic models in cromakalim study.

17.7.3 **Molecular orbitals**

The molecular orbitals of a compound can be calculated using quantum mechanics. For example, **ethene** can be shown to have 12 molecular orbitals. The **highest occupied molecular orbital (HOMO)** and **lowest unoccupied molecular orbital (LUMO)** are shown in Fig. 17.13 (see also Box 17.2).

Test your understanding and practise your molecular modelling with Exercise 17.4 on the Online Resource Centre at www.oxfordtextbooks.co.uk/orc/patrick6e/

17.7.4 **Spectroscopic transitions**

It is possible to calculate the infrared or ultraviolet transitions for a molecule. Although a theoretical infrared spectrum can be generated, it is highly unlikely that it will accurately match the actual infrared spectrum. Nevertheless, the position and identification of specific absorptions can be identified and can be useful in the design of drugs. For example, it is found that the activity

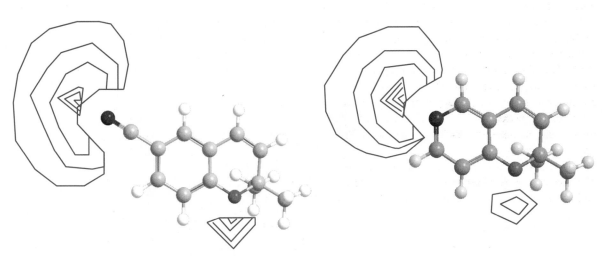

FIGURE 17.12 Molecular electrostatic potentials (MEPs) of bicyclic models (III) and (IV).

HOMO   LUMO

FIGURE 17.13 HOMO and LUMO molecular orbitals for ethene.

BOX 17.2 Study of HOMO and LUMO orbitals

A study of a molecule's HOMO and LUMO orbitals is useful because frontier molecular orbital theory states that these orbitals are the most important in terms of a molecule's reactivity. HOMO and LUMO orbitals can also help to explain drug–receptor interactions. For example, **ketanserin** (Fig. 1) is an antagonist at serotonin receptors, but has a greater binding affinity than would be expected from the more obvious intermolecular interactions.

In order to explain this greater binding affinity, it was proposed that a charge transfer interaction was taking place between the electron-deficient fluorobenzoyl ring system of ketanserin and an electron-rich tryptophan residue which was known to be nearby in the binding site. To check this, HOMO and LUMO energies were calculated for a model complex between the indole system of tryptophan and the fluorobenzoyl system of ketanserin (Fig. 2). This showed that the HOMO for the indole/fluorobenzoyl complex resided on the indole structure, whereas the LUMO was on the fluorobenzoyl moiety, indicating that

FIGURE 1 Ketanserin.

charge transfer was possible. With other antagonists, there was not this same clear-cut separation between the HOMO and LUMO orbitals, with the indole system being involved in both orbitals.

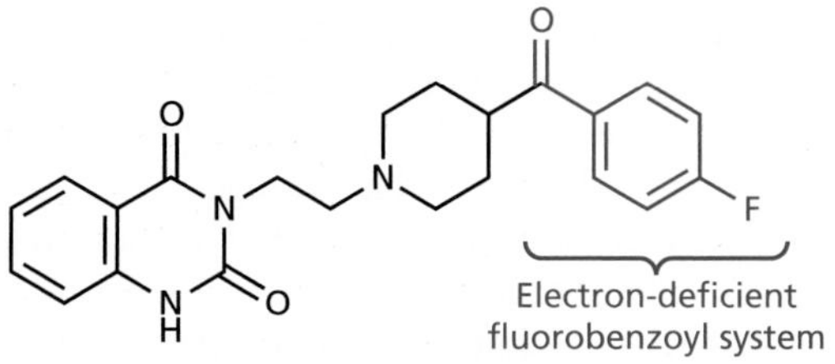

Test your understanding and practise your molecular modelling with Exercise 17.5 on the Online Resource Centre at www.oxfordtextbooks.co.uk/orc/patrick6e/

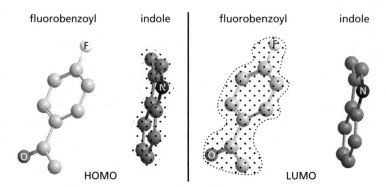

FIGURE 2 HOMO and LUMO molecular orbitals (dot surfaces) for the indole/fluorobenzoyl complex.

of penicillins is related to the position of the β-lactam carbonyl stretching vibration in the infrared spectrum. Calculating the theoretical wavenumber for a range of β-lactam structures can be useful in identifying which ones are likely to have useful activity before synthesizing them.

17.7.5 The use of grids in measuring molecular properties

Grids have become extensively used in measuring molecular properties, and are important in a variety of software programs used both in docking (section 17.12) and 3D QSAR (section 18.10).

There are various properties of a molecule which can be measured as **fields**. A field is defined as the influence that a property has on the space surrounding the molecule. As an analogy, consider a magnet. This creates a magnetic field around it which gets stronger the closer one gets to the magnet. In the same way, it is possible to measure a molecular property by the influence it has on surrounding space. The most commonly measured molecular fields are steric and electrostatic. These can be measured by placing a molecule into a preconstructed 3D lattice or grid (Fig. 17.14). The intersections of this lattice are called lattice (or grid) points and these define the 3D space around the molecule.

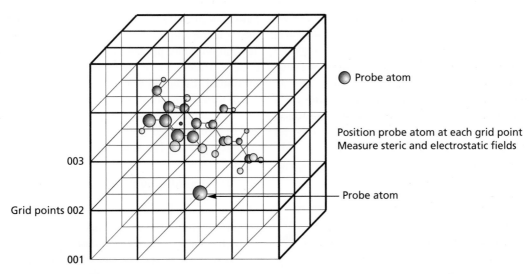

FIGURE 17.14 Measuring fields around a molecule by placing a probe atom at grid points.

Once a molecule has been placed into the lattice, the steric and electrostatic fields around it can be measured. This is done by placing a probe atom, such as a proton or an sp^3 hybridized carbocation, at each of the grid points in turn, and using software to calculate the steric and electrostatic interactions between the probe and the molecule. As far as the steric field is concerned, this will increase as the probe atom gets closer to the molecule. As far as the electrostatic field is concerned, there will be an attraction between the positively charged probe and electron-rich regions of the molecule, and a repulsion between the probe and electron-deficient regions of the molecule.

The steric and electrostatic fields at each grid point are tabulated, and a particular value for the steric energy is chosen which will define the shape of the molecule. The grid points having that value are then connected by contour lines to define the steric field. A similar process is carried out to measure the electrostatic interactions between the positively charged probe atom and the test molecule. Electron-rich and electron-deficient regions for the molecule are then defined by suitable contour lines. It is also possible to use the grid method to measure a hydrophobic field by using a water molecule as a probe.

Grids can also be constructed within the binding sites of enzymes or receptors, and are important in many docking software programs (section 17.12). Various atoms or molecular fragments are used as probes in order to measure interactions with the amino acids that make up the molecular surface of the binding site. The interactions of interest are typical binding interactions such as ionic, van der Waals, and hydrogen bonding. The atoms or fragments used as probes are the typical atoms or fragments that might be found in a drug molecule. For example typical atom probes are C, H, N, and O. Typical fragment

probes are C=O, CO_2^-, N–H, etc. In this way, it is possible to measure whether various binding interactions are possible at the different grid points, as well as their strength. The measurements can then be stored in tables for each atom or molecular fragment. This obviously involves a lot of calculations, but it only has to be done once in order to define the binding characteristics of the binding site. Once this has been done, the binding strengths of different docked molecules can be quickly calculated by identifying which atoms or groups coincide with particular grid points. The relevant entries in the tables are 'looked up' and summed to give the required total. In this way, it is possible to carry out docking studies on hundreds of different molecules within a reasonable time span. The use of grids to measure the binding characteristics of an enzyme's active site was important in the development of the antiflu drug **zanamivir** (section 20.8.3.2).

KEY POINTS

- Several chemical drawing packages include software programs that calculate various physical properties.

- Molecular modelling software makes use of programs based on molecular mechanics and quantum mechanics.

- Molecular mechanics programs use equations based on classical physics to calculate force fields. Atoms are treated as spheres, bonds as springs. Electrons are ignored. This method is suitable for energy minimization and conformational analysis.

- Quantum mechanical methods are *ab initio* or semi-empirical. The former is more rigorous but is restricted to small molecules. These methods are suitable for measuring molecular properties such as molecular orbital energies and coefficients.

- Energy minimization has to be carried out on any molecule constructed with molecular modelling software. The process involves alteration of bond lengths, bond angles, torsion angles, and non-bonded interactions until a stable conformation is obtained.

- Molecular modelling software allows the dimensions of a molecule to be accurately measured, as well as calculations of partial charges, molecular electrostatic potentials, and molecular orbitals.

- Grids and probe atoms are used to measure steric, electrostatic, and hydrophobic fields around molecules.

- Grids can be placed in binding sites in order to identify the nature and strength of potential binding interactions at different locations within the binding site. These can be tabulated and used to measure binding energies of ligands.

17.8 Conformational analysis

17.8.1 Local and global energy minima

In section 17.4, we saw how energy minimization is carried out on a 3D structure to produce a stable conformation. The structure obtained, however, is not necessarily the most stable conformation. This is because energy minimization stops as soon as it reaches the first stable conformation it finds, and that will be the one closest to the starting structure. This is illustrated in Fig. 17.15 where the most stable conformation is separated from a less stable conformation by an energy saddle. If the 3D structure initially created is on the energy curve at the position shown, energy minimization will stop when it reaches the first stable conformation it encounters—a **local energy minimum**. At this point, variations in structure result in low-energy changes and so the minimization will stop. In order to cross the saddle to the more stable conformation, structural variations would

have to be carried out which increase the strain energy of the structure and these will be rejected by the program. The minimization program has no way of knowing that there is a more stable conformation (a **global energy minimum**) beyond the energy saddle. Therefore, in order to identify the most stable conformation, it is necessary to generate different conformations of the molecule and to compare their steric energies. We shall now look at some of the ways in which this can be done.

17.8.2 Molecular dynamics

Molecular dynamics is a molecular mechanics program designed to mimic the movement of atoms within a molecule. The software program works by treating the atoms in the structure as moving spheres. After one femtosecond (1×10^{-15} s) of movement, the position and velocity of each atom in the structure is determined. The forces acting on each atom are then calculated by considering bond lengths, bond angles, torsional terms, and non-bonded interactions with surrounding atoms. The potential energy of each atom is calculated and Newton's laws of motion are then used to determine the acceleration and direction of movement of each atom (the kinetic energy). This allows the program to predict the velocity and position of each atom a femtosecond later. The procedure is then repeated for every femtosecond of the process. The femtosecond duration is important, as it is an order of magnitude less than the rate of a bond-stretching vibration, and so an atom is only allowed to move a fraction of a bond length between each calculation. If this was not the case, and atoms were allowed to move greater distances, one might get the situation where two atoms occupy the same area of space. The calculated forces and potential energies would then be huge leading to atoms moving with excessive velocities and accelerations, and resulting in system failure.

Molecular dynamics can be used to generate a variety of different conformations by 'heating' the molecule to a defined temperature. Of course, this does not mean that the inside of your computer is about to melt! It means that the program allows the structure to undergo bond stretching and bond rotation as if it *was* being heated. As a result, energy barriers between different conformations are overcome, allowing the crossing of energy saddles. In the process, the molecule is 'heated' at a high temperature (900 K) for a certain period (e.g. 5 ps), then 'cooled' to 300 K for another period (e.g. 10 ps) to give a final structure.

The process can be repeated automatically as many times as desired, to give as many different structures as is practical. Each of these structures can then be recovered, energy minimized, and its steric energy measured. By carrying out this procedure, it is usually possible to

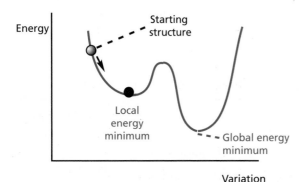

FIGURE 17.15 Local and global energy minima.

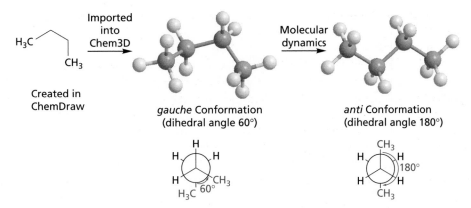

FIGURE 17.16 Use of molecular dynamics to find the most stable conformation.

identify distinct conformations that might be more stable than the initial conformation.

For example, the two-dimensional drawing of **butane** shown in Fig. 17.16 was imported into ChemBio3D and energy minimized. Because of the way the molecule was represented, energy minimization stopped at the first local energy minimum it found—the *gauche* conformation having a steric energy of 3.053 kcal/mol. The molecular dynamics program was run to generate other conformations and successfully produced the fully staggered *anti* conformation which, after minimization, had a steric energy of 2.180 kcal/mol (i.e. it was more stable by about 0.9 kcal/mol).

In fact, this particular problem could be solved more efficiently by the stepwise rotation of bonds described in section 17.8.3. Molecular dynamics is more useful for creating different conformations of molecules which are not conducive to stepwise bond rotation (e.g. cyclic

systems—see Box 17.3), or which would take too long to analyse by that process (large flexible molecules).

Finally, it has to be remembered that biomolecules in the real world are surrounded by water and that this can affect the relative stability of different conformations. Therefore, it is advisable to include water molecules in the modelling system before carrying out molecular dynamics experiments.

17.8.3 Stepwise bond rotation

Although molecular dynamics can be used to generate different conformations, there is no guarantee that it will identify all the conformations that are possible for a structure or find the global minimum. A more systematic process is to generate different conformations by automatically rotating every single bond by a set number of degrees. For example, 12 different conformations

---

BOX 17.3 Finding conformations of cyclic structures by molecular dynamics

The twist boat conformation of **cyclohexane** is not the most stable conformation of cyclohexane, but remains as the twist boat when energy minimization is carried out. 'Heating' the molecule by molecular dynamics produces a variety of different conformations including the more stable chair conformation.

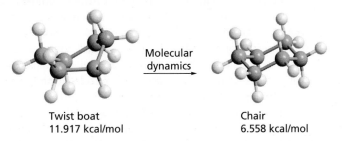

Twist boat
11.917 kcal/mol

Chair
6.558 kcal/mol

FIGURE 1 Generation of the cyclohexane chair conformation by molecular dynamics in Chem 3D.

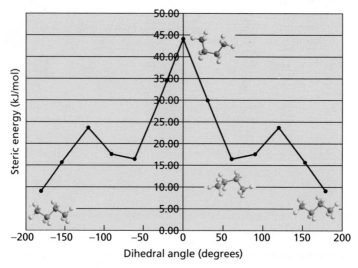

FIGURE 17.17 Graph showing relative stabilities of various butane conformations.

of butane were generated by automatically rotating the central bond in 30° steps. The steric energy of each conformation was calculated and graphed (Fig. 17.17), revealing that the most stable conformation was the fully staggered one, whereas the least stable conformation was the eclipsed one. In this operation, energy minimization is not carried out on each structure, because the aim is to identify both stable and unstable conformations.

Some modelling software packages can automatically identify all the rotatable single bonds in a structure. Bonds to hydrogen or simple substituents are excluded in this analysis, as rotations of these bonds do not generate significantly different conformations. Once the rotatable bonds have been identified, the program generates all the possible conformations which can arise from rotating these bonds by a set amount determined by the operator. The number of conformations generated will depend on the number of rotatable bonds present and the set amount of rotation. For example, a structure with three rotatable bonds could be analysed for conformations resulting from 10° increments at each bond to generate 46 656 conformations. With four rotatable bonds, 30° increments would generate 20 736 conformations.

In general, about 1000 conformations per second can be processed on a standard bench top computer. However, it is important to be as efficient as possible and care should be taken in deciding how much each bond should be rotated at a time to ensure that a representative, but manageable, number of conformations is created.

It is possible to make the process more efficient, depending on the information desired. For example, if we were only interested in identifying stable conformations, the program can automatically filter out conformations which are eclipsed or near eclipsed. It is also possible to filter out 'nonsense' conformations (i.e. conformations

where some atoms occupy the same position in space). Such conformations can arise because bond rotations are being carried out by the program without analysing what is happening elsewhere in the molecule.

Once a series of conformations have been generated, they can be tabulated and sorted into their order of stability. The most stable conformations can then be energy minimized and their structures compared.

Test your understanding and practise your molecular modelling with Exercise 17.6 on the Online Resource Centre at www.oxfordtextbooks.co.uk/orc/patrick6e/

17.8.4 Monte Carlo and the Metropolis method

In molecular dynamics, a search for the most stable conformation involves the generation of random conformations which are all analysed separately. This means that the same amount of processing time is taken up by high-energy and low-energy conformations. The **Monte Carlo method** of conformational analysis introduces a bias towards stable conformations such that more processing time is spent on these—a process known as **importance sampling**. Different conformations are generated by carrying out random bond rotations. This is quite different from molecular dynamics where atoms are shifted in space. As each conformation is generated, it is energy minimized to give a stable conformation, and its steric energy is calculated and compared with the previous structure. If the steric energy of the new conformation is lower (more stable), it is accepted and used as the starting structure for the next conformation. If the steric energy is higher, it may be accepted or rejected depending on a probability formula which takes into account both the energy of the new conformation and the 'temperature' of the

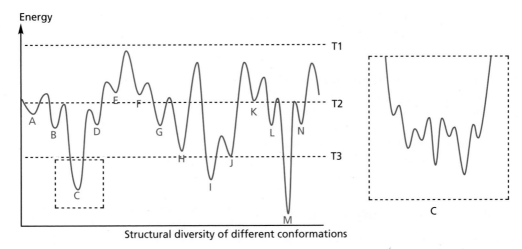

FIGURE 17.18 A 2D representation of conformational space versus structural diversity of conformations.

system. For example, suppose conformation G (Fig. 17.18) is the starting conformation. The new conformation that is generated will be structurally similar to G and could be conformation F or H. Conformation H is more stable and would be immediately accepted and used to generate the next conformation. Conformation F is less stable and so the probability equation would be used to determine whether it is accepted or not. If it is not, another conformation would be generated from conformation G.

The temperature is set by the user and if it is set high enough (for example T1), virtually all the stable conformations that are generated will be accepted and used as the starting point for the next structure. The process is repeated for as long as the user wishes in order to generate a set of different conformations. The advantage of using a high temperature is that it will allow the algorithm to generate a structurally diverse set of conformations such as conformations B, D, F, G, H, J, K, L, and N in Fig. 17.18. The steric energies can be measured, allowing the identification of the most stable conformation present. However, there is no guarantee that the global minimum will be found—in this case conformation M—due to the random nature of the search. This is particularly the case with molecules having several rotatable bonds and a huge number of possible stable conformations.

If this is the case, one might ask why the algorithm could not be run at a lower temperature such that the algorithm only accepts structures that are more stable than the previous one. Unfortunately, this will mean that the system will only focus on a particularly localized area of conformational space. (Conformational space is the term used to describe the various conformations that are possible for a structure.) For example, starting from conformation G, conformations F and H are likely to be generated because they are similar structures, but only H will be accepted. Modifying H may generate conformation

I, but there the process will stop. Although structures C and M are more stable, they will not be generated since the search would have to accept higher energy conformations (such as K) in order to create them. In other words, the algorithm will find the most stable conformation closest in nature to the starting structure.

The **Metropolis method** (also known as **simulated annealing**) is an approach which can be used to increase the chances of finding the global minimum. It involves a number of cycles where the Monte Carlo algorithm is run at different temperatures. In the first cycle, a high temperature is set (T1) and a set of structurally diverse conformations is generated. The most stable conformation is then used as the starting structure for the next run where the temperature is set at a lower value. This process is repeated several times with the probability equation becoming more 'choosy' about which structures are accepted. This slowly 'focuses' the search on a particular area of conformational space which can be searched more rigorously. In this way, there is more chance of finding the global minimum, but there is still no guarantee of success.

For example, an initial run at T1 may generate conformations B, D, F, G, H, J, K, L, and N. Conformation J would be taken as the starting conformation for the next run which would be run at a lower temperature (T2). At this temperature, conformations such as E, F, and K are less likely to be accepted. The second run might then generate conformations A, C, D, G, H, and I but fail to generate conformations L, M, and N since they are structurally different and would only be generated if structure K was generated first. This shows how the search is starting to narrow into specific areas of conformational space. Structure C would be the starting point for the next run at a lower temperature (T3) and this would focus the search into even finer detail where very similar conformations are identified and their steric energies compared, leading to the identification of the most stable conformation in that area of conformational space.

Ideally, the more slowly the temperature is lowered and the more runs that are carried out, the more chance that the global minimum will be identified. For example, just using three runs as described above narrows the search into an area which does not include the global minimum M. By decreasing the temperature in smaller increments, there is more chance that the search will focus into the area of conformational space containing the global minimum.

17.8.5 Genetic and evolutionary algorithms

A search for stable conformations can be carried out using genetic and evolutionary algorithms. As the names suggest, these algorithms are programmed to work using the same principles as biological evolution. Consider, for example, the growth of a bacterial cell. For a cell to divide, DNA has to be copied. However, the copying process is not perfect. Random mutations take place to the genes coding for the bacterial proteins. These mutations may be advantageous or disadvantageous to the individual cell concerned. For example, a mutation may be advantageous if it provides the bacterial cell with immunity against an antibacterial agent. As a result, this cell will survive and the mutation will be passed on to future generations. On the other hand, a mutation might disrupt the function of a vital protein, which results in cell death. Consequently, this mutation is not preserved. When we move to the human level, each member of a new generation receives chromosomes from two parents and this provides variety where particular characteristics are received from one parent or the other.

As far as conformational analysis is concerned, genetic and evolutionary algorithms are designed to create different conformations and to carry out an evolutionary process which will select the most stable conformations. We will now look more closely at how this is done.

First of all, the conformation of a molecule has to be represented in a manner which will allow an evolutionary process of mutation and selection to take place. Quite simply, the torsion angles for the rotatable bonds in the

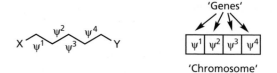

FIGURE 17.19 The representation of torsion angles as genes in chromosomes.

molecule are stored as a sequence of numbers. This sequence corresponds to a 'chromosome', where each 'gene' signifies a torsion angle (Fig. 17.19).

An initial population of 'chromosomes', representing different conformations, is created by randomly choosing values for the different torsion angles. The stability of each conformation is then calculated by molecular mechanics. The next stage in genetic algorithms is to create a new population of chromosomes or conformations. First of all, sets of 'parents' are chosen from the initial population. This is a quasi-random process, but a statistical bias is built into the selection process such that the most stable conformations are chosen as parents. This means that a particularly stable conformation can be involved in several 'relationships'.

The new population of conformations is now generated. The chromosomes from each parent undergo a 'crossover' or recombination process to generate new chromosomes where each chromosome has torsion angles contributed from each parent. In the example shown below (Fig. 17.20), the crossover involves the first two torsion angles of each chromosome. This generates two 'children' conformations, one of which corresponds to a more stable conformation where all the torsion angles are at 180°.

As well as crossovers, random 'single point' mutations are made on individual chromosomes. This corresponds to a random alteration of a single torsion angle within the chromosome. Both crossover processes and mutations generate a new and diverse population of conformations which can now act as parents for the next generation. The process can be repeated for as long as is practical.

There is a risk that a particularly stable conformation might be formed early on in the process and be lost as

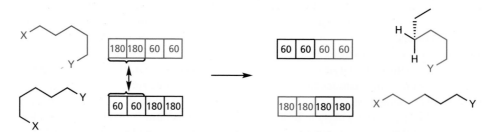

FIGURE 17.20 The crossover or recombination process to generate new conformations.

a result of further crossovers and mutations. To guard against this, most programs carry forward the chromosomes representing the most stable conformations unchanged—a so called **elitist strategy**.

In the above example, the torsion angles in the chromosomes were represented by the actual value of the torsion angle. Many programs store these values as a binary sequence of numbers instead. However, the principle is the same.

Evolutionary programming is slightly different from genetic algorithms in that crossovers do not take place and mutations alone are used to generate the next generation of chromosomes. The first- and second-generation chromosomes then undergo a 'tournament' to see which represent the most stable conformations. This is done by comparing each chromosome against a randomly chosen number of opposing chromosomes. Chromosomes representing stable conformations will score more 'wins' in the tournament and be preserved. Those with poor scores are rejected.

Genetic and evolutionary programs are designed to find a stable conformation in a short time period, but due to their random nature, they cannot be guaranteed to find the global conformation. However, by carrying out several runs, it is possible to find a variety of different stable conformations for a molecule. The method is best used with highly flexible molecules with more than eight rotatable bonds—molecules which would prove difficult to study using systematic searching.

17.9 Structure comparisons and overlays

Using molecular modelling, it is possible to compare the 3D structures of two or more molecules. For example, suppose we wish to compare the structures of the alkaloid **cocaine** with the synthetic agent **procaine**. Both of these compounds have a local anaesthetic property, and structure–activity relationships indicate that the important pharmacophore for local anaesthesia is the presence of an amine, an ester, and an aromatic ring. These functional groups are present in cocaine and procaine, but the pharmacophore also requires the functional groups to be in the same relative positions in space with respect to each other. Looking at the 2D structures of procaine and cocaine, it would be tempting to match up corresponding bonds as shown in Fig. 17.21, but this would place the nitrogen atoms one bond length apart in the overlay.

With molecular modelling, the important atoms of the structures can be matched up, in this case the nitrogens and the aromatic esters of both structures. The software then strives to find the best fit, resulting in the overlay shown in Fig. 17.22. Here, the procaine molecule has been laid across the centre of the bicyclic system in cocaine so that the aromatic esters and nitrogen atoms in both structures are matched up.

How does a software program know when a best fit has been achieved? This is done by calculating the **root mean square distance** (RMSD) between all the atom pairs which are matched up, and finding the relative orientation of the molecules where this value is a minimum. For example, in the overlay between cocaine and procaine, the pairs of atoms to be matched up could be defined as the corresponding nitrogen atoms in each molecule and the oxygen atoms of the two aromatic esters. The distance between the nitrogens in each molecule is measured, as are the distances between each of the corresponding oxygen atoms. The RMSD for all the atom pairs is then calculated. One of the structures is then moved in stages with respect to the other and the calculations repeated until a minimum value of RMSD is obtained, corresponding to the best fit.

Test your understanding and practise your molecular modelling with Exercise 17.7 on the Online Resource Centre at www.oxfordtextbooks.co.uk/orc/patrick6e/

It is important to appreciate that the fitting process described above is carried out on a rigid basis, i.e. the molecules are locked in the one conformation and no bond rotations are permitted. Therefore, it is important that each molecule is in the active conformation before carrying out the fitting process. If the active

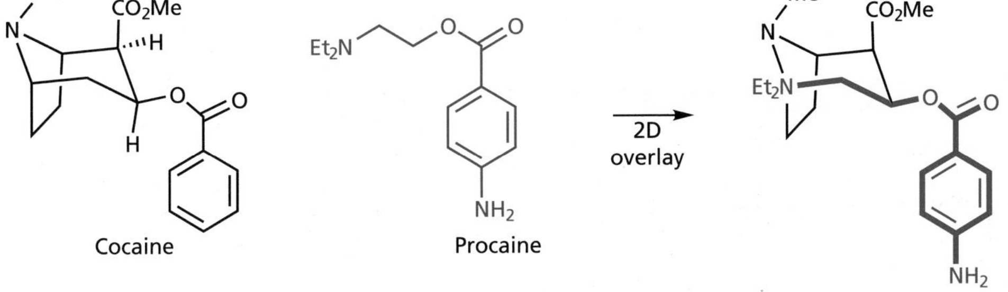

FIGURE 17.21 2D overlay of cocaine and procaine.

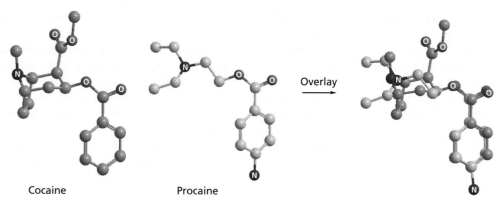

FIGURE 17.22 Overlay of cocaine and procaine using Chem3D.

conformations are not known, it is possible to carry out overlays where one or both molecules can change conformations in order to get the best fit, but this would be more expensive on computer time.

Some modelling software programs have the capacity to automatically overlay two molecules without the operator having to define which atoms should be matched up. The program searches each molecule for centres that are normally involved in binding interactions (i.e. aromatic rings, hydrogen bond donors, hydrogen bond acceptors, positively charged centres, acidic centres, and basic centres). As far as an aromatic ring is concerned, the centre of the ring (the **centroid**) represents the whole ring. For hydrogen bond acceptors (X) or hydrogen bond donors (X–H), the heteroatom (X) is defined as the centre. Of course, the centre for a hydrogen bond donor should be the hydrogen atom, but there is a large uncertainty as to where this atom is located because of bond rotation, and so the heteroatom is defined as the centre of an 'available volume' within which the hydrogen atom is located. Certain functional groups can be defined as being more than one type of centre. For example, the hydroxyl group is considered both as a hydrogen bond donor and a hydrogen bond acceptor centred on the oxygen. A primary amine is considered a hydrogen bond donor, hydrogen bond acceptor, basic centre, and positively charged centre because it could be protonated or non-protonated. Once the centres for each molecule have been identified, the program then strives to overlay them such that equivalent centres are matched up.

KEY POINTS

- Energy minimization produces the nearest stable conformation to the structure presented, and not necessarily the global conformation.

- Molecular dynamics can be carried out on a molecule to generate different conformations which, on energy minimization, give a range of stable conformations. Alternatively,

bonds can be rotated in a stepwise process to generate different conformations.

- Monte Carlo methods allow the random generation of stable conformations. The Metropolis method (simulated annealing) allows the identification of the most stable conformations and may lead to the identification of the global minimum.

- Genetic and evolutionary algorithms are used to generate different conformations and are designed to identify the most stable conformations. They may identify the global minimum, but this cannot be guaranteed.

- Molecular modelling can be used to overlay two molecules in order to assess their similarity.

17.10 Identifying the active conformation

A problem frequently encountered in drug design is trying to decide what shape or conformation a molecule is in when it fits its target binding site—the active conformation. This is particularly true for simple flexible molecules which can adopt a large number of conformations. One might think that the most stable conformation will be the active conformation, as the molecule is most likely to be in that conformation. However, it is possible that a less stable conformation could be the active conformation. This is because the binding interactions with the target result in an energy stabilization which may compensate for the energy required to adopt that conformation.

17.10.1 X-ray crystallography

The easiest way of identifying an active conformation is to study the X-ray crystal structure of a target protein with its ligand (the drug) attached. The crystal structure of the ligand itself can be obtained from the **Cambridge Structural Database** (**CSD**), and the crystal structure of protein–ligand complexes can be obtained from the **Brookhaven National Laboratory Protein Data Bank** (**PDB**) The protein–ligand complex can be downloaded and studied using molecular modelling software, and the active conformation of the ligand identified. Not all proteins can be easily crystallized, however, and other methods of identifying active conformations may have to be used.

17.10.2 Comparison of rigid and non-rigid ligands

Identification of an active conformation is made easier if one of the active compounds interacting with a target binding site is a rigid molecule which has only one possible conformation. The geometry of the pharmacophore (the important binding centres) can thus be determined for the rigid molecule. More flexible molecules can then be analysed to find a conformation which will place the important binding groups in the same relative geometry (see Box 17.4).

BOX 17.4 Identification of an active conformation

The neuromuscular blocking agent **tubocurarine** is a fairly rigid structure where the two quaternary nitrogen atoms represent the pharmacophore. Molecular modelling allows the distance between these atoms to be measured as 11.527 Å (Fig. 1). **Decamethonium** also acts as a neuromuscular blocking agent, but it is an extremely flexible molecule, which means that a large number of conformations are possible.

The most stable conformation is the extended one where the quaternary nitrogens are 14.004 Å apart. Using molecular dynamics, a variety of different conformations for decamethonium can be generated as described in section 17.8.2. One of these conformations has the quaternary nitrogens 11.375 Å apart—a possible candidate for the active conformation (Fig. 2).

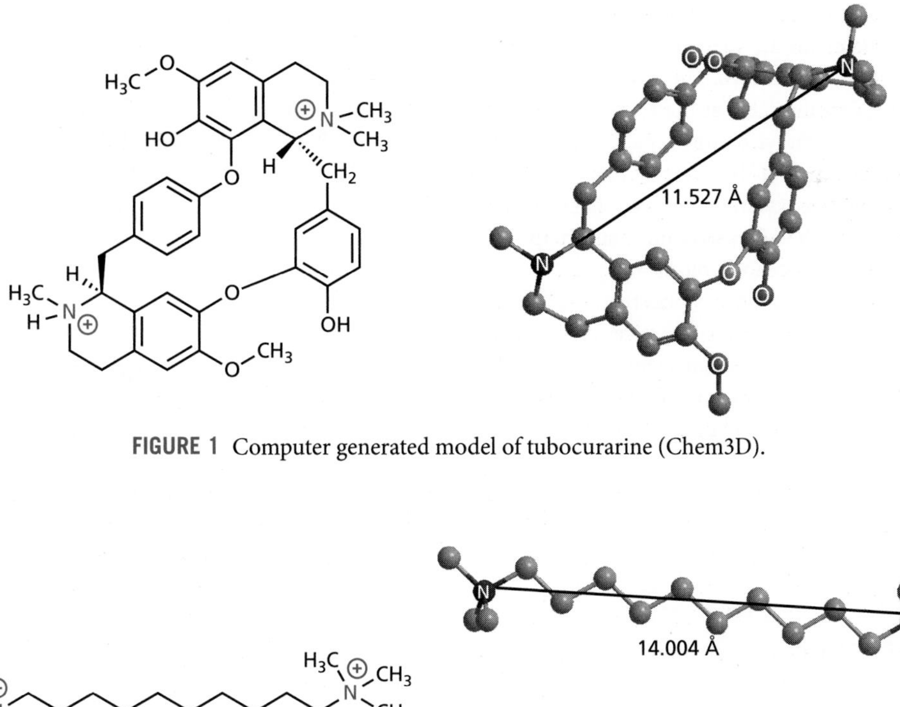

FIGURE 1 Computer generated model of tubocurarine (Chem3D).

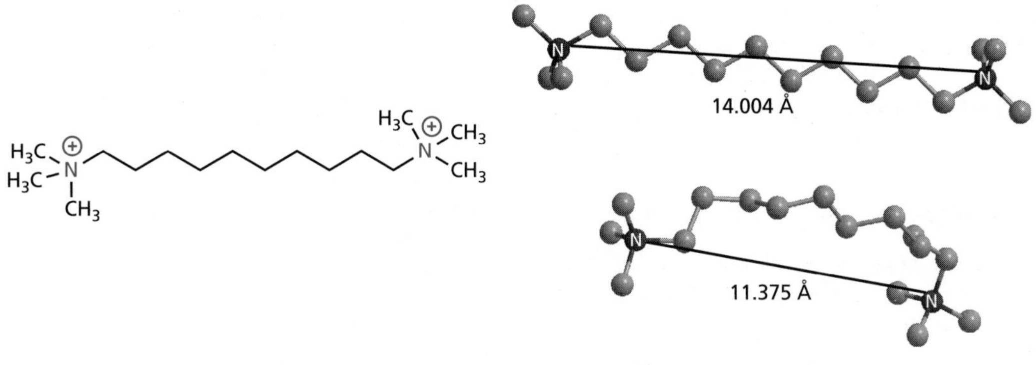

FIGURE 2 Computer generated conformations of decamethonium (Chem3D).

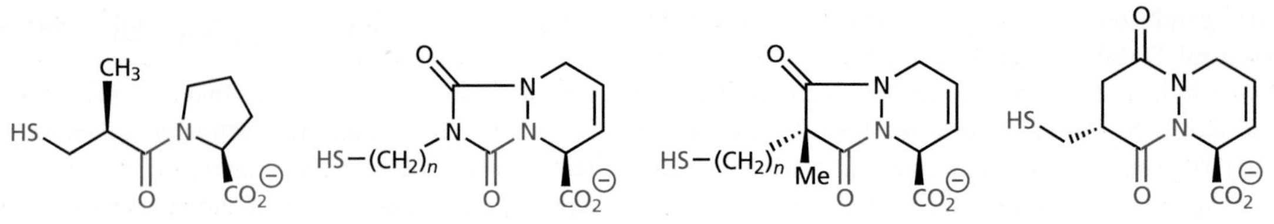

FIGURE 17.23 Captopril and rigid analogues (binding groups coloured).

Captopril   I (n = 1, 2, 3)   II (n = 1, 2)   III

🌐 Test your understanding and practise your molecular modelling with Exercise 17.8 on the Online Resource Centre at www.oxfordtextbooks.co.uk/orc/patrick6e/

Another method of finding active conformations is to consider all the reasonable conformations for a range of active compounds, and then to determine the common volume or space into which the various important binding groups can be placed in order to interact with the binding site. A study such as this was carried out in order to determine the active conformation of the antihypertensive agent **captopril** (Fig. 17.23). Because captopril is flexible, the exact 3D relationship of the important binding groups (i.e. the carboxylate, amide, and thiol groups) in the active conformation is not known. There was also no X-ray crystallographic data available at the time to reveal how captopril binds to its target binding site. To address this problem, rigid analogues (I–III) were synthesized, where the amide and the carboxylate group were fixed in space with respect to each other. Due to bond rotation, the thiol group can still access a sizeable volume of space. The biological activity of these compounds was then measured to identify which analogues were still active. A conformational analysis was then carried out as follows.

The possible conformations for captopril arising from bond rotation around the two bonds shown in Fig. 17.24

were determined using molecular modelling, and a spatial map (A) was generated to show the possible regions in space that were accessible to the thiol group (Fig. 17.25).

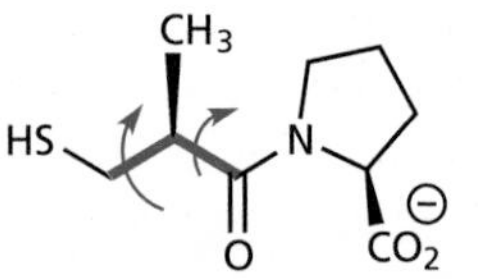

Some of the conformations involved in this analysis are high-energy eclipsed conformations, and these were filtered out of the analysis by programming the software to reject conformations with steric energy greater than 200 kJ mol^{-1}. When this was done, the spatial map (B) showed that the thiol group was restricted to two main regions in space with respect to the other two binding groups.

A spatial map for one of the active rigid analogues was now generated in the same manner and compared with the one generated by captopril. The overlap between the maps was considered to be the most likely location for the thiol group. The process was then repeated for the other active analogues, further narrowing down the possible area that would be occupied by the thiol group. The study identified two 'hot spots' (C) for the thiol group. Conformations of captopril which placed the thiol group in those 'hot spots' were then considered to be likely active conformations.

17.11 3D pharmacophore identification

A 3D pharmacophore represents the relative position of important binding groups in space and disregards the molecular skeleton that holds them there. Thus, the 3D pharmacophore for a particular binding site should be common to all the various ligands which bind to it. Once the 3D pharmacophore has been identified, structures

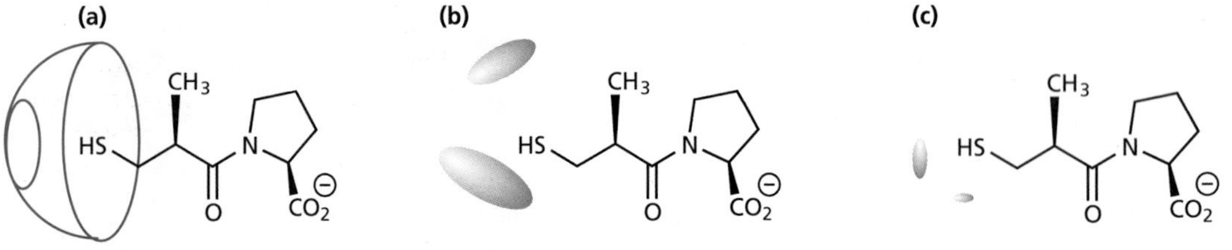

FIGURE 17.24 Bond rotations in captopril.

FIGURE 17.25 Generating spatial maps in conformational analysis. (a) All possible conformations, (b) stable conformations, and (c) after overlaps.

can be analysed to see whether they can adopt a stable conformation which will contain the required pharmacophore. If they do, and there are no steric clashes with the binding site, the structure should be active. There are several 3D chemical databases such as the CSD which can be searched for relevant structures.

17.11.1 X-ray crystallography

The crystal structure of a target protein with its ligand bound to the binding site can be used to identify the 3D pharmacophore. The protein–ligand complex can be downloaded on to the computer and studied to identify the bonding interactions that hold the ligand in the binding site. This is done by measuring the distances between likely binding groups in the drug and complementary binding groups in the binding site to see whether they are within bonding distance. Once the binding groups on the ligand have been identified, their positions can be mapped to produce the pharmacophore. The Brookhaven Protein Databank stores the crystal coordinates of proteins and other large macromolecules with and without bound ligands.

17.11.2 Structural comparison of active compounds

If the structure of the target is unknown, a 3D pharmacophore can be identified from the structures of a range of active compounds. Ideally, the active conformations and the important binding groups of the various compounds should be known. The molecules can then be overlaid as previously described in section 17.9 to ensure that the important binding groups are matched up as closely as possible. It would be rare for the binding groups to match up exactly, so an allowed region in space for each important binding group can be identified for the 3D pharmacophore.

17.11.3 Automatic identification of pharmacophores

It is possible to identify possible 3D pharmacophores for a range of active compounds using some software programs, even if the important binding groups are unknown or uncertain. First of all, the program identifies potential binding centres in a particular molecule. These are the hydrogen bond donors, hydrogen bond acceptors, aromatic rings, acidic groups, and basic groups. It is also possible to search for hydrophobic centres involving hydrocarbon skeletons of three or more carbon atoms. Here, the hydrophobic centre is calculated as the midpoint of the carbon atoms in question.

If dopamine was the structure being analysed, four important binding centres would be identified—the aromatic ring, both phenolic groups (hydrogen bond donors and acceptors), and the amine nitrogen (hydrogen bond donor, hydrogen bond acceptor, base, and a positively charged centre if protonated) (Fig. 17.26).

The program now identifies the various triangles which connect up the important centres. In the case of dopamine, there are apparently four such triangles. Each one is defined by the length of each side and the type of binding centres present, resulting in a set of pharmacophore triangles. Some of the points specified represent more than one type of binding centre, so this means that there will actually be more than four pharmacophore triangles. For example, if one of the points is a phenol then it represents a hydrogen bond donor or a hydrogen bond acceptor. Therefore, any triangle including this point must result in two pharmacophore triangles—one for the hydrogen bond donor and one for the hydrogen bond acceptor.

Of course, this analysis has only been carried out on one conformation of dopamine. The program is now used to generate a range of different conformations as described in section 17.8, and for each conformation another set of pharmacophore triangles is defined. Adding all these together gives the total number of possible pharmacophore triangles for dopamine in all the conformations created.

Another structure with dopamine-like activity is now analysed. Once all of its pharmacophore triangles have been determined, they are compared with those for dopamine, and the pharmacophores that are common to both structures are identified. The process is then repeated for

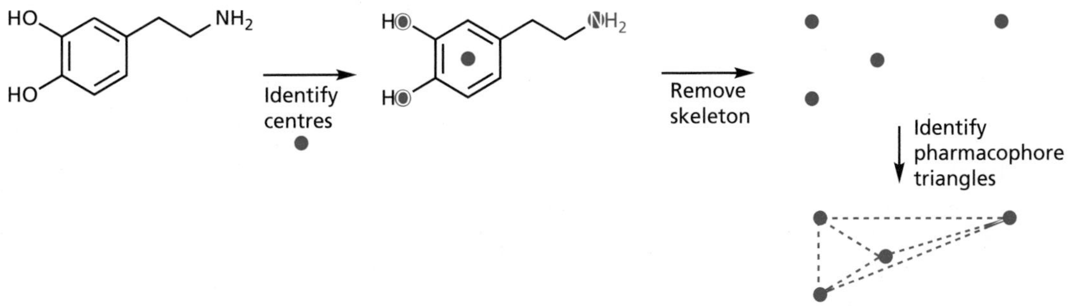

FIGURE 17.26 Pharmacophore identification in dopamine.

all the active compounds until pharmacophore triangles common to all the structures have been identified. These are then plotted on a 3D plot where the *x*-, *y*-, and *z*-axes correspond to the lengths of the three sides of each triangle. This produces a visual display which allows easy identification of distinct pharmacophores. Closely similar pharmacophores can be quickly spotted as they are clustered close together in specific regions of the plot.

For example, the three structures in Fig. 17.27 were analysed and found to have 38 common pharmacophores. When these were plotted, seven distinct groups of pharmacophore were identified (Fig. 17.28). Each pharmacophore present in the grid can be highlighted and revealed. For example, one of the possible active pharmacophores consists of two hydrogen bond acceptors and an aromatic centre. Note that it is advisable to begin this exercise with the most active compound and then proceed through the structures in order of activity.

The above analysis can be simplified enormously if certain groups are known to be essential for binding. The program can then be run such that only triangles containing these centres are included. For example, if the amine of dopamine is known to be an essential binding group, then the triangles connecting the phenolic oxygens and the aromatic ring for each conformation can be omitted.

17.12 Docking procedures

17.12.1 Manual docking

Molecular modelling can be used to dock or fit a molecule into a model of its binding site. If the binding groups on the ligand and the binding site are known, they can be defined by the operator such that each binding group in the ligand is paired with its complementary group in the binding site. The ideal bonding distance for each potential interaction is then defined and the docking procedure is started. The program then moves the molecule around within the binding site to try and get the best fit as defined by the operator. In essence, the procedure is similar to the overlay or fitting process described in section 17.9, only this time the paired groups are not directly overlaid but fitted such that the groups are within preferred bonding distances of each other. Both the ligand and the protein remain in the same conformation throughout the process and so this is a rigid fit. Once a molecule has been successfully docked, fit optimization is carried out. This is essentially the same as energy minimization, but carried out on the ligand–target protein complex. Different conformations of the molecule can be docked in the same way and the interaction energies measured to identify which conformation fits the best.

FIGURE 17.27 Test structures.

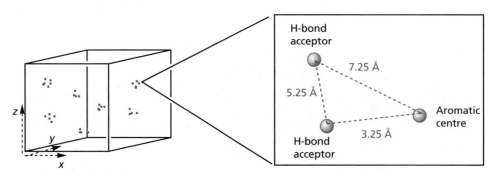

FIGURE 17.28 Pharmacophore plot.

17.12.2 Automatic docking

A variety of docking programs now exist that can automatically dock ligands into a binding site with the minimum of input from an operator. They have the advantage that they do not depend on any preconceived ideas that the operator may have on how a particular ligand should bind, and, as a result, they can reveal unexpected binding modes. They are also amenable to studying many different molecules automatically. Indeed, an important application of automatic docking programs is to carry out virtual screening of hundreds of different molecules with the aim of identifying new lead compounds that will interact with the target. Virtual screening can be seen as complementary to biological screening, in that the former can identify the structures from a chemical 'library' that are most likely to bind to the target. These can then be given priority for biological screening, making the latter more efficient. For virtual screening to be effective, it has to use efficient algorithms which not only dock each molecule realistically, but also give an accurate 'score' of the relative binding energies of the molecules concerned. Moreover, for each molecule studied, the docking program is likely to generate several different orientations or binding modes. It is necessary to score all of these in order to identify the most likely binding mode in terms of how well it fits the space available and how many intermolecular interactions it can form with the binding site.

The calculations required for docking and scoring have to be rapid in order to process the number of molecules involved in a reasonable time period, but they also have to be accurate enough to give a good measure of relative binding energies. This is a difficult compromise to make since increasing the speed at which an algorithm operates involves assumptions or short cuts that inevitably reduce the accuracy of the calculation. As a result, this is an area of intense research interest in the development of new and improved docking programs. For reasons of space, it is not possible to go into the mathematical

details of docking algorithms, and so this section focuses more on the general methods by which automatic docking can be carried out.

The simplest approach to automatic docking is to treat the ligand and the macromolecular target as rigid bodies. This is acceptable if the active conformation of the ligand is known, or if the ligand is a rigid cyclic structure. At the next level of complexity, the target is still considered as a rigid body, but the ligand is allowed to be flexible and can adopt different conformations. The most complex situation is where both the target and the ligand are considered to be flexible. This last situation is extremely expensive in terms of computer time, and most docking studies are carried out by assuming a rigid target.

17.12.3 Defining the molecular surface of a binding site

In order to carry out docking calculations, it is necessary to know the structure of the protein target and the nature of the binding site. This can be obtained from an X-ray crystal structure of the protein which can be downloaded on to the computer. The amino acids lining the binding pocket can then be identified.

The next step is to define the molecular surface of the binding site. One could do this by defining each atom within the binding site by its van der Waals radius, but this results in an extensive surface area, much of which would be inaccessible to a ligand (Fig. 17.29).

A simpler molecular surface can be defined by identifying the parts of the van der Waals surface that are accessible to a solvent molecule. In practice, a probe sphere of radius 1.4–1.5 Å is used to represent a water molecule, and this is 'rolled over' the surface of the binding site (Fig. 17.30). Convex surfaces shown in dark blue are where the probe sphere makes contact with the van der Waals surfaces of a particular atom. Concave surfaces shown in light blue are known as **re-entrants** and represent how far

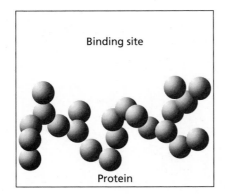

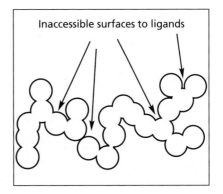

FIGURE 17.29 Defining the surface of the binding site by atoms and van der Waals surfaces.

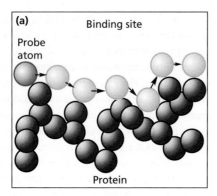
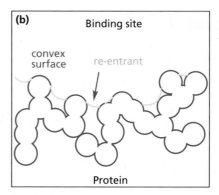

FIGURE 17.30 (a) Defining the Connolly surface of a binding site with a probe atom. (b) The Connolly surface shown in blue.

the probe atom can access the space between the atoms of the binding site. In this area, the probe is in contact with two or three atoms. This kind of molecular surface is also referred to as a **Connolly surface**. The surface is actually represented by a regular distribution of points or dots, and the crucial ones for docking are those on the convex surfaces. Each one of these has a vector associated with it which points into the binding site. The direction of the vector corresponds to the normal of the surface at that point, and so it is a mathematical indication of curvature.

17.12.4 Rigid docking by shape complementarity

The first problem with any docking program is how to position the ligand within the binding site. If you or I were handed real models of the target and the ligand, we would consider the space available in the binding site, eye up the ligand and judge how we could place it into the

binding site before we actually do it. In other words, humans have a spatial awareness which includes the ability to assess the shape of an empty space. This does not come naturally to computers, and the empty space of a binding site has to be defined in a way that a computer program can understand before ligands can be inserted.

The **DOCK program** was one of the earliest programs to tackle this problem. The Connolly surface is first defined, then the empty space of the binding site is defined by identifying a collection of differently sized spheres which will fill up the space available and give a 'negative image' of the binding site (Fig. 17.31). This is achieved as follows.

For each dot representing the molecular surface, spheres are constructed that touch that dot, plus one other dot on the molecular surface. Therefore, if there are n dots representing the molecular surface, $n-1$ spheres will be created at *each* of the dots. This represents a massive number of spheres and so it is necessary to whittle

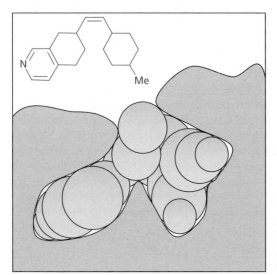
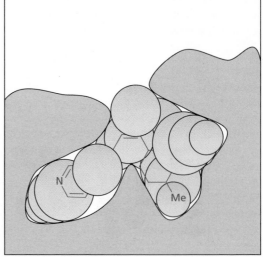

FIGURE 17.31 The DOCK program.

these down. The number of spheres can be significantly reduced as follows;

- Only the sphere of smallest radius touching each dot is chosen. This ensures that none of the spheres chosen intersects the molecular surface.

- There are several dots associated with the surface of a particular atom and each of these now has one sphere associated with it. The next filtering process is to select the sphere with the largest radius. Once this has been completed, the number of spheres left is the same as the number of atoms lining the binding site. Spheres are allowed to overlap and the centre of each sphere accurately defines a unique position of 3D space within the binding site.

Each sphere representing the binding site can be considered as a '**pseudoatom**' and so it is now possible to carry out an overlay operation as described in section 17.9, where ligand atoms are matched with pseudoatoms then overlaid. However, how does the program decide which ligand atom and pseudoatom should be matched up? One could try out every possible combination, but this would take up far too much computer time. Instead, a systematic matching operation takes place called **distance matching** or **clique searching**. First, the distances between each of the ligand atoms are measured. This is repeated for all of the pseudoatoms. These distances are then used to identify which ligand atoms and pseudoatoms can be matched up. The operation takes place like this. A graph is prepared where each ligand atom (1, 2, 3,...) is matched to each of the receptor spheres (A, B, C,...) to give a list of paired atom/pseudoatoms (1A, 1B, 1C..., 2A, 2B, 2C..., 3A, 3B, 3C..., etc.). The next stage is to identify whether two of these pairs are compatible; for example is the pairing 1A possible at the same time as the pairing 2C? This is done by comparing the distance between the ligand atoms 1 and 2, with the distance between the receptor spheres A and C. If the distances are similar, then they are compatible. This process is now repeated for further pairings to see if they are compatible with those already identified. The minimum number of pairings required for an acceptable docking is four. The whole procedure is repeated systematically for each ligand atom to find a variety of matches which will eventually lead to different docking modes.

As an example, consider a ligand represented by atoms 1–10, and a binding site represented by pseudoatoms A–G (Fig. 17.32). The atom/pseudoatom pairs 1A, 6F, 7G, and 10E would be identified as compatible for the docking operation since the distances between the specified ligand atoms match the distances between the specified pseudoatoms.

Once this procedure has been carried out, the actual docking process can take place. Docking then involves an overlay where ligand atoms are fitted onto their paired

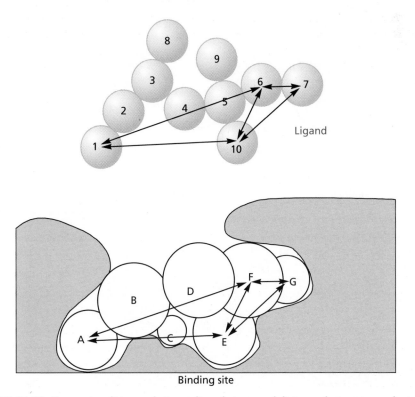

FIGURE 17.32 Comparing distances between ligand atoms and distances between pseudoatoms.

pseudoatoms as described in section 17.9. For example, in Fig. 17.33, ligand atoms 1, 6, 7, and 10 are matched to pseudoatoms A, F, G, and E respectively (Fig. 17.33). This process is repeated for all the other possible matches to give a number of docking or binding modes.

Note that this docking procedure is carried out purely in terms of **steric complementarity** (i.e. whether selected ligand atoms can match up with selected pseudoatoms). It takes no account of possible binding interactions, either favourable or unfavourable. Moreover, since selected ligand atoms are matched up with selected pseudoatoms, it is possible that some of the binding modes obtained are unrealistic. For example, a ligand atom that is not used in the matching operation might be placed in the same space as an atom lining the binding site (Fig. 17.34). Therefore, a filtering process has to be included in the program to remove any such unacceptable binding modes.

If the binding mode is acceptable, an optimization process is carried out which 'fine tunes' the position of the ligand in the binding site. This minimizes unfavourable steric interactions and optimizes intermolecular interactions between the ligand and the binding site. The binding energy of the ligand is now measured and a score is given for that binding mode.

This is repeated for all the possible matches and binding modes. The binding modes with the highest scores are then stored so that they can be analysed further by the operator. In the original version of DOCK, this scoring operation took into account only steric interactions and hydrogen bond interactions, but many other factors have an influence on receptor–ligand binding. These include other types of intermolecular interactions, desolvation, the difference in energy between a ligand's different conformations, and the decrease in entropy resulting from a flexible molecule being bound in a fixed conformation. Later versions of DOCK have tackled these issues, as have other docking programs.

17.12.5 The use of grids in docking programs

A major step forward in the development of docking programs was the use of grids to pre-calculate the binding interactions at different positions within the binding site (section 17.7.5). These values are stored in **'look-up' tables** and automatically accessed for each atom of the

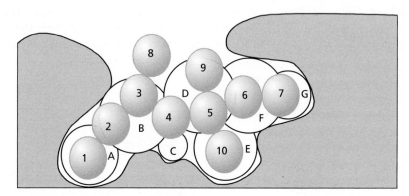

FIGURE 17.33 Docking by overlaying the atom–pseudoatom pairings of 1A, 6F, 7G, and 10E.

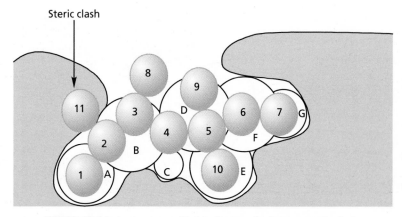

FIGURE 17.34 An unacceptable binding mode due to a steric clash.

ligand based on its particular position within the grid. This means that binding energies can be speedily obtained by adding up the relevant table entries, rather than measuring the interaction between each ligand and the binding site—a process which would take far longer. The first docking programs to implement the use of grids in this way were **AutoDock** and a revised version of **DOCK**. Grids are now used in many other programs.

17.12.6 Rigid docking by matching hydrogen bonding groups

The docking process described in section 17.12.4 is based on whether a ligand has the right shape to fit into the binding site, and takes no account of possible binding interactions. This is ideal for ligands that take up most of the space available in the binding site, but is less satisfactory for ligands which are small in comparison to the size of the binding site.

An alternative method of docking is to use the same 'clique technique' described in section 17.12.4, but this time to match up hydrogen bonding groups in the binding site with complementary hydrogen bonding groups present on the ligand. There are two important factors to take into account. Firstly, a hydrogen bonding group on the ligand must be the correct distance from a hydrogen bonding group in the binding site. Secondly, the two groups concerned must have the correct orientation with respect to each other (section 1.3.2). It is, therefore, necessary to identify positions in space within the binding site where ligand atoms can be positioned to satisfy these criteria. These positions are defined by interaction points as follows.

Firstly, a sphere is created around each hydrogen bonding group in the binding site (Fig. 17.35). The surface of the sphere represents the optimum distance at which a complementary group on the ligand should be placed in order to form a good hydrogen bond. A series of uniformly spaced points is placed over the surface of the sphere to define the surface. These are the interaction points onto which complementary binding groups on the ligand will be positioned during the docking process. However, not all of the points are feasible positions for a good hydrogen bonding interaction, and so a filtering process takes place which:

- removes the points that are not accessible in the binding site;
- removes the points which would not allow a bonding angle (α) greater than 90°.

The interaction points that survive this filtering procedure are now used as 'targets' for the matching operation with suitable ligand atoms.

This method is used in the **Directed Dock** algorithm alongside the matching algorithm based on shape complementarity. This means that it is possible to carry out a docking which takes into account both hydrogen bonding interactions and shape complementarity.

17.12.7 Rigid docking of flexible ligands: the FLOG program

One of the major drawbacks of rigid docking experiments is that they may fail to give a satisfactory answer for flexible ligands. Such ligands can form a variety of different conformations, and unless one knows the active conformation, it is a matter of chance whether the conformation chosen for the docking experiment is the ideal conformation for docking or not. One way round this is to dock as many different conformations of the ligand as possible in order to get the best result. **FLOG (Flexible Ligands Orientated on Grid)** is a docking program that generates conformational libraries called **Flexibases**, which contain 10–20 conformations for each ligand studied. However, there is still a chance that the correct conformation will not be tested, especially for very flexible ligands. The more flexible the ligand, the more conformations that are possible and this can lead to a '**conformational explosion**'. In other words, the number of possible conformations increases exponentially with the number of rotatable bonds present.

17.12.8 Docking of flexible ligands: anchor and grow programs

Various programs have been written to allow the generation of different ligand conformations as part of the docking process. A popular method is to fragment the ligand, identify a rigid anchor fragment which can be

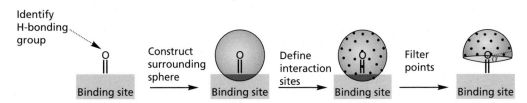

FIGURE 17.35 Identifying interaction sites for hydrogen bonding groups in the binding site.

docked, then reconstruct or grow the molecule back onto the anchor. The following are examples of such programs.

17.12.8.1 Directed Dock and Dock 4.0

Directed Dock and Dock 4.0 use a method where the algorithm identifies the rotatable bonds that are present in a ligand, allowing the identification of rigid and flexible regions. The molecule is then split into molecular components or fragments (Fig. 17.36).

The most rigid fragment is defined as the anchor and is docked by shape complementarity (section 17.12.4). The segments representing the flexible parts of the molecule are then added sequentially to the anchor. As each segment is added, torsion angles are varied in a systematic fashion and this inevitably increases the number of partially built structures (**constructs**). The number of constructs would go through the roof if this was to continue and so, once each segment is added, there is a pruning process which selects a limited number of constructs based on how well they bind and also on how different they are in structure.

The segments are added in 'layers' working outwards. Thus, all the segments in layer 1 are added sequentially before the segments in layer 2 (Fig. 17.36). At each stage of the process, energy minimization of the construct is carried out to relieve any strain arising from the construction process.

17.12.8.2 FlexX

FlexX is a software program that also uses the anchor and grow method, but here the anchor is docked according to chemical complementarity—in other words docking is determined by the intermolecular interactions that can be formed between the anchor and the binding site.

Docking the anchor by chemical complementarity rather than steric complementarity has the advantage that it cuts down the number of possible binding orientations for the anchor.

An interaction surface consisting of interaction points is built round each potential binding group in the binding site (Fig. 17.37; see also section 17.12.6). A matching process now takes place which matches atoms in the anchor to interaction points in the binding site. The distances between atoms in the anchor must match the distance between interaction points in the binding site. This is the same procedure that is carried out in section 17.12.4, but there is the added requirement that the anchor atom and the corresponding interaction point must have binding compatibility. Docking requires the identification of three matched pairs of anchor atoms/interaction points. This corresponds to identifying complementary pharmacophore triangles for the anchor and the binding site.

The matching process is thus brought down to a comparison of the ligand's pharmacophore triangles with the pharmacophore triangles present in the binding site. For a match to occur, a triangle for the ligand must have roughly the same dimensions as a triangle in the binding site. Moreover, the corners of the triangles must have binding compatibility.

The docking is now carried out such that anchor atoms are overlaid with their matched interaction point in the binding site (Fig. 17.38).

The procedure ensures that the angle requirements for hydrogen bonding are fine with respect to the interaction points in the binding site, but angles with respect to the anchor atoms now have to be checked. This is possible since the program builds a set of interaction points round the anchor atoms prior to docking. For example, consider the case where docking matches up an N–H group on the anchor with a carbonyl group in the binding site

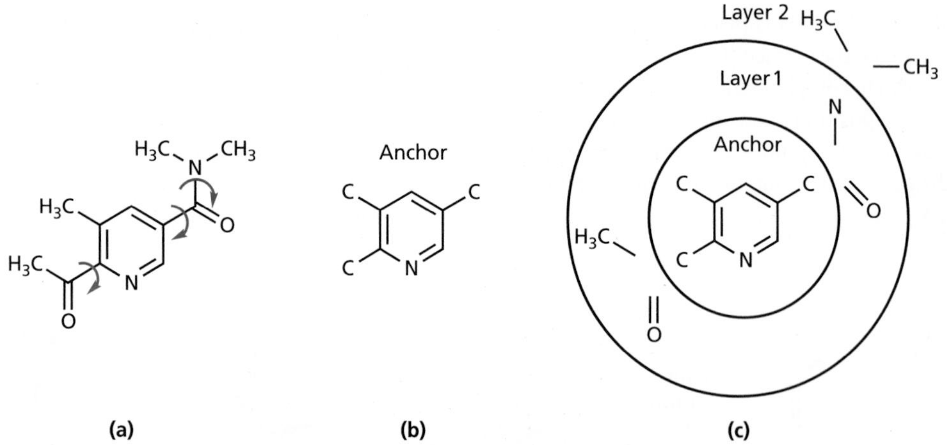

FIGURE 17.36 Anchor and grow algorithm. (a) Identify rotatable bonds. (b) Identify and dock a rigid anchor. (c) Add molecular fragments in layers.

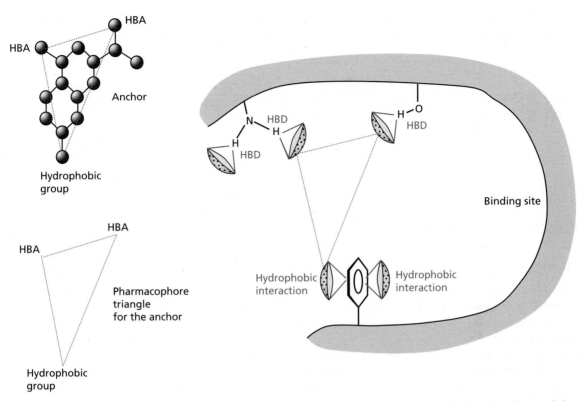

FIGURE 17.37 Docking based on the matching up of complementary pharmacophoric triangles for the anchor and the binding site (HBD = hydrogen bond donor, HBA = hydrogen bond acceptor).

(Fig. 17.39). The docking procedure fits the hydrogen of the N–H group onto an interaction point round the carbonyl group, and so the hydrogen is at the correct orientation and distance with respect to the oxygen. The program now checks to see whether the oxygen is placed on an interaction point round the hydrogen. If so, the oxygen is at the correct distance and orientation with respect to the hydrogen, and a good interaction exists.

There will be several docking solutions for the anchor. These are 'clustered' and a representative binding

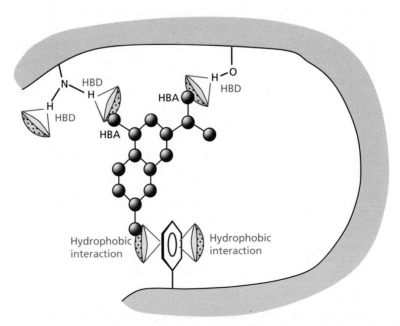

FIGURE 17.38 Docking of the anchor (HBD = hydrogen bond donor, HBA = hydrogen bond acceptor).

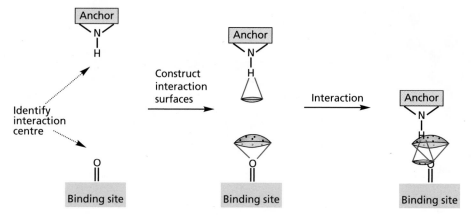

FIGURE 17.39 Assessing whether groups are at the correct distance and orientation with respect to each other for a good hydrogen bond interaction.

orientation is taken from each cluster. The remainder of the molecule is then added. Fragments are attached to the anchor by rejoining the linkers at a discrete set of torsion angles.

One problem associated with the program is that the anchor has to be chosen manually, and this becomes an issue if one wants to do the automatic docking of a series of structures in a database. Another problem is the vast number of different pharmacophore triangles that could be constructed to represent the binding site. One has to remember that each binding group has an interaction surface represented by numerous interaction points, and so the number of triangles which could be built is enormous. There are ways in which the docking algorithm can store the equivalent information in a more efficient manner, but the details of that are beyond the scope of this text.

17.12.8.3 The Hammerhead program

The Hammerhead program is another anchor and grow procedure which was designed to carry out docking studies on a large number of compounds. For example, it has been used to study databases of 10 000–100 000 small molecules in a few days.

Probes are placed into the binding site in order to identify the optimum locations for particular binding interactions (Fig. 17.40). The probes used are hydrogen atoms, as well as C=O and N–H fragments. Each of the probes can be scored as high scoring or low scoring, based on the number of hydrogen bonds it can form or on favourable hydrophobic ring face contacts. Once the probes have been positioned, they act as the targets for the docking procedure. The same kind of matching operations described previously are carried out to match atoms of a molecular fragment with probes, and there is a requirement that docking must involve at least one of the high-scoring probes. Both steric and chemical complementarity is used in the matching process and, once the matches have been identified, the docking operation is carried out.

As far as the ligand is concerned, it is split into different fragments, each of which will have a limited number of rotatable bonds. All the fragments that are formed contain an atom or bond that is shared with another fragment (Fig. 17.41).

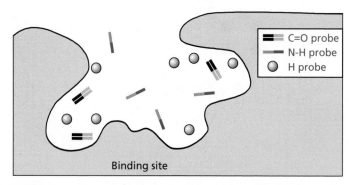

FIGURE 17.40 Positioning probes into a binding site to identify binding interactions.

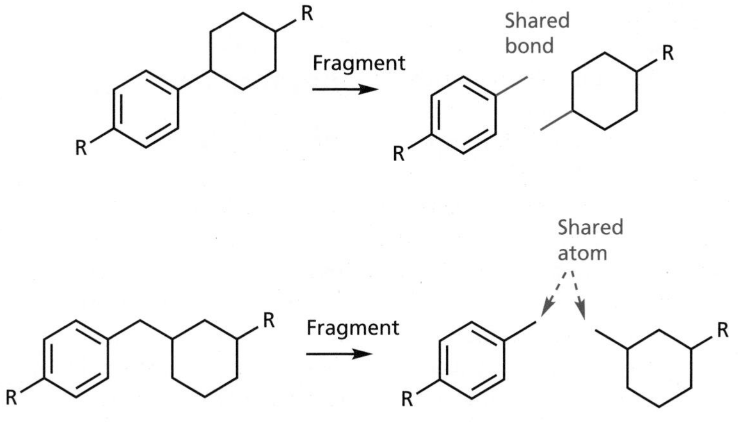

FIGURE 17.41 Fragmenting the ligand.

For each fragment, a number of conformations are generated. The fragments are docked and scored. Fragments that are particularly high scoring are defined as '**heads**' and act as anchors. The remaining fragments are defined as '**tails**'. The reconstruction phase is carried out for each fragment that has been identified as a potential anchor. Fragments that have been identified as 'tails' are then docked one at a time into the area around the anchor head. The first fragment shares an atom or a bond with the anchor and is docked such that it is aligned to the relevant atom or bond on the anchor, as well as to the pocket probes. The two fragments are then merged ('**chaining**') by overlaying the shared atoms or bonds (Figs. 17.42 and 17.43). This involves the tail fragment being moved to the anchor rather than the other way round (i.e. the anchor remains fixed).

An optimization of the construct is carried out after each stage in order to enhance binding interactions with the binding site, and to remove any unfavourable steric interactions that might be present.

This method has advantages over procedures that manually choose a single anchor. Anchors are chosen automatically without any bias from the operator. Moreover, several different anchors are possible, all of which can be investigated. This is important since the anchor with the highest binding score does not necessarily produce the best docking mode for the final ligand.

17.12.9 Docking of flexible ligands: simulated annealing and genetic algorithms

A number of docking programs use simulated annealing or genetic algorithms which incorporate a conformational search as part of the docking process. These methods are viable for the docking of flexible ligands,

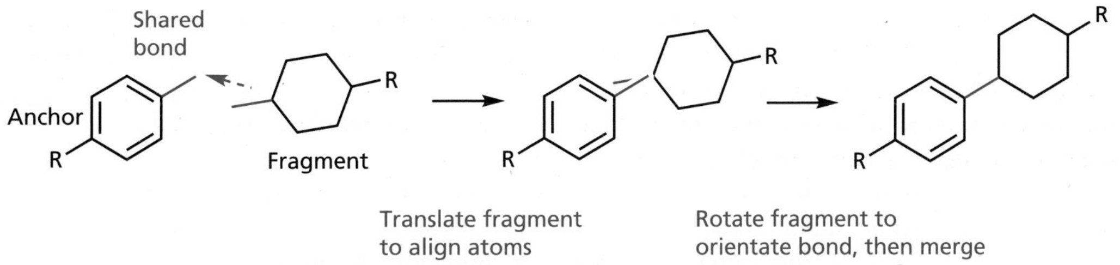

FIGURE 17.42 Merging fragments with a shared bond.

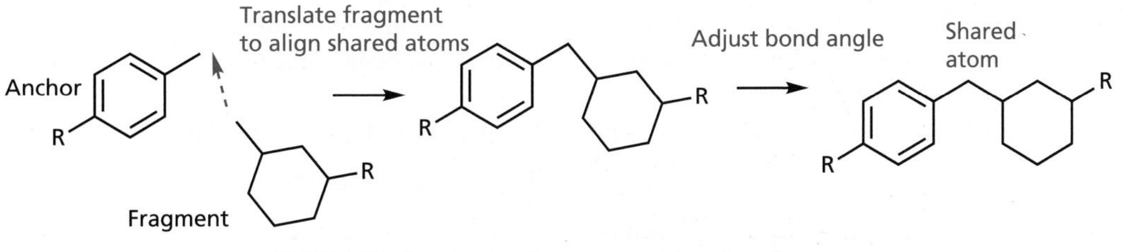

FIGURE 17.43 Merging fragments with a shared atom.

but, compared to previous methods, they are slower and computationally more expensive.

The **Metropolis method** (or **simulated annealing**) involves the use of **Monte Carlo algorithms** and was described in section 17.8.4 as a method of conformational analysis. The same principles are involved in docking studies. The intact ligand is placed randomly in the space close to the binding site. Monte Carlo algorithms are then used to generate different conformations as described in section 17.8.4, but the whole molecule is also translated and rotated such that it 'tumbles' within the binding site. Different conformations are, therefore, generated at different positions and orientations within the binding site. The binding energy of each structure is measured as it is formed and compared with the previous structure. Simulated annealing is carried out in order to identify the best binding modes. The principles of simulated annealing were explained in section 17.8.4 and hold true here—the only difference being that it is the binding energy of different binding modes which is measured, rather than the steric energy of different conformations.

One of the reasons that the procedure is slower and more computationally expensive is the need to measure binding energies for each structure and binding mode as it is formed. This can be speeded up by using 'look-up tables' which are initially prepared by the grid and probe method (section 17.7.5).

A variety of docking programs use Monte Carlo algorithms for docking, including AutoDock, MCDOCK, Prodock, and PRO-LEADS. A disadvantage of this approach is that the quality of results often depends on how the initial structure is placed in the binding site. Some research groups have used a combination of programs to address this problem. For example, DOCK can be used to identify binding modes for a specific ligand conformation based on a rigid fit and steric complementarity. Each of the binding modes identified can then be used as the starting structure for a Monte Carlo based docking program which generates different conformations and orientations in that area of the binding site.

Programs using evolutionary and genetic algorithms have also been used in docking studies. The principles of these programs were described earlier in section 17.8.5 as a method of generating different conformations. The same principles hold for docking. However, chromosomes are set up which not only determine the conformation of a molecule, but also its position and orientation within the binding site. Mutations and crossover procedures change the molecule's conformation and orientation through translation and/or rotation of the whole molecule. Selection of the best docking modes is based on how well each molecule interacts with the binding site. Examples of programs that use genetic algorithms include AutoDock, GOLD, and later versions of DOCK.

17.13 Automated screening of databases for lead compounds and drug design

The automated docking procedures described in section 17.12.2 can be used to screen a variety of different 3D structures to see whether they fit the binding site of a particular target (**electronic screening** or **database mining**). This is useful for a pharmaceutical company wishing to screen its own or other chemical stocks (libraries) for suitable lead compounds.

Screening of databases can also be done purely by searching for suitable pharmacophores. The process is speeded up by a quick filter which eliminates any structure that does not contain the necessary binding centres. The operator has the ability to vary the tolerances involved in the search in order to find pharmacophores which nearly match the desired pharmacophore.

Automated docking and database screening of fragment libraries have also been used in a strategy aimed at modifying the structures of peptide lead compounds. The strategy has been dubbed **REPLACE** and is an alternative approach to the traditional peptidomimetic design strategies described in section 14.9.1. Once the crystal structure of the peptide lead compound bound to its target binding site has been obtained, part of the peptide structure is removed, then databases of small non-peptide molecules are searched to see whether any of them dock into the vacant region of the binding site. Peptide analogues can then be synthesized where the truncated peptide is linked to the docked fragment. For example, the strategy was used on a pentapeptide to replace a highly polar arginine residue with a non-polar moiety that had no peptide characteristics (Fig. 17.44).

17.14 Protein mapping

Drug design is made easier if the structure of the target protein and its binding site are known. The best way of obtaining this information is from X-ray crystallography of protein crystals, preferably with a ligand bound to the binding site. Unfortunately, not all proteins are easily crystallized (e.g. membrane proteins). In cases like these, model proteins and binding sites may be constructed to aid the drug design process.

17.14.1 Constructing a model protein: homology modelling

A model of a protein can be created using molecular modelling if the primary amino acid sequence is known

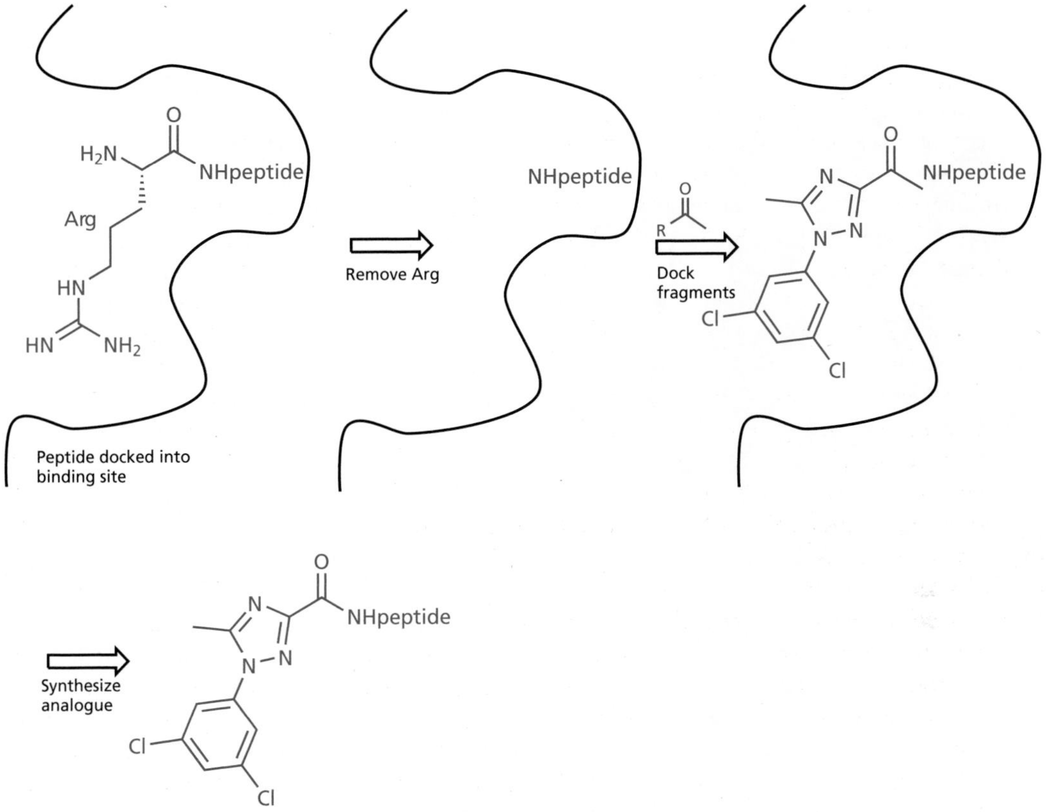

FIGURE 17.44 The REPLACE strategy.

and the X-ray structure of a related protein has been determined. Of historical interest in this respect is the protein **bacteriorhodopsin**, which has been crystallized and its structure determined by X-ray crystallography (Fig. 17.45). Bacteriorhodopsin is structurally similar to **G-protein-coupled receptors** which contain seven transmembrane helices (section 4.7). Many of the important receptors in medicinal chemistry belong to this family of proteins and so the structure of bacteriorhodopsin was used as a template to construct what is termed as **homology models** of these membrane-bound receptors. By identifying the primary amino acid sequence of the target receptor and looking for suitable stretches of hydrophobic amino acids, it is possible to identify the seven transmembrane helices, and then to use bacteriorhodopsin as a template in order to construct the helices in a similar position relative to each other. The linking loops can then be modelled in to give the total 3D structure. Unfortunately, bacteriorhodopsin is not a G-protein-coupled receptor and so it is not the ideal template for constructing these model receptors. In 2000, the crystal structure of the bovine receptor **rhodopsin** was successfully determined by X-ray crystallography. This *is* a G-protein-coupled receptor and provides a better template

for the construction of more accurate receptor models. More recently in 2007, human β_2-adrenoceptor was crystallized in a lipid environment, with an inverse agonist bound to the binding site. An antibody fragment was bound to one of the intracellular loops in order to stabilize the structure. The X-ray crystal structure revealed the structure of the transmembrane helices and the intracellular regions, but the structure of the extracellular regions and the ligand binding site are still to be fully determined.

If a new protein has been discovered, its primary structure is first determined. Suitable software is then used to compare its primary sequence with the primary sequences of other proteins in order to find a closely related protein. This involves comparing the sequences with respect to conserved amino acids, hydrophobic regions, and secondary structure. Once a reference protein of similar structure has been identified, it is used as a template in order to build the peptide backbone of the new protein. First of all, regions which are similar in both the new protein and the template protein are identified. The backbone for the new protein is constructed to match the corresponding region in the template protein. This leaves connecting regions whose structure cannot

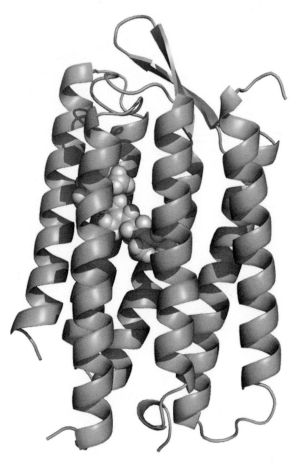

FIGURE 17.45 Bacteriorhodopsin with bound retinal.

be determined from the template. A suitable conformation for these intervening sequences might be found by searching the protein databases for a similar sequence in another protein. Alternatively, a loop may be generated to connect two known regions. Once the backbone has been constructed, the side chains are added in energetically favourable conformations. Energy minimization is carried out and the structure is refined with molecular dynamics in the absence and presence of ligand. Once the model has been constructed, it is tested experimentally. For example, the model would indicate that certain amino acids might be important in the binding site. These could then be mutated to see if this has an effect on ligand binding. Studies such as these have identified amino acids which are important in binding neurotransmitters in a range of G-protein-coupled receptors (Fig. 17.46).

This study shows interesting similarities and differences between the four receptors studied. For example, all four binding sites interact with ligands having a charged nitrogen group, and contain a hydrophobic pocket lined with aromatic residues to receive it. There are several conserved aromatic residues in this pocket at positions

307, 613, and 616. An aspartate residue at position 311 is also present in all cases and is capable of forming an ionic interaction with the charged nitrogen on the ligand.

There are also significant differences between the binding sites, which account for the different ligand selectivities. For example, the amino acids at positions 505 and 508 in the catecholamine receptors are serine, whereas the corresponding amino acids in the cholinergic receptors are alanine. The amino acid at position 617 in the catecholamine receptors is phenylalanine, allowing an interaction with the aromatic portion of the catecholamines, whereas in the cholinergic receptor this amino acid is asparagine, allowing a hydrogen bonding interaction with the ester group of acetylcholine.

17.14.2 Constructing a binding site: hypothetical pseudoreceptors

Rather than construct a complete model protein, it is possible to use molecular modelling to design a model binding site based on the structures of the compounds which bind to it. In order to do this effectively, a range of structurally different compounds with a range of activities is chosen. The active conformations are identified as far as possible and a 3D pharmacophore identified as described previously. The molecules are then aligned with each other such that their pharmacophores are matched. Each molecule is then placed in a potential energy grid and different probes are placed at each grid point in turn to measure interaction energies between the molecule and the probe atom (see section 17.7.5). An aromatic CH probe is used to measure hydrophobic interactions, and an aliphatic OH probe is used to measure polar interactions. The interactions are then displayed by isoenergy contours (typically -6.0 kJ mol^{-1} for hydrophobic interactions and -17.0 kJ mol^{-1} for polar interactions).

The molecules in the study can then be compared to identify common fields. Once these have been identified, suitable amino acids can be positioned to allow the required interaction. For example, an aspartate residue could be used to allow an ionic interaction, and amino acids such as phenylalanine, tryptophan, isoleucine, leucine, or valine could be used for a hydrophobic interaction (see also Box 17.5).

Once built, known compounds can be docked to the model receptor binding site, the complex minimized, and binding energies calculated. These can then be compared with experimental binding affinities to see how well the model agrees with experiment. If the results make sense, the model can then be used for the design and synthesis of new agents.

The procedures described above are an example of a process known as 3D QSAR which is described in

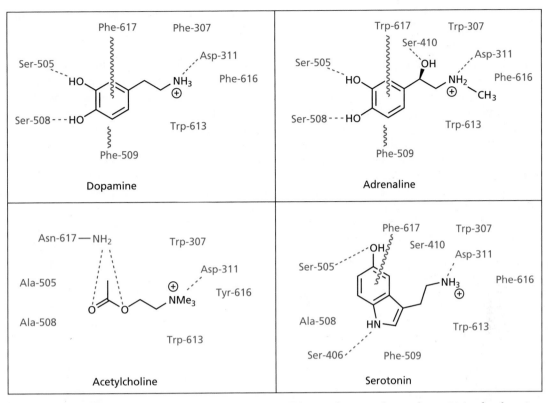

FIGURE 17.46 Important amino acids in various binding sites. (The numbering indicates the position of each amino acid on the seven possible helices of a G-protein-coupled receptor. For example, 311 indicates position 11 on helix 3.)

section 18.10. Examples of specific software programs that generate hypothetical pseudoreceptors are given in section 18.10.5.

- The active conformation and pharmacophore of a molecule can be identified from the X-ray structure of a complex between it and its target protein. Alternatively, the active conformation and pharmacophore can be identified from an active, rigid structure. If no such rigid structures are available, the various conformations of different active compounds can be compared to identify areas of space occupied by binding groups which are shared by all the active structures.

- Docking involves the fitting of a molecule into a binding site. Docking can be carried out on the basis of steric complementarity and/or chemical complementarity.

- Docking is most easily carried out with a rigid ligand and a rigid binding site. It is also possible to carry out docking with a flexible ligand and a rigid binding site.

- The docking of flexible ligands involves programs that use a fragmentation approach followed by the docking of a rigid anchor and reconstruction of the ligand. Alternatively, simulated annealing or genetic/evolutionary algorithms can be used to generate different conformations during the docking process.

- Electronic screening or database mining involves the search of structural databases to identify structures containing a particular pharmacophore.

- The binding sites of a target protein can be constructed by molecular modelling based on the X-ray structure of a protein–ligand complex. Alternatively, a model binding site can be constructed based on the primary sequence of the protein and a structural comparison of known analogous proteins. Another method is to compare a range of active compounds to identify where particular amino acids are likely to be positioned, in order to allow an interaction.

17.15 *De novo* drug design

17.15.1 General principles of *de novo* drug design

De novo drug design involves the design of novel structures based on the structure of the binding site with which they are meant to interact. The structure of the

BOX 17.5 Constructing a receptor map

A range of structures including **altanserin** and **ketanserin** was used to construct a model receptor binding site for the 5-HT$_{2a}$ receptor. Taking ketanserin as the representative structure for these compounds, various potential hydrogen bonding, ionic bonding, and hydrophobic bonding interactions were identified. Structure–activity relationships (SAR) were then used to identify whether any of these proposed interactions were important or not. In this case, SAR indicated that the two carbonyl groups were not important, and so the hydrogen bonding regions derived from these groups probably do not exist in the receptor binding site. Suitable amino acids can now be placed in the relevant positions. The choice of which amino acids should be used is helped by knowing the amino acid sequence of the target protein, along with the structure of a comparable protein. The 5-HT$_{2a}$ receptor belongs to a superfamily of proteins that includes bacteriorhodopsin, whose structure is known. Allying this information with the primary amino acid sequence of the receptor led to the choice of amino acids shown in Fig. 1.

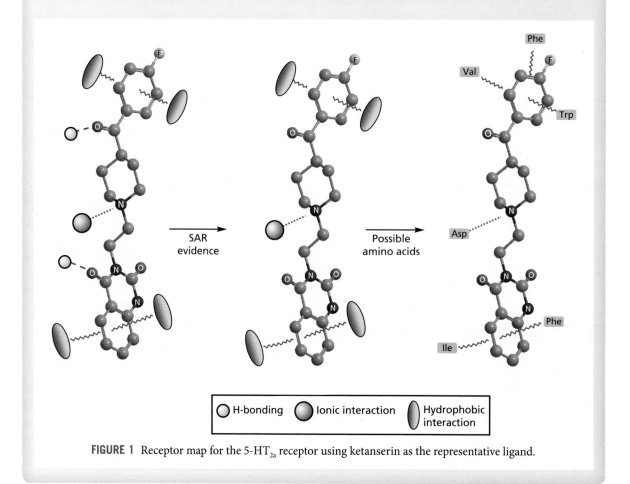

FIGURE 1 Receptor map for the 5-HT$_{2a}$ receptor using ketanserin as the representative ligand.

binding site can be identified from an X-ray crystallographic study of the target protein containing a bound ligand or inhibitor. The position of the ligand identifies where the binding site is located in the protein, and also identifies any induced fit that might have occurred as a result of ligand binding. Once the structure of the protein–ligand complex has been downloaded onto a computer, the ligand can be removed to leave the empty binding site, and *de novo* drug design can then take place. By identifying the amino acids that are present in the binding site, it is possible to identify the binding interactions that are possible within the site. A structure can then be designed which will have the correct size and shape to fit the space available, and will also have the required functional groups to interact with the binding regions. The operator can carry out each of these operations manually. Alternatively, there are several software packages which will carry out the process automatically.

Manual studies allow operators to have full control over the study and to input their own ideas as and when they wish. Such studies have been successful in producing novel active compounds, but they do suffer several disadvantages. For example, the novelty of the structures obtained is inevitably limited to the operator's own imagination and originality. More seriously, manual design is slow and it is really limited to the identification of a single novel structure. Automated design is much faster and can produce large numbers of diverse structures in a short time period.

The early work on *de novo* drug design was carried out manually and an example of one of these studies is given in Case study 5. From these studies, a number of general principles were identified regarding manual *de novo* drug design.

Firstly, it may be tempting to design a molecule which completely fills the space that is available in the binding site. However, this would not be a good idea for the following reasons:

- Normally, the structure of the binding site is identified from X-ray crystallography of the target protein. The position of atoms in the crystal structure is accurate only to 0.2–0.4 Å and allowance should be made for that.
- It is possible that the designed molecule may not bind to the binding site exactly as predicted. If the intended fit is too tight, a slight alteration in the binding mode may prevent the molecule binding at all. It would be better to have a loose-fitting structure in the first instance and to check whether it binds as intended. If it does not, the loose fit gives the molecule a chance to bind in an alternative fashion.
- It is worth leaving scope for variation and elaboration of the molecule. This allows fine tuning of the molecule's binding affinity and pharmacokinetics.

Other important points to take into consideration in *de novo* design are the following:

- Flexible molecules are better than rigid molecules because the former are more likely to find an alternative binding conformation should they fail to bind as expected. This allows modifications to be carried out based on the actual binding mode. If a rigid molecule fails to bind as predicted, it may not bind at all.
- It is pointless designing molecules which are difficult or impossible to synthesize.
- Similarly, it is pointless designing molecules which

need to adopt an unstable conformation in order to bind.

- Consideration of the energy losses involved in water desolvation should be taken into account.
- There may be subtle differences in structure between receptors and enzymes from different species. This is significant if the structure of the binding site used for *de novo* design is based on a protein that is not human in origin.

These principles also hold true for automated *de novo* drug design and some are particularly problematic. For example, automated *de novo* drug design is prone to generating structures which are either difficult or impossible to synthesize. Consequently, efforts have been made to improve the software packages involved, such that they can identify and filter out problem structures, or prevent them being generated in the first place. A second problem with automated *de novo* programs revolves around the scoring functions used to estimate binding affinities. It would be useful to rank the generated structures with respect to their binding strengths, but the results obtained have often been found to be unreliable.

Critics of *de novo* drug design are quick to point out that no clinically useful drug has been designed in this manner. This is true, but it is hard to see how this could be a realistic expectation. The number and variety of structures which could be identified through *de novo* drug design are virtually limitless and so the chances of 'hitting' the ideal structure are poor. Moreover, there is far more to drug design than finding a structure that binds strongly to its target. *De novo* drug design does not identify whether the structures identified will have favourable pharmacokinetic properties or acceptable safety profiles. The real strengths of *de novo* drug design are in stimulating new ideas and identifying novel lead structures that can then be optimized through structure-based drug design (see for example section 20.7.4.4).

17.15.2 Automated *de novo* drug design

Several computer software programs have been written which automatically design novel structures to fit known binding sites. The following are some examples.

17.15.2.1 LUDI

One of the best known *de novo* software programs is called **LUDI**, which works by fitting molecular fragments to different regions of the binding site, then linking the fragments together (Fig. 17.47). There are three stages to the process.

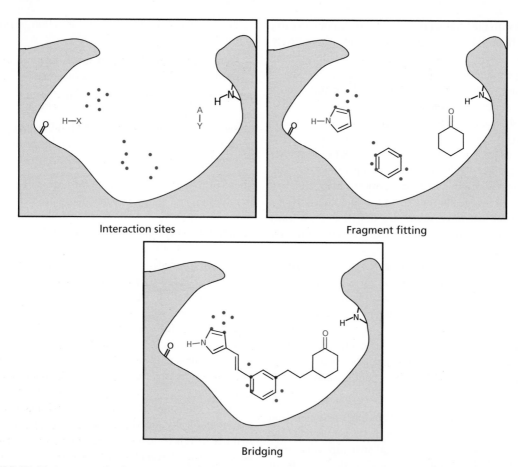

FIGURE 17.47 Stages involved in automated *de novo* drug design using LUDI. H–X, hydrogen bond donor interaction site; A–Y, hydrogen bond acceptor interaction site. The dots indicate aromatic interaction sites.

Stage 1: Identification of interaction sites

First of all, the atoms present in the binding site are analysed to identify those that can take part in hydrogen bonding interactions, and those that can take part in van der Waals interactions. Oxygen atoms and tertiary nitrogen atoms are identified as hydrogen bond acceptors. Any hydrogen attached to oxygen or nitrogen is identified as a hydrogen bond donor. Aromatic and aliphatic carbons are identified as such, and are capable of taking part in van der Waals interactions.

Interaction sites can then be defined. These are positions in the binding site that define where a ligand atom could be placed to interact with any of the above atoms. For example, suppose the binding site contains a methyl group (Fig. 17.48). The program would identify the carbon of that group as an aliphatic carbon capable of taking part in van der Waals interactions. This is a non-directional interaction, so a sphere is constructed around the carbon atom with a radius corresponding to the ideal distance for such an interaction (4 Å). A number of points (normally 14) are then placed evenly over the surface of the sphere to define aliphatic interaction sites.

Regions of the sphere which overlap or come too close to the atoms making up the binding site (i.e. less than 3 Å separation) are rejected, along with any of the 14 points that were on that part of the surface. The remaining points are then used as the aliphatic interaction sites. A similar process is involved in identifying aromatic interaction sites surrounding an aromatic carbon atom.

Identifying interaction sites for hydrogen bonds is carried out in a different fashion. As hydrogen bonds are directional, it is important to define not only the distance between the ligand and the binding region, but the relevant orientation of the atoms as well. This can be done by defining the hydrogen bond interaction site as a vector involving two atoms. The position of these atoms is determined by the ideal bond lengths and bond angles for a hydrogen bond. For example, if the binding site has a carbonyl group present, then there are two possible hydrogen bond interaction sites (X–H) which can be defined (Fig. 17.49).

If the binding site has a hydrogen bond donor present, then interaction sites for a hydrogen bond acceptor would be defined in a similar fashion. For example, Fig. 17.50 shows how the interaction site for a hydrogen bond

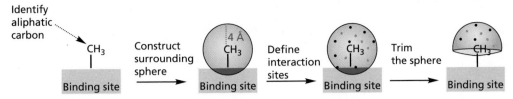

FIGURE 17.48 Identification of aliphatic interaction sites round a methyl group (LUDI).

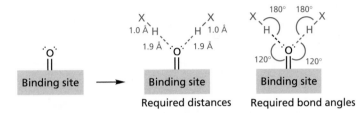

Required distances   Required bond angles

FIGURE 17.49 The interaction sites for a hydrogen bond donor, represented by H–X (LUDI).

acceptor is determined when the binding site has a hydroxyl group present.

There are, in fact, four interaction sites that are normally calculated in this situation. The other three can be visualized if we take a viewpoint along the line of atoms O–H—A and vary the relative position of Y as shown (Fig. 17.51).

As with the van der Waals interaction sites, hydrogen bond interaction sites are checked to ensure that they are no closer than 1.5 Å from any atom present in the binding site. If they are, they are rejected.

Stage 2: Fitting molecular fragments

Once interaction sites have been determined, the LUDI program accesses a library of several hundred molecular

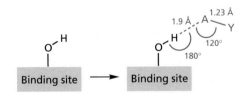

FIGURE 17.50 The interaction site for a hydrogen bond acceptor, represented by A–Y (LUDI). A denotes the hydrogen bond acceptor.

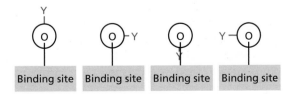

FIGURE 17.51 Four possible interaction sites for A–Y (A is hidden).

fragments, such as those shown in Fig. 17.52. The molecules chosen are typically 5–30 atoms in size and are usually rigid in structure because the fitting procedure assumes rigid fragments. Some fragments are included which *can* adopt different conformations. For these fragments, a selection of different conformations has to be present in the library if they are to be represented fairly in the fitting process. Each conformation is treated as a separate entity during the fitting process.

The atoms which are going to be used in the fitting process have to be predetermined for each fragment. Similarly, the interaction sites on to which each atom can be fitted have to be predetermined. For example, the methyl carbons of an acetone fragment are defined as aliphatic and can only be fitted onto aliphatic interaction sites. The carbonyl group is defined as a hydrogen bond acceptor and can only be fitted on to the corresponding interaction site (Fig. 17.53). The best fit will be the one that matches up the fragment with the maximum number of interaction sites. The program can 'try out' the various fragments in its library and identify those that can be matched up or fitted to the available interaction sites in the binding site.

Stage 3: Fragment bridging

Once fragments have been identified and fitted to the binding site, the final stage is to link them up. The program first identifies the molecular fragments that are closest to each other in the binding site, then identifies the closest hydrogen atoms (Fig. 17.54). These now define the link sites for the bridge. The program now tries out various molecular bridges from a stored library to find out which one fits best. Examples of the types of molecular bridges that are stored are shown in Fig. 17.55. Once a suitable bridge has been found, a final molecule is created.

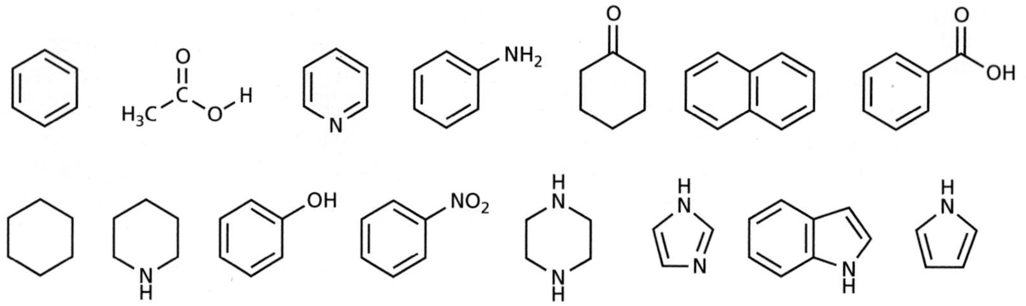

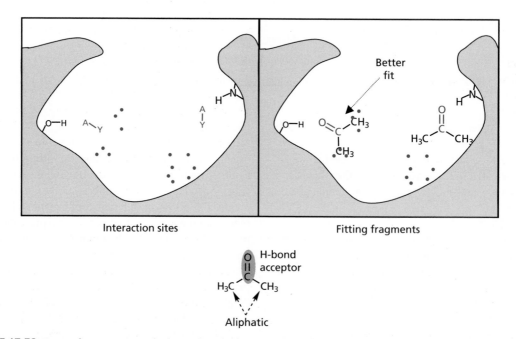

FIGURE 17.52 Examples of molecular fragments used by LUDI.

Interaction sites   Fitting fragments

FIGURE 17.53 Fitting fragments. A–Y, hydrogen bond acceptor interaction site. The dots indicate aliphatic interaction sites.

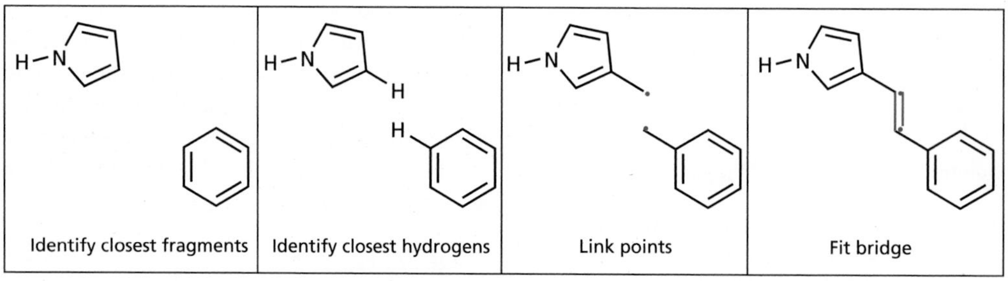

Identify closest fragments | Identify closest hydrogens | Link points | Fit bridge

FIGURE 17.54 The bridging process (LUDI).

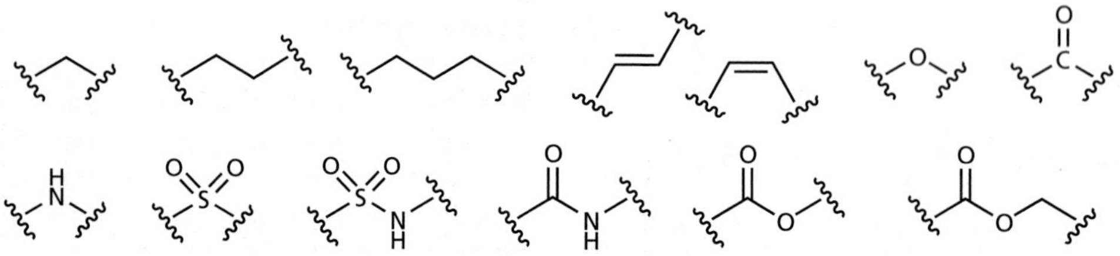

FIGURE 17.55 Examples of molecular bridges (LUDI).

17.15.2.2 SPROUT

Another early example of an automated *de novo* design program is **SPROUT**. Like LUDI, the program fits fragments to interaction sites, but there are interesting differences in the way that the process is carried out. For example, the interaction sites that are used in the program consist of atom-sized spheres. The spheres represent a volume of space within the binding site into which a ligand atom should be placed in order to interact favourably through hydrogen bonding or van der Waals interactions. Alternatively, spheres can be placed into the binding site to ensure that a particular structural feature is present in the final structures—for example an aromatic ring.

As far as the 'building blocks' are concerned, SPROUT uses **templates** to represent molecular fragments. We have already seen examples of molecular fragments used in LUDI. The atoms and bonds are specified in these fragments and there are a huge number of possible fragments which could be considered. Templates, however, are designed to represent several different molecular fragments. Each template is defined by vertices and edges rather than by atoms and bonds. A vertex represents a generalized sp, sp^2, or sp^3 hybridized atom, while an edge represents a single, double, or triple bond, depending on the hybridization of the vertices at either end. For example, the template shown in Fig. 17.56 can represent a large number of different six-membered rings. This approach has the advantage that it radically cuts down the number of different fragments that have to be stored in the program, making the search for novel structures more efficient. However, there is no reason why specific templates cannot be used as well, and the current version of SPROUT allows a mixture of specific molecular fragments and generic templates to be used at the same time.

The generation of the structures takes place in two stages (Fig. 17.57). In the first stage, the emphasis is on generating fragment templates that will fit the binding

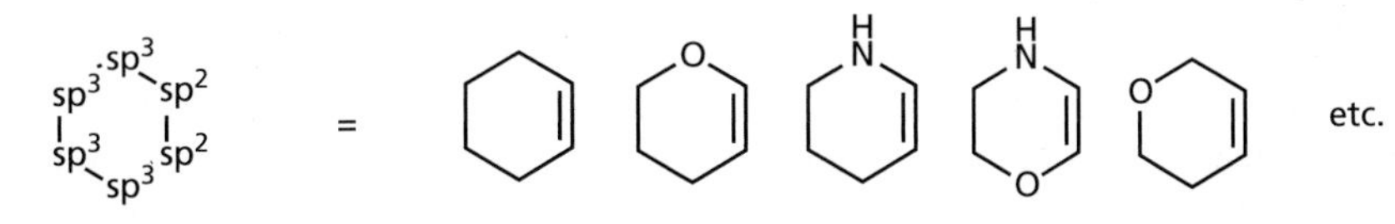

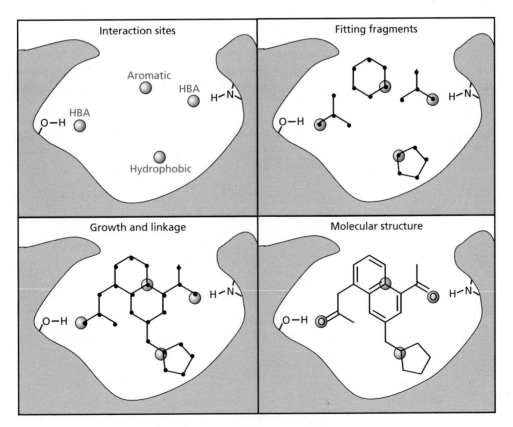

FIGURE 17.56 Examples of structures represented by a template used in SPROUT.

FIGURE 17.57 Generating structures using SPROUT.

site. There is no consideration of binding interactions at this stage, and so there is no need to know what sort of atoms are present in the fragment templates. The program selects a fragment template randomly and positions it into the binding site by placing one of the vertices at the centre of a sphere. In the early versions of SPROUT, further fragment templates were then added sequentially, and the skeleton was 'grown' until it occupied all the other spheres. In the current version, fragment templates are placed at all the spheres and are grown towards each other, until they are finally linked. One advantage of SPROUT is that the 'growth' of fragment templates allows a molecular template to be constructed which bridges interaction sites that are some distance apart. In the LUDI method, single fragments are placed at each interaction point and are then linked. If there is a large separation between the interaction sites, there might not be a sufficiently long linker to connect the fragments.

The second stage in the process is to create specific molecules from the molecular templates that have been produced. This involves replacing the vertices with suitable atoms to allow favourable hydrogen bonding and van der Waals interactions with the binding site. For example, if a vertex is located within a sphere that requires a hydrogen bond acceptor, an oxygen or a nitrogen atom can be added at that position. Since generic templates have been used to generate each skeleton, a large variety of molecular structures can be generated from each molecular template.

SPROUT has the capacity to identify certain structural features that might be unrealistic and then modify them. For example, an OH might be generated during the second stage in order to introduce a hydrogen bond donor, but if the OH is linked to a double bond this results in an enol which would tautomerize to a ketone. The latter would not be able to act as a hydrogen bond donor. The program can identify an enol and modify it to a carboxylic acid which can still act as a hydrogen bond donor (Fig. 17.58).

The program also has the ability to modify structures such that they are more readily synthesized. For example, introducing a heteroatom into a two-carbon link between two rings generates a structure which can be more readily synthesized (Fig. 17.59). In this example, the link could be made synthetically by reacting an alkoxide with an alkyl halide (Fig. 17.60).

The structures that are finally generated by SPROUT are then evaluated *in silico* for a variety of properties, including possible toxicity and pharmacokinetic properties. The program **CAESA** is used to evaluate how easily each structure can be synthesized and to give an indication of likely starting materials for the synthesis. The program does this by carrying out a retrosynthetic analysis of each structure.

More recently, SPROUT has introduced a method of assessing the synthetic feasibility of the partial structures created during the *de novo* construction process. Such an analysis is useful as it allows a pruning process to take

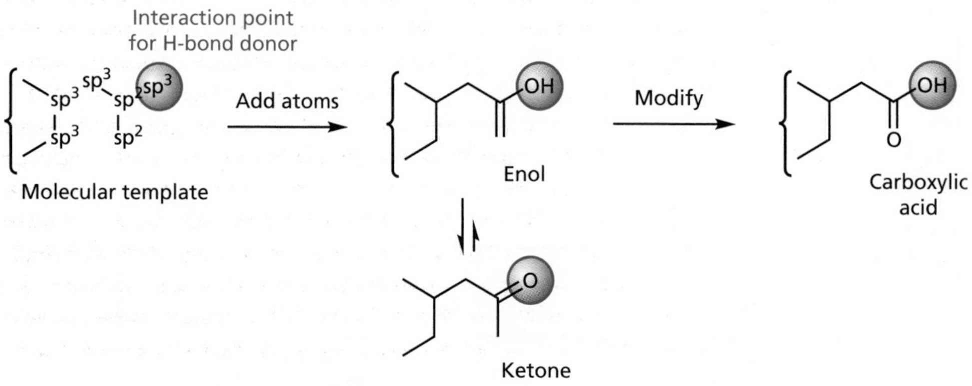

FIGURE 17.58 Modification of an enol to a carboxylic acid by SPROUT.

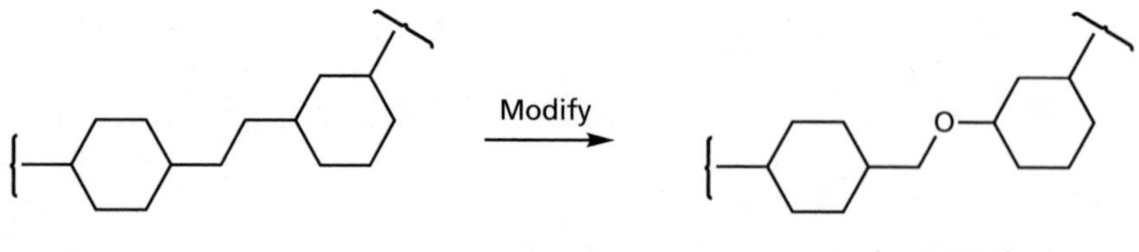

FIGURE 17.59 Modification by SPROUT to generate a more synthetically feasible structure.

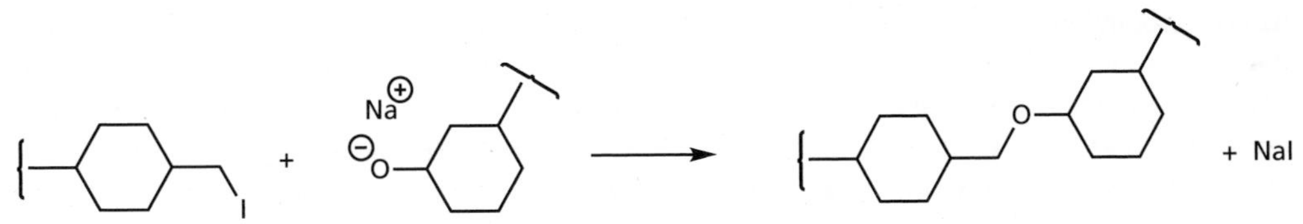

FIGURE 17.60 Possible synthesis allowing the linkage required in Fig. 17.59.

place which rejects partial structures that are not easily synthesized, and directs the program to generate more suitable structures. CAESA itself cannot be used for this purpose since it is relatively slow, taking about a minute per structure. This is acceptable for the analysis of the final structures that are generated, but would slow up the process considerably if it was used to assess the thousands of intermediate structures that are generated during the building process. Therefore, a less accurate but quicker method of analysis is carried out. The method involves the identification of molecular features within each partial structure and identifying how frequently they occur in known structures. The rationale is that if a particular feature commonly exists in known compounds, it is likely that that same feature should be capable of synthesis in the novel structures generated by *de novo* design.

The major structural features within a molecule can be defined as the various sized rings that are present, as well as any connecting chains. The synthetic feasibility of rings and chains is generally dependent on their substitution pattern. For example, the ten most frequently observed substitution patterns for a naphthalene ring amongst a database of known compounds are shown in Fig. 17.61. The analysis of partial structures can be carried out such that structures with uncommon structural features (such as a tetrasubstituted naphthalene ring) are penalized and rejected. A measure of the drug-like character of the partial structures can also be gleaned if the database used in the analysis is restricted to active compounds from drug databases.

17.15.2.3 LEGEND

LEGEND is another long established automated *de novo* drug design program. A grid is set up within the binding site to identify steric and electrostatic interaction energies between each grid point and the binding site (section 17.7.5). These are tabulated for different types of atom and are used to estimate van der Waals interactions for the growing skeletons that are generated by the program, as well as for structure optimization of final structures. The operator has the choice of starting from a single heteroatom, placed in such a position that it can form a hydrogen bond with the binding site. Alternatively, a molecule or molecular fragment can be placed into the binding site to act as a starting structure. This can be useful if one wants to include the partial structure of an active compound within the generated structures. Once the starting atom or fragment has been positioned, the growth stage can commence to generate different skeletons.

Unlike LUDI and SPROUT, LEGEND does not use fragments or templates to generate skeletons. Instead, the skeletons are grown one atom at a time using random choices at each stage of the process. For example, the type of atom to be added is chosen at random, as is the atom to which it will be linked. The latter is termed the **root atom**. The type of bond used to connect the new atom to the root atom is also chosen at random. Finally, a random torsion angle is chosen to position the new atom relative to the existing skeleton. Particular features such as

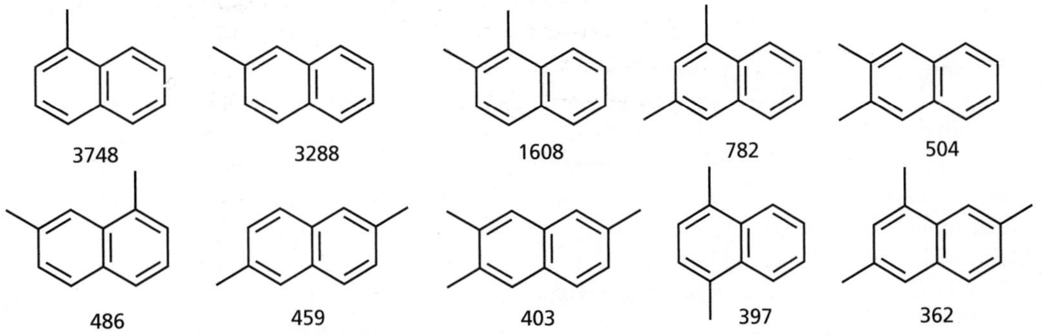

FIGURE 17.61 The ten most frequent substitution patterns for naphthalene in a database of drug-like compounds. Numbers refer to number of occurrences.

aromatic rings, carbonyl groups, and amide groups can be generated since some of the atom types that are used are defined as belonging to these features. For example, if a new atom is defined as being an aromatic carbon, then the final structure must eventually contain an aromatic ring containing that atom. The aromatic ring may not be completely formed when the growth stage is over, but the program will automatically complete the ring.

This approach of adding atoms one by one has the advantage that it can generate a greater diversity of structures than those generated by fragment-based procedures. However, it suffers from the disadvantage that the number of different structures generated can increase dramatically as each atom is joined. For that reason, it is important to evaluate the growing structures at each stage of the process and to carry out pruning operations. Inevitably, this means that the generation of structures is a slower process compared to fragment-based methods. As each atom is added, the structure is checked to ensure that there are no steric clashes within the molecule itself, or between the molecule and the binding site. This is done by measuring the van der Waals interactions using the tabulated values obtained from the grid measurements. If these *are* unfavourable, the structure is rejected and the program backtracks to choose a different root atom. If that also fails to generate an acceptable skeleton, the last atom to be added to the skeleton is removed and a new root atom is chosen.

When a new skeleton is accepted, the position of the new atom is assessed to see if it lies on a grid point associated with a very large electrostatic potential. If this is the case, the program ensures that it is changed to a heteroatom such that a hydrogen bond or ionic bond is possible with the binding site. The skeleton continues to grow until it reaches a size that is pre-determined by the operator. At that stage, hydrogen atoms are added to complete the valencies of each atom. If partially constructed aromatic rings are present, these are also completed. The structure is finally optimized, taking into account both intramolecular and intermolecular interactions. The process is then repeated to generate as many structures as desired.

17.15.2.4 GROW, ALLEGROW, and SYNOPSIS

GROW is a program that uses molecular fragments to generate novel ligands for binding sites. The fragments used represent amino acids, and so the structures that are generated are limited to peptides.

ALLEGROW is a program that can be used to extend a known ligand such that it can access vacant regions of a binding site (Box 17.6).

SYNOPSIS is a *de novo* drug design program that is designed to generate synthetically feasible structures. It does so by incorporating synthetic rules into the structure building process. In other words, fragments can only be linked if there is a known reaction which will allow it. Moreover, the fragments used must be commercially available. This program not only generates synthetically feasible molecules, but also provides a possible synthetic route.

KEY POINTS

- *De novo* design involves the design of a novel ligand, based on the structure of the binding site.

- An X-ray crystal structure of the target protein complexed with a ligand allows identification of the binding site and the binding mode of the ligand.

- A new ligand should initially be loose-fitting and flexible to allow for any alterations in the way binding takes place compared to what is predicted.

- The X-ray structure of the protein complexed with the new ligand will give valuable information on the actual binding mode of the new ligand, and allow modifications to be made which will maximize bonding.

- The new ligand should be capable of synthesis and of interacting with the binding site using a stable conformation.

- Energy losses resulting from desolvation of the ligand should be taken into account when calculating stabilization energies arising from ligand–protein binding.

- Automated programs for *de novo* design identify interaction sites in the binding site, then match molecular fragments to these sites. Bridges are then designed to link the fragments.

17.16 Planning compound libraries

Combinatorial and parallel synthesis (Chapter 16) are methods of rapidly creating a large number of compounds on a small scale using a set reaction scheme. The compounds produced constitute a compound library which could be tested to find active compounds for a set target. A compound library could be created that would include all the possible compounds obtainable from the reaction scheme using available starting materials and reagents. However, molecular modelling can help to focus the study such that a smaller number of structures are made, while maintaining the probability of finding active compounds.

One method of doing this is based on the identification of **pharmacophore triangles** (section 17.11.3). Let us assume that a synthesis is being carried out to generate 1000 compounds with as diverse a range of structures as

BOX 17.6 Designing a non-steroidal glucocorticoid agonist

The arylindazole structure (I) is a non-steroidal agent that acts as an agonist at the glucocorticoid receptor, but it is not possible to dock the structure into the conventional receptor binding site. However, an unusual induced fit has been observed for the steroid **cortivazol** which opens up a new channel in the binding site (Box 8.1). Docking experiments of structure I with this atypical binding site were successful and it was found that the arylpyrazole group (shown in blue) partially occupied the new channel. Using structure

I as a core scaffold, a computational technique called **Alle-Grow** was used to 'grow' the structure into the new channel *in silico*. The program works by adding atoms or small molecular features to the core skeleton, then scoring the resulting structures for binding interactions. Seven thousand virtual structures were created *in silico* by this method, and the most promising of these were synthesized and tested leading to structure II. This was found to have similar activity to the most potent of the clinically used corticosteroids.

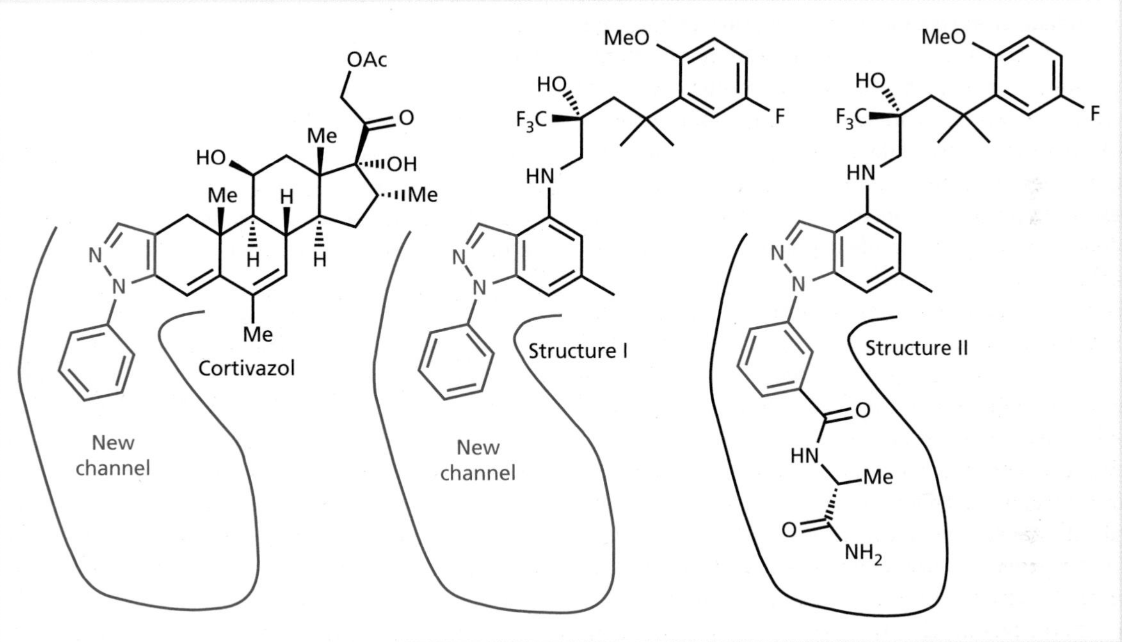

possible. The number of different pharmacophores generated from the 1000 compounds would be an indication of the structural diversity. Therefore, a library of compounds which generates 100 000 different pharmacophores would be superior to a library of similar size which produces only 100 different pharmacophores. An effective way of designing a more focused library is to carry out a pharmacophore search on all the possible products from a reaction scheme, in order to select those products that demonstrate the widest structural diversity. Those compounds would then be the ones chosen for inclusion in the library.

Firstly, all the possible synthetic products are automatically ranked on their level of rigidity. This can be achieved by identifying the number of rotatable bonds. Pharmacophore searching then starts with the most rigid structure, and all the possible pharmacophore triangles

are identified for that structure. If different conformations are possible, these are generated and the various pharmacophore triangles arising from these are added to the total. The next structure is then analysed for all of its pharmacophore triangles. Again, triangles are identified for all the possible conformations. The pharmacophores from the first and second structures are then compared. If more than 10% of the pharmacophores from the second structure are different from those of the first, both structures are added to the list for the intended library. Both sets of pharmacophores are combined, and the next structure is analysed for all of its pharmacophores. These are compared with the total number of pharmacophores from structures 1 and 2 and if there are 10% new pharmacophores represented, the third structure is added to the list and the pharmacophores for all three structures are added together for comparison with the next structure. This process is repeated throughout all the target structures, eliminating all compounds which generate less than 10% of new pharmacophores. In this way, it is possible to cut the number of structures which need to be synthesized by 80–90% with only a 10% drop in the number of pharmacophores generated.

There is a good reason for starting this analysis with a rigid structure. A rigid structure has only a few conformations and there is a good chance that most of these will be represented when the structure interacts with its target. Therefore, one can be confident that the associated pharmacophores are also represented. If the analysis started with a highly flexible molecule having a large number of conformations, there is less chance that all the conformations and their associated pharmacophores will be fairly represented when the structure meets its target binding site. Rigid structures which express some of these conformations more clearly would not be included in the library, as they would be rejected during the analysis. As a result, some pharmacophores which should be present are actually left untested.

It is possible to use modelling software to carry out a substituent search when planning a compound library. Here, one defines the common scaffold created in the synthesis, as well as the number of substituents which are attached and their point of attachment. Next, the general structures of the starting materials used to introduce these substituents are defined. The substituents which can be added to the structure can then be identified by having the computer search databases for commercially available starting materials. The program then generates all the possible structures which can be included in the library, based on the available starting materials. Once these have been identified, they can be analysed for pharmacophore diversity as described above.

Alternatively, the various possible substituents can be clustered into similar groups on the basis of their structural similarity. This allows starting materials to be preselected, choosing a representative compound from each group. The structural similarity of different substituents would be based on a number of criteria, such as the distance between important binding centres, the types of centre present, particular bonding patterns, and functional groups.

17.17 **Database handling**

The development of a drug requires the analysis of large amounts of data. For example, activity against a range of targets has to be measured to ensure that the compounds have good activity against their intended target, and also show selectivity with respect to a range of other targets.

When it comes to rationalizing results, many other parameters have to be considered, such as molecular weight, log P, and pK_a. The handling of such large amounts of data requires dedicated software.

Several software programs are available for the handling of data which allow medicinal chemists to assess biological activity versus physical properties, or to compare the activities of a series of compounds at two different targets. Such programs permit results to be presented in a visual qualitative fashion, allowing a quick identification of any likely correlations between different sets of data.

For example, if one wanted to see whether the log P value of a series of compounds was related to their α- or β-adrenergic activity, a 2D plot could be drawn up comparing α-adrenergic activity versus β-adrenergic activity. The log P value of each compound could then be indicated by a colour code for the various points on the plot. In this way, it would be easy to see whether these three properties were related. Such an analysis might show, for example, that a high log P is associated with compounds having low α-adrenergic activity and high β-adrenergic activity.

Some programs can be used to assess the biological results from a compound library. Firstly, the scaffold used in the library is defined, then the substituents are defined. Once the biological test results are obtained, a tree diagram can be drawn up to assess which substitution point is most important for activity. For example, if there were three substitution points on the scaffold, the program could analyse the data to identify which of the substitution points was the most important in controlling the activity. The data relevant for this particular substituent could then be split into three groups corresponding to good, average, and poor activity. For each of these groups, the program could be used to identify the next most important substitution point, and so on.

KEY POINTS

- Molecular modelling can be used to plan intended combinatorial libraries such that the maximum number of pharmacophores are generated for the minimum number of structures.

- Structures are analysed for their various conformations and resulting pharmacophores, starting with the most rigid structures.

- Each structure is compared with a growing bank of pharmacophores to assess whether it presents a significantly different number of pharmacophores compared with the structures that went before.

QUESTIONS

1. What is meant by energy minimization and how is it carried out?

2. What is meant by the terms local and global energy minima, and what is their relevance to conformational analysis?

3. What two properties should be known about two drugs if they are to be overlaid as a comparison?

4. Is it reasonable to assume that the most stable conformation of a drug is the active conformation?

5. You are carrying out *de novo* drug design to find a ligand for a binding site that contains a hydrogen bonding region and two hydrophobic pockets. Structures I and II below are both suitable candidates. Compare the relevant merits of these structures and decide which one you would synthesize first to test your binding theory.

6. Both structures I and II above show poor water solubility. It is suggested that the phenyl group be replaced by a pyridine ring. What would be the advantages and disadvantages of this idea? Have you any alternative ideas?

7. Assuming that structures I and II both bind to the binding site as predicted, what further modifications might you make to increase binding interactions?

8. Why were such modifications not carried out earlier?

9. The following eight structures have been tested for activity as receptor agonists. Five are active and three are inactive.

Assess the structures and discuss what the pharmacophore might be for agonist activity.

10. How would you go about carrying out overlays of the active structures in Question 9?

🌐 Multiple-choice questions are available on the Online Resource Centre at www.oxfordtextbooks.co.uk/orc/patrick6e/

FURTHER READING

Agrafiotis, D. K., Lobanov, V. S., and Salemme, F. R. (2002) Combinatorial informatics in the post-genomics era. *Nature Reviews Drug Discovery*, 1, 337–46.

Biggadike, K., et al. (2009) Design and x-ray crystal structures of high potency nonsteroidal glucocorticoid agonists exploiting a novel binding site on the receptor. *Proceedings of the New York Academy of Sciences*, 106, 18114–9.

Boda, K., and Johnson, A. P. (2006) Molecular complexity analysis of *de novo* designed ligands. *Journal of Medicinal Chemistry*, 49, 5869–79.

Bourne, P. E., and Wessig, H. (eds) (2003) *Structural bioinformatics*. John Wiley and Sons, New York.

Brooijmans, N., and Kuntz, I. D. (2003) Molecular recognition and docking algorithms. *Annual Review of Biophysics and Biomolecular Structure*, **32**, 335–73.

Kitchen, D. B., et al. (2004) Docking and scoring in virtual screening for drug discovery: methods and applications. *Nature Reviews Drug Discovery*, **3**, 935–49.

Kobilka, B., and Schertler, G. F. X. (2008) New G-protein-coupled receptor crystal structures: insights and limitations. *Trends in Pharmacological Sciences*, **29**, 79–83.

Leach, A. R. (2001) *Molecular modelling: principles and applications*, 2nd edn. Pearson Education, London.

Megget, K. (2011) Idle cures. *Chemistry World,* February, 52–5.

Miller, M. A. (2002) Chemical database techniques. *Nature Reviews Drug Discovery*, **1**, 220–7.

Richards, G. (2002) Virtual screening using grid computing: the screensaver project. *Nature Reviews Drug Discovery*, **1**, 551–5.

Sansom, C. (2010) Model molecules. *Chemistry World*, April, 50–3.

Sansom, C. (2010) Receptive receptors. *Chemistry World*, August, 52–5.

Schlyer, S., and Horuk, R. (2006) I want a new drug: G-protein-coupled receptors in drug development. *Drug Discovery Today*, **11**, 481–93.

Schneider, G., and Fechner, U. (2005). Computer-based *de novo* design of drug-like molecules. *Nature Reviews Drug Discovery*, **4**, 649–63.

Shoichet, B. K. (2004) Virtual screening of chemical libraries. *Nature*, **432**, 862–5.

van Drie, J. H. (2007) Computer-aided drug design: the next 20 years. *Journal of Computer-aided Molecular Design*, **21**, 591–601.

Titles for general further reading are listed on p.845.

Quantitative structure–activity relationships (QSAR)

In Chapters 13 and 14 we studied the various strategies which can be used in the design of drugs. Several of these strategies involved a change in shape such that the new drug had a better 'fit' for its target binding site. Other strategies involved a change in functional groups or substituents such that the drug's pharmacokinetics or binding site interactions were improved. These latter strategies often involved the synthesis of analogues containing a range of substituents on aromatic or heteroaromatic rings or accessible functional groups. The number of possible analogues that could be made is infinite if we were to try and synthesize analogues with every substituent and combination of substituents possible. Therefore, it is clearly advantageous if a rational approach can be followed in deciding which substituents to use. The quantitative structure–activity relationship (QSAR) approach has proved extremely useful in tackling this problem.

The QSAR approach attempts to identify and quantify the physicochemical properties of a drug and to see whether any of these properties has an effect on the drug's biological activity. If such a relationship holds true, an equation can be drawn up which quantifies the relationship and allows the medicinal chemist to say with some confidence which property (or properties) has an important role in the pharmacokinetics or mechanism of action of the drug. It also allows the medicinal chemist some level of prediction. By quantifying physicochemical properties, it should be possible to calculate in advance what the biological activity of a novel analogue might be. There are two advantages to this. Firstly, it allows the medicinal chemist to target efforts on analogues which should have improved activity and, thus, cut down the number of analogues that have to be made. Secondly, if an analogue is discovered which does not fit the equation, it implies that some other feature is important and provides a lead for further development.

What are these physicochemical features that we have mentioned?

Essentially, they could be any structural, physical, or chemical property of a drug. Clearly, any drug will have a large number of such properties and it would be a Herculean task to quantify and relate them all to biological activity at the same time. A simple, more practical approach is to consider one or two physicochemical properties of the drug and to vary these while attempting to keep other properties constant. This is not as simple as it sounds, as it is not always possible to vary one property without affecting another. Nevertheless, there have been numerous examples where the approach has worked.

It is important that the QSAR method is used properly and in relevant situations. Firstly, the compounds studied must be structurally related, act at the same target, and have the same mechanism of action. Secondly, it is crucial that the correct testing procedures are used. *In vitro* tests carried out on isolated enzymes are relevant for a QSAR study since the activities measured for different inhibitors are related directly to how each compound binds to the active site. *In vivo* tests carried out to measure the physiological effects of enzyme inhibitors are not valid, however, since both pharmacodynamic and pharmacokinetic factors come into play. This makes it impossible to derive a sensible QSAR equation.

18.1 Graphs and equations

In the simplest situation, a range of compounds is synthesized in order to vary one physicochemical property (e.g. log P) and to test how this affects the biological activity (log $1/C$) (we will come to the meaning of log $1/C$ and log P in due course). A graph is then drawn to plot the biological activity on the y-axis versus the physicochemical feature on the x-axis (Fig. 18.1).

It is then necessary to draw the best possible line through the data points on the graph. This is done by a procedure known as '**linear regression analysis by the least squares method**'. This is quite a mouthful and can produce a glazed expression on any chemist who is not

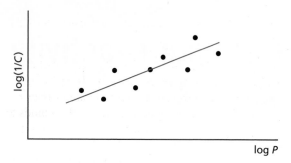

FIGURE 18.1 Biological activity versus log *P*.

mathematically orientated. In fact, the principle is quite straightforward. If we draw a line through a set of data points, most of the points will be scattered on either side of the line. The best line will be the one closest to the data points. To measure how close the data points are, vertical lines are drawn from each point (Fig. 18.2). These verticals are measured and then squared in order to eliminate the negative values. The squares are then added up to give a total (the sum of the squares). The best line through the points will be the line where this total is a minimum. The equation of the straight line will be $y = k_1x + k_2$ where k_1 and k_2 are constants. By varying k_1 and k_2, different equations are obtained until the best line is obtained. This whole process can be speedily done using relevant software.

The next stage in the process is to see whether the relationship is meaningful. As any good politician knows, numbers can be used to 'prove' whatever you want them to prove. Therefore, a proper statistical analysis has to be carried out to assess the validity of any QSAR equation and quantify the goodness of fit for any plot. The **regression** or **correlation coefficient** (*r*) is a measure of how well the physicochemical parameters present in the equation explains the observed variance in activity. An explanation of how *r* is derived is given in Appendix 3. For a perfect fit, *r* = 1, in which case the observed activities would be the same as those calculated by the equation. Such perfection is impossible with biological data and so *r* values greater than 0.9 are considered

acceptable. The regression coefficient is often quoted as r^2, in which case values over 0.8 are considered a good fit. If r^2 is multiplied by 100 it indicates the percentage variation in biological activity that is accounted for by the physicochemical parameters used in the equation. Thus, an r^2 value of 0.85 signifies that 85% of the variation in biological activity is accounted for by the parameters used. There are dangers in putting too much reliance on r, as the value obtained takes no account of the number of compounds (*n*) involved in the study and it is possible to obtain higher values of *r* by increasing the number of compounds tested.

Therefore, another statistical measure for the goodness of fit should be quoted alongside *r*. This is the **standard error of estimate** or the **standard deviation** (*s*). Ideally, *s* should be zero, but this would assume there were no experimental errors in the experimental data or the physicochemical parameters. In reality, *s* should be small, but not smaller than the standard deviation of the experimental data. It is therefore necessary to know the latter to assess whether the value of *s* is acceptably low. Appendix 3 shows how *s* is obtained and demonstrates that the number of compounds (*n*) in the study influences the value of *s*.

Statistical tests called **Fisher's F-tests** are often quoted (Appendix 3). These tests are used to assess the significance of the coefficients *k* for each parameter in the QSAR equation. Normally *p* values (derived from the F-test) should be less than or equal to 0.05 if the parameter is significant. If this is not the case, the parameter should not be included in the QSAR equation.

18.2 Physicochemical properties

Many physical, structural, and chemical properties have been studied by the QSAR approach, but the most common are hydrophobic, electronic, and steric properties. This is because it is possible to quantify these effects. Hydrophobic properties can be easily quantified for complete molecules or for individual substituents. On the other hand, it is more difficult to quantify electronic and steric properties for complete molecules, and this is only really feasible for individual substituents.

Consequently, QSAR studies on a variety of totally different structures are relatively rare and are limited to studies on hydrophobicity. It is more common to find QSAR studies being carried out on compounds of the same general structure, where substituents on aromatic rings or accessible functional groups are varied. The QSAR study then considers how the hydrophobic, electronic, and steric properties of the substituents affect biological activity. The three most studied physicochemical properties are now considered in some detail.

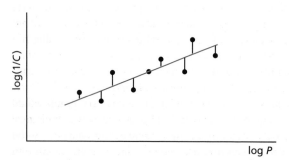

FIGURE 18.2 Proximity of data points to line of best fit.

Test your understanding and practise your molecular modelling with Exercise 18.1 on the Online Resource Centre at www.oxfordtextbooks.co.uk/orc/patrick6e/. You might also find Exercises 25.3 and 25.4 useful at this point.

18.2.1 Hydrophobicity

The hydrophobic character of a drug is crucial to how easily it crosses cell membranes (section 11.3) and may also be important in receptor interactions. Changing substituents on a drug may well have significant effects on its hydrophobic character and, hence, its biological activity. Therefore, it is important to have a means of predicting this quantitatively.

18.2.1.1 The partition coefficient (P)

The hydrophobic character of a drug can be measured experimentally by testing the drug's relative distribution in an *n*-octanol/water mixture. Hydrophobic molecules will prefer to dissolve in the *n*-octanol layer of this two-phase system, whereas hydrophilic molecules will prefer the aqueous layer. The relative distribution is known as the partition coefficient (P) and is obtained from the following equation:

$$P = \frac{\text{Concentration of drug in octanol}}{\text{Concentration of drug in aqueous solution}}$$

Hydrophobic compounds have a high P value, whereas hydrophilic compounds have a low P value.

Varying substituents on the lead compound will produce a series of analogues having different hydrophobicities and, therefore, different P values. By plotting these P values against the biological activity of these drugs, it is possible to see if there is any relationship between the two properties. The biological activity is normally expressed as $1/C$, where C is the concentration of drug required to achieve a defined level of biological activity. The reciprocal of the concentration $(1/C)$ is used, since more active drugs will achieve a defined biological activity at lower concentration.

The graph is drawn by plotting $\log(1/C)$ versus $\log P$. In studies where the range of the $\log P$ values is restricted to a small range (e.g. $\log P = 1$–4), a straight-line graph is obtained (Fig. 18.1) showing that there is a relationship between hydrophobicity and biological activity. Such a line would have the following equation:

$$\log\left(\frac{1}{C}\right) = -k_1 \log P + k_2$$

For example, the binding of drugs to serum albumin is determined by their hydrophobicity and a study of 42 compounds resulted in the following equation:

$$\log\left(\frac{1}{C}\right) = 0.75 \log P + 230 \ (n = 42, r = 0.960, s = 0.159)$$

The equation shows that serum albumin binding increases as $\log P$ increases. In other words, hydrophobic drugs bind more strongly to serum albumin than hydrophilic drugs. Knowing how strongly a drug binds to serum albumin can be important in estimating effective dose levels for that drug. When bound to serum albumin, the drug cannot bind to its receptor, and so the dose levels for the drug should be based on the amount of unbound drug present in the circulation. The equation above allows us to calculate how strongly drugs of similar structure will bind to serum albumin, and gives an indication of how 'available' they will be for receptor interactions. The r value of 0.96 is close to 1, which shows that the line resulting from the equation is a good fit. The value of r^2 is 92%, which indicates that 92% of the variation in serum albumin binding can be accounted for by the different hydrophobicities of the drugs tested. This means that 8% of the variation is unaccounted for, partly as a result of the experimental errors involved in the measurements.

Despite such factors as serum albumin binding, it is generally found that increasing the hydrophobicity of a lead compound results in an increase in biological activity. This reflects the fact that drugs have to cross hydrophobic barriers such as cell membranes in order to reach their target. Even if no barriers are to be crossed (e.g. in *in vitro* studies), the drug has to interact with a target system, such as an enzyme or receptor, where the binding site is more hydrophobic than the surface. Therefore, increasing hydrophobicity also aids the drug in binding to its target site.

This might imply that increasing $\log P$ should increase the biological activity ad infinitum. In fact, this does not happen. There are several reasons for this. For example, the drug may become so hydrophobic that it is poorly soluble in the aqueous phase. Alternatively, it may be 'trapped' in fat depots and never reach the intended site. Finally, hydrophobic drugs are often more susceptible to metabolism and subsequent elimination.

A straight-line relationship between $\log P$ and biological activity is observed in many QSAR studies because the range of $\log P$ values studied is often relatively narrow. For example, the study carried out on serum albumin binding was restricted to compounds having $\log P$ values in the range 0.78–3.82. If these studies were to be extended to include compounds with very high $\log P$ values, then we would see a different picture. The graph would be parabolic, as shown in Fig. 18.3. Here, the biological activity increases as $\log P$ increases until a maximum value is obtained. The value of $\log P$ at the maximum $(\log P^0)$ represents the optimum partition coefficient for

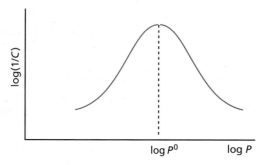

FIGURE 18.3 Parabolic curve of log(1/C) versus log P.

biological activity. Beyond that point, an increase in log P results in a decrease in biological activity.

If the partition coefficient is the only factor influencing biological activity, the parabolic curve can be expressed by the equation:

$$\log\left(\frac{1}{C}\right) = -k_1(\log P)^2 + k_2\log P + k_3$$

Note that the $(\log P)^2$ term has a minus sign in front of it. When P is small, the $(\log P)^2$ term is very small and the equation is dominated by the log P term. This represents the first part of the graph where activity increases with increasing P. When P is large, the $(\log P)^2$ term becomes more significant and eventually 'overwhelms' the log P term. This represents the last part of the graph where activity drops with increasing P. k_1, k_2, and k_3 are constants and can be determined by a suitable software program.

There are relatively few drugs where activity is related to the log P factor alone. Such drugs tend to operate in cell membranes where hydrophobicity is the dominant feature controlling their action. The best examples of drugs which operate in cell membranes are the general anaesthetics. Although they also bind to GABA$_A$ receptors, general anaesthetics are thought to function by entering the central nervous system and 'dissolving' into cell membranes where they affect membrane structure and nerve function. In such a scenario, there are no specific drug–receptor interactions and the mechanism of the drug is controlled purely by its ability to enter cell membranes (i.e. its hydrophobic character). The general anaesthetic activity of a range of ethers was found to fit the following parabolic equation:

$$\log\left(\frac{1}{C}\right) = -0.22\,(\log P)^2 + 1.04\log P + 2.16$$

According to this equation, anaesthetic activity increases with increasing hydrophobicity (P), as determined by the log P factor. The negative $(\log P)^2$ factor shows that the relationship is parabolic and that there is an optimum

value for log P (log P^0) beyond which increasing hydrophobicity causes a decrease in anaesthetic activity. With this equation, it is now possible to predict the anaesthetic activity of other compounds, given their partition coefficients. However, there are limitations. The equation is derived purely for anaesthetic ethers and is not applicable to other structural types of anaesthetics. This is generally true in QSAR studies. The procedure works best if it is applied to a series of compounds which have the same general structure.

However, QSAR studies *have* been carried out on other structural types of general anaesthetics, and a parabolic curve has been obtained in each case. Although, the constants for each equation are different, it is significant that the optimum hydrophobicity (represented by log P^0) for anaesthetic activity is close to 2.3, regardless of the class of anaesthetic being studied. This finding suggests that all general anaesthetics are operating in a similar fashion, controlled by the hydrophobicity of the structure.

Because different anaesthetics have similar log P^0 values, the log P value of any compound can give some idea of its potential potency as an anaesthetic. For example, the log P values of the gaseous anaesthetics **ether, chloroform**, and **halothane** are 0.98, 1.97, and 2.3 respectively. Their anaesthetic activity increases in the same order.

As general anaesthetics have a simple mechanism of action based on the efficiency with which they enter the central nervous system, it implies that log P values should give an indication of how easily any compound can enter the central nervous system. In other words, compounds having a log P value close to 2 should be capable of entering the central nervous system efficiently. This is generally found to be true. For example, the most potent barbiturates for sedative and hypnotic activity are found to have log P values close to 2.

As a rule of thumb, drugs which are to be targeted for the central nervous system should have a log P value of approximately 2. Conversely, drugs which are designed to act elsewhere in the body should have log P values significantly different from 2 in order to avoid possible central nervous system side effects (e.g. drowsiness) (see Box 18.1).

18.2.1.2 The substituent hydrophobicity constant (π)

We have seen how the hydrophobicity of a compound can be quantified using the partition coefficient P. In order to get P, we have to measure it experimentally and that means that we have to synthesize the compounds. It would be much better if we could calculate P theoretically and decide in advance whether the compound is worth synthesizing. QSAR would then allow us to target the most promising-looking structures. For example, if we were planning to synthesize a range of barbiturate

Altering log *P* to remove central nervous system side effects

The cardiotonic agent (I) was found to produce 'bright visions' in some patients, which implied that it was entering the central nervous system. This was supported by the fact that the log *P* value of the drug was 2.59. In order to prevent the drug entering the central nervous system, the 4-OMe group was replaced by a 4-S(O)Me group. This particular group is approximately the same size as the methoxy group, but more hydrophilic. The log *P* value of the new drug (**sulmazole**) was found to be 1.17. The drug was now too hydrophilic to enter the central nervous system and was free of central nervous system side effects.

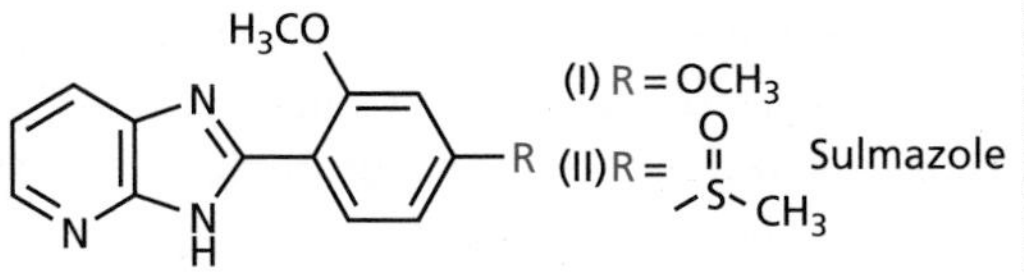

FIGURE 1 Cardiotonic agents.

structures, we could calculate log *P* values for them all and concentrate on the structures which had log *P* values closest to the optimum log P^0 value for barbiturates.

Partition coefficients can be calculated by knowing the contribution that various substituents make to hydrophobicity. This contribution is known as the **substituent hydrophobicity constant** (π) and is a measure of how hydrophobic a substituent is relative to hydrogen. The value can be obtained as follows. Partition coefficients are measured experimentally for a standard compound such as benzene, with and without a substituent (X). The hydrophobicity constant (π_X) for the substituent (X) is then obtained using the following equation:

$$\pi_X = \log P_X - \log P_H$$

where P_H is the partition coefficient for the standard compound and P_X is the partition coefficient for the standard compound with the substituent.

A positive value of π indicates that the substituent is more hydrophobic than hydrogen; a negative value indicates that the substituent is less hydrophobic. The π values for a range of substituents are shown in Table 18.1. These π values are characteristic for the substituent and can be used to calculate how the partition coefficient of a drug would be affected if these substituents

were present. The *P* value for the lead compound would have to be measured experimentally, but, once that is known, the *P* value for analogues can be calculated quite simply.

As an example, consider the log *P* values for benzene (log *P* = 2.13), chlorobenzene (log *P* = 2.84), and benzamide (log *P* = 0.64) (Fig. 18.4). Benzene is the parent compound, and the substituent constants for Cl and $CONH_2$ are 0.71 and −1.49 respectively. Having obtained these values, it is now possible to calculate the theoretical log *P* value for *meta*-chlorobenzamide:

$$\log P_{\text{(chlorobenzamide)}} = \log P_{\text{(benzene)}} + \pi_{Cl} + \pi_{CONH_2}$$
$$= 2.13 + 0.71 + (-1.49)$$
$$= 1.35$$

The observed log *P* value for this compound is 1.51.

It should be noted that π values for aromatic substituents are different from those used for aliphatic substituents. Furthermore, neither of these sets of π values are in fact true constants, and they are accurate only for the structures from which they were derived. They can be used as good approximations when studying other structures, but it is possible that the values will have to be adjusted in order to get accurate results.

In order to distinguish calculated log *P* values from experimental ones, the former are referred to as **Clog *P*** values. There are also software programs which will calculate Clog *P* values for a given structure.

ⓦ Test your understanding and practise your molecular modelling with Exercise 18.2 on the Online Resource Centre at www.oxfordtextbooks.co.uk/orc/patrick6e/

18.2.1.3 *P* versus π

QSAR equations relating biological activity to the partition coefficient *P* have already been described, but there is no reason why the substituent hydrophobicity constant π cannot be used in place of *P* if only the substituents are being varied. The equation obtained would be just as relevant as a study of how hydrophobicity affects biological activity. That is not to say that *P* and π are exactly equivalent—different equations would be obtained with different constants. Apart from that, the two factors have different emphases. The partition coefficient *P* is a measure of the drug's overall hydrophobicity and is, therefore, an important measure of how efficiently a drug is transported to its target and bound to its binding site. The π factor measures the hydrophobicity of a specific region on the drug's skeleton and, if it is present in the QSAR equation, it could emphasize important hydrophobic interactions involving that region of the molecule with the binding site.

TABLE 18.1 Values of π for a range of substituents on aliphatic and aromatic scaffolds

Group	CH$_3$	t-Bu	OH	OCH$_3$	CF$_3$	Cl	Br	F
π (aliphatic scaffolds)	0.50	1.68	−1.16	0.47	1.07	0.39	0.60	−0.17
π (aromatic scaffolds)	0.52	1.68	−0.67	−0.02	1.16	0.71	0.86	0.14

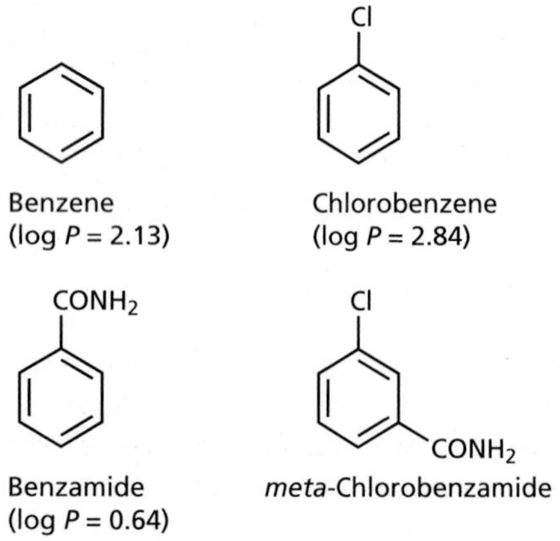

Benzene
(log P = 2.13)

Chlorobenzene
(log P = 2.84)

Benzamide
(log P = 0.64)

meta-Chlorobenzamide

FIGURE 18.4 Values for log P.

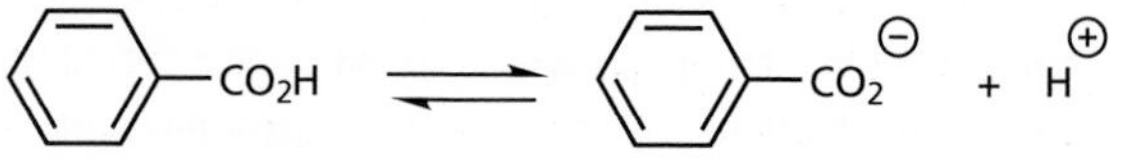

FIGURE 18.5 Ionization of benzoic acid in water.

Most QSAR equations have a contribution from P or from π, but there are examples of drugs for which they have only a slight contribution. For example, a study on antimalarial drugs showed very little relationship between antimalarial activity and hydrophobic character. This finding supports the theory that these drugs act in red blood cells, since previous research has shown that the ease with which drugs enter red blood cells is not related to their hydrophobicity.

18.2.2 Electronic effects

The electronic effects of various substituents will clearly have an effect on a drug's ionization or polarity. This, in turn, may have an effect on how easily a drug can pass through cell membranes or how strongly it can interact with a binding site. It is, therefore, useful to measure the electronic effect of a substituent.

As far as substituents on an aromatic ring are concerned, the measure used is known as the **Hammett substituent constant** (σ). This is a measure of the electron-withdrawing or electron-donating ability of a substituent, and has been determined by measuring the dissociation of a series of substituted benzoic acids compared with the dissociation of benzoic acid itself.

Benzoic acid is a weak acid and only partially ionizes in water (Fig. 18.5). An equilibrium is set up between the ionized and non-ionized forms, where the relative proportion of these species is known as the **equilibrium** or **dissociation constant** K_H (the subscript H signifies that there are no substituents on the aromatic ring).

$$K_H = \frac{[\text{PhCO}_2^-]}{[\text{PhCO}_2\text{H}]}$$

When a substituent is present on the aromatic ring, this equilibrium is affected. Electron-withdrawing groups, such as a nitro group, result in the aromatic ring having a stronger electron-withdrawing and stabilizing influence on the carboxylate anion, and so the equilibrium will shift to the ionized form. Therefore, the substituted benzoic acid is a stronger acid and has a larger K_X value (X represents the substituent on the aromatic ring) (Fig. 18.6).

If the substituent X is an electron-donating group such as an alkyl group, then the aromatic ring is less able to

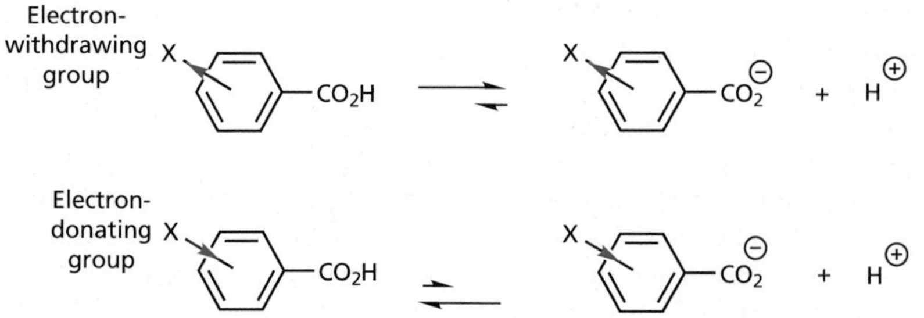

Electron-withdrawing group

Electron-donating group

FIGURE 18.6 Position of equilibrium dependent on substituent group X.

stabilize the carboxylate ion. The equilibrium shifts to the left indicating a weaker acid with a smaller K_X value (Fig. 18.6).

The Hammett substituent constant (σ_X) for a particular substituent (X) is defined by the following equation:

$$\sigma_x = \log\frac{K_x}{K_H} = \log K_x - \log K_H$$

Benzoic acids containing electron-withdrawing substituents will have larger K_X values than benzoic acid itself (K_H) and, therefore, the value of σ_X for an electron-withdrawing substituent will be positive. Substituents such as Cl, CN, or CF_3 have positive σ values.

Benzoic acids containing electron-donating substituents will have smaller K_X values than benzoic acid itself and, hence, the value of σ_X for an electron-donating substituent will be negative. Substituents such as Me, Et, and t-Bu have negative values of σ. The Hammett substituent constant for H is zero.

The Hammett substituent constant takes into account both resonance and inductive effects. Therefore, the value of σ for a particular substituent will depend on whether the substituent is *meta* or *para*. This is indicated by the subscript *m* or *p* after the σ symbol. For example, the nitro substituent has $\sigma_p = 0.78$ and $\sigma_m = 0.71$. In the *meta* position, the electron-withdrawing power is due to the inductive influence of the substituent, whereas at the *para* position inductive and resonance effects both play a part and so the σ_p value is greater (Fig. 18.7).

For the hydroxyl group, $\sigma_m = 0.12$ and $\sigma_p = -0.37$. At the *meta* position, the influence is inductive and electron-withdrawing. At the *para* position, the electron-donating influence due to resonance is more significant than the electron-withdrawing influence due to induction (Fig. 18.8).

Most QSAR studies start off by considering σ, and, if there is more than one substituent present, the σ values can be summed ($\Sigma\sigma$). However, as more compounds are synthesized, it is possible to fine-tune the QSAR equation. As mentioned above, σ is a measure of a substituent's inductive and resonance electronic effects. With more detailed studies, the inductive and resonance effects can be considered separately. Tables of constants are available which quantify a substituent's inductive effect (*F*) and its resonance effect (*R*). In some cases, it might be found that a substituent's effect on activity is due to *F* rather than *R*, and vice versa. It might also be found that a substituent has a more significant effect at a particular position on the ring and this can also be included in the equation.

There are limitations to the electronic constants described so far. For example, Hammett substituent constants cannot be measured for *ortho* substituents as such substituents have an important steric, as well as electronic, effect.

There are very few drugs whose activities are solely influenced by a substituent's electronic effect, and hydrophobicity usually has to be considered as well. Those that do are generally operating by a mechanism whereby they do not have to cross any cell membranes (see Box 18.2). Alternatively, *in vitro* studies on isolated enzymes may result in QSAR equations lacking the hydrophobicity factor, as there are no cell membranes to be considered.

The constants σ, *R*, and *F* can only be used for aromatic substituents and are, therefore, only suitable for drugs containing aromatic rings. However, a series of aliphatic

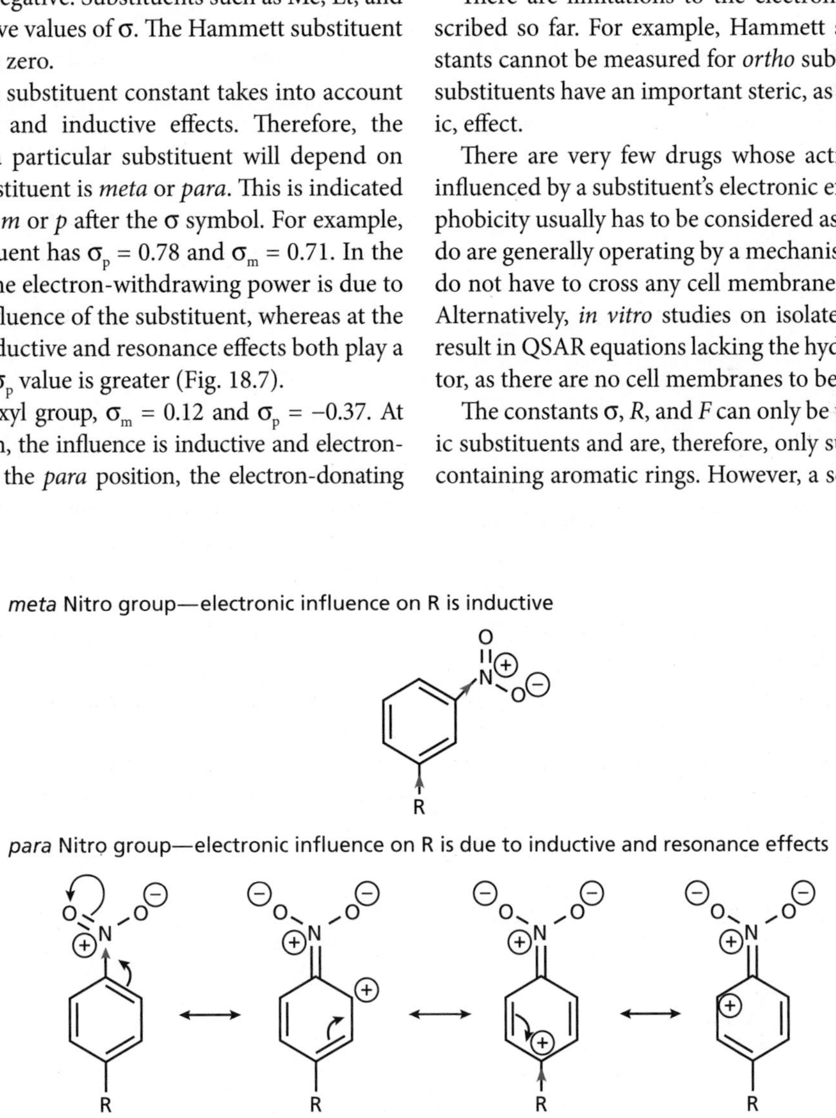

meta Nitro group—electronic influence on R is inductive

para Nitro group—electronic influence on R is due to inductive and resonance effects

FIGURE 18.7 Substituent effects of a nitro group at the *meta* and *para* positions.

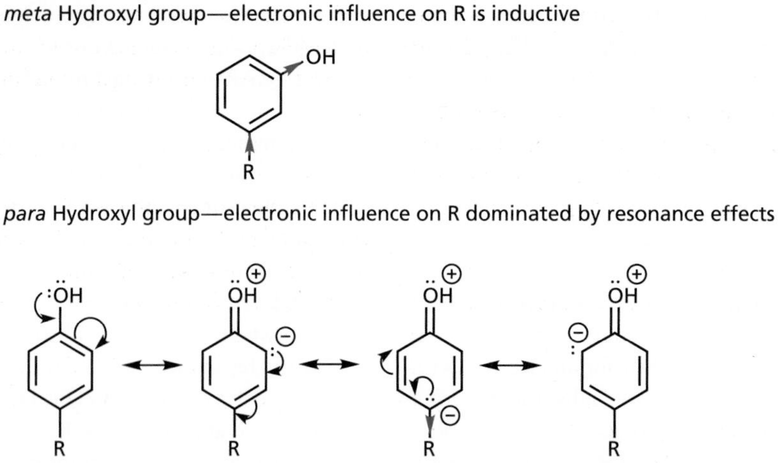

meta Hydroxyl group—electronic influence on R is inductive

para Hydroxyl group—electronic influence on R dominated by resonance effects

FIGURE 18.8 Substituent effects of a phenol at the *meta* and *para* positions.

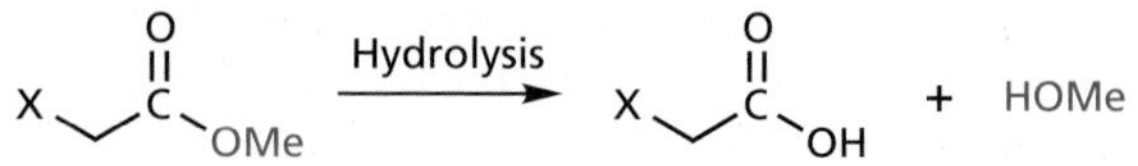

electronic substituent constants are available. These were obtained by measuring the rates of hydrolysis for a series of aliphatic esters (Fig. 18.9). Methyl ethanoate is the parent ester and it is found that the rate of hydrolysis is affected by the substituent X. The extent to which the

FIGURE 18.9 Hydrolysis of an aliphatic ester.

rate of hydrolysis is affected is a measure of the substituent's electronic effect at the site of reaction (i.e. the ester group). The electronic effect is purely inductive and is given the symbol σ_I. Electron-donating groups reduce the rate of hydrolysis and, therefore, have negative values. For example, σ_I values for methyl, ethyl, and propyl are -0.04, -0.07, and -0.36 respectively. Electron-withdrawing groups increase the rate of hydrolysis and have positive values. The σ_I values for NMe_3^+ and CN are 0.93 and 0.53 respectively.

It should be noted that the inductive effect is not the only factor affecting the rate of hydrolysis. The substituent may also have a steric effect. For example, a bulky substituent may shield the ester from attack and lower the rate of hydrolysis. It is, therefore, necessary to separate out these two effects. This can be done by measuring hydrolysis rates under both basic and acidic conditions. Under basic conditions, steric and electronic factors are important, whereas under acidic conditions only steric factors are important. By comparing the rates, values for the electronic effect (σ_I) and the steric effect (E_s) (see section 18.2.3.1) can be determined.

Test your understanding and practise your molecular modelling with Exercise 18.3 on the Online Resource Centre at www.oxfordtextbooks.co.uk/orc/patrick6e/

18.2.3 Steric factors

The bulk, size, and shape of a drug will influence how easily it can approach and interact with a binding site.

BOX 18.2 Insecticidal activity of diethyl phenyl phosphates

The insecticidal activity of diethyl phenyl phosphates is one of the few examples where activity is related to electronic factors alone:

$$\log\left(\frac{1}{C}\right) = 2.282\sigma - 0.348. \quad (r^2\ 0.952, r\ 0.976, s\ 0.286)$$

The equation reveals that substituents with a positive value for σ (i.e. electron-withdrawing groups) will increase activity. The fact that a hydrophobic parameter is not present is a good indication that the drugs do not have to pass into, or through, a cell membrane to have activity. In fact, these drugs are known to act on an enzyme called **acetylcholinesterase** which is situated on the outside of cell membranes (section 22.12).

The value of r is close to 1, which demonstrates that the line is a good fit, and the value of r^2 demonstrates that 95% of the data is accounted for by the σ parameter.

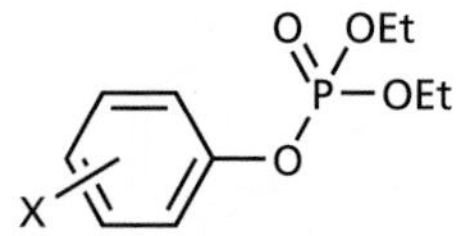

FIGURE 1 Diethyl phenyl phosphates.

A bulky substituent may act like a shield and hinder the ideal interaction between a drug and its binding site. Alternatively, a bulky substituent may help to orientate a drug properly for maximum binding and increase activity. Steric properties are more difficult to quantify than hydrophobic or electronic properties. Several methods have been tried, of which three are described here. It is highly unlikely that a drug's biological activity will be affected by steric factors alone, but these factors are frequently found in Hansch equations (section 18.3).

18.2.3.1 Taft's steric factor (E_s)

Attempts have been made to quantify the steric features of substituents by using Taft's steric factor (E_s). The value for E_s can be obtained by comparing the rates of hydrolysis of substituted aliphatic esters against a standard ester under acidic conditions (Fig. 18.9). Thus,

$$E_s = \log k_x - \log k_0$$

where k_x represents the rate of hydrolysis of an aliphatic ester bearing the substituent X, and k_0 represents the rate of hydrolysis of the reference ester.

The substituents that can be studied by this method are restricted to those which interact sterically with the tetrahedral transition state of the reaction, and not by resonance or internal hydrogen bonding. For example, unsaturated substituents which are conjugated to the ester cannot be measured by this procedure. Examples of E_s values are shown in Table 18.2. Note that the reference ester is X = Me. Substituents such as H and F are smaller than a methyl group and result in a faster rate of hydrolysis ($k_x > k_0$), making E_s positive. Substituents which are larger than methyl reduce the rate of hydrolysis ($k_x < k_0$), making E_s negative. A disadvantage of E_s values is that they are a measure of an *intramolecular* steric effect, whereas drugs interact with target binding sites in an *intermolecular* manner. For example, consider the E_s values for *i*-Pr, *n*-Pr, and *n*-Bu. The E_s value for the branched isopropyl group is significantly greater than that for the linear *n*-propyl group since the bulk of the substituent is closer to the reaction centre. Extending the alkyl chain from *n*-propyl to *n*-butyl has little effect on E_s. The larger *n*-butyl group is extended away from the reaction centre, and so it has little additional steric effect on the rate of

hydrolysis. As a result, the E_s value for the *n*-butyl group undervalues the steric effect which this group might have if it was present on a drug approaching a binding site.

18.2.3.2 Molar refractivity

Another measure of the steric factor is provided by a parameter known as **molar refractivity** (**MR**). This is a measure of the volume occupied by an atom or a group of atoms. The *MR* is obtained from the following equation:

$$MR = \frac{(n^2 - 1)}{(n^2 + 2)} \times \frac{MW}{d}$$

where n is the index of refraction, MW is the molecular weight, and d is the density. The term MW/d defines a volume, and the $(n^2 - 1)/(n^2 + 2)$ term provides a correction factor by defining how easily the substituent can be polarized. This is particularly significant if the substituent has π electrons or lone pairs of electrons.

18.2.3.3 Verloop steric parameter

Another approach to measuring the steric factor involves a computer program called **Sterimol**, which calculates steric substituent values (**Verloop steric parameters**) from standard bond angles, van der Waals radii, bond lengths, and possible conformations for the substituent. Unlike E_s, the Verloop steric parameters can be measured for any substituent. For example, the Verloop steric parameters for a carboxylic acid group are demonstrated in Fig. 18.10. L is the length of the substituent and B_1–B_4 are the radii of the group in different dimensions.

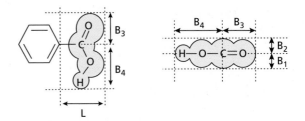

FIGURE 18.10 Verloop parameters for a carboxylic acid group.

TABLE 18.2 Values of E_s for various substituents

Substituent	H	F	Me	Et	*n*-Pr	*n*-Bu	*i*-Pr	*i*-Bu	Cyclopentyl
E_s	1.24	0.78	0	−0.07	−0.36	−0.39	−0.47	−0.93	−0.51

18.2.4 **Other physicochemical parameters**

The physicochemical properties most commonly studied by the QSAR approach have been described above, but other properties have been studied including dipole moments, hydrogen bonding, conformations, and interatomic distances. Difficulties in quantifying these properties limit the use of these parameters, however. Several QSAR formulae have been developed based on the highest occupied and/or the lowest unoccupied molecular orbitals of the test compounds. The energy calculations of these orbitals can be carried out using semi-empirical quantum mechanical methods (section 17.7.3). Indicator variables for different substituents can also be used. These are described in section 18.7.

18.3 **Hansch equation**

In section 18.2, we looked at the physicochemical properties commonly used in QSAR studies and how it is possible to quantify them. In a situation where biological activity is related to only one such property, a simple equation can be drawn up. The biological activity of most drugs, however, is related to a combination of physicochemical properties. In such cases, simple equations involving only one parameter are relevant only if the other parameters are kept constant. In reality, this is not easy to achieve and equations which relate biological activity to a number of different parameters are more common (Box 18.3). These equations are known as Hansch equations and they usually relate biological activity to the most commonly used physicochemical properties (log P, π, σ, and a steric factor). If the range of hydrophobicity values is limited to a small range then the equation will be linear, as follows:

$$\log\left(\frac{1}{C}\right) = k_1 \log P + k_2 \sigma + k_3 E_s + k_4$$

If the log P values are spread over a large range, then the equation will be parabolic for the same reasons described in section 18.2.1.

$$\log\left(\frac{1}{C}\right) = -k_1(\log P)^2 + k_2 \log P + k_3 \sigma + k_4 E_s + k_5$$

The constants k_1–k_5 are determined by computer software in order to get the best fitting equation. Not all the parameters will necessarily be significant. For example, the adrenergic blocking activity of β-halo-arylamines (Fig. 18.11) was related to π and σ and did not include a steric factor. This equation tells us that biological activity increases if the substituents have a positive π value and a negative σ value. In other words, the substituents should be hydrophobic and electron donating.

When carrying out a Hansch analysis, it is important to choose the substituents carefully to ensure that the change in biological activity can be attributed to a particular parameter. There are plenty of traps for the unwary. Take, for example, drugs which contain an amine group. One of the studies most frequently carried out on amines is to synthesize analogues containing a homologous series of alkyl substituents on the nitrogen atom (i.e. Me, Et, n-Pr, n-Bu). If activity increases with the chain length of the substituent, is it due to increasing hydrophobicity, increasing size, or both? If we look at the π and MR values of these substituents, we find that both sets of values increase in a similar fashion across the series and we would not be able to distinguish between them (Table 18.3). In this example, a series of substituents would have to be chosen where π and MR are not correlated. The substituents H, Me, OMe, NHCOCH$_2$, I, and CN would be more suitable.

Test your understanding and practise your molecular modelling with Exercise 18.4 on the Online Resource Centre at www.oxfordtextbooks.co.uk/orc/patrick6e/

18.4 **The Craig plot**

Although tables of π and σ factors are readily available for a large range of substituents, it is often easier to visualize the relative properties of different substituents by considering a plot where the y-axis is the value of the σ factor and the x-axis is the value of the π factor. Such a plot is known as a Craig plot. The example shown in Fig. 18.12 is the Craig plot for the σ and π factors of *para*-aromatic substituents. There are several advantages to the use of such a Craig plot.

- The plot shows clearly that there is no overall relationship between π and σ. The various substituents are scattered around all four quadrants of the plot.

- It is possible to tell at a glance which substituents have positive π and σ parameters, which substituents have negative π and σ parameters, and which substituents have one positive and one negative parameter.

- It is easy to see which substituents have similar π values. For example, the ethyl, bromo, trifluoromethyl, and trifluoromethylsulphonyl groups are all approximately on the same vertical line on the plot. In theory, these groups could be interchangeable on drugs where the principal factor affecting biological activity

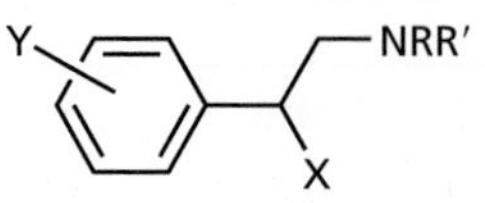

$$\log\left(\frac{1}{C}\right) = 1.22\pi - 1.59\sigma + 7.89$$

$$(n = 22,\ r^2 = 0.841,\ s = 0.238)$$

β-Halo-arylamines

FIGURE 18.11 QSAR equation for β-halo-arylamines.

BOX 18.3 Hansch equation for a series of antimalarial compounds

A series of 102 phenanthrene aminocarbinols was tested for antimalarial activity. In the structure shown, X represents up to four substituents on the left-hand ring while Y represents up to four substituents on the right-hand ring. Experimental log *P* values for the structures were not available and equations were derived which compared the activity with some or all of the following terms:

π_{sum} The π constants for *all* the substituents in the molecule (i.e. all the X and Y substituents, as well as the amino substituents R and R'). This term was used in place of log *P* to represent the overall hydrophobicity for the molecule.

σ_{sum} The σ constants for *all* the substituents in the molecule.

$\sum\pi_X$ The sum of the π constants for all the substituents X in the left-hand ring.

$\sum\pi_Y$ The sum of the π constants for all the substituents Y in the right-hand ring.

$\sum\pi_{X+Y}$ The sum of the π constants for all the substituents X and Y in both the left- and right-hand rings

$\sum\sigma_{X+Y}$ The sum of the σ constants for all the substituents X and Y in both the left- and right-hand rings.

$\sum\sigma_X$ The sum of the σ constants for all the substituents X in the left-hand ring.

$\sum\sigma_Y$ The sum of the σ constants for all the substituents Y in the right-hand ring.

Equations such as equations 1–3 were derived which matched activity against one of the above terms, but none of them had an acceptable value of r^2.

A variety of other equations were derived which included two of the above terms but these were not satisfactory either. Finally, an equation was derived which contained 6 terms and proved satisfactory:

$$\log\left(\frac{1}{C}\right) = -0.015(\pi_{sum})^2 + 0.14\pi_{sum} + 0.27\sum\pi_X$$
$$+ 0.40\sum\pi_Y + 0.65\sum\sigma_X + 0.88\sum\sigma_Y + 2.34$$
$$(n = 102, r = 0.913, r^2 = 0.834, s = 0.258)$$

The equation shows that antimalarial activity increases very slightly as the overall hydrophobicity of the molecule (π_{sum}) increases (the constant 0.14 is low). The $(\pi_{sum})^2$ term shows that there is an optimum overall hydrophobicity for activity and this is found to be 4.44. Activity increases if hydrophobic substituents are present on ring X and, in particular, on ring Y. This could be taken to imply that some form of hydrophobic interaction is involved near both rings. Electron-withdrawing substituents on both rings are also beneficial to activity, more so on ring Y than ring X. The r^2 value is 0.834 which is above the minimum acceptable value of 0.8.

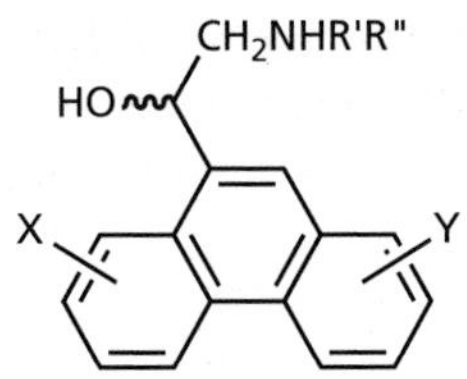

1) $\log\left(\frac{1}{C}\right) = 0.557\sum\pi_{x+y} + 2.699$ ($n = 102, r = 0.768, r^2 = 0.590, s = 0.395$)

2) $\log\left(\frac{1}{C}\right) = 0.017\pi_{sum} + 3.324$ ($n = 102, r = 0.069, r^2 = 0.005, s = 0.616$)

3) $\log\left(\frac{1}{C}\right) = 1.218\sigma_{sum} + 2.721$ ($n = 102, r = 0.814, r^2 = 0.663, s = 0.359$)

FIGURE 1 Phenanthrene aminocarbinols.

is the π factor. Similarly, groups which form a horizontal line can be identified as being isoelectronic or having similar σ values (e.g. CO_2H, Cl, Br, I).

• The Craig plot is useful in planning which substituents should be used in a QSAR study. In order to derive the most accurate equation involving π and σ, analogues should be synthesized with substituents from each quadrant. For example, halogen substituents are useful representatives of substituents with increased hydrophobicity and electron-withdrawing properties (positive

TABLE 18.3 Values for π and *MR* for a series of substituents.

Substituent	H	Me	Et	*n*-Pr	*n*-Bu	OMe	NHCONH$_2$	I	CN
π	0.00	0.56	1.02	1.50	2.13	−0.02	−1.30	1.12	−0.57
MR	0.10	0.56	1.03	1.55	1.96	0.79	1.37	1.39	0.63

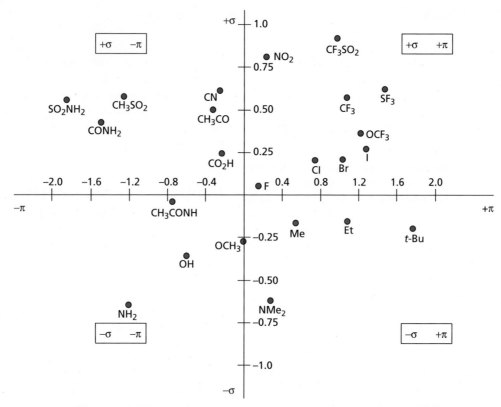

FIGURE 18.12 Craig plot comparing the values of σ and π for various substituents.

π and positive σ), whereas an OH substituent has more hydrophilic and electron-donating properties (negative π and negative σ). Alkyl groups are examples of substituents with positive π and negative σ values, whereas acyl groups have negative π and positive σ values.

- Once the Hansch equation has been derived, it will show whether π or σ should be negative or positive in order to get good biological activity. Further developments would then concentrate on substituents from the relevant quadrant. For example, if the equation shows that positive π and positive σ values are necessary, then further substituents should only be taken from the top right quadrant.

Craig plots can also be drawn up to compare other sets of physicochemical parameters, such as hydrophobicity and *MR*.

18.5 The Topliss scheme

In certain situations, it might not be feasible to make the large range of structures required for a Hansch equation. For example, the synthetic route involved might be so difficult that only a few structures can be made in a limited time. In these circumstances, it would be useful to test compounds for biological activity as they are synthesized and to use these results to determine the next analogue to be synthesized.

A Topliss scheme is a 'flow diagram' which allows such a procedure to be followed. There are two Topliss schemes, one for substituents on an aromatic ring (Fig. 18.13) and one for substituents on aliphatic moieties (Fig. 18.14). The schemes were drawn up by considering the hydrophobicity and electronic factors of various substituents, and are designed such that the optimum substituent can be found as efficiently as possible. They are not meant to be a replacement for a full Hansch analysis, however. Such an analysis would be carried out in due course, once a suitable number of structures have been synthesized.

The Topliss scheme for substituents on an aromatic ring (Fig. 18.13) assumes that the lead compound has been tested for biological activity and contains a monosubstituted aromatic ring. The first analogue in the scheme is the 4-chloro derivative, as this derivative is usually easy to synthesize. The chloro substituent is more hydrophobic and electron-withdrawing than hydrogen and, therefore, π and σ are positive.

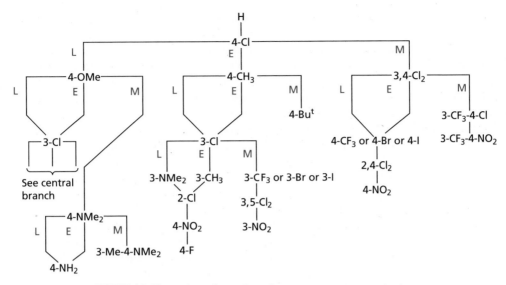

FIGURE 18.13 Topliss scheme for substituents on an aromatic ring.

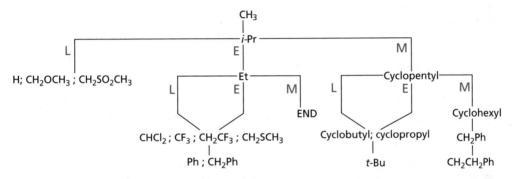

FIGURE 18.14 Topliss scheme for substituents on an aliphatic moiety.

Once the chloro analogue has been synthesized, the biological activity is measured. There are three possibilities. The analogue will have less activity (L), equal activity (E), or more activity (M). The type of activity observed will determine which branch of the Topliss scheme is followed next.

If the biological activity increases, the (M) branch is followed and the next analogue to be synthesized is the 3,4-dichloro-substituted analogue. If, on the other hand, the activity stays the same, then the (E) branch is followed and the 4-methyl analogue is synthesized. Finally, if activity drops, the (L) branch is followed and the next analogue is the 4-methoxy analogue. Biological results from the second analogue now determine the next branch to be followed in the scheme.

What is the rationale behind this?

Let us consider the situation where the 4-chloro derivative increases in biological activity. The chloro substituent has positive π and σ values, which implies that one, or both, of these properties are important to biological activity. If both are important, then adding

a second chloro group should increase biological activity yet further. If it does, substituents are varied to increase the π and σ values even further. If it does not, then an unfavourable steric interaction or excessive hydrophobicity is indicated. Further modifications then test the relative importance of π and steric factors.

Now consider the situation where the 4-chloro analogue drops in activity. This suggests that either negative π and/or σ values are important to activity or a *para* substituent is sterically unfavourable. It is assumed that an unfavourable σ effect is the most likely reason for the reduced activity and so the next substituent is one with a negative σ factor (i.e. 4-OMe). If activity improves, further changes are suggested to test the relative importance of the σ and π factors. On the other hand, if the 4-OMe group does not improve activity, it is assumed that an unfavourable steric factor is at work and the next substituent is a 3-chloro group. Modifications of this group would then be carried out in the same way as shown in the centre branch of Fig. 18.13.

The last scenario is where the activity of the 4-chloro analogue is little changed from the lead compound. This could arise from the drug requiring a positive π value and a negative σ value. As both values for the chloro group are positive, the beneficial effect of the positive π value might be cancelled out by the detrimental effects of a positive σ value. The next substituent to try in that case is the 4-methyl group which has the necessary positive π value and negative σ value. If this still has no beneficial effect, then it is assumed that there is an unfavourable steric interaction at the *para* position and the 3-chloro substituent is chosen next. Further changes continue to vary the relative values of the π and σ factors.

The validity of the Topliss scheme was tested by looking at structure–activity results for various drugs which had been reported in the literature. For example, the biological activities of 19 substituted benzenesulphonamides (Fig. 18.15) have been reported. The second most active compound was the nitro-substituted analogue, which would have been the fifth compound synthesized if the Topliss scheme had been followed.

Another example comes from the anti-inflammatory activities of substituted aryltetrazolylalkanoic acids (Fig. 18.16), of which 28 were synthesized. Using the Topliss scheme, 3 out of the 4 most active structures would have been prepared from the first 8 compounds synthesized.

The Topliss scheme for substituents on aliphatic moieties (Fig. 18.14) was set up following a similar rationale to the aromatic scheme and is used in the same way for substituents attached to a carbonyl, amino, amide, or similar functional group. The scheme attempts to differentiate only between the hydrophobic and electronic effects of substituents, and not their steric properties. Thus, the substituents involved have been chosen to try to minimize any steric differences. It is assumed that the lead compound has a methyl group. The first analogue suggested is the isopropyl analogue. This has an increased π value and, in most cases, would be expected to increase activity. It has been found from experience that the hydrophobicity of most lead compounds is less than optimum.

Let us concentrate first of all on the situation where activity increases. Following this branch, a cyclopentyl group is now used. A cyclic structure is used since it has a larger π value, but keeps any increase in steric factor to a minimum. If activity rises again, more hydrophobic substituents are tried. If activity does not rise, then there could be two explanations. Either the optimum hydrophobicity has been passed or there is an electronic effect (σ_I) at work. Further substituents are then used to determine which is the correct explanation.

Let us now look at the situation where the activity of the isopropyl analogue stays much the same. The most likely explanation is that the methyl group and the isopropyl group are on either side of the hydrophobic optimum. Therefore, an ethyl group is used next, since it has an intermediate π value. If this does not lead to an improvement, it is possible that there is an unfavourable electronic effect. The groups used have been electron-donating, and so electron-withdrawing groups with similar π values are now suggested.

Finally, we shall look at the case where activity drops for the isopropyl group. In this case, hydrophobic and/

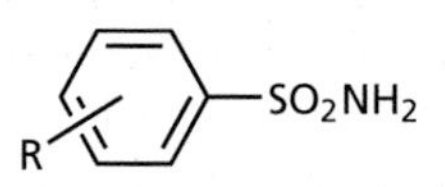

Order of synthesis	R	Biological activity	High potency
1	H	–	
2	4-Cl	More	
3	3,4-Cl$_2$	Less	
4	4-Br	Equal	
5	4-NO$_2$	More	*

FIGURE 18.15 The order of benzenesulphonamide synthesis as directed by the Topliss scheme.

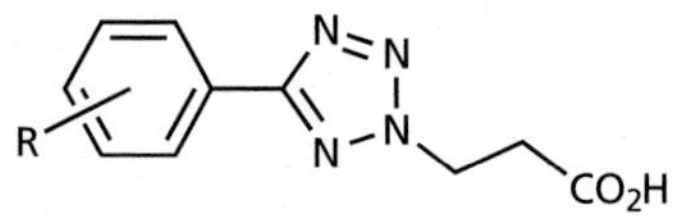

Order of synthesis	R	Biological activity	High potency
1	H	–	
2	4-Cl	Less	
3	4-OMe	Less	
4	3-Cl	More	*
5	3-CF$_3$	Less	
6	3-Br	More	*
7	3-I	Less	
8	3,5-Cl$_2$	More	*

FIGURE 18.16 The order of synthesis for substituted aryltetrazolylalkanoic acids as directed by the Topliss scheme.

or electron-donating groups could be bad for activity and the groups suggested are suitable choices for further development.

18.6 Bio-isosteres

Tables of substituent constants are available for various physicochemical properties. A knowledge of these constants allows the medicinal chemist to identify substituents which may be potential bio-isosteres. Thus, the substituents CN, NO_2, and COMe have similar hydrophobic, electronic, and steric factors, and might be interchangeable. Such interchangeability was observed in the development of **cimetidine** and its analogues (sections 25.2.6 and 25.2.8). The important thing to note is that groups can be bio-isosteric in some situations, but not others. Consider for example the table shown in Fig. 18.17.

This table shows physicochemical parameters for six different substituents. If the most important physicochemical parameter for biological activity is σ_p, then the $COCH_3$ group (0.50) would be a reasonable bio-isostere for the $SOCH_3$ group (0.49). If, on the other hand, the dominant parameter is π, then a more suitable bio-isostere for $SOCH_3$ (−1.58) would be SO_2CH_3 (−1.63).

18.7 The Free–Wilson approach

In the Free–Wilson approach to QSAR, the biological activity of a parent structure is measured then compared with the activities of a range of substituted analogues. An equation is then derived which relates biological activity to the presence or otherwise of particular substituents (X_1–X_n).

$$\text{Activity} = k_1X_1 + k_2X_2 + k_3X_3 + \ldots\ldots + k_nX_n + Z$$

In this equation, X_n is defined as an **indicator variable** and is given the value 1 or 0, depending on whether the substituent (n) is present or not. The contribution that each substituent makes to the activity is determined by

the value of k_n. Z is a constant representing the average activity of the structures studied.

Since the approach considers the overall effect of a substituent on biological activity rather than its various physicochemical properties, there is no need for physicochemical constants and tables, and the method only requires experimental measurements of biological activity. This is particularly useful when trying to quantify the effect of unusual substituents that are not listed in the tables, or when quantifying specific molecular features which cannot be tabulated.

The disadvantage in the approach is the large number of analogues which have to be synthesized and tested to make the equation meaningful. For example, each of the terms k_nX_n refers to a specific substituent at a specific position in the parent structure. Therefore, analogues would not only have to have different substituents, but also have them at different positions of the skeleton.

Another disadvantage is the difficulty in rationalizing the results and explaining why a substituent at a particular position is good or bad for activity. Finally, the effects of different substituents may not be additive. There may be intramolecular interactions which affect activity.

Nevertheless, indicator variables can be useful in certain situations and they can also be used as part of a Hansch equation. An example of this can be seen in the later case study (section 18.9).

18.8 Planning a QSAR study

When starting a QSAR study it is important to decide which physicochemical parameters are going to be studied and to plan the analogues such that the parameters under study are suitably varied. For example, it would be pointless to synthesize analogues where the hydrophobicity and steric volume of the substituents are correlated if these two parameters are to go into the equation.

It is also important to make enough structures to make the results statistically meaningful. As a rule of thumb, five structures should be made for every parameter studied. Typically, the initial QSAR study would involve the

Substituent	$\underset{CH_3}{\overset{\displaystyle O \atop \|}{C}}$	$\underset{CH_3}{\overset{\displaystyle NC\diagdown C \diagup CN \atop \overset{O}{\underset{\|}{\|}} C}}$	$\underset{CH_3}{\overset{\displaystyle O \atop \|}{S}}$	$-\underset{O}{\overset{O}{\underset{\|}{\overset{\|}{S}}}} -CH_3$	$-\underset{O}{\overset{O}{\underset{\|}{\overset{\|}{S}}}} -NHCH_3$	$\underset{NMe_2}{\overset{\displaystyle O \atop \|}{C}}$
π	−0.55	0.40	−1.58	−1.63	−1.82	−1.51
σ_p	0.50	0.84	0.49	0.72	0.57	0.36
σ_m	0.38	0.66	0.52	0.60	0.46	0.35
MR	11.2	21.5	13.7	13.5	16.9	19.2

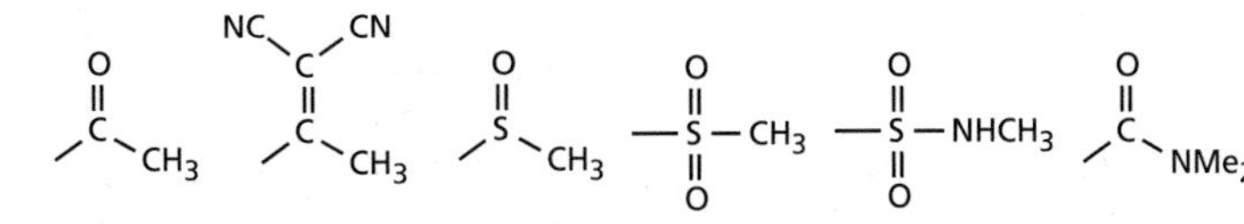

FIGURE 18.17 Physicochemical parameters for six substituents.

two parameters π and σ, and possibly E_s. Craig plots could be used in order to choose suitable substituents.

Certain substituents are worth avoiding in the initial study, as they may have properties other than those being studied. For example, it is best to avoid substituents that might ionize (CO_2H, NH_2, SO_3H) and groups that might be easily metabolized (e.g. esters or nitro groups).

If there are two or more substituents, then the initial equation usually considers the total π and σ contribution.

As more analogues are made, it is often possible to consider the hydrophobic and electronic effect of substituents at specific positions of the molecule. Furthermore, the electronic parameter σ can be split into its inductive and resonance components (F and R). Such detailed equations may show up a particular localized requirement for activity. For example, a hydrophobic substituent may be favoured in one part of the skeleton, while an electron-withdrawing substituent is favoured at another. This, in turn, gives clues about the binding interactions involved between drug and receptor.

18.9 Case study

An example of how a QSAR equation can become more specific as a study develops is demonstrated from work carried out on the anti-allergic activity of a series of pyranenamines (Fig. 18.18). In this study, substituents were varied on the aromatic ring, while the remainder of the molecule was kept constant. Nineteen compounds were synthesized and the first QSAR equation was obtained by considering π and σ:

$$\log\left(\frac{1}{C}\right) = -0.14\sum\pi - 1.35(\sum\sigma)^2 - 0.72$$

$$(n\ 19,\ r^2\ 0.48,\ s\ 0.47,\ F_{2,16}\ 7.3)$$

where $\sum\pi$ and $\sum\sigma$ are the total π and σ values for all the substituents present.

The negative coefficient for the π term shows that activity is inversely proportional to hydrophobicity, which is quite unusual. The $(\sum\sigma)^2$ term is also quite unusual. It was chosen because there was no simple relationship between activity and σ. In fact, it was observed that activity decreased if the substituent was electron-withdrawing *or* electron-donating. Activity was best with neutral substituents. To take account of this, the $(\sum\sigma)^2$ term was introduced. As the coefficient in the equation is negative, activity is lowered if σ is anything other than zero.

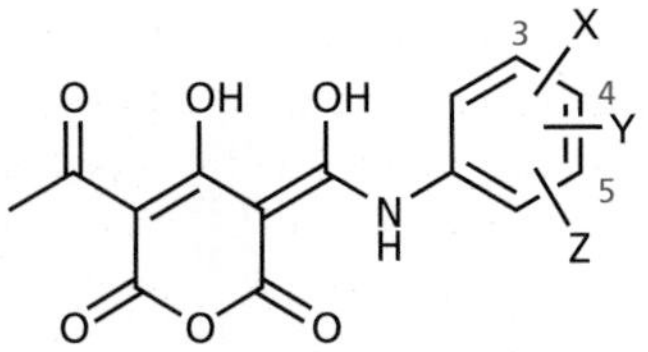

A further range of compounds was synthesized with hydrophilic substituents to test this equation, making a total of 61 structures. This resulted in the following inconsistencies.

- The activities for the substituents 3-NHCOMe, 3-NHCOEt, and 3-NHCOPr were all similar, but according to the equation, these activities should have dropped as the alkyl group got larger as a result of increasing hydrophobicity.

- Activity was greater than expected if there was a substituent such as OH, SH, NH_2, or NHCOR at position 3, 4, or 5.

- The substituent $NHSO_2R$ was bad for activity.

- The substituents 3,5-$(CF_3)_2$ and 3,5-$(NHCOMe)_2$ had much greater activity than expected.

- An acyloxy group at the 4-position resulted in an activity five times greater than predicted by the equation.

These results implied that the initial equation was too simple and that properties other than π and σ were important to activity. At this stage, the following theories were proposed to explain the above results.

- The similar activities for 3-NHCOMe, 3-NHCOEt, and 3-NHCOPr could be due to a steric factor. The substituents had increasing hydrophobicity, which is bad for activity, but they were also increasing in size and it was proposed that this was good for activity. The most likely explanation is that the size of the substituent is forcing the drug into the correct orientation for optimum receptor interaction.

- The substituents which unexpectedly increased activity when they were at positions 3, 4, or 5 are all capable of hydrogen bonding. This suggests an important hydrogen bonding interaction with the receptor. For some reason, the $NHSO_2R$ group is an exception, which implies there is some other unfavourable steric or electronic factor peculiar to this group.

- The increased activity for 4-acyloxy groups was explained by suggesting that these analogues are acting as prodrugs. The acyloxy group is less polar than the hydroxyl group and so these analogues would be expected to cross cell membranes and reach the receptor more efficiently than analogues bearing a free hydroxyl group. Hydrolysis of the ester group would reveal the hydroxyl group which would then take part in hydrogen bonding with the receptor.

FIGURE 18.18 Structure of pyranenamines.

- The structures having substituents 3,5-$(CF_3)_2$ and 3,5-$(NHCOMe)_2$ are the only disubstituted structures where a substituent at position 5 has an electron-withdrawing effect, so this feature was also introduced into the next equation.

The revised QSAR equation was as follows:

$$\log\left(\frac{1}{C}\right) = -0.30\sum\pi - 1.5(\sum\sigma)^2 + 2.0(F\text{-}5)$$
$$+ 0.39(345\text{-}HBD) - 0.63(NHSO_2)$$
$$+ 0.78(M\text{-}V) + 0.72(4\text{-}OCO) - 0.75$$
$$(n\ 61,\ r^2\ 0.77,\ s\ 0.40,\ F_{7,53}\ 25.1)$$

The π and σ parameters are still present, but a number of new parameters have now been introduced.

- The F-5 term represents the inductive effect of a substituent at position 5. The coefficient is positive and large, showing that an electron-withdrawing group substantially increases activity. However, only 2 compounds of the 61 synthesized had a 5-substituent, so there might be quite an error in this result.
- The M-V term represents the volume of any *meta* substituent. The coefficient is positive, indicating that substituents with a large volume at the *meta* position increase activity.
- The advantage of having hydrogen bonding substituents at position 3, 4, or 5 is accounted for by including a hydrogen bonding term (*345-HBD*). The value of this term depends on the number of hydrogen bonding substituents present. If one such group is present, the *345-HBD* term is 1. If two such groups are present, the parameter is 2. Therefore, for each hydrogen bonding substituent present at positions 3, 4, or 5, log (1/C) increases by 0.39. This sort of term is known as an **indicator variable**, which is the basis of the **Free–Wilson approach** described earlier. There is no tabulated value one can use for a hydrogen bonding substituent and so the contribution that this term makes to the biological activity is determined by the value of k, and whether the relevant group is present or not. Indicator variables were also used for the following terms.
- The $NHSO_2$ term was introduced because this group was bad for activity despite being capable of hydrogen bonding. The negative coefficient indicates the drop in activity. A figure of 1 is used for any $NHSO_2R$ substituent present, resulting in a drop of activity by 0.63.
- The *4-OCO* term is 1 if an acyloxy group is present at position 4, and so log (1/C) is increased by 0.72 if this is the case.

A further 37 structures were synthesized to test steric and F-5 parameters as well as exploring further groups capable of hydrogen bonding. Since hydrophilic substituents were good for activity, a range of very hydrophilic substituents were also tested to see if there was an optimum value for hydrophilicity. The results obtained highlighted one more anomaly, in that two hydrogen bonding groups *ortho* to each other were bad for activity. This was attributed to the groups hydrogen bonding with each other rather than to the receptor. A revised equation was obtained as follows:

$$\log\left(\frac{1}{C}\right) = -0.034(\sum\pi)^2 - 0.33(\sum\pi) + 4.3(F\text{-}5)$$
$$+ 1.3(R\text{-}5) - 1.7(\sum\sigma)^2 + 0.73(345\text{-}HBD)$$
$$- 0.86(HB\text{-}INTRA) - 0.69(NHSO_2)$$
$$+ 0.72(4\text{-}OCO) - 0.59$$
$$(n\ 98,\ r^2\ 0.75,\ s\ 0.48,\ F_{9,88}\ 28.7)$$

The main points of interest from this equation are as follows.

- Increasing the hydrophilicity of substituents allowed the identification of an optimum value for hydrophobicity ($\sum\pi = -5$) and introduced the $(\sum\pi)^2$ parameter into the equation. The value of -5 is remarkably low and indicates that the region of the binding site occupied by the drug's aromatic ring is hydrophilic.
- As far as electronic effects are concerned, it is revealed that the resonance effects of substituents at the 5-position also have an influence on activity.
- The unfavourable situation where two hydrogen bonding groups are *ortho* to each other is represented by the *HB-INTRA* parameter. This parameter is given the value 1 if such an interaction is possible, and the negative constant (-0.86) shows that such interactions decrease activity.
- It is interesting to note that the steric parameter is no longer significant and has been removed from the equation.

The compound having the greatest activity has two $NHCOCH(OH)CH_2OH$ substituents at the 3- and 5-positions and is 1000 times more active than the original lead compound. The substituents are very polar and are not ones that would normally be used. They satisfy all the requirements determined by the QSAR study. They are highly polar groups which can take part in hydrogen bonding. They are *meta* with respect to each other, rather than *ortho*, to avoid undesirable intramolecular hydrogen bonding. One of the groups is at the 5-position and has a favourable F-5 parameter. Together, the two groups have a negligible $(\sum\sigma)^2$ value. Such an analogue would certainly not have been obtained by trial and error and this example demonstrates the strengths of the QSAR approach.

All the evidence from this study suggests that the aromatic ring of this series of compounds is fitting into a hydrophilic pocket in the target binding site which contains polar groups capable of hydrogen bonding.

It is further proposed that a positively charged residue such as arginine, lysine, or histidine might be present in the pocket which could interact with an electronegative substituent at position 5 of the aromatic ring (Fig. 18.19).

This example demonstrates that QSAR studies and computers are powerful tools in medicinal chemistry. However, it also shows that the QSAR approach is a long way from replacing the human factor. One cannot put a series of facts and figures into a computer and expect it to magically produce an instant explanation of how a drug works. The medicinal chemist still has to interpret results, propose theories, and test those theories by incorporating the correct parameters into the QSAR equation. Imagination and experience still count for a great deal.

KEY POINTS

- QSAR relates the physicochemical properties of a series of drugs to their biological activity by means of a mathematical equation.

- The commonly studied physicochemical properties are hydrophobicity, electronic factors, and steric factors.

- The partition coefficient is a measure of a drug's overall hydrophobicity. Values of log P are used in QSAR equations, with larger values indicating greater hydrophobicity.

- The substituent hydrophobicity constant is a measure of the hydrophobic character of individual substituents. The values are different for substituents attached to aliphatic and aromatic systems and are only directly relevant to the class of structures

from which the values were derived. Positive values represent substituents more hydrophobic than hydrogen; negative values represent substituents more hydrophilic than hydrogen.

- The Hammett substituent constant is a measure of how electron-withdrawing or electron-donating a substituent is. It is measured experimentally and is dependent on the relative position of the substituent on an aromatic ring. The value takes into account both inductive and resonance effects.

- The parameters F and R are constants quantifying the inductive and resonance effects of a substituent on an aromatic ring.

- The inductive effect of substituents on aliphatic moieties can be measured experimentally and tabulated.

- Steric factors can be measured experimentally or calculated using physical parameters or computer software.

- The Hansch equation is a mathematical equation which relates a variety of physicochemical parameters to biological activity for a series of related structures.

- The Craig plot is a visual comparison of two physicochemical properties for a variety of substituents. It facilitates the choice of substituents for a QSAR study such that the values of each property are not correlated.

- The Topliss scheme is used when structures can only be synthesized and tested one at a time. The scheme is a guide to which analogue should be synthesized next in order to get good activity. There are different schemes for substituents on aromatic and aliphatic systems.

- Indicator variables are used when there are no tabulated or experimental values for a particular property or substituent. The Free–Wilson approach to QSAR only uses indicator variables, whereas the Hansch approach can use a mixture of indicator variables and physicochemical parameters.

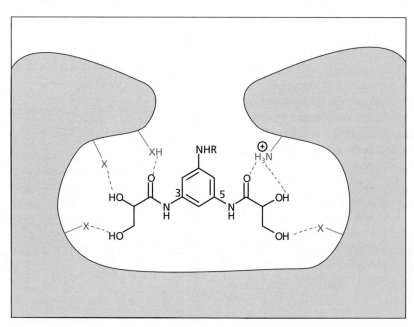

FIGURE 18.19 Hypothetical binding interactions between a pyranenamine and the target binding site.

18.10 **3D QSAR**

In recent years, a method known as 3D QSAR has been developed in which the three-dimensional properties of a molecule are considered as a whole rather than by considering individual substituents or moieties. This has proved remarkably useful in the design of new drugs. Moreover, the necessary software and hardware are readily affordable and relatively easy to use. The philosophy of 3D QSAR revolves around the assumption that the most important features about a molecule are its overall size and shape, and its electronic properties.

If these features can be defined, then it is possible to study how they affect biological properties. There are several approaches to 3D QSAR, but the method which has gained ascendancy was developed by the company Tripos and is known as **CoMFA** (**Comparative Molecular Field Analysis**). CoMFA methodology is based on the assumption that drug–receptor interactions are noncovalent and that changes in biological activity correlate with the changes in the steric and/or electrostatic fields of the drug molecules.

18.10.1 **Defining steric and electrostatic fields**

The steric and electrostatic fields surrounding a molecule can be measured and defined using the grid and probe method described in section 17.7.5. This can be repeated for all the molecules in the 3D QSAR study, but it is crucial that the molecules are all in their **active conformation**, and that they are all positioned within the grid in exactly the same way. In other words, they must all be correctly aligned. Identifying a **pharmacophore** (section 13.2) that is common to all the molecules can assist in this process (Fig. 18.20).

The pharmacophore is placed into the grid and its position is kept constant such that it acts as a reference point when positioning each molecule into the lattice. For each molecule studied, the active conformation and pharmacophore is identified and then the molecule is placed into the lattice such that its pharmacophore matches the reference pharmacophore (Fig. 18.21). Once a molecule has been placed into the lattice, the steric and electrostatic fields around it are measured as described in section 17.7.5.

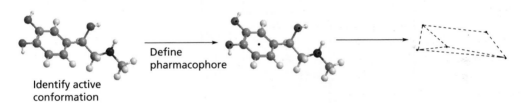

FIGURE 18.20 Identification of the active conformation and pharmacophore.

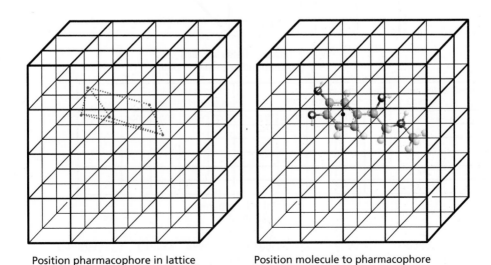

Position pharmacophore in lattice   Position molecule to pharmacophore

FIGURE 18.21 Positioning a pharmacophore and molecules into a lattice.

18.10.2 Relating shape and electronic distribution to biological activity

Defining the steric and electrostatic fields of a series of molecules is relatively straightforward and is carried out automatically by the software program. The next stage is to relate these properties to the biological activity of the molecules. This is less straightforward and differs significantly from traditional QSAR. In traditional QSAR, there are relatively few variables involved. For example, if we consider log P, π, σ, and a steric factor for each molecule, then we have four variables per molecule to compare against biological activity. With 100 molecules in the study, there are far more molecules than variables, and it is possible to come up with an equation relating variables to biological activity as previously described.

In 3D QSAR, the variables for each molecule are the calculated steric and electronic interactions at a couple of thousand lattice points. With 100 molecules in the study, the number of variables now far outweighs the number of structures, and it is not possible to relate these to biological potency by the standard multiple linear regression

analysis described in section 18.1. A different statistical procedure has to be followed using a technique called **partial least squares** (PLS). Essentially, it is an analytical computing process which is repeated over and over again (iterated) to try to find the best formula relating biological potency against the various variables. As part of the process, the number of variables is reduced as the software filters out those which are clearly unrelated to biological activity.

An important feature of the analysis is that a structure is deliberately left out as the computer strives to form some form of relationship. Once a formula has been defined, the formula is tested against the structure which was left out. This is called **cross-validation** and tests how well the formula predicts the biological property for the molecule which was left out. The results of this are fed back into another round of calculations, but now the structure which was left out is included in the calculations and a different structure is left out. This leads to a new improved formula which is once again tested against the compound that was left out, and so the process continues until cross-validation has been carried out against all the structures.

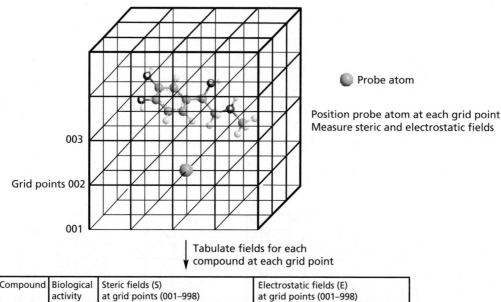

FIGURE 18.22 Measuring steric and electronic fields.

At the end of the process, the final formula is obtained (Fig. 18.22). The predictability of this final equation is quantified by the **cross-validated correlation coefficient** r^2, which is usually referred to as q^2. In contrast to normal QSAR, where r^2 should be greater than 0.8, values of q^2 greater than 0.3 are considered significant. It is more useful, though, to give a graphical representation showing which regions around the molecule are important to biological activity on steric or electronic grounds. Therefore, a steric map shows a series of coloured contours indicating beneficial and detrimental steric interactions around a representative molecule from the set of molecules tested (Fig. 18.23). A similar contour map is created to illustrate electrostatic interactions.

An example of a 3D QSAR study is described in the case study in section 18.10.6.

⊕ For additional material see Web article 5: The design of a serotonin antagonist as a possible anxiolytic agent on the Online Resource Centre at www.oxfordtextbooks.co.uk/orc/patrick6e/

18.10.3 Advantages of CoMFA over traditional QSAR

Some of the problems involved with a traditional QSAR study include the following:

- Only molecules of similar structure can be studied.
- The validity of the numerical descriptors is open to doubt. These descriptors are obtained by measuring reaction rates and equilibrium constants in model reactions and are listed in tables. However, separating one property from another is not always possible in experimental measurement. For example, the Taft steric factor is not purely a measure of the steric factor. This is because the measured reaction rates used to define it are also affected by electronic factors. Also, the *n*-octanol/water partition coefficients which are used to measure log *P* are known to be affected by the hydrogen bonding character of molecules.
- The tabulated descriptors may not include entries for unusual substituents.

FIGURE 18.23 Definition of favourable and unfavourable interactions around a representative molecule.

- It is necessary to synthesize a range of molecules where substituents are varied in order to test a particular property (e.g. hydrophobicity). However, synthesizing such a range of compounds may not be straightforward or feasible.
- Traditional QSAR equations do not directly suggest new compounds to synthesize.

These problems are avoided with CoMFA which has the following advantages:

- Favourable and unfavourable interactions are represented graphically by 3D contours around a representative molecule. A graphical picture such as this is easier to visualize than a mathematical formula.
- In CoMFA, the properties of the test molecules are calculated individually by the computer program. There is no reliance on experimental or tabulated factors. There is no need to confine the study to molecules of similar structure. As long as one is confident that all the compounds in the study share the same pharmacophore and interact in the same way with the target, they can all be analysed in a CoMFA study.
- The graphical representation of beneficial and non-beneficial interactions allows medicinal chemists to design new structures. For example, if a contour map shows a favourable steric effect at one particular location, this implies that the target binding site has space for further extension at that location. This may lead to further favourable receptor–drug interactions.
- Both traditional and 3D QSAR can be used without needing to know the structure of the biological target.

18.10.4 Potential problems of CoMFA

There are several potential problems in using CoMFA:

- It is important to know the active conformation for each of the molecules in the study. Identifying the active conformation is easy for rigid structures such as steroids, but it is more difficult for flexible molecules that are capable of several bond rotations. Therefore, it is useful to have a conformationally restrained analogue which is biologically active and which can act as a guide to the likely active conformation. More flexible molecules can then be constructed on the computer with the conformation most closely matching that of the more rigid analogue. If the structure of the target binding site is known, this can be useful in deciding the likely active conformations of molecules.
- Each molecule in the study must be correctly positioned in the grid so that it is properly aligned with respect to all the other molecules. A common phar-

macophore can be used to aid this process as described earlier. However, it may be difficult to identify the pharmacophore in some molecules. In that case, a pharmacophore mapping exercise could be carried out (section 17.11). This is likely to be successful if there are some rigid active compounds amongst the compounds being studied. An alternative method of alignment is to align the molecules based on their structural similarity. This can be done automatically using what is known as 'topomer' methodology.

- One has to be careful to ensure that all the compounds in the study interact with the target in similar ways. For example, a 3D QSAR study on all possible **acetylcholinesterase** inhibitors is doomed to failure. In the first place, the great diversity of structures involved makes it impossible to align these structures in an unbiased way or to generate a 3D pharmacophore. Secondly, the various inhibitors do not interact with the target enzyme in the same way. X-ray crystallographic studies of enzyme–inhibitor complexes show that the inhibitors **tacrine**, **edrophonium**, and **decamethonium** all have different binding orientations in the active site.

- 3D QSAR provides a summary of how structural changes in a drug affect biological activity, but it is dangerous to assume too much. For example, a 3D QSAR model may show that increasing the bulk of the molecule at a particular location increases activity. This might suggest that there is an accessible hydrophobic pocket allowing extra binding interactions. On the other hand, it is possible that the extra steric bulk causes the molecule to bind in a different orientation from the other molecules in the analysis, and that this is the reason for the increased activity.

- It has been found that slightly different orientations of the grid from one study to another can produce different results for the same set of compounds.

18.10.5 Other 3D QSAR methods

CoMFA continues to be the most popular program for studies into 3D QSAR, but it does suffer a number of disadvantages as described above. Two of the more serious problems are the spurious results that can be obtained if compounds are not properly aligned, or if the orientation of the grid box is slightly different between studies. Users of the program can also choose different spacings between the grid points, and this can give poor results if the grid is too coarse or too fine. The method is also computationally expensive, requiring a lot of calculations for each molecule in the study, and so powerful computers are needed to cope with the huge memory requirement.

Other 3D QSAR programs have been developed in an attempt to address some of these issues. Examples include **HINT** which can be used alongside CoMFA to measure a hydrophobic field, **CoMSIA** which includes hydrogen bonding and hydrophobic fields as well as steric and electrostatic fields, and **CoMASA** which uses fewer calculations.

Some 3D QSAR programs use the intrinsic molecular properties of compounds rather than using a probe to measure the property fields surrounding them. Four examples are **SOMFA**, **HASL**, **CoMMA**, and **MS-Whim**. Other programs are used to model hypothetical pseudoreceptors. These include **Quasar**, **WeP**, and **GRIND**.

KEY POINTS

- CoMFA is an example of a 3D QSAR program which measures steric and electrostatic fields round a series of structures and relates these to biological activity.

- A comparison of the steric and electrostatic fields for different molecules against their biological activity allows the definition of steric and electrostatic interactions which are favourable and unfavourable for activity. These can be displayed visually as contour lines.

- It is necessary to define the active conformation and pharmacophore for each molecule in a CoMFA study. Alignment of the molecules is crucial.

- Unlike conventional QSAR studies, molecules of different structural classes can be compared if they share the same pharmacophore.

- 3D QSAR does not depend on experimentally measured parameters.

- A variety of different 3D QSAR programs have been developed.

18.10.6 Case study: inhibitors of tubulin polymerization

Colchicine (Fig. 18.24) is a lead compound for agents which act as inhibitors of tubulin polymerization (section 2.7.1) and might be useful in the treatment of arthritis. Other lead compounds have been discovered which bind to tubulin at the same binding site, and so a study

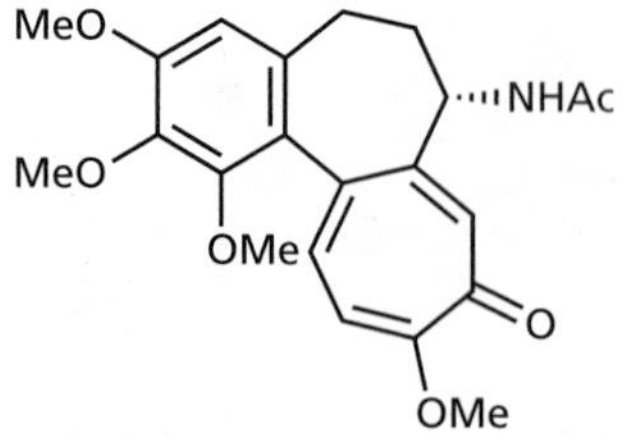

FIGURE 18.24 Colchicine.

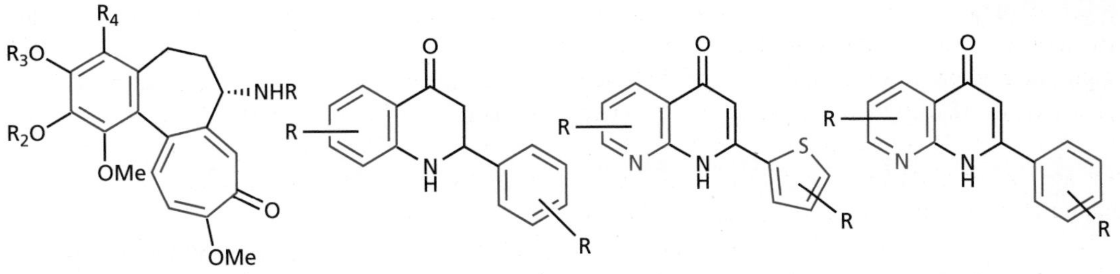

FIGURE 18.25 Structural classes used in the 3D QSAR study.

was carried out to compare the various structural classes interacting in this way. In this 3D QSAR study, 104 such agents were tested, belonging to four distinct families of compounds (Fig. 18.25); 51 compounds were used as a 'training set' for the analysis itself and 53 were used as a 'testing set' to test the predictive value of the results. Both sets contained a mixture of structural classes having both low and high activity.

The first task was to work out how to align these different classes of molecule. Colchicine is the most rigid of the four and also has a high affinity for tubulin. Therefore, it was chosen as the template on to which the other structures would be aligned. The relevant pharmacophore in colchicine was identified as the two aromatic rings. Molecular modelling was now carried out on each of the remaining structures to generate various conformations. Each conformation was compared with colchicine to find the one that would allow the pharmacophores in each structure to be aligned. This was then identified as the active conformation.

Once the active conformations for each structure had been identified, they were fitted into the lattice of grids previously described such that each structure was properly aligned. The steric and electrostatic fields round each molecule were calculated using a probe atom, then the 3D QSAR analysis was carried out to relate the fields to the measured biological activity.

The results of the 3D QSAR analysis are summarized as contour lines round a representative molecule (Fig. 18.26). For the steric interactions, solid contours represent fields that are favourable for activity, and the dashed lines show fields that are unfavourable. For the electrostatic interactions, solid lines are regions where positively charged species improve affinity, and dashed lines indicate regions where negatively charged groups are favourable.

The results revealed that introducing steric bulk round the aromatic ring is more crucial to activity than introducing steric bulk round the bicyclic system. Based on this evidence, the structure shown in Fig. 18.27 was synthesized. The predicted value of pIC_{50} for the compound was 5.62. The actual value was in close agreement at 6.04 ($pIC_{50} = -\log [IC_{50}]$ where IC_{50} is the concentration of inhibitor required to produce 50% enzyme inhibition).

The steric fields of the Tripos CoMFA analysis (Fig. 18.26) were subsequently placed into a model of the binding site. It was found that the bad steric regions were in the same regions as the peptide backbone, whereas the favourable steric areas were in empty spaces.

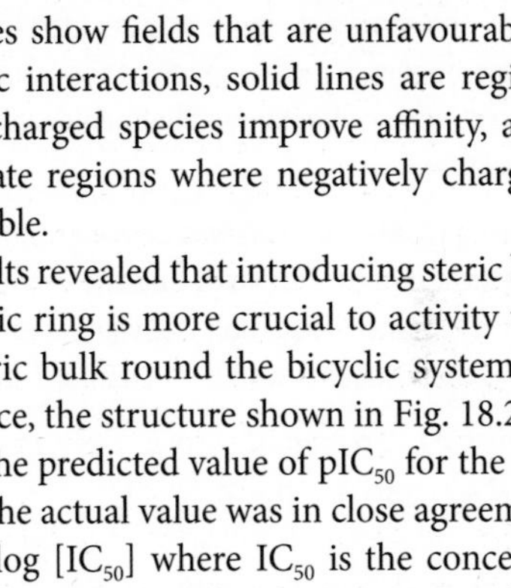

FIGURE 18.27 Novel agent designed on the basis of the 3D QSAR study.

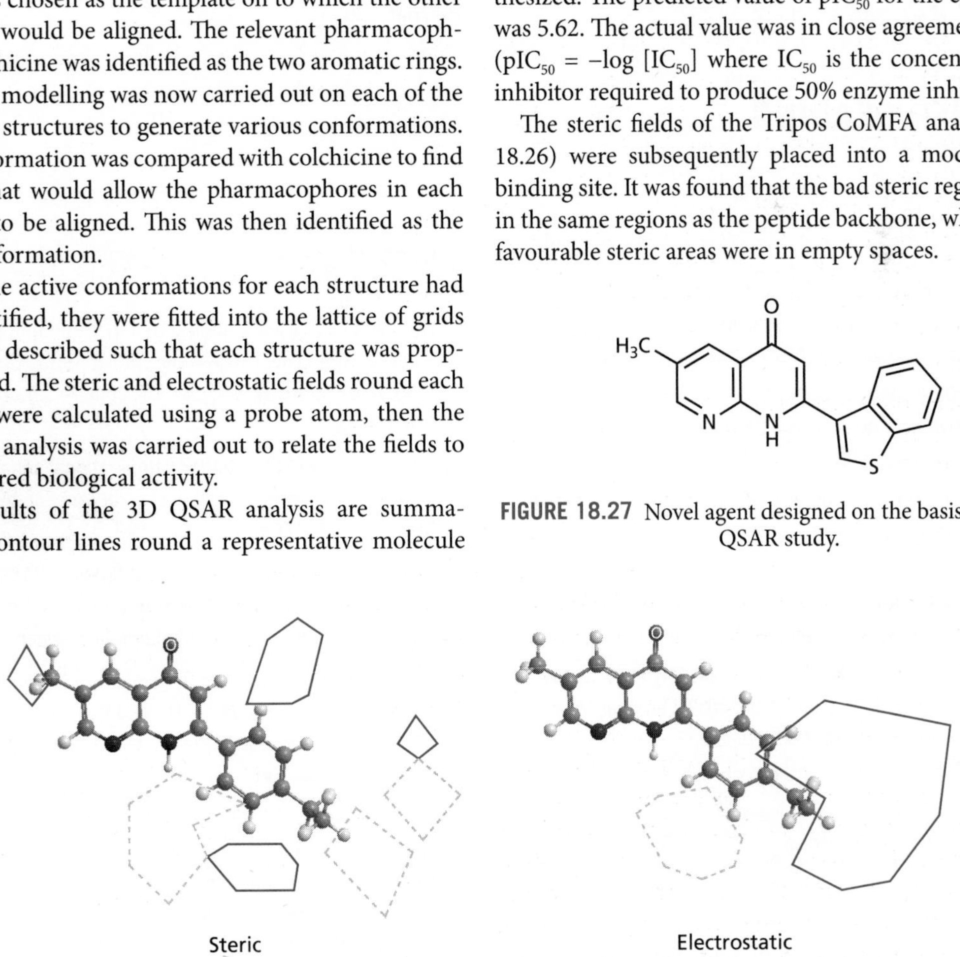

Steric

Electrostatic

FIGURE 18.26 Results of the 3D QSAR analysis ($q^2 = 0.637$).

QUESTIONS

1. Using values from Table 18.1, calculate the log P value for structure (I) (log P for benzene = 2.13).

2. Several analogues of a drug are to be prepared for a QSAR study which will consider the effect of various aromatic substituents on biological activity. You are asked whether the substituents (SO_2NH_2, CF_3, CN, CH_3SO_2, SF_3, $CONH_2$, OCF_3, CO_2H, Br, I) are relevant to the study. What are your thoughts?

3. A lead compound has a monosubstituted aromatic ring present as part of its structure. An analogue was synthesized containing a *para*-chloro substituent which had approximately the same activity. It was decided to synthesize an analogue bearing a methyl group at the *para* position. This showed increased activity. What analogue would you prepare next, and why?

4. The following QSAR equation was derived for the pesticide activity of structure (II). Explain what the various terms mean and whether the equation is a valid one. Identify what kind of substituents would be best for activity.

$$\log 1/C = 1.08\pi_x + 2.41F_x + 1.40\,R_x - 0.072\,MR_x + 5.25$$
$$(n = 16,\ r^2 = 0.840,\ s = 0.59)$$

5. A QSAR equation for the anticonvulsant (III) was derived as follows:

$$\log 1/C = 0.92\,\pi_x - 0.34\,\pi_x^2 + 3.18 \quad (n = 15,\ r^2 = 0.902,\ s = 0.09,\ \pi_0 = 1.35)$$

What conclusions can you draw from this equation? Would you expect activity to be greater if X = CF_3 rather than H or CH_3?

6. The following QSAR equation is related to the mutagenic activity of a series of nitrosoamines; $\log 1/C = 0.92\,\pi + 2.08\,\sigma - 3.26$ ($n = 12$, $r^2 = 0.794$, $s = 0.314$). What sort of substituent is likely to result in high mutagenic activity?

Multiple-choice questions are available on the Online Resource Centre at www.oxfordtextbooks.co.uk/orc/patrick6e/

FURTHER READING

Craig, P. N. (1971) Interdependence between physical parameters and selection of substituent groups for correlation studies. *Journal of Medicinal Chemistry*, **14**, 680–4.

Cramer, R. D., et al. (1979) Application of quantitative structure–activity relationships in the development of the antiallergenic pyranenamines. *Journal of Medicinal Chemistry*, **22**, 714–25.

Cramer, R. D., Patterson, D. E., and Bunce, J. D. (1988) Comparative field analysis (CoMFA). *Journal of the American Chemical Society*, **110**, 5959–67.

Cramer, R. D. (2003) Topomer CoMFA: A design methodology for rapid lead optimization. *Journal of Medicinal Chemistry*, **46**, 374–88.

Hansch, C., and Leo, A. (1995) *Exploring QSAR*. American Chemical Society, Washington, DC.

Kellogg, G. E., and Abraham, D. J. (2000) Hydrophobicity: is Log $P_{o/w}$ more than the sum of its parts? *European Journal of Medicinal Chemistry*, **35**, 651–61.

Kotani, T., and Higashiura, K. (2004) Comparative molecular active site analysis (CoMASA). 1. An approach to rapid evaluation of 3D QSAR. *Journal of Medicinal Chemistry*, **47**, 2732–42.

Kubini, H., Folkers, G., and Martin, Y. C. (eds) (1998) *3D QSAR in drug design*. Kluwer/Escom, Dordrecht.

Martin, Y. C., and Dunn, W. J. (1973) Examination of the utility of the Topliss schemes by analog synthesis. *Journal of Medicinal Chemistry*, **16**, 578–9.

Sutherland, J. J., O'Brien, L. A., and Weaver, D. F. (2004) A comparison of methods for modeling quantitative structure–activity relationships. *Journal of Medicinal Chemistry*, **47**, 5541–54.

Taft, R. W. (1956) Separation of polar, steric and resonance effects in reactivity. In Newman, M. S. (ed.), *Steric effects in organic chemistry*. Chapter 13, John Wiley and Sons, New York.

Topliss, J. G. (1972) Utilization of operational schemes for analog synthesis in drug design. *Journal of Medicinal Chemistry*, **15**, 1006–11.

Verloop, A., Hoogenstraaten, W., and Tipker, J. (1976) Development and application of new steric substituent parameters in drug design. *Medicinal Chemistry*, **11**, 165–207.

van de Waterbeemd, H., Testa, B., and Folkers, G. (eds) (1997) *Computer-assisted lead finding and optimization*. Wiley–VCH, New York.

Zhang, S.-X., et al. (2000) Antitumor agents. 199. Three-dimensional quantitative structure–activity relationship study of the colchicine binding site ligands using comparative molecular field analysis. *Journal of Medicinal Chemistry*, **43**, 167–76.

Titles for general further reading are listed on p.845.

■ CASE STUDY 5
Design of a thymidylate synthase inhibitor

In this case study, we shall look at an early example of how the strategies of *de novo* drug design (section 17.15) and structure-based drug design (section 13.3.11) were used together to develop an active compound that went forward for clinical trials. The research in question involved the design of inhibitors for the enzyme **thymidylate synthase** (section 21.3.2). This enzyme catalyses the methylation of **deoxyuridylate monophosphate (dUMP)** to **deoxythymidylate monophosphate (dTMP)** using **5,10-methylenetetrahydrofolate** as a coenzyme (Fig. CS5.1). Inhibitors of this enzyme have been shown to be antitumour agents which prevent the biosynthesis of one of the required building blocks for DNA. Traditional inhibitors have been modelled on dUMP or the enzyme cofactor 5,10-methylenetetrahydrofolate (Fig. CS5.2), which means that these inhibitors are structurally related to the natural substrate and cofactor. Unfortunately, this increases the possibility of side effects resulting from inhibition of other enzymes and receptors which use these molecules as natural ligands. Therefore, it was decided that *de novo* drug design would be used to design a novel structure which was unrelated to either of the natural substrates.

Before starting the *de novo* design, a good supply of the enzyme was required. Although human thymidylate synthase was not readily available in large quantities, it was possible to obtain good quantities of the bacterial version from *Escherichia coli* by using recombinant DNA technology to clone the gene and then expressing it in fast-growing cells (section 6.4). The bacterial enzyme is not identical to the human version, but it is very similar and so it was considered a reasonable analogue. The enzyme was crystallized along with the known inhibitors

5-fluorodeoxyuridylate and **CB 3717** (Fig. CS5.3). These structures mimic the substrate and the coenzyme, respectively, and bind to the sites normally occupied by these structures. The structure of the enzyme–inhibitor complex was then determined by X-ray crystallography and downloaded on to the computer.

A study of the enzyme–inhibitor complex revealed where the inhibitors were bound and also the binding interactions involved. For CB 3717, the binding interactions around the pteridine portion of the inhibitor were identified as involving hydrogen bonding interactions to two amino acids (the carboxylate ion of Asp-169 and the main chain peptide link next to Ala-263). There was also a hydrogen bonding interaction to a water molecule which acted as a hydrogen bonding bridge to Arg-21 (Fig. CS5.4). Using molecular modelling, the inhibitor was deleted from the binding site to allow further analysis of the empty binding site. Generating the empty binding site from the enzyme–ligand complex is better than studying the empty binding site from the pure enzyme because the latter does not take into account the induced fit that occurs on ligand binding.

A grid was set up within the binding site and an aromatic CH probe was placed at each grid point to measure

FIGURE CS5.2 5,10-Methylenetetrahydrofolate.

FIGURE CS5.1 Reaction catalysed by thymidylate synthase (Ⓟ = Phosphate).

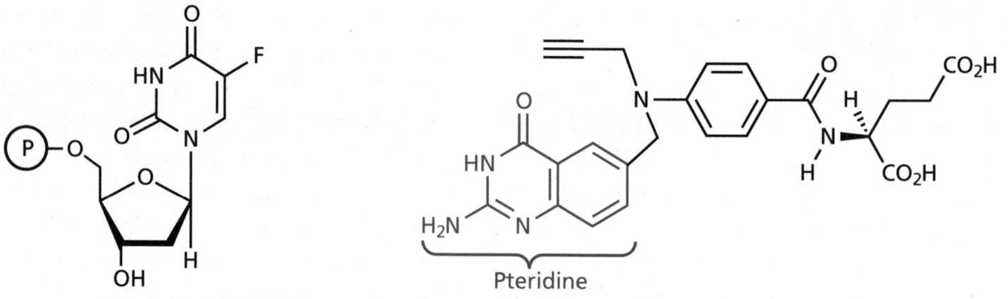

FIGURE CS5.3 Inhibitors of thymidylate synthase.

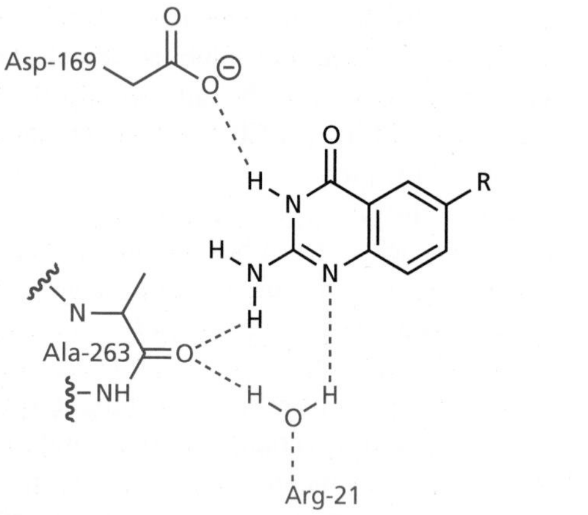

5-Fluorodeoxyuridylate

CB3717

FIGURE CS5.4 Binding interactions holding a pteridine moiety in the active site.

hydrophobic interactions and thus identify hydrophobic regions (see section 17.7.5). From this analysis, it was discovered that the pteridine portion of CB 3717 was positioned in a hydrophobic pocket despite the presence of the hydrogen bonding interactions which held it there. The boundaries of this hydrophobic region were determined and a naphthalene ring was found to be a suitable hydrophobic molecule to fit the pocket, yet still leave

room for the addition of a functional group which would be capable of forming the important hydrogen bonds.

The functional group chosen was a cyclic amide which was fused to the naphthalene scaffold to create a naphthostyryl scaffold (Fig. CS5.5). Modelling suggested that the NH portion of the amide would bind to Asp-169 while the carbonyl group would bind to the water molecule identified above. A substituent was now added to the naphthostyryl scaffold in order to gain access to the space normally occupied by the benzene ring of the cofactor. A dialkylated amine was chosen as the linking unit and was placed at position 5 of the structure. There were several reasons for this. Firstly, adding an amine at this position was easy to carry out synthetically. Secondly, the two substituents on the amine could be easily varied, which would allow fine-tuning of the compound. Lastly, by using an amine, it would be possible to have a branching point which could have different substituents without adding an asymmetric centre. If a carbon atom had been added instead, two different substituents would have led to an asymmetric centre with all the attendant complications that would entail (section 13.3.8).

Modelling demonstrated that a benzyl group was a suitable substituent for the amine in order to access the space normally occupied by the benzene ring of the

Naphthalene

Naphthostyryl scaffold

Substitution

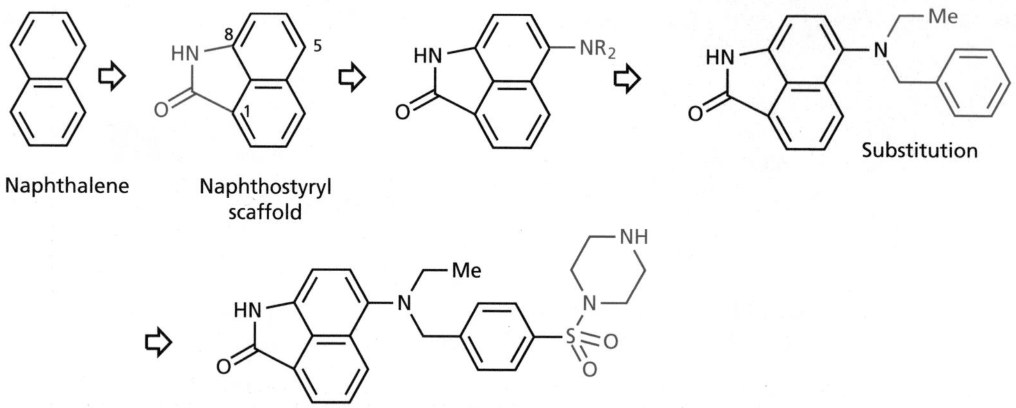

FIGURE CS5.5 Design of an enzyme inhibitor by *de novo* methods.

cofactor. A phenylsulphonyl piperazine group was then added to the *para* position of the aromatic ring in order to make the molecule more water soluble—a necessary property if the synthesized structures were to be bound to the enzyme and crystallized for further X-ray crystallographic studies. The positioning of this group was important because the modelling studies showed that it would protrude from the binding site and still make contact with surrounding water. This meant that the group would not have to be desolvated for the drug to bind—a process which would involve an energy penalty (section 1.3.6).

This structure was now synthesized and was found to inhibit both the bacterial and human versions of the enzyme, with higher activity for the human enzyme. This represents the successful *de novo* design of a novel lead structure.

It was now time to move to structure-based drug design such that the binding interactions and activity of the lead compound could be optimized.

A crystal structure of the novel inhibitor bound to the bacterial enzyme was successfully obtained and studied to see whether the inhibitor had fitted the binding site as expected. In fact, it was found that the naphthalene ring of the inhibitor was wedged deeper into the pocket than expected because of more favourable hydrophobic interactions. As a result, the cyclic amide failed to form the direct hydrogen bond interaction to Asp-169 which had been planned and was hydrogen bonding to a bridging water molecule instead. The lactam carbonyl oxygen was also too close to Ala-263, which had caused this residue to shift 1 Å from its usual position. This, in turn, had displaced the water molecule, which had been the intended target for hydrogen bonding (Fig. CS5.6).

By studying the actual position of the structure in the binding site, it was possible to identify four areas where

extra substituents could fill up empty space and perhaps improve binding. These are shown below in Fig. CS5.7.

Various structures were proposed then overlaid on the lead compound (still docked within the binding site) to see whether they fitted the binding site. Only those which passed this test and were in stable conformations were synthesized and tested for activity (41 in total). The optimum substituent at each position was then identified.

- In region 1 (R^1), modelling showed that this substituent fitted into a hydrophobic pocket that became hydrophilic the deeper one got. This suggested that a hydrogen bonding substituent at the end of an alkyl chain might be worth trying (the *extension strategy*; section 13.3.2) and, indeed, a CH_2CH_2OH group led to an improvement in binding affinity. It was also found that a methyl group was better than the original ethyl group.

- In region 2 (R^2), the carbonyl oxygen was replaced by an amidine group which would be capable of hydrogen bonding to the carbonyl oxygen of Ala-263, rather than repelling it. An added advantage in using a basic amidine group was the fact that there was a good chance that it would become protonated, allowing a stronger ionic interaction with Asp-169, as well as a better hydrogen bonding interaction with Ala-263. When this structure was synthesized, it was found to

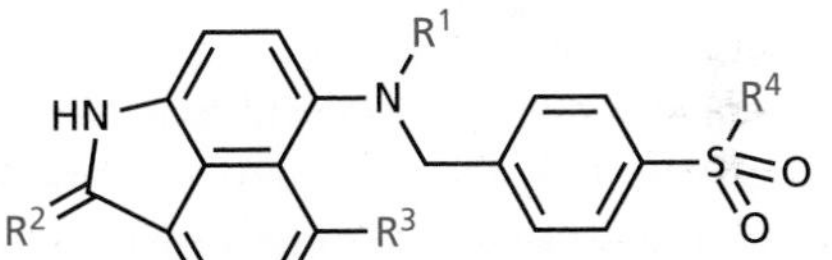

FIGURE CS5.7 Variable positions (in colour).

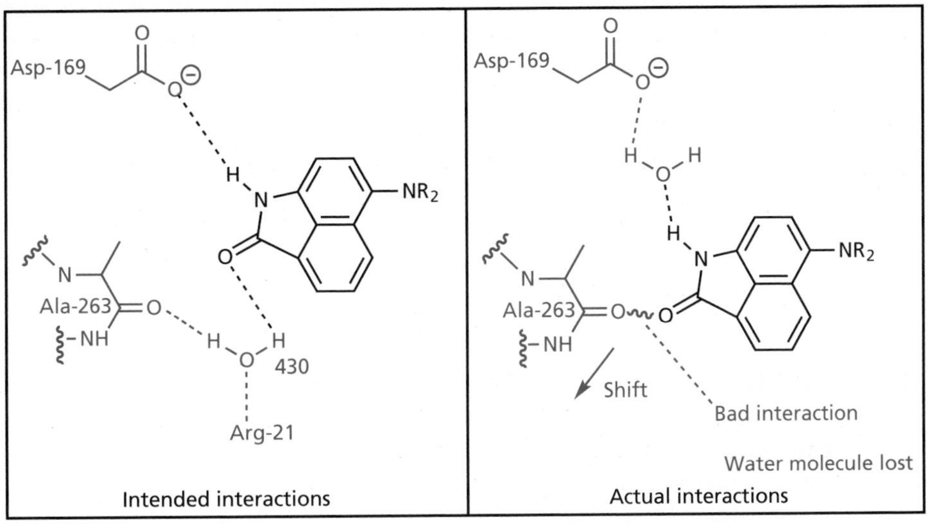

FIGURE CS5.6 Intended versus actual interactions of the inhibitor with the active site.

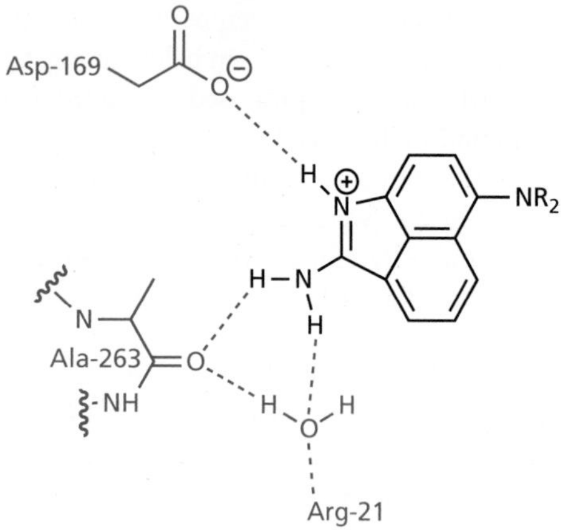

FIGURE CS5.8 Binding interactions of the modified inhibitor with the active site.

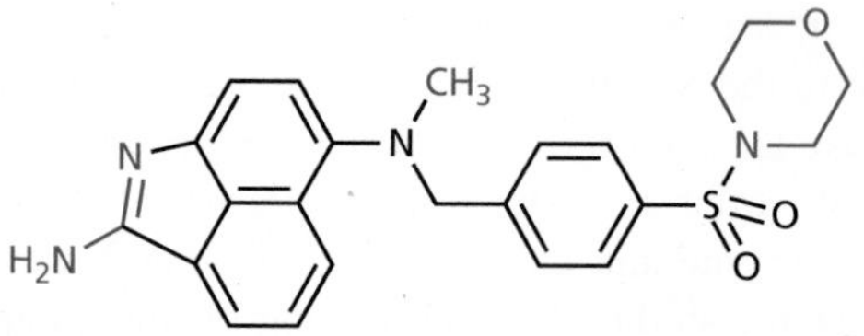

FIGURE CS5.9 Modified inhibitor put forward for clinical trials.

have improved inhibition, and a crystal structure of the enzyme–inhibitor complex showed that the expected interactions were taking place (Fig. CS5.8). Moreover, Ala-263 had returned to its original position, permitting the return of the bridging water molecule.

- In region 3 (R^3), there was room for a small group, such as a chlorine atom or a methyl group; both of these substituents led to an increase in activity.
- Region 4 (R^4) was relatively unimportant for inhibitory activity, since groups at this position protrude out of the active site into the surrounding solvent and have only minimal contact with the enzyme. Nevertheless, the piperazine ring was replaced by a morpholine group because the latter had some advantages with respect to selectivity and pharmacological properties.

Having identified the optimum groups at each position, structures were synthesized combining some or all of these groups. The presence of the amidine resulted in the best improvement in activity and so the presence of this group was mandatory. Interestingly, adding all the optimum groups is not as beneficial as adding some of them.

The modified structure (Fig. CS5.9) was synthesized and was found to be a potent inhibitor, which was 500 times more active than the original amide. A crystal structure of the enzyme–inhibitor complex showed a much better fit, and the compound was put forward for clinical trials as an antitumour agent.

The case study illustrates many of the general principles behind *de novo* design that were described in section 17.15.1. For example, the designed lead compound was fairly flexible and did not fill up all the space available in the binding site. As we have seen, a different binding mode took place from that predicted, and this might not have occurred if a more rigid and more closely fitting structure had been designed. The fact that binding did take place allowed structure-based drug design to be carried out on the information obtained.

The importance of synthetic feasibility was considered throughout the process and was one of the reasons for introducing an amine substituent to the naphthostyryl ring system.

Finally, it is important to appreciate that binding studies using target proteins from different species may produce slightly different results. For example, the computer modelling studies described above were carried out on bacterial thymidylate synthase enzyme rather than the human version. Fortunately, the activities of the designed inhibitors were actually greater for the human enzyme than for the bacterial enzyme, and this was put down to the fact that the hydrophobic space available for the naphthalene ring was larger in the human enzyme than in the bacterial one. Fortunately, most changes carried out had beneficial effects for both enzymes, but with one exception; adding a methyl group at R^3 of the amidine led to an increase in activity for the bacterial enzyme, but not for the human enzyme.

FURTHER READING

Greer, J., Erickson, J. W., Baldwin, J. J., and Varney, M. D. (1994) Application of the three-dimensional structures of protein target molecules in structure-based drug design. *Journal of Medicinal Chemistry*, **37**, 1035–54.

PART E

Selected topics in medicinal chemistry

In Part E, we concentrate on specific topics within medicinal chemistry. The topics which have been included are of the author's choosing, and demonstrate a personal preference or interest. There are many other fascinating topics which could have been included, but sadly there is a limit to what one can include in a textbook of this size.

The different chapters illustrate different methods of classifying drugs, and demonstrate the advantages and disadvantages of such classifications. For example, there are five chapters where drugs are classified by their pharmacological effect, mainly those with antibacterial, antiviral, anticancer, anti-ulcer, and cardiovascular activities. The advantage of this classification is that it gives an overall view of the many different types of drug that can be used to treat these diseases. The disadvantage is the volume of information that has to be imparted. Since these diseases have many causes or mechanisms, there are many different possible targets for drugs. This means that the drugs used in each of these fields are extremely varied in their structure and mechanism of action.

In contrast, the chapters on cholinergic and adrenergic agents concentrate on drugs which interact with specific biological systems, mainly the cholinergic and adrenergic nervous systems. Since the target systems are more focused, there are fewer targets to consider. This has the advantage that it is easier to rationalize the drug structures involved and to study their mechanisms of action. It is also possible to understand why drugs having an effect on these systems can be used in particular fields of medicine, such as in the treatment of asthma or in cardiovascular medicine. The disadvantage in concentrating on a particular biological system is that it ignores drugs that could be anti-asthmatic or cardiovascular by acting on a different biological system.

A study of opioid analgesics is included, where the drugs have been classed by their chemical structure. This is a useful classification for medicinal chemists as the drugs involved have the same pharmacological activity and targets, making their study well focused. The disadvantage with this classification is that analgesics with different structures and mechanisms of action are not included. Similarly, Case study 6 has been included on steroidal anti-inflammatory agents. These all have a common tetracyclic skeleton and so it is possible to compare the different structures to rationalize different activities. However, there are non-steroidal anti-inflammatories which are not covered by this classification.

Topics have also been chosen to include several traditional fields of medicinal chemistry, as well as those which are more recent. For example, the opioid analgesics were discovered more than 100 years ago, while the majority of antibacterial agents were discovered in the mid-twentieth century. In contrast, progress on antiviral agents and the kinase inhibitors used in anticancer therapy has been relatively recent. A comparison of these chapters illustrates the changing face of medicinal chemistry from one of trial and error to a more scientific approach, where diseases are understood at the molecular level and drugs are then designed accordingly.

Case study 7 looks at an example of a research project undertaken to find a novel antidepressant agent, but also includes an overview of clinically useful antidepressants and their mechanism of action. As such, it acts as a link to relevant material within other chapters in the book. Case studies 8 and 9 are new case studies that complement the chapter on cardiovascular drugs. Case study 8 describes the design process leading to the antihypertensive agent aliskiren, while Case study 9 focuses on the design of anticoagulants that inhibit factor X—a key enzyme involved in thrombosis and clot formation. Case study 10 is another new case study that reflects recent work carried out on the design of antiviral agents that target hepatitis C.

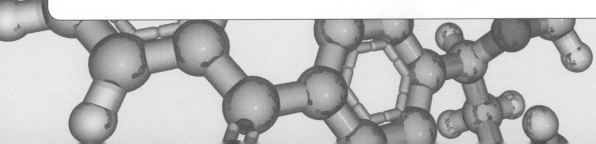

19 Antibacterial agents

The fight against bacterial infections over the last 80 years has been one of the great success stories of medicinal chemistry, yet it remains to be seen whether it will last. Bacteria such as *Staphylococcus aureus* have the worrying ability to gain resistance to known drugs, and so the search for new drugs is never ending. Although deaths from bacterial infection have dropped in the developed world, bacterial infection is still a major cause of death in the developing world. For example, the World Health Organization estimated that tuberculosis was responsible for about 1.5 million deaths in 2014, and that one in three of the world's population was infected. The same organization estimated that, in the year 2000, 1.9 million children died worldwide of respiratory infections, with 70% of these deaths occurring in Africa and Asia. They also estimated that, each year, 1.4 million children died from gut infections and the diarrhoea resulting from these infections. In the developed world, deaths from food poisoning due to virulent strains of *Escherichia coli* have attracted widespread publicity, while tuberculosis has returned as a result of the AIDS epidemic.

The topic of antibacterial agents is a large one and terms are used in this chapter which are unique to this particular field. Rather than clutter the text with explanations and definitions, Appendix 5 contains explanations of such terms as aerobic and anaerobic organisms, antibacterial and antibiotic substances, cocci, bacilli, streptococci, and staphylococci. Appendix 5 also explains briefly the difference between bacteria, algae, protozoa, and fungi. The emphasis in this chapter is on agents that act against bacteria, but some of those described also act against protozoal infections.

19.1 History of antibacterial agents

There is evidence of antibacterial herbs or potions being used for many centuries. For example, the Chinese used mouldy soybean curd to treat carbuncles, boils, and other infections. Greek physicians used wine, myrrh, and inorganic salts. In the middle ages, certain types of honey were used to prevent infections following arrow wounds. Of course in those days, there was no way of knowing that bacteria were the cause of these infections.

Bacteria are single-cell microorganisms which were first identified in the 1670s by van Leeuwenhoek, following his invention of the microscope. However, it was not until the nineteenth century that their link with disease was appreciated. This followed the elegant experiments carried out by the French scientist Pasteur, who demonstrated that specific bacterial strains were crucial to fermentation and that these, and other, microorganisms were more widespread than was previously thought. The possibility that microorganisms might be responsible for disease began to take hold.

An early advocate of a 'germ theory of disease' was the Edinburgh surgeon Lister. Despite the protests of several colleagues who took offence at the suggestion that they might be infecting their own patients, Lister introduced **carbolic acid** as an antiseptic and sterilizing agent for operating theatres and wards. The improvement in surgical survival rates was significant.

During the latter half of the nineteenth century, scientists such as Koch were able to identify the microorganisms responsible for diseases such as tuberculosis, cholera, and typhoid. Methods of vaccination were studied and research was carried out to try and find effective antibacterial agents or antibiotics. The scientist who can lay claim to be the father of chemotherapy—the use of chemicals against infection—was Paul Ehrlich. Ehrlich spent much of his career studying histology, then immunochemistry, and won a Nobel Prize for his contributions to immunology. In 1904, however, he switched direction and entered a field which he defined as chemotherapy. Ehrlich's '**principle of chemotherapy**' was that a chemical could directly interfere with the proliferation of microorganisms at concentrations tolerated by the host. This concept was popularly known as the '**magic bullet**', where the chemical was seen as a bullet which could seek out and destroy the invading microorganism without adversely affecting the host. The process is one of **selective toxicity**, where the chemical shows greater toxicity to microbial cells than to host cells. Such selectivity can be represented by a 'chemotherapeutic index', which compares the minimum effective dose of a drug versus the maximum dose that can be tolerated by the host. This measure of selectivity was eventually replaced by the currently used **therapeutic index**.

By 1910, Ehrlich had successfully developed the first example of a purely synthetic antimicrobial drug. This was the arsenic-containing compound **salvarsan** (Fig. 19.1). Although it was not effective against a wide range of bacterial infections, it did prove effective against the protozoal disease of sleeping sickness (trypanosomiasis) and the spirochaete disease of syphilis. The drug was used until 1945 when it was replaced by penicillin (see also Box 19.20).

Over the next 20 years, progress was made against a variety of protozoal diseases, but little progress was made in finding antibacterial agents until the introduction in 1934 of **proflavine** (Fig. 19.1)—a drug which was used during the Second World War against bacterial infections in deep surface wounds. Unfortunately, it was too toxic to be used against systemic bacterial infections (i.e. those carried in the bloodstream) and there was still an urgent need for agents which would fight those infections.

This need was answered in 1935 when it was discovered that a red dye called **prontosil** was effective against streptococcal infections *in vivo*. As is discussed in section 19.4.1.1, prontosil was eventually recognized as a prodrug for a new class of antibacterial agents—the **sulpha drugs** or **sulphonamides**. The discovery of these drugs was a real breakthrough, as they represented the first drugs to be effective against systemic bacterial infections. In fact, they were the only effective drugs until penicillin became available in the early 1940s.

Although penicillin was discovered in 1928, it was not until 1940 that effective means of isolating it were developed by Florey and Chain. Society was then rewarded with a drug which revolutionized the fight against bacterial infection and proved even more effective than the sulphonamides. Despite penicillin's success, it was not effective against all types of infection and the need for new antibacterial agents still remained. Penicillin is an example of a toxic fungal metabolite that kills bacteria and allows the fungus to compete for nutrients. The realization that fungi might be a source for novel antibiotics spurred scientists on to investigate microbial cultures from all round the globe.

In 1944, the antibiotic **streptomycin** was discovered from a systematic search of soil organisms. It extended the range of chemotherapy to the tubercle bacillus and a variety of Gram-negative bacteria. This compound was the first example of a series of antibiotics known as the **aminoglycoside** antibiotics. After World War II, the search continued leading to the discovery of **chloramphenicol** (1947), the peptide antibiotics (e.g. **bacitracin**; 1945), the tetracycline antibiotics (e.g. **chlortetracycline**; 1948), the macrolide antibiotics (e.g. **erythromycin**; 1952), the cyclic peptide antibiotics (e.g. **valinomycin**), and the first example of a second major group of β-lactam antibiotics, **cephalosporin C** (1955).

As far as synthetic agents were concerned, **isoniazid** was found to be effective against human tuberculosis in 1952, and in 1962 **nalidixic acid** (the first of the quinolone antibacterial agents) was discovered. A second-generation of this class of drugs was introduced in 1987 with **ciprofloxacin**.

Many antibacterial agents are now available and the vast majority of bacterial diseases have been brought under control (e.g. syphilis, tuberculosis, typhoid, bubonic plague, leprosy, diphtheria, gas gangrene, tetanus, and gonorrhoea). This represents a great achievement for medicinal chemistry and it is perhaps sobering to consider the hazards faced by society in the days before penicillin. Septicaemia was a risk faced by mothers during childbirth and could lead to death. Ear infections were common, especially in children, and could lead to deafness. Pneumonia was a frequent cause of death in hospital wards. Tuberculosis was a major problem, requiring special isolation hospitals built away from populated centres. A simple cut or a wound could lead to severe infection requiring the amputation of a limb, while the threat of peritonitis lowered the success rates of surgical operations. This was in the 1930s—still within living memory for many. Perhaps those of us born since the Second World War take the success of antibacterial agents too much for granted.

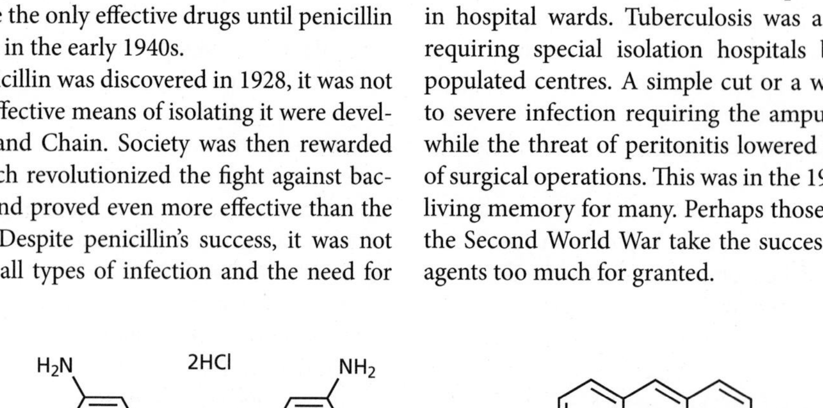

FIGURE 19.1 Salvarsan and proflavine. (The structure of salvarsan shown here is a simplification; it is, in fact, a cyclic trimer with no As=As bonds.)

19.2 **The bacterial cell**

The success of antibacterial agents owes much to the fact that they can act selectively against bacterial cells rather than animal cells. This is largely because bacterial and animal cells differ both in their structure and in their biosynthetic pathways. Let us consider some of the differences between the bacterial cell (defined as **prokaryotic**) (Fig. 19.2) and the animal cell (defined as **eukaryotic**). Differences between bacterial and animal cells:

* eukaryotic cells contain a defined nucleus, whereas prokaryotic cells do not;
* eukaryotic cells contain a variety of structures called organelles (mitochondria, endoplasmic reticulum, etc.), whereas prokaryotic cells are relatively simple;
* the biochemistry of a prokaryotic cell differs significantly from that of a eukaryotic cell. For example, prokaryotic cells synthesize essential vitamins from simple starting materials, whereas eukaryotic cells acquire those vitamins from external sources. Therefore, eukaryotic cells lack the enzymes that are used by prokaryotic cells for vitamin biosynthesis;
* prokaryotic cells have a cell membrane and a cell wall, whereas eukaryotic cells have only a cell membrane.

The cell wall is crucial to a bacterial cell's survival. Bacteria have to survive a wide range of environments and osmotic pressures, whereas plant and animal cells do not. If a bacterial cell lacking a cell wall was placed in an aqueous environment containing a low concentration of salts, water would freely enter the cell due to osmotic pressure. This would cause the cell to swell and eventually burst. The scientific term for this is **lysis**. The cell wall does not stop water flowing into the cell directly, but it does prevent the cell from swelling and so indirectly prevents water entering the cell. Bacteria can be characterized by a staining technique which allows them to be defined as **Gram-positive** or **Gram-negative** (Appendix 5). Bacteria with a thick cell wall (20–40 nm) are stained purple and defined as Gram-positive. Bacteria with a thin cell wall (2–7 nm) are stained pink and are defined as Gram-negative. Although Gram-negative bacteria have a thin cell wall, they have an additional outer membrane not present in Gram-positive bacteria. This outer membrane is made up of lipopolysaccharides—similar in character to the cell membrane. These differences in cell walls and membranes have important consequences for the different vulnerabilities of Gram-positive and Gram-negative bacteria to antibacterial drugs.

19.3 **Mechanisms of antibacterial action**

There are five main mechanisms by which antibacterial agents act (Fig. 19.2).

* *Inhibition of cell metabolism*: antibacterial agents which inhibit cell metabolism are called **antimetabolites**. These compounds inhibit the metabolism of a microorganism, but not the metabolism of the host. They can do this by inhibiting an enzyme-catalysed reaction which is present in the bacterial cell, but not in mammalian cells. The best-known examples of antibacterial agents acting in this way are the sulphonamides. It is also possible for antibacterial agents to show selectivity against enzymes which are present in both the bacterial and mammalian cell, as long as there are significant differences in structure between the two.

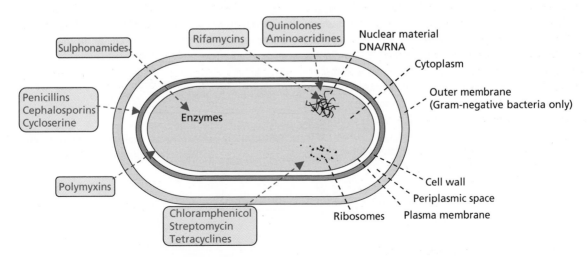

FIGURE 19.2 The bacterial cell and drug targets.

- *Inhibition of bacterial cell wall synthesis* leads to bacterial cell lysis and death. Agents operating in this way include penicillins, cephalosporins, and glycopeptides such as vancomycin. As mammalian cells do not have a cell wall, they are unaffected by such agents.

- *Interactions with the plasma membrane*: some antibacterial agents interact with the plasma membrane of bacterial cells to affect membrane permeability. This has fatal results for the cell. Polymyxins and tyrothricin operate in this way.

- *Disruption of protein synthesis* means that essential proteins and enzymes required for the cell's survival can no longer be made. Agents which disrupt protein synthesis include the rifamycins, aminoglycosides, tetracyclines, and chloramphenicol.

- Inhibition of nucleic acid transcription and replication prevents cell division and/or the synthesis of essential proteins. Agents acting in this way include nalidixic acid and proflavine.

We will now consider these mechanisms in more detail.

19.4 Antibacterial agents which act against cell metabolism (antimetabolites)

19.4.1 Sulphonamides
19.4.1.1 The history of sulphonamides

The best examples of antibacterial agents acting as antimetabolites are the sulphonamides (sometimes called the sulpha drugs). The sulphonamide story began in 1935 when it was discovered that a red dye called **prontosil** (Fig. 19.3) had antibacterial properties *in vivo* (i.e. when given to laboratory animals). Strangely enough, no antibacterial effect was observed *in vitro*. In other words, prontosil could not kill bacteria grown in the test tube. This remained a mystery until it was discovered that prontosil was metabolized by bacteria present in the small intestine of the test animal to give a product called **sulphanilamide** (Fig. 19.3). It was this compound which was the true antibacterial agent. Thus, prontosil was an early example of a **prodrug** (section 14.6).

Sulphanilamide was synthesized in the laboratory and became the first synthetic antibacterial agent found to be active against a wide range of infections. Further developments led to a range of sulphonamides which proved effective against Gram-positive organisms, especially pneumococci and meningococci.

Despite their undoubted benefits, sulpha drugs have proved ineffective against infections such as *Salmonella*—the organism responsible for typhoid. Other problems have resulted from the way these drugs are metabolized, as toxic products are frequently obtained. This led to the sulphonamides being superseded by penicillin.

19.4.1.2 Structure–activity relationships

The synthesis of a large number of sulphonamide analogues (Fig. 19.4) led to the following conclusions:

- the *para*-amino group is essential for activity and must be unsubstituted (i.e. R^1 = H). The only exception is when R^1 = acyl (i.e. amides). The amides themselves are inactive but can be metabolized in the body to regenerate the active compound (Fig. 19.5). Thus amides can be used as sulphonamide prodrugs;

- the aromatic ring and the sulphonamide functional group are both required;

- both the sulphonamide and amino group must be directly attached to the aromatic ring;

- the aromatic ring must be *para*-substituted only. Extra substitution eliminates activity for steric reasons;

- the sulphonamide nitrogen must be primary or secondary;

- R^2 is the only possible site that can be varied in sulphonamides.

19.4.1.3 Sulphanilamide analogues

In sulphanilamide analogues (Fig. 19.4), R^2 is often varied by adding different heterocyclic or aromatic rings to

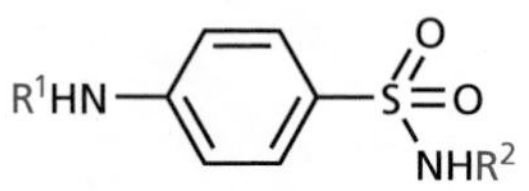

FIGURE 19.4 Sulphonamide analogues used in structure–activity relationship studies.

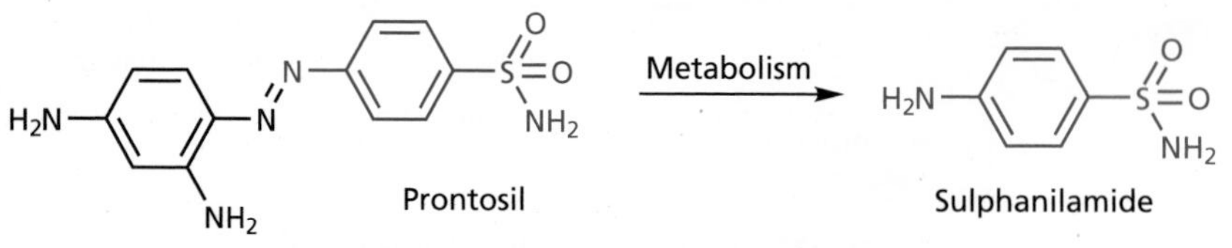

FIGURE 19.3 Metabolism of prontosil.

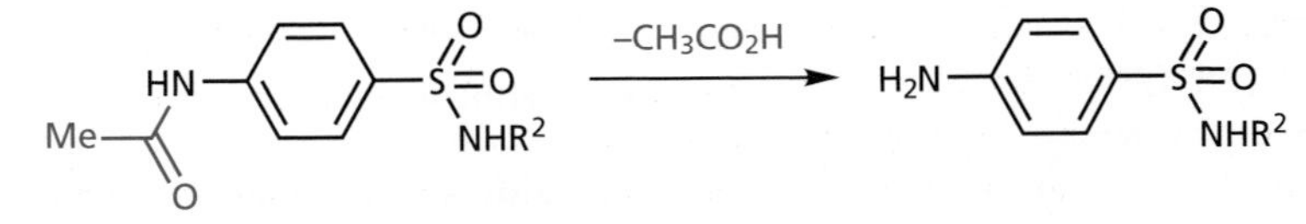

FIGURE 19.5 Metabolism of an *N*-acyl group to regenerate an active sulphonamide.

affect how much of the drug binds to plasma protein. This, in turn, controls the blood levels and lifetime of the drug. Thus, a drug that binds strongly to plasma protein will be longer lasting because it is slowly released into the blood circulation. Varying R^2 can also affect the solubility of sulphonamides. To conclude, variations of R^2 affect the pharmacokinetics of the drug, rather than its mechanism of action (Box 19.1).

19.4.1.4 Applications of sulphonamides

Before the appearance of penicillin, the sulpha drugs were the drugs of choice in the treatment of infectious diseases. Indeed, they played a significant part in world history by saving Winston Churchill's life during World War II. After visiting North Africa for the Casablanca conference in 1943, Churchill became gravely ill with an infection and

---

BOX 19.1 Sulphonamide analogues with reduced toxicity

The primary amino group of sulphonamides is acetylated in the body, and the resulting amides have reduced solubility which can lead to toxic effects. For example, the metabolite formed from **sulphathiazole** (an early sulphonamide) is poorly soluble and can prove fatal if it blocks the kidney tubules (Fig. 1). It is interesting to note that certain populations are more susceptible to this than others. For example, the Japanese and Chinese metabolize sulphathiazole more quickly than the average American and are more susceptible to its toxic effects.

It was discovered that the solubility problem could be overcome by replacing the thiazole ring in sulphathiazole with a pyrimidine ring to give **sulphadiazine** (Fig. 2). The reason for

the improved solubility lies in the acidity of the sulphonamide NH proton. In sulphathiazole, this proton is not very acidic (high pK_a). Therefore, sulphathiazole and its metabolite are mostly un-ionized at blood pH. Replacing the thiazole ring with a more electron-withdrawing pyrimidine ring increases the acidity of the NH proton by stabilizing the resulting anion. Therefore, sulphadiazine and its metabolite are significantly ionized at blood pH. As a consequence, they are more soluble and less toxic. Sulphadiazine was also found to be more active than sulphathiazole and soon replaced it in therapy. Silver sulphadiazine cream is still used topically to prevent infection of burns, although it is really the silver ions which provide the antibacterial effect.

FIGURE 1 Metabolism of sulphathiazole.

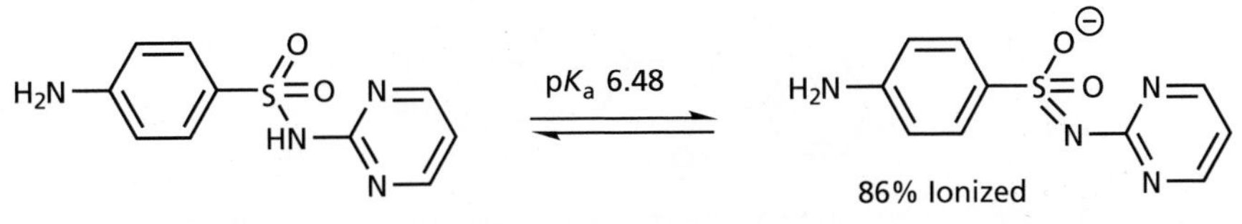

FIGURE 2 Sulphadiazine.

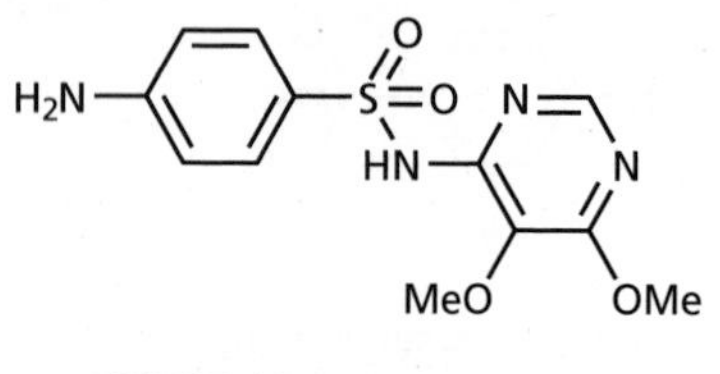

FIGURE 19.6 Sulphadoxine.

was bedridden for several weeks. Fortunately, he responded to the novel sulphonamide drugs of the day.

Penicillins largely superseded sulphonamides and, for a long time, sulphonamides took a back seat. However, there has been a revival of interest with the discovery of a new 'breed' of longer lasting sulphonamides. One example of this new generation is **sulphadoxine** (Fig. 19.6), which is so stable in the body that it need only be taken once a week. The combination of sulphadoxine and **pyrimethamine** is called **Fansidar** and has been used for the treatment of malaria.

The sulpha drugs presently have the following applications in medicine:

- treatment of urinary tract infections;
- eye lotions;

- treatment of mucous membrane infections;
- treatment of gut infections (Box 19.2).

It is also worth noting that sulphonamides have occasionally found uses in other areas of medicine (section 12.4.4.2).

19.4.1.5 Mechanism of action

The sulphonamides act as competitive enzyme inhibitors of **dihydropteroate synthetase** and block the biosynthesis of **tetrahydrofolate** in bacterial cells (Fig. 19.7). Tetrahydrofolate is important in both human and bacterial cells, because it is an enzyme cofactor that provides one carbon units for the synthesis of the pyrimidine nucleic acid bases required for DNA synthesis (section 21.3.1). If pyrimidine and DNA synthesis is blocked, then the cell can no longer grow and divide.

Note that sulphonamides do not actively kill bacterial cells. They do, however, prevent the cells growing and multiplying. This gives the body's own defence systems enough time to gather their resources and wipe out the invader. Antibacterial agents which inhibit cell growth are classed as **bacteriostatic**, whereas agents such as penicillin

---

BOX 19.2 Treatment of intestinal infections

Sulphonamides have been particularly useful against intestinal infections, and can be targeted against these by the use of prodrugs. For example, **succinyl sulphathiazole** is a prodrug of sulphathiazole (Fig. 1). The succinyl moiety contains an acidic group which means that the prodrug is ionized in the intestine. As a result, it is not absorbed into the bloodstream and is retained in

the intestine. Slow enzymatic hydrolysis of the succinyl group then releases the active sulphathiazole where it is needed.

Benzoyl substitution (Fig. 2) on the aniline nitrogen has also given useful prodrugs that are poorly absorbed through the gut wall because they are too hydrophobic (section 11.3). They can be used in the same way.

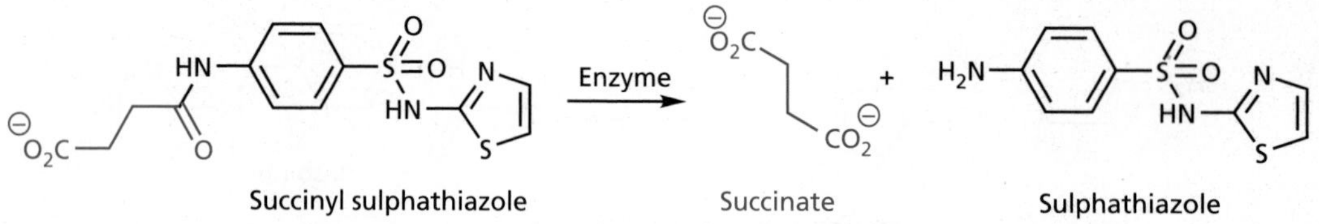

Succinyl sulphathiazole Succinate Sulphathiazole

FIGURE 1 Succinyl sulphathiazole is a prodrug of sulphathiazole.

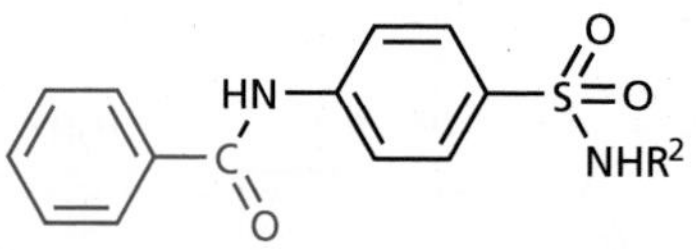

FIGURE 2 Substitution on the aniline nitrogen with benzoyl groups.

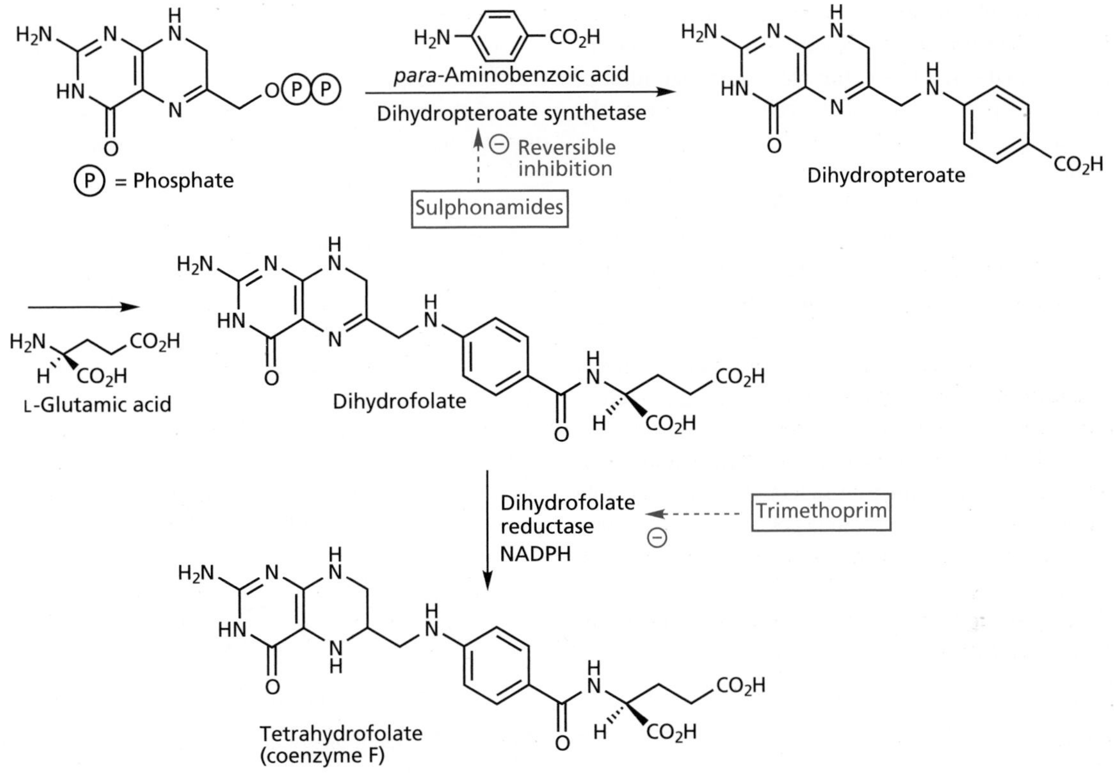

FIGURE 19.7 Mechanism of action of sulphonamides.

which actively kill bacterial cells are classed as **bactericidal**. Because sulphonamides rely on a healthy immune system to complete the job they have started, they are not recommended for patients with a weakened immune system. This includes people with AIDS, as well as patients who are undergoing cancer chemotherapy or have had an organ transplant and are taking immunosuppressant drugs.

Sulphonamides act as inhibitors by mimicking **p-aminobenzoic acid** (PABA)—one of the normal substrates for dihydropteroate synthetase. The sulphonamide molecule is similar enough in structure to PABA that the enzyme is fooled into accepting it into its active site (Fig. 19.8). Once it is bound, the sulphonamide prevents PABA from binding. As a result, dihydropteroate is no longer

synthesized. One might ask why the enzyme does not join the sulphonamide to the other component of dihydropteroate to give a dihydropteroate analogue containing the sulphonamide skeleton. This can, in fact, occur, but the analogue is not accepted by the next enzyme in the biosynthetic pathway.

Sulphonamides are competitive enzyme inhibitors, so inhibition is reversible. This is demonstrated by certain organisms such as staphylococci, pneumococci, and gonococci which can acquire resistance by synthesizing more PABA. The more PABA there is in the cell, the more effectively it can compete with the sulphonamide inhibitor to reach the enzyme's active site. In such cases, the dose levels of sulphonamide have to be increased to obtain

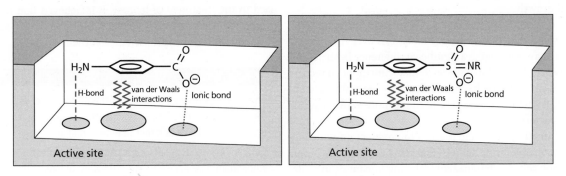

FIGURE 19.8 Sulphonamide prevents PABA from binding by mimicking PABA.

the same level of inhibition. Resistance to sulphonamides can also arise by mutations that modify the target enzyme such that it has less affinity for sulphonamides, or by decreased permeability of the cell membrane to the sulphonamide.

Tetrahydrofolate is clearly necessary for the survival of bacterial cells, but it is also vital for the survival of human cells, so why are the sulpha drugs not toxic to humans? The answer lies in the fact that human cells synthesize tetrahydrofolate in a different manner and do not contain the enzyme dihydropteroate synthetase. In human cells, tetrahydrofolate is synthesized from **folic acid**, which is obtained from the diet as a vitamin and brought across cell membranes by a transport protein.

We could now ask whether bacterial cells could transport folic acid in the same way. In fact, this is not possible because bacterial cells lack the necessary transport protein required to transport folic acid across the bacterial cell membrane.

To sum up, the success of sulphonamides is due to two metabolic differences between mammalian and bacterial cells:

- bacteria have a susceptible enzyme which is not present in mammalian cells;
- bacteria lack the transport protein that would allow them to transport folic acid across their cell membrane.

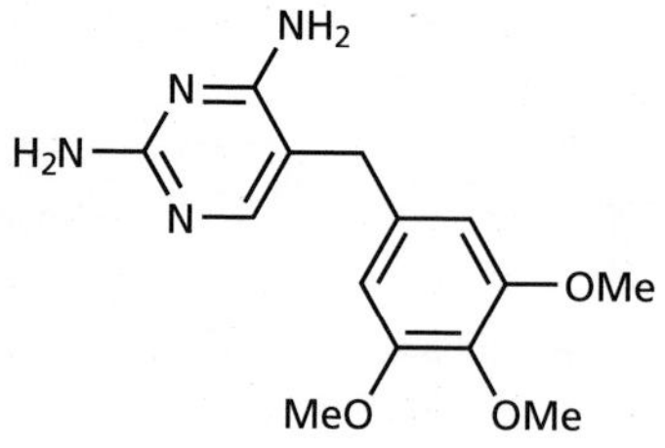

19.4.2 **Examples of other antimetabolites**

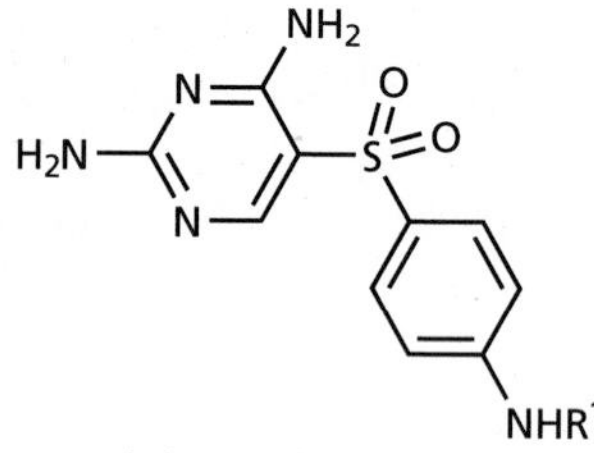

Other antimetabolites in medical use include **trimethoprim** and a group of compounds known as **sulphones** (Fig. 19.9).

19.4.2.1 Trimethoprim

Trimethoprim is an orally active diaminopyrimidine structure, which has proved to be a highly selective antibacterial and antimalarial agent. It acts against **dihydrofolate reductase**—the enzyme which carries out the conversion of dihydrofolate to tetrahydrofolate—leading to the inhibition of DNA synthesis and cell growth.

Dihydrofolate reductase is present in mammalian cells as well as bacterial cells, but mutations over millions of years have resulted in a significant difference in structure between the two enzymes such that trimethoprim recognizes and inhibits the bacterial enzyme more strongly. In fact, trimethoprim is 100 000 times more active against the bacterial enzyme.

Trimethoprim is often given in conjunction with the sulphonamide **sulphamethoxazole** (Fig. 19.9) in a preparation called **cotrimoxazole**. The sulphonamide inhibits the incorporation of PABA into dihydropteroate, while

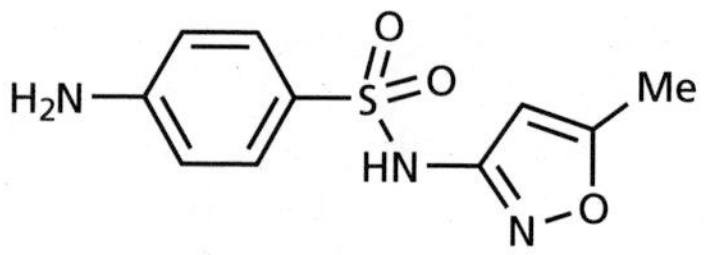

FIGURE 19.9 Examples of antimetabolites in medical use.

trimethoprim inhibits dihydrofolate reductase. Therefore, two enzymes in the one biosynthetic route are inhibited (Fig. 19.7). This is a very effective method of inhibiting a biosynthetic route and has the advantage that the doses of both drugs can be kept down to a safe level. To get the same level of inhibition using a single drug, the dose level would have to be much higher, leading to possible side effects. This approach has been described as '**sequential blocking**'.

Resistance to trimethoprim has been observed in strains of *E. coli* which produce a new form of the target enzyme that has less affinity for the drug.

19.4.2.2 Sulphones

The sulphones (Fig. 19.9) are the most important drugs used in the treatment of leprosy. It is believed that they inhibit the same bacterial enzyme inhibited by the sulphonamides (i.e. dihydropteroate synthetase).

KEY POINTS

- The principle of chemotherapy or the magic bullet involves the design of chemicals which show selective toxicity against bacterial cells rather than mammalian cells.

- Early antibacterial agents were salvarsan, prontosil, and the sulphonamides. Following the discovery of penicillin, several classes of antibiotics were isolated from fungal strains.

- The bacterial cell differs in various respects from mammalian cells, allowing the identification of drug targets which are unique to bacterial cells, or which differ significantly from equivalent targets in mammalian cells.

- Antibacterial agents act on five main targets—cell metabolism, the cell wall, the plasma membrane, protein synthesis, and nucleic acid function.

- Sulphonamides require a primary aromatic amine group and a secondary sulphonamide group for good activity.

- Adding an aromatic or heteroaromatic group to the sulphonamide nitrogen provides a variety of sulphonamides with different pharmacokinetic properties.

- *N*-Acetylation of sulphonamides is a common metabolic reaction.

- Sulphonamides are used to treat infections of the urinary tract, gastrointestinal tract, and mucous membranes. They are also used in eye lotions.

- Sulphonamides are similar in structure to *para*-aminobenzoic acid—a component of dihydropteroate. As a result, they can bind to the bacterial enzyme responsible for dihydropteroate synthesis and act as an inhibitor.

- Mammals synthesize tetrahydrofolate from folic acid acquired from the diet. They lack the enzyme targeted by sulphonamides. Bacteria lack the transport mechanisms required to transport folic acid into their cells.

- Trimethoprim inhibits dihydrofolate reductase—an enzyme which converts folic acid to tetrahydrofolate. It has been used in combination with sulphamethoxazole in a strategy known as sequential blocking.

- Sulphones are used in the treatment of leprosy.

19.5 Antibacterial agents which inhibit cell wall synthesis

19.5.1 Penicillins

19.5.1.1 History of penicillins

In 1877, Pasteur and Joubert discovered that certain moulds could produce toxic substances which killed bacteria. Unfortunately, these substances were also toxic to humans and were of no clinical value. They did demonstrate, however, that moulds could be a potential source of antibacterial agents.

In 1928, Fleming noted that a bacterial culture that had been left several weeks open to the air had become infected by a fungal colony. Of more interest was the fact that there was an area surrounding the fungal colony where the bacterial colonies were dying. He correctly concluded that the fungal colony was producing an antibacterial agent which was spreading into the surrounding area. Recognizing the significance of this, he set out to culture and identify the fungus and showed it to be a relatively rare species of *Penicillium*. It has since been suggested that the *Penicillium* spore responsible for the fungal colony originated from another laboratory in the building, and that the spore was carried by air currents and blown through the window of Fleming's laboratory. This in itself appears to be a remarkable stroke of good fortune. However, a series of other chance events were involved in the story—not least the weather! A period of early cold weather had encouraged the fungus to grow while the bacterial colonies had remained static. A period of warm weather then followed which encouraged the bacteria to grow. These weather conditions were the ideal experimental conditions required for:

- the fungus to produce penicillin during the cold spell;
- the antibacterial properties of penicillin to be revealed during the hot spell.

If the weather had been consistently cold, the bacteria would not have grown significantly and the death of cell colonies close to the fungus would not have been seen. Alternatively, if the weather had been consistently warm, the bacteria would have outgrown the fungus and little penicillin would have been produced. As a final twist to the story, the crucial agar plate had been stacked in a bowl of disinfectant ready for washing up, but was actually placed above the surface of the disinfectant. It says much for Fleming's observational powers that he bothered to take any notice of a discarded culture plate and that he spotted the crucial area of inhibition.

Fleming spent several years investigating the novel antibacterial extract and showed it to have significant antibacterial properties while being remarkably non-toxic to mammals. Unfortunately, Fleming was unable to isolate and purify the active principle, and he came to the conclusion that penicillin was too unstable to be used clinically.

The problem of isolating penicillin was eventually solved in 1938 by Florey and Chain by using processes such as freeze-drying and chromatography, which allowed isolation of the antibiotic under much milder conditions than had previously been available. By 1941, Florey and Chain were able to carry out the first clinical trials on crude extracts of penicillin and achieved spectacular success. Further developments aimed at producing the new agent in large quantities were developed in the USA, and, by 1944, there was enough penicillin to treat casualties arising from the D-Day landings.

Although the use of penicillin was now widespread, the structure of the compound was still not settled and the unusual structures being proposed proved a source

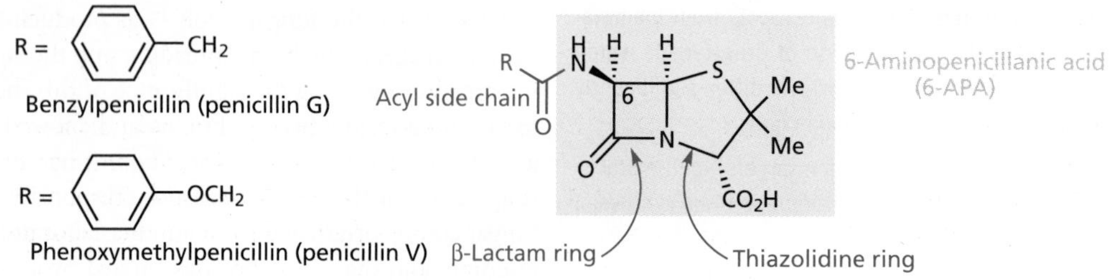

FIGURE 19.10 The structure of penicillin.

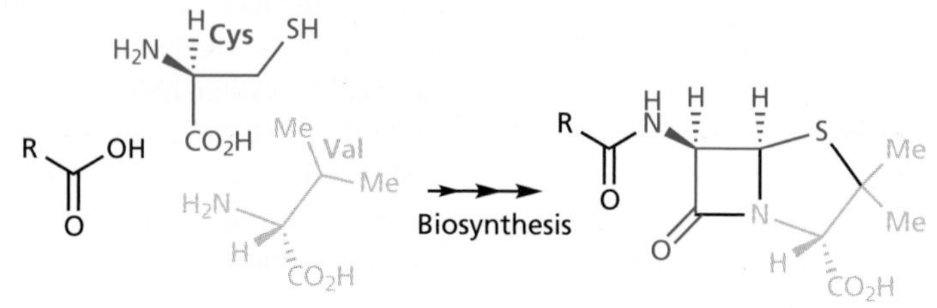

FIGURE 19.11 The biosynthetic precursors of penicillin.

of furious debate. The issue was finally settled in 1945 when Dorothy Hodgkins established the structure by X-ray crystallographic analysis. The structure was quite surprising at the time, as penicillin was clearly a highly strained molecule—which explained why Fleming had been unsuccessful in purifying it.

The full synthesis of such a highly strained molecule presented a huge challenge—one that was met successfully by Sheehan in 1957. Unfortunately, the full synthesis was too involved to be of commercial use, but in the following year Beechams isolated a biosynthetic intermediate of penicillin called **6-aminopenicillanic acid (6-APA)**. This revolutionized the field of penicillins by providing the starting material for a huge range of semi-synthetic penicillins.

Since then, penicillins have been used widely and often carelessly. As a result, penicillin-resistant bacteria have evolved and have become an increasing problem. The fight against penicillin-resistant bacteria was helped in 1976 when Beechams discovered a natural product called **clavulanic acid**, which proved highly effective in protecting penicillins from the bacterial enzymes which attack them (section 19.5.4.1).

19.5.1.2 Structure of benzylpenicillin and phenoxymethylpenicillin

Penicillin (Fig. 19.10) contains a highly unstable-looking bicyclic system consisting of a four-membered β-lactam ring fused to a five-membered thiazolidine ring. The skeleton of the molecule suggests that it is derived from the amino acids cysteine and valine, and this has been established (Fig. 19.11). The overall shape of the molecule is like a half-open book, as shown in Fig. 19.12.

The acyl side chain (R) varies, depending on the components of the fermentation medium. For example, corn steep liquor (the fermentation medium first used for mass production of penicillin) contains high levels of phenylacetic acid ($PhCH_2CO_2H$) and gives **benzylpenicillin (penicillin G**; R = benzyl). A fermentation medium containing phenoxyacetic acid ($PhOCH_2CO_2H$) gives **phenoxymethylpenicillin (penicillin V**; R = $PhOCH_2$) (Fig. 19.10).

Test your understanding and practise your molecular modelling with Exercise 19.1 on the Online Resource Centre at www.oxfordtextbooks.co.uk/orc/patrick6e/

19.5.1.3 Properties of benzylpenicillin

Benzylpenicillin (penicillin G) is active against a range of bacterial infections (Box 19.3) and lacks serious side effects for most patients. However, there are various drawbacks. It cannot be taken orally because it is broken down by stomach acids, it has a narrow spectrum of activity, and there are many bacterial infections against which it has no effect—particularly those where the microorganism produces an enzyme called β-**lactamase**. This is

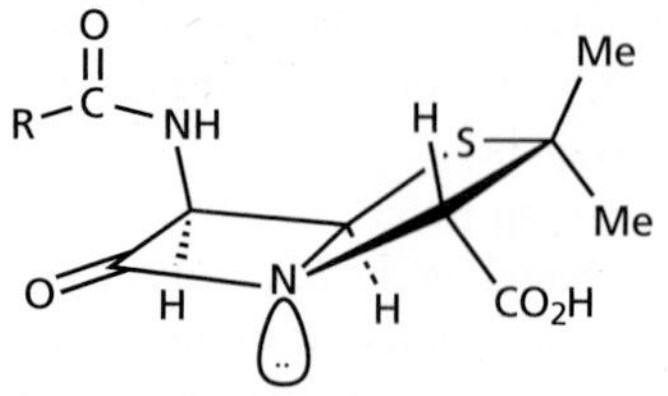

FIGURE 19.12 The three-dimensional shape of penicillin.

BOX 19.3 Clinical properties of benzylpenicillin and phenoxymethylpenicillin

Benzylpenicillin is active against non-β-lactamase producing Gram-positive bacilli (e.g. *Meningitis*, *Gonorrhoea*, and early strains of staphylococci) and several Gram-negative cocci (e.g. *Neisseria*). It is effective for many streptococcal, pneumococcal, gonococcal, and meningococcal infections. It is also used to treat anthrax, diphtheria, gas gangrene, leptospirosis, and Lyme disease in children. It can be effective against tetanus, although **metronidazole** is preferred. Since penicillin is bactericidal, it is most active against rapidly dividing bacteria. There are many bacterial species against which benzylpenicillin shows no activity, in particular Gram-

negative bacteria and those producing β-lactamase enzymes. It is ineffective when taken orally and should be administered by intravenous or intramuscular injection.

Phenoxymethylpenicillin is recommended for the treatment of various problems such as tonsillitis, rheumatic fever, otitis media, and oral infections.

Allergic reactions are suffered by some individuals when they take penicillins, varying from a rash to immediate anaphylactic shock. Anaphylactic reactions occur in 0.2% of patients with a fatality rate of 0.001%. Less serious allergic reactions are more common (1–4%).

an enzyme which hydrolyses the β-lactam ring of benzylpenicillin and makes the structure inactive. Therefore, there is scope for producing analogues with improved properties. Before looking at penicillin analogues, we shall look at penicillin's mechanism of action.

19.5.1.4 Mechanism of action for penicillin

Structure of the cell wall

In order to understand penicillin's mechanism of action, we have to first look at the structure of the bacterial cell wall and the mechanism by which it is formed. Bacteria have cell walls in order to survive a large range of environmental conditions, such as varying pH, temperature, and osmotic pressure. Without a cell wall, water would continually enter the cell as a result of osmotic pressure, causing the cell to swell and burst (lysis). The cell wall is very porous and does not block the entry of water, but it does prevent the cell swelling. Mammalian cells do not

have a cell wall, making it the perfect target for antibacterial agents such as penicillins.

The wall is a peptidoglycan structure (Fig. 19.13). In other words, it is made up of peptide and sugar units. The structure of the wall consists of a parallel series of sugar backbones containing two types of sugar, **N-acetylmuramic acid (NAM)** and **N-acetylglucosamine (NAG)** (Fig. 19.14). Peptide chains are bound to the NAM sugars and it is interesting to note the presence of D-amino acids

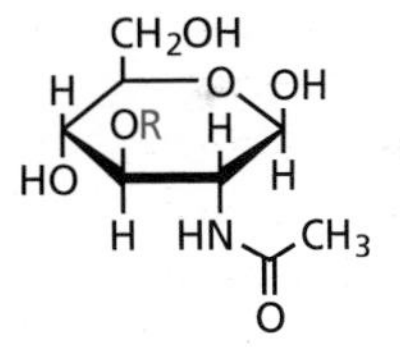

FIGURE 19.14 Sugars contained in the cell wall structure of bacteria. R = H, N-acetylglucosamine (NAG); R = CHMeCO$_2$H, N-acetylmuramic acid (NAM).

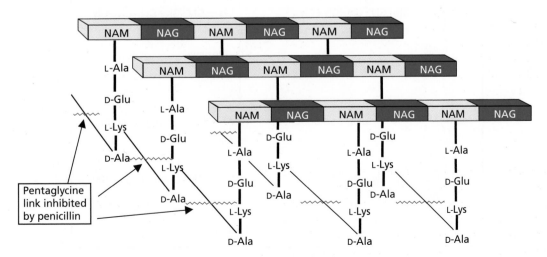

FIGURE 19.13 Peptidoglycan structure of bacterial cell walls.

in these chains. In human biochemistry there are only L-amino acids, whereas bacteria have **racemase** enzymes that can convert L-amino acids into D-amino acids. In the final stage of cell wall biosynthesis, the peptide chains are linked together by the displacement of D-alanine from one chain by glycine in another.

About 30 enzymes are involved in the overall biosynthesis of the cell wall, but it is the final cross-linking reaction which is inhibited by penicillin. This leads to a cell wall framework that is no longer interlinked (Fig. 19.15). As a result, the wall becomes fragile and can no longer prevent the cell from swelling and bursting. The enzyme responsible for the cross-linking reaction is known as the **transpeptidase enzyme**. There are several types of the enzyme which vary in character from one bacterial species to another, but they are all inhibited to various degrees by penicillins.

There are significant differences in the thickness of the cell wall between Gram-positive and Gram-negative bacteria. The cell wall in Gram-positive bacteria consists of 50–100 peptidoglycan layers, whereas in Gram-negative bacteria it consists of only two layers.

The transpeptidase enzyme and its inhibition

The transpeptidase enzyme is bound to the outer surface of the cell membrane and is similar to a class of enzymes called the **serine proteases**, so called because they contain a serine residue in the active site and catalyse the hydrolysis of peptide bonds. In the normal mechanism (see Fig. 19.16(a)), serine acts as a nucleophile to split the peptide bond between the two unusual D-alanine units on a peptide chain. The terminal alanine departs the active site, leaving the peptide chain bound to the active site. The pentaglycyl moiety of another peptide chain now enters the active site and the terminal glycine forms a peptide bond to the alanine group, displacing it from serine and linking the two chains together.

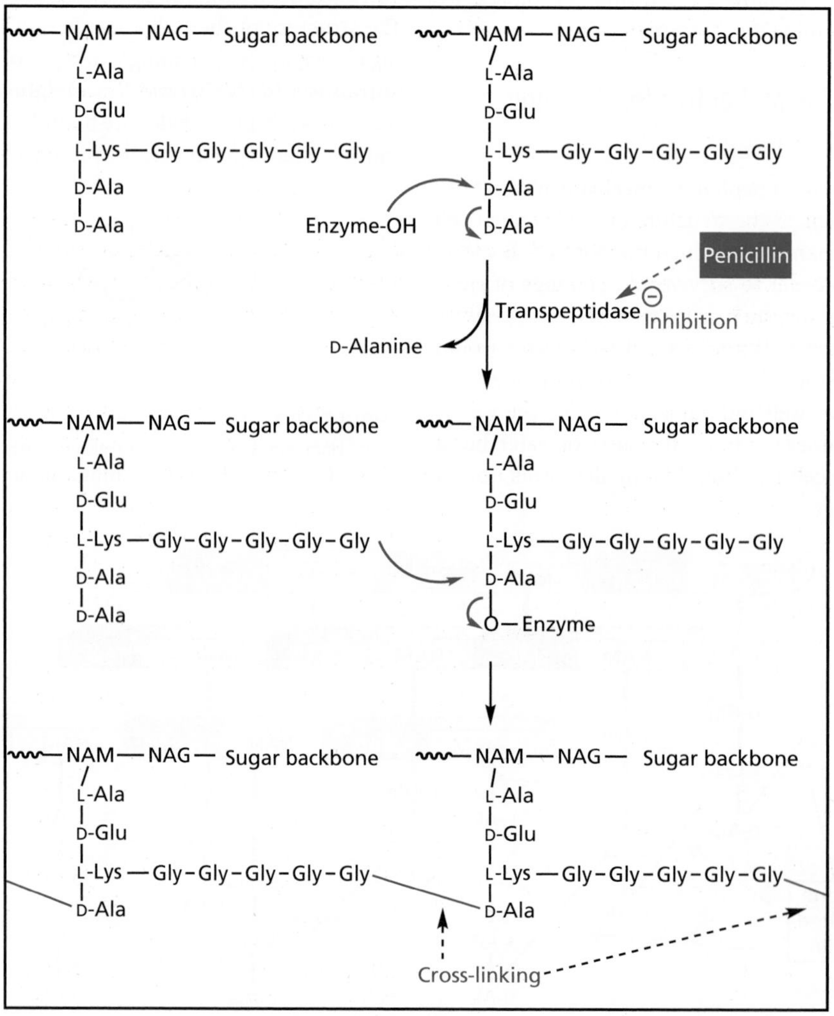

FIGURE 19.15 Cross-linking of bacterial cell walls inhibited by penicillin.

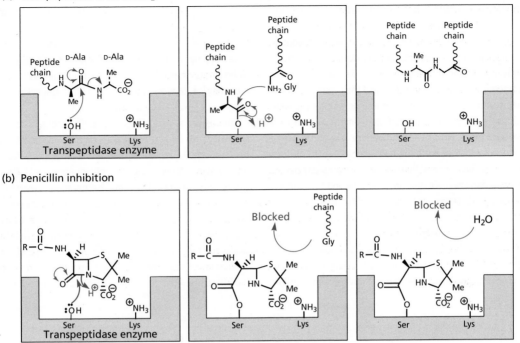

FIGURE 19.16 Mechanisms of transpeptidase cross-linking and penicillin inhibition.

It has been proposed that penicillin has a conformation which is similar to the transition-state conformation taken up by the D-Ala-D-Ala moiety during the cross-linking reaction, and that the enzyme mistakes penicillin for D-Ala-D-Ala and binds it to the active site. Once bound, penicillin is subjected to nucleophilic attack by serine (Fig. 19.16b).

The enzyme can attack the β-lactam ring of penicillin and cleave it in the same way as it did with the peptide bond. However, penicillin is cyclic and so the molecule is not split in two and nothing leaves the active site. Subsequent hydrolysis of the ester group linking the penicillin to the active site does not take place either, since the penicillin structure blocks access to the pentaglycine chain or water.

If penicillin *is* acting as a mimic for a D-Ala-D-Ala moiety, this provides another explanation for its lack of toxicity. Since there are no D-amino acids or D-Ala-D-Ala segments in any human protein, it is unlikely that any of the body's serine protease enzymes would recognize either the segment or penicillin itself. As a result, penicillin is selective for the bacterial transpeptidase enzyme and is ignored by the body's own serine proteases.

This theory has one or two anomalies, though. For example, **6-methylpenicillin** (Fig. 19.17) was thought to be a closer analogue to D-Ala-D-Ala. On that basis, it should fit the active site better and have higher activity. However, when this structure was synthesized, it was found to be inactive. It is now proposed that 6-methoxypenicillin is a closer analogue to acyl-D-Ala-D-Ala than 6-methylpenicillin. Indeed, antibacterial penicillin structures containing a 6-methoxy substituent have been developed, for example **temocillin** (see Fig. 19.27). Molecular modelling studies involving overlays of penicillin analogues (section 17.9) have demonstrated that the methyl group of a 6-methoxy

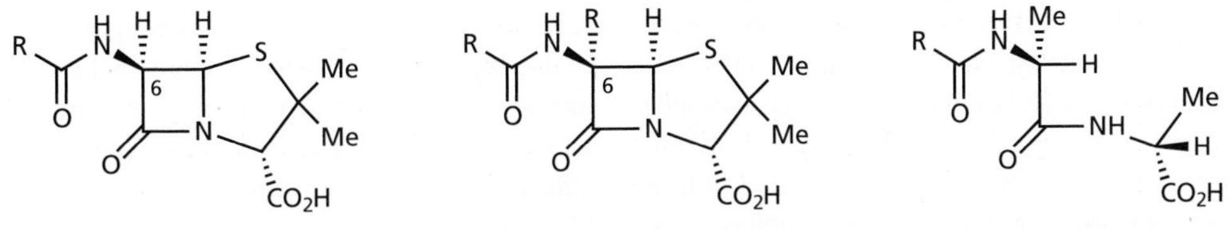

Penicillin

R' = Me; 6-Methylpenicillin
R' = OMe; 6-Methoxypenicillin

Acyl-D-Ala-D-Ala

FIGURE 19.17 Comparison of penicillin, 6-substituted penicillins, and acyl-D-Ala-D-Ala.

BOX 19.4 *Pseudomonas aeruginosa*

Pseudomonas aeruginosa is an example of an **opportunistic pathogen**. Such organisms are not normally harmful to healthy individuals. Indeed, many people carry the organism without being aware of it, because their immune system keeps it under control. Once that immune system is weakened, though, the organism can start multiplying and lead to serious illness. Hospital-bound patients are particularly at risk, especially those suffering from shock or AIDS, or those undergoing cancer chemotherapy. Burn victims are particularly prone to *P. aeruginosa* skin infections and this can lead to septicaemia, which can prove fatal. The organism is also responsible for serious lung infections amongst patients undergoing mechanical ventilation.

The cells of *P. aeruginosa* are rod shaped and can appear blue or green in colour, which is why it was given the name *aeruginosa*. It prefers to grow in moist environments and has been isolated from soil, water, plants, animals, and humans. It can even grow in distilled water and contact lens solutions. In hospitals, there are several possible sources of infection, including respiratory equipment, sinks, uncooked vegetables, and flowers brought by visitors.

Pseudomonas aeruginosa is a difficult organism to treat because it has an intrinsic resistance to a wide variety of antibacterial agents, including many penicillins, cephalosporins, tetracyclines, quinolones, and chloramphenicol. There are two reasons for this. The outer membrane of the cell has a low permeability to drugs and even if a drug does enter the cell, there is an efflux system which can pump it back out again. Nevertheless, there *are* drugs which have proved effective against the organism; in particular aminoglycosides such as tobramycin or gentamicin, and penicillins such as ticarcillin. These are often given in combination with each other.

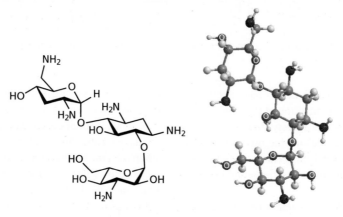

FIGURE 1 Tobramycin.

substituent is more closely aligned to the methyl group of acyl-D-Ala-D-Ala, than a 6-methyl group would be.

19.5.1.5 Resistance to penicillin

Bacterial strains vary in their susceptibility to penicillin. Some species such as streptococci are quite vulnerable, whereas a bacterium like *Pseudomonas aeruginosa* is particularly resistant (see Box 19.4). Other species such as *Staphylococcus aureus* are initially vulnerable, but acquire resistance when they are exposed to penicillin over a period of time. There are several reasons for this varied susceptibility.

Physical barriers

If penicillin is to inhibit the transpeptidase enzyme, it has to reach the outer surface of the bacterial cell membrane where the enzyme is located. Thus, penicillin has to pass through the cell walls of both Gram-positive and Gram-negative bacteria. The cell wall is much thicker in Gram-positive bacteria than in Gram-negative bacteria, so one might think that penicillin would be more effective against Gram-negative bacteria. However, this is not the case. Although the cell wall is a strong, rigid structure, it is also highly porous, which means that small molecules like penicillin can move through it without difficulty. One can imagine the cell wall being like several layers of chicken wire, and the penicillin molecules as small pebbles able to pass through the gaps.

If the cell wall does not prevent penicillin reaching the cell membrane, what does? As far as Gram-positive bacteria are concerned there *is* no barrier, and that is why penicillin G has good activity against these organisms.

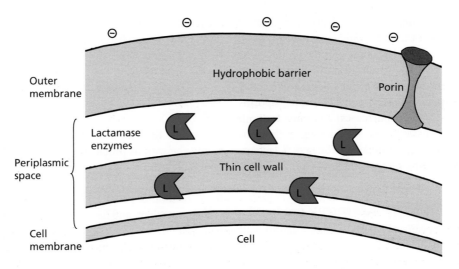

FIGURE 19.18 Outer surface of a Gram-negative bacterial cell.

However, Gram-negative bacteria have an outer lipopolysaccharide membrane surrounding the cell wall which is impervious to water and polar molecules such as penicillin (Fig. 19.18). That can explain why Gram-negative bacteria are generally resistant, but not why some Gram-negative bacteria are susceptible and some are not. Should they not all be resistant?

The answer lies in protein structures called **porins**, which are located in the outer membrane. These act as pores through which water and essential nutrients can pass to reach the cell. Small drugs such as penicillin can also pass this way, but whether they do or not depends on the structure of the porin as well as the characteristics of the penicillin (i.e. its size, structure, and charge). In general, drugs have less chance of passing through the porins if they are large, have a negative charge, and are hydrophobic. In contrast, a small hydrophilic drug that can exist as a zwitterion will move through quickly. Therefore, porins play a crucial role in controlling the amount of penicillin capable of reaching the periplasmic space between the outer membrane and cell membranes. If access is slow, the concentration of penicillin at the transpeptidase enzyme may be insufficient to inhibit it effectively.

Presence of β-lactamase enzymes

The presence of β-lactamase enzymes is the most important mechanism by which bacteria gain resistance to penicillin. β-Lactamases are enzymes which have mutated from transpeptidases and so they are quite similar in nature. For example, they have a serine residue in the active site and can open up the β-lactam ring of penicillin to form an ester link to the structure. Unlike the transpeptidase enzyme, β-lactamases are able to hydrolyse the ester link and shed the ring-opened penicillin. They do this so effectively that 1000 penicillin molecules are hydrolysed per second (Fig. 19.19).

Some Gram-positive bacterial strains are resistant to penicillin because they can release β-lactamase into the surrounding environment such that penicillin is intercepted before it reaches the cell membrane. The enzyme eventually dissipates through the cell wall and is lost, so the bacterium has to keep generating the enzyme to maintain its protection. *Staphylococcus aureus* is a Gram-positive bacterium that used to be susceptible to penicillin, but 95% of *S. aureus* strains now release a β-lactamase which hydrolyses penicillin G.

Most, if not all, Gram-negative bacteria produce β-lactamases which makes them more resistant to penicillins. Moreover, the β-lactamase released is trapped in the periplasmic space between the cell membrane and the outer membrane because it cannot pass through the latter. As a result, any penicillin managing to penetrate the outer membrane encounters a higher concentration of β-lactamase than it would with Gram-positive

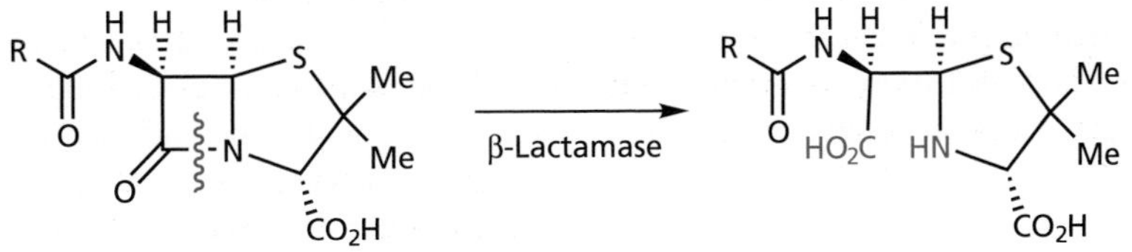

FIGURE 19.19 β-Lactamase deactivation of penicillin.

bacteria. This might suggest again that all Gram-negative bacteria should be resistant to penicillin. However, there are over 1000 types of β-lactamase enzyme produced by both Gram-positive and Gram-negative bacteria, and these vary in their substrate selectivity. Some are selective for penicillins (**penicillinases**), some for cephalosporins (section 19.5.2) (**cephalosporinases**) and some for both penicillins and cephalosporins. The differing levels of enzyme and their differing affinities for different β-lactams account for the varying susceptibilities of Gram-negative bacteria to different β-lactams.

High levels of transpeptidase enzyme produced

In some Gram-negative bacteria, excess quantities of transpeptidase are produced, and penicillin is unable to inhibit all the enzyme molecules present.

Affinity of the transpeptidase enzyme to penicillin

There are several forms of the transpeptidase enzyme present within any bacterial cell and these vary in their affinity for the different β-lactams. Differences in the relative proportions of these enzymes across bacterial species account in part for the variable susceptibility of these bacteria to different penicillins. For example, early strains of *Staphylococcus aureus* contained transpeptidase enzymes which had a high affinity for penicillin and were effectively inhibited. Penicillin-resistant strains of *S. aureus* acquired a transpeptidase enzyme called **penicillin binding protein 2a** (**PBP2a**), which has a much lower affinity to penicillins. The presence of low-affinity transpeptidases is also a problem with enterococci and pneumococci.

Transport back across the outer membrane of Gram-negative bacteria

There are proteins in the outer membrane of some Gram-negative bacteria which are capable of pumping penicillin out of the periplasmic space, thus lowering its concentration and effectiveness. The extent to which this happens varies from species to species and also depends on the structure of the penicillin. This is known as an **efflux** process.

Mutations and genetic transfers

Mutations can occur which will affect any or all of the above mechanisms such that they are more effective in resisting the effects of β-lactams. Small portions of DNA carrying the genes required for resistance can also be transferred from one cell to another by means of genetic vehicles called **plasmids**. These are small pieces of circular bacterial extra-chromosomal DNA. If the transferred DNA contains a gene coding for a β-lactamase enzyme or some other method of improved resistance, then the recipient cell acquires immunity. Genetic material can also be transferred between bacterial cells by viruses and by the uptake of free DNA released by dead bacteria.

19.5.1.6 Methods of synthesizing penicillin analogues

Having studied the mechanism of action of penicillin G and the various problems surrounding resistance, we now look at how analogues of penicillin G can be synthesized which might have improved stability and activity. A method of preparing analogues is required which is cheap, efficient, and flexible. Sheehan's full synthesis of penicillin has too many steps and is too low yielding (1%) for it to be practical, which limits the options to fermentation methods or semi-synthetic procedures.

Fermentation

Originally, the only way to prepare different penicillins was to vary the fermentation conditions. Adding different carboxylic acids to the fermentation medium resulted in penicillins with different acyl side chains (e.g. **phenoxymethylpenicillin**; Fig. 19.10). Unfortunately, there was a limitation to the sort of carboxylic acid which was accepted by the biosynthetic route (i.e. only acids of general formula RCH_2CO_2H). This, in turn, restricted the variety of analogues which could be obtained. The other major disadvantage was the tedious and time-consuming nature of the method.

Semi-synthetic procedure

In 1959, Beechams isolated a biosynthetic intermediate of penicillin from *Penicillium chrysogenum* grown in a fermentation medium which was deficient in a carboxylic acid. The intermediate (**6-aminopenicillanic acid; 6-APA**) proved to be one of Sheehan's synthetic intermediates, and so it was possible to use this to synthesize a huge number of analogues by a semi-synthetic method. Thus, fermentation yielded 6-APA, which could then be treated with a range of acid chlorides (Fig. 19.20).

6-APA is now produced more efficiently by hydrolysing penicillin G or penicillin V with an enzyme (**penicillin acylase**) (Fig. 19.21) or by a chemical method that allows the hydrolysis of the side chain in the presence of the highly strained β-lactam ring. The latter procedure is described in more detail in section 19.5.2.2 where it is used to hydrolyse the side chain from cephalosporins.

We have emphasized the drive to make penicillin analogues with varying acyl side chains, but what is so special about the acyl side chain? Could changes not be made elsewhere in the molecule? In order to answer these questions we need to look at the structure–activity relationships of penicillins.

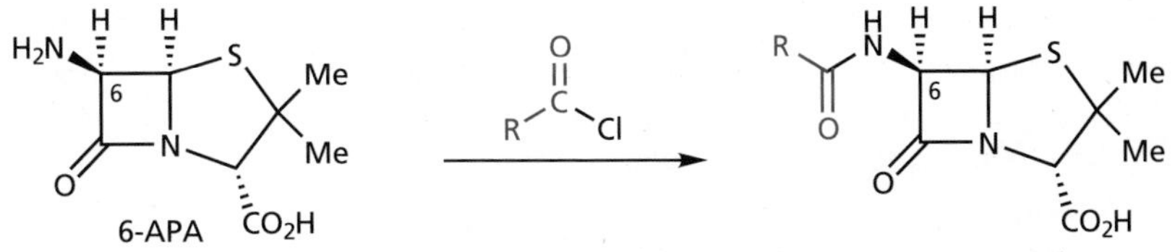

FIGURE 19.20 Penicillin analogues synthesized by acylating 6-APA.

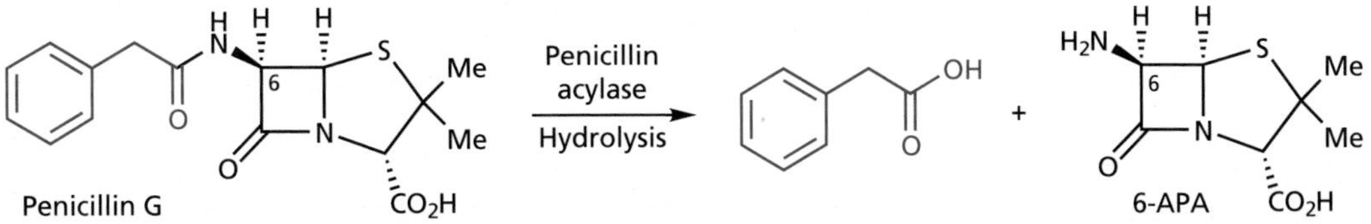

FIGURE 19.21 Synthesis of 6-APA from penicillin G.

19.5.1.7 Structure–activity relationships of penicillins

A large number of penicillin analogues have been synthesized and studied. The results of these studies led to the following SAR conclusions (Fig. 19.22):

- The strained β-lactam ring is essential.
- The free carboxylic acid is essential. This is usually ionized and penicillins are administered as sodium or potassium salts. The carboxylate ion binds to the aminium ion on the side chain of a lysine residue in the binding site.
- The bicyclic system is important. This confers further strain on the β-lactam ring—the greater the strain, the greater the activity, but the greater the instability of the molecule to other factors.
- The 6-acylamino side chain is essential.
- Sulphur is usual but not essential (see section 19.5.3).
- The stereochemistry of the bicyclic ring with respect to the acylamino side chain is important.

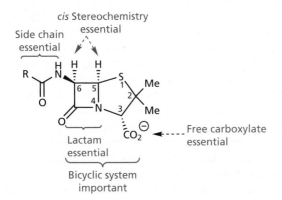

FIGURE 19.22 Structure activity relationships of penicillins.

The results of this analysis led to the inevitable conclusion that very little variation is tolerated by the penicillin nucleus and that any variations are restricted to the R group of the acylamino side chain.

19.5.1.8 Penicillin analogues

In this section we consider the penicillin analogues which proved successful in tackling the problems of acid sensitivity, β-lactamase sensitivity, and limited breadth of activity.

Acid sensitivity of penicillins

There are three reasons for the acid sensitivity of penicillin G.

- *Ring strain*: The bicyclic system in penicillin consists of a four-membered ring fused to a five-membered ring. As a result, penicillin suffers large angle and torsional strains. Acid-catalysed ring opening relieves these strains by breaking open the highly strained β-lactam ring (Fig. 19.23).

- *A highly reactive β-lactam carbonyl group*: The carbonyl group in the β-lactam ring is highly susceptible to nucleophiles and does not behave like a normal tertiary amide. The latter is resistant to nucleophilic attack because the carbonyl group is stabilized by the neighbouring nitrogen atom as shown in Fig. 19.24; the nitrogen can feed its lone pair of electrons into the carbonyl group to form a dipolar resonance structure with bond angles of 120°. This resonance stabilization is impossible for the β-lactam ring because of the increase in angle strain that would result in having a double bond within a four-membered β-lactam ring. The preferred bond angles for a double bond are 120° but the bond angles of the β-lactam ring are constrained to 90°. As a result, the lone pair is localized

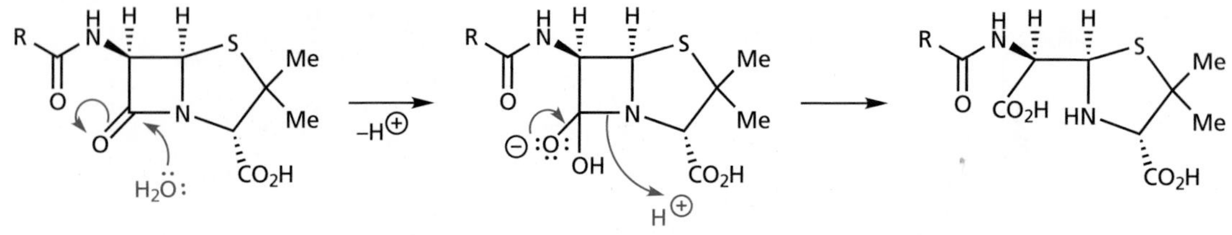

FIGURE 19.23 Ring opening of the β-lactam ring under acidic conditions.

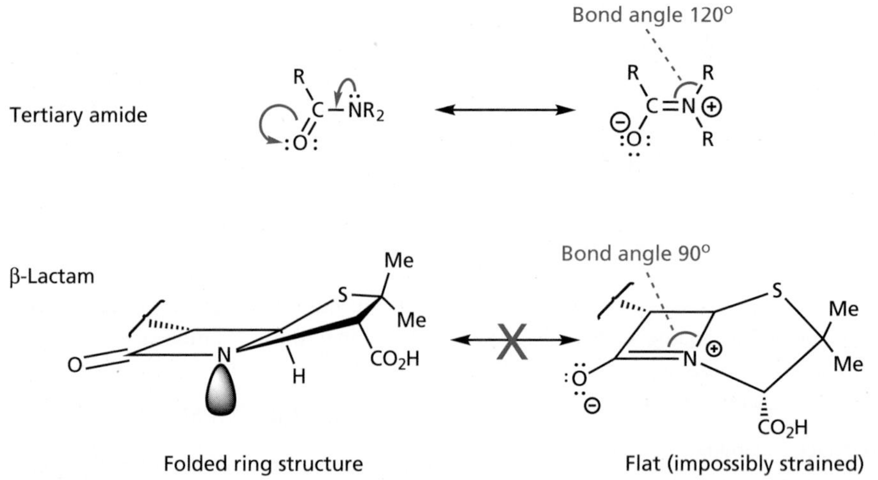

FIGURE 19.24 Comparison of tertiary amide and β-lactam carbonyl groups.

on the nitrogen atom and the carbonyl group is more electrophilic than one would expect for a tertiary amide.

- *Influence of the acyl side chain (neighbouring group participation)*: Figure 19.25 demonstrates how the neighbouring acyl group can actively participate in a mechanism to open up the lactam ring. Thus, penicillin G has a self-destruct mechanism built into its structure.

Acid-resistant penicillins

It can be seen that countering acid sensitivity is a difficult task. Nothing can be done about the first two factors, as the β-lactam ring is vital for antibacterial activity. Therefore, only the third factor can be tackled. The task then becomes one of reducing the amount of neighbouring group participation taking place. This was achieved by placing an electron-withdrawing group in the side chain which could draw electrons away from the carbonyl oxygen and reduce its tendency to act as a nucleophile (Fig. 19.26).

Phenoxymethylpenicillin (penicillin V) has an electronegative oxygen on the acyl side chain with the electron-withdrawing effect required. The molecule has better acid stability than penicillin G and is stable enough to survive the acid in the stomach, so it can be given orally.

Other penicillin analogues with an electron-withdrawing substituent (X) on the α-carbon of the side chain (Fig. 19.26) have also proved resistant to acid hydrolysis and can be given orally (e.g. ampicillin; see Fig. 19.29).

To conclude, the problem of acid sensitivity is fairly easily solved by having an electron-withdrawing group on the acyl side chain.

β-Lactamase-resistant penicillins

The problem of β-**lactamases** (or **penicillinases**) became critical in 1960, when the widespread use of penicillin G led to a dramatic increase of penicillin-resistant *S. aureus* infections. At one point, 80% of all *S. aureus* infections in hospitals were due to virulent, penicillin-resistant strains. Alarmingly, these strains were also resistant to all other available antibiotics. Fortunately, a solution to the problem was just around the corner—the design of β-lactamase-resistant penicillins.

The strategy of steric shields (section 14.2.1) was used successfully to block penicillin from accessing the penicillinase or β-lactamase active site by placing a bulky group on the side chain (Fig. 19.27). However, there was a problem. If the steric shield was *too* bulky, then it also prevented the penicillin from attacking the transpeptidase target enzyme. Therefore, a great deal of work had

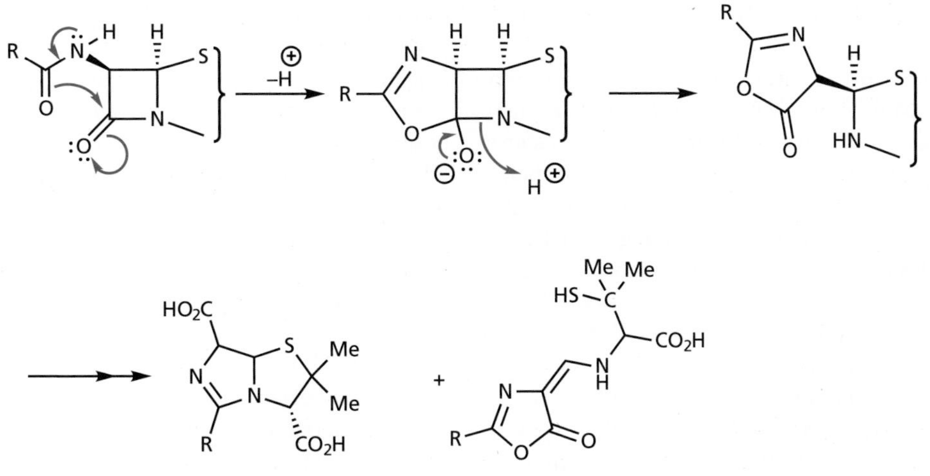

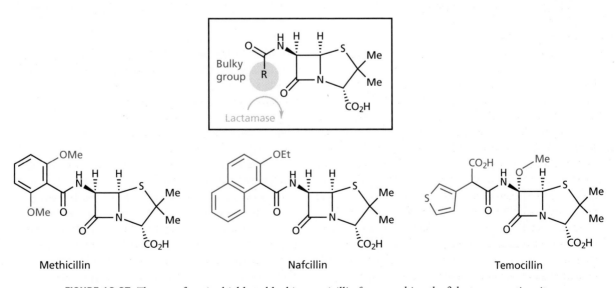

FIGURE 19.25 Influence of the acyl side chain on the acid sensitivity of penicillins.

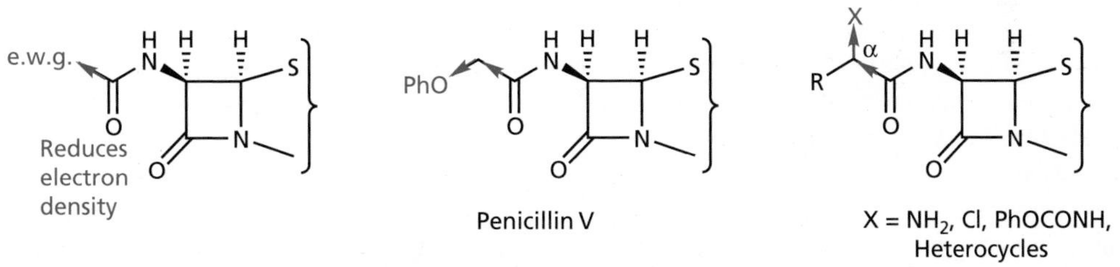

Penillic acids   **Penicillenic acids**

FIGURE 19.26 Reduction of neighbouring group participation with an electron-withdrawing group (e.w.g.).

e.w.g.
Reduces electron density

Penicillin V

X = NH₂, Cl, PhOCONH, Heterocycles

to be done to find the ideal shield—one large enough to ward off the lactamase enzyme, but sufficiently small to allow the penicillin to bind to the target enzyme. The fact that the β-lactam ring interacts with both enzymes in the same way highlights the difficulty in achieving that goal.

Fortunately, shields *were* found which could make that discrimination. **Methicillin** (Fig. 19.27) was the first effective semi-synthetic penicillin with resistance to the *S. aureus* β-lactamase enzyme and reached the clinic just in time to treat the growing *S. aureus* problem. The

Bulky group

Lactamase

Methicillin

Nafcillin

Temocillin

FIGURE 19.27 The use of steric shields to blocking penicillin from reaching the β-lactamase active site.

steric shields are the two *ortho*-methoxy groups on the aromatic ring.

Methicillin is by no means an ideal drug, however. With no electron-withdrawing group on the side chain, it is acid sensitive and has to be injected. It also shows poor activity against many other bacterial strains. Better β-lactamase-resistant agents have since been developed (see Box 19.5), and methicillin is no longer used clinically. **Nafcillin** (Fig. 19.27) is a penicillin that is resistant to β-lactamase enzymes and contains a naphthalene ring which acts as its steric shield. **Temocillin** is another β-lactamase resistant penicillin and is interesting in that it has a 6-methoxy group present (section 19.5.1.4).

In general, β-lactamase-resistant penicillins are kept as 'reserve troops'. They are only introduced into the fray if an infection proves resistant to a broad-spectrum penicillin because of a β-lactamase enzyme being present (e.g. penicillin-resistant *S. aureus* and *S. epidermidis*).

Unfortunately, 95% of *S. aureus* strains detected in hospitals have become resistant to methicillin and the other β-lactamase-resistant penicillins as a result of mutations to the transpeptidase enzyme. These bacteria are referred to as MRSA. The abbreviation stands for methicillin-resistant *S. aureus*, but the term applies to all the β-lactamase-resistant penicillins and not just methicillin.

Broad-spectrum penicillins

There are a variety of factors affecting whether a particular bacterial strain will be susceptible to a penicillin. The spectrum of activity shown by any penicillin depends on its structure, its ability to cross the cell membrane of Gram-negative bacteria, its susceptibility to β-lactamases, its affinity for the transpeptidase target enzyme, and the rate at which it is pumped back out of cells by Gram-negative organisms. All these factors vary in importance across different bacterial species and so there are no clear-cut tactics which can be used

---

BOX 19.5 The isoxazolyl penicillins

The incorporation of an isoxazolyl ring into the penicillin side chain led to orally active compounds which were stable to the β-lactamase enzyme of *S. aureus*. The isoxazolyl ring acts as the steric shield but it is also electron-withdrawing, giving the structure acid stability.

Oxacillin, **cloxacillin**, **flucloxacillin**, and **dicloxacillin** are all useful against *S. aureus* infections. The only difference between them is the type of halogen substitution on the aromatic ring. These substituents affect pharmacokinetic properties such as absorption and plasma protein binding.

Test your understanding and practise your molecular modelling with Exercise 19.3 on the Online Resource Centre at www.oxfordtextbooks.co.uk/orc/patrick6e/

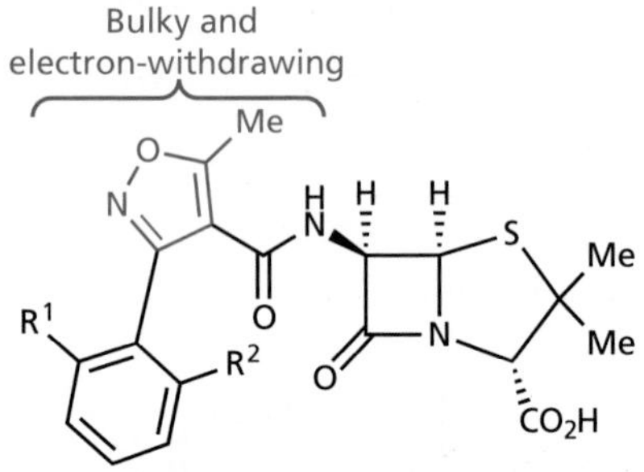

Oxacillin   R^1 = R^2 = H
Cloxacillin   R^1 = Cl, R^2 = H
Flucloxacillin   R^1 = Cl, R^2 = F
Dicloxacillin   R^1 = Cl, R^2 = Cl

FIGURE 1 Incorporation of a five-membered heterocycle into a penicillin side chain.

BOX 19.6 Clinical aspects of β-lactamase-resistant penicillins

Methicillin was useful in the 1960s against penicillin-resistant *S. aureus* infections. However, it is no longer used clinically. **Nafcillin** has more intrinsic activity than methicillin against staphylococci and streptococci, and is administered by injection. **Temocillin** is not active against Gram-positive bacteria, or bacteria with altered penicillin-binding proteins. It should be reserved for the treatment of infections caused by β-lactamase-producing strains of Gram-negative bacteria, including those resistant to third-generation cephalosporins. It is used for the treatment of septicaemia, urinary tract infections, and lower respiratory tract infections caused by susceptible Gram-negative bacteria.

Oxacillin, **cloxacillin**, **flucloxacillin**, and **dicloxacillin** are all useful agents against *S. aureus* infections. Cloxacillin is better absorbed through the gut wall than oxacillin, whereas flucloxacillin is less bound to plasma protein, resulting in higher levels of the free drug in the blood supply. They all show inferior activity to the original penicillins if they are used against bacteria that lack the β-lactamase enzyme. They are also inactive against Gram-negative bacteria. Flucloxacillin is the drug of choice for the treatment of penicillin-resistant staphylococcal infections in the ear. **Co-fluampicil** is a combination of flucloxacillin with ampicillin, which is used against streptococcal or staphylococcal infections.

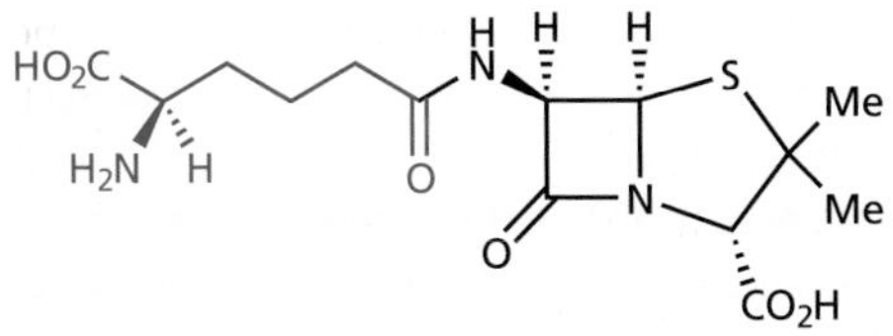

Penicillin N

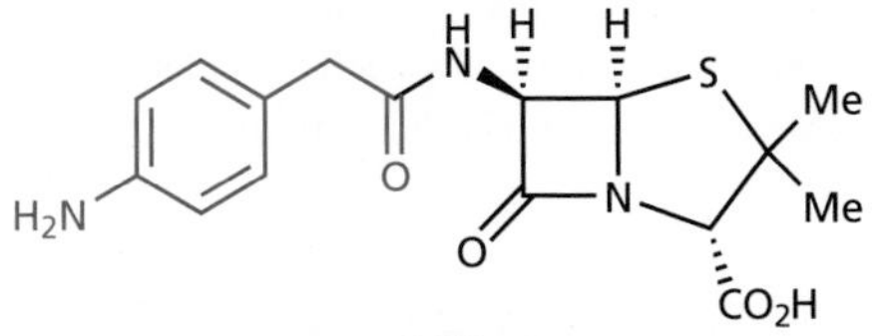

Penicillin T

FIGURE 19.28 Effect of side chain hydrophilic groups on antibacterial activity.

to improve the spectrum of activity. Consequently, the search for broad-spectrum antibiotics was one of trial and error which involved making a huge variety of analogues. These changes were again confined to variations in the side chain and gave the following results:

- Hydrophobic groups on the side chain (e.g. penicillin G) favour activity against Gram-positive bacteria, but result in poor activity against Gram-negative bacteria.
- If the hydrophobic character is increased, there is little effect on Gram-positive activity, but activity drops even further against Gram-negative bacteria.
- Hydrophilic groups on the side chain have little effect on Gram-positive activity (e.g. **penicillin T**) or cause a reduction of activity (e.g. **penicillin N**) (Fig. 19.28). However, they lead to an increase in activity against Gram-negative bacteria.
- Enhancement of Gram-negative activity is found to be greatest if the hydrophilic group (e.g. NH_2, OH, CO_2H) is attached to the carbon that is α to the carbonyl group on the side chain.

Those penicillins having useful activity against both Gram-positive and Gram-negative bacteria are known as **broad-spectrum antibiotics** (Box 19.8). There are three classes of broad-spectrum antibiotics, all of which have an α-hydrophilic group which aids the passage of these penicillins through the porins of the Gram-negative bacterial outer membrane.

Broad-spectrum penicillins: the aminopenicillins

Ampicillin (Fig. 19.29; Beechams 1964) and **amoxicillin** are orally active compounds having a very similar structure, and are commonly used as a first line of defence against infection. Both compounds are acid resistant because of the presence of the electron-withdrawing amino group. There are no steric shields present, and so these agents are sensitive to β-lactamase enzymes. Both structures are poorly absorbed through the gut wall as both the amino group and the carboxylic group are ionized. This problem can be alleviated by using a prodrug where one of the polar groups is masked with a protecting group which can be removed metabolically once the prodrug has been absorbed (Box 19.7).

Broad-spectrum penicillins: the carboxypenicillins

Carbenicillin (Fig. 19.30) was the first example of this class of compounds. It shows a broad spectrum of activity due to the hydrophilic carboxylic acid group (ionized at pH 7) on the side chain. The stereochemistry of this

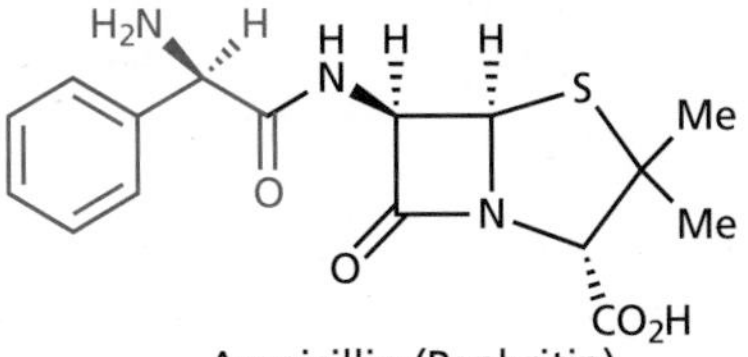

Ampicillin (Penbritin)

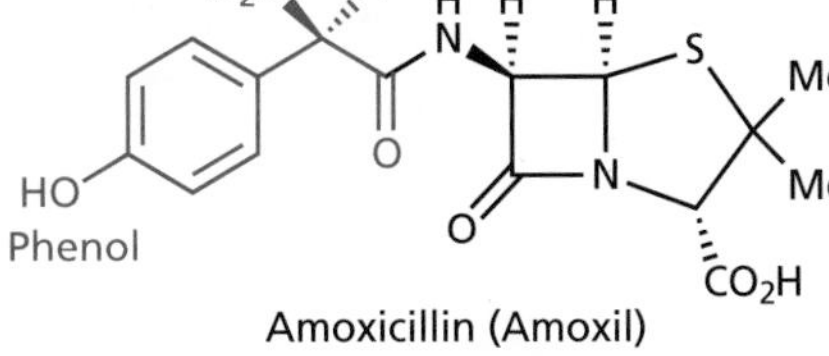

Amoxicillin (Amoxil)

FIGURE 19.29 Broad-spectrum penicillins—the aminopenicillins.

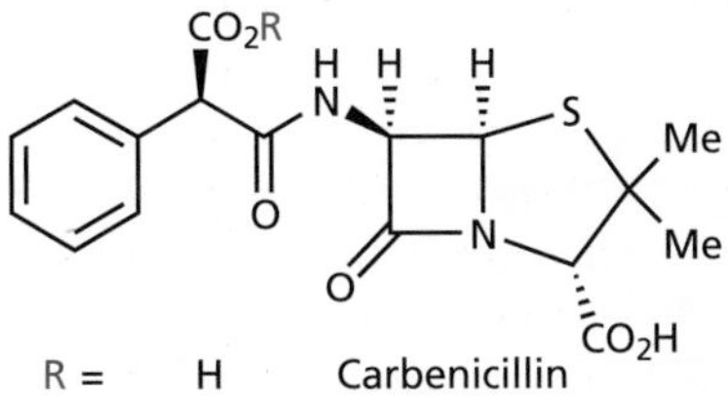

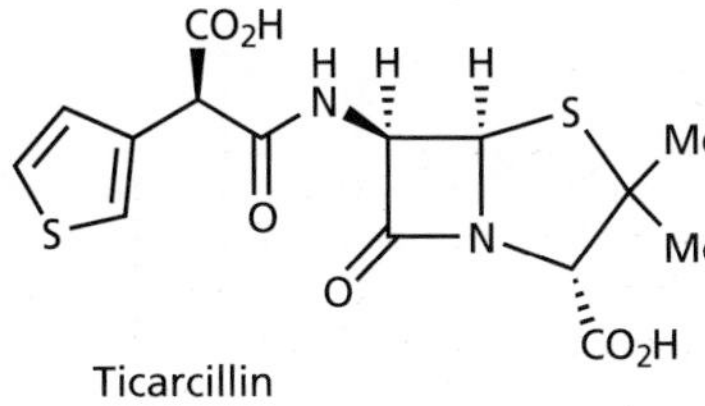

Ticarcillin

R = H Carbenicillin
R = Ph Carfecillin
R = 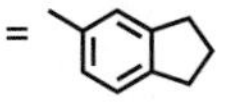 Indanyl carbenicillin

FIGURE 19.30 Carboxypenicillins.

BOX 19.7 Ampicillin prodrugs

Pivampicillin, **talampicillin**, and **bacampicillin** are prodrugs of ampicillin (Fig. 1). In all three cases, the esters used to mask the carboxylic acid group seem rather elaborate and one may ask why a simple methyl ester is not used. The answer is that methyl esters of penicillins are not metabolized in humans. The bulky penicillin skeleton is so close to the ester that it acts as a steric shield and prevents the esterase enzymes that catalyse this reaction from accepting the penicillin ester as a substrate.

Fortunately, acyloxymethyl esters *are* susceptible to esterases. These 'extended' esters contain a second ester group further away from the penicillin nucleus, which is more exposed to attack. The hydrolysis products are inherently unstable and decompose spontaneously to reveal the free carboxylic acid (Fig. 2). The mechanism also releases for-maldehyde or acetaldehyde for pivampicillin and bacampi-cillin respectively. This is not ideal as both aldehydes are toxic. However, these aldehydes are formed naturally in the body from metabolic reactions carried out in the liver, and are subsequently detoxified. The levels produced from the prodrugs described above cause little problem. Moreover, the drugs are only taken for a short duration of time.

Such extended esters can be used to prepare prodrugs of other penicillins, but one has to be careful that one does not go to the other extreme and make the penicillin too lipo-philic. For example, the 1-acyloxyalkyl ester of penicillin G is so lipophilic that it has poor solubility in water. Fortunately, the problem can easily be avoided by making the extended ester more polar (e.g. by attaching valine as in Fig. 3).

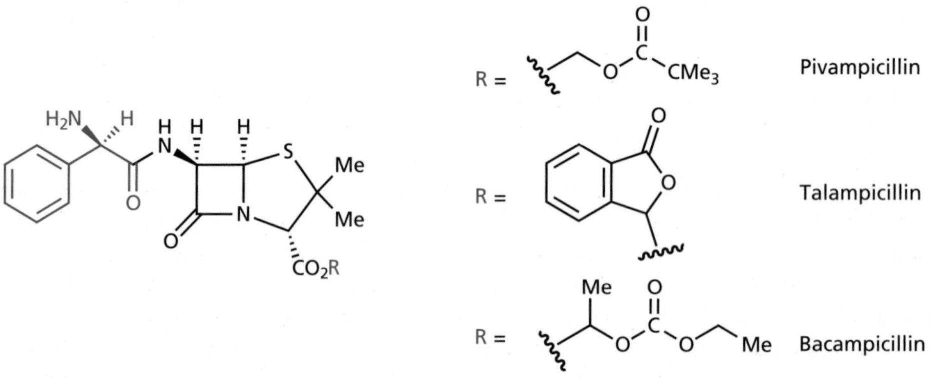

FIGURE 1 Prodrugs used to aid absorption of ampicillin through the gut wall.

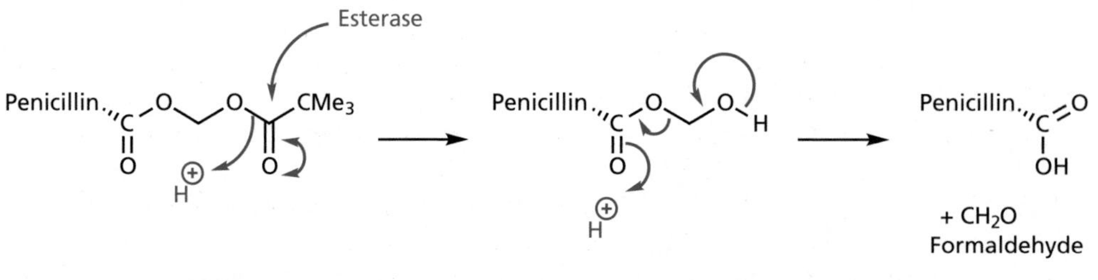

FIGURE 2 Mechanism by which acyloxymethyl esters are hydrolysed.

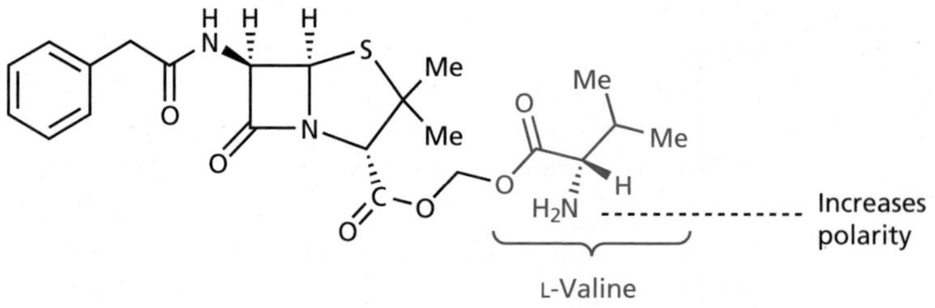

FIGURE 3 Polar extended ester for penicillin G.

group is important and only one of the two enantiomers is active.

Carfecillin and **indanyl carbenicillin** (Fig. 19.30) are prodrugs for carbenicillin and show an improved absorption through the gut wall. Aryl esters are better than alkyl esters since the former are more chemically susceptible to hydrolysis. This is because of the electron-withdrawing inductive effect of the aryl ring. An extended ester is not required in this case as the aryl ester is further from the β-lactam ring and is not shielded (cf. Box 10.7). **Ticarcillin** is similar in structure to carbenicillin, but has a thiophene ring in place of the phenyl group.

Broad-spectrum penicillins: the ureidopenicillins

Ureidopenicillins (Fig. 19.31) are the newest class of broad-spectrum penicillins and have a urea functional group at the α-position. They generally have better properties than the carboxypenicillins and have largely replaced them in the clinic (Box 19.8).

19.5.1.9 Synergism of penicillins with other drugs

There are several examples in medicinal chemistry where the presence of one drug enhances the activity of another. In many cases this can be dangerous, leading to an effective overdose of the enhanced drug. In some cases, though, it can be useful. There are two interesting examples where the activity of penicillin has been enhanced by the presence of another drug.

One of these is the effect of clavulanic acid, described in Section 19.5.4.1. The other is the administration of penicillins with a compound called **probenecid** (Fig. 19.32). Probenecid is a moderately lipophilic carboxylic acid that can block facilitated transport of penicillin through the kidney tubules. In other words, probenecid slows down the rate at which penicillin is excreted by competing with it in the excretion mechanism. Probenecid also competes with penicillin for binding sites on albumin. As

---

BOX 19.8 Clinical aspects of broad-spectrum penicillins

Ampicillin and **amoxicillin** have a similar spectrum of activity to penicillin G, but are more active against Gram-negative cocci and enterobacteria. They are non-toxic and can be taken orally, but they are sensitive to β-lactamases and are inactive against *Pseudomonas aeruginosa*. Some patients get diarrhoea when they take these penicillins. This is a result of poor absorption from the gut, with ampicillin being more poorly absorbed than amoxicillin. If penicillins are used at high doses for prolonged periods, they abolish the normal gut microflora, and this allows the colonization of resistant Gram-negative bacilli or fungi, which cause the intestinal problems. Ampicillin is currently used to treat sinusitis, bronchitis, and a variety of other infections, including oral, ear, and urinary tract infections. Amoxicillin has been used in the treatment of bronchitis, pneumonia, typhoid, gonorrhoea, Lyme disease, and urinary tract infections. Its spectrum of activity is increased when administered with **clavulanic acid** (section 19.5.4.1).

Carbenicillin was the first penicillin to show activity against *P. aeruginosa*. Compared with ampicillin, it is active against a wider range of Gram-negative bacteria, and was used particularly against penicillin-resistant strains. However, it is less active than ampicillin against various other bacterial strains, and requires high dose levels. Toxic side effects are observed and the drug shows a marked reduction in activity against Gram-positive bacteria. It is also acid sensitive and has to be injected. Better penicillins, such as the ureidopenicillins, have since been developed and so the use of carbenicillin is now discouraged.

Carfecillin and **indanyl carbenicillin** proved useful for the treatment of urinary tract infections, but have generally been superseded by fluoroquinolone antibacterial agents (section 19.8.1).

Ticarcillin is administered by injection, and has an identical antibacterial spectrum to carbenicillin. However, it has the advantage that smaller doses can be used. It is also 2–4 times more effective against *P. aeruginosa*, and has fewer side effects. The drug is mainly used against infections due to *Pseudomonas* and *Proteus* species, and is currently administered with clavulanic acid to broaden its spectrum of activity (section 19.5.4.1).

Ureidopenicillins are generally more active than the carboxypenicillins against streptococci and *Haemophilus* species. They show similar activity against Gram-negative aerobic rods such as *P. aeruginosa*, but are generally more active against other Gram-negative bacteria. Unfortunately, they have to be injected. Examples include **azlocillin**, which is 8–16 times more active than carbenicillin against *P. aeruginosa*, and is used primarily for the treatment of infections caused by that organism. It is susceptible to β-lactamases. **Mezlocillin** has a similar spectrum of activity to carbenicillin, but is more active because it has a higher affinity for transpeptidases and can cross the outer membrane of Gram-negative bacteria more effectively. **Piperacillin** is similar to ampicillin in its activity against Gram-positive species. It also has good activity against anaerobic species of both cocci and bacilli, and can be used against a variety of infections. It is more active than ticarcillin against *P. aeruginosa*. Piperacillin can be administered alongside **tazobactam** to widen its spectrum of activity (section 19.5.4.2).

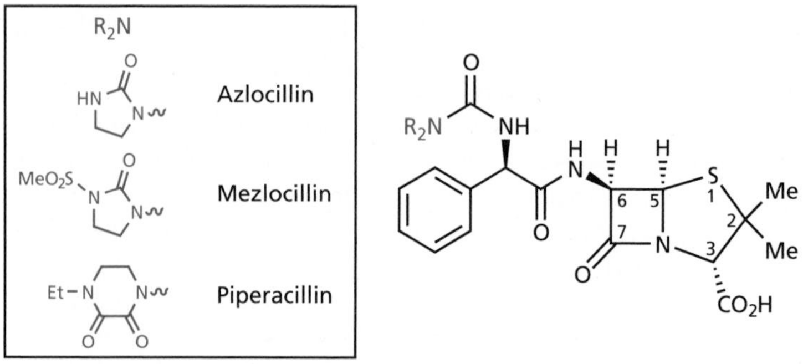

FIGURE 19.31 Ureidopenicillins.

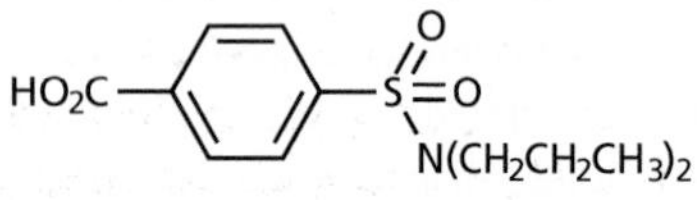

FIGURE 19.32 Probenecid.

a result, penicillin levels in the bloodstream are enhanced and the antibacterial activity increases—a useful tactic if faced with a particularly resistant bacterium.

KEY POINTS

- Penicillins have a bicyclic structure consisting of a β-lactam ring fused to a thiazolidine ring. The strained β-lactam ring reacts irreversibly with the transpeptidase enzyme responsible for the final cross-linking of the bacterial cell wall.

- Penicillin analogues can be prepared by fermentation or by a semi-synthetic synthesis from 6-aminopenicillanic acid. Variation of the penicillin structure is limited to the acyl side chain.

- Penicillins can be made more resistant to acid conditions by incorporating an electron-withdrawing group into the acyl side chain.

- Steric shields can be added to penicillins to protect them from bacterial β-lactamase enzymes.

- Broad-spectrum activity is associated with the presence of an α-hydrophilic group on the acyl side chain of penicillin.

- Prodrugs of penicillins are useful in masking polar groups and improving absorption from the gastrointestinal tract. Extended esters are used which undergo enzyme-catalysed hydrolysis to produce a product which spontaneously degrades to release the penicillin.

- Probenecid can be administered with penicillins to hinder the excretion of penicillins.

19.5.2 **Cephalosporins**

19.5.2.1 Cephalosporin C

Discovery and structure of cephalosporin C

The second major group of β-lactam antibiotics to be discovered were the cephalosporins. The first cephalosporin

(**cephalosporin C**) was derived from a fungus obtained in the mid-1940s from sewer waters on the island of Sardinia. This was the work of an Italian professor who noted that the waters surrounding the sewage outlet periodically cleared of microorganisms. He reasoned that an organism might be producing an antibacterial substance, and so he collected samples and managed to isolate a fungus called *Cephalosporium acremonium* (now called *Acremonium chrysogenum*). The crude extract from this organism was shown to have antibacterial properties, and, in 1948, workers at Oxford University isolated cephalosporin C, but it was not until 1961 that the structure was established by X-ray crystallography.

The structure of cephalosporin C (Fig. 19.33) has similarities to that of penicillin in that it has a bicyclic system containing a four-membered β-lactam ring, but this time the β-lactam ring is fused to a six-membered dihydrothiazine ring. Nevertheless, cephalosporins are derived from the same biosynthetic precursors as penicillin (i.e. cysteine and valine) (Fig. 19.34).

Properties of cephalosporin C

Cephalosporin C is not particularly potent compared to penicillins (1/1000 the activity of penicillin G), but the antibacterial activity that it *does* have is more evenly directed against Gram-negative and Gram-positive bacteria. Another in-built advantage of cephalosporin C is its greater resistance to acid hydrolysis and β-lactamase

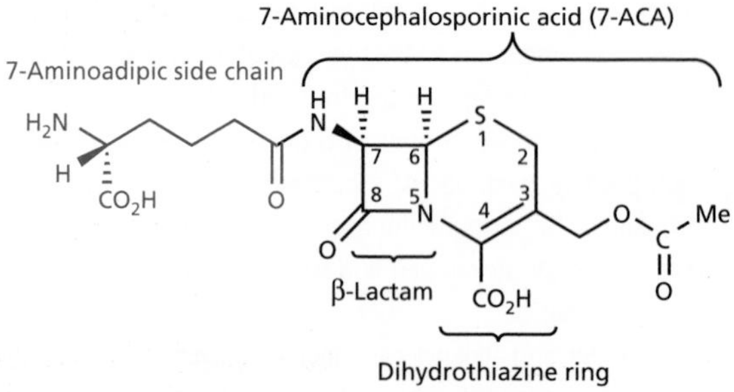

FIGURE 19.33 Cephalosporin C.

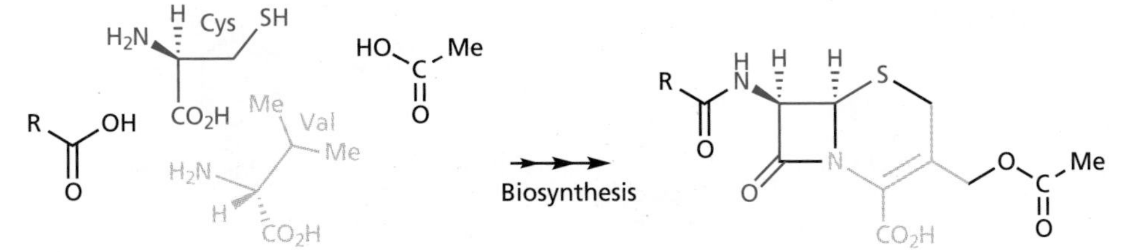

FIGURE 19.34 Biosynthetic precursors of cephalosporin C.

enzymes. It is also less likely to cause allergic reactions. Therefore, cephalosporin C was seen as a useful lead compound for the development of further broad-spectrum antibiotics, hopefully with increased potency.

Structure–activity relationships of cephalosporin C

Many analogues of cephalosporin C have been made which demonstrate the importance of the β-lactam ring within the bicyclic system, an ionized carboxylate group at position 4, and the acylamino side chain at position 7. These results tally closely with those obtained for the penicillins. The strain effect of a six-membered ring fused to a four-membered ring is less than for penicillin, but this is partially offset by the effect of the acetyloxy group at position 3. This can act as a good leaving group in the inhibition mechanism (Fig. 19.35).

There are a limited number of places where modifications can be made (Fig. 19.36), but there are more possibilities than with penicillins. These are as follows:

- variations of the 7-acylamino side chain;
- variations of the 3-acetoxymethyl side chain;
- extra substitution at carbon 7.

19.5.2.2 Synthesis of cephalosporin analogues at position 7

Access to analogues with varied side chains at position 7 initially posed a problem. Unlike penicillins, it proved impossible to obtain cephalosporin analogues by fermentation. Similarly, it was not possible to obtain **7-ACA**

(**7-aminocephalosporinic acid**) either by fermentation or by enzymatic hydrolysis of cephalosporin C, thus preventing the semi-synthetic approach analogous to the preparation of penicillins from 6-APA (section 19.5.1.6).

Therefore, a way had to be found of obtaining 7-ACA from cephalosporin C by chemical hydrolysis. This is no easy task, as a secondary amide has to be hydrolysed in the presence of a highly reactive β-lactam ring. Normal hydrolytic procedures are not suitable and so a special method had to be worked out (Fig. 19.37).

The first step of the procedure requires the formation of an imino chloride by the mechanism shown in Fig. 19.38. This is only possible for the secondary amide group, as ring constraints prevent the β-lactam nitrogen forming a double bond within the β-lactam ring. The imino chloride can then be treated with an alcohol to give an imino ether. This functional group is more susceptible to hydrolysis than the β-lactam ring, and so treatment with aqueous acid successfully gives the desired 7-ACA which can then be acylated to give a range of analogues.

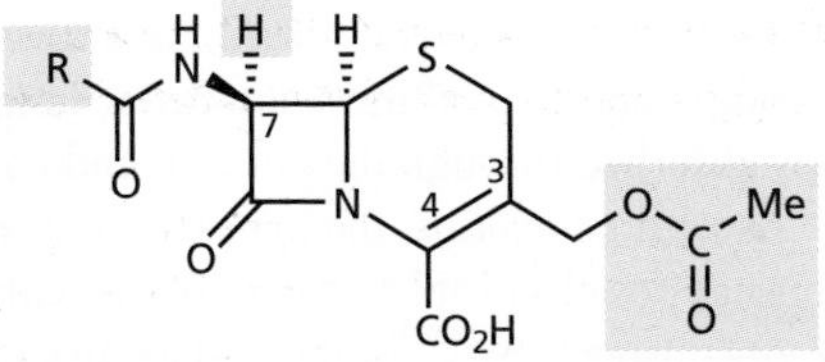

FIGURE 19.36 Positions for possible modification of cephalosporin C. The shading indicates positions which can be varied.

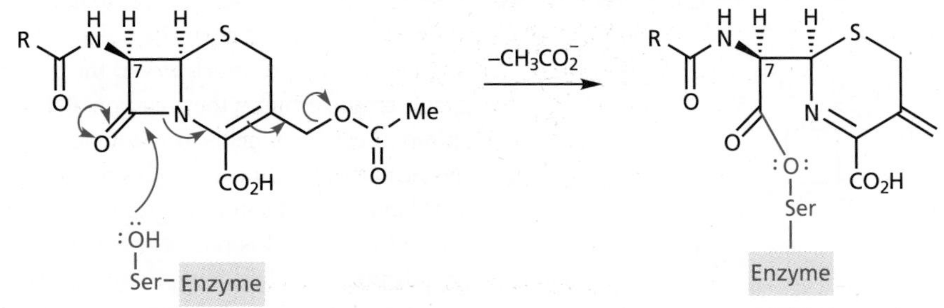

FIGURE 19.35 Mechanism by which cephalosporins inhibit the transpeptidase enzyme.

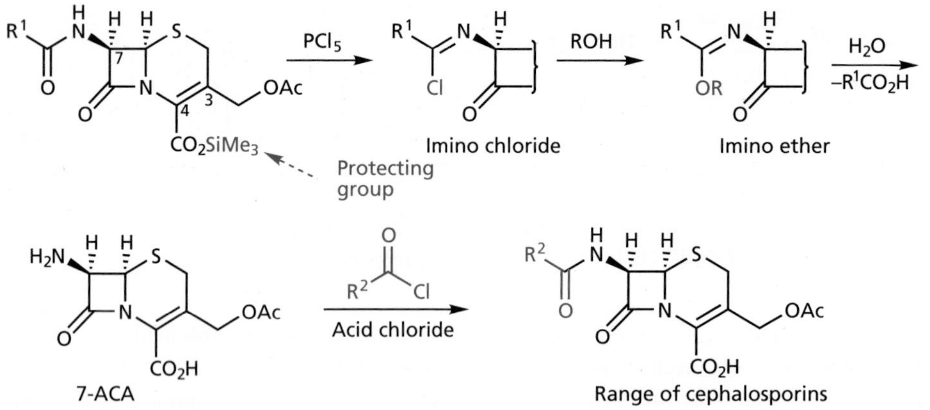

FIGURE 19.37 Synthesis of 7-ACA and cephalosporin analogues.

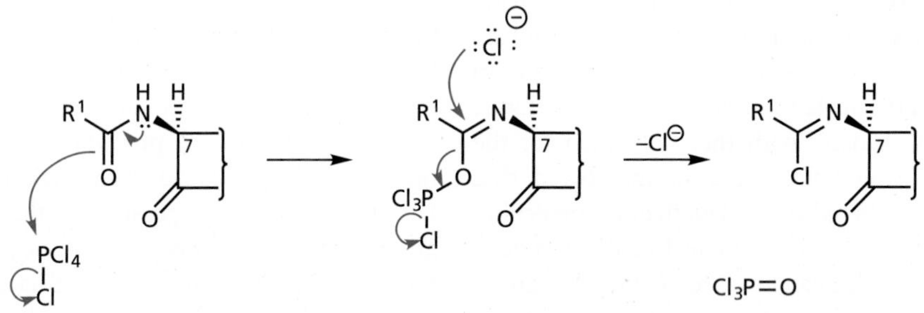

FIGURE 19.38 Mechanism for imino chloride formation.

19.5.2.3 First-generation cephalosporins

Examples of first-generation cephalosporins include **cephalothin**, **cephaloridine**, **cefalexin**, and **cefazolin** (Figs. 19.39–19.42). In general, they have a lower activity than comparable penicillins, but a better range. Most are poorly absorbed through the gut wall and have to be injected. As with penicillins, the appearance of resistant organisms has posed a problem, particularly with Gram-negative organisms. These contain β-lactamases which are more effective than the β-lactamases of Gram-positive organisms. Steric shields are successful in protecting cephalosporins from these β-lactamases, but also prevent them from inhibiting the transpeptidase target enzymes.

One of the most commonly used first-generation cephalosporins was **cephalothin** (Fig. 19.39). A disadvantage

with cephalothin is the fact that the acetyloxy group at position 3 is readily hydrolysed by esterase enzymes to give the less active alcohol (Fig. 19.40). The acetyloxy group is important to the mechanism of inhibition and acts as a good leaving group, whereas the alcohol is a much poorer leaving group. Therefore, it would be useful if this metabolism could be blocked to prolong activity. Replacing the ester with a metabolically stable pyridinium group gives **cephaloridine** (Fig. 19.41). The pyridine can still act as a good leaving group for the inhibition mechanism, but is not cleaved by esterases. Cephaloridine exists as a zwitterion and is soluble in water, but, like most first-generation cephalosporins, it is poorly absorbed through the gut wall and has to be injected.

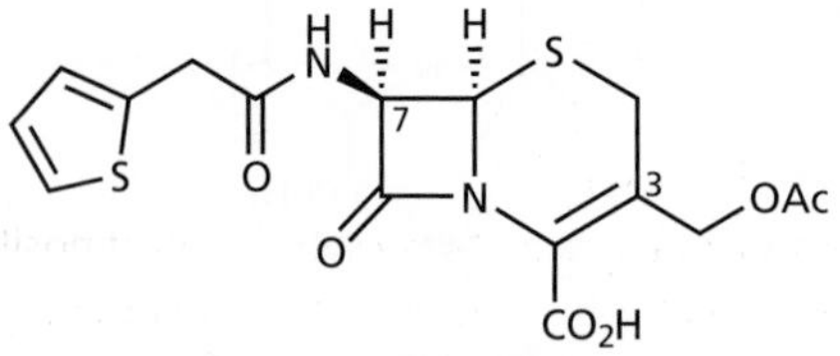

Cefalexin (Fig. 19.41) has a methyl substituent at position 3 (Box 19.9) which appears to help oral absorption. A methyl group would normally be bad for activity as it is not a good leaving group. However, the presence of a hydrophilic amino group at the α-carbon of the 7-acylamino side chain in cefalexin helps to restore activity and cephalexin is one of the few cephalosporins which is orally active. The mechanism of absorption through the gut wall is poorly understood and it is not clear why the 3-methyl group is so advantageous for absorption. **Cefazolin** (Fig. 19.42) is another example of a first-generation cephalosporin.

FIGURE 19.39 Cephalothin.

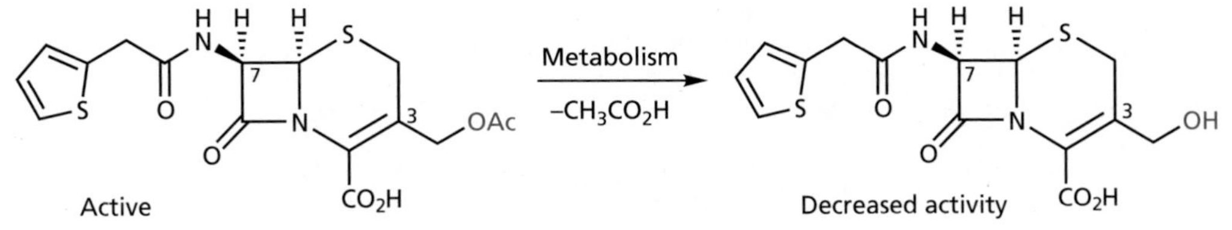

FIGURE 19.40 Metabolic hydrolysis of cephalothin.

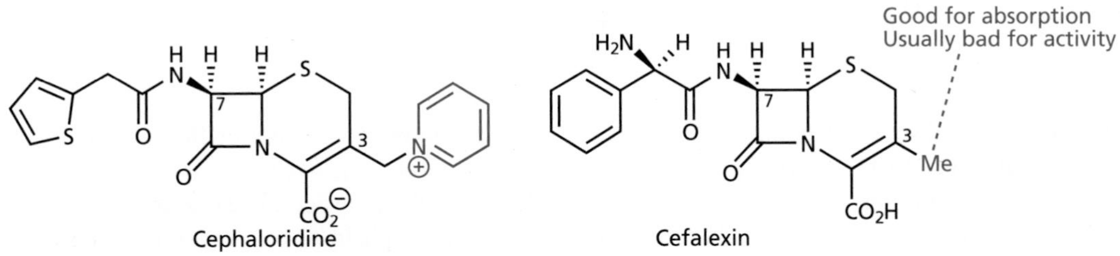

FIGURE 19.41 Cephaloridine and cefalexin.

BOX 19.9 Synthesis of 3-methylated cephalosporins

The synthesis of 3-methylated cephalosporins involves the use of a penicillin starting material as shown below. The synthesis, which was first demonstrated by Eli Lilly Pharmaceuticals, involves oxidation of sulphur followed by an acid-catalysed ring expansion, where the five-membered thiazolidine ring in penicillin is converted to the six-membered dihydrothiazine ring in cephalosporin.

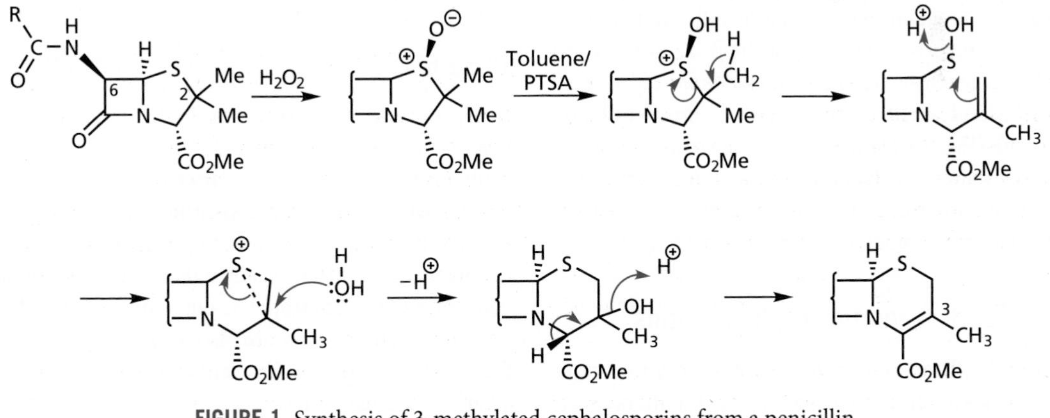

FIGURE 1 Synthesis of 3-methylated cephalosporins from a penicillin.

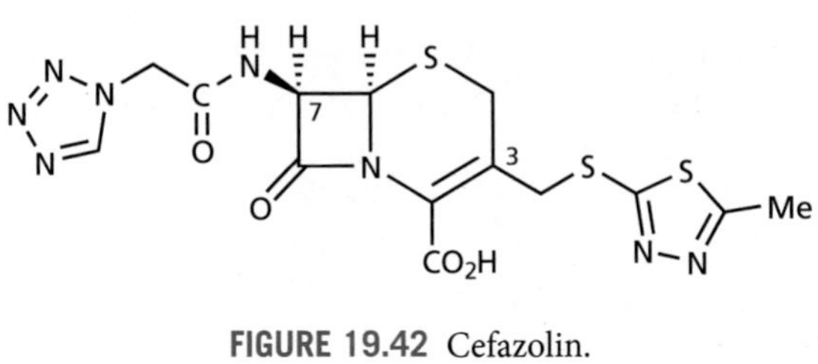

FIGURE 19.42 Cefazolin.

19.5.2.4 Second-generation cephalosporins

Cephamycins

Cephamycins contain a methoxy substituent at position 7, which has proved advantageous. The parent compound **cephamycin C** (Fig. 19.43) was isolated from a culture of *Streptomyces clavuligerus* and was the first β-lactam to be

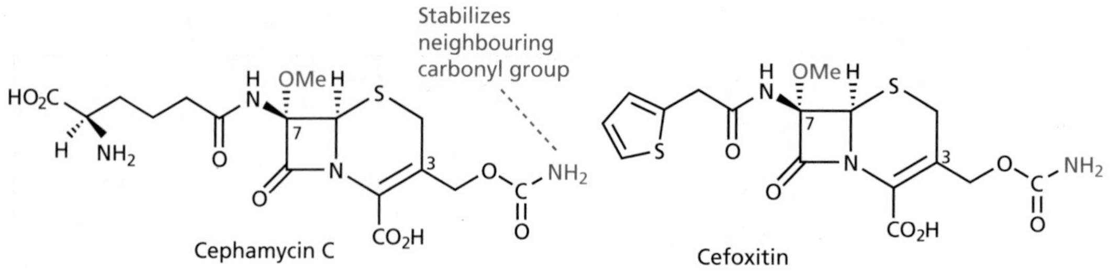

FIGURE 19.43 Cephamycin C and cefoxitin.

isolated from a bacterial source. Modification of the side chain gave **cefoxitin** (Fig. 19.43), which showed a broader spectrum of activity than most first-generation cephalosporins. This is due to greater resistance to β-lactamase enzymes, which may be due to the steric hindrance provided by the methoxy group. Cefoxitin shows good metabolic stability to esterases due to the presence of the urethane group at position 3, rather than an ester (section 14.2.2).

Ⓦ Test your understanding and practise your molecular modelling with Exercise 19.4 on the Online Resource Centre at www.oxfordtextbooks.co.uk/orc/patrick6e/

Oximinocephalosporins

The development of oximinocephalosporins has been a major advance in cephalosporin research. These structures contain an iminomethoxy group at the α-position of the acyl side chain, which significantly increases the stability of cephalosporins against the β-lactamases produced by some organisms (e.g. *Haemophilus influenzae*). The first useful agent in this class of compounds was **cefuroxime** (Fig. 19.44), which, like cefoxitin, has an increased resistance to β-lactamases and mammalian esterases. Unlike cefoxitin, cefuroxime retains activity against streptococci and, to a lesser extent, staphylococci.

19.5.2.5 Third-generation cephalosporins

Replacing the furan ring of the aforesaid oximinocephalosporins with an aminothiazole ring enhances the penetration of cephalosporins through the outer membrane of Gram-negative bacteria, and may also increase affinity for the transpeptidase enzyme. As a result, third-generation cephalosporins containing this ring show a marked increase in activity against these bacteria. A variety of such structures have been prepared, such as **ceftazidime**, **cefotaxime**, **ceftizoxime**, and **ceftriaxone** (Figs. 19.44 and 19.45), with different substituents at position 3 to vary the pharmacokinetic properties. They play a major role in antimicrobial therapy because of their activity against Gram-negative bacteria, many of which are resistant to other β-lactams. As such infections are uncommon outside hospitals, physicians are discouraged from prescribing these drugs routinely and they are viewed as 'reserve troops' to be used for troublesome infections which do not respond to the more commonly prescribed β-lactams.

19.5.2.6 Fourth-generation cephalosporins

Cefepime and **cefpirome** (Fig. 19.45) are oximinocephalosporins which have been classed as fourth-generation cephalosporins. They are zwitterionic compounds having a positively charged substituent at position 3 and a negatively charged carboxylate group at position 4. This property appears to radically enhance the ability of these compounds to penetrate the outer membrane of Gram-negative bacteria. They are also found to have a good affinity for the transpeptidase enzyme and a low affinity for a variety of β-lactamases.

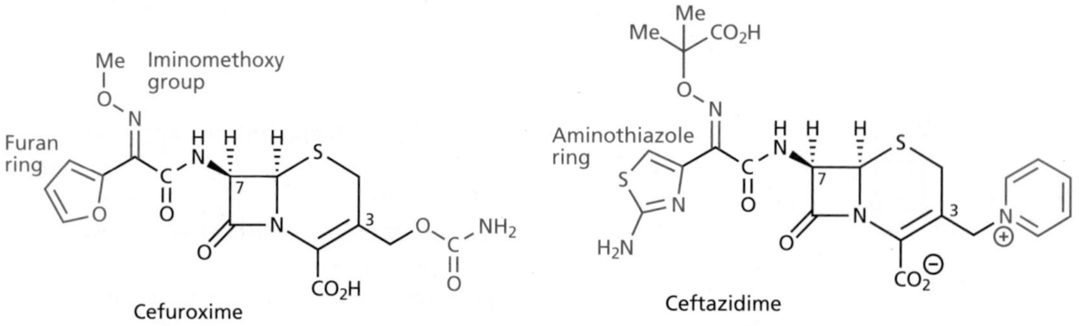

FIGURE 19.44 Oximinocephalosporins.

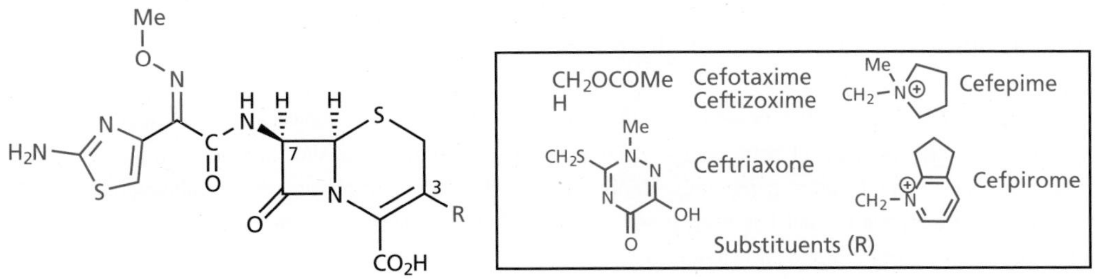

FIGURE 19.45 Third- and fourth-generation oximinocephalosporins.

19.5.2.7 Fifth-generation cephalosporins

Ceftaroline fosamil (Fig. 19.46) is a fifth-generation cephalosporin that has activity against various strains of MRSA and multi-resistant *Streptococcus pneumonia* (MDRSP). It acts as a prodrug for **ceftaroline**, and the 1,3-thiazole ring is thought to be important for its activity against MRSA. **Ceftolozane** was approved in 2015. The basic amino group at position 3 of the pyrazole ring is believed to be important in increasing permeability across the outer membrane of *Pseudomonas aeruginosa* cells, while the basic side chain at position 4 increases activity against strains of *P. aeruginosa* that produce Class C β-lactamases. The urea group partially rigidifies the side chain and this significantly reduces convulsive side effects.

19.5.2.8 Resistance to cephalosporins

The activity of a specific cephalosporin against a particular bacterial cell is dependent on the same factors as those for penicillins, i.e. the ability to reach the transpeptidase enzyme, stability to any β-lactamases which might be present, and the affinity of the antibiotic for the target. For example, most cephalosporins (with the exception of cephaloridine) are stable to the β-lactamase produced by *S. aureus* and can reach the transpeptidase enzyme without difficulty. Therefore, the relative ability of cephalosporins to inhibit *S. aureus* comes down to their affinity for the target transpeptidase enzyme. Agents such as the cephamycins and

ceftazidime have poor affinity, whereas other cephalosporins have a higher affinity. The MRSA organism contains a modified transpeptidase enzyme (**PBP2a**) for which both penicillins and cephalosporins have poor affinity.

KEY POINTS

- Cephalosporins contain a strained β-lactam ring fused to a dihydrothiazine ring.

- In general, first-generation cephalosporins offer advantages over penicillins in that they have greater stability to acid conditions and β-lactamases, and have a good ratio of activity against Gram-positive and Gram-negative bacteria. However, they have poor oral availability and are generally lower in activity.

- Variation of the 7-acylamino side chain alters antimicrobial activity, whereas variation of the side chain at position 3 predominantly alters the metabolic and pharmacokinetic properties of the compound. Introduction of a methoxy substitution at C-7 is possible.

- Semi-synthetic cephalosporins can be prepared from 7-aminocephalosporanic acid (7-ACA).

- 7-ACA is obtained from the chemical hydrolysis of cephalosporins. This requires prior activation of the side chain to make it more reactive than the β-lactam ring.

- Deacetylation of cephalosporins occurs metabolically to produce less active metabolites. Metabolism can be blocked by replacing the susceptible acetoxy group with metabolically stable groups.

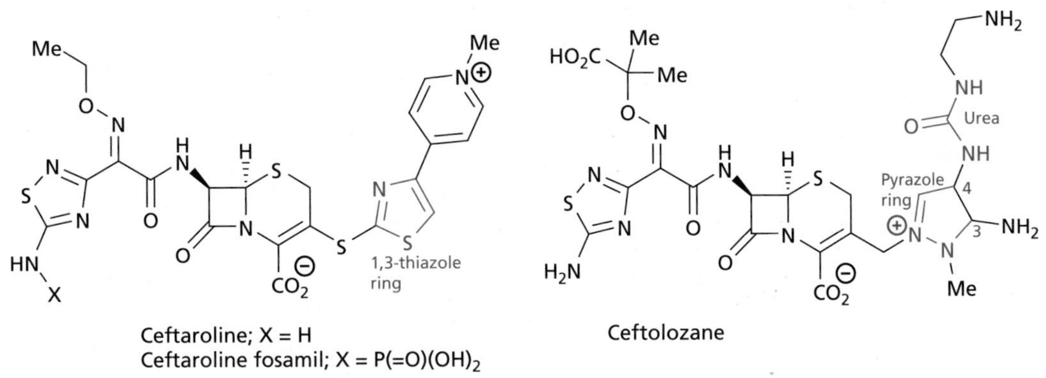

Ceftaroline; X = H
Ceftaroline fosamil; X = P(=O)(OH)$_2$

Ceftolozane

FIGURE 19.46 Fifth-generation cephalosporins.

BOX 19.10 Clinical aspects of cephalosporins

In general, cephalosporins are useful broad-spectrum anti-bacterial agents for the treatment of septicaemia, pneumonia, meningitis, biliary tract infections, peritonitis, and urinary tract infections. **Cephalosporin C** itself has been used in the treatment of urinary tract infections, as it is found to concentrate in the urine and survive the body's hydrolytic enzymes.

First-generation cephalosporins

First-generation cephalosporins have good activity against Gram-positive cocci and they can be used to treat some community-derived Gram-negative infections (i.e. infections not caught in a hospital). They can also be used against *S. aureus* and streptococcal infections when penicillins have to be avoided. **Cephalothin** is more active than penicillin G against some Gram-negative bacteria and is less likely to cause allergic reactions. It can also be used against β-lactamase-producing *S. aureus* strains.

Cefalexin is useful for the treatment of urinary tract infections which do not respond to other drugs or which occur in pregnancy. It is also useful in treating infections of the respiratory tract, ear, skin, and mouth. **Cefazolin** is recommended for use as a prophylactic to prevent infection when surgical procedures are used to implant foreign bodies.

Second-generation cephalosporins

In general, the second-generation cephalosporins have variable activity against Gram-positive cocci, but increased activity against Gram-negative bacteria. **Cefoxitin** is active against bowel flora including *Bacteroides fragilis*, and was once recommended for peritonitis. **Cefuroxime** has a wide spectrum of activity and is useful against organisms which have become resistant to penicillin. However, it is not active against 'difficult' bacteria such as *Pseudomonas aeruginosa*. It is used clinically against *Neisseria gonorrhoeae* and respiratory infections caused by *Haemophilus influenza*, *Moraxella catarrhalis*, and susceptible strains of *S. pneumoniae*. It is also used for surgical prophylaxis, as well as for the treatment of Lyme disease. **Cefotaxime** is used in surgical prophylaxis and for the treatment of gonorrhoea, meningitis, and epiglottitis infections due to *H. influenzae*.

Third-generation cephalosporins

Third-generation cephalosporins have good activity against Gram-negative bacteria, but vary in their activity against Gram-positive cocci. The ability to attack *P. aeruginosa* also varies from structure to structure, and they lack activity against the MRSA organisms and *Enterobacter* species. **Ceftazidime** is an injectable cephalosporin which has excellent activity against *P. aeruginosa,* as well as other Gram-negative bacteria. Because the drug can cross the blood–brain barrier, it can be used to treat meningitis. Compared with the other aminothiazole structures, ceftazidime has good activity against streptococci, but loses activity against strains of methicillin-susceptible *S. aureus*. This is due to a decreased binding affinity for the transpeptidase enzyme present in *S. aureus*. **Ceftriaxone** is used for surgical prophylaxis and as a prophylactic for meningococcal meningitis.

Fourth- and fifth-generation cephalosporins

Fourth-generation cephalosporins have activity against Gram-positive cocci and a broad array of Gram-negative bacteria, including *P. aeruginosa* and many of the enterobacterial species. **Cefpirome** is administered as an intravenous injection or infusion, and has been used against a variety of sensitive Gram-positive and Gram-negative bacteria. **Ceftaroline fosamil** has been licensed for the treatment of bacterial pneumonia and acute bacterial skin infections. **Ceftolozane** was approved in 2014 for the treatment of Gram-negative infections that have gained resistance to traditional antibacterial agents. It is administered alongside tazobactam, particularly for urinary tract infections and intra-abdominal infections. It is also effective against *P. aeruginosa*.

- A methyl substituent at position 3 is good for oral absorption but bad for activity, unless a hydrophilic group is present at the α-position of the acyl side chain.

- 3-Methylated cephalosporins can be synthesized from penicillins.

- Cephamycins are cephalosporins containing a methoxy group at position 7.

- Oximinocephalosporins have resulted in several generations of cephalosporins with increased potency and a broader spectrum of activity, particularly against Gram-negative bacteria.

19.5.3 Other β-lactam antibiotics

Although penicillins and cephalosporins are the best known and most researched β-lactams, there are other β-lactam structures which are of great interest in the antibacterial field.

19.5.3.1 Carbapenems

Thienamycin (Fig. 19.47) was the first example of this class of compounds and was isolated from *Streptomyces*

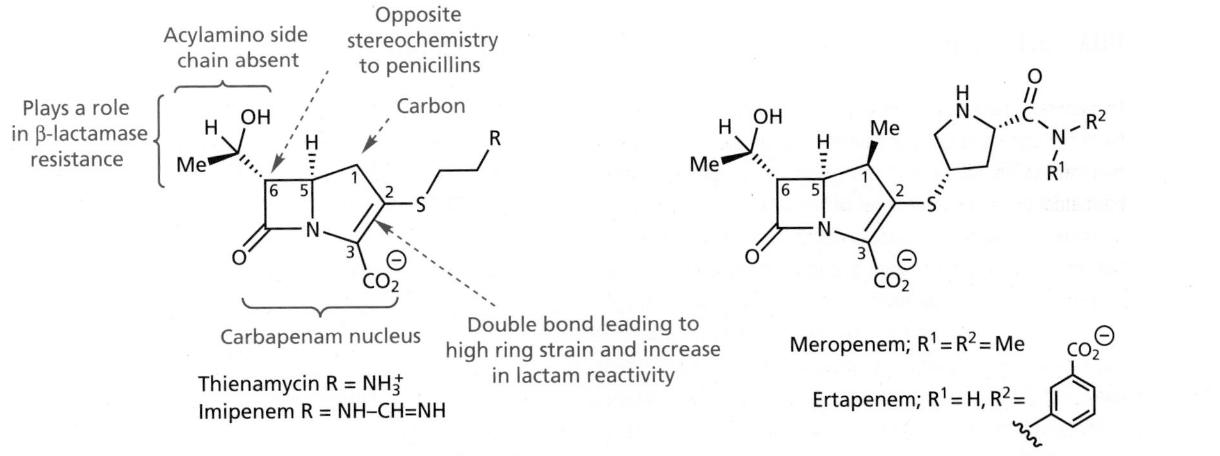

FIGURE 19.47 Carbapenems.

cattleya in 1976. It is potent with an extraordinarily broad range of activity against Gram-positive and Gram-negative bacteria, including *P. aeruginosa*. It has low toxicity and shows a high resistance to β-lactamases. This resistance has been ascribed to the presence of the hydroxyethyl side chain. Unfortunately, it shows poor metabolic and chemical stability, and is not absorbed from the gastrointestinal tract. The surprising features in thienamycin are the missing sulphur atom and acylamino side chain, both of which were thought to be essential to antibacterial activity. Furthermore, the stereochemistry of the side chain at substituent 6 is opposite from the usual stereochemistry in penicillins—another factor contributing to the resistance of this agent to β-lactamases. **Imipenem** and **meropenem** are clinically useful analogues of thienamycin (Box 19.11). Imipenem is susceptible to metabolism by a dehydropeptidase enzyme, whereas meropenem is more resistant because of the substituents at positions 1 and 2. **Ertapenem** was approved in 2002 and is similar in structure to meropenem. It has increased stability against dehydropeptidases, while the presence of an ionized benzoic acid contributes to high protein binding. This prolongs the half-life of the drug such that once daily dosing is feasible. In general, the carbapenems have the broadest spectrum of activity of all the β-lactam antibiotics.

19.5.3.2 Monobactams

Monocyclic β-lactams such as the **nocardicins** (Fig. 19.48) have been isolated from natural sources. At least seven nocardicins have been isolated by the Japanese company Fujisawa. They show moderate activity *in vitro* against a narrow group of Gram-negative bacteria, including *P. aeruginosa*. Surprisingly, they contain a single β-lactam ring, demonstrating that a fused second ring is not always essential for antibacterial activity.

One explanation for this is that nocardicins might have a different mechanism of action from penicillins and cephalosporins—possibly by inhibiting a different enzyme involved in cell wall synthesis. This would help to explain why nocardicins are inactive against Gram-positive bacteria, and generally show a different spectrum of activity from the other β-lactam antibiotics. They also show low levels of toxicity. **Aztreonam** (Fig. 19.48) is an example of a monobactam which has reached the clinic and was developed from a naturally occurring monobactam isolated from *Chromobacterium violaceum*. **BAL30072** is an analogue of aztreonam. It contains a dihydropyridone ring which is intended to act as a siderophore and bind to iron ions. By doing so, it is hoped that the drug will be taken into bacterial cells by iron transport systems. It is currently in phase I of clinical trials.

19.5.4 β-Lactamase inhibitors

The β-lactamases are divided into two major classes—the serine-based β-lactamases (classes A, C, and D) and the metallo-β-lactamases (class B).

There are four clinically approved β-lactamase inhibitors—clavulanic acid, sulbactam, tazobactam, and avibactam. The first three of these are effective against class A serine-based β-lactamases, but are less effective against the other three classes. Avibactam shows activity for class C and some class D β-lactamases.

19.5.4.1 Clavulanic acid

Clavulanic acid (Fig. 19.49) was isolated from *Streptomyces clavuligerus* by Beechams in 1976. It has weak and unimportant antibiotic activity, but it is a powerful and irreversible inhibitor of most β-lactamases, which means that it is used as a sentry drug (section 14.7.1) in

BOX 19.11 Clinical aspects of miscellaneous β-lactam antibiotics

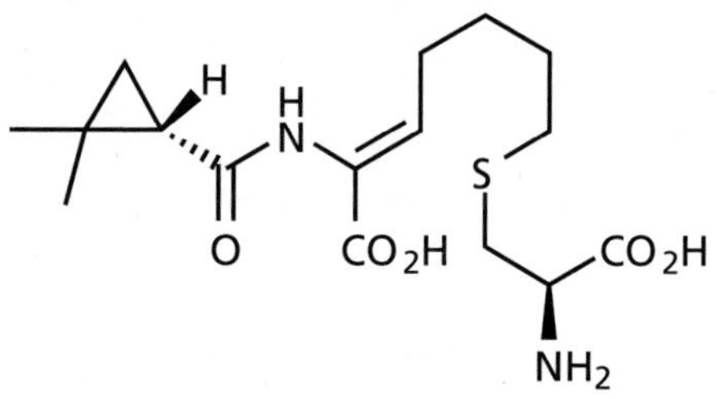

Imipenem is active against a variety of aerobic, anaerobic, Gram-positive and Gram-negative infections, and can be effective against some infections which do not respond to cephalosporins, or infections which have become resistant to more conventional β-lactams. It can be used against hospital-acquired septicaemia, and for surgical prophylaxis. The structure is metabolized by a dehydropeptidase enzyme to produce metabolites that are toxic to the kidney, but this can be alleviated by administrating the drug alongside **cilastatin** (Fig. 1)—a dehydropeptidase inhibitor which protects imipenem from metabolism. Administration is by intramuscular injection or by intravenous infusion. **Meropenem** is also effective against a variety of aerobic, anaerobic, Gram-positive and Gram-negative infections, and is administered by intravenous injection or infusion. Meropenem is slightly less active than imipenem against Gram-positive bacteria, but is more active against Gram-negative bacteria. Unlike imipenem, meropenem is active against *Pseudomonas aeruginosa*, and can be administered on its own because it is more resistant to dehydropeptidases. Both meropenem and imipenem penetrate the outer membrane of Gram-negative bacteria through porins, but meropenem enters more efficiently, resulting in higher activity against these bacteria. The drug has been used to treat pneumonia, meningitis, abdominal infections, and urinary tract infections. **Ertapenem**

FIGURE 1 Cilastatin.

is administered by intravenous infusion and is used for the treatment of abdominal infections, acute gynaecological infections, community-acquired pneumonia, and diabetic foot infections of the skin and soft tissue. It is also used as a prophylactic for colorectal surgery. **Aztreonam** is used against Gram-negative infections including *P. aeruginosa*, *Haemophilus influenzae*, and *Neisseria meningitidis*. It is administered by intravenous injection and can be safely used for patients with allergies to penicillin or cephalosporins. It has no activity against Gram-positive organisms or anaerobic bacteria, because it does not bind to the transpeptidases produced by these organisms. However, it can bind to, and inhibit, the transpeptidases produced by Gram-negative aerobic organisms.

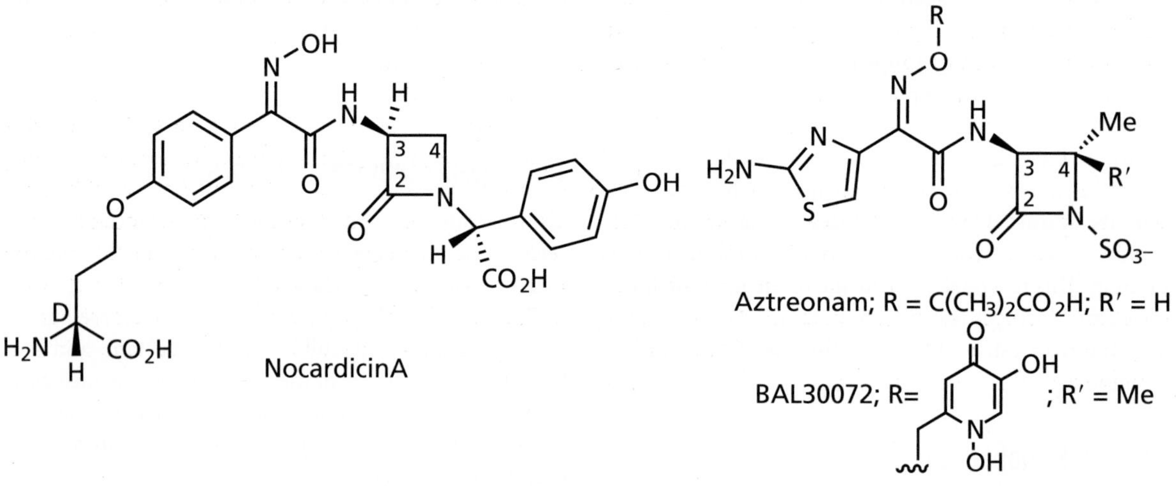

FIGURE 19.48 Monobactams.

combination with traditional penicillins such as amoxicillin (**Augmentin**). This allows the dose levels of amoxicillin to be decreased and also increases its spectrum of activity. However, it should be noted that clavulanic acid is not effective against all classes of β-lactamase enzymes.

Clavulanic acid is also administered intravenously with ticarcillin as **Timentin**.

The structure of clavulanic acid was the first example of a naturally occurring compound where the β-lactam ring was not fused to a sulphur-containing ring; instead

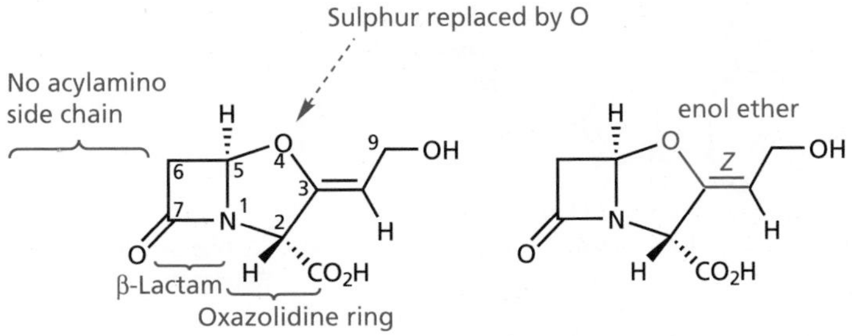

FIGURE 19.49 Clavulanic acid.

it is fused to an oxazolidine ring. The structure is also unusual in that it does not have an acylamino side chain.

Many analogues of clavulanic acid have been made and the essential requirements for β-lactamase inhibition are:

- a strained β-lactam ring;
- the enol ether;
- the Z configuration for the double bond of the enol ether (activity is reduced but not eliminated if the double bond is E);
- no substituent at C-6;
- (R)-stereochemistry at positions 2 and 5;
- the carboxylic acid group.

It is also thought that the 9-hydroxyl group is involved in a hydrogen bonding interaction with the active site of the β-lactamase enzyme. Clavulanic acid is a mechanism-based irreversible inhibitor and can be classed as a **suicide substrate**. The mechanism of inhibition is shown in section 7.5.

19.5.4.2 Penicillanic acid sulphone derivatives

The agents **sulbactam** and **tazobactam** have also been developed as β-lactamase inhibitors and are used in the clinic (Fig. 19.50). They, too, act as suicide substrates for β-lactamase enzymes and have similar properties. Sulbactam has a broader spectrum of activity against β-lactamases than clavulanic acid, but is less potent. It

is combined with ampicillin for intravenous administration in a preparation called **Unasyn**. Tazobactam is similar to sulbactam and has a similar spectrum of activity against β-lactamases. However, its potency is more like clavulanic acid. It is administered intravenously with piperacillin in a preparation called **Tazocin** or **Zosyn**, which has the broadest spectrum of activity of the various combinations described so far.

19.5.4.3 Olivanic acids

The olivanic acids (e.g. **MM 13902**) (Fig. 19.51) were isolated from strains of *Streptomyces olivaceus* and are carbapenem structures like thienamycin. They are very strong inhibitors of β-lactamase—in some cases 1000 times more potent than clavulanic acid. They are also effective against the β-lactamases which break down cephalosporins and are unaffected by clavulanic acid. Unfortunately, olivanic acids lack chemical stability.

19.5.4.4 Avibactam

Avibactam (Fig. 19.51) was approved in 2015 as a combination therapy with ceftazidime for the treatment of antibiotic-resistant infections of the urinary tract. It belongs to a class of compounds called the diazabicyclooctanes, and it is the first new β-lactamase inhibitor to reach the market in over 20 years. Unlike previous β-lactamase inhibitors, it lacks a β-lactam ring, and it inhibits some β-lactamases that are unaffected by tazobactam or clavulanic acid.

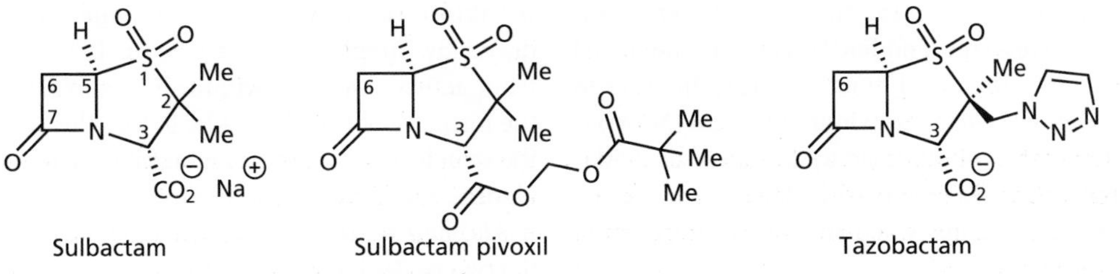

FIGURE 19.50 Penicillanic acid sulphones.

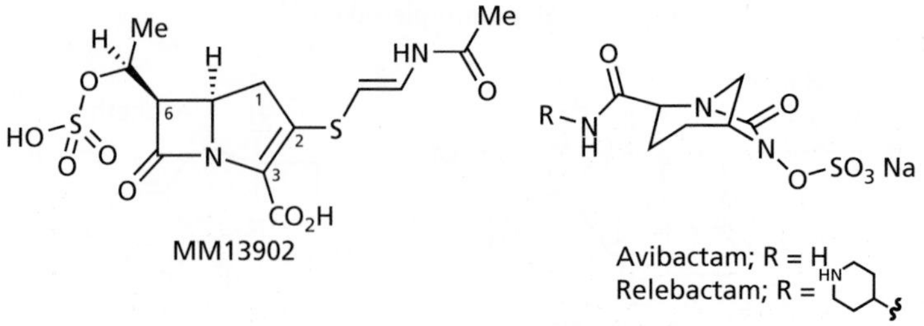

FIGURE 19.51 MM 13902, avibactam, and relebactam.

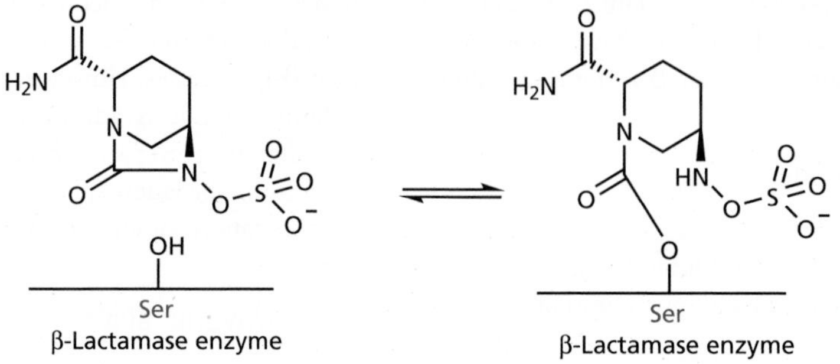

FIGURE 19.52 Reaction of avibactam with the active site of a β-lactamase enzyme.

Avibactam also differs from previous inhibitors by act-ing as a slow, reversible inhibitor. Reaction with a serine residue in the active site opens the five-membered cyclic urea, and a covalent bond is formed between the ring-opened drug and the active site (Fig. 19.52). However, the reaction is reversible, and so the cyclic urea can reform to release the drug from the active site. Nevertheless, the drug is free to re-enter the active site and react once more. This results in a prolonged inhibition. **Relebactam** (Fig. 19.51) is an analogue that is currently undergoing clinical trials as a combination therapy with imipenem and cilastatin.

19.5.5 Other drugs which act on bacterial cell wall biosynthesis

β-Lactams are not the only antibacterial agents that in-hibit cell wall biosynthesis. The antibacterial agents **van-comycin**, **D-cycloserine**, and **bacitracin** also inhibit cell wall biosynthesis, though at different stages. In order to synthesize the cell wall, N-acetylmuramic acid (NAM) is linked to three amino acids, then to the dipeptide D-Ala-D-Ala (Fig. 19.53). The D-Ala-D-Ala dipeptide is derived from two L-alanine units which are first racemized then linked together.

NAM with its pentapeptide chain is then linked to a **C55 carrier lipid** with the aid of a **translocase** enzyme and carried to the outer surface of the cell membrane, where the lipid carrier acts as an anchor to hold the gly-copeptide in place for the subsequent steps. These steps involve the addition of N-acetylglucosamine (NAG) and a pentaglycine chain to give the complete 'building block'. A **transglycosidase** enzyme catalyses the attachment of the disaccharide building block to the growing cell wall and, at the same time, the carrier lipid is released to pick up another molecule of NAM/pentapeptide. Cross-linking between the various chains of the cell wall finally takes place, catalysed by the transpeptidase enzyme as described previously (section 19.5.1.4).

19.5.5.1 D-Cycloserine and bacitracin

D-Cycloserine (Fig. 19.54) is a simple molecule pro-duced by *Streptomyces garyphalus*. It has broad-spec-trum activity and acts within the cytoplasm to prevent the formation of D-Ala-D-Ala. It does this by mimicking the structure of D-alanine and inhibiting the enzymes **L-alanine racemase** (responsible for racemizing L-Ala to D-Ala) and **D-Ala-D-Ala ligase** (responsible for linking the two D-alanine units together).

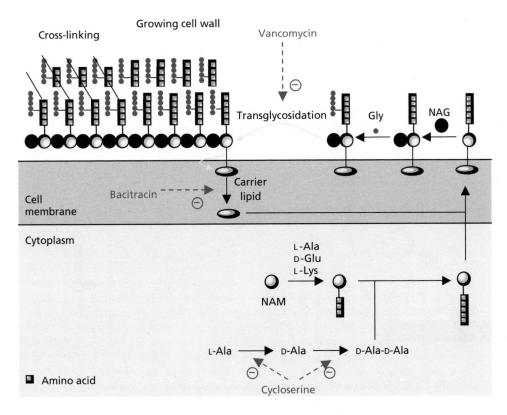

FIGURE 19.53 Cell wall biosynthesis.

Bacitracin is a polypeptide complex produced by *Bacillus subtilis*. It binds to the lipid carrier and prevents it transporting the NAM/pentapeptide unit across the cell membrane.

19.5.5.2 The glycopeptides: vancomycin and vancomycin analogues

Vancomycin (Fig. 19.55) is a narrow-spectrum bactericidal glycopeptide produced by a microorganism called *Streptomyces orientalis* found in Borneo and India. Its name is aptly derived from the verb 'to vanquish'. Vancomycin was introduced in 1956 for the treatment of infections caused by penicillin-resistant *S. aureus*, but was discontinued when methicillin became available. It has since been reintroduced and is now the main stand-by drug for treating MRSA. Vancomycin and related glycopeptides are often the last resort

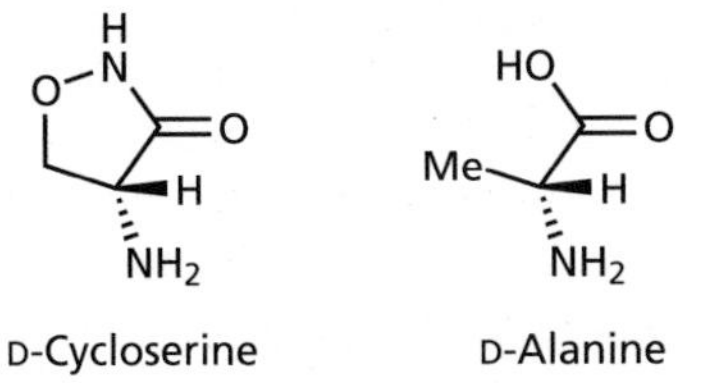

in treating patients with drug-resistant infections. As such, they have become extremely important and a great deal of research is currently being carried out in this area.

Vancomycin is derived biosynthetically from a linear heptapeptide containing five aromatic residues. These undergo oxidative coupling with each other to produce three cyclic moieties within the structure. Chlorination, hydroxylation, and the final addition of two sugar units then complete the structure (Fig. 19.56).

The cyclizations described transform a highly flexible heptapeptide molecule into a rigid structure that holds the peptide backbone in a fixed conformation. Moreover, there is an extra element of rigidity to the structure, which may not be apparent at first sight. The aromatic rings (A–E) cannot rotate and are fixed in space because of hindered single bond rotation. For example, the aromatic rings C and E have a chloro substituent which prevents these rings becoming coplanar with ring D. Similarly, rings A and B have phenol substituents which prevent them becoming coplanar.

The fixed conformation of the hexapeptide chain is important to vancomycin's unique mechanism of action, which involves targeting the cell wall's building blocks rather than a protein or a nucleic acid. To be specific, there is a pocket in the vancomycin structure into which the tail of the building block's pentapeptide moiety can fit. The

D-Cycloserine   D-Alanine

FIGURE 19.54 D-Cycloserine as a mimic for D-alanine.

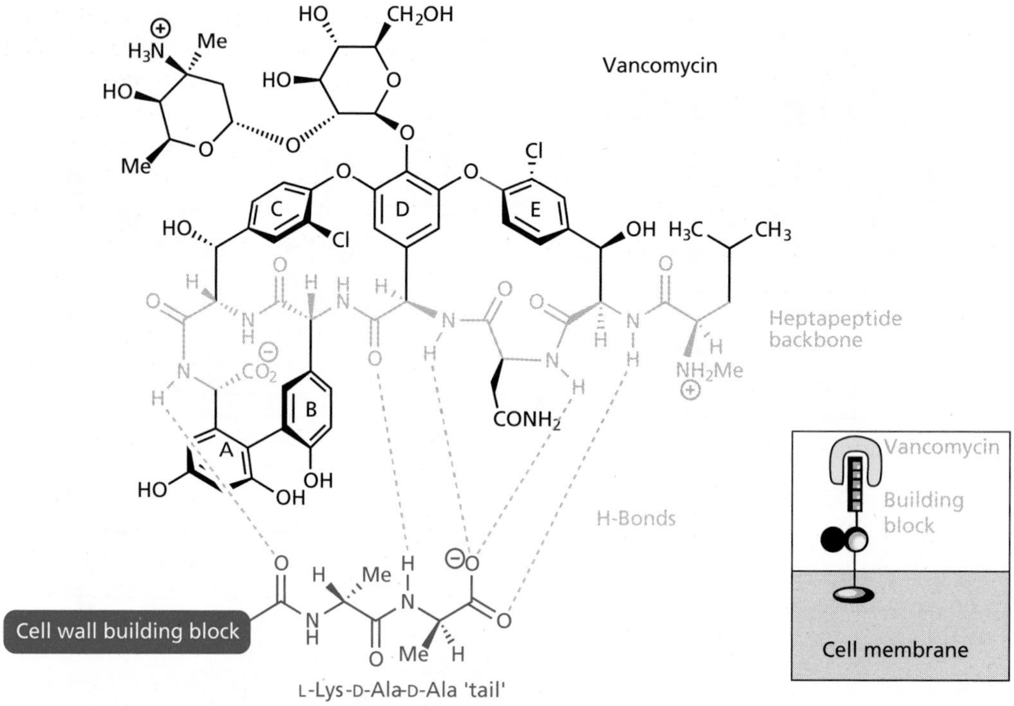

FIGURE 19.55 Vancomycin and its binding interactions to the L-Lys-D-Ala-D-Ala moiety.

pentapeptide is then held there by the formation of five hydrogen bonds between it and the hexapeptide chain of vancomycin (Fig. 19.55). Dimerization can now occur where a highly stable vancomycin dimer is bound to two tails. Because vancomycin is a large molecule, it caps the tails and acts as a steric shield, blocking access to the trans-glycosidase and transpeptidase enzymes (Fig. 19.57).

Dimerization occurs head to tail such that the hep-tapeptide chains of each vancomycin molecule interact through four hydrogen bonds (Fig. 19.58). The sugar and chloro groups also play an important role in this

dimerization, and activity drops if either of these groups is absent.

Because vancomycin is such a large molecule, it is un-able to cross the outer cell membrane of Gram-negative bacteria and consequently lacks activity against those microorganisms. It is also unable to cross the inner cell membrane of Gram-positive bacteria, but this is not re-quired since the construction of the cell wall takes place outside the cell membrane.

Bacterial resistance to vancomycin has been slow to develop, although some hospital strains of *S. aureus* were

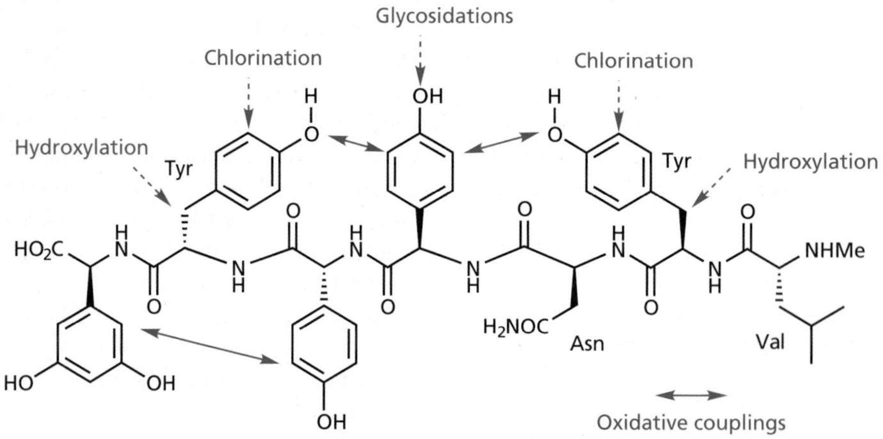

FIGURE 19.56 Reactions involved in the biosynthesis of vancomycin.

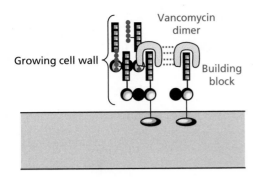

FIGURE 19.57 'Capping' of pentapeptide 'tails' by vancomycin.

identified in 1996 which *do* show resistance (**VRSA**). Of particular concern was the appearance of **vancomycin-resistant enterococci** (VRE) in 1989. These are organisms that can cause life-threatening gut infections in patients whose immune system is weakened. Resistance in the latter microorganisms has arisen from a modification of the cell wall precursors where the terminal D-alanine group in the pentapeptide chain has been replaced by D-lactic acid, resulting in a terminal ester link rather than an amide link (Fig. 19.59). This removes one of the NH groups involved in the hydrogen bonding interaction with vancomycin. It

may not sound like much, but it is sufficient to weaken the binding affinity and make the antibiotic ineffective. The modified building block is still acceptable to the transglycosylase and transpeptidase enzymes. In the latter case, lactate acts as the leaving group rather than D-alanine.

Teicoplanin is a medication that contains five very similar structures which were isolated from a soil microorganism called *Actinoplanes teichomyceticus* and which differ only in the nature of a lengthy alkyl substituent. One example is **teicoplanin A2-5** (Fig. 19.60). The teicoplanins belong to the vancomycin family but they do not dimerize. The long alkyl chain plays an important role in anchoring the antibiotic to the outer surface of the cell membrane where it is perfectly placed to interact with the building blocks for cell wall synthesis (Fig. 19.60). Teicoplanin is used clinically for the treatment of Gram-positive infections and is less toxic than vancomycin. **Dalbavancin** (Fig. 19.61) is an analogue of teicoplanin that was approved in 2014.

Another naturally occurring member of the vancomycin family is **eremomycin** (Fig. 19.62). A biphenyl hydrophobic 'tail' was added to this structure to act as an anchor, resulting in the discovery of **oritavancin**. Oritavancin is 1000 times more active than vancomycin and was approved in 2014. Further modifications involved removal of a tetrahydropyran ring to leave an alcohol

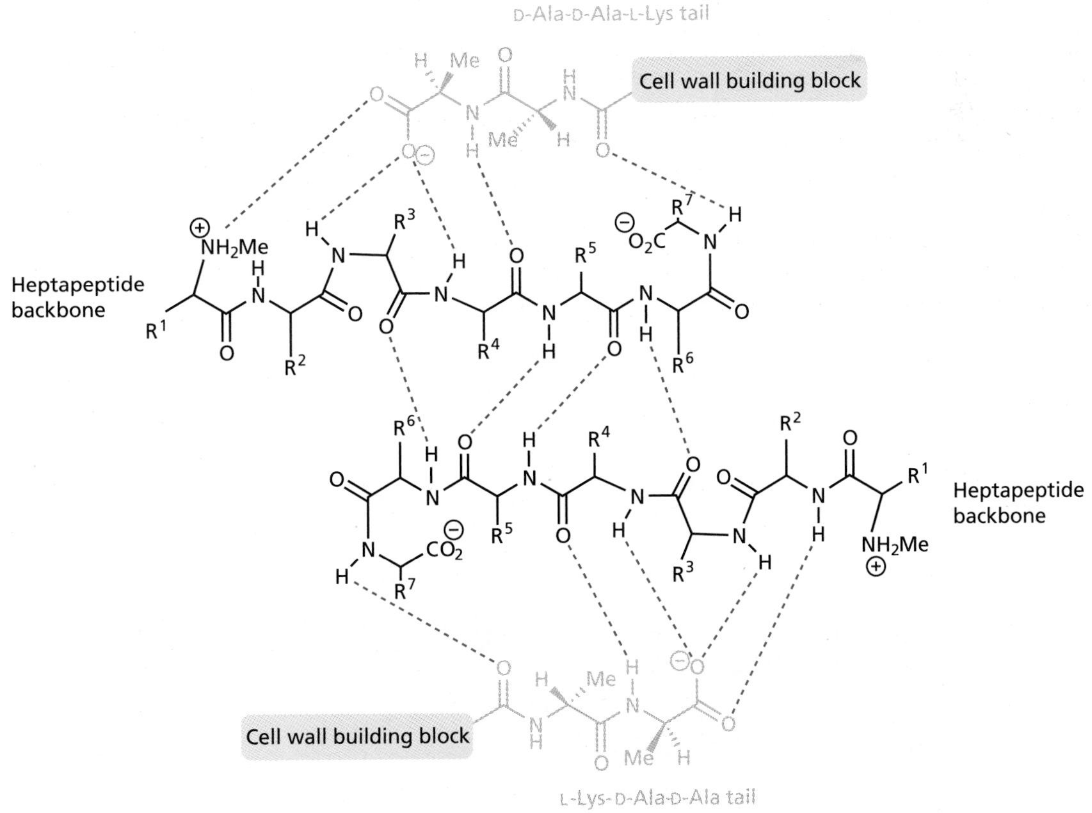

FIGURE 19.58 Dimerization of vancomycin. The dashed lines represent hydrogen bonds.

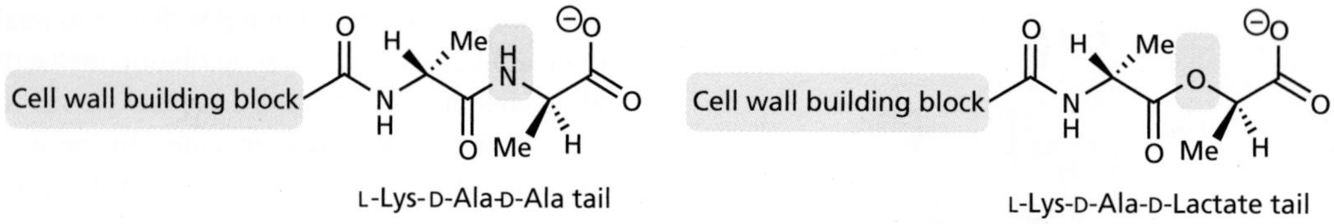

FIGURE 19.59 Modification of the pentapeptide chain leading to resistance.

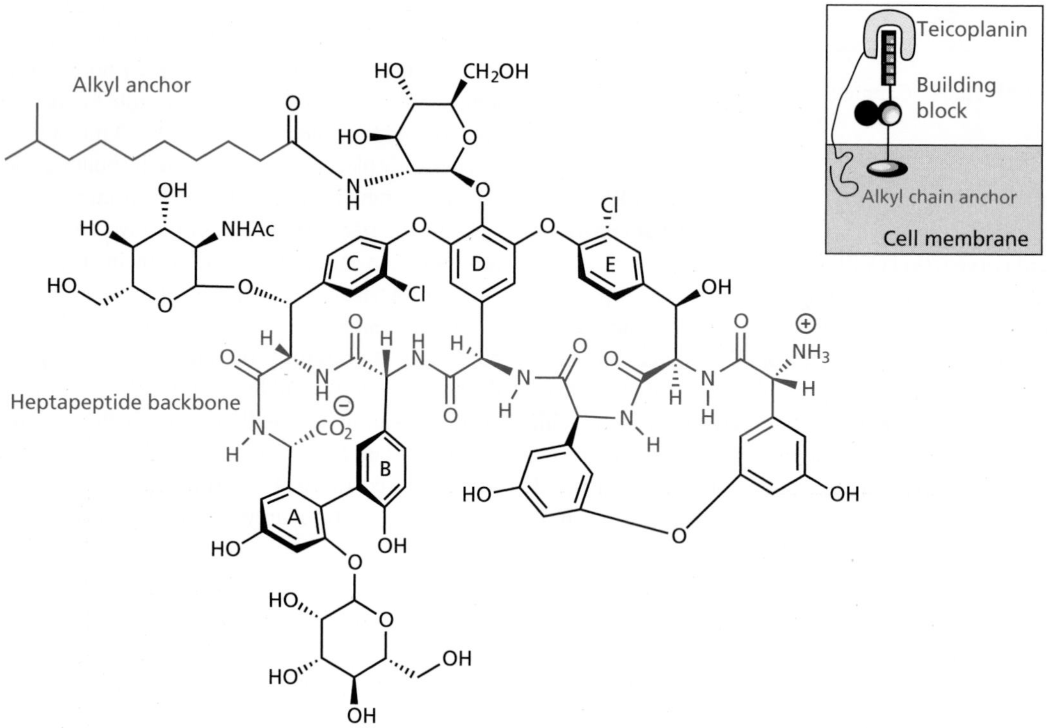

L-Lys-D-Ala-D-Ala tail

L-Lys-D-Ala-D-Lactate tail

FIGURE 19.60 Teicoplanin A$_2$-5.

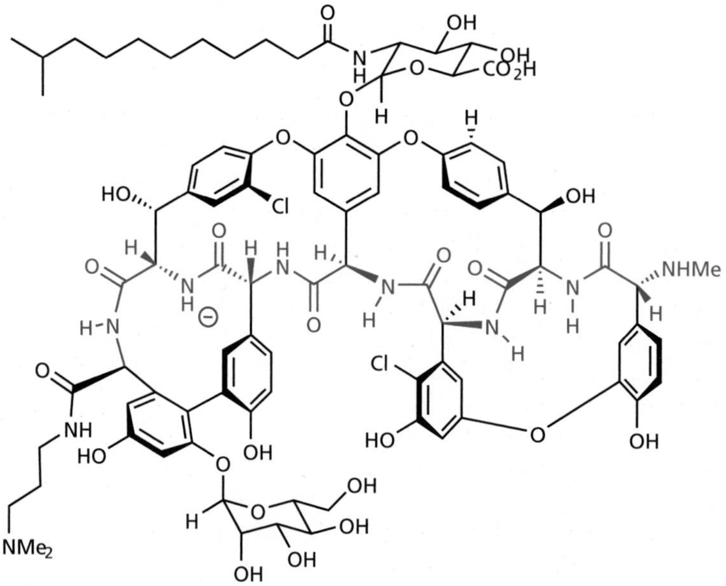

FIGURE 19.61 Dalbavancin.

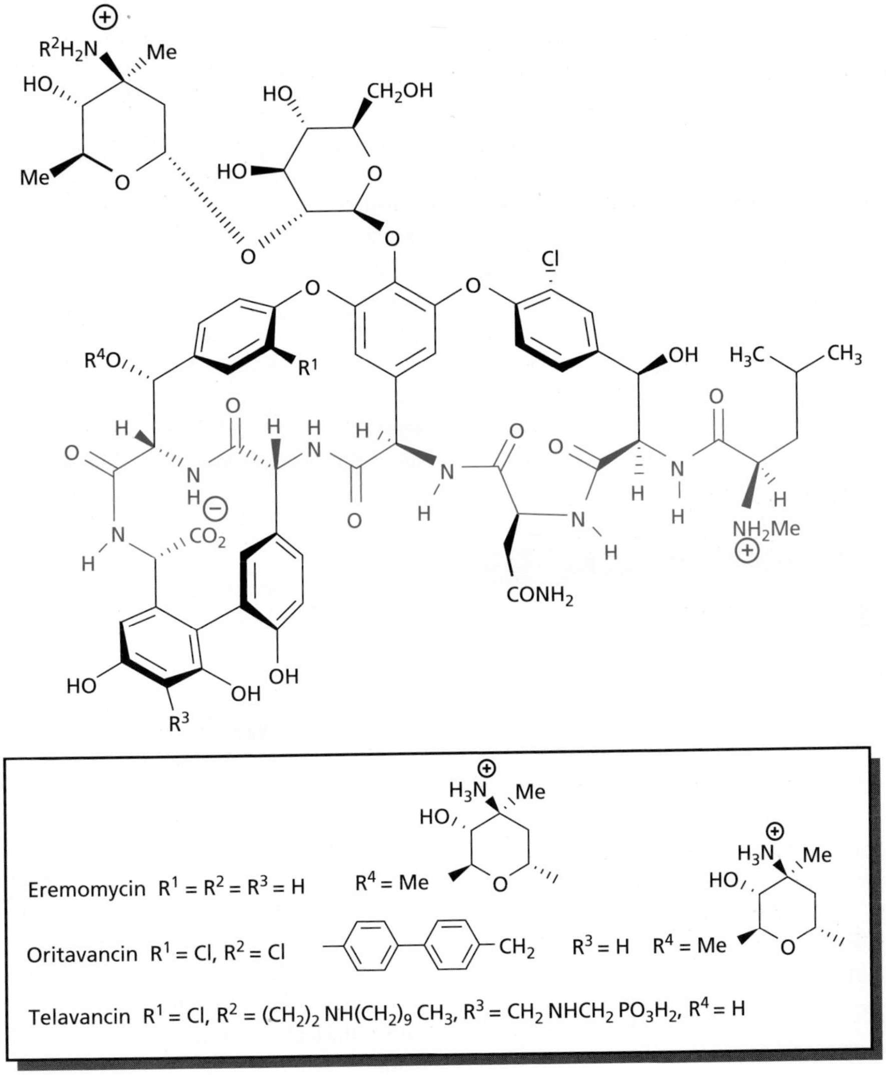

Eremomycin R^1 = R^2 = R^3 = H   R^4 = Me

Oritavancin R^1 = Cl, R^2 = Cl   R^3 = H   R^4 = Me

Telavancin R^1 = Cl, R^2 = (CH$_2$)$_2$ NH(CH$_2$)$_9$ CH$_3$, R^3 = CH$_2$ NHCH$_2$ PO$_3$H$_2$, R^4 = H

FIGURE 19.62 Eremomycin, oritavancin, and telavancin.

group (R^4), modification of the hydrophobic tail (R^2), and addition of a side chain with a phosphate group (R^3) to give **telavancin**, which was approved in 2009.

Although the complexity of the glycopeptides is an advantage in their targeting and selectivity, it is a problem when it comes to synthesizing analogues. Therefore, work has been carried out to try and prepare simplified analogues of vancomycin which are easier to synthesize, yet retain the desired selectivity. Structures such as those shown in Fig. 19.63 have been prepared which are capable of binding to D-Ala-D-Ala and D-Ala-D-Lac. These now represent lead compounds for the development of future antibacterial agents.

There are another two mechanisms by which glycopeptides may have an antibacterial activity. Firstly, it is possible that glycopeptide dimers disrupt the cell membrane structure. This is supported by the fact that glycopeptide

antibiotics enhance the activity of aminoglycosides by increasing their absorption through the cell membrane. Secondly, RNA synthesis is known to be disrupted in the presence of glycopeptides. The possibility of three different mechanisms of action explains why bacteria are slow to acquire resistance to the glycopeptides.

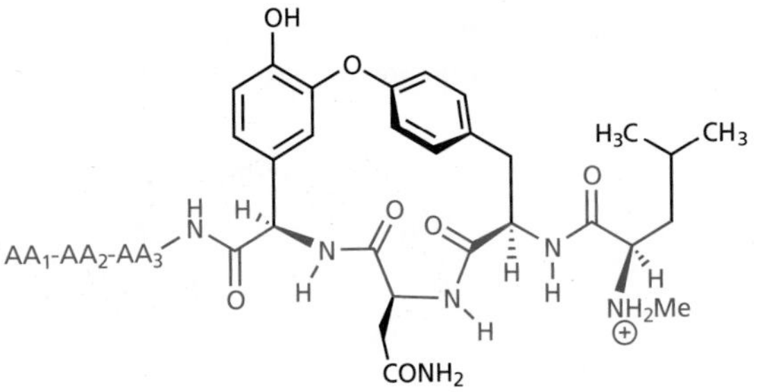

FIGURE 19.63 Simplified analogues of the glycopeptides.

BOX 19.12 Clinical aspects of cycloserine, bacitracin, and vancomycin

ᴅ-**Cycloserine** is administered orally in combination with other drugs to treat tuberculosis that is resistant to first-line drugs. **Bacitracin** is used alongside **polymyxin B sulphate** for the topical treatment of skin infections. The same preparation is also used for the topical treatment of eye infections caused by *Pseudomonas aeruginosa*. **Neomycin sulphate** and bacitracin are used together for the topical treatment of skin infections as a cream or dusting powder.

Vancomycin and **teicoplanin** are bactericidal and are active against aerobic and anaerobic Gram-positive bacteria including MRSA. Vancomycin is not absorbed orally and is administered by intravenous injection every 12 hours for the prophylaxis and

treatment of endocarditis and other serious infections caused by Gram-positive cocci. Vancomycin is also given orally to treat gut infections caused by *Clostridium difficile*. This organism may appear following the use of broad-spectrum antibiotics and produces harmful toxins. Vancomycin has also been used in eye drops. Teicoplanin can be administered once daily and is used for potentially serious Gram-positive infections, including endocarditis, dialysis-associated peritonitis, and serious infections due to *Staphylococcus aureus*. It is also used as a prophylactic in endocarditis and orthopaedic surgery. **Telavancin**, **oritavancin**, and **dalbavancin** have all been approved for the treatment of skin infections, including MRSA.

KEY POINTS

- β-Lactamase inhibitors are β-lactam structures that have negligible antibacterial activity but inhibit β-lactamases. They can be administered alongside penicillins to protect them from β-lactamases and to broaden their spectrum of activity.

- Carbapenems and monobactams are examples of other β-lactam structures with clinically useful antibacterial activity.

- Glycopeptides, such as vancomycin, bind to the building blocks for cell wall synthesis, preventing their incorporation into the cell wall. They also block the cross-linking reaction for those units already incorporated in the wall. The glycopeptides are the drugs of last resort against drug-resistant strains of bacteria.

- Bacitracin binds to the carrier lipid and prevents it carrying the cell wall components across the cell membrane.

- Cycloserine inhibits the synthesis of ᴅ-Ala-ᴅ-Ala.

19.6 Antibacterial agents which act on the plasma membrane structure

19.6.1 Valinomycin and gramicidin A

The peptides valinomycin and gramicidin A both act as ion-conducting antibiotics (ionophores) and

allow the uncontrolled movement of ions across the cell membrane. These agents are described in section 10.6.

19.6.2 Polymyxin B

The polypeptide antibiotic **polymyxin B** (Fig. 19.64) derives from a soil bacterium called *Bacillus polymyxa*. It also operates within the cell membrane and shows a selective toxicity for bacterial cells over animal cells. This appears to be related to the ability of the compound to bind selectively to the different plasma membranes. The mechanism of this selectivity is not fully understood. Polymyxin B acts like valinomycin (section 10.6.2), but it causes the leakage of small molecules such as nucleosides from the cell.

19.6.3 Killer nanotubes

Work is in progress to design cyclic peptides which will self-assemble in the cell membranes of bacteria to form tubules that have been labelled as 'killer nanotubes' (section 10.6.1).

19.6.4 Cyclic lipopeptides

Daptomycin (Fig. 19.65) is a member of a new class of antibacterial agents called the cyclic lipopeptides. It is

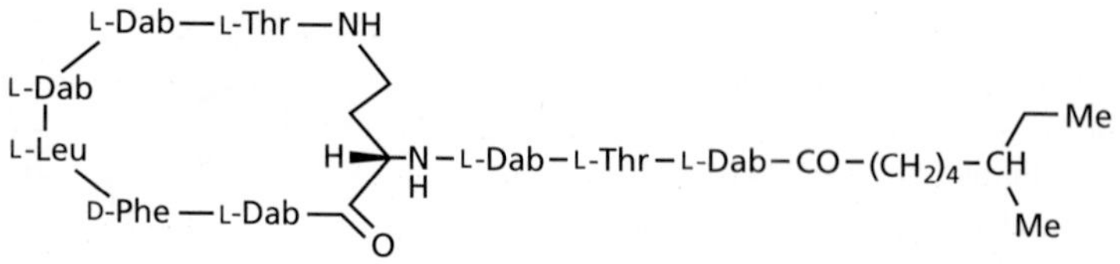

FIGURE 19.64 Polymyxin B (Dab = α,γ-diaminobutyric acid with peptide link involving the α-amino group).

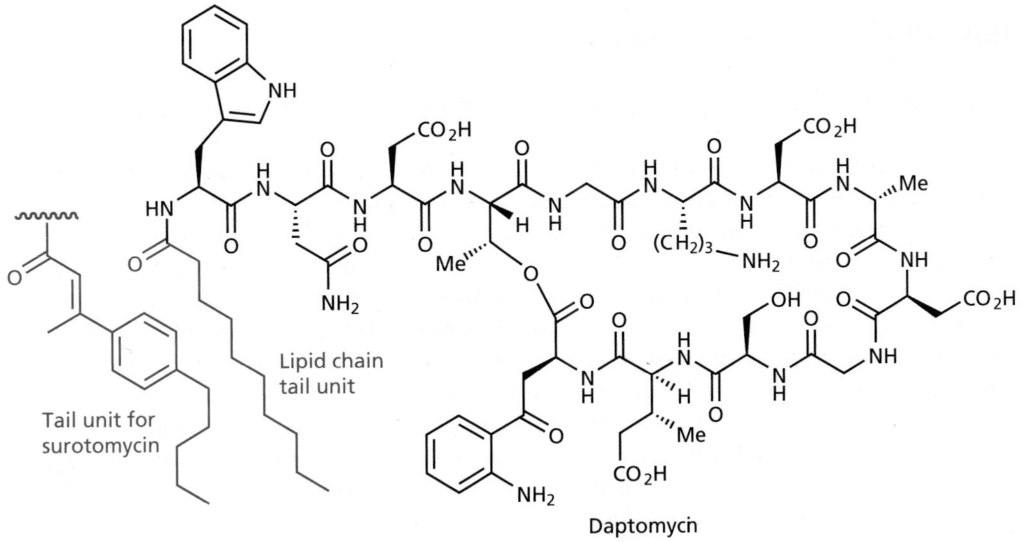

FIGURE 19.65 Daptomycin and surotomycin.

a natural product derived from a bacterial strain called *Streptomyces roseosporus*, and works by disrupting multiple functions of the bacterial cell membrane. The lipid portion of the molecule is derived from decanoic acid and the yield of product obtained is increased if decanoic acid is added to the fermentation medium. Semisynthetic analogues of the structure have been prepared by hydrolysing the amide link to the lipid chain and adding different hydrophobic 'tail units'. **Surotomycin** is one such analogue which is undergoing phase III clinical trials for the treatment of *Clostridium difficile* gut infections.

KEY POINTS

- Ionophores act on the plasma membrane and result in the uncontrolled movement of ions across the cell membrane, leading to cell death.

- Polymyxin B operates selectively on the plasma membrane of bacteria and causes the uncontrolled movement of small molecules across the membrane.

- Cyclic peptides are being designed which will self-assemble to form nanotubes in the cell membranes of bacteria.

- Cyclic lipopeptides are a new class of antibiotic.

BOX 19.13 Clinical aspects of drugs acting on the plasma membrane

Valinomycin and **gramicidin A** show no selective toxicity for bacterial cells over mammalian cells, and are therefore useless as systemic therapeutic agents. However, gramicidin is present as a minor constituent in some topical applications.

Polymyxin B is injected intramuscularly and is useful against *Pseudomonas* strains that are resistant to other antibacterial agents. It can be used topically for the treatment of minor skin infections and has good activity against Gram-negative bacteria. It is less effective against Gram-positive bacteria, as it is difficult for such a big molecule to pass through the thicker cell wall. It is used in combination with **bacitracin** for the treatment of eye and skin infections, or with **dexamethasone** and **neomycin** for the treatment of eye infec-

tions. The ear-drop preparation **Otosporin** contains **hydrocortisone** and polymyxin B, and is used to treat ear infections and inflammation.

Daptomycin was approved in 2003 for the treatment of Gram-positive infections. It is administered by intravenous infusion and has a spectrum of activity similar to vancomycin. In order to guard against the development of drug resistance, the drug is held in reserve for skin and soft tissue infections caused by drug-resistant Gram-positive bacteria such as MRSA. It can be administered alongside other antibacterial agents for mixed infections involving Gram-positive bacteria, Gram-negative bacteria, and some anaerobes.

19.7 Antibacterial agents which impair protein synthesis: translation

The agents described in this section all inhibit protein synthesis by binding to ribosomes and inhibiting different stages of the translation process (Fig. 19.66). Selective toxicity is due to either different diffusion rates through the cell barriers of bacterial versus mammalian cells or to a difference between the ribosomal target structures. The bacterial ribosome is a 70S particle (see section 6.2.2) made up of a 30S subunit and a 50S subunit. The 30S subunit binds messenger RNA (mRNA) and initiates protein synthesis. The 50S subunit combines with the 30S subunit–mRNA complex to form a ribosome, then binds aminoacyl transfer RNA (tRNA) and catalyses the building of the protein chain. There are two main binding sites for the tRNA molecules. The peptidyl site (P-site) binds the tRNA bearing the peptide chain. The acceptor aminoacyl site (A-site) binds the tRNA bearing the next amino acid, to which the peptide chain will be transferred (see also section 6.2.2). The ribosomes of eukaryotic cells are bigger (80S), consisting of 60S and 40S subunits. They are sufficiently different in structure from prokaryotic ribosomes that some drugs can distinguish between them.

19.7.1 Aminoglycosides

Streptomycin (Fig. 19.67) was isolated from the soil microorganism *Streptomyces griseus* in 1944 and is an example of an aminoglycoside—a carbohydrate structure which includes basic amine groups. Streptomycin was the next most important antibiotic to be discovered after penicillin, and a variety of other aminoglycosides have subsequently been isolated from various organisms; for example **gentamicin C1a**, **neomycin**, and **tobramycin**

(Fig. 19.67). In the 1970s, a series of semi-synthetic aminoglycosides were developed, such as **amikacin** (Fig. 19.67), although most of these agents have since been replaced by other types of antibacterial agents. In recent years, however, there has been a resurgence of interest in developing new aminoglycosides because of the growing resistance problem. The fact that aminoglycosides have good activity against Gram-negative bacteria suggests that novel aminoglycosides may prove effective against resistant strains in that category. **Plazomicin** (Fig. 19.67) is one such agent that is currently undergoing clinical trials. Apart from streptomycin, the clinically useful aminoglycosides all contain a 2-deoxystreptamine ring at their core. The semi-synthetic agents include modified amino groups that are no longer susceptible to aminoglycoside modifying enzymes (see later in this section).

The aminoglycosides work best in slightly alkaline conditions. At pH 7.4, they have a positive charge that is beneficial to activity by aiding absorption through the outer membrane of Gram-negative bacteria. An ionic interaction takes place with various negatively charged groups on the outer surface of the cell membrane which displaces magnesium and calcium ions. These ions normally act as bridges between lipopolysaccharides, and their displacement results in rearrangement of cell membrane components to produce pores through which an aminoglycoside can pass. The drug then crosses the cell membrane by an energy-dependent process and is trapped inside the cell where it accumulates to relatively high concentrations.

Binding to bacterial ribosomes now takes place to inhibit protein synthesis. The binding is specifically to the 30S ribosomal subunit and prevents the movement of the ribosome along mRNA. As a result, the triplet code on mRNA can no longer be read. Alternatively, the genetic code is misread resulting in abnormal proteins. In some cases, protein synthesis is terminated and the shortened proteins end up in the cell membrane. This can lead to a

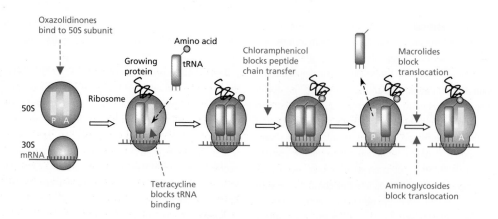

FIGURE 19.66 Stages at which antibacterial agents inhibit translation.

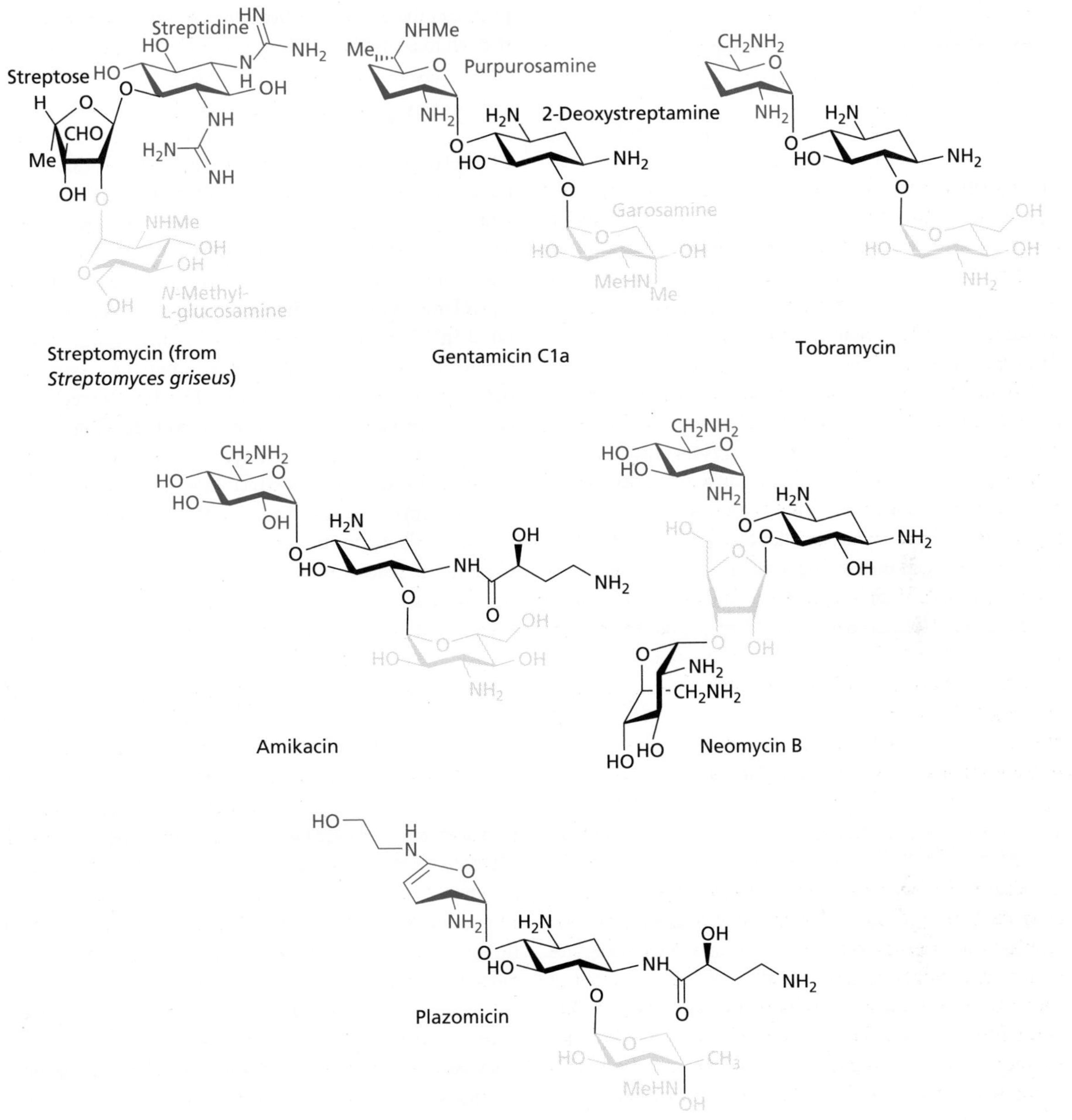

FIGURE 19.67 Aminoglycosides.

further increase in cell permeability, resulting in an even greater uptake of the drug. Aminoglycosides are bactericidal rather than bacteriostatic, and it is thought that their activity may be due to their effects both on the ribosomes and the outer cell membrane.

Because the ribosomes in human cells are different in structure from those in bacterial cells, they have a much lower binding affinity for the aminoglycosides, which explains the selectivity of these drugs.

The binding of streptomycin to ribosomes has been studied by X-ray crystallography, and it has been shown that many of the polar functional groups (alcohols and amines) that are present in streptomycin form hydrogen bonds with the sugar–phosphate backbone of RNA. The other important aminoglycosides bind to a different binding site from streptomycin, but hydrogen bond interactions to RNA are still crucial to their activity.

Resistance can arise to the aminoglycosides by a number of mechanisms. **Aminoglycoside modifying enzymes** (AMEs) are a range of enzymes that modify the structure of aminoglycosides by catalysing the following reactions;

- acetylation of amino groups;
- phosphorylation of hydroxyl groups;
- addition of ADP to hydroxyl groups.

As amino and hydroxyl groups play a key role in the binding interactions of aminoglycosides, such modifications weaken interactions with the target binding site.

Mutations to nucleotides in the target binding site can also lead to resistance. For example, mutation of adenine-1408 results in resistance to neomycin and gentamicin because key hydrogen bond interactions are disrupted. Enzyme-catalysed methylation of key nucleotides also causes resistance.

Resistance can arise from structural changes to the outer membrane of Gram-negative bacteria if that reduces the uptake of aminoglycosides. Resistance can also result from the bacterial cell gaining more effective efflux pumps. Finally, membrane proteases play a role in resistance. These enzymes break down any faulty proteins that enter the cell membrane, such as those produced by the action of aminoglycosides on ribosomes. The more effective the protease enzymes are in dismantling these proteins, the less disruption these proteins can wreak on the cell membrane.

19.7.2 Tetracyclines

The tetracyclines are bacteriostatic antibiotics which have a broad spectrum of activity and are the most widely prescribed form of antibiotic after penicillins. They are also capable of attacking the malarial parasite. One of the best known tetracyclines is **chlortetracycline** (**Aureomycin**) (Fig. 19.68), which was isolated in 1948 from a mud growing microorganism in Missouri called *Streptomyces aureofaciens*—so called because of its golden colour. Further tetracyclines such as **tetracycline** and **doxycycline** (Fig. 19.68) have been isolated from natural sources, or are prepared from naturally occurring tetracyclines in semi-synthetic procedures. Because of the number of chiral centres, substituents, and functional groups that are present in traditional tetracyclines, it has not been economically feasible to prepare them by a full synthesis. However, a synthetic procedure has recently

---

BOX 19.14 Clinical aspects of aminoglycosides

Aminoglycosides are fast acting, but they can also cause ear and kidney problems if the dose levels are not carefully controlled. They are effective in the treatment of infections caused by aerobic Gram-negative bacteria, including *P. aeruginosa*. Indeed, they used to be the only compounds effective against that organism. Some Gram-negative bacteria are resistant to aminoglycosides, due mainly to enzymes which catalyse reactions such as *O*-phosphorylations, *O*-adenylations (addition of an adenine group), and *N*-acylations. Resistance can also occur from alterations of the ribosomes such that they bind aminoglycosides less strongly or by less efficient uptake mechanisms. Because the aminoglycosides are polar in nature, they have to be injected. They are also unable to cross the blood–brain barrier efficiently and so they cannot be used for the treatment of meningitis unless they are injected directly into the central nervous system. The activity of aminoglycosides is increased if they are administered with agents which disrupt cell wall synthesis, as this increases uptake into the cell. However, bacteriostatic agents should not be taken with aminoglycosides, because these inhibit the energy-dependent uptake process by which the aminoglycosides cross the cell membrane.

Streptomycin was the first effective agent used against tuberculosis. However, resistance soon developed and a multi-drug therapy involving streptomycin, **isoniazid**, and **para-ami-**

nosalicylic acid was used until the early 1970s. At that point, **rifampicin** became available, allowing different multi-drug therapies to be developed. Streptomycin is now rarely used for the treatment of tuberculosis, unless there is a known resistance to isoniazid, in which case it is administered by intramuscular injection. Streptomycin is still used to treat enterococcal endocarditis and as an adjunct to **doxycycline** in brucellosis.

Gentamicin is administered by intramuscular or slow intravenous injection for the treatment of a number of infections, including septicaemia, neonatal sepsis, CNS infections (including meningitis), biliary tract infections, acute pyelonephritis or prostatitis, endocarditis, and pneumonia in hospital patients. It can be used topically in drops for the treatment of eye and ear infections. **Amikacin** is more stable to enzyme modification and has been useful for the treatment of infections that have proved resistant to gentamicin. It can also be prescribed as a second line of defence against drug-resistant tuberculosis. **Tobramycin** has similar activity to gentamicin and can be administered by inhalation to treat chronic pulmonary *Pseudomonas aeruginosa* infection in cystic fibrosis. It is also used to treat eye infections. **Neomycin** cannot be used systemically because of its toxicity. However, it can be used to treat infections affecting the skin, ear, eye, or nose.

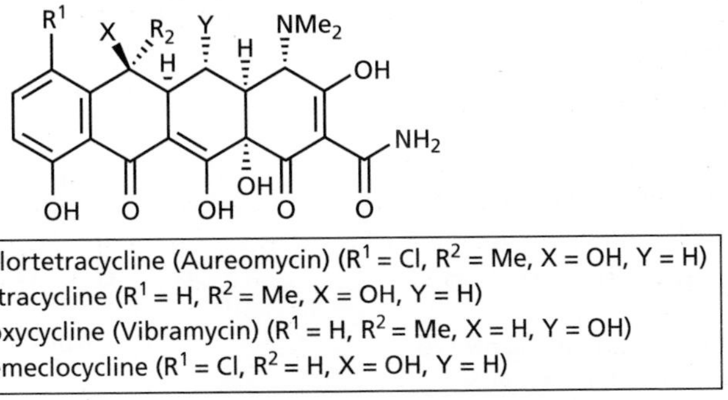

Chlortetracycline (Aureomycin) (R^1 = Cl, R^2 = Me, X = OH, Y = H)
Tetracycline (R^1 = H, R^2 = Me, X = OH, Y = H)
Doxycycline (Vibramycin) (R^1 = H, R^2 = Me, X = H, Y = OH)
Demeclocycline (R^1 = Cl, R^2 = H, X = OH, Y = H)

FIGURE 19.68 Tetracyclines.

been developed that is now being used to prepare a series of novel tetracyclines that are not accessible by any other means.

SAR studies on a variety of tetracyclines have demonstrated that the substituents at positions 6 and 7 of chlortetracycline can be varied or removed without serious loss of activity (Fig. 19.69). Moreover, there is no need to have a chiral centre at position 6. Indeed, activity is retained even if this position is unsubstituted. In contrast, modifications carried out elsewhere in the molecule generally prove detrimental for activity. Similarly, epimerizing any of the chiral centres (other than the one at position 6) proves detrimental. Thus, the pharmacophore for the tetracyclines consists of the tetracycline ring acting as a scaffold for an array of polar functional groups which act as binding groups. The SAR studies also demonstrate that any variation of the structure should be aimed at the northern edge of the molecule—particularly on rings C and D—and that it is possible to introduce substituents at positions 5, 6, 7, 8, and 9 without losing activity.

The southern edge of the tetracycline structure is hydrophilic because of the number of carbonyl and hydroxy groups present, whereas the northern edge is hydrophobic in nature (Fig. 19.70). The fact that so many polar groups are part of the pharmacophore suggests that they are involved in an extensive network of hydrogen bonding interactions with the target binding site.

Tetracyclines have antibacterial activity because they bind to bacterial ribosomes and inhibit protein translation. They do this by preventing aminoacyl-tRNA from binding to the ribosome, which stops further addition of amino acids to the growing protein chain. Protein release is also inhibited. Despite this knowledge, it took several years to identify the exact binding site for tetracyclines. In 2001, a crystal structure was obtained of tetracycline bound to a bacterial ribosome, which revealed that the drug binds to the 30S subunit of the ribosome. The binding interactions observed also explain the SAR results that were obtained 40–50 years earlier (Fig. 19.71).

Tetracycline can actually bind to two different sites on the ribosome, but the more significant binding site

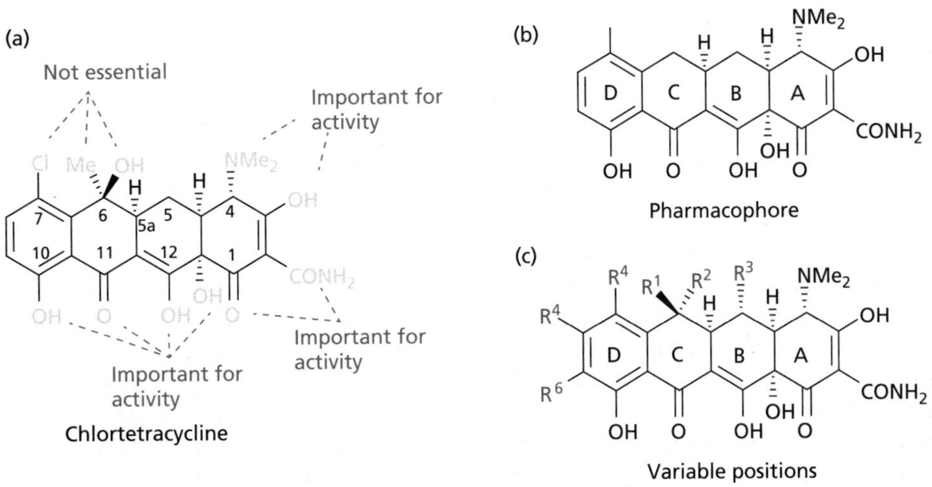

FIGURE 19.69 (a) SAR results on chlortetracycline. (b) Pharmacophore for tetracyclines. (c) Variable positions allowing the introduction of substituents.

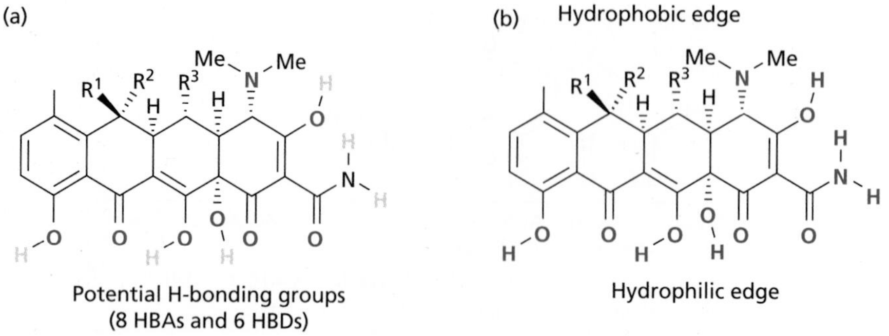

(a)

Potential H-bonding groups
(8 HBAs and 6 HBDs)

(b) Hydrophobic edge

Hydrophilic edge

FIGURE 19.70 (a) Potential hydrogen bonding groups. (b) The hydrophobic and hydrophilic edges of tetracyclines.

involves a series of H-bonds to the sugar–phosphate backbone of ribosomal RNA, as well as interactions involving a bridging magnesium ion. A cytosine ring is also thought to be involved in a pi-pi stacking interaction with ring D of the tetracycline. The binding interactions confirm the importance of the southern edge of the structure to binding and activity, and also help to explain why modifications to the top left region of the molecule can be carried out without losing activity. Modifications to ring D appear to offer the best potential for generating novel tetracycline analogues.

Widespread resistance to traditional tetracyclines has occurred, caused partly by the use of tetracyclines in farming. For example, tetracyclines have been used as food additives to promote the growth of newborn animals. Therefore, a number of novel tetracyclines have been developed that act against resistant strains. Particularly useful modifications have involved the introduction of amino groups at positions 7 and 9 of the D-ring. For example, **minocycline**

(Fig. 19.72) contains a tertiary amine at position 7 and is more potent than any of the early tetracyclines. Moreover, it has a broader range of activity, including activity against tetracycline-resistant staphylococci strains. In general, the presence of the 7-dimethylamino group appears to improve activity against bacterial strains that have gained resistance as a result of efflux pumps. These are proteins within bacterial cell membranes that 'capture' tetracyclines that have entered the cell and transport them back out again.

More recently, tetracyclines with substituents at positions 7 and 9 have proved effective. The most important of these is **tigecycline** (Tigacyl) which was approved by the US FDA in 2005 (Fig. 19.72). This is a broad-spectrum agent which can treat many of the infections that have gained resistance to the older tetracyclines.

As with minocycline, the 7-dimethylamino group increases activity against bacterial strains that have gained resistance as a result of efflux pumps. However,

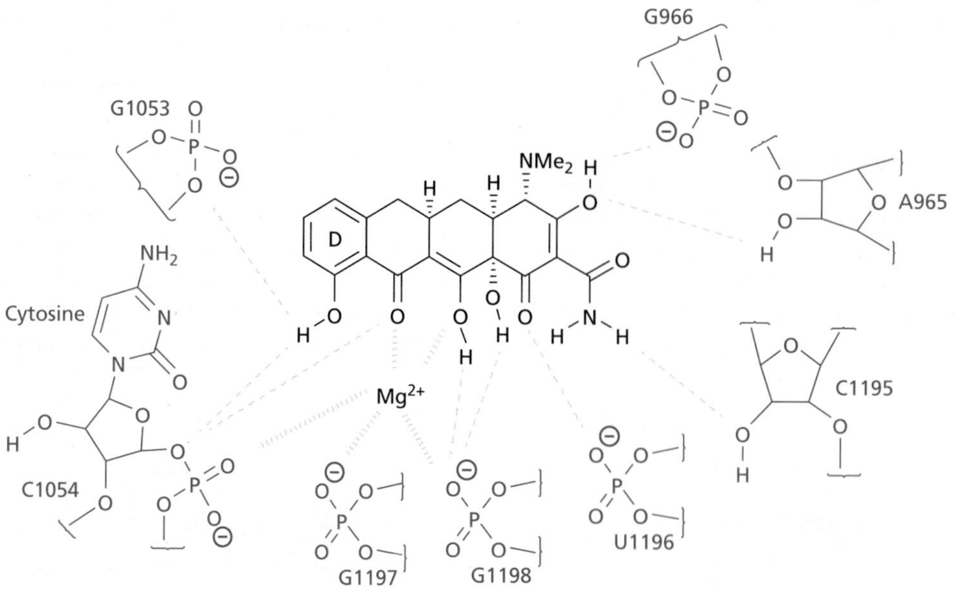

FIGURE 19.71 Binding interactions of tetracycline with the RNA of bacterial ribosomes.

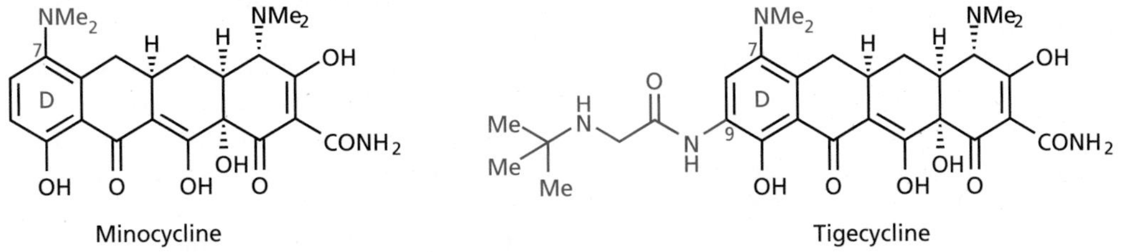

FIGURE 19.72 Clinically important tetracyclines showing activity against tetracycline-resistant bacterial strains.

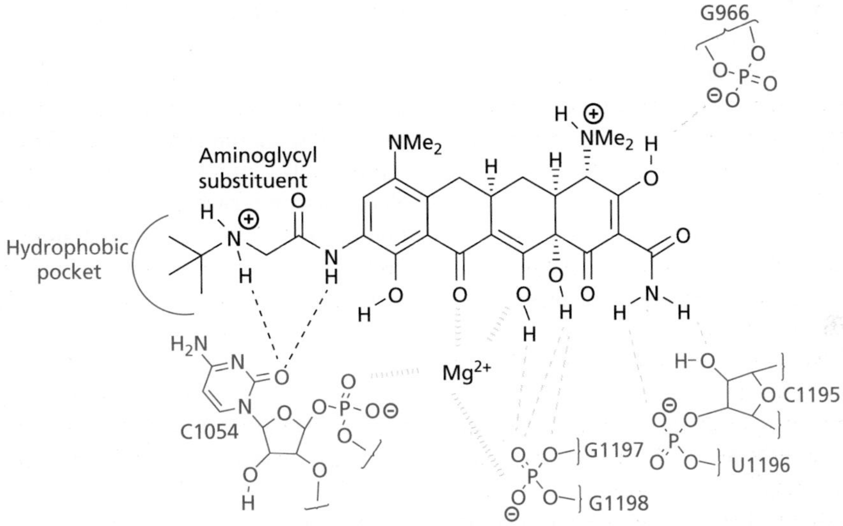

FIGURE 19.73 Binding interactions between tigecycline and the RNA binding site.

tigecycline is also active against bacterial strains that have gained resistance because of mutations to the bacterial ribosome. Such mutations affect the binding site and weaken binding interactions with the older tetracyclines. Molecular modelling studies have suggested that tigecycline is effective against these resistant strains because the side chain at position 9 forms binding interactions that are not possible with tetracycline itself (Fig. 19.73). In particular, there are strong hydrogen bonds between the aminoglycyl substituent and a cytosine base. With tetracycline itself, this cytosine only forms weak pi-pi stacking interactions.

Another group of tetracyclines with useful activity are the aminomethylcyclines represented by **amadacycline** (also called **omadacycline**) (Fig. 19.74). This structure was in phase III clinical trials during 2016.

In recent years, the first economically feasible synthetic route to tetracyclines has been developed and this has resulted in a flurry of novel tetracyclines that were impossible to obtain by semi-synthetic methods. These include agents with a much greater variation of substituents on ring D, as well as structures where the aromatic ring itself has been replaced with a heterocyclic ring. However, the most exciting structures arising from this synthetic breakthrough have been

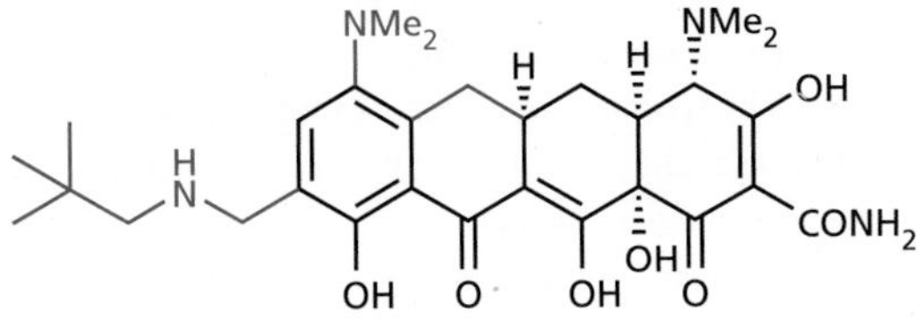

FIGURE 19.74 Amadacycline or omadacycline.

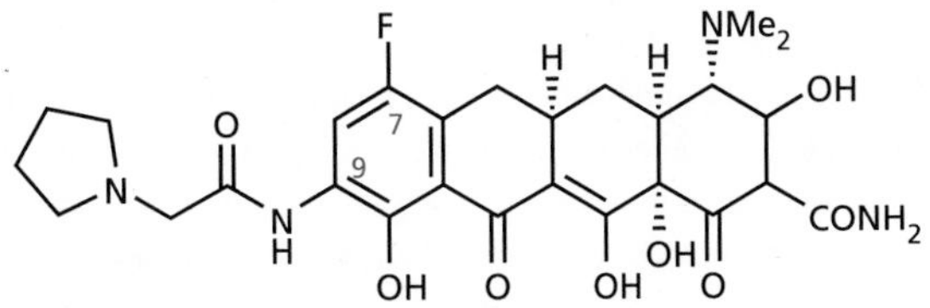

FIGURE 19.75 Eravacycline (TP434).

the **fluorocyclines**—structures that contain a fluorine substituent at position 7 and an amide group at position 9. The most important of these is **eravacycline** (Fig. 19.75) which was in phase III clinical trials during 2016.

The selectivity of tetracyclines is due more to pharmacokinetic factors than selective binding to bacterial ribosomes. In the case of Gram-negative bacteria, tetracyclines cross the outer membrane by passive diffusion through the porins. Passage across the inner membrane is dependent on a pH gradient, which suggests that a proton-driven carrier is involved. Selectivity is due to tetracyclines being concentrated more quickly in bacterial cells than mammalian cells. This is fortunate because tetracyclines are capable of inhibiting protein synthesis in mammalian cells—particularly in mitochondria.

Test your understanding and practise your molecular modelling with Exercise 19.5 on the Online Resource Centre at www.oxfordtextbooks.co.uk/orc/patrick6e/

19.7.3 Chloramphenicol

Chloramphenicol (Fig. 19.76) was originally isolated from a microorganism called *Streptomyces venezuelae* found in a field near Caracas, Venezuela. It is now prepared synthetically and has two asymmetric centres. Only the *R,R*-isomer is active.

Chloramphenicol binds to the A-site of the ribosomal 50S subunit and appears to act by inhibiting the movement of ribosomes along mRNA, probably by inhibiting the peptidyl transferase reaction by which the peptide chain is extended. Since it binds to the same region as macrolides and lincosamides, these drugs cannot be used along with chloramphenicol in combination therapy. The nitrobenzyl group and both alcohol groups are involved in binding interactions. The dichloroacetamide group is also important, but can be replaced by other electronegative groups. Crystal structures of the drug bound to the ribosome have shown that the nitrobenzyl group can form a pi-pi stacking interaction with a cytosine ring, while the primary alcohol group interacts

---

BOX 19.15 Clinical aspects of tetracyclines and chloramphenicol

The tetracyclines are broad-spectrum antibiotics with activity against both Gram-positive and Gram-negative bacteria. Commonly used tetracyclines in the clinic are **tetracycline**, **demeclocycline**, **doxycycline**, **lymecycline**, **minocycline**, and **oxytetracycline**. The use of chlortetracycline has decreased over the years because it kills the intestinal flora that produce vitamin K. However, it is still administered alongside tetracycline and demeclocycline in the preparation **Deteclo**.

In general, the tetracyclines can be divided into short-lasting compounds, such as chlortetracycline, an intermediate group of compounds which includes demeclocycline, and longer-acting compounds which include doxycycline. Minocycline has the broadest spectrum of activity. The tetracyclines were originally used for many types of respiratory infections, but have been largely replaced by β-lactams because of the problems of resistance. However, they are still the agents of choice for the treatment of Lyme disease, rickettsia, and infections caused by chlamydia. They are also used to treat acne and a variety of different infections including respiratory and genital infections. Doxycycline has been found to be useful for the treatment and prophylaxis of malaria, and is cheaper than other antimalarial agents. One drawback is the possibility of skin hypersensitivity to sunlight. The drug can also be used for the treatment of a variety of diseases including syphilis, sinusitis, oral herpes simplex, and acne. It is a

possible agent for the treatment or prophylaxis of anthrax. **Tigecycline** is active against both Gram-positive and Gram-negative infections, including tetracycline-resistant strains. It is also active against MRSA and VRE, but not against *Pseudomonas aeruginosa* or many *Proteus* species. Because of its range of activity, tigecycline is reserved for complicated infections that do not respond to other antibacterial agents.

Tetracyclines should be avoided for young children and pregnant mothers because they can bind to developing teeth and bone leading to tooth discolouration.

Chloramphenicol is a potent broad-spectrum antibiotic. In some regions of the world, it is the drug of choice for the treatment of typhoid when more expensive drugs cannot be afforded. It can also be used to treat severe bacterial infections which are insensitive to other antibacterial agents and is widely used for eye infections. Another application is for ear infections, but the preparation can cause hypersensitivity reactions in about 10% of patients. The drug should only be used in these restricted scenarios as it is quite toxic, especially to bone marrow. The drug is inadequately metabolized in babies leading to a combination of symptoms described as the **grey baby syndrome**, which can be fatal. In adults, the drug undergoes a phase II conjugation reaction to form a glucuronic acid conjugate (section 11.5.5), which is excreted. This reaction fails to take place efficiently in newborn babies and so the drug levels increase to toxic levels.

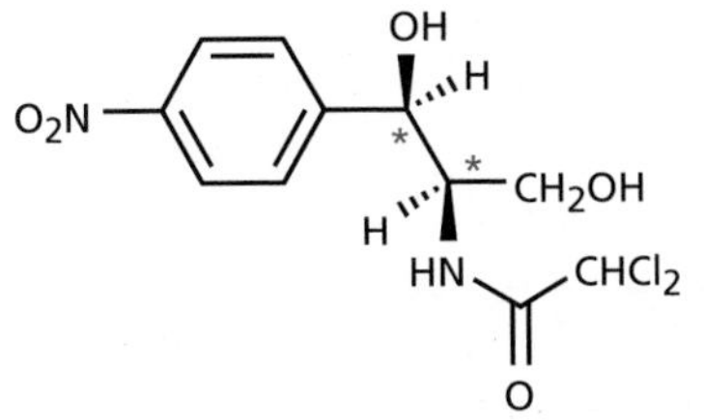

FIGURE 19.76 Chloramphenicol. The asterisks indicate asymmetric centres.

with a ribosome-bound potassium ion. This interaction helps to explain why bacterial strains containing the enzyme **chloramphenicol acetyltransferase** are resistant to chloramphenicol. This enzyme acetylates the primary alcohol group and prevents it interacting with the potassium ion. One of the chlorine substituents can interact with an adenine base, while the amide NH can hydrogen bond to a phosphate group. Chloramphenicol is quite toxic and the nitro substituent is thought to be responsible for this.

19.7.4 Macrolides

Macrolides are bacteriostatic agents. The best-known example of this class of compounds is **erythromycin**—a microbial metabolite isolated in 1952 from the soil microorganism *Streptomyces erythreus* found in the Philippines, and one of the safest antibiotics in clinical use. The structure (Fig. 19.77) consists of a 14-membered macrocyclic lactone ring with a sugar and an aminosugar attached. The alcohol and tertiary amine groups on the aminosugar (desosamine) are crucial for activity.

Crystal structures reveal that erythromycin binds to the 50S subunit of the bacterial ribosome. The macrocyclic ring forms extensive van der Waals interactions with a hydrophobic region of the exit tunnel leading from the ribosomal A-site, while the desosamine hydroxyl group forms a hydrogen bond to an adenine base. This is an

important interaction since mutation of this adenine base to guanine prevents erythromycin from binding. Erythromycin shows selectivity for prokaryotic ribosomes over eukaryotic ribosomes, and a major factor in this is thought to be the nature of nucleotide 2058. This is adenosine in prokaryotic ribosomes and guanosine in eukaryotic ribosomes.

Erythromycin inhibits translocation, but other mechanisms of action also appear likely. Because erythromycin and chloramphenicol bind to the same region of the ribosome, they should not be administered together as they will compete with each other and be less effective.

Erythromycin is unstable to stomach acids, but can be taken orally in a tablet form. The formulation of the tablet involves a coating that is designed to protect the tablet during its passage through the stomach, but which is soluble once it reaches the intestines (enterosoluble). The acid sensitivity of erythromycin is due to the presence of a ketone and two alcohol groups which are set up for the acid-catalysed intramolecular formation of a ketal (Fig. 19.78). One way of preventing this is to protect the hydroxyl groups. For example, **clarithromycin** is a methoxy analogue of erythromycin which is more stable to gastric juices and has improved oral absorption.

Another method of increasing acid stability is to remove the ketone group. There have been two approaches to this. One approach is to convert the ketone to an alkoxyimine as seen in the structure of **roxithromycin**

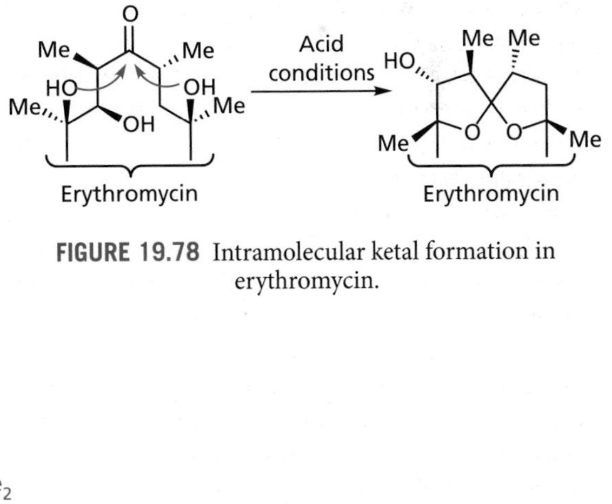

FIGURE 19.78 Intramolecular ketal formation in erythromycin.

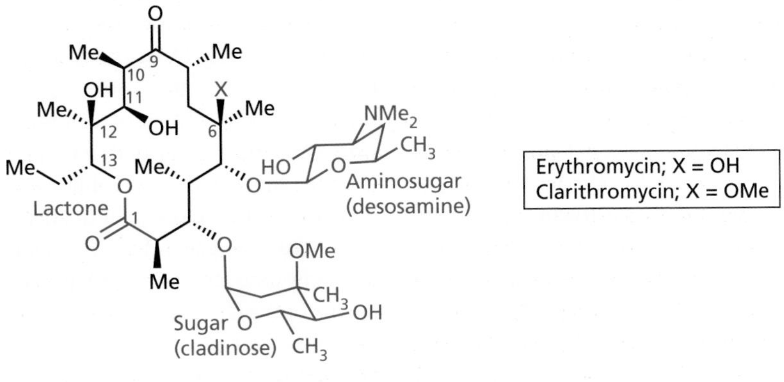

Erythromycin; X = OH
Clarithromycin; X = OMe

FIGURE 19.77 Macrolides.

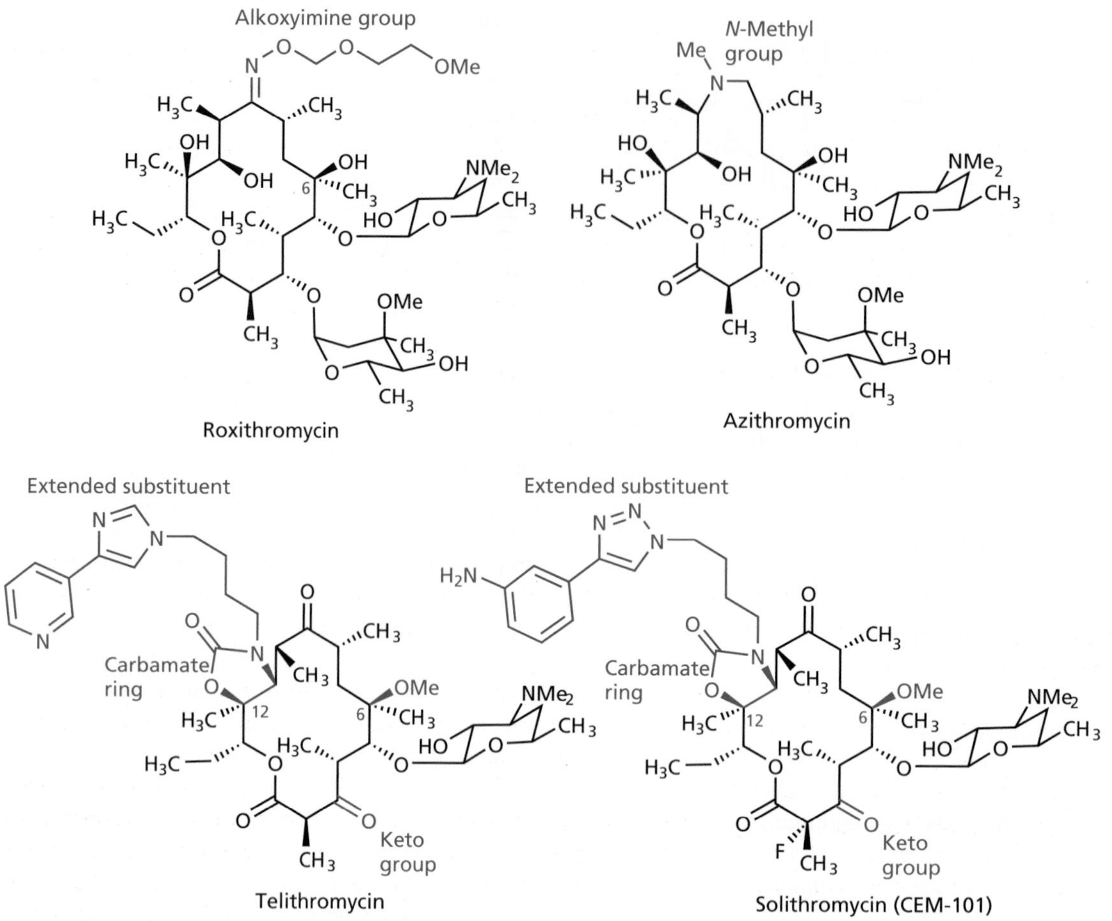

FIGURE 19.79 Examples of macrolides.

(Fig. 19.79). The second approach is seen in **azithromycin**, where an *N*-methyl group has been incorporated into the macrocyclic ring to produce a 15-membered macrocycle. Azithromycin is one of the world's best-selling drugs.

Telithromycin (Fig. 19.79) is a semi-synthetic derivative of erythromycin and reached the European market in 2001. The cladinose sugar in erythromycin has been replaced with a keto-group, and a carbamate ring has been fused to the macrocyclic ring. Because of the introduction of the ketone group, the structure has been classed as a ketolide. The two hydroxyl groups that cause the intramolecular ketal formation in erythromycin have been masked, one as a methoxy group at C6, and the other as part of the carbamate ring at C12. The heteroaromatic rings at the end of the extended substituent are believed to form pi–pi interactions with uracil and adenine bases in the exit tunnel.

Solithromycin is another ketolide structure that is very similar to telithromycin, but shows greater potency due a greater number of binding interactions with the ribosome. It has reached phase III of clinical trials.

19.7.5 Lincosamides

The lincosamide antibiotics (Fig. 19.80) have similar antibacterial properties to the macrolides and act in the same fashion. **Lincomycin** was the first of these agents to be discovered and was isolated in 1962 from a soil organism called *Streptomyces lincolnensis* found near Lincoln, Nebraska. Chemical modification led to the clinically useful **clindamycin** with increased activity. Crystal structures suggest that the propyl substituent interferes with the positioning of tRNA in the A-site of the ribosome, and binds to the same nucleotide (C2452) that interacts with the nitrobenzyl group of chloramphenicol. The sugar ring binds to a similar region of the ribosome as the desosamine sugar in the macrolides, while several hydrogen bonds are possible between the three alcohol groups and surrounding nucleotides. The chlorine substituent of clindamycin occupies a similar position to that occupied by one of the chlorine substituents of chloramphenicol.

Resistance can arise to both macrolides and lincosamides if mutations occur in the region of the ribosome occupied by both agents.

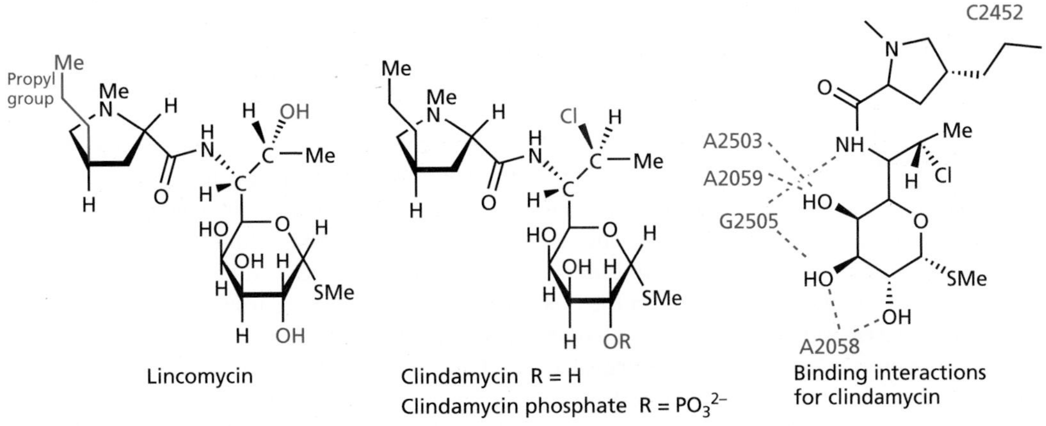

FIGURE 19.80 Lincosamides.

19.7.6 Streptogramins

Pritinamycin is a mixture of macrolactone structures obtained from *Streptomyces pristinaespiralis*. Two of the components (**quinupristin** and **dalfopristin**) have been isolated. These agents bind to different regions of the bacterial ribosome's 50S subunit to form a complex. It is found that the binding of dalfopristin increases the binding affinity for quinupristin, and so the two agents act in synergy with each other. Quinupristin inhibits peptide chain elongation, while dalfopristin interferes with the transfer of the peptide chain from one tRNA to the next.

19.7.7 Oxazolidinones

The oxazolidinones are a relatively new class of synthetic antibacterial agents. They inhibit protein synthesis at a much earlier stage than previous agents, and, consequently, do not suffer the same resistance problems. Before protein synthesis can start, a 70S ribosome has to be formed by the combination of a 30S ribosome with a 50S ribosome. The oxazolidinones bind to the 50S ribosome and prevent this from happening. As a result, translation cannot even start. Other agents that inhibit protein synthesis do so during the translation process itself (Fig. 19.66). **Linezolid** (Fig. 19.81) was the first of this class of compounds to reach the market in 2000, and by 2010, it was netting sales of £716 million per year.

X-ray crystallographic studies have revealed that linezolid binds to the ribosome mostly through van der Waals interactions and pi-pi stacking. This involves the oxazolidinone and aromatic rings of linezolid interacting with nucleic acid bases. There is also a hydrogen bond between the acetamide substituent of linezolid and a phosphate group in the RNA sugar–phosphate backbone. The morpholine ring is the only part of the structure that does not form interactions with the ribosome, and so it can be replaced with other groups. Analogues of linezolid have been developed which bind more strongly. **Radezolid** is one such structure which binds 10 000 times more strongly because of extra binding interactions (*extension strategy*, section13.3.2). It is currently undergoing clinical trials. **Tedizolid** (also known as **torezolid**) is another analogue that was approved in 2014 and is administered as its prodrug **tedizolid phosphate**. It is more potent

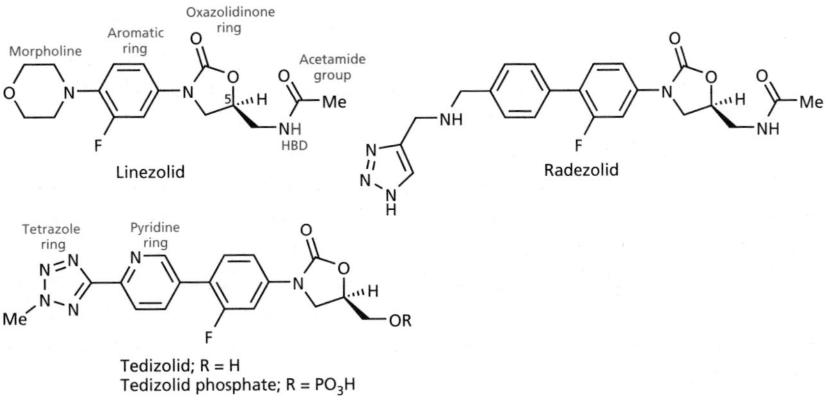

FIGURE 19.81 Oxazolidinones.

than linezolid against a number of infections. Both the pyridine and tetrazole rings provide additional interactions with the binding site.

19.7.8 Pleuromutilins

Pleuromutilin (Fig. 19.82) is a natural product that has been isolated from a fungus (*Clitopilus scyphoides*). It has modest antibacterial activity *in vitro*, and inhibits the start of protein synthesis by binding to the 50S subunit of the bacterial ribosome. The compound has no effect on peptide chain elongation once the process has started. Crystallographic studies have shown that the structure binds to a different binding site of rRNA from other antibacterial agents, and that binding interactions involve nucleosides that are highly conserved and essential for the normal translation process. The carbonyl oxygen of the ester group at position 14 is crucial for activity and acts as a hydrogen bond acceptor, forming two hydrogen bonds to a guanosine nucleotide (G2061). However, most of the binding interactions involve the tricyclic ring system, which forms hydrogen bonds and van der Waals interactions with the target binding site. On binding, an induced fit takes place that involves movement of a couple of uracil nucleotides (U2506 and U2585). This has the effect of sealing the tricyclic ring system into the binding site and allowing efficient binding with surrounding nucleotides. For example, the hydroxyl group at position 11 forms a hydrogen bond to a phosphate group in the sugar–phosphate backbone of RNA. Apart from the ester group, the side chain at position 14 has relatively weak interactions with the binding site, and so it is possible to vary this feature to improve pharmacokinetic properties.

Pleuromutilin itself is not used in the clinic as it has low *in vivo* activity. However, activity can be improved by varying the side chain at position 14 in order to reduce metabolism. The introduction of various thioether substituents proved particularly effective and **retapamulin** was approved in 2007. Two further analogues (**BC-3781** and **BC-7013**) are currently undergoing clinical trials. BC-3781 is the first pleuromutilin to be considered for systemic use.

One potential advantage of these agents is the low risk of rapidly occurring resistance. All the nucleotides involved in binding the drug are essential to the normal functioning of the ribosome, and any mutation would prove detrimental in terms of protein translation and the survival of the bacterial cell. Mutations are more likely to occur with nucleotides that do not interact directly with the drug, and so resistance is only likely to occur if several of those mutations have an indirect effect on binding.

19.8 Agents that act on nucleic acid transcription and replication

19.8.1 Quinolones and fluoroquinolones

The quinolone and fluoroquinolone antibacterial agents are particularly useful in the treatment of urinary tract infections and infections which prove resistant to the more established antibacterial agents.

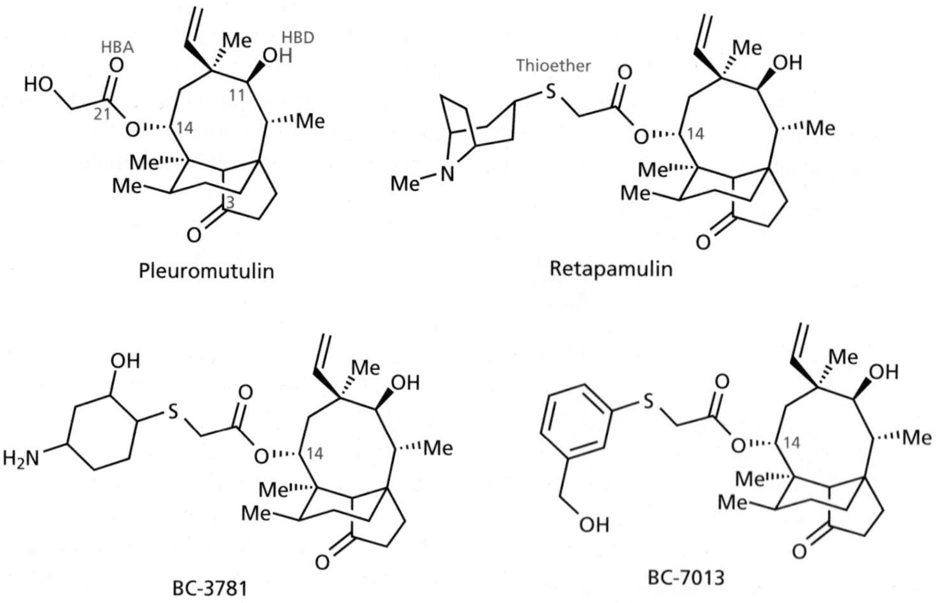

FIGURE 19.82 Pleuromutilins.

BOX 19.16 Clinical aspects of macrolides, lincosamides, streptogramins, oxazolidinones, and pleuromutilins

Macrolides

Erythromycin has an antibacterial spectrum that is similar to penicillins and can be used as an alternative to penicillins for those patients having penicillin allergies. It has been used against penicillin-resistant staphylococci, but newer penicillins are now preferred for these infections owing to increased resistance against erythromycin. It is very useful for the treatment of respiratory infections, including whooping cough and legionnaires' disease. It can also be used to treat syphilis and diphtheria, as well as oral and skin infections. Topically, it can be used for the treatment of acne. **Clarithromycin** has slightly greater activity than erythromycin, with fewer gastrointestinal side effects. Therefore, it is often prescribed instead of erythromycin. Clarithromycin is one of the drugs used in the treatment of ulcers caused by the presence of *Helicobacter pylori* (section 25.4). **Azithromycin** is slightly less active than erythromycin against Gram-positive infections, but is more active against Gram-negative infections, including *Haemophilus influenzae*—against which erythromycin shows poor activity. Azithromycin can also be used for the treatment of Lyme disease. **Telithromycin** has a similar spectrum of activity to other macrolides. It should only be used for specified infections such as pneumonia, tonsillitis, and sinusitis. Resistance to macrolides is due to effective efflux mechanisms which pump the drug back out of the cell. The ribosomal target site may also change in character such that binding is weakened. Enzyme-catalysed modifications can also occur. Recently there has been research into finding novel macrolides which can be effective against respiratory infections due to resistant strains of *Streptococcus pneumoniae*, as well as the organism *H. influenzae*.

Lincosamides

Clindamycin can be taken orally and is active against Gram-positive cocci, including streptococci and penicillin-resistant staphylococci. It is active against peripheral infections involving the anaerobic *B. fragilis*, and is recommended for the treatment of joint and bone infections caused by staphylococci. It is also used topically for the treatment of acne.

Streptogramins

Pritinamycin has been used orally in the treatment of Gram-positive cocci infections, including MRSA. **Quinupristin** and **dalfopristin** are used intravenously in combination (**Synercid**). At present these agents are reserved for life-threatening Gram-positive infections for which there are no alternative therapies; for example hospital-acquired pneumonia, skin and soft tissue infections, and infections due to vancomycin-resistant *Enterococcus faecium*.

Oxazolidinones

The oxazolidinones have a broad spectrum of activity and are active against bacterial strains which have acquired resistance to other antibacterial agents acting against protein synthesis. **Linezolid** has good activity against most clinically important Gram-positive bacteria, including MRSA. It can also be taken orally with 100% uptake from the gastrointestinal tract. Unfortunately, there is a high level of side effects related to its use and, as it is a bacteriostatic agent, there is a greater risk of bacterial resistance developing. **Tedizolid** was approved in 2014 for the treatment of skin infections.

Pleuromutilins

To date, **retapamulin** is the only pleuromutilin to have reached the market. It is used as a topical antibacterial agent and has proved effective against *S. aureus* and *S. pyogenes*—two bacterial strains that are commonly found in skin infections. It is also effective against several bacterial strains that have gained resistance to fusidic acid and mupirocin. However, it is not effective against MRSA.

Nalidixic acid (Fig. 19.83), synthesized in 1962, was the first therapeutically useful agent in this class of compounds. Various analogues were synthesized but offered no great advantage. However, a breakthrough was made in the 1980s with the development of **enoxacin** (Fig. 19.83), which showed improved broad-spectrum activity. The development of enoxacin was based on the discovery that a single fluorine atom at position 6 greatly increased both activity and cellular uptake. A basic substituent such as a piperazinyl ring at position 7 was also beneficial for a variety of pharmacokinetic reasons due to the ability of the basic substituent to form a zwitterion with the carboxylic acid group at position 3. **Norfloxacin** is very similar in structure to enoxacin, but lacks the nitrogen atom at position 8. This modification reduced adverse reactions and increased activity against *S. aureus*. The introduction of a cyclopropyl substituent at position 1 further increased broad-spectrum activity and led to **ciprofloxacin** (Fig. 19.83 and Box 19.17), the most active of the fluoroquinolones against Gram-negative bacteria.

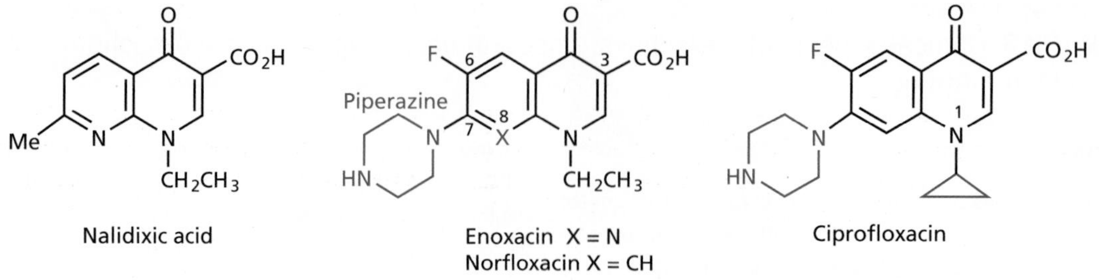

FIGURE 19.83 Quinolones and fluoroquinolones.

The quinolones and fluoroquinolones inhibit the replication and transcription of bacterial DNA by stabilizing the complex formed between DNA and topoisomerases (section 9.2). In Gram-positive bacteria, the stabilized complexes are between DNA and **topoisomerase IV**, with the drugs showing a 1000-fold selectivity for the bacterial enzyme over the corresponding enzyme in human cells. In Gram-negative bacteria, the main target for fluoroquinolones is the complex between DNA and a topoisomerase II enzyme called **DNA gyrase**. It has the same role as topoisomerase IV in reverse and is required when the DNA double helix is being supercoiled after replication and transcription.

A large number of fluoroquinolones have now been synthesized. Those agents having good activity all have a similar bicyclic ring system, which includes a pyridone ring and a carboxylic acid at position 3. A problem with first- and second-generation fluoroquinolones is that they generally show only moderate activity against *S. aureus*, followed by rapidly developing drug resistance. Furthermore, only marginal activity is shown against anaerobes and *Streptococcus pneumoniae*. Third- and fourth-generation fluoroquinolones such as **ofloxacin**, **levofloxacin**, **moxifloxacin**, and **besifloxacin** (Fig. 19.84) began to be developed in the early 1990s to tackle these issues. Ofloxacin has an asymmetric centre and is sold as a racemic mixture of both enantiomers, one of which is active and one of which is not. Levofloxacin is the active enantiomer of ofloxacin and is twice as active as the racemate. **Finafloxacin** was approved in 2014 and shows good activity at acidic pH values where other fluoroquinolones prove inactive. The bulky bicyclic substituent at position 7 of moxifloxacin and finafloxacin is thought to make these agents less susceptible to efflux mechanisms involving a multi-drug transport protein.

19.8.2 Aminoacridines

Aminoacridine agents, such as the yellow-coloured **proflavine** are topical antibacterial agents which were used particularly in the Second World War to treat deep

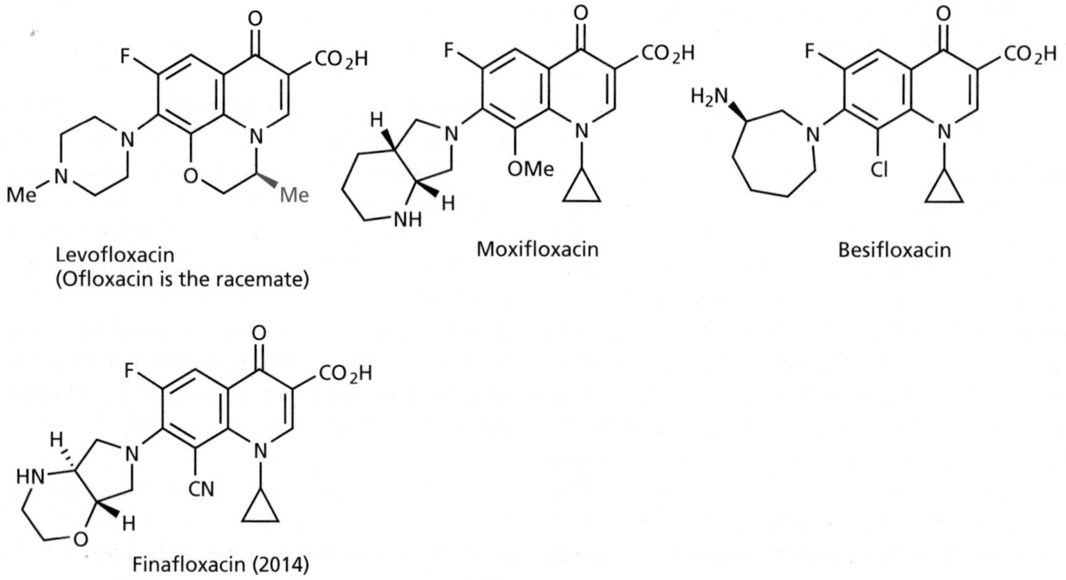

FIGURE 19.84 Third- and fourth-generation fluoroquinolones.

BOX 19.17 Synthesis of ciprofloxacin

The synthesis of ciprofloxacin is a seven-stage route and is applicable to a wide range of fluoroquinolones. It involves the construction of the 'right-hand' pyridone ring onto the fluoro substituted aromatic ring. The cyclopropyl substituent is incorporated just before ring closure and the piperazinyl substituent is added at the final stage of the synthesis.

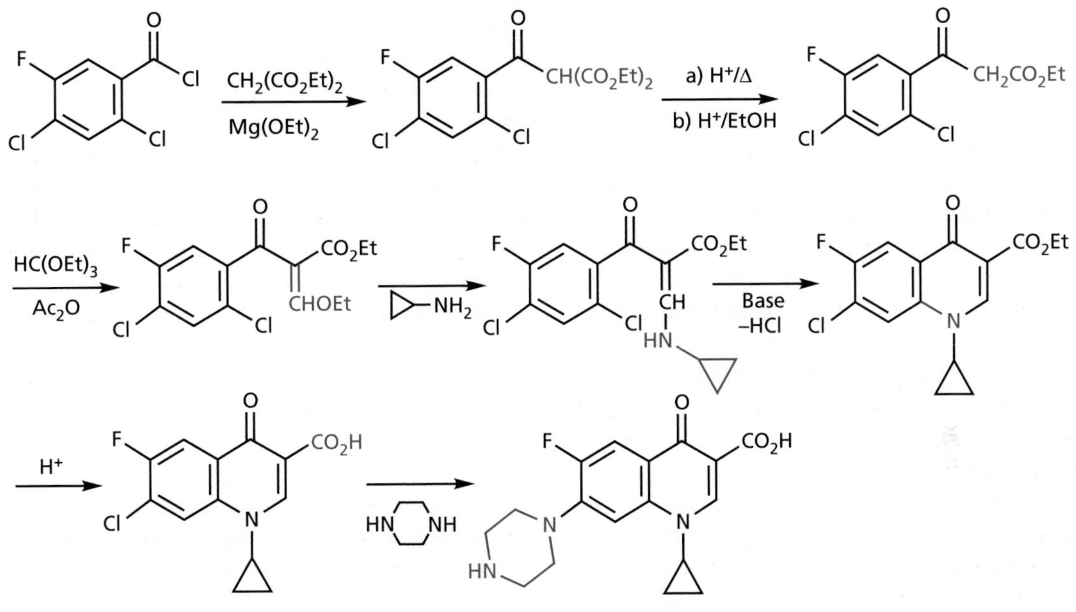

FIGURE 1 Synthesis of ciprofloxacin.

surface wounds. The best agents are completely ionized at pH 7 and they interact directly with bacterial DNA by intercalation (section 9.1). Despite the success of these drugs as topical agents, they are not suitable for the treatment of systemic bacterial infections because they are toxic to host cells.

19.8.3 Rifamycins

Rifampicin (Fig. 19.85) is a semi-synthetic rifamycin made from **rifamycin B**—an antibiotic which was isolated from *Streptomyces mediterranei* in 1957. It inhibits Gram-positive bacteria and works by binding non-covalently to **DNA-dependent RNA polymerase** and inhibiting the start of RNA synthesis. The DNA-dependent RNA polymerases in eukaryotic cells are unaffected because the drug binds to a peptide chain not present in the mammalian RNA polymerase. It is, therefore, highly selective. The flat naphthalene ring and several of the hydroxyl groups are essential for activity and the molecule exists as a zwitterion, giving it good solubility both in lipids and aqueous acid. **Rifaximin** is another

semi-synthetic analogue that was approved in 2004 for the treatment of diarrhoea and *E. coli* infection.

19.8.4 Nitroimidazoles and nitrofurantoin

Metronidazole (Fig. 19.86) is a nitroimidazole structure which was introduced in 1959 as an antiprotozoal agent, but began to be used as an antibacterial agent in the 1970s. The nitro group is reduced when the drug enters the bacterial cell, which lowers the concentration of metronidazole within the cell and sets up a concentration gradient down which more drug can flow. The reduction mechanism also proves toxic to the cell as free radicals are formed which act on DNA. **Nitrofurantoin** also undergoes reduction within bacterial cells to form radical species that act on DNA.

19.8.5 Inhibitors of bacterial RNA polymerase

A recent addition to the arsenal of clinically useful antibiotics is **fidaxomicin** (Fig. 19.87), which is a natural product obtained from a *Dactylosporangium* Gram-positive

BOX 19.18 Clinical aspects of quinolones and fluoroquinolones

Nalidixic acid is active against Gram-negative bacteria and is useful in the short-term therapy of uncomplicated urinary tract infections. It can be taken orally but, unfortunately, bacteria can develop a rapid resistance to it. **Enoxacin** has a greatly increased spectrum of activity against Gram-negative and Gram-positive bacteria. It also shows improved oral absorption, tissue distribution, and metabolic stability, as well as an improvement in the level and spectrum of activity, particularly against Gram-negative bacteria such as *Pseudomonas aeruginosa*. However, it is rarely used now, and the structurally related **norfloxacin** is preferred as a treatment for uncomplicated urinary tract infections. **Ciprofloxacin** is used in the treatment of a large range of infections involving the urinary, respiratory, and gastrointestinal tracts (e.g. travellers' diarrhoea), as well as infections of skin, bone, and joints. It is also used for gonorrhoea and septicaemia, and as part of a cocktail of drugs for anthrax. It has been claimed that ciprofloxacin may be the most active broad-spectrum antibacterial agent on the market. In contrast to nalidixic acid, resistance to the fluoroquinolones is slow to appear, but, when it does appear, it is mainly due to efflux mechanisms which pump the drug back out of the cell. Less common resistance mechanisms include mutations to the

topoisomerase enzymes which reduce their affinity to the agents, and alteration of porins in the outer membrane of Gram-negative organisms to limit access to the cell.

Third-generation fluoroquinolones show improved activity against *Streptococcus pneumoniae*, while maintaining activity against enterobacteria. **Ofloxacin** is administered orally or by intravenous infusion to treat septicaemia, gonorrhoea, and infections of the urinary tract, lower respiratory tract, skin, and soft tissue. **Levofloxacin** has a greater activity against pneumococci than ciprofloxacin and is a second-line treatment for community-acquired pneumonia. It is also used for acute sinusitis, chronic bronchitis, urinary tract infections, skin infections, and soft tissue infections. **Moxifloxacin** also has greater activity against pneumococci than ciprofloxacin. It is used to treat sinusitis and is a second-line treatment of community-acquired pneumonia. **Besifloxacin** is a fourth-generation fluoroquinolone approved in 2009. **Finafloxacin** was approved in 2014 for the treatment of swimmer's ear—an infection caused by *P. aeruginosa* and *Staphylococcus aureus*. Regular swimmers tend to be more prone to ear infections because of residual water in the ear. As the drug shows good activity under acidic conditions, it may also prove useful against infections in acidic environments, such as some urinary tract infections.

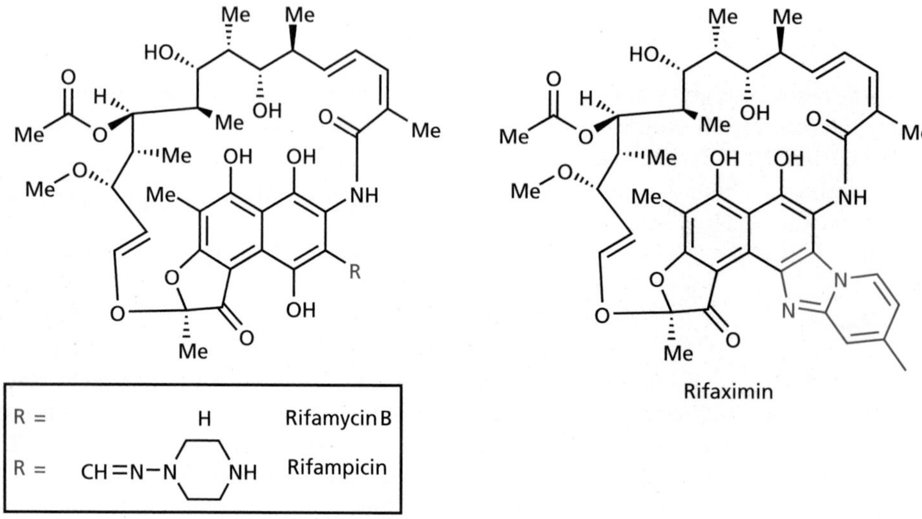

FIGURE 19.85 Rifamycins.

bacterial strain. The agent is a macrocycle and was approved in 2011 as a narrow-spectrum bactericidal agent for the treatment of *Clostridium difficile* infections in the gastrointestinal tract. It inhibits transcription in *C. difficile* by inhibiting RNA polymerase, and has a minimal effect on other gut flora.

19.9 **Miscellaneous agents**

A variety of miscellaneous agents are shown in Fig. 19.88. **Methenamine** is used to treat urinary tract infections where it degrades in acid conditions to give formaldehyde

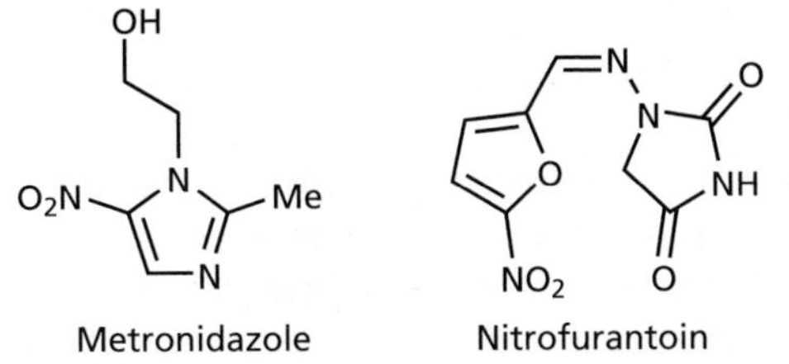

FIGURE 19.86 Metronidazole and nitrofurantoin.

as the active agent (section 14.6.6). **Fusidic acid** is a steroid structure derived from the fungus *Fusidium coccineum* and is used as a topical antibacterial agent. **Isoniazid** is the most widely used drug for the treatment of tuberculosis. It acts by inhibiting the synthetic pathways leading to mycolic acid, an important constituent of mycobacterial cell walls. It is activated in bacterial cells by a catalase-peroxidase enzyme. Resistant strains of tuberculosis (TB) block the action of this enzyme. **Ethambutol**

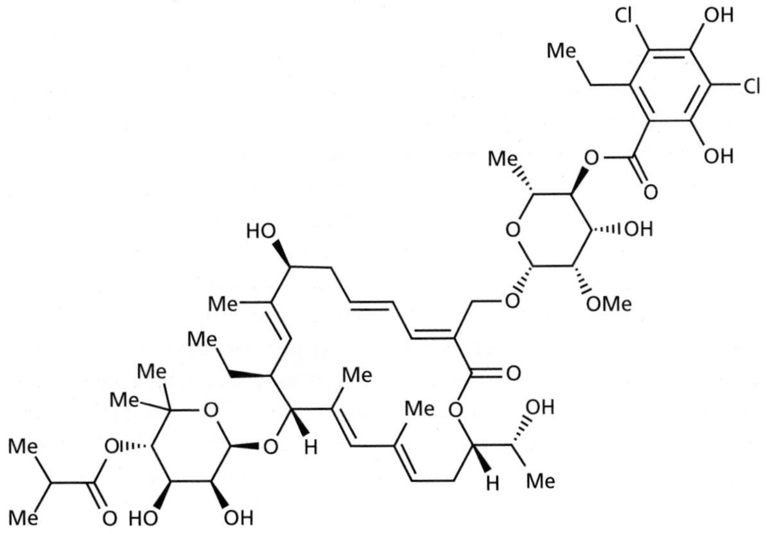

FIGURE 19.87 Fidaxomicin.

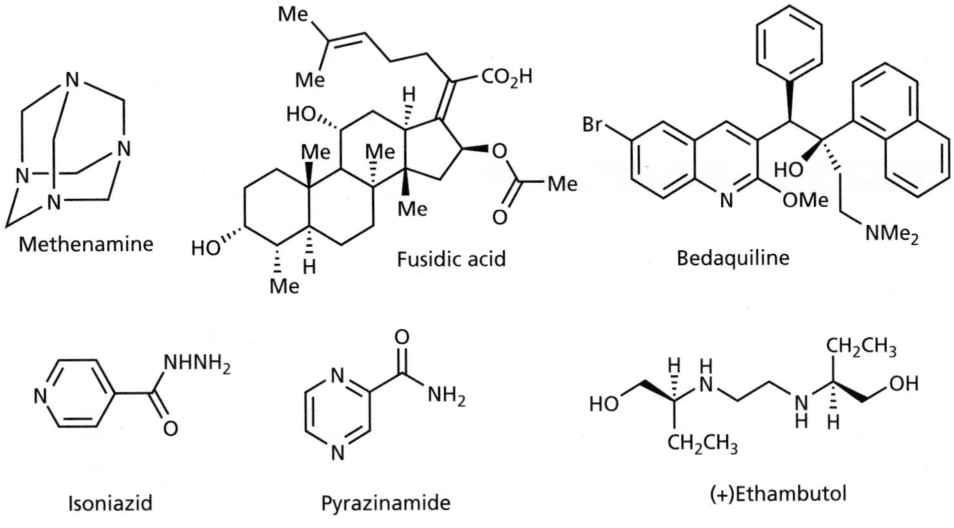

FIGURE 19.88 Miscellaneous agents.

> ## BOX 19.19 Clinical aspects of rifamycins and miscellaneous agents
>
> **Rifampicin** is bactericidal and is mainly used in the treatment of tuberculosis and staphylococci infections that resist penicillin. It is used in combination with **dapsone** in treating leprosy and is also used for the treatment of brucellosis, legionnaires' disease, and serious staphylococcal infection. It is a very useful antibiotic, showing a high degree of selectivity against bacterial cells over mammalian cells. Unfortunately, it is also expensive, which discourages its use against a wider range of infections. Rifampicin is a key component of any antituberculosis regimen, but it poses a special problem when treating tuberculosis in AIDS patients, as it enhances the activity of the cytochrome P450 enzyme family (CYP3A). These enzymes metabolize the HIV-protease inhibitors used in HIV therapy, thus lowering their effectiveness. Increased cytochrome P450 activity also decreases the effect of oral anticoagulants, oral contraceptives, and barbiturates.
>
> **Metronidazole** has good activity in treating infections caused by anaerobic bacteria and protozoa, including difficult-to-treat organisms, such as *Bacteroides fragilis* and *Clostridium difficile*. It is well distributed round the body and crosses the blood–brain barrier, so it can be used for the treatment of brain abscesses and other central nervous system infections involving anaerobic bacteria. Metronidazole is used for the treatment of leg ulcers, bacterial vaginosis, pelvic inflammatory disease, and can also be used as an alternative to penicillins for oral infections including tooth
>
> abscesses. It is administered with amoxicillin (or with tetracycline and bismuth) in the treatment of gastric ulcers involving *H. pylori* (section 25.4). The drug is effective against *Giardia* infections derived from polluted water supplies—a common hazard when visiting the third world. Finally, nitroimidazoles such as metronidazole are commonly combined with cephalosporins or aminoglycosides to treat infections involving both aerobic and anaerobic organisms. Resistance is rare, though not out of the question. **Nitrofurantoin** is used to treat uncomplicated urinary tract infections.
>
> **Methenamine** can be used to treat urinary tract infections, but only if the urine is acidic and the infection is in the lower urinary tract. It can be used as a prophylactic and as a treatment for chronic and recurrent lower urinary tract infections.
>
> **Fusidic acid** is a topical antibacterial agent that is used in eye drops and skin creams. It can penetrate intact and damaged skin, so it is useful for the treatment of boils. It has also been used to eradicate MRSA colonies carried in the nasal passages of hospital patients and health workers. **Isoniazid** is the most widely used drug for the treatment of tuberculosis and is part of a four-drug cocktail which is the first-choice treatment for the initial phase of the disease. **Bedaquiline** was introduced as a treatment for tuberculosis in 2012. As it acts on a different target from previous antituberculosis drugs, it does not suffer cross-resistance with these earlier drugs.

and **pyrazinamide** are synthetic compounds which are both front-line drugs in the treatment of tuberculosis. Ethambutol inhibits **arabinosyl transferase** enzymes that are involved in the biosynthesis of the mycobacterial cell wall. **Bedaquiline** is the most recently approved drug for the treatment of tuberculosis, marking a 40-year gap since the discovery of previous TB drugs. It acts as an enzyme inhibitor of microbial ATP synthase.

KEY POINTS

- Aminoglycosides, tetracyclines, chloramphenicol, streptogramins, lincosamides, and macrolides inhibit protein synthesis by binding to the bacterial ribosomes involved in the translation process.

- Resistance can arise from a variety of mechanisms such as drug efflux, altered binding affinity of the ribosome, altered membrane permeability, and metabolic reactions.

- Oxazolidinones prevent the formation of the bacterial 70S ribosome by binding to the 50S subunit.

- Quinolones and fluoroquinolones inhibit topoisomerase enzymes, resulting in inhibition of replication and transcription.

- Aminoacridines are useful topical antibacterial agents which can intercalate with bacterial DNA and hinder replication and transcription.

- Rifamycins inhibit the enzyme RNA polymerase and prevent RNA synthesis. This, in turn, prevents protein synthesis. Rifampicin is used to treat tuberculosis and staphylococcus infections. Fidaxomicin is a macrocycle which also targets RNA polymerase.

- Nitroimidazoles are used against infections caused by protozoa and anaerobic bacteria.

19.10 Drug resistance

Medicinal chemists are still actively seeking new and improved antibacterial agents to combat the worrying ability of bacteria to acquire resistance to current drugs. For example, 60% of *Streptococcus pneumoniae* strains are resistant to β-lactams, and 60% of *Staphylococcus aureus* strains are resistant to **methicillin**. The last resort in treating *S. aureus* infections is **vancomycin**, but resistance is also beginning to appear to that antibiotic. Some strains of *Enterococcus faecalis* appearing in urinary and wound infections are resistant to all known antibiotics and are untreatable. Other infections that are causing concern because of increasing resistance include those caused by *Clostridium difficile*, *Pseudomonas aeruginosa*, *Escherichia coli*, *Klebsiella pneumoniae*, and *Acinetobacter baumannii*. If antibiotic resistance continues to grow, medicine could be plunged back to the 1930s. Indeed, many of today's advanced surgical procedures would become too risky to carry out due to the risks of infection. Old diseases are already making a comeback. For example, a new antibiotic-resistant strain of tuberculosis (**multi-drug-resistant TB**—MDRTB) appeared in New York and took 4 years and $10 million to bring under control. These strains were resistant to two of the front-line drugs used against TB (isoniazid and rifamycin), and had various levels of resistance against another two (streptomycin and ethambutol). Other examples of bacterial strains acquiring resistance include penicillin-resistant meningococci and pneumococci in South Africa, penicillin-resistant gonococci in Asia and Africa, ampicillin-resistant *H. influenzae* in the USA and Europe, and chloramphenicol-resistant meningococci in France and South East Asia. Resistance to trimethoprim in some of the developing nations has meant that the drug has become ineffective as a treatment of dysentery.

Drug resistance can arise because of a variety of factors described in section 19.5.1.5, but the cell must have the necessary genetic information. This information can be obtained by mutation or by the transfer of genes between cells.

19.10.1 Drug resistance by mutation

Bacteria multiply at such a rapid rate that there is always a chance that a mutation will render a bacterial cell resistant to a particular agent. This feature has been known for a long time and is the reason why patients should complete a full course of antibacterial treatment even though their symptoms may have disappeared well before the end of the course. If this rule is adhered to, the vast majority of invading bacterial cells will be wiped out, leaving the body's own defences to mop up any isolated survivors or resistant cells. If the treatment is stopped too soon, however, then the body's defences struggle to cope with the survivors. Any isolated resistant cell is then given the chance to multiply, resulting in a new infection which will, of course, be completely resistant to the original drug. This was a major factor in the appearance of MDRTB.

Mutations occur naturally and randomly and do not require the presence of a drug. Indeed, it is likely that a drug-resistant cell is present in a bacterial population even before the drug is encountered. This was demonstrated with the identification of **streptomycin-resistant** cells from old cultures of *E. coli*, which had been freeze-dried to prevent multiplication before the introduction of streptomycin into medicine.

19.10.2 Drug resistance by genetic transfer

A second way in which bacterial cells can acquire drug resistance is by gaining that resistance from another bacterial cell. This occurs because it is possible for genetic information to be passed on from one bacterial cell to another. There are two main methods by which this can take place—**transduction** and **conjugation**.

In transduction, small segments of genetic information known as **plasmids** are transferred by means of bacterial viruses (**bacteriophages**) which leave the resistant cell and infect a non-resistant cell. If the plasmid contains the gene required for drug resistance, then the recipient cell will be able to use that information and gain resistance. For example, the genetic information required to synthesize β-**lactamases** can be passed on in this way, rendering bacteria resistant to penicillins. The problem is particularly prevalent in hospitals where currently over 90% of staphylococcal infections are resistant to antibiotics such as penicillin, erythromycin, and tetracycline. It may seem odd that hospitals should be a source of drug-resistant strains of bacteria. In fact, they are the perfect breeding ground. Drugs commonly used in hospitals are present in the air in trace amounts. It has been shown that breathing in these trace amounts kills sensitive bacteria in the nose and allows the nostrils to act as a breeding ground for resistant strains.

In conjugation, bacterial cells pass genetic material directly to each other. This is a method used mainly by Gram-negative, rod-shaped bacteria in the colon, and involves two cells building a connecting bridge of sex pili through which the genetic information can pass.

19.10.3 Other factors affecting drug resistance

The more useful a drug is, the more it will be prescribed and the greater the possibilities of resistant bacterial strains emerging. The original penicillins were used widely in human medicine, but were also commonly used in veterinary medicine. Antibacterial agents have also been used in animal feeding to increase animal weight and this, more than anything else, has resulted in drug-resistant bacterial strains. It is sobering to think that many of the original bacterial strains which were treated so dramatically with penicillin V or penicillin G are now resistant to those early penicillins. In contrast, these two drugs are still highly effective antibacterial agents in poorer, developing African nations, where the use (and abuse) of the drugs has been less widespread.

The ease with which different bacteria acquire resistance varies. For example, *S. aureus* is notorious for its ability to acquire drug resistance owing to the ease with which it can undergo transduction. On the other hand, the microorganism responsible for syphilis seems incapable of acquiring resistance and is still susceptible to the original drugs used against it.

19.10.4 The way ahead

There have been repeated warnings about the ticking time bomb of antibacterial drug resistance, made all the more urgent by the past reluctance of pharmaceutical industries to invest in antibacterial research. This reluctance was understandable due to the poor success rates and limited financial returns that were likely from developing a successful antibacterial drug. After all, novel drugs that prove effective against drug-resistant bacterial strains are likely to be placed on a reserve list of drugs, and only used when necessary. Until recently, there has been little financial incentive for pharmaceutical companies to invest vast amounts of money on antibacterial research. Fortunately, a number of initiatives have appeared that are aimed at encouraging public–private partnerships and providing a different approach to rewarding companies for successful research. The EU has formed the **Innovative Medicines Initiative** which has introduced an initiative called **New Drugs for Bad Bugs**. The US has also provided financial incentives to the pharmaceutical industry which includes more research funding and more generous periods of marketing exclusivity for any successful drugs. As a result, there has been a recent upswing in antibacterial research that has seen a number of novel agents entering clinical trials. The financial problems have been tackled and it is now a case of tackling the scientific problems.

The ability of bacteria to gain resistance to drugs is an ever-present challenge to the medicinal chemist and it is important to continue designing new antibacterial agents. Identifying potential new targets is essential in this never-ending battle. Genomics and proteomics have led to a greater molecular understanding of infectious

agents, leading to the identification of new drug targets. For example, *Mycobacterium tuberculosis*—the causative agent of tuberculosis—has a complex cell wall where three types of polymers are attached to peptidoglycan. The detailed mechanisms by which these polymers are synthesized and incorporated into the cell wall are being investigated to identify new targets, such that novel antibacterial drugs can be designed to disrupt the cell wall structure.

It is also beginning to be appreciated that the drugs with the least susceptibility to resistance are those with several different modes of action. Therefore, designing drugs which act on a number of different targets, rather than one specific target, is more likely to be successful.

Examples of new targets include kinase enzymes. There has already been success in designing kinase inhibitors as anticancer agents (section 21.6.2), and several research groups are now looking at agents that might prove to be selective inhibitors of bacterial kinases. Other potential targets are the enzymes known as **aminoacyl tRNA synthetases**. These enzymes are an ancient group of enzymes responsible for attaching amino acids to tRNA. Because they are ancient, there is a considerable sequence divergence between the bacterial and human enzymes, making selective inhibition possible. **Isoleucyl tRNA synthetase** is one such enzyme which is known to be inhibited by **mupirocin** (Fig. 19.89)—a clinically useful antibiotic isolated from *Pseudomonas fluorescens* with activity against MRSA. Mupirocin is used as a topical agent for skin infections and has also been used to combat the transmission of *S. aureus* within hospitals by treating the nasal passages of patients and hospital staff. Unfortunately, the widespread use of the agent for this purpose has led to strains of *S. aureus* with increasing resistance to the drug. Research is now being carried out to find novel inhibitors for a different aminoacyl tRNA synthetase present in *S. aureus*, namely **tyrosine tRNA synthetase**. The strategy of targeting aminoacyl tRNA synthetases is also proving fruitful in the search for novel antifungal agents. **Tavaborole** inhibits **leucine tRNA synthetase** and was approved in 2014 for the treatment of fungal nail infections.

Other potential targets include the enzymes involved in bacterial fatty acid biosynthesis as there are significant structural differences between these enzymes and the

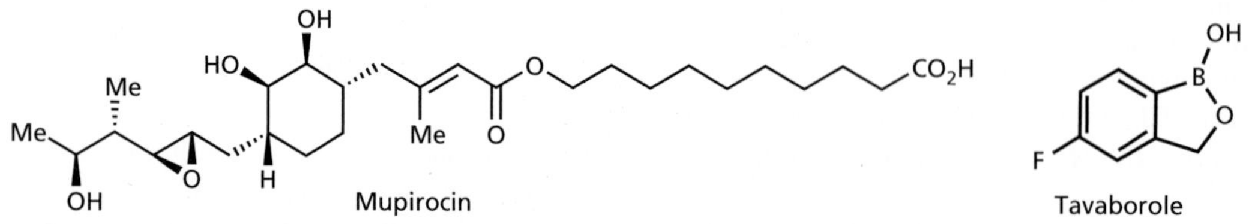

FIGURE 19.89 Inhibitors of aminoacyl tRNA synthetases.

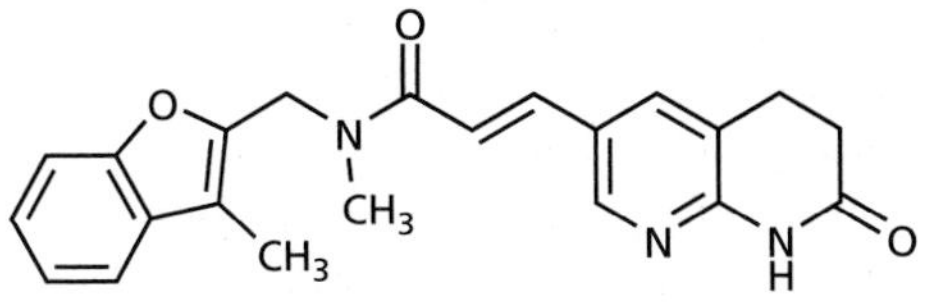

FIGURE 19.90 AFN-1252.

corresponding mammalian enzymes. **AFN-1252** (Fig. 19.90) is currently under investigation as an inhibitor of an enzyme called **FabI**. This is an enoyl-acyl carrier protein (ACP) reductase that catalyses the final rate-limiting step of the elongation cycle in fatty acid biosynthesis.

Another potential approach in countering resistance is to modify antibiotics such that they gain resistance to the mechanisms of resistance used against them! For example, **kanamycin** is an aminoglycoside which is no longer used because resistant bacteria can phosphorylate one of the hydroxyl groups present (Fig. 19.91).

An active analogue has been synthesized which replaces the susceptible alcohol with a ketone (Fig. 19.92). This ketone is in equilibrium with the hydrated gem-diol. When phosphorylation occurs on the diol, the phosphate group thus formed acts as a good leaving group and the ketone is regenerated. *In vitro* tests showed that this agent

was active against strains of bacteria which are resistant to kanamycin.

Another approach is to design molecules with an in-built self-destruct mechanism. One of the problems with antibiotics in medicine or veterinary practice is that much of the active antibiotic is excreted, giving bacteria in the environment the opportunity to gain resistance. This problem could be reduced by incorporating a self-destruct mechanism which kicks in once the antibiotic is excreted. For example, work has been carried out on a cephalosporin containing a protected hydrazine group (Fig. 19.93). The protecting group concerned is *ortho*-nitrobenzylcarbamate which is susceptible to light. Once the antibiotic is excreted and exposed to light, the protecting group is lost allowing the nucleophilic hydrazine moiety to react with the β-lactam ring and deactivate the molecule. This works *in vitro* but has still be tested *in vivo*.

Recent research into drug combinations has shown that there can be a beneficial effect on antibacterial activity *in vitro* if one administers an antibacterial drug with another drug, even if the other drug has no antibacterial activity itself. For example, a small dose of the tetracycline agent **minocycline** showed better activity than expected when it was administered along with the antidiarrhoeal

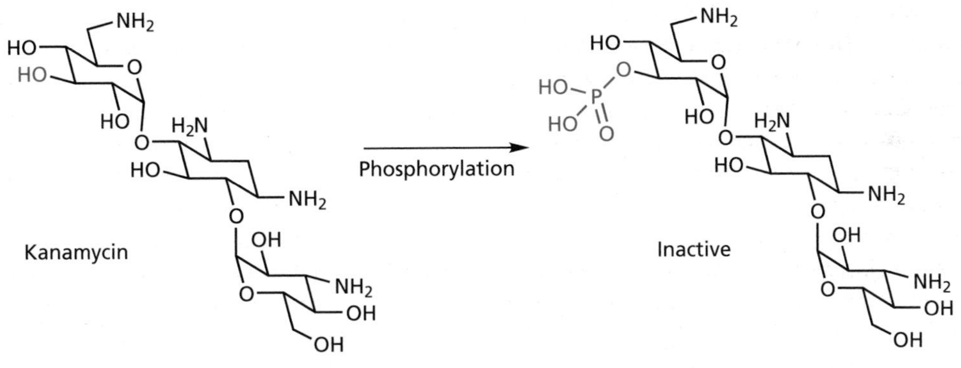

FIGURE 19.91 The phosphorylation reaction causing resistance to kanamycin.

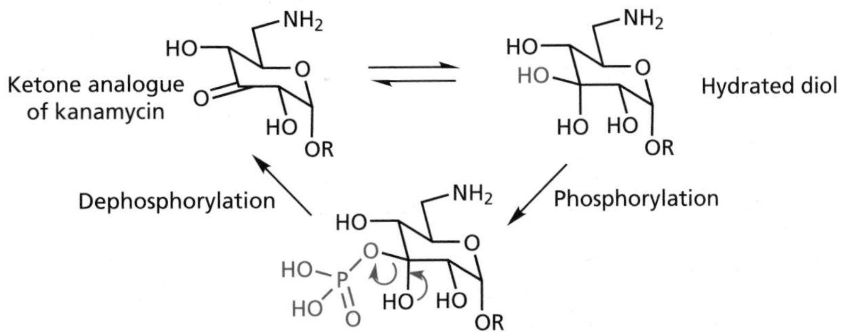

FIGURE 19.92 Analogue of kanamycin which is resistant to phosphorylation.

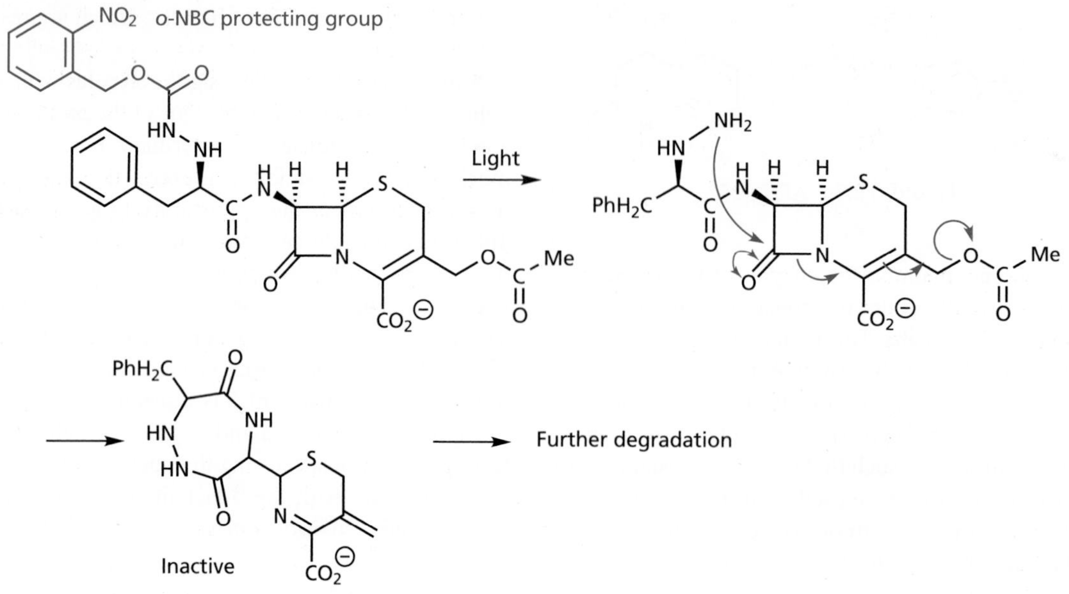

FIGURE 19.93 Self-destruct mechanism.

drug **loperamide** (Box 24.3). Further studies are needed to see if this effect occurs *in vivo*, but it might be another way of tackling drug resistance.

An alternative approach to using drug combinations is to design hybrid drugs that have more than one mechanism of action. For example, **cadazolid** is a hybrid structure which contains a fluoroquinolone and an oxazolidinone (Fig. 19.94). Because cadazolid has two mechanisms of action, it is likely to remain effective even if one of its targets becomes resistant. Currently, cadazolid is undergoing clinical trials as a treatment for infections such as *Clostridium difficile*.

Finally, medicinal chemistry research has come to rely heavily on finding hits and lead compounds from standard chemical libraries. Such libraries have been designed to contain drug-like molecules that obey Lipinski's rules.

However, the success rate from those libraries is relatively poor for antibacterial agents. This is not altogether surprising. Lipinski's rules are a guideline to how well drugs might cross the cell membranes of eukaryotic cells. In contrast, antibacterial drugs have to cross bacterial cell walls and cell membranes, and these are quite different in nature. Therefore, structures designed to fit Lipinski's rules are not necessarily the ideal structures to cross the barriers protecting bacterial prokaryotic cells. It is interesting to note that many of the current antibacterial agents used in medicine are very polar with molecular weights varying from 102 to 1449 Da. Designing compound libraries that contain structures which are more appropriate for the antibacterial field is an important priority if hit compounds are to be identified for novel bacterial targets.

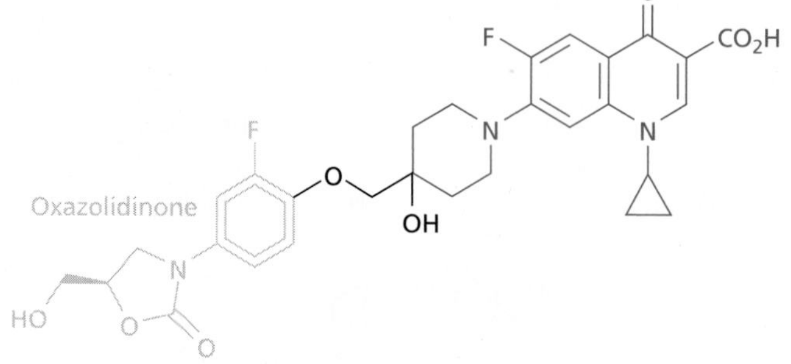

FIGURE 19.94 Cadazolid.

- Bacterial strains vary in their ability to gain resistance to antibacterial drugs. *Staphylococcus aureus* is quick to gain antibacterial resistance. The MRSA strain is a *S. aureus* strain that is resistant to most antibacterials, including methicillin.

- Vancomycin is the antibacterial agent of last resort in the treatment of resistant bacterial strains.

- There are many mechanisms by which bacteria can acquire resistance against antibacterial agents, but they all result from a change in the cell's genetic make-up.

- Drug resistance can result from mutation of a cell's genetic information or from transfer of genetic information from one cell to another. Genetic information can be transferred from one cell to another by transduction or conjugation.

- Care has to be taken to use antibacterial agents in a responsible manner to reduce the chances of resistance developing.

- It is important to identify new targets which can be used for the design of novel antibacterial agents.

BOX 19.20 Organoarsenicals as antiparasitic drugs

The first effective antimicrobial drug to be synthesized was the organoarsenical, **salvarsan** (section 19.1). In the late 1940s, another organoarsenical called **melarsoprol** was introduced into medicine and is the first-choice drug for the treatment of trypanosomiasis and sleeping sickness. This is despite the fact that it has to be injected and can kill one in 20 patients treated with it (Fig. 1).

One of the mechanisms by which melarsoprol might act is through a reaction with the cysteine residues of enzymes involved in glycolysis (Fig. 2). The blocking of glycolysis leads to a loss of cell motility and eventual cell death. Other mechanisms of action have been proposed.

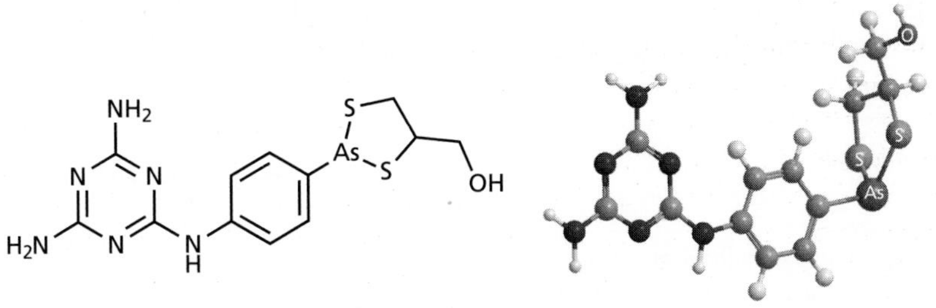

FIGURE 1 Melarsoprol.

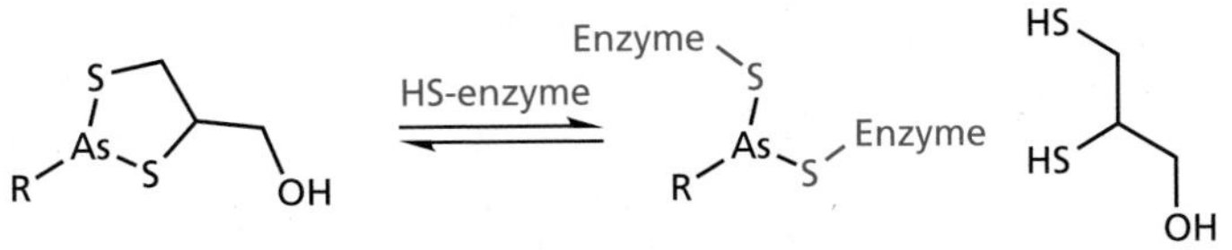

FIGURE 2 Mechanism of action of melarsoprol.

QUESTIONS

1. How would you convert penicillin G to 6-aminopenicillanic acid (6-APA) using chemical reagents? Suggest how you would make ampicillin from 6-APA.

2. Penicillin is produced biosynthetically from cysteine and valine. If the biosynthetic pathway could accept different amino acids, what sort of penicillin analogues might be formed if valine was replaced by alanine, phenylalanine, glycine, or lysine? What sort of penicillin analogue might be formed if cysteine was replaced by serine? (See Appendix 1 for amino acid structures.)

3. Referring to Question 2, why do you think penicillin analogues like this are not formed during the fermentation process?

4. The activity of sulphonamides is decreased if they are taken at the same time as procaine. Suggest why this might be the case.

5. Discuss whether you think the following penicillin analogue would be a useful antibacterial agent.

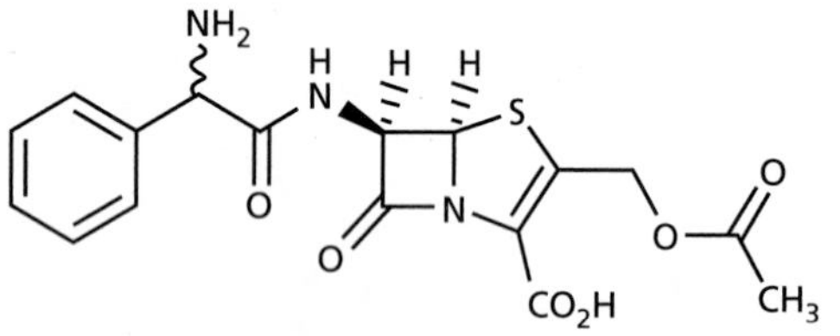

Penicillin analogue

6. Explain what effect replacing the methoxy groups on methicillin with ethoxy groups might have on the properties of the agent.

7. What effect might the bicyclic ring system of cephalosporins have on their chemical and biological properties, compared with the bicyclic ring system of penicillins, and why?

8. The following structure is an analogue of cefoxitin. What sort of properties do you think it might have compared to cefoxitin itself?

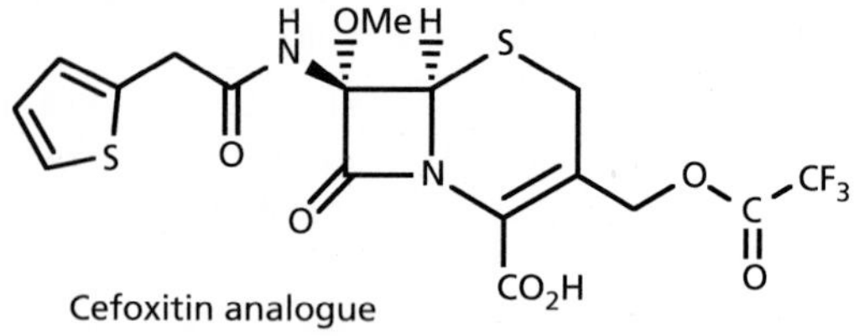

Cefoxitin analogue

9. Show the mechanism by which the prodrug bacampicillin (Box 19.7) is converted to ampicillin. What are the by-products?

10. a. Which of the following fluoroquinolone structures would you expect to have the best antibacterial activity?

    b. Devise a synthesis for the structure chosen in Question 10a.

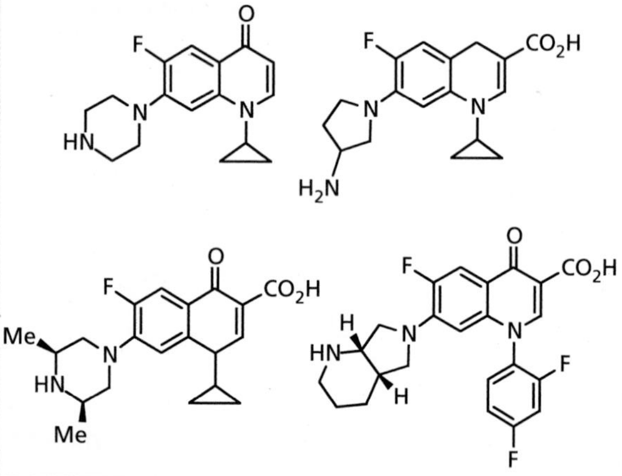

Multiple-choice questions are available on the Online Resource Centre at www.oxfordtextbooks.co.uk/orc/patrick6e/

FURTHER READING

Broadwith, P. (2010) Rousing sleeping sickness research. *Chemistry World*, May, 23.

Brown, D.G., Lister, T., and May-Dracka, T. L. (2014) New natural products as new leads for antibacterial drug discovery. *Bioorganic and Medicinal Chemistry Letters*, **24**, 413–8.

Bush, K. (2015) Investigational agents for the treatment of Gram-negative bacterial infections: a reality check. *ACS Infectious Agents*, **1**, 509–11.

Coates, A., et al. (2002) The future challenges facing the development of new antimicrobial drugs. *Nature Reviews Drug Discovery*, **1**, 895–910.

King, A. (2012) Making light work. *Chemistry World*, April, 52–55 (photodynamic therapy).

Sansom, C. (2012) The latent threat of tuberculosis. *Chemistry World*, September, 48–51.

Singh, S. B. (2014) Confronting the challenges of discovery of novel antibacterial agents. *Bioorganic and Medicinal Chemistry Letters*, **24**, 3683–9.

Skedelj, V., et al. (2011) ATP-binding site of bacterial enzymes as a target for antibacterial drug design. *Journal of Medicinal Chemistry*, 2011, **54**, 915–29.

Lactams and other agents acting on cell walls

Axelsen, P. H., and Li, D. (1998) A rational strategy for enhancing the affinity of vancomycin towards depsipeptide ligands. *Bioorganic and Medicinal Chemistry*, **6**, 877–81.

Nicolaou, K. C., et al. (1999) Chemistry, biology, and medicine of the glycopeptide antibiotics. *Angewandte Chemie, International Edition*, **38**, 2096–152.

Agents acting on the cell membrane

Mann, J. (2001) Killer nanotubes. *Chemistry in Britain*, November, 22.

Linezolid

Ford, C. (2001) First of a kind. *Chemistry in Britain*, March, 22–24.

Quinolones and other agents acting on nucleic acids

Saunders, J. (2000) Quinolones as anti-bacterial DNA gyrase inhibitors. Chapter 10 in: *Top drugs: top synthetic routes*. Oxford University Press, Oxford.

Agents acting against protein synthesis

Becker, B., and Cooper, M.A. (2013) Aminoglycoside antibiotics in the 21st century. *ACS Chemical Biology*, **8**, 105–15.

Novak, R. (2011) Are pleuromutilin antibiotics finally fit for human use? *Annals of the New York Academy of Sciences*, **1241**, 71–81.

Titles for general further reading are listed on p.845.

20.1 Viruses and viral diseases

Viruses are non-cellular, infectious agents which take over a host cell in order to survive and multiply. There are many different viruses capable of infecting bacterial, plant, or animal cells, with more than 400 known to infect humans. Those capable of being transmitted to humans from animals or insects can be particularly dangerous and belong to a class of diseases defined as **zoonoses**. Consequently, both human and veterinary medicine play important roles in the control of such diseases.

Viruses can be transmitted in a variety of ways. Those responsible for diseases such as influenza (flu), chicken pox, measles, mumps, viral pneumonia, rubella, and smallpox can be transmitted through the air by an infected host sneezing or coughing. Other viruses can be transmitted by means of arthropods or ticks leading to diseases such as Colorado tick fever and yellow fever. Some viruses are unable to survive for long outside the host and are transmitted through physical contact. The viruses responsible for AIDS, cold sores, the common cold, genital herpes, certain leukaemias, and rabies are examples of this kind. Finally, food-borne or water-borne viruses can lead to hepatitis A and E, poliomyelitis, and viral gastroenteritis.

Historically, viral infections have proved devastating to human populations. It has been suggested that smallpox was responsible for the major epidemics which weakened the Roman Empire during the periods AD 165–180 and AD 251–266. Smallpox was also responsible for the decimation of indigenous tribes in both North and South America during European colonization. In some areas, it is estimated that 90% of the population died from the disease. Various flu epidemics and pandemics have proved devastating. The number of deaths worldwide due to the flu pandemic of 1918–1919 is estimated to be over 20 million, far larger than the number killed by military action during the First World War. Finally, it is estimated that 30 million people have died as a result of HIV infection since the 1980s.

The African continent has its fair share of lethal viruses, including Ebola and the virus responsible for Lassa fever. In the past, viral diseases such as these occurred in isolated communities and were easily contained. Nowadays, with cheap and readily available air travel, tourists are able to visit remote areas, thus increasing the chances of rare or new viral diseases spreading round the world. Therefore, it is important that world health authorities monitor potential risks and take speedy, appropriate action when required. The outbreak of **severe acute respiratory syndrome** (SARS) in the Far East during 2003 could have had a devastating effect worldwide if it had been ignored. Fortunately, the world community acted swiftly and the disease was brought under control relatively quickly. Nevertheless, the SARS outbreak serves as a timely warning of how dangerous viral infections can be. Scientists have warned of a nightmare scenario involving the possible evolution of a 'supervirus'. Such an agent would have a transmission mode and infection rate equivalent to flu, but a much higher mortality rate. There are already lethal viruses which can be spread rapidly and have a high mortality rate. Fortunately, the latency period between infection and detectable symptoms is short and so it is possible to contain the outbreak, especially if it is in isolated communities. If such viral infections evolved such that the latency period increased to that of AIDS, they could result in devastating pandemics equivalent to the plagues of the Middle Ages.

Considering the potential devastation that viruses can wreak on society, there are fears that terrorists might one day try to release lethal viral strains on civilian populations. This has been termed **bioterrorism**. To date, no terrorist group has carried out such an action, but it would be wrong to ignore the risk.

It is clear that research into effective antiviral drugs is a major priority in medicinal chemistry.

20.2 Structure of viruses

At their simplest, viruses can be viewed as protein packages transmitting foreign nucleic acid between host cells. The type of nucleic acid present depends on the virus concerned. All viruses contain one or more molecules of either RNA or DNA, but not both. They can, therefore, be defined as RNA or DNA viruses. Most RNA viruses contain single-stranded RNA (ssRNA), but some viruses contain double-stranded RNA (dsRNA). If the base sequence of the RNA strand is identical to viral mRNA, it is called the positive (+) strand. If it is complementary, it is called the negative (−) strand. Most DNA viruses contain double-stranded DNA, but a small number contain single-stranded DNA. The size of the nucleic acid varies widely, with the smallest viral genomes coding for 3–4 proteins and the largest coding for over 100 proteins.

The viral nucleic acid is contained and protected within a protein coat called the **capsid**. Capsids are usually made up of protein subunits called **protomers** which are generated in the host cell and can interact spontaneously to form the capsid in a process called **self-assembly**. Once the capsid contains the viral nucleic acid, the whole assembly is known as the **nucleocapsid**. In some viruses, the nucleocapsid may contain viral enzymes which are crucial to its replication in the host cell. For example, the flu virus contains an enzyme called **RNA-dependent RNA polymerase** within its nucleocapsid (Fig. 20.1).

Additional membranous layers of carbohydrates and lipids may surround the nucleocapsid, depending on the virus concerned. These are usually derived from the host cell, but they may also contain viral proteins which have been coded by viral genes.

The complete structure is known as a **virion** and this is the form that the virus takes when it is outside the host cell. The size of a virion can vary from 10 nm to 400 nm. As a result, most viruses are too small to be seen by a light microscope and require the use of an electron microscope.

20.3 **Life cycle of viruses**

The various stages involved in the life cycle of a virus are as follows (Fig. 20.2):

- **Adsorption:** a virion has to first bind to the outer surface of a host cell. This involves a specific molecule on the outer surface of the virion binding to a specific protein or carbohydrate present in the host cell membrane. The relevant molecule on the host cell can thus be viewed as a 'receptor' for the virion. Of course, the host cell has not produced this molecule to be a viral receptor. The molecules concerned are usually glycoproteins which have crucial cellular functions such as the binding of hormones. However, the virion takes advantage of these and once it is bound, the next stage can take place— introduction of the viral nucleic acid into the host cell.

- **Penetration and uncoating:** different viruses introduce their nucleic acid into the host cell by different methods. Some inject their nucleic acid through the cell membrane; others enter the cell intact and are then uncoated. This can also happen in a variety of ways. The viral envelope of some virions fuses with the plasma membrane and the nucleocapsid is then introduced into the cell (Fig. 20.2). Other virions are taken into the cell by endocytosis where the cell membrane wraps itself round the virion and is then pinched off to produce a vesicle called an **endosome** (see for example Fig. 20.40). These vesicles then fuse with **lysosomes**, and host cell enzymes aid the virus in the uncoating process. Low endosomal pH also triggers uncoating. In some cases, the viral envelope fuses with the lysosome membrane, and the nucleocapsid is released into the cell. Whatever the process, the end result is the release of viral nucleic acid into the cell.

- **Replication and transcription:** viral genes can be defined as *early* or *late*. Early genes take over the host cell such that viral DNA and/or RNA are synthesized. The mechanism involved varies from virus to virus. For example, viruses containing negative single-stranded RNA use a viral enzyme called **RNA-dependent RNA polymerase** (or transcriptase) to synthesize mRNA which then codes for viral proteins.

- **Synthesis and assembly of nucleocapsids:** late genes direct the synthesis of capsid proteins and these self-assemble to form the capsid. Viral nucleic acid is then taken into the capsid to form the nucleocapsid.

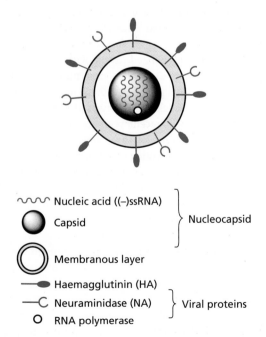

Nucleic acid ((−)ssRNA)
Capsid } Nucleocapsid
Membranous layer
Haemagglutinin (HA)
Neuraminidase (NA) } Viral proteins
RNA polymerase

FIGURE 20.1 Diagrammatic representation of the flu virus.

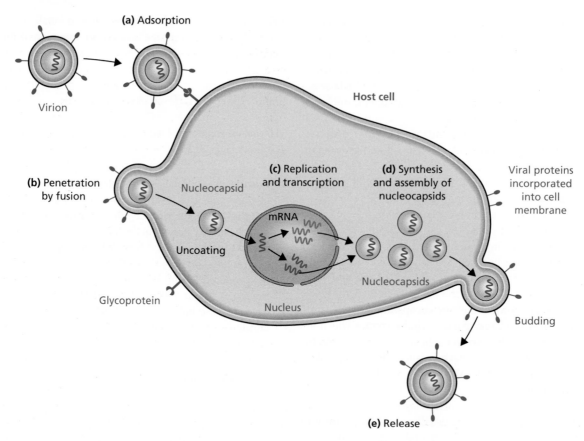

(a) Adsorption

Virion

Host cell

(b) Penetration by fusion

Nucleocapsid

(c) Replication and transcription

(d) Synthesis and assembly of nucleocapsids

Viral proteins incorporated into cell membrane

mRNA

Uncoating

Nucleocapsids

Glycoprotein

Nucleus

Budding

(e) Release

FIGURE 20.2 Life cycle of a DNA virus such as herpes simplex.

- **Virion release:** naked virions (those with no outer layers round the nucleocapsid) are released by cell lysis when the cell is destroyed. In contrast, viruses with envelopes are usually released by a process known as **budding** (Fig. 20.2). Viral proteins are first incorporated into the host cell's plasma membrane. The nucleocapsid then binds to the inner surface of the cell membrane and, at the same time, viral proteins collect at the site and host cell proteins are excluded. The plasma membrane containing the viral proteins then wraps itself round the nucleocapsid and is pinched off from the cell to release the mature virion.

The life cycle stages of herpes simplex, HIV, and flu virus are illustrated in Figs. 20.2, 20.11, and 20.40 respectively.

20.4 Vaccination

Vaccination is the preferred method of protection against viral disease and has proved extremely successful against childhood diseases such as polio, measles, and mumps, as well as historically serious diseases such as smallpox and yellow fever. The first successful vaccination was carried out by Edward Jenner in the eighteenth century. He observed that a milkmaid, who had previously contracted the less virulent cowpox, was immune to smallpox. Therefore, Jenner inoculated people with material from cowpox lesions and discovered that they, too, gained immunity from smallpox. Since then, many other vaccines have been developed. Perhaps the most controversial vaccination in recent years has been the MMR vaccine—a combination of three separate vaccinations that is administered to young children to provide protection against measles, mumps, and rubella. Unfortunately, deep concerns were raised in the UK due to a publication in the *Lancet* that linked the vaccine to an increased risk of autism. This paper was eventually discredited as fraudulent.

Vaccination works by introducing the body to foreign material which bears molecular similarity to some component of the virus, but which lacks its infectious nature or toxic effects. The body then has the opportunity to recognize the molecular fingerprint of the virus (i.e. specific **antigens**), and the immune system is primed to attack the virus should it infect the body. Usually a killed or weakened version of the virus is administered so that it does not lead to infection itself. Alternatively, fragments of the virus (subunit vaccines) can be used if they display a characteristic antigen. Vaccination is a preventive approach and is not usually effective on patients who have already become infected.

Vaccines are currently under investigation for the prevention or treatment of HIV, dengue fever, genital herpes, and haemorrhagic fever caused by the Ebola virus. However, there are difficulties surrounding the HIV and flu viruses, because rapid gene mutation in these viruses results in constant changes to the amino acid composition of glycoproteins normally present on the viral surface. Because these glycoproteins are the important antigens that trigger the immune response, any changes in their structure 'disguise' the virus, and the body's primed immune system fails to recognize it.

Another problem concerning vaccination relates to patients with a weakened immune response. The main categories of patients in this situation are cancer patients undergoing chemotherapy, patients undergoing organ transplants (where the immune system has been deliberately suppressed to prevent organ rejection), and AIDS patients. Vaccination in these patients is less likely to be effective because of the weakened immune response.

20.5 Antiviral drugs: general principles

Antiviral drugs are useful in tackling viral diseases where there is a lack of an effective vaccine, or where infection has already taken place. The life cycle of a virus means that, for most of its time in the body, it is within a host cell and is effectively disguised both from the immune system and from circulating drugs. As it also uses the host cell's own biochemical mechanisms to multiply, the number of potential drug targets that are unique to the virus is more limited than those for invading microorganisms. Thus, the search for effective antiviral drugs has proved more challenging than that for antibacterial drugs. Indeed, the first antiviral agents appeared relatively late on in the 1960s, and only three clinically useful antiviral drugs were in use during the early 1980s. Early antiviral drugs included **idoxuridine** and **vidarabine** for herpes infections, and **amantadine** for influenza A.

Since then, progress has accelerated for two principal reasons: (i) the need to tackle the AIDS pandemic, and (ii) the increased understanding of viral infectious mechanisms resulting from viral genomic research.

In 1981, it was noticed that homosexual men were unusually susceptible to diseases such as pneumonia and fungal infections—ailments which were previously only associated with patients whose immune response had been weakened. The problem soon reached epidemic proportions and it was discovered that a virus (the **human immunodeficiency virus**—HIV) was responsible. It was found that this virus infected **T-cells**—cells which are crucial to the immune response—and was therefore directly attacking the immune system. With a weakened immune system, infected patients proved susceptible to a whole range of opportunistic secondary diseases resulting in the term **acquired immune deficiency syndrome** (AIDS). This discovery led to a major research effort into understanding the disease and counteracting it—an effort which kick-started more general research into antiviral chemotherapy. Fortunately, the tools needed to carry out effective research appeared on the scene at about the same time with the advent of viral genomics. The full genome of any virus can now be quickly determined and compared with those of other viruses, allowing the identification of how the genetic sequence is split into genes. This, in turn, helps to identify viral proteins as potential drug targets. Standard genetic engineering methods permit the production of pure copies of the target protein by inserting the viral gene into a bacterial cell thus providing sufficient quantities of the protein to be isolated and studied (section 6.4).

Good drug targets are proteins which are likely to have the following characteristics:

- They are important to the life cycle of the virus such that their inhibition or disruption has a major effect on infection.

- They bear little resemblance to human proteins, thus increasing the chances of good selectivity and minimal side effects.

- They are common to a variety of different viruses and have a specific region which is identical in its amino acid composition. This makes the chances of developing a drug with broad antiviral activity more likely.

- They are important to the early stages of the viral life cycle, which means that their disruption/inhibition reduces the chances of symptoms and of the virus spreading through the body.

Most antiviral drugs in use today act against HIV, herpesviruses (responsible for a variety of ailments including cold sores and encephalitis), hepatitis B, and hepatitis C. Diseases such as herpes and HIV are chronic in developed countries, and intensive research has been carried out to develop drugs to combat them. In contrast, less research is carried out on viral diseases that are prevalent in developing countries, for example tropical (dengue) and haemorrhagic (Ebola) fevers.

Most antiviral drugs in use today disrupt critical stages of the virus life cycle or the synthesis of virus-specific nucleic acids. Excluding drugs developed for the treatment of HIV, more drugs are available for the treatment of DNA viruses than for RNA viruses. Few drugs show a broad activity against both DNA and RNA viruses.

Studies of the human genome are also likely to be useful for future research. The identification of human proteins which stimulate the body's immune response or the production of antibodies would provide useful leads for the development of drugs that would have an antiviral effect by acting as immunomodulators.

KEY POINTS

- Viruses pose a serious health threat and there is a need for new antiviral agents.

- Viruses consist of a protein coat surrounding nucleic acid, which is either RNA or DNA. Some viruses have an outer membranous coat that is derived from the host cell.

- Viruses are unable to self-multiply and need to enter a host cell in order to do so.

- Vaccination is effective against many viruses, but is less effective against viruses which readily mutate.

- Research into antiviral drugs has increased in recent years as a result of the AIDS epidemic and the need to find drugs to combat it.

- Antiviral research has been aided by advances in viral genomics and genetic engineering, as well as by the use of X-ray crystallography and molecular modelling.

20.6 Antiviral drugs used against DNA viruses

Most of the drugs which are active against DNA viruses have been developed against herpesviruses to combat diseases such as cold sores, genital herpes, chicken pox, shingles, eye diseases, mononucleosis, Burkitt's lymphoma, and Kaposi's sarcoma. Nucleoside analogues have been particularly effective.

20.6.1 Inhibitors of viral DNA polymerase

The clinical aspects of viral DNA polymerase inhibitors are described in Box 20.1. The best known of these inhibitors is **aciclovir**, which was discovered by compound screening and was introduced into the market in 1981. Aciclovir has a nucleoside-like structure and contains the same nucleic acid base as deoxyguanosine. However, it lacks the complete sugar ring. In virally infected cells, it is phosphorylated in three stages to form a triphosphate which is the active agent, and so aciclovir itself is a prodrug (Fig. 20.3).

Nucleotide triphosphates are the building blocks for DNA replication where a new DNA strand is constructed using a DNA template—a process catalysed by the enzyme **DNA polymerase**. Aciclovir triphosphate prevents DNA replication in two ways. Firstly, it is sufficiently similar to the normal deoxyguanosine triphosphate building block (Fig. 20.4) that it can bind to DNA polymerase and inhibit it. Secondly, DNA polymerase can catalyse the attachment of the aciclovir nucleotide to the growing DNA chain. As the sugar unit is incomplete

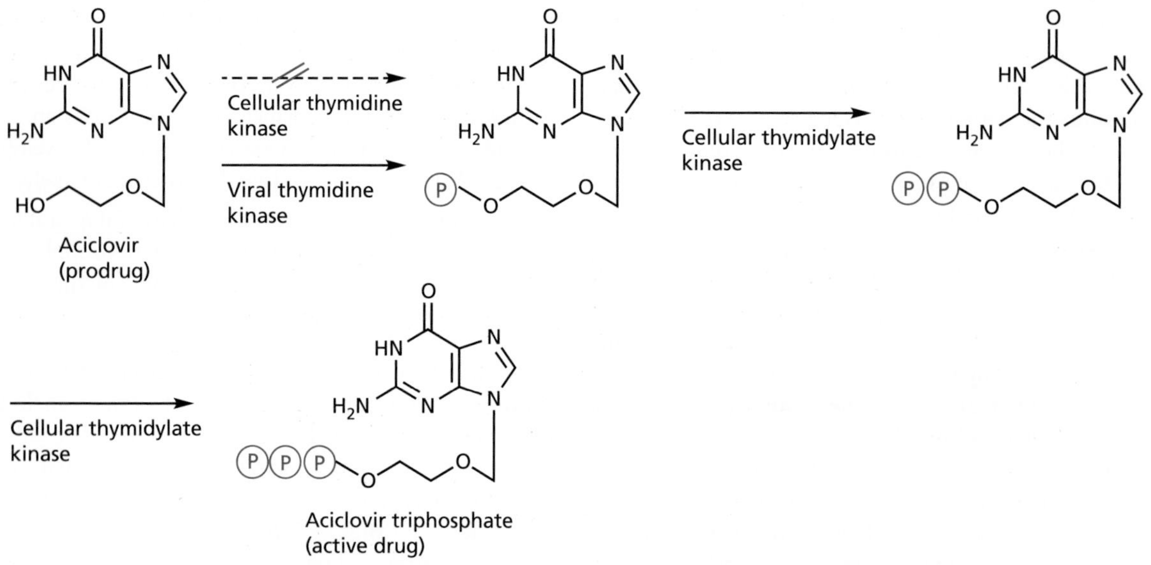

FIGURE 20.3 Activation of aciclovir (Ⓟ represents phosphate groups).

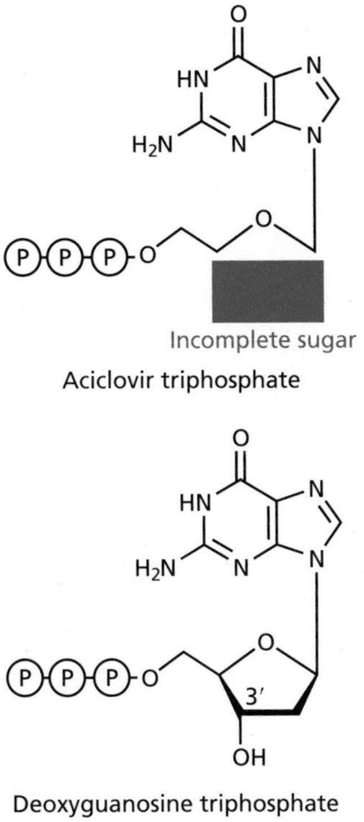

FIGURE 20.4 Comparison of aciclovir triphosphate and deoxyguanosine triphosphate.

and lacks the required hydroxyl group normally present at position 3′ of the sugar ring, the nucleic acid chain cannot be extended any further. Thus, the drug acts as a **chain terminator** (see section 9.5).

However, what is to stop aciclovir triphosphate inhibiting DNA polymerase in normal, uninfected cells? The answer lies in the fact that aciclovir is only converted to the active triphosphate in infected cells. The explanation for this lies in the first phosphorylation reaction catalysed by the enzyme thymidine kinase (Fig. 20.3). Although this enzyme is present in host cells, the herpes

virus carries its own version. It turns out that viral thymidine kinase is 100 times more effective at converting aciclovir to its monophosphate than host cell thymidine kinase. Once formed, the monophosphate is converted to the active triphosphate by cellular enzymes. Therefore, in normal uninfected cells, aciclovir is a poor substrate for cellular thymidine kinase and remains as the prodrug. This, along with the fact that there is a selective uptake of aciclovir by infected cells, explains its excellent activity and much reduced toxicity relative to previous drugs. Another feature which enhances its safety is that aciclovir triphosphate shows a 50-fold selective action against viral DNA polymerases relative to cellular polymerases.

The oral bioavailability of aciclovir is quite low (15–30%) and to overcome this, various prodrugs were developed to increase water solubility. **Valaciclovir** (Fig. 20.5) is an L-valyl ester prodrug that is absorbed from the gut far more effectively than aciclovir itself. However, the prodrug has similar polarity and ionization to aciclovir, and so the prodrug is no more able to cross the cell membranes of the gut wall by passive diffusion than aciclovir itself. Moreover, poorer absorption is observed if D-valine is used for the prodrug instead of L-valine, suggesting that a specific binding interaction is involved in the absorption process. This implies that the prodrug is actively transported by transport proteins in the gut, and that the valine allows the prodrug to be recognized and bound by these proteins. Transport proteins normally responsible for transporting dipeptides across the cell wall have been implicated in this process, i.e. the **human intestinal proton-dependent oligopeptide transporter-1 (hPEPT-1)** and **human intestinal di-/tripeptide transporter-1 (HPT-1)**. Once valaciclovir is absorbed, it is hydrolysed to aciclovir in the liver and gut wall. **Desciclovir** (Fig. 20.5) is a prodrug of aciclovir which lacks the carbonyl group at position 6 of the purine ring and is more water soluble. Once in the blood supply, metabolism by cellular **xanthine oxidase** oxidizes the 6-position to give aciclovir.

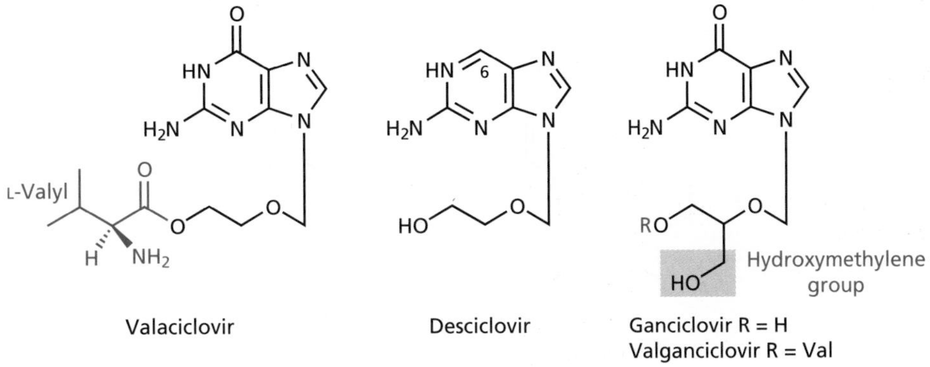

FIGURE 20.5 Prodrugs and analogues of aciclovir.

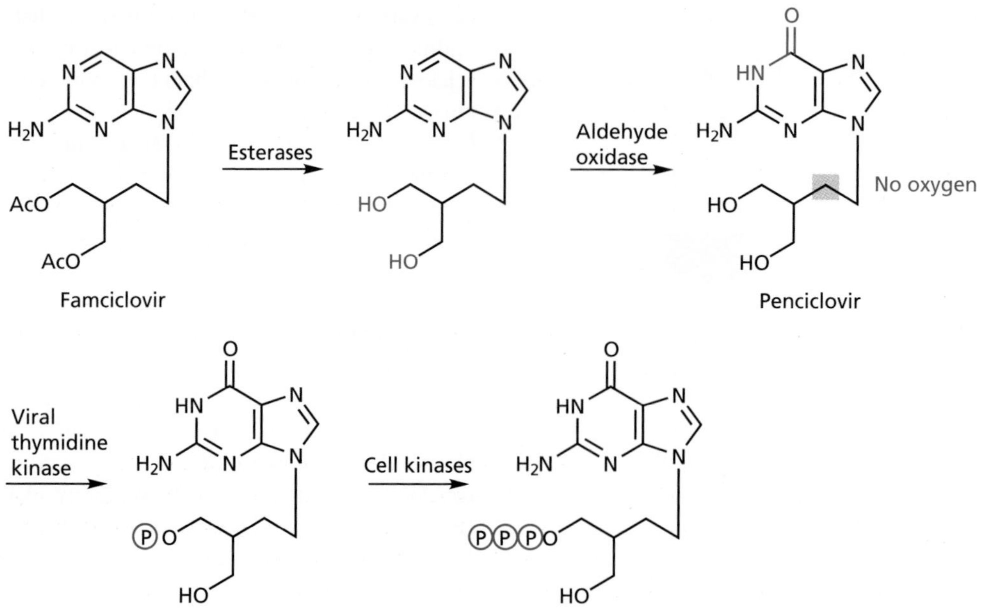

FIGURE 20.6 Penciclovir and famciclovir (Ⓟ represents a phosphate group).

Ganciclovir (Fig. 20.5) is an analogue of aciclovir and bears an extra hydroxymethylene group; **valganciclovir** acts a prodrug for this compound. **Penciclovir** and its prodrug **famciclovir** (Fig. 20.6) are analogues of ganciclovir. In famciclovir, the two alcohol groups of penciclovir are masked as esters making the structure less polar, and resulting in better absorption. Once absorbed, the acetyl groups are hydrolysed by esterases, and the purine ring is oxidized by **aldehyde oxidase** in the liver to generate penciclovir. Phosphorylation reactions then take place in virally infected cells as described previously.

Some viruses are immune from the action of the above antiviral agents because they lack the enzyme thymidine kinase. As a result, phosphorylation fails to take place. **Cidofovir** was designed to combat this problem (Fig. 20.7). It is an analogue of **deoxycytidine 5-monophosphate** where the sugar and phosphate groups have been replaced by an acyclic group and a phosphonomethylene group respectively. The latter group acts as a bio-isostere for the phosphate group and is used because the phosphate group itself would be susceptible to enzymatic hydrolysis. As a phosphate equivalent is now present, the drug does not require thymidine kinase to become activated. Two more phosphorylations can now take place, catalysed by cellular kinases, to convert cidofovir to the active 'triphosphate'.

In contrast to aciclovir, **idoxuridine**, **trifluridine**, and **vidarabine** (Fig. 20.8) are phosphorylated equally well by viral and cellular thymidine kinases, and so there is less selectivity for virally infected cells. As a result, these drugs have more toxic side effects. Idoxuridine, like trifluridine, is an analogue of deoxythymidine and was the first nucleoside-based antiviral agent licensed in the

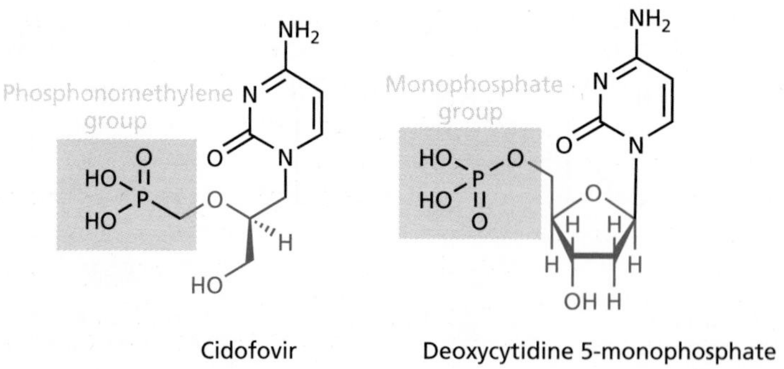

FIGURE 20.7 Comparison of cidofovir and deoxycytidine 5-monophosphate.

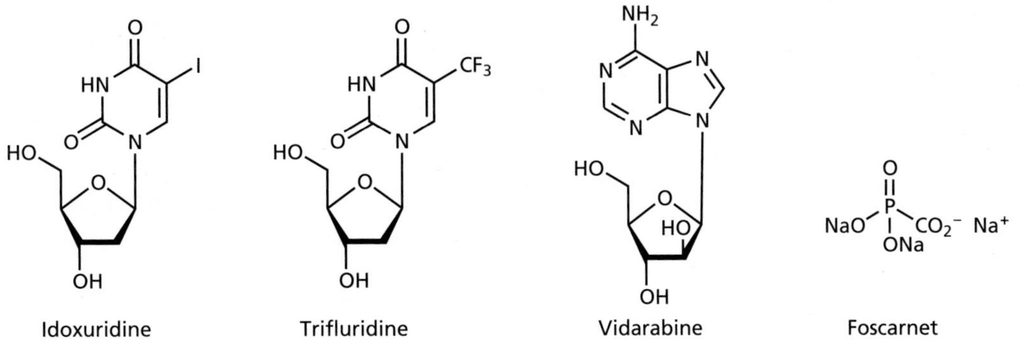

FIGURE 20.8 Miscellaneous antiviral agents.

USA. The triphosphate inhibits viral DNA polymerase as well as thymidylate synthetase. **Vidarabine** (Fig. 20.8) contains an arabinoside sugar ring and was developed from a natural product isolated from a marine sponge.

Foscarnet (Fig. 20.8) was discovered in the 1960s and inhibits viral DNA polymerase. However, it is non-selective and toxic. It also has difficulty crossing cell membranes due to its high charge.

BOX 20.1 Clinical aspects of viral DNA polymerase inhibitors

Aciclovir represented a revolution in the treatment of herpesvirus infections, being the first relatively safe, non-toxic drug to be used systemically. It is used for the treatment of infections due to herpes simplex 1 and 2 (i.e. herpes simplex encephalitis and genital herpes), as well as **varicella-zoster viruses (VZV)** (i.e. chickenpox and shingles). Unfortunately, strains of herpes are appearing which are resistant to aciclovir. This can arise due to mutations, either of the viral thymidine kinase enzyme such that it no longer phosphorylates aciclovir, or of viral DNA polymerase such that it no longer recognizes the activated drug. Aciclovir is not equally effective against all types of herpesvirus. Nine types of herpesvirus are known to infect humans, and aciclovir is most effective against those types described above.

Valaciclovir is a valine prodrug of aciclovir, and is particularly useful in the treatment of VZV infections. When this prodrug is given orally, blood levels of aciclovir are obtained which are equivalent to those obtained by intravenous administration.

Desciclovir is another prodrug for aciclovir but is somewhat more toxic, thus limiting its potential.

Ganciclovir is phosphorylated by thymidine kinases produced by both the α- and β-subfamilies of herpesvirus, and can be used against both viruses. Unfortunately, the drug is not as safe as aciclovir because it can be incorporated into cellular DNA. Nevertheless, it can be used for the treatment of **cytomegalovirus (CMV)** infections. This is a virus which causes eye infections and can lead to blindness. Aciclovir is not effective in this infection because CMV does not encode

a viral thymidine kinase. Ganciclovir, however, can be converted to its monophosphate by kinases other than thymidine kinase. As ganciclovir has a low oral bioavailability, the valine prodrug **valganciclovir** has been introduced for the treatment of CMV infections.

Penciclovir essentially has the same spectrum of activity as aciclovir, but has better potency, faster onset, and longer duration of action. It is used topically for the treatment of cold sores (HSV-1), and intravenously for the treatment of HSV in immunocompromised patients. Like aciclovir, penciclovir has poor oral bioavailability and is poorly absorbed from the gut due to its polarity. Therefore, **famciclovir** is used as a prodrug for better absorption.

Cidofovir is a broad-spectrum antiviral agent which shows selectivity for viral DNA polymerase, and is used to treat retinal inflammation caused by CMV. Unfortunately, the drug is extremely polar and has a poor oral bioavailability (5%). It is also toxic to the kidneys, but this can be reduced by co-administering **probenecid** (section 19.5.1.9).

Idoxuridine can be used for the topical treatment of herpes keratitis, but **trifluridine** is the drug of choice for this disease because it is effective at lower dose frequencies. **Vidarabine** was an early antiviral drug, but aciclovir is now generally preferred.

Foscarnet is used in the treatment of CMV retinitis where it is approximately equal in activity to ganciclovir. It can also be used in immunocompromised patients for the treatment of HSV and VZV strains which prove resistant to aciclovir. It does not undergo metabolic activation.

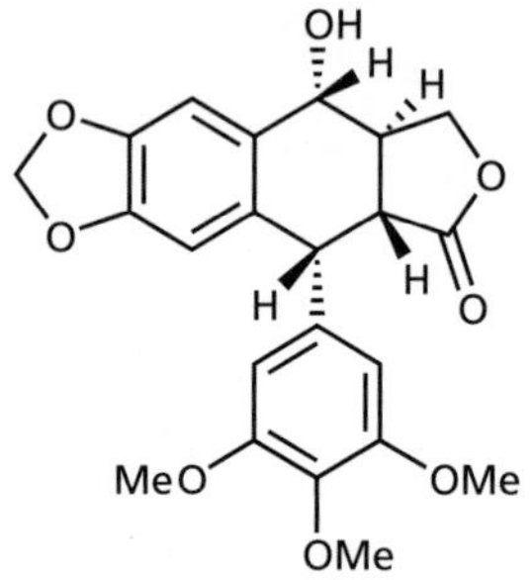

FIGURE 20.9 Podophyllotoxin.

20.6.2 Inhibitors of tubulin polymerization

The plant product **podophyllotoxin** (Fig. 20.9) has been used clinically to treat genital warts caused by the DNA virus **papillomavirus**, but it is not as effective as **imiquimod** (section 20.11.4). It is a powerful inhibitor of tubulin polymerization (sections 2.7.1 and 10.2.2).

20.6.3 Antisense therapy

Fomivirsen (Fig. 20.10) is the first, and so far the only, DNA antisense molecule that has been approved as an antiviral agent. It consists of 21 nucleotides with a phosphonothioate backbone rather than a phosphate backbone to increase the metabolic stability of the molecule (section 14.10). The drug blocks the translation of viral RNA and is used against retinal inflammation caused by CMV in AIDS patients. Because of its high polarity, it is administered as an ocular injection (**intravitreal injection**).

KEY POINTS

- Nucleoside analogues have been effective antiviral agents against DNA viruses, mainly herpesviruses.

- Nucleoside analogues are prodrugs, which are activated by phosphorylation to a triphosphate structure. They have a dual mechanism of action as viral DNA polymerase inhibitors and DNA chain terminators.

- Nucleoside analogues show selectivity for virally infected cells over normal cells when viral thymidine kinase is required to catalyse the first of three phosphorylation steps. They are taken up more effectively into virally infected cells, and their triphosphates show selective inhibition for viral DNA polymerases over cellular DNA polymerases.

- Agents containing a bio-isostere for a phosphate group can be used against DNA viruses lacking thymidine kinase.

- Inhibitors of tubulin polymerization have been used against DNA viruses.

- An antisense molecule has been designed as an antiviral agent.

---

d(*P*-thio)(G-C-G-T-T-T-G-C-T-C-T-T-C-T-T-C-T-T-G-C-G)

FIGURE 20.10 Fomivirsen.

20.7 Antiviral drugs acting against RNA viruses: the human immunodeficiency virus (HIV)

20.7.1 Structure and life cycle of HIV

HIV (Fig. 20.11) is an example of a group of viruses known as the **retroviruses**. There are two variants of HIV. HIV-1 is responsible for AIDS in America, Europe, and Asia, whereas HIV-2 occurs mainly in western Africa. HIV has been studied extensively over the last 20 years and a vast research effort has resulted in a variety of antiviral drugs which have proved successful in slowing down the disease, but not eradicating it. At present, most clinically useful antiviral drugs act against two targets—the viral enzymes **reverse transcriptase** and **protease**. Effective drugs have also been designed against other targets, and a good knowledge of the life cycle of HIV has been essential in identifying those targets (Fig. 20.11).

HIV is an RNA virus which contains two identical strands of (+)ssRNA within its capsid. Also present are the viral enzymes reverse transcriptase and **integrase**, as well as other proteins such as **p7** and **p9**. The capsid is made up of protein units known as **p24**, and surrounding the capsid there is a layer of matrix protein (**p17**). Beyond that, there is a membranous envelope which originates from host cells and which contains the viral glycoproteins **gp120** and **gp41**. Both of these proteins are crucial to the processes of adsorption and penetration. Gp41 traverses the envelope and is bound non-covalently to gp120, which projects from the surface. When the virus approaches the host cell, gp120 interacts and binds with a transmembrane protein called **CD4** which is present on the surface of host T-cells. The gp120 proteins then undergo a conformational change which allows them to bind simultaneously to **chemokine receptors** (**CCR5** and **CXCR4**) on the host cell (not shown). Further conformational changes peel away the gp120 protein allowing the viral protein gp41 to reach the surface of the host cell and anchor the virus to the surface. The gp41 then undergoes a conformational change and pulls the virus and the cell together so that their membranes can fuse.

Once fusion has taken place, the HIV nucleocapsid enters the cell. Disintegration of the protein capsid then takes place, probably aided by the action of a viral enzyme called protease. Viral RNA and viral enzymes are then released into the cell cytoplasm. The released viral

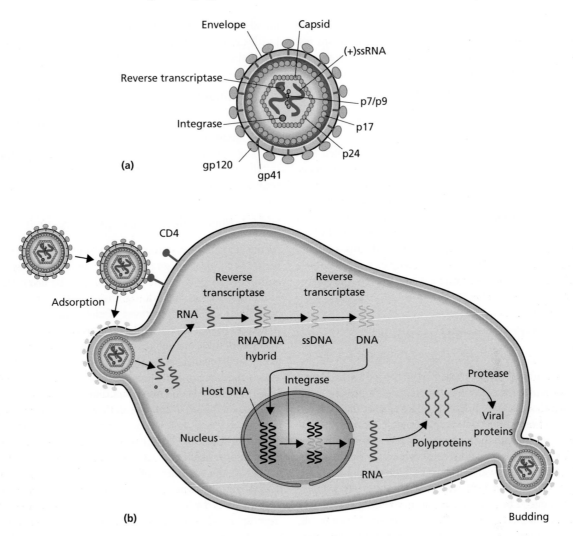

FIGURE 20.11 (a) Structure of HIV particle (p = protein, gp = glycoprotein) and (b) life cycle of the HIV in a host T-cell.

RNA is not capable of coding directly for viral proteins, or of self-replication. Instead, it is converted into DNA and incorporated into the host cell DNA. The conversion of RNA into DNA is not a process that occurs in human cells, so there are no host enzymes to catalyse the process. Therefore, HIV carries its own enzyme—**reverse transcriptase**—to do this. This enzyme is a member of a family of enzymes known as the DNA polymerases, but is unusual in that it can use an RNA strand as a template. The enzyme first catalyses the synthesis of a DNA strand using viral RNA as a template. This leads to a (+)RNA–(−)DNA hybrid. Reverse transcriptase catalyses the degradation of the RNA strand, then uses the remaining DNA strand as a template to catalyse the synthesis of double-stranded DNA (**proviral DNA**). Proviral DNA is now spliced into the host cell's DNA—a process catalysed by the viral protein **integrase**. Once the proviral DNA has been incorporated into host DNA, it is called the

provirus and can remain dormant in host cell DNA until activated by cellular processes. When that occurs, transcription of the viral genes *env*, *gag*, and *pol* takes place to produce viral RNA, some of which will be incorporated into new virions, and the rest of which is used in translation to produce three large non-functional polyproteins, one derived from the *env* gene, one from the *gag* gene, and the other from the *gag-pol* genes. The first of these polyproteins is cleaved by cellular proteinases and produces the viral glycoproteins (gp120 and gp41) which are incorporated into the cell membrane. The remaining two polypeptides (**Pr55** and **Pr160**) remain intact and move to the inner membrane surface. The viral glycoproteins in the cell membrane also concentrate in this area, and cellular proteins are excluded. Budding then takes place to produce an immature membrane-bound virus particle. During the budding process a viral enzyme called protease is released from the gag-pol polypeptide. This is achieved by the protease enzyme autocatalysing the cleavage of susceptible peptide bonds linking it to the rest of the polypeptide. Once released, the protease enzyme dimerizes and cleaves the remaining polypeptide chains to release reverse transcriptase, integrase, and viral structural proteins. The capsid proteins now self-assemble to form new nucleocapsids containing viral RNA, reverse transcriptase, and integrase.

It has been observed that a viral protein called **Vpu** has an important part to play in the budding process. Vpu binds to the host membrane protein CD4 and triggers a host enzyme to tag the CD4 protein with a protein called **ubiquitin**. Proteins that are tagged with ubiquitin are marked out for destruction by the host cell and so the CD4 proteins in the host cell are removed. This is important, as the CD4 proteins could complex with the newly synthesized viral proteins gp120 and prevent the assembly of the new viruses.

20.7.2 **Antiviral therapy against HIV**

Until 1987, no anti-HIV drug was available, but an understanding of the life cycle of HIV has led to the identification of several possible drug targets. At present, most drugs that have been developed act against the viral enzymes reverse transcriptase and protease. However, a serious problem with the treatment of HIV is the fact that the virus undergoes mutations extremely easily. This results in rapid resistance to antiviral drugs. Experience has shown that treatment of HIV with a single drug has a short-term benefit, but, in the long term, the drug serves only to select mutated viruses which are resistant. As a result, current therapy involves combinations of different drugs acting on both reverse transcriptase and protease (Box 20.2). This has been successful in delaying the progression to AIDS and increasing survival rates. Further improvements in HIV treatment have resulted from the development of drugs against other viral targets, such as integrase.

The demands on any HIV drug are immense, especially since it is likely to be taken over long periods of time. It must have a high affinity for its target (in the picomolar range) and it has to be effective in preventing the virus multiplying and spreading. It should show low activity for any similar host targets in the cell, and be safe and well tolerated. It must be active against as large a variety of viral isolates as possible, or else it only serves to select resistant variants. It needs to be synergistic with other drugs used to fight the disease and be compatible with other drugs used to treat opportunistic diseases and infections arising from the weakened immune response. The drug must stay above therapeutic levels within the infected cell and in the circulation. It must be capable of being taken orally and with a minimum frequency of doses, and it should preferably be able to cross the blood–brain barrier in case the virus lurks in the brain. Finally, it must be inexpensive as it is likely to be used for the lifetime of the patient.

20.7.3 **Inhibitors of viral reverse transcriptase**

There are two classes of viral reverse transcriptase inhibitors—nucleoside reverse transcriptase inhibitors (NRTIs) and non-nucleoside reverse transcriptase inhibitors (NNRTIs). The clinical aspects of these inhibitors are described in Box 20.3.

20.7.3.1 Nucleoside reverse transcriptase inhibitors

As the enzyme reverse transcriptase is unique to HIV, it serves as an ideal drug target. Nevertheless, the enzyme is still a DNA polymerase and care has to be taken that inhibitors do not have a significant inhibitory effect on cellular DNA polymerases. Various nucleoside-like structures have proved useful as antiviral agents. The vast majority of these are not active themselves, but are phosphorylated to form an active nucleotide triphosphate. This is the same process previously described in section 20.6.1, but there is one important difference: cellular enzymes are required to catalyse all three phosphorylations because HIV does not produce a viral kinase.

Zidovudine (Fig. 20.12) was originally developed as an anticancer agent but was the first drug to be approved for use in the treatment of AIDS. It is an analogue of deoxythymidine where the sugar 3′-hydroxyl group has been replaced by an azido group. On conversion to the triphosphate, it inhibits reverse transcriptase. Furthermore, the triphosphate is attached to the growing DNA chain. Since the sugar unit has an azide substituent at the 3′ position of the sugar ring, the nucleic acid chain cannot be extended any further.

Didanosine (Fig. 20.12) was the second anti-HIV drug approved for use in the USA (1988). Its activity was unexpected because the nucleic acid base present is inosine—a base which is not naturally incorporated into DNA. However, a series of enzyme reactions converts this compound into 2′,3′-dideoxyadenosine triphosphate which is the active drug. Studies of the target enzyme's active site led to the development of **lamivudine** and **emtricitabine** (Fig. 20.12) (analogues of deoxycytidine where the 3′ carbon has been replaced by sulphur). Other clinically useful NRTIs used against HIV and/or hepatitis B include **abacavir** (the only guanosine analogue), **stavudine**, and **zalcitabine** (Fig. 20.13). **Tenofovir disoproxil** and **adefovir dipivoxil** are prodrugs of modified nucleosides. Both structures contain a monophosphate group protected by two extended esters. Hydrolysis *in vivo* reveals the phosphate group which can then be phosphorylated to the triphosphate as described previously.

BOX 20.2 Clinical aspects of antiviral drugs used against HIV

There is no cure for HIV infection, but anti-HIV drugs can halt or slow the rate at which the disease develops, leading to a significant increase in life expectancy. Unfortunately, the drugs used have toxic side effects, which is particularly significant since patients have to take these drugs for the rest of their lives. This means that patients have to be monitored constantly. A combination of drugs acting against different enzyme targets is required—**highly active antiretroviral therapy** (**HAART**). When choosing which drugs to use, it is important to ensure that they have a synergistic or additive effect, and that they are compatible in terms of their toxic properties.

Currently, **protease inhibitors** (**PIs**) (section 20.7.4) are used with **reverse transcriptase inhibitors** (section 20.7.3) (divergent therapy) or with another PI (convergent therapy). A combination of two **nucleoside reverse transcriptase inhibitors** (**NRTIs**) plus a PI is recommended, but one can also use two PIs with an NRTI, or a **non-nucleoside reverse transcriptase inhibitor** (**NNRTI**) with two NRTIs. For example, the NNRTI efavirenz is used along with the NRTIs emtricitabine and tenofovir.

The NRTIs which can be used against HIV are zidovudine, didanosine, zalcitabine, stavudine, lamivudine, emtricitabine, tenofovir disoproxil, and abacavir. NNRTIs used against HIV are nevirapine, delavirdine, and efavirenz. The

PIs which are used are saquinavir, ritonavir, indinavir, nelfinavir, amprenavir, atazanavir, darunavir, fosamprenavir, lopinavir, and tipranavir.

The fusion inhibitor enfuvirtide has also been approved as an anti-HIV drug, and acts against a different target from the conventional anti-HIV drugs. It can be included alongside conventional drugs if the disease fails to respond to standard HAART therapy. Three integrase inhibitors have also been approved for clinical use—raltegravir, dolutegravir, and elvitegravir.

As a result of HAART, the death rate from AIDS-related deaths appears to be slowing. For example, 1.8 million deaths were recorded in 2009, as compared to 2.4 million in 2004.

One of the problems with combination therapies such as HAART is the number of pills that have to be taken, which can result in patients failing to keep to their medication and thus increase the risks of drug resistance. To tackle this issue, several medications have been combined in a single pill. For example, **Stribild** is a single pill containing elvitegravir, cobicastat, emtricitabine, and tenofovir disoproxil fumarate. Other examples of single pill drug combinations include Combivir, Kaletra, Trizivir, Epzicom/Kivexa, Tuvada, Atripla, Complexa/Eviplera, Triumeq, Evotaz, Prezcobix, Dutrebis, Genvoya, and Descovy.

20.7.3.2 Non-nucleoside reverse transcriptase inhibitors

The NNRTIs (Fig. 20.14) are generally hydrophobic molecules that bind to an allosteric binding site which is hydrophobic in nature. Since the allosteric binding site is separate from the substrate binding site, the NNRTIs are

non-competitive, reversible inhibitors. They include first-generation NNRTIs such as **nevirapine** and **delavirdine**, as well as second-generation drugs such as **efavirenz**, **etravirine**, and **rilpivirine**. X-ray crystallographic studies on inhibitor–enzyme complexes show that the allosteric binding site is adjacent to the substrate binding site. Binding of an NNRTI to the allosteric site results in an induced

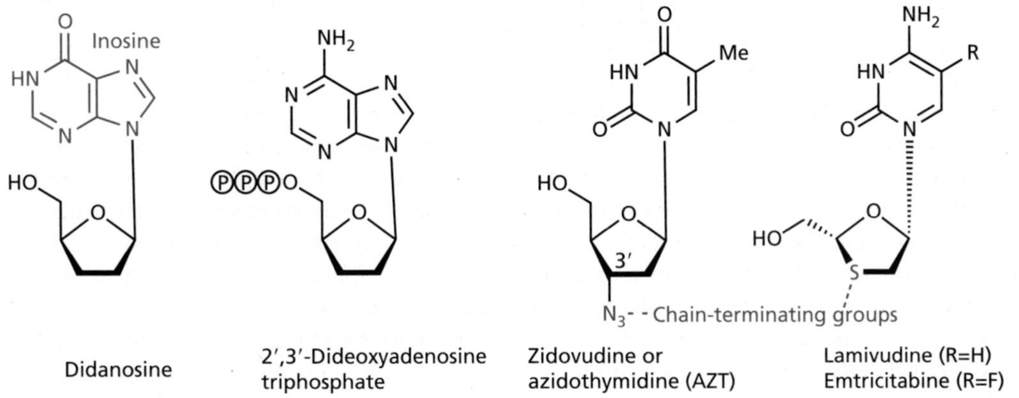

Didanosine

2′,3′-Dideoxyadenosine triphosphate

Zidovudine or azidothymidine (AZT)

Lamivudine (R=H)
Emtricitabine (R=F)

FIGURE 20.12 Inhibitors of viral reverse transcriptase ($\textcircled{P}$ = phosphate).

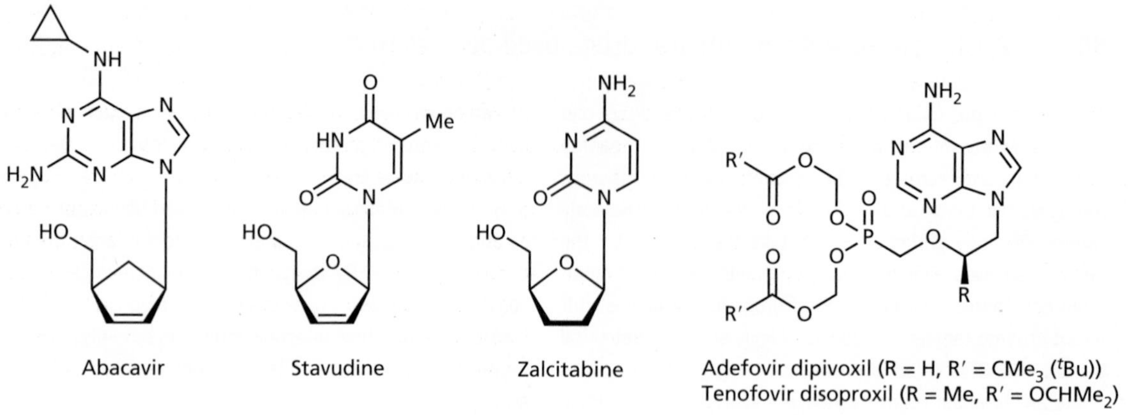

FIGURE 20.13 Further inhibitors of viral reverse transcriptase.

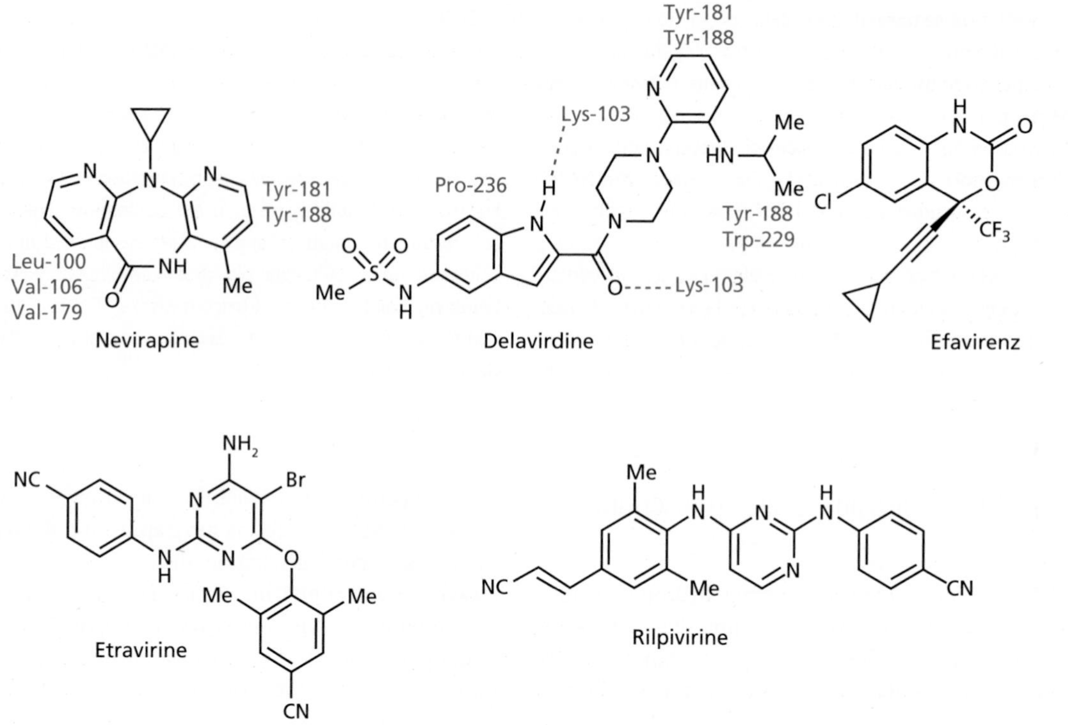

FIGURE 20.14 Non-nucleoside reverse transcriptase inhibitors in clinical use (interactions with amino acids in the binding site are shown in blue).

fit which locks the neighbouring substrate-binding site into an inactive conformation. Unfortunately, rapid resistance emerges due to mutations in the NNRTI binding site, the most common being the replacement of Lys-103 with asparagine. This mutation is called K103N and is defined as a **pan-class resistance mutation**. The resistance problem can be countered by combining an NNRTI with an NRTI from the start of treatment. The two types of drugs can be used together as the binding sites are distinct.

Nevirapine was developed from a lead compound discovered through a random screening programme,

and has a rigid butterfly-like conformation that makes it chiral. One 'wing' interacts through hydrophobic and van der Waals interactions with aromatic residues in the binding site while the other wing interacts with aliphatic residues. The other NNRTI inhibitors bind to the same pocket and appear to function as π electron donors to aromatic side chain residues.

Delavirdine was developed from a lead compound discovered by a screening programme of 1500 structurally diverse compounds. It is larger than other NNRTIs and extends beyond the normal pocket such that it projects into surrounding solvent. The pyridine region and iso-propylamine groups are the most deeply buried parts of the molecule and interact with tyrosine and tryptophan residues. There are also extensive hydrophobic contacts. Unlike other first-generation NNRTIs, there is hydrogen bonding to the main peptide chain next to Lys-103. The indole ring of delavirdine interacts with Pro-236, and mutations involving Pro-236 lead to resistance. Analogues having a pyrrole ring in place of indole might avoid this problem.

Test your understanding and practise your molecular modelling with Exercise 20.1 on the Online Resource Centre at www.oxfordtextbooks.co.uk/orc/patrick6e/

Second-generation NNRTIs were developed specifically to find agents that were active against resistant variants, as well as the wild-type virus. This development has been helped by X-ray crystallographic studies which show how the structures bind to the binding site. It has been shown from sequencing studies that in most of the mutations that cause resistance to first-generation NNRTIs, a large amino acid is replaced by a smaller one, implying that an important binding interaction has been lost. Interestingly, mutations that replace an amino acid with a larger amino acid also appear to be detrimental to

the activity of the enzyme, but no mutations have been found which block NNRTIs sterically from entering the binding site.

Efavirenz is a benzoxazinone structure which has activity against many mutated variants but has less activity against the mutated variant K103N. Nevertheless, activity drops less than for nevirapine, and a study of X-ray structures of each complex revealed that the cyclopropyl group of efavirenz has fewer interactions with Tyr-181 and Tyr-188 than nevirapine. Consequently, mutations of these amino acids have less effect on efavirenz than they do on nevirapine. Efavirenz is also a smaller structure and can shift its binding position when K103N mutation occurs, allowing it to form hydrogen bonds to the main peptide chain of the binding site.

X-ray crystallographic studies of enzyme complexes with several second-generation NNRTIs reveal that these agents contain a non-aromatic moiety which interacts with the aromatic residues Tyr-181, Tyr-188, and Trp-229 at the top of the binding pocket. A relatively small bulk and the ability to form hydrogen bonds to the main peptide chain are important as they allow compounds to change their binding mode when mutations occur. The most recent NNRTIs to be approved are **etravirine** (2008) and **rilpivirine** (2011).

BOX 20.3 Clinical aspects of reverse transcriptase inhibitors

Nucleoside reverse transcriptase inhibitors (NRTIs)

NRTIs are currently used as part of the combination therapy for combating HIV. NRTIs generally have good oral bioavailability, are bound minimally to plasma proteins, and are excreted through the kidneys. They also act against both HIV-1 and HIV-2. However, they are often associated with toxic side effects. **Zidovudine** was the first anti-HIV drug to reach the market, but can cause severe side effects such as anaemia. **Didanosine** was the second anti-HIV drug approved for use and reached the US market in 1988. However, there is a risk of toxicity to the pancreas. **Abacavir** was approved in 1998 and has been used successfully in children, in combination with the protease inhibitors nelfinavir and saquinavir. However, life-threatening hypersensitivity reactions have been reported in some patients. **Tenofovir disoproxil** was approved for HIV-1 treatment in 2001. It remains in infected cells longer than many other antiretroviral drugs, allowing for once-daily dosing, but can have toxic effects on the kidneys. It can be used alongside **emtricitabine** which is relatively free of toxic side effects. Other NRTIs used against HIV include **lamivudine** and **stavudine**. Lamivudine

is less toxic than zidovudine and has also been approved for the treatment of hepatitis B.

Zalcitabine is an NRTI which acts against hepatitis B, but long-term toxicity means that it is unacceptable for the treatment of chronic viral diseases which are not life threatening. **Adefovir dipivoxil** was approved by the US FDA in 2002 for the treatment of chronic hepatitis B. It is also active on viruses such as CMV and herpes.

Non-nucleoside reverse transcriptase inhibitors (NNRTIs)

Compared with the NRTIs, the NNRTIs show a higher selectivity for HIV-1 reverse transcriptase over host DNA polymerases. As a result, NNRTIs are less toxic and have fewer side effects. Unfortunately, rapid resistance emerges if an NNRTI is used on its own, but this does not occur if the NNRTI is combined with an NRTI from the start of treatment. NNRTIs are restricted to HIV-1 activity and are generally metabolized by the liver. They can interact with other drugs and bind more strongly to plasma proteins. **Nevirapine**, **efavirenz**, **delavirdine**, **etravirine**, and **rilpivirine** are NNRTIs currently approved by the FDA for the treatment of HIV.

20.7.4 **Protease inhibitors**

In the mid-1990s, the use of X-ray crystallography and molecular modelling led to the structure-based drug design of a series of inhibitors which act on the viral enzyme HIV protease. Like the reverse transcriptase inhibitors, protease inhibitors (PIs) have a short-term benefit when they are used alone, but resistance soon develops. Consequently, combination therapy is now the accepted method of treating HIV infections. When protease and reverse transcriptase inhibitors are used together, the antiviral activity is enhanced and viral resistance is slower to develop.

Unlike the reverse transcriptase inhibitors, the PIs are not prodrugs and do not need to be activated. Therefore, it is possible to use *in vitro* assays involving virally infected cells in order to test their antiviral activity. The protease enzyme can also be isolated, allowing enzyme assays to be carried out. In general, the latter are used to measure IC_{50} levels as a measure of how effectively novel drugs inhibit the protease enzyme. The IC_{50} is the concentration of drug required to inhibit the enzyme by 50%. Thus, the lower the IC_{50} value, the more potent the inhibitor. However, a good PI does not necessarily mean a good antiviral drug. In order to be effective, the drug has to cross the cell membrane of infected cells, and so *in vitro* whole-cell assays are often used alongside enzyme studies to check cell absorption. EC_{50} values are a measure of antiviral activity and represent the concentration of compounds required to inhibit 50% of the cytopathic effect of the virus in isolated lymphocytes. Another complication is the requirement for anti-HIV drugs to have a good oral bioavailability (i.e. to be orally active). This is a particular problem with the PIs. As we shall see, most PIs are designed from peptide lead compounds. Peptides are well known to have poor pharmacokinetic properties (i.e. poor absorption, metabolic susceptibility, rapid excretion, limited access to the central nervous system, and high plasma protein binding). This is due mainly to high molecular weight, poor water solubility, and susceptible peptide linkages. In the following examples, we will find that potent PIs were discovered relatively quickly, but that these had a high peptide character. Subsequent work was then needed to reduce the peptide character of these compounds in order to retain high antiviral activity, whilst gaining acceptable levels of oral bioavailability and half-life.

Clinically useful PIs (Box 20.4) are generally less well absorbed from the gastrointestinal tract than reverse transcriptase inhibitors, and are also susceptible to first pass metabolic reactions involving the cytochrome P450 isozyme (CYP3A4). This metabolism can result in drug–drug interactions with many of the other drugs given to AIDS patients to combat opportunistic diseases (e.g. rifabutin, ketoconazole, rifampicin, and astemizole).

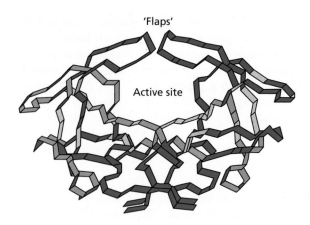

FIGURE 20.15 The HIV protease enzyme.

20.7.4.1 The HIV protease enzyme

The HIV protease enzyme (Fig. 20.15) is an example of an enzyme family called the **aspartyl proteases**—enzymes which catalyse the cleavage of peptide bonds and which contain an aspartic acid in the active site that is crucial to the catalytic mechanism. The enzyme is relatively small and can be obtained by synthesis. Alternatively, it can be cloned and expressed in fast-growing cells, then purified in large quantities. The enzyme can be crystallized with or without an inhibitor bound to the active site, making it an ideal target for structure-based drug design where X-ray crystallographic studies of enzyme–inhibitor complexes allow the design of novel inhibitors.

The HIV protease enzyme is a symmetrical dimer made up of two identical protein units, each consisting of 99 amino acids. The active site is at the interface between the protein units and has a twofold rotational (C2) symmetry. The amino acids Asp-25, Thr-26, and Gly-27 from each monomer are located on the floor of the active site, and each monomer provides a flap to act as the ceiling. The enzyme has a broad substrate specificity and can cleave a variety of peptide bonds in viral polypeptides, but crucially it can cleave bonds between a proline residue and an aromatic residue (phenylalanine or tyrosine) (Fig. 20.16). The cleavage of a peptide bond next to proline is unusual and does not occur with mammalian proteases such as renin, pepsin, or cathepsin D, and so the chances of achieving selectivity against HIV protease over mammalian proteases are good. Moreover, the symmetrical nature of the viral enzyme and its active site is not present in mammalian proteases, again suggesting the possibility of drug selectivity.

There are eight binding subsites in the enzyme, four on each protein unit, located on either side of the catalytic region (Fig. 20.16). These subsites accept the amino

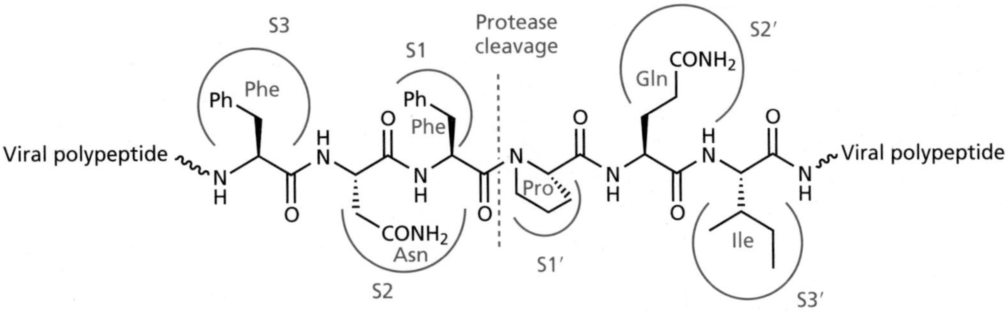

FIGURE 20.16 The aromatic–proline peptide bond that is cleaved by HIV protease (six of the eight binding subsites are shown).

acid side chains of the substrate and are labelled S1–S4 on one side and S1'–S4' on the other side. The relevant side chains on the substrate are labelled P1–P4 and P1'–P4' (Fig. 20.17). Peptide bonds in the substrate are also involved in hydrogen bonding interactions with the active site, as shown in Fig. 20.17. A water molecule is present in the active site which acts as a hydrogen bonding bridge to two isoleucine NH groups on the enzyme flaps. This hydrogen bonding network has the effect of closing the flaps over the active site once the substrate is bound.

There are two variants of HIV protease. The protease enzyme for HIV-2 shares 50% sequence identity with the protease enzyme for HIV-1. The greatest variation occurs outwith the active site, and so inhibitors are found to bind similarly to both enzymes.

The aspartic acids Asp-25 and Asp-25' on the floor of the active site are involved in the catalytic mechanism. Each of these residues is contributed by one of the protein subunits, and the carboxylate side chains interact with a bridging water molecule during the hydrolysis mechanism (Fig. 20.18).

20.7.4.2 Design of HIV protease inhibitors

A similar hydrolytic mechanism to that shown in Fig. 20.18 takes place for a mammalian aspartyl protease called **renin**. This enzyme was studied extensively before the discovery of HIV protease, and a variety of renin inhibitors were designed as antihypertensive agents (section 7.4 and Case study 8). These agents act as **transition-state inhibitors** and many of the strategies resulting from the development of renin inhibitors were adapted to the design of HIV PIs.

Transition-state inhibitors are designed to mimic the transition state of an enzyme-catalysed reaction. The advantage of this approach is that the transition state is likely to be bound to the active site more strongly than either the substrate or product. Therefore, inhibitors resembling the transition state are also likely to be bound more strongly. In the case of the HIV protease-catalysed reaction, the transition state resembles the tetrahedral intermediate shown in Fig. 20.18. As such structures are inherently unstable, it is necessary to design an inhibitor which contains a **transition-state isostere**. Such an

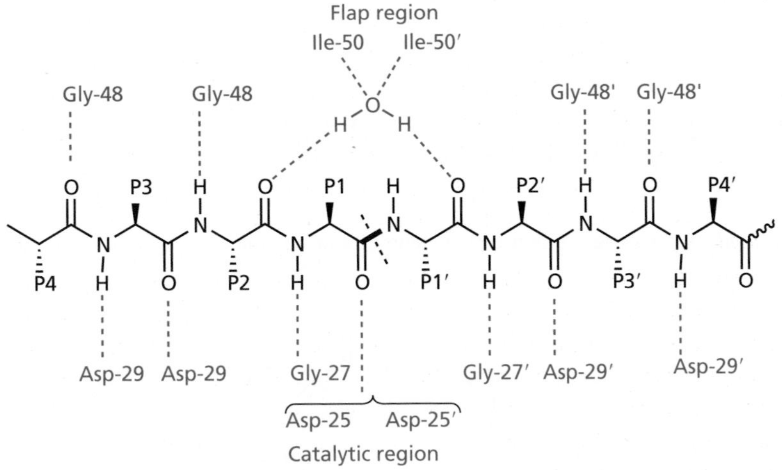

FIGURE 20.17 Interactions between the substrate's peptide backbone and the active site of HIV protease.

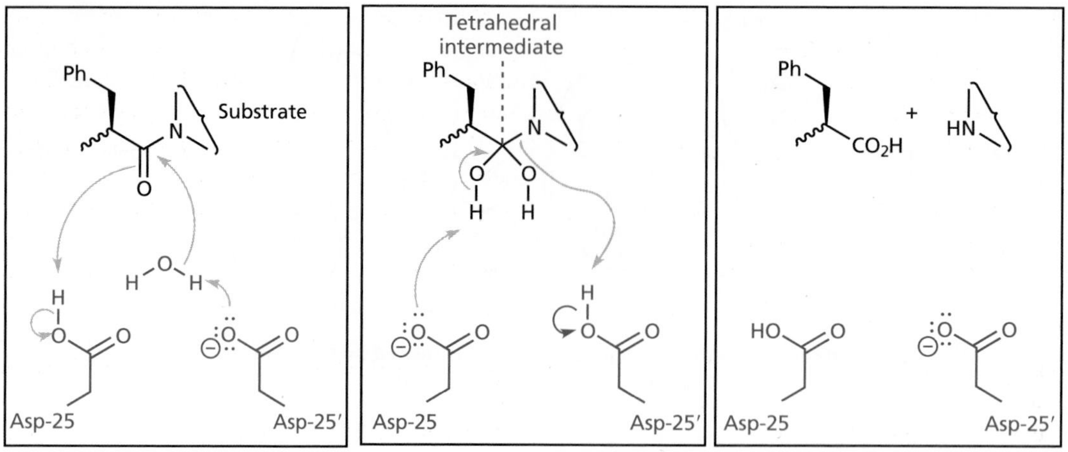

FIGURE 20.18 Mechanism of the reaction catalysed by HIV protease.

isostere would have a tetrahedral centre to mimic the tetrahedral centre of the transition state, yet be stable to hydrolysis. Fortunately, several such isosteres had already been developed in the design of renin inhibitors (Fig. 20.19). Thus, a large number of structures were synthesized incorporating these isosteres, with the hydroxyethylamine isostere proving particularly effective. This isostere has a hydroxyl group which mimics one of the hydroxyl groups of the tetrahedral intermediate and binds to the aspartate residues in the active site. The stereochemistry of this group is also important to activity, with the R-configuration generally being preferred. This preference is determined by the nature of the P1′ group that is present.

Having identified suitable transition-state isosteres, inhibitors were designed based on the enzyme's natural peptide substrates, as these contain amino acid residues which fit the eight subsites and allow a good binding interaction between the substrate and the enzyme. In theory, it might make sense to design inhibitors such that all eight

subsites are filled to allow stronger interactions. However, this leads to structures with a high molecular weight and consequently poor oral bioavailability. Therefore, most of the PIs were designed to have a core unit spanning the S1 to S1′ subsites. Further substituents were then added at either end to fit into the S2/S3 and S2′/S3′ subsites. Early inhibitors, such as saquinavir (see Fig. 20.21), have amino acid side chains that bind to most of the subsites from S3 to S3′. Unfortunately, these compounds have a large molecular weight and a high peptide character leading to poor pharmacokinetic properties. More recent inhibitors have been designed with increased aqueous solubility and oral bioavailability by using a variety of novel P2 and P2′ groups that reduce the molecular weight and peptide character of the compounds. The S2 and S2′ subsites of the protease enzyme appear to contain both polar (Asp-29, Asp-30) and hydrophobic (Val-32, Ile-50, Ile-84) amino acids, allowing the design of drugs that contain hydrophobic P2 groups which are also capable of hydrogen bonding. It has also been possible to design

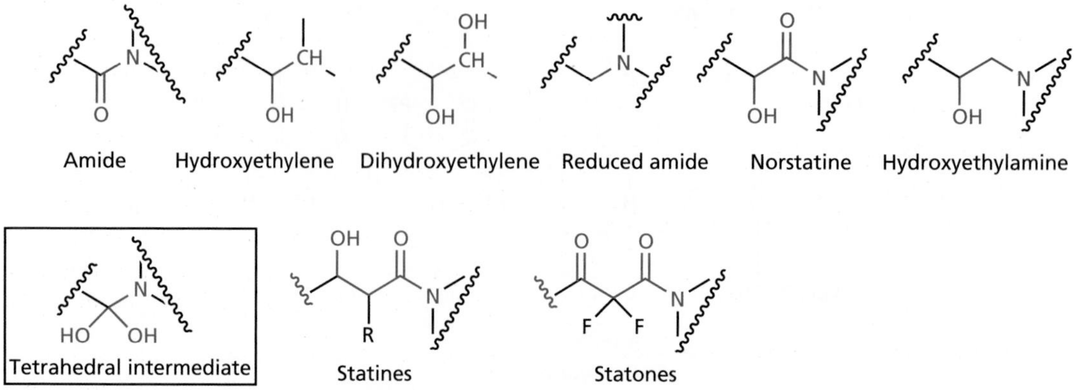

FIGURE 20.19 Transition-state isosteres.

a P1 group that can span both the S1 and S3 subsites, allowing the removal of a P3 moiety, thus lowering the molecular weight. The P2 group is usually attached to P1 by an acyl link, because the carbonyl oxygen concerned acts as an important hydrogen bond acceptor to the bridging water molecule described previously (Fig. 20.17).

We shall now look at how these strategies were used to design individual PIs.

20.7.4.3 Saquinavir

Saquinavir was developed by Roche, and as the first PI to reach the market it serves as the benchmark for all other PIs. The design of saquinavir started by considering a viral polypeptide substrate (pol, see section 20.7.2) and identifying a region of the polypeptide which contains a phenylalanine–proline peptide link. A pentapeptide sequence Leu-Asn-Phe-Pro-Ile was identified and served as the basis for inhibitor design. The peptide link normally hydrolysed in this sequence is between Phe and Pro and so this link was replaced by a hydroxyethylamine transition-state isostere to give a structure which successfully inhibited the enzyme (Fig. 20.20). The amino acid side chains for Leu-Asn-Phe-Pro-Ile are retained in this structure and bind to the five subsites S3-S2′. Despite that, enzyme inhibition is relatively weak. The compound also has high molecular weight and peptide-like character, both of which are detrimental to oral bioavailability.

Consequently, the Roche team set out to identify a smaller inhibitor, starting from the simplest possible substrate for the enzyme—the dipeptide Phe-Pro (Fig. 20.21). The peptide link was replaced by the hydroxylamine transition-state isostere and the resulting *N*- and *C*-protected structure (I) was tested and found to have

weak inhibitory activity. The inclusion of an asparagine group (structure II) to occupy the S2 subsite resulted in a 40-fold increase in activity, which meant that structure II was more active than the pentapeptide analogue (Fig. 20.20). This might seem an unexpected result as the latter occupies more binding subsites. However, it has been found that the crucial interaction of inhibitors is in the core region S2-S2′. If the addition of extra groups designed to bind to other subsites weakens the interaction to the core subsites, it can lead to an overall drop in activity. For example, addition of leucine to structure II resulted in a drop in activity, despite the fact that leucine can occupy the S3 subsite.

Structure II was adopted as the new lead compound and the residues P1 and P2 were varied to find the optimum groups for the S1 and S2 subsites. As it turned out, the benzyl group and the asparagine side chain were already the optimum groups. An X-ray crystallographic study of the enzyme–inhibitor complex revealed that the protecting group (Z) occupied the S3 subsite, which proved to be a large hydrophobic pocket. Therefore, the protecting group was replaced with a larger quinoline ring system which could occupy the subsite more fully. This led to a six-fold increase in activity (structure III). Variations were also carried out on the carboxyl half of the molecule. Proline fits into the S1′ pocket, but it was possible to replace it with a bulkier decahydroisoquinoline ring system. The *t*-butyl ester protecting group was found to occupy the S2′ subsite and could be replaced by a *t*-butylamide group which proved more stable in animal studies. The resulting structure (saquinavir) showed a further 60-fold increase in activity. The *R*-stereochemistry of the transition state hydroxyl group is essential. If the configuration is *S*, all activity is lost.

X-ray crystallography of the enzyme–saquinavir complex (Figs. 20.21 and 20.22) demonstrated the following:

- the substituents on the drug occupy the five subsites S3–S2′;
- the *t*-butylamine nitrogen is positioned in such a way that further *N*-substituents would be incapable of reaching the S3′ subsite;
- there are hydrogen bonding interactions between the hydroxyl group of the hydroxyethylamine moiety and the catalytic aspartates (Asp-25 and Asp-25′);
- the carbonyl groups on either side of the transition-state isostere act as hydrogen bond acceptors to a bridging water molecule. The latter forms hydrogen bonds to the isoleucine groups in the enzyme's flap region in a similar manner to that shown in Fig. 20.17.

Saquinivir is still used clinically but suffers from poor oral bioavailability and susceptibility to drug resistance. Various efforts have been made to design simpler analogues of

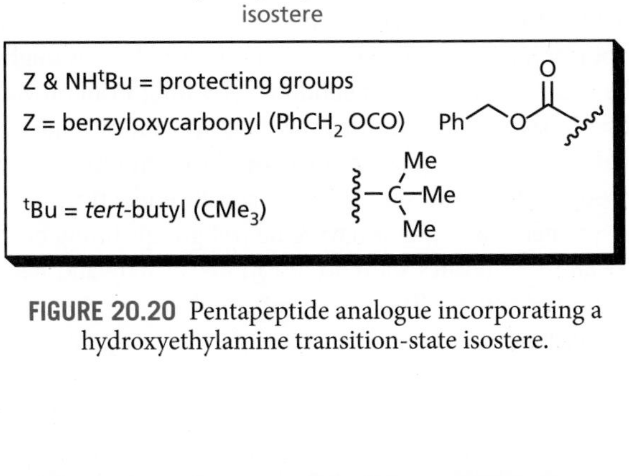

FIGURE 20.20 Pentapeptide analogue incorporating a hydroxyethylamine transition-state isostere.

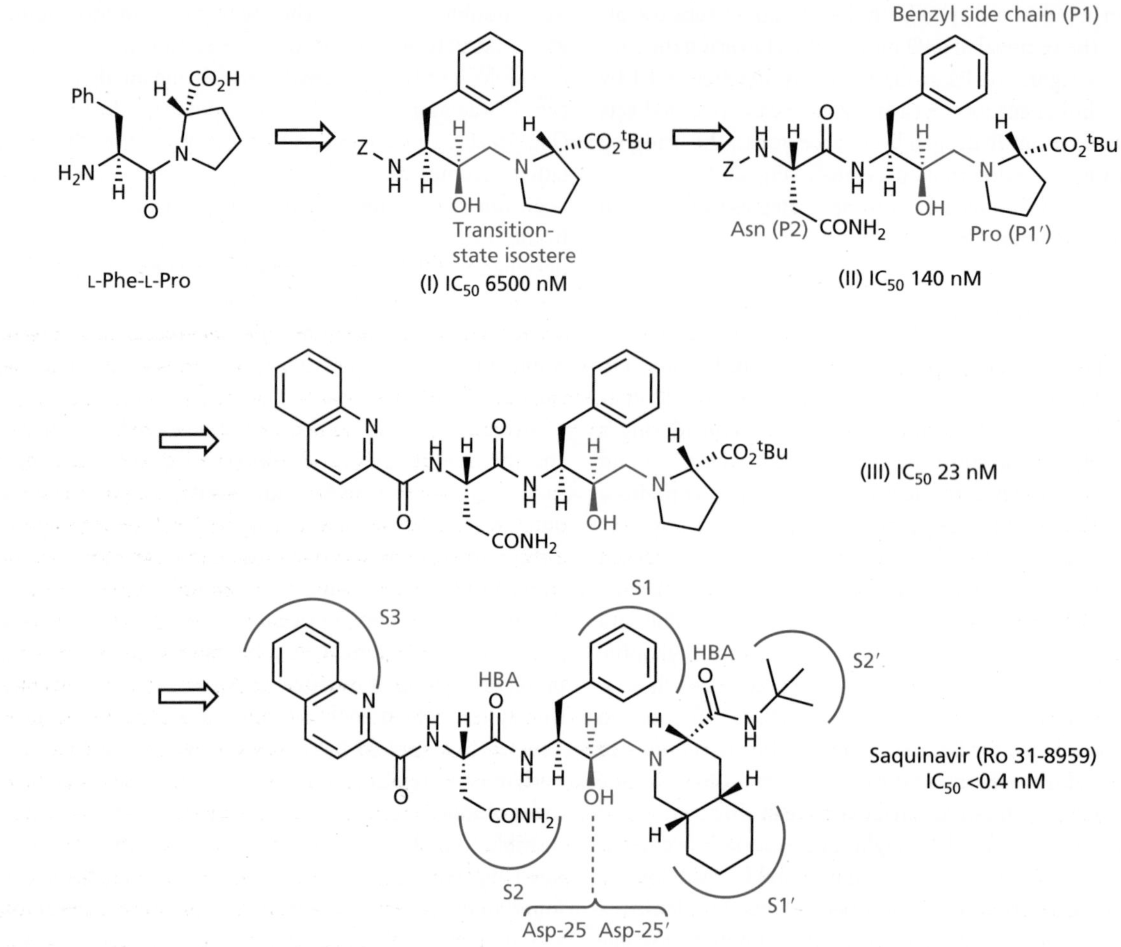

Benzyl side chain (P1)

L-Phe-L-Pro

(I) IC$_{50}$ 6500 nM

Transition-state isostere

Asn (P2)   CONH$_2$ Pro (P1')

(II) IC$_{50}$ 140 nM

(III) IC$_{50}$ 23 nM

S3   S1   HBA   S2'

HBA

Saquinavir (Ro 31-8959)
IC$_{50}$ <0.4 nM

S2   S1'

Asp-25   Asp-25'

FIGURE 20.21 Development of saquinavir (Z = PhCH$_2$OCO).

saquinavir which have lower molecular weight, less peptide character, and consequently better oral bioavailability.

Test your understanding and practise your molecular modelling with Exercise 20.2 on the Online Resource Centre at www.oxfordtextbooks.co.uk/orc/patrick6e/

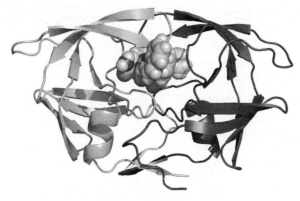

FIGURE 20.22 Saquinavir bound to the active site of HIV protease.

20.7.4.4 Ritonavir and lopinavir

Ritonavir was developed by Abbott Pharmaceuticals to take advantage of the symmetrical properties of the protease enzyme and its active site. Because the active site has C2 symmetry, a substrate is capable of binding 'left to right' or 'right to left' as the binding subsites S1–S4 are identical to subsites S1'–S4'. This implies that it should be possible to design inhibitors having C2 symmetry. There are several advantages to this. Firstly, symmetrical inhibitors should show greater selectivity for the viral protease over mammalian aspartyl proteases, as the active sites of the latter are not symmetrical. Secondly, symmetrical molecules might be less recognizable to peptidases, resulting in improved oral bioavailability. Thirdly, the development of saquinavir showed that a benzyl residue was the optimum binding group for the S1 subsite. As the S1' subsite is identical to S1, a symmetrical inhibitor having benzyl groups fitting both S1 and S1' subsites should bind more strongly and have improved activity. This argument could also be extended for the binding groups fitting the S2/S2' subsites and so on.

As there was no lead compound having C2 symmetry to match the symmetry of the active site, it was necessary to design one. This was done by considering the tetrahedral reaction intermediate derived from the natural substrate. It was assumed that the axis of C2 symmetry for the active site passed through the reaction centre of this intermediate (Fig. 20.23). As the benzyl group was known to be optimum for binding to the S1 subsite, the left-hand portion of the molecule was retained and the right-hand portion was deleted. The left-hand moiety was then rotated such that two benzyl residues were present in the correct orientation for C2 symmetry. The resulting geminal diol is inherently unstable, so one of the alcohols was removed leading to the target alcohol (I; R = H). In order to check whether this target molecule would match the C2 symmetry of the active site when bound, a molecular modelling experiment was carried out whereby the inhibitor was constructed in the active site. The results of this analysis were favourable and so the target alcohol was synthesized. Although it had no antiviral activity, it did show weak activity as an enzyme inhibitor, which meant that it could serve as a lead compound for further development. This represented a success for *de novo* techniques in the design of a lead compound (section 17.15).

The next stage was to extend the molecule to take advantage of the S2 and S2′ subsites. A variety of structures were synthesized and tested, revealing vastly improved enzyme inhibition when valine was added, and further improvement when the valines had *N*-protecting

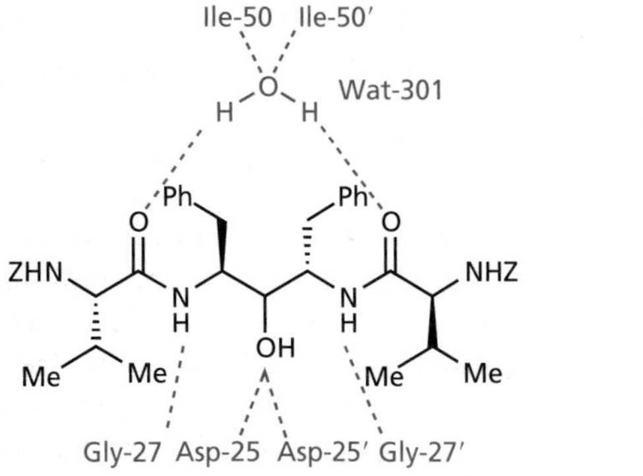

FIGURE 20.25 Binding interactions between the backbone of A 74704 and the active site of HIV protease (Z = PhCH₂OCO).

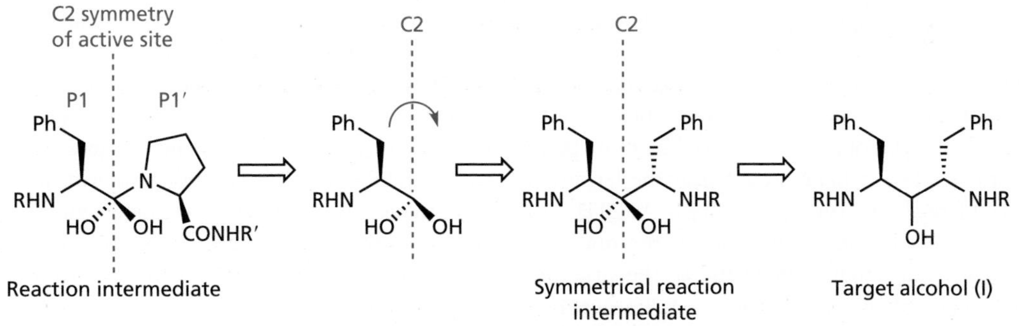

groups (A 74704; Fig. 20.24). A 74704 also showed *in vitro* activity against HIV and was resistant to proteolytic degradation. The structure was co-crystallized with recombinant protease enzyme and studied by X-ray crystallography to reveal a symmetrical pattern of hydrogen bonding between the inhibitor and the enzyme (Fig. 20.25). It was also found that a water molecule (Wat-301) still acted as a hydrogen bonding bridge between the carbonyl groups of P2 and P2′, and the NH groups of Ile-50 and Ile-50′ on the flaps of the enzyme. The C2 symmetry axes of the inhibitor and the active site passed within 0.2 Å of each other and deviated by an angle of only 6°, demonstrating the validity of the design philosophy.

FIGURE 20.23 *De novo* design of a symmetrical lead compound acting as an inhibitor.

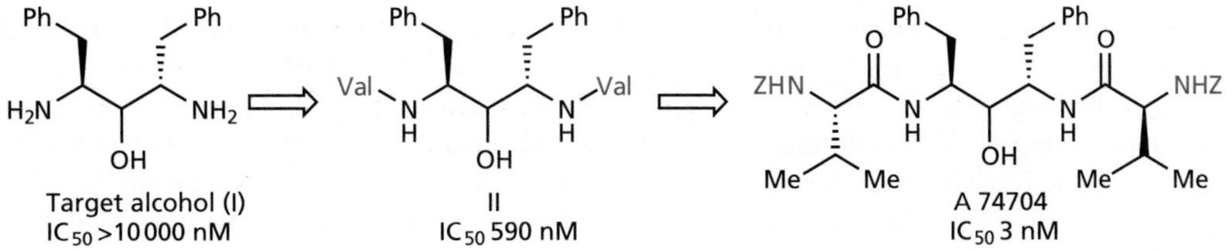

FIGURE 20.24 Development of A 74704 (Z = PhCH₂OCO).

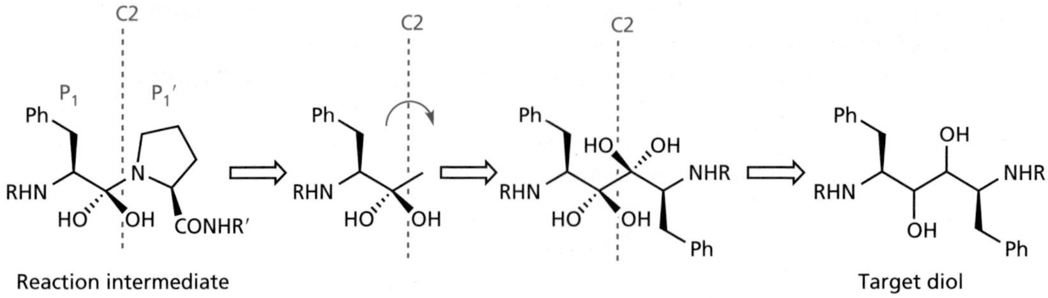

FIGURE 20.26 *De novo* design of a symmetrical diol inhibitor.

Further analysis of the crystal structure suggested that the NH groups on the inhibitor were binding to Gly-27 and Gly-27′ but were too close to each other to allow optimum hydrogen bonding. To address this, it was decided to design symmetrical inhibitors where the relevant NH groups would be separated by an extra bond. In order to achieve this, the axis of C2 symmetry was placed through the centre of the susceptible bond. Accordingly, the design process was repeated to generate the diol shown in Fig. 20.26 as a possible lead compound.

Diol structures analogous to the alcohols previously described were synthesized and tested. Curiously, it was found that the absolute configuration of the diol centres had little effect on activity, and that the activity of the diols was generally better than the corresponding alcohols. For example, the diol equivalent of A 74704 (Fig. 20.27) had a 10-fold better level of activity. Unfortunately, this compound had poor water solubility, indicating that its polarity should be increased. A crystal structure of the enzyme–inhibitor complex revealed that the terminal portions of the molecule were exposed to solvation, which meant that more polar groups could be added at those positions without affecting binding. Consequently, the terminal phenyl groups were replaced by more polar pyridine rings. The urethane groups near the terminals were also replaced by urea groups, leading to A 77003 with improved water solubility. Unfortunately, the oral bioavailability was still unsatisfactory and so the structure entered clinical trials as an intravenous antiviral agent.

Modelling studies of how A 77003 might bind to the active site suggested two possible binding modes, one where each of the diol hydroxyl groups formed symmetrical hydrogen bonds to each of the aspartate residues, and one where only one of the hydroxyl groups hydrogen bonded to both aspartate groups. To investigate this further, X-ray crystallography was carried out on the enzyme–inhibitor complex, revealing that asymmetric binding was taking place, whereby the (R)-OH took part in hydrogen bonding with both aspartate residues, and the (S)-OH was only able to form a single hydrogen bonding interaction. This analysis also showed that the

increased separation of the amide NHs failed to improve the geometry of the hydrogen bonding interactions with Gly-27 and Gly-27′. Thus, the improved activity of the diols over the alcohols was caused by reasons other than those proposed. Results such as these are not totally unexpected when carrying out *de novo* design, as flexible molecules often bind differently from the manner predicted. The better activity for the diols may in fact be due to better binding of the P′ groups to the S′ subsites.

The fact that the (S)-hydroxyl group makes only one hydrogen bonding interaction suggested that it might be worth removing it, as the energy gained from only one hydrogen bonding interaction might be less than the energy required to desolvate the hydroxyl group before binding. This led to A 78791, which had improved activity and was shown by X-ray crystallography to bind in the same manner as A 77003.

A study was then carried out to investigate what effect variations of molecular size, aqueous solubility, and hydrogen bonding would have on the pharmacokinetics and activity of these agents. This led to A 80987, where the P2′ valine was removed, and the urea groups near the ends were replaced by urethane groups. In general, it was found that the presence of *N*-methylureas was good for water solubility and bioavailability, whereas the presence of urethanes (or carbamates) was good for plasma half-life and overall potency. Thus, it was possible to fine-tune these properties by a suitable choice of group at either end of the molecule.

Despite fewer binding interactions, A 80987 retained activity and had improved oral bioavailability. However, it had a relatively short plasma lifetime, was bound strongly to plasma proteins, and it was difficult to maintain therapeutically high levels. Metabolic studies then showed that A 80987 was *N*-oxidized at either or both pyridine rings, and that the resulting metabolites were excreted mainly in the bile. In an attempt to counter this, various design strategies were carried out. Firstly, alkyl groups were placed on the pyridine ring at the vacant position *ortho* to the nitrogen. These were intended to act as a steric shield, but proved ineffective in preventing

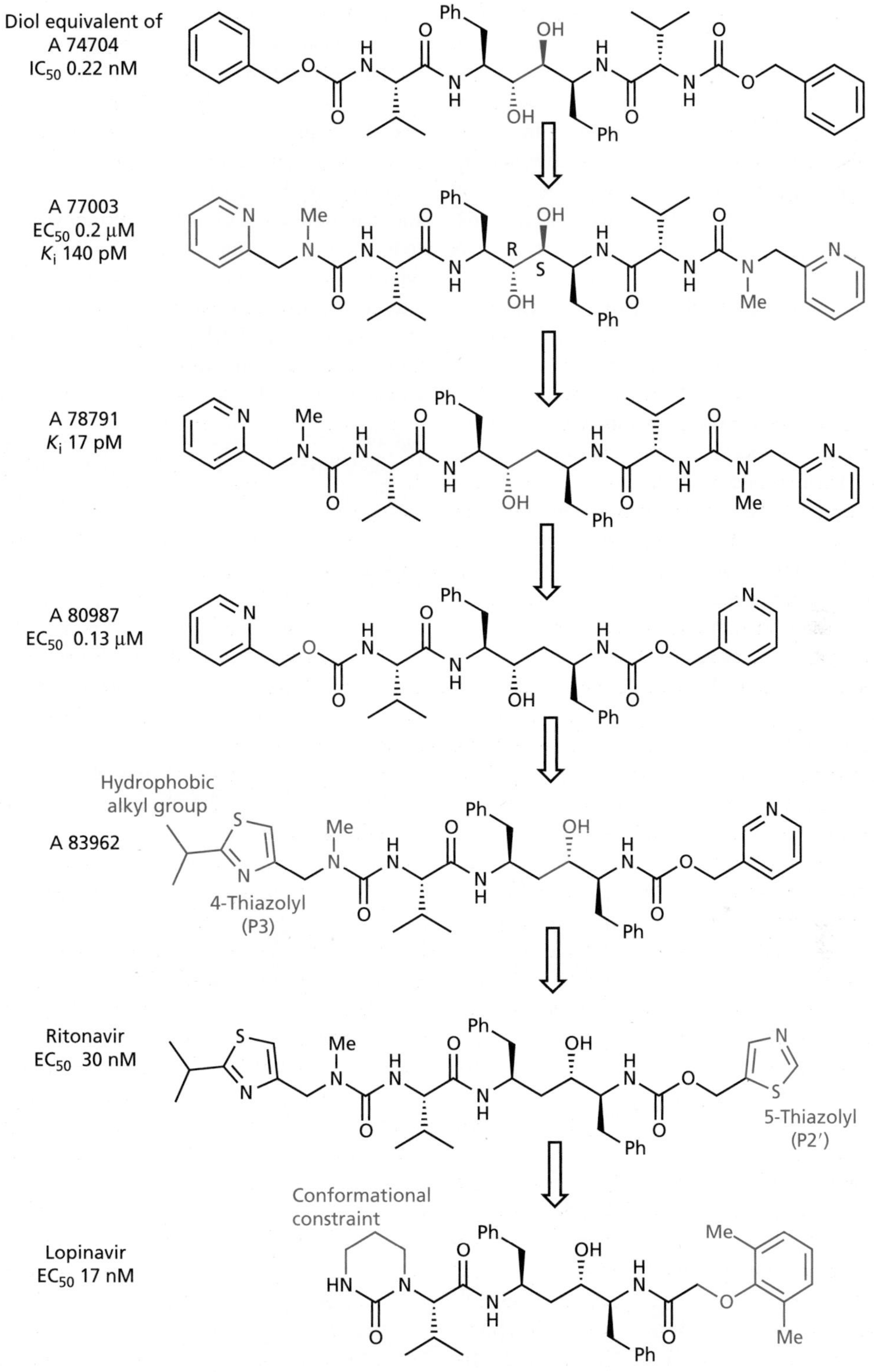

FIGURE 20.27 Development of ritonavir (ABT 538) and lopinavir (ABT 378).

metabolism. It was then proposed that metabolism might be reduced if the pyridine rings were less electron rich, and so methoxy or amino substituents were added as electron-withdrawing groups. However, this too failed to prevent metabolism. Finally, the pyridine ring at P3 was replaced by a variety of heterocycles in an attempt to find a different ring system which would act as a bio-isostere, but would be less susceptible to metabolism. The best results were obtained using the more electron-deficient 4-thiazolyl ring. Although water solubility decreased, it could be restored by reintroducing an *N*-methylurea group in place of one of the urethanes. Further improvements in activity were obtained by placing hydrophobic alkyl groups at the 2-position of the thiazole ring (P3), and by subsequently altering the position of the hydroxyl group in the transition-state isostere. This led to A 83962, which showed an eightfold increase in potency over A 80987.

Attention now turned to the pyridine group at P2′ which was replaced by a 5-thiazolyl group to give ritonavir, which had good activity and oral bioavailability. The good activity observed indicated that a hydrogen bonding interaction was taking place between the thiazolyl N and Asp-30 (specifically the NH of the peptide backbone). This matched a similar hydrogen bonding interaction involving the pyridine N in A 80987. The improved bioavailability is due principally to better metabolic stability (20 times more stable than A 80987), and it was possible to get therapeutic plasma levels of the drug lasting 24 hours following oral administration.

Resistant strains of the virus have developed when ritonavir is used on its own. These arise from a mutation of

valine at position 82 of the enzyme to alanine, threonine, or phenylalanine. X-ray crystallography shows that there is an important hydrophobic interaction between the isopropyl substituent on the P3 thiazolyl group of ritonavir and the isopropyl side chain of Val-82 which is lost as a result of this mutation. Further drug development led to **lopinavir** (Fig. 20.27) where the P3 thiazolyl group was removed and a cyclic urea group was incorporated to introduce conformational constraint. This allowed enhanced hydrogen bonding interactions with the S2 subsite, which balanced out the loss of binding due to the removal of the thiazolyl group. As this structure does not have any interactions with Val-82, it is active against the ritonavir-resistant strain.

Test your understanding and practise your molecular modelling with Exercise 20.3 on the Online Resource Centre at www.oxfordtextbooks.co.uk/orc/patrick6e/

20.7.4.5 Indinavir

The design of **indinavir** included an interesting hybridization strategy (Fig. 20.28). Merck had designed a potent PI that included a hydroxyethylene transition-state

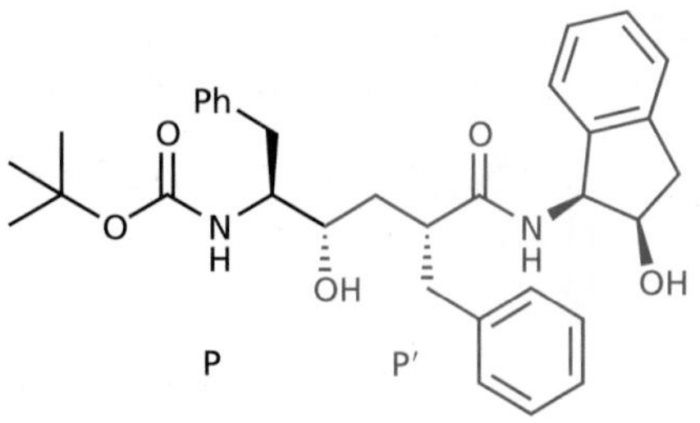

L 685434 IC$_{50}$ 0.3 nM

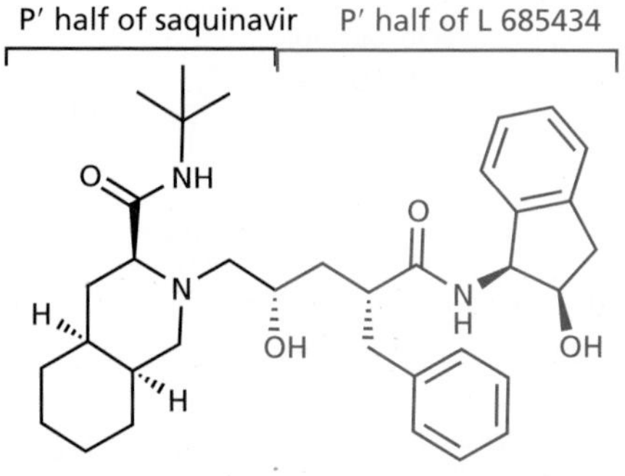

P′ half of saquinavir   P′ half of L 685434

L 704486; IC$_{50}$ 7.6 nM

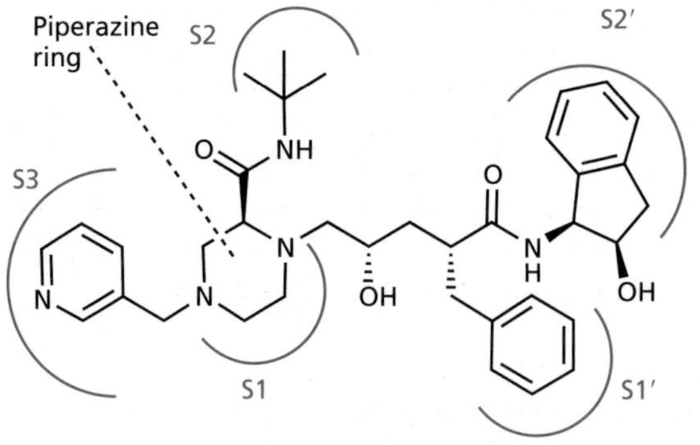

Indinavir IC$_{50}$ 0.56 nM

FIGURE 20.28 Development of indinavir.

isostere (L 685434). Unfortunately, it suffered from poor bioavailability and liver toxicity. At this point, the Merck workers concluded that it might be possible to take advantage of the symmetrical nature of the active site. As the S and S′ subsites are equivalent, it should be possible to combine half of one PI with half of another to give a structurally distinct hybrid inhibitor. A modelling study was carried out to check the hypothesis and the Merck team decided to combine the P′ half of L 685434 with the P′ half of saquinavir. The P′ moiety of saquinavir was chosen for its solubility-enhancing potential, and the P′ moiety of L 685434 is attractive for its lack of peptide character. The resulting hybrid structure (L 704486) was less active as an inhibitor, but was still potent. Moreover, the presence of the decahydroisoquinoline ring system resulted in better water solubility and oral bioavailability (15%), as intended.

Further modifications were aimed at improving binding interactions, aqueous solubility and oral bioavailability. The decahydroisoquinoline ring was replaced by a piperazine ring, the additional nitrogen helping to improve aqueous solubility and oral bioavailability. A pyridine substituent was then added to access the S3 subsite and to improve binding. This resulted in indinavir, which reached the market in 1996.

ⓦ For additional material see Web article 7: The design of indinavir on the Online Resource Centre at www.oxfordtextbooks.co.uk/orc/patrick6e/

20.7.4.6 Nelfinavir

The development of **nelfinavir** was based on work carried out by the Lilly company, aimed at reducing the molecular weight and peptide character of PIs. Structure-based drug design had been used to develop AG1254 (Fig. 20.29), which contains an extended substituent at P1, capable of spanning and binding to both the S1 and S3 subsites of the enzyme. This did away with the need for a separate P3 group and allowed the design of compounds with a lower molecular weight. In addition a new P2 group was designed to replace the asparagine residue that had been present in the lead compound. This group bound effectively to the S2 subsite. Moreover, the peptide character of the compound was reduced because the group was different from any natural amino acid side chain. Unfortunately, the antiviral activity of AG 1254 was not sufficiently high and the compound had poor aqueous solubility.

The company decided to switch direction and see what effect their newly designed substituents would have if they were incorporated into saquinavir, and this led ultimately to nelfinavir. A crystal structure of nelfinavir bound to the enzyme showed that the molecule is bound

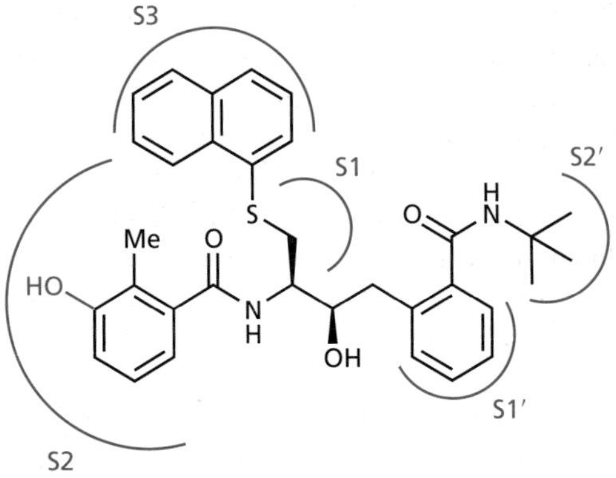

AG 1254
K_i 3 nm

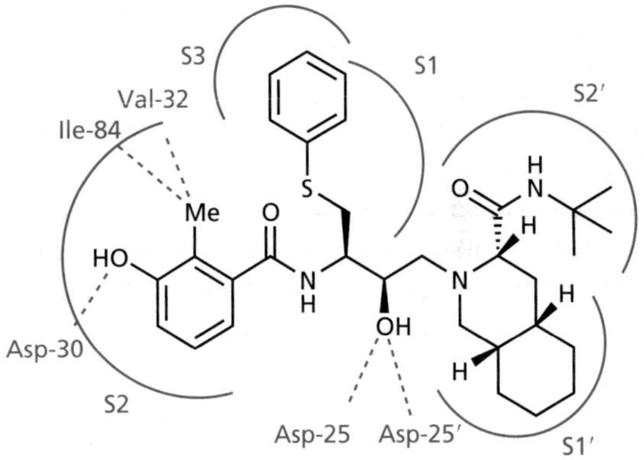

Nelfinavir
K_i 2.0 nM EC$_{50}$ 0.008–0.02 μM

FIGURE 20.29 AG1254 and nelfinavir.

in an extended conformation where the binding interactions involving the molecular backbone are similar to saquinavir. A tightly bound water molecule serves as a hydrogen bonding bridge between the two amide carbonyls of the inhibitor and the flap region of the enzyme, in a similar manner to other enzyme–inhibitor complexes. The crystal structure also showed that the S-phenyl group resides mainly in the S1 site and partially extends into the S3 site. The substituted benzamide group occupies the S2 pocket with the methyl substituent interacting with valine and isoleucine through van der Waals interactions, while the phenol interacts with Asp-30 through hydrogen bonding.

ⓦ Test your understanding and practise your molecular modelling with Exercise 20.4 on the Online Resource Centre at www.oxfordtextbooks.co.uk/orc/patrick6e/

ⓦ For further information see Web article 18: Nelfinavir on the Online Resource Centre.

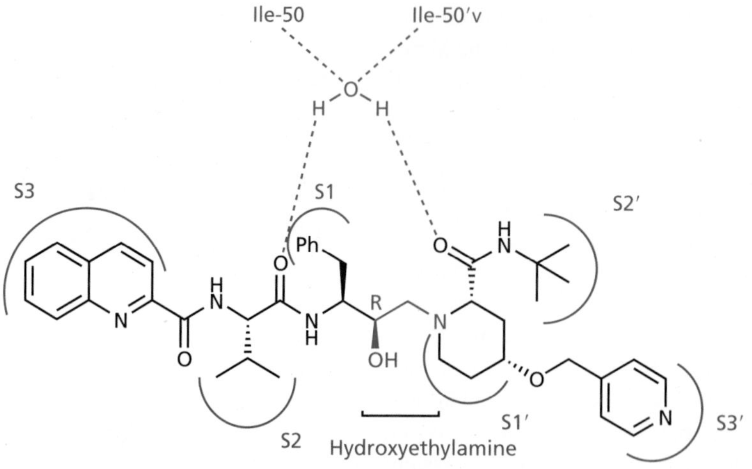

FIGURE 20.30 Palinavir and binding interactions.

20.7.4.7 Palinavir

Palinavir (Fig. 20.30) is a highly potent and specific inhibitor of HIV-1 and HIV-2 proteases. The left-hand or P half of the molecule is similar to saquinavir and the molecule contains the same hydroxyethylamine transition-state mimic. The right-hand (P′) side is different and was designed using the same kind of extension strategy used in nelfinavir. In this case, the P1′ substituent was extended to occupy the S1′ and S3′ subsites. This was achieved by replacing the original proline group at P1′ with 4-hydroxypipecolinic acid and adding a pyridine-containing substituent to access the S3′ subsite.

The crystal structure of the enzyme–inhibitor complex shows that the binding pockets S3–S3′ are all occupied. Two carbonyl groups interact via the bridging water molecule to the isoleucines in the enzyme flaps. The hydroxyl group interacts with both catalytic aspartate residues. Finally, the oxygen and N<u>H</u> atoms of all the amides are capable of hydrogen bonding to complementary groups in the active site. Work is currently in progress to simplify palinavir by introducing a single group that will span two binding subsites (S1 and S3), thus allowing the removal of the separate P3 binding group.

20.7.4.8 Amprenavir and darunavir

Amprenavir (Fig. 20.31) was designed by Vertex Pharmaceuticals as a non-peptide-like PI using saquinavir as the lead compound. Saquinavir suffers from having a high molecular weight and a high peptide character, both of which are detrimental to oral bioavailability. Therefore, it was decided to design a simpler analogue with a lower molecular weight and less peptide character, but retaining good activity. Firstly, the decahydroisoquinoline

group in saquinavir was replaced by an isobutyl sulphonamide group to give structure I. This has the advantage of reducing the number of asymmetric centres from six to three, allowing easier synthesis of analogues. Further simplification and reduction of peptide character was carried out by replacing the P2 and P3 groups with a tetrahydrofuran (THF) carbamate which had previously been found by Merck to be a good binding group for the S2 subsite. Finally an amino group was introduced on the phenylsulphonamide group to increase water solubility and to enhance oral absorption. Fosamprenavir is a phosphate prodrug for amprenavir.

Further work has shown that a fused bis-tetrahydrofuryl ring system is an even better binding group for the hydrophobic S2 pocket than a single THF ring. This is because it fills the pocket more completely and forms hydrogen bonding interactions between the ring oxygens and the peptide backbone of the enzyme. As these interactions are with the protein backbone rather than amino acid side chains, mutations are less likely to lead to drug resistance. **Darunavir** is a second-generation PI which contains this feature, but there are several other compounds currently being studied.

20.7.4.9 Atazanavir

Atazanavir (Fig. 20.32) was approved in June 2003 as the first once-daily HIV-1 PI to be used as part of a combination therapy. It is similar to the early compounds leading towards ritonavir. Current research is looking at the possibility of using a deuterium-labelled analogue of atazanavir which is expected to have a slower rate of metabolism and excretion, and an increased half-life (see also section 14.2.4).

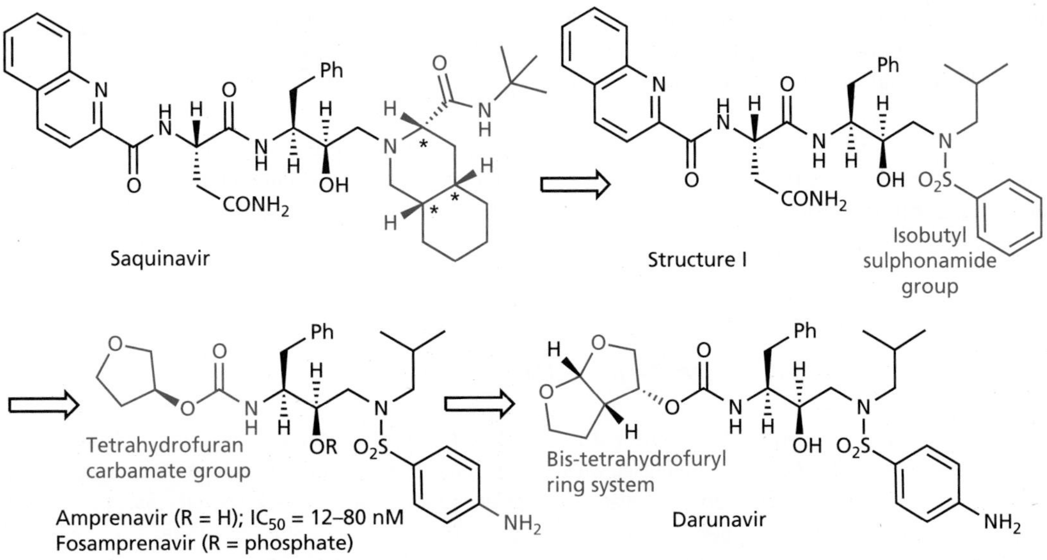

FIGURE 20.31 Development of amprenavir and darunavir.

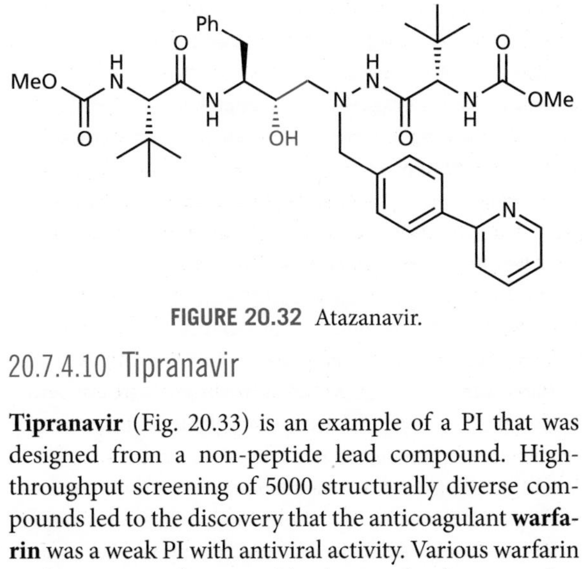

FIGURE 20.32 Atazanavir.

20.7.4.10 Tipranavir

Tipranavir (Fig. 20.33) is an example of a PI that was designed from a non-peptide lead compound. High-throughput screening of 5000 structurally diverse compounds led to the discovery that the anticoagulant **warfarin** was a weak PI with antiviral activity. Various warfarin analogues were then tested leading to the discovery that

phenprocoumon (Fig. 20.33) was a more potent competitive enzyme inhibitor with weak antiviral activity. Both these structures are used therapeutically for other purposes and have high oral bioavailability. Therefore they served as promising lead compounds for non-peptide like antiviral agents with good oral bioavailability.

A crystal structure of the enzyme–inhibitor complex was determined showing that the 4-OH group could form hydrogen bonds with the catalytic aspartate residues, while the two lactone oxygens could form hydrogen bonds directly to the isoleucine groups (Ile-50 and Ile-50′) in the enzyme flaps. Unlike all the previous PIs, there was no bridging water involved in this interaction. Therefore, these compounds represented a new class of inhibitors with a novel pharmacophore of hydrogen bonding interactions. The crystal structure also showed that the ethyl and phenyl groups fitted the S1 and S2 sub-sites respectively, while the benzene ring of the coumarin

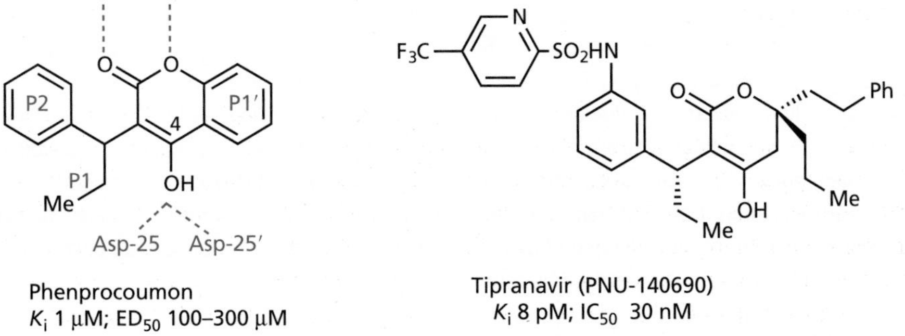

Phenprocoumon
K_i 1 µM; ED$_{50}$ 100–300 µM

Tipranavir (PNU-140690)
K_i 8 pM; IC$_{50}$ 30 nM

FIGURE 20.33 Phenprocoumon and tipranavir.

ring system fitted the S1 subsite. Phenprocoumon was used as the lead compound for further development (see OUP website) and resulted in the discovery of tipranavir.

For further information see the Online Resource Centre at www.oxfordtextbooks.co.uk/orc/patrick6e/

20.7.4.11 Alternative design strategies for antiviral drugs targeting the HIV protease enzyme

An alternative approach to inhibiting the protease enzyme would be to prevent its formation in the first place. Studies are in progress to design protein–protein binding

---

BOX 20.4 Clinical aspects of protease inhibitors

Protease inhibitors (PIs) are an important component of the drug cocktail used to treat HIV. Care has to be taken when administering the agents to haemophiliacs and diabetics since the agents can increase the risk of bleeding, and lower blood sugar levels.

Saquinivir was the first PI to reach the market in 1995. It shows a 100-fold selectivity for both HIV-1 and HIV-2 proteases over human proteases. Approximately 45% of patients develop clinical resistance to the drug over a one-year period, but resistance can be delayed if it is given in combination with reverse transcriptase inhibitors. The oral bioavailability of saquinavir is only 4% in animal studies, although this is improved if the drug is taken with meals. The compound is also highly bound to plasma proteins (98%). As a result, the drug has to be taken in high doses to maintain therapeutically high plasma levels. A curious problem related to saquinavir is that plasma levels are lowered if the patient takes garlic.

Ritonavir reached the market in 1996. It is active against both HIV-1 and HIV-2 proteases and shows selectivity for HIV proteases over mammalian proteases. Despite the fact that ritonavir is highly plasma bound (99%), and has a high molecular weight and peptide-like nature, it has better bioavailability than many other PIs. This is a result of greater stability to drug metabolism, and it is possible to get therapeutic plasma levels of the drug which last 24 hours following oral administration. The metabolic stability of the agent is a result of the drug's ability to act as a potent inhibitor of the cytochrome P450 enzyme CYP3A4, which means that it shuts down its own metabolism. Care has to be taken when drugs affected by CYP3A4 are taken alongside ritonavir, and doses of the latter should be adjusted accordingly. However, ritonavir's ability to inhibit CYP3A4 is useful when it is used alongside other PIs which are normally metabolized by this enzyme (e.g. saquinavir, indinavir, nelfinavir, and amprenavir). As ritonavir inhibits CYP3A4, the lifetime and plasma levels of other PIs can be increased. For this reason, it is often administered in small doses alongside other PIs. If it is intended to be used as an anti-HIV drug in its own right, it is administered with nucleoside reverse transcriptase inhibitors.

Lopinavir is active against ritonavir-resistant strains of HIV, and is administered with ritonavir as a single capsule combination called Kaletra. Each capsule contains 133 mg of lopinavir and 33 mg of ritonavir with the latter serving as a cytochrome P450 inhibitor to increase the levels of lopinavir present in the blood supply.

Indinavir has better oral bioavailability than saquinavir and is less highly bound to plasma proteins (60%). It is usually administered alongside nucleoside reverse transcriptase inhibitors such as didanosine.

Nelfinavir was marketed in 1997 and is used as part of a four-drug combination therapy. Like indinavir and ritonavir, nelfinavir is more potent than saquinavir because of its better pharmacokinetic profile. Compared to saquinavir, it has a lower molecular weight and log P, and an enhanced aqueous solubility, resulting in enhanced oral bioavailability. It can inhibit the metabolic enzyme CYP3A4 and thus affects the plasma levels of other drugs metabolized by this enzyme. It is 98% bound to plasma proteins.

Amprenavir was licensed to GlaxoWellcome and was approved in 1999. It is reasonably specific for the viral protease relative to mammalian proteases, and is about 90% protein bound. It has good oral bioavailability (40–70% in animal studies). **Fosamprenavir** is a phosphate prodrug for amprenavir, and was approved by the US FDA and the European Medicines Agency (EMA) in 2003 and 2004 respectively. The prodrug acts as a slow-release version of amprenavir, reducing the number of pills required. It is usually administered with ritonavir. **Darunavir** is a second-generation PI developed by Tibotec, and was approved by the US FDA in 2006 as the first treatment of drug-resistant HIV. It is usually administered with ritonavir.

Atazanavir was approved in June 2003 as the first once-daily HIV-1 PI to be used as part of a combination therapy. It is usually administered with ritonavir.

Tipranavir is used to treat HIV infections that are resistant to other PIs. However, there have been cases of life-threatening liver toxicity.

inhibitors that will prevent the association of the two protein subunits that make it up (section 10.5).

Another interesting approach is to design prodrugs of toxic compounds that are only activated by HIV protease. The prodrugs would contain a moiety that acts as a substrate for HIV protease, such that the toxin is only released in HIV-infected cells. The toxin would then attack cellular targets and eliminate those cells.

20.7.5 Inhibitors of other targets

Other agents under study for the treatment of HIV include **integrase inhibitors**, and **cell entry inhibitors**. Blocking entry of a virus into a host cell is particularly desirable, because it is so early in the life cycle. **Enfuvirtide** was approved in March 2003 as the first member of a new class of **fusion inhibitors**. It is a polypeptide consisting of 36 amino acids which matches the *C*-terminal end of the viral protein **gp41**. It works by forming an α-helix and binding to a group of three similar α-helices belonging to the gp41 protein. This association prevents the process by which the virus enters the host cell. In order to bring about fusion, the gp41 protein anchors the virus to the cell membrane of the host cell. It then undergoes a conformational change where it builds a grouping of six helices using the three already present as the focus for that grouping (Fig. 20.34). This pulls the membranes of the virion and the host cell together to permit fusion. By binding to the group of three helices, enfuvirtide blocks formation of the required hexamer and prevents fusion.

The manufacture of enfuvirtide involves 106 steps, which makes it expensive and may limit its use. A smaller compound (**BMS 378806**) is being investigated which binds to **gp120** and prevents the initial binding of the virus to **CD4** on the cell surface.

N-**Butyldeoxynojirimycin** (Fig. 20.35) is a carbohydrate that inhibits **glycosidases**—enzymes that catalyse the trimming of carbohydrate moieties linked to viral proteins. If this process is inhibited, too many carbohydrate groups end up attached to a protein, resulting in the protein adopting a different conformation. It is thought that the gp120 protein is affected in this way and cannot be peeled away as described in section 20.7.1 to reveal the gp41 protein.

Bicyclams such as **JM 3100** (Fig. 20.35) block the **CCR5 chemokine receptor** and are under investigation as drugs which will prevent membrane fusion and cell entry.

Maraviroc (Fig. 20.36) was approved as a CCR5 antagonist in 2007, and is the first anti-HIV agent to act on

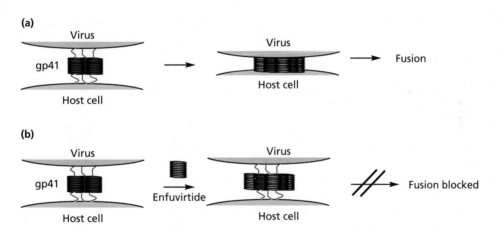

FIGURE 20.34 (a) Normal mechanism of fusion. (b) Enfuvirtide acting as a fusion inhibitor.

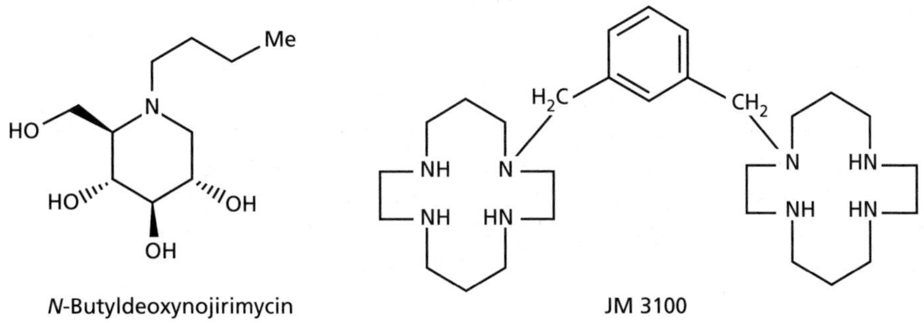

N-Butyldeoxynojirimycin   JM 3100

FIGURE 20.35 Agents that inhibit cell entry.

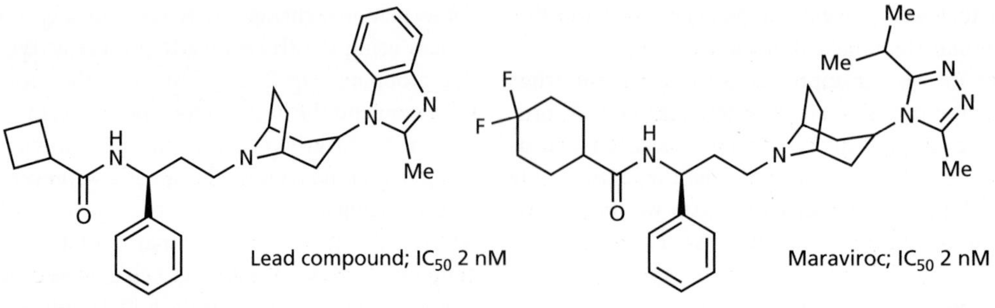

FIGURE 20.36 Comparison of maraviroc and the lead compound from which it was developed.

a molecular target on the host cell rather than the virus. It was developed from a compound that had potent activity, but which blocked HERG ion channels (Box 12.3). Agents which block these channels often have toxic cardiac side effects, and so a large number of analogues were synthesized to find a potent compound which did not block the HERG ion channels. Maraviroc was the result.

It is an example of an agent that works by blocking protein–protein interactions between a viral protein and a host cell protein (section 10.5).

The first integrase inhibitor to reach the market in 2007 was **raltegravir** (Fig. 20.37). The keto–enol system is important for activity as it acts as a chelating group for two magnesium ion cofactors in the enzyme's active site.

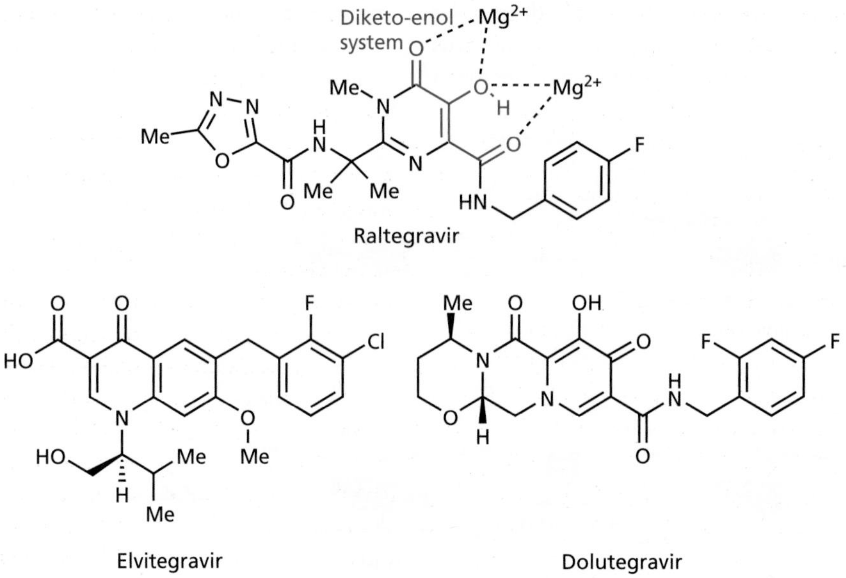

FIGURE 20.37 Clinically approved integrase inhibitors.

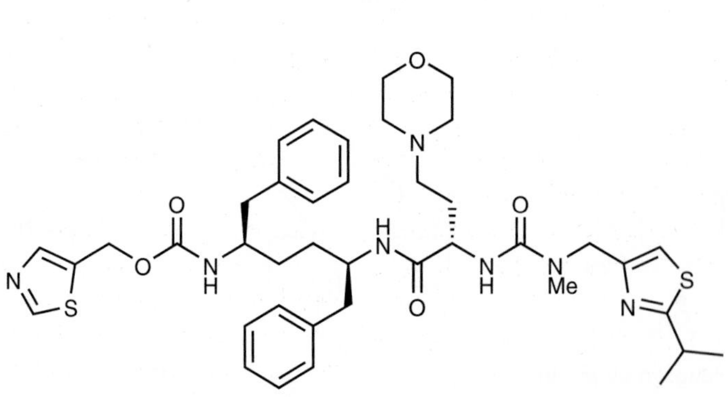

FIGURE 20.38 Cobicistat.

Elvitegravir was the second integrase inhibitor to reach the market in 2012, followed by **dolutegravir** in 2013.

Elvitegravir is administered with another agent called **cobicistat** (Fig. 20.38). Cobicistat is not actually an antiviral agent. Instead, it inhibits the cytochrome P450 enzyme CYP3A4 which normally metabolizes elvitegravir. Therefore, cobicistat is acting as a sentry drug (section 14.7.1) to prolong elvitegravir's half-life.

Cobicistat can also be used to increase the oral absorption of darunavir and atazanavir.

For additional material see Web article 22: Case study on maraviroc, a CCR5 antagonist for HIV treatment on the Online Resource Centre at www.oxfordtextbooks.co.uk/orc/patrick6e/

KEY POINTS

- HIV is a retrovirus containing RNA as its genetic material, and is responsible for AIDS.

- The two main viral targets for anti-HIV drugs are the enzymes reverse transcriptase and protease. A number of drugs acting on other targets have been developed, some of which have been approved. Combination therapy is the favoured approach to treating HIV infection.

- The potency and safety demands for anti-HIV drugs are high, as they are likely to be used for the lifetime of the patient.

- Reverse transcriptase is a DNA polymerase which catalyses the conversion of single-stranded RNA to double-stranded DNA. No such biochemical process occurs in normal cells.

- Nucleoside reverse transcriptase inhibitors are prodrugs that are converted by cellular enzymes to active triphosphates, which act as enzyme inhibitors and chain terminators.

- Non-nucleoside reverse transcriptase inhibitors act as enzyme inhibitors by binding to an allosteric binding site.

- The protease enzyme is a symmetrical dimeric structure consisting of two identical protein subunits. An aspartic acid residue from each subunit is involved in the catalytic mechanism.

- The protease enzyme is distinct from mammalian proteases in being symmetrical and being able to catalyse the cleavage of peptide bonds between proline and aromatic amino acids.

- Protease inhibitors are designed to act as transition-state inhibitors. They contain a transition-state isostere which is tetrahedral but stable to hydrolysis. Suitable substituents are added to fill various binding pockets usually occupied by the amino acid side chains of polypeptide substrates.

- To obtain an orally active PI, it is important to maximize the binding interactions with the enzyme, while minimizing the molecular weight and peptide character of the molecule.

- Cell fusion inhibitors have been developed, one of which has reached the market.

- Three integrase inhibitors have been approved for the market.

20.8 Antiviral drugs acting against RNA viruses: flu virus

20.8.1 Structure and life cycle of the influenza virus

Influenza (or flu) is an airborne, respiratory disease caused by an RNA virus which infects the epithelial cells of the upper respiratory tract. It is a major cause of mortality, especially among the elderly, or among patients with weak immune systems. The most serious pandemic occurred in 1918 with the death of at least 20 million people worldwide caused by the Spanish flu virus. Epidemics then occurred in 1957 (Asian flu), 1968 (Hong Kong flu), and 1977 (Russian flu). Despite the names given to these flus, it is likely that they all derived from China where families live in close proximity to poultry and pigs, increasing the chances of viral infections crossing from one species to another.[1] In 1997, there was an outbreak of flu in Hong Kong which killed 6 out of the 18 people infected. This was contained by slaughtering infected chickens, ducks, and geese which had been the source of the problem. If action had not been swift, it is possible that this flu variant could have become a pandemic and wiped out 30% of the world's population. This emphasizes the need for effective antiviral therapies to combat flu.

The nucleocapsid of the flu virus contains $(-)$ssRNA and a viral enzyme called **RNA polymerase** (see Fig. 20.1). Surrounding the nucleocapsid, there is a membranous envelope derived from host cells which contains two viral glycoproteins called **neuraminidase** (NA) and **haemagglutinin** (HA). The latter acquired its name because it can bind virions to red blood cells and cause haemagglutination. The NA and HA glycoproteins are spike-like objects which project about 10 nm from the surface and are crucial to the infectious process.

In order to reach the epithelial host cells of the upper respiratory tract, the virus has to negotiate a layer of protective mucus, and it is thought that the viral protein NA is instrumental in achieving this. The mucosal secretions are rich in glycoproteins and glycolipids which bear a terminal sugar substituent called **sialic acid** (also called **N-acetylneuraminic acid**). Neuraminidase (also called **sialidase**) is an enzyme which cleaves sialic acid from

---

[1] On the other hand, there has been a recent theory that the 1918 pandemic originated in army transit camps in France. The living conditions in these camps were similar to communities in China in the sense that large numbers of soldiers were camping in close proximity to pigs and poultry used as food stocks. The return of the forces to all parts of the globe following the First World War could explain the rapid spread of the virus.

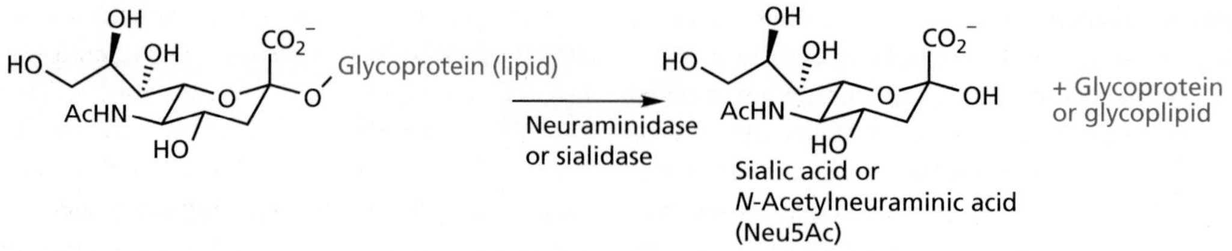

FIGURE 20.39 Action of neuraminidase (sialidase).

these glycoproteins and glycolipids (Fig. 20.39), thus degrading the mucus layer and allowing the virus to reach the surface of epithelial cells.

Once the virus reaches the epithelial cell, adsorption takes place whereby the virus binds to cellular glycoconjugates that are present in the host cell membrane, and which have a terminal sialic acid moiety. The viral protein HA is crucial to this process. Like NA, it recognizes sialic acid but, instead of catalysing the cleavage of sialic acid from the glycoconjugate, HA binds to it (Fig. 20.40). Once the virion has been adsorbed, the cell membrane bulges inwards taking the virion with it to form a vesicle called an **endosome**—a process called **receptor-mediated endocytosis**. The pH in the endosome then decreases, causing HA in the virus envelope to undergo a dramatic conformational change whereby the hydrophobic ends of the protein spring outward and extend towards the endosomal membrane. After contact, fusion occurs and the RNA nucleocapsid is released into the cytoplasm of the host cell. Disintegration of the nucleocapsid releases viral RNA and viral RNA polymerase, which both enter the cell nucleus.

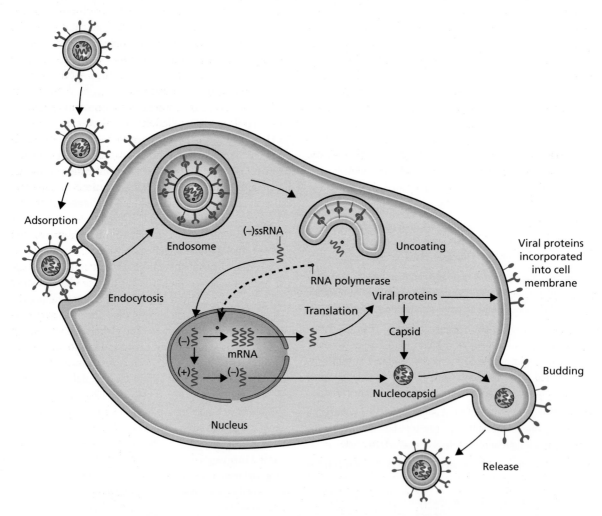

FIGURE 20.40 Life cycle of the influenza virus in a host epithelial cell.

Viral RNA polymerase now catalyses the copying of (−) viral RNA to produce (+) viral RNA, which departs the nucleus and acts as the mRNA required for the translation of viral proteins. Copies of (−) viral RNA are also produced in the nucleus, then exported into the cytoplasm.

Capsid proteins spontaneously self-assemble in the cytoplasm with incorporation of (−) RNA and newly produced RNA polymerase to form new nucleocapsids. Meanwhile, the freshly synthesized viral proteins HA and NA are incorporated into the membrane of the host cell. Newly formed nucleocapsids then move to the cell membrane and attach to the inner surface. HA and NA move through the cell membrane to concentrate at these areas and host cell proteins are excluded. Budding then takes place and a new virion is released. NA aids this release by hydrolysing any interactions that take place between HA on the virus and sialic acid conjugates on the host cell membrane.

There is an important balance between the rate of desialylation by NA (to aid the virion's departure from the host cell) and the rate of attachment by HA to sialylated glycoconjugates (to allow access to the cell). If NA was too active, it would hinder infection of the cell by destroying the receptors recognized by HA. However, if NA was too weak, the newly formed virions would remain adsorbed to the host cell after budding, preventing them from infecting other cells. It is noticeable that the amino acids present in the active site of NA are highly conserved, unlike amino acids elsewhere in the protein. This demonstrates the importance of the enzyme's level of activity.

As HA and NA are on the outer surface of the virion, they can act as antigens (i.e. molecules which can potentially be recognized by antibodies and the body's defence systems). In theory, it should be possible to prepare vaccines which will allow the body to gain immunity from the flu virus. Such vaccinations are available, but they are not totally protective and they lose what protective effect they have with time. This is because the flu virus is adept at varying the amino acids present in HA and NA, thus making these antigens unrecognizable to the antibodies which originally recognized them—a process called **antigenic variation**. The reason it takes place can be traced back to the RNA polymerase enzyme, which is a relatively error-prone enzyme and means that the viral RNA which codes for HA and NA is not consistent. Variations in the code lead to changes in the amino acids present in NA and HA, which results in different types of flu virus based on the antigenic properties of their NA and HA. For example, there are nine antigenic variants of NA.

There are three groups of flu virus, classified as A, B, and C. Antigenic variation does not appear to take place with influenza C, and occurs slowly with influenza B. With influenza A, however, variation occurs almost yearly. If the variation is small, it is called **antigenic drift**. If it is large, it is called **antigenic shift** and it is this that can lead to the more serious epidemics and pandemics. There are two influenza A virus subtypes which are epidemic in humans—those with H1N1 and H3N2 antigens (where H and N stand for HA and NA respectively). A major aim in designing effective antiviral drugs is to find a drug that will be effective against the influenza A virus, and remain effective despite antigenic variations. In general, vaccination is the preferred method of preventing flu, but antiviral drugs have a role in the prevention and treatment of flu when vaccination proves unsuccessful.

20.8.2 Ion channel disrupters: adamantanes

The adamantanes (Fig. 20.41) were discovered by random screening and are the earliest antiviral drugs used clinically against flu, decreasing the incidence of the disease by 50–70%. **Amantadine** and **rimantadine** (Fig. 20.41) are related adamantanes with similar mechanisms of action and can inhibit viral infection in two ways. At low concentration (<1 μg/ml), they inhibit the replication of influenza A viruses by blocking a viral ion channel protein called **matrix (M2) protein**. At higher concentration (>50 μg/ml), the basic nature of the compounds becomes important and they buffer the pH of endosomes to prevent the acidic environment needed for HA to fuse the viral membrane with that of the endosome. These mechanisms inhibit penetration and uncoating of the virus.

Unfortunately, the virus can mutate in the presence of amantadine to form resistant variants. Amantadine binds to a specific region of the M2 ion channel, and resistant variants have mutations which alter the width of the channel. Research carried out to find analogues which might still bind to these mutants proved unsuccessful. Work has also been carried out in an attempt to find an analogue which might affect the ion channel and pH levels at comparable concentrations. This has focused on secondary and tertiary amines with increased basicity, as well as alteration of the structure to reduce activity for the ion channel. The rationale is that resistant flu variants are less likely to be produced if the drug acts on two different targets at the same time. Rimantadine was approved in 1993 as a less toxic alternative to amantadine for the treatment of influenza A. Unfortunately, neither agent is effective against influenza B as this virus does not contain the matrix (M2) protein. Side effects are also a problem, possibly due to effects on host cell ion channels.

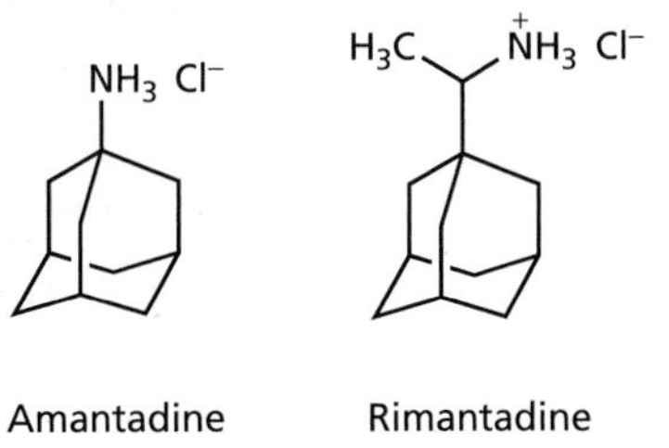

FIGURE 20.41 The adamantanes.

20.8.3 **Neuraminidase inhibitors**

20.8.3.1 Structure and mechanism of neuraminidase

Since neuraminidase (NA) has two crucial roles in the infectious process (section 20.8.1), it is a promising target for potential antiviral agents. Indeed, a screening programme for NA inhibitors was carried out as early as 1966, although without success. Following on from this, researchers set out to design a mechanism-based transition-state inhibitor. This work progressed slowly until the enzyme was isolated and its crystal structure studied by X-ray crystallography and molecular modelling.

Neuraminidase is a mushroom-shaped tetrameric glycoprotein anchored to the viral membrane by a single hydrophobic sequence of some 29 amino acids. As a result, the enzyme can be split enzymatically from the surface and studied without loss of antigenic or enzymic activity. X-ray crystallographic studies have shown that the active site is a deep pocket located centrally on each protein subunit. There are two main types of the enzyme (corresponding to the influenza viruses A and B) and various subtypes. Due to the ease with which mutations occur, there is a wide diversity of amino acids making up the various types and subtypes of the enzyme. However, the 18 amino acids making up the active site itself are constant. As mentioned previously, the level of enzyme activity is crucial to the infectious process, and any variation that affects the active site is likely to affect the activity of the enzyme. This, in turn, will affect the infectious process adversely. Because the active site remains constant, any inhibitor designed to fit it has a good chance of inhibiting all strains of the flu virus. Moreover, it has been observed that the active site is quite different in structure from the active sites of comparable bacterial or mammalian enzymes, so there is a strong possibility that inhibitors can be designed that are selective antiviral drugs.

The enzyme has been crystallized with sialic acid (the product of the enzyme-catalysed reaction) bound to the active site, and the structure determined by X-ray crystallography. A molecular model of the complex was created which resembled the observed crystal structure as closely as possible. From this it was calculated that sialic acid was bound to the active site through a network of hydrogen bonds and ionic interactions, as shown in Fig. 20.42.

The most important interactions involve the carboxylate ion of sialic acid, which is involved in ionic interactions and hydrogen bonds with three arginine residues, particularly with Arg-371. In order to achieve these interactions, the sialic acid has to be distorted from a stable chair conformation (where the carboxylate ion is in the axial position) to a less stable pseudo-boat conformation where the carboxylate ion is equatorial.

There are three other important binding regions or pockets within the active site. The glycerol side chain of sialic acid at C-6 fills one of these pockets, interacting with glutamate residues and a water molecule by hydrogen bonding. The hydroxyl group at C-4 is situated in another binding pocket, interacting with a glutamate residue. Finally, the acetamido substituent at C-5 fits into a hydrophobic pocket which is important for molecular recognition. This pocket includes the hydrophobic residues Trp-178 and Ile-222 which lie close to the methyl carbon (C-11) of sialic acid, as well as the hydrocarbon backbone of the glycerol side chain.

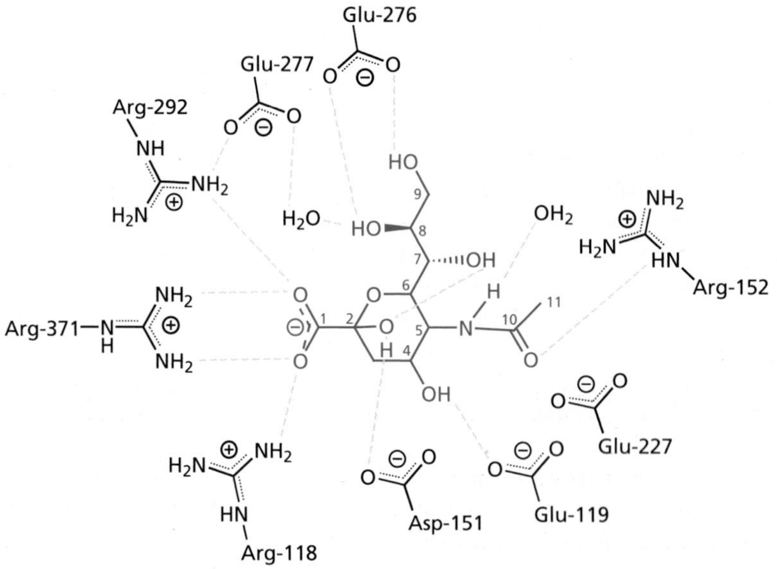

FIGURE 20.42 Hydrogen bonding interactions between sialic acid and the active site of neuraminidase.

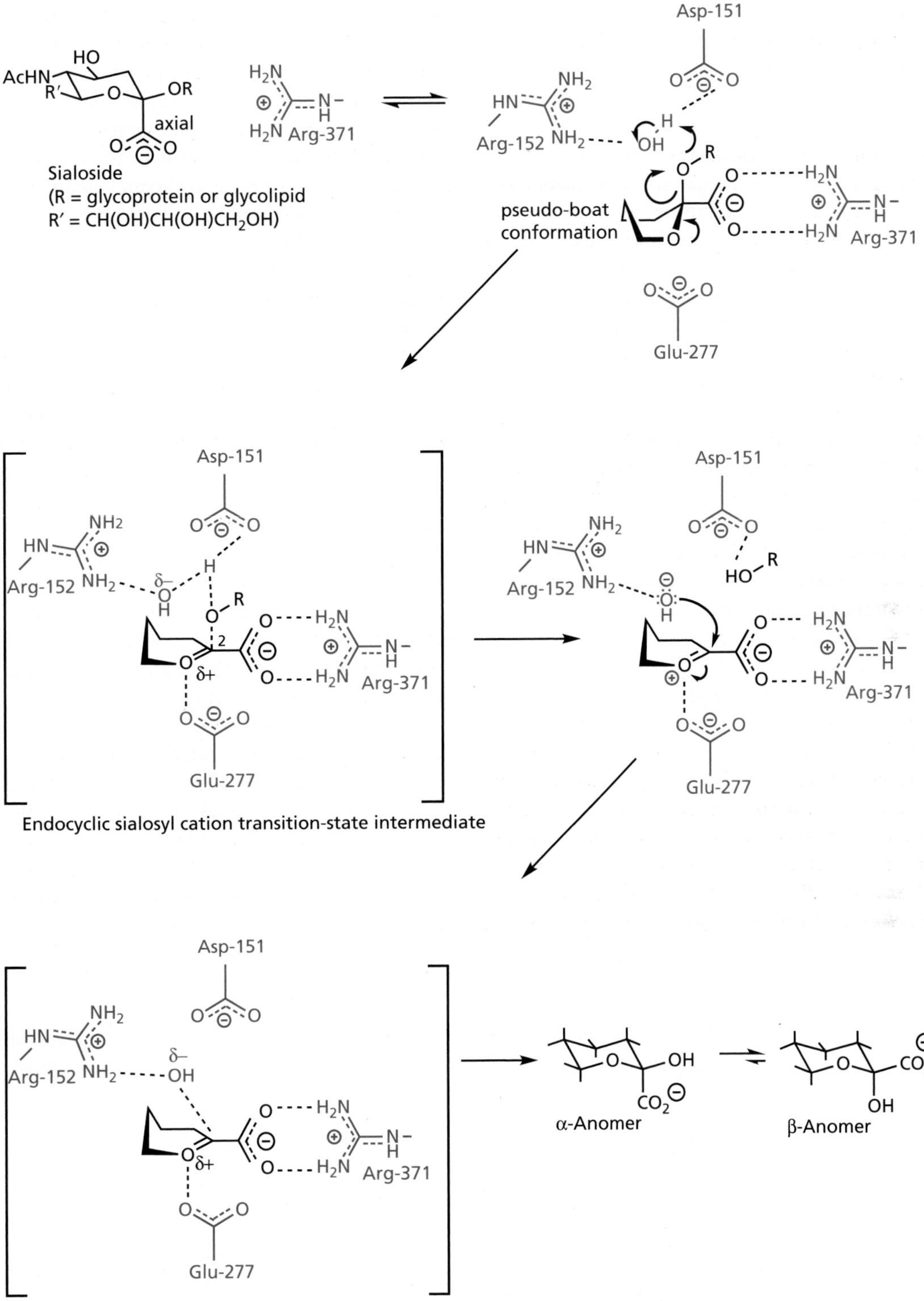

FIGURE 20.43 Proposed mechanism for the enzyme-catalysed hydrolysis of glycoconjugates to sialic acid (substituents are not shown during the mechanism for clarity).

It was further established that the distorted pyranose ring binds to the floor of the active site cavity through its hydrophobic face. The glycosidic OH at C-2 is also shifted from its normal equatorial position to an axial position where it points out of the active site and can form a hydrogen bond to Asp-151, as well as an intramolecular hydrogen bond to the hydroxyl group at C-7.

Based on these results, a mechanism of hydrolysis was proposed which consists of four major steps (Fig. 20.43). The first step involves the binding of the substrate (sialoside) as described above. The second step involves proton donation from an activated water facilitated by the negatively charged Asp-151, and formation of an endocyclic sialosyl cation transition-state intermediate. Glu-277 is proposed to stabilize the developing positive charge on the glycosidic oxygen as the mechanism proceeds.

The final two steps of the mechanism are formation and release of sialic acid. Support for the proposed mechanism comes from kinetic isotope studies which indicate that it is an S_N1 nucleophilic substitution. NMR studies have also been carried out which indicate that sialic acid is released as the α-anomer. This is consistent with an S_N1 mechanism having a high degree of stereofacial selectivity. It is possible that expulsion of the product from the active site is favoured by mutarotation to the more stable β-anomer.

Finally, site-directed mutagenesis studies have shown that the activity of the enzyme is lost if Arg-152 is replaced by lysine, and Glu-277 is replaced by aspartate. These replacement amino acids contain similarly charged, but shorter side chains. As a result, the charged side chains are unable to reach the required area of space in order to stabilize the intermediate.

20.8.3.2 Transition-state inhibitors: development of zanamivir (Relenza)

The transition state shown in Fig. 20.43 has a planar trigonal centre at C-2 and so sialic acid analogues containing a double bond between positions C-2 and C-3 were synthesized to achieve that same trigonal geometry at C-2. This resulted in the discovery of the inhibitor **2-deoxy-2,3-dehydro-*N*-acetylneuraminic acid** (**Neu5Ac2en**) (Fig. 20.44). In order to achieve the required double bond, the hydroxyl group originally present at C-2 of sialic acid had to be omitted, which resulted in lower hydrogen bonding interactions with the active site. However, the inhibitor does not need to distort from a favourable chair shape in

order to bind, and the energy saved by this more than compensates for the loss of one hydrogen bonding interaction. The inhibitor was crystallized with the enzyme and studied by X-ray crystallography and molecular modelling to show that the same binding interactions were taking place with the exception of the missing hydroxyl group at C-2. Unfortunately, this compound also inhibited bacterial and mammalian sialidases and could not be used therapeutically. Moreover, it was inactive *in vivo*.

The search for new inhibitors centred around the use of **GRID** molecular modelling software to evaluate likely binding regions within a model active site. This involved setting up a series of grid points within the active site and placing probe atoms at each point to measure interactions between the probe and amino acid residues (section 17.7.5). Different atomic probes were used to represent various functional groups. These included the oxygen of a carboxylate group, the nitrogen of an aminium ion, the oxygen of a hydroxyl group, and the carbon of a methyl group. Multi-atom probes were also used. These were positioned in the grid such that one atom of the probe was placed at each grid point in turn. Energy calculations were performed for all the atoms within the probe to give a total interaction energy at each grid point. Each time, the probe was rotated to find the orientation for the best possible hydrogen bonding interaction.

The most important result from these studies was the discovery that the binding region normally occupied by the 4-OH of sialic acid could also interact with an aminium or guanidinium ion. As a result, sialic acid analogues, having an amino or guanidinyl group at C-4, instead of a hydroxyl group, were modelled in the active site to study the binding interactions and to check whether there was room for the groups to fit.

These modelling studies were favourable and so the relevant structures were synthesized and tested for activity. **4-Amino-Neu5Ac2en** (Fig. 20.44) contains the aminium group and was found to be more potent than

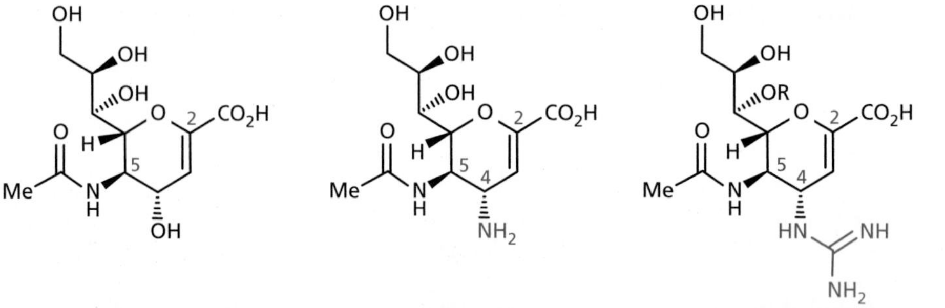

Neu5Ac2en
K_i (M) 4×10^{-6}; IC$_{50}$ 5–10 μM

4-Amino-Neu5Ac2en
K_i (M) 4×10^{-8}

Zanamivir (Relenza); R = H
K_i (M) 3×10^{-11}
Laninamivir (Inavir); R = Me

FIGURE 20.44 Transition-state inhibitors for the enzyme neuraminidase.

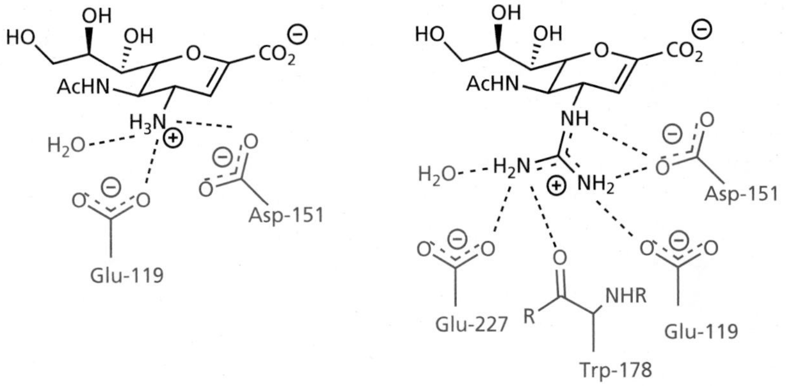

FIGURE 20.45 Binding interactions of aminium and guanidinium moieties at C-4 with the active site of neuraminidase.

Neu5Ac2en. Moreover, it was active in animal studies and showed selectivity against the viral enzyme, implying that the region of the active site which normally binds the 4-hydroxyl group of the substrate is different in the viral enzyme from comparable bacterial or mammalian enzymes. A crystal structure of the inhibitor bound to the enzyme confirmed the binding pattern predicted by the molecular modelling (Fig. 20.45).

Molecular modelling studies had suggested that the larger guanidinium group would be capable of even greater hydrogen bonding interactions, as well as favourable van der Waals interactions. The relevant structure (**zanamivir**; Fig. 20.44) was indeed found to be a more potent inhibitor, having a 100-fold increase in activity. X-ray crystallographic studies of the enzyme–inhibitor complex demonstrated the expected binding interactions (Fig. 20.45). Moreover, the larger guanidinium group was found to expel a water molecule from this binding pocket, which is thought to contribute a beneficial entropic effect. Zanamivir is a slow-binding inhibitor with a high binding affinity to influenza A neuraminidase. It was approved by the US FDA in 1999 for the treatment of influenza A and B, and was marketed by GlaxoWellcome and Biota.

Unfortunately, the polar nature of the molecule means it has poor oral bioavailability (<5%), and it is administered by inhalation. **Laninamivir** (Fig. 20.44) is a closely related structure which was approved in Japan during 2010.

Following on from the success of these studies, **4-epi-amino-Neu5Ac2en** (Fig. 20.46) was synthesized to place the amino group in another binding region predicted by the GRID analysis. This structure proved to be a better inhibitor than Neu5Ac2en, but not as good as zanamivir. The pocket into which this amino group fits is small and there is no room for larger groups.

20.8.3.3 Transition-state inhibitors: 6-carboxamides

A problem with the inhibitors described in section 20.8.3.2 is their polar nature. The glycerol side chain is particularly polar and has important binding interactions with the active site. However, it was found that it could be replaced by a carboxamide side chain with retention of activity (Fig. 20.46).

A series of 6-carboxamide analogues was prepared to explore their structure–activity relationships. Secondary

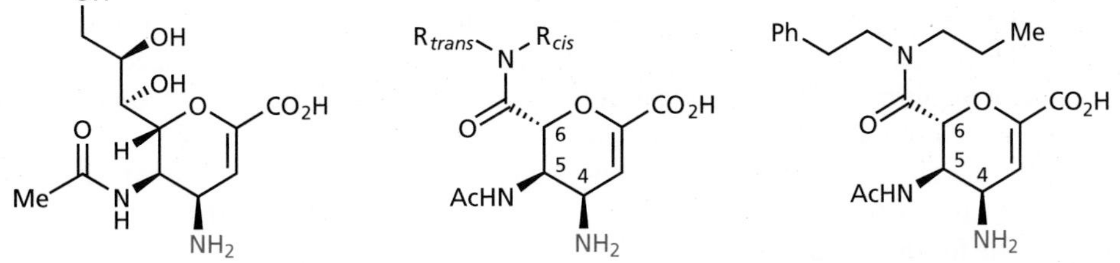

FIGURE 20.46 4-Epi-amino-Neu5Ac2en and carboxamides.

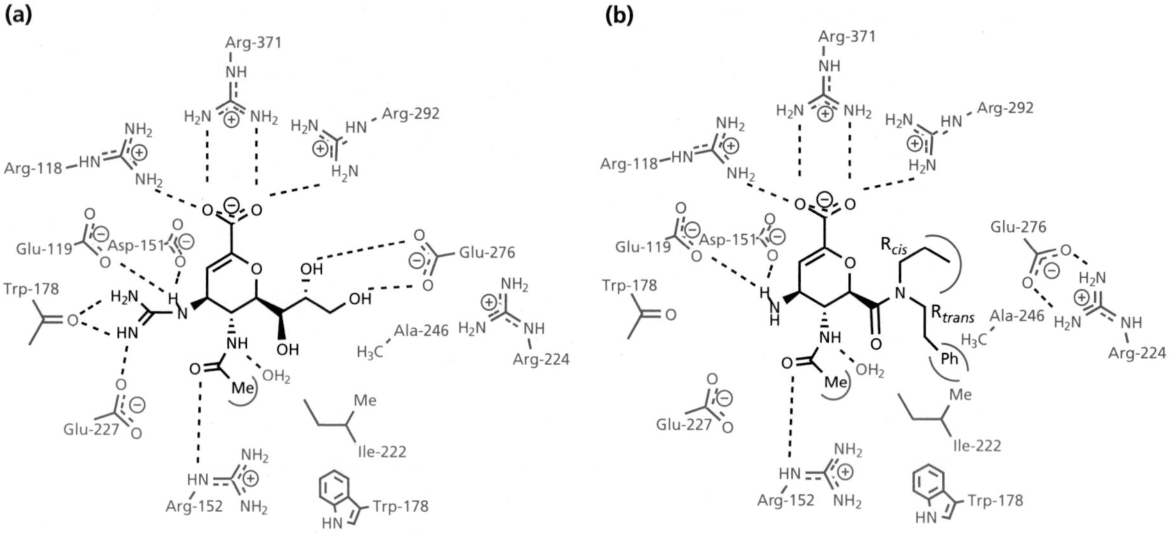

(a)

(b)

FIGURE 20.47 Binding interactions of zanamivir and carboxamides: (a) binding of zanamivir to the active site; (b) binding of carboxamide (I) to the active site.

carboxamides (where R_{cis} = H) showed similar weak inhibition against both A and B forms of the neuraminidase enzyme. Tertiary amides having an alkyl substituent at the *cis* position resulted in a pronounced improvement against the A form of the enzyme, with relatively little effect on the activity against the B form. Thus, tertiary amides showed a marked selectivity of 30- to 1000-fold for the A form of the enzyme. Good activity was related to a variety of different sized R_{trans} substituents larger than methyl, but the size of the R_{cis} group was more restricted, and optimum activity was achieved when R_{cis} was ethyl or *n*-propyl.

The 4-guanidinium analogues are more active than corresponding 4-amino analogues but the improvement is slightly less than that observed for the glycerol series, especially where the 4-amino analogue is already highly active.

Crystal structures of the carboxamide (I in Fig. 20.46) bound to both enzymes A and B were determined by X-ray crystallography (Fig. 20.47). The dihydropyran portion of the carboxamide (I) binds to both the A and B forms of the enzyme in essentially the same manner as observed for zanamivir. The important binding interactions involve the carboxylate ion, the 4-amino group and the 5-acetamido group—the latter occupying a hydrophobic pocket lined by Trp-178 and Ile-222.

However, there is a significant difference in the region occupied by the carboxamide side chain. In the sialic acid analogues, the glycerol side chain forms intermolecular hydrogen bonds to Glu-276. These interactions are not possible for the carboxamide side chain. Instead, the Glu-276 side chain changes conformation and forms a salt bridge with the guanidino side chain of Arg-224, and reveals a lipophilic pocket into which the R_{cis} *n*-propyl substituent can fit. The size of this pocket is optimal for an ethyl or propyl group which matches

the structure–activity (SAR) results. The R_{trans} phenethyl group lies in an extended lipophilic cleft on the enzyme surface formed between Ile-222 and Ala-246. This region can accept a variety of substituents, again consistent with SAR results.

Comparison of the X-ray crystal structures of the native A and B enzymes shows close similarity of position and orientation of the conserved active site residues, except in the region occupied normally by the glycerol side chain. This is particularly the case for Glu-276. Zanamivir can bind to both A and B forms with little or no distortion of the native structures. Binding of the carboxamide (I) to the A form is associated with a change in torsion angles of the Glu-276 side chain such that the residue can form the salt bridge to Arg-224. Very little distortion of the protein backbone is required in order to achieve this. In contrast, when the carboxamide binds to the B form of the enzyme, significant distortion of the protein backbone is required before the salt bridge is formed. Distortion in the B enzyme structure also arises around the phenethyl substituent. This implies that binding of the carboxamide to the B form involves more energy expenditure than to A and this can explain the observed specificity.

Although none of the carboxamides studied reached the market, the information gained from crystal studies on the binding interactions proved relevant in the development of oseltamivir (next section).

20.8.3.4 Carbocyclic analogues: development of oseltamivir (Tamiflu)

The dihydropyran oxygen of Neu5Ac2en and related inhibitors has no important role to play in binding any

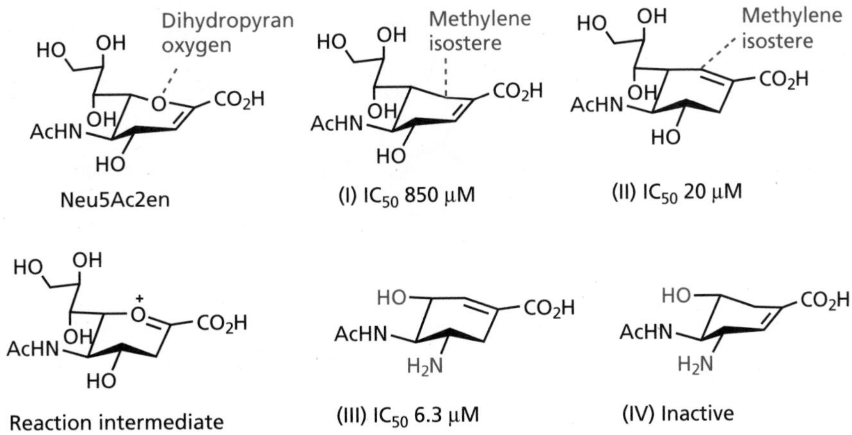

FIGURE 20.48 Comparison of Neu5Ac2en, reaction intermediate, and carbocyclic structures.

of these structures to the active site of NA. Therefore, it should be possible to replace it with a methylene isostere to form carbocyclic analogues such as structure I in Fig. 20.48. This would have the advantage of removing a polar oxygen atom which would increase hydrophobicity and potentially increase oral bioavailability. Moreover, it would be possible to synthesize cyclohexene analogues such as structure II, which more closely match the stereochemistry of the reaction's transition state than previous inhibitors—compare the reaction intermediate in Fig. 20.48 which can be viewed as a transition-state mimic. Such agents might be expected to bind more strongly and be more potent inhibitors.

Structures I and II were synthesized to test this theory, and it was discovered that structure II was 40 times more potent than structure I as an inhibitor. As the substituents are the same, the difference in potencies indicates that the conformation of the ring is crucial for inhibitory activity. Both structures have half chair conformations but these are different due to the position of the double bond.

It was now planned to replace the hydroxyl group on the ring with an amino group to improve binding interactions (compare section 20.8.3.2), and to remove the glycerol side chain to reduce polarity. A hydroxyl group was introduced to replace the glycerol side chain. There were two reasons for this. Firstly, the oxonium double bond in the transition state is highly polarized and electron deficient, whereas the

double bond in the carbocyclic structures is electron rich. Introducing the hydroxyl substituent in place of the glycerol side chain means that the oxygen will have an inductive electron-withdrawing effect on the carbocyclic double bond and reduce its electron density. The second reason for adding the hydroxyl group was that it would be possible to synthesize ether analogues which would allow the addition of hydrophobic groups to fill the binding pocket previously occupied by the glycerol side chain (compare section 20.8.3.3). The resultant structure III was synthesized and proved to be a potent inhibitor. In contrast, the isomer IV failed to show any inhibitory activity.

A series of alkoxy analogues of structure III was now synthesized in order to maximize hydrophobic interactions in the region of the active site previously occupied by the glycerol side chain (Fig. 20.49). For linear alkyl chains, potency increased as the carbon chain length increased from methyl to n-propyl. Beyond that, activity was relatively constant (150–300 nM) up to and including n-nonyl, after which activity dropped. Although longer chains than propyl increase hydrophobic interactions, there is a downside in that there is partial exposure of the side chain to water outside of the active site.

Branching of the optimal propyl group was investigated. There was no increase in activity when methyl branching was at the β-position, but the addition of a methyl group at the α-position increased activity by

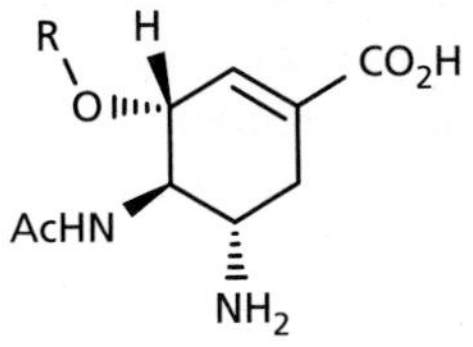

		Linear chains (R)		Branched chains (R)		Miscellaneous chains (R)	
		R	IC$_{50}$ (μM)	R	IC$_{50}$ (μM)	R	IC$_{50}$ (μM)
		Me	3.70	CH_2CHMe_2	0.200	CH_2OMe	2.00
		Et	2.00	$CH(Me)CH_2CH_3$	0.010	$CH_2CH_2CF_3$	0.20
		n-Pr	0.18	$CH(Et)_2$	0.001	$CH_2CH=CH_2$	2.20
		n-Bu	0.30			cyclopentyl	0.02
						cyclohexyl	0.06
						phenyl	0.53

FIGURE 20.49 Alkoxy analogues.

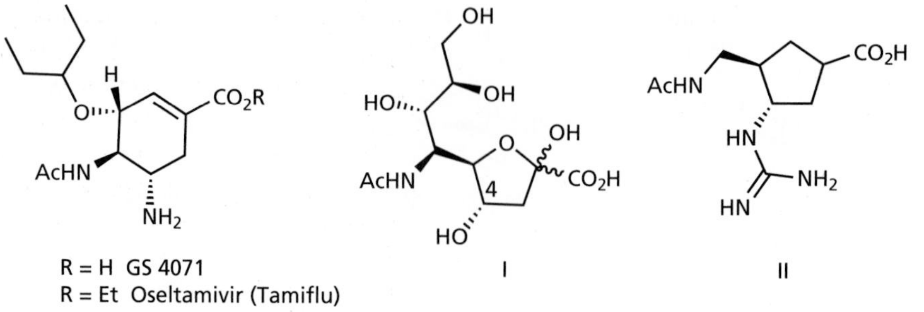

FIGURE 20.50 Oseltamivir and other ring systems.

20-fold. Introduction of an α-methyl group introduces an asymmetric centre, but both isomers were found to have similar activity indicating two separate hydrophobic pockets. The optimal side chain proved to be a pentyloxy side chain ($R = CH(Et)_2$).

The *N*-acetyl group is required for activity and there is a large drop in activity without it. The binding region for the *N*-acetyl group has limitations on the functionality and size of groups which it can accept. Any variations tend to reduce activity. This was also observed with sialic acid analogues.

Replacing the amino group with a guanidine group improves activity, as with the sialic acid series. However, the improvement in activity depends on the type of alkyl group present on the side chain, indicating that individual substituent contributions may not be purely additive.

The most potent of the above analogues was the pentyloxy derivative (GS 4071) (Fig. 20.50). This was co-crystallized with the enzyme and the complex was studied by X-ray crystallography, revealing that the alkoxy side chain makes several hydrophobic contacts in the region of the active site normally occupied by the glycerol side chain. In order to achieve this, the carboxylate group of Glu-276 is forced to orientate outwards from the hydrophobic pocket as observed with the carboxamides. The overall gain in binding energy from these interactions appears to be substantial, as a guanidinium group is not required to achieve low nanomolar inhibition. Interactions elsewhere are similar to those observed with previous inhibitors.

Oseltamivir (Tamiflu) (Fig. 20.50) is the ethyl ester prodrug of GS 4071 and was approved in 1999 for the treatment of influenza A and B. The drug is marketed by Hoffman La Roche and Gilead Sciences. It is taken orally and is converted to GS4071 by esterases in the gastrointestinal tract.

⊕ Test your understanding and practise your molecular modelling with Exercise 20.5 on the Online Resource Centre at www.oxfordtextbooks.co.uk/orc/patrick6e/

20.8.3.5 Other ring systems

Work has been carried out to develop new NA inhibitors where different ring systems act as scaffolds for the important binding groups (Fig. 20.50).

The five-membered tetrahydrofuran (I) is known to inhibit neuraminidase with a potency similar to Neu5Ac2en. It has the same substituents as Neu5Ac2en, although their arrangement on the ring is very different. Nevertheless, a crystal structure of (I) bound to the enzyme shows that the important binding groups (carboxylate, glycerol, acetamido, and C4-OH) can fit into the required pockets. The central ring or scaffold is significantly displaced from the position occupied by the pyranose ring of Neu5Ac2en in order to allow this. This indicates that the position of the central ring is not crucial to activity, and that the relative position of the four important binding groups is more important.

Five-membered carbocyclic rings have also been studied as possible scaffolds. Structure II (Fig. 20.50) was designed such that the guanidine group would fit the negatively charged binding pocket previously described. A crystal structure of the inhibitor with the enzyme showed that the guanidine group occupies the desired pocket and displaces the water molecule originally present. It is involved in charge-based interactions with Asp-151, Glu-119, and Glu-227, analogous to zanamivir.

Modelling studies suggested that the addition of a butyl chain to the structure would allow van der Waals interactions with a small hydrophobic surface in the binding site. The target structure now has four asymmetric centres, and a synthetic route was used which controlled the configuration of two of these. As a result, eight isomers were prepared as a mixture (Fig. 20.51). Neuraminidase crystals were used to select the most active isomer of the mixture by soaking a crystal of the enzyme in the solution of isomers for a day and then collecting X-ray diffraction data from the crystal. This showed the active isomer to be structure I in Fig. 20.52. The structure binds to the active sites of both influenza A and B neuraminidases with the *n*-butyl side chain adopting two different binding modes.

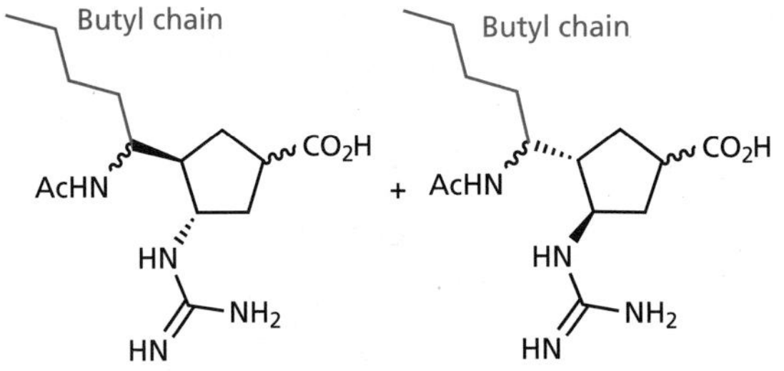

FIGURE 20.51 Mixture of isomers tested for their binding affinity to crystals of the neuraminidase enzyme.

Structure I   Peramivir (BCX 1812)

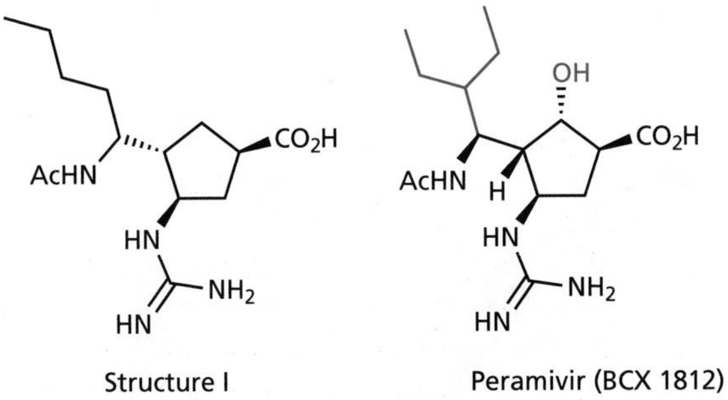

FIGURE 20.52 Development of peramivir (BCX 1812).

In the B version, the side chain is positioned against a hydrophobic surface formed by Ala-246, Ile-222, and Arg-224. In the A version, the chain is in a region formed by the reorientation of the side chain of Glu-276.

Peramivir (Fig. 20.52) was designed to take advantage of both hydrophobic pockets in the active site. It was prepared as a racemic mixture, and a crystal of the neuraminidase enzyme was used to bind the active isomer. Once identified, this was then prepared by a stereospecific synthesis. The relative stereochemistry of the corresponding substituents was the same as in structure I (Fig. 20.52).

In vitro tests of peramivir versus strains of influenza A and B show it to be as active as zanamivir and GS 4071. It is also four orders of magnitude less active against bacterial and mammalian neuraminidases, making it a potent and highly specific inhibitor of flu virus neuraminidase. *In vivo* tests carried out on mice showed it to be orally active and the compound was approved in Japan in 2010.

20.8.3.6 Resistance studies

Studies have been carried out to investigate the likelihood of viruses acquiring resistance to the drugs mentioned above. This is done by culturing the viruses in the presence of the antiviral agents to see if mutation leads to a resistant strain.

Zanamivir has a broad-spectrum efficacy against all type A and B strains tested, and interacts only with conserved residues in the active site of NA. Thus, in order to

gain resistance, one of these important amino acids has to mutate. A variant has been observed where Glu-119 has been mutated to glycine. This has reduced affinity for zanamivir, and the virus can replicate in the presence of the drug. Removing Glu-119 affects the binding interactions with the 4-guanidinium group of zanamivir without affecting interactions with sialic acid. Zanamivir-resistant mutations were also found where a mutation occurred in HA around the sialic acid binding site. This mutation weakened affinity for sialic acid and so lowered binding. Thus, mutant viruses were able to escape more easily from the infected cell after budding. No such mutations have appeared during clinical trials, however.

Another mutation has been observed where Arg-292 is replaced by lysine. In wild-type NA, Arg-292 binds to the carboxylate group of the inhibitor and is partly responsible for distorting the pyranose ring from the chair to the boat conformation. In the mutant structure, the amino group of Lys-292 forms an ionic interaction with Glu-276 which normally binds the hydroxyl groups at positions 8 and 9 of the glycerol side chain. This results in a weaker interaction with inhibitors and substrate alike, leading to a weaker enzyme.

One conclusion that has been made from studies on easily mutatable targets is the desirability to find an inhibitor which is modified as little as possible from the normal substrate, and which uses the same interactions for binding.

KEY POINTS

- The flu virus contains (−)ssRNA and has two glycoproteins called haemagglutinin (HA) and neuraminidase (NA) in its outer membrane.

- HA binds to the sialic acid moiety of glycoconjugates on the outer surface of host cells leading to adsorption and cell uptake.

- NA catalyses the cleavage of sialic acid from glycoconjugates. It aids the movement of the virus through mucus and releases the virus from infected cells after budding.

- HA and NA act as antigens for flu vaccines. However, the influenza A virus readily mutates these proteins, and new flu vaccines are required each year.

- The adamantanes are antiviral agents which inhibit influenza A by blocking a viral ion channel called the matrix (M2) protein. At high concentration they buffer the pH of endosomes. They are ineffective against influenza B which lacks the matrix (M2) protein.

- Neuraminidase has an active site which remains constant for the various types and subtypes of the enzyme and which is different from the active sites of comparable mammalian enzymes.

- There are four important binding pockets in the active site. The sialic acid moiety is distorted from its normal chair conformation when it is bound.

- The mechanism of the enzyme-catalysed reaction is proposed to go through an endocyclic sialosyl cation transition state. Inhibitors were designed to mimic this state by introducing an endocyclic double bond.

- Successful antiviral agents for the treatment of flu have been developed using structure-based drug design.

- Different scaffolds can be used to hold the four important binding groups required for neuraminidase inhibitors.

- There is an advantage in designing drugs which use the same binding interactions as the natural ligand when the target undergoes facile mutations.

20.9 Antiviral drugs acting against RNA viruses: cold virus

The agents used against flu are ineffective against colds, as these infections are caused by a different kind of virus called a rhinovirus. Colds are less serious than flus. Nevertheless, research has taken place to find drugs which can combat them.

There are at least 89 serotypes of **human rhinoviruses** (HRV) and they belong to a group of viruses called the **picornaviruses**. These viruses are responsible for diseases such as polio, hepatitis A, and foot and mouth disease. They are among the smallest of the animal RNA viruses, containing a positive strand of RNA coated by an icosahedral shell made up of 60 copies of 4 distinct proteins, VP1–VP4 (Fig. 20.53). The proteins VP1–VP3 make up the surface of the virion. The smaller VP4 protein lies underneath to form the inner surface and is in contact with the viral RNA. At the junction between each VP1 and VP3 protein, there is a broad canyon 25 Å deep and this is where attachment takes place between the virus and the host cell. On the canyon floor, there is a pore which opens into a hydrophobic pocket within the VP1 protein. This pocket is either empty or occupied by a small molecule called a **pocket factor**. So far the identity of the pocket factor has not been determined but it is known from X-ray crystallographic studies that it is a fatty acid containing seven carbon atoms.

Test your understanding and practise your molecular modelling with Exercise 20.6 on the Online Resource Centre at www.oxfordtextbooks.co.uk/orc/patrick6e/

When the virus becomes attached to the host cell, a receptor molecule on the host cell fits into the canyon and induces conformational changes that cause the VP4 protein and the *N*-terminus of VP1 to move to the exterior of the virus—a process called **externalization**. This is thought to be important to the process by which the virus

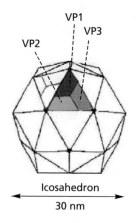

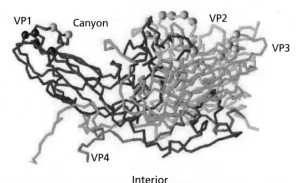

FIGURE 20.53 Structure of human rhinovirus and the proteins VP1–VP4.

is uncoated and releases its RNA into the host cell. It is thought that the pocket factor stabilizes the capsid when it is bound, and prevents the conformational changes that are needed to cause infection.

A variety of drugs having antiviral activity are thought to mimic the pocket factor by displacing it and binding to the same hydrophobic pocket. The drugs concerned are called **capsid-binding agents** and are characteristically long-chain hydrophobic molecules. Like the pocket factor, they stabilize the capsid by locking it into a stable conformation and prevent the conformational changes required for uncoating. They also raise the canyon floor and prevent the host cell receptor from fitting the canyon (Fig. 20.54).

Pleconaril (Fig. 20.55) is one such drug that has undergone phase III clinical trials which demonstrate that it has an effect on the common cold. It is an orally active, broad-spectrum agent which can cross the blood–brain barrier. The drug may also be useful against the enteroviruses that cause diarrhoea, viral meningitis, conjunctivitis, and encephalitis, as these viruses are similar in structure to the rhinoviruses.

The development of pleconaril started when a series of isoxazoles was found to have antiviral activity. This

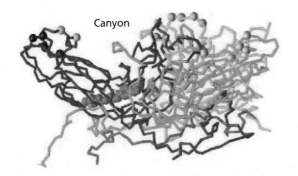

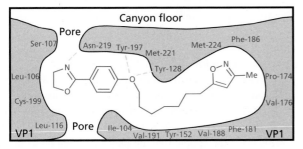

FIGURE 20.54 Binding of disoxaril (possible hydrogen bonds shown as dashed lines).

led to the discovery of **disoxaril** (Fig. 20.55) which entered phase I clinical trials, but proved to be too toxic. X-ray crystallographic studies of VP1–drug complexes involving disoxaril and its analogues showed that the oxazoline and phenyl rings were roughly coplanar and were located in a hydrophilic region of the pocket near the pore leading into the centre of the virion (Fig. 20.54). The hydrophobic isoxazole ring binds into the heart of the hydrophobic pocket and the chain provides sufficient flexibility for the molecule to bend round a corner in the pocket. Ligand binding moves Met-221 which normally seals off the pocket, and this also causes conformational changes in the canyon floor. Structure-based drug design

was carried out to find safer and more effective antiviral agents. For example, the chain cannot be too short or too long, or else there are steric interactions. Placing additional hydrophobic groups on to the phenyl ring improves activity against the HRV2 strain, because increased interactions are possible with a phenylalanine residue at position 116 rather than leucine. The structure **WIN 54954** was developed and entered clinical trials, but results were disappointing because extensive metabolism resulted in 18 different metabolic products due mainly to hydrolysis of the oxazoline ring. Further structure-based drug design led to modifications of the phenyl and oxazoline moieties. This included the introduction of a trifluoromethyl group to block metabolism, resulting in pleconaril, with 70% oral bioavailability.

20.10 Antiviral drugs acting against RNA viruses: hepatitis C

Hepatitis C virus (HCV) is a positive, single-stranded RNA virus that was discovered in 1989. It is a blood-borne virus that affects an estimated 170 million people worldwide, but many of those infected with the agent are unaware of the fact as they do not experience any symptoms. However, in the long term, the virus can cause serious liver damage, cancer, and death. Indeed, HCV is the principle cause for liver transplants in industrialized nations. There are at least six genotypes (gt) of the virus with genotype gt1 being the most common in North and South America, Europe, and Australia.

Until recently, the only therapies available were the broad-spectrum agents **PEGylated alpha interferon (IFN-α)** and **ribavirin** (section 20.11). However, the success rates with these drugs are only about 40%. A substantial research effort over the last 15 years has resulted

FIGURE 20.55 Capsid binding agents.

in a succession of novel drugs reaching the clinic from 2011 onwards (Box 20.5). Success in the hepatitis C field was initially delayed because of the time taken to gain an understanding of the HCV life cycle and the proteins involved, as well as the research required to develop a reliable *in vitro* test. However, drugs have now been successfully designed to act against three different viral targets.

20.10.1 Inhibitors of HCV NS3-4A protease

20.10.1.1 Introduction

Replication of the virus within the host cell includes the synthesis of a 3000-amino acid polyprotein, which is cleaved into individual viral proteins by a viral enzyme called **HCV NS3-4A protease** (NS indicates that the protein is non-structural). This is a serine protease containing a catalytic triad of Asp, His, and Ser (section 3.5.3), which cleaves a peptide bond between cysteine at P1 and either serine or alanine at P1' (Fig. 20.56). The protease is composed of two proteins (NS3 and NS4A). NS3 contains the active site, while NS4A acts as a cofactor and activates NS3. The active site is a long, shallow, and rather featureless groove which is solvent exposed. Substrates generally have 10–11 amino acid residues interacting with various binding pockets (S6 to S5') along the groove (Fig. 20.56). As a result, there are a large number of relatively weak interactions throughout the length of the groove, all of which seem to be important to both the binding affinity and selectivity of the substrate. Thus, it proved a major challenge to design a small, drug-sized inhibitor that would prove both selective and potent, yet bind to only a few of the available binding pockets. However, success was finally achieved in 2011 when **boceprevir** and **telaprevir** both reached the market. Other NS3-4A protease inhibitors followed with the introduction of **simeprevir**, **paritaprevir**, **asunaprevir**, and **grazoprevir**.

20.10.1.2 Design of boceprevir and telaprevir

The development of boceprevir began with the screening of chemical libraries to try and find a small molecule with inhibitory activity that could serve as a lead compound for further development. About four million compounds were screened, but no hits were identified. Therefore, it was decided to use the substrate as the lead compound and design a structure that would bind in a similar manner, but which would not undergo the enzyme-catalysed reaction. A series of peptide structures was studied where the susceptible amide bond was replaced with a ketoamide group, in the expectation that the catalytic serine residue in the active site (Ser-139) would react preferentially with the ketone rather than the amide (Fig. 20.57). As a ketone group undergoes nucleophilic addition rather than nucleophilic substitution, a reversible covalent bond would be formed between the inhibitor and the active site.

A series of peptide structures containing the ketoamide group was screened, leading to the identification of an undecapeptide which reacted as planned and showed good activity as an inhibitor (Fig. 20.58).

This was now adopted as the new lead compound for further development. Similar tactics to those used in the development of the HIV protease inhibitors were employed to design a low molecular weight inhibitor with decreased peptide character and increased binding interactions, resulting in the discovery of **boceprevir** (Fig. 20.59). In the process of designing boceprevir, all the P' binding groups that were present on the lead compound were removed, as well as a couple of P binding groups. This means that boceprevir has substituents that occupy only four of the binding subsites (S1–S4). However, these substituents have been optimized in order to gain the maximum binding interactions with their subsites, and this makes up for the interactions lost with the other seven subsites. It was found that bulky hydrophobic substituents were particularly effective for binding affinity, indicating that subsites S1–S4 were hydrophobic in nature. Bulky substituents were found to fill the subsites more effectively and formed a greater number of van der Waals interactions. The bicyclic structure that replaced proline was particularly effective in filling the S2 pocket, while the cyclobutyl group that fitted the S1 binding pocket was important for selectivity. Hydrogen bonding interactions involving the urea and the two

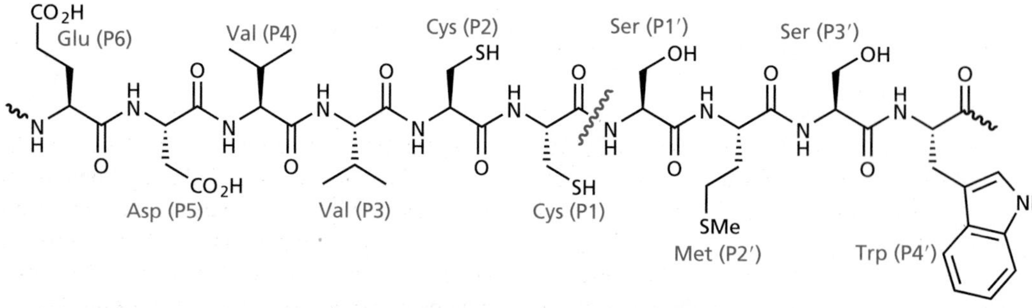

FIGURE 20.56 An example of a substrate for HCV NS3-4A protease.

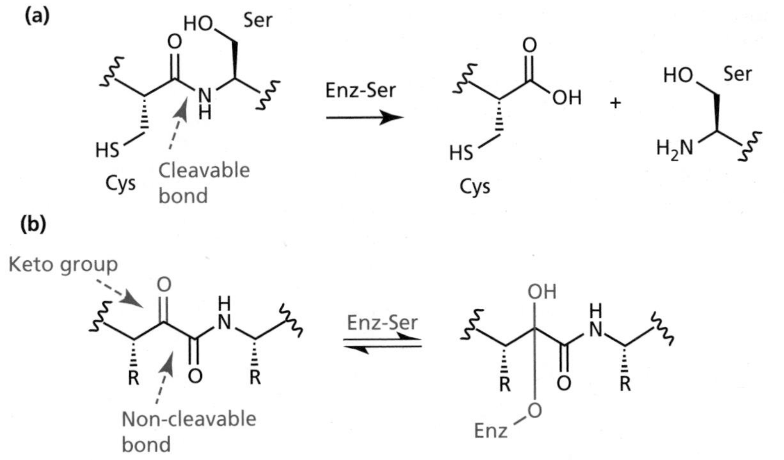

FIGURE 20.57 Design of protease inhibitors for HCV NS3-4A protease. (a) Normal enzyme-catalysed reaction. (b) Interaction of inhibitors with the catalytic serine residue (Ser-139).

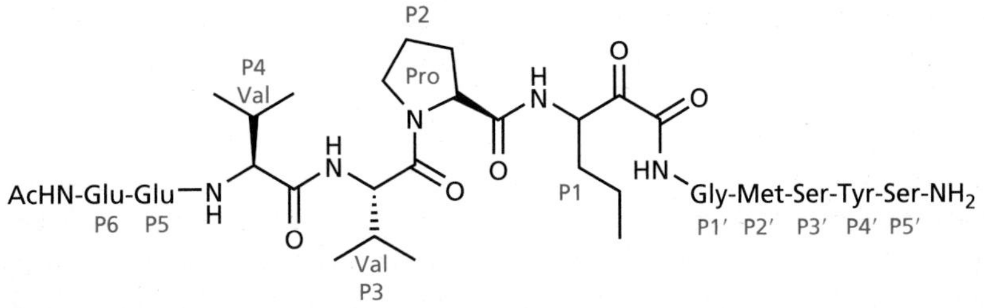

FIGURE 20.58 Structure of the undecapeptide inhibitor that served as the lead compound.

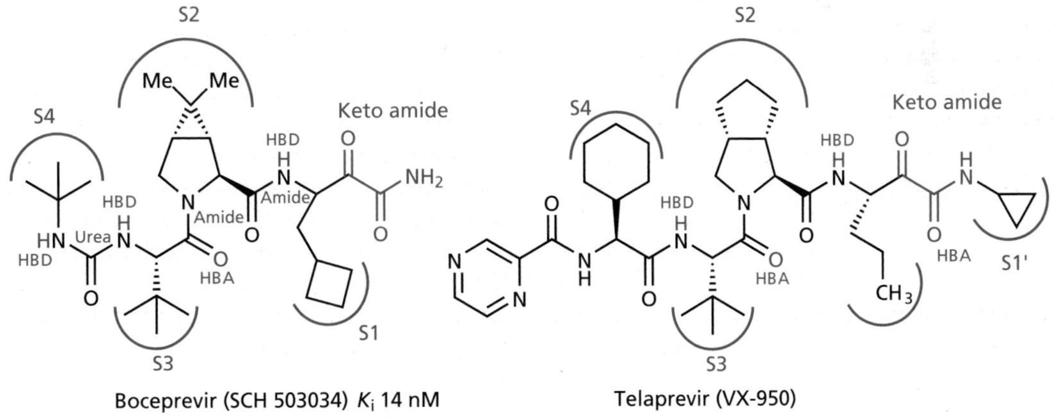

Boceprevir (SCH 503034) K_i 14 nM

Telaprevir (VX-950)

FIGURE 20.59 Boceprevir and telaprevir.

amide groups of boceprevir's backbone also contributed to binding and inhibitory activity. A further hydrogen bonding interaction is possible once the covalent bond is formed between serine and the keto group.

Note that the urea group was introduced to replace a peptide link and so decrease the peptide nature of the structure.

Similar strategies were used in the design of **telaprevir** (Fig. 20.59). It, too, contains bulky, hydrophobic substituents that form efficient van der Waals interactions with the hydrophobic sub pockets S1–S4. It also contains the keto amide group that reacts with Ser-139. However, it contains an additional cyclopropyl group that interacts with the S1′ binding pocket.

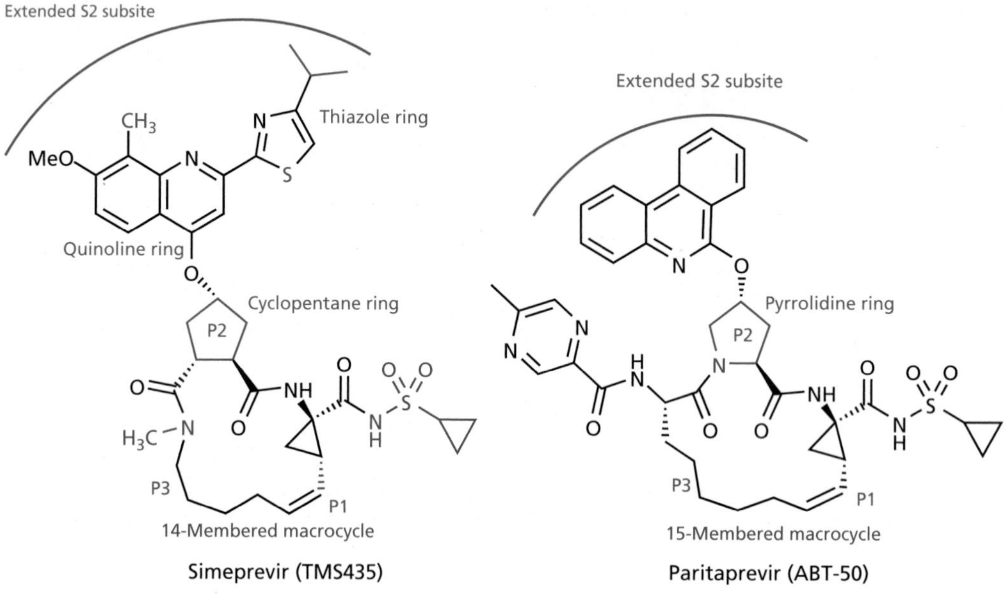

FIGURE 20.60 Simeprevir and paritaprevir.

The introduction of boceprevir and telaprevir significantly improved the treatment of hepatitis C, but in order to avoid the emergence of resistant strains, they have to be administered alongside peg-IFN-alpha and ribavirin. Although this has increased cure rates, the treatment has become more complex as patients are now taking three drugs instead of two. Moreover, side effects have increased and there are resistance problems. Therefore, there was a need for improved inhibitors that could address these issues and possibly replace both interferon and ribavirin.

20.10.1.3 Second-generation protease inhibitors

Following on from boceprevir and telaprevir, a second generation of protease inhibitors were developed that do not require the ketoamide 'warhead'. Instead, they bind purely through intermolecular interactions. This was achieved by further maximizing interactions with the binding subsites. In addition, the strategy of rigidification (section 13.3.9) proved beneficial when a macrocycle was introduced to link the P1 and P3 binding groups (see also Case study 10).

Simeprevir (Fig. 20.60) was the first of these reversible macrocyclic inhibitors to reach the market and was approved by the FDA in 2013. This was followed by **paritaprevir**—a structurally related compound that was approved in 2014. Paritaprevir contains a pyrrolidine ring instead of the cyclopentane ring present in simeprevir. The macrocycle is also 15-membered instead of

14-membered. Both structures take advantage of an extended S2 binding subsite that is not available for substrates, but which opens up when these inhibitors bind.

Unfortunately, HCV strains have emerged that have gained resistance to the protease inhibitors described so far. This is due to mutations that disrupt binding interactions with the inhibitors. To tackle this issue, new structures were designed where the macrocyclic ring linked the P2 and P4 binding groups instead of the P1 and P3 binding groups. This led to the discovery of **grazoprevir** (Fig. 20.61), which has proved effective against HCV strains with mutations to Arg-155 or Asp-168. This is because the extended P2 moiety in grazoprevir no longer interacts with these amino acids. Instead, there are increased interactions with the catalytic triad. The presence of the cyclopropane ring on the macrocyclic ring also makes the structure effective against HCV strains that have gained resistance because of mutations to Ala-156, possibly because the conformation of the macrocycle is slightly altered to move it away from that residue.

The research that led to the discovery of macrocyclic inhibitors has also been instrumental in the development of the linear inhibitor **asunaprevir** (Fig. 20.61). This structure shares many of the features that are present in the macrocyclic inhibitors.

Test your understanding and practise your molecular modelling with Exercise 20.7 on the Online Resource Centre at www.oxfordtextbooks.co.uk/orc/patrick6e/

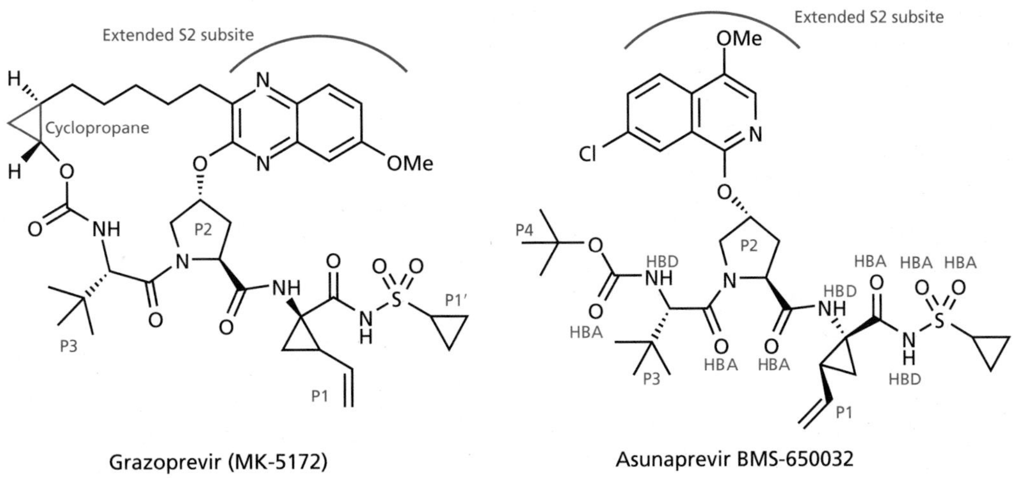

Grazoprevir (MK-5172)

Asunaprevir BMS-650032

FIGURE 20.61 Grazoprevir and asunaprevir.

20.10.2 **Inhibitors of HCV NS5B RNA-dependent RNA polymerase**

An important goal in antiviral therapy is to develop agents that act against different targets, as this will allow combination therapies to be developed that involve different drugs with different mechanisms of action. This would be expected to improve the therapeutic outcome, reduce side effects, and decrease the risk of resistance. To this end, research has been carried out to develop selective agents that will inhibit an HCV enzyme called **NS5B RNA-dependent RNA polymerase**. This enzyme is responsible for catalysing replication of the viral genome, which it does by synthesizing a complementary strand of (−)ssRNA, then using that as the template to generate copies of (+)ssRNA. The enzyme has multiple binding pockets, and various research projects have attempted to develop both competitive and allosteric inhibitors.

Sofosbuvir (Fig. 20.62) is a nucleoside inhibitor of this enzyme that also causes chain termination of the viral

genome, and was approved by the FDA in 2013. Efforts are now being made to design non-nucleoside inhibitors. **Dasabuvir** (Fig. 20.62) is one such agent that has been approved.

20.10.3 **Inhibitors of HCV NS5A protein**

Another target of interest is **non-structural protein 5A** (NS5A). This protein has no catalytic function, but appears crucial to the viral replication cycle. **Daclatasvir** (see Fig. 20.66) is an NS5A inhibitor that gained European approval in 2014. The initial lead compound that led to its discovery was a thiazolidinone structure (BMS-858; Fig. 20.63) which was identified by screening chemical libraries. BMS-858 is a weakly active, but highly selective inhibitor of NS5A. SAR studies demonstrated the importance of the region coloured in blue, and also showed that activity was retained if the alanine moiety was replaced with proline.

It was then discovered that BMS-858 undergoes a radical reaction under biologically conditions to form a dimer,

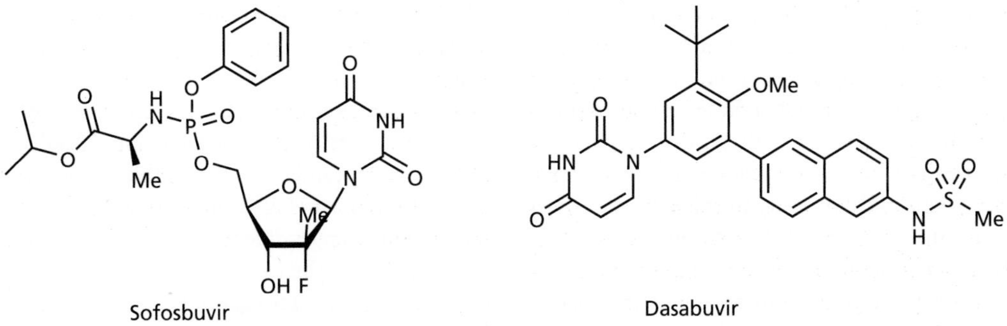

Sofosbuvir

Dasabuvir

FIGURE 20.62 Sofosbuvir and dasabuvir.

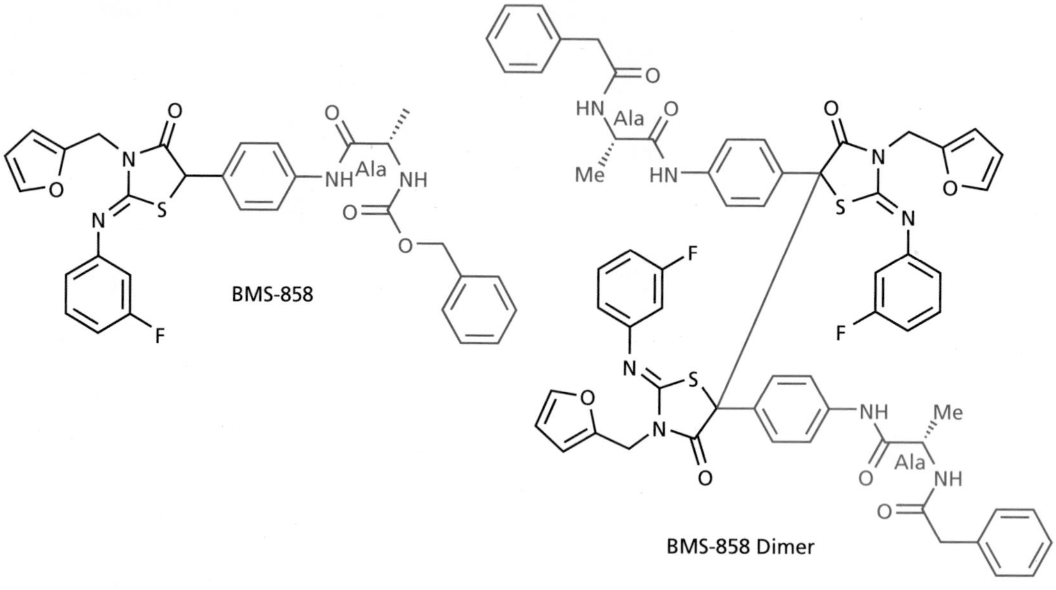

FIGURE 20.63 Initial lead compound.

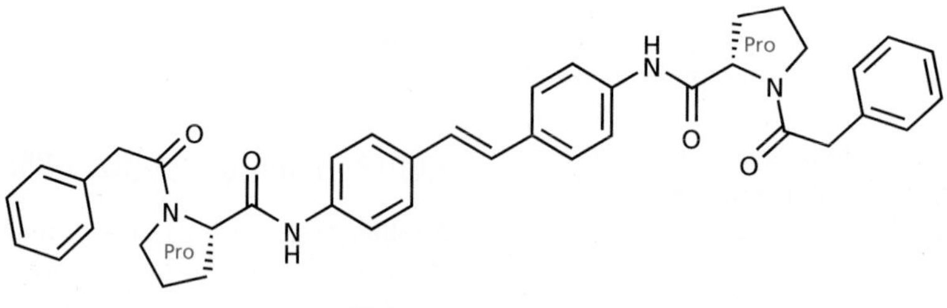

FIGURE 20.64 BMS-346.

and that this was the actual inhibitor. It was decided to simplify the dimer by synthesizing a symmetrical stilbene molecule that incorporated the important regions shown in blue (Fig. 20.63). Replacing the alanine moieties with the more rigid proline residues resulted in a stilbene structure (BMS-346) that was 70 times more potent than BMS-858 (Fig. 20.64). This structure was now adopted as the new lead compound for further development.

Initial SAR studies on a series of stilbenes indicated that shortening the molecule was bad for activity, and that symmetry was important. The importance of symmetry was further supported by studies that showed that the structure interacts with a symmetrical region of the target protein. This region is formed as a result of protein dimerization.

Further work demonstrated that the central alkene group could be replaced with an alkyne group (structure I; Fig. 20.65). Benzimidazole ring systems were then introduced to act as bio-isosteric mimics for the acylated aniline rings (structure II). However, there were concerns that metabolism might still generate toxic aniline ring systems, and so it was decided to remove the alkyne group and link the aromatic rings directly. To maintain the length of the molecule, a bond was inserted between the biphenyl ring

system and the imidazole rings (structure III; Fig. 20.65). The fact that activity was retained during these structural manipulations demonstrated that the central part of the structure is acting as a scaffold, and that the important binding interactions are at each end of the molecule.

The bis-phenylimidazole structure (III) showed good *in vitro* activity, but poor *in vivo* activity. It was proposed that the poor *in vivo* activity was due to poor absorption, and so the phenyl substituents at each end of the structure were replaced with alkyl substituents. This decreased the molecular weight and hydrophobicity of the molecule, as well as increasing aqueous solubility, all of which improved the compound's pharmacokinetic properties. Potent activity was retained by inverting the two chiral centres identified with the stars to give **daclatasvir** (Fig. 20.66), which was granted European approval in 2014, and US approval in 2015. **Ombitasvir** (Fig. 20.67) is another symmetrical inhibitor containing a different central scaffold, and was approved in 2014. The *N*-aryl substituent on the pyrrolidine ring increases potency against HCV gentotypes 1a and 1b by 639- and 23 500-fold respectively.

Further research has shown that symmetry is not essential for good activity. For example, **ledipasvir** (Fig. 20.68)

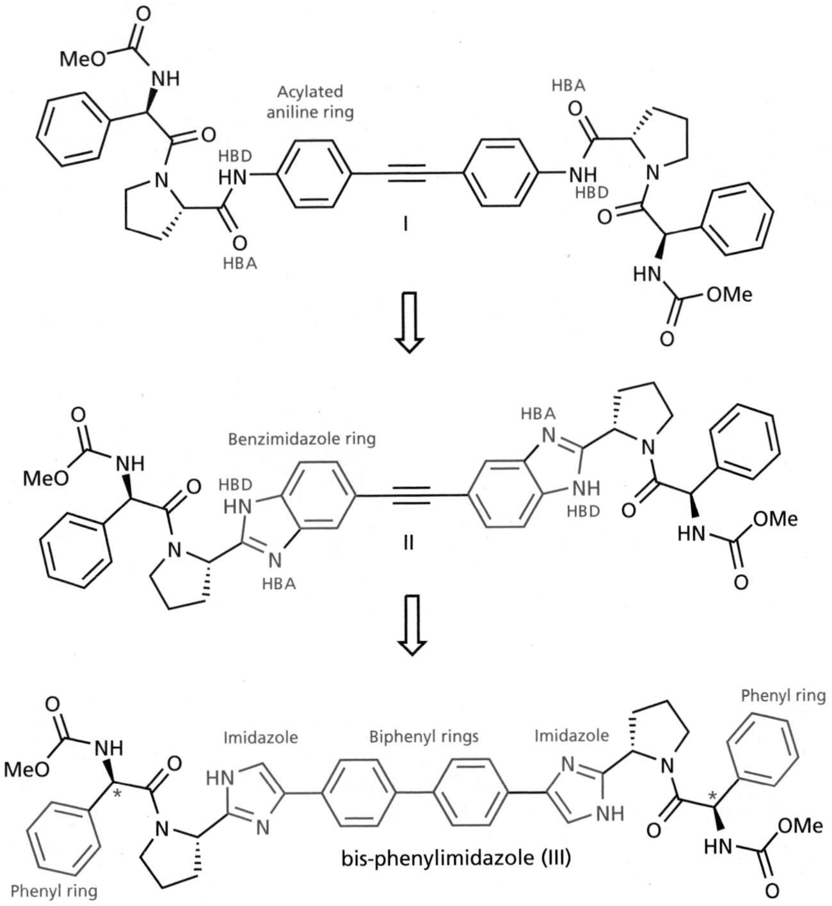

FIGURE 20.65 Development of alkyne and bis-phenylimidazole inhibitors.

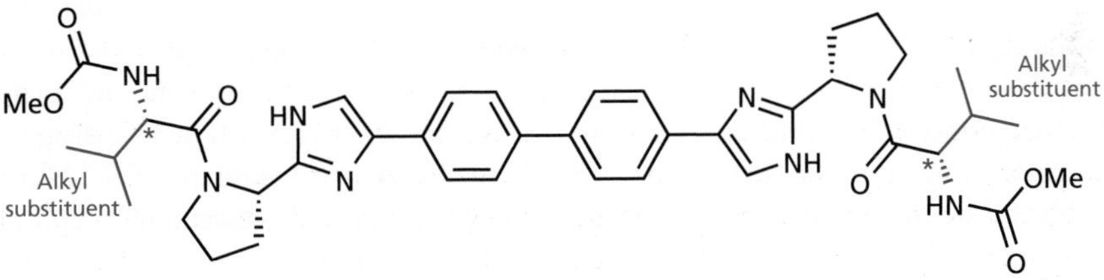

FIGURE 20.66 Daclatasvir (BMS-790,052).

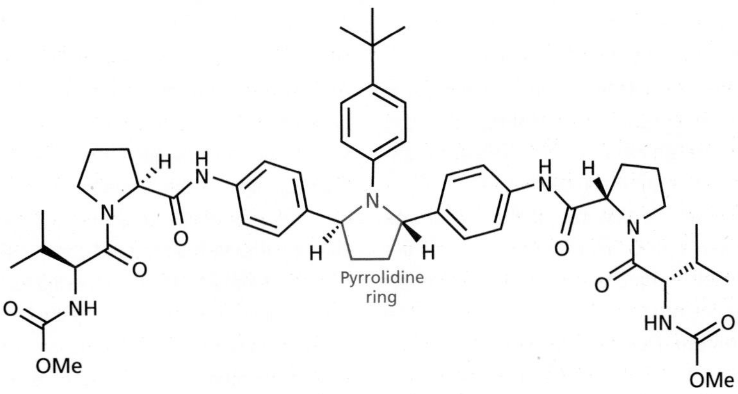

FIGURE 20.67 Ombitasvir (ABT-267).

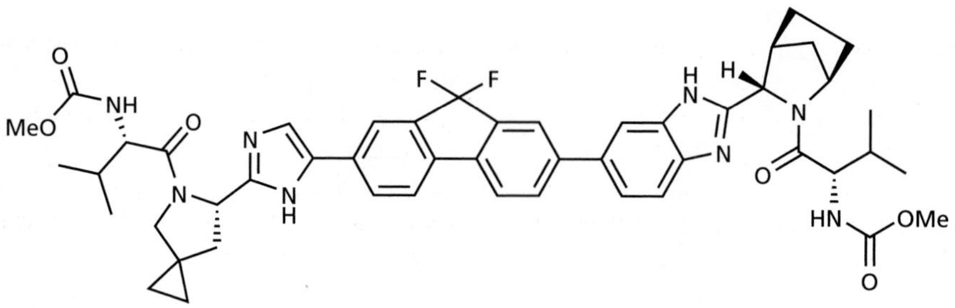

FIGURE 20.68 Ledipasvir (GS-5885).

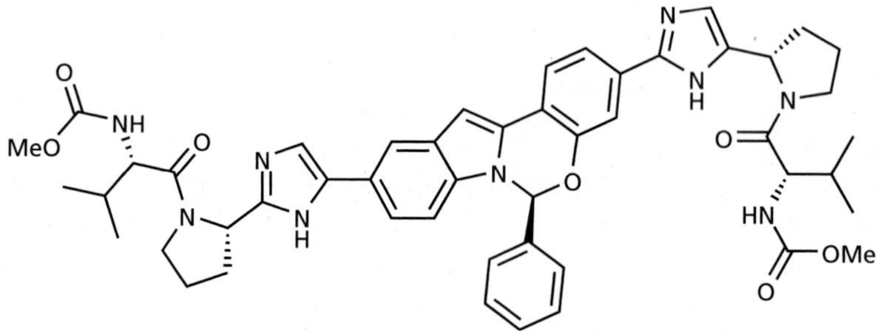

FIGURE 20.69 Elbasvir (MK-8742).

is an unsymmetrical inhibitor that shares many of the features that are present in daclatasvir. It was approved by the FDA in 2014. More recently, **elbasvir** (Fig. 20.69) was approved in January 2016. In both these structures, the central scaffold has been rigidified. This has been achieved with a tricyclic ring system in ledipasvir, and a tetracyclic ring system in elbasvir.

20.10.4 Other targets

A number of research projects have been carried out on agents that target the viral protein NS4B. This is a hydrophobic, membrane-bound protein which has an important role in viral replication. It does so by promoting the aggregation of membranous vesicles from the endoplasmic reticulum to form a membranous web which acts as the site for viral replication.

Studies are also being carried out on agents that act against host targets involved in HCV replication. For example, **cyclosporin A** and its analogues interfere with the interaction between the virus and **cyclophilin A**—a peptidyl *cis-trans* isomerase.

Another compound of interest is **miravirsen**, which is the first micro RNA-targeting drug being developed in humans. Its target is microRNA-122, which binds to the HCV genome and is essential to replication.

BOX 20.5 Clinical aspects of antiviral agents used in the treatment of hepatitis C

The traditional treatment for hepatitis C has been a combination of **ribavarin** with either **interferon alfa** or **peginterferon alfa** (the latter is more effective). Unfortunately, interferon alfa and peginterferon alfa are expensive and have to be administered by injection. The aim in recent years has been to develop orally active agents that can ultimately replace ribavarin and peginterferon alfa. Combination therapies are almost certainly essential to reduce the chances of resistance appearing to any one drug.

Boceprevir and **telaprevir** have proved effective against hepatitis C genotype 1 (gt1), but are less effective against other genotypes. Therefore, it is important to identify the viral genotype present in any infection prior to treatment. Boceprevir or telaprevir have been approved as part of a combination therapy alongside ribavirin and interferon. The disadvantages of this triple combination therapy involve an increased risk of side effects such as anaemia, a more complicated dosing regime, and the possibility of patients being non-compliant. On the other hand, the triple therapy has proved more effective than the traditional dual therapy. The agents are used to treat chronic hepatitis C infection (gt1) in patients with compensated liver disease. Compensated liver disease is where the liver is damaged, but sufficient healthy cells are still present to carry out the liver's normal functions. Decompensated liver disease is where the damage is more serious and the liver fails to function properly.

Sofosbuvir is used in combination with ribavirin, and is effective against hepatitis C infections involving genotypes 1 to 6. A triple therapy can also be used which includes peginterferon alfa. The treatment has been used on patients with compensated liver disease.

Ledipasvir is effective against HCV gentotypes 1a and 1b, and has been approved as a combination therapy with sofosbuvir **(Harvani)**. Ribovarin or peginterferon alfa do not need to be co-administered. One drawback of this dual therapy is the cost, which comes to $1125 per pill.

Elbasvir and **grazoprevir** have been approved as an orally administered combination therapy called **Zepatier**, which can be used with or without ribavarin for the treatment of HCV genotypes 1 and 4. There is no need to include peginterferon alpha in the treatment.

Daclatasvir (Daclinza) has been approved in combination with sofosbuvir for the treatment of gt3 HCV infections. This was the first oral combination therapy for the treatment of gt3 HCV infection that does not require co-administration of ribovarin or peginterferon alfa.

In Japan, **asunaprevir** and **daclatasvir** were approved in 2014 as a dual therapy for the treatment of gt1 HCV infection, including patients with compensated liver disease.

A combination therapy called **Technivie** has been approved for the treatment of HCV genotype 1a and 1b infections. It includes the antiviral agents **paritaprevir**, **ombitasvir**, and **ritonavir**. **Viekira Pak** is a combination therapy that includes these three drugs, as well as **dasabuvir**. Ritonavir is added to inhibit the cytochrome p450 enzyme CYP3A4.

20.11 Broad-spectrum antiviral agents

There are very few clinically useful, broad-spectrum antiviral agents that act on specific targets. The following are some examples.

20.11.1 Agents acting against cytidine triphosphate synthetase

Cytidine triphosphate is an important building block for RNA synthesis, and so blocking its synthesis inhibits the synthesis of viral mRNA. The final stage in the biosynthesis of cytidine triphosphate is the amination of uridine triphosphate—a process that is catalysed by the enzyme cytidine triphosphate synthetase. **Cyclopentenyl cytosine**

(Fig. 20.70) is a carbocyclic nucleoside that is converted in the cell to the triphosphate, which then inhibits this final enzyme in the biosynthetic pathway. The drug has broad antiviral activity against more than 20 RNA and DNA viruses, and has also been studied as an anticancer drug.

20.11.2 Agents acting against S-adenosylhomocysteine hydrolase

The 5′-end of a newly transcribed mRNA is capped with a methyl group in order to stabilize it against phosphatases and nucleases, as well as enhancing its translation. **S-Adenosylhomocysteine hydrolase** is an intracellular enzyme that catalyses this reaction, and many viruses need it to cap their own viral mRNA. **3-Deazaneplanocin A** (Fig. 20.70) is an analogue of cyclopentenyl cytosine, and acts against a range of RNA and DNA viruses by inhibiting S-adenosylhomocysteine hydrolase.

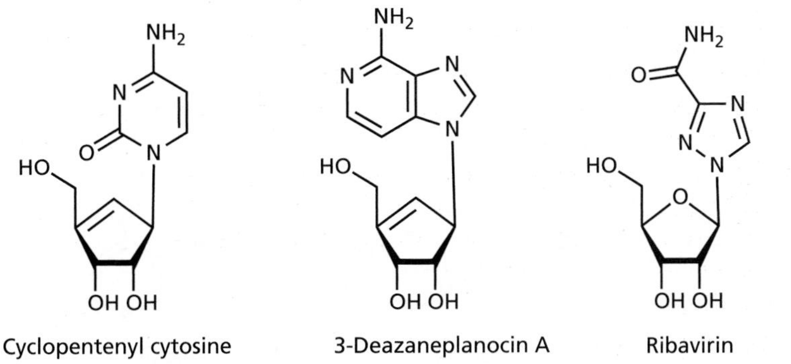

FIGURE 20.70 Broad-spectrum antiviral agents.

20.11.3 Ribavirin

Ribavirin (Fig. 20.70) is a synthetic nucleoside that induces mutations in viral genes and is used against hepatitis C infection (section 20.10). It was the first synthetic, non-interferon-inducing broad-spectrum antiviral nucleoside, and it can inhibit both RNA and DNA viruses by a variety of mechanisms. However, it is only licensed for the treatment of hepatitis C and respiratory syncytial virus. Nevertheless, it has been used in developing countries for the treatment of tropical and haemorrhagic fevers such as Lassa fever when there is no alternative effective treatment. Tests show that it is useful in combination with other drugs such as rimantadine. Its dominant mechanism of action appears to be depletion of intracellular pools of GTP by inhibiting **inosine-5′-monophosphate dehydrogenase**. Phosphorylation of ribavirin results in a triphosphate which inhibits **guanyl transferase** and prevents the 5′ capping of mRNAs. The triphosphate can also inhibit **viral RNA-dependent RNA polymerase**. Because of these multiple mechanisms of action, resistance is rare. The drug's main side effect is anaemia, and it is a suspected teratogen.

20.11.4 Interferons

Interferons are small natural proteins which were discovered in 1957, and are produced by host cells as a response to 'foreign invaders'. Once produced, interferons inhibit protein synthesis and other aspects of viral replication in infected cells. In other words, they shut the cell down. This can be described as an intracellular immune response. Administering interferons to patients has been seen as a possible approach to treating flu, hepatitis, herpes, and colds.

There are several interferons which are named according to their source: α-interferons from lymphocytes, β-interferons from fibroblasts, and γ-interferons from T-cells. α-Interferon (also called **alferon** or **IFN-α**) is the most widely used of the three types. In the past, it was difficult and expensive to isolate interferons from their natural cells, but recombinant DNA techniques allow the production of genetically engineered interferons in larger quantities (section 6.4). Recombinant α-interferon is produced in three main forms. The α-2a and α-2b are natural forms, and **alfacon-1** is the unnatural form. They have proved successful therapeutically, but can have serious toxic side effects. At present, α-interferon is used clinically against hepatitis B infections. It is also used with ribavirin against hepatitis C infections.

Interferon production in the body can also be induced by agents known as **immunomodulators**. One such example is **avridine** (Fig. 20.71) which is used as a vaccine adjuvant for the treatment of animal diseases such as foot and mouth. **Imiquimod** (Fig. 20.71) also induces the production of α-interferon, as well as other cytokines that stimulate the immune system. It is effective against genital warts.

20.11.5 Antibodies and ribozymes

Antibodies that recognize a virion-specific antigen will bind to that antigen and mark the virus out for destruction by the body's immune system. **Palivizumab** is a humanized monoclonal antibody which was approved in 1998 for the treatment of respiratory syncytial infection in babies. It blocks viral spread from cell to cell by targeting a specific protein of the virus. Another monoclonal antibody is being tested for the treatment of hepatitis B.

It has been possible to identify sites in viral RNA that are susceptible to cutting by ribozymes—enzymatic forms of RNA. One such ribozyme is being tested in patients with hepatitis C and HIV. Ribozymes could be generated in the cell by introducing genes into infected cells—a form of gene therapy. Other gene therapy projects are looking at genes that would:

i) code for specialized antibodies capable of seeking out targets inside infected cells, or

ii) code for proteins that would latch on to viral gene sequences within the cell.

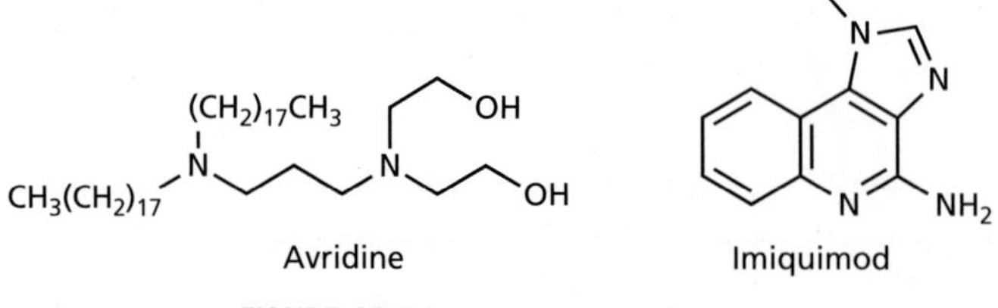

FIGURE 20.71 Immunomodulators.

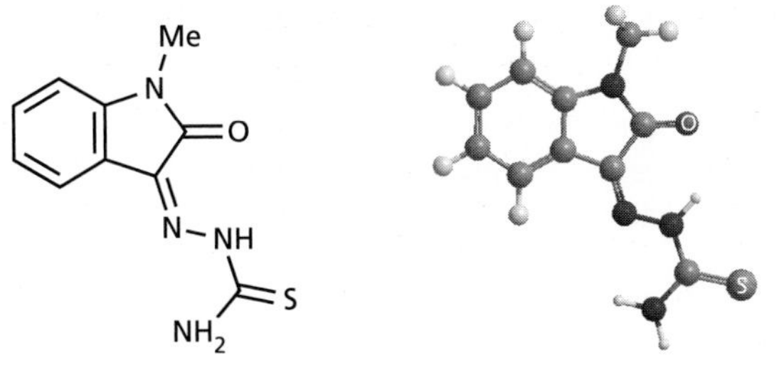

FIGURE 20.72 Methisazone together with ball and stick model.

20.12 Bioterrorism and smallpox

Methisazone (Fig. 20.72) was the first effective antiviral drug to reach the clinic and was used to treat smallpox in the 1960s. The drug was no longer required once the disease was eradicated through worldwide vaccination. In recent years, however, there have been growing worries that terrorists might acquire smallpox and unleash it on a world no longer immunized against the disease. As a result, there has been a regeneration of research into finding novel antiviral agents which are effective against this disease.

KEY POINTS

- There are few broad-spectrum antiviral agents currently available.
- The best broad-spectrum antiviral agents appear to work on a variety of targets, reducing the chances of resistance.
- Interferons are chemicals produced in the body which shut down infected host cells and limit the spread of virus.
- Antibodies and ribozymes are under investigation as antiviral agents.

QUESTIONS

1. Consider the structures of the PIs given in section 20.7.4 and suggest a hybrid structure that might also act as a PI.

2. Consider the structure of the PIs in section 20.7.4 and suggest a novel structure with an extended subsite ligand.

3. What disadvantage might the following structure have as an antiviral agent compared with cidofovir?

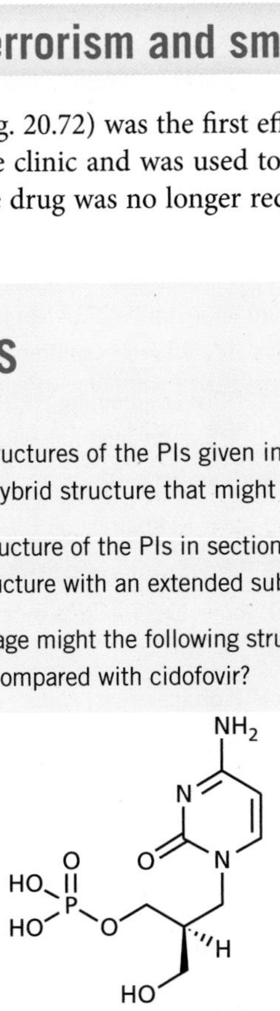

4. Zanamivir has a polar glycerol side chain which forms good hydrogen bonding interactions with a binding pocket, yet carboxamides and oseltamivir have hydrophobic substituents which bind more strongly to this pocket. How is this possible?

5. Show the mechanism by which the prodrugs tenofovir disoproxil and adefovir dipivoxil are converted to their active forms. Why are extended esters used as prodrugs for these compounds?

6. Most PIs bind to the active site with a water molecule acting as a hydrogen bonding bridge to the enzyme flaps. Suggest what relevance this information might have in the design of novel PIs.

7. The following structures were synthesized during the development of L 685434 (Fig. 20.28). Identify the differences between the two structures and suggest why one is more active than the other.

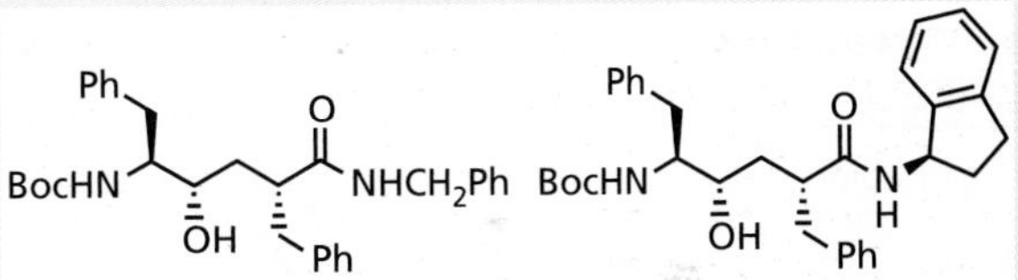

8. Capravirine is a third-generation non-nucleoside reverse transcriptase inhibitor (NNRTI) with a side chain that takes part in important hydrogen bonding to Lys-103 and Pro-236 in the allosteric binding site, yet the side chain has a carbonyl group. Discuss whether this makes the structure prone to enzymatic hydrolysis and inactivation.

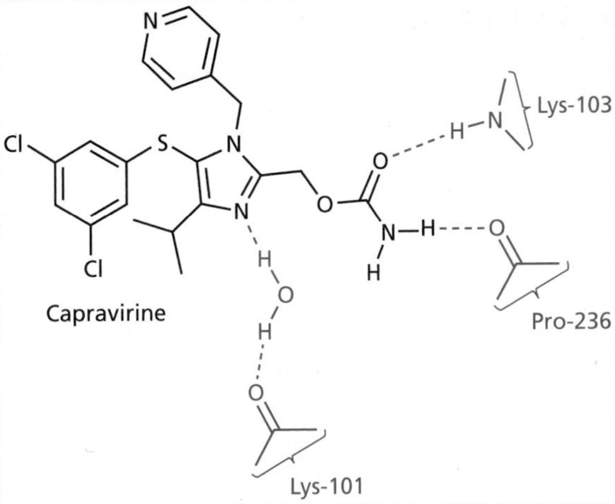

9. The mechanism used by HCV NS3-4A protease to cleave its protein substrate involves a catalytic triad of Ser-139, His-57, and Asp-81. Show how all three amino acids are involved in the catalytic mechanism of hydrolysis.

10. A chiral centre in boceprevir is easily epimerized. Which one is it and why is it so easily epimerized? What implications might this have on antiviral testing? How could this problem of epimerization be prevented?

Multiple-choice questions are available on the Online Resource Centre at www.oxfordtextbooks.co.uk/orc/patrick6e/

FURTHER READING

Carr, A. (2003) Toxicity of antiretroviral therapy and implications for drug development. *Nature Reviews Drug Discovery*, **2**, 624–34.

Coen, D. M., and Schaffer, P. A. (2003) Antiherpesvirus drugs: a promising spectrum of new drugs and drug targets. *Nature Reviews Drug Discovery*, **2**, 278–88.

De Clercq, E. (2002) Strategies in the design of antiviral drugs. *Nature Reviews Drug Discovery*, **1**, 13–25.

Driscoll, J. S. (2002) *Antiviral agents*. Ashgate, Aldershot.

Milroy, D., and Featherstone, J. (2002) Antiviral market overview. *Nature Reviews Drug Discovery*, **1**, 11–12.

Tan, S.-L., et al. (2002) Hepatitis C therapeutics: current status and emerging strategies. *Nature Reviews Drug Discovery*, **1**, 867–81.

Venkatraman, S., et al. (2006) Discovery of a selective, potent, orally bioavailable hepatitis C virus ND3 protease inhibitor: A potential therapeutic agent for the treatment of Hepatitis C infection. *Journal of Medicinal Chemistry*, **49**, 6074–86 (boceprevir).

HIV

Campiani, G., et al. (2002) Non-nucleoside HIV-1 reverse transcriptase (RT) inhibitors: past, present and future perspectives. *Current Pharmaceutical Design*, **8**, 615–57.

De Clercq, E. (2003) The bicyclam AMD3100 story. *Nature Reviews Drug Discovery*, **2**, 581–7.

Dubey, S., Satyanarayana, Y. D., and Lavania, H. (2007) Development of integrase inhibitors for treatment of AIDS: an overview. *European Journal of Medicinal Chemistry*, **42**, 1159–68.

Ghosh, A. K., Dawson, Z. L., and Mitsuya, H. (2007) Darunavir, a conceptually new HIV-1 protease inhibitor for the treatment of drug-resistant HIV. *Bioorganic and Medicinal Chemistry*, **15**, 7576–80.

Ghosh, A. K. (2009) Harnessing nature's insight: design of aspartyl protease inhibitors from treatment of drug-resistant HIV to Alzheimer's disease. *Journal of Medicinal Chemistry*, **52**, 2163–76 (darunavir and amprevanir).

HIV databases: http://www.hiv.lanl.gov/content/index (last accessed April 2012).

Matthews, T., et al. (2004) Enfuvirtide: the first therapy to inhibit the entry of HIV-1 into host CD4 lymphocytes. *Nature Reviews Drug Discovery*, **3**, 215–25.

Miller, J. F., Furfine, E. S., Hanlon, M. H., et al. (2004) Novel arylsulfonamides possessing sub-picomolar HIV protease activities and potent anti-HIV activity against wild-type and drug-resistant viral strains. *Bioorganic and Medical Chemistry Letters*, **14**, 959–63 (amprenavir).

Price, D. A., et al. (2006) Overcoming HERG affinity in the discovery of the CCR5 antagonist maraviroc. *Bioorganic and Medicinal Chemistry Letters*, **16**, 4633–7.

Raja, A., Lebbos, J., and Kirkpatrick, P. (2003) Atazanavir sulphate. *Nature Reviews Drug Discovery*, **2**, 857–8.

Sansom, C. (2009) Molecules made to measure. *Chemistry World*, November, 50–3.

Werber, Y. (2003) HIV drug market. *Nature Reviews Drug Discovery*, **2**, 513–4.

Flu

Ezzell, C. (2001) Magic bullets fly again. *Scientific American*, October, 28–35 (antibodies).

Sansom, C. (2011) Fighting the flu. *Chemistry World*, February, 44–7.

Hepatitis

Manns, M. P., and van Hahn, T. (2013) Novel therapies for hepatitis C—one pill fits all? *Nature Reviews Drug Discovery*, **12**, 595–610.

Cannalire, R., Barreca, M. L., Manfroni, G., and Cecchetti, V. (2016) A journey around the medicinal chemistry of hepatitis C virus inhibitors targeting NS4B: from target to preclinical drug candidates. *Journal of Medicinal Chemistry*, **59**, 16–41.

Titles for general further reading are listed on p.845.

Anticancer agents

21.1 Cancer: an introduction

21.1.1 Definitions

Cancer still remains one of the most feared diseases in the modern world. According to the World Health Organization, it accounted for 8.2 million deaths worldwide in 2012, with 14 million new cases reported that year. It is anticipated that in the next 20 years, the number of new cancer cases will increase to 22 million per year. Lung cancer is the most prevalent cause of death, followed by liver, stomach, colorectal, breast, and oesophageal cancers. Tobacco use is the most common risk factor, and 70% of cancer deaths occur in Africa, Asia, and Central and South America.

Cancer cells are formed when normal cells lose the normal regulatory mechanisms that control growth and multiplication. They become 'rogue cells' and often lose the specialized characteristics that distinguish one type of cell from another (for example a liver cell from a blood cell). This is called a loss of **differentiation**. The term **neoplasm** means new growth and is a more accurate terminology for the disease. The terms **cancer** and **tumour**, however, are more commonly accepted and will be used throughout this chapter. (The word tumour actually means a local swelling.) If the cancer is localized, it is said to be **benign**. If the cancer cells invade other parts of the body and set up secondary tumours—a process known as **metastasis**—the cancer is defined as **malignant**. It is malignant cancer that is life threatening. A major problem in treating cancer is the fact that it is not a single disease. There are more than 200 different cancers resulting from different cellular defects, and so a treatment that is effective in controlling one type of cancer may be ineffective on another.

21.1.2 Causes of cancer

Possibly as many as 30% of cancers are caused by smoking, while another 30% are diet related. Carcinogenic chemicals in smoke, food, and the environment may cause cancer by inducing gene mutations or interfering with normal cell differentiation. The birth of a cancer (**carcinogenesis**) can be initiated by a chemical—usually a **mutagen**—but other triggering events, such as exposure to further mutagens, are usually required before a cancer develops.

Viruses have been implicated in at least six human cancers and are the cause of about 20% of the world's cancer deaths in low and middle-income countries. For example, the Epstein–Barr virus is the cause of Burkitt's lymphoma and nasopharyngeal carcinoma. Human papillomaviruses are sexually transmitted and can lead to cancer of the cervix. Hepatitis B may cause 80% of all liver cancers, and HIV can cause Kaposi's sarcoma and lymphoma. Viruses can bring about cancer in several ways. They may bring oncogenes (see section 21.1.3) into the cell and insert them into the genome. For example, Rous sarcoma virus carries a gene for an abnormal tyrosine kinase. Some viruses carry one or more promoters or enhancers. If these are integrated next to a cellular oncogene, the promoter stimulates its transcription leading to cancer. The bacterium *Helicobacter pylori* is responsible for many stomach ulcers (section 25.4) and is also implicated in stomach cancer.

The treatments used to combat cancer (radiotherapy and chemotherapy) can actually induce a different cancer in surviving patients. For example, 5% of patients cured of Hodgkin's disease developed acute leukaemia. Nevertheless, the risk of a second cancer is outweighed by the benefit of defeating the original one.

Some patients are prone to certain cancers for genetic reasons. Damaged genes can be passed from one generation to another, increasing the risk of cancer in subsequent generations (e.g. certain breast cancers).

21.1.3 Genetic faults leading to cancer: proto-oncogenes and oncogenes

21.1.3.1 Activation of proto-oncogenes

Proto-oncogenes are genes which normally code for proteins involved in the control of cell division and differentiation. If they are mutated, this disrupts the normal function and the cell can become cancerous. The proto-oncogene is then defined as an **oncogene**. The *ras* gene is one example. Normally, it codes for a protein called **Ras** which is involved in the signalling pathway leading to cell division (section 5.4.2). In normal cells, this protein has a self-regulating ability and can 'switch itself off'. If the gene becomes mutated, an abnormal Ras protein is produced which loses this ability and is continually active, leading to continuous cell division. It has been shown that mutation of the *ras* gene is present in 20–30% of human cancers. Oncogenes may also be introduced to the cell by viruses.

21.1.3.2 Inactivation of tumour suppression genes (anti-oncogenes)

If DNA is damaged in a normal cell, there are cellular 'policemen' that can detect the damage and block DNA replication. This gives the cell time to repair the damaged DNA before the next cell division. If repair does not prove possible, the cell commits suicide (**apoptosis**). Tumour suppression genes are genes which code for proteins that are involved in these processes of checking, repair, and suicide. *TP53* is an important example of such a gene and codes for the protein of the same name (**p53**). If the *TP53* gene is damaged, the repair mechanisms become less efficient, defects are carried forward from one cell generation to another, and as the damage increases, the chances of the cell becoming cancerous increase.

21.1.3.3 The consequences of genetic defects

Genetic defects can lead to the following cellular defects, all of which are associated with cancer:

- abnormal signalling pathways;
- insensitivity to growth-inhibitory signals;
- abnormalities in cell cycle regulation;
- evasion of programmed cell death (apoptosis);
- limitless cell division (immortality);
- ability to develop new blood vessels (angiogenesis);
- tissue invasion and metastasis.

It is thought that most, if not all, of these conditions have to be met before a defective cell can spawn a life-threatening malignant growth. Thus, a single defect can be kept under control by a series of safeguards. This can explain why cancers may take many years to develop after exposure to a damaging mutagen, such as asbestos or coal dust. That first exposure may have caused mutations in some cells, but cellular chemistry has the control systems in place to cope and to keep the cells in check. However, a

lifetime's exposure to other damaging mutagens, such as tobacco smoke, results in further genetic damage which overwhelm the safeguards one by one until the abnormal cell finally breaks free of its shackles and becomes cancerous.

The various hurdles and safeguards that a potential cancer cell has to overcome explain why cancers are relatively rare early on in life and are more common in later years. This also helps to explain why cancers are so difficult to treat once they *do* appear. Because so many cellular safeguards have already been overcome, it is unlikely that tackling one specific cellular defect is going to be totally effective. As a result, traditional anticancer drugs have tended to be highly toxic agents that act against a variety of different cellular targets by different mechanisms. Unfortunately, because they are potent cellular poisons, they also affect normal cells and produce serious side effects. Such agents are said to be **cytotoxic** and dose levels have to be chosen which are high enough to affect the tumour, but are bearable to the patient. In recent years, anticancer drugs have been developed which target specific abnormalities in a cancer cell, allowing them to be more selective and have less serious side effects. However, bearing in mind the number of defects in a cancer cell, it is unlikely that a single agent of this kind will be totally effective, and it is more likely that these new agents will be most effective when they are used in combination with other drugs having different mechanisms of action, or with surgery and radiotherapy.

We now look at the various defects that are common in cancer cells.

21.1.4 Abnormal signalling pathways

Whether a normal cell grows and divides depends on the various signals it receives from surrounding cells. The most important of these signals come from hormones called growth factors. These are extracellular chemical messengers which activate protein kinase receptors in the cell membrane (sections 4.8 and 5.4). The receptors concerned trigger a signal transduction pathway which eventually reaches the nucleus and instructs the transcription of the proteins and enzymes required for cell growth and division. Most, if not all, cancers suffer from some defect in this signalling process such that the cell is constantly instructed to multiply. The signalling process is complex, so there are various points at which it can go wrong.

Many cancer cells are capable of growing and dividing in the absence of external growth factors. They can do this by producing the growth factor themselves, then releasing it such that it stimulates its own receptors, often by autophosphorylation. Examples include **platelet-derived growth factor** (PDGF) and **transforming** growth factor α (TGF-α). Other cancer cells can produce abnormal receptors which are constantly switched on despite the lack of growth factors (e.g. **ErB-2 receptors** in breast cancer cells). It is also possible for receptors to be overexpressed. This means that an oncogene is too active and codes for excessive protein receptor. Once these receptors are in the cell membrane, the cell becomes supersensitive to low levels of circulating growth factor.

There are many points where things could go wrong in the signal transduction pathways. For example, the **Ras** protein is a crucial feature in the signal transduction pathways leading to cell growth and division. Abnormal Ras protein is locked in the 'on' position and is constantly active despite the lack of an initial signal from a growth factor.

21.1.5 Insensitivity to growth-inhibitory signals

Several external hormones such as **transforming growth factor β** (TGF-β) counteract the effects of stimulatory growth factors, and signal the inhibition of cell growth and division. Insensitivity to these signals raises the risk of a cell becoming cancerous. This can arise from damage to the genes coding for the receptors for these inhibitory hormones—the tumour suppression genes.

21.1.6 Abnormalities in cell cycle regulation

A cycle of events takes place during cell growth and multiplication which involves four phases known as G_1, S, G_2, and M (Fig. 21.1). As part of this process, decisions have to be made by the cell whether to move from one stage to another, depending on the balance of those chemical signals promoting growth and those inhibiting it.

The G_1 phase (gap 1) is where a cell is actively growing in size and preparing to copy its DNA in response to various growth factors or internal signals. The next phase is the S phase (synthesis) where replication of DNA takes place. Once the cell's chromosomes are copied, there is another interval called the G_2 phase (gap 2) during which the cell readies itself for cell division. This gap, or interval, is crucial, as it gives the cell time to check the copied DNA and to repair any damaged copies. Finally, there is the M phase (mitosis) where cell division takes place to produce two daughter cells, each containing a full set of chromosomes. The daughter cells can then enter the cell cycle again (G_1). Alternatively, they may move into a dormant or resting state (G_0).

Within the cell cycle, there are various decision points which determine whether the cell should continue to the next phase. For example, there is a decision point called the **restriction point** (R) during the G_1 phase which frequently becomes abnormal in tumour cells. There are also various surveillance mechanisms known as checkpoints which assess the integrity of the process. For example, a delay will take place during the G_2 phase if DNA damage is detected. This gives sufficient time for damaged DNA to be repaired or for the cell to commit suicide (**apoptosis**). These checkpoints can also be defective in tumour cells.

Control of the cell cycle involves a variety of proteins called **cyclins** and enzymes called **cyclin-dependent kinases (CDKs)** (Fig. 21.2). There are at least 15 types of cyclin and 9 types of CDK, and each has a role to play at different stages of the cell cycle. Examples are shown in Fig. 21.2. Binding of a cyclin with its associated kinase activates the enzyme and serves to move the cell from one phase of the cell cycle to another. For example, when a cell is in the G_1 phase, a decision has to be made whether to move into the S phase and start copying DNA. This decision is taken depending on the balance of stimulatory versus inhibitory signals being received through

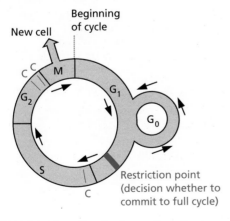

FIGURE 21.1 The cell cycle. G_1, gap 1—cell enlarges and makes new proteins; S, synthesis of DNA; G_2, gap 2—cell prepares to divide; M, mitosis—cell divides; G_0, resting stage—no growth; C, checkpoints.

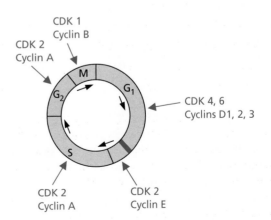

FIGURE 21.2 Control of the cell cycle by cyclins and cyclin-dependent kinases.

signal transduction. If the balance is towards cell growth and division, there is an increase in **cyclin D**. This binds to **CDK4** and **CDK6**. The resulting complexes phosphorylate a powerful growth-inhibitory molecule known as **pRB** which normally binds and inactivates a transcription factor. Phosphorylation alters pRB such that it can no longer bind to the transcription factor and the latter is free to bind to specific regions of DNA. This results in the transcription of specific genes which leads to the production of proteins capable of moving the cell towards the S phase (e.g. **cyclin E** and **thymidine kinase**). Once cyclin E has been produced, it combines with **CDK2** and this complex is responsible for progression from the G_1 phase to the S phase. Other activated cyclin–CDK complexes are important in different phases of the cell cycle. For example, the **cyclin A–CDK2** complex is required for progression through the S phase, and a **cyclin B–CDK1** complex is necessary for mitosis.

Restraining proteins are present which can modify the effect of cyclins (Fig. 21.3). These include **p15** and **p16** which block the activity of the cyclin D–CDK complex. Another is the inhibitory protein **p21** which is controlled by **p53**—an important protein that monitors the health of the cell and the integrity of DNA.

To sum up, progression through the cell cycle is regulated by sequential activation of cyclins and CDKs—a process which can be down-regulated by the CDK inhibitors. The whole process is normally tightly controlled, such that there is an accumulation of a relevant cyclin–CDK complex followed by rapid degradation of the complex once its task is complete.

Overactive cyclins or CDKs have been associated with several cancers. For example, breast cancer cells often produce excess cyclins D and E, and skin melanoma has lost the gene that codes for the inhibitory protein p16. Half of all human tumours lack a proper functioning p53 protein, which means that the level of the inhibitory protein p21 falls. In viral-related cervical cancers both the pRB and p53 proteins are often disabled.

Oncogenic alteration of cyclins, CDKs, **cyclin-dependent kinase inhibitors (CKIs)**, and other components of the pRB pathway have been reported in 90% of human cancers, especially in the G_1 phase. Thus, excessive production of cyclins or CDKs, or insufficient production of CKIs, can lead to a disruption of the normal regulation controls and result in cancer. Efforts have been made to identify how one can restore the control of the cancer cell cycle by targeting molecular abnormalities. These can include CDK inhibition, down-regulation of cyclins, up-regulation of CDK inhibitors, degradation of cyclins, or inhibition of tyrosine kinases that trigger the cell cycle activation in the first place.

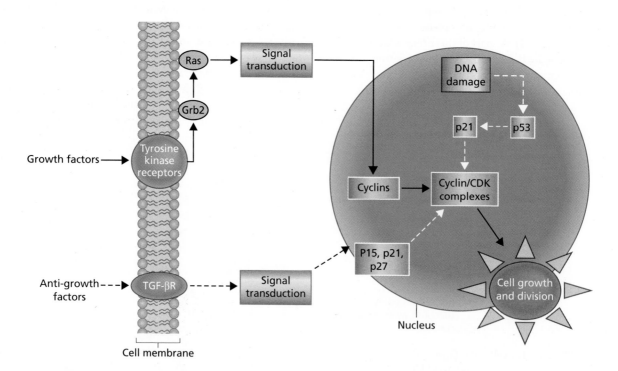

FIGURE 21.3 Cell signalling pathways. Solid arrows indicate pathways that stimulate cell growth and division. Dashed arrows indicate pathways that inhibit cell growth and division.

21.1.7 **Apoptosis and the p53 protein**

There is a built-in cellular destruction process called **apoptosis**, which is the normal way in which the body protects itself against abnormal or faulty cells. Essentially, each cell monitors itself for a series of different chemical signals. Should any of these be absent, a self-destruct mechanism is automatically initiated (Fig. 21.4). Apoptosis is also important in destroying cells that escape from their normal tissue environment. Cancer cells which metastasize have undergone genetic changes that allow them to avoid this process.

Two distinct pathways for apoptosis have been characterized.

- An **extrinsic route** where apoptosis results from external factors. This can take three forms. Firstly, there could be a sustained lack of growth factors or hormones. Secondly, there are proteins called **death activator proteins** which can bind to cell membrane proteins called **tumour necrosis factor receptors** (TNFR). This triggers a signalling process initiating apoptosis. Finally, the immune system produces T-lymphocytes which circulate the body searching for damaged cells. Once found, the lymphocyte perforates the cell membrane of the damaged cell and injects an enzyme called **granzyme**, which initiates apoptosis.

- An **intrinsic pathway** can arise from factors such as DNA damage arising from exposure to chemicals, drugs, or oxidative stress. The cell has monitoring systems which can detect damage and lead to the increased production of the tumour suppressor protein **p53**. At sufficient levels, this protein will trigger apoptosis.

The various signals described above converge on the mitochondria which contain proteins capable of promoting apoptosis, in particular **cytochrome c**. Release of cytochrome c from mitochondria results in the assembly of a large oligomeric protein complex known as an **apoptosome**, which is made up of a scaffolding protein called **Apaf-1**. The apoptosome then recruits and activates an enzyme known as **procaspase 9**, which, in turn, activates **caspases**. Caspases are protease enzymes containing a cysteine residue in the active site which is important to the catalytic mechanism. Because they are proteases, they set about destroying the cell's proteins, which leads to destruction of the cell.

Considering the fatal effect that caspases have on the cell, it is not surprising that there are various checks and balances to ensure that apoptosis does not occur too readily. A family of proteins regulates the process. Proteins such as **Bad** and **Bax** promote it, while others, such as **Bcl-2** and **Bcl-X,** suppress it. The relative levels of

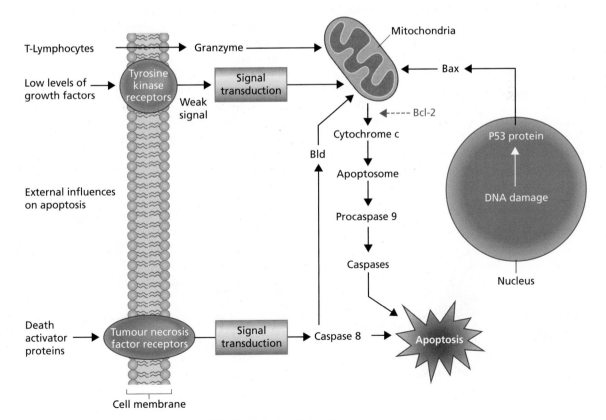

FIGURE 21.4 Signals leading to apoptosis.

these proteins are dependent on the various monitoring procedures within the cell. For example, genetic damage leading to increased levels of p53 induces apoptosis by up-regulating the expression of Bax.

The survival of each cell in the body is, therefore, dependent on the balance of internal and external signals regulating cell growth, as well as the balance of regulatory chemicals promoting or inhibiting apoptosis. A defect in the complex systems leading to apoptosis could inhibit apoptosis, increasing the likelihood of carcinogenesis. For example, it has been found that the gene coding for p53 is the most frequently mutated gene in cancer (30–70%). Damage to this gene means a lack of the apoptosis-inducing p53 protein and an increased chance that the defective cell will survive to become cancerous. The genes coding for the apoptosis suppressors Bcl-2 and Bcl-X$_L$ are also known to be overexpressed in several tumour types. Another genetic defect that has been found in many tumour cells is the overexpression of **HDM2**, a protein that binds to p53 and prevents it from functioning as a transcription factor (section 10.5).

Defects in the apoptosis mechanisms also have serious consequences for radiotherapy and many chemotherapeutic drugs, as both these procedures act by triggering apoptosis. For example, many traditional anticancer drugs damage DNA. This, in itself, may not be fatal to the cell, but the cell's monitoring system detects the damage and goes into self-destruct mode. If the mechanisms involved are defective then apoptosis does not occur, the drugs are not effective, and the cell becomes immortal.

21.1.8 Telomeres

Cancer cells are often described as becoming 'immortal'. This is because there is no apparent limit to the number of times they can divide. The lifetime of normal cells is predetermined by the possible number of times their DNA can be replicated (about 50–60 cell divisions).

Structures called **telomeres** play a key role in this immortalization process. Figure 21.5 shows the structure of a **chromatin**, which consists of a chromosome wrapped round a variety of proteins. The telomere consists of a polynucleotide region at the 3′ end of a chromosome, and contains several thousand repeats of a short (6 base pair) sequence. The purpose of the telomere is to act as a 'splice' for the end of the chromosome and to stabilize and protect the DNA. After each replication process, about 50–100 base pairs are lost from the telomere because **DNA polymerase** is unable to completely replicate the 3′ ends of chromosomal DNA. Eventually, the telomere becomes too short to be effective and the DNA becomes unstable, either unravelling or linking up with another DNA end to end. This proves fatal to the cell and apoptosis is triggered.

It is observed that in the early stages of cancer many cancer cells are also restricted to the number of times they can divide, but eventually a cancer cell develops which breaks free of this restriction and becomes immortal. These cells maintain the length of their telomere by expressing an enzyme called **telomerase**—a member of a group of enzymes called the **RNA-dependent DNA polymerases**. Telomerase has the ability to add hexanucleotide repeats on to the end of telomeric DNA and thus maintain its length. This is an important process during the development of an embryo when telomerase is responsible for creating the telomeres in the first place, but, after birth, the gene encoding the enzyme is suppressed. Immortal cells have found a way of removing that suppression such that the enzyme is expressed once more. The telomerase enzyme is expressed in over 85% of cancers.

Several efforts have been made to design drugs which will inhibit telomerase, but, to date, none have reached the clinic.

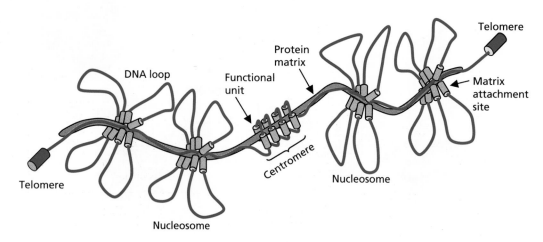

FIGURE 21.5 Chromatin, telomeres, and DNA.

21.1.9 Angiogenesis

As a tumour grows, its cancerous cells require a steady supply of amino acids, nucleic acid bases, carbohydrates, oxygen, and growth factors if they are to continue multiplying. This means that the tumour has to have a good blood supply. As a tumour grows in size, however, its cells become increasingly remote from the blood supply and become starved of these resources. Oxygen levels also fall, resulting in a state of **hypoxia**. This is particularly true for the cells in the centre of the tumour. As a result of this, **hypoxia-inducible factors**, such as **HIF-1**, start to build up within the tumour cells, and these factors serve to up-regulate genes that promote survival in oxygen-starved environments. For example, growth factors are released from the cell, such as **vascular endothelial growth factor (VEGF)** and **fibroblast growth factor (FGF-2)**, which interact with receptors on the endothelial cells of nearby blood vessels and stimulate these cells to divide, leading to the branching and extension of existing capillaries—a process known as **angiogenesis** (Fig. 21.6). Vascular growth factors are present in normal cells and are usually released when tissues have been damaged. The resulting angiogenesis helps in the repair of the injured tissues and is normally controlled by angiogenesis inhibitors such as **angiostatin** and **thrombospondin**. Unfortunately, this balance is disturbed in tumour growth. As a result, tumours are able to receive the increased blood supply required for their survival. Moreover, the chances of cancer cells escaping from the primary source and metastasizing are increased, not only because of the increased availability of blood vessels, but also because the newly developing endothelial cells can release proteins, such as **interleukin-6**, that stimulate metastasis. The blood vessels arising from angiogenesis are abnormal in that they are disorganized in structure, dilated, and leaky. The cells also display molecules called

integrins on their surface which are absent from mature vessels and which protect the new cells from apoptosis. Before angiogenesis can begin, the basement membrane round the blood vessels has to be broken down and this is carried out by enzymes known as **matrix metalloproteinases (MMPs)**. This then allows the endothelial cells to migrate towards the tumour. Dissolution of the matrix also allows angiogenesis factors to be released to encourage angiogenesis.

Inhibiting angiogenesis is a tactic which can help to tackle cancer, and drugs have been developed for that very purpose. Angiogenesis inhibitors are generally safer and less toxic than traditional chemotherapeutic agents, but are unlikely to be used on their own. Instead, they will probably be used alongside standard cancer treatments, such as surgery, chemotherapy, and radiation. Angiogenesis inhibitors appear to 'normalize' the abnormal blood vessels of tumours before they kill them. This normalization can help anticancer agents reach tumours more effectively. In the longer term, it serves to stall tumour growth then shrink it by breaking up abnormal capillaries. As a result, the tumour becomes starved of nutrients and growth should decrease.

Some anticancer treatments take advantage of the leaky blood vessels which result from angiogenesis. Anticancer drugs can be encapsulated into liposomes, nanospheres, and other drug delivery systems which are too big to escape from normal blood vessels, but can escape from the leakier blood vessels supplying the tumour. As a result, the anticancer drug is concentrated at the tumour. Since tumours generally do not develop an effective lymphatic system, the polymeric drug delivery systems tend to be trapped at the tumour site.

Even with angiogenesis, there are regions of a well-developed tumour which fail to receive an adequate blood supply. As a result, cells in the centre of the tumour are starved of oxygen and nutrients, and may well stop

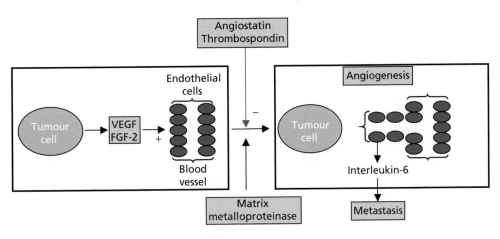

FIGURE 21.6 Angiogenesis.

growing and become dormant. This can pose a serious problem, as most anticancer drugs act best on actively dividing cells. Anticancer therapy may well be successful in halting a cancer and eliminating most of it, but once the treatment is stopped, the dormant cells start multiplying and the tumour reappears. Worryingly, it has been observed that such cells are more likely to metastasize.

Another consequence of an insufficient blood supply and lack of oxygen (hypoxia) is that cells in the centre of the tumour are forced to revert to glycolysis in order to produce energy, which leads to a build-up of acidic by-products within the cell. The cells address this problem by exporting acidic protons into the extracellular space. As a result, the environment around tumours tends to be more acidic than in normal tissues. Several anticancer therapies have attempted to take advantage of this difference in acidities; for example, the selective localization of porphyrins in photodynamic therapy.

Angiogenesis inhibitors are relatively safe, so it may be possible to use these drugs as a prophylactic to prevent the appearance of cancer in susceptible individuals.

Several of the drugs described in the following sections inhibit angiogenesis. These include combretastatins (section 21.5.1), VEGF receptor kinase inhibitors (section 21.6.2.9), matrix metalloproteinase inhibitors (section 21.7.1), thalidomide (section 21.9.1), endostatin and angiostatin (section 21.9.3), and bevacizumab (Box 21.11).

21.1.10 Tissue invasion and metastasis

Not all cancers are life threatening. Benign tumours are growths which remain localized in a particular part of the body and can grow to the size of a football without a fatal result. Malignant cancers, however, are life threatening because the cells involved have the ability to break away from the primary tumour, invade a blood vessel or a lymphatic vessel, travel through the circulation, and set up tumours elsewhere in the body. In order to do this, these cells have to overcome a series of controls that are designed to keep cells in their place.

Cells have a molecular signature on their surface which identifies whether they are in the correct part of the body or not. These are cell adhesion molecules (e.g. **E-cadherin**) which ensure that cells adhere to cells of similar character and to an insoluble meshwork of protein filling the space between them—the extracellular matrix. This is particularly true of epithelial cells—the cell layers forming the outer surface of skin and the lining of the gut, lungs, and other organs.

Adhesion to the extracellular matrix is particularly important, as it is necessary if cells are to survive. Molecules called **integrins** are involved in the anchoring process. If a normal cell becomes detached, it stops growing and

apoptosis is triggered. This prevents cells from one part of the body straying to other parts of the body. Moreover, normal cells can only survive if their adhesion molecules match the relevant extracellular matrix.

Cell adhesion molecules are missing in metastasized cancer cells, allowing them to break away from the primary tumour. Such cells also appear to be anchorage independent: they do not self-destruct once they have become free and can latch on to extracellular matrix in other parts of the body to set up secondary tumours. It is thought that oncogenes in these cells code for proteins which send false messages back to the nucleus implying that the cell is still attached.

It is noticeable that most cancers derive from epithelial cells. Once an epithelial cell has gained the ability to split away from its neighbours, it needs to gain access to the blood supply if it is to spread round the body. However, epithelial cells grow on a basement membrane—a thin layer of extracellular matrix which acts as a physical barrier to the movement of the cells. Cancer cells and white blood cells are the only cells capable of breaching this barrier. White blood cells need to do this in order to reach areas of infection, whereas cancer cells breach the barrier to spread the disease. Both types of cell contain the **matrix metalloproteinase** enzyme that hydrolyses the proteins composing the barrier. Once a cancer cell breaks through the basement barrier, it has to break down a similar barrier surrounding the blood vessel in order to enter the blood supply. It then spreads round the body carried by the blood supply until it finally adheres to the blood vessel and breaks out by the opposite process in order to reach new tissue. It is estimated that fewer than one in 10 000 such cells succeed in setting up a secondary tumour, but it only needs one such cell to do so, and once metastasis has occurred, the prospects of survival are slim. Circulating tumour cells usually get trapped in the first network of capillaries they meet and this is where they are most likely to set up secondary tumours. For most tissues, the focus for secondary tumours will be the lungs. In the case of cells originating from the intestines, it is the liver. Some cancer cells produce factors that cause platelets to initiate blood clotting around them such that they increase in size, become stickier, and stick to the blood vessel wall, allowing them to escape.

21.1.11 Treatment of cancer

There are three traditional approaches to the treatment of cancer—surgery, radiotherapy, and chemotherapy. This chapter is devoted to cancer chemotherapy, but it is important to appreciate that chemotherapy is normally used alongside surgery and radiotherapy. Moreover, it is often the case that combination therapy (the simultaneous use of various anticancer drugs with different mechanisms of action) is more effective than using a single drug. The advantages include increased efficiency of action, decreased toxicity, and evasion of drug resistance.

As cancer cells are derived from normal cells, identifying targets that are unique to cancer cells is not easy. As a result, most traditional anticancer drugs act against targets which are present in both types of cell. Therefore, the effectiveness and selectivity of such drugs is dependent on them becoming more concentrated in cancer cells than normal cells. This often turns out to be the case since cancer cells are generally growing faster than normal cells, and so they accumulate nutrients, synthetic building blocks, and drugs more quickly. Unfortunately, not all cancer cells grow rapidly; cells in the centre of a tumour may be dormant and evade the effects of the drug. Conversely, there are normal cells in the body which grow rapidly, such as bone marrow cells. As a result, they too accumulate anticancer drugs, resulting in bone marrow toxicity—a common side effect of cancer chemotherapy which results in a weakening of the immune response and a decreased resistance to infection. Indeed, many cancer patients are prone to pathogens which would not normally be infectious. Such secondary infections can be difficult to treat, and care has to be taken over which antibacterial drugs are used. For example, bacteriostatic antibacterial agents may not be effective as they rely on the normal functioning of the immune system. Other typical side effects of traditional anticancer drugs are impaired wound healing, loss of hair, damage to the epithelium of the gastrointestinal tract, depression of growth in children, sterility, teratogenicity, nausea, and kidney damage.

Most traditional anticancer drugs work by disrupting the function of DNA and are classed as cytotoxic. Some act on DNA directly; others (antimetabolites) act indirectly by inhibiting the enzymes involved in DNA synthesis. Having said that, cancer chemotherapy is now entering a new era which involves **molecular targeted therapeutics**—highly selective agents which target specific molecular targets that are abnormal or overexpressed in the cancer cell. Progress in this area has arisen from a better understanding of the cellular chemistry involved in particular cancer cells. The development of kinase inhibitors such as **imatinib (Glivec)** is a much-heralded illustration of this approach (section 21.6.2.2). The use of antibodies and gene therapy is another area of research which shows huge potential (section 21.10).

Knowledge of the cell cycle is important in chemotherapy. Some drugs are more effective during one part of the cell cycle than another. For example, drugs which affect microtubules are effective when cells are actively dividing (the M-phase), whereas drugs acting on DNA are more effective if the cells are in the S-phase. Some drugs are effective regardless of the phase; for example, alkylating agents such as **cisplatin.** For this reason, anticancer drugs are most effective against cancers which are rapidly proliferating, since they are more likely to become susceptible when they reach the relevant part of the cell cycle. Conversely, slower growing cancers are less effectively treated.

A better understanding of the molecular mechanisms behind specific cancers is yielding better and more specific treatments, but the importance in detecting cancer early on cannot be overemphasized. Unfortunately, the physical symptoms of most tumours do not become apparent until they are well established. By that time, it may be too late. Therefore, it is preferable to detect actual or potential tumours before symptoms arise. **Personalized medicine** is an approach which is likely to become increasingly important (section 6.1.5). The genetic analysis of tumours in individual patients allows the early detection and identification of cancer, as well as identifying the best treatment to be used for a particular individual. This approach is already used in determining which patients will benefit from the anticancer agents **Herceptin** (Box 21.11) and **imatinib**. Genetic fingerprinting should also identify individuals at risk to particular cancers so that they can be screened regularly.

Although cancer is difficult to treat, there have been notable successes in treating rapidly growing cancers such as Hodgkin's disease, Burkitt's lymphoma, testicular cancer, and several childhood malignancies. Early diagnosis also improves the chances of successful treatment in other cancers. At present, four cancers account for over half of all new cases (lung, breast, colon, and prostate).

Finally, one of the best ways of reducing cancers is to reduce the risk. Public education campaigns are important in highlighting the dangers of smoking, excessive drinking, and hazardous solvents, as well as promoting healthy diets and lifestyles. The benefits of eating high-fibre foods, fruit, and vegetables are clear. Indeed, there have been various research projects aimed at identifying the specific chemicals in these foods which are responsible for this protective property. For example, dithiolthiones are a group of chemicals in broccoli, cauliflower, and cabbage which appear to have protective properties, one of which involves the activation of enzymes in the liver to detoxify carcinogens. **Genistein** (Fig. 21.7) is a protective compound found in soy products used commonly in Asian diets. It is notable that Asian populations have a low incidence of breast, prostate, and colon cancers. **Epigallocatechin gallate**, an antioxidant present in green tea, is another potential protective agent. Synthetic drugs are also being investigated as possible cancer preventives (e.g. **finasteride**, **aspirin**, **ibuprofen**, and **difluoromethylornithine**).

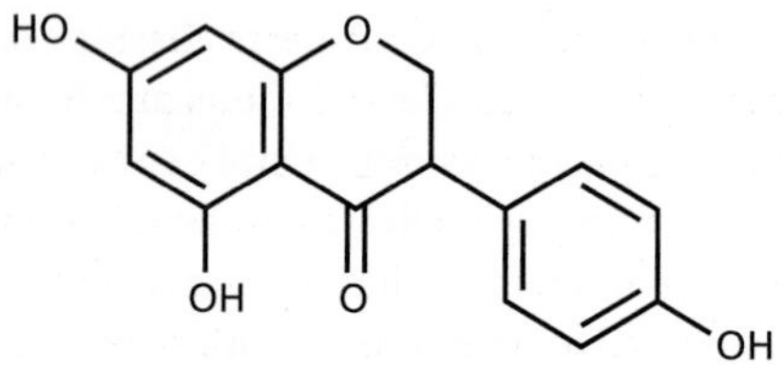

FIGURE 21.7 Genistein.

21.1.12 Resistance

Resistance to anticancer drugs is a serious problem. Resistance can be intrinsic or acquired.

- **Intrinsic resistance** means that the tumour shows little response to an anticancer agent from the very start. This can be due to a variety of possible mechanisms, such as slow growth rate, poor uptake of the drug, or the biochemical/genetic properties of the cell. Tumour cells in the centre of the tumour may be in the resting state and be intrinsically resistant as a result. It has also been proposed that cancer stem cells might exist that are inherently resistant to current anticancer agents. Such stem cells could explain the re-emergence of certain tumours after successful initial treatment.

- When a tumour is initially susceptible to a drug, but becomes resistant, it is said to show **acquired resistance**. This is due to the presence of a mixture of drug-sensitive and drug-resistant cells within the tumour. The drug wipes out the drug-sensitive cells but this only serves to select out and enrich the drug-resistant cells. The survival of even one such cell can lead to failure of the treatment, as that one survivor can spawn a newer drug-resistant tumour. One might ask why a single tumour should contain drug-sensitive and drug-resistant cells as it is likely to have developed from a single cell in the first place. The reason is that cancer cells by their very nature are genetically unstable and so mutations are bound to have occurred during tumour growth, which will result in resistant cells.

There are several molecular mechanisms by which resistance can take place as a result of mutation. For example, resistance can be due to decreased uptake of the drug by the cell or increased synthesis of the drug target. Some drugs need to be activated in the cell and the cancer cell may adapt such that these reactions no longer take place. Alternative metabolic pathways may be found that avoid the effects of antimetabolites. Drugs may be actively expelled from the cell in a process known as **efflux**. A cell membrane carrier protein called **P-glycoprotein** is particularly important in this last mechanism. This protein is a member of a group of energy-dependent transporters

known as **ATP-binding cassette (ABC) transporters**. It expels toxins from normal cells, but mutations in cancer cells can result in increased expression of the protein such that anticancer drugs are efficiently removed as soon as they enter the cell. Unfortunately, the P-glycoprotein can eject a wide diversity of molecules. As a result, cells with excess P-glycoprotein are resistant to a variety of different anticancer drugs, even if they have not been exposed to them before. This is known as **multi-drug resistance (MDR)**. For example, cells acquiring resistance to the **Vinca alkaloids** are also resistant to **dactinomycin** (actinomycin D) and **anthracyclines**.

Efforts have been made to counter this form of resistance by developing drugs which compete for the P-glycoprotein or inhibit it. The calcium ion channel blocker **verapamil** effectively competes for the P-glycoprotein and allows the build-up of an anticancer drug within the cancer cell. Unfortunately, verapamil cannot be used clinically because of its own inherent activity, but there is potential in this approach. **Ciclosporin A** and **quinine** have also been found to inhibit P-glycoprotein and have been investigated in clinical trials, as well as a range of newer agents (e.g. **laniquidar**, **oc144-093**, **zosuquidar**, **elacridar**, **birocodar**, and **tariquidar**). One of the difficulties faced in these studies is finding an agent that will block the transporter protein in cancer cells, but not in normal cells.

To conclude, regardless of the mechanism involved, it is likely that a drug-resistant cell may be present in a cancer. Therefore, it makes sense to use combinations of anticancer drugs with different targets to increase the chances of finding a weakness in every cell, and not just those susceptible to a single drug.

KEY POINTS

- Cancer cells have defects in the normal regulatory controls governing cell growth and division. Such defects arise from mutations resulting in the activation of oncogenes and the inactivation of tumour suppression genes.

- Defects in signalling pathways are commonly found in cancer cells. The pathways stimulating cell growth and division are overactive as a result of the overproduction of a crucial protein in the pathway or the production of an abnormal protein. The proteins involved include growth factors, receptors, signal proteins, and kinases.

- The production of regulatory proteins which suppress cell growth and division is suppressed in many cancers.

- The cell cycle consists of four phases. Progression through the cell cycle is controlled by cyclins and cyclin-dependent kinases, moderated by restraining proteins. Defects in this system have been detected in 90% of cancers.

- Apoptosis is a destructive process leading to cell death. Cells have monitoring systems which check the general health of the cell and trigger the process of apoptosis if there are too many defects. Regulatory proteins have a moderating influence on apoptosis. Defects in apoptosis increase the chances of defective cells developing into cancerous cells and reduce the effectiveness of several drugs.

- Telomeres act as splices to stabilize the ends of DNA. Normally, they decrease in size at each replication until they are too short to be effective, resulting in cell death. Cancer cells activate the expression of an enzyme called telomerase to maintain the telomere and become immortal.

- Angiogenesis is the process by which tumours stimulate the growth of new blood vessels to provide the nutrients required for continued growth. Agents which inhibit angiogenesis are useful in anticancer therapy to inhibit tumour growth and to enhance the effectiveness of other drugs.

- Metastasis is the process by which cancer cells break free of the primary tumour, enter the blood supply, and set up secondary tumours in other tissues. To do this, the regulatory controls which fix cells to a specific environment, and which destroy cells that become detached, are over-ruled.

- Surgery, radiotherapy, and chemotherapy are used to treat cancer. Chemotherapy usually involves combinations of drugs having different targets or mechanisms of action. Traditional anticancer drugs are generally cytotoxic; more modern drugs are selective in their action.

- Cancer cells can have intrinsic or acquired resistance to anticancer drugs. Resistance may be due to poor uptake of the drug, increased production of the target protein, mutations which prevent the drug binding to its target, alternative metabolic pathways, or efflux systems which expel drugs from the cell.

and doxorubicin is one of the most effective anticancer agents ever discovered. The drug intercalates into DNA and is an example of a **topoisomerase II poison** since it stabilizes the complex formed between DNA and topoisomerase II—an enzyme that is crucial to the replication process (sections 9.1 and 6.1.3). It is thought that an excessive number of these stabilized DNA–enzyme complexes triggers apoptosis. Because these enzymes are active during active growth and division, the topoisomerase II poisons are most effective against rapidly proliferating cells. A second mechanism by which doxorubicin can prove harmful to DNA involves the hydroxyquinone moiety, which can chelate iron to form a doxorubicin–DNA–iron complex. Reactive oxygen species are then generated, leading to single-strand breaks in the DNA chain. This mechanism is considered less important than the interaction with topoisomerase II, but it has been implicated in the cardiotoxicity of doxorubicin. A third proposed mechanism involves intercalated doxorubicin inhibiting the **helicases** which unravel DNA into single DNA strands.

A variety of other anthracyclines are used in cancer chemotherapy, mainly **daunorubicin** (also called **cerubidine**, **daunomycin**, or **rubidomycin**), and the second-generation anthracyclines **epirubicin** and **idarubicin** (idamycin) (Fig. 21.8). Idarubicin lacks the methoxy group at R^1, and has an altered metabolism which prolongs its half-life. A third generation of anthracyclines is being studied where the aminium ion on the sugar ring is replaced with an azido group or triazole ring. Such compounds do not appear to be susceptible to the efflux mechanism which leads to drug resistance (Box 21.1). They also show decreased general toxicity.

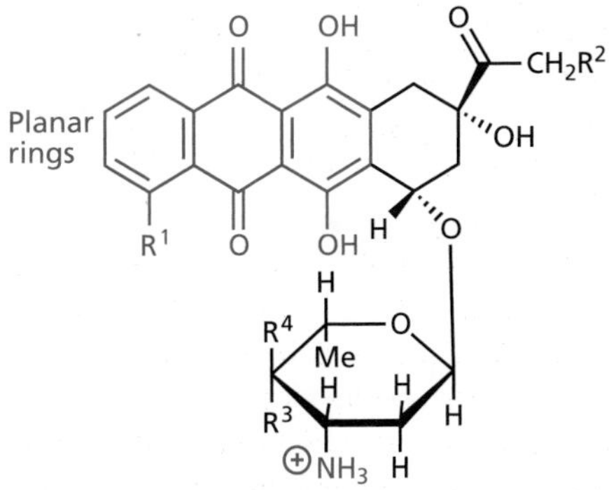

21.2 Drugs acting directly on nucleic acids

21.2.1 Intercalating agents

Intercalating drugs contain a planar aromatic or heteroaromatic ring system which can slip into the double helix of DNA and distort its structure. Once bound, the drug can inhibit the enzymes involved in the replication and transcription processes. Examples include **dactinomycin** and **doxorubicin**, described in section 9.1.

Doxorubicin (previously called **adriamycin**) (Fig. 21.8) belongs to a group of naturally occurring antibiotics called the **anthracyclines,** and was isolated from *Streptomyces peucetius* in 1967. It is very similar in structure to daunorubicin, differing only in one hydroxyl group. However, that has an important effect on activity

Anthracyclines	R^1	R^2	R^3	R^4
Doxorubicin	OMe	OH	OH	H
Epirubicin	OMe	OH	H	OH
Daunorubicin	OMe	H	OH	H
Idarubicin	H	H	OH	H

FIGURE 21.8 The anthracyclines.

BOX 21.1 Clinical aspects of intercalating agents

Most of the anthracyclines are orally inactive and have to be administered by intravenous injection. Another drawback is that they have cardiotoxic side effects which can be irreversible and lead to heart failure, while multi-drug resistance can develop due to amplification of the gene coding for the P-glycoprotein (section 21.1.12). This results in increased efflux of the drug from the tumour cell. Many of the anthracyclines cannot be used alongside radiotherapy owing to enhanced toxic effects, but are widely used otherwise. **Doxorubicin** is used to treat a broad spectrum of solid tumours, as well as acute leukaemias, lymphomas, and childhood tumours. Liposomes (section 11.10) can be useful as carriers to deliver doxorubicin to target tumours, and this approach is associated with less cardiac toxicity. **Daunorubicin** is indicated for acute leukaemias. **Epirubicin** is considered effective against breast cancer. **Idarubicin** is used in the treatment of haematological malignances and can be given orally. Both epirubicin and idarubicin are second-generation anthracyclines with less cardiac toxicity than doxorubicin or

daunorubicin. **Mitoxantrone** is used for the treatment of certain leukaemias and lymphomas, and for advanced breast cancer. It does not have the same level of cardiotoxicity associated with the anthracyclines. **Amsacrine** is given intravenously and is used occasionally for the treatment of acute myeloid leukaemia. **Dactinomycin** is mainly given intravenously to treat paediatric solid tumours, including Wilm's tumour and Ewing's tumour. It has similar side effects to doxorubicin, but has less cardiac toxicity.

Bleomycin is a mixture of bleomycin A_2 and bleomycin B_2, and is used intravenously or intramuscularly in combination therapies for the treatment of certain types of skin cancer, testicular carcinoma, and lymphomas. Unlike most anticancer agents, it produces very little bone marrow depression, but it is quite toxic—particularly to the skin and mucous membranes. The drug is normally inactivated by an enzyme which hydrolyses a primary amide to a carboxylic acid, but this enzyme is present in only very small quantities in the skin. As a result, the active drug can accumulate here to toxic levels.

Mitoxantrone (Fig. 21.9) is a simplified, synthetic analogue of the anthracyclines where the tetracyclic ring system has been 'pruned' back to the planar tricyclic system required for intercalation. There are two identical substituent chains present which make the molecule symmetrical and easier to synthesize. The sugar ring is lacking because it is thought to be responsible for cardiotoxic side effects. However, the amino substituent that is normally present on the sugar is still present within the substituent chains. Structure–activity relationship (SAR) studies on mitoxantrone identify a pharmacophore

involving one of the phenol groups, a carbonyl group, and the amino group in the side chain. Because the molecule is symmetrical, there are two such pharmacophores, but activity remains much the same for analogues containing only one. It was also demonstrated that the amino group linking the side chain to the tricyclic ring system was important to activity. Mitoxantrone intercalates DNA preferentially at guanine–cytosine base pairs such that the side chains lie in the minor groove of DNA, and it is thought to interact with topoisomerase II in a similar fashion to doxorubicin. Other mechanisms of action

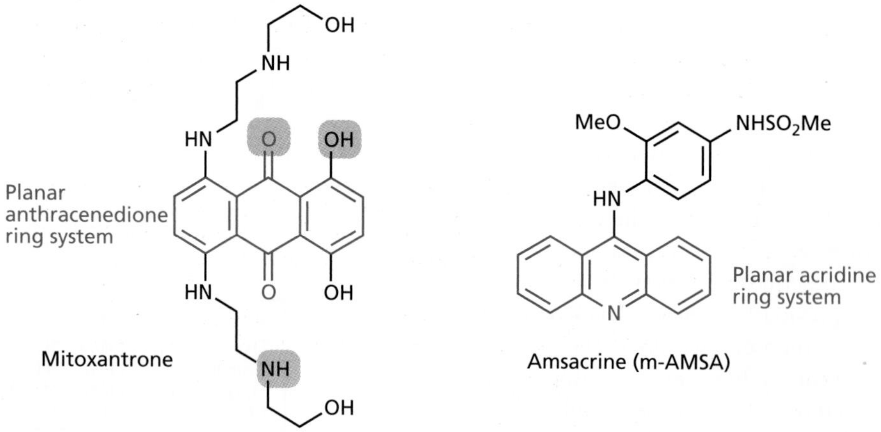

FIGURE 21.9 Mitoxantrone (pharmacophoric groups highlighted in boxes) and amsacrine.

have been proposed, including inhibition of microtubule assembly and inhibition of **protein kinase C. Amsacrine** (Fig. 21.9) contains an acridine tricyclic system capable of intercalating into DNA. It also stabilizes topoisomerase-cleavable complexes.

Another important group of intercalating anticancer agents are the **bleomycins**, which are large, water-soluble glycoproteins derived from *Streptomyces verticillus*. Once they have intercalated with DNA, they are responsible for the production of free radical species that cause oxidative cleavage of DNA strands (section 9.1).

21.2.2 Non-intercalating agents which inhibit the action of topoisomerase enzymes on DNA

21.2.2.1 Podophyllotoxins

Etoposide and **teniposide** (Fig. 9.4; see also Box 21.2) are potent anticancer agents which stabilize the covalent intermediate formed between DNA and topoisomerase II, and are also thought to produce strand breakage by free radical production (section 9.2). The drugs show selectivity for cancer cells, despite the fact that topoisomerase II is present in both cancer cells and normal cells. This is thought to be due to elevated enzyme levels or enzyme activity in the cancer cells. It has also been found that teniposide is more readily taken up by cells than etoposide and has a greater cytotoxic effect. This is thought to be because teniposide is less polar and can cross cell membranes more easily.

Etoposide suffers from poor water solubility but this can be improved by using a phosphate ester prodrug. A range of etoposide analogues has been synthesized in an effort to find agents which have better aqueous solubility, improved activity against drug-resistant cancer cells, and less susceptibility to metabolic inactivation.

21.2.2.2 Camptothecins

Camptothecin (Fig. 21.10) is a naturally occurring cytotoxic alkaloid which was extracted from a Chinese bush (*Camptotheca acuminata*) in 1966. It targets the complex between DNA and **topoisomerase I** (section 9.2). This leads to DNA cleavage and cell death if DNA synthesis is in progress, but it has been observed that these agents are also toxic to cancer cells which are not synthesizing new DNA. This is due to an alternative mechanism of action—possibly the induction of destructive enzymes such as serine proteases and endonucleases.

The camptothecins show selectivity for cancer cells over normal cells when the cancer cells in question show higher levels of topoisomerase I than normal cells. Topoisomerase I can also be more active in certain cancer cells, which may also account for the antitumour selectivity observed.

The lactone group is important for activity, but at blood pH it is in equilibrium with the less active ring-opened carboxylate structure. Introducing substituents into the A and B rings can alter the relative binding affinities of these structures to serum albumin such that the level of the lactone present is altered favourably. Unfortunately, camptothecin itself shows poor aqueous solubility and has unacceptable toxic side effects.

Irinotecan and **topotecan** (Fig. 21.10) are clinically useful, semi-synthetic analogues of camptothecin. They retain the important lactone group and were designed to have aqueous solubility by adding suitable polar functional groups such as alcohols and amines. Irinotecan is a urethane prodrug that is converted to the active phenol (**SN-38**) by carboxylesterases, predominantly in the liver.

21.2.3 Alkylating and metallating agents

Alkylating agents are highly electrophilic compounds that react with nucleophilic groups on DNA to form strong covalent bonds (section 9.3). Drugs with two alkylating groups can cause cross-linking that disrupts replication or transcription. Unfortunately, alkylating agents can alkylate nucleophilic groups on proteins as well, which means they have poor selectivity. Nevertheless, alkylating drugs have been useful in the treatment of cancer. Tumour cells often divide more rapidly than normal cells, and so disruption of DNA function affects these cells more drastically than normal cells. However,

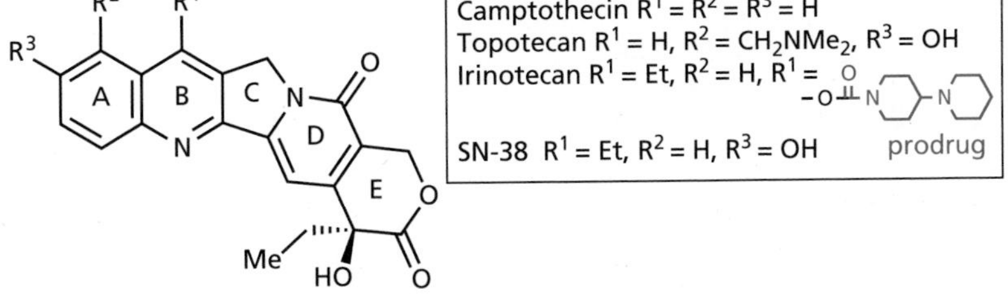

Camptothecin $R^1 = R^2 = R^3 = H$
Topotecan $R^1 = H$, $R^2 = CH_2NMe_2$, $R^3 = OH$
Irinotecan $R^1 = Et$, $R^2 = H$, $R^1 = $

SN-38 $R^1 = Et$, $R^2 = H$, $R^3 = OH$ prodrug

FIGURE 21.10 Camptothecins.

BOX 21.2 Clinical aspects of non-intercalating agents inhibiting the action of topoisomerase enzymes on DNA

Etoposide and **teniposide** are used clinically for a variety of conditions, such as testicular cancer and small cell lung cancer. Resistance can arise due to overexpression of the P-glycoprotein involved in the efflux mechanism, or to mutations in the topoisomerase enzyme which weaken interactions with the drug.

Topotecan is an intravenous drug that is used in the treatment of advanced ovarian cancer when previous treatments have failed. **Irinotecan** is given intravenously and is a prodrug used in combination therapy with **fluorouracil** and **folinic acid (leucovorin)** for the treatment of advanced colorectal cancer. It has a poten-

tial role in treating a variety of other cancers. Unfortunately, the carboxylesterases required to activate the structure are not very efficient, and only 2–5% of an injected dose is actually converted. Gene therapy and ADEPT strategies are being explored to try to improve this process (section 21.10). Resistance to these drugs arises from mutations to the topoisomerase I enzyme.

Severe diarrhoea can be a major side effect of irinotecan which limits its use. This arises from activation of the drug in the intestines by gut flora, resulting in it killing intestinal cells. Research is being carried out to try and find an enzyme inhibitor that will inhibit activation in the intestines.

---

it should be noted that these drugs can be mutagenic and carcinogenic in their own right. This results from the damage that they wreak on DNA in normal, healthy cells. The simple alkylating agents that are used commonly in organic synthesis (e.g. iodomethane and dimethyl sulphate) are considered carcinogenic because they have the capability to alkylate DNA.

21.2.3.1 Nitrogen mustards

Chlormethine (Fig. 21.11) was the first alkylating agent to be used medicinally and was introduced in 1942. The mechanism involves the cross-linking of guanine groups on DNA, as described in section 9.3.1. Chlormethine is highly reactive and can react with water, blood, and

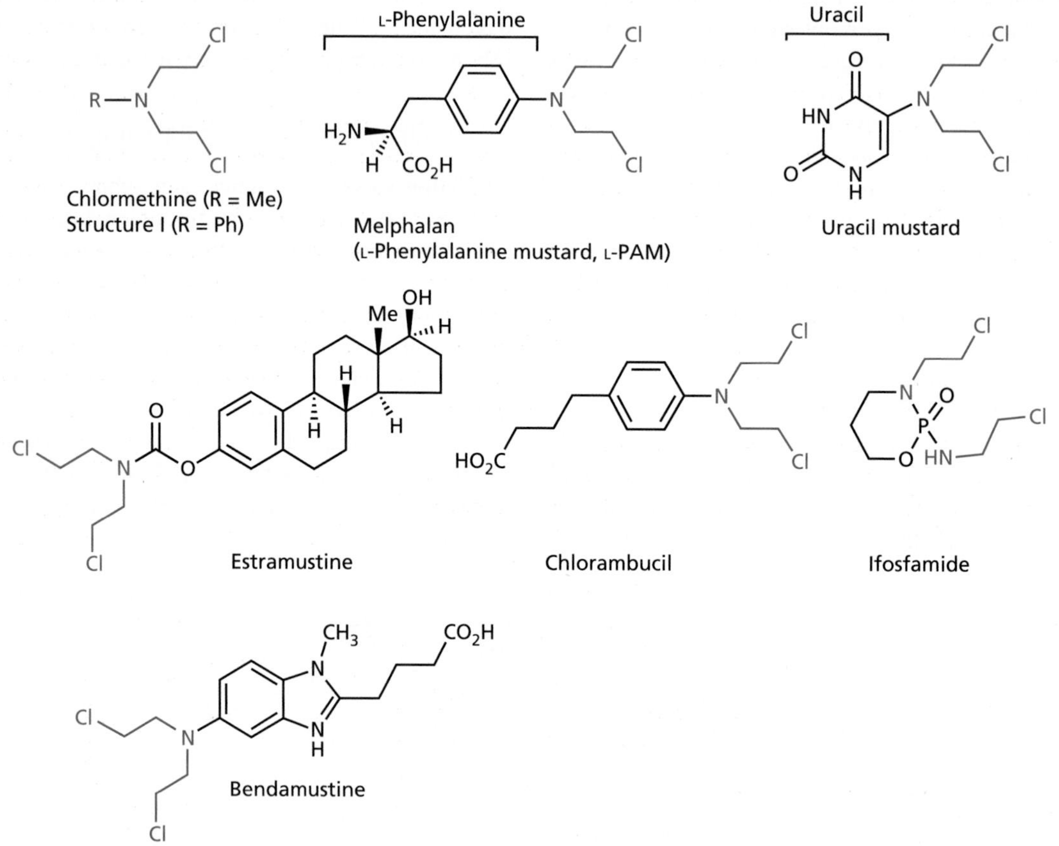

FIGURE 21.11 Mustard-like alkylating agents.

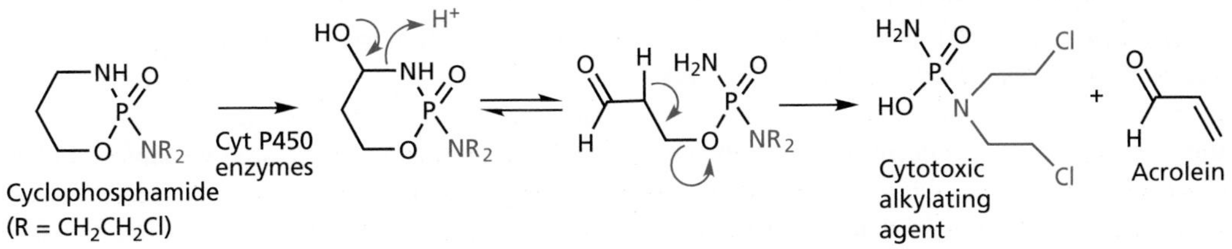

FIGURE 21.12 Phosphoramide mustard from cyclophosphamide.

tissues. It is too reactive to survive the oral route and has to be administered intravenously.

The side reactions mentioned above can be reduced by lowering the reactivity of the alkylating agent. For example, replacing the *N*-methyl group with an *N*-aryl group (I in Fig. 21.11) has such an effect. The lone pair of the nitrogen interacts with the π system of the ring and is less available to displace the chloride ion. As a result, the intermediate aziridinium ion (see Fig. 9.9) is less easily formed and only strong nucleophiles such as guanine will react with it. The alkylating agent **melphalan** (Fig. 21.11) takes advantage of this property and has the added advantage of having a moiety which mimics the amino acid phenylalanine. As a result, the drug is more likely to be recognized as an amino acid and get taken into cells by transport proteins. The increased stability also means that the drug can be given orally. Phenylalanine is a biosynthetic precursor for melanin and it was hoped that this would help to target the drug to skin melanomas. Unfortunately, such targeting has not been particularly significant.

A similar approach has been to attach a nucleic acid building block to the alkylating group. For example, **uracil mustard** (Fig. 21.11) contains the nucleic acid base uracil and shows a certain amount of selectivity for tumour cells over normal cells. Because tumour cells generally divide faster than normal cells, nucleic acid synthesis is faster and so tumour cells need more of the nucleic acid building blocks. The tumour cells scavenge more than their fair share of the building blocks and accumulate the cytotoxic drug more effectively. Unfortunately, this approach has not achieved the high levels of selectivity desired for effective eradication of all relevant tumour cells.

Other examples of alkylating agents include **chlorambucil** and **estramustine** (Fig. 21.11). In the latter drug, the alkylating group is linked to the hormone **estradiol**. As estradiol can cross cell membranes into cells, it carries the alkylating agent with it. The link to the steroid is through a urethane functional group which lowers the nitrogen's nucleophilicity. **Bendamustine** is a more recently approved agent.

Cyclophosphamide (Fig. 21.12) is the most commonly used alkylating agent in cancer chemotherapy. It is a non-toxic prodrug which is converted in the body to the active drug (Fig. 21.12). Metabolism takes place in the liver where cytochrome P450 enzymes oxidize the ring. Ring opening then takes place and a non-enzymatic hydrolysis splits acrolein from the molecule to generate the cytotoxic alkylating agent. The nucleophilicity of the nitrogen is tempered by it being part of a phosphoramide group, and so the active agent is more selective for stronger nucleophiles such as guanine.

Cyclophosphamide itself is relatively non-toxic and can be taken orally without causing damage to the gut wall. It was also hoped that the high level of **phosphoramidase** enzyme present in some tumour cells would lead to a greater concentration of alkylating agent in these cells and result in some selectivity of action. Unfortunately, the acrolein released can sometimes prove toxic to the kidneys and the bladder. One possible explanation is that acrolein alkylates cysteine residues in cell proteins. Certainly, toxicity can be reduced by co-administrating sulphydryl donors such as **N-acetylcysteine** or **sodium-2-mercaptoethane sulphonate** (mesna) ($HSCH_2CH_2SO_3^-$) which interact with the acrolein. **Ifosfamide** (Fig. 21.11) is a

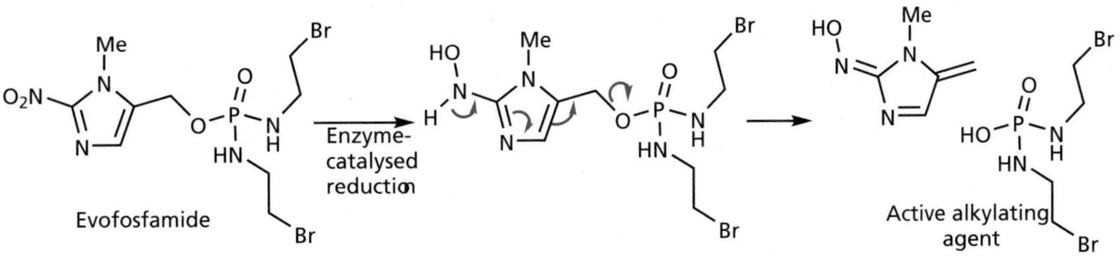

FIGURE 21.13 Activation of the prodrug evofosfamide under hypoxic conditions.

related drug with a similar mechanism and similar problems. **Evofosfamide (TH-302)** (Fig. 21.13) is structurally related to ifosfamide and acts as a hypoxia-activated prodrug. In other words, it is activated in environments where there are low oxygen concentrations, such as the centre of solid tumours. Under such conditions, the prodrug undergoes reduction and degradation to the active alkylating agent. Unfortunately, recent phase III clinical trials have been disappointing.

21.2.3.2 Cisplatin and cisplatin analogues: metallating agents

Cisplatin (Fig. 21.14) is one of the most frequently used anticancer drugs. The structure is activated within cells and produces intrastrand cross-linking (section 9.3.4). **Carboplatin** (Fig. 21.14) is a derivative of cisplatin with reduced side effects.

A range of other platinum drugs (Fig. 21.14) (such as the first orally active compound **satraplatin**) have been developed in an attempt to overcome tumour resistance (Box 21.3). Most of these compounds are still undergoing clinical trials, but **oxaliplatin** was approved in 1999, and is effective against tumours that have gained resistance to cisplatin and carboplatin. This lack of cross-resistance is due to the presence of the diaminocyclohexane ring. The ring is bad for water solubility, but this can be counteracted by introducing an oxalato ligand as the leaving group.

A lot of research is being carried out to try and tackle problems such as side effects and resistance. These include designing prodrugs of metallating agents that will only be activated at target tumour cells. For example, cancer cells have an oxidizing environment and prodrugs are being designed which are activated by hydrogen peroxide. Another approach has been to link the metallating agent to a molecule which will target an overexpressed target in the tumour cells; for example linking the agent to a steroid such that it targets tumour cells which overexpress a steroid receptor.

21.2.3.3 CC 1065 analogues

CC 1065 (Fig. 21.15) is a naturally occurring anticancer agent which binds to the minor groove of DNA then alkylates an adenine base. It is 1000 times more active *in vitro* than doxorubicin and cisplatin. **Adozelesin** is a simplified synthetic analogue and is being considered for use in antibody–drug conjugates (section 21.10.2).

21.2.3.4 Other alkylating agents

There are several other anticancer agents that act as alkylating agents, including **dacarbazine**, **procarbazine**, **lomustine**, **carmustine**, **temozolomide**, **busulfan**, and **mitomycin C**. The mechanisms by which these agents work are described in section 9.3.

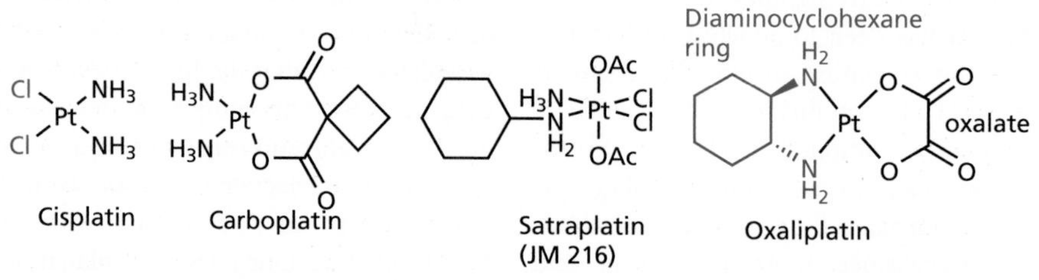

FIGURE 21.14 Platinum-based anticancer drugs.

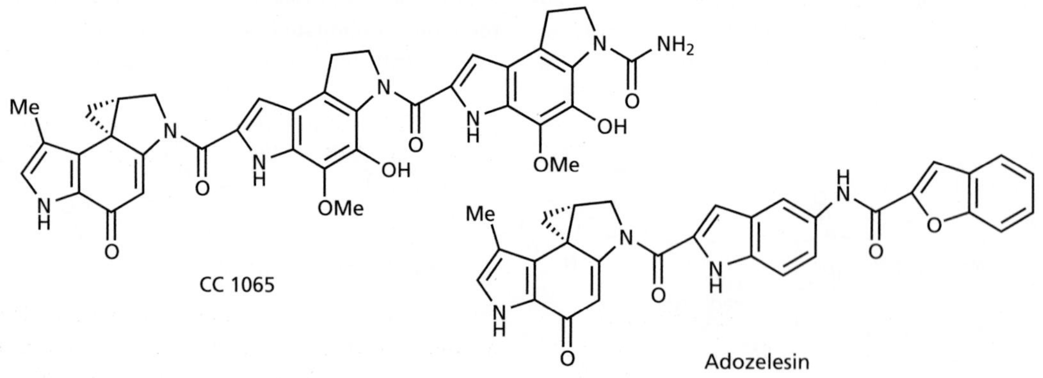

FIGURE 21.15 CC 1065 and adozelesin.

BOX 21.3 Clinical aspects of alkylating and metallating agents

Chlormethine has been used for the treatment of Hodgkin's lymphoma as part of a multi-drug regime. The related structure **melphalan** is currently used in the treatment of multiple myeloma, as well as advanced ovarian and breast cancers. **Uracil mustard** has been used successfully in the treatment of chronic lymphatic leukaemia. **Estramustine** can be given orally and is used predominantly for the treatment of prostate cancer. **Chlorambucil** is an orally active drug used primarily in the treatment of chronic lymphocytic leukaemia and Hodgkin's disease. **Bendamustine** was approved in 2008 for the treatment of chronic lymphatic leukaemia and lymphomas. Resistance to alkylating agents can arise through reaction with cellular thiols and decreased cellular uptake.

Cyclophosphamide is given orally or intravenously, and is widely used for the treatment of leukaemias, lymphomas, soft tissue sarcoma, and solid tumours. Haemorrhagic cystitis is a rare, but serious, side effect which results in inflammation, oedema, bleeding, ulceration, and cell death. This is caused by the metabolite acrolein and can be countered by increased fluid intake or by administering **mesna**. The related drug **ifosfamide** is given intravenously along with mesna.

Lomustine and **carmustine** are lipid soluble and can cross the blood–brain barrier. As a result, they have been used in the treatment of brain tumours and meningeal leukaemia. Lomustine can be given orally, but carmustine is given intravenously because it is rapidly metabolized. Carmustine implants have also been approved. **Streptozotocin** has been used for the treatment of pancreatic islet cell carcinoma. There is a specific uptake of the drug into the pancreas where it carbamoylates proteins.

Busulfan is given orally in the treatment of chronic myeloid leukaemia and may increase the life expectancy of patients by about a year. It is also administered alongside cyclophosphamide prior to stem cell transplantation. It acts selectively on the bone marrow and has little effect on lymphoid tissue or the gastrointestinal tract. However, excessive use may lead to irreversible damage to the bone marrow. Resistance

to busulfan is related to the rapid removal and repair of the DNA cross-links.

Cisplatin is a very useful antitumour agent which is used alone or in combination with other drugs for the intravenous treatment of lung, cervical, bladder, head, neck, testicular, and ovarian tumours. It is also used in various combination therapies to treat other forms of cancer. Unfortunately, cisplatin is associated with very severe nausea and vomiting, but the administration of the 5-HT$_3$ receptor antagonist **ondansetron** (Box 12.2) is effective in combating this problem. **Carboplatin** is now preferred over cisplatin for the intravenous treatment of advanced ovarian tumours, and is also used to treat lung cancers. It is better tolerated than cisplatin and has less severe side effects. **Oxaliplatin** was approved in 1999 for the treatment of colorectal cancer and shows a better safety profile than cisplatin or carboplatin. It is used in combination with **fluorouracil** and **folinic acid**.

Tumour resistance to cisplatin and similar agents has been attributed to a number of factors. Cisplatin requires a transporter protein in order to enter the cell and resistance can occur if there are low levels of the transport protein. The activated species arising from cisplatin (section 9.3.4) reacts easily with cellular thiols, such as glutathione, and resistance can occur if these thiols are present in high concentration. The agent is 'mopped up' before it has a chance to react with DNA. Finally, resistance may arise because of increased efflux of the drug from the cell.

Dacarbazine is used clinically in combination therapies for the treatment of melanoma and soft tissue sarcomas. **Procarbazine** is most often used for the treatment of Hodgkin's disease and is given orally. **Temozolomide** is used for the treatment of certain types of brain tumour, and is administered orally in capsules at least one hour before a meal.

Mitomycin C is used intravenously for the treatment of upper gastrointestinal and breast cancers. It can also be used to treat superficial bladder cancers. It has many side effects and is one of the most toxic anticancer drugs in clinical use. Prolonged use can lead to permanent bone marrow damage.

For additional material see Web article 24: Current metallodrugs in cancer on the Online Resource Centre at www.oxfordtextbooks.co.uk/orc/patrick6e/

21.2.4 Chain cutters

Calicheamicin γ1 is an antitumour agent that was isolated from a bacterium. It binds to DNA and is responsible for generating radical species which lead to the cutting of the DNA chain (section 9.4). It is an extremely potent

agent and is one of the structures being studied in the design of antibody–drug conjugates (section 21.10.2 and Box 21.12).

21.2.5 Antisense therapy

The biopharmaceutical company Genta has developed an antisense drug (section 9.7.2) called **oblimersen** which consists of 18 deoxynucleotides linked by a phosphorothioate backbone (section 14.10). It binds to the initiation codon of the mRNA molecule carrying the genetic instructions for **Bcl-2**. Bcl-2 is a protein which suppresses cell death (apoptosis, section 21.1.7), and so suppressing its synthesis will increase the chances of apoptosis taking place when chemotherapy or radiotherapy is being employed. This is currently being tested in phase III clinical trials in combination with the anticancer drugs **docetaxel** and **irenotecan**. Unfortunately, phase III clinical trials on the treatment of melanoma have proved disappointing. Other possible applications are still being considered, though.

Phosphorothioate oligonucleotides are also being investigated that will target the genetic instructions for **Raf** and **PKCγ**—two proteins which are involved in signal transduction pathways (sections 5.3.2 and 5.4.2).

KEY POINTS

- Intercalating drugs contain planar aromatic or heteroaromatic ring systems which can slide between the base pairs of the DNA double helix.

- Alkylating agents contain electrophilic groups that react with nucleophilic centres on DNA. If two electrophilic groups are present, interstrand and/or intrastrand cross-linking of the DNA is possible.

- Nitrogen mustards react with guanine groups on DNA to produce cross-linking. The reactivity of the agents can be lowered by attaching electron-withdrawing groups to the nitrogen to increase selectivity against DNA over proteins. Incorporation of important biosynthetic building blocks aids the uptake into rapidly dividing cells.

- Cisplatin and its analogues are metallating agents which cause intrastrand cross-linking. They are commonly used for the treatment of testicular and ovarian cancers.

- CC-1065 analogues are highly potent alkylating agents which are being considered for use in antibody–drug conjugates

- Calicheamicin is a natural product which reacts with nucleophiles to produce a diradical species. Reaction with DNA ultimately leads to cutting of the DNA chains.

- Antisense molecules have been designed to inhibit the mRNA molecules that code for the proteins which suppress apoptosis.

21.3 Drugs acting on enzymes: antimetabolites

The drugs described in section 21.2 interact directly with DNA to inhibit its various functions. Another method of disrupting DNA function is to inhibit the enzymes involved in the synthesis of DNA or its nucleotide building blocks. The inhibitors involved are described as **antimetabolites**. The action of antimetabolites leads to the inhibition of DNA function or the synthesis of abnormal DNA, which may trigger the processes leading to apoptosis.

21.3.1 Dihydrofolate reductase inhibitors

Dihydrofolate reductase (DHFR) is an enzyme which is crucial in maintaining levels of the enzyme cofactor **tetrahydrofolate** (FH_4) (Figs. 21.16 and 21.17).

Without this cofactor, the synthesis of the DNA building block (dTMP) would grind to a halt, which, in turn, would slow down DNA synthesis and cell division. The enzyme catalyses the reduction of the vitamin **folic acid** to FH_4 in two steps via **dihydrofolate** (FH_2). Once formed, FH_4 picks up a single carbon unit to form N^5, N^{10}-**methylene** FH_4, which then acts as a source of one-carbon units for various biosynthetic pathways, including the methylation of deoxyuridine monophosphate (dUMP) to form deoxythymidine monophosphate

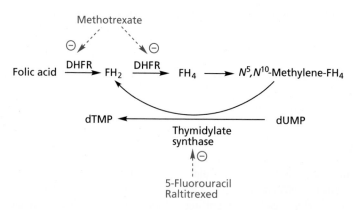

FIGURE 21.16 Reactions catalysed by dihydrofolate reductase and thymidylate synthase.

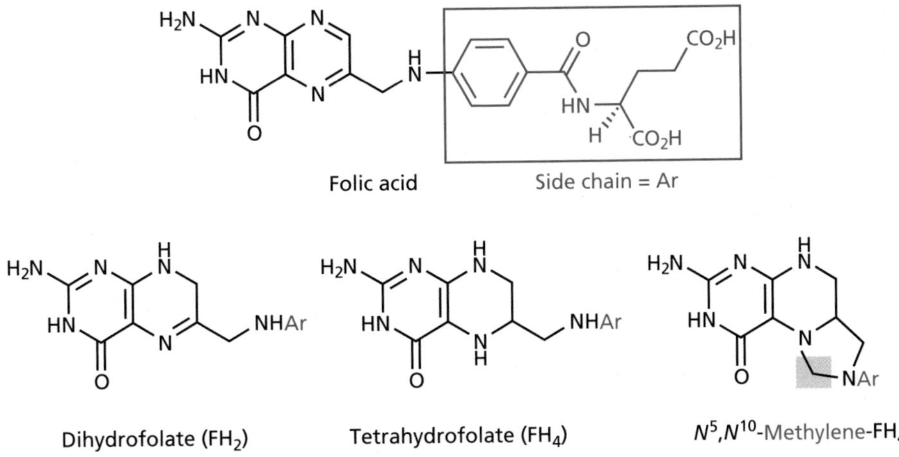

FIGURE 21.17 Structures of folic acid and related cofactors.

(dTMP). N^5,N^{10}-Methylene FH$_4$ is converted back to FH$_2$ in the process and dihydrofolate reductase is vital in restoring the N^5,N^{10}-methylene FH$_4$ for further reaction.

Methotrexate (Fig. 21.18) is one of the most widely used antimetabolites in cancer chemotherapy. It is very similar in structure to the natural folates, differing only in additional amino and methyl groups. It has a stronger binding affinity for DHFR owing to an additional hydrogen bond or ionic bond which is not present when FH$_2$ binds. As a result, methotrexate prevents the binding of FH$_2$ and its conversion to N^5,N^{10}-methylene FH$_4$. Depletion of the cofactor has its greatest effect on the enzyme **thymidylate synthase**, resulting in the lowered synthesis of dTMP.

Methotrexate tends to accumulate in cells as a result of **polyglutamylation**. This is an enzyme-catalysed process which involves the addition of glutamate groups to the glutamate moiety already present in the molecule. This

also happens to natural folates, and the reaction serves to increase the charge and size of the folates such that they are trapped within the cell. **Pemetrexed** and **pralatrexate** are related drugs that were approved in 2004 and 2009 respectively.

21.3.2 Inhibitors of thymidylate synthase

Methotrexate has an indirect effect on thymidylate synthase by lowering the amount of N^5,N^{10}-methylene FH$_4$ cofactor required. **5-Fluorouracil** (Fig. 21.19) is an anticancer drug which inhibits this enzyme directly.

It does so by acting as a prodrug for a **suicide substrate** (section 7.5). 5-Fluorouracil is converted in the body to the fluorinated analogue of 2′-deoxyuridylic acid monophosphate (FdUMP) (Fig. 21.19), which then combines with the enzyme and the cofactor (Fig. 21.20). Up until this point, nothing unusual has happened and the reaction mechanism has been proceeding normally. The tetrahydrofolate

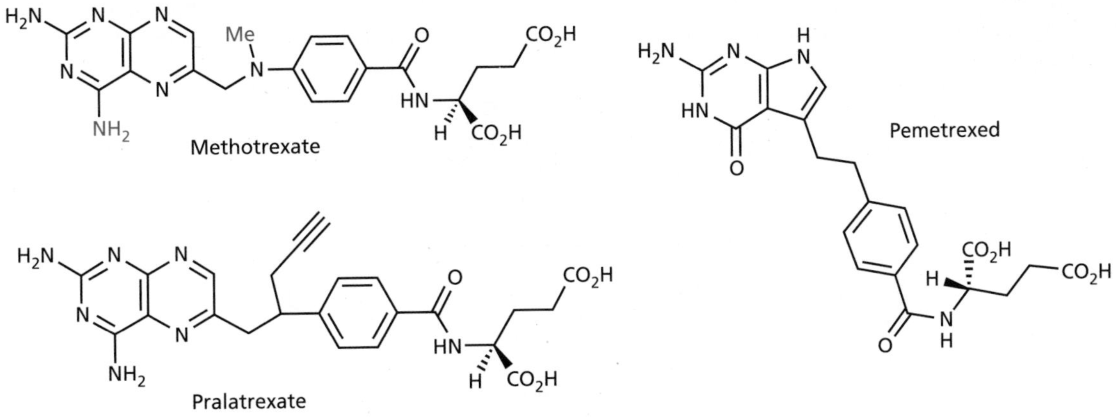

FIGURE 21.18 Methotrexate, pemetrexed, and pralatrexate.

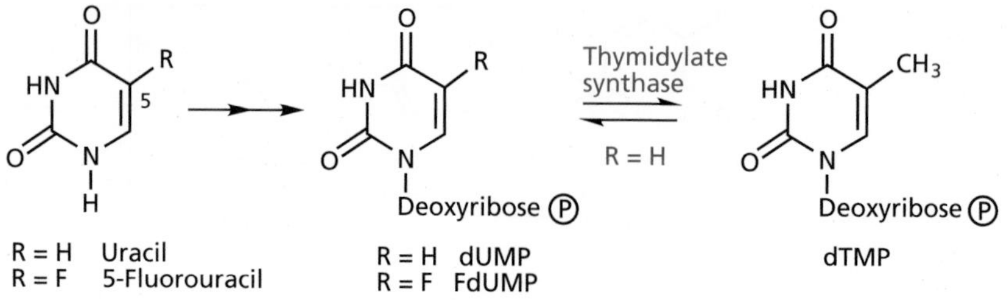

FIGURE 21.19 Biosynthesis of dTMP ($\circled{P}$ = phosphate).

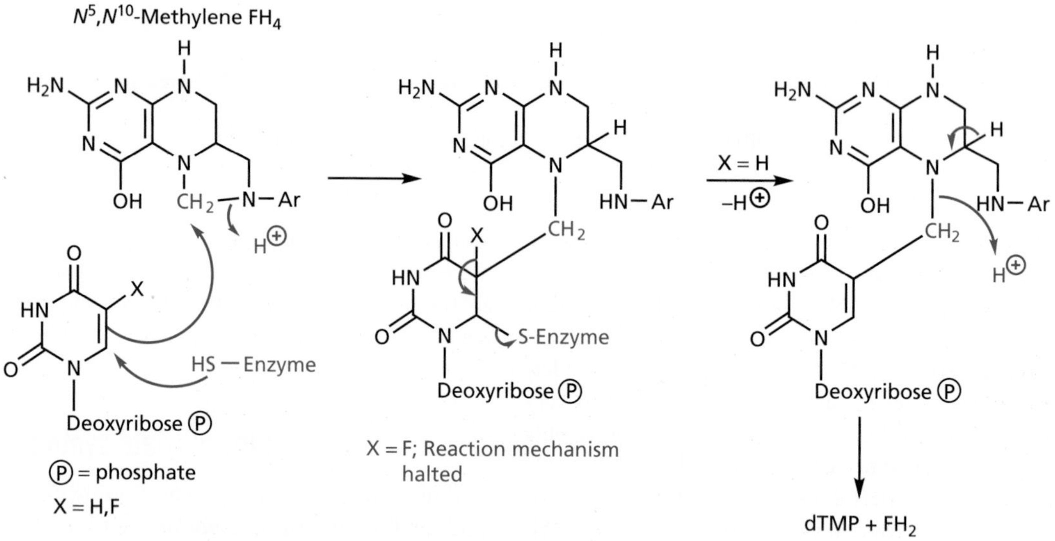

FIGURE 21.20 Use of 5-fluorouracil as a prodrug for a suicide substrate.

has formed a covalent bond to the uracil skeleton via the methylene unit which is usually transferred to uracil, but at this stage, things start to go wrong. A proton is usually lost from position 5 of uracil (X = H). However, 5-fluorouracil has a fluorine atom at that position instead of hydrogen (X = F). Further reaction is impossible, as it would require fluorine to leave as a positive ion. Fluorine is too electronegative for this to occur because it prefers to ionize as the fluoride ion (F$^-$). As a result, the fluorouracil skeleton remains covalently and irreversibly bound to the active

site. The synthesis of thymidine is now terminated, which in turn stops the synthesis of DNA. Consequently, replication and cell division are blocked. **Capecitabine** (Box 21.4) is a prodrug for 5-fluorouracil.

Trifluridine (Fig. 21.21) is very similar to 5-fluorouracil, but contains a trifluoromethyl group at the 5-position instead of fluorine. Like 5-fluorouracil, it acts as a prodrug and is phosphorylated by **thymidylate kinase** to form the active agent **trifluridine monophosphate.** This acts as an irreversible inhibitor of DHFR by reacting with

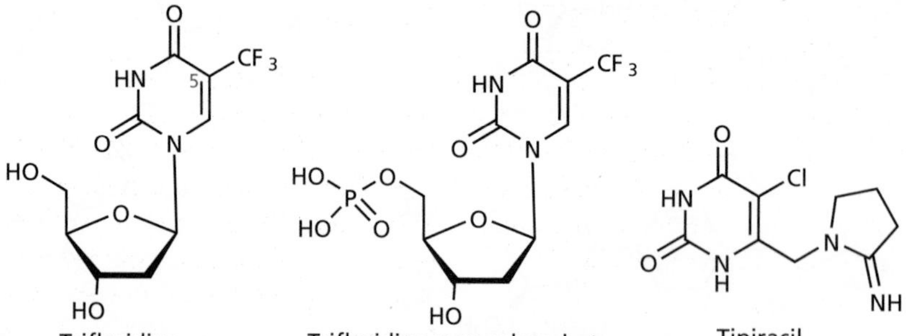

FIGURE 21.21 Trifluridine, trifluridine monophosphate, and tipiracil.

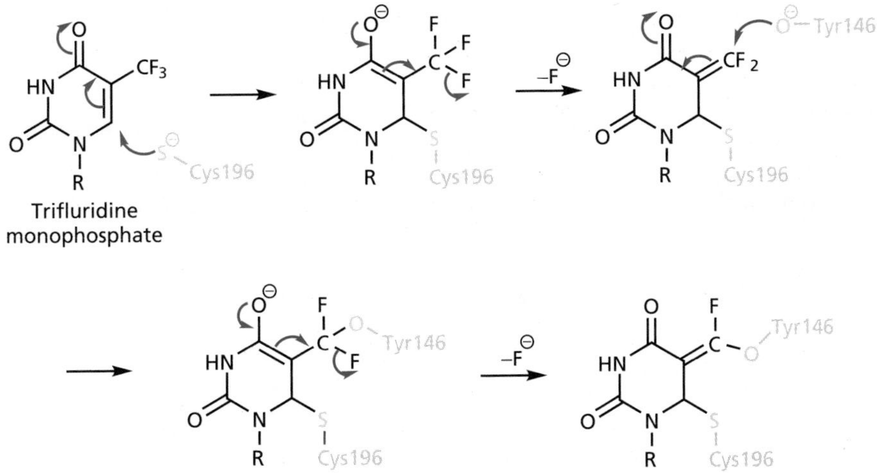

FIGURE 21.22 Mechanism of inhibition for trifluridine monophosphate.

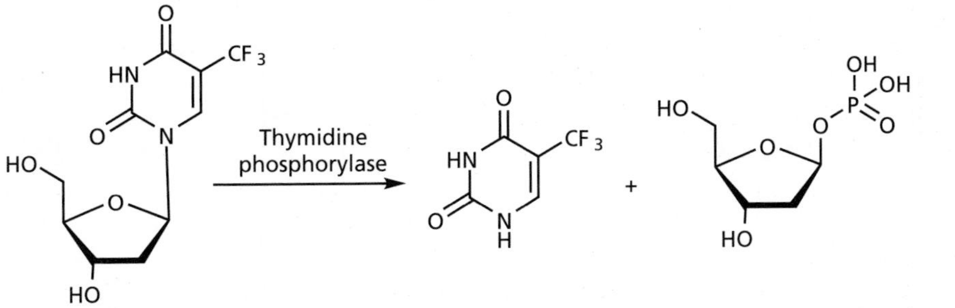

FIGURE 21.23 Metabolic degradation of trifluridine.

Cys-196 and Tyr-146, such that the drug creates a cross-link between the two amino acids (Fig. 21.22).

A second mechanism of action is possible where trifluridine monophosphate is phosphorylated to the triphosphate, and is then incorporated into the growing DNA chain. This induces DNA strand breakage and cell death. Note that trifluridine is also approved as an antiviral agent (section 20.6.1).

One of the drawbacks with trifluridine is its metabolic susceptibility to the enzyme **thymidine phosphorylase** (Fig. 21.23). Therefore, it is administered with an inhibitor of thymidine phosphorylase called **tipiracil**, which serves as a sentry drug (section 14.7.1) to prolong trifluridine's half-life.

The inhibitors described so far bind to the same region of the active site as uracil. Inhibitors which bind to the cofactor binding region have also been developed. **Raltitrexed** (Fig. 21.24) is the first of a new generation of highly specific folate-based thymidylate synthase inhibitors.

@ For additional information see Web article 28: Use of fluorine in medicinal chemistry on the Online Resource Centre at www.oxfordtextbooks.co.uk/orc/patrick6e/

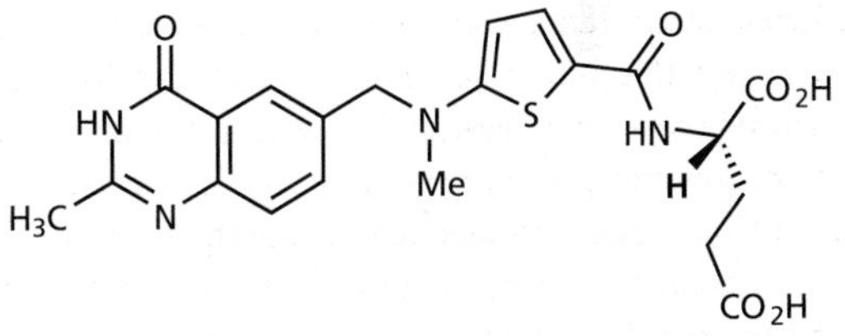

FIGURE 21.24 Raltitrexed (ZD-1694).

21.3.3 Inhibitors of ribonucleotide reductase

Ribonucleotide reductase is responsible for the conversion of ribonucleotide diphosphates to deoxyribonucleotide diphosphates (Fig. 21.25). The enzyme contains an iron cofactor which is crucial to the reaction mechanism. This involves the iron reacting with a tyrosine residue to generate and stabilize a tyrosine free radical, which then abstracts a proton from the substrate and initiates the reaction mechanism. **Hydroxycarbamide** (Fig. 21.25) is a clinically useful agent which inhibits the enzyme by destabilizing the iron centre.

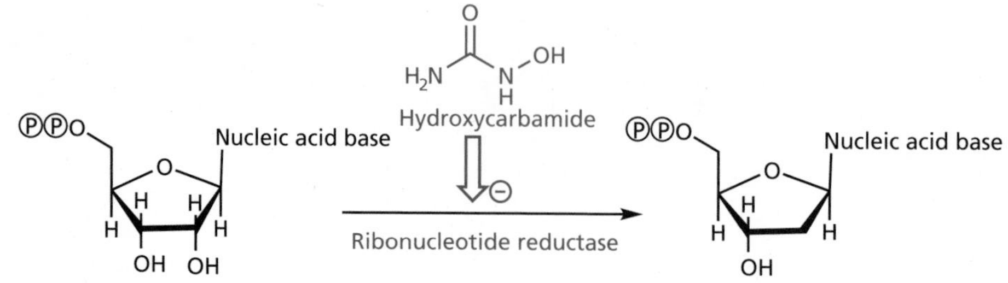

FIGURE 21.25 Reaction catalysed by ribonucleotide reductase ($\mathbb{P}$ = phosphate).

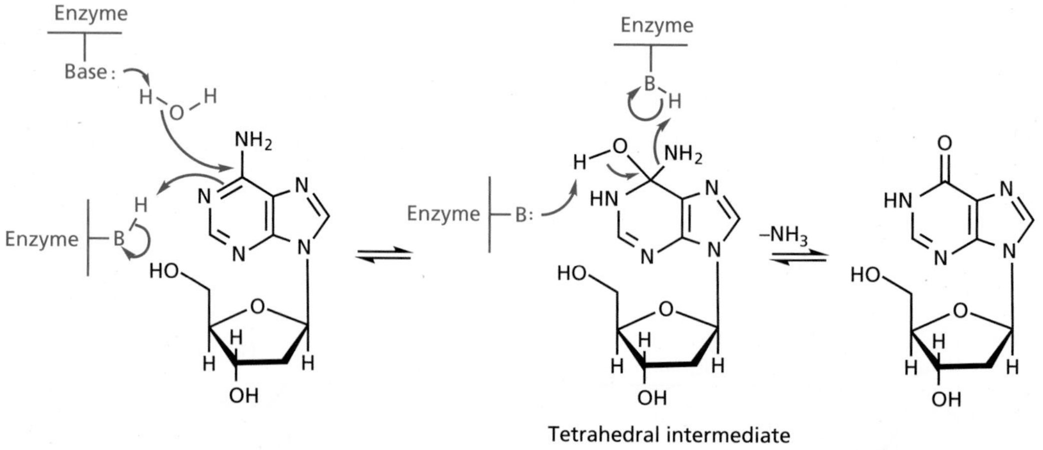

FIGURE 21.26 Mechanism of adenosine deaminase (B = base).

21.3.4 Inhibitors of adenosine deaminase

Ribonucleotide reductase is inhibited directly by hydroxycarbamide, but it can also be inhibited indirectly by increasing the level of natural allosteric inhibitors such as dATP (allosteric inhibitors are described in section 3.6). The enzyme **adenosine deaminase** catalyses the deamination of adenosine to inosine (Fig. 21.26), and it is found that inhibition of the enzyme leads to a build-up of dATP in the cell. This, in turn, inhibits ribonucleotide reductase.

The antileukaemia drug **pentostatin** (Fig. 21.27) is a natural product isolated from *Streptomyces antibioticus*, and is a powerful inhibitor of adenosine deaminase (K_i = 2.5 pM). It acts as a transition-state inhibitor, mimicking the proposed tetrahedral nature of the transition state. The transition state is believed to be similar to the tetrahedral intermediate in Fig. 21.26.

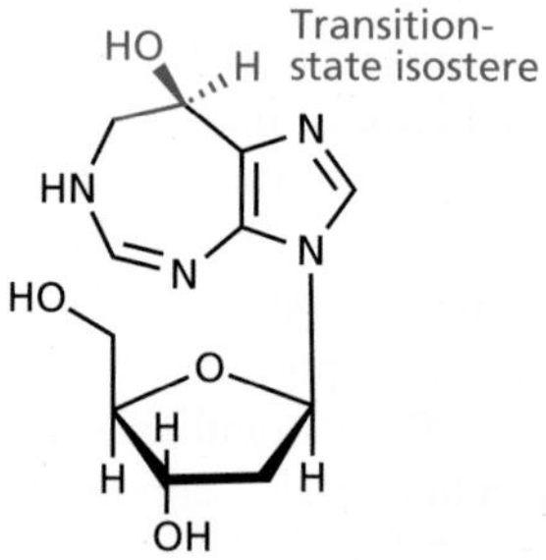

21.3.5 Inhibitors of DNA polymerases

DNA polymerases catalyse the synthesis of DNA using the four deoxyribonucleotide building blocks

FIGURE 21.27 Pentostatin.

dATP, dGTP, dCTP, and dTTP (Chapter 6). The anticancer drug **cytarabine** (Fig. 21.28) is an analogue of 2′-deoxycytidine and acts as a prodrug. It is phosphorylated in cells to the corresponding triphosphate (**ara-CTP**) which acts as a competitive inhibitor. In addition, ara-CTP can act as a substrate for DNA polymerases and become incorporated into the growing DNA chain. This can lead to chain termination or prevent replication of the modified DNA. All of these effects result in the inhibition of DNA synthesis and repair. **Gemcitabine** is

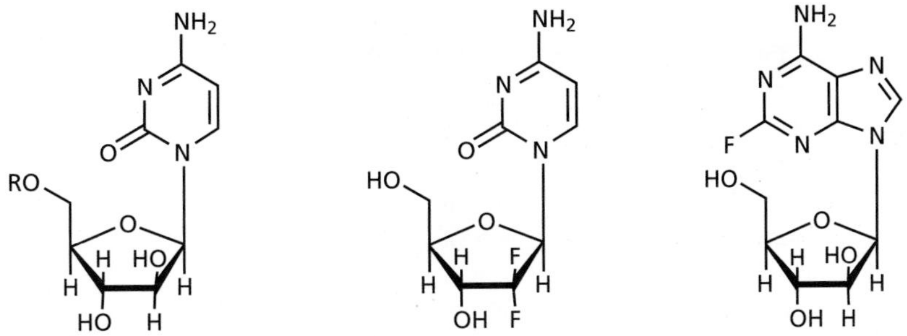

FIGURE 21.28 Inhibitors of DNA polymerase.

R = H; Cytarabine (cytosine arabinoside)
R = Triphosphate; Ara-CTP

Gemcitabine

Fludarabine

an analogue of cytarabine with fewer side effects. The purine analogue **fludarabine** is also metabolized to a triphosphate and has the same mechanism of action as cytarabine. It, too, inhibits transcription and can be incorporated into DNA.

21.3.6 Purine antagonists

The thiopurines **6-mercaptopurine** and **6-tioguanine** (Fig. 21.29) are prodrugs which are converted to their corresponding nucleoside monophosphates by cellular enzymes. The monophosphates then inhibit purine synthesis at a number of points. Both agents are also converted to a common product (**thio-GMP**), which is subsequently converted to **thio-GTP** and **thio-dGTP** before incorporation into RNA and DNA respectively. This leads to complex effects which result in cell death.

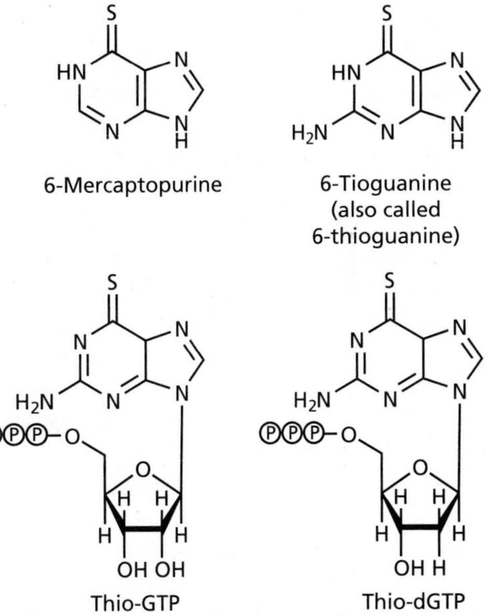

6-Mercaptopurine

6-Tioguanine
(also called
6-thioguanine)

Thio-GTP

Thio-dGTP

FIGURE 21.29 Purine antagonists (Ⓟ = phosphate).

BOX 21.4 Clinical aspects of antimetabolites

Methotrexate can be administered orally and by various other methods. It is used to treat a wide variety of cancers, either alone or in combination with other drugs. Examples include childhood acute lymphoblastic leukaemia, non-Hodgkin's lymphoma, and a number of solid tumours. Resistance to methotrexate can arise from enhanced expression of DHFR, or diminished uptake of methotrexate by the **reduced folate carrier** (RFC)—a membrane transport protein responsible for the cellular uptake of both folates and antifolates. **Pemetrexed** is approved for the treatment of pleural mesothelioma and non-small cell lung cancer. **Pralatrexate** is approved for certain blood tumours.

5-Fluorouracil is usually given intravenously since oral absorption is unpredictable. It is commonly used alongside **folinic acid (leucovorin)** to treat colorectal cancer and is also used for the treatment of various solid tumours, including breast cancer and gastrointestinal tract cancers. Used topically, it is a particularly useful drug for the treatment of skin cancer because it shows a high level of selectivity for cancer cells over normal skin cells. Unfortunately, it has neurotoxic and cardiotoxic side effects. Resistance can occur if the cell produces excess quantities of dUMP to compete with the drug for the active site. **Capecitabine** (Fig. 1) is taken orally and is metabolized to fluorouracil. It can be used as a

(Continued)

BOX 21.4 Clinical aspects of antimetabolites (*Continued*)

monotherapy for metastatic colorectal cancer instead of fluorouracil plus folinic acid. It is also licensed for the first-line treatment of advanced gastric cancer in combination with a platinum agent. The drug can also be useful in the treatment of advanced colon cancer or metastatic breast cancer. A curious side effect in some patients is the elimination of fingerprints as a result of mild inflammation. Patients who have been given an overdose of 5-fluorouracil or capecitabine can be treated with **uradine triacetate** (Fig. 2).

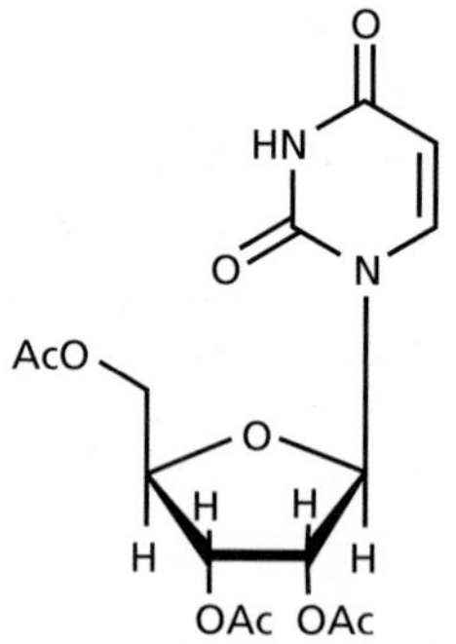

Nelarabine was given accelerated approval by the US FDA in 2005 for the treatment of T-cell acute lymphoblastic leukaemia and T-cell lymphoblastic lymphoma. It acts as a water-soluble prodrug and is demethylated by adenosine deaminase to the less water soluble **ara-G**, which is then phosphorylated by kinases to the active trinucleotide. This is incorporated into DNA, resulting in the triggering of apoptosis.

Raltitrexed is an injectable cytotoxic drug which is used in the treatment of advanced colorectal cancer. Acquired resistance includes impaired cellular uptake, decreased polyglutamation by folylpolyglutamate synthetase (FPGS), or increased thymidylate synthase expression.

Trifluridine was approved in 2015 for the treatment of metastatic colorectal cancer. It is administered with **tipiracil**, which slows the metabolism of trifluridine by inhibiting thymidine phosphorylase. Inhibiting thymidine phosphorylase has the additional effect of slowing angiogenesis by lowering the levels of 2-deoxyribose—a promoter of angiogenesis. **Hydroxycarbamide** is orally administered for the treatment of busulfan-resistant chronic granulocytic leukaemia, and has been used in combination therapy for the treatment of head, neck, and cervical cancers. Resistance can arise due to increased expression of the enzyme.

FIGURE 2 Uridine triacetate—an antidote for capecitabine or 5-fluorouracil overdose.

Pentostatin is a specialist anticancer drug which is given intravenously for the treatment of hairy cell leukaemia.

Cytarabine is used intravenously, subcutaneously, or intrathecally for the treatment of a wide variety of leukaemias. **Gemcitabine** has fewer side effects and is used intravenously to treat pancreatic cancer and non-small-cell lung cancer. It is also administered alongside **cisplatin** for the treatment of advanced bladder cancer or with **paclitaxel** for breast cancer. **Fludarabine** is administered orally or intravenously for the treatment of chronic lymphatic leukaemia, and is available as the 5′ monophosphate prodrug (**Fludara**) to improve solubility.

6-Mercaptopurine and **6-tioguanine** are used primarily for the treatment of acute leukaemias, and are more effective in children than in adults.

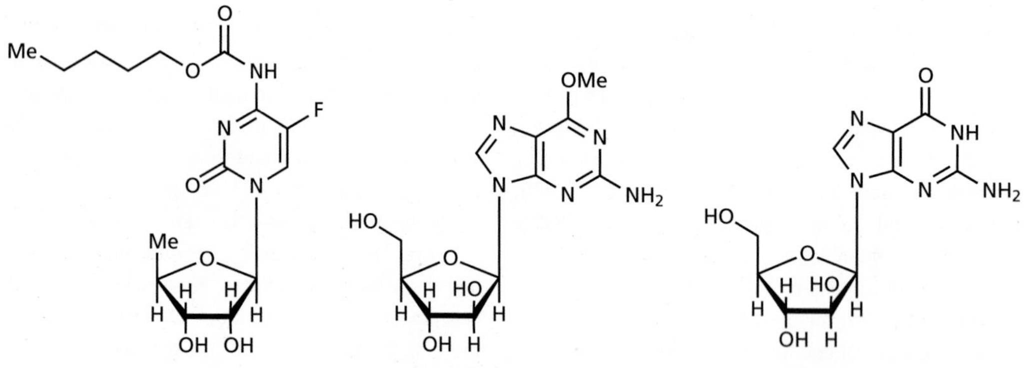

FIGURE 1 Antimetabolites.

- Antimetabolites are agents which inhibit the enzymes involved in the synthesis of DNA or its building blocks.

- Thymidylate synthase catalyses the synthesis of dTMP from dUMP. The cofactor required for this reaction is regenerated by the enzyme dihydrofolate reductase. Inhibition of either enzyme is useful in anticancer therapy.

- Ribonucleotide reductase catalyses the conversion of ribonucleotide diphosphates to deoxyribonucleotide diphosphates. It can be inhibited directly by drugs or indirectly by inhibiting adenosine deaminase. In the latter case, a build-up of dATP results in allosteric inhibition.

- Various nucleosides and purines act as prodrugs and are converted in the cell to agents that inhibit DNA polymerases. The active agents also act as substrates and are incorporated into growing DNA, leading to chain termination or the inhibition of replication.

21.4 Hormone-based therapies

Hormone-based therapies are used for cancers which are hormone dependent. If the cancer cell requires a specific hormone, then a hormone can be administered which has an opposing effect. Alternatively hormone antagonists can be used to block the action of the required hormone. Steroid hormones combine with intracellular receptors to form complexes that act as **nuclear transcription factors**. In other words, they control whether transcription takes place or not (see also Box 8.2 and section 4.9). For clinical aspects of hormone-based therapies see Box 21.5.

21.4.1 Glucocorticoids, estrogens, progestins, and androgens

There are various types of hormones that are used in anticancer therapy, such as the glucocorticoids **prednisolone** and **prednisone** (Fig. 21.30). Prednisone acts as a prodrug and is enzymatically converted to prednisolone in the body.

Estrogens inhibit the production of **luteinizing hormone** (**LH**), and, by doing so, decrease the synthesis of **testosterone**. The most commonly used agents are **ethinylestradiol** (a derivative of **estradiol**) and **diethylstilbestrol** (a non-steroidal estrogen) (Fig. 21.30). **Fosfestrol** is the diphosphate prodrug of diethylstilbestrol.

Progestins used as anticancer agents include **medroxyprogesterone acetate** and **megestrol acetate** (Fig. 21.31).

Androgens are thought to suppress production of LH, resulting in a decrease in estrogen synthesis. The most commonly used agents are **fluoxymesterone** and **testosterone propionate** (Fig. 21.31). The latter is a prodrug which is converted to **dihydrotestosterone**.

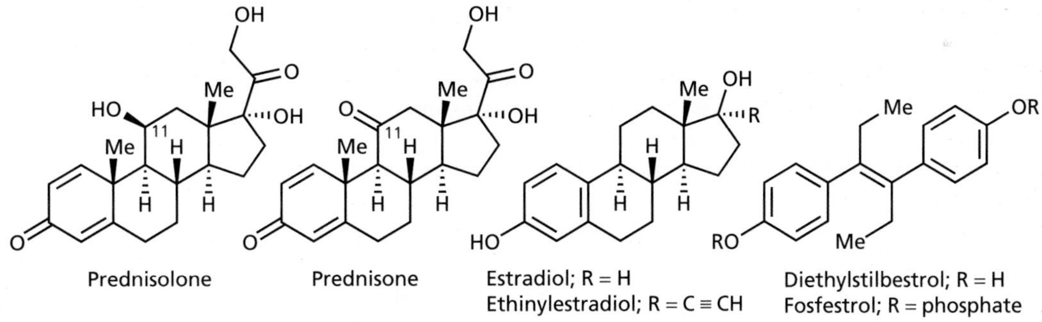

FIGURE 21.30 Glucocorticoids and estrogens.

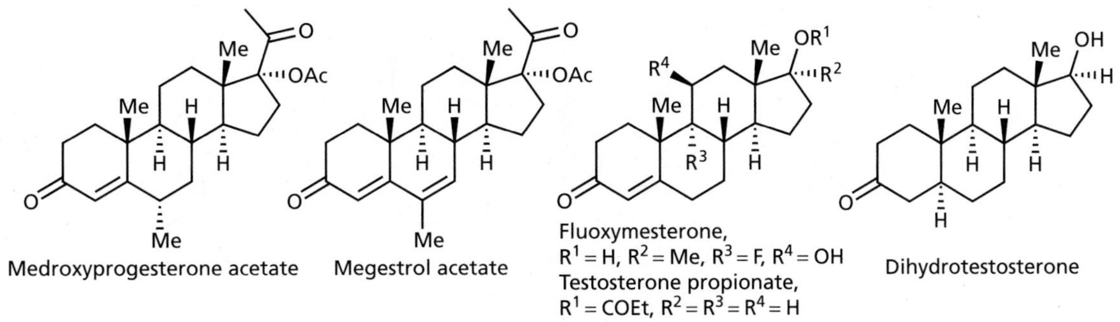

FIGURE 21.31 Progestins and androgens.

21.4.2 Luteinizing hormone-releasing hormone receptor agonists and antagonists

The luteinizing hormone-releasing hormone receptor (LHRH-receptor) is also called the **gonadotrophin-releasing hormone receptor** (GnRH-receptor) and is present in the cell membranes of anterior pituitary cells. It is a G-protein-coupled receptor that is activated by the decapeptide hormone LHRH, and is responsible for raising the levels of luteinizing hormone (LH). This leads, in turn, to increased levels of testosterone. Therefore, agents that reduce the activity of the LHRH-receptor are useful in treating prostate cancers that are sensitive to testosterone levels.

Curiously, most of the agents used in the clinic to reduce the activity of the LHRH-receptor are agonists. These should *increase* receptor activity, but long-term exposure to an agonist desensitizes the receptor, resulting eventually in reduced activity and lowered levels of LH and testosterone. The two agonists most commonly used are **leuprorelin** (also called **leuprolide**) and **goserelin** (Fig. 21.32), both of which are decapeptide analogues of LHRH that are designed to be more resistant to peptidase degradation. This normally takes place next to glycine at position 6, and replacing this amino acid with an unnatural D-amino acid makes this region unrecognizable to the enzyme. Substitution of the glycine residue at position 10 with a suitable group also increases receptor affinity.

Receptor agonists suffer the disadvantage that they initially activate the target receptor and promote tumour growth before desensitization takes place. Therefore, research into receptor antagonists has been carried out to avoid this problem. The only antagonist to be approved so far is **degarelix acetate** (Fig. 21.33), which reached the market in 2009.

21.4.3 Anti-estrogens

Tamoxifen and **raloxifene** (Fig. 21.34) are synthetic agents which antagonize estrogen receptors and prevent estradiol from binding. The mechanism by which these agents work has been extensively studied and is described in Box 8.2. More recent anti—estrogens include **toremifene** and **fulvestrant** (approved in 2002).

21.4.4 Anti-androgens

Flutamide, **bicalutamide**, **enzalutamide**, and **cyproterone acetate** (Fig. 21.35) are used to treat prostate cancer, and work by blocking the action of androgens at their receptors. The lead compound for the design of enzalutamide was an agonist with strong binding affinity for the target receptor. By modifying the structure, it was possible to design structures that lacked agonist activity, but retained good binding affinity as antagonists.

```
        1   2   3   4   5   6   7   8   9   10
pyroGlu-His-Trp-Ser-Tyr-Gly-Leu-Arg-Pro-Gly-NH₂              LHRH
pyroGlu-His-Trp-Ser-Tyr-(D-Leu)-Leu-Arg-Pro-ethylamide       Leuprolide
pyroGlu-His-Trp-Ser-Tyr-(D-(t-Bu)Ser)-Leu-Arg-Pro-Azgly-NH₂  Goserelin
```

FIGURE 21.32 Luteinizing hormone-releasing hormone (LHRH) agonists.

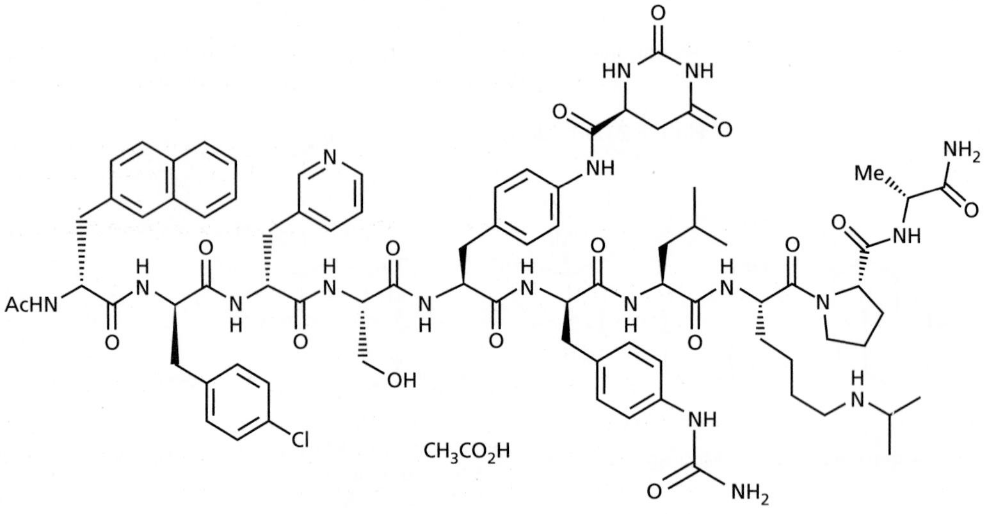

FIGURE 21.33 Degarelix acetate.

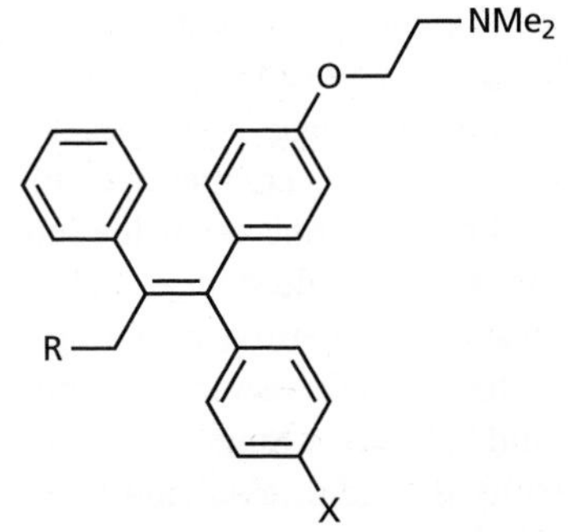

FIGURE 21.34 Anti-estrogens.

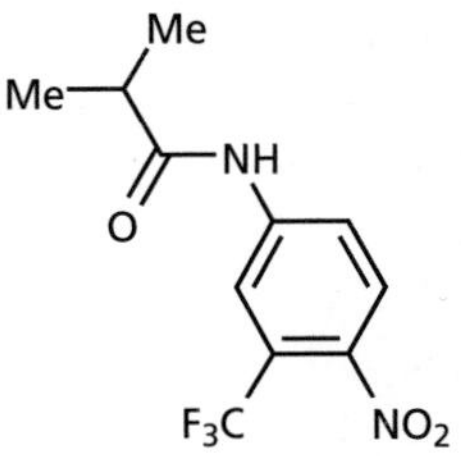

Tamoxifen; X = H, R = Me
4-Hydroxytamoxifen; X = OH, R = Me
Toremifene ; X = H, R = CH₂Cl

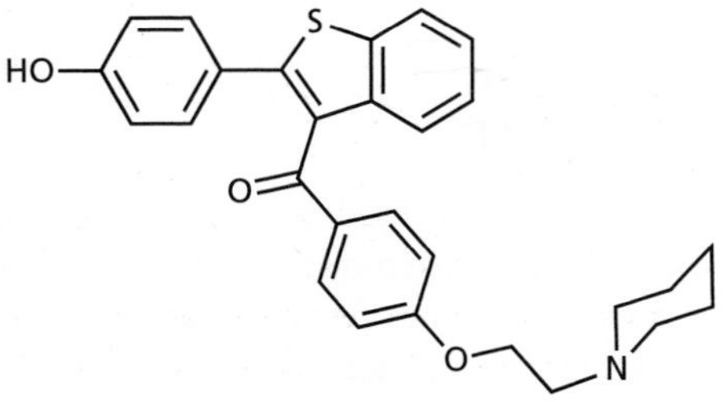

Raloxifene

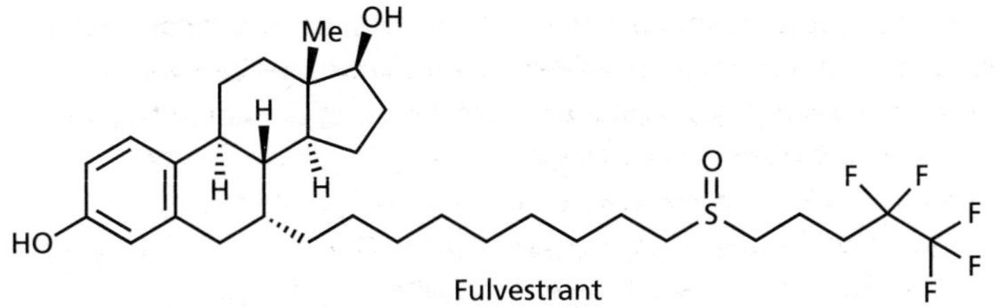

Fulvestrant

Flutamide

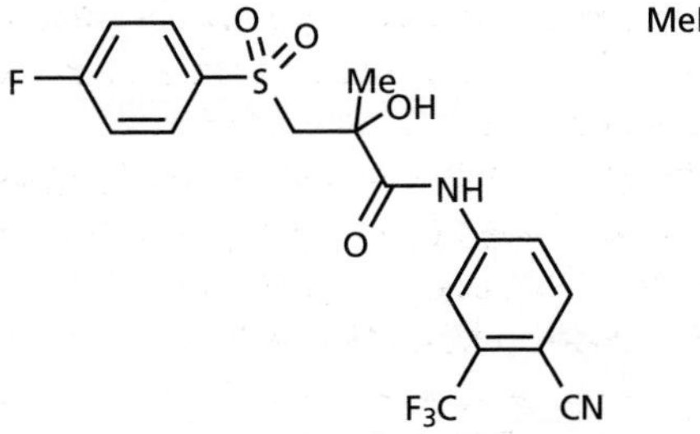

Bicalutamide (1995)

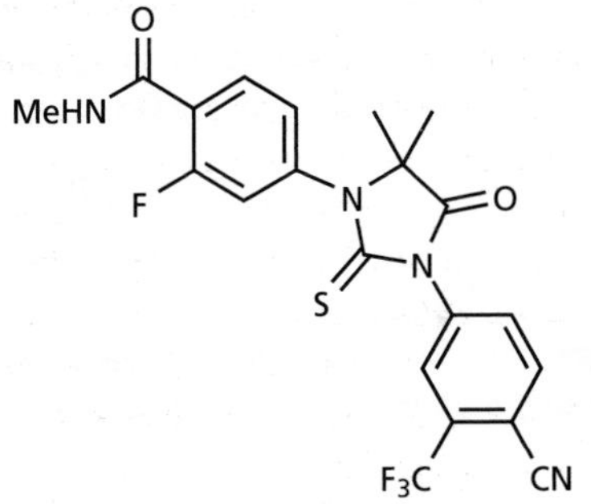

Enzalutamide (2012)

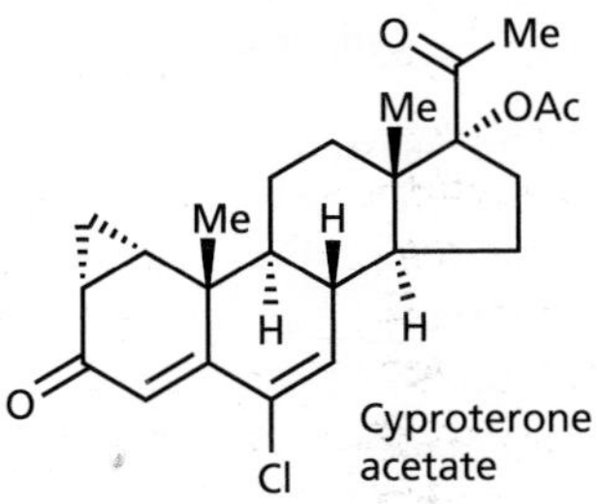

Cyproterone acetate

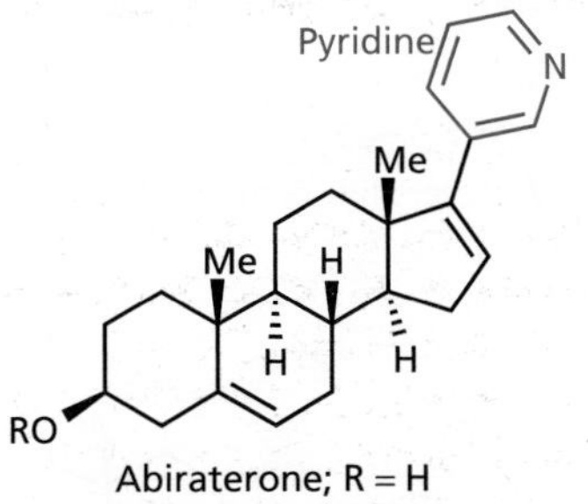

Abiraterone; R = H
Abiraterone acetate: R = MeCO

FIGURE 21.35 Anti-androgens.

Until recently, prostate cancer was treated with a combined therapy of an LHRH agonist and an anti-androgen. A more recent approach is to inhibit a metabolic enzyme called **17α-hydroxylase-17(20)-lyase**. This is a cytochrome P450 enzyme which is involved in the biosynthesis of androgens from cholesterol, and so its inhibition results in lowered androgen levels. **Abiraterone** (Fig. 21.35) is a potent and selective inhibitor of this enzyme, and was approved in 2011 for the treatment of prostate cancer. The pyridine ring plays a key role in its action by interacting with the iron of haem in the enzyme's active site. It is administered as its prodrug, **abiraterone acetate**.

21.4.5 Aromatase inhibitors

Aromatase inhibitors tend to be used as second-line drugs for the treatment of estrogen-dependent breast cancers that prove resistant to tamoxifen. **Aromatase** is a membrane-bound enzyme complex consisting of two proteins; one is a cytochrome P450 enzyme (CYP19A1), and the other is a reductase enzyme that uses NADPH as a cofactor. Aromatase catalyses the formation of an aromatic ring—the last stage in the biosynthesis of estrogens from androgens (Fig. 21.36). The cytochrome enzyme contains haem, which serves to bind both the steroid substrate and oxygen before catalysing the oxidation.

Since the enzyme catalyses the last step of this synthesis, it has been seen as an important target for the design of anti-estrogenic drugs. Two types of inhibitor are used clinically—reversible, competitive inhibitors and irreversible inhibitors acting as suicide substrates.

Aminoglutethimide (Fig. 21.37) is an early example of a reversible, competitive inhibitor, but has disadvantages in that it binds to various cytochrome P450 enzymes and inhibits a range of steroid hydroxylations. This results in undesirable side effects. Using aminoglutethimide as the lead compound, more selective inhibitors have been designed such as **anastrozole** and **letrozole**, both of which are used to treat breast cancer. The *N*-4 nitrogen of the triazole ring interacts with the haem iron of aromatase and prevents binding of the steroid substrate. The anilino nitrogen of aminoglutethimide serves the same purpose.

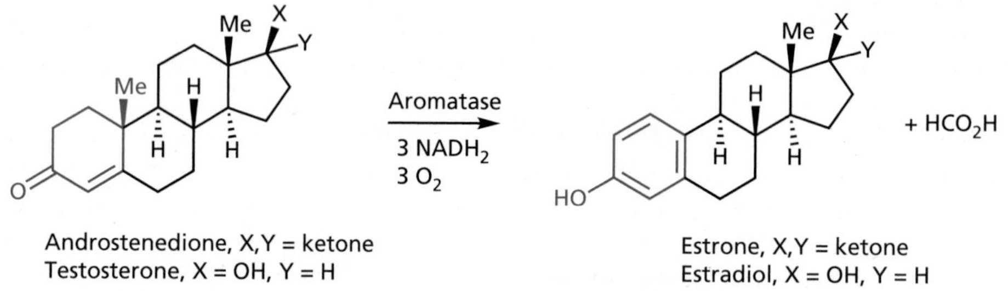

FIGURE 21.36 Reaction catalysed by aromatase.

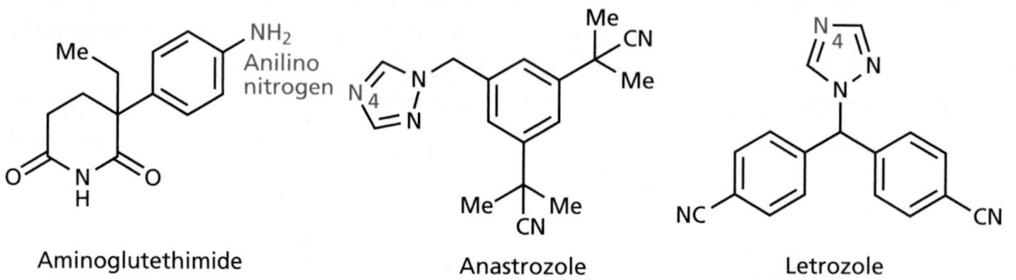

FIGURE 21.37 Reversible, competitive inhibitors of aromatase.

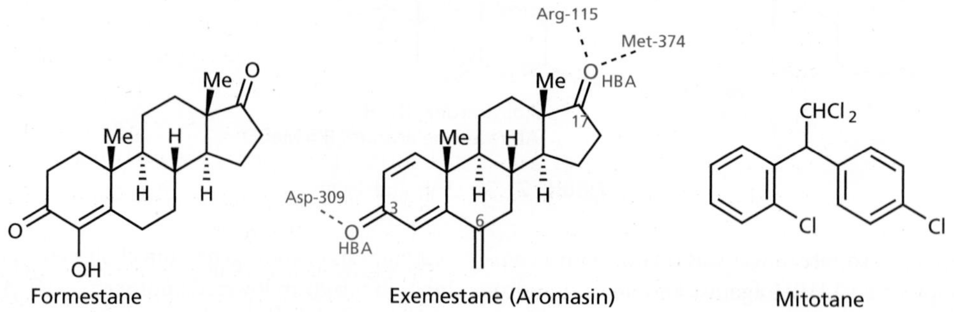

FIGURE 21.38 Formestane, exemestane, and mitotane.

BOX 21.5 Clinical aspects of hormone-based therapies

Prednisolone is used widely for the oral treatment of leukaemias and lymphomas. **Ethinylestradiol** is the most potent estrogen available and is used to treat prostate cancer. **Diethylstilbestrol** is rarely used to treat prostate cancer because of side effects, but is occasionally used for breast cancer. **Fosfestrol** is the diphosphate prodrug of diethylstilbestrol and has been used for the treatment of hormone-resistant metastatic prostate cancer. It is only activated in target cells, where it can reach higher concentrations than using diethylstilbestrol itself. **Progestins** are used primarily to treat advanced endometrial carcinoma that cannot be treated by surgery or radiation. They have also been used as a second-line drug for the treatment of kidney cancers and metastatic breast cancer, but their use in tackling these diseases is now declining. The most commonly used agents are **medroxyprogesterone acetate** and **megestrol acetate**, which can both be administered orally. **Androgens** such as **fluoxymesterone** and **testosterone propionate** are sometimes used to treat metastatic breast cancer. Unfortunately, they have a masculizing effect and so they are only used in a minority of cases. **LHRH agonists** are used to treat advanced prostate and breast cancers. The most commonly used agents are **leuprolide** and **goserelin,** which are administered as their acetates. Leuprolide acetate can be administered daily. Alternatively, it can be inserted into microspheres and administered once monthly, whereupon the drug is released slowly over several weeks. Goserelin acetate can be provided as a slow release implant where the drug is contained within a biodegradable cylindrical polymer rod. This can be implanted into subcutaneous fat every 28 days. Other clinically approved LHRH agonists used in the treatment of prostate and/or breast cancer include **buserelin**, **histrelin**, and **triptorelin**. Administering

LHRH agonists in combination with an anti-androgen (such as cyproterone acetate, flutamide, or bicalutamide) helps to reduce the promotion of tumour growth (tumour flare) in the early stages of treatment.

Degarelix acetate is an LHRH antagonist that has been approved for the treatment of prostate cancer. It avoids the problems of tumour flare, and so it does not need to be administered with an anti-androgen.

Tamoxifen, **toremifine**, and **fulvestrant** are used for the treatment of hormone-dependent breast cancer. The role of **raloxifene** as an anticancer agent is as yet unclear and so it is currently used only for the treatment and prevention of postmenopausal osteoporosis.

Flutamide, **cyproterone acetate**, and **bicalutamide** are used in the treatment of prostate cancer. **Abiraterone acetate** and **enzalutamide** are both approved for the treatment of metastatic castration-resistant prostate cancer that has progressed, despite previous treatment with anticancer drugs such as docetaxel.

Anastrazole, letrozole, and **exemestane** are orally active agents used to treat estrogen-positive breast cancer in postmenopausal women. **Formestane** was also approved for this purpose but has to be administered by intramuscular injection. Therefore, its use has declined in favour of the orally active treatments.

Mitotane (Fig. 21.38) interferes with the synthesis of adrenocortical steroids and is used in the treatment of advanced or inoperable adrenocortical tumours. As it inhibits the activity of the adrenal cortex, corticosteroid replacement therapy is required during its use.

Octreotide is an analogue of somatostatin and is used to treat hormone-secreting tumours of the gastrointestinal tract.

Formestane (Fig. 21.38) is more selective than aminoglutethimide and acts as a suicide substrate to permanently inactivate aromatase. **Exemestane** has also been reported as a suicide substrate and irreversible inhibitor, with the exocyclic double bond at C6 playing a key role in the mechanism of action. X-ray crystal structures reveal that exemestane mimics androstenedione by binding snugly within the hydrophobic binding site, such that its steroidal tetracyclic skeleton forms extensive van der Waals binding interactions. The ketone groups at positions 3 and 17 of both structures also act as hydrogen bond acceptors to the same amino acid residues. Therefore, exemestane can bind effectively by intermolecular interactions before the irreversible reaction takes place.

KEY POINTS

- Hormone-based therapy is used against cancers which are hormone dependent. Hormones can be administered which counteract the offending hormone. Alternatively, antihormonal compounds are administered to prevent the offending hormone from binding to its receptor.

- Glucocorticoids, estrogens, progestins, androgens, and LHRH are used in hormone-based therapy.

- Agents which act as receptor antagonists are used to block estrogens and androgens.

- Enzyme inhibitors are used to block the synthesis of hormones. An important target is the enzyme aromatase which catalyses the last step leading to estrogens.

21.5 Drugs acting on structural proteins

Tubulin is a structural protein which is crucial to cell division (section 2.7.1). The protein acts as a building block for microtubules which are polymerized and depolymerized during cell division. Drugs can block this process by either binding to tubulin to prevent polymerization, or binding to the microtubules to prevent depolymerization. Agents which prevent polymerization do not prevent depolymerization, and so this eventually leads to dissolution of the microtubules and destruction of the mitotic spindle required for cell division. For clinical aspects of the drugs described in this section see Box 21.6.

21.5.1 Agents which inhibit tubulin polymerization

Vincristine, **vinblastine**, **vindesine**, and **vinorelbine** are alkaloids (Fig. 10.3) derived from the Madagascar periwinkle plant (*Catharanthus roseus*, formerly known as *Vinca rosea*), and can bind to tubulin to prevent polymerization. These are discussed in section 10.2.2.

Phyllanthoside (Fig. 21.39) is another natural product that is thought to bind to tubulin and prevent polymerization. It was obtained from the roots of a Costa Rican tree in the early 1970s and entered clinical trials. A variety of other naturally occurring agents have been extracted from marine sources and shown to inhibit microtubule formation. For example, **spongistatin 1** (Fig. 21.39) was extracted from a marine sponge in the Maldives and shows potential as an anticancer agent.

Analogues of the naturally occurring compound **podophyllotoxin** (Fig. 9.4) have already been mentioned in section 21.2.2.1 for their effect on topoisomerase II. Curiously, podophyllotoxin itself has a completely different mechanism of action where it forms a complex with tubulin and prevents the synthesis of microtubules.

Podophyllotoxins belong to a group of compounds called **lignans** and have been isolated from plant sources such as the American mandrake or May apple (*Podophyllum peltatum*), and from the Himalayan plant *Podophyllum emodi*. Extracts of these plants have been used for over 1000 years to treat a variety of diseases, including cancers. For example, it has been recorded that the roots of the wild chervil (*Anthriscus sylvestris*) were used as a treatment for cancer, and it has been shown that these roots contain **deoxypodophyllotoxin**. The crude extract from the above plants is known as **podophyllum** and was shown in 1942 to be effective in the treatment of venereal warts. Podophyllotoxin was eventually isolated from this extract and was used as an anticancer agent for a while. However, its use had to be restricted because of severe side effects. A structural similarity has been noted between podophyllotoxin and **colchicine**—another compound which interacts with tubulin (section 10.2.2).

It is interesting to note that the activity of **epipodophyllotoxin** (section 9.2) against tubulin polymerization is an order of magnitude lower than podophyllotoxin, and when bulky sugar molecules are present (as in **etoposide**—section 21.2.2.1), activity is removed altogether. This implies that the sugar moieties in etoposide form a bad steric interaction with tubulin which prevents binding. **Cryptophycins** (Fig. 21.40) have been isolated from blue-green algae and shown to have an anticancer mechanism which involves the inhibition of microtubule formation. They also inhibit the mechanisms by which

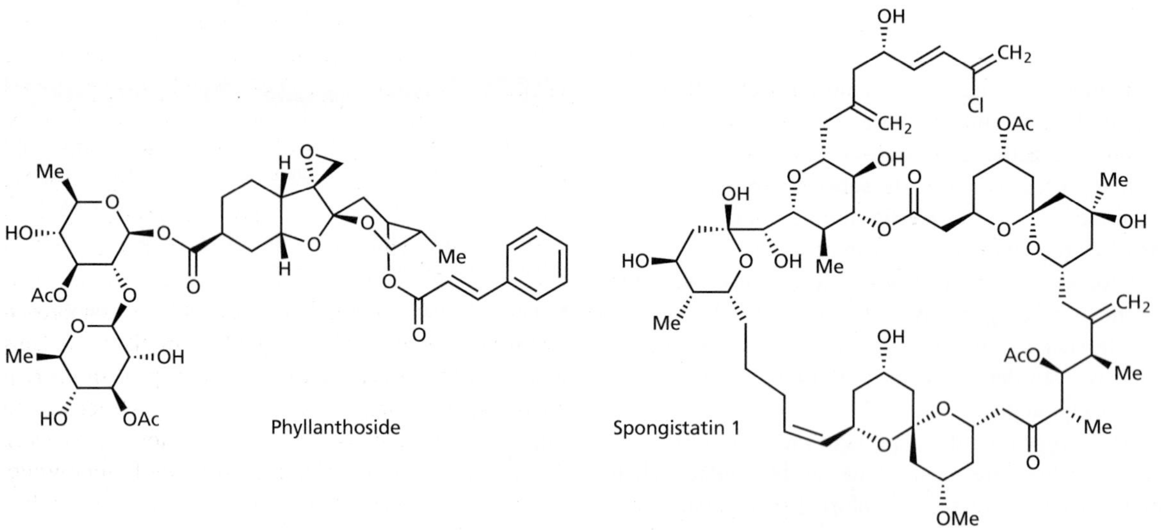

FIGURE 21.39 Natural products inhibiting microtubule formation.

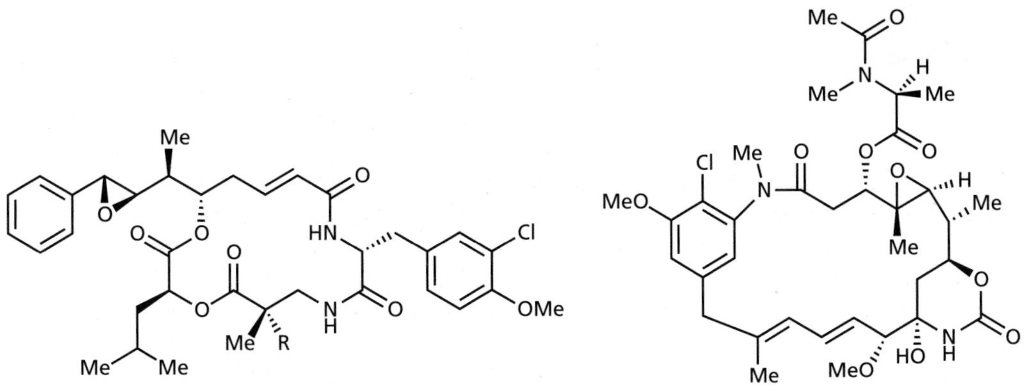

FIGURE 21.40 Cryptophycins and maytansine 1.

microtubules and mitotic spindles function. **Cryptophycin 52** reached phase II clinical trials, but had to be withdrawn because of side effects.

Maytansine 1 (Fig. 21.40) belongs to a group of natural products called the **maytansinoids** which were extracted from an Ethiopian shrub. It has some similarities in structure to the cryptophycins and also inhibits tubulin polymerization, having an activity 1000 times greater than vincristine. Clinical trials had to be abandoned due to its toxic effects and a poor therapeutic window, but it is now being considered as a suitable drug for antibody–drug conjugates (section 21.10.2). **Combretastatins** (Fig. 21.41) are natural products derived from the African bush willow (*Combretum caffrum*), a plant which was used by the Zulus as a medicine and as a charm to ward off enemies. **Combretastatin A-4** is the most active structure in this family and has reached clinical trials as its more water-soluble phosphate prodrug. It shares many of the structural features of other tubulin binding drugs such as colchicine and podophyllotoxin, and binds to tubulin at the same binding region as colchicine. The relative orientation of the two aromatic rings is important and so the *cis*-geometry of the double bond is crucial to activity. The drug has been shown to selectively inhibit the blood supply to tumours and prevent angiogenesis.

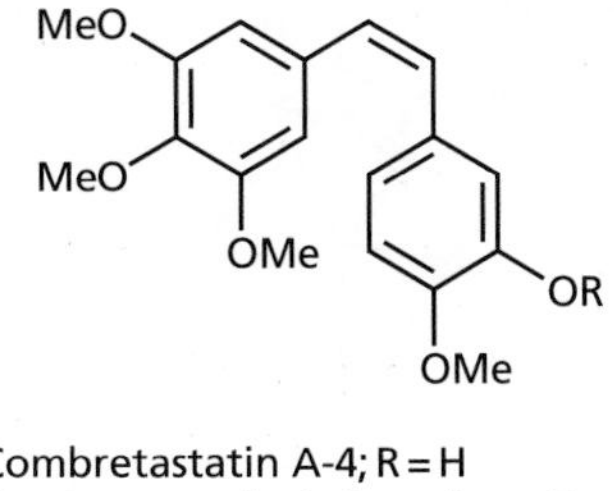

Combretastatin A-4; R = H
Combretastatin A-4 prodrug; R = phosphate

FIGURE 21.41 Combretastatins.

21.5.2 Agents which inhibit tubulin depolymerization

The **taxoids** are an important group of compounds which inhibit tubulin depolymerization and are discussed in section 10.2.2. The best known example is **paclitaxel** (**Taxol**). Semi-synthetic taxoids are currently being investigated to find compounds with better oral bioavailability, improved pharmacological properties, and activity against drug-resistant cancers containing the P-glycoprotein efflux pump.

Since the discovery of paclitaxel, various other natural products have been found to have a similar mechanism of action, and are currently being studied as potential anticancer agents (Fig. 21.42). These include bacterial metabolites called **epothilones**, and marine natural products such as **eleutherobin** isolated from coral. These compounds show several advantages over paclitaxel. Firstly, the epothilones do not appear to be substrates for the P-glycoprotein efflux system, and are potentially effective against drug-resistant cancer cells. Secondly, the epothilones have better aqueous solubility than paclitaxel which may allow the development of better formulations.

A drawback with the epothilones is their metabolic lability, which results from the cleavage of the lactone ring by esterases. They have also been shown to be highly toxic in animal studies. Therefore, research was carried out to find analogues with improved properties. This led to **ixabepilone** (also called **azaepothilone B**), which was approved by the US FDA in 2007 (Fig. 21.42). Ixabepilone contains a lactam ring instead of the lactone ring in the parent epothilone structures. The introduction of the lactam ring improves metabolic stability and reduces toxic side effects. Thus, the lactam acts as a bio-isostere for the lactone group.

These novel agents bind to the same region of tubulin as paclitaxel. A three-dimensional pharmacophore has been developed which encompasses the different

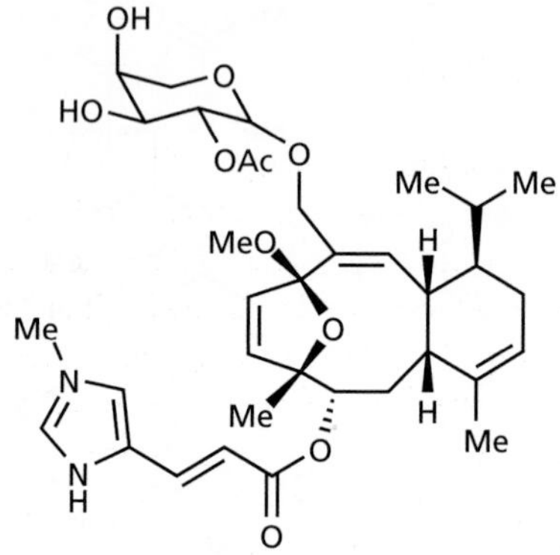

Eleutherobin

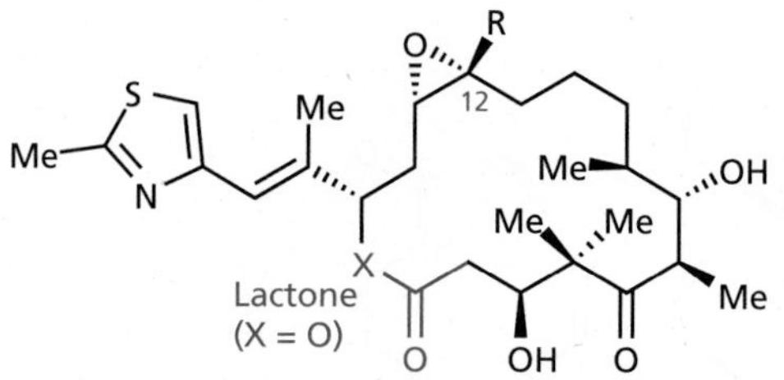

Epothilone A: R = H, X = O
Epothilone B: R = Me, X = O
Ixabepilone (BMS 247550): R = Me, X = NH

FIGURE 21.42 Recently discovered natural products which inhibit microtubule depolymerization.

structures and which is being used as the basis for the design of hybrid molecules that may lead to a third generation of taxoids.

Sarcodictyins (Fig. 21.43) are simplified analogues of eleutherobin and are also active against drug-resistant cancers. Structure–activity relationship (SAR) studies of these compounds have demonstrated the importance of the coloured groups in Fig. 21.43. **Eribulin** is a simplified synthetic analogue of a marine sponge natural product called **halichondrin B**. It was approved in 2010 and is now available in Europe, USA, and Japan.

For additional information see Web article 27: Hybrid drugs in cancer chemotherapy on the Online Resource Centre at www.oxfordtextbooks.co.uk/orc/patrick6e/

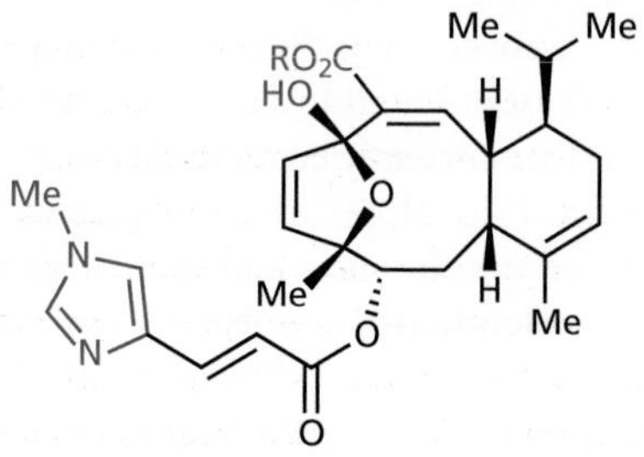

Sarcodictyin A; R = Me
Sarcodictyin B; R = Et

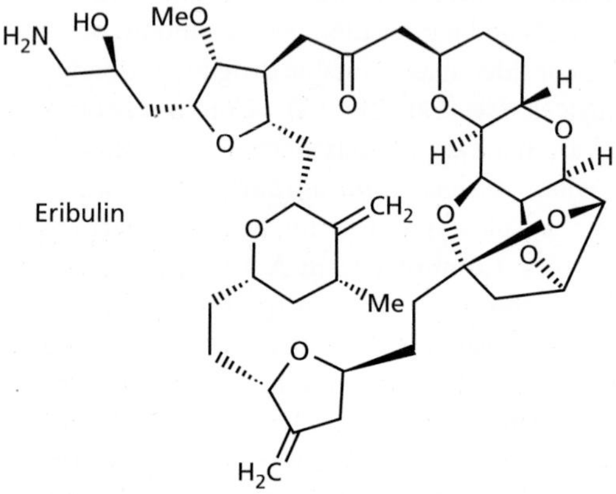

Eribulin

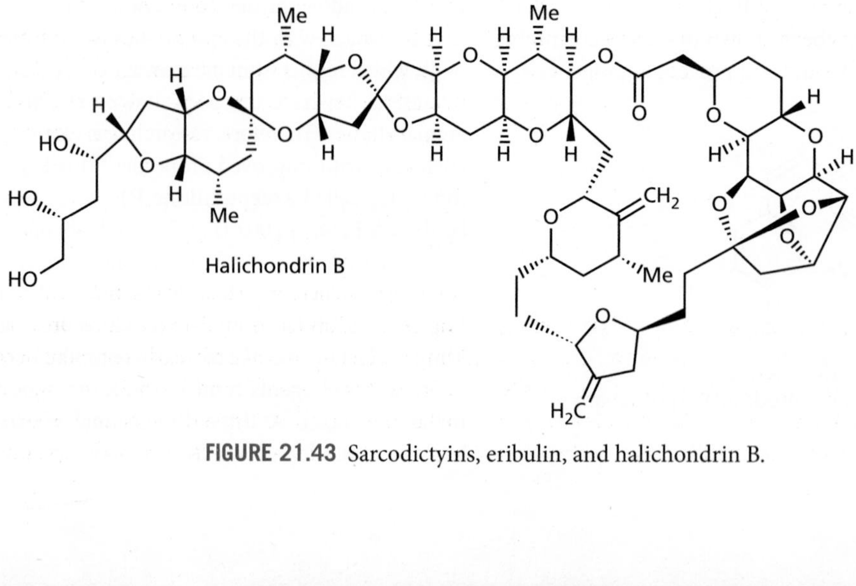

Halichondrin B

FIGURE 21.43 Sarcodictyins, eribulin, and halichondrin B.

BOX 21.6 Clinical aspects of drugs acting on structural proteins

The Vinca alkaloids **vincristine, vinblastine**, and **vindesine** are used intravenously to treat a variety of cancers, including leukaemias, lymphomas, and some solid tumours, such as breast and lung cancer. **Vinorelbine** is widely used for the intravenous treatment of advanced breast cancer and non-small-cell lung carcinomas. Neurotoxicity occurs with all the Vinca alkaloids and may limit their use in some patients. Resistance can arise from overexpression of the P-glycoprotein involved in transporting drugs out of the cell.

Podophyllotoxin is the agent of choice for the treatment of genital warts, but it must be handled with care because of its toxicity. Preparations are available which contain the pure compound or the plant extract (**podophyllum**).

Paclitaxel shows outstanding therapeutic activity against solid tumours, and was approved in 1992 for the treatment of breast and ovarian cancers. **Docetaxel** was approved for the treatment of advanced breast cancer in 1996. Both drugs are in clinical trials for the treatment of a variety of other cancers. They both halt the cell division cycle mainly at the G_2/M stage. Apoptosis then takes place. A problem with the use of taxoids is the fact that they cannot be taken orally and they also have various undesirable side effects. Moreover,

therapy often leads to the development of multi-drug resistance. This involves several mechanisms, including tubulin mutation which results in weaker binding interactions, and overexpression of the P-glycoprotein transport protein which results in faster efflux from the cell. Another problem with paclitaxel is its poor solubility, which makes formulation difficult. Indeed, some patients cannot tolerate the solvents required. Research is being carried out to design solvent-free drug delivery methods such as nanoparticles consisting of albumin-bound paclitaxel. **Cabazitaxel** was introduced in 2012 for the treatment of hormone-refractory prostate cancer, and is used in combination with prednisone. Cabazitaxel is less susceptible to the efflux mechanisms that diminish the effectiveness of paclitaxel and docetaxel.

In 2007, **ixabepilone** was approved by the US FDA (but not by the EMEA) for the treatment of advanced breast cancer. It is currently undergoing clinical trials for the treatment of other types of cancer.

Eribulin binds to the ends of microtubules to prevent depolymerization and trigger apoptosis of cancer cells. It is approved for the treatment of inoperable and recurrent breast cancers.

KEY POINTS

- Agents which inhibit the polymerization or depolymerization of microtubules are important anticancer agents.

- The Vinca alkaloids, podophyllotoxin, the combretastatins, and a variety of other natural products bind to tubulin and inhibit the polymerization process.

- Paclitaxel and its derivatives bind to tubulin and accelerate polymerization by stabilizing the resulting microtubules. Newer analogues are being investigated which show better oral bioavailability, improved pharmacological properties, and activity against drug-resistant cancers.

- A variety of natural products have been discovered which have a similar mechanism of action to paclitaxel.

21.6 Inhibitors of signalling pathways

Most traditional anticancer drugs are cytotoxic both to cancer and normal cells, and any selectivity relies on a greater concentration of the agents within cancer cells. Nowadays, cancer chemotherapy is on the verge of a revolution. Advances in genetics and molecular biology have led to an ever-increasing understanding of the molecular processes behind specific cancers, and the identification of a variety of molecular targets which are either unique to a cancer cell or are overexpressed compared with normal cells. The design of agents that will act on these targets promises the development of more selective anticancer agents with less toxic side effects. Understanding the defects in a cell's signalling pathways and identifying suitable targets have already resulted in clinically useful drugs. Suitable targets include the receptors for growth hormones, and the various signal proteins and kinases in the signal transduction pathways. The following sections illustrate some of the most promising lines of research, but it should be appreciated that there is a vast amount of research being carried out in this area and it is not possible to give a comprehensive coverage of it all.

21.6.1 Inhibition of farnesyl transferase and the Ras protein

It has been observed that an abnormal form of the signalling protein **Ras** (sections 5.4.1–5.4.2) is present in 30% of human cancers, and is particularly prevalent in colonic and pancreatic cancers. Abnormal Ras derives from a mutation of the *ras* gene to form a *ras* oncogene. Ras proteins are an inherent component of the cellular signalling pathways which control cell growth and division. They are small G-proteins which bind GDP when they are in the resting state, and GTP when they are in the active state. Binding to GTP is temporary, as the protein can autocatalyse its hydrolysis back to GDP and return to the resting state. However, mutant Ras proteins have lost the ability to hydrolyse GTP and are constantly active. As Ras is an integral part of the signalling pathways that control cell growth and division, it is believed that this contributes to the development of cancer. Therefore, finding methods of 'neutralizing' Ras could be useful in combating cancer. Direct inhibition of Ras has been attempted by designing inhibitors that are intended to compete with GTP for the GTP binding site. However, GTP has picomolar affinity for Ras, and so designing an effective inhibitor is likely to be extremely challenging. Therefore, other less direct approaches have been attempted.

One of these approaches centres on a zinc metalloenzyme called **farnesyl transferase (FT)**. This enzyme is responsible for attaching a 15-carbon farnesyl group to the Ras protein when it is in the cytoplasm of the cell. The farnesyl group is hydrophobic and can interact with the hydrophobic centre of the cell membrane. As a result, it acts as an anchor to hold the Ras protein to the inner surface of the cell membrane. This is necessary if the Ras protein is to interact with other elements of the signal transduction process. Inhibitors of the FT enzyme have been shown *in vitro* to reverse malignancy in cancer cells containing the *ras* oncogene, without affecting normal cells.

The enzyme mechanism (Fig. 21.44) involves the binding of **farnesyl diphosphate (FPP)** to the active site, followed by the Ras substrate. This order of binding is important, as FPP is actually involved in binding the Ras protein. Farnesylation can then take place where a cysteine residue on the Ras protein displaces a pyrophosphate leaving group from FPP. Magnesium and zinc ions

are present in the active site as cofactors. The former is involved in complexing the negatively charged pyrophosphate group to make it a good leaving group, while the latter interacts with the thiol group of cysteine to enhance its nucleophilicity. Following farnesylation, the terminal tripeptide of Ras is cleaved by the **Ras converting enzyme** (Rce1), and the resulting carboxylic acid is methylated to form a methyl ester by **isoprenylcysteine carboxylmethyltransferase** (ICMT1). Finally, the enzyme **palmitoyl transferase** (PTase) links the prenylated Ras to the cell membrane.

Peptide-based inhibitors of FTase have been developed that mimic the terminal tetrapeptide moiety of the Ras protein. This region is common to different types of Ras protein, and is known as the **CaaX peptide** (C stands for cysteine, a stands for valine, isoleucine, or leucine, and X stands for methionine, glutamine, or serine). Studies on a variety of tetrapeptides demonstrated that placing an aromatic amino acid such as phenylalanine next to X transformed the tetrapeptide from a substrate to an inhibitor. Inhibition could also be enhanced by adding a cysteine residue, since the thiol group in cysteine's side chain can interact with the enzyme's zinc cofactor. A number of peptide-based inhibitors were designed containing these features. However, they had poor pharmacokinetic properties and could not be progressed further.

Better progress was made with the design of nonpeptide inhibitors such as **lonafarnib** and **tipifarnib** (Fig. 21.45), both of which reached late stage clinical trials. Tipifarnib contains an imidazole ring which acts as the ligand for the zinc cofactor. Lonafarnib was developed from a lead compound that was discovered by screening compound libraries, and is 10 000 times more active than the original lead compound. Remarkably, it has no ligand for the zinc cofactor.

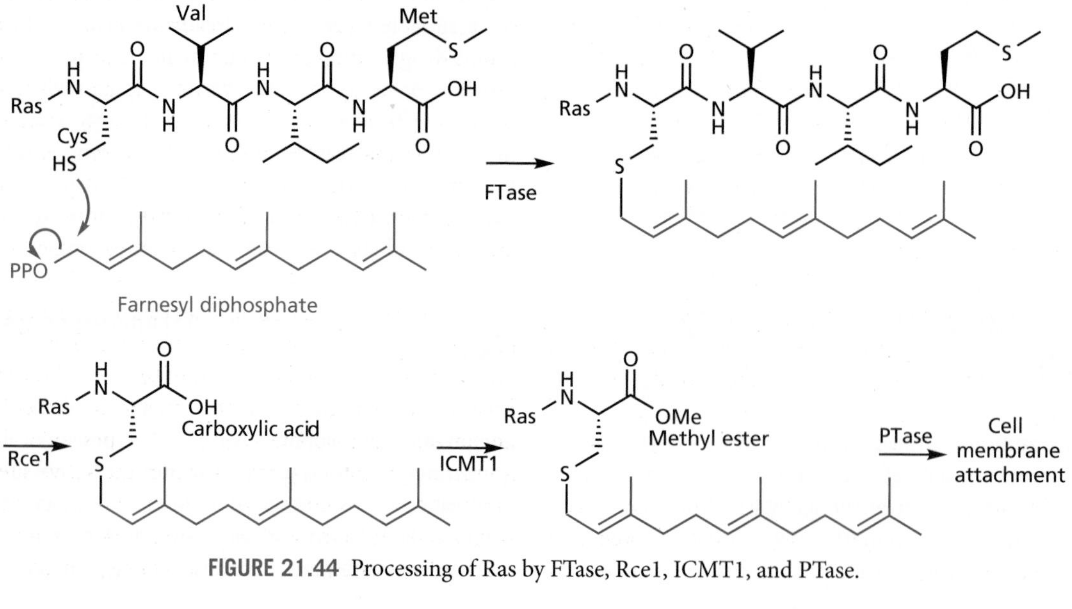

FIGURE 21.44 Processing of Ras by FTase, Rce1, ICMT1, and PTase.

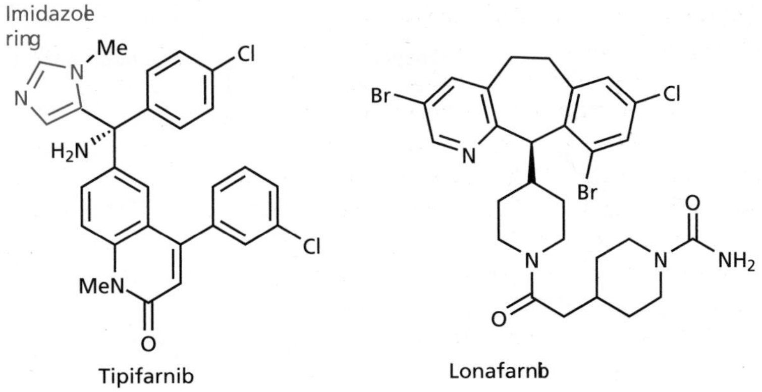

FIGURE 21.45 Tipifarnib and lonafarnib.

Although these farnesyl transferase inhibitors (FTIs) showed promise as anticancer agents, clinical trials were disappointing, and the agents were less effective than anticipated against solid tumours. This has been attributed to alternative methods of prenylation. There are three human Ras proteins (H-Ras, N-Ras, and K-Ras). FTIs inhibit the farnesylation of all three, but can only inhibit the *cellular* functions of H-Ras, as the other two Ras proteins can be prenylated by a related prenylating[1] enzyme called **geranylgeranyltransferase** (**GGTase**). This enzyme catalyses prenylations using a 20-carbon structure called **geranylgeranyl diphosphate** as the prenylating agent. Normally, GGTase prenylates proteins having the CaaX motif where X = Leu, but it can prenylate proteins which are normally prenylated by FT if the latter enzyme is inhibited. This allows these proteins to bind to cell membranes and still be functional, thus bypassing the inhibition of FT. Attempts were made to develop agents that inhibited both enzymes, but these proved too toxic.

Attention has now switched to targeting the later enzymes involved in processing the C-terminus of the Ras protein (Fig. 21.43). Another possible approach is to inhibit the biosynthesis of farnesyl diphosphate by inhibiting the enzyme farnesyl pyrophosphate synthase.

As a final point, it has been observed that **statins** inhibit tumour cell growth, and this has been ascribed to an effect on Ras farnesylation. These structures inhibit the HMGR enzyme (Case study 1) involved in the biosynthetic pathway to both steroids and isoprenoids. Consequently, statins can influence Ras protein farnesylation by lowering the levels of farnesyl diphosphate present in the cell. Other structures are being investigated that can block Ras cell signalling by interacting with Ras itself or with targets involved in the signalling pathway, for example **rasfonin** (section 12.4.1.2).

[1]Prenylation is the term used to describe the formation of a covalent bond between a molecule such as a protein, and a prenyl moiety such as a farnesyl or geranylgeranyl group.

21.6.2 **Protein kinase inhibitors**

Protein kinases are enzymes which phosphorylate specific amino acids in protein substrates. It is estimated that there may be over 500 different types of protein kinase in the body, and a vast amount of research is currently being undertaken on potential inhibitors of these enzymes. Some of these kinases are within the cytoplasm of the cell (sections 5.2 and 5.3), while others (protein kinase receptors) are associated with the cell membrane (sections 4.8 and 5.4). The latter structures have an extracellular binding site to receive an external molecular messenger, and an intracellular kinase active site which is activated when the messenger binds to the receptor's binding site. The chemical messengers involved are a wide variety of growth hormones and growth factors which trigger the start of a signalling cascade that involves the various cytoplasmic protein kinases. This process ultimately controls transcription of specific genes in DNA leading to cell growth and cell division. In many cancers, it has been observed that there is an excess of a particular growth hormone or growth factor, or an excessive quantity of a particular protein kinase or protein kinase receptor. As these structures are intimately involved in the signal transduction processes which drive cell growth and cell division, it is reasonable to assume that protein kinase inhibitors will be useful anticancer agents.

Protein kinases can be divided into two main categories—the tyrosine kinases and the serine-threonine kinases. More recently, histidine kinases have been discovered which phosphorylate the nitrogen of a histidine residue. The tyrosine kinases phosphorylate the phenol group of tyrosine residues, whereas the serine-threonine kinases phosphorylate the alcohol group of serine and threonine residues (section 5.2.2). All the kinases use the cofactor **adenosine triphosphate** (ATP) as the phosphorylating agent, and so there is a region within the active site that binds ATP, and a neighbouring region that binds the substrate. In theory, it should be possible to design inhibitors that bind to one or other of these regions, but, so far, the

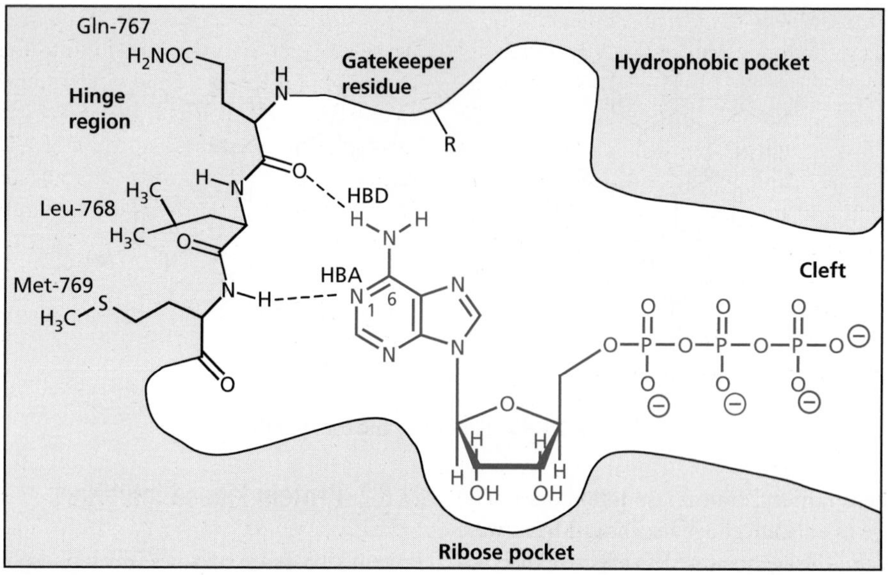

FIGURE 21.46 Binding of ATP to the kinase active site of the epidermal growth factor receptor.

best results have been achieved with inhibitors capable of binding to the cofactor binding region. Considering the fact that there are so many kinases and that they all use ATP as the phosphorylating agent, it was initially thought that achieving selectivity between kinases would be a major problem. This has not turned out to be the case. Crystal structures of protein kinases containing bound ATP reveal that ATP fits quite loosely to the active site, and that there are areas which remain unoccupied. There are also significant differences between kinases with respect to the amino acids present in these unoccupied areas. As a result, it is quite possible to design selective inhibitors.

A knowledge of how ATP is bound to the kinase active site has helped enormously in the design of potent and selective agents. For example, the binding interactions of ATP with the kinase active site of the **epidermal growth factor receptor** (**EGFR**) are shown in Fig. 21.46 and are representative of all the kinase active sites. The purine base is buried deep in the active site and makes two important hydrogen bonding interactions with the protein backbone in a region of the protein known as the **hinge region**, so called because it connects two distinct lobes of the enzyme. The heterocyclic ring also forms van der Waals interactions with the amino acids round about it. The ribose sugar is bound into a **ribose binding pocket** and the triphosphate chain lies along a cleft leading to the surface of the enzyme. The ionized triphosphate interacts with two metal ions and with several amino acids through hydrogen bonding. There are also various areas of unoccupied space, one of which is particularly important and consists of a hydrophobic pocket opposite the ribose binding pocket (**hydrophobic pocket I**). At the entrance to this pocket, there is an important amino acid residue which is called the **gatekeeper residue**. In some kinases, the

gatekeeper residue is large and blocks access to the pocket, whereas in other kinases the gatekeeper residue is small, allowing drugs to be designed that will access and interact with the pocket. The hydrophobic pocket is also lined by different amino acids depending on the kinases involved, which opens up the possibility of designing drugs that can distinguish between the hydrophobic pocket of one kinase active site from the hydrophobic pocket of another.

Kinases exist in an active conformation as well as one or more inactive conformations. The switch by which a kinase changes from an inactive conformation to the active conformation is controlled by an **activation loop**. Activation usually occurs by phosphorylation of residues on this loop, which causes the loop to move position. This has a marked effect on the position of a conserved triad of amino acids (Asp, Phe, and Gly) near the start of the activation loop, whereby Asp and Phe are orientated towards the binding site (**DFG-in**). In inactive forms of the enzyme, an extra hydrophobic region (**hydrophobic pocket II**) is exposed that is not exposed in the active form, and which is close to hydrophobic pocket I. This provides the potential for designing novel kinase inhibitors capable of binding and stabilizing an inactive conformation.

Most kinase inhibitors are classed as **type I** or **type II inhibitors**. In both cases, they compete with ATP for the binding site. Type I inhibitors bind to the active conformation of the enzyme, whereas type II inhibitors bind to the inactive conformation. Both types of inhibitor occupy the region normally occupied by ATP, but type II inhibitors differ from type I inhibitors in that they can form interactions with neighbouring hydrophobic binding regions that are not occupied by ATP when it binds. Protein kinase inhibitors such as **gefitinib**, **erlotinib**, **SU11248**, and **seliciclib** are type I inhibitors, whereas agents such as **imatinib**, **nilotinib**, **sorafenib**, and **vatalanib** are type II inhibitors. **Sunitinib** and **dasatinib** are able to bind to both active and inactive forms of the same kinase enzyme and could be defined as type I or type II. The story is further complicated by the fact that some inhibitors act as a type I inhibitor at one kinase target, and as a type II inhibitor at another.

It has been observed that there is a significant variation of amino acids in hydrophobic region II between different kinases, suggesting that type II inhibitors have the potential to be more selective. However, as the amino acids in the additional hydrophobic region are less conserved, there is a greater possibility of drug resistance caused by random mutations. These could result in a viable kinase that would fail to bind the inhibitor. There is also the problem that Type II inhibitors tend to be larger molecules, which may limit their ability to cross cell membranes.

To date, all clinically important inhibitors have binding interactions that mimic the adenine interactions of the cofactor ATP; namely two or three hydrogen bonds to the hinge region, plus van der Waals interactions with surrounding amino acids. Selectivity is obtained by designing interactions with regions of the active site not occupied by ATP, such as hydrophobic pocket I, or with the gatekeeper residue. In the case of type II inhibitors, additional van der Waals interactions are possible with the extra hydrophobic region II, as well as hydrogen bonds to two conserved amino acids (Glu and Asp) in that same region. The aspartate residue is part of the conserved triad mentioned above.

Another three types of kinase inhibitor have now been identified. **Type III inhibitors** bind to the inactive conformation of the enzyme and bind purely to regions unoccupied by ATP, such as hydrophobic pockets I and II. Therefore, they do not compete with ATP for binding. Such agents have been classed as allosteric inhibitors. **Type IV inhibitors** are also allosteric inhibitors, but bind to a more distant allosteric binding site that is not adjacent to the ATP binding region. Finally, **type V inhibitors** bind in a similar way to type II inhibitors, but do so with the active conformation of the enzyme.

21.6.2.1 Kinase inhibitors of the epidermal growth factor receptor (EGFR)

EGFR is a membrane-bound tyrosine kinase receptor that has an extracellular binding site for epidermal growth factor (EGF) and an intracellular kinase active site (sections 4.8.2–4.8.4). Several agents have been studied as EGFR kinase inhibitors and the first of these to reach the clinic was **gefitinib** (**Iressa**) (Fig. 21.47 and Box 21.7).

Gefitinib was developed by Astra Zeneca and belongs to a group of structures known as the 4-anilinoquinazolines. It was developed from a potent inhibitor (I in Fig. 21.48) which had various important features previously identified by SAR studies, namely a secondary amine, electron-donating substituents at positions 6 and 7, and a small lipophilic substituent on the aromatic ring. The structure had useful *in vitro* activity, but its *in vivo* activity

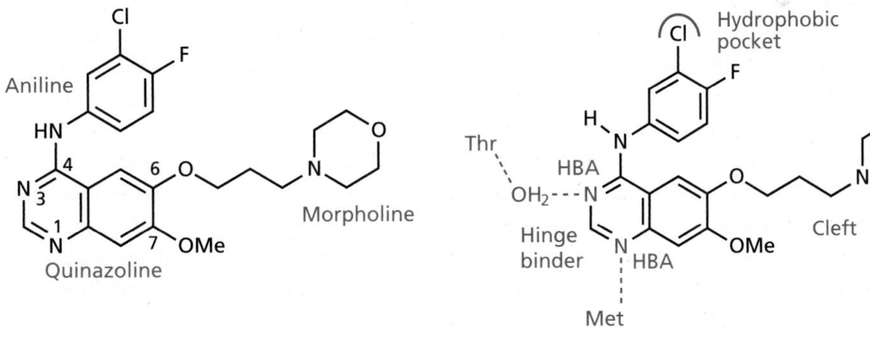

FIGURE 21.47 Structure and binding interactions of gefitinib (Iressa).

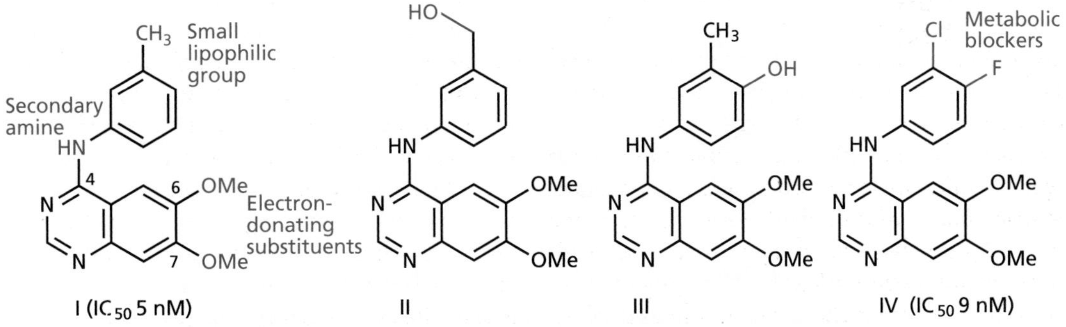

FIGURE 21.48 Design of a metabolically stable analogue of structure I.

was hampered by the fact that it was rapidly metabolized by cytochrome P450 enzymes to give two metabolites. Oxidation of the aromatic methyl group resulted in metabolite II, whereas oxidation of the aromatic *para*-position resulted in metabolite III. Both these features are known to be vulnerable to oxidative metabolism (section 11.5), and so it was decided to modify the structure to block both metabolic routes. In structure IV in Fig. 21.48, the methyl group was replaced by a chloro substituent. This can be viewed as a bio-isostere for the methyl group as it is of similar size and lipophilicity, but it has the advantage that it is resistant to oxidation. A fluoro substituent was chosen to block oxidation of the aromatic *para*-position. Fluorine is essentially the same size as hydrogen, and so there is little risk of any adverse steric effects arising from its introduction. Although the resulting compound was less active *in vitro* as an enzyme inhibitor, it showed better *in vivo* activity since it proved resistant to metabolism. Further modifications were then carried out to optimize the pharmacokinetic properties of the drug. A variety of alkoxy substituents at the 6-position were tried, culminating in the discovery of gefitinib. This contains a morpholine ring, which is often introduced to enhance water solubility. Because the morpholine ring includes a basic nitrogen, it is possible to protonate it and form water-soluble salts of the drug (e.g. hydrochloride or succinate salts). Note that the addition of a water-soluble 'handle' is a common feature in many kinase inhibitors. The group plays no role in target binding, and it is important that it is positioned in such a way that it is in a solvent-exposed region of the drug when the latter is bound to the target binding site. In other words, the group should protrude from the binding site and be exposed to the surrounding aqueous environment. This avoids the energy penalty that would be required if the surrounding solvation coat had to be stripped away from such a polar group (see section 1.3.6). The acidity or basicity of this group also plays an important role in how much of the drug is involved in plasma-protein binding, which affects the drug's distribution and metabolism.

The binding interactions of ATP and gefitinib are worth comparing. Both compounds contain a pyrimidine ring with an amine substituent. In ATP, this substituent binds as an HBD (Fig. 21.46), and it would be tempting to assume that gefitinib binds in a similar manner. In fact, this is not the case. Two hydrogen bonds are still formed to the binding site, but the amine substituent at position 4 of gefitinib is not involved. Instead, both nitrogen atoms in the quinazoline ring act as HBAs (Fig. 21.47). The nitrogen atom at position 1 binds directly to a methionine residue, whereas the nitrogen at position 3 forms a hydrogen bond to a water molecule, which acts as a hydrogen bonding bridge to a threonine residue. This emphasizes the dangers of making assumptions about binding interactions based purely on molecular similarity. The binding site for ATP is quite spacious, so it is perfectly feasible for molecules to bind in different modes, even if they are within the same structural class. Therefore, it is advisable to determine the crystal structure of the enzyme–inhibitor complex in order to establish the binding mode.

A number of other 4-anilinoquinazoline structures have now been developed as reversible EGFR kinase inhibitors. **Erlotinib** (Fig. 21.49) binds like gefitinib, and contains an acetylene group that fits into the hydrophobic pocket guarded by the gatekeeper residue threonine. **Icotonib** is similar in structure, but contains a crown ether as a solubilizing group.

Unlike gefitinib, **lapatinib** (Fig. 21.50) binds to an inactive form of the kinase. This allows the fluorobenzyloxy substituent to interact with a hydrophobic pocket that is not exposed in the active form. The chain containing the amine and the sulphonyl group increases aqueous solubility and is located in a region of the active site that is exposed to solvent. Lapatinib also has potent activity for another receptor tyrosine kinase called **ErbB-2** (**HER2**), which is structurally related to EGFR (section 4.8.4). Thus, lapatinib is a **dual-action inhibitor** that can be used for cancers that overexpress both EGFR and ErbB-2.

Vandetanib was approved in 2011 as an **extended-spectrum agent** capable of inhibiting EGFR, ErbB-2, VEGFR, and RET.

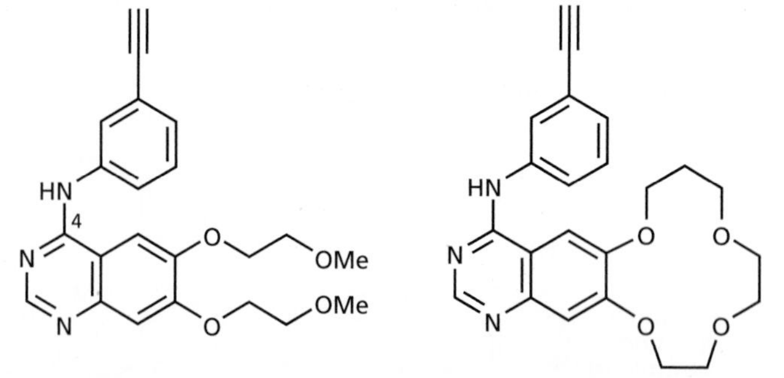

Erlotinib (Tarceva) IC$_{50}$ 2 nM   Icotinib

FIGURE 21.49 Inhibitors of the epidermal growth factor receptor kinase.

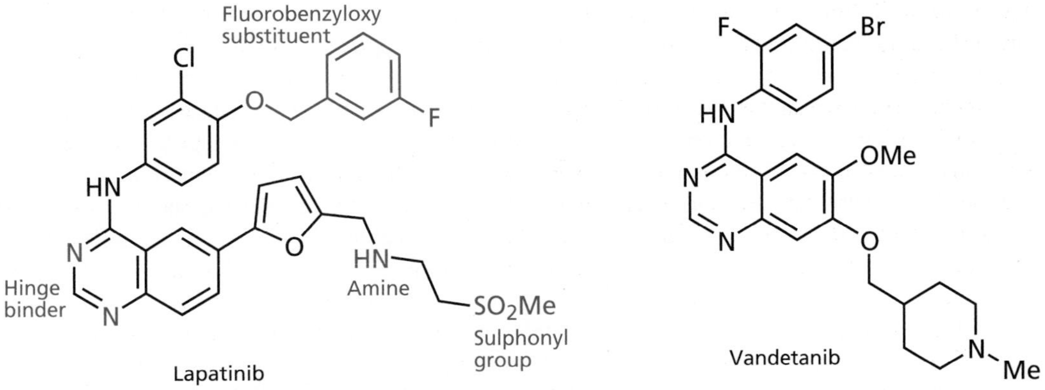

FIGURE 21.50 Inhibitors of the epidermal growth factor receptor kinase.

Not all patients respond to treatment by the aforesaid inhibitors, and the effectiveness of the drugs can diminish after only 6–8 months. In particular, drug resistance can arise as a result of mutation to the gatekeeper residue where Thr-790 is mutated to methionine. The bulkier methionine side chain may prevent the drugs binding on steric grounds, although an alternative theory states that the mutant has an increased affinity for ATP, which allows the latter to compete more effectively with reversible inhibitors. It has also been suggested that receptor internalization may be taking place where the receptor–inhibitor complex ends up in vesicles within the cell. The lower pH of these vesicles may cause the ligand to leave the receptor binding site, allowing free receptor to be returned to the cell membrane.

A number of irreversible kinase inhibitors have recently been approved which have proved more effective against such drug-resistant mutations (Fig. 21.51). These agents contain an electrophilic acrylamide moiety (called the warhead), which reacts with a conserved cysteine residue in EGFR (Cys-797) by means of a Michael addition reaction (Fig. 21.52).

Afatinib is a dual-action, irreversible inhibitor that targets both EGFR and ErbB-2. It was designed to be effective against mutated forms of EGFR where the gatekeeper residue (Thr-790) has been mutated to a methionine residue. Since afatinib is an irreversible inhibitor, dose levels and dosing frequencies are dependent on how quickly cells can synthesize new enzyme. The drawback

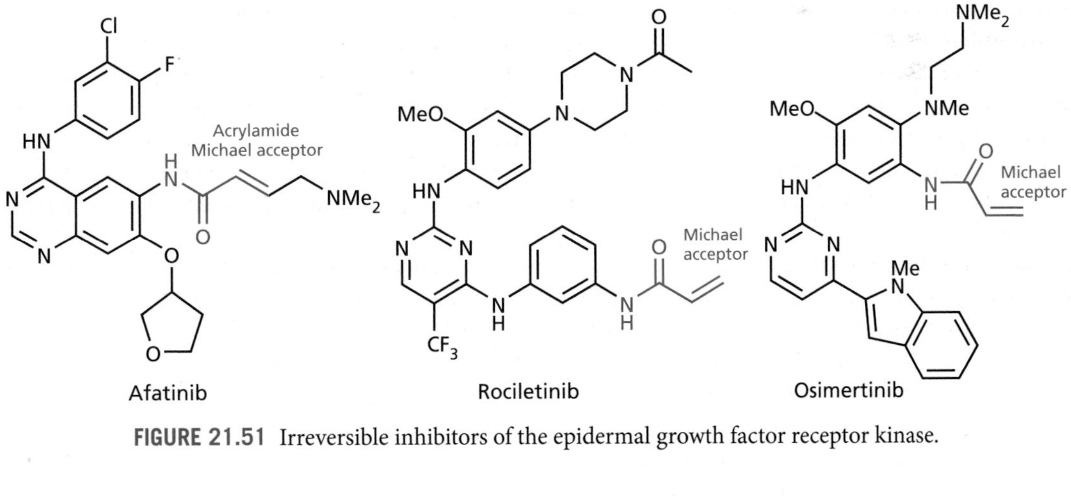

FIGURE 21.51 Irreversible inhibitors of the epidermal growth factor receptor kinase.

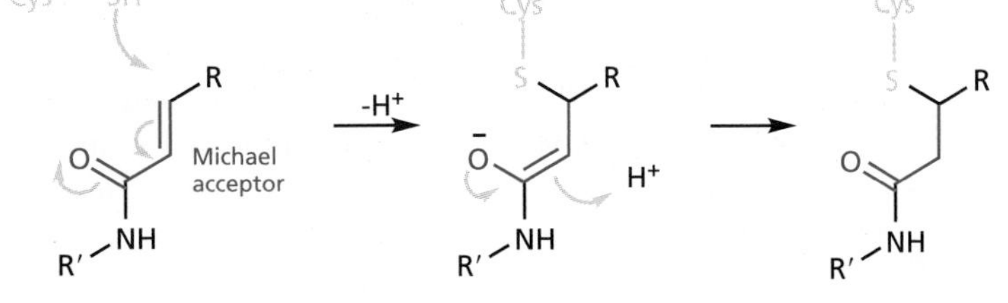

FIGURE 21.52 Michael addition of cysteine to an acrylamide 'warhead'.

BOX 21.7 General synthesis of gefitinib and related analogues

A general synthesis for gefitinib and its analogues starts from a quinazolinone starting material which acts as the central scaffold for the molecule. The synthesis is then a case of introducing the two important substituents. Selective demethylation reveals a phenol which is then protected by an acetate group to prevent it reacting with

subsequent reagents. Chlorination is now carried out on the carbonyl group, and the resulting chloro substituent is substituted by an aniline to introduce the first important substituent. Deprotection of the phenol group and reaction with an alkyl halide introduces the second important substituent.

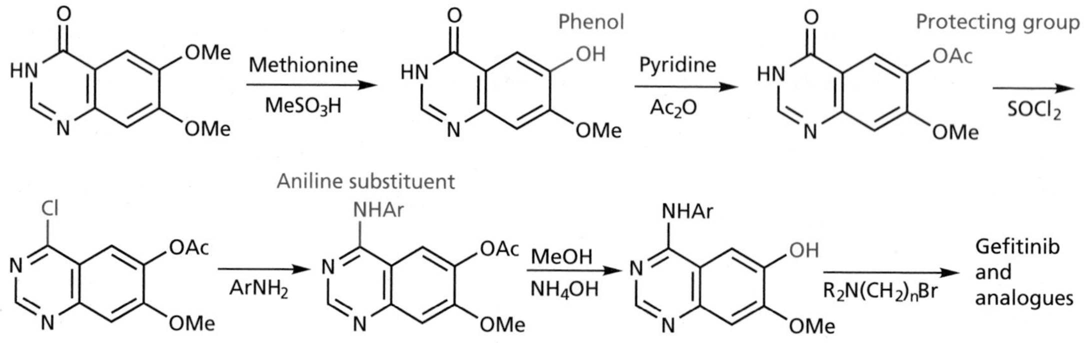

FIGURE 1 Synthesis of gefitinib (Iressa) and related analogues.

of an irreversible inhibitor is the possibility of reaction with other proteins, resulting in toxic side effects. However, the acrylamide moiety shows relatively moderate activity as a Michael acceptor. This means that it is unlikely to react with a nucleophile until the drug is bound to its target binding site through reversible interactions (section 7.1.2). Moreover, the reaction is only likely to occur if a suitable nucleophilic residue is within range, which means that drugs can be designed to target kinase enzymes with cysteine residues at specific positions. Reaction should only take place if reversible binding holds the electrophilic group in the correct position long enough for the reaction to take place.

More selective irreversible inhibitors such as **rociletinib** and **osimertinib** (previously called **mereletinib**) (Fig. 21.51) have been designed to show a greater selectivity for the T790M mutant over wild-type EGFR, thus reducing side effects. Both molecules adopt a U-shaped conformation around a core pyrimidine ring when binding. Drug resistance to these irreversible agents can arise from a mutation that changes Cys-797 to serine.

🌐 Test your understanding and practise your molecular modelling with Exercises 21.3 and 21.4 on the Online Resource Centre at www.oxfordtextbooks.co.uk/orc/patrick6e/

21.6.2.2 Kinase inhibitors of Abelson tyrosine kinase, c-KIT, PDGFR, and SRC

As the first protein kinase inhibitor to reach the market, **imatinib** (**Glivec** or **Gleevec**; Fig. 21.53 and Box 21.8) represents a milestone in anticancer therapy. It was also the first drug designed to target a molecular structure which is unique to a cancer cell. Imatinib acts as a selective inhibitor for a hybrid tyrosine kinase called **Bcr-Abl**, which is active in certain tumour cells. The tyrosine kinase active site resides on the Abl portion of the hybrid protein.

The lead compound (I in Fig. 21.53) used for the development of imatinib was a phenylaminopyrimidine structure identified by random screening of large compound libraries. The original aim of this search was to find inhibitors of a different protein kinase known as protein kinase C (PKC)—a serine-threonine kinase. Strong inhibition of PKC was achieved by adding a pyridyl substituent at the 3′ position of the pyrimidine (II). Adding an amide group to the aromatic ring then led to structures which showed inhibitory activity against tyrosine kinases as well (structure III). For example, structure IV inhibited serine-threonine protein kinases such as PKC-α, and was also a relatively weak inhibitor of tyrosine kinases. A series of chemically related structures was then synthesized to test SAR against a variety of protein kinases and to optimize activity against

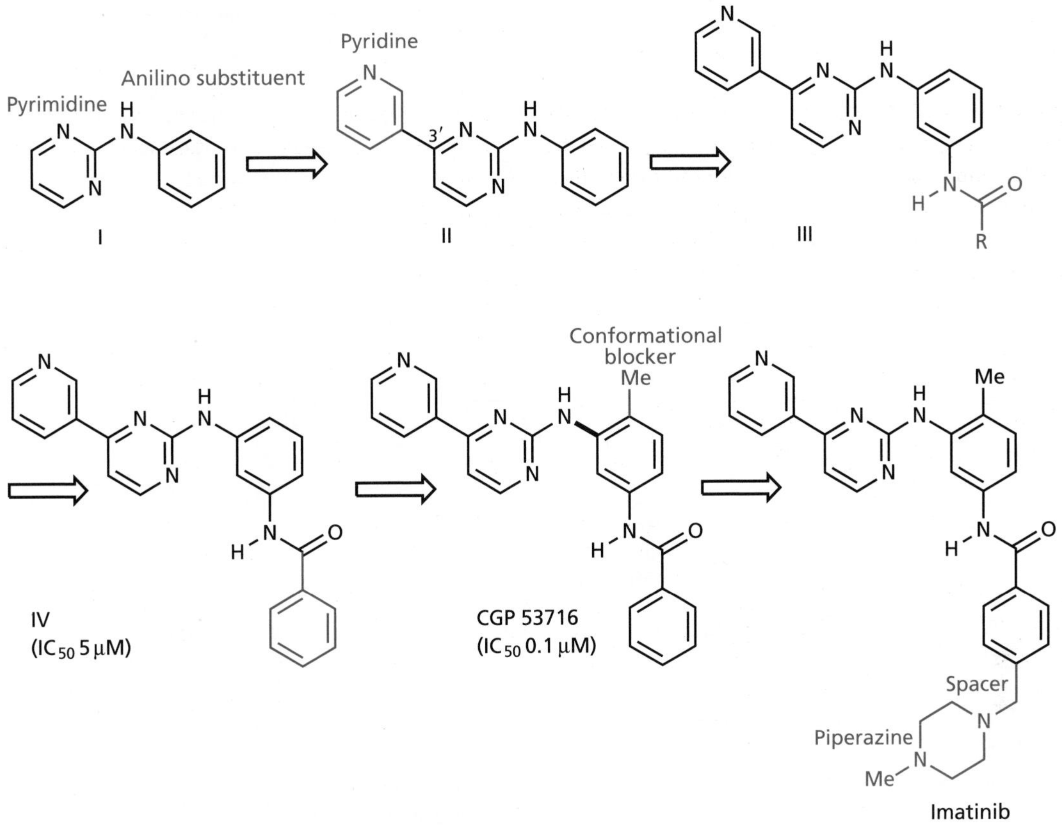

FIGURE 21.53 Development of imatinib.

tyrosine kinases. Introduction of an *ortho* methyl group as a conformational blocker (section 13.3.10) resulted in CGP 53716 which had enhanced activity against tyrosine kinases and no activity against serine-threonine kinases, demonstrating that the molecule had been forced to adopt a conformation which suited binding to tyrosine kinases but not to serine-threonine kinases. The conformational blocker hinders rotation of the Ar–N bond shown in bold (Fig. 21.53) such that the pyridine and pyrimidine rings are positioned away from the conformational blocker. Further modifications were then carried out to maximize activity and selectivity with the addition of a piperazine ring. This ring is also important for aqueous solubility as it contains a basic nitrogen which allows the formation of water-soluble salts. A one-carbon spacer was introduced between the aromatic ring and the piperazine ring, as aniline moieties are known to have mutagenic properties.

The X-ray crystal structure of imatinib bound to an inactive conformation of Abl kinase has been determined. This demonstrates the importance of the amide group within imatinib which serves as an '**anchoring group**' (Fig. 21.54). The amide forms hydrogen bonds to conserved glutamate and aspartate residues. These interactions orientate the molecule allowing either half of

the structure to access hydrophobic pockets which determine target selectivity. There is a hydrogen bonding interaction between an amino group in imatinib and the 'gatekeeper' threonine residue in the active site. The importance of this interaction is emphasized by the loss of activity observed when the amino group is alkylated. The pyridine and pyrimidine rings are located within one of the hydrophobic regions, and the piperazine ring is in the other. Separate modelling studies suggest that the piperazinyl group forms an ionic interaction with a glutamate residue. This residue is conserved in three protein kinases (Abl, c-KIT, and PDGFR, section 4.8.4), and imatinib is an inhibitor of all three. In contrast, the glutamate residue is absent from the tyrosine kinases (EGFR and c-SRC) and these kinases are not inhibited by imatinib. Therefore, this ionic interaction is likely to be important to the selectivity of the agent. Selectivity is also favoured by the *ortho* methyl group that was introduced as a conformational blocker. The methyl group is able to bind to a hydrophobic pocket that would not be accessible if a larger gatekeeper residue was present.

The fact that imatinib is not totally selective and inhibits a number of different kinases led to the concern that it would have serious side effects. Fortunately, this is not the case. It appears that normal cells are able to survive inhibition of

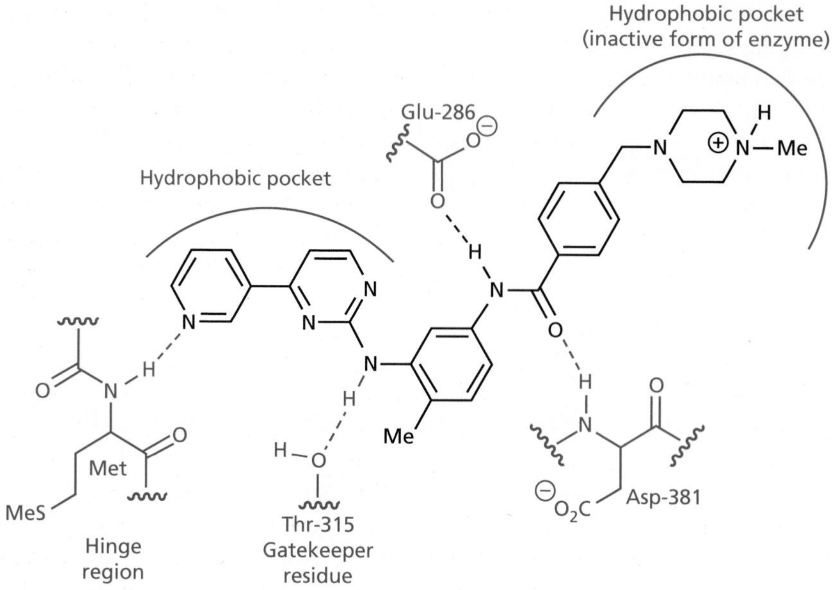

FIGURE 21.54 Binding interactions of imatinib in the active site of Abl kinase.

these kinases, whereas the survival of cancer cells containing Bcr-Abl relies crucially on that protein. Therefore, reliance of a cancer cell on an abnormally functioning protein sensitizes it to agents which target that protein.

Acquired resistance to imatinib has been observed due to mutations in the Abl kinase domain that prevent the drug from binding. Specifically, a mutation that alters the gatekeeper threonine residue to isoleucine has been observed

at position 315 (the T315I mutation). Imatinib forms an important hydrogen bond to Thr-315 which is not possible with an isoleucine residue. Other point mutations have been observed as well. Alternative signalling pathways may be adopted in some resistant cells, and an increased expression of the target receptor may occur in others.

Nilotinib, dasatinib, and **bosutinib** (Fig. 21.55) represent a second generation of Bcr-Abl inhibitors that

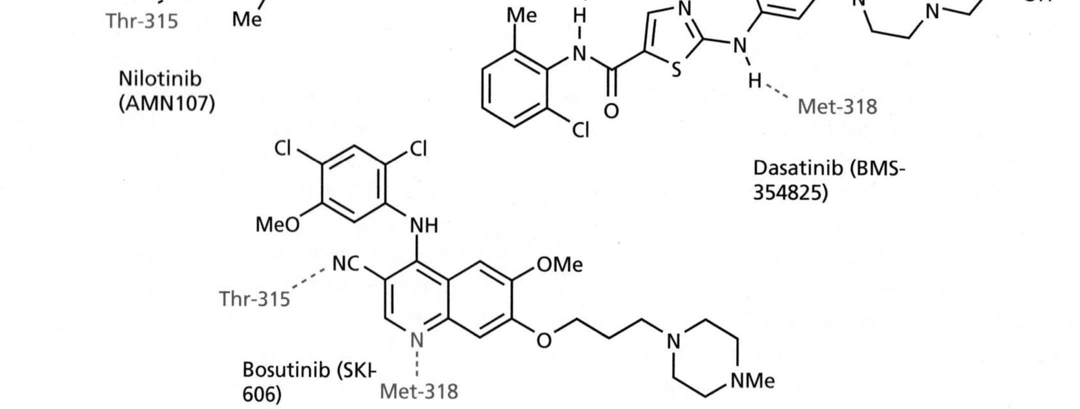

FIGURE 21.55 Nilotinib, dasatinib, and bosutinib.

are active against most imatinib-resistant tumours. The *N*-methylpiperazine ring in imatinib has been replaced by an imidazole ring in nilotinib, increasing affinity for Bcr-Abl by 20–30-fold, while retaining activity for c-KIT and PDGFR. Dasatinib and bosutinib are dual-target inhibitors that bind to the kinase active sites of Abl and **Src**; the latter plays a crucial role in cell movement and proliferation. Dasatinib binds with greater affinity to Bcr-Abl than either nilotinib or imatinib, and can bind to both active and inactive forms of the enzyme. Unfortunately, all these structures are inactive to the T315I mutation since that removes an important hydrogen bond interaction between the ligand and the alcohol group on the side chain of threonine. The side chain of isoleucine is also bigger than the side chain of threonine. This causes steric clashes with the ligand that prevent access and binding to a nearby hydrophobic pocket.

There are several projects aimed at designing kinase inhibitors that are less likely to fall prey to the problems of drug-resistance, especially structures that do not rely on the H-bonding interaction with Thr-315. Such structures should prove effective against the T315I mutant, especially if they take advantage of any unique features in the mutant binding site.

Ponatinib (Fig. 21.56) is one such structure that targets the inactive, DFG-out conformation of the Abl-protein. It is classed as a pan-Bcr-Abl inhibitor because it is effective against the native Abl protein, the T315I mutation, and a wide range of other mutated forms. The alkyne moiety is important for activity against the T315I mutation, because it acts as a simple rod-shaped linker that avoids steric clashes with Ile-315. This allows ponatinib to access the hydrophobic pocket that is guarded by the gatekeeper residue. Moreover, the acetylene linker is able to form van der Waals interactions

with Ile-315 and Phe-382. The presence of a rigid linker also means that the molecule is forced into an extended conformation that allows good binding interactions with the DFG-out mode of the kinase enzyme. Rigidification also cuts down on the number of available conformations, which has a favourable entropic effect on potency (section 13.3.9). The piperazine ring was included to improve solubility as it is likely to be protonated at physiological pH. Moreover, this allows the piperazine ring to form an electrostatic interaction with Asp-400. Van der Waals interactions are also possible between the carbon skeleton of the piperazine ring and a number of hydrophobic amino acid residues. The bicyclic ring binds to the hinge region of the binding site forming a hydrogen bond and a number of van der Waals interactions, while the *N*-arylbenzamide moiety occupies the hydrophobic selectivity pocket that is guarded by the gatekeeper residue. As a result, the structure forms multiple interactions to various parts of the binding site, making it less vulnerable to a single mutation.

Another strategy is to develop allosteric inhibitors of Bcr-Abl. For example, allosteric inhibitors are being studied that stabilize the inactive form of the protein by binding to an autoregulatory binding cleft which is distant from the active site. **GNF-2** (Fig. 21.57) is one such compound which shows extremely good selectivity. Also under study are agents such as **ON012380** (Fig. 21.57), which binds to the substrate binding region rather than the ATP binding region. Many researchers feel that such inhibitors could be more selective and safer to use.

Finally, combination therapies that use drugs capable of targeting different regions of the same protein kinase may be therapeutically important in the future and help to combat resistance against any one drug.

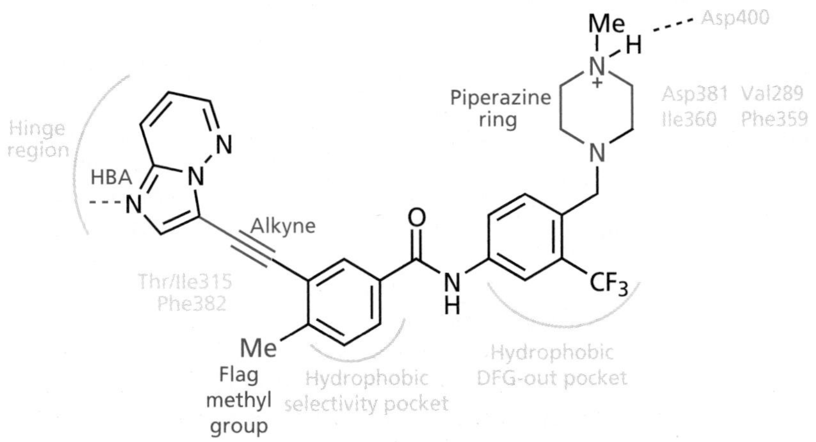

FIGURE 21.56 Ponatinib (AP24534).

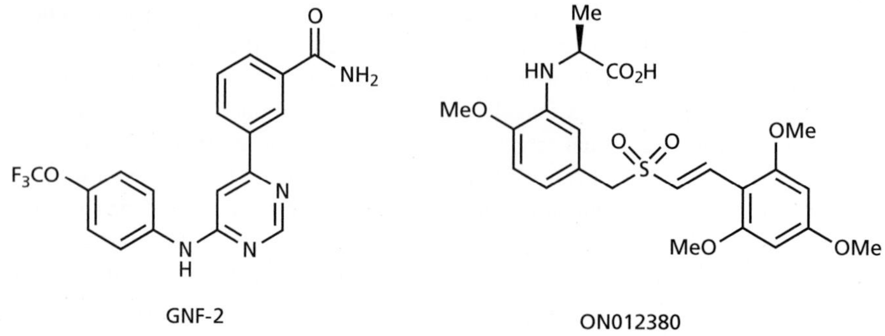

FIGURE 21.57 Structures of GNF-2 and ON012380.

BOX 21.8 General synthesis of imatinib and analogues

The synthesis of imatinib and its analogues involves the pyridine structure (I) as starting material. The central pyrimidine ring is then constructed in two stages to give structure (II). The remaining two steps involve reduction of an aromatic nitro group to an amine, and acylation to give the final product. The route described allows the synthesis of a large variety of amides from intermediate (III).

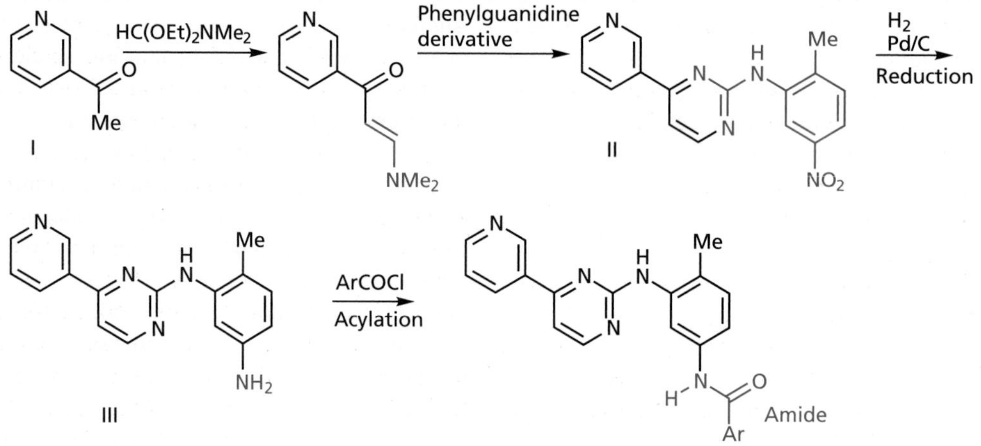

FIGURE 1 General synthesis of imatinib and analogues.

Test your understanding and practise your molecular modelling with Exercises 21.1 and 21.3 on the Online Resource Centre at www.oxfordtextbooks.co.uk/orc/patrick6e/

21.6.2.3 Inhibitors of cyclin-dependent kinases (CDKs)

CDKs are involved in the control of the cell cycle (section 21.1.6), but are overexpressed or overactive in many cancer cells. Since these enzymes are inactive in normal resting cells, drugs that target them should have fewer, and less toxic, side effects than conventional cytotoxic drugs. CDKs are serine-threonine kinases which are activated by **cyclins** and inhibited by **cyclin-dependent kinase inhibitors**

(**CKIs**). The human genome encodes 21 CDKs, seven of which are directly involved in the cell cycle (CDK1–4, 6, 10, and 11). They are typically small proteins of about 300 amino acids. At present, a variety of inhibitors have been identified which compete with ATP for the kinase active site. **Flavopiridol** (also called **alvocidib**) (Fig. 21.58) is one such structure which is undergoing clinical trials and looks promising as part of a combination therapy. It is a semi-synthetic flavone derived from **rohitukine**—a natural product extracted from an Indian plant. One possible problem with flavopiridol is its lack of selectivity between different CDKs, and so analogues which *do* show selectivity may prove beneficial. As far as the binding interactions are concerned, flavopiridol binds to the same region of the active site as ATP. The benzopyran ring lies in the adenine

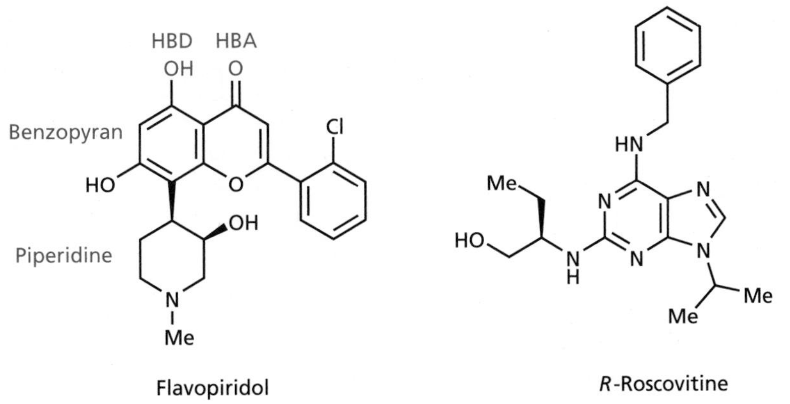

FIGURE 21.58 Flavopiridol and *R*-roscovitine.

binding region such that the ketone acts as a hydrogen bond acceptor and the OH acts as a hydrogen bond donor. The piperidine ring lies in the region normally occupied by the first phosphate moiety of ATP, where it makes several hydrogen bonding interactions with water and nearby amino acid residues. The chlorophenyl group lies over the ribose binding pocket. Flavopiridol is also thought to inhibit the expression of **cyclins D1** and **D3**. It has been found to have an anti-angiogenic effect and can induce apoptosis. **Roscovitine (seliciclib)** (Fig. 21.58) shows selectivity for CDK2 and competes with ATP for the binding site. It induces apoptosis and is currently undergoing clinical trials.

A search for more selective agents resulted in the development of the orally active drug **palbociclib** (Fig. 21.59), which shows selectivity for CDK4 and CDK6. It was approved by the US FDA in 2016 for the treatment of certain types of breast cancer. Another two agents called **abemaciclib** and **ribociclib** were in phase III clinical trials in 2016.

21.6.2.4 Kinase inhibitors of the MAPK signal transduction pathway

The mitogen-activated protein kinase (MAPK) signal transduction pathway (section 5.4.2) is overexpressed in several cancers (section 21.6.1). Key components of the pathway include the signalling protein Ras, the serine-threonine kinases Raf and MAP kinase, and the threonine-tyrosine kinase MEK.

Vemurafenib and **dabrafenib** (Fig. 21.60) are ATP competitive inhibitors that inhibit an abnormal form of B-Raf containing the V600E mutation; in other words, the valine residue at position 600 has been replaced with glutamic acid. Indeed, vemurafenib's name is derived from **V**600 **E**mutated **BRAF** inhibition. As a result of this mutation, Glu-600 forms a salt bridge to Lys-507 that locks the enzyme into the active, DFG-in conformation. Both agents also inhibit the V600K mutation where valine has been replaced with lysine.

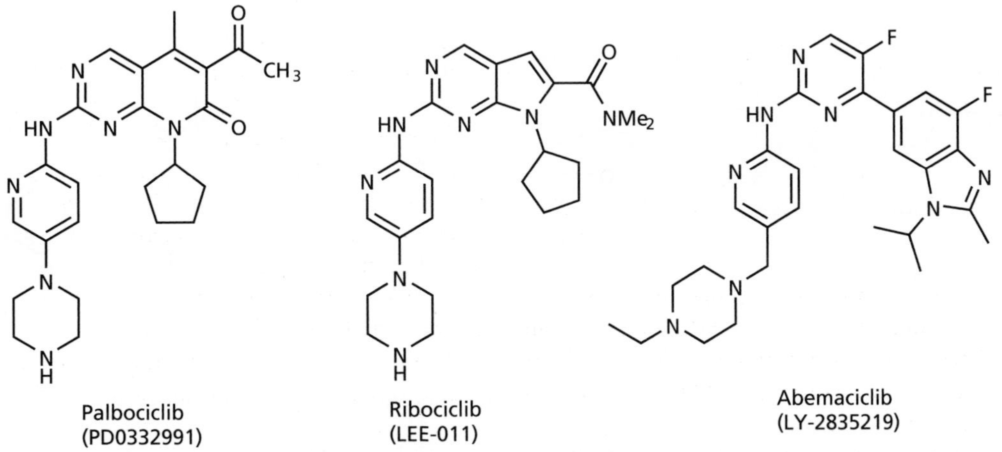

FIGURE 21.59 Kinase inhibitors showing selectivity for CDK4 and CDK6.

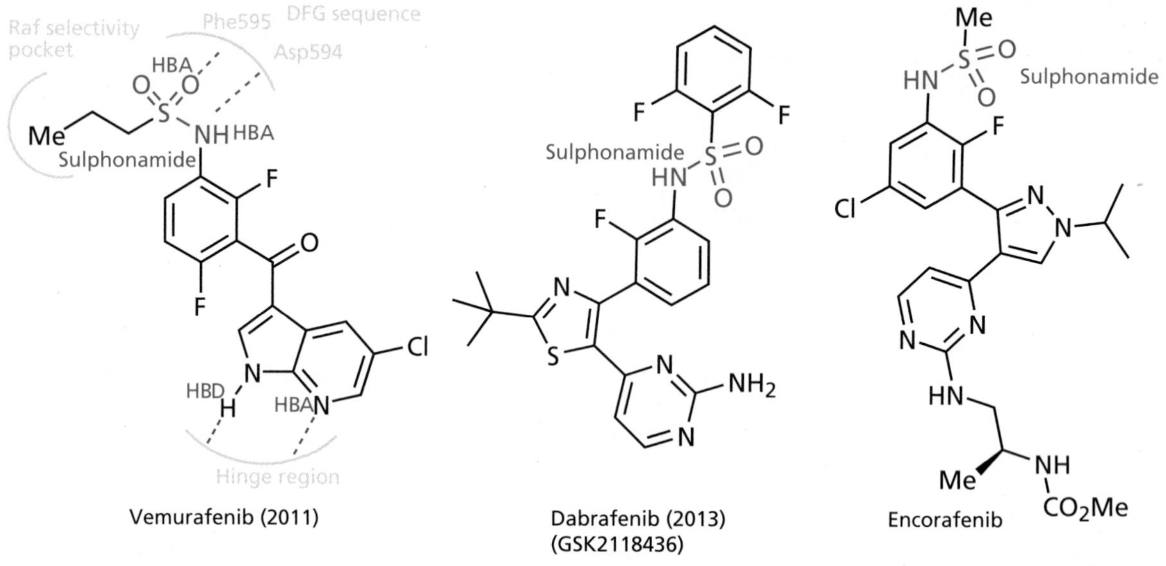

FIGURE 21.60 B-Raf inhibitors.

Both inhibitors interact with Cys-532 in the hinge region, while the sulphonamide moiety interacts with Asp-594 and Phe-595 in the DFG region. It is thought that the sulphonamide is deprotonated and acts as a hydrogen bond acceptor. This interaction directs the propyl chain of vemurafenib into a small hydrophobic pocket that is unique to Raf and is important for selectivity. Second-generation Raf inhibitors under investigation include **encorafenib**.

Cobimetinib and **trametinib** (Fig. 21.61) are MEK inhibitors, and are used in combination therapy with one of the above B-Raf inhibitors. This reduces the risk of drug resistance, which can develop within a year if using a single agent. Both drugs interact with the inactive conformation of the enzyme, but do not bind to the ATP binding region. Instead, they bind to an allosteric

binding region nearby. Cobimetinib contains azetidine and piperidine ring systems that bind to Asp-190, as well as the γ-phosphate group of bound ATP. The iodine substituent forms an interaction with the peptide carbonyl oxygen of Val-127. Several other MEK inhibitors are in clinical trials, including **binimetinib**.

Finally, there are potential advantages in combining agents that target the MAPK signal transduction pathway with those that target the PI3K-mTOR pathway (section 21.6.2.5). The latter is an escape pathway that can cause resistance to B-Raf and MEK inhibitors.

21.6.2.5 Kinase inhibitors of PI3K-PIP$_3$ pathways

The PI3K-PIP$_3$ pathway influences the activity of a number of important downstream kinase enzymes, such as

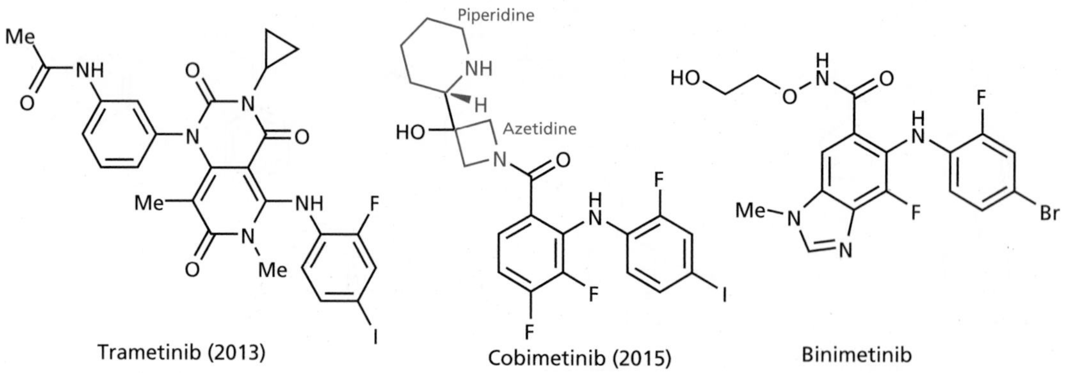

FIGURE 21.61 MEK inhibitors.

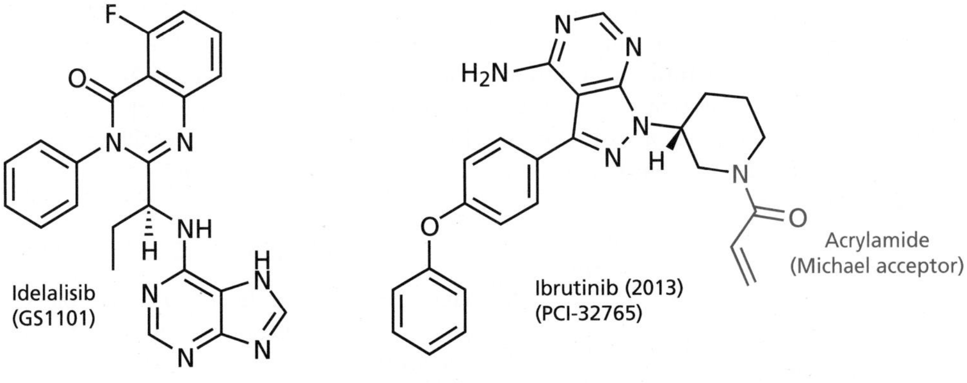

FIGURE 21.62 Idelalisib and ibrutinib.

protein kinase B (Akt), m-TOR, and Bruton's tyrosine kinase (section 5.4.5), all of which regulate cell growth and division. Therefore, there are advantages in targeting various components of this pathway.

Idelalisib (Fig. 21.62) was approved in 2014 and inhibits the P110-δ isoform of phosphoinositide 3-kinase (PI3K). This isoform is found only in spleen, thymus, and blood leukocytes, implying that it plays an important role in the immune system.

Ibrutinib (Fig. 21.62) is an irreversible and selective inhibitor of **Bruton's tyrosine kinase** (BTK), a protein which regulates B-cell proliferation and activation (section 5.4.5). The drug contains an acrylamide group that acts as an electrophile in a Michael addition reaction with the thiol group of a cysteine residue (Cys-481) (compare section 21.6.2.1). Selectivity is largely due to the fact that this cysteine residue is shared with only nine other kinases.

A kinase inhibitor that acts directly on m-TOR has yet to be approved, but m-TOR can be inhibited by inducing protein–protein binding interactions (section 21.6.2.11).

21.6.2.6 Kinase inhibitors of anaplastic lymphoma kinase (ALK)

Anaplastic lymphoma kinase (ALK) (section 4.8.4) is a tyrosine kinase receptor that has been linked to a number of cancers, including anaplastic large cell lymphoma and some lung cancers. ALK appears to have no essential role in mammals, and so it is an attractive drug target. **Crizotinib** is a multi-receptor tyrosine kinase inhibitor, and was the first ALK inhibitor to reach the market (Fig. 21.63) (see also Box 13.4). Crizotinib also inhibits a tyrosine kinase receptor called **c-MET** (section 4.8.4). Overexpression of c-MET has been associated with most solid tumours.

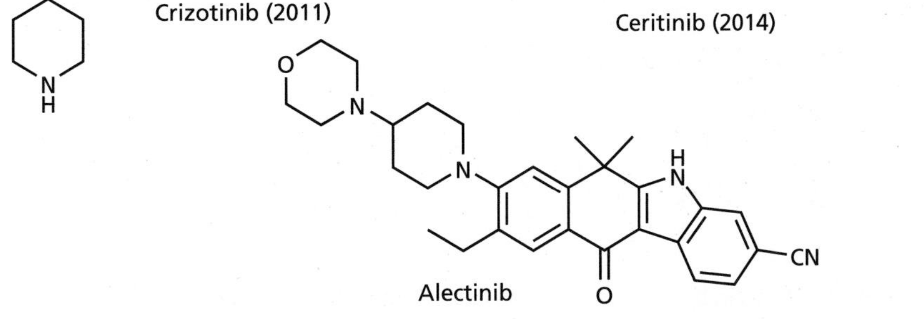

FIGURE 21.63 Crizotinib, ceritinib, and alectinib.

Despite proving highly effective in treating lung cancers, drug resistance has emerged to crizotinib, and so the second-generation inhibitors **ceritinib** and **alectinib** were developed to treat crizotinib-resistant patients (Fig. 21.63) (see also Box 14.3). These have proved effective against most crizotinib-resistant cases, but resistance is developing against these drugs as well. Further research is taking place to find agents that will keep on top of the resistance problem, in particular resistance arising from the G1202R mutation.

21.6.2.7 Kinase inhibitors of RET and KIF5B-RET

Mutations to a tyrosine kinase receptor called RET (section 4.8.4) have been associated with thyroid cancers. The kinase inhibitors **vandetanib** (Fig. 21.50), **sorafenib** (Box 21.10), **cabozantinib**, and **lenvatinib** (Fig. 21.64) have all been approved for the treatment of thyroid cancer because of their ability to inhibit RET. Cabozantinib and lenvatinib are structurally related and contain a quinoline ring system. All four agents are classed as multi-receptor tyrosine kinase inhibitors because they inhibit a number of targets. For example, cabozantinib inhibits the receptor tyrosine kinases RET, c-MET, and VEGFR-2, while lenvatinib inhibits RET, VEGFR-2, and VEGFR-3.

In 2011–12, a KIF5B-RET fusion gene was discovered that appears to promote some cases of lung cancer (NSCLC). It is further proposed that the RET kinase part of the resulting fusion protein has abnormally high catalytic activity. This suggests that inhibitors of the fusion protein might be of potential use in treating NSCLCs. **Ponatinib** (Fig. 21.56) is a multi-receptor tyrosine kinase inhibitor that inhibits RET and is currently being investigated. **Motesanib** (Fig. 21.64) is also under investigation but has not yet been approved. Because all the currently known agents are broad-spectrum kinase inhibitors, there is interest in developing more selective inhibitors for RET. Adverse side effects may be due to the inhibition of VEGFR or EGFR.

21.6.2.8 Kinase inhibitors of Janus kinase

Janus kinases were described in section 5.4.4 and are crucial to the function of several cytokine receptors. **Ruxolitinib** (Fig. 21.65) is a clinically approved agent that inhibits **Janus kinase**. It is believed that the pyrrolopyrimidine ring system of ruxolitinib binds to the hinge region of the active site.

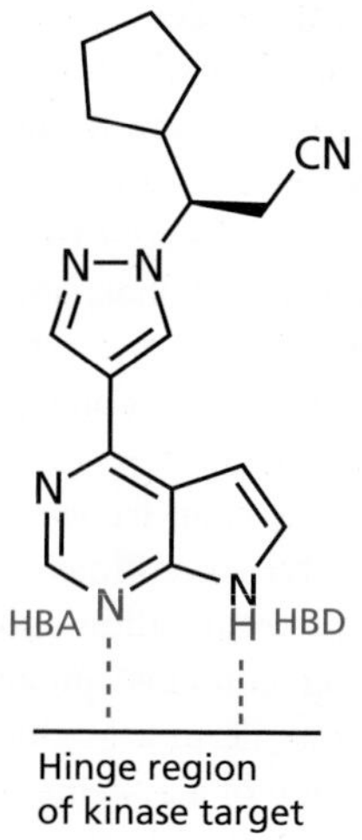

FIGURE 21.65 Ruxolitinib.

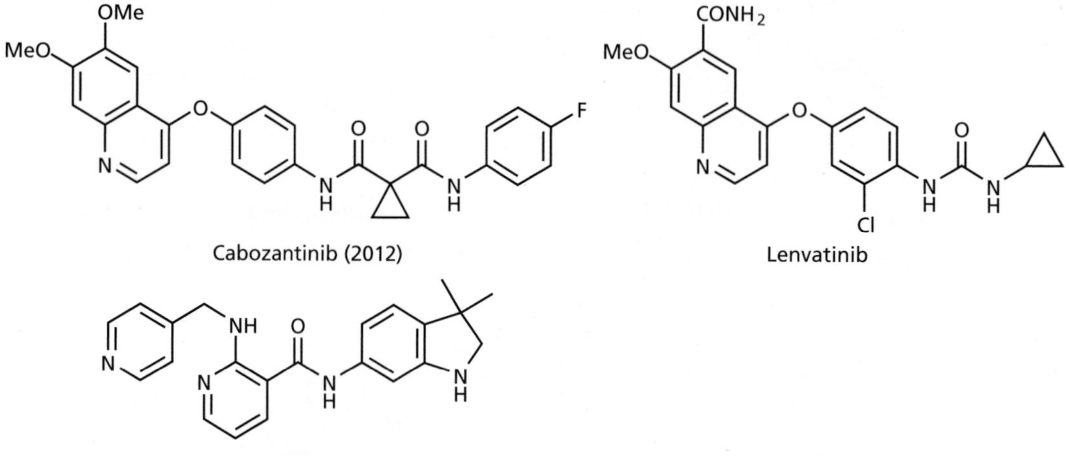

FIGURE 21.64 Cabozantinib, lenvatinib, and motesanib.

21.6.2.9 Kinase inhibitors of vascular endothelial growth factor receptor (VEGFR)

Vascular endothelial growth factor (VEGF) and its receptor (VEGFR) are important in promoting angiogenesis (section 21.1.9). Therefore, tyrosine kinase inhibitors of VEGFR can play an important role in in inhibiting the production of new blood vessels and starving cancer cells of the nutrients they need to grow and spread. One problem with targeting VEGFR is its structural similarity with other tyrosine kinases. Therefore, VEGFR inhibitors tend to be multi-kinase inhibitors. However, inhibiting a variety of targets can be seen as an advantage. The clinically approved multi-receptor tyrosine kinases that include VEGFR as one of their targets are **sorafenib**, **sunitinib**, **pazopanib**, **vandetanib**, **axitinib** (Fig. 21.66), **cabozantinib**, **lenvatinib**, **vatalinib**, and **regorafenib** (a close analogue of sorafenib (Box 21.9)). Lenvatinib is an example of a type V inhibitor (section 21.6.2) in its interactions with the VEGFR-2 receptor.

21.6.2.10 Multi-receptor tyrosine kinase inhibitors

As the title indicates, **multi-receptor tyrosine kinase inhibitors (mRTKIs)** are agents that are designed to

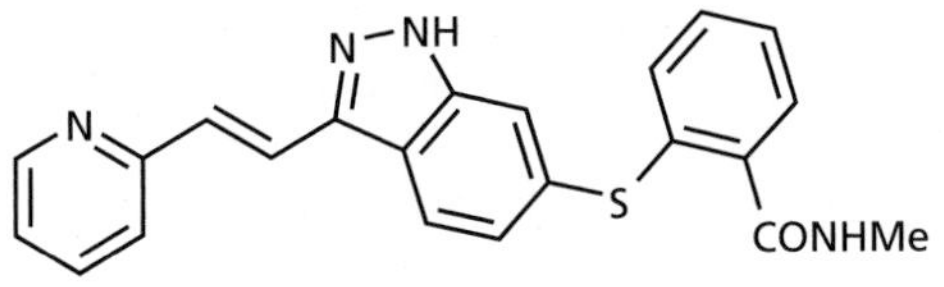

FIGURE 21.66 Axitinib (2012).

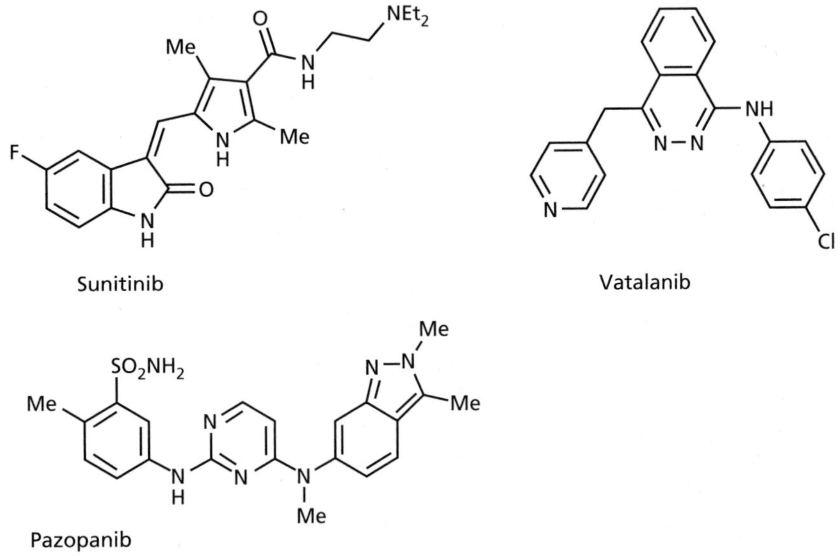

be selective against a number of receptor tyrosine kinase targets, all of which have some bearing on the generation and survival of cancer cells. One author has described them as being selectively non-selective! In other words, they should be non-selective in inhibiting a number of kinases that contribute to a cancer, but selective in the sense that they do not inhibit kinases that would lead to side effects—a difficult goal to achieve. The big advantage of an mRTKI is that drug resistance is less likely to occur. If one of the drug targets mutates and becomes resistant, the other targets are still vulnerable. An mRTKI can be viewed as a combination therapy wrapped up within a single drug. Such drugs are sometimes referred to as promiscuous since they affect a variety of different targets. mRTKIs are likely to be particularly promising agents for the treatment of cancers which are driven by several abnormalities. In truth, none of the kinase inhibitors approved for the market to date are 100% selective for one kinase target and could all, in principle, be defined as mRTKIs. However, it is generally recognized that some inhibitors are more promiscuous than others, and it is those that are defined as mRTKIs. The least promiscuous of the mRTKIs is **lapatinib** (section 21.6.2.1), while the most promiscuous is **sunitinib** described below.

The mRTKIs that have currently reached the market target the kinases associated with angiogenesis, as well as another kinase associated with the tumour itself (for example **c-KIT**). These include **sorafenib**, which was developed from a urea lead compound using a mixture of traditional medicinal chemistry strategies and multiple point variations (Box 21.9). Sunitinib (Fig. 21.67) is another example. It contains an indolinone ring and binds

Sunitinib

Vatalanib

Pazopanib

FIGURE 21.67 Multi-receptor tyrosine kinase inhibitors (mRTKIs).

BOX 21.9 Design of sorafenib

In order to find the lead compound for sorafenib, high-throughput screening of 200 000 compounds was carried out against recombinant **Raf-1 kinase** (also called **c-Raf**). This led to the identification of a urea (I) with micromolar activity (Fig. 1). Substituents and rings were altered in a systematic fashion and it was found that a *para* methyl group on the phenyl ring resulted in a tenfold increase in activity. However, despite the synthesis of many more analogues, no further improvement in activity could be obtained. Up to this point, conventional medicinal chemistry strategies had been followed which involved altering one group at a time. This allows one to rationalize any alterations in activity that result from the change of any ring or substituent. It was now decided to use parallel synthesis to produce 1000 analogues having all possible combinations of the different substituents and rings that had been studied to date. This led to the discovery of a urea (IV) having slightly improved activity over structure (II). The curious thing about this structure is that it deviates from the SAR results obtained by single point modifications. Structure IV has a phenoxy substituent and an isoxazole ring, but neither of these groups would be considered good for activity based on the initial SAR. For example, structure III has the phenoxy substituent while structure VI

has the isoxazole ring, but both structures have low activity compared to the lead compound. Conventionally, this would be taken to imply that neither group is good for activity. However, it does not take into account the synergistic effects that two or more modifications might have. The strategy of multiple point modifications allows the identification of such synergistic effects and demonstrates that there are limitations to simple SAR analyses.

Structure IV was now adopted as the new lead compound. Replacing the phenyl ring with a pyridine ring led to structure (V) and a fivefold increase in activity, as well as improving aqueous solubility and Clog *P*. Conventional optimization strategies then led to sorafenib which is 1000-fold more active than the original lead compound.

The urea functional group serves as an anchor group in a similar manner to the amide group present in imatinib. It forms two hydrogen-bonding interactions to the catalytic aspartate and glutamate residues in the active site, and orientates the molecule such that each half of the molecule is positioned into two selectivity regions. The atoms coloured in blue are involved in important hydrogen bonding interactions.

Regorafenib is a closely related structure that has also been approved.

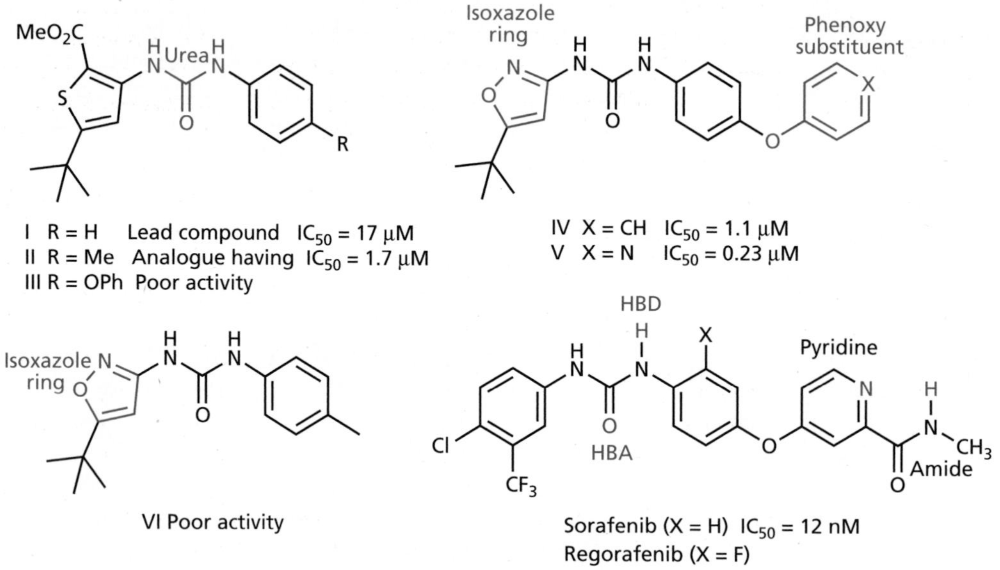

I   R = H   Lead compound   IC$_{50}$ = 17 µM
II R = Me   Analogue having   IC$_{50}$ = 1.7 µM
III R = OPh   Poor activity

IV X = CH   IC$_{50}$ = 1.1 µM
V   X = N   IC$_{50}$ = 0.23 µM

VI Poor activity

Sorafenib (X = H) IC$_{50}$ = 12 nM
Regorafenib (X = F)

FIGURE 1 From lead compound to sorafenib.

mostly to the region of the active site normally occupied by ATP. There is little interaction with the hydrophobic pockets that could confer selectivity for different kinases. **Vatalanib** (Fig. 21.67) is another multi-kinase inhibitor undergoing clinical trials, while **pazopanib** was approved in 2009.

The idea of targeting several targets with the one drug is described as '**polypharmacology**'. Drugs acting in this way may have potentially enhanced antitumour activity relative to more selective kinase inhibitors. An alternative polypharmacology approach is to use a 'cocktail' of different selective kinase inhibitors.

BOX 21.10 Clinical aspects of kinase inhibitors

Kinase inhibitors of the EGF receptor

The EGFR family of kinases has four members—EGFR, HER2, HER3, and HER4. Overexpression of these kinases is associated with a variety of cancers in the breast, lung, brain, prostate, gastrointestinal tract, and ovaries. The first clinically approved kinase inhibitor for the EGFR family was **gefitinib**, which is used for the treatment of refractory lung cancers. This was followed by **erlotinib,** which is approved for the treatment of non-small-cell lung cancer. **Icotinib** was approved in China in 2011 for the same purpose. These agents act on tumours resulting from a mutation in EGFR where Leu-853 is replaced with arginine, resulting in destabilization of inactive conformations of the enzyme and an increased level of the active conformation. The mutation also weakens affinity for ATP while increasing affinity for the inhibitors, such that the latter compete effectively for the active site. Unfortunately, drug-resistant tumours often emerge within a year of initiating treatment, caused by a second mutation where the gatekeeper residue Thr-790 is altered to methionine. This restores the enzyme's affinity for ATP such that competitive inhibitors are much less effective, despite their ability to still bind to the active site. **Lapatinib** was approved in 2007 for the treatment of patients with advanced or metastatic breast tumours which overexpress HER2. Lapatinib is given orally in combination with **capecitabine**. However, it should only be administered if the standard first-line treatment consisting of **trastuzumab** and a taxane has become ineffective because of tumour resistance. Tumours that have resistance to trastuzumab are unlikely to have resistance to lapatinib as the former drug binds to the extracellular region of the receptor and the latter binds to the intracellular kinase active site. Lapatinib is seen to have several advantages over trastuzumab. Whereas trastuzumab inhibits only HER2, lapatinib inhibits both HER2 and EGFR. Inhibition of two proteins is seen as being more effective than inhibiting either one alone. Moreover, drug resistance is less likely to appear if two targets are affected. Lapatinib is also able to cross the blood–brain barrier (unlike trastuzumab) and can combat any breast tumour cells that have reached the brain as a result of metastasis. Finally, lapatinib has less cardiac toxicity compared to trastuzumab. The drug is currently undergoing clinical trials for a range of other cancer treatments such as those involving the head, neck, and kidney. **Afatinib** was approved in 2013 for the treatment of metastatic non-small-cell lung carcinoma (NSCLC) in patients that are EGFR mutation positive. **Osimertinib** was approved in 2015 for the treatment of NSCLC patients having the T790M mutation in EGFR. **Vandetanib** was approved in 2011 for the treatment

of medullary thyroid cancer. It can inhibit both VEGFR and EGFR, thus inhibiting angiogenesis and cell growth respectively. However, its activity against thyroid cancers is more likely to be due to inhibition of RET.

Kinase inhibitors of Bcr-Abl and c-KIT

Imatinib was introduced for the treatment of a rare blood cancer called chronic myeloid leukaemia (CML) which accounts for 15–20% of all cases of adult leukaemia in Western populations. The cancer cells involved contain an abnormal protein kinase which is not found in normal cells. The protein kinase concerned is a member of the tyrosine kinase family and has been named **Bcr-Abl**. This name is derived from the genes which code for the protein (i.e. the c-*abl* and *bcr* genes). In normal cells these genes are distinct and on different chromosomes so they code for separate proteins. In the cancer cells associated with CML, part of one chromosome has been transferred to another resulting in a shortened chromosome called the **Philadelphia chromosome**—a characteristic feature of this type of cancer. The result of this genetic transfer is the formation of a hybrid gene (*bcr-abl*) which is not properly regulated and which codes for excessive levels of the hybrid protein kinase. In turn, this leads to excessive quantities of white blood cells (leukocytes). Imatinib has been successful in 90% of patients, but tumour resistance can often result. Imatinib also inhibits a tyrosine kinase called c-KIT, and has been approved for the treatment of stomach cancers where this kinase is altered or overexpressed. The c-KIT receptor (also called CD117 or c-Kit) is a cytokine receptor expressed on the surface of stem cells, and is activated by stem cell factor. Unfortunately, mutations in c-KIT can result in tumour resistance. Imatinib also inhibits the platelet-derived growth factor receptor (PDGFR), and the drug is currently approved for the treatment of ten different types of cancer.

Nilotinib, **dasatinib**, and **bosutinib** have been approved for the treatment of CML when imatinib therapy is unsuccessful because of drug resistance. Dasatinib is also being considered for metastatic melanoma. The greater binding affinities of these agents mean that they still bind sufficiently strongly if a mutation should result in the loss of one binding interaction. The exception is the T315I mutant, where the interaction with threonine is particularly important. **Ponatinib** was introduced in 2012 as an agent that is effective against the T315I mutation. It is a multi-receptor tyrosine kinase inhibitor that was approved for the treatment of chronic myeloid leukaemia and Philadelphia chromosome-positive acute lymphoblastic leukaemia.

(*Continued*)

BOX 21.10 Clinical aspects of kinase inhibitors (*Continued*)

Inhibitors of cyclin-dependent kinases (CDKs)

Palbociclib was approved by the FDA for the treatment of ER-positive/HER2-negative metastatic breast cancer. It is used in combination with **letrozole** (an aromatase inhibitor).

Kinase inhibitors of the MAPK signal transduction pathway

Vemurafenib and **dabrafenib** were approved in 2011 and 2013 respectively for the treatment of late stage melanoma. They target abnormal B-Raf protein kinases. Dabrafenib is administered alongside the MEK inhibitor **trametinib** in order to combat the emergence of drug resistance. Similarly, vamurafenib is used in combination with the MEK inhibitor **cobimetinib**.

Kinase inhibitors of PI3K-PIP$_3$ pathways

Ibrutinib has gained FDA approval for the treatment of mantle cell lymphoma, chronic lymphocytic leukaemia, and non-Hodgkin's lymphoma. **Idelalisib** was approved in 2014 for the treatment of chronic lymphocytic leukaemia, and is used in combination with rituximab.

Kinase inhibitors of ALK

Crizotinib, **ceritinib**, and **alectinib** have been approved for the treatment of non-small-cell lung cancer, and target an abnormal form of the anaplastic lymphoma kinase receptor (ALK or CD246). Crizotinib was the first of these agents to be introduced, but resistance can develop within a year of treatment. In many cases, this is because the gatekeeper residue Leu-1196 is mutated to methionine. However, other mutations have also been observed to cause resistance (G1269A, 1151T-ins, L1152R, C1156Y, G1202R, F1174L, and S1206Y). Ceritinib and alectinib are effective against several of these mutated forms, but not all. For example, ceritinib is ineffective against G1202R and F1174C/V. Similar resistance is also emerging towards alectinib.

Kinase inhibitors of RET and KIF5B-RET

Vandetanib, **sorafenib**, **cabozantinib,** and **lenvatinib** have all been approved for the treatment of thyroid cancer because of their ability to inhibit RET.

Kinase inhibitors of Janus kinase

Ruxolitinib was approved in 2011 for the treatment of myelofibrosis affecting bone marrow.

Multi-receptor tyrosine kinase inhibitors

Sorafenib has been approved as a treatment of liver, kidney, and thyroid cancers. The agent inhibits the kinase activity of the membrane-bound receptors VEGFR, PDGFR, c-KIT, and RET, as well as an intracellular target (B-Raf). **Sunitinib** was approved in 2006 for the treatment of gastrointestinal stromal tumours (a rare cancer) and advanced renal cell carcinoma (a common kidney cancer). The agent is a simultaneous inhibitor of VEGFR-1, VEGFR-2, and PDGFR-b, and there is evidence that this is more effective than inhibition of either of these targets alone. It also inhibits c-KIT and FLT3. In 2011, it was approved for the treatment of advanced pancreatic neuroendocrine tumours.

Pazopanib was approved in 2009 for the treatment of renal cell carcinoma, and in 2012 for the treatment of soft tissue sarcoma. It inhibits VEGFR-1, VEGFR-2, VEGFR-3, PDGFR, and c-KIT. **Axitinib** is also approved for the treatment of renal cell carcinoma. It inhibits VEGFR, PDGFR, and c-KIT.

Miscellaneous inhibitors

Temsirolimus is an analogue of the antibiotic **rapamycin** and was approved in 2007 for the treatment of advanced renal cell carcinoma. **Everolimus** is a similar structure and was approved in 2009. Finally, **toceranib** is the first anticancer drug approved by the FDA for the treatment of tumours in dogs. It inhibits KIT tyrosine kinase.

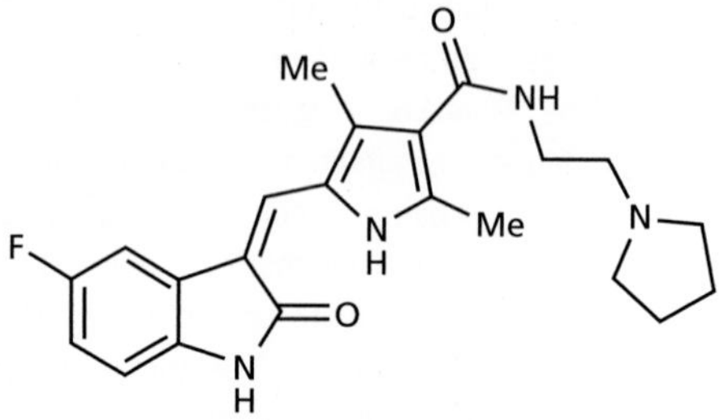

FIGURE 1 Toceranib.

21.6.2.11 Kinase inhibition involving protein–protein binding interactions

Temsirolimus and **everolimus** (Fig. 21.68) are analogues of a natural product called **sirolimus (rapamycin)**—a macrolide which is produced by the bacterium *Streptomyces hygroscopicus*. Sirolimus is used as an immunosuppressant during kidney transplants. However, the more polar temsirolimus and everolimus have been approved for the treatment of certain types of tumour where a kinase enzyme called **mTOR** (or **FRAP**) is excessively active. mTOR triggers a signal transduction process leading to cell growth, and so its inhibition serves to inhibit tumour development. The mechanism of inhibition is rather unusual, compared to those mentioned so far.

The mechanism starts with part of the macrolide structure binding to an immunopholin protein called **FKBP12** (also called the **FK506-binding protein**). Once this drug–protein complex has been formed, a different part of the macrolide ring structure binds to mTOR to form a ternary structure consisting of the two proteins with the drug sandwiched between. In essence, the drug promotes the dimerization of the two proteins, such that they interact with each other. This interaction results in inhibition of mTOR and the signal transduction process that it normally triggers.

It has also been proposed that inhibition of mTOR might reverse the resistance of some tumours to certain other anticancer agents.

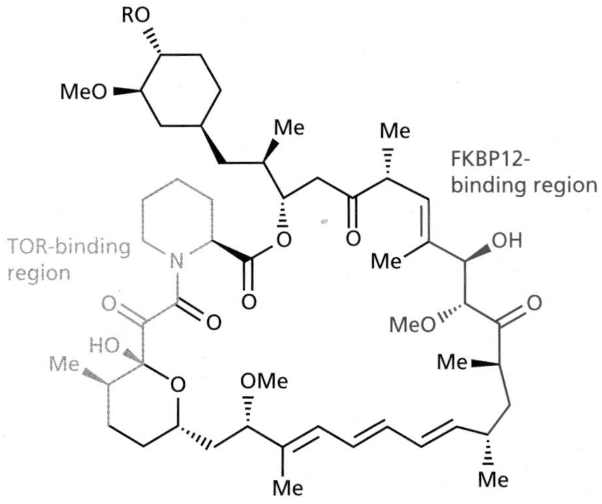

21.6.3 Receptor antagonists of the hedgehog signalling pathway

A number of research projects have been carried out to find antagonists that target a receptor called **smoothened** in the curiously named **hedgehog signalling pathway**—a pathway that is unique to stem cells (section 5.5). **Vismodegib** (Fig. 21.69) was the first of these antagonists to reach the market and was approved in 2012 for the treatment of skin cancers. This was followed by the approval of **sonidegib** in 2015 (Fig. 21.70). The latter was developed from a hit compound identified by screening compound libraries. The hit compound contains a 1,4-diaminophenyl group which could potentially be metabolized to produce toxic metabolites. To counter this possibility, the phenyl ring was replaced with a more electron-deficient pyridine ring. Other modifications were aimed at increasing antagonist activity and optimizing pharmacokinetic properties.

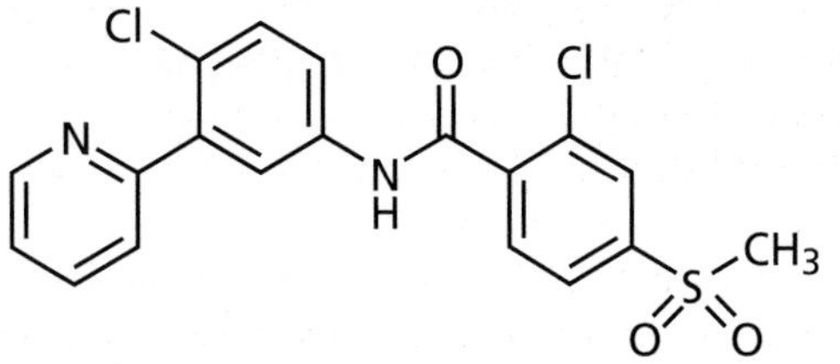

Sirolimus (rapamycin); R = H
Temsirolimus; R = CO-C(Me)(CH₂OH)₂
Everolimus; R = (CH₂)₂OH

FIGURE 21.68 Sirolimus and analogues.

FIGURE 21.69 Vismodegib.

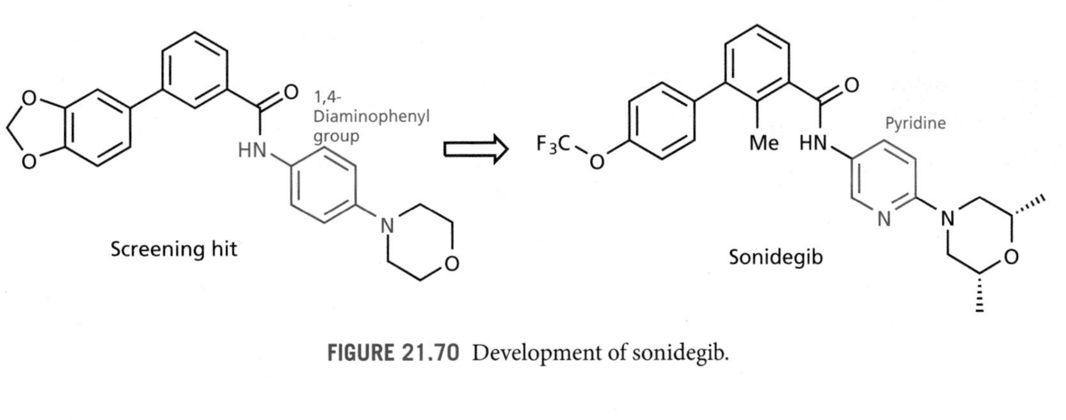

FIGURE 21.70 Development of sonidegib.

- Many cancers produce abnormal or overexpressed proteins that are involved in the signalling pathways which stimulate cell growth and division. Agents which act selectively against these targets are less likely to have the serious side effects associated with traditional cytotoxic agents.

- An abnormal form of the Ras protein is permanently active and is associated with many cancers. Inhibiting the farnesyl transferase (FT) enzyme prevents the Ras protein becoming attached to the cell membrane and prevents it interacting with other elements of the signal transduction process. Unfortunately, FT inhibitors have proved ineffective in clinical trials because of alternative prenylation pathways.

- Protein kinases are enzymes which use ATP to phosphorylate hydroxyl or phenol groups in protein substrates. Protein kinase receptors are proteins in the cell membrane which play a dual role of receptor and enzyme, and which are activated by growth factors. Both the messengers and the receptors have been implicated in various cancers by being overexpressed or abnormal in nature. Anticancer agents have been designed to act as inhibitors of the kinase active site of these proteins. Most bind to the ATP binding region, as well as other regions of the active site. Kinase inhibitors with different selectivities can be designed because there are variations in the amino acids present in the active sites of different kinases.

- Antagonists of the smoothened receptor in the hedgehog signalling pathway have been developed as anticancer agents.

21.7 Miscellaneous enzyme inhibitors

In previous sections, we looked at enzyme inhibitors associated with DNA synthesis and function, as well as enzymes involved in signal transduction. In this section, we look at the inhibition of enzymes which have important roles to play in angiogenesis, metastasis, and apoptosis.

21.7.1 Matrix metalloproteinase inhibitors

Matrix metalloproteinases (**MMPs**) are zinc-dependent enzymes which play an important role in the invasiveness and metastasis of cancer cells—processes that have few anticancer agents acting against them. MMPs catalyse the cleavage of peptide bonds between glycine and isoleucine (or leucine) in protein substrates (Fig. 21.71a). The substrate is thought to bind such that the carbonyl oxygen of glycine coordinates to the zinc cofactor, while

the neighbouring NH moiety acts as a hydrogen bond donor to the peptide backbone of an alanine residue (Fig. 21.71b). A water molecule is held between zinc and a glutamate residue, and acts as the nucleophile that hydrolyses the peptide bond, assisted by the negatively charged glutamate residue and the zinc cofactor (see also section 3.5.5).

The MMPs are extremely destructive enzymes involved in the normal turnover and remodelling of the extracellular matrix or connective tissue (section 21.1.9). This process is usually tightly controlled by natural protein inhibitors. However, excessive activity can result in various problems, including chronic degenerative diseases, inflammation, and tumour invasiveness. There are four main groups of MMPs—**collagenases** (MMP-1, 8, 13, and 18), **gelatinases** (MMP-2 and 9), **stromelysins** (MMP-3, 10, and 11), and **membrane type** (MMP-14, 15, 16, 17, 24, and 25). A number of these are implicated in tumour growth, invasion, metastasis, and angiogenesis. Therefore, **matrix metalloproteinase inhibitors** (**MMPIs**) could be useful in inhibiting the breakdown of the extracellular matrix to make it more difficult for cancer cells to escape and metastasize. They could also be used to inhibit angiogenesis by blocking the release of VEGF from storage depots in the extracellular matrix.

A variety of peptide-based inhibitors were developed that were based on the natural protein substrates for the collagenases. In general, they had the following features:

- Replacement of the susceptible peptide bond with a moiety which is stable to hydrolysis.

- One or more substituents that can fit into the enzyme subsites S1′ and S2′ and form van der Waals interactions. The subsites normally accept the amino acid residues of the substrate.

- At least one functional group capable of forming a hydrogen bond to the enzyme backbone.

- A group capable of strong interactions with the zinc ion cofactor. Typical groups are a thiol, carboxylate, or hydroxamic acid.

These features can be seen in **marimastat** (Fig. 21.72) which is an orally active, synthetic compound that reached phase III clinical trials for breast and prostate cancer. A hydroxamic acid group is present to form a strong bidentate interaction with the zinc cofactor. Substituents (P1′–P3′) are also present to fit three binding pockets in the active site. The NH moiety of the amide bond normally between glycine and isoleucine has been replaced by a hydroxymethylene group which prevents the normal hydrolysis reaction taking place and acts as a transition-state isostere. The hydroxyl group is also beneficial for inhibitory activity and aqueous solubility. Binding studies suggest that the hydroxyl group is directed away from the protein surface and is hydrogen bonded to water. The *t*-butyl substituent (P2′) also serves as a steric shield to protect the terminal amide from hydrolysis.

The nature of the P1′ group can be varied, and determines the activity and selectivity of the inhibitors against the various metalloproteinases. The nature of the P2′ group can also be varied, but it is beneficial to have a bulky group to act as a steric shield. It also serves to desolvate the peptide bonds such that energy is not expended on desolvation prior to binding.

Unfortunately, inhibitors such as marimastat lack selectivity and produce side effects such as tendinitis. They also have poor pharmacokinetic properties owing to

their peptide nature. For example, they have poor aqueous solubility and show susceptibility to peptidases in the gastrointestinal tract. Despite 30 years of research, no MMPI has reached the market and it is clear that MMPIs with much better selectivity are required.

Current research is looking at inhibitors that show selectivity for MMP2, MMP9, and MMP14 as these have been shown to be crucial to angiogenesis and metastasis. However, achieving selectivity is no easy task since there is little distinction between the MMPs. Having said that, the depth and flexibility of the S1′ binding sub pocket varies for different MMPs and this offers some hope of achieving selectivity. Selectivity may also be compromised by using a strong ligand for the zinc cofactor (e.g. a hydroxamic acid group), as this diminishes any distinction that there might be in binding to the S1′ subsite. Therefore, there could be selectivity advantages in using weaker ligands for the zinc cofactor, or no ligand at all. This would mean that any differences in binding to the S1′ pocket of different MMPs would have a more pronounced effect.

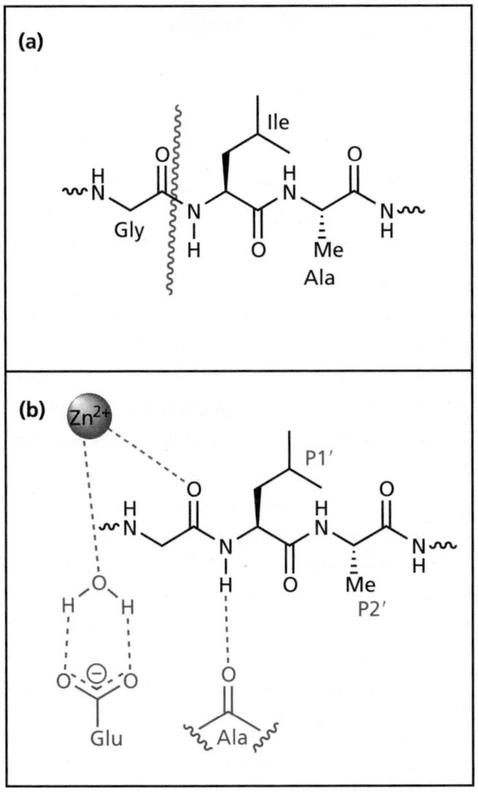

21.7.2 Proteasome inhibitors

The 26S proteasome is a complex structure that can be viewed as the cell's rubbish disposal unit. It is an ATP-dependent multi-catalytic protease and its prime role is to eliminate damaged or misfolded proteins, including key regulatory proteins. The 20S core of the proteasome is a barrel-shaped structure containing four rings, each made up of seven protein subunits (Fig. 21.73). The two outer alpha rings have no catalytic activity, whereas the two inner beta rings each contain three protein subunits with catalytic activity. The β5 subunit is described as being chymotrypsin-like as it splits proteins at similar positions to chymotrypsin itself. Similarly, the β2 and β1 subunits are described as trypsin-like and caspase-like respectively. Unlike chymotrypsin, trypsin, and caspase, the catalytic β subunits all contain an *N*-terminal threonine residue within their active site. This is of significance to the inhibitors described below.

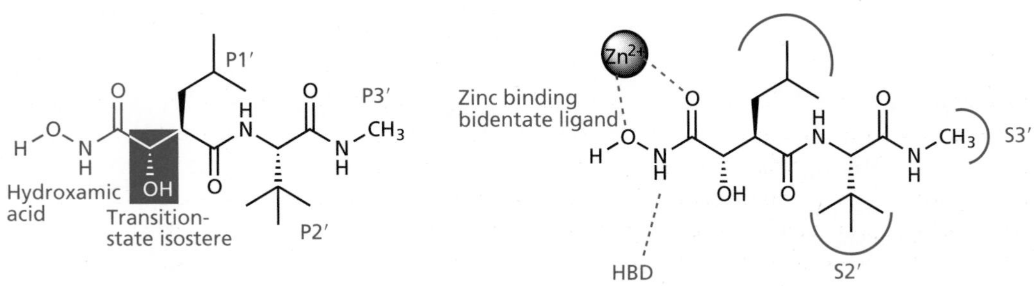

FIGURE 21.71 (a) Peptide bond cleaved by matrix metalloproteinases. (b) Binding interactions.

FIGURE 21.72 Structure and binding interactions of marimastat.

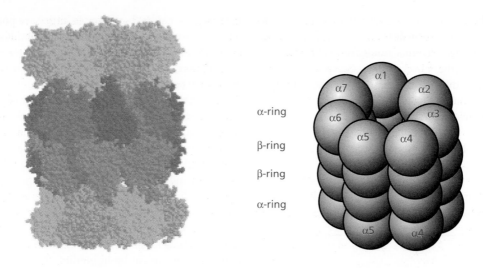

FIGURE 21.73 Structure of the 20S core of the proteasome (pdb 5BXN).

Considering the destructive power of this structure, it is important that it destroys only defective proteins and not normal ones. Therefore, cells mark their defective proteins with a molecular label so that they can be recognized by this protein-killing machine. The molecular label is a protein called **ubiquitin**.

Because the proteasome is so destructive, one might think it would be best to boost its activity in tumour cells. In fact, the opposite strategy is adopted and research is looking at agents that inhibit its action. The rationale lies in the fact that the proteasome removes regulatory proteins which have 'done their job'. Blocking the proteasome will result in an accumulation of various regulatory proteins, which leads to a cellular crisis and triggers

apoptosis. One of the proteins that accumulates as a result of proteasome inhibition is the apoptosis promoter **Bax** (section 21.1.7).

Early research into the design of proteasome inhibitors involved the synthesis of tripeptide structures containing an aldehyde functional group. This group was included such that it would react with the key N-terminal threonine residue to form a hemiacetal group. As a result, the drug would be covalently linked to the residue (Fig. 21.74a). The strategy worked very well and the tripeptides proved to be potent inhibitors *in vitro*. However, it was anticipated that tripeptidyl aldehydes would have poor pharmacokinetic properties. Therefore, other functional groups were investigated as possible replacements

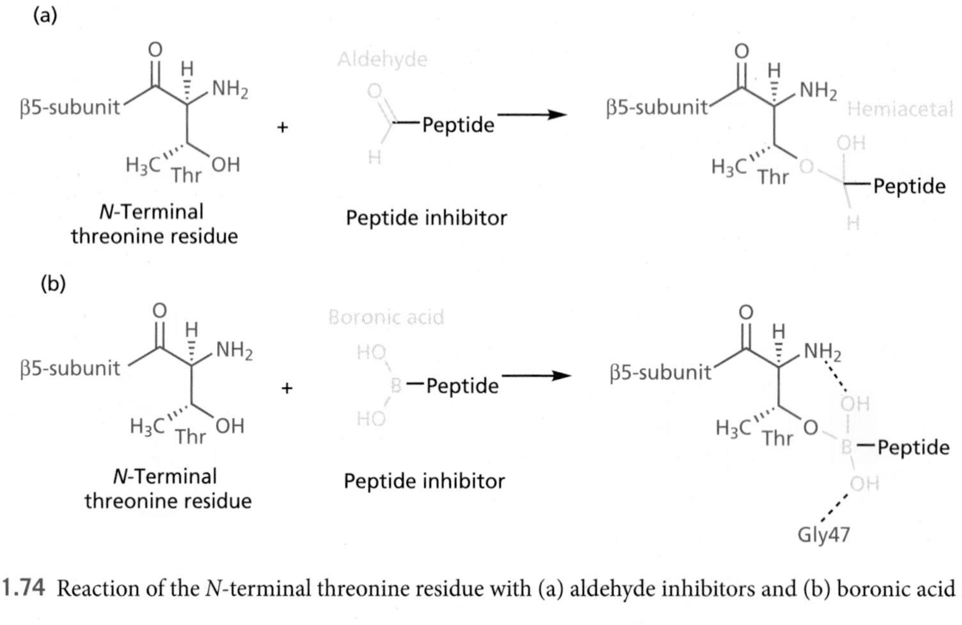

FIGURE 21.74 Reaction of the N-terminal threonine residue with (a) aldehyde inhibitors and (b) boronic acid inhibitors.

for the aldehyde group. These included ketones and keto esters, but the best results were obtained with a boronic acid group, which resulted in compounds having picomolar activities. Boron is electron deficient and contains an empty p-orbital. Because of this, it readily accepts a lone pair of electrons from an oxygen atom to form a strong B–O bond (Fig. 21.74b).

It was now decided to sacrifice some of the high activity observed by shortening the molecule to a dipeptide structure. This served to improve both selectivity and pharmacokinetic properties. Selectivity for the proteasome over serine proteases was improved because dipeptide structures do not bind well with the latter. Serine proteases, such as chymotrypsin and elastase, prefer substrates capable of interacting with four binding subsites (S1–S4), whereas dipeptides can only occupy two of those subsites. The use of a boronic acid inhibitor also increases selectivity for the proteasome with respect to cysteine proteases such as cathepsin B. Boron forms a strong bond with oxygen, but a far weaker bond with sulphur. Therefore, reaction with the thiol group of a cysteine side chain is not favoured.

The research described above led to the development of the reversible inhibitor **bortezomib** (Fig. 21.75), which is particularly effective at binding with the chymotrypsin-like β5 subunit. The presence of the boronic acid group introduces good selectivity as described above. In addition, the presence of a leucine side chain at P1 guarantees good selectivity for the proteasome with respect to thrombin. This is because thrombin prefers to bind substrates with a basic residue at P1. Bortezomib became the first proteasome inhibitor to be approved for the treatment of multiple myeloma. Unlike most anticancer drugs, it is not prone to multi-drug resistance. However, it has to be injected intravenously and it has limited activity against solid tumours. Side effects can also be an issue with some patients.

Ixazomib is a second-generation inhibitor that was approved in 2015 for the treatment of multiple myeloma. It is a prodrug that hydrolyses immediately in plasma to form the boronic acid MLN2238. It, too, shows preferential binding for the β5-subunit. MLN2238 binds more reversibly to proteasomes than borlezomib, and it is thought that this allows more of the drug to reach the proteasomes in tumour cells. A significant proportion of bortezomib is 'mopped up' by proteasomes in non-tumour cells, especially red blood cells and cells receiving a rich blood supply.

Aclarubicin (aclacinomycin A) (Fig. 21.75) is an anthraquinone which also inhibits the proteasome's chymotrypsin-like β5 subunit. The tetracyclic moiety and the three sugar rings are all necessary for activity.

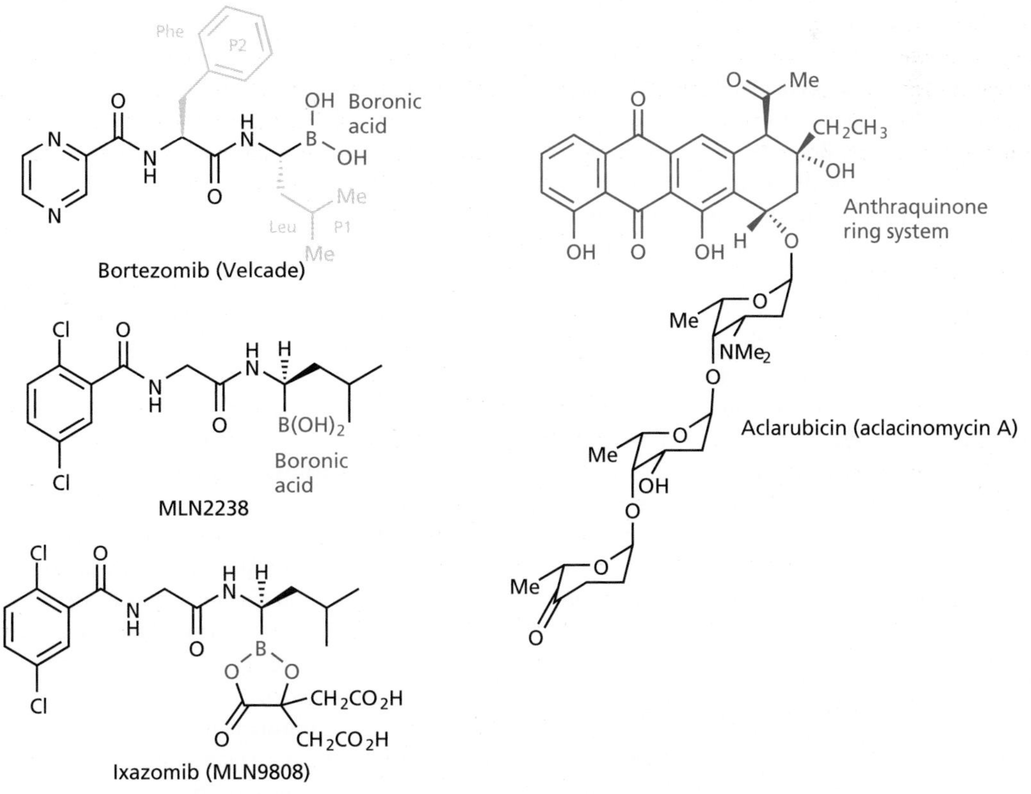

FIGURE 21.75 Proteasome inhibitors of the 20S core of the proteasome.

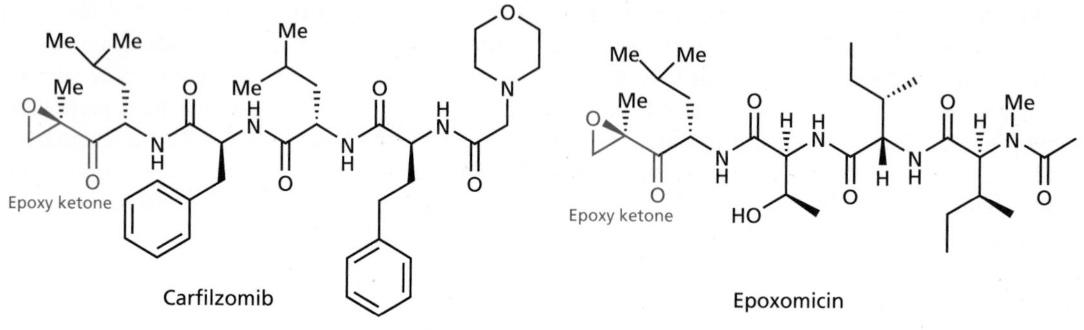

FIGURE 21.76 Irreversible proteasome inhibitors.

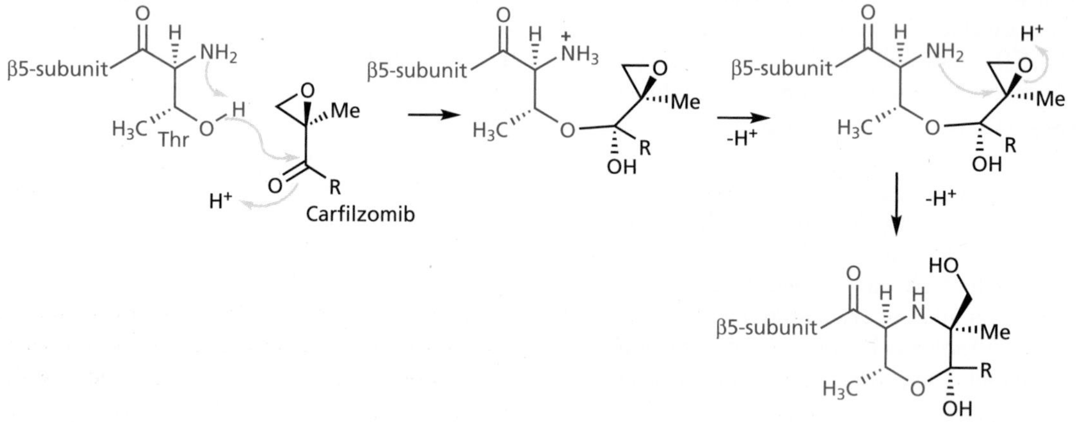

FIGURE 21.77 Mechanism of irreversible inhibition for carfilzomib.

Carfilzomib is an irreversible inhibitor of the β5-subunit, and was developed from a natural product called **epoxomicin** (Fig. 21.76). Both structures show high selectivity for the proteasome β5 subunit over non-proteasome proteases such as trypsin. The epoxy ketone group is crucial both for activity and selectivity since it reacts with catalytic proteins having an N-terminal threonine residue to form a morpholine ring (Fig. 21.77). The reaction involves the primary amino and secondary alcohol groups of threonine reacting with the epoxyketone. The alcohol reacts with the ketone to form a hemiacetal, which is followed by an intramolecular cyclization when the primary amine group reacts with the epoxide ring. Since this reaction can only take place with an N-terminal threonine residue, there is no reaction with proteases containing different N-terminal amino acid residues. Carfilzomib is approved for multiple myeloma patients who have previously taken bortezomib and other therapies.

21.7.3 Histone deacetylase inhibitors

Chromatins (Fig. 21.5) are structures where DNA is wrapped around proteins, most of which are **histones**. The histones assist in DNA packaging and also have a regulatory role. There is a repeating pattern of eight histone proteins along the length of the chromatin structure, with each octet associated with about 200 base pairs of DNA. Each of these repeating units is known as a **nucleosome**.

Histone acetylase is an enzyme that adds acetyl groups to the lysine residues of histone tails which stick out from the chromatin structure. Acetylation neutralizes the positive charge normally associated with the lysine side chain and weakens the ionic interactions between the histones and the negatively charged sugar phosphate backbone of DNA, leading to a less compact structure. The more open structure allows **transcription factors** to access the promoter regions of various genes. **Histone deacetylase** is an enzyme that removes the acetyl groups, leading to a more compact structure and preventing transcription factors accessing the promoter regions. This causes gene silencing, but can also lead to decreased DNA repair, resulting in an increased chance of cancer (see also section 21.9.4).

Several inhibitors of histone deacetylase have been studied (Fig. 21.78). **Romidepsin** (**depsipeptide**) is a natural product derived from a bacterial strain, and was approved in 2009 for the treatment of some lymphomas. The disulphide bond is reduced inside cells to give a dithiol, which

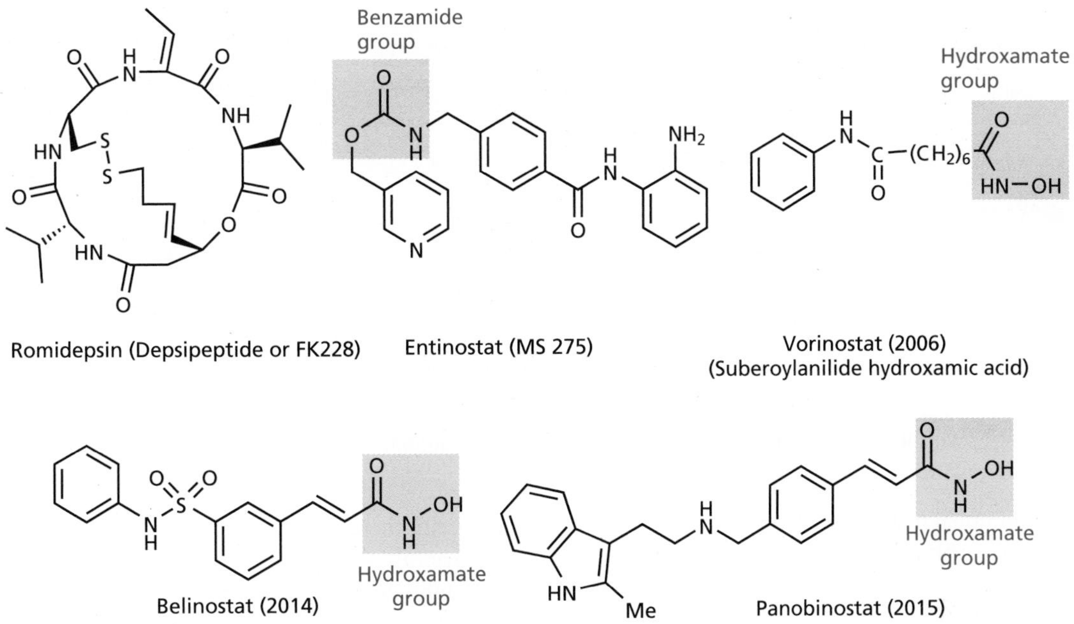

Romidepsin (Depsipeptide or FK228)   Entinostat (MS 275)   Vorinostat (2006)
(Suberoylanilide hydroxamic acid)

Belinostat (2014)   Panobinostat (2015)

FIGURE 21.78 Histone deacetylase inhibitors.

can then bind to a zinc cofactor present in the enzyme. The resulting enzyme inhibition promotes apoptosis and inhibits cell proliferation and angiogenesis. Synthetic agents have also been designed which contain functional groups capable of acting as ligands for the zinc cofactor. For example, **vorinostat**, **belinostat**, and **panobinostat** are clinically approved agents that contain a hydroxamate group that serves that purpose. **Entinostat** contains a benzamide group as a zinc ligand and is undergoing clinical trials. Clinical trials are also being carried out where vorinostat is used in combination with other anticancer drugs—an example of polypharmacology where different drugs are administered to affect different targets.

Crystal structures have revealed that the active site for the target enzyme is tube-shaped, with the zinc cofactor positioned at the bottom (Fig. 21.79). The tubular nature of the active site is significant in terms of the substrates accepted by the enzyme, as only amino acids with long side chains (e.g. lysine) can 'reach down' to the zinc cofactor. The zinc cofactor can then interact with the oxygen of the acetyl group and catalyse its hydrolysis (see also section 3.5.5).

Inhibitors also require a lengthy chain if they are to bind to the zinc cofactor. The general structure for effective inhibitors involves a zinc ligand such as a hydroxamic acid group, connected by a lengthy hydrophobic linker to a polar group. This, in turn, is connected to another hydrophobic group (Fig. 21.80). The polar and hydrophobic groups at the end of the inhibitor interact with binding regions near the entrance to the tube.

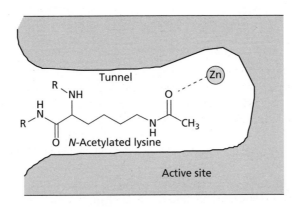

FIGURE 21.79 The tube-shaped nature of the active site in histone deacetylase.

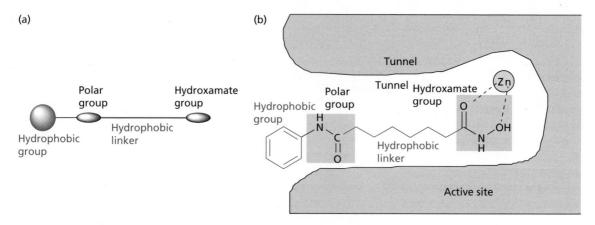

FIGURE 21.80 (a) General structure of hydroxamate inhibitors. (b) Binding of vorinostat.

For additional material see Web article 23: Histone deacetylase inhibitors in medicinal chemistry on the Online Resource Centre at www.oxfordtextbooks.co.uk/orc/patrick6e/

21.7.4 Inhibitors of poly ADP ribose polymerase

A number of research groups are investigating **poly ADP ribose polymerase (PARP) inhibitors** as potential anticancer agents. There are 18 members of the PARP family, and particular interest has been focused on inhibiting PARP-1. This enzyme is present in the nucleus of the cell and detects single- or double-strand breaks in DNA. The enzyme then binds to the affected region and promotes DNA repair. Inhibiting this process would make the cell more susceptible to chemotherapeutic agents that damage DNA and lead to increased cell death. It has also been discovered that some tumour cells are deficient in the tumour-repression proteins **BRCA1** and **BRCA2**. In these cells, there is an increased reliance on PARP-1 for cell survival, which makes the cells more susceptible to PARP inhibitors. Furthermore, PARP-1 appears to have a relatively minor role in DNA repair within normal cells.

When the enzyme binds to DNA, it is activated and catalyses the cleavage of NAD^+ (Fig. 3.11) to form ADP-ribose. A polymerization process is then catalysed that produces chains of poly(ADP-ribose) covalently linked to PARP and other nuclear proteins, such as histones. These protein–polymer structures are thought to play a role in temporarily protecting the damaged DNA. They also promote repair mechanisms by attracting repair proteins.

It has been shown that the nicotinamide ring of NAD^+ forms hydrogen bonds to a serine side chain and a glycine residue in the active site. In addition, pi–pi interactions take place with a tyrosine residue (Fig. 21.81a). Inhibitors have been designed that include a benzamide group that serves to mimic the nicotinamide group and its interactions (Fig. 21.81b).

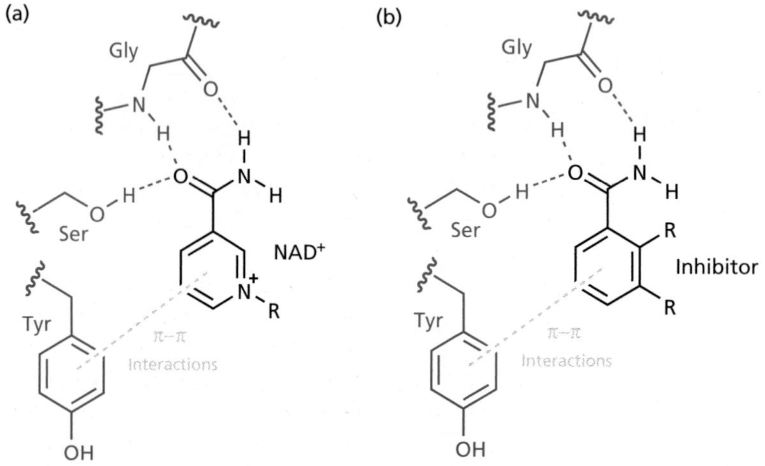

FIGURE 21.81 Binding of NAD^+ to the active site of poly ADP polymerase.

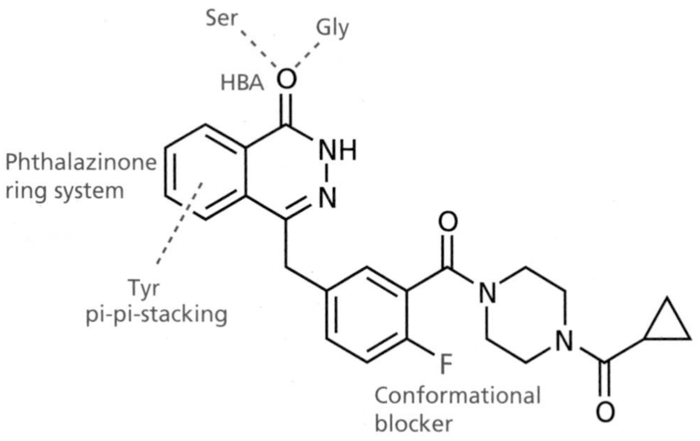

FIGURE 21.82 Olaparib.

However, the first PARP inhibitor to reach the market was **olaparib** (Fig. 21.82). In this structure, the benzamide group is part of a bicyclic phthalazinone ring and is, therefore, conformationally restrained. Docking studies indicate that the phthalazinone ring can form H-bonding interactions with the three key amino acid residues. The presence of the fluorine substituent on the aromatic ring plays an important role in cell permeability, but only if it is *ortho* to the carbonyl group. This suggests that the fluorine substituent acts as a conformational blocker and hinders bond rotation between the aromatic ring and the carbonyl group. This effectively cuts the number of rotatable bonds in the structure from 4 to 3, which is beneficial for membrane permeability (sections 11.3 and 13.3.9). Olaparib was approved in 2014 for the treatment of tumours that are deficient in BRCA1 and BRCA2.

21.7.5 Other enzyme targets

There are many other enzymes which are being studied as potential targets for anticancer agents. For example, inhibiting the **telomerase** enzyme should prevent cells becoming immortal. Several powerful inhibitors have been developed, although no telomerase inhibitor has reached the clinic to date. The inhibition of regulatory enzymes may be useful in shutting down a biosynthetic pathway that is too active. One example is inhibition of **tyrosine hydroxylase** (section 23.12.1). **Methionine aminopeptidase** is an enzyme that plays a key role in endothelial cell proliferation, and blocking it should inhibit angiogenesis. The activation of **caspases** to induce apoptosis is another possible approach to novel anticancer agents.

21.8 Agents affecting apoptosis

Apoptosis (section 21.1.7) is important to the activity of many of the anticancer drugs that create damage within a tumour cell. The process is controlled by a balance of different proteins that can be defined as pro-apoptotic or pro-survival. Bax and Bak are examples of pro-apoptotic proteins that stimulate apoptosis. However, they are kept in check by a range of pro-survival proteins such as Bcl-2 and Bcl-x$_L$. The pro-survival and pro-apoptotic proteins are structurally related and contain regions called the BH motifs. Bax and Bak contain three BH motifs (BH1–BH3), while the pro-survival proteins contain four BH motifs (BH1–BH4). The BH3 region is crucial to the protein–protein interactions that allow a pro-apoptotic protein such as Bax to bind to a pro-survival protein such as Bcl-2. When a cell experiences stress or damage, another group of pro-apoptotic proteins called the BH3-only proteins are stimulated (Fig. 21.83). As the name implies, these proteins only contain the BH3 region. However, this allows the BH3-only proteins to bind to the pro-survival proteins and displace pro-apoptotic proteins such as Bax and Bak. Once released, the pro-apoptotic proteins can trigger apoptosis.

Several cancers have defects to the apoptosis process because they overexpress pro-survival proteins such as BCL-2 and BCL-x$_L$. This allows cancer cells to survive cell damage, and it reduces the effectiveness of many anticancer drugs. Consequently, pro-survival proteins have been studied as possible targets for novel anticancer agents. Research has focused on developing small drug molecules that can mimic the role of the BH3-only proteins by binding to pro-survival proteins, thus preventing them from binding and inhibiting the pro-apoptotic proteins.

In order to achieve this, it is necessary to mimic the protein–protein interactions that occur between the proteins involved in the control of apoptosis. This is no easy task since protein–protein interactions involve several interactions over a wide surface area (see section 10.5). The pro-survival proteins BCL-2 and BCL-x$_L$ contain a hydrophobic groove about 20 Å in length which acts as the binding site for the BH3 motif in pro-apoptotic

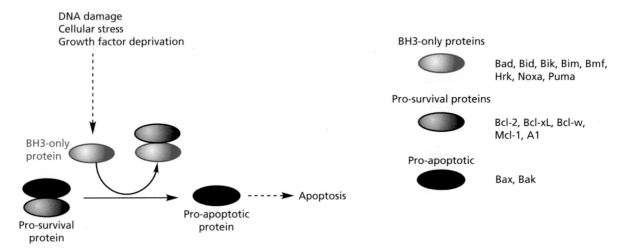

FIGURE 21.83 Interaction of pro-survival and pro-apoptotic proteins in the control of apoptosis.

proteins. This groove contains four hydrophobic pockets (P1–P4), which play a key role in the binding interaction between the two proteins. The BH3 domain of the pro-apoptotic proteins consists of an alpha helix, which fits into the binding groove and allows four conserved hydrophobic residues to slot into the four hydrophobic binding pockets (Fig. 21.84). In addition, the helix contains an aspartate residue that can form a strong ionic interaction with an arginine residue in the binding groove.

Navitoclax and **venetoclax** (Fig. 21.85) were designed to interact with the P2 and P4 pockets in the binding groove and have both reached clinical trials. Navitoclax inhibits both BCL-2 and BCL-x_L, whereas venetoclax was designed to show selectivity for BCL-2. Both agents contain groups that bind to the P2 and P4 pockets. An intramolecular pi–pi interaction is important in determining

the molecule's active conformation, and both aromatic rings involved form further pi–pi interactions with phenylalanine and tyrosine residues in the binding site. This network of pi–pi interactions plays a major factor in these agents having good binding affinity.

21.9 Miscellaneous anticancer agents

The field of anticancer research is a vast one with a wide diversity of novel structures being investigated. The following are examples of various structures which act at different targets, or whose targets have not been identified.

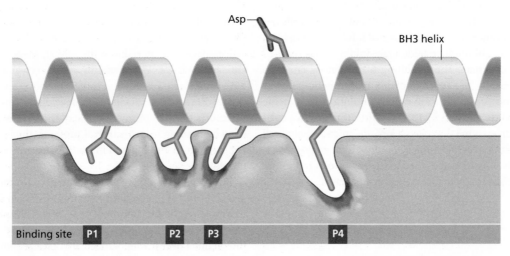

FIGURE 21.84 Binding of the BH3 motif helix to the binding groove present in a pro-survival protein. (Modified from Figure 1 of Lee et al., *Cell Death and Differentiation* (2007), **14**, 1711–13.)

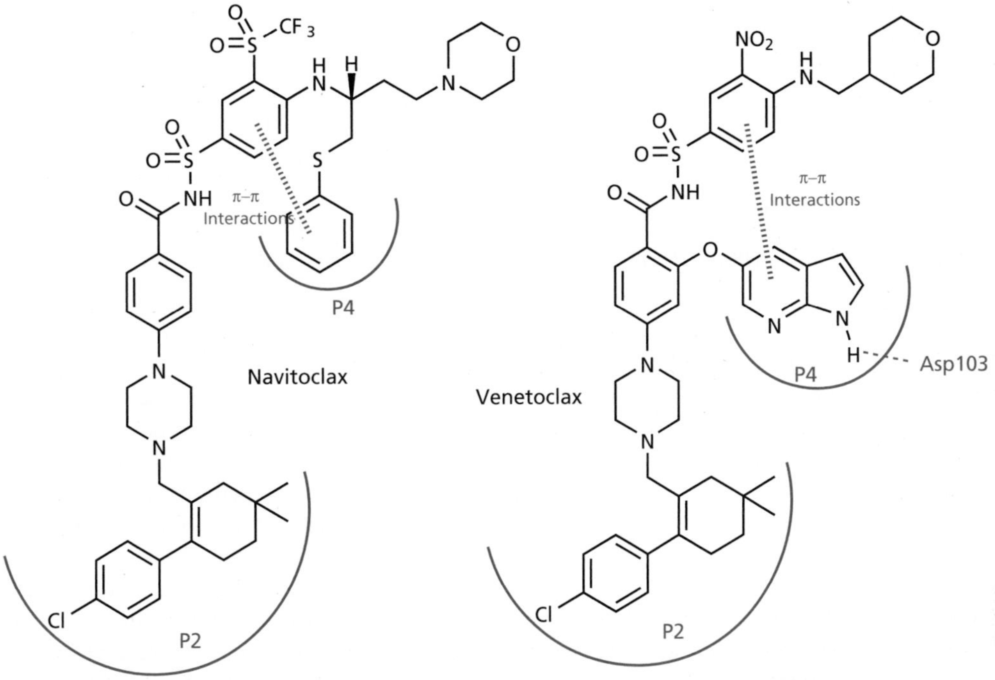

FIGURE 21.85 Navitoclax (ABT-263) and venetoclax (ABT-199).

21.9.1 Synthetic agents

Thalidomide (Fig. 21.86) was originally marketed as a safe, non-toxic sedative and anti-emetic in the 1950s and rapidly became popular to counter the effects of morning sickness during pregnancy. Unfortunately, it was instrumental in one of the major medical disasters of modern times when it produced teratogenic effects in developing fetuses and led to babies being born with stunted limbs and other developmental deformities—the so-called '**thalidomide babies**'. As thalidomide was considered a safe drug at the time, few suspected that it was the cause of the problem and, by the time it was linked to the deformities and withdrawn in 1961, 8000–12 000 babies were affected. Consequently, thalidomide gained a lasting notoriety which was instrumental in a significant tightening of the regulations surrounding the testing of drugs. Despite its notorious past, there has been continuing interest in thalidomide as it has some remarkable properties which indicate a wide variety of clinical uses. Early on, it was recognized that thalidomide has an anti-inflammatory property which was useful in treating leprosy. This activity was eventually linked to thalidomide's ability to inhibit the synthesis of the pro-inflammatory endogenous cytokine TNF-α which is produced by monocytes. In 1998, the drug was approved for the treatment of leprosy. However, thalidomide has a raft of other properties. For example, it is an immunosuppressant and could possibly be used for the treatment of autoimmune diseases or in countering the immune response to allow organ transplants to be accepted by the host. Interest in thalidomide as an anticancer agent started when it was found that it inhibited angiogenesis by an unknown mechanism. Tests showed that thalidomide did, indeed, have anticancer activity and it entered phase III clinical trials for the treatment of renal cancer and multiple myeloma on that basis. Since then, it has been discovered that the anticancer properties of thalidomide are more complex than its effect on angiogenesis alone. In some patients, thalidomide can boost the immune system by a variety of mechanisms rather than suppress it, and this too may account for its anticancer activity. Since thalidomide can suppress or boost the immune system depending on individual circumstances, it is known as an **immunomodulator**. Thalidomide also appears to arrest the growth of cells and promote apoptosis.

Analogues of thalidomide have been synthesized with the aim of removing its teratogenic properties. **Lenalidomide (Revlimid)** and **pomalidomide (Imnovid)** (Fig. 21.86) are two such examples. Both contain an amino substituent on the aromatic ring which is found to be crucial in producing a safer drug. Lenalidomide entered clinical trials in the year 2000 and was given orphan drug status in 2001 for the treatment of the then incurable disease of multiple myeloma. In 2003, it entered phase III clinical trials and was given fast-track status. It is now approved for the treatment of multiple myeloma. In 2013, pomalidomide was also approved for the treatment of multiple myeloma.

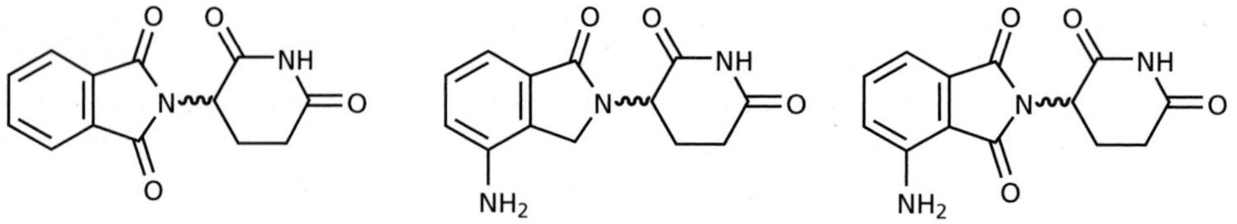

Thalidomide

Lanalidomide (Revlimid)

Pomalidomide (Imnovid)

FIGURE 21.86 Thalidomide and thalidomide analogues.

Arsenic trioxide is an orphan drug used for a variety of leukaemias. It is thought to promote cell suicide by targeting the cell's mitochondria. The compound has been used for many centuries in traditional Chinese medicine.

21.9.2 **Natural products**

Pancratistatin (Fig. 21.87) is a natural product isolated from a plant called *Pancratium littorale* which belongs to the lily family. Records show that extracts from plants in this genus were used by Hippocrates in 200 BC to treat breast cancer. The drug also inhibits angiogenesis, and analogues may have future potential as anticancer agents. The exact mechanism of action is still to be determined. One or more of the hydroxyl groups at positions C-2, C-3, and C-4 are thought to be important.

Bryostatin 1 (Fig. 21.87) is a natural product which was isolated from a marine invertebrate off the coast of California in 1981. It was shown to boost the immune system and make it more effective against cancers. It is undergoing clinical trials in combination with other established anticancer drugs, such as taxol, vincristine, fludarabine, or cisplatin. Bryostatin 1 has been found to affect the activity of protein kinase C—a key component of signal transduction pathways (section 5.3.2).

Dolastatins are natural products which were isolated from the marine sea hare off the island of Mauritius in the Indian Ocean. A full synthesis has been developed to produce **dolastatin 10** and the semi-synthetic analogue **monomethyl auristatin E** (MMAE) (**vedotin**) (Fig. 21.88). Dolostatin 10 entered clinical trials, but was not progressed further. MMAE proved too toxic to be used on its own, but it is the active anticancer agent in the clinically approved antibody–drug conjugate **brentuximab vedotin** (section 21.10.2 and Box 21.11).

Omacetaxine mepesuccinate (Fig. 21.89) is a natural product that was formerly known as **homoharringtone** and is produced by a bush called the Japanese plum yew (*Cephalotaxus harringtonia*). It has been approved for the treatment of chronic myeloid leukaemia (CML) when patients do not respond to kinase inhibitors. The drug

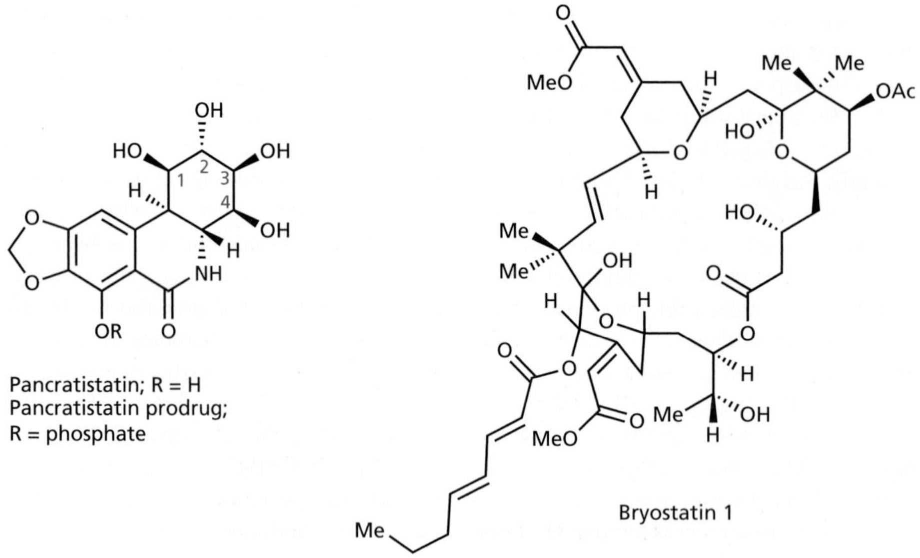

Pancratistatin; R = H
Pancratistatin prodrug;
R = phosphate

Bryostatin 1

FIGURE 21.87 Miscellaneous natural products having anticancer activity.

inhibits the first stage of protein translation by binding to the ribosomal A site, thus preventing aminoacyl t-RNAs from binding properly. The mechanism of action is unique in that it is only the first elongation step of protein translation that is affected. If protein translation has already started, the drug has no effect on the process.

Trabectedin (Fig. 21.89) (also known as **ecteinascidin 743**) was isolated from a sea squirt (*Ecteinascidia turbinata*) in the West Indies during the 1960s. The compound is present in very small quantities and the yields obtained from extraction are very small; only 1 gram of trabectedin is isolated from 1 ton of sea squirt. Therefore, a semi-synthetic process was developed that produces the compound in better yield from a modified microbial antibiotic called **cyanosafracin B**. The compound was approved in 2015 for the treatment of liposarcoma and leiomyosarcoma. It is thought that a variety of mechanisms may be involved in trabectedin's anticancer properties. For example, it is known to bind reversibly to the minor groove of DNA, with rings A and B being involved in this interaction. It also disorganizes the assembly of microtubules, but it is not a tubulin inhibitor.

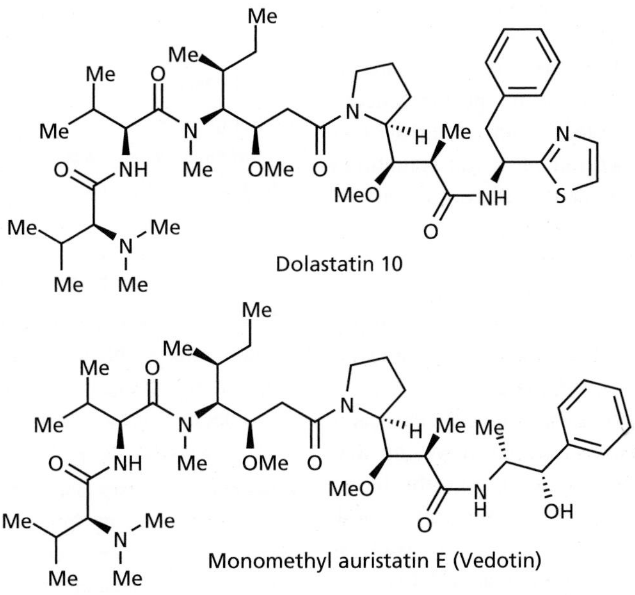

FIGURE 21.88 Dolastatin 10 and monomethyl auristatin E.

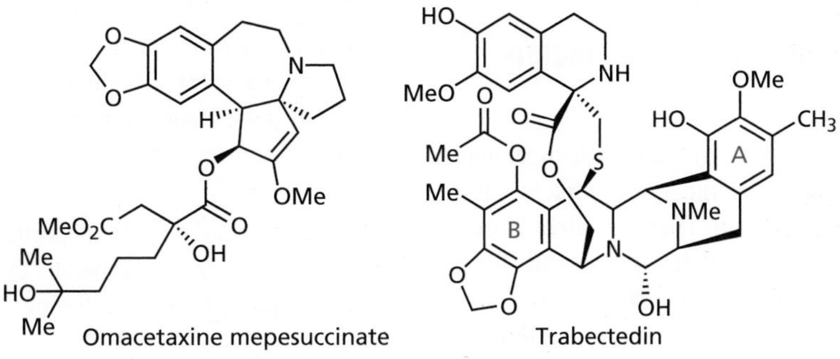

FIGURE 21.89 Omacetaxine mepesuccinate and trabectedin.

21.9.3 Protein therapy

A variety of proteins are being considered as anti-angiogenesis agents. For example, **angiostatin** and **endostatin** are two naturally occurring proteins in the body which inhibit the formation of new blood vessels and are being studied in cancer therapy. **α-Interferon** inhibits the release of growth factors such as VEGF, and can be administered for the treatment of a variety of leukaemias, lymphomas, and melanomas.

Cancer cells are more sensitive than normal cells to a natural death-inducing protein with the catchy title of **tumour necrosis factor-related apoptosis inducing ligand** (**TRAIL**), which stimulates apoptosis. Injecting purified TRAIL *in vivo* might selectively stimulate increased death rates in cancer cells. This is currently being studied in animals.

γ-Interferon is administered as an immunostimulant to reduce the risk of infection in specific cancers. **Aldesleukin** is a recombinant form of **interleukin-2** that can also act as an immunostimulant. It can sometimes cause tumour shrinkage in patients suffering metastatic renal cell carcinoma.

Some cancer cells have lost the ability to carry out normal synthetic routes as a result of gene mutation and the production of inactive enzymes. For example, some leukaemia cells lose the capacity to synthesize the amino acid asparagine and have to obtain it from the blood supply. The enzyme **asparaginase** can catalyse the degradation of asparagine, so providing that enzyme should break down asparagine in the blood supply and starve the cancer cells of this amino acid. A preparation of the asparaginase enzyme (**crisantaspase**) is used to treat acute lymphoblastic leukaemia (see also section 14.8.2).

Aflibercept (also called **zivaflibercept**) is a recombinant fusion protein that combines the binding regions from the VEGF receptor with the Fc portion of an immunoglobulin. It is used to inhibit abnormal growth of blood vessels in the eye by binding to circulating VEGF, and so prevent VEGF binding to its receptor.

21.9.4 Modulation of transcription factor–coactivator interactions

Research is taking place to try and find anticancer agents that work by interacting with transcription factors in order to affect gene transcription. There are already examples of clinically useful agents that act as ligands for nuclear receptor transcription factors. Work is in progress to find small molecules that might disrupt the interaction between a transcription factor and a coactivator protein, and thus prevent the formation of the complex required to signal the start of transcription (Box 10.2).

In a similar vein, drug-like molecules are being investigated as potential ligands for a protein binding region called a **bromodomain**. This is a region that is capable of binding an acetylated lysine residue of another protein, and plays an important role in protein–protein interactions (section 21.7.3). Therefore, drug-like molecules capable of interacting with bromodomains may act as protein–protein binding inhibitors (section 10.5). Contrary to what a chemist might assume, bromodomains do not contain bromine. Instead, their name refers to the *Drosphila* gene *brahma*, the sequence of which first identified the existence of bromodomains.

KEY POINTS

- Matrix metalloproteinases (MMPs) are zinc-dependent enzymes which degrade the extracellular matrix and encourage the processes of angiogenesis, tumour propagation, and metastasis.

- Attempts are being made to design MMP inhibitors that show selectivity between different types of MMP. They contain a group that can bind to the zinc cofactor of the enzyme, as well as a substituent capable of binding to the S1′ binding subsite. The S1′ subsite differs in character between different types of MMP.

- The proteasome is a destructive enzyme complex that breaks down proteins. Inhibition leads to a build-up of conflicting regulatory proteins, which triggers apoptosis. Selective inhibitors take advantage of an *N*-terminal threonine residue which is specific for proteasome enzymes.

- Histone acetylases and deacetylases are involved in the regulation of transcription. Inhibitors of histone deacetylase have been approved as anticancer agents. They contain a ligand capable of interacting with a zinc cofactor which resides at the bottom of a tubular region in the enzyme.

- Inhibitors of poly ADP ribose polymerase (PARP inhibitors) prevent DNA repair and promote cell death. One such inhibitor has been approved as an anticancer agent.

- A variety of other enzymes are potential targets for novel anticancer agents.

- Protein–protein interactions play an important role in the control of apoptosis. Inhibitors have been designed to inhibit protein–protein interactions between pro-apoptotic proteins and pro-survival proteins, such that apoptosis is triggered.

- A large number of synthetic structures and natural products have anticancer properties by unknown mechanisms. Others appear to work by having several different mechanisms.

- Protein therapy has proved useful in the treatment of certain cancers.

21.10 Antibodies, antibody conjugates, and gene therapy

21.10.1 Monoclonal antibodies

Cancer cells have unusual shapes and altered plasma membranes that contain distinctive antigens which have been overexpressed. This allows the possibility of using antibodies against the disease. Although the antigens concerned are likely to be present in some normal cells, they are likely to be present to a greater extent on the cancer cells, making the latter more vulnerable. Monoclonal antibodies (sections 10.7.2 and 14.8.3) have been produced for numerous tumour-associated antigens and a few have reached the clinic as anticancer agents (Box 21.11). These serve to activate the body's immune response to direct killer cells against the tumour. Alternatively, if the antigen is an overexpressed receptor, the antibody may bind to it and block the chemical messenger from binding. In this case, the antibody acts as a receptor antagonist. A new approach has been to develop monoclonal antibodies capable of binding to two different antigens, one on the target tumour cell and one on a normal T-cell of the immune system. In this way, the antibody directs the body's immune system against the tumour. **Blinatumomab** is one such monoclonal antibody that has been approved as an anticancer agent against malignant B-cells (Box 21.11).

BOX 21.11 Clinical aspects of antibodies and antibody–drug conjugates

Antibodies

Antibodies targeting epidermal growth factor receptors

Trastuzumab (Herceptin) is a humanized monoclonal antibody which targets the extracellular region of the HER2 growth factor receptor and inhibits receptor homodimerization. It was approved in 1998 as the standard first-line treatment of HER2 positive metastatic breast cancer in combination with **paclitaxel**. HER2 is a member of the EGFR family of tyrosine kinase receptors, which is overexpressed in 25% of breast cancers. When the antibody binds to the receptor, it induces the immune response to attack the specified cell. It also promotes internalization and degradation of the receptor. Trastuzamab is given by injection. Unfortunately, the drug cannot cross the blood–brain barrier and is ineffective against any tumour cells that have metastasized to the brain. Drug resistance and cardiac toxicity are further problems which can arise. The antibody can also be used in combination with **docetaxel** and **pertuzumab**—another monoclonal antibody that targets the HER2 receptor. Unlike trastuzumab, pertuzumab inhibits heterodimerization of the HER2 receptor with other HER receptor types.

Cetuximab is a chimeric monoclonal antibody that targets the extracellular domain of the EGF receptor and blocks epidermal growth factor from binding. It is used alone, or alongside **irinotecan**, for the treatment of metastatic colorectal tumours which express EGFR, and which have proved resistant to previous chemotherapy that has included irinotecan. The antibody is also used alongside radiotherapy for the treatment of locally advanced squamous cell cancer of the head and neck. For recurrent or metastatic squamous cell cancer of the head and neck, cetuzimab can be used alongside platinum-based drugs. **Panitumumab** is a fully humanized monoclonal antibody that also targets EGFR and has been approved for the treatment of colorectal cancer. **Necitumumab** is another antibody that targets EGFR, and has been approved for the treatment of metastatic squamous non-small cell lung cancers in combination with **gemcitabine** and **cisplatin**.

Antibodies targeting B-cells

Alemtuzumab is a humanized antibody that lyses B-lymphocytes and is used to treat B-cell chronic lymphocytic leukaemia where other therapies have not worked. The antibody binds to a receptor (the CD52 antigen) which is found both on normal and cancerous immune cells (B- and T-lymphocytes). Although the agent shows no selectivity for cancer cells over normal cells, normal cells recover quicker after treatment.

Rituximab is a chimeric antibody targeting the CD20 receptor on B-lymphocytes and was approved in 1997 for the treatment of diffuse B-cell non-Hodgkin's lymphoma and follicular lymphoma. It causes lysis of B-lymphocytes. Patients should be monitored very closely since there are reported fatalities relating to the release of cytokines. In 2010, it was approved by the FDA for the treatment of chronic lymphocytic leukaemia. Another antibody that acts in the same manner is **obinutuzumab** (also called **afutuzumab**), which was approved in 2013. **Ofatumumab** targets a different epitope of the CD20 receptor from rituximab, and was approved in October 2009. It is used to treat leukaemia that cannot be controlled by other forms of chemotherapy.

Blinatumomab was approved in 2014 for the treatment of lymphoblastic leukaemia. It acts as an adaptor that can recognize both T-cells and malignant B-cells. The antibody contains two binding sites. One is for the CD3 site in T-cells, and the other is for the CD19 site of target B-cells. The antibody

(Continued)

BOX 21.11 Clinical aspects of antibodies and antibody-drug conjugates (*Continued*)

can, therefore, bind to both a T-cell and a B-cell. By doing this, it serves as an adaptor that directs and links T-cells to malignant B-cells. As a result, the T-cell attacks the B-cell and destroys it. Antibodies that identify two targets in this way are defined as **bispecific T-cell engagers (BiTEs)**.

Antibodies targeting VEGF and VEGFR

Bevacizumab is a humanized monoclonal antibody that is given intravenously and disables the growth factor VEGF required for angiogenesis. It is used in the treatment of a number of tumours. For example, it is used in the first-line treatment of metastatic colorectal cancer along with **fluorouracil** and **folinic acid (leucovorin)**. It is also used in the first-line treatment of metastatic breast cancer alongside **paclitaxel** or **capecitabine**. For advanced or metastatic renal cell carcinoma, bevacizumab can be used along with interferon alfa-2a, but not as a first-line treatment. Bevacizumab can also be used along with platinum-based drugs for the treatment of non-small-cell lung cancer, and with carboplatin and paclitaxel for tumours of the ovaries or fallopian tubes. **Ramucirumab** targets VEGFR-2 receptors and prevents the binding of VEGF to this receptor. It was approved in 2014 for the treatment of advanced gastric adenocarcinoma, and can be used in combination with paclitaxel. It has also been approved for the treatment of metastatic non-small-cell lung cancers in combination with docetaxel.

Antibodies targeting RANK

Denosumab is a fully humanized monoclonal antibody that targets the ligand for a receptor called **receptor activator of nuclear factor-kappa B (RANK)**. RANK receptors play an important role in bone formation, and so the antibody is approved for the treatment of bone tumours. It is also approved for the treatment of osteoporosis in post-menopausal women because it protects the bone from degradation.

Antibodies targeting melanoma

Ipilimumab was approved in 2011 for the treatment of metastatic melanoma. It activates the immune system by targeting a protein receptor called **CTLA-4** whose normal function is to down-regulate the immune system. By preventing down-regulation, cytotoxic T-lymphocytes can identify and destroy cancer cells. Another antibody which stimulates the immune system is **pembrolizumab**. This antibody targets the **programmed cell death 1 receptor (PD-1 receptor)**, which inhibits the immune response of T-cells against cancer cells. It was approved in 2014/2015 for the treatment of advanced melanoma and metastatic non-small-cell lung cancer. The antibody **nivolumab** also targets the PD-1 receptor, and was approved in 2014 for the treatment of metastatic melanoma and squamous non-small-cell lung cancer. Both pembroli-

zumab and nivolumab block the ability of PD-1 ligands to activate their receptor. It has been shown that many cancer cells produce PD-1 to prevent attack from T-cells.

Elotuzumab is a humanized monoclonal antibody that was approved in 2015 for the treatment of relapsed multiple myeloma, and is used alongside **lenalidomide** and **dexamethasone**. It acts as an immunostimulant and targets the protein **SLAMF7 (CD319)**, which is expressed on the surface of malignant plasma cells.

Daratumumab was approved in 2015 for the treatment of multiple myeloma. It targets **CD38** (cyclic ADP ribose hydrolase), which is a glycoprotein and enzyme present on the surface of plasma cells. CD38 is overexpressed in multiple myeloma.

Antibodies targeting interleukins

Siltuximab is a chimeric monoclonal antibody made from human and mouse proteins. It was approved in 2014 for the treatment of multi-centric **Castleman's disease**. Castleman's disease is a benign cancer which is characterized by enlarged lymph nodes, increased release of cytokines, and increased production of B-cells and T-cells. The antibody targets **interleukin-6** and prevents it binding to interleukin receptors. As a result, the growth of B-lymphocytes and plasma cells is inhibited, as is the secretion of VEGF.

Antibodies targeting glycolipids

Dinutuximab was approved in 2015 for the treatment of neuroblastoma in children. The cancer normally occurs in the adrenal gland, and is the most common cancer identified in infants. The antibody targets an overexpressed glycolipid labelled as **GD2**.

Antibody–drug conjugates

Gemtuzumab ozogamicin (Box 21.12) was approved for the treatment of acute myeloid leukaemia (AML) but was withdrawn in 2010.

Ibritumomab was the first approved drug involving radioimmunotherapy for the treatment of non-Hodgkin's lymphoma.

Tositumomab was approved in 2003 for the treatment of non-Hodgkin's lymphoma that was refractory to rituximab.

Brentuximab vedotin was approved in 2011 for the treatment of Hodgkin's lymphoma and systematic anaplastic large-cell lymphoma. The antibody targets a cell membrane protein called CD30, and is linked to the antitumour agent **monomethyl auristatin E** (MMAE or **vedotin**).

Trastuzumab emantasine has up to eight maytansinoid molecules (DM1) connected by a linker to lysine residues present in trastuzumab (Herceptin). It was approved in 2013 for the treatment of HER2 positive breast cancer in patients who prove resistant to trastuzumab alone.

21.10.2 **Antibody–drug conjugates**

Some monoclonal antibodies have an anticancer activity in the 'naked' form (i.e. without a drug attached), but the level of activity is usually too low to be effective and so a better strategy is to attach an anticancer drug to the antibody (an antibody–drug conjugate) such that the drug is delivered selectively to the cancer cell.

One of the original aims in designing antibody–drug conjugates was to deliver anticancer agents to tumour cells in greater concentrations than was possible by conventional therapy. There is often a narrow therapeutic window between the levels of drug that are effective and the levels leading to unacceptable toxicity. Antibody–drug conjugates were seen as a means of avoiding this problem, as it was anticipated that targeting would lead to higher concentrations of the drug at tumour cells. The first generation of such conjugates involved antibodies linked to anticancer agents such as **methotrexate**, **the Vinca alkaloids**, and **doxorubicin**, but the results were disappointing. The anticancer activity achieved was less than using the drug itself, and yet the toxicity problem remained the same.

It was later realized that the lifetime of the antibody–drug conjugate was substantially greater than that of the drug itself, which contributed to the toxicity problem. Furthermore, delivery to the tumour and subsequent penetration was limited due to the size of the conjugate. This meant that the concentration of antibody–drug reaching the tumour was actually less than when the drug itself was used. As the rate of delivery and penetration is determined by the antibody, there is little that can be done to improve it. Therefore, it was realized that highly potent anticancer agents should be attached to the antibodies in order to be effective at the levels attained at the tumour cell. This rules out anticancer drugs such as **doxorubicin**, **etoposide**, **5-fluorouracil**, and **cisplatin**. More potent anticancer drugs are available, but if they should become detached from the antibody during circulation they are likely to cause severe toxicity. Therefore, it is important that any such drug is attached to the antibody by a stable bond and remains bound until it enters the cancer cell.

The requirements for antibody–drug conjugates now include the following:

- the antibody has to be humanized to avoid an immune response;
- it has to show selectivity for an antigen which is overexpressed in cancer cells rather than normal cells;
- it then needs to be internalized into the cell by receptor-mediated endocytosis such that the antibody–drug can be delivered into the cell;
- the link between the antibody and the drug should be stable until cell entry has taken place and then cleaved to release and activate the drug.

There are various ways in which a drug can be linked to an antibody. For example, there are lysine residues present throughout the whole molecule which contain a nucleophilic primary amino group. A number of drug molecules could be added to the one antibody by acylating or alkylating these groups. There is a problem with this approach, however, as it is quite possible for a molecule to be attached to the region responsible for 'recognizing' the antigen. This would prevent antibody–antigen binding. Moreover, the masking of polar amino groups may lead to precipitation of the antibody–drug complex.

A better approach is to reduce the four intrastrand disulphide links at the hinge region of the antibody (Fig. 21.90) to produce eight thiol groups, and to attach

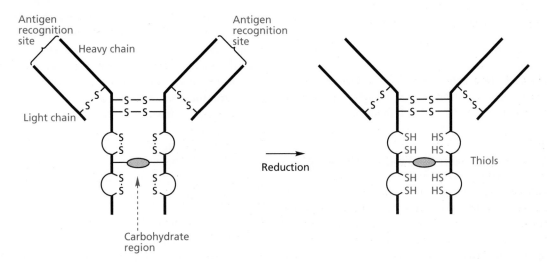

FIGURE 21.90 Reduction of disulphide links.

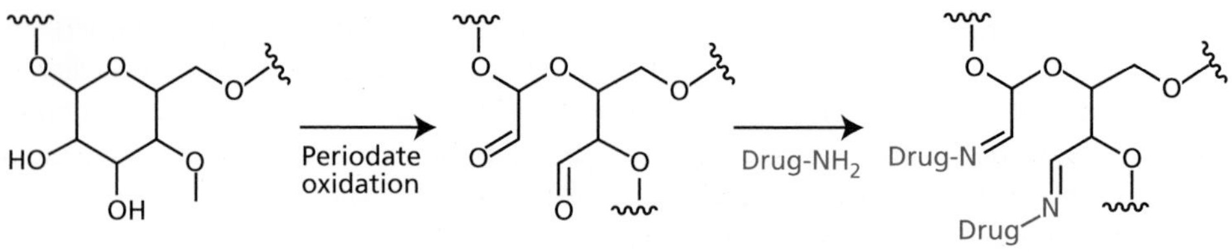

FIGURE 21.91 Linking drugs to carbohydrates.

drugs to these groups by alkylation or via a disulphide linkage. This has the advantage that it can be carried out in a controlled fashion and the drugs do not mask the antigen recognition site. The disadvantage is that a maximum of only eight drugs can be added to any one antibody molecule. One way round this may be to add a linker molecule to the antibody which could itself bear several drug molecules.

Another method of attaching the drug to the antibody is to take advantage of the carbohydrate region between the two heavy chains. Mild oxidation of vicinal diols in the sugar rings produces aldehyde groups to which drugs can be linked through an imine functional group (Fig. 21.91). Further reduction can then be carried out to form more stable amine linkages.

It is important that the linker is cleaved once the antibody–drug complex enters the cancer cell. Various linkers have been tried, such as acid-labile linkers, peptidase-labile linkers, and disulphide linkers. The disulphide linker can be cleaved by disulphide exchange with an intracellular thiol such as glutathione, which has a higher concentration within cells than in plasma.

The drug itself needs to be highly potent ($IC_{50} < 10^{-10}$ M) which involves the use of drugs that are 100–1000 times more cytotoxic than conventional cytotoxic drugs. A variety of agents are being investigated including **radioactive isotopes**, **ricin**, **diphtheria toxin**, *Pseudomonas aeruginosa* **exotoxin A**, **maytansinoids**, **adozelesin**, **calicheamicin γ_1**, **auristatins**, highly potent **taxoids**, and highly potent **doxorubicin analogues**.

Ibritumomab and **tositumomab** are conjugated murine antibodies which carry radioactive isotopes to an antigen called **CD20** on the surface of B-lymphocytes. These cells grow uncontrollably in non-Hodgkin's lymphoma. Ibritumomab carries ^{90}Y and tositumomab carries ^{131}I.

21.10.3 Antibody-directed enzyme prodrug therapy (ADEPT)

Antibody-directed enzyme prodrug therapy (**ADEPT**) involves two steps. The first is the administration of an antibody–enzyme complex. The antibody is raised against tumour-selective antigens and is linked to an enzyme such as **bacterial carboxypeptidase**. This complex then gets bound to the tumour. Unlike antibody–drug conjugates, the antibody–enzyme complex needs to remain attached to the surface of the cell and should not be internalized. A certain period of time needs to elapse to give the complex time to bind to target cells and for unbound complex to be cleared from the blood supply. A prodrug of a cytotoxic drug is then administered. The prodrug is designed such that it will be stable in the blood supply and can only be cleaved and activated by the enzyme complexed to the antibody. This means that the toxic drug is only produced at the tumour and can be administered in higher doses than the parent drug. **CJS 149** is an example of one such prodrug which is activated by a bacterial carboxypeptidase (Fig. 21.92). An advantage of ADEPT over antibody–drug conjugates is that the enzyme is catalytic and can generate a large number of active drug molecules at the site of the tumour. These can then diffuse into the tumour and affect cells which might not have any antibody attached to them.

A lot of research has been carried out on ADEPT using bacterial enzymes such as **carboxypeptidase G2**, **penicillin G acylase**, and **β-lactamase**. The advantage of using a 'foreign' enzyme is that enzymes can be chosen that are not present in the mammalian cell, and so there is no chance of the prodrug being activated by mammalian enzymes during its circulation round the body. It is also possible to use foreign enzymes which have counterparts in the body, as long as the latter are only present in low levels in the blood and/or they are structurally distinct. Prodrugs can be designed that selectively react with the foreign enzyme rather than the mammalian version. Examples of enzymes in this category include **β-glucuronidase** and **nitroreductase**.

Many studies have been carried out on ADEPT. One example is an antibody–β-lactamase complex capable of reacting with the cephalosporin prodrug of an alkylating agent (Fig. 21.93) This takes advantage of the mechanism by which cephalosporins react with β-lactamase to eliminate a leaving group (section 19.5.2.1).

One of the problems associated with ADEPT is the possibility of an immune response to the antibody–enzyme complex since the enzyme is a foreign protein. For this reason, it may be preferable to use human enzymes

BOX 21.12 Gemtuzumab ozogamicin: an antibody–drug conjugate

A humanized antibody called gemtuzumab has been linked to the highly potent anticancer drug **calicheamicin**. The trisulphide group normally present in calicheamicin was first modified to a disulphide with a hydrazide linker attached, while the antibody was treated with periodate to generate aldehyde groups at the carbohydrate region. The two molecules were then linked up by reacting the hydrazine group on the drug with the aldehyde groups on the antibody (Fig. 1). The resulting conjugate is called **gemtuzumab ozogamicin** and has been approved for the treatment of acute myeloid leukaemia (AML) in patients older than 60 years.

When the antibody–drug complex reaches the target leukemic cell, the antibody attaches itself to a **CD33 antigen** and the antibody–drug complex is then taken into the cell by endocytosis. It is thought that the drug is then released from the antibody in lysosomes or endosomes by acidic hydrolysis of the hydrazone, and that reduction of the disulphide group occurs later in the cell nucleus to produce the active thiol (Fig. 2).

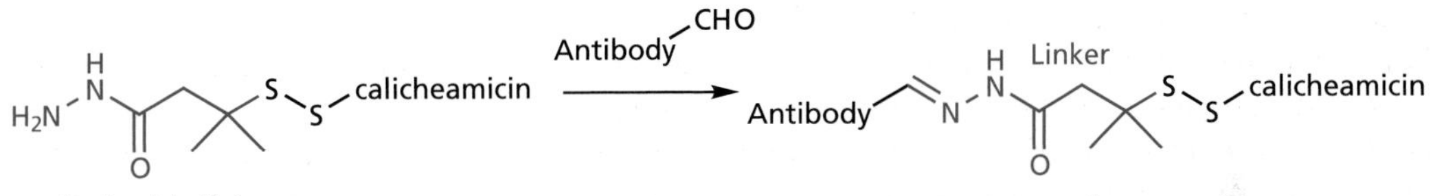

FIGURE 1 Linking calicheamicin to an antibody.

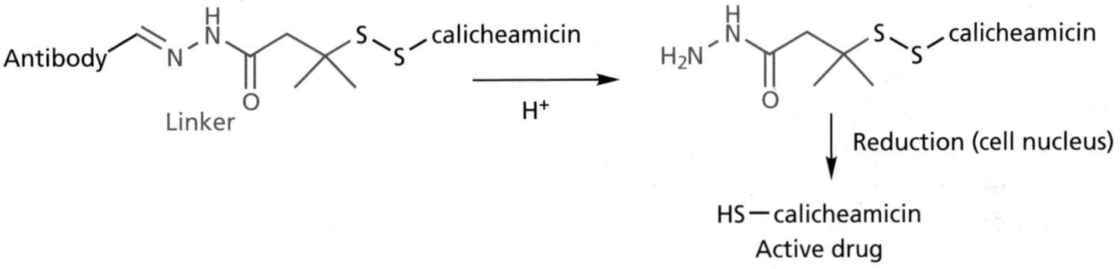

FIGURE 2 Release and activation of calicheamicin.

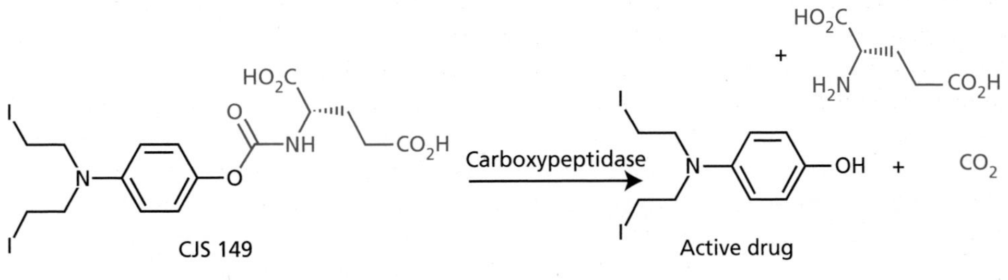

FIGURE 21.92 Activation of a prodrug by carboxypeptidase.

along with prodrugs that are already approved for anticancer use. Research has been carried out on human enzymes such as **alkaline phosphatase**, **carboxypeptidase A**, and **β-glucuronidase**. The advantage of using a human enzyme is the decreased chance of an immune response, but the disadvantage is the increased risk of prodrug activation occurring during circulation in the blood supply.

Another problem may be insufficient enzyme activity. For example, the activation of **irinotecan** has been achieved using a particularly active isozyme of **human carboxylesterase** enzyme isolated from the liver. The

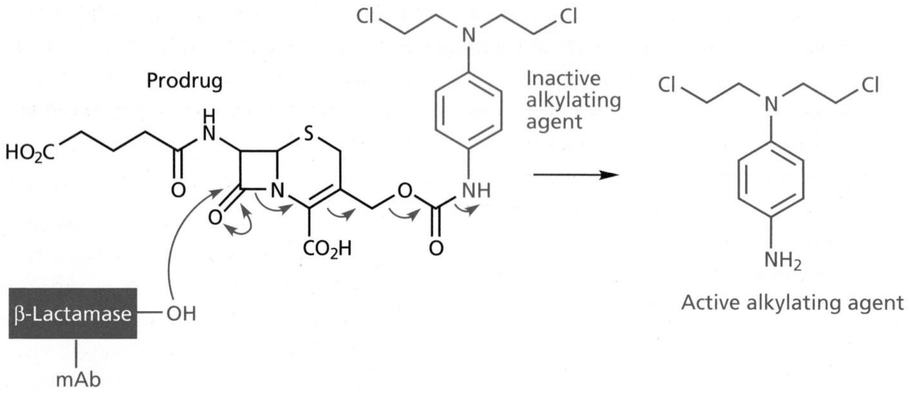

FIGURE 21.93 ADEPT strategy to release an alkylating agent.

isozyme concerned (hCE-2) was 26 times more active than another isozyme hCE-1, but was still too low to be effective for ADEPT. Nevertheless, the isozyme may be suitable for gene therapy (section 21.10.5) where greater concentrations of the isozyme could be achieved within the cell than could be brought to the cell by antibodies.

The time gap between the administration of the antibody–enzyme complex and the prodrug is critical. Enough time must be provided to ensure that unbound complex has dropped to low levels, otherwise the prodrug will be activated in the blood supply. However, the longer the time gap, the more chance the levels of the antibody–enzyme complex will drop at the tumour. One way to tackle this problem is a three-stage ADEPT strategy. The antibody–enzyme complex is administered as before. Sufficient time is given for the complex to concentrate at the tumour, then a second antibody is administered which targets the conjugate and speeds up its clearance from the blood supply. The second antibody can be galactosylated to speed up its clearance rate such that it only has time to target circulating conjugate and does not survive long enough to penetrate the tumour. Finally, the prodrug is added as before.

21.10.4 Antibody-directed abzyme prodrug therapy (ADAPT)

Abzymes are antibodies which have a catalytic property. It is possible that prodrugs could be designed that act as antigens for these antibodies and are activated by the abzyme's catalytic properties. This can be done by immunizing mice with a transition-state analogue of the reaction that is desired, followed by isolation of the monoclonal antibodies by hybridization techniques. Since the antibody targets the prodrug rather than antigens on the cancer cell, this fails to target drugs

to cancer cells. However, it should be possible to construct hybrid antibodies where one arm recognizes antigens on cancer cells while the other arm recognizes the prodrug and activates it. This approach is still in its early stages, but it has several potential advantages over ADEPT. For example, it should be possible to design catalytic mechanisms that do not occur naturally, allowing highly selective activation of prodrugs at tumours. It also removes the risk of an immune response due to foreign enzymes. At present, the catalytic activity of abzymes is too low to be useful and much more research has to be carried out.

21.10.5 Gene-directed enzyme prodrug therapy (GDEPT)

Gene-directed enzyme prodrug therapy involves the delivery of a gene to the cancer cell. Once delivered, the gene codes for an enzyme capable of transforming a prodrug into an active drug. As the enzyme will be produced inside the cell, the prodrug is required to enter the cell.

The main challenge in GDEPT is to deliver the gene selectively to tumour cells. In one method, the gene is packaged inside a virus, such as a retrovirus or adenovirus. In the case of adenoviruses, the desired genes could be spliced into the viral DNA such that the virus inserts them into host cell DNA on infection. The virus is also genetically modified such that it is no longer virulent and can do no harm to normal cells. Non-viral vectors have also been tried, such as cationic lipids and peptides. So far, it has not been possible to achieve the required selectivity for cancer cells over normal cells, and so the delivery vector has to be administered directly to the tumour.

The enzymes which are ultimately produced by the introduced genes should not be present in normal cells, so that prodrug activation only occurs in tumour cells. One advantage of GDEPT over ADEPT is the fact that foreign enzymes could be generated inside cancer cells and hidden from the immune response. The **thymidine kinase** enzyme produced by herpes simplex virus has been studied intensively in GDEPT. This enzyme activates the antiviral drugs **aciclovir** and **ganciclovir** (section 20.6.1). As these drugs are poor substrates for mammalian thymidine kinase, activation will only be significant in the tumour cells containing the viral form of the enzyme. Several clinical trials have been carried out using this approach.

One problem associated with GDEPT is that it is unlikely that all tumour cells will receive the necessary gene to activate the prodrug. It is therefore important that the anticancer drug is somehow transferred between cells in the tumour—a so-called **'bystander' effect**. This may occur by a variety of means, such as release of the activated drug from the infected cell, direct transfer through intercellular gap junctions, or by the release of drug-carrying vesicles following cell death.

GDEPT has been used to introduce the genes for the bacterial enzymes **nitroreductase** and **carboxypeptidase G2** into cancer cells. Prodrugs were then administered which were converted to alkylating agents by the resulting enzymes. One of the problems with carboxypeptidase G2 is the difficulty some of the prodrugs have in crossing cell membranes. In order to overcome this problem, the gene was modified such that the resulting enzyme was incorporated into the cell membrane with the active site revealed on the outer surface of the cell.

Gene therapy aimed at activating the prodrug **irinotecan** is being explored to try and improve the process by which the urethane is hydrolysed to the active drug (section 21.2.2.2). This could involve the introduction of a gene encoding a more active carboxypeptidase enzyme into tumour cells. For example, **rabbit livercarboxypeptidase** is 100–1000 times more efficient than the human form of the enzyme.

21.10.6 **Other forms of gene therapy**

Gene therapy could also be used to introduce the genes coding for regulatory proteins which have been suppressed in cancer cells. For example, attempts have been made to introduce the gene for the **p53 protein** via a virus vector.

KEY POINTS

- Monoclonal antibodies have been targeted against antigens which are over expressed in certain cancer cells. They are useful in the treatment of breast cancers, colorectal cancer, and lymphomas.

- Antibody–drug conjugates involve the linking of a highly potent drug or radioisotope to an antibody. The conjugate is designed to target specific cancer cells and then be enveloped by the cell such that the drug can be released inside the cell.

- Antibodies should be humanized to avoid the immune response.

- Drugs can be attached to antibodies via lysine residues. Alternatively, the antibody can be modified to produce thiol or aldehyde groups to which drugs can be attached.

- ADEPT involves an antibody–enzyme conjugate which is targeted to specific cancer cells. Once the antibody has become attached to the outer surface of cancer cells, a prodrug is administered which is activated by the enzyme at the tumour site.

- ADAPT involves an antibody which has catalytic activity designed to activate a prodrug. At present, the activity of such abzymes is too low to be useful.

- GDEPT involves the delivery of a gene into a cancer cell. The gene codes for an enzyme capable of activating an anticancer prodrug.

21.11 **Photodynamic therapy**

Conventional prodrugs are inactive compounds which are normally metabolized in the body to their active form. A variation of the prodrug approach is the concept of a sleeping agent. This is an inactive compound which is converted to the active drug by some form of external influence. The best example of this approach is the use of photosensitizing agents such as **porphyrins** or **chlorins** in cancer treatment—photodynamic therapy (PDT). Porphyrins occur naturally in chlorophyll in plants and haemoglobin in red blood cells. They usually complex a metal ion in the centre of the molecule (magnesium in chlorophyll and iron in haemoglobin). In this form, they are non-toxic. However, if they lack the central ion, they have the potential to do great damage. Given intravenously, these agents accumulate within cells and have some selectivity for tumour cells. By themselves, the agents have little effect, but if the cancer cells are irradiated with red light or a red laser, the porphyrins are converted to an excited state and react with molecular oxygen to produce highly toxic singlet oxygen. Singlet oxygen can then attack proteins and unsaturated lipids in the cell membrane, leading to the formation of hydroxyl radicals which further react with DNA and cause cell destruction. **Temoporfin (Foscan)** (Fig. 21.94) is an example of a chlorin photosensitizing agent which is used to treat advanced head and neck tumours that do not respond to other treatments.

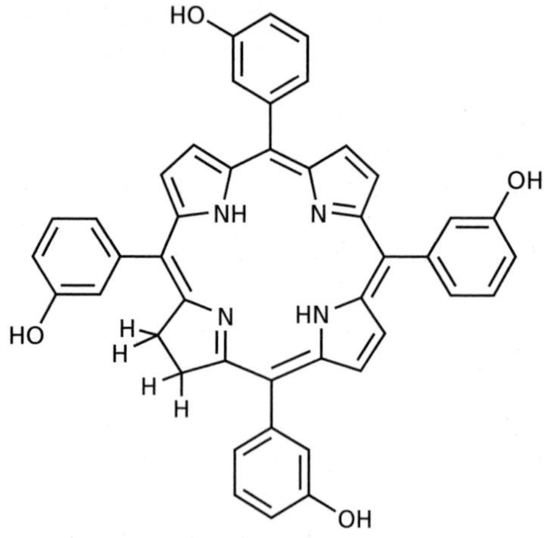

FIGURE 21.94 Temoporfin.

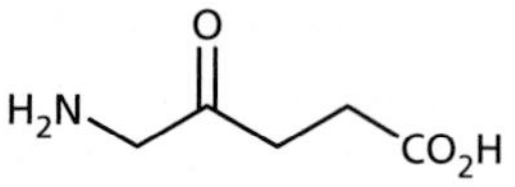

FIGURE 21.95 5-Aminolevulinic acid.

Problems such as photosensitivity have limited the application of PDT, but research is underway to find improved methods of delivering the agent.

5-Aminolevulinic acid (Fig. 21.95) is used as a photosensitizer to treat skin blemishes that may turn cancerous. The compound is a biosynthetic precursor for porphyrins and is applied to the blemishes several hours before photodynamic therapy is carried out. This gives sufficient time for a build-up of porphyrins in the affected tissue.

KEY POINTS

- Photodynamic therapy involves the irradiation of tumours containing porphyrin photosensitizers. This produces reactive oxygen species which are fatal to the cell.
- Photosensitivity is a serious problem, as the porphyrins can accumulate in the eyes and skin where they become activated by daylight.

21.12 **Viral therapy**

Talimogene laherparepvec (T-VEC) is a virus with anticancer properties that was approved in 2015 for the treatment of advanced, inoperable melanomas. It is the first such virus to reach the market and is a genetically modified version of herpes simplex virus 1—the virus that normally causes cold sores. Genetic modifications were carried out to ensure that the virus cannot produce cold sores. Other modifications ensured that it replicates in tumour cells but not in normal cells, which means that it has cytotoxic selectivity for cancer cells. Finally, the virus was modified to secrete a cytokine which stimulates the immune system to attack cancer cells.

The virus is injected directly into a tumour, where it replicates within tumour cells and destroys them. Once a tumour cell has been destroyed, further viruses are released to invade neighbouring tumour cells. The cytokine that is released attracts immune cells to the site of the tumour where they identify tumour antigens. T-cells are then programmed to identify that antigen, and an immune response is triggered that attacks cancer cells that have metastasized and spread to other parts of the body.

Unfortunately, the porphyrin structures used for PDT are inherently hydrophobic, which makes them difficult to formulate. Encapsulation using liposomes, oils, or polymeric micelles is one method of avoiding this problem, and has the advantage that tumours engulf and retain macromolecules more readily than would be the case with normal tissue. This is because the blood vessels nourishing tumours are leaky (see section 21.1.9) and release larger molecules than would be released from normal blood vessels.

Despite this, problems still remain. For example, the liposomes which carry the agent can be engulfed and destroyed by cells of the reticuloendothelial system. The most serious disadvantage with PDT, however, is photosensitivity. Once the drug has been released from liposomes and activated, it is free to circulate round the body and accumulate in the eyes and skin, leading to phototoxic side effects which render the patient highly sensitive to light. Indeed, it is this property that first highlighted the possibility of using porphyrins in PDT. **Porphyria** is a disease where porphyrins accumulate in the skin and result in photosensitization and disfigurement. Victims are unable to tolerate sunlight, and disfigurements can include erosion of the gums to reveal red, fang-like teeth. It is likely that sufferers from this disease may have inspired the medieval vampire legends. Indeed, it is interesting to note that victims would have been averse to garlic, as components of garlic exacerbate the symptoms and cause an agonizing reaction. It was the observation that porphyrins could break down cells that led to the idea that these agents could be used to break down cancer cells.

QUESTIONS

1. Do you think sulphonamides would be suitable antibacterial agents to treat opportunistic infections in cancer patients?

2. Since esters are commonly used as prodrugs, esterases would be suitable enzymes to use in ADEPT. What are your thoughts on this statement?

3. Staurosporin is a kinase inhibitor that shows no selectivity, but is a useful lead compound for potential antitumour agents. A simplification strategy resulted in arcyriaflavin, which is selective for PKC. There are three reasons why this molecule is simpler. Explain what they are and why simplification is desirable.

4. CGP 52411 is a further simplification of arcyriaflavin A. Remarkably, this compound is inactive against PKC and is selective for the kinase active site of the epidermal growth factor receptor. Suggest why there might be such a drastic change in selectivity.

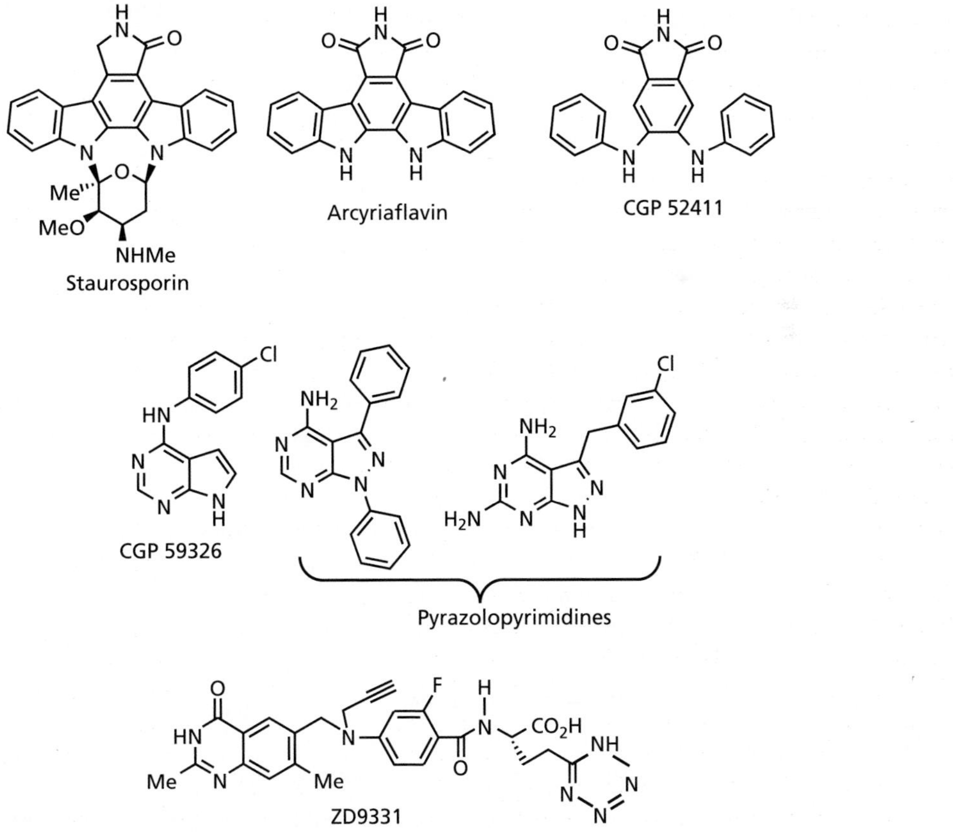

5. Further studies showed that CGP 52411 was bound to the ATP binding site of the kinase active site (Fig. 21.46). Suggest how this structure might bind. SAR studies show that substitution on any of the NH groups or the aromatic rings is bad for activity.

6. CGP 59326 and the pyrazolopyrimidine structures shown are also useful inhibitors of the EGFR kinase active site. Suggest how they might be bound to the active site.

7. In the development of imatinib, a conformational blocker was introduced (Fig. 21.53). Suggest a conformation which would be feasible in the lead compound that would be prevented by the conformational blocker.

8. Imatinib has a pyrimidine ring where one of the nitrogens is involved in an important hydrogen bond interaction. It has been suggested that it should be possible to produce an analogue where the pyrimidine ring is replaced by a pyridine ring. What are your thoughts on this suggestion?

9. Suggest a mechanism by which CC-1065 and adozelesin act as alkylating agents.

10. ZD 9331 is an analogue of raltitrexed (Fig. 21.24) that was studied as a possible anticancer agent. It contains a tetrazole ring as part of its structure. What purpose does this ring serve?

Staurosporin

Arcyriaflavin

CGP 52411

CGP 59326

Pyrazolopyrimidines

ZD9331

11. Identify whether belinostat and panobinostat fit the general structure expected for a histone deacetylase inhibitor.

12. It has been proposed that a fluorine substituent in olaparib is acting as a conformational blocker. However, a fluorine atom is not much bigger than a hydrogen atom. Why should it act as a conformational blocker? Identify where the rotatable bonds are located in the structure.

13. It has been suggested that the sulphonamide group of vemurafenib is ionized when it binds to its target enzyme. Why should this group be prone to ionization?

Multiple-choice questions are available on the Online Resource Centre at www.oxfordtextbooks.co.uk/orc/patrick6e/

FURTHER READING

Atkins, J. H., and Gershell, L. J. (2002) Selective anticancer drugs. *Nature Reviews Drug Discovery*, **1**, 491–2.

Elsayed, Y. A., and Sausville, E. A. (2001) Selected novel anticancer treatments targeting cell signalling proteins. *The Oncologist*, **6**, 517–37.

Featherstone, J., and Griffiths, S. (2002) Drugs that target angiogenesis. *Nature Reviews Drug Discovery*, **1**, 413–4.

Goldberg, A. L., Elledge, S. J., and Harper, J. W. (2001) The cellular chamber of doom. *Scientific American*, January, 68–73.

Hanahan, D., and Weinberg, R. A. (2000) The hallmarks of cancer. *Cell*, **100**, 57–70.

Jain, R. K., and Carmeliet, P. F. (2001) Vessels of death. *Scientific American*, December, 27–33 (angiogenesis).

Jordan, V. C. (2003) Tamoxifen: a most unlikely pioneering medicine. *Nature Reviews Drug Discovery*, **2**, 205–13.

Neidle, S., and Parkinson, G. (2002) Telomere maintenance as a target for anticancer drug discovery. *Nature Reviews Drug Discovery*, **1**, 383–93.

Ojima, I., Vite, G. D., and Altmann, K.-H. (eds) (2001) *Anticancer agents*. ACS Symposium Series 796, American Chemical Society, Washington, DC.

Opalinska, J. B., and Gewirtz, A. M. (2002) Nucleic-acid therapeutics: basic principles and recent applications. *Nature Reviews Drug Discovery*, **1**, 503–14.

Pecorino, L. (2008) *Molecular biology of cancer: mechanisms, targets and therapeutics,* 2nd edn. Oxford University Press, Oxford.

Reed, J. C. (2002) Apoptosis-based therapies. *Nature Reviews Drug Discovery*, **1**, 111–21.

Sansom, C. (2009) Temozolomide—birth of a blockbuster. *Chemistry World*, July, 48–51.

Szakacs, G., et al. (2006) Targeting multidrug resistance in cancer. *Nature Reviews Drug Discovery*, **5**, 219–34.

Thurston, D. E. (2006) *Chemistry and pharmacology of anticancer drugs*. CRC Press.

Wayt Gibbs, W. (2003) Untangling the roots of cancer. *Scientific American*, July, 48–57.

Weissman, K. (2003) Life and cell death. *Chemistry in Britain*, August, 19–22.

Zhang, J. Y. (2002) Apoptosis-based anticancer drugs. *Nature Reviews Drug Discovery*, **1**, 101–2.

Topoisomerase poisons

Fang, L., et al. (2006) Discovery of a daunorubicin analogue that exhibits potent antitumour activity and overcomes P-gp-meditated drug resistance. *Journal of Medicinal Chemistry*, **49**, 932–41.

Fortune, J. M., and Osheroff, N. (2000) Topoisomerase II as a target for anticancer drugs. *Progress in Nucleic Acid Research*, **64**, 221–53.

Platinum-based agents

Kelland, L. (2007) The resurgence of platinum-based cancer chemotherapy. *Nature Reviews Cancer*, **7**, 573–84.

Wang, D., and Lippard, S. J. (2005) Cellular processing of platinum anticancer drugs. *Nature Reviews Drug Discovery*, **4**, 307–20.

Wheate, N. J., et al. (2010) The status of platinum anticancer drugs in the clinic and in clinical trials. *Dalton Transactions*, **39**, 8113–27.

Hormonal therapy

Jordan, C. (2003) Antiestrogens and selective estrogen receptor modulators as multifunctional medicines. *Journal of Medicinal Chemistry*, **46**, 1081–111.

Mann, J. (2009) Design for life. *Chemistry World,* November, 54–7 (abiraterone).

Photodynamic therapy

Lane, N. (2003) New light on medicine. *Scientific American*, January, 26–33.

Farnesyltransferase inhibitors

Bell, I. M. (2004) Inhibitors of farnesyltransferase: a rational approach to cancer chemotherapy? *Journal of Medicinal Chemistry*, **47**, 1–10.

Cox, A. D., Fesik, S. W., Kimmelman, A. C., et al. (2014) Drugging the undruggable RAS: Mission Possible? *Nature Reviews Drug Discovery*, **13**, 828–51.

Protein kinase inhibitors

Atkins, M., Jones, C. A., and Kirkpatrick, P. (2006) Sunitinib maleate. *Nature Reviews Drug Discovery*, **5**, 279–80.

Barker, A. J., et al. (2001) Studies leading to the identification of ZD1839 (Iressa). *Bioorganic and Medical Chemistry Letters*, **11**, 1911–4.

Capdeville, R., et al. (2002) Glivec (STI571, imatinib), a rationally developed, targeted anticancer drug. *Nature Reviews Drug Discovery*, **1**, 493–502.

Collins, I., and Workman, P. (2006) Design and development of signal transduction inhibitors for cancer treatment: experience and challenges with kinase targets. *Current Signal Transduction Therapy*, **1**, 13–23.

Cui, J. J., et al. (2011) Structure based drug design of crizotinib. *Journal of Medicinal Chemistry*, **54**, 6342–63.

Cui, J. J. (2014) A new challenging and promising era of tyrosine kinase inhibitors. *ACS Medicinal Chemistry Letters*, **5**, 272–4.

Dancey, J., and Sausville, E. A. (2003) Issues and progress with protein kinase inhibitors for cancer treatment. *Nature Reviews Drug Discovery*, **2**, 296–313.

Houlton, S. (2011) Stemming the tide. *Chemistry World*, September, 57–9.

Janin, Y. L. (2005) Heat shock protein 90 inhibitors. A text book example of medicinal chemistry? *Journal of Medicinal Chemistry*, **48**, 7503–12.

Morphy, R. (2010) Selectively nonselective kinase inhibition: striking the right balance. *Journal of Medicinal Chemistry*, **53**, 1413–37.

Moy, B., Kirkpatrick, P., and Goss, P. (2007) Lapatinib. *Nature Reviews Drug Discovery*, **6**, 431–2.

Quintas-Cardama, A., Kantarjian, H., and Cortes, J. (2007) Flying under the radar: the new wave of BCR-ABL inhibitors. *Nature Reviews Drug Discovery*, **6**, 1–15.

Rini, B., Kar, S., and Kirkpatrick, P. (2007) Temsirolimus. *Nature Reviews Drug Discovery*, **6**, 599–600.

Zaiac, M. (2002) Taking aim at cancer. *Chemistry in Britain*, November, 44–6.

Miscellaneous enzyme inhibitors and other agents

Bartlett, J. B., Dredge, K., and Dalgleish, A. G. (2004) The evolution of thalidomide and its IMiD derivatives as anticancer agents. *Nature Reviews Cancer*, **4**, 314–22.

Jarvis, L. M. (2007) Living on the edge. *Chemical and Engineering News*, February 26, 15–23.

Johnstone, R. W. (2002) Histone-deacetylase inhibitors: novel drugs for the treatment of cancer. *Nature Reviews Drug Discovery*, **1**, 287–99.

McLaughlin, F., Finn, P., and La Thangue, N. B. (2003) The cell cycle, chromatin and cancer. *Drug Discovery Today*, **8**, 793–802.

Sanchez-Serrano, I. (2006) Success in translational research: lessons from the development of bortezomib. *Nature Reviews Drug Discovery*, **5**, 107–14.

Yoo, C, B., and Jones, P. A. (2006) Epigenetic therapy of cancer: past, present and future. *Nature Reviews Drug Discovery*, **5**, 37–50.

Antibodies and gene therapy

Ezzell, C. (2001) Magic bullets fly again. *Scientific American*, October, 28–35 (antibodies).

Schrama, D., Reisfeld, R. A., and Becker, J. C. (2006) Antibody targeted drugs as cancer therapeutics. *Nature Reviews Drug Discovery*, **5**, 147–59.

Senter, P. D., and Springer, C. J. (2001) Selective activation of anticancer prodrugs by monoclonal antibody–enzyme conjugates. *Advanced Drug Delivery Reviews*, **53**, 247–64.

Titles for general further reading are listed on p.845.

Cholinergics, anticholinergics, and anticholinesterases

In this chapter, we shall concentrate on drugs that have an effect on the cholinergic nervous system. There are several clinically important drugs in this category which act on the peripheral and/or the central nervous system (CNS).

22.1 The peripheral nervous system

The peripheral nervous system is so called because it is peripheral to the CNS (the brain and spinal column). There are many divisions and subdivisions of the peripheral system that can lead to confusion. The first distinction to make is between **sensory** and **motor nerves**:

- sensory nerves take messages from the body to the CNS;
- motor nerves carry messages from the CNS to the rest of the body.

An individual nerve cell is called a **neuron** (Appendix 4) and neurons must communicate with each other in order to relay messages. However, the neurons are not physically connected. Instead, there are gaps which are called **synapses** (Fig. 22.1). If a neuron is to communicate its message to another neuron (or a target organ), it can only do so by releasing a chemical that crosses the synaptic gap and binds to receptors on the target cell. This interaction between chemical and receptor can then stimulate other processes, which, in the case of a second neuron, continues the message. Because these

chemicals effectively carry a message from a neuron, they are known as chemical messengers or **neurotransmitters**. There are a large number of neurotransmitters in the body, but the important ones in the peripheral nervous system are **acetylcholine** and **noradrenaline** (Fig. 22.2). The very fact that neurotransmitters are chemicals allows the medicinal chemist to design and synthesize organic compounds which can mimic (**agonists**) or block (**antagonists**) their action.

22.2 Motor nerves of the peripheral nervous system

In this chapter, we are concerned primarily with drugs that influence the activity of motor nerves. Motor nerves take messages from the CNS to various parts of the body, such as skeletal muscle, smooth muscle, cardiac muscle, and glands (Figs. 4.1 and 22.3). The message travelling along a single neuron is often compared to an electrical pulse, but the analogy with electricity should not be taken too far since the pulse is a result of ion flow across the membranes of neurons and not a flow of electrons (see Appendix 4).

It should be evident that the workings of the human body depend crucially on an effective motor nervous system. Without it, we would not be able to operate our muscles and we would end up as flabby blobs, unable to move or breathe. We would not be able to eat, digest, or excrete our food because the smooth muscle activity of

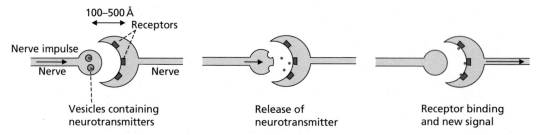

FIGURE 22.1 Signal transmission at a synapse.

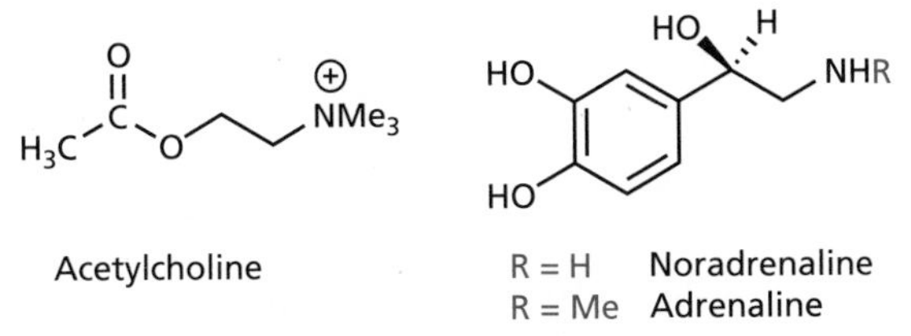

FIGURE 22.2 Acetylcholine, noradrenaline, and adrenaline.

the gastrointestinal tract (GIT) and the urinary tract is controlled by motor nerves. We would not be able to control body temperature, as the smooth muscle controlling the diameter of our peripheral blood vessels would cease to function. Finally, our heart would resemble a wobbly jelly rather than a powerful pump. In short, if the motor nerves failed to function, we would be in a mess! Let us now look at the motor nerves in more detail.

The motor nerves of the peripheral nervous system have been classified into three subsystems: the **somatic motor nervous system**, **autonomic motor nervous system**, and **enteric nervous system**. These are considered in the following sections.

22.2.1 The somatic motor nervous system

The somatic motor nerves carry messages from the CNS to the skeletal muscles. There are no synapses en route and the neurotransmitter at the neuromuscular junction is **acetylcholine**. Acetylcholine binds to cholinergic receptors within the cell membranes of muscle cells and the final result is contraction of skeletal muscle.

22.2.2 The autonomic motor nervous system

The autonomic motor nerves carry messages from the CNS to smooth muscle, cardiac muscle, and the adrenal medulla. This system can be divided into the **sympathetic** and **parasympathetic nervous systems**.

Sympathetic neurons leave the CNS, and synapse almost immediately with a second neuron using

acetylcholine as neurotransmitter. The second neuron then proceeds to various tissues and organs around the body. **Noradrenaline** is the neurotransmitter released from the second neuron, and this interacts with adrenergic receptors present in target cells and organs. At the heart, the action of noradrenaline leads to contraction of cardiac muscle and an increase in heart rate. Elsewhere, it relaxes smooth muscle and reduces the contractions of the gastrointestinal and urinary tracts. It also reduces salivation and the dilatation of the peripheral blood vessels. In general, the sympathetic nervous system promotes the **fight or flight** response by shutting down the body's housekeeping roles (digestion, defecation, urination, etc.), while stimulating the heart.

There are some neurons in the sympathetic nervous system which do not synapse with a second neuron, but go directly to a gland called the **adrenal medulla**. Acetylcholine is the neurotransmitter released by these neurons and it stimulates the adrenal medulla to release the hormone **adrenaline**, which then circulates through the blood system. **Adrenaline** reinforces the actions of noradrenaline by activating adrenergic receptors throughout the body, whether they are directly supplied with nerves or not.

Parasympathetic neurons leave the CNS, travel some distance, then synapse with a second neuron using acetylcholine as neurotransmitter. The second neuron then proceeds to synapse with the same target tissues and organs as the sympathetic neurons. However, acetylcholine acts as the neurotransmitter, rather than noradrenaline, and activates cholinergic receptors on the target cells. The resulting effects are the opposite from those caused by activation of adrenergic receptors. For example, cardiac muscle is relaxed, whereas the smooth muscle of the digestive and urinary tracts is contracted.

As the sympathetic and parasympathetic nervous systems oppose each other in their actions, they can be looked upon as acting like a brake and an accelerator on the different tissues and organs around the body. The analogy is not quite apt because both systems are always operating and the overall result depends on which effect is the stronger.

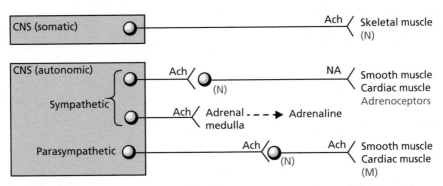

FIGURE 22.3 Motor nerves of the peripheral nervous system. N = nicotinic receptor; M = musarinic receptor; AcH = acetylcholine; NA = noradrenaline

22.2.3 The enteric system

The third constituent of the peripheral nervous system is the enteric system, which is located in the walls of the gastrointestinal tract (GIT). It receives messages from sympathetic and parasympathetic nerves, but it also responds to local effects to provide local reflex pathways which are important in the control of GIT function. A large variety of neurotransmitters are involved, including **serotonin**, **neuropeptides**, and **ATP**. **Nitric oxide (NO)** is also involved as a chemical messenger.

22.2.4 Defects in motor nerve transmission

Defects in motor nerve transmission would clearly lead to a large variety of ailments involving the heart, skeletal muscle, gastrointestinal tract, urinary tract, and many other organs. Such defects might be the result of either a deficit or an excess of neurotransmitter. Therefore, treatment involves the administration of drugs which can act as agonists or antagonists depending on the problem. There is a difficulty with this approach, however. Usually, the problem we wish to tackle occurs at a certain location where there might, for example, be a lack of neurotransmitter. Application of an agonist to make up for low levels of neurotransmitter at the heart might solve the problem there, but would lead to problems elsewhere in the body where the levels of neurotransmitter would be normal. At those areas, the agonist would cause too much activity and cause unwanted side effects. Therefore, drugs showing selectivity for different parts of the body would clearly be preferred. This selectivity has been achieved to a great extent with both cholinergic and adrenergic agents. In this chapter, we concentrate on cholinergic agents (adrenergic agents are covered in Chapter 23).

22.3 The cholinergic system

22.3.1 The cholinergic signalling system

Let us look first at what happens at synapses involving acetylcholine as the neurotransmitter. Figure 22.4 shows the synapse between two neurons and the events involved when a message is transmitted from one neuron to another. The same general process takes place when a message is passed from a neuron to a muscle cell.

1. The first stage involves the biosynthesis of acetylcholine (Fig. 22.5). Acetylcholine is synthesized from **choline** and **acetyl coenzyme A** at the end of the presynaptic neuron. The reaction is catalysed by the enzyme **choline acetyltransferase**.

2. Acetylcholine is incorporated into membrane-bound vesicles by means of a specific transport protein.

3. The arrival of a nerve signal leads to an opening of calcium ion channels and an increase in intracellular calcium concentration. This induces the vesicles to fuse with the cell membrane and release the transmitter into the synaptic gap.

4. Acetylcholine crosses the synaptic gap and binds to the cholinergic receptor, resulting in stimulation of the second neuron.

5. Acetylcholine moves to an enzyme called **acetylcholinesterase**, which is situated on the postsynaptic neuron, and which catalyses the hydrolysis of acetylcholine to produce choline and acetic acid (ethanoic acid).

6. Choline is taken up into the presynaptic neuron by a transport protein to continue the cycle.

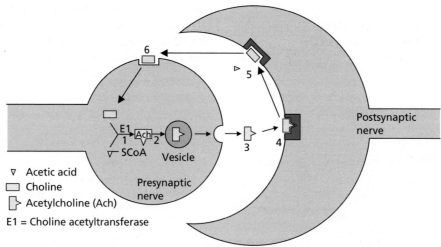

FIGURE 22.4 Synapse with acetylcholine acting as the neurotransmitter.

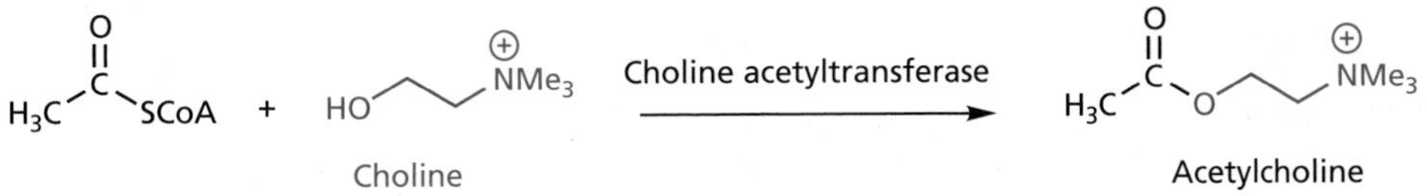

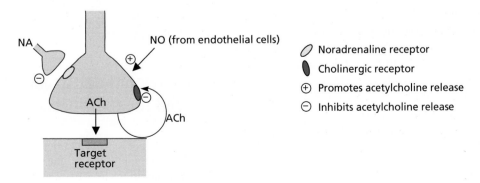

Choline acetyltransferase

Choline

Acetylcholine

FIGURE 22.5 Biosynthesis of acetylcholine.

The most important thing to note is that there are several stages where it is possible to use drugs to either promote or inhibit the overall process. The greatest success so far has been with drugs targeting stages 4 and 5 (i.e. the cholinergic receptor and the acetylcholinesterase enzyme). These are considered in more detail in subsequent sections.

22.3.2 Presynaptic control systems

Cholinergic receptors (called **autoreceptors**) are present at the terminus of the presynaptic neuron (Fig. 22.6). The purpose of these receptors is to provide a means of local control over nerve transmission. When acetylcholine is released from the neuron, some of it will find its way to these autoreceptors and switch them on. This has the effect of inhibiting further release of acetylcholine.

The presynaptic neuron also contains receptors for **noradrenaline**, which act as another control system for acetylcholine release. Branches from the sympathetic nervous system lead to the cholinergic synapses and when the sympathetic nervous system is active, noradrenaline is released and binds to these receptors. Once again, the effect is to inhibit acetylcholine release. This indirectly enhances the activity of noradrenaline at target organs by lowering cholinergic activity.

The chemical messenger **nitric oxide (NO)** can also influence acetylcholine release, but in this case it promotes release. A large variety of other chemical messengers including **cotransmitters** (section 23.3.3) are also implicated in presynaptic control. The important thing to appreciate is that presynaptic receptors offer another possible drug target to influence the cholinergic nervous system.

22.3.3 Cotransmitters

Cotransmitters are messenger molecules released along with acetylcholine. The particular cotransmitter released depends on the location and target cell of the neurons. Each cotransmitter interacts with its own receptor on the postsynaptic cell. Cotransmitters have a variety of structures and include peptides such as **vasoactive intestinal peptide** (VIP), **gonadotrophin-releasing hormone** (GnRH), and **substance P**. The role of these agents appears to be as follows:

- they are longer lasting and reach more distant targets than acetylcholine, resulting in longer-lasting effects;
- the balance of cotransmitters released varies under different circumstances (e.g. presynaptic control) and so can produce different effects.

22.4 Agonists at the cholinergic receptor

One point might have occurred to you. If there is a lack of acetylcholine acting at a certain part of the body, why not just administer more acetylcholine? After all, it is easy enough to make in the laboratory (Fig. 22.7).
There are three reasons why this is not feasible.

- Acetylcholine is easily hydrolysed in the stomach by acid catalysis and cannot be given orally.

NA

NO (from endothelial cells)

⊕

ACh

⊖

ACh

Target receptor

◯ Noradrenaline receptor
⬤ Cholinergic receptor
⊕ Promotes acetylcholine release
⊖ Inhibits acetylcholine release

FIGURE 22.6 Presynaptic control systems.

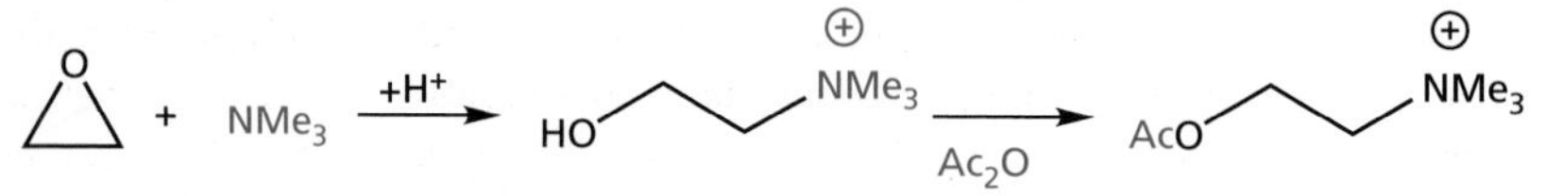

FIGURE 22.7 Synthesis of acetylcholine.

- Acetylcholine is easily hydrolysed in the blood by enzymes called esterases.
- There is no selectivity of action. Additional acetylcholine will switch on all cholinergic receptors in the body.

Therefore, we need analogues of acetylcholine that are more stable to hydrolysis and more selective with respect to where they act in the body. We shall look at selectivity first.

There are two ways in which selectivity can be achieved. Firstly, some drugs may be distributed more efficiently to one part of the body than another. Secondly, there are different types of cholinergic receptor, which vary in the way they are distributed in tissues. It is possible to design synthetic agents that show selectivity for these receptors and, hence, have tissue selectivity.

This is not just a peculiarity of cholinergic receptors. Differences have been observed for other types of receptors, such as those for dopamine, noradrenaline, and serotonin, and there are many types and subtypes of receptor for each chemical messenger (see section 4.3).

The first indications that different types of cholinergic receptor existed came from the action of natural compounds. It was discovered that the compounds **nicotine** (present in tobacco) and **muscarine** (the active principle of a poisonous mushroom) (Fig. 22.8) were both cholinergic agonists, but that they had different physiological effects.

Nicotine showed selectivity for cholinergic receptors present on skeletal muscle or at the synapses between different neurons, whereas muscarine showed selectivity for cholinergic receptors present on smooth muscle and cardiac muscle. From these results, it was concluded that there was one type of cholinergic receptor on skeletal muscles and at nerve synapses (the **nicotinic receptor**), and a different type of cholinergic receptor on smooth muscle and cardiac muscle (the **muscarinic receptor**) (Fig. 22.3).

Muscarine and nicotine were the first compounds to indicate that receptor selectivity was possible, but they are unsuitable as medicines because they have undesirable side effects resulting from their interactions with other receptors. In the search for a good drug, it is important to gain selectivity for one class of receptor over another (e.g. the cholinergic receptor in preference to an adrenergic receptor), and selectivity between receptor types (e.g. the muscarinic receptor in preference to a nicotinic receptor). It is also preferable to gain selectivity for particular subtypes of a receptor. For example, not every muscarinic receptor is the same throughout the body. At present, five subtypes of the muscarinic receptor have been discovered (M1–M5) and ten subtypes of the nicotinic receptor ($\alpha1$–$\alpha10$).

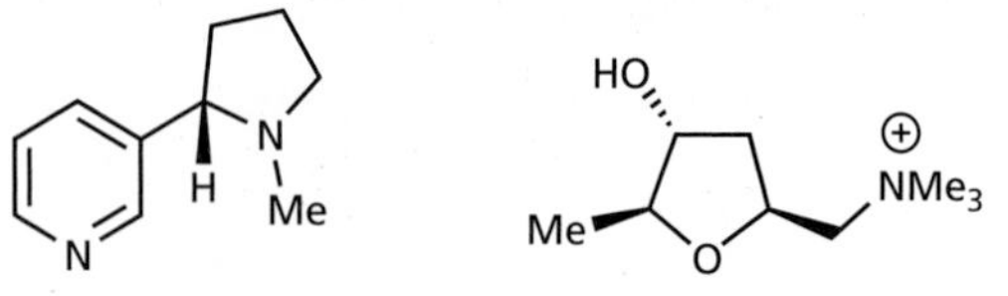

The principle of selectivity was proven with nicotine and muscarine, and so the race was on to design novel drugs which had the selectivity of nicotine or muscarine, but not the side effects.

KEY POINTS

- The cholinergic nervous system involves nerves which use the neurotransmitter acetylcholine as a chemical messenger. These include the motor nerves which innervate skeletal muscle, nerves which synapse with other nerves in the peripheral nervous system, and the parasympathetic nerves innervating cardiac and smooth muscle.

- There are two types of cholinergic receptor. Muscarinic receptors are present in smooth and cardiac muscle. Nicotinic receptors are present in skeletal muscle and in synapses between neurons.

- Acetylcholine is hydrolysed by the enzyme acetylcholinesterase when it departs the cholinergic receptor. The hydrolytic product choline is taken up into presynaptic neurons and acetylated back to acetylcholine. The cholinergic receptor and the enzyme acetylcholinesterase are useful drug targets.

- Acetylcholine cannot be used as a drug, because it is hydrolysed rapidly by stomach acids and enzymes. It shows no selectivity for different types and subtypes of cholinergic receptor.

22.5 **Acetylcholine: structure, SAR, and receptor binding**

The first stage in any drug development is to study the lead compound and to find out which parts of the molecule are important to activity so that they can be retained in future analogues (i.e. structure–activity

FIGURE 22.8 Nicotine and muscarine.

relationships—SAR). These results also provide information about what the binding site of the cholinergic receptor looks like and help decide what changes are worth making in new analogues.

In this case, the lead compound is acetylcholine itself. The results described below are valid for both the nicotinic and muscarinic receptors and were obtained by the synthesis of a large range of analogues.

- The positively charged nitrogen atom is essential to activity. Replacing it with a neutral carbon atom eliminates activity.
- The distance from the nitrogen to the ester group is important.
- The ester functional group is important.
- The overall size of the molecule cannot be altered much. Bigger molecules have poorer activity.
- The ethylene bridge between the ester and the nitrogen atom cannot be extended (Fig. 22.9).
- There must be two methyl groups on the nitrogen. A larger, third alkyl group is tolerated, but more than one large alkyl group leads to loss of activity.
- Bigger ester groups lead to a loss of activity.

Clearly, there is a tight fit between acetylcholine and its binding site, which leaves little scope for variation. The above findings tally with a receptor binding site as shown in Fig. 22.10.

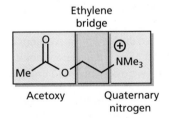

FIGURE 22.9 Acetylcholine.

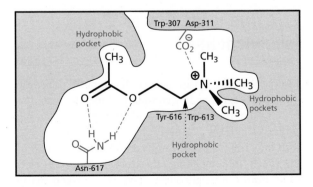

FIGURE 22.10 Muscarinic receptor binding site.

It is proposed that important hydrogen bonding interactions exist between the ester group of acetylcholine and an asparagine residue. It is also thought that a small hydrophobic pocket exists which can accommodate the methyl group of the ester, but nothing larger. This interaction is thought to be more important in the muscarinic receptor than the nicotinic receptor.

The evidence suggests that the NMe_3^+ group is placed in a hydrophobic pocket lined with three aromatic amino acids. It is also thought that the pocket contains two smaller hydrophobic pockets, which are large enough to accommodate two of the three methyl substituents on the NMe_3^+ group. The third methyl substituent on the nitrogen is positioned in an open region of the binding site and so it is possible to replace it with other groups. A strong ionic interaction has been proposed between the charged nitrogen atom and the anionic side group of an aspartate residue. The existence of this ionic interaction represents the classical view of the cholinergic receptor, but there is an alternative suggestion which states that there may be an induced dipole interaction between the NMe_3^+ group and the aromatic residues in the hydrophobic pocket.

There are several reasons for this. Firstly, the positive charge on the NMe_3^+ group is not localized on the nitrogen atom, but is spread over the three methyl groups (compare section 17.7.1). Such a diffuse charge is less likely to be involved in a localized ionic interaction and it has been shown by model studies that NMe_3^+ groups can be stabilized by binding to aromatic rings. It might seem strange that a hydrophobic aromatic ring should be capable of stabilizing a positively charged group, but it has to be remembered that aromatic rings are electron rich, as shown by the fact they can undergo reaction with electrophiles. It is thought that the diffuse positive charge on the NMe_3^+ group is capable of distorting the π electron cloud of aromatic rings to induce a dipole moment (section 1.3.4). Induced ion–dipole interactions between the NMe_3^+ group and an aromatic residue such as tyrosine would then account for the binding. The fact that three aromatic amino acids are present in the pocket adds weight to the argument.

Of course, it is possible that both types of binding interactions are taking place, which will please both parties!

A large amount of effort has been expended trying to identify the active conformation of acetylcholine, i.e. the shape adopted by the neurotransmitter when it binds to the cholinergic receptor. This has been no easy task, as acetylcholine is a highly flexible molecule (Fig. 22.11)

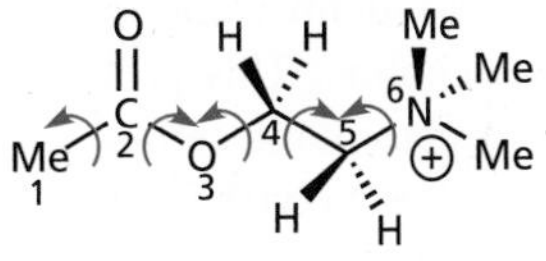

FIGURE 22.11 Bond rotations in acetylcholine leading to different conformations.

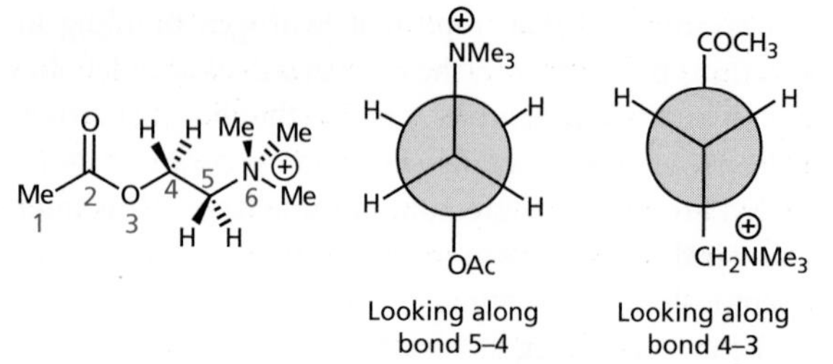

FIGURE 22.12 The sawhorse and Newman projections of acetylcholine.

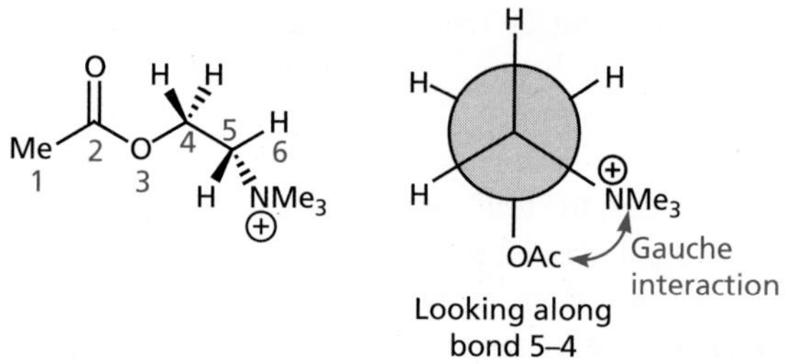

FIGURE 22.13 A gauche conformation for acetylcholine.

Muscarine

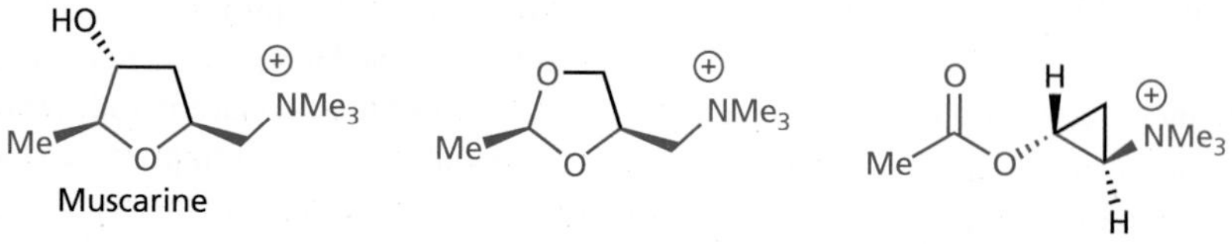

FIGURE 22.14 Rigid molecules incorporating the acetylcholine skeleton (C—C—O—C—C—N).

where bond rotation along the length of its chain can lead to many possible stable conformations (or shapes).

In the past, it was assumed that a flexible neurotransmitter would adopt its most stable conformation when binding. In the case of acetylcholine, that would be the conformation represented by the sawhorse and Newman projections shown in Fig. 22.12. However, there is not a massive energy difference between this and alternative stable conformations such as the gauche conformation shown in Fig. 22.13. The stabilization energy gained from binding interactions within the binding site could more than compensate for any energy penalties involved in adopting a slightly less stable conformation.

In order to try and establish the active conformation of acetylcholine, rigid cyclic molecules have been studied which contain the skeleton of acetylcholine within their structure; for example muscarine and the analogues shown in Fig. 22.14. In these structures, the portion of the acetylcholine skeleton which is included in a ring is locked into a particular conformation because bonds within rings cannot rotate freely. If such molecules bind to the cholinergic receptor, this indicates that this particular conformation is 'allowed' for activity.

Many such structures have been prepared, but it has not been possible to identify one *specific* active conformation for acetylcholine. This probably indicates that the cholinergic receptor has a certain amount of latitude and can recognize the acetylcholine skeleton within the rigid analogues, even when it is not in the ideal active conformation. Nevertheless, such studies have shown that the separation between the ester group and the

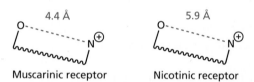

FIGURE 22.15 Pharmacophore of acetylcholine.

quaternary nitrogen is important for binding, and that this distance differs for the muscarinic and the nicotinic receptor (Fig. 22.15).

Having identified the binding interactions and pharmacophore of acetylcholine, we shall now look at how acetylcholine analogues were designed with improved stability.

Test your understanding and practise your molecular modelling with Exercises 22.1 and 22.2 on the Online Resource Centre at www.oxfordtextbooks.co.uk/orc/patrick6e/

22.6 The instability of acetylcholine

As described previously, acetylcholine is prone to hydrolysis. This is explained by considering one of the conformations that the molecule can adopt (Fig. 22.16). In this conformation, the positively charged nitrogen interacts with the carbonyl oxygen and has an electron-withdrawing effect. To compensate, the oxygen atom pulls electrons from the neighbouring carbon atom and makes that carbon atom electron deficient and more prone to

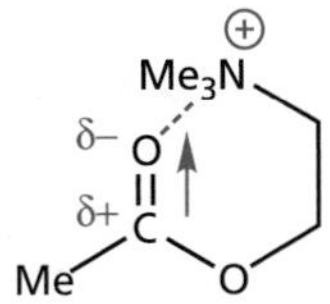

FIGURE 22.16 Neighbouring group participation. The arrow indicates the inductive pull of oxygen which increases the electrophilicity of the carbonyl carbon.

nucleophilic attack. Water is a poor nucleophile, but, because the carbonyl group is more electrophilic, hydrolysis takes place relatively easily. This influence of the nitrogen ion is known as **neighbouring group participation** or **anchimeric assistance**.

We shall now look at how the problem of hydrolysis was overcome, but it should be appreciated that we are doing this with the benefit of hindsight. At the time the problem was tackled, the SAR studies were incomplete and the format of the cholinergic receptor binding site was unknown.

22.7 Design of acetylcholine analogues

There are two possible approaches to tackling the inherent instability of acetylcholine: steric shields and electronic stabilization.

22.7.1 Steric shields

The principle of steric shields was described in section 14.2.1 and can be demonstrated with **methacholine** (Fig. 22.17). Here, an extra methyl group has been placed on

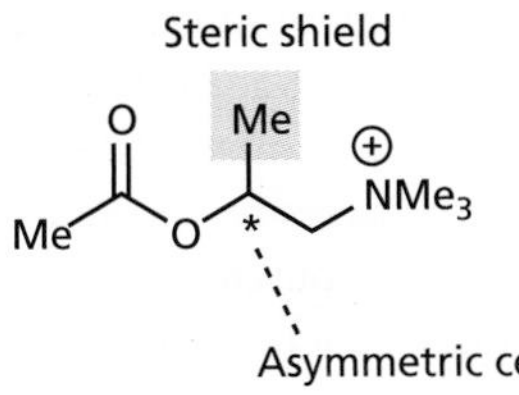

FIGURE 22.17 Methacholine (racemic mixture).

the ethylene bridge as a steric shield to protect the carbonyl group. The shield hinders the approach of any potential nucleophile and also hinders binding to esterase enzymes, thus slowing down chemical and enzymatic hydrolysis. As a result, methacholine is three times more stable to hydrolysis than acetylcholine.

The obvious question now is, why not put on a bigger alkyl group like an ethyl group or a propyl group? Alternatively, why not put a bulky group on the acyl half of the molecule, as this would be closer to the carbonyl centre and have a greater shielding effect?

In fact, these approaches were tried. They certainly increased stability but they lowered cholinergic activity. We should already know why—the fit between acetylcholine and its receptor is so tight that there is little scope for enlarging the molecule. The extra methyl group is acceptable, but larger substituents hinder the molecule binding to the cholinergic receptor and decrease its activity.

Introducing a methyl steric shield has another useful effect. It was discovered that methacholine has significant muscarinic activity, but very little nicotinic activity. Therefore, methacholine shows good selectivity for the muscarinic receptor. This is perhaps more important than the gain in stability.

Selectivity for the muscarinic receptor can be explained if we compare the proposed active conformation of methacholine with muscarine (Fig. 22.18), as the methyl group of methacholine occupies the same position as a methylene group in muscarine. This is only possible for the *S*-enantiomer of methacholine and when the two enantiomers of methacholine were separated, it was found that the *S*-enantiomer was, indeed, the more active enantiomer. However, it is not used therapeutically.

22.7.2 Electronic effects

The use of electronic factors to stabilize functional groups has been described in sections 14.2.2–14.2.3, and was used in the design of **carbachol** (Fig. 22.19)—a long-acting cholinergic agent which is resistant to hydrolysis. Here, the acyl methyl group has been replaced by NH_2 which means that the ester has been replaced by a urethane or carbamate group. This functional group is more resistant to hydrolysis because the lone pair of electrons on nitrogen can interact with the neighbouring carbonyl group and lower its electrophilic character (Fig. 22.20).

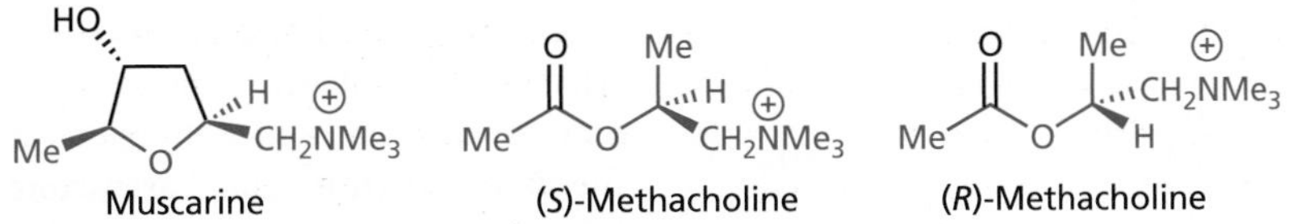

FIGURE 22.18 Comparison of muscarine and the *R*- and *S*-enantiomers of methacholine.

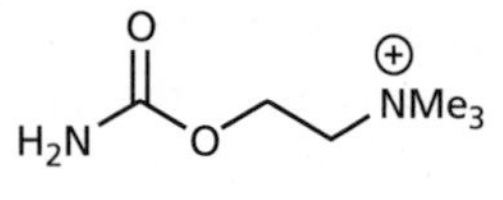

FIGURE 22.19 Carbachol.

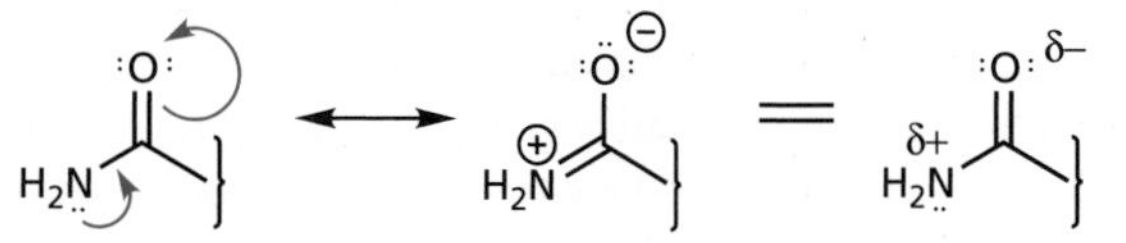

FIGURE 22.20 Resonance structures of carbachol.

The tactic worked, but it was by no means a foregone conclusion that it would. Although the NH_2 group is equivalent in size to the methyl group, the former is polar and the latter is hydrophobic, and it was by no means certain that a polar NH_2 group would be accepted into a hydrophobic pocket in the binding site. Fortunately, it is accepted and activity is retained, which means that the amino group acts as a **bio-isostere** for the methyl group. A bio-isostere is a group which can replace another group without affecting the pharmacological activity of interest (sections 13.3.7 and 14.2.2). Thus, the amino group is a bio-isostere for the methyl group as far as the cholinergic receptor is concerned, but not as far as the esterase enzymes are concerned.

The inclusion of the electron-donating amino group greatly increases chemical and enzymatic stability. Unfortunately, carbachol shows very little selectivity between the muscarinic and nicotinic receptors. Nevertheless, it is used clinically for the treatment of glaucoma where it can be applied locally, thus avoiding the problems of receptor selectivity. Glaucoma arises when the aqueous contents of the eye cannot be drained. This raises the pressure on the eye and can lead to blindness. Agonists cause the eye muscles to contract and allow drainage, thus relieving the pressure.

22.7.3 Combining steric and electronic effects

We have seen that the β-methyl group of methacholine increases stability and introduces receptor selectivity. Therefore, it made sense to add a β-methyl group to carbachol. The resulting compound is **bethanechol** (Fig. 22.21) which is both stable to hydrolysis and selective in its action. It is occasionally used therapeutically in stimulating the GIT and urinary bladder after surgery. Both these organs are 'shut down' with drugs during surgery (section 22.9).

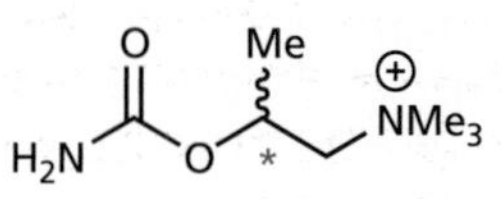

FIGURE 22.21 Bethanechol.

22.8 Clinical uses for cholinergic agonists

22.8.1 Muscarinic agonists

A possible future use for muscarinic agonists is in the treatment of Alzheimer's disease. Current clinical uses include:

- treatment of glaucoma;
- 'switching on' the GIT and urinary tract after surgery;
- treatment of certain heart defects by decreasing heart muscle activity and heart rate.

Pilocarpine (Fig. 22.22) is an example of a muscarinic agonist which is used in the treatment of glaucoma. It is an alkaloid obtained from the leaves of shrubs belonging to the genus *Pilocarpus*. Although there is no quaternary ammonium group present in pilocarpine, it is assumed that the drug is protonated before it interacts with the muscarinic receptor. Molecular modelling shows that pilocarpine can adopt a conformation having the correct pharmacophore for the muscarine receptor, i.e. a separation between nitrogen and oxygen of 4.4 Å.

Pilocarpine is also being considered for the treatment of Alzheimer's disease, as are other muscarinic agonists such as **oxotremorine** and various **arecoline** analogues (Fig. 22.22). At present, anticholinesterases are used clinically for the treatment of this disease (section 22.15).

Test your knowledge and practise your molecular modelling with Exercise 22.7 on the Online Resource Centre at www.oxfordtextbooks.co.uk/orc/patrick6e/

22.8.2 Nicotinic agonists

Nicotinic agonists are used in the treatment of myasthenia gravis. This is an autoimmune disease where the body has produced antibodies against its own cholinergic receptors. As a result, the number of available receptors drops and so fewer messages reach the muscle cells. In turn, this leads to severe muscle weakness and fatigue. Administering an agonist increases the chance of activating what few receptors remain. An example of a selective nicotinic agonist is the first structure shown in Fig. 22.23. This agent is very similar in structure to methacholine, and differs only in the position of the methyl substituent. This is sufficient, however, to completely alter receptor selectivity. Despite that, this particular compound is not used clinically and anticholinesterases (section 22.15) are the preferred treatment. **Varenicline** *is* used clinically, however. It is a partial agonist at nicotinic receptors and was approved in 2006 as an aid to stop smoking.

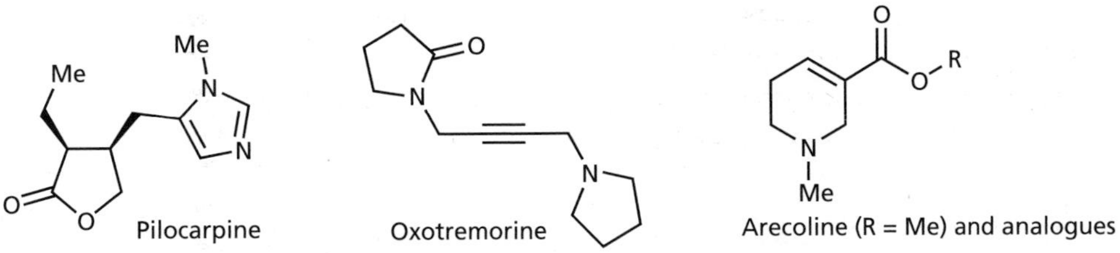

FIGURE 22.22 Examples of muscarinic agonists.

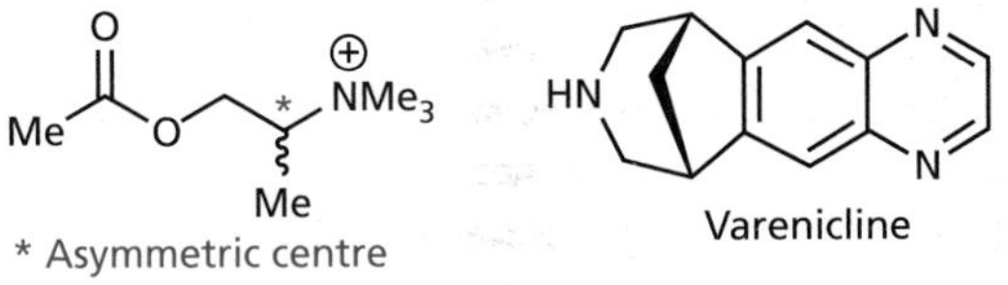

* Asymmetric centre

FIGURE 22.23 Examples of selective nicotinic agonists.

22.9 Antagonists of the muscarinic cholinergic receptor

22.9.1 Actions and uses of muscarinic antagonists

Antagonists of the cholinergic receptor are drugs which bind to the receptor but do not 'switch it on'. By binding to the receptor, an antagonist acts like a plug at the receptor binding site and prevents acetylcholine from binding (Fig. 22.24). The overall effect on the body is the same as if there was a lack of acetylcholine. Therefore, antagonists have the opposite clinical effect from agonists.

The antagonists described in this section act only at the muscarinic receptor, and, therefore, affect nerve transmissions to glands, the CNS, and the smooth muscle of the GIT, urinary tract, and airways. The physiological

effects and clinical uses of these antagonists reflect this (see also Boxes 22.1 and 22.2).

The physiological effects of muscarinic antagonists are:

- reduction of saliva and gastric secretions;
- relaxation of smooth muscle in the GIT, urinary tract, and airways;
- dilatation of eye pupils;
- CNS effects.

The clinical uses are:

- shutting down the GIT and urinary tract during surgery;
- treatment of incontinence;
- ophthalmic examinations;
- relief of peptic ulcers;
- treatment of Parkinson's disease;
- treatment of anticholinesterase poisoning;
- treatment of motion sickness;
- treatment of chronic obstructive pulmonary disease.

A potential use for M2 antagonists is in the treatment of Alzheimer's disease.

22.9.2 Muscarinic antagonists

The first antagonists to be discovered were natural products—in particular alkaloids (nitrogen-containing compounds derived from plants).

22.9.2.1 Atropine and hyoscine

Atropine (Fig. 22.25) is present in the roots of *Atropa belladonna* (deadly nightshade) and is included in a root extract which was once used by Italian women to dilate their eye pupils. This was considered a sign of beauty—hence the name belladonna. Clinically, atropine has been used to decrease gastrointestinal motility and to counteract anticholinesterase poisoning.

Atropine has an asymmetric centre but exists as a racemate. Usually, natural products exist exclusively as

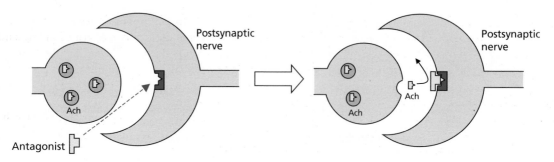

FIGURE 22.24 Action of an antagonist to block a receptor.

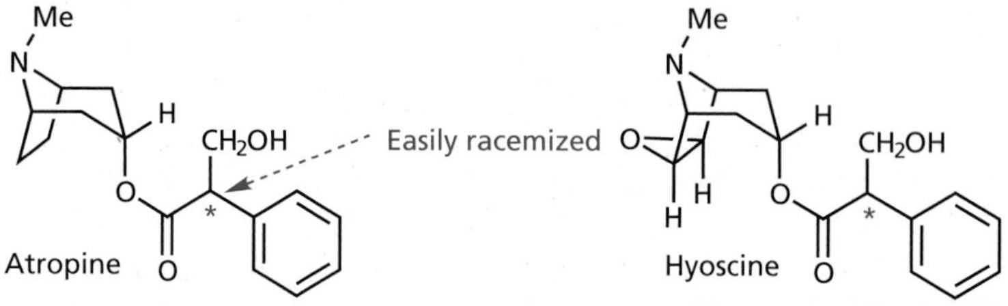

FIGURE 22.25 Atropine and hyoscine.

one enantiomer. This is also true for atropine, which is present in the plants of the family Solanaceae as a single enantiomer called **hyoscyamine**. As soon as the natural product is extracted into solution, however, racemization takes place. The asymmetric centre in atropine is easily racemized as it is next to a carbonyl group and an aromatic ring. This makes the proton attached to the asymmetric centre acidic and easily removed.

Hyoscine (or **scopolamine**) (Fig. 22.25) is obtained from the thorn apple (*Datura stramonium*) and is very similar in structure to atropine. It has been used in the treatment of motion sickness.

These two compounds bind to the cholinergic receptor, but at first sight they do not look anything like acetylcholine. If we look more closely though, we can see that a basic nitrogen and an ester group are present, and if we superimpose the acetylcholine skeleton on to the atropine skeleton, the distances between the ester and the nitrogen groups are similar in both molecules (Fig. 22.26). There is, of course, the problem that the nitrogen in atropine is uncharged, whereas the nitrogen in acetylcholine has a full positive charge. This implies that the nitrogen atom in atropine must be protonated and charged when it binds to the cholinergic receptor.

Therefore, atropine has two important binding features shared with acetylcholine—a charged nitrogen when protonated and an ester group. It is able to bind to the receptor, but why is it unable to switch it on? Because atropine is a larger molecule than acetylcholine, it is capable of binding to other binding regions within the binding site which are not used by acetylcholine itself. As

a result, it interacts differently with the receptor and does not induce the same conformational changes (induced fit) as acetylcholine. This means that the receptor is not activated.

As both atropine and hyoscine are tertiary amines rather than quaternary salts, they are able to cross the blood–brain barrier as the free base. Once they are in the brain, they can become protonated and antagonize muscarinic receptors to produce CNS effects; for example, hallucinogenic activity is brought on with high doses, and both hyoscine and atropine were used by witches in past centuries to produce that very effect. Other CNS effects observed in atropine poisoning are restlessness, agitation, and hyperactivity.

In recent times, the disorientating effect of scopolamine has seen it being used as a truth drug for the interrogation of spies and so it is no surprise to find it

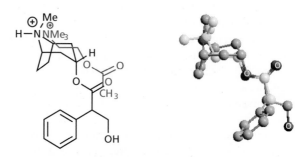

FIGURE 22.26 Acetylcholine skeleton superimposed on to the atropine skeleton.

cropping up in various novels. An interesting application for scopolamine was described in Jack Higgins' novel *Day of Judgement* where it was used in association with **suxamethonium** (see Fig. 22.34) to torture one hapless victim. Suxamethonium was applied to the conscious victim in order to create initial convulsive muscle spasms, followed by paralysis, inability to breathe, agonizing pain, and a living impression of death. Scopolamine was then used to erase the memory of this horror, so that the impact would be just as bad when the process was repeated!

Test your understanding and practise your molecular modelling with Exercise 22.3 on the Online Resource Centre at www.oxfordtextbooks.co.uk/orc/patrick6e/

22.9.2.2 Structural analogues of atropine and hyoscine

Quaternary salts of atropine and hyoscine are used in the clinic to avoid CNS side effects. This is because fully charged structures are unable to cross the blood–brain barrier. Commonly used salts include **atropine methonitrate**, **ipratropium**, and **hyoscine butylbromide** (Fig. 22.27). There are also a number of structural analogues for both atropine and hyoscine. **Trospium chloride** is one such example and contains two branched aromatic rings and a tertiary alcohol group—a common feature in many muscarinic antagonists. These features are also

present in **tiotropium bromide**—an analogue of hyoscine with a prolonged period of activity. Binding studies on tiotropium bromide have shown that the tertiary alcohol group is a crucial factor in prolonging activity because it can form a hydrogen bond to an arginine residue in the binding site.

A large number of atropine analogues have been synthesized to investigate their structure–activity relationships. These reveal the importance of the aromatic ring, the ester group, and the basic nitrogen (which is ionized).

22.9.2.3 Simplified analogues of atropine

Further research has shown that the complex ring system of atropine is not necessary for antagonist activity. For example, **propiverine** (Fig. 22.28) is a clinically useful antagonist that has a simple piperidine ring instead of the bicyclic ring system of atropine. Ring contraction is observed in **glycopyrronium bromide**, which contains a five-membered pyrrolidine ring. Further simplification can be seen in antagonists where the ring has been removed altogether and replaced with a two-carbon linker between the amine and the ester. Examples include **dicycloverine**, **cyclopentolate**, and **propantheline bromide**. It is also possible to link the amine and the ester by means of an alkyne group, as seen in **oxybutynin**.

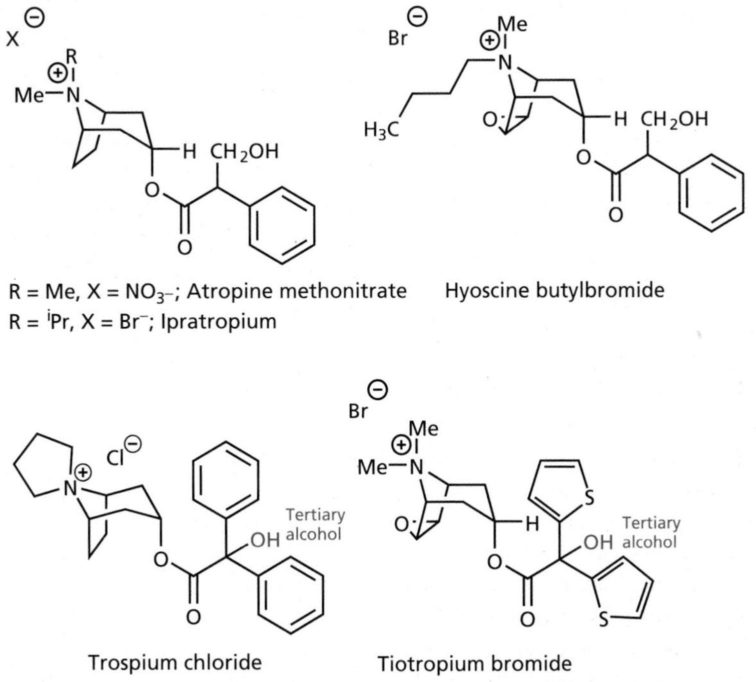

R = Me, X = NO_3^-; Atropine methonitrate
R = ^{i}Pr, X = Br^-; Ipratropium

Hyoscine butylbromide

Trospium chloride

Tiotropium bromide

FIGURE 22.27 Structural analogues of atropine and hyoscine.

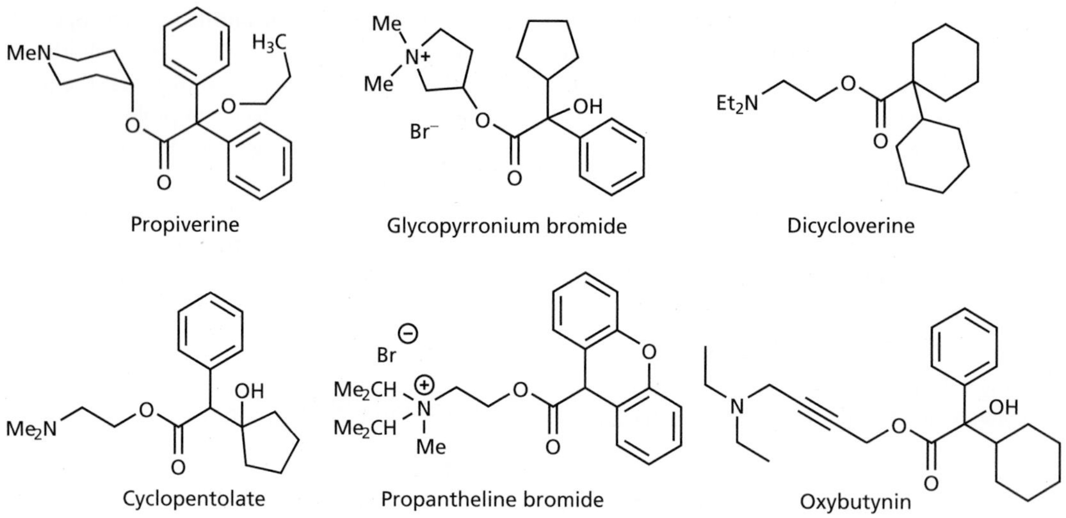

Propiverine

Glycopyrronium bromide

Dicycloverine

Cyclopentolate

Propantheline bromide

Oxybutynin

FIGURE 22.28 Simplified analogues of atropine.

A number of generalizations can be made about the structural requirements for muscarinic antagonists based on the structures described so far:

- The alkyl groups (R) on nitrogen can be larger than methyl (in contrast to agonists).
- The nitrogen can be tertiary or quaternary, whereas agonists must have a quaternary nitrogen. Note, however, that the tertiary nitrogen is probably charged when it interacts with the receptor.
- An alcohol group is commonly present, but is not essential.
- Very large acyl groups are allowed (R^1 and R^2 = aromatic, heteroaromatic, or alicyclic rings). This is in contrast to agonists where only the acetyl group is permitted.

This last point appears to be the most crucial in determining whether a compound will act as an antagonist or not. The acyl group has to be bulky, but it also has to have that bulk arranged in a certain manner: in other words, there must be some sort of branching in the acyl group.

The conclusion that can be drawn from these results is that there must be hydrophobic binding regions next to the normal acetylcholine binding site. The overall shape of the acetylcholine binding site plus the extra binding regions would have to be T-shaped or Y-shaped in order to explain the importance of branching in antagonists (Fig. 22.29). A structure such as **propantheline**, which

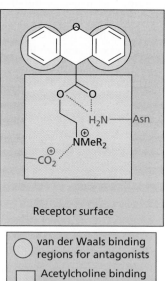

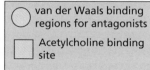

FIGURE 22.29 The binding of propantheline to the muscarinic receptor.

contains the complete acetylcholine skeleton as well as the hydrophobic acyl side chain, binds more strongly to the receptor than acetylcholine itself. The extra binding interactions mean that the conformational changes induced in the receptor will be different from those induced by acetylcholine and will fail to induce the secondary

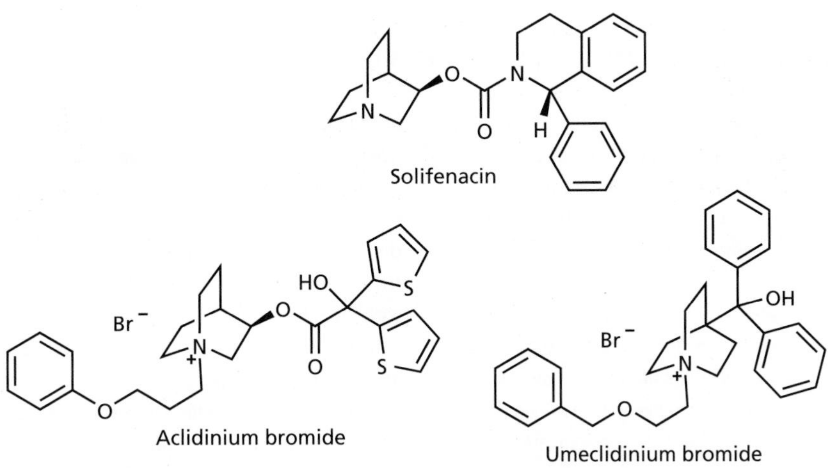

Solifenacin

Aclidinium bromide

Umeclidinium bromide

FIGURE 22.30 Muscarinic agents containing a quinuclidine ring.

biological response. As long as the antagonist is bound, acetylcholine is unable to bind and pass on its message.

For additional material see Web article 8: Photoaffinity labelling on the Online Resource Centre at www.oxfordtextbooks.co.uk/orc/patrick6e/

22.9.2.4 Quinuclidine muscarinic agents

Another strategy that has been successfully carried out on muscarinic antagonists is ring variation, where the tropane ring present in in atropine is replaced with a quinuclidine ring. This has resulted

in clinically useful agents such as **solifenacin**, **aclidinium bromide**, and **umeclidinium bromide** (Fig. 22.30; see also Box 22.2).

22.9.2.5 Other muscarinic antagonists

Most of the structures described so far contain an ester group. However, umeclidinium bromide is an exception to the rule and demonstrates that the ester group is not essential for activity. Other examples of muscarinic agents that lack the ester group are **tolterodine**, **fesoterodine**, **darifenacin**, **tropicamide**, **trihexyphenidyl**, and **pirenzepine** (Fig. 22.31).

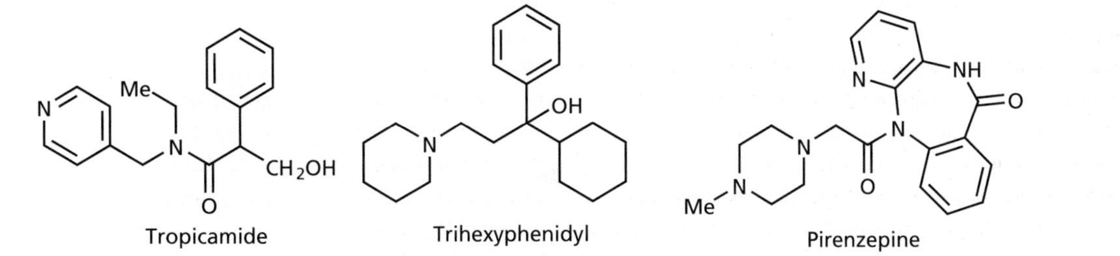

Tolterodine

Fesoterodine

Darifenacin

Tropicamide

Trihexyphenidyl

Pirenzepine

FIGURE 22.31 Muscarinic agents lacking an ester group.

BOX 22.1 Clinical applications for muscarinic antagonists

Muscarinic antagonists have proved useful medicines in a number of ways. Many show selectivity for specific organs, although this selectivity owes more to the distribution of the agents than to their receptor selectivity. In other words, the compounds can reach some parts of the body more easily than others.

Several muscarinic antagonists relax urinary smooth muscle and are useful in treating incontinence and bladder problems. These include **oxybutynin**, **tolterodine**, **flavoxate** (Fig. 1), **darifenacin**, **fesoterodine**, **propiverine**, **solifenacin**, and **trospium chloride**. Solifenacin and darifenacin are selective M3 antagonists, while fesoteridine is a prodrug. The ester that is present in fesoterodine is hydrolysed by esterase enzymes to reveal a phenol group in the active compound.

Muscarinic antagonists can also be used to relieve spasm by relaxing intestinal smooth muscle. **Atropine** and **atropine**

methonitrate are used for this purpose, as are **dicycloverine**, **hyoscine butylbromide**, and **propantheline bromide**.

Trihexyphenidyl and **benzatropine** (Fig. 1) are used centrally to counteract movement disorders caused by Parkinson's disease, while **pirenzepine** is used in some countries for the treatment of peptic ulcers. It is a selective M_1 antagonist with no activity for M_2 receptors.

Choosing the most suitable method of administration is another important factor in achieving selectivity of action. For example, **tropicamide** and **cyclopentolate** are used in eye drops to dilate pupils for ophthalmic examination. Similarly, atropine, cyclopentolate, and tropicamide can be applied as eye drops for the treatment of glaucoma. Inhaled muscarinic antagonists are taken to relax smooth muscle in the airways as a treatment for chronic obstructive pulmonary disease (Box 22.2).

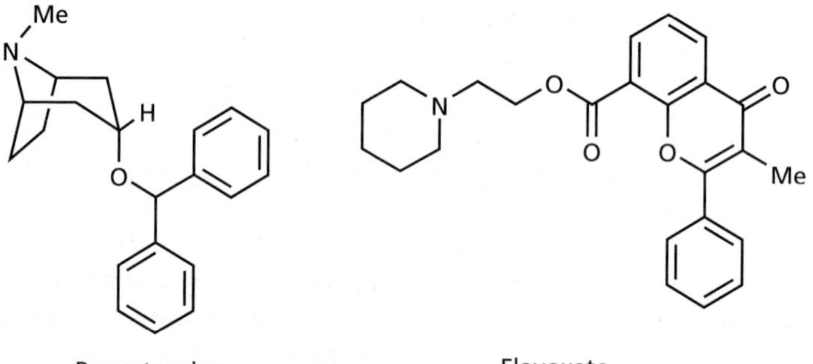

Benzatropine   Flavoxate

FIGURE 1 Benzatropine and flavoxate.

BOX 22.2 Muscarinic antagonists for the treatment of COPD

Chronic obstructive pulmonary disease (COPD) is a progressive disease characterized by reduced airflow to the lungs. It is caused by a number of factors such as smoking or exposure to chemicals. The World Health Organization (WHO) has predicted that COPD could become the fourth most common cause of death by the year 2030, and the third most common cause of chronic disability by 2020. There is no cure, but inhaled muscarinic antagonists can alleviate the symptoms by acting on M_3 receptors in the airways to relax smooth muscle and produce bronchodilation. **Ipratropium** is one such agent, but has to be taken four times a day. Therefore, there is an advantage in designing longer-lasting

agents that require less frequent doses. **Tiotropium bromide** and **glycopyrronium bromide** are two examples of longer-lasting agents that are taken once daily. Unfortunately, they suffer from a number of side effects caused by anticholinergic effects elsewhere in the body. Although these agents are taken by inhalation, up to 90% of the dose can be swallowed, then absorbed across the gut wall into the blood supply to cause systemic side effects. All the agents described have a quaternary nitrogen atom with a permanent positive charge which helps to limit the amount of drug absorbed from the gastrointestinal tract, but that does not prevent absorption totally.

Aclidinium bromide is a more recently approved agent (2012) that is taken twice daily. The agent has fewer side effects because its ester group is hydrolysed quickly by esterase enzymes in the blood supply. SAR studies have shown that the thiophene rings and the quaternary nitrogen are important factors in this susceptibility to enzyme-catalysed hydrolysis. Both these features are also responsible for strong binding and prolonged activity to the M_3 receptors in the airways. A related structure, **umeclidinium bromide**, was approved in 2013 and is taken once daily.

For more advanced stages of the disease, a combination therapy consisting of a muscarinic antagonist with a β_2-adrenergic agonist is often recommended. Examples involve the adrenergic agonist **indacaterol** taken with tiotropium bromide or glycopyrronium bromide. Alternatively, the adrenergic agonist **vilanterol** can be taken with umeclidinium bromide. Triple therapy, which includes a corticosteroid, is also recommended in some cases.

In recent years, there have been efforts to design a dual-action agent that combines muscarinic antagonism and β_2-adrenergic agonism within the same molecule. One example which has reached clinical trials is **batefenterol** (Fig. 1). The feature providing adrenergic activity is similar to indacaterol (Fig. 23.20).

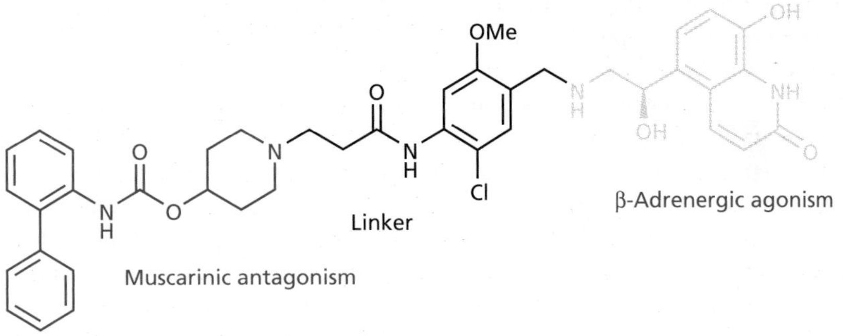

FIGURE 1 Batefenterol.

22.10 Antagonists of the nicotinic cholinergic receptor

22.10.1 Applications of nicotinic antagonists

Nicotinic receptors are present in nerve synapses at ganglia, as well as at the neuromuscular synapse. However, drugs are able to show a level of selectivity between these two sites, mainly because of the distinctive routes which have to be taken to reach them. Antagonists of ganglionic nicotinic receptor sites are not therapeutically useful because they cannot distinguish between the ganglia of the sympathetic nervous system and the ganglia of the parasympathetic nervous system (both use nicotinic receptors) (Fig. 22.3). Consequently, they have many side effects. However, antagonists of the neuromuscular junction are therapeutically useful and are known as **neuromuscular blocking agents**.

22.10.2 Nicotinic antagonists

22.10.2.1 Curare and tubocurarine

Curare was first identified in the sixteenth century when Spanish soldiers in South America found themselves under attack by indigenous people using poisoned arrows. It was discovered that the Indians were using a crude, dried extract from a plant called *Chondrodendron tomentosum*, which stopped the heart and also caused paralysis. Curare is a mixture of compounds, but the active principle is a cholinergic antagonist that blocks nerve transmissions from nerve to muscle.

It might seem strange to consider such a compound for medicinal use, but at the right dose levels and under proper control, there are useful applications for this sort of action. The main application is in the relaxation of abdominal muscles in preparation for surgery. This allows the surgeon to use lower levels of general anaesthetic than would otherwise be required and increase the safety margin for operations.

As mentioned above, curare is actually a mixture of compounds, and it was not until 1935 that the active principle (**tubocurarine**) was isolated. The determination of the structure took even longer, and it was not established until 1970 (Fig. 22.32). Tubocurarine was used clinically as a neuromuscular blocker, but it had undesirable side effects because it also acted as an antagonist at the nicotinic receptors of the autonomic nervous system (Fig. 22.2). Better agents are now available.

The structure of tubocurarine presents a problem to our theory of receptor binding. Although it has a couple

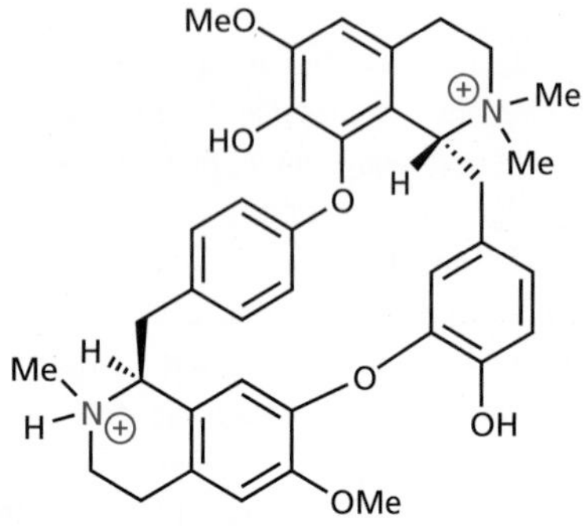

FIGURE 22.32 Tubocurarine.

of charged nitrogen centres, there is no ester to interact with the acetyl binding region. Studies on the compounds discussed so far show that the positively charged nitrogen on its own is not sufficient for good binding, so why should tubocurarine bind to the nicotinic receptor?

The answer lies in the fact that the molecule has *two* positively charged nitrogen atoms (one tertiary, which is protonated, and one quaternary). Originally, it was believed that the distance between the two centres (1.15 nm) might be equivalent to the distance between two separate cholinergic receptors, and that the tubocurarine molecule could bridge the two binding sites to act as a steric shield for both. However pleasing that theory may be, the dimensions of the nicotinic receptor make this impossible. The nicotinic receptor is a protein dimer made up of two identical protein complexes separated by 9–10 nm—far too large to be bridged by the tubocurarine molecule (Fig. 22.33 and section 22.11).

Another possibility is that the tubocurarine molecule bridges two acetylcholine binding sites within the one protein complex. As there are two such sites within the complex, this appears to be an attractive theory. However, the two sites are more than 1.15 nm apart and so this too has to be ruled out. It has now been proposed that one of the positively charged nitrogens on tubocurarine binds to the anionic binding region of an acetylcholine binding site, while the other binds to a nearby cysteine residue 0.9–1.2 nm away (Fig. 22.33).

Despite the uncertainty surrounding the binding interactions of tubocurarine, it seems highly probable that two ionic binding regions are involved. Such an interaction is extremely strong and would more than make up for the lack of the ester binding interaction. It is also clear that the distance between the two positively charged nitrogen atoms is crucial to activity. Therefore, analogues that retain this distance should also be good antagonists. Strong evidence for this comes from the fact that the simple molecule decamethonium is a good antagonist (section 22.10.2.2).

ω Test your understanding and practise your molecular modelling with Exercise 22.4 on the Online Resource Centre at www.oxfordtextbooks.co.uk/orc/patrick6e/

22.10.2.2 Decamethonium and suxamethonium

Decamethonium (Fig. 22.34) is as simple an analogue of tubocurarine as one could imagine. It is a flexible, straight-chain molecule that is capable of a large number of conformations. The fully extended conformation places the nitrogen atoms 1.4 nm apart, but there are other more folded conformations that position the nitrogen centres 1.14 nm apart, which compares well with the equivalent distance in tubocurarine (1.15 nm) (see also Box 17.4 and molecular modelling exercise 22.4).

The drug binds strongly to cholinergic receptors and has proved a useful clinical agent, but it suffers from several disadvantages. For example, when it binds initially to nicotinic receptors, it acts as an agonist rather than an antagonist. In other words, it switches on receptors, such that sodium ion channels open up to depolarize muscle cell membranes and cause brief contractions of the muscle. Because the drug is not rapidly hydrolysed in the same way as acetylcholine, it remains bound to the receptor, leading to persistent depolarization and subsequent desensitization of the end plate. At that stage, it can be viewed as an antagonist since it no longer stimulates muscle contraction and blocks access to acetylcholine. (A theory of how

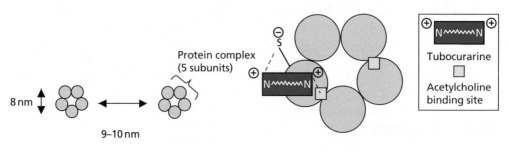

(a) Receptor dimer

(b) Interaction with tubocurarine

FIGURE 22.33 Tubocurarine binding to the cholinergic receptor.

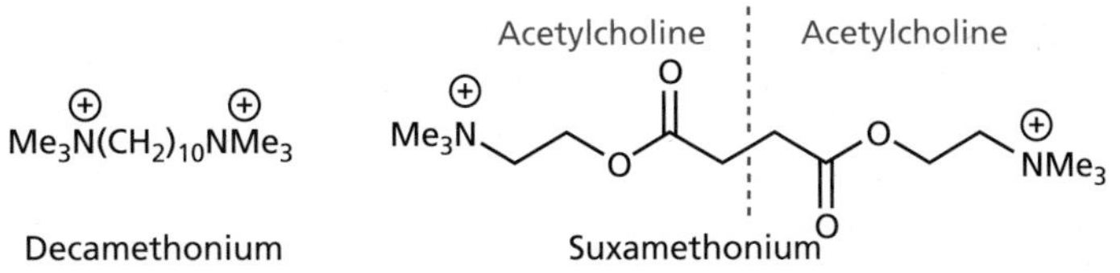

FIGURE 22.34 Decamethonium and suxamethonium.

such an effect might take place is described in section 8.6.) Another disadvantage is that it binds too strongly, so patients take a long time to recover from its effects.

We now face the opposite problem from the one faced when designing cholinergic agonists. Instead of stabilizing a molecule, we need to introduce some instability—a sort of timer control whereby the molecule can be inactivated more quickly. Success was first achieved with **suxamethonium** (Fig. 22.34) where two ester groups are incorporated into the chain in such a way that the distance between the charged nitrogens remains the same. The ester groups are susceptible to chemical and enzymatic hydrolysis and the resulting products are inactive since they can no longer bridge the two binding regions on the receptor. The ester groups are also introduced such that suxamethonium mimics two acetylcholine molecules linked end on. Suxamethonium has a fast onset and short duration of action (5–10 minutes), but suffers from various side effects. Furthermore, about 1 person in every 2000 lacks the plasma cholinesterase enzyme which hydrolyses suxamethonium. Nevertheless, it is still used clinically in short surgical procedures, such as the insertion of tracheal tubes.

Both decamethonium and suxamethonium are classed as depolarizing neuromuscular blockers and have effects on the autonomic ganglia, which explains some of their side effects. Decamethonium also lacks total selectivity for the neuromuscular junction and has an effect on cholinergic receptors in the heart. This leads to an increased heart rate and a fall in blood pressure.

22.10.2.3 Steroidal neuromuscular blocking agents

The design of **pancuronium**, **vecuronium**, and **rocuronium** (Fig. 22.35) was based on tubocurarine, but involved a steroid nucleus acting as a spacer between the two nitrogen groups. The distance between the quaternary nitrogens is 1.09 nm, compared to 1.15 nm in tubocurarine. Acyl groups were also added to introduce one or two acetylcholine skeletons into the molecule in order to improve affinity for the receptor sites. These compounds have a faster onset of action than tubocurarine and do not affect blood pressure. They are not as rapid in onset as suxamethonium and have a longer duration of action (45 minutes), but their main advantage is that they have fewer side effects and so they are widely used clinically. Unlike decamethonium and suxamethonium, these agents have no agonist activity and act as pure antagonists, so they have no depolarizing effect on target muscle cells. The neuromuscular blocking activity of rocuronium can be reversed with a cyclodextrin called **Sugammadex** (Box 10.3).

22.10.2.4 Atracurium and mivacurium

The design of **atracurium** (Fig. 22.36) was based on the structures of tubocurarine and suxamethonium. It is superior to both as it lacks cardiac side effects and is rapidly broken down in blood. This rapid breakdown allows the drug to be administered as an intravenous drip.

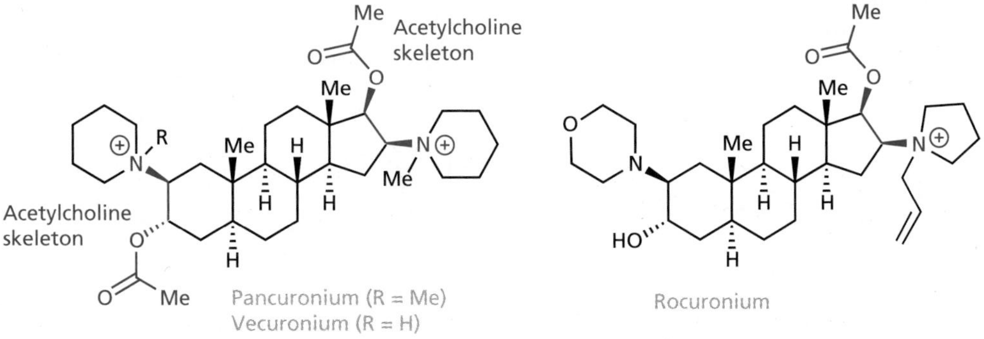

FIGURE 22.35 Steroidal neuromuscular blocking agents.

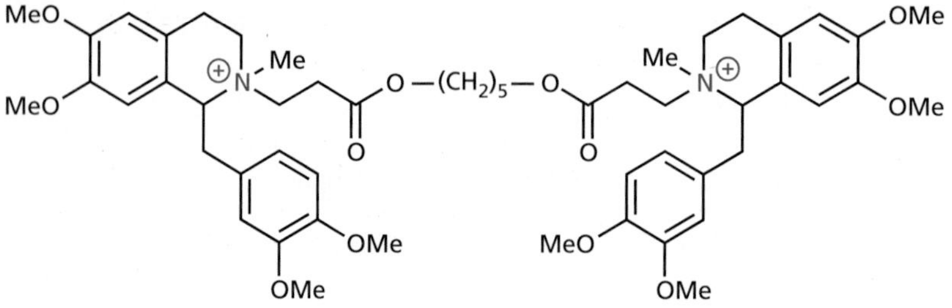

FIGURE 22.36 Atracurium.

The rapid breakdown is due to a self-destruct mechanism. At the slightly alkaline pH of blood (pH = 7.4), the molecule can undergo a **Hofmann elimination** (Fig. 22.37). Once this happens, the compound is inactivated because the positive charge on the nitrogen is lost and the molecule is split in two. It is a particularly clever example of drug design in that the very element responsible for the molecule's biological activity promotes its deactivation.

The important features of atracurium are:

- *The spacer*—a 13-atom chain connects the two quaternary centres.
- *The blocking units*—the cyclic structures at either end of the molecule which block the binding site from acetylcholine.
- *The quaternary centres*—these are essential for receptor binding. If one is lost through Hofmann elimination, the binding interaction is too weak and the antagonist leaves the binding site.
- *The Hofmann elimination*—the ester groups within the spacer chain are crucial to the rapid deactivation process. Hofmann eliminations normally require strong alkaline conditions and high temperatures—hardly normal physiological conditions. However, if a good electron-withdrawing group is present on the carbon that is *beta* to the quaternary nitrogen centre, it allows the reaction to proceed under the much milder alkaline conditions present in blood (pH 7.4). The electron-withdrawing ester group increases the acidity of the hydrogen on the *beta*-carbon such that it is easily lost. The Hofmann elimination does not occur at acid pH, and so the drug is stable in solution at a pH of 3–4 and can be stored safely in a refrigerator.

Because the drug acts very briefly (approx. 30 minutes), it is added intravenously for as long as it is needed. As soon as surgery is over, the intravenous drip is stopped and antagonism ceases almost instantaneously. Another major advantage is that the drug does not require enzymes to become deactivated and so deactivation occurs at a constant rate between patients. With previous neuromuscular blockers, deactivation depended on metabolic mechanisms involving enzymic deactivation and/or excretion. The efficiency of these processes varies from patient to patient and is particularly poor for patients with kidney failure or with low levels of plasma esterases.

Mivacurium (Fig. 22.38) is a newer drug which is similar to atracurium and is rapidly inactivated by plasma enzymes as well as by the Hofmann elimination. It has a faster onset (about 2 minutes) and shorter duration of action (about 15 minutes), although the duration is longer if the patients have liver disease or enzyme deficiencies.

22.10.2.5 Other nicotinic antagonists

Local anaesthetics and **barbiturates** appear to prevent the changes in ion permeability which would normally result from the interaction of acetylcholine with the nicotinic receptor. They do not, however, bind to the cholinergic binding site. It is believed that they bind instead to the part of the receptor which is on the inside of the cell membrane, perhaps binding to the ion channel itself and blocking it.

Certain snake toxins have been found to bind irreversibly to the nicotinic receptor, thus blocking cholinergic transmissions. These include toxins such as

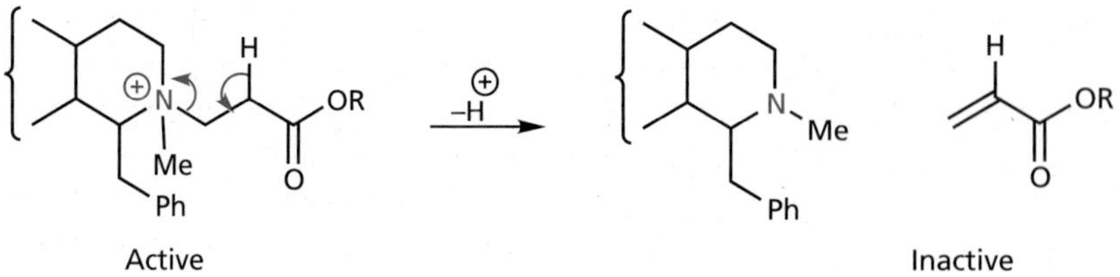

Active   Inactive

FIGURE 22.37 Hofmann elimination of atracurium.

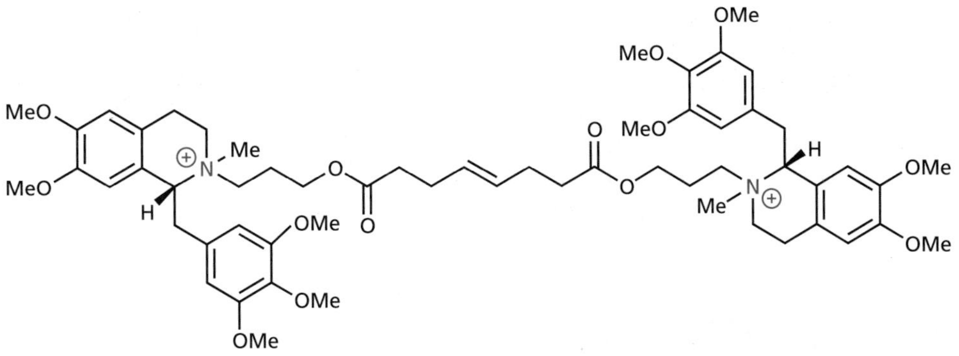

FIGURE 22.38 Mivacurium.

α-**bungarotoxin** from the Indian cobra. The toxin is a polypeptide containing 70 amino acids which cross-links the α- and β-subunits of the cholinergic receptor (section 22.11).

Finally, the antidepressant and antismoking drug **bupropion** (section 23.12.4) has been shown to be a nicotinic antagonist, as well as a reuptake inhibitor of noradrenaline and dopamine. It is possible that the drug's effectiveness as an antismoking aid may be related to its blockage of neuronal nicotinic receptors in the brain.

KEY POINTS

- Cholinergic antagonists bind to cholinergic receptors but fail to activate them. They block binding of acetylcholine and have a variety of clinical uses.

- Muscarinic antagonists normally contain a tertiary or quaternary nitrogen, a functional group involving oxygen, and a branch point containing two hydrophobic ring substituents.

- Nicotinic antagonists are useful as neuromuscular blockers in surgery.

- The pharmacophore for a nicotinic antagonist consists of two charged nitrogen atoms separated by a spacer molecule such that the centres are a specific distance apart.

- One of the charged nitrogens binds to the cholinergic binding site; the other interacts with a nucleophilic group neighbouring the binding site.

- Neuromuscular blockers should have a fast onset of action, minimal side effects, and a short duration of action to allow fast recovery. The lifetime of neuromuscular blockers can be decreased by introducing ester groups which are susceptible to enzymatic hydrolysis.

- Neuromuscular blockers which chemically degrade by means of the Hofmann elimination are not dependent on metabolic reactions and are more consistent from patient to patient.

22.11 Receptor structures

The **nicotinic receptor** has been successfully isolated from the electric ray (*Torpedo marmorata*)—a fish found in the Atlantic and the Mediterranean—allowing the receptor to be carefully studied. As a result, a great deal is known about its structure and operation. It is a protein complex made up of five subunits, two of which are the same. The five subunits (two α, one β, one γ, and one δ) form a cylindrical or barrel shape which traverses the cell membrane (section 4.6.2). The centre of the cylinder acts as an ion channel for sodium, and a gating or lock system is controlled by the interaction of the nicotinic receptor with acetylcholine. In the absence of acetylcholine, the gate is shut. When acetylcholine binds, the gate is opened. The binding site for acetylcholine is situated mainly on the α-subunit and there are two binding sites per ion channel complex. It is usually found that nicotinic receptors occur in pairs, linked together by a disulphide bridge between the δ-subunits.

This is the make-up of the nicotinic receptor at neuromuscular junctions. The nicotinic receptors at ganglia and in the CNS are more diverse in nature involving different α- and β-subunits. This allows drugs to act selectively on neuromuscular, rather than neuronal, receptors. For example, **decamethonium** is only a weak antagonist at autonomic ganglia, whereas **epibatidine** (extracted from a South American frog) is a selective agonist for neuronal receptors. The snake toxin α-**bungarotoxin** is specific for receptors at neuromuscular junctions.

Muscarinic receptors belong to the superfamily of G-protein-coupled receptors (section 4.7) which operate by activation of a signal transduction process (sections 5.1–5.3). Five subtypes of muscarinic receptors have been identified and are labelled M_1–M_5. These subtypes tend to be concentrated in specific tissues. For example, M_2 receptors occur mainly in the heart, whereas M_4 receptors are found mainly in the CNS. M_2 receptors are also used as the autoreceptors on presynaptic cholinergic neurons (section 22.3.2).

The M_1, M_3, and M_5 receptors are associated with a signal transduction process involving the secondary messenger **inositol triphosphate** (IP_3) (section 5.3). The M_2 and M_4 receptors involve a process which inhibits the production of the secondary messenger **cyclic AMP** (section 5.2). Lack of M_1 activity is thought to be associated with dementia.

KEY POINTS

- The nicotinic receptor is an ion channel consisting of five protein subunits. There are two binding sites for each ion channel.
- The muscarinic receptor is a G-protein-coupled receptor. Various subtypes of muscarinic receptor predominate in different tissues.

22.12 Anticholinesterases and acetylcholinesterase

22.12.1 Effect of anticholinesterases

Anticholinesterases are inhibitors of **acetylcholinesterase**—the enzyme that hydrolyses acetylcholine (section 22.3.1). If acetylcholine is not destroyed, it can return to reactivate the cholinergic receptor and increase cholinergic effects (Fig. 22.39). Therefore, an acetylcholinesterase inhibitor will have the same biological effect as a cholinergic agonist.

22.12.2 Structure of the acetylcholinesterase enzyme

The acetylcholinesterase enzyme has a fascinating tree-like structure (Fig. 22.40). The trunk of the tree is a collagen molecule which is anchored to the cell membrane.

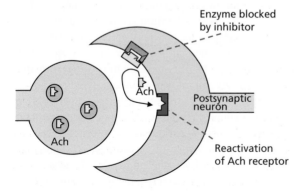

FIGURE 22.39 Effect of anticholinesterases (Ach = acetylcholine).

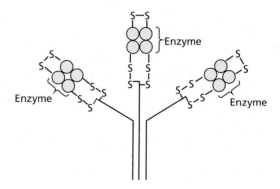

FIGURE 22.40 The acetylcholinesterase enzyme.

There are three branches with disulphide bridges that lead off from the trunk, each of which holds the acetylcholinesterase enzyme above the surface of the membrane. The enzyme itself is made up of four protein subunits, each of which has an active site. Therefore, each enzyme tree has twelve active sites. The trees are rooted immediately next to the cholinergic receptors such that they efficiently capture acetylcholine as it departs the receptor. In fact, the acetylcholinesterase enzyme is one of the most efficient enzymes known. A soluble cholinesterase enzyme called **butyrylcholinesterase** is also present in various tissues and plasma. This enzyme has a broader substrate specificity than acetylcholinesterase and can hydrolyse a variety of esters. Its physiological function is not totally clear, but it has been found to catalyse the hydrolysis of toxic esters, such as **cocaine**, and appears to have a non-catalytic role in cell differentiation and development. It is also more effective than acetylcholinesterase at hydrolysing high levels of acetylcholine when the acetylcholinesterase enzyme itself becomes substrate-inhibited.

22.12.3 The active site of acetylcholinesterase

The design of anticholinesterases depends on the shape of the enzyme's active site, the binding interactions involved with acetylcholine, and the mechanism of hydrolysis. The active site is at the foot of a narrow gorge (Fig. 22.41a), while a peripheral binding site is located at the entrance to the gorge. It is believed that the latter site plays a crucial role in recognizing acetylcholine as the substrate. One of the key interactions is a weak pi-cation interaction between the heteroaromatic ring of a tryptophan residue and the charged quaternary nitrogen of acetylcholine (Fig. 22.41b). After acetylcholine has been 'captured', it is rapidly transferred down the gorge to the active site (Fig. 22.41c). This process is aided by the fact that the gorge is lined with fourteen conserved aromatic residues, which can also form pi-cation interactions with acetylcholine, and thus channel the substrate down the

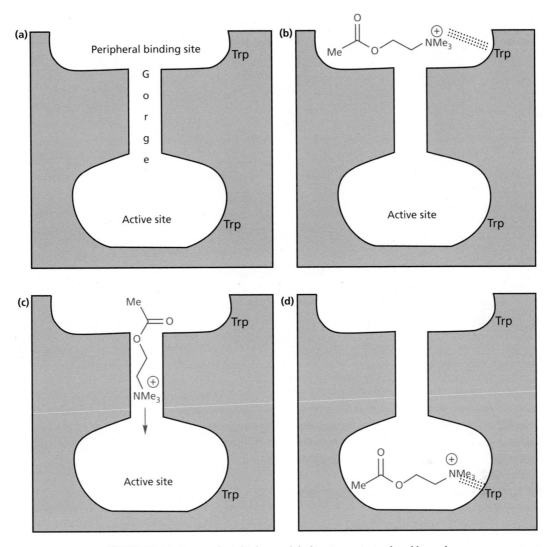

FIGURE 22.41 Process by which acetylcholine is recognized and bound.

gorge into the active site. Once acetylcholine enters the active site, another tryptophan residue forms yet another pi-cation interaction (Fig. 22.41d). An electrostatic gradient running down the gorge encourages the movement of acetylcholine. The gradient is due to several negatively charged amino acid residues in the active site, which create a dipole that points down the gorge to serve as an electronic steering mechanism for the positively charged substrate. The tryptophan residues in the peripheral binding site and the active site are 12 Å apart and this is significant when it comes to designing potential **dual-action drugs** (section 22.15.2).

22.12.3.1 Crucial amino acids within the active site

The important amino acids within the active site are those which bind acetylcholine, as well as those involved in the mechanism of hydrolysis. As far as binding is concerned,

several amino acids are thought to be involved, but a key interaction is the interaction between a tryptophan residue and the quaternary nitrogen atom (Fig. 22.42). The key amino acid residues involved in the catalytic mechanism are serine, histidine, and glutamate.

22.12.3.2 Mechanism of hydrolysis

The histidine residue acts as an acid–base catalyst throughout the mechanism, while serine acts as a nucleophile. This is not a particularly good role for serine, as an aliphatic alcohol is a poor nucleophile and is unable to hydrolyse an ester, but the acid–base catalysis provided by histidine overcomes that disadvantage. The glutamate residue interacts with the histidine residue and serves to orientate and activate the ring (compare chymotrypsin— section 3.5.3). There are several stages to the mechanism (Fig. 22.43):

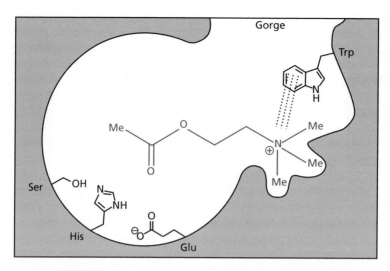

FIGURE 22.42 Key amino acid residues within the active site.

1. Acetylcholine approaches and binds to the active site. Serine acts as a nucleophile and uses a lone pair of electrons to form a bond to the ester of acetylcholine. Nucleophilic addition to the ester takes place and opens up the carbonyl group.

2. The histidine residue catalyses this reaction by acting as a base and removing a proton, thus making serine more nucleophilic.

3. Histidine now acts as an acid catalyst and protonates the alkoxy (OR) portion of the intermediate, turning it into a much better leaving group.

4. The carbonyl group reforms and expels the alcohol portion of the ester (i.e. choline).

5. The acyl portion of acetylcholine is now covalently bound to the active site. Choline leaves the active site and is replaced by water.

6. Water acts as a nucleophile and uses a lone pair of electrons on oxygen to attack the acyl group.

7. Water is normally a poor nucleophile, but histidine aids the process again by acting as a basic catalyst and removing a proton.

8. Histidine acts as an acid catalyst by protonating the intermediate.

9. The carbonyl group is reformed and the serine residue is released. Because it is now protonated, it is a much better leaving group.

10. Ethanoic acid leaves the active site and the cycle can be repeated.

The enzymatic process is remarkably efficient owing to the close proximity of the glutamate residue (not shown), the serine nucleophile, and the histidine acid–base catalyst. As a result, hydrolysis by acetylcholinesterase is 10^8 (one hundred million) times faster than in its absence. The process is so efficient that acetylcholine is hydrolysed within 100 μs of reaching the enzyme.

22.13 Anticholinesterase drugs

Anticholinesterase drugs act as inhibitors of the enzyme acetylcholinesterase. This inhibition can be either reversible or irreversible depending on how the drug interacts with the active site. Two main groups of acetylcholinesterases are considered here—carbamates and organophosphorus agents.

22.13.1 Carbamates

22.13.1.1 Physostigmine

As in so many fields of medicinal chemistry, it was a natural product that provided the lead for the carbamate inhibitors. The natural product was **physostigmine** (Fig. 22.44; also called **eserine**) which was discovered in 1864 as a product of the poisonous calabar bean (the ordeal bean, *Physostigma venenosum*) from West Africa. Extracts of these beans were fed to people accused of crimes to assess whether they were guilty or innocent. Death indicated a guilty verdict. The structure was established in 1925 and physostigmine is still used clinically to treat glaucoma.

SAR studies of physostigmine demonstrate that:

- the carbamate group is essential to activity;
- the benzene ring is important;
- the pyrrolidine nitrogen is important and is ionized at blood pH.

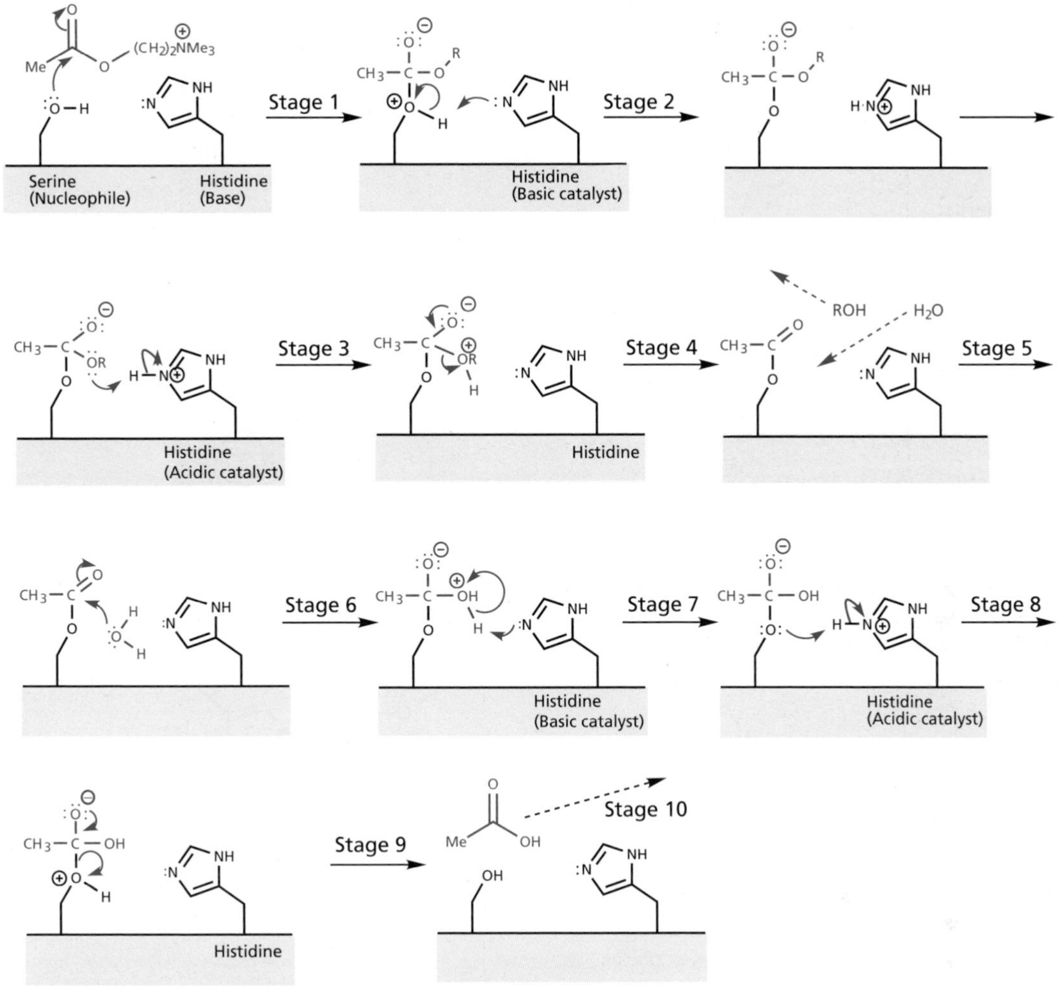

FIGURE 22.43 Mechanism of hydrolysis for the acetylcholinesterase enzyme (the glutamate component of the catalytic triad is not shown).

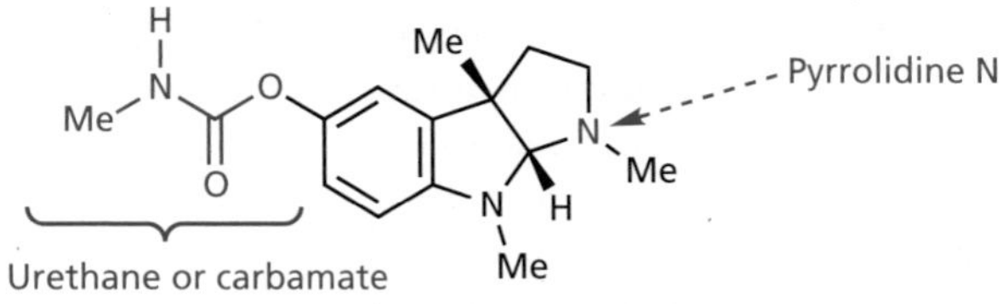

Urethane or carbamate

FIGURE 22.44 Physostigmine.

Working backwards, the positively charged pyrrolidine nitrogen is important because it binds to the anionic binding region of the enzyme. The benzene ring may be involved in some extra hydrophobic bonding with the active site. Alternatively, it may be important in the mechanism of inhibition as it provides a good leaving group. The carbamate group is the crucial group responsible for physostigmine's inhibitory properties. To understand

why, we must look at what happens when physostigmine acts as the substrate for acetylcholinesterase (Fig. 22.45).

The first four stages proceed as normal, with histidine catalysing the nucleophilic attack of the serine residue on physostigmine (stages 1 and 2). The leaving group (this time a phenol) is expelled with the aid of acid catalysis from histidine (stages 3 and 4), and departs the active site to be replaced by a water molecule.

The next stage turns out to be extremely slow. Despite the fact that histidine can still act as a basic catalyst, water finds it difficult to attack the carbamoyl intermediate. This step becomes the rate-determining step for the whole process and the overall rate of hydrolysis of physostigmine is 4×10^7 times slower than that of acetylcholine. As a result, the cholinesterase active site becomes blocked and is unable to react with acetylcholine.

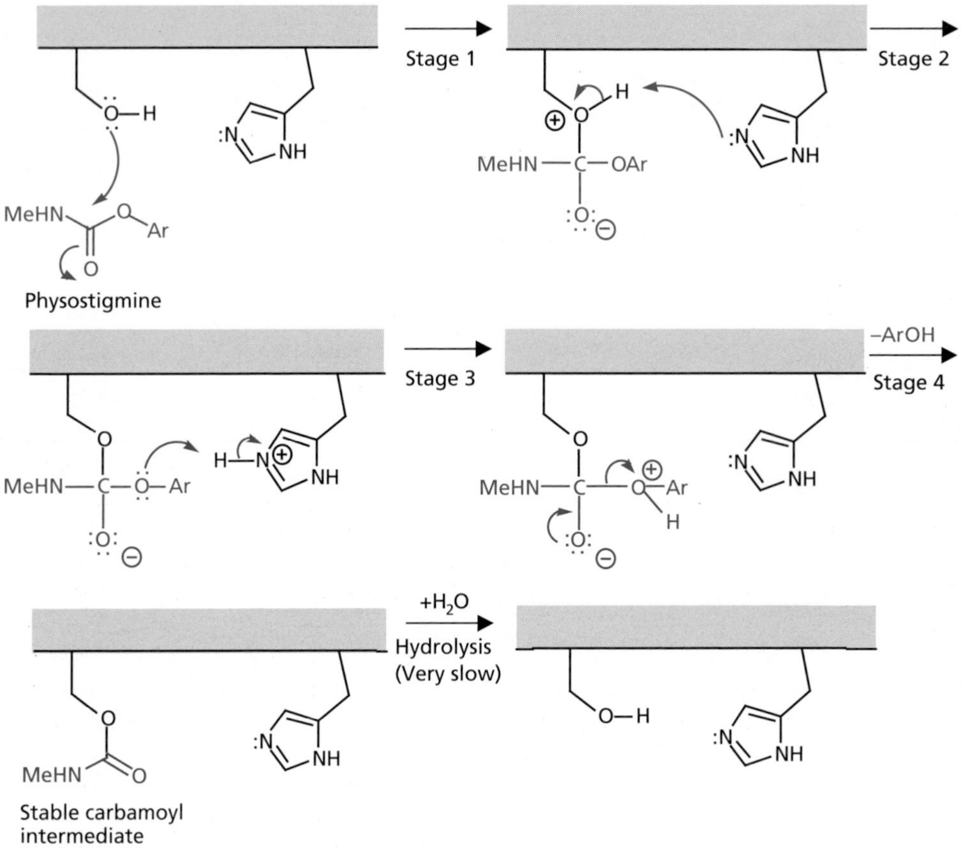

FIGURE 22.45 Mechanism of inhibition by physostigmine (Ar represents the tricyclic system of physostigmine).

The final stage is slow because of the stability of the carbamoyl–enzyme complex. This is because the nitrogen can feed a lone pair of electrons into the carbonyl group and drastically reduce its electrophilic character (Fig. 22.46; compare section 22.7.2).

22.13.1.2 Analogues of physostigmine

Physostigmine has limited medicinal use because of serious side effects, and it has only been used in the treatment of glaucoma or as an antidote for atropine poisoning. Simpler analogues, however, have been used in the

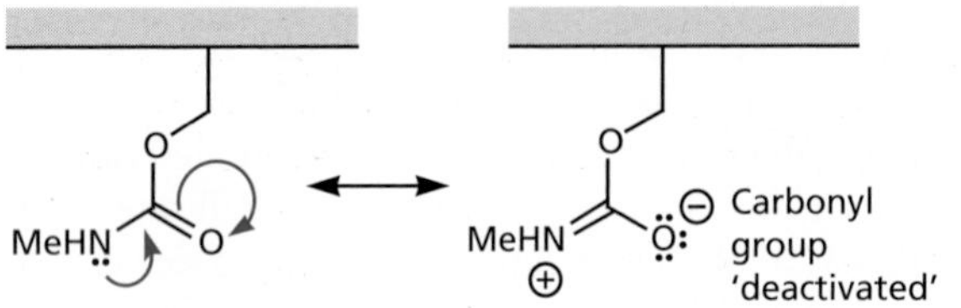

FIGURE 22.46 Stabilization of the carbamoyl–enzyme intermediate.

treatment of myasthenia gravis and as an antidote to curare poisoning.

Miotine (Fig. 22.47) still has the necessary carbamate, aromatic, and tertiary aliphatic nitrogen groups. It is active as an antagonist but it also has disadvantages: it is susceptible to chemical hydrolysis and it can cross the blood–brain barrier (section 11.4.5) as the free base, resulting in side effects due to its action in the CNS.

Neostigmine and **pyridostigmine** (Fig. 22.47) were designed to deal with both these problems. Firstly, a quaternary nitrogen atom is present and so there is no chance of the free base being formed. As the molecule is permanently charged, it cannot cross the blood–brain barrier and so the drug is free of CNS side effects. Increased stability is achieved by using a dimethylcarbamate group rather than a methylcarbamate group.

Two further points to note about neostigmine are the following:

- the quaternary nitrogen is 4.7 Å away from the carbamate group;
- the direct bonding of the quaternary centre to the aromatic ring reduces the number of conformations that

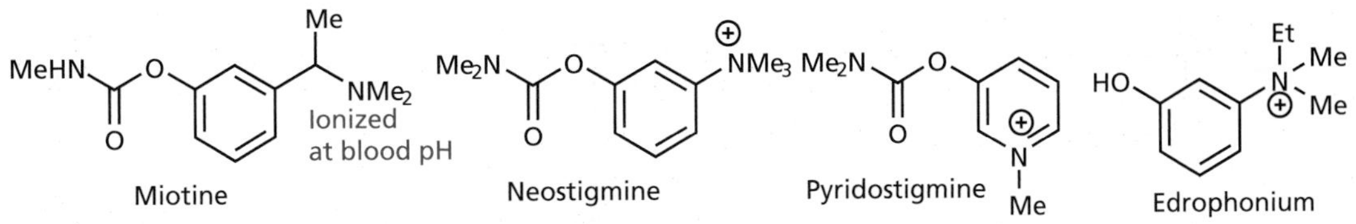

FIGURE 22.47 Analogues of physostigmine. Miotine is a chiral molecule, which has been studied as a racemate.

the molecule can adopt. This is an advantage if the active conformation is retained because the molecule is more likely to be in the active conformation when it approaches the active site.

Both neostigmine and pyridostigmine are in use today. They are given intravenously to reverse the actions of neuromuscular blockers or used orally in the treatment of myasthenia gravis. Pyridostigmine was one of the drugs used in the chemical cocktail provided to allied troops in Iraq during **Operation Desert Shield**. The agent was present to help protect against possible exposure to **organophosphate nerve gases**. **Edrophonium** is a similar agent used to reverse neuromuscular blocking, and is also used as a treatment of myasthenia gravis.

22.13.2 **Organophosphorus compounds**

The potential of organophosphorus agents as nerve agents was first recognized by German scientists in the 1920s and 30s, and research was carried out to investigate their potential as weapons of war. When World War II broke out, governments in the UK, USA, Sweden, and Russia recognized the danger of Germany perfecting these weapons and started their own research efforts during the 1940s. In the UK, this was carried out at the **Porton Down Defence Centre**. Fortunately, these agents were never used, but researchers in different countries continued work to find suitable antidotes that would protect troops from a possible attack. It has not been proved whether organophosphate nerve gases have ever been used in combat, but many believe that they were part of

the chemical weapons arsenal that was used against the Kurds by the Iraqi government. It has also been proposed that **sarin** (Fig. 22.48) may have been released when Iraqi chemical plants and ammunition dumps were bombed during 1990–91, and that this might be a possible cause of the mystery illness that afflicted many of the veterans of that war—**Gulf War syndrome**. Bosnians, Serbs, and Croats have also been accused of using nerve agents during the break-up of Yugoslavia in the 1990s. Certainly, nerve agents have been used by terrorist groups: the most notorious example was the release of sarin in the Tokyo subway during 1995.

The organophosphate nerve agents are examples of the weapons of mass destruction which several Western countries feared might be used by Iraq on its neighbours or supplied to extremist groups. The invasion of Iraq in 2003 was designed to combat this threat, but subsequent searches failed to reveal any such weapons.

It would be wrong to give the impression that the only use for organophosphates is as weapons of war and terror. They have also been extremely important insecticides used in agriculture and animal husbandry, and have a variety of uses in medicine. We shall consider these aspects in the following sections.

22.13.2.1 Nerve agents

The nerve gases **dyflos** and **sarin** (GB) (Fig. 22.48) were discovered and perfected long before their mode of action was known. Both agents inhibit acetylcholinesterase by irreversibly phosphorylating the serine residue at the active site (Fig. 22.49).

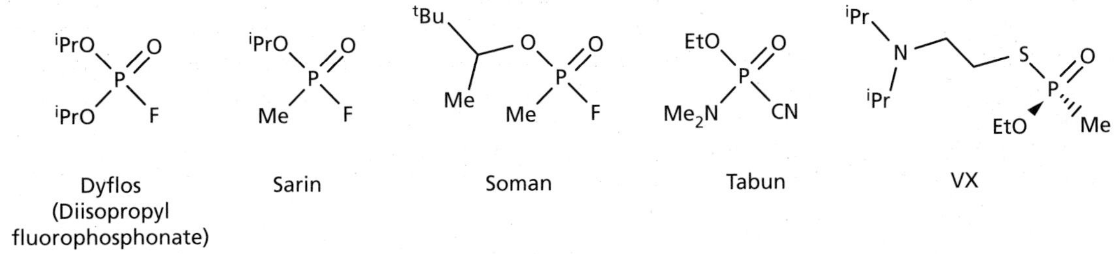

FIGURE 22.48 Examples of nerve agents (^{i}Pr = CHMe$_2$; ^{t}Bu = CMe$_3$).

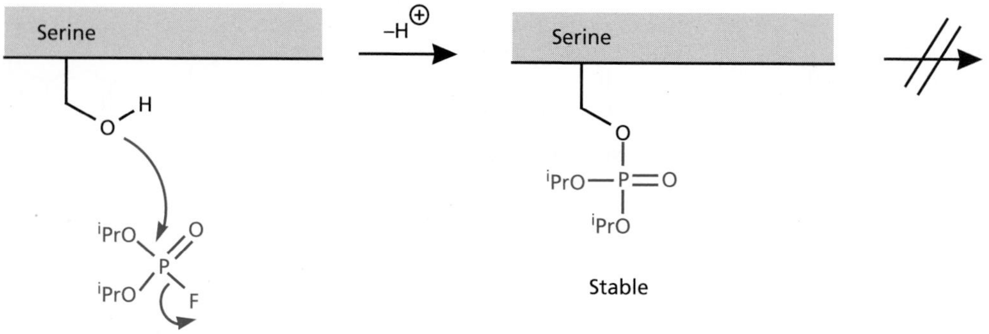

FIGURE 22.49 Simplified mechanism of action of dyflos at the active site of acetylcholinesterase.

The early part of the mechanism is similar to the normal mechanism, but the phosphorylated adduct which is formed is extremely resistant to hydrolysis. Consequently, the enzyme is permanently inactivated. As acetylcholine cannot be hydrolysed, the cholinergic system is continually stimulated. This results in permanent contraction of skeletal muscle, followed by death.

Other nerve agents include **tabun** (GA), **soman** (GD) and **VX**. VX is the most toxic of the nerve agents, having an LD_{50} of 10 mg through skin contact. It was discovered at Porton Down in the UK in 1954, then traded to the USA in exchange for technological information on nuclear weapons. The USA produced several tons of the material for its chemical warfare programme, but decided to dispose of its stockpiles in the late 1960s—a process that was only completed in 2008. Much of the nerve agent now lies at the bottom of the Atlantic Ocean.

22.13.2.2 Medicines

Once the mechanism of action of nerve agents was discovered, compounds such as **ecothiopate** (Fig. 22.50) were designed to fit the active site more effectively by including a quaternary amine to bind with the anionic region. This meant that lower doses would be more effective. Ecothiopate is used medicinally in the form of eye drops for the treatment of glaucoma, and has advantages over dyflos which has also been used in this way. Unlike dyflos, ecothiopate slowly hydrolyses from the enzyme over a matter of days.

22.13.2.3 Insecticides

The insecticides **parathion**, **malathion**, and **chlorpyrifos** (Fig. 22.50) are good examples of how a detailed knowledge of biosynthetic pathways can be useful in drug design. These agents are relatively non-toxic compared with nerve gases because the P=S double bond prevents inhibition of the acetylcholinesterase enzymes. In contrast, the equivalent compounds containing a P=O double bond are highly lethal.

Fortunately, there are no metabolic pathways in mammals which can convert the P=S double bond to a P=O double bond. In insects, however, the insecticides act as prodrugs and are metabolized by oxidative desulphurization. The resulting anticholinesterases prove lethal. In mammals, the same compounds are metabolized in a different way to give inactive compounds which are then excreted (Fig. 22.51). Despite this, organophosphate insecticides are not totally safe and prolonged exposure to them can cause serious side effects if they are not handled with care. Parathion has high lipid solubility and is easily absorbed through mucous membranes, and can also be absorbed through the skin. Preparations of malathion are used medicinally for the treatment of head lice, crab lice, and scabies, but should not be used too frequently or over prolonged periods.

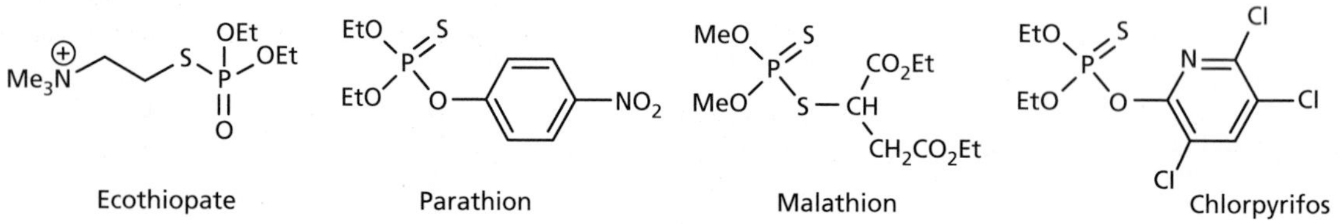

FIGURE 22.50 Organophosphates used as medicines and insecticides.

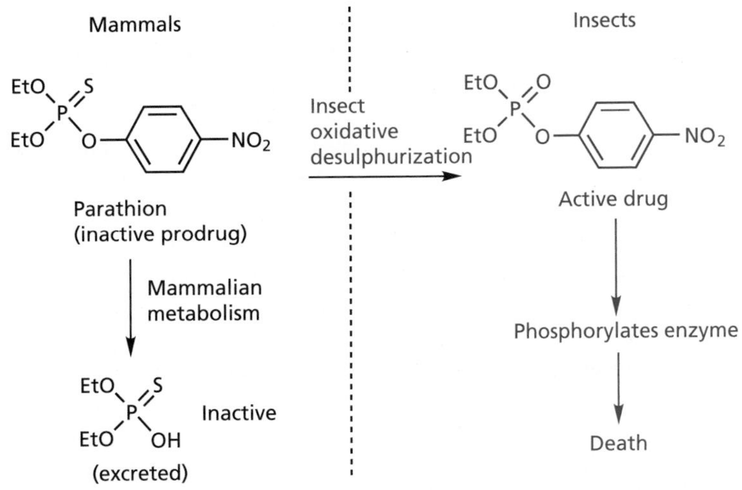

FIGURE 22.51 Metabolism of insecticides in mammals and insects.

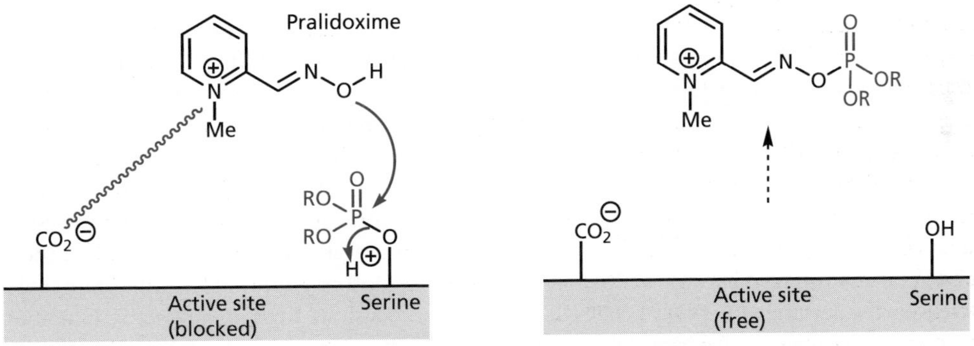

FIGURE 22.52 Pralidoxime as an antidote for organophosphate poisoning.

22.14 Pralidoxime: an organophosphate antidote

Pralidoxime (Fig. 22.52) is an antidote to organophosphate poisoning and represents one of the early examples of rational drug design. Any antidote for organophosphate poisoning has to displace the organophosphate moiety from serine by hydrolysing the phosphate–serine bond. However, this is a strong bond and not easily broken. Therefore, a strong nucleophile is required.

The literature revealed that phosphates can be hydrolysed with hydroxylamine (Fig. 22.53). This proved too toxic a compound to be used on humans, so the next stage was to design an equally reactive nucleophilic group which would specifically target the acetylcholinesterase enzyme. If such a compound could be designed, then there was less chance of the antidote taking part in toxic side reactions.

The designers' job was made easier by the knowledge that the organophosphate group does not fill the active site, and the anionic binding region is vacant. The obvious thing to do was to find a suitable group to bind to this anionic centre and attach a hydroxylamine moiety to it. Once positioned in the active site, the hydroxylamine group could react with the phosphate ester (Fig. 22.52).

Pralidoxime was the result. The positive charge is provided by a methylated pyridine ring and the nucleophilic side group is attached to the *ortho* position, as it was calculated that this would place the nucleophilic hydroxyl group in exactly the correct position to react with the phosphate ester. The results were spectacular, with

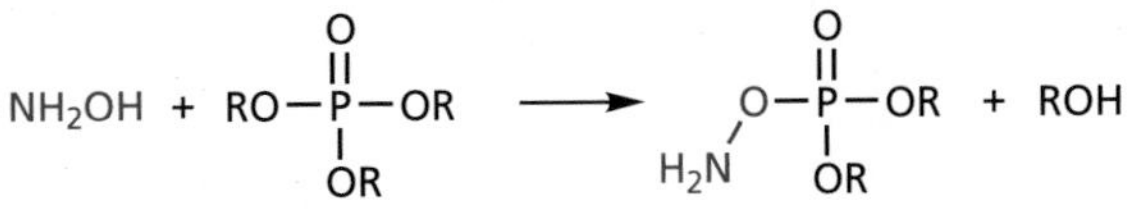

FIGURE 22.53 Hydrolysis of phosphates.

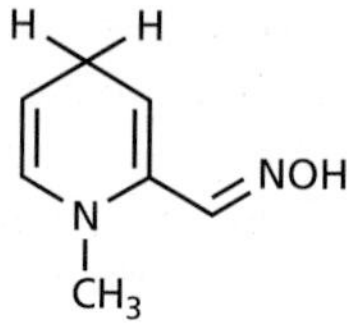

FIGURE 22.54 Pro-2-PAM.

pralidoxime showing a potency as an antidote 10^6 times greater than hydroxylamine.

Because pralidoxime has a quaternary nitrogen, it is fully charged and cannot pass through the blood–brain barrier into the CNS. This means that the antidote cannot work on any enzymes that have been inhibited in the brain. **Pro-2-PAM** (Fig. 22.54) is a prodrug of pralidoxime which avoids this problem. As a tertiary amine it can pass through the blood–brain barrier, and is oxidized to pralidoxime once it has entered the CNS.

22.15 Anticholinesterases as 'smart drugs'

22.15.1 Acetylcholinesterase inhibitors

Acetylcholine is an important neurotransmitter in the CNS, as well as in the peripheral nervous system. It has been proposed that the memory loss, intellectual deterioration, and personality changes associated with Alzheimer's disease may, in part, be due to the destruction of cholinergic nerves in the brain. Such damage is associated with the appearance of extracellular protein plaques and intracellular protein tangles in nerve fibres. These aberrant protein structures are neurotoxic and responsible for the destruction of neurons.

Although Alzheimer's disease is primarily a disease of the elderly, it can strike victims as young as 30 years of age and is the fourth leading cause of death in the developed world, affecting nearly 50% of those aged 85 years or more. It has been predicted that there will be 70 million sufferers worldwide by 2050, representing 1.2% of the total population.

The destruction of cholinergic nerves results in a drop in both cholinergic receptors and acetylcholine levels in the brain. Therefore, research has been carried out into the use of anticholinesterases for the treatment of Alzheimer's disease—the so-called 'smart drugs'. There is no evidence that these compounds assist general memory improvement, and so they are not a student's answer to exam cramming! The treatment does not offer a cure for Alzheimer's disease either, but it can alleviate the symptoms by prolonging the action of acetylcholine at the remaining cholinergic receptors. Unlike anticholinesterases acting in the periphery, 'smart drugs' have to cross the blood–brain barrier and so structures containing quaternary nitrogen atoms are not suitable. Tests with **physostigmine** were carried out in 1979, but the compound was not ideal as it does not enter the brain sufficiently well and shows short-lived, non-selective inhibition. The first drug to be approved for the treatment of Alzheimer's was **tacrine** (Fig. 22.55) in 1993. However, this is an extremely toxic drug and is only beneficial for about a year. Other

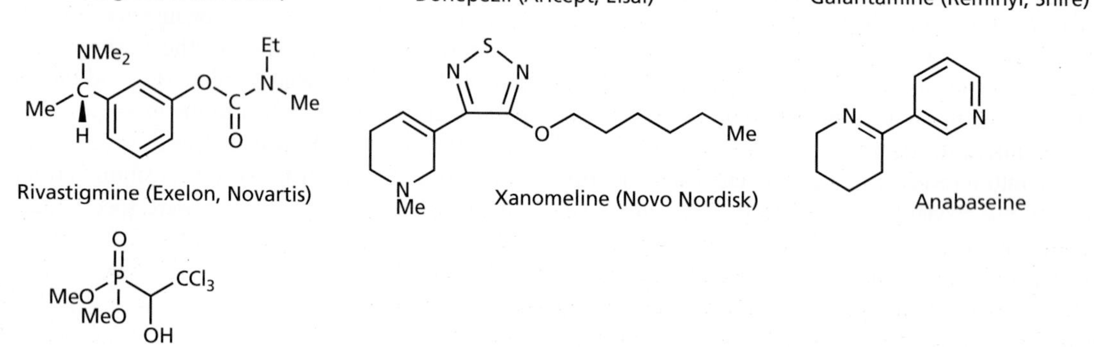

Tacrine (Cognex, Parke-Davis)

Donepezil (Aricept, Eisai)

Galantamine (Reminyl, Shire)

Rivastigmine (Exelon, Novartis)

Xanomeline (Novo Nordisk)

Anabaseine

FIGURE 22.55 'Smart drugs'.

agents which have subsequently been introduced include **donepezil** in 1997, **rivastigmine** in 2000, and **galantamine** (obtained from daffodils or snowdrop bulbs) in 2001. Rivastigmine (an analogue of physostigmine) was the first drug to be approved in all countries of the European Union. It shows selectivity for the brain and has beneficial effects on cognition, memory, concentration, and functional abilities involving day-to-day tasks or hobbies. The drug has a short half-life, reducing the risk of accumulation or drug–drug interactions. **Metrifonate** (an organophosphate) and **anabaseine** (from ants and marine worms) have also been tested for the treatment of Alzheimer's. Herbal medicines have been used in the past to treat the symptoms of Alzheimer's disease, and may provide useful lead compounds for further research (Box 22.2).

The anticholinesterase drugs have been shown to be beneficial in the early stages of Alzheimer's disease, but are of less benefit when the disease has become advanced. One disadvantage with the long-term use of these agents is the fact that they increase acetylcholine levels all round the body and not just in the brain; this leads to gastrointestinal side effects. Another problem is that the increased acetylcholine levels result in an increased activation of presynaptic cholinergic receptors which act as a feedback control to lower the amounts of acetylcholine released. Consequently, research has been carried out to find selective cholinergic agonists that could be used to treat the symptoms of the disease.

22.15.2 Dual-action agents acting on the acetylcholinesterase enzyme

In recent years, it has been discovered that the acetylcholinesterase enzyme appears to do more than just catalyse the hydrolysis of acetylcholine. Under normal conditions, the enzyme plays a non-catalytic role in neural development, cell adhesion, and differentiation. Protein–protein interactions involving the interaction of the peripheral binding site of acetylcholinesterase with other proteins promote these processes, with the tryptophan residue previously described (section 22.12.3) playing a crucial role.

Unfortunately, it has also been discovered that the enzyme can play an active role in promoting the deposits of aberrant protein that are found in the brain of Alzheimer's sufferers. Studies have shown that the peripheral binding site of the enzyme is capable of binding β-**amyloid protein**, which is normally soluble and has an antioxidant role. However, on binding to acetylcholinesterase, the protein undergoes a conformational change which causes it to become insoluble, leading to the appearance of the protein plaques and tangles associated with Alzheimer's disease. The enzyme has been described

as a **pathological chaperone** for this process and becomes associated with the protein deposits. Moreover, soluble oligomers of the protein are also formed within cells, which disrupt mitochondria function and increase oxidative stress, resulting in cell toxicity and cell death. These, indeed, may be more relevant to the disease than the visible extracellular plaques.

There is an exciting possibility that drugs might be developed that could halt the progression of the disease by preventing the binding of β-amyloid protein to the peripheral binding site of acetylcholinesterase. Research is currently in progress aimed at designing dual-action drugs that are capable of inhibiting this process, as well as acting as acetylcholinesterase inhibitors. **Donepezil** (Fig. 22.55) is one currently used inhibitor that can span the gorge to interact with both the peripheral binding site and the active site. It has also been shown to have an inhibitory effect on protein aggregation. However, much of the early work has looked at tacrine dimers. **Tacrine** (Fig. 22.55) is believed to enter the active site of the enzyme in a similar manner to acetylcholine; in other words it is protonated and binds initially to the peripheral binding site. It is then transferred down the gorge into the active site (Fig. 22.41). A dimer was designed where two tacrine molecules were linked by a hydrocarbon chain of sufficient length to allow one tacrine moiety to bind to the active site while the other interacted simultaneously with the peripheral binding site. Different lengths of linker were tried and it was found that a seven-carbon chain was ideal—**bis(7)-tacrine** (Fig. 22.56). This compound was found to be 150–1000 times more potent as an enzyme inhibitor depending on the source of enzyme studied. Studies have shown that the key tryptophan residues in the active site and the peripheral binding site can form pi-cation interactions with each of the tacrine components. The linker

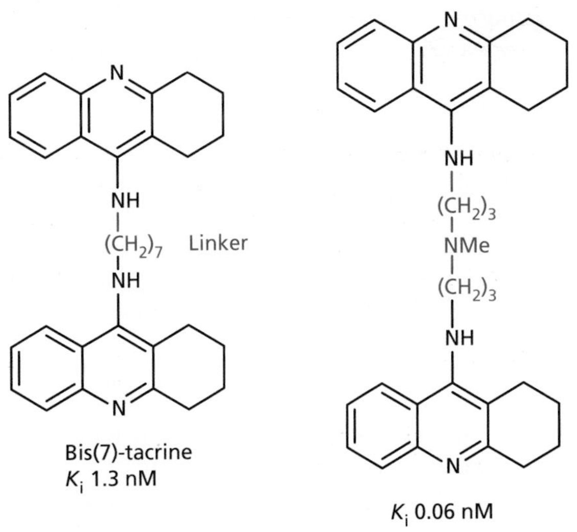

Bis(7)-tacrine
K_i 1.3 nM

K_i 0.06 nM

FIGURE 22.56 Tacrine dimers as dual-action agents.

can also form van der Waals interactions with the gorge and there is an entropy gain achieved by the displacement of water from the gorge. On the other hand, there is an entropy penalty resulting from the restriction in flexibility of the linker once it is constrained within the gorge. The tricyclic hydrophobic nature of the tacrine moieties is also important as there is only a small desolvation penalty involved when the structure binds. Stronger pi-cation interactions would be possible if one of the tacrine ring systems was replaced with a simpler amine, but the latter would be strongly solvated, and binding would involve a higher desolvation penalty.

The introduction of an N-methyl group into the linker resulted in further binding interactions and increased potency. The N-methyl group is protonated when the dimer binds and so it can form pi-cation interactions with the aromatic residues lining the gorge. Following on from this work, a large number of structures were synthesized, including homodimers of **galantamine** and **huperzine B** (see Box 22.3), as well as heterodimers containing two different acetylcholinesterase inhibitors. Other dual-action structures have been prepared consisting of a standard acetylcholinesterase inhibitor linked to a moiety designed to bind more effectively with the peripheral binding site. Many of these have been shown to inhibit both the catalytic activity of acetylcholinesterase and protein aggregation. Nevertheless, none of these compounds has entered the clinic to date.

🌐 Test your understanding and practise your molecular modelling with Exercises 22.5 and 22.6 on the Online Resource Centre at www.oxfordtextbooks.co.uk/orc/patrick6e/

There is a good chance that a dual-action agent will eventually reach the clinic. However, there are many factors involved in Alzheimer's disease and so drugs interacting with acetylcholinesterase alone are unlikely to provide a total cure. Attention is now turning to treatments that can address more than one of the various targets implicated in Alzheimer's disease. These treatments could involve a cocktail of different drugs acting at different targets. An alternative approach is to use an agent that can interact with different targets in a predictable way (**multiple-target directed ligands**) (see also section 13.3.14). For example, dual-action agents that inhibit the acetylcholinesterase enzyme have been designed which have one or more of the following properties:

- antioxidant activity and/or the ability to chelate metals;
- the ability to inhibit enzymes such as butyrylcholinesterase, monoamine oxidase, or BACE1;
- antagonist activity at α_2-adrenoceptors, 5HT$_3$ receptors, N-methyl-D-aspartate (NMDA) receptors, muscarinic (M$_2$) receptors, or H$_3$ receptors;

- inhibition of serotonin reuptake from nerve synapses;
- the blockade of calcium ion channels.

22.15.3 Multi-targeted agents acting on the acetylcholinesterase enzyme and the muscarinic M$_2$ receptor

As an example of one area of research into multi-targeted directed ligands, we shall consider agents that have been designed to act as dual-action agents that target the acetylcholinesterase enzyme (AChE), as well as being antagonists of the M$_2$ receptor. The M$_2$ receptor is an autoreceptor present on presynaptic cholinergic neurons. Activation of the autoreceptor inhibits the release of acetylcholine from the presynaptic neuron (sections 22.3.2 and 22.11), and so M$_2$ antagonists will increase acetylcholine release and help to raise acetylcholine levels. The lead compound for this work was a polyamine structure called **benextramine** (Fig. 22.57). This is an irreversible α-adrenoceptor antagonist, but it also shows activity as an anticholinesterase and M$_2$ receptor antagonist. Polyamines have been identified as good lead compounds for multi-targeted directed ligands because the protonated nitrogens present have the capability of forming pi-cation interactions with aromatic residues in virtually any protein target. Moreover, the flexible linear structure allows the polyamine to adopt a huge number of different conformations, making it more likely that suitable conformations are present that allow interaction with different targets. Such compounds are defined as **promiscuous ligands** (section 12.2.7).

Studies showed that the 2-methoxybenzyl group was important to activity, but not the disulphide bridge. Varying the chain length led to **methoctramine**, which had improved M$_2$ activity, whilst retaining good AChE activity. It was also shown that a diamine diamide backbone retained affinity for M$_2$, and so the two 'internal' amines were replaced with amides in order to improve lipophilicity. This decreased affinity for M$_2$ receptors and the butylcholinesterase enzyme (BuChE), but increased affinity for AChE. N-Methylation further increased affinity for AChE resulting in the discovery of **caproctamine**.

Compared with benextramine, caproctamine was found to be 42 times more active as an AChE inhibitor and 2 times less active as a BuChE inhibitor, while retaining affinity for the M$_2$ receptor. It was also demonstrated that the structure could bind simultaneously to the tryptophan residues in the active and peripheral binding sites, while the linker formed hydrophobic interactions with aromatic residues in the gorge. However, caproctamine showed very little ability to inhibit AChE-induced Aβ aggregation, which demonstrated that the ability to interact with the peripheral binding site does not necessarily block protein aggregation.

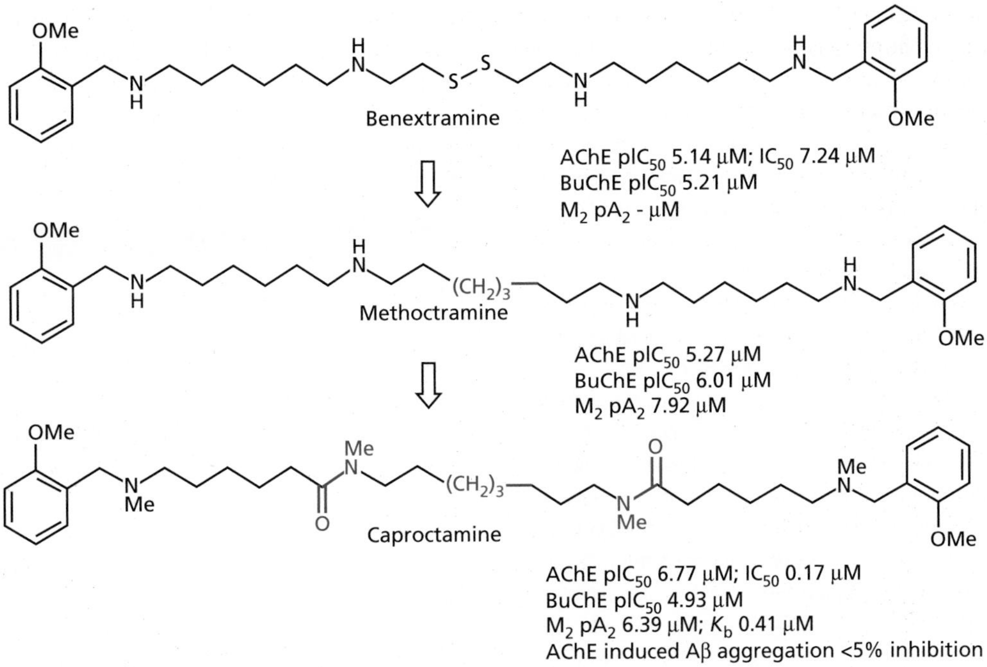
Benextramine
AChE pIC50 5.14 μM; IC50 7.24 μM
BuChE pIC50 5.21 μM
M2 pA2 - μM
Methoctramine
AChE pIC50 5.27 μM
BuChE pIC50 6.01 μM
M2 pA2 7.92 μM
Caproctamine
AChE pIC50 6.77 μM; IC50 0.17 μM
BuChE pIC50 4.93 μM
M2 pA2 6.39 μM; Kb 0.41 μM
AChE induced Aβ aggregation <5% inhibition

FIGURE 22.57 Development of caproctamine from benextramine.

Further work was done to introduce some rigidity into the linker chain by introducing piperidine rings (Fig. 22.58). This resulted in increased anticholinesterase activity and M_2 antagonism, as well as inhibition of AChE-induced Aβ aggregation.

A more substantial aromatic system was introduced into the middle of the linker because it was believed that this would form pi–pi interactions with aromatic residues in the gorge of AChE, and would also allow the structure to interact directly with Aβ proteins to inhibit self-induced aggregation of the protein. This led to the structure shown in Fig. 22.59, which proved to have nanomolar activity as an AChE inhibitor. The structure also proved more active in its ability to inhibit AChE-induced

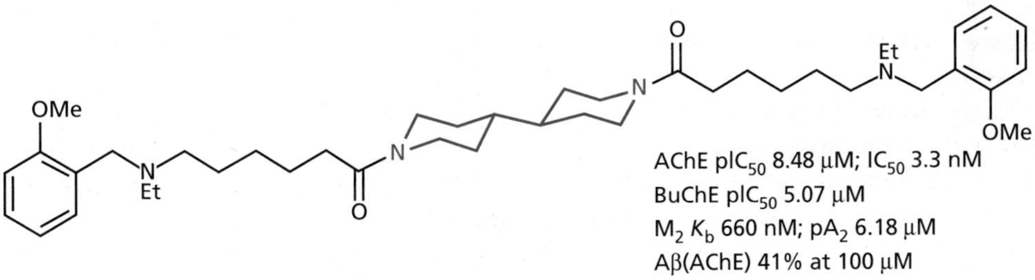
AChE pIC50 8.48 μM; IC50 3.3 nM
BuChE pIC50 5.07 μM
M2 Kb 660 nM; pA2 6.18 μM
Aβ(AChE) 41% at 100 μM

FIGURE 22.58 Rigidification of caproctamine.

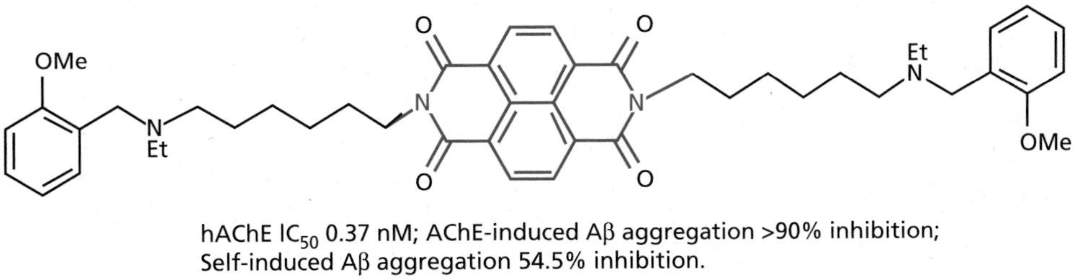
hAChE IC50 0.37 nM; AChE-induced Aβ aggregation >90% inhibition;
Self-induced Aβ aggregation 54.5% inhibition.

FIGURE 22.59 Further rigidified analogue of caproctamine.

BOX 22.3 Mosses play it smart

An extract from the club moss *Huperzia serrata* has been used for centuries in Chinese herbal medicine to treat ailments varying from confusion in Alzheimer's disease to schizophrenia. The extract contains a novel alkaloid called **huperzine A**, which acts as an anticholinesterase. Binding is very specific and so the drug can be used in small doses, thus minimizing the risk of side effects. Huperzine A has been approved for clinical use in China and has been shown to have memory-enhancing effects.

A synthetic route to the natural product has been worked out which has allowed the synthesis of different analogues, but none of these is as active as the natural product. The tricyclic ring system seems to be necessary for good activity, ruling out the possibility of significant simplification. All the functional groups in the molecule are also required for good activity.

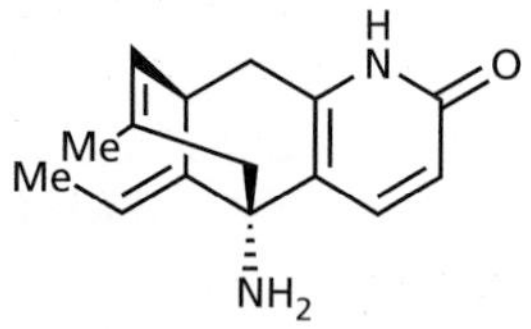

FIGURE 1 Huperzine A.

ring system interacts by pi–pi or van der Waals interactions with aromatic residues in the gorge. Hydrogen bonding is possible between the methoxy groups and a tyrosine residue in the active site. It remains to be seen whether further development of this compound will result in a clinically useful agent.

KEY POINTS

- Anticholinesterases inhibit the enzyme acetylcholinesterase and have the same clinical effects as cholinergic agonists.

- The active site for acetylcholinesterase includes a catalytic triad of amino acids—histidine, serine, and glutamate.

- Histidine acts as an acid–base catalyst, while serine acts as a nucleophile during the hydrolytic mechanism. Glutamate orientates and activates histidine.

- The carbamate inhibitors are derived from the lead compound physostigmine. They react with acetylcholinesterase to produce a stable carbamoyl-bound intermediate which is slow to hydrolyse.

- Organophosphorus agents have been used as nerve gases, medicines, and insecticides. They irreversibly phosphorylate serine in the active site.

- Pralidoxime was designed as an antidote for organophosphate poisoning. It can bind to the active site of phosphorylated enzymes and displace the phosphate group from serine.

- Anticholinesterases have been used as smart drugs in the treatment of Alzheimer's disease. They have to cross the blood–brain barrier and cannot be permanently charged.

- Dual-action agents have been designed as potential drugs for the treatment of Alzheimer's disease. They are designed to bind simultaneously to the active site and the peripheral binding site.

- Multi-target agents have been designed that target acetylcholinesterase and other targets that have been implicated in Alzheimer's disease.

Aβ aggregation. Its activity as an M_2 antagonist was not reported, however.

Docking experiments indicated that the structure could bind to the key tryptophan residues in the catalytic and peripheral binding sites, while the central tetracyclic

QUESTIONS

1. Based on the binding site described in Fig. 22.10, suggest whether the following structures are likely to act as agonists or not.

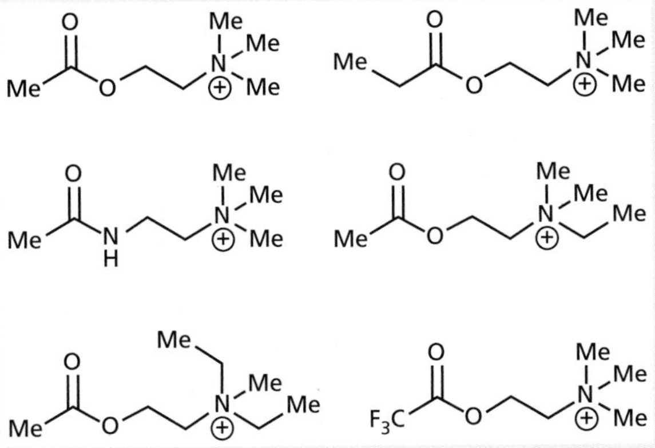

2. Suggest a mechanism by which atropine is racemized.

3. A fine balance of binding interactions is required of a neurotransmitter. What do you think is meant by this and what consequences does it have for drug design?

4. Suggest how the binding interactions holding acetylcholine to the active site of acetylcholinesterase might aid in the hydrolysis of acetylcholine.

5. Explain how the following diester could act as a prodrug for pilocarpine.

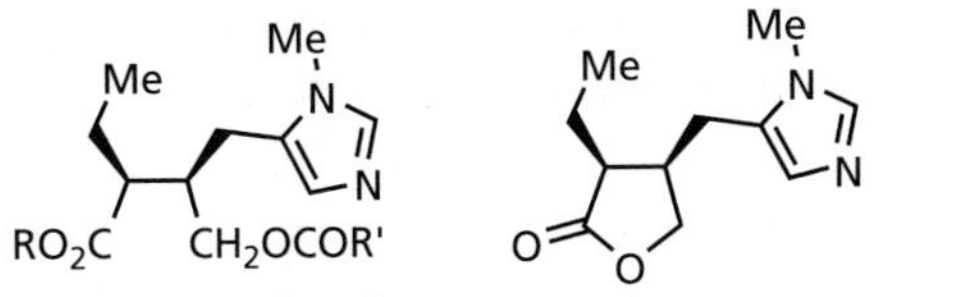

Diester prodrug for pilocarpine   Pilocarpine analogue

6. What advantage do you think the pilocarpine analogue shown might have over pilocarpine itself, and why?

7. Arecoline has been described as a cyclic 'reverse ester' bioisostere of acetylcholine. What is meant by this, and what similarity is there, if any, between arecoline and acetylcholine?

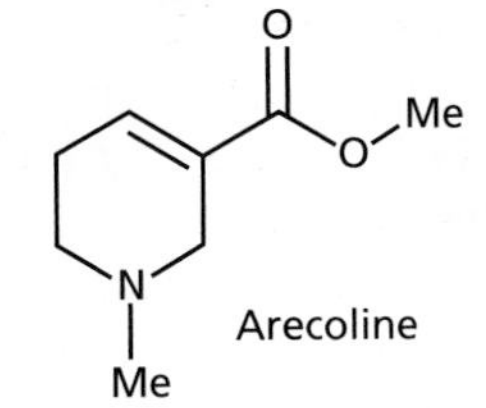

Arecoline

8. Arecoline has a very short duration of action. Why do you think this is?

9. Suggest analogues of arecoline that might have better properties, such as a longer duration of action.

10. Neuromuscular blocking activity for tubocurarine is associated with a pharmacophore where the distance between two charged nitrogen atoms is 1.15 nm. Decamethonium can adopt a folded conformation where the N–N separation is 1.14 nm. Octamethonium is an analogue of decamethonium which contains an 8-carbon bridge between the charged nitrogens. The fully extended conformation is the most stable conformation and corresponds to an N–N distance of 1.157 nm apart. Discuss whether octamethonium is likely to be more active than decamethonium.

11. An electrostatic gradient has been proposed that guides acetylcholine into the active site of the acetylcholinesterase enzyme. Can you foresee any problems associated with the presence of such a gradient? It has also been proposed that there may be a 'back door' into the active site. What do you think this means, how could it occur, and why would it be necessary?

12. Research is being carried out to design Alzheimer's drugs that will inhibit both acetylcholinesterase and butyrylcholinesterase, despite the fact that the former enzyme is more effective at catalysing the hydrolysis of acetylcholine. Why do you think this approach is considered relevant? What might be the disadvantages of such an approach?

⊕ Multiple-choice questions are available on the Online Resource Centre at www.oxfordtextbooks.co.uk/orc/patrick6e/

FURTHER READING

Roberts, S. M., and Price, B. J. (eds) (1985) Atracurium design and function. in *Medicinal chemistry—the role of organic research in drug research*, Chapter 8. Academic Press, London.

Teague, S. J. (2003) Implications of protein flexibility for drug discovery. *Nature Reviews Drug Discovery*, **2**, 527–41.

Titles for general further reading are listed on p.845.

Drugs acting on the adrenergic nervous system

23.1 The adrenergic nervous system

23.1.1 Peripheral nervous system

In Chapter 22, we studied the cholinergic system and the important role it plays in the peripheral nervous system. **Acetylcholine** is the crucial neurotransmitter in the cholinergic system and has specific actions at various synapses and tissues. The other important player in the peripheral nervous system (sections 22.1–22.2) is the adrenergic system, which makes use of the chemical messengers **adrenaline** and **noradrenaline**. Noradrenaline (also called **norepinephrine**) is the neurotransmitter released by the sympathetic nerves which feed smooth muscle and cardiac muscle, whereas adrenaline (**epinephrine**) is a hormone released along with noradrenaline from the **adrenal medulla**.

The action of noradrenaline at various tissues is the opposite to that of acetylcholine, which means that tissues are under a dual control. For example, if noradrenaline has a stimulant activity at a specific tissue, acetylcholine has an inhibitory activity at that same tissue. Both the cholinergic and adrenergic systems have a 'background' activity, so the situation is analogous to driving a car with one foot on the brake and one foot on the accelerator. The overall effect on the tissue depends on which effect is predominant.

The adrenergic nervous system has a component that the cholinergic system does not have—the facility to release adrenaline during times of danger or stress. This is known as the '**fight or flight**' response. Adrenaline is carried by the blood supply round the body and activates adrenergic receptors in preparation for immediate physical action, whether that be to fight the perceived danger or to flee from it. This means that the organs required for physical activity are activated, while those

that are not important are suppressed. For example, adrenaline stimulates the heart and dilates the blood vessels to muscles so that the muscles are supplied with sufficient blood for physical activity. At the same time, smooth muscle activity in the gastrointestinal tract is suppressed as digestion is not an immediate priority. This 'fight or flight' response is clearly an evolutionary advantage and stood early humans in good stead when faced with an unexpected encounter with a grumpy old bear. Nowadays, it is unlikely that you will meet a grizzly bear on your way to the supermarket, but the 'fight or flight' response is still functional when you are faced with modern dangers such as crazy drivers. It also functions in any situation of stress such as an imminent exam, important football game, or public performance. In general, the effects of noradrenaline are the same as those of adrenaline, although noradrenaline constricts blood vessels to skeletal muscle rather than dilates them.

23.1.2 Central nervous system

There are also adrenergic receptors in the central nervous system (CNS) and noradrenaline is important in many functions of the CNS including sleep, emotion, temperature regulation, and appetite. However, the emphasis in this chapter is on the peripheral role of adrenergic agents.

23.2 Adrenergic receptors

23.2.1 Types of adrenergic receptor

In Chapter 22, we saw that there are two types of cholinergic receptor, with subtypes of each. The same holds true for adrenergic receptors. The two main types of adrenergic receptor are called the α- **and** β-**adrenoceptors**. Both the α- and the β-adrenoceptors are G-protein-coupled receptors (section 4.7), but differ in the type of G-protein with which they couple (G_o for α-adrenoceptors; G_s for β-adrenoceptors).

For each type of receptor, there are various receptor subtypes with slightly different structures. The α-adrenoceptor consists of α_1- and α_2-subtypes, which differ in the type of secondary message produced. The α_1-receptors activate **inositol triphosphate** (IP_3) and **diacylglycerol** (DG) as secondary messengers (section 5.3), whereas the α_2-receptors inhibit the production of the secondary messenger **cyclic AMP** (section 5.2.3). The β-adrenoceptor consists of β_1-, β_2-, and β_3-subtypes, all of which activate the formation of cyclic AMP. To complicate matters slightly further, both the α_1- and α_2-adrenoceptors have further subcategories (α_{1A}, α_{1B}, α_{1D}, α_{2A}, α_{2B}, α_{2C}).

All of these adrenergic receptor types and subtypes are 'switched on' by adrenaline and noradrenaline, but the fact that they have slightly different structures means that it should be possible to design selective agonists that can distinguish between them. This is crucial in developing drugs that have minimal side effects and act at specific organs in the body, for as we shall see, the various adrenoceptors are not evenly distributed in different tissues. By the same token, it should be possible to design selective antagonists with minimal side effects that switch off particular types and subtypes of adrenoceptor.

23.2.2 Distribution of receptors

The various adrenoceptor types and subtypes vary in their distribution, with certain tissues containing more of one type of adrenoceptor than another. Table 23.1 describes various tissues, the types of adrenoceptor which predominate in these tissues, and the effect of activating these receptors (see also Box 23.1).

A few points are worth highlighting here:

- Activation of α-receptors generally contracts smooth muscle (except in the gut), whereas activation of β-receptors generally relaxes smooth muscle. This latter effect is mediated through the most common of the β-adrenoceptors—the β_2-receptor. In the heart, the β_1-adrenoceptors predominate and activation results in contraction of muscle.

- Different types of adrenoceptor explain why adrenaline can have different effects at different parts of the body. For example, the blood vessels supplying skeletal muscle have mainly β_2-adrenoceptors and are dilated by adrenaline, whereas the blood vessels elsewhere have mainly α-adrenoceptors and are constricted by adrenaline. As more blood vessels are constricted than dilated in the system, the overall effect of adrenaline is to increase the blood pressure, while at the same time providing sufficient blood for the muscles in the fight or flight response.

TABLE 23.1 Distribution and effects of adrenoceptors in different parts of the body

Organ or tissue	Predominant adrenoceptors	Effect of activation	Physiological effect
Heart muscle	β_1	Muscle contraction	Increased heart rate and force
Bronchial smooth muscle	α_1	Smooth muscle contraction	Closes airways
	β_2	Smooth muscle relaxation	Dilates and opens airways
Arteriole smooth muscle (not supplying muscles)	α	Smooth muscle contraction	Constricts arterioles and increases blood pressure (hypertension)
Arteriole smooth muscle (supplying muscles)	β_2	Smooth muscle relaxation	Dilates arterioles and increases blood supply to muscles
Veins	α	Smooth muscle contraction	Constricts veins and increases blood pressure (hypertension)
	β_2	Smooth muscle relaxation	Dilates veins and decreases blood pressure (hypotension)
Liver	α_1 and β_2	Activates enzymes which metabolize glycogen and deactivates enzymes which synthesize glycogen	Breakdown of glycogen to produce glucose
Gastrointestinal tract smooth muscle	α_1, α_2, and β_2	Relaxation	'Shuts down' digestion
Kidney	β_2	Increases renin secretion	Increases blood pressure
Fat cells	β_3	Activates enzymes	Fat breakdown

BOX 23.1 Clinical aspects of adrenergic agents

The main clinical use for adrenergic agonists is in the treatment of asthma. Activation of β_2-adrenoceptors causes the smooth muscles of the bronchi to relax, thus widening the airways. Agonists acting selectively on α_1-adrenoceptors cause vasoconstriction and can be used alongside local anaesthetics in dentistry to localize and prolong the effect of the anaesthetic at the site of injection. They are also used as nasal decongestants and for the treatment of hypotension (low blood pressure). Selective α_2-agonists are used in the treatment of glaucoma, hypertension, and pain.

The main uses for adrenergic antagonists are in treating angina and hypertension. Agents which act on the

α-receptors of blood vessels cause relaxation of smooth muscle, dilatation of the blood vessels, and a drop in blood pressure. Selective α_1-antagonists are now preferred for the treatment of hypertension, and are also being investigated as potential agents for the treatment of benign prostatic hyperplasia. Selective α_2-antagonists are being studied for the treatment of depression. Agents that block β_1-receptors in the heart (β-blockers) slow down the heart rate and reduce the force of contractions. β-Blockers also have a range of effects in other parts of the body, which combine to lower blood pressure.

23.3 Endogenous agonists for the adrenergic receptors

The term **endogenous** refers to any chemical which is naturally present in the body. As far as the adrenergic system is concerned, the body's endogenous chemical messengers are the neurotransmitter noradrenaline and the hormone adrenaline. Both act as agonists and switch on adrenoceptors. They belong to a group of compounds called the **catecholamines**—so called because they have an alkylamine chain linked to a **catechol** ring (the 1,2-benzenediol ring) (Fig. 23.1).

💧 Test your understanding and practice your molecular modelling with Exercise 23.1 on the Online Resource Centre at www.oxfordtextbooks.co.uk/orc/patrick6e/

23.4 Biosynthesis of catecholamines

The biosynthesis of noradrenaline and adrenaline starts from the amino acid **L-tyrosine** (Fig. 23.2). The enzyme **tyrosine hydroxylase** catalyses the introduction

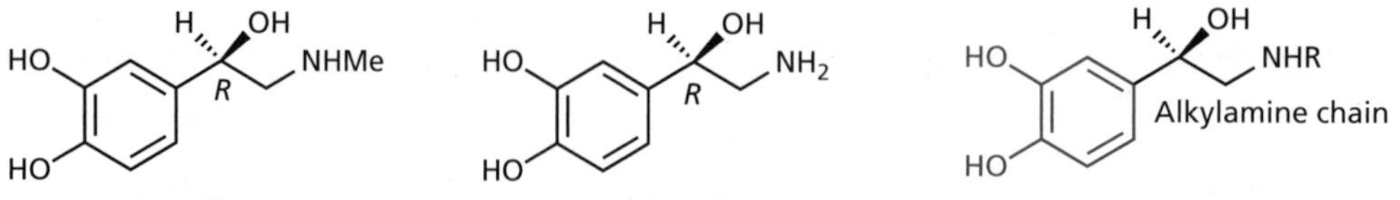

FIGURE 23.1 Endogenous adrenergic agonists.

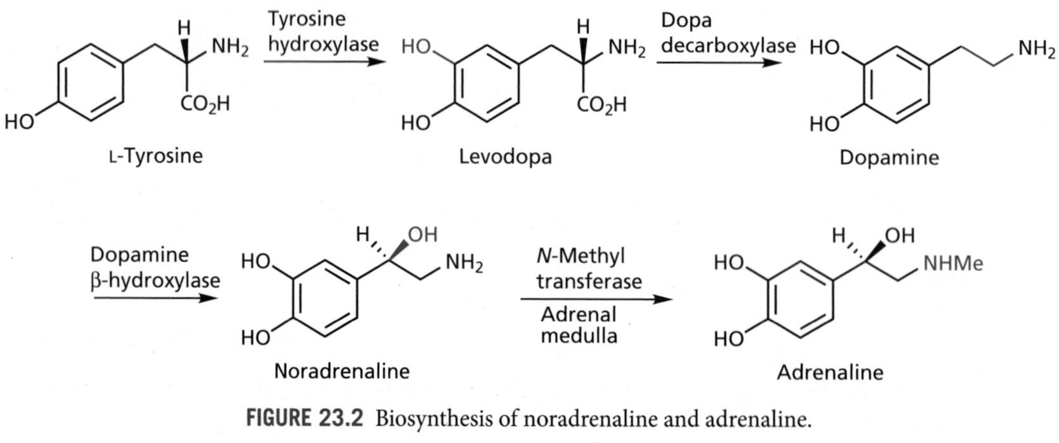

FIGURE 23.2 Biosynthesis of noradrenaline and adrenaline.

of a second phenol group to form **levodopa** (l-dopa) which is then decarboxylated by **aromatic L-amino-acid decarboxylase** (**dopa decarboxylase**) to give **dopamine**—an important neurotransmitter in its own right. Dopamine is then hydroxylated to **noradrenaline**, which is the end product in adrenergic neurons. In the adrenal medulla, however, noradrenaline is *N*-methylated to form **adrenaline**. The biosynthesis of the catecholamines is controlled by regulation of **tyrosine hydroxylase**—the first enzyme in the pathway. This enzyme is inhibited by noradrenaline—the end product of biosynthesis, thus allowing self-regulation of catecholamine synthesis and control of catecholamine levels.

23.5 Metabolism of catecholamines

Metabolism of catecholamines in the periphery takes place within cells and involves two enzymes—**monoamine oxidase** (**MAO**) and **catechol *O*-methyltransferase** (**COMT**). MAO converts catecholamines to their corresponding aldehydes. These compounds are inactive as adrenergic agents and undergo further metabolism (as shown in Fig. 23.3 for noradrenaline). The final carboxylic acid is polar and excreted in the urine.

An alternative metabolic route is possible which results in the same product. This time the enzyme COMT catalyses the methylation of one of the phenolic groups of the catecholamine. The methylated product is oxidized by MAO then converted to the final carboxylic acid and excreted (Fig. 23.4).

Metabolism in the CNS is slightly different, but still involves MAO and COMT as the initial enzymes.

23.6 Neurotransmission

23.6.1 The neurotransmission process

The mechanism of neurotransmission is shown in Fig. 23.5 and applies to adrenergic neurons innervating smooth or cardiac muscle, as well as synaptic connections within the CNS.

Noradrenaline is biosynthesized in a presynaptic neuron then stored in membrane-bound vesicles. When a nerve impulse arrives at the terminus of a neuron, it stimulates the opening of calcium ion channels and promotes the fusion of the vesicles with the cell membrane to release noradrenaline. The neurotransmitter then diffuses to adrenergic receptors on the target cell where it binds and activates the receptor, leading to the signalling process which will eventually result in a cellular response. After the message has been received, noradrenaline departs the receptor and is taken back into the presynaptic neuron by a transport protein. Once in the cell, noradrenaline is repackaged into the vesicles. Some of the noradrenaline is metabolized before it is repackaged, but this is balanced out by noradrenaline biosynthesis.

23.6.2 Cotransmitters

The process of adrenergic neurotransmission is actually more complex than that illustrated in Fig. 23.5. For example, noradrenaline is not the only neurotransmitter released during the process. **Adenosine triphosphate** (**ATP**) and a protein called **chromogranin A** are released from the vesicles along with noradrenaline, and act as cotransmitters. They interact with their own specific

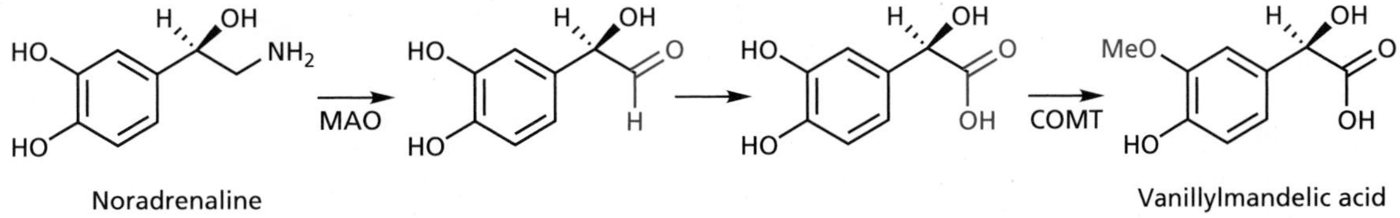

FIGURE 23.3 Metabolism of noradrenaline with monoamine oxidase (MAO) then catechol *O*-methyltransferase (COMT).

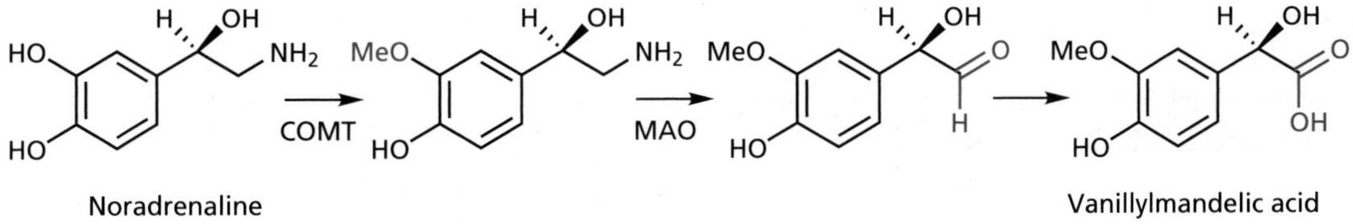

FIGURE 23.4 Metabolism of noradrenaline with catechol *O*-methyltransferase (COMT) then monoamine oxidase (MAO).

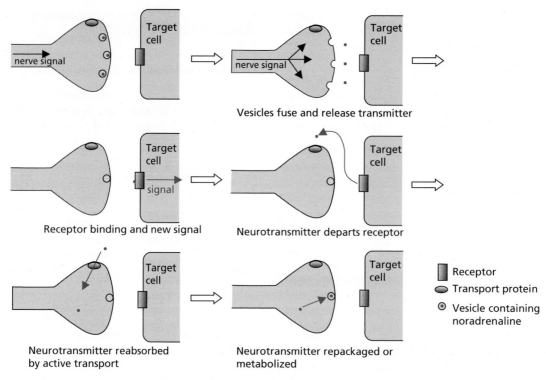

FIGURE 23.5 Neurotransmission process for noradrenaline.

receptors on the target cell and allow a certain variation in the speed and type of message which the target cell receives. For example, ATP leads to a fast response in smooth muscle contraction.

23.6.3 Presynaptic receptors and control

A further feature of the neurotransmission process not shown in Fig. 23.5 is the existence of presynaptic receptors which have a controlling effect on noradrenaline release (Fig. 23.6).

There are a variety of these receptors, each of which responds to a specific chemical messenger. For example, there is an adrenergic receptor (the α_2-**adrenoceptor**) which interacts with released noradrenaline and has an inhibitory effect on further release of noradrenaline. Thus, noradrenaline acts to control its own release by a negative feedback system.

There are receptors specific for **prostaglandins** released from the target cell. For example, the prostaglandin **PGE$_2$** appears to inhibit transmission, whereas **PGF$_{2\alpha}$** appears to facilitate it. Thus, the target cell itself can have some influence on the adrenergic signals coming to it.

There are presynaptic muscarinic receptors that are specific for **acetylcholine** and serve to inhibit release of noradrenaline. These receptors respond to side branches of the cholinergic nervous system which synapse on to the adrenergic neuron. This means that when the

cholinergic system is active, it sends signals along its side branches to inhibit adrenergic transmission. Therefore, as the cholinergic activity to a particular tissue increases, the adrenergic activity decreases, both of which enhance the overall cholinergic effect (cf. section 22.3.2).

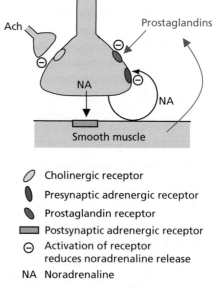

FIGURE 23.6 Presynaptic receptors on a neuron releasing noradrenaline as neurotransmitter.

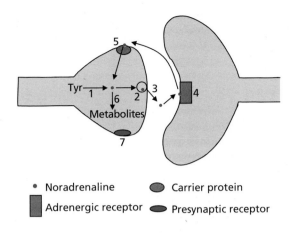

• Noradrenaline   ● Carrier protein

▮ Adrenergic receptor   ⬭ Presynaptic receptor

FIGURE 23.7 Drug targets affecting noradrenaline neurotransmission.

23.7 Drug targets

Having studied the nerve transmission process, it is now possible to identify several potential drug targets which will affect the process (Fig. 23.7):

1. The biosynthetic enzymes involved in the synthesis of noradrenaline within presynaptic neurons (section 23.4).

2. The vesicle carriers which package noradrenaline within the presynaptic neuron prior to release.

3. The process of exocytosis where vesicles fuse with the cell membrane and release noradrenaline into the synaptic gap when the neuron is active.

4. Adrenergic receptors in the postsynaptic neuron which are activated by noradrenaline to generate a signal in that neuron.

5. The transport proteins which are responsible for the reuptake of noradrenaline from the synaptic gap.

6. The metabolic enzymes which metabolize noradrenaline (section 23.5).

7. The presynaptic adrenergic receptors which regulate noradrenaline release (section 23.6.3).

In the next section, we concentrate on the adrenergic receptors. In later sections, we will consider some of the other possible drug targets.

KEY POINTS

• The neurotransmitter involved in the adrenergic nervous system is noradrenaline. Adrenaline is a hormone which is released by the adrenal medulla at times of stress and activates adrenergic receptors.

• The sympathetic nerves innervating smooth muscle and cardiac muscle release noradrenaline.

• Adrenergic receptors are G-protein-coupled receptors. There are two main types: the α- and the β-adrenoceptors. There are various subtypes of each.

• The different types and subtypes of adrenoceptor predominate in different tissues. Drugs which show receptor selectivity also show tissue selectivity.

• The major use of adrenergic agonists is in the treatment of asthma. The major use of adrenergic antagonists is in cardiovascular medicine.

• Adrenaline, noradrenaline, and dopamine are catecholamines.

• The biosynthesis of catecholamines starts from tyrosine and involves levodopa as an intermediate.

• Catecholamines are metabolized by monoamine oxidase and catechol O-methyltransferase.

• Noradrenaline is synthesized in presynaptic neurons, and packaged in vesicles prior to release. Once released, it activates receptors on target cells. It is then is taken up into presynaptic neurons by a transport protein and repacked into vesicles. A certain percentage of noradrenaline is metabolized.

• Adrenergic receptors are the main targets for adrenergic drugs.

23.8 The adrenergic binding site

The adrenergic receptors are G-protein-linked receptors which consist of seven transmembrane (TM) helices (section 4.7). In order to study the binding site of a receptor, one would ideally crystallize it with a ligand bound to the binding site. X-ray crystallography would then be used to determine the crystal structure and identify how the ligand binds. Unfortunately, membrane-bound receptors are very difficult to crystallize, and it was only in 2007 that the β_2-adrenoceptor was crystallized (section 17.14.1). However, the crystal structure obtained does not reveal how an agonist binds to the ligand binding site. Therefore, our knowledge of the binding site is based on mutagenesis studies and molecular modelling. Mutagenesis studies involve mutating amino acids to see which ones are crucial for ligand binding, while molecular modelling involves the construction of a model binding site based on the structures of similar proteins whose structures are known (section 17.14.1). From these studies, it has been proposed that three of the transmembrane helices (TM3, TM5, and TM6) are involved in the binding site, illustrated for the β-adrenoceptor in Fig. 23.8. Mutagenesis studies have indicated the importance of an aspartic acid residue (Asp-113), a phenylalanine residue (Phe-290), and two serine residues (Ser-207 and Ser-204). Modelling studies indicate that these groups can bind to

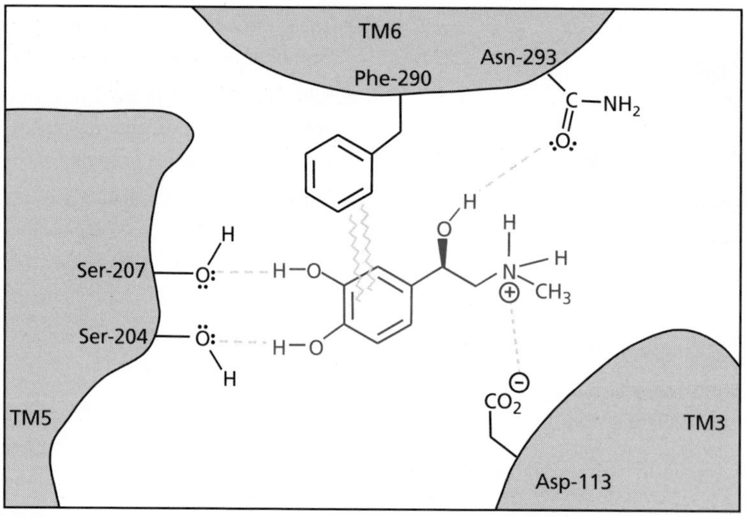

FIGURE 23.8 Adrenergic binding site.

adrenaline or noradrenaline as shown in the figure. The serine residues interact with the phenolic groups of the catecholamine via hydrogen bonding. The aromatic ring of Phe-290 interacts with the catechol ring by van der Waals interactions, while Asp-113 interacts with the protonated nitrogen of the catecholamine by ionic bonding. There is also a proposed hydrogen bonding interaction between Asn-293 and the alcohol function of the catecholamine.

23.9 Structure–activity relationships

23.9.1 Important binding groups on catecholamines

Support for the above binding site interactions is provided by studies of structure–activity relationships (SAR) on catecholamines. These emphasize the importance of

having the alcohol group, the intact catechol ring system with both phenolic groups unsubstituted, and the ionized amine (Fig. 23.9).

Some of the evidence supporting these conclusions is as follows:

- **The alcohol group**—the *R*-enantiomer of noradrenaline is more active than the *S*-enantiomer, indicating that the secondary alcohol is involved in a hydrogen bonding interaction. Compounds lacking the hydroxyl group (e.g. dopamine) have a greatly reduced interaction. Some of the activity *is* retained, indicating that the alcohol group is important, but not essential.

- **The amine** is normally protonated and ionized at physiological pH. This is important since replacing nitrogen with carbon results in a large drop in activity. Activity is also affected by the number of substituents on the nitrogen. Primary and secondary amines have good adrenergic activity, whereas tertiary amines and quaternary ammonium salts do not.

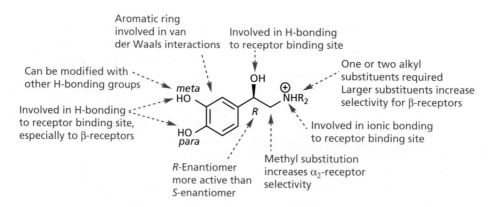

FIGURE 23.9 Important binding groups for adrenergic agents.

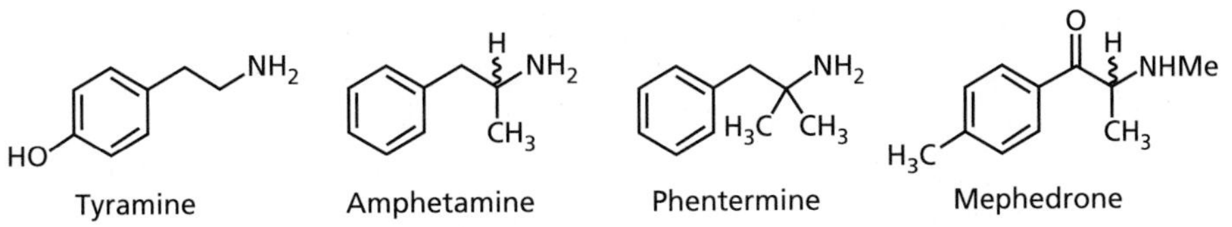

Tyramine   **Amphetamine**   **Phentermine**   **Mephedrone**

FIGURE 23.10 Agents that have no affinity for the adrenergic receptor.

- **Both phenol substituents** are important. For example, **tyramine**, **amphetamine**, **phentermine**, and the banned substance **mephedrone** (Fig. 23.10) have little, or no, affinity for adrenoceptors, although they do have an effect on the adrenergic system through other mechanisms (section 23.12.4). Having said that, the phenol groups can be replaced by other groups capable of interacting with the binding site by hydrogen bonding. This is particularly true for the *meta* phenol group which can be replaced by groups such as CH_2OH, CH_2CH_2OH, NH_2, NHMe, NHCOR, NMe_2, and $NHSO_2R$.

- **Alkyl substitution** on the side chain linking the aromatic ring to the amine decreases activity at both α- and β-adrenergic receptors. This may be a steric effect which blocks hydrogen bonding to the alcohol or which prevents the molecule adopting the active conformation.

23.9.2 **Selectivity for α- versus β-adrenoceptors**

SAR studies demonstrate certain features which introduce a level of receptor selectivity between the α- and β-adrenoceptors:

- *N*-**Alkyl substitution**: it was discovered that adrenaline has the same potency for both types of adrenoceptor, whereas noradrenaline has a greater potency for α-adrenoceptors than for β-adrenoceptors. This indicates that an *N*-alkyl substituent has a role to

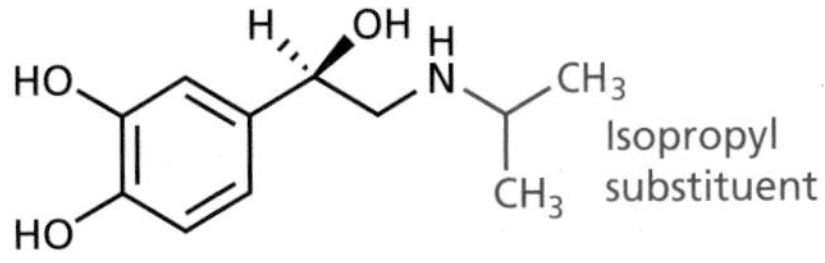

FIGURE 23.11 (*R*)-Isoprenaline.

play in receptor selectivity. Further work demonstrated that increasing the size of the *N*-alkyl substituent resulted in loss of potency at the α-receptor but an increase in potency at β-receptors. For example, the synthetic analogue **isoprenaline** (Fig. 23.11) is a powerful β-stimulant devoid of α-agonist activity. The presence of a bulky *N*-alkyl group, such as isopropyl or butyl, is particularly good for β-adrenoceptor activity. These results indicate that the β-adrenoceptor has a hydrophobic pocket into which a bulky alkyl group can fit, whereas the α-adrenoceptor does not (Fig. 23.12).

- **Phenol groups** seem particularly important for β-receptors. If they are absent, activity drops more significantly for β-receptors than for α-receptors.

- α-**Methyl substitution**: addition of an α-methyl group (e.g. α-methylnoradrenaline; Fig. 23.13) increases α_2-receptor selectivity.

- **Extension**: as mentioned earlier, isopropyl or *t*-butyl substituents on the amine nitrogen are particularly good for β-selectivity. Increasing the length of the alkyl chain offers no advantage, but if a polar functional group is placed at the end of the alkyl group, the situation changes. In particular, adding a phenol group to the end of a C_2 alkyl chain results in a dramatic rise in

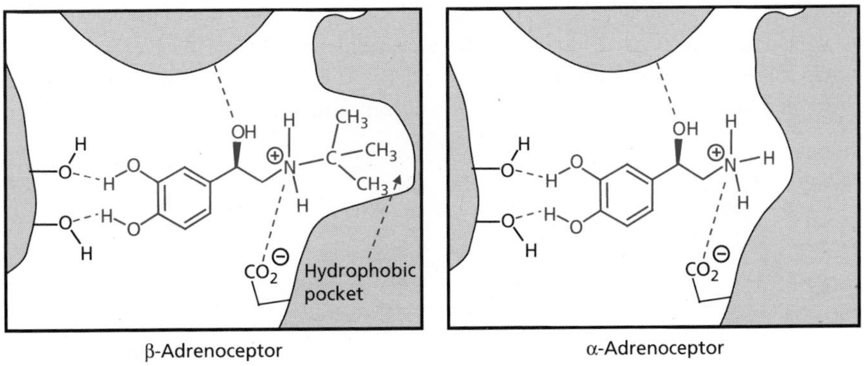

β-Adrenoceptor   α-Adrenoceptor

FIGURE 23.12 Comparison of β- and α-adrenoceptor binding sites.

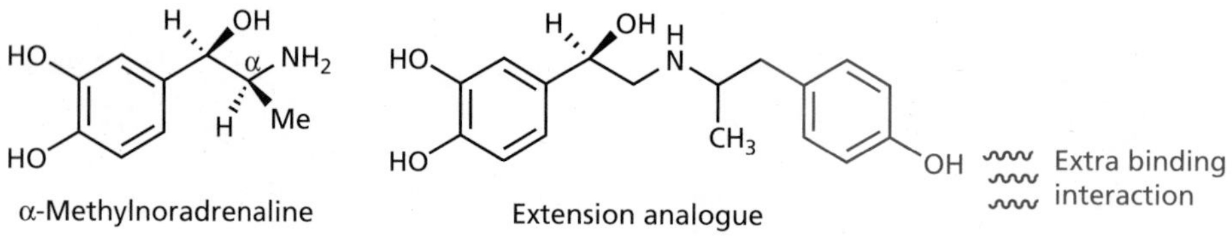

FIGURE 23.13 α-Methylnoradrenaline and extension analogue of noradrenaline.

activity, demonstrating that an extra polar binding region has been accessed which can take part in hydrogen bonding. For example, the activity of the extension analogue shown in Fig. 23.13 is increased by a factor of 800.

23.10 Adrenergic agonists

23.10.1 General adrenergic agonists

Adrenaline itself is an obvious agonist for the overall adrenergic system and it is frequently used in emergency situations such as cardiac arrest or anaphylactic reactions. The latter can be caused by hypersensitivity to certain foodstuffs (e.g. nuts) or foreign chemicals (e.g. a bee sting or penicillin). Individuals who have a high risk of suffering a severe anaphylactic reaction should carry a pre-assembled syringe carrying adrenaline which can be injected intramuscularly (**Anapen** or **Epipen**). Adrenaline is also administered with local anaesthetics in order to constrict blood vessels and prolong local anaesthetic activity at the site of injection.

Adrenaline is fast acting which makes it ideal for emergency situations, but it has a short duration of action and is rapidly cleared from the system. Moreover, it switches on all possible adrenergic receptors, leading to a whole range of side effects, including nausea, tachycardia, arrhythmias, hypertension, palpitations, anxiety, tremor, headache, restlessness, sweating, and dizziness. Therefore, if long-term medication is required, it is preferable to have agonists which are selective for specific adrenoceptors.

Ephedrine (Fig. 23.14) is a natural product present in various plants which have been used in folk medicine for many years. There are two asymmetric centres, and ephedrine exists as a racemate of the *R, S* and *S, R*

stereoisomers. It activates both α- and β-adrenoceptors and has been used extensively in non-prescription preparations as a bronchodilator. It has also been used as a vasopressor and cardiac stimulant. As it lacks the phenolic groups of adrenaline, it is not susceptible to metabolism by catechol *O*-methyltransferase. It is also more lipophilic, and so it can cross the blood–brain barrier and act as a stimulant. Ephedrine is the active constituent of herbal remedies that contain the dried plant material *ma-huang*.

Pseudoephedrine (Fig. 23.14) occurs naturally in certain plant species and is the *S,S* diastereomer of ephedrine. It is used as a nasal decongestant in preparations such as **Sudafed**, **Benylin**, and **Lemsip**. Unfortunately, it can be used in the illicit manufacture of amphetamines and so many pharmaceutical firms are starting to replace it with alternative decongestants.

23.10.2 α$_1$-, α$_2$-, β$_1$-, and β$_3$-Agonists

Compared to the β$_2$-agonists, there are fewer applications for agonists at the α$_1$-, α$_2$-, β$_1$-, and β$_3$-receptors (Fig. 23.15). The β$_1$-agonist **dobutamine** is used to treat cardiogenic shock, while the β$_3$-agonist **mirabegron** was approved in 2012 as a treatment for overactive bladders. **Vibegron** is another β$_3$-agonist which is currently undergoing clinical trials for the treatment of overactive bladders. Activation of β$_3$-receptors in the bladder relaxes smooth muscle and eases the urge to pass urine (see also Box 22.1). β$_3$-Agonists were also developed as potential anti-obesity agents or for the treatment of diabetes. Unfortunately, clinical results proved disappointing.

Agonists acting on both α-adrenoceptors constrict blood vessels, raise blood pressure, and can cause cardiovascular problems. However, α-agonists such as **ephedrine** (Fig. 23.14), **metaraminol** (Fig. 23.34), and **noradrenaline** are useful in treating hypotension.

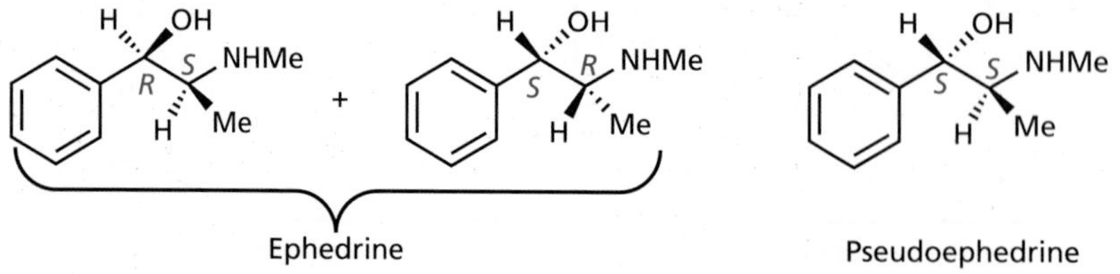

FIGURE 23.14 Ephedrine and pseudoephedrine.

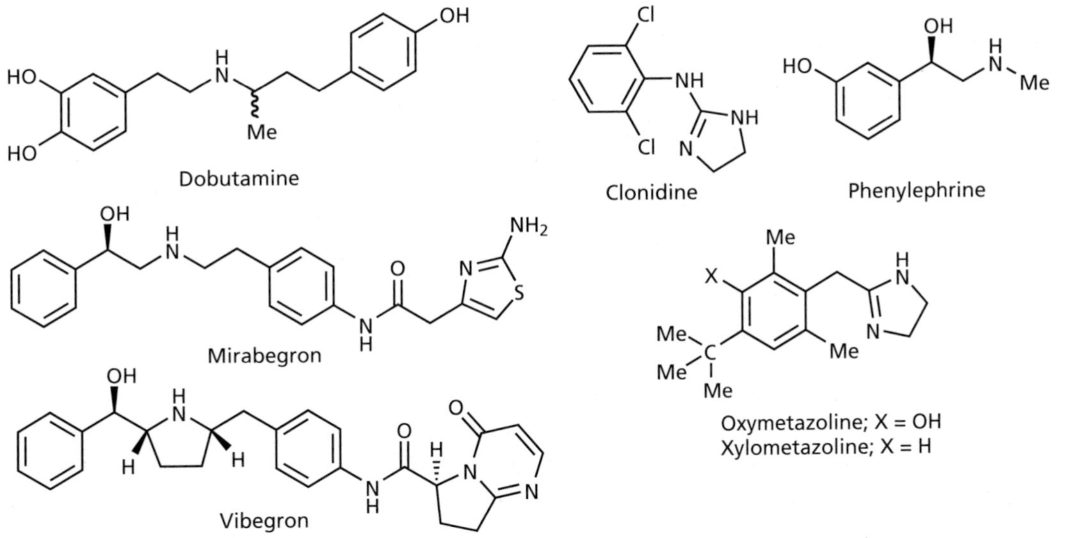

FIGURE 23.15 Adrenergic agonists.

Selective α_1- and α_2-agonists have found a number of uses as described in Box 23.1. **Clonidine** is a selective α_2-agonist which is used for the treatment of hypertension. There is also strong evidence that it acts as an analgesic, especially if it is injected directly into the spinal cord. Selective α_1-agonists, such as **oxymetazoline, xylometazoline**, and **phenylephrine**, act as vasoconstrictors, and are used widely as topical medicines for the treatment of nasal congestion and bloodshot eyes.

23.10.3 β_2-Agonists and the treatment of asthma

The most useful adrenergic agonists in medicine today are the β_2-agonists. These can be used to relax smooth muscle in the uterus to delay premature labour, but they are more commonly used for the treatment of asthma. Activation of the β_2-adrenoceptor results in smooth muscle relaxation and, since β_2-receptors predominate in bronchial smooth muscle, this leads to dilatation of the airways.

Adrenaline is often used to dilate the airways in emergency situations, but it is not suitable for long-term use because of its short duration of action and cardiovascular side effects (section 23.10.1). These side effects result from adrenaline interacting with all available adrenergic receptors, and so a more selective agent for β_2-receptors is preferable.

Isoprenaline (Fig. 23.11) was used for some time as an anti-asthmatic agent and shows some selectivity for β-receptors over α-receptors because of its bulky N-alkyl substituent. However, it produces cardiovascular side effects since it lacks selectivity between the different subtypes of β-receptors and activates β_1-receptors of the heart. The search was then on to find a selective agonist for β_2-receptors which could be inhaled and have a long duration of action. Further research demonstrated that selectivity between different types of β-receptors could be obtained by introducing alkyl substituents to the side chain linking the aromatic ring and the amine, and/or varying the alkyl substituents on the nitrogen. For example, **isoetharine** (Fig. 23.16) was shown to be selective for β_2-receptors. Unfortunately, it was short lasting.

This short duration of action occurs because drugs such as isoetharine and adrenaline are taken up by tissues and methylated by the metabolic enzyme **catechol-O-methyltransferase** (COMT) to form an inactive ether. In order to prevent this, attempts were made to modify the *meta* phenol group and make it more resistant to metabolism (Fig. 23.17). This was no easy task as the phenolic

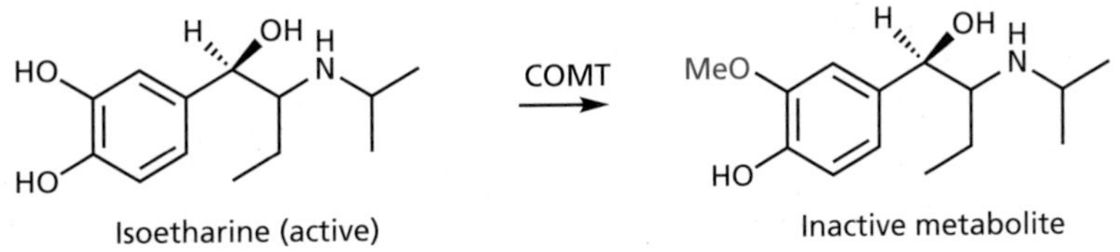

FIGURE 23.16 Metabolism of isoetharine (COMT = catechol O-methyltransferase).

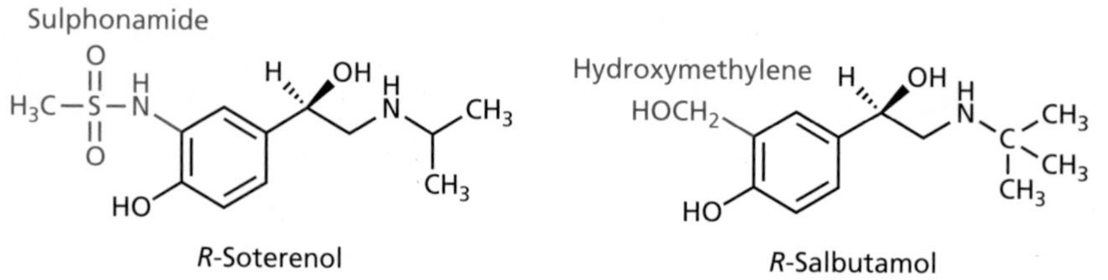

FIGURE 23.17 Selective β_2-agonists.

group is important to activity, so it was necessary to replace it with a group which could still bind to the receptor and retain biological activity, but would not be recognized by the metabolic enzyme.

Various functional groups were tried at the *meta* position with a sulphonamide group (MeSO$_2$NH) proving successful. This resulted in a long-lasting selective β_2-agonist called **soterenol** (Fig. 23.17). However, this compound was never used clinically because a better compound was obtained in **salbutamol** (known as **albuterol** in the USA) (Box 23.2). Here, the *meta* phenol group of the catecholamine skeleton was replaced by a hydroxymethylene group—an example of a *group shift*

strategy (section 14.2.6). Salbutamol has the same potency as isoprenaline, but is 2000 times less active on the heart. It has a duration of 4 hours and is not taken up by transport proteins or metabolized by COMT. Instead, it is more slowly metabolized to a phenolic sulphate. Salbutamol was marketed as a racemate and soon became a market leader in 26 countries for the treatment of asthma. The *R* enantiomer is 68 times more active than the *S* enantiomer. Furthermore, the *S* enantiomer accumulates to a greater extent in the body and produces side effects. Consequently, the pure *R* enantiomer (**levalbuterol**) was eventually marketed—an example of **chiral switching** (section 15.2.1).

BOX 23.2 Synthesis of salbutamol

Salbutamol is an important anti-asthmatic agent that can be synthesized from aspirin. **Fries rearrangement** of aspirin produces a ketoacid which is then esterified. A bromoketone is then prepared which allows the introduction of an amino

group by nucleophilic substitution. The methyl ester and ketone are then reduced, and, finally, the *N*-benzyl protecting group is removed by hydrogenolysis.

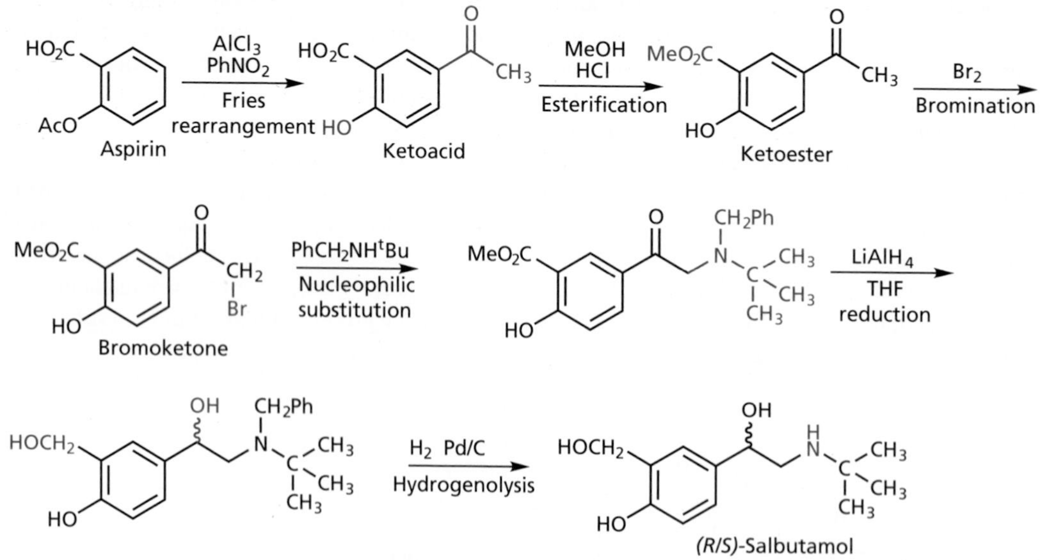

FIGURE 1 Synthesis of salbutamol.

Several analogues of salbutamol have been synthesized to test whether the *meta* CH$_2$OH group could be modified further. These demonstrated the following requirements for the *meta* substituent:

- it has to be capable of taking part in hydrogen bonding—substituents such as MeSO$_2$NHCH$_2$, HCONHCH$_2$, and H$_2$NCONHCH$_2$ permit this;
- substituents with an electron-withdrawing effect on the ring have poor activity (e.g. CO$_2$H);
- bulky *meta* substituents are bad for activity because they prevent the substituent adopting the necessary conformation for hydrogen bonding;
- the CH$_2$OH group can be extended to CH$_2$CH$_2$OH but no further.

Having identified the advantages of a hydroxymethyl group at the *meta* position, attention turned to the *N*-alkyl substituents. Salbutamol itself has a bulky *t*-butyl group. *N*-Arylalkyl substituents were added which would be capable of reaching the polar region of the binding site described earlier (*extension strategy*; section 23.9.2). For example, **salmefamol** (Fig. 23.18) is 1.5 times more active than salbutamol and has a longer duration of action (6 hours). The drug is given by inhalation, but in severe attacks it may be given intravenously.

Further developments were carried out to find a longer lasting agent in order to cope with nocturnal asthma—a condition which usually occurs at about 4 a.m. (commonly called the **morning dip**). It was decided to increase the lipophilicity of the drug because it was believed that a more lipophilic drug would bind more strongly to the tissue in the vicinity of the adrenoceptor and be available to act for a longer period. Increased lipophilicity was achieved by increasing the length of the *N*-substituent with a further hydrocarbon chain and aromatic ring. This led to **salmeterol** (Fig. 23.19), which has twice the potency of salbutamol and an extended action of 12 hours. Another theory for salmeterol's long-lasting activity is

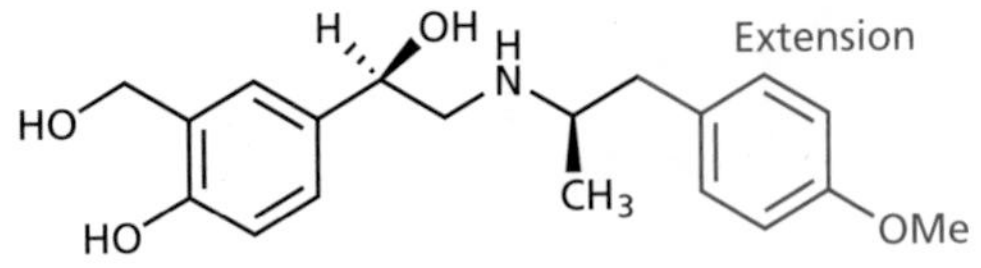

FIGURE 23.18 Salmefamol.

that its hydrophobic groups might be interacting with the adrenergic receptor itself at another binding site called the **exosite**, such that the drug is held in close proximity to the catechol binding region. The ether oxygen atom within the hydrophobic region is important for activity and might be acting as a hydrogen bond acceptor to that exosite.

Formoterol is another agent with long-lasting activity and is taken twice a day.

In recent years, a number of long-lasting β$_2$-agonists that only need to be taken once a day have been approved for the treatment of chronic obstructive pulmonary disease (Fig. 23.20) (see also Box 22.2). **Indacaterol** was approved in Europe in 2009, and in the USA in 2011. The left-hand part of the structure is responsible for adrenergic activity and contains a bicyclic quinolone ring system that has been shown to mimic a catechol ring. In other words, it is a bio-isostere for a catechol ring. The indan ring on the right was introduced as a hydrophobic moiety to prolong activity. The level of hydrophobicity can be fine-tuned by varying the substituents present on the indan ring system, with ethyl groups proving optimal. **Olodaterol** was approved in Europe in 2013, and in the US in 2014. It also contains a hydrophobic region and a bicyclic bio-isostere for the catechol ring. It is interesting to speculate whether the oxygen of the methoxy group serves the same purpose as the ether oxygen present in salmeterol. Placing the phenol group at the *meta* position relative to the main chain improved selectivity for the β$_2$-receptor. **Vilanterol** is similar in structure to salmeterol and was approved in 2013.

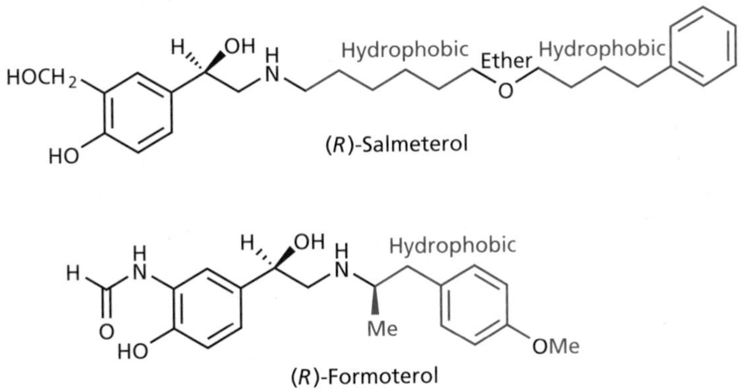

FIGURE 23.19 Salmeterol and formoterol.

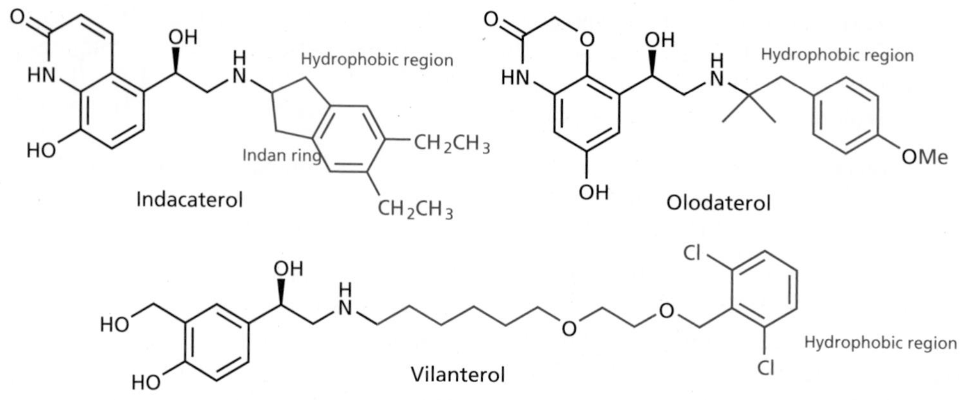

FIGURE 23.20 β_2-Agonists used in the treatment of COPD.

KEY POINTS

- The important binding groups in catecholamines are the two phenolic groups, the aromatic ring, the secondary alcohol, and the ionized amine.

- Placing a bulky alkyl group on the amine leads to selectivity for β-receptors over α-receptors.

- Extending the *N*-alkyl substituent to include a hydrogen-bonding group increases affinity for β-receptors.

- Agents which are selective for β_2-adrenoceptors are useful anti-asthmatic agents.

- Early β_2-agonists were metabolized by catechol-*O*-methyltransferase. Replacing the susceptible phenol group with a hydroxymethylene group prevented metabolism while retaining receptor interactions.

- Longer-lasting anti-asthmatics have been obtained by increasing the lipophilic character of the compounds.

23.11 **Adrenergic receptor antagonists**

23.11.1 **General α/β-blockers**

Carvedilol and **labetalol** are agents which act as antagonists at both the α- and β-adrenoceptors (Fig. 23.21). They have both been used as antihypertensives. Carvedilol has also been used to treat cardiac failure.

23.11.2 α-**Blockers**

Selective α_1-antagonists have been used to treat hypertension or to control urinary output. **Prazosin** (Fig. 23.22) was the first α_1-selective antagonist to be used for the treatment of hypertension, but it is short acting. Longer lasting drugs such as **doxazosin** and **terazosin** are better because

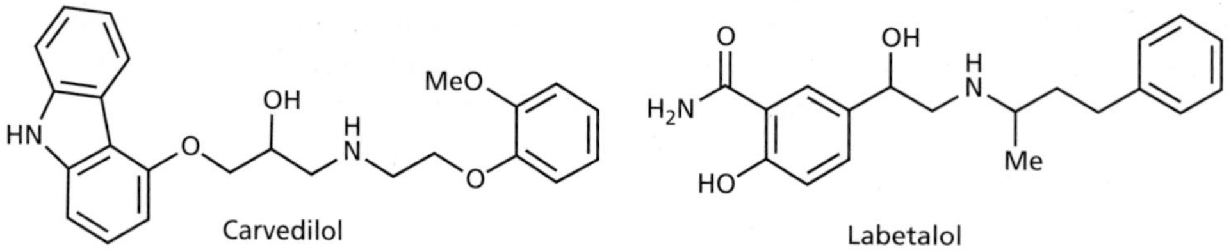

FIGURE 23.21 General α/β-blockers.

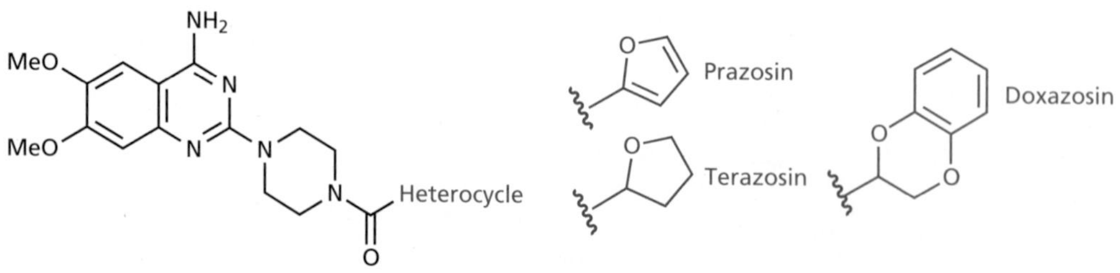

FIGURE 23.22 α_1-Selective antagonists.

they are given as once-daily doses. These agents relieve hypertension by blocking the actions of noradrenaline or adrenaline at the α_1-receptors of smooth muscle in blood vessels. This results in relaxation of the smooth muscle and dilatation of the blood vessels, leading to a lowering in blood pressure. These drugs have also been used for the treatment of patients with an enlarged prostate—a condition known as **benign prostatic hyperplasia**. The enlarged prostate puts pressure on the urinary tract and it becomes difficult to pass urine. The α_1-blockers prevent activation of the α_1-adrenoceptors that are responsible for smooth muscle contraction of the prostate gland, prostate urethra, and the neck of the bladder. This leads to smooth muscle relaxation at these areas, reducing the pressure on the urinary tract and helping the flow of urine. The agents are not a cure for the problem, but they relieve the symptoms.

α_2-Antagonists are being considered as antidepressants. Depression is associated with decreased release of noradrenaline and serotonin in the CNS, and antidepressants work by increasing the levels of one or both of these neurotransmitters. It may sound odd then, to consider an adrenergic antagonist as an antidepressant agent, but it makes sense when it is appreciated that the α_2-receptors are presynaptic adrenergic receptors or **autoreceptors** (section 23.6.3). Activation of these results in a decrease of noradrenaline released from the neuron, so blocking the autoreceptor will actually increase noradrenaline levels.

Mirtazepine (Fig. 23.23) is an antidepressant agent which blocks the α_2-receptor and increases the amount of noradrenaline released. α_2-Receptors also control the release of serotonin from serotonin nerve terminals, and so mirtazepine increases serotonin levels as well. It is not known for certain whether the antidepressant activity observed is due to increased noradrenaline levels or serotonin levels or both. Current work is looking at the design of dual-action drugs which include the ability to block α_2-adrenoceptors (Case study 7).

Older antidepressants that are designed to increase noradrenaline and serotonin levels by different mechanisms can take 2–6 weeks before they have an effect. This delay in action is due to feedback control involving the α_2-receptors. When taken initially, the drugs certainly cause noradrenaline levels to increase, but feedback control counteracts this effect. It is only when the presynaptic

receptors become desensitized that neurotransmitter levels increase sufficiently to have a clinical effect.

23.11.3 β-**Blockers as cardiovascular drugs**

23.11.3.1 First-generation β-blockers

The most important adrenergic antagonists used in medicine today are the β-blockers, which were originally designed to act as antagonists at the β_1-receptors of the heart.

The first goal in the development of these agents was to achieve selectivity for β-receptors over α-receptors. **Isoprenaline** (Fig. 23.24) was chosen as the lead compound. Although this is an agonist, it is active at β-receptors and not α-receptors. Therefore, the goal was to take advantage of this inherent specificity and modify the molecule to convert it from an agonist to an antagonist.

The phenolic groups are important for agonist activity, but this does not necessarily mean that they are essential for antagonist activity as antagonists can often block receptors by binding in a different way. Therefore, one of the early experiments was to replace the phenol groups with other substituents. Replacing the phenolic groups of isoprenaline with chloro substituents produced **dichloroisoprenaline** (Fig. 23.24). This compound is a partial agonist. In other words, it has some agonist activity, but it is weaker than a pure agonist. Nevertheless, dichloroisoprenaline blocks natural chemical messengers from binding and can therefore be viewed as an antagonist because it lowers adrenergic activity.

The next stage was to try to remove the partial agonist activity. A common method of converting an agonist into an antagonist is to add an extra aromatic ring. This can sometimes result in an extra hydrophobic interaction with the receptor which is not involved when the

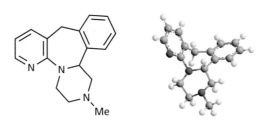

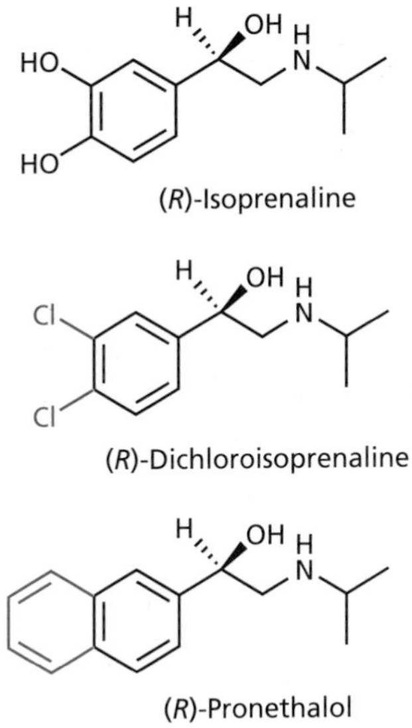

(R)-Isoprenaline

(R)-Dichloroisoprenaline

(R)-Pronethalol

FIGURE 23.24 Partial β-agonists.

FIGURE 23.23 Mirtazepine.

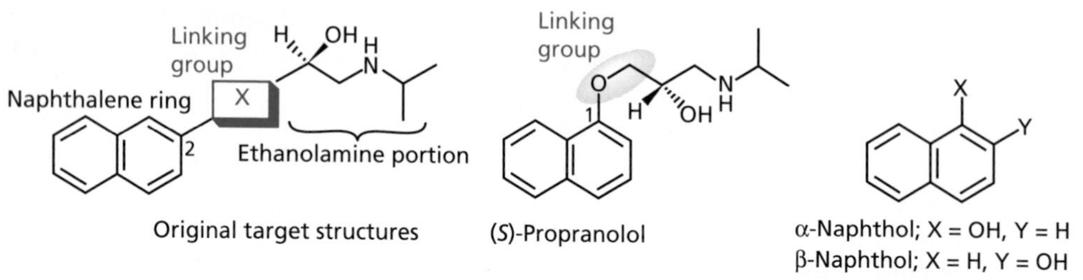

FIGURE 23.25 Chain extension tactics and the discovery of propranolol.

agonist binds. This, in turn, means a different induced fit between the ligand and the binding site, such that the ligand binds without activating the receptor. Therefore, the chloro groups of dichloroisoprenaline were replaced by an extra benzene ring to give a naphthalene ring system. The product obtained (**pronethalol**; Fig. 23.24) was still a partial agonist, but was the first β-blocker to be used clinically for angina, arrhythmia, and high blood pressure.

Further research was carried out to see what effect extending the length of the chain between the aromatic ring and the amine would have. One of these projects involved the introduction of various linking groups between the naphthalene ring and the ethanolamine portion of the molecule (Fig. 23.25). At this stage, a chance event occurred. The researchers wanted to use β-naphthol as a starting material in order to introduce a linking group of X = O–CH$_2$ (Fig. 23.25). However, the stores had run out of the reagent and so α-naphthol was used instead to prepare the structure now known as **propranolol** (Fig. 23.25). In this structure, the chain was at the 1-position of the naphthalene ring rather than the 2-position, and nobody expected it to be active. To everyone's astonishment, propranolol was found to be a pure antagonist, having 10–20 times greater activity than pronethalol. It was

introduced into the clinic for the treatment of angina and is now the benchmark against which all β-blockers are rated. Its contribution to medicine was so significant that its inventor James Black got the Nobel Prize in 1988. The S-enantiomer is the active form although propranolol is used clinically as a racemate. When the original target structure from β-naphthol was eventually synthesized, it was found to have similar properties to pronethalol.

23.11.3.2 Structure–activity relationships of aryloxypropanolamines

Propranolol is an example of an aryloxypropanolamine structure (see Box 23.3). A large number of aryloxypropanolamines have been synthesized and tested, demonstrating the following SAR (Fig. 23.26):

- branched bulky N-alkyl substituents such as isopropyl and t-butyl groups are good for β-antagonist activity, suggesting an interaction with a hydrophobic pocket in the binding site (compare β-agonists);

- variation of the aromatic ring system is possible and heteroaromatic rings can be introduced, such as those in pindolol and timolol (Fig. 23.27);

BOX 23.3 Synthesis of aryloxypropanolamines

Propranolol is a first-generation β-blocker and acts as an antagonist at β-adrenoceptors. The synthesis of propranolol is relatively simple and can be easily adapted to produce a large number of analogues. A phenol is reacted with 2-chloromethyloxirane such that nucleophilic substitution of the alkyl chloride takes place. The resulting product is then treated with an amine to ring-open the epoxide. This introduces the

amine and generates the secondary alcohol at the same time. Because of the nature of the synthetic route, a huge variety of phenols and amines can be used to produce different analogues. There is an asymmetric centre in the final product, but it is only possible to synthesize the racemate using this route. A different and more expensive route would have to be used to synthesize only one of the specific enantiomers.

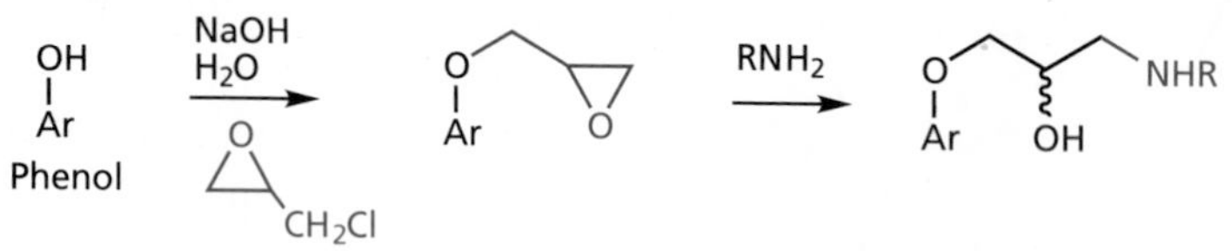

FIGURE 1 Synthesis of aryloxypropanolamines.

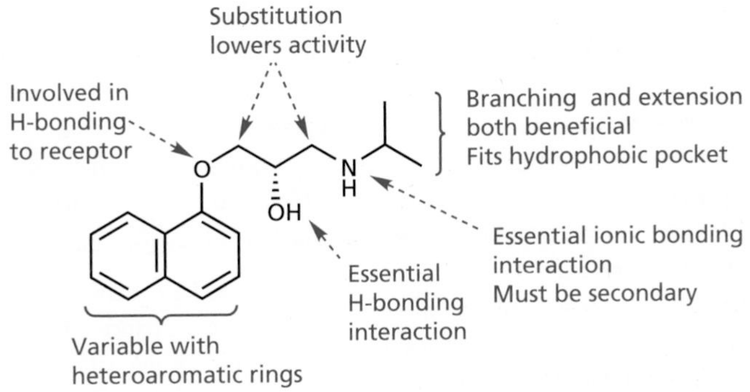

FIGURE 23.26 Structure–activity relationships of aryloxypropanolamines.

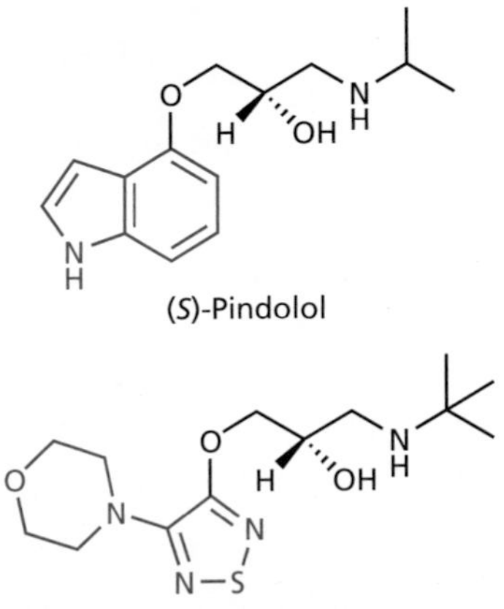

(S)-Pindolol

(S)-Timolol

FIGURE 23.27 β_1-Antagonists containing heteroaromatic ring systems.

NHCOCH$_3$

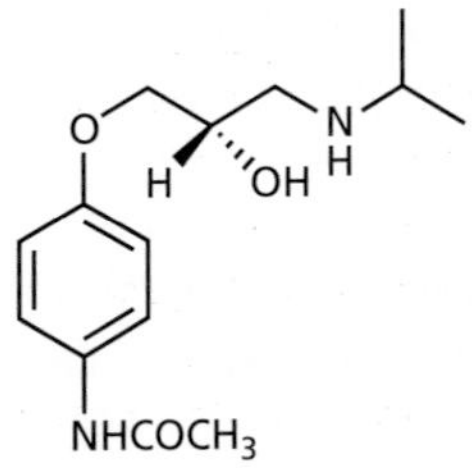

FIGURE 23.28 (S)-Practolol.

- substitution on the side chain methylene group increases metabolic stability but lowers activity;

- the alcohol group on the side chain is essential for activity;

- replacing the ether oxygen on the side chain with S, CH$_2$, or NMe is detrimental, although a tissue-selective β-blocker has been obtained replacing O with NH;

- N-alkyl substituents longer than isopropyl or t-butyl are less effective (but see next point);

- adding an N-arylethyl group such as –CHMe$_2$–CH$_2$Ph or –CHMe–CH$_2$Ph is beneficial (*extension*);

- the amine must be secondary.

23.11.3.3 Selective β$_1$-blockers (second-generation β-blockers)

Propranolol is a non-selective β-antagonist which acts as an antagonist at β$_2$-receptors, as well as β$_1$-receptors. Normally, this is not a problem, but it is serious if the patient is asthmatic since the propranolol could initiate an asthmatic attack by antagonizing the β$_2$-receptors in bronchial smooth muscle. This leads to contraction of bronchial smooth muscle and closure of the airways.

Practolol (Fig. 23.28) is not as potent as propranolol, but it is a selective cardiac β$_1$-antagonist which does not block vascular or bronchial β$_2$-receptors. It is much safer for asthmatic patients and, because it is more polar than propranolol, it has much fewer CNS effects.

Practolol was marketed as the first cardioselective β$_1$-blocker for the treatment of angina and hypertension, but after a few years it had to be withdrawn because of unexpected, but serious, side effects in a very small number of patients. These side effects included skin rashes, eye problems, and peritonitis.

Further investigations were carried out and it was demonstrated that the amido group had to be in the *para* position of the aromatic ring rather than the *ortho* or *meta* positions if the structure was to retain selectivity for the cardiac β$_1$-receptors. This implied that there was an extra hydrogen bonding interaction taking place with β$_1$-receptors (Fig. 23.29) which was not taking place with β$_2$-receptors.

Replacement of the acetamido group with other groups capable of hydrogen bonding led to a series of cardioselective β$_1$-blockers which included **acebutolol**, **atenolol**, **metoprolol**, and **betaxolol** (Fig. 23.30).

23.11.3.4 Short-acting β-blockers

Most clinically useful β-blockers should have a reasonably long duration of action such that they need only be

BOX 23.4 Clinical aspects of β-blockers

β-Blockers are used for the treatment of angina, myocardial infarction, arrhythmias, and hypertension. The effects of **propranolol** and other first-generation β-blockers depend on how active the patient is. At rest, propranolol causes little change in heart rate, output, or blood pressure. On the other hand, if the patient exercises or becomes excited, propranolol reduces the resulting effects of circulating adrenaline. The β-blockers were originally intended for use in angina, but they also had an unexpected antihypertensive activity (i.e. they lowered blood pressure). Indeed, the β-blockers are now more commonly used as antihypertensives rather than for the treatment of angina. The antihypertensive activity arises from the following effects:

- action at the heart to reduce cardiac output;
- action at the kidneys to reduce renin release; renin catalyses formation of angiotensin I, which is quickly converted to angiotensin II—a potent vasoconstrictor (Case study 2 and section 26.3);
- action in the CNS to lower the overall activity of the sympathetic nervous system.

These effects override the fact that β-blockers block the β-receptors on blood vessels and would normally cause vasoconstriction.

First-generation β-blockers have various side effects such as the following:

- bronchoconstriction in asthmatics—this is a dangerous side effect and the β-blockers are not recommended for patients with asthma;
- fatigue and tiredness of limbs due to reduced cardiac output;
- CNS effects (dizziness, nightmares, and sedation), especially with lipophilic β-blockers, such as **propranolol**, **pindolol**, and

oxprenolol, all of which can cross the blood–brain barrier. More water-soluble agents, such as **nadolol** (Fig. 1), are less likely to have such side effects;

- coldness of the extremities;
- heart failure for patients on the verge of a heart attack—the β-blockers produce a fall in the resting heart rate and this may push some patients over the threshold;
- inhibition of noradrenaline release at synapses.

The second-generation β-blockers are more cardioselective and have fewer side effects. However, they still have some effect on bronchial smooth muscle, and so they should only be used on asthmatic patients when there is no alternative treatment. Water-soluble β-blockers such as **atenolol** are less likely to enter the brain, and so there is less risk of sleep disturbance or nightmares. β-Blockers which act as partial agonists (e.g. **acebutolol**) tend to cause less bradycardia and may also cause less coldness of the extremities. **Esmolol** is a short-acting β-blocker with a rapid onset of action. It is administered by slow intravenous injection during surgical procedures in order to treat any tachycardia (rapid heart rates) that might occur.

β-Blockers have a range of other clinical uses apart from cardiovascular medicine. They are used to counteract overproduction of catecholamines resulting from an enlarged thyroid gland or tumours of the adrenal gland. They can also be used to alleviate the trauma of alcohol and drug withdrawal, as well as relieving the stress associated with situations such as exams, public speaking, and public performances. There are some studies which suggest that propranolol might be a useful treatment for post-traumatic stress disorder and for the removal of traumatic memories. **Timolol** and **betaxolol** are used in the treatment of glaucoma (although their mechanism of action is not clear), while propranolol is used to treat anxiety and migraine.

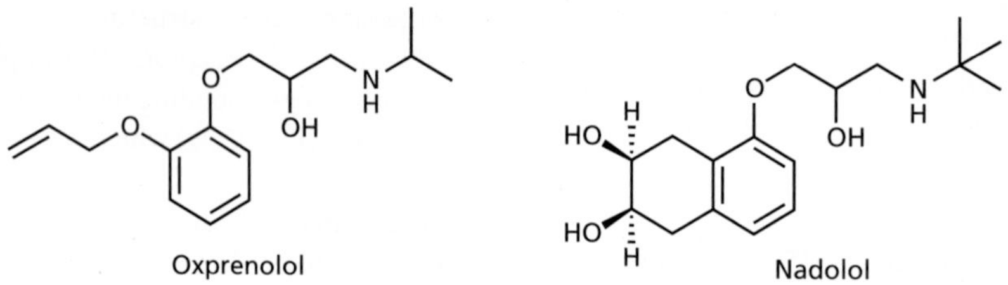

FIGURE 1 Oxprenolol and nadolol.

OH
HN
CH3
O
H
X
H3C
N
H
H2
para Substitution
Extra H-bonding interaction
meta Substitution

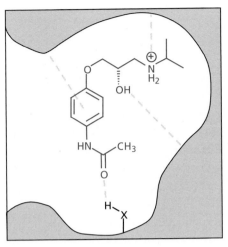

FIGURE 23.29 Binding interactions of antagonists with β_1-receptors.

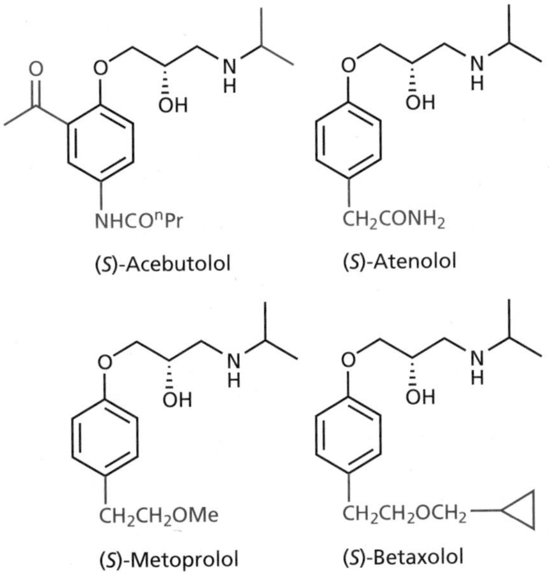

NHCO^{n}Pr
(*S*)-Acebutolol
CH$_2$CONH$_2$
(*S*)-Atenolol
CH$_2$CH$_2$OMe
(*S*)-Metoprolol
CH$_2$CH$_2$OCH$_2$
(*S*)-Betaxolol

FIGURE 23.30 Second-generation β-blockers.

taken once or twice a day. However, there is an advantage in having a very short-acting agent with a half-life measured in minutes rather than hours, because they can be administered during surgical procedures to treat any cardiac problems that may arise during the operation. **Esmolol** (Fig. 23.31) is one such agent. It has a rapid onset of action and is administered if the heart starts to beat too rapidly. Because it is a short-acting agent, its actions are quickly reversed once administration has been stopped.

Practolol was the lead compound used in the development of esmolol. The amide group was replaced with an ester, with the expectation that the ester would act as a bio-isostere for the amide. Moreover, it was anticipated that the ester group would prove susceptible to esterase enzymes and be rapidly hydrolysed to an inactive metabolite. The aryl ester was indeed active as a β-blocker, but was not hydrolysed rapidly enough to be clinically useful. It was concluded that the aromatic ring was acting as a steric shield to the esterase enzymes, and so linker

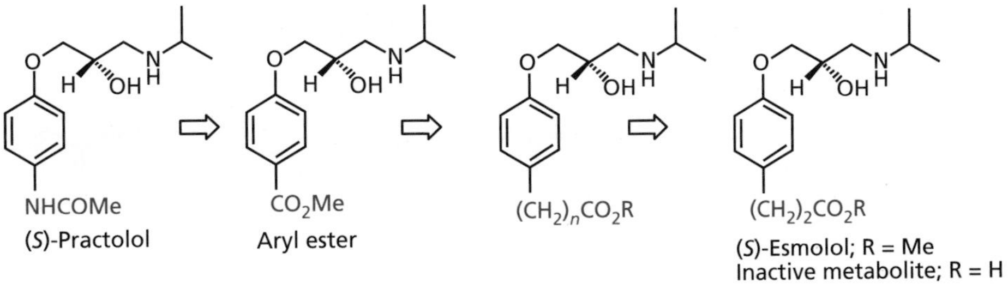

NHCOMe
(*S*)-Practolol
CO$_2$Me
Aryl ester
(CH$_2$)$_n$CO$_2$R
(CH$_2$)$_2$CO$_2$R
(*S*)-Esmolol; R = Me
Inactive metabolite; R = H

FIGURE 23.31 Development of short-acting β-blockers.

chains were inserted between the aromatic ring and the ester group to make the ester more 'exposed'. An ethylene linker proved ideal resulting in the discovery of esmolol. The structure is slightly more potent than practolol, and is significantly more cardioselective. Once administration has been stopped, it takes 12 minutes to reach 80% recovery and 20 minutes to reach full recovery. The inactive carboxylic acid metabolite that is formed is rapidly conjugated and excreted.

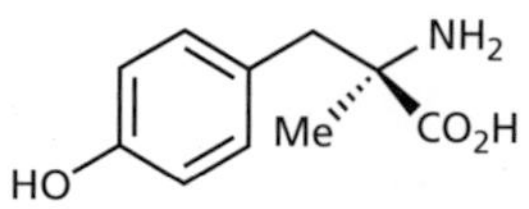

FIGURE 23.32 α-Methyltyrosine.

KEY POINTS

- Antagonists of β-adrenoceptors are known as β-blockers.
- Replacing the catechol ring with a naphthalene ring changes an agonist into a partial agonist.
- Variation of the linking group between naphthalene and the ethanolamine moiety resulted in the first β-antagonists.
- SAR of aryloxypropanolamines reveal the importance of the ionized amine, the side chain alcohol, and the ether linkage. Substituents on the nitrogen can be varied. The naphthalene ring can be replaced by various heterocyclic rings.
- First-generation β-blockers inhibit all β-receptors and can induce asthma in susceptible patients.
- Second-generation β-blockers show selectivity for β_1-receptors over β_2-receptors. Aryloxypropanolamines bearing a hydrogen-bonding group at the *para* position of an aromatic ring show β_1-selectivity.
- Third-generation β-blockers bear an extended *N*-substituent, which includes a hydrogen-bonding group capable of an extra interaction with the β_1-adrenoceptor.

23.12 Other drugs affecting adrenergic transmission

In the previous sections, we discussed drugs which act as agonists or antagonists at adrenergic receptors. However, there are various other drug targets involved in the adrenergic transmission process which are important in controlling adrenergic activity. In this section, we briefly cover some of the most important aspects of these.

23.12.1 Drugs that affect the biosynthesis of adrenergics

In section 23.4, we identified **tyrosine hydroxylase** as the regulatory enzyme for catecholamine biosynthesis. This makes it a potential drug target. For example, **α-methyltyrosine** (Fig. 23.32) inhibits tyrosine hydroxylase and is sometimes used clinically to treat tumour cells which overproduce catecholamines.

It is sometimes possible to 'fool' the enzymes of a biosynthetic process into accepting an unnatural substrate such that a false transmitter is produced and stored in the storage vesicles. For example, **α-methyldopa** is converted and stored in vesicles as **α-methylnoradrenaline** (Fig. 23.33), where it displaces noradrenaline. Such false transmitters are less active than noradrenaline, so this is another way of down-regulating the adrenergic system. The drug has serious side effects, however, and is limited to the treatment of hypertension in late pregnancy.

A similar example is the use of **α-methyl-*m*-tyrosine** in the treatment of shock. This unnatural amino acid is accepted by the enzymes of the biosynthetic pathway and converted to **metaraminol** (Fig. 23.34).

23.12.2 Drugs inhibiting the uptake of noradrenaline into storage vesicles

The uptake of noradrenaline into storage vesicles can be inhibited by drugs. The natural product **reserpine** binds to the transport protein responsible for transporting noradrenaline into the vesicles, and so noradrenaline accumulates in the cytoplasm where it is metabolized by MAO. As noradrenaline levels drop, adrenergic activity drops. Reserpine was once prescribed as an antihypertensive agent, but it has serious side effects (e.g. depression). Therefore, it is no longer used.

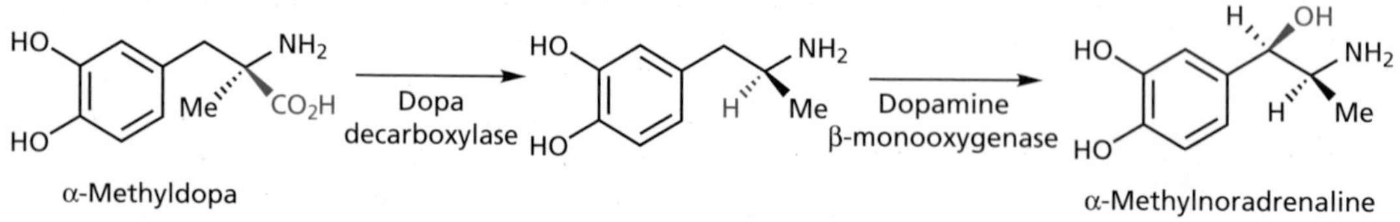

FIGURE 23.33 A false transmitter—α-methylnoradrenaline.

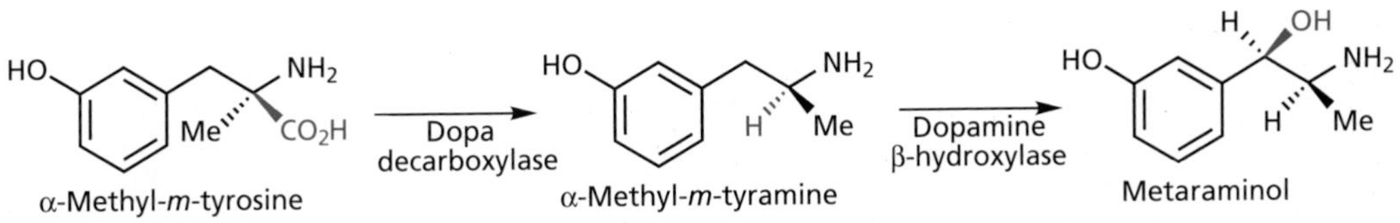

The chemical structures for α-Methyl-*m*-tyrosine, α-Methyl-*m*-tyramine, and Metaraminol with the enzymatic transformations.

FIGURE 23.34 A false transmitter—metaraminol.

23.12.3 Release of noradrenaline from storage vesicles

The storage vesicles are also the targets for the drugs **guanethidine** and **bretylium** (Fig. 23.35). Guanethidine is taken up into presynaptic neurons and storage vesicles by the same transport proteins as noradrenaline, and it displaces noradrenaline in the same way as reserpine. The drug also prevents exocytosis of the vesicle and so prevents release of the vesicle's contents into the synaptic gap. Guanethidine is an effective antihypertensive agent, but it is no longer used in the clinic because of side effects resulting from non-specific inhibition of adrenergic nerve transmission. Bretylium works in the same way as guanethidine and is sometimes used to treat irregular heart rhythms.

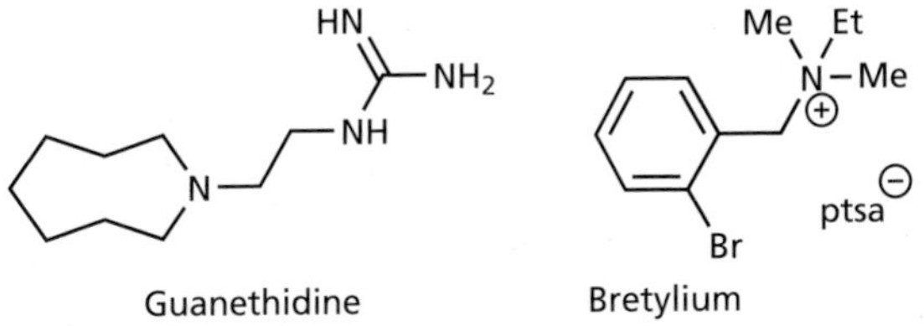

23.12.4 Reuptake inhibitors of noradrenaline into presynaptic neurons

Once noradrenaline has interacted with its receptor, it is normally taken back into the presynaptic neuron by a transport protein. This transport protein is an important target for various drugs which inhibit noradrenaline uptake and thus prolong adrenergic activity. The tricyclic

The chemical structures of Guanethidine and Bretylium.

FIGURE 23.35 Agents that affect adrenergic activity (ptsa = *para*-toluenesulphonate).

antidepressants **desipramine**, **imipramine**, and **amitriptyline** (Fig. 23.36) work in this fashion in the CNS and were the principal treatment for depression from the 1960s to the 1980s.

It has been proposed that the tricyclic antidepressants (TCAs) are able to act as inhibitors because they are partly superimposable on noradrenaline. This can be seen in Fig. 23.37 where the aromatic ring and the nitrogen atoms of noradrenaline are overlaid with the nitrogen atom and one of the aromatic rings of desipramine.

Test your understanding and practice your molecular modelling with Exercise 23.2 on the Online Resource Centre at www.oxfordtextbooks.co.uk/orc/patrick6e/

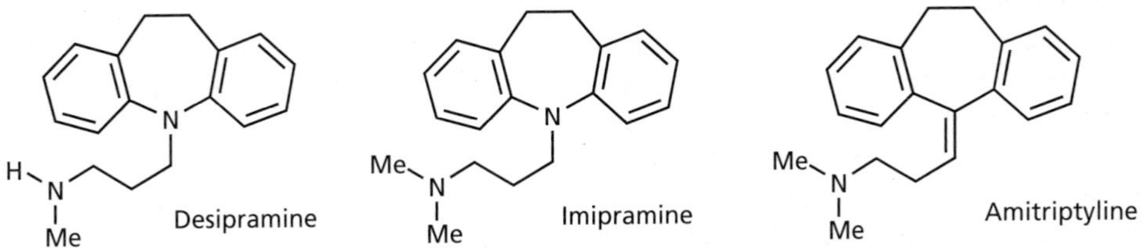

The chemical structures of Desipramine, Imipramine, and Amitriptyline.

FIGURE 23.36 Tricyclic antidepressants.

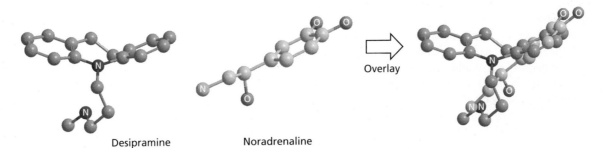

FIGURE 23.37 Overlay of desipramine and noradrenaline.

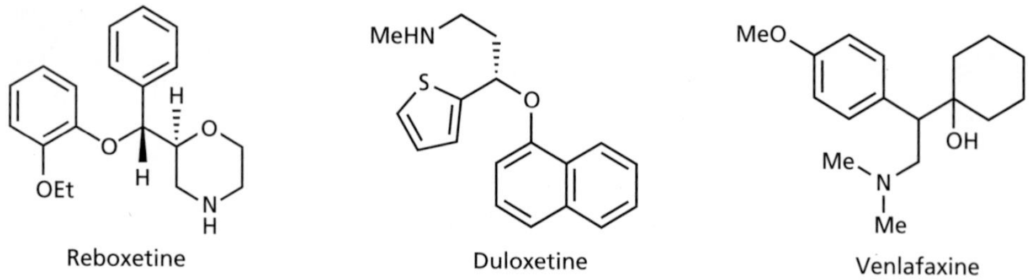

FIGURE 23.38 Reuptake inhibitors.

Note that the tricyclic system of desipramine is V-shaped, so that when the molecules are overlaid the second aromatic ring is held above the plane of the noradrenaline structure. Planar tricyclic structures would be expected to be less active as inhibitors, because the second aromatic ring would then occupy the space required for the amine nitrogen.

Unfortunately, the TCAs are not selective and interact with a variety of other targets, such as the reuptake protein for serotonin, the sodium and calcium ion channels in the heart, and the receptors for histamine, acetylcholine, and noradrenaline (mainly H_1, M_1, and α_1 respectively). Blockage of the transport protein for serotonin is beneficial to antidepressant activity, but interaction with ion channels and receptors results in various side effects including cardiotoxicity. Those agents containing tertiary amines (e.g. imipramine and amitriptyline) have the greatest side effects on the cholinergic system.

Newer antidepressant agents with better selectivity have now been developed and are termed **selective noradrenaline reuptake inhibitors** (SNRIs). **Reboxetine** (Fig. 23.38) is one such example and was marketed in 2003. It selectively inhibits noradrenaline uptake and has no appreciable action on cholinergic or α_1-adrenergic receptors. It also rapidly desensitizes presynaptic α_2-adrenergic receptors, which further enhances its activity and speeds up its onset of action. Dual noradrenaline and serotonin reuptake inhibitors such as **duloxetine** and **venlafaxine** (Fig. 23.38) are clinical agents which block the transport proteins for both noradrenaline and serotonin, but are more selective than the classical TCAs.

Bupropion (**Zyban**; Fig. 23.39) inhibits the reuptake of both noradrenaline and dopamine. It has been used for the treatment of depression, and as an aid to giving up smoking (see also section 22.10.2.5). It is also being considered for the treatment of obesity in combination with the opioid antagonist naltrexone. This represents a massive market. The World Health Organization stated that 1.9 billion adults were overweight in 2014, with 600 million of those classified as obese.

Stimulants acting as noradrenaline reuptake inhibitors have been used for the treatment of **attention deficit hyperactivity disorder**. This is the most commonly diagnosed childhood behavioural disorder and is associated with inattention, hyperactivity, and impulsivity. **Methylphenidate** (**Ritalin**; Fig. 23.39) is the most commonly prescribed medication for this disorder, while **atomoxetine** (Fig. 23.39) was approved in 2002. Both agents lead to increased levels of noradrenaline and dopamine in the brain.

Cocaine also inhibits noradrenaline uptake when it is chewed from coca leaves, but this time the inhibition is in the peripheral nervous system rather than the CNS. Chewing coca leaves was well known to the Incas as a means of increasing endurance and suppressing hunger, and they would chew the leaves whenever they were faced with situations requiring long periods of physical effort or stamina. When coca leaves are chewed, cocaine is absorbed into the systemic blood supply and acts predominantly on peripheral adrenergic receptors to increase adrenergic activity. Nowadays, cocaine abusers prefer to smoke or snort the drug, which allows it to enter the CNS more efficiently. There, it inhibits the uptake of dopamine rather than noradrenaline, resulting in its CNS effects.

Some amines such as **tyramine**, **amphetamine**, and **ephedrine** (Figs. 23.10 and 23.14) closely resemble noradrenaline in structure and are transported into the nerve cell by noradrenaline's transport proteins. Once in the cell, they are taken up into the vesicles. Because these amines are competing with noradrenaline for transport proteins, noradrenaline is more slowly reabsorbed into the nerve

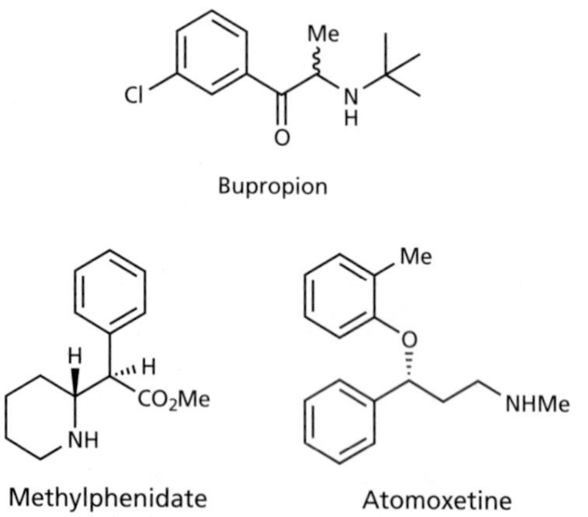

FIGURE 23.39 Adrenergic agents acting in the central nervous system.

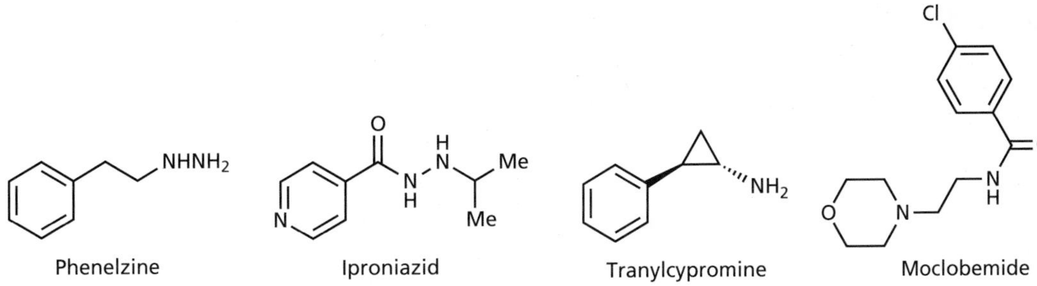

FIGURE 23.40 Monoamine oxidase inhibitors.

cells. Moreover, as the foreign amines are transported into the nerve cell, noradrenaline is transported out by those same transport proteins. Both of these facts means that more noradrenaline is available to interact with its receptors. Therefore, amphetamines and similar amines have an indirect agonist effect on the adrenergic system.

Phentermine (Fig. 23.10) is very similar in structure to amphetamine and causes increased levels of adrenaline and noradrenaline that result in hunger suppression. Consequently, it was approved in 1959 to suppress the appetite of obese patients. A combination of phentermine with the anticonvulsant and antimigraine drug **topiramate** is currently being considered as a treatment for obesity.

23.12.5 Inhibition of metabolic enzymes

Inhibition of the enzymes responsible for the metabolism of noradrenaline should prolong noradrenaline activity. We have seen how amines such as tyramine, amphetamine, and ephedrine inhibit the reuptake of noradrenaline into the presynaptic neuron. These amines also inhibit MAO, one of the important enzymes involved in the metabolism of noradrenaline. This, in turn, leads to a build-up in noradrenaline levels and an increase in adrenergic activity.

Monoamine oxidase inhibitors (**MAOIs**) such as **phenelzine**, **iproniazid**, and **tranylcypromine** (Fig. 23.40) have been used clinically as antidepressants, but other classes of compound such as the **tricyclic antidepressants** and **selective serotonin reuptake inhibitors** (SSRIs) are now more favoured as they have fewer side effects. It is important to realize that the MAOIs affect the levels of all neurotransmitters that are normally metabolized by these enzymes—in particular, noradrenaline, dopamine, and serotonin. As a result of these widespread effects, it is difficult to be sure what mechanism is most involved in the antidepressant activity of these agents.

Another serious problem associated with MAOIs is their interaction with other drugs and food. A well-known example of this is the **cheese reaction**. Ripe cheese contains **tyramine** which is normally metabolized by MAOs in the gut wall and the liver, and so never enters the systemic circulation. If the MAOs are inhibited

by MAOIs, tyramine is free to circulate round the body, enhancing the adrenergic system and leading to acute hypertension and severe headaches.

Better agents such as **moclobemide** (Fig. 23.40) have been designed to act selectively on one of the isozymes of MAO (MAO-A; Box 7.4). They have also been designed to be reversible rather than irreversible in their action. This has the advantage that high levels of ingested tyramine will displace the inhibitor from MAO-A in the gut, allowing the enzyme to metabolize tyramine and prevent the high blood levels that would lead to toxic effects.

In recent years, there has been interest in using MAO inhibitors as part of the treatment for Alzheimer's disease, as blocking MAO would lower the levels of free radical species present in the brain. A hybrid molecule with the ability to inhibit MAO and the cholinesterase enzymes has reached clinical trials (sections 13.3.14 and 22.15).

KEY POINTS

- Inhibitors of catecholamine biosynthesis affect adrenergic activity.

- Drugs that are similar to tyrosine may be converted by the catecholamine biosynthetic pathway to structures that act as false transmitters and lower adrenergic activity.

- The uptake and release of noradrenaline from storage vesicles can be inhibited by certain drugs.

- The tricyclic antidepressants inhibit the reuptake of noradrenaline into presynaptic neurons by blocking transport proteins. Adrenergic activity is increased in the CNS.

- Cocaine increases peripheral adrenergic activity by blocking noradrenaline reuptake. In the CNS it inhibits the reuptake of dopamine.

- Amphetamines compete with noradrenaline for the transport proteins responsible for transporting noradrenaline back into the presynaptic neuron. Adrenergic activity is increased in the CNS.

- Monoamine oxidase inhibitors inhibit the metabolic enzyme monoamine oxidase and result in increased levels of noradrenaline and other catecholamines.

QUESTIONS

1. How would you synthesize the following structures to test their adrenergic agonist activity?

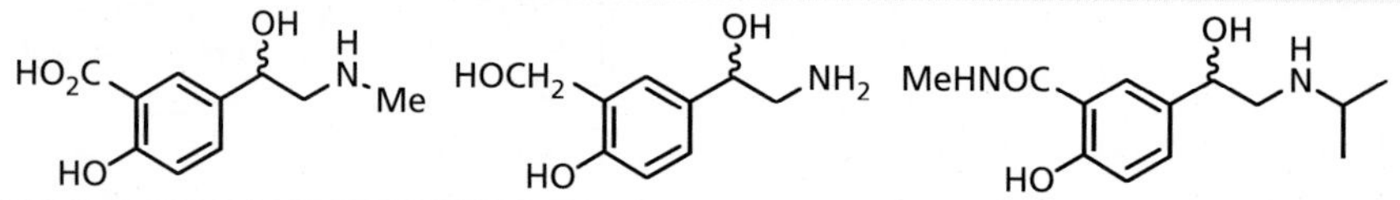

2. Suggest how you might synthesize the adrenergic antagonist, pindolol (Fig. 23.27).

3. Suggest whether the structures in the box below are likely to have good or bad activity as β-blockers.

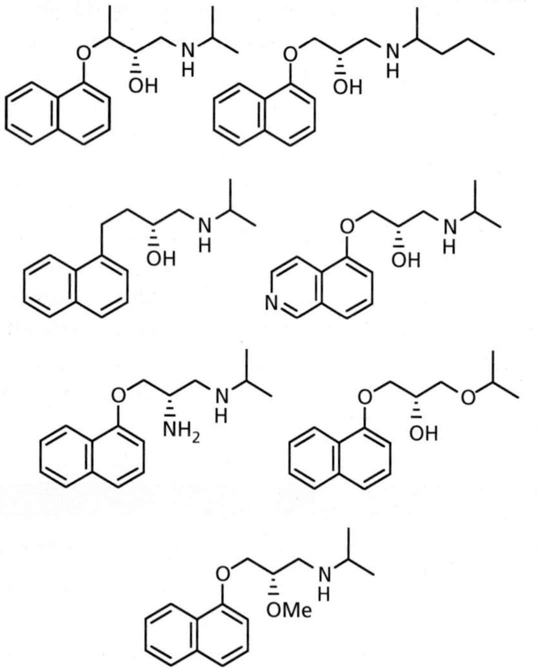

4. The catechol system is important for the binding of adrenergic agonists, yet is not required for adrenergic antagonists. Why should this be the case?

5. How would α-substitution affect the metabolism of adrenergic agents and why?

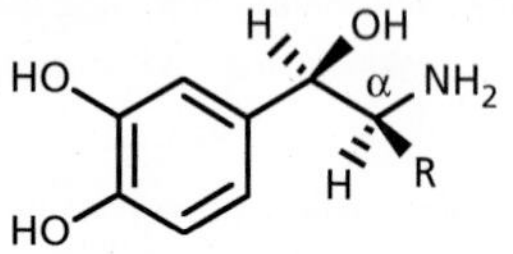

6. What synthetic complication arises from introducing an α-substituent as described in Question 5?

7. The active enantiomer of aryloxypropanolamines is the *S*-form, whereas the active enantiomer of arylethanolamines is the *R*-form. Does this imply that the two agents are binding differently to the binding site?

8. Variation of the substituents on the 2-aminoindan ring of indacaterol allows the hydrophobic nature of the molecule to be varied, without affecting important binding interactions with the target receptor. Therefore, the analogue shown below should be just as effective as indacaterol. Suggest why indacaterol was favoured over the analogue.

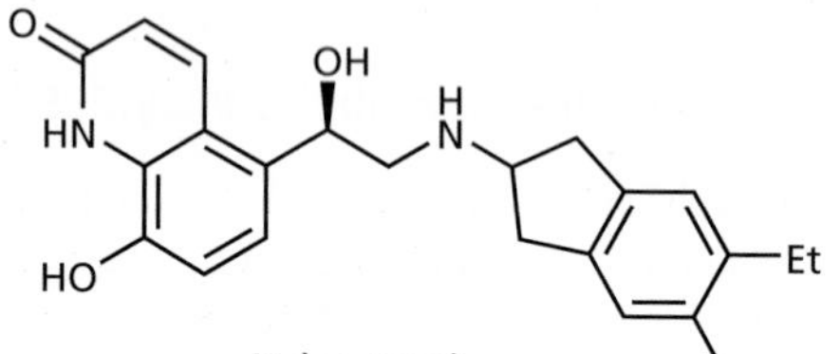

Indacaterol

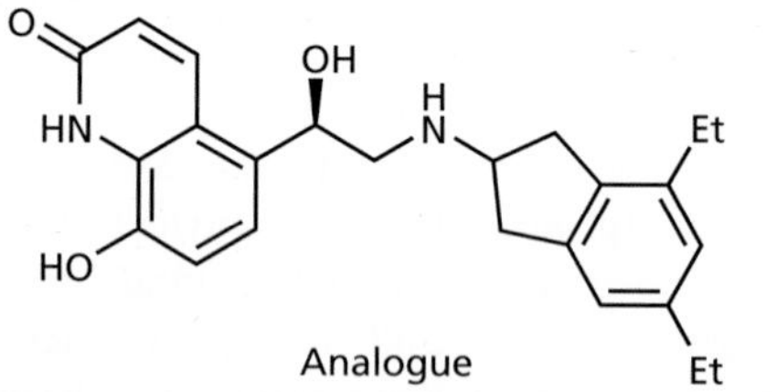

Analogue

Multiple-choice questions are available on the Online Resource Centre at www.oxfordtextbooks.co.uk/orc/patrick6e/

FURTHER READING

Abraham, D. J. (ed.) (2003) Adrenergics and adrenergic-blocking agents. in *Burger's medicinal chemistry and drug discovery*, 6th edn. **Volume 6**, Chapter 1, John Wiley and Sons, New York.

Bolognesi, M. L., et al. (2009) Alzheimer's disease: new approaches to drug discovery. *Current Opinion in Chemical Biology*, **13**, 303–8.

Furse, K. E., and Lybrand, T. P. (2003) Three-dimensional models for β-adrenergic receptor complexes with agonists and antagonists. *Journal of Medicinal Chemistry*, **46**, 4450–62.

Ganellin, C. R., and Roberts, S. M. (eds.) (1994) Salbutamol: a selective β$_2$-stimulant bronchodilator. in *Medicinal chemistry—the role of organic research in drug research*, 2nd edn. Chapter 11, Academic Press, New York.

Ganellin, C. R., and Roberts, S. M. (eds.) (1994) Beta blockers. in *Medicinal chemistry—the role of organic research in drug research*, 2nd edn. Chapter 10, Academic Press, New York.

Kobilka, B., and Schertler, G. F. X. (2008), New G-protein-coupled receptor crystal structures: insights and limitations. *Trends in Pharmacological Sciences*, **29**, 79–83.

Megget, K. (2010) Roadblock on memory lane. *Chemistry World*, July, 46–50.

O'Driscoll, C. (2001) Attack on asthma. *Chemistry in Britain*, September, 40–2.

Williams, D. A., and Lemke, T. L. (eds.) (2002) Drugs affecting adrenergic neurotransmission. in *Foye's principles of medicinal chemistry*, 5th edn. Chapter 11, Lippincott Williams and Wilkins, Baltimore.

Titles for general further reading are listed on p.845.

24 The opioid analgesics

24.1 History of opium

The search for a safe, orally active, and non-addictive analgesic based on the opiate structure is one of the oldest fields in medicinal chemistry, yet one where true success has proved elusive. The term 'opiates' refers to narcotic analgesics that are structurally related to morphine, whereas 'opioids' is the term used to cover all the synthetic, semi-synthetic, naturally occurring, and endogenous compounds that interact with opioid receptors in the body.

It is important to appreciate that the opioids are not the only compounds which are of use in the relief of pain: there are several other classes of analgesic, including **aspirin**. However, these compounds operate by different mechanisms from those used by the opioids and are effective against different types of pain.

The first opioids were extracted from opium—the sticky exudate obtained from the opium poppy (*Papaver somniferum*). Opium is perhaps the oldest herbal medicine known to humanity and has been used for myriad afflictions. It was particularly effective as a sedative and a treatment for diarrhoea. By the eighteenth and nineteenth centuries, preparations of opium known as **laudanum** had become extremely popular in Europe, not least in the British Royal Navy where the concoction was used by ships' surgeons as an analgesic and sedative. Laudanum also proved to be one of the first examples of a drug taken for 'recreational purposes'. A number of famous nineteenth century authors and poets are known to have taken the drug on a regular basis, with several becoming addicted. Opium was also smoked in opium dens, which became widespread around the world—especially among Chinese communities. A growing realization of the long-term problems associated with taking opium led eventually to laws being introduced in the twentieth century that restricted its use to medical and scientific purposes.

For additional material see Web article 9: History of opium on the Online Resource Centre at www.oxford-textbooks.co.uk/orc/patrick6e/

24.2 The active principle: morphine

24.2.1 Isolation of morphine

Opium contains a complex mixture of over 20 alkaloids. The principle alkaloid in the mixture, and the one responsible for opium's analgesic and sedative activity, is **morphine** (Fig. 24.1). Pure morphine was first isolated in 1803, but it was not until 1833 that chemists at the Edinburgh firm of Macfarlane and Co. (now Macfarlane-Smith) were able to isolate and purify it on a commercial scale. Because morphine is poorly absorbed orally, it was little used in medicine until the hypodermic syringe was invented in 1853. Injecting the drug directly into

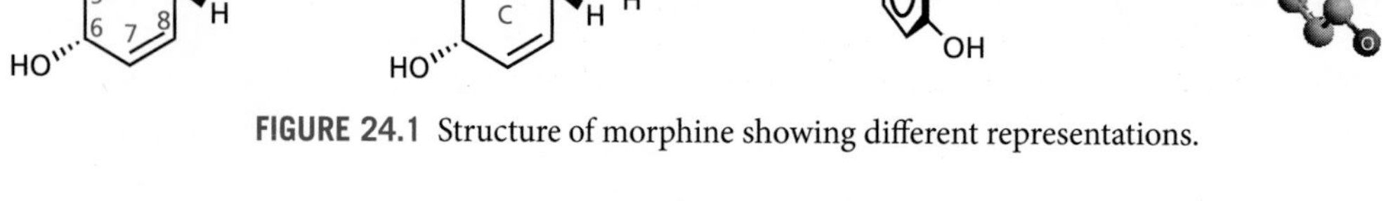

FIGURE 24.1 Structure of morphine showing different representations.

the blood supply revealed that morphine was a potent analgesic and sedative, and was far more effective than opium. However, there was a price to be paid. The risks of addiction, tolerance, and respiratory depression (Box 24.1) were also greatly increased.

24.2.2 Structure and properties

Morphine is an extremely complex molecule by nineteenth-century standards, and identifying its structure posed a huge challenge to chemists. By 1881, the functional groups on morphine had been identified, but it took many more years to establish the full structure. In those days, the only way to find out the structure of a complicated molecule was to degrade the compound into simpler molecules that were already known and could be identified. For example, the degradation of morphine with strong base produced methylamine, which established that there was an N–CH$_3$ fragment in the molecule. From such evidence, chemists would propose a structure. This would be like trying to work out the structure of a bombed cathedral from the rubble. Once a structure had been proposed, chemists would attempt to synthesize it. If the properties of the synthesized compound were the same as those of the natural compound, then the structure was proven. This was a long drawn-out process, made all the more difficult because nineteenth-century chemists had few of the synthetic reagents or procedures available today. As a result, it was not until 1925 that Sir Robert Robinson proposed the correct structure of morphine. A full synthesis was achieved in 1952 and the structure was finally proved by X-ray crystallography in 1968 (164 years after the original isolation). The molecule contains five rings labelled A–E and has a pronounced T shape. It is basic because of the tertiary amino group, but it also contains a phenol, alcohol, aromatic ring, ether bridge, and alkene double bond. The nitrogen atom of the amine can undergo inversion, which means that the N-methyl group can slowly interconvert between the axial and the equatorial positions.

Test your understanding and practise your molecular modelling with Exercise 24.1 on the Online Resource Centre at www.oxfordtextbooks.co.uk/orc/patrick6e/

24.3 Structure–activity relationships

Following the discovery of morphine, it was natural for chemists to use the known reactions of the day to synthesize various analogues and to see whether these had analgesic activity or not. Different tests have been used to assess analgesic activity which complicates the picture (see web article 11). Nevertheless, some conclusions can be made regarding the importance or otherwise of different functional groups. For example, **heterocodeine**,

---

BOX 24.1 Clinical aspects of morphine

Morphine is still one of the most effective painkillers available to medicine and is currently the drug of choice in the treatment of severe pain. It is especially good for treating dull, constant pain, rather than sharp, periodic pain. It acts mainly in the brain and appears to work by elevating the pain threshold, thus decreasing the brain's awareness of pain. Unfortunately, it has a large number of side effects, which include depression of the respiratory centre, constipation, excitation, euphoria, nausea, vomiting, itching, pupil constriction, tolerance, and dependence.

Some side effects can be advantageous. For example, the observation that morphine causes constipation has led to the design of opioids which are used in the treatment of diarrhoea. Euphoria can be a useful side effect when treating pain in terminally ill patients. However, the effect is not observed in patients suffering severe pain. Moreover, the euphoric effects of morphine in healthy individuals can encourage people to take the drug for the wrong reasons. Other side effects, such as constipation, itching, and nausea may not appear serious, but they can become so uncomfortable that treatment has to be stopped.

The dangerous side effects of morphine are those of tolerance and dependence, allied with the effects morphine can have on breathing. In fact, the most common cause of death from a morphine overdose is suffocation. This is caused by morphine decreasing the sensitivity of the respiratory centre in the brain to carbon dioxide. The properties of tolerance and dependence in a drug are particularly dangerous and lead to severe withdrawal symptoms when the drug is no longer taken. Withdrawal symptoms associated with morphine include anorexia, weight loss, pupil dilation, chills, excessive sweating, abdominal cramps, muscle spasms, hyperirritability, lacrimation, tremor, increased heart rate, and increased blood pressure. No wonder addicts find it hard to kick the habit!

For additional material see Web article 10: Clinical applications of opioids on the Online Resource Centre at www.oxfordtextbooks.co.uk/orc/patrick6e/

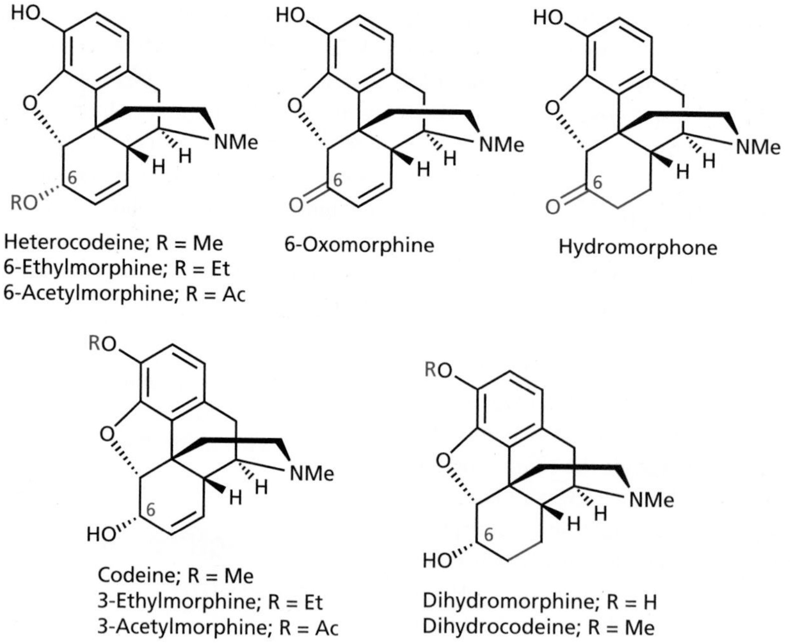

Heterocodeine; R = Me
6-Ethylmorphine; R = Et
6-Acetylmorphine; R = Ac

6-Oxomorphine

Hydromorphone

Codeine; R = Me
3-Ethylmorphine; R = Et
3-Acetylmorphine; R = Ac

Dihydromorphine; R = H
Dihydrocodeine; R = Me

FIGURE 24.2 Analogues of morphine.

6-ethylmorphine, **6-acetylmorphine**, **6-oxomorphine**, **hydromorphone**, and **dihydromorphine** (Fig. 24.2) are examples of structures where the alkene or 6-hydroxy groups have been modified or removed. Analgesic activity is retained in these structures, indicating that neither of these groups is crucial to activity. However, analgesic activity drops significantly for **codeine**, **dihydrocodeine**, and **3-ethylmorphine**, indicating the importance of the phenolic group.

For additional material see Web article 11: Testing methods on the Online Resource Centre at www.oxford-textbooks.co.uk/orc/patrick6e/

These, and other, results led to the conclusion that the important functional groups for analgesic activity are the phenol OH group, the aromatic ring, and the tertiary amine which is protonated and ionized when the drug interacts with its target binding site. These functional groups play an important role in binding the drug to the binding site by the intermolecular bonding forces indicated in Fig. 24.3.

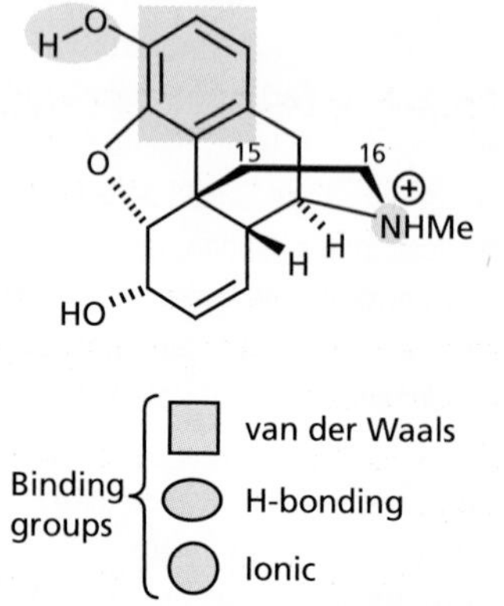

At this stage, it is worth making some observations on the stereochemistry of morphine. Morphine is a chiral molecule containing several asymmetric centres and exists naturally as a single stereoisomer. When morphine was first synthesized, it was made as a racemic mixture of the naturally occurring enantiomer plus its mirror-image enantiomer (Fig. 24.4). It was noticeable that the activity of synthetic morphine was half that of natural morphine, and separation of the enantiomers revealed that the unnatural enantiomer had no analgesic activity. This should come as no surprise as the macromolecules targeted by drugs are themselves asymmetric and are able to distinguish between the enantiomers of a chiral drug.

Epimerization of a single asymmetric centre is not beneficial for activity either, as changing the stereochemistry of even one asymmetric centre can result in a drastic

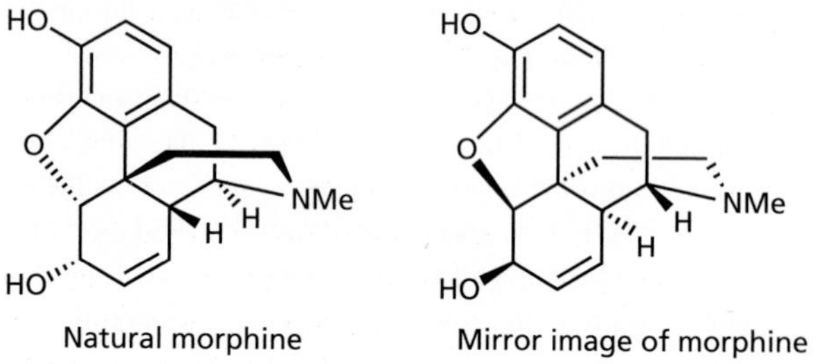

Binding groups
☐ van der Waals
⬭ H-bonding
◯ Ionic

FIGURE 24.3 Important functional groups for analgesic activity in morphine.

Natural morphine

Mirror image of morphine

FIGURE 24.4 Morphine and its mirror image.

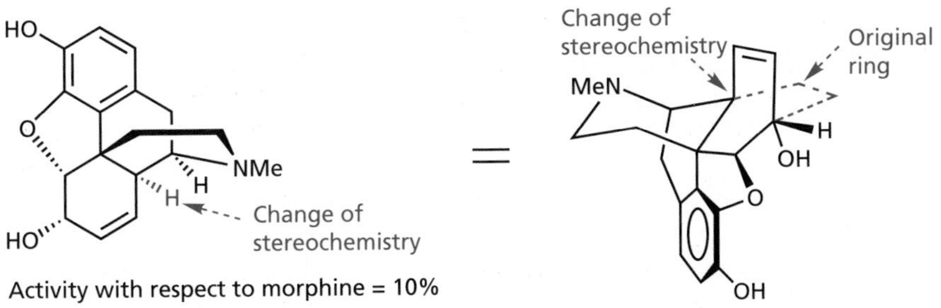

FIGURE 24.5 Epimerization of a single asymmetric centre.

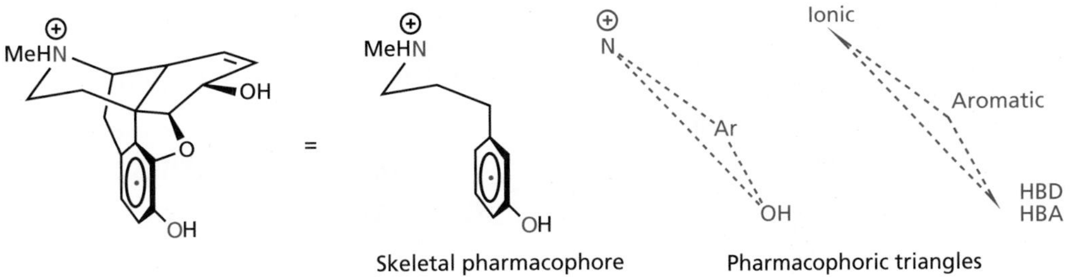

FIGURE 24.6 Opioid pharmacophores for morphine and related opioids.

change of shape that could affect how the molecule binds to its target binding site. For example, epimerization of the asymmetric centre at position 14 results in a stereoisomer that has only 10% the activity of morphine (Fig. 24.5).

To sum up, analgesic activity is not only related to the presence of the important functional groups defined earlier, but also to their relative position with respect to each other—the **pharmacophore** (section 13.2). Opioid pharmacophores can be defined in different ways, either by defining a simple skeleton that links the important functional groups, or by pharmacophoric triangles where the corners correspond to functional groups or binding interactions (Fig. 24.6).

Finally, a word of caution regarding the importance of the phenol group. There is no doubt that the phenol group is an important part of the opioid pharmacophore for receptor binding, but it is not necessarily as important when one considers the analgesic activity of different opioid structures *in vivo*. That is because pharmacokinetic factors also have an important role in the level of analgesic activity observed. As we will see, there are examples of opioid analgesics where the phenol group is masked or missing altogether. This has the advantage of making the molecule less susceptible to metabolism by phase II conjugation reactions (section 11.5.5). Moreover, the masking or absence of the phenol group increases hydrophobicity, such that the molecule is absorbed from the gastrointestinal tract more easily, and/or can cross the blood–brain barrier more efficiently. Consequently, the increased levels of opioid reaching the target receptors

can compensate for weaker binding interactions. Some opioids with a masked phenol group also act as prodrugs, where the masking group is removed by metabolic enzymes. There are even instances where the phenol group is no longer required as part of the binding pharmacophore. Simpler, more flexible opioids are believed to interact with opioid receptors in a different way from morphine such that the phenol group becomes redundant (see sections 24.6.3.4 and 24.6.3.5). Alternatively, more complex opioids such as the orvinols contain additional binding groups that can compensate for the masking of the phenol group (section 24.6.4).

Test your understanding and practise your molecular modelling with Exercises 24.2 and 24.3 on the Online Resource Centre at www.oxfordtextbooks.co.uk/orc/patrick6e/

KEY POINTS

- Morphine is extracted from opium and is one of the oldest drugs used in medicine.

- Morphine is a powerful analgesic but has various side effects, the most serious being respiratory depression, tolerance, and dependence.

- The structure of morphine consists of five rings forming a T-shaped molecule.

- The important binding groups on morphine are the phenol, the aromatic ring, and the ionized amine.

24.4 The molecular target for morphine: opioid receptors

Although morphine was isolated in the 19th century, it took many years to discover how it produced its analgesic effect. It is now known that morphine activates analgesic receptors in the central nervous system and that this leads to a reduction in the transmission of pain signals to the brain. There are three main types of analgesic or opioid receptor that are activated by morphine: the mu (μ), kappa (κ), and delta (δ) receptors. All of them are G-protein-coupled receptors which activate G_i or G_o signal proteins (section 4.7 and Chapter 5). Morphine acts as an agonist at all three types of receptor and activation leads to a variety of cellular effects depending on the type of receptor involved. These include the opening of potassium ion channels, the closing of calcium ion channels, or the inhibition of neurotransmitter release—all of which reduce the transmission of pain signals from one nerve cell to another. A newer form of terminology has now been introduced where the μ, κ, and δ receptors are called the **MOR, KOR**, and **DOR** receptors, respectively. Nevertheless, we will continue to use the original nomenclature in this chapter as it is still more prevalent.

There are differences between the three opioid receptors in terms of their effects. Activation of the μ receptor results in sedation and the strongest analgesic effect, but this receptor is also associated with the strongest and most dangerous side effects of respiratory depression, euphoria, and addiction. Activation of the δ and κ receptors does not produce the same level of analgesia, but there are less serious side effects. For example, the δ receptor does not cause sedation, euphoria, or physical dependence, while the κ receptor has no effect on breathing, is free of euphoric effects, and has a low risk of physical dependence. The κ receptor is considered the safest of the three types of receptor and a lot of research has been carried out to develop κ-selective opioids. Unfortunately, it has been discovered that activation of the κ receptor can lead to sedation and psychological side effects, such as anxiety, depression, and psychosis. As a result, these agents have failed to fulfil their original promise.

A fourth opioid receptor was later identified in the 1990s which shows a lot of structural similarity to the classical opioid receptors. It was, therefore, referred to as the **opioid receptor-like receptor (ORL1)**, but is now known as the **NOR** receptor. It was originally classed as an **orphan receptor** because its endogenous ligand was not known, but an endogenous ligand has now been identified as a polypeptide structure called **nociceptin**. Activation of the ORL1 receptor can either increase or decrease the sensitivity to pain depending on the location of the receptor and the method by which agonists are administered.

Morphine and most of its analogues bind strongly to the μ receptor, and less strongly to the κ or δ receptors. This explains why it has been so difficult to find a safe, powerful analgesic as the receptor with which they bind most strongly produces the most serious side effects.

Recently, it has been demonstrated that opioid receptors can occur as homomeric and heteromeric dimers. This has important consequences for drug design (section 24.9.2).

For additional material see Web article 12: Opioid dimers and receptors on the Online Resource Centre at www.oxfordtextbooks.co.uk/orc/patrick6e/

24.5 Morphine: pharmacodynamics and pharmacokinetics

Pharmacodynamics refers to the manner in which a drug binds to its target and produces a pharmacological effect. The functional groups that are important to the activity of morphine act as binding groups in the following manner (Fig. 24.7):

- the amine nitrogen is protonated and charged, allowing it to form an ionic bond with a negatively charged region of the binding site;
- the phenol acts as a hydrogen bond donor and forms a hydrogen bond to a hydrogen bond acceptor in the binding site;
- the rigid structure of morphine means that its aromatic ring has a defined orientation with respect to the rest of the molecule, allowing van der Waals interactions with a suitable hydrophobic location in the binding site.

Pharmacokinetics refers to the ability of a drug to reach its target and to survive in the body. Morphine is relatively polar and is poorly absorbed from the gut, and so it is normally given by intravenous injection. However, only a small percentage of the dose administered actually reaches the analgesic receptors in the central nervous system (CNS) due to the blood–brain barrier (section 11.4.5). This acts as a barrier to polar drugs and effectively prevents any ionized drug from crossing into the CNS. For example the *N*-methyl quaternary salt of morphine (Fig. 24.8) is inactive when it is administered by intravenous injection because it is blocked by the blood–brain barrier. If this same compound is injected directly into the brain, however, it has a similar analgesic activity to morphine. If morphine was fully charged, it would not be able to enter the brain either. However, the amine group is a weak base and so morphine can exist both as the free base and ionized form. This means that morphine can cross the blood–brain barrier as the free base then ionize in order to interact with the opioid receptors. The pK_a values of useful analgesics should be 7.8–8.9 such that there is an approximately equal chance of the amine being ionized or unionized at physiological pH.

The extent to which different structures cross the blood–brain barrier plays an important role in analgesic activity. For example, **normorphine** (Fig. 24.8) has only 25% the activity of morphine. The secondary NH group is more polar than the original tertiary group, and so normorphine is less efficient at crossing the blood–brain barrier, leading to a drop in activity.

It is possible to get increased levels of morphine in the brain by using prodrugs (section 14.6) where some of the polar functional groups are masked. It is interesting to compare the activities of morphine, **6-acetylmorphine** (Fig. 24.2), and **diamorphine** (heroin) (Fig. 24.8). The most active (and the most dangerous) compound of the three is 6-acetylmorphine, which is four times more active than morphine. Diamorphine is also more active than morphine by a factor of two, but less active than 6-acetylmorphine. How do we explain this?

6-Acetylmorphine is less polar than morphine and will cross the blood–brain barrier into the CNS more quickly and in greater concentrations. The phenolic group is free and therefore it will interact immediately with the analgesic receptors.

Diamorphine has two masked polar groups, and so it is the most efficient compound of the three in crossing the blood–brain barrier. However, the 3-acetyl group has to be removed by esterases in the CNS before the structure can bind to an opioid receptor. This means that diamorphine is more powerful than morphine because of the ease with which it crosses the blood–brain barrier, but less powerful than 6-acetylmorphine because the 3-acetyl group has to be hydrolysed.

Diamorphine and 6-acetylmorphine are both more potent analgesics than morphine. Unfortunately, they also have greater side effects, as well as severe tolerance and dependence characteristics. Diamorphine is still used in Canada and the UK to treat terminally ill patients suffering chronic pain, but 6-acetylmorphine is considered so dangerous that its synthesis is banned in many countries.

The lifetime of morphine in the blood supply is quite short, with 90% of each dose being metabolized and excreted within 24 hours. The presence of the alcohol and phenol groups means that the molecule readily undergoes phase II conjugation reactions (section 11.5.5) and the resulting polar conjugates are quickly excreted.

Drug metabolism also plays an important role in the activity of different opioid structures. For example, **codeine** (Fig. 24.2) is the 3-methyl ether of morphine and has a binding affinity for the opioid receptor which is only 0.1% of morphine. It also has no analgesic activity when it is injected directly into the brain. This is not surprising as methylation of the phenol group would be expected to disrupt its ability to act as a binding group (section 13.1.1). What *is* surprising is the fact that codeine has an analgesic effect which is 20% that of morphine—much

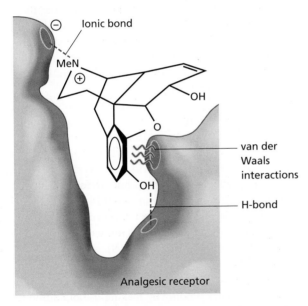

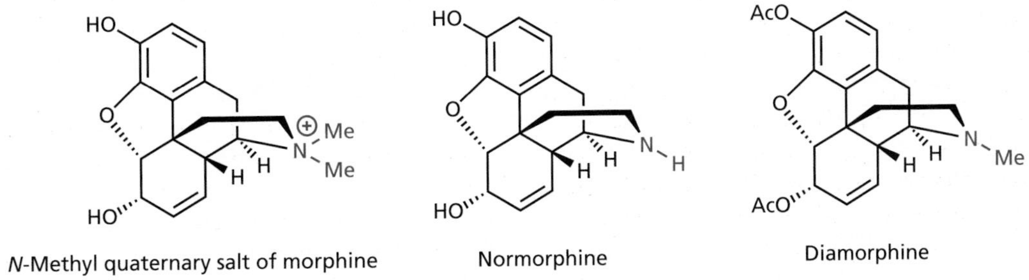

FIGURE 24.7 Binding interactions of morphine with the hypothetical binding site of an opioid receptor.

FIGURE 24.8 Analogues of morphine with differing abilities to cross the blood–brain barrier.

better than expected. Why is this? The answer lies in the fact that codeine is metabolized by *O*-demethylation in the liver to give morphine. Thus, codeine can be viewed as a prodrug for morphine. Codeine is present in opium and is used for treating moderate pain, coughs, and diarrhoea (see also section 11.5.6).

KEY POINTS

- The important binding interactions between morphine and opioid receptors are a hydrogen bonding interaction via a phenol group, an ionic interaction via a charged amine, and van der Waals interactions involving the aromatic ring.

- There are three different analgesic receptors (μ, κ, and δ) with which morphine interacts. All require the presence of a pharmacophore involving the phenol, aromatic ring, and ionized amine.

- Morphine binds most strongly to the μ receptor. This receptor is responsible for the serious side effects associated with morphine.

- The κ receptor is responsible for analgesia and sedation, and lacks serious side effects. However, activation causes psychological side effects which have prevented κ-selective opioids reaching the market.

- The δ receptor is favoured by the enkephalins.

- The opioid receptors are G-protein-linked receptors.

- The ability of opioids to cross the blood–brain barrier plays an important role in analgesic activity.

- Some analgesics such as codeine and diamorphine act as prodrugs for morphine.

24.6 Morphine analogues

Considering the problems associated with morphine, there is a need for novel analgesic agents which retain the analgesic activity of morphine, but which have fewer side effects and can be administered orally. The following sections illustrate how many of the classical drug

design strategies described in Chapter 13 were effective in obtaining novel analgesic structures.

24.6.1 Variation of substituents

A series of alkyl substituents were placed on the phenolic group, but the resulting compounds were inactive or poorly active. We have already identified that the phenol group must be free for good analgesic activity.

The removal of the *N*-methyl group to give normorphine allowed a series of alkyl chains to be added to the basic centre (Box 24.2). These results are discussed in the next section.

24.6.2 Drug extension

The strategy of drug extension described in section 13.3.2 involves the addition of extra functional groups to a lead compound in order to probe for extra binding regions in a binding site. Many analogues of morphine containing extra functional groups have been prepared, but have rarely shown any improvement. However, there are two exceptions. The introduction of a hydroxyl group at position 14 (Fig. 24.9) increases activity for structures such as **oxymorphone** and **oxycodone**, and suggests that there might be an extra hydrogen bond interaction taking place with the binding site.

The other exception involves the variation of alkyl substituents on the nitrogen atom. As the alkyl group is increased in size from a methyl to a butyl group, the activity drops to zero (Fig. 24.10). With a larger group, such as a pentyl or a hexyl group, activity recovers slightly. None of this is particularly exciting, but when a phenethyl group is attached, the activity increases 14-fold relative to morphine—a strong indication that a hydrophobic binding region has been located which interacts favourably with the new aromatic ring (Fig. 24.9).

To conclude, the size and nature of the group on nitrogen is important to the activity spectrum. Drug extension can lead to better binding by making use of additional binding interactions.

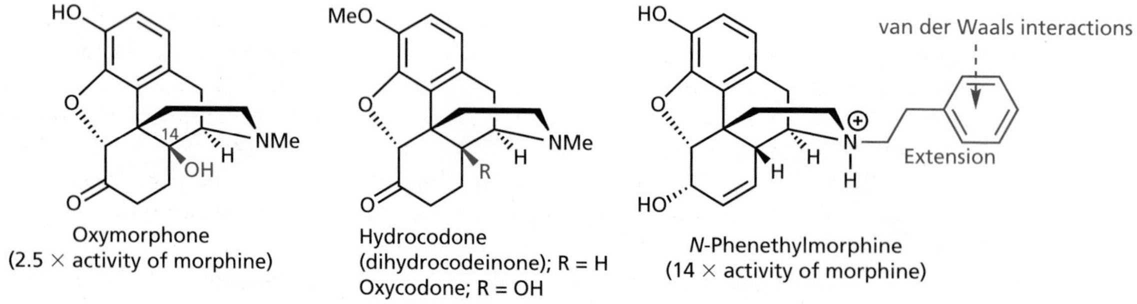

Oxymorphone
(2.5 × activity of morphine)

Hydrocodone
(dihydrocodeinone); R = H
Oxycodone; R = OH

N-Phenethylmorphine
(14 × activity of morphine)

FIGURE 24.9 Extended analogues of morphine.

BOX 24.2 Synthesis of *N*-alkylated morphine analogues

The synthesis of *N*-alkylated morphine analogues is easily achieved by removing the *N*-methyl group from morphine to give **normorphine**, then alkylating the amino group with an alkyl halide. Removal of the *N*-methyl group was achieved originally by a von Braun degradation with cyanogen bromide, but is now more conveniently carried out using a chloroformate reagent such as vinyloxycarbonyl chloride. The final alkylation step can sometimes be profitably replaced by a two-step process involving an acylation to give an amide, followed by reduction.

Demethylation and alkylation of the basic centre present in morphine.

Before leaving this subject, it is worth describing important results which occurred when an allyl or a cyclopropylmethyl group was attached to nitrogen (Fig. 24.11) (see also section 24.7). **Naloxone** and **naltrexone** have no analgesic activity at all, and **nalorphine** retains only weak analgesic activity. Not very exciting, you might think. What *is* important is that they act as antagonists to morphine, i.e. they bind to the analgesic receptors without 'switching them on' and then block morphine from binding. As a result, morphine can no longer act as an analgesic. One might be hard pushed to see an advantage in this, and with good reason. If we are just considering analgesia, there is none. However, the fact that morphine is blocked from all its receptors means that none of its side effects are produced either, and it is the blocking of these effects that makes antagonists extremely useful. For example, accident victims have sometimes been given an overdose of morphine. If this is not treated quickly, then the casualty may die of suffocation. Administering nalorphine means that the antagonist can block morphine from binding to opioid receptors and lead to recovery.

The opioid antagonists have also proved useful in treating addictions. Naltrexone is eight times more active than naloxone as an antagonist and is given to drug addicts who have been weaned off morphine or heroin. Naltrexone blocks the opioid receptors, preventing the effects that addicts seek if they are tempted to restart their habit. As a result, they are more likely to remain abstinent. Naltrexone in combination with **bupropion** (section 23.12.4) is also being considered for the treatment of obesity. **Nalmefene** (Fig. 24.11) is a close analogue which is currently undergoing clinical trials as an oral treatment for alcoholism. It binds more strongly than naltrexone to opioid receptors and blocks the effects of natural opioids released as a result of drinking.

There is another interesting observation related to these antagonists. For many years, chemists had been trying to find a morphine analogue without serious side effects. There had been so little success in this search that many believed it would be impossible to separate the analgesic effects from the side effects. The fact that the antagonist naloxone blocks both the analgesic and side effects of morphine did nothing to change that view. However, the properties of nalorphine offered a glimmer of hope.

Nalorphine acts as an antagonist at the μ receptor and as a weak agonist at the κ receptor. Therefore, the slight analgesia observed with nalorphine is due

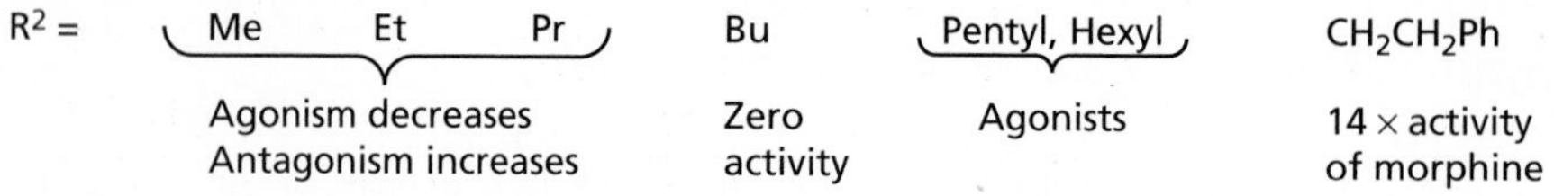

$R^2 =$	Me Et Pr	Bu	Pentyl, Hexyl	CH_2CH_2Ph
	Agonism decreases Antagonism increases	Zero activity	Agonists	14 × activity of morphine

FIGURE 24.10 Change in activity with respect to alkyl group size.

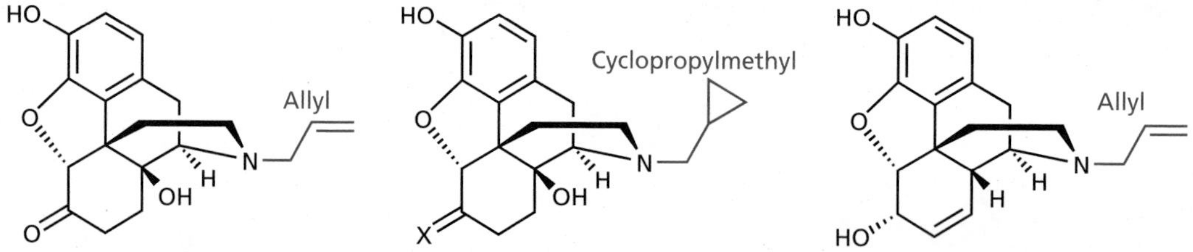

Naloxone

Naltrexone (X = O)
Nalmefene (X = CH$_2$)

Nalorphine

FIGURE 24.11 Antagonists to morphine.

to partial activation of the κ receptor. Moreover, this activity appears to be free of the undesired side effects associated with morphine. This was the first sign that a non-addictive, safe analgesic might be possible if structures were made that were selective for the κ receptor. Unfortunately, nalorphine has hallucinogenic and psychological side effects, which result from activation of the κ receptor.

24.6.3 Simplification or drug dissection

We turn now to more drastic alterations of the morphine structure and ask whether the complete carbon skeleton is really necessary. If the molecule could be simplified, it would be easier to synthesize analogues (section 13.3.8). The structure of morphine has five rings and five chiral centres (Fig. 24.12) and analogues were made to see whether structures with fewer rings and chiral centres were still active.

24.6.3.1 Removing ring E

Removing ring E leads to complete loss of activity. This emphasizes the importance of the basic nitrogen to analgesic activity.

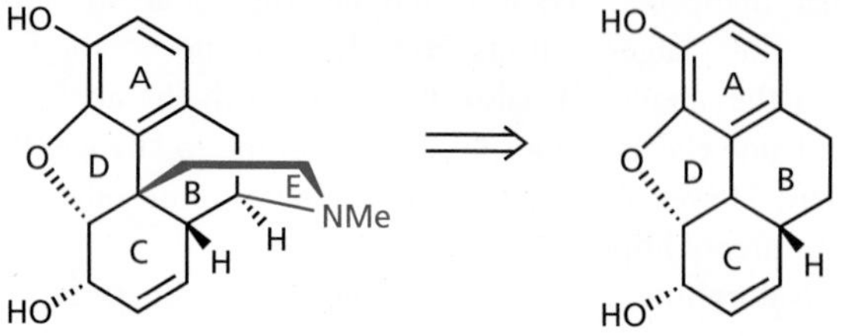

FIGURE 24.12 Removing ring E from morphine.

24.6.3.2 Removing ring D

Removing the oxygen bridge, as well as the alcohol and alkene functional groups gives a series of tetracyclic compounds called the **morphinans** (Fig. 24.13) which have useful analgesic activity. This demonstrates that the oxygen bridge is not essential. The structures shown in Fig. 24.13 also have three asymmetric centres, rather than five.

N-Methylmorphinan was the first such compound tested and is only 20% as active as morphine, but since the phenolic group is missing, this is not surprising. The more relevant **levorphanol** structure is five times more active than morphine and, although side effects are also increased, levorphanol has a massive advantage over morphine in that it can be taken orally and lasts much longer in the body. This is because levorphanol is not metabolized in the liver to the same extent as morphine. As might be expected, the mirror image of levorphanol (**dextrorphan**) has insignificant analgesic activity.

The same strategy of drug extension already described for the morphine structures was tried on the morphinans with similar results. For example, adding an allyl substituent on the nitrogen gives antagonists. Adding a phenethyl group to the nitrogen greatly increases potency. Adding a 14-hydroxyl group also increases activity.

To conclude:

- morphinans are more potent and longer acting than their morphine counterparts, but they also have higher toxicity and comparable dependence characteristics;
- modifications carried out on the morphinans have the same SAR results as they do with morphine—this im-

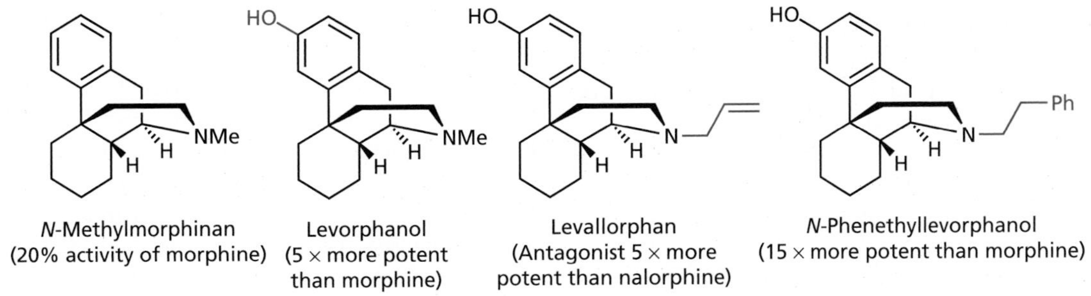

N-Methylmorphinan
(20% activity of morphine)

Levorphanol
(5 × more potent
than morphine)

Levallorphan
(Antagonist 5 × more
potent than nalorphine)

N-Phenethyllevorphanol
(15 × more potent than morphine)

FIGURE 24.13 Examples of morphinans.

plies that morphine and morphinans are binding to the same receptors in the same way;

* the morphinans are easier to synthesize because they are simpler molecules with fewer rings and chiral centres.

24.6.3.3 Removing rings C and D

Removing both rings C and D gives an interesting group of compounds called the **benzomorphans** (Fig. 24.14) which retain analgesic activity. One of the simplest of these structures is **metazocine**, which has the same analgesic activity as morphine. Notice that the two methyl groups in metazocine are *cis* with respect to each other and represent the remnants of the C ring. It is important that these methyl groups are retained in order to obtain good activity.

The same chemical modifications were carried out on the benzomorphans as described for the morphinans and morphine to produce the same biological effects, implying a similar interaction with the analgesic receptors. For example, replacing the N-methyl group of metazocine

with a phenethyl group gives **phenazocine**, which is four times more active than morphine and was the first compound discovered to have a useful level of analgesia without dependence properties.

Further developments led to **pentazocine** (Fig. 24.14), which has proved to be a useful long-term analgesic with a very low risk of addiction. Like nalorphine, pentazocine acts as an antagonist at the μ receptor but, unlike nalorphine, it is a full agonist at the κ receptor rather than a partial agonist. Pentazocine also acts as a weak agonist at the δ receptor. Unfortunately, the compound has hallucinogenic and psychotomimetic side effects as a result of activating the κ receptor. A newer compound (**bremazocine**) has a longer duration, has 200 times the activity of morphine, appears to have no addictive properties, and does not depress breathing.

To conclude:

* rings C and D are not essential to analgesic activity;

* analgesia and addiction are not necessarily coexistent;

* 6,7-benzomorphans are clinically useful compounds with reasonable analgesic activity, less addictive liability, and less tolerance;

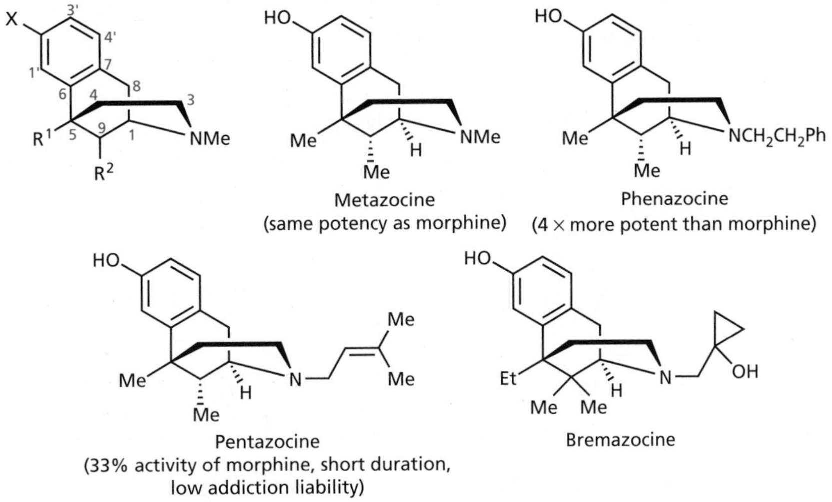

Metazocine
(same potency as morphine)

Phenazocine
(4 × more potent than morphine)

Pentazocine
(33% activity of morphine, short duration,
low addiction liability)

Bremazocine

FIGURE 24.14 Benzomorphans.

- benzomorphans are simpler to synthesize than morphine and morphinans;
- benzomorphans bind to opioid receptors in the same manner as morphine and morphinans.

24.6.3.4 Removing rings B, C, and D

Removing rings B, C, and D gives a series of compounds known as **4-phenylpiperidines**. The analgesic activity of these compounds was discovered by chance in the 1940s when chemists were studying analogues of cocaine for antispasmodic properties. Their structural relationship to morphine was only identified when they were found to be analgesics—this is evident if the structure is drawn as shown in Fig. 24.15. Activity can be increased sixfold by introducing the phenolic group and altering the ester to a ketone to give **ketobemidone**.

Pethidine (meperidine) is a weaker analgesic than morphine, but shares the same undesirable side effects. On the plus side, it has a rapid onset and a shorter duration of action. As a result, it has been used as an analgesic in childbirth. The rapid onset and short duration of action mean that there is less chance of the drug depressing the baby's breathing once it is born. The structure was discovered in 1939 and was the first fully synthetic opioid analgesic to enter clinical practice.

The piperidines are more easily synthesized than any of the previous groups and a large number of analogues have been studied. There is some doubt as to whether they act in the same way as morphine at analgesic receptors, as some of the chemical adaptations we have already described do not lead to comparable biological results. For example, adding allyl or cyclopropyl groups does not give antagonists. The replacement of the methyl group of pethidine with a cinnamic acid residue increases the activity by 30-fold, whereas putting the same group on morphine eliminates activity (Fig. 24.16).

These results might have something to do with the fact that the phenylpiperidines are more flexible molecules than the previous structures, and are likely to have different binding modes with opioid receptors (see section 24.8.3).

Fentanyl and its analogues (Fig. 24.17) represent a class of opioids known as the **4-anilinopiperidines** and are among the most potent agonists known for the μ receptor. These drugs lack a phenolic group and are very lipophilic. As a result, they can cross the blood–brain barrier more efficiently. Fentanyl itself is up to 100 times more active than morphine as a sedative and analgesic, and it is thought that the Russian authorities used it in an attempt to incapacitate a group of terrorists during the infamous cinema siege of recent years. Apparently, the drug was introduced as a gas through the ventilation system into the auditorium and succeeded in rendering

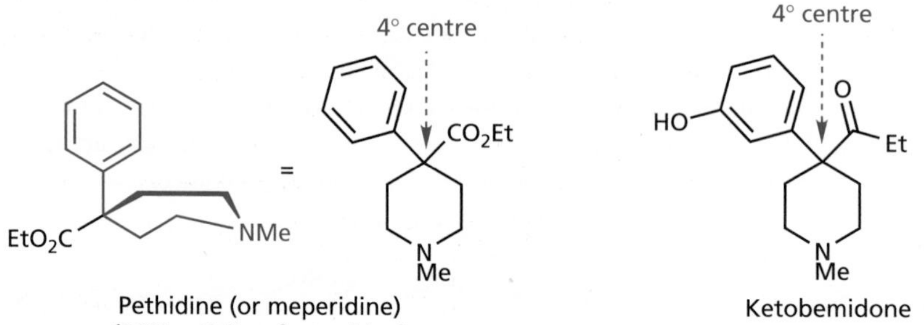

Pethidine (or meperidine)
(20% activity of morphine)

Ketobemidone

FIGURE 24.15 4-Phenylpiperidines.

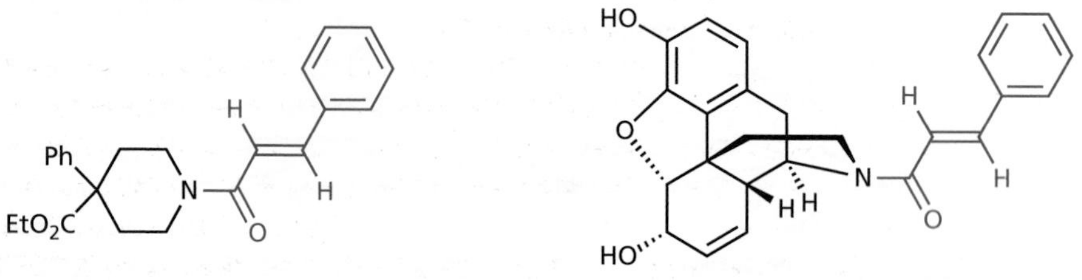

N-Cinnamoyl analogue of pethidine
30 × more potent than pethidine

N-Cinnamoyl analogue of morphine
Zero activity

FIGURE 24.16 Effect of addition of a cinnamic acid residue (in blue) on meperidine and morphine.

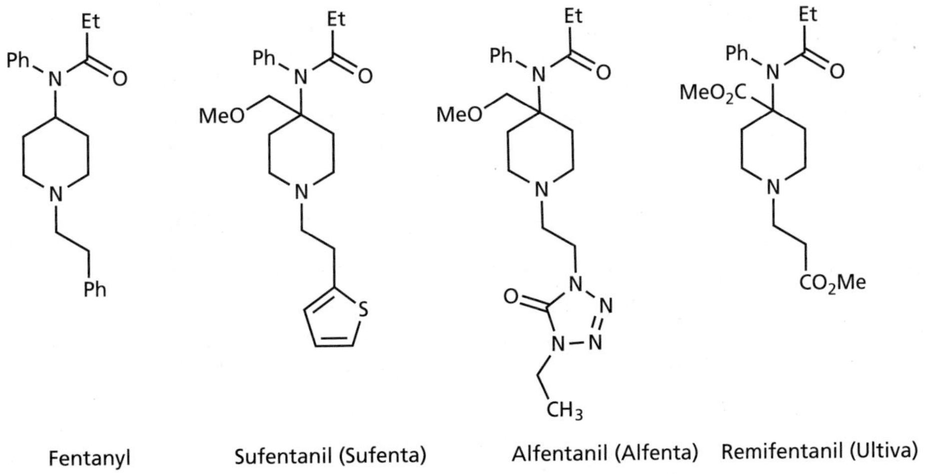

FIGURE 24.17 Fentanyl and analogues.

both terrorists and hostages unconscious. Unfortunately, the authorities waited too long to enter the building and many innocent people died as a result of suffocation. Like morphine, an overdose of fentanyl can stop breathing by depressing the respiratory centre in the brain.

Fentanyl and the shorter lasting **alfentanil** and **remifentanil** are used during surgery for analgesia and to enhance anaesthesia. Remifentanil was designed to have a very short duration of action by introducing ester groups which are rapidly metabolized by non-specific esterase enzymes. It can be administered as an intravenous drip and does not accumulate in the body because of its rapid metabolism. This reduces the risk of serious side effects, such as depression of the respiratory centre.

To conclude:

- rings C, D, and E are not essential for analgesic activity;
- piperidines retain side effects, such as addiction and depression of the respiratory centre, because they are agonists at the μ receptor;
- piperidine analgesics are faster acting and have a shorter duration of action than morphine;
- the quaternary centre present in piperidines is usually necessary (fentanyl and its analogues are exceptions);

- the aromatic ring and basic nitrogen are essential to activity, but the phenol group is not;
- piperidine analgesics appear to bind with analgesic receptors in a different manner to previous structural classes.

For additional material see Web article 13: 4-Anilinopiperidines

Test your understanding and practise your molecular modelling with Exercises 24.4–24.7 on the Online Resource Centre at www.oxfordtextbooks.co.uk/orc/patrick6e/

24.6.3.5 Removing rings B, C, D, and E

The analgesic **methadone** (Fig. 24.18) was discovered in Germany during the Second World War and is comparable in activity to morphine. It is orally active and has less severe emetic and constipation effects. Side effects such as sedation, euphoria, and withdrawal symptoms are also less severe, and so the compound has been given to drug addicts as a substitute for morphine or heroin in order to wean them off these drugs. This is not a complete cure, as it merely swaps an addiction to heroin or morphine for an addiction to methadone. This is considered less dangerous, though.

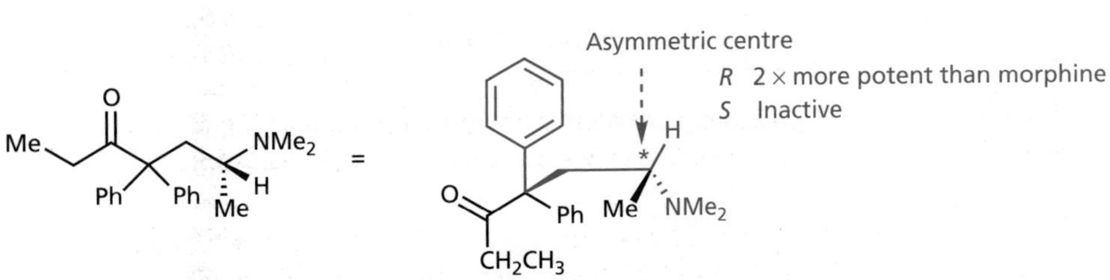

FIGURE 24.18 Methadone.

BOX 24.3 Opioids as antidiarrhoeal agents

One of the main aims in drug design is to find agents that have minimal side effects, but, it is sometimes possible to take advantage of a side effect. For example, one of the side effects of opioid analgesics is constipation. This is not very comfortable, but it is a useful property if you wish to counteract diarrhoea. The aim then is to design a drug such that the original side effect becomes the predominant feature. **Loperamide** is a successful antidiarrhoeal agent which was first synthesized in 1969, approved by the US Food and Drugs Administration (FDA) in 1976, and marketed as **Imodium**. It can be viewed as a hybrid molecule involving a 4-phenylpiperidine and a methadone-like structure. The compound is lipophilic, slowly absorbed, and prone to metabolism, meaning that it acts as a selective agonist on opioid receptors in the gastrointestinal tract. It is also free from any euphoric effect, as it cannot cross the blood–brain barrier. All these features make it a safe medicine, free from the addictive properties of the opioid analgesics. **Diphenoxylate** is a structurally related agent that is also used in the treatment of diarrhoea.

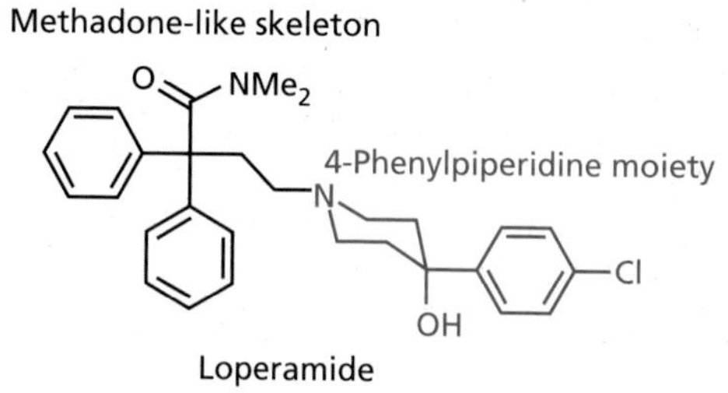

The molecule is a **diphenylpropylamine** structure containing a single asymmetric centre. When the molecule is drawn in the same manner as morphine, we would expect the *R*-enantiomer to be the more active enantiomer. This proves to be the case with the *R*-enantiomer being twice as potent as morphine, whereas the *S*-enantiomer is inactive. This is quite a dramatic difference. Because the *R*- and *S*-enantiomers have identical physical properties and lipid solubility, they should both reach analgesic receptors to the same extent, and so the difference in activity is most probably due to receptor–ligand interactions.

Many analogues of methadone have been synthesized, such as **dipipanone**, which is an oral analgesic, and L-α-acetylmethadol (LAAM) (Fig. 24.19). The latter has been used as a longer acting alternative for maintenance therapy in opioid dependence (see also buprenorphine, section 24.6.4). A methadone-like structure has also been linked to the 4-phenylpiperidine skeleton to produce a useful agent for the treatment of diarrhoea (Box 24.3).

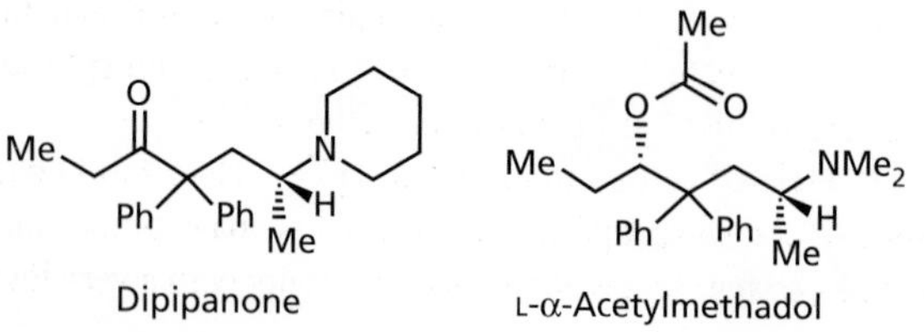

24.6.4 Rigidification

The strategy of rigidification is used to limit the number of conformations that a molecule can adopt. The aim is to retain the active conformation for the desired target and eliminate alternative conformations that might fit different targets (section 13.3.9). This should increase activity, improve selectivity, and decrease side effects. The best examples of this tactic in the analgesic field are the **orvinols** (or **oripavines**), which often show remarkably high activity. A comparison of these structures with morphine shows that an extra ring sticks out from what used to be the crossbar of the T-shaped morphine skeleton (Fig. 24.20).

Some remarkably powerful orvinols have been obtained (Box 24.4). **Etorphine** (Fig. 24.21), for example,

FIGURE 24.19 Dipipanone and L-α-acetylmethadol (LAAM).

For additional material see Web article 14: Orvinols on the Online Resource Centre at www.oxfordtextbooks.co.uk/orc/patrick6e/

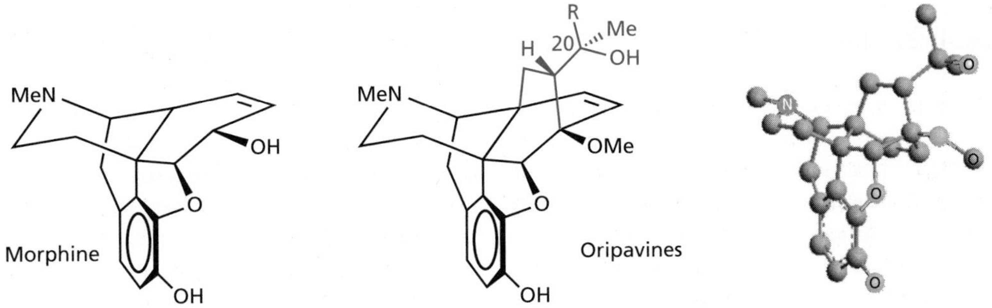

FIGURE 24.20 Comparison of morphine and orvinols.

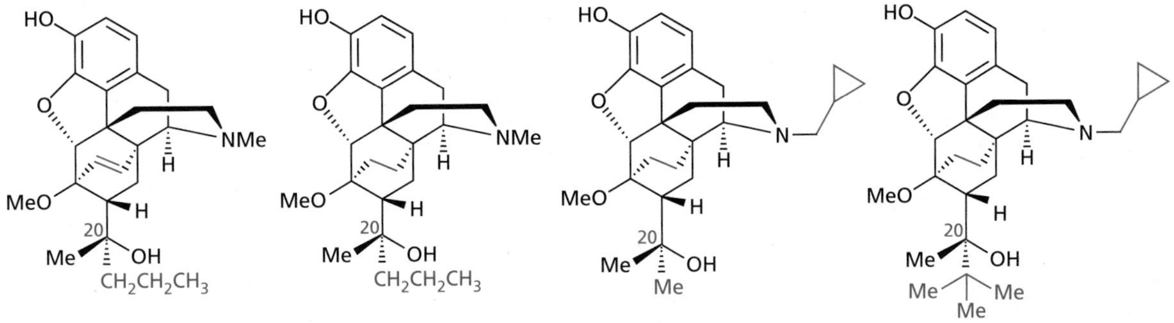

FIGURE 24.21 Etorphine and related structures

is 10 000 times more potent than morphine. This is due to a combination of its high hydrophobicity, which allows it to cross the blood–brain barrier 300 times more easily than morphine, allied to 20 times higher affinity for the analgesic receptor because of better binding interactions. At slightly higher doses than those required for analgesia, it can act as a 'knock-out' drug or sedative. It has a considerable margin of safety and is used to immobilize large animals, such as elephants. As the compound is so active, only very small doses are required and these can be dissolved in such small volumes (1 ml) that they can be placed in darts which can be fired into the hide of the animal. Reducing the double bond of etorphine increases activity over tenfold, and the resulting structure (**dihydroetorphine**) is one of the most potent analgesics ever reported. Dihydroetorphine is used in China as a strong analgesic and as a treatment for opioid addiction.

The presence of lipophilic groups at C20 (R in Fig. 24.20) is found to improve activity dramatically, indicating the

existence of an extra hydrophobic binding region in the receptor binding site.[1] The group best able to interact with this region is a phenethyl substituent, and the product containing this group is even more active than etorphine. As one might imagine, these highly active compounds have to be handled very carefully in the laboratory.

Because of their rigid structures, these compounds are highly selective agents for the analgesic receptors. Unfortunately, the increased analgesic activity is also accompanied by unacceptable side effects due to strong interactions with the μ receptor. Therefore, it was decided to see whether *N*-substituents, such as an allyl or cyclopropyl group, would give the oripavine equivalent of a pentazocine or a nalorphine—an agent acting as an antagonist at the μ receptor and an agonist at the κ receptor.

Adding a cyclopropyl group gives a very powerful antagonist called **diprenorphine** (Fig. 24.21), which is 100 times more potent than nalorphine and can be used to reverse the immobilizing effects of etorphine. Diprenorphine has no analgesic activity.

The related compound **buprenorphine** (Fig. 24.21) has similar clinical properties to drugs like nalorphine and pentazocine in that it has analgesic activity with a very low

Test your knowledge and practise your molecular modelling with Exercises 24.8–24.11 on the Online Resource Centre at www.oxfordtextbooks.co.uk/orc/patrick6e/

[1]It has been proposed that the phenylalanine aromatic ring on enkephalins interacts with this same binding region (see section 24.8.3).

BOX 24.4 Synthesis of the orvinols

The orvinols are synthesized from an alkaloid called **thebaine**, which is extracted from opium along with codeine and morphine. Although similar in structure to both these compounds, thebaine has no analgesic activity and is extremely toxic. There is a diene group present in ring C and when thebaine is treated with methyl vinyl ketone, a Diels–Alder reaction takes place to give an extra ring and increased rigidity to the structure (Fig. 1). Because a ketone group has been introduced, it is now possible to try the strategy of drug

extension by adding various groups to the ketone via a Grignard reaction. It is noteworthy that this reaction is stereospecific. The Grignard reagent complexes to both the 6-methoxy group and the ketone, and an alkyl group is then delivered to the less-hindered face of the ketone in an asymmetric reaction (Fig. 2). The final stage in the synthesis involves treatment with KOH and ethylene glycol to demethylate the methyl ether at position 3 without demethylating the methyl ether at position 6.

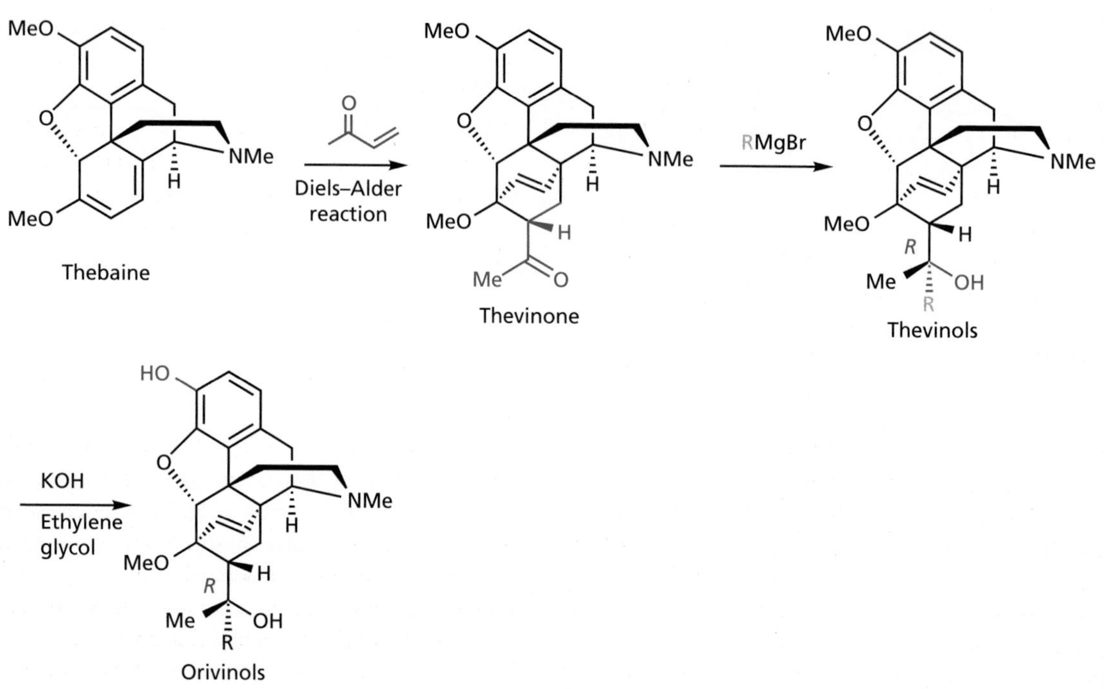

FIGURE 1 Formation of orvinols.

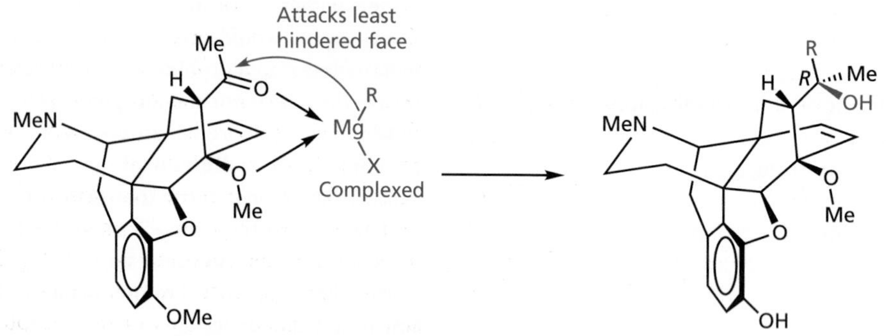

FIGURE 2 Grignard reaction leads to an asymmetric centre.

risk of addiction. It is a particularly safe drug because it has very little effect on respiration, and what little effect it does have actually decreases at high doses. Therefore, the risks of suffocation from a drug overdose are much smaller than with morphine. Buprenorphine has been used in hospitals to treat patients recovering from surgery, as well as those suffering from cancer. It has also been used as an alternative to methadone for weaning addicts off heroin. Its drawbacks include side effects such as nausea and vomiting, as well as the fact that it cannot be taken orally.

Buprenorphine has unusual receptor binding properties with respect to other opioids. It has a strong affinity for the μ receptor where it acts as a partial agonist, whereas it acts as an antagonist at the κ and δ receptors. Normally, one would expect compounds that act as antagonists at the μ receptor and agonists at the κ receptor to be safer analgesics, and so the clinical properties of buprenorphine are quite surprising. It is thought that the lack of serious side effects is related in some way to the rate at which buprenorphine interacts with the receptor. It is slow to bind, but once it has bound, it binds strongly and is slow to leave. As the effects of binding are gradual, it means that there are no sudden changes in transmitter levels. Buprenorphine is the most lipophilic compound in the orvinol series of compounds and enters the brain very easily, and so the slow onset of binding has nothing to do with how easily it reaches the receptor. Because buprenorphine binds very strongly, less of it is required to interact with a certain percentage of analgesic receptors than morphine. On the other hand, buprenorphine is only a partial agonist and is less efficient at activating analgesic receptors. This means that it is unable to reach the maximum level of analgesia which can be acquired by morphine. Thus, buprenorphine can produce analgesia at lower doses than morphine, but if the pain levels are high, buprenorphine is not as effective as morphine. Nevertheless, buprenorphine provides another example of an opioid analogue where analgesia has been separated from dangerous side effects.

KEY POINTS

- The addition of a 14-hydroxyl group or an *N*-phenethyl group usually increases activity as a result of interactions with extra binding regions.

- *N*-Alkylated analogues of morphine are easily synthesized by demethylating morphine to normorphine, then alkylating with alkyl halides.

- The addition of suitable *N*-substituents results in compounds which act as antagonists or partial agonists. Such compounds can be used as antidotes to morphine overdose, as treatment for addiction, or as safer analgesics.

- The morphinans and benzomorphans are analgesics which have a simpler structure than morphine and interact with analgesic receptors in a similar fashion.

- The 4-phenylpiperidines are a group of analgesic compounds which contain part of the analgesic pharmacophore present in morphine. They may bind to analgesic receptors slightly differently from analgesics of more complex structure.

- Methadone is a synthetic agent which contains part of the analgesic pharmacophore present in morphine. It is administered to drug addicts to wean them off heroin.

- Thebaine is an alkaloid derived from opium which lacks analgesic activity. It is the starting material for a three-stage synthesis of orvinols.

- Orvinols are extremely potent compounds due to enhanced receptor interactions and an increased ability to cross the blood–brain barrier.

- The addition of *N*-cycloalkyl groups to the orvinols results in powerful antagonists or partial agonists, which can be used to reverse the actions of etorphine, as antidotes for opioid overdose or addiction, or as safer analgesics.

24.7 **Agonists and antagonists**

We return now to look at a particularly interesting problem regarding the agonist/antagonist properties of morphine analogues. Why should such a small change as replacing an *N*-methyl group with an *N*-allyl group result in such a dramatic change in biological activity, such that an agonist becomes an antagonist? Why should a molecule such as nalorphine act as an agonist at one analgesic receptor and an antagonist at another? How can different receptors distinguish between such subtle changes in a molecule?

We shall consider one possible explanation. Let us assume that an opioid receptor exists in an active or an inactive conformation (Fig. 24.22a). The active conformation is capable of binding G-proteins and triggering signal transduction, while the inactive conformation is not. Let us further assume that an equilibrium exists between the two conformations and that the equilibrium shifts depending on what type of ligand is bound. If the active conformation binds an agonist (Fig. 24.22b), the equilibrium shifts to the active form leading to increased signal transduction. If an antagonist binds to the inactive conformation, the opposite happens (Fig. 24.22c).

This argument assumes that the binding sites of the active and inactive forms of the receptor are capable of distinguishing between the structures of an agonist and an antagonist. This is quite feasible as the binding sites are likely to have different conformations. We shall assume that the binding regions required to bind

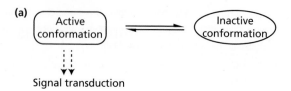

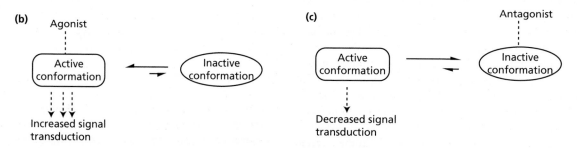

FIGURE 24.22 (a) Equilibrium between two receptor conformations. (b) Effect on adding an agonist. (c) Effect on adding an antagonist (see also Figure 8.16).

the opioid pharmacophore are identically positioned in both binding sites (the blue regions in Figs 24.23 and 24.24), but that an additional hydrophobic region is positioned closer to the ionic binding region in the inactive binding site than it is in the active binding

site. Let us now consider the binding of the agonist *N*-phenethylmorphine (Fig. 24.23). Like morphine, it binds using its phenol, aromatic, and amine functional groups. The aromatic ring of the phenethyl group is quite far from the amine group. Therefore, it overlaps

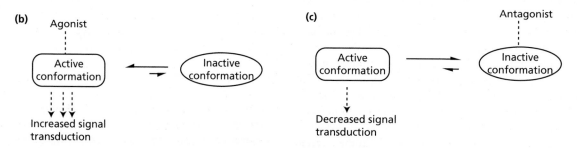

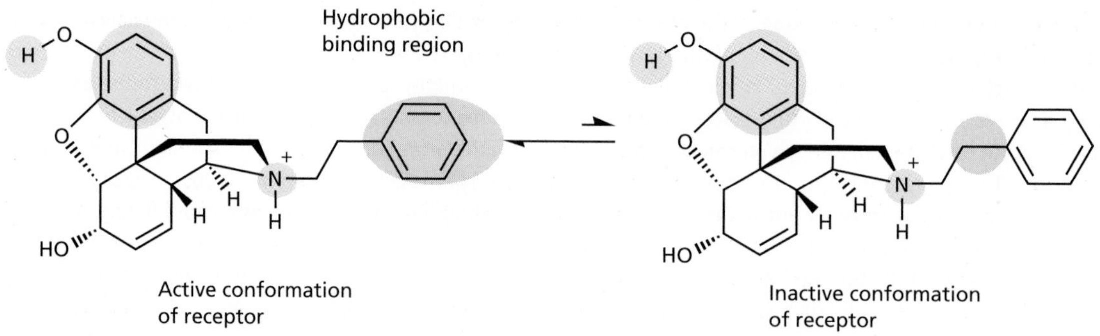

FIGURE 24.23 *N*-Phenethylmorphine binding interactions with the active and inactive conformations of the opioid receptor.

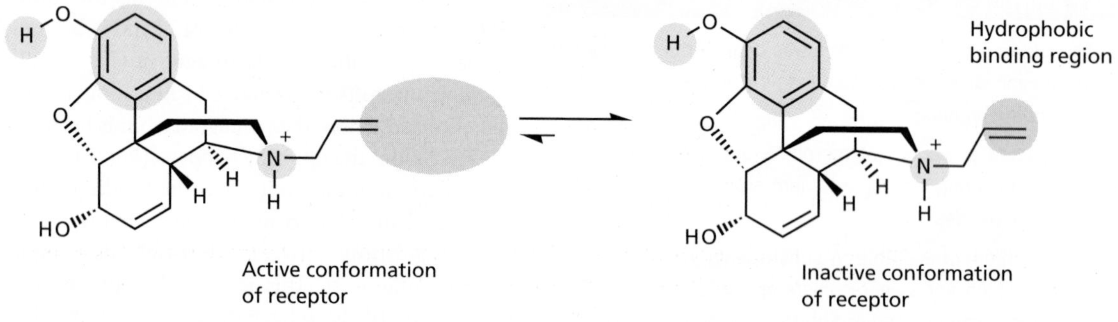

FIGURE 24.24 *N*-Allylmorphine binding interactions with the active and inactive conformations of the opioid receptor.

more effectively with the more distant hydrophobic region, causing the equilibrium to shift to the active form of the receptor.

Now consider what happens if the phenethyl group is replaced by an allyl group (Fig. 24.24). The allyl group is much closer to the amine and interacts better with a closer hydrophobic region. Therefore, the equilibrium would shift to the inactive conformation.

How then do we explain the fact that some opioids act as an agonist with one type of opioid receptor, and as an antagonist at another? We could propose that the relative positions of the extra hydrophobic regions are different in the different types of receptor. In Fig. 24.24, the allyl group is almost overlapping with the hydrophobic binding region of the active conformation. If this binding region was slightly closer in a different type of receptor, it would permit the allyl group to form a better interaction and have agonist activity.

24.8 Endogenous opioid peptides and opioids

24.8.1 Endogenous opioid peptides

Morphine relieves pain by binding to analgesic receptors in the CNS, which implies that endogenous chemicals must exist that interact with these receptors. The search for these natural analgesics took many years, but led, ultimately, to the discovery of the **enkephalins**. The term enkephalin is derived from the Greek meaning 'in the head', and that is exactly where the enkephalins are produced. There are two enkephalins called **Met-enkephalin** and **Leu-enkephalin** (Fig. 24.25). Both of the enkephalins are pentapeptides and have a slight preference for the δ receptor (Box 24.5). It has been proposed that

BOX 24.5 A comparison of opioids and their effects on opioid receptors

Table 24.1 shows the relative activities of different opioids as agonists, partial agonists, and antagonists at different opioid receptors. A plus sign indicates that the compound acts as an agonist, whereas a minus sign means that it acts as an antagonist. The number of plus signs or minus signs indicates the binding affinity. Plus signs in brackets indicate partial agonist activity.

The search for κ-selective agents has resulted in the clinically useful agents nalbuphine and butorphanol (Fig. 1). Unfortunately, many of the κ-selective agents are limited in their utility because they are partial agonists and are not potent enough to treat severe pain. Moreover, activation of the κ receptor has been associated with hallucinations and psychotomimetic side effects.

TABLE 24.1 Relative activities of opioids at opioid receptors.

Receptor	Mor	Meth	Peth	Etor	Fent	Pent	Nal	Bup	Nalo	Nalt	Lenk	End	Dyn
mu	+++	+++	++	+++	+++	−	−−	(+++)	−−−	−−−	+	+++	++
kappa	+		+	+++		++	(++)	−−	−−	−−−		+++	+++
delta	+		+	+++	+	+			−	−	+++	+++	+

Mor (morphine); Meth (methadone); Peth (pethidine); Etor (etorphine); Fent (fentanyl); Pent (pentazocine); Nal (nalorphine); Bup (buprenorphine); Nalo (naloxone); Nalt (naltrexone); Lenk (Leu-enkephalin); End (β-endorphin); Dyn (dynorphin)

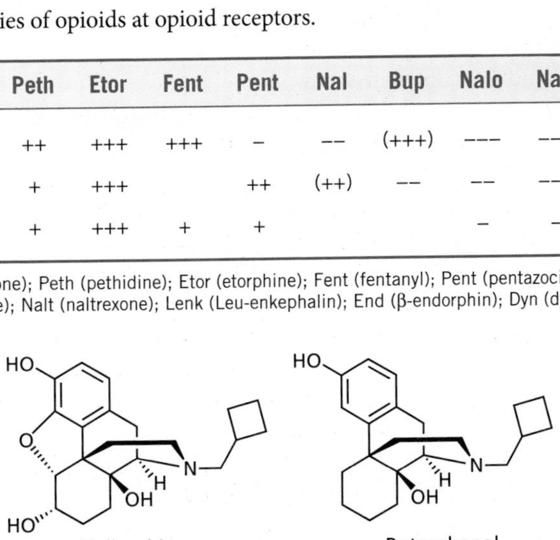

Nalbuphine

Butorphanol

FIGURE 1 Nalbuphine has the same activity as morphine, low addiction liability, no psychotomimetic activity, but is orally inactive. Butorphanol is also orally inactive.

H-Tyr-Gly-Gly-Phe-Met-OH   H-Tyr-Gly-Gly-Phe-Leu-OH
Met-enkephalin   Leu-enkephalin

FIGURE 24.25 Enkephalins.

Proenkephalin   Endorphins
Prodynorphin   ⟶ + enkephalins
Pro-opiomelanocortin   + dynorphins

FIGURE 24.26 Production of the body's natural painkillers.

enkephalins are responsible for the analgesic effects of acupuncture.

At least 15 endogenous peptides have now been discovered (the enkephalins, **dynorphins,** and the **endorphins**), varying in length from 5 to 33 amino acids. These compounds are thought to be neurotransmitters or neurohormones in the brain, and operate as the body's natural painkillers. They are mostly derived from three inactive precursor proteins—**proenkephalin, prodynorphin**, and **pro-opiomelanocortin** (Fig. 24.26).

All of these compounds have either the Met- or the Leu-enkephalin skeleton at their N-terminus, which emphasizes the importance of this pentapeptide structure towards analgesic activity. It has also been shown that tyrosine is essential to activity, and much has been made of the fact that there is a tyrosine skeleton in the morphine skeleton (Fig. 24.27).

If the crucial part of these molecules is the N-terminal pentapeptide, why should there be so many different peptides carrying out the same task? One suggestion is that the remaining peptide chain of each molecule is responsible for targeting each peptide to particular types of analgesic receptor. It is known that enkephalins show preference for the δ receptor, whereas dynorphins show selectivity for the κ receptor, and β-endorphins show selectivity to both the μ and δ receptors. This has led to a theory called the **message-address concept**, which proposes that part of a molecule is responsible for its pharmacological activity (the message) and another part is responsible for its target selectivity (the address) (see also sections 24.9.1 and 24.10).

H-Tyr-Pro-Trp-Phe-NH₂   H-Tyr-Pro-Phe-Phe-NH₂
Endomorphin-1   Endomorphin-2

FIGURE 24.28 Endomorphins.

The most recent endogenous opioid ligand was discovered in 1995 by two groups and was named **nociceptin** or **orphanin-FQ**. It is a heptadecapeptide derived from the protein **pronociceptin/orphanin FQ** and is a ligand for the **ORL₁-receptor** (section 24.4). Curiously, the N-terminal amino acid is phenylalanine rather than tyrosine and it appears that this plays a crucial role in receptor selectivity. The endogenous opioids such as the enkephalins, endorphins, and dynorphins have tyrosine at the N-terminus and have no affinity for the ORL₁-receptor, whereas nociceptin/orphanin-FQ has negligible affinity for the μ, κ, and δ receptors.

The **endomorphins** (Fig. 24.28) have also been recently discovered and are unlike previous opioid peptides. For a start, they are tetrapeptides, whereas all other opioid peptides are pentapeptides or larger. Secondly, the second and third amino acids in their skeleton differ from glycine, another break from convention. Finally, they have a primary amide functional group at the C-terminus. They do, however, have the mandatory tyrosine and phenylalanine residues that are present in other opioid peptides. The endomorphins have a strong affinity and selectivity for the μ receptor. However, there is some doubt as to whether these are truly endogenous opioids or whether they are merely breakdown products resulting from the extraction process used to isolate proteins.

For additional material see Web article 15: The message-address concept on the Online Resource Centre at www.oxfordtextbooks.co.uk/orc/patrick6e/

24.8.2 Analogues of enkephalins and δ-selective opioids

SAR studies on the enkephalins have shown the importance of the phenol ring and amino group of the tyrosine residue. Without either, activity is lost. If tyrosine is replaced

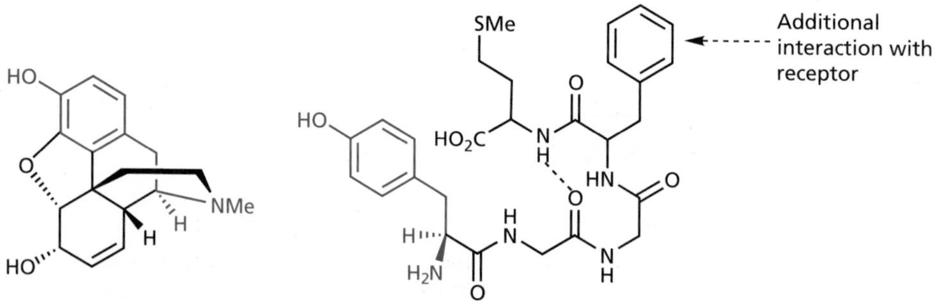

FIGURE 24.27 Comparison of morphine and Met-enkephalin (dashed line is a H-bond).

by another amino acid, activity is also lost—the only exception being D-serine.

It has been found that the enkephalins are inactivated by peptidase enzymes *in vivo*, with the most labile bond being the peptide link between tyrosine and glycine. Efforts have been made to synthesize analogues which are resistant towards this hydrolysis. It is possible to replace either, or both, of the glycine units with unnatural D-amino acids, such as D-alanine. Since D-amino acids do not occur naturally in the human body, peptidases do not recognize the structure and the peptide bond is not attacked. The alternative tactic of replacing L-tyrosine with D-tyrosine is not possible as it completely alters the relative orientation of the tyrosine aromatic ring with respect to the rest of the molecule. As a result, the analogue is unable to bind to the analgesic receptor and is inactive.

N-Methylating the peptide link also blocks peptidase hydrolysis. Another tactic is to use unusual amino acids which are not recognized by peptidases, or prevent the molecule from fitting the peptidase active site (Fig. 24.29). Unfortunately, the enkephalins also have some activity at the μ receptor and so the search for selective agents continues.

The first non-peptide structure to show selectivity for the δ receptor was the antagonist **naltrindole** (Box 24.6). Several selective non-peptide agonists have since been developed, such as **SB 213698** and **SNC-80** (Fig. 24.30).

24.8.3 Binding theories for enkephalins

It is clear from SAR studies on enkephalins that the tyrosine residue and the aromatic ring of phenylalanine are important for analgesic activity, which suggests that they act as important binding groups in their interaction with opioid receptors. This, in turn, implies that the receptor binding site contains two hydrophobic binding regions— one which interacts with the phenol ring of tyrosine (the T-binding region) and one which interacts with the aromatic ring of phenylalanine (the P-binding region) (Fig. 24.31). The T-binding region is distinct from the P-binding region in terms of its position and the fact that it must contain a group capable of forming a hydrogen bond to the phenol group of the ligand.

It has also been suggested that the two hydrophobic binding regions may be approximately equidistant from the ionic binding region. This is supported by various studies on the conformations of enkephalins, which indicate that the Gly-Gly segment introduces a bend into the peptide backbone of the molecule such that it adopts a folded conformation. Assuming that the active conformation is similar in nature, this means that the T-binding region is likely to be closer to the P-binding region than one might have imagined.

The possibility of two hydrophobic binding regions roughly equidistant from the ionic binding region provides a possible explanation for the different SAR results obtained for simple opioids such as pethidine, compared to rigid opioids such as morphine. It makes sense that morphine should mimic the tyrosine residue of the enkephalins and interact with the T-binding region (Fig. 24.32). An alternative binding mode with the P-binding region might not be possible because of bad steric interactions.

In contrast, pethidine is a smaller, more flexible molecule and may well bind more easily to the P-binding region

H-L-Tyr-Gly-Gly-L-Phe-L-Met-OH   Met-enkephalin—δ agonist and some μ activity.
H-L-Tyr-D-AA-Gly-NMe-L-Phe-L-Met-OH   Resistant to peptidase. Orally active.
N,*N*-Diallyl-L-Tyr-aib-aib-L-Phe-L-Leu-OH   Antagonist to δ receptor (aib = α-aminobutyric acid).
Longer enkephalins/endorphins   Increase in κ activity. Slight increase in μ activity.

FIGURE 24.29 Tactics to stabilize the bond between the tyrosine and glycine residues.

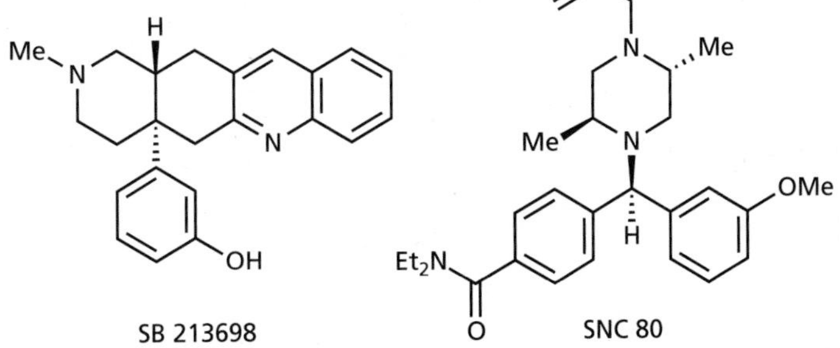

FIGURE 24.30 Non-peptide agonists that are selective for the δ receptor.

BOX 24.6 Design of naltrindole

The message-address concept has been extremely useful in designing selective opioids. **Leu-enkephalin** shows selectivity for the δ receptor, and it has been shown that the tyrosine residue acts as the analgesic message, while the aromatic ring of phenylalanine acts as the address for the δ receptor (Fig. 1). It is believed that the observed selectivity is due to the aromatic ring interacting with a binding region that is unique to the δ receptor.

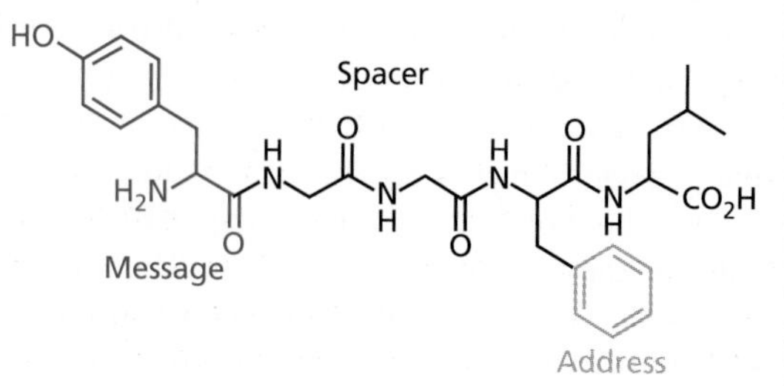

In an attempt to obtain non-peptide, δ-selective opioids, an aromatic ring was now fused to morphine-like structures to see whether it could act as an address segment. The position of the aromatic ring relative to the opioid message would be crucial, and success was achieved by fusing the aromatic ring to the C-ring of **naltrexone** to give **naltrindole**. Whereas naltrexone is a non-selective antagonist, naltrindole is a highly potent, δ-selective antagonist with 240 times more potency at the δ receptor. A molecular dynamics simulation demonstrated that Leu-enkephalin could adopt a conformation where the relative positions of the aromatic rings of Tyr and Phe were reasonably similar to the corresponding rings in naltrindole (Fig. 2).

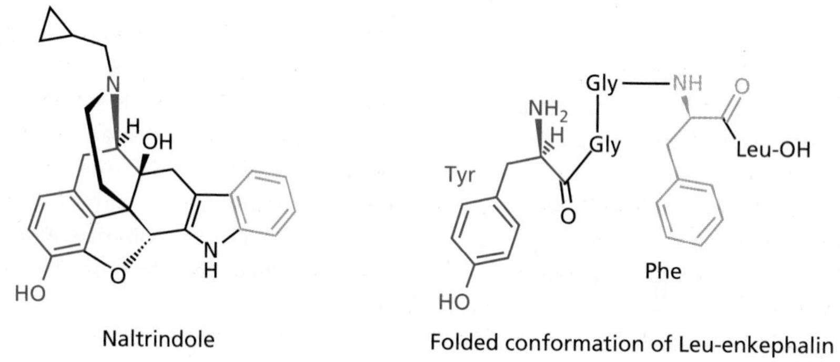

FIGURE 1 Leu-enkephalin.

Naltrindole

Folded conformation of Leu-enkephalin

FIGURE 2 Comparison of naltrindole with Leu-enkephalin.

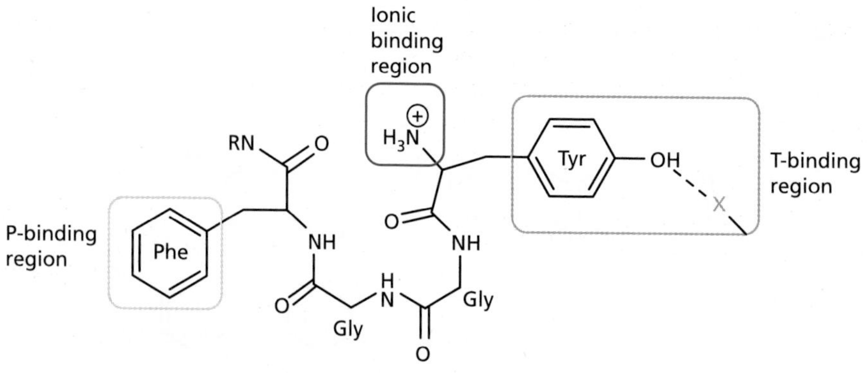

FIGURE 24.31 Proposed binding interactions of an enkephalin with its receptor binding site.

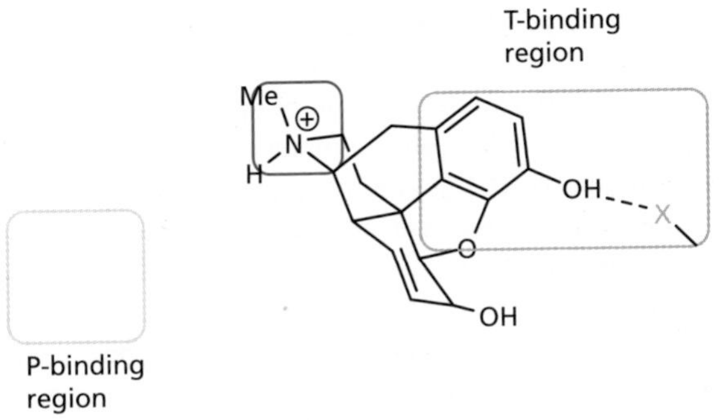

FIGURE 24.32 Interaction of morphine with proposed binding site.

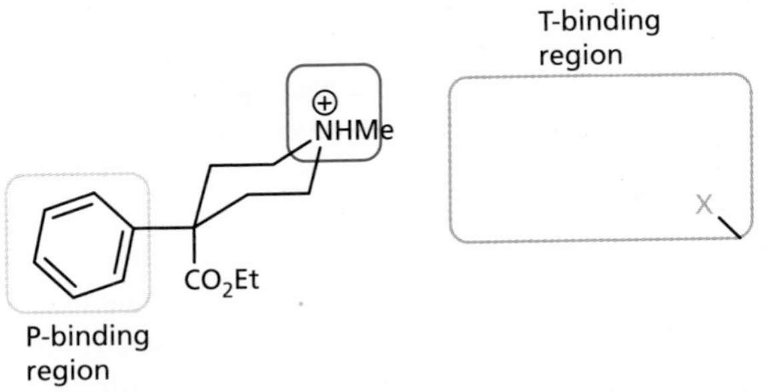

FIGURE 24.33 Interaction of pethidine with proposed binding site.

(Fig. 24.33). If so, this would explain why the activity of phenylpiperidines is not dependent on a phenol group, as the P-binding region lacks the necessary group to interact with it. The different binding mode would also explain why certain *N*-substituents on phenylpiperidines do not produce the same pharmacological results observed with rigid opioids. By interacting with the P-binding region, the phenylpiperidines would be differently orientated. This would mean that their *N*-substituents would occupy a different region of space in the binding site and be unable to make the same kind of interactions.

24.8.4 Inhibitors of peptidases

An alternative approach to pain relief is to enhance the activity of natural enkephalins by inhibiting the peptidase enzymes which metabolize them (**enkephalinases**). Studies have shown that the enzyme responsible for metabolism has a zinc ion present in the active site, as well as a hydrophobic pocket which normally accepts the phenylalanine side chain present in enkephalins. A dipeptide (L-Phe-Gly) was chosen as the lead compound and a thiol group was incorporated to act as a binding group for the zinc ion (a similar strategy was used in the design of the ACE inhibitor captopril—Case study 2). The result was a structure called **thiorphan** (Fig. 24.34), which was shown to have analgesic activity. It remains to be seen whether agents such as these will prove useful as analgesics in the clinic. However, an enkephalinase inhibitor called **racecadotril** (or **acetorphan**) (Fig. 24.35) is used in some countries for the treatment of diarrhoea. The agent is actually a prodrug for thiorphan, which is formed after hydrolysis of the ester and thioester groups.

24.8.5 Endogenous morphine

For many years, it was assumed that morphine itself could not possibly be an endogenous compound since the structure is an alkaloid produced by the poppy plant. Remarkably, morphine has now been identified as being present in tissues and body fluids, as have thebaine and codeine. It has also been demonstrated that human cells are capable of synthesizing morphine via a biosynthetic route similar to that used in the poppy plant. The levels of morphine are low and it is not yet clear what role it plays.

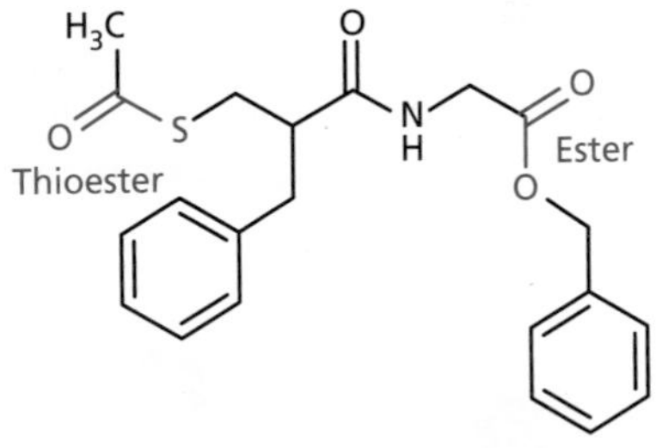

FIGURE 24.35 Racecadotril.

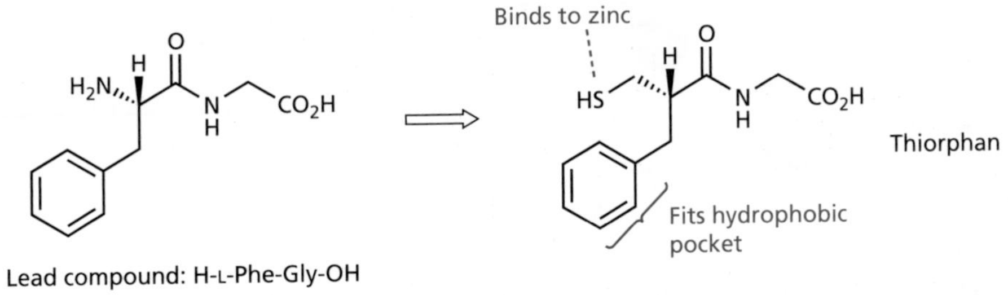

FIGURE 24.34 Development of thiorphan.

For additional material see Web article 16: Morphine biosynthesis on the Online Resource Centre at www.oxfordtextbooks.co.uk/orc/patrick6e/

24.9 The future

24.9.1 The message-address concept

There has been a lot of research in recent years aimed at developing opioids that show selectivity for a particular type of opioid receptor. The message-address concept has been extremely useful in guiding this research. The basis behind the concept is that opioids have a pharmacophore (the message) that is responsible for its activity, whether that be as an agonist or an antagonist. In addition, selective agents have a feature (the address) that is responsible for its receptor selectivity. This feature could be a functional group that interacts with a binding region that is unique to one type of receptor, resulting in increased affinity for that receptor. Alternatively, it could be a feature that acts as a steric shield and prevents the molecule from binding to some receptor types but not others. The message-address concept was first applied to endogenous opioids, such as the enkephalins (section 24.8.3), and has since been applied to the design of novel opioids (Box 24.6 and section 24.10).

For additional material see Web article 15: The message-address concept on the Online Resource Centre at www.oxfordtextbooks.co.uk/orc/patrick6e/

24.9.2 Receptor dimers

The opioids may be some of the oldest drugs used in medicine, but they are also some of the least understood. Investigations are still ongoing to try and discover the Holy Grail of opioid research—an opioid analgesic that is potent, orally active, and devoid of serious side effects. A better understanding of opioid receptors and the manner in which they interact with each other will help immensely in this ambitious goal.

There is now good evidence that opioid receptors form dimers in specific tissues (Fig. 24.36). These can exist as homodimers involving two identical opioid receptor types or as heterodimers where the receptor types are different. It is thought that the transmembrane (TM) regions of the component receptors are intertwined, resulting in the equivalent of two hybrid receptors. Each hybrid receptor would have the same overall arrangement of seven TM regions as in the monomeric receptor, but five of the TM regions would be contributed by one of the receptor proteins, while the remaining two TM regions would be contributed from the other receptor. Therefore, it is quite possible that the binding of a ligand to one of the hybrid receptors in a homodimer will affect the ability of the other hybrid receptor to bind a second ligand. For example, it is thought that an antagonist binding to one of the hybrid receptor binding sites will cause a conformational change in the dimeric complex that distorts the binding site of the second hybrid and prevent binding.

The picture becomes more complex when one considers heterodimeric receptors. Here, the hybrid receptors are not identical and so selective opioids will show selectivity for one of the hybrid receptors over the other, depending on which part of the receptor is most important in binding selectivity. For example, in κ receptors, there is a glutamic acid residue in the sixth TM region that is important in binding κ-selective opioids. This means that κ-selective opioids will bind to receptor hybrid A in the complex shown in Fig. 24.36 if the light spheres represent a δ receptor and the dark spheres represent a κ receptor. Similarly, the extracellular loop 3 in the δ receptor is important in binding

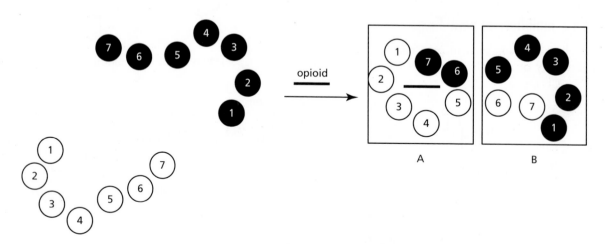

FIGURE 24.36 Formation of a receptor dimer with bound ligand.

δ-selective opioids, and so these agents would prefer to bind to receptor hybrid B. In either case, the binding of a selective antagonist to one of the hybrid receptors can result in antagonism at the other if the binding results in a conformational change over the whole complex. This can explain why selective opioid ligands appear to give contradictory results when tested on different tissues. For example, **norbinaltorphimine** (see Fig. 24.39) is a selective κ-antagonist when tested on some types of tissue, but a δ-antagonist when tested on others. This used to be explained by proposing that different receptor subtypes were present in different tissues, but the same results can be achieved by proposing the presence of κ-δ receptor heterodimers in some tissues but not others.

Heterodimeric receptors are of current interest because it is believed that the tolerance and dependence effects associated with μ-agonists might be caused by their interaction with δ-μ heterodimers, rather than by interaction with unassociated μ receptors. A bivalent ligand (**MDAN-21**) (Fig. 24.37) consisting of the μ-selective agonist **oxymorphone** linked to the δ-selective antagonist **naltrindole** has been found to have 50-fold more potency than morphine, without causing tolerance or dependence. This has exciting potential for the development of a new generation of safer opioid analgesics with fewer side effects.

For additional material see Web article 12: Opioid dimers and receptors on the Online Resource Centre at www.oxfordtextbooks.co.uk/orc/patrick6e/

24.9.3 Selective opioid agonists versus multi-targeted opioids

The early hopes of finding a highly selective κ-agonist with minimal side effects were dashed when it was found that psychotomimetic and dysphoric side effects were associated with activation of the κ receptor. Research into designing δ-selective agonists is still in progress. The main problem is to design structures that are potent and can cross the blood–brain barrier. If this could be achieved, it would allow an understanding of whether the δ receptor has any role in addiction and whether safe δ-selective agonists are feasible. However, it may be more advantageous to design opioids that have controlled activities at a combination of opioid and non-opioid receptors.

Test your understanding and molecular modelling with Exercises 24.12–24.18 on the Online Resource Centre at www.oxfordtextbooks.co.uk/orc/patrick6e/

24.9.4 Peripheral-acting opioids

Another approach has been to design opioids that act on the peripheral nervous system rather than the central nervous system. **Nalfurafine** (TRK 820) is one such agent that is used in Japan as a κ-agonist for the treatment of uremic pruritus (section 24.10). Another agent is **naloxegol** which is a PEGylated opioid antagonist prepared from naloxone (Fig. 24.38). It was approved in 2014 for the treatment of opioid-induced constipation. PEGylation prevents the drug crossing the blood–brain barrier into the CNS.

24.10 Case study: design of nalfurafine

The message-address concept has been a very useful guideline in designing opioid structures showing selectivity for different opioid receptors. In this case study, we will look at how a non-selective antagonist was developed into

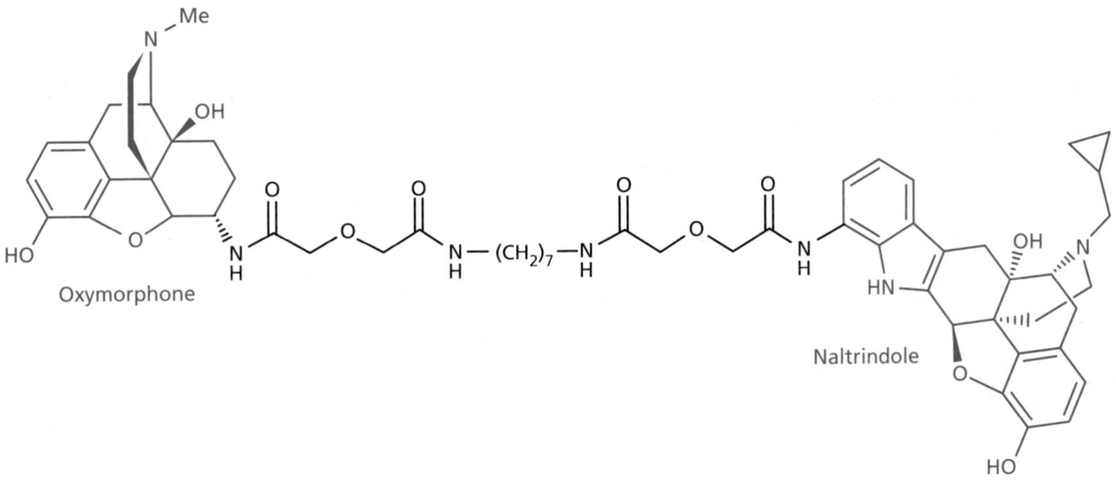

FIGURE 24.37 The bivalent ligand MDAN-21.

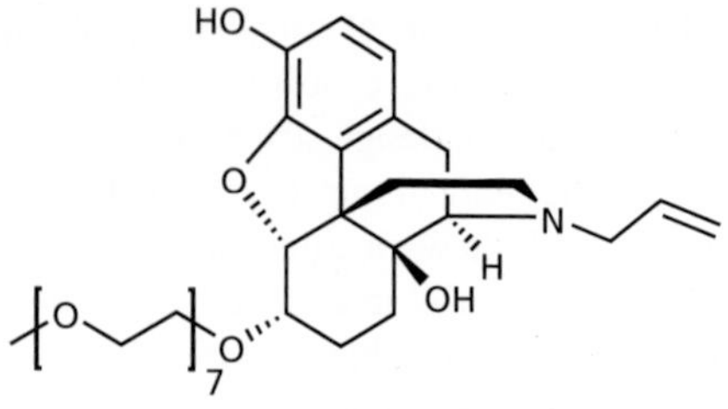

FIGURE 24.38 Naloxegol.

a κ-selective antagonist, and then a κ-selective agonist. The story begins with the design of an opioid dimer that was meant to bind simultaneously to the two κ receptors making up a κ,κ-receptor homodimer (section 24.9.2). There is evidence that the separation between the κ receptors in the κ,κ-homodimer is smaller than for other types of homodimers. Therefore, an opioid dimer with a short linker unit between the two opioid structures should show selectivity. Dimers of the non-selective antagonist **naltrexone** (Fig. 24.11) were synthesized leading to the discovery of **norbinaltorphimine (nor-BNI)** (Fig. 24.39), which is a κ-selective antagonist used frequently in pharmacological studies. Despite the apparent success of the strategy, it soon became clear that the separation between the two opioid moieties is too short to allow simultaneous interactions with both components of a receptor dimer. Moreover, SAR studies have demonstrated that only one of the full opioid pharmacophores is required for activity. The second opioid structure certainly plays a role in the high potency and selectivity observed for nor-BNI, but the intact pharmacophore is not required. It was therefore concluded that the dimer was interacting with a single κ receptor, such that one of the opioid components binds to the receptor binding site and acts as the message, while the second opioid component contains a specific feature that serves as the address and interacts with a binding region that is unique to the κ receptor. This extra interaction

would explain both the selectivity and the increased potency. Further work demonstrated that the basic amine group is the crucial feature in the address half of the molecule and that it interacts with a unique glutamate residue. Simplification of the structure led to an analogue with an octahydroisoquinoline ring system, which also acts as a κ-selective antagonist.

This antagonist was now used as the lead compound for the design of a κ-selective agonist. There are several examples of projects where an agonist lead compound has been modified to obtain an antagonist, but relatively few where an antagonist has been modified to obtain an agonist. In the former case, the normal strategy is to add an extra functional group in order to form an extra binding interaction with the target receptor, such that the resulting induced fit differs from that caused by the binding of an agonist. In order to modify an antagonist to an agonist, the opposite strategy is required—it is necessary to identify an extra interaction that is causing antagonist activity, and then remove the group that is involved.

It was thought that the feature responsible for antagonism might be the bulky, hydrophobic octahydroisoquinoline ring system. Therefore, a further simplification was carried out, replacing this ring with a less bulky, flexible, acyclic chain of sufficient length to match the original bicyclic ring (Fig. 24.40). The flexibility of the chain was considered important as this would increase the chances of it adopting the correct active conformation for κ-agonist activity. Various chains were studied. For example, a thioester group with a pentyl side chain was essentially inactive. However, replacing the thioester with an amide group resulted in agonist activity with κ-selectivity. The side chain was then modified by reintroducing some rigidity in the form of a double bond and an aromatic ring. This resulted in improved selectivity and fewer psychotomimetic side effects. A variety of analogues were prepared containing different heteroaromatic rings instead of the aromatic ring; the best of these was **nalfurafine**.

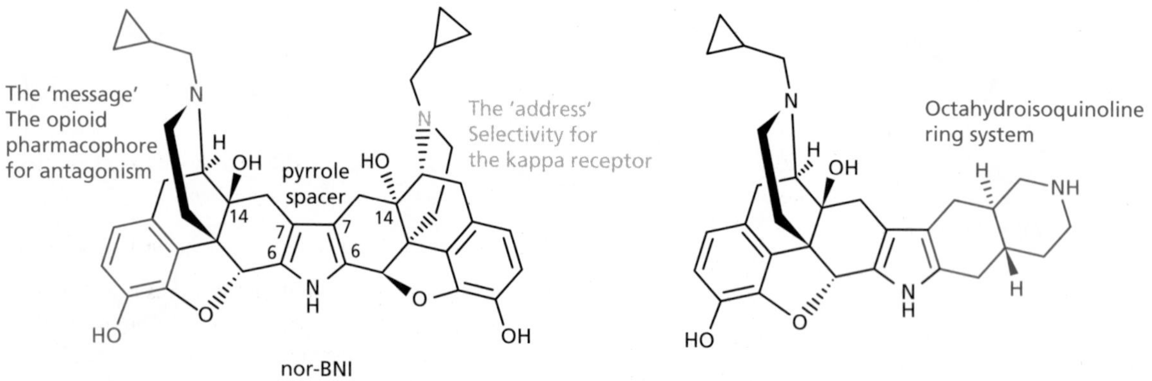

FIGURE 24.39 Norbinaltorphimine and a simplified analogue.

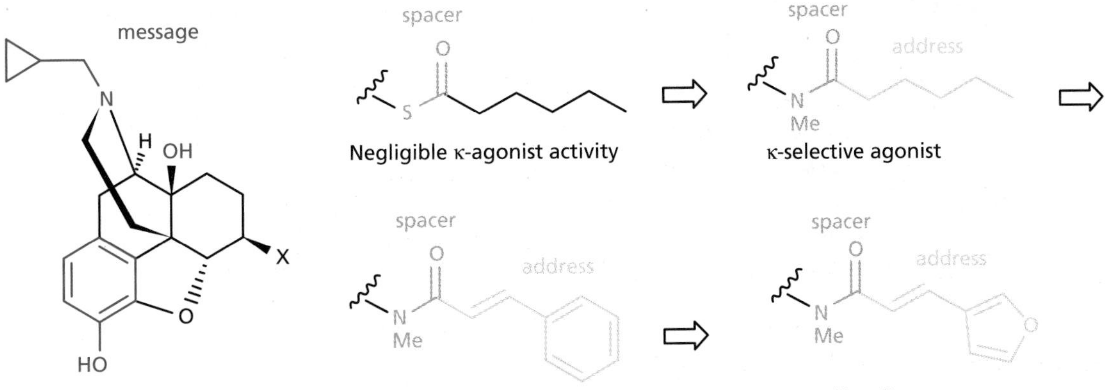

FIGURE 24.40 Design of nalfurafine.

It is interesting to note that κ-selectivity in this structure appears to be related to the presence of the hydrophobic heteroaromatic ring. This appears to contradict the earlier finding with antagonists where an ionized basic group is required for selectivity. However, given the increased flexibility of the modified address segment, it is possible that the heteroaromatic ring is interacting with a different amino acid residue in the binding site.

Nalfurafine is free of the serious side effects of morphine, as well as some of the common side effects of other κ-selective agonists, for example psychotomimetic and dysphoric effects. It was originally proposed as an analgesic in surgery, but sedative effects meant that it was not approved. However, low doses have been found to inhibit the itching associated with the injection of morphine. The compound was subsequently approved as an anti-itching (antipruritic) medication in 2009 for patients undergoing dialysis, and was the only κ-selective agent clinically approved at that time.

KEY POINTS

- It is proposed that there are two accessory hydrophobic binding regions in the receptor binding site. An agent will act as an agonist or antagonist depending on which of these regions it can access.

- Enkephalins, dynorphins, endomorphins, and endorphins are peptides which act as the body's natural painkillers. The presence of an N-terminal tyrosine is crucial to activity.

- Analogues of enkephalins have been designed to be more stable to peptidases by the inclusion of unnatural amino acids, D-amino acids, or N-methylated peptide links.

- Enkephalinase inhibitors may have a future role as analgesics by inhibiting the metabolism of enkephalins.

- The existence of homodimeric and heterodimeric opioid receptors has an important role in understanding the activity of opioids and in designing novel opioids.

- The message-address concept has been used to design opioids that are selective for a particular type of opioid receptor.

QUESTIONS

1. Morphine is an example of a plant alkaloid. Alkaloids tend to be secondary metabolites that are not crucial to a plant's growth and are produced when the plant is mature. If that is the case, what role do you think these compounds have in plants, if any?

2. The synthesis in Box 24.2 shows that N-alkylated analogues can be synthesized by N-alkylation directly or by a two-stage process involving N-acylation. Why might a two-stage process be preferred to direct N-alkylation? What

sort of products could not be synthesized by the two-stage process? Is this likely to be a problem?

3. Show how you would synthesize nalorphine (Fig. 24.11).

4. Pethidine has been used in childbirths as it is short acting and less hazardous than morphine in the newborn baby. Several drugs taken by the mother before giving birth can prove hazardous to a newly born child, but less so before birth. Why is this?

5. Show how you would synthesize diprenorphine and buprenorphine.

6. Why is buprenorphine considered the most lipophilic of the oripavine series of compounds?

7. Identify the potential hydrogen bond donors and acceptors in morphine. Structure–activity relationships reveal that one functional group in morphine is important as a hydrogen bond donor or as a hydrogen bond acceptor. Which group is that?

8. Propose the likely analgesic activity of 3-acetyl morphine relative to morphine, heroin, and 6-acetylmorphine.

9. Describe how you would synthesize the *N*-phenethyl analogue of morphine.

10. The *N*-phenethyl analogue of morphine is a semi-synthetic product. What does this mean?

11. Explain whether you think the following structures would act as agonists or antagonists.

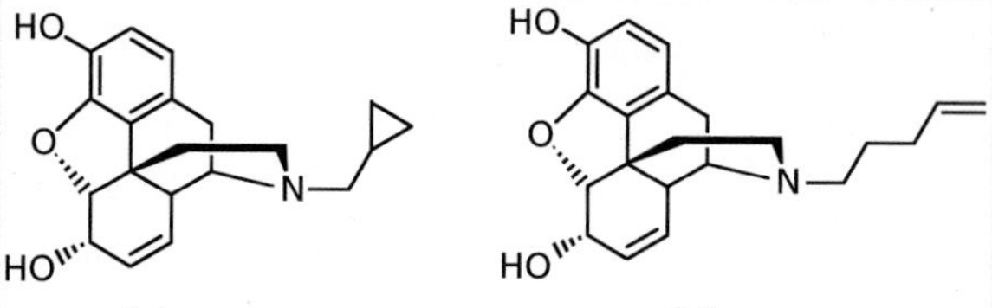

(a)   (b)

12. Thebaine has no analgesic activity. Suggest why this might be so.

13. Morphine is the active principle of opium. What is meant by an active principle?

14. Identify the asymmetric centres in morphine.

15. How could heroin be synthesized from morphine? What problems does this pose for drug regulation authorities?

16. Propose a synthesis of naloxegol from naloxone.

Multiple-choice questions are available on the Online Resource Centre at www.oxfordtextbooks.co.uk/orc/patrick6e/

FURTHER READING

Abraham, D. J. (ed.) (2003) Narcotic analgesics. in *Burger's medicinal chemistry and drug discovery,* 6th edn. Chapter 7, John Wiley and Sons, New York.

Corbett, A. D., et al. (2006) 75 Years of opioid research: the exciting but vain quest for the Holy Grail. *British Journal of Pharmacology,* **147,** S153–62.

Feinberg, A. P., Creese, I., and Snyder, S. H. (1976) The opiate receptor: a model explaining structure-activity relationships of opiate agonists and antagonists. *Proceedings of the National Academy of Sciences of the USA,* **73,** 4215–9.

Hruby, V. J. (2002) Designing peptide receptor agonists and antagonists. *Nature Reviews Drug Discovery,* **1,** 847–58.

Kreek, M. J., LaForge, K. S., and Butelman, E. (2002) Pharmacotherapy of addictions. *Nature Reviews Drug Discovery,* **1,** 710–25.

Pouletty, P. (2002) Drug addictions: towards socially accepted and medically treatable diseases. *Nature Reviews Drug Discovery,* **1,** 731–6.

Roberts, S. M., and Price, B. J. (eds.) (1985) Discovery of buprenorphine, a potent antagonist analgesic. in *Medicinal chemistry—the role of organic research in drug research.* Chapter 7, Academic Press, New York.

Williams, D. A., and Lemke, T. L. (eds.) (2002) Opioid analgesics. in *Foye's principles of medicinal chemistry,* 5th edn. Chapter 19, Lippincott Williams and Wilkins, Philadelphia.

Titles for general further reading are listed on p.845.

Anti-ulcer agents

25.1 Peptic ulcers

25.1.1 Definition

Peptic ulcers are localized erosions of the mucous membranes of the stomach or duodenum. The pain associated with ulcers is caused by irritation of exposed surfaces by the stomach acids. Before the appearance of effective anti-ulcer drugs in the 1960s, ulcer sufferers often suffered intense pain for many years and, if left untreated, the ulcer could result in severe bleeding and even death. For example, the film star Rudolph Valentino died from a perforated ulcer in 1926 at the age of 31. It is also believed that William Pitt the Younger (Britain's youngest Prime Minister) died of a perforated ulcer at the age of 46.

25.1.2 Causes

The causes of ulcers have been disputed. Stress, alcohol, and diet have been considered important factors, but there is no clear evidence for this. Scientific evidence indicates that the two main culprits are the use of non-steroidal anti-inflammatories (NSAIDS) or the presence of a bacterium called *Helicobacter pylori*. As far as the NSAIDS are concerned, agents such as **aspirin** inhibit the enzyme **cyclooxygenase 1 (COX-1)**. This enzyme is responsible for the synthesis of prostaglandins that inhibit acid secretion and protect the gastric mucosa. Once an ulcer has erupted, the presence of gastric acid aggravates the problem and delays recovery.

25.1.3 Treatment

Anti-ulcer therapy has been a huge money spinner for the pharmaceutical industry with drugs such as **cimetidine**, **ranitidine**, and the **proton pump inhibitors (PPIs)**. None of these drugs were available until the 1960s, however, and it is perhaps hard for us now to appreciate how dangerous ulcers could be before that. In the early 1960s, the conventional treatment was to try to neutralize gastric acid in the stomach by administering antacids. These were bases such as sodium bicarbonate or calcium carbonate. The dose levels required for neutralization were large and caused unpleasant side effects. Relief was only temporary and patients were often advised to stick to rigid diets, such as strained porridge and steamed fish. Ultimately, the only answer to severe ulcers was a surgical operation to remove part of the stomach.

The first effective anti-ulcer agents were the **H_2 histamine antagonists** which appeared in the 1960s. These were followed in the 1980s by the PPIs. The discovery of *H. pylori* then led to the use of antibacterial agents in anti-ulcer therapy. The current approach for treating ulcers caused by *H. pylori* is to use a combination of drugs, which includes a PPI such as **omeprazole**, and two antibiotics such as **amoxicillin** and **metronidazole**.

25.1.4 Gastric acid release

Gastric juices consist of digestive enzymes and hydrochloric acid designed to break down food. Hydrochloric acid is secreted from **parietal cells**, and the stomach secretes a layer of mucus to protect itself from its own gastric juices. Bicarbonate ions are also released and are trapped in the mucus to create a pH gradient within the mucus layer.

The H_2 antagonists and PPIs both work by reducing the amount of gastric acid released into the stomach by the parietal cells lining the stomach wall (Fig. 25.1). These parietal cells are innervated with nerves (not shown on the diagram) from the autonomic nervous system (sections 22.1–22.2). When the autonomic nervous system is stimulated, a signal is sent to the parietal cells culminating in the release of the neurotransmitter **acetylcholine** at the nerve termini. Acetylcholine activates the cholinergic receptors of the parietal cells leading to the release of gastric acid into the stomach. The trigger for this process is provided by the sight, smell, or even the thought of food. Thus, gastric acid is released before food has even entered the stomach.

Nerve signals also stimulate a region of the stomach called the **antrum**, which contains hormone-producing

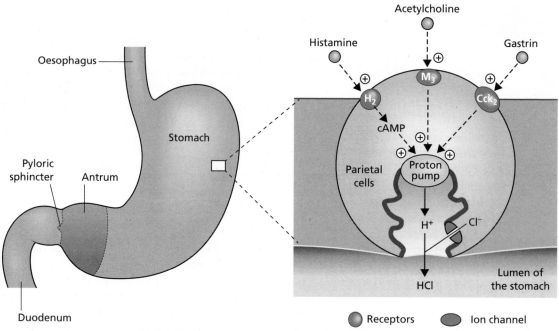

FIGURE 25.1 Factors influencing the release of gastric acid.

cells known as G cells. The hormone released is a peptide called **gastrin** (Fig. 25.2), which is also released when food is present in the stomach. Gastrin moves into the blood supply and travels to the parietal cells, further stimulating the release of gastric acid. Release of gastric acid should, therefore, be inhibited by antagonists blocking either the cholinergic receptor or the receptor for gastrin.

Agents which block the cholinergic receptor are known as **anticholinergic drugs** (section 22.9). These agents certainly block the cholinergic receptor in parietal cells and inhibit release of gastric acid. Unfortunately, they also inhibit cholinergic receptors at other parts of the body and cause unwanted side effects.

The local hormone **histamine** also stimulates the release of gastric acid by interacting with a specific type of histamine receptor called the **H$_2$ receptor**. Thus, histamine antagonists have proved to be important anti-ulcer drugs, although they have now largely been superseded by the PPIs, which block the mechanism by which hydrochloric acid is released from parietal cells.

25.2 H$_2$ antagonists

The first breakthrough in anti-ulcer therapy came with the design of the H$_2$ antagonist **cimetidine** (**Tagamet**) (see Fig. 25.32) produced by the company Smith Kline and French (SKF). The cimetidine programme started in 1964 and was one of the early examples of rational drug design. Up until that time, many of the successes in medicinal chemistry involved the fortuitous discovery of useful pharmaceutical agents from natural sources, and the study of analogues often synthesized on a trial-and-error basis. Although this approach yielded a large range of medicinal compounds, it was wasteful in terms of the time and effort involved. Nowadays, the emphasis is on rational drug design using the tools of X-ray crystallography, molecular modelling, and genetic engineering (Chapters 13 and 17). Unfortunately, such tools were not available in the 1960s and the story of cimetidine is a good example of how to carry out rational drug design when the target has not been identified or isolated.

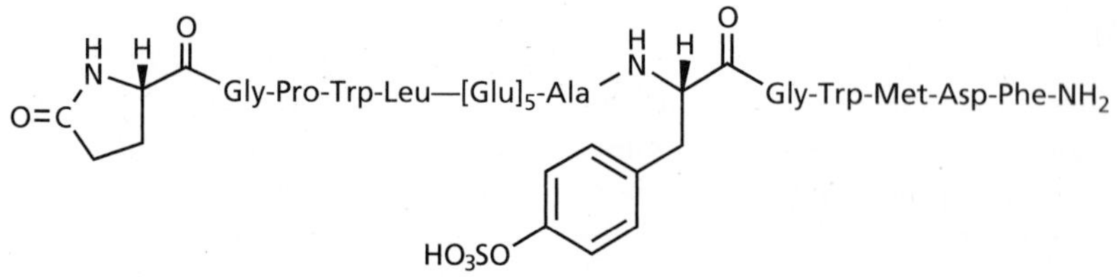

FIGURE 25.2 Gastrin.

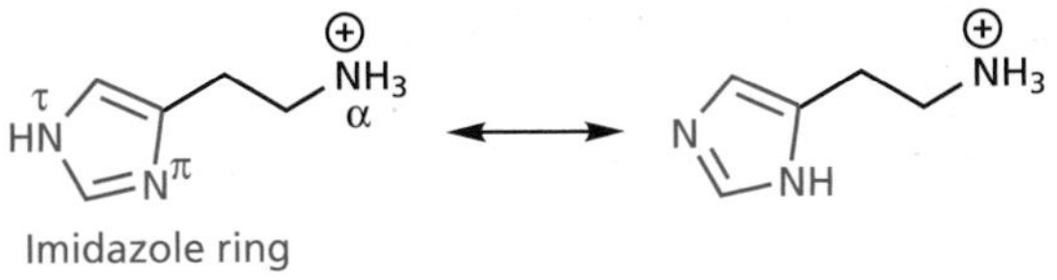

FIGURE 25.3 Histamine.

The remarkable aspect of the cimetidine story is that at the onset of the project there were no lead compounds available, and it was not even known if the target histamine receptor existed! In 1964, the best hope of achieving an anti-ulcer agent appeared to be in finding a drug which would block the hormone gastrin. Several research teams were active in this field, but the research team at SKF decided to follow a different tack altogether.

It was known experimentally that histamine (Fig. 25.3) stimulated gastric acid release *in vitro*, so the SKF team proposed that an antihistamine agent might be effective in treating ulcers. At the time, this was a highly speculative proposal as it was by no means certain that histamine played any significant role *in vivo*. Many workers at the time discounted the importance of histamine, especially when it was found that conventional antihistamines failed to inhibit gastric acid release. This suggested the absence of histamine receptors in the parietal cells. The fact that histamine had a stimulatory effect was explained away by suggesting that histamine coincidentally switched on the gastrin or cholinergic receptors. Even if a histamine receptor was present, opponents argued that blocking it would have little effect as the receptors for acetylcholine and gastrin would remain unaffected and could still be activated by their respective messengers. Initiating a project which had no known target and no known lead compound was unprecedented, and represented a massive risk. Indeed, for a long time little progress was made and it is said that company accountants demanded that the project be terminated. It says much for the scientists involved that they stuck to their guns and eventually confounded their critics. Why did the SKF team persevere in their search for an effective antihistamine? What was their reasoning? Before answering that, let us look at histamine itself and the antihistamines available at that time.

25.2.1 Histamine and histamine receptors

Histamine contains an imidazole ring which can exist in two tautomeric forms, as shown in Fig. 25.3. Attached to the imidazole ring is a two-carbon chain with a terminal α-amino group. The pK_a of this amino group is 9.80, which means that at a plasma pH of 7.4, the side chain of histamine is 99.6% ionized. The pK_a of the imidazole ring is 5.74 and so the ring is mostly un-ionized at pH 7.4 (Fig. 25.4). Note that the lower the pK_a value, the more acidic the proton. It is also useful to remember that 50% ionization takes place when the pH is the same value as the pK_a (section 11.3).

Whenever cell damage occurs, histamine is released and stimulates the dilatation and increased permeability of small blood vessels. This allows defensive cells such as white blood cells to be released from the blood supply into an area of tissue damage and to combat any potential infection. Unfortunately, the release of histamine can also be a problem. Allergic reactions and irritations are caused by release of histamine when it is not really needed.

The early antihistamine drugs were therefore designed to treat conditions such as hay fever, rashes, insect bites, or asthma. Two examples of these early antihistamines are **mepyramine** and **diphenhydramine** ('**Benadryl**') (Fig. 25.5), neither of which has any effect on gastric acid release.

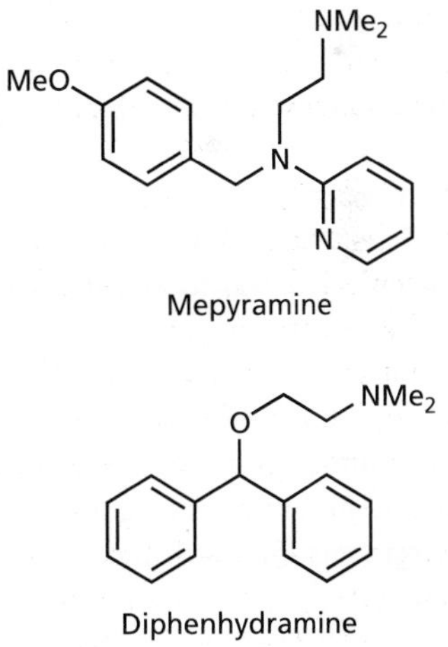

Mepyramine

Diphenhydramine

FIGURE 25.5 Early antihistamines.

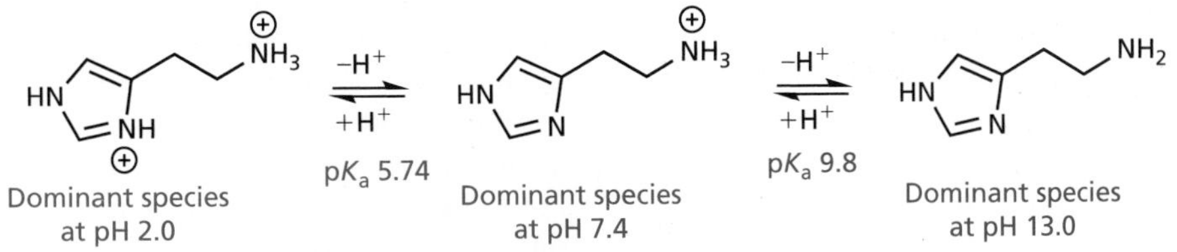

Dominant species at pH 2.0 $-H^+$ / $+H^+$ pK_a 5.74   Dominant species at pH 7.4 $-H^+$ / $+H^+$ pK_a 9.8   Dominant species at pH 13.0

FIGURE 25.4 Ionization of histamine.

Bearing this in mind, why did the SKF team persevere with the antihistamine approach? The main reason was the fact that conventional antihistamines failed to inhibit *all* the then-known actions of histamine. For example, they failed to fully inhibit the dilatation of blood vessels induced by histamine. The SKF scientists, therefore, proposed that there might be two different types of histamine receptor, analogous to the two types of cholinergic receptor mentioned in Chapter 22. Histamine—the natural messenger—would switch both on equally effectively and would not distinguish between them, whereas suitably designed antagonists might be capable of making that distinction. By implication, this meant that the conventional antihistamines known in the early 1960s were already selective in inhibiting the histamine receptors involved in the inflammation process (classified as H_1 receptors), rather than the proposed histamine receptors responsible for gastric acid secretion (classified as H_2 receptors).

It was an interesting theory, but the fact remained that there was no known antagonist for the proposed H_2 receptors. Until such a compound was found, it could not be certain that the H_2 receptors even existed.

Test your understanding and practise your molecular modelling with Exercise 25.1 on the Online Resource Centre at www.oxfordtextbooks.co.uk/orc/patrick6e/

25.2.2 Searching for a lead

25.2.2.1 Histamine

The SKF team obviously had a problem. They had a theory but no lead compound. How could they make a start?

Their answer was to start from histamine itself. If the hypothetical H_2 receptor existed, then histamine must bind to it. The task then was to vary the structure of histamine in such a way that it would bind as an antagonist rather than an agonist.

This meant exploring how histamine itself bound to its receptors. Structure–activity relationship (SAR) studies on histamine and histamine analogues revealed that the binding requirements for histamine to the H_1 receptors were as follows:

- the side chain had to have a positively charged nitrogen atom with at least one attached proton—quaternary ammonium salts which lacked such a proton were extremely weak in activity;
- there had to be a flexible chain between the above cation and a heteroaromatic ring;
- the heteroaromatic ring did not have to be imidazole, but it did have to contain a nitrogen atom with a lone pair of electrons, *ortho* to the side chain.

For the proposed H_2 receptor, SAR studies were carried out experimentally by determining whether histamine

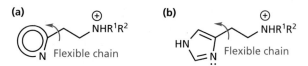

FIGURE 25.6 Summary of structure–activity relationship (SAR) results: (a) SAR for agonists at the H_1 receptor; (b) SAR for agonists at the proposed H_2 receptor.

analogues could stimulate gastric acid release in stomach tissue. The essential SAR requirements were the same as for the H_1 receptor except that the heteroaromatic ring had to contain an amidine unit (HN–CH=N:).

The results are summarized in Fig. 25.6 and appear to show that the terminal α-amino group is involved in a binding interaction with both types of receptor via ionic and/or hydrogen bonding, while the nitrogen atom(s) in the heteroaromatic ring interact(s) via hydrogen bonding, as shown in Fig. 25.7.

25.2.2.2 $N^α$-Guanylhistamine

Having gained a knowledge of the SAR for histamine, the task was now to design a molecule that would be recognized by the proposed H_2 receptor, but would not activate it. In other words, an agonist had to be converted to an antagonist. This meant altering the way in which the molecule bound to the receptor.

Pictorially, one can imagine histamine fitting into its binding site and stabilizing a change in shape which 'switches on' the receptor (Fig. 25.8). An antagonist can often be found by adding a functional group that binds to an extra binding region in the binding site and prevents the change in shape required for activation.

This was one of several strategies tried out by the SKF workers. To begin with, the structural differences between agonists and antagonists in other areas of medicinal chemistry were identified and similar alterations were tried on histamine. Analogues were tested to see whether they stimulated or blocked gastric acid release—the assumption being that an H_2 receptor would be responsible for such an effect.

Fusing an aromatic ring on to noradrenaline had been a successful tactic used in the design of adrenergic antagonists (see section 23.11.3). This same tactic was tried with histamine to give analogues such as the one shown in Fig. 25.9, but none of these compounds proved to be an antagonist.

Another approach which had been used successfully in the development of anticholinergic agents (section 22.9.2) had been the addition of non-polar, hydrophobic substituents. Similar substituents were attached to various locations of the histamine skeleton, but none proved to be antagonists.

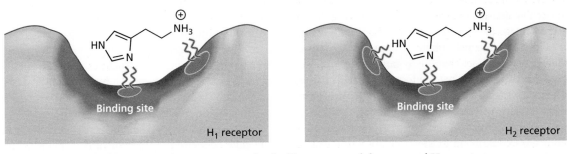

FIGURE 25.7 Binding interactions for the H₁ receptor and the proposed H₂ receptor.

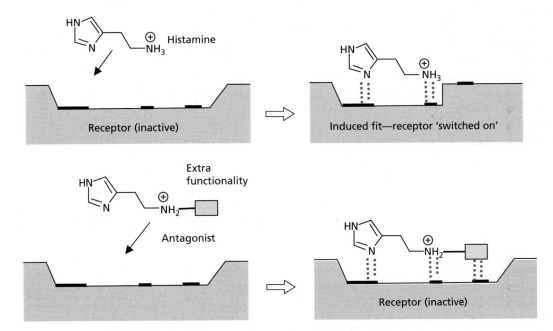

FIGURE 25.8 Possible receptor interactions of histamine and an antagonist.

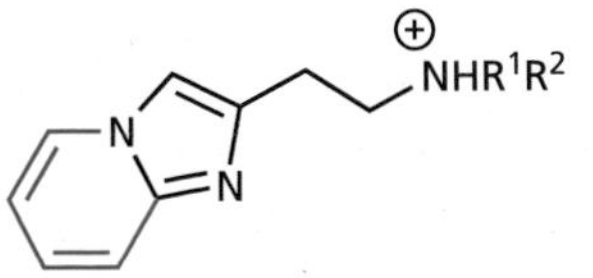

FIGURE 25.9 Histamine analogue with no antagonist activity.

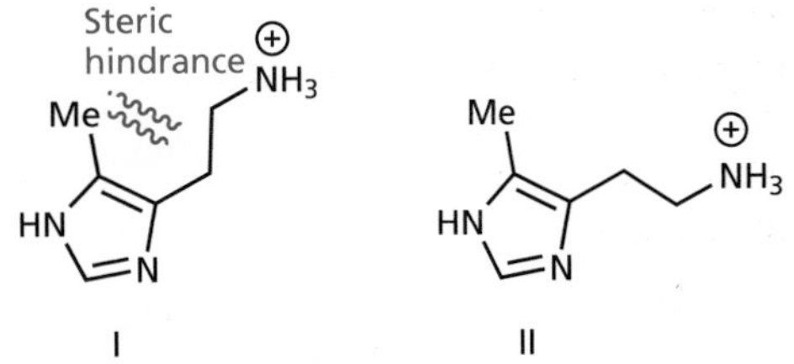

FIGURE 25.10 4-Methylhistamine.

Nevertheless, there was one interesting result which proved relevant to later studies. It was discovered that **4-methylhistamine** (Fig 25.10) was a highly selective H₂ agonist. In other words, it stimulated gastric acid release in the test assay, but had weak activity for all the other actions of histamine. How could this be?

4-Methylhistamine (like histamine) is a highly flexible molecule because of its side chain, but structural studies show that some of its conformations are less stable than others. In particular, conformation I in Fig. 25.10 is not favoured because of a large steric interaction between

the 4-methyl group and the side chain. This means that the 4-methyl group is acting as a conformational blocker (section 13.3.10). The selectivity observed suggests that 4-methylhistamine (and by inference histamine) has to adopt two different conformations in order to fit the H₁ and putative H₂ receptor. As 4-methylhistamine is more active at the hypothetical H₂ receptor, it implies that conformation II is required for the H₂ receptor and conformation I is required for the H₁ receptor.

Despite this interesting result, the SKF workers were no closer to an H$_2$ antagonist. Two hundred compounds had been synthesized and not one had shown a hint of being an antagonist. Research up until this stage had concentrated on adding hydrophobic groups to search for an additional hydrophobic binding region in the proposed receptor binding site. Now the focus switched to study the effect of varying polar groups on the molecule. In particular, the terminal α-NH$_3^+$ group was replaced by different polar functional groups, the reasoning being that such groups could bond to the same binding region as the NH$_3^+$ group, but that the geometry of bonding might be altered sufficiently to produce an antagonist. This led to the first crucial breakthrough, with the discovery that *N*$^\alpha$-**guanylhistamine** (Fig. 25.11) was a weak antagonist of gastric acid release.

This structure had, in fact, been synthesized very early on in the project, but had not been recognized as an antagonist. This is not too surprising since it acts as an agonist! It was not until later pharmacological studies were carried out that it was realized that *N*$^\alpha$-guanylhistamine was acting as a partial agonist (section 8.4). This means that *N*$^\alpha$-guanylhistamine activates the H$_2$ receptor, but not to the same extent as histamine. As a result, the amount of gastric acid released is lower. More importantly, as long as *N*$^\alpha$-guanylhistamine is bound to the receptor, it prevents histamine from binding and thus prevents

complete receptor activation. This was the first indication of any sort of antagonism to histamine, but still did not prove the existence of the H$_2$ receptor.

The question now arose as to which parts of the *N*$^\alpha$-guanylhistamine skeleton were really necessary for this effect. Various guanidine structures were synthesized that lacked the imidazole ring, but none had the desired antagonist activity, demonstrating that both the imidazole ring and the guanidine group were required.

The structures of *N*$^\alpha$-guanylhistamine and histamine were now compared. Both structures contain an imidazole ring and a positively charged group linked by a two-carbon bridge. The guanidine group is basic and protonated at pH 7.4, so the analogue has a positive charge, similar to histamine. However, the charge on the guanidine group can be spread around a planar arrangement of three nitrogens which means that it can be further away from the imidazole ring (Fig. 25.11). This leads to the possibility that the analogue could be interacting with an extra polar binding region on the receptor which is 'out of reach' of histamine. This is demonstrated in Figs. 25.12 and 25.13. Two alternative binding regions might be available for the cationic group—an agonist region where binding leads to activation of the receptor, and an antagonist region where binding does not activate the receptor. In Fig. 25.13, histamine is only able to reach the agonist region, whereas the analogue with

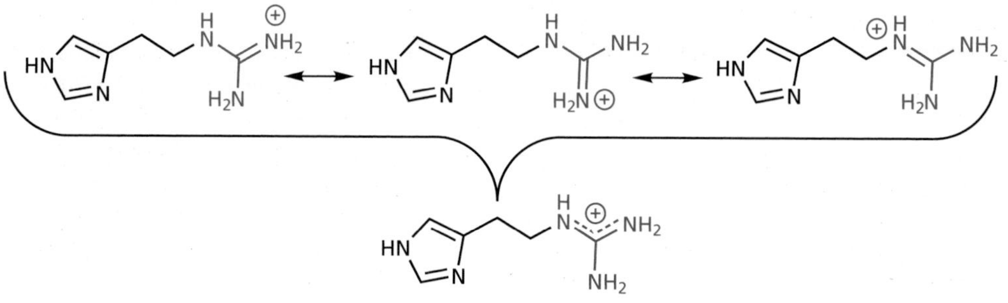

FIGURE 25.11 *N*$^\alpha$-Guanylhistamine.

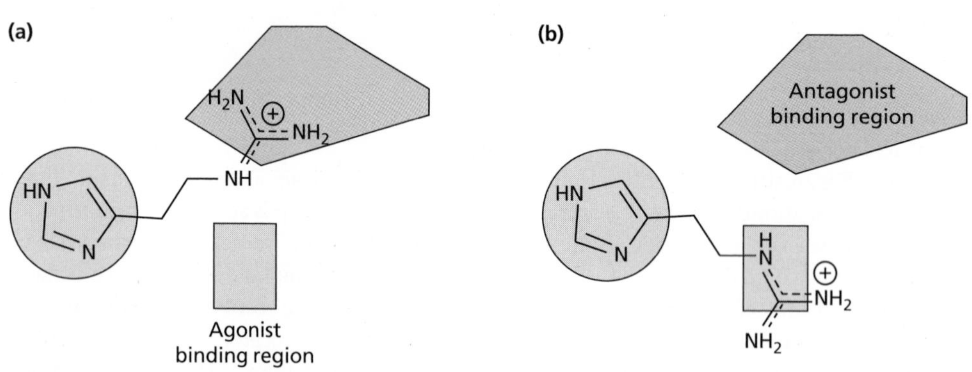

FIGURE 25.12 Possible binding modes for *N*$^\alpha$-guanylhistamine as (a) an antagonist and (b) an agonist.

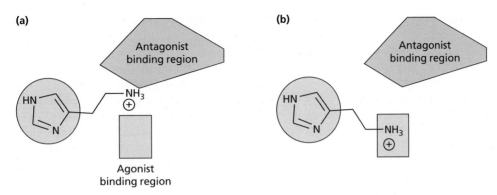

FIGURE 25.13 Binding of histamine: (a) no binding to the antagonist binding region; (b) binding to the agonist binding region.

its extended functionality is capable of reaching either region (Fig. 25.12).

If most of the analogue molecules bind to the agonist region and the remainder bind to the antagonist region, then this could explain the partial agonist activity. Regardless of the mode of binding, histamine would be prevented from binding and an antagonism would be observed due to the fraction of N^{α}-guanylhistamine bound to the antagonist region.

25.2.3 Developing the lead: a chelation bonding theory

The task was now to find an analogue which would bind to the antagonist region only. The isothiourea (Fig. 25.14a) was synthesized as the positive charge would be restricted to the terminal portion of the chain and should interact more strongly with the more distant antagonist binding region. Antagonist activity did increase, but the compound was still a partial agonist, showing that binding was still possible to the agonist region.

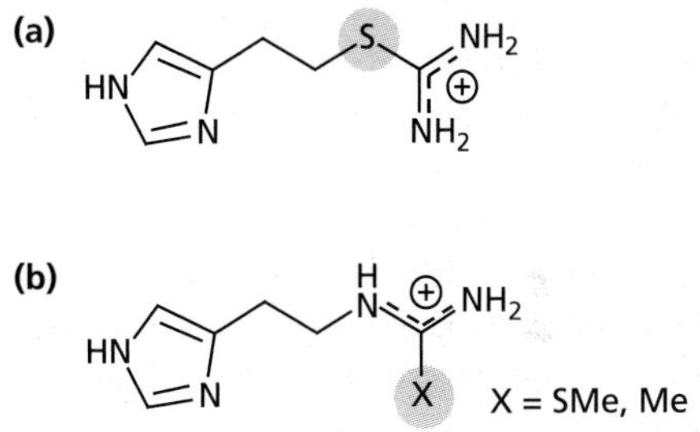

FIGURE 25.14 (a) An isothiourea. (b) Other analogues.

Two other analogues were synthesized where one of the terminal amino groups in the guanidine group was replaced by a methylthio group or a methyl group (Fig. 25.14b). Both these structures were partial agonists, but with poorer antagonist activity.

From these results, it was concluded that both terminal amino groups were required for binding to the antagonist binding site. It was proposed that the charged guanidine group was interacting with a charged carboxylate residue on the receptor via two hydrogen bonds (Fig. 25.15). If

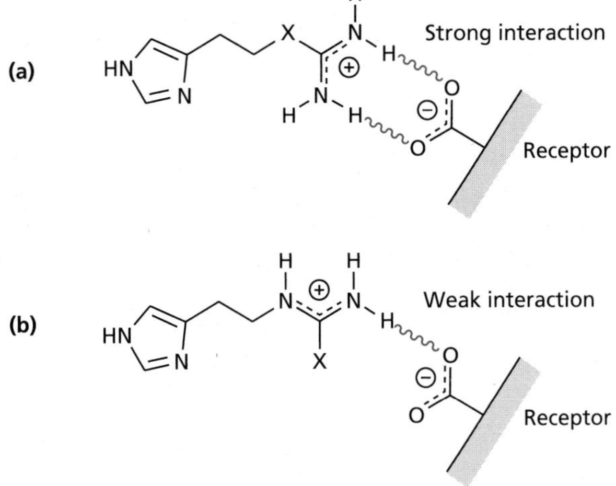

FIGURE 25.15 Proposed hydrogen bonding interactions for (a) a structure with two terminal amino groups (X = NH or S) and (b) an analogue with one terminal amino group (X = Me or SMe).

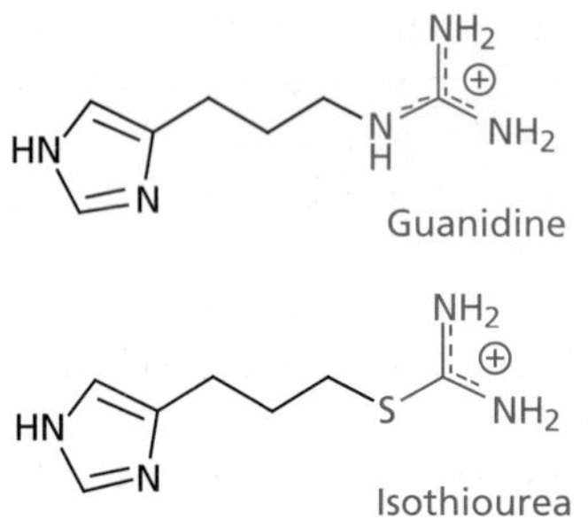

FIGURE 25.16 Guanidine and isothiourea structures with a 3-C linker.

either of these terminal amino groups was absent, then binding would be weaker, resulting in a lower level of antagonism.

The chain was now extended from a two-carbon unit to a three-carbon unit to see what would happen if the guanidine group was moved further away from the imidazole ring. The antagonist activity increased for the guanidine structure (Fig. 25.16), but, strangely enough, decreased for the isothiourea structure (Fig. 25.16). Therefore, it was proposed that with a chain length of two carbon units, hydrogen bonding to the receptor involved the terminal NH_2 groups, but with a chain length of three carbon units, hydrogen bonding to the same carboxylate group involved one terminal NH_2 group along with the NH group within the chain (Fig. 25.17). Support for this theory was provided by the fact that replacing one of the terminal NH_2 groups in the guanidine analogue with SMe or Me (Fig. 25.18) did not affect antagonist activity adversely. This was completely different from the results obtained when similar changes were carried out on the C_2 bridged compound. These bonding interactions are represented pictorially in Figs. 25.19 and 25.20.

25.2.4 From partial agonist to antagonist: the development of burimamide

The problem was now to completely remove the agonist activity to get a pure antagonist. This meant designing a structure which would differentiate between the agonist and antagonist binding regions. At first sight this looks impossible, as both regions appear to involve the same type of bonding. Histamine's activity as an agonist depends on the imidazole ring and the charged amino function, with the two groups taking part in hydrogen and ionic bonding respectively. The antagonist activity of the partial agonists described so far also appears to depend on a hydrogen-bonding imidazole ring and an ionic-bonding guanidine group.

Fortunately, a distinction can be made between the charged groups.

The structures which show antagonist activity are all capable of forming a chelated bonding structure, as shown in Fig. 25.17. This interaction involves two hydrogen bonds between two charged species, but is it really necessary for the chelating group to be charged? Could a neutral group also chelate to the antagonist region by hydrogen bonding alone? If so, it might be possible to distinguish between the agonist and antagonist regions, especially since ionic bonding appears mandatory for the agonist region.

Therefore, it was decided to see what would happen if the strongly basic guanidine group was replaced by a neutral group, capable of interacting with the receptor by two hydrogen bonds. There are many such groups, but the SKF workers limited the options by adhering to a principle which they followed throughout their research programme. Whenever they wished to alter a specific physical or chemical property, they strove to ensure that other properties were changed as little as possible. Only in this way could they rationalize any observed improvement

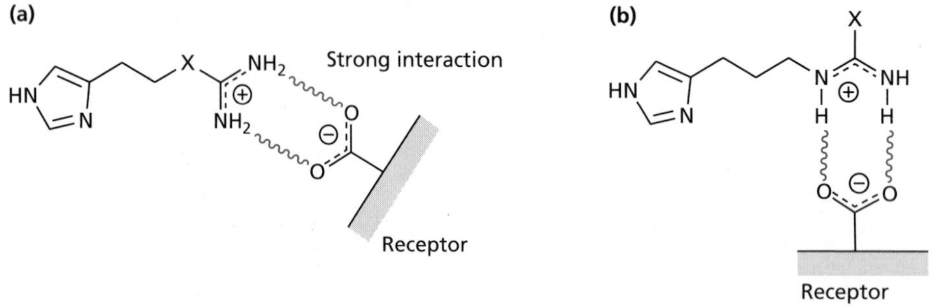

FIGURE 25.17 Proposed binding interactions for analogues of different chain length: (a) H-bonding involving two terminal amino groups for the three-atom chain (X = NH or S); (b) H-bonding involving a terminal and internal amino group for a four-atom chain (X = NH_2, SMe, or Me).

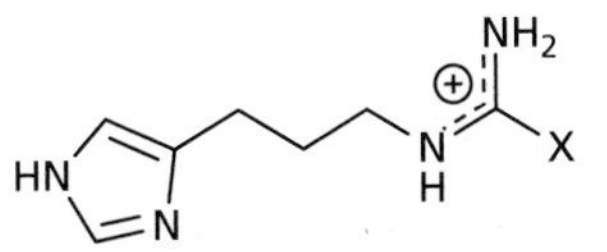

FIGURE 25.18 Guanidine analogue with X = SMe or Me.

in activity. Thus, it was necessary to ensure that the new group was similar to guanidine in terms of size, shape, and hydrophobicity.

Several functional groups were tried, but success was ultimately achieved by using a thiourea group to give **SKF 91581** (Fig. 25.21). The thiourea group is neutral at physiological pH because the C=S group has an electron-withdrawing effect on the neighbouring nitrogens, making them non-basic and more like amide nitrogens. Apart from basicity, the properties of the thiourea group are very similar to the guanidine group. Both groups are planar, similar in size, and can take part in hydrogen bonding. This means that the alteration in biological activity can reasonably be attributed to the differences in basicity between the two groups.

SKF 91581 proved to be a weak antagonist with no agonist activity, establishing that the agonist binding region involves ionic bonding, whereas the antagonist region involves hydrogen bonding.

Further chain extension and the addition of an *N*-methyl group led to **burimamide** (Fig. 25.21) which was found to have enhanced antagonist activity, suggesting that the thiourea group has been moved closer to the antagonist binding region. The beneficial addition of the *N*-methyl group is due to an increase in hydrophobicity and a possible explanation for this will be described in section 25.2.8.2 (desolvation).

Burimamide is a highly specific competitive histamine antagonist at H₂ receptors, and is 100 times more potent than N^α-guanylhistamine in inhibiting gastric acid release induced by histamine. Its discovery gave the SKF researchers far greater evidence for the existence of H₂ receptors.

Test your understanding and practise your molecular modelling with Exercise 25.2 on the Online Resource Centre at www.oxfordtextbooks.co.uk/orc/patrick6e/

25.2.5 Development of metiamide

Despite this success, burimamide was not suitable for clinical trials because its activity was still too low for oral administration. Attention was now directed to the imidazole ring of burimamide and, in particular, to its possible tautomeric and protonated forms. It was argued that if one of these forms was preferred for binding with the H₂ receptor, then activity might be enhanced by modifying the burimamide structure to favour that form.

At pH 7.4, it is possible for the imidazole ring to equilibrate between the two tautomeric forms (I) and (II) via the protonated intermediate (III) (Fig. 25.22). The necessary proton for this process is supplied by water or by an

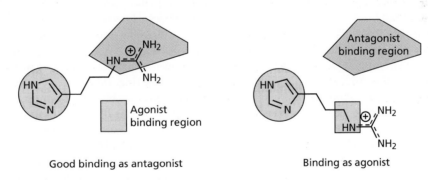

Good binding as antagonist   Binding as agonist

FIGURE 25.19 Proposed binding interactions for the 3-C bridged guanidine analogue.

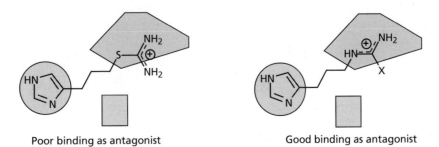

Poor binding as antagonist   Good binding as antagonist

FIGURE 25.20 Proposed binding effect at the antagonist region if the guanidine group is modified.

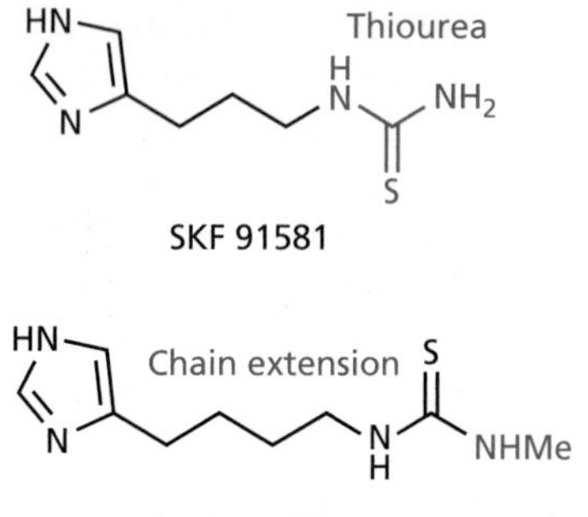

FIGURE 25.21 SKF 91581 and burimamide.

exchangeable proton on a suitable amino acid residue in the binding site. If the exchange is slow, then it is possible that the drug will enter and leave the receptor at a faster rate than the equilibration between the two tautomeric forms. If bonding involves only one of the tautomeric forms or the protonated form, then, clearly, antagonism would be increased if the structure was varied to prefer that form over the other. Our model hypothesis for receptor binding shows that the imidazole ring is important for the binding of both agonists and antagonists. Therefore, it is reasonable to assume that the preferred imidazole form is the same for both agonists and antagonists. If so, then the preferred form for a strong agonist such as histamine should also be the preferred form for a strong antagonist.

Figure 25.22 shows that the imidazole ring can exist as two un-ionized tautomers and one protonated form. Is the protonated form likely?

We have already seen that the pK_a for the imidazole ring in histamine is 5.74, meaning that the ring is a weak base and mostly un-ionized at physiological pH. The pK_a value for imidazole itself is 6.80 and for the imidazole ring in burimamide it is 7.25, showing that these rings are more basic and more likely to be ionized. Why should this be so?

The explanation lies in the side chains, which have an electronic effect affecting the basicity of the imidazole ring. A measure of the electronic effect of the side chain can be worked out by the Hammett equation (section 18.2.2):

$$pK_{a(R)} = pK_{a(H)} + \rho\sigma_R$$

where $pK_{a(R)}$ is the pK_a of the imidazole ring bearing a side chain R, $pK_{a(H)}$ is the pK_a of the unsubstituted imidazole ring, ρ is a constant, and σ_R is the Hammett substituent constant for the side chain R.

From the pK_a values, the value of the Hammett substituent constant can be calculated to show whether the side chain R is electron-withdrawing or electron-donating. In burimamide, the side chain is slightly electron-donating (of the same order as a methyl group). Therefore, the imidazole ring in burimamide is more likely to be ionized than in histamine, where the side chain is electron-withdrawing. At pH 7.4, 40% of the imidazole ring in burimamide is ionized compared with approximately 3% in histamine. This represents quite a difference between the two structures and, as the binding of the imidazole ring is important for both antagonist and agonist activity, it suggests that a pK_a value closer to that of histamine might lead to better binding and to better antagonist activity.

Therefore, it was necessary to make the side chain electron-withdrawing rather than electron-donating. This can be done by inserting an electronegative atom into the side chain—preferably one which causes minimum disturbance to the rest of the molecule. In other words, an isostere for a methylene group is required—one which has an electronic effect, but which has approximately the same size and properties as the methylene group.

The first isostere to be tried was a sulphur atom. Sulphur is quite a good isostere for the methylene unit, as both groups have similar van der Waals radii and similar bond angles. However, the C–S bond is slightly longer than a C–C bond, leading to a slight extension (15%) of the structure.

The methylene group replaced was next but one to the imidazole ring. This site was chosen not for any strategic reasons, but because a synthetic route was readily available to carry out that particular transformation. As hoped, the resulting compound—**thiaburimamide** (Fig. 25.23)—had a significantly lower pK_a of 6.25 and was found to have enhanced antagonistic activity, supporting the theory that the un-ionized form is preferred over the protonated, ionized form.

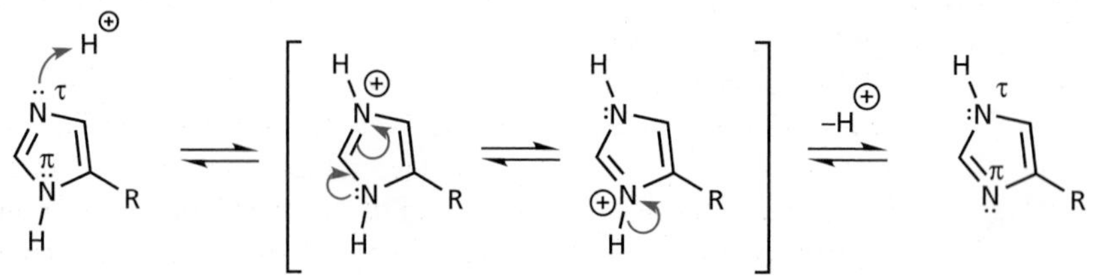

FIGURE 25.22 Imidazole ring can equilibrate between tautomeric forms (I and II) via the protonated intermediate (III).

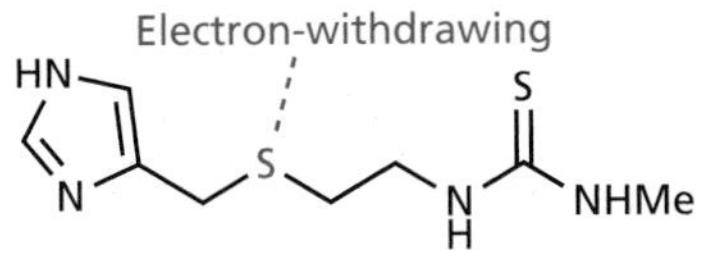

FIGURE 25.23 Thiaburimamide.

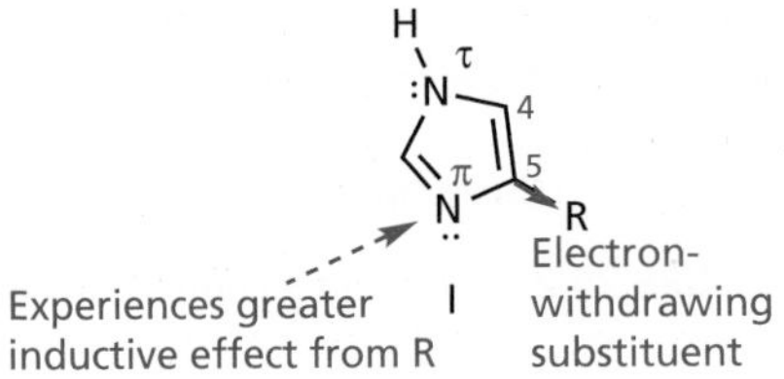

FIGURE 25.24 Inductive effect of the side chain on the imidazole nitrogens.

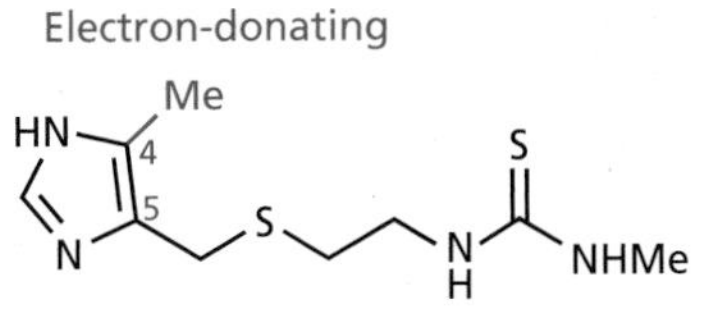

FIGURE 25.25 Metiamide.

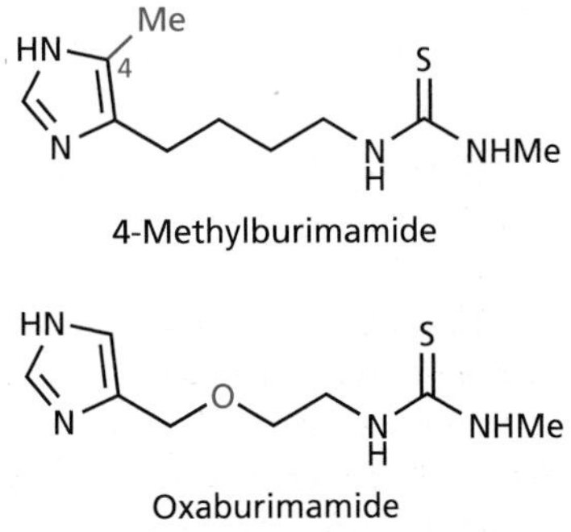

FIGURE 25.26 4-Methylburimamide and oxaburimamide.

Thiaburimamide favours the un-ionized imidazole ring over the ionized ring, but there are two possible un-ionized tautomers. The next question is whether either of these is preferred for receptor binding.

Let us return to histamine. If one of the un-ionized tautomers is preferred over the other, it would be reasonable to assume that the preferred tautomer is the favoured tautomer for receptor binding, as it is more likely to be present. The preferred tautomer for histamine is tautomer I (Fig. 25.22), where Nτ is protonated and Nπ is not. This implies that Nτ in tautomer II is more basic than Nπ in tautomer I. This might not appear obvious, but we can rationalize it as follows. If Nτ in tautomer II is more basic than Nπ in tautomer I, it is more likely to become protonated to form the ionized intermediate (III). Moreover, deprotonation of III is more likely to give the weaker base which would be Nπ in tautomer I. Therefore the equilibrium should shift to favour tautomer I.

This is all very well, but why should Nτ (tautomer II) be more basic than Nπ (tautomer I)? The answer lies in the side chain R. The side chain on histamine has a positively charged terminal amino group, which means that the side chain has an electron-withdrawing effect on the imidazole ring. Because this effect is inductive, the strength of the effect will decrease with distance round the ring, which means that the nitrogen atom closest to the side chain (Nπ) experiences a greater electron-withdrawing effect than the one further away (Nτ). As a result, the closer nitrogen (Nπ) is less basic, and is less likely to bond to hydrogen (Fig. 25.24). As the side chain in thiaburimamide is also electron-withdrawing, then tautomer I will also be favoured here.

It was now argued that tautomer I could be further enhanced if an electron-*donating* group was placed at position 4 of the imidazole ring. At this position, the inductive effect would be felt most strongly at the neighbouring nitrogen (Nτ), further enhancing its basic character over

Nπ. At the same time, it was important to choose a group that would not interfere with the normal receptor binding interactions. For example, a large substituent might be too bulky and prevent the analogue fitting the binding site. A methyl group was chosen because it was known that 4-methylhistamine was an agonist that was highly selective for the H₂ receptor (section 25.2.2.2). This resulted in **metiamide** (Fig. 25.25), which was found to have enhanced antagonist activity, supporting the proposed theory.

It is interesting to note that the percentage increase in tautomer I outweighs an undesirable rise in pK_a. By adding an electron-donating methyl group, the pK_a of the imidazole ring rises to 6.80 compared with 6.25 for thiaburimamide. Coincidentally, this is the same pK_a as for imidazole itself, which shows that the electronic effects of the methyl group and the side chain cancel each other out as far as pK_a is concerned. A pK_a of 6.80 means that 20% of metiamide exists as the protonated form (III), but this is still lower than the corresponding 40% for burimamide. More importantly, the beneficial effect on activity due to the increase in tautomer (I) outweighs the detrimental effect caused by the increase in the protonated form (III).

4-Methylburimamide (Fig. 25.26) was also synthesized for comparison. Here, the introduction of the 4-methyl group does not lead to an increase in activity. The pK_a is increased to 7.80, resulting in the population of ionized imidazole ring rising to 72%. This demonstrates the importance of rationalizing structural changes. Adding the 4-methyl group to thiaburimamide is advantageous, but adding it to burimamide is not.

The design and synthesis of metiamide followed a rational approach aimed at favouring one specific tautomer. Such a study is known as a **dynamic structure-activity analysis**.

Strangely enough, it has since transpired that the improvement in antagonism may have resulted from conformational effects. X-ray crystallography studies have indicated that the longer thioether linkage in the chain increases the flexibility of the side chain and that the 4-methyl substituent in the imidazole ring may help to orientate the imidazole ring correctly for receptor binding. It is significant that the oxygen analogue **oxaburimamide** (Fig. 25.26) is less potent than burimamide, despite the fact that the electron-withdrawing effect of the oxygen-containing chain on the ring is similar to the sulphur-containing chain. The bond lengths and angles of the ether link are similar to the methylene unit and, in this respect, it is a better isostere than sulphur. This is because the oxygen atom is substantially smaller than sulphur. However, this does not imply that it will be a better bio-isostere, as other properties might be detrimental to activity. For example, the oxygen atom is significantly more basic and more hydrophilic than either sulphur or methylene. In fact, oxaburimamide's lower activity might be due to a variety of reasons. The oxygen may not allow the same flexibility permitted by the sulphur atom. Alternatively, the oxygen may be involved in a hydrogen bonding interaction with the binding site that is detrimental to activity. Another possibility is the fact that oxygen is more likely to be solvated than sulphur and there is an energy penalty involved in desolvating the group before binding.

Metiamide is 10 times more active than burimamide and showed promise as an anti-ulcer agent. Unfortunately, a number of patients suffered from kidney damage and granulocytopenia—a condition which results in the reduction of circulating white blood cells and makes patients susceptible to infection. Further developments were now required to find an improved drug without these side effects.

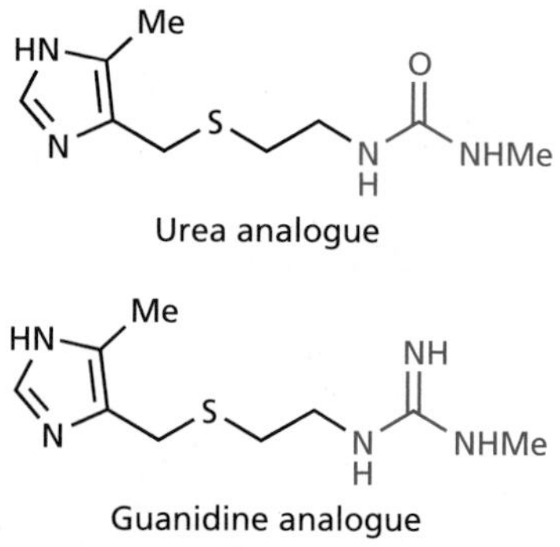

FIGURE 25.27 Urea and guanidine analogues.

25.2.6 Development of cimetidine

It was proposed that metiamide's side effects were associated with the thiourea group—a group which is not particularly common in human biochemistry. Therefore, consideration was given to replacing the thiourea with a group which had similar properties, but which would be more acceptable in a biochemical context. The urea analogue (Fig. 25.27) was found to be less active. The guanidine analogue (Fig. 25.27) was also less active, but it was interesting to note that this compound had no agonist activity. This contrasts with the C_3-bridged guanidine (Fig. 25.16), which is a partial agonist. Therefore, the guanidine analogue (Fig 25.27) was the first example of a guanidine-containing structure having pure antagonist activity.

One possible explanation for this is that the longer four-atom chain extends the guanidine binding group beyond the reach of the agonist binding region (Fig. 25.28), whereas the shorter three-atom chain still allows binding to both agonist and antagonist regions (Fig. 25.29).

The antagonist activity for the guanidine analogue (Fig. 25.27) is weak, but it was decided to look more closely at this compound, as it was thought that the guanidine unit would lack the toxic side effects of the thiourea unit. This is a reasonable assumption as the guanidine unit is present naturally in the amino acid

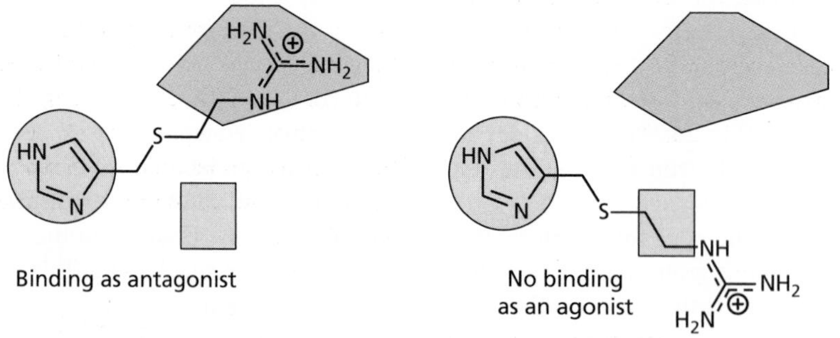

FIGURE 25.28 Binding of the guanidine analogue with a four-atom linker.

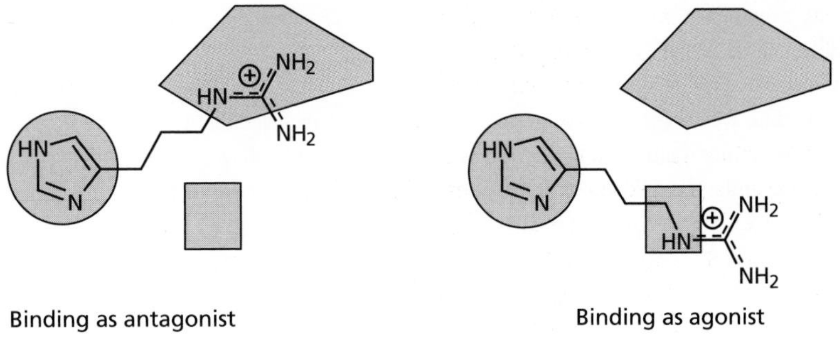

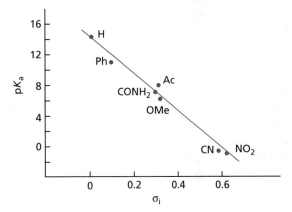

FIGURE 25.29 Binding of the guanidine analogue with a three-atom linker.

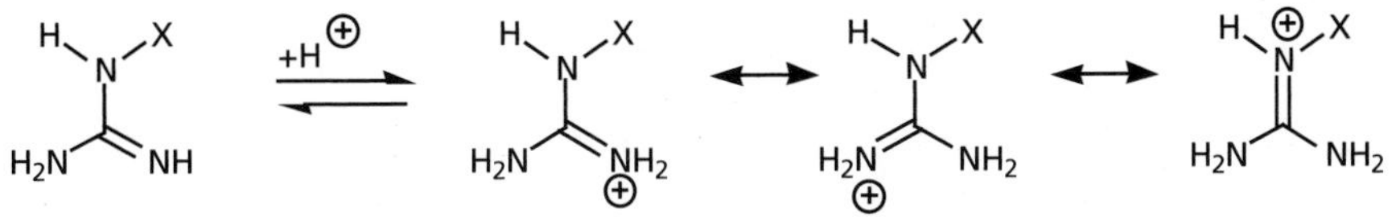

Binding as antagonist   Binding as agonist

FIGURE 25.30 Ionization of monosubstituted guanidines.

arginine (Appendix 1). The problem was how to retain the guanidine unit while increasing activity. It seemed likely that the low activity observed was because the basic guanidine group would be essentially fully protonated and ionized at pH 7.4. The challenge was now to make this group non-basic—no easy task as guanidine is one of the strongest neutral organic bases in organic chemistry.

Nevertheless, a search of the literature revealed a useful study on the ionization of monosubstituted guanidines (Fig. 25.30). A comparison of the pK_a values of these compounds with the inductive substituent constants σ_i (section 18.2.2) for the substituents X gave a straight line, as shown in Fig. 25.31, showing that pK_a is inversely proportional to the electron-withdrawing power of the substituent. Thus, strongly electron-withdrawing substituents make the guanidine group less basic and less ionized. The nitro and cyano groups are particularly strong electron-withdrawing groups. The pK_as for cyanoguanidine and nitroguanidine are 0.4 and 0.9, respectively (Fig. 25.31)—similar values to the pK_a for thiourea itself (−1.2).

Both the nitroguanidine and cyanoguanidine analogues of metiamide were synthesized and found to have comparable antagonist activities to metiamide. The cyanoguanidine analogue (**cimetidine**; Fig. 25.32) was the more potent analogue and was chosen for clinical studies. Its synthesis is described in Box 25.1.

FIGURE 25.31 pK_a versus inductive substituent constants (σ_i).

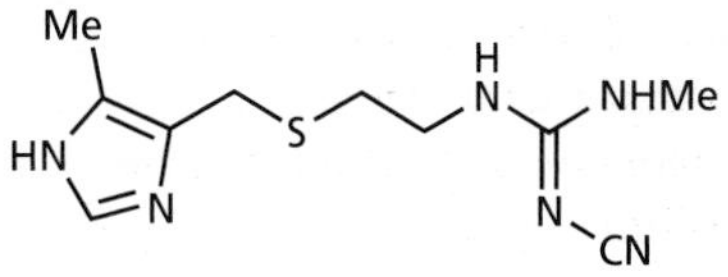

FIGURE 25.32 Cimetidine.

effects observed for metiamide, and has been shown to be slightly more active. It has also been found to inhibit **pentagastrin** (Fig. 25.33) from stimulating release of gastric acid. Pentagastrin is an analogue of gastrin, and the fact that cimetidine inhibits it suggests some relationship between histamine and gastrin in the release of gastric acid.

25.2.7 **Cimetidine**

25.2.7.1 Biological activity

Cimetidine inhibits gastric acid release by acting as an antagonist at H₂ receptors. It does not show the toxic side

(*t*-Boc)N-β-Ala-Trp-Met-Asp-Phe-NH₂

FIGURE 25.33 Pentagastrin.

BOX 25.1 Synthesis of cimetidine

The synthesis of cimetidine was originally carried out as a four-step process, where lithium aluminium hydride was used as the reagent for the initial reduction step. Subsequent research revealed that this reduction could be carried out more cheaply and safely using sodium in liquid ammonia, and so this became the method used in the manufacturing process.

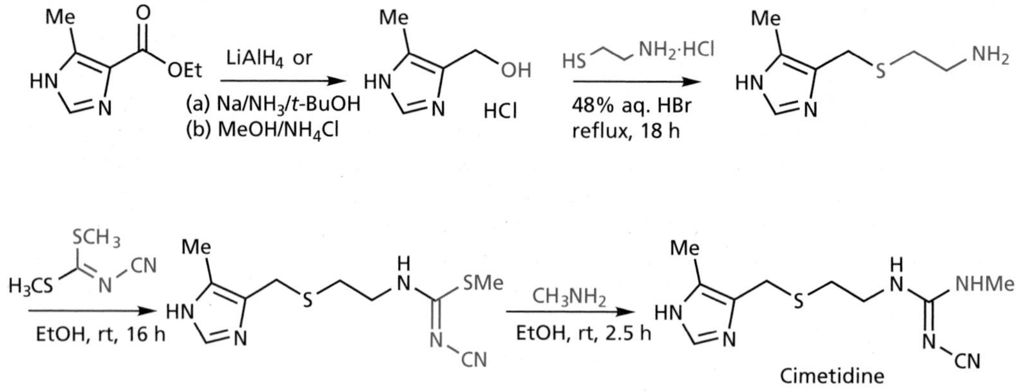

FIGURE 1 Synthesis of cimetidine.

Cimetidine was first marketed in the UK in 1976 under the trade name of **Tagamet** (derived from antagonist and cimetidine). It was the first really effective anti-ulcer drug, doing away with the need for surgery. For several years, it was the world's biggest selling prescription product until it was pushed into second place by **ranitidine** in 1988 (section 25.2.9.1).

25.2.7.2 Structure and activity

The finding that metiamide and cimetidine are both good H$_2$ antagonists of similar activity shows that the cyanoguanidine group is a good bio-isostere for the thiourea group. Three tautomeric forms (Fig. 25.34) are possible for the guanidine group, with the imino tautomer (II) being the preferred one. This is because the cyano group has a stronger electron-withdrawing effect on the neighbouring nitrogen compared with the two nitrogens further away. As a result, the neighbouring nitrogen is less basic and less likely to be protonated. Moreover, tautomer II has an extra stabilization because of the conjugation of the double bond and the cyano group.

As tautomer II is favoured, the guanidine group bears a close structural similarity to the thiourea group. Both groups have a planar π-electron system with similar geometries (equal C–N distances and angles). They are polar and hydrophilic, with high dipole moments and low partition coefficients. They are weakly basic and also weakly acidic such that they are un-ionized at pH 7.4.

25.2.7.3 Metabolism

It is important to study the metabolism of a new drug in case the metabolites have biological activity in their own right. Any such activity might lead to undesirable side effects. Alternatively, a metabolite might have enhanced activity of the type desired and give clues to further development. Cimetidine itself is metabolically stable and

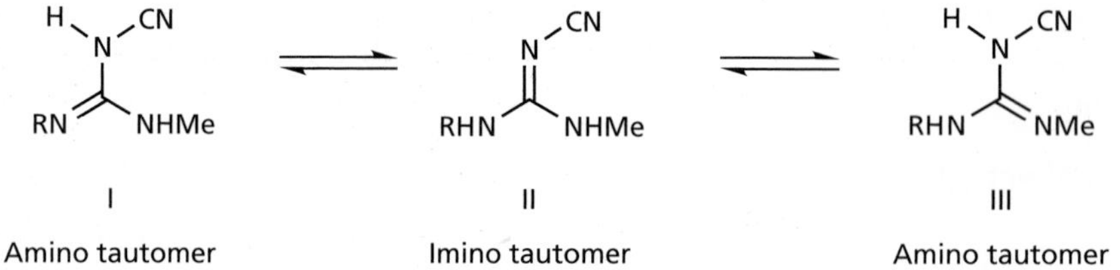

FIGURE 25.34 Three tautomeric forms of guanidine unit.

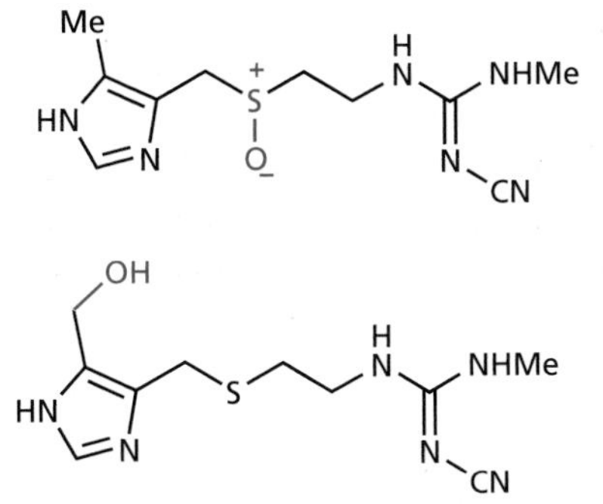

FIGURE 25.35 Metabolites of cimetidine.

is excreted largely unchanged. The only metabolites that have been identified are due to oxidation of the sulphur link or oxidation of the ring methyl group (Fig. 25.35).

It has been found that cimetidine inhibits the cytochrome P-450 enzymes in the liver (section 11.5.2). These enzymes are involved in the metabolism of several clinically important drugs, and inhibition by cimetidine may result in toxic side effects as a result of increased blood levels of these drugs. In particular, caution is required when cimetidine is taken with drugs such as **diazepam**, **lidocaine**, **warfarin**, or **theophylline**.

25.2.8 **Further studies of cimetidine analogues**

25.2.8.1 Conformational isomers

A study of the various stable conformations of the guanidine group in cimetidine led to a rethink of the type of bonding taking place at the antagonist binding region. Up until this point, the favoured theory had been a bidentate hydrogen bonding interaction, as shown in the top diagram of Fig. 25.15, where the two hydrogens involved in hydrogen bonding are pointing in the same direction. In order to achieve this kind of bonding, the guanidine group in cimetidine would have to adopt the Z,Z conformation shown in Fig. 25.36. (The Z and E nomenclature is relevant here, as there is double bond character in the N–C bonds of the guanidine unit.)

However, X-ray and NMR studies have shown that cimetidine exists as an equilibrium mixture of the E,Z and Z,E conformations. Neither the Z,Z nor the E,E form is favoured because of steric interactions. If either the E,Z or Z,E form is the active conformation then it implies that the chelation type of hydrogen bonding described previously is not taking place. An alternative possibility is that the guanidine unit is hydrogen bonding to two distinct hydrogen bonding regions, rather than to a single carboxylate group (Fig. 25.37). Further support for this theory is provided by the weak activity observed for the urea analogue (Fig. 25.27). This compound is known to prefer the Z,Z conformation over the Z,E or E,Z conformations, and would, therefore, be unable to bind to two distinct hydrogen bonding regions.

If this bonding theory is correct and the active conformation is the E,Z or Z,E form, restricting the group to adopt one or other of these forms may lead to more active

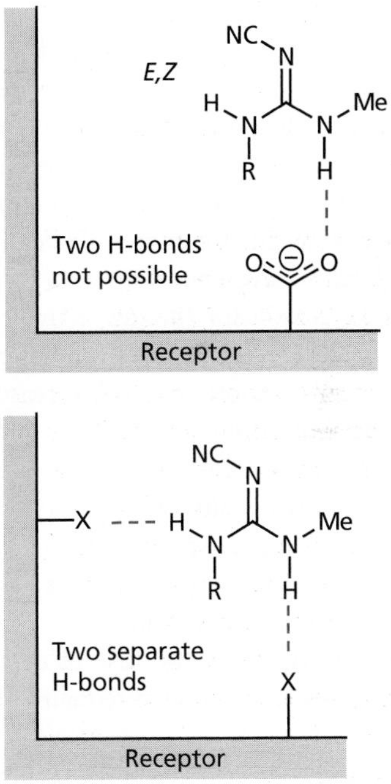

FIGURE 25.37 Alternative theory for cimetidine binding at the antagonist region.

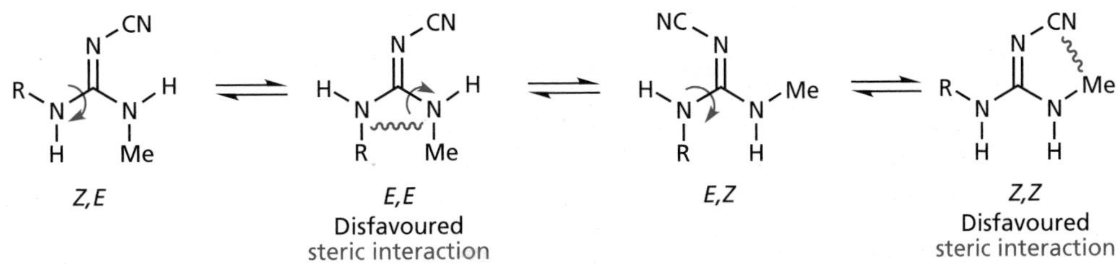

FIGURE 25.36 Conformations of the guanidine group in cimetidine.

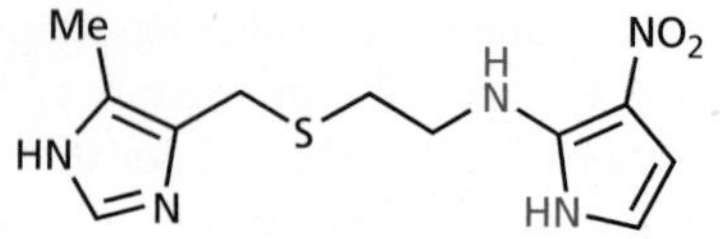

FIGURE 25.38 Nitropyrrole derivative of cimetidine.

compounds and the identification of the active conformation. This can be achieved by incorporating part of the guanidine unit within a ring—a strategy of rigidification (section 13.3.9). For example, the nitropyrrole derivative (Fig. 25.38) has been shown to be the strongest antagonist in the cimetidine series, implying that the *E,Z* conformation is the active conformation.

The isocytosine ring (Fig. 25.39) has also been used to 'lock' the guanidine group, limiting the number of conformations available. The ring allows for further substitution and development, as described in the following sections.

25.2.8.2 Desolvation

The guanidine and thiourea groups, used so successfully in the development of H$_2$ antagonists, are polar and hydrophilic. This implies that they are likely to be highly solvated and surrounded by a 'water coat'. Before hydrogen bonding can take place to the receptor, this water coat has to be removed. The more solvated the group is, the more difficult that will be.

One possible reason for the low activity of the urea derivative (Fig. 25.27) has already been described in section 25.2.8.1 above. Another possible reason could be the fact that the urea group is more hydrophilic than the thiourea or cyanoguanidine groups, and is more highly solvated. The energy penalty involved in desolvating the urea group might explain why this analogue has a lower activity than cimetidine, despite having a lower partition coefficient and greater water solubility.

Leading on from this, if the ease of desolvation is a factor in antagonist activity, then reducing the solvation

of the polar group should increase activity. One way of achieving this would be to increase the hydrophobic character of the polar binding group.

A study was carried out on a range of cimetidine analogues containing different planar aminal systems (Z) (Fig. 25.40) to see whether there was any relationship between antagonist activity and the hydrophobic character of the aminal system (HZ).

This study showed that antagonist activity was proportional to the hydrophobicity (log *P*) of the aminal unit HZ (Fig. 25.41) and supported the desolvation theory. The relationship could be quantified as follows:

$$\log(\text{activity}) = 2.0 \log P + 7.4$$

Further studies on hydrophobicity were carried out by adding hydrophobic substituents to the isocytosine analogue (Fig. 25.39). These studies showed that there was an optimum hydrophobicity for activity corresponding to the equivalent of a butyl or pentyl substituent. A benzyl substituent was particularly good for activity, but proved to have toxic side effects. These could be decreased by adding alkoxy substituents to the aromatic ring and this led to the synthesis of **oxmetidine** (Fig. 25.42), which had enhanced activity over cimetidine. Oxmetidine was considered for clinical use, but was eventually withdrawn as it still retained undesirable side effects.

Test your understanding and practise your molecular modelling with Exercises 25.3 and 25.4 on the Online Resource Centre at www.oxfordtextbooks.co.uk/orc/patrick6e/

25.2.8.3 Development of the nitroketeneaminal binding group

As we have seen, antagonist activity increases if the hydrophobicity of the polar binding group is increased. Therefore, it was decided to see what would happen if

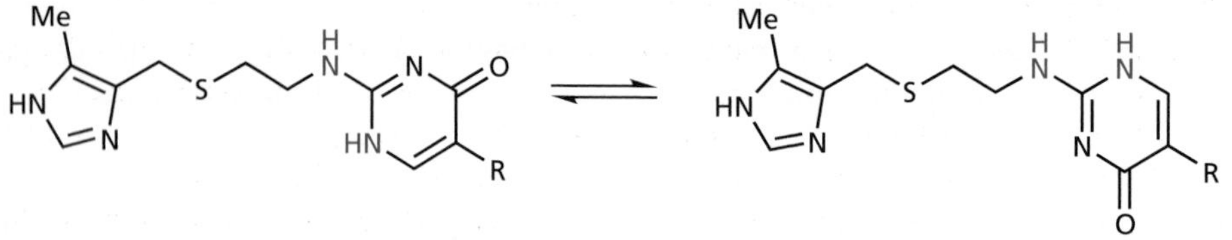

FIGURE 25.39 Isocytosine ring.

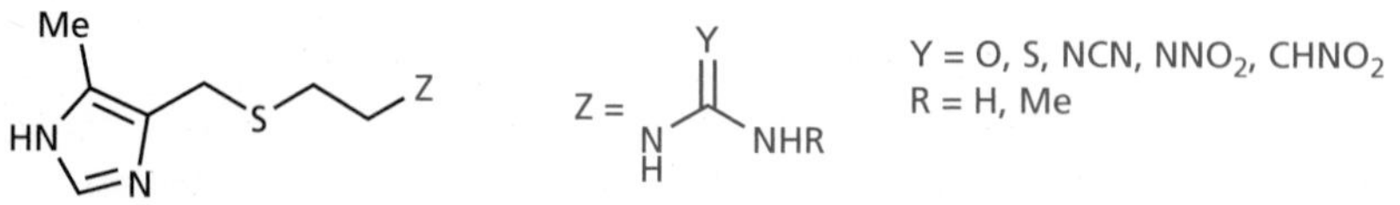

Y = O, S, NCN, NNO$_2$, CHNO$_2$
R = H, Me

FIGURE 25.40 Cimetidine analogue with planar aminal system (Z).

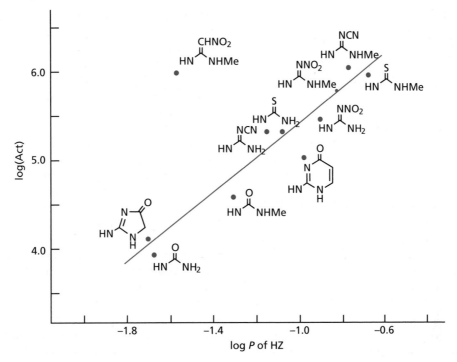

FIGURE 25.41 Antagonist activity is proportional to the hydrophobicity (log P) of the aminal unit Z.

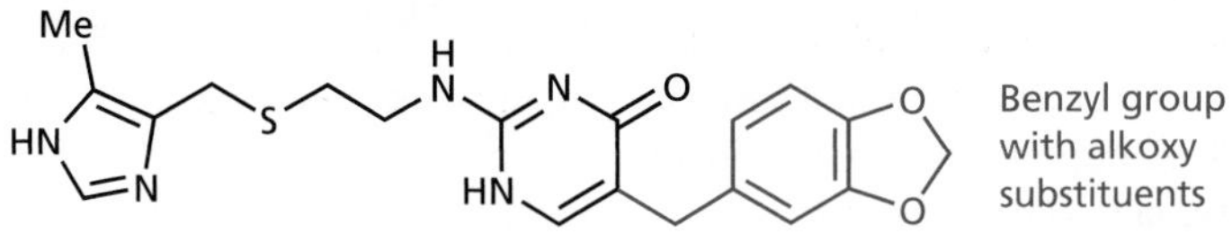

FIGURE 25.42 Oxmetidine.

Benzyl group with alkoxy substituents

the polar imino nitrogen of cimetidine was replaced by a non-polar carbon atom. This would result in a keteneaminal group, as shown in Fig. 25.43. Unfortunately, keteneaminals are more likely to exist as their amidine tautomers unless a strongly electronegative group (e.g. NO_2) is attached to the carbon atom.

A nitroketeneaminal group was, therefore, used to give the structure shown in Fig. 25.44. Surprisingly, there was no great improvement in activity, but, when the structure was studied in detail, it was discovered that it was far more hydrophilic than expected. This explained why the activity had not increased, but it highlighted a different puzzle. The compound was *too* active. Based on its hydrophilicity, it should have been a much weaker antagonist (Fig. 25.41).

It was clear that this compound did not fit the pattern followed by previous compounds as its antagonist activity was 30 times higher than predicted. Nor was the nitroketeneaminal the only analogue to deviate from the expected pattern. The imidazolinone analogue (Fig. 25.44), which is relatively hydrophobic, had a much lower activity than would have been predicted from the equation. Findings like these are particularly exciting, as any deviation from the normal pattern suggests that some other factor is at work, which may give a clue to future development.

In this case, it was concluded that the polarity of the group might be important in some way. In particular, the orientation of the dipole moment appeared to be crucial. In Fig. 25.45, the orientation of the dipole moment is defined by φ—the angle between the dipole moment

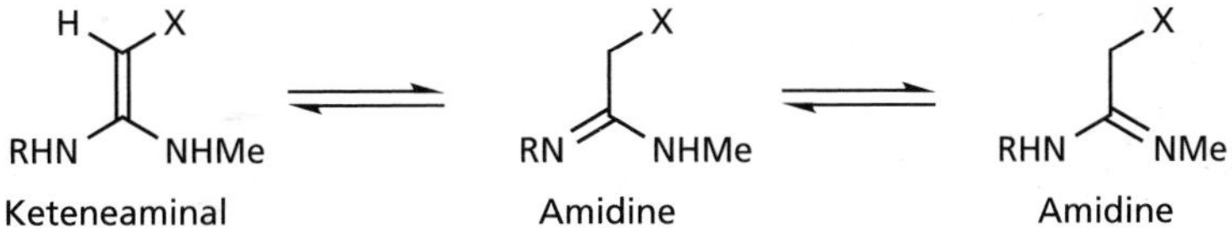

FIGURE 25.43 The keteneaminal group and amidine tautomers.

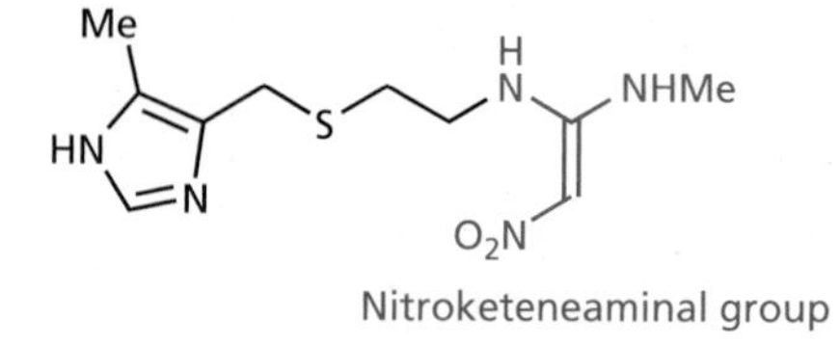

Nitroketeneaminal group

Imidazolinone group

FIGURE 25.44 Cimetidine analogues with unexpected levels of activity.

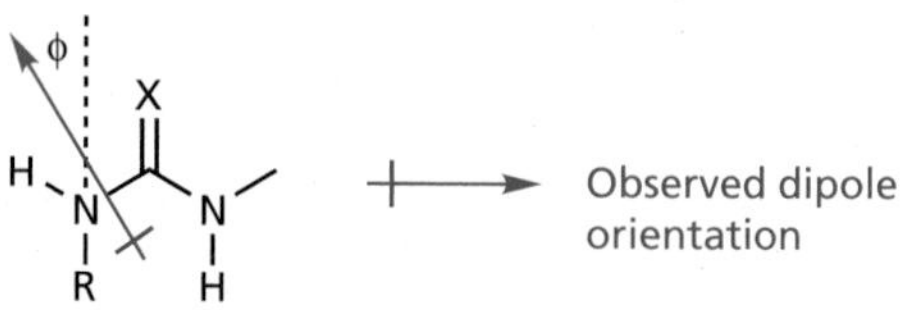

Observed dipole orientation

FIGURE 25.45 Orientation of dipole moment.

and an extension of the NR bond. The cyanoguanidine, nitroketeneaminal, and nitropyrrole groups all have high antagonist activity and have dipole moment orientations of 13°, 33°, and 27° respectively (Fig. 25.46). The isocytosine and imidazolinone groups have lower activity and have dipole orientations of 2° and −6° respectively. The strength of the dipole moment (μ) does not appear to be crucial.

Why should the orientation of a dipole moment be important? One possible explanation is as follows. As the drug approaches the receptor, its dipole interacts with a dipole on the receptor surface such that the dipole moments are aligned. This orientates the drug in a specific way before hydrogen bonding takes place and determines how strong the subsequent hydrogen bonding will be (Fig. 25.47). If the dipole moment is correctly orientated, as in the keteneaminal analogue, the group is correctly positioned for strong hydrogen bonding, and high activity will result. If the orientation is wrong, as in the imidazolinone analogue, then the bonding is less efficient and activity is weaker.

QSAR studies (Chapter 18) were carried out to determine what the optimum angle ϕ should be for activity.

This resulted in an ideal angle for ϕ of 30°. A correlation was worked out between the dipole moment orientation, partition coefficient, and activity as follows:

$$\log A = 9.12 \cos \theta + 0.6 \log P - 2.71$$
$$(n = 13, r = 0.91, s = 0.41)$$

where A is the antagonist activity, P is the partition coefficient, and θ is the deviation in angle of the dipole moment from the ideal orientation of 30° (Fig. 25.48).

The equation shows that activity increases with increasing hydrophobicity (P). The $\cos \theta$ term shows that activity drops if the orientation of the dipole moment varies from the ideal angle of 30°. At the ideal angle, θ is 0° and $\cos \theta$ is 1. If the orientation of the dipole moment deviates from 30°, then $\cos \theta$ will be less than 1 and will lower the calculated activity. The nitroketeneaminal group did not result in a more powerful cimetidine analogue, but we shall see it again in ranitidine (section 25.2.9.1).

Test your understanding and practise your molecular modelling with Exercises 25.5 and 25.6 on the Online Resource Centre at www.oxfordtextbooks.co.uk/orc/patrick6e/

25.2.9 **Further H$_2$ antagonists**

25.2.9.1 Ranitidine

Further studies on cimetidine analogues showed that the imidazole ring could be replaced by other nitrogen-containing heterocyclic rings. Glaxo moved one step further, however, and replaced the imidazole ring with a furan ring bearing a nitrogen-containing substituent. This led to the introduction of **ranitidine** (**Zantac**) (Fig. 25.49). Ranitidine has fewer side effects than cimetidine, a longer duration of action, and is 10 times more active. SAR results for ranitidine include the following:

- the nitroketeneaminal group is optimum for activity, but can be replaced by other planar π systems capable of hydrogen bonding;
- replacing the sulphur atom with a methylene atom leads to a drop in activity;
- placing the sulphur next to the ring lowers activity;

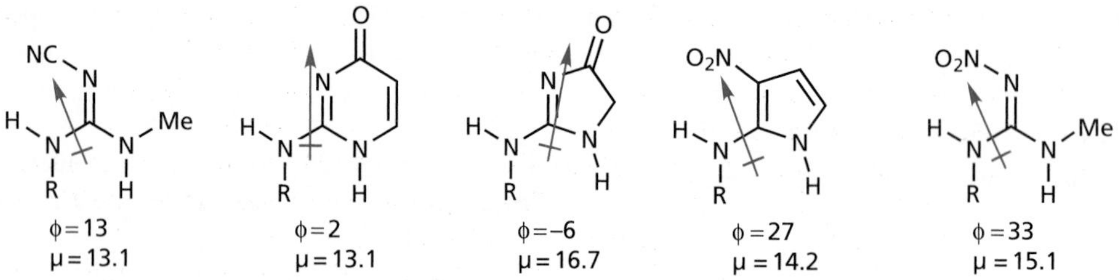

| $\phi = 13$ | $\phi = 2$ | $\phi = -6$ | $\phi = 27$ | $\phi = 33$ |
| $\mu = 13.1$ | $\mu = 13.1$ | $\mu = 16.7$ | $\mu = 14.2$ | $\mu = 15.1$ |

FIGURE 25.46 Dipole moments of various antagonistic groups.

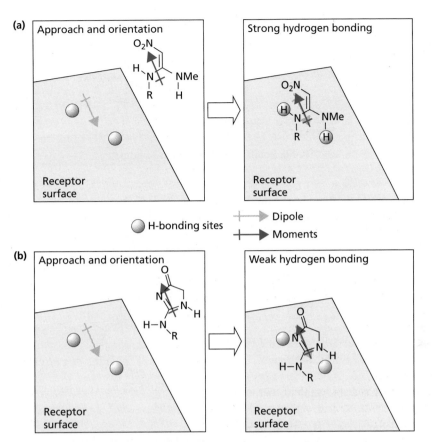

FIGURE 25.47 Dipole–dipole interactions and their effects on orientation and receptor binding: (a) strong binding of the nitroketeneaminal group; (b) weak binding of the imidazolinone group.

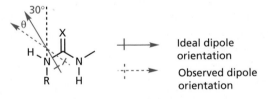

FIGURE 25.48 Definition of the angle θ.

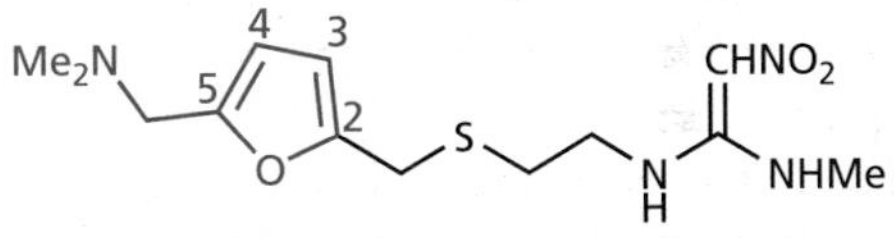

FIGURE 25.49 Ranitidine.

- replacing the furan ring with more hydrophobic rings such as phenyl or thiophene reduces activity;

- 2,5-disubstitution is the best substitution pattern for the furan ring;

- the methyl substituents of the dimethylamino group can be varied, showing that the basicity and hydrophobicity of this group are not crucial to activity;

- methyl substitution at carbon-3 of the furan ring eliminates activity, whereas the equivalent substitution in the imidazole series increases activity;

- methyl substitution at carbon-4 of the furan ring increases activity.

The last two results imply that the heterocyclic rings for cimetidine and ranitidine are not interacting in the same way with the H₂ receptor. This is supported by the fact that a corresponding dimethylaminomethylene group attached to cimetidine leads to a drop in activity. Ranitidine was introduced to the market in 1981, and by 1988 it had taken over from cimetidine as the world's biggest selling prescription drug. Over a 10-year period, it earned Glaxo profits of around £4 billion ($7 billion) and at one time was earning profits of £4 million ($7 million) per day.

25.2.9.2 Famotidine and nizatidine

Over the period 1985–87, two new anti-ulcer drugs were introduced to the market—**famotidine** and **nizatidine** (Fig. 25.50).

Famotidine (Pepcid) is 30 times more active than cimetidine *in vitro*. The side chain contains a

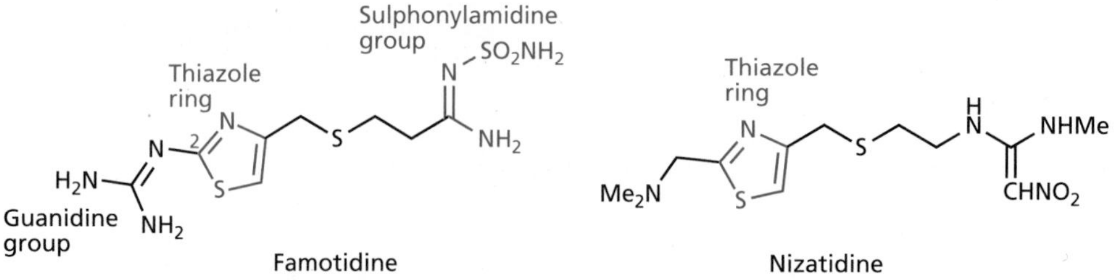

FIGURE 25.50 Famotidine and nizatidine.

sulphonylamidine group, and the heterocyclic imidazole ring of cimetidine has been replaced by a 2-guanidino-thiazole ring. SAR studies gave the following results:

- the sulphonylamidine binding group is not essential and can be replaced by a variety of structures as long as they are planar, have a dipole moment, and are capable of interacting with the receptor by hydrogen bonding;

- a low pK_a is not essential, which allows a larger variety of planar groups to be used than is possible for cimetidine;

- activity is optimum for a chain length of four or five units;

- replacement of sulphur by a CH_2 group *increases* activity;

- modification of the chain is possible with, for example, inclusion of an aromatic ring;

- a methyl substituent on the heterocyclic ring, *ortho* to the chain, leads to a drop in activity (unlike the cimetidine series);

- three of the four hydrogens in the two guanidine NH_2 groups are required for activity.

There are several results here which are markedly different from cimetidine, implying that famotidine and cimetidine are not interacting in the same way with the H_2 receptor. Further evidence for this is the fact that replacing the guanidine of cimetidine analogues with a sulphonylamidine group leads to very low activity.

Nizatidine (Fig. 25.50) was introduced into the UK in 1987 by the Lilly Corporation and is equipotent with ranitidine. The furan ring in ranitidine is replaced by a thiazole ring.

25.2.9.3 H_2 antagonists with prolonged activity

Glaxo carried out further development on ranitidine by placing the oxygen of the furan ring exocyclic to a phenyl ring and replacing the dimethylamino group with a piperidine ring to give a series of novel structures (I in Fig. 25.51). The most promising of these compounds

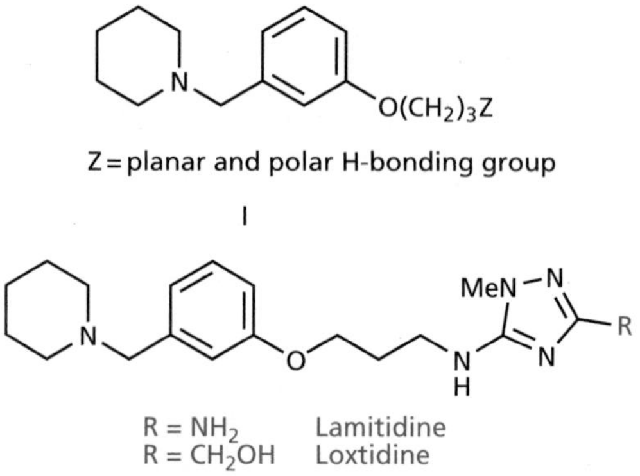

FIGURE 25.51 Long lasting anti-ulcer agents.

were **lamitidine** and **loxtidine** (Fig. 25.51) which were 5–10 times more potent than ranitidine and three times longer lasting. Unfortunately, these compounds showed toxicity in long-term animal studies with the possibility that they caused gastric cancer, so they were subsequently withdrawn from clinical study. The relevance of these studies has been disputed.

25.2.10 **Comparison of H_1 and H_2 antagonists**

The structures of the H_2 antagonists are markedly different from the classical H_1 antagonists, so it is not surprising that H_1 antagonists failed to antagonize the H_2 receptor. H_1 antagonists, like H_1 agonists, possess an ionic amino group at the end of a flexible chain. Unlike the agonists, they possess two aryl or heteroaryl rings in place of the imidazole ring (Fig. 25.52). Because of the aryl rings, H_1 antagonists are hydrophobic molecules having high partition coefficients.

In contrast, H_2 antagonists are polar, hydrophilic molecules having high dipole moments and low partition coefficients. At the end of the flexible chain they have a polar, π-electron system which is amphoteric and un-ionized at pH 7.4. This binding group appears to be the key feature leading to antagonism of H_2 receptors (Fig. 25.52). The heterocycle generally contains a nitrogen atom or, in the case of furan or phenyl, a nitrogen-containing side chain. The hydrophilic character of H_2

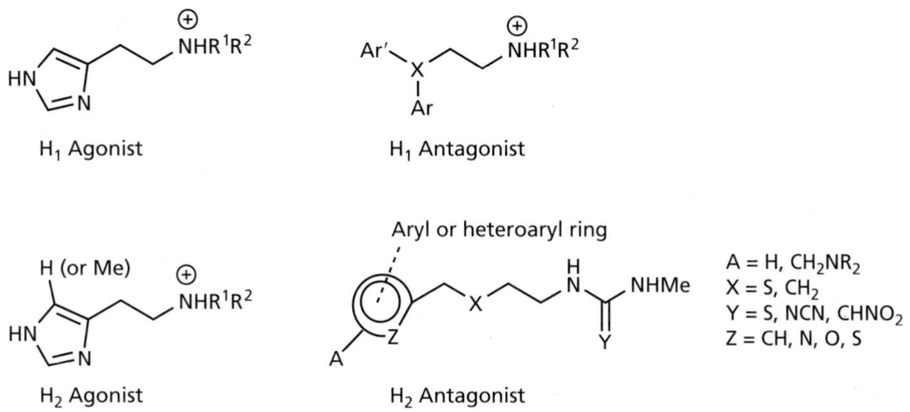

FIGURE 25.52 Comparison of H_1 agonists, H_1 antagonists, H_2 agonists, and H_2 antagonists.

antagonists helps to explain why H_2 antagonists are less likely to have the central nervous system side effects often associated with H_1 antagonists.

25.2.11 H_2 receptors and H_2 antagonists

H_2 receptors are present in a variety of organs and tissues, but their main role is in acid secretion. As a result, H_2 antagonists are remarkably safe and mostly free of side effects. The four most used agents on the market are cimetidine, ranitidine, famotidine, and nizatidine. They inhibit all aspects of gastric secretion and are absorbed rapidly from the gastrointestinal tract with half-lives of 1–2 hours. About 80% of ulcers are healed after 4–6 weeks. Attention must be given to possible drug interactions when using cimetidine because of inhibition of drug metabolism (section 25.2.7.3). The other three H_2 antagonists mentioned do not inhibit the P450 cytochrome oxidase system and are less prone to such interactions.

KEY POINTS

- Peptic ulcers are localized erosions of the mucous membranes which occur in the stomach and duodenum. The hydrochloric acid present in gastric juices results in increased irritation, and so drugs which inhibit the release of hydrochloric acid act as anti-ulcer agents. Such agents relieve the symptoms rather than treat the cause.

- The chemical messengers histamine, acetylcholine, and gastrin stimulate the release of hydrochloric acid from stomach parietal cells by acting on their respective receptors.

- H_2 antagonists are anti-ulcer drugs that act on H_2 receptors present on parietal cells. They reduce the amount of acid released.

- The design of H_2 antagonists was based on the natural agonist histamine as a lead compound. Chain extension accessed an antagonist binding region, and the replacement

of an ionized terminal group with a polar, un-ionized group capable of hydrogen bonding led to pure antagonists.

- The design of improved H_2 antagonists was aided by dynamic structure–activity analysis where changes were made to favour one tautomer over another.

- The alignment of dipole moments between a drug and its binding site plays a role in the binding and activity of H_2 antagonists. Desolvation of polar groups also has an important effect on binding affinity.

25.3 Proton pump inhibitors

Although the H_2 antagonists have been remarkably successful in the treatment of ulcers, they have been largely superseded by the proton pump inhibitors (PPIs). These work by irreversibly inhibiting an enzyme complex called the proton pump and have been found to be superior to the H_2 antagonists. They are used on their own to treat ulcers that are caused by NSAIDs, and in combination with antibacterial agents to treat ulcers caused by the bacterium *H. pylori* (section 25.4).

25.3.1 Parietal cells and the proton pump

When the parietal cells are actively secreting hydrochloric acid into the stomach, they form invaginations called **canaliculi** (Fig. 25.53). Each canaliculus can be viewed as a sheltered channel or inlet that flows into the overall 'ocean' of the stomach lumen. Being a channel, it is not part of the cell, but it penetrates 'inland' and increases the amount of 'coastline' (surface area) across which the cell can release its hydrochloric acid. The protons required for the hydrochloric acid are generated from water and carbon dioxide, catalysed by an enzyme called **carbonic anhydrase** (Fig. 25.54). Once the protons have been generated, they have to be exported out of the cell rather than

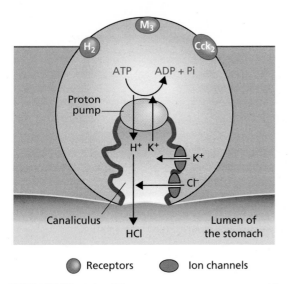

FIGURE 25.53 Role of the proton pump in secreting HCl.

$$H_2O + CO_2 \underset{}{\overset{\text{Carbonic}}{\underset{\text{anhydrase}}{\rightleftharpoons}}} H_2CO_3 \rightleftharpoons H^+ + HCO_3^-$$

FIGURE 25.54 Enzyme-catalysed generation of protons in the parietal cell.

stored. There are two reasons for this. Firstly, a build-up of acid within the cell would prove harmful to the cell. Secondly, the enzyme-catalysed reaction which generates the protons is reversible, and so a build-up of protons within the cell would encourage the reverse reaction and slow the production down. The export of protons from the parietal cell is achieved by an enzyme complex called the **proton pump** or **H$^+$/K$^+$-ATPase**.

The proton pump is only present in the canalicular membranes of parietal cells and is crucial to the mechanism by which hydrochloric acid is released into the stomach. It is called an H$^+$/K$^+$-ATPase because it pumps protons out of the cell into the canaliculus at the same time as it pumps potassium ions back in. Energy is required for this process, as both the protons and the potassium ions are being moved against their concentration

gradients. In fact, the ratio of protons inside the cell to protons in the canaliculus is 1 to 10^6! The energy required to carry out this exchange is obtained by the hydrolysis of ATP (Fig. 25.55)—hence the term ATPase.

The pump is not responsible for the efflux of chloride ions; these depart the cell through separate chloride ion channels. This outflow closely matches the efflux of protons such that a chloride ion is released for every proton that is pumped out. As a result, hydrochloric acid is formed in the canaliculus, rather than inside the parietal cell.

As each chloride ion departs the cell, it is accompanied by a potassium ion which flows through its own ion channel. No energy is required for this outflow, because the potassium ion is flowing down a concentration gradient. The potassium ion acts as a counterion for the chloride ion and, once it is in the canaliculus, it is pumped back into the cell by the proton pump as described previously. Consequently, potassium ions undergo a cyclic movement in and out of the cell.

25.3.2 Proton pump inhibitors

There are four PPIs in clinical use: **omeprazole**, **lansoprazole**, **pantoprazole**, and **rabeprazole** (Fig. 25.56). The S-enantiomer of omeprazole has also been recently approved. All the PPIs have a pyridyl methylsulphinyl benzamidazole skeleton and act as prodrugs, since they are activated when they reach the acidic canaliculi of parietal cells. Once activated, they bind irreversibly to exposed cysteine residues of the proton pump and 'block' the pump, preventing further release of hydrochloric acid.

There is a strategic advantage in inhibiting the proton pump, rather than blocking histamine or cholinergic receptors. For example, H$_2$ antagonists block histamine receptors and block the stimulatory effect of histamine, but this does not block the receptors for acetylcholine or gastrin, and so it is still possible for the parietal cell to be activated towards secretion. The proton pump is 'downstream' of all these targets and operates the final stage of hydrochloric acid release. Blocking it prevents the release of hydrochloric acid regardless of what mechanisms are involved in stimulating hydrochloric acid secretion.

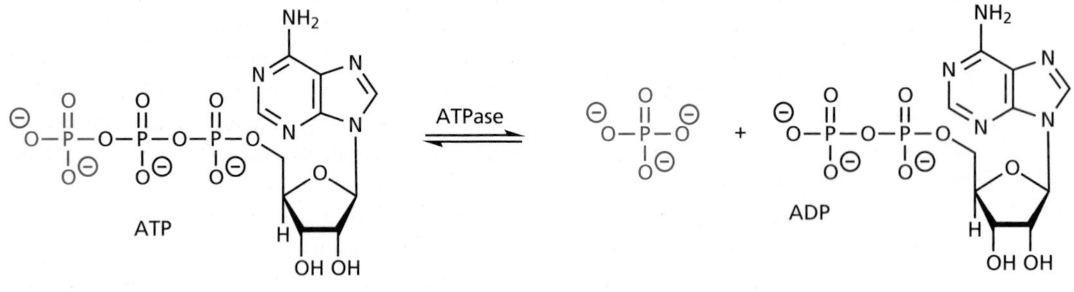

FIGURE 25.55 Enzyme-catalysed hydrolysis of ATP.

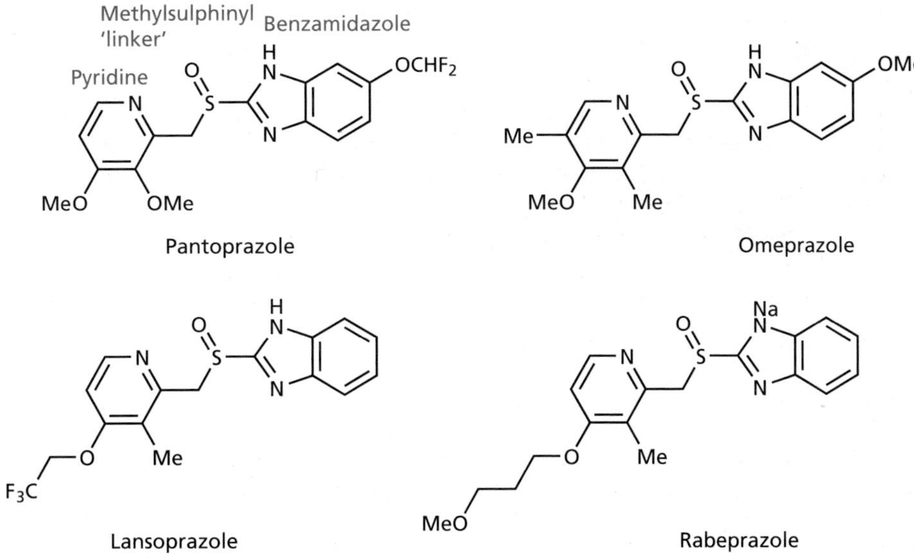

Methylsulphinyl 'linker'

Benzamidazole

Pyridine

Pantoprazole

Omeprazole

Lansoprazole

Rabeprazole

FIGURE 25.56 Proton pump inhibitors (PPIs).

25.3.3 Mechanism of inhibition

The PPIs are weak bases having a pK_a of about 4.0. As a result, they are free bases at blood pH (7.4) and are only ionized in strongly acidic environments where the pH is less than 4. These are conditions found only in the secretory canaliculus of the parietal cell, where the pH is 2 or less. The drugs are taken orally and are absorbed into the blood supply where they are carried round the body as fairly innocuous passengers until they reach the parietal cells. Because they are un-ionized weak bases at this stage, and are also lipophilic in nature, they are able to cross the cell membrane of the parietal cell into the strongly acidic conditions of the canaliculi. Here, the drugs undergo a personality change and become particularly vicious! The canaliculus is highly acidic, so the drug becomes protonated. The consequences of this are twofold:

- the ionized drug is too polar to cross back into the cell through the cell membrane. This leads to a 1000-fold accumulation of the drug in the canaliculi where it is intended to act;

- protonation triggers an acid-catalysed conversion, as shown in Fig. 25.57, which activates the drug.

Protonation takes place on the benzimidazole ring of the drug. The nitrogen of the pyridine ring then acts as a nucleophile and uses its lone pair of electrons to form a bond to the electron-deficient 2-carbon of the benzimidazole ring to form a spiro structure. By doing so, the aromatic character of the imidazole portion of the ring is lost and so there is a high tendency for this ring to re-aromatize. This can be achieved by a lone pair of electrons

from nitrogen reforming the double bond and cleaving the S–C bond to form a sulphenic acid. Sulphenic acids are highly reactive to nucleophiles and so a rapid reaction takes place involving an intramolecular attack by the NH group of the benzimidazole on the sulphenic acid to displace the hydroxyl group. A cationic, tetracyclic pyridinium sulphenamide is formed, which acts as an irreversible enzyme inhibitor (Fig. 25.57). It does so by forming a covalent bond to an accessible cysteine residue on the proton pump. There are three such accessible cysteine residues (Cys-813, Cys-892, and Cys-821) and it has been found that the specific cysteine residues attacked depend on which PPI is involved. For example, omeprazole reacts with two of the accessible cysteine residues (Cys-813 and Cys-892), lansoprazole reacts with all three, and pantoprazole only reacts with one (either Cys-813 or Cys-892). Cys-813 is the only cysteine residue which appears to react with all the PPIs.

As acid conditions are required to activate the PPIs, they are most active when parietal cells are actively secreting hydrochloric acid, and they show little activity when the parietal cells are in a resting state. Since a covalent disulphide bond is formed between the inhibitor and the proton pump, inhibition is irreversible, and so PPIs have a long duration of action. The duration depends on how quickly new pumps are generated by the cell.

PPIs also have very few side effects because of their selectivity of action. This is a result of several factors:

- the target enzyme (H^+/K^+-ATPase) is only present in parietal cells;

- the canaliculi of the parietal cells are the only compartments in the body which have such a low pH (1–2);

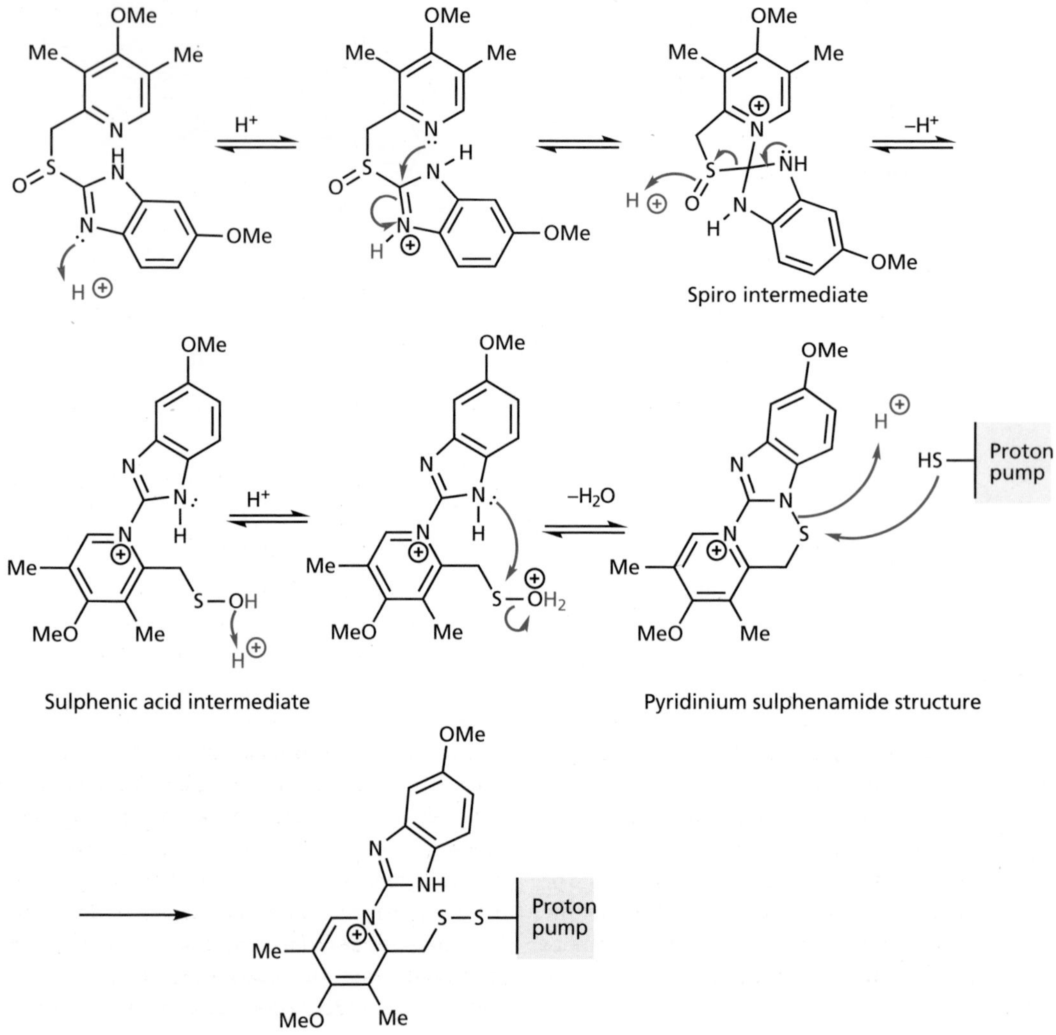

FIGURE 25.57 Mechanism of inhibition by proton pump inhibitors.

- the drug is concentrated at the target site due to protonation and is unable to return to the parietal cell or to the general circulation;

- the drug is rapidly activated close to the target;

- once activated, the drug reacts rapidly with the target;

- the drug is inactive at neutral pH.

25.3.4 Metabolism of proton pump inhibitors

PPIs are metabolized by cytochrome P450 enzymes, particularly *S*-mephenytoin hydroxylase (**CYP2C19**) and nifedipine hydroxylase (**CYP3A4**). As a result of genetic variations, about 3% of white people of European descent are slow metabolizers of PPIs. Pantoprazole, in contrast to omeprazole and lansoprazole, is also metabolized by the conjugating enzyme **sulphotransferase**.

25.3.5 Design of omeprazole and esomeprazole

Omeprazole was the first PPI to be approved and was marketed as **Losec** in 1988. In 1996, it became the biggest selling pharmaceutical ever. The story of how omeprazole was developed can be traced back to the 1970s. The lead compound for the project was a thiourea structure (CMN 131 in Fig. 25.58). This had originally been investigated as an antiviral drug, but general pharmacological tests showed that it could inhibit acid secretion. Unfortunately, toxicology tests showed that the compound was toxic to the liver, which was attributed to the presence of the thioamide group. Various analogues were made to try to modify or disguise this group, which included incorporating the thiourea group within a ring. This led eventually to the discovery of H 77/67, which was also found to inhibit acid secretion. A variety of analogues having

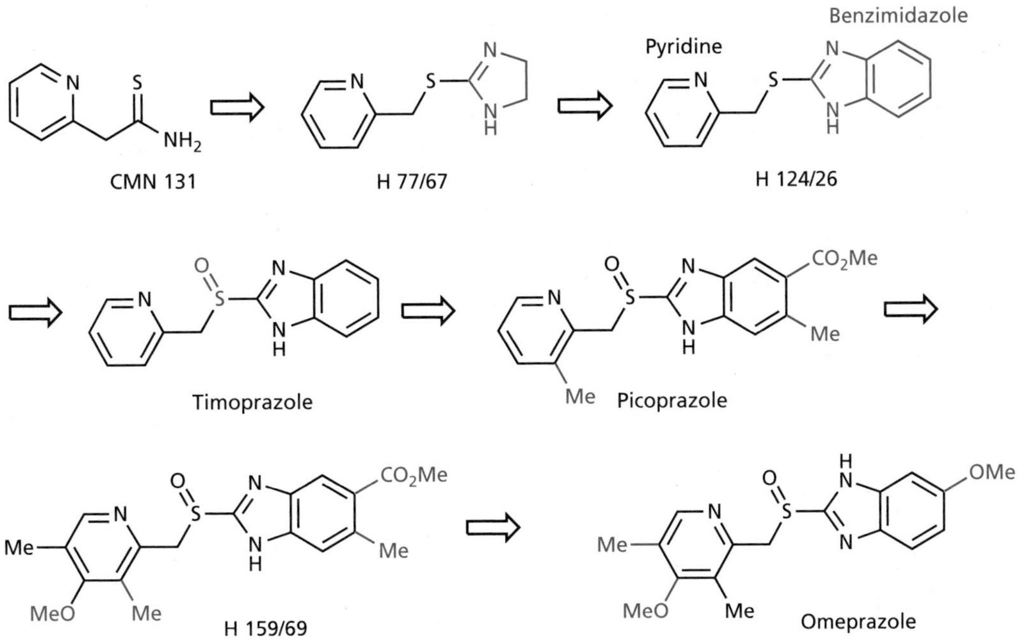

FIGURE 25.58 Development of omeprazole.

the general structure (heterocycle–X–Y–heterocycle) were synthesized which demonstrated that the pyridine ring and the bridging CH_2–S group already present in H 77/67 were optimal for activity. However, activity was increased by replacing the imidazole ring of H 77/67 with a benzimidazole group to give H 124/26. At this stage, drug metabolism studies revealed that a sulphoxide metabolite of H 124/26 was formed *in vivo*, which was more active than the original structure. The metabolite was called **timoprazole** and was the first example of a pyridinylmethylsulphinyl benzimidazole structure. It went forward for preclinical trials, but toxicological studies revealed that it inhibited iodine uptake by the thyroid gland and so it could not go on to clinical trials.

Analogues were now synthesized to find a structure which retained the antisecretory properties, but did not inhibit iodine uptake. Eventually, it was found that the two effects could be separated by placing suitable substituents on the two heterocyclic rings. This led to **picoprazole**, which showed potent antisecretory properties over a long period without the toxic side effect on the thyroid. Animal toxicology studies showed no other toxic effects and the drug went forward for clinical trials, where it was found to be the most effective antisecretory compound ever tested in humans. At this point (1977) the proton pump was discovered and identified as the target for picoprazole. Further development was carried out with the aim of getting a more potent drug by varying the substituents on the pyridine ring.

It was discovered that substituents which increased the basicity of the pyridine ring were good for activity.

This fits in with the mechanism of activation (Fig. 25.57) where the nitrogen of the pyridine ring acts as a nucleophile. In order to increase the nucleophilicity of the pyridine ring, a methoxy group was placed at the *para* position relative to the nitrogen, and two methyl groups were placed at the *meta* positions. The latter have an inductive effect which is electron-donating and increases the electron density of the ring. The methoxy substituent was added at the *para* position to increase electron density on the pyridine nitrogen by the resonance mechanism shown in Fig. 25.59.

It is noticeable that all the PPIs shown in Fig. 25.56 have an alkoxy substituent at the *para* position of the pyridine. The position of the substituent is important. If the substituent was at the *meta* position, none of the possible resonance structures would place the negative charge on the nitrogen atom (Fig. 25.60). Finally, if the methoxy substituent was at the *ortho* position it would be likely to have a bad steric effect and hinder the mechanism.

The introduction of two methyl groups and a methoxy group led to H 159/69 (Fig. 25.58), which was extremely potent but too chemically labile. Further analogues were synthesized where substituents round the benzimidazole ring were varied in order to get the right balance of potency, chemical stability, and synthetic accessibility. Finally, omeprazole was identified as the structure having the best balance of these properties.

Omeprazole was launched in 1988 and became the world's biggest-selling drug, earning its makers vast profits. For example, worldwide sales in 2000 were $6.2 billion (£3.6 billion). The patents on omeprazole ran out in

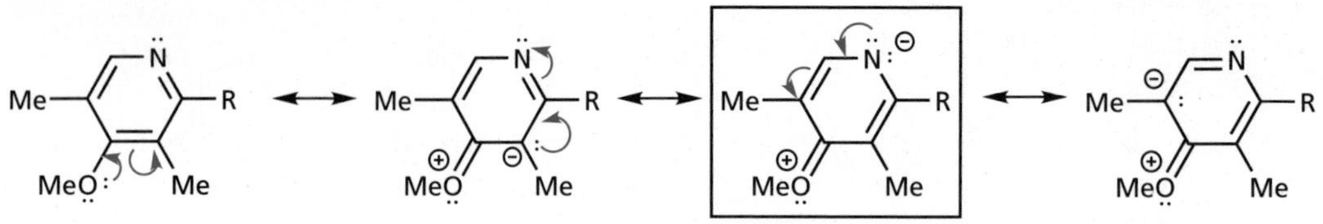

FIGURE 25.59 Influence of the methoxy substituent on the pyridine ring.

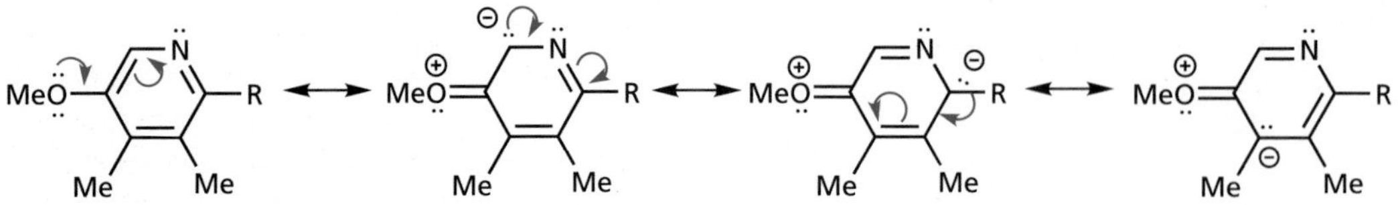

FIGURE 25.60 Possible resonance structures for methoxy substitution at the *meta* position.

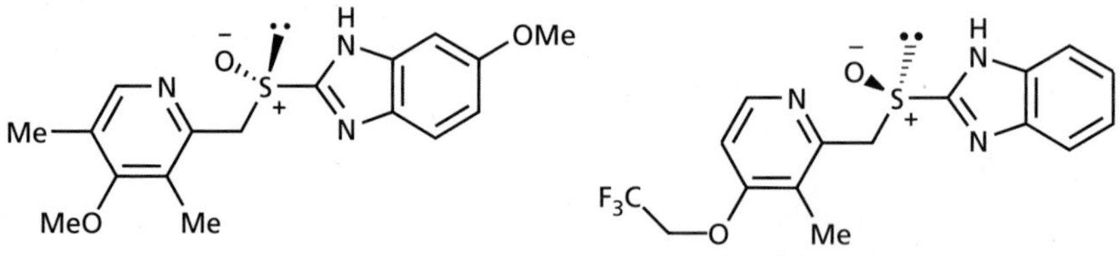

FIGURE 25.61 Esomeprazole and dexlansoprazole.

Europe in 1999 and in the USA in 2001, but its makers (Astra) had already started a programme to find an even better compound. In particular, they were looking for a compound with better bioavailability.

Substitution was varied on both the pyridine and benzimidazole rings, but the best compound was eventually found to be the *S*-enantiomer of omeprazole—**esomeprazole** (**Nexium**; Fig. 25.61). At first sight, it may not be evident that omeprazole has an asymmetric centre. In fact, the sulphur atom is an asymmetric centre as it has a lone pair of electrons and is tetrahedral. Unlike the nitrogen atoms of amines, sulphur atoms do not undergo pyramidal inversion and so it is possible to isolate both enantiomers. The *S*-enantiomer of omeprazole was found to be superior to the *R*-enantiomer in terms of its pharmacokinetic profile, and was launched in Europe in 2000 and in the USA in 2001. The story of esomeprazole is an example of **chiral switching** (section 15.2.1) where a racemic drug is replaced on the market with a single enantiomer. There is no difference between the two enantiomers of omeprazole as far as the mechanism of action is concerned, but it is possible to use double the dose levels of esomeprazole compared to omeprazole, resulting in greater activity. Esomeprazole is metabolized

mainly by CYP2C19 in the liver, to form the hydroxy and desmethyl metabolites shown in Fig. 25.62. However, it undergoes less hydroxylation than the *R*-isomer and has a lower clearance rate. Because of these differences in metabolism and excretion, higher plasma levels of the *S*-enantiomer are achieved compared with the *R*-enantiomer. The synthesis of omeprazole and esomeprazole is described in Box 25.2. **Dexlansoprazole**, the *R*-enantiomer of lansoprazole, was also approved by the FDA in 2009.

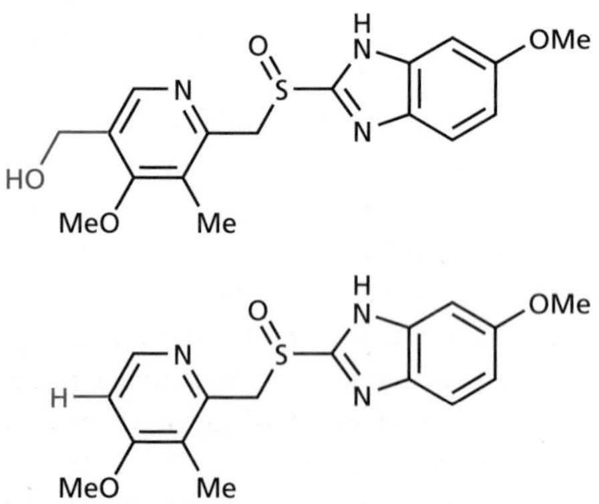

FIGURE 25.62 Metabolites of esomeprazole.

BOX 25.2 Synthesis of omeprazole and esomeprazole

The synthesis of omeprazole appears relatively simple, involving the linkage of the two halves of the molecule through a nucleophilic substitution reaction. The benzimidazole half of the molecule has a thiol substituent which is treated with sodium hydroxide to give a thiolate. On reaction with the chloromethylpyridine, the thiolate group displaces the chloride ion to link up the two halves of the molecule. Subsequent oxidation of the sulphur atom with *meta*-chloroperbenzoic acid gives omeprazole. What is not obvious from the scheme is the effort required to synthesize the required chloromethylpyridine starting material. This is not the sort of molecule that is easily bought off the shelf and its synthesis involves six steps.

The same route can be used for the synthesis of esomeprazole (the *S*-enantiomer of omeprazole) by employing asymmetric conditions for the final sulphoxidation step. Early attempts to carry out this reaction involved the Sharpless reagent formed from Ti(O-*i*Pr)$_4$, the oxidizing agent cumene hydroperoxide (Ph(CH$_3$)$_2$OOH), and the chiral auxiliary

(*S,S*)-diethyl tartrate. Although sulphoxidation took place, it required almost stoichiometric quantities of the titanium reagent and there was little enantioselectivity. The reaction conditions were modified in three ways to improve enantioselectivity to over 94% enantiomeric excess, using less of the titanium reagent (4–30 mol%).

• Formation of the titanium complex was carried out in the presence of the sulphide starting material.

• The solution of the titanium complex was equilibrated at an elevated temperature for a prolonged time period.

• The oxidation was carried out in the presence of an amine such as *N,N*-diisopropylethylamine. The role of the amine is not fully understood, but it may participate in the titanium complex.

The enantiomeric excess can be enhanced further by preparing a metal salt of the crude product and carrying out a crystallization which boosts the enantiomeric excess to more than 99.5%.

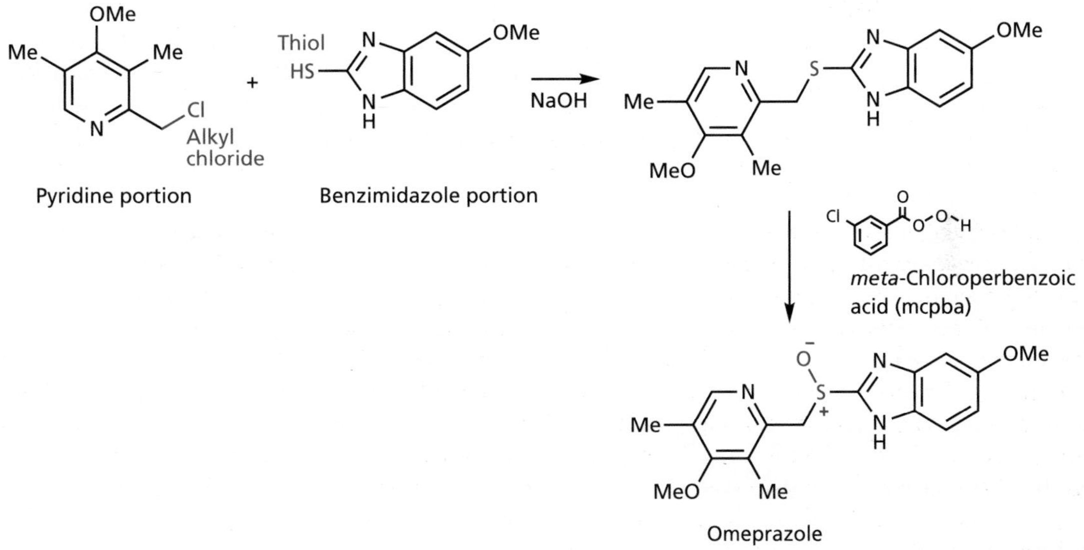

FIGURE 1 Synthesis of omeprazole.

25.3.6 Other proton pump inhibitors

The other PPIs shown in Fig. 25.56 retain the pyridinylmethylsulphinyl benzimidazole structure of omeprazole. They also share the alkoxy substituent at the *para* position of the pyridine ring. Variation has been limited to the other substituents present on the heterocyclic rings. These play a role in determining the lipophilic character of

the drug, as well as its stability. As far as the latter is concerned, there has to be a balance between the drug being sufficiently stable and un-ionized at neutral pH to reach its target unchanged, and its ability to undergo rapid acid-induced conversion into the active sulphenamide when it reaches the target. Stability to mild acid is important to avoid activation in other cellular compartments, such as lysosomes and chromaffin granules. Drugs which undergo the acid-induced conversion extremely easily are more active, but they are less stable and are more likely to undergo transformation in the blood supply before they reach their target. Drugs which are too stable are less reactive under acid conditions and react slower with the target.

The various PPIs all work by the same mechanism, but have slightly different properties. For example, pantoprazole is chemically more stable than omeprazole or lansoprazole under neutral to mildly acidic conditions (3.5–7.4), but it is a weaker, irreversible inhibitor under strong acid conditions. Rabeprazole is the least stable at neutral pH and is the most active inhibitor.

25.4 *Helicobacter pylori* and the use of antibacterial agents

25.4.1 Discovery of *Helicobacter pylori*

One of the problems relating to anti-ulcer therapy, both with the H_2 antagonists and the PPIs, is the high rate of ulcer recurrence once the therapy is finished. The reappearance of ulcers has been attributed to the presence of a microorganism called *Helicobacter pylori*, which is naturally present in the stomachs of many people, and can cause inflammation of the stomach wall. As a result, patients who are found to have *H. pylori* are currently given a combination of three drugs—a PPI to reduce gastric acid secretion, and two antibacterial agents (such as **nitroimidazole**, **clarithromycin**, **amoxicillin**, or **tetracycline**) to eradicate the organism.

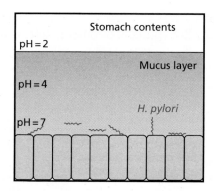

FIGURE 25.63 *Helicobacter pylori* attached to stomach cells.

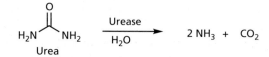

FIGURE 25.64 Action of urease.

It was once considered unthinkable that a bacterium could survive the acid conditions of the stomach. However, in 1979, it was shown that *H. pylori* can do just that. The microorganism is able to attach to a sugar molecule on the surface of the cells that line the stomach wall and use the mucus layer which protects the stomach wall from gastric juices as its own protection. As there is a pH gradient across the mucus layer, the microorganisms can survive at the surface of the mucus cells where the pH is closer to neutral (Fig. 25.63). *H. pylori* is a spiral, curved bacterium which is highly motile and grows best in oxygen concentrations of 5%, matching those of the mucus layer. The bacterium also produces large amounts of the enzyme **urease** which catalyses the hydrolysis of urea to ammonia and carbon dioxide, thus neutralizing any acid in the local environment (Fig. 25.64). The bacterial cells can contribute to the formation of stomach ulcers because they secrete proteins and toxins that interact with the stomach's epithelial cells, leading to inflammation and cell damage. It is also thought that the microorganism increases the risk of gastric cancers.

25.4.2 Treatment

As mentioned earlier, *H. pylori* is treated with a triple therapy of a PPI and at least two antibacterial agents. A PPI is administered because the antibiotics used work best at higher pH levels than those normally present in the stomach. The combination of **omeprazole**, **amoxicillin**, and **metronidazole** is frequently used, but combinations involving other antibacterial agents, such as **clarithromycin** and **tetracycline** are also possible. **Bismuth chelate** (bismuth subcitrate and tripotassium dicitratobismuthate) is present in some combination therapies. This preparation has a toxic effect on *H. pylori* and may help to prevent adherence to the mucosa. Other protective properties include an enhancement of local prostaglandin synthesis, a coating of the ulcer base, and an adsorption of pepsin.

Combination therapy has been shown to eradicate *H. pylori* in over 90% of duodenal ulcers and significantly reduce ulcer recurrence. Similar treatment is recommended for *H. pylori*-related stomach ulcers.

Eradication of *H. pylori* can be difficult because of the emergence of resistant strains and the difficulty in delivering the antibacterial agents at the required therapeutic concentration. *H. pylori* can also assume a resting coccoid form that is more resistant to therapy.

It has been found that PPIs have an inherent anti-*H. pylori* action and it has been suggested that they inhibit urease, possibly by linking to exposed cysteine residues. However, the PPIs also inhibit strains of *H. pylori* which do not have urease, so this is not the full story. This antibacterial activity is sufficient to suppress the organism but not eradicate it, so traditional antibacterial agents are still required.

Research has been carried out into the design of drugs which act as sugar decoys to prevent *H. pylori* binding with stomach cells in the first place.

25.5 Traditional and herbal medicines

Several herbal remedies have been used for the treatment of ulcers.

Liquorice has been reported to have a variety of medicinal properties and has been used as a medicine for several thousand years. It is reported to have anti-ulcer properties and this has been attributed to a component called **glycyrrhetinic acid**—the aglycone of **glycyrrhizin**. **Carbenoxolone** is a derivative of glycyrrhetinic acid and has been used in ulcer therapy. It has some antibacterial action against *H. pylori* and is thought to have a mucosal protective role by increasing mucus production.

Silymarin is a mixture of compounds (mainly **silibinin**, **silichristin**, and **siliianin**) obtained from the fruit of the milk thistle (*Silybum marianum*). It has anti-ulcer activity and has been shown to reduce histamine concentrations in rats.

Extracts from the **neem tree** (*Azadirachta indica*) have been used extensively in India as a medicine for a variety of ailments. The aqueous extract of the neem bark has been reported to have anti-ulcer effects. Possible mechanisms include proton pump inhibition or antioxidant effects in the scavenging of OH radicals.

Other herbal medicines include **comfrey** and **marshmallow**.

KEY POINTS

- *Helicobacter pylori* is a bacterium which is responsible for many ulcers. It can survive at the surface of mucus cells and produce proteins and toxins which damage epithelial cells.

- Ulcers which are caused by *H. pylori* are treated with a combination of drugs which includes a PPI and at least two antibiotics.

- Several traditional and herbal remedies are used in the treatment of ulcers.

QUESTIONS

1. Omeprazole is administered orally as a galenic formulation to protect it from being activated during its journey through the acidic contents of the stomach. Once it is released in the intestines, it is absorbed into the blood supply and carried to the parietal cells where it crosses the cell membrane into the canaliculi and is activated. As the canaliculi lead directly into the lumen of the stomach, why is omeprazole not orally administered directly to the stomach?

2. In the development of omeprazole, the methoxy and methyl groups were added to the pyridine ring to increase the pK_a.

Subsequently, it was found that analogue (I) with only one of the methyl groups had a higher pK_a than omeprazole. Suggest why this might be the case.

3. Suggest whether you think structure (I) would be a better PPI than omeprazole.

4. The acid-catalysed activation of PPIs requires pyridine to be nucleophilic, which is why two methyl groups and a methoxy group are present in omeprazole. Suggest whether the addition of an extra methyl group (structure II) would lead to a more potent PPI.

5. The phenol (III) is a very difficult compound to synthesize and is unstable at neutral pH. Suggest why this might be the case.

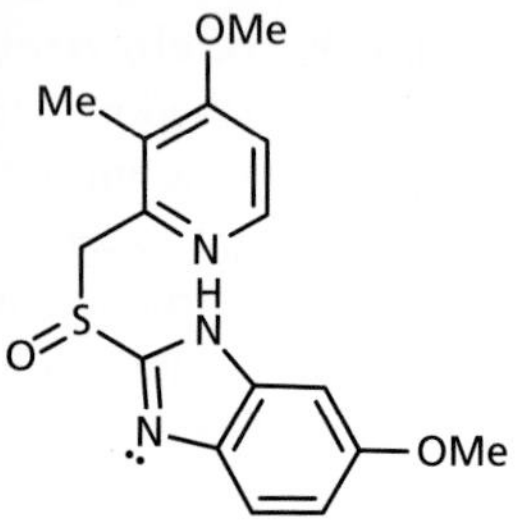

I

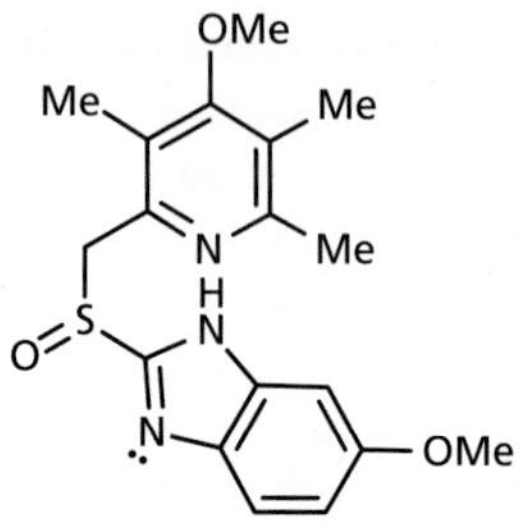

II

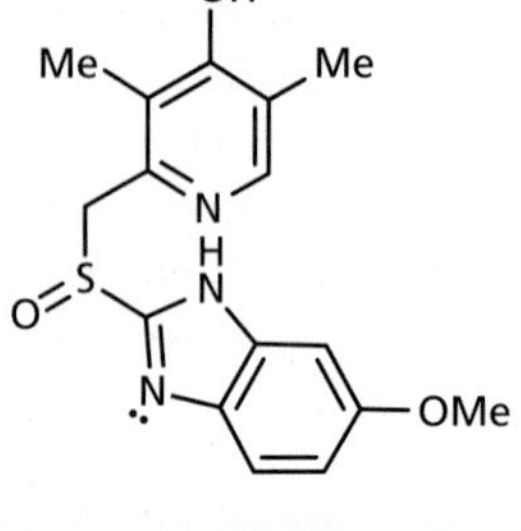

III

6. Suggest what types of metabolite might be possible from omeprazole.

7. One of the metabolic reactions that takes place on cimetidine is oxidation of the methyl substituent on the imidazole ring (Fig. 25.35). A common strategy to prevent such a metabolic reaction occurring is to replace a susceptible methyl group with a chloro substituent. Why is a chloro substituent commonly used for this purpose? Do you think an analogue of cimetidine with a 4-chloro substituent would be an improvement over cimetidine itself?

8. The acidic contents of the stomach encourage the digestion of food and the destruction of cells. Why are the cells lining the stomach not digested in that case?

⚫ Multiple-choice questions are available on the Online Resource Centre at www.oxfordtextbooks.co.uk/orc/patrick6e/

FURTHER READING

Agranat, I., Caner, H., and Caldwell, J. (2002) Putting chirality to work: the strategy of chiral switches. *Nature Reviews Drug Discovery*, **1**, 753–68.

Carlsson, E., Lindberg, P., and von Unge, S. (2002) Two of a kind. *Chemistry in Britain*, May, 42–5 (PPIs).

Ganellin, C. R., and Roberts, S. M. (eds.) (1994) Discovery of cimetidine, ranitidine and other H₂-receptor histamine antagonists. in *Medicinal chemistry—the role of organic research in drug research,* 2nd edn. Chapter 12, Academic Press, New York.

Hall, N. (1997) A landmark in drug design. *Chemistry in Britain*, December, 25–27 (cimetidine).

O'Brien, D. P., et al. (2006) The role of decay-accelerating factor as a receptor for *Helicobacter pylori* and a mediator of gastric inflammation. *Journal of Biological Chemistry*, **281**, 13317–23.

Olbe, L., Carlsson, E., and Lindberg, P. (2003) A proton-pump inhibitor expedition: the case histories of omeprazole and esomeprazole. *Nature Reviews Drug Discovery*, **2**, 132–9.

Saunders, J. (2000) Antagonists of histamine receptors (H₂) as anti-ulcer remedies. in *Top drugs: top synthetic routes.* Chapter 4, Oxford Science Publications, Oxford.

Saunders, J. (2000) Proton pump inhibitors as gastric acid secretion inhibitors. in *Top drugs: top synthetic routes.* Chapter 5, Oxford Science Publications, Oxford.

Titles for general further reading are listed on p.845.

26 Cardiovascular drugs

26.1 Introduction

Cardiovascular disease is the leading cause of death in industrial nations, and involves disorders that affect the heart and the circulatory system. Significant risk factors include hypertension, diabetes, obesity, smoking, and lack of exercise. The World Health Organization stated that in 2008, there were 17 327 000 deaths worldwide and that this could rise to 23.6 million per year by 2030. Statistics reveal that the Russian Federation and Ukraine top the international 'league tables' for death rates from cardiovascular disease. The total number of US deaths per year from cardiovascular disease is still increasing, although death *rates* (the number of deaths per 100 000 people) have decreased. This has been attributed to the introduction of drugs such as the statins, and public health initiatives aimed at discouraging smoking. Nevertheless, cardiovascular disease is still a major cause of death in western societies.

Considering the impact of cardiovascular disease on society, it is not surprising that cardiovascular drugs include some of the most profitable therapeutics on the market. Recent statistics showed that NHS England spent £6.8 billion on cardiovascular drugs for the year 2012–13.

A wide variety of biological pathways and mediators control the cardiovascular system, providing many different molecular targets for drug design. For example, the cholinergic and adrenergic systems described in Chapters 22 and 23 play a vital role in controlling blood pressure, heart rate, and the strength of heart contractions. Several cholinergic and adrenergic agents have cardiovascular activity, but the most important are the beta-blockers described in section 23.11. In this chapter, we consider other cardiovascular drugs that are used to treat hypertension (high blood pressure), cardiac problems, atherosclerosis, and thrombosis.

26.2 The cardiovascular system

The cardiovascular system (Fig. 26.1) is the system by which blood is circulated round the body. The heart acts as the pump and contains four chambers—the left and right atria, and the left and right ventricles. The atria are the collecting chambers for blood entering the heart, and the ventricles pump blood out of the heart. This is achieved by contraction of cardiac muscle cells, which are quite distinct from skeletal muscle cells.

The left ventricle is responsible for pumping blood out of the heart and round the body (**systemic circulation**). When the ventricle contracts, blood is pumped into the **aorta**—the main artery leaving the heart. The aorta branches to form **arteries**, which carry blood to the various tissues and organs of the body. When the arteries reach these target sites, they branch into narrower blood

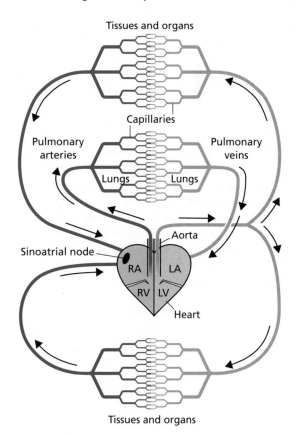

FIGURE 26.1 The cardiovascular system (LA = left atrium, RA = right atrium, LV = left ventricle, RV = right ventricle).

vessels called **arterioles**, which continue to divide into even narrower blood vessels called **capillaries**. This can be viewed like the branching of a tree, where the aorta is the main trunk, the arteries are the main branches, the arterioles are smaller branches, and the capillaries are the twigs.

The blood in the capillaries supplies oxygen and nutrients to the body's cells and takes up carbon dioxide and waste products from those same cells. The capillaries then merge back together to form small veins (venules), which, in turn, merge to form veins. These carry the blood back to the right atrium of the heart to complete the systemic circulation.

The next part of the process is **pulmonary circulation**. The right atrium contracts to force blood into the right ventricle through a one-way valve. Once that has been achieved, the right ventricle contracts and blood is forced through the **pulmonary arteries** towards the lungs. The arteries divide to form capillaries within the lung, and carbon dioxide diffuses from the blood in the capillaries into the air sacs of the lungs (alveoli). At the same time, oxygen diffuses in the opposite direction to oxygenate the blood. The capillaries now merge to form the pulmonary veins which carry the oxygenated blood to the left atrium of the heart. The blood is pumped from the left atrium into the left ventricle, which pumps it back into the systemic circulation.

It is important to appreciate that all the blood departing the left ventricle travels through the lungs, whereas the blood departing the right ventricle travels to different organs and tissues, and so only a certain percentage of the blood travelling round the body reaches a specific tissue or organ. This includes the heart itself, which is supplied with oxygen and nutrients by blood in the **coronary arteries**, rather than the blood that is pumped through it.

The amount of blood pumped by each half of the heart is about 5 litres/min for an average male, but this can increase fivefold during exercise due to an increase in both heart rate and contractility (force of contraction). Various factors control the heart rate and contractility. First of all, the cardiac muscles in the heart have to contract in a coordinated fashion if the heart is to act as an efficient pump. This is achieved by a collection of natural pacemaker cells in a region of the heart called the **sinoatrial node**, which is present in the right atrium. When the pacemaker cells contract, they stimulate neighbouring cardiac muscle cells to contract. Therefore, the signal initiated by the sinoatrial node spreads over the left and right atria, such that the atria contract before the ventricles. There is a time delay before the ventricles contract, by which time the atria have relaxed.

Heart rate is regulated by neural and hormonal influences on the sinoatrial node. Parasympathetic neurons release acetylcholine to slow the heart rate, whereas sympathetic neurons release noradrenaline to speed it up. The hormone adrenaline also speeds up the heart rate. Other influences on heart rate include temperature, plasma electrolyte concentrations, and other hormones.

The heart's force of contraction (contractility) is also influenced by various factors. If there is an increased flow of blood into the heart, the ventricle is distended and this automatically increases contractility. Noradrenaline released from the sympathetic nervous system also increases contractility, as does circulating adrenaline. In contrast, the parasympathetic nervous system has little influence on contractility.

The amount of blood that reaches different organs and tissues varies determining on circumstances, and is determined by the dilation/constriction of the arterioles supplying those tissues. For example, when exercising, blood vessels supplying skeletal muscle and the heart dilate to increase the amount of blood received. At the same time, the blood vessels supplying the kidneys and abdomen constrict such that these areas receive a reduced blood supply. Dilation or constriction of arterioles is affected by both local and neural factors. An increase in metabolic activity for any tissue produces decreased levels of oxygen and increased levels of carbon dioxide, hydrogen ions, potassium ions and metabolites. These changes stimulate dilation of arterioles to the tissues concerned.

The activity of the sympathetic nervous system also influences the constriction and dilation of arterioles. If the sympathetic nervous system increases in activity, more noradrenaline is released resulting in the constriction of most arterioles. If the sympathetic nervous system decreases in activity, the opposite occurs and most arterioles dilate. One exception involves the arterioles supplying skeletal muscle, which dilate when the sympathetic nervous system increases in activity.

A number of important hormones have a powerful effect on the dilation or constriction of arterioles. For example, histamine dilates arterioles following tissue damage. Other potent vasodilators are nitric oxide, prostacyclin, bradykinin, and natriuretic factors. Hormones with powerful vasoconstricting properties include angiotensin II and endothelin.

Blood pressure is a key indicator of a person's health and is affected by a number of factors, namely heart rate and contractility, blood volume, and the peripheral resistance to blood flow though the blood vessels. Regulation of blood pressure is controlled by the **medullary cardiovascular centre**, which determines the activity of the sympathetic and parasympathetic nervous systems acting on the cardiovascular system. This centre receives signals from arterial baroreceptors that are sensitive to blood pressure. The higher the blood pressure, the more the arterioles stretch, and the larger the signal from the baroreceptors to the medullary cardiovascular centre. If blood pressure is high, the control centre reduces the activity of the sympathetic nervous system and increases the activity of the parasympathetic nervous system. This produces a decrease in heart rate and contractility, and dilation of blood vessels. The medullary cardiovascular centre also receives information from many other sources such as peripheral sensory receptors and arterial chemoreceptors that are sensitive to concentrations of oxygen, carbon dioxide, and hydrogen ions.

If these feedback systems fail to bring blood pressure back to normal levels, then chronic high blood pressure (hypertension) results. This is a major cause of cardiovascular disease, and so antihypertensive agents are a key part of cardiovascular medicine.

26.3 Antihypertensives affecting the activity of the RAAS system

26.3.1 Introduction

Hypertension is a key risk factor for cardiovascular diseases such as stroke, heart attack, kidney failure, and atherosclerosis, and so reducing blood pressure with antihypertensive agents is one of the most effective strategies for lowering that risk. Several biochemical processes can lead to increased blood pressure, and one of the most important is the **Renin–Angiotensin–Aldosterone System (RAAS)** (Fig. 26.2). The key component

of this process is **angiotensin II**, which is a peptide hormone with potent vasoconstrictive properties. In other words, it causes blood vessels to constrict and produce an increase in blood pressure. Angiotensin II also promotes the release of another hormone called **aldosterone** from the adrenal cortex. Aldosterone promotes increased fluid retention in the kidneys—another factor that increases blood pressure because of increased blood volume. Problems of hypertension occur when the RAAS cascade becomes too active, and so a number of antihypertensive agents have been designed which block the cascade at various points. For example, agents that inhibit the enzymes **renin** and **angiotensin-converting enzyme** (**ACE**) reduce the levels of angiotensin II produced. Another approach is to design agents that block the **AT$_1$ receptors** that respond to angiotensin II. Finally, aldosterone inhibitors block the effects of aldosterone. We shall now look at these agents in more detail.

26.3.2 Renin inhibitors

Renin is an enzyme that is produced in the kidneys, and which catalyses the cleavage of a protein substrate called **angiotensinogen** to form **angiotensin I** (Fig. 26.2).

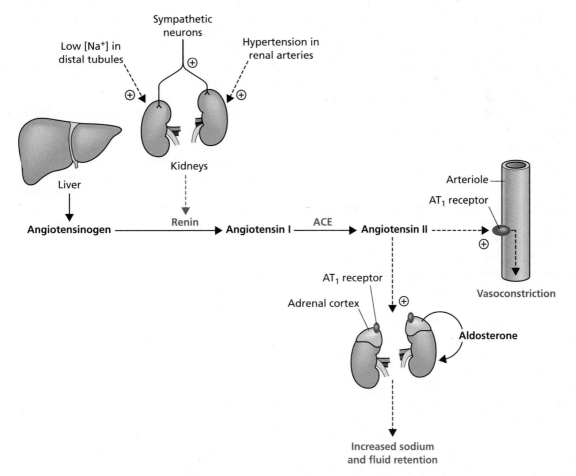

FIGURE 26.2 Biosynthesis and actions of angiotensin II. Biosynthesis is indicated by solid arrows. Actions are indicated by dashed arrows. Enzymes are coloured blue.

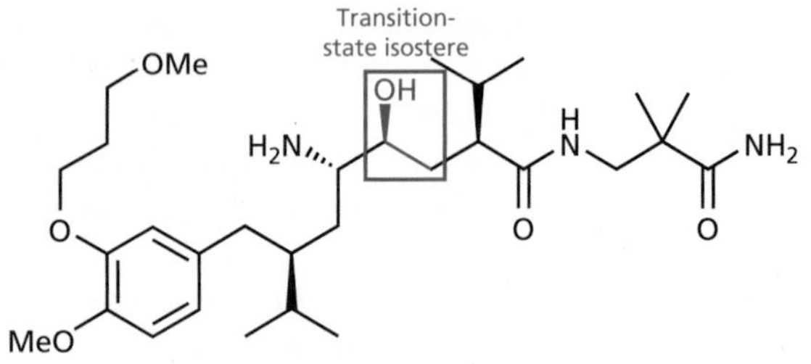

FIGURE 26.3 Aliskiren.

Angiotensin I has no vasoconstrictive properties, but it acts as the substrate for the next enzyme in the RAAS cascade—angiotensin-converting enzyme (ACE) (section 26.3.3). Therefore, blocking the synthesis of angiotensin I will also lower angiotensin II levels.

There are several advantages in targeting renin. It is the first enzyme in the RAAS cascade, and it catalyses the rate-limiting step for the process. Therefore, renin inhibition should be more efficient than inhibiting ACE. Blocking renin also avoids the build-up of angiotensin I that results from inhibiting ACE (section 26.3.3). Finally, renin only accepts angiotensinogen as a substrate, and appears to have no other biochemical role. Therefore, enzyme inhibitors showing selectivity for renin should only affect the RAAS cascade, but not other biochemical processes. This will minimize side effects.

A substantial research effort into renin inhibitors led eventually to the discovery of **aliskiren** (Fig. 26.3; Case study 8), which was approved by the US FDA in 2007. The structure of aliskiren includes a transition-state isostere (section 13.3.7) that mimics the tetrahedral transition state of the enzyme-catalysed reaction and enhances binding affinity. Despite the successful approval of aliskiren, no other renin inhibitor has reached the market.

26.3.3 ACE inhibitors

The **angiotensin-converting enzyme** (ACE) is a membrane-bound enzyme which catalyses the conversion of **angiotensin I** to the potent vasoconstrictor **angiotensin II** (Fig. 26.2). Research into ACE inhibitors as potential antihypertensive agents took place as early as the 1970s,

resulting in the design of **captopril** (Fig. 26.4), which was approved in 1981 as the first clinically effective agent to target the RAAS cascade (Case study 2). Captopril contains a thiol group that binds to the zinc ion cofactor present in the active site. It also includes a carboxylic acid which is ionized under physiological conditions, and forms an ionic interaction with a lysine residue. The methyl group (labelled P1′) fits the enzyme's S1′ binding subsite and forms van der Waals interactions.

Enalaprilat (Fig. 26.4) was the next ACE inhibitor to be designed, and includes an additional phenethyl substituent that binds to the S1 binding subsite. A second carboxylic acid was added to replace the thiol group as the binding group for zinc, but this increases the polarity of the drug and results in poor oral absorption. Therefore, the drug is administered as its ethyl ester prodrug (**enalapril**). **Lisinopril** is similar in structure, varying only in the P1′ substituent.

A flurry of further ACE inhibitors were approved in the 1990s when the proline ring was replaced with a variety of bicyclic rings—**moexipril, quinapril, perindopril, trandolapril, ramipril,** and **benzazepril** (Fig. 26.5). Such variations were possible because the S2′ subsite in ACE can accept a wide variety of amino acid side chains. Aromatic side chains are particularly favoured, which is why an aromatic ring is included in the bicyclic ring systems of moexipril, quinapril, and benzazepril. Benzazepril contains a level of conformational restraint (section 13.3.9) because the P1′ methyl group and neighbouring peptide bond are encapsulated within a seven-membered ring. All of these bicyclic agents are ester prodrugs that are hydrolysed *in vivo* to generate the active ACE inhibitor.

Fosinopril (Fig. 26.5) is a prodrug which contains an extended ester (section 14.6.1.1). Enzymatic hydrolysis of this ester generates an unstable intermediate which spontaneously decomposes to generate **fosinoprilat**, which contains a phosphinic acid group. The phosphinic acid group ($HOPOCH_2-$) was introduced as a replacement for a peptide group ($-CONH-$), and acts as a mimic for the transition state involved in peptide bond hydrolysis. Fosinoprilat also takes advantage of the spacious S2′ pocket by having a cyclohexyl substituent on the proline ring.

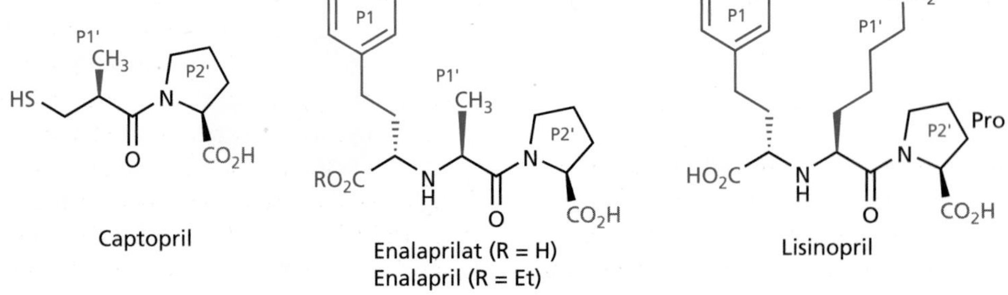

FIGURE 26.4 Early examples of ACE inhibitors.

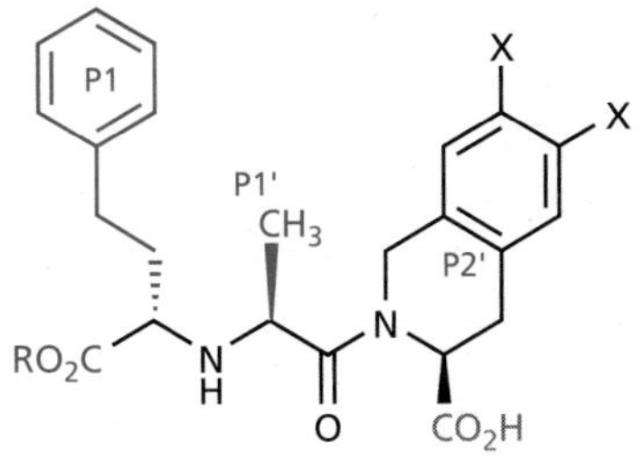

FIGURE 26.5 Further examples of ACE inhibitors.

The ACE inhibitors are successful antihypertensive agents, but there are side effects associated with prolonged use. The resulting drop in angiotensin II levels triggers a feedback mechanism that increases the levels of renin. This, in turn, increases levels of angiotensin I. Although ACE is inhibited, some of that angiotensin I is converted to angiotensin II by the action of other enzymes such as **chymases** and **cathepsin G**. Thus, the antihypertensive effects of ACE inhibitors decrease with time. In addition, ACE has an important role in catalysing the hydrolysis of a peptide inflammatory mediator called **bradykinin**. Therefore, ACE inhibition results in increased bradykinin levels. Although bradykinin can dilate blood vessels and contribute to the antihypertensive effect, it can also act on the airways and trigger persistent coughs in some patients. Another side effect is angioedema (e.g. facial swelling).

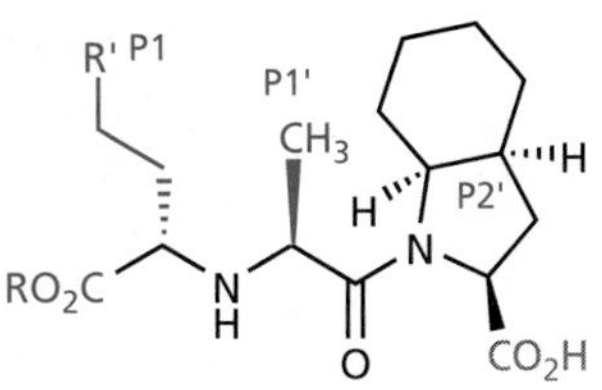

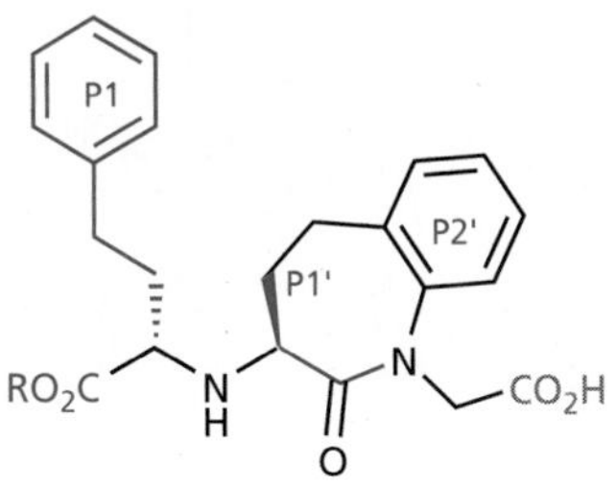

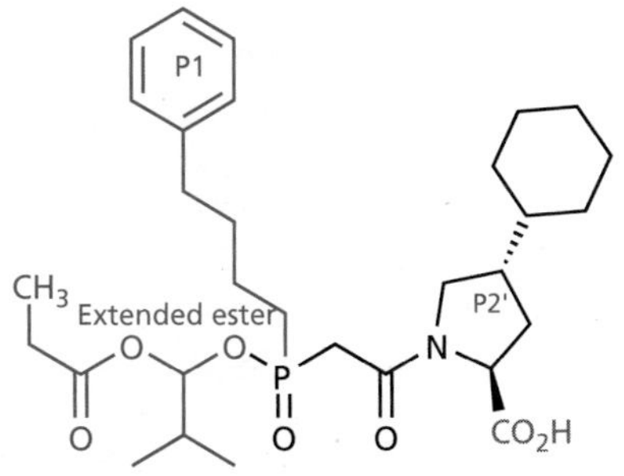

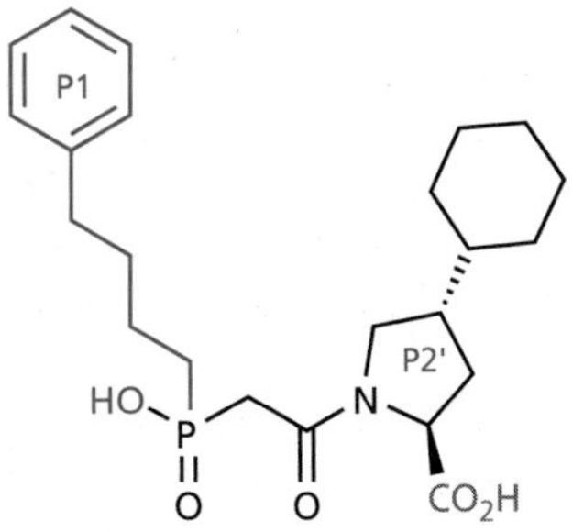

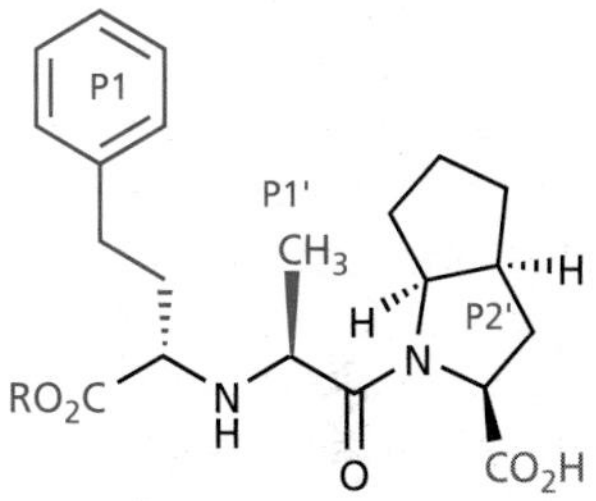

26.3.4 Angiotensin receptor antagonists

Angiotensin II produces its various effects by activating **angiotensin II receptors**, which are examples of membrane-bound G-protein-coupled receptors (sections 4.7 and 5.1–5.3). There are four subtypes of angiotensin II receptor (AT_1 to AT_4), and it is the AT_1 receptor that is responsible for vasoconstriction. In the 1990s, a group of

selective AT_1 antagonists known as the 'sartans' reached the clinic. The first of these was **losartan** (Fig. 26.6)**,** which was launched by Du Pont. The precursor for losartan was a biphenyl structure that had good antihypertensive activity, but had to be injected due to poor oral absorption. This problem was overcome by replacing the polar carboxylic acid with a less polar tetrazole ring. The tetrazole ring is commonly used as a bio-isostere for a carboxylic acid (sections 13.3.7 and 14.1.6). It was later discovered that losartan undergoes oxidative metabolism to form **EXP3174** (Fig. 26.7), which proved to be a more potent antagonist and established the importance of the resulting carboxylic acid.

SAR studies have demonstrated the importance of three key features in EXP3174 for good antagonist activity. Strong binding is associated with the acidic tetrazole ring and a hydrophobic substituent on the imidazole ring (Fig. 26.8). These features serve as an 'address' that directs the structure to the AT_1 receptor. The third feature is the carboxylic acid, which is ionized under physiological conditions. This group is crucial for antagonist activity and acts as the 'message' for receptor antagonism. It is thought that the carboxylate group forms an ionic interaction with a lysine residue to produce an induced fit that forces the receptor into an inactive conformation. The imidazole and biphenyl rings mainly serve as a scaffold to orientate the three key binding groups, allowing them

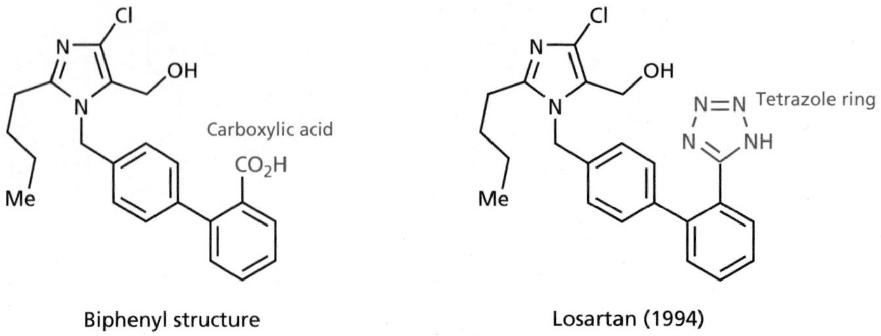

FIGURE 26.6 Development of losartan.

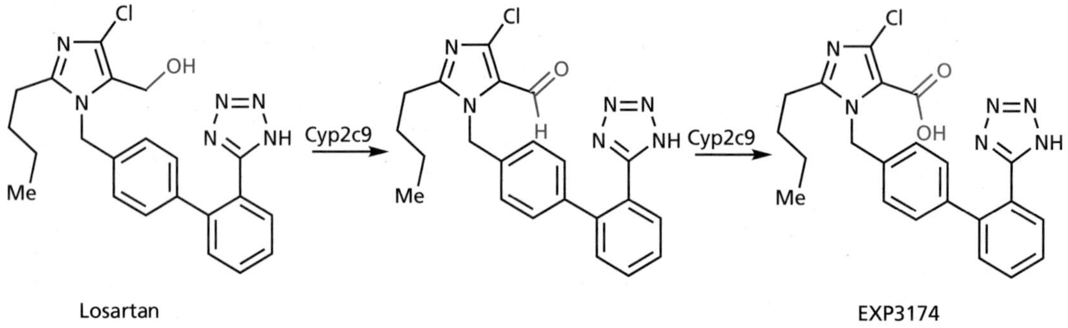

FIGURE 26.7 Metabolism of losartan.

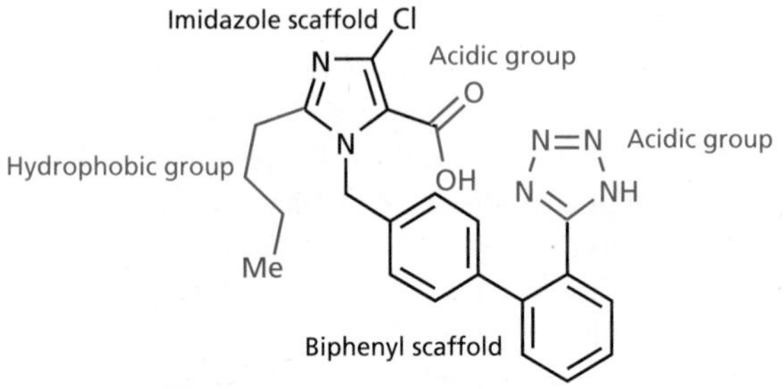

FIGURE 26.8 Key structural features in EXP3174.

to interact simultaneously with complementary binding regions in the binding site.

Other examples of sartans include **valsartan, candesartan, irbesartan, telmisartan, eprosartan, azilsartan,** and **olmesartan** (Fig. 26.9). Some sartans (candesartan, olmesartan, and valsartan) act as inverse agonists rather than pure antagonists by stabilizing the inactive state of the receptor (section 8.5).

All the sartans contain the hydrophobic substituent and the two acidic groups required for good binding and activity, apart from irbesartan which lacks one of the acidic groups. The tetrazole feature acts as one of the acidic groups in several of the sartans, but alternative bioisosteres are possible. For example, azilsartan contains an oxadiazole ring. Although the binding affinity of azilsartan is weaker than the closely related candesartan, its *in vivo* activity is superior due to increased bioavailability.

This has been attributed to the fact that the oxadiazole ring is less polar than the tetrazole ring and allows better oral absorption.

The nature of the hydrophobic substituent varies in the different structures.

Many of the sartans contain a scaffold that includes a biphenyl ring and an imidazole or benzimidazole ring. However, valsartan and eprosartan are exceptions, and contain only one of these features.

The presence of two acidic groups in sartans can result in poor oral absorption, and so a number of extended ester prodrugs have been designed. For example, **candesartan cilexetil, olmesartan medoxomil,** and **azilsartan medoxomil** (Fig. 26.10) are prodrugs for candesartan, olmesartan, and azilsartan respectively. Simple methyl or ethyl esters were found to be unsuitable as prodrugs for the sartans because they proved to be poor substrates for the body's esterase enzymes. This is likely to be due to steric issues, and so extended esters similar to those used in the design of penicillin prodrugs are used instead (see also Box 19.7). The more exposed ester undergoes enzymatic hydrolysis to form an unstable product that undergoes a spontaneous chemical transformation to 'release' the sartan.

AT_1 antagonists have proved highly effective, but they can lead to elevated levels of angiotensin II which then stimulates AT_2 and AT_4 receptors. Moreover, the feedback system by which increased levels of angiotensin II inhibit the release of renin is dependent on the activation

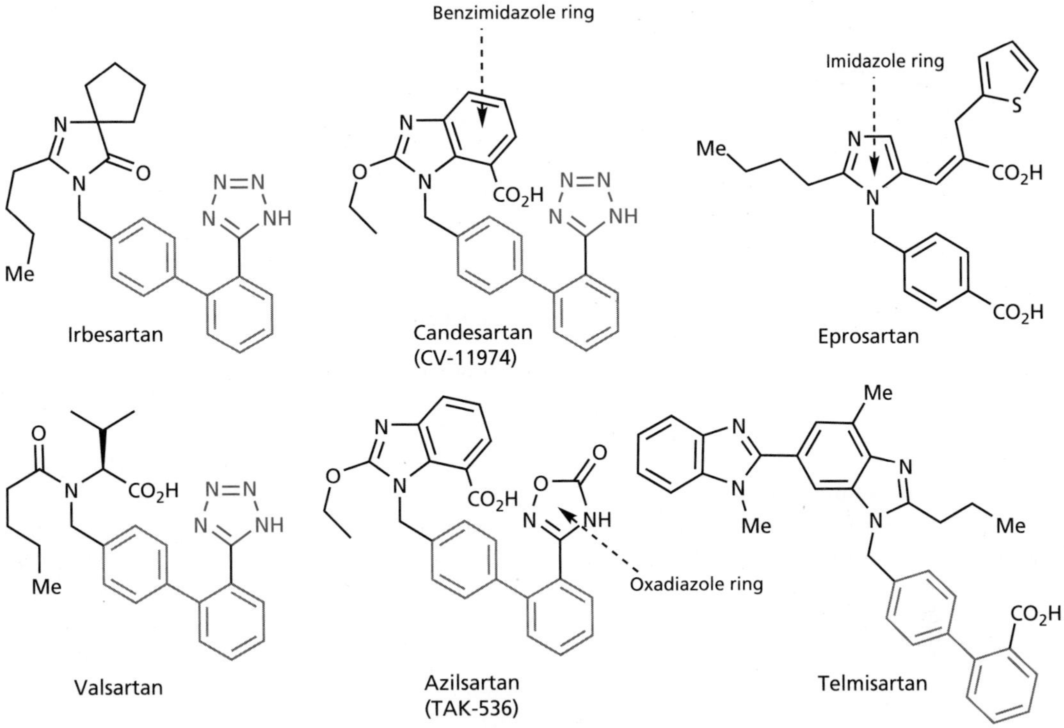

FIGURE 26.9 Examples of other 'sartans'.

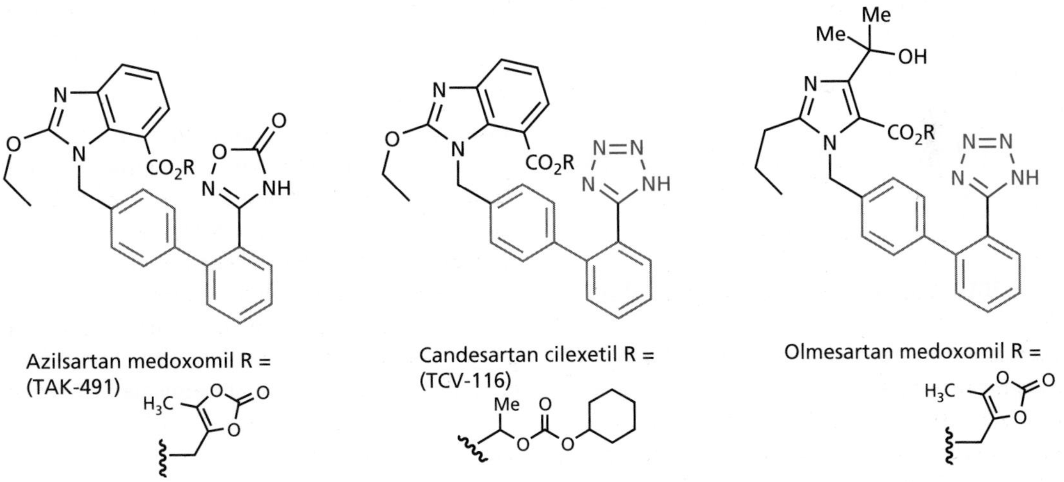

FIGURE 26.10 Sartan prodrugs.

of AT_1 receptors. Therefore, AT_1 antagonists will result in increased renin levels.

26.3.5 Mineralocorticoid receptor antagonists

Angiotensin II promotes the release of **aldosterone** (Fig. 26.11), which is the most potent of the endogenous mineralocorticoids responsible for increasing fluid and sodium ion retention in the kidneys. As a result, blood volume and blood pressure increase. Aldosterone acts on intracellular mineralocorticoid receptors in order to produce these effects, and so mineralocorticoid receptor antagonists have been studied as antihypertensive agents. **Spironolactone** is one such agent that reached the clinic in 1959, and acts as a prodrug for the active antagonist **canrenone** (Fig. 26.11). However, the agent is not totally selective and undesirable side effects can result from antagonism of progesterone and androgen receptors.

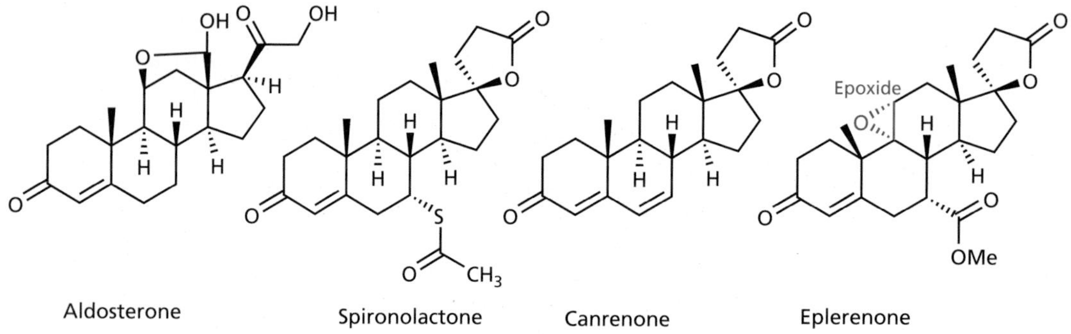

FIGURE 26.11 Aldosterone and mineralocorticoid receptor antagonists.

Eplerenone is a more selective agent, with the epoxide ring proving a crucial feature in this respect. Spironolactone and eplerenone are currently used as diuretics in patients suffering from heart failure.

26.3.6 Dual-action agents

Blocking the RAAS cascade is not guaranteed to be totally effective in treating hypertension, as there are other biological pathways that can play a role in increasing blood pressure. Therefore, there are potential advantages in designing dual-action agents that not only block the RAAS cascade, but also affect one of those other pathways. Examples of such agents are given in sections 26.4.3, 26.5.1, and 26.5.3.

26.4 Endothelin receptor antagonists as antihypertensive agents

26.4.1 Endothelins and endothelin receptors

Endothelins are endogenous peptides that contain 21 amino acids and have powerful vasoconstrictive activity.

There are three different endothelins (ET-1, ET-2, and ET-3), but ET-1 is the most important in terms of hypertension. It is 10 times more potent than angiotensin II.

The endothelins produce their physiological effects by activating endothelin receptors—members of the G-protein-coupled receptor family (section 4.7.3). There are four types of endothelin receptor (ET_A, ET_{B1}, ET_{B2}, and ET_C), with ET_A receptors predominating in the smooth muscles of pulmonary arteries. Therefore, endothelin antagonists that show good selectivity for ET_A receptors are beneficial for the treatment of pulmonary arterial hypertension. Nevertheless, antagonists acting at ET_{B2} receptors can also be beneficial in treating hypertension.

26.4.2 Endothelin antagonists

Bosentan, **sitaxsentan**, **ambrisentan**, and **macitentan** are endothelin receptor antagonists that have been approved for the treatment of hypertension (Figs. 26.12 and 13). However, sitaxsentan was withdrawn by Pfizer in 2010 due to a number of fatalities. Darusentan was in clinical trials in 2009.

Apart from sitaxsentan, the clinically approved ET antagonists contain a pyrimidine ring as a central core or scaffold for the various substituents that are present.

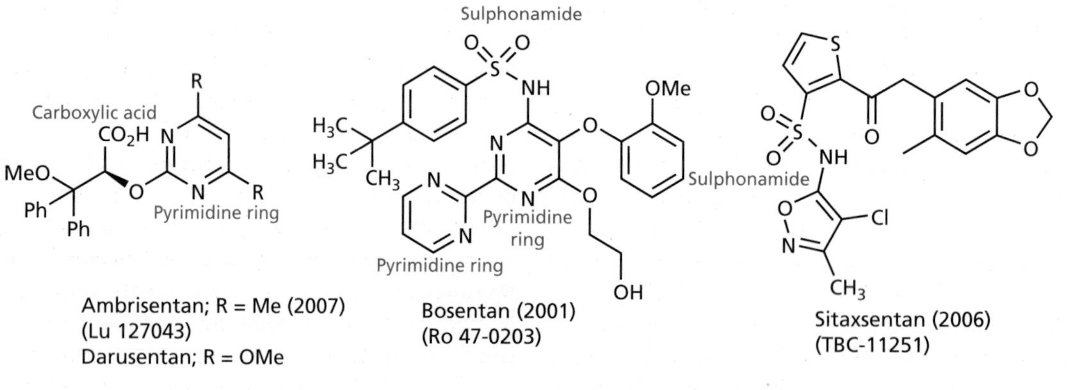

FIGURE 26.12 Endothelin antagonists.

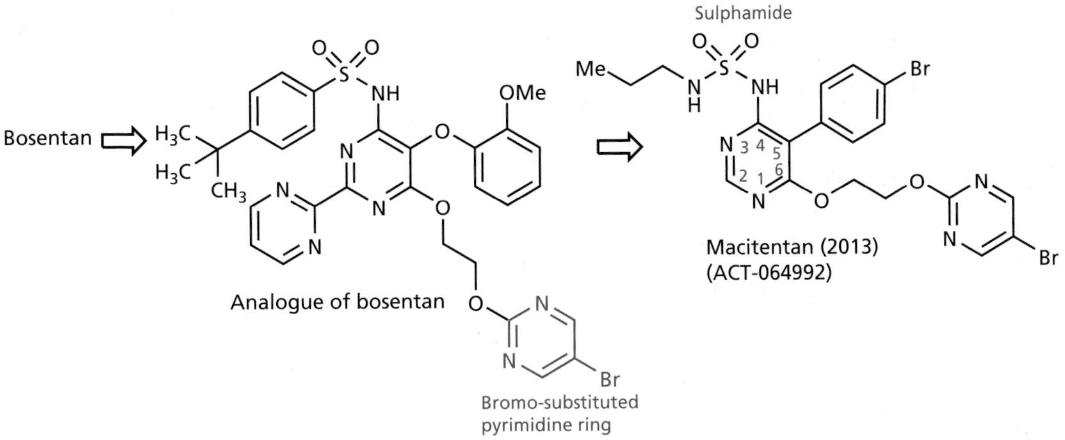

FIGURE 26.13 Development of macitentan.

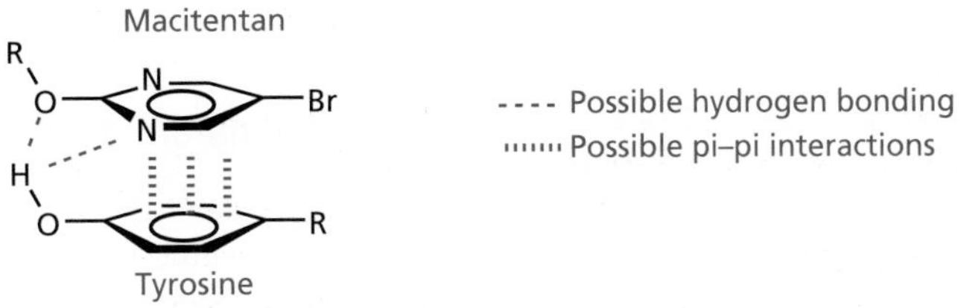

FIGURE 26.14 Possible binding interactions.

These substituents include an acidic group (in the form of a carboxylic acid, sulphonamide, or sulphamide), which may be interacting with a basic amino acid in the binding site. The structures also contain at least three aromatic or heteroaromatic rings that may bind to hydrophobic regions of the binding site.

Bosentan was the first agent to be approved. It interacts with both ET_A and ET_B receptors, showing a slight selectivity for the ET_A receptor. Later antagonists have been designed with greater selectivity for the ET_A receptor, although it is not yet proven whether this is beneficial or not.

Bosentan acted as the lead compound for the development of macitentan, which was approved in 2013 (Fig. 26.13). Adding a bromo-substituted pyrimidine ring to the alcohol group of bosentan resulted in increased binding affinity, which was attributed to hydrogen bonding and pi–pi interactions with a tyrosine residue (Fig. 26.14). This is an example of the strategy of drug extension where additional groups are added in order to achieve extra binding interactions (section 13.3.2). However, the resulting structure contained five aromatic/heteroaromatic rings and had a molecular weight above 600 g mol^{-1}, both of which are likely to be detrimental to oral activity (section 11.3). Therefore, further development

focused on optimizing the substituents around the pyrimidine core in order to reduce the molecular weight and the number of aromatic rings. The introduction of a novel sulphamide functional group was a significant step forward in achieving that goal, and it was possible to remove the substituent at position 2 altogether.

26.4.3 Dual-action agents

Dual-action agents are being developed that act as antagonists at both endothelin and AT_1 receptors. There are some structural similarities between these two receptors, and similar SAR results were obtained for peptide ligands acting on both targets. Therefore, Merck screened a library of AT_1 receptor antagonists to identify agents that also acted as ET antagonists. This led to the discovery of **L-746072** (Fig. 26.15) with nanomolar affinity for both kinds of receptor.

Coming from the other direction, Bristol-Myers Squibb studied known endothelin antagonists to see whether any had AT_1 antagonist activity. This led to the development of **PS-433540** (**BMS-346567**) (Fig. 26.15), which shows good activity against both the AT_1 and ET_B receptors. This compound reached clinical trials in 2007.

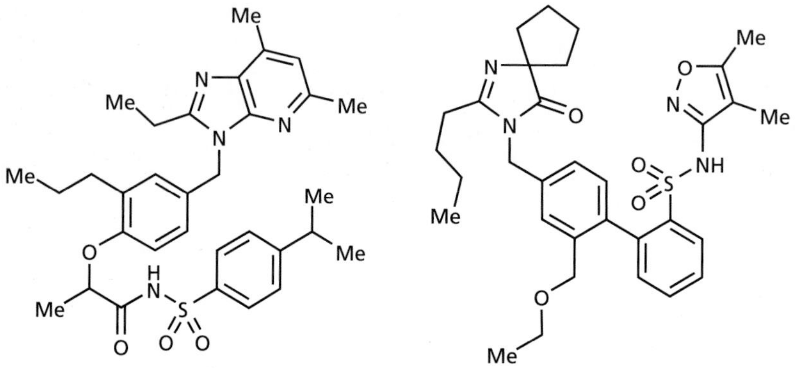

FIGURE 26.15 Dual-action agents that act as endothelin and AT_1 antagonists.

KEY POINTS

- The RAAS cascade produces angiotensin II, which is a potent vasoconstricting hormone.

- Renin and angiotensin-converting enzyme (ACE) are key enzymes in the RAAS cascade, and are important targets for antihypertensive agents.

- Angiotensin II produces its vasoconstrictive effects by activating the angiotensin II receptor (AT_1), making this receptor an important target for antihypertensive agents.

- Aliskiren is the only approved renin inhibitor for treating hypertension. It blocks the synthesis of angiotensin I.

- A large variety of ACE inhibitors have been brought to the market. They all contain a functional group that binds to the zinc ion cofactor of ACE.

- Most ACE inhibitors are administered as prodrugs to lower their polarity and to increase absorption.

- The sartans are antihypertensive agents that act on the AT_1 receptor as either antagonists or inverse agonists. The presence of two acidic groups and a hydrophobic substituent is important for binding and activity.

- Mineralocorticoid receptor antagonists that block the effects of aldosterone have antihypertensive effects.

- Endothelins are endogenous peptides with potent vasoconstrictive properties. Endothelin antagonists have been successfully developed as antihypertensive agents.

26.5 **Vasodilators**

In the previous section, we looked at agents that block the actions of endogenous hormones with vasoconstrictive properties. Vasodilation results because the levels of those hormones are lowered. In this section, we consider agents that actively promote the dilation of blood vessels (vasodilation). Such agents can be used in the treatment of angina or hypertension.

26.5.1 **Modulators of soluble guanylate cyclase**

Soluble guanylate cyclase (sGC) is an enzyme that catalyses the synthesis of **cyclic GMP** (section 5.4.3)—an important secondary messenger involved in the signal transduction pathways leading to vasodilation. The enzyme is activated by the gaseous signalling molecule **nitric oxide**, which binds to the ferrous ion of a penta-coordinated haem complex within the enzyme (Fig. 26.16). The resulting hexa-coordinated complex then loses histidine as a ligand to form a penta-coordinated nitrosyl haem complex, which produces a 200-fold activation of the enzyme.

Medications such as **glyceryl trinitrate, isosorbide mononitrate**, and **isosorbide dinitrate** (Fig. 26.17) are used to provide relief from angina by dilating blood vessels and easing the work load on the heart. They act by providing a source of nitrate ion, which is then converted *in vivo* to nitric oxide. The nitric oxide then activates the sGC enzyme. **Sodium nitroprusside** (Fig. 26.17) is an inorganic compound that acts as a source of nitric oxide and is used in emergencies to treat severe hypertension.

Various research teams have now designed dual-action agents that release nitrate in addition to their primary role. These include calcium channel activators and AT_1 receptor antagonists. The latter category includes analogues of losartan and telmisartan (Fig. 26.18). However, none of these agents have reached the clinic to date.

One disadvantage with the nitrate-releasing agents is their reliance on metabolic reactions to generate nitric oxide. This is because the effects can vary depending on how efficiently these metabolic reactions take place in different individuals. In addition, some patients can gain a tolerance to the agents over prolonged use. Finally,

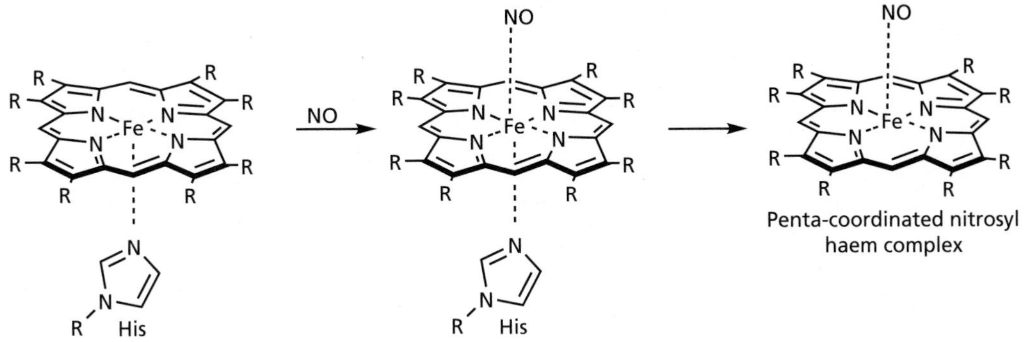

FIGURE 26.16 Activation of soluble guanylate cyclase by nitric oxide.

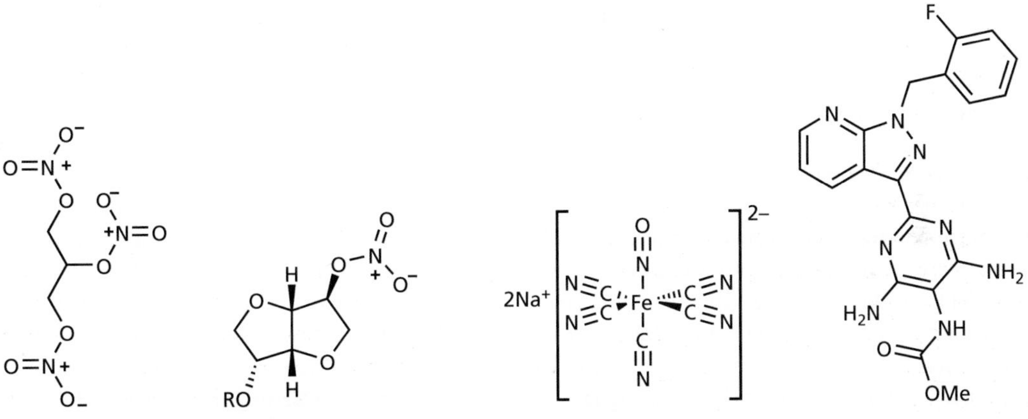

Glyceryl trinitrate
(nitroglycerine)

Isosorbide mononitrate (R = H)
Isosorbide dinitrate (R = NO₂)

Sodium nitroprusside

Riociguat
(BAY 63-2521)

FIGURE 26.17 Examples of vasodilators.

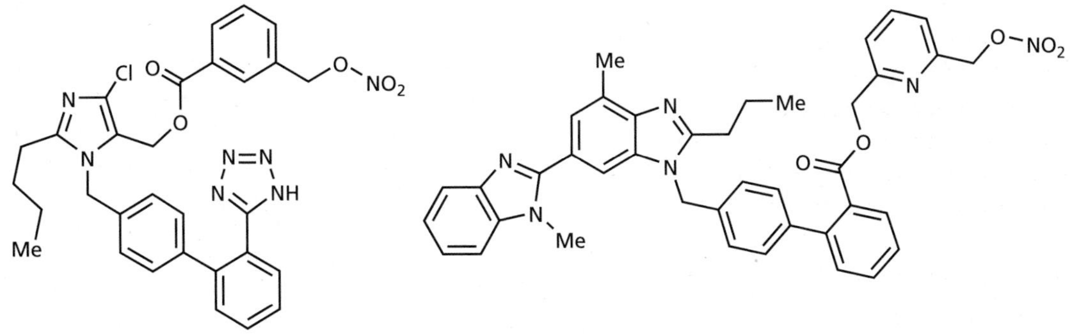

Analogue of losartan

Analogue of telmisartan (WB1106)

FIGURE 26.18 Dual-action agents that release nitrate ions.

the NO that is generated is non-specific in its actions, and can affect any biological pathway that includes an enzyme with iron in its active site. Therefore, research has recently focused on developing drugs that interact directly with the cyclase enzyme.

Riociguat (Fig. 26.19) is one such example, and was approved by the US FDA in 2013 for the treatment of

pulmonary hypertension—a condition where the blood vessels between the heart and the lungs are constricted. It is thought that the drug binds to an allosteric binding site of sGC and serves to stabilize the nitrosyl–haem complex of the enzyme. As a result, riociguat stimulates the enzyme to bind NO more efficiently, and helps to keep the enzyme in its active conformation for longer periods.

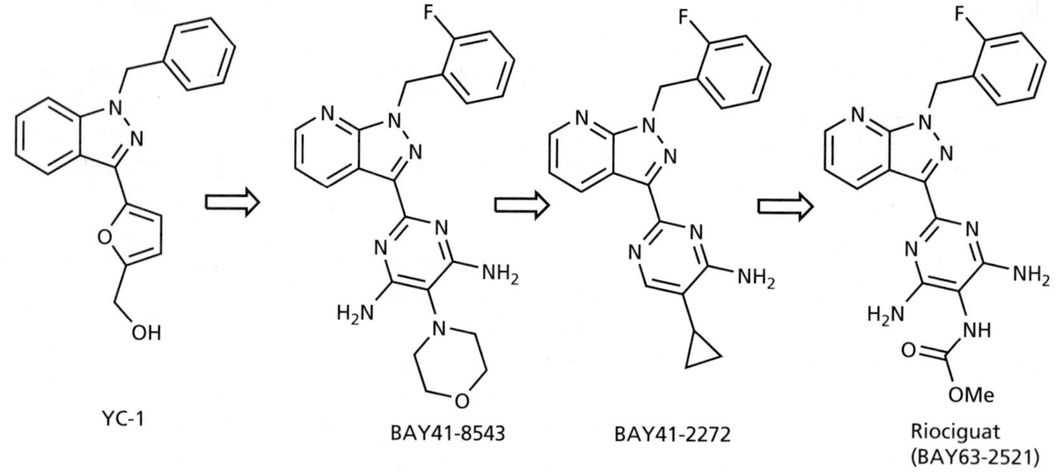

FIGURE 26.19 Development of riociguat.

Riociguat was developed at Bayer Health Care AG, starting from the lead compound YC-1 (Fig. 26.19). Analogues were synthesized in which the rings and their substituents were varied, leading to the development of **BAY41-8543** and **BAY41-2272**. These showed good activity, but adverse pharmacokinetic properties. Further modifications were carried out to improve pharmacokinetic properties leading to the discovery of riociguat.

26.5.2 Phosphodiesterase type 5 inhibitors

Cyclic nucleotide phosphodiesterases are enzymes that catalyse the hydrolysis of cyclic GMP to GMP, or cyclic AMP to AMP. By inhibiting these enzymes, the half-life and activity of these cyclic nucleotides is prolonged. Phosphodiesterase type 5 (PDE5) is of particular interest as a drug target in cardiovascular medicine, because it is present in vascular smooth muscle and specifically hydrolyses cyclic GMP. By inhibiting PDE5, the half-life of cyclic GMP is increased, which prolongs its activity as a vasodilator. Two PDE5 inhibitors have been approved for the treatment of pulmonary hypertension. **Sildenafil** (Fig. 12.13) was approved by the FDA in 2005, and **tadalafil** (Fig. 26.20) was approved in 2009.

The design of tadalafil started with the screening of compounds containing a β-carboline scaffold since

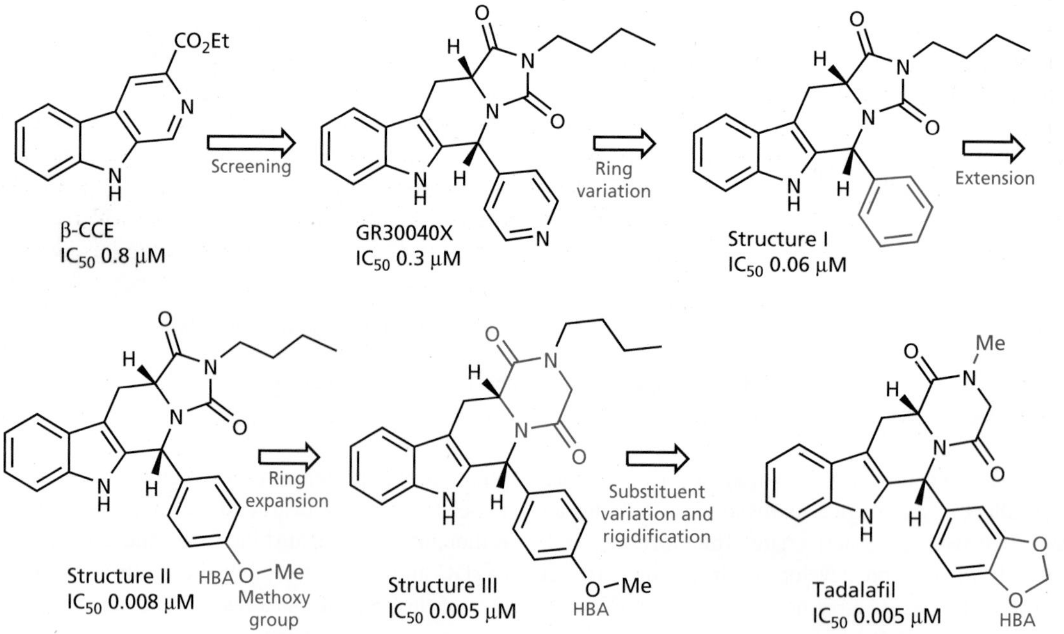

FIGURE 26.20 Phosphodiesterase inhibitors.

β-carboline-3-carboxylate (β-CCE) was known to be a modest PDE5 inhibitor. From the various hits obtained, GR30040X was adopted as the lead compound. A variety of drug design strategies were carried out which involved varying rings and substituents. A key modification was the introduction of a methoxy substituent which resulted in a significant increase in activity. It is likely that this is caused by an extra hydrogen bond with the binding site, where the oxygen of the methoxy group acts as a hydrogen bond acceptor. The methoxy group is replaced by a methylendioxy group in tadalafil itself, but the crucial oxygen atom is still present.

26.5.3 Neprilysin inhibitors

Neprilysin is a membrane-bound zinc metallopeptidase, and is also known as **neutral endopeptidase** (**NEP**). It hydrolyses endogenous peptide hormones such as **bradykinin**, **atrial natriuretic factor**, and **brain natriuretic factor**. These hormones act as vasodilators, and so inhibiting neprilysin prolongs their half-life and antihypertensive effects.

An example of a neprilysin inhibitor is **sacubitril** (Fig. 26.21), which was approved in 2015 as a treatment for heart failure. The drug is used in combination with valsartan (section 26.3.4). Sacubitril is an ester prodrug for the active agent **sacubitrilat**.

Dual-action agents have also been designed that include the ability to inhibit neprilysin. **Omapatrilat** (Fig. 26.21) is an ACE inhibitor that also inhibits neprilysin—a so-called **vasopeptidase inhibitor**. The agent has strong antihypertensive properties, but can cause side effects in some patients due to the resulting increase in bradykinin levels.

Therefore, research has switched to designing AT_1 receptor antagonists that can also inhibit neprilysin. Various dual-action agents have been developed that contain an aryl acid capable of binding to the AT_1 receptor, and another group capable of interacting with the zinc ion cofactor present in neprilysin. However, none have reached the market to date.

26.5.4 Prostacyclin agonists

Prostacyclin (PGI$_2$) (Fig. 26.22) is a prostaglandin which acts as a potent vasodilator. It is the natural ligand for the prostacyclin receptor—a G-protein-coupled receptor that interacts with G$_s$ proteins and leads to the activation of adenylate cyclase (section 5.2). Prostacyclin has been synthesized under the name of **epoprostenol** and can be administered by intravenous injection for the treatment of pulmonary hypertension. However, it is both chemically and metabolically unstable, resulting in a short half-life of only 3 minutes. This means that it has to be administered by continuous intravenous transfusion. Therefore, research has focused on finding stable analogues. **Iloprost** is one such analogue, but is prone to oxidative metabolism if taken orally. Because of this, it has to be administered by inhalation 6–9 times per day. **Beraprost sodium** is structurally related to iloprost, and is an orally active agonist that is approved in Japan and South Korea. However, it has a short half-life due to rapid elimination. **Selexipag** is the most recent innovation and was approved in 2015 as an orally active agent. It is actually a prodrug and is slowly metabolized by carboxylesterase enzymes to form the active agent **ACT333679**. This is an important factor in the drug's prolonged activity. Selexipag is structurally distinct from previous prostacyclin agonists, but shares the carboxylic acid which is crucial for agonist activity. SAR studies on selexipag also demonstrate the importance of the tertiary amine and its isopropyl substituent. The ether group in selexipag was introduced to enhance metabolic stability by blocking metabolic oxidation at that position.

26.5.5 Miscellaneous vasodilators

Hydralazine (Fig. 26.23) relieves hypertension by relaxing the smooth muscle of arteries and arterioles, but its exact mechanism of action is yet to be established. **Minoxidil** is thought to be a prodrug which is metabolized to form **minoxidil sulphate** as the active agent. This

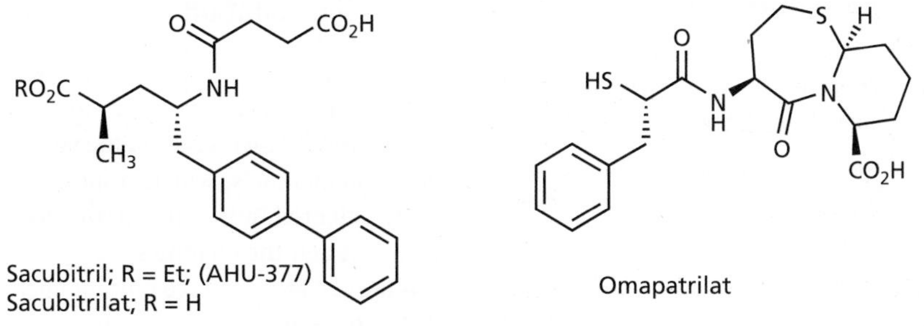

Sacubitril; R = Et; (AHU-377)
Sacubitrilat; R = H

Omapatrilat

FIGURE 26.21 Neprilysin inhibitors.

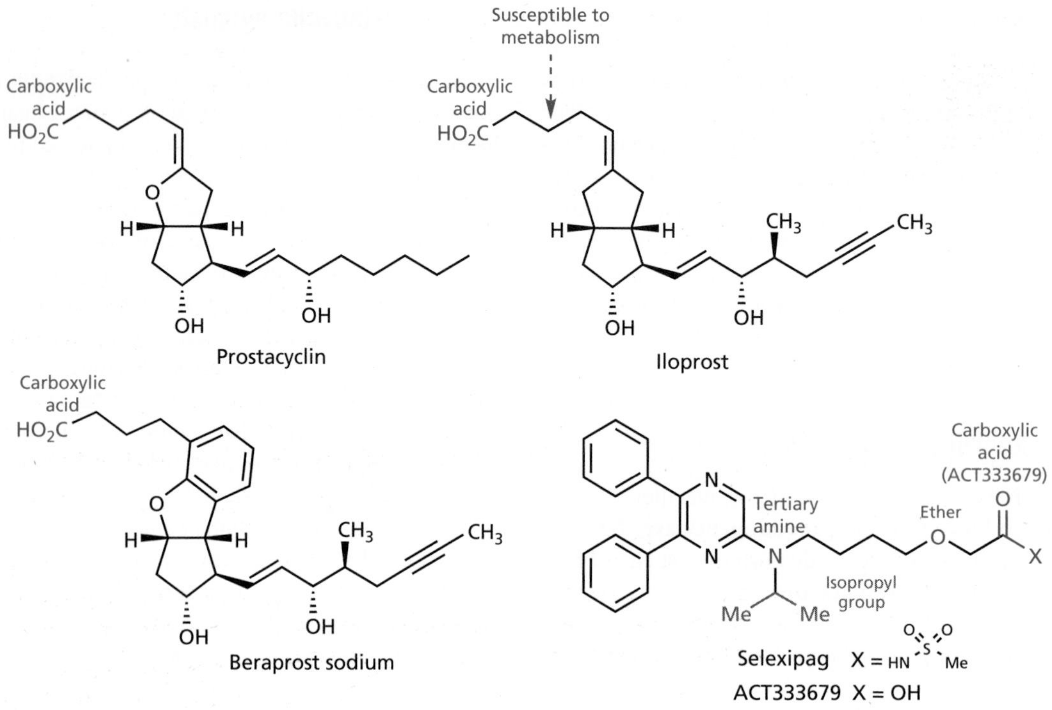

FIGURE 26.22 Prostacyclin and prostacyclin agonists.

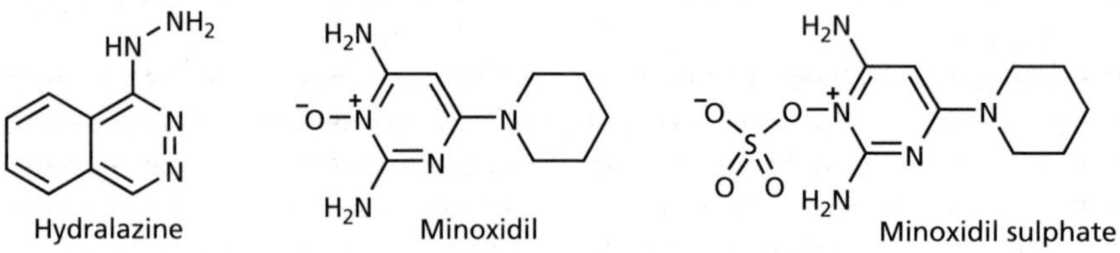

FIGURE 26.23 Hydralazine and minoxidil.

promotes the opening of potassium ion channels, causing hyperpolarization of cell membranes. Minoxidil is also used to slow hair loss.

KEY POINTS

- Vasodilators relieve hypertension by dilating blood vessels and reducing blood pressure.
- Cyclic GMP is a secondary messenger that promotes the dilation of blood vessels.
- Endogenous peptide hormones such as bradykinin and the natriuretic factors promote vasodilation.
- Levels of cyclic GMP can be increased by enhancing the activity of soluble guanylate cyclase. This enzyme catalyses the synthesis of cyclic GMP.
- Levels of cyclic GMP can be enhanced by inhibiting phosphodiesterase type 5 (PDE5). PDE5 catalyses the hydrolysis of cyclic GMP.
- Levels of bradykinin and the natriuretic factors can be increased by inhibiting neprilysin—the enzyme responsible for hydrolysing these hormones.

26.6 Calcium entry blockers

26.6.1 Introduction

Various types of calcium ion channel control the passage of calcium ions into cells. The most important in terms of cardiovascular medicine are the voltage-gated L-type calcium ion channels, which respond to changes in membrane potential rather than to the presence of an external ligand. When the channels are open, calcium ions flood into the cell and produce physiological effects depending on the type of cell involved. In vascular smooth muscle this corresponds to muscle contraction, which leads to

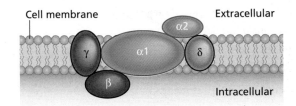

FIGURE 26.24 Protein subunits present in the L-type calcium ion channel. Only skeletal muscle cells contain ion channels with the gamma subunit.

constriction of the blood vessels and an increase in blood pressure. In heart muscle, the effects are an increase in heart rate and a stronger force of contraction.

The structure of the ion channel consists of a multiprotein complex involving four or five subunits (α_1, α_2, β, β, and γ) (Fig. 26.24). The α_1 subunit controls the flow of calcium ions through the ion channel, and it is thought

that glutamate residues are involved in binding the calcium ions during the transport process. The α_1 subunit also provides the target binding sites for the calcium ion channel blockers.

The calcium ion channel blockers are drugs that prevent calcium ions crossing the ion channel, and are used mainly for the treatment of hypertension and angina pectoris. However, they can also be used for other cardiovascular diseases.

There are three main structural classes—the dihydropyridines, the phenylalkylamines, and the benzothiazepines. The different structural classes bind to distinct binding sites on the α_1 subunit. However, the binding sites are very close together, and there is evidence that some of the amino acids involved in binding the benzothiazepines are also involved in binding the phenylalkylamines. Moreover, the binding sites are said to be allosterically linked, which means that a drug binding to one of the binding sites influences the ability of a drug to

---

BOX 26.1 Synthesis of dihydropyridines

The **Hantzsch condensation** is a particularly good method for synthesizing dihydropyridines. The reaction can be carried out with three reagents—ammonia, an aldehyde, and two equivalents of a β-keto ester. For example, the calcium channel blocker **nifedipine** was synthesized in this manner (Fig. 1).

Although it is possible to synthesize nifedipine in one step as shown in Fig. 1, better yields are obtained when the synthesis is carried out in two stages (Fig. 2).

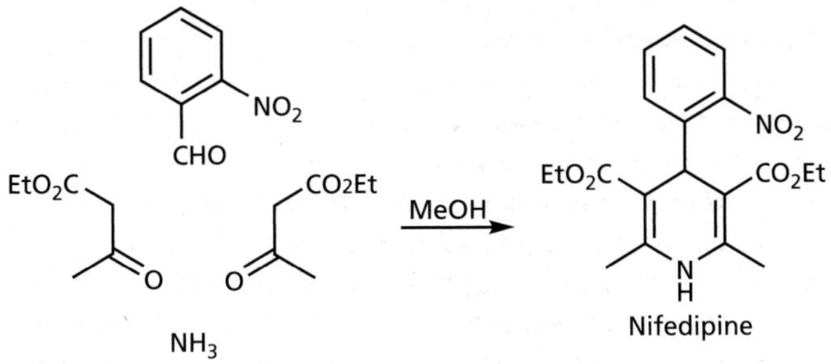

FIGURE 1 Synthesis of nifedipine.

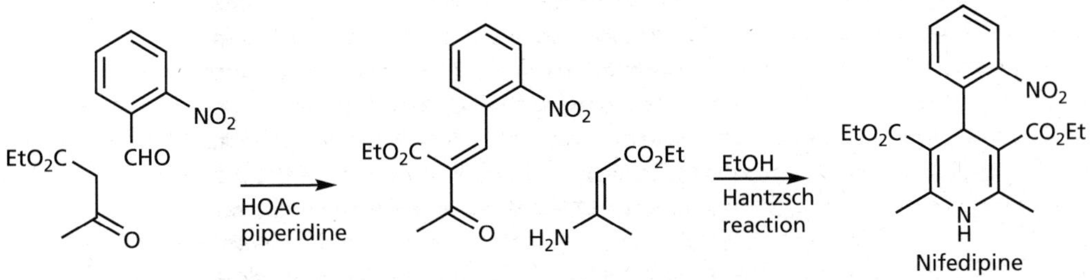

FIGURE 2 Synthesis of nifedipine in two stages.

bind to a neighbouring binding site. For example, when a phenylalkylamine binds to its binding site, it inhibits dihydropyridines and benzothiazepines from binding to their binding sites. Similarly, when a benzothiazepine or dihydropyridine binds to its respective binding site, it inhibits the binding of phenylalkylamines. On the other hand, when a benzothiazepine is bound, it *promotes* the binding of a dihydropyridine.

26.6.2 Dihydropyridines

The dihydropyridines approach their binding site from the extracellular side of the ion channel, and show selectivity for the ion channels in vascular smooth muscle. As a result, they are used mainly for the treatment of hypertension.

A large number of dihydropyridine analogues have been synthesized, and it has been found that their ability to act as calcium ion channel blockers is dependent on the following factors (Fig. 26.25):

- the structures are generally hydrophobic in nature suggesting that they interact with a hydrophobic binding site;
- the nitrogen in the dihydropyridine ring must be unsubstituted;

- small hydrophobic alkyl groups are preferred at positions 2 and 6;
- ester groups are preferred at positions 3 and 5;
- there must be an aryl substituent at position 4;
- the aryl ring normally contains a substituent at the *ortho* or *meta* position. However, a substituent at the *para* position is bad for activity.

The dihydropyridine ring is essential for activity. For example, if a pyridine ring is present, activity is lost. It has been shown that the preferred conformation of the dihydropyridine ring is a slightly flattened boat structure, with the aryl substituent in a pseudoaxial position (Fig. 26.26). The aromatic ring itself is perpendicular to the dihydropyridine ring such that one of the *ortho* protons hangs over the face of the dihydropyridine ring. Any substituent on the aromatic ring points away from the dihydropyridine ring. This is thought to correspond to the active conformation. In contrast, a pyridine ring is planar. This has a significant effect on the orientation of the substituents around the ring system, and, in particular, the orientation of the aromatic ring. The nature of the substituent on the aromatic ring has a role in the activity of the dihydropyridines. Larger substituents are generally better for activity and appear to have a greater flattening effect on the dihydropyridine ring.

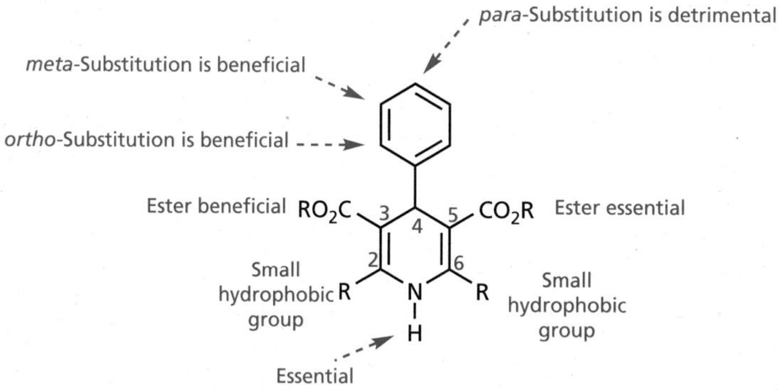

FIGURE 26.25 SAR results for dihydropyridines as calcium-ion channel blockers.

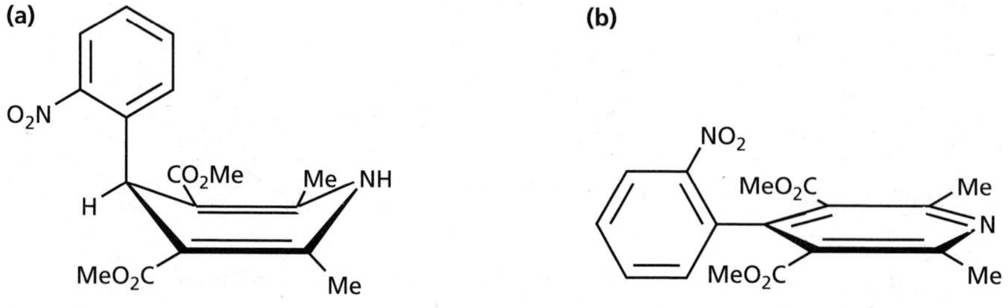

FIGURE 26.26 (a) Preferred conformation of nifedipine. (b) A comparable pyridine structure.

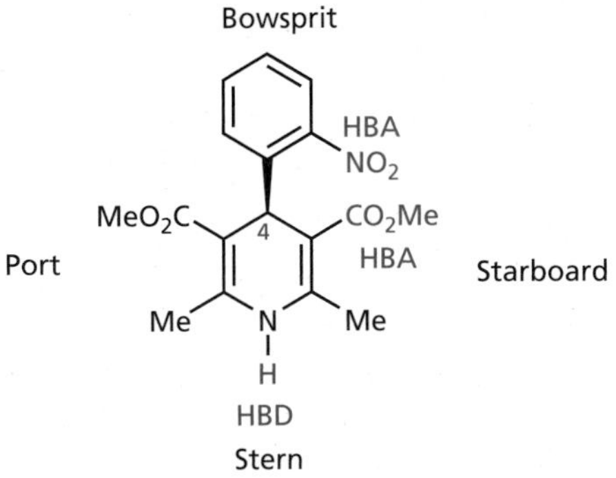

FIGURE 26.27 Groups in nifedipine that are proposed to form hydrogen bonds with the binding site.

Unsymmetrical dihydropyridines contain an asymmetric centre at position 4, and it is found that one enantiomer is more active than the other. In such structures, a distinction can be made between the two halves of the dihydropyridine ring, which can be labelled port and starboard. It has been demonstrated that the ester group on the port side is not essential, although it plays a role in whether the agent acts as an agonist or an antagonist.

Modelling studies carried out on a homology model of the L-type calcium ion channel have also provided evidence that three hydrogen bonding interactions are possible with the binding site (Fig. 26.27).

In addition to nifedipine, various other dihydropyridines have been approved (Fig. 26.28). Dual-action dihydropyridines have also been designed that serve as a source of nitric oxide, as well as being calcium ion channel blockers.

The dihydropyridine ring has been defined as a **privileged scaffold** as it is found in various drugs having a wide range of different activities. The type of activity observed depends on the substituents that are present around the ring.

26.6.3 Phenylalkylamines

The phenylalkylamines are so called because they contain hydrophobic aromatic rings and an aliphatic tertiary amine. They block the calcium ion channel by binding to an intracellular binding site, and it is believed that the structures have to cross the cell membrane in order to reach that binding site. Although the tertiary amine is

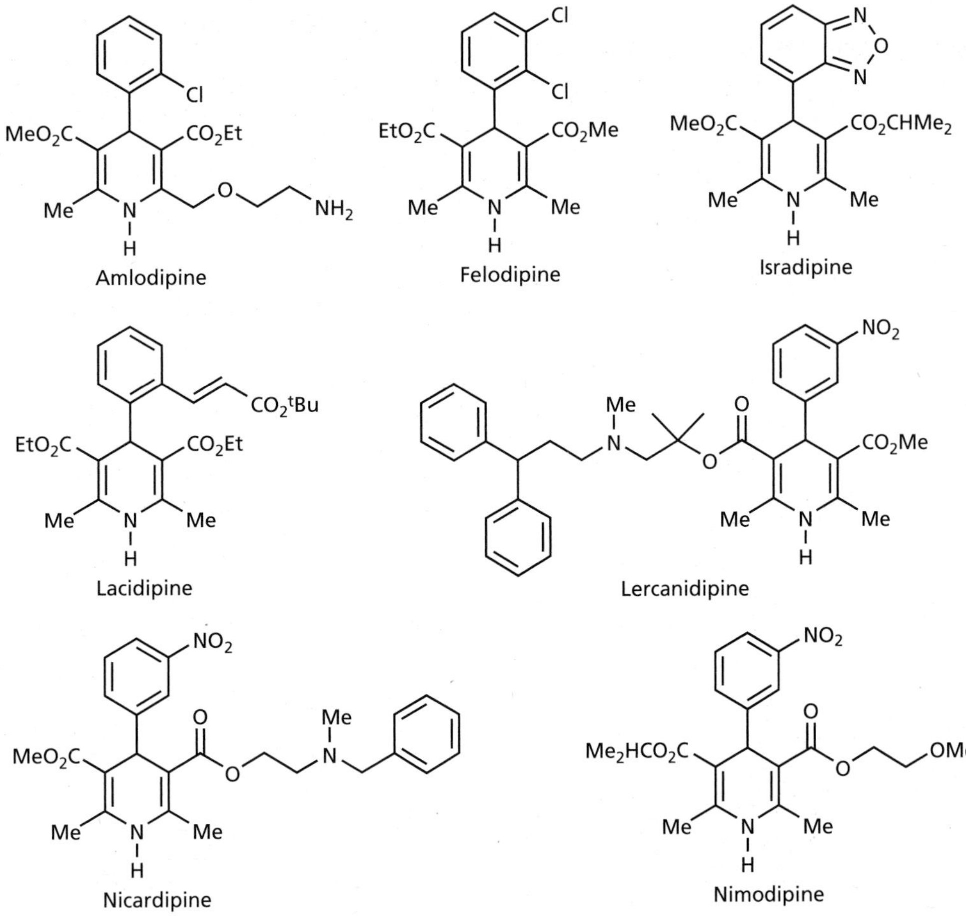

FIGURE 26.28 Examples of clinically important dihydropyridines.

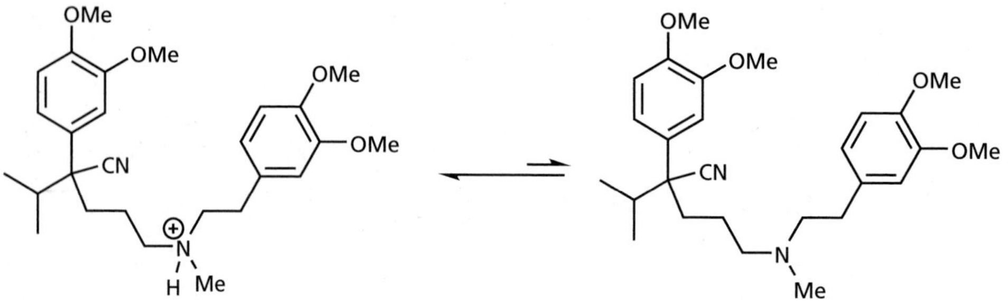

FIGURE 26.29 Verapamil in its ionized and non-ionized forms.

mostly protonated at physiological pH, a small percentage of the drug exists as the free base, and it is in this form that phenylalkylamines can cross cell membranes. Evidence for this comes from methylating the amine to form a quaternary salt, which results in loss of activity. Because the quaternary salt has a permanent positive charge, it is unable to cross the cell membrane to reach its binding site. However, if the salt is injected into the cell, activity *is* observed, proving that the positive charge is important for binding. It is thought that ionic interactions are formed with the negatively charged glutamate residues involved in calcium ion transport. The rest of the structure is mainly hydrophobic and is likely to form van der Waals interactions with hydrophobic regions of the ion channel.

Verapamil (Fig. 26.29) is the only example of a phenylalkylamine that is currently used in the clinic. It is used to treat hypertension by blocking calcium ion channels in vascular smooth muscle. It is also useful in treating arrhythmias by blocking calcium ion channels in cardiac muscle.

26.6.4 Benzothiazepines

The benzothiazepines access the calcium ion channel from the extracellular side of the side membrane. **Diltiazem** (Fig. 26.30) is the only example of a benzothiazepine that is currently approved, and it is used to treat hypertension and angina.

Diltiazem has two chiral centres, which means that there are four possible stereoisomers. However, only the 2S,3S isomer is active.

SAR studies on diltiazem-like structures have demonstrated the following features (Fig. 26.30):

- The basic tertiary amine of the alkylamine chain is crucial for activity and is likely to be protonated when it binds to the target binding site. This allows both ionic and hydrogen bonding interactions.

- *Ortho* and *meta* substituents on the phenyl group are not well tolerated, suggesting that the aromatic ring is in a rather confined pocket.

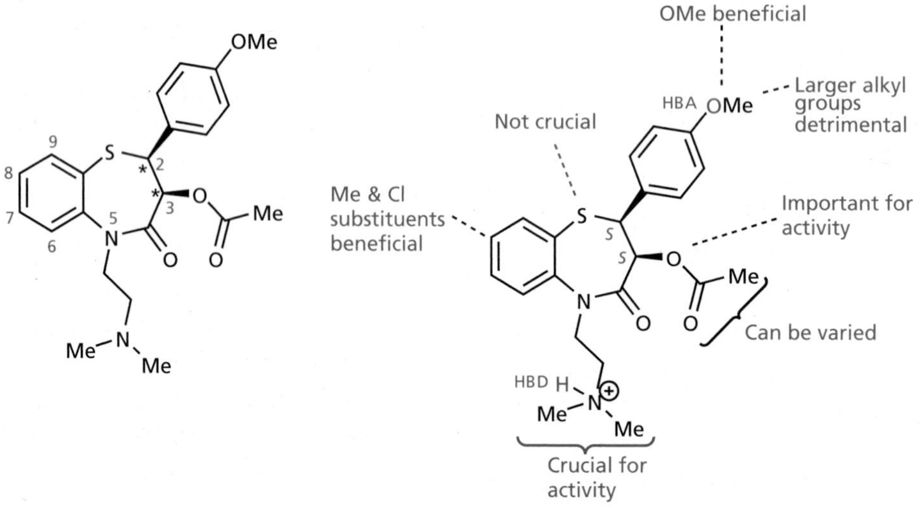

FIGURE 26.30 Diltiazem.

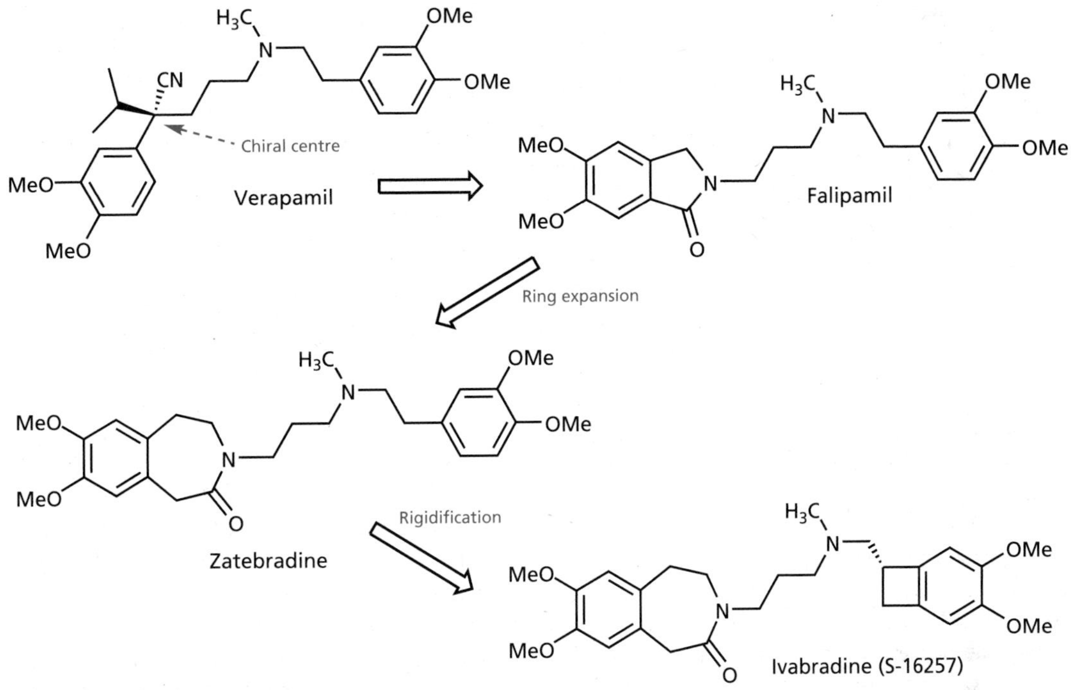

FIGURE 26.31 Key structures in the design and development of ivabradine.

- A methoxy substituent at the *para* position of the phenyl group increases activity 15-fold, which suggests that the oxygen acts as a hydrogen bond acceptor. Activity is lost if larger alkoxy groups are present, which supports the theory that the aromatic ring fits a tight binding pocket. Activity is lost if a phenol group is present rather than a methoxy group.

- The presence of a hydrophobic methyl or chloro substituent at position 8 is beneficial, which suggests that the substituent occupies a small hydrophobic pocket. The same increase in activity is not observed if the chloro substituent is placed at position 7 or 9.

- The sulphur atom is not essential for activity, but binding affinity drops over tenfold if it is replaced by CH_2. The hydrophobic nature and high electron density of the sulphur atom is thought to contribute to binding.

The two major metabolic reactions affecting diltiazem involve hydrolysis of the methyl ester group and demethylation of the tertiary amine.

26.7 **Funny ion channel inhibitors**

Funny ion channels (If) are ion channels that are present in the heart's pacemaker cells, and which affect the heart rate. They are unusual in that they can be activated both by voltage effects and a ligand (cyclic AMP). **Ivabradine** (Fig. 26.31) is a selective If inhibitor which has recently been approved by the FDA for the treatment of angina. Unlike the beta blockers and calcium ion channel blockers, ivabradine reduces heart rate without affecting blood pressure or the strength of heart contractions. Therefore, it is classed as a **specific bradycardic agent**.

The lead compound for the development of ivabradine was **verapamil** (section 26.6.3). The chiral centre present in verapamil was removed by introducing a bicyclic ring system, leading eventually to the discovery of **falipamil**, which proved to be the first specific bradycardic agent. Ring expansion then led to the discovery of **zatebradine**, which proved more selective and 20 times more potent. Finally, a second bicyclic ring system was introduced to give ivabradine. This has the effect of rigidifying the structure by locking one of the previously rotatable bonds into the new ring.

KEY POINTS

- Calcium entry blockers are drugs that block voltage-gated L-type calcium ion channels. They are used to treat hypertension and angina.

- The three structural classes of calcium entry blockers are dihydropyridines, phenylalkylamines, and benzothiazepines.

- The three structural classes of calcium entry blockers bind to distinct binding sites in the α_1 subunit of the calcium ion channel.

- The dihydropyridines are used mainly to treat hypertension.
- The dihydropyridines generally contain five substituents, which include an aromatic ring and two ester groups. The nitrogen atom is unsubstituted.
- The phenylalkylamines bind to the calcium ion channel by approaching from the intracellular side of the ion channel. Therefore, they must cross the cell membrane to be effective.
- Verapamil is the only phenylalkylamine that has been approved for the clinic.
- The benzothiazepines bind to the calcium ion channel by approaching from the extracellular side of the ion channel.
- Diltiazem is the only benzothiazepine that has been approved for the clinic, and is used for the treatment of hypertension and angina.
- Ivabradine inhibits the funny ion channel and has been approved for the treatment of angina.

26.8 Lipid-regulating agents

Excess levels of cholesterol and other lipids can be a major risk factor for cardiovascular disease. The most important drugs used to lower cholesterol levels are the statins, with the fibrates offering an alternative therapy.

26.8.1 Statins

Statins are cholesterol-lowering agents that block the enzyme involved in the rate-limiting step of cholesterol biosynthesis (**3-hydroxy-3-methylglutaryl-coenzyme A reductase**) (Case study 1). All statins contain the same polar 'head group', linked to a variety of different hydrophobic groups (Fig. 26.32). The hydrophobic moiety for **type I statins** includes a decalin ring, whereas a variety of rings and substituents make up the hydrophobic moiety of **type II statins**.

The polar head group of the statins mimics the natural substrate for the target enzyme, and forms the same binding interactions with the active site. However, the statins bind more strongly than the natural substrate because of their hydrophobic moiety. A more detailed description is given in Case study 1.

Statins have been used in combination with drugs that target proteins not directly involved in cholesterol biosynthesis. For example, **Inegy** or **Vytorin** is a preparation that includes **simvastatin** and a cholesterol absorption inhibitor called **ezetimibe** (Fig. 26.33). The latter agent lowers the levels of cholesterol absorbed from the gastrointestinal tract.

26.8.2 Fibrates

The fibrates were discovered in the 1930s and are currently used to treat patients who have high blood levels of triglycerides, or for patients who are unable to take statins. The agents increase HDL cholesterol levels and lower triglyceride levels. They do so by activating intracellular receptors called **peroxisome proliferator-activated alpha receptors** (**PPARαs**), which are mainly found in the liver. The natural agonists for these nuclear hormone receptors (section 4.9) are fatty acids and their metabolites. When activated, the PPARs control gene transcription, which, in turn, increases the levels of enzymes involved in metabolizing fats and carbohydrates. For

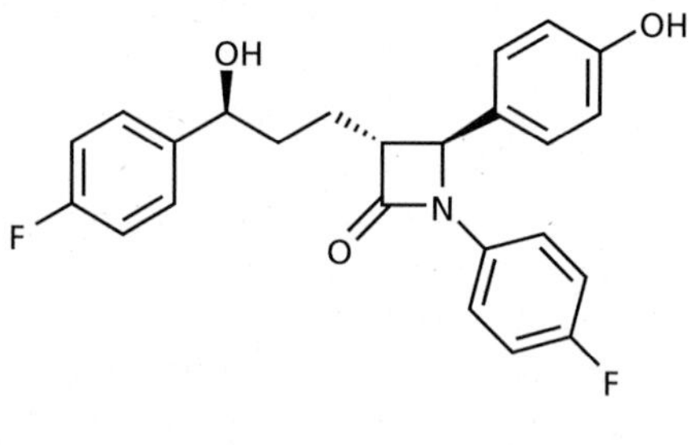

FIGURE 26.33 Ezetimibe.

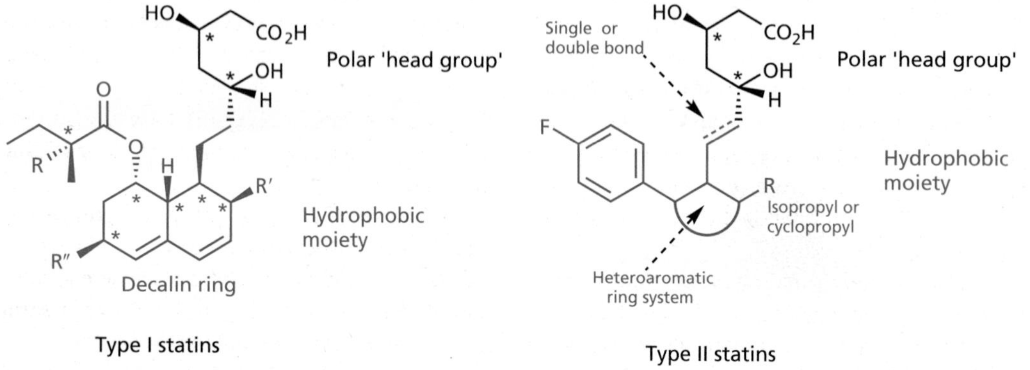

FIGURE 26.32 General structure for the type I and type II statins (* represents an asymmetric centre).

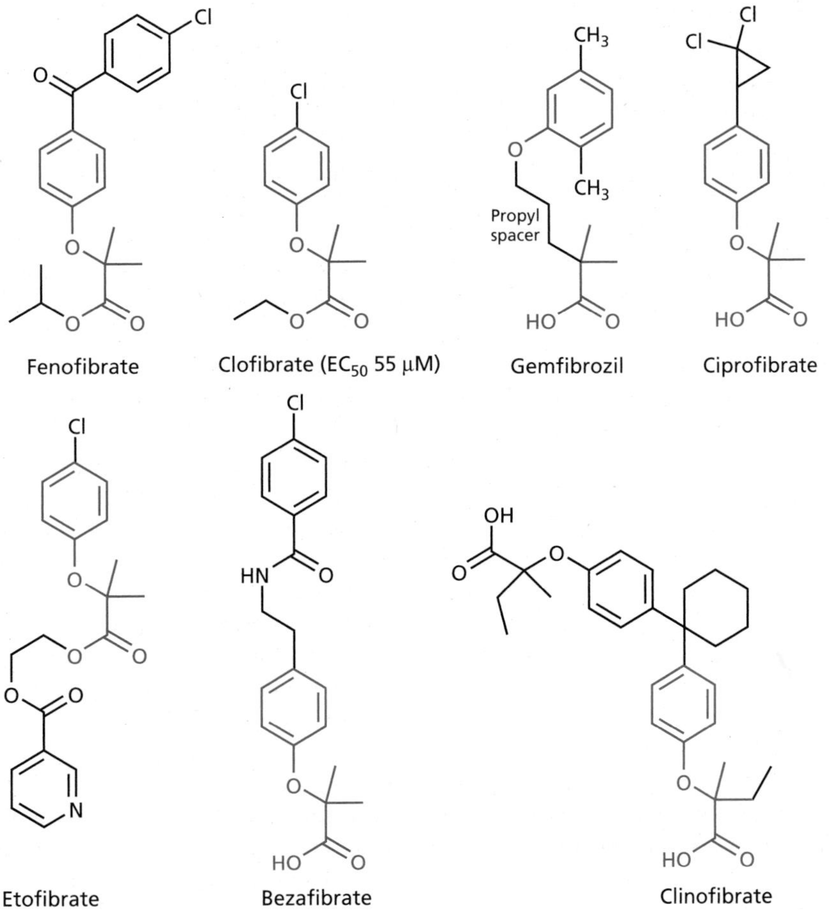

FIGURE 26.34 Examples of fibrates. The blue region represents the common pharmacophore.

example, lipoprotein lipase levels are increased, resulting in an increased clearance of triglycerides.

Clofibrate (Fig. 26.34) was the first fibrate to reach the market in the 1960s. However, it is rarely used nowadays because of serious side effects. Commonly prescribed fibrates include **bezafibrate**, **ciprofibrate**, **gemfibrozil**, **fenofibrate**, and **clinofibrate**.

The key pharmacophore for most of the fibrates is the fibric acid moiety (2-phenoxyisobutyric acid). Gemfibrozil is an exception, in that it has a propyl spacer between the phenoxy ring and the isobutyric acid moiety. The carboxylic acid group is essential for activity, and is predominantly ionized at blood pH. Those fibrates that contain an ester group instead of a carboxylic acid act as prodrugs. Esterases in the blood catalyse the hydrolysis of the ester group to reveal the free acid.

Several of the fibrates contain a *para*-chloro-substituted aromatic ring. These fibrates have longer half-lives, and it is likely that chloro substitution blocks oxidative metabolism at the exposed *para*-position of the aromatic ring (section 11.5.2).

Current research is focused on designing fibrates with increased binding affinity and selectivity for PPARα receptors.

26.8.3 Dual- and pan-PPAR agonists

There are three subtypes of PPAR receptor. The PPARα receptor is the target for fibrates (section 26.8.2). PPARγ receptors are found mostly in adipose tissue with smaller levels in the liver and spleen, whereas the PPARβ receptors are more widely distributed. Agents that show selectivity for the PPARγ receptors have been used for the treatment of diabetes—namely the **thiazolidinediones** (or **glitazones**).

Dual PPARα/PPARγ agonists called **glitazars** are being studied to treat patients suffering from insulin resistance and high levels of triglycerides. Several of these agents reached clinical trials, but were not approved. However, **saroglitazar** (Fig. 26.35) was approved in India during 2013.

Elafibranor (Fig. 26.35) is a dual acting PPARα/ PPARβ agonist that is currently in clinical trials. Agents

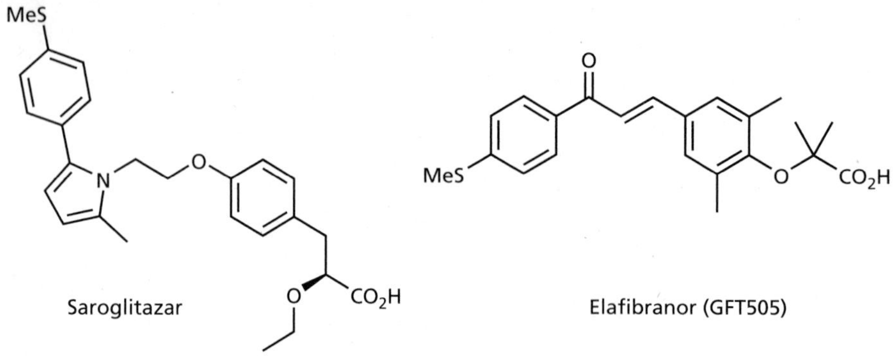

FIGURE 26.35 Examples of dual-acting PPAR agonists.

targeting PPARβ receptors could possibly be used to treat obesity and insulin resistance.

Bezafibrate (Fig. 26.34) has served as the starting point in the development of pan-PPAR agonists that act on all three PPAR subtypes. Such agents might be of use in treating type 2 diabetes and associated cardiovascular complications.

26.8.4 Antisense drugs

Mipomersen is an antisense oligonucleotide drug that was approved by the FDA in 2013 as a cholesterol-reducing drug. The backbone contains phosphorothioate links for better resistance to nuclease enzymes (sections 9.7.2 and 14.10). The drug does not directly target cholesterol or its synthesis. Instead, it binds to the messenger RNA for **apolipoprotein B** and prevents the translation of that protein. Apolipoprotein B is a key structural component of low-density lipoproteins (so called 'bad cholesterol') where it serves to bind lipids such as cholesterol.

26.8.5 Inhibitors of transfer proteins

The **microsomal triglyceride transfer protein** (MTP) acts as a lipid carrier and plays an important role in the formation of very low-density lipoproteins (VLDL). Inhibitors of this protein have been shown to lower levels of plasma cholesterol and triglycerides.

Lomitapide (Fig. 26.36) is one such inhibitor that was approved by the FDA in 2012 as an orphan drug to reduce cholesterol levels. The structure was developed from a hybrid molecule, containing structural features of two hit compounds that had been identified from screening programmes. Further optimization included the introduction of a fluorinated *N*-substituent which significantly improved *in vivo* activity. The bicyclic indoline ring was also replaced with a simpler benzamide group, which allowed substantial numbers of analogues to be synthesized. This led to the discovery of lomitapide.

Another strategy is to inhibit the **cholesteryl ester transfer protein** (CETP). This is a plasma protein that aids the transfer of triglcyerides and cholesterol between LDLs and HDLs. Studies on CETP inhibitors indicate that these can raise HDL levels and lower LDL levels in the presence of statins. **Anacetrapib** (Fig. 26.37) is one such inhibitor undergoing clinical trials.

26.8.6 Antibodies as lipid-lowering agents

Alirocumab and **evolocumab** are fully humanized monoclonal antibodies that were approved for the treatment of hyperlipidaemia in 2014 and 2015 respectively. They target a protein called **PCSK9**, which is responsible for promoting the degradation of LDL receptors. By inhibiting this protein, the antibodies increase the number of LDL receptors available in the liver. As a result, the removal of LDL-cholesterol from the blood supply increases. The antibodies are used for patients who experience high LDL-cholesterol levels despite normal drug therapies.

KEY POINTS

- Statins are the most important agents currently prescribed for the lowering of cholesterol levels.

- Statins inhibit the enzyme that controls the rate-limiting step of cholesterol biosynthesis.

- Statins contain a polar 'head group' that mimics the natural substrate for the target enzyme, plus a hydrophobic region that forms additional binding interactions.

- Fibrates lower triglyceride levels by activating peroxisome proliferator-activated alpha receptors (PPARαs) in the liver.

- Mipomersen is an antisense drug that inhibits the translation of apolipoprotein B—a key component of low-density lipoprotein particles.

- Lomitapide lowers levels of cholesterol and triglycerides by inhibiting a triglyceride transfer protein.

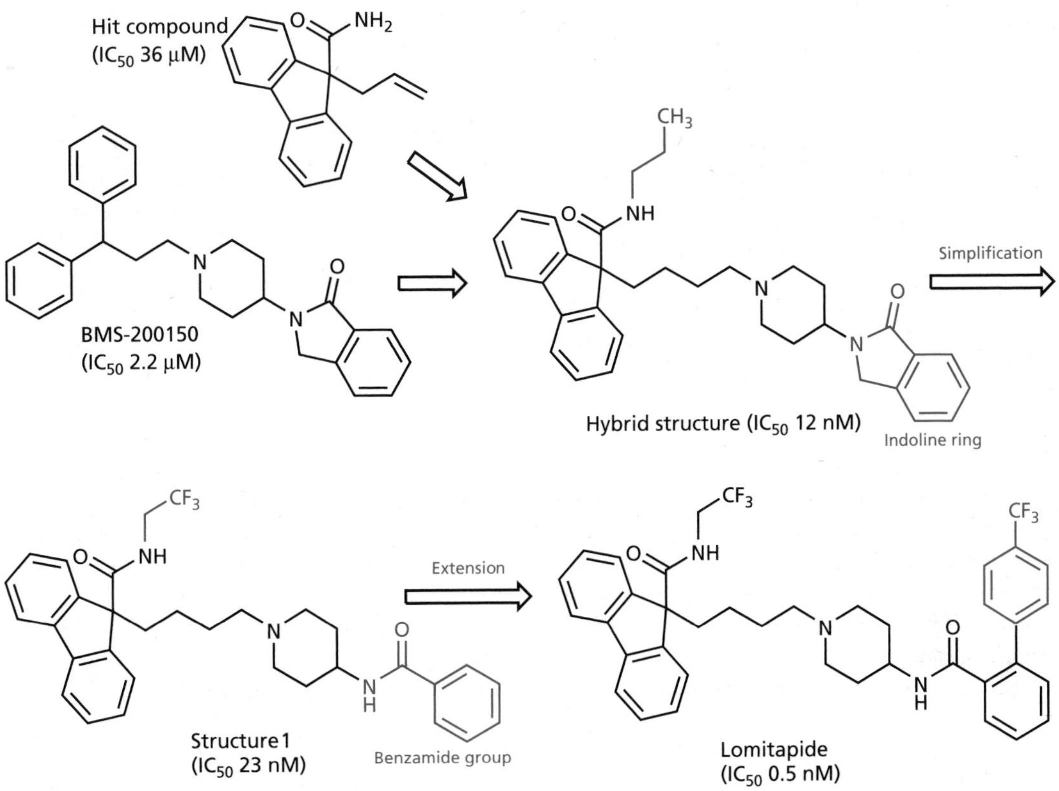

FIGURE 26.36 Development of lomitapide (BMS-201038).

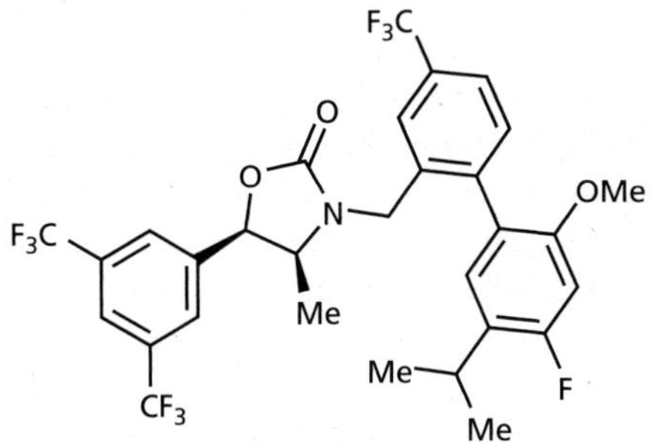

FIGURE 26.37 Anacetrapib.

- Antibodies targeting a protein called PCSK9 prevent the degradation of LDL receptors. This increases the removal of LDL-cholesterol from the blood supply.

26.9 Antithrombotic agents

Antithrombotic agents counter the effects of thrombosis—a condition that can result in a variety of cardiovascular disorders such as acute myocardial infarction or stroke. Thrombosis is the process by which thrombi form in the circulatory system. If this occurs in the veins, then it can lead to deep vein thrombosis resulting in leg pain, swelling, and ulcers. Moreover, if the thrombi are dislodged, it can cause pulmonary embolism.

The process by which a thrombus is formed involves enzymes such as **factor Xa** and **thrombin** (Fig. 26.38). Factor Xa is the active form of factor X and catalyses the conversion of **prothrombin** to **thrombin**. Thrombin then catalyses the cleavage of a protein substrate called **fibrinogen** to form **fibrin**. Thrombin also activates other

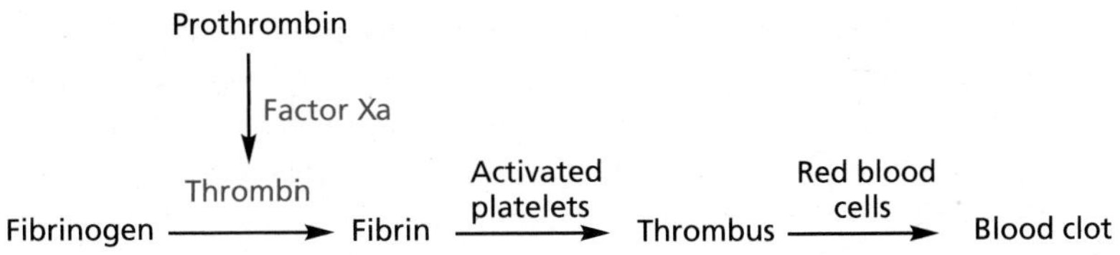

FIGURE 26.38 Key features involved in clot formation.

blood factors that promote the cross-linking of fibrin and inhibit fibrinolysis. Once fibrin becomes cross-linked, it combines with activated platelets to form a thrombus, which can then trap red blood cells to form a clot.

In recent years, a vast amount of research has gone into designing antithrombotic agents that counteract the formation of thrombi. They can be classed as anticoagulants, antiplatelet agents, or fibrinolytic agents. Anticoagulants influence the levels or activity of thrombin in the circulatory system, antiplatelet agents prevent the activation and aggregation of platelets, and fibrinolytic agents break up clots that have already formed.

26.9.1 Anticoagulants
26.9.1.1 Introduction

Anticoagulants are used to prevent a thrombus developing in the venous system, where the blood flow is slower than the arterial system. They can protect against deep-vein thrombosis and pulmonary embolism.

Traditional anticoagulants such as **heparin** and **hirudin** bind to thrombin and inhibit its enzymatic activity. Heparin is a polymeric carbohydrate which is fast acting, but has a short duration of action and needs to be injected. Low molecular weight heparins (the 'parins') have a longer duration of action and are preferred for prophylactic treatments. They include **bemiparin**, **dalteparin**, **enoxaparin**, and **tinzaparin**. Hirudin is the anticoagulant protein released by leeches when they feed from a host. Because of the difficulty of extracting naturally occurring hirudin, the hirudins used in medicine are recombinant proteins such as **bivalirudin** and **lepirudin**. Like heparin, they have to be administered by injection.

The best known oral anticoagulant is **warfarin** (Fig. 26.39). Warfarin was originally marketed as a rat killer in 1948, but was introduced as a therapeutic agent in 1954. Since then, it has been the most important oral anticoagulant used in medicine. Warfarin acts by preventing the recycling of **vitamin K**—an important cofactor required by the enzymes that activate blood coagulation proteins such as prothrombin and factor VII. During this activation process, vitamin K is converted into its oxidized form, and, if it is to continue acting as a cofactor, it has to be converted back to its reduced form. An enzyme called **vitamin K epoxide reductase** carries out this task, and it

is this enzyme that is inhibited by warfarin. In the presence of warfarin, the reduced form of vitamin K is gradually used up and the activation process becomes starved of an essential cofactor. This explains why there is a substantial time lag between administering warfarin and the onset of anticoagulant activity. The drug only becomes effective once the vitamin K already present in the body has been converted to its oxidized form. It also explains why foods that are rich in vitamin K (such as green vegetables) counteract the activity of warfarin.

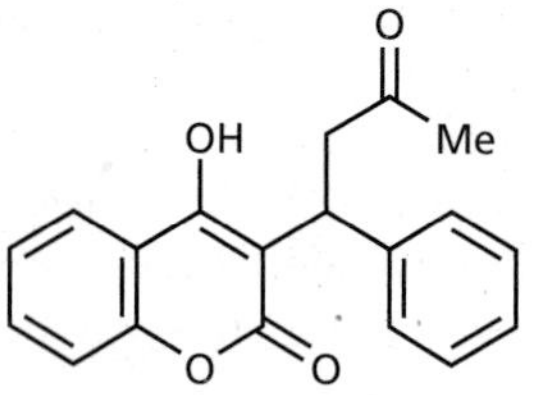

Another drawback with warfarin is the problem of drug–drug interactions. Several commonly prescribed drugs affect how quickly warfarin is metabolized, which, in turn, affects the levels of warfarin present (section 11.5.6). This is a particular problem because of warfarin's low therapeutic index, and so there has been substantial research into finding alternative oral anticoagulants that are safer to use and are quicker acting.

26.9.1.2 Direct thrombin inhibitors

Direct thrombin inhibitors bind to thrombin in order to inhibit its activity. These include heparin and the hirudins (section 26.9.1.1). **Dabigatran** also acts as a direct thrombin inhibitor and can be administered orally as the 'double' prodrug—**dabigatran etexilate** (Fig. 26.40), where both the amidine and carboxylic acid groups are masked.

The development of dabigatran began by choosing a trisubstituted benzimidazole ring as the central scaffold. Known binding groups for thrombin were then added as substituents to give a weakly active compound that served as a lead compound for further development (Fig. 26.41). Various analogues were synthesized, where different aryl groups and linkers were employed, ending up with the discovery of dabigatran.

Dabigatran interacts with the binding site mainly through van der Waals interactions, but there is an important ionic interaction between the amidine group and an aspartate residue (Asp189) (Fig. 26.42). The *N*-methyl group of dabigatran fits into a hydrophobic pocket and forms van der Waals interactions. The carboxylic acid group is not involved in binding, and was introduced in order to make the drug less hydrophobic. This means that less of the drug becomes bound to plasma proteins, resulting in increased bioavailability. The carboxylic acid group was positioned in the molecule such that it would protrude from the binding site and be solvent exposed, thus avoiding the energy penalty associated with desolvation (see also section 1.3.6).

The anticoagulant properties of dabigatran can be reversed by administering **idarucizumab**. This is a monoclonal antibody that acts within minutes and was introduced to the market in 2015.

FIGURE 26.39 Warfarin.

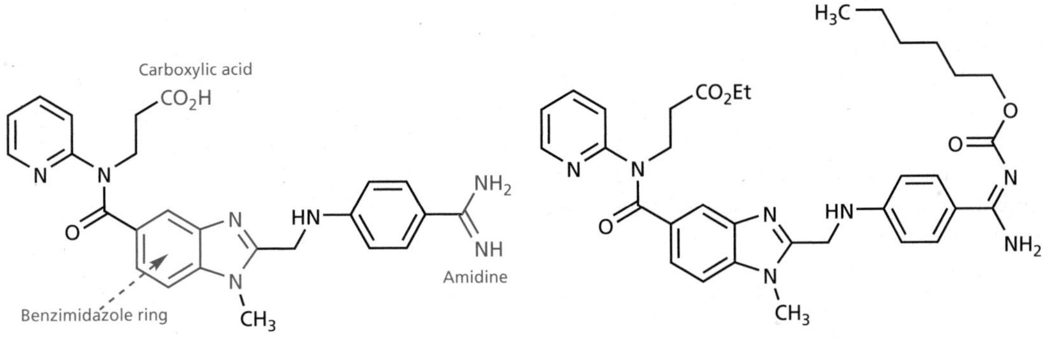

Dabigatran (BIBR 953) Dabigatran etexilate (BIBR 1048)

FIGURE 26.40 Dabigatran and dabigatran etexilate.

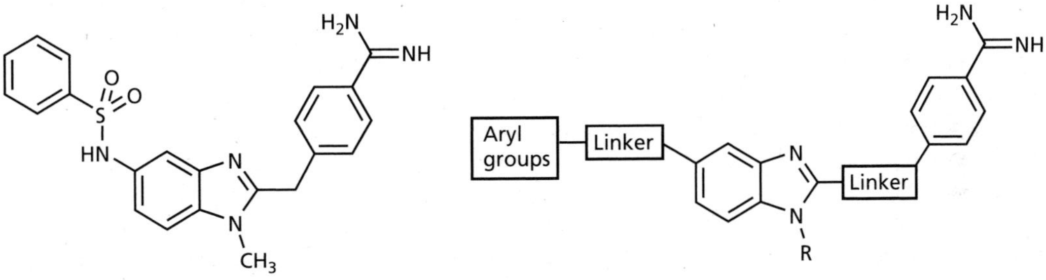

Lead compound (IC_{50} 1.5 μM) Structural variations

FIGURE 26.41 Development of dabigatran.

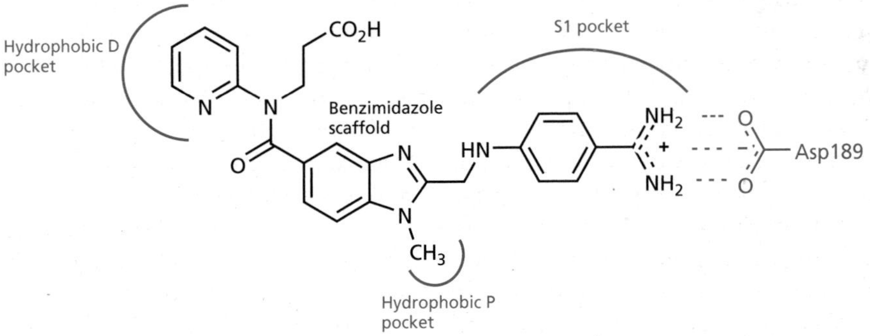

FIGURE 26.42 Binding interactions of dabigatran. P is for the proximal binding pocket, D is for the distal pocket, and S1 is the specificity pocket.

26.9.1.3 Factor Xa inhibitors

Factor Xa (the activated form of factor X) is the enzyme that catalyses the conversion of prothrombin to thrombin (Fig. 26.38). One advantage of targeting factor Xa is that it does not appear to have any other significant physiological role. Therefore, selective factor Xa inhibitors should be relatively free of side effects. Moreover, inhibiting factor Xa should be more efficient than inhibiting thrombin. Since one molecule of factor Xa generates several molecules of thrombin, a much smaller dose of a factor Xa inhibitor should be able to produce the same anticoagulant effect as a thrombin inhibitor.

Initial research into factor Xa inhibitors focused on dibasic structures that could form strong interactions with the S1 and S4 subsites of the enzyme. Although these proved potent *in vitro*, their polarity meant that they had poor oral absorption and were rapidly cleared from the circulation.

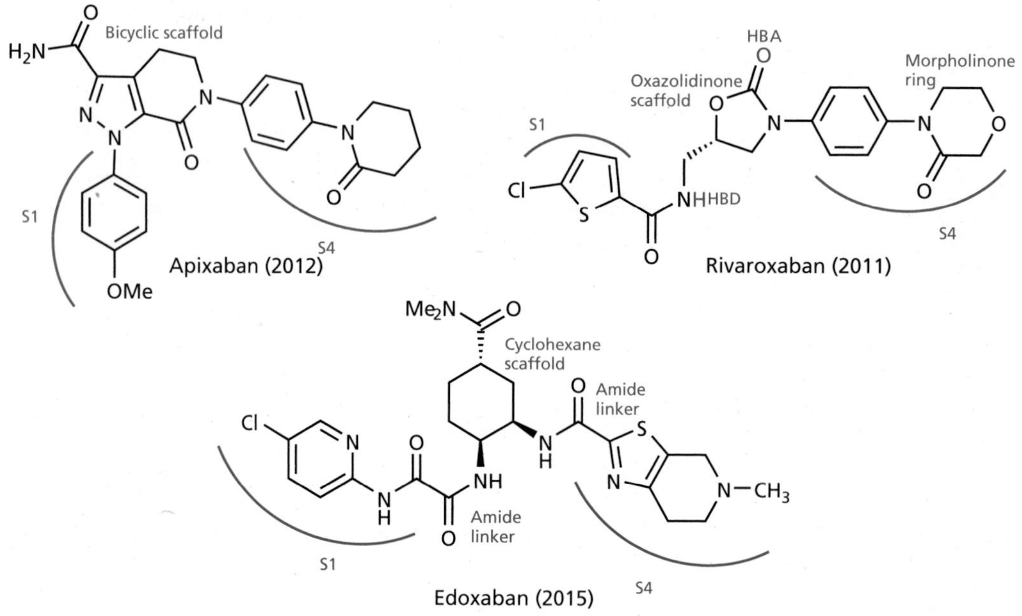

FIGURE 26.43 Factor Xa inhibitors.

A breakthrough was made when it was discovered that the presence of two basic groups was not as crucial as originally thought. Indeed, strong binding affinities could be achieved with non-basic groups. This was an important discovery since it led to the development of clinically useful agents where the predominant interactions between ligands and the binding pockets are hydrophobic in nature. Moreover, it was discovered that these interactions displaced water molecules from hydrophobic regions of the binding site to produce a positive entropic effect that enhanced binding affinity (section 1.3.6).

Three orally active factor Xa inhibitors have successfully received approval in recent years (Fig. 26.43; Case study 9). These agents show high target selectivity, and have proved as effective as warfarin, However, they suffer the disadvantage that there is no antidote available to their anticoagulant activity should a patient suffer serious bleeding. Each of the compounds contains a central scaffold that serves to direct two substituents into the S1 and S4 subsites of the target enzyme.

Test your understanding and practice your molecular modelling with Exercise 26.1 on the Online Resource Centre at www.oxfordtextbooks.co.uk/orc/patrick6e/

26.9.2 **Antiplatelet agents**

26.9.2.1 Introduction

When an atherosclerotic plaque ruptures, platelets form a clot at the site of the injury. The process by which this happens involves activation of platelet receptors by a number of agents such as thrombin, ADP, thromboxane A2, epinephrine, and collagen. Thrombin is the most potent of these.

Activation triggers the expression of **glycoprotein IIb/IIIa (GpIIb/IIIa) receptors**, which bind **fibrin** (itself formed by the action of thrombin on fibrinogen). Platelet aggregation now takes place, and the resulting **thrombus** traps red blood cells and other plasma particles to form a clot that can cause angina or myocardial infarction.

Most antiplatelet agents inhibit the activation of platelets by acting on the relevant platelet receptors. For example, **clopidogrel** (see Fig. 26.47) acts as an antagonist at ADP receptors. However, **aspirin** has antiplatelet activity because it inhibits **cyclooxygenase 1**—one of the enzymes responsible for the formation of thromboxane A2.

26.9.2.2 PAR-1 antagonists

Thrombin is an enzyme that activates platelet G-protein-coupled receptors called **protease activated receptors (PARs)**. It does so by binding to the extracellular N-terminal chain of the receptor, then cleaving part of that chain. The remaining part of the N-terminal chain then acts as a 'tethered ligand' and activates the receptor (Fig. 26.44). There are 4 subtypes of PAR receptor. PAR-1 is the most relevant of these in terms of platelet activation and is also known as the **thrombin receptor**.

Vorapaxar (Fig. 26.45) is a PAR-1 antagonist that was approved in 2014 and is believed to compete for the binding site normally occupied by the tethered ligand. Since vorapaxar does not inhibit fibrin generation, it is less likely to have the bleeding side effects of traditional antithrombotics.

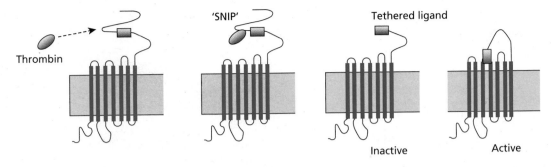

FIGURE 26.44 Activation of PAR-1 by thrombin.

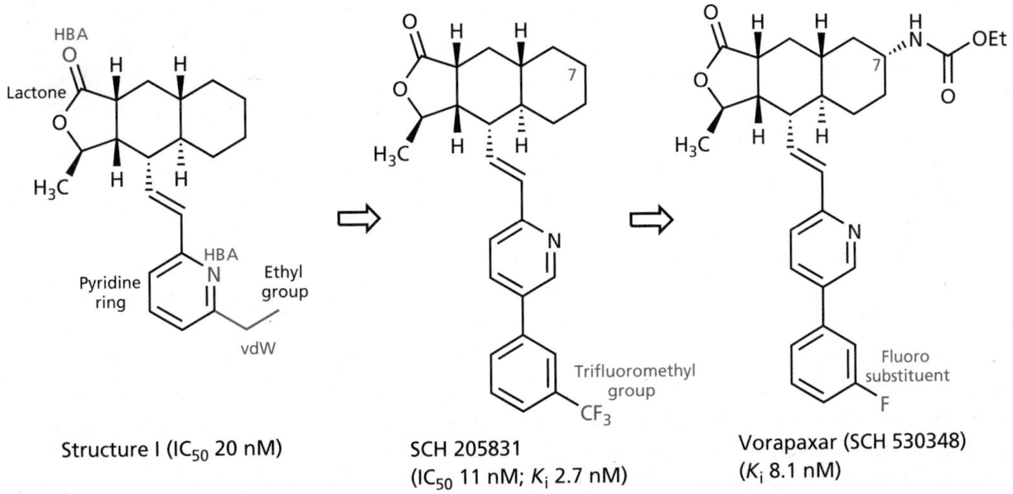

Structure I (IC$_{50}$ 20 nM)

SCH 205831
(IC$_{50}$ 11 nM; K_i 2.7 nM)

Vorapaxar (SCH 530348)
(K_i 8.1 nM)

FIGURE 26.45 Development of vorapaxar.

The development of vorapaxar began with a screening program that identified a synthetic compound (Structure I; Fig. 26.45) as having *in vitro* activity. This compound had previously been synthesized as an analogue of a natural product called (+)-**himbacine** (Fig. 26.46). Structure I was originally tested as a racemate, and when it was subsequently resolved, it was found that the chiral centres present in the more potent enantiomer had the opposite configurations to those present in himbacine. Indeed, himbacine itself has no activity as a PAR-1 antagonist.

Structure I showed good *in vitro* activity, and SAR studies demonstrated that the nitrogen atom of the pyridine ring and the carbonyl oxygen of the lactone were both important hydrogen bond acceptors. The ethyl substituent was also beneficial for activity. Unfortunately, the compound was not orally active, and so further modifications led to the orally active, biaryl structure SCH 205831. It was found that the ethyl substituent that had been beneficial for activity in structure I proved detrimental if it was introduced into **SCH 205831**. It is likely that the ethyl group acts as a conformational blocker in the biaryl system and forces the aryl rings into an orthogonal arrangement that results in an unfavourable binding conformation. The trifluoromethyl group that is present in SCH205831 is thought to help protect the aromatic ring against metabolism. In vorapaxar, this group was replaced with a fluoro substituent to make the molecule less hydrophobic. In addition, a substituent was added at C7 to protect this position against metabolism.

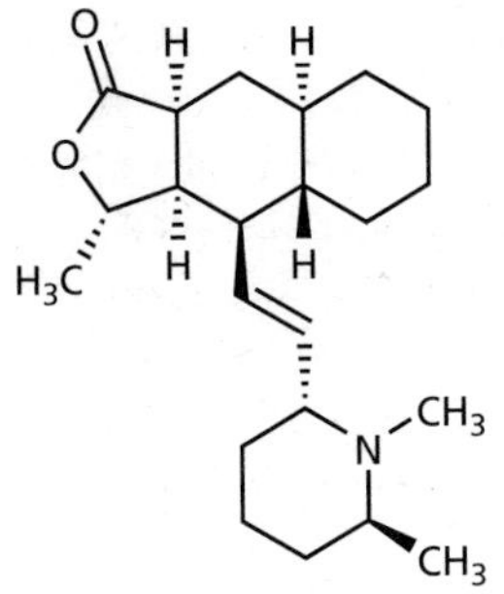

26.9.2.3 P2Y$_{12}$ antagonists

The nucleotide ADP is released from damaged cells and promotes platelet activation by activating G-protein-coupled receptors called P2Y$_{12}$ receptors. These ADP

FIGURE 26.46 Himbacine.

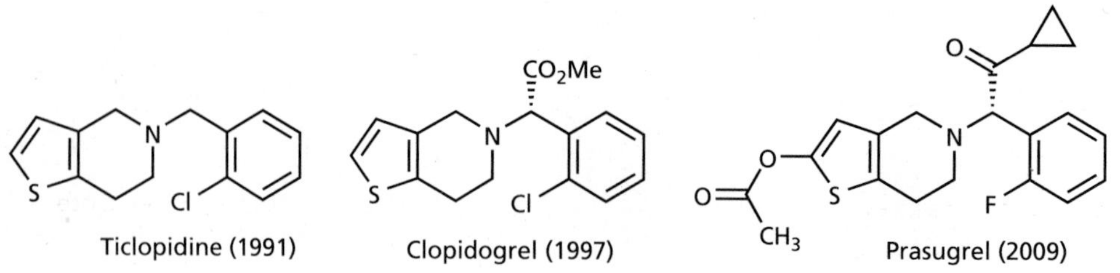

FIGURE 26.47 Irreversible antagonists of the P2Y$_{12}$ ADP receptor.

receptors promote thrombin generation and thrombus stabilization. Therefore, P2Y$_{12}$ antagonists have an anti-platelet activity. The first orally active antagonists to be discovered were **ticlopidine** and **clopidogrel** (Fig. 26.47). These agents are converted by a two-stage metabolic oxidation to a thiol, which binds irreversibly to the receptor. However, this is a disadvantage for patients who lack the enzyme responsible for this transformation, namely **cytochrome P$_{450}$ 2C19**. Moreover, the agents have a slow onset of action because of the two-stage metabolic process.

Prasugrel was designed to be activated by one metabolic oxidation rather than two. As a result, it is faster acting than clopidogrel and has greater potency. On the debit side, it has greater risks of causing major bleeding.

Reversible antagonists have also been designed. The P2Y$_{12}$ receptor contains an allosteric binding site that accepts ATP as a natural allosteric antagonist. Therefore,

it made sense to adopt ATP as the lead compound in the design of synthetic allosteric antagonists (Fig. 26.48). **Cangrelor** was designed as a metabolically stable analogue of ATP and is administered by IV injection. It is a fast-acting, short-lasting agent with a half-life of 2–3 minutes, making it suitable for continuous infusion.

Further research led to **ticagrelor**, which is orally active and was approved in Europe in 2010, and in the US in 2011. A triazolopyrimidine bicyclic ring system was introduced as a bio-isostere for the purine ring system, which resulted in an unexpected 100-fold increase in binding affinity. The ribose sugar has also been replaced with a cyclopentane ring. Initially, attempts were made to replace the labile triphosphate side chain with a more stable substituent that would act as an acidic mimic. However, it was found that the triphosphate chain was not crucial for binding because of the additional binding affinity introduced by the triazolopyrimidine ring.

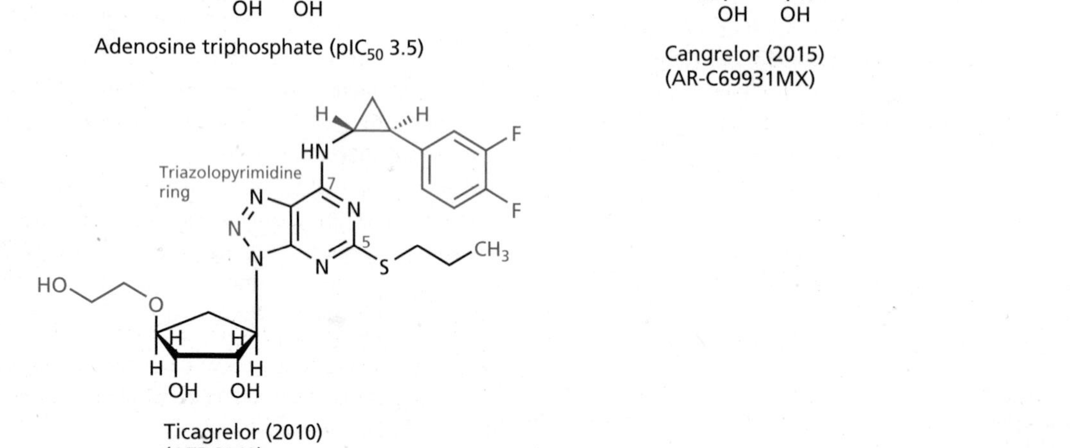

Adenosine triphosphate (pIC$_{50}$ 3.5)

Cangrelor (2015)
(AR-C69931MX)

Ticagrelor (2010)
(AZD6140)

FIGURE 26.48 Allosteric antagonists of the P2Y$_{12}$ ADP receptor

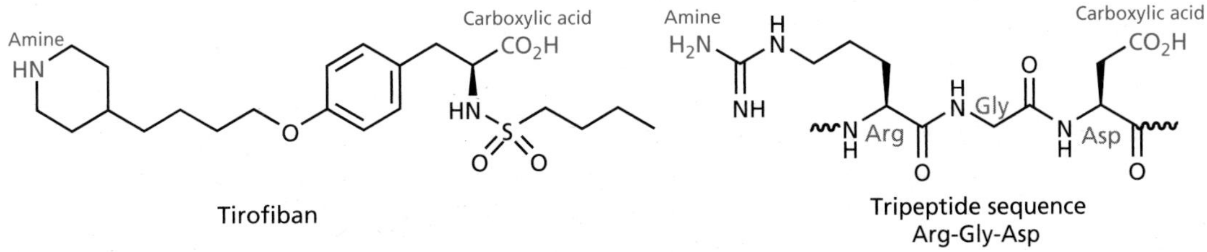

FIGURE 26.49 Tirofiban and the tripeptide sequence Arg-Gly-Asp.

Further modification of the substituents at positions 5 and 7 also increased binding affinity.

26.9.2.4 GpIIb/IIIa antagonists

The final step in platelet aggregation is the binding of fibrinogen to an activated membrane-bound glycoprotein complex called **GoIIb/IIIa**. Fibrinogen is a hexavalent ligand that can bind to GpIIb/IIIa complexes on adjacent platelets, and provide cross-linking between the platelets. Tripeptide sequences in fibrinogen involving Arg-Gly-Asp are key to these binding interactions. GpIIb/IIIa antagonists are available that block this aggregation process, but have to be taken intravenously. They include the monoclonal antibody **abciximab** and a cyclic heptapeptide called **eptifibatide**, which is related to a protein found in rattlesnake venom. **Tirofiban** (Fig. 26.49) is a small-molecular-weight drug that was designed to mimic the tripeptide sequence Arg-Gly-Asp. It includes an amine and a carboxylic acid that mimic the basic and acidic functional groups of the tripeptide, and which are separated by a similar distance.

26.9.3 Fibrinolytic drugs

Thrombi are broken down naturally by the actions of an enzyme called **plasmin**. Plasmin catalyses the degradation of fibrin, and is formed when its precursor protein **plasminogen** is cleaved by serine protease enzymes known as **tissue plasminogen activators** (tPAs).

The fibrinolytic drugs **tenecteplase**, **reteplase**, and **alteplase** are genetically modified versions of natural human tPA that are produced by recombinant DNA technology. **Urokinase** and **streptokinase** are naturally occurring enzymes that are also approved as fibrinolytics. Urokinase is present in human urine, and streptokinase is present in some streptococcal strains of bacteria.

KEY POINTS

- Antithrombotic agents help to prevent blood clots in the circulatory system, thus reducing the risks of heart attack and stroke.

- Antithrombotic agents can be classified as anticoagulants, antiplatelet agents, and fibrinolytic agents.

- Anticoagulants inhibit the formation of thrombi.

- Heparins, hirudin, and the parins are injectable agents that bind to the enzyme thrombin, and inhibit its action.

- Warfarin is an oral anticoagulant that inhibits vitamin K epoxide reductase, but suffers from a low therapeutic index and drug–drug interactions.

- Dabigatran is a direct thrombin inhibitor that was developed as a safer, orally active anticoagulant.

- Idarucizumab is a monoclonal antibody that reverses the effects of dabigatran.

- Factor Xa inhibitors have been developed as orally active anticoagulants which block the conversion of prothrombin to thrombin. However, no antidote has been developed.

- Most antiplatelet agents block platelet aggregation by binding as antagonists to platelet receptors.

- Aspirin has antiplatelet activity because it inhibits cyclooxygenase 1—an enzyme that catalyses the synthesis of thromboxane A2.

- Protease activated receptors (PARs) are platelet receptors which are activated by thrombrin.

- Vorapaxar is an antiplatelet agent that acts as an antagonist of PAR-1.

- $P2Y_{12}$ receptors are platelet receptors that are activated by ADP and antagonized by ATP.

- ATP was used as the lead compound for the development of cangrelor—a $P2Y_{12}$ antagonist that is approved as an antiplatelet agent.

- When activated, glycoprotein IIb/IIIa receptors bind fibrin to cause platelet aggregation and the formation of a thrombus.

- Antagonists of glycoprotein IIb/IIIa receptors block platelet aggregation.

- The fibrinolytic drugs currently used in cardiovascular medicine are enzymes which convert plasminogen to plasmin. Plasmin catalyses the degradation of fibrin, leading to the breakdown of a thrombus.

QUESTIONS

1. Fosinopril (Fig. 26.5) contains an extended ester that undergoes enzymatic hydrolysis to form an intermediate that spontaneously decomposes to generate fosinoprilat. Identify the intermediate and propose a mechanism by which it decomposes. Identify the other product from the decomposition.

2. Explain why the phosphinic acid group in fosinoprilat can be considered to be a transition state mimic for peptide hydrolysis.

3. Lomitapide (Fig. 26.36) contains two trifluoromethyl groups. Suggest why these groups might have been introduced into the structure and why this would increase *in vivo* activity.

4. Factor Xa inhibitors were originally designed to include two polar, basic groups that would be ionized such that they could form strong interactions with the S1 and S4 subsites.

   i) What sort of interactions must be taking place, and what kind of amino acids must be present to allow such an interaction?

   ii) If polar interactions were so important between the early factor Xa inhibitors and the S1 and S4 subsites, why should more recent inhibitors containing P1 and P4 hydrophobic groups have even greater binding affinity?

5. When orally active factor Xa inhibitors were being developed, one of the priorities was to achieve an extremely high selectivity for factor Xa over the protease enzyme trypsin. Why was this considered such a priority?

6. The structure of cangrelor (Fig. 26.48) includes a dichloromethylene group in place of an oxygen atom. Why has this modification been carried out, and why has a dichloromethylene group (CCl_2) been incorporated, rather than CH_2 or CMe_2?

Multiple-choice questions are available on the Online Resource Centre at www.oxfordtextbooks.co.uk/orc/patrick6e/

FURTHER READING

RAAS

Zaman, M. A., Oparil, S., and Calhoun, D. A. (2002) Drugs targeting the renin–angiotensin–aldosterone system. *Nature Reviews Drug Discovery*, **1**, 621–36.

Renin

Jensen, C., Herold, P., and Brunner, H. R. (2008) Aliskiren: the first renin inhibitor for clinical treatment. *Nature Reviews Drug Discovery*, **7**, 399–410.

Cohen, N. C. (2007) Structure-based drug design and the discovery of aliskiren (Tekturna[R]): perseverance and creativity to overcome a R&D pipeline challenge. *Chemical Biology & Drug Design*, **70**, 557–65.

ATII

Kurtz, T. W., and Klein, U. (2009) Next generation multifunctional angiotensin receptor blockers. *Hypertension Research*, **32**, 826–34.

Endothelin antagonists

Bolli, M. H., Boss, C., and Binkert, C. (2012) The discovery of N-[5-(4-bromophenyl)-6-[2-[(5-bromo-2-pyrimidinyl)oxy]ethoxy]-4-pyrimidinyl]-N'-propylsulfamide (Macitentan), an orally active, potent dual endothelin receptor antagonist. *Journal of Medicinal Chemistry*, **55**, 7849–61.

Neprilysin

Sahli, S., Stump, B., Welti, T., et al. (2004) Structure-based design, synthesis and in vitro evaluation of nonpeptidic neprilysin inhibitors. *ChemBioChem*, **5**, 996–1000.

Phosphodiesterases

Daugan, A., Grondin, P., Ruault, C., et al. (2003) The discovery of tadalafil: a novel and highly selective PDE5 inhibitor. 1: 5,6,11,11a-tetrahydro-1*H*-imidazo[1',5':1,6]pyrido[3,4-b]indole-1,3(2*H*)-dione analogues. *Journal of Medicinal Chemistry*, **46**, 4525–32.

Daugan, A., Grondin, P., Ruault, C., et al. (2003) The discovery of tadalafil: a novel and highly selective PDE5 inhibitor. 2: 2,3,6,7,12,12a-hexahydropyrazion[1',2':1,6][3,4-b]indole-1,4-dione analogues. *Journal of Medicinal Chemistry*, **46**, 4533–42.

Calcium channel blockers

Budriesi, R., Cosimelli, B., Ioan, P., Carosati, E., Ugenti, M. P., and Spisani, R. (2007) Diltiazem analogues: the last ten years on structure activity relationships. *Current Medicinal Chemistry*, **14**, 279–87.

Triggle, D. J. (2006) L-Type calcium channels. *Current Pharmaceutical Design*, **12**, 443–57.

Funny ion channel

Reiffen, M., Eberlein, W., Muller, P., et al. (1990) Specific bradycardic agents. 1. Chemistry, pharmacology, and structure-activity relationships of substituted benzaepinones, a new class of compounds exerting anti-ischaemic properties. *Journal of Medicinal Chemistry*, **33**, 1496–504 (zatebradine).

PPARS

Pirat, C., Farce, A., Lebegue, N., et al. (2012) Targeting peroxisome proliferator-activated receptors (PPARs): development of modulators. *Journal of Medicinal Chemistry*, **55**, 4027–61.

sGC

Evgenov, O. V., Pacher, P., Schmidt, P. M., et al. (2006) NO-independent stimulators and activators of soluble guanylate cyclase: discovery and therapeutic potential. *Nature Reviews Drug Discovery*, **5**, 755–68.

Microsomal triglyceride transfer protein inhibitors

Wetterau, J. R., Gregg, R. E., Harrity, T. W., et al. (1998) An MTP inhibitor that normalizes atherogenic lipoprotein levels in WHHL rabbits. *Science*, **282**, 751–4 (lomitapide).

Direct thrombin inhibitors

Hauel, N. H., Nar, H., Priepke, H., et al. (2002) Structure-based design of novel potent nonpeptide thrombin inhibitors. *Journal of Medicinal Chemistry*, **45**, 1757–66 (dabigatran).

Factor Xa inhibitors

Pinto, D. J. P., Smallheer, J. M., Cheney, D. L., et al. (2010) Factor Xa inhibitors: next generation antithrombotic agents. *Journal of Medicinal Chemistry*, **53**, 6243–74.

Antiplatelet agents
PAR-1 antagonist

Chackalamannil, S., Wang, Y., Greenlee, W. J., et al. (2008) Discovery of a novel, orally active himbacine-based thrombin receptor antagonist (SCH 530348) with potent antiplatelet activity. *Journal of Medicinal Chemistry*, **51**, 3061–4 (vorapaxar).

P2Y$_{12}$ antagonists

Springthorpe, B., Bailey, A., Barton, P., et al. (2007) From ATP to AZD6140: the discovery of an orally active P2T$_{12}$ receptor antagonist for the prevention of thrombosis. *Bioorganic and Medicinal Chemistry Letters*, **17**, 6013–8 (ticagrelor).

Ingall, A. H., Dixon, J., Bailey, A., et al. (1999) Antagonists of the platelet P2T receptor: a novel approach to antithrombotic therapy. *Journal of Medicinal Chemistry*, **42**, 213–20 (cangrelor).

Titles for general further reading are listed on p.845.

CASE STUDY 6
Steroidal anti-inflammatory agents

CS6.1 Introduction to steroids

Steroids are important endogenous hormones found in many life forms. They all share a common tetracyclic structure as shown in Fig. CS6.1, but they vary in the substituents and functional groups that are present. The stereochemistry of the rings in fully saturated steroids is identical in mammalian steroids, where the three six-membered rings have chair conformations. There are several asymmetric centres present, but only one stereoisomer occurs naturally for any particular steroid. For example, **cortisol** has seven asymmetric centres, but only the stereoisomer shown in Fig. CS6.1 exists naturally.

Test your understanding and practise your molecular modelling with Exercises CS6.1 and CS6.2 on the Online Resource Centre at www.oxfordtextbooks.co.uk/orc/patrick6e/

Some of the terminology used in the nomenclature of steroids is worth explaining at this point. Substituents are often described as being alpha (α) or beta (β). α-Substituents are below the general 'plane' of the steroid as represented in Fig. CS6.1, and are represented by hatched wedges in 2D diagrams, whereas β-substituents are above the plane and are represented by solid wedges. For example, in cortisol, the axial methyl groups (C18 and C19) are β-substituents, whereas the axial hydrogens at positions 9 and 14 are in the α position.

The position of double bonds in steroids is usually identified by the symbol delta (Δ). For example, Δ^4 signifies the double bond between C4 and C5 in cortisol. If there is any ambiguity, then the numbers of both carbons are indicated. For example, cholesterol has a double bond between C5 and C6, rather than C5 and C10, so this is indicated as $\Delta C^{5(6)}$.

Steroids are hydrophobic compounds because of their extensive hydrocarbon skeleton. This is an important characteristic as the hormonal steroids have to cross cell membranes in order to interact with intracellular steroid receptors (see section 4.9 and Box 8.2). All of the important endogenous steroids have polar functional groups, such as alcohols, phenols, and ketones. These play a crucial role in the binding of steroids to their target receptors, but their presence does not alter the hydrophobic nature of the molecule as a whole. Because most steroids are hormones, they are present in very small quantities in the body (less than 1 mg). The exception is cholesterol, which is present in much larger quantities (250 g) and has a number of non-hormonal roles (Case study 1).

In this Case study, we will be concentrating on those steroids released from the adrenal cortex of the adrenal gland—the **adrenocorticoids**. There are two types of adrenocorticoids—the **glucocorticoids** and the **mineralocorticoids**. The former act on carbohydrate, fat, and protein metabolism, mainly in the liver, muscle, and brain cells. They also have an important anti-inflammatory effect which is separate from their metabolic effects. The mineralocorticoids regulate electrolyte balance through sodium ion retention in kidney cells. The major endogenous glucocorticoids are **corticosterone, cortisone**, and **cortisol** (also known as **hydrocortisone**) (Figs CS6.1 and CS6.2). **Aldosterone** is the major endogenous mineralocorticoid. An imbalance of these steroids can

(a)

Axial bonds are in grey
equatorial bonds are in blue

(b)

Cortisol (hydrocortisone)

FIGURE CS6.1 (a) General tetracyclic structure of a steroid with numbering; (b) structure of cortisol (asymmetric centres indicated by stars).

FIGURE CS6.2 Adrenocorticoids.

lead to certain diseases. For example, an excess of glucocorticoids causes **Cushing's syndrome**, whereas a deficit results in **Addison's disease**. An excess of mineralocorticoids leads to **Conn's syndrome**.

The glucocorticoids have an important clinical role in replacement therapy for Addison's disease, and have also been used as anti-inflammatories and immunosuppressants in the treatment of a number of conditions, such as asthma, hypersensitivity, rheumatoid arthritis, cancer, and diseases which have an autoimmune or inflammatory effect. The adrenocorticoids are examples of steroids having a **pregnane** skeleton—steroids containing a two-carbon side chain at position 17 of the tetracyclic steroid skeleton (Fig. CS6.2).

One of the most important applications of glucocorticoids in medicine is as anti-inflammatory agents. Unfortunately, the endogenous glucocorticoids suffer from the fact that they have mineralocorticoid and immunosuppressant effects, which can cause oedema and increased susceptibility to infection. Moreover, the endogenous glucocorticoids affect a large number of enzymes in different cell types in order to control metabolism. This means that they have a large number of undesired side effects if they are taken as drugs to control inflammation. Consequently, glucocorticoids are best used as topical anti-inflammatory agents. A lot of research has gone into designing glucocorticoids that act locally at the site of administration and are metabolized rapidly in the blood supply such that they cannot act on other targets. Having said that, there are some glucocorticoids which *can* be administered orally and which have been designed to have fewer side effects.

CS6.2 **Orally active analogues of cortisol**

In 1947, it was found that cortisone could relieve the symptoms of rheumatoid arthritis. However, cortisone is readily converted in the liver to cortisol, and it is now thought that the effects of cortisone are actually due to cortisol. A large number of analogues have been synthesized which have identified the features of cortisol that are important for corticosteroid activity. In essence, all the functional groups are important, and the removal of any of these groups either reduces or eliminates activity.

However, further studies have shown that the introduction of extra substituents can increase activity, which allows the removal of one of the original functional groups.

Introducing a 9α-fluoro substituent to give **fludrocortisone** (Fig. CS6.3) increased activity tenfold, but it also increased mineralocorticoid activity 300–600 times. In contrast, the introduction of an extra double bond at the Δ^1 position increased activity fourfold without increasing mineralocorticoid activity—see **prednisolone** and **prednisone** (Fig. CS6.3). Introducing substituents such as methyl or fluorine at the 6α-position has also been found to be beneficial because these groups serve to block metabolism at that position. For example, **methylprednisolone** has a 6α-methyl group.

A methyl group was introduced at C-16 to see whether it would block the metabolic reduction of the C-20 keto group of hydrocortisone analogues—a reaction that is known to lead to inactive metabolites. There is no evidence that such protection actually occurs and there is no obvious increase in glucocorticoid activity, but the presence of the methyl group does suppress the mineralocorticoid properties of sodium and water retention. It is thought that the 16-methyl substituent blocks the ability of these analogues to bind to the mineralocorticoid receptor. Further research revealed that the introduction of C-16 substituents, such as a methyl or hydroxyl group, counteracted the mineralocorticoid effect of a 9-fluoro substituent. This resulted in the development of **triamcinolone**, **dexamethasone**, **betamethasone**, and **flumetasone pivalate** (Fig. CS6.3), all of which have increased glucocorticoid activity and negligible mineralocorticoid side effects.

Test your understanding and practise your molecular modelling with Exercises CS6.2 and CS6.3 on the Online Resource Centre at www.oxfordtextbooks.co.uk/orc/patrick6e/

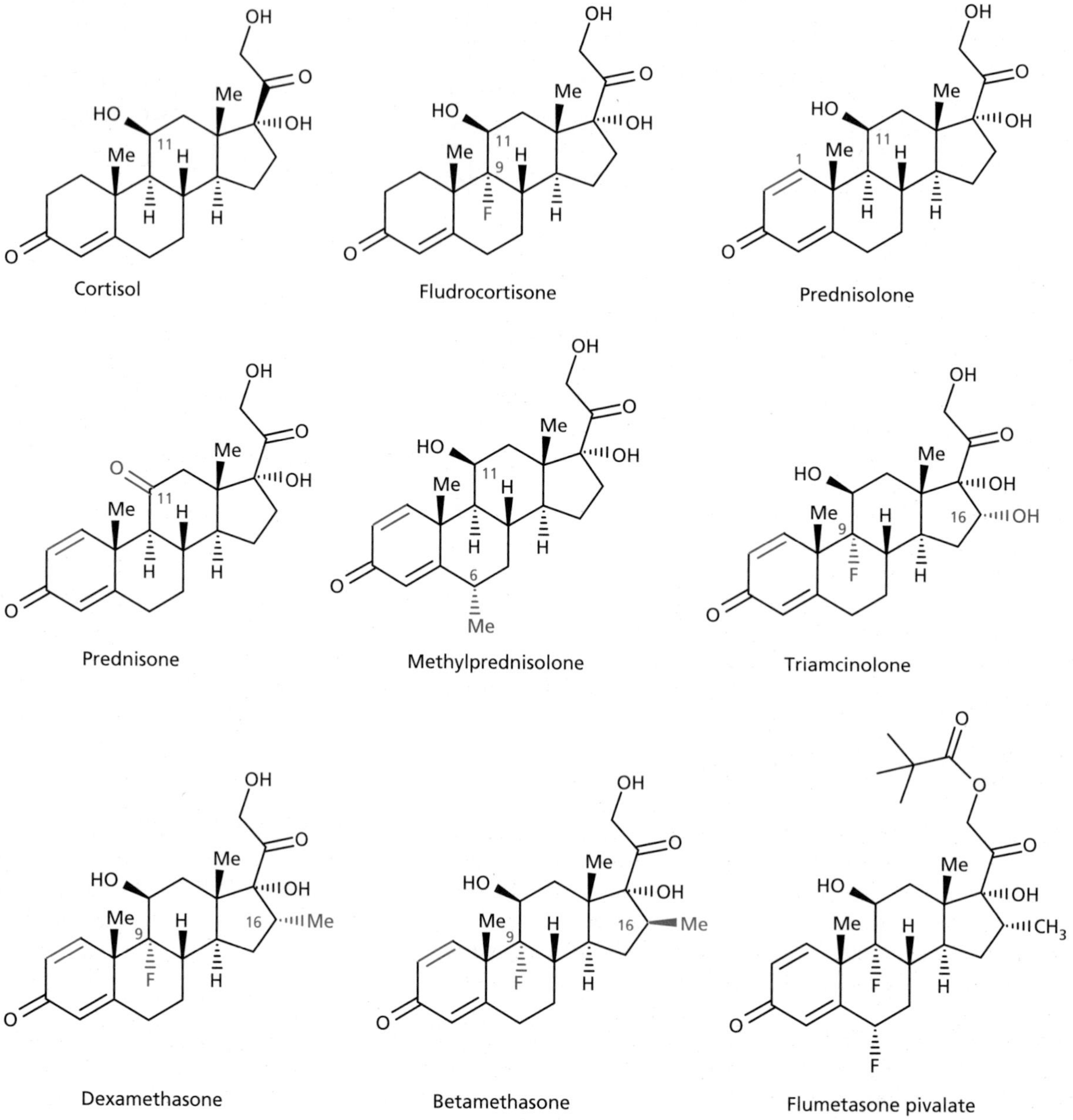

FIGURE CS6.3 Analogues of cortisol.

CS6.3 **Topical glucocorticoids as anti-inflammatory agents**

CS6.3.1 **Cortisol analogues**

Glucocorticoids are often applied topically to treat skin inflammations. **Triamcinolone acetonide** (Fig. CS6.4) is one such agent. The acetonide group links the alcohol substituents at C16 and C17 of triamcinoline, thus reducing the polarity of the molecule. This leads to better skin absorption and a 1000-fold increase in activity compared with triamcinolone itself. If the compounds are injected

under the skin, they have equal activity. It is not yet clear whether the acetonide is acting as a prodrug and is rapidly metabolized once it reaches the tissues, or whether the acetonide group increases binding to a hydrophobic region in the glucocorticoid receptor. **Fluocinolone acetonide, fluocinonide,** and **flunisolide** (Fig. CS6.4) are clinical agents that contain the same acetonide group (see also **fludroxycortide**; Box CS6.1).

Good skin absorption can also be achieved by esterifying one or more alcohol groups. The corresponding phosphate esters were less active, providing further evidence that lipophilicity is important to the activity of topically applied anti-inflammatories. Glaxo used this strategy to develop the

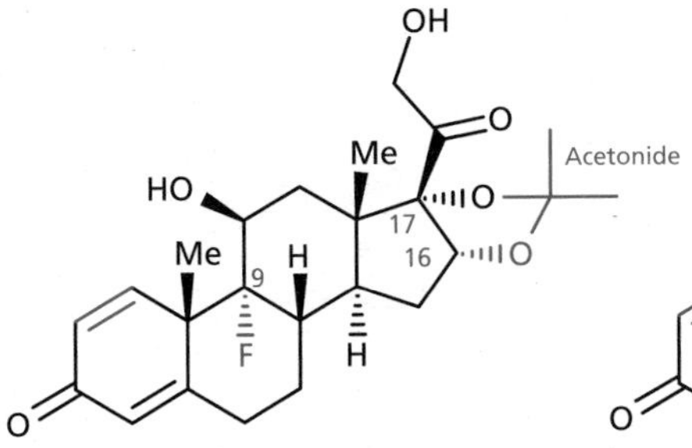

FIGURE CS6.4 Steroid acetonides used as topical agents.

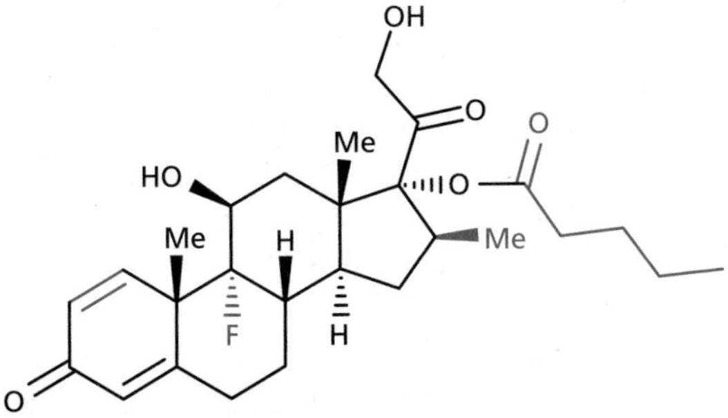

Triamcinolone acetonide

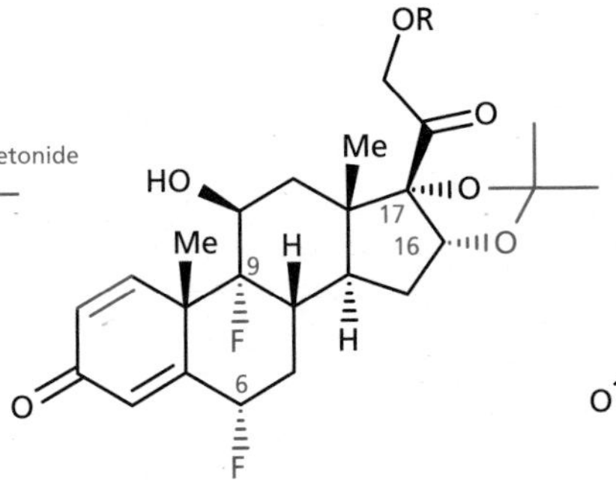

Fluocinolone acetonide; R = H
Fluocinonide; R = Ac

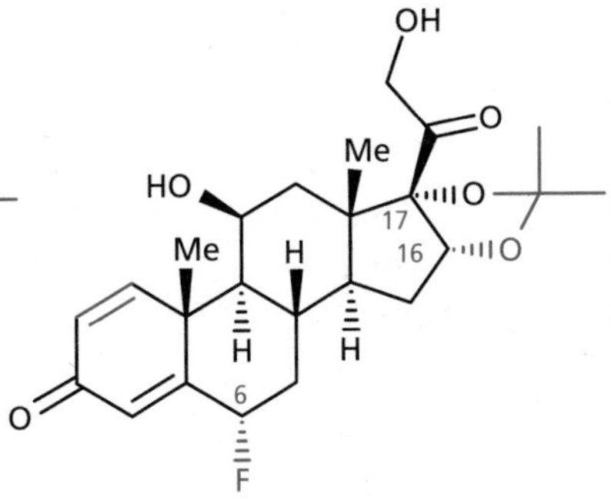

Flunisolide

Betamethasone 17-valerate

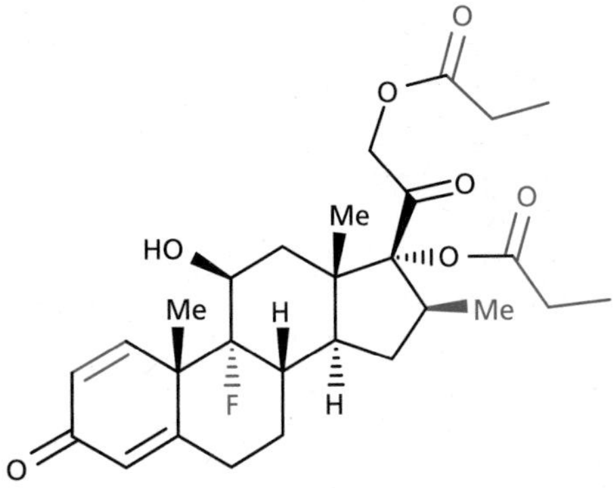

Betamethasone 17,21-dipropionate

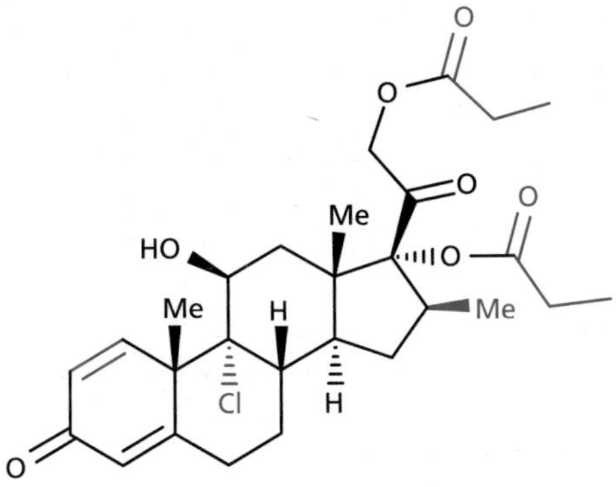

Beclometasone 17,21-dipropionate

FIGURE CS6.5 Clinically useful esters and analogues of betamethasone.

clinically useful agents **betamethasone 17-valerate**, **betamethasone dipropionate**, and **beclometasone dipropionate** (formerly beclomethasone dipropionate) (Fig. CS6.5).

CS6.3.2 **21-Deoxysteroids**

Removal of the 21-OH group from cortisol eliminates activity, but activity can be restored by adding similar substituents to the ones described in section CS6.3.1. Thus, the introduction of an extra double bond in the A ring, along with substituents at C-6 and C-9, results in **fluorometholone** (Fig. CS6.6).

Esterification of the 17-OH group results in better skin absorption and increased topical activity, for example, **21-deoxybetamethasone 17-propionate** (Fig. CS6.6).

Introducing a halogen at position 21 was particularly beneficial for the 17-esters. The best activity was obtained

FIGURE CS6.6 21-Deoxysteroids with glucocorticoid activity.

using F or Cl, with short chain esters at C-17. The best compound arising from these studies was **clobetasol propionate** (Fig. CS6.6).

CS6.3.3 **11-Ketosteroids**

In general, replacing the 11β-OH group of cortisol with a keto group results in a drop in activity, and it is believed that the ketone group has to be reduced *in vivo* for the compound to be active. However, activity can be restored by introducing suitable substituents elsewhere. Halogens at positions C-9 and C-21 are particularly important in this respect, for example **clobetasone butyrate** (Fig. CS6.7).

CS6.3.4 **Analogues with modified C-17 side chains**

The two-carbon chain at C-17 is generally important for activity, but it was found that activity could be retained if the side chain was replaced with a carboxylic acid as long as both it and the 17-OH group were esterified. If only one or other of the functional groups was esterified, then there was no activity. This was an important discovery as it meant that the diesters would be active at the site of administration but would be hydrolysed to inactive compounds as soon as they reached the blood

Clobetasone 17-butyrate

FIGURE CS6.7 Clobetasone 17-butyrate.

circulation, thus reducing the chances of unwanted side effects elsewhere in the body. A variety of esters were synthesized which demonstrated that the 17α-propionate and 17β-fluoromethyl esters were ideal (structure I, Fig. CS6.8). Further variations led to the discovery that the 17β-fluoromethyl thioester was also beneficial, leading to the clinically important **fluticasone propionate** (Fig. CS6.8). This agent has a high affinity for target receptors, high potency, and low oral bioavailability (1%) because of low solubility and rapid metabolism in the liver.

CS6.3.5 **Glucocorticoids used in asthma treatment**

Glucocorticoids are used as anti-inflammatory agents in the treatment of asthma, and are administered by inhalation in order to reduce the risks of side effects caused by their presence in the blood supply. However, it is not possible to completely prevent these agents reaching the blood supply; a certain percentage of inhaled glucocorticoid is swallowed and absorbed orally. However, most of the glucocorticoids used in asthma treatment are rapidly metabolized in the liver. Of more significance is the proportion of inhaled dose that gets absorbed into the blood supply through the lungs. Therefore, it is important that glucocorticoids used in asthma treatment are susceptible to metabolic deactivation in the blood, for example by esterases.

Beclometasone dipropionate (Fig. CS6.5) represented a breakthrough in asthma treatment and is currently used as an inhaler, as are **budesonide, ciclesonide, mometasone furoate**, and **fluticasone propionate** (Figs. CS6.8 and CS6.9). Budesonide is an example of a new generation of non-halogenated glucocorticoids. One would actually expect a drop in activity as a result of the lack of halogen substituents, but the nature of the acetal is key in providing high topical anti-inflammatory activity. The acetal group increases the hydrophobic nature of the compound leading to prolonged residence in lung tissue.

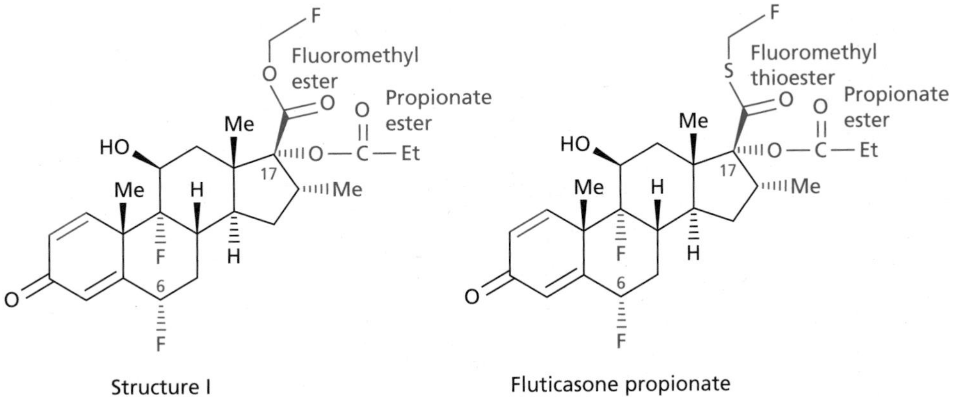

FIGURE CS6.8 Development of fluticasone propionate.

Budesonide has been found to have high receptor affinity and a higher anti-inflammatory potency than fluticasone propionate. In contrast, its systemic glucocorticoid activity is 4–7 times lower because of extensive first pass metabolism in the liver by the cytochrome P450 enzyme (CYP3A4) to much less potent metabolites. **Ciclesonide** is the latest in this series and is an example of a **soft steroid**. The structure acts as a prodrug and is activated by esterases in lung tissue which hydrolyse the C-21 ester to reveal a free alcohol group. This is the active compound and has a prolonged duration of action in lung tissue. However, it has negligible activity elsewhere in the body despite it being able to reach the circulatory system. This is because it is rapidly metabolized by cytochrome P450 enzymes to inactive metabolites.

The use of heterocyclic esters at C-17 also results in high topical anti-inflammatory activity, as in mometasone furoate.

CS6.3.6 Glucocorticoids used in ophthalmology

A number of steroids have been used as topical anti-inflammatory agents in ophthalmology, such as dexamethasone (Fig. CS6.3), **fluorometholone** (Fig. CS6.6), **betamethasone sodium phosphate, hydrocortisone acetate, prednisolone acetate, prednisolone sodium phosphate**, and **rimexolone** (Fig. CS6.10). Rimexolone is surprisingly short of many of the features that are present in other anti-inflammatory glucocorticoids. For example, it lacks the 17α-OH group, as well as halogen substituents.

Unfortunately, glucocorticoids can cause side effects such as glaucoma and cataract formation. The latter is thought to be associated with the C-20 keto group forming Schiff bases with lysine residues on proteins, followed by a rearrangement reaction involving the C-21 hydroxyl group to give amine-linked adducts. Indeed, efficacy appears to go hand in hand with toxicity.

To tackle this problem, it was decided to design soft drugs which would quickly metabolize in the circulation to inactive compounds. The ideal drug would be one which is metabolized at a reasonable rate in the blood supply, but survives long enough to act as an anti-inflammatory agent at its intended target. This requires the correct balance of activity, solubility, lipophilicity, tissue distribution, protein binding, and rate of metabolic deactivation.

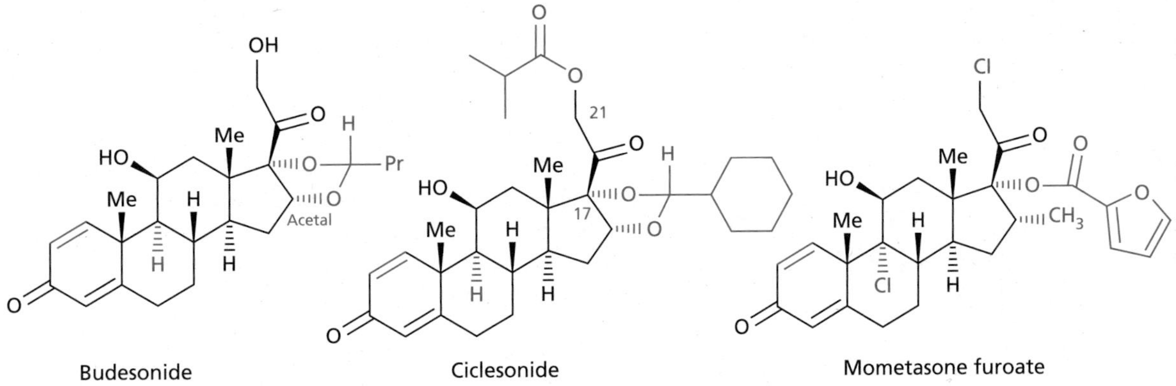

FIGURE CS6.9 Glucocorticoids used in the treatment of asthma.

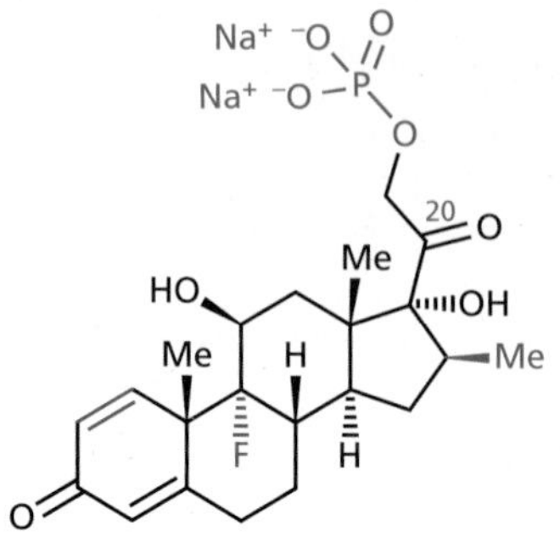

Betamethasone sodium phosphate

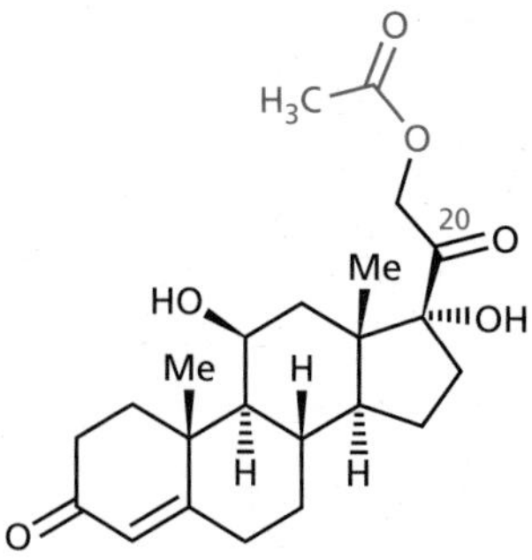

Hydrocortisone acetate

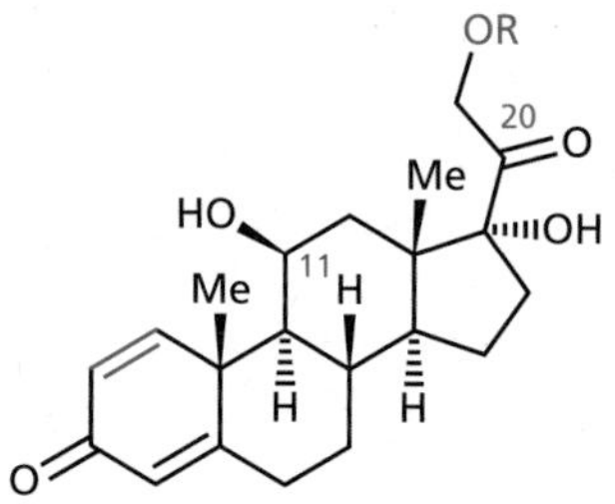

Prednisolone acetate, R = Ac
Prednisolone sodium phosphate,
R = PO₃Na₂

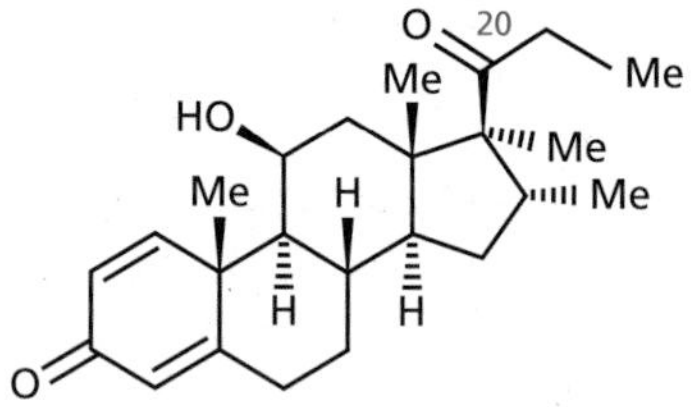

Rimexolone

FIGURE CS6.10 Glucocorticoids used in ophthalmology.

The lead compound for the design of these compounds was **cortienic acid**, which was known to be an inactive metabolite of hydrocortisone resulting from oxidation of the dihydroxyacetone side chain (Fig. CS6.11).

The aim was to now restore activity by adding suitable esters to the functional groups at C-17. As the esters would be susceptible to hydrolysis by esterases in the blood, any activity introduced in this manner would be lost completely after hydrolysis had taken place. Other features that were known to be beneficial to anti-inflammatory activity were also included in various analogues, such as an extra double bond in the A ring or fluorination at C-6 or C-9. A first generation of compounds was synthesized that illustrated the following important features for activity:

- a fluoromethyl or chloromethyl ester at C-17β;
- a carbonate or ether group at C-17α.

This led to the discovery of **loteprednol etabonate** (Fig. CS6.12), which has a much better therapeutic ratio than the traditional corticosteroids. The compound contains the extra double bond in the A ring which is good for activity, as well as two hydrolysable esters. As predicted, it is metabolized in two stages to the Δ^1 analogue of cortienic acid. Ester hydrolysis occurs first to give an inactive metabolite, followed by hydrolysis of the less reactive carbonate ester.

The use of a carbonate ester over a normal ester at C-17α was a deliberate strategy to prevent the possibility of the intramolecular reaction shown in Fig. CS6.13, which would result in toxic anhydrides being formed.

Etiprednol dicloacetate (Fig. CS6.14) is a second-generation soft drug where two normal esters have been employed. The two chloro groups increase the rate of hydrolysis of the 17α-ester, which means that this ester

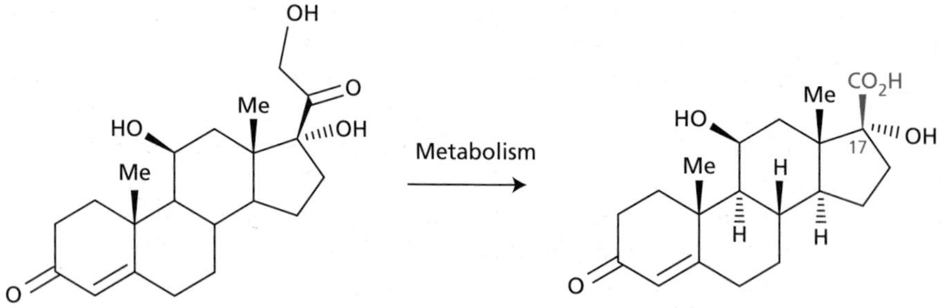

Cortisol (hydrocortisone)

Cortienic acid

FIGURE CS6.11 Metabolism of cortisol to cortienic acid.

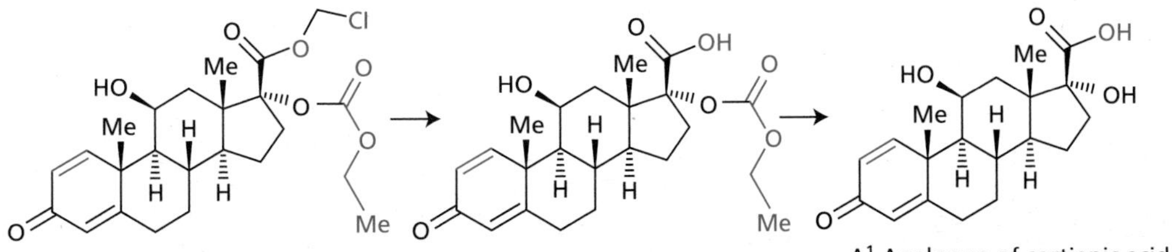

FIGURE CS6.12 Metabolism of loteprednol etabonate.

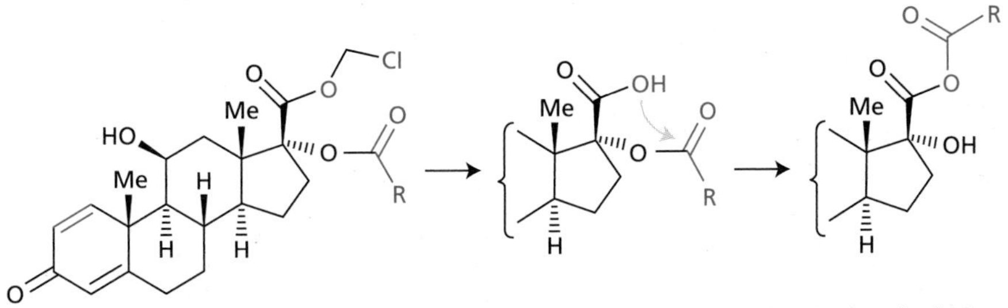

FIGURE CS6.13 Intramolecular reaction leading to toxic anhydrides.

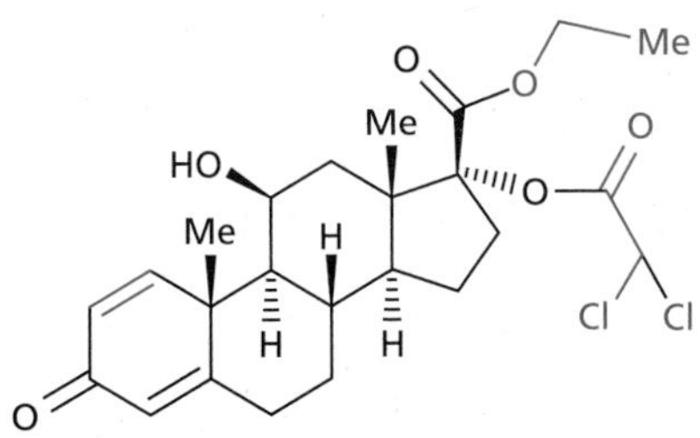

FIGURE CS6.14 Etiprednol dicloacetate.

is hydrolysed first instead of the 17β-ester, thus avoiding the risk of anhydride formation.

The absence of the chlorine substituent from the 17β-ester is potentially a problem as this is part of the pharmacophore for activity. However, molecular modelling studies demonstrated that one of the two chlorine substituents on the 17α-ester could occupy the same position in space as the original chlorine substituent.

Soft drugs containing a lactone group are of potential interest as anti-asthmatic agents (Fig. CS6.15). The lactone in the figure displays sufficient activity and stability in lung tissue to be effective. However, when it reaches the plasma, it undergoes rapid hydrolysis to form inactive metabolites. This is due to the enzyme **serum paraoxonase**, which is present in plasma and the liver, but not in lung tissue.

CS6.3.7 Sustained release of topical anti-inflammatory agents

An interesting example of a pro-soft drug approach in drug design involves the design of a sustained chemical release system for hydrocortisone (Fig. CS6.16). When the spirothiazolidine derivative of hydrocortisone is applied topically, it undergoes a spontaneous ring opening to form an imine and a thiol. The latter group reacts with the thiol group of cysteine residues in proteins, and becomes tethered to local tissue via a disulphide bond. Eventually, the imine is hydrolysed to release the drug. The compound has been found to be more active than hydrocortisone itself, and less of it crosses the dermis into the blood supply.

Test your understanding and practise your molecular modelling with Exercises CS6.1–CS6.8 on the Online Resource Centre at www.oxfordtextbooks.co.uk/orc/patrick6e/

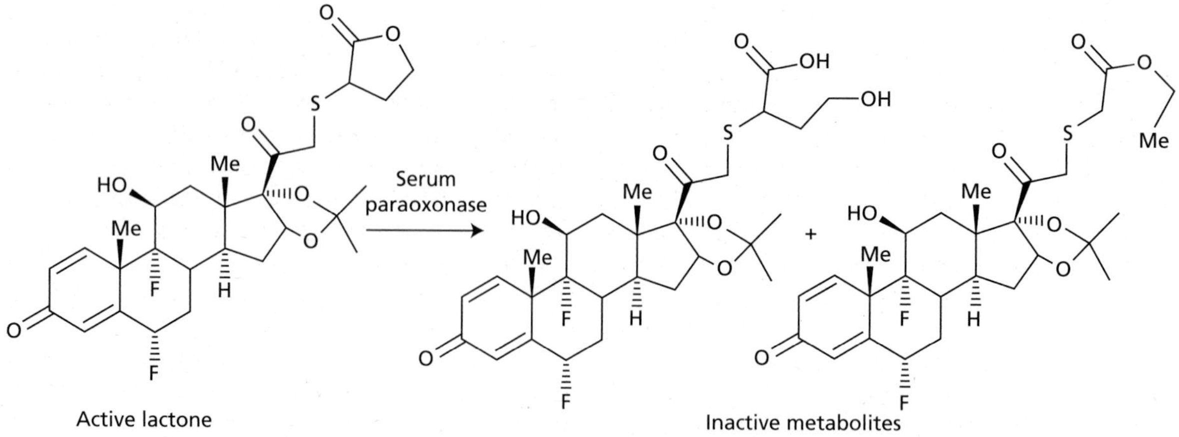

FIGURE CS6.15 Inactivation of an active lactone by the enzyme serum paraoxonase.

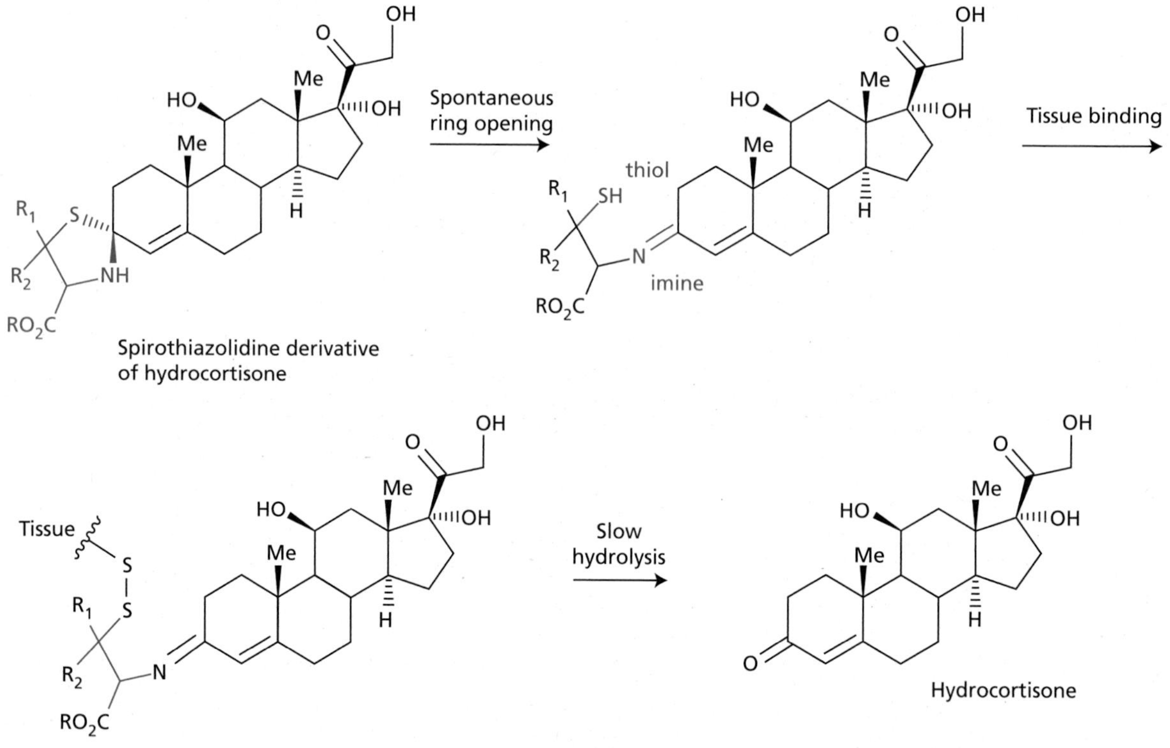

FIGURE CS6.16 Sustained release of hydrocortisone.

BOX CS6.1 Clinical aspects of glucocorticoids

The main clinical application for glucocorticoids is in the treatment of inflammation associated with conditions such as rheumatoid arthritis, asthma, and allergies. The agents used should have a low to negligible mineralocorticoid side effect. Ideally, they should be administered topically, whether that be as a cream or ointment for skin inflammations; drops for inflammations of the eye, ear, and nose; or aerosols for the prophylaxis and treatment of asthma. However, there are occasions when oral administration is acceptable and, in certain emergency situations, they can be injected, for example in severe asthma or anaphylactic shock. They can also be injected directly into joints or soft tissue for the treatment of joint inflammations. Long-term use of glucocorticoids is discouraged because it can lead to growth suppression in children, susceptibility to infection (especially chicken pox and measles), and suppression of the pituitary/adrenal glands. The last effect can result in serious medical problems if the treatment is stopped suddenly and so a steroid treatment card should be carried by any patients taking glucocorticoids on a long-term basis. Systemic administration can also result in a wide range of psychiatric conditions varying from nightmares to depression and suicidal tendencies, especially with patients having a history of mental disorders. High doses can lead to Cushing's syndrome, but this is usually reversible when the treatment is gradually withdrawn.

Orally active glucocorticoids currently used in the clinic include **cortisol**, **cortisone acetate**, **deflazacort**, **dexamethasone**, **methylprednisolone**, **prednisolone**, **riamcinolone acetonide**, and the ester prodrugs of **betamethasone** and **dexamethasone**.

There are a large variety of topical agents used as creams, drops, or sprays, including **alclometasone dipropionate**, **beclometasone dipropionate**, **budesonide**, **cortisol**, **dexamethasone**, **diflucortolone valerate**, **fludroxycortide**, **flumetasone pivalate**, **fluorometholone**, **flunisolide**, **fluocinolone acetonide**, **fluocinonide**, **fluticasone propionate**, **halobetasol propionate**, **loteprednol etabonate**, **mometasone furoate**, **rimexolone**, and **triamcinolone acetonide**. Ester prodrugs of **betamethasone**, **clobetasol**, **cortisol**, **dexamethasone**, **fluocortolone**, and **prednisolone** are also available.

Preparations used for injections include **triamcinolone acetonide** and ester prodrugs of **betamethasone**, **cortisol**, **dexamethasone**, **methylprednisolone**, and **prednisolone**.

Agents used in the prophylaxis of asthma include **budesonide**, **ciclesonide**, **fluticasone propionate**, and **mometasone furoate**.

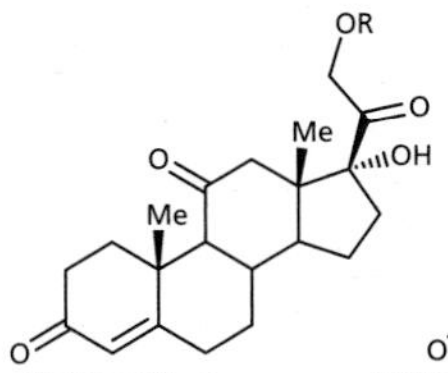

Cortisone; R = H
Cortisone acetate; R = Ac

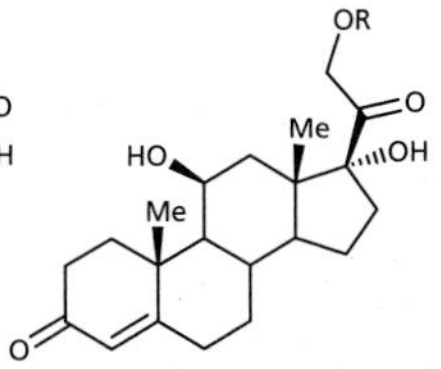

Hydrocortisone; R = H
Hydrocortisone acetate; R = Ac
Hydrocortisone phosphate; R = Phosphate
Hydrocortisone succinate; R = CO(CH$_2$)$_2$CO$_2$H

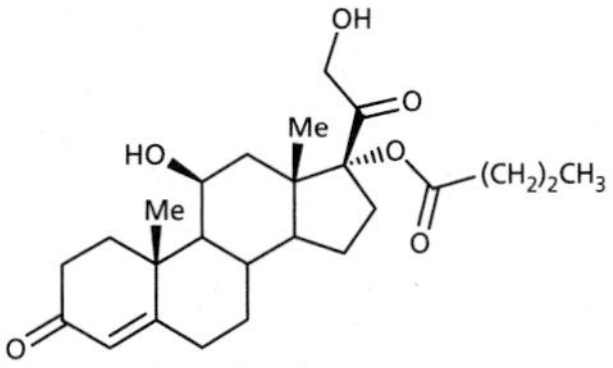

Hydrocortisone butyrate

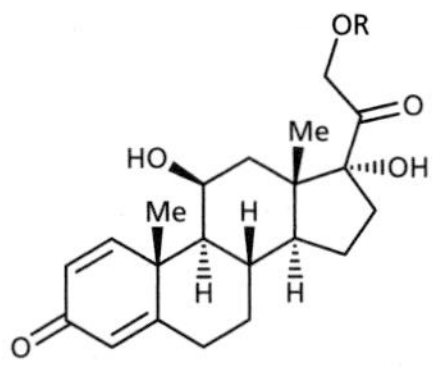

Prednisolone; R = H
Prednisolone acetate; R = Ac
Prednisolone sodium phosphate; R = PO$_3$Na$_2$
Prednisolone hexanoate; R = CO(CH$_2$)$_4$CH$_3$
Prednisolone metasulphobenzoate
R =

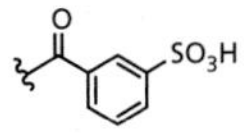

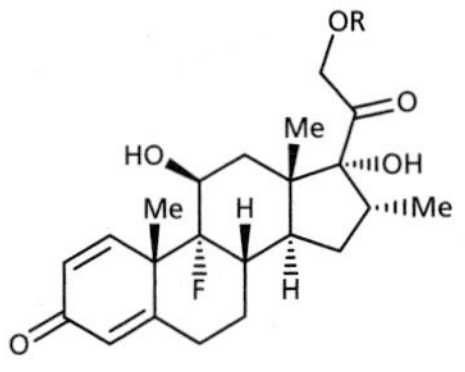

Dexamethasone; R = H
Dexamethasone acetate; R = Ac
Dexamethasone phosphate; R = PO$_3^{2-}$
Dexamethasone metasulphobenzoate;
R =

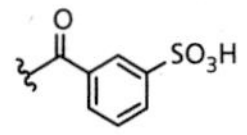

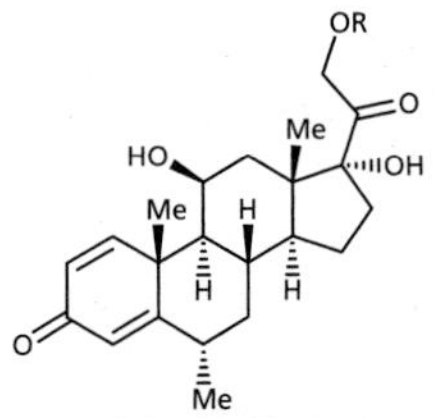

Methylprednisolone; R = H
Methylprednisolone acetate; R = Ac
Methylprednisolone succinate;
R = CO(CH$_2$)CO$_2^-$ Na$^+$

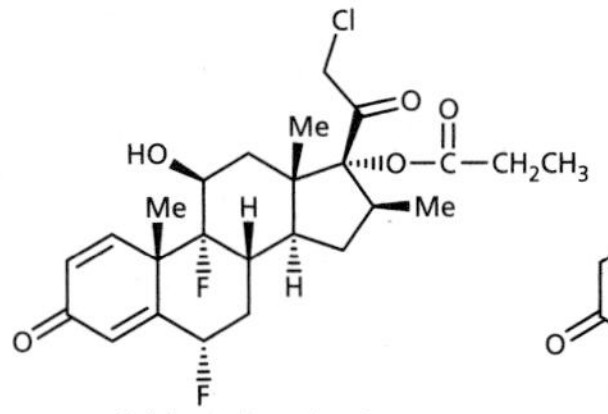

Halobetasol propionate

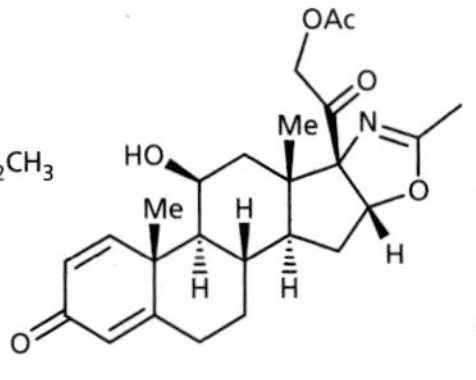

Deflazacort

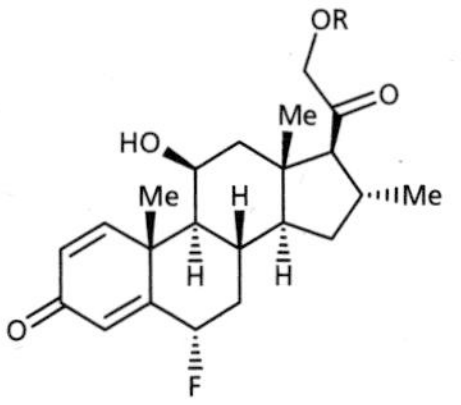

Fluocortolone; R = H
Fluocortolone pivalate; R = COCMe$_3$
Fluocortolone caproate; R = CO(CH$_2$)$_4$CH$_3$

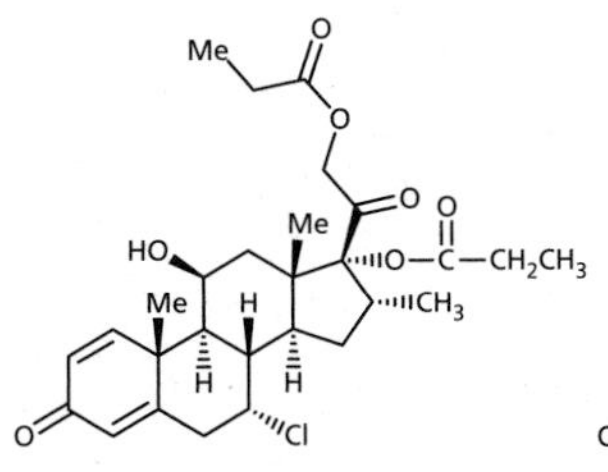

Alclometasone dipropionate

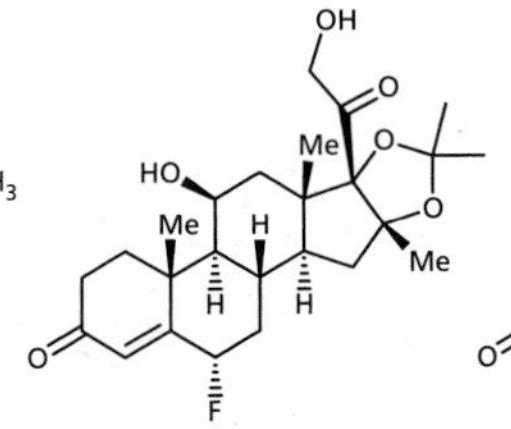

Fludroxycortide

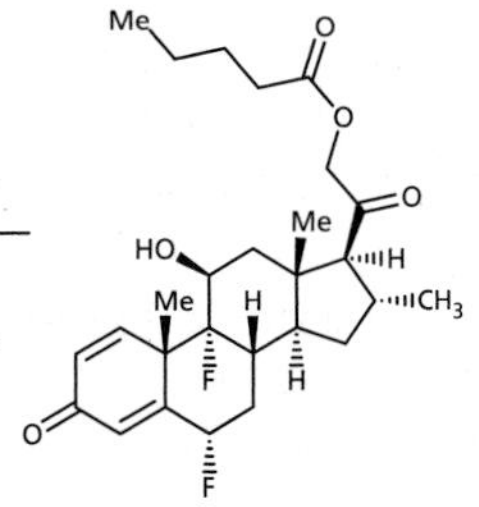

Diflucortolone valerate

CASE STUDY 7
Current research into antidepressant agents

CS7.1 Introduction

I am worn out with grief;
every night my bed is damp from my weeping;
my pillow is soaked with tears.
I can hardly see;
my eyes are so swollen;
from the weeping caused by my enemies.

Psalm 6, verses 6 & 7

Major depression is a common ailment that affects up to 10% of the population. It is estimated that 18 million people suffer from it in the USA and 340 million worldwide. The World Health Organization believes that by the year 2020, depression could be the second leading ailment in the world after heart disease. Depression is common in the elderly, and it is estimated that 21% of women and 13% of men will suffer major depression at some point in their lives. Symptoms include misery, apathy, pessimism, low self-esteem, feelings of guilt, inability to concentrate or work, loss of libido, poor sleep patterns, loss of motivation, and loss of appetite. Sufferers of long-term depression are more prone to other diseases and their lifespan can be shortened.

The causes of depression are many and varied. Some people are genetically predisposed to depression, but, in many cases, a stressful life-changing event precipitates the condition. Such events include loss of employment, divorce, bereavement, rejection, victimization, false accusation, and slander. Often, the sufferer has no control or redress over what has taken place, and the sense of helplessness and hopelessness that results exacerbates the situation.

Those suffering severe depression describe each day as a living nightmare. The same distressing thoughts whirl round in their minds pulling them deeper and deeper into a bottomless psychological whirlpool from which there seems to be no escape. Each day is an ordeal to be endured and, for some, it can be too much. Some turn to alcohol or illicit drugs for a temporary oblivion; a few turn to suicide for a permanent oblivion. Those who have never suffered depression have no concept of the disease, and telling the sufferer to 'snap out of it' or 'pull yourself together' is worse than useless.

CS7.2 The monoamine hypothesis

The pharmacological processes that cause depression are still an area of a research, but the accepted theory proposes that a deficit of monoamine neurotransmitters in certain parts of the brain causes the condition. This is known as the **monoamine** or **monoaminergic hypothesis**. The principle neurotransmitters believed to be involved are **dopamine**, **noradrenaline**, and **serotonin** (also known as **5-hydroxytryptamine**, 5-HT). There are various lines of evidence which support this. For example, the antihypertensive agent **reserpine** lowers monoamine levels in the brain and is known to cause depression as a side effect. Moreover, the clinically important antidepressant agents are known to increase monoamine levels by a variety of mechanisms. However, there are anomalies which indicate that there is more to the story than an increase in monoamine levels. For example, **amphetamine** and **cocaine** are agents that increase noradrenaline and serotonin transmission, but are ineffective as antidepressants. There is also evidence that a wide range of endogenous hormones and neurotransmitters play a role in depression; substance P, corticotrophin-releasing factor, arginine, vasopressin, neuropeptide Y, melanin-concentrating hormone, acetylcholine, glutamic acid, gamma-aminobutyric acid, glucocorticoids, cytokines, enkephalins, and anandamide. Nevertheless, most clinically useful agents in use today are responsible for raising monoamine levels.

CS7.3 Current antidepressant agents

First-generation antidepressants were introduced about 50 years ago, and include the **monoamine oxidase inhibitors** (MAOIs), which are discussed in section 23.12.5, and the **tricyclic antidepressants** (TCAs), which are described in section 23.12.4. Unfortunately, these drugs have low target selectivity and many side effects.

Second-generation antidepressants were introduced in the 1980s and are represented by agents known as **selective serotonin reuptake inhibitors** (SSRIs) (Box

10.1). These represented a major step forward in treatment because they are more selective and have fewer side effects. However, like the TCAs and MAOIs, they have a slow onset of action and it can take 2–6 weeks before patients feel any benefit. Another problem with their use is their negative effect on libido.

Third-generation antidepressant agents include **selective noradrenaline reuptake inhibitors** (section 23.12.4), and dual action **serotonin and noradrenaline reuptake inhibitors** (SNRIs) (section 23.12.4).

CS7.4 **Current areas of research**

Currently, there is research into novel agents designed to interact with the following targets:

- transport proteins for dopamine, serotonin, and noradrenaline;
- adrenergic receptors such as the α_2-adrenoceptor;
- serotonin receptors such as the 5-HT$_{1A}$, 5-HT$_{2A}$, 5-HT$_{2C}$, 5-HT$_6$, and 5-HT$_7$ receptors.

Dual action agents that act on two of the above targets are of particular interest. Examples include agents that:

- block the reuptake of both noradrenaline and serotonin;
- block α_2-adrenoceptors (section 23.11.2) and activate 5-HT receptors;
- block serotonin reuptake and are antagonists for the 5-HT$_{1A}$ receptor. The 5-HT$_{1A}$ receptor is an autoreceptor present on the presynaptic neurons that release serotonin. When activated, this receptor inhibits the release of serotonin from the neuron, and so an antagonist should counteract this effect;
- block serotonin reuptake and act as antagonists for the 5-HT$_{2A}$ receptor. This receptor is responsible for the sexual dysfunction side effect associated with SSRIs.

In this Case study, we shall look at a research project aimed at discovering antagonists for the 5-HT$_7$ receptor.

CS7.5 **Antagonists for the 5-HT$_7$ receptor**

There are seven main types of serotonin receptors (5-HT$_1$–5-HT$_7$) and several subtypes of these. The 5-HT$_7$ receptor is the most recent serotonin receptor to be discovered and appears to play an important role in psychiatric disorders, such as depression. It has been shown that antagonists of this receptor have an antidepressant activity in animal studies, although the mechanism by which this takes place is unclear. At first sight, it may seem odd that a serotonin antagonist should have an antidepressant activity, as antidepressant activity is normally associated with increased serotonin levels and increased activation of serotonin receptors. However, it should be borne in mind that different receptors for the same neurotransmitter serve different purposes and some act as autoreceptors to provide a negative feedback control for neurotransmitter release. For example, the α_2-adrenergic receptor is a presynaptic autoreceptor which has the effect of inhibiting noradrenaline release (sections 23.6.3 and 23.11.2). It is conceivable that activation of 5-HT$_7$ receptors might lead to a drop in serotonin levels by a similar manner. Therefore, an antagonist that is selective for this receptor over other serotonin receptors could be advantageous.

Workers at SmithKline Beecham carried out high-throughput screening of their compound bank for structures having affinity for the 5-HT$_7$ receptor and identified the sulphonamide (I; Fig. CS7.1) as a lead compound with slight selectivity. The structure has two asymmetric centres and was tested as a mixture of the two possible diastereomers. As there are two enantiomers for each diastereomer, this means that there are four possible stereoisomers (R,R; S,S; R,S; and S,R). All four stereoisomers

FIGURE CS7.1 Identification of a lead compound.

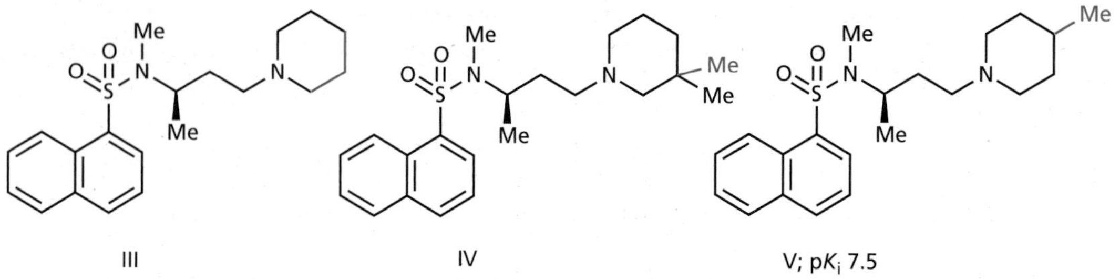

FIGURE CS7.2 Methods of removing the asymmetric centre in the piperidine ring.

were tested separately and the *R,R* isomer (II) was found to have the best affinity.

The affinity for the *R,S*-diastereomer was still 6.2, which indicated that the stereochemistry of the asymmetric centre in the piperidine ring was not essential. Therefore, it was decided to remove this asymmetric centre as this would simplify the synthesis of analogues (*simplification*; section 13.3.8) and avoid the need to separate and purify diastereomers for each analogue produced. The obvious way of removing the asymmetric centre was to remove the methyl substituent, but the resulting structure III (Fig. CS7.2) had no affinity. This indicated the importance of the methyl group, which suggests that it might be interacting with a hydrophobic pocket in the binding site. Another method of removing the asymmetric centre was to add a second methyl substituent at the same position. However, the resulting structure IV had no affinity either, implying that the second methyl group might be bad for steric reasons. The problem was eventually solved by shifting the methyl group to position 4 of the piperidine ring, which not only removed the asymmetric centre but improved affinity (*simplification* and *group shift*; sections 13.3.8 and 14.2.6).

A conformational analysis of the flexible chain linking the two ring systems was now carried out (*conformational analysis*; section 17.8). This revealed that all the bonds are relatively free to rotate apart from the bonds

shown in bold (Fig. CS7.3). Concentrating on conformations involving these bonds, an energy minimum was found when the two methyl substituents are gauche with respect to each other, corresponding to a dihedral angle of 60°.

As the gauche conformation is an energy minimum, it represents a stable conformation and the molecule will spend a greater amount of time in this conformation than in others. Therefore, there is a possibility that this might correspond to the active conformation (*active conformation*; section 13.2). If this is the case, locking the molecule into this conformation should increase binding affinity (*rigidification*; section 13.3.9).

Rigidification can be carried out by introducing a ring that incorporates both methyl groups and the connecting bonds, for example structures VI and VII where the ring is six-membered and five-membered respectively (Figs. CS7.4 and CS7.5). Before synthesizing these structures, docking experiments were carried out using a 5-HT$_7$ receptor homology model (*docking*; section 17.12, *homology models*; section 17.14.1). These experiments predicted that the *R*-enantiomer of structure VI would have greater binding affinity than the *S*-enantiomer. Both enantiomers were duly synthesized and the *R*-enantiomer had a 25-fold better affinity as predicted. It also had slightly better affinity than structure V. Structure VII containing the five-membered ring was then synthesized

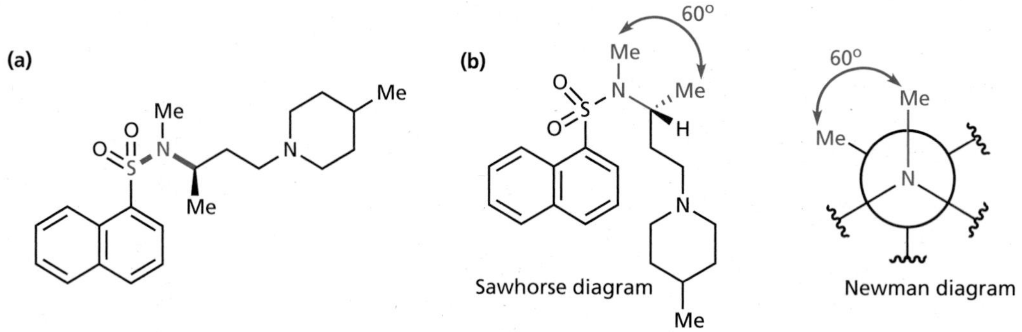

FIGURE CS7.3 Conformational analysis shows that (a) the bonds shown in blue have restricted rotation, and (b) there is a stable conformation having a torsion angle of 60°.

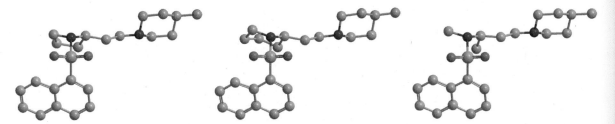

FIGURE CS7.4 Three-dimensional representations of the lead compound and rigidified analogues.

(*ring contraction*; section 13.3.4), which resulted in an increase in affinity.

The naphthalene ring system is not essential for activity and it was possible to replace it with a single aromatic ring to give structure VIII (*simplification* or *ring variation*; sections 13.3.8 and 13.3.5). A number of different aromatic substituents were tested at different positions (*variation of aromatic substituents*; section 13.3.1.2) and it was found that a phenolic group was best for activity, giving SB 269970. It is possible that this group is participating in a hydrogen bonding interaction with the binding site, as a methoxy substituent has less affinity. This was confirmed by docking the structure into the model binding site and identifying a possible hydrogen bonding interaction.

The selectivity of SB 269970 was tested against various receptors, and it was found to have greater than 250-fold selectivity over 13 other receptors, as well as a 50-fold selectivity over 5-HT$_{5A}$. Further testing with a

commercial screening package (Cerep) showed that it had a 100-fold selectivity over a total of 50 other receptors, enzymes or ion channels. The compound has been shown to be an inverse agonist (section 8.5).

Because SB 269970 contains a phenolic group, it is prone to phase II conjugation reactions (section 11.5.5), which leads to rapid excretion. The phenolic group is involved in an important binding interaction and so, rather than removing it entirely, it was replaced with a metabolically stable bio-isostere (*bio-isosteres*; sections 13.3.7 and 14.1.6) that would still be capable of forming the important hydrogen bond. This was achieved by fusing a five-membered heterocycle onto the aromatic ring such that an NH group would be placed at the same position as the original phenol. Various heterocycles were tried with an indole ring system being the best (structure IX, Fig. CS7.5).

Unfortunately, the compound was rapidly cleared from the blood and had zero bioavailability when tested

FIGURE CS7.5 The design process from lead compound to structure IX.

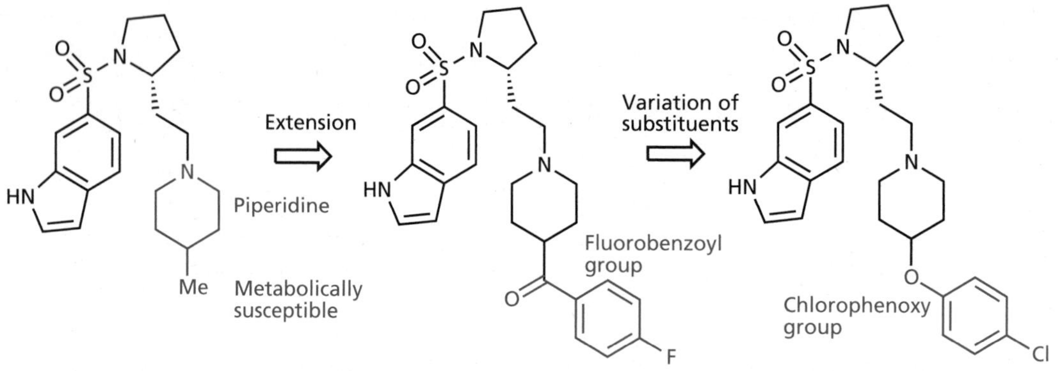

FIGURE CS7.6 Development of SB 656104.

in rats, and so attention now turned to the methyl substituent on the piperidine ring, as this was also likely to be susceptible to metabolism (section 11.5.2). Molecular modelling showed that it might be possible to replace the methyl group with a substituent that would extend into a large hydrophobic pocket close by in the binding site. It was decided to try a substituent containing an aromatic ring. This would not only remove the susceptible methyl group, but offer the possibility of increased binding with the hydrophobic pocket (*extension*; section 13.3.2). Various substituents were tried and a fluorobenzoyl

substituent was one of the best (structure X; Fig. CS7.6). Unfortunately, structure X had increased affinity for the α_{1B}-adrenoceptor as well as the 5HT$_7$ receptor. Variation of the substituents (section 13.3.1.2) at either end of the aromatic ring showed that the chlorophenoxy group had much better selectivity (SB 656104; Fig. CS7.6). Although binding affinity for the 5HT$_7$ receptor had dropped, this structure had the best balance of properties. Crucially it lasted far longer than SB 269970 in the blood supply and had an oral bioavailability of 16%. This compound was taken forward as the basis for further studies.

FURTHER READING

Forbes, I. T., et al. (2002) SB-656104-A: a novel 5-HT$_7$ receptor antagonist with improved in vivo properties. *Bioorganic and Medicinal Chemistry Letters*, **12**, 3341–4.

Leopoldo, M. (2004) Serotonin7 receptors (5-HT$_7$Rs) and their ligands. *Current Medicinal Chemistry*, **11**, 629–61.

Pacher, P., and Kecskemeti, V. (2004) Trends in the development of new antidepressants. Is there a light at the end of the tunnel? *Current Medicinal Chemistry*, **11**, 925–43.

Stromgaard, K. (2009) Recognising antidepressants. *Chemistry World*, July, 33.

The design and development of aliskiren

CS8.1 Introduction

Aliskiren is an antihypertensive agent that inhibits **renin**—the enzyme that catalyses the initial reaction at the start of the renin–angiotensin–aldosterone system (RAAS; section 26.3.2). Inhibiting renin has the effect of blocking the RAAS cascade and preventing the production of angiotensin II—a potent vasoconstrictor that increases blood pressure. The development of aliskiren involved a substantial research effort spanning 35 years (1972–2007), and demonstrates many of the strategies and methods involved in designing an effective therapeutic agent from a peptide lead compound.

CS8.2 Reaction catalysed by renin

Renin is an aspartyl protease enzyme that catalyses the hydrolysis of a protein substrate called **angiotensinogen**. The specific peptide bond that is hydrolysed is situated between a leucine and a valine residue, and results in the release of a decapeptide called **angiotensin I** from the *N*-terminus of the protein substrate (Fig. CS8.1).

The mechanism of action by which renin cleaves angiotensinogen involves two aspartyl residues and a bridging water molecule (Fig. CS8.2). In the first stage of the mechanism, a tetrahedral intermediate is formed which contains two hydroxyl groups, one of which is supplied by the water molecule. In the second part of the mechanism, the carbonyl group is reformed and the peptide link is split. In both parts of the mechanism, the aspartate residues act as acid–base catalysts. One of the aspartate residues acts as a base to take up a proton, while the other acts as an acid to provide a proton.

CS8.3 From lead compound to peptide inhibitors

It is common practice to consider an enzyme's substrate as the lead compound when designing an enzyme inhibitor

H2N—Asp-Arg-Val-Tyr-Ile-His-Pro-Phe-His-Leu│Val-Ile-〜〜〜   Angiotensinogen

Renin ↓

H2N—Asp-Arg-Val-Tyr-Ile-His-Pro-Phe-His-Leu-CO2H   Angiotensin I

FIGURE CS8.1 Action of renin on angiotensinogen.

FIGURE CS8.2 Mechanism of renin-catalysed hydrolysis.

(section 12.4.5.2). This is because the substrate is recognized by the enzyme and can bind to the active site, prior to the enzyme-catalysed reaction. In this case, the substrate is a protein, and so only a small section is actually bound to renin. In fact, it has been shown that seven amino acid residues are involved in the substrate's binding interactions with renin, and that those residues are on either side of the susceptible peptide bond. Therefore, a heptapeptide having the same amino acid sequence as that portion of the protein would be a sensible starting point for drug design (Fig. CS8.3).

The trick is then to design a similar molecule that will bind to the active site, but will not undergo the enzyme-catalysed reaction. This can be done by replacing the susceptible peptide bond between Leu and Val with a functional group that cannot be hydrolysed. A particularly effective strategy is to introduce a group that acts as a transition-state isostere (section 13.3.7) and mimics the high-energy transition state involved in the first stage of the enzyme-catalysed reaction. Compounds containing a transition-state isostere are known as **transition-state analogues or inhibitors**, and should bind more strongly than either the substrate or the product. In this reaction, the transition state resembles the tetrahedral intermediate, rather than the planar peptide bond of the substrate. Therefore, the transition-state isostere should also be tetrahedral.

The hydroxyethylene group (Fig. CS8.4) has the required tetrahedral geometry, as well as one of the two hydroxyl groups required for good binding. It is also stable to hydrolysis, because there is no leaving group present.

A number of peptide structures were synthesized by Ciba-Geigy that included this transition-state isostere, leading to the discovery of an effective inhibitor called **CGP29287** in 1982 (Fig. CS8.5). Note that the left-hand portion of the structure contains the side chains corresponding to the four amino acid residues—Pro, Phe, His,

Cleavage site

HN—Pro-Phe-His-Leu⋮Val-Ile-His-OH

FIGURE CS8.3 The heptapeptide section of angiotensinogen that binds to renin.

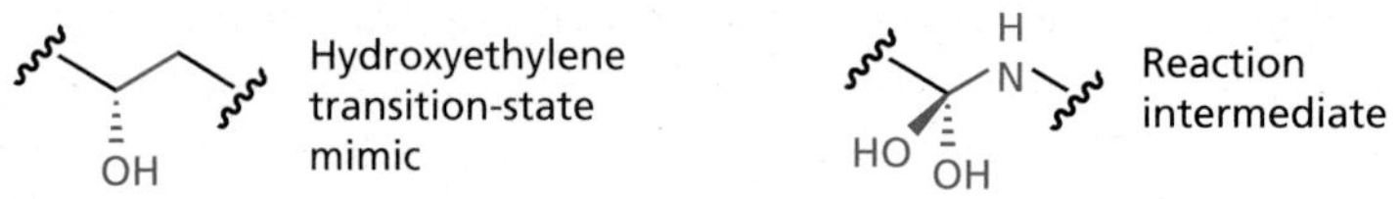

FIGURE CS8.4 The hydroxyethylene transition-state isostere.

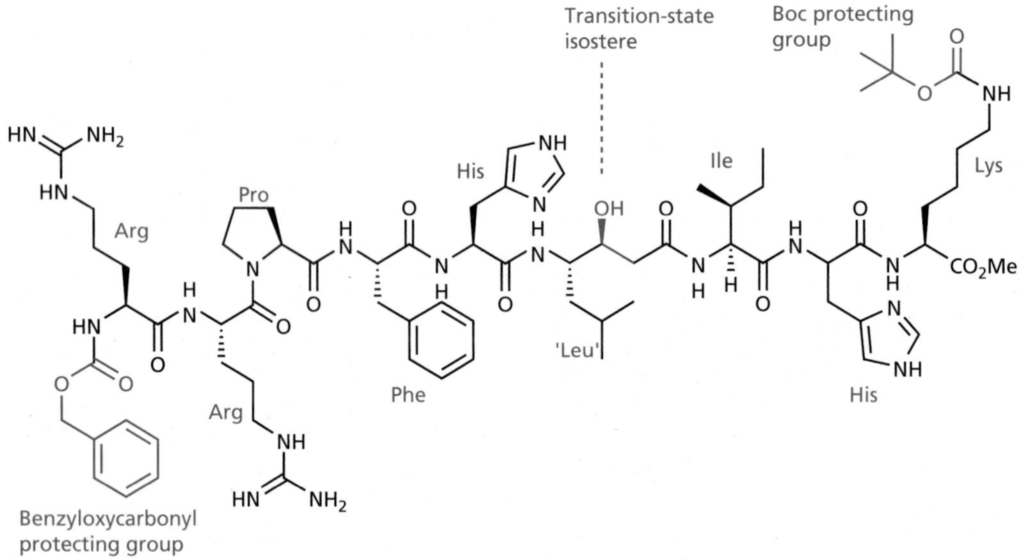

FIGURE CS8.5 CGP29287 (IC$_{50}$ = 7 nM). 'Leu' indicates that the side chain of Leu is present.

and Leu. These are present in the substrate and fit the four binding subsites S1–S4. On the right-hand side, the amino acids Ile and His are retained for subsites S2′ and S3′. The valine side chain that would normally fit the S1′ subsite has been omitted because of the inclusion of the transition-state isostere.

CS8.4 Peptidomimetic strategies

Unfortunately, none of the peptide inhibitors that were discovered were suitable for clinical studies due to poor oral absorption and metabolic susceptibility. Therefore, a number of peptidomimetic strategies (section 14.9) were carried out to produce smaller structures with a lower molecular weight and less peptide character. For example, shorter peptides were studied by removing amino acid residues from both the *N* and *C* termini of CGP29287. This led to the discovery of **CGP38560** (Fig. CS8.6), which retains the transition-state isostere and one of the histidine residues present in CGP29287. It also retains the benzyl group that mimics the side chain of phenylalanine in CGP29287. On the right-hand side, a peptide bond has been removed which allows the side chain for valine to be incorporated. This can fit into the S1′ binding pocket.

The structure now contains three peptide bonds rather than eight. It also contains an unnatural side chain in the form of a cyclohexane ring—another strategy commonly used in peptidomimetic design. CGP38560 proved even more potent than the original first-generation drugs, and went forward to clinical trials. However, it, too, suffered from poor oral absorption and rapid excretion in bile. It was concluded that the molecular weight of the structure was still too high for good oral absorption.

CS8.5 Design of non-peptide inhibitors

At this stage, it was decided that a radical change of approach was needed in order to design a structure that was completely non-peptide in nature, but would still mimic the binding interactions of CGP38560. This required identifying the active conformation and the crucial binding interactions involved when CGP38560 binds to renin.

Ideally, this would have involved crystallizing renin with CGP38560 bound to the active site, then studying the enzyme–ligand complex by X-ray crystallography. However, it was not possible to crystallize renin at the time, and so a homology model of the enzyme was built (see section 17.14.1). Docking experiments were then carried out which demonstrated that the ligand could bind in an extended conformation with the side chains fitting into six of the enzyme's binding subsites (Fig. CS8.7). Important hydrogen bonding interactions were also identified between the ligand's backbone and the active site, including an interaction between the transition-state isostere and the catalytic aspartate residues.

The modelling studies also indicated that the binding subsites S1 and S3 adjoined each other and formed an extended hydrophobic binding region. This suggested that it might be possible to design a smaller molecule that linked the groups occupying the S1 and S3 subsites. The rationale was that linking the two groups would increase hydrophobic interactions with the S1 and S3 subsites, which would compensate for the binding interactions lost by removing groups elsewhere in the molecule. In particular, it was hoped to remove the groups occupying

FIGURE CS8.6 CGP38560 ($IC_{50} = 0.7$ nM). The labels 'AA' indicate that the side chains of these amino acids are present rather than the intact amino acid residues.

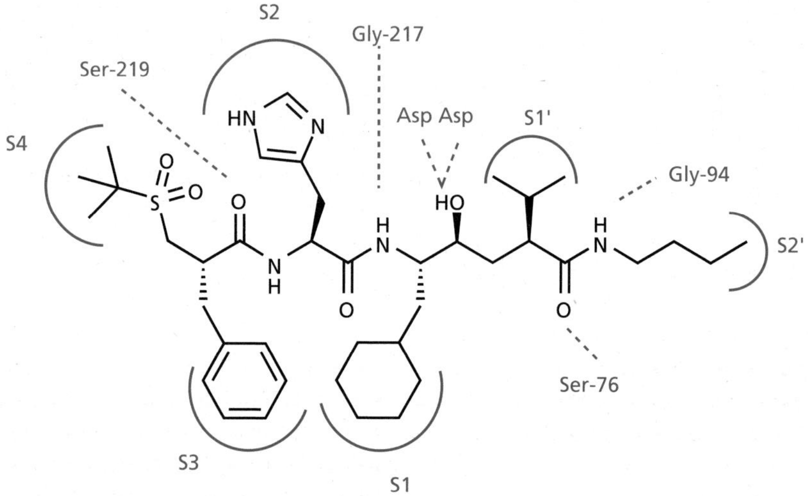

FIGURE CS8.7 Binding interactions proposed for CGP38560 as a result of docking experiments.

subsites S2 and S4, along with the peptide chain holding them in place. This would significantly reduce the molecular weight and peptide character of the resulting structure. In order to design such a molecule, the key binding fragments of CGP38560 were identified (Fig. CS8.8).

The goal was now to link up these fragments with a non-peptide scaffold that would ideally include a hydrogen bond acceptor capable of interacting with Ser-219. In addition, the chain linking the P1 and P3 binding groups should be hydrophobic to allow additional interactions with that region of the binding site.

A large number of structures with different scaffolds were studied by docking them into the model binding site. Only synthetically feasible structures capable of adopting a stable conformation when docked were considered for further investigation. Moreover, it was important that the

active conformation should be reasonably close in stability to the most stable conformation. Otherwise, the relative population of the active conformation would prove too low for good activity. For that reason, it was determined that there should be no more than 1 kcal mol^{-1} difference between the steric energies of the active conformation and the most stable conformation.

One of the early structures investigated was a truncated structure bearing the cyclohexyl ring as a P1 group (Fig. CS8.9). Not surprisingly, there was a large drop in activity from 0.7 nM to 30 000 nM. This is because the cyclohexyl group can only occupy the S1 binding site. Modelling showed that a propyl linker connecting the cyclohexyl group with an aromatic ring would allow these groups to occupy the S1 and S3 pockets respectively. However, the structure that was synthesized contained a tertiary butyl

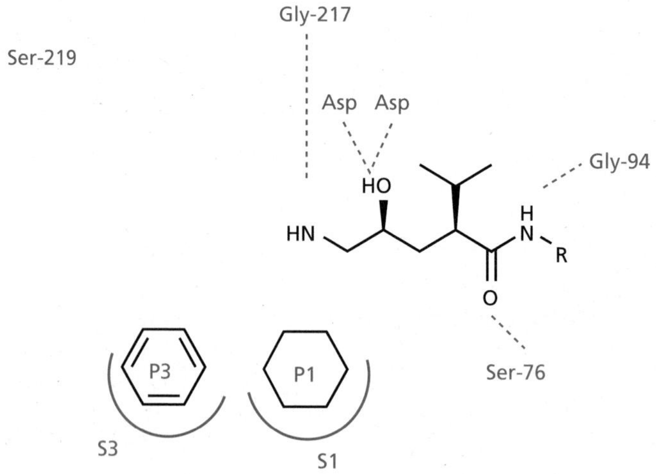

FIGURE CS8.8 Key binding fragments of CGP38560. S indicates a binding subsite. P indicates the group occupying a binding subsite.

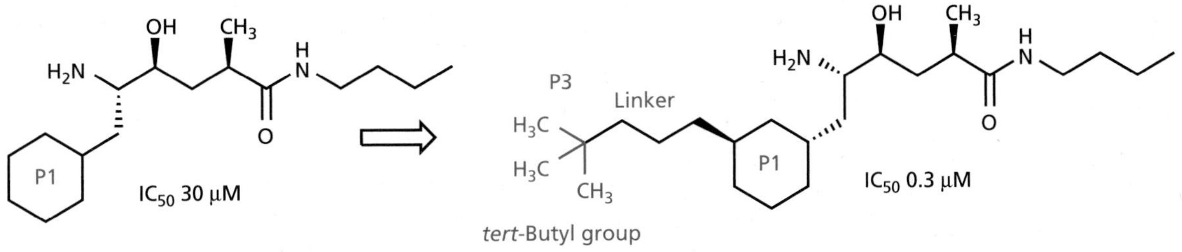

FIGURE CS8.9 Creating a new scaffold.

group instead of the aromatic ring, since the butyl group was expected to fill the S3 pocket more efficiently. A 100-fold increase in activity was observed, which justified the approach.

CS8.6 Optimization of the structure

Unfortunately, two extra chiral centres were now present on the ring, which involved the added complication of having to test four different stereoisomers. Therefore, it was decided to replace the cyclohexane ring with alternative hydrophobic substituents.

An alternative series of compounds (Fig. CS8.10) was now studied where a methyl group was added as the P1 group, rather than the cyclohexane ring. A rigid linker was introduced that included an aromatic ring and allowed enhanced hydrophobic interactions with the binding site. At the end of the linker, a phenyl or *tert*-butyl substituent was included as the P3 group. Altering the P1 methyl group to a larger isopropyl group led to a 20-fold increase in van der Waals interactions, and a dramatic increase in potency was observed when an extended ester substituent was added to the aromatic ring in the linker. Originally, it was thought that the ester substituent was forming

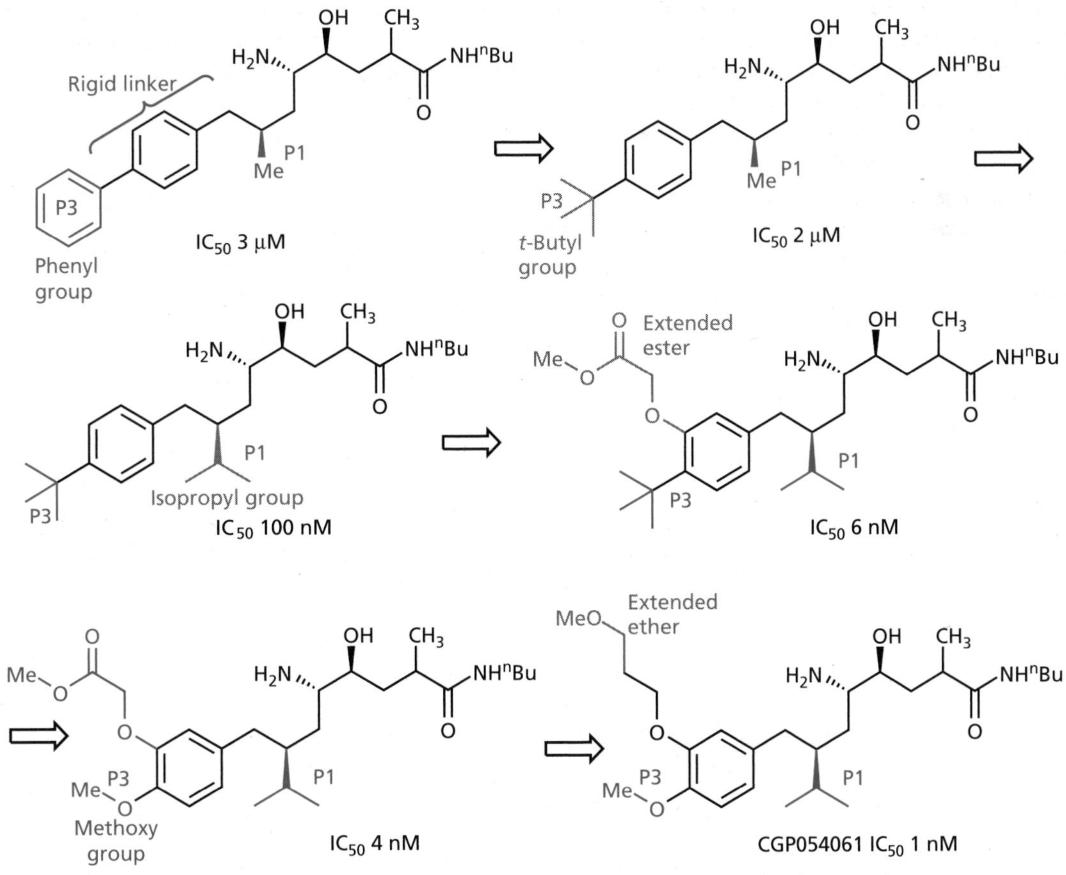

FIGURE CS8.10 Optimizing P1 and P3 (^{n}Bu = $(CH_2)_3CH_3$).

a hydrogen bond with Ser-219. However, later studies demonstrated that the substituent fits into a small pocket (labelled the S3^{sp} pocket) that is not revealed when the normal substrate binds.

At this stage it was observed that the potency of these compounds was decreased in the presence of plasma. This was attributed to the hydrophobic nature of the compounds, and so the *tert*-butyl group was replaced with a more polar methoxy group (Fig. CS8.10). Further modification of the extended ester side chain led to an extended ether with improved activity regardless of whether plasma was present or not. This side chain proved the optimum binding group for the S3^{sp} pocket, which not only resulted in increased potency, but also enhanced selectivity for renin with respect to other aspartyl proteases.

An X-ray crystal structure confirmed that the extended ether side chain fits the narrow S3^{sp} sub pocket and is not binding to Ser-219 (Fig. CS8.11). The terminal ether oxygen can form a hydrogen bond to a tyrosine residue. However, hydrophobic interactions between the ether chain and the S3^{sp} sub pocket are more important.

The P1 and P3 binding groups are positioned in their respective binding pockets. However, the hydroxyl group of the transition-state isostere is shifted slightly from what had been expected, which means that it forms a hydrogen bond to Asp-32, but not Asp-215. The basic amino group forms a hydrogen bond with Gly-217, but the P1' methyl group does not fit the S1' subsite as well as it could do.

Further modifications to the 'right-hand' part of the chain were now carried out to optimize potency and pharmacokinetic properties. The P1' methyl group was altered to an isopropyl group in order to fill the hydrophobic S1' pocket more efficiently. This represents a return to the original side chain of valine. The P2' butyl chain was also modified by adding two methyl substituents and a primary amide group, resulting in the structure of **aliskiren** (Fig. CS8.12).

Aliskiren is a potent inhibitor with a high selectivity for renin, showing a 10 000-fold lower affinity for other aspartyl peptidases. It is also significantly more hydrophilic (log P 2.45) than the peptide-like renin inhibitors, and has good water solubility. This is one factor behind its better oral bioavailability. An X-ray crystal structure demonstrated that aliskiren binds to the target enzyme as predicted. A water molecule acts as a hydrogen bonding bridge between the aryl ether oxygen and Ser-219, but this does not appear to make a large contribution to the binding affinity. The hydroxyl group of the transition-state isostere is able to interact with both aspartates, in contrast to the earlier structure. The primary amino group can hydrogen bond to both Gly-217 and Asp-215, while the P1' isopropyl group fits snugly into the S1' subsite. The terminal primary amide forms hydrogen bonds to a number of amino acids in the S2' pocket, while the methyl substituents form van der Waals interactions that are not available to the original *n*-butyl chain.

Aliskiren was approved by the US FDA in 2007 and remains the only renin inhibitor on the market.

FIGURE CS8.11 Expected binding interactions versus actual (^{n}Bu = $(CH_2)_3CH_3$).

FIGURE CS8.12 Aliskiren ($IC_{50} = 0.6$ nM).

FURTHER READING

Cohen, N. C. (2007) Structure-based drug design and the discovery of aliskiren (Tekturna): perseverance and creativity to overcome a R&D pipeline challenge. *Chemical Biology & Drug Design*, **6**, 1589–94.

Goschke, R., Cohen, N. C., et al. (1997) Design and synthesis of novel 2,7-dialkyl substituted 5(*S*)-amino-4(*S*) hydroxy-8-phenyl-octanecarboxamides as *in vitro* potent peptidomimetic inhibitors of human renin. *Bioorganic and Medicinal Chemistry Letters*, **7**, 2735–40.

Rahuel, J., Rasetti, V., et al. (2000) Structure-based drug design: the discovery of novel nonpeptide orally active inhibitors of human renin. *Chemistry and Biology*, **7**, 493–504.

Rasetti, V., Cohen, N. C., et al. (1996) Bioactive hydroxyethylene dipeptide isosteres with hydrophobic (P3-P1)-moieties. A novel strategy towards small non-peptide renin inhibitors. *Bioorganic and Medicinal Chemistry Letters*, **6**, 1589–94.

CASE STUDY 9
Factor Xa inhibitors

CS9.1 Introduction

Factor Xa is a key enzyme involved in thrombosis and clot formation (Fig. CS9.1). Therefore, factor Xa inhibitors are potential anticoagulants for the treatment of deep-vein thrombosis and pulmonary embolism. In recent years, three such anticoagulants have been approved—**apixaban**, **rivaroxaban**, and **edoxaban**. These drugs are the first factor Xa inhibitors to reach the market, and this Case study describes the design process that led to their discovery.

CS9.2 The target

Factor Xa is a serine protease enzyme that converts **prothrombin** to **thrombin** by catalysing the hydrolysis of two peptide bonds in the substrate. The active site contains a catalytic triad of Ser-195, His-57, and Asp-102, which play a key role in the catalytic reaction. The mechanism by which this takes place is the same as the one described for chymotrypsin in section 3.5.3. Serine provides the nucleophilic group that reacts with the susceptible peptide bonds, histidine acts as an acid–base catalyst, and aspartate activates and orientates the histidine residue.

The enzyme also contains binding subsites that accept the side chains of amino acid residues on either side of the susceptible peptide bonds. The S1 subsite is a deep, hydrophobic cleft containing Asp-189 and Tyr-228, and is important for both binding affinity and selectivity (Fig. CS9.2). Strong binding interactions are possible between the negatively charged carboxylate group of Asp-189 and an arginine residue in the substrate (Fig. CS9.3). Selectivity arises from the fact that the amino acid residues at positions 190–192 of the subsite vary for different protease enzymes. For example, factor Xa has an alanine residue at position 190, whereas the digestive enzyme trypsin has a larger serine residue. Consequently, the S1 pocket in trypsin is significantly smaller than that of factor Xa.

Because of the strong binding interactions and selectivity provided by the S1 subsite, an early priority in the design of factor Xa inhibitors was to include a basic group that could interact with the aspartate residue in that subsite.

Early research also demonstrated that potent inhibitors were obtained if they could interact with both the S1 and S4 subsites. The S4 subsite is a narrow, hydrophobic channel that contains aromatic residues—Tyr-99, Phe-174, and Trp-215 (Fig. CS9.2). These residues can form good interactions with hydrophobic groups, but they can also interact with positively charged groups through pi-cation interactions (see also section 1.3.4). The S4 pocket is also good for selectivity

FIGURE CS9.1 Key features involved in clot formation.

FIGURE CS9.2 Key amino acid residues in the S1 and S4 subsites of factor Xa.

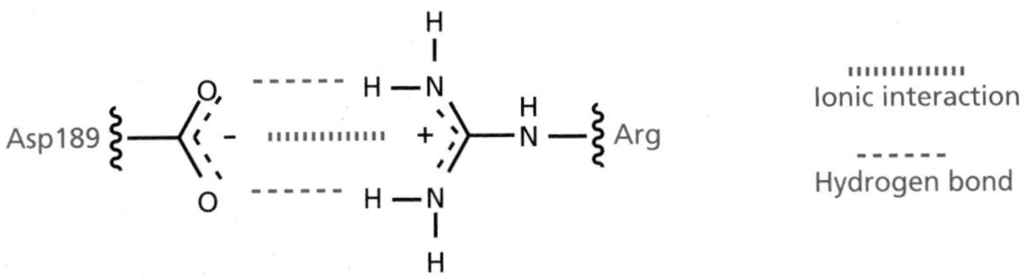

FIGURE CS9.3 The interaction between Asp-189 and a basic arginine residue.

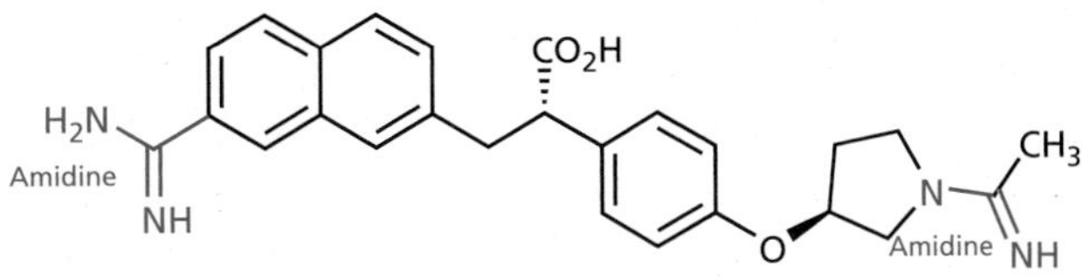

DX-9065a (K_i 41 nM)

FIGURE CS9.4 Example of an early dibasic factor Xa inhibitor.

because of significant differences in the amino acids that are present in different protease enzymes. For example, in thrombin, the amino acid residues at positions 99, 174, and 215 are leucine, isoleucine, and tryptophan. Therefore, the S4 subsite in thrombin lacks two of the aromatic residues that are present in factor Xa.

CS9.3 General strategies in the design of factor Xa inhibitors

Factor Xa inhibitors were initially designed to have a central scaffold containing two substituents capable of interacting with the S1 and S4 subsites. The relative position of these substituents on the scaffold was important since that determined whether the scaffold could direct each group towards its respective binding subsite. The substituents themselves contained basic functional groups that are ionized under physiological conditions and can form strong binding interactions with each subsite. Thus, the group fitting the S1 pocket forms ionic and hydrogen bonds with Asp-189, while the group fitting the S4 pocket forms pi-cation interactions with the aromatic residues there.

An amidine functional group is one of the most basic functional groups in organic chemistry, and so early inhibitors were designed to include two of these groups—one for each subsite. This resulted in a number of very potent and selective inhibitors, such as **DX-9065a** (Fig. CS9.4). This compound entered clinical trials, but the presence of two highly polar groups meant that it

had poor oral absorption—a story that was repeated with other dibasic factor Xa inhibitors. Real progress in designing clinically effective agents only became possible when it was discovered that potent inhibitors could still be obtained using hydrophobic binding groups for the S1 and S4 subsites. The development of apixaban illustrates the changing emphasis of these strategies.

CS9.4 Apixaban: from hit structure to lead compound

Apixaban was developed by Bristol-Myers Squibb as an anticoagulant and antithrombotic agent. The company started off by screening chemical libraries in order to find a compound that could bind to the target protein. This can often be a hit-and-miss affair. However, the odds of success can be improved by factoring in what is known about the target binding site and the substrate, and then choosing chemical libraries that contain the most relevant structures. In this case, it was decided to screen a library of compounds that had been prepared for a previous project and which included structures that were intended to mimic the tripeptide Arg-Gly-Asp. It was thought that this library might prove productive because a similar tripeptide sequence (Glu-Gly-Asp) is present in prothrombin. Since this sequence contributes to the binding of prothrombin to factor Xa, some of the library structures might also be 'recognized'. Moreover, the library structures included a strong basic group that could mimic the side chain of arginine. Such a group could bind strongly to either the S1 or S4 subsite and would further increase the chances of finding a hit compound. Once the screening was carried out, a hit structure was, indeed, identified with weak micromolar activity (Fig. CS9.5).

The next task was to increase binding affinity. It was already known that the presence of two basic groups was good for potency if they could fit into the S1 and S4 subsites of the enzyme, and so a second benzamidine group was added to replace the aspartate group and the tetrahydroisoquinoline ring of the hit structure. This could be viewed as an example of an extension strategy where an additional binding group is added to the structure (section 13.3.2). The position of the amidine substituents on each of the benzamidine groups was now varied to find the optimum positions (section 13.3.1.2), resulting in the identification of a compound with increased activity (structure IV; Fig. CS9.5). A chain contraction strategy (section 13.3.3) was now carried out and involved the removal of a methylene group to see whether the two binding groups were the ideal distance apart. This resulted in a five-fold increase in activity for structure V.

Molecular modelling studies on structure V indicated that the *meta*-benzamidine group could fit the S1 subsite to interact with Asp-189, while the *para*-benzamidine ring fitted the S4 subsite to form pi-cation interactions

with Phe-174 and Tyr-99. A hydrogen bond was also possible between the amide carbonyl group and Gly-218.

The modelling studies also revealed that there was space in the binding site for the addition of an ester substituent that would provide an additional hydrogen bonding interaction—an example of how molecular modelling can help in the design of more potent compounds, and another example of the extension strategy (section 13.3.2). This led to the discovery of structure VI and a three-fold increase in activity. Structure VI was now adopted as a new lead structure for further modifications.

CS9.5 Apixaban: from lead compound to final structure

The presence of two basic amidine groups proved excellent for binding and *in vitro* activity, but was detrimental for *in vivo* activity due to the poor oral absorption associated with a molecule containing two ionized groups. Therefore, it was decided to modify the structure such that the group fitting the S4 subsite would be hydrophobic rather than ionic. This made sense since van der Waals

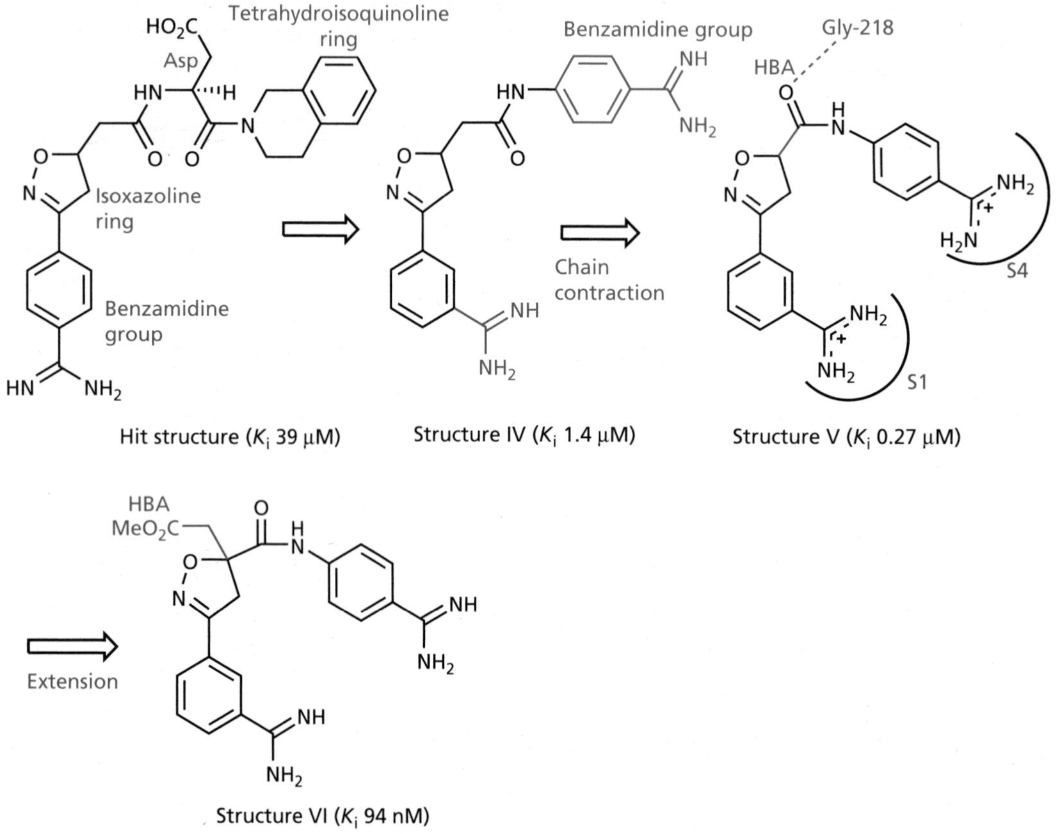

FIGURE CS9.5 From hit structure to lead compound. All the structures were synthesized and tested as racemates.

interactions would be possible with the three aromatic residues present in that subsite (Fig. CS9.2). Consequently, the P4 amidine group was replaced with an aromatic ring to form a biphenyl ring system (Structure VII; Fig. CS9.6). This ring system was deliberately chosen because earlier work on peptide-based inhibitors had demonstrated that a biphenyl ring could occupy the region of the binding site between the S1 and S4 subsites. Although activity dropped slightly for structure VII, it could be restored by adding a substituent to the terminal aromatic ring. A number of different substituents were tried out at different positions (section 13.3.1), demonstrating that a substituent at the *ortho* position was good for activity if it could act as a hydrogen bond acceptor. A sulphonamide substituent was particularly effective and resulted in a substantial increase in activity for **SF303**. Further molecular modelling studies indicated that the enhanced activity was likely to be due to a hydrogen bond between one of the sulphonamide oxygen atoms and the phenol OH of Tyr-99. It was also observed that the terminal ring of the biphenyl group could form edge-to-face interactions with Trp-215. SF303 was the first monobasic factor Xa inhibitor to be discovered with good potency and selectivity (Fig. CS9.6). It also demonstrated that a polar, ionized group was not essential for good binding with the S4 subsite.

At this stage, it was decided that the methyl ester in SF303 was a potential problem since it could be hydrolysed by esterases to form a carboxylic acid. Therefore,

a number of different substituents were tested including a tetrazole ring—a common bio-isostere for a carboxylic acid. Although many of these analogues showed improved activity, more fruitful results were achieved by a completely different strategy. This involved replacing the isoxazoline ring with an aromatic isoxazole ring—an example of a ring variation strategy (section 13.3.5). This resulted in **SA862**. Although activity dropped due to the loss of the methyl ester, SA862 had the advantage of lacking a chiral centre, thus avoiding the complication of having to test different enantiomers. Various substituents (R) were now introduced to see whether they would mimic the interactions of the ester substituent in SF303. However, the unsubstituted structure proved to have the best activity.

Further ring variations led to the highly potent compound **SN429**, which contains a pyrazole ring. This demonstrated that the oxygen present in the isoxazole ring was not important for activity, and that the five-membered ring served mainly as a scaffold to orientate its two major substituents into the S1 and S4 pockets. The pyrazole ring also provided the opportunity to introduce a substituent—in this case a methyl group—at a different position of the molecule.

Molecular modelling demonstrated that the P1 benzamidine substituent still formed a bidentate interaction with the aspartate residue in the S1 pocket, and several hydrogen bonding groups were also identified

Structure VI (K_i 94 nM)

Structure VII (X = H; K_i 220 nM)
SF303 (X = SO$_2$NH$_2$; K_i 6.3 nM)

SA862 (R = H; K_i 0.15 nM)

SN429 K_i 0.013 nM

FIGURE CS9.6 Development of SN429.

(Fig. CS9.6). The two rings in the biphenyl moiety were orthogonal to each other allowing the terminal ring to form interactions with a tryptophan residue. This also orientated the sulphonamide group such that it could interact with Tyr-99. Finally, the methyl substituent could form van der Waals interactions with a hydrophobic pocket.

SN429 was extremely potent *in vitro*, but had low *in vivo* activity due to poor oral absorption and a short half-life. It was now clear that the remaining benzamidine group was proving bad for pharmacokinetic properties. For the first time, consideration was given to replacing the amidine group with a non-basic group. Previously, this would have been considered unthinkable due to the strong interactions that the amidine group forms with the S1 subsite. However, picomolar levels of *in vitro* activity had now been achieved with SN429, and so it might be possible to sacrifice some of that activity in order to improve pharmacokinetic properties. Therefore, the amidine group was replaced with a *para*-methoxy group to give structure VIII (Fig. CS9.7). *In vitro* activity certainly dropped, by 850-fold, but the compound was still potent with an activity of 11 nM. Moreover, it had much better pharmacokinetic properties than previous structures, reflected in improved oral absorption and decreased clearance. As an added bonus, there was an improvement in selectivity.

At this stage, concerns were raised over the amide linker since metabolic hydrolysis could conceivably generate a toxic biaryl aniline structure (section 11.5.2; Fig. CS9.8). One of the strategies used to solve this problem was to incorporate the amide group into a bicyclic ring system. A couple of further optimizations were then carried out. The hydrophobic methyl group on the pyrazole ring was replaced with a more polar primary amide. This had the effect of decreasing plasma protein binding and increasing oral bioavailability. The phenylsulphonamide ring was also replaced with a lactam ring to produce the final structure—apixaban (Fig. CS9.8). The *ortho* position of the carbonyl group on this ring encourages the lactam ring to be orthogonal to the neighbouring aromatic ring, as with the previous biphenyl structures—an example of conformational restraint (section 13.3.10). This, in turn, is beneficial for binding interactions with the S4 subsite.

A crystal structure of the protein–drug complex has established the binding mode of apixaban (Fig. CS9.9). Because the lactam ring of the drug is orthogonal to the neighbouring aromatic ring, it is in the correct orientation to slide between the three aromatic residues that are present in the S4 subsite. This allows it to form stacking interactions with the aromatic rings of Phe-174 and Tyr-99, as well as edge-to-face interactions with Trp-215 (Fig. CS9.10).

In the S1 subsite, the methoxy group forms van der Waals interactions with the side chain of Val-213. The methoxy group is sufficiently lipophilic to allow such an interaction. The anisole ring is also responsible for a 20 000-fold increase in selectivity for factor Xa over trypsin, since it clashes with the Ser-190 residue that is present in trypsin. For similar reasons, the anisole ring increases selectivity for factor Xa over thrombin.

Elsewhere, three important hydrogen bonding interactions are formed that involve the primary amide group and the bicyclic scaffold.

Finally, the relative rigidity of apixaban is an important factor in its picomolar activity. A conformational analysis

FIGURE CS9.7 Development of apixaban.

FIGURE CS9.8 Possible metabolic reaction.

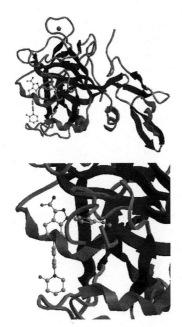

FIGURE CS9.9 Crystal structure of apixaban bound to factor X (accessed from the protein data bank, pdb 2P16)

has revealed that the drug can only form eight stable conformations. This means that there is only a small entropic penalty to be paid when the structure binds and adopts the active conformation.

Apixaban was approved in Europe in 2012 for the treatment of deep-vein thrombosis and pulmonary embolism, with US FDA approval following on in 2014.

CS9.6 The development of rivoraxaban

Rivoraxaban was developed by Bayer (Fig. CS9.11). It contains an oxazolidinone ring and has structural similarities to the antibacterial agent linezolid (section 19.7.7). However, rivoraxaban has no antibacterial activity.

The structure was developed from two lead compounds identified through high-throughput screening. In the first lead compound, the bicyclic ring was modified to an isoindoline ring system in structure IX to remove three chiral centres—a simplification strategy (section 13.3.8). The aromatic side chain was also modified by replacing the highly basic amidine group with a pyridine ring and an amine. The left-hand side was left unmodified since SAR studies demonstrated that the chlorothiophene ring and amide group were essential for good activity. However, none of the compounds investigated had acceptable pharmacokinetic properties.

At this stage, it was decided to adopt a new lead compound based on an oxazolidinone structure (X) that had shown very weak activity in the original screening process. One of the reasons for considering this structure was the presence of the thiophene ring. Considering the benefits of the chlorothiophene ring in the previous lead compound, a chlorine substituent was now added to the thiophene ring of the new lead compound. This led to structure XI and a remarkable 200-fold increase in activity, demonstrating again that highly basic groups are not essential for good binding with the S1 subsite. Optimization of the P4 region resulted in rivaroxaban.

Crystal structures show that the oxazolidinone ring acts as a scaffold, and directs its two substituents into the S1 and S4 pockets of the enzyme by forming an L-shaped active conformation. The aromatic and morphilinone rings are positioned in the S4 channel, whereas the chlorothiophene ring fits into the S1 pocket, allowing the chlorine substituent to form an important interaction with the aromatic ring of a tyrosine residue. This interaction compensates for the lack of the interactions that would be possible between a basic group and an aspartate residue. There are two important hydrogen bonds

FIGURE CS9.10 Binding interactions of apixaban.

FIGURE CS9.11 Development of rivaroxaban.

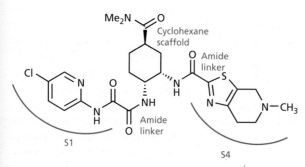

FIGURE CS9.12 Binding interactions for edoxaban (DU-176b).

CS9.7 **The development of edoxaban**

Edoxaban (Fig. CS9.12) was developed by Daiichi Sankyo and gained approval in Japan in 2011. FDA approval was granted in 2015. The structure contains a central cyclohexane ring which acts as the scaffold for the two main substituents. These are linked through amide groups to neighbouring positions of the scaffold. As with the previous agents, the scaffold ring serves to position and orientate the two halves of the molecule into the S1 and S4 binding pockets.

identified in Fig. CS9.11, but it is the chlorine group that is crucial for the drug's high potency and good oral bioavailability.

Test your understanding and practice your molecular modelling with Exercise 26.1 on the Online Resource Centre at www.oxfordtextbooks.co.uk/orc/patrick6e/

FURTHER READING

Pinto, D. J. P., Smallheer, J. M., Cheney, D. L., et al. (2010) Factor Xa inhibitors: next generation antithrombotic agents. *Journal of Medicinal Chemistry*, **53**, 6243–74.

Pinto, D. J. P., Orwat, M. J., Koch, S., et al. (2007) Discovery of 1-(4-methoxyphenyl)-7-oxo-6-(4-(2-oxopiperidin-1-yl)phenyl)-4,5,6,7-tetrahydro-1*H*-pyrazolo[3,4-c]pyridine-3-carboxamide (Apixaban, BMS-562247), a highly potent selective, efficacious and orally bioavailable inhibitor of blood coagulation factor Xa. *Journal of Medicinal Chemistry*, **50**, 5339–56.

Roehrig, S., Straub, A., Pohlmann, J., et al. (2005) Discovery of the novel antithrombotic agent 5-chloro-*N*-({(5*S*)-2-oxo-3-[4-(3-oxomorpholin-4-yl)phenyl]-1,3-oxazolidin-5-yl}methyl) thiophene-2-carboxamide (BAY 59-7939): an oral, direct factor Xa inhibitor. *Journal of Medicinal Chemistry*, **48**, 5900–8 (rivoraxaban).

■ CASE STUDY 10
Reversible inhibitors of HCV NS3-4A protease

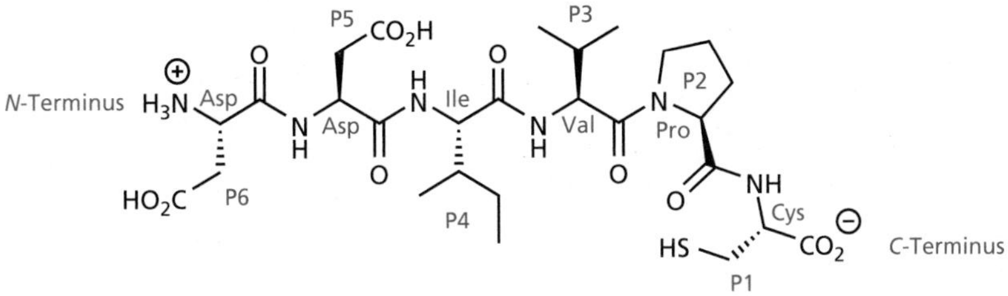

FIGURE CS10.1 Hexapeptide lead compound (K_i 79 mM).

CS10.1 Introduction

In recent years, there has been remarkable progress in the number of antiviral agents approved for the treatment of hepatitis C, starting with the approval of **boceprevir** and **telaprevir** in 2009. Both of these agents contain a ketoamide group that reacts with the catalytic serine residue of a viral protease enzyme labelled HCV NS3-4A to form a reversible covalent bond (section 20.10.1). Since then, a number of reversible inhibitors have been approved which bind through intermolecular bonds alone. **Simeprevir** was the first of these to reach the market and was approved by the FDA in 2013. This Case study describes some of the research that led to the discovery of these reversible inhibitors.

CS10.2 Identification of a lead compound

The lead compound for the development of reversible inhibitors was one of the peptide products resulting from the enzyme-catalysed reaction of a larger peptide substrate. It was found that that this product remained bound to the active site following the enzyme-catalysed reaction and acted as a weak inhibitor. The structure in question was a hexapeptide that could interact with six enzyme binding subsites S1–S6 (Fig. CS10.1). Therefore, it was adopted as the starting point for the development of more potent inhibitors.

The hexapeptide contains a cysteine residue at the C-terminus which forms a number of important binding interactions. For example, the carboxylate group forms strong ionic and hydrogen bonds with a catalytic histidine residue, as well as hydrogen bonds with peptide bonds close to the **oxyanion hole** (Fig. CS10.2); the oxyanion hole is a region occupied by the oxyanion during the enzyme-catalysed reaction (section 3.5.4). In addition, the thiol group of cysteine's side chain interacts with the aromatic ring of Phe-154 in the S1 binding subsite. Crystal structures of the protease enzyme reveal that this subsite is a shallow hydrophobic groove which is quite distinct from the S1 pocket of other proteases. This implies that there is a good chance of designing selective inhibitors that take advantage of this distinctiveness.

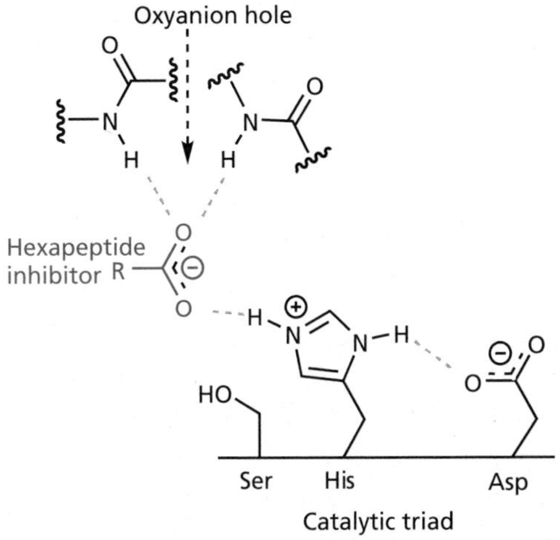

FIGURE CS10.2 Binding interactions between the active site and the carboxylate ion of the hexapeptide inhibitor.

CS10.3 Modifications of the lead compound

Although the thiol group of the lead compound forms useful binding interactions with the S1 subsite, an early priority was to replace it with an alternative binding group. This is because the thiol group is chemically reactive and could cause side effects. A number of hydrophobic groups were investigated which demonstrated that a propyl side chain was the best option. The primary amine group at the N-terminus of the hexapeptide was also considered detrimental because of its chemical reactivity,

and so it was capped with an acetyl group (Ac) (Structure I; Fig. CS10.3).

A significant improvement in activity was obtained by adding a substituent to the proline ring (P2). Adding a phenethyl or benzyloxy substituent led to a 21-fold increase in activity (Structure II; Fig. CS10.3), while a naphthyl ring increased activity a further 18-fold (Structure III; Fig. CS10.3). These results were surprising since the S2 subsite should not be not large enough to accept such groups. However, it was discovered that the presence of these groups produces an induced fit in the enzyme that opens up an additional binding region next to the S2 binding subsite. This additional region (the extended S2 binding subsite) allows a drug extension strategy to

FIGURE CS10.3 Early modifications of the hexapeptide lead compound.

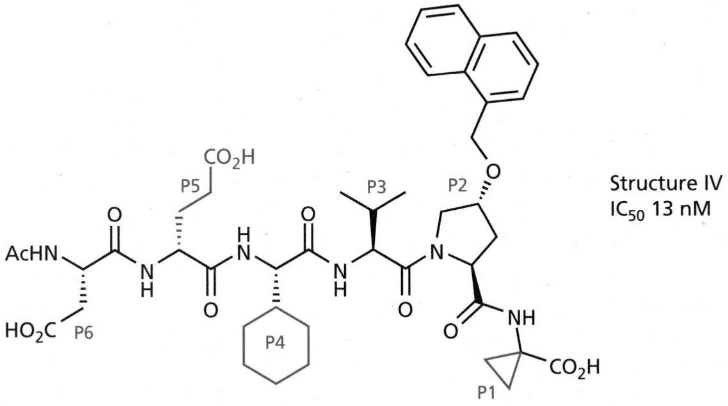

FIGURE CS10.4 An inhibitor with low nanomolar activity.

be carried out whereby additional groups are added to obtain additional binding interactions. With normal substrates, this region is not accessible.

A further increase in binding interactions was now obtained by modifying the binding groups that occupy the S1, S4, and S5 subsites. This resulted in structure IV, which was active in the low nanomolar range (Fig. CS10.4).

Further modifications of the P1–P3 side chains were carried out to optimize binding interactions with the S1–S3 subsites, resulting in an extremely potent compound with picomolar activity (Structure V; Fig. CS10.5). This corresponds to a million-fold increase in potency relative to the original lead compound.

CS10.4 From hexapeptide to tripeptide

Despite the extremely high potency of structure V, it proved to have unfavourable pharmacokinetic properties, which meant that there was a 25 000-fold difference between its *in vitro* and *in vivo* activities. This is not too surprising, considering the structure's peptide nature and high molecular weight. Moreover, it contains three ionizable carboxylic acids which are extremely polar. All these characteristics are known to be detrimental to oral absorption and *in vivo* activity (section 11.3). Therefore, the two amino acids at the *N*-terminus were removed to produce a tetrapeptide with only one of the original three carboxylic acids. Not surprisingly, activity dropped dramatically to micromolar levels (1–4 μM) due to the loss of the binding interactions associated with the P5 and P6 groups. However, further modifications to the P1 and P2 side chains successfully restored activity to nanomolar levels (Structure VI; Fig. CS10.6). Further increases in potency were observed by adding a methoxy group to the quinoline ring and a *tert*-butyl group at P3 (Structure VII). Because of this increased activity, it was now possible to shorten the structure yet further to a tripeptide, whilst retaining good activity at nanomolar levels (Structure VIII; Fig. CS10.7).

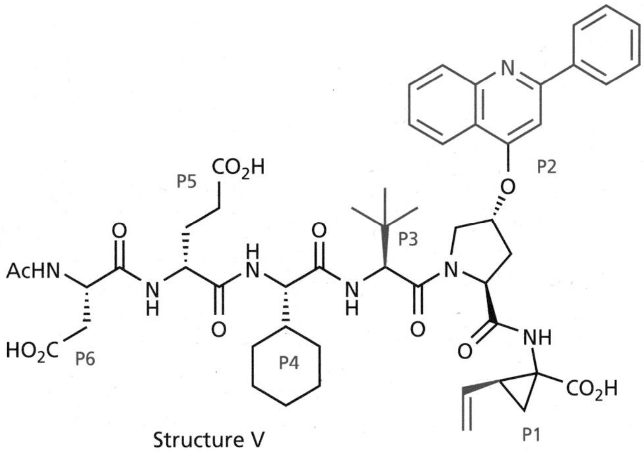

FIGURE CS10.5 An inhibitor with picomolar activity.

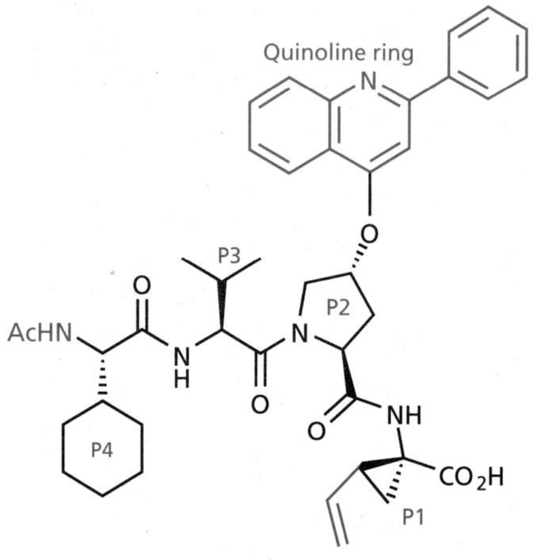

FIGURE CS10.6 Tetrapeptide inhibitors with nanomolar activity.

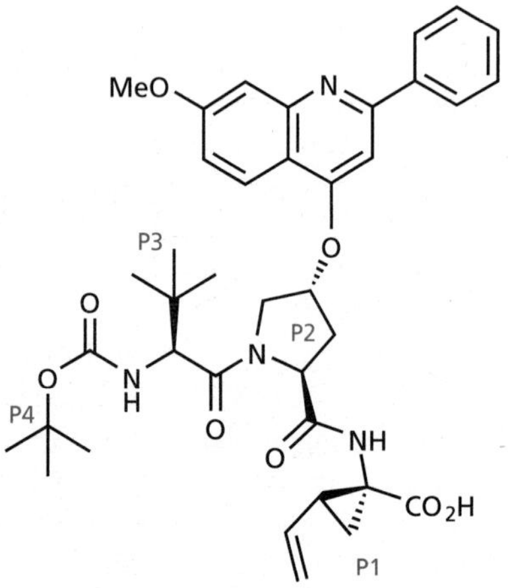

Structure VI (IC$_{50}$ 13 nM)

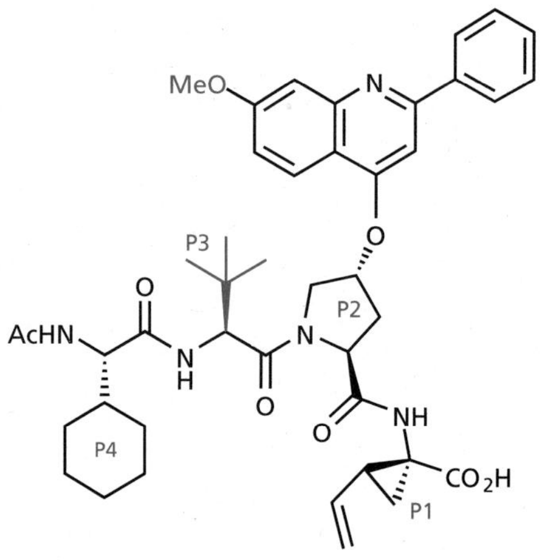

Structure VII (IC$_{50}$ 1 nM)

Structure VIII (IC$_{50}$ 29 nM)

FIGURE CS10.7 The tripeptide inhibitor (IC$_{50}$ 29nM).

CS10.5 **From tripeptide to macrocycle (BILN-2061)**

At this stage, NMR studies were carried out to determine what the active conformation of the inhibitors might be when they are bound to the target enzyme. These studies revealed that the inhibitors were bound in an extended conformation, with the peptide bond between P2 and P3 existing in the *trans*-configuration rather than the *cis* configuration. It was also found that the P1 and P3 side chains lie relatively close together in the active conformation (Fig. CS10.8). This led to the idea that the P1 and P3 side chains of the tripeptide could be linked up to form a macrocycle which would rigidify the molecule and trap that portion of the molecule in the correct conformation

for binding (section 13.3.9). The level of rigidification that would result from this would significantly reduce the number of possible conformations for the structure, since five rotatable bonds in the tripeptide would be incorporated within the macrocycle and would no longer be capable of full rotation. Cutting down the number of possible conformations also reduces the entropy penalty that results when a highly flexible molecule has to adopt a single conformation for binding. Incorporating rotatable bonds into a ring system is a common strategy of rigidification, but there is a risk that it might prevent the molecule adopting the active conformation. However, that risk is minimized by using a macrocycle. Large rings are more flexible than smaller ring systems, and this makes it easier for the molecule to find an optimum binding conformation.

Another advantage of rigidification is that a smaller number of rotatable bonds would be expected to improve oral absorption (section 11.3).

Modelling studies were carried out to determine suitable linkers between the P1 and P3 binding groups. These studies indicated that a hydrocarbon bridge would be ideal since it would not only link the groups, but would form additional van der Waals interactions with the S1–S3 binding pockets. A propyl linker proved the best option, resulting in a 15-membered macrocycle with enhanced activity (Structure IX; Fig. CS10.8). Unfortunately, the structure still had poor pharmacokinetic properties, including poor bioavailability and rapid clearance. Further modifications were carried out on the P2 extension and the *N*-terminal cap (P4) to try and improve pharmacokinetic properties and *in vivo* activity. This led to **BILN-2061** (Fig. CS10.9), which had nanomolar potency in both *in vitro* and *in vivo* tests (IC$_{50}$ 3 nM, EC$_{50}$ 1.2 nM respectively). The agent proved orally active, and

Active conformation for linear tripeptide (IC_{50} 29 nM)

Macrocyclic analogue of the tripeptide Structure IX (IC_{50} 11 nM)

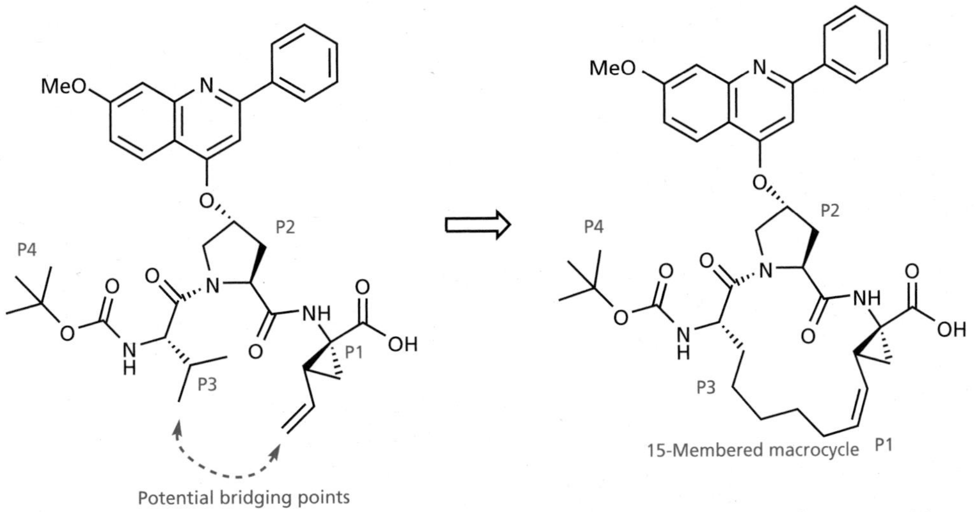

FIGURE CS10.8 Design of a macrocyclic inhibitor.

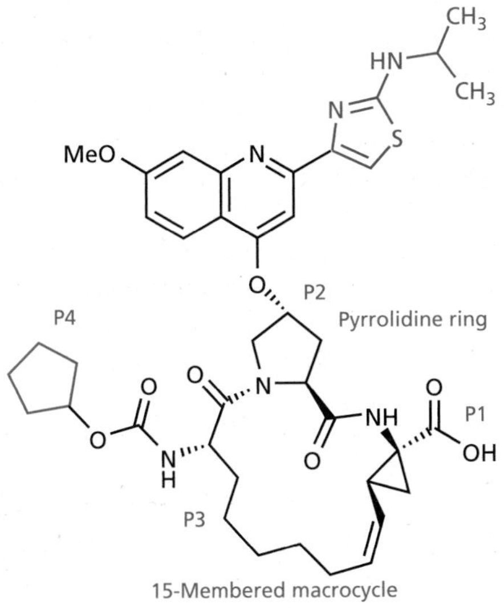

FIGURE CS10.9 BILN-2061 (IC_{50} 3 nM).

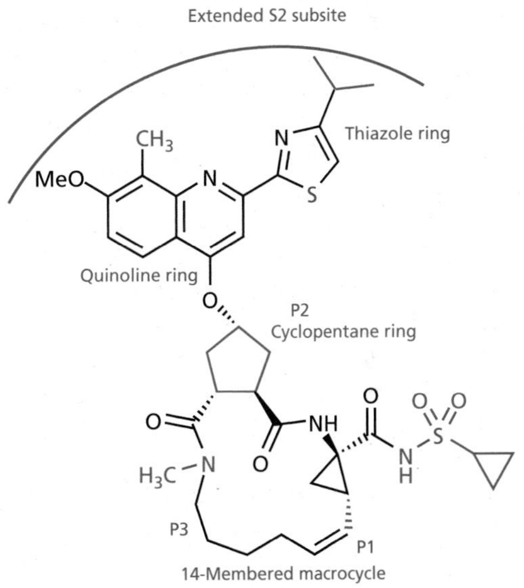

Simeprevir (TMS435)

FIGURE CS10.10 Simeprevir.

was highly selective—showing negligible activity against host proteases such as elastase and cathepsin.

CS10.6 From BILN-2061 to simeprevir

BILN-2061 was the first HCV protease inhibitor to enter clinical trials, but had to be withdrawn due to toxicity in animal studies. However, further modifications of the structure led to **simeprevir** (Fig. CS10.10) which was approved in 2013. One of the key modifications was the replacement of the 'proline ring' with a cyclopentane ring, while another important change was ring contraction of the macrocycle to a 14-membered ring. A third important modification was the discovery that *in vivo* activity could be improved by using an acylsulphonamide group as a bio-isostere for the important carboxylate group (section 13.3.7). The acylsulphonamide mimics the carboxylic acid by containing an acidic proton, which means that the group can become ionized. However the resulting negative charge is spread over more oxygen atoms, which proves beneficial for crossing cell membranes. The presence of the cyclopropyl group on sulphur also allows an extra binding interaction with the S1' binding pocket. The incorporation of the acylsulphonamide bio-isostere resulted in increased binding affinity and better cell permeability, which meant that it was possible to remove the P4 moiety and reduce the size of the molecule.

Binding studies carried out on simeprevir reveal that the hydrocarbon region of the macrocyclic ring occupies the S1 and S3 subsites as planned and forms several van der Waals interactions with hydrophobic amino acid residues. The cyclopentane ring occupies the S2 pocket, while the quinoline and thiazole rings occupy the region defined as the extended S2 subsite. This allows the quinoline ring to form an important pi–pi interaction with the guanidine group of Arg-155. Two hydrogen bonds are also formed to Arg-155 and Ala-157. The acylsulphonamide group mimics the original carboxylic acid by forming hydrogen bonds to the catalytic region and the oxyanion hole.

Appendix 1

Essential amino acids

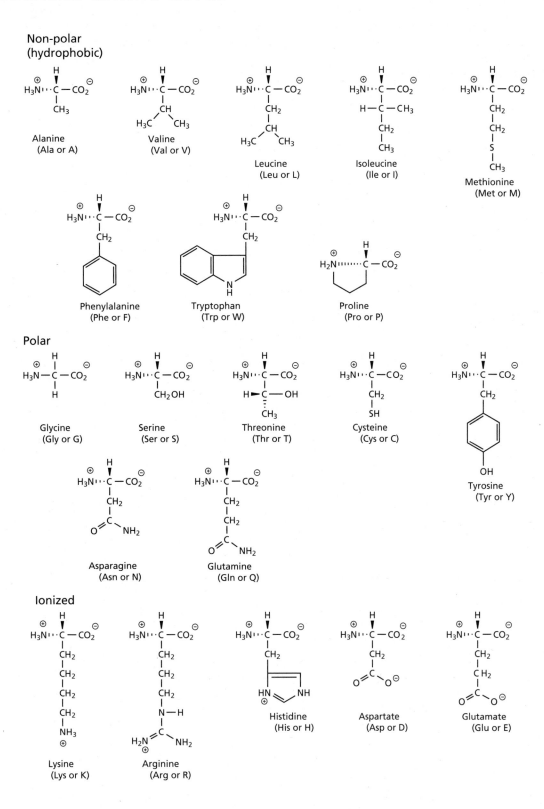

Non-polar
(hydrophobic)

Alanine
(Ala or A)

Valine
(Val or V)

Leucine
(Leu or L)

Isoleucine
(Ile or I)

Methionine
(Met or M)

Phenylalanine
(Phe or F)

Tryptophan
(Trp or W)

Proline
(Pro or P)

Polar

Glycine
(Gly or G)

Serine
(Ser or S)

Threonine
(Thr or T)

Cysteine
(Cys or C)

Tyrosine
(Tyr or Y)

Asparagine
(Asn or N)

Glutamine
(Gln or Q)

Ionized

Lysine
(Lys or K)

Arginine
(Arg or R)

Histidine
(His or H)

Aspartate
(Asp or D)

Glutamate
(Glu or E)

Appendix 2

The standard genetic code

UUU	Phe	**UCU**	Ser	**UAU**	Tyr	**UGU**	Cys
UUC	Phe	**UCC**	Ser	**UAC**	Tyr	**UGC**	Cys
UUA	Leu	**UCA**	Ser	**UAA**	Stop	**UGA**	Stop
UUG	Leu	**UCG**	Ser	**UAG**	Stop	**UGG**	Trp
CUU	Leu	**CCU**	Pro	**CAU**	His	**CGU**	Arg
CUC	Leu	**CCC**	Pro	**CAC**	His	**CGC**	Arg
CUA	Leu	**CCA**	Pro	**CAA**	Gln	**CGA**	Arg
CUG	Leu	**CCG**	Pro	**CAG**	Gln	**CGG**	Arg
AUU	Ile	**ACU**	Thr	**AAU**	Asn	**AGU**	Ser
AUC	Ile	**ACC**	Thr	**AAC**	Asn	**AGC**	Ser
AUA	Ile	**ACA**	Thr	**AAA**	Lys	**AGA**	Arg
AUG	Met	**ACG**	Thr	**AAG**	Lys	**AGG**	Arg
GUU	Val	**GCU**	Ala	**GAU**	Asp	**GGU**	Gly
GUC	Val	**GCG**	Ala	**GAC**	Asp	**GGC**	Gly
GUA	Val	**GCA**	Ala	**GAA**	Glu	**GGA**	Gly
GUG	Val	**GCG**	Ala	**GAG**	Glu	**GGG**	Gly

Appendix 3

Statistical data for QSAR

To illustrate how statistical terms such as r, s, and F are derived and interpreted, the numerical data in Table A3.1 will be used. There are six compounds in the study ($n = 6$). Y_{exp} is the logarithm of the observed activity for each of the compounds and X is a physicochemical parameter. The QSAR equation derived from the data is

$$\log(\text{activity}) = Y_{calc} = k_1 X + k_2 = -0.47\,X - 0.022$$

The slope of the line is −0.47 and the intercept with the y-axis is −0.022.

The correlation coefficient r for the above QSAR equation is calculated using the following equation:

$$r^2 = 1 - \frac{SS_{calc}}{SS_{mean}}$$

SS_{calc} is a measure of how much the experimental activity of the compounds varies from the calculated value. For each compound, the difference between the experimental activity and the calculated activity is $Y_{exp} - Y_{calc}$ (Fig. A3.1). This is then squared and the values are added together to give the sum of the squares (SS_{calc}).

SS_{mean} is a measure of how much the experimental activity varies from the mean of all the experimental activities and represents the situation where no correlation with X has been attempted (Fig. A3.1).

If there is a correlation between the activity (Y) and the parameter (X), the line of the equation should pass closer to the data points than the line representing the mean. This means that SS_{calc} should be less than SS_{mean}. For a perfect correlation, the calculated values for the activity would be the same as the experimental ones and so SS_{calc} would be zero. This would make $r^2 = 1$.

For the figures shown in table A3.1, the value of r works out as follows:

$$r^2 = 1 - \frac{SS_{calc}}{SS_{mean}} = 1 - \frac{0.1912}{0.5279} = 1 - 0.3622 = 0.638$$

This indicates that only 64% of the variability in activity is due to the parameter X. This is much lower than the minimum acceptable figure of 80% and so the equation is not a particularly good one. Nevertheless, it is possible that X may have some influence on the activity. To check whether there is any significance to the equation

TABLE A3.1

Compound ($n = 6$)	Physicochemical parameter (X)	Log(act.)$_{exp}$ Y_{exp}	Log(act.)$_{calc}$ Y_{calc}	$Y_{exp} - Y_{calc}$	Square of $Y_{exp} - Y_{calc}$	$Y_{exp} - Y_{mean}$	Square of $Y_{exp} - Y_{mean}$
1	0.23	0.049	−0.129	0.178	0.0317	0.263	0.0692
2	0.23	0.037	−0.129	0.166	0.0276	0.251	0.0630
3	−0.17	0	0.057	−0.057	0.0032	0.214	0.0458
4	0	−0.155	−0.022	−0.133	0.0177	0.059	0.0035
5	1.27	−0.468	−0.613	0.145	0.0210	−0.254	0.0645
6	0.91	−0.745	−0.445	−0.3	0.0900	−0.531	0.2820
		Mean value Y_{mean}			Sum of squares SS_{calc}		Sum of squares SS_{mean}
		−0.214			0.1912		0.5279

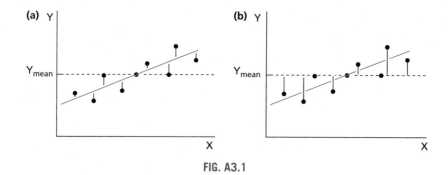

FIG. A3.1

a statistical test called an *F*-test can be carried out. The equation used for this specific example is as follows:

$$F_{p_2-p_1,\,n-p_2} = \frac{SS_{mean} - SS_{calc}}{SS_{calc}} \times \frac{n-p_2}{p_2-p_1}$$

where p_2 is the number of parameters involved in the derived QSAR equation (Y and X) and p_1 represents the number of parameters involved in the reference equation (Y only in this example). N, SS_{mean}, and SS_{calc} are as described above. This gives the following:

$$F_{2-1,6-2} = \frac{0.528 - 01912}{0.1912} \times \frac{6-2}{2-1}$$

$$\text{or } F_{1,4} = \frac{0.528 - 0.1912}{0.19112} \times \frac{6-2}{2-1} = 1.7615 \times 4 = 7.05$$

$F_{1,4}$ is now compared against tables of *F* values which indicate the probability level of a significant correlation. For $F_{1,4}$ the tables show that a value of 4.54 would indicate a probability level of 0.9, whereas 7.71 represents

a probability level of 0.95. A value of 21.2 represents a probability level of 0.99. The higher the value of F_{14}, the closer the probability level approaches 1. The calculated value of 7.05 shows that the probability level is between 0.9 and 0.95.

The standard deviation (*s*) for the equation is calculated by the following equation and is dependent on the number of compounds (*n*) tested.

$$s^2 = \frac{SS_{calc}}{n-2}$$

This gives a value of 0.218 for the data provided in Table A3.1. The value of *s* should be as small as possible, but not smaller than the standard deviation of the experimental data.

A QSAR equation could now be derived to see whether the biological activity matches a different physicochemical parameter. Table A3.2 shows values for a different parameter (Z). In this case, the derived equation is:

$$Y_{calc} = 0.33\,Z - 0.62$$

TABLE A3.2

Compound ($n=6$)	Physicochemical parameter (Z)	Log(act.)$_{exp}$ Y_{exp}	Log(act.)$_{calc}$ Y_{calc}	$Y_{exp} - Y_{calc}$	Square of $Y_{exp} - Y_{calc}$	$Y_{exp} - Y_{mean}$	Square of $Y_{exp} - Y_{mean}$
1	2.03	0.049	0.0499	−0.0009	0.0000	0.263	0.0692
2	1.83	0.037	−0.0161	0.0531	0.0028	0.251	0.0630
3	1.38	0.000	−0.1646	0.1646	0.0271	0.214	0.0458
4	0.90	−0.155	−0.323	0.1680	0.0282	0.059	0.0035
5	1.40	−0.468	−0.158	−0.3100	0.0961	−0.254	0.0645
6	−0.26	−0.745	−0.7058	−0.0392	0.0015	−0.531	0.2820
		Mean value			Sum of squares		Sum of squares
		Y_{mean}			SS_{calc}		SS_{mean}
		−0.214			0.1558		0.5279

TABLE A3.3

Compound (n = 6)	Physicochemical parameter (X)	Physicochemical parameter (Z)	Log(act.)$_{exp}$ Y_{exp}	Log(act.)$_{calc}$ Y_{calc}	$Y_{exp} - Y_{calc}$	Square of $Y_{exp} - Y_{calc}$	$Y_{exp} - Y_{mean}$	Square of $Y_{exp} - Y_{mean}$
1	0.23	2.03	0.049	0.0493	−0.0003	0.0000	0.263	0.0692
2	0.23	1.83	0.037	−0.0007	0.0377	0.0014	0.251	0.0630
3	−0.17	1.38	0.000	0.0228	−0.0228	0.0005	0.214	0.0458
4	0.00	0.90	−0.155	−0.1550	0.0000	0.0000	0.059	0.0035
5	1.27	1.40	−0.468	−0.4618	−0.0062	0.0000	−0.254	0.0645
6	0.91	−0.26	−0.745	−0.7544	0.0094	0.0001	−0.531	0.2820
			Mean value			Sum of squares		Sum of squares
			Y_{mean}			SS_{calc}		SS_{mean}
			−0.214			0.0021		0.5279

The statistical analysis of this gives the following:

$$n = 6; r = 0.840, s = 0.199, F_{1,4} = 9.6$$

All these results are better than the previous ones, showing that the parameter Z is more important than X in explaining the variation in activity. r is still less than 0.9, however, and further improvements are necessary.

If both of the above parameters are included in the analysis, the equation becomes:

$$Y_{calc} = -0.34X + 0.25Z - 0.38$$

The corresponding table of results is shown in Table A3.3. The statistical results are $n = 6$, $r = 0.998$, $s = 0.028$, and $F_{1,3} = 230.3$. Note that there are three parameters in the QSAR equation and so the F term is $F_{1,3}$ rather than $F_{1,4}$. Comparison with tabulated $F_{1,3}$ values shows that the probability level for this equation is 0.999.

A final check has to be made to ensure that the values for the two parameters (X and Z) are not related in any way. An equation attempting to relate X and Z is derived and assessed statistically. For the values shown, $r^2 = 0.122$, which shows that there is little correlation between X and Z. The final equation is therefore validated.

QSAR equations may also include terms in parenthesis. For example, taking the previous equation:

$$Y_{calc} = -0.34(\pm0.08)X + 0.25(\pm0.05)Z - 0.38(\pm0.09)$$

The numbers in parenthesis represent the 95% confidence limits for the various parameters. For example, there is 95% confidence that the coefficient for Z lies between the values 0.20 and 0.30. If the number in parenthesis is smaller than the coefficient, it means the parameter is statistically significant in the F-test.

Appendix 4

The action of nerves

The structure of a typical nerve cell or neuron is shown in Fig. A4.1. The nucleus of the cell is found in the large cell body situated at one end of the neuron. Small arms (dendrites) radiate from the cell body and receive messages from other neurons. These messages either stimulate or destimulate the neuron. The cell body 'collects' the sum total of these messages.

Ion channels are selective for different ions. There are cationic ion channels for Na^+, K^+, and Ca^{2+} ions. When these channels are open, they are generally excitatory and lead to depolarization of the cell.

It is worth emphasizing that the cell body of a neuron receives messages not just from one other neuron, but from a range of different neurons. These pass on different messages (neurotransmitters). Therefore, a message received from a single neuron is unlikely to stimulate a neuron signal by itself, unless other neurons are acting in sympathy.

Assuming that the overall stimulation is large enough, an electrical signal is fired down the length of the neuron (the axon). The axon is covered with sheaths of lipid (myelin sheaths), which act to insulate the signal as it passes down the axon.

The axon leads to a knob-shaped swelling (**synaptic button**) if the neuron is communicating with another neuron. Alternatively, if the neuron is communicating with a muscle cell, the axon leads to what is known as a **neuromuscular endplate**, where the end plate is spread like an amoeba over an area of the muscle cell.

Within the synaptic button or neuromuscular endplate there are small globules (**vesicles**) containing the neurotransmitter. When a signal is received from the axon, the vesicles merge with the cell membrane and release their neurotransmitter into the gap between the neuron and the target cell (**synaptic gap**). The neurotransmitter binds to a receptor as described in Chapter 4, and passes on its message. Once the message has been received, the neurotransmitter leaves the receptor and is either broken down enzymatically (e.g. acetylcholine) or taken up intact by the presynaptic neuron (e.g. noradrenaline). Either way, the neurotransmitter is removed from the synaptic gap and is unable to bind with its receptor a second time.

To date, we have talked about nerves 'firing' and the generation of 'electrical signals' without really considering the mechanism of these processes. The secret behind nerve transmission lies in the movement of ions across cell membranes, but there is an important difference in what happens in the cell body of a neuron compared with the axon. We shall consider what happens in the cell body first.

All cells contain sodium, potassium, calcium, and chloride ions, and it is found that the concentration of these ions is different inside the cell compared with outside. The concentration of potassium inside the cell is larger than the surrounding medium, whereas the concentration of sodium and chloride ions is smaller. Thus, a concentration gradient exists across the membrane.

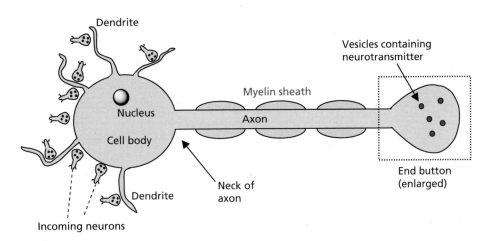

FIG. A4.1 Structure of a typical nerve cell (neuron).

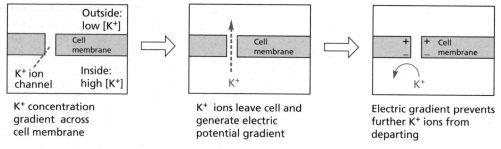

FIG. A4.2 Generation of electric potential across a cell membrane.

Potassium is able to move down its concentration gradient (i.e. out of the cell), as it can pass through the potassium ion channels (Fig. A4.2). But, if potassium ions can move out of the cell, why does the potassium concentration inside the cell not fall to equal that of the outside? The answer lies in the fact that potassium is a positively charged ion and, as it leaves the cell, an electric potential is set up across the cell membrane. This would not happen if a negatively charged counterion could leave with the potassium ion. However, the counterions in question are large proteins which cannot pass through the cell membrane. As a result, a few potassium ions are able to escape through the ion channels out of the cell and an electric potential builds up across the cell membrane such that the inside of the cell membrane is more negative than the outside. This electric potential (50–80 mV) opposes, and eventually prevents, the flow of potassium ions.

But what about the sodium ions? Could they flow into the cell along their concentration gradient to balance the charged potassium ions that are departing? The answer is that they cannot because they are too big for the potassium ion channels. This appears to be a strange argument, as sodium ions are smaller than potassium ions, but it has to be remembered that we are dealing with an aqueous environment where the ions are solvated (i.e. they have a 'coat' of water molecules). Sodium, being a smaller ion than potassium, has a greater localization of charge and is able to bind its solvating water molecules more strongly. As a result, sodium, along with its water coat, is bigger than a potassium ion with or without its water coat.

Ion channels for sodium do exist and these channels are capable of removing the water coat around sodium and letting it through. However, the sodium ion channels are mostly closed when the neuron is in the resting state. As a result, the flow of sodium ions across the membrane is very small compared to potassium. Nevertheless, the presence of sodium ion channels is crucial to the transmission of a nerve signal.

To conclude, the movement of potassium across the cell membrane sets up an electric potential across the cell membrane which opposes this flow. Charged protein structures are unable to move across the membrane, while sodium ions cross very slowly, so an equilibrium is established. The cell membrane is polarized and the electric potential at equilibrium is known as the resting potential.

The number of potassium ions required to establish that potential is of the order of a few million compared to the several hundred billion present in the cell. Therefore the effect on concentration is negligible.

As mentioned above, potassium ions are able to flow out of potassium ion channels, but not all of these channels are open in the resting state. What would happen if more were to open? The answer is that more potassium ions would flow out of the cell and the electric potential across the cell membrane would become more negative to counter this increased flow. This is known as **hyperpolarization** and the effect is to destimulate the neuron (Fig. A4.3).

Suppose instead that a few sodium ion channels were to open up. In this case, sodium ions would flow into the cell and, as a result, the electric potential would become less negative. This is known as **depolarization** and results in a stimulation of the neuron.

If chloride ion channels are opened, chloride ions flow into the cell, and the cell membrane becomes hyperpolarized, destimulating the neuron.

Ion channels do not open or close by chance. They are controlled by the neurotransmitters released by communicating neurons. The neurotransmitters bind with their receptors and this leads to the opening or closing of ion channels. Such ion channels are known as **ligand-gated ion channels**. For example, acetylcholine controls the sodium ion channel, whereas γ-aminobutyric acid (GABA) and glycine control chloride ion channels. The resulting flow of ions leads to a localized hyperpolarization or depolarization in the area of the ion channel. The cell body collects and sums all this information such that the neck of the axon experiences an overall depolarization or hyperpolarization depending on the sum total of the various excitatory or inhibitory signals received.

We shall now consider what happens at the axon of the neuron (Fig. A4.4). The cell membrane of the axon also has sodium and potassium ion channels, but they are different in character from those in the cell body. The axon

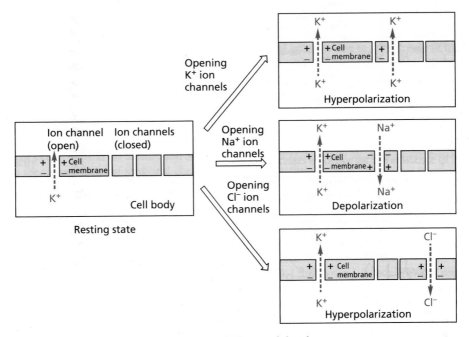

FIG. A4.3 Hyperpolarization and depolarization.

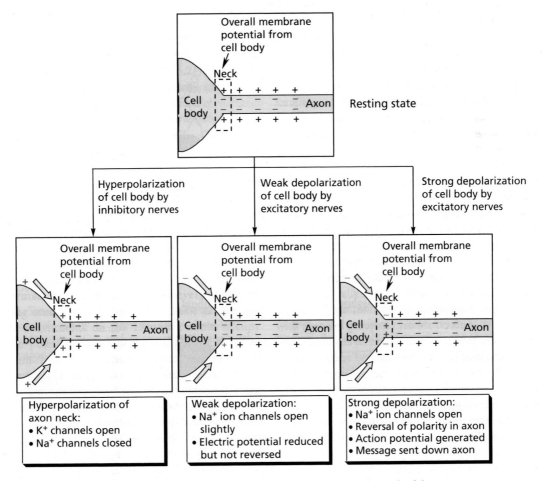

FIG. A4.4 Hyperpolarization and depolarization effects at the neck of the axon.

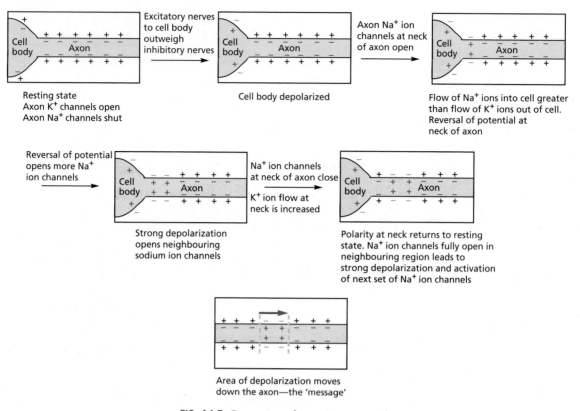

FIG. A4.5 Generation of an action potential.

ion channels are not controlled by neurotransmitters, but by the electric potential of the cell membrane. Therefore, they are known as **voltage-gated ion channels.**

The sodium ion channels located at the junction of the nerve axon with the cell body are the crucial channels as they are the first ion channels to experience whether the cell body has been depolarized or hyperpolarized.

If the cell body is strongly depolarized then a signal is fired along the neuron. A specific threshold value has to be reached before this happens, however. If the depolarization from the cell body is weak, only a few sodium channels open up and the depolarization at the neck of the axon does not reach that threshold value. The sodium channels then reclose and no signal is sent.

With stronger depolarization, more sodium channels open up until the flow of sodium ions entering the axon becomes greater than the flow of potassium ions leaving it. This results in a rapid increase in depolarization, which, in turn, opens up more sodium channels, resulting in very strong depolarization at the neck of the axon. The flow of sodium ions into the cell increases dramatically and becomes greater than the flow of potassium ions out of the axon. As a result, the electric potential

across the membrane is reversed, such that it is positive inside the cell and negative outside the cell. This process lasts less than a millisecond before the sodium channels reclose and sodium permeability returns to its normal state. More potassium channels then open and permeability to potassium ions increases for a while to speed up the return to the resting state.

The process is known as an action potential and can only take place in the axon of the neuron. The cell membrane of the axon is said to be excitable, unlike the membrane of the cell body. The important point to note is that once an action potential has fired at the neck of the axon, it has reversed the polarity of the membrane at that point. This, in turn, has an effect on the neighbouring area of the axon and depolarizes it beyond the critical threshold level. It, too, fires an action potential and so the process continues along the whole length of the axon (Fig. A4.5). The number of ions involved in this process is minute, such that the concentrations are unaffected. Once the action potential reaches the synaptic button or the neuromuscular endplate, it causes an influx of calcium ions into the cell and an associated release of neurotransmitter into the synaptic gap. The mechanism of this is not well understood.

Appendix 5

Microorganisms

Bacterial nomenclature

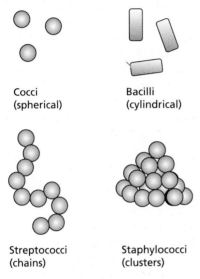

Cocci
(spherical)

Bacilli
(cylindrical)

Streptococci
(chains)

Staphylococci
(clusters)

FIG. A5.1 Bacterial nomenclature.

Some clinically important bacteria

The Gram stain

The Gram stain is a staining procedure of great value in the identification of bacteria.

The staining technique involves the addition of a purple dye followed by washing with acetone. Bacteria with a thick cell wall (20–40 nm) absorb the dye and are defined as Gram-positive because they are stained purple. Bacteria with a thin cell wall (2–7 nm) absorb only a small amount of dye, and the excess dye is washed out with acetone. These bacteria are then stained pink with a second dye and are said to be Gram-negative.

- Gram-negative bacteria—these cells have a thin cell wall and are coloured pink.
- Gram-positive bacteria—these cells have a thick cell wall and are coloured purple.

Classifications

Bacteria can be classified as being Gram-positive or Gram-negative depending on what colour they retain on treatment with the Gram-stain procedure. They can also

TABLE A5.1

Organism	Gram	Infections
Staphylococcus aureus	Positive	Skin and tissue infections, septicaemia, endocarditis; accounts for about 25% of all hospital infections
Streptococcus	Positive	Several types—commonly cause sore throats, upper respiratory tract infections, and pneumonia
Escherichia coli	Negative	Urinary tract and wound infections, common in the gastrointestinal tract, and often causes problems after surgery; accounts for about 25% of hospital infections
Proteus species	Negative	Urinary tract infections
Salmonella species	Negative	Food poisoning and typhoid
Shigella species	Negative	Dysentery
Enterobacter species	Negative	Urinary tract and respiratory tract infections, septicaemia
Pseudomonas aeruginosa	Negative	An opportunist pathogen, can cause very severe infections in burn victims and other compromised patients, e.g. cancer patients; commonly causes chest infections in patients with cystic fibrosis
Haemophilus influenzae	Negative	Chest and ear infections, occasionally meningitis in young children
Bacteroides fragilis	Negative	Septicaemia following gastrointestinal surgery

be classed as **aerobic** or **anaerobic** depending on their dependency on oxygen. Aerobic organisms grow in the presence of oxygen, whereas anaerobic organisms do not.

Definitions of different microorganisms

Bacteria are unicellular organisms that have a prokaryotic cell structure. They are diverse in nature and some can carry out photosynthesis. Examples of typical infections are given in Table A5.1.

Blue-green algae are made up of prokaryotic cells that can form multicellular filaments and carry out photosynthesis in the same manner as the eukaryotic algae.

Algae, with the exception of the blue-green algae, are made up of eukaryotic cells and can perform oxygen-evolving photosynthesis. Some are unicellular and some are multicellular. The latter have little or no cell differentiation, which sets them apart from higher multicellular organisms such as plants and animals.

Protozoa are unicellular eukaryotic organisms that are unable to carry out photosynthesis. They are responsible for diseases such as malaria, African sleeping sickness, Chagas' disease, leishmaniosis, and amoebic dysentery.

Fungi are multicellular eukaryotic organisms with little or no cell differentiation. They can form long filaments of interconnected cells called mycelia. They, too, are unable to carry out photosynthesis. Fungi are responsible for infections such as athlete's foot, ringworm, aspergillosis, candidiasis, and histoplasmosis.

Appendix 6

Trade names and drugs

This appendix shows the drugs present in various drug trade names. If you wish to identify the trade name associated with a specific drug, go to the index where you will find the trade name in brackets after each drug.

Abthrax raxibacumab
Accupril quinapril
Accupro quinapril
Accuretic quinapril with hydrochlorothiazide
Aceon perindopril erbumine
Aclacin aclarubicin
Aclaplastin aclarubicin
Acnamino minocycline
Actemra tocilizumab
Actilyse alteplase
Activase alteplase
Actiq fentanyl
Adagen pegademase
Adalat nifedipine
Adcetris brentuximab vedotin
Adcirca tadalafil
Adempas riociguat
Afrin oxymetazoline
Agenerase amprenavir
Aggrastat tirofiban
Airomir salbutamol
Alecensa alectinib
Aldactone spironolactone
Aldara imiquimod
Aldomet methyldopa
Alimta pemetrexed
Alkeran melphalan
Allegra fexofenadine
Altabax retapamulin
Altace ramipril
Altargo retapamulin
Altromid-S clofibrate
Alvesco ciclesonide
Amias candesartan cilexetil
Amikin amikacin
Amoxil amoxicillin
Amsidine amsacrine
Anectine suxamethonium
Angiomax bivalirudin
Angiox bivalirudin
Angitak isosorbide dinitrate
Antabuse disulfiram

Apresoline hydralazine hydrochloride
Aprovel irbesartan
Apsin phenoxymethylpenicillin
Aramine metaraminol
Aricept donepezil
Arimidex anastrozole
Aromasin exemestane
Arzerra ofatumumab
Asmabec Clickhaler beclometasone dipropionate
Asmanex mometasone furoate
Asmasal Clickhaler salbutamol
Atacand candesartan cilexetil
Atripla emtricitabine, tenofovir disproxil and efavirenz
Atrovent ipratropium bromide
Augmentin amoxicillin with clavulanic acid
Augmentin clavulanic acid with amoxicillin
Augmentin co-amoxiclav
Aureocort triamcinolone acetonide and chlortetracyline
Avamys fluticasone furoate
Avastin bevacixumab
Avalox moxifloxacin
Avapro irbesartan
Avelox moxifloxacin
Avloclor chloroquine
Avycaz avibactam and ceftazidime
Axid nizatidine
Azactam aztreonam
Bactroban mupirocin
Baycol cerivastatin
BCNU carmustine
Becodisks beclometasone dipropionate
Beconase beclometasone dipropionate
Belbuca buprenorphine
Beleodaq belinostat
Benadryl diphenhydramine
Benlysta belimumab
Benuryl probenicid
Besivance besifloxacin
Betacap betamethasone valerate
Betaloc metoprolol
Betanis mirabegron
Betesil betamethasone valerate
Betim timolol
Betmiga mirabegron
Betnelan betamethasone
Betnesol betamethasone sodium phosphate

Betnesol-N betamethasone sodium phosphate and neomycin
Betnovate betamethasone valerate
Betnovate-C betamethasone valerate and clioquinol
Betnovate-N betamethasone valerate and neomycin sulphate
Betoptic betaxolol
Bettamousse betamethasone valerate
Bextra valdecoxib
Bexxar tositumomab
Bezalip bezafibrate
BiCNU carmustine
Biocef cefalexin
Blincyto blinatumomab
Bosulif bosutinib
Brevibloc esmolol
Bridion sugammadex
Brilinta ticagrelor
Brilique ticagrelor
Broflex trihexyphenidyl
Bronkaid ephedrine
Budelin Novolizer budenoside
Buscopan hyoscine butylbromide
Busilvex busulfan
Butropina atropine methonitrate and secbutabarbital
Calcort deflazacort
Camcolit lithium carbonate
Campto irinotecan
Camptosar irinotecan
Canesten hydrocortisone and clomitrazole
Capoten captopril
Caprelsa vandetanib
Carace lisinopril
Carbometyx cabozantinib
Cardene nicardipine
Cardura doxazosin
Carelux clobetasol propionate
Casodex bicalutamide
Catapres clonidine
CCNU lomustine
Cefrom cefpirome
Celebrix celecoxib
Celsentri maraviroc
Ceporex cefalexin
Cerezyme imiglucerase
Certican everolimus
Chantix varenicline
Chirocaine levobupivicaine
Chloromycetin chloramphenicol
Cicatrin bacitracin and neomycin sulphate
Cidomycin gentamicin
Cialis tadalafil
Cinqair reslizumab
Cipralex escitalopram

Cipramil citalopram
Cipro ciprofloxacin
Ciprobay ciprofloxacin
Ciproxan ciprofloxacin
Ciproxin ciprofloxacin
Claforan cefotaxime
Clarosip clarithromycin
Clenil Modulite beclometasone dipropionate
Clexane enoxaparin
Clozaril clozapine
CoAprovel irbesartan with hydrochlorothiazide
Co-Diovan valsartan with hydrochlorothiazide
Co-fluampicil ampicillin and flucloxacillin
Cogentin benzatropine
Cognex tacrine
Combivir lapivudine and ziodvudine
Complera emtricitabine, tenofovir disproxil, and rilpivirine
Condyline podophyllotoxin
Conmana icotonib
Contrave bupropion and naltrexone
Copegus ribavirin
Co-phenotrope diphenoxylate and atropine sulphate
Cordilox verapamil
Corgard nadolol
Corlan hydrocortisone
Corlanor ivabradine
Corvitol metoprolol
Cosentyx secukinumab
Cosmegen Lyovac dactinomycin
Cotellic cobimetinib
Coumadin warfarin
Coversum perindopril arginine
Coversyl perindopril arginine
Cozaar losartan
Crestor rosuvastatin
Crixivan indinavir
Crystapen benzylpenicillin
Cubicin daptomycin
Cutivate fluticasone propionate
Cymbalta duloxetine
Cymevene ganciclovir
Cyprostat cyproterone acetate
Cyramza ramucirumab
Cysticide praziquantel
Cystrin oxybutynin
Cytadren aminoglutethimide
Descovy emtricitabine and tenofovir disproxil
DF118 Forte dihydrocodeine
DHC Continus dihydrocodeine
Daktacort hydrocortisone and miconazole nitrate
Dalacin C clindamycin
Daklinza daclatasvir
Dalvance dabavancin

Darzalex daratumumab
DaunoXome daunorubicin
Demerol pethidine
Depo-Medrone methylprednisolone
Deprenyl selegiline
Derbac-M malathion
Dermovate clobetasol propionate
Dermovate-NN clobetasol propionate, neomycin sulphate, and nystatin
Deteclo chlortetracycline, demeclocycline, and tetracycline
Detrol tolterodine
Detrunorm propiverine
Detrusitol tolterodine
Dexilant dexlansoprazole
Diconal dipipanone
Dificlir fidaxomicin
Diflucan fluconazole
Dioderm hydrocortisone
Diovan valsartan
Diprosalic betamethasone dipropionate and salicylic acid
Diprosone betamethasone dipropionate
Ditropan oxybutynin
Dixarit clonidine
Dobutrex dobutamine
Doxil doxorubicin
Drixine oxymetazoline
Drogenil flutamide
DTIC-Dome dacarbazine
Durogesic fentanyl
Dutrebis lamivudine and raltegravir
DynaCirc isradipine
Dyspamet cimetidine
Easi-Breathe beclometasone dipropionate
Edarbi azilsartan medoximil
Edronax reboxetine
Edurant rilpivirine
Efcortesol hydrocortisone
Effient prasugrel
Efudix fluorouracil
Eklira Genuair aclidinium bromide
Eldepryl selegiline
Eldisine vindesine
Elelyso taliglucerase alfa
Eliquis apixaban
Elitek rasburicase
Elocon mometasone furoate
Eloxatin oxaliplatin
Empliciti elotuzumab
Emselex darifenacin
Emtriva emtricitabine
Enablex darifenacin
Enbrel etanercept

Endoxana cyclophosphamide
Entresto sacubitril and valsartan
Epivir lamivudine
Epzicom abacavir and lamivudine
Erbitux cetuximab
Erivedge vismodegib
Erwinase crisantaspase
Erymax erythromycin
Erythrocin erythromycin
Erythroped erythromycin
Esmeron rocuronium
Estracyt estramustine
Etopophos etoposide
Etrivex clobetasol propionate
Eulexin flutamide
Eumovate clobetasone butyrate
Eviplera emtricitabine, tenofovir disproxil, and rilpivirine
Evista raloxifene
Evotaz atazanavir and cobicastat
Exelon rivastigmine
Exemptia adalimumab
Exforge amlodipine with valsartan
Eylea aflibercept
Fabrazyme agalsidase beta
Famvir famciclovir
Fansidar sulfadoxine with pyrimethamine
Fareston toremifene
Farlutal medroxyprogesterone acetate
Farydak panobinostat
Faslodex fulvestrant
Fasturtec rasburicase
Faverin fluvoxamine
Femera letrozole
Firmagon degarelix acetate
Flagyl metronidazole
Flixonase fluticasone propionate
Flixotide fluticasone propionate
Floxapen flucloxacillin
Floxin ofloxacin
Fluanxol venlafaxine
Fludara fludarabine
FML fluorometholone
Folotyn pralatrexate
Foradil formoterol
Fortovase saquinavir
Fortral pentazocine
Fortum ceftazidime
Foscan temoporfin
Foscavir foscarnet
Fostair beclometasone dipropionate and formoterol fumarate
Fragmin dalteparin
Fucibet betamethasone valerate and fusidic acid
Fucidin fusidic acid

Fucidin H hydrocortisone acetate and fusidic acid
Fungilin amphotericin
Fungizone amphotericin
Furadantin nitrofurantoin
Fuzeon enfuvirtide
Galexos simeprevir
Galseud pseudoephedrine
Gastrozepin pirenzepine
Gazyva obinutuzumab
Gemzar gemcitabine
Genasense oblimersen
Genticin gentamicin
Gentisone hydrocortisone and gentamicin
Genvoya elvitegravir, cobicastat, emtricitabine, and
  tenofovir disproxil
Genzyme alglucosidase alfa
Gilotrif afatinib
Giotrif afatinib
Gliadel carmustine
Glivec imatinib
Haelan fludroxycortide
Halaven eribulin
Harvoni ledipasvir and sofosbuvir
Hepsera adefovir dipivoxil
Herceptin trastuzumab
Herpid idoxuridine
Hidrasec racecadotril or acetorphan
Hiprex methenamine
Hivid zalcitabine
Humira adalimumab
Hycamptin topotecan
Hycamtin topotecan
Hycodan hydrocodone (dihydrocodeinone)
Hypovase prazosin
Hytrin terazosin
Ibrance palboclicib
Iclusig ponatinib
Imbruvica ibrutinib
Imigran sumatriptan
Imlygic talimogene laherparepvec
Immukin γ-interferon
Imnovid pomalidomide
Imodium loperamide
Imuran azathioprine
Inavir laninamivir
Incivek telaprevir
Incruse Ellipta umeclidinium bromide
Inderal propranolol
Inlyta axitinib
Innohep tinzaparin
Innovace enalapril
Inspra eplerenone
Integrilin eptifibatide
Istin amlodipine

Intelence etravirine
IntronA γ-interferon
Invanz ertapenem
Invirase saquinavir
Ipratropium Steri-Neb ipratropium
Iressa gefitinib
Isentress raltegravir
Ismelin guanethidine
Ismo isosorbide mononitrate
Isoket isosorbide dinitrate
Istodax romidepsin or depsipeptide
Jakafi ruxolitinib
Jantoven warfarin
Jetrea ocriplasmin
Jevtana cabazitaxel
Joicela lumiracoxib and biomarker
Juxtapid lomitapide
Kadcyla trastuzumab emantasine
Kaletra lopinavir with ritinavir
Kanuma sebelipase alfa
Karvea irbesartan
Katek telithromycin
Kefadim ceftazidime
Keflex cefalexin
Keflin cefalothin
Keftab cefalexin
Kefurox cefuroxime
Kemicetine chloramphenicol
Kenalog triamcinolone acetonide
Kengreal cangrelor
Kengrexal cangrelor
Kentera oxybutynin
Keytruda pembrolizumab
Kivexa abacavir and lamivudine
Klaricid clarithromycin
Kolanticon dicycloverine
Krystexxa pegloticase
Kwells hyoscine hydrobromide
Kynamro mipomersen
Kytril granisetron
Lacipil lacidipine
Lanoxin digoxin
Lanvis tioguanine
Largactil chlorpromazine
Lentaron formestane
Lenvima lenvatinib
Lescol fluvastatin
Letairis ambrisentan
Leukeran chlorambucil
Levaquin levofloxacin
Levulan 5-aminolevulanic acid
Lexiva fosamprenavir
Librium chlordiazepoxide
Lipaglyn saroglitazar

Lipantil fenofibrate
Lipitor atorvastatin
Lipoclin clinofibrate
Liskonum lithium carbonate
Lixiana edoxaban
Locoid hydrocortisone butyrate
Locorten-Vioform flumetasone pivalate and clioquinol
Lojuxta lomitapide
Loniten minoxidil
Lonsurf tipiracil and trifluridine
Lopresor metoprolol
Lopid gemfibrozil
Losec omeprazole
Lotemax loteprednol etabonate
Lotensin benzazepril
Lotriderm betamethasone dipropionate and
  clotrimazole
Lucentis ranibizumab
Lumizyme alglucosidase alfa
Lustral sertraline
Lynparza olaparib
Lyrinel oxybutynin
Lysovir amantadine
MabCampath alemtuzumab
MabThera rituximab
Macrobid nitrofurantoin
Macrodantin nitrofurantoin
Marcaine bupivacaine
Macugen pegaptanib
Manerix moclobemide
Mansil oxamniquine
Marevan warfarin
Mavik trandolapril
Maxidex dexamethasone
Maxitrol dexamethasone, neomycin sulphate, and
  polymyxin B sulphate
Maxolon metoclopramide
Medrone methylprednisolone
Mefoxin cefoxitin
Mekinist trametinib
Megace megestrol acetate
Meronem meropenem
Mestinon pyridostigmine
Metalyse tenecteplase
Metaramin metaraminol
Methadose methadone
Metosyn fluocinonide
Metrolyl metronidazole
Mevacor lovastatin
Mevastatin compactin
Micardis telmisartan
Micardis Plus telmisartan with hydrochlorothiazide
Mictral nalidixic acid
Mildison hydrocortisone

Minocin minocycline
Mircel D lucanthone
Mitoxana ifosfamide
Mivacron mivacurium
Modalim ciprofibrate
Modecate fluphenazine decanoate
Modrasone alclometasone dipropionate
Monopril fosinopril
Monotrim trimethoprim
Morcap morphine
Morphgesic morphine
Motens lacidipine
Movantik naloxegol
Moventig naloxegol
MST Cintinus morphine
MXL morphine
Mydriacyl tropicamide
Mydriasert phenylephrine with tropicamide
Mydrilate cyclopentolate
Myelostat omacetaxine mepesuccinate
Myleran busulfan
Mylotarg gemtuzumab
Myotonine bethanechol
Myozyme alglucosidase alfa
Myrbetriq mirabegron
Nalorex naltrexone
Narcan naloxone
Nardil phenelzine
Nasacort triamcinolone acetonide
Nasofan fluticasone propionate
Nasonex mometasone furoate
Navelbine vinorelbine
Nebcin tobramycin
Negaban temocillin
Negram nalidixic acid
Neoral ciclosporin
Nerisone diflucortolone valerate
Neulasta pegfilgrastim
Neupogen filgrastim
Neurontin gabapentin
Nexavar sorafenib
Nexium esomeprazole
Nicorette nicotine
Nicotinell nicotine
Nifedipress nifedipine
Nimotop nimodipine
Ninlaro ixazomib
Nipent pentostatin
Nipride sodium nitroprusside
NiQuitin CQ nicotine
Nitromin glyceryl trinitrate
Nitropress sodium nitroprusside
Nivaquine chloroquine
Nolvadex tamoxifen

Norcuron vecuronium
Norvasc amlodipine
Norvir ritonavir
Novantrone mitoxantrone
Nucala mepolizumab
Nuelin theophylline
Nulojix belatacept
Nuvigil armodafinal
Nystaform-HC hydrocortisone, nystatin, and chlorhexidine
Ocuclear oxymetazoline
Odomzo sonidegib
Olmecip olmesartan medoximil
Olmetec olmesartan medoximil
Olysio simeprevir
Onbrez indacaterol
Oncaspar pegaspargase (PEG-asparaginase)
Oncovin vincristine
Onivyde irinotecan
Onkotrone mitoxantrone
Opdivo nivolumab
Opsumit macitentan
Oramorph morphine
Orbactiv oritavancin
Orimeten aminoglutethimide
Otomize dexamethasone and neomycin sulphate
Otosporin hydrocortisone, neomycin sulphate, and polymyxin B sulphate
Otradrops xylometazoline
Otravin xylometazoline
Otrivine xylometazoline
Oxecta oxycodone
Oxis formoterol
Palladia toceranib
Palladone hydromorphone
Pamergan P100 pethidine
Paraplatin carboplatin
Pariet rabeprazole
Peg-Intron peginterferon a2b
Pegasys peginterferon a2a
Penbritin ampicillin
Pepcid famotidine
Perdix moexipril
Perforomist formoterol
Perjeta pertuzumab
Pharmorubicin epirubicin
Phenergan promethazine
Picovir pleconaril
Pilogel pilocarpine
Pitressin vasopressin
Plavix clopidogrel
Plendil felodipine
Polyfax bacitracin and polymyxin B sulphate
Pomalyst pomalidomide

Portrazza necitumumab
Posiject dobutamine
Possia ticagrelor
Pradaxa dabigatran etexilate
Praluent alirocumab
Pravachol pravastatin
Prazaxa dabigatran etexilate
Praxbind idarucizumab
Pred Forte prednisolone acetate
Predsol prednisolone sodium phosphate
Predsol-N prednisolone sodium phosphate with neomycin sulphate
Prescal isradipine
Pressonex metaraminol
Prestalia perindopril arginine and amlodopine besylate
Preterax perindopril and indapamide
Prexige lumiracoxib
Prezcobix darunavir and cobicastat
Prezita darunavir
Priadel lithium carbonate
Prilosec omeprazole
Primatene ephedrine
Primaxin cilastatin with imipenem
Prinivil lisinopril
Prioderm malathion
Pro-Banthine propantheline bromide
Probecid probenicid
Procardia nifedipine
Procoralan ivabradine
Proleukin aldesleukin
Prolia denosumab
Protium pantoprazole
Provera medroxyprogesterone acetate
Provigil modafinil
Prozac fluoxetine
Pulmicort budenoside
Puri-Nethol mercaptopurine
Purivase pegloticase
Pyrogastrone carbenoxolone
Qnexa phentermine and topiramate
Quellada M malathion
Questran colestyramine
Qvar beclometasone dipropionate
Raplacta peramivir
Rebetol ribavirin
Refludin lepirudin
Regurin trospium chloride
Relenza zanamivir
Relvar Elipta vilanterol with fluticasone furoate
Rapilysin reteplase
Remicade infliximab
Reminyl galantamine
Remitch nalfurafine
ReoPro abciximab

Repatha evolocumab
Rescriptor delavirdine
Respontin ipratropium
Retevase reteplase
Retrovir zidovudine
Revex nalmefene
Revlimid lanalidomide
Reyataz atazanavir
Rhinocort Aqua budenoside
Rifadin rifampicin
Rifater rifampicin
Rifinah rifampicin
Rimacid indometacin
Rimactane rifampicin
Rimactazid rifampicin
Risperdal risperidone
Ritalin methylphenidate
Rituxan rituximab
RoActemra tocilizumab
Rocephin ceftriaxone
Roferon-A α-interferon
Rogaine minoxidil
Rubex doxorubicin
Salamol Easi-Breathe salbutamol
Sanctura trospium chloride
Sandimmun ciclosporin
Sativex nabiximols
Savaysa edoxaban
Scopoderm TTS hyoscine
Sebomin minocycline
Sectral acebutolol
Securon verapamil
Selincro nalmefene
Septrin co-trimoxazole
Seretide fluticasone propionate and salmeterol
Serevent salmetrol
Seroxat paroxetine
Sevikar olmesartan medoximil with amlodipine besilate
Sevikar HCT olmesartan medoximil with amlodipine besilate and hydrochlorothiazide
Sevredol morphine
Sinemet co-careldopa
Sirturo bedaquiline
Sivextro tedizolid phosphate
Slo-Phyllin theophylline
Sofradex dexamethasone sodium metasulphobenzoate, framycetin sulphate, and gramicidin
Solu-Cortef hydrocortisone
Solu-Medrone methylprednisolone
Somavert pegvisomant
Sovaldi sofosbuvir
Sovriad simeprevir
Spiriva tiotropium bromide

Sprycel dasatinib
Stalevo co-careldopa
Stiolto Respimat tiotropium bromide with olodaterol
Stivarga regorafenib
Stratter atomoxetine
Strensiq asfotase alfa
Streptase streptokinase
Stribild emtricitabine, tenofovir disproxil, cobicastat, and elvitegravir
Striverdi Respimat olodaterol
Sublimaze fentanyl
Subutex buprenorphine
Sudafed pseudoephedrine
Suleo-M malathion
Sunvepra asunaprevir
Supralip fenofibrate
Sustanon 250 testosterone propionate
Sustiva efavirenz
Sutent sunitinib
Sylvant siltuximab
Symbicort budenoside and formoterol fumarate
Symmetrel amantadine
Synagis palivizumab
Synalar fluocinolone acetonide
Synalar C fluocinolone acetonide and clioquinol
Synalar N fluocinolone acetonide and neomycin sulphate
Synartis flunisolide
Synercid dalfopristin with quinupristine
Syner-kinase urokinase
Synribo omacetaxine mepesuccinate
Syntocinon oxytocin
Tafinlar dabrafenib
Tagamet cimetidine
Tagrisso osimertinib
Tamiflu oseltamivir
Tarceva erlotinib
Targocid teicoplanin
Tarivid ofloxacin
Tasigna nilotinib
Tavanic levofloxacin
Taxol paclitaxel
Taxotere docetaxel
Tazocin tazobactam with piperacillin
Technivie paritaprevir, ombitasvir, and ritonavir
Teflaro ceftaroline fosamil
Tekturna aliskiren
Telfast fexofenadine
Telzir fosamprenavir
Temgesic buprenorphine
Temodal temozolomide
Temodar temozolomide
Tenormin atenolol
Tensipine nifedipine

Teveten eprosartan
Thelin sitaxsentan
Ticlid ticlopidine
Tildiem diltiazem
Timentin ticarcillin with clavulanic acid
Timodine hydrocortisone, nystatin, and benzalkonium chloride
Tiova tiotropium bromide
Tivicay dolutegravir
TNKase tenecteplase
Tobi tobramycin
Tobradex dexamethasone and tobramycin
Tofranil imipramine
Tomudex raltitrexed
Torisel temsirolimus
Toviaz fesoterodine
Tracleer bosentan
Tracrium atracurium
Transtec buprenorphine
Trasicor oxprenolol
Treanda bendamustine
Triapin ramipril with felodipine
Trimopan trimethoprim
Trimovate clobetasone butyrate and oxytetracycline
Trisenox arsenic trioxide
Tritace ramipril
Triumeq abacavir, lamivudine, and dolutegravir
Trizivir abacavir, lamivudine, and zidovudine
Trosyl tioconazole
Truvada emtricitabine and tenofovir disoproxil
Tybost cobicistat
Tygacil tigecycline
Tykerb lapatinib
Tysabri natalizumab
Ultralanum Plain fluocortolone
Ultravate halobetasol propionate
Uniphyllin Continus theophylline
Unituxin dinutuximab
Univasc moexipril
Uptravi selexipag
Uriben nalidixic acid
Urispas 200 flavoxate
Uromitexan mesna
Utibron Neohaler indacaterol and glycopyrrolate
Valcyte valganciclovir
Valium diazepam
Valni nifedipine
Valtrex valaciclovir
Vancocin vancomycin
Vansil oxamniquine
Vascace cilazapril
Veasnoid tretinoin
Vectavir penciclovir
Vectibix panitumumab

Velbe vinblastine
Velcade bortezomib
Ventavis iloprost
Ventmax salbutamol
Ventodisks salbutamol
Ventolin salbutamol
Vepesid etoposide
Vesicare solifenacin
Viagra sildenafil
Vibramycin doxycycline
Videx didanosine
Viekira Pak dasabuvir, paritaprevir, ombitasvir, and ritonavir
Vigamox moxifloxacin
Vimizim elsosulfase alfa
Vioxx rofecoxib
Victrelis boceprevir
Viracept nelfinavir
Viraferon α-interferon
Viramune nevirapine
Virazole ribavirin
Viread tenofovir
Viroptic trifluridine
Virormone testosterone propionate
Virovir aciclovir
Virulex Forte atropine methonitrate, moroxydine, paracetamol, and scopolamine
Vistamethasone betamethasone sodium phosphate
Vistide cidofovir
Vistogard uridine triacetate
Vitravene fomivirsen
Volibris ambrisentan
Volmax salbutamol
Voraxaze glucarpidase
Votrient pazopanib
VPRIV velaglucerase alfa
Vumon teniposide
Vytorin ezetimibe and simvastatin
Warticon podophyllotoxin
Xalkori crizotinib
Xarelto rivaroxaban
Xeloda lapatinib
Xgeva denosumab
Xifaxin rifaximin
Xiaflex collagenase clostridium histolyticum
Xiapex collagenase clostridium histolyticum
Xolair omalizumab
Xopenex levalbuterol
Xtandi enzalutamide
Xtoro finafloxacin
Xylocaine lidocaine
Yentreve duloxetine
Yervoy ipilimumab
Yondelis trabectedin

Zactima vandetanib
Zaltrap aflibercept
Zanidip lercanidipine
Zantac ranitidine
Zarnestra tipifarnib
Zavedos idarubicin
Zeffix lamivudine
Zelapar selegiline
Zelboraf vemurafenib
Zenapax daclizumab
Zepatier elbasvir and grazoprevir
Zerbaxa ceftolozane with tazobactam
Zerit stavudine
Zestril lisinopril
Zetia ezetimibe
Zevalin ibritumomab
Zeftera/Zevtera ceftobiprole
Zelboraf vemurafenib
Ziagen abacavir
Zinacef cefuroxime

Zinnat cefuroxime
Zispin mirtazepine
Zithromax azithromycin
Zocor simvastatin
Zofran ondansetron
Zoladex goserelin
Zolinza vorinostat
Zomorph morphine
Zontivity vorapaxar
Zortress everolimus
Zosyn tazobactam with piperacillin
Zoton lansoprazole
Zovirax aciclovir
Zyban bupropion
Zydelig idelalisib
Zykadia ceritinib
Zyloric allopurinol
Zyprexa olanzapine
Zytiga abiraterone
Zyvox linezolid

Appendix 7

Hydrogen bonding interactions

The following table summarizes the possible hydrogen bonding interactions for selected functional groups (see also sections 1.3 and 13.1). The number of hydrogen bond donors (HBDs) and acceptors (HBAs) present in each functional group is given beneath each structure. In medicinal chemistry, the number of HBDs and HBAs correspond to the number of atoms capable of forming such interactions. Weak HBAs are not included in this text; for example nitrogen atoms that are part of an amide or aniline structure, nitrogen atoms where the lone pair is part of an aromatic sextet, or sp³ hybridized oxygen atoms that are linked to an sp² hybridized centre.

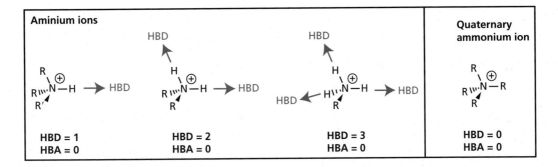

Aminium ions

HBD = 1
HBA = 0

HBD = 2
HBA = 0

HBD = 3
HBA = 0

Quaternary ammonium ion

HBD = 0
HBA = 0

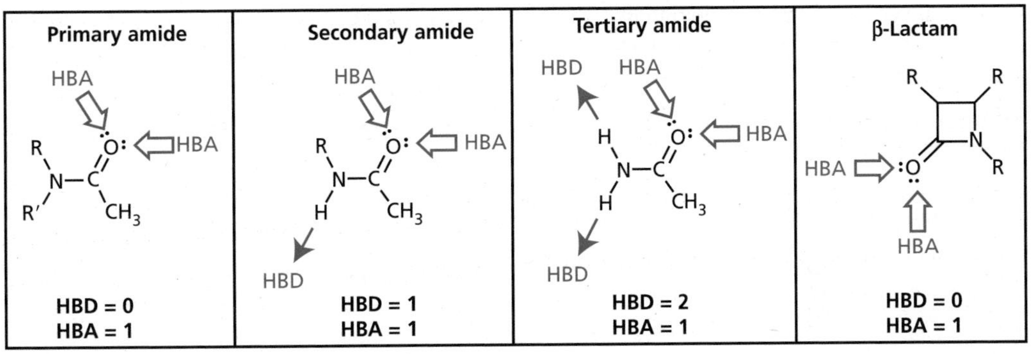

Primary amide

HBD = 0
HBA = 1

Secondary amide

HBD = 1
HBA = 1

Tertiary amide

HBD = 2
HBA = 1

β-Lactam

HBD = 0
HBA = 1

Glossary

3D QSAR QSAR studies which relate the biological activities of a series of compounds to their steric and electrostatic fields determined by molecular modelling software.

Abzyme An antibody with catalytic properties.

ACE inhibitors Drugs which inhibit the enzyme angiotensin-converting enzyme. Inhibition prevents the synthesis of a powerful vasoconstrictor and so ACE inhibitors are used as antihypertensive agents.

Acetylcholine A neurotransmitter that is present in both the peripheral and central nervous systems.

Acetylcholinesterase An enzyme that hydrolyses the neurotransmitter acetylcholine.

Acquired resistance Resistance that a microorganism acquires to a drug to which it was previously susceptible.

Acromegaly A long-term condition where the body is producing too much growth hormone. It results in increased tissue growth.

Action potential Refers to the reversal in membrane potential as a signal travels along the axon of a nerve.

Activation energy The energy required for a reaction to reach its transition state.

Active conformation The conformation adopted by a compound when it binds to its target binding site.

Active principle The single chemical in a mixture of compounds which is chiefly responsible for that mixture's biological activity.

Active site The binding site of an enzyme where a reaction is catalysed by the enzyme.

Active transport The process by which a substance is transported across a membrane against its concentration gradient. Energy in the form of ATP hydrolysis is required to drive the process.

ADAPT Antibody-directed abzyme prodrug therapy.

Addiction Addiction can be defined as a habitual form of behaviour. It need not be harmful. For example, one can be addicted to eating chocolate or watching television without suffering more than a bad case of toothache or a surplus of soap operas.

Adenoviruses Icosahedral-shaped viruses containing double-stranded DNA. They are responsible for respiratory infections.

ADEPT Antibody-directed enzyme prodrug therapy.

ADME Refers to drug absorption, drug distribution, drug metabolism, and drug excretion.

Adrenal medulla A gland that produces adrenaline.

Adrenaline A catecholamine that acts as a hormone and neurotransmitter, and which plays a crucial part in the 'fight or flight' response. It is also called epinephrine.

Adrenergics Refers to compounds that interact with the receptors targeted by adrenaline and noradrenaline.

Adrenoceptors Receptors that are activated by adrenaline and noradrenaline.

Adrenocorticoids Those steroids released from the adrenal cortex of the adrenal gland.

Adsorption Refers to the situation where a molecule or a structure adheres to a surface. In virology it refers to a virus binding to the surface of a host cell.

Aerobic bacteria Bacteria that grow in the presence of oxygen.

Affinity Affinity is a measure of how strongly a ligand binds to its target binding site.

Affinity constant A measure of the bonding affinity between two molecules at equilibrium. It is the reverse of the dissociation constant.

Affinity screening A method of screening compounds based on their binding affinity to a target.

Agonist A drug that produces the same response at a receptor as the natural messenger.

AIDS Acquired immune deficiency syndrome.

Alchemy A molecular modelling software package.

Alkaloids Natural products extracted from plants. They contain an amine functional group.

Alkylating agents Agents which act as electrophiles and form irreversible covalent bonds with macromolecular targets. These agents are classed as cytotoxic and are used as anticancer agents.

Allosteric Refers to a protein binding site other than the one used by the normal ligand, and which affects the activity of the protein. An allosteric inhibitor binding to an allosteric binding site induces a change of shape in the protein which disguises the normal binding site from its ligand.

Alveoli The small sacs in lung tissue where gaseous exchange takes place between the contents of the lungs and the blood vessels surrounding the sacs.

Ames test A biological test used to assess whether potential drugs are mutagenic.

Aminoacridines A group of synthetic antibacterial agents that target bacterial DNA.

Aminoacyl tRNA synthetases Enzymes that catalyse the attachment of an amino acid to tRNA. Potentially useful targets in antibacterial therapy.

Aminoglycosides A group of antibacterial agents that contain sugar components and a basic amino function.

Aminopeptidases Enzymes that catalyse the hydrolysis of amino acids from the *N*-terminus of a peptide or protein.

Anaerobic bacteria Bacteria that grow in the absence of oxygen.

Analgesics A group of compounds used clinically as painkillers.

Anaphylactic reaction A serious allergic response to a chemical stimulus. It can be life threatening.

Anchimeric assistance The process by which a functional group in a molecule can accelerate a reaction at another functional group by participating in the reaction mechanism. Also known as neighbouring group participation.

Anchor *see* Linker

Androgens Hormones that are used in anticancer therapy.

Angiogenesis The process by which new blood vessels are formed.

Angioplasty A process by which a narrowed blood vessel is mechanically widened by the use of a balloon catheter and stent.

Angiostatin An endogenous compound that inhibits angiogenesis.

Angiotensin-converting enzyme An enzyme that catalyses the conversion of angiotensin I to the hypertensive hormone angiotensin II.

Angiotensin receptor A receptor that is activated by the hypertensive agent angiotensin II.

Angiotensinogen The protein substrate for the enzyme renin. The enzyme-catalysed reaction produces angiotensin I.

Anomers Cyclic stereoisomers of sugars that differ only in their configurations at the hemiacetal (anomeric) carbon.

Antacid A substance that is taken orally to lower the acidity of the stomach contents.

Antagonist A drug which binds to a receptor without activating it, and which prevents an agonist or a natural messenger from binding.

Antedrugs *see* Soft drugs

Anthracyclines A group of antibiotics that are important in anticancer therapy.

Anti-androgens Anticancer agents that block the action of androgens at their receptors.

Anti-angiogenesis agents Agents used in anticancer therapy that inhibit the growth of new blood vessels.

Antibacterial agent A synthetic or naturally occurring agent which can kill or inhibit the growth of bacterial cells.

Antibiotic An antibacterial agent derived from a natural source.

Antibody A Y-shaped glycoprotein generated by the body's immune system to interact with an antigen present on a foreign molecule. Marks the foreign molecule for destruction.

Antibody–drug conjugates Refers to antibodies with drugs covalently linked to their structure.

Anticholinesterases Agents which inhibit the enzyme acetylcholinesterase.

Anticodon A set of three nucleic acid bases on tRNA that base pair with a triplet of nucleic acid bases on mRNA during translation. The amino acid linked to tRNA is determined by the anticodon that is present.

Antidiuretic An agent used to reduce the level of urination and increase water retention.

Anti-emetic A drug used to prevent nausea and vomiting.

Anti-estrogens Agents which bind to estrogen receptors and block the binding of estrogen. Used in anticancer therapy.

Antigen A region of a molecule that is 'recognized' by the body's immune system and which will interact with antibodies targeted against it.

Antigenic drift The process by which antigens gradually vary in nature.

Antigenic shift Refers to a large alteration in the nature of antigens.

Antigenic variation A property of some viruses which are able to vary the chemical structure of antigens on their surface through rapid mutations.

Antihypertensives Agents used to lower blood pressure by dilating blood vessels.

Antimetabolites Agents which inhibit enzymes that are crucial to the normal metabolism of the cell. Used in antibacterial and anticancer therapy.

Anti-oestrogens *see* Anti-estrogens

Anti-oncogenes Genes that code for proteins which check the 'normality' of a cell and which induce cell death if abnormalities are present. Important in preventing the birth of a cancer cell.

Antisense therapy The design of molecules which will bind to specific regions of mRNA and prevent mRNA acting as a code for protein synthesis.

Antrum Part of the pyloric region situated at the bottom part of the stomach.

Anxiolytic A drug that relieves anxiety.

Aorta The principal artery carrying blood away from the heart.

Apaf-1 A scaffolding protein for apoptosomes.

Apoptosis The process by which a cell commits suicide.

Apoptosome A protein complex that destroys the cell's proteins and leads to apoptosis.

Aptamers Oligonucleotides or peptides that bind to a target molecule.

Aquaporins Membrane-bound proteins containing a pore that allows water to pass through the membrane.

Area under the plasma drug concentration curve (AUC) Represents the total amount of drug that is available in the blood supply during a dosing regimen.

Aromatase An enzyme that catalyses an aromatization reaction in oestrogen synthesis. Aromatase inhibitors are used as anticancer agents.

Arteries Blood vessels taking blood away from the heart.

Aspartyl proteases Enzymes that catalyse the hydrolysis of peptide bonds in protein substrates and which contain aspartate residues in the active site that take part in the hydrolysis mechanism.

Aspergillosis A fungal infection caused by the *Aspergillus* fungus.

Asymmetric centre An atom with four different substituents that frequently results in asymmetry for the whole molecule.

Asymmetric synthesis A synthesis which shows selectivity for a particular enantiomer or diastereomer of an asymmetric compound.

Attention deficit hyperactivity disorder A disorder associated with children—the name speaks for itself.

AUC *see* Area under the plasma drug concentration curve

Autonomic motor nervous system Nerves carrying messages from the central nervous system to smooth muscle, cardiac muscle, and the adrenal medulla.

Autoreceptors Presynaptic receptors that are involved in a feedback control whereby the ligand released by the presynaptic neuron binds and inhibits further ligand release.

Bacilli Bacterial cells that are rod shaped.

Bacteriorhodopsin A protein present in Archaea microorganisms that captures light energy and acts as a proton pump to pump protons across the cell membrane.

Bacteriostatic Bacteriostatic drugs inhibit the growth and multiplication of bacteria, but do not directly kill them.

Bactericidal Bactericidal drugs actively kill bacterial cells.

Bacteriophage A virus that invades bacterial cells.

Bad A protein that promotes apoptosis.

Barbiturates A series of synthetic compounds with sedative properties.

Bax A protein that promotes apoptosis.

Bcl-2 and Bcl-X Proteins that suppress apoptosis.

Bcr-Abl protein A protein that is formed as a result of a chromosomal defect called the Philadelphia chromosome. It is related to the disease known as chronic myelogenous leukaemia.

Benign cancer or tumour A localized tumour that is not life threatening.

Benign prostatic hyperplasia The term for an enlarged prostate, resulting in difficulty in passing urine.

Benzodiazepines A structural class of compounds that are used as hypnotics, anxiolytics, sedatives, anticonvulsants, and muscle relaxants.

Beta-blockers Compounds that block or antagonize β-adrenoceptors. Particularly useful in cardiovascular medicine.

Beta-lactamase inhibitors Agents which inhibit the beta lactamase enzymes.

Beta lactamases *see* Lactamases

Beta-lactams Structures that contain a four-membered β-lactam ring and are commonly used as antibacterial agents.

Bile duct A duct leading from the liver to the intestines. Some drugs and drug metabolites are excreted through the bile duct, but can be reabsorbed from the intestines.

Binding region A region within a binding site that is capable of interacting with a drug or an endogenous molecule by intermolecular bonding.

Binding site The location where an endogenous molecule or drug binds on a macromolecule. Normally a hollow or cleft in the surface of the macromolecule.

Bioavailability Refers to the fraction of drug that is available in the blood supply following administration.

Bioequivalence studies Studies carried out to ensure that the bioavailability of a drug remains the same should there be any alteration to the manufacture or formulation of the drug.

Bio-isostere A chemical group which can replace another chemical group without affecting the biological activity of the drug.

Bioterrorism The use of toxic infectious agents by terrorist groups.

Bleomycins A group of naturally occurring glycoproteins used as anticancer agents.

Blood–brain barrier Blood vessels in the brain are less porous than blood vessels in the periphery. They also have a fatty coating. Drugs entering the brain have to be lipophilic in order to cross this barrier.

Boc A shorthand term for the protecting group, *t*-butyloxycarbonyl.

Bromodomain A binding region in one protein that is capable of binding an acetylated lysine residue of another protein to allow protein–protein interactions.

Bronchodilator An agent which dilates the airways and can combat asthma.

Cache A molecular modelling software package.

Calmodulin A calcium-binding protein that activates several protein kinases.

Camptothecins A group of naturally occurring alkaloids and semi-synthetic derivatives used as anticancer agents.

Canaliculae Invaginations or channels formed by parietal cells which connect with the lumen of the stomach.

Capillaries Small blood vessels.

Carboxypenicillins A family of penicillins having a carboxylic acid substituent at the α-position. They have been largely superseded by newer agents.

Capsid A protein coat that encapsulates the nucleic acid of a virus.

Capsid-binding agents Antiviral drugs that stabilize the capsid of the human rhinovirus by binding to a hydrophobic pocket normally occupied by a pocket factor.

Carbapenems A group of β-lactam antibacterial agents, so called because they lack a sulphur atom.

Carboxypeptidases Enzymes that hydrolyse the peptide link at the C-terminus of a peptide chain.

Carcinogenesis The birth of a cancer.

Carrier protein A protein in the membrane of a cell which is capable of transporting specific polar molecules across the membrane. The molecules transported are too polar to cross the membrane themselves and are crucial to the survival and functions of the cell.

Caspases Enzymes which have important roles to play in the ageing process of cells.

Catecholamines Compounds that contain a basic amino group and a catechol ring. The catechol ring consists of an aromatic ring with two phenolic groups, *ortho* to each other.

Catechol-O-methyltransferase A metabolic enzyme that catalyses the methylation of a phenol group in catecholamines such as noradrenaline and adrenaline.

Cation–pi interaction The interaction of a positively charged group with a pi-electron system to produce an induced dipole that results in a binding interaction between the dipole and the ion.

Cell cycle Refers to recognizable phases of cell growth, DNA synthesis, and cell division.

Cell membrane A phospholipid bilayer surrounding all cells. It acts as a hydrophobic barrier.

Central nervous system The nervous tissue of the brain and the spinal column.

Centroid A dummy atom used in molecular modelling to define the centre of an aromatic or heteroaromatic ring. Has also been used to mean the scaffold of a molecule.

Cephalosporinases *see* Lactamases

Cephalosporins A group of β-lactam semi-synthetic antibacterial agents that target the bacterial transpeptidase enzymes.

Cephamycins A family of cephalosporins having a methoxy substituent at the 7-position.

Chain cutters Agents that interact with DNA, leading to the splitting of the DNA backbone. Generally operate by producing radicals. Used as anticancer agents.

Chain contraction/extension strategy The variation of chain length in a drug to optimize the separation between different binding groups.

Chem3D A molecular modelling software package.

ChemDraw A chemical drawing software package.

ChemWindow A chemical drawing software package.

Chemokine receptors G-protein-coupled receptors that are activated by small proteins called chemokines. Activation results in movement of the cell to a particular location within the organism.

Chemotherapeutic index A comparison of the minimum effective dose of a drug with the maximum dose which can be tolerated by the host.

Chimeric antibodies Antibodies that are part human and part mouse (or other species) in nature.

Chirality The property of asymmetry where the mirror images of a molecule are non-superimposable.

Chiral switching The replacement of a racemic drug on the market with its more active enantiomer.

Chloramphenicol acetyltransferase An enzyme present in chloramphenicol-resistant bacteria that catalyses the acylation of hydroxyl groups present in the drug.

CHO cells Chinese Hamster ovarian cells. Commonly used to express a cloned receptor on their surface for *in vitro* tests.

Choline acetyltransferase An enzyme that catalyses the synthesis of acetylcholine.

Cholinergics Refers to compounds that interact with cholinergic receptors.

Cholinergic receptors Receptors that are activated by acetylcholine.

Chromatin A structure consisting of DNA wrapped round proteins such as histone.

Cloning The process by which identical copies of a DNA molecule or a gene are obtained.

CMV *see* Cytomegalovirus

Coactivator protein A protein that interacts with a transcription factor to form a protein complex that either activates or represses transcription.

Cocci Bacterial cells that are spherical in shape.

Coenzyme A small organic molecule that acts as a co-factor.

Cofactor An ion or small organic molecule (other than the substrate) which is bound to the active site of an enzyme and takes part in the enzyme-catalysed reaction.

Combinatorial libraries A store of compounds that have been synthesized by combinatorial synthesis.

Combinatorial synthesis A method of synthesizing large quantities of compounds in small scale using automated or semi-automated processes. Normally carried out as solid phase syntheses.

Combretastatins Naturally occurring anticancer agents that inhibit tubulin polymerization.

Comparative molecular field analysis (CoMFA) A method of carrying out 3D QSAR that was developed by the company Tripos.

Competitive inhibitors Reversible inhibitors that compete with the normal substrate for an enzyme's active site.

Compound banks or libraries A store of synthetic compounds that have been produced by traditional methods or by combinatorial syntheses.

Conformational analysis A study of the various conformations permitted for a molecule.

Conformations are different three-dimensional shapes arising from single bond rotations.

Conformational blockers Groups that are added to molecules to prevent them adopting certain conformations.

Conformational space The three-dimensional space surrounding the scaffold of a molecule.

Conjugation In the chemical sense it refers to interacting systems of π bonds. In the microbiological sense, it refers to the process by which bacterial cells pass genetic information directly between each other.

Conjugation reactions *see* Phase II reactions.

Constitutional activity Some receptors (e.g. the GABA*, serotonin, and dihydropyridine receptors) are found to have an inherent activity even in the absence of the chemical messenger. They are said to be constitutionally active.

Convergent evolution In a biochemical sense, refers to receptors which are in different branches of the receptor evolutionary tree, but which have converged to recognize the same endogenous ligand.

Correlation coefficient *see* Regression coefficient

Cotransmitters Chemical messengers which are released from nerves along with the major neurotransmitter. They have a fine-tuning effect on the signal received.

Craig plot A plot which compares the values of two physicochemical parameters for different substituents.

Cross-validated correlation coefficient (q^2) A measure of the predictability of a 3D QSAR equation.

Cross-validation Used in 3D QSAR in order to obtain a QSAR equation. Tests how an equation predicts the activity of a test compound that has not been included in the derivation of the equation.

Cryptophycins Naturally occurring anticancer agents that inhibit tubulin polymerization.

Cyclases Enzymes that catalyse cyclization reactions such as the formation of cyclic AMP from ATP.

Cyclin dependent kinases Enzymes that are activated by cyclins and which catalyse phosphorylation reactions that control the cell cycle.

Cyclins A group of proteins that are important in the control and regulation of the cell cycle.

Cyclodextrins Cyclic structures made up of sugar molecules.

Cyclooxygenases Enzymes that are important in the production of prostaglandins.

CYP Shorthand terminology for cytochrome P450 enzymes; for example CYP3A4.

Cytochrome C Released by mitochondria to promote apoptosis.

Cytochrome P450 enzymes Enzymes that are extremely important in the metabolism of drugs. They catalyse oxidation reactions.

Cytomegalovirus A virus that causes eye infections and blindness.

Cytoplasm The contents of a cell.

Cytotoxic agents Anticancer agents that are generally toxic to cells by a number of mechanisms.

Database mining The use of computers to automatically search databases of compounds for structures containing specified pharmacophores.

De novo drug design The design of a drug or lead compound based purely on molecular modelling studies of a binding site.

Death activator proteins Chemical messengers that trigger a cell to commit suicide.

Deconvolution The isolation and identification of an active compound in a mixture of compounds obtained from a combinatorial synthesis.

Dependence A compulsive urge to take a drug for psychological or physical needs. The psychological need is usually why the drug was taken in the first place (to change one's mood) but physical needs are often associated with this. This shows up when the drug is no longer taken leading to psychological withdrawal symptoms (feeling miserable) and physical withdrawal symptoms (headaches, shivering, etc.). Dependence need not be a serious matter if it is mild and the drug is non-toxic (e.g. dependence on coffee). However, it is serious if the drug is toxic and/or shows tolerance. Examples: opiates, alcohol, barbiturates, and diazepams.

Desensitization The process by which a receptor becomes less sensitive to the continued presence of an agonist.

Desolvation A process that involves the removal of surrounding water from molecules before they can interact with another; for example a drug with its binding site. Energy is required to break the intermolecular interactions involved.

Diacylglycerol A secondary messenger that is generated by the action of the enzyme phospholipase C on phosphatidylinositol diphosphate.

Differentiation The ability of cells to become specialized in a multicellular organism.

Dihydrofolate reductase An enzyme involved in generating tetrahydrofolate—an important enzyme cofactor. Dihydrofolate reductase inhibitors prevent the synthesis of nucleic acids and are used as antibacterial and anticancer agents.

Dihydropteroate synthetase A bacterial enzyme that catalyses the synthesis of dihydropteroate. It is the molecular target for the sulphonamide antibacterial agents.

Dipole–dipole interactions Interactions between two separate dipoles. A dipole is a directional property and can be represented by an arrow between an electron-rich part of a molecule and an electron-deficient part of a molecule. Different dipoles align such that an electron-rich area interacts with an electron-deficient area.

Discovery Studio Pro A molecular modelling software package.

Displacer A test compound that competes with a radioligand for the binding site of a receptor.

Divergent evolution Receptors that diverged early in evolution have greater differences in their binding sites and ligand preferences.

DNA Deoxyribonucleic acid.

DNA ligase An enzyme that repairs breaks in the DNA chain.

DNA polymerases Enzymes that catalyse the synthesis of DNA from a DNA template.

DNA viruses Viruses that contain DNA as their nucleic acid.

DOCK A software program used for docking molecules into target binding sites.

Docking The in silico process by which a molecular modelling program fits a molecule into a target binding site.

Dose ratio The agonist concentration required to produce a specified level of effect when no antagonist is present, compared to the agonist concentration required to produce the same level in the presence of an antagonist.

Drug–drug interactions Related to the effect one drug has on the activity of another if both drugs are taken together.

Drug load The ratio of active drug in the total contents of a dose.

Drug metabolism The reactions undergone by a drug when it is in the body. Most metabolic reactions are catalysed by enzymes, especially in the liver.

dsDNA Double-stranded DNA. A term used in virology.

dsRNA Double-stranded RNA. A term used in virology.

Dual-action inhibitor An agent that inhibits two separate targets or two separate regions of the same target.

Dummy atom *see* Centroid

Dynamic combinatorial chemistry The generation of a mixture of products from a mixture of starting materials in the presence of a target. Products are in equilibrium with starting materials and the equilibrium shifts to products binding to the target.

Dynamic structure–activity analysis The design of drugs based on which tautomer is preferred for activity.

Dynorphins Endogenous polypeptides that act as analgesics.

E_s *see* Taft's steric factor

EC$_{50}$ The concentration of drug required to produce 50% of the maximum possible effect.

ED$_{50}$ The mean effective dose of a drug necessary to produce a therapeutic effect in 50% of the test sample.

Efficacy A measure of how effectively an agonist activates a receptor. It is possible for a drug to have high affinity for a receptor (i.e. strong binding interactions) but have low efficacy.

Efflux A process by which drugs are expelled from a cell through the action of cell membrane carrier proteins.

Electronic screening *see* Database mining

Electrostatic interactions *see* Ionic interactions

EMEA *see* European Agency for the Evaluation of Medicinal Projects

Enantiomers The mirror image forms of an asymmetric molecule.

Endocytosis The process by which a segment of cell membrane folds inwards and is 'nipped off' to form a vesicle within the cell.

Endogenous compounds Chemicals which are naturally present in the body.

Endomorphins Endogenous tetrapeptides that act as analgesics.

Endorphins Endogenous polypeptides that act as analgesics.

Endoplasmic reticulum Folds of membrane within eukaryotic cells. Endoplasmic reticulum can be defined as smooth or rough according to its appearance under the electron microscope. Rough endoplasmic reticulum has ribosomes attached to it and is where protein synthesis takes place.

Endosome A membrane-bound vesicle within eukaryotic cells.

Endpoint Some form of measurable effect. Used in clinical trials to determine whether a drug is successful or not.

Energy minimization An operation carried out by molecular modelling software to find a stable conformation of a molecule.

Enkephalinases Enzymes which hydrolyse enkephalins.

Enkephalins Endogenous peptides which act as analgesics.

Enteric nervous system Located in the walls of the intestine. Responds to the autonomic nervous system and local hormones.

Enzyme A protein that acts as a catalyst for a reaction.

Epimerization The inversion of an asymmetric centre.

Epitopes Small molecules that bind to part of a binding site and do not produce a biological effect as a result of binding.

Epothilones Naturally occurring anticancer agents that inhibit tubulin depolymerization.

Ergosterol A fungal steroid that is an important constituent of the fungal cell membrane.

Estradiol A female sex hormone with estrogenic activity.

Estrogens Compounds that are important to the estrous cycle in humans or animals. The natural estrogens are steroids and act as female sex hormones.

Eukaryotic cell The cells that are present in plants, animals, and multicellular organisms. They contain a membrane-bound nucleus and organelles.

European Patent Convention (EPC) A group of European countries for which patents can be drawn up based on a European patent.

European Agency for the Evaluation of Medicinal Projects (EMEA) The European regulatory authority for the testing and approval of drugs.

European Patent Office (EPO) Issues European patents.

Exocytosis The process by which vesicles within a cell fuse with a cell membrane and release their contents out of the cell.

Exons The ends of an mRNA molecule that are spliced together after the removal of an intron during post-transcriptional modifications.

Extension strategies The addition of functional groups to a drug with the aim of achieving a further binding interaction with another binding region in the binding site.

F A symbol used in pharmacokinetic equations to represent oral bioavailability. Alternatively, a symbol used in QSAR equations to represent the inductive effect of a substituent.

Farnesyl transferase An enzyme that attaches a farnesyl group to the Ras protein to allow membrane attachment.

Fast tracking A method of pushing a drug through clinical trials and the regulatory process as quickly as possible. Applied to drugs that show distinct advantages over current drugs in the treatment of life-threatening diseases, or for drugs that can be used to treat diseases that have no current treatment.

FDA *see* Food and Drugs Administration

Feedback control The process by which the product of an enzymatic reaction or a series of enzymatic reactions controls the level of its own production.

FGF *see* Fibroblast growth factor

Fibroblast growth factor A growth factor that stimulates angiogenesis.

Fight or flight response Refers to the reaction of the body to situations of stress or danger, and which involves the release of adrenaline and other chemical messengers that prepare the body for physical effort.

First pass effect The extent to which an orally administered drug is metabolized during its first passage through the gut wall and the liver.

Fischer's lock and key hypothesis *see* Lock and key hypothesis

Fisher's F-test A statistical test used to assess the significance of coefficients in a QSAR equation.

Flagellum A tail-like structure used by some microorganisms as a method of propulsion.

Fluoroquinolones A group of synthetic antibacterial agents.

Fmoc A shorthand term for the protecting group, 9-fluorenylmethoxycarbonyl.

Folic acid A vitamin that is converted to an important enzyme cofactor

Food and Drugs Administration (FDA) The drugs regulatory authority in the US.

Force field Relevant to molecular modelling. Refers to the calculation of the interactions and energies between different atoms resulting from bond stretching, angle bending, torsional angles, and non-bonded interactions.

Free–Wilson approach A QSAR equation which uses indicator variables rather than physicochemical parameters.

Fusion inhibitors Agents that inhibit the fusion of HIV with the cell membrane of host cells.

G-proteins Membrane-bound proteins consisting of three subunits which are important in the signal transduction process from activated G-protein-coupled receptors.

G-protein-coupled receptors Membrane-bound receptors that interact with G-proteins when they are activated by a ligand.

Gastrointestinal tract Consists of the mouth, throat, stomach, and the upper and lower intestines.

Gating The mechanism by which ion channels are opened or closed.

GCP *see* Good Clinical Practice

GDEPT Gene-directed enzyme prodrug therapy.

Genetic polymorphism The variation in DNA sequence for a particular gene amongst different individuals.

Genomics The study of the genetic code for an organism.

Glomerulus A knotted arrangement of blood vessels which fits into the opening of a nephron and from which water and small molecules are filtered into the nephron.

Global energy minimum The most stable conformation of a molecule.

GLP *see* Good Laboratory Practice

Glucagon A peptide hormone that is released by the pancreas and promotes a rise in blood sugar levels.

Glucocorticoids Hormones that are used in anticancer therapy and as anti-inflammatory agents.

Gluconeogenesis The biochemical process by which glucose is produced in the body from non-sugar substrates.

Glycoconjugate The general term for macromolecules that are linked to carbohydrates.

Glycolipid A lipid molecule linked to one or more carbohydrates.

Glycomics The study of carbohydrates.

Glycopeptides and glycoproteins Peptides and proteins that are linked to one or more carbohydrates.

Glycopeptide antibacterial agents Glycopeptides with antibacterial activity, the most important being vancomycin.

Glycosidases Enzymes that catalyse the hydrolysis of the glycosidic bond between carbohydrate groups.

Glycosphingolipids Glycoconjugates which are thought to be important in the regulation of cell growth. Includes the molecules responsible for labelling blood cells.

GMP *see* Good Manufacturing Practice

Gonadotrophin-releasing hormone *see* Luteinizing hormone-releasing hormone

Good Clinical Practice (GCP) Scientific codes of practice that apply to clinical trials and which are monitored by regulatory authorities.

Good Laboratory Practice (GLP) Scientific codes of practice that apply to a pharmaceutical company's research laboratories and which are monitored by regulatory authorities.

Good Manufacturing Practice (GMP) Scientific codes of practice that apply to a pharmaceutical company's production plants and which are monitored by regulatory authorities.

Granzyme An enzyme introduced into defective cells by T-lymphocytes, and which induces apoptosis.

GRID A molecular modelling software program that maps the nature of binding regions within a binding site.

Group shifts The transposition of a group within a molecule to make it unidentifiable to metabolic enzymes but not to target binding sites.

Growth factors Hormones that activate membrane-bound receptors and trigger a signal transduction pathway leading to cell growth and division.

GTPase activating proteins (GAPs) Regulatory proteins that bind to activated small G-proteins and promote the autocatalytic process by which G-proteins hydrolyse bound GTP to GDP. This terminates the G-protein's activity as a signalling protein.

Guanine nucleotide exchange factors (GEF) Regulatory proteins that enhance signalling by small G-proteins. They facilitate the exchange of bound GDP for bound GTP.

HAART *see* Highly active antiretroviral therapy.

Haemagglutinin A glycoprotein on the surface of the flu virus that is crucial to the infection process.

Half-life The time taken for the plasma concentration of a drug to fall by half.

HAMA response Human anti-mouse antibodies are antibodies that are produced against monoclonal antibodies which have been derived from a mouse source, and are recognized as foreign by the body's immune system.

Hammett substituent constant (σ) A measure of whether a substituent is electron-withdrawing or electron-donating and to what extent.

Hansch equation A QSAR equation involving various parameters.

Hard drugs Drugs that are resistant to metabolism.

HBA *see* Hydrogen bond acceptor

HBD *see* Hydrogen bond donor

Helicases Enzymes that catalyse the coiling and uncoiling of DNA.

Helicobacter pylori An organism that can survive in the stomach and cause damage to the stomach lining, leading to ulcers.

Henderson–Hasselbalch equation An equation that is used to determine the extent of ionization of an ionizable drug at a particular pH.

Herpes viruses Viruses responsible for cold sores and other herpes infections.

High-throughput screening An automated method of carrying out a large number of *in vitro* assays on small scale.

Highly active antiretroviral therapy (HAART) A therapy used in the treatment of HIV which involves a combination of antiviral drugs.

Histone acetylase and histone deacetylase Enzymes that acetylate and deacetylate the lysine residues of the structural protein, histone. Important in the control of gene expression.

HIV Human immunodeficiency virus.

HOMO Highest occupied molecular orbital.

Homology models A term used in molecular modelling for the construction of a model protein or binding site based on the structure of known proteins or binding sites.

Hormones Endogenous chemicals that act as chemical messengers. They are typically released from glands and travel in the blood supply to reach their targets. Some hormones are local hormones and are released from cells to act in the immediate area around the cell.

HRV *see* Human rhinoviruses

Human genome project The sequencing of human DNA.

Human intestinal di-/tripeptide transporter-1 A transport protein that transports dipeptides across the gut wall.

Human intestinal proton-dependent oligopeptide transporter-1 A transport protein that transports dipeptides across the gut wall.

Human rhinoviruses RNA viruses responsible for the common cold.

Hybridization The mixing of atomic orbitals to form hybridized atomic orbitals. With atoms such as carbon, nitrogen, and oxygen, it involves the mixing of 2s and 2p orbitals to produce sp, sp^2, or sp^3 hybridized orbitals. This is important in determining whether the atoms concerned can form π bonds.

Hybridomas Cells that are formed from the fusion of B-lymphocytes with immortal B-lymphocytes in the production of monoclonal antibodies.

Hydrogen bond A non-covalent bond that takes place between an electron-deficient hydrogen and an electron-rich atom, particularly oxygen and nitrogen.

Hydrogen bond acceptor A functional group that provides the electron-rich atom required to interact with a hydrogen in a hydrogen bond.

Hydrogen bond donor A functional group that provides the hydrogen required for a hydrogen bond.

Hydrolases Enzymes that catalyse hydrolysis reactions.

Hydrophilic Refers to compounds that are polar and water soluble. Literally means water loving.

Hydrophobic Refers to compounds that are non-polar and water insoluble. Literally means water hating.

Hydrophobic interactions Refers to the stabilization that is gained when two hydrophobic regions of a molecule or molecules interact and shed the ordered water 'coat' surrounding them. The water molecules concerned become less ordered resulting in an increase in entropy.

17α-Hydroxylase-17(20)-lyase A cytochrome P450 enzyme which is involved in the biosynthesis of androgens from cholesterol.

Hyperchem A molecular modelling software package.

Hypoglycaemia Lowered glucose levels in the blood.

Hypoxia A lack of oxygen.

Hypoxia-inducible factors (HIF) Transcription factors that respond to low levels of oxygen to up-regulate genes that promote cell survival in oxygen-starved environments

IC$_{50}$ The concentration of an inhibitor required to inhibit an enzyme by 50%.

IND *see* Investigational Exemption to a New Drug Application

Immunomodulators Agents that either suppress or enhance the immune system.

Immunosuppressants Drugs that inhibit the immune response. Useful in the treatment of autoimmune disease and in reducing the chances of rejection following organ transplants.

Impurity profiling The study of drug batches to identify and quantify any impurities that might be present.

Indicator variables A variable used in QSAR equations which is given the value of 1 or 0 depending on whether a substituent is present or not.

Induced dipole interactions The situation where a charge or a dipole on one molecule induces a dipole in another molecule to allow an ion–dipole interaction or a dipole–dipole interaction respectively. An induced dipole normally requires the presence of π electrons.

Induced fit The alteration in shape that arises in a macromolecule such as a receptor or an enzyme when a ligand binds to its binding site.

Inhibition constant A measure of the equilibrium between an enzyme–inhibitor complex and the uncomplexed enzyme and inhibitor.

Inhibitor An agent that binds to an enzyme and inhibits its activity.

Inositol triphosphate A secondary messenger that is generated by the action of the enzyme phospholipase C on phosphatidylinositol diphosphate.

In silico Refers to procedures that are carried out on a computer.

Institutional Review Board (IRB) A regulatory body in the USA that grants approval to clinical trials at a particular site.

Integrase An HIV enzyme that catalyses the insertion of viral DNA into human DNA.

Integrase inhibitors Antiviral agents that inhibit the HIV enzyme integrase.

Integrins Molecules that are involved in anchoring cells to the extracellular matrix.

Intron The middle portion of an mRNA molecule that is excised during a post-transcriptional splicing operation.

Investigational Exemption to a New Drug Application A document required by the FDA before clinical trials on a drug can begin.

Iontophoresis A means of encouraging topical absorption of a drug by applying a painless pulse of electricity to increase skin permeability.

Intercalating agents Agents containing a planar moiety that is capable of slipping between the base pairs of DNA. Important anticancer and antibacterial agents.

Interferons Endogenous proteins that are part of the body's defence system against viral infections. They work by inhibiting the metabolism of infected cells.

Interleukin-6 A protein that stimulates metastasis.

Intermolecular bonds Bonding interactions that take place between two separate molecules.

International Preliminary Examination Report (IPER) A report on a patent application that can be used when applying for patents to individual countries.

International Search Report (ISR) A report on a patent application that can be used when applying for patents to individual countries.

Intramolecular bonds Bonding interactions other than covalent bonds that take place within the same molecule.

Intramuscular injection The administration of a drug by injection into muscle.

Intraperitoneal injection The administration of a drug by injection into the abdominal cavity.

Intrathecal injection The administration of a drug by injection into the spinal column.

Intravenous injection The administration of a drug by injection into a vein.

Inverse agonist A compound which acts as an antagonist, but which also decreases the 'resting' activity of target receptors (i.e. those receptors which are active in the absence of agonist).

In vitro **procedures** Testing procedures carried out on isolated macromolecules, whole cells, or tissue samples.

In vivo **procedures** Studies carried out on animals or humans.

Ion channels Protein complexes in the cell membrane which allow the passage of specific ions across the cell membrane.

Ion channel disrupters A term used to describe a group of antiviral agents that act against the flu virus by disrupting ion channels.

Ion–dipole interactions A non-covalent bonding interaction that takes place between a charged atom and a dipole moment, such as the interaction of a positive charge with the negative end of the dipole.

Ionic interaction A non-covalent bonding interaction between two molecular regions having opposite charges.

Ionophores Agents which act on a cell membrane to produce an uncontrollable passage of ions across the membrane.

Irreversible inhibitor An enzyme inhibitor that binds so strongly to the enzyme that it cannot be displaced.

IsisDraw A chemical drawing software package.

Isomerases Enzymes that catalyse isomerizations and intramolecular group transfers.

Isostere A chemical group which can be considered to be equivalent in physical and chemical properties to another chemical group.

Isozymes A series of enzymes that catalyse the same chemical reaction but which differ in their amino acid composition.

K_d The dissociation binding constant.

K_i The inhibitory or affinity constant.

Kinases Enzymes which catalyse the phosphorylation of alcoholic or phenolic groups present in a substrate. The substrate is normally a protein.

Koshland's theory of induced fit *see* Induced fit

β-Lactamase inhibitors Agents which inhibit the β-lactamase enzymes.

Lactamases Bacterial enzymes that hydrolyse the β-lactam ring of penicillins and cephalosporins.

Lactate dehydrogenase An enzyme that catalyses the conversion of lactic acid to pyruvic acid and vice versa.

LD_{50} The mean lethal dose of a drug required to kill 50% of the test sample.

Lead compound A compound showing a desired pharmacological property which can be used to initiate a medicinal chemistry project.

LHRH *see* Luteinizing hormone-releasing hormone

Ligand Any molecule capable of binding to a binding site.

Ligand-gated ion channels Ion channels that are under the control of a chemical messenger or ligand.

Ligases Enzymes that join two substrates together at the expense of ATP hydrolysis.

Lignans Plant compounds which are estrogen-like and have antioxidant properties

Lincosamides A group of antibiotics acting against protein synthesis.

Lineweaver–Burk plots Plots which can be used to determine whether an enzyme inhibitor is competitive or non-competitive.

Linker A term used in combinatorial chemistry for a molecule that is covalently linked to a solid phase support and contains a functional group to which another molecule can be attached for the start of a synthesis.

Lipinski's rule of five A set of rules obeyed by the majority of orally active drugs. The rules take into account the molecular weight, the number of hydrogen bonding groups, and the hydrophobic character of the drug.

Lipolysis The process by which lipids are broken down by hydrolysis to free fatty acids.

Lipophilic Refers to compounds that are fatty and non-polar in character. Literally means fat loving.

Liposomes Small vesicles consisting of a phospholipid bilayer membrane. Used to encapsulate drugs for drug delivery.

Local energy minimum Refers to the nearest stable conformation reached when energy minimization is carried out on a molecule by molecular modelling software.

Lock and key hypothesis The now redundant theory that a ligand fits its binding site like a key fitting a lock.

log P *see* Partition coefficient

LUDI A software program used for *de novo* drug design.

LUMO Lowest unoccupied molecular orbital.

Luteinizing hormone A hormone that is important to ovulation and development of the corpus luteum in females, and in the production of testosterone in males.

Luteinizing hormone-releasing hormone Hormone that is used in anticancer therapy.

Lyases Enzymes that catalyse the addition or removal of groups to form double bonds.

Lysis The process where a cell loses its contents due to weakening of a cell wall or cell membrane.

Lysosomes Membrane-bound structures within eukaryotic cells that contain destructive enzymes.

MAA *see* Marketing Authorisation Application

Macrolides Macrocyclic structures that act as antibacterial agents. Erythromycin is the best known example of this class of agents.

Macromolecule A molecule of high molecular weight such as a protein, carbohydrate, lipid, or nucleic acid.

Magic bullet *see* Principle of chemotherapy

Marketing Authorisation Application (MAA) A document provided to the EMEA in order to receive marketing approval for a new drug.

Malignant cancers or tumours Life-threatening tumours that are undergoing metastasis and setting up secondary tumours elsewhere in the body.

Matrix metalloproteinases Enzymes that catalyse the hydrolysis of the proteins making up basement membranes. A target for new anticancer drugs called matrix metalloproteinase inhibitors.

Maytansinoids A group of natural products extracted from an Ethiopian shrub.

MDRTB Multi-drug resistant tuberculosis.

Membrane potential The electric potential difference between the outer and inner surfaces of a membrane.

Merrifield resin A resin used in solid phase peptide synthesis.

Message-address concept A concept used in opioid research which states that one part of an opioid is responsible for the pharmacological activity of the agent, while another part is responsible for its selectivity for different opioid receptors.

Messenger RNA (mRNA) Carries the genetic code required for the synthesis of a specific protein.

'Me too' drugs Drugs which have been modelled as variations of an existing drug.

Metabolic blockers Groups added to a drug to block metabolism at a particular part of the skeleton.

Metalloproteinases Enzymes that catalyse the hydrolysis of peptide bonds in protein substrates and which contain a metal ion as a cofactor in the active site.

Metastasis Refers to the breaking away of individual cancer cells from an established tumour such that they enter the blood supply and start up new tumours elsewhere in the body.

Methylene shuffle A strategy used to alter the hydrophobicity of a molecule. One alkyl chain is shortened by one carbon unit, while another is lengthened by a one carbon unit.

Michaelis constant The Michaelis constant equals the substrate concentration when the reaction rate of an enzyme-catalysed reaction is half of its maximum value.

Microfluidics The manipulation of tiny volumes of liquids in a confined space.

Micro RNA (miRNA) Short segments of double stranded mRNA molecules.

miRNP (micro-RNA-protein) A protein complex that binds miRNA, unwinds it, and discards one of the strands to produce bound siRNA. Subsequent binding with a target mRNA suppresses translation.

Microspheres Small spheres made up of a biologically degradable polymer. Used in drug delivery.

Microtubules Small tubules that are formed in cells by the polymerization of a structural protein called tubulin. Important for cell division and as targets for anticancer drugs.

Mineralocorticoids Steroids released from the adrenal cortex that regulate electrolyte balance.

Mitochondria Organelles within eukaryotic cells that can be viewed as the cell's energy generators. They also play a role in cell apoptosis.

Mitogen-activated protein kinase An enzyme that phosphorylates and activates proteins called transcription factors.

Mitosis The process of cell division.

Mix and split The procedure involved when synthesizing mixtures of compounds by combinatorial synthesis.

MMR vaccine A combination of three vaccinations that provides protection against measles, mumps, and rubella.

Modulator An agent that binds to the allosteric binding site of a target and modulates the activity of that target.

Molar refractivity (*MR*) A measure of a substituent's steric influence in a QSAR equation.

Molecular dynamics A molecular mechanics program that mimics the movement of atoms within a molecule.

Molecular targeted therapeutics The administration of highly selective agents that target specific molecular targets which are abnormal or overexpressed in a cancer cell.

Monoamine oxidase A metabolic enzyme that catalyses the oxidation of monoamines such as noradrenaline to give an aldehyde.

Monoamine oxidase inhibitors Compounds which inhibit the metabolic enzyme monoamine oxidase. Have been used as antidepressants but are less favoured now since they have side effects.

Monoclonal antibodies Refers to antibodies that are cloned and are identical in nature.

Monosaccharides The carbohydrate or sugar monomers that make up a polysaccharide.

Motor nerves Nerves carrying messages from the central nervous system to the periphery.

MR *see* Molar refractivity

MRSA Stands for methicillin-resistant *Staphylococcus aureus*: strains of *S. aureus* that have acquired resistance to methicillin (a penicillin).

Multi-drug resistance Refers to the situation where a cancer cell acquires resistance to a range of drugs other than the one it was exposed to. Related to the overexpression of P-glycoprotein which expels drugs from the cell.

Multi-target directed ligand An agent that has been designed to interact with different molecular targets in a predictable fashion.

Murine antibodies Refers to monoclonal antibodies that were originally isolated from mice.

Mutation An alteration in the nucleic acid base sequence making up a gene. May result in a different amino acid in the resultant protein.

Muscarinic receptors One of the two main types of cholinergic receptor.

Mutagen A chemical or substance that induces a mutation in DNA.

NCE *see* New Chemical Entity

NDA *see* New Drug Application

Neighbouring group participation A mechanism by which a functional group in a molecule assists a reaction without being altered itself.

Nanotubes Tubular structures on the molecular scale which are being considered as possible antibacterial agents.

Neoplasm The proper term for a cancer or tumour. Means new growth.

Nephrons Tubes that collect water and small molecules from the glomeruli and carry these towards the bladder. Much of the water, along with hydrophobic molecules, is reabsorbed into the blood supply from the nephrons and does not reach the bladder.

Neuraminidase An enzyme present in the flu virus that catalyses the hydrolysis of a sialic acid molecule from host glycoconjugates and which is crucial to the infection process.

Neuromuscular blocking agents Agents that block the action of acetylcholine at nicotinic receptors, resulting in the relaxation of skeletal muscle.

Neuropeptides Peptides that act as neurotransmitters.

Neurotransmission The process by which nerves communicate with other cells.

Neurotransmitter A chemical released by a nerve ending that acts as a chemical messenger by interacting with a receptor on a target cell.

New Chemical Entity (NCE) A novel drug structure.

New Drug Application (NDA) A document provided to the FDA in order to receive marketing approval for a new drug.

New Molecular Entity *see* New Chemical Entity

Nicotinic receptors One of the two main types of cholinergic receptor.

Nitric oxide synthase An enzyme that catalyses the generation of nitric oxide from L-arginine.

Nitrogen mustards Alkylating agents used in anticancer therapy.

NME *see* New Molecular Entity

Nocardicins Monocyclic β-lactams with antibacterial activity that were isolated from natural sources.

Non-nucleoside reverse transcriptase inhibitors (NNRTI) A group of antiviral agents that target an allosteric binding site on the viral enzyme reverse transcriptase.

Noradrenaline A catecholamine that acts as a neurotransmitter. It is also called norepinephrine.

NRTI *see* Nucleoside reverse transcriptase inhibitors

Nuclear hormone or transcription receptors *see* Transcription factors

Nucleases Enzymes that hydrolyse oligonucleotides and nucleic acids.

Nucleic acids RNA or DNA macromolecules made up of nucleotide units. Each nucleotide is made up of a nucleic acid base, sugar, and phosphate group.

Nucleocapsid Consists of a viral capsid and its nucleic acid contents. Viral enzymes may be present.

Nucleoside A building block for RNA or DNA that consists of a nucleic acid base linked to a sugar molecule.

Nucleoside reverse transcriptase inhibitors A group of antiviral agents that mimic nucleosides and target the viral enzyme reverse transcriptase.

Nucleosomes Repeating units of histone proteins within a chromatin structure.

Nucleotide A molecule consisting of a nucleoside linked to one, two, or three phosphate groups.

NVOC The nitroveratryloxycarbonyl protecting group.

Oligonucleotides A series of nucleotides linked together by phosphate bonds. Smaller versions of nucleic acid.

Olivanic acids A group of agents which inhibit β-lactamases.

Oncogenes Genes which normally code for proteins involved in the control of cell growth and division, but which have undergone a mutation such that they code for rogue proteins, resulting in the uncontrolled growth and division of cells.

Opportunistic pathogens Pathogens which are normally harmless but which cause serious infection when the immune system is weakened.

Organelles Identifiable structures within the cytoplasm of a eukaryotic cell.

Oripavines Complex multicyclic analogues of morphine which have powerful analgesic and sedative properties.

Organophosphates Agents that inhibit the acetylcholinesterase enzyme and which are used as nerve gases, medicines, and insecticides.

Orphan drugs Drugs that are effective against rare diseases. Special financial incentives are given to pharmaceutical industries to develop such drugs.

Orphan receptors Novel receptors for which the endogenous ligand is unknown.

Oxazolidinones A group of synthetic antibacterial agents that act against protein synthesis.

Oxidases Enzymes that catalyse oxidation reactions.

Oximinocephalosporins A group of second- and third-generation cephalosporins.

P1 or P1′ Nomenclature used to label the substituents of a substrate that can fit into the binding subsites of an enzyme. P1, P2, P3, etc. refer to substituents on one side of the reaction centre, while P1′, P2′, P3′ etc. refer to substituents on the other side.

p53 protein An important protein that monitors the health of the cell and the integrity of its DNA. Important to the apoptosis process.

Pancreatic lipase An enzyme responsible for catalysing the digestion of fats in the gut.

Papillomavirus A DNA virus responsible for genital warts.

Parasympathetic nerves Nerves of the autonomic motor nervous system that use acetylcholine as neurotransmitter.

Parietal cells Cells lining the stomach which release hydrochloric acid into the stomach.

Partial agonist A drug which acts like an antagonist by blocking an agonist, but which retains some agonist activity of itself.

Partition coefficient (*P*) A measure of a drug's hydrophobic character. Usually quoted as a value of log *P*.

Patent Cooperation Treaty (PCT) A treaty to which about 122 countries have signed up.

PEGylation Covalently linking molecules of polyethylene glycol to macromolecules.

Penicillin binding protein 2a A transpeptidase enzyme present in penicillin-resistant strains of *Staphylococcus aureus*.

Penicillins Natural and semi-synthetic antibacterial agents that are bactericidal in nature.

Peptidases Enzymes which hydrolyse peptide bonds.

Peptidomimetics Agents that have been developed from peptide lead compounds such that their peptide nature is removed or disguised in order to improve their pharmacokinetic properties.

Peptoids Peptides which are partly or wholly made up of non-naturally occurring amino acids. As such, they may no longer be recognized as peptides by the body's protease enzymes.

Personalized medicine The treatment of a patient based on a knowledge of the patient's genetic make-up, and their likely susceptibility to specific drugs.

P-glycoprotein A protein that expels toxins and drugs from cells. Plays an important role in drug resistance in the anticancer field when cancer cells mutate and produce increased levels of the protein.

Phage *see* Bacteriophage.

Pharmacodynamics The study of how ligands interact with their target binding site.

Pharmacokinetics The study of drug absorption, drug distribution, drug metabolism, and drug excretion.

Pharmacophore The atoms and functional groups required for a specific pharmacological activity, and their relative positions in space.

Pharmacophore triangle A triangle connecting three of the important binding centres making up the overall pharmacophore of a molecule.

Partial charges A measure of the partial charge on each atom of a molecule calculated by molecular modelling software.

Partial least squares A statistical method of reaching a QSAR equation in 3D QSAR.

PDGF Platelet-derived growth factor.

Penicillanic acid sulphone derivatives A group of agents which inhibit beta-lactamases.

Penicillinases *see* Lactamases.

Phase I metabolism Reactions undergone by a drug which normally result in the introduction or unmasking of a polar functional group. Most phase I reactions are oxidations.

Phase II metabolism Conjugation reactions where a polar molecule is attached to a functional group that has often been introduced by a phase I reaction.

Phosphatase An enzyme that catalyses the hydrolysis of phosphate bonds.

Phosphatidylinositol diphosphate A cell membrane component that acts as the substrate for the enzyme phospholipase C to generate the secondary messengers inositol triphosphate and diacylglycerol.

Phosphodiesterases Enzymes which are responsible for hydrolysing the secondary messengers cyclic AMP and cyclic GMP.

Phosphorylase An enzyme that catalyses the hydrolysis of phosphate bonds.

Photodynamic therapy The use of light to activate a prodrug in the body. Used in cancer therapy.

Photolithography A method of combinatorial synthesis involving the synthesis of products on a solid surface. Reactions only occur on those areas of the surface where photolabile protecting groups have been removed by exposure to light.

Pi (π) bond A weak covalent bond resulting from the 'side on' overlap of p-orbitals. Only occurs when the atoms concerned are sp or sp^2 hybridized, and when the bond between the atoms is a double bond or a triple bond.

Pi (π) bond cooperativity A situation which can arise in conjugated systems where a hydrogen bond donor and a hydrogen bond acceptor enhance their respective hydrogen bonding strengths by a resonance mechanism involving π bonds.

Picornaviruses A family of viruses that include polio, hepatitis A, cold viruses, and foot and mouth viruses.

Pinocytosis A method by which molecules can enter cells without passing through cell membranes. The molecule is 'engulfed' by the cell membrane and taken into the cell in a membrane-bound vesicle.

pK_a A measure of the acid–base strength for a drug or a functional group.

Placebo A preparation that contains no active drug, but should look and taste as similar as possible to the preparation of the actual drug. Used to test for the placebo effect where a patient's condition improves because he or she believes they have been given a useful drug, regardless of whether they received it or not.

Placental barrier Membranes that separate a mother's blood from the blood of her fetus. Some drugs can pass through this barrier.

Plasma proteins Proteins in the plasma of the blood. Drugs which bind to plasma proteins are unavailable to reach their target.

Plasmid Segments of circular DNA that are transferred naturally between bacterial cells. Useful in cloning and genetic engineering.

Podophyllotoxins A group of natural and semi-synthetic agents used as anticancer agents.

Poly ADP ribose polymerase An enzyme that repairs single-strand breaks in DNA.

Polyglutamylation An enzyme-catalysed process which involves addition of glutamate residues to a glutamate moiety already present in a molecule.

Polymerases Enzymes that catalyse the polymerization of molecular units to form macromolecules.

Polypharmacology The administration of different drugs to interact with different targets.

Porins Protein structures that create pores in the outer membrane of Gram-negative bacteria through which essential nutrients can pass. Some drugs can pass through these pores if they have the correct physical properties.

Potency The amount of drug required to achieve a defined biological effect.

pRB A powerful growth-inhibitory molecule that binds to a transcription factor to inactivate it.

Presynaptic control systems Receptors on the ends of presynaptic nerves that affect the release of neurotransmitter from the nerve.

Principle of chemotherapy The principle where a drug shows selective toxicity towards a target cell but not a normal cell.

Privileged scaffolds Scaffolds that are commonly present in established drugs.

Procaspase 9 An enzyme that activates caspase enzymes to produce apoptosis.

Prodrug A molecule that is inactive in itself, but which is converted to the active drug in the body, normally by an enzymatic reaction. Used to avoid problems related to the pharmacokinetics of the active drug, and for targeting.

Progestins Hormones that are used in anticancer therapy.

Prokaryotic cells Simple bacterial cells that contain no organelles or well defined nucleus.

Promiscuous ligands Ligands that interact with a range of different molecular targets.

Prostaglandins Endogenous chemicals that play an important role as chemical messengers.

Prosthetic group A cofactor which is covalently linked to the active site of an enzyme.

Protease inhibitors A group of antiviral agents which inhibit the HIV protease enzyme.

Proteases Enzymes which hydrolyse peptide bonds.

Protein A macromolecule made up of amino acid monomers. Includes enzymes, receptors, carrier proteins, ion channels, hormones, and structural proteins.

Protein kinases *see* Kinases

Protein-protein binding inhibitors (PPBIs) Drugs designed to inhibit the binding interactions between different proteins.

Proteoglycan A molecule consisting of a protein and a carbohydrate.

Protomers The protein subunits that make up a viral capsid.

Proteomics A study of the structure and function of novel proteins discovered from genomic studies.

Proto-oncogenes Genes which code for proteins involved in the control of cell growth and division, but which can cause cancer if they undergo mutation to form oncogenes.

Proton pump inhibitors A series of drugs which inhibit the proton pump responsible for releasing hydrochloric acid into the stomach.

q^2 *see* Cross-validated correlation coefficient

Quantitative structure–activity relationships (QSAR) Studies which relate the physicochemical properties of compounds with their pharmacological activity.

Quinolones A group of synthetic antibacterial agents, largely replaced by fluoroquinolones.

R A symbol used in QSAR equations to represent the electronic influence of a substituent due to resonance effects. Also used in stereochemistry to define the absolute configuration of a chiral centre.

Racemase A bacterial enzyme capable of racemizing a chiral centre.

Racemate or racemic mixture A mixture of the various stereoisomers of a molecule. A molecule having one asymmetric centre would be present as both possible enantiomers.

Racemization A reaction which affects the absolute configuration of asymmetric centres to produce a racemic mixture.

Radioligand labelling The use of a radioactively labelled irreversible inhibitor to label a macromolecular target.

Ras protein A small G-protein that plays an important role in the signal transduction pathways leading to cell growth and division.

Receptor A protein with which a chemical messenger or drug can interact to produce a biological response.

Receptor-mediated endocytosis Refers to the process by which a virus binds to a host cell glycoprotein and induces endocytosis to enter the cell.

Recombinant DNA technology The process by which DNA is manipulated to produce new DNA. Involves the controlled splitting of DNA from different sources, followed by the formation or recombination of hybrid DNA.

Recursive deconvolution A method of identifying the constituents in a combinatorial synthetic mixture. The method requires the storage of intermediate mixtures.

Reductases Enzymes that catalyse reduction reactions.

Regression coefficient A measure of how well a QSAR equation explains the variance in biological activity of a series of drugs.

Relaxation time The time taken for excited nuclei to return to their resting state in NMR spectroscopy.

Renal Relating to the kidney.

Replication The process by which a copy of DNA is produced.

Restriction enzymes Enzymes that are used in recombinant DNA technology to split DNA chains in a controlled fashion.

Restriction point A point within the cell cycle where a decision is taken whether to progress to the next stage or not.

Retroviruses RNA viruses that use a viral reverse transcriptase enzyme to generate viral DNA from viral RNA within a host cell.

Reverse transcriptase A viral enzyme present in HIV that catalyses DNA from an RNA template.

Reverse transcriptase inhibitors A group of antiviral compounds that inhibit the viral enzyme reverse transcriptase.

Reversible inhibitors Enzyme inhibitors that compete with the substrate for the enzyme's active site and which can be displaced by increasing the concentration of substrate.

Ribosomes Structures consisting of rRNA and protein which bind mRNA and catalyse the synthesis of the protein coded by mRNA.

Ribosomal RNA (rRNA) Present in ribosomes as the major structural and catalytic component.

Ribozymes RNA molecules with an enzymatic property.

Rifamycins A group of antibiotics and semi-synthetic agents used as antibacterial agents.

Rigidification strategies Strategies used to limit the number of conformations that a drug can adopt whilst retaining the active conformation.

Ring contraction/expansion strategy The variation of ring size in a drug to optimize the relative positions of different binding groups.

Ring fusion or extension strategy The fusion of one ring onto another to enhance a drug's binding interactions.

Ring variation strategies The replacement of an aromatic, heteroaromatic, or saturated ring with a different ring system to obtain different structural classes of a drug.

Rink resin A resin used in combinatorial chemistry.

RNA Ribonucleic acid.

RNA-dependent RNA polymerase An enzyme that catalyses the synthesis of RNA from an RNA template.

RNA induced silencing complex (RISC) A complex that catalyses the unravelling of the strands of micro-RNA to produce single-stranded segments of RNA called small interfering or small inhibitory RNAs (siRNA).

RNA viruses Viruses that contain RNA as their nucleic acid.

S1 or S1′ Nomenclature used to label binding subsites of an enzyme. The subsites accept the amino acid residues of a peptide substrate. S1, S2, S3, etc. refer to subsites on one side of the reaction centre, while S1′, S2′, S3′, etc. refer to subsites on the other side.

Safety-catch linker An example of a linker in combinatorial chemistry on which two molecules can be constructed, one the target molecule and the other a tagging molecule.

SAR *see* Structure–activity relationships

Sarcodictyins Naturally occurring anticancer agents that inhibit tubulin depolymerization.

SARS Severe acute respiratory syndrome. A viral infection.

Scaffolds The molecular core of a drug to which the important binding groups are attached as substituents.

Scatchard plot A plot used to measure the affinity of a drug for its binding site.

Schild analysis Used to determine the dissociation constant of competitive antagonists.

Screening A procedure by which compounds are tested for biological activity.

Scintillation proximate assay A visual method of detecting whether a ligand binds to a target by its ability to compete with a radiolabelled ligand that emits light in the presence of scintillant.

Secondary messenger A natural chemical which is produced by the cell as a result of receptor activation, and which carries the chemical message from the cell membrane to the cytoplasm.

Secondary metabolites Natural products that are not crucial to cell growth and division. Generally produced in mature cells.

Selective noradrenaline reuptake inhibitors (SNRIs) Agents that inhibit the reuptake of noradrenaline from nerve synapses. The agents show selectivity for the transport proteins that uptake noradrenaline.

Selective serotonin reuptake inhibitors (SSRIs) Agents that inhibit the reuptake of serotonin from nerve synapses. The agents show selectivity for the transport proteins that uptake serotonin.

Self-assembly The process by which molecular units assemble into a structure without the aid of enzymes or other structures; for example the assembly of protomers to form a viral capsid.

Self-destruct drugs Drugs which are designed to be inactivated in the body through chemical or enzymatic mechanisms.

Semi-synthetic product A product that has been synthesized from a naturally occurring compound.

Sensitization The process by which a cell adapts to the continued presence of an antagonist, resulting in increased receptor sensitivity or the production of more receptors.

Sequential blocking Describes the situation where two agents inhibit two different enzymes in a biosynthetic pathway. Allows each agent to be administered in lower and safer doses.

Serine proteases Enzymes that catalyse the hydrolysis of peptide bonds in protein substrates. A serine residue in the active site acts as a nucleophilic group during the reaction mechanism.

Serine-threonine kinases Enzymes which catalyse the phosphorylation of serine and threonine residues in protein substrates.

Sialidase An enzyme that catalyses the cleavage of sialic acid from glycoproteins and glycolipids. Also called neuraminidase.

Sigma (σ) bond A strong covalent bond taking place between two atoms. It involves strong overlap between two atomic orbitals whose lobes point towards each other.

Signal transduction The mechanism by which an activated receptor transmits a message into the cell, resulting in a cellular response.

Simplification strategies The simplification of a drug to remove functional groups, asymmetric centres, and skeletal frameworks that are not required for activity.

Small G-proteins Proteins that have an important role in signal transduction pathways. So called because they are similar to G-proteins but are a single protein.

Small inhibitory RNAs (siRNA) Single-stranded segments of RNA which are attached to a protein called RISC and can bind to mRNA containing complementary base pairs. The enzyme complex then destroys the mRNA molecule.

Small nuclear RNA Small molecules of RNA that are in the nucleus and are a constituent of a spliceosome. They are important to the modification and splicing of mRNA following transcription.

'Smart drugs' Anticholinesterases that act in the central nervous system to increase levels of acetylcholine. They relieve the symptoms of Alzheimer's disease.

Soft drugs Drugs that are designed to undergo metabolism in a predictable manner to produce non-toxic, inactive metabolites that are excreted.

Somatic gene therapy The use of a carrier virus to smuggle a gene into a human cell which has a defective form of the gene.

Somatic motor nervous system Motor nerves carrying messages to skeletal muscle.

Specifications The tests that have to be carried out on a manufactured drug, and the standards of purity required.

Spider scaffolds Scaffolds which have binding group substituents placed round the whole scaffold.

Spindle The arrangement of microtubules that is formed in order to separate cells during cell division.

Spliceosome A structure made up of protein and small nuclear RNA. Serves to modify and splice mRNA following transcription.

ssDNA Single-stranded DNA. A term used in virology.

ssRNA Single-stranded RNA. A term used in virology.

Statins Drugs that inhibit the enzyme 3-hydroxy-3-methylglutaryl-coenzyme A reductase and lower cholesterol levels in the blood supply

Steady state concentration The concentration of a drug that is maintained in the blood supply following regular administrations.

Steric shields Groups that are added to molecules to protect vulnerable groups by nature of their size.

Streptogramins A group of macrocyclic antibiotics acting against protein synthesis.

Structure–activity relationships (SAR) Studies carried out to determine those atoms or functional groups which are important to a drug's activity.

Structure-based drug design The design of drugs based on a study of their target binding interactions with the aid of X-ray crystallography and molecular modelling.

Subcutaneous injection The administration of a drug by injection under the surface of the skin.

Subsites Often refers to enzymes that accept peptides or proteins as substrates. The subsites are binding pockets that accept amino acid residues from the substrate.

Substituent hydrophobicity constant (π) A measure of a substituent's hydrophobic character.

Substrate A chemical which undergoes a reaction that is catalysed by an enzyme.

Suicide substrates Enzyme inhibitors which have been designed to be activated by an enzyme catalysed reaction, and which will bind irreversibly to the active site as a result.

Sulphonamides Synthetic antibacterial drugs that are bacteriostatic in nature.

Sulphotransferases Enzymes that catalyse conjugation reactions involving sulphate groups.

Supercoiling The process by which DNA coils into a compact shape.

Suppositories Drug preparations that are administered rectally.

Surface plasmon resonance An optical method of detecting the binding of a ligand with its target.

Sybyl A molecular modelling software package.

Sympathetic nerves Nerves of the autonomic motor nervous system that use noradrenaline as a neurotransmitter at target cells and which use acetylcholine as a neurotransmitter between nerves.

Synapse The small gap between a nerve and a target cell, across which a neurotransmitter has to travel in order to reach its receptor.

Synergy An effect where the presence of one drug enhances the activity of another.

Tadpole scaffold A scaffold where substituents acting as binding groups are located at one region of the scaffold.

Taft's steric factor (E_s) A measure of a substituent's steric influence in QSAR equations.

Tagging A method of identifying what structures are being synthesized on a resin bead during a combinatorial synthesis. The tag is a peptide or nucleotide sequence which is constructed in parallel with the synthesis.

Tautomers The different structures that a conjugated system can adopt arising from the rearrangement of double bonds and hydrogen atoms.

Taxoids Naturally occurring and semi-synthetic anticancer agents that inhibit tubulin depolymerization.

Telomerase An enzyme that catalyses the construction of telomeres.

Telomeres Polynucleotide structures at the 3′ ends of chromosomes that stabilize DNA.

Tetracyclines Tetracyclic antibiotics that are bacteriostatic in their action.

Teratogen A compound that produces abnormalities in a developing fetus.

TGF Transforming growth factor.

Therapeutic index or ratio The ratio of a drug's undesirable effects with respect to its desirable effects. The larger the therapeutic index, the safer the drug. The therapeutic index compares the drug dose levels which lead to toxic effects in 50% of cases studied to the dose levels leading to maximum therapeutic effects in 50% of cases studied.

Therapeutic window The range of a drug's plasma concentration between its therapeutic level and its toxic level.

Thymidylate synthase Catalyses the synthesis of an important building block for DNA. Inhibitors are used as anticancer agents.

Thrombospondin An endogenous compound that inhibits angiogenesis.

TNF and TNFR Tumour necrosis factors and tumour necrosis factor receptors. Play a role in apoptosis or cell death.

Transdermal absorption Refers to the absorption of a drug through the skin.

Tolerance Repeat doses of a drug may result in smaller biological results. The drug may block or antagonize its own action and larger doses are needed for the same pharmacological effect. Alternatively, the body may 'learn' how to metabolize the drug more efficiently. Again, larger doses are needed for the same pharmacological effect, increasing the chances of toxic side effects.

Topliss scheme A scheme used to determine which substituents should be introduced in order to get more active drugs. Useful when analogues are synthesized and tested one at a time.

Topoisomerases Enzymes that catalyse transient breaks in one or both strands of DNA to allow coiling and uncoiling of the molecule. These act as targets for several antibacterial and anticancer drugs.

Transcription The process by which a segment of DNA is copied to mRNA.

Transcription factors Complexes which bind to DNA and control the expression of specific genes.

Transduction The process by which plasmids are exchanged between bacterial cells.

Transfer RNA (tRNA) An RNA molecule that bears an amino acid which is specific for a particular triplet of nucleic acid bases.

Transferases Enzymes that catalyse transfer reactions.

Transgenic animals Animals that have been genetically modified such that they can be used for the *in vivo* testing of drugs.

Transglycosidase A bacterial enzyme that catalyses the attachment of a disaccharide building block to the growing sugar chain of a new cell wall.

Translation The process by which proteins are synthesized based on the genetic code present in mRNA.

Translocase A bacterial enzyme that links a building block for the bacterial cell wall to a C55 carrier lipid located within the cell membrane.

Translocation Part of the translation process where a tRNA molecule departs the P binding site of a ribosome, and the ribosome shifts along mRNA to reveal the next triplet.

Transition state A high-energy intermediate that must be formed during an enzyme-catalysed reaction. The energy required to reach the transition state determines the rate of reaction. It is proposed that an enzyme binds the transition state more strongly than the substrate or the product resulting in a stabilization of the transition state.

Transition-state analogues or inhibitors Enzyme inhibitors which have been designed to mimic the transition state of an enzyme-catalysed reaction.

Transition-state isostere An arrangement of atoms that mimics the arrangement of atoms in a transition state, but which is more stable.

Transpeptidases Important bacterial enzymes that catalyse the final cross-linking of the bacterial cell wall. Targeted by penicillins and cephalosporins.

Transport proteins *see* Carrier proteins

Tricyclic antidepressants A series of tricyclic compounds that have antidepressant activity by blocking the uptake of noradrenaline from nerve synapses back into the presynaptic nerve.

Triplet code Refers to the fact that the genetic code is read in sets of three nucleic acid bases at a time. Each triplet codes for a specific amino acid.

Tumour necrosis factor-related apoptosis inducing ligand (TRAIL) A death-inducing protein which stimulates cell death.

Tumour suppression genes *see* Anti-oncogenes

Tyrosine kinases Enzymes which catalyse the phosphorylation of tyrosine residues in protein substrates.

Tyrosine kinase receptors Membrane-bound receptors that are activated by external ligands, resulting in subsequent intracellular kinase activity that phosphorylates tyrosine residues in protein substrates.

Ubiquitin A small regulatory protein that is attached to proteins and marks them out for destruction.

Ureidopenicillins A group of penicillins bearing a urea group at the α-position.

Vaccination The introduction of foreign antigens to prime the immune system such that it will work more effectively against later infections.

van der Waals interactions Weak interactions that occur between two hydrophobic regions and which involve interactions between transient dipoles. The dipoles arise from uneven electron distributions with time.

Varicella zoster viruses Viruses responsible for chickenpox and shingles.

Vascular endothelial growth factor A growth factor that stimulates angiogenesis.

Vasopressin A hormone that is responsible for increasing water retention in the kidneys and increasing blood pressure.

Vectors A process by which a molecule can be taken into a cell. Particularly important to gene therapy.

VEGF *see* Vascular endothelial growth factor

Veins Blood vessels carrying blood back to the heart.

Verloop steric parameter A measure of a substituent's steric properties. Used in QSAR equations.

Vesicle A membrane-bound 'bubble' within the cell. Neurotransmitters are stored within vesicles prior to release.

Vinca alkaloids Naturally occurring compounds that inhibit tubulin polymerization and are used as anticancer agents.

Virion The form that a virus takes when it is not within a host cell.

Viruses Non-cellular infectious agents consisting of DNA or RNA wrapped in a protein coat. Require a host cell to multiply.

Voltage-gated ion channels Ion channels that are controlled by the potential difference across the cell membrane. Important to the mechanism of transmission in nerves.

VRE Vancomycin-resistant enterococci.

VRSA Vancomycin-resistant *Staphylococcus aureus*.

VZV *see* Varicella zoster viruses

Wang resin A resin used in combinatorial chemistry.

Withdrawal symptoms The symptoms that arise when a drug associated with physical dependence is no longer taken.

Z A shorthand term for the protecting group benzyloxycarbonyl.

Zinc finger domains Refers to a region of a steroid receptor that is rich in cysteine residues and zinc co-factors. Involved in binding to DNA when the receptor is part of a transcription factor.

Zoonose A disease caused by an infectious agent (for example virus, bacterium, fungus, or protozoa) that can be transmitted from one species to another. It includes transmission between different species of mammals, as well as between insects and mammals.

General further reading

It is recommended that you read round various topics as much as you can. There are strengths and weaknesses in every publication and if you find an explanation difficult to follow in one textbook or article, you may find a clearer one in a different publication. Textbooks also differ in the breadth and detail of coverage given to the various topics of medicinal chemistry. For example, this textbook has chapters concentrating in some detail on topics such as antibacterial, antiviral, and anticancer agents, but does not cover antifungal or antimigraine agents to the same extent. The following are useful reference and general texts. You will also find references to articles and books at the end of each chapter which are more specific to the topics covered in that chapter.

REFERENCE WORKS

Abraham, D. J. and Rotella, D. P. (ed.) (2010) *Burger's medicinal chemistry, drug discovery and development*, 7th edn, Vols 1–8. John Wiley and Sons, New York.
Association of the British Pharmaceutical Industry (2008) *Medicines compendium*. Datapharm Publications Ltd., London.
British Medical Association and Royal Pharmaceutical Society of Great Britain. *British national formulary* (BNF). Pharmaceutical Press, London (twice-yearly publication).
Triggle, D. J. and Taylor, J. B. (eds.) (2006) *Comprehensive medicinal chemistry*, Vols 1–8. Elsevier Science.
O'Neil, M. J. (ed.) (2013) *The Merck index: an encyclopaedia of chemistry, drugs and biologicals*, 15th edn. Royal Society of Chemistry, Cambridge.

GENERAL TEXTBOOKS ON MEDICINAL CHEMISTRY

Drews, J. (2013) *In quest of tomorrow's medicines*. Springer-Verlag, New York.
Ganellin, C. R., and Roberts, S. M. (eds.) (1994) *Medicinal chemistry—the role of organic research in drug research*, 2nd edn. Academic Press, London.
Brunton, L., Chabner, B., and Knollman, B. (eds.) (2011) *Goodman and Gilmans' the pharmacological basis of therapeutics*, 12th edn. McGraw-Hill Prof Med/Tech, New York.
Jie, J. L. (2015) *Top drugs: history, pharmacology, syntheses*. Oxford University Press, Oxford.
King, F. D. (ed.) (2002) *Medicinal chemistry, principles and practice*, 2nd edn. Royal Society of Chemistry, Cambridge.
Krogsgaard-Larsen, P. (eds) (2009) *Textbook of drug design and development*, 4th edn. CRC Press.
Le Fanu, J. (2011) *The rise and fall of modern medicine*. Abacus Press, Berkhamsted.
Patrick, G. L. (2015) *An introduction to drug synthesis*. Oxford University Press, Oxford.
Patrick, G. (2001) *Instant notes medicinal chemistry*. Bios Scientific, Oxford.
Silverman, R. B., and Holladay, M. W. (2014) *The organic chemistry of drug design and action*, 3rd edn. Academic Press, San Diego.
Sneader, W. (2005) *Drug discovery: a history*. John Wiley and Sons, Chichester.
Thomas, G. (2007) *Medicinal chemistry: an introduction*, 2nd edn. John Wiley and Sons, Chichester.
Wermuth, C. G., Aldous, D., Raboisson, P., and Rognan, D. (eds.) (2015) *The practice of medicinal chemistry*, 4th edn. Academic Press, London.
Lemke, T. L., Williams, D. A., Roche, V. F. and Zito, S. W. (eds.) (2012) *Foye's principles of medicinal chemistry*, 7th edn. Lippincott, Williams and Wilkins, Philadelphia.

GENERAL TEXTBOOKS ON RELATED AREAS

Berg, J. M., Tymoczko, J. L., Gatto, G. J., and Stryer, L. (2015) *Biochemistry*, 8th edn. W. H. Freeman and Co., New York.
Cairns, D. (2012) *Essentials of pharmaceutical chemistry*, 4th edn. Pharmaceutical Press, London.
Page, C. P., Hoffman, B., Curtis, M., and Walker, M. (2006) *Integrated pharmacology*, 3rd edn. Mosby, St. Louis.
Rang, H. P., Ritter, J. M., Flower, R. J., and Henderson, G. (2015) *Pharmacology*, 8th edn. Elsevier, Churchill Livingstone, Edinburgh.

JOURNALS

Advances in Drug Research
Advances in Medicinal Chemistry
Annual Reports in Medicinal Chemistry
Antimicrobial Agents and Chemotherapy
Bioorganic and Medicinal Chemistry
Bioorganic and Medicinal Chemistry Letters
Chemical and Pharmaceutical Bulletins
Chemistry in Britain
Current Medicinal Chemistry
Current Opinion in Drug Discovery and Development

Drug Design and Delivery
Drug Discovery Today
Drug News and Perspectives
Drugs
Drugs of the Future
Drugs Today
European Journal of Medicinal Chemistry
Journal of Combinatorial Chemistry
Journal of Computational Chemistry
Journal of Computer-aided Molecular Design
Journal of Medicinal Chemistry

Medicinal Chemistry Research
Medicinal Research Reviews
Nature
Nature Reviews Drug Discovery
Pharmacochemistry Library
Progress in Drug Research
Progress in Medicinal Chemistry
QSAR
Science
Scientific American
Trends in Pharmacological Sciences

Index

Trade names are indicated in blue. Appendix 6 also provides a list of trade names not included in the index.

A

A74704, 509–11
A77003, 510–11
A78791, 510–11
A80987, 510–12
A83962, 511–12
ab initio quantum mechanics, 350
abacavir (Ziagen), 500–3
abatacept, 275
Abbott pharmaceuticals, 12, 508
abciximab (ReoPro), 149, 763
abdominal infections, treatment of, 456
abdominal muscles, relaxation, 635
Abelson tyrosine kinase inhibitors, 582–5
    resistance, 585
abemaciclib, 587
abiraterone (Zytiga), 569–70
abiraterone acetate, 570–1
ABT199, 605
ABT263, 605
ABT267, 537
ABT378, 511
ABT538, 511
abzymes, 614
7-ACA, see 7-aminocephalosporinic acid
ACE, see angiotensin-converting enzyme
acebutolol (Sectral), 669–71
acetaldehyde, 216
acetorphan (Hidrasec), 699
acetyl coenzyme A, 172, 622–3
acetylation
    aminoglycoside modifying enzymes, 468
    aspirin synthesis, 295–6
    by aspirin, 230
    drug metabolism, 171–2, 174
    histone acetylase, 600
    of proteins, 25
N-acetylation, 25
acetylcholine, 44–6, 620–30, 654, 808, see
    also acetylcholinesterase
    active conformation, 625–6
    binding interactions with cholinergic
        binding site, 380, 625
    biosynthesis, 622–3
    control of gastric acid release, 705–6
    hydrolysis by acetylcholinesterase, 640–2
    interaction with presynaptic
        receptors, 623, 658
    pharmacophore, 626
    regulation of heart rate, 736
    role in depression, 776
acetylcholinesterase, 102, 250, 252, 402,
    622, 640–3
    inhibitors, see also anticholinesterases
N-acetylcysteine, 557

N-acetylgalactosamine-6-sulphatase, 275
N-acetylglucosamine, 435, 458
L-α-acetylmethadol, 690
3-acetylmorphine, 680
6-acetylmorphine, 680, 683
N-acetylmuramic acid, 435, 458
N-acetylneuraminic acid, 519–20
N-acetylneuraminic acid, see sialic acid
acetylsalicylic acid, 296, see also aspirin
aciclovir (Virovir, Zovirax), 101, 137–8,
    494–5, 497, 615
    prodrugs, 495
aciclovir triphosphate, 138, 494–5
Acinetobacter baumannii, resistance, 483
aclacinomycin A, see aclarubicin
aclarubicin (Aclacin, Aclaplastin), 599
aclidinium bromide (Eklira Genuair), 633–5
acne, treatment of, 472, 477
acquired immunodeficiency syndrome, 179,
    425, 490, 493, 498, see also HIV
    appetite stimulation in patients, 214
    clinical trials of HIV drugs, 288–9
    patient susceptibility to Pseudomonas
        aeruginosa, 438
    treatment, 141, 152, 500, 503–4
    treatment of other diseases in AIDS
        patients, 431, 482, 498
    vaccination, 493
acquired resistance, 552
Acremonium chrysogenum, 448
acrolein, 557–9
acromegaly, treatment of, 276
acrylamides, 95–6
    Michael addition acceptor, 589
    warheads in irreversible inhibitors, 581
ACT064992, 743
ACT333679, 747–8
actinomycin D, see dactinomycin
Actinoplanes teichomyceticus, 461
action potentials, 204, 810
activation energy, 30
active conformation, 235, 247–8, 364–6,
    413, 415
    acetylcholine, 625–6
    captopril, 366
    cimetidine, 719–20
    decamethonium, 365
    dihydropyridines, 750
    HCV protease inhibitors, 798
    kinases, 579
    muscarine, 627
    receptors, 119
    renin inhibitors, 784
    serotonin antagonists, 778
    tubulin polymerization inhibitors, 417
active principle, 207, 340

active site, 23, 31–33
    acetylcholinesterase, 640–2
    angiotensin-converting enzyme, 304–6
    carboxypeptidase, 303
    epidermal growth factor receptor, 578
    neuraminidase, 522
    PARP enzyme, 602
    thymidylate synthase, 420
acupuncture, 696
acute bacterial skin infections, treatent
    of, 454
acute lymphoblastic leukaemia, treatment
    of, 608
acute myeloid leukaemia, treatment of, 613
acute myocardial infarction, 757
acyclovir, see aciclovir
acyl CoA synthetase, 173
acylases, see penicillin acylase
acylating agent, 230
N-acylsulphonamide, bioisostere for
    carboxylate group, 244, 799–800
N-acyltransferase, 174
adalimumab (Humira, Exemptia), 277
adamanolol, 149–50
adamantanes, 521
ADAPT, see antibody-directed abzyme
    therapy
addiction, 1–2, 274–5, see also drug addiction
Addison's disease, 767
adefovir dipivoxil (Hepsera), 500, 502–3
adenine, 77–8
    binding interactions, 232
    nucleophilic groups, 132
adenosine, 38, 46, 53, 564
    adenosine 5'-diphosphate (ADP), 57,
        72–4, 760–1
    aminoglycoside modifying enzymes, 468
adenosine 5'-monophosphate (AMP), 38, 746
adenosine 5'-triphosphate (ATP), 56–7,
    63–5, 72–4, 577–80, 622, 657–8, 726
adenosine deaminase, 102, 276, 564, 566
    inhibitors, 564
adenosine receptors, 59
S-adenosyl methionine, 172, 174
S-adenosylhomocysteine hydrolase, 539
adenovirus, 614
adenylate cyclase, 62–6, 68, 747
adenylyl cyclase, see adenylate cyclase
ADEPT strategy, see antibody-directed
    enzyme prodrug therapy
ADME, 11, 162, 256
adozelesin, 558, 612
ADP, see adenosine 5'-diphosphate
ADP receptors, see P2Y12 receptors
    antagonist, 760
ADP-ribose, 602

adrenal cortex, 737, 766
adrenal gland tumours, treatment of, 670
adrenal medulla, 621, 654, 656–7
adrenaline, 621, 654–6, 662–3, *see also* adrenergic receptor agonists, adrenergic receptor antagonists, and adrenergic receptors
   as a lead compound, 214
   as a vasoconstrictor, 179, 274, 662
   binding site interactions, 660–1
   biosynthesis, 656–7
   effects on the peripheral nervous system, 621–2
   fight or flight response, 46, 621, 654
   metabolism, 663
   regulation of heart rate and contractility, 736
   relaxation of heart muscle, 66
   role in fat metabolism, 66
   role in glycogen breakdown, 40, 64–6
   structure and properties, 350–3
   treatment of asthma, 663
adrenergic agents, clinical aspects, 656
adrenergic nervous system, 654
adrenergic receptors, 52, 654–5
   agonists, 123, 662–6
   antagonists, 242–3, 666–72, *see also* β-blockers
   binding site, 53, 659–62
   distribution, 655
   evolutionary tree, 53–4
   in the peripheral nervous system, 621
   neurotransmission process, 657–9
   physiological effects, 655, 663
   presynaptic control systems, 623, 658
   role in fat metabolism, 66
   role in glycogen metabolism, 64–6
   role in smooth muscle contraction, 66
   signal transduction, 61, 64–6
   target for antidepressants, 777
   types and subtypes, 47, 54, 123, 654–5
   X-ray crystal structure, 55, 379, 659
adrenoceptors, 200, *see* adrenergic receptors
adrenocortical tumours, treatment of, 571
adrenocorticoids, 766
adriamycin, 553, *see* doxorubicin
adsorption, 491–2, 520
   HIV, 498
aerobic bacteria, 812
afatinib (**Giotrif Gilotrif**), 581–2, 593
affinity, 124
affinity constant (*K*i), 125
affinity screening, 205
aflibercept (**Eylea, Zaltrap**), 608
AFN1252, 485
African bush willow, 573
African clawed frog, 209
afutuzumab, 609, *see* obinutuzumab
AG1254, 513
agalsidase beta (**Fabrazyme**), 275
age-related vision loss, treatment of, 277
agonists, 109–13, 120
AHU377, 747
AIDS, *see* acquired immunodeficiency syndrome

Akira Endo, 190
Akt protein, 73–4
D-Ala-D-Ala ligase, 458
D-alanine, 435–6, 458, 461, 697
L-alanine, 17, 22, 157–8, 801
   ACE inhibitors, 305
   cholinergic binding site, 380
   decapeptide, 351
   matrix metalloproteinase active site, 596–7
   racemization, 458
   resistance to HIV protease inhibitors, 512
   yeast alanine tRNA, 83
L-alanine racemase, 458
D-alanyl-D-alanine, 205, 437–8, 458–60
albinism, 85
albumin, 165–6, 397, 447
albuterol, 664, *see* salbutamol
   formulation, 182
alclometasone dipropionate (**Modrasone**), 775
alcohol, 2, 166, 178, 216
   aversion therapy, 270
   cure for antifreeze poisoning, 94
alcohol dehydrogenases, 94–5, 169, 170–2
alcoholism, treatment of, 95, 216, 685
alcohols, binding role, 224–5
aldehyde in proteasome inhibitors, 598
aldehyde dehydrogenases, 170
aldehyde oxidase, 495–6
aldehydes, binding role, 226
aldesleukin (**Proleukin**), 608
aldometasone dipropionate, 775
aldosterone, 737, 741–2, 766–7
alectinib (**Alecensa**), 589–90, 594
alemtuzumab (**MabCampath**), 609
alfacon-1, 540
alfentanil, 689
alferon, 540
algae, 812
alglucosidase alfa (**Myozyme, Lumizyme, Genzyme**), 275
algorithms, 349
aliphatic electronic substituent constants, 402
alirocumab (**Praluent**), 756
aliskiren (**Tekturna**), 98, 738, 781–7
alkaline phosphatase, 613
alkenes, binding role, 226
alkyl fluorides, 231
alkyl halides, binding role, 230
alkyl substituents, *see also* variation of alkyl substituents
   binding role, 231
   variation in drug design, 236–9
N-alkylated N- morphine analogues, synthesis, 685
alkylating agents, 95, 131–35, 551, 555–9, 614
   oxamniquine, 319
alkylation by alkyl halides, 230–1
allantoin, 275–6
Allegra, 175, 813
AlleGrow, 391
allergic reactions to penicillin, 435
allergies, treatment of, *see* anti-allergic agents

allopurinol (**Zyloric**), 102
allosteric antagonist for PY12 receptor, 762
allosteric binding sites, 38–9, 59, 96, 118
   P2Y12 receptor, 762
   soluble guanylate cyclase, 745
allosteric inhibitors, 501, 564, 585
allosteric modulators, 59, 113, 117–8
N-allylmorphine, 694
aloe plant, 221
altanserin, 382
alteplase (**Actilyse, Activase**), 763
alvocidib, 586
Alzheimer's disease, treatment of, 102, 648–9, 675
amadacycline, 471
amantadine (**Lysovir, Symmetrel**), 493, 521
ambrisentan (**Volibris, Letairis**), 742
American mandrake, 572
Ames test, 202
amides
   binding role, 227–9
   synthesis, 334, 337–8
amidine as a binding group, 789
amikacin (**Amikin**), 466–8
amines, binding role, 226–7
amino acid conjugates, 173
amino acid N-acyltransferase, 173
aminoacridines, 128, 478–9
aminoacyl tRNA, 466
aminoacyl tRNA synthetases, 484
para-aminobenzoic acid, 431
γ-aminobutyric acid (GABA), 46, 53, 776, 808
γ-aminobutyric acid receptors, 59, 113, 119, 215, 398
7-aminocephalosporinic acid, 449–50
aminoglutethimide (**Orimeten, Cytadren**), 216, 570
aminoglycoside modifying enzymes, 467
aminoglycosides, 208, 426, 466–8, 485
   resistance mechanisms, 467–8
aminolevulinic acid (**Levulan**), 616
4-amino-Neu5Ac2en, 524–5
6-aminopenicillanic acid, 434, 440–1
aminopenicillins, 445
aminopeptidase enzyme, 271
para-aminosalicylic acid, 468
amitriptyline, 673–4
amlodipine (**Istin, Norvasc**), 751
   with valsartan (**Exforge**) 815
   amlodopine besylate with perindopril arginine (**Prestalia**) 818
AMN107, 584
amoebic dysentery, 812
amoxicillin (**Amoxil**), 445, 456
   clinical aspects, 447, 482
   treatment of *Helicobacter pylori*, 705, 732
   with clavulanic acid (**Augmentin**) 813
amoxycillin, *see* amoxicillin
AMP, *see* adenosine 5'-monophosphate
amphetamines, 181, 661–2, 674–6
amphotericin (**Fungilin, Fungizone**), 152–4
ampicillin (**Penbritin**), 272, 442, 444–7, 457
   with flucloxacillin (**Co-fluampicil**) 814

ampicillin-resistant *H. influenza*, 483
amprenavir (**Agenerase**), 501, 514–6
amsacrine (**Amsidine**), 554–5
β-amyloid protein, 649
anabaseine, 648–9
anacetrapib, 756–7
anaerobic bacteria, 812
analgesic receptors, *see* opioid receptors
analgesics, 123, 210, 263, 270, 274, 663, *see also* opioids
anandamide, 214, 776
Anapen, 662
anaphylactic reactions, treatment of, 662
anaplastic lymphoma kinase, 58, 589
    inhibitors, 589–90, 594
anastrozole (**Arimidex**), 570–1
anchimeric assistance, 132, 627
anchor and grow programs, 373–77
anchors, 156, 327–8
androgen receptors, 741
androgens, 567–8, 570–1
androstenedione, 570
angina, 760
    treatment of, 216, 656, 668–70, 744, 753
angina pectoris, treatment of, 749
angiogenesis, 544, 549–50, 596
    inhibitors, 549–50, 573, 605–6
angioplasty, 180
angiostatin, 549, 608
angiostatin II receptors, 200
angiotensin II receptors, 54, 739
    antagonists, 739–41
angiotensin-converting enzyme, 97, 200, 302, 737–8
    crystal structure with lisinopril, 306
    inhibitors, 2, 102, 240, 302–7, 738–9, 747, *see also* captopril, cilazapril, enalaprilate and lisinopril
    model binding site, 304–6
angiotensinogen, 97, 737, 781–2
angiotensins, 46, 53, 97, 302, 305, 307, 670, 736–41, 781
anhydrases, *see* carbonic anhydrase
anilinopiperidines, 688
annealing, 87
Antabuse, 95, 216, 813
antacids, 705
antagonists, 109, 114–20
antedrugs, 176
anthracyclines, 129, 553–4
    resistance, 552
anthrax, treatment of, 277, 435, 472, 480
Anthriscus sylvestris, 572
anti-allergic agents, 123, 212, 410
anti-androgens, 568–70
anti-angiogenesis agents, 608
anti-arthritic agent, 263
anti-asthmatics, 264, 277, 663–66
    administration, 179
antibacterial agents, 257, 425–87, 705, 732
antibodies, 158, 160, 276–7
    abzymes, 614
    anticancer therapy, 609–14
    antiviral agents, 540

as lipid-lowering agents, 756
chimeric, 277
fusion protein, 275
humanized, 277
hybrid, 614
inhibiting protein-protein
    interactions, 148–50
linking drugs, 613
monoclonal, 265, 276
murine, 277
reduction of disulphide links, 611
targeting drugs, 183–4, 265
antibody-directed abzyme prodrug therapy
    ADAPT, 614
antibody-directed enzyme prodrug therapy
    ADEPT, 612–4
antibody-drug conjugates, 183, 609–13
antibody-enzyme complex, 612
anticancer agents, 102, 129–37, 251, 276, 281, 290, 543–616
anticholinergic agents, 210, 706, 708, *see also* cholinergic antagonists
anticholinesterase poisoning, treatment
    of, 629
anticholinesterases, 252, 416, 640–52
anticoagulants, 150, 176, 212, 220, 290, 515, 758–60, 788–94
    effect of aspirin, 181
anticodon, 83–5
antidepressants, 212, 639, 776–7, *see also* monoamine oxidase inhibitors, selective noradrenaline reuptake inhibitors, selective serotonin reuptake inhibitor, and tricyclic antidepressants
    adrenergic antagonists, 667
    dopamine antagonists, 123, 197
    drug-drug interactions, 181
    dual-action agents, 777
    reuptake inhibitors, 144–5
    serotonin agonist, 213
    serotonin antagonists, 296, 777–80
antidiabetic agents, 212, 261
antidiarrhoeal agents, *see* diarrhoea
    treatments
antidiuretic hormone, 178
antidotes
    anticholinesterase poisoning, 629
    atropine poisoning, 644
    curare poisoning, 644
    to lewisite, 103
    to morphine overdose, 123
    to organophosphates, 647–8
anti-emetics, 123, 200
anti-epileptics, *see* epilepsy treatments
anti-estrogens, 568–70
antifreeze poisoning, 94
antifungal agents, 199, 242, 257, 262, 266, 285, 484
antigenic drift, 521
antigenic variation, 521
antigens, 158, 160, 184, 265, 492, 521, 609
anti-growth factors, 546
antihistamines, 175, 201, 213, 216
anti-HIV drugs, *see* HIV

antihypertensive agents, 211, 672–3, 737–53, 776, 781–7, *see also* angiotensin-converting enzyme inhibitors, adrenergic antagonists, beta-blockers, clonidine, cromkalim, losartan, and renin inhibitors
anti-impotence drugs, 212
anti-inflammatory agents, 102, 230, 289–90, 605, 766–75
antimalarial agents, 212, 269, 309–14, 405
    doxycycline, 472
antimetabolites, 428, 560–66
antimicrobial agents, *see* antibacterial agents, antifungal agents, and antiprotozoal agents
antimigraine agents, 123
anti-obesity drugs
    adrenergic agonists, 123, 297, 662
    lipase inhibitors, 95–6, 208
    melanin concentrating hormone receptor antagonists, 202
    reuptake inhibitors, 144, 674
anti-oncogenes, 544
antiparasitic drugs, 487
antiplatelet drugs, 758, 760–3
antiprotozoal agents, 2–3, 426, 479, 482, *see also* antimalarial agents and oxamniquine
antipsychotics, 122–3, 199, 244, 269
antirheumatic agent, 259
antisense drugs, 756
antisense therapy, 139–41, 184, 280, 498, 559–60
antiseptic, 425
antismoking drugs, 212, 639, 674
antithrombotic agents, 258, 757–63, 788–94
antituberculosis agents, *see* tuberculosis treatments
antitumour agents, *see* anticancer agents
anti-ulcer agents, 123, 477, 482, 629, 634, 705–33, *see also* ulcers, treatment of
antiviral agents, 242, 285, 490–541
antrum, 705
ants, 649
anxiety, treatment of, 208, 670
aorta, 165, 735
AP24534, 585
6-APA. *See* 6-aminopenicillanic acid
Apaf-1 protein, 547
apixaban (**Eliquis**), 760, 789–93
apoliprotein B, 756
apomorphine, 352
apoptosis, 151, 544, 547–8, 603–4
    agents affecting apoptosis, 603–5
apoptosome, 547
appetite, control of, 208
appetite stimulants, 214
apricots, 3
aptamers, 152, 280
aquaporins, 177
arabinosyl transferase enzymes, 482
2-arachidonyl glycerol, 214
arachidonylethanolamine, 214

ara-CTP, 564–5
ara-G, 566
area under the plasma drug concentration curve (AUC), 182
arecoline, 628
Arf protein, 71
arginine, 17, 21–2, 717, 801
    ACE binding site, 303–6
    binding of statins, 193
    carboxypeptidase binding site, 302–3
    neuraminidase binding site, 522–4, 526
    replace strategy, 378–9
    role in depression, 776
    synthesis of nitric oxide, 39
argonaute protein, 141
armodafinil (Nuvigil), 292–3
aromatase, 570
    inhibitors, 570–1
aromatic L-aminoacid decarboxylase, 657
aromatic rings, binding role, 225–6
aromatic substituents, variation in drug design, 237–9, see also variation of aromatic substituents
arrhythmia, treatment of, 668, 670, 752
arsenic, 2
arsenic trioxide (Trisenox), 606
arsenite, 103
Artabotrys uncinatus, 314
arteether, 310, 313
arteflene, 314
artemether, 310, 313
Artemisia annua, 309–10
artemisinin, 207–8, 210, 309–13
Artemisinin Combination Therapy, 313
arterial baroreceptors, 736
arterial chemoreceptors, 737
arteries, 735
arterioles, 736
arthritis, 27
    treatment, 99, 221, 275, 277, 289, 417
arthropods, 490
artificial viruses, 89, 184
aryl halides, binding role, 231
aryloxypropanolamines, 668–9
asbestos, 544
asfatase alfa (Strensiq), 276
Asian flu, 519
asparaginase, 276, 608
asparagine, 17, 801
    cholinergic binding site, 380, 625
    glycoproteins, 25
    in leukaemia cells, 608
    in protease inhibitors, 507, 509, 513
    mutation in reverse transcriptase, 502
    rigidification, 248
aspartate (aspartic acid), 17, 21–2, 801
    acetylcholinesterase active site, 642
    acid base catalysis, 781
    adrenergic receptor binding site, 659–60
    as an activating group, 788, 795
    BH3 domain, 604
    binding group in factor Xa, 788–9
    binding of protease inhibitors, 504–10, 512–6

binding site of pro-survival proteins, 604
binding to estrogen antagonists, 117
catalytic triad, 34, 96
cholinergic binding site, 625
HIV-protease active site, 504–10, 512–6
HMG-CoA reductase active site, 188–90
in fibrinogen, 150
kinase active site, 583–5, 592
lipase active site, 96
mechanism of renin, 97
mimic, 198
neuraminidase active site, 523–6, 528
proton acceptor, 33
proton donor, 33
receptor binding sites, 380
substrates for caspases, 198
thymidylate kinase active site, 419–22
aspartic acid, see aspartate
aspartyl proteases, 504, see also renin
aspergillosis, 812
Aspergillus alliaceus, 208
Aspergillus sclerotiorum, 320
Aspergillus terreus, 190
asperlicin, 208, 245
aspirin, 3, 220, 288, 551, 678, 760
    as a prodrug, 270
    as an acylating agent, 230
    drug-drug interactions, 180–1
    interaction with cyclooxygenases, 102, 230, 705
    role in causing ulcers, 705
    synthesis of, 295–6
Association of British Pharmaceutical Industry, 294
astemizole, 212, 504
asthma treatment, 200, 221, 277, 656, 707, 770–1, see also antiasthmatics and salbutamol
Astra, 296, 730
AstraZeneca, 579
asunaprevir (Sunvepra), 534–5, 539
asymmetric compounds, 111
asymmetric synthesis, 111
AT$_1$ receptors, 737
    antagonists, 747
atazanavir (Reyataz), 501, 514–6, 519
atenolol (Tenormin), 669–71
atherosclerotic plaque, 760
athletes foot, treatment of, 152
atomoxetine (Strattera), 674
atorvastatin (Lipitor), 163, 187, 191, 193
ATP, see adenosine 5'-triphosphate
ATPase, 726
ATP-binding cassette (ABC) transporters, 552
atracurium (Tracrium), 637–8
atria, 735
atrial natriuretic factor, 747
α-atrial natriuretic peptide, 72
Atripla, 501, 813
Atropa belladonna, 629
atropine, 210, 629–30, 634
    antidote, 644
    quaternary salts, 631

atropine methonitrate, 631, 634
    with secbutabarbital (Butropina) 814
attention deficit hyperactivity disorder, treatment of, 674
AUC, see area under the plasma drug concentration curve
Augmentin, 98, 456, 813
aureomycin, 468
AutoDock program, 373, 378
autoimmune disease, 628
    treatment of, 198
autonomic motor nervous system, 621
autoreceptors, 12, 623, 667, 777
avibactam, 455, 457–8
    with ceftazidime (Avycaz) 813
avridine, 540
axitinib (Inlyta), 591, 594
axon, 807
Azadirachta indica, 733
azaepothilone B, 573
azathioprine (Imuran), 269
azidothymidine, 501, see zidovudine
azilsartan, 740–1
azilsartan medoxomil (Edarbi), 740–1
azithromycin (Zithromax), 474, 477
azlocillin, 447–8
AZT, see zidovudine
Aztec, 210
aztreonam (Azactam), 455–6

B

bacampicillin, 446
Bacillus polymyxa, 464
Bacillus subtilis, 459
bacitracin, 426, 458–9, 464–5
    with neomycin sulphate (Cicatrin) 814
    with polymyxin B sulphate (Polyfax) 818
bacteria, 812
bacterial carboxypeptidase, 612
bacterial cell, 427
bacterial pneumonia, treatment of, 454
bacterial RNA polymerase inhibitors, 479–80
bactericidal, definition, 431
bacteriophages, 88, 483
bacteriorhodopsin, 379–80, 382
bacteriostatic, definition, 430
Bacteroides fragilis, 454, 811
    treatment of, 477, 482
Bad protein, 547
Bak protein, 603–4
BAL30072, 455–6
bar coding, 343
barbiturates, 1, 165–6, 398, 638
Barlos resin, 327–8
base pairing, 78–80, 82, 232–3
    abnormal for guanine, 132
basil, 3
batefenterol, 635
Bax protein, 547, 598, 603–4
BAY41-2272, 746
BAY41-8543, 746
BAY59-7939, 794

BAY63-2521, 745–6
Bayer, 793
Bayer Health Care AG, 746
bazinaprine, 213
BC3781, 476
BC7013, 476
B-cell activating factor, 277
B-cells as targets for antibodies, 609
    Bcl-2 protein, 280, 547–8, 560, 603–4
        inhibitors, 604
Bcl-x protein, 151–2, 547–8
    Bcl-xL protein, 603
        inhibitors, 604
Bcr-Abl kinase inhibitors, 593
BCX1812, 529
beclometasone 17,21-dipropionate
        (Becodisks, Beconase), 769–70, 775
    with formoterol fumarate (Fostair) 815
beclometasone dipropionate, see
        beclometasone 17,21-dipropionate
bedaquiline (Sirturo), 481–2
bee sting, 662
Beechams Pharmaceuticals, 434, 440, 445, 455
belatacept (Nulojix), 275
belimumab (Benlysta), 277
belinostat (Beleodaq), 601
belladonna, 629
bemiparin, 758
Benadryl, 707, 813
bendamustine (Treanda), 556–7, 559
benextramine, 650–1
benign prostatic hyperplasia, treatment
        of, 667
benign tumours, 543, 550
Benylin, 662
benzamide group
        as a mimic of nicotinamide, 602
        as a zinc ligand, 601
benzatropine (Cogentin), 634
benzazepril (Lotensin), 738–9
benzazeprilat, 739
benzhexol see trihexyphenidyl
benzodiazepines, 113, 215
    solid phase synthesis, 329
benzoic acid, ionisation, 400
benzomorphans, 687–8
benzothiazepines, 749–50, 752–3
benztropine see benzatropine
benzylpenicillin (Crystapen), 434–5,
        438–42, 483–4
L-benzylsuccinic acid, 214–5, 302–4
beraprost sodium, 747–8
Bergman cyclization, 136
Besifloxacin (Besivance), 477–8, 480
beta-blockers, see β-blockers
beta-lactams, see β-lactams
betamethasone (Betnelan), 767–9, 775
betamethasone 17,21-dipropionate
        (Diprosone), 769
    with salicylic acid (Diprosalic) 815
    with clotrimazole (Lotriderm) 817
betamethasone sodium phosphate
        (Betnesol, Vistamethasone), 771–2
    with neomycin (Betnesol-N) 814

betamethasone 17-valerate, 769 (Betacap,
        Betesil, Betnovate, Bettamousse)
    with clioquinol (Betnovate-C) 814
    with fusidic acid (Fucibet) 815
    with neomycin sulphate (Betnovate-N) 814
betaxolol (Betoptic), 669–71
bethanechol (Myotonine), 628
bevacizumab (Avastin), 152, 277, 610
bezafibrate (Bezalip), 755–6
BH motif, 603
BH3-only proteins, 603–4
BIAcore, 205
BIBR1048, 759
BIBR953, 759
bicalutamide (Casodex), 568, 571
bicyclams, 517
bile duct, 176
bilharzia, 315
biliary tract infections, treatment of, 468
Bill and Melinda Gates Foundation, 197
BILN-2061, 798–9
binding efficiency, 219
binding groups, 5, 48, 109–12
binding regions, 5, 32, 48
binding sites, 4–5, 47, see also allosteric
        binding sites
    adrenergic receptors, 659–60
    catecholamine receptors, 380
    cholinergic receptor, 380
    construction of, 380
    estrogen receptor, 116–7
    G-protein coupled receptors, 381
    model, 382
    muscarinic receptor, 625
binimetinib, 588
bioassays, 203–7
bioavailability, 182, 288
bio-equivalence studies, 288
bio-isosteres, 243–4, 259, 278, 409, 511
    acylsulphonamide, 799
    benzimidazoles, 536
    tetrazole ring, 739, 791
    for alpha, beta unsaturated esters, 268
    for carboxylic acid, 258–9
    for catechol ring, 665
    for lactone, 573
    for methyl group, 580, 628
    for phenols, 259
    for phosphate, 496
    for purine ring, 762
    for thiourea, 718
    serotonin antagonists, 779
biomarker, 290
Biota pharmaceuticals, 525
bioterrorism, 490, 541
birocodar, 552
bis(7)-tacrine, 649–50
bismuth, 482
bismuth chelate, 732
bismuth subcitrate, 732
bi-specific T-cell engagers, 610
bivalirudin (Angiox, Angiomax), 758
Black, James, 668
bladder cancer, treatment of, 559, 566

bladder problems, treatment of, 634
Bld protein, 547
bleomycins, 129–30, 136, 554–5
blinatumomab (Blincyto), 609
β-blockers, 123, 289, 666–72, see also
        adrenergic receptor antagonists
    short-acting, 669–72
blood clotting, inhibition, 149
blood pressure, 736
blood-brain barrier, 166, 203
blood clot, 788
bloodshot eyes, treatment of, 663
blue green algae, 572, 812
BMS184476, 146–7
BMS188797, 146–7
BMS 247550, 574
BMS 378806, 517
BMS200150, 757
BMS201038, 757
BMS346, 536
BMS346567, 743
BMS790052, 537
BMS858, 535–6
boceprevir (Victrelis), 532–4, 538, 795
boils, treatment of, 425, 482
bond pi cooperativity, 233
bone infections, treatment of, 477, 480
bone tumours, treatment of, 594, 610
boronic acid in proteasome inhibitors, 598–9
bortezomib (Velcade), 102, 599
bosentan (Tracleer), 742–3
bosutinib (Bosulif), 584–5, 593
botulism, 210
bradykinin, 46, 736, 739, 747
    receptor, 53
B-Raf kinase inhibitors, 587–8, 594
brain abscesses, treatment of, 482
brain natriuretic factor, 747
brain natriuretic peptide, 72
brain tumours, treatment of, 559
brand names, see trade names
Brazilian viper, 210
BRCA protein, 602
breast cancer, 543, 545–6, 551
    treatment of, 571, 575, 587, 594, 610
    treatment with alkylating agents, 559
    treatment with antibodies, 609–10
    treatment with antimetabolites, 565–6
    treatment with aromatase inhibitors, 570
    treatment with drugs acting on
        tubulin, 575
    treatment with hormone-based
        therapies, 571
    treatment with intercalating agents, 554
    treatment with kinase inhibitors, 593
    treatment with pancratistatin, 606
    treatment with raloxifene, 117
    treatment with tamoxifen, 123
breast milk, 176
bremazocine, 687
brentuximab vedotin (Adcetris), 606, 610
bretylium, 673
Bristol-Myers Squibb, 743, 789
British Approved Name, 12

British National Formulary for Children, 290
broad-spectrum antiviral agents, 539–40
broad-spectrum penicillins, 444–48
  clinical aspects, 447
broccoli, 2, 551
bromodomain, 608
bronchitis, treatment of, 447, 480
bronchodilator, 662
Brookhaven National Laboratory Protein Data Bank, 20, 365, 367
brucellosis, treatment of, 468, 482
Brussels sprouts, 174
Bruton's tyrosine kinase, 74, 589
bryostatins, 209, 606
bubonic plague, 426
budding, 492
budesonide, 770–1, 775
bungarotoxin, 210, 639
bupivacaine (Marcaine), 292
buprenorphine (Belbuca, Subutex, Temgesic, Transtec), 691, 693, 695
bupropion (Zyban), 212, 639, 674, 685
  with naltrexone (Contrave) 814
burimamide, 214, 712–5
Burkitt's lymphoma, 543
  treatment of, 494, 551
burns, treatment of, 429
buserelin, 571
busulfan (Busilvex, Myleran), 132–4, 559
busulphan, see busulfan
butorphanol, 695
N-butyldeoxynojirimycin, 517
butyrylcholinesterase, 640
bystander effect, 615

C

C55 carrier lipid, 458
CaaX peptide, 576
cabazitaxel (Jevtana), 146–7, 575
cabbage, 2, 551
cabozantinib (Carbometyx), 590–1, 594
Caco-2 cell monolayer absorption model, 203
cadazolid, 486
cadherins, 550
CAESA, 388
caffeine, 2, 288
calabar bean, 642
calcitonin, 275
calcium carbonate, 705
calcium entry blockers, 748
calcium ion channels, 62, 69, 200, 748–50
  blockers, 552, 748–53
calcium ions, 69, 71
calcium-dependent protein kinases, 69
calcium-sensing receptor, 113
calicheamicin, 136–7, 559, 612–3
calmodulin, 69, 151
Cambridge Structural Database, 351, 365
cAMP, see cyclic AMP
Camptotheca acuminata, 130, 555
camptothecin, 130–1, 555

canaliculus, 725–6
cancer, 543–52
cancer chemotherapy, see anticancer agents
cancer stem cells, 552
candesartan, 740–1
candesartan cilexetil (Amias, Atacand), 740–1
candidiasis, 812
candoxatril, 269, 273
candoxatrilat, 269
cangrelor (Kengreal, Kengrexal), 762
cannabinoid receptors, 214
  antagonists, 123
cannabis, 1–2, 179
canrenone, 741
capecitabine (Xeloda), 562, 565–6, 593, 610
capillaries, 735–6
caproctamine, 650–1
capsid, 144, 491
capsid-binding agents, 530–1
capsules, 12, 178, 182, 287–8, 516
captopril (Capoten), 102, 332, 738
  active conformation, 366
  as a lead compound, 211–2
  design of, 303–5
  rigidification of, 248
  side effects, 304
  synthesis of, 307
carbachol, 627–8
carbapenems, 454–5
carbenicillin, 445, 447
carbenoxolone (Pyrogastrone), 733
carbidopa, 273–4
  with levodopa, see co-careldopa
carbohydrates, 25, 157–60, see also aminoglycosides, anthracyclines, N-butyldeoxynojirimycin, cyclodextrins, deoxyribose, glucose, glycoconjugates, macrolides, peptidoglycan, podophyllotoxins, ribose, sialic acid, vancomycin
  in antibodies, 612
carbolic acid, 425
β-carboline-3-carboxylate, 746–7
carbon buckyballs, 89
carbonic anhydrase, 42, 213, 345, 725–6
carboplatin (Paraplatin), 198, 558–9, 610
carboxamides, 525–6
carboxylate ion
  as a zinc ligand, 596
  binding role, 229
carboxylation, 25
carboxylesterase, 747
carboxylesterases, 555–6, 613–4
carboxylic acids, binding role, 229
carboxypenicillins, 445, 447
carboxypeptidase, 302–3, 612–3, 615
  inhibitor, 303
carbuncles, treatment of, 425
carcinogenesis, 543
cardiac arrest, treatment of, 662
cardiac failure, treatment of, 666
cardiac muscle, 735
cardiac stimulant, 662

cardiogenic shock, treatment of, 662
cardiotonic agents, 399
cardiovascular drugs, 628, 667–8, 735–63, see also antihypertensive agents
cardiovascular system, 735–7
carfecillin, 445, 447
carfilzomib, 600
carmustine (BCNU, BiCNU, Gliadel), 132, 134, 559
carrier lipids, see C55 carrier lipid
carrier proteins, see transport proteins
carvedilol, 666
caspases, 103, 198, 547, 603
  inhibitors, 198
Castleman's disease, treatment of, 610
catalase-peroxidase enzyme, 481
catalytic triad, 34, 96, 642–3
  factor Xa, 788
  HCV NS3-4A protease, 795
catch and release, 336–7
catechol O-methyltransferase, 172, 262, 657, 662–3
catecholamine receptors, 380, see also adrenergic and dopamine receptors
catecholamines, 656–7, see also adrenaline, adrenergic agonists, dopamine and noradrenaline
catechols, 352, 656, see also catecholamines
Catharanthus roseus, 571
cathepsin B, 599
cathepsin D, 504
cathepsin G, 739
cation-pi interaction, 9
cauliflower, 2, 551
CB3717, 419–20
CC1065, 558
CCK, see cholecystokinin
CCR5 receptor, 498, 517
  antagonists, 517–8
CD117 receptor, 593
CD20 antigen, 612
CD20 receptor, 609
CD319 protein, 610
CD33 antigen, 613
CD38 protein, 610
CD4 protein, 498–500, 517
CD52 antigen, 609
CDKs, see cyclin-dependent kinases
cefalexin (Ceporex), 261, 450–1, 454
cefalothin, see cephalothin
cefazolin, 450–1, 454
cefepime, 452–3
cefotaxime (Claforan), 452–3
cefoxitin (Mefoxin), 260, 452, 454
cefpirome (Cefrom), 452–4
ceftaroline, 453
ceftaroline fosamil (Teflaro), 453–4
ceftazidime (Fortum, Kefadim), 452–4, 457
ceftizoxime, 452–3
ceftolozane, 453–4
  with tazobactam (Zerbaxa) 821
ceftriaxone (Rocephin), 452–4
cefuroxime (Zinacef, Zinnat, Kefurox), 452, 454

celecoxib (Celebrix), 99, 101
cell adhesion molecules, 550
cell cycle regulation, 545–6
cell death, see apoptosis and caspases
cell entry inhibitors, 517
cell membrane, 3
cell wall structure, 435–6
cellulose, 157
central dogma, 83
central nervous system, 44–5
central nervous system infections, treatment
    of, 482
centroid, 235, 364
    scaffold, 330
centromere, 548
cephalexin, see cefalexin
cephaloridine, 261, 450–1, 453
cephalosporin C, 426, 448–9, 454
cephalosporinases, 440, see also
    β-lactamases
cephalosporins, 208, 448–54, 485–6
    prodrug, 612
Cephalosporium acremonium, 448
cephalostatins, 152, 209
Cephalotaxus harringtonia, 606
cephalothin (Keflin), 450–1, 454
cephamycins, 451–3
Cerep, 779
ceritinib (Zykadia), 267, 589–90, 594
cerivastatin (Baycol), 191, 289–90
cerubidine, 553
cervical cancer, 543
    treatment of, 559, 566
cetuximab (Erbitux), 609
cGMP, see cyclic GMP
CGP52411, 260–1
CGP53353, 261
CGP53716, 583
CGP29287, 782–3
CGP38560, 783–4
CGP53716, 583
Chain, see Florey and Chain
chain contraction, 239, 241, 251
chain cutters, 136, 559
chain extension, 239, 241, 668, 713
    ACE inhibitors, 305
    oxamniquine, 316
chain terminators, 128, 137–8, 148, 495, 564
chaperones, 28
cheese reaction, 675
ChemBio3D program, 350
ChemDraw program, 350
chemical development, 295–97
chemical libraries for antibacterial
    agents, 486
chemokine receptors, 498, 517
chemotherapeutic index, 426
chemotherapy, 425, 543
ChemWindow program, 350
cherries, 3
chickenpox, 490
    treatment of, 494, 497
chimeric antibodies, see antibodies
chimeric drug, 252

Chinese Hamster Ovarian cells, 203
chiral compounds, 111
chiral switching, 145, 292, 664, 730
chlamydia, treatment of, 472
chlorambucil (Leukeran), 556–7, 559
chloramphenicol (Kemicetine,
    Chloromycetin), 208, 270, 290, 426,
    466, 472–3
    binding interactions, 472–3
    resistance, 473
chloramphenicol acetyltransferase, 473
chloramphenicol palmitate, 270
chloramphenicol succinate, 270–1
chloramphenicol-resistant
    meningococci, 483
chlordiazepoxide, see Librium
chloride ion channels, 808, see also
    γ-aminobutyric acid receptors and
    glycine receptors
chlorins, 273, 615
chlormethine, 132–3, 556–7, 559
chloroform, 398
chlorophyll, 615
chloroquine (Avloclor, Nivaquine), 309
chlorothiazide, 213
chlorpromazine (Largactil), 213, 216
chlorpropamide, 261
chlorpyrifos, 646
chlortetracycline, 426, 468–9, 472
    with demeclocycline and tetracycline
    (Deteclo) 815
CHO cells, see Chinese hamster ovarian cells
cholecystokinin, 208, 245
cholera, 425
cholesterol, 187–8, 220, 766
    biosynthesis, 188
cholesterol absorption inhibitor, 754
cholesterol conjugates, 172
cholesterol-lowering agents, 102, 141, 178,
    246, 289, 754–7, see also statins
cholesteryl ester transfer protein, 756
cholestyramine, see colestyramine
choline, 622–3, 642
choline acetyltransferase, 622–3
cholinergic nervous system, 622–23, 658
cholinergic receptors, 47, 621, 624, 706, see
    also muscarinic and nicotinic receptors
    agonists, 623–8
    antagonists, see anticholinergics, and
        muscarinic and nicotinic receptor
        antagonists
Chondrodendron tomentosum, 635
chromatin, 82, 548, 600
Chromobacterium violaceum, 455
chromogranin A, 657
chromosomes, algorithms, 362–3
chronic myeloid leukaemia, treatment
    of, 593, 606
chronic obstructive pulmonary disease, 634
    treatment of, 629, 634–5, 665–6
Churchill, Winston, 429
chymases, 739
chymotrypsin, 19, 34, 599
Ciba-Geigy, 782

ciclesonide (Alvesco), 770–1, 775
ciclosporin (Neoral, Sandimmun), 208–9,
    216, 552
    absorption, 163
    administration, 280
    metabolism, 175
cidofovir (Vistide), 496–7
cigarette, 179
cigarette smoke, 174
cilastatin, 456, 458
    with imipenem (Primaxin) 818
cilazapril (Vascace), 210, 212, 248
cilazaprilat, 240–1
cimetidine (Dyspamet, Tagamet), 176,
    705–7, 716–9, 725
cinacalcet, 113–4
cinchona bark, 210
ciprofibrate (Modalim), 755
ciprofloxacin (Ciproxin, Ciprobay, Cipro,
    Ciproxan), 426, 477–80
cisapride, 201
cisplatin, 133–5, 551, 558–9, 566, 606, 609, 611
citalopram (Cipramil), 145
citrulline, 39
CJS 149, 612
CKIs, see cyclin-dependent kinase inhibitors
c-Kit receptor, 58, 593
    kinase inhibitors, 582–5, 591, 593–4
clarithromycin (Clarosip, Klaricid), 473,
    477, 732
clavulanic acid, 98–9, 434, 447, 455–7
    with amoxicillin (Augmentin) 813
click chemistry in situ, 218–9
clindamycin (Dalacin C), 271, 474–5, 477
clindamycin phosphate, 271, 475
clinical trials, 287–90
clinofibrate (Lipoclin), 755
clique searching, 371
Clitopilus scyphoides, 476
clobetasol, 775
clobetasol 17-propionate (Carelux,
    Dermovate, Etrivex), 770
    with neomycin sulphate and nystatin
        (Dermovate-NN) 815
clobetasone 17-butyrate (Eumovate), 770
    with oxytetracycline (Trimovate) 820
clobetasone butyrate, 770
clofibrate (Altromid-S), 755
Clog P, 256, 399
clonidine (Catapres, Dixarit), 178, 216, 663
clopidogrel (Plavix), 760, 762
clorgiline, 101–2
Clostridium botulinum, 210
Clostridium difficile,
    resistance, 483
    treatment of, 464–5, 480, 482, 486
Clostridium histolyticum, 275
clot formation, 757–8, 760, 788
cloxacillin, 444
clozapine (Clozaril), 123
club moss, 652
c-MET kinase inhibitors, 589–90
c-MET receptor, 58
CML, treatment of, 593

CMN 131, 728–9
CMV, *see* cytomegalovirus
CMV retinitis, treatment of, 497
CNS infections, treatment of, 468
coactivator proteins, 59, 116–7, 149, 608
coagulation factors, 86–7
coal dust, 544
co-amoxiclav, *see* Augmentin
cobicistat (Tybost), 518–9
cobimetinib (Cotellic), 588, 594
cobras, 210
coca bush, 210
coca leaves, 674
cocaine, 2, 688, 776
    ability to cross placental barrier, 166
    administration of, 178–9
    lead compound for local anaesthetics, 245
    mode of action, 144, 674
    overlay with procaine, 363–4
    source, 207, 210
co-careldopa (Sinemet, Stalevo), 273
cocoa, 2
codeine, 170, 174, 680, 683–4, 699
coenzyme A, 188–9, 192–3, *see also* acetyl
    coenzyme A
coenzyme F, 431
coenzymes, 35–6
cofactors, 20, 35–37, *see also* adenosine
    5'-triphosphate, dihydrolipoate,
    S-adenosyl methionine,
    5,10-methylenetetrahydrofolate,
    magnesium, NAD$^+$, NADH, NADPH,
    NS4A, 3'-phosphoadenosine
    5'-phosphosulphate, tetrahydrofolate,
    vitamin K, and zinc
    competition with inhibitors, 94
coffee, 2
co-fluampicil, 444, 814
colchicine, 145–6, 252, 416–7, 572–3
cold sores, 490
    treatment of, 494, 497
cold virus, 530
colestyramine (Questran), 178
collagen, 760
    degradation, 275
collagenase clostridium histolyticum
    (Xiapex, Xiaflex), 275
collagenases, 596
colon cancer, 551, 575
    treatment of, 566
colony-stimulating factors, 56, 275
Colorado tick fever, 490
colorectal cancer, treatment of, 556, 559,
    565–6, 609–10
CoMASA, 416
combination therapies, 201
    MEK and B-Raf inhibitors, 588
    treatment of COPD, 635
    treatment of HCV, 538
combinatorial libraries, 341
combinatorial synthesis, 325, 340–6
    planning, 392
Combivir, 501, 814
combretastatins, 573

Combretum caffrum, 573
CoMFA, 354, 413–7
comfrey, 733
CoMMA, 416
common cold, 490
compactin (Mevastatin), 190
compensated liver disease, 538–9
competitive inhibition, 93
competitive inhibitors, 94
Complexa/Eviplera, 501, 814
compound libraries, 325, 330, 341
    planning, 390–2
computer-aided design of lead
    compounds, 215
CoMSIA, 416
conformational analysis, 358–63, 778
    apixaban, 792–3
conformational blockers, 248, 250, 709, *see*
    also conformational restraint
    imatinib, 583
    olaparib, 603
    oxamniquine, 317–8
conformational explosion, 373
conformational restraint, 792, *see also*
    conformational blockers
    benzazepril, 738
    olaparib, 603
conformational space, 330
conjugation, 483
conjugation reactions, 167, 171
Conn's syndrome, 767
Connolly surface, 370
conotoxin, 210
constipation, treatment of, 701
constructs, 374
contraception, 123
contraceptives, 113, 208, 260
    drug-drug interactions, 176, 482
convergent evolution, 54
coronary arteries, 736
coronary heart disease, 187
correlation coefficient, 396
corticosterone, 766–7
corticotrophin-releasing factor, 776
cortienic acid, 772
cortisol, 113, 766, 768, 772, 775
cortisone, 766–7, 775
cortisone acetate, 775
cortivazol, 113, 391
co-transmitters, 623, 657–8
co-trimoxazole (Septrin), 432
coughs, treatment of, 684
cowpox, 492
Craig plot, 404–6
crisantaspase (Erwinase), 608
Crixivan, 12, 814
crizotinib (Xalkori), 234, 251, 589–90, 594
Crohn's disease, treatment of, 277
cromakalim, 264, 286, 354–5
cross-linking, 132–6, 555, 558
cross-validated correlation coefficient, 415
cross-validation, 414
crown ether solubilising group, 580
cryptophycins, 572–3

crystal structure
    factor Xa and apixaban, 792–3
CrystalLEAD, 218
c-Src tyrosine kinase, 583
CTLA-4 receptor, 610
curacin A, 209
curare, 2, 635
curare poisoning, antidote, 644
Cushing's syndrome, 767
CXCR4 receptor, 498
cyanosafracin B, 607
cyclases, 39, *see also* adenylate cyclase and
    soluble guanylate cyclase
cyclic ADP ribose hydrolase, 610
cyclic AMP, 62–6, 640, 655, 746
    ligand for funny ion channels, 753
cyclic GMP, 39, 62, 71–2, 744, 746
cyclic lipopeptides, 464–5
cyclic nucleotide phosphodiesterases, 746
cyclic peptide antibiotics, 426
cyclic peptides, 154
cyclin-dependent kinases, 19–20, 545–6, 586
    inhibitors, 546, 586–7, 594
cyclins, 545–6, 586–7
cyclodextrins, 89, 159–60
cycloguanil pamoate, 269–70
cyclohexane, 359
cyclooxygenases, 99, 230, 290, 705
    inhibitors, 101–2, 760
cyclopentenyl cytosine, 539
cyclopentolate (Mydrilate), 631–2, 634
cyclophilin A, 538
cyclophosphamide (Endoxana), 132, 270,
    557, 559
cyclopropanecarboxylic acid esters as
    prodrugs, 268
cyclopropyl group as a bioisostere, 243
D-cycloserine, 458–9, 464
cyclosporin A, 538, *see* ciclosporin
cyproterone acetate (Cyprostat), 568–9, 571
cysteine, 17, 801
    alkylation, 96
    as a zinc ligand, 576
    as nucleophile with irreversible
        inhibitors, 589
    biosynthetic building block for
        β-lactams, 434, 448–9
    caspase active site, 198, 547
    cytochrome P450, 100
    disulphide bonds, 21–3
    farnesyl transferase inhibitors, 576
    farnesylation, 576
    glutathione conjugates, 174
    glycolysis enzymes, 487
    intracellular receptors, 59
    mercury poisoning, 102
    nicotinic receptor, 636
    nucleophile, 34, 95
    PEGylation of antibodies, 277
    proton pump, 726–8
    reaction with acrolein, 558
    target for irreversible inhibitors, 581–2
    urease, 733
cysteine proteases, 599, *see also* cathepsins

cytarabine, 564–6
cytidine triphosphate (CTP), 539
cytidine triphosphate synthetase, 539
cytochrome c, 547
cytochrome P450 enzymes, 36, 134, 167–70, 173, 174–6, *see also* S-mephenytoin hydroxylase and nifedipine hydroxylase
    cytochrome P450 2C19, 762
    inhibitors, 519, 539
    irreversible inhibition, 100
cytokine receptors, 56, 72–3, 593
    targets for antibodies, 277
cytokines, 56, 58, 605, 616, 776
    inhibition, 277
cytomegalovirus (CMV) infections, treatment of, 497
cytoplasm, 3
cytosine, 77–8
    alkylation of, 134
    as a nucleophile, 563
    nucleophilic groups, 132
cytosine arabinoside, 565

D

D1927, 259
dabigatran, 758–9
dabigatran etexilate (Pradaxa, Prazaxa), 758–9
dabrafenib (Tafinlar), 587–8, 594
dacarbazine (DTIC-Dome), 134–5, 559
daclatasvir, 539
daclatasvir (Daklinza), 535–7
daclizumab (Zenapax), 149
dactinomycin (Cosmegen Lyovac), 129, 553–4
    resistance, 552
daffodils, 649
Daiichi Sankyo, 794
dalbavancin (Dalvance), 461–2, 464
dalfopristin, 475, 477
    with quinupristin (Synercid) 819
dalteparin (Fragmin), 758
dantron, 210
dapsone, 482
daptomycin (Cubicin), 152, 464–5
daratumumab (Darzalex), 610
darifenacin (Emselex, Enablex), 633–4
darunavir (Prezita), 501, 514–6, 519
darusentan, 742
dasabuvir, 535, 539
    with ombitasvir, paritaprevir, and ritonavir (Viekira Pak) 820
dasatinib (Sprycel), 579, 584–5, 593
database handling, 392
database mining, 217, 378
dATP, *see* deoxyadenosine triphosphate
Datura stramonium, 630
daunomycin, 553
daunorubicin (DaunoXome), 553–4
dCTP, *see* deoxycytosine triphosphate
DDT, 309
de novo drug design, 249, 381–90, 419–2, 509–10
deacetylases, *see* histone deacetylase

10-deacetylbaccatin III, 299
deacetylcortivazol, 113
deadly nightshade, 629
dealkylation, 237
deaminases, *see* adenosine deaminase
death activator proteins, 547
death-inducing protein, 608
3-deazaneplanocin A, 539
decamethonium, 365, 416, 636–7, 639
decarboxylases, *see* aromatic L-aminoacid decarboxylase, dopa decarboxylase
decompensated liver disease, 539
deep-vein thrombosis, 757–8
    treatment of, 788, 793
deflazacort (Calcort), 775
degarelix acetate (Firmagon), 568, 571
degree of inhibition, 105
dehydrogenases, *see* alcohol dehydrogenase, aldehyde dehydrogenase, 17β-hydroxysteroid dehydrogenase, inosine-5'-monophosphate dehydrogenase, lactate dehydrogenase
dehydropeptidase, 455–6
    inhibitor, 456
delavirdine (Rescriptor), 501–3
delta-selective opioids, 696–8
demeclocycline, 469, 472
    with chlortetracycline and tetracycline (Detecto) 815
dementia, 640
demethylation, 228, 685
    of methyl ethers, 231
N-demethylation, 231, 268
dendrites, 807
dengue fever, treatment of, 493
denosumab (Prolia, Xgeva), 277, 610
deoxoartemisinin, 311
deoxodeoxyartemisinin, 311
2-deoxy-2,3-dehydro-N-acetylneuraminic acid (Neu5Ac2en), 524–7
deoxyadenosine, 77
deoxyadenosine triphosphate, 564
    allosteric inhibitor, 564
deoxyartemisinin, 311
21-deoxybetamethasone 17-propionate, 769–70
deoxycytidine, 77, 500, 564
deoxycytosine monophosphate, 496
deoxycytosine triphosphate, 564
deoxyguanosine, 77, 138
deoxyguanosine triphosphate, 495
deoxypodophyllotoxin, 572
2-deoxyribose, 77, 82, 566
deoxythymidine, 77, 496, 500
deoxythymidine monophosphate (dTMP), *see* deoxythymidylate monophosphate
deoxythymidine triphosphate (dTTP), 564
deoxythymidylate monophosphate (dTMP), 419, 560–2
deoxyuridine monophosphate, *see* deoxyuridylate monophosphate
deoxyuridylate monophosphate (dUMP), 419, 560, 562
    fluorinated analogue, 561

dependence, 121–2
depolarization, 808–9
depression, 667, 776
    treatment of, 176, 674
depsipeptide (Istodax), 600–1
desciclovir, 495, 497
Descovy, 501, 814
desensitization, 119, 121, 568, 667
desert hedgehog, 74
desipramine, 174, 197–8, 673–4
Detecto, 472, 815
deuterium as a metabolic blocker, 261, 514
devazepide, 245
dexamethasone (Maxidex), 465, 610, 767–8, 771, 775
    with neomycin sulphate (Otomize) 818
    with neomycin sulphate, and polymyxin B sulphate (Maxitrol) 817
    with sodium metasulphobenzoate, framycetin sulphate, and gramicidin (Sofradex) 819
    with tobramycin (Tobradex) 820
dexamethasone acetate, 775
dexamethasone metasulphobenzoate, 775
dexamethasone phosphate, 775
dexlansoprazole (Dexilant), 730
dextrorphan, 686
dGTP, 564
diabetes, 735
    treatment of, 180, 297, 755
diacylglycerol, 62, 68–71, 74, 655
1,4-diaminosubstituted aromatic ring, toxicity, 267
diamorphine, 1, 683
diarrhoea, treatment of, 479, 678–9, 684, 690, 699
diazepam (Valium), 181, 270–1, 719
dicer, 141
dichloroisoprenaline, 667–8
dicloxacillin, 444
dicycloverine (Kolanticon), 631–2, 634
didanosine (Videx), 500–1, 503, 516
dideoxyadenosine triphosphate, 500–1
Diels-Alder reaction, 692
diethyl ether, 398
diethylstilbestrol, 567, 571
differentiation, 543
diflucortolone valerate (Nerisone) 775
difluoromethylornithine, 551
digitalin, 210
digitalis, 207, 210
digitonin, 210
digitoxin, 210
dihydroartemisinin, 310–1, 313
dihydrocodeine (DF118 Forte, DHC Continus), 680
dihydroetorphine, 691
dihydrofolate, 431–2, 560–2
dihydrofolate reductase, 102, 199, 431–2, 560–1
    inhibitors, 560
dihydrolipoate, 103
dihydromorphine, 680
dihydropteroate, 431–2

dihydropteroate synthetase, 430–2
dihydropyran-derivatized resin, 327–9
dihydropyridine receptor, 119
dihydropyridines, 749–51
dihydrotestosterone, 567
diiminoquinone, toxicity, 267
diltiazem (Tildiem), 752–3
2,3-dimercaptopropanol, 103
dinutuximab (Unituxin), 610
diphenhydramine (Benadryl), 707
diphenoxylate, 690
    with atropine sulphate (Co-phenotrope) 814
diphenylpropylamine, 690
diphtheria, treatment of, 435, 477
diphtheria toxin, 612
dipipanone (diconal), 690
dipole-dipole interactions, 8–9
diprenorphine, 691
Directed Dock, 373–4
dirty drugs, 202, 253
Discovery Studio Pro, 350–1
disoxaril, 531
displacement or inhibition curve, 125
displacer, 124–5
dissociation binding constant (Kd), 124
dissociation constant, 400
distance matching, 371
disulfiram (Antabuse), 95, 216
disulphide bonds, 21–2
dithiolthiones, 551
diuretics, 100, 213, 289, 742
divergent evolution, 54
DNA, 77–82, 184
    as a drug target, 128–39
DNA gyrase, 477
DNA ligase, 129, 136
DNA polymerases, 102, 499–500, 548, 564,
    see also viral DNA polymerases
    inhibitors, 495, 564–5
    RNA-dependent, 548
DNA-dependent RNA polymerase, 479
dobutamine (Dobutrex, Posiject), 214,
    662–3
docetaxel (Taxotere), 146–7, 281, 560, 575,
    609–10
DOCK program, 370–74
docking, 368–78, 778
    renin inhibitors, 783–4
dolastatins, 209, 606–7
dolutegravir (Tivicay), 501, 518–9
donepezil (Aricept), 102, 648–9
dopa decarboxylase, 273–4, 656–7, 672–3
dopamine, 46
    as a drug, 274
    biosynthesis, 656–7
    interaction with adrenergic receptors, 660
    metabolism of, 101
    pharmacophore, 367–8
    prodrugs, 269, 273–4
    reuptake inhibitors, 639, 674
    role in depression, 776
    role in Parkinsons disease, 200
    role in treating depression, 144, 675
    structure, 46

dopamine receptors, 47, 52–4, 123, 200
    agonists, 113, 123, 213, 200, 213
    antagonists, 123, 197, 199–200, 202, 244,
        248, 260, 266
    regulation, 59
dopamine β-hydroxylase, 656, 673
dopamine β-monooxygenase, 672
dose ratio, 125
double-blind placebo-controlled
    studies, 288–9
doxazosin (Cardura), 666–7
Doxil, 184, 815
doxorubicin (Rubex, Doxil), 129, 184,
    553–4, 610–1
doxorubicin analogues, 612
doxycycline (Vibramycin), 468–9, 472
drosha, 140–1
Drosophila subatrata, 476
drug absorption, 162–64
drug addiction, treatment of, 685, 689, 693
drug administration, 177–80
drug alliances, 273–74
drug delivery, 183–84
drug distribution, 165–66
drug dosing, 180–82
drug excretion, 176–77
drug half-life, 181
drug load, 287
drug metabolism, 167–76
drug metabolism studies, 285–7
drug scavengers, 159
drug specifications, 296
drug targeting, see targeting drugs
drug-drug interactions, 166, 175–6, 180–1,
    203, 289, 516, 719
    protease inhibitors, 504
dTMP, see deoxythymidylate
    monophosphate
dTTP, see deoxythymidine triphosphate
DU122290, 243
Du Pont pharmaceuticals, 259, 739
Du176b, 794
dual-action agents, 252, 742, 747, see also
    dual-target agents
    acting on acetylcholinesterase, 649–50
    afatinib, 581
    antagonists at endothelin and AT1
        receptors, 743–4
    batefenterol, 635
    dihydropyridines, 751
    lapatinib, 580
    release of nitrate ions, 744–5
    reuptake inhibitors, 674
    serotonin and noradrenaline reuptake
        inhibitors (SNRIs), 674, 777
dual- and pan-PPAR agonists, 755–6
dual-target agents, 585, see also dual-action
    agents
duloxetine (Cymbalta, Yentreve), 674
dUMP, see deoxyuridylate monophosphate
DuP 697, 242
duplex regions, 83
Dutrebis, 501, 815
DX9065a, 789

dyflos, 645–6
dynamic combinatorial synthesis, 343–6
dynamic structure-activity analysis, 716
dynorphins, 695–6
dysentery, treatment of, 483
dystrophin, 87

E

ear drops, 465
ear infections, treatment of, 444, 447, 454,
    465, 468, 472
ebalzotan, 296
Ebola virus, 490, 493
EC_{50} value, 125
ecothiopate, 646
Ecteinascidia turbinata, 607
ecteinascidin 743, see trabectidin
Ecuadorian poison frog, 209
ED_{50}, 284
edoxaban (Savaysa, Lixiana), 760, 794
edrophonium, 416, 645
efavirenz (Sustiva), 501–3
efficacy, 124–5
efflux, 440
efflux pumps, 146–7, 468, 470, 573
EGF, see epidermal growth factor
Ehrlich, Paul, 425–6
elacridar, 552
elafibranor, 755–6
elastase, 599
elbasvir, 538–9
    with grazoprevir (Zepatier) 821
electric potential, 808
electric ray, 639
electronic fields, 414
electronic screening, 378
electrophilic functional groups, 95
electrostatic bonds, 5, 21
electrostatic fields, 413
eleutherobin, 209, 573–4
elitist strategy, 363
elotuzumab (Empliciti), 610
elsosulfase alfa (Vimizim), 275
elvitegravir, 501, 518–9
    with emtricitabine, tenofovir disproxil,
        and cobicastat (Stribald) 819
emetine, 210
Emisphere Technologies Inc, 184
emtricitabine (Emtriva), 500–1, 503
    with tenofovir (Travuda) 820
    with tenofovir and efavirenz (Atripla) 813
enalapril (Innovace), 102, 212, 305–6, 332,
    738
enalaprilat, 239, 305–7, 738
    synthesis, 307
enantiomers, 111
enantiospecific reactions, 112
encorafenib, 588
end bouton, 807
endocarditis, treatment of, 464, 468
endocytosis, 121, 520
endogenous compounds as drugs, 274–77
endogenous opioids, 53, 695–6

endometrial carcinoma, treatment of, 571
endomorphins, 696
endonucleases, 141, 555, *see also* drosha
endoplasmic reticulum, 83
endorphins, 46, 52, 198, 214, 695–6
endosomes, 265, 520
endostatin, 608
endothelin receptors, 742
    antagonists, 742–3
endothelins, 736, 742
endozepines, 215
endpoint, 289
energy minimization, 351, 358
    apomorphine, 352
enfuvirtide (Fuzeon), 145, 280, 501, 517
enkephalinases, 102, 214, 699
    inhibitors, 699
enkephalins, 198, *see also* Leu-enkephalin
        and Met-enkephalin
    analogues, 696–7
    as lead compounds for enzyme
        inhibitors, 214
    binding interactions, 691
    binding theories, 697–9
    discovery of, 695–6
    inactivation, 697
    oral activity, 332
    production of, 696
    role in depression, 776
    structure-activity relationships, 696–7
    target receptors, 52, 695–7
enlarged thyroid gland, treatment of, 670
enoxacin, 477–8, 480
enoxaparin (Clexane, Lovenox, Oksapar,
        Xaparin), 758
enoyl-acyl carrier protein reductase, 485
enteric nervous system, 622
Enterococcus faecalis, resistance, 483
Enterococcus faecium, treatment of, 477
entinostat, 601
enzalutamide (Xtandi), 568, 571
enzymes, 4, 30–42, *see also* aminoglycoside
        modifying enzymes, anhydrases,
        cyclises, deacetylases, deaminases,
        decarboxylases, dehydrogenases,
        endonucleases, esterases, glycosidases,
        helicases, hydrolases, integrase,
        isomerases, kinases, lactamases,
        ligases, lipases, lyases, metalloenzymes,
        oxygenases, nucleases, oxidoreductases,
        peptidases, phosphodiesterases,
        phosphoramidases, phosphorylases,
        phosphatases, polymerases,
        proteases proton pump, racemases,
        recombinant enzymes, reductases,
        restriction enzymes, serum
        paraoxonase, synthases, transferases,
        transglycosidases, translocases, urease
    as drug targets, 93–106, *see also*
        individual enzymes listed above
    enzyme kinetics, 104–6
    enzyme modulators, 103
ephedrine (Bronkaid, Primatene), 662, 674–5
4-epi-amino-Neu5Ac2en, 525

epibatidine, 209, 639
epidermal growth factor (EGF), 56, 72
epidermal growth factor receptors
        (EGFR), 56–7, 70–2, 578, 583, 593
    as targets for antibodies, 609
    inhibition, 609
    kinase inhibitors, 579–82, 593
    resistance to kinase inhibitors, 581–2
epigallocatechin gallate, 551
epilepsy, treatment of, 216, 268
epinephrine, 654, 760, *see also* adrenaline
Epipen, 662
epipodophyllotoxin, 130, 572
epirubicin (Pharmorubicin), 553–4
epitope mapping, 218
epitopes, 217–8, 250
eplerenone (Inspra), 742
epoprostenol, 747
epothilones, 573–4
epoxide hydrolase, 169–70
epoxomicin, 600
eprosartan (Teveten), 740–1
Epstein-Barr virus, 543
eptifibatide (Integrilin), 763
Epzicom/Kivexa, 501, 815
equilibrium constant, 400
eravacycline, 471–2
ErbB2 receptor tyrosine kinase, 57, 545, 580
erectile dysfunction and sexual
        impotence, 212
eremomycin, 461, 463
ergosterol, 153–4
eribulin (Halaven), 574–5
ERK, 71
erlotinib (Tarceva), 578, 580, 593
ertapenem (Invanz), 455–6
erythromycin (Erymax, Erythrocin,
        Erythroped), 164, 426, 473, 477, 483
    binding interactions, 473
    structure-activity relationships, 473
erythropoietin, 56, 275
Escherichia coli, 88, 203, 275, 425, 811
    resistance, 483
    treatment of, 479
escitalopram (Cipralex), 145
eserine, 642
esmolol (Brevibloc), 670–2
esomeprazole (Nexium), 730–1
esterases, 170, 172, 230, 264, *see*
        also acetylcholinesterase,
        butyrylcholinesterase, carboxylesterases
        and phosphodiesterases
    activation of prodrugs, 266, 268, 306,
        495, 528, 683
    drug susceptibility, 573, 689
    resistance of penicillin methyl esters, 446
    susceptibility of cephalosporins, 450–2
esters
    as prodrugs, 267–9
    binding role, 230
estradiol, 34, 116–7, 240, 557, 567–8, 570
estramustine (Estracyt), 556–7, 559
estrogen receptor, 59, 113, 116–7, 123
estrogens, 178, 567, 570

estrone, 34, 240, 272, 570
    lysine ester, 272
etanercept (Enbrel), 275
ethambutol, 481–3
ethers, binding role, 230–1
ethinylestradiol, 567, 571
ethylene glycol, 94
3-ethylmorphine, 680
6-ethylmorphine, 680
etiprednol dicloacetate, 773
etofibrate, 755
etoposide (Etopophos, Vepesid), 130,
        555–6, 572, 611
etorphine, 690–1, 695
etravirine (Intelence), 501–3
eukaryotic cells, 427
European Agency for the Evaluation of
        Medicinal Products (EMEA), 293
European Patent Convention, 291
European Patent Office, 291
evanescent wave, 206
everolimus (Zortress, Certican), 152, 594–5
evofosfamide, 557–8
evolocumab (Repatha), 756
evolutionary programs, 362–3
Evotaz, 501, 815
Ewing's tumour, treatment of, 554
excess sleepiness, treatment of, 292
exemestane (Aromasin), 571
exocytosis, 659
exons, 85–6
exotoxin A, 612
EXP3174, 739–40
extended esters, 446
    as prodrugs, 268
    fosinopril, 738
extended-spectrum agent, 580
extension strategy, 421, 514, 661, 669
    ACE inhibitors, 303–4
    factor Xa inhibitors, 790
    HCV protease inhibitors, 796–7
    in drug design, 239
    macitentan, 743
    opioids, 684–5
    serotonin antagonists, 780
    tadalafil, 746
externalization, 530
eye diseases, treatment of, 494
eye drops, 464, 646
    fusidic acid, 482
    pupil dilation, 634
eye infections
    treatment of, 464–5, 468, 472, 497
eye lotions, 430
ezetimibe (Zetia), 754
    with simvastatin (Vytorin) 820

F

F13640, 263
F15599, 263
FabI enzyme, 485
Fabry disease, treatment of, 275
facilitated transport, 177

factor VII, 758
factor Xa, 757, 759, 788
  inhibitors, 759–60, 788
fail-fast fail-cheap strategy, 203
falipamil, 753
false negatives, 333
false transmitters, 672–3
famciclovir (Famvir), 496–7
famotidine (Pepcid), 723–5
Fansidar, 430, 815
farnesyl diphosphate, 576–7
farnesyl pyrophosphate synthase, 577
farnesyl transferase, 102, 576
  inhibitors, 575–7
fast tracking, 294
fatty acid biosynthesis, 484–5
FdUMP, 562
feedback control, 38–9
feedback system, RAAS cascade, 739–41
felodipine (Plendil), 751
fenofibrate (Lipantil, Supralip), 755
fenozan, 313
fentanyl (Actiq, Durogesic, Sublimaze), 178, 181, 688–9, 695
fermentation of penicillins, 440
fesoterodine (Toviaz), 633–4
fevers, treatment of, 314
fexofenadine (Allegra, Telfast), 175–6
fialuridine, 285
fibrates, 754–5
fibrin, 757–8, 760, 763, 788
fibrinogen, 150, 757, 760, 763, 788
fibrinolytic agents, 758, 763
fibroblast growth factor, 549
fidaxomicin (Dificlir), 479–80
fields, 356–7
fight or flight response, 621, 654
filgrastim (Neupogen), 276
finafloxacin (Xtoro), 477–8, 480
finasteride, 551
first pass effect, 176
Fischer's lock and key hypothesis, 32
Fishers F-tests, 396
FK228, 601
FK506 binding protein, 217, 595
FKBP12, 595
flagellum, 199
flavin-containing monooxygenases, 170–1
flavopiridol, 586–7
flavoxate (Urispas 200), 634
Fleming, Alexander, 220, 433–4
Flexibases, 373
Flexible Ligands Orientated on Grid, 373
FlexX, 374–6
FLOG, 373
Florey and Chain, 426, 433
flu epidemics, 490
flu virus, 491
flucloxacillin (Floxapen), 444
  with ampicillin (Co-fluampicil) 814
fluconazole (Diflucan), 199, 257, 262, 266, 285
Fludara, 566, 815
fludarabine (Fludara), 565–6, 606

fludrocortisone, 767–8
fludroxycortide (Haelan), 775
flumetasone pivalate, 767–8, 775
  with clioquinol (Locorten-Vioform) 817
flunisolide (Synartis), 768–9, 775
fluocinolone acetonide (Synalar), 768–9, 775
  with clioquinol (Synalar C) 819
  with neomycin sulphate (Synalar N) 819
fluocinonide (Metosyn), 768–9, 775
fluocortolone (Ultralanum Plain), 775
fluocortolone caproate, 775
fluocortolone pivalate, 775
fluorine
  as a metabolic blocker, 260–1, 580
  as an isostere for hydrogen, 243
  role in fluorouracil, 562
fluorocyclines, 472
5-fluorodeoxyuridylate, 419–20
fluorometholone (FML), 769–71, 775
fluoroquinolones, 131, 476–80
fluorouracil (Efudix), 5, 243, 559–62, 565, 611
  clinical aspects, 556, 610
  transport into cells, 164
fluorous solid phase extraction, 335
fluoxetine (Prozac), 145, 197–8
fluoxymesterone, 567, 571
fluphenazine, 269
fluphenazine decanoate (Modecate), 270
flutamide (Drogenil, Eulexin), 568–9, 571
fluticasone furoate (Avamys)
fluticasone propionate (Cutivate, Flixonase, Flixotide), 770–1, 775
  with salmeterol (Seretide) 819
fluvastatin (Lescol), 191
fluvoxamine (Faverin), 145
folic acid, 432, 560–1
folinic acid, 556, 559, 565, 610
folylpolyglutamate synthetase, 566
fomivirsen (Vitravene), 140, 498
Food and Drug Administration (FDA), 293
foot and mouth disease, 530
  treatment of, 540
foot infections, treatment of, 456
force fields, 349
formaldehyde, 271
formestane (Lentaron), 571
formoterol (Foradil, Oxis), 665
formulation, 182, 287
Fortovase, 12, 815
fosamprenavir (Lexiva, Telzir), 501, 514–6
Foscan, 615, 815
foscarnet (Foscavir), 497
fosfestrol, 567, 571
fosinopril (Monopril), 738–9
fosinoprilat, 738–9
foxglove, 210
fragment-based lead discovery, 217–19
fragment evolution, 218
fragment self-assembly, 218
FRAP, inhibition of, 595
Free-Wilson approach to QSAR, 409, 411

frog, 639
fructose, 157
fruit fly, 476
F-test, 804
Fujisawa, 455
fulvestrant (Faslodex), 568–9, 571
fungal nail infections, treatment of, 484
fungi, 812
funny ion channels, 753
  inhibitors, 753
fusidic acid (Fucidin), 481–2
Fusidium coccineum, 481
fusion inhibitors, 501, 517
fusion protein, 275
Fuzeon, 280, 816

G

G cells, 706
GABA, see γ-aminobutyric acid
galantamine (Reminyl), 113, 207, 250, 648–9
  dimers, 650
galanthamine, see galantamine
ganciclovir (Cymevene), 495–7, 615
GAP, 71, see GTPase activating proteins
garlic, 616
gas gangrene, treatment of, 435
gastric acid, 705–7, 717
gastric adenocarcinoma, treatment with antibodies, 610
gastric cancers, 732
  treatment of, 566
gastric motility, stimulation, 274
gastrin, 706–7, 717
gastrin receptor, 706
gastrointestinal stromal tumours, treatment of, 594
gastrointestinal tract, 162–3
  stimulation, 123, 628
  suppressant, 629
gastrointestinal tract cancers, treatment of, 559, 565, 571
gastrointestinal tract infections, treatment of, 265, 430, 480
gatekeeper residue, 578, 583–4
gating, 51
Gaucher's disease, treatment of, 275
G-coupled receptors, dimerization, 55
GD2 glycolipid, 610
GDP, see guanosine diphosphate
gefitinib (Iressa), 257, 261, 578–80, 582, 593
gel electrophoresis, 26
gelatinases, 596
gemcitabine (Gemzar), 564–6, 609
gemfibrozil (Lopid), 755
gemtuzumab (Mylotarg), 610, 613
gene silencing, 600
gene therapy, 87, 184, 540, 609, 614–5
gene transcription, control, 59
gene-directed enzyme prodrug therapy GDEPT, 614–5
general anaesthetics, 1, 152, 176, 179, 398
General Medical Council, 294

genetic algorithms, 362–3, 377–8
genetic diseases, 37, 85–7, 140
genetic engineering, 87–9
genetic fingerprinting, 290, 551
genetic polymorphism, 37, 60, 82
genistein, 551–2
genital herpes, 490
    treatment of, 493–4, 497
genital infections, treatment of, 472
genital warts, treatment of, 498, 540, 575
genomic research, viral diseases, 493
genomics, 26
Genta, 280, 559
gentamicin (**Cidomycin, Genticin**), 438,
    466–8
Genvoya, 501, 816
geranylgeranyl diphosphate, 577
geranylgeranyltransferase, 577
GFT505, 756
Gilead Sciences, 528
Ginko, 220
glaucoma, 178, 628
    treatment of, 123, 628, 634, 642, 646,
    656, 670
Glaxo pharmaceuticals, 722, 724
GlaxoWellcome pharmaceuticals, 516, 525
Gleevec, 582
Gliadel, 180, 816
glitazars, 755
glitazones, 755
Glivec, 551, 582, 816
global energy minima, 358
glomerulus, 177
glucagon, 66
glucarpidase (**Voraxaze**), 275
glucocerebrosidase, 275
glucocorticoid receptor, 113
    agonist, 391
glucocorticoids, 113, 567, 766–76
glucose, 38, 40, 64–5, 157, 332, 655
glucose-1-phosphate, 38, 40, 64–5
α-glucosidase, 275
β-glucuronidase, 612–3
glucuronidation, 173
C-glucuronides, 171
N-glucuronides, 171
O-glucuronides, 171
S-glucuronides, 171
glucuronyltransferase, 172
glutamate (glutamic acid), 17, 21–2, 801
    acetylcholinesterase active site, 641–2
    acid-catalyst in enzyme mechanisms, 33,
    189
    calcium ion channels, 749, 752
    carboxylation of prothrombin, 25
    kinase active site, 583–4, 592
    matrix metalloproteinase active site, 596–7
    neuraminidase active site, 522–4, 526,
    528–9
    neurotransmitter, 46
    role in depression, 776
glutamate receptors, 52–3
glutamic acid, see glutamate
glutamine, 17, 171, 576, 801

glutathione, 171, 269
glutathione conjugates, 171, 173–4
glutathione S-transferase, 170–1, 173
glyceryl trinitrate (**Nitromin**), 178, 216,
    744–5
glycine, 801, 808
    active site of PARP enzyme, 602
    as neurotransmitter, 46
glycine receptor, 50
glycoconjugates, 157
glycogen, 38, 40, 64–5, 157, 655
glycogen synthase, 40, 64–5
glycogen-1-phosphate, 40
glycolipids, 157
    as targets for antibodies, 610
glycomics, 157
glycoprotein IIb/IIIa receptors, 760
glycoproteins, 4, 157, see also bleomycins,
    gp40, gp120, P-glycoprotein,
    neuraminidase and haemagglutinin
glycopyrronium bromide, 631–2, 634
glycosidases, 517, see also β-glucuronidase
glycosylase, 461
glycosylation, 25
glycrrhetinic acid, 733
glycyrrhizin, 733
GMP, 746
GNF-2, 585–6
GOLD, 378
gonadotrophin-releasing hormone, 623
gonadotrophin-releasing hormone
    receptor, 568
gonorrhoea, treatment of, 447, 454, 480
Good Clinical Practice (GCP), 294
Good Laboratory Practice (GLP), 294
Good Manufacturing Practice (GMP), 294
goserelin (**Zoladex**), 280, 568, 571
gout, treatment of, 102, 145, 275–6
GP IIb/IIIa, 763
    antagonists, 763
gp120 glycoprotein, 498–500, 517
gp41 glycoprotein, 498–9, 517
G-protein-coupled receptors, 52–6, 59,
    61, see also adrenergic receptors,
    angiotensin II receptors, calcium
    sensing receptor, dopamine receptors,
    endothelin receptors, histamine
    receptors, luteinising hormone-
    releasing hormone receptor, muscarinic
    receptors, opioid receptors, P2Y12
    receptors, prostacyclin receptor,
    protease activated receptors, rhodopsin,
    serotonin receptors
    dimerization, 55–6
    signal transduction, 61–70
G-proteins, 52–3, 61–8, see also small
    G-proteins
GR30040X, 746–7
graft survival, 275
gramicidin, 154, 465
gramicidin S, 265
Gram-negative bacteria, 811
Gram-positive bacteria, 811
Gram-stain, 811

granisetron (**Kytril**), 200, 214
granzyme, 547
grapefruit juice, 174
grazoprevir, 534–5, 539
    with elbasvir (**Zepatier**) 821
Grb2 protein, 71–2
grey baby syndrome, 290, 472
grid program, 356–7, 372–3, 413, 416, 419,
    524–5
Grignard reaction, 692
GRIND, 416
group shift, 261–2
    serotonin antagonists, 778
GROW, 390
growth factors, 546–7
growth hormone, 56
growth hormone receptor, 56–7
GS4071, 528
GS5885, 538
GSK2118436, 588
GTPase activating proteins, 71
guanethidine (**Ismelin**), 673
guanine, 77–8
    alkylation of, 134
    metallation of, 134
    nucleophilic groups, 132
guanine nucleotide exchange factors, 71
guanosine diphosphate, 61–4, 68, 71–2
guanosine triphosphate, 39, 61–4, 68, 71–2
guanyl transferase, 540
guanylate cyclase, 71–2
N^α-guanylhistamine, 710
Guardia infections, treatment of, 482
Gulf War syndrome, 645
gut infections, 425
    treatment of, 430, 464–5
gynaecological infections, treatment of, 456

H

H124/26, 729
H159/69, 729
H77/67, 728–9
H^+/K^+-ATPase, see proton pump
haem, 570
    in soluble guanylate cyclase, 744–5
haemagglutinin, 491, 519–21
haematological malignancies, treatment
    of, 554
haemoglobin, 23, 312, 615
haemophilias, 86
Haemophilus epiglottis, treatment of, 454
Haemophilus influenzae, 452, 454, 811
    treatment of, 456, 477
haemoproteins, 167
haemorrhagic cystitis, 559
haemorrhagic fevers, treatment of, 493, 540
haemorrhoids, treatment of, 309
hair loss treatment, 748
halichondrin B, 209, 574
halobetasol propionate (**Ultravate**), 775
haloperidol, 174
halothane, 398
HAMA response, 277

Hammerhead program, 376–77
Hammett equation, 714
Hammett substituent constant, 400–2, 714
Hansch equation, 404–5
Hantzsch condensation, 749
hard drugs, 176
HASL, 416
hay fever, treatment of, 707
HCV NS3-4A protease, 532–3
   inhibitors, 533, 795–800
HCV NS4B protein, 538
HDLs, 756
HDM2 protein, 151–2, 548
head cancers, treatment of, 559, 566,
   609, 615
heart, 735
heart failure, treatment of, 747
heart irregularities, treatment of, 123
heavy metals, 102
hedgehog signalling pathway, 74–5
   receptor antagonists, 595
helicases, 81
   inhibition, 553
Helicobacter pylori, 705, 732–3
   cause of cancer, 543
α-helix, 18–9
heme, 312
Henderson-Hasselbalch equation, 163
heparin, 758
hepatitis, 179
hepatitis A, 490, 530
hepatitis B, 87
   cause of cancer, 543
   treatment of, 493, 500, 503, 540
hepatitis C, 87, 531
   antiviral agents, 276, 493, 531–40,
   795–800
   in vitro tests, 203
hepatitis E, 490
hepatocyte growth factor, 58
hepatocyte growth factor receptor, 58
hepatocytes, 203
HER receptors, 57, 593
HER-2 growth factor receptor, 580, 593
   inhibition, 609
herbal medicines, 220–1, 649, 652, 733
Herceptin, 551, 609–10, 816
HERG potassium ion channels, 201, 285
heroin, 1, 179, 683, 685
herpes infections, treatment of, 497
herpes keratitis, treatment of, 497
herpes simplex virus, 615
   encephalitis, treatment of, 497
   genetically modified variant, 616
   life cycle, 492
   treatment of, 472
herpesviruses, 494, 497
hetacillin, 272
heterocodeine, 679–80
heterocycles, binding interactions, 232–3
hexamine, see methenamine
hexobarbitone, 268
hexyl-insulin monoconjugate 2, 184
Higgins, Jack, 631

high blood pressure, treatment of, 668
high density lipoproteins, 187
high throughput screening, 205, 333
highly active antiretroviral therapy
   (HAART, 501
high-throughput screening, 204
himbacine, 761
HINT, 416
Hippocrates, 606
hirudins, 758
histamine, 353–4, 706–11, 714–5, 717, 736
histamine receptors, 52, 123, 706–8
   antagonists, 705–25, see also
      antihistamines
   evolution, 54–5
histidine, 17, 21–2, 189, 801
   acid/base catalyst, 33–7, 189, 198, 641–3,
     788, 795
   Ames test, 202
   as iron ligand, 744–5
   catalytic triad, 34–5, 96
   chymotrypsin, 34–5
   lipase active site, 96
histidine kinases, 577
histone acetylase, 600
histone deacetylase, 102, 600
   active site, 601–2
   inhibitors, 600–2
   mechanism, 36–7
histones, 82, 600, 602
histrelin, 571
HIV, 493, 498–9
   antiviral therapy, 500–1
   cause of cancer, 543
   life cycle, 498–500
   structure, 498–500
   treatment of, 493
HIV protease, 20, 28, 152, 504–6
   cloning, 203
   role in the viral life cycle, 498–500
   substrates, 505
   target in HIV therapy, 500–1
HIV protease inhibitors, 503–17
   drug-drug interactions, 482
   in HIV therapy, 501
HMG-CoA, 33, 102, 188–9, 192
HMG-CoA reductase, 33, 102, 188–90,
   193–4, 577, 745
   active site, 193
HMGR, see HMG-CoA reductase
Hodgkin's disease, 543
   treatment of, 551, 559
Hodgkin's lymphoma, treatment of, 559, 610
Hodgkins, Dorothy, 434
Hofmann elimination, 638
homoharringtone, 606
homology modelling, 378–80, 778
   L-type calcium ion channel, 751
   renin, 783
honey, 425
Hong Kong flu, 519
hormone-based therapies, 567–71
hormones, 44, 46
   as drugs, 275–6

HR780, 246
HSP90, 28
HSV, treatment of, 497
human ether-a-go-go related gene, 201
human genome project, 26, 88
human granulocyte-colony stimulating
   factor, 276
human growth factor, 88, 150, 275
human growth hormone, 184
   antagonist, 276
human immune deficiency virus, see HIV
human intestinal di-/tripeptide
   transporter, 495
human intestinal proton-dependent
   oligopeptide transporter, 495
human parathyroid hormone, 275
human rhinovirus, 530
hunger suppression, 675
Huperzia serrata, 652
huperzine A, 652
huperzine B dimers, 650
hybrid drugs, 252, 486
hybridomas, 276
hycanthone, 319–20
hydantoins, solid phase synthesis, 330
hydralazine (**Apresoline**), 747–8
hydrocortisone (Corlan, Dioderm,
   Efcortesol, Mildison), 465, 766, 773–5
   sustained release, 773–4
   with clomitrazole (**Canesten**) 814
   with gentamicin (**Gentisone**) 816
   with miconazole nitrate (**Daktacort**) 814
   with neomycin sulphate and polymyxin B
     sulphate (**Otosporin**) 818
   with nystatin and benzalkonium chloride
     (**Timodine**) 820
   with nystatin and chlorhexidine
     (**Nystaform-HC**) 818
hydrocortisone acetate, 771–2, 775
   with fusidic acid (**Fucidin H**) 815
hydrocortisone butyrate (**Locoid**), 775
hydrocortisone phosphate, 775
hydrocortisone succinate, 775
hydrogen bond acceptor, 6
hydrogen bond donor, 6
hydrogen bond flip-flop, 6
hydrogen bonding, 6–8, 21–2
hydrolases, 37, see also
         S-adenosylhomocysteine hydrolase,
         cyclic ADP ribose hydrolase, epoxide
         hydrolase, neuraminidase, penicillin
         acylase
hydromorphone (**Palladone**), 680
   formulation, 182
hydrophobic interactions, 10, 22
hydrophobicity, 397
hydroxamic acid as a zinc ligand, 596, 601
3-hydroxy-3-methyl glutaryl coenzyme
   A, 188
3-hydroxy-3-methylglutaryl-coenzyme A
   reductase, see HMG-Co reductase
hydroxycarbamide, 102, 563–4, 566
17α-hydroxylase-17(20)-lyase, inhibition
   of, 569

hydroxylation
    drug metabolism, 170, 175, 730
    metabolic inhibitors, 570
    of proline, 25
    vancomycin biosynthesis, 459–60
4-hydroxyminaprine, 213
N-(2-hydroxypropyl)methacrylamide, 183
17β-hydroxysteroid dehydrogenase
    type 134, 214, 240, 252
4-hydroxytamoxifen, 569
5-hydroxytryptamine, 214, see serotonin
hyoscine (Scopoderm TTS), 210, 629–30
    quaternary salts, 631
hyoscine butylbromide (Buscopan), 631, 634
hyoscine hydrobromide (Kwells)
hyoscyamine, 630
Hyperchem program, 350
hyperpolarization, 808–9
hypertension, 200, 735, 737
    treatment of, 656, see also
        antihypertensive agents
hypodermic syringe, 678
hypoglycaemia, 212
hypotension, treatment of, 656, 662
hypoxia, 550
hypoxia-activated prodrug, 558
hypoxia-inducible factors, 549

I

ibritumomab (Zevalin), 610, 612
ibrutinib (Imbruvica), 589, 594
ibuprofen, 220, 551
IC$_{50}$ value, 125
ICI D7114, 297–8
icotinib (Conmana), 580, 593
idamycin, 553
idarubicin (Zavedos), 553–4
idarucizumab (Praxbind), 758
idelalisib (Zydelig), 589, 594
idoxuridine (Herpid), 493, 496–7
IFN-alpha, 538, 540
ifosfamide (Mitoxana), 556–9
iloprost (Ventavis), 747–8
imatinib (Glivec), 239, 294, 551, 579,
    582–4, 586, 593
imidazole, as zinc ligand, 576
imiglucerase (Cerezyme), 275
imipenem, 455–6, 458
    with cilastatin (Primaxin) 818
imipramine (Tofranil), 216, 673–4
imiquimod (Aldara), 498, 540
immunoglobulin E, 277
immunomodulators, 494, 540, 605
immunostimulants, 608, 610
immunosuppressants, 149, 216, 269, 275,
    277, 605
Imnovid, 605–6, 816
Imodium, 690, 816
implants, 180
importance sampling, 360
impurity profiling, 297
in vitro tests, 203
in vivo tests, 203–4

Incas, 178, 674
incontinence, treatment of, 629, 634
indacaterol (Onbrez), 635, 665–6
    with glycopyrrolate (Utibron Neohaler) 820
indanyl carbenicillin, 445, 447
Indian cobra, 639
Indian hedgehog, 74
indicator variable, 409, 411
indinavir (Crixivan), 12, 175, 501, 512–3, 516
    metabolism of, 175
indometacin (Rimacid), 99, 101
indomethacin, see indometacin
induced dipole interactions, 9, 229
induced fit, 47
Inegy, 754
infliximab (Remicade), 277
influenza, 490
influenza A, treatment of, 493
influenza virus
    structure and life cycle, 519–21
inhibition constant, 105, 125
Innovative Medicines Initiative, 484
inosine, 564
inosine-5'-monophosphate
    dehydrogenase, 540
inositol, 70
inositol triphosphate (IP3), 62, 68–71, 74,
    640, 655
insect bites, treatment of, 707
insecticides, 402, 646–7
Institutional Review Board, 294
insulin, 28, 275
    administration, 162, 180
    control of glycogen synthesis, 40
    crossing the blood brain barrier, 166
    dosing regimes, 180
    oral delivery system, 184
    production of, 88
    resistance, treatment of, 756
insulin receptor, 56–7
integrase, 498–500
    inhibitors, 501, 517–8
integrins, 149–50, 549–50
    inhibition, 277
intercalating agents, 128–29, 553–5
interferons (IntronA, Immukin,
    Viraferon), 56, 275–6, 540, 608, 610
interleukin 6 receptor, 277
interleukins, 17, 56, 549, 608
    as targets for antibodies, 610
    inhibition, 277
intermolecular bonds, 4
International Preliminary Examination
    Report, 291
International Search Report, 291
intestinal infections, see gastrointestinal
    infections
intestinal spasm, treatment of, 634
intra-abdominal infections, treatment of, 454
intracellular receptors, 59
intramolecular bonds, 4, 21
intramuscular injection, 179
intraperitoneal injection, 180
intrathecal injection, 180

intravenous drip, 179
intravenous injection, 179
intravitreal injection, 498
intron, 85–6
inverse agonists, 119–20, 740
Investigational Exemption to a New Drug
    Application (IND), 293
Invirase, 12, 816
ion carriers, 155
ion channel disrupters, 521
ion channels, 49–51, see also funny ion
    channels, calcium ion channels
ion-dipole interactions, 8–9
ionic bonding, 5, 21–2
ionophores, 156
iontophoresis, 178–9
ipecacuanha, 210
IP3, see inositol triphosphate
ipilimumab (Yervoy), 610
ipratropium (Atrovent, Ipratropium Steri-
    Neb, Respontin), 631, 634
iproniazid, 675
irbesartan (Aprovel, Avapro, Karvea), 740–1
    with hydrochlorothiazide (CoAprovel) 814
Iressa, 579, 582, 816
irinotecan (Campto, Camptosar), 281,
    555–6, 560, 609, 613, 615
iron ions
    activation of artemisinin, 311–2
    as cofactors, 36, 167, 312, 563, 570, 615,
        744–5
    interaction with doxorubicin, 553
    transport system, 455
irregular heart rhythms, treatment of, 673
irreversible antagonists of the P2Y12
    receptor, 762
irreversible inhibitors, 94–6, 571
    ibrutinib, 589
    kinase inhibitors, 581
    of the proteasome, 600
    trifluridine, 562
isoetharine, 663
isoleucine, 17, 22, 801
    HIV protease, 505, 507, 509, 513–5
    kinase mutation, 584
    mimic, 149
    Ras protein, 576
isoleucyl tRNA synthetase, 484
isomerases, 37, see also topoisomerases
isoniazid, 174, 211, 216, 426, 468, 481–3
isonicotinaldehyde thiosemicarbazone, 211
isoprenaline, 237, 661, 663, 667
isoprenylcysteine
    carboxylmethyltransferase, 576
isosorbide dinitrate (Angitak, Isoket), 744–5
isosorbide mononitrate (Ismo), 744–5
isosteres, 233–4, 243–4, 260
isothermal titration calorimetry, 206
isoxazolyl penicillins, 444
isozymes, 40, 99, 101, 199
isradipine, 751
ivabradine (Procoralan, Corlanor), 753
ixabepilone, 573–5
ixazomib (Ninlaro), 599

J

JAK-STAT signal transduction pathway, 72–3
Janus kinases, 72–3, 104
    inhibitors, 590, 594
Japanese plum yew, 606
Jenner, 492
JM216, 558
JM3100, 517
John Hopkins Clinical Compound
    Library, 211
joint infections, treatment of, 477, 480
Joubert, 433

K

Kaletra, 273, 501, 516, 816
kanamycin, 485
Kaposi's sarcoma, 543
    treatment of, 494
ketanserin, 123, 356, 382
ketobemidone, 688
ketoconazole, 504
ketones, binding role, 226
kidney cancer, treatment of, 571, 594
kidneys, 176–7, 737
KIF5B-RET fusion gene, 590
killer nanotubes, 154–5
kinases, see cyclin-dependent kinases,
    kinase-linked receptors, protein
    kinases, serine threonine kinases,
    thymidine kinase, tyrosine kinases
kinase-linked receptors, 55–58
    signal transduction, 70–4
kinase inhibitors, 104, see also protein
    kinases
Kivexa, 501, 816
Klebsiella pneumoniae, resistance, 483
Koch, 425
Koshland's theory of induced fit, 32

L

L685434, 512–3
L704486, 512–3
L787257, 263
L791456, 263
L746072, 743–4
L-dopa, see levodopa
L-PAM, see L-phenylalanine mustard
labetalol, 666
lacidipine (Motens, Lacipil), 751
lactamases, 434–5, 439–40, 483, 612, 614
    inhibitors, 98, 455–57
lactams
    as an acylating agent, 229
    binding role, 229
lactate dehydrogenase, 30, 32, 35–6, 40
lactic acid, 30, 36, 40
ladostigil, 252
lamitidine, 724
lamivudine (Epivir, Zeffix), 500–1, 503
lanalidomide, 606
laninamivir (Inavir), 525

laniquidar, 552
lansoprazole (Zoton), 102, 726–8, 732
lapatinib (Tykerb), 580–1, 591, 593
lasalocid A, 156
Lassa fever, 490
    treatment of, 540
laudanum, 678
laxative, 210
LD$_{50}$ value, 284
LDL receptors, 756
LDLs, 756
LDZ, 271
lead compounds, 207
    enzyme substrate, 781
    for apixaban, 790
    for dabigatran, 759
    for ivabradine, 753
    for macitentan, 743
    for tadalafil, 747
    P2Y12 allosteric antagonists, 762
    product from enzyme reaction, 795
lecithin, 3
ledipasvir, 536, 538–9
    with sofosbuvir (Harvoni) 816
LEE011, 587
leeches, 758
leg ulcers, treatment of, 482
LEGEND program, 389–90
legionnaires disease, treatment of, 477, 482
leiomyosarcoma, treatment of, 607
Lemsip, 662
lenalidomide (Revlimid), 605, 610
lenvatinib (Lenvima), 590–1, 594
lepirudin (Refludin), 758
leprosy, treatment of, 432, 482, 605
leptospirosis, treatment of, 435
lercanidipine (Zanidip), 751
letrozole (Femera), 570–1, 594
leucine, 17, 22, 801
    interaction with capsid binding
        agents, 531
    interaction with statins, 193
    mimic, 149
    Ras protein, 576
leucine tRNA synthetase, 484
leucovorin, 556, 565, 610
Leu-enkephalin, 183, 695–6, 698
leukaemia, 490, 543, 608
    leukaemia treatments, 593–4, 608
    alkylating agents, 559
    antibodies, 609–10
    antibody-drug conjugates, 613
    antimetabolites, 564–6
    arsenic trioxide, 606
    asparaginase, 276, 608
    hormone-based, 571
    intercalating agents, 554
    kinase inhibitors, 593
    6-mercaptopurine, 96, 566
    methotrexate, 180, 565
    pegasparagase and pegademase, 276
    vinca alkaloids, 575
leuprolide, 568, 571
leuprorelin, 568

levalbuterol (Xopenex), 664
levallorphan, 687
levobupivacaine (Chirocaine), 292–3
levodopa, 101, 164, 268–9, 273–4, 656–7
    with carbidopa, see co-careldopa
levofloxacin (Levaquin, Tavanic), 477–8,
    480
levorphanol, 686–7
levothyroxine sodium, 178
lewisite, 103
LHRH, see luteinizing hormone-releasing
    hormone
Librium, 269, 816
lice, treatment of, 646
lidocaine (Xylocaine), 179, 260, 719
ligand efficiency, 219–20
ligand-gated ion channels, 808
ligases, 37, 87–8, see also D-Ala-D-Ala ligase
    and DNA ligase
lignans, 572
lignocaine, see lidocaine
Lilly Pharmaceuticals, 451, 513, 724
lincomycin, 474
lincosamides, 472, 474–5, 477
linear regression analysis, 395–6
linezolid (Zyvox), 475–7
linkers, 326–9
linoleate, 187
lipase enzymes, 65, see also phospholipases
lipid carrier, 459
lipid regulating agents, 754–7
Lipinski's rule of five, 163–4, 219, 331–2
lipophilic efficiency, 234
lipoprotein lipase, 755
liposarcoma, treatment of, 607
liposomes, 183–4, 549, 554, 616
5-lipoxygenase, 102
lipstatin, 208–9
liquorice, 733
lisinopril (Carace, Prinivil, Zestril), 164,
    212, 306–7, 738
    crystal structure, 306
Lister, 425
lithium salts (Camcolit, Liskonum,
    Priadel), 70
liver, 737
liver cancers, 543
liver microsomal fractions, 287
local anaesthetics
    ability to cross cell membranes, 165
    administration, 162, 179
    development of, 245, 260
    localization of action, 274, 656, 662
    long lasting, 292
    molecular target, 51, 638
    structure comparisons, 363
    testing methods, 204
local energy minimum, 358
log P, 256, 397–9
lollipop phase separator, 335–6
lomitapide (Lojuxta, Juxtapid), 756–7
lomustine (CCNU), 132, 134, 559
lonafarnib, 102, 576–7